Atomic Numbers and Atomic Masses of the Elements

Based on $^{12}_{6}C$. Numbers in parentheses are the mass numbers of the most stable isotopes of radioactive elements.

Element	Symbol	Atomic Number	Atomic Mass	Element	Symbol	Atomic Number	Atomic Mass
Actinium	Ac	89	(227)	Mendelevium	Md	101	(258)
Aluminum	Al	13	26.98	Mercury	Hg	80	200.59
Americium	Am	95	(243)	Molybdenum	Mo	42	95.94
Antimony	Sb	51	121.76	Neodymium	Nd	60	144.24
Argon	Ar	18	39.95	Neon	Ne	10	20.18
Arsenic	As	33	74.92	Neptunium	Np	93	(237)
Astatine	At	85	(210)	Nickel	Ni	28	58.69
Barium	Ba	56	137.33	Niobium	Nb	41	92.91
Berkelium	Bk	97	(247)	Nitrogen	N	7	14.01
Beryllium	Be	4	9.01	Nobelium	No	102	(259)
Bismuth	Bi	83	208.98	Osmium	Os	76	190.23
Bohrium	Bh	107	(267)	Oxygen	O	8	16.00
Boron	B	5	10.81	Palladium	Pd	46	106.42
Bromine	Br	35	79.90	Phosphorus	P	15	30.97
Cadmium	Cd	48	112.41	Platinum	Pt	78	195.08
Calcium	Ca	20	40.08	Plutonium	Pu	94	(244)
Californium	Cf	98	(251)	Polonium	Po	84	(209)
Carbon	C	6	12.01	Potassium	K	19	39.10
Cerium	Ce	58	140.12	Praseodymium	Pr	59	140.91
Cesium	Cs	55	132.91	Promethium	Pm	61	(145)
Chlorine	Cl	17	35.45	Protactinium	Pa	91	(231)
Chromium	Cr	24	52.00	Radium	Ra	88	(226)
Cobalt	Co	27	58.93	Radon	Rn	86	(222)
Copper	Cu	29	63.55	Rhenium	Re	75	186.21
Curium	Cm	96	(247)	Rhodium	Rh	45	102.91
Darmstadtium	Ds	110	(281)	Roentgenium	Rg	111	(280)
Dubnium	Db	105	(262)	Rubidium	Rb	37	85.47
Dysprosium	Dy	66	162.50	Ruthenium	Ru	44	101.07
Einsteinium	Es	99	(252)	Rutherfordium	Rf	104	(263)
Erbium	Er	68	167.26	Samarium	Sm	62	150.36
Europium	Eu	63	151.96	Scandium	Sc	21	44.96
Fermium	Fm	100	(257)	Seaborgium	Sg	106	(266)
Fluorine	F	9	19.00	Selenium	Se	34	78.96
Francium	Fr	87	(223)	Silicon	Si	14	28.09
Gadolinium	Gd	64	157.25	Silver	Ag	47	107.87
Gallium	Ga	31	69.72	Sodium	Na	11	22.99
Germanium	Ge	32	72.64	Strontium	Sr	38	87.62
Gold	Au	79	196.97	Sulfur	S	16	32.07
Hafnium	Hf	72	178.49	Tantalum	Ta	73	180.95
Hassium	Hs	108	(277)	Technetium	Tc	43	(98)
Helium	He	2	4.00	Tellurium	Te	52	127.60
Holmium	Ho	67	164.93	Terbium	Tb	65	158.93
Hydrogen	H	1	1.01	Thallium	Tl	81	204.38
Indium	In	49	114.82	Thorium	Th	90	(232)
Iodine	I	53	126.90	Thulium	Tm	69	168.93
Iridium	Ir	77	192.22	Tin	Sn	50	118.71
Iron	Fe	26	55.85	Titanium	Ti	22	47.87
Krypton	Kr	36	83.80	Tungsten	W	74	183.84
Lanthanum	La	57	138.91	Uranium	U	92	(238)
Lawrencium	Lr	103	(262)	Vanadium	V	23	50.94
Lead	Pb	82	207.19	Xenon	Xe	54	131.29
Lithium	Li	3	6.94	Ytterbium	Yb	70	173.04
Lutetium	Lu	71	174.97	Yttrium	Y	39	88.91
Magnesium	Mg	12	24.31	Zinc	Zn	30	65.41
Manganese	Mn	25	54.94	Zirconium	Zr	40	91.22
Meitnerium	Mt	109	(276)				

General, Organic, and Biological Chemistry

General, Organic, and Biological Chemistry

FIFTH EDITION

H. STEPHEN STOKER

Weber State University

BROOKS/COLE
CENGAGE Learning™

Australia • Brazil • Japan • Korea • Mexico • Singapore • Spain • United Kingdom • United States

**General, Organic, and Biological Chemistry,
Fifth Edition**
H. Stephen Stoker

Acquisitions Editor: Kilean Kennedy

Senior Development Editor: Rebecca Berardy
Schwartz

Associate Editor: Stephanie Van Camp

Assistant Editor: Elizabeth Woods

Editorial Assistant: Jon Olafsson

Senior Media Editor: Rebecca Berardy Schwartz

Senior Marketing Manager: Amee Mosley

Marketing Coordinator: Kevin Carroll

Senior Marketing Communications Manager:
Linda Yip

Project Manager, Editorial Production: Andrea
Cava

Art & Design Manager: Jill Haber

Manufacturing Coordinator: Miranda Klapper

Text Designer: Nesbitt Graphics, Inc./Lisa
Adamitis

Art Editor: Jessyca Broekman

Photo Researcher: Naomi Kornhauser

Copy Editor: Peggy J. Flanagan

Illustrators: Laura Brown McEntee, Rossi
Illustration & Design

Cover Designer: Leonard Massiglia

Cover Image: © Masterfile Royalty Free

Compositor: Aptara, Inc.

For product information and technology assistance, contact us at
Cengage Learning Customer & Sales Support, 1-800-354-9706

For permission to use material from this text or product,
submit all requests online at **www.cengage.com/permissions**
Further permissions questions can be emailed to
permissionrequest@cengage.com

Library of Congress Control Number: 2008934779

ISBN-13: 978-0-547-15281-3

ISBN-10: 0-547-15281-7

Brooks/Cole
10 Davis Drive
Belmont, CA 94002
USA

Cengage Learning is a leading provider of customized learning solutions with
office locations around the globe, including Singapore, the United Kingdom,
Australia, Mexico, Brazil, and Japan. Locate your local office at
www.cengage.com/global

Cengage Learning products are represented in Canada by Nelson Education, Ltd.

To learn more about Brooks/Cole, visit **www.cengage.com/brookscole**

Purchase any of our products at your local college store or at our preferred
online store **www.ichapters.com**

Printed in the United States of America
3 4 5 6 7 12 11 10 09

Brief Contents

Contents

Chapter 6 Chemical Calculations: Formula Masses, Moles, and Chemical Equations 137

Chapter 7 Gases, Liquids, and Solids 163

Chapter 8 Solutions 192

Chapter 9 Chemical Reactions 223

Preface

The positive responses of instructors and students who used the previous four editions of this text have been gratifying—and have lead to the new fifth edition that you hold in your hands. This new edition represents a renewed commitment to the goals I initially set out to meet when writing the first edition. These goals have not changed with the passage of time. My initial and still ongoing goals are to write a text in which

- The needs are simultaneously met for the many students in the fields of nursing, allied health, biological sciences, agricultural sciences, food sciences, and public health who are required to take such a course.
- The development of chemical topics always starts out at ground level. The students who will use this text often have little or no background in chemistry and hence approach the course with a good deal of trepidation. This "ground level" approach addresses this situation.
- The amount and level of mathematics is purposefully restricted. Clearly, some chemical principles cannot be divorced entirely from mathematics and, when this is the case, appropriate mathematical coverage is included.
- The early chapters focus on fundamental chemical principles and the later chapters, built on these principles, develop the concepts and applications central to the fields of organic chemistry and biochemistry.

FOCUS ON BIOCHEMISTRY   Most students taking this course have a greater interest in the biochemistry portion of the course than the preceding two parts. But biochemistry, of course, cannot be understood without a knowledge of the fundamentals of organic chemistry, and understanding organic chemistry in turn depends on knowing the key concepts of general chemistry. Thus, in writing this text, I essentially started from the back and worked forward. I began by determining what topics would be considered in the biochemistry chapters and then tailored the organic and then general sections to support that presentation. Users of the previous editions confirm that this approach ensures an efficient but thorough coverage of the principles needed to understand biochemistry.

EMPHASIS ON VISUAL SUPPORT   I believe strongly in visual reinforcement of key concepts in a textbook; thus, this book uses art and photos wherever possible to teach key concepts. Artwork is used to make connections and highlight what is important for the student to know. Reaction equations use color to emphasize the portions of a molecule that undergo change. Colors are likewise assigned to things like valence shells and classes of compounds to help students follow trends. Computer-generated, three-dimensional molecular models accompany many discussions in the organic and biochemistry sections of the text. Color photographs show applications of chemistry to help make concepts real and more readily remembered.

Visual summary features, called *Chemistry at a Glance,* pull together material from several sections of a chapter to help students see the larger picture. For example, Chapter 3 features a *Chemistry at a Glance* on the shell–subshell–orbital interrelationships; Chapter 10 presents buffer solutions; Chapter 13 includes IUPAC nomenclature for alkanes, alkenes, and alkynes; and Chapter 22 summarizes DNA replication. The *Chemistry at a Glance* feature serves both as an overview for the student reading the material for the first time and as a review tool for the student preparing for exams. Given the popularity of the *Chemistry at a Glance* summaries in the previous editions, several new ones have been added and several existing ones have been updated or expanded. New topics selected for *Chemistry at a Glance* boxes include

- Acid–base definitions
- Structural representations for alkanes
- Structural characteristics and naming of alcohols, thiols, ethers, and thioethers
- Structural components of nucleic acids

COMMITMENT TO STUDENT LEARNING   In addition to the study help *Chemistry at a Glance* offers, the text is built on a strong foundation of learning aids designed to help students master the course material.

- **Problem-solving pedagogy.** Because problem solving is often difficult for students in this course to master, I have taken special care to provide support to help students build their skills. Within the chapters, worked-out *Examples* follow the explanation of many concepts. These examples walk students through the thought processes involved in problem solving, carefully outlining all the steps involved. Each is immediately followed by a *Practice Exercise,* to reinforce the information just presented.
- **Diversity of worked-out Examples.** The number of worked-out Examples has been dramatically increased in this new edition, with most of the increase in the biochemistry chapters of the text. Worked-out Examples are a standard feature in the general chemistry portions of all textbooks for this market. This relates primarily to the mathematical nature of many general chemistry topics. In most texts, including earlier editions of this text, fewer worked-out Examples appear in the organic chemistry chapters and still fewer (almost none) in the biochemistry portion because of the mathematical demands decrease. This new edition changes that perspective. Thirty-two new worked-out Examples are found within the biochemistry scope of the text. With the increasing detail that now accompanies biochemistry discussions about the human body, such Examples are warranted and can be very helpful for students.
- **Chemical Connections.** In every chapter *Chemical Connections* show chemistry as it appears in everyday life. These boxes focus on topics that are relevant to students' future careers in the health and environmental fields and on those that are important for informed citizens to understand. Many of the health-related *Chemical Connections* have been updated to include the latest research findings, and include new boxes on metallic elements and the human body, iron (the most abundant transition element in the human body), carbon monoxide toxicity and the human body, the chemistry of nicotine addiction, and changing sugar consumption patterns (decreased sucrose, increased fructose).
- **Margin notes.** Liberally distributed throughout the text, *margin notes* provide tips for remembering and distinguishing between concepts, highlight links across chapters, and describe interesting historical background information.
- **Defined terms.** All definitions are highlighted in the text when they are first presented, using boldface and italic type. Each defined term appears as a complete sentence; students are never forced to deduce a definition from context. In addition, the definitions of all terms appear in the combined *Index/Glossary* found at the end of the text. A major emphasis in this new edition has been "refinements" in the defined terms arena. All defined terms were reexamined to see if they could be stated with greater clarity. The result was a "rewording" of many defined terms.
- **Review aids.** Several review aids appear at the ends of the chapters. *Concepts to Remember* and *Key Reactions and Equations* provide concise review of the material presented in the chapter. A *Key Terms Review* lists all the key terms in the chapter alphabetically and cross-references the section of the chapter in which they appear. These aids help students prepare for exams.
- **End-of-chapter problems.** An extensive set of end-of-chapter problems complements the worked examples within the chapters. Each end-of-chapter problem set is divided into two sections: *Exercises and Problems* and *Additional Problems*. The *Exercises*

and Problems are organized by topic and paired, with each pair testing similar material and the answer to the odd-numbered member of the pair at the back of the book. These problems always involve only a single concept. The *Additional Problems* involve more than one concept and are more difficult than the Exercises and Problems.

A new feature of this edition is the placement of 67 new "visual concept" problems in the general chemistry problem sets. This new problem type, which is integrated throughout the problem sets, is designed to facilitate learning through the use of molecular models, pictorial representations of chemical systems, graphs, and other visual portrayals.

■ **Multiple-choice practice tests.** Practice tests at the end of each chapter act as a cumulative overview and self-study tool.

CONTENT CHANGES   Coverage of a number of topics has been expanded in this edition. The two driving forces in expanded coverage considerations were (1) the requests of users and reviewers of the previous editions and (2) my desire to incorporate new research findings, particularly in the area of biochemistry, into the text. Topics with expanded coverage include

- Percent composition by mass
- Colloidal dispersion and suspensions
- Equivalent and milliequivalent concentration units
- Nuclear stability
- Nuclear medicine
- Alkane–base polymers
- Cyclic esters (lactones)
- Polyester type polymers
- Acyl transfer reactions
- Amine salts
- Polyamide and polyurethane type polymers
- Protein isoelectric points
- Protein classification by molecular shape and by function
- Regulation of enzyme activity
- Ribosome structure
- Post-translational phase of protein synthesis
- Recombinant DNA technology
- Glycogenolysis
- Ketogeneis
- Lipogenesis

SUPPORTING MATERIALS

■ For the Instructor

Supporting instructor materials are available to qualified adopters. Please consult your local Cengage Learning, Brooks/Cole representative for details. Visit **www.cengage.com/chemistry/stoker** to

- See samples of materials
- Request a desk copy
- Locate your local representative
- Download electronic files of the *Lab Manual Instructor's Resource Manual* and other helpful materials for instructors and students.

POWERLECTURE WITH DIPLOMA TESTING AND JOININ™ INSTRUCTOR'S DVD. PowerLecture (ISBN-10: 0-495-83160-3; ISBN-13: 978-0-495-83160-0) is a one-stop digital library and presentation tool that includes

- Prepared *Microsoft® PowerPoint® Lecture Slides* that cover all key points from the text in a convenient format with art and photographs. You can enhance the slides with your own materials or with additional interactive video and animations on the DVD for personalized, media-enhanced lectures.
- *Image libraries in PowerPoint and JPEG formats* that contain electronic files for all text art, most photographs, and all numbered tables in the text. These files can be used to create your own transparencies or PowerPoint lectures.
- *JoinIn "clicker" Slides* written specifically for the use of Chemistry with the classroom response system of your choice that allows you to seamlessly display student answers.
- The *Complete Solutions Manual* (H. Stephen Stoker), which contains answers to all end-of-chapter exercises.
- Sample chapters from the *Study Guide with Solutions to Selected Problems*.
- The *Instructor's Resource Guide* for the textbook and for *Experimental Chemistry*.
- The Test Bank in printable *Word* and *PDF* documents, which provide an easy way to view all questions and answers from Diploma® Testing.
- *Diploma® Testing,* which combines a flexible test-editing program with comprehensive gradebook functions for easy administration and tracking. With *Diploma Testing*, instructors can administer tests via print, network server, or the Web. Questions can be selected based on their chapter/section, level of difficulty, question format, algorithmic functionality, topic, learning objective, and five levels of key words. With *Diploma® Testing* you can

> Choose from the 2,500 test items designed to measure the concepts and principles covered in the text.
>
> Ensure that each student gets a different version of the problem by selecting from preprogrammed algorithmic questions.
>
> Edit or author algorithmic or static questions that integrate into the existing bank, becoming part of the question database for future use.
>
> Choose problems designated as single-skill (easy), multi-skill (medium), and challenging and multi-skill (hard).
>
> Customize tests to assess the specific content from the text.

- Create several forms of the same test where questions and answers are scrambled.

ⓞWL Online Web-based Learning **OWL: ONLINE WEB-BASED LEARNING** OWL is authored by Roberta Day, Beatrice Botch, and David Gross of the University of Massachusetts, Amherst; William Vining of the State University of New York at Oneonta; and Susan Young of Hartwick College:

> OWL Instant Access (two semesters): ISBN-10: 0-495-11105-8; ISBN-13:978-0-495-11105-4
> Instant Access to OWL e-Book (two semesters): ISBN-10: 0-495-83162-X; ISBN-13: 978-0-495-83162-4

Developed at the University of Massachusetts, Amherst, and class tested by tens of thousands of chemistry students, OWL is a fully customizable and flexible web-based learning system. OWL supports mastery learning and offers numerical, chemical, and contextual parameterization to produce thousands of problems correlated to this text. The OWL system also features a database of simulations, tutorials, and exercises, as well as end-of-chapter problems from the text. With OWL, you get the most widely used online learning system available for chemistry with unsurpassed reliability and dedicated training and support.

The optional **e-Book in OWL** includes the complete electronic version of the text, fully integrated and linked to OWL homework problems. Most e-books in OWL are interactive and offer highlighting, notetaking, and bookmarking features that can all be saved. To view an OWL demo and for more information, visit **www.cengage.com/owl** or contact your Cengage Learning, Brooks/Cole representative.

LAB MANUAL INSTRUCTORS RESOURCE MANUAL Available on PowerLecture DVD and on the instructor companion site, this guide, by G. Lynn Carlson, Senior Lecturer

Emeritus University of Wisconsin-Parkside, includes additional information related to the experiments in the Lab Manual. Additional safety notes, references, and web resources enhance the experience of the Lab for both the instructor and the students.

CENGAGE LEARNING CUSTOM SOLUTIONS This allows you to develop personalized text solutions to meet your course needs. Match your learning materials to your syllabus and create the perfect learning solution—your customized text will contain the same thought-provoking, scientifically sound content, superior authorship, and stunning art that you've come to expect from Cengage Learning, Brooks/Cole texts, yet in a more flexible format. Visit **www.cengage.com/custom.com** to start building your book today.

■ For the Student

Visit the *student* website at **www.cengage.com/chemistry/stoker** to see samples of select student supplements. Students can purchase any Cengage Learning product at your local college store or at our preferred online store **www.ichapters.com.**

STUDENT COMPANION WEBSITE Accessible from **www.cengage.com/chemistry/ stoker,** this site provides online study tools including practice tests, flashcards, and Careers in Chemistry.

OWL Online Web-based Learning OWL FOR GENERAL, ORGANIC, AND BIOLOGICAL CHEMISTRY See the above description in the instructor support materials section.

STUDY GUIDE WITH SOLUTIONS TO SELECTED PROBLEMS By Danny V. White of American River College and Joanne A. White, this useful resource (ISBN: 0-547-16808-X; ISBN 13: 978-0-547-16808-1) will reinforce your skills with activities and practice problems for each chapter. After completing the end-of-chapter exercises, you'll be able to check your answers for the odd-numbered questions.

LAB MANUAL The **Lab Manual** (ISBN-10: 0-547-16793-8; ISBN-13: 978-0-547-16793-0) to accompany this textbook, by G. Lynn Carlson, Senior Lecturer Emeritus University of Wisconsin-Parkside, includes 42 experiments that were selected to match the topics in your textbook. Each experiment has an introduction, a procedure, a page of pre-lab exercises about the concepts the lab illustrates, and a report form. Some have a scenario that places the experiment in a real-world context. In addition, each experiment has a link to a set of references and on-line resources that might help you succeed with the experiment.

ESSENTIAL ALGEBRA FOR CHEMISTRY STUDENTS, SECOND EDITION This short book by David W. Ball, Cleveland State University (ISBN-10: 0-495-01327-7; ISBN-13 978-0-495-01327-3) is intended for students who lack confidence or competency in their essential mathematics skills necessary to survive in general chemistry. Each chapter focuses on a specific type of skill and has worked-out examples to show how these skills translate to chemical problem solving. It includes references to OWL, our web-based tutorial program that offers students access to online algebra skills exercises.

SURVIVAL GUIDE FOR GENERAL CHEMISTRY WITH MATH REVIEW AND PROFICIENCY QUESTIONS, SECOND EDITION Intended to help you practice for exams, this survival guide by Charles H. Atwood, University of Georgia (ISBN-10: 0-495-38751-7; ISBN-13 978-0-495-38751-0) shows you how to solve difficult problems by dissecting them into manageable chunks. The guide includes three levels of proficiency questions—A, B, and minimal—to quickly build confidence as you master the knowledge you need to succeed in your course.

■ For the Laboratory

CENGAGE LEARNING, BROOKS/COLE LAB MANUALS   We offer a variety of printed manuals to meet all your general chemistry laboratory needs. Instructors can visit the chemistry site at **www.cengage.com/chemistry** for a full listing and description of these laboratory manuals and laboratory notebooks. All Cengage Learning lab manuals can be customized for your specific needs.

SIGNATURE LABS . . . FOR THE CUSTOMIZED LABORATORY   Signature Labs combines the resources of Brooks/Cole, CER, and OuterNet Publishing to provide you unparalleled service in creating your ideal customized lab program. Select the experiments and artwork you need from our collection of content and imagery to find the perfect labs to match your course. Visit **www.signaturelabs.com** or contact your Cengage Learning representative for more information.

ACKNOWLEDGMENTS

I would like to gratefully acknowledge the helpful comments of reviewers.

■ Reviewers of the Fifth Edition

Teresa Brown, *Rochester Community and Technical College*; Karen Frindell, *Santa Rosa Junior College*; Irene Gerow, *East Carolina University*; Kevin Gratton, *Johnson County Community College*; Sherell Hickman, *Brevard Community College*; Martina Kaledin, *Kennesaw State University*; Allen W. Leung, *Rio Hondo Community College*; Michael J. Muhitch, *Rochester University*; Anthony Oertling, *Eastern Washington University*; James R. Paulson, *University of Wisconsin—Oshkosh*; Paul Sampson, Kent State University; Heather Sklenicka, *Rochester Community and Technical College*; Bobby Stanton, *University of Georgia*; Richard B. Triplett, *Des Moines Area Community College*; David A. Tramontozzi, *Macomb Community College; Paolos Yohannes, Georgia Perimeter College.*

■ Reviewers of the Fourth Edition

Jennifer Adamski, *Old Dominion University*; M. Reza Asdjodi, *University of Wisconsin—Eau Claire*; Irene Gerow, *East Carolina University*; Ernest Kho, *University of Hawaii at Hilo*; Larry L. Land, *University of Florida*; Michael Myers, *California State University—Long Beach*; H. A. Peoples, *Las Positas College*; Shashi Rishi, *Greenville Technical College*; Steven M. Socol, *McHenry County College.*

Special thanks go to Richard B. Triplett, *Des Moines Area Community College;* David Vanderlinden, *Des Moines Area Community College;* Barry Ganong, *Mansfield University*; and Michelle B. Moore, *Weber State University* for their help in ensuring this book's accuracy. Thanks to Richard Gurney, *Simmons College,* for his contribution of PowerPoint Lecture Outline content.

I also give special thanks to the people at Brooks/Cole, Cengage Learning, who guided the revision through various stages of development and production: Charles Hartford, Publisher, Chemistry; Senior Development Editor, Rebecca Berardy Schwartz; Andrea Cava, Project Editor; Naomi Kornhauser, Photo Researcher; and Stephanie VanCamp, Associate Editor.

H. Stephen Stoker
Weber State University

EMPHASIS ON VISUAL SUPPORT

Visual reinforcement is integral to the approach of this textbook. Art and photos are used whenever possible to teach key concepts.

Artwork is used to make connections between macroscale and microscale phenomena.

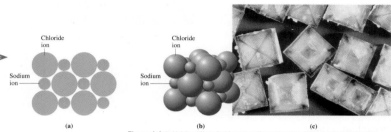

Figure 4.4 (a, b) A two-dimensional cross-section and a three-dimensional view of sodium chloride (NaCl), an ionic solid. Both views show an alternating array of positive and negative ions. (c) Sodium chloride crystals.

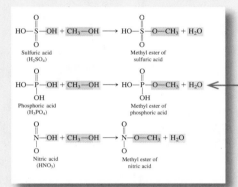

Reaction equations use color to emphasize the portions of a molecule that undergo change.

In the same manner that carboxylic acids are acidic (Section 16.8), phosphoric acid, diphosphoric acid, and triphosphoric acid are also acidic. The phosphoric acids are, however, polyprotic rather than monoprotic acids. The hydrogen atom in each of the —OH groups possesses acidic properties. All three phosphoric acids undergo esterification reactions with alcohols, producing species such as

Color is often used to highlight differences between related structures.

Carbohydrate Metabolism

24

Chapter Outline

Carbohydrates are the major energy source for human beings.

In this chapter we explore the relationship between carbohydrate metabolism and energy production in cells. The molecule glucose is the focal point of carbohydrate metabolism. Commonly called blood sugar, glucose is supplied to the body via the circulatory system and, after being absorbed by a cell, can be either oxidized to yield energy or stored as glycogen for future use. When sufficient oxygen is present, glucose is totally oxidized to CO_2 and H_2O. However, in the absence of oxygen, glucose is only partially oxidized to lactic acid. Besides supplying energy needs, glucose and other six-carbon sugars can be converted into a variety of different sugars (C_3, C_4, C_5, and C_7) needed for biosynthesis. Some of the oxidative steps in carbohydrate metabolism also produce NADH and NADPH, sources of reductive power in cells.

Throughout the text, an exciting **photo program** helps students see the everyday applications of the chemistry they are learning.

CHEMISTRY AT GLANCE

Chemistry at a Glance diagrams demonstrate interrelationships among concepts.

Chemistry at a Glance pulls together material from a group of sections or a whole chapter to help students see the larger picture through a visual summary.

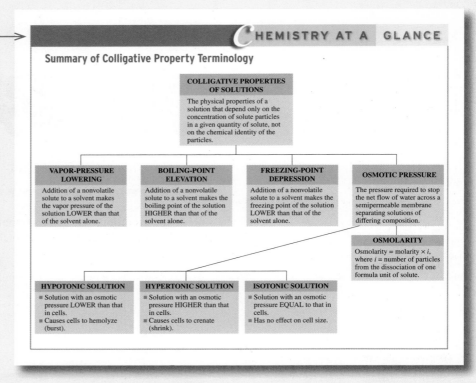

Many *Chemistry at a Glance* features have been revised and several new ones have been added.

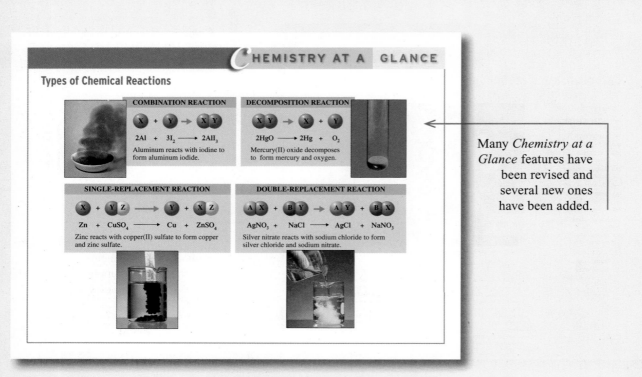

CONNECTING TO CHEMISTRY

In addition to Chemistry at a Glance, the text is built on a strong foundation of learning aids designed to help students master the course material.

 CHEMICAL Connections

Iron: The Most Abundant Transition Element in the Human Body

Small amounts of nine transition metals are necessary for the proper functioning of the human body. They include all of the Period 4 transition metals except scandium and titanium plus the Period 5 transition metal molybdenum, as shown in the following transition metal portion of the periodic table.

Transition Metals

Period 4		V	Cr	Mn	Fe	Co	Ni	Cu	Zn
Period 5		Mo							
Period 6									

Iron is the most abundant, from a biochemical standpoint, of these transition metals; zinc is the second most abundant.

Most of the body's iron is found as a component of the proteins hemoglobin and myoglobin, where it functions in the transport and storage of oxygen. Hemoglobin is the oxygen carrier in red blood cells, and myoglobin stores oxygen in muscle cells. Iron-deficient blood has less oxygen-carrying capacity and often cannot completely meet the body's energy needs. Energy deficiency—tiredness and apathy—is one of the symptoms of iron deficiency.

Iron deficiency is a worldwide problem. Millions of people are unknowingly deficient. Even in the United States and Canada, about 20% of women and 3% of men have this problem; some 8% of women and 1% of men are anemic, experiencing fatigue, weakness, apathy, and headaches.

Inadequate intake of iron, either from malnutrition or from high consumption of the wrong foods, is the usual cause of iron deficiency. In the Western world, the cause is often displacement of iron-rich foods by foods high in sugar and fat.

About 80% of the iron in the body is in the blood, so iron losses are greatest whenever blood is lost. Blood loss from menstruation makes a woman's need for iron nearly twice as great as a man's. Also, women usually consume less food than men do. These two factors—lower intake and higher loss—cause iron deficiency to be likelier in women than in men. The iron RDA (recommended dietary allowance) is 8 mg per day for adult males and older women. For women of childbearing age, the RDA is 18 mg. This amount is necessary to replace menstrual loss and to provide the extra iron needed during pregnancy.

Iron deficiency may also be caused by poor absorption of ingested iron. A normal, healthy person absorbs about 2%–10% of the iron in vegetables and about 10%–30% in meats. About 40% of the iron in meat, fish, and poultry is bound into molecules of heme, the iron-containing part of hemoglobin and myoglobin. Heme iron is much more readily absorbed (23%) than nonheme iron (2%–10%). (See the accompanying charts.)

Cooking utensils can enhance the amount of iron delivered by the diet. The iron content of 100 g of spaghetti sauce simmered in a glass dish is 3 mg, but it is 87 mg when the sauce is cooked in an unenameled iron skillet. Even in the short time it takes to scramble eggs, their iron content can be tripled by cooking them in an iron pan.

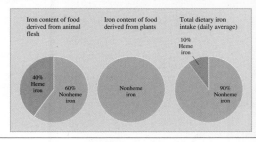

Iron content of food derived from animal flesh: 40% Heme iron, 60% Nonheme iron

Iron content of food derived from plants: Nonheme iron

Total dietary iron intake (daily average): 10% Heme iron, 90% Nonheme iron

Chemical Connections boxes show chemistry as it appears in everyday life. Topics are relevant to students' future careers in the health and environmental fields and are important for informed citizens to understand.

 1.8 NAMES AND CHEMICAL SYMBOLS OF THE ELEMENTS

Each element has a unique name that, in most cases, was selected by i wide variety of rationales for choosing a name have been applied. Som geographical names; germanium is named after the native country of coverer, and the elements francium and polonium are named after Fran The elements mercury, uranium, and neptunium are all named for p gets its name from the Greek word *helios*, for "sun," because it was spectroscopically in the sun's corona during an eclipse. Some elemer that reflect specific properties of the element or of the compounds Chlorine's name is derived from the Greek *chloros*, denoting "greeni color of chlorine gas. Iridium gets its name from the Greek *iris*, mea this alludes to the varying colors of the compounds from which it was

Abbreviations called chemical symbols also exist for the names of **chemical symbol** *is a one- or two-letter designation for an element a element's name.* These chemical symbols are used more frequently tha names. Chemical symbols can be written more quickly than the names, less space. A list of the known elements and their chemical symbols is gi The chemical symbols and names of the more frequently encountered elem in color in this table.

Note that the first letter of a chemical symbol is always capitalized is not. Two-letter chemical symbols are often, but not always, the first tv element's name.

Eleven elements have chemical symbols that bear no relationship t English-language name. In ten of these cases, the symbol is derived name of the element; in the case of the element tungsten, a German name source. Most of these elements have been known for hundreds of years

---

Looking forward/looking back margin notes show students the connections of concepts both in what they've learned previously and in what's to come.

 Learning the chemical symbols of the more common elements is an important key to success in studying chemistry. Knowledge of chemical symbols is essential for writing chemical formulas (Section 1.10) and chemical equations (Section 6.6).

 1.3 PROPERTIES OF MATTER

Various kinds of matter are distinguished from each other by their properties. A **property** *is a distinguishing characteristic of a substance that is used in its identification and description.* Each substance has a unique set of properties that distinguishes it from all other substances. Properties of matter are of two general types: physical and chemical.

A **physical property** *is a characteristic of a substance that can be observed without changing the basic identity of the substance.* Common physical properties include color, odor, physical state (solid, liquid, or gas), melting point, boiling point, and hardness.

During the process of determining a physical property, the physical appearance of a substance may change, but the substance's identity does not. For example, it is impossible to measure the melting point of a solid without changing the solid into a liquid. Although the liquid's appearance is much different from that of the solid, the substance is still the same; its chemical identity has not changed. Hence melting point is a physical property.

A **chemical property** *is a characteristic of a substance that describes the way the substance undergoes or resists change to form a new substance.* For example, copper objects

Margin notes summarize key information, give tips for remembering or distinguishing between similar ideas, and provide additional details and links between concepts.

Chemical properties describe the ability of a substance to form new substances, either by reaction with other substances or by decomposition. Physical properties are properties associated with a substance's physical existence. They can be determined without reference to any other substance, and determining them causes no change in the identity of the substance.

PROBLEM-SOLVING PEDAGOGY

Learning how to solve problems is a key concept in chemistry, as it is in life. The Examples
support students as they build these skills.

Within the chapters worked-out **Examples** follow the explanation of many concepts. These examples walk students through the thought process involved in problem solving, carefully outlining all the steps involved.

The value of the ideal gas constant (R) varies with the units chosen for pressure and volume. With pressure in atmospheres and volume in liters, R has the value

$$R = \frac{PV}{nT} = 0.0821 \frac{\text{atm} \cdot \text{L}}{\text{mole} \cdot \text{K}}$$

The value of R is the same for all gases under normally encountered conditions of temperature, pressure, and volume.

If three of the four variables in the ideal gas law equation are known, then the fourth can be calculated using the equation. Example 7.4 illustrates the use of the ideal gas law.

 **EXAMPLE 7.4**

Using the Ideal Gas Law to Calculate the Volume of a Gas

The colorless, odorless, tasteless gas carbon monoxide, CO, is a by-product of incomplete combustion of any material that contains the element carbon. Calculate the volume, in liters, occupied by 1.52 moles of this gas at 0.992 atm pressure and a temperature of 65°C.

Solution

This problem deals with only one set of conditions, so the ideal gas equation is applicable. Three of the four variables in the ideal gas equation (P, n, and T) are given, and the fourth (V) is to be calculated.

$$P = 0.992 \text{ atm} \qquad n = 1.52 \text{ moles}$$
$$V = ? \text{ L} \qquad T = 65°C = 338 \text{ K}$$

Rearranging the ideal gas equation to isolate V on the left side of the equation gives

$$V = \frac{nRT}{P}$$

Because the pressure is given in atmospheres and the volume unit is liters, the R value 0.0821 is valid. Substituting known numerical values into the equation gives

$$V = \frac{(1.52 \text{ moles}) \times \left(0.0821 \dfrac{\text{atm} \cdot \text{L}}{\text{mole} \cdot \text{K}}\right)(338 \text{ K})}{0.992 \text{ atm}}$$

Note that all the parts of the ideal gas constant unit cancel except for one, the volume part. Doing the arithmetic yields the volume of CO.

$$V = \left(\frac{1.52 \times 0.0821 \times 338}{0.992}\right) \text{L} = 42.5 \text{ L}$$

Practice Exercise 7.4

Calculate the volume, in liters, occupied by 3.25 moles of Cl_2 gas at 1.54 atm pressure and a temperature of 213°C.

Answer: 84.2 L Cl_2

7.8 DALTON'S LAW OF PARTIAL PRESSURES

In a mixture of gases that do not react with one another, each type of molecule moves around in the container as though the other kinds were not there. This type of behavior is possible because a gas is mostly empty space, and attractions between molecules in the gaseous state are negligible at most temperatures and pressures. Each gas in the mixture occupies the entire volume of the container; that is, it distributes itself uniformly throughout the container. The molecules of each type strike the walls of the container as frequently and with the same energy as though they were the only gas in the mixture. Consequently, the pressure exerted by each gas in a mixture is the same as it would be if the gas were alone in the same container under the same conditions.

The English scientist John Dalton (Figure 7.13) was the first to notice the independent behavior of gases in mixtures. In 1803, he published a summary statement concerning this behavior that is now known as Dalton's law of partial pressures. **Dalton's law of partial**

Examples are immediately followed by a **Practice Exercise** to reinforce the information just presented.

Figure 7.13 John Dalton (1766-1844) throughout his life had a particular interest in the study of weather. From "weather" he turned his attention to the nature of the atmosphere and then to the study of gases in general.

A sample of clean air is the most common example of a mixture of gases that do not react with one another.

COMPREHENSIVE END-OF-CHAPTER REVIEW AND PRACTICE

End-of-chapter review material and an extensive set of problems geared for different types of learners provide ample opportunity for mastery.

Concepts to Remember and **Key Reactions and Equations** provide concise review of the material presented in the chapter, helping students prepare for exam.

Concepts to Remember

Chemical bonds. Chemical bonds are the attractive forces that hold atoms together in more complex units. Chemical bonds result from the transfer of valence electrons between atoms (ionic bond) or from the sharing of electrons between atoms (covalent bond) (Section 4.1).

Valence electrons. Valence electrons, for representative elements, are the electrons in the outermost electron shell, which is the shell with the highest shell number. These electrons are particularly important in determining the bonding characteristics of a given atom (Section 4.2).

Octet rule. In compound formation, atoms of representative elements lose, gain, or share electrons in such a way that their electron configurations become identical to those of the noble gas nearest them in the periodic table (Section 4.3).

Ionic compounds. Ionic compounds commonly involve a metal atom and a nonmetal atom. Metal atoms lose one or more electrons, producing positive ions. Nonmetal atoms acquire the electrons lost by the metal atoms, producing negative ions. The oppositely charged ions attract one another, creating ionic bonds (Section 4.4).

Charge magnitude for ions. Metal atoms containing one, two, or three valence electrons tend to lose such electrons, producing ions of +1, +2, or +3 charge, respectively. Nonmetal atoms containing five, six, or seven valence electrons tend to gain electrons, producing ions of −3, −2, or −1 charge, respectively (Section 4.5).

Chemical formulas for ionic compounds. The ratio in which positive and negative ions combine is the ratio that causes the total amount of positive and negative charges to add up to zero (Section 4.7).

Structure of ionic compounds. Ionic solids consist of positive and negative ions arranged in such a way that each ion is surrounded by ions of the opposite charge (Section 4.8).

Binary ionic compound nomenclature. Binary ionic compounds are named by giving the full name of the metallic element first, followed by a separate word containing the stem of the nonmetallic element name and the suffix -ide. A Roman numeral specifying ionic charge is appended to the name of the metallic element if it is a metal that exhibits variable ionic charge (Section 4.9).

Polyatomic ions. A polyatomic ion is a group of covalently bonded atoms that has acquired a charge through the loss or gain of electrons. Polyatomic ions are very stable entities that generally maintain their identity during chemical reactions (Section 4.10).

Key Reactions and Equations

1. Number of valence electrons for representative elements (Section 4.2)

    Number of valence electrons = periodic-table group number

2. Charges on metallic monatomic ions (Section 4.5)

    Group IA metals form 1^+ ions.

    Group IIA metals form 2^+ ions.

    Group IIIA metals form 3^+ ions.

3. Charges on nonmetallic monatomic ions (Section 4.5)

    Group VA nonmetals form 3^- ions.

    Group VIA nonmetals form 2^- ions.

    Group VIIA nonmetals form 1^- ions.

EXERCISES *and* PROBLEMS

The members of each pair of problems in this section test similar material.

Types of Chemical Bonds (Section 4.1)

4.1 Contrast the two general types of chemical *bonds* in terms of the mechanism by which they form.

4.2 Contrast the two general types of chemical *compounds* in terms of their general physical properties.

Valence Electrons (Section 4.2)

4.3 How many valence electrons do atoms with the following electron configurations have?
a. $1s^2 2s^2$
b. $1s^2 2s^2 2p^6 3s^2$
c. $1s^2 2s^2 2p^6 3s^2 3p^1$
d. $1s^2 2s^2 2p^6 3s^2 3p^6 4s^2 3d^{10} 4p^2$

4.4 How many valence electrons do atoms with the following electron configurations have?
a. $1s^2 2s^2 2p^6$
b. $1s^2 2s^2 2p^6 3s^2 3p^1$
c. $1s^2 2s^2 2p^6 3s^1$
d. $1s^2 2s^2 2p^6 3s^2 3p^6 4s^2 3d^{10} 4p^5$

4.5 Give the periodic-table group number and the number of valence electrons present for each of the following representative elements.
a. $_3$Li   b. $_{10}$Ne   c. $_{20}$Ca   d. $_{53}$I

4.6 Give the periodic-table group number and the number of valence electrons present for each of the following representative elements.
a. $_{12}$Mg   b. $_{19}$K   c. $_{15}$P   d. $_{35}$Br

4.7 Write the complete electron configuration for each of the following representative elements.
a. Period 2 element with four valence electrons
b. Period 2 element with seven valence electrons
c. Period 3 element with two valence electrons
d. Period 3 element with five valence electrons

4.8 Write the complete electron configuration for each of the following representative elements.
a. Period 2 element with one valence electron
b. Period 2 element with six valence electrons
c. Period 3 element with seven valence electrons
d. Period 3 element with three valence electrons

Lewis Symbols for Atoms (Section 4.2)

4.9 Draw Lewis symbols for atoms of each of the following elements.
a. $_{12}$Mg   b. $_{19}$K   c. $_{15}$P   d. $_{36}$Kr

4.10 Draw Lewis symbols for atoms of each of the following elements.
a. $_{13}$Al   b. $_{20}$Ca   c. $_{17}$Cl   d. $_4$Be

4.11 Each of the following Lewis symbols represents a Period 2 element. Determine each element's identity.
a. X ·   b. : X :   c. Ẋ ·   d. · Ẍ·

Extensive and varied **Exercises and Problems** at the end of each chapter are organized by topic and paired. These problems always involve only a single concept. Answers to selected problems can be found at the back of the book.

4.53 What is a *formula unit* of an ionic compound?

4.54 In general terms, how many *formula units* are present in a crystal of an ionic compound?

4.55 The following drawings represent solid-state ionic compounds, with red spheres denoting positive ions and blue spheres denoting negative ions.

I II III IV

Which of these drawings could be used as a representation for each of the following ionic compounds?
a. Al_2S_3 b. MgF_2 c. K_3N d. CaO

New visual problem types are integrated throughout these problem sets and are designed to facilitate learning through the use of molecular models, pictorial representations of chemical systems, graphs, and other visual portrayals.

END-OF-CHAPTER REVIEW AND PRACTICE

(continued)

ADDITIONAL PROBLEMS

4.87 Fill in the blanks to complete the following table.

Positive Ion	Negative Ion	Chemical Formula	Name
			Magnesium hydroxide
		$BaBr_2$	
Zn^{2+}	NO_3^-		
			Iron(III) chlorate
		PbO_2	
Co^{2+}	PO_4^{3-}		
K^+	I^-		
		Cu_2SO_4	
			Lithium nitride
Al^{3+}	S^{2-}		

4.88 Fill in the blanks to complete the following table.

Chemical Symbol	Number of Protons	Number of Neutrons	Number of Electrons	Net Charge
$_{26}^{59}Fe^{3+}$				
	28	31	25	
	77	120		2+
		9	10	1-

4.89 What would be the chemical symbol for an ion with each of the following characteristics?
a. A sodium ion with ten electrons
b. A fluorine ion with ten electrons
c. A sulfur ion with two fewer protons than electrons
d. A calcium ion with two more protons than electrons

Additional Problems involve more than one concept and are more difficult than the Exercises and Problems.

4.90 Write the formula of the ionic compound that could form from the elements X and Z if
a. X has two valence electrons and Z has seven valence electrons
b. X has one valence electron and Z has six valence electrons
c. X has three valence electrons and Z has five valence electrons
d. X has six valence electrons and Z has two valence electrons

4.91 Identify the Period 3 element that most commonly produces each of the following ions.
a. X^{2-} b. X^{2+} c. X^{3-} d. X^{3+}

4.92 Indicate whether each of the following compounds contains (1) only monatomic ions, (2) only polyatomic ions, (3) both monatomic and polyatomic ions, or (4) no ions.
a. CaF_2 b. $NaNO_3$ c. NH_4CN d. AlP

4.93 Write chemical formulas (symbol and charge) for both kinds of ions present in each of the following compounds.
a. KCl b. CaS c. BeF_2 d. Al_2S_3 .

4.94 Give the chemical formula for, and the name of the compound formed from, each of the following pairs of ions.
a. Na^+ and N_3^- b. K^+ and NO_3^-
c. Mg^{2+} and O^{2-} d. NH_4^+ and PO_4^{3-}

4.95 Name each compound in the following pairs of binary ionic compounds.
a. $SnCl_4$ and $SnCl_2$ b. FeS and Fe_2S_3
c. Cu_3N and Cu_3N_2 d. NiI_2 and NiI_3

4.96 In which of the following pairs of binary ionic compounds do both members of the pair contain positive ions with the same charge?
a. Co_2O_3 and $CoCl_3$ b. Cu_2O and CuO
c. K_2O and Al_2O_3 d. MgS and NaI

4.97 Name each compound in the following pairs of polyatomic-ion-containing compounds.
a. $CuNO_3$ and $Cu(NO_3)_2$ b. $Pb_3(PO_4)_2$ and $Pb_3(PO_4)_4$
c. $Mn(CN)_3$ and $Mn(CN)_2$ d. $Co(ClO_3)_2$ and $Co(ClO_3)_3$

4.98 Write chemical formulas for the following compounds.
a. Sodium sulfide b. Sodium sulfate
c. Sodium sulfite d. Sodium thiosulfate

*M*ultiple-Choice Practice Test

The **Multiple-Choice Practice Test** at the end each chapter provides a cumulative review.

4.99 For which of the following elements is the listed number of valence electrons *correct*?
a. Mg (2 valence electrons) b. N (3 valence electrons)
c. F (1 valence electron) d. S (2 valence electrons)

4.100 Which of the following is an *incorrect* statement about the number of electrons lost or gained by a representative element during ion formation?
a. The number usually does not exceed three.
b. The number is governed by the octet rule.
c. The number is related to the position of the element in the periodic table.
d. The number is the same as the number of valence electrons present.

4.101 Which of the following is a *correct* statement concerning the mechanism for ionic bond formation?
a. Electrons are transferred from nonmetallic atoms to metallic atoms.
b. Protons are transferred from the nuclei of metallic atoms to the nuclei of nonmetallic atoms.
c. Sufficient electrons are transferred to form ions of equal but opposite charge.
d. Electron loss is always equal to electron gain.

4.102 In which of the following pairings is the chemical formula *not* consistent with the ions shown?
a. M^{2+} and X^{3-} (M_3X_2)
b. M^{2+} and X^- (MX_2)
c. M^+ and X^{3-} (MX_3)
d. M^{2+} and X^{2-} (MX)

4.103 The correct chemical formula for the ionic compound formed between Mg and O is
a. MgO b. Mg_2O_2 c. MgO_2 d. Mg_2O

4.104 In which of the following pairs of ionic compounds do both members of the pair contain positive ions with a +1 charge?
a. KCl and CaO b. Na_3N and Li_2S
c. $AlCl_3$ and MgF_2 d. BaI_2 and $BeBr_2$

4.105 The correct chemical formula for the compound aluminum nitride is
a. AlN b. AlN_2 c. Al_2N_3 d. Al_3N_2

4.106 In which of the following pairs of metals are both members of the pair variable-charge metals?
a. Na and Al b. Au and Ag
c. Cu and Zn d. Fe and Ni

4.107 In which of the following pairs of polyatomic ions do both members of the pair have the same charge?
a. ammonium and phosphate
b. sulfate and nitrate
c. cyanide and hydroxide
d. hydrogen carbonate and carbonate

4.108 Which of the following ionic compounds contains 4 atoms per formula unit?
a. Lithium nitride
b. Potassium sulfide
c. Copper(II) iodide
d. Sodium cyanide

General, Organic, and Biological Chemistry

Basic Concepts About Matter

1

Chapter Outline

Numerous physical and chemical changes in matter occur during a volcanic eruption.

In this chapter we address the question, "What exactly is chemistry about?" In addition, we consider common terminology associated with the field of chemistry. Much of this terminology is introduced in the context of the ways in which matter is classified. Like all other sciences, chemistry has its own specific language. It is necessary to restrict the meanings of some words so that all chemists (and those who study chemistry) can understand a given description of a chemical phenomenon in the same way.

 CHEMISTRY: THE STUDY OF MATTER

Chemistry *is the field of study concerned with the characteristics, composition, and transformations of matter.* What is matter? **Matter** *is anything that has mass and occupies space.* The term *mass* refers to the amount of matter present in a sample.

Matter includes all things—both living and nonliving—that can be seen (such as plants, soil, and rocks), as well as things that cannot be seen (such as air and bacteria). Various forms of energy such as heat, light, and electricity are not considered to be matter. However, chemists must be concerned with energy as well as with matter because nearly all changes that matter undergoes involve the release or absorption of energy.

The scope of chemistry is extremely broad, and it touches every aspect of our lives. An iron gate rusting, a chocolate cake baking, the diagnosis and treatment of a heart attack, the propulsion of a jet airliner, and the digesting of food all fall within the realm of chemistry. The key to understanding such diverse processes is an understanding of the fundamental nature of matter, which is what we now consider.

The universe is composed entirely of matter and energy.

1

Figure 1.1 (a) A solid has a definite shape and a definite volume. (b) A liquid has an indefinite shape—it takes the shape of its container—and a definite volume. (c) A gas has an indefinite shape and an indefinite volume—it assumes the shape and volume of its container.

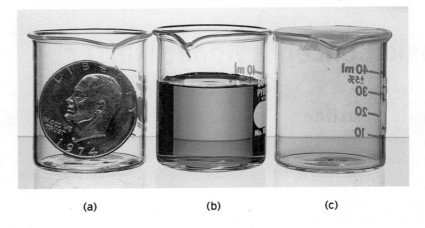

(a)   (b)   (c)

 PHYSICAL STATES OF MATTER

Three physical states exist for matter: solid, liquid, and gas. The classification of a given matter sample in terms of physical state is based on whether its shape and volume are definite or indefinite.

The volume of a sample of matter is a measure of the amount of space occupied by the sample.

Solid *is the physical state characterized by a definite shape and a definite volume.* A dollar coin has the same shape and volume whether it is placed in a large container or on a table top (Figure 1.1a). For solids in powdered or granulated forms, such as sugar or salt, a quantity of the solid takes the shape of the portion of the container it occupies, but each individual particle has a definite shape and definite volume. **Liquid** *is the physical state characterized by an indefinite shape and a definite volume.* A liquid always takes the shape of its container to the extent that it fills the container (Figure 1.1b). **Gas** *is the physical state characterized by an indefinite shape and an indefinite volume.* A gas always completely fills its container, adopting both the container's volume and its shape (Figure 1.1c).

The state of matter observed for a particular substance depends on its temperature, the surrounding pressure, and the strength of the forces holding its structural particles together. At the temperatures and pressures normally encountered on Earth, water is one of the few substances found in all three of its physical states: solid ice, liquid water, and gaseous steam (Figure 1.2). Under laboratory conditions, states other than those commonly observed can be attained for almost all substances. Oxygen, which is nearly always thought of as a gas, becomes a liquid at $-183°C$ and a solid at $-218°C$. The metal iron is a gas at extremely high temperatures (above $3000°C$).

Figure 1.2 Water can be found in the solid, liquid, and vapor (gaseous) forms simultaneously, as shown here at Yellowstone National Park.

 PROPERTIES OF MATTER

Various kinds of matter are distinguished from each other by their properties. A **property** *is a distinguishing characteristic of a substance that is used in its identification and description.* Each substance has a unique set of properties that distinguishes it from all other substances. Properties of matter are of two general types: physical and chemical.

A **physical property** *is a characteristic of a substance that can be observed without changing the basic identity of the substance.* Common physical properties include color, odor, physical state (solid, liquid, or gas), melting point, boiling point, and hardness.

During the process of determining a physical property, the physical appearance of a substance may change, but the substance's identity does not. For example, it is impossible to measure the melting point of a solid without changing the solid into a liquid. Although the liquid's appearance is much different from that of the solid, the substance is still the same; its chemical identity has not changed. Hence melting point is a physical property.

A **chemical property** *is a characteristic of a substance that describes the way the substance undergoes or resists change to form a new substance.* For example, copper objects

Chemical properties describe the ability of a substance to form new substances, either by reaction with other substances or by decomposition. Physical properties are properties associated with a substance's physical existence. They can be determined without reference to any other substance, and determining them causes no change in the identity of the substance.

"Good" Versus "Bad" Properties for a Chemical Substance

It is important not to judge the significance or usefulness of a chemical substance on the basis of just one or two of the many chemical and physical properties it exhibits. Possession of a "bad" property, such as toxicity or a strong noxious odor, does not mean that a chemical substance has nothing to contribute to the betterment of human society.

A case in point is the substance carbon monoxide. Everyone knows that it is a gaseous air pollutant present in automobile exhaust and cigarette smoke and that it is toxic to human beings. For this reason, some people automatically label carbon monoxide a "bad" substance, a substance we do not need or want.

Indeed, carbon monoxide is toxic to human beings. It impairs human health by reducing the oxygen-carrying capacity of the blood. Carbon monoxide does this by interacting with the hemoglobin in red blood cells in a way that prevents the hemoglobin from distributing oxygen throughout the body. Someone who dies from carbon monoxide poisoning actually dies from lack of oxygen.

The fact that carbon monoxide is colorless, odorless, and tasteless is very significant. Because of these properties, carbon monoxide gives no warning of its initial presence. There are several other common air pollutants that are more toxic than carbon monoxide. However, they have properties that give warning of their presence and hence they are not considered as "dangerous" as carbon monoxide.

Despite its toxicity, carbon monoxide plays an important role in the maintenance of the high standard of living we now enjoy. Its contribution lies in the field of iron metallurgy and the production of steel. The isolation of iron from iron ores, necessary for the production of steel, involves a series of high-temperature reactions, carried out in a blast furnace, in which the iron content of molten iron ores reacts with carbon monoxide. These reactions release the iron from its ores. The carbon monoxide needed in steel making is obtained by reacting coke (a product derived by heating coal to a high temperature without air being present) with oxygen.

The industrial consumption of the metal iron, both in the United States and worldwide, is approximately ten times greater than that of all other metals combined. Steel production accounts for nearly all of this demand for iron. Without steel, our standard of living would drop dramatically, and carbon monoxide is necessary for the production of steel.

Is carbon monoxide a "good" or a "bad" chemical substance? The answer to this question depends on the context in which the carbon monoxide is encountered. In terms of air pollution, it is a "bad" substance. In terms of steel making, it is a "good" substance. A similar "good–bad" dichotomy exists for almost every chemical substance.

---

turn green when exposed to moist air for long periods of time (Figure 1.3); this is a chemical property of copper. The green coating formed on the copper is a new substance that results from the copper's reaction with oxygen, carbon dioxide, and water present in air. The properties of this new substance (the green coating) are very different from those of metallic copper. On the other hand, gold objects resist change when exposed to air for long periods of time. The lack of reactivity of gold with air is a chemical property of gold.

Most often the changes associated with chemical properties result from the interaction (reaction) of a substance with one or more other substances. However, the presence of a second substance is not an absolute requirement. Sometimes the presence of energy (usually heat or light) can trigger the change called *decomposition*. That hydrogen peroxide, in the presence of either heat or light, decomposes into the substances water and oxygen is a chemical property of hydrogen peroxide.

When we specify chemical properties, we usually give conditions such as temperature and pressure because they influence the interactions between substances. For example, the gases oxygen and hydrogen are unreactive toward each other at room temperature, but they interact explosively at a temperature of several hundred degrees.

● E X A M P L E 1.1
Classifying Properties as Physical or Chemical

▶ Classify each of the following properties for selected metals as a *physical* property or a *chemical* property.

a. Iron metal rusts in an atmosphere of moist air.
b. Mercury metal is a liquid at room temperature.
c. Nickel metal dissolves in acid to produce a light green solution.
d. Potassium metal has a melting point of 63°C.

(continued)

Figure 1.3 The green color of the Statue of Liberty results from the reaction of the copper skin of the statue with the components of air. That copper will react with the components of air is a chemical property of copper.

Solution

a. Chemical property. The interaction of iron metal with moist air produces a new substance (rust).
b. Physical property. Visually determining the physical state of a substance does not produce a new substance.
c. Chemical property. A change in color indicates the formation of a new substance.
d. Physical property. Measuring the melting point of a substance does not change the substance's composition.

Practice Exercise 1.1

Classify each of the following properties for selected metals as a *physical* property or a *chemical* property.

a. Titanium metal can be drawn into thin wires.
b. Silver metal shows no sign of reaction when placed in hydrochloric acid.
c. Copper metal possesses a reddish brown color.
d. Beryllium metal, when inhaled in a finely divided form, can produce serious lung disease.

Answers: a. physical property; **b.** chemical property; **c.** physical property; **d.** chemical property

1.4 CHANGES IN MATTER

Changes in matter are common and familiar occurrences. Changes take place when food is digested, paper is burned, and a pencil is sharpened. Like properties of matter, changes in matter are classified into two categories: physical and chemical.

A **physical change** *is a process in which a substance changes its physical appearance but not its chemical composition.* A new substance is never formed as a result of a physical change.

A change in physical state is the most common type of physical change. Melting, freezing, evaporation, and condensation are all changes of state. In any of these processes, the composition of the substance undergoing change remains the same even though its physical state and appearance change. The melting of ice does not produce a new substance; the substance is water both before and after the change. Similarly, the steam produced from boiling water is still water.

A **chemical change** *is a process in which a substance undergoes a change in chemical composition.* Chemical changes always involve conversion of the material or materials under consideration into one or more new substances, each of which has properties and composition distinctly different from those of the original materials. Consider, for example, the rusting of iron objects left exposed to moist air (Figure 1.4). The reddish brown substance (the rust) that forms is a new substance with chemical properties that are obviously different from those of the original iron.

Figure 1.4 As a result of chemical change, bright steel girders become rusty when exposed to moist air.

 **EXAMPLE 1.2**

Correct Use of the Terms *Physical* and *Chemical* in Describing Changes

Complete each of the following statements about changes in matter by placing the word *physical* or *chemical* in the blank.

a. The fashioning of a piece of wood into a round table leg involves a _____ change.
b. The vigorous reaction of potassium metal with water to produce hydrogen gas is a _____ change.
c. Straightening a bent piece of iron with a hammer is an example of a _____ change.
d. The ignition and burning of a match involve a _____ change.

Solution

a. *Physical.* The table leg is still wood. No new substances have been formed.
b. *Chemical.* A new substance, hydrogen, is produced.
c. *Physical.* The piece of iron is still a piece of iron.
d. *Chemical.* New gaseous substances, as well as heat and light, are produced as the match burns.

Physical changes need not involve a change of state. Pulverizing an aspirin tablet into a powder and cutting a piece of adhesive tape into small pieces are physical changes that involve only the solid state.

Practice Exercise 1.2

Complete each of the following statements about changes in matter by placing the word *physical* or *chemical* in the blank.

a. The destruction of a newspaper through burning involves a _____ change.
b. The grating of a piece of cheese is a _____ change.
c. The heating of a blue powdered material to produce a white glassy substance and a gas is a _____ change.
d. The crushing of ice cubes to make ice chips is a _____ change.

Answers: a. chemical; **b.** physical; **c.** chemical; **d.** physical

Chemists study the nature of changes in matter to learn how to bring about favorable changes and prevent undesirable ones. The control of chemical change has been a major factor in attainment of the modern standard of living now enjoyed by most people in the developed world. The many plastics, synthetic fibers, and prescription drugs now in common use are derived via controlled chemical change.

The Chemistry at a Glance feature below reviews the ways in which the terms *physical* and *chemical* are used to describe the properties of substances and the changes

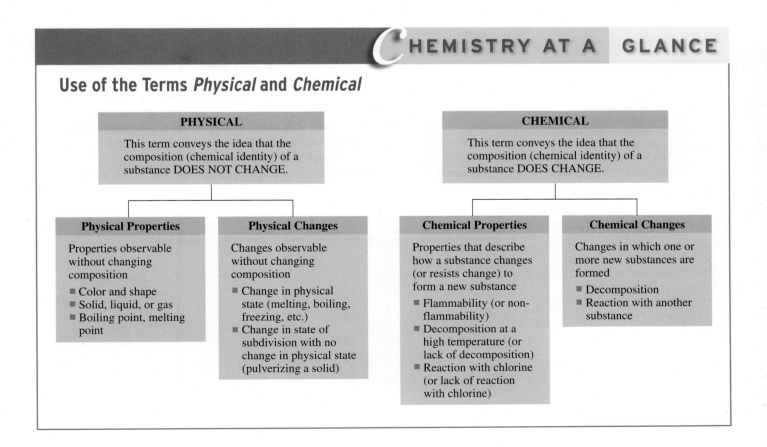

*C*HEMISTRY AT A GLANCE

Use of the Terms *Physical* and *Chemical*

PHYSICAL	**CHEMICAL**
This term conveys the idea that the composition (chemical identity) of a substance DOES NOT CHANGE.	This term conveys the idea that the composition (chemical identity) of a substance DOES CHANGE.

Physical Properties	**Physical Changes**	**Chemical Properties**	**Chemical Changes**
Properties observable without changing composition	Changes observable without changing composition	Properties that describe how a substance changes (or resists change) to form a new substance	Changes in which one or more new substances are formed
▪ Color and shape ▪ Solid, liquid, or gas ▪ Boiling point, melting point	▪ Change in physical state (melting, boiling, freezing, etc.) ▪ Change in state of subdivision with no change in physical state (pulverizing a solid)	▪ Flammability (or non-flammability) ▪ Decomposition at a high temperature (or lack of decomposition) ▪ Reaction with chlorine (or lack of reaction with chlorine)	▪ Decomposition ▪ Reaction with another substance

that substances undergo. Note that the term *physical,* used as a modifier, always conveys the idea that the composition (chemical identity) of a substance did not change, and that the term *chemical,* used as a modifier, always conveys the idea that the composition of a substance did change.

 1.5   PURE SUBSTANCES AND MIXTURES

In addition to its classification by physical state (Section 1.2), matter can also be classified in terms of its chemical composition as a pure substance or as a mixture. A **pure substance** *is a single kind of matter that cannot be separated into other kinds of matter by any physical means.* All samples of a pure substance contain only that substance and nothing else. Pure water is water and nothing else. Pure sucrose (table sugar) contains only that substance and nothing else.

A pure substance always has a definite and constant composition. This invariant composition dictates that the properties of a pure substance are always the same under a given set of conditions. Collectively, these definite and constant physical and chemical properties constitute the means by which we identify the pure substance.

A **mixture** *is a physical combination of two or more pure substances in which each substance retains its own chemical identity.* Components of a mixture retain their identity because they are physically mixed rather than chemically combined. Consider a mixture of small rock salt crystals and ordinary sand. Mixing these two substances changes neither the salt nor the sand in any way. The larger, colorless salt particles are easily distinguished from the smaller, light-gray sand granules.

One characteristic of any mixture is that its components can be separated by using physical means. In our salt–sand mixture, the larger salt crystals could be—though very tediously—"picked out" from the sand. A somewhat easier separation method would be to dissolve the salt in water, which would leave the undissolved sand behind. The salt could then be recovered by evaporation of the water. Figure 1.5a shows a heterogeneous mixture of potassium dichromate (orange crystals) and iron filings. A magnet can be used to separate the components of this mixture (Figure 1.5b).

Another characteristic of a mixture is variable composition. Numerous different salt–sand mixtures, with compositions ranging from a slightly salty sand mixture to a slightly sandy salt mixture, could be made by varying the amounts of the two components.

Mixtures are subclassified as heterogeneous or homogeneous. This subclassification is based on visual recognition of the mixture's components. A **heterogeneous mixture** *is a mixture that contains visibly different phases (parts), each of which has different properties.* A nonuniform appearance is a characteristic of all heterogeneous mixtures. Examples include chocolate chip cookies and blueberry muffins. Naturally occurring heterogeneous mixtures include rocks, soils, and wood.

A **homogeneous mixture** *is a mixture that contains only one visibly distinct phase (part), which has uniform properties throughout.* The components present in a homogeneous mixture cannot be visually distinguished. A sugar–water mixture in which all of the

Substance is a general term used to denote any variety of matter. *Pure substance* is a specific term that is applied to matter that contains only a single substance.

All samples of a pure substance, no matter what their source, have the same properties under the same conditions.

Most naturally occurring samples of matter are mixtures. Gold and diamond are two of the few naturally occurring pure substances. Despite their scarcity in nature, numerous pure substances are known. They are obtained from natural mixtures by using various types of separation techniques or are synthesized in the laboratory from naturally occurring materials.

Figure 1.5 (a) A magnet (on the left) and a mixture consisting of potassium dichromate (the orange crystals) and iron filings. (b) The magnet can be used to separate the iron filings from the potassium dichromate.

(a)

(b)

Figure 1.6 Matter falls into two basic classes: pure substances and mixtures. Mixtures, in turn, may be homogeneous or heterogeneous.

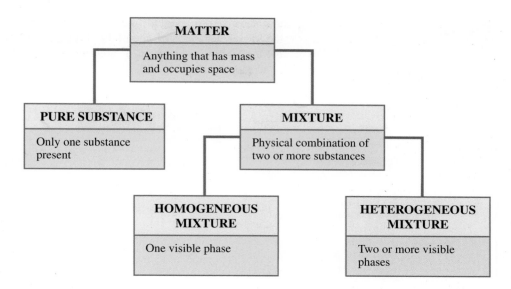

sugar has dissolved has an appearance similar to that of pure water. Air is a homogeneous mixture of gases; motor oil and gasoline are multicomponent homogeneous mixtures of liquids; and metal alloys such as 14-karat gold (a mixture of copper and gold) are examples of homogeneous mixtures of solids. The homogeneity present in solid-state metallic alloys is achieved by mixing the metals while they are in the molten state.

Figure 1.6 summarizes what we have learned thus far about various classifications of matter.

 ## 1.6 ELEMENTS AND COMPOUNDS

Chemists have isolated and characterized an estimated 9 million pure substances. A very small number of these pure substances, 117 to be exact, are different from all of the others. They are elements. All of the rest, the remaining millions, are compounds. What distinguishes an element from a compound?

An **element** *is a pure substance that cannot be broken down into simpler pure substances by chemical means such as a chemical reaction, an electric current, heat, or a beam of light.* The metals gold, silver, and copper are all elements.

A **compound** *is a pure substance that can be broken down into two or more simpler pure substances by chemical means.* Water is a compound. By means of an electric current, water can be broken down into the gases hydrogen and oxygen, both of which are elements. The ultimate breakdown products for any compound are elements. A compound's properties are always different from those of its component elements, because the elements are chemically rather than physically combined in the compound (Figure 1.7).

Both elements and compounds are pure substances.

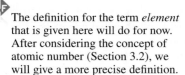

The definition for the term *element* that is given here will do for now. After considering the concept of atomic number (Section 3.2), we will give a more precise definition.

Figure 1.7 A pure substance can be either an element or a compound.

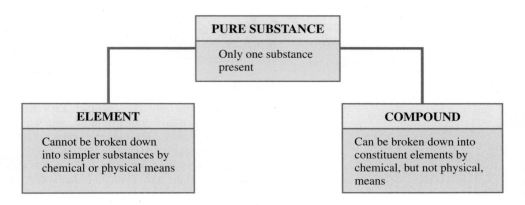

CHEMISTRY AT A GLANCE

Classes of Matter

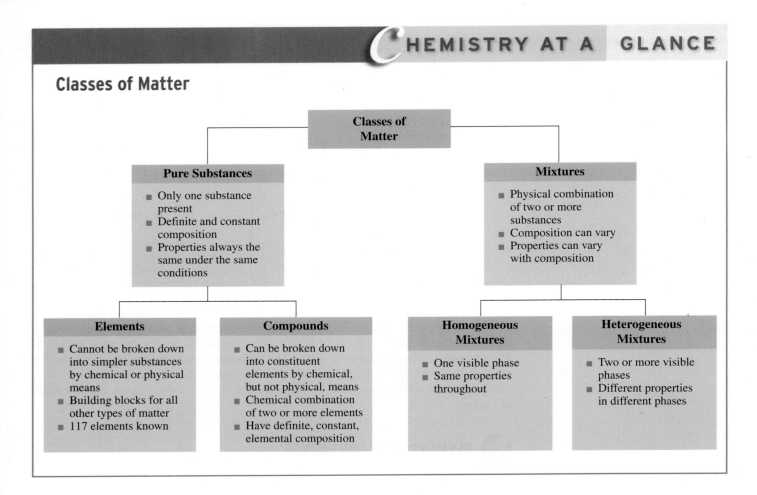

Classes of Matter

Pure Substances
- Only one substance present
- Definite and constant composition
- Properties always the same under the same conditions

Mixtures
- Physical combination of two or more substances
- Composition can vary
- Properties can vary with composition

Elements
- Cannot be broken down into simpler substances by chemical or physical means
- Building blocks for all other types of matter
- 117 elements known

Compounds
- Can be broken down into constituent elements by chemical, but not physical, means
- Chemical combination of two or more elements
- Have definite, constant, elemental composition

Homogeneous Mixtures
- One visible phase
- Same properties throughout

Heterogeneous Mixtures
- Two or more visible phases
- Different properties in different phases

Every known compound is made up of some combination of two or more of the 117 known elements. In any given compound, the elements are combined chemically in fixed proportions by mass.

Even though two or more elements are obtained from decomposition of compounds, compounds are not mixtures. Why is this so? Remember, substances can be combined either physically or chemically. Physical combination of substances produces a mixture. Chemical combination of substances produces a compound, a substance in which combining entities are *bound* together. No such binding occurs during physical combination. Example 1.3, which involves two comparisons involving locks and their keys, nicely illustrates the difference between compounds and mixtures.

The Chemistry at a Glance feature above summarizes what we have learned thus far about the subdivisions of matter called pure substances, elements, compounds, and mixtures.

● EXAMPLE 1.3

The "Composition" Difference Between a Mixture and a Compound

Consider two boxes with the following contents: the first contains 10 locks and 10 keys that fit the locks; the second contains 10 locks with each lock's key inserted into the cylinder. Which box has contents that would be an analogy for a mixture and which box has contents that would be an analogy for a compound?

Solution

The box containing the locks with their keys inserted in the cylinder represents the compound. Two objects withdrawn from this box will always be the same; each will be a lock with its associated key. Each item in the box has the same "composition."

The box containing separated locks and keys represents the mixture. Two objects withdrawn from this box need not be the same; results could be two locks, two keys, or a lock and a key. All items in the box do not have the same "composition."

There are three major property distinctions between compounds and mixtures.

1. Compounds have properties distinctly different from those of the substances that combined to form the compound. The components of mixtures retain their individual properties.
2. Compounds have a definite composition. Mixtures have a variable composition.
3. Physical methods are sufficient to separate the components of a mixture. The components of a compound cannot be separated by physical methods; chemical methods are required.

A student who attended a university in the year 1700 would have been taught that 13 elements existed. In 1750 he or she would have learned about 16 elements, in 1800 about 34, in 1850 about 59, in 1900 about 82, and in 1950 about 98. Today's total of 117 elements was reached in 2006.

Figure 1.8 Questions used in classifying matter into various categories.

Figure 1.8 summarizes the thought processes a chemist goes through in classifying a sample of matter as a heterogeneous mixture, a homogeneous mixture, an element, or a compound. This figure is based on the following three questions about a sample of matter:

1. Does the sample of matter have the same properties throughout?
2. Are two or more different substances present?
3. Can the pure substance be broken down into simpler substances?

1.7 DISCOVERY AND ABUNDANCE OF THE ELEMENTS

The discovery and isolation of the 117 known elements, the building blocks for all matter, have taken place over a period of several centuries. Most of the discoveries have occurred since 1700, the 1800s being the most active period.

Eighty-eight of the 117 elements occur naturally, and 29 have been synthesized in the laboratory by bombarding samples of naturally occurring elements with small particles. Figure 1.9 shows samples of selected naturally occurring elements. The synthetic (laboratory-produced) elements are all unstable (radioactive) and usually revert quickly back to naturally occurring elements (see Section 11.5).

The naturally occurring elements are not evenly distributed on Earth and in the universe. What is startling is the nonuniformity of the distribution. A small number of elements account for the majority of elemental particles (atoms). (An atom is the smallest particle of an element that can exist; see Section 1.9.)

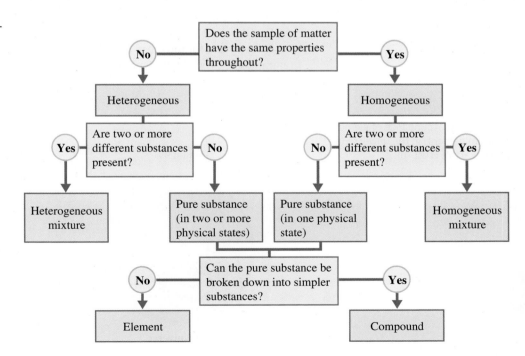

Figure 1.9 Outward physical appearance of selected naturally occurring elements. *Center:* Sulfur. *From upper right, clockwise:* Arsenic, iodine, magnesium, bismuth, and mercury.

Any increase in the number of known elements from 117 will result from the production of additional synthetic elements. Current chemical theory strongly suggests that all naturally occurring elements have been identified. The isolation of the last of the known naturally occurring elements, rhenium, occurred in 1925.

Studies of the radiation emitted by stars enable scientists to estimate the elemental composition of the universe (Figure 1.10a). Results indicate that two elements, hydrogen and helium, are absolutely dominant. All other elements are mere "impurities" when their abundances are compared with those of these two dominant elements. In this big picture, in which Earth is but a tiny microdot, 91% of all elemental particles (atoms) are hydrogen, and nearly all of the remaining 9% are helium.

If we narrow our view to the chemical world of humans—Earth's crust (its waters, atmosphere, and outer solid surface)—a different perspective emerges. Again, two elements dominate, but this time they are oxygen and silicon. Figure 1.10b provides information on elemental abundances for Earth's crust. The numbers given are atom percents—that is, the percentage of total atoms that are of a given type. Note that the eight elements listed (the only elements with atom percents greater than 1%) account for over 98% of total atoms in Earth's crust. Note also the dominance of oxygen and silicon; these two elements account for 80% of the atoms that make up the chemical world of humans.

Oxygen, the most abundant element in Earth's crust, was isolated in pure form for the first time in 1774 by the English chemist and theologian Joseph Priestly (1733–1804). Discovery years for the other "top five" elements of Earth's crust are 1824 (silicon), 1827 (aluminum), 1766 (hydrogen), and 1808 (calcium).

Figure 1.10 Abundance of elements (in atom percent) in the universe (a) and in Earth's crust (b).

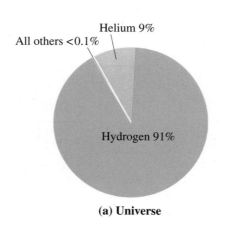

Helium 9%
All others <0.1%
Hydrogen 91%

(a) Universe

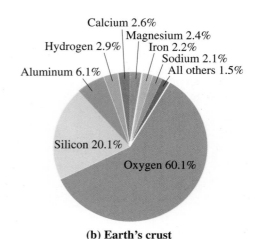

Calcium 2.6%
Magnesium 2.4%
Hydrogen 2.9%
Iron 2.2%
Sodium 2.1%
Aluminum 6.1%
All others 1.5%
Silicon 20.1%
Oxygen 60.1%

(b) Earth's crust

Elemental Composition of the Human Body

The distribution of elements in the human body and other living systems is very different from that found in Earth's crust. This distribution is the result of living systems *selectively* taking up matter from their external environment rather than simply accumulating matter representative of their surroundings. Food intake constitutes the primary selective intake process.

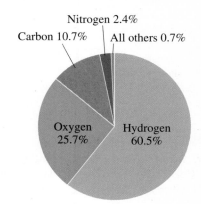

Nitrogen 2.4%
Carbon 10.7%   All others 0.7%
Oxygen 25.7%   Hydrogen 60.5%

Only four elements are found in the human body at atom percent levels greater than 1%.

Hydrogen, carbon, and nitrogen are all much more abundant than in Earth's crust (Figure 1.10b), and oxygen is significantly less abundant than in Earth's crust.

The dominance of hydrogen and oxygen in the human body reflects its high water content. Hydrogen is over twice as abundant as oxygen, largely because water contains hydrogen and oxygen in a 2-to-1 atom ratio.

Carbohydrates, fats, and proteins, nutrients required by the human body in large amounts, are all sources of carbon, hydrogen, and oxygen. Proteins are the body's primary nitrogen source.

	Hydrogen	Oxygen	Carbon	Nitrogen
Water				
Carbohydrate				
Fat				
Protein				

 ## 1.8 NAMES AND CHEMICAL SYMBOLS OF THE ELEMENTS

Each element has a unique name that, in most cases, was selected by its discoverer. A wide variety of rationales for choosing a name have been applied. Some elements bear geographical names; germanium is named after the native country of its German discoverer, and the elements francium and polonium are named after France and Poland. The elements mercury, uranium, and neptunium are all named for planets. Helium gets its name from the Greek word *helios,* for "sun," because it was first observed spectroscopically in the sun's corona during an eclipse. Some elements carry names that reflect specific properties of the element or of the compounds that contain it. Chlorine's name is derived from the Greek *chloros,* denoting "greenish-yellow," the color of chlorine gas. Iridium gets its name from the Greek *iris,* meaning "rainbow"; this alludes to the varying colors of the compounds from which it was isolated.

Abbreviations called chemical symbols also exist for the names of the elements. A **chemical symbol** *is a one- or two-letter designation for an element derived from the element's name.* These chemical symbols are used more frequently than the elements' names. Chemical symbols can be written more quickly than the names, and they occupy less space. A list of the known elements and their chemical symbols is given in Table 1.1. The chemical symbols and names of the more frequently encountered elements are shown in color in this table.

Note that the first letter of a chemical symbol is always capitalized and the second is not. Two-letter chemical symbols are often, but not always, the first two letters of the element's name.

Eleven elements have chemical symbols that bear no relationship to the element's English-language name. In ten of these cases, the symbol is derived from the Latin name of the element; in the case of the element tungsten, a German name is the symbol's source. Most of these elements have been known for hundreds of years and date back to the time when Latin was the language of scientists. Elements whose chemical symbols are derived from non-English names are marked with an asterisk in Table 1.1.

Learning the chemical symbols of the more common elements is an important key to success in studying chemistry. Knowledge of chemical symbols is essential for writing chemical formulas (Section 1.10) and chemical equations (Section 6.6).

TABLE 1.1
The Chemical Symbols for the Elements
The names and symbols of the more frequently encountered elements are shown in color.

Ac	actinium	Gd	gadolinium	Po	polonium
Ag	silver*	Ge	germanium	Pr	praseodymium
Al	aluminum	H	hydrogen	Pt	platinum
Am	americium	He	helium	Pu	plutonium
Ar	argon	Hf	hafnium	Ra	radium
As	arsenic	Hg	mercury*	Rb	rubidium
At	astatine	Ho	holmium	Re	rhenium
Au	gold*	Hs	hassium	Rf	rutherfordium
B	boron	I	iodine	Rg	roentgenium
Ba	barium	In	indium	Rh	rhodium
Be	beryllium	Ir	iridium	Rn	radon
Bh	bohrium	K	potassium*	Ru	ruthenium
Bi	bismuth	Kr	krypton	S	sulfur
Bk	berkelium	La	lanthanum	Sb	antimony*
Br	bromine	Li	lithium	Sc	scandium
C	carbon	Lr	lawrencium	Se	selenium
Ca	calcium	Lu	lutetium	Sg	seaborgium
Cd	cadmium	Md	mendelevium	Si	silicon
Ce	cerium	Mg	magnesium	Sm	samarium
Cf	californium	Mn	manganese	Sn	tin*
Cl	chlorine	Mo	molybdenum	Sr	strontium
Cm	curium	Mt	meitnerium	Ta	tantalum
Co	cobalt	N	nitrogen	Tb	terbium
Cr	chromium	Na	sodium*	Tc	technetium
Cs	cesium	Nb	niobium	Te	tellurium
Cu	copper*	Nd	neodymium	Th	thorium
Db	dubnium	Ne	neon	Ti	titanium
Ds	darmstadtium	Ni	nickel	Tl	thallium
Dy	dysprosium	No	nobelium	Tm	thulium
Er	erbium	Np	neptunium	U	uranium
Es	einsteinium	O	oxygen	V	vanadium
Eu	europium	Os	osmium	W	tungsten*
F	fluorine	P	phosphorus	Xe	xenon
Fe	iron*	Pa	protactinium	Y	yttrium
Fm	fermium	Pb	lead*	Yb	ytterbium
Fr	francium	Pd	palladium	Zn	zinc
Ga	gallium	Pm	promethium	Zr	zirconium

Only 111 elements are listed in this table. The remaining six elements, discovered (synthesized) between 1996 and 2006, are yet to be named.

* These elements have symbols that were derived from non-English names.

Figure 1.11 A computer reconstruction of the surface of a crystal as observed with a scanning tunneling microscope. The image reveals a regular pattern of individual atoms. The color was added to the image by the computer and is used to show that two different kinds of atoms are present.

1.9   ATOMS AND MOLECULES

Consider the process of subdividing a sample of the element gold (or any other element) into smaller and smaller pieces. It seems reasonable that eventually a "smallest possible piece" of gold would be reached that could not be divided further and still be the element gold. This smallest possible unit of gold is called a gold atom. An **atom** is *the smallest particle of an element that can exist and still have the properties of the element.*

A sample of any element is composed of atoms of a single type, those of that element. In contrast, a compound must have two or more types of atoms present, because by definition at least two elements must be present (Section 1.6).

No one ever has seen or ever will see an atom with the naked eye; they are simply too small for such observation. However, sophisticated electron microscopes, with magnification factors in the millions, have made it possible to photograph "images" of individual atoms (Figure 1.11).

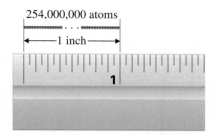

254,000,000 atoms

1 inch

1

Figure 1.12 254 million atoms arranged in a straight line would extend a distance of approximately 1 inch.

Reasons for the tendency of atoms to assemble into molecules and information on the binding forces involved are considered in Chapter 4.

The Latin word *mole* means "a mass." The word *molecule* denotes "a little mass."

The concept that heteroatomic molecules are the building blocks for *all* compounds will have to be modified when certain solids, called ionic solids, are considered in Section 4.8.

Atoms are incredibly small particles. Atomic dimensions, although not directly measurable, can be calculated from measurements made on large-size samples of elements. The diameter of an atom is approximately four-billionths of an inch. If atoms of such diameter were arranged in a straight line, it would take 254 million of them to extend a distance of 1 inch (see Figure 1.12).

Free atoms are rarely encountered in nature. Instead, under normal conditions of temperature and pressure, atoms are almost always found together in aggregates or clusters ranging in size from two atoms to numbers too large to count. When the group or cluster of atoms is relatively small and bound together tightly, the resulting entity is called a molecule. A **molecule** *is a group of two or more atoms that functions as a unit because the atoms are tightly bound together.* This resultant "package" of atoms behaves in many ways as a single, distinct particle would.

A **diatomic molecule** *is a molecule that contains two atoms.* It is the simplest type of molecule that can exist. Next in complexity are triatomic molecules. A **triatomic molecule** *is a molecule that contains three atoms.* Continuing on numerically, we have *tetraatomic* molecules, *pentatomic* molecules, and so on.

The atoms present in a molecule may all be of the same kind, or two or more kinds may be present. On the basis of this observation, molecules are classified into two categories: *homoatomic* and *heteroatomic.* A **homoatomic molecule** *is a molecule in which all atoms present are of the same kind.* A substance containing homoatomic molecules must be an element. The fact that homoatomic molecules exist indicates that *individual* atoms are not always the preferred structural unit for an element. The gaseous elements hydrogen, oxygen, nitrogen, and chlorine exist in the form of diatomic molecules. There are four atoms present in a gaseous phosphorus molecule and eight atoms present in a gaseous sulfur molecule (see Figure 1.13).

A **heteroatomic molecule** *is a molecule in which two or more kinds of atoms are present.* Substances that contain heteroatomic molecules must be compounds because the presence of two or more kinds of atoms reflects the presence of two or more kinds of elements. The number of atoms present in the heteroatomic molecules associated with compounds varies over a wide range. A water molecule contains 3 atoms: 2 hydrogen atoms and 1 oxygen atom. The compound sucrose (table sugar) has a much larger molecule: 45 atoms are present, of which 12 are carbon atoms, 22 are hydrogen atoms, and 11 are oxygen atoms. Figure 1.14 shows general models for four simple types of heteroatomic molecules. Comparison of parts (c) and (d) of this figure shows that molecules with the same number of atoms need not have the same arrangement of atoms.

A molecule is the smallest particle of a compound capable of a stable independent existence. Continued subdivision of a quantity of table sugar to yield smaller and smaller amounts would ultimately lead to the isolation of one single "unit" of table sugar: a molecule of table sugar. This table sugar molecule could not be broken down any further and still exhibit the physical and chemical properties of table sugar. The table sugar molecule could be broken down further by chemical (not physical) means to produce atoms, but if that occurred, we would no longer have table sugar. The *molecule* is the limit of *physical* subdivision. The *atom* is the limit of *chemical* subdivision.

Figure 1.13 Molecular structure of (a) chlorine molecule, (b) phosphorus molecule, and (c) sulfur molecule.

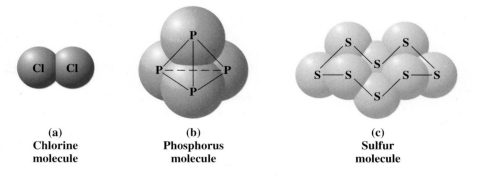

(a)
Chlorine molecule

(b)
Phosphorus molecule

(c)
Sulfur molecule

Figure 1.14 Depictions of various simple heteroatomic molecules using models. Spheres of different sizes and colors represent different kinds of atoms.

(a) A diatomic molecule containing one atom of A and one atom of B

(b) A triatomic molecule containing two atoms of A and one atom of B

(c) A tetraatomic molecule containing two atoms of A and two atoms of B

(d) A tetraatomic molecule containing three atoms of A and one atom of B

 EXAMPLE 1.4

Classifying Molecules on the Basis of Numbers of and Types of Atoms

Classify each of the following molecules as (1) *diatomic, triatomic,* etc., (2) *homoatomic* or *heteroatomic,* and (3) representing an *element* or a *compound.*

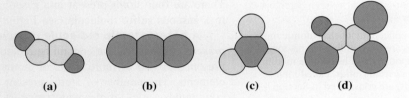

(a)   (b)   (c)   (d)

Solution

a. Tetraatomic (four atoms); heteroatomic (two kinds of atoms); a compound (two kinds of atoms)

b. Triatomic (three atoms); homoatomic (only one kind of atom); an element (one kind of atom)

c. Tetraatomic (four atoms); heteroatomic (two kinds of atoms); a compound (two kinds of atoms)

d. Hexatomic (six atoms); heteroatomic (three kinds of atoms); a compound (three kinds of atoms)

Practice Exercise 1.4

Classify each of the following molecules as (1) *diatomic, triatomic,* etc., (2) *homoatomic* or *heteroatomic,* and (3) representing an *element* or a *compound.*

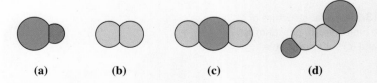

(a)   (b)   (c)   (d)

Answers: **a.** diatomic (two atoms); heteroatomic (two kinds of atoms); compound (two kinds of atoms); **b.** diatomic (two atoms); homoatomic (one kind of atom); element (one kind of atom); **c.** triatomic (three atoms); heteroatomic (two kinds of atoms); compound (two kinds of atoms); **d.** tetraatomic (four atoms); heteroatomic (three kinds of atoms); compound (three kinds of atoms)

1.10 CHEMICAL FORMULAS

Information about compound composition can be presented in a concise way by using a chemical formula. A **chemical formula** *is a notation made up of the chemical symbols of the elements present in a compound and numerical subscripts (located to the right of each chemical symbol) that indicate the number of atoms of each element present in a molecule of the compound.*

The chemical formula for the compound aspirin is $C_9H_8O_4$. This chemical formula conveys the information that an aspirin molecule contains three different elements—carbon (C), hydrogen (H), and oxygen (O)—and 21 atoms—9 carbon atoms, 8 hydrogen atoms, and 4 oxygen atoms.

When only one atom of a particular element is present in a molecule of a compound, that element's symbol is written without a numerical subscript in the formula for the compound. The formula for rubbing alcohol, C_3H_8O, reflects this practice for the element oxygen.

In order to write formulas correctly, one must follow the capitalization rules for elemental symbols (Section 1.8). Making the error of capitalizing the second letter of an element's symbol can dramatically alter the meaning of a chemical formula. The formulas $CoCl_2$ and $COCl_2$ illustrate this point; the symbol Co stands for the element cobalt, whereas CO stands for one atom of carbon and one atom of oxygen.

Sometimes chemical formulas contain parentheses; an example is $Al_2(SO_4)_3$. The interpretation of this formula is straightforward; in a formula unit, there are present 2 aluminum (Al) atoms and 3 SO_4 groups. The subscript following the parentheses always indicates the number of units in the formula of the polyatomic entity inside the parentheses. In terms of atoms, the formula $Al_2(SO_4)_3$ denotes 2 aluminum (Al) atoms, $3 \times 1 = 3$ sulfur (S) atoms, and $3 \times 4 = 12$ oxygen (O) atoms. Example 1.5 contains further comments about chemical formulas that contain parentheses.

Further information about the use of parentheses in chemical formulas (when and why) will be presented in Section 4.11. The important concern now is being able to interpret chemical formulas that contain parentheses in terms of total atoms present.

EXAMPLE 1.5

Interpreting Chemical Formulas

For each of the following chemical formulas, determine how many atoms of each element are present in one molecule of the compound.

a. HCN—hydrogen cyanide, a poisonous gas
b. $C_{18}H_{21}NO_3$—codeine, a pain-killing drug
c. $Ca_{10}(PO_4)_6(OH)_2$—hydroxyapatite, present in tooth enamel

Solution

a. One atom each of the elements hydrogen, carbon, and nitrogen is present. Remember that the subscript 1 is implied when no subscript is written.
b. This formula indicates that 18 carbon atoms, 21 hydrogen atoms, 1 nitrogen atom, and 3 oxygen atoms are present in one molecule of the compound.
c. There are 10 calcium atoms. The amounts of phosphorus, hydrogen, and oxygen are affected by the subscripts outside the parentheses. There are 6 phosphorus atoms and 2 hydrogen atoms present. Oxygen atoms are present in two locations in the formula. There are a total of 26 oxygen atoms: 24 from the PO_4 subunits (6×4) and 2 from the OH subunits (2×1).

Practice Exercise 1.5

For each of the following chemical formulas, determine how many atoms of each element are present in one molecule of the compound.

a. H_2SO_4—sulfuric acid, an industrial acid
b. $C_{17}H_{20}N_4O_6$—riboflavin, a B vitamin
c. $Ca(NO_3)_2$—calcium nitrate, used in fireworks to give a reddish color

Answers: a. 2 hydrogen atoms, 1 sulfur atom, and 4 oxygen atoms; **b.** 17 carbon atoms, 20 hydrogen atoms, 4 nitrogen atoms, and 6 oxygen atoms; **c.** 1 calcium atom, 2 nitrogen atoms, and 6 oxygen atoms

CONCEPTS TO REMEMBER

Chemistry. Chemistry is the field of study that is concerned with the characteristics, composition, and transformations of matter (Section 1.1).

Matter. Matter, the substances of the physical universe, is anything that has mass and occupies space. Matter exists in three physical states: solid, liquid, and gas (Section 1.2).

Properties of matter. Properties, the distinguishing characteristics of a substance that are used in its identification and description, are of two types: physical and chemical. Physical properties are properties that can be observed without changing a substance into another substance. Chemical properties are properties that matter exhibits as it undergoes or resists changes in chemical composition. The failure of a substance to undergo change in the presence of another substance is considered a chemical property (Section 1.3).

Changes in matter. Changes that can occur in matter are classified into two types: physical and chemical. A physical change is a process that does not alter the basic nature (chemical composition) of the substance under consideration. No new substances are ever formed as a result of a physical change. A chemical change is a process that involves a change in the basic nature (chemical composition) of the substance. Such changes always involve conversion of the material or materials under consideration into one or more new substances that have properties and composition distinctly different from those of the original materials (Section 1.4).

Pure substances and mixtures. All specimens of matter are either pure substances or mixtures. A pure substance is a form of matter that has a definite and constant composition. A mixture is a physical combination of two or more pure substances in which the pure substances retain their identity (Section 1.5).

Types of mixtures. Mixtures can be classified as heterogeneous or homogeneous on the basis of the visual recognition of the components present. A heterogeneous mixture contains visibly different parts or phases, each of which has different properties. A homogeneous mixture contains only one phase, which has uniform properties throughout (Section 1.5).

Types of pure substances. A pure substance can be classified as either an element or a compound on the basis of whether it can be broken down into two or more simpler substances by chemical means. Elements cannot be broken down into simpler substances. Compounds yield two or more simpler substances when broken down. There are 117 pure substances that qualify as elements. There are millions of compounds (Section 1.6).

Chemical symbols. Chemical symbols are a shorthand notation for the names of the elements. Most consist of two letters; a few involve a single letter. The first letter of a chemical symbol is always capitalized, and the second letter is always lowercase (Section 1.8).

Atoms and molecules. An atom is the smallest particle of an element that can exist and still have the properties of the element. Free isolated atoms are rarely encountered in nature. Instead, atoms are almost always found together in aggregates or clusters. A molecule is a group of two or more atoms that functions as a unit because the atoms are tightly bound together (Section 1.9).

Types of molecules. Molecules are of two types: homoatomic and heteroatomic. Homoatomic molecules are molecules in which all atoms present are of the same kind. A pure substance containing homoatomic molecules is an element. Heteroatomic molecules are molecules in which two or more different kinds of atoms are present. Pure substances that contain heteroatomic molecules must be compounds (Section 1.9).

Chemical formulas. Chemical formulas are used to specify compound composition in a concise manner. They consist of the symbols of the elements present in the compound and numerical subscripts (located to the right of each symbol) that indicate the number of atoms of each element present in a molecule of the compound (Section 1.10).

EXERCISES *and* PROBLEMS

The members of each pair of problems in this section test similar material.

● Chemistry: The Study of Matter (Section 1.1)

1.1   What are the two general characteristics that all types of matter possess?

1.2   What are the three aspects of matter that are of particular interest to chemists?

1.3   Classify each of the following as matter or energy (nonmatter).
a. Air   b. Pizza   c. Sound
d. Light   e. Gold   f. Virus

1.4   Classify each of the following as matter or energy (nonmatter).
a. Electricity   b. Bacteria   c. Silver
d. Cake   e. Water   f. Magnetism

● Physical States of Matter (Section 1.2)

1.5   Give a characteristic that distinguishes
a. liquids from solids   b. gases from liquids

1.6   Give a characteristic that is the same for
a. liquids and solids   b. gases and liquids

1.7   Indicate whether each of the following substances does or does not take the shape of its container and also whether it has a definite volume.

a. Copper wire   b. Oxygen gas
c. Granulated sugar   d. Liquid water

1.8   Indicate whether each of the following substances does or does not take the shape of its container and also whether it has an indefinite volume.
a. Aluminum powder   b. Carbon dioxide gas
c. Clean air   d. Gasoline

● Properties of Matter (Section 1.3)

1.9   The following are properties of the substance magnesium. Classify each property as physical or chemical.
a. Solid at room temperature
b. Ignites upon heating in air
c. Hydrogen gas is produced when it is dissolved in acids
d. Has a density of 1.738 g/cm^3 at 20°C

1.10 The following are properties of the substance magnesium. Clasify each property as physical or chemical.
a. Silvery-white in color
b. Does not react with cold water
c. Melts at 651°C
d. Finely divided form burns in oxygen with a dazzling white flame

1.11 Indicate whether each of the following statements describes a physical or a chemical property.
   a. Silver salts discolor the skin by reacting with skin protein.
   b. Hemoglobin molecules have a red color.
   c. Beryllium metal vapor is extremely toxic to humans.
   d. Aspirin tablets can be pulverized with a hammer.

1.12 Indicate whether each of the following statements describes a physical or a chemical property.
   a. Diamonds are very hard substances.
   b. Gold metal does not react with nitric acid.
   c. Lithium metal is light enough to float on water.
   d. Mercury is a liquid at room temperature.

⬤ **Changes in Matter (Section 1.4)**

1.13 Classify each of the following changes as physical or chemical.
   a. Crushing a dry leaf
   b. Hammering a metal into a thin sheet
   c. Burning your chemistry textbook
   d. Slicing a ham

1.14 Classify each of the following changes as physical or chemical.
   a. Evaporation of water from a lake
   b. "Scabbing over" of a skin cut
   c. Cutting a string into two pieces
   d. Melting of some candle wax

1.15 Classify each of the following changes as physical or chemical.
   a. A match burns
   b. "Rubbing alcohol" evaporates
   c. A copper object turns green over time
   d. A pan of water boils

1.16 Classify each of the following changes as physical or chemical.
   a. A newspaper page turns yellow over time
   b. A rubber band breaks
   c. A firecracker explodes
   d. Dry ice "disappears" over time

1.17 Correctly complete each of the following sentences by placing the word *chemical* or *physical* in the blank.
   a. The freezing over of a pond's surface is a _____ process.
   b. The crushing of some ice to make ice chips is a _____ procedure.
   c. The destruction of a newspaper through burning it is a _____ process.
   d. Pulverizing a hard sugar cube using a mallet is a _____ procedure.

1.18 Correctly complete each of the following sentences by placing the word *chemical* or *physical* in the blank.
   a. The reflection of light by a shiny metallic object is a _____ process.
   b. The heating of a blue powdered material to produce a white glassy-type substance and a gas is a _____ procedure.
   c. A burning candle produces light by _____ means.
   d. The grating of a piece of cheese is a _____ technique.

⬤ **Pure Substances and Mixtures (Section 1.5)**

1.19 Classify each of the following statements as true or false.
   a. All heterogeneous mixtures must contain three or more substances.
   b. Pure substances cannot have a variable composition.
   c. Substances maintain their identity in a heterogeneous mixture but not in a homogeneous mixture.
   d. Pure substances are seldom encountered in the "everyday" world.

1.20 Classify each of the following statements as true or false.
   a. All homogeneous mixtures must contain at least two substances.
   b. Heterogeneous mixtures, but not homogeneous mixtures, can have a variable composition.
   c. Pure substances cannot be separated into other kinds of matter by physical means.
   d. The number of known pure substances is less than 100,000.

1.21 Assign each of the following descriptions of matter to one of the following categories: *heterogeneous mixture, homogeneous mixture,* or *pure substance.*
   a. Two substances present, two phases present
   b. Two substances present, one phase present
   c. One substance present, two phases present
   d. Three substances present, three phases present

1.22 Assign each of the following descriptions of matter to one of the following categories: *heterogeneous mixture, homogeneous mixture,* or *pure substance.*
   a. Three substances present, one phase present
   b. One substance present, three phases present
   c. One substance present, one phase present
   d. Two substances present, three phases present

1.23 Classify each of the following as a *heterogeneous mixture,* a *homogeneous mixture,* or a *pure substance.* Also indicate how many phases are present, assuming all components are present in the same container.
   a. Water and dissolved salt
   b. Water and sand
   c. Water, ice, and oil
   d. Carbonated water (soda water) and ice

1.24 Classify each of the following as a *heterogeneous mixture,* a *homogeneous mixture,* or a *pure substance.* Also indicate how many phases are present, assuming all components are present in the same container.
   a. Water and dissolved sugar
   b. Water and oil
   c. Water, wax, and pieces of copper metal
   d. Salt water and sugar water

⬤ **Elements and Compounds (Section 1.6)**

1.25 From the information given, classify each of the pure substances A through D as elements or compounds, or indicate that no such classification is possible because of insufficient information.
   a. Analysis with an elaborate instrument indicates that substance A contains two elements.
   b. Substance B decomposes upon heating.
   c. Heating substance C to 1000°C causes no change in it.
   d. Heating substance D to 500°C causes it to change from a solid to a liquid.

1.26 From the information given, classify each of the pure substances A through D as elements or compounds, or indicate that no such classification is possible because of insufficient information.
   a. Substance A cannot be broken down into simpler substances by chemical means.
   b. Substance B cannot be broken down into simpler substances by physical means.
   c. Substance C readily dissolves in water.
   d. Substance D readily reacts with the element chlorine.

1.27 From the information given in the following equations, classify each of the pure substances A through G as elements or compounds, or indicate that no such classification is possible because of insufficient information.
   a. A + B → C   b. D → E + F + G

1.28 From the information given in the following equations, classify each of the pure substances A through G as elements or compounds, or indicate that no such classification is possible because of insufficient information.
  a. A → B + C   b. D + E → F + G

1.29 Indicate whether each of the following statements is true or false.
  a. Both elements and compounds are pure substances.
  b. A compound results from the physical combination of two or more elements.
  c. In order for matter to be heterogeneous, at least two compounds must be present.
  d. Compounds, but not elements, can have a variable composition.

1.30 Indicate whether each of the following statements is true or false.
  a. Compounds can be separated into their constituent elements by chemical means.
  b. Elements can be separated into their constituent compounds by physical means.
  c. A compound must contain at least two elements.
  d. A compound is a physical mixture of different elements.

● **Discovery and Abundance of the Elements (Section 1.7)**

1.31 Indicate whether each of the following statements about elements is true or false.
  a. Elements that do not occur in nature have been produced in a laboratory setting.
  b. At present 108 elements are known.
  c. Current chemical theory suggests there are more naturally occurring elements yet to be discovered.
  d. More laboratory-produced elements exist than naturally occuring elements.

1.32 Indicate whether each of the following statements about elements is true or false.
  a. The majority of the known elements have been discovered since 1990.
  b. New naturally occuring elements have been identified within the past 10 years.
  c. More than 25 laboratory-produced elements are known.
  d. All laboratory-produced elements are unstable.

1.33 Indicate whether each of the following statements about elemental abundances is true or false.
  a. Silicon is the second most abundant element in Earth's crust.
  b. Hydrogen is the most abundant element in the universe but not in Earth's crust.
  c. Oxygen and hydrogen are the two most abundant elements in the universe.
  d. One element accounts for over one-half of the atoms in Earth's crust.

1.34 Indicate whether each of the following statements about elemental abundances is true or false.
  a. Hydrogen is the most abundant element in both Earth's crust and the universe.
  b. Oxygen and silicon are the two most abundant elements in the universe.
  c. Helium is the second most abundant element in Earth's crust.
  d. Two elements account for over three-fourths of the atoms in Earth's crust.

1.35 With the help of Figure 1.10 indicate whether the first listed element in each of the given pairs of elements is more abundant or less abundant in Earth's crust, in terms of atom percent, than the second listed element.
  a. Silicon and aluminum   b. Calcium and hydrogen
  c. Iron and oxygen   d. Sodium and potassium

1.36 With the help of Figure 1.10 indicate whether the first listed element in each of the given pairs of elements is more abundant or less abundant in Earth's crust, in terms of atom percent, than the second listed element.
  a. Oxygen and hydrogen   b. Iron and aluminum
  c. Calcium and magnesium   d. Copper and sodium

● **Names and Chemical Symbols of the Elements (Section 1.8)**

1.37 Give the name of the element denoted by each of the following chemical symbols.
  a. N   b. Ni   c. Pb   d. Sn

1.38 Give the name of the element denoted by each of the following chemical symbols.
  a. Li   b. He   c. F   d. Zn

1.39 Give the chemical symbol for each of the following elements.
  a. Aluminum   b. Neon
  c. Hydrogen   d. Uranium

1.40 Give the chemical symbol for each of the following elements.
  a. Mercury   b. Chlorine
  c. Gold   d. Beryllium

1.41 Write the chemical symbol for each member of the following pairs of elements.
  a. Sodium and sulfur   b. Magnesium and manganese
  c. Calcium and cadmium   d. Arsenic and argon

1.42 Write the chemical symbol for each member of the following pairs of elements.
  a. Copper and cobalt   b. Potassium and phosphorus
  c. Iron and iodine   d. Silicon and silver

1.43 In which of the following sequences of elements do all the elements have two-letter symbols?
  a. Magnesium, nitrogen, phosphorus
  b. Bromine, iron, calcium
  c. Aluminum, copper, chlorine
  d. Boron, barium, beryllium

1.44 In which of the following sequences of elements do all the elements have symbols that start with a letter that is *not* the first letter of the element's English name?
  a. Silver, gold, mercury
  b. Copper, helium, neon
  c. Cobalt, chromium, sodium
  d. Potassium, iron, lead

● **Atoms and Molecules (Section 1.9)**

1.45 Classify the substances represented by the following models as *homoatomic* or *heteroatomic* molecules.
  a.

  b.

  c.

  d.

1.46 Classify the substances represented by the following models as *homoatomic* or *heteroatomic* molecules.

a.

b.

c.

d.

1.47 Classify the substances represented by the models in Problem 1.45 as to the number of atoms present per molecule, that is, as *diatomic, triatomic, tetraatomic,* etc.

1.48 Classify the substances represented by the models in Problem 1.46 as to the number of atoms present per molecule, that is, as *diatomic, triatomic, tetraatomic,* etc.

1.49 Classify the substances represented by the models in Problem 1.45 as to type of pure substance, that is, as *element* or *compound*.

1.50 Classify the substances represented by the models in Problem 1.46 as to type of pure substance, that is, as *element* or *compound*.

1.51 Indicate whether each of the following statements is true or false. If a statement is false, change it to make it true. (Such a rewriting should involve more than merely converting the statement to the negative of itself.)

a. The atom is the limit of chemical subdivision for both elements and compounds.

b. Triatomic molecules must contain at least two kinds of atoms.

c. A molecule of a compound must be heteroatomic.

d. Only heteroatomic molecules may contain three or more atoms.

1.52 Indicate whether each of the following statements is true or false. If a statement is false, change it to make it true. (Such a rewriting should involve more than merely converting the statement to the negative of itself.)

a. A molecule of an element may be homoatomic or heteroatomic, depending on which element is involved.

b. The limit of chemical subdivision for a compound is a molecule.

c. Heteroatomic molecules do not maintain the properties of their constituent elements.

d. Only one kind of atom may be present in a homoatomic molecule.

1.53 Draw a diagram of each of the following molecules using circular symbols of your choice to represent atoms.

a. A diatomic molecule of a compound

b. A molecule that is triatomic and homoatomic

c. A molecule that is tetraatomic and contains three kinds of atoms

d. A molecule that is triatomic, is symmetrical, and contains two elements

1.54 Draw a diagram of each of the following molecules using circular symbols of your choice to represent atoms.

a. A triatomic molecule of an element

b. A molecule that is diatomic and heteroatomic

c. A molecule that is triatomic and contains three elements

d. A molecule that is triatomic, is not symmetrical, and contains two kinds of atoms

● **Chemical Formulas (Section 1.10)**

1.55 Write chemical formulas for the substances represented by the molecular models given in Problem 1.45.

1.56 Write chemical formulas for the substances represented by the molecular models given in Problem 1.46.

1.57 On the basis of its chemical formula, classify each of the following substances as an element or a compound.

a. $LiClO_3$ b. CO c. Co d. S_8

1.58 On the basis of its chemical formula, classify each of the following substances as an element or a compound.

a. $CoCl_2$ b. $COCl_2$ c. SN d. Sn

1.59 List the element(s) present in each of the substances whose chemical formulas were given in Problem 1.57.

1.60 List the element(s) present in each of the substances whose chemical formulas were given in Problem 1.58.

1.61 Write chemical formulas for compounds in which the combining ratios of atoms are as follows:

a. Table sugar: carbon:hydrogen:oxygen, 12:22:11

b. Caffeine: carbon:hydrogen:nitrogen:oxygen, 8:10:4:2

1.62 Write chemical formulas for compounds in which the combining ratios of atoms are as follows:

a. Vitamin C: carbon:hydrogen:oxygen, 6:8:6

b. Nicotine: carbon:hydrogen:nitrogen, 10:14:2

1.63 Write a chemical formula for each of the following substances based on the information given about a molecule of the substance.

a. A molecule of hydrogen cyanide is triatomic and contains the elements hydrogen, carbon, and nitrogen.

b. A molecule of sulfuric acid is heptaatomic and contains two atoms of hydrogen, one atom of sulfur, and the element oxygen.

1.64 Write a chemical formula for each of the following substances based on the information given about a molecule of the substance.

a. A molecule of nitrous oxide contains twice as many atoms of nitrogen as of oxygen and is triatomic.

b. A molecule of nitric acid is pentaatomic and contains three atoms of oxygen and the elements hydrogen and nitrogen.

1.65 Write chemical formulas for the following compounds by using the given "verbally transmitted" information.

a. BA (pause) CL2

b. H (pause) N (pause) O3

c. NA3 (pause) P (pause) O4

d. MG (pause) OH taken twice

1.66 Write chemical formulas for the following compounds by using the given "verbally transmitted" information.

a. NA (pause) BR

b. H2 (pause) S (pause) O4

c. ZN (pause) CL2

d. FE (pause) CN taken three times

ADDITIONAL PROBLEMS

1.67   In the following diagrams, the different colored spheres represent atoms of different elements.

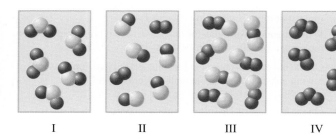

Select the diagram or diagrams that represent each of the listed situations. (Note that there may be more than one correct answer for a given situation and that answers may be used more than once or not at all.)
   a. Which diagram(s) represent(s) a compound whose molecules are triatomic?
   b. Which diagram(s) represent(s) a mixture of two compounds?
   c. Which diagram(s) represent(s) a mixture that contains two different types of diatomic molecules?
   d. Which diagram(s) represent(s) a pure substance?

1.68   In the following diagrams, different colored spheres represent atoms of different elements.

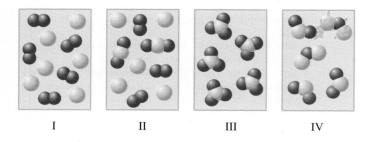

Select the diagram or diagrams that represent each of the listed situations. (Note that there may be more than one correct answer for a given situation and that answers may be used more than once or not at all.)
   a. Which diagram(s) represent(s) a compound whose molecules are tetraatomic?
   b. Which diagram(s) represent(s) a mixture of two substances?
   c. Which diagram(s) represent(s) a mixture of two elements?
   d. Which diagram(s) represent(s) a pure substance?

1.69   Assign each of the following descriptions of matter to one of the following categories: *element, compound,* or *mixture.*
   a. One substance present, one phase present, substance cannot be decomposed by chemical means
   b. One substance present, three elements present
   c. Two substances present, two phases present
   d. Two elements present, composition is definite and constant

1.70   Indicate whether each of the following samples of matter is a *heterogeneous mixture,* a *homogeneous mixture,* a *compound,* or an *element.*
   a. A colorless gas, only part of which reacts with hot iron
   b. A "cloudy" liquid that separates into two layers upon standing for two hours

   c. A green solid, all of which melts at the same temperature to produce a liquid that decomposes upon further heating
   d. A colorless gas that cannot be separated into simpler substances using physical means and that reacts with copper to produce both a copper-nitrogen compound and a copper-oxygen compound

1.71   Assign each of the following descriptions of matter to one of the following categories: *element, compound,* or *mixture.*
   a. One substance present, one phase present, one kind of homoatomic molecule present
   b. Two substances present, two phases present, all molecules are heteroatomic
   c. One phase present, two kinds of homoatomic molecules present
   d. One phase present, all molecules are triatomic, all molecules are heteroatomic, all molecules are identical

1.72   In the following diagram, the different colored spheres represent atoms of different elements. Four changes, denoted by the four numbered arrows, are shown.

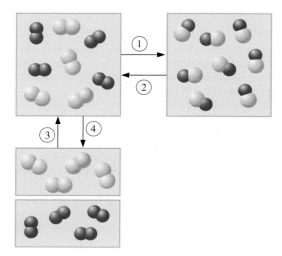

Select the change, by listing the arrow number, that represents each of the listed situations. (Note that there may be more than one correct answer for a given situation and that answers may be used more than once or not at all.)
   a. Which change(s) is a (are) physical change(s)?
   b. Which change(s) is a (are) chemical change(s)?
   c. Which change(s) is a (are) change(s) in which a compound is decomposed into its constituent elements?
   d. Which change(s) is a (are) change(s) in which two elements combine to form a compound?

1.73   Certain words can be viewed whimsically as sequential combinations of symbols of elements. For example, the given name Stephen is made up of the following sequence of chemical symbols: S-Te-P-He-N. Analyze each of the following given names in a similar manner.
   a. Barbara   b. Eugene
   c. Heather   d. Allan

1.74   In each of the following pairs of chemical formulas, indicate whether the first chemical formula listed denotes *more total atoms, the same number of total atoms,* or *fewer total atoms* than the second chemical formula listed.

a. HN_3 and NH_3
b. $CaSO_4$ and $Mg(OH)_2$
c. $NaClO_3$ and $Be(CN)_2$
d. $Be_3(PO_4)_2$ and $Mg(C_2H_3O_2)_2$

1.75 On the basis of the given information, determine the numerical value of the subscript x in each of the following chemical formulas.
a. BaS_2O_x: formula unit contains 6 atoms
b. $Al_2(SO_x)_3$; formula unit contains 17 atoms
c. SO_xCl_x; formula unit contains 5 atoms
d. $C_xH_{2x}Cl_x$; formula unit contains 8 atoms

1.76 A mixture contains the following five pure substances: N_2, N_2H_4, NH_3, CH_4, and CH_3Cl.
a. How many different kinds of molecules that contain four or fewer atoms are present in the mixture?
b. How many different kinds of atoms are present in the mixture?
c. How many total atoms are present in a mixture sample containing five molecules of each component?
d. How many total hydrogen atoms are present in a mixture sample containing four molecules of each component?

Multiple-Choice Practice Test

1.77 Which of the following is a property of *both* liquids and solids?
a. Definite shape
b. Definite volume
c. Indefinite shape
d. Indefinite volume

1.78 In which of the following pairs of properties are both of the properties *physical* properties?
a. Freezes at 10°C, red in color
b. Decomposes at 75°C, reacts with oxygen
c. Good conductor of electricity, flammable
d. Has a low density, nonflammable

1.79 Which of the following is *always* a characteristic of a *chemical change*?
a. Heat is absorbed.
b. Light is emitted.
c. One or more new substances are produced.
d. A change of state occurs.

1.80 Which of the following statements is *incorrect*?
a. Some, but not all, pure substances contain homoatomic molecules.
b. Some, but not all, pure substances contain heteroatomic molecules.
c. Some, but not all, compounds are pure substances.
d. Some, but not all, compounds contain three or more elements.

1.81 A pure substance A is found to change upon heating into two new pure substances B and C. Substance B, but not substance C, undergoes reaction with oxygen. Based on this information we definitely know that
a. A is a compound and C is an element.
b. Both A and B are compounds.
c. B is a compound and C is an element.
d. A is a compound.

1.82 In which of the following listings of elements do each of the elements have a two-letter chemical symbol?
a. Tin, nitrogen, zinc
b. Potassium, fluorine, carbon
c. Lead, hydrogen, iodine
d. Magnesium, silicon, chlorine

1.83 Which of the following statements concerning elemental abundances is *incorrect*?
a. One element accounts for over one-half of all atoms present in Earth's crust.
b. Two elements account for over three-fourths of all atoms present in Earth's crust.
c. Elemental abundances in Earth's crust closely parallel elemental abundances in the universe as a whole.
d. Hydrogen is the most abundant type of atom in the universe as a whole.

1.84 Which of the following pairings of terms is *incorrect*?
a. Element—a single type of atom
b. Pure substance—variable composition
c. Heterogeneous mixture—two or more regions with different properties
d. Compound—two or more elements present

1.85 Which of the following classifications of matter could *not* contain molecules that are both heteroatomic and diatomic?
a. Heterogeneous mixture
b. Homogeneous mixture
c. Compound
d. Element

1.86 In which of the following pairs of chemical formulas do the two members of the pair contain the same number of atoms per molecule?
a. $NaSCN$ and H_2CO_3
b. $CoCl_2$ and $COCl_2$
c. H_3N and HN_3
d. $Mg(OH)_2$ and SO_3

Measurements in Chemistry

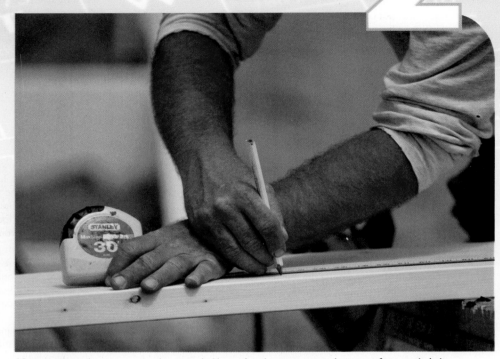

Measurements can never be exact; there is always some degree of uncertainty.

It would be extremely difficult for a carpenter to build cabinets without being able to use tools such as hammers, saws, drills, tape measures, rulers, straight edges, and T-squares. They are the tools of a carpenter's trade. Chemists also have "tools of the trade." The tool they use most is called *measurement*. Understanding measurement is indispensable in the study of chemistry. Questions such as "How much . . . ?," "How long . . . ?," and "How many . . . ?" simply cannot be answered without resorting to measurements. This chapter will help you learn what you need to know to deal properly with measurement. Much of the material in the chapter is mathematical. This is necessary; measurements require the use of numbers.

2.1 MEASUREMENT SYSTEMS

We all make measurements on a routine basis. For example, measurements are involved in following a recipe for making brownies, in determining our height and weight, and in fueling a car with gasoline. **Measurement** *is the determination of the dimensions, capacity, quantity, or extent of something.* In chemical laboratories, the most common types of measurements are those of mass, volume, length, time, temperature, pressure, and concentration.

Two systems of measurement are in use in the United States: (1) the English system of units and (2) the metric system of units. Common measurements of commerce, such as those used in a grocery store, are made in the *English system*. The units of this system include the inch, foot, pound, quart, and gallon. The *metric system* is used in scientific work. The units of this system include the gram, meter, and liter.

The word *metric* is derived from the Greek word *metron,* which means "measure."

The United States is in the process of voluntary conversion to the metric system for measurements of commerce. Metric system units now appear on numerous consumer products. Soft drinks now come in 1-, 2-, and 3-liter containers. Road signs in some states display distances in both miles and kilometers (Figure 2.1). Canned and packaged goods such as cereals and mixes on grocery store shelves now have the masses of their contents listed in grams as well as in pounds and ounces.

The metric system is superior to the English system. Its superiority stems from the interrelationships between units of the same type (volume, length, etc.). Metric unit interrelationships are less complicated than English unit interrelationships because the metric system is a decimal unit system. In the metric system, conversion from one unit size to another can be accomplished simply by moving the decimal point to the right or left an appropriate number of places. Thus, the metric system is simply more convenient to use.

Figure 2.1 Metric system units are becoming increasingly evident on highway signs.

The modern version of the metric system is called the International System, or SI (the abbreviation is taken from the French name, *le Système International*).

2.2 METRIC SYSTEM UNITS

In the metric system, there is one base unit for each type of measurement (length, mass, volume, and so on). The names of fractional parts of the base unit and multiples of the base unit are constructed by adding prefixes to the base unit. These prefixes indicate the size of the unit relative to the base unit. Table 2.1 lists common metric system prefixes, along with their symbols or abbreviations and mathematical meanings. The prefixes in color are the ones most frequently used.

The meaning of a metric system prefix is independent of the base unit it modifies and always remains constant. For example, the prefix *kilo-* always means 1000; a *kilo*second is 1000 seconds, a *kilo*watt is 1000 watts, and a *kilo*calorie is 1000 calories. Similarly, the prefix *nano-* always means one-billionth; a *nano*meter is one-billionth of a meter, a *nano*gram is one-billionth of a gram, and a *nano*liter is one-billionth of a liter.

● **E X A M P L E 2.1**

Recognizing the Mathematical Meanings of Metric System Prefixes

The use of numerical prefixes should not be new to you. Consider the use of the prefix *tri-* in the words *tri*angle, *tri*cycle, *tri*o, *tri*nity, and *tri*ple. Each of these words conveys the idea of three of something. The metric system prefixes are used in the same way.

▶ Using Table 2.1, write the name of the metric system prefix associated with the listed power of 10 or the power of 10 associated with the listed metric system prefix.

  a. nano-   **b.** micro-   **c.** deci-   **d.** 10^3 **e.** 10^6 **f.** 10^9

Solution

  a. The prefix nano- denotes 10^{-9} (one-billionth).
  b. The prefix micro- denotes 10^{-6} (one-millionth).
  c. The prefix deci- denotes 10^{-1} (one-tenth).
  d. 10^3 (one thousand) is denoted by the prefix *kilo-*.
  e. 10^6 (one million) is denoted by the prefix *mega-*.
  f. 10^9 (one billion) is denoted by the prefix *giga-*.

Practice Exercise 2.1

Using Table 2.1, write the name of the metric system prefix associated with the listed power of 10 or the power of 10 associated with the listed metric system prefix.

  a. milli-   **b.** pico-   **c.** mega-   **d.** 10^{-6} **e.** 10^{-2} **f.** 10^{-1}

Answers: a. 10^{-3}; **b.** 10^{-12}; **c.** 10^6; **d.** micro-; **e.** centi-; **f.** deci-

Metric Length Units

The **meter** (m) *is the base unit of length in the metric system.* It is about the same size as the English yard; 1 meter equals 1.09 yards (Figure 2.2a). The prefixes listed in Table 2.1 enable us to derive other units of length from the meter. The kilometer (km) is 1000 times larger than the meter; the centimeter (cm) and millimeter (mm) are, respectively, one-hundredth

Length is measured by determining the distance between two points.

TABLE 2.1
Common Metric System Prefixes
with Their Symbols and
Mathematical Meanings

	Prefix*	Symbol	Mathematical Meaning†
Multiples	giga-	G	1,000,000,000 (10^9, billion)
	mega-	M	1,000,000 (10^6, million)
	kilo-	k	1000 (10^3, thousand)
Fractional Parts	deci-	d	0.1 (10^{-1}, one-tenth)
	centi-	c	0.01 (10^{-2}, one-hundredth)
	milli-	m	0.001 (10^{-3}, one-thousandth)
	micro-	μ (Greek mu)	0.000001 (10^{-6}, one-millionth)
	nano-	n	0.000000001 (10^{-9}, one-billionth)
	pico-	p	0.000000000001 (10^{-12}, one-trillionth)

*Other prefixes also are available but are less commonly used.
†The power-of-10 notation for denoting numbers is considered in Section 2.6.

and one-thousandth of a meter. Most laboratory length measurements are made in centimeters rather than meters because of the meter's relatively large size.

Metric Mass Units

The **gram** (g) *is the base unit of mass in the metric system.* It is a very small unit compared with the English ounce and pound (Figure 2.2b). It takes approximately 28 grams to equal 1 ounce and nearly 454 grams to equal 1 pound. Both grams and milligrams (mg) are commonly used in the laboratory, where the kilogram (kg) is generally too large.

The terms *mass* and *weight* are often used interchangeably in measurement discussions; technically, however, they have different meanings. **Mass** *is a measure of the total quantity of matter in an object.* **Weight** *is a measure of the force exerted on an object by gravitational forces.*

The mass of a substance is a constant; the weight of an object varies with the object's geographical location. For example, matter weighs less at the equator than it would at the North Pole because the pull of gravity is less at the equator. Because Earth is not a perfect sphere, but bulges at the equator, the magnitude of gravitational attraction is less at the equator. An object would weigh less on the moon than on Earth because of the smaller

Mass is measured by determining the amount of matter in an object.

Students often erroneously think that the terms *mass* and *weight* have the same meaning. *Mass* is a measure of the amount of material present in a sample. *Weight* is a measure of the force exerted on an object by the pull of gravity.

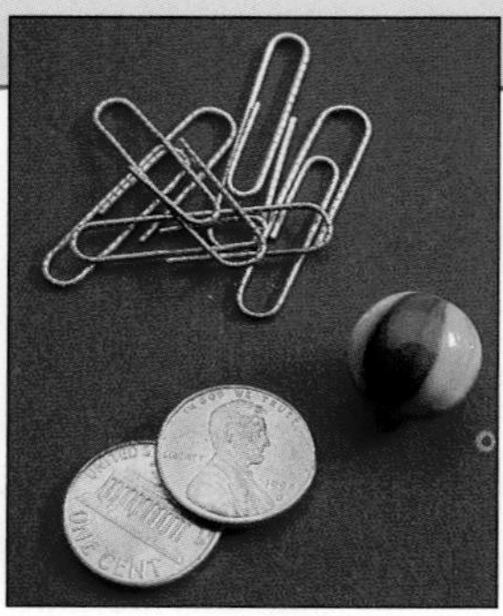

Figure 2.2 Comparisons of the base metric system units of length (meter), mass (gram), and volume (liter) with common objects.

(a) Length	(b) Mass	(c) Volume
A meter is slightly larger than a yard.	A gram is a small unit compared to a pound.	A liter is slightly larger than a quart.
1 meter = 1.09 yards.	1 gram = 1/454 pound.	1 liter = 1.06 quarts.
A baseball bat is about 1 meter long.	Two pennies, five paper-clips, and a marble have masses of about 5, 2, and 5 grams, respectively.	Most beverages are now sold by the liter rather than by the quart.

Volume is measured by determining the amount of space occupied by a three-dimensional object.

Total volume of large cube = 1000 cm³ = 1 L

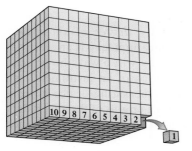

1 cm³ = 1 mL

Figure 2.3 A cube 10 cm on a side has a volume of 1000 cm³, which is equal to 1 L. A cube 1 cm on a side has a volume of 1 cm³, which is equal to 1 mL.

Another abbreviation for the unit cubic centimeter, used in medical situations, is cc.

$$1 \text{ cm}^3 = 1 \text{ cc}$$

Figure 2.4 The use of the concentration unit milligrams per deciliter (mg/dL) is common in clinical laboratory reports dealing with the composition of human body fluids.

size of the moon and the correspondingly lower gravitational attraction. Quantitatively, a 22.0-lb mass weighing 22.0 lb at Earth's North Pole would weigh 21.9 lb at Earth's equator and only 3.7 lb on the moon. In outer space, an astronaut may be weightless but never massless. In fact, he or she has the same mass in space as on Earth.

Metric Volume Units

The **liter** (L) *is the base unit of volume in the metric system.* The abbreviation for liter is a capital L rather than a lower-case l because a lower-case l is easily confused with the number 1. A liter is a volume equal to that occupied by a cube that is 10 centimeters on each side. Because the volume of a cube is calculated by multiplying length times width times height (which are all the same for a cube), we have

$$1 \text{ liter} = \text{volume of a cube with 10 cm edges}$$
$$= 10 \text{ cm} \times 10 \text{ cm} \times 10 \text{ cm}$$
$$= 1000 \text{ cm}^3$$

A liter is also equal to 1000 milliliters; the prefix *milli-* means one-thousandth. Therefore,

$$1000 \text{ mL} = 1000 \text{ cm}^3$$

Dividing both sides of this equation by 1000 shows that

$$1 \text{ mL} = 1 \text{ cm}^3$$

Consequently, the units milliliter and cubic centimeter are the same. In practice, mL is used for volumes of liquids and gases, and cm³ for volumes of solids. Figure 2.3 shows the relationship between 1 mL (1 cm³) and its parent unit, the liter, in terms of cubic measurements.

A liter and a quart have approximately the same volume; 1 liter equals 1.06 quarts (Figure 2.2c). The milliliter and deciliter (dL) are commonly used in the laboratory. Deciliter units are routinely encountered in clinical laboratory reports detailing the composition of body fluids (Figure 2.4). A deciliter is equal to 100 mL (0.100 L).

Healthy, I.M.		9/23/07	9/23/07	9/24/07
M 37	Your Doctor Anywhere, U.S.A.			05169
000-00-000				032136

Test Name	Result	Units	Normal Reference Range
CHEM-SCREEN PROFILE			
CALCIUM	9.70	mg/dL	9.00-10.40
PHOSPHATE (as PHOSPHORUS)	3.00	mg/dL	2.20-4.30
BUN	16.00	mg/dL	9.00-23.0
CREATININE	1.30	mg/dL	0.80-1.30
BUN/CREAT RATIO	12.31		12-20
URIC ACID	7.50	mg/dL	3.60-8.30
GLUCOSE	114.00	mg/dL	65.0-130
TOTAL PROTEIN	7.90	g/dL	6.50-8.00
ALBUMIN	5.10	g/dL	3.90-4.90
GLOBULIN	2.80	g/dL	2.10-3.50
ALB/GLOB RATIO	1.82		1.20-2.20
TOTAL BILIRUBIN	0.55	mg/dL	0.30-1.40
DIRECT BILIRUBIN	0.18	mg/dL	0.04-0.20
CHOLESTEROL	203.00	mg/dL	140-233
CHOLESTEROL PERCENTILE	50	PERCENTILE	1-74
HDL CHOLESTEROL	71	mg/dL	
CHOL./HDL CHOLESTEROL	*(01)-2.77		
TRIGLYCERIDES	148.00	mg/dL	50.0-200

(01) THE RESULT OBTAINED FOR THE CHOLESTEROL/HDL CHOLESTEROL RATIO FOR THIS PATIENT'S SAMPLE IS ASSOCIATED WITH THE LOWEST CORONARY HEART DISEASE (CHD) RISK.

EXACT AND INEXACT NUMBERS

In scientific work, numbers are grouped in two categories: *exact numbers and inexact numbers*. An **exact number** *is a number whose value has no uncertainty associated with it—that is, it is known exactly.* Exact numbers occur in definitions (for example, there are exactly 12 objects in a dozen, not 12.01 or 12.02); in counting (for example, there can be 7 people in a room, but never 6.99 or 7.03); and in simple fractions (for example, 1/3, 3/5, and 5/9).

An **inexact number** *is a number whose value has a degree of uncertainty associated with it.* Inexact numbers result any time a measurement is made. It is impossible to make an *exact* measurement; some uncertainty will always be present. Flaws in measuring device construction, improper calibration of an instrument, and the skills (or lack of skills) possessed by a person using a measuring device all contribute to error (uncertainty). Section 2.4 considers further the origins of the uncertainty associated with measurements and also the methods used to "keep track" of such uncertainty.

> All measurements have some degree of uncertainty associated with them. Thus, the numerical value associated with a measurement is always an inexact number.

 UNCERTAINTY IN MEASUREMENT AND SIGNIFICANT FIGURES

As noted in the previous section, because of the limitations of the measuring device and the limited powers of observation of the individual making the measurement, every measurement carries a degree of uncertainty or error. Even when very elaborate measuring devices are used, some degree of uncertainty is always present.

Origin of Measurement Uncertainty

To illustrate how measurement uncertainty arises, let us consider how two different rulers, shown in Figure 2.5, are used to measure a given length. Using ruler A, we can say with certainty that the length of the object is between 3 and 4 centimeters. We can further say that the actual length is closer to 4 centimeters and estimate it to be 3.7 centimeters. Ruler B has more subdivisions on its scale than ruler A. It is marked off in tenths of a centimeter instead of in centimeters. Using ruler B, we can definitely say that the length of the object is between 3.7 and 3.8 centimeters and can estimate it to be 3.74 centimeters.

Note how both length measurements (ruler A and ruler B) contain some digits (all those except the last one) that are exactly known and one digit (the last one) that is estimated. It is this last digit, the estimated one, that produces uncertainty in a measurement. Note also that the uncertainty in the second length measurement is less than that in the first one—an uncertainty in the hundredths place compared with an uncertainty in the tenths place. We say that the second measurement is more *precise* than the first one; that is, it has less uncertainty than the first measurement.

Only one estimated digit is ever recorded as part of a measurement. It would be incorrect for a scientist to report that the length of the object in Figure 2.5 is 3.745 centimeters as read by using ruler B. The value 3.745 contains two estimated digits, the 4 and the 5, and indicates a measurement with less uncertainty than what is actually obtainable with that particular measuring device. Again, only one estimated digit is ever recorded as part of a measurement.

Because measurements are never exact, two types of information must be conveyed whenever a numerical value for a measurement is recorded: (1) the magnitude of the measurement and (2) the uncertainty of the measurement. The magnitude is indicated by the digit values. Uncertainty is indicated by the number of significant figures recorded. **Significant figures** *are the digits in a measurement that are known with certainty plus one digit that is estimated.* To summarize, in equation form,

Number of significant figures = all certain digits + one estimated digit

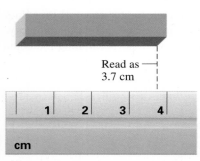

Ruler A

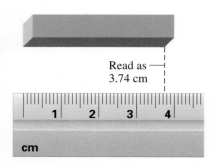

Ruler B

Figure 2.5 The scale on a measuring device determines the magnitude of the uncertainty for the recorded measurement. Measurements made with ruler A will have greater uncertainty than those made with ruler B.

Guidelines for Determining Significant Figures

Recognizing the number of significant figures in a measured quantity is easy for measurements we make ourselves, because we know the type of instrument we are using and its limitations. However, when someone else makes the measurement, such information is often not available. In such cases, we follow a set of guidelines for determining the number of significant figures in a measured quantity.

The term *significant figures* is often verbalized in shortened form as "sig figs."

1. In any measurement, all nonzero digits are significant.
2. *Zeros* may or may not be significant because zeros can be used in two ways: (1) to position a decimal point and (2) to indicate a measured value. Zeros that perform the first function are not significant, and zeros that perform the second function are significant. When zeros are present in a measured number, we follow these rules:
    a. *Leading zeros,* those at the beginning of a number, are never significant.

    > 0.0141 has three significant figures.
    >
    > 0.0000000048 has two significant figures.

    b. *Confined zeros,* those between nonzero digits, are always significant.

    > 3.063 has four significant figures.
    >
    > 0.001004 has four significant figures.

    c. *Trailing zeros,* those at the end of a number, are significant if a decimal point is present in the number.

    > 56.00 has four significant figures.
    >
    > 0.05050 has four significant figures.

    d. *Trailing zeros,* those at the end of a number, are not significant if the number lacks an explicitly shown decimal point.

    > 59,000,000 has two significant figures.
    >
    > 6010 has three significant figures.

It is important to remember what is "significant" about significant figures. The number of significant figures in a measurement conveys information about the uncertainty associated with the measurement. The "location" of the last significant digit in the numerical value of a measurement specifies the measurement's uncertainty: Is this last significant digit located in the hundredths, tenths, ones, or tens position, etc.? Consider the following measurement values (with the last significant digit "boxed" for emphasis).

4620.0 (five significant figures) has an uncertainty of tenths.
4620 (three significant figures) has an uncertainty in the tens place.
462,000 (three significant figures) has an uncertainty in the thousands place.

The Chemistry at a Glance feature on page 28 reviews the rules that govern which digits in a measurement are significant.

● EXAMPLE 2.2

Determining the Number of Significant Figures in a Measurement and the Uncertainty Associated with the Measurement

For each of the following measurements, give the number of significant figures present and the uncertainty associated with the measurement.

a. 5623.00   **b.** 0.0031   **c.** 97,200   **d.** 637

Solution

a. Six significant figures are present because trailing zeros are significant when a decimal point is present. The uncertainty is in the hundredths place (± 0.01), the location of the last significant digit.

(continued)

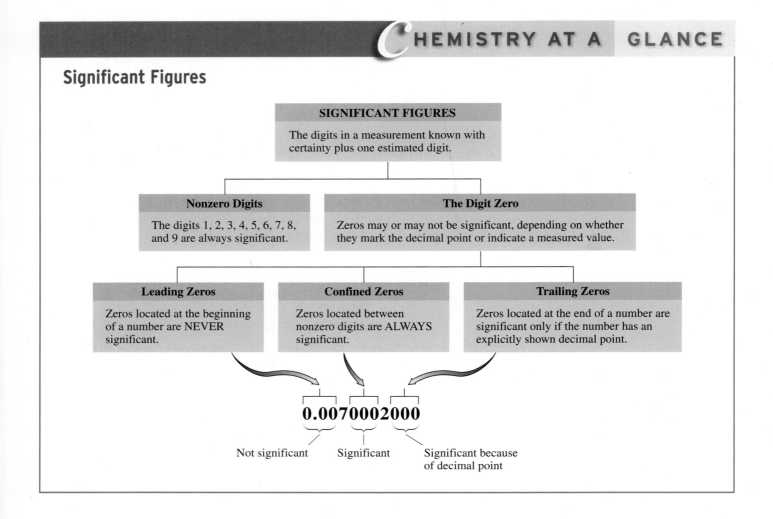

CHEMISTRY AT A GLANCE

Significant Figures

SIGNIFICANT FIGURES

The digits in a measurement known with certainty plus one estimated digit.

Nonzero Digits

The digits 1, 2, 3, 4, 5, 6, 7, 8, and 9 are always significant.

The Digit Zero

Zeros may or may not be significant, depending on whether they mark the decimal point or indicate a measured value.

Leading Zeros

Zeros located at the beginning of a number are NEVER significant.

Confined Zeros

Zeros located between nonzero digits are ALWAYS significant.

Trailing Zeros

Zeros located at the end of a number are significant only if the number has an explicitly shown decimal point.

0.0070002000

Not significant   Significant   Significant because of decimal point

b. Two significant figures are present because leading zeros are never significant. The uncertainty is in the ten-thousandths place (±0.0001), the location of the last significant digit.

c. Three significant figures are present because the trailing zeros are not significant (no explicit decimal point is shown). The uncertainty is in the hundreds place (±100).

d. Three significant figures are present, and the uncertainty is in the ones place (±1).

Practice Exercise 2.2

For each of the following measurements, give the number of significant figures present and the uncertainty associated with the measurement.

a. 727.23   **b.** 0.1031   **c.** 47,230   **d.** 637,000,000

Answers: a. 5, ±0.01; **b.** 4, ±0.0001; **c.** 4, ±10; **d.** 3, $\pm1,000,000$

 2.5 SIGNIFICANT FIGURES AND MATHEMATICAL OPERATIONS

When measurements are added, subtracted, multiplied, or divided, consideration must be given to the number of significant figures in the computed result. Mathematical operations should not increase (or decrease) the uncertainty of experimental measurements.

Figure 2.6 The digital readout on an electronic calculator usually shows more digits than are warranted. Calculators are not programmed to account for significant figures.

Hand-held electronic calculators generally "complicate" uncertainty considerations because they are not programmed to take significant figures into account. Consequently, the digital readouts display more digits than are warranted (Figure 2.6). It is a mistake to record these extra digits, because they are not significant figures and hence are meaningless.

Rounding Off Numbers

When we obtain calculator answers that contain too many digits, it is necessary to delete (drop) the nonsignificant digits, a process that is called rounding off. **Rounding off** *is the process of deleting unwanted (nonsignificant) digits from calculated numbers.* There are two rules for rounding off numbers.

1. *If the first digit to be deleted is 4 or less, simply drop it and all the following digits.* For example, the number 3.724567 becomes 3.72 when rounded to three significant figures.
2. *If the first digit to be deleted is 5 or greater, that digit and all that follow are dropped, and the last retained digit is increased by one.* The number 5.00673 becomes 5.01 when rounded to three significant figures.

These rounding rules must be modified slightly when digits to the left of the decimal point are to be dropped. To maintain the inferred position of the decimal point in such situations, zeros must replace all the dropped digits that are to the left of the inferred decimal point. Parts (c) and (d) of Example 2.3 illustrate this point.

 EXAMPLE 2.3

Rounding Numbers to a Specified Number of Significant Figures

Round off each of the following numbers to two significant figures.

a. 25.7   **b.** 0.4327   **c.** 432,117   **d.** 13,500

Solution

a. Rule 2 applies. The last retained digit (the 5) is increased in value by one unit.

$$25.7 \text{ becomes } 26$$

b. Rule 1 applies. The last retained digit (the 3) remains the same, and all digits that follow it are simply dropped.

$$0.4327 \text{ becomes } 0.43$$

c. Since the first digit to be dropped is a 2, rule 1 applies

$$432,117 \text{ becomes } 430,000$$

Note that to maintain the position of the inferred decimal point, zeros must replace all of the dropped digits. This will always be the case when digits to the left of the inferred decimal place are dropped.

d. This is a rule 2 situation because the first digit to be dropped is a 5. The 3 is rounded up to a 4 and zeros take the place of all digits to the left of the inferred decimal place that are dropped.

$$13,500 \text{ becomes } 14,000$$

Practice Exercise 2.3

Round off each of the following numbers to three significant figures.

a. 432.55   **b.** 0.03317   **c.** 162,700   **d.** 65,234

Answers: a. 433; **b.** 0.0332; **c.** 163,000; **d.** 65,200

Operational Rules

Significant-figure considerations in mathematical operations that involve measured numbers are governed by two rules, one for multiplication and division and one for addition and subtraction.

1. *In multiplication and division, the number of significant figures in the answer is the same as the number of significant figures in the measurement that contains the fewest significant figures.* For example,

Four significant   Three significant
figures   figures

$$6.038 \times 2.57 = 15.51766 \qquad \text{(calculator answer)}$$
$$= 15.5 \qquad \text{(correct answer)}$$

Three significant
figures

The calculator answer is rounded to three significant figures because the measurement with the fewest significant figures (2.57) contains only three significant figures.

2. *In addition and subtraction, the answer has no more digits to the right of the decimal point than are found in the measurement with the fewest digits to the right of the decimal point.* For example,

$$9.333 \qquad \leftarrow \text{Uncertain digit (thousandths)}$$
$$+1.4 \qquad \leftarrow \text{Uncertain digit (tenths)}$$
$$10.733 \qquad \text{(calculator answer)}$$
$$10.7 \qquad \text{(correct answer)}$$

Uncertain digit (tenths)

The calculator answer is rounded to the tenths place because the uncertainty in the number 1.4 is in the tenths place.

> Concisely stated, the significant-figure operational rules are
>
> × or ÷ : Keep smallest number of significant figures in answer.
>
> + or − : Keep smallest number of decimal places in answer.

Note the contrast between the rule for multiplication and division and the rule for addition and subtraction. In multiplication and division, significant figures are counted; in addition and subtraction, decimal places are counted. It is possible to gain or lose significant figures during addition or subtraction, but *never* during multiplication or division. In our previous sample addition problem, one of the input numbers (1.4) has two significant figures, and the correct answer (10.7) has three significant figures. This is allowable in addition (and subtraction) because we are counting decimal places, not significant figures.

● **EXAMPLE 2.4**

Expressing Answers to the Proper Number of Significant Figures

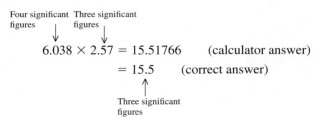

Perform the following computations, expressing your answers to the proper number of significant figures.

a. 6.7321×0.0021 **b.** $\dfrac{16.340}{23.42}$ **c.** 6.000×4.000

d. $8.3 + 1.2 + 1.7$ **e.** $3.07 \times (17.6 - 13.73)$

Solution

a. The calculator answer to this problem is

$$6.7321 \times 0.0021 = 0.01413741$$

The input number with the least number of significant figures is 0.0021.

$$6.7321 \times 0.0021$$

Five significant   Two significant
figures   figures

Thus the calculator answer must be rounded to two significant figures.

$$0.01413741 \quad \text{becomes} \quad 0.014$$
Calculator answer   Correct answer

b. The calculator answer to this problem is

$$\frac{16{,}340}{23.42} = 697.69427$$

Both input numbers contain four significant figures. Thus the correct answer will also contain four significant figures.

$$697.69427 \quad \text{becomes} \quad 697.7$$
Calculator answer   Correct answer

c. The calculator answer to this problem is

$$6.000 \times 4.000 = 24$$

Both input numbers contain four significant figures. Thus the correct answer must also contain four significant figures, and

$$24 \quad \text{becomes} \quad 24.00$$
(calculator answer)   (correct answer)

Note here how the calculator answer had too few significant figures. Most calculators cut off zeros after the decimal point even if those zeros are significant. Using too few significant figures in an answer is just as wrong as using too many.

d. The calculator answer to this problem is

$$8.3 + 1.2 + 1.7 = 11.2$$

All three input numbers have uncertainty in the tenths place. Thus the last retained digit in the correct answer will be that of tenths. (In this particular problem, the calculator answer and the correct answer are the same, a situation that does not occur very often.)

e. This problem involves the use of both multiplication and subtraction significant-figure rules. We do the subtraction first.

$$17.6 - 13.73 = 3.87 \quad \text{(calculator answer)}$$
$$= 3.9 \quad \text{(correct answer)}$$

This answer must be rounded to tenths because the input number 17.6 involves only tenths. We now do the multiplication.

$$3.07 \times 3.9 = 11.973 \quad \text{(calculator answer)}$$
$$= 12 \quad \text{(correct answer)}$$

The number 3.9 limits the answer to two significant figures.

> Calculators are not programmed to take significant figures into account, which means that students must always adjust their calculator answer to the correct number of significant figures. Sometimes this involves deleting a number of digits through rounding, and other times it involves adding zeros to increase the number of significant figures.

Practice Exercise 2.4

Perform the following computations, expressing your answers to the proper number of significant figures. Assume that all numbers are measured numbers.

a. 5.4430×1.203 **b.** $\dfrac{17.4}{0.0031}$

c. $7.4 + 20.74 + 3.03$ **d.** $4.73 \times (2.2 + 8.9)$

Answers: **a.** 6.548; **b.** 5600; **c.** 31.2; **d.** 52.5

Some numbers used in computations may be *exact numbers* rather than measured numbers. Because exact numbers (Section 2.3) have no uncertainty associated with them, they possess an unlimited number of significant figures. Therefore, such numbers never limit the number of significant figures in a computational answer.

SCIENTIFIC NOTATION

Up to this point in the chapter, we have expressed all numbers in decimal notation, the everyday method for expressing numbers. Such notation becomes cumbersome for very large and very small numbers (which occur frequently in scientific work). For example, in one drop of blood, which is 92% water by mass, there are approximately

$$1,600,000,000,000,000,000,000 \text{ molecules}$$

of water, each of which has a mass of

$$0.000000000000000000000030 \text{ gram}$$

Recording such large and small numbers is not only time-consuming but also open to error; often, too many or too few zeros are recorded. Also, it is impossible to multiply or divide such numbers with most calculators because they can't accept that many digits. (Most calculators accept either 8 or 10 digits.)

A method called *scientific notation* exists for expressing in compact form multidigit numbers that involve many zeros. **Scientific notation** *is a numerical system in which numbers are expressed in the form A × 10ⁿ, where A is a number with a single nonzero digit to the left of the decimal place and n is a whole number.* The number A is called the *coefficient*. The number 10^n is called the *exponential term*. The coefficient is always multiplied by the exponential term.

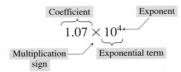

The two previously cited numbers that deal with molecules of water are expressed in scientific notation as

$$1.6 \times 10^{21} \text{ molecules}$$

and

$$3.0 \times 10^{-23} \text{ gram}$$

Obviously, scientific notation is a much more concise way of expressing numbers. Such scientific notation is compatible with most calculators.

Converting from Decimal to Scientific Notation

The procedure for converting a number from decimal notation to scientific notation has two parts.

1. *The decimal point in the decimal number is moved to the position behind (to the right of) the first nonzero digit.*
2. *The exponent for the exponential term is equal to the number of places the decimal point has been moved.* The exponent is positive if the original decimal number is 10 or greater and is negative if the original decimal number is less than 1. For numbers between 1 and 10, the exponent is zero.

The following two examples illustrate the use of these procedures:

$$93,000,000 = 9.3 \times 10^7$$
Decimal point is moved 7 places

$$0.0000037 = 3.7 \times 10^{-6}$$
Decimal point is moved 6 places

Scientific notation is also called exponential notation.

Significant Figures and Scientific Notation

The decimal and scientific notation forms of a number *always* contain the same number of significant figures.

How do significant-figure considerations affect scientific notation? The answer is simple. *Only significant figures become part of the coefficient.* The numbers 63, 63.0, and 63.00, which respectively have two, three, and four significant figures, when converted to scientific notation become, respectively,

$$6.3 \times 10^1 \qquad \text{(two significant figures)}$$
$$6.30 \times 10^1 \qquad \text{(three significant figures)}$$
$$6.300 \times 10^1 \qquad \text{(four significant| figures)}$$

Multiplication and Division in Scientific Notation

Multiplication and division of numbers expressed in scientific notation are common procedures. For these two types of operations, the coefficients, which are decimal numbers, are combined in the usual way. The rules for handling the exponential terms are

1. To multiply exponential terms, *add* the exponents.
2. To divide exponential terms, *subtract* the exponents.

● EXAMPLE 2.5
Multiplication and Division in Scientific Notation

Carry out the following mathematical operations involving numbers that are expressed in scientific notation.

a. $(2.33 \times 10^3) \times (1.55 \times 10^4)$ **b.** $\dfrac{8.42 \times 10^6}{3.02 \times 10^4}$

Solution

a. Multiplying the two coefficients gives

$$2.33 \times 1.55 = 3.6115 \qquad \text{(calculator answer)}$$
$$= 3.61 \qquad \text{(correct answer)}$$

Remember that the coefficient obtained by multiplication can have only three significant figures in this case, the same number as in both input numbers for the multiplication.

Multiplication of the two powers of 10 to give the exponential term requires that we add the exponents.

$$10^3 \times 10^4 = 10^{3+4} = 10^7$$

Combining the new coefficient with the new exponential term gives the answer.

$$3.61 \times 10^7$$

b. Performing the indicated division of the coefficients gives

$$\frac{8.42}{3.02} = 2.7880794 \qquad \text{(calculator answer)}$$
$$= 2.79 \qquad \text{(correct answer)}$$

Because both input numbers have three significant figures, the answer also has three significant figures.

The division of exponential terms requires that we subtract the exponents.

$$\frac{10^6}{10^4} = 10^{(+6)-(+4)} = 10^2$$

Combining the new coefficient and the new exponential term gives

$$2.79 \times 10^2$$

(continued)

Practice Exercise 2.5

Carry out the following mathematical operations involving numbers that are expressed in scientific notation.

a. $(4.057 \times 10^3) \times (2.001 \times 10^7)$ **b.** $\dfrac{4.1 \times 10^{-10}}{3.112 \times 10^{-7}}$

Answers: a. 8.118×10^{10}; **b.** 1.3×10^{-3}

Uncertainty and Scientific Notation

The uncertainty associated with a measurement whose value is expressed in scientific notation cannot be obtained directly from the coefficient in the scientific notation. The coefficient decimal point location is not the true location for the decimal point. The value of the exponent in the exponential term must be taken into account in determining the uncertainty.

The uncertainty associated with a scientific notation number is obtained by determining the uncertainty associated with the coefficient and then multiplying this value by the exponential term. For the number 3.72×10^{-3}, we have

$$10^{-2} \quad \times \quad 10^{-3} \quad = \quad 10^{-5}$$

Uncertainty in coefficient	Exponential term	Uncertainty in number

That the uncertainty is, indeed, 1×10^{-5} can be readily seen by rewriting the number in decimal notation.

$$3.72 \times 10^{-3} = 0.00372$$

The uncertainty for the decimal number is in the ten-thousandths place (10^{-5}).

 2.7 CONVERSION FACTORS

With both the English unit and metric unit systems in common use in the United States, we often must change measurements from one system to their equivalent in the other system. The mathematical tool we use to accomplish this task is a general method of problem solving called *dimensional analysis*. Central to the use of dimensional analysis is the concept of conversion factors. A **conversion factor** *is a ratio that specifies how one unit of measurement is related to another unit of measurement.*

Conversion factors are derived from equations (equalities) that relate units. Consider the quantities "1 minute" and "60 seconds," both of which describe the same amount of time. We may write an equation describing this fact.

$$1 \text{ min} = 60 \text{ sec}$$

This fixed relationship is the basis for the construction of a pair of conversion factors that relate seconds and minutes.

$$\frac{1 \text{ min}}{60 \text{ sec}} \quad \text{and} \quad \frac{60 \text{ sec}}{1 \text{ min}} \qquad \text{These two quantities are the same}$$

Note that conversion factors always come in pairs, one member of the pair being the reciprocal of the other. Also note that the numerator and the denominator of a conversion factor always describe the same amount of whatever we are considering. One minute and 60 seconds denote the same amount of time.

Conversion Factors Within a System of Units

Most students are familiar with and have memorized numerous conversion factors within the English system of measurement (English-to-English conversion factors). Some of these factors, with only one member of a conversion factor pair being listed, are

$$\frac{12 \text{ in.}}{1 \text{ ft}} \qquad \frac{3 \text{ ft}}{1 \text{ yd}} \qquad \frac{4 \text{ qt}}{1 \text{ gal}} \qquad \frac{16 \text{ oz}}{1 \text{ lb}}$$

> In order to avoid confusion with the word *in,* the abbreviation for inches, in., includes a period. This is the only unit abbreviation in which a period appears.

Such conversion factors contain an unlimited number of significant figures because the numbers within them arise from definitions.

Metric-to-metric conversion factors are similar to English-to-English conversion factors in that they arise from definitions. Individual conversion factors are derived from the meanings of the metric system prefixes (Section 2.2). For example, the set of conversion factors involving kilometer and meter come from the equality

$$1 \text{ kilometer} = 10^3 \text{ meters}$$

and those relating microgram and gram come from the equality

$$1 \text{ microgram} = 10^{-6} \text{ gram}$$

> In order to obtain metric-to-metric conversion factors, you need to know the meaning of the metric system prefixes in terms of powers of 10 (see Table 2.1).

The two pairs of conversion factors are

$$\frac{10^3 \text{ m}}{1 \text{ km}} \quad \text{and} \quad \frac{1 \text{ km}}{10^3 \text{ m}} \qquad \frac{1 \text{ } \mu\text{g}}{10^{-6} \text{ g}} \quad \text{and} \quad \frac{10^{-6} \text{ g}}{1 \text{ } \mu\text{g}}$$

Note that the numerical equivalent of the prefix is always associated with the base (unprefixed) unit in a metric-to-metric conversion factor.

The number 1 always goes with the *prefixed* unit. ⟶

$$\frac{1 \text{ mL}}{10^{-3} \text{ L}}$$

The power of 10 always ↗ goes with the *unprefixed* unit.

Conversion Factors Between Systems of Units

Conversion factors that relate metric units to English units and vice versa are not defined quantities because they involve two different systems of measurement. The numbers associated with these conversion factors must be determined experimentally (see Figure 2.7). Table 2.2 lists commonly encountered relationships between metric system and English system units. These few conversion factors are sufficient to solve most of the problems we will encounter.

Metric-to-English conversion factors can be specified to differing numbers of significant figures. For example,

$$1.00 \text{ lb} = 454 \text{ g}$$
$$1.000 \text{ lb} = 453.6 \text{ g}$$
$$1.0000 \text{ lb} = 453.59 \text{ g}$$

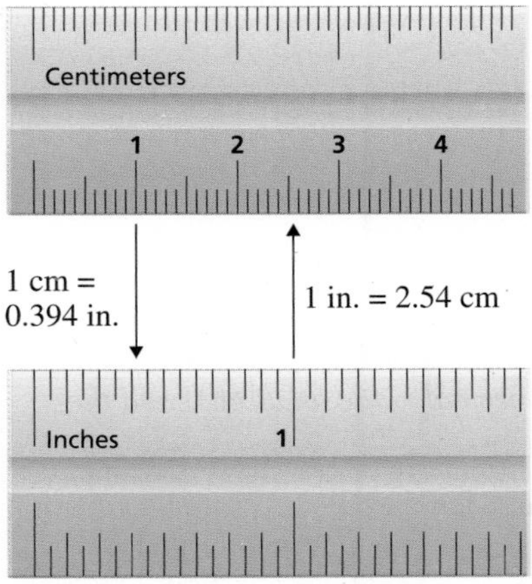

In a problem-solving context, which "version" of a conversion factor is used depends on how many significant figures there are in the other numbers of the problem. Conversion factors should never limit the number of significant figures in the answer to a problem. The conversion factors in Table 2.2 are given to three significant figures, which is sufficient for the applications we will make of them.

The Chemistry at a Glance feature on page 38 summarizes what we have discussed in this section about conversion factors.

Centimeters

1 cm = 0.394 in.

1 in. = 2.54 cm

Inches   1

Figure 2.7 It is experimentally determined that 1 inch equals 2.54 centimeters, or 1 centimeter equals 0.394 inch.

TABLE 2.2
Equalities and Conversion Factors
That Relate the English and Metric
Systems of Measurement

	Metric to English	English to Metric
Length		
1.00 inch = 2.54 centimeters	$\dfrac{1.00 \text{ in.}}{2.54 \text{ cm}}$	$\dfrac{2.54 \text{ cm}}{1.00 \text{ in.}}$
1.00 meter = 39.4 inches	$\dfrac{39.4 \text{ in.}}{1.00 \text{ m}}$	$\dfrac{1.00 \text{ m}}{39.4 \text{ in.}}$
1.00 kilometer = 0.621 mile	$\dfrac{0.621 \text{ mi}}{1.00 \text{ km}}$	$\dfrac{1.00 \text{ km}}{0.621 \text{ mi}}$
Mass		
1.00 pound = 454 grams	$\dfrac{1.00 \text{ lb}}{454 \text{ g}}$	$\dfrac{454 \text{ g}}{1.00 \text{ lb}}$
1.00 kilogram = 2.20 pounds	$\dfrac{2.20 \text{ lb}}{1.00 \text{ kg}}$	$\dfrac{1.00 \text{ kg}}{2.20 \text{ lb}}$
1.00 ounce = 28.3 grams	$\dfrac{1.00 \text{ oz}}{28.3 \text{ g}}$	$\dfrac{28.3 \text{ g}}{1.00 \text{ oz}}$
Volume		
1.00 quart = 0.946 liter	$\dfrac{1.00 \text{ qt}}{0.946 \text{ L}}$	$\dfrac{0.946 \text{ L}}{1.00 \text{ qt}}$
1.00 liter = 0.265 gallon	$\dfrac{0.265 \text{ gal}}{1.00 \text{ L}}$	$\dfrac{1.00 \text{ L}}{0.265 \text{ gal}}$
1.00 milliliter = 0.0338 fluid ounce	$\dfrac{0.0338 \text{ fl oz}}{1.00 \text{ mL}}$	$\dfrac{1.00 \text{ mL}}{0.0338 \text{ fl oz}}$

DIMENSIONAL ANALYSIS

Dimensional analysis *is a general problem-solving method in which the units associated with numbers are used as a guide in setting up calculations.* In this method, units are treated in the same way as numbers; that is, they can be multiplied, divided, or canceled. For example, just as

$$5 \times 5 = 5^2 \qquad (5 \text{ squared})$$

we have

$$\text{cm} \times \text{cm} = \text{cm}^2 \qquad (\text{cm squared})$$

Also, just as the 3s cancel in the expression

$$\frac{\cancel{3} \times 5 \times 7}{\cancel{3} \times 2}$$

the centimeters cancel in the expression

$$\frac{(\cancel{\text{cm}}) \times (\text{in.})}{(\cancel{\text{cm}})}$$

"Like units" found in the numerator and denominator of a fraction will always cancel, just as like numbers do.

The following steps show how to set up a problem using dimensional analysis.

Step 1: *Identify the known or given quantity (both numerical value and units) and the units of the new quantity to be determined.*

This information will always be found in the statement of the problem. Write an equation with the given quantity on the left and the units of the desired quantity on the right.

Step 2: *Multiply the given quantity by one or more conversion factors in such a manner that the unwanted (original) units are canceled, leaving only the desired units.*

The general format for the multiplication is

$$(\text{Information given}) \times (\text{conversion factors}) = (\text{information sought})$$

The number of conversion factors used depends on the individual problem.

Step 3: *Perform the mathematical operations indicated by the conversion factor setup.*

When performing the calculation, double-check to make sure all units except the desired set have canceled.

EXAMPLE 2.6
Unit Conversions Within the Metric System

A standard aspirin tablet contains 324 mg of aspirin. How many grams of aspirin are in a standard aspirin tablet?

Solution

Step 1: The given quantity is 324 mg, the mass of aspirin in the tablet. The unit of the desired quantity is grams.

$$324 \text{ mg} = ? \text{ g}$$

Step 2: Only one conversion factor will be needed to convert from milligrams to grams, one that relates milligrams to grams. The two forms of this conversion factor are

$$\frac{1 \text{ mg}}{10^{-3} \text{ g}} \quad \text{and} \quad \frac{10^{-3} \text{ g}}{1 \text{ mg}}$$

The second factor is the one needed because it allows for cancellation of the milligram units, leaving us with grams as the new units.

$$324 \text{ m\cancel{g}} \times \left(\frac{10^{-3} \text{ g}}{1 \text{ m\cancel{g}}}\right) = ? \text{ g}$$

Step 3: Combining numerical terms as indicated generates the final answer.

$$\left(324 \times \frac{10^{-3}}{1}\right) \text{ g} = 0.324 \text{ g}$$

Number from first factor   Numbers from second factor

The answer is given to three significant figures because the given quantity in the problem, 324 mg, has three significant figures. The conversion factor used arises from a definition and thus does not limit significant figures in any way.

Practice Exercise 2.6

Analysis shows the presence of 203 μg of cholesterol in a sample of blood. How many grams of cholesterol are present in this blood sample?

Answer: 2.03×10^{-4} g cholesterol

EXAMPLE 2.7
Unit Conversions Between the Metric and English System

Capillaries, the microscopic vessels that carry blood from small arteries to small veins, are on the average only 1 mm long. What is the average length of a capillary in inches?

Solution

Step 1: The given quantity is 1 mm, and the units of the desired quantity are inches.

$$1 \text{ mm} = ? \text{ in.}$$

Step 2: The conversion factor needed for a one-step solution, millimeters to inches, is not given in Table 2.2. However, a related conversion factor, meters to inches, is given. Therefore, we first convert millimeters to meters and then use the meters-to-inches conversion factor in Table 2.2.

$$\text{mm} \longrightarrow \text{m} \longrightarrow \text{in.}$$

(continued)

The correct conversion factor setup is

$$1 \text{ mm} \times \left(\frac{10^{-3} \text{ m}}{1 \text{ mm}}\right) \times \left(\frac{39.4 \text{ in.}}{1.00 \text{ m}}\right) = ? \text{ in.}$$

All of the units except for inches cancel, which is what is needed. The information for the first conversion factor was obtained from the meaning of the prefix *milli-*.

This setup illustrates the fact that sometimes the given units must be changed to intermediate units before common conversion factors, such as those found in Table 2.2, are applicable.

Step 3:   Collecting the numerical factors and performing the indicated math gives

$$\left(\frac{1 \times 10^{-3} \times 39.4}{1 \times 1.00}\right) \text{in.} = 0.0394 \text{ in.} \qquad \text{(calculator answer)}$$

$$= 0.04 \text{ in.} \qquad \text{(correct answer)}$$

The calculator answer must be rounded to one significant figure because 1 mm, the given quantity, contains only one significant figure.

Practice Exercise 2.7

Blood analysis reports often give the amounts of various substances present in the blood in terms of milligrams per deciliter. What is the measure, in quarts, of 1.00 deciliter?

Answer: 0.106 qt

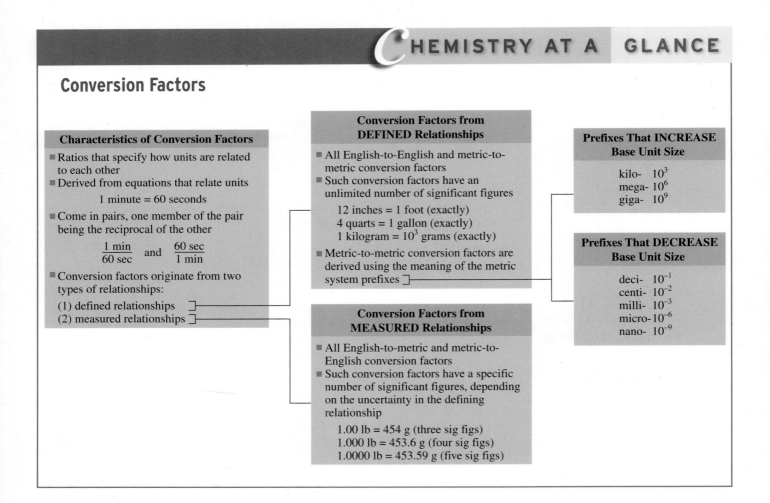

CHEMISTRY AT A GLANCE

Conversion Factors

Characteristics of Conversion Factors

- Ratios that specify how units are related to each other
- Derived from equations that relate units

    1 minute = 60 seconds

- Come in pairs, one member of the pair being the reciprocal of the other

    $\dfrac{1 \text{ min}}{60 \text{ sec}}$ and $\dfrac{60 \text{ sec}}{1 \text{ min}}$

- Conversion factors originate from two types of relationships:

    (1) defined relationships
    (2) measured relationships

Conversion Factors from DEFINED Relationships

- All English-to-English and metric-to-metric conversion factors
- Such conversion factors have an unlimited number of significant figures

    12 inches = 1 foot (exactly)
    4 quarts = 1 gallon (exactly)
    1 kilogram = 10^3 grams (exactly)

- Metric-to-metric conversion factors are derived using the meaning of the metric system prefixes

Conversion Factors from MEASURED Relationships

- All English-to-metric and metric-to-English conversion factors
- Such conversion factors have a specific number of significant figures, depending on the uncertainty in the defining relationship

    1.00 lb = 454 g (three sig figs)
    1.000 lb = 453.6 g (four sig figs)
    1.0000 lb = 453.59 g (five sig figs)

Prefixes That INCREASE Base Unit Size

kilo- 10^3
mega- 10^6
giga- 10^9

Prefixes That DECREASE Base Unit Size

deci- 10^{-1}
centi- 10^{-2}
milli- 10^{-3}
micro- 10^{-6}
nano- 10^{-9}

Density is a *physical property* peculiar to a given substance or mixture under fixed conditions.

DENSITY

Density *is the ratio of the mass of an object to the volume occupied by that object.*

$$\text{Density} = \frac{\text{mass}}{\text{volume}}$$

People often speak of a substance as being heavier or lighter than another substance. What they actually mean is that the two substances have different densities; a specific volume of one substance is heavier or lighter than the same volume of the second substance. Equal masses of substances with different densities occupy different volumes; the contrast in volume is often very striking (see Figure 2.8).

A correct density expression includes a number, a mass unit, and a volume unit. Although any mass and volume units can be used, densities are generally expressed in grams per cubic centimeter (g/cm^3) for solids, grams per milliliter (g/mL) for liquids, and grams per liter (g/L) for gases. Table 2.3 gives density values for a number of substances. Note that temperature must be specified with density values because substances expand and contract with changes in temperature. For the same reason, the pressure of gases is also given with their density values.

An object placed in a liquid either floats on the liquid's surface, sinks to the bottom of the liquid, or remains at some intermediate position in which it has been placed in the liquid (neither floating nor sinking), depending on how its density compares to that of the liquid. A floating object has a density that is less than that of the liquid (see Figure 2.9), a sinking object has a density that is greater than that of the liquid, and a stationary object (neither floats nor sinks) has a density that is the same as that of the liquid.

Figure 2.8 Both of these items have a mass of 23 grams, but they have very different volumes; therefore, their densities are different as well.

Figure 2.9 The penny is less dense than the mercury it floats on.

TABLE 2.3
Densities of Selected Substances

Solids (25°C)			
gold	19.3 g/cm^3	table salt	2.16 g/cm^3
lead	11.3 g/cm^3	bone	1.7–2.0 g/cm^3
copper	8.93 g/cm^3	table sugar	1.59 g/cm^3
aluminum	2.70 g/cm^3	wood (pine)	0.30–0.50 g/cm^3
Liquids (25°C)			
mercury	13.55 g/mL	water	0.997 g/mL
milk	1.028–1.035 g/mL	olive oil	0.92 g/mL
blood plasma	1.027 g/mL	ethyl alcohol	0.79 g/mL
urine	1.003–1.030 g/mL	gasoline	0.56 g/mL
Gases (25°C and 1 atmosphere pressure)			
chlorine	3.17 g/L	nitrogen	1.25 g/L
carbon dioxide	1.96 g/L	methane	0.66 g/L
oxygen	1.42 g/L	hydrogen	0.08 g/L
air (dry)	1.29 g/L		

CHEMICAL Connections   Body Density and Percent Body Fat

More than half the adult population of the United States is over-weight. But what does "overweight" mean? In years past, people were considered overweight if they weighed more for their height than called for in standard height/mass charts. Such charts are now considered outdated. Today, we realize that body composition is more important than total body mass. The proportion of fat to total body mass—that is, the percent of body fat—is the key to defining *overweight*. A very muscular person, for example, can be overweight according to height/mass charts although he or she has very little body fat. Some athletes fall into this category. Body composition ratings, tied to percent body fat, are listed here.

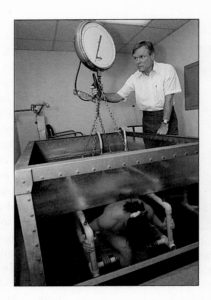

Body Composition Rating	Percent Body Fat	
	Men	**Women***
excellent	less than 13	less than 18
good	13–17	18–22
average	18–21	23–26
fair	22–30	27–35
poor	greater than 30	greater than 35

*Women are genetically predisposed to maintain a higher percentage of body fat.

The percentage of fat in a person's body can be determined by hydrostatic (underwater) weighing. Fat cells, unlike most other human body cells and fluids, are less dense than water. Consequently, a person with a high percentage of body fat is more buoyed up by water than a lean person. The hydrostatic-weighing technique for determining body fat is based on this difference in density. A person is first weighed in air and then weighed again submerged in water. The difference between these two masses (with a correction for residual air in the lungs and for the temperature of the water) is used to calculate body density. The higher the density of the body, the lower the percent of body fat. Sample values relating body density and percent body fat are given here.

Body Density (g/mL)	Percent Body Fat
1.070	12.22
1.062	15.25
1.052	19.29
1.036	25.35
1.027	29.39

 EXAMPLE 2.8

Calculating Density Given a Mass and a Volume

A student determines that the mass of a 20.0-mL sample of olive oil is 18.4 g. What is the density of the olive oil in grams per milliliter?

Solution

To calculate density, we substitute the given mass and volume values into the defining formula for density.

$$\text{Density} = \frac{\text{mass}}{\text{volume}} = \frac{18.4 \text{ g}}{20.0 \text{ mL}} = 0.92 \frac{\text{g}}{\text{mL}} \quad \text{(calculator answer)}$$

$$= 0.920 \frac{\text{g}}{\text{mL}} \quad \text{(correct anwer)}$$

Because both input numbers contain three significant figures, the density is specified to three significant figures.

Practice Exercise 2.8

A sample of table sugar (sucrose) with a mass of 2.500 g occupies a volume of 1.575 cm³. What is the density, in grams per cubic centimeter, of this sample of table sugar?

Answer: 1.587 g/cm³

Density as a Conversion Factor

Density can be used as a conversion factor that relates the volume of a substance to its mass. This use of density enables us to calculate the volume of a substance if we know its mass. Conversely, the mass can be calculated if the volume is known.

Density conversion factors, like all other conversion factors, have two reciprocal forms. For a density of 1.03 g/mL, the two conversion factor forms are

$$\frac{1.03 \text{ g}}{1 \text{ mL}} \quad \text{and} \quad \frac{1 \text{ mL}}{1.03 \text{ g}}$$

Note that the number 1 always goes in front of a "naked" unit in a conversion factor; that is, a density given as 5.2 g/mL means 5.2 grams per 1 mL.

> Density may be used as a conversion factor to convert from mass to volume or vice versa.

● **EXAMPLE 2.9**

Converting from Mass to Volume by Using Density as a Conversion Factor

Blood plasma has a density of 1.027 g/mL at 25°C. What volume, in milliliters, does 125 g of plasma occupy?

Solution

Step 1:   The given quantity is 125 g of blood plasma. The units of the desired quantity are milliliters. Thus our starting point is

$$125 \text{ g} = ? \text{ mL}$$

Step 2:   The conversion from grams to milliliters can be accomplished in one step because the given density, used as a conversion factor, directly relates grams to milliliters. Of the two conversion factor forms

$$\frac{1.027 \text{ g}}{1 \text{ mL}} \quad \text{and} \quad \frac{1 \text{ mL}}{1.027 \text{ g}}$$

we will use the latter because it allows for cancellation of gram units, leaving milliliters.

$$125 \text{ g} \times \left(\frac{1 \text{ mL}}{1.027 \text{ g}}\right) = ? \text{ mL}$$

Step 3:   Doing the necessary arithmetic gives us our answer:

$$\left(\frac{125 \times 1}{1.027}\right) \text{mL} = 121.71372 \text{ mL} \quad \text{(calculator answer)}$$
$$= 122 \text{ mL} \quad \text{(correct answer)}$$

Even though the given density contained four significant figures, the correct answer is limited to three significant figures. This is because the other given number, the mass of blood plasma, had only three significant figures.

Practice Exercise 2.9

If your blood has a density of 1.05 g/mL at 25°C, how many grams of blood would you lose if you made a blood bank donation of 1.00 pint (473 mL) of blood?

Answer: 497 g blood

 **2.10 TEMPERATURE SCALES**

Heat is a form of energy. Temperature is an indicator of the tendency of heat energy to be transferred. Heat energy flows from objects of higher temperature to objects of lower temperature.

Figure 2.10 The relationships among the Celsius, Kelvin, and Fahrenheit temperature scales are determined by the degree sizes and the reference point values. The reference point values are not drawn to scale.

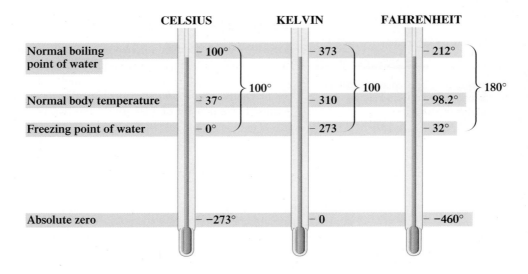

Zero on the Kelvin scale is known as *absolute zero*. It corresponds to the lowest temperature allowed by nature. How fast particles (molecules) move depends on temperature. The colder it gets, the more slowly they move. At absolute zero, movement stops. Scientists in laboratories have been able to attain temperatures as low as 0.0001 K, but a temperature of 0 K is impossible.

Three different temperature scales are in common use: Celsius, Kelvin, and Fahrenheit (Figure 2.10). Both the Celsius and the Kelvin scales are part of the metric measurement system; the Fahrenheit scale belongs to the English measurement system. Degrees of different size and different reference points are what produce the various temperature scales.

The *Celsius scale* is the scale most commonly encountered in scientific work. The normal boiling and freezing points of water serve as reference points on this scale, the former having a value of 100° and the latter 0°. Thus there are 100 "degree intervals" between the two reference points.

The *Kelvin scale* is a close relative of the Celsius scale. Both have the same size degree, and the number of degrees between the freezing and boiling points of water is the same. The two scales differ only in the numbers assigned to the reference points. On the Kelvin scale, the boiling point of water is 373 kelvins (K) and the freezing point of water is 273 K. The choice of these reference points makes all temperature readings on the Kelvin scale positive values. Note that the degree sign (°) is not used with the Kelvin scale. For example, we say that an object has a temperature of 350 K (*not* 350°K).

The *Fahrenheit scale* has a smaller degree size than the other two temperature scales. On this scale, there are 180 degrees between the freezing and boiling points of water as contrasted to 100 degrees on the other two scales. Thus the Celsius (and Kelvin) degree size is almost two times ($\frac{9}{5}$) larger than the Fahrenheit degree. Reference points on the Fahrenheit scale are 32° for the freezing point of water and 212° for the normal boiling point of water.

Besides the boiling and freezing point for water, a third reference point is shown in Figure 2.10 for each of the temperature scales—normal human body temperature. The Chemical Connections feature on page 44 gives further information concerning this reference point.

Conversions Between Temperature Scales

Because the size of the degree is the same, the relationship between the Kelvin and Celsius scales is very simple. No conversion factors are needed; all that is required is an adjustment for the differing numerical scale values. The adjustment factor is 273, the number of degrees by which the two scales are offset from one another.

$$K = °C + 273$$
$$°C = K - 273$$

The relationship between the Fahrenheit and Celsius scales can also be stated in an equation format.

$$°F = \frac{9}{5}(°C) + 32 \qquad \text{or} \qquad °C = \frac{5}{9}(°F - 32)$$

EXAMPLE 2.10

Converting from One Temperature
Scale to Another

Body temperature for a person with a high fever is found to be 104°F. To what is this temperature equivalent on the following scales?

a. Celsius scale   **b.** Kelvin scale

Solution

a. We substitute 104 for °F in the equation

$$°C = \frac{5}{9}(°F - 32)$$

Then solving for °C gives

$$°C = \frac{5}{9}(104 - 32) = \frac{5}{9}(72) = 40°$$

b. Using the answer from part **a** and the equation

$$K = °C + 273$$

we get, by substitution,

$$K = 40 + 273 = 313$$

Practice Exercise 2.10

In the human body, heat stroke occurs at a temperature of 41°C. To what is this temperature equivalent on the following scales?

a. Fahrenheit scale   **b.** Kelvin scale

Answers: a. 106°F; **b.** 314 K

TABLE 2.4
Specific Heats of Selected Common Substances

Substance	Specific Heat (cal/g · °C)*
water, liquid	1.00
ethyl alcohol	0.58
olive oil	0.47
wood	0.42
aluminum	0.21
glass	0.12
silver	0.057
gold	0.031

*The unit notation cal/g · °C means calories per gram per degree Celsius.

In discussions involving nutrition, the energy content of foods, and dietary tables, the term *Calorie* (spelled with a capital C) is used. The dietetic Calorie is actually 1 kilocalorie (1000 calories). The statement that an oatmeal raisin cookie contains 60 Calories means that 60 kcal (60,000 cal) of energy is released when the cookie is metabolized (undergoes chemical change) within the body.

Temperature Readings and Significant Figures

Standard operating procedure in reading a thermometer is to estimate the temperature to the closest degree, giving a degree reading having an uncertainty in the "ones place." This means that Celsius or Fahrenheit temperatures of 10°, 20°, 30°, etc., are considered to have two significant figures even though no decimal point is explicitly shown after the zero. A temperature reading of 100°C or 100°F is considered to possess three significant figures.

 2.11 HEAT ENERGY AND SPECIFIC HEAT

The form of energy most often required for or released by chemical reactions and physical changes is *heat energy*. A commonly used unit for the measurement of heat energy is the calorie. A **calorie** (cal) *is the amount of heat energy needed to raise the temperature of 1 gram of water by 1 degree Celsius.* For large amounts of heat energy, the measurement is usually expressed in kilocalories.

$$1 \text{ kilocalorie} = 1000 \text{ calories}$$

Another unit for heat energy that is used with increasing frequency is the joule (J). The relationship between the joule (which rhymes with *pool*) and the calorie is

$$1 \text{ calorie} = 4.184 \text{ joules}$$

Heat energy values in calories can be converted to joules by using the conversion factor

$$\frac{4.184 \text{ J}}{1 \text{ cal}}$$

Specific heat *is the quantity of heat energy, in calories, necessary to raise the temperature of 1 gram of a substance by 1 degree Celsius.* Specific heats for a number of substances in various states are given in Table 2.4.

HEMICAL Connections

Normal Human Body Temperature

Studies show that "normal" human body temperature varies from individual to individual. For oral temperature measurements, this individual variance spans the range from 96°F to 101°F.

Furthermore, individual body temperatures vary with exercise and with the temperature of the surroundings. When excessive heat is produced in the body by strenuous exercise, oral temperature can rise as high as 103°F. On the other hand, when the body is exposed to cold, oral temperature can fall to values considerably below 96°F. A rapid fall in temperature of 2°F to 3°F produces uncontrollable shivering.

Each individual also has a characteristic pattern of temperature variation during the day, with differences of as much as 1°F to 3°F between high and low points. Body temperature is typically lowest in the very early morning, after several hours of sleep, when one is inactive and not digesting food. During the day, body temperature rises to a peak and begins to fall again. "Morning people"—people who are most productive early in the day—have a body temperature peak at midmorning or midday. "Night people"—people who feel as though they are just getting started as evening approaches and who work best late at night—have a body temperature peak in the evening.

What, then, is the average (normal) human body temperature? Reference books list the value 98.6°F (37.0°C) as the answer to this question. The source for this value is a study in-

volving over 1 million human body temperature readings that was published in 1868, over 140 years ago.

A 1992 study, published in the *Journal of the American Medical Association,* questions the validity of this average value (98.6°F). This new study notes that the 1868 study was carried out using thermometers that were more difficult to get accurate readings from than modern thermometers. The 1992 study is based on oral temperature readings obtained using electronic thermometers. Findings of this new study include the following:

1. The range of temperatures was 96.0°F to 100.8°F.
2. The mean (average) temperature was 98.2°F (36.8°C).
3. At 6 A.M., the temperature 98.9°F is the upper limit of the normal temperature range.
4. In late afternoon (4 P.M.), the temperature 99.9°F is the upper limit of the normal temperature range.
5. Women have a slightly higher average temperature than men (98.4°F versus 98.1°F).
6. Over the temperature range 96°F to 101°F, there is an average increase in heart rate of 2.44 beats per minute for each 1°F rise in temperature.

As a result of this study, future reference books will probably use 98.2°F rather than 98.6°F as the value for average (normal) human body temperature.

Water has the highest specific heat of all common substances.

The property of specific heat varies *slightly* with temperature and pressure. We will ignore such variations in this text.

The higher the specific heat of a substance, the less its temperature will change as it absorbs a given amount of heat. Water has a relatively high specific heat; it is thus a very effective coolant. The moderate climates of geographical areas near large bodies of water—the Hawaiian Islands, for example—are related to the ability of water to absorb large amounts of heat without undergoing drastic temperature changes. Desert areas (obviously lacking in water) experience low- and high-temperature extremes.

Specific heat is an important quantity because it can be used to calculate the number of calories required to heat a known mass of a substance from one temperature to another. It can also be used to calculate how much the temperature of a substance increases when it absorbs a known number of calories of heat. The equation used for such calculations is

$$\text{Heat absorbed} = \text{specific heat} \times \text{mass} \times \text{temperature change}$$

If any three of the four quantities in this equation are known, the fourth quantity can be calculated. If the units for specific heat are cal/g · °C, the units for mass are grams, and the units for temperature change are °C; then the heat absorbed has units of calories.

● EXAMPLE 2.11

Calculating the Amount of Heat Released as the Result of a Temperature Decrease

▶ If a hot-water bottle contains 1200 g of water at 65°C, how much heat, in calories, will it have supplied to a person's "aching back" by the time it has cooled to 37°C (assuming all of the heat energy goes into the person's back)?

Solution

We will substitute known quantities into the equation

Heat released = specific heat × mass × temperature change

Table 2.4 shows that the specific heat of liquid water is 1.00 cal/g · °C. The mass of the water is given as 1200 g. The temperature change in going from 65°C to 37°C is 28°C. Substituting these values into the preceding equation gives

$$\text{Heat released} = \left(\frac{1.00 \text{ cal}}{\cancel{g} \cdot \cancel{°C}}\right) \times (1200 \cancel{g}) \times (28 \cancel{°C})$$

$$= 33{,}600 \text{ cal} \qquad \text{(calculator answer)}$$

$$= 34{,}000 \text{ cal} \qquad \text{(correct answer)}$$

The given quantity of 1200 g and the temperature difference of 28°C, both of which have only two significant figures, limit the answer to two significant figures.

Practice Exercise 2.11

How much heat energy, in calories, must be absorbed by 125.0 g of water to raise its temperature by 12°C?

Answer: 1500 cal

CONCEPTS TO REMEMBER

The metric system. The metric system, the measurement system preferred by scientists, is a decimal system in which larger and smaller units of a quantity are related by factors of 10. Prefixes are used to designate relationships between the basic unit and larger or smaller units of a quantity. Units in the metric system include the gram (mass), liter (volume), and meter (length) (Section 2.2).

Exact and inexact numbers. Numbers are of two kinds: exact and inexact. An exact number has a value that has no uncertainty associated with it. Exact numbers occur in definitions, in counting, and in simple fractions. An inexact number has a value that has a degree of uncertainty associated with it. Inexact numbers are generated any time a measurement is made (Section 2.3).

Significant figures. Significant figures in a measurement are those digits that are certain, plus a last digit that has been estimated. The maximum number of significant figures possible in a measurement is determined by the design of the measuring device (Section 2.4).

Calculations and significant figures. Calculations should never increase (or decrease) the precision of experimental measurements. In multiplication and division, the number of significant figures in the answer is the same as that in the measurement containing the fewest significant figures. In addition and subtraction, the answer has no more digits to the right of the decimal point than are found in the measurement with the fewest digits to the right of the decimal point (Section 2.5).

Scientific notation. Scientific notation is a system for writing decimal numbers in a more compact form that greatly simplifies the mathematical operations of multiplication and division. In this system, numbers are expressed as the product of a number between 1 and 10 and 10 raised to a power (Section 2.6).

Conversion factors. A conversion factor is a ratio that specifies how one unit of measurement is related to another. They are derived from equations (equalities) that relate units and they always come in reciprocal pairs (Section 2.7).

Dimensional analysis. Dimensional analysis is a general problem-solving method in which the units associated with numbers are used as a guide in setting up calculations. A given quantity is multiplied by one or more conversion factors in such a manner that the unwanted (original) units are canceled, leaving only the desired units (Section 2.8).

Density. Density is the ratio of the mass of an object to the volume occupied by that object. A correct density expression includes a number, a mass unit, and a volume unit (Section 2.9).

Temperature scales. The three major temperature scales are the Celsius, Kelvin, and Fahrenheit scales. The size of the degree for the Celsius and Kelvin scales is the same; they differ only in the numerical values assigned to the reference points. The Fahrenheit scale has a smaller degree size than the other two temperature scales (Section 2.10).

Heat energy and specific heat. The most commonly used units of measurement for heat energy are the calorie and the joule. A calorie is the amount of heat energy needed to raise the temperature of 1 gram of water by 1 degree Celsius. A calorie is equivalent to 4.184 joules. The specific heat of a substance is the quantity of heat energy, in calories, that is necessary to raise the temperature of 1 gram of the substance by 1 degree Celsius (Section 2.11).

kEY REACTIONS AND EQUATIONS

1. Density of a substance (Section 2.9)

$$\text{Density} = \frac{\text{mass}}{\text{volume}}$$

2. Conversion of temperature readings from one scale to another (Section 2.10)

$$K = {}^\circ C + 273 \qquad\qquad {}^\circ C = K - 273$$

$${}^\circ F = \frac{9}{5}({}^\circ C) + 32 \qquad\qquad {}^\circ C = \frac{9}{5}({}^\circ F - 32)$$

3. Relationship between a calorie and a joule (Section 2.11)

$$1 \text{ calorie} = 4.184 \text{ joules}$$

4. Heat energy absorbed by a substance (Section 2.11)

$$\begin{array}{c}\text{Heat energy} \\ \text{absorbed}\end{array} = \begin{array}{c}\text{specific} \\ \text{heat}\end{array} \times \text{mass} \times \begin{array}{c}\text{temperature} \\ \text{change}\end{array}$$

EXERCISES *and* PROBLEMS

The members of each pair of problems in this section test similar material.

⬤ Measurement Systems (Section 2.1)

2.1 What is the main reason scientists prefer to use the metric system of measurement rather than the English system of measurement?

2.2 List the more common types of measurements made in chemical laboratories.

⬤ Metric System Units (Section 2.2)

2.3 Write the name of the metric system prefix associated with each of the following mathematical meanings.
a. 10^3 b. 10^{-3} c. 10^{-6} d. $1/10$

2.4 Write the name of the metric system prefix associated with each of the following mathematical meanings.
a. 10^{-2} b. 10^{-9} c. 10^6 d. $1/1000$

2.5 Write out the names of the metric system units that have the following abbreviations.
a. cm b. kL c. μL d. ng

2.6 Write out the names of the metric system units that have the following abbreviations.
a. mg b. pg c. Mm d. dL

2.7 Arrange each of the following from smallest to largest.
a. Milligram, centigram, nanogram
b. Gigameter, megameter, kilometer
c. Microliter, deciliter, picoliter
d. Milligram, kilogram, microgram

2.8 Arrange each of the following from smallest to largest.
a. Milliliter, gigaliter, microliter
b. Centigram, megagram, decigram
c. Micrometer, picometer, kilometer
d. Nanoliter, milliliter, centiliter

⬤ Exact and Inexact Numbers (Section 2.3)

2.9 A person is told that there are 60 minutes in an hour and also that a section of fence is 60 feet long. What is the difference relative to uncertainty between the value of 60 in these two pieces of information?

2.10 A person is told that there are 27 people in attendance at a meeting and also that the speed of a bicyclist is 27 miles per hour. What is the difference relative to uncertainty between the value of 27 in these two pieces of information?

2.11 Indicate whether the number in each of the following statements is an *exact* or an *inexact* number.
a. A classroom contains 32 chairs.
b. There are 60 seconds in a minute.
c. A bowl of cherries weighs 3.2 pounds.
d. A newspaper article contains 323 words.

2.12 Indicate whether the number in each of the following statements is an *exact* or an *inexact* number.
a. A classroom contains 63 students.
b. The car is traveling at a speed of 56 miles per hour.
c. The temperature on the back porch is $-3{}^\circ$F.
d. There are 3 feet in a yard.

2.13 Indicate whether each of the following quantities would involve an *exact* number or an *inexact* number.
a. The length of a swimming pool
b. The number of gummi bears in a bag
c. The number of quarts in a gallon
d. The surface area of a living room rug

2.14 Indicate whether each of the following quantities would involve an *exact* number or an *inexact* number.
a. The number of pages in a chemistry textbook
b. The number of teeth in a bear's mouth
c. The distance from Earth to the sun
d. The temperature of a heated oven

⬤ Uncertainty in Measurement (Section 2.4)

2.15 What is the magnitude of the uncertainty associated with each of the following measured numbers?
a. 2.730 b. 2345 c. 0.2222 d. 280

2.16 What is the magnitude of the uncertainty associated with each of the following measured numbers?
a. 24.35 b. 0.4006 c. 330 d. 3756.3

2.17 Indicate to what decimal position readings should be recorded (nearest 0.1, 0.01, etc.) for measurements made with the following devices.
a. A thermometer with a smallest scale marking of 1°C
b. A graduated cylinder with a smallest scale marking of 0.1 mL
c. A volumetric device with a smallest scale marking of 10 mL
d. A ruler with a smallest scale marking of 1 mm

2.18 Indicate to what decimal position readings should be recorded (nearest 0.1, 0.01, etc.) for measurements made with the following devices.
   a. A ruler with a smallest scale marking of 1 cm
   b. A device for measuring angles with a smallest scale marking of 1°
   c. A thermometer with a smallest scale marking of 0.1°F
   d. A graduated cylinder with a smallest scale marking of 10 mL

2.19 The number of people present at a college football game was estimated by police to be 50,000. How many people were present at the football game if you assume that this estimate has each of the following uncertainties?
   a. 10,000   b. 1000   c. 100   d. 10

2.20 The number of people present at an outdoor rock concert was estimated by police to be 38,000. How many people were present at the concert if you assume that this estimate has each of the following uncertainties?
   a. 10,000   b. 1000   c. 100   d. 10

2.21 Consider the following rulers as instruments for the measurement of length.

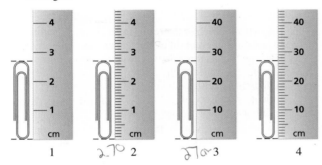

1   2.70 2   27 or 3   4

What would the uncertainty be in measurements made using the following?
   a. Ruler 1   b. Ruler 4

2.22 Using the rulers given in Problem 2.21, what would the uncertainty be in measurements made using the following?
   a. Ruler 2   b. Ruler 3

2.23 Using the rulers given in Problem 2.21, what is the length of the paper clip shown by the side of the following?
   a. Ruler 2   b. Ruler 3

2.24 Using the rulers given in Problem 2.21, what is the length of the paper clip shown by the side of the following?
   a. Ruler 1   b. Ruler 4

2.25 With which of the rulers in Problem 2.21 was each of the following measurements made, assuming that you cannot use a ruler multiple times in making the measurement? (It is possible that there may be more than one correct answer.)
   a. 20.4 cm   b. 2.3 cm   c. 3.74 cm   d. 32 cm

2.26 With which of the rulers in Problem 2.21 was each of the following measurements made, assuming that you cannot use a ruler multiple time in making the measurement? (It is possible that there may be more than one correct answer.)
   a. 3.2 cm   b. 3.22 cm   c. 22 cm   d. 3 cm

⬤ **Significant Figures and Measurement (Section 2.4)**

2.27 Determine the number of significant figures in each of the following measured values.
   a. 6.000   b. 0.0032   c. 0.01001
   d. 65,400   e. 76.010   f. 0.03050

2.28 Determine the number of significant figures in each of the following measured values.
   a. 23,009   b. 0.00231   c. 0.3330
   d. 73,000   e. 73.000   f. 0.40040

2.29 In which of the following pairs of numbers do both members of the pair contain the same number of significant figures?
   a. 11.01 and 11.00   b. 2002 and 2020
   c. 0.000066 and 660,000   d. 0.05700 and 0.05070

2.30 In which of the following pairs of numbers do both members of the pair contain the same number of significant figures?
   a. 345,000 and 340,500   b. 2302 and 2320
   c. 0.6600 and 0.66   d. 936 and 936,000

2.31 Identify the *estimated digit* in each of the measured values in Problem 2.27.

2.32 Identify the *estimated digit* in each of the measured values in Problem 2.28.

2.33 What is the magnitude of the uncertainty (±10, ±0.1, etc.) associated with each of the measured values in Problem 2.27?

2.34 What is the magnitude of the uncertainty (±10, ±0.1, etc.) associated with each of the measured values in Problem 2.28?

⬤ **Significant Figures and Mathematical Operations (Section 2.5)**

2.35 Round off each of the following numbers to the number of significant figures indicated in parentheses.
   a. 0.350763 (three)   b. 653,899 (four)
   c. 22.55555 (five)   d. 0.277654 (four)

2.36 Round off each of the following numbers to the number of significant figures indicated in parentheses.
   a. 3883 (two)   b. 0.0003011 (two)
   c. 4.4050 (three)   d. 2.1000 (three)

2.37 Without actually solving, indicate the number of significant figures that should be present in the answers to the following multiplication and division problems.
   a. $10.300 \times 0.30 \times 0.300$ b. $3300 \times 3330 \times 333.0$
   c. $\dfrac{6.0}{33.0}$ d. $\dfrac{6.000}{33}$

2.38 Without actually solving, indicate the number of significant figures that should be present in the answers to the following multiplication and division problems.
   a. $3.00 \times 0.0003 \times 30.00$ b. $0.3 \times 0.30 \times 3.0$
   c. $\dfrac{6.00}{33,000}$ d. $\dfrac{6.00000}{3}$

2.39 Carry out the following multiplications and divisions, expressing your answer to the correct number of significant figures. Assume that all numbers are measured numbers.
   a. $2.0000 \times 2.00 \times 0.0020$ b. 4.1567×0.00345
   c. $0.0037 \times 3700 \times 1.001$ d. $\dfrac{6.00}{33.0}$
   e. $\dfrac{530,000}{465.300}$ f. $\dfrac{4670 \times 3.00}{2.450}$

2.40 Carry out the following multiplications and divisions, expressing your answer to the correct number of significant figures. Assume that all numbers are measured numbers.
   a. $2.000 \times 0.200 \times 0.20$ b. 3.6750×0.04503
   c. $0.0030 \times 0.400 \times 4.00$ d. $\dfrac{6.0000}{33.00}$
   e. $\dfrac{45,000}{1.2345}$ f. $\dfrac{3.00 \times 6.53}{13.567}$

2.41 Carry out the following additions and subtractions, expressing your answer to the correct number of significant figures. Assume that all numbers are measured numbers.
   a. $12 + 23 + 127$ b. $3.111 + 3.11 + 3.1$
   c. $1237.6 + 23 + 0.12$ d. $43.65 - 23.7$

2.42 Carry out the following additions and subtractions, expressing your answer to the correct number of significant figures. Assume that all numbers are measured numbers.
   a. $237 + 37.0 + 7.0$ b. $4.000 + 4.002 + 4.20$
   c. $235.45 + 37 + 36.4$ d. $3.111 - 2.07$

● Scientific Notation (Section 2.6)

2.43 Express the following numbers in scientific notation.
   a. 120.7   b. 0.0034   c. 231.00
   d. 23,000   e. 0.200   f. 0.1011

2.44 Express the following numbers in scientific notation.
   a. 37.06   b. 0.00571   c. 437.0
   d. 4370   e. 0.20340   f. 230,000

2.45 Which number in each pair of numbers is the larger of the two numbers?
   a. 1.0×10^{-3} or 1.0×10^{-6}
   b. 1.0×10^{3} or 1.0×10^{-2}
   c. 6.3×10^{4} or 2.3×10^{4}
   d. 6.3×10^{-4} or 1.2×10^{-4}

2.46 Which number in each pair of numbers is the larger of the two numbers?
   a. 2.0×10^{2} or 2.0×10^{-2} b. 1.0×10^{6} or 3.0×10^{6}
   c. 4.4×10^{-4} or 4.4×10^{-5} d. 9.7×10^{3} or 8.3×10^{2}

2.47 How many significant figures are present in each of the following measured numbers?
   a. 1.0×10^{2} b. 5.34×10^{6}
   c. 5.34×10^{-4} d. 6.000×10^{3}

2.48 How many significant figures are present in each of the following measured numbers?
   a. 1.01×10^{2} b. 1.00×10^{2}
   c. 6.6700×10^{8} d. 6.050×10^{-3}

2.49 Carry out the following multiplications and divisions, expressing your answer in scientific notation to the correct number of significant figures.
   a. $(3.20 \times 10^{7}) \times (1.720 \times 10^{5})$
   b. $(3.71 \times 10^{-4}) \times (1.117 \times 10^{2})$
   c. $(1.00 \times 10^{3}) \times (5.00 \times 10^{3}) \times (3.0 \times 10^{-3})$
   d. $\dfrac{3.0 \times 10^{-5}}{1.5 \times 10^{2}}$ e. $\dfrac{4.56 \times 10^{7}}{3.0 \times 10^{-4}}$
   f. $\dfrac{(2.2 \times 10^{6}) \times (2.3 \times 10^{-6})}{(1.2 \times 10^{-3}) \times (3.5 \times 10^{-3})}$

2.50 Carry out the following multiplications and divisions, expressing your answer in scientific notation to the correct number of significant figures.
   a. $(4.0 \times 10^{4}) \times (1.32 \times 10^{8})$
   b. $(2.23 \times 10^{-6}) \times (1.230 \times 10^{-2})$
   c. $(3.200 \times 10^{7}) \times (1.10 \times 10^{-2}) \times (2.3 \times 10^{-7})$
   d. $\dfrac{6.0 \times 10^{-5}}{3.0 \times 10^{3}}$ e. $\dfrac{5.132 \times 10^{7}}{1.12 \times 10^{3}}$
   f. $\dfrac{(3.2 \times 10^{2}) \times (3.31 \times 10^{6})}{(4.00 \times 10^{-3}) \times (2.0 \times 10^{6})}$

2.51 What is the uncertainty, in terms of a power of ten, associated with each of the following measured values?
   a. 3.60×10^{4} b. 3.60×10^{6}
   c. 3.6×10^{5} d. 3.6×10^{-3}

2.52 What is the uncertainty, in terms of a power of ten, associated with each of the following measured values?
   a. 4.30×10^{-2} b. 4.300×10^{-1}
   c. 4.3×10^{3} d. 4.300×10^{5}

● Conversion Factors (Section 2.7)

2.53 Give the two forms of the conversion factor that relate each of the following pairs of units.
   a. Days and hours   b. Decades and centuries
   c. Feet and yards   d. Quarts and gallons

2.54 Give the two forms of the conversion factor that relate each of the following pairs of units.
   a. Days and weeks   b. Years and centuries
   c. Inches and feet   d. Pints and quarts

2.55 Give the two forms of the conversion factor that relate each of the following pairs of units.
   a. kL and L   b. mg and g
   c. m and cm   d. μsec and sec

2.56 Give the two forms of the conversion factor that relate each of the following pairs of units.
   a. ng and g   b. dL and L
   c. m and Mm   d. psec and sec

2.57 Indicate whether each of the following equations relating units would generate an *exact* set of conversion factors or an *inexact* set of conversion factors relative to significant figures.
   a. 1 dozen = 12 objects
   b. 1 kilogram = 2.20 pounds
   c. 1 minute = 60 seconds
   d. 1 millimeter = 10^{-3} meter

2.58 Indicate whether each of the following equations relating units would generate an *exact* set of conversion factors or an *inexact* set of conversion factors relative to significant figures.
   a. 1 gallon = 16 cups   b. 1 week = 7 days
   c. 1 pint = 0.4732 liter   d. 1 mile = 5280 feet

● Dimensional Analysis (Section 2.8)

2.59 Using dimensional analysis, convert each of the following measurements to meters.
   a. 1.6×10^{3} dm   b. 24 nm
   c. 0.003 km   d. 3.0×10^{8} mm

2.60 Using dimensional analysis, convert each of the following measurements to meters.
   a. 2.7×10^{3} mm   b. 24 μm
   c. 0.003 pm   d. 4.0×10^{5} cm

2.61 The human stomach produces approximately 2500 mL of gastric juice per day. What is the volume, in liters, of gastric juice produced?

2.62 A typical loss of water through sweating per day for a human is 450 mL. What is the volume, in liters, of sweat produced per day?

2.63 The mass of premature babies is customarily determined in grams. If a premature baby weighs 1550 g, what is its mass in pounds?

2.64 The smallest bone in the human body, which is in the ear, has a mass of 0.0030 g. What is the mass of this bone in pounds?

2.65 What volume of water, in gallons, would be required to fill a 25-mL container?

2.66 What volume of gasoline, in milliliters, would be required to fill a 17.0-gal gasoline tank?

2.67 An individual weighs 83.2 kg and is 1.92 m tall. What are the person's equivalent measurements in pounds and feet?

2.68 An individual weighs 135 lb and is 5 ft 4 in. tall. What are the person's equivalent measurements in kilograms and meters?

Density (Section 2.9)

2.69 A sample of mercury is found to have a mass of 524.5 g and to have a volume of 38.72 cm^3. What is its density in grams per cubic centimeter?

2.70 A sample of sand is found to have a mass of 12.0 g and to have a volume of 2.69 cm^3. What is its density in grams per cubic centimeter?

2.71 Acetone, the solvent in nail polish remover, has a density of 0.791 g/mL. What is the volume, in milliliters, of 20.0 g of acetone?

2.72 Silver metal has a density of 10.40 g/cm^3. What is the volume, in cubic centimeters, of a 100.0-g bar of silver metal?

2.73 The density of homogenized milk is 1.03 g/mL. How much does 1 cup (236 mL) of homogenized milk weigh in grams?

2.74 Nickel metal has a density of 8.90 g/cm^3. How much does 15 cm^3 of nickel metal weigh in grams?

2.75 Water has a density of 1.0 g/cm^3 at room temperature. State whether each of the following will sink or float when placed in water.
a. Paraffin wax (density = 0.90 g/cm^3)
b. Limestone (density = 2.8 g/cm^3)

2.76 Air has a density of 1.29 g/L at room temperature. State whether each of the following will rise or sink in air.
a. Helium gas (density = 0.18 g/L)
b. Argon gas (density = 1.78 g/L)

Temperature Scales (Section 2.10)

2.77 An oven for baking pizza operates at approximately 525°F. What is this temperature in degrees Celsius?

2.78 A comfortable temperature for bathtub water is 95°F. What temperature is this in degrees Celsius?

2.79 Mercury freezes at −38.9°C. What is the coldest temperature, in degrees Fahrenheit, that can be measured using a mercury thermometer?

2.80 The body temperature for a hypothermia victim is found to have dropped to 29.1°C. What is this temperature in degrees Fahrenheit?

2.81 Which is the higher temperature, −10°C or 10°F?

2.82 Which is the higher temperature, −15°C or 4°F?

Heat Energy and Specific Heat (Section 2.11)

2.83 A substance has a specific heat of 0.63 cal/g · °C. What is its specific heat in J/g · °C?

2.84 A substance has a specific heat of 0.24 cal/g · °C. What is its specific heat in J/g · °C?

2.85 If it takes 18.6 cal of heat to raise the temperature of 12.0 g of a substance by 10.0°C, what is the specific heat of the substance?

2.86 If it takes 35.0 cal of heat to raise the temperature of 25.0 g of a substance by 12.0°C, what is the specific heat of the substance?

2.87 How many calories of heat energy are required to raise the temperature of 42.0 g of each of the following substances from 20.0°C to 40.0°C?
a. Silver   b. Liquid water   c. Aluminum

2.88 How many calories of heat energy are necessary to raise the temperature of 20.0 g of each of the following substances from 25°C to 55°C?
a. Gold   b. Ethyl alcohol   c. Olive oil

ADDITIONAL PROBLEMS

2.89 A person is told that there are 12 inches in a foot and also that a piece of rope is 12 inches long. What is the fundamental difference between the value of 12 in these two pieces of information?

2.90 Round off the number 4.7205059 to the indicated number of significant figures.
a. Six   b. Five   c. Four   d. Two

2.91 Write each of the following numbers in scientific notation to the number of significant figures indicated in parentheses.
a. 0.00300300 (three)   b. 936,000 (two)
c. 23.5003 (three)   d. 450,000,001 (six)

2.92 For each of the pairs of units listed, indicate whether the first unit is larger or smaller than the second unit, and then indicate how many times larger or smaller it is.
a. Milliliter, liter   b. Kiloliter, microliter
c. Nanoliter, deciliter   d. Centiliter, megaliter

2.93 Indicate how each of the following conversion factors should be interpreted in terms of significant figures present.
a. $\dfrac{2.540 \text{ cm}}{1.000 \text{ in.}}$ b. $\dfrac{453.6 \text{ g}}{1.000 \text{ lb}}$
c. $\dfrac{2.113 \text{ pt}}{1.00 \text{ L}}$ d. $\dfrac{10^{-9} \text{ m}}{1 \text{ nm}}$

2.94 A one-gram sample of a powdery white solid is found to have a volume of two cubic centimeters. Calculate the solid's density using the following uncertainty specifications, and express your answers in scientific notation.
a. 1.0 g and 2.0 cm^3 b. 1.000 g and 2.00 cm^3
c. 1.0000 g and 2.0000 cm^3 d. 1.000 g and 2.0000 cm^3

2.95 Calculate the volume, in *milliliters*, for each of the following.
a. 75.0 g of gasoline (density = 0.56 g/mL)
b. 75.0 g of sodium metal (density = 0.93 g/cm^3)
c. 75.0 g of ammonia gas (density = 0.759 g/L)
d. 75.0 g of mercury (density = 13.6 g/mL)

2.96 Which quantity of heat energy in each of the following pairs of heat energy values is the larger?
a. 2.0 joules or 2.0 calories
b. 1.0 kilocalorie or 92 calories
c. 100 Calories or 100 calories
d. 2.3 Calories or 1000 kilocalories

2.97 The concentration of salt in a salt solution is found to be 4.5 mg/mL. What is the salt concentration in each of the following units?
a. mg/L   b. pg/mL   c. g/L   d. kg/m^3

$\mathcal{M}$ultiple-Choice Practice Test

2.98 In which of the following pairings of metric system prefix and power of ten is the pairing *incorrect*?
a. Kilo- and 10^{-3} b. Micro- and 10^{-6}
c. Deci- and 10^{-1} d. Mega and 10^{6}

2.99 Which of the following statements about the "significance" of zeros in recorded measurements is *incorrect*?
a. Leading zeros are never significant.
b. Confined zeros are always significant.
c. Trailing zeros are never significant.
d. Trailing zeros may or may not be significant.

2.100 The *estimated* digit in the measurement 65,430 seconds is
a. the zero.   b. the three.   c. the four.   d. the five.

2.101 When rounded to three significant figures, the number 43267 becomes
a. 432   b. 433   c. 43200   d. 43300

2.102 The uncertainty associated with the measurement 0.3030 lies in the
a. tenths place (0.1).
b. hundredths place (0.01).
c. thousandths place (0.001).
d. ten-thousandths place (0.0001).

2.103 The number 273.00, when expressed in scientific notation, becomes

a. 2.73×10^{-2} b. 2.7300×10^{-2}
c. 2.73×10^{2} d. 2.7300×10^{2}

2.104 The calculator answer obtained by multiplying the measurements 53.534 and 5.00 is 267.67. This answer
a. is correct as written.   b. should be rounded to 267.7.
c. should be rounded to 268.   d. should be rounded to 270.

2.105 The calculator answer obtained by adding the measurements 8.1, 2.19, and 3.123 is 13.413. This answer
a. is correct as written.
b. should be rounded to two significant figures.
c. should be rounded to 13.41.
d. should be rounded to 13.4.

2.106 What is the volume, in milliliters, of 50.0 g of a liquid if its density is 1.20 g/mL?
a. 32.1 mL   b. 41.7 mL   c. 60.0 mL   d. 75.0 mL

2.107 Which of the following statements concerning the three major temperature scales is *incorrect*?
a. Kelvin scale temperatures can never have negative values.
b. A Celsius degree and a Kelvin are equal in size.
c. The addition of 273 to a Fahrenheit scale reading will convert it to a Kelvin scale reading.
d. The freezing point of water has a lower numerical value on the Celsius scale than on the Fahrenheit scale.

Atomic Structure and the Periodic Table

Chapter Outline

Music consists of a series of tones that build octave after octave. Similarly, elements have properties that recur period after period.

In Chapter 1 we learned that all matter is made up of small particles called atoms and that 117 different types of atoms are known, each type of atom corresponding to a different element. Furthermore, we found that compounds result from the chemical combination of different types of atoms in various ratios and arrangements.

Until the last two decades of the nineteenth century, scientists believed that atoms were solid, indivisible spheres without an internal structure. Today, this model of the atom is known to be incorrect. Evidence from a variety of sources indicates that atoms are made up of even smaller particles called *subatomic particles*. In this chapter we consider the fundamental types of subatomic particles, how they arrange themselves within an atom, and the relationship between an atom's subatomic makeup and its chemical identity.

3.1 INTERNAL STRUCTURE OF AN ATOM

Atoms possess internal structure; that is, they are made up of even smaller particles, which are called subatomic particles. A **subatomic particle** *is a very small particle that is a building block for atoms.* Three types of subatomic particles are found within atoms: electrons, protons, and neutrons. Key properties of these three types of particles are summarized in Table 3.1. An **electron** *is a subatomic particle that possesses a negative (−) electrical charge.* It is the smallest, in terms of mass, of the three types of subatomic particles. A **proton** *is a subatomic particle that possesses a positive (+) electrical charge.* Protons and electrons carry the *same amount* of charge; the charges, however, are opposite

TABLE 3.1
Charge and Mass Characteristics of Electrons, Protons, and Neutrons

	Electron	Proton	Neutron
Charge	−1	+1	0
Actual mass (g)	9.109×10^{-28}	1.673×10^{-24}	1.675×10^{-24}
Relative mass (based on the electron being 1 unit)	1	1837	1839

Atoms of all 117 elements contain the same three types of subatomic particles. Different elements differ only in the numbers of the various subatomic particles they contain.

The radius of a nucleus is approximately 10,000 times smaller than the radius of an entire atom.

(positive versus negative). A **neutron** *is a subatomic particle that has no charge associated with it; that is, it is neutral.* Both protons and neutrons are massive particles compared to electrons; they are almost 2000 times heavier.

Arrangement of Subatomic Particles Within an Atom

The arrangement of subatomic particles within an atom is not haphazard. *All* protons and *all* neutrons present are found at the center of an atom in a very tiny volume called the *nucleus* (Figure 3.1). The **nucleus** *is the small, dense, positively charged center of an atom.* A nucleus is always positively charged because it contains positively charged protons. Because the nucleus houses the heavy subatomic particles (protons and neutrons), almost all (over 99.9%) of the mass of an atom is concentrated in its nucleus. The small size of the nucleus, coupled with its large amount of mass, causes nuclear material to be extremely dense.

The outer (extranuclear) region of an atom contains all of the electrons. In this region, which accounts for most of the volume of an atom, the electrons move rapidly about the nucleus. The electrons are attracted to the positively charged protons of the nucleus by the forces that exist between particles of opposite charge. The motion of the electrons in the extranuclear region determines the volume (size) of the atom in the same way that the blade of a fan determines a volume by its circular motion. The volume occupied by the electrons is sometimes referred to as the *electron cloud.* Because electrons are negatively charged, the electron cloud is also negatively charged. Figure 3.1 contrasts the nuclear and extranuclear regions of an atom. The attractive force between the nucleus (positively charged) and the electrons (negatively charged) keeps the electrons within the extranuclear region of the atom. By analogy, the attractive force of gravity keeps the planets in their positions about the sun.

Closely resembling the term *nucleus* is the term *nucleon.* A **nucleon** *is any subatomic particle found in the nucleus of an atom.* Thus both protons and neutrons are nucleons, and the nucleus can be regarded as containing a collection of nucleons (protons and neutrons).

Charge Neutrality of an Atom

An atom as a whole is electrically neutral; that is, it has no *net* electrical charge. For this to be the case, the same number of positive and negative charges must be present in the atom. Equal numbers of positive and negative charges give a *net* electrical charge of zero. Thus equal numbers of protons and electrons are present in an atom.

Number of protons = number of electrons

Size Relationships Within an Atom

To help you visualize the size relationships among the parts of an atom, imagine enlarging (magnifying) the nucleus until it is the size of a baseball (about 2.9 inches in diameter). If the nucleus were this large, the whole atom would have a diameter of approximately 2.5 miles. The electrons would still be smaller than the periods used to end sentences in this text, and they would move about at random within that 2.5-mile region.

The concentration of nearly all of the mass of an atom in the nucleus can also be illustrated by using an imaginary example. If a coin the same size as a copper penny contained copper nuclei (copper atoms stripped of their electrons) rather than copper atoms (which are mostly empty space), the coin would weigh 190,000,000 tons! Nuclei are indeed very dense matter.

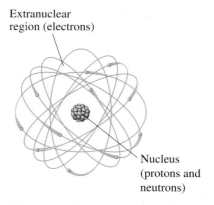

Extranuclear region (electrons)

Nucleus (protons and neutrons)

Figure 3.1 The protons and neutrons of an atom are found in the central nuclear region, or nucleus, and the electrons are found in an electron cloud outside the nucleus. Note that this figure is not drawn to scale; the correct scale would be comparable to a penny (the nucleus) in the center of a baseball field (the atom).

Despite the existence of subatomic particles, we will continue to refer to atoms as the fundamental building blocks for all types of matter. Subatomic particles do not lead an independent existence for any appreciable length of time; they gain stability by joining together to form atoms.

 3.2 ATOMIC NUMBER AND MASS NUMBER

Atomic number and mass number are always *whole* numbers because they are obtained by counting whole objects (protons, neutrons, and electrons).

An **atomic number** *is the number of protons in the nucleus of an atom.* Because an atom has the same number of electrons as protons (Section 3.1), the atomic number also specifies the number of electrons present.

Atomic number = number of protons = number of electrons

The symbol Z is used as a general designation for atomic number.

A **mass number** *is the sum of the number of protons and the number of neutrons in the nucleus of an atom.* Thus the mass number gives the number of subatomic particles present in the nucleus.

Mass number = number of protons + number of neutrons

The mass of an atom is almost totally accounted for by the protons and neutrons present—hence the term *mass number*. The symbol A is used as a general designation for mass number.

The number and identity of subatomic particles present in an atom can be calculated from its atomic and mass numbers in the following manner.

The *sum* of the mass number and the atomic number for an atom ($A + Z$) corresponds to the total number of subatomic particles present in the atom (protons, neutrons, and electrons).

Number of protons = atomic number = Z

Number of electrons = atomic number = Z

Number of neutrons = mass number − atomic number = $A - Z$

Note that neutron count is obtained by subtracting atomic number from mass number.

● **EXAMPLE 3.1**

Determining the Subatomic Particle Makeup of an Atom Given Its Atomic Number and Mass Number

▶ An atom has an atomic number of 9 and a mass number of 19.

  a. Determine the number of protons present.
  b. Determine the number of neutrons present.
  c. Determine the number of electrons present.

Solution

  a. There are 9 protons because the atomic number is always equal to the number of protons present.
  b. There are 10 neutrons because the number of neutrons is always obtained by subtracting the atomic number from the mass number ($19 - 9 = 10$).

$$\underbrace{(\text{Protons} + \text{neutrons})}_{\text{Mass number}} - \underbrace{\text{protons}}_{\substack{\text{Atomic} \\ \text{number}}} = \text{neutrons}$$

  c. There are 9 electrons because the number of protons and the number of electrons are always the same in an atom.

Practice Exercise 3.1

An atom has an atomic number of 11 and a mass number of 23.

  a. Determine the number of protons present.
  b. Determine the number of neutrons present.
  c. Determine the number of electrons present.

Answers: a. 11 protons; **b.** 12 neutrons; **c.** 11 electrons

An alphabetical list of the 117 known elements, with their atomic numbers as well as other information, is found on the inside front cover of this text. If you check the atomic number column in this tabulation, you will find an entry for each of the numbers in the sequence 1 through 116 plus the number 118. Scientists interpret this continuous atomic number sequence 1 through 116 as evidence that there are no "missing elements" yet to be discovered in nature. The naturally occurring element with the highest atomic number is element 92 (uranium). Elements 93 through 116 and 118 are all laboratory produced (Section 1.7); to date, laboratory experiments to produce element 117 have not been successful.

The mass and atomic numbers of a given atom are often specified using the notation

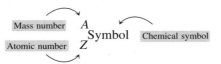

In this notation, often called *complete chemical symbol notation,* the atomic number is placed as a subscript in front of the chemical symbol and the mass number is placed as a superscript in front of the chemical symbol. Examples of such notation for actual atoms include

$$^{19}_{9}\text{F} \qquad ^{23}_{11}\text{Na} \qquad ^{197}_{79}\text{Au}$$

The first of these notations specifies a fluorine atom that has an atomic number of 9 and a mass number of 19.

Electrons and Chemical Properties

The chemical properties of an atom, which are the basis for its identification, are determined by the number and arrangement of the electrons about the nucleus. When two atoms interact, the outer part (electrons) of one interacts with the outer part (electrons) of the other. The small nuclear centers never come in contact with each other in a chemical reaction. The number of electrons about a nucleus may be considered to be determined by the number of protons in the nucleus; charge balance requires an equal number of the two (Section 3.1). Hence the number of protons (which is the atomic number) characterizes an atom. All atoms with the same atomic number have the same chemical properties and are atoms of the same element.

In Section 1.6, an element was defined as a pure substance that cannot be broken down into simpler substances by ordinary chemical means. Although this is a good historical definition for an element, we can now give a more rigorous definition by using the concept of atomic number. An **element** *is a pure substance in which all atoms present have the same atomic number.*

 ## 3.3 ISOTOPES AND ATOMIC MASSES

Charge neutrality (Section 3.1) requires the presence in an atom of an equal number of protons and electrons. However, because neutrons have no electrical charge, their numbers in atoms do not have to be the same as the number of protons or electrons. Most atoms contain more neutrons than either protons or electrons.

Studies of atoms of various elements also show that the number of neutrons present in atoms of an element is not constant; it varies over a small range. This means that not all atoms of an element have to be identical. They must have the same number of protons and electrons, but they can differ in the number of neutrons.

Isotopes

Atoms of an element that differ in neutron count are called isotopes. **Isotopes** *are atoms of an element that have the same number of protons and the same number of electrons but different numbers of neutrons.* Different isotopes always have the same atomic number and different mass numbers.

The word *isotope* comes from the Greek *iso,* meaning "equal," and *topos,* meaning "place." Isotopes occupy an equal place (location) in listings of elements because all isotopes of an element have the same atomic number.

CHEMICAL Connections

Protium, Deuterium, and Tritium: The Three Isotopes of Hydrogen

Measurable differences in physical properties are found among isotopes for elements with low atomic numbers. This results from differences in mass among isotopes being relatively large compared to the masses of the isotopes themselves. The situation is greatest for the element hydrogen, the element with the lowest atomic number.

Three isotopes of hydrogen exist: ^1H, ^2H, and ^3H. With a single proton and no neutrons in its nucleus, hydrogen-1 is by far the most abundant isotope (99.985%). Hydrogen-2, with a neutron in addition to a proton in its nucleus, has an abundance of 0.015%. The presence of the additional neutron in ^2H doubles its mass compared to that of ^1H. Hydrogen-3 has two neutrons and a proton in its nucleus and has a mass triple that of ^1H. Only minute amounts of ^3H, which is radioactive (unstable: see Section 11.1), occur naturally.

In discussions involving hydrogen isotopes, special names and symbols are given to the isotopes—something that does not occur for any other element. Hydrogen-1 is usually called hydrogen but is occasionally called *protium*. Hydrogen-2 has the name *deuterium* (symbol D), and hydrogen-3 is called *tritium* (symbol T).

(a) Protium   (b) Deuterium   (c) Tritium
Names for the three isotopes of hydrogen.

The following table contrasts the properties of H_2 and D_2.

Isotope	Melting Point	Boiling Point	Density (at 0°C and 1 atmosphere pressure)
H_2	−259°C	−253°C	0.090 g/L
D_2	−253°C	−250°C	0.18 g/L

Water in which both hydrogen atoms are deuterium (D_2O) is called "heavy water." The properties of heavy water are measurably different from those of "ordinary" H_2O.

Compound	Melting Point	Boiling Point	Density (at 0°C and 1 atmosphere pressure)
H_2O	0.0°C	100.0°C	0.99987 g/mL
D_2O	3.82°C	101.4°C	1.1047 g/mL

Heavy water (D_2O) can be obtained from natural water by distilling a sample of natural water, because the D_2O has a slightly higher boiling point than H_2O. Pure deuterium (D_2) is produced by decomposing the D_2O. Heavy water is used in the operation of nuclear power plants (to slow down free neutrons present in the reactor core).

Tritium, the heaviest hydrogen isotope, is used in nuclear weapons. Because of the minute amount of naturally occurring tritium, it must be synthesized in the laboratory using bombardment reactions (Section 11.5).

Isotopes of an element have the same chemical properties, but their physical properties are often slightly different. Isotopes of an element have the same chemical properties because they have the same number of electrons. They have slightly different physical properties because they have different numbers of neutrons and therefore different masses.

Most elements found in nature exist in isotopic forms, with the number of naturally occurring isotopes ranging from two to ten. For example, all silicon atoms have 14 protons and 14 electrons. Most silicon atoms also contain 14 neutrons. However, some silicon atoms contain 15 neutrons and others contain 16 neutrons. Thus three different kinds of silicon atoms exist.

When it is necessary to distinguish between isotopes of an element, complete chemical symbol notation (Section 3.2) is used. The three silicon isotopes are designated, respectively, as

$$^{28}_{14}Si, \qquad ^{29}_{14}Si, \qquad \text{and} \qquad ^{30}_{14}Si$$

Names for isotopes include the mass number. $^{28}_{14}Si$ is called silicon-28, and $^{29}_{14}Si$ is called silicon-29. The atomic number is not included in the name because it is the same for all isotopes of an element.

The various isotopes of a given element are of varying abundance; usually one isotope is predominant. Silicon is typical of this situation. The percentage abundances for its three isotopes are 92.21% ($^{28}_{14}Si$), 4.70% ($^{29}_{14}Si$), and 3.09% ($^{30}_{14}Si$). Percentage abundances are number percentages (numbers of atoms) rather than mass percentages. A sample of 10,000 silicon atoms contains 9221 $^{28}_{14}Si$ atoms, 470 $^{29}_{14}Si$ atoms, and 309 $^{30}_{14}Si$ atoms.

There are a few elements for which all naturally occurring atoms have the same number of neutrons—that is, for which all atoms are identical. They include the elements Be, F, Na, Al, P, and Au.

A mass number, in contrast to an atomic number, lacks uniqueness. Atoms of different elements can have the same mass number. For example, carbon-14 and nitrogen-14 have the same mass numbers. Atoms of different elements, however, cannot have the same atomic number.

There are 286 isotopes that occur naturally. In addition, over 2000 more have been synthesized in the laboratory via nuclear rather than chemical reactions (Section 11.5). All these synthetic isotopes are unstable (radioactive). Despite their instability, many are used in chemical research, as well as in medicine.

● **E X A M P L E 3 . 2**

Determining the Subatomic Particle Characteristics of an Atom Given Its Complete Chemical Symbol

▶ Determine the following for an atom whose complete chemical symbol is $^{26}_{12}Mg$.

a. The total number of subatomic particles present in the atom
b. The total number of subatomic particles present in the nucleus of the atom
c. The total number of nucleons present in the atom
d. The total charge (including sign) associated with the nucleus of the atom

Solution

a. The mass number gives the combined number of protons and neutrons present. The atomic number gives the number of electrons present. Adding these two numbers together gives the total number of subatomic particles present. There are 38 subatomic particles present ($26 + 12 = 38$).
b. The nucleus contains all protons and all neutrons. The mass number (protons + neutrons), thus, gives the total number of subatomic particles present in the nucleus of an atom. There are 26 subatomic particles present in the nucleus.
c. A nucleon is any subatomic particle present in the nucleus. Thus, both protons and neutrons are nucleons. There are 26 such particles present in the nucleus. Parts **b** and **c** of this example are, thus, asking the same thing using different terminology.
d. The charge associated with a nucleus originates from the protons present. It will always be positive because protons are positively charged particles. The atomic number, 12, indicates that 12 protons are present. Thus, the nuclear charge is $+12$.

Practice Exercise 3.2

Determine the following for an atom whose complete chemical symbol is $^{63}_{29}Cu$.

a. The total number of subatomic particles present in the atom
b. The total number of subatomic particles present in the nucleus of the atom
c. The total number of nucleons present in the atom
d. The total charge (including sign) associated with the nucleus of the atom

Answers: **a.** 92 subatomic particles; **b.** 63 subatomic particles; **c.** 63 nucleons; **d.** $+29$ charge

Atomic Masses

An analogy involving isotopes and identical twins may be helpful: Identical twins need not weigh the same, even though they have identical "gene packages." Likewise, isotopes, even though they have different masses, have the same number of protons.

The terms *atomic mass* and *atomic weight* are often used interchangeably. *Atomic mass,* however, is the correct term.

The existence of isotopes means that atoms of an element can have several different masses. For example, silicon atoms can have any one of three masses because there are three silicon isotopes. Which of these three silicon isotopic masses is used in situations in which the mass of the element silicon needs to be specified? The answer is none of them. Instead we use a *weighted-average mass* that takes into account the existence of isotopes and their relative abundances.

The *weighted-average mass* of the isotopes of an element is known as the element's atomic mass. An **atomic mass** *is the calculated average mass for the isotopes of an element, expressed on a scale where $^{12}_{6}C$ serves as the reference point.* What we need to calculate an atomic mass are the masses of the various isotopes on the $^{12}_{6}C$ reference scale and the percentage abundance of each isotope.

The $^{12}_{6}C$ reference scale mentioned in the definition of *atomic mass* is a scale scientists have set up for comparing the masses of atoms. On this scale, the mass of a $^{12}_{6}C$ atom is defined to be exactly 12 atomic mass units (amu). The masses of all other atoms are then determined relative to that of $^{12}_{6}C$. For example, if an atom is twice as heavy as $^{12}_{6}C$, its mass is 24 amu, and if an atom weighs half as much as an atom of $^{12}_{6}C$, its mass is 6 amu.

Example 3.3 shows how an atomic mass is calculated using the amu ($^{12}_{6}C$) scale, the percentage abundances of isotopes, and the number of isotopes of an element.

EXAMPLE 3.3

Calculation of an Element's Atomic Mass

Naturally occurring chlorine exists in two isotopic forms, $_{17}^{35}Cl$ and $_{17}^{37}Cl$. The relative mass of $_{17}^{35}Cl$ is 34.97 amu, and its abundance is 75.53%; the relative mass of $_{17}^{37}Cl$ is 36.97 amu, and its abundance is 24.47%. What is the atomic mass of chlorine?

Solution

An element's atomic mass is calculated by multiplying the relative mass of each isotope by its fractional abundance and then totaling the products. The fractional abundance for an isotope is its percentage abundance converted to decimal form (divided by 100).

$$_{17}^{35}Cl: \qquad \left(\frac{75.53}{100}\right) \times 34.97\ \text{amu} = (0.7553) \times 34.97\ \text{amu} = 26.41\ \text{amu}$$

$$_{17}^{37}Cl: \qquad \left(\frac{24.47}{100}\right) \times 36.97\ \text{amu} = (0.2447) \times 36.97\ \text{amu} = 9.047\ \text{amu}$$

$$\text{Atomic mass of Cl} = (26.41 + 9.047)\ \text{amu}$$

$$= 35.46\ \text{amu}$$

This calculation involved an element containing just two isotopes. A similar calculation for an element having three isotopes would be carried out the same way, but it would have three terms in the final sum; an element possessing four isotopes would have four terms in the final sum.

Practice Exercise 3.3

Naturally occurring copper exists in two isotopic forms, $_{29}^{63}Cu$ and $_{29}^{65}Cu$. The relative mass of $_{29}^{63}Cu$ is 62.93 amu, and its abundance is 69.09%; the relative mass of $_{29}^{65}Cu$ is 64.93 amu, and its abundance is 30.91%. What is the atomic mass of copper?

Answer: 63.55 amu

The alphabetical list of the known elements printed inside the front cover of this text gives the calculated atomic mass for each of the elements; it is the last column of numbers. Table 3.2 gives isotopic data for the elements with atomic numbers 1 through 12.

TABLE 3.2
Isotopic Data for Elements with Atomic Numbers 1 Through 12
Information given for each isotope includes mass number, isotopic mass in terms of amu, and percentage abundance.

1	Hydrogen	2	Helium	3	Lithium
$_1^1H$ 1.008 amu 99.985% $_1^2H$ 2.014 amu 0.015% $_1^3H$ 3.016 amu trace		$_2^3He$ 3.016 amu trace $_2^4He$ 4.003 amu 100%		$_3^6Li$ 6.015 amu 7.42% $_3^7Li$ 7.016 amu 92.58%	
4	Beryllium	5	Boron	6	Carbon
$_4^9Be$ 9.012 amu 100%		$_5^{10}B$ 10.013 amu 19.6% $_5^{11}B$ 11.009 amu 80.4%		$_6^{12}C$ 12.000 amu 98.89% $_6^{13}C$ 13.003 amu 1.11% $_6^{14}C$ 14.003 amu trace	
7	Nitrogen	8	Oxygen	9	Fluorine
$_7^{14}N$ 14.003 amu 99.63% $_7^{15}N$ 15.000 amu 0.37%		$_8^{16}O$ 15.995 amu 99.759% $_8^{17}O$ 16.999 amu 0.037% $_8^{18}O$ 17.999 amu 0.204%		$_9^{19}F$ 18.998 amu 100%	
10	Neon	11	Sodium	12	Magnesium
$_{10}^{20}Ne$ 19.992 amu 90.92% $_{10}^{21}Ne$ 20.994 amu 0.26% $_{10}^{22}Ne$ 21.991 amu 8.82%		$_{11}^{23}Na$ 22.990 amu 100%		$_{12}^{24}Mg$ 23.985 amu 78.70% $_{12}^{25}Mg$ 24.986 amu 10.13% $_{12}^{26}Mg$ 25.983 amu 11.17%	

Percent Composition by Mass

There are two methods for expressing the percent elemental composition of a substance: (1) percent by number of atoms (used in atomic mass calculations) and (2) percent by mass. These two percent methods for elemental composition produce quite different numerical values. Compare the approximate percent elemental composition for the human body, as given in Table 3.3, using these two methods.

By mass, our bodies are mostly oxygen (65%) because of the large amount of water (H_2O) in our bodies and because oxygen is 16 times heavier than hydrogen. Carbon is the second most abundant element, by mass, in the human body, and

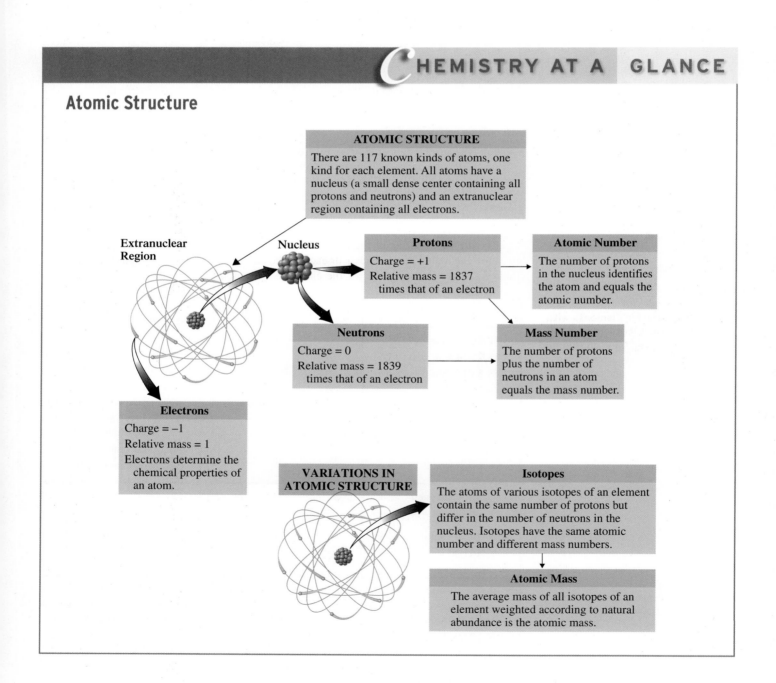

CHEMISTRY AT A GLANCE

Atomic Structure

ATOMIC STRUCTURE

There are 117 known kinds of atoms, one kind for each element. All atoms have a nucleus (a small dense center containing all protons and neutrons) and an extranuclear region containing all electrons.

Extranuclear Region

Nucleus

Protons

Charge = +1
Relative mass = 1837 times that of an electron

Atomic Number

The number of protons in the nucleus identifies the atom and equals the atomic number.

Neutrons

Charge = 0
Relative mass = 1839 times that of an electron

Mass Number

The number of protons plus the number of neutrons in an atom equals the mass number.

Electrons

Charge = −1
Relative mass = 1
Electrons determine the chemical properties of an atom.

VARIATIONS IN ATOMIC STRUCTURE

Isotopes

The atoms of various isotopes of an element contain the same number of protons but differ in the number of neutrons in the nucleus. Isotopes have the same atomic number and different mass numbers.

Atomic Mass

The average mass of all isotopes of an element weighted according to natural abundance is the atomic mass.

TABLE 3.3
Approximate Percent Elemental Composition of the Human Body

Element	Percent by Mass	Percent by Number of Atoms
Oxygen	65	26
Carbon	18	11
Hydrogen	10	60
Nitrogen	3	2
All other elements	4	1

hydrogen ranks third. Based on the number of atoms, hydrogen is the most abundant element (60%), with oxygen second and carbon third. There are twice as many hydrogen atoms as oxygen atoms in water, which is the most abundant substance in the human body.

Water's contribution to human body mass is approximately 70%, a value that is relatively easy to determine. It would be extremely difficult to determine the percentage of molecules in the human body that are water molecules because of the many different types of complex biochemical molecules, most of which are present in small numbers.

The Chemistry at a Glance feature on page 58 summarizes what has been said about atoms in Sections 3.1 through 3.3.

3.4 THE PERIODIC LAW AND THE PERIODIC TABLE

During the early part of the nineteenth century, scientists began to look for order in the increasing amount of chemical information that had become available. They knew that certain elements had properties that were very similar to those of other elements, and they sought reasons for these similarities in the hope that these similarities would suggest a method for arranging or classifying the elements.

In 1869, these efforts culminated in the discovery of what is now called the *periodic law*, proposed independently by the Russian chemist Dmitri Mendeleev (Figure 3.2) and the German chemist Julius Lothar Meyer. Given in its modern form, the **periodic law** *states that when elements are arranged in order of increasing atomic number, elements with similar chemical properties occur at periodic (regularly recurring) intervals.*

A periodic table is a visual representation of the behavior described by the periodic law. A **periodic table** *is a tabular arrangement of the elements in order of increasing atomic number such that elements having similar chemical properties are positioned in vertical columns.* The most commonly used form of the periodic table is shown in Figure 3.3 (see also the inside front cover of the text). Within the table, each element is represented by a rectangular box that contains the symbol, atomic number, and atomic mass of the element. Elements within any given column of the periodic table exhibit similar chemical behavior.

Figure 3.2 Dmitri Ivanovich Mendeleev (1834–1907). Mendeleev constructed a periodic table as part of his effort to systematize chemistry. He received many international honors for his work, but his reception at home in czarist Russia was mixed. Element 101 carries his name.

Using the information on a periodic table, you can quickly determine the number of protons and electrons for atoms of an element. However, no information concerning neutrons is available from a periodic table; mass numbers are not part of the information given because they are not unique to an element.

Groups and Periods of Elements

The location of an element within the periodic table is specified by giving its period number and group number.

A **period** *is a horizontal row of elements in the periodic table.* For identification purposes, the periods are numbered sequentially with Arabic numbers, starting at the top of the periodic table. In Figure 3.3, period numbers are found on the left side of the table. The elements Na, Mg, Al, Si, P, S, Cl, and Ar are all members of Period 3, the

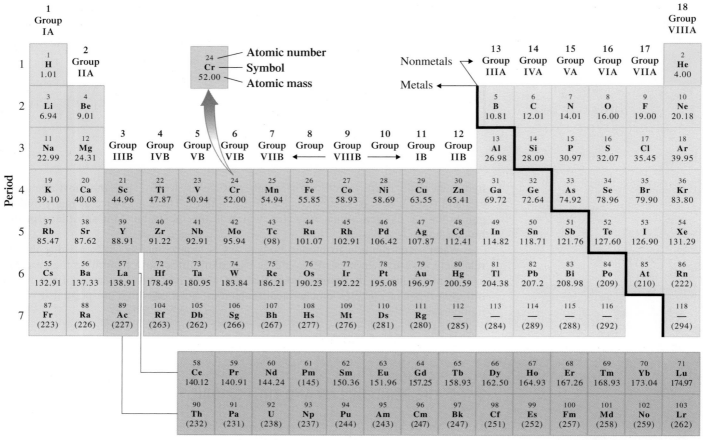

Figure 3.3 The periodic table of the elements is a graphical way to show relationships among the elements. Elements with similar chemical properties fall in the same vertical column.

third row of elements. Period 4 is the fourth row of elements, and so on. There are only two elements in Period 1, H and He.

A **group** *is a vertical column of elements in the periodic table.* There are two notations in use for designating individual periodic-table groups. In the first notation, which has been in use for many years, groups are designated by using Roman numerals and the letters A and B. In the second notation, which an international scientific commission has recommended, the Arabic numbers 1 through 18 are used. Note that in Figure 3.3 both group notations are given at the top of each group. The elements with atomic numbers 8, 16, 34, 52, and 84 (O, S, Se, Te, and Po) constitute Group VIA (old notation) or Group 16 (new notation).

Four groups of elements also have common (non-numerical) names. On the extreme left side of the periodic table are found the *alkali metals* (Li, Na, K, Rb, Cs, Fr) and the *alkaline earth metals* (Be, Mg, Ca, Sr, Ba, Ra). **Alkali metal** *is a general name for any element in Group IA of the periodic table, excluding hydrogen.* The alkali metals are soft, shiny metals that readily react with water. **Alkaline earth metal** *is a general name for any element in Group IIA of the periodic table.* The alkaline earth metals are also soft, shiny metals but they are only moderately reactive toward water. On the extreme right of the periodic table are found the *halogens* (F, Cl, Br, I, At) and the *noble gases* (He, Ne, Ar, Kr, Xe, Rn). **Halogen** *is a general name for any element in Group VIIA of the periodic table.* The halogens are reactive elements that are gases at room temperature or become such at temperatures slightly above room temperature. **Noble gas** *is a general name for any element in Group VIIIA of the periodic table.* Noble gases are unreactive gases that undergo few, if any, chemical reactions.

The elements within a given periodic-table group show numerous similarities in properties, the degree of similarity varying from group to group. In no case are the group members "clones" of one another. Each element has some individual characteristics not found in other elements of the group. By analogy, the members of a human family often bear many resemblances to each other, but each member also has some (and often much) individuality.

The location of any element in the periodic table is specified by giving its group number and its period number. The element gold, with an atomic number of 79, belongs to Group IB (or 11) and is in Period 6. The element nitrogen, with an atomic number of 7, belongs to Group VA (or 15) and is in Period 2.

EXAMPLE 3.4

Identifying Groups, Periods, and Specially Named Families of Elements

▶ What is the chemical symbol of the element that fits each of the following descriptions based on periodic table location?

a. Located in both Period 3 and Group IVA
b. The Period 4 noble gas
c. The Period 2 alkaline earth metal
d. The Period 3 halogen

Solution

a. Period 3 is the third row of elements and Group IVA is the fifth column from the right side of the periodic table. The element that has this column–row (period–group) location is Si (silicon).
b. The noble gases are the elements of Group VIIIA (the right-most column in the periodic table). The Period 4 (fourth row) noble gas is Kr (krypton).
c. The alkaline earth metals are the elements of Group IIA (the second column from the left side of the periodic table). The Period 2 (second row) alkaline earth metal is Be (beryllium).
d. The halogens are the elements of Group VIIA (the second column from the right side of the periodic table). The Period 3 (third row) halogen is Cl (chlorine).

Practice Exercise 3.4

What is the chemical symbol of the element that fits each of the following descriptions based on periodic table location?

a. Located in both Period 4 and Group VIA
b. The Period 3 alkali metal
c. The Period 2 noble gas
d. The Period 4 halogen

Answers: a. Se (selenium); **b.** Na (sodium); **c.** Ne (neon); **d.** Br (bromine)

The Shape of the Periodic Table

When the phrase "the first ten elements" is used, it means the first ten elements in the periodic table, the elements with atomic numbers 1 through 10.

Within the periodic table of Figure 3.3, the practice of arranging the elements according to increasing atomic number is violated in Groups IIIB and IVB. Element 72 follows element 57, and element 104 follows element 89. The missing elements, elements 58 through 71 and 90 through 103 are located in two rows at the bottom of the periodic table. Technically, the elements at the bottom of the table should be included in the body of the table, as shown in Figure 3.4. However, in order to have a more compact table, we place them at the bottom of the table as shown in Figure 3.3.

1																	2
3	4											5	6	7	8	9	10
11	12											13	14	15	16	17	18
19	20	21				22	23	24	25	26	27	28	29	30	31	32	33	34	35	36
37	38	39				40	41	42	43	44	45	46	47	48	49	50	51	52	53	54
55	56	57	58 59 60 61 62 63 64 65 66 67 68 69 70 71	72	73	74	75	76	77	78	79	80	81	82	83	84	85	86
87	88	89	90 91 92 93 94 95 96 97 98 99 100 101 102 103	104	105	106	107	108	109	110	111	112	113	114	115	116		118

Figure 3.4 In this periodic table, elements 58 through 71 and 90 through 103 (in color) are shown in their proper positions.

(a) Metals　　　　　**(b) Nonmetals**

Figure 3.5 (a) Some familiar metals (clockwise, starting on left) are aluminum, lead, tin, and zinc. (b) Some familiar nonmetals are sulfur (yellow), phosphorus (dark red), and bromine (reddish-brown liquid).

 METALS AND NONMETALS

In the previous section, we noted that the Group IA and IIA elements are known, respectively, as the alkali metals and the alkaline earth metals. Both of these designations contain the word *metal.* What is a metal?

On the basis of selected physical properties, elements are classified into the categories metal and nonmetal. A **metal** *is an element that has the characteristic properties of luster, thermal conductivity, electrical conductivity, and malleability.* With the exception of mercury, all metals are solids at room temperature (25°C). Metals are good conductors of heat and electricity. Most metals are ductile (can be drawn into wires) and malleable (can be rolled into sheets). Most metals have high luster (shine), high density, and high melting points. Among the more familiar metals are the elements iron, aluminum, copper, silver, gold, lead, tin, and zinc (see Figure 3.5a).

A **nonmetal** *is an element characterized by the absence of the properties of luster, thermal conductivity, electrical conductivity, and malleability.* Many of the nonmetals, such as hydrogen, oxygen, nitrogen, and the noble gases, are gases. The only nonmetal found as a liquid at room temperature is bromine. Solid nonmetals include carbon, iodine, sulfur, and phosphorus (Figure 3.5b). In general, the nonmetals have lower densities and lower melting points than metals. Table 3.4 contrasts selected physical properties of metals and nonmetals.

Metals generally are malleable, ductile, and lustrous and are good thermal and electrical conductors. *Nonmetals* tend to lack these properties. In many ways, the general properties of metals and nonmetals are opposites.

TABLE 3.4
Selected Physical Properties of Metals and Nonmetals

Metals	Nonmetals
1. High electrical conductivity that decreases with increasing temperature	1. Poor electrical conductivity (except carbon in the form of graphite)
2. High thermal conductivity	2. Good heat insulators (except carbon in the form of diamond)
3. Metallic gray or silver luster*	3. No metallic luster
4. Almost all are solids[†]	4. Solids, liquids, or gases
5. Malleable (can be hammered into sheets)	5. Brittle in solid state
6. Ductile (can be drawn into wires)	6. Nonductile

*Except copper and gold.
[†]Except mercury; cesium and gallium melt on a hot summer day (85°F) or when held in a person's hand.

Figure 3.6 This portion of the periodic table shows the dividing line between metals and nonmetals. All elements that are not shown are metals.

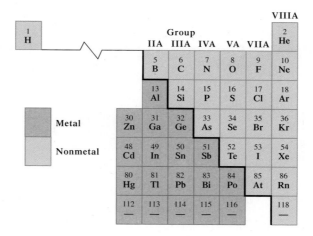

Periodic Table Locations for Metals and Nonmetals

The majority of the elements are metals. Only 22 elements are nonmetals. It is not necessary to memorize which elements are nonmetals and which are metals; this information is obtainable from a periodic table (Figure 3.6). The steplike heavy line that runs through the right third of the periodic table separates the metals on the left from the nonmetals on the right. Note also that the element hydrogen is a nonmetal.

The fact that the vast majority of elements are metals in no way indicates that metals are more important than nonmetals. Most nonmetals are relatively common and are found in many important compounds. For example, water (H_2O) is a compound involving two nonmetals.

An analysis of the abundance of the elements in Earth's crust (Figure 1.10) in terms of metals and nonmetals shows that the two most abundant elements, which account for 80.2% of all atoms, are nonmetals—oxygen and silicon. The four most abundant elements in the human body (see the Chemical Connections feature on page 10 in Chapter 1), which comprise over 99% of all atoms in the body, are nonmetals—hydrogen, oxygen, carbon, and nitrogen. Besides these four nonmetals, a number of other elements (including several metals), in small amounts, are also important in the functioning of the human body (see the Chemical Connections feature on page 64).

ELECTRON ARRANGEMENTS WITHIN ATOMS

As electrons move about an atom's nucleus, they are restricted to specific regions within the extranuclear portion of the atom. Such restrictions are determined by the amount of energy the electrons possess. Furthermore, electron energies are limited to certain values, and a specific "behavior" is associated with each allowed energy value.

The space in which electrons move rapidly about a nucleus is divided into subspaces called *shells, subshells,* and *orbitals.*

Electron Shells

Electrons within an atom are grouped into main energy levels called electron shells. An **electron shell** *is a region of space about a nucleus that contains electrons that have approximately the same energy and that spend most of their time approximately the same distance from the nucleus.*

Electron shells are numbered 1, 2, 3, and so on, outward from the nucleus. Electron energy increases as the distance of the electron shell from the nucleus increases. An electron in shell 1 has the minimum amount of energy that an electron can have.

The maximum number of electrons that an electron shell can accommodate varies; the higher the shell number (n), the more electrons that can be present. In higher-energy shells, the electrons are farther from the nucleus, and a greater volume of space is available for them; hence more electrons can be accommodated. (Conceptually, electron shells

Electrons that occupy the first electron shell are closer to the nucleus and have a lower energy than electrons in the second electron shell.

Chemical Connections   Metallic Elements and the Human Body

Four elements—hydrogen, oxygen, carbon, and nitrogen—supply 99% of the atoms in the human body and give rise to 96% of the body's mass (Section 3.3). These four elements, often called the *building-block elements,* are all nonmetals. Hydrogen and oxygen, besides being present in water, are also present along with carbon in carbohydrates and fats. These same three elements along with nitrogen are found in proteins. Does this mean that metals, which constitute the vast majority of the elements, are unimportant in the proper functioning of the human body? The answer is an unqualified no.

Another group of elements essential to proper human body function, which includes several metals, are the *minerals,* elements needed in small amounts that must be obtained from food. There are the *major minerals* and the *trace minerals*, with the former being required in larger amounts than the latter.

The major minerals, seven in number, include four metals and three nonmetals. The four metals, all located on the left side of the periodic table, are sodium, potassium, magnesium, and calcium. The three nonmetals, all Period 3 nonmetals, are phosphorus, sulfur, and chlorine. The relative amounts of the major minerals present in a human body are given in the top part of the accompanying graph. Note that these minerals are not present in the body in elemental form, but rather as constituents of compounds; for example, sodium is not present as sodium metal but as the compound sodium chloride (table salt).

Trace minerals are needed in much smaller quantities than the major minerals. The least abundant major mineral is over ten times more abundant than the most abundant trace mineral, as shown in the bottom part of the accompanying graph. This graph only shows the six most abundant of the trace minerals, four of which are metals. Other trace minerals that are metals include zinc, copper, cobalt, molybdenum, and chromium. One of the purposes of many dietary supplements of the multivitamin type is to ensure that adequate amounts of trace metals are part of a person's dietary intake (see accompanying dietary supplement label). The biological importance of iron, the most abundant of the trace minerals, is considered in a Chemical Connections feature later in this chapter.

As knowledge concerning the biological functions of trace minerals increases as the result of research endeavors, the way doctors and nutritionists think about diet and health changes. For example, it is now known that a combined supplement of manganese, copper, and zinc, in combination with calcium, improves

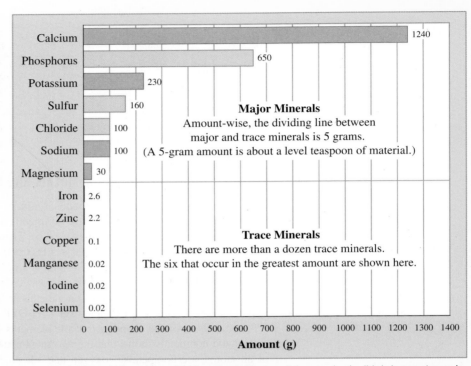

Major Minerals
Amount-wise, the dividing line between major and trace minerals is 5 grams. (A 5-gram amount is about a level teaspoon of material.)

Trace Minerals
There are more than a dozen trace minerals. The six that occur in the greatest amount are shown here.

Amounts of minerals found in a 65 kilogram (143 pound) human body. (Metals are shown in gray and nonmetals are shown in blue.)

Supplement Facts
Serving Size 1 tablet

Amount Per Tablet	% DV	Amount Per Tablet	% DV
Vitamin A 3,000 I.U.	60%	Iodine 150 mcg	100%
Vitamin C 120 mg	200%	Magnesium 100 mg	25%
Vitamin D 400 I.U.	100%	Zinc 15 mg	100%
Vitamin E 50 I.U.	167%	Selenium 25 mcg	36%
Vitamin K 25 mcg	31%	Copper 2 mg	100%
Thiamin 1.5 mg	100%	Manganese 2 mg	100%
Riboflavin 1.7 mg	100%	Chromium 120 mcg	100%
Niacin 20 mg	100%	Molybdenum 25 mcg	33%
Vitamin B6 2 mg	100%	Chloride 36 mg	1%
Folic Acid 400 mcg	100%	Potassium 40 mg	1%
Vitamin B12 6 mcg	100%	Boron 150 mcg	*
Biotin 30 mcg	10%	Nickel 5 mcg	*
Pantothenic Acid 10 mg	100%	Silicon 2 mg	*
Calcium 100 mg	10%	Tin 10 mcg	*
Iron 9 mg	50%	Vanadium 10 mcg	*
Phosphorus 77 mg	8%	Lutein 250 mcg	*
*Daily Value (DV) not established.			

bone health to a greater degree than a calcium supplement alone. Likewise, trace amounts of copper are needed for the proper absorption and mobilization of iron in the body.

may be considered to be nested one inside another, somewhat like the layers of flavors inside a jawbreaker or similar type of candy.)

The lowest-energy shell ($n = 1$) accommodates a maximum of 2 electrons. In the second, third, and fourth shells, 8, 18, and 32 electrons, respectively, are allowed. The relationship among these numbers is given by the formula $2n^2$, where n is the shell number. For example, when $n = 4$, the quantity $2n^2 = 2(4)^2 = 32$.

Electron Subshells

Within each electron shell, electrons are further grouped into energy sublevels called electron subshells. An **electron subshell** *is a region of space within an electron shell that contains electrons that have the same energy.* We can draw an analogy between the relationship of shells and subshells and the physical layout of a high-rise apartment complex. The shells are analogous to the floors of the apartment complex, and the subshells are the counterparts of the various apartments on each floor.

The number of subshells within a shell is the same as the shell number. Shell 1 contains one subshell, shell 2 contains two subshells, shell 3 contains three subshells, and so on.

Subshells within a shell differ in size (that is, the maximum number of electrons they can accommodate) and energy. The higher the energy of the contained electrons, the larger the subshell.

Subshell size (type) is designated using the letters *s, p, d,* and *f.* Listed in this order, these letters denote subshells of increasing energy and size. The lowest-energy subshell within a shell is always the *s* subshell, the next highest is the *p* subshell, then the *d* subshell, and finally the *f* subshell. An *s* subshell can accommodate 2 electrons, a *p* subshell 6 electrons, a *d* subshell 10 electrons, and an *f* subshell 14 electrons.

Both a number and a letter are used in identifying subshells. The number gives the shell within which the subshell is located, and the letter gives the type of subshell. Shell 1 has only one subshell—the 1*s.* Shell 2 has two subshells—the 2*s* and 2*p.* Shell 3 has three subshells—the 3*s,* 3*p,* and 3*d,* and so on. Figure 3.7 summarizes the relationships between electron shells and electron subshells for the first four shells.

The letters used to label the different types of subshells come from old spectroscopic terminology associated with the lines in the spectrum of the element hydrogen. These lines were denoted as *s*harp, *p*rincipal, *d*iffuse, and *f*undamental. Relationships exist between such lines and the arrangement of electrons in an atom.

Figure 3.7 The number of subshells within a shell is equal to the shell number, as shown here for the first four shells. Each individual subshell is denoted with both a number (its shell) and a letter (the type of subshell it is in).

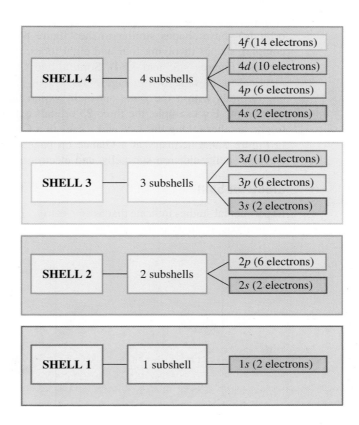

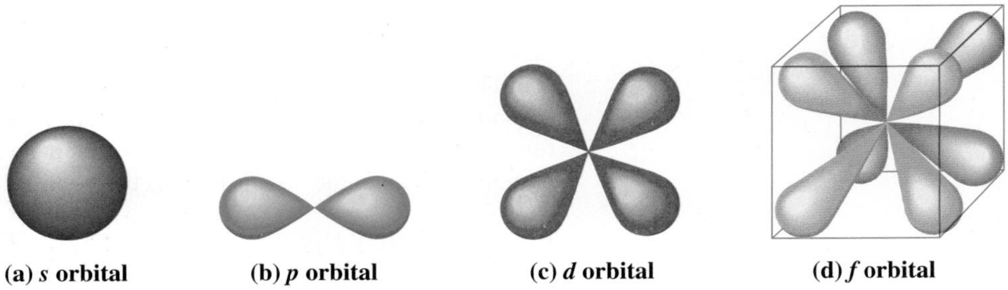

Figure 3.8 An s orbital has a spherical shape, a p orbital has two lobes, a d orbital has four lobes, and an f orbital has eight lobes. The f orbital is shown within a cube to illustrate that its lobes are directed toward the corners of a cube. Some d and f orbitals have shapes related to, but not identical to, those shown.

(a) s orbital　　**(b) p orbital**　　**(c) d orbital**　　**(d) f orbital**

The four subshell types (s, p, d, and f) are sufficient when dealing with shells of higher number than shell 4 because in such shells any additional subshells present are not needed to accommodate electrons. For example, in shell 5 there are five subshell types (5s, 5p, 5d, 5f, and a fifth one that is never used). The reason why some subshells are not needed involves consideration of the order of filling of subshells with electrons, which is the topic of Section 3.7.

Electron Orbitals

An electron orbital is also often called an atomic orbital.

Electron subshells have within them a certain, definite number of locations (regions of space), called electron orbitals, where electrons may be found. In our apartment complex analogy, if shells are the counterparts of floor levels and subshells are the apartments, then electron orbitals are the rooms of the apartments. An **electron orbital** *is a region of space within an electron subshell where an electron with a specific energy is most likely to be found.*

An electron orbital, independent of all other considerations, can accommodate a maximum of 2 electrons. Thus an s subshell (2 electrons) contains one orbital, a p subshell (6 electrons) contains three orbitals, a d subshell (10 electrons) contains five orbitals, and an f subshell (14 electrons) contains seven orbitals.

Orbitals have distinct shapes that are related to the type of subshell in which they are found. Note that we are talking not about the shape of an electron, but rather about the shape of the region in which the electron is found. An orbital in an s subshell, which is called an s orbital, has a spherical shape (Figure 3.8a). Orbitals found in p subshells—p orbitals—have shapes similar to the "figure 8" of an ice skater (Figure 3.8b). More complex shapes involving four and eight lobes, respectively, are associated with d and f orbitals (Figures 3.8c and 3.8d). Some d and f orbitals have shapes related to, but not identical to, those shown in Figure 3.8.

Orbitals within the same subshell, which have the same shape, differ mainly in orientation. For example, the three 2p orbitals extend out from the nucleus at 90° angles to one another (along the x, y, and z axes in a Cartesian coordinate system), as is shown in Figure 3.9. Chemistry at a Glance on page 67 shows key interrelationships among electron shells, electron subshells, and electron orbitals.

Electron Spin

Experimental studies indicate that as an electron "moves about" within an orbital, it spins on its own axis in either a clockwise or a counterclockwise direction. Furthermore, when two

Figure 3.9 Orbitals within a sub-shell differ mainly in orientation. For example, the three p orbitals within a p subshell lie along the x, y, and z axes of a Cartesian coordinate system.

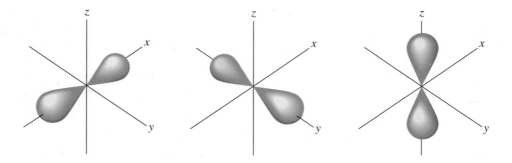

CHEMISTRY AT A GLANCE

Shell-Subshell-Orbital Interrelationships

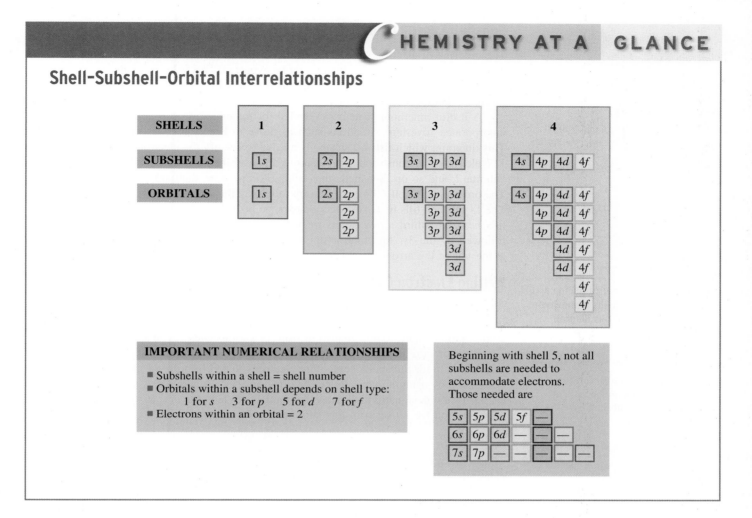

IMPORTANT NUMERICAL RELATIONSHIPS

- Subshells within a shell = shell number
- Orbitals within a subshell depends on shell type:
  1 for *s*   3 for *p*   5 for *d*   7 for *f*
- Electrons within an orbital = 2

Beginning with shell 5, not all subshells are needed to accommodate electrons. Those needed are

electrons are present in an orbital, they always have opposite spins; that is, one is spinning clockwise and the other counterclockwise. This situation of opposite spins is energetically the most favorable state for two electrons in the same orbital. We will have more to say about electron spin when we discuss orbital diagrams in Section 3.7.

3.7 ELECTRON CONFIGURATIONS AND ORBITAL DIAGRAMS

Electron shells, subshells, and orbitals describe "permissible" locations for electrons — that is, where electrons *can* be found. We are now ready to discuss *actual* locations of the electrons in specific atoms.

There are many orbitals about the nucleus of an atom. Electrons do not occupy these orbitals in a random, haphazard fashion; a very predictable pattern exists for electron orbital occupancy. There are three rules, all quite simple, for assigning electrons to various shells, subshells, and orbitals.

1. *Electron subshells are filled in order of increasing energy.*
2. *Electrons occupy the orbitals of a subshell such that each orbital acquires one electron before any orbital acquires a second electron. All electrons in such singly occupied orbitals must have the same spin.*
3. *No more than two electrons may exist in a given orbital — and then only if they have opposite spins.*

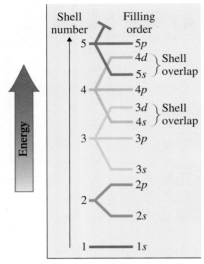

Figure 3.10 The order of filling of various electron subshells is shown on the right-hand side of this diagram. Above the 3p subshell, subshells of different shells "overlap."

All electrons in a given subshell have the same energy because all orbitals within a subshell have the same energy.

An *electron configuration* specifies *subshell* occupancy for electrons, and an *orbital diagram* specifies *orbital* occupancy for electrons.

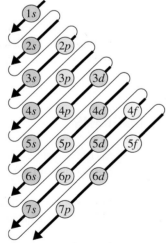

Figure 3.11 The order for filling electron subshells with electrons follows the order given by the arrows in this diagram. Start with the arrow at the top of the diagram and work toward the bottom of the diagram, moving from the bottom of one arrow to the top of the next-lower arrow.

Subshell Energy Order

The ordering of electron subshells in terms of increasing energy, which is experimentally determined, is more complex than might be expected. This is because the energies of subshells in different shells often "overlap," as shown in Figure 3.10. This diagram shows, for example, that the 4s subshell has lower energy than the 3d subshell.

A useful mnemonic (memory) device for remembering subshell filling order, which incorporates "overlap" situations such as those in Figure 3.10, is given in Figure 3.11. This diagram, which lists all subshells needed to specify the electron arrangements for all 117 elements, is constructed by locating all *s* subshells in column 1, all *p* subshells in column 2, and so on. Subshells that belong to the same shell are found in the same row. The order of subshell filling is given by following the diagonal arrows, starting at the top. The 1s subshell fills first. The second arrow points to (goes through) the 2s subshell, which fills next. The third arrow points to both the 2p and the 3s subshells. The 2p fills first, followed by the 3s. Any time a single arrow points to more than one subshell, we start at the tail of the arrow and work to its tip to determine the proper filling sequence.

Writing Electron Configurations and Orbital Diagrams

An **electron configuration** *is a statement of how many electrons an atom has in each of its electron subshells.* Because subshells group electrons according to energy, electron configurations indicate how many electrons of various energies an atom has.

Electron configurations are not written out in words; rather, a shorthand system with symbols is used. Subshells containing electrons, listed in order of increasing energy, are designated by using number–letter combinations (1s, 2s, and 2p). A superscript following each subshell designation indicates the number of electrons in that subshell. The electron configuration for nitrogen in this shorthand notation is

$$1s^2 2s^2 2p^3$$

Thus a nitrogen atom has an electron arrangement of two electrons in the 1s subshell, two electrons in the 2s subshell, and three electrons in the 2p subshell.

An **orbital diagram** *is a notation that shows how many electrons an atom has in each of its occupied electron orbitals.* Note that electron configurations deal with *subshell* occupancy and that orbital diagrams deal with *orbital* occupancy. The orbital diagram for the element nitrogen is

1s	2s	2p
↑↓	↑↓	↑ ↑ ↑

This diagram indicates that both the 1s and the 2s orbitals are filled, each containing two electrons of opposite spin. In addition, each of the three 2p orbitals contains one electron. Electron spin is denoted by the direction (up or down) in which an arrow points. For two electrons of opposite spin, which is the case in a fully occupied orbital, one arrow must point up and the other down. Let us now systematically consider electron configurations and orbital diagrams for the first few elements in the periodic table.

Hydrogen (atomic number = 1) has only one electron, which goes into the 1s subshell; this subshell has the lowest energy of all subshells. Hydrogen's electron configuration is written as $1s^1$, and its orbital diagram is

1s

H: ↑

Helium (atomic number = 2) has two electrons, both of which occupy the 1s subshell. (Remember, an *s* subshell contains one orbital, and an orbital can accommodate two electrons.) Helium's electron configuration is $1s^2$, and its orbital diagram is

1s

He: ↑↓

The two electrons present are of opposite spin.

Lithium (atomic number = 3) has three electrons, and the third electron cannot enter the 1s subshell because its maximum capacity is two electrons. (All *s* subshells are completely filled with two electrons.) The third electron is placed in the next-highest-energy subshell, the 2s. The electron configuration for lithium is $1s^2 2s^1$, and its orbital diagram is

$$\text{Li:} \quad \begin{array}{cc} 1s & 2s \\ \boxed{\uparrow\downarrow} & \boxed{\uparrow} \end{array}$$

For *beryllium* (atomic number = 4), the additional electron is placed in the 2s subshell, which is now completely filled, giving beryllium the electron configuration $1s^2 2s^2$. The orbital diagram for beryllium is

$$\text{Be:} \quad \begin{array}{cc} 1s & 2s \\ \boxed{\uparrow\downarrow} & \boxed{\uparrow\downarrow} \end{array}$$

For *boron* (atomic number = 5), the 2p subshell, which is the subshell of next highest energy (Figures 3.10 and 3.11), becomes occupied for the first time. Boron's electron configuration is $1s^2 2s^2 2p^1$, and its orbital diagram is

$$\text{B:} \quad \begin{array}{ccc} 1s & 2s & 2p \\ \boxed{\uparrow\downarrow} & \boxed{\uparrow\downarrow} & \boxed{\uparrow\ \ \ } \end{array}$$

The 2p subshell contains three orbitals of equal energy. It does not matter which of the 2p orbitals is occupied because they are of equivalent energy.

With the next element, *carbon* (atomic number = 6), we come to a new situation. We know that the sixth electron must go into a 2p orbital. However, does this new electron go into the 2p orbital that already has one electron or into one of the others? Rule 2, at the start of this section, covers this situation. Electrons will occupy equal-energy orbitals singly to the maximum extent possible before any orbital acquires a second electron. Thus, for carbon, we have the electron configuration $1s^2 2s^2 2p^2$ and the orbital diagram

$$\text{C:} \quad \begin{array}{ccc} 1s & 2s & 2p \\ \boxed{\uparrow\downarrow} & \boxed{\uparrow\downarrow} & \boxed{\uparrow\ |\ \uparrow\ \ } \end{array}$$

A *p* subshell can accommodate six electrons because there are three orbitals within it. The 2p subshell can thus accommodate the additional electrons found in the elements with atomic numbers 7 through 10: *nitrogen* (N), *oxygen* (O), *fluorine* (F), and *neon* (Ne). The electron configurations and orbital diagrams for these elements are

			1s	2s	2p
N:	$1s^2 2s^2 2p^3$	N:	$\boxed{\uparrow\downarrow}$	$\boxed{\uparrow\downarrow}$	$\boxed{\uparrow\,|\,\uparrow\,|\,\uparrow}$
O:	$1s^2 2s^2 2p^4$	O:	$\boxed{\uparrow\downarrow}$	$\boxed{\uparrow\downarrow}$	$\boxed{\uparrow\downarrow\,|\,\uparrow\,|\,\uparrow}$
F:	$1s^2 2s^2 2p^5$	F:	$\boxed{\uparrow\downarrow}$	$\boxed{\uparrow\downarrow}$	$\boxed{\uparrow\downarrow\,|\,\uparrow\downarrow\,|\,\uparrow}$
Ne:	$1s^2 2s^2 2p^6$	Ne:	$\boxed{\uparrow\downarrow}$	$\boxed{\uparrow\downarrow}$	$\boxed{\uparrow\downarrow\,|\,\uparrow\downarrow\,|\,\uparrow\downarrow}$

With *sodium* (atomic number = 11), the 3s subshell acquires an electron for the first time. Sodium's electron configuration is

$$1s^2 2s^2 2p^6 3s^1$$

Note the pattern that is developing in the electron configurations we have written so far. Each element has an electron configuration that is the same as the one just before it except for the addition of one electron.

The symbols $1s^2$, $2s^2$, and $2p^3$ are read as "one s two," "two s two," and "two p three," not as "one s squared," "two s squared," and "two p cubed."

The sum of the superscripts in an electron configuration equals the total number of electrons present and hence must equal the atomic number of the element.

Electron configurations for other elements are obtained by simply extending the principles we have just illustrated. A subshell of lower energy is always filled before electrons are added to the next highest subshell; this continues until the correct number of electrons have been accommodated.

For a few elements in the middle of the periodic table, the actual distribution of electrons within subshells differs slightly from that obtained by using the procedures outlined in this section. These exceptions are caused by very small energy differences between some subshells and are not important in the uses we shall make of electron configurations.

 **EXAMPLE 3.5**

Writing an Electron Configuration

Write the electron configurations for the following elements.

a. Strontium (atomic number = 38)
b. Lead (atomic number = 82)

Solution

a. The number of electrons in a strontium atom is 38. Remember that the atomic number gives the number of electrons (Section 3.2). We will need to fill subshells, in order of increasing energy, until 38 electrons have been accommodated.

The $1s$, $2s$, and $2p$ subshells fill first, accommodating a total of 10 electrons among them.

$$1s^2 2s^2 2p^6 \ldots$$

Next, according to Figures 3.10 and 3.11, the $3s$ subshell fills and then the $3p$ subshell.

$$1s^2 2s^2 2p^6 \boxed{3s^2 3p^6} \ldots$$

We have accommodated 18 electrons at this point. We still need to add 20 more electrons to get our desired number of 38.

The $4s$ subshell fills next, followed by the $3d$ subshell, giving us 30 electrons at this point.

$$1s^2 2s^2 2p^6 3s^2 3p^6 \boxed{4s^2 3d^{10}} \ldots$$

Note that the maximum electron population for d subshells is 10 electrons.

Eight more electrons are needed, which are added to the next two higher subshells, the $4p$ and the $5s$. The $4p$ subshell can accommodate 6 electrons, and the $5s$ can accommodate 2 electrons.

$$1s^2 2s^2 2p^6 3s^2 3p^6 4s^2 3d^{10} \boxed{4p^6 5s^2}$$

To double-check that we have the correct number of electrons, 38, we add the superscripts in our final electron configuration.

$$2 + 2 + 6 + 2 + 6 + 2 + 10 + 6 + 2 = 38$$

The sum of the superscripts in any electron configuration should add up to the atomic number if the configuration is for a neutral atom.

b. To write this configuration, we continue along the same lines as in part **a,** remembering that the maximum electron subshell populations are $s = 2$, $p = 6$, $d = 10$, and $f = 14$.

Lead, with an atomic number of 82, contains 82 electrons, which are added to subshells in the following order. (The line of numbers beneath the electron configuration is a running total of added electrons and is obtained by adding the superscripts up to that point. We stop when we have 82 electrons.)

$$1s^2 2s^2 2p^6 3s^2 3p^6 4s^2 3d^{10} 4p^6 5s^2 4d^{10} 5p^6 6s^2 4f^{14} 5d^{10} 6p^2$$

2   4   10   12   18   20   30   36   38   48   54   56   70   80   82

Running total of electrons added

Note in this electron configuration that the $6p$ subshell contains only 2 electrons, even though it can hold a maximum of 6. We put only 2 electrons in this subshell because that is sufficient to give 82 total electrons. If we had completely filled this subshell, we would have had 86 total electrons, which is too many.

Practice Exercise 3.5

Write the electron configurations for the following elements.

a. Manganese (atomic number = 25)
b. Xenon (atomic number = 54)

Answers: a. $1s^2 2s^2 2p^6 3s^2 3p^6 4s^2 3d^5$; **b.** $1s^2 2s^2 2p^6 3s^2 3p^6 4s^2 3d^{10} 4p^6 5s^2 4d^{10} 5p^6$

3.8 THE ELECTRONIC BASIS FOR THE PERIODIC LAW AND THE PERIODIC TABLE

For many years, there was no explanation available for either the periodic law or why the periodic table has the shape that it has. We now know that the theoretical basis for both the periodic law and the periodic table is found in electronic theory. As we saw earlier in the chapter (Section 3.2), when two atoms interact, it is their electrons that interact. Thus the number and arrangement of electrons determine how an atom reacts with other atoms—that is, what its chemical properties are. The properties of the elements repeat themselves in a periodic manner because the arrangement of electrons about the nucleus of an atom follows a periodic pattern, as we saw in Section 3.7.

Electron Configurations and the Periodic Law

The periodic law (Section 3.4) points out that the properties of the elements repeat themselves in a regular manner when the elements are arranged in order of increasing atomic number. The elements that have similar chemical properties are placed under one another in vertical columns (groups) in the periodic table.

Groups of elements have similar chemical properties because of similarities in their electron configuration. *Chemical properties repeat themselves in a regular manner among the elements because electron configurations repeat themselves in a regular manner among the elements.*

To illustrate this correlation between similar chemical properties and similar electron configurations, let us look at the electron configurations of two groups of elements known to have similar chemical properties.

We begin with the elements lithium, sodium, potassium, and rubidium, all members of Group IA of the periodic table. The electron configurations for these elements are

$$_3\text{Li: } 1s^2 \,\boxed{2s^1}$$
$$_{11}\text{Na: } 1s^2 2s^2 2p^6 \,\boxed{3s^1}$$
$$_{19}\text{K: } 1s^2 2s^2 2p^6 3s^2 3p^6 \,\boxed{4s^1}$$
$$_{37}\text{Rb: } 1s^2 2s^2 2p^6 3s^2 3p^6 4s^2 3d^{10} 4p^6 \,\boxed{5s^1}$$

> The electron arrangement in the outermost shell is the same for elements in the same group. This is why elements in the same group have similar chemical properties.

Note that each of these elements has one electron in its outermost shell. (The outermost shell is the shell with the highest number.) This similarity in outer-shell electron arrangements causes these elements to have similar chemical properties. In general, elements with similar outer-shell electron configurations have similar chemical properties.

Let us consider another group of elements known to have similar chemical properties: fluorine, chlorine, bromine, and iodine of Group VIIA of the periodic table. The electron configurations for these four elements are

$$_9\text{F: } 1s^2 \,\boxed{2s^2 2p^5}$$
$$_{17}\text{Cl: } 1s^2 2s^2 2p^6 \,\boxed{3s^2 3p^5}$$
$$_{35}\text{Br: } 1s^2 2s^2 2p^6 \, 3s^2 3p^6 \,\boxed{4s^2}\, 3d^{10} \,\boxed{4p^5}$$
$$_{53}\text{I: } 1s^2 2s^2 2p^6 3s^2 3p^6 \, 4s^2 \, 3d^{10} 4p^6 \,\boxed{5s^2}\, 4d^{10} \,\boxed{5p^5}$$

Figure 3.12 Electron configurations and the positions of elements in the periodic table. The periodic table can be divided into four areas that are 2, 6, 10, and 14 columns wide. The four areas contain elements whose distinguishing electron is located, respectively, in *s*, *p*, *d*, and *f* subshells. The extent of filling of the subshell that contains an element's distinguishing electron can be determined from the element's position in the periodic table.

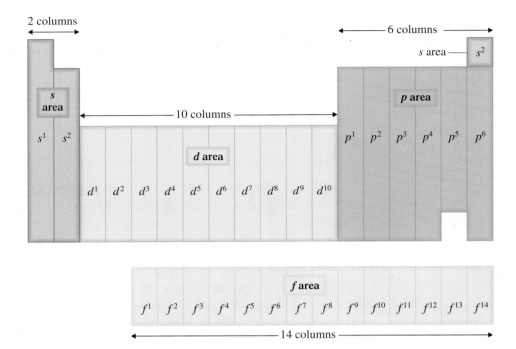

Once again, similarities in electron configuration are readily apparent. This time, the repeating pattern involves an outermost *s* and *p* subshell containing a combined total of seven electrons (shown in color). Remember that for Br and I, shell numbers 4 and 5 designate, respectively, electrons in the outermost shells.

Electron Configurations and the Periodic Table

One of the strongest pieces of supporting evidence for the assignment of electrons to shells, subshells, and orbitals is the periodic table itself. The basic shape and structure of this table, which were determined many years before electrons were even discovered, are consistent with and can be explained by electron configurations. Indeed, the specific location of an element in the periodic table can be used to obtain information about its electron configuration.

As the first step in linking electron configurations to the periodic table, let us analyze the general shape of the periodic table in terms of columns of elements. As shown in Figure 3.12, on the extreme left of the table, there are 2 columns of elements; in the center there is a region containing 10 columns of elements; to the right there is a block of 6 columns of elements; and in the two rows at the bottom of the table, there are 14 columns of elements.

The number of columns of elements in the various regions of the periodic table— 2, 6, 10, and 14—is the same as the maximum number of electrons that the various types of subshells can accommodate. We will see shortly that this is a very significant observation; the number matchup is no coincidence. The various columnar regions of the periodic table are called the *s* area (2 columns), the *p* area (6 columns), the *d* area (10 columns), and the *f* area (14 columns), as shown in Figure 3.12.

The concept of *distinguishing electrons* is the key to obtaining electron configuration information from the periodic table. A **distinguishing electron** *is the last electron added to the electron configuration for an element when electron subshells are filled in order of increasing energy.* This last electron is the one that causes an element's electron configuration to differ from that of the element immediately preceding it in the periodic table.

For all elements located in the *s* area of the periodic table, the distinguishing electron is always found in an *s* subshell. All *p* area elements have distinguishing electrons in *p* subshells. Similarly, elements in the *d* and *f* areas of the periodic table have distinguishing electrons located in *d* and *f* subshells, respectively. Thus the area location of an element in the periodic table can be used to determine the type of subshell that contains the distinguishing electron. Note that the element helium belongs to the *s* rather than the *p* area of the periodic table, even though its periodic table position is on the right-hand side. (The reason for this placement of helium will be explained in Section 4.3.)

The extent to which the subshell containing an element's distinguishing electron is filled can also be determined from the element's position in the periodic table. All elements in the first column of a specific area contain only one electron in the subshell; all elements in the second column contain two electrons in the subshell; and so on. Thus all elements in the first column of the *p* area (Group IIIA) have an electron configuration ending in p^1. Elements in the second column of the *p* area (Group IVA) have electron configurations ending in p^2; and so on. Similar relationships hold in other areas of the table, as shown in Figure 3.12.

3.9 CLASSIFICATION OF THE ELEMENTS

The elements can be classified in several ways. The two most common classification systems are

1. A system based on selected physical properties of the elements, in which they are described as metals or nonmetals. This classification scheme was discussed in Section 3.5.
2. A system based on the electron configurations of the elements, in which elements are described as *noble-gas, representative, transition,* or *inner transition elements.*

The classification scheme based on electron configurations of the elements is depicted in Figure 3.13.

A **noble-gas element** *is an element located in the far right column of the periodic table.* These elements are all gases at room temperature, and they have little tendency to form chemical compounds. With one exception, the distinguishing electron for a noble gas completes the *p* subshell; therefore, noble gases have electron configurations ending

The electron configurations of the noble gases will be an important focal point when we consider chemical bonding theory in Chapters 4 and 5.

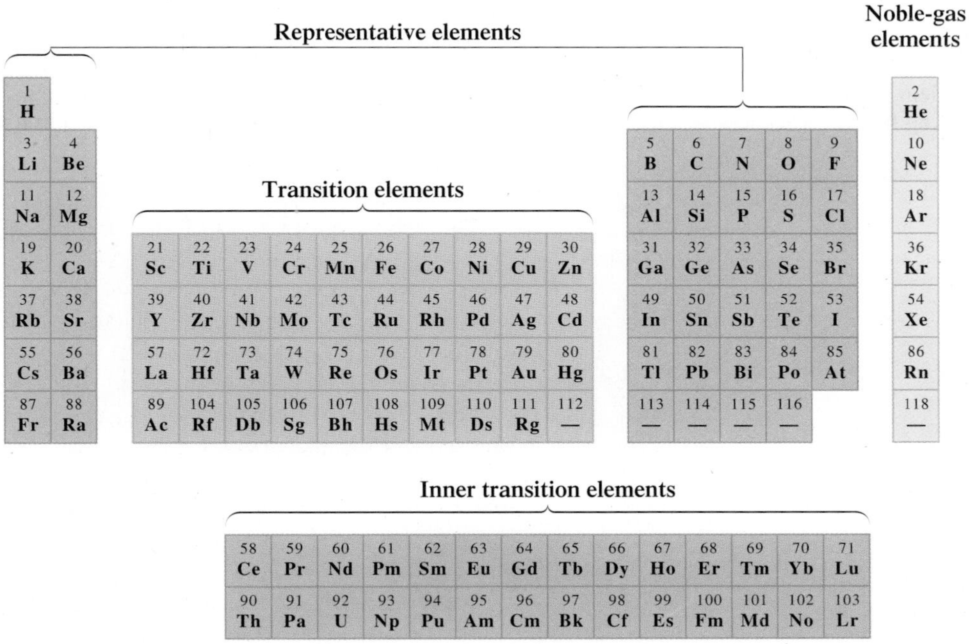

Figure 3.13 A classification scheme for the elements based on their electron configurations. Representative elements occupy the *s* area and most of the *p* area shown in Figure 3.12. The noble-gas elements occupy the last column of the *p* area. The transition elements are found in the *d* area, and the inner transition elements are found in the *f* area.

in p^6. The exception is helium, in which the distinguishing electron completes the first shell—a shell that has only two electrons. Helium's electron configuration is $1s^2$.

A **representative element** *is an element located in the s area or the first five columns of the p area of the periodic table.* The distinguishing electron in these elements partially or completely fills an *s* subshell or partially fills a *p* subshell. Some representative elements are nonmetals while others are metals. The four most abundant elements in the human body—hydrogen, oxygen, carbon, and nitrogen—are nonmetallic representative elements.

Iron: The Most Abundant Transition Element in the Human Body

Small amounts of nine transition metals are necessary for the proper functioning of the human body. They include all of the Period 4 transition metals except scandium and titanium plus the Period 5 transition metal molybdenum, as shown in the following transition metal portion of the periodic table.

Transition Metals

Period 4		V	Cr	Mn	Fe	Co	Ni	Cu	Zn
Period 5			Mo						
Period 6									

Iron is the most abundant, from a biochemical standpoint, of these transition metals; zinc is the second most abundant.

Most of the body's iron is found as a component of the proteins hemoglobin and myoglobin, where it functions in the transport and storage of oxygen. Hemoglobin is the oxygen carrier in red blood cells, and myoglobin stores oxygen in muscle cells. Iron-deficient blood has less oxygen-carrying capacity and often cannot completely meet the body's energy needs. Energy deficiency—tiredness and apathy—is one of the symptoms of iron deficiency.

Iron deficiency is a worldwide problem. Millions of people are unknowingly deficient. Even in the United States and Canada, about 20% of women and 3% of men have this problem; some 8% of women and 1% of men are anemic, experiencing fatigue, weakness, apathy, and headaches.

Inadequate intake of iron, either from malnutrition or from high consumption of the wrong foods, is the usual cause of iron deficiency. In the Western world, the cause is often displacement of iron-rich foods by foods high in sugar and fat.

About 80% of the iron in the body is in the blood, so iron losses are greatest whenever blood is lost. Blood loss from menstruation makes a woman's need for iron nearly twice as great as a man's. Also, women usually consume less food than men do. These two factors—lower intake and higher loss—cause iron deficiency to be likelier in women than in men. The iron RDA (recommended dietary allowance) is 8 mg per day for adult males and older women. For women of childbearing age, the RDA is 18 mg. This amount is necessary to replace menstrual loss and to provide the extra iron needed during pregnancy.

Iron deficiency may also be caused by poor absorption of ingested iron. A normal, healthy person absorbs about 2%–10% of the iron in vegetables and about 10%–30% in meats. About 40% of the iron in meat, fish, and poultry is bound into molecules of heme, the iron-containing part of hemoglobin and myoglobin. Heme iron is much more readily absorbed (23%) than nonheme iron (2%–10%). (See the accompanying charts.)

Cooking utensils can enhance the amount of iron delivered by the diet. The iron content of 100 g of spaghetti sauce simmered in a glass dish is 3 mg, but it is 87 mg when the sauce is cooked in an unenameled iron skillet. Even in the short time it takes to scramble eggs, their iron content can be tripled by cooking them in an iron pan.

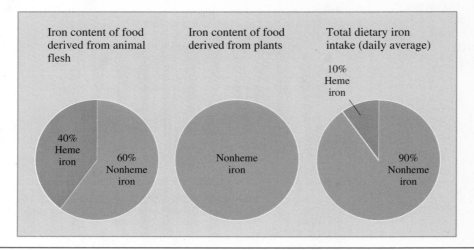

A **transition element** *is an element located in the* d *area of the periodic table.* Each has its distinguishing electron in a *d* subshell. All of the transition elements are metals. The most abundant transition element in the human body is iron. The Chemical Connections feature on page 74 considers several aspects of the biochemical role of iron in the human body.

An **inner transition element** *is an element located in the* f *area of the periodic table.* Each has its distinguishing electron in an *f* subshell. All of the inner transition elements are metals. Many of them are laboratory-produced elements rather than naturally occurring elements (Section 11.5).

The Chemistry at a Glance feature below contrasts the three element classification schemes that have been considered so far in this chapter: by physical properties (Section 3.5), by electronic properties (Section 3.9), and by non-numerical periodic table group names (Section 3.4).

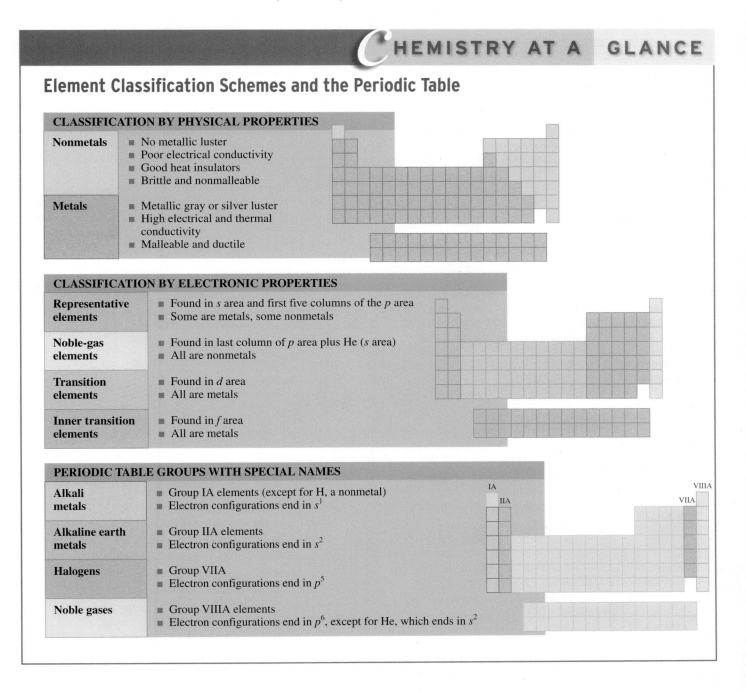

CHEMISTRY AT A GLANCE

Element Classification Schemes and the Periodic Table

CLASSIFICATION BY PHYSICAL PROPERTIES

| Nonmetals | ■ No metallic luster
■ Poor electrical conductivity
■ Good heat insulators
■ Brittle and nonmalleable |
| Metals | ■ Metallic gray or silver luster
■ High electrical and thermal conductivity
■ Malleable and ductile |

CLASSIFICATION BY ELECTRONIC PROPERTIES

Representative elements	■ Found in *s* area and first five columns of the *p* area ■ Some are metals, some nonmetals
Noble-gas elements	■ Found in last column of *p* area plus He (*s* area) ■ All are nonmetals
Transition elements	■ Found in *d* area ■ All are metals
Inner transition elements	■ Found in *f* area ■ All are metals

PERIODIC TABLE GROUPS WITH SPECIAL NAMES

Alkali metals	■ Group IA elements (except for H, a nonmetal) ■ Electron configurations end in s^1
Alkaline earth metals	■ Group IIA elements ■ Electron configurations end in s^2
Halogens	■ Group VIIA ■ Electron configurations end in p^5
Noble gases	■ Group VIIIA elements ■ Electron configurations end in p^6, except for He, which ends in s^2

CONCEPTS TO REMEMBER

Subatomic particles. Subatomic particles, the very small building blocks from which atoms are made, are of three major types: electrons, protons, and neutrons. Electrons are negatively charged, protons are positively charged, and neutrons have no charge. All neutrons and protons are found at the center of the atom in the nucleus. The electrons occupy the region about the nucleus. Protons and neutrons have much larger masses than electrons (Section 3.1).

Atomic number and mass number. Each atom has a characteristic atomic number and mass number. The atomic number is equal to the number of protons in the nucleus of the atom. The mass number is equal to the total number of protons and neutrons in the nucleus (Section 3.2).

Isotopes. Isotopes are atoms that have the same number of protons and electrons but have different numbers of neutrons. The isotopes of an element always have the same atomic number and different mass numbers. Isotopes of an element have the same chemical properties (Section 3.3).

Atomic mass. The atomic mass of an element is a calculated average mass. It depends on the percentage abundances and masses of the naturally occurring isotopes of the element (Section 3.3).

Periodic law and periodic table. The periodic law states that when elements are arranged in order of increasing atomic number, elements with similar chemical properties occur at periodic (regularly recurring) intervals. The periodic table is a graphical representation of the behavior described by the periodic law. In a modern periodic table, vertical columns contain elements with similar chemical properties. A group in the periodic table is a vertical column of elements. A period in the periodic table is a horizontal row of elements (Section 3.4).

Metals and nonmetals. Metals exhibit luster, thermal conductivity, electrical conductivity, and malleability. Nonmetals are characterized by the absence of the properties associated with metals. The majority of the elements are metals. The steplike heavy line that runs through the right third of the periodic table separates the metals on the left from the nonmetals on the right (Section 3.5).

Electron shell. An electron shell contains electrons that have approximately the same energy and spend most of their time approximately the same distance from the nucleus (Section 3.6).

Electron subshell. An electron subshell contains electrons that all have the same energy. The number of subshells in a particular shell is equal to the shell number. Each subshell can hold a specific maximum number of electrons. These values are 2, 6, 10, and 14 for s, p, d, and f subshells, respectively (Section 3.6).

Electron orbital. An electron orbital is a region of space about a nucleus where an electron with a specific energy is most likely to be found. Each subshell consists of one or more orbitals. For s, p, d, and f subshells there are 1, 3, 5, and 7 orbitals, respectively. No more than two electrons may occupy any orbital (Section 3.6).

Electron configuration. An electron configuration is a statement of how many electrons an atom has in each of its subshells. The principle that electrons normally occupy the lowest-energy subshell available is used to write electron configurations (Section 3.7).

Orbital diagram. An orbital diagram is a notation that shows how many electrons an atom has in each of its orbitals. Electrons occupy the orbitals of a subshell such that each orbital within the subshell acquires one electron before any orbital acquires a second electron. All electrons in such singly occupied orbitals must have the same spin (Section 3.7).

Electron configurations and the periodic law. Chemical properties repeat themselves in a regular manner among the elements because electron configurations repeat themselves in a regular manner among the elements (Section 3.8).

Electron configurations and the periodic table. The groups of the periodic table consist of elements with similar electron configurations. Thus the location of an element in the periodic table can be used to obtain information about its electron configuration (Section 3.8).

Classification system for the elements. On the basis of electron configuration, elements can be classified into four categories: noble gases (far right column of the periodic table); representative elements (s and p areas of the periodic table, with the exception of the noble gases); transition elements (d area of the periodic table); and inner transition elements (f area of the periodic table) (Section 3.9).

KEY REACTIONS AND EQUATIONS

1. Relationships involving atomic number and mass number for a neutral atom (Section 3.2)

   Atomic number = number of protons = number of electrons

   Mass number = number of protons + number of neutrons

   Mass number = total number of subatomic particles in the nucleus

   Mass number − atomic number = number of neutrons

   Mass number + atomic number = total number of subatomic particles

2. Relationships involving electron shells, electron subshells, and electron orbitals (Section 3.6)

   Number of subshells in a shell = shell number

   Maximum number of electrons in an s subshell = 2

   Maximum number of electrons in a p subshell = 6

   Maximum number of electrons in a d subshell = 10

   Maximum number of electrons in an f subshell = 14

   Maximum number of electrons in an orbital = 2

3. Order of filling of subshells in terms of increasing energy (Section 3.7)

   1s, 2s, 2p, 3s, 3p, 4s, 3d, 4p, 5s, 4d, 5p, 6s, 4f, 5d, 6p, 7s, 5f, 6d, 7p

EXERCISES *and* PROBLEMS

The members of each pair of problems in this section test similar material.

⬤ Internal Structure of an Atom (Section 3.1)

3.1 Indicate which subatomic particle (proton, neutron, or electron) correctly matches each of the following phrases. More than one particle can be used as an answer.
a. Possesses a negative charge
b. Has no charge
c. Has a mass slightly less than that of a neutron
d. Has a charge equal to, but opposite in sign from, that of an electron

3.2 Indicate which subatomic particle (proton, neutron, or electron) correctly matches each of the following phrases. More than one particle can be used as an answer.
a. Is not found in the nucleus
b. Has a positive charge
c. Can be called a nucleon
d. Has a relative mass of 1837 if the relative mass of an electron is 1

3.3 Indicate whether each of the following statements about the nucleus of an atom is true or false.
a. The nucleus of an atom is neutral.
b. The nucleus of an atom contains only neutrons.
c. The number of nucleons present in the nucleus is equal to the number of electrons present outside the nucleus.
d. The nucleus accounts for almost all the mass of an atom.

3.4 Indicate whether each of the following statements about the nucleus of an atom is true or false.
a. The nucleus of an atom contains all of the "heavy" subatomic particles.
b. The nucleus of an atom accounts for almost all of the volume of the atom.
c. The nucleus of an atom has an extremely low density compared to that of the atom as a whole.
d. The nucleus of an atom can be positively or negatively charged, depending on the identity of the atom.

⬤ Atomic Number and Mass Number (Section 3.2)

3.5 Determine the atomic number and mass number for atoms with the following subatomic makeups.
a. 2 protons, 2 neutrons, and 2 electrons
b. 4 protons, 5 neutrons, and 4 electrons
c. 5 protons, 4 neutrons, and 5 electrons
d. 28 protons, 30 neutrons, and 28 electrons

3.6 Determine the atomic number and mass number for atoms with the following subatomic makeups.
a. 1 proton, 1 neutron, and 1 electron
b. 10 protons, 12 neutrons, and 10 electrons
c. 12 protons, 10 neutrons, and 12 electrons
d. 50 protons, 69 neutrons, and 50 electrons

3.7 Determine the number of protons, neutrons, and electrons present in atoms with the following characteristics.
a. Atomic number = 8 and mass number = 16
b. Mass number = 18 and $Z = 8$
c. Atomic number = 20 and $A = 44$
d. $A = 257$ and $Z = 100$

3.8 Determine the number of protons, neutrons, and electrons present in atoms with the following characteristics.
a. Atomic number = 10 and mass number = 20
b. Mass number = 110 and $Z = 48$
c. $A = 11$ and atomic number = 5
d. $Z = 92$ and $A = 238$

3.9 Indicate whether the *atomic number*, the *mass number*, or *both the atomic number and the mass number* are needed to determine the following.
a. Number of protons in an atom
b. Number of neutrons in an atom
c. Number of nucleons in an atom
d. Total number of subatomic particles in an atom

3.10 What information about the subatomic particles present in an atom is obtained from each of the following?
a. Atomic number
b. Mass number
c. Mass number − atomic number
d. Mass number + atomic number

3.11 Determine the following information for an atom whose complete chemical symbol is $^{39}_{19}K$.
a. Atomic number
b. Mass number
c. Number of protons present
d. Number of electrons present

3.12 Determine the following information for an atom whose complete chemical symbol is $^{31}_{15}P$.
a. Atomic number
b. Mass number
c. Number of protons present
d. Number of electrons present

3.13 Using the information inside the front cover, identify the element X based on the given complete chemical symbol for X.
a. $^{15}_{7}X$ b. $^{27}_{13}X$ c. $^{139}_{56}X$ d. $^{197}_{79}X$

3.14 Using the information inside the front cover, identify the element X based on the given complete chemical symbol for X.
a. $^{16}_{8}X$ b. $^{19}_{9}X$ c. $^{45}_{21}X$ d. $^{63}_{29}X$

3.15 Using the information inside the front cover, and the given information, write the complete chemical symbol for the following atoms.

a. (7p, 8n)   b. (14p, 14n)   c. (18p, 22n)   d. (22p, 26n)

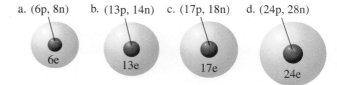

3.16 Using the information inside the front cover, and the given information, write the complete chemical symbol for the following atoms.

a. (6p, 8n)   b. (13p, 14n)   c. (17p, 18n)   d. (24p, 28n)

3.17 Arrange the following atoms in the orders specified.

$$^{32}_{16}S \quad ^{40}_{18}Ar \quad ^{35}_{17}Cl \quad ^{37}_{19}K$$

a. Order of increasing atomic number
b. Order of decreasing mass number
c. Order of increasing number of electrons
d. Order of increasing number of neutrons

3.18 Arrange the following atoms in the orders specified.

$$^{14}_{6}C \quad ^{17}_{8}O \quad ^{13}_{7}N \quad ^{19}_{9}F$$

a. Order of decreasing atomic number
b. Order of increasing mass number
c. Order of decreasing number of neutrons
d. Order of increasing number of nucleons

3.19 Determine the following information for an atom whose complete chemical symbol is $^{23}_{11}Na$.
a. The total number of subatomic particles present
b. The total number of subatomic particles present in the nucleus of the atom
c. The total number of nucleons present
d. The total charge (including sign) associated with the nucleus of the atom

3.20 Determine the following information for an atom whose complete chemical symbol is $^{37}_{17}Cl$.
a. The total number of subatomic particles present
b. The total number of subatomic particles present in the nucleus of the atom
c. The total number of nucleons present
d. The total charge (including sign) associated with the nucleus of the atom

Isotopes and Atomic Masses (Section 3.3)

3.21 The atomic number of the element carbon (C) is 6. Write the complete chemical symbols for each of the following carbon isotopes: carbon-12, carbon-13, and carbon-14.

3.22 The atomic number of the element sulfur (S) is 16. Write the complete chemical symbols for each of the following sulfur isotopes: sulfur-32, sulfur-33, sulfur-34, and sulfur-36.

3.23 Using the information given in the following table, indicate whether each of the following pairs of atoms are isotopes.

	Atom A	Atom B	Atom C	Atom D
Number of protons	9	10	10	9
Number of neutrons	10	9	10	9
Number of electrons	9	10	10	9

a. Atom A and atom B
b. Atom A and atom C
c. Atom A and atom D

3.24 Using the information given in the table in Problem 3.23, indicate whether each of the following pairs of atoms are isotopes.
a. Atom B and atom C
b. Atom B and atom D
c. Atom C and atom D

3.25 Indicate whether each of the following statements about sodium isotopes is true or false.
a. $^{23}_{11}Na$ has one more electron than $^{24}_{11}Na$.
b. $^{23}_{11}Na$ and $^{24}_{11}Na$ contain the same number of neutrons.
c. $^{23}_{11}Na$ has one less subatomic particle than $^{24}_{11}Na$.
d. $^{23}_{11}Na$ and $^{24}_{11}Na$ have the same atomic number.

3.26 Indicate whether each of the following statements about magnesium isotopes is true or false.

a. $^{24}_{12}Mg$ has one more proton than $^{25}_{12}Mg$.
b. $^{24}_{12}Mg$ and $^{25}_{12}Mg$ contain the same number of subatomic particles.
c. $^{24}_{12}Mg$ has one less neutron than $^{25}_{12}Mg$.
d. $^{24}_{12}Mg$ and $^{25}_{12}Mg$ have different mass numbers.

3.27 The following are selected properties for the most abundant isotope of a particular element. Which of these properties would also be the same for the second-most-abundant isotope of the element?
a. Mass number is 70
b. 31 electrons are present
c. Isotopic mass is 69.92 amu
d. Isotope reacts with chlorine to give a green compound

3.28 The following are selected properties for the most abundant isotope of a particular element. Which of these properties would also be the same for the second-most-abundant isotope of the element?
a. Atomic number is 31
b. Does not react with the element gold
c. 40 neutrons are present
d. Density is 1.03 g/mL

3.29 Calculate the atomic mass of each of the following elements using the given data for the percentage abundance and mass of each isotope.
a. Lithium: 7.42% ^6Li (6.01 amu) and 92.58% ^7Li (7.02 amu)
b. Magnesium: 78.99% ^{24}Mg (23.99 amu), 10.00% ^{25}Mg (24.99 amu), and 11.01% ^{26}Mg (25.98 amu)

3.30 Calculate the atomic mass of each of the following elements using the given data for the percentage abundance and mass of each isotope.
a. Silver: 51.82% ^{107}Ag (106.9 amu) and 48.18% ^{109}Ag (108.9 amu)
b. Silicon: 92.21% ^{28}Si (27.98 amu), 4.70% ^{29}Si (28.98 amu), and 3.09% ^{30}Si (29.97 amu)

3.31 Using information available on the inside front cover, determine the atomic mass associated with the listed elements or the element name associated with the listed atomic masses.
a. Iron   b. Nitrogen   c. 40.08 amu   d. 126.90 amu

3.32 Using information available on the inside front cover, determine the atomic mass associated with the listed elements or the element name associated with the listed atomic masses.
a. Phosphorus   b. Nickel   c. 101.07 amu   d. 20.18 amu

The Periodic Law and the Periodic Table (Section 3.4)

3.33 Give the symbol of the element that occupies each of the following positions in the periodic table.
a. Period 4, Group IIA   b. Period 5, Group VIB
c. Group IA, Period 2   d. Group IVA, Period 5

3.34 Give the symbol of the element that occupies each of the following positions in the periodic table.
a. Period 1, Group IA   b. Period 6, Group IB
c. Group IIIB, Period 4   d. Group VIIA, Period 3

3.35 Using the periodic table, determine the following.
a. The atomic number of the element carbon
b. The atomic mass of the element silicon
c. The atomic number of the element with an atomic mass of 88.91 amu
d. The atomic mass of the element located in Period 2 and Group IIA

3.36 Using the periodic table, determine the following.
a. The atomic number of the element magnesium
b. The atomic mass of the element nitrogen

c. The atomic mass of the element with an atomic number of 10

d. The atomic number of the element located in Group IIIA and Period 3

3.37 Based on periodic table position, select the two elements in each set of elements that would be expected to have similar chemical properties.

a. $_{19}K$, $_{29}Cu$, $_{37}Rb$, $_{41}Nb$ b. $_{13}Al$, $_{14}Si$, $_{15}P$, $_{33}As$

c. $_9F$, $_{40}Zr$, $_{50}Sn$, $_{53}I$ d. $_{11}Na$, $_{12}Mg$, $_{54}Xe$, $_{55}Cs$

3.38 Based on periodic table position, select the two elements in each set of elements that would be expected to have similar chemical properties.

a. $_{11}Na$, $_{14}Si$, $_{23}V$, $_{55}Cs$ b. $_{13}Al$, $_{19}K$, $_{32}Ge$, $_{50}Sn$

c. $_{37}Rb$, $_{38}Sr$, $_{54}Xe$, $_{56}Ba$ d. $_2He$, $_6C$, $_8O$, $_{10}Ne$

3.39 The following statements either define or are closely related to the terms *periodic law, period,* and *group.* Match each statement with the appropriate term.

a. This is a vertical arrangement of elements in the periodic table.

b. The properties of the elements repeat in a regular way as atomic numbers increase.

c. The chemical properties of elements 12, 20, and 38 demonstrate this principle.

d. Carbon is the first member of this arrangement.

3.40 The following statements either define or are closely related to the terms *periodic law, period,* and *group.* Match each statement with the appropriate term.

a. This is a horizontal arrangement of elements in the periodic table.

b. Element 19 begins this arrangement in the periodic table.

c. Elements 24 and 33 belong to this arrangement.

d. Elements 10, 18, and 36 belong to this arrangement.

3.41 Identify each of the following elements by name.

a. Period 2 halogen   b. Period 3 alkali metal

c. Period 4 noble gas   d. Period 5 alkaline earth metal

3.42 Identify each of the following elements by name.

a. Period 2 alkali metal   b. Period 3 noble gas

c. Period 4 alkaline earth metal   d. Period 5 halogen

3.43 How many elements exist with an atomic number less than 40 that are

a. Halogens   b. Noble gases

c. Alkali metals   d. Alkaline earth metals

3.44 How many elements exist with an atomic number greater than 20 that are

a. Halogens   b. Noble gases

c. Alkali metals   d. Alkaline earth metals

3.45 Determine the following for the "highlighted" elements in the given periodic table. (Specify your answer by giving the "color" of the element.)

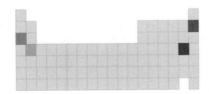

a. Which of the highlighted elements is a halogen?

b. Which of the highlighted elements is an alkaline earth metal?

c. Which of the highlighted elements is in Group IIA?

d. Which of the highlighted elements is in Period 3?

3.46 Determine the following for the "highlighted" elements in the periodic table given in Problem 3.45. (Specify your answer by giving the "color" of the element.)

a. Which of the highlighted elements is an alkali metal?

b. Which of the highlighted elements is a noble gas?

c. Which of the highlighted elements is in Group VIIA?

d. Which of the highlighted elements is in Period 2?

● **Metals and Nonmetals (Section 3.5)**

3.47 In which of the following pairs of elements are both members of the pair metals?

a. $_{17}Cl$ and $_{35}Br$ b. $_{13}Al$ and $_{14}Si$

c. $_{29}Cu$ and $_{42}Mo$ d. $_{30}Zn$ and $_{83}Bi$

3.48 In which of the following pairs of elements are both members of the pair metals?

a. $_7N$ and $_{34}Se$ b. $_{16}S$ and $_{48}Cd$

c. $_3Li$ and $_{26}Fe$ d. $_{50}Sn$ and $_{53}I$

3.49 Identify the nonmetal in each of the following sets of elements.

a. S, Na, K   b. Cu, Li, P

c. Be, I, Ca   d. Fe, Cl, Ga

3.50 Identify the nonmetal in each of the following sets of elements.

a. Al, H, Mg   b. C, Sn, Sb

c. Ti, V, F   d. Sr, Se, Sm

3.51 Classify each of the following general physical properties as a property of metallic elements or of nonmetallic elements.

a. Ductile   b. Low electrical conductivity

c. High thermal conductivity   d. Good heat insulator

3.52 Classify each of the following general physical properties as a property of metallic elements or of nonmetallic elements.

a. Nonmalleable   b. High luster

c. Low thermal conductivity   d. Brittle

3.53 Determine the following for the "highlighted" elements in the given periodic table.

a. Is the element highlighted in red a metal or nonmetal?

b. Is the element highlighted in green a metal or nonmetal?

c. Is the element highlighted in yellow a good or poor conductor of electricity?

d. Is the element highlighted in blue a good or poor conductor of heat?

3.54 Determine the following, using the periodic table given in Problem 3.53.

a. Is the element highlighted in yellow a metal or nonmetal?

b. Is the element highlighted in blue a metal or nonmetal?

c. Is the element highlighted in red a good or poor conductor of electricity?

d. Is the element highlighted in green a good or poor conductor of heat?

● **Electron Arrangements Within Atoms (Section 3.6)**

3.55 The following statements define or are closely related to the terms *electron shell, electron subshell,* and *electron orbital.* Match each statement with the appropriate term.

a. In terms of electron capacity, this unit is the smallest of the three.

b. This unit can contain a maximum of two electrons.

c. This unit is designated just by a number.

d. The term *energy level* is closely associated with this unit.

3.56 The following statements define or are closely related to the terms *electron shell, electron subshell,* and *electron orbital.* Match each statement with the appropriate term.

a. This unit can contain as many electrons as, or more electrons than, either of the other two.

b. The term *energy sublevel* is closely associated with this unit.

c. Electrons that occupy this unit do not need to have identical energies.

d. The unit is designated in the same way as the orbitals contained within it.

3.57 Indicate whether each of the following statements is true or false.

a. An orbital has a definite size and shape, which are related to the energy of the electrons it could contain.

b. All the orbitals in a subshell have the same energy.

c. All subshells accommodate the same number of electrons.

d. A $2p$ subshell and a $3p$ subshell contain the same number of orbitals.

3.58 Indicate whether each of the following statements is true or false.

a. All the subshells in a shell have the same energy.

b. An s orbital has a shape that resembles a four-leaf clover.

c. The third shell can accommodate a maximum number of 18 electrons.

d. All orbitals accommodate the same number of electrons.

3.59 Give the maximum number of electrons that can occupy each of the following electron-accommodating units.

a. One of the orbitals in the $2p$ subshell

b. One of the orbitals in the $3d$ subshell

c. The $4p$ subshell

d. The third shell

3.60 Give the maximum number of electrons that can occupy each of the following electron-accommodating units.

a. One of the orbitals in the $4d$ subshell

b. One of the orbitals in the $5f$ subshell

c. The $3d$ subshell

d. The second shell

Electron Configurations and Orbital Diagrams (Section 3.7)

3.61 Write complete electron configurations for atoms of each of the following elements.

a. $_6$C   b. $_{11}$Na   c. $_{16}$S   d. $_{18}$Ar

3.62 Write complete electron configurations for atoms of each of the following elements.

a. $_{10}$Ne   b. $_{13}$Al   c. $_{19}$K   d. $_{22}$Ti

3.63 On the basis of the total number of electrons present, identify the elements whose electron configurations are

a. $1s^2 2s^2 2p^4$ b. $1s^2 2s^2 2p^6$

c. $1s^2 2s^2 2p^6 3s^2 3p^1$ d. $1s^2 2s^2 2p^6 3s^2 3p^6 4s^2$

3.64 On the basis of the total number of electrons present, identify the elements whose electron configurations are

a. $1s^2 2s^2 2p^2$ b. $1s^2 2s^2 2p^6 3s^1$

c. $1s^2 2s^2 2p^6 3s^2 3p^5$ d. $1s^2 2s^2 2p^6 3s^2 3p^6 4s^2 3d^{10} 4p^3$

3.65 Write *complete* electron configurations for atoms whose electron configurations *end* as follows.

a. $3p^5$ b. $4d^7$ c. $4s^2$ d. $3d^1$

3.66 Write *complete* electron configurations for atoms whose electron configurations *end* as follows.

a. $4p^2$ b. $3d^{10}$ c. $5s^1$ d. $4p^6$

3.67 Draw the orbital diagram associated with each of the following electron configurations.

a. $1s^2 2s^2 2p^2$ b. $1s^2 2s^2 2p^6 3s^2$

c. $1s^2 2s^2 2p^6 3s^2 3p^3$ d. $1s^2 2s^2 2p^6 3s^2 3p^6 4s^2 3d^7$

3.68 Draw the orbital diagram associated with each of the following electron configurations.

a. $1s^2 2s^2 2p^5$ b. $1s^2 2s^2 2p^6 3s^1$

c. $1s^2 2s^2 2p^6 3s^2 3p^1$ d. $1s^2 2s^2 2p^6 3s^2 3p^6 4s^2 3d^5$

3.69 How many unpaired electrons are present in the orbital diagram for each of the following elements?

a. $_7$N   b. $_{12}$Mg   c. $_{17}$Cl   d. $_{25}$Mn

3.70 How many unpaired electrons are present in the orbital diagram for each of the following elements?

a. $_9$F   b. $_{16}$S   c. $_{20}$Ca   d. $_{30}$Zn

Electron Configurations and the Periodic Law (Section 3.8)

3.71 Indicate whether the elements represented by the given pairs of electron configurations have similar chemical properties.

a. $1s^2 2s^1$ and $1s^2 2s^2$

b. $1s^2 2s^2 2p^6$ and $1s^2 2s^2 2p^6 3s^2 3p^6$

c. $1s^2 2s^2 2p^3$ and $1s^2 2s^2 2p^6 3s^2 3p^6 4s^2 3d^3$

d. $1s^2 2s^2 2p^6 3s^2 3p^4$ and $1s^2 2s^2 2p^6 3s^2 3p^6 4s^2 3d^{10} 4p^4$

3.72 Indicate whether the elements represented by the given pairs of electron configurations have similar chemical properties.

a. $1s^2 2s^2 2p^4$ and $1s^2 2s^2 2p^5$

b. $1s^2 2s^2$ and $1s^2 2s^2 2p^2$

c. $1s^2 2s^1$ and $1s^2 2s^2 2p^6 3s^2 3p^6 4s^1$

d. $1s^2 2s^2 2p^6$ and $1s^2 2s^2 2p^6 3s^2 3p^6 4s^2 3d^6$

3.73 Specify the location of each of the following elements in the periodic table in terms of s area, p area, d area, or f area.

a. Magnesium   b. Copper   c. Bromine   d. Iron

3.74 Specify the location of each of the following elements in the periodic table in terms of s area, p area, d area, or f area.

a. Aluminum   b. Potassium   c. Sulfur   d. Gold

3.75 For each of the following elements, specify the extent to which the subshell containing the distinguishing electron is filled (s^2, p^3, p^5, d^4, etc.).

a. $_{13}$Al   b. $_{23}$V   c. $_{20}$Ca   d. $_{36}$Kr

3.76 For each of the following elements, specify the extent to which the subshell containing the distinguishing electron is filled (s^2, p^3, p^5, d^4, etc.).

a. $_{10}$Ne   b. $_{19}$K   c. $_{33}$As   d. $_{30}$Zn

3.77 Determine the following for the "highlighted" elements in the given periodic table.

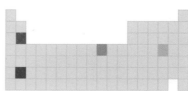

a. Is the element highlighted in red in the s area or p area of the periodic table?

b. Is the element highlighted in green in the d area or p area of the periodic table?

c. Is the element highlighted in yellow a p^2 element or a p^4 element?

d. Is the element highlighted in blue a d^2 or an s^2 element?

3.78 Determine the following, using the periodic table given in Problem 3.77.

a. Is the element highlighted in yellow in the p area or d area of the periodic table?

b. Is the element highlighted in blue in the s area or d area of the periodic table?

c. Is the element highlighted in red an s^2 or a d^2 element?

d. Is the element highlighted in green a d^6 or a d^8 element?

● **Classification of the Elements (Section 3.9)**

3.79 Classify each of the following elements as a noble gas, representative element, transition element, or inner transition element.

a. $_{15}P$ b. $_{18}Ar$ c. $_{79}Au$ d. $_{92}U$

3.80 Classify each of the following elements as a noble gas, representative element, transition element, or inner transition element.

a. $_1H$ b. $_{44}Ru$ c. $_{51}Sb$ d. $_{86}Rn$

3.81 Classify the element with each of the following electron configurations as a representative element, transition element, noble gas, or inner transition element.

a. $1s^2 2s^2 2p^6$

b. $1s^2 2s^2 2p^6 3s^2 3p^4$

c. $1s^2 2s^2 2p^6 3s^2 3p^6 4s^2 3d^1$

d. $1s^2 2s^2 2p^6 3s^2 3p^6 4s^2$

3.82 Classify the element with each of the following electron configurations as a representative element, transition element, noble gas, or inner transition element.

a. $1s^2 2s^2 2p^6 3s^1$

b. $1s^2 2s^2 2p^6 3s^2 3p^6$

c. $1s^2 2s^2 2p^6 3s^2 3p^6 4s^2 3d^7$

d. $1s^2 2s^2 2p^6 3s^2 3p^6 4s^2 3d^{10} 4p^5$

3.83 Determine the following for the "highlighted" elements in the given periodic table.

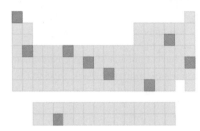

a. How many of the highlighted elements are representative elements?

b. How many of the highlighted elements are noble gases?

c. How many of the highlighted elements are nonmetallic representative elements?

d. How many of the highlighted elements are metals?

3.84 Determine the following for the "highlighted" elements in the periodic table given in Problem 3.83.

a. How many of the highlighted elements are inner transition elements?

b. How many of the highlighted elements are transition elements?

c. How many of the highlighted elements are metallic representative elements?

d. How many of the highlighted elements are nonmetals?

ADDITIONAL PROBLEMS

3.85 With the help of the periodic table, complete the following table.

Element	Complete Chemical Symbol	Atomic Number	Mass Number	Number of Protons	Number of Neutrons
(a)	^3_2He				
(b) nickel			60		
(c)		18	37		
(d)			90		52
(e)	$^{235}_{92}U$				
(f)				17	20
(g)			232	94	
(h)	$^{32}_{16}S$				
(i) iron			56		
(j) calcium					20

3.86 With the help of the periodic table, write complete chemical symbols for atoms with the following characteristics.

a. Contains 20 electrons and 24 neutrons

b. Radon atom with a mass number of 211

c. Silver atom that contains 157 subatomic particles

d. Beryllium atom that contains 9 nucleons

3.87 Characterize each of the following pairs of atoms as containing (1) the same number of neutrons, (2) the same number of electrons, or (3) the same total number of subatomic particles.

a. $^{13}_6C$ and $^{14}_7N$ b. $^{18}_8O$ and $^{19}_9F$

c. $^{37}_{17}Cl$ and $^{36}_{18}Ar$ d. $^{35}_{17}Cl$ and $^{37}_{17}Cl$

3.88 Indicate whether each of the following numbers are the same or different for two isotopes of an element.

a. Atomic number   b. Mass number

c. Number of neutrons   d. Number of electrons

3.89 Write the complete chemical symbol ($^A_Z E$) for the isotope of chromium with each of the following characteristics.

a. Two more neutrons than $^{55}_{24}Cr$

b. Two fewer subatomic particles than $^{52}_{24}Cr$

c. The same number of neutrons as $^{60}_{29}Cu$

d. The same number of subatomic particles as $^{60}_{29}Cu$

3.90 How many electrons are present in nine molecules of the compound $C_{12}H_{22}O_{11}$ (table sugar)?

3.91 Which of the six elements nitrogen, beryllium, argon, aluminum, silver, and gold belong(s) in each of the following classifications?
a. Period and Roman numeral group numbers are numerically equal
b. Readily conducts electricity and heat
c. Has an atomic mass greater than its atomic number
d. All atoms have a nuclear charge greater than +20

3.92 The electron configuration of the isotope $^{16}_{8}O$ is $1s^2 2s^2 2p^4$. What is the electron configuration for the isotope $^{18}_{8}O$?

3.93 Write electron configurations for the following elements.
a. The Group IIIA element in the same period as $_4$Be
b. The Period 3 element in the same group as $_5$B
c. The lowest-atomic-numbered metal in Group IA
d. The Period 3 element that has three unpaired electrons

3.94 Referring only to the periodic table, determine the element of lowest atomic number whose electron configuration contains each of the following.
a. Three completely filled orbitals
b. Three completely filled subshells
c. Three completely filled shells
d. Three completely filled s subshells

$\mathcal{M}$ultiple-Choice Practice Test

3.95 Which of the following listings correctly orders subatomic particles in terms of increasing mass?
a. Electron, proton, neutron
b. Electron, neutron, proton
c. Proton, neutron, electron
d. Neutron, proton, electron

3.96 Which of the following statements concerning the *nucleus* of an atom is *correct*?
a. Contains only neutrons
b. Contains all protons and all electrons
c. Is always positively charged
d. Accounts for most of the total volume of an atom

3.97 The number of protons, neutrons, and electrons, respectively, in an atom of $^{60}_{27}Co$ is
a. 27, 27, 33   b. 27, 33, 27
c. 33, 27, 33   d. 27, 60, 27

3.98 All atoms of a given element have the same
a. mass number.   b. number of nucleons.
c. number of neutrons.   d. number of protons.

3.99 The atomic number of an oxygen atom containing 10 neutrons is
a. 8   b. 10   c. 18   d. 20

3.100 The correct electron configuration for the element $_{20}Ca$ is
a. $1s^2 2s^2 2p^6 3s^2$
b. $1s^2 2s^2 3s^2 4s^2$
c. $1s^2 2s^2 2p^6 3s^2 3p^6 4s^2$
d. $1s^2 2s^2 2p^6 3s^2 3p^6 3d^{10} 4s^2$

3.101 Which of the following statements is consistent with the electron configuration $1s^2 2s^2 2p^6 3s^2 3p^4$?
a. There are 4 electrons present in a $3p$ orbital.
b. There are 4 electrons present in a $3p$ subshell.
c. There are 4 electrons present in a $3p$ shell.
d. There are 4 electrons present in the third shell.

3.102 Which of the following elements is located in Period 3 and Group IVA of the periodic table?
a. N   b. Si   c. Ge   d. In

3.103 In which of the following pairs of elements is one element a metal and the other element a nonmetal?
a. $_{30}Zn$ and $_{31}Ga$ b. $_{16}S$ and $_{17}Cl$
c. $_9F$ and $_{53}I$ d. $_{15}P$ and $_{53}Bi$

3.104 Which of the following statements concerning types of elements is *correct*?
a. There are more noble gas elements than transition elements.
b. There are more "s area" elements than "p area" elements.
c. There are more nonmetals than metals.
d. There are more representative elements than inner transition elements.

Chemical Bonding: The Ionic Bond Model

4

Chapter Outline

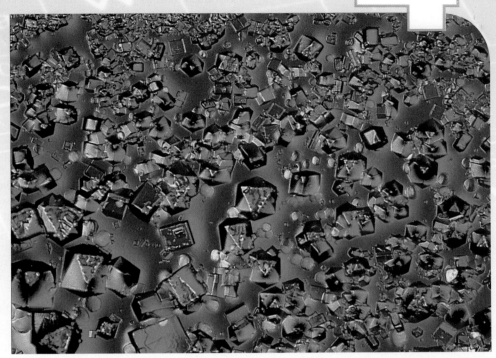

Magnification of crystals of sodium chloride (table salt), one of the most commonly encountered ionic compounds. Color has been added to the image by computer.

As scientists study living organisms and the world in which we live, they rarely encounter free isolated atoms. Instead, under normal conditions of temperature and pressure, they nearly always find atoms associated in aggregates or clusters ranging in size from two atoms to numbers too large to count. In this chapter, we will explain why atoms tend to join together in larger units, and we will discuss the binding forces (chemical bonds) that hold them together.

As we examine the nature of attractive forces between atoms, we will discover that both the tendency and the capacity of an atom to be attracted to other atoms are dictated by its electron configuration.

4.1 CHEMICAL BONDS

Chemical compounds are conveniently divided into two broad classes called *ionic compounds* and *molecular compounds*. Ionic and molecular compounds can be distinguished from each other on the basis of general physical properties. Ionic compounds tend to have high melting points ($500°C - 2000°C$) and are good conductors of electricity when they are in a molten (liquid) state or in solution. Molecular compounds, on the other hand, generally have much lower melting points and tend to be gases, liquids, or low-melting point solids. They do not conduct electricity in the molten state. Ionic compounds, unlike molecular compounds, do not have molecules as their basic structural unit. Instead, an extended array of positively and negatively charged particles called *ions* is present (Section 4.8).

Some combinations of elements produce ionic compounds, whereas other combinations of elements form molecular compounds. What determines whether the interaction of two elements produces ions (an ionic compound) or molecules (a molecular compound)? To answer this question, we need to learn about chemical bonds. A **chemical bond** *is the attractive force that holds two atoms together in a more complex unit.* Chemical bonds form as a result of interactions between electrons found in the combining atoms. Thus the nature of chemical bonds is closely linked to electron configurations (Section 3.7).

Corresponding to the two broad categories of chemical compounds are two types of chemical attractive forces (chemical bonds): ionic bonds and covalent bonds. An **ionic bond** *is a chemical bond formed through the transfer of one or more electrons from one atom or group of atoms to another atom or group of atoms.* As its name suggests, the ionic bond model (electron transfer) is used in describing the attractive forces in ionic compounds. An **ionic compound** *is a compound in which ionic bonds are present.*

A **covalent bond** *is a chemical bond formed through the sharing of one or more pairs of electrons between two atoms.* The covalent bond model (electron sharing) is used in describing the attractions between atoms in molecular compounds. A **molecular compound** *is a compound in which covalent bonds are present.*

Even before we consider the details of these two bond models, it is important to emphasize that the concepts of ionic and covalent bonds are actually "convenience concepts." Most bonds are not 100% ionic or 100% covalent. Instead, most bonds have some degree of both ionic and covalent character—that is, some degree of both the transfer and the sharing of electrons. However, it is easiest to understand these intermediate bonds (the real bonds) by relating them to the pure or ideal bond types called ionic and covalent.

Two concepts fundamental to understanding both the ionic and the covalent bonding models are

1. Not all electrons in an atom participate in bonding. Those that do are called *valence electrons.*
2. Certain arrangements of electrons are more stable than others, as is explained by the *octet rule.*

Section 4.2 addresses the concept of valence electrons, and Section 4.3 discusses the octet rule.

Another designation for *molecular compound* is *covalent compound.* The two designations are used interchangeably. The modifier *molecular* draws attention to the basic structural unit present (the molecule), and the modifier *covalent* focuses on the mode of bond formation (electron sharing).

Purely ionic bonds involve a complete transfer of electrons from one atom to another. *Purely* covalent bonds involve equal sharing of electrons. Experimentally, it is found that most actual bonds have some degree of both ionic and covalent character. The exceptions are bonds between identical atoms; here, the bonding is purely covalent.

The term *valence* is derived from the Latin word *valentia,* which means "capacity" (to form bonds).

In a Lewis symbol, the chemical symbol represents the nucleus and all of the nonvalence electrons. The valence electrons are then shown as "dots."

4.2   VALENCE ELECTRONS AND LEWIS SYMBOLS

Certain electrons called *valence* electrons are particularly important in determining the bonding characteristics of a given atom. A **valence electron** *is an electron in the outermost electron shell of a representative element or noble-gas element.* Note the restriction on the use of this definition; it applies only to representative elements and noble-gas elements. For such elements, valence electrons are always found in either *s* or *p* subshells. (We will not consider in this text the more complicated valence electron definitions for transition elements or inner transition elements; here, the presence of incompletely filled *d* or *f* subshells is a complicating factor.)

The number of valence electrons in an atom of a representative element can be determined from the atom's electron configuration, as is illustrated in Example 4.1.

Scientists have developed a shorthand system for designating the number of valence electrons present in atoms of an element. This system involves the use of Lewis symbols. A **Lewis symbol** *is the chemical symbol of an element surrounded by dots equal in number to the number of valence electrons present in atoms of the element.* Figure 4.1 gives the Lewis symbols for the first 20 elements, all of which are representative elements or noble gases. Lewis symbols, named in honor of the American chemist Gilbert N. Lewis (Figure 4.2), who first introduced them, are also frequently called *electron-dot structures.*

Figure 4.1 Lewis symbols for selected representative and noble-gas elements.

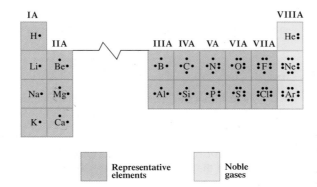

● **E**XAMPLE **4**.**1**

Determining the Number of Valence Electrons in an Atom

▶ Determine the number of valence electrons in atoms of each of the following elements.

a. $_{12}Mg$ **b.** $_{14}Si$ **c.** $_{33}As$

Solution

a. Atoms of the element magnesium have two valence electrons, as can be seen by examining magnesium's electron configuration.

Number of valence electrons

$$1s^2 2s^2 2p^6 ③s^②$$

Highest value of the electron shell number

The highest value of the electron shell number is $n = 3$. Only two electrons are found in shell 3: the two electrons in the $3s$ subshell.

b. Atoms of the element silicon have four valence electrons

Number of valence electrons

$$1s^2 2s^2 2p^6 ③s^② ③p^②$$

Highest value of the electron shell number

Electrons in two different subshells can simultaneously be valence electrons. The highest shell number is 3, and both the $3s$ and the $3p$ subshells belong to this shell. Hence all of the electrons in both of these subshells are valence electrons.

c. Atoms of the element arsenic have five valence electrons.

Number of valence electrons

$$1s^2 2s^2 2p^6 3s^2 3p^6 ④s^② 3d^{10} ④p^③$$

Highest value of the electron shell number

The $3d$ electrons are not counted as valence electrons because the $3d$ subshell is in shell 3, and this shell does not have maximum n value. Shell 4 is the outermost shell and has maximum n value.

Practice Exercise 4.1

Determine the number of valence electrons in atoms of each of the following elements.

a. $_{11}Na$ **b.** $_{16}S$ **c.** $_{35}Br$

Answers: a. 1 valence electron; **b.** 6 valence electrons; **c.** 7 valence electrons

The general practice in writing Lewis symbols is to place the first four "dots" separately on the four sides of the chemical symbol and then begin pairing the dots as further dots are added. It makes no difference on which side of the symbol the process of adding dots begins. The following notations for the Lewis symbol of the element calcium are all equivalent.

$$\text{Ca}\cdot \qquad \text{Ca}\cdot \qquad \cdot\text{Ca} \qquad \cdot\text{Ca}\cdot$$

Three important generalizations about valence electrons can be drawn from a study of the Lewis symbols shown in Figure 4.1.

1. *Representative elements in the same group of the periodic table have the same number of valence electrons.* This should not be surprising. Elements in the same group in the periodic table have similar chemical properties as a result of their similar outer-shell electron configurations (Section 3.8). The electrons in the outermost shell are the valence electrons.
2. *The number of valence electrons for representative elements is the same as the Roman numeral periodic-table group number.* For example, the Lewis symbols for oxygen and sulfur, which are both members of Group VIA, have six dots. Similarly, the Lewis symbols of hydrogen, lithium, sodium, and potassium, which are all members of Group IA, have one dot.
3. *The maximum number of valence electrons for any element is eight.* Only the noble gases (Section 3.9), beginning with neon, have the maximum number of eight electrons. Helium, which has only two valence electrons, is the exception in the noble-gas family. Obviously, an element with a total of two electrons cannot have eight valence electrons. Although electron shells with n greater than 2 are capable of holding more than eight electrons, they do so only when they are no longer the outermost shell and thus are not the valence shell. For example, arsenic has 18 electrons in its third shell; however, shell 4 is the valence shell for arsenic.

Figure 4.2 Gilbert Newton Lewis (1875–1946), one of the foremost chemists of the twentieth century, made significant contributions in other areas of chemistry besides his pioneering work in describing chemical bonding. He formulated a generalized theory for describing acids and bases and was the first to isolate deuterium (heavy hydrogen).

● **E X A M P L E   4 . 2**

Writing Lewis Symbols for Elements

▶ Write Lewis symbols for the following elements.

a. O, S, and Se   **b.** B, C, and N

Solution

a. These elements are all Group VIA elements and thus possess six valence electrons. (The number of valence electrons and the periodic-table group number will always match for representative elements.) The Lewis symbols, which all have six "dots," are

$$\cdot\ddot{\underset{\cdot\cdot}{\text{O}}}: \qquad \cdot\ddot{\underset{\cdot\cdot}{\text{S}}}: \qquad \cdot\ddot{\underset{\cdot\cdot}{\text{Se}}}:$$

b. These elements are sequential elements in Period 2 of the periodic table; B is in Group IIIA (three valence electrons), C is in Group IVA (four valence electrons), and N is in Group VA (five valence electrons). The Lewis symbols for these elements are

$$\cdot\overset{\cdot}{\text{B}}\cdot \qquad \cdot\overset{\cdot}{\underset{\cdot}{\text{C}}}\cdot \qquad :\overset{\cdot}{\text{N}}\cdot$$

Practice Exercise 4.2

Write Lewis symbols for the following elements.

a. Be, Mg, and Ca   **b.** P, S, and Cl

Answers: a. $\cdot\text{Be}$ $\cdot\text{Mg}$ $\cdot\text{Ca}$ **b.** $:\overset{\cdot}{\text{P}}\cdot$ $:\overset{\cdot}{\underset{\cdot\cdot}{\text{S}}}\cdot$ $:\overset{\cdot}{\underset{\cdot\cdot}{\text{Cl}}}:$

 THE OCTET RULE

A key concept in elementary bonding theory is that certain arrangements of valence electrons are more stable than others. The term *stable* as used here refers to the idea that a system, which in this case is an arrangement of electrons, does not easily undergo spontaneous change.

The valence electron configurations of the noble gases (helium, neon, argon, krypton, xenon, and radon) are considered the *most stable of all valence electron configurations.* All of the noble gases except helium possess eight valence electrons, which is the maximum number possible. Helium's valence electron configuration is $1s^2$. All of the other noble gases possess ns^2np^6 valence electron configurations, where n has the maximum value found in the atom.

He: $\boxed{1s^2}$

Ne: $1s^2\boxed{2s^22p^6}$

Ar: $1s^22s^22p^6\boxed{3s^23p^6}$

Kr: $1s^22s^22p^63s^23p^6\boxed{4s^2}3d^{10}\boxed{4p^6}$

Xe: $1s^22s^22p^63s^23p^64s^23d^{10}4p^6\boxed{5s^2}4d^{10}\boxed{5p^6}$

Rn: $1s^22s^22p^63s^23p^64s^23d^{10}4p^65s^24d^{10}5p^6\boxed{6s^2}4f^{14}5d^{10}\boxed{6p^6}$

Except for helium, all the noble-gas valence electron configurations have the outermost s and p subshells *completely filled.*

The conclusion that an ns^2np^6 configuration ($1s^2$ for helium) is the most stable of all valence electron configurations is based on the chemical properties of the noble gases. The noble gases are the *most unreactive* of all the elements. They are the only elemental gases found in nature in the form of individual uncombined atoms. There are no known compounds of helium and neon, and only a few compounds of argon, krypton, xenon, and radon are known. The noble gases have little or no tendency to form bonds to other atoms.

Atoms of many elements that lack the very stable noble-gas valence electron configuration tend to acquire it through chemical reactions that result in compound formation. This observation is known as the **octet rule:** *In forming compounds, atoms of elements lose, gain, or share electrons in such a way as to produce a noble-gas electron configuration for each of the atoms involved.*

4.4 THE IONIC BOND MODEL

Electron transfer between two or more atoms is central to the ionic bond model. This electron transfer process produces charged particles called ions. An **ion** *is an atom (or group of atoms) that is electrically charged as a result of the loss or gain of electrons.* An atom is neutral when the number of protons (positive charges) is equal to the number of electrons (negative charges). Loss or gain of electrons destroys this proton–electron balance and leaves a net charge on the atom.

If an atom *gains* one or more electrons, it becomes a *negatively* charged ion; excess negative charge is present because electrons outnumber protons. If an atom *loses* one or more electrons, it becomes a *positively* charged ion; more protons are present than electrons. There is excess positive charge (Figure 4.3). Note that the excess positive charge associated with a positive ion is never caused by proton gain but always by electron loss. If the number of protons remains constant and the number of electrons decreases, the result is net positive charge. The number of protons, which determines the identity of an element, never changes during ion formation.

The charge on an ion depends on the number of electrons that are lost or gained. Loss of one, two, or three electrons gives ions with $+1$, $+2$, or $+3$ charges, respectively. A gain of one, two, or three electrons gives ions with -1, -2, or -3 charges, respectively. (Ions that have lost or gained more than three electrons are very seldom encountered.)

The notation for charges on ions is a superscript placed to the right of the chemical symbol. Some examples of ion symbols are

Positive ions: Na^+, K^+, Ca^{2+}, Mg^{2+}, Al^{3+}

Negative ions: Cl^-, Br^-, O^{2-}, S^{2-}, N^{3-}

Note that we use a single plus or minus sign to denote a charge of 1, instead of using the notation $^{1+}$ or $^{1-}$. Also note that in multicharged ions, the number precedes the charge sign; that is, the notation for a charge of plus two is $^{2+}$ rather than $^{+2}$.

The *outermost* electron shell of an atom is also called the *valence* electron shell.

Some compounds exist whose formulation is not consistent with the octet rule, but the vast majority of simple compounds have formulas that are consistent with its precepts.

The word *ion* is pronounced "eye-on."

An atom's nucleus *never* changes during the process of ion formation. The number of neutrons and protons remains constant.

A loss of electrons by an atom always produces a positive ion. A gain of electrons by an atom always produces a negative ion.

Figure 4.3 Loss of an electron from a sodium atom leaves it with one more proton than electrons, so it has a net electrical charge of +1. When chlorine gains an electron, it has one more electron than protons, so it has a net electrical charge of −1.

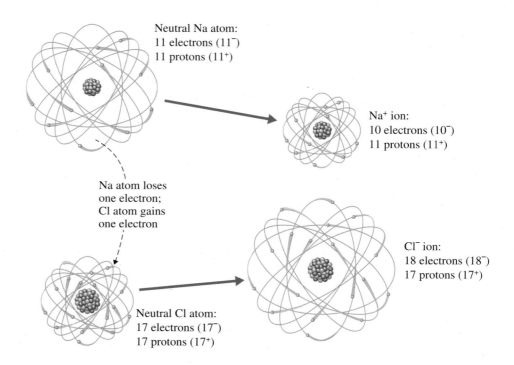

Neutral Na atom:
11 electrons (11⁻)
11 protons (11⁺)

Na⁺ ion:
10 electrons (10⁻)
11 protons (11⁺)

Na atom loses
one electron;
Cl atom gains
one electron

Cl⁻ ion:
18 electrons (18⁻)
17 protons (17⁺)

Neutral Cl atom:
17 electrons (17⁻)
17 protons (17⁺)

EXAMPLE 4.3

Writing Chemical
Symbols for Ions

Give the chemical symbol for each of the following ions.

a. The ion formed when a barium atom loses two electrons
b. The ion formed when a phosphorus atom gains three electrons

Solution

a. A neutral barium atom contains 56 protons and 56 electrons because barium has an atomic number of 56. The barium ion formed by the loss of 2 electrons would still contain 56 protons but would have only 54 electrons because 2 electrons were lost.

$$56 \text{ protons} = 56 + \text{charges}$$
$$\underline{54 \text{ electrons} = 54 - \text{charges}}$$
$$\text{Net charge} = 2+$$

The chemical symbol of the barium ion is thus Ba^{2+}.

b. The atomic number of phosphorus is 15. Thus 15 protons and 15 electrons are present in a neutral phosphorus atom. A gain of 3 electrons raises the electron count to 18.

$$15 \text{ protons} = 15 + \text{charges}$$
$$\underline{18 \text{ electrons} = 18 - \text{charges}}$$
$$\text{Net charge} = 3-$$

The chemical symbol for the ion is P^{3-}.

Practice Exercise 4.3

Give the chemical symbol for each of the following ions.

a. The ion formed when cesium loses one electron
b. The ion formed when selenium gains two electrons

Answers: a. Cs^+; **b.** Se^{2-}

The chemical properties of a particle (atom or ion) depend on the particle's electron arrangement. Because an ion has a different electron configuration (fewer or more electrons) from the atom from which it was formed, it has different chemical properties as well. For example, the drug many people call lithium, which is used to treat mental illness (manic-depressive symptoms), does not involve lithium (Li, the element) but rather lithium ions (Li^+). The element lithium, if ingested, would be poisonous and possibly fatal. The lithium ion, ingested in the form of lithium carbonate, has entirely different effects on the human body.

4.5 THE SIGN AND MAGNITUDE OF IONIC CHARGE

The octet rule provides a very simple and straightforward explanation for the charge magnitude associated with ions of the representative elements. *Atoms tend to gain or lose electrons until they have obtained an electron configuration that is the same as that of a noble gas.* The element sodium has the electron configuration

$$1s^2 2s^2 2p^6 3s^1$$

One valence electron is present. Sodium can attain a noble-gas electron configuration by losing this valence electron (to give it the electron configuration of neon) or by gaining seven electrons (to give it the electron configuration of argon).

$$Na\ (1s^2 2s^2 2p^6 3s^1) \begin{array}{c} \xrightarrow{\text{Loss of 1 } e^-} Na^+ \quad (1s^2 2s^2 2p^6) \\ \text{Electron configuration of neon} \\ \xrightarrow{\text{Gain of 7 } e^-} Na^{7-} \quad (1s^2 2s^2 2p^6 3s^2 3p^6) \\ \text{Electron configuration of argon} \end{array}$$

The electron loss or gain that involves the fewest electrons will always be the more favorable process from an energy standpoint and will be the process that occurs. Thus for sodium the loss of one electron to form the Na^+ ion is the process that occurs.

The element chlorine has the electron configuration

$$1s^2 2s^2 2p^6 3s^2 3p^5$$

Seven valence electrons are present. Chlorine can attain a noble-gas electron configuration by losing seven electrons (to give it the electron configuration of neon) or by gaining one electron (to give it the electron configuration of argon). The latter occurs for the reason we previously cited.

$$Cl\ (1s^2 2s^2 2p^6 3s^2 3p^5) \begin{array}{c} \xrightarrow{\text{Loss of 7 } e^-} Cl^{7+} \quad (1s^2 2s^2 2p^6) \\ \text{Electron configuration of neon} \\ \xrightarrow{\text{Gain of 1 } e^-} Cl^- \quad (1s^2 2s^2 2p^6 3s^2 3p^6) \\ \text{Electron configuration of argon} \end{array}$$

The considerations we have just applied to sodium and chlorine lead to the following generalizations:

1. Metal atoms containing one, two, or three valence electrons (the metals in Groups IA, IIA, and IIIA of the periodic table) tend to lose electrons to acquire a noble-gas electron configuration. The noble gas involved is the one preceding the metal in the periodic table.
   Group IA metals form 1^+ ions.
   Group IIA metals form 2^+ ions.
   Group IIIA metals form 3^+ ions.
2. Nonmetal atoms containing five, six, or seven valence electrons (the nonmetals in Groups VA, VIA, and VIIA of the periodic table) tend to gain electrons to acquire a

The positive charge on metal ions from Groups IA, IIA, and IIIA has a magnitude equal to the metal's periodic-table group number.

Fresh Water, Seawater, Hard Water, and Soft Water: A Matter of Ions

Water is the most abundant compound on the face of Earth. We encounter it everywhere we go: as water vapor in the air; as a liquid in rivers, lakes, and oceans; and as a solid (ice and snow) both on land and in the oceans.

All water as it occurs in nature is impure in a chemical sense. The impurities present include suspended matter, microbiological organisms, dissolved gases, and dissolved minerals. Minerals dissolved in water produce ions. For example, rock salt (NaCl) dissolves in water to produce Na^+ and Cl^- ions.

The major distinction between *fresh water* and *seawater* (salt water) is the number of ions present. On a relative scale, where the total concentration of ions in fresh water is assigned a value of 1, seawater has a value of approximately 500; that is, seawater has a concentration of dissolved ions 500 times greater than that of fresh water.

The dominant ions in fresh water and seawater are not the same. In seawater, Na^+ ion is the dominant positive ion and Cl^- ion is the dominant negative ion. This contrasts with fresh water, where Ca^{2+} and Mg^{2+} ions are the most abundant positive ions and HCO_3^- (a polyatomic ion; Section 4.10) is the most abundant negative ion.

When fresh water is purified for drinking purposes, suspended particles, disease-causing agents, and objectionable odors are removed. Dissolved ions are not removed. At the concentrations at which they are normally present in fresh water, dissolved ions are not harmful to health. Indeed, some of the taste of water is caused by the ions present; water without any ions present would taste "unpleasant" to most people.

Hard water is water that contains Ca^{2+}, Mg^{2+}, and Fe^{2+} ions. The presence of these ions does not affect the drinkability of water, but it does affect other uses for the water. The hard-water ions form insoluble compounds with soap (producing scum) and lead to the production of deposits of scale in steam boilers, tea kettles, and hot water pipes.

The most popular method for obtaining *soft water* from hard water involves the process of "ion exchange." In this process, the offending hard-water ions are exchanged for Na^+ ions. Sodium ions do not form insoluble soap compounds or scale. People with high blood pressure or kidney problems are often advised to avoid drinking soft water because of its high sodium content.

Nonmetals from Groups VA, VIA, and VIIA form negative ions whose charge is equal to the group number minus 8. For example, S, in Group VIA, forms S^{2-} ions $(6 - 8 = -2)$.

noble-gas electron configuration. The noble gas involved is the one following the nonmetal in the periodic table.

Group VIIA nonmetals form 1^- ions.
Group VIA nonmetals form 2^- ions.
Group VA nonmetals form 3^- ions.

3. Elements in Group IVA occupy unique positions relative to the noble gases. They would have to gain or lose four electrons to attain a noble-gas structure. Theoretically, ions with charges of $+4$ or -4 could be formed by carbon, but in most cases the bonding that results is more adequately described by the covalent bond model discussed in Chapter 5.

Isoelectronic Species

An ion formed in the preceding manner with an electron configuration the same as that of a noble gas is said to be *isoelectronic* with the noble gas. **Isoelectronic species** *are an atom and ion, or two ions, that have the same number and configuration of electrons.* An atom and an ion or two ions may be isoelectronic. The following is a list of ions that are isoelectronic with the element Ne; all, like Ne, have the electron configuration $1s^2 2s^2 2p^6$.

$$N^{3-} \quad O^{2-} \quad F^- \quad Na^+ \quad Mg^{2+} \quad Al^{3+}$$

It should be emphasized that an ion that is isoelectronic with a noble gas does not have the properties of the noble gas. It has not been converted into the noble gas. The number of protons in the nucleus of the isoelectronic ion is different from that in the noble gas. This point is emphasized by the comparison in Table 4.1 between Mg^{2+} ion and Ne, the noble gas with which Mg^{2+} is isoelectronic.

TABLE 4.1
Comparison of the Characteristics of the Isoelectronic Species Mg^{2+} and Ne

	Ne Atom	Mg^{2+} Ion
Atomic number	10	12
Protons (in the nucleus)	10	12
Electrons (around the nucleus)	10	10
Charge	0	+2

4.6 LEWIS STRUCTURES FOR IONIC COMPOUNDS

No atom can lose electrons unless another atom is available to accept them.

Ion formation through the loss or gain of electrons by atoms is not an isolated, singular process. In reality, electron loss and electron gain are always partner processes; if one occurs, the other also occurs. Ion formation requires the presence of two elements: a metal that can donate electrons and a nonmetal that can accept electrons. The electrons lost by the metal are the same ones gained by the nonmetal. The positive and negative ions simultaneously formed from such *electron transfer* attract one another. The result is the formation of an ionic compound.

Lewis structures are helpful in visualizing the formation of simple ionic compounds. A **Lewis structure** *is a combination of Lewis symbols that represents either the transfer or the sharing of electrons in chemical bonds.* Lewis *symbols* involve individual elements. Lewis *structures* involve compounds. The reaction between the element sodium (with one valence electron) and chlorine (with seven valence electrons) is represented as follows with a Lewis structure:

$$\text{Na} \cdot \ddot{\ddot{\text{Cl}}} \colon \longrightarrow [\text{Na}]^{+} \left[\colon \ddot{\ddot{\text{Cl}}} \colon \right]^{-} \longrightarrow \text{NaCl}$$

The loss of an electron by sodium empties its valence shell. The next inner shell, which contains eight electrons (a noble-gas configuration), then becomes the valence shell. After the valence shell of chlorine gains one electron, it has the needed eight valence electrons.

When sodium, which has one valence electron, combines with oxygen, which has six valence electrons, the oxygen atom requires the presence of two sodium atoms to acquire two additional electrons.

$$\begin{array}{c} \text{Na} \cdot \\[4pt] + \ddot{\text{O}} \colon \\[4pt] \text{Na} \cdot \end{array} \longrightarrow \begin{array}{c} [\text{Na}]^{+} \\[2pt] [\text{Na}]^{+} \end{array} \left[\colon \ddot{\ddot{\text{O}}} \colon \right]^{2-} \longrightarrow \text{Na}_2\text{O}$$

Note that because oxygen has room for two additional electrons, two sodium atoms are required per oxygen atom—hence the formula Na_2O.

An opposite situation to that in Na_2O occurs in the reaction between calcium, which has two valence electrons, and chlorine, which has seven valence electrons. Here, two chlorine atoms are necessary to accommodate electrons transferred from one calcium atom because a chlorine atom can accept only one electron. (It has seven valence electrons and needs only one more.)

$$\begin{array}{c} \cdot \ddot{\ddot{\text{Cl}}} \colon \\[6pt] \text{Ca} \colon + \\[6pt] \cdot \ddot{\ddot{\text{Cl}}} \colon \end{array} \longrightarrow [\text{Ca}]^{2+} \begin{array}{c} \left[\colon \ddot{\ddot{\text{Cl}}} \colon \right]^{-} \\[6pt] \left[\colon \ddot{\ddot{\text{Cl}}} \colon \right]^{-} \end{array} \longrightarrow \text{CaCl}_2$$

● E X A M P L E   4 . 4

Using Lewis Structures to Depict Ionic Compound Formation

Show the formation of the following ionic compounds using Lewis structures.

a. Na_3N **b.** MgO **c.** Al_2S_3

Solution

a. Sodium (a Group IA element) has one valence electron, which it would "like" to lose. Nitrogen (a Group VA element) has five valence electrons and would thus "like" to acquire three more. Three sodium atoms are needed to supply enough electrons for one nitrogen atom.

$$\begin{array}{c} \text{Na} \cdot \\[4pt] \text{Na} \cdot \ddot{\text{N}} \colon \\[4pt] \text{Na} \cdot \end{array} \longrightarrow \begin{array}{c} [\text{Na}]^{+} \\[2pt] [\text{Na}]^{+} \\[2pt] [\text{Na}]^{+} \end{array} \left[\colon \ddot{\text{N}} \colon \right]^{3-} \longrightarrow \text{Na}_3\text{N}$$

(continued)

b. Magnesium (a Group IIA element) has two valence electrons, and oxygen (a Group VIA element) has six valence electrons. The transfer of the two magnesium valence electrons to an oxygen atom results in each atom having a noble-gas electron configuration. Thus these two elements combine in a one-to-one ratio.

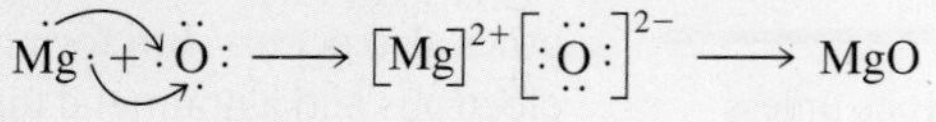

$$Mg + \ddot{O}: \longrightarrow [Mg]^{2+}[:\ddot{O}:]^{2-} \longrightarrow MgO$$

c. Aluminum (a Group IIIA element) has three valence electrons, all of which need to be lost through electron transfer. Sulfur (a Group VIA element) has six valence electrons and thus needs to acquire two more. Three sulfur atoms are needed to accommodate the electrons given up by two aluminum atoms.

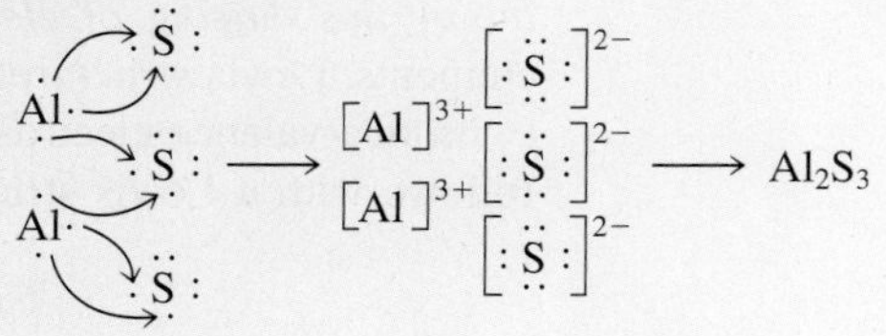

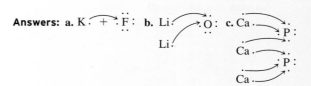

Practice Exercise 4.4

Show the formation of the following ionic compounds using Lewis structures.

a. KF   **b.** Li$_2$O   **c.** Ca$_3$P$_2$

Answers: a.   **b.**   **c.**

$\text{K} \cdot + \cdot \ddot{\text{F}}:$ **b.** Li ... $\ddot{\text{O}}:$ **c.** Ca ... $\ddot{\text{P}}:$
$\qquad\qquad\qquad \text{Li}$ $\qquad \text{Ca}$
$\qquad\qquad\qquad\qquad\qquad\qquad\quad \ddot{\text{P}}:$
$\qquad\qquad\qquad\qquad\qquad\qquad\quad \text{Ca}$

4.7 CHEMICAL FORMULAS FOR IONIC COMPOUNDS

Electron loss always equals electron gain in an electron transfer process. Consequently, ionic compounds are always neutral; no net charge is present. The total positive charge present on the ions that have lost electrons always is exactly counterbalanced by the total negative charge on the ions that have gained electrons. Thus *the ratio in which positive and negative ions combine is the ratio that achieves charge neutrality for the resulting compound.* This generalization can be used instead of Lewis structures to determine ionic compound formulas. Ions are combined in the ratio that causes the positive and negative charges to add to zero.

The correct combining ratio when K^+ ions and S^{2-} ions combine is two to one. Two K^+ ions (each of $+1$ charge) are required to balance the charge on a single S^{2-} ion.

$$2(K^+): \qquad (2 \text{ ions}) \times (\text{charge of } +1) = +2$$
$$S^{2-}: \qquad \underline{(1 \text{ ion}) \times (\text{charge of } -2) = -2}$$
$$\text{Net charge} = \quad 0$$

The formula of the compound formed is thus K$_2$S.

There are three rules to remember when writing chemical formulas for ionic compounds.

1. The symbol for the positive ions is always written first.
2. The charges on the ions that are present are *not* shown in the formula. You need to know the charges to determine the formula; however, the charges are not explicitly shown in the formula.
3. The numbers in the formula (the subscripts) give the combining ratio for the ions.

EXAMPLE 4.5

Using Ionic Charges to Determine the Chemical Formula of an Ionic Compound

Determine the chemical formula for the compound that is formed when each of the following pairs of ions interact.

a. Na^+ and P^{3-} **b.** Be^{2+} and P^{3-}

Solution

a. The Na^+ and P^{3-} ions combine in a three-to-one ratio because this combination causes the charges to add to zero. Three Na^+ ions give a total positive charge of 3. One P^{3-} ion results in a total negative charge of 3. Thus the chemical formula for the compound is Na_3P.

b. The numbers in the charges for these ions are 2 and 3. The lowest common multiple of 2 and 3 is 6 ($2 \times 3 = 6$). Thus we need 6 units of positive charge and 6 units of negative charge. Three Be^{2+} ions are needed to give the 6 units of positive charge, and two P^{3-} ions are needed to give the 6 units of negative charge. The combining ratio of ions is three to two, and the chemical formula is Be_3P_2.

The strategy of finding the lowest common multiple of the numbers in the charges of the ions always works, and it is a faster process than that of drawing the Lewis structures.

Practice Exercise 4.5

Determine the chemical formula for the compound that is formed when each of the following pairs of ions interact.

a. Ca^{2+} and F^- **b.** Al^{3+} and O^{2-}

Answers: a. CaF_2; **b.** Al_2O_3

4.8 THE STRUCTURE OF IONIC COMPOUNDS

An ionic compound, in the solid state, consists of positive and negative ions arranged in such a way that each ion is surrounded by nearest neighbors of the opposite charge. Any given ion is bonded by electrostatic (positive–negative) attractions to all the other ions of opposite charge immediately surrounding it. Figure 4.4 shows a two-dimensional cross section and a three-dimensional view of the arrangement of ions in the ionic

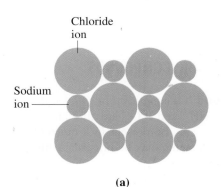

Chloride ion

Sodium ion

(a)

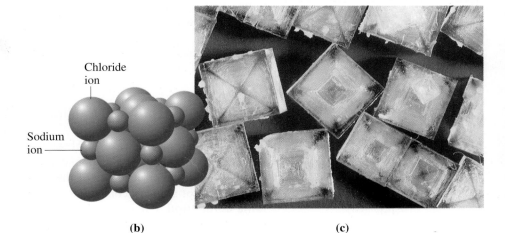

Chloride ion

Sodium ion

(b)   (c)

Figure 4.4 (a, b) A two-dimensional cross-section and a three-dimensional view of sodium chloride (NaCl), an ionic solid. Both views show an alternating array of positive and negative ions. (c) Sodium chloride crystals.

Figure 4.5 Cross-section of the structure of the ionic solid NaCl. No molecule can be distinguished in this structure. Instead, a basic *formula unit* is present that is repeated indefinitely.

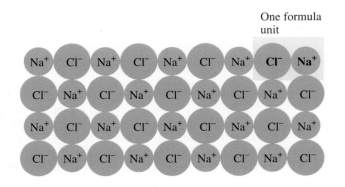

One formula unit

In Section 1.9 the molecule was described as the smallest unit of a pure substance that is capable of a stable, independent existence. Ionic compounds, with their formula units, are exceptions to this generalization.

(a)

(b)

Figure 4.6 Ionic compounds usually have crystalline forms in the solid state, such as those associated with (a) fluorite and (b) ruby.

compound sodium chloride (NaCl). Note in these structural representations that no given ion has a single partner. A given sodium ion has six immediate neighbors (chloride ions) that are equidistant from it. A chloride ion in turn has six immediate sodium ion neighbors.

The alternating array of positive and negative ions present in an ionic compound means that discrete molecules do not exist in such compounds. Therefore, the chemical formulas of ionic compounds cannot represent the composition of molecules of these substances. Instead, such chemical formulas represent the simplest combining ratio for the ions present. The chemical formula for sodium chloride, NaCl, indicates that sodium and chloride ions are present in a one-to-one ratio in this compound. Chemists use the term *formula unit,* rather than *molecule,* to refer to the smallest unit of an ionic compound. A **formula unit** *is the smallest whole-number repeating ratio of ions present in an ionic compound that results in charge neutrality.* A formula unit is "hypothetic," because it does not exist as a separate entity; it is only "a part" of the extended array of ions that constitute an ionic solid (see Figure 4.5).

Although the chemical formulas for ionic compounds represent only ratios, they are used in equations and chemical calculation in the same way as are the chemical formulas for molecular species. Remember, however, that they cannot be interpreted as indicating that molecules exist for these substances; they merely represent the simplest ratio of ions present.

The ions present in an ionic solid adopt an arrangement that maximizes attractions between ions of opposite charge and minimizes repulsions between ions of like charge. The specific arrangement that is adopted depends on ion sizes and on the ratio between positive and negative ions. Arrangements are usually very symmetrical and result in crystalline solids—that is, solids with highly regular shapes. Crystalline solids usually have flat surfaces or faces that make definite angles with one another, as is shown in Figure 4.6.

The Chemistry at a Glance feature on page 95 reviews the general concepts we have considered so far about ionic compounds.

4.9 RECOGNIZING AND NAMING BINARY IONIC COMPOUNDS

The term *binary* means "two." A **binary compound** *is a compound in which only two elements are present.* The compounds NaCl, CO_2, NH_3, and P_4O_{10} are all binary compounds. Any number of atoms of the two elements may be present in a molecule or formula unit of a binary compound, but only two elements may be present. A **binary ionic compound** *is an ionic compound in which one element present is a metal and the other element present is a nonmetal.* The metal is always present as the positive ion, and the nonmetal is always present as the negative ion. The joint presence of a metal and a nonmetal in a binary compound is the "recognition key" that the compound is an ionic compound.

CHEMISTRY AT A GLANCE

Ionic Bonds and Ionic Compounds

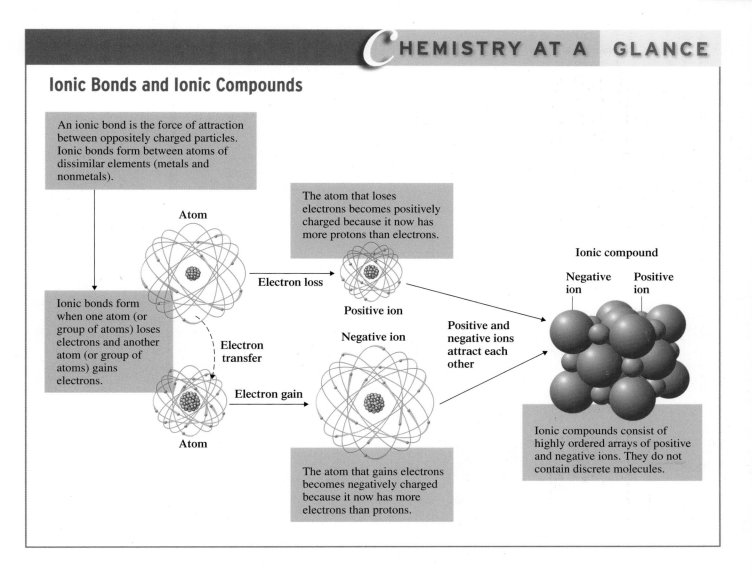

An ionic bond is the force of attraction between oppositely charged particles. Ionic bonds form between atoms of dissimilar elements (metals and nonmetals).

Ionic bonds form when one atom (or group of atoms) loses electrons and another atom (or group of atoms) gains electrons.

Atom

Electron loss

Electron transfer

Electron gain

Atom

The atom that loses electrons becomes positively charged because it now has more protons than electrons.

Positive ion

Negative ion

The atom that gains electrons becomes negatively charged because it now has more electrons than protons.

Positive and negative ions attract each other

Ionic compound

Negative ion   Positive ion

Ionic compounds consist of highly ordered arrays of positive and negative ions. They do not contain discrete molecules.

EXAMPLE 4.6

Recognizing a Binary Ionic Compound on the Basis of Its Chemical Formula

Which of the following binary compounds would be expected to be an ionic compound?

a. Al_2S_3 **b.** H_2O **c.** KF   **d.** NH_3

Solution

a. Ionic; a metal (Al) and a nonmetal (S) are present.
b. Not ionic; two nonmetals are present.
c. Ionic; a metal (K) and a nonmetal (F) are present.
d. Not ionic; two nonmetals are present.

The two compounds that are not ionic are *molecular* compounds (Section 4.1). Chapter 5 includes an extended discussion of molecular compounds. In general, molecular compounds contain just nonmetals.

Practice Exercise 4.6

Which of the following binary compounds would be expected to be an ionic compound?

a. CO_2 **b.** $MgCl_2$ **c.** Fe_2O_3 **d.** PF_3

Answers: a. Not ionic; **b.** Ionic; **c.** Ionic; **d.** Not ionic

TABLE 4.2
Names of Selected Common Nonmetallic Ions

Element	Stem	Name of Ion	Formula of Ion
bromine	brom-	bromide	Br^-
carbon	carb-	carbide	C^{4-}
chlorine	chlor-	chloride	Cl^-
fluorine	fluor-	fluoride	F^-
hydrogen	hydr-	hydride	H^-
iodine	iod-	iodide	I^-
nitrogen	nitr-	nitride	N^{3-}
oxygen	ox-	oxide	O^{2-}
phosphorus	phosph-	phosphide	P^{3-}
sulfur	sulf-	sulfide	S^{2-}

Binary ionic compounds are named using the following rule: *The full name of the metallic element is given first, followed by a separate word containing the stem of the nonmetallic element name and the suffix* -ide. Thus, in order to name the compound NaF, we start with the name of the metal (sodium), follow it with the stem of the name of the nonmetal (fluor-), and then add the suffix *-ide*. The name becomes *sodium fluoride.*

The stem of the name of the nonmetal is the name of the nonmetal with its ending chopped off. Table 4.2 gives the stem part of the name for each of the most common nonmetallic elements. The name of the metal ion is always exactly the same as the name of the metal itself; the metal's name is never shortened. Example 4.7 illustrates the use of the rule for naming binary ionic compounds.

EXAMPLE 4.7

Naming Binary Ionic Compounds

Name the following binary ionic compounds.

a. MgO   **b.** Al_2S_3 **c.** K_3N **d.** $CaCl_2$

Solution

The general pattern for naming binary ionic compounds is

Name of metal + stem of name of nonmetal + *-ide*

a. The metal is magnesium and the nonmetal is oxygen. Thus the compound's name is *magnesium oxide.*
b. The metal is aluminum and the nonmetal is sulfur; the compound's name is *aluminum sulfide.* Note that no mention is made of the subscripts present in the formula—the 2 and the 3. The name of an ionic compound never contains any reference to formula subscript numbers. There is only one ratio in which aluminum and sulfur atoms combine. Thus, just telling the names of the elements present in the compound is adequate nomenclature.
c. Potassium (K) and nitrogen (N) are present in the compound, and its name is *potassium nitride.*
d. The compound's name is *calcium chloride.*

Practice Exercise 4.7

Name the following binary ionic compounds.

a. Na_2S **b.** BeO   **c.** Li_3P **d.** BaI_2

Answers: a. Sodium sulfide; **b.** Beryllium oxide; **c.** Lithium phosphide; **d.** Barium iodide

All the inner transition elements (*f* area of the periodic table), most of the transition elements (*d* area), and a few representative metals (*p* area) exhibit variable ionic charge behavior.

Thus far in our discussion of ionic compounds, it has been assumed that the only behavior allowable for an element is that predicted by the octet rule. This is a good assumption for nonmetals and for most representative element metals. However, there are other metals that exhibit a less predictable behavior because they are able to form

Figure 4.7 Copper(II) oxide (CuO) is black, whereas copper(I) oxide (Cu$_2$O) is reddish brown. Iron(II) chloride (FeCl$_2$) is green, whereas iron(III) chloride (FeCl$_3$) is bright yellow.

Copper (I) Oxide

Iron (III) Chloride

Copper (II) Oxide

Iron (II) Chloride

An older method for indicating the charge on metal ions uses the suffixes *-ic* and *-ous* rather than the Roman numeral system. It is mentioned here because it is still sometimes encountered. In this system, when a metal has two common ionic charges, the suffix *-ous* is used for the ion of lower charge and the suffix *-ic* for the ion of higher charge. The metal's Latin name is also used. In this older system, iron(II) ion is called ferrous ion, and iron(III) ion is called ferric ion.

more than one type of ion. For example, iron forms both Fe^{2+} ions and Fe^{3+} ions, depending on chemical circumstances.

When we name compounds that contain metals with variable ionic charges, the charge on the metal ion must be incorporated into the name. This is done by using Roman numerals. For example, the chlorides of Fe^{2+} and Fe^{3+} (FeCl$_2$ and FeCl$_3$, respectively) are named iron(II) chloride and iron(III) chloride (Figure 4.7). Likewise, CuO is named copper(II) oxide. If you are uncertain about the charge on the metal ion in an ionic compound, use the charge on the nonmetal ion (which does not vary) to calculate it. For example, in order to determine the charge on the copper ion in CuO, you can note that the oxide ion carries a -2 charge because oxygen is in Group VIA. This means that the copper ion must have a $+2$ charge to counterbalance the -2 charge.

 **EXAMPLE 4.8**

Using Roman Numerals in the Naming of Binary Ionic Compounds

Name the following binary ionic compounds, each of which contains a metal whose ionic charge can vary.

a. AuCl   **b.** Fe$_2$O$_3$

Solution

We will need to indicate the magnitude of the charge on the metal ion in the name of each of these compounds by means of a Roman numeral.

a. To calculate the metal ion charge, use the fact that total ionic charge (both positive and negative) must add to zero.

$$\text{(Gold charge)} + \text{(chlorine charge)} = 0$$

The chloride ion has a -1 charge (Section 4.5). Therefore,

$$\text{(Gold charge)} + (-1) = 0$$

Thus,

$$\text{Gold charge} = +1$$

Therefore, the gold ion present is Au$^+$, and the name of the compound is *gold(I) chloride*.

b. For charge balance in this compound we have the equation

$$2\text{(iron charge)} + 3\text{(oxygen charge)} = 0$$

Note that we have to take into account the number of each kind of ion present (2 and 3 in this case). Oxide ions carry a -2 charge (Section 4.5). Therefore,

$$2\text{(iron charge)} + 3(-2) = 0$$
$$2\text{(iron charge)} = +6$$
$$\text{Iron charge} = +3$$

(continued)

Here, we are interested in the charge on a single iron ion ($+3$) and not in the total positive charge present ($+6$). The compound is named *iron(III) oxide* because Fe^{3+} ions are present. As is the case for all ionic compounds, the name does not contain any reference to the numerical subscripts in the compound's formula.

Practice Exercise 4.8

Name the following binary ionic compounds, each of which contains a metal whose ionic charge can vary.

a. PbO_2 **b.** Cu_2S

Answers: a. Lead(IV) oxide; **b.** Copper(I) sulfide

The fixed-charge metals are those in Group IA ($+1$ ionic charge), those in Group IIA ($+2$ ionic charge), and five others (Al^{3+}, Ga^{3+}, Zn^{2+}, Cd^{2+}, and Ag^+).

In order to know when to use Roman numerals in binary ionic compound names, you must know which metals exhibit variable ionic charge and which have a fixed ionic charge. There are many more of the former (Roman numeral required) than of the latter (no Roman numeral required). Thus you should learn the identity of the metals that have a fixed ionic charge (the short list); any metal not on the short list must exhibit variable charge. Figure 4.8 shows the metals that always form a single type of ion in ionic compound formation. Ionic compounds that contain these metals are the only ones without Roman numerals in their names.

 4.10 POLYATOMIC IONS

There are two categories of ions: monatomic and polyatomic. A **monatomic ion** *is an ion formed from a single atom through loss or gain of electrons.* All of the ions we have discussed so far have been monatomic (Cl^-, Na^+, Ca^{2+}, N^{3-}, and so on).

A **polyatomic ion** *is an ion formed from a group of atoms (held together by covalent bonds) through loss or gain of electrons.* An example of a polyatomic ion is the sulfate ion, $SO_4{}^{2-}$. This ion contains four oxygen atoms and one sulfur atom, and the whole group of five atoms has acquired a -2 charge. The whole sulfate group is the ion

Figure 4.8 A periodic table in which the metallic elements that exhibit a fixed ionic charge are highlighted.

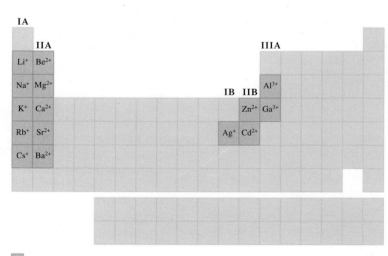

Fixed ionic charge metals

TABLE 4.3
Formulas and Names of Some Common Polyatomic Ions

Key Element Present	Formula	Name of Ion
nitrogen	NO_3^-	nitrate
	NO_2^-	nitrite
	NH_4^+	ammonium
	N_3^-	azide
sulfur	SO_4^{2-}	sulfate
	HSO_4^-	bisulfate or hydrogen sulfate
	SO_3^{2-}	sulfite
	HSO_3^-	bisulfite or hydrogen sulfite
	$S_2O_3^{2-}$	thiosulfate
phosphorus	PO_4^{3-}	phosphate
	HPO_4^{2-}	hydrogen phosphate
	$H_2PO_4^-$	dihydrogen phosphate
	PO_3^{3-}	phosphite
carbon	CO_3^{2-}	carbonate
	HCO_3^-	bicarbonate or hydrogen carbonate
	$C_2O_4^{2-}$	oxalate
	$C_2H_3O_2^-$	acetate
	CN^-	cyanide
chlorine	ClO_4^-	perchlorate
	ClO_3^-	chlorate
	ClO_2^-	chlorite
	ClO^-	hypochlorite
hydrogen	H_3O^+	hydronium
	OH^-	hydroxide
metals	MnO_4^-	permanganate
	CrO_4^{2-}	chromate
	$Cr_2O_7^{2-}$	dichromate

rather than any one atom within the group. Covalent bonding, discussed in Chapter 5, holds the sulfur and oxygen atoms together.

There are numerous ionic compounds in which the positive or negative ion (sometimes both) is polyatomic. Polyatomic ions are very stable and generally maintain their identity during chemical reactions.

Note that polyatomic ions are not molecules. They never occur alone as molecules do. Instead, they are always found associated with ions of opposite charge. Polyatomic ions are *charged pieces* of compounds, not compounds. Ionic compounds require the presence of both positive and negative ions and are neutral overall.

Table 4.3 lists the names and formulas of some of the more common polyatomic ions. The following generalizations concerning polyatomic ion names and charges emerge from consideration of the ions listed in Table 4.3.

1. Most of the polyatomic ions have a negative charge, which can vary from −1 to −3. Only two positive ions are listed in the table: NH_4^+ (ammonium) and H_3O^+ (hydronium).
2. Two of the negatively charged polyatomic ions, OH^- (hydroxide) and CN^- (cyanide), have names ending in *-ide,* and the rest of them have names ending in either *-ate* or *-ite.*
3. A number of *-ate, -ite* pairs of ions exist, as in SO_4^{2-} (sulfate) and SO_3^{2-} (sulfite). The *-ate* ion always has one more oxygen atom than the *-ite* ion. Both the *-ate* and *-ite* ions of a pair carry the same charge.
4. A number of pairs of ions exist wherein one member of the pair differs from the other by having a hydrogen atom present, as in CO_3^{2-} (carbonate) and HCO_3^- (hydrogen carbonate or bicarbonate). In such pairs, the charge on the ion that contains hydrogen is always 1 less than that on the other ion.

Learning the names of the common polyatomic ions is a memorization project. There is no shortcut. The charges and formulas for the various polyatomic ions *cannot* be easily related to the periodic table, as was the case for many of the monatomic ions.

The prefix *bi-* in polyatomic ion names means *hydrogen* rather than the number *two.*

CHEMICAL Connections

Tooth Enamel: A Combination of Monatomic and Polyatomic Ions

The hard outer covering of a tooth, its enamel, is made up of a three-dimensional network of calcium ions (Ca^{2+}), phosphate ions (PO_4^{3-}), and hydroxide ions (OH^-) arranged in a regular pattern. The formula for this material is $Ca_{10}(PO_4)_6(OH)_2$, and its name is hydroxyapatite. Fibrous protein is dispersed in the spaces between the ions. (See the accompanying figure.)

Hydroxyapatite continually dissolves and reforms within the mouth. Tooth enamel is continually dissolving to a slight extent, to give a water solution in saliva of Ca^{2+}, PO_4^{3-}, and OH^- ions. This process is called *demineralization*. At the same time, however, the ions in the saliva solution are recombining to deposit enamel back on the teeth. This process is called *mineralization*. As long as demineralization and mineralization occur at equal rates, no net loss of tooth enamel occurs.

$$Ca_{10}(PO_4)_6(OH)_2 \underset{\text{mineralization}}{\overset{\text{demineralization}}{\rightleftharpoons}} 10Ca^{2+} + 6PO_4^{3-} + 2OH^-$$

Tooth decay results when chemical factors within the mouth cause the rate of demineralization to exceed the rate of mineralization. The acidic H^+ ion is the chemical species that most often causes the demineralization process to dominate. The continuation of this process over an extended period of time results in the formation of pits or cavities in tooth enamel. Eventually, the pits break through the enamel, allowing bacteria to enter the tooth structure and cause decay.

When fluoride ion (F^-) exchanges with hydroxide ion in the hydroxyapatite structure, tooth enamel is strengthened.

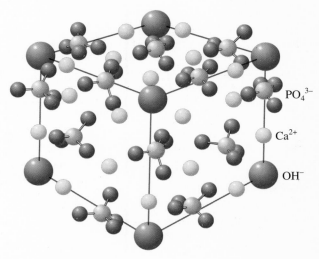

PO_4^{3-}
Ca^{2+}
OH^-

$$Ca_{10}(PO_4)_6(OH)_2 + 2\ F^- \longrightarrow Ca_{10}(PO_4)_6F_2 +\ 2\ OH^-$$

Hydroxyapatite   Fluoroapatite

This replacement of hydroxide by fluoride in the apatite crystal produces an enamel that is less soluble in acidic medium—hence the effectiveness of fluoride mouthwashes and fluoride-containing toothpastes.

The Chemical Connections feature above considers the structure of tooth enamel, a substance that contains the polyatomic ions phosphate (PO_4^{3-}) and hydroxide (OH^-).

 ## 4.11 CHEMICAL FORMULAS AND NAMES FOR IONIC COMPOUNDS CONTAINING POLYATOMIC IONS

Chemical formulas for ionic compounds that contain polyatomic ions are determined in the same way as those for ionic compounds that contain monatomic ions (Section 4.7). The positive and negative charges present must add to zero.

Two conventions not encountered previously in chemical formula writing often arise when we write chemical formulas containing polyatomic ions.

1. When more than one polyatomic ion of a given kind is required in a chemical formula, the polyatomic ion is enclosed in parentheses, and a subscript, placed outside the parentheses, is used to indicate the number of polyatomic ions needed. An example is $Fe(OH)_3$.

2. So that the identity of polyatomic ions is preserved, the same elemental symbol may be used more than once in a chemical formula. An example is the formula NH_4NO_3, where the chemical symbol for nitrogen (N) appears in two locations because both the NH_4^+ and NO_3^- ions contain N.

Example 4.9 illustrates the use of both of these new conventions.

EXAMPLE 4.9

Writing Chemical Formulas
for Ionic Compounds Containing
Polyatomic Ions

Determine the chemical formulas for the ionic compounds that contain these pairs of ions.

a. Na^+ and SO_4^{2-} **b.** Mg^{2+} and NO_3^- **c.** NH_4^+ and CN^-

Solution

a. In order to equalize the total positive and negative charge, we need two sodium ions ($+1$ charge) for each sulfate ion (-2 charge). We indicate the presence of two Na^+ ions with the subscript 2 following the symbol of this ion. The chemical formula of the compound is Na_2SO_4. The convention that the positive ion is always written first in a chemical formula still holds when polyatomic ions are present.

b. Two nitrate ions (-1 charge) are required to balance the charge on one magnesium ion ($+2$ charge). Because more than one polyatomic ion is needed, the chemical formula contains parentheses, $Mg(NO_3)_2$. The subscript 2 outside the parentheses indicates two of what is inside the parentheses. If parentheses were not used, the chemical formula would appear to be $MgNO_{32}$, which is not intended and conveys false information.

c. In this compound, both ions are polyatomic, which is a perfectly legal situation. Because the ions have equal but opposite charges, they combine in a one-to-one ratio. Thus the chemical formula is NH_4CN. No parentheses are necessary because we need only one polyatomic ion of each type in a formula unit. The appearance of the symbol for the element nitrogen (N) at two locations in the chemical formula could be prevented by combining the two nitrogens, resulting in N_2H_4C. But the chemical formula N_2H_4C does not convey the message that NH_4^+ and CN^- ions are present. Thus, when writing chemical formulas that contain polyatomic ions, we always maintain the identities of these ions, even if it means having the same elemental symbol at more than one location in the formula.

Practice Exercise 4.9

Determine the chemical formulas for the ionic compounds that contain the following pairs of ions.

a. K^+ and CO_3^{2-} **b.** Ca^{2+} and OH^- **c.** NH_4^+ and HPO_4^{2-}

Answers: a. K_2CO_3; **b.** $Ca(OH)_2$; **c.** $(NH_4)_2HPO_4$

The names of ionic compounds containing polyatomic ions are derived in a manner similar to that for binary ionic compounds (Section 4.9). The rule for naming binary ionic compounds is as follows: Give the name of the metallic element first (including, when needed, a Roman numeral indicating ion charge), and then give a separate word containing the stem of the nonmetallic name and the suffix -*ide*.

For our present situation, *if the polyatomic ion is positive, its name is substituted for that of the metal. If the polyatomic ion is negative, its name is substituted for the nonmetal stem plus* -ide. Where both positive and negative ions are polyatomic, dual substitution occurs, and the resulting name includes just the names of the polyatomic ions.

EXAMPLE 4.10

Naming Ionic Compounds
in Which Polyatomic Ions
Are Present

Name the following compounds, which contain one or more polyatomic ions.

a. $Ca_3(PO_4)_2$ **b.** $Fe_2(SO_4)_3$ **c.** $(NH_4)_2CO_3$

Solution

a. The positive ion present is the calcium ion (Ca^{2+}). We will not need a Roman numeral to specify the charge on a Ca^{2+} ion because it is always $+2$. The negative ion is the polyatomic phosphate ion (PO_4^{3-}). The name of the compound is *calcium phosphate*. As in naming binary ionic compounds, subscripts in the formula are not incorporated into the name.

(continued)

CHEMISTRY AT A GLANCE

Nomenclature of Ionic Compounds

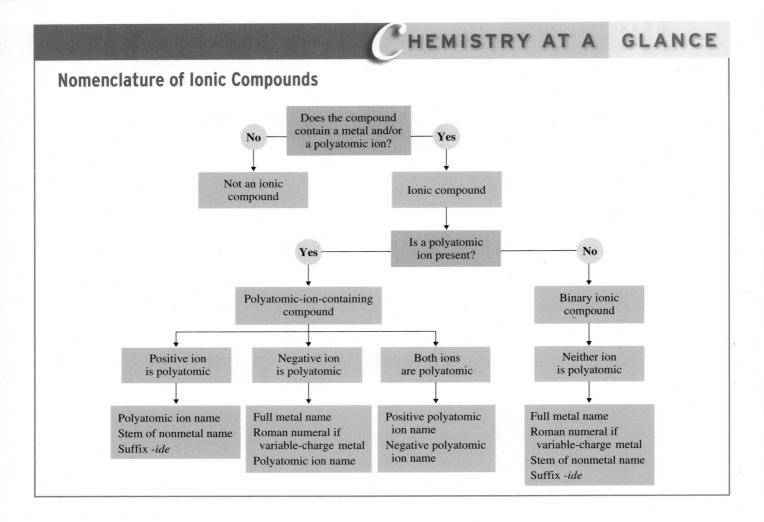

b. The positive ion present is iron(III). The negative ion is the polyatomic sulfate ion (SO_4^{2-}). The name of the compound is *iron(III) sulfate*. The determination that iron is present as iron(III) involves the following calculation dealing with charge balance:

$$2(\text{iron charge}) + 3(\text{sulfate charge}) = 0$$

The sulfate charge is -2. (You had to memorize that.) Therefore,

$$2(\text{iron charge}) + 3(-2) = 0$$
$$2(\text{iron charge}) = +6$$
$$\text{Iron charge} = +3$$

c. Both the positive and the negative ions in this compound are polyatomic—the ammonium ion (NH_4^+) and the carbonate ion (CO_3^{2-}). The name of the compound is simply the combination of the names of the two polyatomic ions: *ammonium carbonate*.

Practice Exercise 4.10

Name the following compounds, which contain one or more polyatomic ions.

a. $Ba(NO_3)_2$ **b.** Cu_3PO_4 **c.** $(NH_4)_2SO_4$

Answers: a. Barium nitrate; **b.** Copper(I) phosphate; **c.** Ammonium sulfate

The Chemistry at a Glance feature above summarizes the "thought processes" involved in naming ionic compounds, both those with monatomic ions and those with polyatomic ions.

CONCEPTS TO REMEMBER

Chemical bonds. Chemical bonds are the attractive forces that hold atoms together in more complex units. Chemical bonds result from the transfer of valence electrons between atoms (ionic bond) or from the sharing of electrons between atoms (covalent bond) (Section 4.1).

Valence electrons. Valence electrons, for representative elements, are the electrons in the outermost electron shell, which is the shell with the highest shell number. These electrons are particularly important in determining the bonding characteristics of a given atom (Section 4.2).

Octet rule. In compound formation, atoms of representative elements lose, gain, or share electrons in such a way that their electron configurations become identical to those of the noble gas nearest them in the periodic table (Section 4.3).

Ionic compounds. Ionic compounds commonly involve a metal atom and a nonmetal atom. Metal atoms lose one or more electrons, producing positive ions. Nonmetal atoms acquire the electrons lost by the metal atoms, producing negative ions. The oppositely charged ions attract one another, creating ionic bonds (Section 4.4).

Charge magnitude for ions. Metal atoms containing one, two, or three valence electrons tend to lose such electrons, producing ions of

+1, +2, or +3 charge, respectively. Nonmetal atoms containing five, six, or seven valence electrons tend to gain electrons, producing ions of −3, −2, or −1 charge, respectively (Section 4.5).

Chemical formulas for ionic compounds. The ratio in which positive and negative ions combine is the ratio that causes the total amount of positive and negative charges to add up to zero (Section 4.7).

Structure of ionic compounds. Ionic solids consist of positive and negative ions arranged in such a way that each ion is surrounded by ions of the opposite charge (Section 4.8).

Binary ionic compound nomenclature. Binary ionic compounds are named by giving the full name of the metallic element first, followed by a separate word containing the stem of the nonmetallic element name and the suffix *-ide*. A Roman numeral specifying ionic charge is appended to the name of the metallic element if it is a metal that exhibits variable ionic charge (Section 4.9).

Polyatomic ions. A polyatomic ion is a group of covalently bonded atoms that has acquired a charge through the loss or gain of electrons. Polyatomic ions are very stable entities that generally maintain their identity during chemical reactions (Section 4.10).

KEY REACTIONS AND EQUATIONS

1. Number of valence electrons for representative elements (Section 4.2)

    Number of valence electrons = periodic-table group number

2. Charges on metallic monatomic ions (Section 4.5)

    Group IA metals form 1^+ ions.

    Group IIA metals form 2^+ ions.

    Group IIIA metals form 3^+ ions.

3. Charges on nonmetallic monatomic ions (Section 4.5)

    Group VA nonmetals form 3^- ions.

    Group VIA nonmetals form 2^- ions.

    Group VIIA nonmetals form 1^- ions.

EXERCISES *and* PROBLEMS

The members of each pair of problems in this section test similar material.

● Types of Chemical Bonds (Section 4.1)

4.1 Contrast the two general types of chemical *bonds* in terms of the mechanism by which they form.

4.2 Contrast the two general types of chemical *compounds* in terms of their general physical properties.

● Valence Electrons (Section 4.2)

4.3 How many valence electrons do atoms with the following electron configurations have?
   a. $1s^2 2s^2$
   b. $1s^2 2s^2 2p^6 3s^2$
   c. $1s^2 2s^2 2p^6 3s^2 3p^1$
   d. $1s^2 2s^2 2p^6 3s^2 3p^6 4s^2 3d^{10} 4p^2$

4.4 How many valence electrons do atoms with the following electron configurations have?
   a. $1s^2 2s^2 2p^6$
   b. $1s^2 2s^2 2p^6 3s^2 3p^1$
   c. $1s^2 2s^2 2p^6 3s^1$
   d. $1s^2 2s^2 2p^6 3s^2 3p^6 4s^2 3d^{10} 4p^5$

4.5 Give the periodic-table group number and the number of valence electrons present for each of the following representative elements.
   a. $_3$Li   b. $_{10}$Ne   c. $_{20}$Ca   d. $_{53}$I

4.6 Give the periodic-table group number and the number of valence electrons present for each of the following representative elements.
   a. $_{12}$Mg   b. $_{19}$K   c. $_{15}$P   d. $_{35}$Br

4.7 Write the complete electron configuration for each of the following representative elements.
   a. Period 2 element with four valence electrons
   b. Period 2 element with seven valence electrons
   c. Period 3 element with two valence electrons
   d. Period 3 element with five valence electrons

4.8 Write the complete electron configuration for each of the following representative elements.
   a. Period 2 element with one valence electron
   b. Period 2 element with six valence electrons
   c. Period 3 element with seven valence electrons
   d. Period 3 element with three valence electrons

● Lewis Symbols for Atoms (Section 4.2)

4.9 Draw Lewis symbols for atoms of each of the following elements.
   a. $_{12}$Mg   b. $_{19}$K   c. $_{15}$P   d. $_{36}$Kr

4.10 Draw Lewis symbols for atoms of each of the following elements.
   a. $_{13}$Al   b. $_{20}$Ca   c. $_{17}$Cl   d. $_4$Be

4.11 Each of the following Lewis symbols represents a Period 2 element. Determine each element's identity.
   a. X·   b. :X: c. X·   d. ·X·

4.12 Each of the following Lewis symbols represents a Period 3 element. Determine each element's identity.

a. $\cdot \overset{\cdot}{X} \cdot$ b. $\cdot \overset{\cdot}{\underset{\cdot}{X}} \cdot$ c. $\cdot \overset{\cdot\cdot}{\underset{\cdot}{X}} :$ d. $: \overset{\cdot\cdot}{\underset{\cdot\cdot}{X}} :$

● The Octet Rule (Section 4.3)

4.13 What is the chemical property of the noble gases that leads to the conclusion that they possess extremely stable electron arrangements?

4.14 What is the maximum number of valence electrons that a representative element/noble gas can possess, and which group of elements possesses this maximum number of valence electrons?

4.15 What does the octet rule indicate happens to elements that lack a noble-gas electron configuration?

4.16 Which noble gas is an exception to the rule that noble-gas elements possess eight valence electrons?

● Notation for Ions (Section 4.4)

4.17 Give the chemical symbol for each of the following ions.
a. An oxygen atom that has gained two electrons
b. A magnesium atom that has lost two electrons
c. A fluorine atom that has gained one electron
d. An aluminum atom that has lost three electrons

4.18 Give the chemical symbol for each of the following ions.
a. A chlorine atom that has gained one electron
b. A sulfur atom that has gained two electrons
c. A potassium atom that has lost one electron
d. A beryllium atom that has lost two electrons

4.19 What would be the chemical symbol for an ion with each of the following numbers of protons and electrons?
a. 20 protons and 18 electrons b. 8 protons and 10 electrons
c. 11 protons and 10 electrons d. 13 protons and 10 electrons

4.20 What would be the chemical symbol for an ion with each of the following numbers of protons and electrons?
a. 15 protons and 18 electrons b. 17 protons and 18 electrons
c. 12 protons and 10 electrons d. 19 protons and 18 electrons

4.21 Calculate the number of protons and electrons in each of the following ions.
a. P^{3-} b. N^{3-} c. Mg^{2+} d. Li^+

4.22 Calculate the number of protons and electrons in each of the following ions.
a. S^{2-} b. F^- c. K^+ d. H^+

4.23 Fill in the blanks to complete the following table.

Chemical Symbol	Ion Formed	Number of Electrons in Ion	Number of Protons in Ion
Ca	Ca^{2+}		
	Be^{2+}	2	
		54	53
Al			13

4.24 Fill in the blanks to complete the following table.

Chemical Symbol	Ion Formed	Number of Electrons in Ion	Number of Protons in Ion
F	F^-		
S		18	
Br			35
	Ba^{2+}		56

4.25 Write the complete chemical symbol for the following atoms ($_A^Z E$) or ions ($_A^Z E^n$).

a. (6p, 8n) b. (13p, 14n) c. (17p, 18n) d. (24p, 28n)

6e   10e   18e   24e

4.26 Write the complete chemical symbol for the following atoms ($_A^Z E$) or ions ($_A^Z E^n$).

a. (7p, 8n) b. (14p, 14n) c. (18p, 22n) d. (20p, 20n)

10e   14e   18e   18e

● Ionic Charge Sign and Magnitude (Section 4.5)

4.27 What is the charge on the monatomic ion formed by each of the following elements?
a. $_{12}Mg$ b. $_7N$ c. $_{19}K$ d. $_9F$

4.28 What is the charge on the monatomic ion formed by each of the following elements?
a. $_3Li$ b. $_{15}P$ c. $_{16}S$ d. $_{13}Al$

4.29 Indicate the number of electrons lost or gained when each of the following atoms forms an ion.
a. $_4Be$ b. $_{35}Br$ c. $_{38}Sr$ d. $_{34}Se$

4.30 Indicate the number of electrons lost or gained when each of the following atoms forms an ion.
a. $_{37}Rb$ b. $_{53}I$ c. $_8O$ d. $_{11}Na$

4.31 Which noble gas has an electron configuration identical to that of each of the following ions?
a. O^{2-} b. P^{3-} c. Ca^{2+} d. K^+

4.32 Which noble gas has an electron configuration identical to that of each of the following ions?
a. F^- b. Al^{3+} c. Si^{4+} d. C^{4-}

4.33 Which noble gas is isoelectronic with each of the ions in Problem 4.31?

4.34 Which noble gas is isoelectronic with each of the ions in Problem 4.32?

4.35 In what group in the periodic table would representative elements that form ions with the following charges most likely be found?
a. $+2$ b. -2 c. -3 d. $+1$

4.36 In what group in the periodic table would representative elements that form ions with the following charges most likely be found?
a. $+3$ b. -4 c. $+4$ d. -1

4.37 Write the electron configuration of the following.
a. An aluminum atom   b. An aluminum ion

4.38 Write the electron configuration of the following.
a. An oxygen atom   b. An oxygen ion

● Lewis Structures for Ionic Compounds (Section 4.6)

4.39 Using Lewis structures, show how ionic compounds are formed by atoms of
a. Be and O b. Mg and S c. K and N d. F and Ca

4.40 Using Lewis structures, show how ionic compounds are formed by atoms of
a. Na and F b. Li and S c. Be and S d. P and K

4.41 The following Lewis symbols for ions have the charges omitted. Determine the number of extra electrons that each structure is showing and give the proper charge for each ion.

a. $\left[\, :\!\overset{\displaystyle ..}{\underset{\displaystyle ..}{S}}\!: \,\right]$ b. $\left[\, :\!\overset{\displaystyle ..}{\underset{\displaystyle ..}{F}}\!: \,\right]$ c. $\left[\, :\!\overset{\displaystyle ..}{N}\!: \,\right]$ d. $\left[\, :\!\overset{\displaystyle ..}{\underset{\displaystyle ..}{Se}}\!: \,\right]$

4.42 The following Lewis symbols for ions have the charges omitted. Determine the number of extra electrons that each structure is showing and give the proper charge for each ion.

a. $\left[\, :\!\overset{\displaystyle ..}{P}\!: \,\right]$ b. $\left[\, :\!\overset{\displaystyle ..}{\underset{\displaystyle ..}{O}}\!: \,\right]$ c. $\left[\, :\!\overset{\displaystyle ..}{\underset{\displaystyle ..}{Br}}\!: \,\right]$ d. $\left[\, :\!\overset{\displaystyle ..}{\underset{\displaystyle ..}{I}}\!: \,\right]$

● Chemical Formulas for Ionic Compounds (Section 4.7)

4.43 Write the chemical formula for an ionic compound formed from Ba^{2+} ions and each of the following ions.
a. Cl^- b. Br^- c. N^{3-} d. O^{2-}

4.44 Write the chemical formula for an ionic compound formed from K^+ ions and each of the following ions.
a. Cl^- b. Br^- c. N^{3-} d. O^{2-}

4.45 Write the chemical formula for an ionic compound formed from F^- ions and each of the following ions.
a. Mg^{2+} b. Be^{2+} c. Li^+ d. Al^{3+}

4.46 Write the chemical formula for an ionic compound formed from S^{2-} ions and each of the following ions.
a. Mg^{2+} b. Be^{2+} c. Li^+ d. Al^{3+}

4.47 Write the chemical formula for an ionic compound formed from the following ions.
a. Na^+ and S^{2-} b. Ca^{2+} and I^- c. Li^+ and N^{3-} d. Al^{3+} and Br^-

4.48 Write the chemical formula for an ionic compound formed from the following ions.
a. Li^+ and O^{2-} b. Al^{3+} and N^{3-} c. K^+ and Cl^- d. Mg^{2+} and I^-

4.49 The component elements for four binary ionic compounds are shown with different colors on the following periodic table.

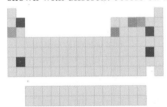

What is the likely chemical formula for the following?
a. Red compound   b. Blue compound
c. Yellow compound   d. Green compound

4.50 The component elements for four binary ionic compounds are shown with different colors on the following periodic table.

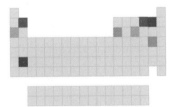

What is the likely chemical formula for the following?
a. Red compound   b. Blue compound
c. Yellow compound   d. Green compound

● Structure of Ionic Compounds (Section 4.8)

4.51 Describe the general structure of a solid-state ionic compound.

4.52 Explain why ionic compounds do not contain individual molecules.

4.53 What is a *formula unit* of an ionic compound?

4.54 In general terms, how many *formula units* are present in a crystal of an ionic compound?

4.55 The following drawings represent solid-state ionic compounds, with red spheres denoting positive ions and blue spheres denoting negative ions.

I   II   III   IV

Which of these drawings could be used as a representation for each of the following ionic compounds?
a. Al_2S_3 b. MgF_2 c. K_3N d. CaO

4.56 Use the solid-state ionic drawings in Problem 4.55, but with blue spheres denoting positive ions and red spheres denoting negative ions. Which of these drawings could be used as a representation for each of the following compounds?
a. Ba_3N_2 b. $NaCl$ c. AlF_3 d. K_2O

● Binary Ionic Compound Nomenclature (Section 4.9)

4.57 Which of the following pairs of elements would be expected to form a binary ionic compound?
a. Sodium and oxygen   b. Magnesium and sulfur
c. Nitrogen and chlorine   d. Copper and fluorine

4.58 Which of the following pairs of elements would be expected to form a binary ionic compound?
a. Potassium and sulfur   b. Calcium and nitrogen
c. Carbon and chlorine   d. Iron and iodine

4.59 Which of the following binary compounds would be expected to be an ionic compound?
a. Al_2O_3 b. H_2O_2 c. K_2S d. N_2H_4

4.60 Which of the following binary compounds would be expected to be an ionic compound?
a. Cu_2O b. CO c. $NaBr$ d. Be_3P_2

4.61 Name the following binary ionic compounds, each of which contains a fixed-charge metal.
a. KI b. BeO c. AlF_3 d. Na_3P

4.62 Name the following binary ionic compounds, each of which contains a fixed-charge metal.
a. $CaCl_2$ b. Ca_2C c. Be_3N_2 d. K_2S

4.63 Calculate the charge on the metal ion in the following binary ionic compounds, each of which contains a variable-charge metal.
a. Au_2O b. CuO c. SnO_2 d. SnO

4.64 Calculate the charge on the metal ion in the following binary ionic compounds, each of which contains a variable-charge metal.
a. Fe_2O_3 b. FeO c. $SnCl_4$ d. Cu_2S

4.65 Name the following binary ionic compounds, each of which contains a variable-charge metal.
a. FeO b. Au_2O_3 c. CuS d. $CoBr_2$

4.66 Name the following binary ionic compounds, each of which contains a variable-charge metal.
a. PbO b. $FeCl_3$ c. SnO_2 d. NiI_2

4.67 Name each of the following binary ionic compounds.
a. $AuCl$ b. KCl c. $AgCl$ d. $CuCl_2$

4.68 Name each of the following binary ionic compounds.
a. NiO b. FeN c. AlN d. BeO

4.69 Write chemical formulas for the following binary ionic compounds.
 a. Potassium bromide　　 b. Silver oxide
 c. Beryllium fluoride　　 d. Barium phosphide

4.70 Write chemical formulas for the following binary ionic compounds.
 a. Gallium nitride　　　 b. Zinc chloride
 c. Magnesium sulfide　　 d. Aluminum nitride

4.71 Write chemical formulas for the following binary ionic compounds.
 a. Cobalt(II) sulfide　　 b. Cobalt(III) sulfide
 c. Tin(IV) iodide　　　 d. Lead(II) nitride

4.72 Write chemical formulas for the following binary ionic compounds.
 a. Iron(III) oxide　　　 b. Iron(II) oxide
 c. Nickel(III) sulfide　　 d. Copper(I) bromide

● **Compounds Containing Polyatomic Ions (Sections 4.10 and 4.11)**

4.73 With the help of Table 4.3, write chemical formulas (including charge) for each of the following polyatomic ions.
 a. Sulfate　 b. Chlorate　 c. Hydroxide　 d. Cyanide

4.74 With the help of Table 4.3, write chemical formulas (including charge) for each of the following polyatomic ions.
 a. Ammonium b. Nitrate c. Perchlorate　 d. Phosphate

4.75 With the help of Table 4.3, write chemical formulas (including charge) for each of the following pairs of polyatomic ions.
 a. Phosphate and hydrogen phosphate
 b. Nitrate and nitrite
 c. Hydronium and hydroxide
 d. Chromate and dichromate

4.76 With the help of Table 4.3, write chemical formulas (including charge) for each of the following pairs of polyatomic ions.
 a. Chlorate and perchlorate
 b. Hydrogen phosphate and dihydrogen phosphate
 c. Carbonate and bicarbonate
 d. Sulfate and hydrogen sulfate

4.77 Write chemical formulas for the compounds formed between the following positive and negative ions.
 a. Na^+ and ClO_4^-　　　 b. Fe^{3+} and OH^-
 c. Ba^{2+} and NO_3^-　　　 d. Al^{3+} and CO_3^{2-}

4.78 Write chemical formulas for the compounds formed between the following positive and negative ions.
 a. K^+ and CN^-　　　 b. NH_4^+ and SO_4^{2-}
 c. Co^{2+} and $H_2PO_4^-$　 d. Ca^{2+} and PO_4^{3-}

4.79 Fill in the blanks to complete the following table of chemical formulas for polyatomic-ion-containing compounds. For each compound the positive ion present is listed on the left side of the table and the negative ion present is listed at the top.

	CN^-	NO_3^-	HCO_3^-	SO_4^{2-}
NH_4^+				
Al^{3+}				
Ag^+		$AgNO_3$		
Ca^{2+}				

4.80 Fill in the blanks to complete the following table of chemical formulas for polyatomic-ion-containing compounds. For each compound the positive ion present is listed on the left side of the table and the negative ion present is listed at the top.

	OH^-	PO_4^{3-}	CO_3^{2-}	HSO_4^-
Li^+				
Cu^{2+}				
Ba^{2+}			$BaCO_3$	
Fe^{3+}				

4.81 Name the following compounds, all of which contain polyatomic ions and fixed-charge metals.
 a. $MgCO_3$　 b. $ZnSO_4$　 c. $Be(NO_3)_2$　 d. Ag_3PO_4

4.82 Name the following compounds, all of which contain polyatomic ions and fixed-charge metals.
 a. $LiOH$　 b. $Al(CN)_3$　 c. $Ba(ClO_3)_2$　 d. $NaNO_3$

4.83 Name the following compounds, all of which contain polyatomic ions and variable-charge metals.
 a. $Fe(OH)_2$　 b. $CuCO_3$　 c. $AuCN$　 d. $Mn_3(PO_4)_2$

4.84 Name the following compounds, all of which contain polyatomic ions and variable-charge metals.
 a. $Fe(NO_3)_3$　 b. $Co_2(CO_3)_3$　 c. Cu_3PO_4　 d. $Pb(SO_4)_2$

4.85 Write formulas for the following compounds, all of which contain polyatomic ions.
 a. Potassium bicarbonate　　 b. Gold(III) sulfate
 c. Silver nitrate　　　　　 d. Copper(II) phosphate

4.86 Write formulas for the following compounds, all of which contain polyatomic ions.
 a. Aluminum nitrate　　　 b. Iron(III) sulfate
 c. Calcium cyanide　　　 d. Lead(IV) hydroxide

ADDITIONAL PROBLEMS

4.87 Fill in the blanks to complete the following table.

Positive Ion	Negative Ion	Chemical Formula	Name
			Magnesium hydroxide
		$BaBr_2$	
Zn^{2+}	NO_3^-		
			Iron(III) chlorate
		PbO_2	
Co^{2+}	PO_4^{3-}		
K^+	I^-		
		Cu_2SO_4	
			Lithium nitride
Al^{3+}	S^{2-}		

4.88 Fill in the blanks to complete the following table.

Chemical Symbol	Number of Protons	Number of Neutrons	Number of Electrons	Net Charge
$^{59}_{26}Fe^{3+}$				
	28	31	25	
	77	120		$2+$
		9	10	$1-$

4.89 What would be the chemical symbol for an ion with each of the following characteristics?
 a. A sodium ion with ten electrons
 b. A fluorine ion with ten electrons
 c. A sulfur ion with two fewer protons than electrons
 d. A calcium ion with two more protons than electrons

4.90 Write the formula of the ionic compound that could form from the elements X and Z if
   a. X has two valence electrons and Z has seven valence electrons
   b. X has one valence electron and Z has six valence electrons
   c. X has three valence electrons and Z has five valence electrons
   d. X has six valence electrons and Z has two valence electrons

4.91 Identify the Period 3 element that most commonly produces each of the following ions.
   a. X^{2-} b. X^{2+} c. X^{3-} d. X^{3+}

4.92 Indicate whether each of the following compounds contains (1) only monatomic ions, (2) only polyatomic ions, (3) both monatomic and polyatomic ions, or (4) no ions.
   a. CaF_2 b. $NaNO_3$ c. NH_4CN d. AlP

4.93 Write chemical formulas (symbol and charge) for both kinds of ions present in each of the following compounds.
   a. KCl b. CaS c. BeF_2 d. Al_2S_3

4.94 Give the chemical formula for, and the name of the compound formed from, each of the following pairs of ions.
   a. Na^+ and N_3^- b. K^+ and NO_3^-
   c. Mg^{2+} and O^{2-} d. NH_4^+ and PO_4^{3-}

4.95 Name each compound in the following pairs of binary ionic compounds.
   a. $SnCl_4$ and $SnCl_2$ b. FeS and Fe_2S_3
   c. Cu_3N and Cu_3N_2 d. NiI_2 and NiI_3

4.96 In which of the following pairs of binary ionic compounds do both members of the pair contain positive ions with the same charge?
   a. Co_2O_3 and $CoCl_3$ b. Cu_2O and CuO
   c. K_2O and Al_2O_3 d. MgS and NaI

4.97 Name each compound in the following pairs of polyatomic-ion-containing compounds.
   a. $CuNO_3$ and $Cu(NO_3)_2$ b. $Pb_3(PO_4)_2$ and $Pb_3(PO_4)_4$
   c. $Mn(CN)_3$ and $Mn(CN)_2$ d. $Co(ClO_3)_2$ and $Co(ClO_3)_3$

4.98 Write chemical formulas for the following compounds.
   a. Sodium sulfide   b. Sodium sulfate
   c. Sodium sulfite   d. Sodium thiosulfate

$\mathcal{M}$ultiple-Choice Practice Test

4.99 For which of the following elements is the listed number of valence electrons *correct*?
   a. Mg (2 valence electrons)   b. N (3 valence electrons)
   c. F (1 valence electron)   d. S (2 valence electrons)

4.100 Which of the following is an *incorrect* statement about the number of electrons lost or gained by a representative element during ion formation?
   a. The number usually does not exceed three.
   b. The number is governed by the octet rule.
   c. The number is related to the position of the element in the periodic table.
   d. The number is the same as the number of valence electrons present.

4.101 Which of the following is a *correct* statement concerning the mechanism for ionic bond formation?
   a. Electrons are transferred from nonmetallic atoms to metallic atoms.
   b. Protons are transferred from the nuclei of metallic atoms to the nuclei of nonmetallic atoms.
   c. Sufficient electrons are transferred to form ions of equal but opposite charge.
   d. Electron loss is always equal to electron gain.

4.102 In which of the following pairings is the chemical formula *not* consistent with the ions shown?
   a. M^{2+} and X^{3-} (M_3X_2)
   b. M^{2+} and X^- (MX_2)
   c. M^+ and X^{3-} (MX_3)
   d. M^{2+} and X^{2-} (MX)

4.103 The correct chemical formula for the ionic compound formed between Mg and O is
   a. MgO b. Mg_2O_2 c. MgO_2 d. Mg_2O

4.104 In which of the following pairs of ionic compounds do both members of the pair contain positive ions with a $+1$ charge?
   a. KCl and CaO b. Na_3N and Li_2S
   c. $AlCl_3$ and MgF_2 d. BaI_2 and $BeBr_2$

4.105 The correct chemical formula for the compound aluminum nitride is
   a. AlN b. AlN_2 c. Al_2N_3 d. Al_3N_2

4.106 In which of the following pairs of metals are both members of the pair variable-charge metals?
   a. Na and Al   b. Au and Ag
   c. Cu and Zn   d. Fe and Ni

4.107 In which of the following pairs of polyatomic ions do both members of the pair have the same charge?
   a. ammonium and phosphate
   b. sulfate and nitrate
   c. cyanide and hydroxide
   d. hydrogen carbonate and carbonate

4.108 Which of the following ionic compounds contains 4 atoms per formula unit?
   a. Lithium nitride
   b. Potassium sulfide
   c. Copper(II) iodide
   d. Sodium cyanide

Chemical Bonding: The Covalent Bond Model

5

Chapter Outline

The pleasant odor of flowers is produced by a mixture of volatile molecular compounds emitted from the flower blooms. Such molecular compounds contain covalent bonds.

The forces that hold atoms in compounds together as a unit are of two general types: (1) ionic bonds (which involve electron transfer) and (2) covalent bonds (which involve electron sharing). The ionic bond model was the subject of Chapter 4. We now consider the covalent bond model.

5.1 THE COVALENT BOND MODEL

We begin our discussion of covalent bonding and the molecular compounds that result from such bonding by listing several key differences between ionic and covalent bonding and the resulting ionic and molecular compounds.

1. Ionic bonds form between atoms of dissimilar elements (a metal and a nonmetal). Covalent bond formation occurs between *similar* or even *identical* atoms. Most often two nonmetals are involved.
2. Electron transfer is the mechanism by which ionic bond formation occurs. Covalent bond formation involves *electron sharing*.
3. Ionic compounds do not contain discrete molecules. Instead, such compounds consist of an extended array of alternating positive and negative ions. In covalently bonded compounds, the basic structural unit is a molecule. Indeed, such compounds are called molecular compounds.
4. All ionic compounds are solids at room temperature. Molecular compounds may be solids (glucose), liquids (water), or gases (carbon dioxide) at room temperature.

Figure 5.1 Electron sharing can occur only when electron orbitals from two different atoms overlap.

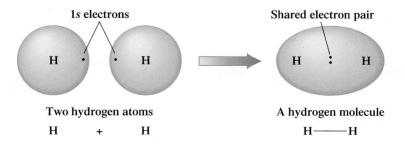

1*s* electrons

Two hydrogen atoms

H   +   H

Shared electron pair

A hydrogen molecule

H———H

5. An ionic solid, if soluble in water, forms an aqueous solution that conducts electricity. The electrical conductance is related to the presence of ions (charged particles) in the solution. A molecular compound, if soluble in water, usually produces a nonconducting aqueous solution.

Among the millions of compounds that are known, those that have covalent bonds are dominant. Almost all compounds encountered in the fields of organic chemistry and biochemistry contain covalent bonds.

A **covalent bond** *is a chemical bond resulting from two nuclei attracting the same shared electrons.* Consideration of the hydrogen molecule (H_2), the simplest of all molecules, provides initial insights into the nature of the covalent bond and its formation. When two hydrogen atoms, each with a single electron, are brought together, the orbitals that contain the valence electrons *overlap* to create an orbital common to both atoms. This overlapping is shown in Figure 5.1. The two electrons, one from each H atom, now move throughout this new orbital and are said to be *shared* by the two nuclei.

Once two orbitals overlap, the most favorable location for the shared electrons is the area directly between the two nuclei. Here the two electrons can simultaneously interact with (be attracted to) both nuclei, a situation that produces increased stability. This concept of increased stability can be explained by using an analogy. Consider the nuclei of the two hydrogen atoms in H_2 to be "old potbellied stoves" and the two electrons to be running around each of the stoves trying to keep warm. When the two nuclei are together (an H_2 molecule) the electrons have two sources of heat. In particular, in the region between the nuclei (the overlap region) the electrons can keep both front and back warm at the same time. This is a better situation than when each electron has only one "stove" (nucleus) as a source of heat.

Covalent bonds result from a common attraction of two nuclei for one or more shared pairs of electrons.

In terms of Lewis notation, this sharing of electrons by the two hydrogen atoms is diagrammed as follows:

 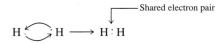

Shared electron pair

H   H ⟶ H : H

The two shared electrons do double duty, helping each hydrogen atom achieve a helium noble-gas configuration.

5.2 LEWIS STRUCTURES FOR MOLECULAR COMPOUNDS

Using the octet rule (Section 4.3), which applies to both electron transfer and electron sharing, and Lewis symbols (Section 4.2), let us now consider the formation of selected simple covalently bonded molecules that contain the element fluorine. Fluorine, located in Group VIIA of the periodic table, has seven valence electrons. Its Lewis symbol is

Fluorine needs only one electron to achieve the octet of electrons that enables it to have a noble-gas electron configuration. When fluorine bonds to other nonmetals, the octet of electrons is completed by means of electron sharing. The molecules HF, F_2, and BrF, whose Lewis structures follow, are representative of this situation.

$$H \overset{\frown}{\underset{\smile}{}} \ddot{\underset{..}{F}} : \longrightarrow H : \ddot{\underset{..}{F}} :$$

$$: \ddot{\underset{..}{F}} \overset{\frown}{\underset{\smile}{}} \ddot{\underset{..}{F}} : \longrightarrow : \ddot{\underset{..}{F}} : \ddot{\underset{..}{F}} :$$

$$: \ddot{\underset{..}{Br}} \overset{\frown}{\underset{\smile}{}} \ddot{\underset{..}{F}} : \longrightarrow : \ddot{\underset{..}{Br}} : \ddot{\underset{..}{F}} :$$

The HF and BrF molecules illustrate the point that the two atoms involved in a covalent bond need not be identical (as is the case with H_2 and F_2).

A common practice in writing Lewis structures for covalently bonded molecules is to represent the *shared* electron pairs with dashes. Using this notation, the H_2, HF, F_2, and BrF molecules are written as

$$H\!-\!H \qquad H\!-\!\ddot{\underset{..}{F}}: \qquad :\ddot{\underset{..}{F}}\!-\!\ddot{\underset{..}{F}}: \qquad :\ddot{\underset{..}{Br}}\!-\!\ddot{\underset{..}{F}}:$$

The atoms in covalently bonded molecules often possess both *bonding* and *nonbonding* electrons. **Bonding electrons** *are pairs of valence electrons that are shared between atoms in a covalent bond.* Each of the fluorine atoms in the molecules HF, F_2, and BrF possesses one pair of bonding electrons. **Nonbonding electrons** *are pairs of valence electrons on an atom that are not involved in electron sharing.* Each of the fluorine atoms in HF, F_2, and BrF possesses three pairs of nonbonding electrons, as does the bromine atom in BrF.

Nonbonding electron pairs are often also referred to as *unshared electron pairs* or *lone electron pairs* (or simply *lone pairs*).

In Section 5.8 we will learn that nonbonding electrons play an important role in determining the shape (geometry) of molecules when three or more atoms are present.

Nonbonding electrons

$$H\overset{/}{\underset{\diagdown}{}}H \qquad H:\ddot{\underset{..}{F}}: \qquad :\ddot{\underset{..}{F}}:\ddot{\underset{..}{F}}: \qquad :\ddot{\underset{..}{Br}}:\ddot{\underset{..}{F}}:$$

Bonding electrons

Bonding electrons (black)
Nonbonding electrons (blue)

The preceding four examples of Lewis structures involved diatomic molecules, the simplest type of molecule. The "thinking pattern" used to draw these diatomic Lewis structures easily extends to triatomic and larger molecules. Consider the molecules H_2O, NH_3, and CH_4, molecules in which two, three, and four hydrogen atoms are attached, respectively, to the O, N, and C atoms. The hydrogen content of these molecules is correlated directly with the fact that oxygen, nitrogen, and carbon have six, five, and four valence electrons, respectively, and therefore need to gain two, three, and four electrons, respectively, through electron sharing in order for the octet rule to be obeyed. The electron-sharing patterns and Lewis structures for these three molecules are as follows:

$$H\!-\!\ddot{O}\!-\!H$$
Water, H_2O

(a) O, with six valence electrons, forms two convalent bonds.

$$H\!-\!\ddot{N}\!-\!H$$
$$\underset{\displaystyle H}{|}$$
Ammonia, NH_3

(b) N, with five valence electrons, forms three covalent bonds.

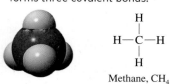

$$\underset{\displaystyle H}{\overset{\displaystyle H}{H\!-\!C\!-\!H}}$$
Methane, CH_4

(c) C, with four valence electrons, forms four covalent bonds.

Figure 5.2 The number of covalent bonds formed by a nonmetallic element is directly correlated with the number of electrons it must share in order to obtain an octet of electrons.

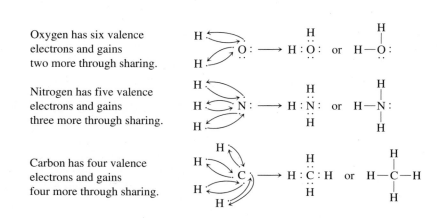

Oxygen has six valence electrons and gains two more through sharing.

Nitrogen has five valence electrons and gains three more through sharing.

Carbon has four valence electrons and gains four more through sharing.

Thus we see here that just as the octet rule was useful in determining the ratio of ions in ionic compounds (Section 4.6), it can be used to predict chemical formulas for molecular compounds. Figure 5.2 and Example 5.1 illustrate further the use of the octet rule to determine chemical formulas for molecular compounds.

E X A M P L E 5.1

Using the Octet Rule to Predict the
Formulas of Simple Molecular
Compounds

Draw Lewis structures for the simplest binary compounds that can be formed from the following pairs of nonmetals.

a. Nitrogen and iodine   **b.** Sulfur and hydrogen

Solution

a. Nitrogen is in Group VA of the periodic table and has five valence electrons. It will need to form three covalent bonds to achieve an octet of electrons. Iodine, in Group VIIA of the periodic table, has seven valence electrons and will need to form only one covalent bond in order to have an octet of electrons. Therefore, three iodine atoms will be needed to meet the needs of one nitrogen atom. The Lewis structure for this molecule is

Each atom in NI_3 has an octet of electrons; these octets are circled in color in the following diagram.

b. Sulfur has six valence electrons and hydrogen has one valence electron. Thus, sulfur will form two covalent bonds (6 + 2 = 8), and hydrogen will form one covalent bond (1 + 1 = 2). Remember that for hydrogen, an "octet" is two electrons; the noble gas that hydrogen mimics is helium, which has only two valence electrons.

Practice Exercise 5.1

Draw Lewis structures for the simplest binary compounds that can be formed from the following pairs of nonmetals.

a. Phosphorus and hydrogen   **b.** Oxygen and chlorine

Answers:

5.3 SINGLE, DOUBLE, AND TRIPLE COVALENT BONDS

A **single covalent bond** *is a covalent bond in which two atoms share one pair of electrons.* All of the bonds in all of the molecules considered in the previous section were *single* covalent bonds.

Single covalent bonds are not adequate to explain covalent bonding in all molecules. Sometimes two atoms must share two or three pairs of electrons in order to provide a complete octet of electrons for each atom involved in the bonding. Such bonds are called *double* covalent bonds and *triple* covalent bonds. A **double covalent bond** *is a covalent bond in which two atoms share two pairs of electrons.* A double covalent bond between two atoms is approximately twice as strong as a single covalent bond between the same two atoms; that is, it takes approximately twice as much energy to break the double bond as it does the single bond. A **triple covalent bond** *is a covalent bond in which two atoms share three pairs of electrons.* A triple covalent bond is approximately three times as strong as a single covalent bond between the same two atoms. The term *multiple covalent bond* is a designation that applies to both double and triple covalent bonds.

One of the simplest molecules possessing a multiple covalent bond is the N_2 molecule, which has a triple covalent bond. A nitrogen atom has five valence electrons and needs three additional electrons to complete its octet.

$$\cdot \ddot{N} \cdot$$

In order to acquire a noble-gas electron configuration, each nitrogen atom must share three of its electrons with the other nitrogen atom.

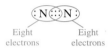

Note that all three shared electron pairs are placed in the area between the two nitrogen atoms in the Lewis structure. Just as one line is used to denote a single covalent bond, three lines are used to denote a triple covalent bond.

When you are "counting" electrons in a Lewis structure to make sure that all atoms in the molecule have achieved their octet of electrons, *all* electrons in a double or triple bond are considered to belong to *both* of the atoms involved in that bond. The "counting" for the N_2 molecule would be

Eight   Eight
electrons   electrons

Each of the circles around a nitrogen atom contains eight valence electrons. Circles are never drawn to include just some of the electrons in a double or triple bond.

A slightly more complicated molecule containing a triple covalent bond is the molecule C_2H_2 (acetylene). A carbon–carbon triple covalent bond is present as well as two carbon–hydrogen single bonds. The arrangement of valence electrons in C_2H_2 is as follows:

$$H \overset{\frown}{C} \cdot \longrightarrow \quad \cdot C \overset{\frown}{H} \longrightarrow H : C ::: C : H \quad \text{or} \quad H—C \equiv C—H$$

The two atoms in a triple covalent bond are commonly the same element. However, they do not have to be. The molecule HCN (hydrogen cyanide) contains a heteroatomic triple covalent bond.

$$H : C ::: N : \quad \text{or} \quad H—C \equiv N :$$

A common molecule that contains a double covalent bond is CO_2 (carbon dioxide). In fact, there are two carbon–oxygen double covalent bonds present in CO_2.

$$: \ddot{O} \overset{\frown}{} C \overset{\frown}{} \ddot{O} : \longrightarrow : \ddot{O} :: C :: \ddot{O} : \quad \text{or} \quad : \ddot{O} = C = \ddot{O} :$$

Note for the CO_2 Lewis structure how the circles are drawn for the octet of electrons about each of the atoms.

5.4 VALENCE ELECTRONS AND NUMBER OF COVALENT BONDS FORMED

Not all elements can form double or triple covalent bonds. There must be at least two vacancies in an atom's valence electron shell prior to bond formation if it is to participate in a double bond, and at least three vacancies are necessary for triple-bond formation. This

A single line (dash) is used to denote a single covalent bond, two lines to denote a double covalent bond, and three lines to denote a triple covalent bond.

requirement eliminates Group VIIA elements (fluorine, chlorine, bromine, iodine) and hydrogen from participating in such bonds. The Group VIIA elements have seven valence electrons and one vacancy, and hydrogen has one valence electron and one vacancy. All covalent bonds formed by these elements are single covalent bonds.

Double bonding becomes possible for elements that need two electrons to complete their octet, and triple bonding becomes possible when three or more electrons are needed to complete an octet. Note that the word *possible* was used twice in the previous sentence. Multiple bonding does not have to occur when an element has two, three, or four vacancies in its octet; single covalent bonds can be formed instead. When more than one behavior is possible, the "bonding behavior" of an element is determined by the element or elements to which it is bonded.

Let us consider the possible "bonding behaviors" for O (six valence electrons, two octet vacancies), N (five valence electrons, three octet vacancies), and C (four valence electrons, four octet vacancies).

To complete its octet by electron sharing, an oxygen atom can form either two single bonds or one double bond.

$$: \overset{\displaystyle |}{\underset{\displaystyle \cdot\cdot}{O}} — \qquad\qquad : \overset{\cdot\cdot}{O} =$$

Two single bonds   One double bond

Nitrogen is a very versatile element with respect to bonding. It can form single, double, or triple covalent bonds as dictated by the other atoms present in a molecule.

$$— \overset{\cdot\cdot}{\underset{|}{N}} — \qquad\qquad — \overset{\cdot\cdot}{N} = \qquad\qquad : N \equiv$$

Three single bonds   One single and   One triple bond
                        one double bond

Note that the nitrogen atom forms three bonds in each of these bonding situations. A double bond counts as two bonds, a triple bond as three. Because nitrogen has only five valence electrons, it must form three covalent bonds to complete its octet.

Carbon is an even more versatile element than nitrogen with respect to variety of types of bonding, as illustrated by the following possibilities. In each case, carbon forms four bonds.

$$— \overset{\displaystyle |}{\underset{\displaystyle |}{C}} — \qquad — \overset{\displaystyle |}{C} = \qquad = C = \qquad — C \equiv$$

Four single bonds   Two single bonds and   Two double bonds   One single bond and
                     one double bond   one triple bond

> There is a strong tendency for atoms of nonmetallic elements to form a specific number of covalent bonds. The number of bonds formed is equal to the number of electrons the nonmetallic atom must share to obtain an octet of electrons.

5.5 COORDINATE COVALENT BONDS

In the covalent bonds we have considered so far (single, double, and triple), the two participating atoms in the bond contributed the same number of electrons to the bond. There is another, *less common* way in which a covalent bond can form. It is possible for one atom to supply two electrons and the other atom none to a shared electron pair. A **coordinate covalent bond** *is a covalent bond in which both electrons of a shared pair come from one of the two atoms involved in the bond.* Coordinate covalent bonding enables an atom that has two or more vacancies in its valence shell to share a pair of nonbonding electrons that are located on another atom.

The element oxygen, with two vacancies in its valence octet, quite often forms coordinate covalent bonds. Consider the Lewis structures of the molecules HOCl (hypochlorous acid) and HClO₂ (chlorous acid).

> An "ordinary" covalent bond can be thought of as a "Dutch-treat" bond; each atom "pays" its part of the bill. A coordinate covalent bond can be thought of as a "you-treat" bond; one atom pays the whole bill.

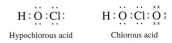

H : Ö : Cl :   H : Ö : Cl : Ö :

Hypochlorous acid   Chlorous acid

Figure 5.3 (a) A "regular" covalent single bond is the result of overlap of two half-filled orbitals. (b) A coordinate covalent single bond is the result of overlap of a filled and a vacant orbital.

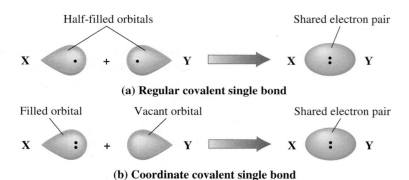

(a) Regular covalent single bond

(b) Coordinate covalent single bond

In hypochlorous acid, all the bonds are "ordinary" covalent bonds. In chlorous acid, which differs from hypochlorous acid in that a second oxygen atom is present, the "new" chlorine–oxygen bond is a coordinate covalent bond. The second oxygen atom with six valence electrons (denoted by x's) needs two more for an octet. It shares one of the nonbonding electron pairs present on the chlorine atom. (The chlorine atom does not need any of the oxygen's electrons because it already has an octet.)

> Atoms participating in coordinate covalent bonds generally do not form their normal number of covalent bonds.

Atoms participating in coordinate covalent bonds generally deviate from the common bonding pattern (Section 5.4) expected for that type of atom. For example, oxygen normally forms two bonds; yet in the molecules N_2O and CO, which contain coordinate covalent bonds, oxygen forms one and three bonds, respectively.

> Once a coordinate covalent bond forms, it is indistinguishable from other covalent bonds in a molecule.

Once a coordinate covalent bond is formed, there is no way to distinguish it from any of the other covalent bonds in a molecule; all electrons are identical regardless of their source. The main use of the concept of coordinate covalency is to help rationalize the existence of certain molecules and polyatomic ions whose electron-bonding arrangement would otherwise present problems. Figure 5.3 contrasts the formation of a "regular" covalent bond with that of a coordinate covalent bond.

5.6 SYSTEMATIC PROCEDURES FOR DRAWING LEWIS STRUCTURES

Could you generate the Lewis structures of HOCl, $HClO_2$, N_2O, and CO given in the preceding section without any help? Drawing Lewis structures for diatomic molecules is usually straightforward and uncomplicated. However, with triatomic and even larger molecules, students often have trouble. Here is a stepwise procedure for distributing valence electrons as bonding and nonbonding pairs within a Lewis structure.

Let us apply this stepwise procedure to the molecule SO_2, a molecule in which two oxygen atoms are bonded to a central sulfur atom (see Figure 5.4).

Step 1: *Calculate the total number of valence electrons available in the molecule by adding together the valence electron counts for all atoms in the molecule.* The periodic table is a useful guide for determining this number.

An SO_2 molecule has 18 valence electrons available for bonding. Sulfur (Group VIA) has 6 valence electrons, and each oxygen (also Group VIA) has 6 valence electrons. The total number is therefore $6 + 2(6) = 18$.

Step 2: *Write the chemical symbols of the atoms in the molecule in the order in which they are bonded to one another, and then place a single covalent bond, involving two electrons, between each pair of bonded atoms.* For SO_2, the S atom is the central atom. Thus we have

Figure 5.4 The sulfur dioxide (SO_2) molecule. A computer-generated model.

O:S:O

Determining which atom is the *central atom*—that is, which atom has the most other atoms bonded to it—is the key to determining the arrangement of atoms in a molecule or polyatomic ion. Most other atoms present will be bonded to the central atom. For common binary molecular compounds, the molecular formula can help us determine the identity of the central atom. The central atom is the atom that appears only once in the formula; for example, S is the central atom in SO_3, O is the central atom in H_2O, and P is the central atom in PF_3. In molecular compounds containing hydrogen, oxygen, and an additional element, that additional element is the central atom; for example, N is the central atom in HNO_3, and S is the central atom in H_2SO_4. In compounds of this type, the oxygen atoms are bonded to the central atom, and the hydrogen atoms are bonded to the oxygens. Carbon is the central atom in nearly all carbon-containing compounds. Neither hydrogen nor fluorine is ever the central atom.

Step 3: *Add nonbonding electron pairs to the structure such that each atom bonded to the central atom has an octet of electrons. Remember that for hydrogen, an "octet" is only 2 electrons.*

For SO_2, addition of the nonbonding electrons gives

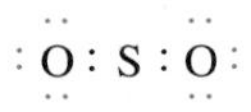

$$\ddot{\text{:}}\ddot{\text{O}}\text{:}\,\text{S}\,\text{:}\ddot{\text{O}}\ddot{\text{:}}$$

At this point, 16 of the 18 available electrons have been used.

Step 4: *Place any remaining electrons on the central atom of the structure.*

Placing the two remaining electrons on the S atom gives

$$\text{:}\ddot{\ddot{\text{O}}}\text{:}\,\ddot{\text{S}}\,\text{:}\ddot{\ddot{\text{O}}}\text{:}$$

Step 5: *If there are not enough electrons to give the central atom an octet, then use one or more pairs of nonbonding electrons on the atoms bonded to the central atom to form double or triple bonds.*

The S atom has only 6 electrons. Thus a nonbonding electron pair from an O atom is used to form a sulfur–oxygen double bond.

This structure now obeys the octet rule.

Step 6: *Count the total number of electrons in the completed Lewis structure to make sure it is equal to the total number of valence electrons available for bonding, as calculated in Step 1.* This step serves as a "double-check" on the correctness of the Lewis structure.

For SO_2, there are 18 valence electrons in the Lewis structure of Step 5, the same number we calculated in Step 1.

EXAMPLE 5.2
Drawing a Lewis Structure Using Systematic Procedures

Draw Lewis structures for the following molecules.

a. PF_3, a molecule in which P is the central atom and all F atoms are bonded to it (see Figure 5.5).

b. HCN, a molecule in which C is the central atom (see Figure 5.6).

Solution

a. *Step 1:*   Phosphorus (Group VA) has 5 valence electrons, and each of the fluorine atoms (Group VIIA) has 7 valence electrons. The total electron count is $5 + 3(7) = 26$.

(continued)

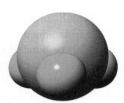

Figure 5.5 A computer-generated model of the phosphorus trifluoride (PF_3) molecule.

Step 2:   Drawing the molecular skeleton with single covalent bonds (2 electrons) placed between all bonded atoms gives

$$F : P : F$$
$$F$$

Step 3:   Adding nonbonding electrons to the structure to complete the octets of all atoms bonded to the central atom gives

$$: \ddot{F} : P : \ddot{F} :$$
$$: \ddot{F} :$$

At this point, we have used 24 of the 26 available electrons.

Step 4:   The central P atom has only 6 electrons; it needs 2 more. The 2 remaining available electrons are placed on the P atom, completing its octet. All atoms now have an octet of electrons.

$$: \ddot{F} : \ddot{P} : \ddot{F} :$$
$$: \ddot{F} :$$

Step 5:   This step is not needed; the central atom already has an octet of electrons.

Step 6:   There are 26 electrons in the Lewis structure, the same number of electrons we calculated in Step 1.

b. *Step 1:*   Hydrogen (Group IA) has 1 valence electron, carbon (Group IVA) has 4 valence electrons, and nitrogen (Group VA) has 5 valence electrons. The total number of electrons is 10.

Step 2:   Drawing the molecular skeleton with single covalent bonds between bonded atoms gives

$$H : C : N$$

Figure 5.6 A computer-generated model of the hydrogen cyanide (HCN) molecule.

Step 3:   Adding nonbonding electron pairs to the structure such that the atoms bonded to the central atom have "octets" gives

$$H : C : \ddot{N} :$$

Remember that hydrogen needs only 2 electrons.

Step 4:   The structure in Step 3 has 10 valence electrons, the total number available. Thus there are no additional electrons available to place on the carbon atom to give it an octet of electrons.

Step 5:   To give the central carbon atom its octet, 2 nonbonding electron pairs on the nitrogen atom are used to form a carbon-nitrogen triple bond.

$$H : C :\!\!\curvearrowright\!\! \ddot{N} : \longrightarrow H : C ::: N :$$

Step 6:   The Lewis structure has 10 electrons, as calculated in Step 1.

Practice Exercise 5.2

Draw Lewis structures for the following molecules.

a. $SiCl_4$, a molecule in which Si is the central atom and all Cl atoms are bonded to it.

b. H_2CS, a molecule in which C is the central atom and the other three atoms are bonded to it.

Answers:

$$\text{a.} \quad : \ddot{Cl} : \\ : \ddot{Cl} : \ddot{Si} : \ddot{Cl} : \\ : \ddot{Cl} :$$

$$\text{b.} \quad \ddot{S} : \\ H : C : H$$

Nitric Oxide: A Molecule Whose Bonding Does Not Follow "The Rules"

The bonding in most, *but not all,* simple molecules is easily explained using the systematic procedures for drawing Lewis structures described in Section 5.6. The molecule NO (nitric oxide) is an example of a simple molecule whose bonding does not conform to the standard rules for bonding. The presence of an odd number of valence electrons (11) in nitric oxide (5 from nitrogen and 6 from oxygen) makes it impossible to write a Lewis structure in which all electrons are paired as required by the octet rule. Thus an unpaired electron is present in the Lewis structure of NO.

Despite the "nonconforming" nature of the bonding in nitric oxide, it is an abundant and important molecule within our environment. This colorless, odorless, nonflammable gas is generated by numerous natural and human-caused processes, including (1) lightning passing through air, which causes the N_2 and O_2 of air to react (to a small extent) with each other to produce NO, (2) automobile engines, within which the hot walls of the cylinders again cause N_2 and O_2 of air to become slightly reactive toward each other, and (3) a burning cigarette.

The fact that NO is produced in the preceding ways has been known for many years. The environmental effects of such NO have also been well documented. The NO serves as a precursor for the formation of both acid rain and smog.

During the early 1990s, it was found that NO is also an important biochemical that is naturally present in the human body. The body generates its own NO, usually from amino acids, and once formed, the NO has a life of 10 seconds or less. Its biochemical functions within the human body include (1) helping maintain blood pressure by dilating blood vessels, (2) helping kill foreign invading molecules as part of the body's immune system response, and (3) serving as a biochemical messenger in the brain for processes associated with long-term memory.

5.7 BONDING IN COMPOUNDS WITH POLYATOMIC IONS PRESENT

Ionic compounds containing polyatomic ions (Section 4.10) present an interesting combination of both ionic and covalent bonds: covalent bonding *within* the polyatomic ion and ionic bonding *between* it and ions of opposite charge.

Polyatomic ion Lewis structures, which show the covalent bonding within such ions, are drawn using the same procedures as for molecular compounds (Section 5.6), with the accommodation that the total number of electrons used in the structure must be adjusted (increased or decreased) to take into account ion charge. The number of electrons is increased in the case of negatively charged ions and decreased in the case of positively charged ions.

In the Lewis structure for an *ionic compound* that contains a polyatomic ion, the positive and negative ions are treated separately to show that they are individual ions not linked by covalent bonds. The Lewis structure of potassium sulfate, K_2SO_4, is written as

> Students often erroneously assume that the charge associated with a polyatomic ion is assigned to a particular atom within the ion. Polyatomic ion charge is not localized on a particular atom but rather is associated with the ion as a whole.

> When we write the Lewis structure of an ion (monatomic or polyatomic), it is customary to use brackets and to show ionic charge outside the brackets.

Correct structure   Incorrect structure

● **EXAMPLE 5.3**
Drawing Lewis Structures for Polyatomic Ions

▶ Draw a Lewis structure for $SO_4{}^{2-}$, a polyatomic ion in which S is the central atom and all O atoms are bonded to the S atom (see Figure 5.7).

Solution

Step 1: Both S and O are Group VIA elements. Thus each of the atoms has 6 valence electrons. Two extra electrons are also present, which accounts for the −2 charge on the ion. The total electron count is $6 + 4(6) + 2 = 32$.

(continued)

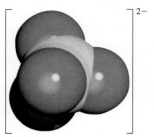

Figure 5.7 A computer-generated model of the sulfate ion (SO_4^{2-}).

Step 2: Drawing the molecular skeleton with single covalent bonds between bonded atoms gives

$$\left[\begin{array}{c} \text{O} \\ \text{O} : \text{S} : \text{O} \\ \text{O} \end{array}\right]^{2-}$$

Step 3: Adding nonbonding electron pairs to give each oxygen atom an octet of electrons yields

$$\left[\begin{array}{c} :\ddot{\text{O}}: \\ :\ddot{\text{O}} : \text{S} : \ddot{\text{O}}: \\ :\ddot{\text{O}}: \end{array}\right]^{2-}$$

Step 4: The Step 3 structure has 32 electrons, the total number available. No more electrons can be added to the structure, and indeed, none need to be added because the central S atom has an octet of electrons. There is no need to proceed to Step 5.

Practice Exercise 5.3

Draw a Lewis structure for BrO_3^-, a polyatomic ion in which Br is the central atom and all O atoms are bonded to it.

Answer: $\left[\begin{array}{c} :\ddot{\text{O}} :\ddot{\text{Br}}: \ddot{\text{O}}: \\ :\ddot{\text{O}}: \end{array}\right]^-$

5.8 MOLECULAR GEOMETRY

Lewis structures show the numbers and types of bonds present in molecules. They do not, however, convey any information about molecular geometry—that is, molecular shape. **Molecular geometry** *is a description of the three-dimensional arrangement of atoms within a molecule.* Indeed, Lewis structures falsely imply that all molecules have flat, two-dimensional shapes. This is not the case, as can be seen from the previously presented computer-generated models for the molecules SO_2, PF_3, and HCN (Figures 5.4 through 5.6).

Molecular geometry is an important factor in determining the physical and chemical properties of a substance. Dramatic relationships between geometry and properties are often observed in research associated with the development of prescription drugs. A small change in overall molecular geometry, caused by the addition or removal of atoms, can enhance drug effectiveness and/or decrease drug side effects. Studies also show that the human senses of taste and smell depend in part on the geometries of molecules.

For molecules that contain only a few atoms, molecular geometry can be predicted by using the information present in a molecule's Lewis structure and a procedure called valence shell electron pair repulsion (VSEPR) theory. **VSEPR theory** *is a set of procedures for predicting the molecular geometry of a molecule using the information contained in the molecule's Lewis structure.*

The central concept of VSEPR theory is that electron pairs in the valence shell of an atom adopt an arrangement in space that minimizes the repulsions between the like-charged (all negative) electron pairs. The specific arrangement adopted by the electron pairs depends on the number of electron pairs present. The electron pair arrangements about a *central atom* in the cases of two, three, and four electron pairs are as follows:

1. Two electron pairs, to be as far apart as possible from one another, are found on opposite sides of a nucleus—that is, at $180°$ angles to one another (Figure 5.8a). Such an electron pair arrangement is said to be *linear.*

180°

Central atom

(a) Linear

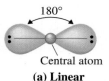

120°

(b) Trigonal planar

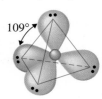

109°

(c) Tetrahedral

Figure 5.8 Arrangements of valence electron pairs about a central atom that minimize repulsions between the pairs.

The preferred arrangement of a given number of valence electron pairs about a central atom is the one that maximizes the separation among them. Such an arrangement minimizes repulsions between electron pairs.

2. Three electron pairs are as far apart as possible when they are found at the corners of an equilateral triangle. In such an arrangement, they are separated by 120° angles, giving a *trigonal planar* arrangement of electron pairs (Figure 5.8b).
3. A *tetrahedral* arrangement of electron pairs minimizes repulsions among four sets of electron pairs (Figure 5.8c). A tetrahedron is a four-sided solid in which all four sides are identical equilateral triangles. The angle between any two electron pairs is 109°.

Electron Groups

Before we use VSEPR theory to predict molecular geometry, an expansion of the concept of an "electron pair" to that of an "electron group" is needed. This will enable us to extend VSEPR theory to molecules in which double and triple bonds are present. A **VSEPR electron group** *is a collection of valence electrons present in a localized region about the central atom in a molecule.* A VSEPR electron group may contain two electrons (a single covalent bond), four electrons (a double covalent bond) or six electrons (a triple covalent bond). VSEPR electron groups that contain four and six electrons repel other VSEPR electron groups in the same way electron pairs do. This makes sense. The four electrons in a double bond or the six electrons in a triple bond are localized in the region between two bonded atoms in a manner similar to the two electrons of a single bond.

Let us now apply VSEPR theory to molecules in which two, three, and four VSEPR electron groups are present about a central atom. Our operational rules will be

1. Draw a Lewis structure for the molecule and identify the specific atom for which geometrical information is desired. (This atom will usually be the central atom in the molecule.)
2. Determine the number of VSEPR electron groups present about the central atom. The following conventions govern this determination:
   a. No distinction is made between bonding and nonbonding electron groups. Both are counted.
   b. Single, double, and triple bonds are all counted equally as "one electron group" because each takes up only one region of space about a central atom.
3. Predict the VSEPR electron group arrangement about the atom by assuming that the electron groups orient themselves in a manner that minimizes repulsions (see Figure 5.8).

The acronym VSEPR is pronounced "vesper."

Molecules with Two VSEPR Electron Groups

All molecules with two VSEPR electron groups are *linear.* Two common molecules with two VSEPR electron groups are carbon dioxide (CO_2) and hydrogen cyanide (HCN), whose Lewis structures are

$$\ddot{O}=C=\ddot{O}: \qquad H-C\equiv N:$$

In CO_2, the central carbon atom's two VSEPR electron groups are the two double bonds. In HCN, the central carbon atom's two VSEPR electron groups are a single bond and a triple bond. In both molecules, the VSEPR electron groups arrange themselves on opposite sides of the carbon atom, which produces a linear molecule.

Molecules with Three VSEPR Electron Groups

Molecules with three VSEPR electron groups have two possible molecular structures: *trigonal planar* and *angular.* The former occurs when all three VSEPR electron groups are bonding and the latter when one of the three VSEPR electron groups is nonbonding. The molecules H_2CO (formaldehyde) and SO_2 (sulfur dioxide) illustrate these two possibilities. Their Lewis structures are

Trigonal planar   Angular

VSEPR electron group arrangement and molecular geometry are not the same when a central atom possesses nonbonding electron pairs. The word used to describe the molecular geometry in such cases does not include the positions of the nonbonding electron groups.

In both molecules, the VSEPR electron groups are found at the corners of an equilateral triangle.

The shape of the SO_2 molecule is described as *angular* rather than *trigonal planar*, because molecular geometry describes only *atom positions*. The positions of nonbonding electron groups are not taken into account in describing molecular geometry. Do not interpret this to mean that nonbonding electron groups are unimportant in molecular geometry determinations; indeed, in the case of SO_2, it is the presence of the nonbonding electron group that makes the molecule angular rather than linear.

Molecules with Four VSEPR Electron Groups

Molecules with four VSEPR electron groups have three possible molecular geometries: *tetrahedral* (no nonbonding electron groups present), *trigonal pyramidal* (one nonbonding electron group present), and *angular* (two nonbonding electron groups present). The molecules CH_4 (methane), NH_3 (ammonia), and H_2O (water) illustrate this sequence of molecular geometries.

"Dotted line" and "wedge" bonds can be used to indicate the directionality of bonds, as shown below.

Bond behind page

Bonds in the plane of the page

Bond in front of page

In all three molecules, the VSEPR electron groups arrange themselves at the corners of a tetrahedron. Again, note that the word used to describe the geometry of the molecule does not take into account the positioning of nonbonding electron groups.

Molecules with More Than One Central Atom

The molecular shape of molecules that contain more than one central atom can be obtained by considering each central atom separately and then combining the results. Let us apply this principle to the molecules C_2H_2 (acetylene), H_2O_2 (hydrogen peroxide), and HN_3 (hydrogen azide), all of which have a four-atom "chain" structure. Their Lewis structures and VSEPR electron group counts are as follows:

Acetylene

H—C≡C—H

2 VSEPR electron groups 2 VSEPR electron groups

Linear C center   Linear C center

Hydrogen peroxide

H—O—O—H

4 VSEPR electron groups 4 VSEPR electron groups

Angular O center   Angular O center

Hydrogen azide

H—N=N=N:

3 VSEPR electron groups 2 VSEPR electron groups

Angular N center   Linear N center

These three molecules thus have, respectively, zero bends, two bends, and one bend in their four-atom chain.

H—C≡C—H

Zero bends in the chain

Two bends in the chain

One bend in the chain

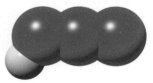

(a) The acetylene (C_2H_2) molecule.

(b) The hydrogen peroxide (H_2O_2) molecule.

(c) The hydrogen azide (HN_3) molecule.
Figure 5.9 Computer-generated models of (a) C_2H_2, (b) H_2O_2, and (c) HN_3.

Computer-generated three-dimensional models for these three molecules are given in Figure 5.9.

CHEMISTRY AT A GLANCE

The Geometry of Molecules

PREDICTING MOLECULAR GEOMETRY USING VSEPR THEORY	**Operational rules** 1. Draw a Lewis structure for the molecule. 2. Count the number of VSEPR electron groups about the central atom in the Lewis structure. 3. Assign a geometry based on minimizing repulsions between electron groups.

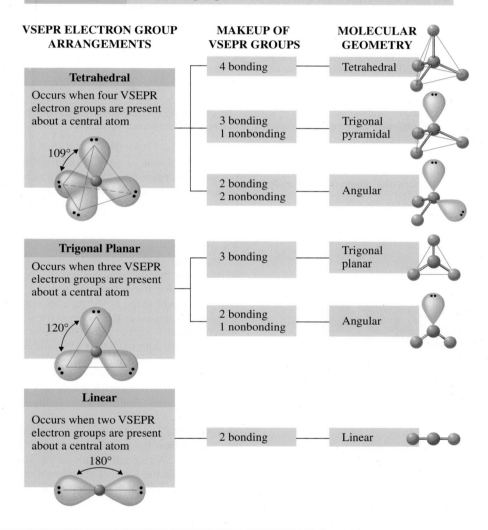

VSEPR ELECTRON GROUP ARRANGEMENTS	MAKEUP OF VSEPR GROUPS	MOLECULAR GEOMETRY
Tetrahedral Occurs when four VSEPR electron groups are present about a central atom 109°	4 bonding	Tetrahedral
	3 bonding 1 nonbonding	Trigonal pyramidal
	2 bonding 2 nonbonding	Angular
Trigonal Planar Occurs when three VSEPR electron groups are present about a central atom 120°	3 bonding	Trigonal planar
	2 bonding 1 nonbonding	Angular
Linear Occurs when two VSEPR electron groups are present about a central atom 180°	2 bonding	Linear

The Chemistry at a Glance feature above summarizes the key concepts involved in using VSEPR theory to predict molecular geometry.

 5.9 **ELECTRONEGATIVITY**

The ionic and covalent bonding models seem to represent two very distinct forms of bonding. Actually, the two models are closely related; they are the extremes of a broad continuum of bonding patterns. The close relationship between the two bonding models becomes

CHEMICAL Connections

The Chemical Senses of Smell and Taste

Whether a substance has an odor depends on whether it can excite the olfactory nerve endings in the nose. Factors that determine whether excitation of nerve endings in the nose occurs include:

1. *Molecular volatility.* Odoriferous substances must be either gases or easily vaporized liquids or solids at room temperature; otherwise, the molecules of such substances would never reach the nose.
2. *Molecular solubility.* Olfactory nerve endings are covered with *mucus,* an aqueous solution that contains dissolved proteins and carbohydrates. Odoriferous molecules must be at least slightly soluble in this mucus.
3. *Molecular geometry and polarity.* Olfactory nerve endings have receptor sites with distinct shapes. To be "accepted" at a given receptor site, an incoming molecule must have a shape that is complementary to that of the receptor site (see the accompanying diagram). The interaction that causes excitation is similar to that between a key and a lock; the key must have a particular shape in order to open the lock.

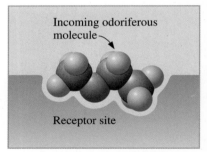

An incoming odoriferous molecule must have a shape complementary to that of the receptor site.

The current theory in use for explaining the sense of smell (olfaction), first proposed in the 1950s, suggests that every known odor can be made from a combination of seven primary odors (see the accompanying table) in much the same way that all colors can be made from the three primary colors of red, yellow, and blue. Associated with each of the primary odors is a particular molecular polarity and/or molecular geometry.

Primary Odor	Familiar Substance with This Odor	Molecular Characteristics
ethereal	dry cleaning fluid	rodlike shape
peppermint	mint candy	wedgelike shape
musk	some perfumes	disclike shape
camphoraceous	moth balls	spherical shape
floral	rose	disc with tail (kite) shape
pungent*	vinegar	negative polarity interaction
putrid*	skunk odor	positive polarity interaction

*These odors are less specific than are the other five odors, which indicates that they involve different interactions.

If two substances have similar molecular shapes that fit the same receptor site, they should produce the same odor. This is true in many, but not all, cases. If different parts of a large molecule fit different receptor sites, the combined interactions should produce a "mixed" odor, different from any of the primary odors. Such is the case in many, but not all, situations.

The occurrence of exceptions to the simple generalizations of this "odor theory" indicates that the theory is oversimplified. One reason for the exceptions may be that some molecules have a different shape in the gaseous state than when they are in solution (mucus) at the receptor sites. Also, more recent research indicates a greater diversity of receptor sites than previously thought. Approximately 1000 different kinds of receptor sites have now been identified, which magnifies greatly the possibilities for multiple-site interactions. Tests show that humans have the ability to distinguish among over 10,000 different odors, which obviously implies a very sophisticated operational system.

The "taste" of a food is actually more related to a person's nose than his or her tongue; that is, "flavor" is sensed by the nose. With each breath of air, some of the incoming air goes from the back of the mouth up into the nasal passages. When food first enters the mouth, the most volatile molecules present in the food are carried via air into the nose, where they are "smelled." As food is chewed, more volatile molecules are released from the food and enter the nasal passages.

apparent when the concepts of *electronegativity* (discussed in this section) and *bond polarity* (discussed in the next section) are considered.

The electronegativity concept has its origins in the fact that the nuclei of various elements have differing abilities to attract shared electrons (in a bond) to themselves. Some elements are better electron attractors than other elements. **Electronegativity**

Figure 5.10 Linus Carl Pauling (1901–1994). Pauling received the Nobel Prize in chemistry in 1954 for his work on the nature of the chemical bond. In 1962 he received the Nobel Peace Prize in recognition of his efforts to end nuclear weapons testing.

Figure 5.11 Abbreviated periodic table showing Pauling electronegativity values for selected representative elements.

is a measure of the relative attraction that an atom has for the shared electrons in a bond.

Linus Pauling (Figure 5.10), whose contributions to chemical bonding theory earned him a Nobel Prize in chemistry, was the first chemist to develop a *numerical* scale of electronegativity. Figure 5.11 gives Pauling electronegativity values for the more frequently encountered representative elements. The higher the electronegativity value for an element, the greater the attraction of atoms of that element for the shared electrons in bonds. The element fluorine, whose Pauling electronegativity value is 4.0, is the most electronegative of all elements; that is, it possesses the greatest electron-attracting ability for electrons in a bond.

As Figure 5.11 shows, electronegativity values increase from left to right across periods and from bottom to top within groups of the periodic table. These two trends result in nonmetals generally having higher electronegativities than metals. This fact is consistent with our previous generalization (Section 4.5) that metals tend to lose electrons and nonmetals tend to gain electrons when an ionic bond is formed. Metals (low electronegativities, poor electron attractors) give up electrons to nonmetals (high electronegativities, good electron attractors).

Note that the electronegativity for an element is not a directly measurable quantity. Rather, electronegativity values are calculated from bond energy information and other related experimental data. Values differ from element to element because of differences in atom size, nuclear charge, and number of inner-shell (nonvalence) electrons.

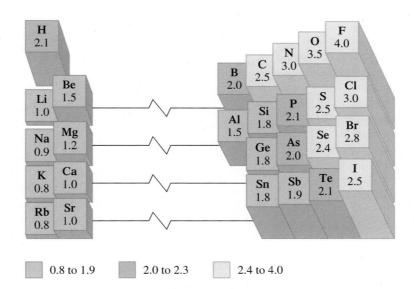

| 0.8 to 1.9 | 2.0 to 2.3 | 2.4 to 4.0 |

● **EXAMPLE 5.4**

Predicting Electronegativity Relationships Using General Periodic Table Trends

▶ Predict which member of each of the following pairs of elements has the greater electronegativity value based on periodic table electronegativity relationships.

a. Cl or Br   **b.** C or O   **c.** Li or K   **d.** Al or N

Solution

a. Cl and Br are in the same periodic table group, with Cl being above Br. Therefore Cl is the more electronegative element, as electronegativity decreases going down a group.

b. C and O are in the same periodic table period, with O being farther to the right. Therefore O has the greater electronegativity, as electronegativity increases going across a period.

c. Li and K are in the same group, with Li being two positions above K. Therefore Li is the more electronegative element, as electronegativity decreases going down a group.

(continued)

d. Al is in Period 3 and N is in Period 2. The element P can be used as a "link" between the two elements. It is in the same period as Al and the same group as N. The periodic table period trend predicts that P is more electronegative than Al and the periodic table group trend predicts that N is more electronegative than P. Therefore, N is more electronegative than Al.

Practice Exercise 5.4

Predict which member of each of the following pairs of elements has the greater electronegativity value based on periodic table electronegativity trends.

a. Be or C   **b.** O or S   **c.** Na or Cl   **d.** Mg or N

Answers: a. C; **b.** O; **c.** Cl; **d.** N

 5.10 **BOND POLARITY**

When two atoms of equal electronegativity share one or more pairs of electrons, each atom exerts the same attraction for the electrons, which results in the electrons being *equally* shared. This type of bond is called a nonpolar covalent bond. A **nonpolar covalent bond** *is a covalent bond in which there is equal sharing of electrons between two atoms.*

When the two atoms involved in a covalent bond have different electronegativities, the electron-sharing situation is more complex. The atom that has the higher electronegativity attracts the electrons more strongly than the other atom, which results in an *unequal* sharing of electrons. This type of covalent bond is called a polar covalent bond. A **polar covalent bond** *is a covalent bond in which there is unequal sharing of electrons between two atoms.* Figure 5.12 pictorially contrasts a nonpolar covalent bond and a polar covalent bond using the molecules H_2 and HCl.

The significance of unequal sharing of electrons in a polar covalent bond is that it creates fractional positive and negative charges on atoms. Although both atoms involved in a polar covalent bond are initially uncharged, the unequal sharing means that the electrons spend more time near the more electronegative atom of the bond (producing a fractional negative charge) and less time near the less electronegative atom of the bond (producing a fractional positive charge). The presence of such fractional charges on atoms within a molecule often significantly affects molecular properties (Section 5.11).

The fractional charges associated with atoms involved in a polar covalent bond are always values less than 1 because complete electron transfer does not occur. Complete electron transfer, which produces an ionic bond, would produce charges of $+1$ and -1. A notation that involves the lower-case Greek letter delta (δ) is used to denote fractional charge. The symbol δ^-, meaning "fractional negative charge," is placed above the more electronegative atom of the bond, and the symbol δ^+, meaning "fractional positive charge," is placed above the less electronegative atom of the bond.

The δ^+ and δ^- symbols are pronounced "delta plus" and "delta minus." Whatever the magnitude of δ^+, it must be the same as that of δ^- because the sum of δ^+ and δ^- must be zero.

Figure 5.12 (a) In the nonpolar covalent bond present in H_2 (H—H), there is a symmetrical distribution of electron density between the two atoms; that is, equal sharing of electrons occurs. (b) In the polar covalent bond present in HCl (H—Cl), electron density is displaced toward the Cl atom because of its greater electronegativity; that is, unequal sharing of electrons occurs.

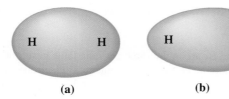

(a)   (b)

With delta notation, the direction of polarity of the bond in hydrogen chloride (HCl) is depicted as

$$\overset{\delta^+}{H}\!-\!\overset{\delta^-}{\ddot{\underset{\cdot\cdot}{Cl}}}\!:$$

Chlorine is the more electronegative of the two elements; it dominates the electron-sharing process and draws the shared electrons closer to itself. Hence the chlorine end of the bond has the δ^- designation (the more electronegative element always has the δ^- designation).

The direction of polarity of a polar covalent bond can also be designated by using an arrow with a cross at one end (+→). The cross is near the end of the bond that is "positive," and the arrowhead is near the "negative" end of the bond. Using this notation, we would denote the bond in the molecule HCl as

$$\overset{\longmapsto}{H}\!-\!\overset{\cdot\cdot}{\underset{\cdot\cdot}{Cl}}\!:$$

An extension of the reasoning used in characterizing the covalent bond in the HCl molecule as polar leads to the generalization that most chemical bonds are not 100% covalent (equal sharing) or 100% ionic (no sharing). Instead, most bonds are somewhere in between (unequal sharing).

Bond polarity *is a measure of the degree of inequality in the sharing of electrons between two atoms in a chemical bond.* The numerical value of the electronegativity difference between two bonded atoms gives an approximate measure of the polarity of the bond. The greater the numerical difference, the greater the inequality of electron sharing and the greater the polarity of the bond. As the polarity of the bond increases, the bond is increasingly ionic.

The existence of *bond polarity* means that there is no natural boundary between ionic and covalent bonding. Most bonds are a mixture of pure ionic and pure covalent bonds; that is, unequal sharing of electrons occurs. Most bonds have both ionic and covalent character. Nevertheless, it is still convenient to use the terms *ionic* and *covalent* in describing chemical bonds, based on the following arbitary but useful (though not infallible) guidelines, which relate to electronegativity difference between bonded atoms.

1. Bonds that involve atoms with the same or very similar electronegativities are called *nonpolar covalent bonds.* "Similar" here means an electronegativity difference of 0.4 or less.

   Technically, the only purely nonpolar covalent bonds are those between identical atoms. However, bonds with a small electronegativity difference behave very similarly to purely nonpolar covalent bonds.
2. Bonds with an electronegativity difference greater that 0.4 but less than 1.5 are called *polar covalent bonds.*
3. Bonds with an electronegativity difference greater than 2.0 are called *ionic bonds.*
4. Bonds with an electronegativity difference between 1.5 and 2.0 are considered *ionic* if the bond involves a metal and a nonmetal, and *polar covalent* if the bond involves two nonmetals. In the 1.5–2.0 range of electronegativity difference some compounds exhibit characteristics associated with ionic compounds (see Sections 4.1 and 5.1). This rule helps in dealing with this "borderline" area.

⬤ **EXAMPLE 5.5**
Using Electronegativity Difference to Predict Bond Polarity and Bond Type

▶ Consider the following bonds

N—Cl   Ca—F   C—O   B—H   N—O

a. Rank the bonds in order of increasing polarity.
b. Determine the direction of polarity for each bond.
c. Classify each bond as nonpolar covalent, polar covalent, or ionic.

(continued)

Solution

First, the electronegativity difference for each of the bonds is calculated using the electronegativity values in Figure 5.11.

$$N—Cl: \quad 3.0 - 3.0 = 0.0$$
$$Ca—F: \quad 4.0 - 1.0 = 3.0$$
$$C—O: \quad 3.5 - 2.5 = 1.0$$
$$B—H: \quad 2.1 - 2.0 = 0.1$$
$$N—O: \quad 3.5 - 3.0 = 0.5$$

a. Bond polarity increases as electronegativity difference increases. Using the mathematical symbol $<$, which means "is less than," we can rank the bonds in terms of increasing bond polarity as follows:

$$N—Cl \; < \; B—H \; < \; N—O \; < \; C—O \; < \; Ca—F$$
$$\quad 0.0 \qquad\quad 0.1 \qquad\quad 0.5 \qquad\quad 1.0 \qquad\quad 3.0$$

b. The direction of bond polarity is from the least electronegative atom to the most electronegative atom. The more electronegative atom bears the fractional negative charge (δ^-).

$$N—Cl \qquad B—H \qquad N—O \qquad C—O \qquad Ca—F$$

c. Nonpolar covalent bonds require a difference in electronegativity of 0.4 or less, and an electronegativity difference of 2.0 or greater corresponds to ionic bonds. Bonds with an electronegativity difference greater than 0.4 but less than 1.5 are polar covalent. If the electronegativity difference is between 1.5 and 2.0, the bond is polar covalent if it involves two nonmetals, but is considered ionic if the bond involves a metal and a nonmetal. Using these guidelines the bond classifications are:

Nonpolar covalent: N—Cl, B—H

Polar covalent: N—O, C—O

Ionic: Ca—F

Practice Exercise 5.5

Consider the following bonds:

$$N—S \qquad H—H \qquad Na—F \qquad K—Cl \qquad F—Cl$$

a. Rank the bonds in order of increasing polarity.
b. Determine the direction of polarity for each bond.
c. Classify each bond as nonpolar covalent, polar covalent, or ionic.

Answers: a. $H—H \; < \; N—S \; < \; F—Cl \; < \; K—Cl \; < \; Na—F$

b. H—H   N—S   F—Cl   K—Cl   Na—F

c. Nonpolar covalent: H—H

Polar covalent: N—S, F—Cl

Ionic: K—Cl, Na—F

The Chemistry at a Glance feature on page 127 summarizes much that we have said about chemical bonds in this chapter.

5.11   MOLECULAR POLARITY

Molecules, as well as bonds (Section 5.10), can have polarity. **Molecular polarity** *is a measure of the degree of inequality in the attraction of bonding electrons to various locations within a molecule.* In terms of electron attraction, if one part of a molecule is

CHEMISTRY AT A GLANCE

Covalent Bonds and Molecular Compounds

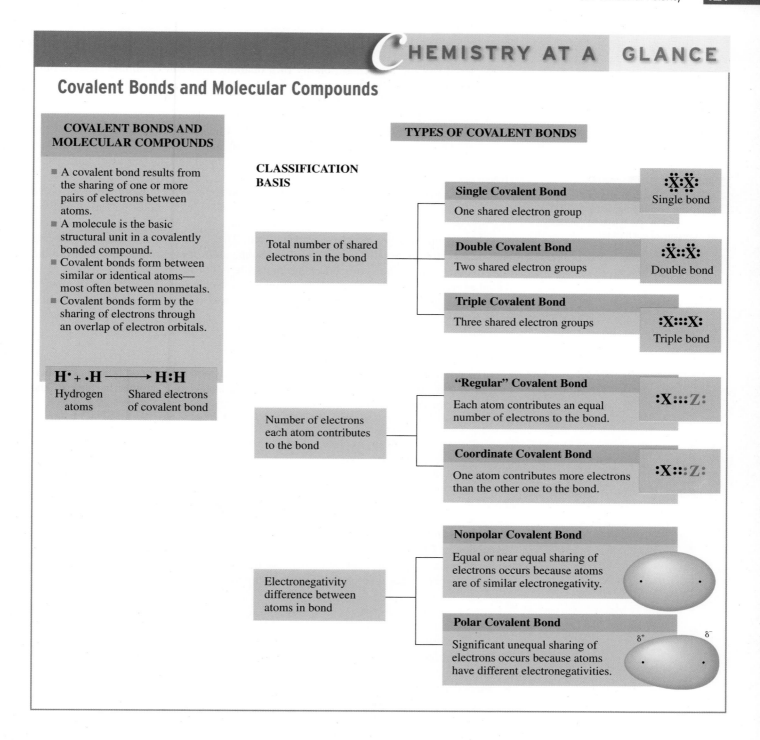

COVALENT BONDS AND MOLECULAR COMPOUNDS

- A covalent bond results from the sharing of one or more pairs of electrons between atoms.
- A molecule is the basic structural unit in a covalently bonded compound.
- Covalent bonds form between similar or identical atoms—most often between nonmetals.
- Covalent bonds form by the sharing of electrons through an overlap of electron orbitals.

H· + ·H ⟶ H:H

Hydrogen atoms | Shared electrons of covalent bond

TYPES OF COVALENT BONDS

CLASSIFICATION BASIS

Total number of shared electrons in the bond

Single Covalent Bond
One shared electron group

:X:X:
Single bond

Double Covalent Bond
Two shared electron groups

:X::X:
Double bond

Triple Covalent Bond
Three shared electron groups

:X:::X:
Triple bond

Number of electrons each atom contributes to the bond

"Regular" Covalent Bond
Each atom contributes an equal number of electrons to the bond.

:X:::Z:

Coordinate Covalent Bond
One atom contributes more electrons than the other one to the bond.

:X:::Z:

Electronegativity difference between atoms in bond

Nonpolar Covalent Bond
Equal or near equal sharing of electrons occurs because atoms are of similar electronegativity.

Polar Covalent Bond
Significant unequal sharing of electrons occurs because atoms have different electronegativities.

δ^+ δ^-

A prerequisite for determining molecular polarity is a knowledge of molecular geometry.

favored over other parts, then the molecule is *polar*. A **polar molecule** *is a molecule in which there is an unsymmetrical distribution of electronic charge.* In a polar molecule, bonding electrons are more attracted to one part of the molecule than to other parts. A **nonpolar molecule** *is a molecule in which there is a symmetrical distribution of electron charge.* Attraction for bonding electrons is the same in all parts of a *nonpolar* molecule. Molecular polarity depends on two factors: (1) bond polarities and (2) molecular geometry (Section 5.8). In molecules that are symmetrical, the effects of polar bonds may cancel each other, resulting in the molecule as a whole having no polarity.

Determining the molecular polarity of a diatomic molecule is simple because only one bond is present. If that bond is nonpolar, then the molecule is nonpolar; if the bond is polar, then the molecule is polar.

Determining molecular polarity for triatomic molecules is more complicated. Two different molecular geometries are possible: linear and angular. In addition, the symmetrical nature of the molecule must be considered. Let us consider the polarities of three specific triatomic molecules: CO_2 (linear), H_2O (angular), and HCN (linear).

> Molecules in which all bonds are polar can be nonpolar if the bonds are so oriented in space that the polarity effects cancel each other.

In the linear CO_2 molecule, both bonds are polar (oxygen is more electronegative than carbon). Despite the presence of these polar bonds, CO_2 molecules are *nonpolar*. The effects of the two polar bonds are canceled as a result of the oxygen atoms being arranged symmetrically around the carbon atom. The shift of electronic charge toward one oxygen atom is exactly compensated for by the shift of electronic charge toward the other oxygen atom. Thus one end of the molecule is not negatively charged relative to the other end (a requirement for polarity), and the molecule is nonpolar. This cancellation of individual bond polarities, with crossed arrows used to denote the polarities, is diagrammed as follows:

$$\overset{\longleftarrow\ \ \longrightarrow}{O\!=\!C\!=\!O}$$

The nonlinear (angular) triatomic H_2O molecule is polar. The bond polarities associated with the two hydrogen–oxygen bonds do not cancel one another because of the nonlinearity of the molecule.

$$\underset{H \qquad\quad H}{\overset{\nearrow\ \ O\ \ \nwarrow}{}}$$

As a result of their orientation, both bonds contribute to an accumulation of negative charge on the oxygen atom. The two bond polarities are equal in magnitude but are not opposite in direction.

The generalization that linear triatomic molecules are nonpolar and nonlinear triatomic molecules are polar, which you might be tempted to make on the basis of our discussion of CO_2 and H_2O molecular polarities, is not valid. The linear molecule HCN, which is polar, invalidates this statement. Both bond polarities contribute to nitrogen's acquiring a partial negative charge relative to hydrogen in HCN.

$$\overset{\longrightarrow\ \ \longrightarrow}{H\!-\!C\!\equiv\!N}$$

(The two polarity arrows point in the same direction because nitrogen is more electronegative than carbon, and carbon is more electronegative than hydrogen.)

Molecules that contain four and five atoms commonly have trigonal planar and tetrahedral geometries, respectively. Such molecules in which all of the atoms attached to the central atom are identical, such as SO_3 (trigonal planar) and CH_4 (tetrahedral), are *nonpolar*. The individual bond polarities cancel as a result of the highly symmetrical arrangement of atoms around the central atom.

If two or more kinds of atoms are attached to the central atom in a trigonal planar or tetrahedral molecule, the molecule is polar. The high degree of symmetry required for cancellation of the individual bond polarities is no longer present. For example, if one of the hydrogen atoms in CH_4 (a nonpolar molecule) is replaced by a chlorine atom, then a polar molecule results, even though the resulting CH_3Cl is still a tetrahedral molecule. A carbon–chlorine bond has a greater polarity than a carbon–hydrogen bond; chlorine has an electronegativity of 3.0, and hydrogen has an electronegativity of only 2.1. Figure 5.13 contrasts the polar CH_3Cl and nonpolar CH_4 molecules. Note that the direction of polarity of the carbon–chlorine bond is opposite to that of the carbon–hydrogen bonds.

Figure 5.13 (a) Methane (CH_4) is a nonpolar tetrahedral molecule. (b) Methyl chloride (CH_3Cl) is a polar tetrahedral molecule. Bond polarities cancel in the first case, but not in the second.

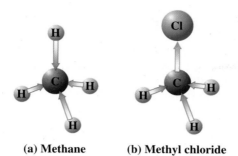

(a) Methane   **(b) Methyl chloride**

EXAMPLE 5.6

Predicting the Polarity of Molecules Given Their Molecular Geometry

Predict the polarity of each of the following molecules.

a. PCl_3 (trigonal pyramidal)   **b.** SCl_2 (angular)

c. $SiBr_4$ (tetrahedral)   **d.** C_2Cl_2 (linear)

Cl—C≡C—Cl

Solution

Knowledge of a molecule's geometry, which is given for each molecule in this example, is a prerequisite for predicting molecular polarity.

a. Noncancellation of the individual bond polarities in the trigonal pyramidal PF_3 molecule results in its being a *polar* molecule.

Net polarity toward fluorine

The bond polarity arrows all point toward fluorine atoms because fluorine is more electronegative than phosphorus.

b. For the bent SCl_2 molecule, the shift in electron density in the polar sulfur–chlorine bonds will be toward the chlorine atoms because chlorine is more electronegative than sulfur. The SCl_2 molecule as a whole is *polar* because of the noncancellation of the individual sulfur–chlorine bond polarities.

Net polarity toward chlorine

c. The $SiBr_4$ molecule is a tetrahedral molecule and all four atoms attached to the central atom (Si) are the same. This, and the highly symmetrical nature of a tetrahedral geometry (all bond angles are the same), means that the Si–Br bond polarities cancel each other and the molecule as a whole is *nonpolar*.

No net polarity

d. The carbon–carbon bond is nonpolar. The two carbon–chlorine bonds are polar and are "equal but opposite" in terms of effect; that is, they cancel. The C_2Cl_2 molecule as a whole is thus *nonpolar*.

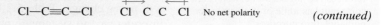

Cl—C≡C—Cl   Cl C C Cl   No net polarity

(continued)

Practice Exercise 5.6

Predict the polarity of each of the following molecules.

a. NH_3 (trigonal pyramidal)

b. H_2S (angular)

c. N_2O (linear)

$:N≡N—\ddot{O}:$

d. C_2H_4 (trigonal planar about C atoms)

Answers: **a.** polar; **b.** polar; **c.** polar; **d.** nonpolar

5.12 NAMING BINARY MOLECULAR COMPOUNDS

A **binary molecular compound** *is a molecular compound in which only two nonmetallic elements are present.* The names of binary molecular compounds are derived by using a rule very similar to that used for naming binary ionic compounds (Section 4.9). However, one major difference exists. Names for binary molecular compounds always contain numerical prefixes that give the number of each type of atom present in addition to the names of the elements present. This is in direct contrast to binary ionic compound nomenclature, where formula subscripts are never mentioned in the names.

Here is the basic rule to use when constructing the name of a binary molecular compound: *The full name of the nonmetal of lower electronegativity is given first, followed by a separate word containing the stem of the name of the more electronegative nonmetal and the suffix -ide. Numerical prefixes, giving numbers of atoms, precede the names of both nonmetals.*

Prefixes are necessary because several different compounds exist for most pairs of nonmetals. For example, all of the following nitrogen–oxygen compounds exist: NO, NO_2, N_2O, N_2O_3, N_2O_4, and N_2O_5. The compounds N_2O, N_2O_3, and N_2O_4 are named dinitrogen monoxide, dinitrogen trioxide, and dinitrogen tetroxide, respectively. Such diverse behavior between two elements is related to the fact that single, double, and triple covalent bonds exist. The prefixes used are always the standard numerical prefixes, which are given for the numbers 1 through 10 in Table 5.1. Example 5.7 shows how these prefixes are used in nomenclature for binary covalent compounds; it also includes special instructions concerning use of the prefix *mono-*.

There is one standard exception to the use of numerical prefixes when naming binary molecular compounds. Compounds in which hydrogen is the first listed element in the formula are named without numerical prefixes. Thus the compounds H_2S and HCl are hydrogen sulfide and hydrogen chloride, respectively.

A few binary molecular compounds have names that are completely unrelated to the rules we have been discussing. They have common names that were coined prior to the development of systematic rules. At one time, in the early history of chemistry, all compounds had common names. With the advent of systematic nomenclature, most common names were discontinued. A few, however, have persisted and are now officially accepted. The most "famous" example is the compound H_2O, which has the systematic name hydrogen oxide, a name that is never used. The compound H_2O is *water,* a name that will never change. Table 5.2 lists other compounds for which common names are used in preference to systematic names.

Numerical prefixes are used in naming binary molecular compounds. They are *never* used, however, in naming binary ionic compounds.

When an element name begins with a vowel, an *a* or *o* at the end of the Greek prefix is dropped for phonetic reasons, as in pentoxide instead of pentaoxide.

In Section 10.3, we will learn that placing hydrogen first in a formula conveys the message that the compound behaves as an acid in aqueous solution.

Classification of a compound as ionic or molecular determines which set of nomenclature rules is used. For *nomenclature purposes,* binary compounds in which a metal and a nonmetal are present are considered ionic, and binary compounds that contain two nonmetals are considered covalent. Electronegativity differences are *not* used in classifying a compound as ionic or molecular for nomenclature purposes.

TABLE 5.1
Common Numerical Prefixes for the Numbers 1 Through 10

Prefix	Number
mono-	1
di-	2
tri-	3
tetra-	4
penta-	5
hexa-	6
hepta-	7
octa-	8
nona-	9
deca-	10

TABLE 5.2
Selected Binary Molecular Compounds That Have Common Names

Compound Formula	Accepted Common Name
H_2O	water
H_2O_2	hydrogen peroxide
NH_3	ammonia
N_2H_4	hydrazine
CH_4	methane
C_2H_6	ethane
PH_3	phosphine
AsH_3	arsine

● **E X A M P L E   5 . 7**

Naming Binary Molecular Compounds

Name the following binary molecular compounds.

a. S_2Cl_2 **b.** CS_2 **c.** P_4O_{10} **d.** CBr_4

Solution

The names of each of these compounds will consist of two words. These words will have the following general formats:

$$\text{First word: (prefix)} + \left(\begin{array}{c}\text{full name of least}\\\text{electronegative nonmetal}\end{array}\right)$$

$$\text{Second word: (prefix)} + \left(\begin{array}{c}\text{stem of name of more}\\\text{electronegative nonmetal}\end{array}\right) + \text{(ide)}$$

a. The elements present are sulfur and chlorine. The two portions of the name (including prefixes) are *disulfur* and *dichloride,* which are combined to give the name *disulfur dichloride.*

b. When only one atom of the first nonmetal is present, it is customary to omit the initial prefix *mono-.* Thus the name of this compound is *carbon disulfide.*

c. The prefix for four atoms is *tetra-* and for ten atoms is *deca-.* This compound has the name *tetraphosphorus decoxide,* and the structure shown in Figure 5.14.

d. Omitting the initial *mono-* (see part **b**), we name this compound *carbon tetrabromide.*

Practice Exercise 5.7

Name the following binary molecular compounds.

a. PF_3 **b.** SO_2 **c.** P_4S_{10} **d.** N_2O_4

Answers: a. Phosphorus trifluoride; **b.** Sulfur dioxide; **c.** Tetraphosphorus decasulfide; **d.** Dinitrogen tetroxide

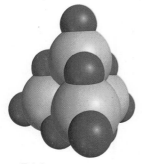

Figure 5.14 A computer-generated molecular model of the tetraphosphorus decoxide (P_4O_{10}) molecule.

CONCEPTS TO REMEMBER

Molecular compounds. Molecular compounds usually involve two or more nonmetals. The covalent bonds within molecular compounds involve electron sharing between atoms. The covalent bond results from the common attraction of the two nuclei for the shared electrons (Section 5.1).

Bonding and nonbonding electron pairs. Bonding electrons are pairs of valence electrons that are shared between atoms in a covalent bond. Nonbonding electrons are pairs of valence electrons about an atom that are not involved in electron sharing (Section 5.2).

Types of covalent bonds. One shared pair of electrons constitutes a single covalent bond. Two or three pairs of electrons may be shared between atoms to give double and triple covalent bonds. Most often, both atoms of the bond contribute an equal number of electrons to the bond. In a few cases, however, both electrons of a shared pair come from the same atom; this is a coordinate covalent bond (Sections 5.3 and 5.5).

Number of covalent bonds formed. There is a strong tendency for nonmetals to form a particular number of covalent bonds. The number of valence electrons the nonmetal has and the number of covalent bonds it forms give a sum of eight (Section 5.4).

Molecular geometry. Molecular geometry describes the way atoms in a molecule are arranged in space relative to one another. VSEPR theory is a set of procedures used to predict molecular geometry from a compound's Lewis structure. VSEPR theory is based on the concept that valence shell electron groups about an atom (bonding or nonbonding) orient themselves as far away from one another as possible (to minimize repulsions) (Section 5.8).

Electronegativity. Electronegativity is a measure of the relative attraction that an atom has for the shared electrons in a bond. Electronegativity values are useful in predicting the type of bond that forms (ionic or covalent) (Section 5.9).

Bond polarity. When atoms of like electronegativity participate in a bond, the bonding electrons are equally shared and the bond is nonpolar. When atoms of differing electronegativity participate in a bond, the bonding electrons are unequally shared and the bond is polar. In a polar bond, the more electronegative atom dominates the sharing process. The greater the electronegativity difference between two bonded atoms, the greater the polarity of the bond (Section 5.10).

Molecular polarity. Molecules as a whole can have polarity. If individual bond polarities do not cancel because of the symmetrical nature of a molecule, then the molecule as a whole is polar (Section 5.11).

Binary molecular compound nomenclature. Names for binary molecular compounds usually contain numerical prefixes that give the number of each type of atom present per molecule in addition to the names of the elements (Section 5.12).

KEY REACTIONS AND EQUATIONS

1. Molecular geometry and central atom VSEPR electron group count (Section 5.8)

Four VSEPR electron groups none of which is nonbonding = tetrahedral geometry

Four VSEPR electron groups one of which is nonbonding = trigonal pyramidal geometry

Four VSEPR electron groups two of which are nonbonding = angular geometry

Three VSEPR electron groups none of which is nonbonding = trigonal planar geometry

Three VSEPR electron groups one of which is nonbonding = angular geometry

Two VSEPR electron groups none of which is nonbonding = linear geometry

2. Bond characterization and electronegativity difference (Section 5.10)

Electronegativity difference of 0 to 0.4 = nonpolar covalent bond

Electronegativity difference greater than 0.4 but less than 1.5 = polar covalent bond

Electronegativity difference from 1.5 to 2.0 = ionic bond or polar covalent bond depending on the atoms involved in the bond

Electronegativity difference greater than 2.0 = ionic bond

EXERCISES and PROBLEMS

The members of each pair of problems in this section test similar material.

The Covalent Bond Model (Section 5.1)

5.1 Contrast the types of atoms involved in ionic and covalent bonds.

5.2 Contrast the mechanisms by which ionic and covalent bonds form.

5.3 Contrast the basic structural unit in ionic and molecular compounds.

5.4 Contrast the general physical state at room temperature and room pressure for ionic and molecular compounds.

Lewis Structures for Molecular Compounds (Section 5.2)

5.5 Draw Lewis structures to illustrate the covalent bonding in the following diatomic molecules.
a. Br_2 b. HI   c. IBr   d. BrF

5.6 Draw Lewis structures to illustrate the covalent bonding in the following diatomic molecules.
a. I_2 b. HF   c. IF   d. F_2

5.7 How many nonbonding electron pairs are present in each of the following Lewis structures?
a. :F:S:F: b. H:O:O:H   c. H:C:H (with H above and below)   d. :Cl:C:Cl: (with H above and below)

5.8 How many nonbonding electron pairs are present in each of the following Lewis structures?
a. H:O:Cl: b. :Cl:O:Cl: c. H:N:H (with H below)   d. H:N:N:H (with H below each N)

5.9 What would be the predicted chemical formula for the simplest molecular compound formed between the following pairs of elements?

a. Nitrogen and fluorine   b. Chlorine and oxygen
c. Hydrogen and sulfur   d. Carbon and hydrogen

5.10 What would be the predicted chemical formula for the simplest molecular compound formed between the following pairs of elements?

a. Nitrogen and hydrogen   b. Oxygen and fluorine
c. Sulfur and bromine   d. Carbon and chlorine

● Single, Double, and Triple Covalent Bonds (Section 5.3)

5.11 Specify the number of single, double, and triple covalent bonds present in molecules represented by the following Lewis structures.

a. $:N:::N:$ b. $H:O:O:H$

c. $H:C:H$ d. $H:C::C:H$
$\quad:O$ $\quad H \quad H$

5.12 Specify the number of single, double, and triple covalent bonds present in molecules represented by the following Lewis structures.

a. $:C:::O:$ b. $H:N:N:H$
$\qquad\qquad\qquad\qquad H \quad H$

c. $:O:S:O:$ d. $H:C:::C:H$
$\quad:O$

5.13 Convert each of the Lewis structures in Problem 5.11 into the form in which lines are used to denote shared electron pairs. Include nonbonding electron pairs in the rewritten structures.

5.14 Convert each of the Lewis structures in Problem 5.12 into the form in which lines are used to denote shared electron pairs. Include nonbonding electron pairs in the rewritten structures.

● Valence Electrons and Number of Covalent Bonds Formed (Section 5.4)

5.15 Indicate whether each of the following is or is not a normally expected bonding pattern for the element shown.

a. $—\overset{..}{\underset{..}{O}}—$ b. $:N≡$ c. $:\overset{.}{C}=$ d. $:\overset{..}{O}=$

5.16 Indicate whether each of the following is or is not a normally expected bonding pattern for the element shown.

a. $—\overset{..}{N}=$ b. $:O≡$ c. $—C≡$ d. $=C=$

5.17 Identify the Period 2 nonmetal that would normally be expected to exhibit each of the following bonding capabilities.
a. Forms three single bonds
b. Forms two double bonds
c. Forms one single bond and one double bond
d. Forms two single bonds and one double bond

5.18 Identify the Period 3 nonmetal that would normally be expected to exhibit each of the following bonding capabilities.
a. Forms one triple bond
b. Forms one single bond and one triple bond
c. Forms four single bonds
d. Forms one double bond

● Coordinate Covalent Bonds (Section 5.5)

5.19 What aspect of the following Lewis structure gives you a "hint" that the concept of coordinate covalency is needed to explain the bonding in the molecule?

$$:C:::O:$$

5.20 What aspect of the following Lewis structure gives you a "hint" that the concept of coordinate covalency is needed to explain the bonding in the molecule?

$$:N:::N:\overset{..}{\underset{..}{O}}:$$

5.21 Identify the coordinate covalent bond(s) present, if any, in each of the following molecules by listing the two atoms involved in the bond. Name the atom on the left or the bottom in the bond first.

a. $:N≡N—\overset{..}{\underset{..}{O}}:$ b. $H—\overset{..}{\underset{..}{O}}—\overset{..}{\underset{..}{F}}:$

c. $:\overset{..}{\underset{..}{O}}—\overset{..}{\underset{..}{Cl}}—\overset{..}{\underset{..}{O}}—H$ d. $:\overset{..}{\underset{..}{O}}—\overset{..}{\underset{..}{Br}}—\overset{..}{\underset{..}{O}}—H$
$\qquad\qquad\qquad\qquad\qquad\qquad\qquad\quad |$
$\qquad\qquad\qquad\qquad\qquad\qquad\qquad :\overset{..}{\underset{..}{O}}:$

5.22 Identify the coordinate covalent bond(s) present, if any, in each of the following molecules by listing the two atoms involved in the bond. Name the atom on the left or the bottom in the bond first.

a. $:S=S—\overset{..}{\underset{..}{O}}:$ b. $H—\overset{..}{\underset{..}{O}}—\overset{..}{\underset{..}{Br}}:$

c. $:\overset{..}{\underset{..}{O}}—\overset{..}{\underset{..}{I}}—\overset{..}{\underset{..}{O}}—H$ d. $:\overset{..}{\underset{..}{O}}—\overset{..}{\underset{..}{Cl}}—\overset{..}{\underset{..}{O}}—H$
$\qquad\qquad\qquad\qquad\qquad\qquad\qquad\quad |$
$\qquad\qquad\qquad\qquad\qquad\qquad\qquad :\overset{..}{\underset{..}{O}}:$

● Drawing Lewis Structures (Section 5.6)

5.23 Without actually drawing the Lewis structure, determine the total number of "dots" present in the Lewis structure of each of the following molecules. That is, determine the total number of valence electrons available for bonding in each of the molecules.
a. Cl_2O b. H_2S c. NH_3 d. SO_3

5.24 Without actually drawing the Lewis structure, determine the total number of "dots" present in the Lewis structure of each of the following molecules. That is, determine the total number of valence electrons available for bonding in each of the molecules.
a. PCl_3 b. H_2O_2 c. SF_2 d. HCl

5.25 Draw Lewis structures to illustrate the covalent bonding in the following polyatomic molecules. The first atom in each formula is the central atom to which all other atoms are bonded.
a. PH_3 b. PCl_3 c. $SiBr_4$ d. OF_2

5.26 Draw Lewis structures to illustrate the covalent bonding in the following polyatomic molecules. The first atom in each formula is the central atom to which all other atoms are bonded.
a. AsH_3 b. $AsCl_3$ c. CBr_4 d. SCl_2

5.27 Draw Lewis structures for the simplest molecular compound likely to form between these pairs of elements.
a. Sulfur and fluorine   b. Carbon and iodine
c. Nitrogen and bromine   d. Selenium and hydrogen

5.28 Draw Lewis structures for the simplest molecular compound likely to form between these pairs of elements.
a. Nitrogen and chlorine   b. Bromine and hydrogen
c. Phosphorus and fluorine   d. Selenium and bromine

5.29 Draw Lewis structures to illustrate the bonding in the following molecules. In each case, there will be at least one multiple bond present in a molecule.
a. C_3H_4: A central carbon atom has two other carbon atoms bonded to it. Each of the noncentral carbon atoms also has two hydrogen atoms bonded to it.

b. N_2F_2: The two nitrogen atoms are bonded to one another, and each nitrogen atom also has a fluorine atom bonded to it.

c. C_2H_3N: The two carbon atoms are bonded to each other. One of the carbon atoms has a nitrogen atom bonded to it, and the other carbon atom has three hydrogen atoms bonded to it.

d. C_3H_4: A central carbon atom has two other carbon atoms bonded to it. One of the noncentral carbon atoms also has one hydrogen atom bonded to it, and the other one has three hydrogen atoms bonded to it.

5.30 Draw Lewis structures to illustrate the bonding in the following molecules. In each case, there will be at least one multiple bond present in a molecule.

a. $COCl_2$: Both chlorine atoms and the oxygen atom are bonded to the carbon atom.

b. $C_2H_2Br_2$: The two carbon atoms are bonded to one another. Each carbon atom also has a bromine atom and a hydrogen atom bonded to it.

c. C_2N_2: The two carbon atoms are bonded to one another, and each carbon atom also has a nitrogen bonded to it.

d. CH_2N_2: A central carbon atom has both nitrogen atoms bonded to it. Both hydrogen atoms are bonded to one of the two nitrogen atoms.

● Lewis Structures for Polyatomic Ions (Section 5.7)

5.31 Draw Lewis structures for the following polyatomic ions.
a. OH^- b. BeH_4^{2-} c. $AlCl_4^-$ d. NO_3^-

5.32 Draw Lewis structures for the following polyatomic ions.
a. CN^- b. PF_4^+ c. BH_4^- d. ClO_3^-

5.33 Draw Lewis structures for the following compounds that contain polyatomic ions.
a. NaCN   b. K_3PO_4

5.34 Draw Lewis structures for the following compounds that contain polyatomic ions.
a. KOH   b. NH_4Br

● Molecular Geometry (VSEPR Theory) (Section 5.8)

5.35 Specify the molecular geometry of each of the following molecules using the terms *linear, angular, trigonal pyramidal,* and *tetrahedral.*

a.   b.   c.   d.

5.36 Specify the molecular geometry of each of the following molecules using the terms *linear, angular, trigonal pyramidal,* and *tetrahedral.*

a.   b.

c.   d.

5.37 Using VSEPR theory, predict whether each of the following triatomic molecules is linear or angular (bent).

a. H:S̈:H   b. H:Ö:C̈l:

c. :Ö::O:Ö: d. :N̈::N::Ö:

5.38 Using VSEPR theory, predict whether each of the following triatomic molecules is linear or angular (bent).

a. H:C⁝⁝N: b. :N̈::S̈:F̈:

c. :F̈:S̈:F̈:   d. :C̈l:Ö:C̈l:

5.39 Using VSEPR theory, predict the molecular geometry of the following molecules.

a. :F̈:N̈:F̈:   b. :C̈l:C:C̈l:
    :F̈:   :Ö:

c.   :Ö: d.   H
   :C̈l:P:C̈l: :C̈l:C:C̈l:
    :C̈l: :C̈l:

5.40 Using VSEPR theory, predict the molecular geometry of the following molecules.

a. H:P̈:H   b. :C̈l:C::Ö:
   :C̈l: H

c.   H   d.   :F̈:
  :C̈l:C:C̈l: H:Si:H
    H   :C̈l:

5.41 Using VSEPR theory, predict the molecular geometry of the following molecules.
a. NCl_3 b. $SiCl_4$ c. H_2Se d. SBr_2

5.42 Using VSEPR theory, predict the molecular geometry of the following molecules.
a. HOBr   b. H_2Te c. NBr_3 d. SiF_4

5.43 Using VSEPR theory, predict the molecular geometry of the following molecules.

a. H:C::C:H   b.   H
     H   H   H:C:Ö:H
                           H

5.44 Using VSEPR theory, predict the molecular geometry of the following molecules.

a.   :Ö: b.   H Ö:
   :O::N:Ö:H   H:C:C:H
                            H

● Electronegativity (Section 5.9)

5.45 Using a periodic table, but not a table of electronegativity values, arrange each of the following sets of atoms in order of increasing electronegativity.
a. Na, Al, P, Mg   b. Cl, Br, I, F
c. S, P, O, Al   d. Ca, Mg, O, C

5.46 Using a periodic table, but not a table of electronegativity values, arrange each of the following sets of atoms in order of increasing electronegativity.
a. Be, N, O, B   b. Li, C, B, K
c. S, Te, Cl, Se   d. S, Mg, K, Ca

5.47 Use the information in Figure 5.11 as a basis for answering the following questions.
a. Which elements have electronegativity values that exceed that of the element carbon?
b. Which elements have electronegativity values of less than 1.0?
c. What are the four most electronegative elements listed in Figure 5.11?
d. By what constant amount do the electronegativity values for sequential Period 2 elements differ?

5.48 Use the information in Figure 5.11 as a basis for answering the following questions.
   a. Which elements have electronegativity values that exceed that of the element sulfur?
   b. What are the three least electronegative elements listed in Figure 5.11?
   c. Which three elements in Figure 5.11 have an electronegativity of 2.5?
   d. How does the electronegativity of the element hydrogen compare to that of the Period 2 elements?

● **Bond Polarity (Section 5.10)**

5.49 Place δ^+ above the atom that is relatively positive and δ^- above the atom that is relatively negative in each of the following bonds. Try to answer this question without referring to Figure 5.11.
   a. B—N   b. Cl—F   c. N—C   d. F—O

5.50 Place δ^+ above the atom that is relatively positive and δ^- above the atom that is relatively negative in each of the following bonds. Try to answer this question without referring to Figure 5.11.
   a. Cl—Br   b. Al—S   c. Br—S   d. O—N

5.51 Rank the following bonds in order of increasing polarity.
   a. H—Cl, H—O, H—Br   b. O—F, P—O, Al—O
   c. H—Cl, Br—Br, B—N   d. P—N, S—O, Br—F

5.52 Rank the following bonds in order of increasing polarity.
   a. H—Br, H—Cl, H—S   b. N—O, Be—N, N—F
   c. N—P, P—P, P—S   d. B—Si, Br—I, C—H

5.53 Classify each of the following bonds as nonpolar covalent, polar covalent, or ionic on the basis of electronegativity differences.
   a. C—O   b. Na—Cl   c. C—I   d. Ca—S

5.54 Classify each of the following bonds as nonpolar covalent, polar covalent, or ionic on the basis of electronegativity differences.
   a. Cl—F   b. P—H   c. C—H   d. Ca—O

● **Molecular Polarity (Section 5.11)**

5.55 Indicate whether each of the following hypothetical triatomic molecules is polar or nonpolar. Assume that A, X, and Y have different electronegativities.
   a. A linear X—A—X molecule
   b. A linear X—X—A molecule
   c. An angular A—X—Y molecule
   d. An angular X—A—Y molecule

5.56 Indicate whether each of the following hypothetical triatomic molecules is polar or nonpolar. Assume that A, X, and Y have different electronegativities.
   a. A linear X—A—Y molecule
   b. A linear A—Y—A molecule
   c. An angular X—A—X molecule
   d. An angular X—X—X molecule

5.57 Indicate whether each of the following triatomic molecules is polar or nonpolar. The molecular geometry is given in parentheses.
   a. CS_2 (linear with C in the center position)
   b. H_2Se (angular with Se in the center position)
   c. FNO (angular with N in the center position)
   d. N_2O (linear with N in the center position)

5.58 Indicate whether each of the following triatomic molecules is polar or nonpolar. The molecular geometry is given in parentheses.
   a. SCl_2 (angular with S in the center position)
   b. OF_2 (angular with O in the center position)
   c. SO_2 (angular with S in the center position)
   d. H_2S (angular with S in the center position)

5.59 Indicate whether each of the following molecules is polar or nonpolar. The molecular geometry is given in parentheses.
   a. NF_3 (trigonal pyramid with N at the apex)
   b. H_2Se (angular with Se in the center position)
   c. CS_2 (linear with C in the center position)
   d. $CHCl_3$ (tetrahedral with C in the center position)

5.60 Indicate whether each of the following molecules is polar or nonpolar. The molecular geometry is given in parentheses.
   a. PH_2Cl (trigonal pyramid with P at the apex)
   b. SO_3 (trigonal planar with S in the center position)
   c. CH_2Cl_2 (tetrahedral with C in the center position)
   d. CCl_4 (tetrahedral with C in the center position)

● **Naming Binary Molecular Compounds (Section 5.12)**

5.61 Name the following binary molecular compounds.
   a. SF_4 b. P_4O_6 c. ClO_2 d. H_2S

5.62 Name the following binary molecular compounds.
   a. Cl_2O b. CO   c. PI_3 d. HI

5.63 Each of the following diagrams depicts a collection of sulfur–oxygen molecules, where the yellow atoms are sulfur and the red atoms are oxygen.

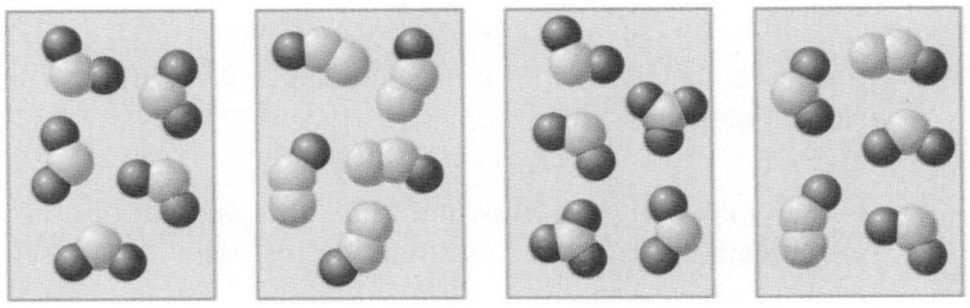

I   II   III   IV

Which diagram depicts the following?
   a. Sulfur dioxide molecules
   b. Disulfur monoxide molecules
   c. A mixture of sulfur dioxide and disulfur monoxide molecules
   d. A mixture of sulfur dioxide and sulfur trioxide molecules

5.64 Each of the following diagrams depicts a collection of nitrogen–oxygen molecules, where the blue atoms are nitrogen and the red atoms are oxygen.

 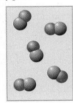 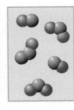 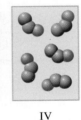

I   II   III   IV

Which diagram depicts the following?
   a. Dinitrogen monoxide molecules
   b. Nitrogen monoxide molecules
   c. A mixture of nitrogen dioxide and dinitrogen monoxide molecules
   d. A mixture of dinitrogen monoxide, nitrogen dioxide, and nitrogen monoxide molecules

5.65 Write chemical formulas for the following binary molecular compounds.
   a. Iodine monochloride   b. Dinitrogen monoxide
   c. Nitrogen trichloride   d. Hydrogen bromide

5.66 Write chemical formulas for the following binary molecular compounds.
   a. Bromine monochloride   b. Tetrasulfur dinitride
   c. Sulfur trioxide   d. Dioxygen difluoride

5.67   Write chemical formulas for the following binary molecular compounds.
    a. Hydrogen peroxide   b. Methane
    c. Ammonia   d. Phosphine

5.68   Write chemical formulas for the following binary molecular compounds.
    a. Ethane   b. Water
    c. Hydrazine   d. Arsine

ADDITIONAL PROBLEMS

5.69   How many electron dots should appear in the Lewis structure for each of the following molecules or polyatomic ions?
    a. O_2F_2 b. $C_2H_2Br_2$ c. S_2^{2-} d. NH_4^+

5.70   In which of the following pairs of diatomic species do both members of the pair have bonds of the same multiplicity (single, double, triple)?
    a. HCl and HF   b. S_2 and Cl_2
    c. CO and NO^+ d. OH^- and HS^-

5.71   Specify the reason why each of the following Lewis structures is incorrect using the following choices: (1) not enough electron dots, (2) too many electron dots, or (3) improper placement of a correct number of electron dots.
    a. $: O{\equiv}O :$ b. $H : \overset{..}{O} : Cl$
    c. $H : \overset{..}{O} ::: \overset{..}{O} : H$ d. $\left[: \overset{..}{N} :: \overset{..}{O} : \right]^+$

5.72   Specify both the VSEPR electron group geometry about the central atom and the molecular geometry for each of the following species.
    a. SiH_4 b. NH_4^+ c. ClNO   d. NO_3^-

5.73   Classify each of the following molecules as polar or nonpolar, or indicate that no such classification is possible because of insufficient information.
    a. A molecule in which all bonds are polar
    b. A molecule in which all bonds are nonpolar
    c. A molecule with two bonds, both of which are polar
    d. A molecule with two bonds, one that is polar and one that is nonpolar

5.74   Indicate which molecule in each of the following pairs of molecules is *more* polar.
    a. BrCl and BrI   b. CO_2 and SO_2
    c. SO_3 and NF_3 d. H_3CF and Cl_3CF

5.75   Four hypothetical elements, A, B, C, and D, have electronegativities A = 3.8, B = 3.3, C = 2.8, and D = 1.3. These elements form the compounds BA, DA, DB, and CA. Arrange these compounds in order of increasing *ionic* bond character.

5.76   Successive substitution of F atoms for H atoms in the molecule CH_4 produces the molecules CH_3F, CH_2F_2, CHF_3, and CF_4.
    a. Draw Lewis structures for each of the five molecules.
    b. Using VSEPR theory, predict the geometry of each of the five molecules.
    c. Give the polarity (polar or nonpolar) of each of the five molecules.

5.77   The chemical formula for a compound containing two nitrogen atoms and one oxygen atom is written as N_2O and the compound's name is dinitrogen monoxide. The chemical formula is not written as ON_2 and the compound is not named oxygen dinitride. Explain.

5.78   The correct name for the compound $NaNO_3$ is not sodium nitrogen trioxide. Explain.

5.79   Name each of the following binary compounds. (*Caution:* At least one of the compounds is ionic.)
    a. NaCl   b. BrCl   c. K_2S d. Cl_2O

*M*ultiple-Choice Practice Test

5.80   In which of the following pairs of compounds are both members of the pair *molecular* compounds?
    a. PCl_3 and LiBr   b. CCl_4 and KOH
    c. NaH and CaF_2 d. CO_2 and NH_3

5.81   Eighteen electrons are present in the Lewis structure of which of the following molecules?
    a. CO_2 b. N_2O c. SO_2 d. HCN

5.82   Two nonbonding electrons are present in the Lewis structure of which of the following molecules?
    a. CH_4 b. HCN   c. SO_2 d. H_2O

5.83   An angular molecular geometry is associated with molecules in which the central atom has which of the following?
    a. Three bonding groups and one nonbonding group
    b. Two bonding groups and two nonbonding groups
    c. Two bonding groups and zero nonbonding groups
    d. Three bonding groups and zero nonbonding groups

5.84   In which of the following pairs of molecules do both members of the pair have the same molecular geometry?
    a. SO_2 and CO_2 b. H_2S and HCN
    c. NH_3 and SO_3 d. H_2O and OF_2

5.85   In which of the following pairs of bonds is the first listed bond *more* polar than the second listed bond?
    a. N—N and N—F   b. P—Cl and S—Cl
    c. C—N and C—F   d. N—Cl and C—H

5.86   If the electronegativity difference between two atoms is 1.0, the bond between the two atoms would be classified as which of the following?
    a. Ionic   b. Nonpolar covalent
    c. Polar covalent   d. Coordinate covalent

5.87   Which of the following statements concerning molecular polarity is *correct*?
    a. All diatomic molecules are polar.
    b. All diatomic molecules are nonpolar.
    c. Some diatomic molecules are polar.
    d. No diatomic molecules are nonpolar.

5.88   Which of the following is a molecular compound that contains five atoms per molecule?
    a. Phosphorus trichloride   b. Disulfur monoxide
    c. Dinitrogen trioxide   d. Carbon dioxide

5.89   Which of the following molecular compounds is paired with an incorrect name?
    a. NH_3, ammonia
    b. H_2S, hydrogen sulfide
    c. N_2O_5, dinitrogen pentoxide
    d. NO_2, mononitrogen dioxide

Chemical Calculations: Formula Masses, Moles, and Chemical Equations

6

Chapter Outline

A half-carat diamond contains approximately 5 × 10²¹ carbon atoms. In this chapter we learn how to calculate the number of atoms in a particular amount of substance.

In this chapter we discuss "chemical arithmetic," the quantitative relationships between elements and compounds. Anyone who deals with chemical processes needs to understand at least the simpler aspects of this topic. All chemical processes, regardless of where they occur—in the human body, at a steel mill, on top of the kitchen stove, or in a clinical laboratory setting—are governed by the same mathematical rules.

We have already presented some information about chemical formulas (Section 1.10). In this chapter we discuss chemical formulas again, and here we look beyond describing the composition of compounds in terms of constituent atoms. A new unit, the mole, will be introduced and its usefulness discussed. Chemical equations will be considered for the first time. We will learn how to represent chemical reactions by using chemical equations and how to derive quantitative relationships from these chemical equations.

Many chemists use the term *molecular mass* interchangeably with *formula mass* when dealing with substances that contain discrete molecules. It is incorrect, however, to use the term *molecular mass* when dealing with ionic compounds because such compounds do not have molecules as their basic structural unit (Section 4.8).

FORMULA MASSES

Our entry point into the realm of "chemical arithmetic" is a discussion of the quantity called formula mass. A **formula mass** *is the sum of the atomic masses of all the atoms represented in the chemical formula of a substance.* Formula masses, like the atomic masses from which they are calculated, are relative masses based on the $^{12}_{6}C$ relative-mass scale (Section 3.3). Example 6.1 illustrates how formula masses are calculated.

EXAMPLE 6.1

Using a Compound's Chemical Formula and Atomic Masses to Calculate Its Formula Mass

Calculate the formula mass of each of the following substances.

a. SnF_2 (tin(II) fluoride, a toothpaste additive)
b. $Al(OH)_3$ (aluminum hydroxide, a water purification chemical)

Solution

Formula masses are obtained simply by adding the atomic masses of the constituent elements, counting each atomic mass as many times as the symbol for the element occurs in the chemical formula.

a. A formula unit of SnF_2 contains three atoms: one atom of Sn and two atoms of F. The formula mass, the collective mass of these three atoms, is calculated as follows:

$$1 \text{ atom Sn} \times \left(\frac{118.71 \text{ amu}}{1 \text{ atom Sn}} \right) = 118.71 \text{ amu}$$

$$2 \text{ atoms F} \times \left(\frac{19.00 \text{ amu}}{1 \text{ atom F}} \right) = 38.00 \text{ amu}$$

$$\text{Formula mass} = 156.71 \text{ amu}$$

We derive the conversion factors in the calculation from the atomic masses listed on the inside front cover of the text. Our rules for the use of conversion factors are the same as those discussed in Section 2.7.

Conversion factors are usually not explicitly shown in a formula mass calculation, as they are in the preceding calculation; the calculation is simplified as follows:

$$\begin{aligned} \text{Sn:} & \quad 1 \times 118.71 \text{ amu} = 118.71 \text{ amu} \\ \text{F:} & \quad 2 \times 19.00 \text{ amu} = \underline{38.00 \text{ amu}} \\ & \quad \text{Formula mass} = 156.71 \text{ amu} \end{aligned}$$

b. The chemical formula for this compound contains parentheses. Improper interpretation of parentheses (see Section 4.11) is a common error made by students doing formula mass calculations. In the formula $Al(OH)_3$, the subscript 3 outside the parentheses affects both of the symbols inside the parentheses. Thus we have

$$\begin{aligned} \text{Al:} & \quad 1 \times 26.98 \text{ amu} = 26.98 \text{ amu} \\ \text{O:} & \quad 3 \times 16.00 \text{ amu} = 48.00 \text{ amu} \\ \text{H:} & \quad 3 \times 1.01 \text{ amu} = \underline{3.03 \text{ amu}} \\ & \quad \text{Formula mass} = 78.01 \text{ amu} \end{aligned}$$

In this text, we will always use atomic masses rounded to the hundredths place, as we have done in this example. This rule allows us to use, without rounding, the atomic masses given inside the front cover of the text. A benefit of this approach is that we always use the same atomic mass for a given element and thus become familiar with the atomic masses of the common elements.

Practice Exercise 6.1

Calculate the formula mass of each of the following substances.

a. $Na_2S_2O_3$ (sodium thiosulfate, a photographic chemical)
b. $(NH_2)_2CO$ (urea, a chemical fertilizer for crops)

Answers: **a.** 158.12 amu; **b.** 60.07 amu

Figure 6.1 Oranges may be bought in units of mass (4-lb bag) or units of amount (3 oranges).

 6.2 THE MOLE: A COUNTING UNIT FOR CHEMISTS

The quantity of material in a sample of a substance can be specified either in terms of units of mass or in terms of units of *amount*. Mass is specified in terms of units such as grams, kilograms, and pounds. The amount of a substance is specified by indicating the number of objects present—3, 17, or 437, for instance.

Figure 6.2 A basic process in chemical laboratory work is determining the mass of a substance.

How large is the number 6.02×10^{23}? It would take a modern computer that can count 100 million times a second 190 million years to count 6.02×10^{23} times. If each of the 6 billion people on Earth were made a millionaire (receiving 1 million dollar bills), we would still need 100 million other worlds, each inhabited with the same number of millionaires, in order to have 6.02×10^{23} dollar bills in circulation.

Why the number 6.02×10^{23}, rather than some other number, was chosen as the counting unit of chemists is discussed in Section 6.3. A more formal definition of the mole will also be presented in that section.

We all use both units of mass and units of amount on a daily basis. For example, when buying oranges at the grocery store, we can decide on quantity in either mass units (4-lb bag or 10-lb bag) or amount units (three oranges or eight oranges) (Figure 6.1). In chemistry, as in everyday life, both mass and amount methods of specifying quantity are used. In laboratory work, practicality dictates working with quantities of known mass (Figure 6.2). Counting out a given number of atoms for a laboratory experiment is impossible because we cannot see individual atoms.

When we perform chemical calculations after the laboratory work has been done, it is often useful and even necessary to think of the quantities of substances present in terms of numbers of atoms or molecules instead of mass. When this is done, very large numbers are always encountered. Any macroscopic-sized sample of a chemical substance contains many trillions of atoms or molecules.

In order to cope with this large-number problem, chemists have found it convenient to use a special unit when counting atoms and molecules. Specialized counting units are used in many areas—for example, a *dozen* eggs or a *ream* (500 sheets) of paper (Figure 6.3).

The chemist's counting unit is the *mole*. What is unusual about the mole is its magnitude. A **mole** *is* 6.02×10^{23} *objects*. The extremely large size of the mole unit is necessitated by the extremely small size of atoms and molecules. To the chemist, *one mole* always means 6.02×10^{23} objects, just as *one dozen* always means 12 objects. Two moles of objects is two times 6.02×10^{23} objects, and five moles of objects is five times 6.02×10^{23} objects.

Avogadro's number *is the name given to the numerical value* 6.02×10^{23}. This designation honors Amedeo Avogadro (Figure 6.4), an Italian physicist whose pioneering work on gases later proved valuable in determining the number of particles present in given volumes of substances. When we solve problems dealing with the number of objects (atoms or molecules) present in a given number of moles of a substance, Avogadro's number becomes part of the conversion factor used to relate the number of objects present to the number of moles present.

From the definition

$$1 \text{ mole} = 6.02 \times 10^{23} \text{ objects}$$

two conversion factors can be derived:

$$\frac{6.02 \times 10^{23} \text{ objects}}{1 \text{ mole}} \quad \text{and} \quad \frac{1 \text{ mole}}{6.02 \times 10^{23} \text{ objects}}$$

Example 6.2 illustrates the use of these conversion factors in solving problems.

 **E X A M P L E 6.2**

Calculating the Number of Objects in a Molar Quantity

How many objects are there in each of the following quantities?

a. 0.23 mole of aspirin molecules   **b.** 1.6 moles of oxygen atoms

Solution

Dimensional analysis (Section 2.8) will be used to solve each of these problems. Both of the problems are similar in that we are given a certain number of moles of substance and want to find the number of objects present in the given number of moles. We will need Avogadro's number to solve each of these moles-to-particles problems.

Moles of Substance	Conversion factor involving Avogadro's number	Particles of Substance

a. The objects of concern are molecules of aspirin. The given quantity is 0.23 mole of aspirin molecules, and the desired quantity is the number of aspirin molecules.

$$0.23 \text{ mole aspirin molecules} = ? \text{ aspirin molecules}$$

(continued)

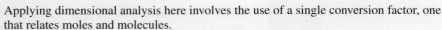

Applying dimensional analysis here involves the use of a single conversion factor, one that relates moles and molecules.

$$0.23 \text{ mole aspirin molecules} \times \left(\frac{6.02 \times 10^{23} \text{ aspirin molecules}}{1 \text{ mole aspirin molecules}} \right)$$

$$= 1.4 \times 10^{23} \text{ aspirin molecules}$$

b. This time we are dealing with atoms instead of molecules. This switch does not change the way we work the problem. We will need the same conversion factor.

The given quantity is 1.6 moles of oxygen atoms, and the desired quantity is the actual number of oxygen atoms present.

$$1.6 \text{ moles oxygen atoms} = ? \text{ oxygen atoms}$$

The setup is

$$1.6 \text{ moles oxygen atoms} \times \left(\frac{6.02 \times 10^{23} \text{ oxygen atoms}}{1 \text{ mole oxygen atoms}} \right)$$

$$= 9.6 \times 10^{23} \text{ oxygen atoms}$$

Figure 6.3 Everyday counting units–a dozen, a pair, and a ream.

Practice Exercise 6.2

How many objects are there in each of the following quantities?

a. 0.46 mole of vitamin C molecules   **b.** 1.27 moles of copper atoms

Answers: a. 2.8×10^{23} vitamin C molecules; **b.** 7.65×10^{23} copper atoms

Figure 6.4 Amedeo Avogadro (1776-1856) was the first scientist to distinguish between atoms and molecules. His name is associated with the number 6.02×10^{23}, the number of particles (atoms or molecules) in a mole.

The mass value below each symbol in the periodic table is both an atomic mass in atomic mass units and a molar mass in grams. For example, the mass of one nitrogen atom is 14.01 amu, and the mass of 1 mole of nitrogen atoms is 14.01 g.

6.3 THE MASS OF A MOLE

How much does a mole weigh? Are you uncertain about the answer to that question? Let us consider a similar but more familiar question first: "How much does a dozen weigh?" Your response is now immediate: "A dozen what?" The mass of a dozen identical objects obviously depends on the identity of the object. For example, the mass of a dozen elephants is greater than the mass of a dozen peanuts. The mass of a mole, like the mass of a dozen, depends on the identity of the object. Thus the mass of a mole, or *molar mass,* is not a set number; it varies and is different for each chemical substance (see Figure 6.5). This is in direct contrast to the *molar number,* Avogadro's number, which is the same for all chemical substances.

The **molar mass** *is the mass, in grams, of a substance that is numerically equal to the substance's formula mass.* For example, the formula mass (atomic mass) of the element sodium is 22.99 amu; therefore, 1 mole of sodium weighs 22.99 g. In Example 6.1, we calculated that the formula mass of tin(II) fluoride is 156.71 amu; therefore, 1 mole of tin(II) fluoride weighs 156.71 g. We can obtain the actual mass in grams of 1 mole of any substance by computing its formula mass (atomic mass for elements) and writing "grams" after it. Thus, when we add atomic masses to get the formula mass (in amu's) of a compound, we are simultaneously finding the mass of 1 mole of that compound (in grams).

It is not a coincidence that the molar mass of a substance, in grams, and its formula mass or atomic mass, in amu, match numerically. The value selected for Avogadro's number is the one that validates this relationship; only for this unique value is the relationship valid. The numerical match between molar mass and atomic or formula mass makes calculating the mass of any given number of moles of a substance a very simple procedure. When you solve problems of this type, the numerical value of the molar mass becomes part of the conversion factor used to convert from moles to grams.

Figure 6.5 The mass of a mole is not a set number of grams; it depends on the substance. For the substances shown, the mass of 1 mole (clockwise from sulfur, the yellow solid) is as follows: sulfur, 32.07 g; zinc, 65.41 g; carbon, 12.01 g; magnesium, 24.31 g; lead, 207.2 g; silicon, 28.09 g; copper, 63.55 g; and, in the center, mercury, 200.59 g.

For example, for the compound CO_2, which has a formula mass of 44.01 amu, we can write the equality

$$44.01 \text{ g } CO_2 = 1 \text{ mole } CO_2$$

From this statement (equality), two conversion factors can be written:

$$\frac{44.01 \text{ g } CO_2}{1 \text{ mole } CO_2} \quad \text{and} \quad \frac{1 \text{ mole } CO_2}{44.01 \text{ g } CO_2}$$

Example 6.3 illustrates the use of gram-to-mole conversion factors like these in solving problems.

 EXAMPLE 6.3

Calculating the Mass of a Molar Quantity of Compound

Acetaminophen, the pain-killing ingredient in Tylenol formulations, has the formula $C_8H_9O_2N$. Calculate the mass, in grams, of a 0.30-mole sample of this pain reliever.

Solution

We will use dimensional analysis to solve this problem. The relationship between molar mass and formula mass will serve as a conversion factor in the setup of this problem.

Moles of Substance	Conversion factor involving molar mass →	Grams of Substance

Molar masses are conversion factors between grams and moles for any substance. Because the periodic table is the usual source of the atomic masses needed to calculate molar masses, the periodic table can be considered to be a useful source of conversion factors.

The given quantity is 0.30 mole of $C_8H_9O_2N$, and the desired quantity is grams of this same substance.

$$0.30 \text{ mole } C_8H_9O_2N = ? \text{ grams } C_8H_9O_2N$$

The calculated formula mass of $C_8H_9O_2N$ is 151.18 amu. Thus,

$$151.18 \text{ grams } C_8H_9O_2N = 1 \text{ mole } C_8H_9O_2N$$

With this relationship in the form of a conversion factor, the setup for the problem becomes

$$0.30 \text{ mole } C_8H_9O_2N \times \left(\frac{151.18 \text{ g } C_8H_9O_2N}{1 \text{ mole } C_8H_9O_2N} \right) = 45 \text{ g } C_8H_9O_2N$$

Practice Exercise 6.3

Carbon monoxide (CO) is an air pollutant that enters the atmosphere primarily in automobile exhaust. Calculate the mass in grams of a 2.61-mole sample of this air pollutant.

Answer: 73.1 g CO

The molar mass of an *element* is unique. No two natural elements have the same molar mass. The molar mass of a *compound* lacks uniqueness. More than one compound can have the same molar mass. For example, the compounds carbon dioxide (CO_2), nitrous oxide (N_2O), and propane (C_3H_8) all have a molar mass of 44.0 g. Despite having like molar masses, these compounds have very different chemical properties. Chemical properties are related to electron arrangements of atoms (Section 3.6) and to the bonding that results when the atoms interact in compound formation. Molar mass is a physical rather than a chemical property of a substance.

The atomic mass unit (amu) and the grams (g) unit are related to one another through Avogadro's number.

$$6.02 \times 10^{23} \text{ amu} = 1.00 \text{ g}$$

That this is the case can be deduced from the following equalities:

$$\text{Atomic mass of N} = \text{mass of 1 N atom} = 14.01 \text{ amu}$$
$$\text{Molar mass of N} = \text{mass of } 6.02 \times 10^{23} \text{ N atoms} = 14.01 \text{ g}$$

Because the second equality involves 6.02×10^{23} times as many atoms as the first equality and the masses come out numerically equal, the gram unit must be 6.02×10^{23} times larger than the amu unit.

In Section 6.2 we defined the mole simply as

$$1 \text{ mole} = 6.02 \times 10^{23} \text{ objects}$$

Although this statement conveys correct information (the value of Avogadro's number to three significant figures is 6.02×10^{23}), it is not the officially accepted definition for the mole. The official definition, which is based on mass, is as follows: The **mole** *is the amount of a substance that contains as many elementary particles (atoms, molecules, or formula units) as there are atoms in exactly 12 grams of* $^{12}_{6}C$. The value of Avogadro's number is an experimentally determined quantity (the number of atoms in exactly 12 g of $^{12}_{6}C$ atoms) rather than a defined quantity. Its value is not even mentioned in the preceding definition. The most up-to-date experimental value for Avogadro's number is 6.0221415×10^{23}, which is consistent with our previous definition (Section 6.2).

6.4 CHEMICAL FORMULAS AND THE MOLE CONCEPT

A chemical formula has two meanings or interpretations: a microscopic-level interpretation and a macroscopic-level interpretation. At a microscopic level, a chemical formula indicates the number of atoms of each element present in one molecule or formula unit of a substance (Section 1.10). *The numerical subscripts in a chemical formula give the number of atoms of the various elements present in 1 formula unit of the substance.* The formula N_2O_4, interpreted at the microscopic level, conveys the information that two atoms of nitrogen and four atoms of oxygen are present in one molecule of N_2O_4 (see Figure 6.6).

Now that the mole concept has been introduced, a macroscopic interpretation of chemical formulas is possible. At a macroscopic level, a chemical formula indicates the number of moles of atoms of each element present in one mole of a substance. *The numerical subscripts in a chemical formula give the number of moles of atoms of the various elements present in 1 mole of the substance.* The designation *macroscopic* is given to this molar interpretation because moles are laboratory-sized quantities of atoms. The formula N_2O_4, interpreted at the macroscopic level, conveys the information that 2 moles of nitrogen atoms and 4 moles of oxygen atoms are present in 1 mole of N_2O_4 molecules. Thus the subscripts in a formula always carry a dual meaning: atoms at the microscopic level and moles of atoms at the macroscopic level.

When it is necessary to know the number of moles of a particular element *within* a compound, the subscript of that element's symbol in the chemical formula becomes part of the conversion factor used to convert from moles of compound to moles of element *within* the compound. Using N_2O_4 as our chemical formula, we can write the following conversion factors:

The numerical relationship between the amu unit and the grams unit is

$$6.02 \times 10^{23} \text{ amu} = 1.00 \text{ g}$$

or

$$1 \text{ amu} = 1.66 \times 10^{-24} \text{ g}$$

This second equality is obtained from the first by dividing each side of the first equality by 6.02×10^{23}.

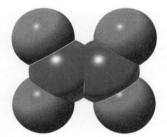

Figure 6.6 A computer-generated model of the molecular structure of the compound N_2O_4.

The molar (macroscopic-level) interpretation of a chemical formula is used in calculations where information about a *particular element within a compound* is needed.

Conversion factors that relate a component of a substance to the substance as a whole are dependent on the formula of the substance. By analogy, the relationship of body parts of an animal to the animal as a whole is dependent on the animal's identity. For example, in 1 mole of elephants there would be 4 moles of elephant legs, 2 moles of elephant ears, 1 mole of elephant tails, and 1 mole of elephant trunks.

$$\text{For N:} \quad \frac{2 \text{ moles N atoms}}{1 \text{ mole N}_2\text{O}_4 \text{ molecules}} \quad \text{or} \quad \frac{1 \text{ mole N}_2\text{O}_4 \text{ molecules}}{2 \text{ moles N atoms}}$$

$$\text{For O:} \quad \frac{4 \text{ moles O atoms}}{1 \text{ mole N}_2\text{O}_4 \text{ molecules}} \quad \text{or} \quad \frac{1 \text{ mole N}_2\text{O}_4 \text{ molecules}}{4 \text{ moles O atoms}}$$

Example 6.4 illustrates the use of this type of conversion factor in problem solving.

EXAMPLE 6.4

Calculating Molar Quantities of Compound Components

Lactic acid, the substance that builds up in muscles and causes them to hurt when they are worked hard, has the formula $C_3H_6O_3$. How many moles of carbon atoms, hydrogen atoms, and oxygen atoms are present in a 1.2-mole sample of lactic acid?

Solution

One mole of $C_3H_6O_3$ contains 3 moles of carbon atoms, 6 moles of hydrogen atoms, and 3 moles of oxygen atoms. We obtain the following conversion factors from this statement:

$$\left(\frac{3 \text{ moles C atoms}}{1 \text{ mole C}_3\text{H}_6\text{O}_3}\right) \quad \left(\frac{6 \text{ moles H atoms}}{1 \text{ mole C}_3\text{H}_6\text{O}_3}\right) \quad \left(\frac{3 \text{ moles O atoms}}{1 \text{ mole C}_3\text{H}_6\text{O}_3}\right)$$

Using the first conversion factor, we calculate the moles of carbon atoms present as follows:

$$1.2 \text{ moles } C_3H_6O_3 \times \left(\frac{3 \text{ moles C atoms}}{1 \text{ mole } C_3H_6O_3}\right) = 3.6 \text{ moles C atoms}$$

Similarly, from the second and third conversion factors, the moles of hydrogen and oxygen atoms present are calculated as follows:

$$1.2 \text{ moles } C_3H_6O_3 \times \left(\frac{6 \text{ moles H atoms}}{1 \text{ mole } C_3H_6O_3}\right) = 7.2 \text{ moles H atoms}$$

$$1.2 \text{ moles } C_3H_6O_3 \times \left(\frac{3 \text{ moles O atoms}}{1 \text{ mole } C_3H_6O_3}\right) = 3.6 \text{ moles O atoms}$$

Practice Exercise 6.4

The compound deoxyribose, whose chemical formula is $C_5H_{10}O_4$, is an important component of DNA molecules, the molecules responsible for the transfer of genetic information from one generation to the next in living organisms. How many moles of carbon atoms, hydrogen atoms, and oxygen atoms are present in a 0.456-mole sample of deoxyribose?

Answers: 2.28 moles C atoms, 4.56 moles H atoms, and 1.82 moles O atoms

6.5 THE MOLE AND CHEMICAL CALCULATIONS

In this section, we will combine the things we have learned about moles to produce a general approach to problem solving that is applicable to a variety of chemical situations. In Section 6.2, we learned that *Avogadro's number* provides a relationship between the number of particles of a substance and the number of moles of that same substance:

Particles of Substance → Conversion factor involving Avogadro's number → Moles of Substance

In Section 6.3, we learned that *molar mass* provides a relationship between the number of grams of a substance and the number of moles of that substance:

Grams of Substance → Conversion factor involving molar mass → Moles of Substance

Figure 6.7 In solving chemical-formula-based problems, the only "transitions" allowed are those between quantities (boxes) connected by arrows. Associated with each arrow is the concept on which the required conversion factor is based.

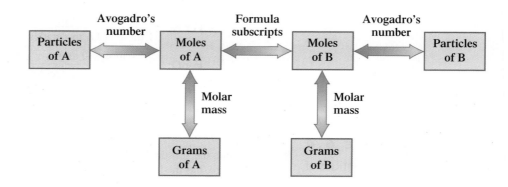

In Section 6.4, we learned that the *molar interpretation of chemical formula subscripts* provides a relationship between the number of moles of a substance and the number of moles of its components:

$$\boxed{\text{Moles of Compound}} \xrightarrow[\text{chemical formula subscripts}]{\text{Conversion factor involving}} \boxed{\text{Moles of Element within Compound}}$$

The preceding three concepts can be combined into a single diagram that is very useful in problem solving. This diagram, Figure 6.7, can be viewed as a road map from which conversion factor sequences (pathways) may be obtained. It gives all the relationships we need for solving two general types of problems:

1. Calculations where information (moles, particles, or grams) is given about a particular substance, and additional information (moles, particles, or grams) is needed concerning the same substance.
2. Calculations where information (moles, particles, or grams) is given about a particular substance, and information is needed concerning a *component* of that same substance.

For the first type of problem, only the left side of Figure 6.7 (the "A" boxes) is needed. For problems of the second type, both sides of the diagram (both "A" and "B" boxes) are used.
   The thinking pattern needed to use Figure 6.7 is very simple.

1. Determine which box in the diagram represents the *given* quantity in the problem.
2. Locate the box that represents the *desired* quantity.
3. Follow the indicated pathway that takes you from the given quantity to the desired quantity. This involves simply following the arrows. There will always be only one pathway possible for the needed transition.

Examples 6.5 and 6.6 illustrate two of the types of problems that can be solved by using the relationships shown in Figure 6.7.

   **EXAMPLE 6.5**

Calculating the Number of Particles in a Given Mass of Compound

Vitamin C has the formula $C_6H_8O_6$. Calculate the number of vitamin C molecules present in a 0.250-g tablet of pure vitamin C.

Solution

We will solve this problem by using the three steps of dimensional analysis (Section 2.8) and Figure 6.7.

Step 1:   The given quantity is 0.250 g of $C_6H_8O_6$, and the desired quantity is molecules of $C_6H_8O_6$.

$$0.250 \text{ g } C_6H_8O_6 = ? \text{ molecules } C_6H_8O_6$$

In terms of Figure 6.7, this is a "grams of A" to "particles of A" problem. We are given grams of a substance, A, and desire to find molecules (particles) of that same substance.

Step 2: Figure 6.7 gives us the pathway we need to solve this problem. Starting with "grams of A," we convert to "moles of A" and finally reach "particles of A." The arrows between the boxes along our path give the type of conversion factor needed for each step.

$$\boxed{\text{Grams of A}} \xrightarrow[\text{mass}]{\text{Molar}} \boxed{\text{Moles of A}} \xrightarrow[\text{number}]{\text{Avogadro's}} \boxed{\text{Particles of A}}$$

Using dimensional analysis, the setup for this sequence of conversion factors is

$$0.250 \text{ g } C_6H_8O_6 \times \left(\frac{1 \text{ mole } C_6H_8O_6}{176.14 \text{ g } C_6H_8O_6} \right) \times \left(\frac{6.02 \times 10^{23} \text{ molecules } C_6H_8O_6}{1 \text{ mole } C_6H_8O_6} \right)$$

$$\text{g } C_6H_8O_6 \longrightarrow \text{moles } C_6H_8O_6 \longrightarrow \text{molecules } C_6H_8O_6$$

The number 176.14 that is used in the first conversion factor is the formula mass of $C_6H_8O_6$. It was not given in the problem but had to be calculated by using atomic masses and the method for calculating formula masses shown in Example 6.1.

Step 3: The solution to the problem, obtained by doing the arithmetic, is

$$\frac{0.250 \times 1 \times 6.02 \times 10^{23}}{176.14 \times 1} \text{ molecules } C_6H_8O_6$$

$$= 8.54 \times 10^{20} \text{ molecules } C_6H_8O_6$$

Practice Exercise 6.5

The compound lithium carbonate, used to treat manic depression, has the formula Li_2CO_3. Calculate the number of formula units of lithium carbonate present in a 0.500-g sample of lithium carbonate.

Answer: 4.07×10^{21} Li_2CO_3 formula units

EXAMPLE 6.6

Calculating the Mass of an Element Present in a Given Mass of Compound

How many grams of nitrogen are present in a 0.10-g sample of caffeine, the stimulant in coffee and tea? The formula of caffeine is $C_8H_{10}N_4O_2$.

Solution

Step 1: There is an important difference between this problem and the preceding one; here we are dealing with not one but *two* substances, caffeine and nitrogen. The given quantity is grams of caffeine (substance A), and we are asked to find the grams of nitrogen (substance B). This is a "grams of A" to "grams of B" problem.

$$0.10 \text{ g } C_8H_{10}N_4O_2 = ? \text{ g N}$$

Step 2: The appropriate set of conversions for a "grams of A" to "grams of B" problem, from Figure 6.7, is

$$\boxed{\text{Grams of A}} \xrightarrow[\text{mass}]{\text{Molar}} \boxed{\text{Moles of A}} \xrightarrow[\text{subscripts}]{\text{Formula}} \boxed{\text{Moles of B}} \xrightarrow[\text{mass}]{\text{Molar}} \boxed{\text{Grams of B}}$$

The conversion factor setup is

$$0.10 \text{ g } C_8H_{10}N_4O_2 \times \left(\frac{1 \text{ mole } C_8H_{10}N_4O_2}{194.26 \text{ g } C_8H_{10}N_4O_2} \right) \times \left(\frac{4 \text{ moles N}}{1 \text{ mole } C_8H_{10}N_4O_2} \right) \times \left(\frac{14.01 \text{ g N}}{1 \text{ mole N}} \right)$$

The number 194.26 that is used in the first conversion factor is the formula mass for caffeine. The conversion from "moles of A" to "moles of B" (the second conversion factor) is made by using the information contained in the formula $C_8H_{10}N_4O_2$. One mole of caffeine contains 4 moles of nitrogen. The number 14.01 in the final conversion factor is the molar mass of nitrogen.

(continued)

Step 3: Collecting the numbers from the various conversion factors and doing the arithmetic give us our answer.

$$\left(\frac{0.10 \times 1 \times 4 \times 14.01}{194.26 \times 1 \times 1}\right) \text{g N} = 0.029 \text{ g N}$$

Practice Exercise 6.6

How many grams of oxygen are present in a 0.10-g sample of adrenaline, a hormone secreted into the bloodstream in times of stress? The formula of adrenaline is $C_9H_{13}NO_3$.

Answer: 0.026 g O

6.6 WRITING AND BALANCING CHEMICAL EQUATIONS

A **chemical equation** *is a written statement that uses chemical symbols and chemical formulas instead of words to describe the changes that occur in a chemical reaction.* The following example shows the contrast between a word description of a chemical reaction and a chemical equation for the same reaction.

Word description: Calcium sulfide reacts with water to produce calcium oxide and hydrogen sulfide.

Chemical equation: $CaS + H_2O \longrightarrow CaO + H_2S$

In the same way that chemical symbols are considered the *letters* of chemical language, and chemical formulas are considered the *words* of the language, chemical equations can be considered the *sentences* of chemical language.

The substances present at the start of a chemical reaction are called *reactants.* A **reactant** *is a starting material in a chemical reaction that undergoes change in the chemical reaction.* As a chemical reaction proceeds, reactants are consumed (used up) and new materials with new chemical properties, called *products,* are produced. A **product** *is a substance produced as a result of the chemical reaction.*

Conventions Used in Writing Chemical Equations

Four conventions are used to write chemical equations.

> In a chemical equation, the *reactants* are always written on the left side of the equation, and the *products* are always written on the right side of the equation.

1. *The correct formulas of the **reactants** are always written on the **left** side of the equation.*

$$\boxed{CaS} + \boxed{H_2O} \longrightarrow CaO + H_2S$$

2. *The correct formulas of the **products** are always written on the **right** side of the equation.*

$$CaS + H_2O \longrightarrow \boxed{CaO} + \boxed{H_2S}$$

3. *The reactants and products are separated by an arrow pointing toward the products.*

$$CaS + H_2O \boxed{\longrightarrow} CaO + H_2S$$

This arrow means "to produce."

4. *Plus signs are used to separate different reactants or different products.*

$$CaS \oplus H_2O \longrightarrow CaO \oplus H_2S$$

Plus signs on the reactant side of the equation mean "reacts with," and plus signs on the product side mean "and."

A *valid* chemical equation must satisfy two conditions:

1. *It must be consistent with experimental facts.* Only the reactants and products that are actually involved in a reaction are shown in an equation. An accurate formula must be used for each of these substances. Elements in solid and liquid states are represented in equations by the chemical symbol for the element. Elements that are gases at room temperature are represented by the molecular formula denoting the form in which they actually occur in nature. The following monatomic, diatomic, and tetratomic elemental gases are known.

The diatomic elemental gases are the elements whose names end in -*gen* (hydrogen, oxygen, and nitrogen) or -*ine* (fluorine, chlorine, bromine, and iodine).

Monatomic:	He, Ne, Ar, Kr, Xe
Diatomic:	H_2, O_2, N_2, F_2, Cl_2, Br_2 (vapor), I_2 (vapor)
Tetraatomic:	P_4 (vapor), As_4 (vapor)*

2. *There must be the same number of atoms of each kind on both sides of the chemical equation.* Chemical equations that satisfy this condition are said to be balanced. A **balanced chemical equation** *is a chemical equation that has the same number of atoms of each element involved in the reaction on each side of the equation.* Because the conventions previously listed for writing equations do not guarantee that an equation will be balanced, we now consider procedures for balancing equations.

Atoms are neither created nor destroyed in a chemical reaction. The production of new substances in a chemical reaction results from the rearrangement of the existent groupings of atoms into new groupings. Because only rearrangement occurs, the products always contain the same number of atoms of each kind as do the reactants. This generalization is often referred to as the *law of conservation of mass.* The mass of the reactants and the mass of the products are the same, because both contain exactly the same number of atoms of each kind present.

Guidelines for Balancing Chemical Equations

An unbalanced chemical equation is brought into balance by adding *coefficients* to the equation to adjust the number of reactant or product molecules present. An **equation coefficient** *is a number that is placed to the left of a chemical formula in a chemical equation; it changes the amount, but not the identity, of the substance.* In the notation $2H_2O$, the 2 on the left is a coefficient; $2H_2O$ means two molecules of H_2O, and $3H_2O$ means three molecules of H_2O. Thus equation coefficients tell how many molecules or formula units of a given substance are present.

The following is a balanced chemical equation, with the coefficients shown in color.

$$4NH_3 + 3O_2 \longrightarrow 2N_2 + 6H_2O$$

This balanced equation tells us that four NH_3 molecules react with three O_2 molecules to produce two N_2 molecules and six H_2O molecules.

The coefficients of a balanced equation represent numbers of molecules or formula units of various species involved in the chemical reaction.

A coefficient of 1 in a balanced equation is not explicitly written; it is considered to be understood. Both Na_2SO_4 and Na_2S have "understood coefficients" of 1 in the following balanced equation:

$$Na_2SO_4 + 2C \longrightarrow Na_2S + 2CO_2$$

An equation coefficient placed in front of a formula applies to the whole formula. By contrast, subscripts, which are also present in formulas, affect only parts of a formula.

Coefficient (affects both H and O)

$2H_2O$

Subscript (affects only H)

In balancing a chemical equation, formula subscripts are *never changed.* You must use the formulas just as they are given. *The only thing you can do is add coefficients.*

The preceding notation denotes two molecules of H_2O; it also denotes a total of four H atoms and two O atoms.

Let's look at the mechanics involved in determining the coefficients needed to balance a chemical equation. Suppose we want to balance the chemical equation

$$FeI_2 + Cl_2 \longrightarrow FeCl_3 + I_2$$

Step 1: *Examine the equation and pick one element to balance first.* It is often convenient to start with the compound that contains the greatest number of atoms, whether a reactant or a product, and focus on the element in that compound that has the greatest number of atoms. Using this guideline, we select $FeCl_3$ and the element chlorine within it.

---

*The four elements listed as vapors are not gases at room temperature but vaporize at slightly higher temperatures. The resultant vapors contain molecules with the formulas indicated.

We note that there are three chlorine atoms on the right side of the equation and two atoms of chlorine on the left (in Cl_2). For the chlorine atoms to balance, we will need six on each side; 6 is the lowest number that both 3 and 2 will divide into evenly. In order to obtain six atoms of chlorine on each side of the equation, we place the coefficient 3 in front of Cl_2 and the coefficient 2 in front of $FeCl_3$.

$$FeI_2 + ③Cl_2 \longrightarrow ②FeCl_3 + I_2$$

We now have six chlorine atoms on each side of the equation.

$$3Cl_2: \qquad 3 \times 2 = 6$$
$$2FeCl_3: \qquad 2 \times 3 = 6$$

Step 2: *Now pick a second element to balance.* We will balance the iron next. The number of iron atoms on the right side has already been set at 2 by the coefficient previously placed in front of $FeCl_3$. We will need two iron atoms on the reactant side of the equation instead of the one iron atom now present. This is accomplished by placing the coefficient 2 in front of FeI_2.

$$②FeI_2 + 3Cl_2 \longrightarrow 2FeCl_3 + I_2$$

It is always wise to pick, as the second element to balance, one whose amount is already set on one side of the equation by a previously determined coefficient. If we had chosen iodine as the second element to balance instead of iron, we would have run into problems. Because the coefficient for neither FeI_2 nor I_2 had been determined, we would have had no guidelines for deciding on the amount of iodine needed.

Step 3: *Now pick a third element to balance.* Only one element is left to balance—iodine. The number of iodine atoms on the left side of the equation is already set at four ($2FeI_2$). In order to obtain four iodine atoms on the right side of the equation, we place the coefficient 2 in front of I_2.

$$2FeI_2 + 3Cl_2 \longrightarrow 2FeCl_3 + ②I_2$$

The addition of the coefficient 2 in front of I_2 completes the balancing process; all the coefficients have been determined.

Step 4: *As a final check on the correctness of the balancing procedure, count atoms on each side of the equation.* The following table can be constructed from our balanced equation.

$$2FeI_2 + 3Cl_2 \longrightarrow 2FeCl_3 + 2I_2$$

Atom	Left Side	Right Side
Fe	$2 \times 1 = 2$	$2 \times 1 = 2$
I	$2 \times 2 = 4$	$2 \times 2 = 4$
Cl	$3 \times 2 = 6$	$2 \times 3 = 6$

All elements are in balance: two iron atoms on each side, four iodine atoms on each side, and six chlorine atoms on each side (see Figure 6.8).

Figure 6.8 When 16.90 g of the compound CaS (left photo) is decomposed into its constituent elements, the Ca and S produced (right photo) has an identical mass of 16.90 grams. Because atoms are neither created nor destroyed in a chemical reaction, the masses of reactants and products in a chemical reaction are always equal.

$$CaS \longrightarrow Ca + S$$

Note that the elements chlorine and iodine in the preceding equation are written in the form of diatomic molecules (Cl_2 and I_2). This is in accordance with the guideline given at the start of this section on the use of molecular formulas for elements that are gases at room temperature.

In Example 6.7 we will balance another chemical equation.

EXAMPLE 6.7
Balancing a Chemical Equation

Balance the following chemical equation.

$$C_2H_6O + O_2 \longrightarrow CO_2 + H_2O$$

Solution

The element oxygen appears in four different places in this chemical equation. This means we do not want to start the balancing process with the element oxygen. Always start the balancing process with an element that appears only once on both the reactant and product sides of the equation.

Step 1: *Balancing of H atoms.* There are six H atoms on the left and two H atoms on the right. Placing the coefficient 1 in front of C_2H_6O and the coefficient 3 in front of H_2O balances the H atoms at six on each side.

$$1C_2H_6O + O_2 \longrightarrow CO_2 + 3H_2O$$

Step 2: *Balancing of C atoms.* An effect of balancing the H atoms at six (Step 1) is the setting of the C atoms on the left side at two. Placing the coefficient 2 in front of CO_2 causes the carbon atoms to balance at two on each side of the chemical equation.

$$1C_2H_6O + O_2 \longrightarrow 2CO_2 + 3H_2O$$

Step 3: *Balancing of O atoms.* The oxygen content of the right side of the chemical equation is set at seven atoms: four oxygen atoms from $2CO_2$ and three oxygen atoms from $3H_2O$. To obtain seven oxygen atoms on the left side of the chemical equation, we place the coefficient 3 in front of O_2; $3O_2$ gives six oxygen atoms, and there is an additional O in $1C_2H_6O$. The element oxygen is present in all four formulas in the chemical equation.

$$1C_2H_6O + 3O_2 \longrightarrow 2CO_2 + 3H_2O$$

Step 4: *Final check.* The equation is balanced. There are two carbon atoms, six hydrogen atoms, and seven oxygen atoms on each side of the chemical equation.

$$C_2H_6O + 3O_2 \longrightarrow 2CO_2 + 3H_2O$$

Practice Exercise 6.7

Balance the following chemical equation.

$$C_4H_{10}O + O_2 \longrightarrow CO_2 + H_2O$$

Answer: $C_4H_{10}O + 6O_2 \longrightarrow 4CO_2 + 5H_2O$

Some additional comments and guidelines concerning chemical equations in general, and the process of balancing in particular, are given here.

1. The coefficients in a balanced chemical equation are always the *smallest set of whole numbers* that will balance the equation. We mention this because more than one set of coefficients will balance a chemical equation. Consider the following three equations:

$$2H_2 + O_2 \longrightarrow 2H_2O$$
$$4H_2 + 2O_2 \longrightarrow 4H_2O$$
$$8H_2 + 4O_2 \longrightarrow 8H_2O$$

All three of these chemical equations are mathematically correct; there are equal numbers of hydrogen and oxygen atoms on both sides of the equation. However, the first equation is considered the correct form because the coefficients used there are the smallest set of whole numbers that will balance the equation. The coefficients in the second equation are two times those in the first equation, and the third equation has coefficients that are four times those of the first equation.

2. At this point, you are not expected to be able to write down the products for a chemical reaction when given the reactants. After learning how to balance chemical equations, students sometimes get the mistaken idea that they ought to be able to write down equations from scratch. This is not so. You will need more chemical knowledge before attempting this task. At this stage, you should be able to balance simple equations, given *all* of the reactants and *all* of the products.

3. It is often useful to know the physical state of the substances involved in a chemical reaction. We specify physical state by using the symbols (*s*) for solid, (*l*) for liquid, (*g*) for gas, and (*aq*) for aqueous solution (a substance dissolved in water). Two examples of such symbol use in chemical equations are

$$2Fe_2O_3(s) + 3C(s) \longrightarrow 4Fe(s) + 3CO_2(g)$$

$$2HNO_3(aq) + 3H_2S(aq) \longrightarrow 2NO(g) + 3S(s) + 4H_2O(l)$$

6.7 CHEMICAL EQUATIONS AND THE MOLE CONCEPT

The coefficients in a balanced chemical equation, like the subscripts in a chemical formula (Section 6.4), have two levels of interpretation—a microscopic level of meaning and a macroscopic level of meaning. The microscopic level of interpretation was used in the previous two sections. *The coefficients in a balanced chemical equation give the numerical relationships among formula units consumed or produced in the chemical reaction.* Interpreted at the microscopic level, the chemical equation

$$N_2 + 3H_2 \longrightarrow 2NH_3$$

conveys the information that one molecule of N_2 reacts with three molecules of H_2 to produce two molecules of NH_3.

At the macroscopic level of interpretation, chemical equations are used to relate mole-sized quantities of reactants and products to each other. At this level, *the coefficients in a balanced chemical equation give the fixed molar ratios between substances consumed or produced in the chemical reaction.* Interpreted at the macroscopic level, the chemical equation

$$N_2 + 3H_2 \longrightarrow 2NH_3$$

conveys the information that 1 mole of N_2 reacts with 3 moles of H_2 to produce 2 moles of NH_3.

The coefficients in a balanced chemical equation can be used to generate mole-based conversion factors to be used in solving problems. Several pairs of conversion factors are obtainable from a single balanced chemical equation. Consider the following balanced chemical equation:

$$4Fe + 3O_2 \longrightarrow 2Fe_2O_3$$

Three mole-to-mole relationships are obtainable from this chemical equation:

4 moles of Fe produce 2 moles of Fe_2O_3.

3 moles of O_2 produce 2 moles of Fe_2O_3.

4 moles of Fe react with 3 moles of O_2.

From each of these macroscopic-level relationships, two conversion factors can be written. The conversion factors for the first relationship are

$$\left(\frac{4 \text{ moles Fe}}{2 \text{ moles Fe}_2O_3} \right) \quad \text{and} \quad \left(\frac{2 \text{ moles Fe}_2O_3}{4 \text{ moles Fe}} \right)$$

Conversion factors that relate two different substances to one another are valid only for systems governed by the chemical equation from which they were obtained.

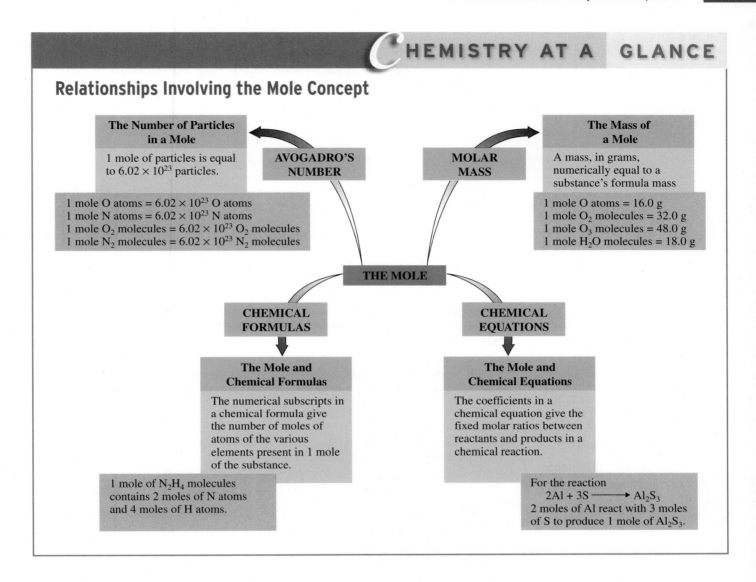

CHEMISTRY AT A GLANCE

Relationships Involving the Mole Concept

The Number of Particles in a Mole

1 mole of particles is equal to 6.02×10^{23} particles.

AVOGADRO'S NUMBER

1 mole O atoms = 6.02×10^{23} O atoms
1 mole N atoms = 6.02×10^{23} N atoms
1 mole O_2 molecules = 6.02×10^{23} O_2 molecules
1 mole N_2 molecules = 6.02×10^{23} N_2 molecules

MOLAR MASS

The Mass of a Mole

A mass, in grams, numerically equal to a substance's formula mass

1 mole O atoms = 16.0 g
1 mole O_2 molecules = 32.0 g
1 mole O_3 molecules = 48.0 g
1 mole H_2O molecules = 18.0 g

THE MOLE

CHEMICAL FORMULAS

CHEMICAL EQUATIONS

The Mole and Chemical Formulas

The numerical subscripts in a chemical formula give the number of moles of atoms of the various elements present in 1 mole of the substance.

1 mole of N_2H_4 molecules contains 2 moles of N atoms and 4 moles of H atoms.

The Mole and Chemical Equations

The coefficients in a chemical equation give the fixed molar ratios between reactants and products in a chemical reaction.

For the reaction
$$2Al + 3S \longrightarrow Al_2S_3$$
2 moles of Al react with 3 moles of S to produce 1 mole of Al_2S_3.

All balanced chemical equations are the source of numerous conversion factors. The more reactants and products there are in the chemical equation, the greater the number of derivable conversion factors. The next section details how conversion factors such as those in the preceding illustration are used in solving problems.

The Chemistry at a Glance feature above reviews the relationships that involve the mole.

 6.8 CHEMICAL CALCULATIONS USING CHEMICAL EQUATIONS

When the information contained in a chemical equation is combined with the concepts of molar mass (Section 6.3) and Avogadro's number (Section 6.2), several useful types of chemical calculations can be carried out. A typical chemical-equation-based calculation gives information about one reactant or product of a reaction (number of grams, moles, or particles) and requests similar information about another reactant or product of the same reaction. The substances involved in such a calculation may both be reactants or products or may be a reactant and a product.

The conversion factor relationships needed to solve problems of this general type are given in Figure 6.9. This diagram should seem very familiar to you; it is almost identical to Figure 6.7, which you used in solving problems based on chemical formulas. There is

The quantitative study of the relationships among reactants and products in a chemical reaction is called *chemical stoichiometry*. The word *stoichiometry*, pronounced stoy-key-om-eh-tree, is derived from the Greek *stoicheion* ("element") and *metron* ("measure"). The stoichiometry of a chemical reaction always involves the *molar relationships* between reactants and products and thus is given by the coefficients in the balanced equation for the chemical reaction.

Figure 6.9 In solving chemical-equation-based problems, the only "transitions" allowed are those between quantities (boxes) connected by arrows. Associated with each arrow is the concept on which the required conversion factor is based.

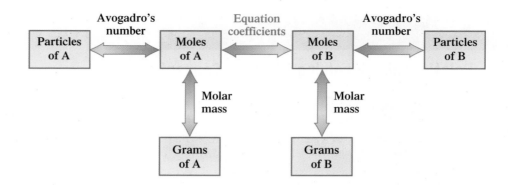

only one difference between the two diagrams. In Figure 6.7, the subscripts in a chemical formula are listed as the basis for relating "moles of A" to "moles of B." In Figure 6.9, the same two quantities are related by using the coefficients of a balanced chemical equation.

The most common type of chemical-equation-based calculation is a "grams of A" to "grams of B" problem. In this type of problem, the mass of one substance involved in a chemical reaction (either reactant or product) is given, and information is requested about the mass of another substance involved in the reaction (either reactant or product). This type of problem is frequently encountered in laboratory settings. For example, a chemist may have a certain number of grams of a chemical available and may want to know how many grams of another substance can be produced from it or how many grams of a third substance are needed to react with it. Examples 6.8 and 6.9 illustrate this type of problem.

EXAMPLE 6.8

Calculating the Mass of a Product in a Chemical Reaction

The human body converts the glucose, $C_6H_{12}O_6$, contained in foods to carbon dioxide, CO_2, and water, H_2O. The chemical equation for the chemical reaction is

$$C_6H_{12}O_6 + 6O_2 \longrightarrow 6CO_2 + 6H_2O$$

Assume a person eats a candy bar containing 14.2 g (1/2 oz) of glucose. How many grams of water will the body produce from the ingested glucose, assuming all of the glucose undergoes reaction?

Solution

Step 1:   The given quantity is 14.2 g of glucose. The desired quantity is grams of water.

$$14.2 \text{ g } C_6H_{12}O_6 = ? \text{ g } H_2O$$

In terms of Figure 6.9, this is a "grams of A" to "grams of B" problem.

Step 2:   Using Figure 6.9 as a road map, we determine that the pathway for this problem is

$$\boxed{\text{Grams of A}} \xrightarrow[\text{mass}]{\text{Molar}} \boxed{\text{Moles of A}} \xrightarrow[\text{coefficients}]{\text{Equation}} \boxed{\text{Moles of B}} \xrightarrow[\text{mass}]{\text{Molar}} \boxed{\text{Grams of B}}$$

The mathematical setup for this problem is

$$14.2 \text{ g } C_6H_{12}O_6 \times \left(\frac{1 \text{ mole } C_6H_{12}O_6}{180.18 \text{ g } C_6H_{12}O_6} \right) \times \left(\frac{6 \text{ moles } H_2O}{1 \text{ mole } C_6H_{12}O_6} \right) \times \left(\frac{18.02 \text{ g } H_2O}{1 \text{ mole } H_2O} \right)$$

$$\text{g } C_6H_{12}O_6 \longrightarrow \text{moles } C_6H_{12}O_6 \longrightarrow \text{moles } H_2O \longrightarrow \text{g } H_2O$$

The 180.18 g in the first conversion factor is the molar mass of glucose, the 6 and 1 in the second conversion factor are the coefficients, respectively, of H_2O and $C_6H_{12}O_6$ in the balanced chemical equation, and the 18.02 g in the third conversion factor is the molar mass of H_2O.

Step 3:   The solution to the problem, obtained by doing the arithmetic after all the numerical factors have been collected, is

$$\left(\frac{14.2 \times 1 \times 6 \times 18.02}{180.18 \times 1 \times 1} \right) \text{g } H_2O = 8.52 \text{ g } H_2O$$

Practice Exercise 6.8

Silicon carbide, SiC, which is used as an abrasive on sandpaper, is prepared using the chemical reaction

$$SiO_2 + 3C \longrightarrow SiC + 2CO$$

How many grams of SiC can be produced from 15.0 g of C?

Answer: 16.7 g SiC

Carbon Monoxide Air Pollution: A Case of Combining Ratios

Experimental conditions—such as temperature, pressure, and the relative amounts of different reactants present—are often key factors in determining the products of a chemical reaction. Under one set of experimental conditions two reactants may produce a certain set of products, and under a different set of experimental conditions these same two reactants may produce another set of products. Such is the case when methane gas (CH_4) reacts with oxygen gas from the air.

When methane is burned in an excess of oxygen the products are carbon dioxide and water.

$$CH_4(g) + 2O_2(g) \longrightarrow CO_2(g) + 2H_2O(g)$$

These same two reactants, with less oxygen present, undergo combustion to produce carbon monoxide and water.

$$2CH_4(g) + 3O_2(g) \longrightarrow 2CO(g) + 4H_2O(g)$$

The difference between the two reactions is the molar combining ratio for the two reactants; 1 to 2 (twice as many moles of oxygen) in the first case and 2 to 3 (1.5 times as many moles of oxygen) in the second case.

Methane gas, the reactant in both of the preceding reactions, is the major component present in the natural gas used in home heating during the winter season. Natural gas furnaces are designed to operate under conditions that favor the first reaction and at the same time minimize the second reaction. With a properly operating gas furnace, the products of combustion are, thus, predominately CO_2 and H_2O (with a small amount of CO), all of which leave the home through an external venting system. However, if the gas burners are out of adjustment (improper oxygen/fuel ratio) or the venting system becomes obstructed, CO levels within the home can build up to levels that are toxic to humans.

Within the human body, inhaled CO reacts with the hemoglobin (Hb) present in red blood cells to form the substance carboxyhemoglobin (COHb). Such COHb formation reduces the ability of hemoglobin to transport oxygen from the lungs to the tissues of the body.

The important factor concerning the effect of CO on the human body is the amount of COHb present in the blood. The higher the percentage of COHb present, the more serious the effect, as indicated in the following table.

Health Effects of COHb Blood Levels

COHb Blood Level (%)	Demonstrated Effects
Less than 1.0	No apparent effect.
1.0 to 2.0	Some evidence of effect on behavioral performance.
2.0 to 5.0	Central nervous system effects. Impairment of time-interval discrimination, visual acuity, brightness discrimination, and certain other psychomotor functions.
Greater than 5.0	Cardiac and pulmonary functional changes.
10.0 to 80.0	Headache, fatigue, drowsiness, coma, respiratory failure, death.

The CO concentration of inhaled air determines COHb levels in the blood. The normal or background level of blood COHb is about 0.5%. This is partly the result of CO produced by the body during the destructive metabolism of heme, a component of hemoglobin (to be discussed in Section 26.7). The remainder comes from low levels of CO in ambient air. The following table gives the relationship between ambient air CO concentrations and blood carboxyhemoglobin levels.

Blood COHb and Ambient Air CO Concentrations

Ambient Air CO Concentration (ppm)	Concentration of Blood COHb (%)
10	2.1
20	3.7
30	5.3
50	8.5
70	11.7

Carbon monoxide levels in ambient air, from both natural sources and human-generated air pollution, are seldom dangerous, ranging from 0.1 part per million (ppm) in rural environments to 3 to 4 ppm in urban areas with large numbers of automobiles and coal-burning industrial complexes.

Cigarette smoking is a form of individualized CO air pollution. Cigarette smoke contains a CO concentration greater than

(continued)

20,000 ppm, the result of the oxygen-deficient conditions (smoldering) under which the cigarette burns. During inhalation, this high CO concentration is diluted to a level of about 400–500 ppm. This "diluted" CO concentration is still sufficiently high to produce elevated COHb levels in the blood of smokers, as shown in the following table.

Blood COHb Levels of Smokers

Category of Smoker	Blood Level of COHb (%)
Never smoked	1.3
Ex-smoker	1.4
Pipe and/or cigar smoker only	1.7
Light cigarette smoker (½ pack or less/day) noninhaler	2.3
Light cigarette smoker (½ pack or less/day) inhaler	3.8
Moderate cigarette smoker (½ to 2 packs/day) inhaler	5.9
Heavy cigarette smoker (2 or more packs/day) inhaler	6.9

Indoor CO air pollution caused by cigarette smoking is significant enough that many states have banned smoking in public gathering areas ranging from airports to restaurants. Studies indicate that nonsmokers present for an extended time in areas where smoking occurs have elevated levels of COHb. The elevated levels are not as high, however, as those of the smokers themselves.

 **EXAMPLE 6.9**

Calculating the Mass of a Reactant Taking Part in a Chemical Reaction

The active ingredient in many commercial antacids is magnesium hydroxide, $Mg(OH)_2$, which reacts with stomach acid (HCl) to produce magnesium chloride ($MgCl_2$) and water. The equation for the reaction is

$$Mg(OH)_2 + 2HCl \longrightarrow MgCl_2 + 2H_2O$$

How many grams of $Mg(OH)_2$ are needed to react with 0.30 g of HCl?

Solution

Step 1: This problem, like Example 6.8, is a "grams of A" to "grams of B" problem. It differs from the previous problem in that both the given and the desired quantities involve reactants.

$$0.30 \text{ g HCl} \longrightarrow ? \text{ g } Mg(OH)_2$$

Step 2: The pathway used to solve it will be the same as in Example 6.8.

$$\boxed{\text{Grams of A}} \xrightarrow[\text{mass}]{\text{Molar}} \boxed{\text{Moles of A}} \xrightarrow[\text{coefficients}]{\text{Equation}} \boxed{\text{Moles of B}} \xrightarrow[\text{mass}]{\text{Molar}} \boxed{\text{Grams of B}}$$

The dimensional-analysis setup is

$$0.30 \text{ g HCl} \times \left(\frac{1 \text{ mole HCl}}{36.46 \text{ g HCl}} \right) \times \left(\frac{1 \text{ mole } Mg(OH)_2}{2 \text{ moles HCl}} \right) \times \left(\frac{58.33 \text{ g } Mg(OH)_2}{1 \text{ mole } Mg(OH)_2} \right)$$

$$\text{g HCl} \longrightarrow \text{ moles HCl} \longrightarrow \text{ moles } Mg(OH)_2 \longrightarrow \text{ g } Mg(OH)_2$$

The balanced chemical equation for the reaction is used as the bridge that enables us to go from HCl to $Mg(OH)_2$. The numbers in the second conversion factor are coefficients from this equation.

Step 3: The solution obtained by combining all of the numbers in the manner indicated in the setup is

$$\left(\frac{0.30 \times 1 \times 1 \times 58.33}{36.46 \times 2 \times 1} \right) \text{ g } Mg(OH)_2 = 0.24 \text{ g } Mg(OH)_2$$

To put our answer in perspective, we note that a common brand of antacid tablets has tablets containing 0.10 g of $Mg(OH)_2$.

Practice Exercise 6.9

The chemical equation for the photosynthesis reaction in plants is

$$6CO_2 + 6H_2O \longrightarrow C_6H_{12}O_6 + 6O_2$$

How many grams of H_2O are consumed at the same time that 20.0 g of CO_2 is consumed?

Answer: 8.19 g H_2O

"Grams of A" to "grams of B" problems (Examples 6.8 and 6.9) are not the only type of problem for which the coefficients in a balanced equation can be used to relate the quantities of two substances. As a further example of the use of equation coefficients in problem solving, consider Example 6.10 (a "particles of A" to "moles of B" problem).

● EXAMPLE 6.10

Calculating the Amount of a Substance Taking Part in a Chemical Reaction

Figure 6.10 Testing apparatus for measuring the effects of airbag deployment.

Automotive airbags inflate when sodium azide, NaN_3, rapidly decomposes to its constituent elements. The equation for the chemical reaction is

$$2NaN_3(s) \longrightarrow 2Na(s) + 3N_2(g)$$

The gaseous N_2 so generated inflates the airbag (see Figure 6.10). How many moles of NaN_3 would have to decompose in order to generate 253 million (2.53×10^8) molecules of N_2?

Solution

Although a calculation of this type does not have a lot of practical significance, it tests your understanding of the problem-solving relationships discussed in this section of the text.

Step 1: The given quantity is 2.53×10^8 molecules of N_2, and the desired quantity is moles of NaN_3.

$$2.53 \times 10^8 \text{ molecules } N_2 = ? \text{ moles } NaN_3$$

In terms of Figure 6.9, this is a "particles of A" to "moles of B" problem.

Step 2: Using Figure 6.9 as a road map, we determine that the pathway for this problem is

$$\boxed{\text{Particles of A}} \xrightarrow[\text{number}]{\text{Avogadro's}} \boxed{\text{Moles of A}} \xrightarrow[\text{coefficients}]{\text{Equation}} \boxed{\text{Moles of B}}$$

The mathematical setup is

$$2.53 \times 10^8 \text{ molecules } N_2 \times \left(\frac{1 \text{ mole } N_2}{6.02 \times 10^{23} \text{ molecules } N_2}\right) \times \left(\frac{2 \text{ moles } NaN_3}{3 \text{ moles } N_2}\right)$$

Avogadro's number is present in the first conversion factor. The 2 and 3 in the second conversion factor are the coefficients, respectively, of NaN_3 and N_2 in the balanced chemical equation.

Step 3: The solution to the problem, obtained by doing the arithmetic after all the numerical factors have been collected, is

$$\left(\frac{2.53 \times 10^8 \times 1 \times 2}{6.02 \times 10^{23} \times 3}\right) \text{ mole } NaN_3 = 2.80 \times 10^{-16} \text{ mole } NaN_3$$

Practice Exercise 6.10

Decomposition of $KClO_3$ serves as a convenient laboratory source of small amounts of oxygen gas. The reaction is

$$2KClO_3 \longrightarrow 2KCl + 3O_2$$

How many moles of $KClO_3$ must be decomposed to produce 64 billion (6.4×10^{10}) O_2 molecules?

Answer: 7.1×10^{-14} mole $KClO_3$

CHEMICAL Connections

Chemical Reactions on an Industrial Scale: Sulfuric Acid

The various calculations in this chapter can be considered to be "laboratory-based" calculations. Chemical substance amounts are always specified in grams, the common laboratory unit for mass. These gram-sized laboratory amounts are very small, almost "infinitesimal," when compared with industrial production figures for various "high-volume" chemicals, which are specified in terms of billions of pounds per year. About 50 of the millions of compounds known are produced in amounts exceeding 1 billion pounds per year in the United States.

The number-one chemical in the United States, in terms of production amount, is sulfuric acid (H_2SO_4), with an annual production approaching 100 billion pounds. Its production amount is almost twice that of any other chemical. So important is sulfuric acid production in the United States (and the world) that some economists use sulfuric acid production as a measure of a nation's industrial strength.

Why is so much sulfuric acid produced in the United States? What are its uses? What are its properties? Where do we encounter it in our everyday life?

Pure sulfuric acid is a colorless, corrosive, oily liquid. It is usually marketed as a concentrated (96% by mass) aqueous solution. People rarely have direct contact with this strong acid because it is seldom part of finished consumer products. The closest encounter most people have with the acid (other than in a chemical laboratory) is involvement with automobile batteries. The acid in a standard automobile battery is a 38%-by-mass aqueous solution of sulfuric acid. However, less than 1% of annual sulfuric acid production ends up in car batteries.

Approximately two-thirds of sulfuric acid production is used in the manufacture of chemical fertilizers. These fertilizer compounds are an absolute necessity if the food needs of an ever-increasing population are to be met. The connection between sulfuric acid and fertilizer revolves around the element phosphorus, which is necessary for plant growth. The starting material for phosphate fertilizer production is phosphate rock, a highly insoluble material containing calcium phosphate, $Ca_3(PO_4)_2$. The treatment of phosphate rock with H_2SO_4 results in the formation of phosphoric acid, H_3PO_4.

$$Ca_3(PO_4)_2 + 3H_2SO_4 \longrightarrow 3CaSO_4 + 2H_3PO_4$$

The phosphoric acid so produced is then used to produce soluble phosphate compounds that plants can use as a source of

The source of phosphorus for ammonium phosphate fertilizer is phosphate rock.

phosphorus. The major phosphoric acid fertilizer derivative is ammonium hydrogen phosphate [$(NH_4)_2HPO_4$].

The raw materials needed to produce sulfuric acid are simple: sulfur, air, and water. In the first step of production, elemental sulfur is burned to give sulfur dioxide gas.

$$S + O_2 \longrightarrow SO_2$$

Some SO_2 is also obtainable as a by-product of metallurgical operations associated with zinc and copper production. Next, the SO_2 gas is combined with additional O_2 (air) to produce sulfur trioxide gas.

$$2SO_2 + O_2 \longrightarrow 2SO_3$$

The SO_3 is then dissolved in water, which yields sulfuric acid as the product.

$$SO_3 + H_2O \longrightarrow H_2SO_4$$

Reactions similar to the last two steps in commercial H_2SO_4 production can also occur naturally in the atmosphere. The H_2SO_4 so produced is a major contributor to the phenomenon called acid rain (see the Chemical Connection feature on page 272 in Chapter 10).

CONCEPTS TO REMEMBER

Formula mass. The formula mass of a substance is the sum of the atomic masses of the atoms in its chemical formula (Section 6.1).

The mole concept. The mole is the chemist's counting unit. One mole of any substance—element or compound—consists of 6.02×10^{23} formula units of the substance. Avogadro's number is the name given to the numerical value 6.02×10^{23} (Section 6.2).

Molar mass. The molar mass of a substance is the mass in grams that is numerically equal to the substance's formula mass. Molar mass is not a set number; it varies and is different for each chemical substance (Section 6.3).

The mole and chemical formulas. The numerical subscripts in a chemical formula give the number of moles of atoms of the various elements present in 1 mole of the substance (Section 6.4).

Chemical equation. A chemical equation is a written statement that uses symbols and formulas instead of words to represent how reactants undergo transformation into products in a chemical reaction (Section 6.5).

Balanced chemical equation. A balanced chemical equation has the same number of atoms of each element involved in the reaction on each side of the equation. An unbalanced chemical equation is brought into balance through the use of equation coefficients. An equation coefficient is a number that is placed to the left of the formula of a substance in a chemical equation and that changes the amount, but not the identity, of the substance (Section 6.6).

The mole and chemical equations. The equation coefficients in a balanced chemical equation give the molar ratios between substances consumed or produced in the chemical reaction described by the equation (Section 6.7).

$\mathcal{K}$EY REACTIONS AND EQUATIONS

1. Calculation of formula mass (Section 6.1)

   Formula mass = sum of atomic masses of all components

2. The mole (Section 6.2)

   $$1 \text{ mole} = 6.02 \times 10^{23} \text{ objects}$$

3. Avogadro's number (Section 6.2)

   $$\text{Avogadro's number} = 6.02 \times 10^{23} \text{ objects}$$

4. Mass of a mole (Section 6.3)

   $$\text{Molar mass} = \frac{\text{mass, in grams, numerically equal to}}{\text{a substance's formula mass}}$$

5. Balanced chemical equation (Section 6.6)

   $$\text{Balanced chemical equation} = \frac{\text{same number of atoms of each}}{\text{kind on each side of the equation}}$$

EXERCISES *and* PROBLEMS

The members of each pair of problems in this section test similar material.

● Formula Masses (Section 6.1)

6.1 Calculate, to two decimal places, the formula mass of each of the following substances. Obtain the needed atomic masses from the inside front cover of the text.
   a. $C_{12}H_{22}O_{11}$ (sucrose, table sugar)
   b. C_7H_{16} (heptane, a component of gasoline)
   c. $C_7H_5NO_3S$ (saccharin, an artificial sweetener)
   d. $(NH_4)_2SO_4$ (ammonium sulfate, a lawn fertilizer)

6.2 Calculate, to two decimal places, the formula mass of each of the following substances. Obtain the needed atomic masses from the inside front cover of the text.
   a. $C_{20}H_{30}O$ (vitamin A)
   b. $C_{14}H_9Cl_5$ (DDT, formerly used as an insecticide)
   c. $C_8H_{10}N_4O_2$ (caffeine, a central nervous system stimulant)
   d. $Ca(NO_3)_2$ (calcium nitrate, gives fireworks their red color)

● The Mole as a Counting Unit (Section 6.2)

6.3 Indicate the number of objects present in each of the following molar quantities.
   a. Number of apples in 1.00 mole of apples
   b. Number of elephants in 1.00 mole of elephants
   c. Number of atoms in 1.00 mole of Zn atoms
   d. Number of molecules in 1.00 mole of CO_2 molecules

6.4 Indicate the number of objects present in each of the following molar quantities.
   a. Number of oranges in 1.00 mole of oranges
   b. Number of camels in 1.00 mole of camels
   c. Number of atoms in 1.00 mole of Cu atoms
   d. Number of molecules in 1.00 mole of CO molecules

6.5 How many atoms are present in the following molar quantities of various elements?
   a. 1.50 moles Fe   b. 1.50 moles Ni
   c. 1.50 moles C   d. 1.50 moles Ne

6.6 How many atoms are present in the following molar quantities of various elements?
   a. 1.20 moles Au   b. 1.20 moles Ag
   c. 1.20 moles Be   d. 1.20 moles Si

6.7 Select the quantity that contains the greater number of atoms in each of the following pairs of substances.
   a. 0.100 mole C atoms or 0.200 mole Al atoms
   b. Avogadro's number of C atoms or 0.750 mole Al atoms
   c. 6.02×10^{23} C atoms or 1.50 moles Al atoms
   d. 6.50×10^{23} C atoms or Avogadro's number of Al atoms

6.8 Select the quantity that contains the greater number of atoms in each of the following pairs of substances.
   a. 0.100 mole N atoms or 0.300 mole P atoms
   b. 6.18×10^{23} N atoms or Avogadro's number of P atoms
   c. Avogadro's number of N atoms or 1.20 moles of P atoms
   d. 6.18×10^{23} N atoms or 2.00 moles P atoms

● Molar Mass (Section 6.3)

6.9 How much, in grams, does 1.00 mole of each of the following substances weigh?
   a. CO (carbon monoxide)   b. CO_2 (carbon dioxide)
   c. NaCl (table salt)   d. $C_{12}H_{22}O_{11}$ (table sugar)

6.10 How much, in grams, does 1.00 mole of each of the following substances weigh?

a. H_2O (water)
b. H_2O_2 (hydrogen peroxide)
c. NaCN (sodium cyanide)
d. KCN (potassium cyanide)

6.11 What is the mass, in grams, of each of the following quantities of matter?
a. 0.034 mole of gold atoms
b. 0.034 mole of silver atoms
c. 3.00 moles of oxygen atoms
d. 3.00 moles of oxygen molecules (O_2)

6.12 What is the mass, in grams, of each of the following quantities of matter?
a. 0.85 mole of copper atoms
b. 0.85 mole of nickel atoms
c. 2.50 moles of nitrogen atoms
d. 2.50 moles of nitrogen molecules (N_2)

6.13 How many moles of specified particles are present in a sample of each of the following substances if each sample weighs 5.00 g?
a. CO molecules   b. CO_2 molecules
c. B_4H_{10} molecules   d. U atoms

6.14 How many moles of specified particles are present in a sample of each of the following substances if each sample weighs 7.00 g?
a. N_2O molecules   b. NO_2 molecules
c. P_4O_{10} molecules   d. V atoms

Chemical Formulas and the Mole Concept (Section 6.4)

6.15 Write the six mole-to-mole conversion factors that can be derived from each of the following chemical formulas.
a. H_2SO_4 b. $POCl_3$

6.16 Write the six mole-to-mole conversion factors that can be derived from each of the following chemical formulas.
a. HNO_3 b. $C_2H_4Br_2$

6.17 How many moles of each type of atom are present in each of the following molar quantities?
a. 2.00 moles SO_2 molecules
b. 2.00 moles SO_3 molecules
c. 3.00 moles NH_3 molecules
d. 3.00 moles N_2H_4 molecules

6.18 How many moles of each type of atom are present in each of the following molar quantities?
a. 4.00 moles NO_2 molecules
b. 4.00 moles N_2O molecules
c. 7.00 moles H_2O molecules
d. 7.00 moles H_2O_2 molecules

6.19 How many *total moles* of atoms are present in each of the following molar quantities?
a. 4.00 moles SO_3 b. 2.00 moles H_2SO_4
c. 1.00 mole $C_{12}H_{22}O_{11}$ d. 3.00 moles $Mg(OH)_2$

6.20 How many *total moles* of atoms are present in each of the following molar quantities?
a. 3.00 moles N_2O_4 b. 4.00 moles HNO_3
c. 0.500 mole C_2H_6O d. 5.00 moles $(NH_4)_2S$

6.21 Based on the chemical formula H_3PO_4, write the conversion factor that would be needed to do each of the following one-step conversions.
a. Moles of H_3PO_4 to moles of H atoms
b. Moles of H_3PO_4 to moles of O atoms
c. Moles of H_3PO_4 to total moles of atoms
d. Moles of P atoms to moles of O atoms

6.22. Based on the chemical formula H_2CO_3, write the conversion factor that would be needed to do each of the following one-step conversions.
a. Moles of H_2CO_3 to moles of C atoms
b. Moles of H_2CO_3 to moles of H atoms
c. Moles of H_2CO_3 to total moles of atoms
d. Moles of H atoms to moles of O atoms

Calculations Based on Chemical Formulas (Section 6.5)

6.23 Determine the number of atoms in each of the following quantities of an element.
a. 10.0 g B   b. 32.0 g Ca   c. 2.0 g Ne   d. 7.0 g N

6.24 Determine the number of atoms in each of the following quantities of an element.
a. 10.0 g S   b. 39.1 g K   c. 3.2 g U   d. 7.0 g Be

6.25 Determine the mass, in grams, of each of the following quantities of substance.
a. 6.02×10^{23} copper atoms   b. 3.01×10^{23} copper atoms
c. 557 copper atoms   d. 1 copper atom

6.26 Determine the mass, in grams, of each of the following quantities of substance.
a. 6.02×10^{23} silver atoms   b. 3.01×10^{23} silver atoms
c. 1.00×10^6 silver atoms   d. 1 silver atom

6.27 Determine the number of moles of substance present in each of the following quantities.
a. 10.0 g He   b. 10.0 g N_2O
c. 4.0×10^{10} atoms P   d. 4.0×10^{10} atoms Be

6.28 Determine the number of moles of substance present in each of the following quantities.
a. 25.0 g N   b. 25.0 g Li
c. 8.50×10^{15} atoms S   d. 8.50×10^{15} atoms Cl

6.29 Determine the number of atoms of sulfur present in each of the following quantities.
a. 10.0 g H_2SO_4 b. 20.0 g SO_3
c. 30.0 g Al_2S_3 d. 2.00 moles S_2O

6.30 Determine the number of atoms of nitrogen present in each of the following quantities.
a. 10.0 g N_2H_4 b. 20.0 g HN_3
c. 30.0 g $LiNO_3$ d. 4.00 moles N_2O_5

6.31 Determine the number of grams of sulfur present in each of the following quantities.
a. 3.01×10^{23} S_2O molecules
b. 3 S_4N_4 molecules
c. 2.00 moles SO_2 molecules
d. 4.50 moles S_8 molecules

6.32 Determine the number of grams of oxygen present in each of the following quantities.
a. 4.50×10^{22} SO_3 molecules   b. 7 P_4O_{10} molecules
c. 3.00 moles H_2SO_4 molecules   d. 1.50 moles O_3 molecules

Writing and Balancing Chemical Equations (Section 6.6)

6.33 Indicate whether each of the following chemical equations is balanced.
a. $SO_3 + H_2O \rightarrow H_2SO_4$
b. $CuO + H_2 \rightarrow Cu + H_2O$
c. $CS_2 + O_2 \rightarrow CO_2 + SO_2$
d. $AgNO_3 + KCl \rightarrow KNO_3 + AgCl$

6.34 Indicate whether each of the following chemical equations is balanced.
a. $H_2 + O_2 \rightarrow H_2O$
b. $NO + O_2 \rightarrow NO_2$
c. $C + O_2 \rightarrow CO_2$
d. $HNO_3 + NaOH \rightarrow NaNO_3 + H_2O$

6.35 For each of the following balanced chemical equations, indicate how many atoms of each element are present on the reactant and product sides of the chemical equation.
a. $2N_2 + 3O_2 \rightarrow 2N_2O_3$
b. $4NH_3 + 6NO \rightarrow 5N_2 + 6H_2O$
c. $PCl_3 + 3H_2 \rightarrow PH_3 + 3HCl$
d. $Al_2O_3 + 6HCl \rightarrow 2AlCl_3 + 3H_2O$

6.36 For each of the following balanced chemical equations, indicate how many atoms of each element are present on the reactant and product sides of the chemical equation.
a. $4Al + 3O_2 \rightarrow 2Al_2O_3$
b. $2Na + 2H_2O \rightarrow 2NaOH + H_2$
c. $2Co + 3HgCl_2 \rightarrow 2CoCl_3 + 3Hg$
d. $H_2SO_4 + 2NH_3 \rightarrow (NH_4)_2SO_4$

6.37 Balance the following chemical equations.
a. $Na + H_2O \rightarrow NaOH + H_2$
b. $Na + ZnSO_4 \rightarrow Na_2SO_4 + Zn$
c. $NaBr + Cl_2 \rightarrow NaCl + Br_2$
d. $ZnS + O_2 \rightarrow ZnO + SO_2$

6.38 Balance the following chemical equations.
a. $H_2S + O_2 \rightarrow SO_2 + H_2O$
b. $Ni + HCl \rightarrow NiCl_2 + H_2$
c. $IBr + NH_3 \rightarrow NH_4Br + NI_3$
d. $C_2H_6 + O_2 \rightarrow CO_2 + H_2O$

6.39 Balance the following chemical equations.
a. $CH_4 + O_2 \rightarrow CO_2 + H_2O$
b. $C_6H_6 + O_2 \rightarrow CO_2 + H_2O$
c. $C_4H_8O_2 + O_2 \rightarrow CO_2 + H_2O$
d. $C_5H_{10}O + O_2 \rightarrow CO_2 + H_2O$

6.40 Balance the following chemical equations.
a. $C_2H_4 + O_2 \rightarrow CO_2 + H_2O$
b. $C_6H_{12} + O_2 \rightarrow CO_2 + H_2O$
c. $C_3H_6O + O_2 \rightarrow CO_2 + H_2O$
d. $C_5H_{10}O_2 + O_2 \rightarrow CO_2 + H_2O$

6.41 Balance the following chemical equations.
a. $PbO + NH_3 \rightarrow Pb + N_2 + H_2O$
b. $Fe(OH)_3 + H_2SO_4 \rightarrow Fe_2(SO_4)_3 + H_2O$

6.42 Balance the following chemical equations.
a. $SO_2Cl_2 + HI \rightarrow H_2S + H_2O + HCl + I_2$
b. $Na_2CO_3 + Mg(NO_3)_2 \rightarrow MgCO_3 + NaNO_3$

6.43 The following diagrams represent the reaction of A_2 (red spheres) with B_2 (blue spheres) to give specific products. Write a balanced equation for each reaction based on the information in the diagram.

a.
Reactants   Products
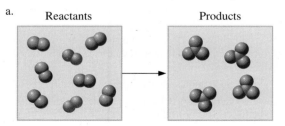

b.
Reactants   Products
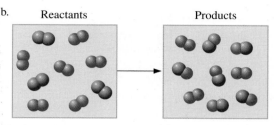

6.44 The following diagrams represent the reaction of A_2 (red spheres) with B_2 (blue spheres) to give specific products. Write a balanced equation for each reaction based on the information in the diagram.

a.
Reactants   Products
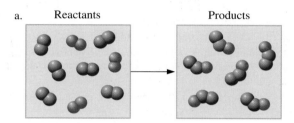

b.
Reactants   Products
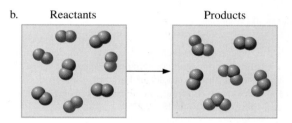

6.45 Diagram I represents the reactant mixture for a chemical reaction. Select from diagrams II through IV the product mixture that is consistent with both diagram I and the concepts associated with a balanced chemical equation.

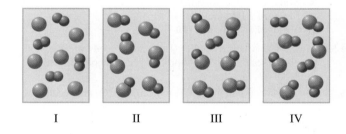

I   II   III   IV

6.46 Diagram I represents the reactant mixture for a chemical reaction. Select from diagrams II through IV the product mixture that is consistent with both diagram I and the concepts associated with a balanced chemical equation.

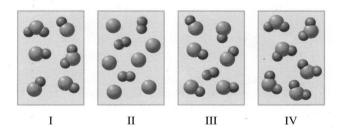

I   II   III   IV

● **Chemical Equations and the Mole Concept (Section 6.7)**

6.47 Write the 12 mole-to-mole conversion factors that can be derived from the following balanced chemical equation.

$$2Ag_2CO_3 \longrightarrow 4Ag + 2CO_2 + O_2$$

6.48 Write the 12 mole-to-mole conversion factors that can be derived from the following balanced chemical equation.

$$N_2H_4 + 2H_2O_2 \longrightarrow N_2 + 4H_2O$$

6.49 Using each of the following chemical equations, calculate the number of moles of CO_2 that can be obtained from 2.00 moles of the first listed reactant with an excess of the other reactant.
a. $C_7H_{16} + 11O_2 \rightarrow 7CO_2 + 8H_2O$
b. $2HCl + CaCO_3 \rightarrow CaCl_2 + CO_2 + H_2O$
c. $Na_2SO_4 + 2C \rightarrow Na_2S + 2CO_2$
d. $Fe_3O_4 + CO \rightarrow 3FeO + CO_2$

6.50 Using each of the following chemical equations, calculate the number of moles of CO_2 that can be obtained from 3.50 moles of the first listed reactant with an excess of the other reactant.
a. $FeO + CO \rightarrow Fe + CO_2$
b. $3O_2 + CS_2 \rightarrow CO_2 + 2SO_2$
c. $2C_8H_{18} + 25O_2 \rightarrow 16CO_2 + 18H_2O$
d. $C_6H_{12}O_6 + 6O_2 \rightarrow 6CO_2 + 6H_2O$

6.51 For the chemical reaction

$$Sb_2S_3 + 6HCl \longrightarrow 2SbCl_3 + 3H_2S$$

write the conversion factor that would be needed to do each of the following one-step conversions.
a. Moles of $SbCl_3$ to moles of H_2S
b. Moles of Sb_2S_3 to moles of HCl
c. Moles of H_2S to moles of HCl
d. Moles of Sb_2S_3 to moles of $SbCl_3$

6.52 For the chemical reaction

$$UF_6 + 2H_2O \longrightarrow UO_2F_2 + 4HF$$

write the conversion factor that would be needed to do each of the following one-step conversions.
a. Moles of UF_6 to moles of HF
b. Moles of UO_2F_2 to moles of H_2O
c. Moles of HF to moles of UO_2F_2
d. Moles of H_2O to moles of UF_6

6.53 How many water molecules (H_2O) are needed to react with 8 ethene molecules (C_2H_4) to produce ethyl alcohol molecules (C_2H_5OH)?

6.54 How many carbon monoxide molecules (CO) are needed to react with 8 hydrogen molecules (H_2) to produce methyl alcohol molecules (CH_3OH)?

6.55 The following diagram represents the high-temperature reaction between CH_4 and O_2.

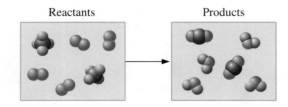

| Reactants | Products |

Use this diagram to answer the following questions.
a. What are the chemical formulas of the products?
b. How many moles of each product can be obtained if 6.0 moles of CH_4 reacts?

6.56 The following diagram represents the high-temperature reaction between CH_4 and H_2O.

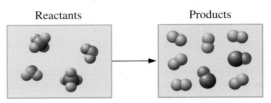

| Reactants | Products |

Use this diagram to answer the following questions.
a. What are the chemical formulas of the products?
b. How many moles of each product can be obtained if 6.0 moles of CH_4 reacts?

● **Calculations Based on Chemical Equations (Section 6.8)**

6.57 How many grams of the first reactant in each of the following chemical equations would be needed to produce 20.0 g of N_2 gas?
a. $4NH_3 + 3O_2 \rightarrow 2N_2 + 6H_2O$
b. $(NH_4)_2Cr_2O_7 \rightarrow N_2 + 4H_2O + Cr_2O_3$
c. $N_2H_4 + 2H_2O_2 \rightarrow N_2 + 4H_2O$
d. $2NH_3 \rightarrow N_2 + 3H_2$

6.58 How many grams of the first reactant in each of the following chemical equations would be needed to produce 20.0 g of H_2O?
a. $N_2H_4 + 2H_2O_2 \rightarrow N_2 + 4H_2O$
b. $H_2O_2 + H_2S \rightarrow 2H_2O + S$
c. $2HNO_3 + NO \rightarrow 3NO_2 + H_2O$
d. $3H_2 + WO_3 \rightarrow W + 3H_2O$

6.59 The principal constituent of natural gas is methane, which burns in air according to the reaction

$$CH_4 + 2O_2 \longrightarrow CO_2 + 2H_2O$$

How many grams of O_2 are needed to produce 3.50 g of CO_2?

6.60 Tungsten (W) metal, which is used to make incandescent bulb filaments, is produced by the reaction

$$WO_3 + 3H_2 \longrightarrow 3H_2O + W$$

How many grams of H_2 are needed to produce 1.00 g of W?

6.61 The catalytic converter that is standard equipment on American automobiles converts carbon monoxide (CO) to carbon dioxide (CO_2) by the reaction

$$2CO + O_2 \longrightarrow 2CO_2$$

What mass of O_2, in grams, is needed to react completely with 25.0 g of CO?

6.62 A mixture of hydrazine (N_2H_4) and hydrogen peroxide (H_2O_2) is used as a fuel for rocket engines. These two substances react as shown by the equation

$$N_2H_4 + 2H_2O_2 \longrightarrow N_2 + 4H_2O$$

What mass of N_2H_4, in grams, is needed to react completely with 35.0 g of H_2O_2?

6.63 Both water and sulfur dioxide are products from the reaction of sulfuric acid (H_2SO_4) with copper metal, as shown by the equation

$$2H_2SO_4 + Cu \longrightarrow SO_2 + 2H_2O + CuSO_4$$

How many grams of H_2O will be produced at the same time that 10.0 g of SO_2 is produced?

6.64 Potassium thiosulfate ($K_2S_2O_3$) is used to remove any excess chlorine from fibers and fabrics that have been bleached with that gas. The reaction is

$$K_2S_2O_3 + 4Cl_2 + 5H_2O \longrightarrow 2KHSO_4 + 8HCl$$

How many grams of HCl will be produced at the same time that 25.0 g of $KHSO_4$ is produced?

ADDITIONAL PROBLEMS

6.65 The compound 1-propanethiol, which is the eye irritant released when fresh onions are chopped up, has a formula mass of 76.18 amu and the formula C_3H_yS. What number does y stand for in the formula?

6.66 Select the quantity that has the greater number of atoms in each of the following pairs of quantities. Make your selection using the periodic table but without performing an actual calculation.
   a. 1.00 mole S or 1.00 mole S_8
   b. 28.0 g Al or 1.00 mole Al
   c. 28.1 g Si or 30.0 g Mg
   d. 2.00 g Na or 6.02×10^{23} atoms He

6.67 What amount or mass of each of the following substances would be needed to obtain 1.000 g of Si?
   a. Moles of SiH_4
   b. Grams of SiO_2
   c. Molecules of $(CH_3)_3SiCl$
   d. Atoms of Si

6.68 How many grams of Si would contain the same number of atoms as there are in 2.10 moles of Ar?

6.69 After the following chemical equation was balanced, the name of one of the reactants was substituted for its formula.
$$2 \text{ butyne} + 11O_2 \longrightarrow 8CO_2 + 6H_2O$$
Using only the information found within the chemical equation, determine the molecular formula of butyne.

6.70 Ammonium dichromate decomposes according to the following reaction.
$$(NH_4)_2Cr_2O_7 \longrightarrow N_2 + 4H_2O + Cr_2O_3$$
How many grams of each of the products can be formed from the decomposition of 75.0 g of ammonium dichromate?

6.71 Black silver sulfide can be produced from the reaction of silver metal with sulfur.
$$2Ag + S \longrightarrow Ag_2S$$
How many grams of Ag and how many grams of S are needed to produce 125 g of Ag_2S?

6.72 How many grams of beryllium (Be) are needed to react completely with 45.0 g of nitrogen (N_2) in the synthesis of Be_3N_2?

$m$ultiple-Choice Practice Test

6.73 Which of the following are the values of the formula masses, respectively, of the compounds H_2O and CO_2?
   a. 10.00 amu and 22.00 amu
   b. 17.01 amu and 30.01 amu
   c. 18.02 amu and 30.01 amu
   d. 18.02 amu and 44.01 amu

6.74 Which statement concerning Avogadro's number is *correct*?
   a. It has the value 6.02×10^{26}.
   b. It denotes the number of molecules in 1 mole of any molecular compound.
   c. It is the mass, in grams, of 1 mole of any substance.
   d. It denotes the number of atoms in 1 mole of any substance.

6.75 Which set of quantities is needed to calculate the mass of 1 mole of a substance?
   a. Chemical formula and Avogadro's number
   b. Chemical formula and atomic masses
   c. Atomic masses and Avogadro's number
   d. Atomic numbers and Avogadro's number

6.76 Which of the following are the values of the molar masses, respectively, of Na and K?
   a. 11.00 amu and 19.00 amu
   b. 11.00 grams and 19.00 grams
   c. 22.99 amu and 39.10 amu
   d. 22.99 grams and 39.10 grams

6.77 Which compound sample contains the greatest number of atoms?
   a. 4.0 moles NH_3 b. 3.0 moles SO_3
   c. 6.0 moles CO   d. 4.0 moles CO_2

6.78 Which of the following statements is true for all balanced chemical equations?
   a. The total number of molecules on each side of the equation must be equal.
   b. The total number of atoms on each side of the equation must be equal.
   c. The sum of the subscripts on each side of the equation must be equal.
   d. The sum of the coefficients on each side of the equation must be equal.

6.79 When the chemical equation $NH_3 \longrightarrow N_2 + H_2$ is correctly balanced, the coefficients are
   a. 1, 2, 3   b. 2, 1, 3   c. 3, 1, 2   d. 1, 1, 3

6.80 In which of the following is the first listed quantity less than the second listed quantity?
   a. Mass of 1 mole of CO_2, mass of 1 mole of CO
   b. Moles in 28.0 g of CO_2, moles in 28.0 g of CO
   c. Molecules in 2 moles of CO_2, molecules in 2 moles of CO
   d. Atoms in 2 moles of CO_2, atoms in 2 moles of CO

6.81 Which of the following is the correct "setup" for the problem, "How many grams of S are present in 50.0 g of S_4N_4?"

   a. $50.0 \text{ g } S_4N_4 \times \dfrac{1 \text{ mole } S_4N_4}{184.32 \text{ g } S_4N_4} \times \dfrac{4 \text{ moles } S}{1 \text{ mole } S_4N_4} \times \dfrac{32.07 \text{ g } S}{4 \text{ moles } S}$

   b. $50.0 \text{ g } S_4N_4 \times \dfrac{1 \text{ mole } S_4N_4}{184.32 \text{ g } S_4N_4} \times \dfrac{4 \text{ moles } S}{1 \text{ mole } S_4N_4} \times \dfrac{32.07 \text{ g } S}{1 \text{ mole } S}$

c. $50.0 \text{ g } S_4N_4 \times \dfrac{1 \text{ mole } S_4N_4}{184.32 \text{ g } S_4N_4} \times \dfrac{1 \text{ mole } S}{1 \text{ mole } S_4N_4} \times \dfrac{32.07 \text{ g } S}{1 \text{ mole } S}$

d. $50.0 \text{ g } S_4N_4 \times \dfrac{1 \text{ mole } S_4N_4}{184.32 \text{ g } S_4N_4} \times \dfrac{1 \text{ mole } S}{4 \text{ moles } S_4N_4} \times \dfrac{32.07 \text{ g } S}{1 \text{ mole } S}$

6.82 Which of the following is the correct "setup" for the problem, "How many grams of H_2O form when 3.2 moles of O_2 react according to the following reaction?"

$$2H_2S + 3O_2 \longrightarrow 2H_2O + 2SO_2$$

a. $3.2 \text{ moles } O_2 \times \dfrac{18.02 \text{ g } H_2O}{2 \text{ moles } H_2O}$

b. $3.2 \text{ moles } O_2 \times \dfrac{32.00 \text{ g } O_2}{1 \text{ mole } O_2} \times \dfrac{18.02 \text{ g } H_2O}{32.00 \text{ g } O_2}$

c. $3.2 \text{ moles } O_2 \times \dfrac{2 \text{ moles } H_2O}{3 \text{ moles } O_2} \times \dfrac{18.02 \text{ g } H_2O}{1 \text{ mole } H_2O}$

d. $3.2 \text{ moles } O_2 \times \dfrac{32.00 \text{ g } O_2}{1 \text{ mole } O_2} \times \dfrac{2 \text{ moles } H_2O}{3 \text{ moles } O_2}$

Gases, Liquids, and Solids

7

Chapter Outline

During a volcanic eruption many interconversions occur among the three states of matter. Products of the eruption include gaseous substances, liquid (molten) substances, and solid substances.

In Chapters 3, 4, and 5, we considered the structure of matter from a submicroscopic point of view—in terms of molecules, atoms, protons, neutrons, and electrons. In this chapter, we are concerned with the macroscopic characteristics of matter as represented by the physical states—solid, liquid, and gas. Of particular concern are the properties exhibited by matter in the various physical states and a theory that correlates these properties with molecular behavior.

7.1 THE KINETIC MOLECULAR THEORY OF MATTER

Solids, liquids, and gases (Section 1.2) are easily distinguished by using four common physical properties of matter: (1) volume and shape, (2) density, (3) compressibility, and (4) thermal expansion. We discussed the property of density in Section 2.9. **Compressibility** *is a measure of the change in volume of a sample of matter resulting from a pressure change.* **Thermal expansion** *is a measure of the change in volume of a sample of matter resulting from a temperature change.* These distinguishing characteristics are compared in Table 7.1 for the three states of matter. The physical characteristics of the solid, liquid, and gaseous states listed in Table 7.1 can be explained by kinetic molecular theory, which is one of the fundamental theories of chemistry. The **kinetic molecular theory of matter** *is a set of five statements used to explain the physical behavior of the three states of matter (solids, liquids, and gases).* The basic idea of this theory is that the particles (atoms, molecules, or ions) present in a substance, independent of the physical state of the substance, are always in motion.

The word *kinetic* comes from the Greek *kinesis,* which means "movement." The kinetic molecular theory deals with the movement of particles.

Figure 7.1 The water in the lake behind the dam has potential energy as a result of its position. When the water flows over the dam, its potential energy becomes kinetic energy that can be used to turn the turbines of a hydroelectric plant.

The energy released when gasoline is burned represents potential energy associated with chemical bonds.

For gases, the attractions between particles (statement 3) are minimal and as a first approximation are considered to be zero (see Section 7.2).

Two consequences of the elasticity of particle collisions (statement 5) are that (1) the energy of any given particle is continually changing, and (2) particle energies for a system are not all the same; a range of particle energies is always encountered.

The five statements of the kinetic molecular theory of matter follow.

Statement 1:   *Matter is ultimately composed of tiny particles (atoms, molecules, or ions) that have definite and characteristic sizes that do not change.*

Statement 2:   *The particles are in constant random motion and therefore possess kinetic energy.*

Kinetic energy *is energy that matter possesses because of particle motion.* An object that is in motion has the ability to transfer its kinetic energy to another object upon collision with that object.

Statement 3:   *The particles interact with one another through attractions and repulsions and therefore possess potential energy.*

Potential energy *is stored energy that matter possesses as a result of its position, condition, and/or composition* (Figure 7.1). The potential energy of greatest importance when considering the differences among the three states of matter is that which originates from electrostatic interactions among particles. An **electrostatic interaction** *is an attraction or repulsion that occurs between charged particles.* Particles of opposite charge (one positive and the other negative) attract one another, and particles of like charge (both positive or both negative) repel one another. Further use of the term *potential energy* in this text will mean potential energy of electrostatic origin.

Statement 4:   *The kinetic energy (velocity) of the particles increases as the temperature is increased.*

The *average* kinetic energy (velocity) of all particles in a system depends on the temperature; kinetic energy increases as temperature increases.

Statement 5:   *The particles in a system transfer energy to each other through elastic collisions.*

In an elastic collision, the total kinetic energy remains constant; no kinetic energy is lost. The difference between an *elastic* and an *inelastic* collision is illustrated by comparing the collision of two hard steel spheres with the collision of two masses of putty. The collision of spheres approximates an elastic collision (the spheres bounce off one another and continue moving, as in Figure 7.2); the putty collision has none of these characteristics (the masses "glob" together with no resulting movement).

The differences among the solid, liquid, and gaseous states of matter can be explained by the relative magnitudes of kinetic energy and potential energy (in this case, electrostatic attractions) associated with the physical state. Kinetic energy can be considered a *disruptive force* that tends to make the particles of a system increasingly independent of one another. This is because the particles tend to move away from one another as a result of the energy of motion. Potential energy of attraction can be considered a *cohesive force* that tends to cause order and stability among the particles of a system.

TABLE 7.1
Distinguishing Properties of Solids, Liquids, and Gases

Property	Solid State	Liquid State	Gaseous State
volume and shape	definite volume and definite shape	definite volume and indefinite shape; takes the shape of its container to the extent that it is filled	indefinite volume and indefinite shape; takes the volume and shape of the container that it completely fills
density	high	high, but usually lower than corresponding solid	low
compressibility	small	small, but usually greater than corresponding solid	large
thermal expansion	very small: about 0.01% per °C	small: about 0.10% per °C	moderate: about 0.30% per °C

How much kinetic energy a chemical system has depends on its temperature. Kinetic energy increases as temperature increases (statement 4 of the kinetic molecular theory of matter). Thus the higher the temperature, the greater the magnitude of disruptive influences within a chemical system. Potential energy magnitude, or cohesive force magnitude, is essentially independent of temperature. The fact that one of the types of forces depends on temperature (disruptive forces) and the other does not (cohesive forces) causes temperature to be the factor that determines in which of the three physical states a given sample of matter is found. We will discuss the reasons for this in Section 7.2.

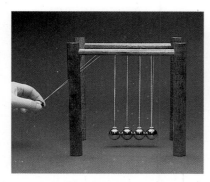

Figure 7.2 Upon release, the steel ball on the left transmits its kinetic energy through a series of elastic collisions to the ball on the right.

7.2 KINETIC MOLECULAR THEORY AND PHYSICAL STATES

A **solid** *is the physical state characterized by a dominance of potential energy (cohesive forces) over kinetic energy (disruptive forces).* The particles in a solid are drawn close together in a regular pattern by the strong cohesive forces present (Figure 7.3a). Each particle occupies a fixed position, about which it vibrates because of disruptive kinetic energy. With this model, the characteristic properties of solids (Table 7.1) can be explained as follows:

1. *Definite volume and definite shape.* The strong, cohesive forces hold the particles in essentially fixed positions, resulting in definite volume and definite shape.
2. *High density.* The constituent particles of solids are located as close together as possible (touching each other). Therefore, a given volume contains large numbers of particles, resulting in a high density.
3. *Small compressibility.* Because there is very little space between particles, increased pressure cannot push the particles any closer together; therefore, it has little effect on the solid's volume.
4. *Very small thermal expansion.* An increased temperature increases the kinetic energy (disruptive forces), thereby causing more vibrational motion of the particles. Each particle occupies a slightly larger volume, and the result is a slight expansion of the solid. The strong, cohesive forces prevent this effect from becoming very large.

A **liquid** *is the physical state characterized by potential energy (cohesive forces) and kinetic energy (disruptive forces) of about the same magnitude.* The liquid state consists of particles that are randomly packed but relatively near one another (Figure 7.3b). The molecules are in constant, random motion; they slide freely over one another but do not move with enough energy to separate. The fact that the particles freely slide over each other indicates the influence of disruptive forces; however, the fact that the particles do

Figure 7.3 (a) In a solid, the particles (atoms, molecules, or ions) are close together and vibrate about fixed sites. (b) The particles in a liquid, though still close together, freely slide over one another. (c) In a gas, the particles are in constant random motion, each particle being independent of the others present.

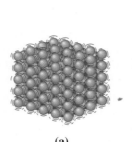

(a)

(b)

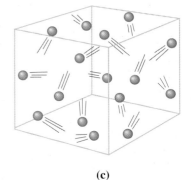

(c)

Figure 7.4 Gas molecules can be compared to billiard balls in random motion, bouncing off one another and off the sides of the pool table.

not separate indicates fairly strong cohesive forces. With this model, the characteristic properties of liquids (Table 7.1) can be explained as follows:

1. *Definite volume and indefinite shape.* The attractive forces are strong enough to restrict particles to movement within a definite volume. They are not strong enough, however, to prevent the particles from moving over each other in a random manner that is limited only by the container walls. Thus liquids have no definite shape except that they maintain a horizontal upper surface in containers that are not completely filled.
2. *High density.* The particles in a liquid are not widely separated; they are still touching one another. Therefore, there will be a large number of particles in a given volume— a high density.
3. *Small compressibility.* Because the particles in a liquid are still touching each other, there is very little empty space. Therefore, an increase in pressure cannot squeeze the particles much closer together.
4. *Small thermal expansion.* Most of the particle movement in a liquid is vibrational because a particle can move only a short distance before colliding with a neighbor. The increased particle velocity that accompanies a temperature increase results only in increased vibrational amplitudes. The net effect is an increase in the effective volume a particle occupies, which causes a slight volume increase in the liquid.

A **gas** *is the physical state characterized by a complete dominance of kinetic energy (disruptive forces) over potential energy (cohesive forces).* Attractive forces among particles are very weak and, as a first approximation, are considered to be zero. As a result, the particles of a gas move essentially independently of one another in a totally random manner (Figure 7.3c). Under ordinary pressure, the particles are relatively far apart, except when they collide with one another. In between collisions with one another or with the container walls, gas particles travel in straight lines (Figure 7.4).

The kinetic theory explanation of the properties of gases follows the same pattern that we saw earlier for solids and liquids.

1. *Indefinite volume and indefinite shape.* The attractive (cohesive) forces between particles have been overcome by high kinetic energy, and the particles are free to travel in all directions. Therefore, gas particles completely fill their container, and the shape of the gas is that of the container.
2. *Low density.* The particles of a gas are widely separated. There are relatively few particles in a given volume (compared with liquids and solids), which means little mass per volume (a low density).
3. *Large compressibility.* Particles in a gas are widely separated; essentially, a gas is mostly empty space. When pressure is applied, the particles are easily pushed closer together, decreasing the amount of empty space and the volume of the gas (see Figure 7.5).
4. *Moderate thermal expansion.* An increase in temperature means an increase in particle velocity. The increased kinetic energy of the particles enables them to push back whatever barrier is confining them into a given volume, and the volume increases. You will note that the size of the particles is not changed during expansion or compression of gases, solids, or liquids; they merely move either farther apart or closer together. It is the space between the particles that changes.

Gas at low pressure   Gas at higher pressure

Figure 7.5 When a gas is compressed, the amount of empty space in the container is decreased. The size of the molecules does not change; they simply move closer together.

CHEMICAL Connections

The Importance of Gas Densities

In the gaseous state, particles are approximately 10 times farther apart than in the solid or liquid state at a given temperature and pressure. Consequently, gases have densities much lower than those of solids and liquids. The fact that gases have low densities is a major factor in explaining many commonly encountered phenomena.

Popcorn pops because of the difference in density between liquid and gaseous water (1.0 g/mL versus 0.001 g/mL). As the corn kernels are heated, water within the kernels is converted into steam. The steam's volume, approximately 1000 times greater than that of the water from which it was generated, causes the kernels of corn to "blow up."

Changes in density that occur as a solid is converted to gases via a chemical reaction are the basis for the operation of automobile air bags and the effects of explosives. Automobile air bags are designed to inflate rapidly (in a fraction of a second) in the event of a crash and then to deflate immediately. Their

activation involves mechanical shock causing a steel ball to compress a spring that electronically ignites a detonator cap, which in turn causes solid sodium azide (NaN_3) to decompose. The decomposition reaction is

$$2NaN_3(s) \longrightarrow 2Na(l) + 3N_2(g)$$

The nitrogen gas so generated inflates the air bag. A small amount of NaN_3 (high density) will generate over 50 L of N_2 gas at 25°C. Because the air bag is porous, it goes limp quickly as the generated N_2 gas escapes.

Millions of hours of hard manual labor are saved annually by the use of industrial explosives in quarrying rock, constructing tunnels, and mining coal and metal ores. The active ingredient in dynamite, a heavily used industrial explosive, is nitroglycerin, whose destructive power comes from the generation of large volumes of gases at high temperatures. The reaction is

$$4C_3H_5O_3(NO_2)_3(s) \longrightarrow$$
Nitroglycerin
$$12CO_2(g) + 10H_2O(g) + 6N_2(g) + O_2(g)$$

At the temperature of the explosion, about 5000°C, there is an approximately 20,000-fold increase in volume as the result of density changes. No wonder such explosives can blow materials to pieces!

The density difference associated with temperature change is the basis for the operation of hot air balloons. Hot air, which is less dense than cold air, rises.

Weather balloons and blimps are filled with helium, a gas less dense than air. Thus, such objects rise in air.

Water vapor is less dense than air. Thus, moist air is less dense than dry air. Decreasing barometric pressure (from lower-density moist air) is an indication that a storm front is approaching.

7.3 GAS LAW VARIABLES

The behavior of a gas can be described reasonably well by *simple* quantitative relationships called *gas laws*. A **gas law** is *a generalization that describes in mathematical terms the relationships among the amount, pressure, temperature, and volume of a gas.*

Gas laws involve four variables: amount, pressure, temperature, and volume. Three of these four variables (amount, volume, and temperature) have been previously discussed (Sections 6.2, 2.2, and 2.10, respectively). Amount is usually specified in terms of *moles* of gas present. The units *liter* and *milliliter* are generally used in specifying gas volume. Only one of the three temperature scales discussed in Section 2.10, the *Kelvin scale*, can be used in gas law calculations if the results are to be valid. We have not yet discussed pressure, the fourth gas law variable. The remainder of this section consists of a discussion of pressure. **Pressure** is *the force applied per unit area on an object—that is, the total force on a surface divided by the area of that surface.* The mathematical equation for pressure is

$$P(\text{pressure}) = \frac{F(\text{force})}{A(\text{area})}$$

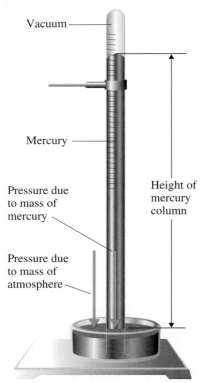

Figure 7.6 The essential components of a mercury barometer are a graduated glass tube, a glass dish, and liquid mercury.

"Millimeters of mercury" is the pressure unit most often encountered in clinical work in allied health fields. For example, oxygen and carbon dioxide pressures in respiration are almost always specified in millimeters of mercury.

Blood pressure is measured with the aid of an apparatus known as a sphygmomanometer, which is essentially a barometer tube connected to an inflatable cuff by a hollow tube. A typical blood pressure is 120/80; this ratio means a systolic pressure of 120 mm Hg above atmospheric pressure and a diastolic pressure of 80 mm Hg above atmospheric pressure.

For a gas, the force that creates pressure is that which is exerted by the gas molecules or atoms as they constantly collide with the walls of their container. Barometers, manometers, and gauges are the instruments most commonly used to measure gas pressures.

The air that surrounds Earth exerts pressure on every object it touches. A **barometer** *is a device used to measure atmospheric pressure.* The essential components of a simple barometer are shown in Figure 7.6. Atmospheric pressure is expressed in terms of the height of the barometer's mercury column, usually *in millimeters of mercury* (mm Hg). Another name for millimeters of mercury is *torr,* used in honor of Evangelista Torricelli (1608–1647), the Italian physicist who invented the barometer.

$$1 \text{ mm Hg} = 1 \text{ torr}$$

Atmospheric pressure varies with the weather and the altitude. It averages about 760 mm Hg at sea level, and it decreases by approximately 25 mm Hg for every 1000-ft increase in altitude. The pressure unit *atmosphere* (atm) is defined in terms of this average pressure at sea level. By definition,

$$1 \text{ atm} = 760 \text{ mm Hg} = 760 \text{ torr}$$

Another commonly used pressure unit is *pounds per square inch* (psi or lb/in^2). One atmosphere is equal to 14.7 psi.

$$1 \text{ atm} = 14.7 \text{ psi}$$

Pressure Readings and Significant Figures

Standard procedure in obtaining pressures that are based on the height of a column of mercury (barometric readings) is to estimate the column height to the closest millimeter. Thus such pressure readings have an uncertainty in the "ones place," that is, to the closest millimeter of mercury. The preceding operational procedure means that millimeter of mercury pressure readings such as 750, 730, and 650 are considered to have three significant figures even though no decimal point is explicitly shown after the zero (Section 2.4). Likewise, a pressure reading of 700 mm Hg or 600 mm Hg is considered to possess three significant figures.

7.4   BOYLE'S LAW: A PRESSURE-VOLUME RELATIONSHIP

Of the several relationships that exist among gas law variables, the first to be discovered relates gas pressure to gas volume. It was formulated over 300 years ago, in 1662, by the British chemist and physicist Robert Boyle (Figure 7.7). **Boyle's law** states that *the volume of a fixed amount of a gas is inversely proportional to the pressure applied to the gas if the temperature is kept constant.* This means that if the pressure on the gas increases, the volume decreases proportionally; conversely, if the pressure decreases, the volume increases. Doubling the pressure cuts the volume in half; tripling the pressure reduces the volume to one-third its original value; quadrupling the pressure reduces the volume to one-fourth its original value; and so on. Figure 7.8 illustrates Boyle's law.

The mathematical equation for Boyle's law is

$$P_1 \times V_1 = P_2 \times V_2$$

where P_1 and V_1 are the pressure and volume of a gas at an initial set of conditions, and P_2 and V_2 are the pressure and volume of the same sample of gas under a new set of conditions, with the temperature and amount of gas remaining constant.

EXAMPLE 7.1

Using Boyle's Law to Calculate the New Volume of a Gas

▶ A sample of O_2 gas occupies a volume of 1.50 L at a pressure of 735 mm Hg and a temperature of 25°C. What volume will it occupy, in liters, if the pressure is increased to 770 mm Hg with no change in temperature?

Solution

A suggested first step in working gas law problems that involve two sets of conditions is to analyze the given data in terms of initial and final conditions.

$$P_1 = 735 \text{ mm Hg} \qquad P_2 = 770 \text{ mm Hg}$$
$$V_1 = 1.50 \text{ L} \qquad V_2 = ? \text{ L}$$

When we know any three of the four quantities in the Boyle's law equation, we can calculate the fourth, which is usually the final pressure, P_2, or the final volume, V_2. The Boyle's law equation is valid only if the temperature and amount of the gas remain constant.

We know three of the four variables in the Boyle's law equation, so we can calculate the fourth, V_2. We will rearrange Boyle's law to isolate V_2 (the quantity to be calculated) on one side of the equation. This is accomplished by dividing both sides of the Boyle's law equation by P_2.

$$P_1V_1 = P_2V_2 \qquad \text{(Boyle's law)}$$

$$\frac{P_1V_1}{P_2} = \frac{P_2V_2}{P_2} \qquad \begin{array}{l}\text{(Divide each side of} \\ \text{the equation by } P_2.)\end{array}$$

$$V_2 = V_1 \times \frac{P_1}{P_2}$$

Substituting the given data into the rearranged equation and doing the arithmetic give

$$V_2 = 1.50 \text{ L} \times \left(\frac{735 \text{ mm Hg}}{770 \text{ mm Hg}}\right) = 1.43 \text{ L}$$

Practice Exercise 7.1

A sample of H_2 gas occupies a volume of 2.25 L at a pressure of 628 mm Hg and a temperature of 35°C. What volume will it occupy, in liters, if the pressure is decreased to 428 mm Hg with no change in temperature?

Answer: 3.30 L H_2

Figure 7.7 Robert Boyle (1627–1691), like most men of the seventeenth century who devoted themselves to science, was self-taught. It was through his efforts that the true value of experimental investigation was first recognized.

Boyle's law is consistent with kinetic molecular theory. The pressure that a gas exerts results from collisions of the gas molecules with the sides of the container. If the volume of a container holding a specific number of gas molecules is increased, the total wall area of the container will also increase, and the number of collisions in a given area (the pressure) will decrease because of the greater wall area. Conversely, if the volume of the container is decreased, the wall area will be smaller and there will be more collisions within a given wall area. Figure 7.9 illustrates this concept.

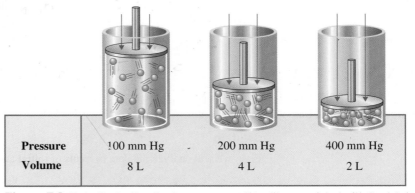

Pressure	100 mm Hg	200 mm Hg	400 mm Hg
Volume	8 L	4 L	2 L

Figure 7.8 Data illustrating the inverse proportionality associated with Boyle's law.

Figure 7.9 When the volume of a gas at constant temperature decreases by half (a), the average number of times a molecule hits the container walls is doubled (b).

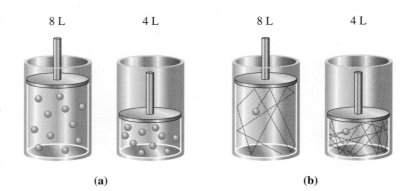

(a)   (b)

Boyle's law explains the process of breathing. Breathing in occurs when the diaphragm flattens out (contracts). This contraction causes the volume of the thoracic cavity to increase and the pressure within the cavity to drop (Boyle's law) below atmospheric pressure. Air flows into the lungs and expands them, because the pressure is greater outside the lungs than within them. Breathing out occurs when the diaphragm relaxes (moves up), decreasing the volume of the thoracic cavity and increasing the pressure (Boyle's law) within the cavity to a value greater than the external pressure. Air flows out of the lungs. The air flow direction is always from a high-pressure region to a low-pressure region.

Filling a medical syringe with a liquid demonstrates Boyle's law. As the plunger is drawn out of the syringe (see Figure 7.10), the increase in volume inside the syringe chamber results in decreased pressure there. The liquid, which is at atmospheric pressure, flows into this reduced-pressure area. This liquid is then expelled from the chamber by pushing the plunger back in. This ejection of the liquid does not involve Boyle's law; a liquid is incompressible, and mechanical force pushes it out.

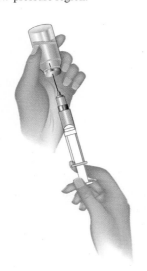

Figure 7.10 Filling a syringe with a liquid is an application of Boyle's law.

7.5 CHARLES'S LAW: A TEMPERATURE-VOLUME RELATIONSHIP

The relationship between the temperature and the volume of a gas at constant pressure is called *Charles's law* after the French scientist Jacques Charles (Figure 7.11). This law was discovered in 1787, over 100 years after the discovery of Boyle's law. **Charles's law** states that *the volume of a fixed amount of gas is directly proportional to its Kelvin temperature if the pressure is kept constant* (Figure 7.12). Whenever a *direct* proportion exists between two quantities, one increases when the other increases and one decreases when the other decreases. The direct-proportion relationship of Charles's law means that if the temperature increases, the volume will also increase and that if the temperature decreases, the volume will also decrease.

A balloon filled with air illustrates Charles's law. If the balloon is placed near a heat source such as a light bulb that has been on for some time, the heat will cause the balloon to increase visibly in size (volume). Putting the same balloon in the refrigerator will cause it to shrink.

Charles's law, stated mathematically, is

$$\frac{V_1}{T_1} = \frac{V_2}{T_2}$$

where V_1 is the volume of a gas at a given pressure, T_1 is the Kelvin temperature of the gas, and V_2 and T_2 are the volume and Kelvin temperature of the gas under a new set of conditions, with the pressure remaining constant.

 E X A M P L E   7.2

Using Charles's Law to Calculate the New Volume of a Gas

A sample of the gaseous anesthetic cyclopropane, with a volume of 425 mL at a temperature of 27°C, is cooled at constant pressure to 20°C. What is the new volume, in milliliters, of the sample?

Solution

First, we will analyze the data in terms of initial and final conditions.

$$V_1 = 425 \text{ mL} \qquad\qquad V_2 = ? \text{ mL}$$
$$T_1 = 27°C + 273 = 300 \text{ K} \qquad T_2 = 20°C + 273 = 293 \text{ K}$$

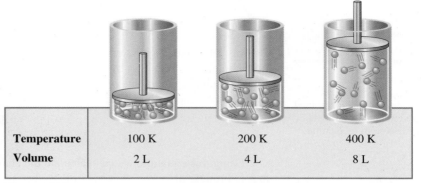

Temperature	100 K	200 K	400 K
Volume	2 L	4 L	8 L

Figure 7.12 Data illustrating the direct proportionality associated with Charles's law.

Figure 7.11 Jacques Charles (1746–1823), a French physicist, in the process of working with hot-air balloons, made the observations that ultimately led to the formulation of what is now known as Charles's law.

When you use the mathematical form of Charles's law, the temperatures used *must be* Kelvin scale temperatures.

Charles's law predicts that gas volume will become smaller and smaller as temperature is reduced, until eventually a temperature is reached at which gas volume becomes zero. This "zero-volume" temperature is calculated to be $-273°C$ and is known as *absolute zero* (see Section 2.10). Absolute zero is the basis for the Kelvin temperature scale. In reality, gas volume never vanishes. As temperature is lowered, at some point before absolute zero, the gas condenses to a liquid, at which point Charles's law is no longer valid.

Note that both of the given temperatures have been converted to Kelvin scale readings. This change is accomplished by simply adding 273 to the Celsius scale value (Section 2.10).

We know three of the four variables in the Charles's law equation, so we can calculate the fourth, V_2. We will rearrange Charles's law to isolate V_2 (the quantity desired) by multiplying each side of the equation by T_2.

$$\frac{V_1}{T_1} = \frac{V_2}{T_2} \qquad \text{(Charles's law)}$$

$$\frac{V_1 T_2}{T_1} = \frac{V_2 \cancel{T_2}}{\cancel{T_2}} \qquad \text{(Multiply each side by } T_2.\text{)}$$

$$V_2 = V_1 \times \frac{T_2}{T_1}$$

Substituting the given data into the equation and doing the arithmetic give

$$V_2 = 425 \text{ mL} \times \left(\frac{293 \text{ K}}{300 \text{ K}} \right) = 415 \text{ mL}$$

Practice Exercise 7.2

A sample of dry air, with a volume of 125 mL at a temperature of 53°C, is heated at constant pressure to 95°C. What is the new volume, in milliliters, of the sample?

Answer: 141 mL air

Charles's law is consistent with kinetic molecular theory. When the temperature of a gas increases, the kinetic energy (velocity) of the gas molecules increases. The speedier particles hit the container walls harder. In order for the pressure of the gas to remain constant, the container volume must increase. This will result in fewer particles hitting a unit area of wall at a given instant. A similar argument applies when the temperature of a gas is lowered. This time the velocity of the molecules decreases, and the wall area (volume) must also decrease in order to increase the number of collisions in a given area in a given time.

Charles's law is the principle used in the operation of a convection heater. When air comes in contact with the heating element, it expands (its density becomes less). The hot, less dense air rises, causing continuous circulation of warm air. This same principle has ramifications in closed rooms that lack effective air circulation. The warmer and less dense air stays near the top of the room. This is desirable in the summer but not in the winter.

7.6 THE COMBINED GAS LAW

Boyle's and Charles's laws can be mathematically combined to give a more versatile equation than either of the laws by themselves. The **combined gas law** states that *the product of the pressure and volume of a fixed amount of gas is directly proportional to its Kelvin temperature*. The mathematical equation for the combined gas law is

$$\frac{P_1 V_1}{T_1} = \frac{P_2 V_2}{T_2}$$

Using this equation, we can calculate the change in pressure, temperature, or volume that is brought about by changes in the other two variables.

> Any time a gas law contains temperature terms, as is the case for both Charles's law and the combined gas law, these temperatures must be specified on the Kelvin temperature scale.

 EXAMPLE 7.3

Using the Combined Gas Law to Calculate the New Volume of a Gas

A sample of O_2 gas occupies a volume of 1.62 L at 755 mm Hg pressure and a temperature of 0°C. What volume, in liters, will this gas sample occupy at 725 mm Hg pressure and 50°C?

Solution

First, we analyze the data in terms of initial and final conditions.

$$P_1 = 755 \text{ mm Hg} \qquad P_2 = 725 \text{ mm Hg}$$
$$V_1 = 1.62 \text{ L} \qquad V_2 = ? \text{ L}$$
$$T_1 = 0°C + 273 = 273 \text{ K} \quad T_2 = 50°C + 273 = 323 \text{ K}$$

We are given five of the six variables in the combined gas law, so we can calculate the sixth one, V_2. Rearranging the combined gas law to isolate the variable V_2 on a side by itself gives

$$V_2 = \frac{V_1 P_1 T_2}{P_2 T_1}$$

Substituting numerical values into this "version" of the combined gas law gives

$$V_2 = 1.62 \text{ L} \times \frac{755 \text{ mm Hg}}{725 \text{ mm Hg}} \times \frac{323 \text{ K}}{273 \text{ K}} = 2.00 \text{ L}$$

Practice Exercise 7.3

A helium-filled weather balloon, when released, has a volume of 10.0 L at 27°C and a pressure of 663 mm Hg. What volume, in liters, will the balloon occupy at an altitude where the pressure is 96 mm Hg and the temperature is −30.0°C?

Answer: 56 L He

7.7 THE IDEAL GAS LAW

The three gas laws so far considered in this chapter are used to describe gaseous systems where change occurs. Two sets of conditions, with one unknown variable, is the common feature of the systems they describe. It is also useful to have a gas law that describes a gaseous system where no changes in condition occur. Such a law exists and is known as the *ideal gas law*. The **ideal gas law** *is a gas law that describes the relationships among the four variables temperature, pressure, volume, and molar amount for a gaseous substance at a given set of conditions*.

Mathematically, the ideal gas law has the form

$$PV = nRT$$

In this equation, pressure, temperature, and volume are defined in the same manner as in the gas laws we have already discussed. The symbol n stands for the *number of moles* of gas present in the sample. The symbol R represents the *ideal gas constant*, the proportionality constant that makes the equation valid.

> The ideal gas law is used in calculations when *one* set of conditions is given with one missing variable. The combined gas law (Section 7.6) is used when *two* sets of conditions are given with one missing variable.

The value of the ideal gas constant (R) varies with the units chosen for pressure and volume. With pressure in atmospheres and volume in liters, R has the value

$$R = \frac{PV}{nT} = 0.0821 \frac{atm \cdot L}{mole \cdot K}$$

The value of R is the same for all gases under normally encountered conditions of temperature, pressure, and volume.

If three of the four variables in the ideal gas law equation are known, then the fourth can be calculated using the equation. Example 7.4 illustrates the use of the ideal gas law.

EXAMPLE 7.4

Using the Ideal Gas Law to Calculate the Volume of a Gas

The colorless, odorless, tasteless gas carbon monoxide, CO, is a by-product of incomplete combustion of any material that contains the element carbon. Calculate the volume, in liters, occupied by 1.52 moles of this gas at 0.992 atm pressure and a temperature of 65°C.

Solution

This problem deals with only one set of conditions, so the ideal gas equation is applicable. Three of the four variables in the ideal gas equation (P, n, and T) are given, and the fourth (V) is to be calculated.

$$P = 0.992 \text{ atm} \qquad n = 1.52 \text{ moles}$$
$$V = ? \text{ L} \qquad T = 65°C = 338 \text{ K}$$

Rearranging the ideal gas equation to isolate V on the left side of the equation gives

$$V = \frac{nRT}{P}$$

Because the pressure is given in atmospheres and the volume unit is liters, the R value 0.0821 is valid. Substituting known numerical values into the equation gives

$$V = \frac{(1.52 \text{ moles}) \times \left(0.0821 \frac{atm \cdot L}{mole \cdot K} \right)(338 \text{ K})}{0.992 \text{ atm}}$$

Note that all the parts of the ideal gas constant unit cancel except for one, the volume part. Doing the arithmetic yields the volume of CO.

$$V = \left(\frac{1.52 \times 0.0821 \times 338}{0.992} \right) L = 42.5 \text{ L}$$

Practice Exercise 7.4

Calculate the volume, in liters, occupied by 3.25 moles of Cl_2 gas at 1.54 atm pressure and a temperature of 213°C.

Answer: 84.2 L Cl_2

7.8 DALTON'S LAW OF PARTIAL PRESSURES

In a mixture of gases that do not react with one another, each type of molecule moves around in the container as though the other kinds were not there. This type of behavior is possible because a gas is mostly empty space, and attractions between molecules in the gaseous state are negligible at most temperatures and pressures. Each gas in the mixture occupies the entire volume of the container; that is, it distributes itself uniformly throughout the container. The molecules of each type strike the walls of the container as frequently and with the same energy as though they were the only gas in the mixture. Consequently, the pressure exerted by each gas in a mixture is the same as it would be if the gas were alone in the same container under the same conditions.

The English scientist John Dalton (Figure 7.13) was the first to notice the independent behavior of gases in mixtures. In 1803, he published a summary statement concerning this behavior that is now known as Dalton's law of partial pressures. **Dalton's law of partial**

Figure 7.13 John Dalton (1766–1844) throughout his life had a particular interest in the study of weather. From "weather" he turned his attention to the nature of the atmosphere and then to the study of gases in general.

A sample of clean air is the most common example of a mixture of gases that do not react with one another.

Figure 7.14 A set of four identical containers can be used to illustrate Dalton's law of partial pressures. The pressure in the fourth container (the mixture of gases) is equal to the sum of the pressures in the first three containers (the individual gases).

| P_A | + | P_B | + | P_C | = | P_{Total} |
| 1 | + | 3 | + | 2 | = | 6 |

pressures states that *the total pressure exerted by a mixture of gases is the sum of the partial pressures of the individual gases present.* A **partial pressure** *is the pressure that a gas in a mixture of gases would exert if it were present alone under the same conditions.*

Expressed mathematically, Dalton's law states that

$$P_{Total} = P_1 + P_2 + P_3 + \cdots$$

where P_{Total} is the total pressure of a gaseous mixture and P_1, P_2, P_3, and so on are the partial pressures of the individual gaseous components of the mixture.

As an illustration of Dalton's law, consider the four identical gas containers shown in Figure 7.14. Suppose we place amounts of three different gases (represented by *A, B,* and *C*) into three of the containers and measure the pressure exerted by each sample. We then place all three samples in the fourth container and measure the pressure exerted by this mixture of gases. We find that

$$P_{Total} = P_A + P_B + P_C$$

 **EXAMPLE 7.5**

Using Dalton's Law to Calculate a Partial Pressure

The total pressure exerted by a mixture of the three gases oxygen, nitrogen, and water vapor is 742 mm Hg. The partial pressures of the nitrogen and oxygen in the sample are 581 mm Hg and 143 mm Hg, respectively. What is the partial pressure of the water vapor present in the mixture?

Solution

Dalton's law says that

$$P_{Total} = P_{N_2} + P_{O_2} + P_{H_2O}$$

The known values for variables in this equation are

$$P_{Total} = 742 \text{ mm Hg}$$
$$P_{N_2} = 581 \text{ mm Hg}$$
$$P_{O_2} = 143 \text{ mm Hg}$$

Rearranging Dalton's law to isolate P_{H_2O} on the left side of the equation gives

$$P_{H_2O} = P_{Total} - P_{N_2} - P_{O_2}$$

Substituting the known numerical values into this equation and doing the arithmetic give

$$P_{H_2O} = 742 \text{ mm Hg} - 581 \text{ mm Hg} - 143 \text{ mm Hg} = 18 \text{ mm Hg}$$

Practice Exercise 7.5

A gaseous mixture contains the three noble gases He, Ar, and Kr. The total pressure exerted by the mixture is 1.57 atm, and the partial pressures of the He and Ar are 0.33 atm and 0.39 atm, respectively. What is the partial pressure of the Kr in the mixture?

Answer: 0.85 atm for Kr

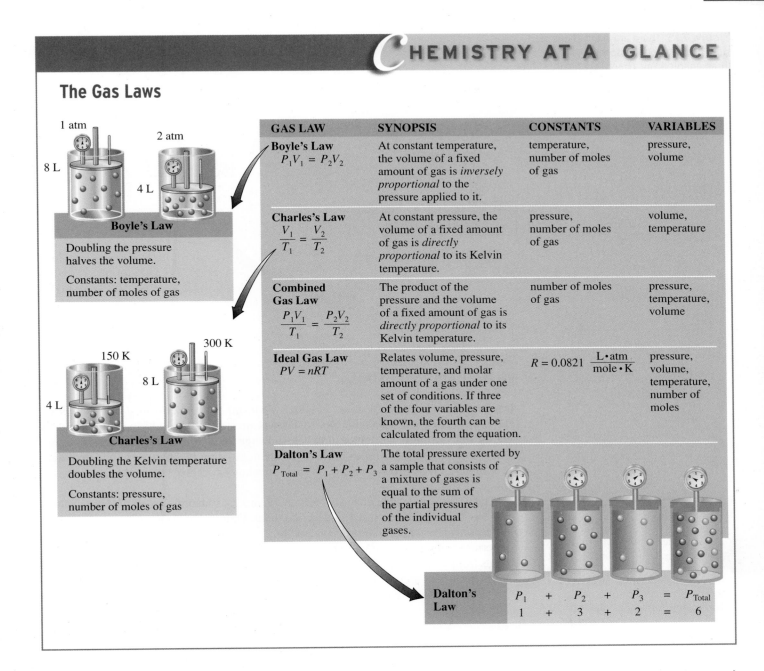

CHEMISTRY AT A GLANCE

The Gas Laws

Boyle's Law

1 atm — 8 L

2 atm — 4 L

Doubling the pressure halves the volume.

Constants: temperature, number of moles of gas

Charles's Law

150 K — 4 L

300 K — 8 L

Doubling the Kelvin temperature doubles the volume.

Constants: pressure, number of moles of gas

GAS LAW	SYNOPSIS	CONSTANTS	VARIABLES
Boyle's Law $P_1V_1 = P_2V_2$	At constant temperature, the volume of a fixed amount of gas is *inversely proportional* to the pressure applied to it.	temperature, number of moles of gas	pressure, volume
Charles's Law $\dfrac{V_1}{T_1} = \dfrac{V_2}{T_2}$	At constant pressure, the volume of a fixed amount of gas is *directly proportional* to its Kelvin temperature.	pressure, number of moles of gas	volume, temperature
Combined Gas Law $\dfrac{P_1V_1}{T_1} = \dfrac{P_2V_2}{T_2}$	The product of the pressure and the volume of a fixed amount of gas is *directly proportional* to its Kelvin temperature.	number of moles of gas	pressure, temperature, volume
Ideal Gas Law $PV = nRT$	Relates volume, pressure, temperature, and molar amount of a gas under one set of conditions. If three of the four variables are known, the fourth can be calculated from the equation.	$R = 0.0821 \dfrac{L \cdot atm}{mole \cdot K}$	pressure, volume, temperature, number of moles
Dalton's Law $P_{Total} = P_1 + P_2 + P_3$	The total pressure exerted by a sample that consists of a mixture of gases is equal to the sum of the partial pressures of the individual gases.		

Dalton's Law

P_1	+	P_2	+	P_3	=	P_{Total}
1	+	3	+	2	=	6

Using the actual gauge pressure values given in Figure 7.14, we see that

$$P_{Total} = 1 + 3 + 2 = 6$$

Dalton's law of partial pressures is important when we consider the air of our atmosphere, which is a mixture of numerous gases. At higher altitudes, the total pressure of air decreases, as do the partial pressures of the individual components of air. An individual going from sea level to a higher altitude usually experiences some tiredness because his or her body is not functioning as efficiently at the higher altitude. At higher elevation, the red blood cells absorb a smaller amount of oxygen because the oxygen partial pressure at the higher altitude is lower. A person's body acclimates itself to the higher altitude after a period of time as additional red blood cells are produced by the body.

The Chemistry at a Glance feature above summarizes key concepts about the gas laws we have considered in this chapter.

Figure 7.15 There are six changes of state possible for substances. The three endothermic changes, which require the input of heat, are melting, evaporation, and sublimation. The three exothermic changes, which release heat, are freezing, condensation, and deposition.

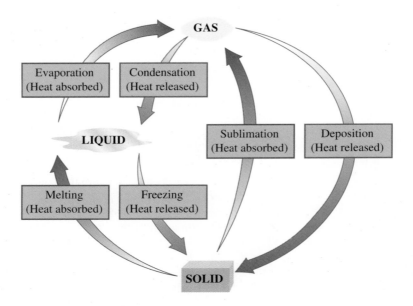

7.9 CHANGES OF STATE

A **change of state** *is a process in which a substance is transformed from one physical state to another physical state.* Changes of state are usually accomplished by heating or cooling a substance. Pressure change is also a factor in some systems. Changes of state are examples of physical changes—that is, changes in which chemical composition remains constant. No new substances are ever formed as a result of a change of state.

There are six possible changes of state. Figure 7.15 identifies each of these changes and gives the terminology used to describe them. Four of the six terms used in describing state changes are familiar: freezing, melting, evaporation, and condensation. The other two terms—sublimation and deposition—are not so common. *Sublimation* is the direct change from the solid to the gaseous state; *deposition* is the reverse of this, the direct change from the gaseous to the solid state (Figure 7.16).

Changes of state are classified into two categories based on whether heat (thermal energy) is given up or absorbed during the change process. An **endothermic change of state** *is a change of state in which heat energy is absorbed.* The endothermic changes of

Although the processes of sublimation and deposition are not common, they are encountered in everyday life. Dry ice sublimes, as do mothballs placed in a clothing storage area. It is because of sublimation that ice cubes left in a freezer get smaller as time passes. Ice or snow forming in clouds (from water vapor) during the winter season is an example of deposition.

Figure 7.16 Sublimation and deposition of iodine. (a) The beaker contains iodine crystals, I_2; a dish of ice rests on top of the beaker. (b) Iodine has an appreciable vapor pressure even below its melting point (114°C); thus, when heated carefully, the solid sublimes without melting. The vapor deposits crystals on the cool underside of the dish, the process of deposition.

(a)   (b)

state are melting, sublimation, and evaporation. An **exothermic change of state** *is a change of state in which heat energy is given off.* Exothermic changes of state are the reverse of endothermic changes of state; they are freezing, condensation, and deposition.

7.10 EVAPORATION OF LIQUIDS

Evaporation *is the process by which molecules escape from the liquid phase to the gas phase.* We are all aware that water left in an open container at room temperature slowly disappears by evaporation.

Evaporation can be explained using kinetic molecular theory. Statement 4 of this theory (Section 7.1) indicates that not all the molecules in a liquid (or solid or gas) possess the same kinetic energy. At any given instant, some molecules will have above-average kinetic energies and others will have below-average kinetic energies as a result of collisions between molecules. A given molecule's energy constantly changes as a result of collisions with neighboring molecules. When molecules that happen to be considerably above average in kinetic energy at a given moment are on the liquid surface and are moving in a favorable direction relative to the surface, they can overcome the attractive forces (potential energy) holding them in the liquid and escape.

Evaporation is a surface phenomenon. Surface molecules are subject to fewer attractive forces because they are not completely surrounded by other molecules; thus escape is much more probable. Liquid surface area is an important factor to consider when determining the rate at which evaporation occurs. Increased surface area results in an increased evaporation rate because a greater fraction of the total molecules are on the surface.

> For a liquid to evaporate, its molecules must gain enough kinetic energy to overcome the attractive forces among them.

Rate of Evaporation and Temperature

Water evaporates faster from a glass of hot water than from a glass of cold water, because a certain minimum kinetic energy is required for molecules to escape from the attractions of neighboring molecules. As the temperature of a liquid increases, a larger fraction of the molecules present acquire this minimum kinetic energy. Consequently, the rate of evaporation always increases as liquid temperature increases.

The escape of high-energy molecules from a liquid during evaporation affects the liquid in two ways: The amount of liquid decreases, and the liquid temperature is lowered. The lower temperature reflects the loss of the most energetic molecules. (Analogously, when all the tall people are removed from a classroom of students, the average height of the remaining students decreases.) A lower average kinetic energy corresponds to a lower temperature (statement 4 of the kinetic molecular theory); hence a cooling effect is produced.

> Evaporative cooling is important in many processes. Our own bodies use evaporation to maintain a constant temperature. We perspire in hot weather, and evaporation of the perspiration cools our skin. The cooling effect of evaporation is quite noticeable when one first comes out of an outdoor swimming pool on a hot, breezy day.

The molecules that escape from an evaporating liquid are often collectively referred to as vapor, rather than gas. A **vapor** *is a gas that exists at a temperature and pressure at which it ordinarily would be thought of as a liquid or solid.* For example, at room temperature and atmospheric pressure, the normal state for water is the liquid state. Molecules that escape (evaporate) from liquid water at these conditions are frequently called *water vapor.*

7.11 VAPOR PRESSURE OF LIQUIDS

The evaporative behavior of a liquid in a closed container is quite different from its behavior in an open container. Some liquid evaporation occurs in a closed container; this is indicated by a drop in liquid level. However, unlike the liquid level in an open-container system, the liquid level in a closed-container system eventually ceases to drop (becomes constant).

Blood Pressure and the Sodium Ion/Potassium Ion Ratio

In a manner similar to gases, liquids also exert a pressure on the walls of their container. Thus, blood exerts a pressure on the body's blood vessels as it moves throughout the body. Such pressure, generated by a contracting heart, is necessary to move the blood to all parts of the body. The pressure that blood exerts within blood vessels is an important indicator of health. If the pressure is too low, dizziness from a shortage of oxygen-carrying blood to the brain can result. If it is too high, the risk of kidney damage, stroke, and heart failure increases.

Blood pressure readings are reported as a ratio of two numbers, such as 120/80. Such numbers represent pressures in terms of the height of a column of mercury (in millimeters) that the pressure can support. The higher of the two numbers in a blood pressure reading (the systolic pressure) represents pressure when the heart contracts, pushing blood into the arteries. The smaller number (the diastolic pressure) represents pressure when the heart is "resting" between contractions. Normal range systolic values are 100–120 mm Hg for young adults and 115–135 mm Hg for older adults. The corresponding normal diastolic ranges are 60–80 mm Hg and 75–85 mm Hg, respectively.

High blood pressure, or hypertension, occurs in an estimated one-third of the U.S. population. While high blood pressure by itself doesn't make a person feel sick, it is the most common risk factor for heart disease. And heart disease is the leading cause of death in the United States. Hypertension forces the heart to work too hard, and it damages blood vessels.

Besides taking medication, factors known to help reduce high blood pressure include increasing physical activity, losing weight, decreasing the consumption of alcohol, and limiting the intake of sodium.

The major dietary source of sodium is sodium chloride (NaCl, table salt), the world's most common food additive. Most people find its taste innately appealing. Salt use tends to enhance other flavors, probably by suppressing the bitter flavors. In general, processed foods contain the most sodium chloride, and unprocessed foods, such as fresh fruits and vegetables, contain the least. Studies on the sodium content of foods show that as much as 75% of it is added during processing and manufacturing, 15% comes from salt added during cooking and at the table, and only 10% is naturally present in the food.

Recent research indicates that sodium's contribution to hypertension may be more complex than was originally thought. More important than total sodium intake (in the form of Na^+ ion) is the dietary sodium ion/potassium ion (Na^+/K^+) ratio. Ideally, this ratio should be about 0.6, meaning significantly more potassium compared to sodium is needed. The Na^+/K^+ ratio in a typical American diet is about 1.05.

Increasing potassium in our diet and at the same time decreasing sodium has a positive effect on reducing hypertension. The following two tables list low sodium ion/high potassium ion foods (desirable) and high sodium ion/low potassium ion foods (undesirable).

Low Sodium Ion/High Potassium Ion Foods (Desirable)

Food Category	Examples
Fruit and fruit juices	Pineapple, grapefruit, pears, strawberries, watermelon, raisins, bananas, apricots, oranges
Low-sodium cereals	Oatmeal (unsalted), shredded wheat
Nuts (unsalted)	Hazelnuts, macadamia nuts, almonds, peanuts, cashews, coconut
Vegetables	Summer squash, zucchini, eggplant, cucumber, onions, lettuce, green beans, broccoli
Beans (dry, cooked)	Great Northern beans, lentils, lima beans, red kidney beans

High Sodium Ion/Low Potassium Ion Foods (Undesirable)

Food Category	Examples
Fats	Butter, margarine, salad dressings
Soups	Onion, mushroom, chicken noodle, tomato, split pea
Breakfast cereals	Many varieties; consult the label for specific nutritional information.
Breads	Most varieties
Processed meats	Most varieties
Cheese	Most varieties

Kinetic molecular theory explains these observations in the following way. The molecules that evaporate in a closed container do not leave the system as they do in an open container. They find themselves confined in a fixed space immediately above the liquid (see Figure 7.17a). These trapped vapor molecules undergo many random collisions with the container walls, other vapor molecules, and the liquid surface. Molecules that collide with the liquid surface are recaptured by the liquid. Thus two processes, evaporation (escape) and condensation (recapture), take place in a closed container (see Fig. 7.17b).

For a short time, the rate of evaporation in a closed container exceeds the rate of condensation, and the liquid level drops. However, as more of the liquid evaporates, the number of vapor molecules increases; the chance of their recapture through striking the

Figure 7.17 In the evaporation of a liquid in a closed container (a), the liquid level drops for a time (b) and then becomes constant (ceases to drop). At that point a state of equilibrium has been reached in which the rate of evaporation equals the rate of condensation (c).

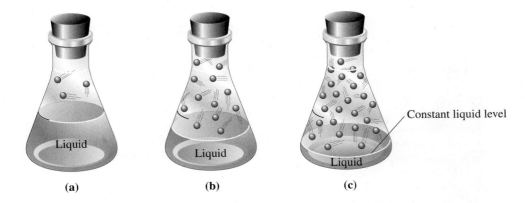

Constant liquid level

(a)   (b)   (c)

Remember that for a system at equilibrium, change at the molecular level is still occurring even though you cannot see it.

liquid surface also increases. Eventually, the rate of condensation becomes equal to the rate of evaporation, and the liquid level stops dropping (see Figure 7.17c). At this point, the number of molecules that escape in a given time is the same as the number recaptured; a steady-state situation has been reached. The amounts of liquid and vapor in the container do not change, even though both evaporation and condensation are still occurring.

This steady-state situation, which will continue as long as the temperature of the system remains constant, is an example of physical equilibrium. **Equilibrium** *is a condition in which two opposite processes take place at the same rate.* For systems in a state of equilibrium, no net macroscopic changes can be detected. However, the system is dynamic; the forward and reverse processes are occurring at equal rates.

When there is a liquid–vapor equilibrium in a closed container, the vapor in the fixed space immediately above the liquid exerts a constant pressure on both the liquid surface and the walls of the container. This pressure is called the *vapor pressure* of the liquid. **Vapor pressure** *is the pressure exerted by a vapor above a liquid when the liquid and vapor are in equilibrium with each other.*

The magnitude of a vapor pressure depends on the nature and temperature of the liquid. Liquids that have strong attractive forces between molecules have lower vapor pressures than liquids that have weak attractive forces between particles. Substances that have high *vapor pressures* (weak attractive forces) evaporate readily—that is, they are *volatile.* A **volatile substance** *is a substance that readily evaporates at room temperature because of a high vapor pressure.* Gasoline is a substance whose components are very volatile.

The vapor pressure of all liquids increases with temperature because an increase in temperature results in more molecules having the minimum kinetic energy required for evaporation. Table 7.2 shows the variation in vapor pressure, as temperature increases, of water.

TABLE 7.2
Vapor Pressure of Water at Various Temperatures

Temperature (°C)	Vapor Pressure (mm Hg)	Temperature (°C)	Vapor Pressure (mm Hg)
0	4.6	50	92.5
10	9.2	60	149.4
20	17.5	70	233.7
25*	23.8	80	355.1
30	31.8	90	525.8
37†	37.1	100	760.0
40	55.3		

*Room temperature
†Body temperature

Figure 7.18 Bubbles of vapor form within a liquid when the temperature of the liquid reaches the liquid's boiling point.

 BOILING AND BOILING POINT

In order for a molecule to escape from the liquid state, it usually must be on the surface of the liquid. **Boiling** *is a form of evaporation where conversion from the liquid state to the vapor state occurs within the body of the liquid through bubble formation.* This phenomenon begins to occur when the vapor pressure of a liquid, which steadily increases as the liquid is heated, reaches a value equal to that of the prevailing external pressure on the liquid; for liquids in open containers, this value is atmospheric pressure. When these two pressures become equal, bubbles of vapor form around any speck of dust or around any irregularity associated with the container surface (Figure 7.18). These vapor bubbles quickly rise to the surface and escape because they are less dense than the liquid itself. We say the liquid is boiling.

A **boiling point** *is the temperature at which the vapor pressure of a liquid becomes equal to the external (atmospheric) pressure exerted on the liquid.* Because the atmospheric pressure fluctuates from day to day, the boiling point of a liquid does also. Thus, in order for us to compare the boiling points of different liquids, the external pressure must be the same. The boiling point of a liquid that is most often used for comparison and tabulation purposes is called the *normal* boiling point. A **normal boiling point** *is the temperature at which a liquid boils under a pressure of 760 mm Hg.*

Conditions That Affect Boiling Point

At any given location, the changes in the boiling point of a liquid caused by *natural* variations in atmospheric pressure seldom exceed a few degrees; in the case of water, the maximum is about 2°C. However, variations in boiling points *between* locations at different elevations can be quite striking, as shown in Table 7.3.

The boiling point of a liquid can be increased by increasing the external pressure. This principle is used in the operation of a pressure cooker. Foods cook faster in pressure cookers because the elevated pressure causes water to boil above 100°C. An increase in temperature of only 10°C will cause food to cook in approximately half the normal time (see Figure 7.19). Table 7.4 gives the boiling temperatures reached by water under several household pressure cooker conditions. Hospitals use this same principle to sterilize instruments and laundry in autoclaves, where sufficiently high temperatures are reached to destroy bacteria.

Liquids that have high normal boiling points or that undergo undesirable chemical reactions at elevated temperatures can be made to boil at low temperatures by reducing the external pressure. This principle is used in the preparation of numerous food products, including frozen fruit juice concentrates. Some of the water in a fruit juice is boiled away at a reduced pressure, thus concentrating the juice without heating it to a high temperature (which spoils the taste of the juice and reduces its nutritional value).

Figure 7.19 The converse of the pressure cooker "phenomenon" is that food cooks more slowly at reduced pressures. The pressure reduction associated with higher altitudes, and the accompanying reduction in boiling points of liquids, mean that food cooked over a campfire in the mountains requires longer cooking times.

TABLE 7.3
Boiling Point of Water at Various Locations That Differ in Elevation

Location	Feet Above Sea Level	Atmospheric Pressure (mm Hg)	Boiling Point (°C)
Top of Mt. Everest, Tibet	29,028	240	70
Top of Mt. McKinley, Alaska	20,320	340	79
Leadville, Colorado	10,150	430	89
Salt Lake City, Utah	4,390	650	96
Madison, Wisconsin	900	730	99
New York City, New York	10	760	100
Death Valley, California	−282	770	100.4

TABLE 7.4
Boiling Point of Water at Various Pressure Cooker Settings When Atmospheric Pressure Is 1 Atmosphere

Pressure Cooker Setting (additional pressure beyond atmospheric, lb/in.²)	Internal Pressure in Cooker (atm)	Boiling Point of Water (°C)
5	1.34	108
10	1.68	116
15	2.02	121

7.13 INTERMOLECULAR FORCES IN LIQUIDS

Boiling points vary greatly among substances. The boiling points of some substances are well below 0°C; for example, oxygen has a boiling point of −183°C. Numerous other substances do not boil until the temperature is much higher. An explanation of this variation in boiling points involves a consideration of the nature of the intermolecular forces that must be overcome in order for molecules to escape from the liquid state into the vapor state. An **intermolecular force** *is an attractive force that acts between a molecule and another molecule.*

Intermolecular forces are similar in one way to the previously discussed *intramolecular* forces (forces *within* molecules) that are involved in covalent bonding (Sections 5.3 and 5.4); they are electrostatic in origin; that is, they involve positive–negative interactions. A major difference between inter- and intramolecular forces is their strength. Intermolecular forces are weak compared to intramolecular forces (true chemical bonds). Generally, their strength is less than one-tenth that of a single covalent bond. However, intermolecular forces are strong enough to influence the behavior of liquids, and they often do so in very dramatic ways.

There are three main types of intermolecular forces: dipole–dipole interactions, hydrogen bonds, and London forces.

Dipole-Dipole Interactions

A **dipole-dipole interaction** *is an intermolecular force that occurs between polar molecules.* A polar molecule (Section 5.10) has a negative end and a positive end; that is, it has a *dipole* (two poles resulting from opposite charges being separated from one another). As a consequence, the positive end of one molecule attracts the negative end of another molecule, and vice versa. This attraction constitutes a dipole–dipole interaction. The greater the polarity of the molecules, the greater the strength of the dipole–dipole interactions. And the greater the strength of the dipole–dipole interactions, the higher the boiling point of the liquid. Figure 7.20 shows the many dipole–dipole interactions that are possible for a random arrangement of polar chlorine monofluoride (ClF) molecules.

Hydrogen Bonds

Unusually strong dipole–dipole interactions are observed among hydrogen-containing molecules in which hydrogen is covalently bonded to a highly electronegative element of small atomic size (fluorine, oxygen, and nitrogen). Two factors account for the extra strength of these dipole–dipole interactions.

1. The highly electronegative element to which hydrogen is covalently bonded attracts the bonding electrons to such a degree that the hydrogen atom is left with a significant δ^+ charge.

$$\overset{\delta^+ \;\; \delta^-}{H—F} \qquad \overset{\delta^+ \;\; \delta^-}{H—O} \qquad \overset{\delta^+ \;\; \delta^-}{H—N}$$

Indeed, the hydrogen atom is essentially a "bare" nucleus because it has no electrons besides the one attracted to the electronegative element—a unique property of hydrogen.

2. The small size of the "bare" hydrogen nucleus allows it to approach closely, and be strongly attracted to, a lone pair of electrons on the electronegative atom of another molecule.

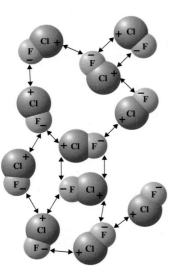

Figure 7.20 There are many dipole-dipole interactions possible between randomly arranged ClF molecules. In each interaction, the positive end of one molecule is attracted to the negative end of a neighboring molecule.

Figure 7.21 Depiction of hydrogen bonding among water molecules. The dotted lines are the hydrogen bonds.

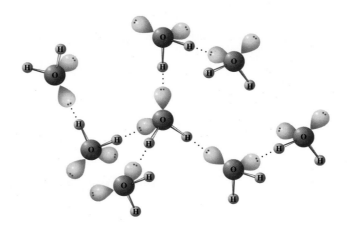

Dipole–dipole interactions of this type are given a special name, hydrogen bonds. A **hydrogen bond** *is an extra-strong dipole–dipole interaction between a hydrogen atom covalently bonded to a small, very electronegative atom (F, O, or N) and a lone pair of electrons on another small, very electronegative atom (F, O, or N).*

Water (H_2O) is the most commonly encountered substance wherein hydrogen bonding is significant. Figure 7.21 depicts the process of hydrogen bonding among water molecules. Note that each oxygen atom in water can participate in two hydrogen bonds—one involving each of its nonbonding electron pairs.

The two molecules that participate in a hydrogen bond need not be identical. Hydrogen bond formation is possible whenever two molecules, the same or different, have the following characteristics.

1. One molecule has a hydrogen atom attached by a covalent bond to an atom of nitrogen, oxygen, or fluorine.
2. The other molecule has a nitrogen, oxygen, or fluorine atom present that possesses one or more nonbonding electron pairs.

Figure 7.22 gives additional examples of hydrogen bonding involving simple molecules.

The three elements that have significant hydrogen-bonding ability are fluorine, oxygen, and nitrogen. They are all very electronegative elements of small atomic size. Chlorine has the same electronegativity as nitrogen, but its larger atomic size causes it to have little hydrogen-bonding ability.

A series of dots is used to represent a hydrogen bond, as in the notation

—X—H ⋯ Y—

X and Y represent small, highly electronegative elements (fluorine, oxygen, or nitrogen).

● **EXAMPLE 7.6**

Predicting Whether Hydrogen Bonding Will Occur Between Molecules

Hydrogen bonding plays an important role in many biochemical systems because biomolecules contain many oxygen and nitrogen atoms that can participate in hydrogen bonding. This type of bonding is particularly important in determining the structural characteristics and functionality of proteins (Chapter 20) and nucleic acids (Chapter 22).

Indicate whether hydrogen bonding should occur between two molecules of each of the following substances.

a. Ethyl amine

b. Methyl alcohol

c. Diethyl ether

Solution

a. Hydrogen bonding should occur because we have an N—H bond and a nitrogen atom with a nonbonding electron pair.
b. Hydrogen bonding should occur because we have an O—H bond and an oxygen atom with nonbonding electron pairs.
c. Hydrogen bonding should not occur. We have an oxygen atom with nonbonding electron pairs, but no N—H, O—H, or F—H bond is present.

Practice Exercise 7.6

Indicate whether hydrogen bonding should occur between two molecules of each of the following substances.

a. Nitrogen trifluoride   **b.** Ethyl alcohol   **c.** Formaldehyde

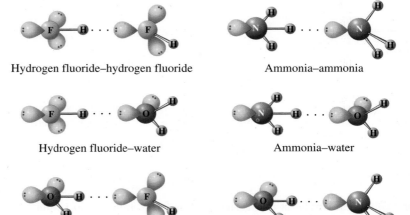

Answers: a. No; no hydrogen atoms are present; **b.** Yes; a hydrogen atom attached to an oxygen atom is present; **c.** No; hydrogen is present, but it is attached to a carbon atom.

Figure 7.22 Diagrams of hydrogen bonding involving selected simple molecules. The solid lines represent covalent bonds; the dotted lines represent hydrogen bonds.

Hydrogen fluoride–hydrogen fluoride

Ammonia–ammonia

Hydrogen fluoride–water

Ammonia–water

Water–hydrogen fluoride

Water–ammonia

The vapor pressures (Section 7.11) of liquids that have significant hydrogen bonding are much lower than those of similar liquids wherein little or no hydrogen bonding occurs. This is because the presence of hydrogen bonds makes it more difficult for molecules to escape from the condensed state; additional energy is needed to overcome the hydrogen bonds. For this reason, boiling points are much higher for liquids in which hydrogen bonding occurs. The effect that hydrogen bonding has on boiling point can be seen by comparing water's boiling point with those of other hydrogen compounds of Group VIA elements—H_2S, H_2Se, and H_2Te (see Figure 7.23). Water is the only compound in this series where significant hydrogen bonding occurs.

Figure 7.23 If there were no hydrogen bonding between water molecules, the boiling point of water would be approximately −80°C; this value is obtained by extrapolation (extension of the line connecting the three heavier compounds). Because of hydrogen bonding, the actual boiling point of water, 100°C, is nearly 200°C higher than predicted. Indeed, in the absence of hydrogen bonding, water would be a gas at room temperature, and life as we know it on Earth would not be possible.

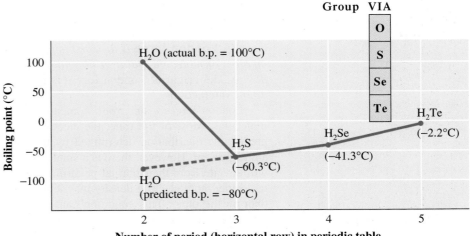

CHEMICAL Connections — Hydrogen Bonding and the Density of Water

The density pattern that liquid water exhibits as its temperature is lowered is different from that of nearly all other liquids. For most liquids, density increases with decreasing temperature and reaches a maximum for the liquid at its freezing point. Water's maximum density is reached at a temperature of 4°C rather than at its freezing point (see the accompanying graph).

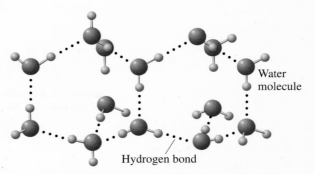
Water molecule

Hydrogen bond

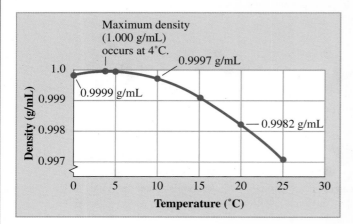

Maximum density (1.000 g/mL) occurs at 4°C.

0.9997 g/mL

0.9999 g/mL

0.9982 g/mL

Temperature (°C)

Density (g/mL)

water's oxygen atom. The net result is that when water molecules are hydrogen-bonded, they are farther apart than when they are not hydrogen-bonded. The accompanying diagram shows the hydrogen-bonding pattern that is characteristic of ice.

When natural bodies of water are gradually cooled as winter approaches, the surface water eventually reaches a temperature of 4°C, the temperature of water's highest density. Such water "sinks" to the bottom. Over time, this process results in a stratification (layering) that creates temperature zones. The "heaviest" water, at 4°C, is on the bottom; "lighter" water of lower temperatures comes next, with ice at the surface. The fact that ice is less dense than liquid water explains why lakes freeze from top to bottom, a phenomenon that allows aquatic life to continue to exist for extended periods of time in bodies of water that are frozen over.

This "abnormality"—that water at its freezing point is less dense than water at slightly higher temperatures—is a consequence of hydrogen bonding between water molecules. Furthermore, at 0°C, solid water (ice) is significantly less dense than liquid water (0.9170 g/mL versus 0.9999 g/mL) because of hydrogen bonding.

Hydrogen bonds can form only between water molecules that are positioned at certain angles to each other. These angles are dictated by the location of the nonbonding pairs of electrons of

Because ice is less dense than water, ice floats in liquid water; also, liquid water expands upon freezing. Such expansion is why antifreeze is used in car radiators in the winter in cold climates. During the winter season, the weathering of rocks and concrete and the formation of potholes in streets are hastened by the expansion of freezing water in cracks.

London Forces

The third type of intermolecular force, and the weakest, is the London force, named after the German physicist Fritz London (1900–1954), who first postulated its existence. A **London force** *is a weak temporary intermolecular force that occurs between an atom or molecule (polar or nonpolar) and another atom or molecule (polar or nonpolar).* The origin of London forces is more difficult to visualize than that of dipole–dipole interactions.

London forces result from momentary (temporary) uneven electron distributions in molecules. Most of the time, the electrons in a molecule can be considered to have a predictable distribution determined by their energies and the electronegativities of the atoms present. However, there is a small statistical chance (probability) that the electrons will deviate from their normal pattern. For example, in the case of a nonpolar diatomic molecule, more electron density may temporarily be located on one side of the molecule than on the other. This condition causes the molecule to become polar for an instant. The negative side of this *instantaneously* polar molecule tends to repel electrons of

CHEMISTRY AT A GLANCE

Intermolecular Forces

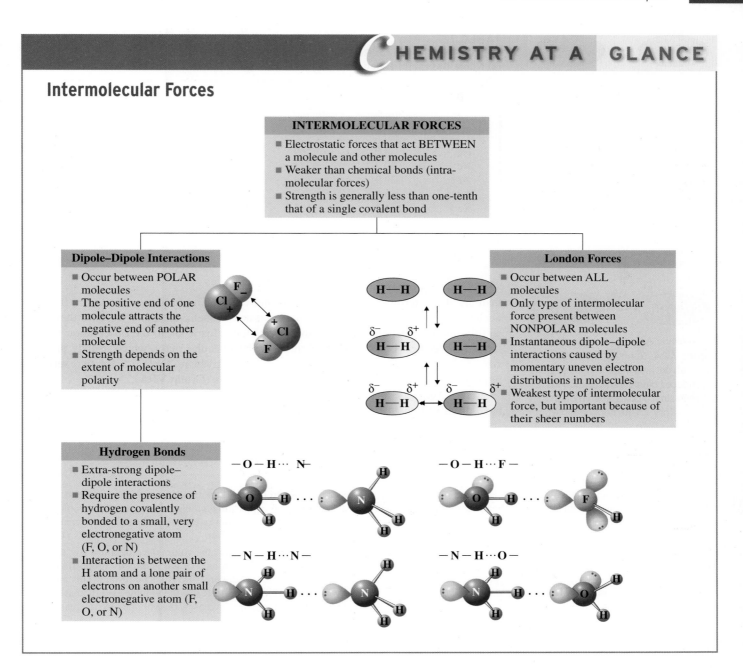

INTERMOLECULAR FORCES
- Electrostatic forces that act BETWEEN a molecule and other molecules
- Weaker than chemical bonds (intramolecular forces)
- Strength is generally less than one-tenth that of a single covalent bond

Dipole–Dipole Interactions
- Occur between POLAR molecules
- The positive end of one molecule attracts the negative end of another molecule
- Strength depends on the extent of molecular polarity

London Forces
- Occur between ALL molecules
- Only type of intermolecular force present between NONPOLAR molecules
- Instantaneous dipole–dipole interactions caused by momentary uneven electron distributions in molecules
- Weakest type of intermolecular force, but important because of their sheer numbers

Hydrogen Bonds
- Extra-strong dipole–dipole interactions
- Require the presence of hydrogen covalently bonded to a small, very electronegative atom (F, O, or N)
- Interaction is between the H atom and a lone pair of electrons on another small electronegative atom (F, O, or N)

adjoining molecules and causes these molecules also to become polar (*induced polarity*). The original polar molecule and all of the molecules with induced polarity are then attracted to one another. This happens many, many times per second throughout the liquid, resulting in a net attractive force. Figure 7.24 depicts the situation that prevails when London forces exist.

As an analogy for London forces, consider what happens when a bucket filled with water is moved. The water will "slosh" from side to side. This is similar to the movement of electrons. The "sloshing" from side to side is instantaneous; a given "slosh" quickly disappears. "Uneven" electron distribution is likewise a temporary situation.

The strength of London forces depends on the ease with which an electron distribution in a molecule can be distorted (polarized) by the polarity present in another molecule. In large molecules, the outermost electrons are necessarily located farther from the nucleus

The boiling points of substances with similar molar masses increase in this order: nonpolar molecules < polar molecules with no hydrogen bonding < polar molecules with hydrogen bonding.

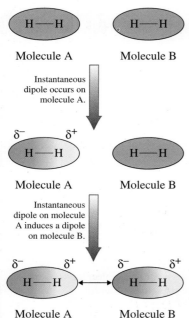

Figure 7.24 Nonpolar molecules such as H_2 can develop instantaneous dipoles and induced dipoles. The attractions between such dipoles, even though they are transitory, create London forces.

TABLE 7.5

Boiling Point Trends for Related Series of Nonpolar Molecules: (a) Noble Gases, (b) Halogens

(a) Noble Gases (Group VIIIA Elements)			(b) Halogens (Group VIIA Elements)		
Substance	Molecular Mass (amu)	Boiling Point (°C)	Substance	Molecular Mass (amu)	Boiling Point (°C)
He	4.0	−269	F_2	38.0	−187
Ne	20.2	−246	Cl_2	70.9	−35
Ar	39.9	−186	Br_2	159.8	+59
Kr	83.8	−153			
Xe	131.3	−107			
Rn	222.0	−62			

than are the outermost electrons in small molecules. The farther electrons are from the nucleus, the weaker the attractive forces that act on them, the more freedom they have, and the more susceptible they are to polarization. This leads to the observation that for *related* molecules, boiling points increase with molecular mass, which usually parallels size. This trend is reflected in the boiling points given in Table 7.5 for two series of related substances: the noble gases and the halogens (Group VIIA).

The Chemistry at a Glance feature on page 185 provides a summary of what we have discussed about intermolecular forces.

CONCEPTS TO REMEMBER

Kinetic molecular theory. The kinetic molecular theory of matter is a set of five statements that explain the physical behavior of the three states of matter (solids, liquids, and gases). The basic idea of this theory is that the particles (atoms, molecules, or ions) present in a substance are in constant motion and are attracted or repelled by each other (Section 7.1).

The solid state. A solid is characterized by a dominance of potential energy (cohesive forces) over kinetic energy (disruptive forces). As a result, the particles of solids are held in a rigid three-dimensional arrangement in which the particle kinetic energy is manifest as vibrational movement of particles (Section 7.2).

The liquid state. A liquid is characterized by neither potential energy (cohesive forces) nor kinetic energy (disruptive forces) being dominant. As a result, particles of liquids are randomly arranged but are relatively close to each other and are in constant random motion, sliding freely over each other but without enough kinetic energy to become separated (Section 7.2).

The gaseous state. A gas is characterized by a complete dominance of kinetic energy (disruptive forces) over potential energy (cohesive forces). As a result, particles move randomly, essentially independently of each other. Under ordinary pressure, the particles of a gas are separated from each other by relatively large distances, except when they collide (Section 7.2).

Gas laws. Gas laws are generalizations that describe, in mathematical terms, the relationships among the amount, pressure, temperature, and volume of a specific quantity of gas. When these relationships are used, it is necessary to express the temperature on the Kelvin scale. Pressure is usually expressed in atm, mm Hg, or torr (Section 7.3).

Boyle's law. Boyle's law, the pressure–volume law, states that the volume of a fixed amount of a gas is inversely proportional to the

pressure applied to the gas if the temperature is kept constant. This means that when the pressure on the gas increases, the volume decreases proportionally; conversely, when the volume decreases, the pressure increases (Section 7.4).

Charles's law. Charles's law, the volume–temperature law, states that the volume of a fixed amount of gas is directly proportional to its Kelvin temperature if the pressure is kept constant. This means that when the temperature increases, the volume also increases and that when the temperature decreases, the volume also decreases (Section 7.5).

The combined gas law. The combined gas law is an expression obtained by mathematically combining Boyle's and Charles's laws. A change in pressure, temperature, or volume that is brought about by changes in the other two variables can be calculated by using this law (Section 7.6).

Ideal gas law. The ideal gas law has the form $PV = nRT$, where R is the ideal gas constant (0.0821 atm · L/mole · K). With this equation, any one of the characteristic gas properties (P, V, T, or n) can be calculated, given the other three (Section 7.7).

Dalton's law of partial pressures. Dalton's law of partial pressures states that the total pressure exerted by a mixture of gases is the sum of the partial pressures of the individual gases. A partial pressure is the pressure that a gas in a mixture would exert if it were present alone under the same conditions (Section 7.8).

Changes of state. Most matter can be changed from one physical state to another by heating, cooling, or changing pressure. The state changes that release heat are called exothermic (condensation, deposition, and freezing), and those that absorb heat are called endothermic (melting, evaporation, and sublimation) (Section 7.9).

Vapor pressure. The pressure exerted by vapor in equilibrium with its liquid is the vapor pressure of the liquid. Vapor pressure increases as liquid temperature increases (Section 7.11).

Boiling and boiling point. Boiling is a form of evaporation in which bubbles of vapor form within the liquid and rise to the surface. The boiling point of a liquid is the temperature at which the vapor pressure of the liquid becomes equal to the external (atmospheric) pressure exerted on the liquid. The boiling point of a liquid increases or decreases as the prevailing atmospheric pressure increases or decreases (Section 7.12).

Intermolecular forces. Intermolecular forces are forces that act between a molecule and another molecule. The three principal types of intermolecular forces in liquids are dipole–dipole interactions, hydrogen bonds, and London forces (Section 7.13).

Hydrogen bonds. A hydrogen bond is an extra-strong dipole–dipole interaction between a hydrogen atom covalently bonded to a very electronegative atom (F, O, or N) and a lone pair of electrons on another small, very electronegative atom (F, O, or N) (Section 7.13).

kEY REACTIONS AND EQUATIONS

1. Boyle's law (Section 7.4)
$$P_1V_1 = P_2V_2 \qquad (n, T \text{ constant})$$

2. Charles's law (Section 7.5)
$$\frac{V_1}{T_1} = \frac{V_2}{T_2} \qquad (n, P \text{ constant})$$

3. Combined gas law (Section 7.6)
$$\frac{P_1V_1}{T_1} = \frac{P_2V_2}{T_2} \qquad (n \text{ constant})$$

4. Ideal gas law (Section 7.7)
$$PV = nRT$$

5. Ideal gas constant (Section 7.7)
$$R = 0.0821 \text{ atm} \cdot \text{L/mole} \cdot \text{K}$$

6. Dalton's law of partial pressures (Section 7.8)
$$P_{\text{Total}} = P_1 + P_2 + P_3 + \cdots$$

EXERCISES *and* PROBLEMS

The members of each pair of problems in this section test similar material.

● **Kinetic Molecular Theory of Matter (Section 7.1)**

7.1 Using kinetic molecular theory concepts, answer the following questions.
  a. What type of energy is related to cohesive forces?
  b. What effect does temperature have on the magnitude of disruptive forces?
  c. What is the general effect of cohesive forces on a system of particles?
  d. What type of potential energy is particularly important when considering the physical states of matter?

7.2 Using kinetic molecular theory concepts, answer the following questions.
  a. What type of energy is related to disruptive forces?
  b. What effect does temperature have on the magnitude of cohesive forces?
  c. What is the general effect of disruptive forces on a system of particles?
  d. How do molecules transfer energy from one to another?

● **Kinetic Molecular Theory and Physical States (Section 7.2)**

7.3 Indicate the state of matter to which each of the following characterizations applies. There may be more than one correct answer for some characterizations.
  a. Disruptive forces and cohesive forces are of about the same magnitude.
  b. Disruptive forces are significantly less than cohesive forces.
  c. Kinetic energy dominates over potential energy.
  d. Kinetic energy is significantly less than potential energy.

7.4 Indicate the state of matter to which each of the following characterizations applies. There may be more than one correct answer for some characterizations.
  a. Disruptive forces are significantly greater than cohesive forces.
  b. Cohesive forces are significantly less than disruptive forces.
  c. Potential energy dominates over kinetic energy.
  d. Kinetic energy and potential energy are of about the same magnitude.

7.5 With which of the states of matter is each of the following relative characterizations best associated? There may be more than one correct answer for some characterizations.
  a. Indefinite shape and indefinite volume
  b. Very small thermal expansion
  c. Small thermal expansion
  d. High density

7.6 With which of the states of matter is each of the following relative characterizations best associated? There may be more than one correct answer for some characterizations.
  a. Moderate thermal expansion
  b. Definite volume and indefinite shape
  c. Small compressibility
  d. Low density

● **Gas Law Variables (Section 7.3)**

7.7 Identify the gas law variable with which each of the following measurements is associated.
  a. 7.4 moles   b. 3.70 L   c. 173 mm Hg   d. 32.3 K

7.8 Identify the gas law variable with which each of the following measurements is associated.
  a. 343 K   b. 6.7 mL   c. 673 torr   d. 0.23 mole

7.9 Carry out the following pressure unit conversions using the dimensional-analysis method of problem solving.
   a. 735 mm Hg to atmospheres
   b. 0.530 atm to millimeters of mercury
   c. 0.530 atm to torr
   d. 12.0 psi to atmospheres

7.10 Carry out the following pressure unit conversions using the dimensional-analysis method of problem solving.
   a. 73.5 mm Hg to atmospheres
   b. 1.75 atm to millimeters of mercury
   c. 735 torr to atmospheres
   d. 1.61 atm to pounds per square inch

● Boyle's Law (Section 7.4)

7.11 At constant temperature, a sample of 6.0 L of O_2 at 3.0 atm pressure is compressed until the volume decreases to 2.5 L. What is the new pressure, in atmospheres?

7.12 At constant temperature, a sample of 6.0 L of N_2 at 2.0 atm pressure is allowed to expand until the volume reaches 9.5 L. What is the new pressure, in atmospheres?

7.13 A sample of ammonia (NH_3), a colorless gas with a pungent odor, occupies a volume of 3.00 L at a pressure of 655 mm Hg and a temperature of 25°C. What volume, in liters, will this NH_3 sample occupy at the same temperature if the pressure is increased to 725 mm Hg?

7.14 A sample of nitrogen dioxide (NO_2), a toxic gas with a reddish-brown color, occupies a volume of 4.00 L at a pressure of 725 mm Hg and a temperature of 35°C. What volume, in liters, will this NO_2 sample occupy at the same temperature if the pressure is decreased to 125 mm Hg?

7.15 Diagram I depicts a gas, in a cylinder of variable volume, at the conditions specified in the diagram. Which of the diagrams II through IV depicts the result of doubling the pressure at constant temperature and constant number of moles of gas?

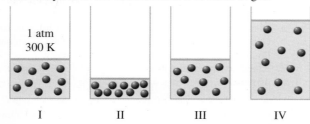

I   II   III   IV

7.16 Using the diagrams given in Problem 7.15, which of the diagrams II through IV correctly depicts the result of halving the pressure at constant temperature and constant number of moles of gas?

● Charles's Law (Section 7.5)

7.17 At atmospheric pressure, a sample of H_2 gas has a volume of 2.73 L at 27°C. What volume, in liters, will the H_2 gas occupy if the temperature is increased to 127°C and the pressure is held constant?

7.18 At atmospheric pressure, a sample of O_2 gas has a volume of 55 mL at 27°C. What volume, in milliliters, will the O_2 gas occupy if the temperature is decreased to 0°C and the pressure is held constant?

7.19 A sample of N_2 gas occupies a volume of 375 mL at 25°C and a pressure of 2.0 atm. Determine the temperature, in degrees Celsius, at which the volume of the gas would be 525 mL at the same pressure.

7.20 A sample of Ar gas occupies a volume of 1.2 L at 125°C and a pressure of 1.0 atm. Determine the temperature, in degrees Celsius, at which the volume of the gas would be 1.0 L at the same pressure.

7.21 Diagram I depicts a gas, in a cylinder of variable volume, at the conditions specified in the diagram. Which of the diagrams II through IV depicts the results of decreasing the Kelvin temperature by a factor of two at constant pressure and constant number of moles of gas?

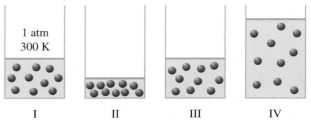

I   II   III   IV

7.22 Using the diagrams given in Problem 7.21, which of the diagrams II through IV correctly depicts the result of increasing the Kelvin temperature by a factor of two at constant pressure and constant number of moles of gas?

● Combined Gas Law (Section 7.6)

7.23 Rearrange the standard form of the combined gas law equation so that each of the following variables is by itself on one side of the equation.
   a. T_1 b. P_2 c. V_1

7.24 Rearrange the standard form of the combined gas law equation so that each of the following variables is by itself on one side of the equation.
   a. V_2 b. T_2 c. P_1

7.25 A sample of carbon dioxide (CO_2) gas has a volume of 15.2 L at a pressure of 1.35 atm and a temperature of 33°C. Determine the following for this gas sample.
   a. Volume, in liters, at $T = 35°C$ and $P = 3.50$ atm
   b. Pressure, in atmospheres, at $T = 42°C$ and $V = 10.0$ L
   c. Temperature, in degrees Celsius, at $P = 7.00$ atm and $V = 0.973$ L
   d. Volume, in milliliters, at $T = 97°C$ and $P = 6.70$ atm

7.26 A sample of carbon monoxide (CO) gas has a volume of 7.31 L at a pressure of 735 mm Hg and a temperature of 45°C. Determine the following for this gas sample.
   a. Pressure, in millimeters of mercury, at $T = 357°C$ and $V = 13.5$ L
   b. Temperature, in degrees Celsius, at $P = 1275$ mm Hg and $V = 0.800$ L
   c. Volume, in liters, at $T = 45°C$ and $P = 325$ mm Hg
   d. Pressure, in atmospheres, at $T = 325°C$ and $V = 2.31$ L

7.27 Diagram I depicts a gas, in a cylinder of variable volume, at conditions given in the diagram. Which of the diagrams II through IV correctly depicts the result of doubling the pressure and doubling the Kelvin temperature of the gas while keeping the number of moles of gas constant?

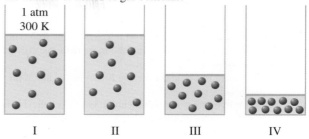

I   II   III   IV

7.28 Using the diagrams given in Problem 7.27, which of the diagrams II through IV correctly depicts the result of doubling the pressure and halving the Kelvin temperature of the gas while keeping the number of moles of gas constant?

Ideal Gas Law (Section 7.7)

7.29 What is the temperature, in degrees Celsius, of 5.23 moles of helium (He) gas confined to a volume of 5.23 L at a pressure of 5.23 atm?

7.30 What is the temperature, in degrees Celsius, of 1.50 moles of neon (Ne) gas confined to a volume of 2.50 L at a pressure of 1.00 atm?

7.31 Calculate the volume, in liters, of 0.100 mole of O_2 gas at 0°C and 2.00 atm pressure.

7.32 Calculate the pressure, in atmospheres, of 0.100 mole of O_2 gas in a 2.00-L container at a temperature of 75°C.

7.33 Determine the following for a 0.250-mole sample of CO_2 gas.
a. Volume, in liters, at 27°C and 1.50 atm
b. Pressure, in atmospheres, at 35°C in a 2.00-L container
c. Temperature, in degrees Celsius, at 1.20 atm pressure in a 3.00-L container
d. Volume, in milliliters, at 125°C and 0.500 atm pressure

7.34 Determine the following for a 0.500-mole sample of CO gas.
a. Pressure, in atmospheres, at 35°C in a 1.00-L container
b. Temperature, in degrees Celsius, at 5.00 atm pressure in a 5.00-L container
c. Volume, in liters, at 127°C and 3.00 atm
d. Pressure, in millimeters of mercury, at 25°C in a 2.00-L container

Dalton's Law of Partial Pressures (Section 7.8)

7.35 The total pressure exerted by a mixture of O_2, N_2, and He gases is 1.50 atm. What is the partial pressure, in atmospheres, of the O_2, given that the partial pressures of the N_2 and He are 0.75 and 0.33 atm, respectively?

7.36 The total pressure exerted by a mixture of He, Ne, and Ar gases is 2.00 atm. What is the partial pressure, in atmospheres, of Ne, given that the partial pressures of the other gases are both 0.25 atm?

7.37 A gas mixture contains O_2, N_2, and Ar at partial pressures of 125, 175, and 225 mm Hg, respectively. If CO_2 gas is added to the mixture until the total pressure reaches 623 mm Hg, what is the partial pressure, in millimeters of mercury, of CO_2?

7.38 A gas mixture contains He, Ne, and H_2S at partial pressures of 125, 175, and 225 mm Hg, respectively. If all of the H_2S is removed from the mixture, what will be the partial pressure, in millimeters of mercury, of Ne?

7.39 The following diagram depicts a gaseous mixture of neon (red spheres), argon (black spheres), and krypton (yellow spheres).

If the total pressure in the container is 6.0 atm, what is the partial pressure, in atm, of the following?
a. Neon   b. Argon   c. Krypton

7.40 The following diagram depicts a gaseous mixture of O_2 (red spheres), N_2 (blue spheres), and Cl_2 (yellow spheres).

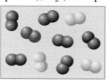

Assuming that the gases do not react with each other, what is the total pressure, in atm, in the container if the partial pressure of the O_2 gas is 2.0 atm?

Changes of State (Section 7.9)

7.41 Indicate whether each of the following is an exothermic or an endothermic change of state.
a. Sublimation   b. Melting   c. Condensation

7.42 Indicate whether each of the following is an exothermic or an endothermic change of state.
a. Freezing   b. Evaporation   c. Deposition

7.43 Indicate whether the liquid state is involved in each of the following changes of state.
a. Sublimation   b. Melting   c. Condensation

7.44 Indicate whether the solid state is involved in each of the following changes of state.
a. Freezing   b. Deposition   c. Evaporation

Evaporation of Liquids (Section 7.10)

7.45 What are the two ways in which the escape of high-energy molecules from the surface of a liquid during the process of evaporation affects the liquid?

7.46 What are the two general requirements that must be met before a molecule can escape from the surface of a liquid during the evaporation process?

7.47 How does an increase in the temperature of a liquid affect the rate of evaporation of the liquid?

7.48 How does an increase in the surface area of a liquid affect the rate of evaporation of the liquid?

Vapor Pressure of Liquids (Section 7.11)

7.49 Indicate whether or not each of the following statements about an equilibrium state is true or false.
a. For systems in a state of equilibrium, two opposite processes take place at equal rates.
b. A liquid–vapor equilibrium situation is an example of a physical equilibrium state.

7.50 Indicate whether or not each of the following statements about an equilibrium state is true or false.
a. For systems in a state of equilibrium, no net macroscopic changes can be detected.
b. For a liquid–vapor equilibrium situation the two opposite processes occurring are evaporation and condensation.

7.51 What is the relationship between the temperature of a liquid and the vapor pressure of that liquid?

7.52 What is the relationship between the strength of the attractive forces in a liquid and the liquid's vapor pressure?

7.53 What term is used to describe a substance that readily evaporates at room temperature because of a high vapor pressure?

7.54 What term is used to describe gaseous molecules of a substance at a temperature and pressure at which we ordinarily would think of the substance as a liquid or solid?

● **Boiling and Boiling Point (Section 7.12)**

7.55 Indicate whether each of the following statements concerning boiling or boiling point is true or false.
  a. The process of boiling is a form of evaporation.
  b. The boiling point of a liquid heated in an open container is determined by atmospheric pressure.
  c. The normal boiling point of a liquid is the temperature at which it boils when the liquid is at sea level.
  d. At a pressure of 760 mm Hg, all liquids boil at the same temperature.

7.56 Indicate whether each of the following statements concerning boiling and boiling point is true or false.
  a. A liquid can be made to boil at temperatures higher than its normal boiling point.
  b. A liquid can be made to boil at temperatures lower than its normal boiling point.
  c. In a boiling liquid, vapor formation occurs within the body of the liquid.
  d. To compare the boiling points of two different liquids the external pressure should be the same.

7.57 What is the relationship between external pressure and the boiling point of a liquid?

7.58 What is the relationship between location (elevation) and the boiling point of a liquid?

● **Intermolecular Forces in Liquids (Section 7.13)**

7.59 Describe the molecular conditions necessary for the existence of a dipole–dipole interaction.

7.60 Describe the molecular conditions necessary for the existence of a London force.

7.61 In liquids, what is the relationship between boiling point and the strength of intermolecular forces?

7.62 In liquids, what is the relationship between vapor pressure magnitude and the strength of intermolecular forces?

7.63 For liquid-state samples of the following diatomic substances, classify the dominant intermolecular forces present as London forces, dipole–dipole interactions, or hydrogen bonds.
  a. H_2 b. HF   c. CO   d. F_2

7.64 For liquid-state samples of the following diatomic substances, classify the dominant intermolecular forces present as London forces, dipole–dipole interactions, or hydrogen bonds.
  a. O_2 b. HCl   c. Cl_2 d. BrCl

7.65 In which of the following substances, in the pure liquid state, would hydrogen bonding occur?

7.66 In which of the following substances, in the pure liquid state, would hydrogen bonding occur?

7.67 How many hydrogen bonds can form between a single water molecule and other water molecules?

7.68 How many hydrogen bonds can form between a single ammonia molecule (NH_3) and other ammonia molecules?

ADDITIONAL PROBLEMS

7.69 A sample of NO_2 gas in a 575-mL container at a pressure of 1.25 atm and a temperature of 125°C is transferred to a new container with a volume of 825 mL.
  a. What is the new pressure, in atmospheres, if no change in temperature occurs?
  b. What is the new temperature, in degrees Celsius, if no change in pressure occurs?
  c. What is the new temperature, in degrees Celsius, if the pressure is increased to 2.50 atm?

7.70 A sample of NO_2 gas in a nonrigid container at a temperature of 24°C occupies a certain volume at a certain pressure. What will be its temperature, in degrees Celsius, in each of the following situations?
  a. Both pressure and volume are doubled.
  b. Both pressure and volume are cut in half.
  c. The pressure is doubled and the volume is cut in half.
  d. The pressure is cut in half and the volume is tripled.

7.71 Match each of the listed restrictions on variables to the following gas laws: *Boyle's law, Charles's law,* and *the combined gas law.* More than one answer may be correct in a given situation.
  a. The number of moles is constant.
  b. The pressure is constant.
  c. The temperature is constant.
  d. Both the number of moles and the temperature are constant.

7.72 Suppose a helium-filled balloon used to carry scientific instruments into the atmosphere has a volume of 1.00×10^6 L at 25°C and a pressure of 752 mm Hg at the time it is launched. What will be the volume of the balloon, in liters, when, at a height of 37 km, it encounters a temperature of −33°C and a pressure of 75.0 mm Hg?

7.73 What is the pressure, in atmospheres, inside a 4.00-L container that contains the following amounts of O_2 gas at 40.0°C?
  a. 0.72 mole   b. 4.5 moles   c. 0.72 g   d. 4.5 g

7.74 How many molecules of hydrogen sulfide (H_2S) gas are contained in 2.00 L of H_2S at 0.0°C and 1.00 atm pressure?

7.75 A 1.00-mole sample of dry ice (solid CO_2) is placed in a flexible sealed container and allowed to sublime. After complete sublimation, what will be the container volume, in liters, at 23°C and 0.983 atm pressure?

7.76 A piece of Ca metal is placed in a 1.00-L container with pure N_2. The N_2 is at a pressure of 1.12 atm and a temperature of 26°C. One hour later, the pressure has dropped to 0.924 atm and the temperature has dropped to 24°C. Calculate the number of grams of N_2 that reacted with the Ca.

7.77 A gas mixture containing He, Ne, and Ar exerts a pressure of 3.00 atm. What is the partial pressure of each gas present in the mixture under the following conditions?
   a. There is an equal number of moles of each gas present.
   b. There is an equal number of atoms of each gas present.
   c. The partial pressures of He, Ne, and Ar are in a 3:2:1 ratio.
   d. The partial pressure of He is one-half that of Ne and one-third that of Ar.

7.78 Under which of the following "pressure situations" will a liquid boil?
   a. Vapor pressure and atmospheric pressure are equal.
   b. Vapor pressure is less than atmospheric pressure.
   c. Vapor pressure = 635 mm Hg, and atmospheric pressure = 735 mm Hg.
   d. Vapor pressure = 735 torr, and atmospheric pressure = 1.00 atm.

7.79 The vapor pressure of PBr_3 reaches 400 mm Hg at 150°C. The vapor pressure of PI_3 reaches 400 mm Hg at 57°C.
   a. Which substance should evaporate at the slower rate at 100°C?
   b. Which substance should have the lower boiling point?
   c. Which substance should have the weaker intermolecular forces?

7.80 In each of the following pairs of molecules, indicate which member of the pair would be expected to have the higher boiling point.
   a. Cl_2 and Br_2 b. H_2O and H_2S
   c. O_2 and CO   d. C_3H_8 and CO_2

*M*ultiple-Choice Practice Test

7.81 Which of the following statements is correct, according to kinetic molecular theory?
   a. Solids have small compressibilities because there is very little space between particles.
   b. In the gaseous state, attractive forces between particles are of about the same magnitude as disruptive forces.
   c. An increase in temperature means an increase in potential energy.
   d. Gases have high densities because the particles are widely separated.

7.82 How many times larger or smaller in size is the mm Hg pressure unit compared to the atmosphere pressure unit?
   a. 100 times larger   b. 100 times smaller
   c. 760 times larger   d. 760 times smaller

7.83 Charles's law involves which of the following?
   a. An inverse proportion   b. A constant volume
   c. A constant temperature   d. A constant pressure

7.84 A sample of 20.0 liters of nitrogen gas is under a pressure of 20.0 atm. If the volume of this gas is decreased to 5.00 liters at constant temperature, what will the new pressure be?
   a. 80.0 atm   b. 40.0 atm
   c. 10.0 atm   d. 5.00 atm

7.85 Which of the following gas laws has the mathematical form $PV = nRT$?
   a. Dalton's law of partial pressures
   b. Ideal gas law
   c. Combined gas law
   d. Boyle's law

7.86 In which of the following pairs of state changes is the final state (solid, liquid, or gas) the same for both members of the pair?
   a. Sublimation and evaporation
   b. Condensation and freezing
   c. Deposition and melting
   d. Sublimation and condensation

7.87 Molecules of a liquid can pass into the vapor phase only under which of the following conditions?
   a. Atmospheric pressure is less than 640 mm Hg.
   b. The vapor pressure of the liquid is greater than the atmospheric pressure.
   c. The temperature of the liquid exceeds the liquid's normal boiling point.
   d. The molecules have sufficient kinetic energy to overcome the intermolecular forces in the liquid.

7.88 Liquids boil at lower temperatures at higher elevations because
   a. The intermolecular forces are weaker.
   b. The intramolecular forces are weaker.
   c. The atmospheric pressure is greater.
   d. The vapor pressure at which boiling occurs is lower.

7.89 Which of the following is an intermolecular force that occurs between all molecules?
   a. Hydrogen bonding
   b. Weak dipole–dipole interaction
   c. Strong dipole–dipole interaction
   d. London force

7.90 Which of the following statements concerning intermolecular forces is *correct*?
   a. Dipole–dipole interaction strength increases with molecular polarity.
   b. Hydrogen bonding occurs only between nonpolar hydrogen-containing molecules.
   c. London forces are extra-strong dipole–dipole interactions.
   d. Substances in which hydrogen bonding is present usually have high vapor pressures.

Solutions

Chapter Outline

Ocean water is a solution in which many different substances are dissolved.

Solutions are common in nature, and they represent an abundant form of matter. Solutions carry nutrients to the cells of our bodies and carry away waste products. The ocean is a solution of water, sodium chloride, and many other substances (even gold). A large percentage of all chemical reactions take place in solution, including most of those discussed in later chapters in this text.

8.1 CHARACTERISTICS OF SOLUTIONS

All samples of matter are either *pure substances* or *mixtures* (Section 1.5). Pure substances are of two types: *elements* and *compounds*. Mixtures are of two types: *homogeneous* (uniform properties throughout) and *heterogeneous* (different properties in different regions).

Where do solutions fit in this classification scheme? The term *solution* is just an alternative way of saying *homogeneous mixture*. A **solution** *is a homogeneous mixture of two or more substances with each substance retaining its own chemical identity.*

It is often convenient to call one component of a solution the solvent and other components that are present solutes (Figure 8.1). A **solvent** *is the component of a solution that is present in the greatest amount.* A solvent can be thought of as the medium in which the other substances present are dissolved. A **solute** *is a component of a solution that is present in a lesser amount relative to that of the solvent.* More than one solute can be present in the same solution. For example, both sugar and salt (two solutes) can be dissolved in a container of water (solvent) to give salty sugar water.

"All solutions are mixtures" is a valid statement. However, the reverse statement, "All mixtures are solutions," is not valid. Only those mixtures that are *homogeneous* are solutions.

Figure 8.1 The colored crystals are the solute, and the clear liquid is the solvent. Stirring produces the solution.

In most of the situations we will encounter, the solutes present in a solution will be of more interest to us than the solvent. The solutes are the active ingredients in the solution. They are the substances that undergo reaction when solutions are mixed.

The general properties of a solution (homogeneous mixture) were outlined in Section 1.5. These properties, restated using the concepts of solvent and solute, are as follows:

1. A solution contains two or more components: a solvent (the substance present in the greatest amount) and one or more solutes.
2. A solution has a variable composition; that is, the ratio of solute to solvent may be varied.
3. The properties of a solution change as the ratio of solute to solvent is changed.
4. The dissolved solutes are present as individual particles (molecules, atoms, or ions). Intermingling of components at the particle level is a requirement for homogeneity.
5. The solutes remain uniformly distributed throughout the solution and will not settle out with time. Every part of a solution has exactly the same properties and composition as every other part.
6. The solute(s) generally can be separated from the solvent by physical means such as evaporation.

Generally, solutions are *transparent;* that is, you can see through them. A synonym for *transparent* is *clear.* Clear solutions may be colorless or colored. A solution of potassium dichromate is a clear yellow-orange solution.

Most solutes are more soluble in hot solvent than in cold solvent.

Solutions used in laboratories and clinical settings are most often liquids, and the solvent is nearly always water. However, gaseous solutions (dry air), solid solutions (metal alloys—see Figure 8.2), and liquid solutions in which water is not the solvent (gasoline, for example) are also possible and are relatively common.

 SOLUBILITY

In addition to *solvent* and *solute,* several other terms are used to describe characteristics of solutions. **Solubility** *is the maximum amount of solute that will dissolve in a given amount of solvent under a given set of conditions.* Many factors affect the numerical value of a solute's solubility in a given solvent, including the nature of the solvent itself, the temperature, and, in some cases, the pressure and presence of other solutes. Solubility is commonly expressed as grams of solute per 100 grams of solvent.

Effect of Temperature on Solubility

Most solids become more soluble in water with increasing temperature. The data in Table 8.1 illustrate this temperature–solubility pattern. Here, the solubilities of selected ionic solids in water are given at three different temperatures.

In contrast to the solubilities of solids, gas solubilities in water decrease with increasing temperature. For example, both N_2 and O_2, the major components of air, are less soluble in hot water than in cold water. The Chemical Connections feature on page 195 considers further the topic of *temperature* and gas solubility.

Figure 8.2 Jewelry often involves solid solutions in which one metal has been dissolved in another metal.

TABLE 8.1
Solubilities of Various Compounds in Water at 0°C, 50°C, and 100°C

Solute	Solubility (g solute/100 g H₂O)		
	0°C	50°C	100°C
lead(II) bromide (PbBr₂)	0.455	1.94	4.75
silver sulfate (Ag₂SO₄)	0.573	1.08	1.41
copper(II) sulfate (CuSO₄)	14.3	33.3	75.4
sodium chloride (NaCl)	35.7	37.0	39.8
silver nitrate (AgNO₃)	122	455	952
cesium chloride (CsCl)	161.4	218.5	270.5

Respiratory therapy procedures take advantage of the fact that increased pressure increases the solubility of a gas. Patients with lung problems who are unable to get sufficient oxygen from air are given an oxygen-enriched mixture of gases to breathe. The larger oxygen partial pressure in the enriched mixture translates into increased oxygen uptake in the patient's lungs.

Effect of Pressure on Solubility

Pressure has little effect on the solubility of solids and liquids in water. However, it has a major effect on the solubility of gases in water. The pressure–solubility relationship for gases was first formalized by the English chemist William Henry (1775–1836) and is now known as *Henry's law.* **Henry's law** states that *the amount of gas that will dissolve in a liquid at a given temperature is directly proportional to the partial pressure of the gas above the liquid.* In other words, as the pressure of a gas above a liquid increases, the solubility of the gas increases; conversely, as the pressure of the gas decreases, its solubility decreases. The Chemical Connections feature on page 195 considers further the topic of *pressure* and gas solubility.

Saturated, Supersaturated, and Unsaturated Solutions

A **saturated solution** *is a solution that contains the maximum amount of solute that can be dissolved under the conditions at which the solution exists.* A saturated solution containing excess undissolved solute is an equilibrium situation where an amount of undissolved solute is continuously dissolving while an equal amount of dissolved solute is continuously crystallizing.

When the amount of dissolved solute in a solution corresponds to the solute's solubility in the solvent, the solution formed is a saturated solution.

Consider the process of adding table sugar (sucrose) to a container of water. Initially, the added sugar dissolves as the solution is stirred. Finally, as we add more sugar, we reach a point where no amount of stirring will cause the added sugar to dissolve. The last-added sugar remains as a solid on the bottom of the container; the solution is saturated. Although it appears to the eye that nothing is happening once the saturation point is reached, this is not the case on the molecular level. Solid sugar from the bottom of the container is continuously dissolving in the water, and an equal amount of sugar is coming out of solution. Accordingly, the net number of sugar molecules in the liquid remains the same. The equilibrium situation in the saturated solution is somewhat similar to the evaporation of a liquid in a closed container (Section 7.10). Figure 8.3 illustrates the dynamic equilibrium process occurring in a saturated solution that contains undissolved excess solute.

Sometimes it is possible to exceed the maximum solubility of a compound, producing a *supersaturated* solution. A **supersaturated solution** *is an unstable solution that temporarily contains more dissolved solute than that present in a saturated solution.* An indirect rather than a direct procedure is needed to prepare a supersaturated solution; it involves the slow cooling, without agitation of any kind, of a high-temperature saturated solution in which no excess solid solute is present. Even though solute solubility decreases as the temperature is reduced, the excess solute often remains in solution. A supersaturated solution is an unstable situation; with time, excess solute will crystallize out, and the solution will revert to a saturated solution. A supersaturated solution will produce crystals rapidly, often in a dramatic manner, if it is slightly disturbed or if it is "seeded" with a tiny crystal of solute.

An **unsaturated solution** *is a solution that contains less than the maximum amount of solute that can be dissolved under the conditions at which the solution exists.* Most solutions we encounter fall into this category.

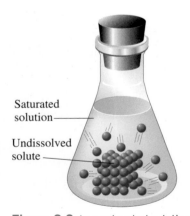

Saturated solution

Undissolved solute

Figure 8.3 In a saturated solution, the dissolved solute is in dynamic equilibrium with the undissolved solute. Solute enters and leaves the solution at the same rate.

CHEMICAL Connections

Factors Affecting Gas Solubility

Both temperature and pressure affect the solubility of a gas in water. The effects are opposite. Increased temperature decreases gas solubility, and increased pressure increases gas solubility. The following table quantifies such effects for carbon dioxide (CO_2), a gas we often encounter dissolved in water.

Solubility of CO_2 (g/100 mL water)

Temperature Effect (at 1 atm pressure)		Pressure Effect (at 0°C)	
0°C	0.348	1 atm	0.348
20°C	0.176	2 atm	0.696
40°C	0.097	3 atm	1.044
60°C	0.058		

The effect of temperature on gas solubility has important environmental consequences because of the use of water from rivers and lakes for industrial cooling. Water used for cooling and then returned to its source at higher than ambient temperatures contains less oxygen and is less dense than when it was diverted. This lower-density "oxygen-deficient" water tends to "float" on colder water below, which blocks normal oxygen adsorption processes. This makes it more difficult for fish and other aquatic forms to obtain the oxygen they need to sustain life. This overall situation is known as *thermal pollution.*

Thermal pollution is sometimes unrelated to human activities. On hot summer days, the temperature of shallow water sometimes reaches the point where dissolved oxygen levels are insufficient to support some life. Under these conditions, suffocated fish may be found on the surface.

A flat taste is often associated with boiled water. This is due in part to the removal of dissolved gases during the boiling process. The removal of dissolved carbon dioxide particularly affects the taste.

The effect of pressure on gas solubility is observed every time a can or bottle of carbonated beverage is opened. The fizzing that occurs results from the escape of gaseous CO_2. The atmospheric pressure associated with an open container is much lower than the pressure used in the bottling process.

Pressure is a factor in the solubility of gases in the bloodstream. In hospitals, persons who are having difficulty obtaining oxygen are given supplementary oxygen. The result is an oxygen pressure greater than that in air. Hyperbaric medical procedures involve the use of pure oxygen. Oxygen pressure is sufficient to cause it to dissolve directly into the bloodstream, bypassing the body's normal mechanism for oxygen uptake (hemoglobin). Treatment of carbon monoxide poisoning is a situation where hyperbaric procedures are often needed.

Deep-sea divers can experience solubility-pressure problems. For every 30 feet that divers descend, the pressure increases by 1 atm. As a result, the air they breathe (particularly the N_2 component) dissolves to a greater extent in the blood. If a diver returns to the surface too quickly after a deep dive, the dissolved gases form bubbles in the blood (in the same way CO_2 does in a freshly opened can of carbonated beverage). This bubble formation may interfere with nerve impulse transmission and restrict blood flow. This painful condition, known as the *bends,* can cause paralysis or death.

Divers can avoid the bends by returning to the surface slowly and by using a helium–oxygen gas mixture instead of air in their breathing apparatus. Helium is less soluble in blood than N_2 and, because of its small atomic size, can escape from body tissues. Nitrogen must be removed via normal respiration.

Carbon dioxide escaping from an opened bottle of a carbonated beverage.

The terms *concentrated* and *dilute* are also used to convey qualitative information about the degree of saturation of a solution. A **concentrated solution** *is a solution that contains a large amount of solute relative to the amount that could dissolve.* A concentrated solution does not have to be a saturated solution (see Figure 8.4). A **dilute solution** *is a solution that contains a small amount of solute relative to the amount that could dissolve.*

When the term *solution* is used, it is generally assumed that "aqueous solution" is meant, unless the context makes it clear that the solvent is not water.

Aqueous and Nonaqueous Solutions

Another set of solution terms involves the modifiers *aqueous* and *nonaqueous.* An **aqueous solution** *is a solution in which water is the solvent.* The presence of water is not a prerequisite

Figure 8.4 Both solutions contain the same amount of solute. A concentrated solution (left) contains a relatively large amount of solute compared with the amount that could dissolve. A dilute solution (right) contains a relatively small amount of solute compared with the amount that could dissolve.

The fact that water molecules are polar is very important in the dissolving of an ionic solid in water.

for a solution, however. A **nonaqueous solution** *is a solution in which a substance other than water is the solvent.* Alcohol-based solutions are often encountered in a medical setting.

 SOLUTION FORMATION

In a solution, solute particles are uniformly dispersed throughout the solvent. Considering what happens at the molecular level during the solution process will help us understand how this is accomplished.

In order for a solute to dissolve in a solvent, two types of interparticle attractions must be overcome: (1) attractions between solute particles (solute–solute attractions) and (2) attractions between solvent particles (solvent–solvent attractions). Only when these attractions are overcome can particles in both pure solute and pure solvent separate from one another and begin to intermingle. A new type of interaction, which does not exist prior to solution formation, arises as a result of the mixing of solute and solvent. This new interaction is the attraction between solute and solvent particles (solute–solvent attractions). These attractions are the primary driving force for solution formation.

An important type of solution process is one in which an ionic solid dissolves in water. Let us consider in detail the process of dissolving sodium chloride, a typical ionic solid, in water (Figure 8.5). The polar water molecules become oriented in such a way that the negative oxygen portion points toward positive sodium ions and the positive hydrogen portions point toward negative chloride ions. As the polar water molecules begin to surround ions on the crystal surface, they exert sufficient attraction to cause these ions to break away from the crystal surface. After leaving the crystal, an ion retains its surrounding group of water molecules; it has become a *hydrated ion.* As each hydrated ion leaves the surface, other ions are exposed to the water, and the crystal is picked apart ion by ion. Once in solution, the hydrated ions are uniformly distributed either by stirring or by random collisions with other molecules or ions.

The random motion of solute ions in solutions causes them to collide with one another, with solvent molecules, and occasionally with the surface of any undissolved solute. Ions undergoing the latter type of collision occasionally stick to the solid surface and thus leave the solution. When the number of ions in solution is low, the chances for collision with the undissolved solute are low. However, as the number of ions in solution increases, so do the chances for collisions, and more ions are recaptured by the undissolved solute. Eventually, the number of ions in solution reaches a level where ions return to the undissolved solute at the same rate at which other ions leave. At this point, the solution is saturated, and the equilibrium process discussed in the previous section is in operation.

Factors Affecting the Rate of Solution Formation

The rate at which a solution forms is governed by how rapidly the solute particles are distributed throughout the solvent. Three factors that affect the rate of solution formation are

Figure 8.5 When an ionic solid, such as sodium chloride, dissolves in water, the water molecules *hydrate* the ions. The positive ions are bound to the water molecules by their attraction for the fractional negative charge on the water's oxygen atom, and the negative ions are bound to the water molecules by their attraction for the fractional positive charge on the water's hydrogen atoms.

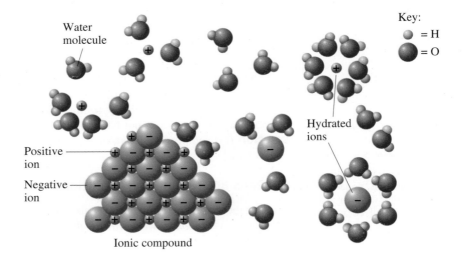

1. *The state of subdivision of the solute.* A crushed aspirin tablet will dissolve in water more rapidly than a whole aspirin tablet. The more compact whole aspirin tablet has less surface area, and thus fewer solvent molecules can interact with it at a given time.
2. *The degree of agitation during solution preparation.* Stirring solution components disperses the solute particles more rapidly, increasing the possibilities for solute–solvent interactions. Hence the rate of solution formation is increased.
3. *The temperature of the solution components.* Solution formation occurs more rapidly as the temperature is increased. At a higher temperature, both solute and solvent molecules move more rapidly (Section 7.1) so more interactions between them occur within a given time period.

8.4 SOLUBILITY RULES

In this section, we will present some rules for qualitatively predicting solute solubilities. These rules are based on polarity considerations—specifically, on the magnitude of the difference between the polarity of the solute and solvent. In general, it is found that the greater the difference in solute–solvent polarity, the less soluble is the solute. This means that *substances of like polarity tend to be more soluble in each other than substances that differ in polarity.* This conclusion is often expressed as the simple phrase *"like dissolves like."* Polar substances, in general, are good solvents for other polar substances but not for nonpolar substances (see Figure 8.6). Similarly, nonpolar substances exhibit greater solubility in nonpolar solvents than in polar solvents.

The generalization "like dissolves like" is a useful tool for predicting solubility behavior in many, but not all, solute–solvent situations. Results that agree with this generalization are nearly always obtained in the cases of gas-in-liquid and liquid-in-liquid solutions and for solid-in-liquid solutions in which the solute is not an ionic compound. For example, NH_3 gas (a polar gas) is much more soluble in H_2O (a polar liquid) than is O_2 gas (a nonpolar gas).

In the common case of solid-in-liquid solutions in which the solute is an ionic compound, the rule "like dissolves like" is not adequate. Their polar nature would suggest that all ionic compounds are soluble in a polar solvent such as water, but this is not the case. The failure of the generalization for ionic compounds is related to the complexity of the factors involved in determining the magnitude of the solute–solute (ion–ion) and solvent–solute (solvent–ion) interactions. Among other things, both the charge and the size of the ions in the solute must be considered. Changes in these factors affect both types of interactions, but not to the same extent.

Some guidelines concerning the solubility of ionic compounds in water, which should be used in place of "like dissolves like," are given in Table 8.2.

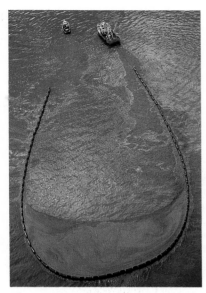

Figure 8.6 Oil spills can be contained to some extent by using trawlers and a boom apparatus because oil and water, having different polarities, are relatively insoluble in each other. The oil, which is of lower density, floats on top of the water.

The generalization "like dissolves like" is not adequate for predicting the solubilities of *ionic compounds* in water. More detailed solubility guidelines are needed (see Table 8.2).

TABLE 8.2
Solubility Guidelines for Ionic Compounds in Water

Soluble Compounds	Important Exceptions
Compounds containing the following ions are soluble with exceptions as noted.	
group IA (Li^+, Na^+, K^+, etc.)	none
ammonium (NH_4^+)	none
acetate ($C_2H_3O_2^-$)	none
nitrate (NO_3^-)	none
chloride (Cl^-), bromide (Br^-), and iodide (I^-)	Ag^+, Pb^{2+}, Hg_2^{2+}
sulfate (SO_4^{2-})	Ca^{2+}, Sr^{2+}, Ba^{2+}, Pb^{2+}
Insoluble Compounds*	**Important Exceptions**
Compounds containing the following ions are insoluble with exceptions as noted.	
carbonate (CO_3^{2-})	group IA and NH_4^+
phosphate (PO_4^{3-})	group IA and NH_4^+
sulfide (S^{2-})	groups IA and IIA and NH_4^+
hydroxide (OH^-)	group IA, Ca^{2+}, Sr^{2+}, Ba^{2+}

*All ionic compounds, even the least soluble ones, dissolve to some slight extent in water. Thus the "insoluble" classification really means ionic compounds that have very limited solubility in water.

 **E X A M P L E 8 . 1**

Predicting Solute Solubility Using
Solubility Rules

With the help of Table 8.2, predict the solubility of each of the following solutes in the solvent indicated.

a. CH_4 (a nonpolar gas) in water
b. Ethyl alcohol (a polar liquid) in chloroform (a polar liquid)
c. AgCl (an ionic solid) in water
d. Na_2SO_4 (an ionic solid) in water
e. $AgNO_3$ (an ionic solid) in water

Solution

a. Insoluble. They are of unlike polarity because water is polar.
b. Soluble. Both substances are polar, so they should be relatively soluble in one another—like dissolves like.
c. Insoluble. Table 8.2 indicates that all chlorides except those of silver, lead, and mercury(I) are soluble. Thus AgCl is one of the exceptions.
d. Soluble. Table 8.2 indicates that all ionic sodium-containing compounds are soluble.
e. Soluble. Table 8.2 indicates that all compounds containing the nitrate ion (NO_3^-) are soluble.

Practice Exercise 8.1

With the help of Table 8.2, predict the solubility of each of the following solutes in the solvent indicated.

a. NO_2 (a polar gas) in water
b. CCl_4 (a nonpolar liquid) in benzene (a nonpolar liquid)
c. NaBr (an ionic solid) in water
d. $MgCO_3$ (an ionic solid) in water
e. $(NH_4)_3PO_4$ (an ionic solid) in water

Answers: a. soluble; **b.** soluble; **c.** soluble; **d.** insoluble; **e.** soluble

The Chemical Connections feature on page 199 considers further the topic of polarity and solubility as it relates to those substances known as vitamins.

 ## 8.5 SOLUTION CONCENTRATION UNITS

Because solutions are mixtures (Section 8.1), they have a variable composition. Specifying what the composition of a solution is involves specifying solute concentrations. A **concentration** *is the amount of solute present in a specified amount of solution.* Many methods of expressing concentration exist, and certain methods are better suited for some purposes than others. In this section we consider two methods: *percent concentration* and *molarity.*

Percent Concentration

There are three different ways of representing percent concentration:

1. Percent by mass (or mass–mass percent)
2. Percent by volume (or volume–volume percent)
3. Mass–volume percent

Percent by mass (or mass–mass percent) is the percentage unit most often used in chemical laboratories. **Percent by mass** *is the mass of solute in a solution divided by the total mass of solution, multiplied by 100 (to put the value in terms of percentage).*

$$\text{Percent by mass} = \frac{\text{mass of solute}}{\text{mass of solution}} \times 100$$

CHEMICAL Connections

Solubility of Vitamins

Polarity plays an important role in the solubility of many substances in the fluids and tissues of the human body. Consider vitamin solubilities. The 13 known vitamins fall naturally into two classes: fat-soluble and water-soluble. The fat-soluble vitamins are A, D, E, and K. Water-soluble vitamins are vitamin C and the eight B vitamins (thiamine, riboflavin, niacin, vitamin B_6, folic acid, vitamin B_{12}, pantothenic acid, and biotin). Water-soluble vitamins have polar molecular structures, as does water. By contrast, fat-soluble vitamins have nonpolar molecular structures that are compatible with the nonpolar nature of fats.

Certain vegetables, including mushrooms, broccoli, and asparagus, are good B vitamin sources, although cooking them in water can leach out much of their vitamin B content.

Vitamin E, which is soluble in fatty tissue because of its nonpolar structure, helps to protect the lungs against air pollutants, especially when a person is breathing rapidly during strenuous exercise.

Vitamin C is water-soluble. Because of this, vitamin C is not stored in the body and must be ingested in our daily diet. Unused vitamin C is eliminated rapidly from the body via bodily fluids. Vitamin A, however, is fat-soluble. It can be, and is, stored by the body in fat tissue for later use. If vitamin A is consumed in excess quantities (from excessive vitamin supplements), illness can result. Because of its limited water solubility, vitamin A cannot be rapidly eliminated from the body by bodily fluids.

The water-soluble vitamins can be easily leached out of foods as they are prepared. As a rule of thumb, you should eat foods every day that are rich in the water-soluble vitamins. Taking megadose vitamin supplements of water-soluble vitamins is seldom effective. The extra amounts of these vitamins are usually picked up by the extracellular fluids, carried away by blood, and excreted in the urine. As one person aptly noted, "If you take supplements of water-soluble vitamins, you may have the most expensive urine in town."

The solute and solution masses must be measured in the same unit, which is usually grams. The mass of the solution is equal to the mass of the solute plus the mass of the solvent.

$$\text{Mass of solution} = \text{mass of solute} + \text{mass of solvent}$$

A solution whose mass percent concentration is 5.0% would contain 5.0 g of solute per 100.0 g of solution (5.0 g of solute and 95.0 g of solvent). Thus percent by mass directly gives the number of grams of solute in 100 g of solution. The percent-by-mass concentration unit is often abbreviated as %(m/m).

> The concentration of butterfat in milk is expressed in terms of percent by mass. When you buy 1% milk, you are buying milk that contains 1 g of butterfat per 100 g of milk.

● EXAMPLE 8.2
Calculating the Percent-by-Mass Concentration of a Solution

▶ What is the percent-by-mass, %(m/m), concentration of sucrose (table sugar) in a solution made by dissolving 7.6 g of sucrose in 83.4 g of water?

Solution

Both the mass of solute and the mass of solvent are known. Substituting these numbers into the percent-by-mass equation

(continued)

$$\%(\text{m/m}) = \frac{\text{mass of solute}}{\text{mass of solution}} \times 100$$

gives

$$\%(\text{m/m}) = \frac{7.6 \text{ g sucrose}}{7.6 \text{ g sucrose} + 83.4 \text{ g water}} \times 100$$

Remember that the denominator of the preceding equation (mass of solution) is the combined mass of the solute and the solvent.

Doing the mathematics gives

$$\%(\text{m/m}) = \frac{7.6 \text{ g}}{91.0 \text{ g}} \times 100 = 8.4\%$$

Practice Exercise 8.2

What is the percent-by-mass, %(m/m), concentration of Na_2SO_4 in a solution made by dissolving 7.6 g of Na_2SO_4 in enough water to give 87.3 g of solution?

Answer: 8.7%(m/m) Na_2SO_4

The second type of percentage unit, percent by volume (or volume–volume percent), which is abbreviated %(v/v), is used as a concentration unit in situations where the solute and solvent are both liquids or both gases. In these cases, it is more convenient to measure volumes than masses. **Percent by volume** *is the volume of solute in a solution divided by the total volume of solution, multiplied by 100.*

$$\text{Percent by volume} = \frac{\text{volume of solute}}{\text{volume of solution}} \times 100$$

Solute and solution volumes must always be expressed in the same units when you use percent by volume.

When the numerical value of a concentration is expressed as a percent by volume, it directly gives the number of milliliters of solute in 100 mL of solution. Thus a 100-mL sample of a 5.0%(v/v) alcohol-in-water solution contains 5.0 mL of alcohol dissolved in enough water to give 100 mL of solution. Note that such a 5.0%(v/v) solution could not be made by adding 5 mL of alcohol to 95 mL of water, because the volumes of two liquids are not usually additive. Differences in the way molecules are packed, as well as differences in distances between molecules, almost always result in the volume of the solution being less than the sum of the volumes of solute and solvent (see Figure 8.7). For example, the final volume resulting from the addition of 50.0 mL of ethyl alcohol to 50.0 mL of water is 96.5 mL of solution (see Figure 8.8). Working problems involving percent by volume entails using the same procedures as those used for problems involving percent by mass.

The proof system for specifying the alcoholic content of beverages is twice the percent by volume. Hence 40 proof is 20%(v/v) alcohol; 100 proof is 50%(v/v) alcohol.

Figure 8.7 When volumes of two different liquids are combined, the volumes are not additive. This process is somewhat analogous to pouring marbles and golf balls together. The marbles can fill in the spaces between the golf balls. This results in the "mixed" volume being less than the sum of the "premixed" volumes.

The third type of percentage unit in common use is mass–volume percent; it is abbreviated %(m/v). This unit, which is often encountered in clinical and hospital settings, is particularly convenient to use when you work with a solid solute, which is easily weighed, and a liquid solvent. Solutions of drugs for internal and external use, intravenous and intramuscular injectables, and reagent solutions for testing are usually labeled in mass–volume percent.

Mass-volume percent *is the mass of solute in a solution (in grams) divided by the total volume of solution (in milliliters), multiplied by 100.*

$$\text{Mass–volume percent} = \frac{\text{mass of solute (g)}}{\text{volume of solution (mL)}} \times 100$$

Note that in the definition of mass–volume percent, specific mass and volume units are given. This is necessary because the units do not cancel, as was the case with mass percent and volume percent.

Mass–volume percent indicates the number of grams of solute dissolved in each 100 mL of solution. Thus a 2.3%(m/v) solution of any solute contains 2.3 g of solute in each 100 mL of solution, and a 5.4%(m/v) solution contains 5.4 g of solute in each 100 mL of solution.

Using Percent Concentrations as Conversion Factors

It is often necessary to prepare solutions of a specific percent concentration. Such a preparation requires knowledge about the amount of solute needed and/or the final volume of solution. These quantities are easily calculated using percent concentration as a conversion factor. Table 8.3 shows the relationship between the definitions for the three types of percent concentrations and the conversion factors that can be derived from them. Examples 8.3 and 8.4 illustrate how these definition-derived conversion factors are used in a problem-solving context.

For dilute aqueous solutions, where the density is close to 1.00 g/mL, %(m/m) and %(m/v) are almost the same because mass in grams of the solution equals the volume in milliliters of the solution.

Figure 8.8 Identical volumetric flasks are filled to the 50.0-mL mark with ethanol and with water. When the two liquids are poured into a 100-mL volumetric flask, the volume is seen to be less than the expected 100.0 mL; it is only 96.5 mL.

EXAMPLE 8.3

Calculating the Mass of Solute Needed to Produce a Solution of a Given Percent-by-Mass Concentration

How many grams of sucrose must be added to 375 g of water to prepare a 2.75%(m/m) solution of sucrose?

Solution

Usually, when a solution concentration is given as part of a problem statement, the concentration information is used in the form of a conversion factor when you solve the problem. That will be the case in this problem.

(continued)

The given quantity is 375 g of H_2O (grams of solvent), and the desired quantity is grams of sucrose (grams of solute).

$$375 \text{ g } H_2O = ? \text{ g sucrose}$$

The conversion factor relating these two quantities (solvent and solute) is obtained from the given concentration. In a 2.75%-by-mass sucrose solution, there are 2.75 g of sucrose for every 97.25 g of water.

$$100.00 \text{ g solution} - 2.75 \text{ g sucrose} = 97.25 \text{ g } H_2O$$

The relationship between grams of solute and grams of solvent (2.75 to 97.25) gives us the needed conversion factor.

$$\frac{2.75 \text{ g sucrose}}{97.25 \text{ g } H_2O}$$

The problem is set up and solved, using dimensional analysis, as follows:

$$375 \text{ g } H_2O \times \left(\frac{2.75 \text{ g sucrose}}{97.25 \text{ g } H_2O} \right) = 10.6 \text{ g sucrose}$$

Practice Exercise 8.3

How many grams of $LiNO_3$ must be added to 25.0 g of water to prepare a 5.00%(m/m) solution of $LiNO_3$?

Answer: 1.32 g $LiNO_3$

TABLE 8.3
Conversion Factors Obtained from Percent Concentration Units

Percent Concentration	Meaning in Words	Conversion Factors	
15%(m/m) NaCl solution	There are 15 g of NaCl in 100 g of solution.	$\dfrac{15 \text{ g NaCl}}{100 \text{ g solution}}$ and	$\dfrac{100 \text{ g solution}}{15 \text{ g NaCl}}$
6%(v/v) methanol solution	There are 6 mL of methanol in 100 mL of solution.	$\dfrac{6 \text{ mL ethanol}}{100 \text{ mL solution}}$ and	$\dfrac{100 \text{ mL solution}}{6 \text{ mL ethanol}}$
25%(m/v) sucrose solution	There are 25 g of sucrose in 100 mL of solution.	$\dfrac{25 \text{ g sucrose}}{100 \text{ mL solution}}$ and	$\dfrac{100 \text{ mL solution}}{25 \text{ g sucrose}}$

 **EXAMPLE 8.4**

Calculating the Mass of Solute Needed to Produce a Solution of a Given Mass–Volume Percent Concentration

Normal saline solution that is used to dissolve drugs for intravenous use is 0.92%(m/v) NaCl in water. How many grams of NaCl are required to prepare 35.0 mL of normal saline solution?

Solution

The given quantity is 35.0 mL of solution, and the desired quantity is grams of NaCl.

$$35.0 \text{ mL solution} = ? \text{ g NaCl}$$

The given concentration, 0.92%(m/v), which means 0.92 g of NaCl per 100 mL of solution, is used as a conversion factor to go from milliliters of solution to grams of NaCl. The setup for the conversion is

$$35.0 \text{ mL solution} \times \left(\frac{0.92 \text{ g NaCl}}{100 \text{ mL solution}} \right)$$

Doing the arithmetic after canceling the units gives

$$\left(\frac{35.0 \times 0.92}{100} \right) \text{g NaCl} = 0.32 \text{ g NaCl}$$

When a percent concentration is given without specifying which of the three types of percent concentration it is (not a desirable situation), it is assumed to mean percent by mass. Thus a 5% NaCl solution is assumed to be a 5%(m/m) NaCl solution.

Practice Exercise 8.4

How many grams of glucose ($C_6H_{12}O_6$) are needed to prepare 500.0 mL of a 4.50%(m/v) glucose–water solution?

Answer: 22.5 g glucose

Molarity

Molarity *is the moles of solute in a solution divided by the liters of solution.* The mathematical equation for molarity is

$$\text{Molarity (M)} = \frac{\text{moles of solute}}{\text{liters of solution}}$$

Note that the abbreviation for molarity is a capital M. A solution containing 1 mole of KBr in 1 L of solution has a molarity of 1 and is said to be a 1 M (1 *molar*) solution.

The molarity concentration unit is often used in laboratories where chemical reactions are being studied. Because chemical reactions occur between molecules and atoms, use of the mole—a unit that counts particles—is desirable. Equal volumes of two solutions of the same molarity contain the same number of solute molecules.

In order to find the molarity of a solution, we need to know the solution volume in liters and the number of moles of solute present. An alternative to knowing the number of moles of solute is knowing the number of grams of solute present and the solute's formula mass. The number of moles can be calculated by using these two quantities (Section 6.4).

 **EXAMPLE 8.5**

Calculating the Molarity of a Solution

Determine the molarities of the following solutions

a. 4.35 moles of $KMnO_4$ are dissolved in enough water to give 750 mL of solution.
b. 20.0 g of NaOH is dissolved in enough water to give 1.50 L of solution.

Solution

a. The number of moles of solute is given in the problem statement.

$$\text{Moles of solute } (KMnO_4) = 4.35 \text{ moles}$$

The volume of the solution is also given in the problem statement, but not in the right units. Molarity requires liters for the volume units, and we are given milliliters of solution. Making the unit change yields

$$750 \text{ mL} \times \left(\frac{10^{-3} \text{ L}}{1 \text{ mL}}\right) = 0.750 \text{ L}$$

The molarity of the solution is obtained by substituting the known quantities into the equation

$$M = \frac{\text{moles of solute}}{\text{liters of solution}}$$

which gives

$$M = \frac{4.35 \text{ moles } KMnO_4}{0.750 \text{ L solution}} = 5.80 \frac{\text{moles } KMnO_4}{\text{L solution}}$$

Note that the units for molarity are always moles per liter.

(continued)

b. This time, the volume of solution is given in liters.

$$\text{Volume of solution} = 1.50 \text{ L}$$

The moles of solute must be calculated from the grams of solute (given) and the solute's molar mass, which is 40.00 g/mole (calculated from atomic masses).

$$20.0 \text{ g NaOH} \times \left(\frac{1 \text{ mole NaOH}}{40.00 \text{ g NaOH}} \right) = 0.500 \text{ mole NaOH}$$

Substituting the known quantities into the defining equation for molarity gives

$$\text{M} = \frac{0.500 \text{ mole NaOH}}{1.50 \text{ L solution}} = 0.333 \frac{\text{mole NaOH}}{\text{L solution}}$$

Practice Exercise 8.5

Determine the molarities of the following solutions.

a. 2.37 moles of KNO_3 are dissolved in enough water to give 650.0 mL of solution.
b. 40.0 g of KCl is dissolved in enough water to give 0.850 L of solution.

Answers: a. 3.65 M KNO_3; **b.** 0.631 M KCl

> When you perform molarity concentration calculations, you need the *identity* of the solute. You cannot calculate moles of solute without knowing the chemical identity of the solute. When you perform percent concentration calculations, the *identity* of the solute is not used in the calculation; all you need is the *amount* of solute.

> In preparing 100 mL of a solution of a specific molarity, enough solvent is added to a weighed amount of solute to give a *final* volume of 100 mL. The weighed solute is not added to a *starting* volume of 100 mL; this would produce a final volume greater than 100 mL because the solute volume increases the total volume.

Using Molarity as a Conversion Factor

The mass of solute present in a known volume of solution is an easily calculable quantity if the molarity of the solution is known. When we do such a calculation, molarity serves as a conversion factor that relates liters of solution to moles of solute. In a similar manner, the volume of solution needed to supply a given amount of solute can be calculated by using the solution's molarity as a conversion factor. Examples 8.6 and 8.7 show, respectively, these uses of molarity as a conversion factor.

 **EXAMPLE 8.6**

Calculating the Amount of Solute Present in a Given Amount of Solution

How many grams of sucrose (table sugar, $C_{12}H_{22}O_{11}$) are present in 185 mL of a 2.50 M sucrose solution?

Solution

The given quantity is 185 mL of solution, and the desired quantity is grams of $C_{12}H_{22}O_{11}$.

$$185 \text{ mL of solution} = ? \text{ g } C_{12}H_{22}O_{11}$$

The pathway used to solve this problem is

$$\text{mL solution} \longrightarrow \text{L solution} \longrightarrow \text{moles } C_{12}H_{22}O_{11} \longrightarrow \text{g } C_{12}H_{22}O_{11}$$

The given molarity (2.50 M) serves as the conversion factor for the second unit change; the formula mass of sucrose (which is not given and must be calculated) is used to accomplish the third unit change.

The dimensional-analysis setup for this pathway is

$$185 \text{ mL solution} \times \left(\frac{10^{-3} \text{ L solution}}{1 \text{ mL solution}} \right) \times \left(\frac{2.50 \text{ moles } C_{12}H_{22}O_{11}}{1 \text{ L solution}} \right)$$
$$\times \left(\frac{342.34 \text{ g } C_{12}H_{22}O_{11}}{1 \text{ mole } C_{12}H_{22}O_{11}} \right)$$

Canceling the units and doing the arithmetic, we find that

$$\left(\frac{185 \times 10^{-3} \times 2.50 \times 342.34}{1 \times 1 \times 1} \right) \text{ g } C_{12}H_{22}O_{11} = 158 \text{ g } C_{12}H_{22}O_{11}$$

Practice Exercise 8.6

How many grams of silver nitrate ($AgNO_3$) are present in 375 mL of 1.50 M silver nitrate solution?

Answer: 95.6 g $AgNO_3$

EXAMPLE 8.7

Calculating the Amount of Solution Needed to Supply a Given Amount of Solute

A typical dose of iron (II) sulfate ($FeSO_4$) used in the treatment of iron-deficiency anemia is 0.35 g. How many milliliters of a 0.10 M iron(II) sulfate solution would be needed to supply this dose?

Solution

The given quantity is 0.35 g of $FeSO_4$; the desired quantity is milliliters of $FeSO_4$ solution.

$$0.35 \text{ g } FeSO_4 = ? \text{ mL } FeSO_4 \text{ solution}$$

The pathway used to solve this problem is

$$\text{g } FeSO_4 \longrightarrow \text{moles } FeSO_4 \longrightarrow \text{L } FeSO_4 \text{ solution} \longrightarrow \text{mL } FeSO_4 \text{ solution}$$

We accomplish the first unit conversion by using the formula mass of $FeSO_4$ (which must be calculated) as a conversion factor. The second unit conversion involves the use of the given molarity as a conversion factor.

$$0.35 \text{ g } FeSO_4 \times \left(\frac{1 \text{ mole } FeSO_4}{151.92 \text{ g } FeSO_4} \right) \times \left(\frac{1 \text{ L solution}}{0.10 \text{ mole } FeSO_4} \right) \times \left(\frac{1 \text{ mL solution}}{10^{-3} \text{ L solution}} \right)$$

Canceling units and doing the arithmetic, we find that

$$\left(\frac{0.35 \times 1 \times 1 \times 1}{151.92 \times 0.10 \times 10^{-3}} \right) \text{mL solution} = 23 \text{ mL solution}$$

Practice Exercise 8.7

How many milliliters of a 0.100 M NaOH solution would be needed to provide 15.0 g of NaOH for a chemical reaction?

Answer: 3750 mL NaOH solution

The Chemical Connections feature on page 206 discusses how drug-concentration levels can be controlled in the delivery of medication to the human body.

The Chemistry at a Glance feature on page 207 reviews the ways in which solution concentrations are specified.

8.6 DILUTION

A common activity encountered when working with solutions is that of diluting a solution of known concentration (usually called a stock solution) to a lower concentration. **Dilution** *is the process in which more solvent is added to a solution in order to lower its concentration.* The same amount of solute is present, but it is now distributed in a larger amount of solvent (the original solvent plus the added solvent).

Often, we prepare a solution of a specific concentration by adding a predetermined volume of solvent to a specific volume of stock solution (see Figure 8.9). A simple

Figure 8.9 Frozen orange juice concentrate is diluted with water prior to drinking.

CHEMICAL Connections

Controlled-Release Drugs: Regulating Concentration, Rate, and Location of Release

In the use of both prescription and over-the-counter drugs, body concentration levels of the drug are obviously of vital importance. All drugs have an optimum concentration range where they are most effective. Below this optimum concentration range, a drug is ineffective, and above it the drug may have adverse side effects. Hence, the much-repeated warning "Take as directed."

Ordinarily, in the administration of a drug, the body's concentration level of the drug rapidly increases toward the higher end of the effective concentration range and then gradually declines and falls below the effective limit. The period of effectiveness of the drug can be extended by using the drug in a controlled-release form. This causes the drug to be released in a regulated, continuous manner over a longer period of time. The accompanying graph contrasts "ordinary-release" and "controlled-release" modes of drug action.

The use of controlled-release medication began in the early 1960s with the introduction of the decongestant Contac. Contac's controlled-release mechanism, which is now found in many drugs and used by all drug manufacturers, involves drug particles encapsulated within a slowly dissolving coating that *varies in thickness* from particle to particle. Particles of the drug with a thinner coating dissolve fast. Those particles with a thicker coating dissolve more slowly, extending the period of drug release. The number of particles of various thicknesses, within a formulation, is predetermined by the manufacturer.

When drugs are taken orally, they first encounter the acidic environment of the stomach. Two problems can occur here:

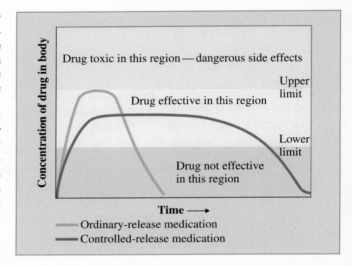

(1) The drug itself may damage the stomach lining. (2) The drug may be rendered inactive by the gastric acid present in the stomach. Controlled-release techniques are useful in overcoming these problems. Drug particle coatings are now available that are acid-resistant; that is, they do not dissolve in acidic solution. Drugs with such coatings pass from the stomach into the small intestine in undissolved form. Within the nonacidic (basic) environment of the small intestine, the dissolving process then begins.

relationship exists between the volumes and concentrations of the diluted and stock solutions. This relationship is

$$\left(\begin{array}{c}\text{Concentration of}\\\text{stock solution}\end{array}\right) \times \left(\begin{array}{c}\text{volume of}\\\text{stock solution}\end{array}\right) = \left(\begin{array}{c}\text{concentration of}\\\text{diluted solution}\end{array}\right) \times \left(\begin{array}{c}\text{volume of}\\\text{diluted solution}\end{array}\right)$$

or

$$C_s \times V_s = C_d \times V_d$$

Use of this equation in a problem-solving context is shown in Example 8.8.

 **EXAMPLE 8.8**

Calculating the Amount of Solvent That Must Be Added to a Stock Solution to Dilute It to a Specified Concentration

A nurse wants to prepare a 1.0%(m/v) silver nitrate solution from 24 mL of a 3.0%(m/v) stock solution of silver nitrate. How much water should be added to the 24 mL of stock solution?

Solution

The volume of water to be added will be equal to the difference between the final and initial volumes. The initial volume is known (24 mL). The final volume can be calculated by using the equation

$$C_s \times V_s = C_d \times V_d$$

Once the final volume is known, the difference between the two volumes can be obtained.

CHEMISTRY AT A GLANCE

Solutions

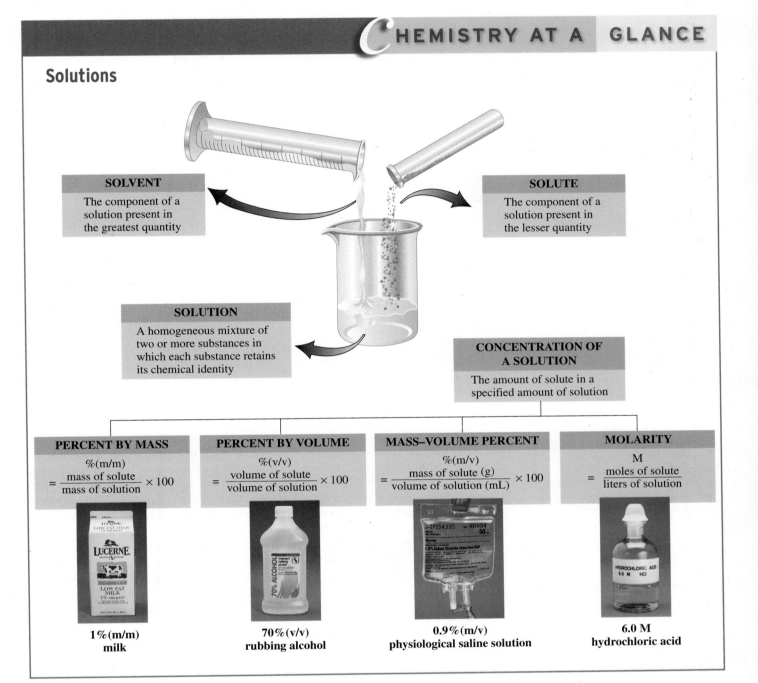

SOLVENT
The component of a solution present in the greatest quantity

SOLUTE
The component of a solution present in the lesser quantity

SOLUTION
A homogeneous mixture of two or more substances in which each substance retains its chemical identity

CONCENTRATION OF A SOLUTION
The amount of solute in a specified amount of solution

PERCENT BY MASS
$$\%(m/m) = \frac{\text{mass of solute}}{\text{mass of solution}} \times 100$$

1%(m/m) milk

PERCENT BY VOLUME
$$\%(v/v) = \frac{\text{volume of solute}}{\text{volume of solution}} \times 100$$

70%(v/v) rubbing alcohol

MASS–VOLUME PERCENT
$$\%(m/v) = \frac{\text{mass of solute (g)}}{\text{volume of solution (mL)}} \times 100$$

0.9%(m/v) physiological saline solution

MOLARITY
$$M = \frac{\text{moles of solute}}{\text{liters of solution}}$$

6.0 M hydrochloric acid

Substituting the known quantities into the dilution equation, which has been rearranged to isolate V_d on the left side, gives

$$V_d = \frac{C_s \times V_s}{C_d} = \frac{3.0\% \ (m/v) \times 24 \text{ mL}}{1.0\% \ (m/v)} = 72 \text{ mL}$$

The solvent added is

$$V_d - V_s = (72 - 24) \text{ mL} = 48 \text{ mL}$$

(continued)

8.7 COLLOIDAL DISPERSIONS AND SUSPENSIONS

Colloidal dispersions are mixtures that have many properties similar to those of solutions, although they are not true solutions. In a broad sense, colloidal dispersions may be thought of as mixtures in which a material is *dispersed* rather than *dissolved.* A **colloidal dispersion** *is a homogeneous mixture that contains dispersed particles that are intermediate in size between those of a true solution and those of an ordinary heterogeneous mixture.* The terms *solute* and *solvent* are not used to indicate the components of a colloidal dispersion. Instead, the particles dispersed in a colloidal dispersion are called the *dispersed phase,* and the material in which they are dispersed is called the *dispersing medium.*

Particles of the dispersed phase in a colloidal dispersion are so small that (1) they are not usually discernible by the naked eye, (2) they do not settle out under the influence of gravity, and (3) they cannot be filtered out using filter paper that has relatively large pores. In these respects, the dispersed phase behaves similarly to a solute in a solution.

However, the dispersed-phase particle size is sufficiently large to make the dispersion nonhomogeneous to light. When we shine a beam of light through a true solution, we cannot see the track of the light. However, a beam of light passing through a colloidal dispersion can be observed because the light is scattered by the dispersed phase (Figure 8.10). This scattered light is reflected into our eyes. This phenomenom, first described by the Irish physicist John Tyndall (1820–1893), is called the *Tyndall effect.* The **Tyndall effect** *is the light-scattering phenomenon that causes the path of a beam of visible light through a colloidal dispersion to be observable.*

Many different biochemical colloidal dispersions occur within the human body. Foremost among them is blood, which has numerous components that are colloidal in size. Fat is transported in the blood and lymph systems as colloidal-sized particles.

The diameters of the dispersed particles in a colloidal dispersion are in the range of 10^{-7} cm to 10^{-5} cm. This compares with diameters of less than 10^{-7} cm for particles such as ions, atoms, and molecules. Thus colloidal particles are up to 1000 times larger

Some chemists use the term *colloid* instead of colloidal dispersion.

Milk is a colloidal dispersion. If you shine a flashlight through a glass of salt water and a glass of milk, you can duplicate the experiment illustrated in Figure 8.10. (For the best effect, dilute the milk with some water until it just looks cloudy.)

Particle size for the dispersed phase in a colloidal dispersion is larger than that for solutes in a true solution.

Figure 8.10 A beam of light travels through a true solution (the yellow liquid) without being scattered—that is, its path cannot be seen. This is not the case for a colloidal dispersion (the red liquid), where scattering of light by the dispersed phase makes the light pathway visible.

TABLE 8.4   Property Comparison for Solutions, Colloidal Dispersions, and Suspensions

Property	Solution	Colloidal Dispersion	Suspension
type of mixture	homogeneous	homogeneous	heterogeneous
type of particles	atoms, ions, and small molecules	groups of small particles or individual larger molecules	very large particles, which are often visible
effect of light	transparent	scatters light (Tyndall effect)	not transparent
settling properties	particles do not settle	particles do not settle	particles settle rapidly
filtration properties	particles cannot be filtered out	particles cannot be filtered out	particles can be filtered out

Several medications are intended to be taken as suspensions. They always carry the instructions "shake well before using." Included among suspension medications are Kaopectate, calamine lotion, and milk of magnesia.

than those present in a true solution. The dispersed particles are usually aggregates of molecules, but this is not always the case. Some protein molecules are large enough to form colloidal dispersions that contain single molecules in suspension.

Colloidal dispersions that contain particles with diameters larger than 10^{-5} cm are usually not encountered. Suspended particles of this size usually settle out under the influence of gravity. For example, within a short period of time, the particles in a sample of stirred "muddy water" will settle to the bottom of a container, producing a clear liquid above the sediment. Stirred muddy water is an example of a *suspension* rather than a colloidal dispersion. A **suspension** *is a heterogeneous mixture that contains dispersed particles that are heavy enough that they settle out under the influence of gravity.* Most filters will remove the dispersed particles from a suspension, which is in direct contrast to the effect of filtration on a colloidal dispersion Table 8.4 compares the properties of "true" solutions, colloidal dispersions, and suspensions.

COLLIGATIVE PROPERTIES OF SOLUTIONS

Adding a solute to a pure solvent causes the solvent's physical properties to change. A special group of physical properties that change when a solute is added are called colligative properties. A **colligative property** *is a physical property of a solution that depends only on the number (concentration) of solute particles (molecules or ions) present in a given quantity of solvent and not on their chemical identities.* Examples of colligative properties include vapor-pressure lowering, boiling-point elevation, freezing-point depression, and osmotic pressure. The first three of these colligative properties are discussed in this section. The fourth, osmotic pressure, will be considered in Section 8.9.

Adding a nonvolatile solute to a solvent *lowers* the vapor pressure of the resulting solution below that of the pure solvent at the same temperature. (A nonvolatile solute is one that has a low vapor pressure and therefore a low tendency to vaporize; Section 7.11.) This lowering of vapor pressure is a direct consequence of some of the solute molecules or ions occupying positions on the surface of the liquid. Their presence decreases the probability of solvent molecules escaping; that is, the number of surface-occupying solvent molecules has been decreased. Figure 8.11 illustrates the decrease in surface concentration of solvent molecules when a solute is added. As the *number* of solute particles increases, the reduction in vapor pressure also increases; thus vapor pressure is a colligative property. What is important is not the identity of the solute molecules but the fact that they take up room on the surface of the liquid.

Adding a nonvolatile solute to a solvent *raises* the boiling point of the resulting solution above that of the pure solvent. This is logical when we remember that the vapor pressure of the solution is lower than that of pure solvent and that the boiling point is

Figure 8.11 Close-ups of the surface of a liquid solvent (a) before and (b) after solute has been added. There are fewer solvent molecules on the surface of the liquid after solute has been added. This results in a decreased vapor pressure for the solution compared with pure solvent.

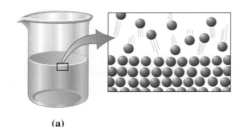

(a)

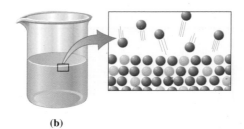

(b)

Figure 8.12 A water-antifreeze mixture has a higher boiling point and a lower freezing point than pure water.

In the making of homemade ice cream, the function of the rock salt added to the ice is to depress the freezing point of the ice–water mixture surrounding the ice cream mix sufficiently to allow the mix (which contains sugar and other solutes and thus has a freezing point below 0°C) to freeze.

The term *osmosis* comes from the Greek *osmos*, which means "push."

An osmotic semipermeable membrane contains very small pores (holes)—too small to see—that are big enough to let small solvent molecules through but not big enough to let larger solute molecules pass through.

dependent on vapor pressure (Section 7.12). A higher temperature will be needed to raise the depressed vapor pressure of the solution to atmospheric pressure; this is the condition required for boiling.

A common application of the phenomenon of boiling point elevation involves automobiles. The coolant ethylene glycol (a nonvolatile solute) is added to car radiators to prevent boilover in hot weather (see Figure 8.12). The engine may not run any cooler, but the coolant–water mixture will not boil until it reaches a temperature well above the normal boiling point of water.

Adding a nonvolatile solute to a solvent *lowers* the freezing point of the resulting solution below that of the pure solvent. The presence of the solute particles within the solution interferes with the tendency of solvent molecules to line up in an organized manner, a condition necessary for the solid state. A lower temperature is necessary before the solvent molecules will form the solid.

Applications of freezing-point depression are even more numerous than those for boiling-point elevation. In climates where the temperature drops below 0°C in the winter, it is necessary to protect water-cooled automobile engines from freezing. This is done by adding antifreeze (usually ethylene glycol) to the radiator. The addition of this nonvolatile material causes the vapor pressure and freezing point of the resulting solution to be much lower than those of pure water. Also in the winter, a salt, usually NaCl or $CaCl_2$, is spread on roads and sidewalks to melt ice or prevent it from forming. The salt dissolves in the water to form a solution that will not freeze until the temperature drops much lower than 0°C, the normal freezing point of water.

8.9 OSMOSIS AND OSMOTIC PRESSURE

The process of osmosis and the colligative property of osmotic pressure are extremely important phenomena when we consider biochemical solutions. These phenomena govern many of the processes important to a functioning human body.

Osmosis

Osmosis *is the passage of a solvent through a semipermeable membrane separating a dilute solution (or pure solvent) from a more concentrated solution.* The simple apparatus shown in Figure 8.13a is helpful in explaining, at the molecular level, what actually occurs during the osmotic process. The apparatus consists of a tube containing a concentrated salt–water solution that has been immersed in a dilute salt–water solution. The immersed end of the tube is covered with a semipermeable membrane. A **semipermeable membrane** *is a membrane that allows certain types of molecules to pass through it but prohibits the passage of other types of molecules.* The selectivity of a semipermeable membrane is based on size differences between molecules. The particles that are allowed to pass through (usually just solvent molecules like water) are relatively small. Thus, the membrane functions somewhat like a sieve. Using the experimental setup of Figure 8.13a, we can observe a net flow of solvent from the dilute to the concentrated solution over the course of time. This is indicated by a rise

Figure 8.13 (a) Osmosis, the flow of solvent through a semipermeable membrane from a dilute to a more concentrated solution, can be observed with this apparatus. (b) The liquid level in the tube rises until equilibrium is reached. At equilibrium, the solvent molecules move back and forth across the membrane at equal rates.

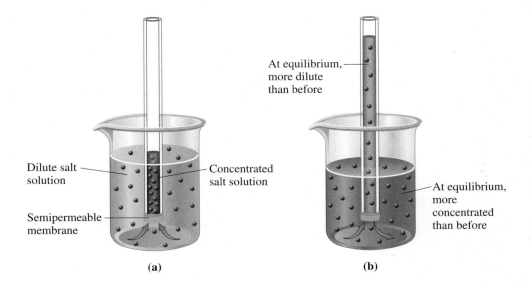

(a)   **(b)**

A process called *reverse osmosis* is used in the desalination of seawater to make drinking water. Pressure greater than the osmotic pressure is applied on the salt water side of the membrane to force solvent water across the membrane from the salt water side to the "pure" water side.

in the level of the solution in the tube and a drop in the level of the dilute solution, as shown in Figure 8.13b.

What is actually happening on a molecular level as the process of osmosis occurs? Water is flowing in both directions through the membrane. However, the rate of flow into the concentrated solution is greater than the rate of flow in the other direction (see Figure 8.14). Why? The presence of solute molecules diminishes the ability of water molecules to cross the membrane. The solute molecules literally get in the way; they occupy some of the surface positions next to the membrane. Because there is a greater concentration of solute molecules on one side of the membrane than on the other, the flow rates differ. The flow rate is diminished to a greater extent on the side of the membrane where the greater concentration of solute is present.

The net transfer of solvent across the membrane continues until (1) the concentrations of solute particles on both sides of the membrane become equal or (2) the hydrostatic pressure on the concentrated side of the membrane (from the difference in liquid levels) becomes sufficient to counterbalance the greater escaping tendency of molecules from the dilute side. From here on, there is an equal flow of solvent in both directions across the membrane, and the volume of liquid on each side of the membrane remains constant.

Osmotic Pressure

Osmotic pressure *is the pressure that must be applied to prevent the net flow of solvent through a semipermeable membrane from a solution of lower solute concentration to a*

Figure 8.14 Enlarged views of a semi-permeable membrane separating (a) pure water and a salt-water solution, and (b) a dilute salt-water solution and a concentrated salt-water solution. In both cases, water moves from the area of lower solute concentration to the area of higher solute concentration.

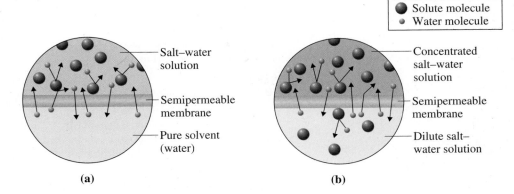

(a)   **(b)**

P (osmotic pressure)

No net flow into the tube because of the applied pressure

Figure 8.15 Osmotic pressure is the amount of pressure needed to prevent the solution in the tube from rising as a result of the process of osmosis.

Osmolarity is greater for ionic solutes than for molecular solutes (solutes that do not separate into ions, such as glucose and sucrose), if the concentrations of the solutions are equal, because ionic solutes dissociate to form more than 1 mole of particles per mole of compound.

The concept of osmolarity also applies to freezing-point depression and to boiling-point elevation (Section 8.8). The freezing-point depression for a 0.1 M NaCl solution ($i = 2$) is twice that for a 0.1 M glucose solution ($i = 1$).

solution of higher solute concentration. In terms of Figure 8.13, osmotic pressure is the pressure required to prevent water from rising in the tube. Figure 8.15 shows how this pressure can be measured. The greater the concentration difference between the separated solutions, the greater the magnitude of the osmotic pressure.

Cell membranes in both plants and animals are semipermeable in nature. The selective passage of fluid materials through these membranes governs the balance of fluids in living systems (see Figure 8.16). Thus osmotic-type phenomena are of prime importance for life. We say "osmotic-type phenomena" instead of "osmosis" because the semipermeable membranes found in living cells usually permit the passage of small solute molecules (nutrients and waste products) in addition to solvent. The term *osmosis* implies the passage of solvent only. The substances prohibited from passing through the membrane in osmotic-type processes are colloidal-sized molecules and insoluble suspended materials (see Section 8.10).

It is because of an osmotic-type process that plants will die if they are watered with salt water. The salt solution outside the root membranes is more concentrated than the solution in the root, so water flows out of the roots; then the plant becomes dehydrated and dies. This same principle is the reason for not drinking excessive amounts of salt water, even if you are stranded on a raft in the middle of the ocean. When salt water is taken into the stomach, water flows out of the stomach wall membranes and into the stomach; then the tissues become dehydrated. Drinking seawater will cause greater thirst because the body will lose water rather than absorb it.

Osmolarity

The osmotic pressure of a solution depends on the number of solute particles present. This in turn depends on the solute concentration and on whether the solute forms ions once it is in solution. Note that two factors are involved in determining osmotic pressure.

The fact that some solutes dissociate into ions in solution is of utmost importance in osmotic pressure considerations. For example, the osmotic pressure of a 1 M NaCl solution is twice that of a 1 M glucose solution, despite the fact that both solutions have equal concentrations (1 M). Sodium chloride is an ionic solute, and it dissociates in solution to give two particles (a Na^+ and a Cl^- ion) per formula unit; however, glucose is a molecular solute and does not dissociate. It is the number of particles present that determines osmotic pressure.

The concentration unit *osmolarity* is used to compare the osmotic pressures of solutions. **Osmolarity** *is the product of a solution's molarity and the number of particles produced per formula unit if the solute dissociates.* The equation for osmolarity is

$$\text{Osmolarity} = \text{molarity} \times i$$

where i is the number of particles produced from the dissociation of one formula unit of solute. The abbreviation for osmolarity is osmol.

● EXAMPLE 8.9

Calculating the Osmolarity of Various Solutions

What is the osmolarity of each of the following solutions?

a. 2 M NaCl   **b.** 2 M $CaCl_2$ **c.** 2 M glucose
d. 2 M in both NaCl and glucose
e. 2 M in NaCl and 1 M in glucose

Solution

The general equation for osmolarity will be applicable in each of the parts of the problem.

$$\text{Osmolarity} = \text{molarity} \times i$$

a. Two particles per dissociation are produced when NaCl dissociates in solution.

$$NaCl \longrightarrow Na^+ + Cl^-$$

The molarity of a 5.0%(m/v) glucose solution is 0.31 M. The molarity of a 0.92%(m/v) NaCl solution is 0.16 M. Despite the differing molarities, these two solutions have the same osmotic pressure. The concept of osmolarity explains why these solutions of different concentration can exhibit the same osmotic pressure. For glucose $i = 1$ and for NaCl $i = 2$.

Figure 8.16 The dissolved substances in tree sap create a more concentrated solution than the surrounding ground water. Water enters membranes in the roots and rises in the tree, creating an osmotic pressure that can exceed 20 atm in extremely tall trees.

The pickling of cucumbers and salt curing of meat are practical applications of the concept of crenation. A concentrated salt solution (brine) is used to draw water from the cells of the cucumber to produce a pickle. Salt on the surface of the meat preserves the meat by crenation of bacterial cells.

The value of i is 2, and the osmolarity is twice the molarity.

$$\text{Osmolarity} = 2\,M \times 2 = 4\ \text{osmol}$$

b. For $CaCl_2$, the value of i is 3, because three ions are produced from the dissociation of one $CaCl_2$ formula unit.

$$CaCl_2 \longrightarrow Ca^{2+} + 2Cl^-$$

The osmolarity will therefore be triple the molarity:

$$\text{Osmolarity} = 2\,M \times 3 = 6\ \text{osmol}$$

c. Glucose is a nondissociating solute. Thus the value of i is 1, and the molarity and osmolarity will be the same—two molar and two osmolar.

d. With two solutes present, we must consider the collective effects of both solutes. For NaCl, $i = 2$; and for glucose, $i = 1$. The osmolarity is calculated as follows:

$$\text{Osmolarity} = \underbrace{2\,M \times 2}_{\text{NaCl}} + \underbrace{2\,M \times 1}_{\text{glucose}} = 6\ \text{osmol}$$

e. This problem differs from the previous one in that the two solutes are not present in equal concentrations. This does not change the way we work the problem. The i values are the same as before, and the osmolarity is

$$\text{Osmolarity} = \underbrace{2\,M \times 2}_{\text{NaCl}} + \underbrace{1\,M \times 1}_{\text{glucose}} = 5\ \text{osmol}$$

Practice Exercise 8.9

What is the osmolarity of each of the following solutions?

a. 3 M $NaNO_3$ **b.** 3 M $Ca(NO_3)_2$ **c.** 3 M sucrose
d. 3 M in both $Ca(NO_3)_2$ and sucrose
e. 3 M in both $NaNO_3$ and $Ca(NO_3)_2$

Answers: a. 6 osmol; **b.** 9 osmol; **c.** 3 osmol; **d.** 12 osmol; **e.** 15 osmol

Solutions of equal osmolarity have equal osmotic pressures. If the osmolarity of one solution is three times that of another, then the osmotic pressure of the first solution is three times that of the second solution. A solution with high osmotic pressure will take up more water than a solution of lower osmotic pressure; thus more pressure must be applied to prevent osmosis.

Hypotonic, Hypertonic, and Isotonic Solutions

The terms *hypotonic solution, hypertonic solution,* and *isotonic solution* pertain to osmotic-type phenomena that occur in the human body. A consideration of what happens to red blood cells when they are placed in three different liquids will help us understand the differences in meaning of these three terms. The liquid media are distilled water, concentrated sodium chloride solution, and physiological saline solution.

When red blood cells are placed in pure water, they swell up (enlarge in size) and finally rupture (burst); this process is called *hemolysis* (Figure 8.17a). Hemolysis is caused by an increase in the amount of water entering the cells compared with the amount of water leaving the cells. This is the result of cellular fluid having a greater osmotic pressure than pure water.

When red blood cells are placed in a concentrated sodium chloride solution, a process opposite to hemolysis occurs. This time, water moves from the cells to the solution, causing the cells to shrivel (shrink in size); this process is called *crenation* (Figure 8.17b). Crenation occurs because the osmotic pressure of the concentrated salt solution surrounding the red cells is greater than that of the fluid within the cells. Water always moves in the direction of greater osmotic pressure.

(a)

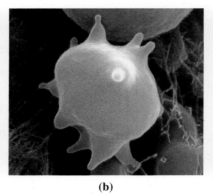
(b)

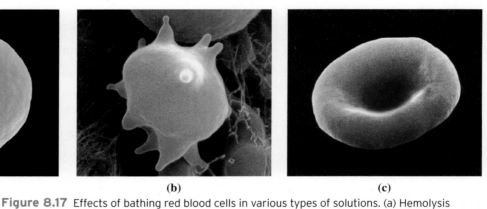
(c)

Figure 8.17 Effects of bathing red blood cells in various types of solutions. (a) Hemolysis occurs in pure water (a hypotonic solution). (b) Crenation occurs in concentrated sodium chloride solution (a hypertonic solution). (c) Cells neither swell nor shrink in physiological saline solution (an isotonic solution).

The word *tonicity* refers to the tone, or firmness, of a biological cell.

The terminology "D5W," often heard in television shows involving doctors and paramedics, refers to a 5%(m/v) solution of glucose (also called dextrose [D] in water (W).

The use of 5%(m/v) glucose solution for intravenous feeding has a shortcoming. A patient can accommodate only about 3 L of water in a day. Three liters of 5%(m/v) glucose water will supply only about 640 kcal of energy, an inadequate amount of energy. A resting patient requires about 1400 kcal/day.

This problem is solved by using solutions that are about 6 times as concentrated as isotonic solutions. They are administered, through a tube, directly into a large blood vessel leading to the heart (the superior vena cava) rather than through a small vein in the arm or leg. The large volume of blood flowing through this vein quickly dilutes the solution to levels that do not upset the osmotic balance in body fluids. Using this technique, patients can be given up to 5000 kcal/day of nourishment.

Finally, when red blood cells are placed in physiological saline solution, a 0.9%(m/v) sodium chloride solution, water flow is balanced and neither hemolysis nor crenation occurs (Figure 8.17c). The osmotic pressure of physiological saline solution is the same as that of red blood cell fluid. Thus the rates of water flow into and out of the red blood cells are the same.

We will now define the terms *hypotonic, hypertonic,* and *isotonic.* A **hypotonic solution** *is a solution with a lower osmotic pressure than that within cells.* The prefix *hypo-* means "under" or "less than normal." Distilled water is hypotonic with respect to red blood cell fluid, and these cells will hemolyze when placed in it (Figure 8.17a). A **hypertonic solution** *is a solution with a higher osmotic pressure than that within cells.* The prefix *hyper-* means "over" or "more than normal." Concentrated sodium chloride solution is hypertonic with respect to red blood cell fluid, and these cells undergo crenation when placed in it (Figure 8.17b). An **isotonic solution** *is a solution with an osmotic pressure that is equal to that within cells.* Red blood cell fluid, physiological saline solution, and 5%(m/v) glucose water are all isotonic with respect to one another. The processes of replacing body fluids and supplying nutrients to the body intravenously require the use of isotonic solutions such as physiological saline and glucose water. If isotonic solutions were not used, the damaging effects of hemolysis or crenation would occur.

It is sometimes necessary to introduce a hypotonic or hypertonic solution, under controlled conditions, into the body to correct an improper "water balance" in a patient. A hypotonic solution can be used to cause water to flow from the blood into surrounding tissue; blood pressure can be decreased in this manner. A hypertonic solution will cause the net transfer of water from tissues to blood; then the kidneys will remove the water. Some laxatives, such as Epsom salts, act by forming hypertonic solutions in the intestines. Table 8.5 summarizes the differences in meaning among the terms *hypotonic, hypertonic,* and *isotonic.*

The Chemistry at a Glance on page 215 summarizes this chapter's discussion of colligative properties of solutions.

TABLE 8.5   Characteristics of Hypotonic, Hypertonic, and Isotonic Solutions

	Type of Solution		
	Hypotonic	Hypertonic	Isotonic
osmolarity relative to body fluids	less than	greater than	equal
osmotic pressure relative to body fluids	less than	greater than	equal
osmotic effect on cells	net flow of water into cells	net flow of water out of cells	equal water flow into and out of cells

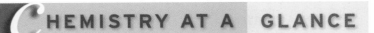

CHEMISTRY AT A GLANCE

Summary of Colligative Property Terminology

COLLIGATIVE PROPERTIES OF SOLUTIONS

The physical properties of a solution that depend only on the concentration of solute particles in a given quantity of solute, not on the chemical identity of the particles.

VAPOR-PRESSURE LOWERING

Addition of a nonvolatile solute to a solvent makes the vapor pressure of the solution LOWER than that of the solvent alone.

BOILING-POINT ELEVATION

Addition of a nonvolatile solute to a solvent makes the boiling point of the solution HIGHER than that of the solvent alone.

FREEZING-POINT DEPRESSION

Addition of a nonvolatile solute to a solvent makes the freezing point of the solution LOWER than that of the solvent alone.

OSMOTIC PRESSURE

The pressure required to stop the net flow of water across a semipermeable membrane separating solutions of differing composition.

OSMOLARITY

Osmolarity = molarity × i, where i = number of particles from the dissociation of one formula unit of solute.

HYPOTONIC SOLUTION

- Solution with an osmotic pressure LOWER than that in cells.
- Causes cells to hemolyze (burst).

HYPERTONIC SOLUTION

- Solution with an osmotic pressure HIGHER than that in cells.
- Causes cells to crenate (shrink).

ISOTONIC SOLUTION

- Solution with an osmotic pressure EQUAL to that in cells.
- Has no effect on cell size.

In *osmosis*, only solvent passes through the membrane. In *dialysis*, both solvent and small solute particles (ions and small molecules) pass through the membrane.

 DIALYSIS

Dialysis is closely related to osmosis. It is the osmotic-type process that occurs in living systems. Osmosis, you recall (Section 8.9), occurs when a solution and a solvent are separated by a semipermeable membrane that allows solvent but not solute to pass through it.

Figure 8.18 In dialysis, there is a net movement of ions from a region of higher concentration to a region of lower concentration. (a) Before dialysis. (b) After dialysis.

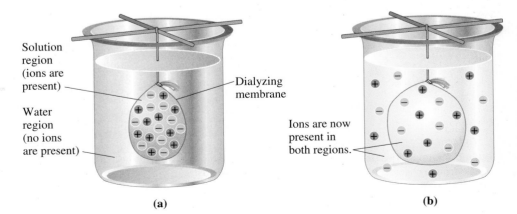

Solution region (ions are present)

Water region (no ions are present)

Dialyzing membrane

Ions are now present in both regions.

(a)   (b)

CHEMICAL Connections

The Artificial Kidney: A Hemodialysis Machine

Individuals who once would have died of kidney failure can now be helped through the use of artificial-kidney machines, which clean the blood of toxic waste products. In these machines, the blood is pumped through tubing made of dialyzing membrane. The tubing passes through a water bath that collects the impurities from the blood. Blood proteins and other important large molecules remain in the blood. This procedure to cleanse the blood is called *hemodialysis.*

In hemodialysis, a catheter is attached to a major artery of one arm, and the patient's blood is passed through a collection of tiny tubes with a carefully selected pore size. These tubes are immersed in a bath (dialyzing solution) that is isotonic in the normal components of blood. The isotonic solution consists of 0.6%(m/v) NaCl, 0.04%(m/v) KCl, 0.2%(m/v) $NaHCO_3$, and 1.5%(m/v) glucose. This solution does not contain urea or other wastes, which diffuse from the blood through the membrane and into the dialyzing solution.

The accompanying figure illustrates a typical hollow-fiber dialysis device.

Typically, an artificial-kidney patient must receive dialysis treatment two or three times a week, for 4 hours per treatment, in order to maintain proper health. A kidney transplant is preferable to many years on hemodialysis. However, kidney transplants are possible only when the donor kidneys are close tissue matches to the recipient.

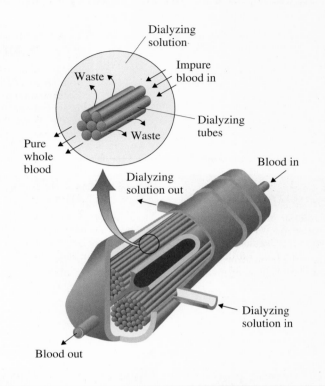

The food taken into our bodies consists primarily of molecules too large to cross cellular membranes. Digestion of food converts these large molecules into smaller molecules that can cross the membranes of cells in the intestinal wall, enter the bloodstream, and then enter cells throughout the body where they are used to produce the energy needed to "run" the body.

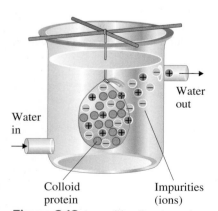

Figure 8.19 Impurities (ions) can be removed from a colloidal dispersion by using a dialysis procedure.

There is a net transfer of solvent from the dilute solution (or pure solvent) into the more concentrated solution. **Dialysis** *is the process in which a semipermeable membrane allows the passage of solvent, dissolved ions, and small molecules but blocks the passage of colloidal-sized particles and large molecules.* Thus dialysis allows for the separation of small particles from colloids and large molecules. Many plant and animal membranes function as dialyzing membranes.

Consider the placement of an aqueous solution of sodium chloride in a dialyzing bag that is surrounded by water (Figure 8.18a). What happens? Sodium ions and chloride ions move through the dialyzing membrane into the water; that is, there is a net movement of ions from a region of high concentration to a region of low concentration. This will occur until both sides of the membrane have equal concentrations of ions (Figure 8.18b).

Dialysis can be used to purify a colloidal solution containing protein molecules and solute. The smaller solute molecules pass through the dialyzing membrane and leave the solution. The larger protein molecules remain behind. The result is a purified protein colloidal dispersion (Figure 8.19).

The human kidneys are a complex dialyzing system that is responsible for removing waste products from the blood. The removed products are then eliminated in urine. When the kidneys fail, these waste products build up and eventually poison the body.

When a person goes into shock, there is a sudden increase in the permeability of the membranes of the blood capillaries. Large colloidally dispersed molecules, such as proteins, leave the bloodstream and leak into the space between cells. This damage disrupts the normal chemistry of the blood. If a patient in shock is left untreated, death can occur.

CONCEPTS TO REMEMBER

Solution components. The component of a solution that is present in the greatest amount is the *solvent*. A *solute* is a solution component that is present in a small amount relative to the solvent (Section 8.1).

Solution characteristics. A solution is a homogeneous (uniform) mixture. Its composition and properties are dependent on the ratio of solute(s) to solvent. Dissolved solutes are present as individual particles (molecules, atoms, or ions) (Section 8.1).

Solubility. The solubility of a solute is the maximum amount of solute that will dissolve in a given amount of solvent. The extent to which a solute dissolves in a solvent depends on the structure of solute and solvent, the temperature, and the pressure. Molecular polarity is a particularly important factor in determining solubility. A *saturated* solution contains the maximum amount of solute that can be dissolved under the conditions at which the solution exists (Section 8.2).

Solution concentration. Solution concentration is the amount of solute present in a specified amount of solution. Percent solute and molarity are commonly encountered concentration units. Percent concentration units include percent by mass, percent by volume, and mass–volume percent. Molarity gives the moles of solute per liter of solution (Section 8.5).

Dilution. Dilution involves adding solvent to an existing solution. Although the amount of solvent increases, the amount of solute remains the same. The net effect of dilution is a decrease in the concentration of the solution (Section 8.6).

Colloidal dispersions and suspensions. Colloidal dispersions are homogeneous mixtures in which the particles in the dispersed phase are small enough that they do not settle out under the influence of gravity but yet large enough to scatter light (Tyndall effect), even though they cannot be seen with the naked eye. Colloidal dispersions are often encountered within the human body. A suspension is a heterogeneous mixture in which the particles in the dispersed phase are large enough to settle out under the influence of gravity (Section 8.7).

Colligative properties of solutions. Properties of a solution that depend on the number of solute particles in solution, not on their identity, are called colligative properties. Vapor-pressure lowering, boiling-point elevation, freezing-point depression, and osmotic pressure are all colligative properties (Section 8.8).

Osmosis and osmotic pressure. Osmosis involves the passage of a solvent from a dilute solution (or pure solvent) through a semipermeable membrane into a more concentrated solution. Osmotic pressure is the amount of pressure needed to prevent the net flow of solvent across the membrane in the direction of the more concentrated solution (Section 8.9).

Dialysis. Dialysis is the process in which a semipermeable membrane permits the passage of solvent, dissolved ions, and small molecules but blocks the passage of large molecules. Many plant and animal membranes function as dialyzing membranes (Section 8.10).

KEY REACTIONS AND EQUATIONS

1. Percent by mass (Section 8.5)

$$\%(m/m) = \frac{\text{mass of solute}}{\text{mass of solution}} \times 100$$

2. Percent by volume (Section 8.5)

$$\%(v/v) = \frac{\text{volume of solute}}{\text{volume of solution}} \times 100$$

3. Mass–volume percent (Section 8.5)

$$\%(m/v) = \frac{\text{mass of solute (g)}}{\text{volume of solution (mL)}} \times 100$$

4. Molarity (Section 8.5)

$$M = \frac{\text{moles of solute}}{\text{liters of solution}}$$

5. Dilution of stock solution to make less-concentrated solution (Section 8.6)

$$C_s \times V_s = C_d \times V_d$$

6. Osmolarity (Section 8.9)

$$\text{osmol} = M \times i$$

EXERCISES *and* PROBLEMS

The members of each pair of problems in this section test similar material.

● Solution Characteristics (Section 8.1)

8.1 Indicate whether each of the following statements about the general properties of solutions is true or false.
  a. A solution may contain more than one solute.
  b. All solutions are homogeneous mixtures.
  c. Every part of a solution has exactly the same properties as every other part.
  d. The solutes present in a solution will "settle out" with time if the solution is left undisturbed.

8.2 Indicate whether each of the following statements about the general properties of solutions is true or false.
  a. All solutions have a variable composition.
  b. For solution formation to occur, the solute and solvent must chemically react with each other.
  c. Solutes are present as individual particles (molecules, atoms, or ions) in a solution.
  d. A general characteristic of all solutions is the liquid state.

8.3 Identify the *solute* and the *solvent* in solutions composed of the following:
a. 5.00 g of sodium chloride (table salt) and 50.0 g of water
b. 4.00 g of sucrose (table sugar) and 1000 g of water
c. 2.00 mL of water and 20.0 mL of ethyl alcohol
d. 60.0 mL of methyl alcohol and 20.0 mL of ethyl alcohol

8.4 Identify the *solute* and the *solvent* in solutions composed of the following:
a. 5.00 g of NaBr and 200.0 g of water
b. 50.0 g of $AgNO_3$ and 1000 g of water
c. 50.0 mL of water and 100.0 mL of methyl alcohol
d. 50.0 mL of isopropyl alcohol and 20.0 mL of ethyl alcohol

● **Solubility (Section 8.2)**

8.5 For each of the following pairs of solutions, select the solution for which solute solubility is greatest.
a. Ammonia gas in water with P = 1 atm and T = 50°C
Ammonia gas in water with P = 1 atm and T = 90°C
b. Carbon dioxide gas in water with P = 2 atm and T = 50°C
Carbon dioxide gas in water with P = 1 atm and T = 50°C
c. Table salt in water with P = 1 atm and T = 60°C
Table salt in water with P = 1 atm and T = 50°C
d. Table sugar in water with P = 2 atm and T = 40°C
Table sugar in water with P = 1 atm and T = 70°C

8.6 For each of the following pairs of solutions, select the solution for which solute solubility is greatest.
a. Oxygen gas in water with P = 1 atm and T = 10°C
Oxygen gas in water with P = 1 atm and T = 20°C
b. Nitrogen gas in water with P = 2 atm and T = 50°C
Nitrogen gas in water with P = 1 atm and T = 70°C
c. Table salt in water with P = 1 atm and T = 40°C
Table salt in water with P = 1 atm and T = 70°C
d. Table sugar in water with P = 3 atm and T = 30°C
Table sugar in water with P = 1 atm and T = 80°C

8.7 Classify each of the following solutions as *saturated, unsaturated,* or *supersaturated* based on the following observations.
a. Agitation of the solution produces a large amount of solid crystals.
b. Heating the solution causes excess undissolved solute present to dissolve.
c. Excess undissolved solute is present at the bottom of the solution container.
d. The amount of solute dissolved is less than the maximum amount that could dissolve under the conditions at which the solution exists.

8.8 Classify each of the following solutions as *saturated, unsaturated,* or *supersaturated* based on the following observations about adding a small piece of solid solute to the solution.
a. The added solute rapidly dissolves.
b. The added solute falls to the bottom of the container where it remains without any decrease in size.
c. The added solute falls to the bottom of the container where it decreases in size for several hours and thereafter its size remains constant.
d. The added solute causes the production of a large amount of solid white crystals.

8.9 Use Table 8.1 to determine whether each of the following solutions is *saturated* or *unsaturated*.
a. 1.94 g of $PbBr_2$ in 100 g of H_2O at 50°C
b. 34.0 g of NaCl in 100 g of H_2O at 0°C
c. 75.4 g of $CuSO_4$ in 200 g of H_2O at 100°C
d. 0.540 g of Ag_2SO_4 in 50 g of H_2O at 50°C

8.10 Use Table 8.1 to determine whether each of the following solutions is *saturated* or *unsaturated*.
a. 175 g of CsCl in 100 g of H_2O at 100°C
b. 455 g of $AgNO_3$ in 100 g of H_2O at 50°C
c. 2.16 g of Ag_2SO_4 in 200 g of H_2O at 50°C
d. 0.97 g of $PbBr_2$ in 50 g of H_2O at 50°C

8.11 Use Table 8.1 to determine whether each of the following solutions is *dilute* or *concentrated*.
a. 0.20 g of $CuSO_4$ in 100 g of H_2O at 100°C
b. 1.50 g of $PbBr_2$ in 100 g of H_2O at 50°C
c. 61 g of $AgNO_3$ in 100 g of H_2O at 50°C
d. 0.50 g of Ag_2SO_4 in 100 g of H_2O at 0°C

8.12 Use Table 8.1 to determine whether each of the following solutions is *dilute* or *concentrated*.
a. 255 g of $AgNO_3$ in 100 g of H_2O at 100°C
b. 35.0 g of NaCl in 100 g of H_2O at 0°C
c. 1.87 g of $PbBr_2$ in 100 g of H_2O at 50°C
d. 1.87 g of $CuSO_4$ in 100 g of H_2O at 50°C

● **Solution Formation (Section 8.3)**

8.13 Match each of the following statements about the dissolving of the ionic solid NaCl in water with the term *hydrated ion, hydrogen atom,* or *oxygen atom*.
a. A Na^+ ion surrounded with water molecules
b. A Cl^- ion surrounded with water molecules
c. The portion of a water molecule that is attracted to a Na^+ ion
d. The portion of a water molecule that is attracted to a Cl^- ion

8.14 Match each of the following statements about the dissolving of the ionic solid KBr in water with the term *hydrated ion, hydrogen atom,* or *oxygen atom*.
a. A K^+ ion surrounded with water molecules
b. A Br^- ion surrounded with water molecules
c. The portion of a water molecule that is attracted to a K^+ ion
d. The portion of a water molecule that is attracted to a Br^- ion

8.15 Indicate whether each of the following actions will *increase* or *decrease* the rate of the dissolving of a sugar cube in water.
a. Cooling the sugar cube–water mixture
b. Stirring the sugar cube–water mixture
c. Breaking the sugar cube up into smaller "chunks"
d. Crushing the sugar cube to give a granulated form of sugar

8.16 Indicate whether each of the following actions will *increase* or *decrease* the rate of the dissolving of table salt in water.
a. Heating the table salt–water mixture
b. Shaking the table salt–water mixture
c. Heating the table salt prior to adding it to the water
d. Heating the water prior to its receiving the table salt

● **Solubility Rules (Section 8.4)**

8.17 Predict whether the following solutes are *very soluble* or *slightly soluble* in water.
a. O_2 (a nonpolar gas)   b. CH_3OH (a polar liquid)
c. CBr_4 (a nonpolar liquid)   d. AgCl (an ionic solid)

8.18 Predict whether the following solutes are *very soluble* or *slightly soluble* in water.
a. NH_3 (a polar gas)   b. N_2 (a nonpolar gas)
c. C_6H_6 (a nonpolar liquid)   d. Na_3PO_4 (an ionic solid)

8.19 Ethanol is a polar solvent and carbon tetrachloride is a nonpolar solvent. In which of these two solvents are each of the following solutes more likely to be soluble?
a. NaCl, ionic   b. Cooking oil, nonpolar
c. Sugar, polar   d. $LiNO_3$, ionic

8.20 Methanol is a polar solvent and heptane is a nonpolar solvent. In which of these two solvents are each of the following solutes more likely to be soluble?
- a. KCl, ionic
- b. Rubbing alcohol, polar
- c. Gasoline, nonpolar
- d. $NaNO_3$, ionic

8.21 Classify each of the following types of ionic compounds in the solubility categories *soluble, soluble with exceptions, insoluble,* or *insoluble with exceptions.*
- a. Chlorides and sulfates
- b. Nitrates and ammonium–ion containing
- c. Carbonates and phosphates
- d. Sodium–ion containing and potassium–ion containing

8.22 Classify each of the following types of ionic compounds in the solubility categories *soluble, soluble with exceptions, insoluble,* or *insoluble with exceptions.*
- a. Nitrates and sodium–ion containing
- b. Chlorides and bromides
- c. Hydroxides and phosphates
- d. Sulfates and iodides

8.23 In each of the following sets of ionic compounds, identify the members of the set that are soluble in water.
- a. $NaCl$, Na_2SO_4, $NaNO_3$, Na_2CO_3
- b. $AgNO_3$, KNO_3, $Ca(NO_3)_2$, $Cu(NO_3)_2$
- c. $CaBr_2$, $Ca(OH)_2$, $CaCl_2$, $CaSO_4$
- d. $NiSO_4$, $Ni_3(PO_4)_2$, $Ni(OH)_2$, $NiCO_3$

8.24 In each of the following sets of ionic compounds, identify the members of the set that are soluble in water.
- a. K_2SO_4, KOH, KI, K_3PO_4
- b. $NaCl$, $AgCl$, $BeCl_2$, $CuCl_2$
- c. $Ba(OH)_2$, $BaSO_4$, $BaCO_3$, $Ba(NO_3)_2$
- d. $CoBr_2$, $CoCl_2$, $Co(OH)_2$, $CoSO_4$

● **Percent Concentration Units (Section 8.5)**

8.25 The following diagrams show varying amounts of the same solute (the red spheres) in varying amounts of solution.

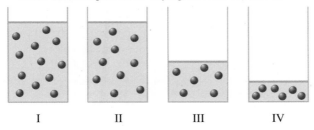

I	II	III	IV

- a. In which of the diagrams is the solution concentration the largest?
- b. In which two of the diagrams are the solution concentrations the same?

8.26 The following diagrams show varying amounts of the same solute (the red spheres) in varying amounts of solution.

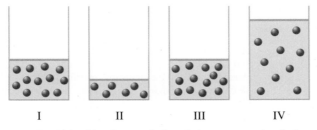

I	II	III	IV

- a. In which of the diagrams is the solution concentration the largest?
- b. In which two of the diagrams are the solution concentrations the same?

8.27 Calculate the mass percent of solute in the following solutions.
- a. 6.50 g of NaCl dissolved in 85.0 g of H_2O
- b. 2.31 g of LiBr dissolved in 35.0 g of H_2O
- c. 12.5 g of KNO_3 dissolved in 125 g of H_2O
- d. 0.0032 g of NaOH dissolved in 1.2 g of H_2O

8.28 Calculate the mass percent of solute in the following solutions.
- a. 2.13 g of $AgNO_3$ dissolved in 30.0 g of H_2O
- b. 135 g of CsCl dissolved in 455 g of H_2O
- c. 10.3 g of K_2SO_4 dissolved in 93.7 g of H_2O
- d. 10.3 g of KBr dissolved in 125 g of H_2O

8.29 How many grams of glucose must be added to 275 g of water in order to prepare each of the following percent-by-mass concentrations of aqueous glucose solution?
- a. 1.30%
- b. 5.00%
- c. 20.0%
- d. 31.0%

8.30 How many grams of lactose must be added to 655 g of water in order to prepare each of the following percent-by-mass concentrations of aqueous lactose solution?
- a. 0.50%
- b. 2.00%
- c. 10.0%
- d. 25.0%

8.31 Calculate the mass, in grams, of K_2SO_4 needed to prepare 32.00 g of 2.000%(m/m) K_2SO_4 solution.

8.32 Calculate the mass, in grams, of KCl needed to prepare 200.0 g of 5.000%(m/m) KCl solution.

8.33 How many grams of water must be added to 20.0 g of NaOH in order to prepare a 6.75%(m/m) solution?

8.34 How many grams of water must be added to 10.0 g of $Ca(NO_3)_2$ in order to prepare a 12.0%(m/m) solution?

8.35 Calculate the volume percent of solute in each of the following solutions.
- a. 20.0 mL of methyl alcohol in enough water to give 475 mL of solution
- b. 4.00 mL of bromine in enough carbon tetrachloride to give 87.0 mL of solution

8.36 Calculate the volume percent of solute in each of the following solutions.
- a. 60.0 mL of water in enough ethylene glycol to give 970.0 mL of solution
- b. 455 mL of ethyl alcohol in enough water to give 1375 mL of solution

8.37 What is the percent by volume of isopropyl alcohol in an aqueous solution made by diluting 22 mL of pure isopropyl alcohol with water to give a volume of 125 mL of solution?

8.38 What is the percent by volume of acetone in an aqueous solution made by diluting 75 mL of pure acetone with water to give a volume of 785 mL of solution?

8.39 Calculate the mass–volume percent of $MgCl_2$ in each of the following solutions.
- a. 5.0 g of $MgCl_2$ in enough water to give 250 mL of solution
- b. 85 g of $MgCl_2$ in enough water to give 580 mL of solution

8.40 Calculate the mass–volume percent of $NaNO_3$ in each of the following solutions.
- a. 1.00 g of $NaNO_3$ in enough water to give 75.0 mL of solution
- b. 100.0 g of $NaNO_3$ in enough water to give 1250 mL of solution

8.41 How many grams of Na_2CO_3 are needed to prepare 25.0 mL of a 2.00%(m/v) Na_2CO_3 solution?

8.42 How many grams of $Na_2S_2O_3$ are needed to prepare 50.0 mL of a 5.00%(m/v) $Na_2S_2O_3$ solution?

8.43 How many grams of NaCl are present in 50.0 mL of a 7.50%(m/v) NaCl solution?

8.44 How many grams of glucose are present in 250.0 mL of a 10.0%(m/v) glucose solution?

● Molarity (Section 8.5)

8.45 Calculate the molarity of the following solutions.
  a. 3.0 moles of potassium nitrate (KNO_3) in 0.50 L of solution
  b. 12.5 g of sucrose ($C_{12}H_{22}O_{11}$) in 80.0 mL of solution
  c. 25.0 g of sodium chloride (NaCl) in 1250 mL of solution
  d. 0.00125 mole of baking soda ($NaHCO_3$) in 2.50 mL of solution

8.46 Calculate the molarity of the following solutions.
  a. 2.0 moles of ammonium chloride (NH_4Cl) in 2.50 L of solution
  b. 14.0 g of silver nitrate ($AgNO_3$) in 1.00 L of solution
  c. 0.025 mole of potassium chloride (KCl) in 50.0 mL of solution
  d. 25.0 g of glucose ($C_6H_{12}O_6$) in 1.25 L of solution

8.47 Calculate the number of grams of solute in each of the following solutions.
  a. 2.50 L of a 3.00 M HCl solution
  b. 10.0 mL of a 0.500 M KCl solution
  c. 875 mL of a 1.83 M $NaNO_3$ solution
  d. 75 mL of a 12.0 M H_2SO_4 solution

8.48 Calculate the number of grams of solute in each of the following solutions.
  a. 3.00 L of a 2.50 M HCl solution
  b. 50.0 mL of a 12.0 M HNO_3 solution
  c. 50.0 mL of a 12.0 M $AgNO_3$ solution
  d. 1.20 L of a 0.032 M Na_2SO_4 solution

8.49 Calculate the volume, in milliliters, of solution required to supply each of the following.
  a. 1.00 g of sodium chloride (NaCl) from a 0.200 M sodium chloride solution
  b. 2.00 g of glucose ($C_6H_{12}O_6$) from a 4.20 M glucose solution
  c. 3.67 moles of silver nitrate ($AgNO_3$) from a 0.400 M silver nitrate solution
  d. 0.0021 mole of sucrose ($C_{12}H_{22}O_{11}$) from an 8.7 M sucrose solution

8.50 Calculate the volume, in milliliters, of solution required to supply each of the following.
  a. 4.30 g of lithium chloride (LiCl) from a 0.089 M lithium chloride solution
  b. 429 g of lithium nitrate ($LiNO_3$) from an 11.2 M lithium nitrate solution
  c. 2.25 moles of potassium sulfate (K_2SO_4) from a 0.300 M potassium sulfate solution
  d. 0.103 mole of potassium hydroxide (KOH) from an 8.00 M potassium hydroxide solution

● Dilution (Section 8.6)

8.51 What is the molarity of the solution prepared by diluting 25.0 mL of 0.220 M NaCl to each of the following final volumes?
  a. 30.0 mL   b. 75.0 mL   c. 457 mL   d. 2.00 L

8.52 What is the molarity of the solution prepared by diluting 35.0 mL of 1.25 M $AgNO_3$ to each of the following final volumes?
  a. 50.0 mL   b. 95.0 mL   c. 975 mL   d. 3.60 L

8.53 For each of the following solutions, how many milliliters of water should be added to yield a solution that has a concentration of 0.100 M?
  a. 50.0 mL of 3.00 M NaCl   b. 2.00 mL of 1.00 M NaCl
  c. 1.45 L of 6.00 M NaCl   d. 75.0 mL of 0.110 M NaCl

8.54 For each of the following solutions, how many milliliters of water should be added to yield a solution that has a concentration of 0.125 M?
  a. 25.0 mL of 1.00 M $AgNO_3$
  b. 5.00 mL of 10.0 M $AgNO_3$
  c. 2.50 L of 2.50 M $AgNO_3$
  d. 75.0 mL of 0.130 M $AgNO_3$

8.55 Determine the final concentration of each of the following solutions after 20.0 mL of water has been added.
  a. 30.0 mL of 5.0 M NaCl solution
  b. 30.0 mL of 5.0 M $AgNO_3$ solution
  c. 30.0 mL of 7.5 M NaCl solution
  d. 60.0 mL of 2.0 M NaCl solution

8.56 Determine the final concentration of each of the following solutions after 30.0 mL of water has been added.
  a. 20.0 mL of 5.0 M NaCl solution
  b. 20.0 mL of 5.0 M $AgNO_3$ solution
  c. 20.0 mL of 0.50 M NaCl solution
  d. 60.0 mL of 3.0 M NaCl solution

8.57 The following diagrams show various amounts of the same solute (the red spheres) in varying amounts of solution. If one-half of the solution in diagram I is withdrawn and then diluted by a factor of 4, which of the other diagrams (II–IV) represent the newly formed solution?

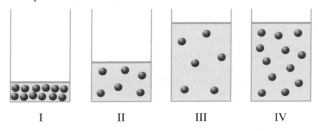

I   II   III   IV

8.58 The following diagrams show various amounts of the same solute (the red spheres) in varying amounts of solution. If one-half of the solution in diagram I is withdrawn and then diluted by a factor of 2, which of the other diagrams (II–IV) represent the newly formed solution?

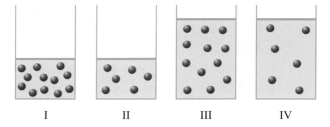

I   II   III   IV

● Colloidal Dispersions and Suspensions (Section 8.7)

8.59 Indicate whether each of the following characterizations applies to a *true solution*, a *colloidal dispersion*, or a *suspension*. There may be more than one correct answer for a given characterization.
  a. Dispersed phase can be trapped by filter paper
  b. Dispersed phase can often be seen with the naked eye
  c. Appearance is that of a homogeneous mixture
  d. Dispersed phase is transparent to light

8.60 Indicate whether each of the following characterizations applies to a *true solution*, a *colloidal dispersion*, or a *suspension*. There may be more than one correct answer for a given characterization.
  a. Dispersed phase cannot be trapped by filter paper

b. Dispersed phase rapidly settles out under the influence of gravity
c. Dispersed phase scatters light
d. Dispersed phase contains very large particles

8.61 Indicate whether each of the following statements about colloidal dispersions and suspensions are *true* or *false*.
  a. The dispersed particles in a colloidal dispersion can be seen with the naked eye.
  b. With time, the particles in a colloidal dispersion settle out under the influence of gravity.
  c. Milk is an example of a suspension.
  d. The dispersed phase in a suspension can be trapped by filter paper.

8.62 Indicate whether each of the following statements about colloidal dispersions and suspensions are *true* or *false*.
  a. Blood is an example of a colloidal dispersion.
  b. The dispersed particles in a colloidal dispersion are several times larger than those in a suspension.
  c. A beam of light is scattered by the dispersed phase in a colloidal dispersion.
  d. With time, the particles in a suspension settle out under the influence of gravity.

● **Colligative Properties of Solutions (Section 8.8)**

8.63 Why is the vapor pressure of a solution that contains a non-volatile solute always less than that of pure solvent?

8.64 How are the boiling point and freezing point of water affected by the addition of solute?

8.65 Why does seawater evaporate more slowly than fresh water at the same temperature?

8.66 How does the freezing point of seawater compare with that of fresh water?

8.67 Assume that you have identical volumes of two liquids; the first is pure water and the second is 0.1 M sucrose (table sugar) solution. Consider the following diagrams, where red is the pure water and blue is the sucrose solution.

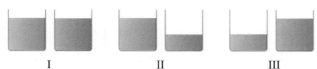

I   II   III

Which one of these diagrams best represents the two liquids after they have stood uncovered for a few days and some evaporation of liquid has occurred?

8.68 Assume that you have identical volumes of two liquids; the first is 0.3 M glucose solution and the second is 0.1 M glucose solution. Based on the diagrams in Problem 8.67, where red is the 0.3 M glucose and blue is the 0.1 M glucose, which one of the diagrams best represents the two liquids after they have stood uncovered for a few days and some evaporation of liquid has occurred?

● **Osmosis and Osmotic Pressure (Section 8.9)**

8.69 Indicate whether the osmotic pressure of a 0.1 M NaCl solution will be *less than, the same as,* or *greater than* that of each of the following solutions.
  a. 0.1 M NaBr   b. 0.050 M $MgCl_2$
  c. 0.1 M $MgCl_2$ d. 0.1 M glucose

8.70 Indicate whether the osmotic pressure of a 0.1 M $NaNO_3$ solution will be *less than, the same as,* or *greater than* that of each of the following solutions.
  a. 0.1 M NaCl   b. 0.1 M KNO_3
  c. 0.1 M Na_2SO_4 d. 0.1 M glucose

8.71 What is the ratio of the osmotic pressures of 0.30 M NaCl and 0.10 M $CaCl_2$?

8.72 What is the ratio of the osmotic pressures of 0.20 M NaCl and 0.30 M $CaCl_2$?

8.73 Would red blood cells *swell, remain the same size,* or *shrink* when placed in each of the following solutions?
  a. 0.9%(m/v) glucose solution
  b. 0.9%(m/v) NaCl solution
  c. 2.3%(m/v) glucose solution
  d. 5.0%(m/v) NaCl solution

8.74 Would red blood cells *swell, remain the same size,* or *shrink* when placed in each of the following solutions?
  a. Distilled water
  b. 0.5%(m/v) NaCl solution
  c. 3.3%(m/v) glucose solution
  d. 5.0%(m/v) glucose solution

8.75 Will red blood cells *crenate, hemolyze,* or *remain unaffected* when placed in each of the solutions in Problem 8.73?

8.76 Will red blood cells *crenate, hemolyze,* or *remain unaffected* when placed in each of the solutions in Problem 8.74?

8.77 Classify each of the solutions in Problem 8.73 as hypotonic, hypertonic, or isotonic.

8.78 Classify each of the solutions in Problem 8.74 as hypotonic, hypertonic, or isotonic.

8.79 Consider two solutions, A and B, separated by an osmotic semi-permeable membrane that allows only water to pass through, as shown in the following diagram.

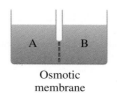

Osmotic
membrane

Based on each of the following identities for solutions A and B, indicate whether the liquid level in compartment A, with time, will *increase, decrease,* or *not change.*
  a. A = 1.0 M NaCl solution and B = 2.0 M NaCl solution
  b. A = 5.0%(m/v) glucose solution and B = 4.0%(m/v) glucose solution
  c. A = 2.0 M KCl solution and B = 2.0 M KNO_3 solution
  d. A = 1.0 M glucose solution and B = 1.0 M NaCl solution

8.80 Consider two solution, A and B, separated by an osmotic semi-permeable membrane that allows only water to pass through, as shown in the diagram in Problem 8.79. Based on each of the following identities for solutions A and B, indicate whether the liquid level in compartment A, with time, will *increase, decrease,* or *not change.*
  a. A = 1.0 M glucose solution and B = 2.0 M glucose solution
  b. A = 5.0%(m/v) NaCl solution and B = 4.0%(m/v) NaCl solution
  c. A = 2.0 M Na_2SO_4 solution and B = 3.0 M KNO_3 solution
  d. A = 2.0 M glucose solution and B = 1.0 M NaCl solution

● **Dialysis (Section 8.10)**

8.81 What happens in each of the following situations?
  a. A dialyzing bag containing a 1 M solution of potassium chloride is immersed in pure water.
  b. A dialyzing bag containing colloidal-sized protein, 1 M potassium chloride, and 1 M glucose is immersed in pure water.

8.82 What happens in each of the following situations?

a. A dialyzing bag containing a 1 M solution of potassium chloride is immersed in a 1 M sodium chloride solution.

b. A dialyzing bag containing colloidal-sized protein is immersed in a 1 M glucose solution.

ADDITIONAL PROBLEMS

8.83 With the help of Table 8.2, determine in which of the following pairs of compounds both members of the pair have like solubility in water (both soluble or both insoluble).

a. $(NH_4)_2CO_3$ and $AgNO_3$

b. $ZnCl_2$ and $Mg(OH)_2$

c. BaS and $NiCO_3$

d. $AgCl$ and $Al(OH)_3$

8.84 How many grams of solute are dissolved in the following amounts of solution?

a. 134 g of 3.00%(m/m) KNO_3 solution

b. 75.02 g of 9.735%(m/m) $NaOH$ solution

c. 1576 g of 0.800%(m/m) HI solution

d. 1.23 g of 12.0%(m/m) NH_4Cl solution

8.85 What volume of water, in quarts, is contained in 3.50 qt of a 2.00%(v/v) solution of water in acetone?

8.86 How many liters of 0.10 M solution can be prepared from 60.0 g of each of the following solutes?

a. $NaNO_3$ b. HNO_3 c. KOH d. $LiCl$

8.87 What is the molarity of the solution prepared by concentrating, by evaporation of solvent, 2212 mL of 0.400 M potassium sulfate (K_2SO_4) solution to each of the following final volumes?

a. 1875 mL   b. 1.25 L   c. 853 mL   d. 553 mL

8.88 After all the water is evaporated from 10.0 mL of a CsCl solution, 3.75 g of CsCl remains. Express the original concentration of the CsCl solution in each of the following units.

a. mass–volume percent   b. molarity

8.89 Find the molarity of a solution obtained when 352 mL of 4.00 M sodium bromide (NaBr) solution is mixed with

a. 225 mL of 4.00 M NaBr solution

b. 225 mL of 2.00 M NaBr solution

8.90 Which of the following aqueous solutions would give rise to a greater osmotic pressure?

a. 8.00 g of NaCl in 375 mL of solution or 4.00 g of NaBr in 155 mL of solution

b. 6.00 g of NaCl in 375 mL of solution or 6.00 g of $MgCl_2$ in 225 mL of solution

𝑀ultiple-Choice Practice Test

8.91 Which of the following statements about solutions is *incorrect*?

a. A solution is a homogeneous mixture.

b. Solutions in which both solute and solvent are solids are possible.

c. Solutions readily separate into solute and solvent if left undisturbed for 24 hours.

d. The substance present in the greatest amount is considered to be the solvent.

8.92 Which of the following statements is true for an *unsaturated* solution?

a. Undissolved solute must be present.

b. No undissolved solute may be present.

c. The solubility limit for the solute has been reached.

d. Solid crystallizes out if the solution is stirred.

8.93 Which of the following statements is most closely related to Henry's law?

a. Most solid solutes become more soluble in water with increasing temperature.

b. Most solid solutes become less soluble in water with increasing pressure.

c. Gaseous solutes become less soluble in water with increasing temperature.

d. Gaseous solutes become more soluble in water with increasing pressure.

8.94 Solubility in water is a general characteristic of which of the following types of ionic compounds?

a. Phosphates   b. Nitrates

c. Carbonates   d. Hydroxides

8.95 What is the concentration, in mass percent, of a solution that contains 20.0 g of NaCl dissolved in 250.0 g of water?

a. 6.76% by mass   b. 7.41% by mass

c. 8.00% by mass   d. 8.25% by mass

8.96 For which of the following solutions is the concentration 1.0 molar?

a. 0.050 mole of solute in 25.0 mL of solution

b. 2.0 moles of solute in 500.0 mL of solution

c. 3.0 moles of solute in 1.5 L of solution

d. 0.50 mole of solute in 500.0 mL of solution

8.97 Which of the following is a correct characterization for the particles present in the dispersed phase of a colloidal dispersion?

a. Large enough that they can be seen by the naked eye

b. Small enough that they do not settle out under the influence of gravity

c. Large enough that they can be filtered out using filter paper

d. Small enough that they do not scatter a beam of light

8.98 Which of the following is not a colligative property?

a. Osmotic pressure   b. Boiling-point elevation

c. Freezing-point depression   d. Density

8.99 Which of the following solutions has an osmolarity of 3.0?

a. 1.5 M glucose   b. 2.0 M sucrose

c. 1.0 M $CaCl_2$ d. 3.0 M NaCl

8.100 The osmotic pressure of a *hypotonic solution* is which of the following?

a. The same as that in cells

b. Lower than that in cells

c. Double that in cells

d. Higher than that in cells

Chemical Reactions

Chapter Outline

A fireworks display involves numerous different chemical reactions occurring at the same time.

In the previous two chapters we considered the properties of matter in various pure and mixed states. Nearly all of the subject matter dealt with interactions and changes of a *physical* nature. We now concern ourselves with the *chemical* changes that occur when various types of matter interact.

We first consider several types of chemical reactions and then discuss important fundamentals common to all chemical changes. Of particular concern to us will be how fast chemical changes occur (chemical reaction rates) and the extent to which chemical changes occur (chemical equilibrium).

9.1 TYPES OF CHEMICAL REACTIONS

A **chemical reaction** *is a process in which at least one new substance is produced as a result of chemical change.* An almost inconceivable number of chemical reactions are possible. The majority of chemical reactions (but not all) fall into five major categories: *combination* reactions, *decomposition* reactions, *single-replacement* reactions, *double-replacement* reactions, and *combustion* reactions.

Combination Reactions

A **combination reaction** *is a chemical reaction in which a single product is produced from two (or more) reactants.* The general equation for a combination reaction involving two reactants is

$$X + Y \longrightarrow XY$$

Figure 9.1 When a hot nail is stuck into a pile of zinc and sulfur, a fiery combination reaction occurs and zinc sulfide forms.

$$Zn + S \longrightarrow ZnS$$

B ◄► F

In organic chemistry (Chapters 12–17), combination reactions are called *addition reactions*. One reactant, usually a small molecule, is considered to be added to a larger reactant molecule to produce a single product.

In organic chemistry, decomposition reactions are often called *elimination reactions*. In many reactions, including some metabolic reactions that occur in the human body, either H_2O or CO_2 is eliminated from a molecule (a decomposition).

In such a combination reaction, two substances join together to form a more complicated product (see Figure 9.1). The reactants X and Y can be elements or compounds or an element and a compound. The product of the reaction (XY) is always a compound. Some representative combination reactions that have elements as the reactants are

$$Ca + S \longrightarrow CaS$$
$$N_2 + 3H_2 \longrightarrow 2NH_3$$
$$2Na + O_2 \longrightarrow Na_2O_2$$

Some examples of combination reactions in which compounds are involved as reactants are

$$SO_3 + H_2O \longrightarrow H_2SO_4$$
$$2NO + O_2 \longrightarrow 2NO_2$$
$$2NO_2 + H_2O_2 \longrightarrow 2HNO_3$$

Decomposition Reactions

A **decomposition reaction** *is a chemical reaction in which a single reactant is converted into two (or more) simpler substances (elements or compounds).* Thus a decomposition reaction is the opposite of a combination reaction. The general equation for a decomposition reaction in which there are two products is

$$XY \longrightarrow X + Y$$

Although the products may be elements or compounds, the reactant is *always* a compound.

At sufficiently high temperatures, all compounds can be broken down (decomposed) into their constituent elements. Examples of such reactions include

$$2CuO \longrightarrow 2Cu + O_2$$
$$2H_2O \longrightarrow 2H_2 + O_2$$

At lower temperatures, compound decomposition often produces other compounds as products.

$$CaCO_3 \longrightarrow CaO + CO_2$$
$$2KClO_3 \longrightarrow 2KCl + 3O_2$$
$$4HNO_3 \longrightarrow 4NO_2 + 2H_2O + O_2$$

Decomposition reactions are easy to recognize because they are the only type of reaction in which there is only one reactant.

Single-Replacement Reactions

A **single-replacement reaction** *is a chemical reaction in which an atom or molecule replaces an atom or group of atoms from a compound.* There are always two reactants and two products in a single-replacement reaction. The general equation for a single-replacement reaction is

$$X + YZ \longrightarrow Y + XZ$$

A common type of single-replacement reaction is one in which an element and a compound are reactants, and an element and a compound are products. Examples of this type of single-replacement reaction include

$$Fe + CuSO_4 \longrightarrow Cu + FeSO_4$$
$$Mg + Ni(NO_3)_2 \longrightarrow Ni + Mg(NO_3)_2$$
$$Cl_2 + NiI_2 \longrightarrow I_2 + NiCl_2$$
$$F_2 + 2NaCl \longrightarrow Cl_2 + 2NaF$$

The first two equations illustrate one metal replacing another metal from its compound. The latter two equations illustrate one nonmetal replacing another nonmetal from its compound. A more complicated example of a single-replacement reaction, in which all reactants and products are compounds, is

$$4PH_3 + Ni(CO)_4 \longrightarrow 4CO + Ni(PH_3)_4$$

Double-Replacement Reactions

A **double-replacement reaction** *is a chemical reaction in which two substances exchange parts with one another and form two different substances.* The general equation for a double-replacement reaction is

$$AX + BY \longrightarrow AY + BX$$

Such reactions can be thought of as involving "partner switching." The AX and BY partnerships are disrupted, and new AY and BX partnerships are formed in their place.

When the reactants in a double-replacement reaction are ionic compounds in solution, the parts exchanged are the positive and negative ions of the compounds present.

$$AgNO_3(aq) + NaCl(aq) \longrightarrow AgCl(s) + NaNO_3(aq)$$
$$2KI(aq) + Pb(NO_3)_2(aq) \longrightarrow 2KNO_3(aq) + PbI_2(s)$$

In most reactions of this type, one of the product compounds is in a different physical state (solid or gas) from that of the reactants (see Figure 9.2). Insoluble solids formed from such a reaction are called *precipitates;* AgCl and PbI_2 are precipitates in the foregoing reactions.

Combustion Reactions

Combustion reactions are a most common type of chemical reaction. A **combustion reaction** *is a chemical reaction between a substance and oxygen (usually from air) that proceeds with the evolution of heat and light (usually from a flame).* Hydrocarbons—binary compounds of carbon and hydrogen (of which many exist)—are the most common type of compound that undergoes combustion. In hydrocarbon combustion, the carbon of the hydrocarbon combines with the oxygen of air to produce carbon dioxide (CO_2). The hydrogen of the hydrocarbon also interacts with the oxygen of air to give water (H_2O) as a product. The relative amounts of CO_2 and H_2O produced depend on the composition of the hydrocarbon.

$$2C_2H_2 + 5O_2 \longrightarrow 4CO_2 + 2H_2O$$
$$C_3H_8 + 5O_2 \longrightarrow 3CO_2 + 4H_2O$$
$$C_4H_8 + 6O_2 \longrightarrow 4CO_2 + 4H_2O$$

Examples of combustion reactions in which products other than CO_2 and H_2O are produced include

$$CS_2 + 3O_2 \longrightarrow CO_2 + 2SO_2$$
$$2H_2S + 3O_2 \longrightarrow 2SO_2 + 2H_2O$$
$$4NH_3 + 5O_2 \longrightarrow 4NO + 6H_2O$$
$$2ZnS + 3O_2 \longrightarrow 2ZnO + 2SO_2$$

In organic chemistry, replacement reactions (both single and double) are often called *substitution* reactions. Substitution reactions of the single-replacement type are seldom encountered. However, double-replacement reactions are common in organic chemistry.

Figure 9.2 A double-replacement reaction involving solutions of potassium iodide and lead(II) nitrate (both colorless solutions) produces yellow, insoluble lead(II) iodide as one of the products.

$$2KI(aq) + Pb(NO_3)_2(aq) \longrightarrow$$
$$2KNO_3(aq) + PbI_2(s)$$

Hydrocarbon combustion reactions are the basis of an industrial society, making possible the burning of gasoline in cars, of natural gas in homes, and of coal in factories. Gasoline, natural gas, and coal all contain hydrocarbons. Unlike most other chemical reactions, hydrocarbon combustion reactions are carried out for the energy they produce rather than for the material products.

 **E X A M P L E   9 . 1**

Classifying Chemical Reactions

Classify each of the following chemical reactions as a *combination, decomposition, single-replacement, double-replacement,* or *combustion* reaction.

a. $2KNO_3 \rightarrow 2KNO_2 + O_2$
b. $Zn + 2AgNO_3 \rightarrow Zn(NO_3)_2 + 2Ag$
c. $Ni(NO_3)_2 + 2NaOH \rightarrow Ni(OH)_2 + 2NaNO_3$
d. $3Mg + N_2 \rightarrow Mg_3N_2$

(continued)

Solution

a. Decomposition. Two substances are produced from a single substance.
b. Single-replacement. An element and a compound are reactants, and an element and a compound are products.
c. Double-replacement. Two compounds exchange parts with each other; the nickel ion (Ni^{2+}) and sodium ion (Na^+) are "swapping partners."
d. Combination. Two substances combine to form a single substance.

Practice Exercise 9.1

Classify each of the following chemical reactions as a *combination, decomposition, single-replacement, double-replacement,* or *combustion* reaction.

a. $CH_4 + 2O_2 \rightarrow CO_2 + 2H_2O$
b. $N_2 + 3H_2 \rightarrow 2NH_3$
c. $Ni + Cu(NO_3)_2 \rightarrow Cu + Ni(NO_3)_2$
d. $CuCO_3 \rightarrow CuO + O_2$

Answers: a. Combustion; **b.** Combination; **c.** Single-replacement; **d.** Decomposition

Combination reactions in which oxygen reacts with another element to form a single product are also combustion reactions. Two such reactions are

$$S + O_2 \longrightarrow SO_2$$
$$2Mg + O_2 \longrightarrow 2MgO$$

Many, *but not all,* chemical reactions fall into one of the five categories we have discussed in this section. Even though this classification system is not all-inclusive, it is still very useful because of the many reactions it does help correlate. The Chemistry at a Glance feature on page 228 summarizes pictorially reaction types that we have considered in this section.

 REDOX AND NONREDOX CHEMICAL REACTIONS

Chemical reactions can also be classified, in terms of whether transfer of electrons occurs, as either oxidation–reduction (redox) or nonoxidation–reduction (nonredox) reactions. An **oxidation-reduction (redox) chemical reaction** *is a chemical reaction in which there is a transfer of electrons from one reactant to another reactant.* A **nonoxidation-reduction (nonredox) chemical reaction** *is a chemical reaction in which there is no transfer of electrons from one reactant to another reactant.*

A "bookkeeping system" known as oxidation numbers is used to identify whether electron transfer occurs in a chemical reaction. An **oxidation number** *is a number that represents the charge that an atom appears to have when the electrons in each bond it is participating in are assigned to the more electronegative of the two atoms involved in the bond.*

There are several rules for determining oxidation numbers.

1. *The oxidation number of an element in its elemental state is zero.* For example, the oxidation number of copper in Cu is zero, and the oxidation number of chlorine in Cl_2 is zero.
2. *The oxidation number of a monatomic ion is equal to the charge on the ion.* For example, the Na^+ ion has an oxidation number of $+1$, and the S^{2-} ion has an oxidation number of -2.
3. *The oxidation numbers of Groups IA and IIA metals in compounds are always $+1$, and $+2$, respectively.*
4. *The oxidation number of hydrogen is $+1$ in most hydrogen-containing compounds.*

Oxidation numbers are also sometimes called *oxidation states.*

CHEMICAL Connections

Combustion Reactions, Carbon Dioxide, and Global Warming

Most fuels used in our society, including coal, petroleum, and natural gas, are carbon-hydrogen-containing substances. When such fuels are burned (combustion; Section 9.1), carbon dioxide is one of the combustion products. For example, equations for the combustion of methane (CH_4) and propane (C_3H_8) are

$$CH_4 + 2O_2 \longrightarrow CO_2 + 2H_2O$$

$$C_3H_8 + 5O_2 \longrightarrow 3CO_2 + 4H_2O$$

Almost all combustion-generated CO_2 enters the atmosphere. Significant amounts of this atmospheric CO_2 are absorbed into the oceans because of this compound's solubility in water, and plants also remove CO_2 from the atmosphere via the process of photosynthesis. However, these removal mechanisms are not sufficient to remove all the combustion-generated CO_2; it is being generated faster than it can be removed. Consequently, atmospheric concentrations of CO_2 are slowly increasing, as the following graph shows.

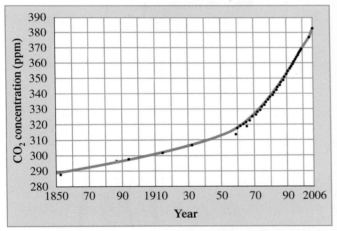

The concentration unit on the vertical axis is parts per million (ppm)—the number of molecules of CO_2 per million molecules present in air. Data for periods before 1958 were derived from analysis of air trapped in bubbles in glacial ice.

Increasing atmospheric CO_2 levels pose an environmental concern because within the atmosphere, CO_2 acts as a heat-trapping agent. During the day, Earth receives energy from the sun, mostly in the form of *visible light*. At night, as Earth cools, it re-radiates the energy it received during the day in the form of *infrared light* (heat energy). Carbon dioxide does not absorb visible light, but it has the ability to absorb infrared light. The CO_2 thus traps some of the heat energy re-radiated by the surface of the Earth as it cools, preventing this energy from escaping to outer space. Because this action is similar to that of glass in a greenhouse, CO_2 is called a *greenhouse gas*. The warming caused by CO_2 as it prevents heat loss from Earth is called the *greenhouse effect* or *global warming*.

Some scientists believe that the presence of increased concentrations of CO_2 (and other greenhouse gases present in the atmosphere in lower concentrations than CO_2) is beginning to cause a change in our climate as the result of a small increase in the average temperature of Earth's surface. Some computer models predict an average global temperature increase of 1 to 3°C toward the end of the twenty-first century if atmospheric CO_2 concentrations continue to increase at their current rate. Because numerous other factors are also involved in determining climate, however, predictions cannot be made with certainty. Much research concerning this situation is in progress, and many governments around the world are now trying to reduce the amount of combustion-generated CO_2 that enters the atmosphere.

The other greenhouse gases besides CO_2 include CH_4, N_2O, and CFCs (chlorofluorocarbons). Atmospheric concentrations of these other greenhouse gases are lower than that of CO_2. However, because they are more effective absorbers of infrared radiation than CO_2 is, they make an appreciable contribution to the overall greenhouse effect. Further information about CFCs is given in the Chemical Connections feature "Chlorofluorocarbons and the Ozone Layer" in Chapter 12. Estimated contributions of various greenhouse gases to global warming are as follows:

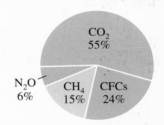

Contributions of various gases to the greenhouse effect.

5. *The oxidation number of oxygen is −2 in most oxygen-containing compounds.*

6. *In binary molecular compounds, the more electronegative element is assigned a negative oxidation number equal to its charge in binary ionic compounds. For example, in CCl_4 the element Cl is the more electronegative, and its oxidation number is −1 (the same as in the simple Cl^- ion).*

7. *For a compound, the sum of the individual oxidation numbers is equal to zero; for a polyatomic ion, the sum is equal to the charge on the ion.*

Example 9.2 illustrates the use of these rules.

CHEMISTRY AT A GLANCE

Types of Chemical Reactions

COMBINATION REACTION

$$X + Y \rightarrow XY$$

$$2Al + 3I_2 \longrightarrow 2AlI_3$$

Aluminum reacts with iodine to form aluminum iodide.

DECOMPOSITION REACTION

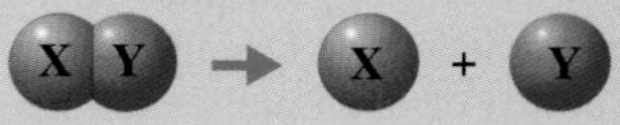

$$XY \rightarrow X + Y$$

$$2HgO \longrightarrow 2Hg + O_2$$

Mercury(II) oxide decomposes to form mercury and oxygen.

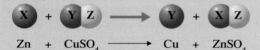

SINGLE-REPLACEMENT REACTION

$$X + YZ \longrightarrow Y + XZ$$

$$Zn + CuSO_4 \longrightarrow Cu + ZnSO_4$$

Zinc reacts with copper(II) sulfate to form copper and zinc sulfate.

DOUBLE-REPLACEMENT REACTION

$$AX + BY \rightarrow AY + BX$$

$$AgNO_3 + NaCl \longrightarrow AgCl + NaNO_3$$

Silver nitrate reacts with sodium chloride to form silver chloride and sodium nitrate.

● **EXAMPLE 9.2**

Assigning Oxidation Numbers to Elements in a Compound or Polyatomic Ion

▶ Assign an oxidation number to each element in the following compounds or polyatomic ions.

a. P_2O_5 **b.** $KMnO_4$ **c.** NO_3^-

Solution

a. The sum of the oxidation numbers of all the atoms present must add to zero (rule 7).

$$2(\text{oxid. no. P}) + 5(\text{oxid. no. O}) = 0$$

The oxidation number of oxygen is -2 (rule 5 or rule 6). Substituting this value into the previous equation enables us to calculate the oxidation number of phosphorus.

$$2(\text{oxid. no. P}) + 5(-2) = 0$$

$$2(\text{oxid. no. P}) = +10$$

$$(\text{oxid no. P}) = +5$$

Thus the oxidation numbers for the elements involved in this compound are

$$P = +5 \quad \text{and} \quad O = -2$$

Note that the oxidation number of phosphorus is not $+10$; that is the calculated charge associated with two phosphorus atoms. Oxidation number is always specified on a *per-atom* basis.

b. The sum of the oxidation numbers of all the atoms present must add to zero (rule 7).

$$(\text{oxid. no. K}) + (\text{oxid. no. Mn}) + 4(\text{oxid. no. O}) = 0$$

The oxidation number of potassium, a Group IA element, is +1 (rule 3), and the oxidation number of oxygen is −2 (rule 5). Substituting these two values into the rule 7 equation enables us to calculate the oxidation number of manganese.

$$(+1) + (\text{oxid. no. Mn}) + 4(-2) = 0$$

$$(\text{oxid. no. Mn}) = 8 - 1 = +7$$

Thus the oxidation numbers for the elements involved in this compound are

$$K = +1 \qquad Mn = +7 \qquad \text{and} \qquad O = -2$$

Note that all the oxidation numbers add to zero when it is taken into account that there are four oxygen atoms.

$$(+1) + (+7) + 4(-2) = 0$$

c. The species NO_3^- is a polyatomic ion rather than a neutral compound. Thus the second part of rule 7 applies: The oxidation numbers must add to −1, the charge on the ion.

$$(\text{oxid. no. N}) + 3(\text{oxid. no. O}) = -1$$

The oxidation number of oxygen is −2 (rule 5). Substituting this value into the sum equation gives

$$(\text{oxid. no. N}) + 3(-2) = -1$$

$$(\text{oxid. no. N}) = -1 + 6 = +5$$

Thus the oxidation numbers for the elements involved in the polyatomic ion are

$$N = +5 \qquad \text{and} \qquad O = -2$$

Practice Exercise 9.2

Assign oxidation numbers to each element in the following compounds or polyatomic ions.

a. N_2O_4 **b.** $K_2Cr_2O_7$ **c.** NH_4^+

Answers: a. N = +4, O = −2; **b.** K = +1, Cr = +6, O = −2; **c.** N = −3, H = +1

Many elements display a range of oxidation numbers in their various compounds. For example, nitrogen exhibits oxidation numbers ranging from −3 to +5. Selected examples are

$$\underset{-3}{NH_3} \qquad \underset{+1}{N_2O} \qquad \underset{+2}{NO} \qquad \underset{+3}{N_2O_3} \qquad \underset{+4}{NO_2} \qquad \underset{+5}{HNO_3}$$

As shown in this listing of nitrogen-containing compounds, the oxidation number of an atom is written *underneath* the symbol of that atom in the chemical formula. This convention is used to avoid confusion with the charge on an ion.

To determine whether a reaction is a redox reaction or a nonredox reaction, we look for changes in the oxidation number of elements involved in the reaction. Changes in oxidation number are a requirement for a redox reaction. The reaction between calcium metal and chlorine gas (see Figure 9.3) is a redox reaction.

$$\underset{0}{Ca} + \underset{0}{Cl_2} \longrightarrow \underset{+2 \; -1}{CaCl_2}$$

The oxidation number of Ca changes from zero to +2, and the oxidation number of Cl changes from zero to −1.

The decomposition of calcium carbonate is a nonredox reaction.

$$\underset{+2 \; +4 \; -2}{CaCO_3} \longrightarrow \underset{+2 \; -2}{CaO} + \underset{+4 \; -2}{CO_2}$$

It is a nonredox reaction because there are no changes in oxidation number.

Figure 9.3 The burning of calcium metal in chlorine is a redox reaction. The burning calcium emits a red-orange flame.

EXAMPLE 9.3

Using Oxidation Numbers to Determine Whether a Chemical Reaction Is a Redox Reaction

By using oxidation numbers, determine whether the following reaction is a redox reaction or a nonredox reaction.

$$4NH_3 + 3O_2 \longrightarrow 2N_2 + 6H_2O$$

Solution

For the reactant NH_3, H has an oxidation number of $+1$ (rule 4) and N an oxidation number of -3 (rule 7). The other reactant, O_2, is an element and thus has an oxidation number of zero (rule 1). The product N_2 also has an oxidation number of zero because it is an element. In H_2O, the other product, H has an oxidation number of $+1$ (rule 4) and oxygen an oxidation number of -2 (rule 5).

The overall oxidation number analysis is

$$4NH_3 + 3O_2 \longrightarrow 2N_2 + 6H_2O$$
$$\underset{-3 +1}{} \quad \underset{0}{} \quad \underset{0}{} \quad \underset{+1 \; -2}{}$$

This reaction is a redox reaction because the oxidation numbers of both N and O change.

Practice Exercise 9.3

By using oxidation numbers, determine whether the following reaction is a redox reaction or a nonredox reaction.

$$SO_3 + H_2O \longrightarrow H_2SO_4$$

Answer: Nonredox reaction; the oxidation numbers of S ($+6$), O (-2), and H ($+1$) are the same on both sides of the equation.

9.3 TERMINOLOGY ASSOCIATED WITH REDOX PROCESSES

Four key terms used in describing redox processes are *oxidation, reduction, oxidizing agent,* and *reducing agent.* The definitions for these terms are closely tied to the concepts of "electron transfer" and "oxidation number change"—concepts considered in Section 9.2. It is electron transfer that links all redox processes together. Change in oxidation number is a direct consequence of electron transfer.

In a redox reaction, one reactant undergoes oxidation, and another reactant undergoes reduction. **Oxidation** *is the process whereby a reactant in a chemical reaction loses one or more electrons.* **Reduction** *is the process whereby a reactant in a chemical reaction gains one or more electrons.*

Oxidation and reduction are complementary processes that always occur together. When electrons are lost by one species, they do not disappear: rather, they are always gained by another species. Thus electron transfer always involves both oxidation and reduction.

Electron loss (oxidation) always leads to an increase in oxidation number. Conversely, electron gain (reduction) always leads to a decrease in oxidation number. These generalizations are consistent with the rules for monatomic ion formation (Section 4.5); electron loss produces positive ions (increase in oxidation number), and electron gain produces negative ions (decrease in oxidation number). Figure 9.4 summarizes the relationship between change in oxidation number and the processes of oxidation and reduction.

Oxidation involves the *loss* of electrons, and reduction involves the *gain* of electrons. Students often have trouble remembering which is which. Two helpful mnemonic devices follow.

LEO the lion says *GER*

Loss of *E*lectrons: *O*xidation.
Gain of *E*lectrons: *R*eduction.

OIL RIG

*O*xidation *I*s *L*oss (of electrons).
*R*eduction *I*s *G*ain (of electrons).

Figure 9.4 An increase in oxidation number is associated with the process of oxidation, a decrease with the process of reduction.

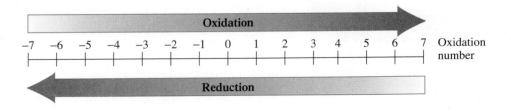

TABLE 9.1
Summary of Redox Terminology in Terms of Electron Transfer

Term	Electron Transfer
oxidation	loss of electron(s)
reduction	gain of electron(s)
oxidizing agent (substance reduced)	electron(s) gained
reducing agent (substance oxidized)	electron(s) lost

Oxidizing Agents and Reducing Agents

There are two different ways of looking at the reactants in a redox reaction. First, the reactants can be viewed as being "acted on." From this perspective, one reactant is *oxidized* (the one that loses electrons), and one is *reduced* (the one that gains electrons). Second, the reactants can be looked on as "bringing about" the reaction. In this approach, the terms *oxidizing agent* and *reducing agent* are used. An **oxidizing agent** *is the reactant in a redox reaction that causes oxidation of another reactant by accepting electrons from it.* This acceptance of electrons means that the oxidizing agent itself is reduced. Similarly, a **reducing agent** *is the reactant in a redox reaction that causes reduction of another reactant by providing electrons for the other reactant to accept.* Thus the reducing agent and the substance oxidized are one and the same, as are the oxidizing agent and the substance reduced:

> The terms *oxidizing agent* and *reducing agent* sometimes cause confusion, because the oxidizing agent is not oxidized (it is reduced) and the reducing agent is not reduced (it is oxidized). A simple analogy is that a travel agent is not the one who takes a trip; he or she is the one who plans (causes) the trip that is taken.

$$\text{Substance oxidized} = \text{reducing agent}$$
$$\text{Substance reduced} = \text{oxidizing agent}$$

Table 9.1 summarizes the redox terminology presented in this section in terms of electron transfer.

 **EXAMPLE 9.4**

Identifying the Oxidizing Agent and Reducing Agent in a Redox Reaction

For the redox reaction

$$FeO + CO \longrightarrow Fe + CO_2$$

identify the following.
a. The substance oxidized
b. The substance reduced
c. The oxidizing agent
d. The reducing agent

Solution

Oxidation numbers are calculated using the methods illustrated in Example 9.2.

$$FeO + CO \longrightarrow Fe + CO_2$$
$$\overset{+2\ -2}{}\quad \overset{+2\ -2}{}\qquad \overset{0}{}\quad \overset{+4\ -2}{}$$

a. Oxidation involves an increase in oxidation number. The oxidation number of C has increased from $+2$ to $+4$. Therefore, the reactant that contains C, which is CO, is the substance that has been oxidized.
b. Reduction involves a decrease in oxidation number. The oxidation number of Fe has decreased from $+2$ to zero. Therefore, the reactant that contains Fe, which is FeO, is the substance that has been reduced.
c. The oxidizing agent and the substance reduced are always one and the same. Therefore, FeO is the oxidizing agent.
d. The reducing agent and the substance oxidized are always one and the same. Therefore, CO is the reducing agent.

(continued)

Practice Exercise 9.4

For the redox reaction

$$3MnO_2 + 4Al \longrightarrow 2Al_2O_3 + 3Mn$$

identify the following.

a. The substance oxidized   **b.** The substance reduced
c. The oxidizing agent   **d.** The reducing agent

Answers: a. Al; **b.** MnO_2; **c.** MnO_2; **d.** Al

CHEMICAL Connections

"Undesirable" Oxidation-Reduction Processes: Metallic Corrosion

One of the biggest problems in society related to the use of metals is that of corrosion. *Corrosion* is the deterioration of a metal as a result of naturally occurring oxidation–reduction processes. In corrosion processes, metals are converted to compounds of the metals. Let us consider corrosion as it relates to the metals iron, copper, silver, aluminum, and gold.

Iron. Iron, the most used of all metals, is the primary ingredient in all types of steel. It is estimated that as much as one-seventh of annual iron production simply replaces that lost by corrosion. The iron corrosion process is called *rusting,* a process that requires the presence of both moisture and oxygen. Rusting involves a three-step series of reactions. The iron metal is first converted to iron(II) hydroxide [$Fe(OH)_2$], then to iron(III) hydroxide [$Fe(OH)_3$], and finally to rust, the hydrated oxide $Fe_2O_3 \cdot H_2O$. The overall reaction, the sum of the three steps, is

$$4Fe(s) + 3O_2(g) + 2H_2O(l) \longrightarrow 2Fe_2O_3 \cdot H_2O(s)$$

The rust so produced, a reddish-brown solid, does not adhere to the surface of the metal and protect it from further reaction, but instead "flakes" off. This exposes a fresh surface of iron, and the rusting process continues.

Copper. The corrosion product for metallic copper is green—the familiar green coating found on many statues and buildings. Copper corrosion requires the presence of oxygen, water, and carbon dioxide. All three of these substances are normally present in air. The overall reaction is

$$2Cu(s) + O_2(g) + CO_2(g) + H_2O(l) \longrightarrow Cu(OH)_2 \cdot CuCO_3(s)$$

The green copper hydroxide–copper carbonate coating associated with copper corrosion is a tough film that adheres to the copper surface. This protects the copper from further corrosion.

Silver. Silver is not oxidized by oxygen in the air at ordinary temperatures, but it does tarnish quickly in the presence of sulfur-containing air pollutants such as hydrogen sulfide (H_2S) and sulfur-containing foods such as eggs and mustard. The corrosion product, which is often called silverware tarnish, is a thin layer of black silver sulfide (Ag_2S). The equation for its formation from H_2S, in the presence of air, is

$$4Ag(s) + 2H_2S(g) + O_2(g) \longrightarrow 2Ag_2S(s) + 2H_2O(g)$$

Tarnished and untarnished pieces of silverware.

Aluminum. Aluminum, the second most used of all metals, readily undergoes corrosion. Freshly cut aluminum has a *bright* silvery appearance. Its surface quickly changes to a dull silver-white as a thin film of aluminum oxide (Al_2O_3) forms through atmospheric oxidation. When household aluminum objects are "cleaned" with scouring pads or abrasive chemicals, the oxide coating is usually removed, giving the aluminum a shinier appearance. The cleaning is, however, in vain; a new oxide coating quickly forms, which prevents the aluminum from undergoing further oxidation.

$$4Al(s) + 3O_2(g) \longrightarrow 2Al_2O_3(s)$$

Gold. Gold is completely resistant to atmospheric corrosion. In its earliest uses, it was valued more for its beauty (color and luster) due to its corrosion resistance. Today, in addition to being a component of jewelry, it is valued as a medium of exchange and used as a basis for monetary systems.

COLLISION THEORY AND CHEMICAL REACTIONS

What causes a chemical reaction, either redox or nonredox, to take place? A set of three generalizations, developed after the study of thousands of different reactions, helps answer this question. Collectively these generalizations are known as collision theory. **Collision theory** *is a set of statements that give the conditions necessary for a chemical reaction to occur.* Central to collision theory are the concepts of molecular collisions, activation energy, and collision orientation. The statements of collision theory are

1. *Molecular collisions.* Reactant particles must interact (that is, collide) with one another before any reaction can occur.
2. *Activation energy.* Colliding particles must possess a certain minimum total amount of energy, called the activation energy, if the collision is to be effective (that is, result in reaction).
3. *Collision orientation.* Colliding particles must come together in the proper orientation unless the particles involved are single atoms or small, symmetrical molecules.

Let's look at these statements in the context of a reaction between two molecules or ions.

Molecular Collisions

When reactions involve two or more reactants, collision theory assumes (statement 1) that the reactant molecules, ions, or atoms must come in contact (collide) with one another in order for any chemical change to occur. The validity of this statement is fairly obvious. Reactants cannot react if they are separated from each other.

Most reactions are carried out either in liquid solution or in the gaseous phase, wherein reacting particles are more free to move around, and thus it is easier for the reactants to come in contact with one another. Reactions in which reactants are solids can and do occur; however, the conditions for molecular collisions are not as favorable as they are for liquids and gases. Reactions of solids usually take place only on the solid surface and thus include only a small fraction of the total particles present in the solid. As the reaction proceeds and products dissolve, diffuse, or fall from the surface, fresh solid is exposed. Thus the reaction eventually consumes all of the solid. The rusting of iron is an example of this type of process.

Activation Energy

The collisions between reactant particles do not always result in the formation of reaction products. Sometimes, reactant particles rebound unchanged from a collision. Statement 2 of collision theory indicates that in order for a reaction to occur, particles must collide with a certain minimum energy; that is, the kinetic energies of the colliding particles must add to a certain minimum value. **Activation energy** *is the minimum combined kinetic energy that colliding reactant particles must possess in order for their collision to result in a chemical reaction.* Every chemical reaction has a different activation energy. In a slow reaction, the activation energy is far above the average energy content of the reacting particles. Only those few particles with above-average energy undergo collisions that result in reaction; this is the reason for the overall slowness of the reaction.

It is sometimes possible to start a reaction by providing activation energy and then have the reaction continue on its own. Once the reaction is started, enough energy is released to activate other molecules and keep the reaction going. The striking of a kitchen match is an example of such a situation (Figure 9.5). Activation energy is initially provided by rubbing the match head against a rough surface; heat is generated by friction. Once the reaction is started, the match continues to burn.

Figure 9.5 Rubbing a match head against a rough surface provides the activation energy needed for the match to ignite.

Figure 9.6 In the reaction of NO_2 with CO to produce NO and CO_2, the most favorable collision orientation is one that puts an O atom from NO_2 in close proximity to the C atom of CO.

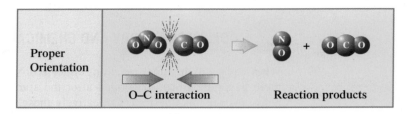

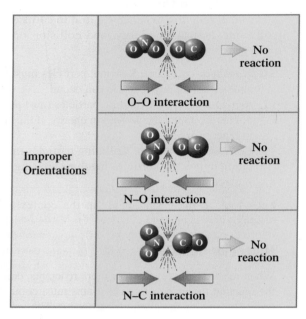

Collision Orientation

Reaction rates are sometimes very slow because reactant molecules must be oriented in a certain way in order for collisions to lead successfully to products. For nonspherical molecules and nonspherical polyatomic ions, orientation relative to one another at the moment of collision is a factor that determines whether a collision produces a reaction.

As an illustration of the importance of proper collision orientation, consider the chemical reaction between NO_2 and CO to produce NO and CO_2.

$$NO_2(g) + CO(g) \longrightarrow NO(g) + CO_2(g)$$

Many reactions in the human body do not occur unless specialized proteins called *enzymes* (Chapter 21) are present. One of the functions of these enzymes is to hold reactant molecules in the orientation required for a reaction to occur.

In this reaction, an O atom is transferred from an NO_2 molecule to a CO molecule. The collision orientation most favorable for this to occur is one that puts an O atom from NO_2 near a C atom from CO at the moment of collision. Such an orientation is shown in Figure 9.6 (top). Figure 9.6 (bottom) shows three undesirable NO_2–CO orientations, where the likelihood of successful reaction is very low.

9.5 EXOTHERMIC AND ENDOTHERMIC CHEMICAL REACTIONS

Exothermic means energy is released; energy is a "product" of the chemical reaction. *Endothermic* means energy is absorbed; energy is a "reactant" in the reaction.

In Section 7.9, the terms *exothermic* and *endothermic* were used to classify changes of state. Melting, sublimation, and evaporation are endothermic changes of state, and freezing, condensation, and deposition are exothermic changes of state. The terms *exothermic* and *endothermic* are also used to classify chemical reactions. An **exothermic chemical reaction** *is a chemical reaction in which energy is released as the reaction occurs.* The burning of a fuel (reaction of the fuel with oxygen) is an exothermic process. An **endothermic chemical reaction** *is a chemical reaction in which a continuous input of energy is needed for the reaction to occur.* The photosynthesis process that occurs in plants is an example of an endothermic reaction. Light is the energy source for photosynthesis. Light energy must be continuously supplied in order for photosynthesis to occur; a green plant that is kept in the dark will die.

Figure 9.7 Energy diagram graphs showing the difference between an exothermic and an endothermic reaction. (a) In an exothermic reaction, the average energy of the reactants is higher than that of the products, indicating that energy has been released in the reaction. (b) In an endothermic reaction, the average energy of the reactants is less than that of the products, indicating that energy has been absorbed in the reaction.

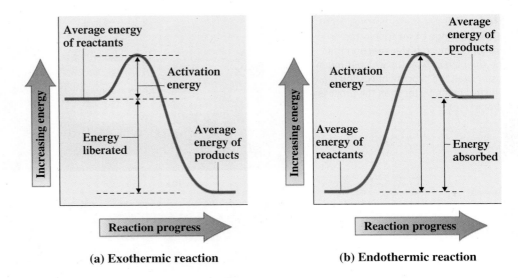

(a) Exothermic reaction

(b) Endothermic reaction

What determines whether a chemical reaction is exothermic or endothermic? The answer to this question is related to the strength of chemical bonds—that is, the energy stored in chemical bonds. Different types of bonds, such as oxygen–hydrogen bonds and fluorine–nitrogen bonds, have different energies associated with them. In a chemical reaction, bonds are broken within reactant molecules, and new bonds are formed within product molecules. The energy balance between this bond-breaking and bond-forming determines whether there is a net loss or a net gain of energy.

An exothermic reaction (release of energy) occurs when the energy required to break bonds in the reactants is less than the energy released by bond formation in the products. The opposite situation applies for an endothermic reaction. There is more energy stored in product molecule bonds than in reactant molecule bonds. The necessary additional energy must be supplied from external sources as the reaction proceeds. Figure 9.7 illustrates the energy relationships associated with exothermic and endothermic chemical reactions. Note that both of these diagrams contain a "hill" or "hump." The height of this "hill" corresponds to the activation energy needed for reaction between molecules to occur. This activation energy is independent of whether a given reaction is exothermic or endothermic.

 ## FACTORS THAT INFLUENCE CHEMICAL REACTION RATES

A **chemical reaction rate** *is the rate at which reactants are consumed or products produced in a given time period in a chemical reaction.* Natural processes have a wide range of reaction rates (see Figure 9.8). In this section we consider four different factors that affect reaction rate: (1) the physical nature of the reactants, (2) reactant concentrations, (3) reaction temperature, and (4) the presence of catalysts.

Physical Nature of Reactants

The physical nature of reactants includes not only the physical state of each reactant (solid, liquid, or gas) but also the particle size. In reactions where reactants are all in the same physical state, the reaction rate is generally faster between liquid-state reactants than between solid-state reactants and is fastest between gaseous-state reactants. Of the three states of matter, the gaseous state is the one where there is the most freedom of movement; hence, collisions between reactants are the most frequent in this state.

In the solid state, reactions occur at the boundary surface between reactants. The reaction rate increases as the amount of boundary surface area increases. Subdividing a solid into smaller particles increases surface area and thus increases reaction rate.

When the particle size of a solid is extremely small, reaction rates can be so fast that an explosion results. Although a lump of coal is difficult to ignite, the spontaneous ignition of coal dust is a real threat to underground coal-mining operations.

For reactants in the solid state, reaction rate increases as subdivision of the solid increases.

Figure 9.8 Natural processes occur at a wide range of reaction rates. A fire (a) is a much faster reaction than the ripening of fruit (b), which is much faster than the process of rusting (c), which is much faster than the process of aging (d).

(a)   (b)

(c)   (d)

Reactant Concentrations

An increase in the concentration of a reactant causes an increase in the rate of the reaction. Combustible substances burn much more rapidly in pure oxygen than in air (21% oxygen). A person with a respiratory problem such as pneumonia or emphysema is often given air enriched with oxygen because an increased partial pressure of oxygen facilitates the absorption of oxygen in the alveoli of the lungs and thus expedites all subsequent steps in respiration.

Increasing the concentration of a reactant means that there are more molecules of that reactant present in the reaction mixture; thus collisions between this reactant and other reactant particles are more likely. An analogy can be drawn to the game of billiards. The more billiard balls there are on the table, the greater the probability that a moving cue ball will strike one of them.

When the concentration of reactants is increased, the actual quantitative change in reaction rate is determined by the specific reaction. The rate usually increases, but not to the same extent in all cases. Sometimes the rate doubles with a doubling of concentration, but not always.

> Reaction rate increases as the concentration of reactants increases.

Reaction Temperature

The effect of temperature on reaction rates can also be explained by using the molecular-collision concept. An increase in the temperature of a system results in an increase in the average kinetic energy of the reacting molecules. The increased molecular speed causes more collisions to take place in a given time. Because the average energy of the colliding molecules is greater, a larger fraction of the collisions will result in reaction from the point of view of activation energy. As a rule of thumb, chemists have found that for the temperature ranges we normally encounter, the rate of a chemical reaction doubles for every 10°C increase in temperature.

> Reaction rate increases as the temperature of the reactants increases.

Presence of Catalysts

A **catalyst** *is a substance that increases a chemical reaction rate without being consumed in the chemical reaction.* Catalysts enhance reaction rates by providing alternative reaction pathways that have lower activation energies than the original, uncatalyzed pathway. This lowering of activation energy is diagrammatically shown in Figure 9.9.

> Catalysts lower the activation energy for a reaction. Lowered activation energy increases the rate of a reaction.

Figure 9.9 Catalysts lower the activation energy for chemical reactions. Reactions proceed more rapidly with the lowered activation energy.

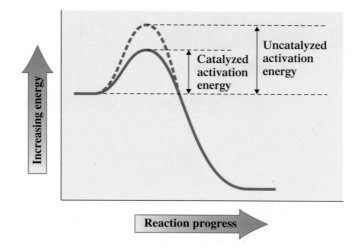

Catalysts exert their effects in varying ways. Some catalysts provide a lower-energy pathway by entering into a reaction and forming an "intermediate," which then reacts further to produce the desired products and regenerate the catalyst. The following equations, where C is the catalyst, illustrate this concept.

Catalysts are extremely important for the proper functioning of the human body and other biochemical systems. Enzymes, which are proteins, are the catalysts within the human body (Chapter 21). They cause many reactions to take place rapidly under mild conditions and at body temperature. Without these enzymes, the reactions would proceed very slowly and then only under harsher conditions.

Uncatalyzed reaction: $X + Y \longrightarrow XY$

Catalyzed reaction: *Step 1:* $X + C \longrightarrow XC$

Step 2: $XC + Y \longrightarrow XY + C$

Solid catalysts often act by providing a surface to which reactant molecules are physically attracted and on which they are held with a particular orientation. These "held" reactants are sufficiently close to and favorably oriented toward one another that the reaction takes place. The products of the reaction then leave the surface and make it available to catalyze other reactants.

The Chemistry at a Glance feature on page 238 summarizes the factors that increase reaction rates.

9.7 CHEMICAL EQUILIBRIUM

In our discussions of chemical reactions up to this point, we have assumed that chemical reactions go to completion; that is, reactions continue until one or more of the reactants are used up. This assumption is valid as long as product concentrations are not allowed to build up in the reaction mixture. If one or more products are gases that can escape from the reaction mixture or insoluble solids that can be removed from the reaction mixture, no product buildup occurs.

When product buildup does occur, reactions do not go to completion. This is because product molecules begin to react with one another to re-form reactants. With time, a steady-state situation results wherein the rate of formation of products and the rate of re-formation of reactants are equal. At this point, the concentrations of all reactants and all products remain constant, and a state of *chemical equilibrium* is reached. **Chemical equilibrium** *is the state in which forward and reverse chemical reactions occur simultaneously at the same rate.* We discussed equilibrium situations in Sections 7.11 (vapor pressure) and 8.2 (saturated solutions), but the previous examples involved physical equilibrium rather than chemical equilibrium.

A chemical reaction is in a state of chemical equilibrium when the rates of the forward and reverse reactions are equal. At this point, the concentrations of reactants and products no longer change.

The conditions that exist in a system in a state of chemical equilibrium can best be seen by considering an actual chemical reaction. Suppose equal molar amounts of gaseous H_2 and I_2 are mixed together in a closed container and allowed to react to produce gaseous HI.

$$H_2 + I_2 \longrightarrow 2HI$$

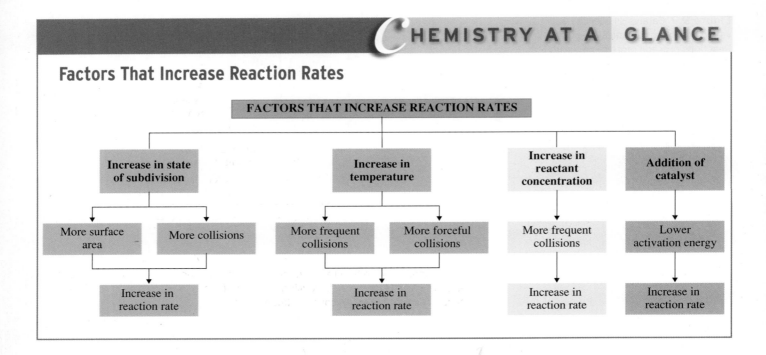

Initially, no HI is present, so the only reaction that can occur is that between H_2 and I_2. However, as the HI concentration increases, some HI molecules collide with one another in a way that causes a reverse reaction to occur:

$$2HI \longrightarrow H_2 + I_2$$

The initially low concentration of HI makes this reverse reaction slow at first, but as the concentration of HI increases, the reaction rate also increases. At the same time that the reverse-reaction rate is increasing, the forward-reaction rate (production of HI) is decreasing as the reactants are used up. Eventually, the concentrations of H_2, I_2, and HI in the reaction mixture reach a level at which the rates of the forward and reverse reactions become equal. At this point, a state of chemical equilibrium has been reached.

Figure 9.10a illustrates the behavior of reaction rates over time for both the forward and reverse reactions in the H_2–I_2–HI system. Figure 9.10b illustrates the important point that the reactant and product concentrations are usually not equal at the point at which equilibrium is reached.

> At chemical equilibrium, forward and reverse reaction rates are equal. Reactant and product concentrations, although constant, do not have to be equal.

The equilibrium involving H_2, I_2, and HI could have been established just as easily by starting with pure HI and allowing it to change into H_2 and I_2 (the reverse reaction). The final position of equilibrium does not depend on the direction from which equilibrium is approached.

It is normal procedure to represent an equilibrium by using a single equation and two half-headed arrows pointing in opposite directions. Thus the reaction between H_2 and I_2 at equilibrium is written as

$$H_2 + I_2 \rightleftharpoons 2HI$$

The half-headed arrows denote a chemical system at equilibrium.

> Theoretically, all reactions are reversible (can go in either direction). Sometimes, the reverse reaction is so slight, however, that we say the reaction has "gone to completion" because no detectable reactants remain.

The term *reversible* is often used to describe a reaction like the one we have just discussed. A **reversible reaction** is *a chemical reaction in which the conversion of reactants to products (the forward reaction) and the conversion of products to reactants (the reverse reaction) occur simultaneously.* When the half-headed arrow notation is used in an equation, it means that a reaction is reversible.

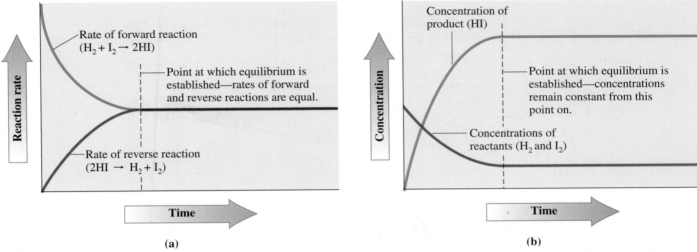

Figure 9.10 Graphs showing how reaction rates and reactant concentrations vary with time for the chemical system H_2-I_2-HI. (a) At equilibrium, rates of reaction are equal. (b) At equilibrium, concentrations of reactants remain constant but are not equal.

9.8 EQUILIBRIUM CONSTANTS

As noted in Section 9.7, the concentrations of reactants and products are constant (not changing) in a system at chemical equilibrium. This constancy allows us to describe the extent of reaction in a given equilibrium system by a single number called an equilibrium constant. An **equilibrium constant** *is a numerical value that characterizes the relationship between the concentrations of reactants and products in a system at chemical equilibrium.*

An equilibrium constant is obtained by writing an *equilibrium constant expression* and then evaluating it numerically. For a hypothetical chemical reaction, where A and B are reactants, C and D are products, and *w*, *x*, *y*, and *z* are equation coefficients,

$$w\mathrm{A} + x\mathrm{B} \rightleftharpoons y\mathrm{C} + z\mathrm{D}$$

the equilibrium constant expression is

$$K_{eq} = \frac{[\mathrm{C}]^y[\mathrm{D}]^z}{[\mathrm{A}]^w[\mathrm{B}]^x}$$

Note the following points about this general equilibrium constant expression:

1. The square brackets refer to molar (moles/liter) concentrations.
2. Product concentrations are always placed in the numerator of the equilibrium constant expression.
3. Reactant concentrations are always placed in the denominator of the equilibrium constant expression.
4. The coefficients in the balanced chemical equation for the equilibrium system determine the powers to which the concentrations are raised.
5. The abbreviation K_{eq} is used to denote an equilibrium constant.

> In equilibrium constants, square brackets mean that concentrations are expressed in molarity units.

An additional convention in writing equilibrium constant expressions, which is not apparent from the equilibrium constant definition, is that only *concentrations of gases and substances in solution* are written in an equilibrium constant expression. The reason for this convention is that other substances (pure solids and pure liquids) have constant concentrations. These constant concentrations are incorporated into the equilibrium constant itself. For example, pure water in the liquid state has a concentration of 55.5 moles/L. It does not matter whether we have 1, 50, or 750 mL of liquid water. The concentration will be the same. In the liquid state, pure water is pure water, and it has only one concentration. Similar reasoning applies to other pure liquids and pure solids. All such substances have constant concentrations.

> The concentrations of *pure liquids* and *pure solids,* which are constants, are never included in an equilibrium constant expression.

The only information we need to write an equilibrium constant expression is a balanced chemical equation, which includes information about physical state. Using the preceding generalizations about equilibrium constant expressions, for the reaction

$$4NH_3(g) + 7O_2(g) \rightleftharpoons 4NO_2(g) + 6H_2O(g)$$

we write the equilibrium constant expression as

Coefficient of NO_2 ↘ ↙ Coefficient of H_2O

$$K_{eq} = \frac{[NO_2]^4[H_2O]^6}{[NH_3]^4[O_2]^7}$$

Coefficient of NH_3 ↑ ↑ Coefficient of O_2

Chemical Connections: Stratospheric Ozone: An Equilibrium Situation

Ozone is oxygen that has undergone conversion from its normal diatomic form (O_2) to a triatomic form (O_3). The presence of ozone in the *lower atmosphere* is considered undesirable because its production contributes to air pollution; it is the major "active ingredient" in smog.

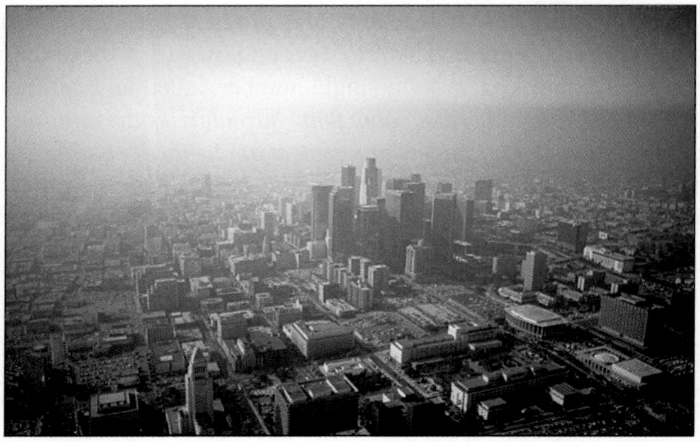

Los Angeles smog.

In the *upper atmosphere* (stratosphere), ozone is a naturally occurring species whose presence is not only desirable but absolutely essential to the well-being of humans on Earth. Stratospheric ozone screens out 95% to 99% of the ultraviolet radiation that comes from the sun. It is ultraviolet light that causes sunburn and that can be a causative factor in some types of skin cancer. The upper region of the stratosphere, where ozone concentrations are greatest, is often called the ozone layer. This ozone maximization occurs at altitudes of 25 to 30 miles (see the accompanying graph).

Within the ozone layer, ozone is continually being consumed and formed through the equilibrium process

$$3O_2(g) \rightleftharpoons 2O_3(g)$$

The source for ozone is thus diatomic oxygen. It is estimated that on any given day, 300 million tons of stratospheric ozone is formed and an equal amount destroyed in this equilibrium process.

Since the mid-1970s, scientists have observed a seasonal thinning (depletion) of ozone in the stratosphere above Antarctica. This phenomenon, which is commonly called the *ozone hole,* occurs in September and October of each year, the beginning of the Antarctic spring. Up to 70% of the ozone above Antarctica is lost during these two months. (A similar, but smaller, manifestation of this same phenomenon also occurs in the north pole region.)

Winter conditions in Antarctica include extreme cold (it is the coldest location on Earth) and total darkness. When sunlight appears in the spring, it triggers the chemical reactions that lead to ozone depletion. By the end of November, weather conditions are such that the ozone-depletion reactions stop. Then the ozone hole disappears as air from nonpolar areas flows into the polar region, replenishing the depleted ozone levels.

Chlorofluorocarbons (CFCs), synthetic compounds that have been developed primarily for use as refrigerants, are considered a causative factor for this ozone hole phenomenon. How their presence in the atmosphere contributes to this situation is considered in the Chemical Connections feature "Chlorofluorocarbons and the Ozone Layer" in Chapter 12.

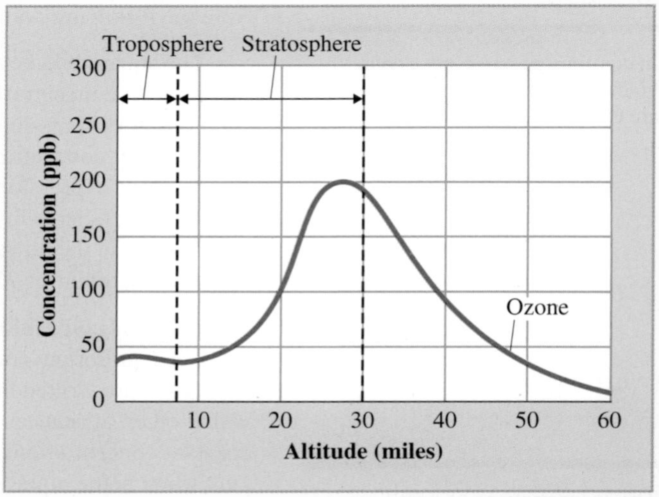

Atmospheric ozone concentration as a function of altitude.

**EXAMPLE 9.5**

Writing the Equilibrium Constant
Expression for a Chemical Reaction
from the Chemical Equation
for the Reaction

Write the equilibrium constant expression for each of the following reactions.

a. $I_2(g) + Cl_2(g) \rightleftharpoons 2ICl(g)$ **b.** $C(s) + H_2O(g) \rightleftharpoons CO(g) + H_2(g)$

Solution

a. All of the substances involved in this reaction are gases. Therefore, each reactant and product will appear in the equilibrium constant expression.

The numerator of an equilibrium constant expression always contains product concentrations. There is only one product, ICl. Write its concentration in the numerator and square it, because the coefficient of ICl in the equation is 2.

$$[ICl]^2$$

Next, place the concentrations of the reactants in the denominator. Their powers will be an understood (not written) 1, because the coefficient of each reactant is 1.

$$K_{eq} = \frac{[ICl]^2}{[I_2][Cl_2]}$$

The equilibrium constant expression is now complete.

b. The reactant carbon (C) is a solid and thus will not appear in the equilibrium constant expression. Therefore,

$$K_{eq} = \frac{[CO][H_2]}{[H_2O]}$$

Note that all of the powers in this expression are 1 as a result of all the coefficients in the balanced equation being equal to unity.

Practice Exercise 9.5

Write the equilibrium constant expression for each of the following reactions.

a. $2Cl_2(g) + 2H_2O(g) \rightleftharpoons 4HCl(g) + O_2(g)$ **b.** $NH_4Cl(s) \rightleftharpoons HCl(g) + NH_3(g)$

Answers: a. $K_{eq} = \dfrac{[O_2][HCl]^4}{[H_2O]^2[Cl_2]^2}$; **b.** $K_{eq} = [NH_3][HCl]$

If the concentrations of all reactants and products are known at equilibrium, the numerical value of the equilibrium constant can be calculated by using the equilibrium constant expression.

EXAMPLE 9.6

Calculating the Value of an
Equilibrium Constant from
Equilibrium Concentrations

Calculate the value of the equilibrium constant for the equilibrium system

$$2NO(g) \rightleftharpoons N_2(g) + O_2(g)$$

at 1000°C, given that the equilibrium concentrations are 0.0026 M for NO, 0.024 M for N_2, and 0.024 M for O_2.

Solution

First, write the equilibrium constant expression.

$$K_{eq} = \frac{[N_2][O_2]}{[NO]^2}$$

Next, substitute the equilibrium concentrations into the equilibrium constant expression and solve the equation.

$$K_{eq} = \frac{[0.024][0.024]}{[0.0026]^2}$$

$$K_{eq} = 85$$

In doing the mathematics, remember that the number 0.0026 must be squared.

(continued)

Practice Exercise 9.6

Calculate the value of the equilibrium constant for the equilibrium system

$$N_2(g) + 3H_2(g) \rightleftharpoons 2NH_3(g)$$

at 532°C, given that the equilibrium concentrations are 0.079 M for N_2, 0.12 M for H_2, and 0.0051 M for NH_3.

Answer: $K_{eq} = 0.19$

Temperature Dependence of Equilibrium Constants

The value of K_{eq} for a reaction depends on the reaction temperature. If the temperature changes, the value of K_{eq} also changes, and thus differing amounts of reactants and products will be present. Note that the equilibrium constant calculated in Example 9.6 is for a temperature of 1000°C. The equilibrium constant for this reaction would have a different value at a lower or a higher temperature.

Does the value of an equilibrium constant increase or decrease when reaction temperature is increased? For reactions where the forward reaction is *exothermic,* the equilibrium constant *decreases* with increasing temperature. For reactions where the forward reaction is *endothermic,* the equilibrium constant *increases* with increasing temperature (Section 9.5).

Equilibrium Constant Values and Reaction Completeness

The magnitude of an equilibrium constant value conveys information about how far a reaction has proceeded toward completion. If the equilibrium constant value is large (10^3 or greater), the equilibrium system contains more products than reactants. Conversely, if the equilibrium constant value is small (10^{-3} or less), the equilibrium system contains more reactants than products. Table 9.2 further compares equilibrium constant values and the extent to which a chemical reaction has occurred.

Equilibrium position *is a qualitative indication of the relative amounts of reactants and products present when a chemical reaction reaches equilibrium.* As shown in the last column of Table 9.2, the terms *far to the right, to the right, neither to the right nor to the left, to the left,* and *far to the left* are used in describing equilibrium position. In equilibrium situations where the concentrations of products are greater than those of reactants, the equilibrium position is said to lie to the *right* because products are always listed on the right side of a chemical equation. Conversely, when reactants dominate at equilibrium, the equilibrium position lies to the *left*. The terminology *neither to the right nor to the left* indicates that significant amounts of both reactants and products are present in an equilibrium mixture.

TABLE 9.2
Equilibrium Constant Values and the Extent to Which a Chemical Reaction Has Taken Place

Value of K_{eq}	Relative Amounts of Products and Reactants	Description of Equilibrium Position
very large (10^{30})	essentially all products	far to the right
large (10^{10})	more products than reactants	to the right
near unity (between 10^3 and 10^{-3})	significant amounts of both reactants and products	neither to the right nor to the left
small (10^{-10})	more reactants than products	to the left
very small (10^{-30})	essentially all reactants	far to the left

Equilibrium position can also be indicated by varying the length of the arrows in the half-headed arrow notation for a reversible reaction. The longer arrow indicates the direction of the predominant reaction. For example, the arrow notation in the equation

$$CO_2 + H_2O \rightleftharpoons H_2CO_3$$

indicates that the equilibrium position lies to the right.

9.9 ALTERING EQUILIBRIUM CONDITIONS: LE CHÂTELIER'S PRINCIPLE

Products are written on the right side of a chemical equation. A *shift to the right* means more products are produced. Conversely, because reactants are written on the left side of an equation, a *shift to the left* means more reactants are produced.

The surname Le Châtelier is pronounced "le-SHOT-lee-ay."

Figure 9.11 Henri Louis Le Châtelier (1850-1936), although most famous for the principle that bears his name, was amazingly diverse in his interests. He worked on metallurgical processes, cements, glasses, fuels, and explosives and was also noted for his skills in industrial management.

A chemical system at equilibrium is very susceptible to disruption from outside forces. A change in temperature or a change in pressure can upset the balance within the equilibrium system. Changes in the concentrations of reactants or products also upset an equilibrium.

Disturbing an equilibrium has one of two results: Either the forward reaction speeds up (to produce more products), or the reverse reaction speeds up (to produce additional reactants). Over time, the forward and reverse reactions again become equal, and a new equilibrium, different from the previous one, is established. If more products have been produced as a result of the disruption, the equilibrium is said to have *shifted to the right*. Similarly, when disruption causes more reactants to form, the equilibrium has *shifted to the left*.

An equilibrium system's response to disrupting influences can be predicted by using a principle introduced by the French chemist Henry Louis Le Châtelier (Figure 9.11). **Le Châtelier's principle** *states that if a stress (change of conditions) is applied to a system in equilibrium, the system will readjust (change the equilibrium position) in the direction that best reduces the stress imposed on the system.* We will use this principle to consider how four types of change affect equilibrium position. The changes are (1) concentration changes, (2) temperature changes, (3) pressure changes, and (4) addition of catalysts.

Concentration Changes

Adding a reactant or product to, or removing it from, a reaction mixture at equilibrium always upsets the equilibrium. If an additional amount of any reactant or product has been *added* to the system, the stress is relieved by shifting the equilibrium in the direction that *consumes* (uses up) some of the added reactant or product. Conversely, if a reactant or product is *removed* from an equilibrium system, the equilibrium shifts in a direction that *produces* more of the substance that was removed.

Let us consider the effect that concentration changes will have on the gaseous equilibrium

$$N_2(g) + 3H_2(g) \rightleftharpoons 2NH_3(g)$$

Suppose some additional H_2 is added to the equilibrium mixture. The stress of "added H_2" causes the equilibrium to shift to the right; that is, the forward reaction rate increases in order to use up some of the additional H_2.

Stress: Too much H_2 $N_2(g) + 3H_2(g) \rightleftharpoons 2NH_3(g)$
Response: Use up "extra" H_2 Shift to the right →

$[N_2]$ decreases $[H_2]$ decreases $[NH_3]$ increases

As the H_2 reacts, the amount of N_2 also decreases (it reacts with the H_2) and the amount of NH_3 increases (it is formed as H_2 and N_2 react).

With time, the equilibrium shift to the right caused by the addition of H_2 will cease because a new equilibrium condition (not identical to the original one) has been reached. At this new equilibrium condition, most (but not all) of the added H_2 will have been converted to NH_3. Necessary accompaniments to this change are a decreased N_2 concentration (some of it reacted with the H_2) and an increased NH_3 concentration (produced from the N_2–H_2 reaction). Figure 9.12 quantifies the changes that occur in

Figure 9.12 Concentration changes that result when H_2 is added to an equilibrium mixture involving the system

$$N_2(g) + 3H_2(g) \rightleftharpoons 2NH_3(g)$$

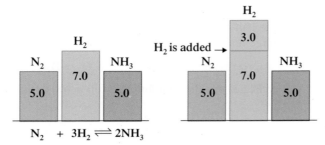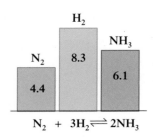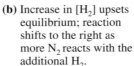

(a) Original equilibrium conditions

(b) Increase in $[H_2]$ upsets equilibrium; reaction shifts to the right as more N_2 reacts with the additional H_2.

(c) New equilibrium conditions. Compared with the original equilibrium in (a): $[N_2]$ has decreased. $[H_2]$ has increased because of addition. (Note that $[H_2]$ is actually decreased from conditions at (b) because some of it has reacted with N_2 to form more NH_3.) $[NH_3]$ has increased.

the N_2–H_2–NH_3 equilibrium system when it is upset by the addition of H_2 for a specific set of concentrations.

Consider again the reaction between N_2 and H_2 to form NH_3.

$$N_2(g) + 3H_2(g) \rightleftharpoons 2NH_3(g)$$

Le Châtelier's principle applies in the same way to removing a reactant or product from the equilibrium mixture as it does to adding a reactant or product at equilibrium. Suppose that at equilibrium we remove some NH_3. The equilibrium position shifts to the right to replenish the NH_3.

Within the human body, numerous equilibrium situations exist that shift in response to a concentration change. Consider, for example, the equilibrium between glucose in the blood and stored glucose (glycogen) in the liver:

$$\text{Glucose in blood} \rightleftharpoons \text{stored glucose} + H_2O$$

> Thousands of chemical equilibria simultaneously exist in biochemical systems. Many of them are interrelated. When the concentration of one substance changes, many equilibria are affected.

Strenuous exercise or hard work causes our blood glucose level to decrease. Our bodies respond to this stress (not enough glucose in the blood) by the liver converting glycogen into glucose. Conversely, when an excess of glucose is present in the blood (after a meal), the liver converts the excess glucose in the blood into its storage form (glycogen).

Temperature Changes

Le Châtelier's principle can be used to predict the influence of temperature changes on an equilibrium, provided we know whether the reaction is exothermic or endothermic. For *exothermic reactions,* heat can be treated as one of the *products;* for *endothermic reactions,* heat can be treated as one of the *reactants.*

Consider the exothermic reaction

$$H_2(g) + F_2(g) \rightleftharpoons 2HF(g) + \text{heat}$$

Heat is produced when the reaction proceeds to the right. Thus if we add heat to an exothermic system at equilibrium (by raising the temperature), the system will shift to the left in an "attempt" to decrease the amount of heat present. When equilibrium is reestablished, the concentrations of H_2 and F_2 will be higher, and the concentration of HF will have decreased. Lowering the temperature of an exothermic reaction mixture causes the reaction to shift to the right as the system acts to replace the lost heat (Figure 9.13).

Figure 9.13 Effect of temperature change on the equilibrium mixture

$$CoCl_4^{2-} + 6H_2O \rightleftharpoons$$
Blue
$$Co(H_2O)_6^{2+} + 4Cl^- + heat$$
Pink

At room temperature, the equilibrium mixture is blue from $CoCl_4^{2-}$. When cooled by the ice bath, the equilibrium mixture turns pink from $Co(H_2O)_6^{2+}$. The temperature decrease causes the equilibrium position to shift to the right.

Increasing the pressure associated with an equilibrium system by adding an inert gas (a gas that is not a reactant or a product in the reaction) does not affect the position of the equilibrium.

The behavior, with temperature change, of an equilibrium system involving an endothermic reaction, such as

$$Heat + 2CO_2(g) \rightleftharpoons 2CO(g) + O_2(g)$$

is opposite to that of an exothermic reaction because a shift to the left produces heat. Consequently, an increase in temperature will cause the equilibrium to shift to the right (to decrease the amount of heat present), and a decrease in temperature will produce a shift to the left (to generate more heat).

Pressure Changes

Pressure changes affect systems at equilibrium only when gases are involved—and then only in cases where the chemical reaction is such that a change in the total number of moles in the gaseous state occurs. This latter point can be illustrated by considering the following two gas-phase reactions:

$$\underbrace{2H_2(g) + O_2(g)}_{\text{3 moles of gas}} \longrightarrow \underbrace{2H_2O(g)}_{\text{2 moles of gas}}$$

$$\underbrace{H_2(g) + Cl_2(g)}_{\text{2 moles of gas}} \longrightarrow \underbrace{2HCl(g)}_{\text{2 moles of gas}}$$

In the first reaction, the total number of moles of gaseous reactants and products decreases as the reaction proceeds to the right. This is because 3 moles of reactants combine to give only 2 moles of products. In the second reaction, there is no change in the total number of moles of gaseous substances present as the reaction proceeds. This is because 2 moles of reactants combine to give 2 moles of products. Thus a pressure change will shift the equilibrium position in the first reaction but not in the second.

Pressure changes are usually brought about through volume changes. A pressure increase results from a volume decrease, and a pressure decrease results from a volume increase (Section 7.4). Le Châtelier's principle correctly predicts the direction of the equilibrium position shift resulting from a pressure change only when the pressure change is caused by a change in volume. It does not apply to pressure increases caused by the addition of a nonreactive (inert) gas to the reaction mixture. This addition has no effect on the equilibrium position. The partial pressure (Section 7.8) of each of the gases involved in the reaction remains the same.

According to Le Châtelier's principle, the stress of increased pressure is relieved by decreasing the number of moles of gaseous substances in the system. This is accomplished by the reaction shifting in the direction of fewer moles; that is, it shifts to the side of the equation that contains the fewer moles of gaseous substances. For the reaction

$$2NO_2(g) + 7H_2(g) \rightleftharpoons 2NH_3(g) + 4H_2O(g)$$

an increase in pressure would shift the equilibrium position to the right because there are 9 moles of gaseous reactants and only 6 moles of gaseous products. On the other hand, the stress of decreased pressure causes an equilibrium system to produce more moles of gaseous substances.

⬤ **EXAMPLE 9.7**

Using Le Châtelier's Principle to Predict How Various Changes Affect an Equilibrium System

▶ How will the gas-phase equilibrium

$$CH_4(g) + 2H_2S(g) + heat \rightleftharpoons CS_2(g) + 4H_2(g)$$

be affected by each of the following?

a. The removal of $H_2(g)$
b. The addition of $CS_2(g)$
c. An increase in the temperature
d. An increase in the volume of the container (a decrease in pressure)

(continued)

Solution

a. The equilibrium will *shift to the right,* according to Le Châtelier's principle, in an "attempt" to replenish the H_2 that was removed.
b. The equilibrium will *shift to the left* in an attempt to use up the extra CS_2 that has been placed in the system.
c. Raising the temperature means that heat energy has been added. In an attempt to minimize the effect of this extra heat, the position of the equilibrium will *shift to the right,* the direction that consumes heat; heat is one of the reactants in an endothermic reaction.
d. The system will *shift to the right,* the direction that produces more moles of gaseous substances (an increase of pressure). In this way, the reaction produces 5 moles of gaseous products for every 3 moles of gaseous reactants consumed.

Practice Exercise 9.7

How will the gas-phase equilibrium

$$CO(g) + 3H_2(g) \rightleftharpoons CH_4(g) + H_2O(g) + heat$$

be affected by each of the following?

a. The removal of $CH_4(g)$
b. The addition of $H_2O(g)$
c. A decrease in the temperature
d. A decrease in the volume of the container (an increase in pressure)

Answers: a. Shift to the right; **b.** Shift to the left; **c.** Shift to the right; **d.** Shift to the right

Addition of Catalysts

Catalysts cannot change the position of an equilibrium. A catalyst functions by lowering the activation energy for a reaction. It speeds up both the forward and the reverse reactions, so it has no net effect on the position of the equilibrium. However, the lowered activation energy allows equilibrium to be established more quickly than if the catalyst were absent.

CONCEPTS TO REMEMBER

Chemical reaction. A process in which at least one new substance is produced as a result of chemical change (Section 9.1).

Combination reaction. A chemical reaction in which a single product is produced from two or more reactants (Section 9.1).

Decomposition reaction. A chemical reaction in which a single reactant is converted into two or more simpler substances (elements or compounds) (Section 9.1).

Single-replacement reaction. A chemical reaction in which an atom or a molecule replaces an atom or a group of atoms from a compound (Section 9.1).

Double-replacement reaction. A chemical reaction in which two substances exchange parts with one another and form two different substances (Section 9.1).

Combustion reaction. A chemical reaction in which oxygen (usually from air) reacts with a substance with evolution of heat and usually the presence of a flame (Section 9.1).

Redox reaction. A chemical reaction in which there is a transfer of electrons from one reactant to another reactant (Section 9.2).

Nonredox reaction. A chemical reaction in which there is no transfer of electrons from one reactant to another reactant (Section 9.2).

Oxidation number. A number that represents the charge that an atom appears to have when the electrons in each bond it is participating in are assigned to the more electronegative of the two atoms involved in the bond. Oxidation numbers are used to identify the electron transfer that occurs in a redox reaction (Section 9.2).

Oxidation-reduction terminology. Oxidation is the loss of electrons; reduction is the gain of electrons. An oxidizing agent causes oxidation by accepting electrons from the other reactant. A reducing agent causes reduction by providing electrons for the other reactant to accept (Section 9.3).

Collision theory. Collision theory summarizes the conditions required for a chemical reaction to take place. The three basic tenets of collision theory are: (1) Reactant molecules must collide with each other, (2) The colliding reactants must possess a certain minimum of energy, and (3) In some cases, colliding reactants must be oriented in a specific way if reaction is to occur (Section 9.4).

Exothermic and endothermic chemical reactions. An exothermic chemical reaction releases energy as the reaction occurs. An endothermic chemical reaction requires an input of energy as the reaction occurs (Section 9.5).

Chemical reaction rates. A chemical reaction rate is the speed at which reactants are converted to products. Four factors affect the rates of all reactions: (1) the physical nature of the reactants, (2) reactant concentrations, (3) reaction temperature, and (4) the presence of catalysts (Section 9.6).

Chemical equilibrium. Chemical equilibrium is the state wherein the rate of the forward reaction is equal to the rate of the reverse reaction. Equilibrium is indicated in chemical equations by writing half-headed arrows pointing in both directions between reactants and products (Section 9.7).

Equilibrium constant. The equilibrium constant relates the concentrations of reactants and products at equilibrium. The value of an equilibrium constant is obtained by writing an equilibrium constant expression and then numerically evaluating it. Equilibrium constant expressions can be obtained from the balanced chemical equations for reactions (Section 9.8).

Equilibrium position. The relative amounts of reactants and products present in a system at equilibrium define the equilibrium position. The equilibrium position is toward the right when a large amount of product is present and is toward the left when a large amount of reactant is present (Section 9.8).

Le Châtelier's principle. Le Châtelier's principle states that when a stress (change of conditions) is applied to a system in equilibrium, the system will readjust (change the equilibrium position) in the direction that best reduces the stress imposed on it. Stresses that change an equilibrium position include (1) changes in amount of reactants and/or products, (2) changes in temperature, and (3) changes in pressure (Section 9.9).

KEY REACTIONS AND EQUATIONS

1. Combination reaction (Section 9.1)
$$X + Y \longrightarrow XY$$

2. Decomposition reaction (Section 9.1)
$$XY \longrightarrow X + Y$$

3. Single-replacement reaction (Section 9.1)
$$X + YZ \longrightarrow Y + XZ$$

4. Double-replacement reaction (Section 9.1)
$$AX + BY \longrightarrow AY + BX$$

5. Equilibrium constant expression equation for a general reaction (Section 9.8)
$$wA + xB \rightleftharpoons yC + zD$$

$$K_{eq} = \frac{[C]^y[D]^z}{[A]^w[B]^x}$$

EXERCISES *and* PROBLEMS

The members of each pair of problems in this section test similar material.

● **Types of Chemical Reactions (Section 9.1)**

9.1 What is the general chemical equation for each of the following types of chemical reactions?
a. Single-replacement reaction b. Combination reaction

9.2 What is the general chemical equation for each of the following types of chemical reactions?
a. Double-replacement reaction b. Decomposition reaction

9.3 Classify each of the following reactions as a combination, decomposition, single-replacement, double-replacement, or combustion reaction.
a. $3CuSO_4 + 2Al \rightarrow Al_2(SO_4)_3 + 3Cu$
b. $K_2CO_3 \rightarrow K_2O + CO_2$
c. $2AgNO_3 + K_2SO_4 \rightarrow Ag_2SO_4 + 2KNO_3$
d. $2P + 3H_2 \rightarrow 2PH_3$

9.4 Classify each of the following reactions as a combination, decomposition, single-replacement, double-replacement, or combustion reaction.
a. $2NaHCO_3 \rightarrow Na_2CO_3 + CO_2 + H_2O$
b. $2Ag_2CO_3 \rightarrow 4Ag + 2CO_2 + O_2$

c. $2C_2H_6 + 7O_2 \rightarrow 4CO_2 + 6H_2O$
d. $Mg + 2HCl \rightarrow MgCl_2 + H_2$

9.5 Indicate to which of the following types of reactions each of the statements listed applies: *combination, decomposition, single-replacement, double-replacement,* and *combustion.* More than one answer is possible for a given statement.
a. An element may be a reactant.
b. An element may be a product.
c. A compound may be a reactant.
d. A compound may be a product.

9.6 Indicate to which of the following types of reactions each of the statements listed applies: *combination, decomposition, single-replacement, double-replacement,* and *combustion.* More than one answer is possible for a given statement.
a. Two reactants are required.
b. Only one reactant is present.
c. Two products are present.
d. Only one product is present.

● **Redox and Nonredox Chemical Reactions (Section 9.2)**

9.7 Determine the oxidation number of each of the following.
a. Ba in Ba^{2+}
b. S in SO_3
c. F in F_2
d. P in PO_4^{3-}

9.8 Determine the oxidation number of each of the following.
a. Al in Al^{3+}
b. N in NO_2
c. O in O_3
d. S in SO_4^{2-}

9.9 Determine the oxidation number of Cr in each of the following chromium-containing species.
a. Cr_2O_3
b. CrO_2
c. CrO_3
d. Na_2CrO_4
e. $BaCrO_4$
f. $BaCr_2O_7$
g. $Na_2Cr_2O_7$
h. CrF_5

9.10 Determine the oxidation number of Cl in each of the following chlorine-containing species.
a. $BeCl_2$
b. $Ba(ClO)_2$
c. ClF_4^+
d. Cl_2O_7
e. NCl_3
f. $AlCl_4^-$
g. ClF
h. ClO^-

9.11 What is the oxidation number of each element in each of the following substances?
a. PF_3
b. NaOH
c. Na_2SO_4
d. CO_3^{2-}

9.12 What is the oxidation number of each element in each of the following substances?
a. H_2S
b. H_2
c. N^{3-}
d. MnO_4^-

9.13 Classify each of the following reactions as a redox reaction or a nonredox reaction.
a. $2Cu + O_2 \rightarrow 2CuO$
b. $K_2O + H_2O \rightarrow 2KOH$
c. $2KClO_3 \rightarrow 2KCl + 3O_2$
d. $CH_4 + 2O_2 \rightarrow CO_2 + 2H_2O$

9.14 Classify each of the following reactions as a redox reaction or a nonredox reaction.
a. $2NO + O_2 \rightarrow 2NO_2$
b. $CO_2 + H_2O \rightarrow H_2CO_3$
c. $Zn + 2AgNO_3 \rightarrow Zn(NO_3)_2 + 2Ag$
d. $HNO_3 + NaOH \rightarrow NaNO_3 + H_2O$

● **Terminology Associated with Redox Processes (Section 9.3)**

9.15 Identify which substance is oxidized and which substance is reduced in each of the following redox reactions.
a. $N_2 + 3H_2 \rightarrow 2NH_3$
b. $Cl_2 + 2KI \rightarrow 2KCl + I_2$
c. $Sb_2O_3 + 3Fe \rightarrow 2Sb + 3FeO$
d. $3H_2SO_3 + 2HNO_3 \rightarrow 2NO + H_2O + 3H_2SO_4$

9.16 Identify which substance is oxidized and which substance is reduced in each of the following redox reactions.
a. $2Al + 3Cl_2 \rightarrow 2AlCl_3$
b. $Zn + CuCl_2 \rightarrow ZnCl_2 + Cu$
c. $2NiS + 3O_2 \rightarrow 2NiO + 2SO_2$
d. $3H_2S + 2HNO_3 \rightarrow 3S + 2NO + 4H_2O$

9.17 Identify which substance is the oxidizing agent and which substance is the reducing agent in each of the redox reactions of Problem 9.15.

9.18 Identify which substance is the oxidizing agent and which substance is the reducing agent in each of the redox reactions of Problem 9.16.

● **Collision Theory and Chemical Reactions (Section 9.4)**

9.19 What are the three central concepts associated with collision theory?

9.20 Why are most chemical reactions carried out either in liquid solution or in the gaseous phase?

9.21 What two factors determine whether a collision between two reactant molecules will result in a reaction?

9.22 What happens to the reactants in an ineffective molecular collision?

● **Exothermic and Endothermic Chemical Reactions (Section 9.5)**

9.23 Which of the following reactions are endothermic, and which are exothermic?
a. $C_2H_4 + 3O_2 \rightarrow 2CO_2 + 2H_2O + heat$
b. $N_2 + 2O_2 + heat \rightarrow 2NO_2$
c. $2H_2O + heat \rightarrow 2H_2 + O_2$
d. $2KClO_3 \rightarrow 2KCl + 3O_2 + heat$

9.24 Which of the following reactions are endothermic, and which are exothermic?
a. $CaCO_3 + heat \rightarrow CaO + CO_2$
b. $N_2 + 3H_2 \rightarrow 2NH_3 + heat$
c. $CO + 3H_2 + heat \rightarrow CH_4 + H_2O$
d. $2N_2 + 6H_2O + heat \rightarrow 4NH_3 + 3O_2$

9.25 In a certain chemical reaction, the average energy of the products is 35 kcal lower than the average energy of the reactants.
a. Is the chemical reaction exothermic or endothermic?
b. Is energy released or absorbed in the chemical reaction?

9.26 Indicate whether each of the following is a characteristic of an endothermic or an exothermic chemical reaction.
a. There is more energy stored in product molecule bonds than in reactant molecule bonds.
b. The energy required to break bonds in the reactants is less than the energy released by bond formation in the products.

9.27 Sketch an energy diagram graph representing an exothermic reaction, and label the following.
a. Average energy of reactants
b. Average energy of products
c. Activation energy
d. Amount of energy liberated during the reaction

9.28 Sketch an energy diagram graph representing an endothermic reaction, and label the following.
a. Average energy of reactants
b. Average energy of products
c. Activation energy
d. Amount of energy absorbed during the reaction

● **Factors That Influence Chemical Reaction Rates (Section 9.6)**

9.29 Using collision theory, indicate why each of the following factors influences the rate of a reaction.
a. Temperature of reactants
b. Presence of a catalyst

9.30 Using collision theory, indicate why each of the following factors influences the rate of a reaction.
a. Physical nature of reactants
b. Reactant concentrations

9.31 Substances burn more rapidly in pure oxygen than in air. Explain why.

9.32 Milk will sour in a couple of days when left at room temperature, yet it can remain unspoiled for 2 weeks when refrigerated. Explain why.

9.33 Will each of the changes listed increase or decrease the rate of the following chemical reaction?

$$2CO + O_2 \longrightarrow 2CO_2$$

a. Adding some O_2 to the reaction mixture
b. Lowering the temperature of the reaction mixture
c. Adding a catalyst to the reaction mixture
d. Removing some CO from the reaction mixture

9.34 Will each of the changes listed increase or decrease the rate of the following chemical reaction?

$$N_2 + 3H_2 \longrightarrow 2NH_3$$

a. Adding some N_2 to the reaction mixture
b. Raising the temperature of the reaction mixture
c. Removing a catalyst present in the reaction mixture
d. Removing some H_2 from the reaction mixture

9.35 Draw an energy diagram graph for an exothermic reaction where no catalyst is present. Then draw an energy diagram graph for the same reaction when a catalyst is present. Indicate the similarities and differences between the two diagrams.

9.36 Draw an energy diagram graph for an endothermic reaction where no catalyst is present. Then draw an energy diagram graph for the same reaction when a catalyst is present. Indicate the similarities and differences between the two diagrams.

9.37 The characteristics of four reactions, each of which involves only two reactants, are given.

Reaction	Activation Energy	Temperature	Concentration of Reactants
1	low	low	1 mole/L of each
2	high	low	1 mole/L of each
3	low	high	1 mole/L of each
4	low	low	1 mole/L of first reactant and 4 moles/L of second reactant

For each of the following pairs of the preceding reactions, compare the reaction rates when the two reactants are first mixed by indicating which reaction is faster.
a. 1 and 2   b. 1 and 3   c. 1 and 4   d. 2 and 3

9.38 The characteristics of four reactions, each of which involves only two reactants, are given.

Reaction	Activation Energy	Temperature	Concentration of Reactants
1	high	low	1 mole/L of each
2	high	high	1 mole/L of each
3	low	low	1 mole/L of first reactant and 4 moles/L of second reactant
4	low	low	4 moles/L of each

For each of the following pairs of the preceding reactions, compare the reaction rates when the two reactants are first mixed by indicating which reaction is faster.

a. 1 and 2   b. 1 and 3
c. 1 and 4   d. 3 and 4

Chemical Equilibrium (Section 9.7)

9.39 What condition must be met in order for a system to be in a state of chemical equilibrium?

9.40 What relationship exists between the rates of the forward and reverse reactions for a system in a state of chemical equilibrium?

9.41 Consider the following equilibrium system.

$$N_2(g) + O_2(g) \rightleftharpoons 2NO(g)$$

a. Write the chemical equation for the forward reaction.
b. Write the chemical equation for the reverse reaction.

9.42 Consider the following equilibrium system.

$$N_2(g) + 3H_2(g) \rightleftharpoons 2NH_3(g)$$

a. Write the chemical equation for the forward reaction.
b. Write the chemical equation for the reverse reaction.

9.43 Sketch a graph showing how the concentrations of the reactants and products of a typical reversible chemical reaction vary with time.

9.44 Sketch a graph showing how the rates of the forward and reverse reactions for a typical reversible chemical reaction vary with time.

9.45 The following series of diagrams represent the reaction $X \rightleftharpoons Y$ followed over a period of time. The X molecules are red and the Y molecules are green.

Increasing time

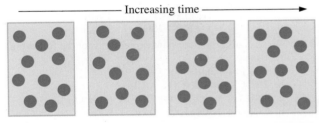

At the end of the time period depicted, has the reaction system reached equilibrium? Justify your answer with a one-sentence explanation.

9.46 The following series of diagrams represent the reaction $X \rightleftharpoons Y$ followed over a period of time. The X molecules are red and the Y molecules are green.

Increasing time

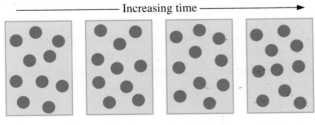

At the end of the time period depicted, has the reaction system reached equilibrium? Justify your answer with a one-sentence explanation.

9.47 For the reaction $A_2 + 2B \rightarrow 2AB$, diagram I depicts an initial reaction mixture, where A_2 molecules are red and B atoms are green. Which of the diagrams II through IV is a possible equilibrium state for the reaction system? There may be more than one correct answer.

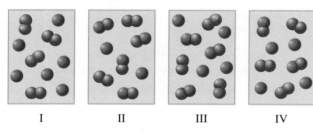

I   II   III   IV

9.48 For the reaction $A_2 + B_2 \rightarrow 2AB$, diagram I depicts an initial reaction mixture, where A_2 molecules are red and B_2 molecules are green. Which of the diagrams II through IV is a possible equilibrium state for the reaction system? There may be more than one correct answer.

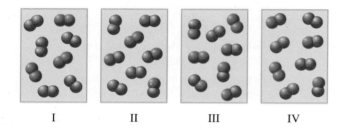

I   II   III   IV

● Equilibrium Constants (Section 9.8)

9.49 Write equilibrium constant expressions for the following reactions.
   a. $N_2O_4(g) \rightleftharpoons 2NO_2(g)$
   b. $COCl_2(g) \rightleftharpoons CO(g) + Cl_2(g)$
   c. $CS_2(g) + 4H_2(g) \rightleftharpoons CH_4(g) + 2H_2S(g)$
   d. $2SO_2(g) + O_2(g) \rightleftharpoons 2SO_3(g)$

9.50 Write equilibrium constant expressions for the following reactions.
   a. $3O_2(g) \rightleftharpoons 2O_3(g)$
   b. $2NOCl(g) \rightleftharpoons 2NO(g) + Cl_2(g)$
   c. $4NH_3(g) + 5O_2(g) \rightleftharpoons 4NO(g) + 6H_2O(g)$
   d. $CO(g) + H_2O(g) \rightleftharpoons CO_2(g) + H_2(g)$

9.51 Write equilibrium constant expressions for the following reactions.
   a. $H_2SO_4(l) \rightleftharpoons SO_3(g) + H_2O(l)$
   b. $2Ag(s) + Cl_2(g) \rightleftharpoons 2AgCl(s)$
   c. $BaCl_2(aq) + Na_2SO_4(aq) \rightleftharpoons 2NaCl(aq) + BaSO_4(s)$
   d. $2Na_2O(s) \rightleftharpoons 4Na(l) + O_2(g)$

9.52 Write equilibrium constant expressions for the following reactions.
   a. $2KClO_3(s) \rightleftharpoons 2KCl(s) + 3O_2(g)$
   b. $PCl_5(s) \rightleftharpoons PCl_3(l) + Cl_2(g)$
   c. $AgNO_3(aq) + NaCl(aq) \rightleftharpoons AgCl(s) + NaNO_3(aq)$
   d. $2FeBr_3(s) \rightleftharpoons 2FeBr_2(s) + Br_2(g)$

9.53 Calculate the value of the equilibrium constant for the reaction
$$N_2O_4(g) \rightleftharpoons 2NO_2(g)$$
if the concentrations of the species at equilibrium are $[N_2O_4] = 0.213$ and $[NO_2] = 0.0032$.

9.54 Calculate the value of the equilibrium constant for the reaction
$$N_2(g) + 2O_2(g) \rightleftharpoons 2NO_2(g)$$
if the concentrations of the species at equilibrium are $[N_2] = 0.0013$, $[O_2] = 0.0024$, and $[NO_2] = 0.00065$.

9.55 Use the given K_{eq} value and the terminology in Table 9.2 to describe the relative amounts of reactants and products present in each of the following equilibrium situations.

   a. $H_2(g) + Br_2(g) \rightleftharpoons 2HBr(g)$ $K_{eq}(25°C) = 2.0 \times 10^9$
   b. $2HCl(g) \rightleftharpoons H_2(g) + Cl_2(g)$ $K_{eq}(25°C) = 3.2 \times 10^{-34}$
   c. $SO_2(g) + NO_2(g) \rightleftharpoons NO(g) + SO_3(g)$ $K_{eq}(460°C) = 85.0$
   d. $COCl_2(g) \rightleftharpoons CO(g) + Cl_2(g)$ $K_{eq}(395°C) = 0.046$

9.56 Use the given K_{eq} value and the terminology in Table 9.2 to describe the relative amounts of reactants and products present in each of the following equilibrium situations.
   a. $2NO(g) \rightleftharpoons N_2(g) + O_2(g)$ $K_{eq}(25°C) = 1 \times 10^{30}$
   b. $N_2(g) + 3H_2(g) \rightleftharpoons 2NH_3(g)$ $K_{eq}(25°C) = 1 \times 10^9$
   c. $PCl_5(g) \rightleftharpoons PCl_3(g) + Cl_2(g)$ $K_{eq}(127°C) = 1 \times 10^{-2}$
   d. $2Na_2O(s) \rightleftharpoons 4Na(l) + O_2(g)$ $K_{eq}(427°C) = 1 \times 10^{-25}$

9.57 The following four diagrams represent gaseous equilibrium mixtures for the reaction $A_2 + B_2 \rightarrow 2AB$ at four different temperatures. For which of the diagrams is the numerical value of the equilibrium constant the largest? (A atoms are red and B atoms are green in the various diagrams.)

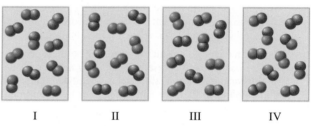

I   II   III   IV

9.58 Based on the diagrams, chemical reaction, and reaction conditions depicted in Problem 9.57, for which of the diagrams is the numerical value of the equilibrium constant the smallest?

9.59 The following four diagrams represent gaseous reaction mixtures for the chemical reaction $A_2 + B_2 \rightarrow 2AB$. If the numerical value of the equilibrium constant for the reaction is 64, which of the diagrams represents the equilibrium mixture? (A atoms are red and B atoms are green in the various diagrams.)

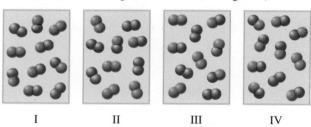

I   II   III   IV

9.60 Based on the diagrams, chemical reaction, and reaction conditions depicted in Problem 9.59, which of the diagrams represents the equilibrium mixture if the numerical value of the equilibrium constant is 9.0?

● Le Châtelier's Principle (Section 9.9)

9.61 For the reaction
$$2Cl_2(g) + 2H_2O(g) \rightleftharpoons 4HCl(g) + O_2(g)$$
determine in what direction the equilibrium will be shifted by each of the following changes.
   a. Increase in Cl_2 concentration
   b. Increase in O_2 concentration
   c. Decrease in H_2O concentration
   d. Decrease in HCl concentration

9.62 For the reaction
$$2Cl_2(g) + 2H_2O(g) \rightleftharpoons 4HCl(g) + O_2(g)$$
determine in what direction the equilibrium will be shifted by each of the following changes.

a. Increase in H_2O concentration
b. Increase in HCl concentration
c. Decrease in O_2 concentration
d. Decrease in Cl_2 concentration

9.63 For the reaction

$$C_6H_6(g) + 3H_2(g) \rightleftharpoons C_6H_{12}(g) + \text{heat}$$

determine in what direction the equilibrium will be shifted by each of the following changes.
a. Increasing the concentration of C_6H_{12}
b. Decreasing the concentration of C_6H_6
c. Increasing the temperature
d. Decreasing the pressure by increasing the volume of the container

9.64 For the reaction

$$C_6H_6(g) + 3H_2(g) \rightleftharpoons C_6H_{12}(g) + \text{heat}$$

determine in what direction the equilibrium will be shifted by each of the following changes.
a. Decreasing the concentration of H_2
b. Increasing the concentration of C_6H_6
c. Decreasing the temperature
d. Increasing the pressure by decreasing the volume of the container

9.65 Consider the following chemical system at equilibrium.

$$CO(g) + H_2O(g) + \text{heat} \rightleftharpoons CO_2(g) + H_2(g)$$

For each of the following adjustments of conditions, indicate the effect (shifts left, shifts right, or no effect) on the position of equilibrium.
a. Refrigerating the equilibrium mixture
b. Adding a catalyst to the equilibrium mixture
c. Adding CO to the equilibrium mixture
d. Increasing the size of the reaction container

9.66 Consider the following chemical system at equilibrium.

$$CO(g) + H_2O(g) + \text{heat} \rightleftharpoons CO_2(g) + H_2(g)$$

For each of the following adjustments of conditions, indicate the effect (shifts left, shifts right, or no effect) on the position of equilibrium.

a. Heating the equilibrium mixture
b. Increasing the pressure on the equilibrium mixture by adding a nonreactive gas
c. Adding H_2 to the equilibrium mixture
d. Decreasing the size of the reaction container

9.67 The following two diagrams represent the composition of an equilibrium mixture for the reaction $A_2 + B_2 \longrightarrow 2AB$ at two different temperatures. Based on the diagrams, is the chemical reaction endothermic or exothermic? Explain your answer using Le Châtelier's principle. (A atoms are red and B atoms are green in the diagrams.)

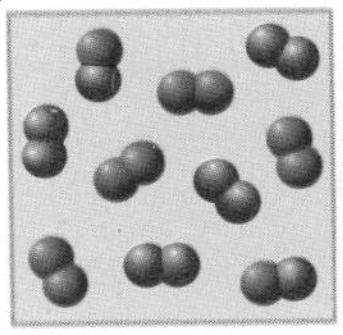

$T = 150°C$ 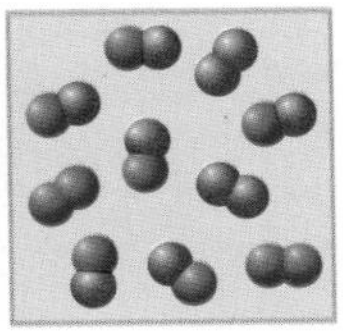 $T = 200°C$

9.68 The following two diagrams represent the composition of an equilibrium mixture for the reaction $A_2 + B_2 \longrightarrow 2AB$ at two different temperatures. Based on the diagrams, is the chemical reaction endothermic or exothermic? Explain your answer using Le Châtelier's principle. (A atoms are red and B atoms are green in the diagrams.)

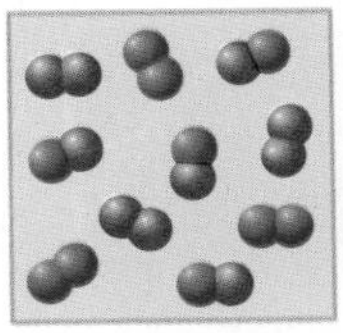

$T = 250°C$ 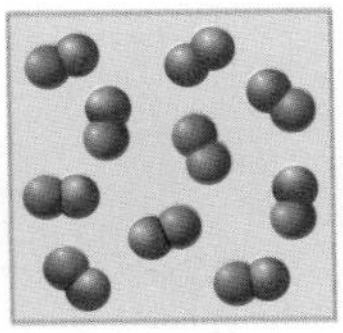 $T = 400°C$

ADDITIONAL PROBLEMS

9.69 Characterize each of the following reactions using one selection from the choices *redox* and *nonredox* combined with one selection from the choices *combination, decomposition, single-replacement, double-replacement,* and *combustion.*
a. $Zn + Cu(NO_3)_2 \rightarrow Zn(NO_3)_2 + Cu$
b. $CH_4 + 2O_2 \rightarrow CO_2 + 2H_2O$
c. $2CuO \rightarrow 2Cu + O_2$
d. $NaCl + AgNO_3 \rightarrow AgCl + NaNO_3$

9.70 Classify each of the following reactions as (1) a redox reaction, (2) a nonredox reaction, or (3) "can't classify" because of insufficient information.
a. A combination reaction in which one reactant is an element and the other is a compound
b. A decomposition reaction in which the products are all elements
c. A decomposition reaction in which one of the products is an element
d. A single-replacement reaction in which both of the reactants are compounds

9.71 In each of the following statements, choose the word in parentheses that best completes the statement.
a. The process of reduction is associated with the (loss, gain) of electrons.
b. The oxidizing agent in a redox reaction is the substance that undergoes (oxidation, reduction).
c. Reduction always results in an (increase, decrease) in the oxidation number of an element.
d. A reducing agent in a redox reaction is the substance that contains the element that undergoes an (increase, decrease) in oxidation number.

9.72 Indicate whether each of the following substances loses or gains electrons in a redox reaction.
a. The oxidizing agent
b. The reducing agent
c. The substance undergoing oxidation
d. The substance undergoing reduction

9.73 Indicate whether each of the following substances undergoes an increase in oxidation number or a decrease in oxidation number in a redox reaction.
a. The oxidizing agent
b. The reducing agent
c. The substance undergoing oxidation
d. The substance undergoing reduction

9.74 Which of the following changes would affect the *value* of a system's equilibrium constant?
a. Removal of a reactant or product from an equilibrium mixture
b. Decrease in the system's total pressure
c. Increase in the system's temperature
d. Addition of a catalyst to the equilibrium mixture

9.75 Write a balanced chemical equation for the totally gaseous equilibrium system that would lead to the following expression for the equilibrium constant.

$$K_{eq} = \frac{[CH_4][H_2S]^2}{[CS_2][H_2]^4}$$

9.76 For which of the following reactions is product formation favored by high temperature?
a. $N_2(g) + 2O_2(g) + heat \rightleftharpoons 2NO_2(g)$
b. $2N_2(g) + 6H_2O(g) + heat \rightleftharpoons 4NH_3(g) + 3O_2(g)$
c. $C_2H_4(g) + 3O_2(g) \rightleftharpoons 2CO_2(g) + 2H_2O(g) + heat$
d. $2KClO_3(s) + heat \rightleftharpoons 2KCl(s) + 3O_2(g)$

9.77 Predict the direction in which each of the following equilibria will shift if the pressure within the system is increased by reducing volume, using the choices *left*, *right*, and *no effect*.
a. $H_2(g) + C_2N_2(g) \rightleftharpoons 2HCN(g)$
b. $CO(g) + Br_2(g) \rightleftharpoons COBr_2(g)$
c. $CS_2(g) + 4H_2(g) \rightleftharpoons CH_4(g) + 2H_2S(g)$
d. $Ni(s) + 4CO(g) \rightleftharpoons Ni(CO)_4(g)$

*M*ultiple-Choice Practice Test

9.78 Which of the following general equations is a representation of a single-replacement reaction?
a. $X + Y \rightarrow XY$
b. $XY \rightarrow X + Y$
c. $X + YZ \rightarrow Y + XZ$
d. $AX + BY \rightarrow AY + BX$

9.79 In which of the following compounds does Cl have an oxidation number of +5?
a. KCl   b. $KClO_2$ c. $KClO_3$ d. HClO

9.80 Which substance is oxidized in the following redox reaction?

$$2H_2S + O_2 \rightarrow 2H_2O + 2S$$

a. H_2S b. O_2 c. H_2O d. S

9.81 In a redox reaction, which of the following is true for the substance that is reduced?
a. It is also the reducing agent.
b. It always gains electrons.
c. It never contains oxygen.
d. It must contain hydrogen.

9.82 For a collision between molecules to result in reaction, the molecules must possess a certain minimum energy and also undergo which of the following?
a. Exchange electrons
b. Interact with a catalyst
c. Have a favorable orientation relative to each other when they collide
d. Have the same activation energy

9.83 Increasing the temperature at which a chemical reaction occurs will also do which of the following?
a. Increase the activation energy
b. Cause more reactant collisions to take place in a given time
c. Increase the energy of the system, thus decreasing the reaction rate
d. Decrease the energy of the system, thus increasing the reaction rate

9.84 Which of the following conditions characterizes a system in a state of chemical equilibrium?
a. The concentrations of reactants and products are equal.
b. The rate of the forward reaction has dropped to zero.
c. Reactants are being consumed at the same rate at which products are converted to reactants.
d. Reactant molecules no longer react with each other.

9.85 In writing an equilibrium constant expression, which of the following is *incorrect*?
a. Concentrations are always expressed in molarity units.
b. Product concentrations are always placed in the numerator of the expression.
c. Reactant concentrations are always placed in the denominator of the expression.
d. Concentrations of pure solids and pure liquids are always placed in the denominator of the expression.

9.86 According to Le Châtelier's principle, which of these effects will occur if NH_3 is *removed* from an equilibrium mixture governed by the following chemical equation?

$$N_2(g) + 3H_2(g) \rightarrow 2NH_3(g) + heat$$

a. Concentration of N_2 will increase.
b. Heat will be generated.
c. Concentration of H_2 will remain the same.
d. Concentration of N_2 will decrease and that of H_2 will increase.

9.87 Which of these changes will cause the equilibrium position to shift to the left for the following chemical reaction?

$$4NH_3(g) + 3O_2(g) \rightarrow 2N_2(g) + 6H_2O(g) + heat$$

a. Adding more NH_3
b. Decreasing the temperature
c. Adding a catalyst
d. Increasing the pressure by decreasing the volume

10 Acids, Bases, and Salts

Chapter Outline

The tartness of an oil-vinegar salad dressing is caused by the acetic acid present in the vinegar.

Acids, bases, and salts are among the most common and important compounds known. In the form of aqueous solutions, these compounds are key materials in both biochemical systems and the chemical industry. A major ingredient of gastric juice in the stomach is hydrochloric acid. Quantities of lactic acid are produced when the human body is subjected to strenuous exercise. The lye used in making homemade soap contains the base sodium hydroxide. Bases are ingredients in many stomach antacid formulations. The white crystals you sprinkle on your food to make it taste better represent only one of many hundreds of salts that exist.

 10.1 ARRHENIUS ACID–BASE THEORY

In 1884 the Swedish chemist Svante August Arrhenius (1859–1927) proposed that acids and bases be defined in terms of the chemical species they form when they dissolve in water. An **Arrhenius acid** *is a hydrogen-containing compound that, in water, produces hydrogen ions (H^+ ions).* The acidic species in Arrhenius theory is thus the hydrogen ion. An **Arrhenius base** *is a hydroxide-containing compound that, in water, produces hydroxide ions (OH^- ions).* The basic species in Arrhenius theory is thus the hydroxide ion. For this reason, Arrhenius bases are also called *hydroxide bases.*

Two common examples of Arrhenius acids are HNO_3 (nitric acid) and HCl (hydrochloric acid).

$$HNO_3(l) \xrightarrow{H_2O} H^+(aq) + NO_3^-(aq)$$

$$HCl(g) \xrightarrow{H_2O} H^+(aq) + Cl^-(aq)$$

Figure 10.1 The difference between the aqueous solution processes of ionization (Arrhenius acids) and dissociation (Arrhenius bases). Ionization is the production of ions from a molecular compound that has been dissolved in solution. Dissociation is the production of ions from an ionic compound that has been dissolved in solution.

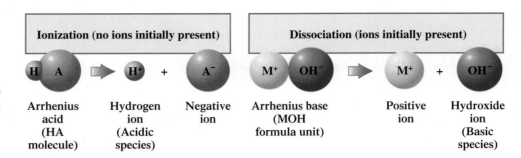

When Arrhenius acids are in the pure state (not in solution), they are covalent compounds; that is, they do not contain H^+ ions. This ion is formed through an interaction between water and the acid when they are mixed. **Ionization** *is the process in which individual positive and negative ions are produced from a molecular compound that is dissolved in solution.*

Two common examples of Arrhenius bases are NaOH (sodium hydroxide) and KOH (potassium hydroxide).

$$NaOH(s) \xrightarrow{H_2O} Na^+(aq) + OH^-(aq)$$

$$KOH(s) \xrightarrow{H_2O} K^+(aq) + OH^-(aq)$$

In direct contrast to acids, Arrhenius bases are ionic compounds in the pure state. When these compounds dissolve in water, the ions separate to yield the OH^- ions. **Dissociation** *is the process in which individual positive and negative ions are released from an ionic compound that is dissolved in solution.* Figure 10.1 contrasts the processes of ionization (acids) and dissociation (bases).

Arrhenius acids have a sour taste, change blue litmus paper to red (see Figure 10.2), and are corrosive to many materials. Arrhenius bases have a bitter taste, change red litmus paper to blue, and are slippery (soapy) to the touch. (The bases themselves are not slippery, but they react with the oils in the skin to form new slippery compounds.)

Figure 10.2 Litmus is a vegetable dye obtained from certain lichens found principally in the Netherlands. Paper treated with this dye turns from blue to red in acids (left) and from red to blue in bases (right).

10.2 BRØNSTED–LOWRY ACID–BASE THEORY

Although it is widely used, Arrhenius acid–base theory has some shortcomings. It is restricted to aqueous solution, and it does not explain why compounds like ammonia (NH_3), which do not contain hydroxide ion, produce a basic water solution.

In 1923, Johannes Nicolaus Brønsted (1879–1947), a Danish chemist, and Thomas Martin Lowry (1874–1936), a British chemist, independently and almost simultaneously proposed broadened definitions for acids and bases—definitions that applied in both aqueous and nonaqueous solutions and that also explained how some nonhydroxide-containing substances, when added to water, produce basic solutions.

A **Brønsted-Lowry acid** *is a substance that can donate a proton (H^+ ion) to some other substance.* A **Brønsted-Lowry base** *is a substance that can accept a proton (H^+ ion) from some other substance.* In short, a Brønsted–Lowry acid is a *proton donor* (or hydrogen ion donor), and a Brønsted–Lowry base is a *proton acceptor* (or hydrogen ion acceptor). The terms *proton* and *hydrogen ion* are used interchangeably in acid–base discussions. Remember that a H^+ ion is a hydrogen atom (proton plus electron) that has lost its electron; hence it is a proton.

Any chemical reaction involving a Brønsted–Lowry acid must also involve a Brønsted–Lowry base. You cannot have one without the other. Proton donation (from an acid) cannot occur unless an acceptor (a base) is present.

Brønsted–Lowry acid–base theory also includes the concept that hydrogen ions in an aqueous solution do not exist in the free state but, rather, react with water to form

The terms *hydrogen ion* and *proton* are used synonymously in acid–base discussions. Why? The predominant hydrogen isotope, ^1_1H, is unique in that no neutrons are present; it consists of a proton and an electron. Thus the ion $^1_1H^+$, a hydrogen atom that has lost its only electron, is simply a proton.

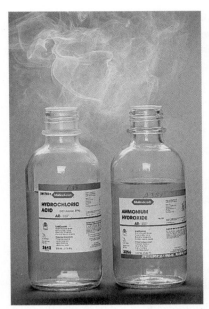

Figure 10.3 A white cloud of finely divided solid NH_4Cl is produced by the acid-base reaction that results when the colorless gases HCl and NH_3 mix. (The gases escaped from the concentrated solutions of HCl and NH_3.)

A Brønsted–Lowry base, a proton acceptor, must contain an atom that possesses a pair of nonbonding electrons that can be used in forming a coordinate covalent bond to an incoming proton (from a Brønsted–Lowry acid).

hydronium ions. The attraction between a hydrogen ion and polar water molecules is sufficiently strong to bond the hydrogen ion to a water molecule to form a hydronium ion (H_3O^+). The bond between them is a coordinate covalent bond (Section 5.5) because both electrons are furnished by the oxygen atom.

$$H^+ + :\ddot{O}-H \longrightarrow \left[H:\ddot{O}:H \right]^+$$

Coordinate covalent bond

Hydronium ion

When gaseous hydrogen chloride dissolves in water, it forms hydrochloric acid. This is a simple Brønsted–Lowry acid–base reaction. The chemical equation for this process is

$$\underset{\substack{\text{Base:}\\ H^+ \text{ acceptor}}}{H_2O(l)} + \underset{\substack{\text{Acid:}\\ H^+ \text{ donor}}}{HCl(g)} \longrightarrow H_3O^+(aq) + Cl^-(aq)$$

The hydrogen chloride behaves as an acid by donating a proton to a water molecule. Because the water molecule accepts the proton, to become H_3O^+, it is the base.

It is not necessary that a water molecule be one of the reactants in a Brønsted–Lowry acid–base reaction; the reaction does not have to take place in the liquid state. Brønsted–Lowry acid–base theory can be used to describe gas-phase reactions. The white solid haze that often covers glassware in a chemistry laboratory results from the gas-phase reaction between HCl and NH_3:

$$\underset{\substack{\text{Base:}\\ H^+ \text{ acceptor}}}{NH_3(g)} + \underset{\substack{\text{Acid:}\\ H^+ \text{ donor}}}{HCl(g)} \longrightarrow NH_4^+(g) + Cl^-(g)$$

This is a Brønsted–Lowry acid–base reaction because the HCl molecules donate protons to the NH_3, forming NH_4^+ and Cl^- ions. These ions instantaneously combine to form the white solid NH_4Cl (see Figure 10.3).

All acids and bases included in Arrhenius theory are also acids and bases according to Brønsted–Lowry theory. However, the converse is not true; some substances that are not considered Arrhenius bases are Brønsted–Lowry bases.

Conjugate Acid-Base Pairs

For most Brønsted–Lowry acid–base reactions, 100% proton transfer does not occur. Instead, an equilibrium situation (Section 9.7) is reached in which a forward reaction and a reverse reaction occur at the same rate.

The equilibrium mixture for a Brønsted–Lowry acid–base reaction always has two acids and two bases present. Consider the acid–base reaction involving hydrogen fluoride and water:

$$HF(aq) + H_2O(l) \rightleftharpoons H_3O^+(aq) + F^-(aq)$$

For the forward reaction, the HF molecules donate protons to water molecules. Thus the HF is functioning as an acid, and the H_2O is functioning as a base.

$$\underset{\text{Acid}}{HF(aq)} + \underset{\text{Base}}{H_2O(l)} \longrightarrow H_3O^+(aq) + F^-(aq)$$

For the reverse reaction, the one going from right to left, a different picture emerges. Here, H_3O^+ is functioning as an acid (by donating a proton), and F^- behaves as a base (by accepting the proton).

$$\underset{\text{Acid}}{H_3O^+(aq)} + \underset{\text{Base}}{F^-(aq)} \longrightarrow HF(aq) + H_2O(l)$$

Conjugate means "coupled" or "joined together" (as in a pair).

Every Brønsted–Lowry acid has a conjugate base, and every Brønsted–Lowry base has a conjugate acid. In general terms, these relationships can be diagrammed as follows:

$$HA + B \rightleftharpoons HB^+ + A^-$$

Acid · · · · Base · · · · Conjugate acid · · · · Conjugate base

The two acids and two bases involved in a Brønsted–Lowry acid–base equilibrium mixture can be grouped into two conjugate acid–base pairs. A **conjugate acid-base pair** *is two species, one an acid and one a base, that differ from each other through the loss or gain of a proton (H^+ ion).* The two conjugate acid–base pairs in our example are HF and F, and H_3O^+ and H_2O.

Conjugate pair

$$HF(aq) + H_2O(l) \rightleftharpoons H_3O^+(aq) + F^-(aq)$$

Acid · · · · Base · · · · Acid · · · · Base

Conjugate pair

The notation for specifying a conjugate acid–base pair is "acid/base." Using this notation, the two conjugate acid–base pairs in the preceding example are HF/F and H_3O^+/H_2O.

For any given conjugate acid–base pair

1. The *acid* in the acid–base pair always has one *more* H atom and one *fewer* negative charge than the base. Note this relationship for the HF/F$^-$ conjugate acid–base pair.
2. The *base* in the acid–base pair always has one *fewer* H atom and one *more* negative charge than the acid. Note this relationship for the HF/F$^-$ conjugate acid–base pair.

The acid in a conjugate acid–base pair is called the *conjugate acid* of the base, and the base in the conjugate acid–base pair is called the *conjugate base* of the acid. **A conjugate acid** *is the species formed when a proton (H^+ ion) is added to a Brønsted–Lowry base.* The H_3O^+ ion is the conjugate acid of a H_2O molecule. **A conjugate base** *is the species that remains when a proton (H^+ ion) is removed from a Brønsted–Lowry acid.* The H_2O molecule is the conjugate base of the H_3O^+ ion.

● **EXAMPLE 10.1**

Determining the Formula of One Member of a Conjugate Acid-Base Pair When Given the Other Member

▶ Write the chemical formula of each of the following.

a. The conjugate base of HSO_4^- b. The conjugate acid of NO_3^-
c. The conjugate base of H_3PO_4 d. The conjugate acid of $HC_2O_4^-$

Solution

a. A conjugate base can always be found by removing one H^+ from a given acid. Removing one H^+ (both the atom and the charge) from HSO_4^- leaves SO_4^{2-}. Thus SO_4^{2-} is the conjugate base of HSO_4^-.

b. A conjugate acid can always be found by adding one H^+ to a given base. Adding one H^+ (both the atom and the charge) to NO_3^- produces HNO_3. Thus HNO_3 is the conjugate base of NO_3^-.

c. Proceeding as in part **a**, the removal of a H^+ ion from H_3PO_4 produces the $H_2PO_4^-$ ion. Thus $H_2PO_4^-$ is the conjugate base of H_3PO_4.

d. Proceeding as in part **b**, the addition of a H^+ ion to $HC_2O_4^-$ produces the $H_2C_2O_4$ molecule. Thus $H_2C_2O_4$ is the conjugate acid of $HC_2O_4^-$.

Practice Exercise 10.1

Write the chemical formula of each of the following.

a. The conjugate acid of ClO_3^- b. The conjugate base of NH_3
c. The conjugate acid of PO_4^{3-} d. The conjugate base of HS^-

Answers: a. $HClO_3$; b. NH_2^-; c. HPO_4^{2-}; d. S^{2-}

Amphiprotic Substances

Some molecules or ions are able to function as either Brønsted–Lowry acids or bases, depending on the kind of substance with which they react. Such molecules are said to be amphiprotic. An **amphiprotic substance** *is a substance that can either lose or accept a proton and thus can function as either a Brønsted–Lowry acid or a Brønsted–Lowry base.*

Water is the most common amphiprotic substance. Water functions as a base in the first of the following two reactions and as an acid in the second.

$$HNO_3(aq) + H_2O(l) \rightleftharpoons H_3O^+(aq) + NO_3^-(aq)$$

Acid   Base

$$NH_3(aq) + H_2O(l) \rightleftharpoons NH_4^+(aq) + OH^-(aq)$$

Base   Acid

The Chemistry at a Glance feature on page 258 summarizes the terminology associated with defining what acids and bases are.

10.3 MONO-, DI-, AND TRIPROTIC ACIDS

Acids can be classified according to the number of hydrogen ions they can transfer per molecule during an acid–base reaction. A **monoprotic acid** *is an acid that supplies one proton (H^+ ion) per molecule during an acid–base reaction.* Hydrochloric acid (HCl) and nitric acid (HNO_3) are both monoprotic acids.

A **diprotic acid** *is an acid that supplies two protons (H^+ ions) per molecule during an acid–base reaction.* Carbonic acid (H_2CO_3) is a diprotic acid. The transfer of protons for a diprotic acid always occurs in steps. For H_2CO_3, the two steps are

$$H_2CO_3(aq) + H_2O(l) \rightleftharpoons H_3O^+(aq) + HCO_3^-(aq)$$
$$HCO_3^-(aq) + H_2O(l) \rightleftharpoons H_3O^+(aq) + CO_3^{2-}(aq)$$

A few triprotic acids exist. A **triprotic acid** *is an acid that supplies three protons (H^+ ions) per molecule during an acid–base reaction.* Phosphoric acid, H_3PO_4, is the most common triprotic acid. The three proton-transfer steps for this acid are

$$H_3PO_4(aq) + H_2O(l) \rightleftharpoons H_3O^+(aq) + H_2PO_4^-(aq)$$
$$H_2PO_4^-(aq) + H_2O(l) \rightleftharpoons H_3O^+(aq) + HPO_4^{2-}(aq)$$
$$HPO_4^{2-}(aq) + H_2O(l) \rightleftharpoons H_3O^+(aq) + PO_4^{3-}(aq)$$

A **polyprotic acid** *is an acid that supplies two or more protons (H^+ ions) during an acid–base reaction.* Both diprotic and triprotic acids are examples of polyprotic acids.

The number of hydrogen atoms present in one molecule of an acid *cannot* always be used to classify the acid as mono-, di-, or triprotic. For example, a molecule of acetic acid contains four hydrogen atoms, and yet it is a monoprotic acid. Only one of the hydrogen atoms in acetic acid is *acidic;* that is, only one of the hydrogen atoms leaves the molecule when it is in solution.

Whether a hydrogen atom is acidic is related to its location in a molecule—that is, to which other atom it is bonded. From a structural viewpoint, the acidic behavior of acetic acid can be represented by the equation

$$H-\underset{\underset{H}{|}}{\overset{\overset{H}{|}}{C}}-\overset{\overset{O}{\|}}{C}-O-H + H_2O \rightleftharpoons H_3O^+ + \left[H-\underset{\underset{H}{|}}{\overset{\overset{H}{|}}{C}}-\overset{\overset{O}{\|}}{C}-O \right]^-$$

Note that one hydrogen atom is bonded to an oxygen atom and the other three hydrogen atoms are bonded to a carbon atom. The hydrogen atom bonded to the oxygen atom is the acidic hydrogen atom; the hydrogen atoms that are bonded to carbon atoms are too tightly held to be removed by reaction with water molecules. Water has very little effect on a carbon–hydrogen bond because that bond is only slightly polar. On the other hand, the hydrogen bonded to oxygen is involved in a very polar bond because of oxygen's large electronegativity (Section 5.9). Water, which is a polar molecule, readily attacks this bond.

The term *amphiprotic* is related to the Greek *amphoteres*, which means "partly one and partly the other." Just as an amphibian is an animal that lives partly on land and partly in the water, an amphiprotic substance is sometimes an acid and sometimes a base.

If the double arrows in the equation for a system at equilibrium are of unequal length, the longer arrow indicates the direction in which the equilibrium is displaced.

$\rightleftharpoons$ Equilibrium displaced toward reactants

$\rightleftharpoons$ Equilibrium displaced toward products

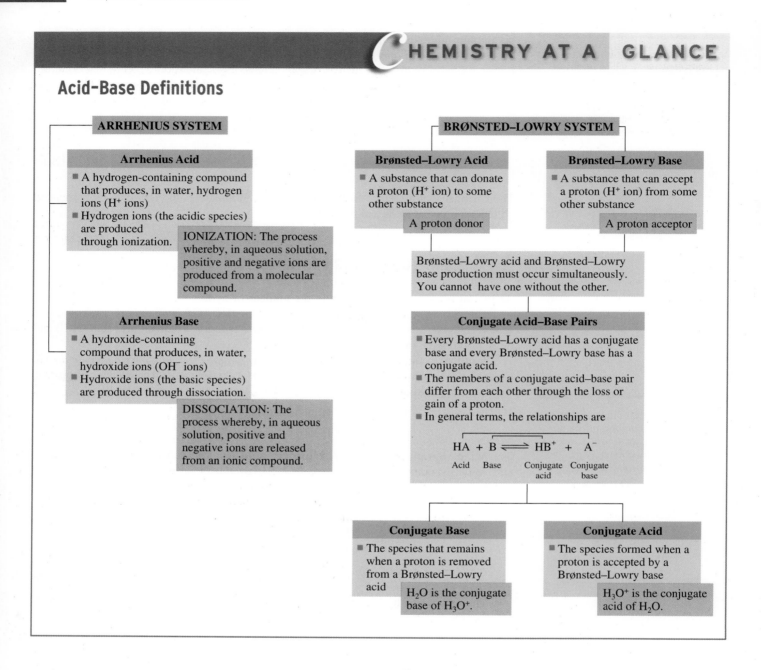

CHEMISTRY AT A GLANCE

Acid-Base Definitions

ARRHENIUS SYSTEM

Arrhenius Acid
- A hydrogen-containing compound that produces, in water, hydrogen ions (H^+ ions)
- Hydrogen ions (the acidic species) are produced through ionization.

IONIZATION: The process whereby, in aqueous solution, positive and negative ions are produced from a molecular compound.

Arrhenius Base
- A hydroxide-containing compound that produces, in water, hydroxide ions (OH^- ions)
- Hydroxide ions (the basic species) are produced through dissociation.

DISSOCIATION: The process whereby, in aqueous solution, positive and negative ions are released from an ionic compound.

BRØNSTED–LOWRY SYSTEM

Brønsted–Lowry Acid
- A substance that can donate a proton (H^+ ion) to some other substance

A proton donor

Brønsted–Lowry Base
- A substance that can accept a proton (H^+ ion) from some other substance

A proton acceptor

Brønsted–Lowry acid and Brønsted–Lowry base production must occur simultaneously. You cannot have one without the other.

Conjugate Acid–Base Pairs
- Every Brønsted–Lowry acid has a conjugate base and every Brønsted–Lowry base has a conjugate acid.
- The members of a conjugate acid–base pair differ from each other through the loss or gain of a proton.
- In general terms, the relationships are

$$HA + B \rightleftharpoons HB^+ + A^-$$

Acid   Base   Conjugate   Conjugate
               acid   base

Conjugate Base
- The species that remains when a proton is removed from a Brønsted–Lowry acid

H_2O is the conjugate base of H_3O^+.

Conjugate Acid
- The species formed when a proton is accepted by a Brønsted–Lowry base

H_3O^+ is the conjugate acid of H_2O.

Figure 10.4 The sour taste of limes and other citrus fruit is due to the citric acid present in the fruit juice.

Writing the formula for acetic acid as $HC_2H_3O_2$ instead of $C_2H_4O_2$ indicates that there are two different kinds of hydrogen atoms present. One of the hydrogen atoms is acidic, and the other three are not. When some hydrogen atoms are acidic and others are not, we write the acidic hydrogens first, thus separating them from the other hydrogen atoms in the formula. Citric acid, the principal acid in citrus fruits (see Figure 10.4), is another example of an acid that contains both acidic and nonacidic hydrogens. Its formula, $H_3C_6H_5O_7$, indicates that three of the eight hydrogen atoms present in a molecule are acidic.

 ## 10.4 STRENGTHS OF ACIDS AND BASES

Brønsted–Lowry acids vary in their ability to transfer protons and produce hydronium ions in aqueous solution. Acids can be classified as strong or weak on the basis of the extent to which proton transfer occurs in aqueous solution. A **strong acid** *is an acid that transfers 100%, or*

TABLE 10.1
Commonly Encountered Strong Acids

HCl	hydrochloric acid
HBr	hydrobromic acid
HI	hydroiodic acid
HNO_3	nitric acid
$HClO_3$	chloric acid
$HClO_4$	perchloric acid
H_2SO_4	sulfuric acid

Learn the names and formulas of the seven commonly encountered strong acids, and then assume that all other acids you encounter are weak unless you are told otherwise.

It is important not to confuse the terms *strong* and *weak* with the terms *concentrated* and *dilute*. *Strong* and *weak* apply to the *extent of proton transfer,* not to the concentration of acid or base. *Concentrated* and *dilute* are relative concentration terms. Stomach acid (gastric juice) is a dilute (not weak) solution of a strong acid (HCl); it is 5% by mass hydrochloric acid.

TABLE 10.2
Commonly Encountered Strong Hydroxide Bases

Group IA Hydroxides	Group IIA Hydroxides
LiOH	—
NaOH	—
KOH	$Ca(OH)_2$
RbOH	$Sr(OH)_2$
CsOH	$Ba(OH)_2$

very nearly 100%, of its protons (H^+ ions) to water in an aqueous solution. Thus if an acid is strong, nearly all of the acid molecules present give up protons to water. This extensive transfer of protons produces many hydronium ions (the acidic species) within the solution. A **weak acid** *is an acid that transfers only a small percentage of its protons (H^+ ions) to water in an aqueous solution.* The extent of proton transfer for weak acids is usually less than 5%.

The extent to which an acid undergoes ionization depends on the molecular structure of the acid; molecular polarity and the strength and polarity of individual bonds are particularly important factors in determining whether an acid is strong or weak. The vast majority of acids are weak rather than strong.

Only seven commonly encountered acids are strong. Their chemical formulas and names are given in Table 10.1.

The difference between a strong acid and a weak acid can also be stated in terms of equilibrium position (Section 9.7). Consider the reaction wherein HA represents the acid and H_3O^+ and A^- are the products from the proton transfer to H_2O. For strong acids, the equilibrium lies far to the right (100% or almost 100%):

$$HA + H_2O \rightleftharpoons H_3O^+ + A^-$$

For weak acids, the equilibrium position lies far to the left:

$$HA + H_2O \rightleftharpoons H_3O^+ + A^-$$

Thus, in solutions of strong acids, the predominant species are H_3O^+ and A^-. In solutions of weak acids, the predominant species is HA; very little proton transfer has occurred. The differences between strong and weak acids, in terms of species present in solution, are illustrated in Figure 10.5.

Just as there are strong acids and weak acids, there are also strong bases and weak bases. As with acids, there are only a few strong bases. Strong bases are limited to the hydroxides of Groups IA and IIA listed in Table 10.2. Of the strong bases, only NaOH and KOH are commonly encountered in a chemical laboratory.

Only one of the many weak bases that exist is fairly common—aqueous ammonia. In a solution of ammonia gas (NH_3) in water, small amounts of OH^- ions are produced through the reaction of NH_3 molecules with water.

$$NH_3(g) + H_2O(l) \rightleftharpoons NH_4^+(aq) + OH^-(aq)$$

A solution of aqueous ammonia is sometimes erroneously called ammonium hydroxide. Aqueous ammonia is the preferred designation because most of the NH_3 present has not reacted with water; the equilibrium position lies far to the left. Only a few ammonium ions (NH_4^+) and hydroxide ions (OH^-) are present.

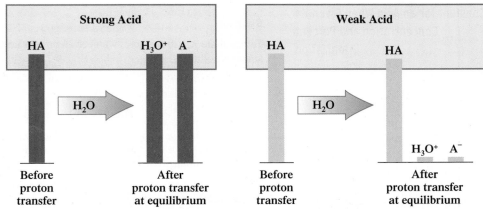

Figure 10.5 A comparison of the number of H_3O^+ ions (the acidic species) present in strong acid and weak acid solutions of the same concentration.

TABLE 10.3
Ionization Constant Values (K_a) and Percent Ionization Values for 1.0 M Solutions, at 24°C, of Selected Weak Acids

Name	Formula	K_a	Percent Ionization
phosphoric acid	H_3PO_4	7.5×10^{-3}	8.3
hydrofluoric acid	HF	6.8×10^{-4}	2.6
nitrous acid	HNO_2	4.5×10^{-4}	2.1
acetic acid	$HC_2H_3O_2$	1.8×10^{-5}	0.42
carbonic acid	H_2CO_3	4.3×10^{-7}	0.065
dihydrogen phosphate ion	$H_2PO_4^-$	6.2×10^{-8}	0.025
hydrocyanic acid	HCN	4.9×10^{-10}	0.0022
hydrogen carbonate ion	HCO_3^-	5.6×10^{-11}	0.00075
hydrogen phosphate ion	HPO_4^{2-}	4.2×10^{-13}	0.000065

10.5 IONIZATION CONSTANTS FOR ACIDS AND BASES

The strengths of various acids and bases can be quantified by use of ionization constants, which are forms of equilibrium constants (Section 9.8).

An **acid ionization constant** *is the equilibrium constant for the reaction of a weak acid with water.* For an acid with the general formula HA, the acid ionization constant is obtained by writing the equilibrium constant expression for the reaction

$$HA(aq) + H_2O(l) \rightleftharpoons H_3O^+(aq) + A^-(aq)$$

which is

$$K_a = \frac{[H_3O^+][A^-]}{[HA]}$$

The concentration of water is not included in the equilibrium constant expression because water is a pure liquid (Section 9.8). The symbol K_a is used to denote an acid ionization constant.

Table 10.3 gives K_a values and percent ionization values (which can be calculated from K_a values) for selected weak acids. Acid strength increases as the K_a value increases. The actual value of K_a for a given acid must be determined by experimentally measuring the concentrations of HA, H_3O^+, and A^- in the acid solution and then using these values to calculate K_a. Example 10.2 shows how an acid ionization constant value (K_a) can be calculated by using concentration (molarity) and percent ionization data for an acid.

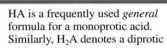

HA is a frequently used *general* formula for a monoprotic acid. Similarly, H_2A denotes a diprotic acid.

Note the following relationships among acid strength, percent ionization, and K_a magnitude.

- Acid strength increases as percent ionization increases.
- Acid strength increases as the magnitude of K_a increases.
- Percent ionization increases as the magnitude of K_a increases.

 E XAMPLE 10.2

Calculating the Acid Ionization Constant for an Acid When Given Its Concentration and Percent Ionization

 A 0.0100 M solution of an acid, HA, is 15% ionized. Calculate the acid ionization constant for this acid.

Solution

To calculate K_a for the acid, we need the molar concentrations of H_3O^+, A^-, and HA in the aqueous solution.

The concentration of H_3O^+ will be 15% of the original HA concentration. Thus the concentration of hydronium ion is

$$H_3O^+ = (0.15) \times (0.0100 \text{ M}) = 0.0015 \text{ M}$$

The ionization of a monoprotic acid produces hydronium ions and the conjugate base of the acid (A^- ions) in a 1:1 ratio. Thus the concentration of A^- will be the same as that of hydronium ion—that is, 0.0015 M.

The concentration of HA is equal to the original concentration diminished by that which ionized (15%, or 0.0015 M):

$$HA = 0.0100 \text{ M} - 0.0015 \text{ M} = 0.0085 \text{ M}$$

Substituting these values in the equilibrium expression gives

$$K_a = \frac{[H_3O^+][A^-]}{[HA]} = \frac{[0.0015][0.0015]}{[0.0085]} = 2.6 \times 10^{-4}$$

Practice Exercise 10.2

A 0.100 M solution of an acid, HA, is 6.0% ionized. Calculate the acid ionization constant for this acid.

Answer: $K_a = 3.8 \times 10^{-4}$

The acids HNO_3, $HClO_4$, $HClO_3$, and H_2SO_4 are strong acids. The acids HNO_2, $HClO_2$, $HClO$, H_2SO_3, H_2CO_3, and H_3PO_4 are weak acids. A "generalization" exists relative to this situation. For simple oxyacids (H, O, and one other nonmetal), if the number of oxygen atoms present exceeds the number of acidic hydrogen atoms present by two or more, the acid strength is strong. For oxyacids where the oxygen–hydrogen difference is less than two, the acid strength is weak.

In Section 10.3, we noted that dissociation of a polyprotic acid occurs in a stepwise manner. In general, each successive step of proton transfer for a polyprotic acid occurs to a lesser extent than the previous step. For the dissociation series

$$H_2CO_3(aq) + H_2O(l) \rightleftharpoons H_3O^+(aq) + HCO_3^-(aq)$$
$$HCO_3^-(aq) + H_2O(l) \rightleftharpoons H_3O^+(aq) + CO_3^{2-}(aq)$$

the second proton is not as easily transferred as the first because it must be pulled away from a negatively charged particle, HCO_3^-. (Remember that particles with opposite charge attract one another.) Accordingly, HCO_3^- is a weaker acid than H_2CO_3. The K_a values for these two acids (Table 10.3) are 5.6×10^{-11} and 4.3×10^{-7}, respectively.

Base strength follows the same principle as acid strength. Here, however, we deal with a base ionization constant, K_b. A **base ionization constant** *is the equilibrium constant for the reaction of a weak base with water.* The general expression for K_b is

$$K_b = \frac{[BH^+][OH^-]}{[B]}$$

where the reaction is

$$B(aq) + H_2O(l) \rightleftharpoons BH^+(aq) + OH^-(aq)$$

For the reaction involving the weak base NH_3,

$$NH_3(aq) + H_2O(l) \rightleftharpoons NH_4^+(aq) + OH^-(aq)$$

the base ionization constant expression is

$$K_b = \frac{[NH_4^+][OH^-]}{[NH_3]}$$

The K_b value for NH_3, the only common weak base, is 1.8×10^{-5}.

 SALTS

To a nonscientist, the term *salt* denotes a white granular substance that is used as a seasoning for food. To the chemist, the term *salt* has a much broader meaning; sodium chloride (table salt) is only one of thousands of salts known to a chemist. A **salt** *is an ionic compound containing a metal or polyatomic ion as the positive ion and a nonmetal or polyatomic ion (except hydroxide) as the negative ion.* (Ionic compounds that contain hydroxide ion are bases rather than salts.)

Much information about salts has been presented in previous chapters, although the term *salt* was not explicitly used in these discussions. Formula writing and nomenclature for binary ionic compounds (salts) were covered in Sections 4.7 and 4.9. Many salts contain polyatomic ions such as nitrate and sulfate; these ions were discussed in Section 4.10. The solubility of ionic compounds (salts) in water was the topic of Section 8.4.

All common soluble salts are *completely* dissociated into ions in solution (Section 8.3). Even if a salt is only slightly soluble, the small amount that does dissolve completely dissociates. Thus the terms *weak* and *strong,* which are used to denote qualitatively the percent ionization/dissociation of acids and bases, are not applicable to salts. The terms *weak salt* and *strong salt* are not used.

Acids, bases, and salts are related in that a salt is one of the products that results from the chemical reaction of an acid with a hydroxide base. This particular type of reaction will be discussed in Section 10.7.

10.7 ACID–BASE NEUTRALIZATION CHEMICAL REACTIONS

When acids and hydroxide bases are mixed, they react with one another and their acidic and basic properties disappear; we say they have neutralized each other. An **acid–base neutralization chemical reaction** *is the chemical reaction between an acid and a hydroxide base in which a salt and water are the products.* The neutralization process can be viewed as either a double-replacement reaction or a proton transfer reaction.

From a *double-replacement viewpoint* (Section 9.1),

$$AX + BY \longrightarrow AY + BX$$

we have, for the HCl–KOH neutralization,

$$HCl + KOH \longrightarrow HOH + KCl$$

Acid   Base   Water   Salt

The salt that is formed contains the negative ion from the acid ionization and the positive ion from the base dissociation (see Figure 10.6).

From a *proton transfer viewpoint,* the formation of water results from the transfer of protons from H_3O^+ ions (the acidic species in aqueous solution) to OH^- ions (the basic species) (Figure 10.7).

Any time an acid is completely reacted with a base, neutralization occurs. It does not matter whether the acid and base are strong or weak. Sodium hydroxide (a strong base) and nitric acid (a strong acid) react as follows:

$$HNO_3 + NaOH \longrightarrow NaNO_3 + H_2O$$

The equation for the reaction of potassium hydroxide (a strong base) with hydrocyanic acid (a weak acid) is

$$HCN + KOH \longrightarrow KCN + H_2O$$

Note that in both reactions, the products are a salt ($NaNO_3$ in the first reaction and KCN in the second) and water.

Figure 10.6 The acid-base reaction between sulfuric acid and barium hydroxide produces the insoluble salt barium sulfate.

Figure 10.7 Formation of water by the transfer of protons from H_3O^+ ions to OH^- ions.

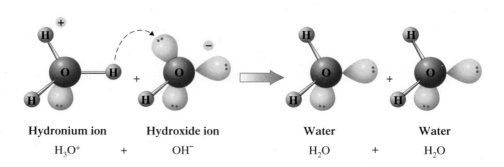

Hydronium ion		Hydroxide ion	Water		Water
H_3O^+	+	OH^-	H_2O	+	H_2O

Balancing Acid–Base Neutralization Equations

In any acid–base neutralization reaction, the amounts of H^+ ion and OH^- ion that react are equal. These two ions always react in a one-to-one ratio to form water.

$$H^+ + OH^- \longrightarrow H_2O \text{ (HOH)}$$

This constant reaction ratio between the two ions enables us to balance chemical equations for neutralization reactions quickly.

Let us consider the neutralization reaction between H_2SO_4 and KOH.

$$H_2SO_4 + KOH \longrightarrow \text{salt} + H_2O$$

Because the acid H_2SO_4 is diprotic and the base KOH contains only one OH^- ion, we will need twice as many base molecules as acid molecules. Thus we place the coefficient 2 in front of the formula for KOH in the chemical equation; this gives two H^+ ions reacting with two OH^- ions to produce two H_2O molecules.

$$H_2SO_4 + 2KOH \longrightarrow \text{salt} + 2H_2O$$

The salt formed is K_2SO_4; there are two K^+ ions and one SO_4^{2-} ion on the left side of the equation, which combine to give the salt. The balanced equation is

$$H_2SO_4 + 2KOH \longrightarrow K_2SO_4 + 2H_2O$$

 SELF-IONIZATION OF WATER

Although we usually think of water as a covalent substance, experiments show that an *extremely small* percentage of water molecules in pure water interact with one another to form ions, a process that is called *self-ionization* (Figure 10.8). This interaction can be thought of as the transfer of protons between water molecules (Brønsted–Lowry theory, Section 10.2):

$$H_2O + H_2O \rightleftharpoons H_3O^+ + OH^-$$

The net effect of this transfer is the formation of *equal amounts* of hydronium and hydroxide ion. Such behavior for water should not seem surprising; we have already discussed the fact that water is an amphiprotic substance (Section 10.2)—one that can either gain or lose protons. We have already seen several reactions in which H_2O acts as an acid and others wherein it acts as a base.

At any given time, the number of H_3O^+ and OH^- ions present in a sample of pure water is always extremely small. At equilibrium and 24°C, the H_3O^+ and OH^- concentrations are 1.00×10^{-7} M (0.000000100 M).

The smallness of the amount of H^+ ion present in water (1.00×10^{-7} M) is illustrated by this example: If once every second one molecule (or ion) were removed from a liter of water, examined, and then returned, a H^+ ion would be encountered once every 17.4 years.

Ion Product Constant for Water

The constant concentration of H_3O^+ and OH^- ions present in pure water at 24°C can be used to calculate a very useful number called the ion product constant for water. The **ion product constant for water** *is the numerical value* 1.00×10^{-14}, *obtained by multiplying*

Figure 10.8 Self-ionization of water through proton transfer between water molecules.

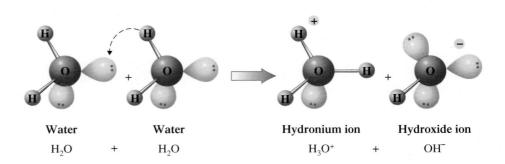

Water	Water	Hydronium ion	Hydroxide ion
H_2O +	H_2O	H_3O^+ +	OH^-

together the molar concentrations of H_3O^+ ion and OH^- ion present in pure water at 24°C. We have the following equation for the ion product constant for water:

$$\text{Ion product constant for water} = [H_3O^+] \times [OH^-]$$
$$= (1.00 \times 10^{-7}) \times (1.00 \times 10^{-7})$$
$$= 1.00 \times 10^{-14}$$

Remember that square brackets mean concentration in moles per liter (molarity).

The ion product constant expression for water is valid not only in water but also in water with solutes present. At all times, the product of the hydronium ion and hydroxide ion molarities in an aqueous solution at 24°C must equal 1.00×10^{-14}. Thus, if $[H_3O^+]$ is increased by the addition of an acidic solute, then $[OH^-]$ must decrease so that their product will still be 1.00×10^{-14}. Similarly, if additional OH^- ions are added to the water, then $[H_3O^+]$ must correspondingly decrease.

We can easily calculate the concentration of either H_3O^+ ion or OH^- ion present in an aqueous solution, if we know the concentration of the other ion, by simply rearranging the ion product expression $[H_3O^+] \times [OH^-] = 1.00 \times 10^{-14}$.

$$[H_3O^+] = \frac{1.00 \times 10^{-14}}{[OH^-]} \quad \text{or} \quad [OH^-] = \frac{1.00 \times 10^{-14}}{[H_3O^+]}$$

EXAMPLE 10.3

Calculating the Hydroxide Ion Concentration of a Solution from a Given Hydronium Ion Concentration

Sufficient acidic solute is added to a quantity of water to produce a solution with $[H_3O^+] = 4.0 \times 10^{-3}$. What is the $[OH^-]$ in this solution?

Solution

$[OH^-]$ can be calculated by using the ion product expression for water, rearranged in the form

$$[OH^-] = \frac{1.00 \times 10^{-14}}{[H_3O^+]}$$

Substituting into this expression the known $[H_3O^+]$ and doing the arithmetic give

$$[OH^-] = \frac{1.00 \times 10^{-14}}{4.0 \times 10^{-3}} = 2.5 \times 10^{-12}$$

If we know $[H_3O^+]$, we can always calculate $[OH^-]$, and vice versa, because of the ion product constant for water:

$$[H_3O^+] \times [OH^-] = 1.00 \times 10^{-14}$$

Practice Exercise 10.3

Sufficient acidic solute is added to a quantity of water to produce a solution with $[H_3O^+] = 5.7 \times 10^{-6}$. What is the $[OH^-]$ in this solution?

Answer: $[OH^-] = 1.8 \times 10^{-9}$

Neither $[H_3O^+]$ nor $[OH^-]$ is ever zero in an aqueous solution.

The relationship between $[H_3O^+]$ and $[OH^-]$ is that of an inverse proportion; when one increases, the other decreases. If $[H_3O^+]$ increases by a factor of 10^2, then $[OH^-]$ decreases by the same factor, 10^2. A graphic portrayal of this increase–decrease relationship for $[H_3O^+]$ and $[OH^-]$ is given in Figure 10.9.

Acidic, Basic, and Neutral Solutions

A basic solution is also often referred to as an *alkaline solution*.

Small amounts of both H_3O^+ ion and OH^- ion are present in all aqueous solutions. What, then, determines whether a given solution is acidic or basic? It is the relative amounts of these two ions present. An **acidic solution** *is an aqueous solution in which the concentration of H_3O^+ ion is higher than that of OH^- ion.* A **basic solution** *is an aqueous solution in which the concentration of the OH^- ion is higher than that of the H_3O^+ ion.* It is possible to have an aqueous solution that is neither acidic nor basic but is, rather, a neutral solution. A **neutral solution** *is an aqueous solution in which the concentrations of H_3O^+ ion and*

Excessive Acidity Within the Stomach: Antacids and Acid Inhibitors

Gastric juice, an acidic digestive fluid, secreted by glands in the mucous membrane that lines the stomach, is produced at the rate of 2–3 liters per day in an average adult. It contains hydrochloric acid (HCl), a substance necessary for the proper digestion of food, at a concentration of about 0.03 M.

Overeating and emotional factors can cause the stomach to produce too much HCl. This leads to hyperacidity, the condition we often call "acid indigestion" or "heartburn." Ordinarily, the stomach and digestive tract themselves are protected from the corrosive effect of excess stomach acid by the stomach's mucosal lining. Constant excess acid can, however, damage this lining to the extent that swelling, inflammation, and bleeding (symptoms of ulcers) occur.

There are two approaches to combating the problem of excess stomach acid: (1) removing the excess acid through neutralization and (2) decreasing the production of stomach acid. The first approach involves the use of *antacids,* and the second approach involves the use of *acid inhibitors.*

An *antacid* is an over-the-counter drug containing one or more basic substances that are capable of neutralizing the HCl present in gastric juice. Neutralizing agents present in selected brand-name antacids are shown in the table opposite.

Magnesium hydroxide and aluminum hydroxide neutralize HCl to produce a salt and water as follows:

$$2HCl + Mg(OH)_2 \longrightarrow MgCl_2 + 2H_2O$$
$$3HCl + Al(OH)_3 \longrightarrow AlCl_3 + 3H_2O$$

Neutralization involving sodium bicarbonate and calcium carbonate produce the gas carbon dioxide in addition to a salt and water.

Brand Name	Neutralizing Agent(s)
Alka-Seltzer	$NaHCO_3$
BiSoDol	$NaHCO_3$
DiGel	$Mg(OH)_2, Al(OH)_3$
Gaviscon	$Al(OH)_3, NaHCO_3$
Gelusil	$Mg(OH)_2, Al(OH)_3$
Maalox	$Mg(OH)_2, Al(OH)_3$
Milk of Magnesia	$Mg(OH)_2$
Mylanta	$Mg(OH)_2, Al(OH)_3$
Riopan	$AlMg(OH)_5$
Rolaids	$NaAl(OH)_2CO_3$
Tums	$CaCO_3$

$$HCl + NaHCO_3 \longrightarrow NaCl + CO_2 + H_2O$$
$$2HCl + CaCO_3 \longrightarrow CaCl_2 + CO_2 + H_2O$$

The CO_2 released by these reactions increases the gas pressure in the stomach, causing a person to belch often.

Brand-name over-the-counter *acid inhibitors* include Pepcid, Tagamet, and Zantac. These substances inhibit gastric acid production by blocking the action of histamine, a gastric acid secretion regulator, at receptor sites in the gastric-acid-secreting cells of the stomach lining. The net effect is decreased amounts of gastric secretion in the stomach. This lowered acidity allows for healing of ulcerated tissue.

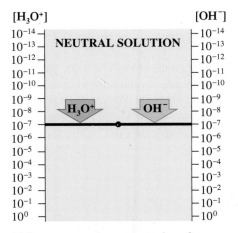

(a) In pure water the concentration of hydronium ions, [H_3O^+], and that of hydroxide ions, [OH^-], are equal. Both are 1.00×10^{-7} M at 24°C.

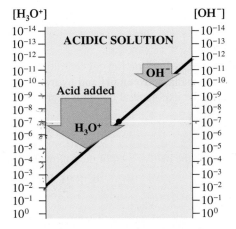

(b) If [H_3O^+] is increased by a factor of 10^5 (from 10^{-7} M to 10^{-2} M), then [OH^-] is decreased by a factor of 10^5 (from 10^{-7} M to 10^{-12} M).

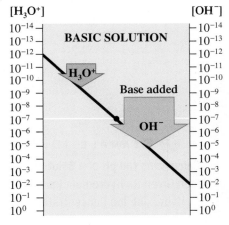

(c) If [OH^-] is increased by a factor of 10^5 (from 10^{-7} M to 10^{-2}M), then [H_3O^+] is decreased by a factor of 10^5 (from 10^{-7} M to 10^{-12} M).

Figure 10.9 The relationship between [H_3O^+] and [OH^-] in aqueous solution is an inverse proportion; when [H_3O^+] is increased, [OH^-] decreases, and vice versa.

TABLE 10.4
Relationship Between [H₃O⁺] and [OH⁻] in Neutral, Acidic, and Basic Solutions

Type of Solution	Relationship Between [H₃O⁺] and [OH⁻]
neutral solution	$[H_3O^+] = [OH^-] = 1.00 \times 10^{-7}$
acidic solution $[H_3O^+] > [OH^-]$	$[H_3O^+]$ is greater than 1.00×10^{-7} $[OH^-]$ is less than 1.00×10^{-7}
basic solution $[OH^-] > [H_3O^+]$	$[H_3O^+]$ is less than 1.00×10^{-7} $[OH^-]$ is greater than 1.00×10^{-7}

OH⁻ ion are equal. Table 10.4 summarizes the relationships between [H₃O⁺] and [OH⁻] that we have just considered.

10.9 THE pH CONCEPT

The pH scale is a compact method for representing solution acidity.

Hydronium ion concentrations in aqueous solution range from relatively high values (10 M) to extremely small ones (10^{-14} M). It is inconvenient to work with numbers that extend over such a wide range; a hydronium ion concentration of 10 M is 1,000 trillion times larger than a hydronium ion concentration of 10^{-14} M. The pH scale was developed as a more practical way to handle such a wide range of numbers. The **pH scale** *is a scale of small numbers that is used to specify molar hydronium ion concentration in an aqueous solution.*

The *p* in pH comes from the German word *potenz*, which means "power," as in "power of 10."

The calculation of pH scale values involves the use of logarithms. The **pH** *is the negative logarithm of an aqueous solution's molar hydronium ion concentration.* Expressed mathematically, the definition of pH is

$$pH = -\log[H_3O^+]$$

(The letter *p*, as in pH, means "negative logarithm of.")

Integral pH Values

The rule for the number of significant figures in a logarithm is: The number of digits after the decimal place in a logarithm is equal to the number of significant figures in the original number.

$$[H_3O^+] = \underline{6.3} \times 10^{-5}$$
Two significant figures

$$pH = 4.\underline{20}$$
Two digits

For any hydronium ion concentration expressed in exponential notation in which the coefficient is 1.0, the pH is given directly by the negative of the exponent value of the power of 10:

$$[H_3O^+] = 1.0 \times 10^{-x}$$
$$pH = x$$

Thus, if the hydronium ion concentration is 1.0×10^{-9}, then the pH will be 9.00. This simple relationship between pH and hydronium ion concentration is valid only when the coefficient in the exponential notation expression for the hydronium ion concentration is 1.0.

● **E X A M P L E 10.4**

Calculating the pH of a Solution When Given Its Hydronium Ion or Hydroxide Ion Concentration

Calculate the pH for each of the following solutions.

a. $[H_3O^+] = 1.0 \times 10^{-6}$ **b.** $[OH^-] = 1.0 \times 10^{-6}$

Solution

a. Because the coefficient in the exponential expression for the molar hydronium ion concentration is 1.0, the pH can be obtained from the relationships

$$[H_3O^+] = 1.0 \times 10^{-x}$$
$$pH = x$$

The power of 10 is −6 in this case, so the pH will be 6.00.

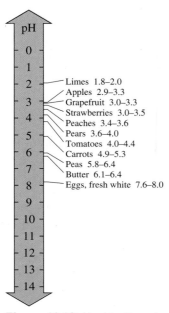

Figure 10.10 Most fruits and vegetables are acidic. Tart or sour taste is an indication that such is the case. Nonintegral pH values for selected foods are as shown here.

b. The given quantity involves hydroxide ion rather than hydronium ion. Thus we must calculate the hydronium ion concentration first and then calculate the pH.

$$[H_3O^+] = \frac{1.00 \times 10^{-14}}{1.0 \times 10^{-6}} = 1.0 \times 10^{-8}$$

A solution with a hydronium ion concentration of 1.0×10^{-8} M will have a pH of 8.00.

Practice Exercise 10.4

Calculate the pH for each of the following solutions.

a. $[H_3O^+] = 1.0 \times 10^{-3}$ **b.** $[OH^-] = 1.0 \times 10^{-8}$

Answers: a. 3.00; **b.** 6.00

Nonintegral pH Values

If the coefficient in the exponential expression for the molar hydronium ion concentration is *not* 1.0, then the pH will have a nonintegral value; that is, it will not be a whole number. For example, consider the following nonintegral pH values.

$$[H_3O^+] = 6.3 \times 10^{-5} \quad pH = 4.20$$
$$[H_3O^+] = 5.3 \times 10^{-5} \quad pH = 4.28$$
$$[H_3O^+] = 2.2 \times 10^{-4} \quad pH = 3.66$$

Figure 10.10 gives nonintegral pH values for selected fruits and vegetables.

The easiest way to obtain nonintegral pH values such as these involves using an electronic calculator that allows for the input of exponential numbers and that has a base-10 logarithm key (LOG).

In using such an electronic calculator, you can obtain logarithm values simply by pressing the LOG key after having entered the number whose log is desired. For pH, you must remember that after obtaining the log value, you must change signs because of the negative sign in the defining equation for pH.

● **E X A M P L E 10.5**

Calculating the pH of a Solution When Given Its Hydronium Ion Concentration

▶ Calculate the pH for each of the following solutions.

a. $[H_3O^+] = 7.23 \times 10^{-8}$ **b.** $[H_3O^+] = 5.70 \times 10^{-3}$

Solution

a. Using an electronic calculator, first enter the number 7.23×10^{-8} into the calculator. Then use the LOG key to obtain the logarithm value, -7.1408617. Changing the sign of this number (because of the minus sign in the definition of pH) and adjusting for significant figures yields a pH value of 7.141.

b. Entering the number 5.70×10^{-3} into the calculator and then using the LOG key give a logarithm value of -2.2441251. This value translates into a pH value, after rounding, of 2.244.

Practice Exercise 10.5

Calculate the pH for each of the following solutions.

a. $[H_3O^+] = 4.44 \times 10^{-11}$ **b.** $[H_3O^+] = 8.92 \times 10^{-6}$

Answers: a. 10.353; **b.** 5.050

pH Values and Hydronium Ion Concentration

It is often necessary to calculate the hydronium ion concentration for a solution from its pH value. This type of calculation, which is the reverse of that illustrated in Examples 10.4 and 10.5, is shown in Example 10.6.

● EXAMPLE 10.6

Calculating the Molar Hydronium Ion Concentration of a Solution from the Solution's pH

The pH of a solution is 6.80. What is the molar hydronium ion concentration for this solution?

Solution

From the defining equation for pH, we have

$$pH = -\log{[H_3O^+]} = 6.80$$

$$\log{[H_3O^+]} = -6.80$$

To find $[H_3O^+]$, we need to determine the *antilog* of -6.80.

How an antilog is obtained using a calculator depends on the type of calculator. Many calculators have an antilog function (sometimes labeled INV log) that performs this operation. If this key is present, then

1. Enter the number -6.80. Note that it is the *negative* of the pH that is entered into the calculator.
2. Press the INV log key (or an inverse key and then a log key). The result is the desired hydronium ion concentration.

$$\log{[H_3O^+]} = -6.80$$

$$\text{antilog}{[H_3O^+]} = 1.5848931 \times 10^{-7}$$

Rounded off, this value translates into a hydronium ion concentration of 1.6×10^{-7} M.

Some calculators use a 10^x key to perform the antilog operation. Use of this key is based on the mathematical identity

$$\text{antilog } x = 10^x$$

In our case, this means

$$\text{antilog } -6.80 = 10^{-6.80}$$

If the 10^x key is present, then

1. Enter the number -6.80 (the negative of the pH).
2. Press the function key 10^x. The result is the desired hydronium ion concentration.

$$[H_3O^+] = 10^{-6.80} = 1.6 \times 10^{-7}$$

Practice Exercise 10.6

The pH of a solution is 3.44. What is the molar hydronium ion concentration for this solution?

Answer: 3.6×10^{-4} M

Interpreting pH Values

Identifying an aqueous solution as acidic, basic, or neutral based on pH value is a straightforward process. A **neutral solution** *is an aqueous solution whose pH is 7.0.* An **acidic solution** *is an aqueous solution whose pH is less than 7.0.* A **basic solution** *is an aqueous solution whose pH is greater than 7.0.* The relationships among $[H_3O^+]$, $[OH^-]$, and pH are summarized in Figure 10.11. Note the following trends from the information presented in this figure.

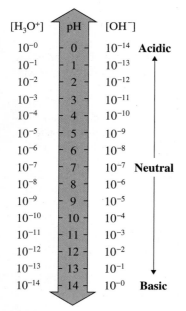

Figure 10.11 Relationships among pH values, $[H_3O^+]$, and $[OH^-]$ at 24°C.

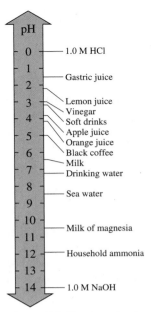

pH

0	1.0 M HCl
1	
2	Gastric juice
3	Lemon juice
	Vinegar
4	Soft drinks
5	Apple juice
	Orange juice
6	Black coffee
	Milk
7	Drinking water
8	
	Sea water
9	
10	
	Milk of magnesia
11	
12	Household ammonia
13	
14	1.0 M NaOH

Figure 10.12 The pH values of selected common liquids. The lower the numerical value of the pH, the more acidic the substance is.

Solutions of low pH are more acidic than solutions of high pH; conversely, solutions of high pH are more basic than solutions of low pH.

TABLE 10.5
The Normal pH Range of Selected Body Fluids

Type of Fluid	pH Value
bile	6.8–7.0
blood plasma	7.3–7.5
gastric juices	1.0–3.0
milk	6.6–7.6
saliva	6.5–7.5
spinal fluid	7.3–7.5
urine	4.8–8.4

Like pH, pK_a is a positive number. The lower the pK_a value, the stronger the acid.

● **E X A M P L E   1 0 . 7**

Calculating the pK_a of an Acid from the Acid's K_a Value

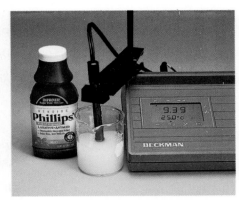

Figure 10.13 A pH meter gives an accurate measurement of pH values. The pH of vinegar is 2.32 (left). The pH of milk of magnesia in water is 9.39 (right).

1. The higher the concentration of hydronium ion, the lower the pH value. Another statement of this same trend is that lowering the pH always corresponds to increasing the hydronium ion concentration.
2. A change of 1 unit in pH always corresponds to a tenfold change in hydronium ion concentration. For example,

$$\text{Difference of 1} \begin{cases} \text{pH} = 1.0, \text{ then } [H_3O^+] = 0.1 \text{ M} \\ \text{pH} = 2.0, \text{ then } [H_3O^+] = 0.01 \text{ M} \end{cases} \text{tenfold difference}$$

In a laboratory, solutions of any pH can be created. The range of pH values that are displayed by natural solutions is more limited than that of prepared solutions, but solutions corresponding to most pH values can be found (see Figure 10.12). A pH meter (Figure 10.13) helps chemists determine accurate pH values.

The pH values of several human body fluids are given in Table 10.5. Most human body fluids except gastric juices and urine have pH values within one unit of neutrality. Both blood plasma and spinal fluid are always slightly basic.

The Chemistry at a Glance feature on page 271 summarizes what we have said about acids and acidity.

(10.10) THE pK_a METHOD FOR EXPRESSING ACID STRENGTH

In Section 10.5 ionization constants for acids and bases were introduced. These constants give an indication of the strengths of acids and bases. An additional method for expressing the strength of acids is in terms of pK_a units. The definition for pK_a is

$$pK_a = -\log K_a$$

The pK_a for an acid is calculated from K_a in exactly the same way that pH is calculated from hydronium ion concentration.

Determine the pK_a for acetic acid, $HC_2H_3O_2$, given that K_a for this acid is 1.8×10^{-5}.

Solution

Because the K_a value is 1.8×10^{-5} and p$K_a = -\log K_a$, we have

$$pK_a = -\log(1.8 \times 10^{-5}) = 4.74$$

The logarithm value 4.74 was obtained using an electronic calculator, as explained in Example 10.5.

(continued)

Practice Exercise 10.7

Determine the pK_a for hydrocyanic acid, HCN, given that K_a for this acid is 4.4×10^{-10}.

Answer: p$K_a = 9.36$

 THE pH OF AQUEOUS SALT SOLUTIONS

The addition of an acid to water produces an acidic solution. The addition of a base to water produces a basic solution. What type of solution is produced when a salt is added to water? Because salts are the products of acid–base neutralizations, a logical supposition would be that salts dissolve in water to produce neutral (pH = 7.0) solutions. Such is the case for a *few* salts. Aqueous solutions of *most* salts, however, are either acidic or basic rather than neutral. Let us consider why this is so.

When a salt is dissolved in water, it completely ionizes; that is, it completely breaks up into the ions of which it is composed (Section 8.3). For many salts, one or more of the ions so produced are reactive toward water. The ensuing reaction, which is called hydrolysis, causes the solution to have a non-neutral pH. A **hydrolysis reaction** *is the reaction of a salt with water to produce hydronium ion or hydroxide ion or both.*

> The term *hydrolysis* comes from the Greek *hydro,* which means "water," and *lysis,* which means "splitting."

Types of Salt Hydrolysis

Not all salts hydrolyze. Which ones do and which ones do not? Of those salts that do hydrolyze, which produce acidic solutions and which produce basic solutions? The following guidelines, based on the neutralization "parentage" of a salt—that is, on the acid and base that produce the salt through neutralization—can be used to answer these questions.

1. The salt of a *strong acid* and a *strong base* does not hydrolyze, so the solution is neutral.
2. The salt of a *strong acid* and a *weak base* hydrolyzes to produce an acidic solution.
3. The salt of a *weak acid* and a *strong base* hydrolyzes to produce a basic solution.
4. The salt of a *weak acid* and a *weak base* hydrolyzes to produce a slightly acidic, neutral, or slightly basic solution, depending on the relative weaknesses of the acid and base.

These guidelines are summarized in Table 10.6.

The first prerequisite for using these guidelines is the ability to classify a salt into one of the four categories mentioned in the guidelines. This classification is accomplished by writing the neutralization equation (Section 10.7) that produces the salt and then specifying the strength (strong or weak) of the acid and base involved. The "parent" acid

TABLE 10.6
Neutralization "Parentage" of Salts and the Nature of the Aqueous Solutions They Form

Type of Salt	Nature of Aqueous Solution	Examples
strong acid–strong base	neutral	$NaCl$, KBr
strong acid–weak base	acidic	NH_4Cl, NH_4NO_3
weak acid–strong base	basic	$NaC_2H_3O_2$, K_2CO_3
weak acid–weak base	depends on the salt	$NH_4C_2H_3O_2$, NH_4NO_2

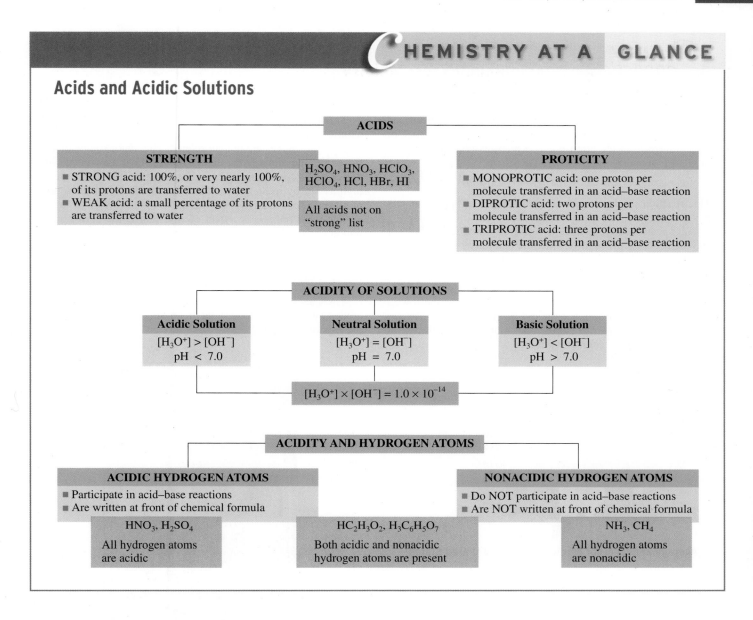

CHEMISTRY AT A GLANCE

Acids and Acidic Solutions

ACIDS

STRENGTH
- STRONG acid: 100%, or very nearly 100%, of its protons are transferred to water
- WEAK acid: a small percentage of its protons are transferred to water

H_2SO_4, HNO_3, $HClO_3$, $HClO_4$, HCl, HBr, HI

All acids not on "strong" list

PROTICITY
- MONOPROTIC acid: one proton per molecule transferred in an acid–base reaction
- DIPROTIC acid: two protons per molecule transferred in an acid–base reaction
- TRIPROTIC acid: three protons per molecule transferred in an acid–base reaction

ACIDITY OF SOLUTIONS

Acidic Solution
$[H_3O^+] > [OH^-]$
pH < 7.0

Neutral Solution
$[H_3O^+] = [OH^-]$
pH = 7.0

Basic Solution
$[H_3O^+] < [OH^-]$
pH > 7.0

$[H_3O^+] \times [OH^-] = 1.0 \times 10^{-14}$

ACIDITY AND HYDROGEN ATOMS

ACIDIC HYDROGEN ATOMS
- Participate in acid–base reactions
- Are written at front of chemical formula

HNO_3, H_2SO_4

All hydrogen atoms are acidic

$HC_2H_3O_2$, $H_3C_6H_5O_7$

Both acidic and nonacidic hydrogen atoms are present

NONACIDIC HYDROGEN ATOMS
- Do NOT participate in acid–base reactions
- Are NOT written at front of chemical formula

NH_3, CH_4

All hydrogen atoms are nonacidic

and base for the salt are identified by pairing the negative ion of the salt with H^+ (to form the acid) and pairing the positive ion of the salt with OH^- (to form the base). The following two equations illustrate the overall procedure.

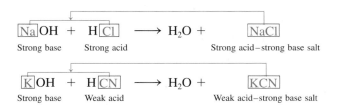

$Na\text{OH} + H\text{Cl} \longrightarrow H_2O + NaCl$
Strong base   Strong acid   Strong acid–strong base salt

$K\text{OH} + H\text{CN} \longrightarrow H_2O + KCN$
Strong base   Weak acid   Weak acid–strong base salt

Knowing which acids and bases are strong and which are weak (Section 10.4) is a necessary part of the classification process. Once the salt has been so classified, the guideline that is appropriate for the situation is easily selected.

CHEMICAL Connections

Acid Rain: Excess Acidity

Rainfall, even in a pristine environment, has always been and will always be acidic. This acidity results from the presence of carbon dioxide in the atmosphere, which dissolves in water to produce carbonic acid (H_2CO_3), a weak acid.

$$CO_2(g) + H_2O(l) \rightarrow H_2CO_3(aq)$$

This reaction produces rainwater with a pH of approximately 5.6–5.7.

Acid rain is a generic term used to describe rainfall (or snowfall) whose pH is lower than the naturally produced value of 5.6. Acid rain has been observed with increasing frequency in many areas of the world. Rainfall with pH values between 4 and 5 is now common, and occasionally rainfall with a pH as low as 2 is encountered. Within the United States, the lowest acid-rain pH values are encountered in the northeastern states (see the accompanying map). The maritime provinces of Canada have also been greatly affected.

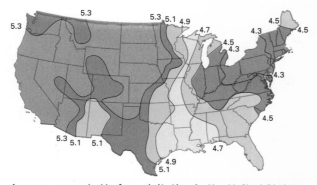

Average annual pH of precipitation in the United States

Acid rain originates from the presence of sulfur oxides (SO_2 and SO_3) and, to a lesser extent, nitrogen oxides (NO and NO_2) in the atmosphere. After being discharged into the atmosphere, these pollutants can be converted into sulfuric acid (H_2SO_4) and nitric acid (HNO_3) through oxidation processes. Several complicated pathways exist by which these two strong acids are produced. Which pathway is actually taken depends on numerous factors, including the intensity of sunlight and the amount of ammonia present in the atmosphere.

The high solubility of sulfur oxides in water is a major factor contributing to atmospheric sulfuric acid production; SO_2 is approximately 70 times more soluble in water than CO_2 and 2600 times more soluble in water than O_2.

Small amounts of sulfur oxides and nitrogen oxides arise naturally from volcanic activity, lightning, and forest fires, but their major sources are human-related. The major source of sulfur oxide emissions is the combustion of coal associated with power plant operations. (The sulfur content of coal can be as high as 5% by mass.) Automobile exhaust is the major source of nitrogen oxides.

The most observable effect of acid rain is the corrosion of building materials. Sulfuric acid (acid rain) readily attacks carbonate-based building materials (limestone, marble); the calcium carbonate is slowly converted into calcium sulfate.

$$CaCO_3(s) + H_2SO_4(aq) \rightarrow CaSO_4(s) + CO_2(g) + H_2O(l)$$

The $CaSO_4$, which is more soluble than $CaCO_3$, is gradually eroded away. Many stone monuments show distinctly discernible erosion damage.

Erosion damage to an 1869 grave marker.

How does the acidity of acid rain compare to the acidity of other common acidic substances? A soft drink is 10 times as acidic as commonly encountered acid rain (pH = 4.5), vinegar is 15 times more acidic, lemon juice is 50 times more acidic, and automobile battery acid (sulfuric acid) is 3,200 times more acidic.

● **EXAMPLE 10.8**

Predicting Whether a Salt's Aqueous Solution Will Be Acidic, Basic, or Neutral

▶ Determine the acid–base "parentage" of each of the following salts, and then use this information to predict whether each salt's aqueous solution is acidic, basic, or neutral.

a. Sodium acetate, $NaC_2H_3O_2$ **b.** Ammonium chloride, NH_4Cl

c. Potassium chloride, KCl   **d.** Ammonium fluoride, NH_4F

Solution

a. The ions present are Na^+ and $C_2H_3O_2^-$. The "parent" base of Na^+ is NaOH, a strong base. The "parent" acid of $C_2H_3O_2^-$ is $HC_2H_3O_2$, a weak acid. Thus the acid–base neutralization that produces this salt is

$$NaOH + HC_2H_3O_2 \longrightarrow H_2O + NaC_2H_3O_2$$

Strong base Weak acid Weak acid–strong base salt

The solution of a weak acid–strong base salt (guideline 3) produces a basic solution.

b. The ions present are NH_4^+ and Cl^-. The "parent" base of NH_4^+ is NH_3, a weak base. The "parent" acid of Cl^- is HCl, a strong acid. This "parentage" will produce a strong acid–weak base salt through neutralization. Such a salt gives an acidic solution upon hydrolysis (guideline 2).

c. The ions present are K^+ and Cl^-. The "parent" base is KOH (a strong base), and the "parent" acid is HCl (a strong acid). The salt produced from neutralization involving this acid–base pair will be a strong acid–strong base salt. Such salts do not hydrolyze. The aqueous solution is neutral (guideline 1).

d. The ions present are NH_4^+ and F^-. Both ions are of weak "parentage"; NH_3 is a weak base, and HF is a weak acid. Thus NH_4F is a weak acid–weak base salt. This is a guideline 4 situation. In this situation, you cannot predict the effect of hydrolysis unless you know the relative strengths of the weak acid and weak base (which is the weaker of the two). HF has a K_a of 6.8×10^{-4} (Table 10.3). NH_3 has a K_b of 1.8×10^{-5} (Section 10.5). Thus, NH_3 is the weaker of the two and will hydrolyze to the greater extent, causing the solution to be acidic.

Practice Exercise 10.8

Predict whether solutions of each of the following salts will be acidic, basic, or neutral.

a. Sodium bromide, NaBr
b. Potassium cyanide, KCN
c. Ammonium iodide, NH_4I
d. Barium chloride, $BaCl_2$

Answers: a. Neutral; **b.** Basic; **c.** Acidic; **d.** Neutral

Chemical Equations for Salt Hydrolysis Reactions

Salt hydrolysis reactions are Brønsted–Lowry acid–base (proton transfer) reactions (Section 10.2). Such reactions are of the following two general types.

1. *Basic hydrolysis:* The reaction of the *negative ion* from a salt with water to produce the ion's conjugate acid and hydroxide ion.

Conjugate acid–base pair

$$CN^- + H_2O \longrightarrow HCN + OH^-$$

Proton acceptor   Proton donor   Weak acid   Makes solution basic

Conjugate acid–base pair

$$F^- + H_2O \longrightarrow HF + OH^-$$

Proton acceptor   Proton donor   Weak acid   Makes solution basic

The only *negative ions* that undergo hydrolysis are those of "weak-acid parentage." The driving force for the reaction is the formation of the weak-acid "parent."

2. *Acidic hydrolysis:* The reaction of the *positive ion* from a salt with water to produce the ion's conjugate base and hydronium ion. The most common ion to undergo this type of reaction is the NH_4^+ ion.

Blood Plasma pH and Hydrolysis

Blood plasma has a slightly basic pH (7.35–7.45), as shown in Table 10.5. The reason for this is related to salt hydrolysis and becomes apparent when the identity of the ions present in blood plasma is specified.

The most abundant positive ion present in blood plasma is Na^+, an ion associated with a strong base (NaOH). Thus it does not hydrolyze. The predominant negative ion present is Cl^-, an ion that comes from a strong acid (HCl). Thus it also does not hydrolyze. Together, these two ions, Na^+ and Cl^-, produce a neutral solution because neither hydrolyzes.

The third most abundant ion in blood plasma is the hydrogen carbonate ion, HCO_3^-, which comes from the weak acid H_2CO_3. Hydrolysis of this ion produces hydroxide ion.

$$HCO_3^- + H_2O \longrightarrow H_2CO_3 + OH^-$$

Thus blood plasma has a slightly basic pH value. Other negative ions present in blood plasma, such as HPO_4^{2-} ion (from the triprotic acid H_3PO_4), also hydrolyze and add to the basic character.

$$HPO_4^{2-} + H_2O \longrightarrow H_2PO_4^- + OH^-$$

However, because of their lower concentrations, their effect on the pH is not as great as that of HCO_3^- ion. The concentration of HCO_3^- ion in blood plasma is 16 times greater than that of HPO_4^{2-} ion (see the Chemical Connections feature on page 283).

The pH of numerous other body fluids besides blood plasma is also directly influenced by hydrolysis reactions.

The only *positive ions* that undergo hydrolysis are those of "weak-base parentage." The driving force for the reaction is the formation of the weak-base "parent."

Conjugate acid–base pair

$$NH_4^+ + H_2O \longrightarrow NH_3 + H_3O^+$$

Proton donor   Proton acceptor   Weak base   Makes solution acidic

Hydrolysis reactions do not go 100% to completion. They occur only until equilibrium conditions are reached (Section 9.7). At the equilibrium point, solution pH change can be significant—differing from neutrality by two to four units. Table 10.7 shows the range of pH values encountered for selected 0.1 M aqueous salt solutions after hydrolysis has occurred.

10.12 BUFFERS

A **buffer** *is an aqueous solution containing substances that prevent major changes in solution pH when small amounts of acid or base are added to it.* Buffers are used in a laboratory setting to maintain optimum pH conditions for chemical reactions. Many commercial products contain buffers, which are needed to maintain optimum pH conditions for product behavior. Examples include buffered aspirin (Bufferin) and pH-controlled hair shampoos. Most human body fluids are highly buffered. For example, a buffer system maintains blood's pH at a value close to 7.4, an optimum pH for oxygen transport.

A less common type of buffer involves a weak base and its conjugate acid. We will not consider this type of buffer here.

Buffers contain two active chemical species: (1) a substance to react with and remove added base, and (2) a substance to react with and remove added acid. Typically, a buffer system is composed of a weak acid *and* its conjugate base—that is, a conjugate acid–base

---

TABLE 10.7
Approximate pH of Selected 0.1 M Aqueous Salt Solutions at 24°C

Name of Salt	Formula of Salt	pH	Category of Salt
ammonium nitrate	NH_4NO_3	5.1	strong acid–weak base
ammonium nitrite	NH_4NO_2	6.3	weak acid–weak base
ammonium acetate	$NH_4C_2H_3O_2$	7.0	weak acid–weak base
sodium chloride	NaCl	7.0	strong acid–strong base
sodium fluoride	NaF	8.1	weak acid–strong base
sodium acetate	$NaC_2H_3O_2$	8.9	weak acid–strong base
ammonium cyanide	NH_4CN	9.3	weak acid–weak base
sodium cyanide	NaCN	11.1	weak acid–strong base

● EXAMPLE 10.9

Recognizing Pairs of Chemical Substances That Can Function as a Buffer in Aqueous Solution

▶ Predict whether each of the following pairs of substances could function as a buffer system in aqueous solution.

a. HCl and NaCl
b. HCN and KCN
c. HCl and HCN
d. NaCN and KCN

Solution

Buffer solutions contain either a weak acid and a salt of that weak acid or a weak base and a salt of that weak base.

a. No. We have an acid and the salt of that acid. However, the acid is a strong acid rather than a weak acid.
b. Yes. HCN is a weak acid, and KCN is a salt of that weak acid.
c. No. Both HCl and HCN are acids. No salt is present.
d. No. Both NaCN and KCN are salts. No weak acid is present.

Practice Exercise 10.9

Predict whether each of the following pairs of substances could function as a buffer system in aqueous solution.

a. HCl and NaOH
b. $HC_2H_3O_2$ and $KC_2H_3O_2$
c. NaCl and NaCN
d. HCN and $HC_2H_3O_2$

Answers: **a.** No; **b.** Yes; **c.** No; **d.** No

pair (Section 10.2). Conjugate acid–base pairs that are commonly employed as buffers include $HC_2H_3O_2/C_2H_3O_2^-$, $H_2PO_4^-/HPO_4^{2-}$, and H_2CO_3/HCO_3^-.

As an illustration of buffer action, consider a buffer solution containing approximately equal concentrations of acetic acid (a weak acid) and sodium acetate (a salt of this weak acid). This solution resists pH change by the following mechanisms:

1. When a small amount of a strong acid such as HCl is added to the solution, the newly added H_3O^+ ions react with the acetate ions from the sodium acetate to give acetic acid.

$$H_3O^+ + C_2H_3O_2^- \longrightarrow HC_2H_3O_2 + H_2O$$

Most of the added H_3O^+ ions are tied up in acetic acid molecules, and the pH changes very little.

2. When a small amount of a strong base such as NaOH is added to the solution, the newly added OH^- ions react with the acetic acid (neutralization) to give acetate ions and water.

$$OH^- + HC_2H_3O_2 \longrightarrow C_2H_3O_2^- + H_2O$$

Most of the added OH^- ions are converted to water, and the pH changes only slightly.

The reactions that are responsible for the buffering action in the acetic acid/acetate ion system can be summarized as follows:

$$C_2H_3O_2^- \underset{OH^-}{\overset{H_3O^+}{\rightleftharpoons}} HC_2H_3O_2$$

Note that one member of the buffer pair (acetate ion) removes excess H_3O^+ ion and that the other (acetic acid) removes excess OH^- ion. The buffering action always results in the active species being converted to its partner species.

To resist both increases and decreases in pH effectively, a weak-acid buffer must contain significant amounts of both the weak acid and its conjugate base. If a solution has a large amount of weak acid but very little conjugate base, it will be unable to consume much added acid. Consequently, the pH tends to drop significantly when acid is added. Conversely, a solution that contains a large amount of conjugate base but very little weak acid will provide very little protection against added base. Addition of just a little base will cause a big change in pH.

EXAMPLE 10.10

Writing Equations for Reactions That Occur in a Buffered Solution

Write an equation for each of the following buffering actions.

a. The response of $H_2PO_4^-/HPO_4^{2-}$ buffer to the addition of H_3O^+ ions
b. The response of HCN/CN^- buffer to the addition of OH^- ions

Solution

a. The base in a conjugate acid–base pair is the species that responds to the addition of acid. (Recall, from Section 10.2, that the base in a conjugate acid–base pair always has one less hydrogen than the acid.) The base for this reaction is HPO_4^{2-}. The equation for the buffering action is

$$H_3O^+ + HPO_4^{2-} \longrightarrow H_2PO_4^- + H_2O$$

In the buffering response, the base is always converted into its conjugate acid.

b. The acid in a conjugate acid–base pair is the species that responds to the addition of base. The acid for this reaction is HCN. The equation for the buffering action is

$$HCN + OH^- \longrightarrow CN^- + H_2O$$

Water will always be one of the products of buffering action.

Practice Exercise 10.10

Write an equation for each of the following buffering actions.

a. The response of H_2CO_3/HCO_3^- buffer to the addition of H_3O^+ ions
b. The response of $H_2PO_4^-/HPO_4^{2-}$ buffer to the addition of OH^- ions

Answers: a. $H_3O^+ + HCO_3^- \rightarrow H_2CO_3 + H_2O$; **b.** $H_2PO_4^- + OH^- \rightarrow HPO_4^{2-} + H_2O$

A common misconception about buffers is that a buffered solution is always a neutral (pH 7.0) solution. This is false. One can buffer a solution at any desired pH. A pH 7.4 buffer will hold the pH of the solution near pH 7.4, whereas a pH 9.3 buffer will tend to hold the pH of a solution near pH 9.3. The pH of a buffer is determined by the degree of weakness of the weak acid used and by the concentrations of the acid and its conjugate base.

A false idea about buffers is that they will hold the pH of a solution *absolutely* constant. The addition of even small amounts of a strong acid or a strong base to any solution, buffered or not, will lead to a change in pH. The important concept is that the shift in pH will be much less when an effective buffer is present (see Figure 10.14).

Buffer systems have their limits. If large amounts of H_3O^+ or OH^- are added to a buffer, the buffer capacity can be exceeded; then the buffer system is overwhelmed and the pH changes. For example, if large amounts of H_3O^+ were added to the acetate/acetic acid buffer previously discussed, the H_3O^+ ion would react with acetate ion until the acetate was depleted. Then the pH would begin to drop as free H_3O^+ ions accumulated in the solution.

Additional insights into the workings of buffer systems are obtained by considering buffer action within the framework of Le Châtelier's principle and an equilibrium system.

Figure 10.14 A comparison of pH changes in buffered and unbuffered solutions. When 0.01 mole of strong acid and 0.01 mole of strong base are added to 1.0 L of pure water and to 1.0 L of 0.1 M HPO_4^{2-} ion/0.1 M $H_2PO_4^-$ ion buffer, the pH of the water varies between 2.0 and 12.0, while the pH of the buffer stays in the narrow range of 7.1 to 7.3.

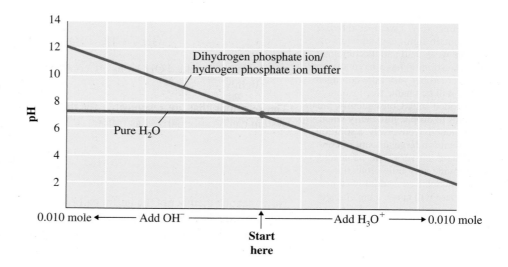

Let us again consider an acetic acid/acetate ion buffer system. An equilibrium is established in solution between the acetic acid and the acetate ion.

$$HC_2H_3O_2(aq) + H_2O(l) \rightleftharpoons H_3O^+(aq) + C_2H_3O_2^-(aq)$$

This equilibrium system functions in accordance with *Le Châtelier's principle* (Section 9.9), which states that an equilibrium system, when stressed, will shift its position in such a way as to counteract the stress. Stresses for the buffer will be (1) addition of base (hydroxide ion) and (2) addition of acid (hydronium ion). Further details concerning these two stress situations are as follows.

Addition of base [OH⁻ ion] *to the buffer.* The addition of base causes the following changes to occur in the solution:

1. The added OH⁻ ion reacts with H_3O^+ ion, producing water (neutralization).
2. The neutralization reaction produces the stress of *not enough* H_3O^+ ion because H_3O^+ ion was consumed in the neutralization.
3. The equilibrium shifts to the right, in accordance with Le Châtelier's principle, to produce more H_3O^+ ion, which maintains the pH close to its original level.

Addition of acid [H_3O^+ ion] *to the buffer.* The addition of acid causes the following changes to occur in the solution:

1. The added H_3O^+ ion increases the overall amount of H_3O^+ ion present.
2. The stress on the system is *too much* H_3O^+ ion.
3. The equilibrium shifts to the left, in accordance with Le Châtelier's principle, consuming most of the excess H_3O^+ ion and resulting in a pH close to the original level.

The Chemistry at a Glance feature on page 278 reviews important concepts about buffer systems.

10.13 THE HENDERSON-HASSELBALCH EQUATION

Buffers may be prepared from any ratio of concentrations of a weak acid and the salt of its conjugate base. However, a buffer is most effective in counteracting pH change when the acid-to-conjugate-base ratio is 1:1. If a buffer contains considerably more acid than the conjugate base, it is less efficient in handling an acid. Conversely, a buffer with considerably more of the conjugate base than the acid is less efficient in handling added base.

When the concentrations of an acid and its conjugate base are equal in a buffer solution, the solution's hydronium ion concentration is equal to the acid ionization constant of the weak acid—or, stated more concisely, the pH of the solution is equal to the pK_a of the weak acid. The mathematical basis for this equality is as follows:

For the weak acid,

$$K_a = \frac{[H_3O^+][A^-]}{[HA]}$$

If HA and A⁻ are equal, then they cancel from the equation, and we have

$$K_a = [H_3O^+]$$

Taking the negative logarithm of both sides of this equation gives

$$pK_a = pH$$

The relationship between pK_a and pH for buffer solutions where the conjugate acid–base pair concentration ratio is something other than 1:1 is given by the equation

$$pH = pK_a + \log \frac{[A^-]}{[HA]}$$

This equation is called the *Henderson–Hasselbalch equation.* The Henderson–Hasselbalch equation indicates that if there is more A⁻ than HA in a solution, the pH is higher than pK_a; and if there is more HA than A⁻, the pH is lower than pK_a.

CHEMISTRY AT A GLANCE

Buffer Systems

BUFFER SOLUTION

- A solution that resists major change in pH when small amounts of strong acid or strong base are added
- A typical buffer system contains a weak acid and its conjugate base

Common biochemical buffer systems are H_2CO_3/HCO_3^- $H_2PO_4^-/HPO_4^{2-}$

Weak Acid

- The buffer component that reacts with added base
- Reaction converts it into its conjugate base

Conjugate Base of Weak Acid

- The buffer component that reacts with added acid
- Reaction converts it into its conjugate acid

DIAGRAMS OF BUFFER ACTION

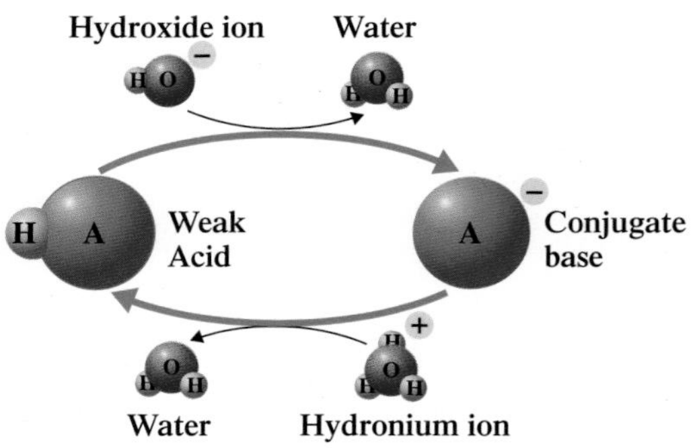

EXAMPLE 10.11

Calculating the pH of a Buffer Solution Using the Henderson–Hasselbalch Equation

What is the pH of a buffer solution that is 0.5 M in formic acid ($HCHO_2$) and 1.0 M in sodium formate ($NaCHO_2$)? The pK_a for formic acid is 3.74.

Solution

The concentrations for the buffering species are

$$HCHO_2 = 0.5\ M \qquad CHO_2^- = 1.0\ M$$

Substituting these values into the Henderson–Hasselbalch equation gives

$$pH = pK_a + \log\frac{[CHO_2^-]}{[HCHO_2]} = 3.74 + \log\frac{1.0}{0.5}$$

$$= 3.74 + \log 2 = 3.74 + 0.30$$

$$= 4.04$$

Practice Exercise 10.11

What is the pH of a buffer solution that is 0.6 M in acetic acid ($HC_2H_3O_2$) and 1.5 M in sodium acetate ($NaC_2H_3O_2$)? The pK_a for acetic acid is 4.74.

Answer: pH = 5.14

10.14 ELECTROLYTES

Aqueous solutions in which ions are present are good conductors of electricity, and the greater the number of ions present, the better the solution conducts electricity. Acids, bases, and soluble salts all produce ions in solution; thus they all produce solutions that conduct electricity. All three types of compounds are said to be electrolytes. An **electrolyte** *is a substance whose aqueous solution conducts electricity.* The presence of ions (charged particles) explains the electrical conductivity.

Buffering Action in Human Blood

Metabolic processes normally maintain blood pH within the narrow range of 7.35–7.45. Even small departures from blood's normal pH range can cause serious illness, and death can result from pH variations that exceed a few tenths of a unit. The most immediate threat to the survival of a person with severe injury or burns is a change in blood pH; individuals in such a situation are said to be in *shock*. Paramedics immediately administer intravenous fluids in such cases to combat changes in blood pH.

The body's primary buffer system for controlling blood pH is the carbonic acid/hydrogen carbonate ion system. Any excess acid formed in the blood reacts with the HCO_3^- ion, and any excess base reacts with H_2CO_3.

$$H_3O^+ + HCO_3^- \longrightarrow H_2CO_3 + H_2O$$

$$OH^- + H_2CO_3 \longrightarrow HCO_3^- + H_2O$$

The ratio of $[H_2CO_3]$ to $[HCO_3^-]$ in blood is approximately 1 to 10, which means this buffer has a greater ability to interact with acid than with base. Significant amounts of acids (up to 10 moles a day) are produced in the human body as a result of normal metabolic reactions. For example, lactic acid ($HC_3H_5O_2$) is produced in muscle tissue during exercise.

This 1-to-10 ratio of buffering species is also needed to maintain the blood at a pH of 7.4. A 1-to-1 ratio buffer would produce a pH of 6.4. The 1-to-10 ratio is easily maintained.

Carbonic acid concentration is controlled by respiration. Excess H_2CO_3 decomposes to CO_2 and H_2O and is removed from the blood by the lungs.

$$H_2CO_3 \longrightarrow CO_2 + H_2O$$

Hydrogen carbonate ion concentration is controlled by the kidneys. Excess HCO_3^- ion is eliminated from the body through urine.

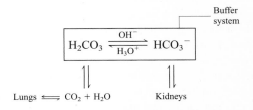

Under certain stress conditions, the blood's buffer systems can be overwhelmed. **Acidosis** *is a body condition in which the pH of blood drops from its normal value of 7.4 to 7.1–7.2.* **Alkalosis** *is a body condition in which the pH of blood increases from its normal value of 7.4 to a value of 7.5.* Both can be life-threatening if not properly taken care of; both can be caused by either metabolic processes or changes in breathing patterns (respiration).

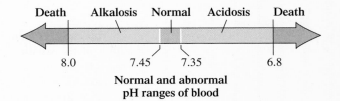

Normal and abnormal
pH ranges of blood

Metabolic acidosis is seen in diabetics, who accumulate acidic substances from the metabolism of fats. Excessive loss of bicarbonate ion in cases of severe diarrhea is another cause. A temporary metabolic acidosis condition can result from prolonged intensive exercise. Exercise generates lactic acid (a weak acid) in the muscles. Some of the lactic acid ionizes, and this produces an influx of H_3O^+ ions into the bloodstream.

Metabolic alkalosis is less common than metabolic acidosis. It results from elevated HCO_3^- ion levels. Causes include prolonged vomiting and the side effects of certain drugs that change the concentrations of sodium, potassium, and chloride ions in the blood.

Respiratory acidosis results from higher than normal levels of CO_2 in the blood; inefficient CO_2 removal is usually the origin of this problem. Hypoventilation (a lowered breathing rate), caused by lung diseases such as emphysema and asthma or obstructed air passages, produces respiratory acidosis.

Respiratory alkalosis is caused by hyperventilation (an elevated breathing rate). Causes include hysteria and anxiety (occasioned, for example, by chemistry tests) and the rapid breathing associated with extremely high fevers.

Some substances, such as table sugar (sucrose), glucose, and isopropyl alcohol, do not produce ions in solution. These substances are called nonelectrolytes. A **nonelectrolyte** *is a substance whose aqueous solution does not conduct electricity.*

Electrolytes can be divided into two groups—strong electrolytes and weak electrolytes. A **strong electrolyte** *is a substance that completely (or almost completely) ionizes/dissociates into ions in aqueous solution.* Strong electrolytes produce strongly conducting solutions. All strong acids and strong bases and all soluble salts are strong electrolytes. A **weak electrolyte** *is a substance that incompletely ionizes/dissociates into ions in aqueous solution.* Weak electrolytes produce solutions that are intermediate between those containing strong electrolytes and those containing nonelectrolytes in their ability to conduct an electric current. Weak acids and weak bases constitute the weak electrolytes.

The differences between the processes of *ionization* and *dissociation* were considered in Section 10.1.

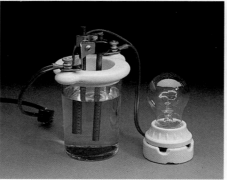

Figure 10.15 This simple device can be used to distinguish among strong electrolytes, weak electrolytes, and nonelectrolytes. The light bulb glows strongly for strong electrolytes (left), weakly for weak electrolytes (center), and not at all for nonelectrolytes (right).

You can determine whether a substance is an electrolyte in solution by testing the ability of the solution to conduct an electric current. A device such as that shown in Figure 10.15 can be used to distinguish among strong electrolytes, weak electrolytes, and nonelectrolytes. If the medium between the electrodes (the solution) is a conductor of electricity, the light bulb glows. A strong glow indicates a strong electrolyte. A faint glow occurs for a weak electrolyte, and there is no glow for a nonelectrolyte.

10.15 EQUIVALENTS AND MILLIEQUIVALENTS OF ELECTROLYTES

The solution that results from dissolving equal molar amounts of the soluble salts KNO_3 and NaCl in water contains four ions in equal concentrations; they are the K^+, Na^+, NO_3^-, and Cl^- ions (see Section 10.14). An identical solution to this one could be made from dissolving the same molar amounts of the soluble salts $NaNO_3$ and KCl in water; again, the ions K^+, Na^+, NO_3^-, and Cl^- are present. In solutions such as these, trying to assign specific positive ions to specific anions, or to talk about specific ionic compounds being present, has lost meaning. We have a solution containing individual ions of four types; the "original partners" for the ions are no longer important. What is important is the identity of the ions present and the total number of each type of ion present.

A similar situation occurs in the human body. All body fluids are electrolyte solutions that contain several positive and negative ions. The ions present usually have more than one source. In discussing such mixtures of ions, the focus is on ion identity and amount of ion present, rather than on the compounds from which the ions were originally supplied.

For solutions that contain electrolytes the concentrations of the ions present are usually specified using the *equivalent* unit. An **equivalent (Eq)** *of an ion is the molar amount of that ion needed to supply one mole of positive or negative charge.* Thus, 1 mole of Na^+ ion and 1 mole of Cl^- ion each equal one equivalent because they supply one mole of electrical charge. For all ions with a $+1$ or -1 charge, one equivalent is equal to 1 mole of the ion. For ions with a $+2$ or -2 charge, there are two equivalents per one mole of ion, since each ion carries 2 units of charge. Similar considerations apply to ions with a -3 charge, such as the PO_4^{3-} ion.

$$1 \text{ mole } Cl^- \text{ ion} = 1 \text{ equivalent}$$
$$1 \text{ mole } Ca^{2+} \text{ ion} = 2 \text{ equivalents}$$
$$1 \text{ mole } PO_4^{3-} \text{ ion} = 3 \text{ equivalents}$$

Used even more frequently than the *equivalent* unit in specifying electrolyte concentrations is the smaller *milliequivalent* unit. This is because of the relatively low concentrations of ions present in body fluids.

$$1 \text{ milliequivalent} = 10^{-3} \text{ equivalent}$$

TABLE 10.8
Concentrations of Major Electrolytes in Blood Plasma

Positive Ions		Negative Ions	
Ion Identity	Concentration (mEq/L)	Ion Identity	Concentration (mEq/L)
Na^+	142	Cl^-	100
Ca^{2+}	5	HCO_3^-	24
K^+	4	HPO_4^{2-}	2
Mg^{2+}	2	SO_4^{2-}	1

As representative of electrolyte concentrations in body fluids, in mEq/L, consider the blood plasma electrolyte values given in Table 10.8.

Examples 10.12 and 10.13 illustate how electrolyte concentrations, specified in mEq/L, are used in a calculational context.

EXAMPLE 10.12

Converting Electrolyte Concentrations from mEq/L to Moles/L

A typical concentration of Cl^- in blood is 106 mEq/L. At this concentration, how many moles of Cl^- ion are present in 1.00 L of blood?

Solution

Step 1: The given quantity is 1.00 L of blood and the desired quantity is moles of Cl^- ion.

$$1.00 \text{ L blood} = ? \text{ moles } Cl^- \text{ ion}$$

Step 2: The pathway for solving the problem is

$$\boxed{\text{L blood}} \rightarrow \boxed{\text{mEq } Cl^-} \rightarrow \boxed{\text{Eq } Cl^-} \rightarrow \boxed{\text{moles } Cl^-}$$

The mathematical setup, using dimensional analysis and conversion factors (Chapter 2), is

$$1.00 \text{ L blood} \times \frac{106 \text{ mEq } Cl^-}{1 \text{ L blood}} \times \frac{10^{-3} \text{ Eq } Cl^-}{1 \text{ mEq } Cl^-} \times \frac{1 \text{ mole } Cl^-}{1 \text{ Eq } Cl^-}$$

$$= 0.106 \text{ mole } Cl^- \text{ ion}$$

The first conversion factor in the setup is the given concentration of Cl^-. The second conversion factor effects the change from mEq to Eq units and the final conversion factor effects the change from Eq to moles. This final factor comes from the definition for an

equivalent given earlier in this section.

Practice Exercise 10.12

A typical concentration of Na^+ in blood is 141 mEq/L. How many moles of Na^+ ion are

EXAMPLE 10.13

Calculating the Mass of an Electrolyte Present in a Given Volume of Solution

The concentration of Mg^{2+} ion present in a blood sample is 4.2 mEq/L. How many milligrams of Mg^{2+} ion are present in 250.0 mL of the blood?

Solution

Step 1: The given quantity is 250.0 mL of solution and the desired quantity is milligrams of Mg^{2+} ion.

$$250.0 \text{ mL blood} = ? \text{ mg } Mg^{2+} \text{ ion}$$

(continued)

Step 2: The pathway for solving the problem is

$$\boxed{\text{mL blood}} \rightarrow \boxed{\text{L blood}} \rightarrow \boxed{\text{mEq Mg}^{2+}} \rightarrow \boxed{\text{Eq Mg}^{2+}} \rightarrow \boxed{\text{moles Mg}^{2+}}$$

$$\rightarrow \boxed{\text{g Mg}^{2+}} \rightarrow \boxed{\text{mg Mg}^{2+}}$$

The mathematical setup, in terms of conversion factors, is

$$250.0 \text{ mL blood} \times \frac{10^{-3} \text{ L blood}}{1 \text{ mL blood}} \times \frac{4.2 \text{ mEq Mg}^{2+}}{1 \text{ L blood}} \times \frac{10^{-3} \text{ Eq Mg}^{2+}}{1 \text{ mEq Mg}^{2+}} \times \frac{1 \text{ mole Mg}^{2+}}{2 \text{ Eq Mg}^{2+}}$$

$$\times \frac{24.31 \text{g Mg}^{2+}}{1 \text{ mole Mg}^{2+}} \times \frac{1 \text{ mg Mg}^{2+}}{10^{-3} \text{ g Mg}^{2+}}$$

$$= 13 \text{ mg Mg}^{2+}\text{ion}$$

Note the value 2 in the fourth conversion factor. The 2 is needed because of the +2 charge on the Mg^{2+} ion; there are 2 equivalents per mole. The next-to-last conversion factor, relating grams and moles, is based on the atomic mass of magnesium. One mole of magnesium has a mass of 24.31 grams (Section 6.3). [Note that the mass of a Mg^{2+} ion and that of a Mg atom are considered to be the same; their difference in mass is that of 2 electrons and that mass difference is negligible in terms of significant figures (Section 2.5).]

Practice Exercise 10.13

The concentration of Ca^{2+} ion present in a blood sample is found to be 4.3 mEq/L. How many milligrams of Ca^{2+} ion are present in 500.0 mL of the blood?

Answer: 43 mg Ca^{2+} ion

10.16 ACID–BASE TITRATIONS

The analysis of solutions to determine the concentration of acid or base present is performed regularly in many laboratories. Such activity is different from determining a solution's pH value. The pH of a solution gives information about the *concentration* of hydronium ions in solution. Only ionized molecules influence the pH value. The concentration of an acid or a base gives information about the *total number* of acid or base molecules present; both dissociated and undissociated molecules are counted. Thus acid or base concentration is a measure of total acidity or total basicity.

The procedure most frequently used to determine the concentration of an acid or a base solution is an acid–base titration. An **acid–base titration** *is a neutralization reaction in which a measured volume of an acid or a base of known concentration is completely reacted with a measured volume of a base or an acid of unknown concentration.* Note that the chemical reaction that occurs in an acid–base titration is that of *neutralization* (Section 10.7).

Suppose we want to determine the concentration of an acid solution by means of titration. We first measure out a known volume of the acid solution into a beaker or flask. Then we slowly add a solution of base of known concentration to the flask or beaker by means of a buret (Figure 10.16). We continue to add base until all the acid has completely reacted with the added base. The volume of base needed to reach this point is obtained from the buret readings. When we know the original volume of acid, the concentration of the base, and the volume of added base, we can calculate the concentration of the acid, as will be shown in Example 10.14.

In order to complete a titration successfully, we must be able to detect when the reaction between acid and base is complete. Neither the acid nor the base gives any outward sign that the reaction is complete. Thus an indicator is always added to the reaction mixture (Figure 10.17). An **acid–base indicator** *is a compound that exhibits*

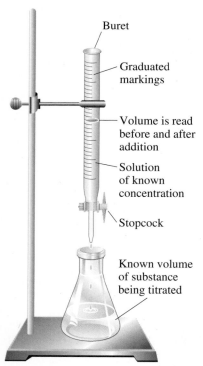

Buret

Graduated markings

Volume is read before and after addition

Solution of known concentration

Stopcock

Known volume of substance being titrated

Figure 10.16 A schematic diagram showing the setup used for titration procedures.

CHEMICAL Connections

Electrolytes and Body Fluids

Structurally speaking, there are three types of body fluids: blood plasma, which is the liquid part of the blood; interstitial fluid, which is the fluid in tissues between and around cells; and intracellular fluids, which are the fluids within cells. For every kilogram of its mass, the body contains about 400 mL of intracellular fluid, 160 mL of interstitial fluid, and 40 mL of plasma.

Water is the main component of any type of body fluid. In addition, all body fluids contain electrolytes. It is the electrolytes present in body fluids that govern numerous body processes. The chemical makeup of the three types of body fluids, in terms of electrolytes (ions present), is shown in the accompanying figure.

Chemically, two of the body fluids (plasma and interstitial fluid) are almost identical. Intracellular fluid, on the other hand, shows striking differences. For example, K^+ is the dominant positive ion in intracellular fluid, and Na^+ dominates in the other two fluids. A similar situation occurs with negative ions. A different ion dominates in intracellular fluid (HPO_4^{2-}) than in the other two fluids (Cl^-).

The electrolytes present in body fluids (1) govern the movement of water between body fluid compartments and (2) maintain acid–base balance within the body fluids. Osmotic pressure, a major factor in controlling water movement, is directly related to electrolyte concentration gradients.

The fact that the presence of ions causes a solution to conduct electricity is of extreme biochemical significance. For example, messages to and from the brain are sent in the form of electrical signals. Ions in intracellular and interstitial fluids are often the carriers of these signals. The presence of electrolytes (ions) is essential to the proper functioning of the human body.

In a hospital setting, patients routinely receive fluids in an intravenous manner. The specific composition of the solutions used, all of which contain electrolytes, depends on the nutritional and fluid needs of the individual. If a patient's system is out of balance from an electrolyte perspective, the solution used is one designed to restore balance. The following table lists commonly used intravenous electrolyte solutions, the conditions they are used for, and the concentrations of the electrolytes present.

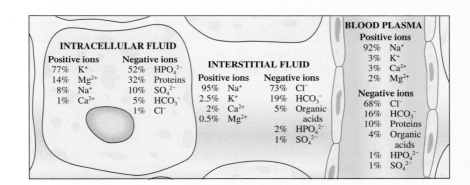

Commonly Used Intravenous Replacement Solutions

Condition Being Treated	Solution Used	Electrolytes (mEq/L)
replacement of fluid loss	sodium chloride (0.9%)	Na^+ 154, Cl^- 154
replacement of electrolytes in extracellular fluids	replacement solution (extracellular)	Na^+ 140, K^+ 10, Ca^{2+} 5, Mg^{2+} 3, Cl^- 103, acetate$^-$ 47, citrate^{3-} 8
replacement of fluids and electrolytes lost through dehydration	ringer's solution	Na^+ 147, K^+ 4, Ca^{2+} 4, Cl^- 155
maintenance of fluid and electrolyte levels	maintenance solution with 5% glucose	Na^+ 40, K^+ 35, Cl^- 40, lactate$^-$ 20, HPO_4^{2-} 15
treatment of malnutrition (low potassium levels)	potassium chloride with 5% glucose	K^+ 40, Cl^- 40

Figure 10.17 An acid-base titration using an indicator that is yellow in acidic solution and red in basic solution.

different colors depending on the pH of its solution. Typically, an indicator is one color in basic solutions and another color in acidic solutions. An indicator is selected that changes color at a pH that corresponds as nearly as possible to the pH of the solution when the titration is complete. If the acid and base are both strong, the pH at that point is 7.0. However, because of hydrolysis (Section 10.11), the pH is not 7.0 if a weak acid or weak base is part of the titration system.

Titration of a weak acid by a strong base requires an indicator that changes color above pH 7.0 because the salt formed in the titration will hydrolyze to form a basic solution. Conversely, titration of a weak base by a strong acid requires an indicator that changes color below pH 7.0.

Example 10.14 shows how titration data are used to calculate the molarity of an acid solution of unknown concentration.

● **EXAMPLE 10.14**

Calculating an Unknown Molarity from Acid–Base Titration Data

In a sulfuric acid (H_2SO_4)–sodium hydroxide (NaOH) acid–base titration, 17.3 mL of 0.126 M NaOH is needed to neutralize 25.0 mL of H_2SO_4 of unknown concentration. Find the molarity of the H_2SO_4 solution, given that the neutralization reaction that occurs is

$$H_2SO_4(aq) + 2NaOH(aq) \rightarrow Na_2SO_4(aq) + 2H_2O(l)$$

Solution

First, we calculate the number of moles of H_2SO_4 that reacted with the NaOH. The pathway for this calculation, using dimensional analysis (Section 6.8), is

mL of NaOH $\longrightarrow$ L of NaOH $\longrightarrow$ moles of NaOH $\longrightarrow$ moles of H_2SO_4

The sequence of conversion factors that effects this series of unit changes is

$$17.3 \text{ mL NaOH} \times \left(\frac{10^{-3}\text{L NaOH}}{1 \text{ mL NaOH}}\right) \times \left(\frac{0.126 \text{ mole NaOH}}{1 \text{ L NaOH}}\right) \times \left(\frac{1 \text{ mole } H_2SO_4}{2 \text{ moles NaOH}}\right)$$

The first conversion factor derives from the definition of a milliliter, the second conversion factor derives from the definition of molarity (Section 8.5), and the third conversion factor uses the coefficients in the balanced chemical equation for the titration reaction (Section 6.7).

The number of moles of H_2SO_4 that react is obtained by combining all the numbers in the dimensional analysis setup in the manner indicated.

$$\left(\frac{17.3 \times 10^{-3} \times 0.126 \times 1}{1 \times 1 \times 2}\right) \text{mole } H_2SO_4 = 0.00109 \text{ mole } H_2SO_4$$

Now that we know how many moles of H_2SO_4 reacted, we calculate the molarity of H_2SO_4 solution using the definition for molarity.

$$\text{Molarity } H_2SO_4 = \frac{\text{moles } H_2SO_4}{\text{L } H_2SO_4 \text{ solution}} = \frac{0.00109 \text{ mole}}{0.0250 \text{ L}}$$

$$= 0.0436 \frac{\text{mole}}{\text{L}}$$

Note that the units in the denominator of the molarity equation must be liters (0.0250) rather than milliliters (25.0).

Practice Exercise 10.14

In a nitric acid (HNO_3)–potassium hydroxide (KOH) acid–base titration, 32.4 mL of 0.352 M KOH is required to neutralize 50.0 mL of HNO_3 of unknown concentration. Find the molarity of the HNO_3 solution, given that the neutralization reaction that occurs is

$$HNO_3(aq) + KOH(aq) \rightarrow KNO_3(aq) + H_2O(l)$$

Answer: 0.228 M HNO_3

CONCEPTS TO REMEMBER

Arrhenius acid-base theory. An Arrhenius acid is a hydrogen-containing compound that, in water, produces hydrogen ions. An Arrhenius base is a hydroxide-containing compound that, in water, produces hydroxide ions (Section 10.1).

Brønsted-Lowry acid-base theory. A Brønsted–Lowry acid is any substance that can donate a proton (H^+) to some other substance. A Brønsted–Lowry base is any substance that can accept a proton from some other substance. Proton donation (from an acid) cannot occur unless an acceptor (a base) is present (Section 10.2).

Conjugate acids and bases. A conjugate acid–base pair consists of two species that differ by one proton. The conjugate base of an acid is the species that remains when the acid loses a proton. The conjugate acid of a base is the species formed when the base accepts a proton (Section 10.2).

Polyprotic acids. Polyprotic acids are acids that can transfer two or more hydrogen ions during an acid–base reaction (Section 10.3).

Strengths of acids and bases. Acids can be classified as strong or weak based on the extent to which proton transfer occurs in aqueous solution. A strong acid completely transfers its protons to water. A weak acid transfers only a small percentage of its protons to water (Section 10.4).

Acid ionization constant. The acid ionization constant quantitatively describes the degree of ionization of an acid. It is the equilibrium constant expression that corresponds to the ionization of the acid (Section 10.5).

Salts. Salts are ionic compounds containing a metal or polyatomic ion as the positive ion and a nonmetal or polyatomic ion (except hydroxide ion) as the negative ion. Ionic compounds containing hydroxide ion are bases rather than salts (Section 10.6).

Acid-base neutralization. Acid–base neutralization is the chemical reaction between an acid and a hydroxide base to form a salt and water (Section 10.7).

Self-ionization of water. In pure water, a small number of water molecules (1.0×10^{-7} mole/L) donate protons to other water molecules to produce small concentrations (1.0×10^{-7} mole/L) of hydronium and hydroxide ions (Section 10.8).

The pH scale. The pH scale is a scale of small numbers that are used to specify molar hydronium ion concentration in an aqueous solution. Mathematically, the pH is the negative logarithm of the hydronium ion concentration. Solutions with a pH lower than 7.0 are acidic, those with a pH higher than 7.0 are basic, and those with a pH equal to 7.0 are neutral (Section 10.9).

Hydrolysis of salts. Salt hydrolysis is a chemical reaction in which a salt interacts with water to produce an acidic or a basic solution. Only salts that contain the conjugate base of a weak acid and/or the conjugate acid of a weak base hydrolyze (Section 10.11).

Buffer solutions. A buffer solution is a solution that resists pH change when small amounts of acid or base are added to it. The resistance to pH change in most buffers is caused by the presence of a weak acid and a salt of its conjugate base (Section 10.12).

Electrolytes. An electrolyte is a substance that forms a solution in water that conducts electricity. Strong acids, strong bases, and soluble salts are strong electrolytes. Weak acids and weak bases are weak electrolytes (Section 10.14).

Equivalents and milliequivalents. Body fluids contain small amounts of many different electrolytes, whose concentrations are expressed in equivalents or milliequivalents. An equivalent is the amount of an electrolyte that carries one mole of positive or negative charge (Section 10.15).

Acid-base titrations. An acid–base titration is a procedure in which an acid–base neutralization reaction is used to determine an unknown concentration. A measured volume of an acid or a base of known concentration is exactly reacted with a measured volume of a base or an acid of unknown concentration (Section 10.16).

KEY REACTIONS AND EQUATIONS

1. Weak-acid equilibrium and acid ionization constant (K_a) expression (Section 10.5)

$$HA(aq) + H_2O(l) \rightleftharpoons H_3O^+(aq) + A^-(aq)$$

$$K_a = \frac{[H_3O^+][A^-]}{[HA]}$$

2. Weak-base equilibrium and base ionization constant (K_b) expression (Section 10.5)

$$B(aq) + H_2O(l) \rightleftharpoons BH^+(aq) + OH^-(aq)$$

$$K_b = \frac{[BH^+][OH^-]}{[B]}$$

3. Ion product constant for water (Section 10.8)

$$[H_3O^+][OH^-] = 1.0 \times 10^{-14}$$

4. Relationship between $[H_3O^+]$ and pH (Section 10.9)

$$pH = -\log[H_3O^+]$$

5. Henderson–Hasselbalch equation (Section 10.13)

$$pH = pK_a + \log\frac{[A^-]}{[HA]}$$

6. Equivalent concentration unit (Section 10.15)

1 equivalent = molar amount of an ion that carries 1 mole of electrical charge

EXERCISES and PROBLEMS

The members of each pair of problems in this section test similar material.

● Arrhenius Acid-Base Theory (Section 10.1)

10.1 In Arrhenius acid–base theory, what ion is responsible for the properties of
a. acidic solutions   b. basic solutions

10.2 What term is used to describe the formation of ions, in aqueous solution, from
a. a molecular compound   b. an ionic compound

10.3 Classify each of the following as a property of an Arrhenius acid or the property of an Arrhenius base.
a. Has a sour taste   b. Has a bitter taste

10.4 Classify each of the following as a property of an Arrhenius acid or the property of an Arrhenius base.
a. Changes the color of blue litmus paper to red
b. Changes the color of red litmus paper to blue

10.5 Write equations depicting the behavior of the following Arrhenius acids and bases in water.
a. HI (hydroiodic acid)   b. HClO (hypochlorous acid)
c. LiOH (lithium hydroxide)   d. CsOH (cesium hydroxide)

10.6 Write equations depicting the behavior of the following Arrhenius acids and bases in water.
a. HBr (hydrobromic acid)
b. HCN (hydrocyanic acid)
c. RbOH (rubidium hydroxide)
d. KOH (potassium hydroxide)

● Brønsted–Lowry Acid-Base Theory (Section 10.2)

10.7 In each of the following reactions, decide whether the underlined species is functioning as a Brønsted–Lowry acid or base.
a. $\underline{HF} + H_2O \rightarrow H_3O^+ + F^-$
b. $\underline{H_2O} + S^{2-} \rightarrow HS^- + OH^-$
c. $H_2O + \underline{H_2CO_3} \rightarrow H_3O^+ + HCO_3^-$
d. $\underline{HCO_3^-} + H_2O \rightarrow H_3O^+ + CO_3^{2-}$

10.8 In each of the following reactions, decide whether the underlined species is functioning as a Brønsted–Lowry acid or base.
a. $\underline{HClO_2} + H_2O \rightarrow H_3O^+ + ClO_2^-$
b. $\underline{OCl^-} + H_2O \rightarrow HOCl + OH^-$
c. $\underline{NH_3} + HNO_2 \rightarrow NH_4^+ + NO_2^-$
d. $HCl + \underline{H_2PO_4^-} \rightarrow H_3PO_4 + Cl^-$

10.9 Write equations to illustrate the acid–base reactions that can take place between the following Brønsted–Lowry acids and bases.
a. Acid: HClO; base: H_2O
b. Acid: $HClO_4$; base: NH_3
c. Acid: H_3O^+; base: OH^-
d. Acid: H_3O^+; base: NH_2^-

10.10 Write equations to illustrate the acid–base reactions that can take place between the following Brønsted–Lowry acids and bases.
a. Acid: $H_2PO_4^-$; base: NH_3
b. Acid: H_2O; base: ClO_4^-
c. Acid: HCl; base: OH^-
d. Acid: $HC_2H_3O_2$; base: H_2O

10.11 Write the formula of each of the following.
a. Conjugate base of H_2SO_3
b. Conjugate acid of CN^-
c. Conjugate base of $HC_2O_4^-$
d. Conjugate acid of HPO_4^{2-}

10.12 Write the formula of each of the following.
a. Conjugate base of NH_4^+
b. Conjugate acid of OH^-
c. Conjugate base of H_2S
d. Conjugate acid of NO_2^-

10.13 For each of the following amphiprotic substances, write the two equations needed to describe its behavior in aqueous solution.
a. HS^- b. HPO_4^{2-} c. NH_3 d. OH^-

10.14 For each of the following amphiprotic substances, write the two equations needed to describe its behavior in aqueous solution.
a. $H_2PO_4^-$ b. HSO_4^- c. $HC_2O_4^-$ d. PH_3

● Polyprotic Acids (Section 10.3)

10.15 Classify the following acids as monoprotic, diprotic, or triprotic.
a. $HClO_4$ (perchloric acid)
b. $H_2C_2O_4$ (oxalic acid)
c. $HC_2H_3O_2$ (acetic acid)
d. H_2SO_4 (sulfuric acid)

10.16 Classify the following acids as monoprotic, diprotic, or triprotic.
a. $HC_4H_7O_2$ (butyric acid)
b. H_3PO_4 (phosphoric acid)

c. HNO_3 (nitric acid)

d. $H_2C_4H_4O_4$ (succinic acid)

10.17 Write equations showing all steps in the dissociation of citric acid ($H_3C_6H_5O_7$).

10.18 Write equations showing all steps in the dissociation of arsenic acid (H_3AsO_4).

10.19 How many acidic hydrogen atoms and how many nonacidic hydrogen atoms are present in each of the following molecules?

a. HNO_3 (nitric acid)

b. $H_2C_4H_4O_4$ (succinic acid)

c. $HC_4H_7O_2$ (butyric acid)

d. CH_4 (methane)

10.20 How many acidic hydrogen atoms and how many nonacidic hydrogen atoms are present in each of the following molecules?

a. H_2CO_3 (carbonic acid)

b. $H_2C_3H_2O_4$ (malonic acid)

c. NH_3 (ammonia)

d. $HC_3H_5O_2$ (propanoic acid)

10.21 The formula for lactic acid is preferably written as $HC_3H_5O_3$ rather than as $C_3H_6O_3$. Explain why.

10.22 The formula for tartaric acid is preferably written as $H_2C_4H_4O_6$ rather than as $C_4H_6O_6$. Explain why.

10.23 Pyruvic acid, which is produced in metabolic reactions, has the structure

$$H-\underset{\underset{H}{|}}{\overset{\overset{H}{|}}{C}}-\overset{\overset{O}{\|}}{C}-\overset{\overset{O}{\|}}{C}-O-H$$

Would you predict that this acid is a mono-, di-, tri-, or tetraprotic acid? Give your reasoning.

10.24 Oxaloacetic acid, which is produced in metabolic reactions, has the structure

$$H-O-\overset{\overset{O}{\|}}{C}-\overset{\overset{O}{\|}}{C}-\underset{\underset{H}{|}}{\overset{\overset{H}{|}}{C}}-\overset{\overset{O}{\|}}{C}-O-H$$

Would you predict that this acid is a mono-, di-, tri-, or tetra-protic acid? Give your reasoning.

Strengths of Acids and Bases (Section 10.4)

10.25 Classify each of the acids in Problem 10.15 as a strong acid or a weak acid.

10.26 Classify each of the acids in Problem 10.16 as a strong acid or a weak acid.

10.27 What is the difference between a strong acid and a weak acid in terms of equilibrium position?

10.28 What is the difference between a strong base and a weak base in terms of equilibrium position?

10.29 The HCl in a 0.10 M HCl solution is 100% dissociated. What are the molar concentrations of HCl, H_3O^+, and Cl^- in the solution?

10.30 The HNO_3 in a 0.50 M HNO_3 solution is 100% dissociated. What are the molar concentrations of HNO_3, H_3O^+, and NO_3^- in the solution?

10.31 The following four diagrams represent aqueous solutions of four different acids with the general formula HA. Which of the four acids is the strongest acid?

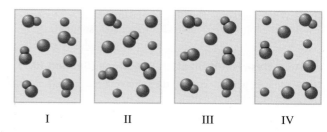

I   II   III   IV

10.32 Using the diagrams shown in Problem 10.31, which of the four acids is the weakest acid?

Ionization Constants for Acids and Bases (Section 10.5)

10.33 Write the acid ionization constant expression for the ionization of each of the following monoprotic acids.

a. HF (hydrofluoric acid)   b. $HC_2H_3O_2$ (acetic acid)

10.34 Write the acid ionization constant expression for the ionization of each of the following monoprotic acids.

a. HCN (hydrocyanic acid)   b. $HC_6H_7O_6$ (ascorbic acid)

10.35 Write the base ionization constant expression for the ionization of each of the following bases. In each case, the nitrogen atom accepts the proton.

a. NH_3 (ammonia)   b. $C_6H_5NH_2$ (aniline)

10.36 Write the base ionization constant expression for the ionization of each of the following bases. In each case, the nitrogen atom accepts the proton.

a. CH_3NH_2 (methylamine)   b. $C_2H_5NH_2$ (ethylamine)

10.37 Using the acid ionization constant information given in Table 10.3, indicate which acid is the stronger in each of the following acid pairs.

a. H_3PO_4 and HNO_2 b. HCN and HF

c. H_2CO_3 and HCO_3^- d. HNO_2 and HCN

10.38 Using the acid ionization constant information given in Table 10.3, indicate which acid is the stronger in each of the following acid pairs.

a. H_3PO_4 and $H_2PO_4^-$ b. H_3PO_4 and H_2CO_3

c. HPO_4^{2-} and $H_2PO_4^-$ d. $HC_2H_3O_2$ and HCN

10.39 A 0.00300 M solution of an acid is 12% ionized. Calculate the acid ionization constant K_a.

10.40 A 0.0500 M solution of a base is 7.5% ionized. Calculate the base ionization constant K_b.

Salts (Section 10.6)

10.41 Classify each of the following substances as an acid, a base, or a salt.

a. HBr   b. NaI   c. NH_4NO_3 d. $Ba(OH)_2$

10.42 Classify each of the following substances as an acid, a base, or a salt.

a. $AlPO_4$ b. KOH   c. HNO_3 d. $HC_2H_3O_2$

10.43 Classify each of the following substances as an acid, a base, or a salt.

a. NaOH   b. NH_4I c. H_2SO_4 d. $Ba_3(PO_4)_2$

10.44 Classify each of the following substances as an acid, a base, or a salt.

a. $Ca(OH)_2$ b. HCN   c. KBr   d. HCl

10.45 Write a balanced equation for the dissociation into ions of each of the following soluble salts in aqueous solution.

a. $Ba(NO_3)_2$ b. Na_2SO_4 c. $CaBr_2$ d. K_2CO_3

10.46 Write a balanced equation for the dissociation into ions of each of the following soluble salts in aqueous solution.
a. CaS b. $BeSO_4$ c. $MgCl_2$ d. $NaC_2H_3O_2$

● **Acid–Base Neutralization Reactions (Section 10.7)**

10.47 Indicate whether each of the following reactions is an acid–base neutralization reaction.
a. $NaCl + AgNO_3 \longrightarrow AgCl + NaNO_3$
b. $HNO_3 + NaOH \longrightarrow NaNO_3 + H_2O$
c. $HBr + KOH \longrightarrow KBr + H_2O$
d. $H_2SO_4 + Pb(NO_3)_2 \longrightarrow PbSO_4 + 2HNO_3$

10.48 Indicate whether each of the following reactions is an acid–base neutralization reaction.
a. $H_2S + CuSO_4 \longrightarrow H_2SO_4 + CuS$
b. $HCN + LiOH \longrightarrow LiCN + H_2O$
c. $H_2SO_4 + Ba(OH)_2 \longrightarrow BaSO_4 + 2H_2O$
d. $Ni + 2HCl \longrightarrow NiCl_2 + H_2$

10.49 Without writing an equation, specify the molecular ratio in which each of the following acid–base pairs will react.
a. HNO_3 and $NaOH$ b. H_2SO_4 and $NaOH$
c. H_2SO_4 and $Ba(OH)_2$ d. HNO_3 and $Ba(OH)_2$

10.50 Without writing an equation, specify the molecular ratio in which each of the following acid–base pairs will react.
a. HCl and KOH b. H_2CO_3 and KOH
c. HCl and $Ca(OH)_2$ d. H_2CO_3 and $Ca(OH)_2$

10.51 Write a balanced chemical equation to represent each of the following acid–base neutralization reactions.
a. HCl and $NaOH$ b. HNO_3 and KOH
c. H_2SO_4 and $LiOH$ d. $Ba(OH)_2$ and H_3PO_4

10.52 Write a balanced chemical equation to represent each of the following acid–base neutralization reactions.
a. HCl and $LiOH$ b. HNO_3 and $Ba(OH)_2$
c. H_2SO_4 and $NaOH$ d. KOH and H_3PO_4

10.53 Write a balanced molecular equation for the preparation of each of the following salts, using an acid–base neutralization reaction.
a. Li_2SO_4 (lithium sulfate)
b. $NaCl$ (sodium chloride)
c. KNO_3 (potassium nitrate)
d. $Ba_3(PO_4)_2$ (barium phosphate)

10.54 Write a balanced molecular equation for the preparation of each of the following salts, using an acid–base neutralization reaction.
a. $LiNO_3$ (lithium nitrate)
b. $BaCl_2$ (barium chloride)
c. K_3PO_4 (potassium phosphate)
d. Na_2SO_4 (sodium sulfate)

● **Hydronium Ion and Hydroxide Ion Concentrations (Section 10.8)**

10.55 Calculate the molar H_3O^+ ion concentration of a solution if the OH^- ion concentration is
a. 3.0×10^{-3} M   b. 6.7×10^{-6} M
c. 9.1×10^{-8} M   d. 1.2×10^{-11} M

10.56 Calculate the molar H_3O^+ ion concentration of a solution if the OH^- ion concentration is
a. 5.0×10^{-4} M   b. 7.5×10^{-7} M
c. 2.3×10^{-12} M   d. 1.1×10^{-10} M

10.57 Indicate whether each of the following solutions is acidic, basic, or neutral.
a. $[H_3O^+] = 1.0 \times 10^{-3}$ b. $[H_3O^+] = 3.0 \times 10^{-11}$
c. $[OH^-] = 4.0 \times 10^{-6}$ d. $[OH^-] = 2.3 \times 10^{-10}$

10.58 Indicate whether each of the following solutions is acidic, basic, or neutral.
a. $[H_3O^+] = 2.0 \times 10^{-4}$
b. $[H_3O^+] = 2.0 \times 10^{-8}$
c. $[OH^-] = 1.0 \times 10^{-7}$
d. $[OH^-] = 5.0 \times 10^{-9}$

10.59 Selected information about four solutions, each at 24°C, is given in the following table. Complete the table by filling in the missing information.

$[H_3O^+]$	$[OH^-]$	Acidic or Basic
2.2×10^{-2}		
	3.3×10^{-3}	
6.8×10^{-8}		
	7.2×10^{-8}	

10.60 Selected information about four solutions, each at 24°C, is given in the following table. Complete the table by filling in the missing information.

$[H_3O^+]$	$[OH^-]$	Acidic or Basic
	7.7×10^{-2}	
6.3×10^{-8}		
4.2×10^{-6}		
	3.3×10^{-10}	

● **pH Scale (Section 10.9)**

10.61 Calculate the pH of the following solutions.
a. $[H_3O^+] = 1.0 \times 10^{-4}$
b. $[H_3O^+] = 1.0 \times 10^{-11}$
c. $[OH^-] = 1.0 \times 10^{-3}$
d. $[OH^-] = 1.0 \times 10^{-7}$

10.62 Calculate the pH of the following solutions.
a. $[H_3O^+] = 1.0 \times 10^{-6}$
b. $[H_3O^+] = 1.0 \times 10^{-2}$
c. $[OH^-] = 1.0 \times 10^{-9}$
d. $[OH^-] = 1.0 \times 10^{-5}$

10.63 Calculate the pH of the following solutions.
a. $[H_3O^+] = 2.1 \times 10^{-8}$
b. $[H_3O^+] = 4.0 \times 10^{-8}$
c. $[OH^-] = 7.2 \times 10^{-11}$
d. $[OH^-] = 7.2 \times 10^{-3}$

10.64 Calculate the pH of the following solutions.
a. $[H_3O^+] = 3.3 \times 10^{-5}$
b. $[H_3O^+] = 7.6 \times 10^{-5}$
c. $[OH^-] = 8.2 \times 10^{-10}$
d. $[OH^-] = 8.2 \times 10^{-4}$

10.65 What is the molar hydronium ion concentration in solutions with each of the following pH values?
a. 2.0   b. 6.0   c. 8.0   d. 10.0

10.66 What is the molar hydronium ion concentration in solutions with each of the following pH values?
a. 3.0   b. 5.0   c. 9.0   d. 12.0

10.67 What is the molar hydronium ion concentration in solutions with each of the following pH values?
a. 3.67   b. 5.09   c. 7.35   d. 12.45

10.68 What is the molar hydronium ion concentration in solutions with each of the following pH values?
a. 2.05   b. 4.88   c. 6.75   d. 11.33

10.69 Selected information about four solutions, each at 24°C, is given in the following table. Complete the table by filling in the missing information.

$[H_3O^+]$	$[OH^-]$	pH	Acidic or Basic
		7.21	
7.2×10^{-10}			
		5.30	
	7.2×10^{-10}		

10.70 Selected information about four solutions, each at 24°C, is given in the following table. Complete the table by filling in the missing information.

$[H_3O^+]$	$[OH^-]$	pH	Acidic or Basic
		8.73	
	7.2×10^{-5}		
6.3×10^{-3}			
		2.00	

● pK_a Values (Section 10.10)

10.71 Calculate the pK_a value for each of the following acids.
a. Nitrous acid (HNO_2), $K_a = 4.5 \times 10^{-4}$
b. Carbonic acid (H_2CO_3), $K_a = 4.3 \times 10^{-7}$
c. Dihydrogen phosphate ion ($H_2PO_4^-$), $K_a = 6.2 \times 10^{-8}$
d. Sulfurous acid (H_2SO_3), $K_a = 1.5 \times 10^{-2}$

10.72 Calculate the pK_a value for each of the following acids.
a. Phosphoric acid (H_3PO_4), $K_a = 7.5 \times 10^{-3}$
b. Hydrofluoric acid (HF), $K_a = 6.8 \times 10^{-4}$
c. Hydrogen phosphate ion (HPO_4^{2-}), $K_a = 4.2 \times 10^{-13}$
d. Propanoic acid ($HC_3H_5O_2$), $K_a = 1.3 \times 10^{-5}$

10.73 Acid A has a pK_a value of 4.23, and acid B has a pK_a value of 3.97. Which of the two acids is the stronger?

10.74 Acid A has a pK_a value of 5.71, and acid B has a pK_a value of 5.30. Which of the two acids is the weaker?

● Hydrolysis of Salts (Section 10.11)

10.75 Classify each of the following salts as a "strong acid–strong base salt," a "strong acid–weak base salt," a "weak acid–strong base salt," or a "weak acid–weak base salt."
a. NaCl   b. $KC_2H_3O_2$
c. NH_4Br d. $Ba(NO_3)_2$

10.76 Classify each of the following salts as a "strong acid–strong base salt," a "strong acid–weak base salt," a "weak acid–strong base salt," or a "weak acid–weak base salt."
a. K_3PO_4 b. $NaNO_3$
c. KCl   d. $Na_2C_2O_4$

10.77 Identify the ion (or ions) present in each of the salts in Problem 10.75 that will undergo hydrolysis in aqueous solution.

10.78 Identify the ion (or ions) present in each of the salts in Problem 10.76 that will undergo hydrolysis in aqueous solution.

10.79 Predict whether solutions of each of the salts in Problem 10.75 will be acidic, basic, or neutral.

10.80 Predict whether solutions of each of the salts in Problem 10.76 will be acidic, basic, or neutral.

● Buffers (Section 10.12)

10.81 Predict whether each of the following pairs of substances could function as a buffer system in aqueous solution.
a. HNO_3 and $NaNO_3$ b. HF and NaF
c. KCl and KCN   d. H_2CO_3 and $NaHCO_3$

10.82 Predict whether each of the following pairs of substances could function as a buffer system in aqueous solution.
a. HNO_3 and HCl   b. HNO_2 and KNO_2
c. $NaC_2H_3O_2$ and $KC_2H_3O_2$ d. $HC_2H_3O_2$ and $NaNO_3$

10.83 Identify the two "active species" in each of the following buffer systems.
a. HCN and KCN   b. H_3PO_4 and NaH_2PO_4
c. H_2CO_3 and $KHCO_3$ d. $NaHCO_3$ and K_2CO_3

10.84 Identify the two "active species" in each of the following buffer systems.
a. HF and LiF   b. Na_2HPO_4 and KH_2PO_4
c. K_2CO_3 and $KHCO_3$ d. $NaNO_2$ and HNO_2

10.85 Write an equation for each of the following buffering actions.
a. The response of a HF/F^- buffer to the addition of H_3O^+ ions
b. The response of a H_2CO_3/HCO_3^- buffer to the addition of OH^- ions
c. The response of a HCO_3^-/CO_3^{2-} buffer to the addition of H_3O^+ ions
d. The response of a $H_3PO_4/H_2PO_4^-$ buffer to the addition of OH^- ions

10.86 Write an equation for each of the following buffering actions.
a. The response of a HPO_4^{2-}/PO_4^{3-} buffer to the addition of OH^- ions
b. The response of a HF/F^- buffer to the addition of OH^- ions
c. The response of a HCN/CN^- buffer to the addition of H_3O^+ ions
d. The response of a $H_3PO_4/H_2PO_4^-$ buffer to the addition of H_3O^+ ions

10.87 The following four diagrams represent aqueous solutions containing a weak acid (HA) and/or its conjugate base (A^-). Which of the four solutions is a buffer solution? There may be more than one correct answer.

I   II   III   IV

10.88 Using the diagrams shown in Problem 10.87, which of the solutions would have the greatest buffer capacity, that is, greatest protection against pH change, when the following occurs?
a. A strong acid is added to the solution.
b. A strong base is added to the solution.

● The Henderson-Hasselbalch Equation (Section 10.13)

10.89 What is the pH of a buffer that is 0.230 M in a weak acid and 0.500 M in the acid's conjugate base? The pK_a for the acid is 6.72.

10.90 What is the pH of a buffer that is 0.250 M in a weak acid and 0.260 M in the acid's conjugate base? The pK_a for the acid is 5.53.

10.91 What is the pH of a buffer that is 0.150 M in a weak acid and 0.150 M in the acid's conjugate base? The acid's ionization constant is 6.8×10^{-6}.

10.92 What is the pH of a buffer that is 0.175 M in a weak acid and 0.200 M in the acid's conjugate base? The acid's ionization constant is 5.7×10^{-4}.

Electrolytes (Section 10.14)

10.93 Classify each of the following compounds as a strong electrolyte or a weak electrolyte.
a. H_2CO_3 b. KOH   c. NaCl   d. H_2SO_4

10.94 Classify each of the following compounds as a strong electrolyte or a weak electrolyte.
a. H_3PO_4 b. HNO_3 c. KNO_3 d. NaOH

10.95 Indicate whether solutions of each of the following substances contain ions, molecules, or both.
a. Acetic acid, a weak acid
b. Sucrose, a nonelectrolyte
c. Sodium sulfate, a soluble salt
d. Hydrofluoric acid, a weak electrolyte

10.96 Indicate whether solutions of each of the following substances contain ions, molecules, or both.
a. Hydrochloric acid, a strong acid
b. Sodium nitrate, a soluble salt
c. Potassium chloride, a strong electrolyte
d. Ethanol, a nonelectrolyte

10.97 How many ions, per formula unit, are produced when each of the following soluble salts (strong electrolytes) dissolves in water?
a. NaCl   b. $Mg(NO_3)_2$ c. K_2S d. NH_4CN

10.98 How many ions, per formula unit, are produced when each of the following soluble salts (strong electrolytes) dissolves in water?
a. KNO_3 b. Na_2CO_3 c. $MgCl_2$ d. K_3N

10.99 Write a balanced chemical equation for the dissociation in water of each of the salts listed in Problem 10.97.

10.100 Write a balanced chemical equation for the dissociation in water of each of the salts listed in Problem 10.98.

10.101 Four different substances of the generalized formula HA were dissolved in water with the results shown in the diagrams. Which of the diagrams represents the substance that is the strongest electrolyte?

I   II   III   IV

10.102 Which of the diagrams in Problem 10.101 represents the substance that is the weakest electrolyte?

Equivalents and Milliequivalents of Electrolytes (Section 10.15)

10.103 Indicate the number of equivalents in each of the following molar quantities of ions.
a. 1 mole Na^+ b. 1 mole NO_3^-
c. 1 mole Mg^{2+} d. 1 mole HCO_3^-

10.104 Indicate the number of equivalents in each of the following molar quantities of ions.
a. 1 mole K^+ b. 1 mole $H_2PO_4^-$
c. 1 mole HPO_4^{2-} d. 1 mole Ca^{2+}

10.105 Indicate the number of equivalents in each of the following molar quantities of ions.
a. 2 moles K^+ b. 3 moles $H_2PO_4^-$
c. 2 moles HPO_4^{2-} d. 7 moles Ca^{2+}

10.106 Indicate the number of equivalents in each of the following molar quantities of ions.
a. 4 moles Na^+ b. 2 moles NO_3^-
c. 3 moles Mg^{2+} d. 5 moles HCO_3^-

10.107 A physiological solution contains 47 mEq/L of Cl^- ion. How many moles of Cl^- ion are present in 2.00 L of the solution?

10.108 A physiological solution contains 49 mEq/L of Na^+ ion. How many moles of Na^+ ion are present in 1.50 L of the solution?

10.109 How many milligrams of Ca^{2+} ion are present in 250.0 mL of a solution whose Ca^{2+} ion concentration is 4.1 mEq/L?

10.110 How many milligrams of Mg^{2+} ion are present in 350.0 mL of a solution whose Mg^{2+} ion concentration is 3.8 mEq/L?

Titration Calculations (Section 10.16)

10.111 Determine the molarity of a NaOH solution when each of the following amounts of acid neutralizes 25.0 mL of the NaOH solution.
a. 5.00 mL of 0.250 M HNO_3
b. 20.00 mL of 0.500 M H_2SO_4
c. 23.76 mL of 1.00 M HCl
d. 10.00 mL of 0.100 M H_3PO_4

10.112 Determine the molarity of a KOH solution when each of the following amounts of acid neutralizes 25.0 mL of the KOH solution.
a. 5.00 mL of 0.500 M H_2SO_4
b. 20.00 mL of 0.250 M HNO_3
c. 13.07 mL of 0.100 M H_3PO_4
d. 10.00 mL of 1.00 M HCl

ADDITIONAL PROBLEMS

10.113 In which of the following pairs of substances do the two members of the pair constitute a conjugate acid–base pair?
a. HN_3 and N_3^- b. H_2SO_4 and SO_4^{2-}
c. H_2CO_3 and $HClO_3$ d. NH_3 and NH_2^-

10.114 For which of the following pairs of acids are both members of the pair of "like strength" — that is, both strong or both weak?
a. HNO_3 and HNO_2 b. HCl and HBr
c. H_3PO_4 and $HClO_4$ d. H_2CO_3 and $H_2C_2O_4$

10.115 Solution A has a pH of 3.20, solution B a pH of 12.50, solution C a pH of 7.00, and solution D a pH of 4.44. Arrange the four solutions in order of the following:
a. Decreasing acidity   b. Increasing $[H_3O^+]$
c. Decreasing $[OH^-]$ d. Increasing basicity

10.116 What would be the pH of a solution that contains 0.1 mole of each of the solutes NaCl, NaOH, and HCl in enough water to give 2.00 L of solution?

10.117 Arrange the following 0.1 M aqueous solutions in order of increasing pH: HCl, HCN, NaOH, and KCl.

10.118 Identify the buffer system(s)—the conjugate acid–base pair(s)—present in solutions that contain equal molar amounts of the following.
a. HCN, KCN, NaBr, and NaCl
b. HF, HCl, $NaC_2H_3O_2$, and NaF

10.119 It is possible to make two completely different buffers that involve the dihydrogen phosphate ion, $H_2PO_4^-$. Characterize each of the buffers by specifying the conjugate acid–base pair that is present.

10.120 Both ions in each of the salts NH_4CN and $NH_4C_2H_3O_2$ undergo hydrolysis in aqueous solution. Upon hydrolysis, the first listed salt gives a basic solution and the second a neutral solution. Explain how this is possible.

10.121 How many grams of NaOH are needed to make 875 mL of a solution with a pH of 10.00?

$\mathcal{M}$ultiple-Choice Practice Test

10.122 In an Arrhenius acid–base context, the compounds HCl, HNO_3, and NaOH, when dissolved in water, are which of the following, respectively?
a. Acid, acid, and base   b. Base, base, and acid
c. Base, acid, and base   d. Acid, base, and acid

10.123 Which statement is *correct* for the Brønsted–Lowry acid–base reaction

$$HCN + HCO_3^- \rightleftharpoons CN^- + H_2CO_3$$

a. HCO_3^- is the conjugate base of HCN.
b. HCN is the conjugate acid of HCO_3^-.
c. HCO_3^- is the Brønsted–Lowry acid for the forward reaction.
d. CN^- is the Brønsted–Lowry base for the reverse reaction.

10.124 In which of the following pairs of acids are both members of the pair *weak* acids?
a. HNO_3 and HCl   b. HCN and HF
c. H_2SO_4 and H_3PO_4 d. HBr and HI

10.125 Which of the following is produced in the *second* step of the dissociation of the polyprotic acid H_3PO_4?
a. H_3PO_3 b. $H_2PO_4^-$
c. HPO_4^{2-} d. PO_4^{3-}

10.126 Which of the following chemical equations represents an acid–base neutralization reaction?
a. $H_2SO_4 + Zn \longrightarrow ZnSO_4 + H_2$
b. $HNO_3 + KOH \longrightarrow KNO_3 + H_2O$
c. $Ba(OH)_2 + Na_2SO_4 \longrightarrow BaSO_4 + 2NaOH$
d. $2H_2 + O_2 \longrightarrow 2H_2O$

10.127 Which of the following is a *correct* statement concerning an aqueous solution with a pH of 8.00?
a. The hydroxide ion concentration is greater than the hydronium ion concentration.
b. The hydronium ion concentration is 8.00 M.
c. The hydroxide ion concentration is 1.0×10^{-8} M.
d. The hydronium ion concentration is 8.0×10^{-8} M.

10.128 In which of the following salts would both the positive ion and the negative ion hydrolyze when the salt is dissolved in water?
a. NaCl   b. NaCN   c. NH_4Cl d. NH_4CN

10.129 Which of the following combinations of substances would produce a buffer?
a. A strong acid and a salt of the strong acid
b. A weak acid and a salt of the weak acid
c. A salt of a strong acid and a salt of a weak acid
d. A strong acid and a weak acid

10.130 Which of the following statements concerning electrolytes is *correct*?
a. All strong acids are strong electrolytes.
b. All salts are weak electrolytes.
c. Some molecular substances are strong electrolytes and others are weak electrolytes.
d. All bases are weak electrolytes.

10.131 Determining the concentration of an acid using an acid–base titration always involves which of the following?
a. Reacting a strong acid with a weak base
b. Reacting an acid of known concentration with an indicator
c. Reacting a 1.0 M acid solution with a 1.0 M base solution
d. An acid–base neutralization reaction

Nuclear Chemistry

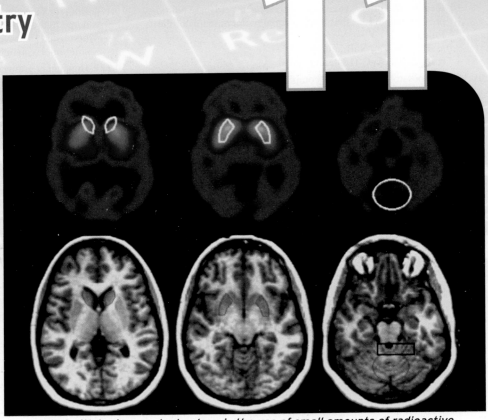

Associated with brain-scan technology is the use of small amounts of radioactive substances.

In this chapter we consider nuclear reactions. It is in the study of such reactions that we encounter the terms *radioactivity, nuclear power plants, nuclear weapons,* and *nuclear medicine.* The electricity produced by a nuclear power plant is generated through the use of heat energy obtained from nuclear reactions. In modern medicine, nuclear reactions are used in the diagnosis and treatment of numerous diseases. Despite some controversy concerning the use of nuclear reactions in weapons and power plants, it is important to remember that it is far more likely that your life will be extended by nuclear medicine than that your life will be taken by nuclear weapons.

 11.1 STABLE AND UNSTABLE NUCLIDES

A **nuclear reaction** *is a reaction in which changes occur in the nucleus of an atom.* Nuclear reactions are not considered to be ordinary chemical reactions. The governing principles for ordinary chemical reactions deal with the rearrangement of electrons; this rearrangement occurs as the result of electron transfer or electron sharing (Section 4.1). In nuclear reactions, it is nuclei rather than electron arrangements that undergo change.

Specific atoms, in nuclear chemistry discussions, are called *nuclides* rather than *isotopes.* The term *isotopes* refers to atoms of the same element that have different mass numbers (Section 3.3). The term *nuclide,* a much more general term, refers to atoms of either the same or different elements. A **nuclide** *is an atom with a specific atomic number and a specific mass number.* In practice, the designation *nuclides* is used to describe atoms of different elements and the designation *isotopes* is used to describe different types of atoms of the same element. The species $^{12}_{6}C$ and $^{16}_{8}O$ are *nuclides* of different elements, while the species $^{12}_{6}C$ and $^{13}_{6}C$ are *isotopes* of the same element.

To identify a nucleus or atom uniquely, both its atomic number and its mass number must be specified. Three different notation systems exist for doing this. Consider a nuclide of nitrogen that has seven protons and eight neutrons. This nuclide can be denoted as $^{15}_{7}N$ or nitrogen-15 or N-15. In the first notation, the superscript is the mass number and the subscript is the atomic number. In the second and third notations, the mass number is placed immediately after the name or chemical symbol of the element. All three types of notation will be used in this chapter. Note that all notations give the mass number.

Nuclides (atoms) may be divided into two types based on nuclear stability. Some nuclides have nuclei that are stable and others possess nuclei that are unstable. A **stable nuclide** *is a nuclide with a stable nucleus, a nucleus that does not readily undergo change.* Conversely, an **unstable nuclide** *is a nuclide with an unstable nucleus, a nucleus that spontaneously undergoes change.* The spontaneous change that unstable nuclei undergo involves emission of radiation from the nucleus, a process by which the unstable nucleus can become more stable. The radiation emitted from unstable nuclei is called *radioactivity.* **Radioactivity** *is the radiation spontaneously emitted from an unstable nucleus.* Nuclides that possess unstable nuclei are said to be radioactive. A **radioactive nuclide** *is a nuclide with an unstable nucleus from which radiation is spontaneously emitted.* The term *radioactive nuclide* is often shortened to simply *radionuclide.*

Naturally occurring radionuclides exist for 29 of the 88 elements that occur in nature (Section 1.7). Radionuclides are known for *all* 117 elements, however, even though they occur naturally for only the above-mentioned 29. This is because laboratory procedures have been developed by which scientists can convert nonradioactive nuclides (stable nucleus) into radioactive nuclides (unstable nucleus). Such procedures are considered in Section 11.5.

No simple rule exists for predicting whether a particular nuclide is radioactive. However, considering some observations about those nuclides that are *stable* is helpful in understanding why some nuclides are stable and others are not. Two generalizations are readily apparent from a study of the properties of naturally occurring stable nuclides.

1. *There is a correlation between nuclear stability and the total number of nucleons found in a nucleus.* All nuclei with 84 or more protons are unstable. The largest stable nucleus known is that of $^{209}_{83}Bi$, a nucleus that contains 209 nucleons. It thus appears that there is a limit to the number of nucleons that can be packed into a stable nucleus.

2. *There is a correlation between nuclear stability and neutron-to-proton ratio in a nucleus.* The number of neutrons found in a stable nucleus increases as the number of protons increases. For elements of low atomic number, neutron-to-proton ratios are very close to 1. For heavier elements, stable nuclei have higher neutron-to-proton ratios, and the ratio reaches approximately 1.5 for the heaviest stable elements. These observations suggest that neutrons are at least partially responsible for the stability of a nucleus. It should be remembered that like charges repel each other and that most nuclei contain many protons (with identical positive charges) squeezed together into a very small volume. As the number of protons increases, the forces of repulsion between protons sharply increase. Therefore, a greater number of neutrons is necessary to counteract the increasing repulsions. Finally, at element 84, the repulsive forces become sufficiently great that the nuclei are unstable regardless of the number of neutrons present.

 ## 11.2 THE NATURE OF RADIOACTIVE EMISSIONS

The fact that unstable nuclei spontaneously emit radiation was accidentally discovered by the French physicist and engineer Antoine Henri Becquerel (1852–1908) in 1896. While working on an experiment involving rocks that phosphoresce, Becquerel discovered that a particular uranium-containing rock gave off radiation. Soon other scientists, such as the French chemists Marie (1867–1934; see Figure 11.1) and Pierre (1859–1906) Curie and

The term *radioactive isotope* is sometimes used in place of *radioactive nuclide.* We will use *radioactive nuclide* in this text.

Figure 11.1 Marie Curie, one of the pioneers in the study of radioactivity, is the first person to have been awarded two Nobel Prizes for scientific work. In 1903, she, her husband Pierre, and Henri Becquerel were corecipients of the Nobel Prize in physics. In 1911, she received the Nobel Prize in chemistry. In 1934, Marie, now respectfully called Madame Curie, died of leukemia caused by overexposure to radiation.

Figure 11.2 The effect of an electromagnetic field on alpha, beta, and gamma radiation. Alpha and beta particles are deflected in opposite directions, whereas gamma radiation is not affected.

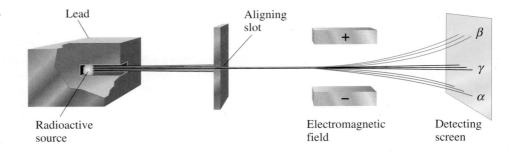

Lead

Aligning slot

Radioactive source

Electromagnetic field

Detecting screen

the British chemist Ernest Rutherford (1871–1937), began their own investigations into this strange phenomenon—a phenomenon that Marie Curie named radioactivity.

The first information concerning the nature of the radiation emanating from naturally radioactive materials was obtained by Rutherford in 1898–1899. Using an apparatus similar to that shown in Figure 11.2, he found that if radiation from uranium is passed between electrically charged plates, it is split into three components. This finding indicates the presence of three different types of emissions from naturally radioactive materials. A closer analysis of Rutherford's experiment reveals that one radiation component is positively charged (it is attracted to the negative plate); a second component is negatively charged (it is attracted to the positive plate); and the third component carries no charge (it is unaffected by either charged plate). Rutherford chose to call the three radiation components alpha rays (α rays)—the positive component; beta rays (β rays)—the negative component; and gamma rays (γ rays)—the uncharged component. (Alpha, beta, and gamma are the first three letters of the Greek alphabet.) We mention Rutherford's nomenclature system because it "stuck"; we still use these Greek-letter designations. Today, however, we speak of alpha particles, beta particles, and gamma rays. Further research has shown that both alpha particles and beta particles have specific masses and that gamma radiation has no mass—that is, it is a form of energy.

The complete characterization of the three types of natural radioactive emissions required many years. Early work in the field was hampered by the fact that many of the details concerning atomic structure were not yet known. For example, the neutron was not discovered until 1932, 36 years after the discovery of radioactivity. In terms of modern-day scientific knowledge, Rutherford's three types of "radiation" are characterized as follows:

An **alpha particle** *is a particle in which two protons and two neutrons are present that is emitted by certain radioactive nuclei.* The notation used to represent an alpha particle is $^4_2\alpha$. The numerical subscript indicates that the charge on the particle is $+2$ (from the two protons). The numerical superscript indicates a mass of 4 amu. Alpha particles are identical to the nuclei of helium-4 ($^4_2$He) atoms; because of this, an alternative designation for an alpha particle is $^4_2$He.

A **beta particle** *is a particle whose charge and mass are identical to those of an electron that is emitted by certain radioactive nuclei.* Beta particles are not extranuclear electrons; they are particles that have been produced inside the nucleus and then ejected. We will discuss this formation process in Section 11.3. The symbol used to represent a beta particle is $^0_{-1}\beta$. The numerical subscript indicates that the charge on the beta particle is -1; it is the same as that of an electron. The use of the superscript zero for the mass of a beta particle should be interpreted as meaning not that a beta particle has no mass but, rather, that the mass is very close to zero amu. The actual mass of a beta particle is 0.00055 amu.

A **gamma ray** *is a form of high-energy radiation without mass or charge that is emitted by certain radioactive nuclei.* Gamma rays are similar to X rays except that gamma rays have higher energy. The symbol for gamma rays is $^0_0\gamma$.

There are two key concepts for understanding the phenomenon of radioactivity: (1) certain nuclides possess unstable nuclei, and (2) nuclides with unstable nuclei spontaneously emit energy (radiation).

Alpha, beta, and gamma radiation are designated with the notations $^4_2\alpha$, $^0_{-1}\beta$, and $^0_0\gamma$, respectively.

EQUATIONS FOR RADIOACTIVE DECAY

Alpha, beta, and gamma emissions come from the nucleus of an atom. These spontaneous emissions alter nuclei; obviously, if a nucleus loses an alpha particle (two protons and two neutrons), it will not be the same as it was before the departure of the particle. In the case of alpha and beta emissions, the nuclear alteration causes the identity of atoms to change, forming a new element. Thus nuclear reactions differ dramatically from ordinary chemical reactions, where the identities of the elements are always maintained. **Radioactive decay** *is the process whereby a radionuclide is transformed into a nuclide of another element as a result of the emission of radiation from its nucleus.* The terms *parent nuclide* and *daughter nuclide* are often used in descriptions of radioactive decay processes. A **parent nuclide** *is the nuclide that undergoes decay in a radioactive decay process.* A **daughter nuclide** *is the nuclide that is produced in a radioactive decay process.*

> In a *chemical* reaction, element identity is maintained. Atoms are rearranged to form new substances involving the same elements. In a *nuclear* reaction, element identity is not maintained. An element changes into a different element.

Alpha Particle Decay

Alpha particle decay *is the radioactive decay process in which an alpha particle is emitted from an unstable nucleus.* It always results in the formation of a nuclide of a different element. The product nucleus has an atomic number that is 2 less than that of the original nucleus and a mass number that is 4 less than that of the original nucleus. We can represent alpha particle decay in general terms by the equation

> Loss of an alpha particle from an unstable nucleus results in (1) a decrease of 4 units in the mass number (A) and (2) a decrease of 2 units in the atomic number (Z).

$$^{A}_{Z}\text{X} \longrightarrow {}^{4}_{2}\alpha + {}^{A-4}_{Z-2}\text{Y}$$

where X is the chemical symbol for the nucleus of the original element undergoing decay and Y is the chemical symbol of the nucleus formed as a result of the decay.

Specific radioactive decay processes are represented using *nuclear equations.* A **nuclear equation** *is an equation in which the chemical symbols present represent atomic nuclei rather than atoms.* Both $^{211}_{83}\text{Bi}$ and $^{238}_{92}\text{U}$ are radionuclides that undergo alpha particle decay. The nuclear equations for these two decay processes are

$$^{211}_{83}\text{Bi} \longrightarrow {}^{4}_{2}\alpha + {}^{207}_{81}\text{Tl}$$
$$^{238}_{92}\text{U} \longrightarrow {}^{4}_{2}\alpha + {}^{234}_{90}\text{Th}$$

In the first equation, $^{211}_{83}\text{Bi}$ is the parent nuclide and $^{207}_{81}\text{Tl}$ is the daughter nuclide; in the second equation, $^{238}_{92}\text{U}$ is the parent nuclide and $^{234}_{90}\text{Th}$ is the daughter nuclide.

Nuclear equations differ from ordinary chemical equations in three important ways:

> Note that the symbols in nuclear equations stand for *nuclei* rather than atoms. We do not worry about electrons when writing nuclear equations.

1. The symbols in nuclear equations stand for nuclei rather than atoms. (We do not worry about electrons when writing nuclear equations.)
2. Mass numbers and atomic numbers (nuclear charge) are always specifically included in nuclear equations.
3. The elemental symbols on both sides of the equation frequently are not the same in nuclear equations.

The procedures for balancing nuclear equations are different from those used for ordinary chemical equations. A **balanced nuclear equation** *is a nuclear equation in which the sum of the subscripts (atomic numbers or particle charges) on both sides of the equation are equal, and the sum of the superscripts (mass numbers) on both sides of the equation are equal.* Both of the preceding nuclear equations are balanced. In the alpha decay of $^{211}_{83}\text{Bi}$, the subscripts on both sides total 83, and the superscripts total 211. For the alpha decay of $^{238}_{92}\text{U}$, the subscripts total 92 on both sides, and the superscripts total 238.

> The rules for balancing nuclear equations are
>
> 1. The sum of the subscripts must be the same on both sides of the equation.
> 2. The sum of the superscripts must be the same on both sides of the equation.

Beta Particle Decay

Beta particle decay *is the radioactive decay process in which a beta particle is emitted from an unstable nucleus.* Beta particle decay, like alpha particle decay, always produces a nuclide of a different element. The mass number of the new nuclide is the same as that of

Loss of a beta particle from an unstable nucleus results in (1) no change in the mass number (A) and (2) an increase of 1 unit in the atomic number (Z).

the parent nuclide. However, the atomic number has increased by 1 unit. The general equation for beta decay is

$$_Z^A X \longrightarrow {}_{-1}^{0}\beta + {}_{Z+1}^{A}Y$$

Specific examples of beta particle decay are

$$_4^{10}Be \longrightarrow {}_{-1}^{0}\beta + {}_5^{10}B$$

$$_{90}^{234}Th \longrightarrow {}_{-1}^{0}\beta + {}_{91}^{234}Pa$$

Both of these nuclear equations are balanced; superscripts and subscripts add to the same sums on both sides of the equation.

At this point in the discussion, you may be wondering how a nucleus, which is composed only of neutrons and protons, ejects a negative particle (beta particle) when no such particle is present in the nucleus. Explained simply, a neutron in the nucleus is transformed into a proton and a beta particle through a complex series of steps; that is,

$$\text{Neutron} \longrightarrow \text{proton} + \text{beta particle}$$

$$_0^1 n \longrightarrow {}_1^1 p + {}_{-1}^{0}\beta$$

Once it is formed within the nucleus, the beta particle is ejected with a high velocity. Note the symbols used to denote a neutron ($_0^1 n$; no charge and a mass of 1 amu) and a proton ($_1^1 p$; a +1 charge and a mass of 1 amu).

Gamma Ray Emission

Gamma rays are to nuclear reactions what heat is to ordinary chemical reactions.

Gamma ray emission *is the radioactive decay process in which a gamma ray is emitted from an unstable nucleus.* For naturally occurring radionuclides, gamma ray emission always takes place in conjunction with an alpha or a beta decay process; it never occurs independently. These gamma rays are often not included in the nuclear equation because they do not affect the balancing of the equation or the identity of the daughter nuclide. This can be seen from the following two nuclear equations.

Among *synthetically* produced radionuclides (Section 11.5), pure "gamma emitters," radionuclides that give off gamma rays but no alpha or beta particles, occur. These radionuclides are important in diagnostic nuclear medicine (Section 11.11). Pure "gamma emitters" are not found among naturally occurring radionuclides.

$$_{88}^{226}Ra \longrightarrow {}_{86}^{222}Rn + {}_2^4\alpha + {}_0^0\gamma$$

Balanced nuclear equation with gamma radiation included

$$_{88}^{226}Ra \longrightarrow {}_{86}^{222}Rn + {}_2^4\alpha$$

Balanced nuclear equation with gamma radiation omitted

The fact that gamma rays are often left out of balanced nuclear equations should not be interpreted to mean that such rays are not important in nuclear chemistry. On the contrary, gamma rays are more important than alpha and beta particles when the effects of external radiation exposure on living organisms are considered (Section 11.8).

● **E X A M P L E 11.1**

Writing Balanced Nuclear Equations, Given the Parent Nuclide and Its Mode of Decay

▶ Write a balanced nuclear equation for the decay of each of the following radioactive nuclides. The mode of decay is indicated in parentheses.

a. $_{31}^{70}Ga$ (beta emission)

b. $_{60}^{144}Nd$ (alpha emission)

c. $_{100}^{248}Fm$ (alpha emission)

d. $_{47}^{113}Ag$ (beta emission)

Solution

In each case, the atomic and mass numbers of the daughter nucleus are obtained by writing the symbols of the parent nucleus and the particle emitted by the nucleus (alpha or beta). Then the equation is balanced.

a. Let X represent the daughter nuclide, the product of the radioactive decay. Then

$$_{31}^{70}Ga \longrightarrow {}_{-1}^{0}\beta + X$$

The sum of the superscripts on both sides of the equation must be equal, so the superscript for X must be 70. In order for the sum of the subscripts on both sides of the equation to be equal, the subscript for X must be 32. Then $31 = (-1) + (32)$. As soon as we determine the subscript of X, we can obtain the identity of X by looking at a periodic table. The element with an atomic number of 32 is Ge (germanium). Therefore,

$$^{70}_{31}\text{Ga} \longrightarrow \, ^{\,0}_{-1}\beta + ^{70}_{32}\text{Ge}$$

b. Letting X represent the daughter nuclide, we have, for the alpha decay of $^{144}_{60}\text{Nd}$,

$$^{144}_{60}\text{Nd} \longrightarrow \, ^{4}_{2}\alpha + \text{X}$$

We balance the equation by making the superscripts on each side of the equation total 144 and the subscripts total 60. We get

$$^{144}_{60}\text{Nd} \longrightarrow \, ^{4}_{2}\alpha + ^{140}_{58}\text{Ce}$$

c. Similarly, we write

$$^{248}_{100}\text{Fm} \longrightarrow \, ^{4}_{2}\alpha + \text{X}$$

Balancing superscripts and subscripts, we get

$$^{248}_{100}\text{Fm} \longrightarrow \, ^{4}_{2}\alpha + ^{244}_{98}\text{Cf}$$

In alpha emission, the atomic number of the daughter nuclide always decreases by 2, and the mass number of the daughter nuclide always decreases by 4.

d. Finally, we write

$$^{113}_{47}\text{Ag} \longrightarrow \, ^{\,0}_{-1}\beta + \text{X}$$

In beta emission, the atomic number of the daughter nuclide always increases by 1, and the mass number does not change from that of the parent. The balancing procedure gives us the result

$$^{113}_{47}\text{Ag} \longrightarrow \, ^{\,0}_{-1}\beta + ^{113}_{48}\text{Cd}$$

Practice Exercise 11.1

Write a balanced nuclear equation for the decay of the following radioactive nuclides. The mode of decay is indicated in parentheses.

a. $^{245}_{97}\text{Bk}$ (alpha emission)   **b.** $^{89}_{38}\text{Sr}$ (beta emission)

c. $^{230}_{90}\text{Th}$ (alpha emission)   **d.** $^{40}_{19}\text{K}$ (beta emission)

Answers: a. $^{245}_{97}\text{Bk} \rightarrow \, ^{4}_{2}\alpha + ^{241}_{95}\text{Am}$; **b.** $^{89}_{38}\text{Sr} \rightarrow \, ^{\,0}_{-1}\beta + ^{89}_{39}\text{Y}$;
c. $^{230}_{90}\text{Th} \rightarrow \, ^{4}_{2}\alpha + ^{226}_{88}\text{Ra}$; **d.** $^{40}_{19}\text{K} \rightarrow \, ^{\,0}_{-1}\beta + ^{40}_{20}\text{Ca}$

11.4 RATE OF RADIOACTIVE DECAY

Radioactive nuclides do not all decay at the same rate. Some decay very rapidly; others undergo disintegration at extremely low rates. This indicates that radionuclides are not all equally unstable. The greater the decay rate, the lower the stability of the nuclide.

The greater the decay rate for a radionuclide, the shorter its half-life.

The concept of *half-life* is used to express nuclear stability quantitatively. A **half-life** ($t_{1/2}$) *is the time required for one-half of a given quantity of a radioactive substance to undergo decay.* For example, if a radionuclide's half-life is 12 days and you have a 4.00-g sample of it, then after 12 days (1 half-life), only 2.00 g of the sample (one-half of the original amount) will remain undecayed; the other half will have decayed into some other substance. Similarly, during the next half-life, one-half of the 2.00 g remaining will decay, leaving one-fourth of the original atoms (1.00 g) unchanged. After three half-lives, one-eighth ($1/2 \times 1/2 \times 1/2$) of the original sample will remain undecayed. Figure 11.3 illustrates the radioactive decay curve for a radionuclide.

Most radionuclides used in diagnostic medicine have short half-lives. This limits to a short time interval the exposure of the human body to radiation.

Figure 11.3 Decay of 80.0 mg of ^{131}I, which has a half-life of 8.0 days. After each half-life period, the quantity of material present at the beginning of the period is reduced by half.

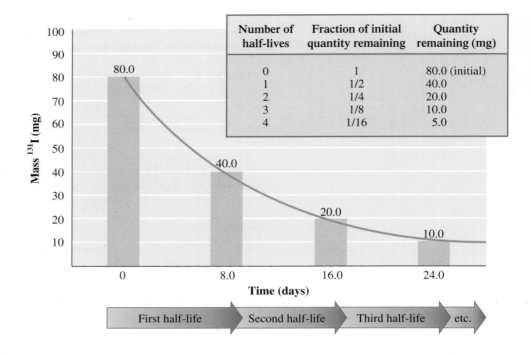

Number of half-lives	Fraction of initial quantity remaining	Quantity remaining (mg)
0	1	80.0 (initial)
1	1/2	40.0
2	1/4	20.0
3	1/8	10.0
4	1/16	5.0

The half-life for a radionuclide is independent of external conditions such as temperature, pressure, and state of chemical combination.

TABLE 11.1

Range of Half-lives Found for Naturally Occurring Radionuclides

Element	Half-life ($t_{1/2}$)
vanadium-50	6×10^{15} yr
platinum-190	6.9×10^{11} yr
uranium-238	4.5×10^{9} yr
uranium-235	7.1×10^{8} yr
thorium-230	7.5×10^{4} yr
lead-210	22 yr
bismuth-214	19.7 min
polonium-212	3.0×10^{-7} sec

There is a wide range of half-lives for radionuclides. Half-lives as long as billions of years and as short as a fraction of a second have been determined (Table 11.1). Most naturally occurring radionuclides have long half-lives. However, some radionuclides with short half-lives are also found in nature. Naturally occurring mechanisms exist for the continual production of the short-lived species.

The decay rate (half-life) of a radionuclide is constant. It is independent of physical conditions such as temperature, pressure, and state of chemical combination. It depends only on the identity of the radionuclide. For example, radioactive sodium-24 decays at the same rate whether it is incorporated in NaCl, NaBr, Na_2SO_2, or $NaC_2H_3O_2$. If a nuclide is radioactive, nothing will stop it from decaying and nothing will increase or decrease its decay rate.

Calculations involving amounts of radioactive material decayed, amounts remaining undecayed, and time elapsed can be carried out by using the following equation:

$$\left(\begin{matrix} \text{Amount of radionuclide} \\ \text{undecayed after } n \text{ half-lives} \end{matrix} \right) = \left(\begin{matrix} \text{original amount} \\ \text{of radionuclide} \end{matrix} \right) \times \left(\frac{1}{2^n} \right)$$

The Chemistry at a Glance feature on page 299 summarizes important concepts about radioactive decay that we have considered so far in this chapter.

● **E X A M P L E 11.2**

Using Half-life to Calculate the Amount of Radioisotope That Remains Undecayed After a Certain Time

▶ Iodine-131 is a radionuclide that is frequently used in nuclear medicine. Among other things, it is used to detect fluid buildup in the brain. The half-life of iodine-131 is 8.0 days. How much, in grams, of a 0.16-g sample of iodine-131 will remain undecayed after a period of 32 days?

Solution

First, we must determine the number of half-lives that have elapsed.

$$32 \text{ days} \times \left(\frac{1 \text{ half-life}}{8.0 \text{ days}} \right) = 4 \text{ half-lives}$$

Knowing the number of elapsed half-lives and the original amount of radioactive iodine present, we can use the equation

$$\left(\begin{matrix}\text{Amount of radionuclide}\\\text{undecayed after } n \text{ half-lives}\end{matrix}\right) = \left(\begin{matrix}\text{original amount}\\\text{of radionuclide}\end{matrix}\right) \times \left(\frac{1}{2^n}\right)$$

$$= 0.16 \text{ g} \times \frac{1}{2^4} \quad \longleftarrow \quad \text{4 half-lives}$$

$$= 0.16 \text{ g} \times \frac{1}{16} = 0.010 \text{ g}$$

Constructing a tabular summary of the amount of sample remaining after each of the elapsed half-lives yields

Half-lives	0	1	2	3	4
Number of days	0	8	16	24	32
Amount remaining	0.16 g	0.080 g	0.040 g	0.020 g	0.010 g

Practice Exercise 11.2

The half-life of cobalt-60 is 5.3 years. If 2.0 g of cobalt-60 is allowed to decay for a period of 15.9 years, how many grams of cobalt-60 remain?

Answer: 0.25 g

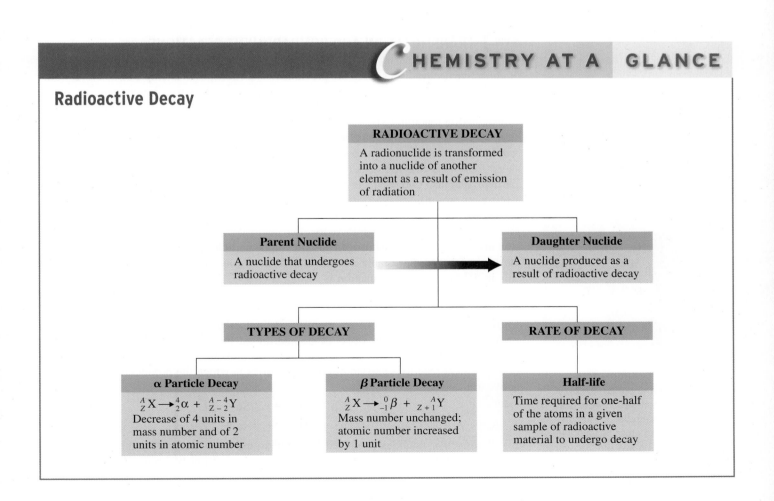

CHEMISTRY AT A GLANCE

Radioactive Decay

RADIOACTIVE DECAY

A radionuclide is transformed into a nuclide of another element as a result of emission of radiation

Parent Nuclide

A nuclide that undergoes radioactive decay

Daughter Nuclide

A nuclide produced as a result of radioactive decay

TYPES OF DECAY

RATE OF DECAY

α Particle Decay

$$^A_Z X \longrightarrow \, ^4_2\alpha + \, ^{A-4}_{Z-2}Y$$

Decrease of 4 units in mass number and of 2 units in atomic number

β Particle Decay

$$^A_Z X \longrightarrow \, ^{\,0}_{-1}\beta + \, ^A_{Z+1}Y$$

Mass number unchanged; atomic number increased by 1 unit

Half-life

Time required for one-half of the atoms in a given sample of radioactive material to undergo decay

EXAMPLE 11.3

Using Half-life to Calculate the Time Needed to Reduce Radioactivity to a Specific Level

Strontium-90 is a nuclide found in radioactive fallout from nuclear weapon explosions. Its half-life is 28.0 years. How long, in years, will it take for 94% (15/16) of the strontium-90 atoms present in a sample of material to undergo decay?

Solution

If 15/16 of the sample has decayed, then 1/16 of the sample remains undecayed. In terms of $1/2^n$, 1/16 is equal to $1/2^4$; that is,

$$\frac{1}{2} \times \frac{1}{2} \times \frac{1}{2} \times \frac{1}{2} = \frac{1}{2^4} = \frac{1}{16}$$

Thus 4 half-lives have elapsed in reducing the amount of strontium-90 to 1/16 of its original amount.

The half-life of strontium-90 is 28 years, so the total time elapsed will be

$$4 \text{ half-lives} \times \left(\frac{28.0 \text{ years}}{1 \text{ half-life}}\right) = 112 \text{ years}$$

Practice Exercise 11.3

Iodine-135 is a nuclide found in radioactive fallout from nuclear weapon explosions. Its half-life is 6.70 hr. How long, in hours, will it take for 75.0% (three-quarters) of the iodine-135 atoms in a "fallout" sample to undergo decay?

Answer: 13.4 hr

11.5 TRANSMUTATION AND BOMBARDMENT REACTIONS

Radioactive decay, discussed in the previous two sections, is an example of a natural transmutation reaction. A **transmutation reaction** *is a nuclear reaction in which a nuclide of one element is changed into a nuclide of another element.* It is also possible to cause a transmutation reaction to occur in a laboratory setting by means of a bombardment reaction. A **bombardment reaction** *is a nuclear reaction brought about by bombarding stable nuclei with small particles traveling at very high speeds.* In bombardment reactions, there are always two reactants (the target nuclide and the small, high-energy bombarding particle) and also two products (the daughter nuclide and another small particle such as a neutron or proton).

The first successful bombardment reaction was carried out in 1919 by Ernest Rutherford (see Figure 11.4) 25 years after the discovery of radioactive decay. The reaction involved bombarding nitrogen gas with alpha particles from a natural source (radium). In this process, a new stable nuclide was formed: oxygen-17. The nuclear equation for this initial bombardment reaction is

$$^{14}_{7}\text{N} + {}^{4}_{2}\alpha \longrightarrow {}^{17}_{8}\text{O} + {}^{1}_{1}\text{p}$$

Further research carried out by many investigators has shown that numerous nuclei undergo change under the stress of bombardment by small, high-energy particles. In most cases, the new nuclide that is produced is radioactive (unstable). Two examples of bombardment reactions carried out in laboratories in which the product nuclide is radioactive are

$$^{44}_{20}\text{Ca} + {}^{1}_{1}\text{p} \longrightarrow {}^{44}_{21}\text{Sc} + {}^{1}_{0}\text{n}$$
$$^{23}_{11}\text{Na} + {}^{2}_{1}\text{H} \longrightarrow {}^{21}_{10}\text{Ne} + {}^{4}_{2}\alpha$$

Radioactive nuclides produced by bombardment reactions obey the same laws as naturally occurring radionuclides. In many cases, the previously discussed alpha and beta modes of decay occur (Section 11.3).

Figure 11.4 Ernest Rutherford (1871–1937), the first person to carry out a bombardment reaction, was a "world-class" researcher. Earlier, he discovered that an atom has a nucleus (Section 3.1), and he was the discoverer of the alpha and beta radiation associated with radioactivity.

TABLE 11.2
The Transuranium Elements

Name	Symbol	Atomic Number	Mass Number of Most Stable Nuclide	Half-life of Most Stable Nuclide	Discovery Year for First Isotope
neptunium	Np	93	237	2.14×10^6 yr	1940
plutonium	Pu	94	244	7.6×10^7 yr	1940
americium	Am	95	243	8.0×10^3 yr	1944
curium	Cm	96	247	1.6×10^7 yr	1944
berkelium	Bk	97	247	1400 yr	1950
californium	Cf	98	251	900 yr	1950
einsteinium	Es	99	252	472 days	1952
fermium	Fm	100	257	100 days	1953
mendelevium	Md	101	258	52 days	1955
nobelium	No	102	259	58 min	1958
lawrencium	Lr	103	262	3.6 hr	1961
rutherfordium	Rf	104	263	10 min	1969
dubnium	Db	105	262	34 sec	1970
seaborgium	Sg	106	266	21 sec	1974
bohrium	Bh	107	267	17 sec	1980
hassium	Hs	108	277	11 min	1984
meitnerium	Mt	109	276	0.72 sec	1982
darmstadtium	Ds	110	281	11.1 sec	1994
roentgenium	Rg	111	280	3.6 sec	1994
element 112	—	112	285	34 sec	1996
element 113	—	113	284	0.48 sec	2004
element 114	—	114	289	2.6 sec	1999
element 115	—	115	288	87 msec	2004
element 116	—	116	292	0.61 msec	2006
element 118	—	118	294	0.89 msec	2006

Production of the small, *high-energy* bombarding particles needed to effect a bombardment reaction requires use of a cyclotron or a linear accelerator (both very expensive pieces of equipment). Both use magnetic fields to accelerate charged particles to velocities at which the energy is sufficient to allow the particle to penetrate the nucleus and induce a nuclear reaction.

All nuclides of all elements beyond bismuth ($Z = 83$) in the periodic table are radioactive.

Synthetic Elements

Over 2000 bombardment-produced radionuclides that do not occur naturally are now known. This number is seven times greater than the number of naturally occurring nuclides (Section 3.3). In this total is at least one radionuclide of every naturally occurring element. In addition, nuclides of 29 elements that do not occur in nature have been produced in small quantities as the result of bombardment reactions. Four of these "synthetic" elements, produced between 1937 and 1941, filled gaps in the periodic table for which no naturally occurring element had been found. These four elements are technetium (Tc, element 43), an element with numerous uses in nuclear medicine (Section 11.11); promethium (Pm, element 61); astatine (At, element 85); and francium (Fr, element 87). The remainder of the "synthetic" elements, elements 93 to 116 and 118, are called the *transuranium elements* because they occur immediately following uranium in the periodic table. (Uranium is the naturally occurring element with the highest atomic number.) All nuclides of all of the transuranium elements are radioactive. Table 11.2 gives information about the stability of the transuranium elements. Note the extremely short half-lives of the more recently produced elements.

Most radioisotopes used in the field of medicine are "synthetic" radionuclides. For example, the synthetic radionuclides cobalt-60, yttrium-90, iodine-131, and gold-198 are used in radiotherapy treatments for cancer. Section 11.11 provides more information about the medical uses for radionuclides. The synthetic element americium (element 95) is present in nearly all standard smoke detectors (see Figure 11.5).

In a decay series such as the one involving uranium-238, gamma rays are emitted at each step (even though they are not shown) in addition to the alpha or beta particle. Such gamma rays are very important in the effects of radiation exposure on health (Section 11.8).

Figure 11.6 In the $^{238}_{92}$U decay series, each nuclide except $^{206}_{82}$Pb (the stable end product) is unstable; the successive transformations continue until this stable product is formed.

11.6 RADIOACTIVE DECAY SERIES

Radioactive nuclides with high atomic numbers attain nuclear stability through a series of decay steps. When such nuclides decay, they produce daughter nuclei that are also radioactive. These daughter nuclei in turn decay to a third radioactive product, and so on. Eventually, a stable nucleus is produced. Such a sequence of decay products is called a radioactive decay series. A **radioactive decay series** *is a series of radioactive decay processes beginning with a long-lived radionuclide and ending with a stable nuclide of lower atomic number.*

Uranium-238, the most abundant isotope of uranium (99.2%), is the beginning nuclide for an important naturally occurring decay series. As shown in Figure 11.6, 14 steps are needed for uranium-238 to reach lead-206, its stable end product. Note from Figure 11.6 that both alpha and beta emissions are part of the decay sequence and that there is no simple pattern as to which is emitted when.

In the uranium-238 decay series, all the intermediate products are solids except one. Radon-222 is a gas at normal temperatures and is therefore a very mobile species. Its presence has been detected in both aqueous and atmospheric environments. Exposure to radon-222 constitutes the major source of radiation exposure for the average American (Section 11.10).

11.7 CHEMICAL EFFECTS OF RADIATION

The very energetic radiations produced from radioactive decay travel outward from their nuclear sources into the material surrounding the radioactive source. There, they interact with the atoms and molecules of the material, which dissipates their excess energy. Numerous interactions (collisions) between atoms or molecules and a "particle" of radiation are required before the energy of the radiation is reduced to the level of surrounding materials. At this point the radiation is "harmless." Let us consider in closer detail the interactions that do occur between radiation and atoms or molecules.

It is the electrons of molecules that are most directly affected by radiation, whether that radiation is from a radioactive material or some other source. In general, two things can happen to an electron subjected to radiation: excitation or ionization. *Excitation* occurs when radiation, through energy release, excites an electron from an occupied orbital into an empty, higher-energy orbital. *Ionization* occurs when the radiation carries enough energy to remove an electron from an atom or molecule.

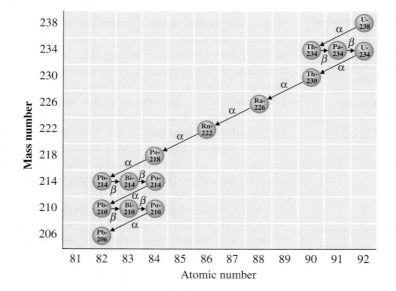

Tobacco Radioactivity and the Uranium-238 Decay Series

The adverse health effects from cigarette smoking account for 440,000 deaths, or nearly 1 in 5 deaths, each year in the United States. Approximately one-third of these deaths are from lung cancer and an almost equal number from cardiovascular disease. The risk of dying from lung cancer is 22 times higher among men who smoke cigarettes compared with nonsmokers. Cigarette smokers are 2–4 times more likely to develop coronary heart disease than nonsmokers. Cigarette smoking approximately doubles a person's risk of stroke. One's life is shortened 14 minutes for every cigarette smoked. A 30- to 40-year cigarette smoker who smokes two packs of cigarettes per day loses an estimated 8 years of life.

The link between cigarette smoke and cancer is definitely established. The causative agents for the cancer involve many of the more than 2000 compounds identified in cigarette tar. And radioactivity has also been implicated.

The link between radioactivity and tobacco involves the following sequence of events. The soil in which tobacco is grown is heavily treated with phosphate fertilizers. The source for phosphate fertilizers is ultimately phosphate rock. Nearly all phosphate rock contains small amounts of uranium and its decay products as impurities. Hence small amounts of radioactive nuclides are present in fertilized tobacco-growing soil (as well as in many other crop soils).

Radon-222, one of the intermediate products from the uranium-238 decay series (Section 11.6), is present at relatively high concentrations in soil gas and in the surface air layer under the vegetation canopy provided by a field of growing tobacco plants. The decay products from radon-222 (see Figure 11.6), which include the solids polonium-218 and bismuth-214, often become firmly attached to the surface and interior of tobacco leaves. These short-lived isotopes decay further to lead-210, which has a half-life of 20.4 years. Gradually, lead-210 levels build in tobacco leaves.

During the burning of a cigarette, small particles (particulates) are produced in addition to gaseous products. Many of these particulates, some of which have a lead-210 content, are inhaled and deposited in the respiratory tract of the smoker and are eventually transported to storage sites in the liver, spleen, and bone marrow. With time (years of smoking), lead-210 concentrations (and decay products) continue to build within the body. The results are constant added exposure of organs and bone marrow to alpha and beta particles and an increased probability of cancer development in the smoker as compared with the nonsmoker.

Based on its effects on electrons, radiation of various types is classified into the categories *nonionizing radiation* and *ionizing radiation*. **Nonionizing radiation** *is radiation with insufficient energy to remove an electron from an atom or molecule.* Radio waves, microwaves, infrared light, and visible light are forms of nonionizing radiation. The first three possess insufficient energy to excite electrons to higher energy states. Such radiation can, however, cause increased movement, vibration, and rotation of molecules with a resultant increase in the temperature of a material. Visible light, the fourth type of nonionizing radiation, possesses sufficient energy to excite electrons to higher energy states. Electrons that undergo such excitation return, with time, to their normal states.

Ionizing radiation *is radiation with sufficient energy to remove an electron from an atom or molecule.* Most radiation associated with radioactive decay processes is ionizing radiation. Cosmic rays, X rays, and ultraviolet light are also forms of ionizing radiation. (Cosmic rays are energetic particles coming from interstellar space; they are made up primarily of protons, alpha particles, and beta particles.)

The result of the interaction of ionizing radiation with matter is *ion pair formation*. In ion pair formation, the incoming radiation transfers sufficient energy into a molecule to knock an electron out of it, converting the molecule into a positive ion (see Figure 11.7); that is, ionization occurs and an ion pair is formed. An **ion pair** *is the electron and positive ion that are produced during an interaction between an atom or molecule and ionizing radiation.* This ionization process is not the normal "voluntary" transfer of electrons that occurs during ionic compound formation (Section 4.6) but rather the involuntary, nonchemical removal of an electron from a molecule to form an ion. Many ion pairs are produced by a single "particle" of radiation because such a particle must undergo many collisions before its energy is reduced to the level of surrounding material. The electrons ejected from atoms or molecules during ion pair formation frequently have enough energy to bombard neighboring molecules and cause additional ionizations.

Figure 11.7 Ion pair formation. When radiation interacts with an atom, electrons are often knocked away from the atom. The atoms that lose electrons become ions. An ion so produced and its "free electron" constitute an ion pair.

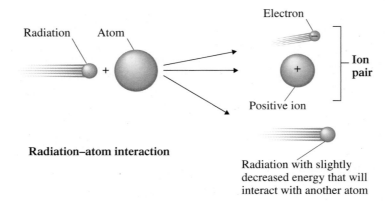

Free-radical formation, either directly or indirectly, usually accompanies ion pair formation. A **free radical** *is an atom, molecule, or ion that contains an unpaired electron.* The presence of the unpaired electron in a free radical usually causes it to be a very reactive species. (Recall from Section 5.2 that electrons normally occur in pairs in molecules.) Free radicals can rapidly react with other chemical species nearby, often precipitating a series of totally undesirable chemical reactions inside a living cell. Once formed, a free radical can combine with another free radical to form a molecule in which electrons are paired, or it can react with another molecule to produce a new free radical. The latter is a common occurrence. It is such production of new free radicals in a "chainlike" manner that causes major problems within a living cell. Such free-radical production is what makes the injury from radiation exposure far greater in magnitude than that expected merely on the basis of the energy of the incoming radiation.

Because water is the most abundant molecule in living organisms, the effects of ionizing radiation on water is of prime importance in assessing the effects of radiation exposure on health and life. The first step in the interaction of ionizing radiation with water is usually ion pair formation.

$$H_2O + \text{radiation} \longrightarrow \underbrace{H_2O^+ + e^-}_{\text{Ion pair}}$$

The H_2O^+ ion formed in this interaction should not be confused with the H_3O^+ ion produced when acids dissolve in water (Section 10.2). The H_2O^+ ion is a free radical and is extremely reactive; the H_3O^+ ion is not a free radical. The Lewis structure for the water free radical (H_2O^+) is

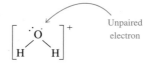

Just as the water free radical (H_2O^+) is not to be confused with the acidic species H_3O^+, the hydroxyl free radical (OH) is not to be confused with the hydroxide ion (OH^-); the latter is the basic species in aqueous solution. The difference between these two species is one electron, as comparing their Lewis structures illustrates:

$: \ddot{O} - H$ $\left[: \ddot{O} - H \right]^-$

Hydroxyl free   Hydroxide ion
radical

The highly reactive ·H_2O^+, a species not normally present in biochemical systems, can then react with another water molecule, causing further free-radical formation.

$$·H_2O^+ + H_2O \longrightarrow H_3O^+ + ·OH$$
Free   New free
radical   radical

The ·OH free radicals produced in this manner then interact with many different biomolecules to produce new free radicals, which in turn can react further. The result often devastatingly upsets cellular activity.

11.8 BIOCHEMICAL EFFECTS OF RADIATION

The three types of naturally occurring radioactive emissions—alpha particles, beta particles, and gamma rays—differ in their ability to penetrate matter and cause ionization. Consequently, the extent of the biochemical effects of radiation depends on the type of radiation involved.

Alpha Particle Effects

Alpha particles are the most massive and also the slowest particles involved in natural radioactive decay processes. Maximum alpha particle velocities are on the order of one-tenth of the speed of light. For a given alpha-emitting radionuclide, all alpha particles have the same energy; different alpha-emitting radionuclides, however, produce alpha particles of differing energies.

The speed of light, 3.0×10^8 m/sec (186,000 miles/sec), is the maximum limit of velocity. Objects cannot travel faster than the speed of light.

Because of their "slowness," alpha particles have low penetrating power and cannot penetrate the body's outer layers of skin. The major damage from alpha radiation occurs when alpha-emitting radionuclides are ingested—for example, in contaminated food. There are no protective layers of skin within the body.

Beta Particle Effects

Unlike alpha particles, which are all emitted with the same discrete energy from a given radionuclide, beta particles emerge from a beta-emitting substance with a continuous range of energies up to a specific limit that is characteristic of the particular radionuclide. Maximum beta particle velocities are on the order of nine-tenths of the speed of light.

With their greater velocity, beta particles can penetrate much deeper than alpha particles and can cause severe skin burns if their source remains in contact with the skin for an appreciable time. Because of their much smaller size, they do not ionize molecules (Section 11.7) as readily as alpha particles do. An alpha particle is approximately 8000 times heavier than a beta particle. A typical alpha particle travels about 6 cm in air and produces 40,000 ion pairs, and a typical beta particle travels 1000 cm in air and produces about 2000 ion pairs. Internal exposure to beta radiation is as serious as internal alpha exposure.

Gamma Radiation Effects

Gamma radiation is released at a velocity equal to the speed of light. Gamma rays readily penetrate deeply into organs, bone, and tissue.

X rays and gamma rays are similar except that X rays are of lower energy. X rays used in diagnostic medicine have energies approximately 10% of that of gamma rays.

Figure 11.8 contrasts the abilities of alpha, beta, and gamma radiations to penetrate paper, aluminum foil, and a thin layer of a lead–concrete mixture.

The minimum radiation dosage that causes human injury is unknown. However, the effects of larger doses have been studied (Table 11.3). As you can see, very serious damage or death can result from large doses of ionizing radiation.

Figure 11.8 Alpha, beta, and gamma radiation differ in penetrating ability.

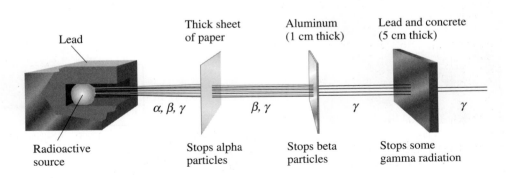

Lead

Thick sheet of paper

Aluminum (1 cm thick)

Lead and concrete (5 cm thick)

α, β, γ β, γ γ γ

Radioactive source

Stops alpha particles

Stops beta particles

Stops some gamma radiation

TABLE 11.3
The Effects of Short-Term
Whole-Body Radiation Exposure
on Humans

Dose (rems)*	Effects
0–25	no detectable clinical effects.
25–100	slight short-term reduction in number of some blood cells; disabling sickness not common.
100–200	nausea and fatigue, vomiting if dose is greater than 125 rems; longer-term reduction in number of some blood cells.
200–300	nausea and vomiting first day of exposure; up to a 2-week latent period followed by appetite loss, general malaise, sore throat, pallor, diarrhea, and moderate emaciation. Recovery in about 3 months, unless complicated by infection or injury.
300–600	nausea, vomiting, and diarrhea in first few hours. Up to a 1-week latent period followed by loss of appetite, fever, and general malaise in the second week, followed by hemorrhage, inflammation of mouth and throat, diarrhea, and emaciation. Some deaths in 2 to 6 weeks. Eventual death for 50% if exposure is above 450 rems; others recover in about 6 months.
600 or more	nausea, vomiting, and diarrhea in first few hours. Rapid emaciation and death as early as second week. Eventual death of nearly 100%.

*A rem is the quantity of ionizing radiation that must be absorbed by a human to produce the same biological effect as 1 roentgen of high-penetration X rays. A roentgen is the quantity of high-penetration X rays that produces approximately 2×10^9 ion pairs per cubic centimeter of dry air at 0°C and 1 atm.

11.9 DETECTION OF RADIATION

You cannot hear, feel, taste, see, or smell low levels of radiation. However, there are numerous methods for detecting its presence. Becquerel's initial discovery of radioactivity (Section 11.2) was a result of the effect of radiation on photographic plates. Radiation affects photographic film as ordinary light does; it exposes the film. Technicians and others who work around radiation usually wear film badges (see Figure 11.9) to record the extent of their exposure to radiation. When the film from the badge is developed, the degree of darkening of the film negative indicates the extent of radiation exposure. Different filters are present in the badge, so various parts of the film register exposures to the different types of radiations (alpha, beta, gamma, and X rays).

Radiation can also be detected by making use of the fact that it ionizes atoms and molecules (Section 11.7). The Geiger counter operates on this principle. The basic components of a Geiger counter are shown in Figure 11.10. The detection part of such a

Figure 11.9 Film badges, such as the one worn by this technician, are used to determine a person's exposure to radiation.

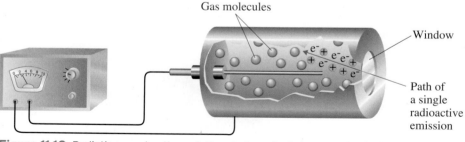

Figure 11.10 Radiation passing through the window of a Geiger counter ionizes one or more gas atoms, producing ion pairs. The electrons from the ion pairs are attracted to the central wire, and the positive ions are drawn to the metal tube. This constitutes a pulse of electric current, which is amplified and displayed on a meter or other readout.

CHEMICAL Connections

Preserving Food Through Food Irradiation

In some parts of the world, spoilage of food can claim up to 50% of a year's harvest. This is not so in the United States and many other countries, where food preservation, which takes many forms, is routinely carried out. Accepted modes of food preservation include freezing, canning, refrigeration, the use of chemical additives (such as antioxidants and mold inhibitors) and packaging (to keep out pests).

Food irradiation with gamma rays from ^{60}Co or ^{137}Cs sources is a newer form of food preservation. It is used extensively in Europe but has been "slow to catch on" in the United States even though it has been endorsed by the World Health Organization, the American Medical Association, and the U.S. Food and Drug Administration.

Spoilage of food is a biochemical process that usually involves bacteria, molds, and yeasts. Gamma radiation either kills or retards the growth of such species, the effects being determined by the irradiation dosage. Treatment levels can be grouped into three general categories:

1. "Low"-dose irradiation is used to delay physiological processes, including the ripening of fresh fruits and the sprouting of vegetables (such as onions and potatoes) and to control insects and parasites in foods. For example, irradiated strawberries stay unspoiled for up to three weeks, compared with three to five days for untreated berries.
2. "Medium"-dose irradiation is used to reduce spoilage and pathogenic microorganisms, to improve technological properties of food (such as reduced cooking time for dehydrated vegetables), and to extend the shelf life of many foods. Salmonella and other food-borne pathogens in meat, fish, and poultry are reduced by "medium"-dose irradiation. Such irradiation kills the parasite in pork that results in trichinosis.
3. "High"-dose irradiation is used to sterilize meat, poultry, and seafood and to kill insects in spices and seasonings.

Irradiation does not make foods radioactive, just as an airport luggage scanner does not make luggage radioactive. Nor does it cause harmful chemical changes. Some small loss of vitamin activity and nutrients may occur, but these losses are usually several orders of magnitude smaller than that which occurs with heat treatment (cooking). Scientists have repeatedly concluded from animal-feeding studies that there are no toxic effects associated with irradiated foods.

The Food and Drug Administration has approved irradiation of meat and poultry and allows its use for a variety of other foods, including spices and fresh fruits and vegetables. Federal rules require irradiated foods to be labeled as such to distinguish them from nonirradiated foods. Studies show that consumers are getting less "leery" of irradiated foods as more information becomes available about the safety of this technology. Irradiated foods sold to date have cost slightly more than their conventional counterparts. The estimated increase is two to three cents a pound for fruits and vegetables and three to five cents a pound for meat and poultry products.

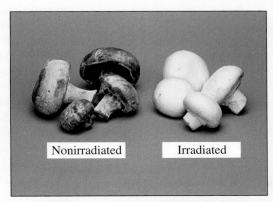

Irradiation increases the "shelf-life" of many types of food, including mushrooms.

counter is a metal tube filled with a gas (usually argon). The tube has a thin-walled window made of a material that can be penetrated by alpha, beta, or gamma rays. In the center of the tube is a wire attached to the positive terminal of an electrical power source. The metal tube is attached to the negative terminal of the same source. Radiation entering the tube ionizes the gas, which allows a pulse of electricity to flow. This pulse of electricity is then amplified and displayed on a meter or some other type of readout display.

11.10 SOURCES OF RADIATION EXPOSURE

Most of us will never come in contact with the radiation dosage necessary to cause the effects listed in Table 11.3. Nevertheless, *low-level* exposure to ionizing radiation is something we constantly encounter. In fact, there is no way we can totally avoid this low-level exposure because much of it results from naturally occurring environmental processes.

Figure 11.11 Components of the estimated annual radiation exposure (in millirems) of an average American. Individual exposures vary widely, but most such radiation comes from natural sources, the largest single contributor being radon gas.

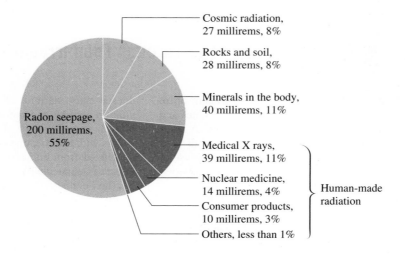

Cosmic radiation, 27 millirems, 8%

Rocks and soil, 28 millirems, 8%

Minerals in the body, 40 millirems, 11%

Medical X rays, 39 millirems, 11%

Nuclear medicine, 14 millirems, 4%

Consumer products, 10 millirems, 3%

Others, less than 1%

Human-made radiation

Radon seepage, 200 millirems, 55%

Radiation that comes from natural environmental sources is called *background radiation*. **Background radiation** *is radiation that comes from natural sources to which living organisms are exposed on a continuing basis.* Numerous sources for background radiation exist, including the following:

1. *Cosmic radiation*: Earth, and all things on it, are constantly bombarded by radiation from outer space. Some cosmic radiation, but not all of it, comes from the sun; sources outside the solar system also exist.
2. *Rocks and minerals*: Trace to larger amounts of thorium and uranium minerals are present in almost all rocks and soil. Building materials, including brick and concrete, will contain small amounts of these substances. Consequently, such radiation exists in all types of buildings, including work environments, schools, and homes.
3. *Food and drink*: The main radioactive substance present in food and drink is potassium-40. Potassium-40 is a naturally occurring radioactive isotope of potassium (0.012% abundance; Section 3.3) and is therefore found in all potassium-containing foods. It is estimated that, on average, 30 mg of potassium-40 is present in a human body.
4. *Radon seepage in buildings*: Radon-222 exposure, which is an airborne exposure, is the greatest background radiation exposure source for most individuals. The Chemical Connections feature on page 309 considers this topic in greater detail.

Exposure to cosmic radiation increases with altitude. At higher elevations there are fewer molecules present in the atmosphere and thus there are fewer molecules available to absorb incoming radiation. The passenger in an airplane receives increased exposure to cosmic radiation during the flight. People who live at an altitude of 5000 feet receive approximately twice the exposure to cosmic radiation as those who live at sea level.

Figure 11.11 quantifies exposure amounts for various sources of background radiation and also gives data for radiation exposure arising from human activities, which include (1) medical X rays, (2) nuclear medicine, (3) consumer products, and (4) miscellaneous sources, including occupational exposure, nuclear fallout from weapons testing, and nuclear power plants. The estimates of per capita radiation exposure given in Figure 11.11 are averages for all Americans; the actual exposure of individuals varies according to where they live and work, their medical history, and other factors.

Comparing the values in Figure 11.11 with those of Table 11.3, taking into account that the units in the former are millirems and those in the latter are rems, shows that the current dosage levels received by the general population are very small compared with those known to cause serious radiation sickness.

With low-level radiation exposure, chromosome damage rather than cell death can occur. If the damaged genetic material repairs itself improperly, then new, abnormal cells are produced when the cells replicate.

Cells that reproduce at a rapid rate, such as those in bone marrow, lymph nodes, and embryonic tissue, are the most sensitive to radiation damage. The sensitivity of embryonic tissue to radiation damage is the reason why pregnant women need to be protected from radiation exposure. One of the first signs of overexposure to radiation is a drop in red blood count. This is a direct consequence of the sensitivity of bone marrow, the site of red blood cell formation, to radiation.

The Indoor Radon-222 Problem

Over 80% of the radiation exposure that an average American experiences comes from naturally occurring environmental processes. Foremost among these processes, both in terms of amount of and seriousness of radiation exposure, is the generation of radon gas. Radon gas accounts for over half the background radioactivity on Earth (see Figure 11.11).

The element radon, located in group VIIIA of the periodic table, is a member of the noble gas family, and thus is a very unreactive substance. Substances that are chemically unreactive are not ordinarily thought of as posing any health risks. However, radon's nuclear properties—its radioactivity—causes it to be a health risk. More than 20 isotopes of radon exist, all of which are radioactive. Radon-222, the radon isotope with the longest half-life (3.82 days), is the isotope of most concern relative to human exposure. Its source is uranium ores and minerals. Radon-222 is one of the intermediate decay products in the uranium-238 decay series (see Figure 11.6).

Because uranium compounds are present in trace amounts in many types of rocks and soils, radon-222 (and its decay products) are found almost everywhere in our environment. What distinguishes radon-222 from other decay products in the uranium-238 decay series is the fact that it is a gas while other decay products are solids. In the gaseous state, radon-222 readily migrates from soil and rocks into the surrounding air and sometimes into water sources. Radon gas is sparingly soluble in water.

In outdoor situations, radon-222 gas is not considered a major health hazard because it dissipates into the air. However, indoor exposure to radon-222, where ventilation is restricted, is a serious hazard because the radon can be *directly* inhaled by those living or working in the indoor space. If a person inhales air containing radon and then exhales before the radon-222 undergoes radioactive decay, no harm is done. If the radon-222, however, undergoes decay while in the lungs, an alpha particle is emitted and the radon is changed to polonium-218.

$$^{222}_{86}Rn \longrightarrow {}^{4}_{2}\alpha + {}^{218}_{84}Po$$

Polonium-218 is a solid rather than a gas and can become attached to lung tissue, where it will undergo further alpha particle decay.

$$^{218}_{84}Po \longrightarrow {}^{4}_{2}\alpha + {}^{214}_{82}Pb$$

Additional radiation exposure related to radon-222 can *indirectly* occur by breathing air in which radon-222 decay has

already occurred. The solid radioactive products from the radon-222 decay adhere to airborne dust and smoke, which are inhaled into the lungs and deposited in the respiratory tract where the solids undergo further alpha particle decay.

Serious radon-222 contamination of homes has been found to occur in several areas in the United States, where the natural uranium content of the soil is high. The radon-222 seeps into the homes through cracks in the cement foundation or through other openings and then accumulates in basements. Because of this problem, commercially available kits for testing radon in the home are now readily available in stores. As radon-222 awareness increases, an increasing number of home buyers ask for a determination of radon levels before buying a house.

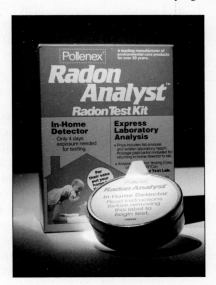

A commercially available kit to test for radon gas in the home.

The decay products of radon-222 that accumulate in the lungs, because of their alpha-particle emissions, are able to irradiate tissue, damage cells, and possibly lead to lung cancer. The U.S. Environmental Protection Agency (EPA) estimates that 10% of all lung cancer deaths are related to radon-222 exposure. Radon-222 exposure is believed to be the leading cause of lung cancer among nonsmokers (30% of deaths).

 **NUCLEAR MEDICINE**

Nuclear medicine *is a field of medicine in which radionuclides are used for diagnostic and therapeutic purposes.* In diagnostic applications, technicians use small amounts of radionuclides whose progress through the body or localization in specific organs can be followed. Larger quantities of radionuclides are used in therapeutic applications.

An additional use for radionuclides in medicine, besides diagnostic and therapeutic uses, is as a source of power (Section 11.12). Cardiac pacemakers powered by plutonium-238 can remain in a patient for longer periods than those powered by chemical batteries, *and* the additional surgery required to replace batteries is not needed.

Radionuclides used in diagnostic nuclear medicine are often called *radioactive tracers*. Techniques in which radioactive tracers are used are called *nuclear imaging* procedures.

Diagnostic Uses for Radioisotopes

The fundamental chemical principle behind the use of radionuclides in diagnostic medical work is the fact that a radioactive nuclide of an element has the same chemical properties as a nonradioactive nuclide of the element. Thus body chemistry is not upset by the presence of a small amount of a radioactive substance whose nonradioactive form is already present in the body.

The criteria used in selecting radionuclides for diagnostic procedures include the following:

1. At low concentrations (to minimize radiation damage), the radionuclide must be detectable by instrumentation placed outside the body. Nearly all diagnostic radionuclides are gamma emitters because the penetrating power of alpha and beta particles is too low.
2. The radionuclide must have a short half-life so that the intensity of the radiation is sufficiently great to be detected. A short half-life also limits the time period of radiation exposure.
3. The radionuclide must have a known mechanism for elimination from the body so that the material does not remain in the body indefinitely.
4. The chemical properties of the radionuclide must be such that it is compatible with normal body chemistry. It must be able to be selectively transmitted to the part or system of the body that is under study.

The following examples illustrate the diverse uses of radionuclides in procedures for diagnosis of disease or malfunction in the human body.

- *Determination of Blood Volume.* A person's blood volume can be found by injecting a known quantity of red blood cells labeled with radioactive chromium-51 into the bloodstream and measuring the dilution factor. The labeled red blood cells are distributed throughout the body and a new blood sample taken after a period of time has elapsed can be used to calculate the blood volume. The concentration of labeled cells in the new blood sample is compared to the concentration of labeled cells previously injected in the body.
- *Location of Sites of Infection.* Abscesses and additional sites of infection can be located using gallium-67, a gamma ray emitter. The gallium-67 is incorporated into a compound that binds to white blood cells. The tagged white blood cells, as well as untagged white blood cells, migrate to sites of infection. The tagged white blood cells enable the infection site to be located because of the gamma ray emission that occurs from the gallium-67.
- *Diagnosis of Impaired Heart Muscle.* The radionuclide thallium-201 has been found to be effective in helping to diagnose heart disease. This radionuclide, when injected into the blood, is found to have particular affinity for heart muscle. Only heart muscle tissue that receives a normal flow of blood will bond to the thallium-201. This leads to the detection of heart tissue that lacks sufficient blood because of narrowed arteries.
- *Location of Impaired Circulation.* Sodium-24 is used to follow the circulation of blood in the body. A small amount of this radionuclide is injected into the bloodstream in the form of a sodium chloride solution. The movement of the sodium-24 through the circulatory system is followed with radiation detection equipment. If it takes longer than normal for the radionuclide to show up at a particular spot in the body, this is an indication that the circulation is impaired at that spot.
- *Assessment of Thyroid Activity.* Administering iodine-123, used in the form of a sodium iodide solution, gives information about the functioning of the thyroid gland. The radioactive iodide is absorbed by the thyroid at a rate related to the activity of the gland. If a hypothyroid condition exists, then the amount accumulated is less than normal; if a hyperthyroid condition exists, then a greater-than-average amount accumulates. Figure 11.12 shows the contrast between a normal and an abnormal thyroid scan.

Figure 11.12 Iodine-123 is the radio-nuclide involved in obtaining these thyroid gland scans. The scan on the left, which shows uniform iodine-123 uptake, is considered a normal scan. The scan on the right shows a thyroid gland in which the right lobe is not functioning properly.

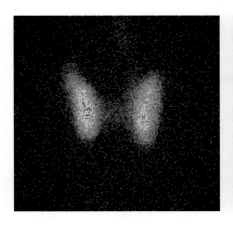
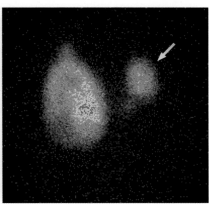

■ *Determination of Tumor Size and Shape.* The size and shape of organs, as well as the presence of tumors, can be determined in some situations by scanning the organ in which a radionuclide tends to concentrate. Iodine-131 and technetium-99 are used to generate thyroid and brain scans, respectively. In the brain, technetium-99, in the form of a polyatomic ion (TcO_4^-), concentrates in brain tumors more than in normal brain tissue; this helps radiologists determine the presence, size, and location of brain tumors.

Table 11.4 lists a number of radionuclides that are used in diagnostic procedures. The half-life of the radionuclide, the body locations wherein it concentrates, and its diagnostic function are also given.

TABLE 11.4
Selected Radionuclides Used in Diagnostic Procedures

Nuclide	Half-life	Part of Body Affected	Use in Diagnosis
barium-131	11.6 days	bone	detection of bone tumors
chromium-51	27.8 days	blood kidney	determination of blood volume and red blood cell lifetime assessment of kidney activity
gallium-67	3.2 days	blood	detection of sites of infection
iodine-123	13 hours	thyroid	assessment of thyroid gland activity
iron-59	45 days	blood	evaluation of iron metabolism in blood
phosphorus-32	14.3 days	blood breast	blood studies assessment of breast carcinoma
potassium-42	12.4 hours	tissue	determination of intercellular spaces in fluids
sodium-24	15.0 hours	blood	detection of circulatory problems; assessment of peripheral vascular disease
technetium-99m[*,†]	6.0 hours	brain spleen thyroid lung	detection of brain tumors, hemorrhages, or blood clots measurement of size and shape of spleen measurement of size and shape of thyroid location of blood clots
thallium-201	3.0 days	blood	assessment of normal flow of blood

[*]The form in which the technetium-99m is administered, such as a phosphate or a chloride, determines its target location; e.g., technetium-99m as a phosphate is adsorbed on the surface of bone.
[†]The designations for some radionuclides have a small *m*, which means *metastable,* associated with their mass number, such as technetium-99m. Such nuclides, which are formed at high energies, decay by emitting gamma radiation without a change to a new element. For technetium-99m the decay equation is
$$^{99m}Tc \longrightarrow {}^{99}Tc + {}^{0}_{0}\gamma \text{ (gamma radiation)}.$$

TABLE 11.5
Some Radionuclides Used
in Radiation Therapy

Nuclide	Half-life	Type of Emitter	Use in Therapy
cobalt-60	5.3 years	gamma	external source of radiation in treatment of cancer
iodine-131	8 days	beta, gamma	cancer of thyroid
phosphorus-32	14.3 days	beta, gamma	treatment of some types of leukemia and widespread carcinomas
radium-226	1620 years	alpha, gamma	used in implantation cancer therapy
radon-222	3.8 days	alpha, gamma	used in treatment of uterine, cervical, oral, and bladder cancers
yttrium-90	64 hours	beta, gamma	implantation therapy

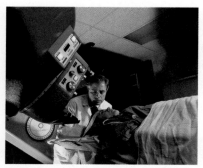

Figure 11.13 Cobalt-60 is used as a source of gamma radiation in radiation therapy.

Abnormal cells are more susceptible to radiation damage than normal cells because abnormal cells divide more frequently.

Therapeutic Uses for Radioisotopes

The objectives in therapeutic radionuclide use are entirely different from those for diagnostic procedures. The main objective in the therapeutic use of radionuclides is to *selectively destroy* abnormal (usually cancerous) cells. The radionuclide is often, but not always, placed within the body. Therapeutic radionuclides implanted in the body are usually alpha or beta emitters because an intense dose of radiation in a small localized area is needed.

A commonly used implantation radionuclide that is effective in the localized treatment of tumors is yttrium-90, a beta emitter with a half-life of 64 hours. Yttrium-90 salts are implanted by inserting small, hollow needles into the tumor.

External, high-energy beams of gamma radiation are also extensively used in the treatment of certain cancers. Cobalt-60 is frequently used for this purpose; a beam of radiation is focused on the small area of the body where the tumor is located (see Figure 11.13). This therapy usually causes some radiation sickness because normal cells are also affected, but to a lesser extent. The operating principle here is that abnormal cells are more susceptible to radiation damage than normal cells. Radiation sickness is the price paid for abnormal-cell destruction. Table 11.5 lists some radionuclides that are used in therapy.

NUCLEAR FISSION AND NUCLEAR FUSION

Our glimpse into the world of nuclear chemistry would not be complete without a brief mention of two additional types of nuclear reactions that are used as sources of energy: nuclear fission and nuclear fusion.

Fission Reactions

Nuclear fission *is a nuclear reaction in which a large nucleus (high atomic number) splits into two medium-sized nuclei with the release of several free neutrons and a large amount of energy.* The most important nucleus that undergoes fission is uranium-235. Bombardment of this nucleus with neutrons causes it to split into two fragments. Characteristics of the uranium-235 fission reaction include the following:

1. There is no unique way in which the uranium-235 nucleus splits. Thus, many different, lighter elements are produced during uranium-235 fission reactions. The following are examples of the ways in which this fission process proceeds.

$$^{235}_{92}\text{U} + ^{1}_{0}\text{n}$$

$$^{135}_{53}\text{I} + ^{97}_{39}\text{Y} + 4\,^{1}_{0}\text{n}$$
$$^{139}_{56}\text{Ba} + ^{94}_{36}\text{Kr} + 3\,^{1}_{0}\text{n}$$
$$^{131}_{50}\text{Sn} + ^{103}_{42}\text{Mo} + 2\,^{1}_{0}\text{n}$$
$$^{139}_{54}\text{Xe} + ^{95}_{38}\text{Sr} + 2\,^{1}_{0}\text{n}$$

2. Very large amounts of energy, which are many times greater than that released by ordinary radioactive decay, are emitted during the fission process. It is this large release of energy that makes nuclear fission of uranium-235 the important process that it is. In general, the term *nuclear energy* is used to refer to the energy released during a nuclear fission process. An older term for this energy is *atomic energy*.

3. Neutrons, which are reactants in the fission process, are also produced as products. The number of neutrons produced per fission depends on the way in which the nucleus splits; it ranges from 2 to 4 (as can be seen from the foregoing fission equations). On the average, 2.4 neutrons are produced per fission. The significance of the neutrons that are produced is that they can cause the fission process to continue by colliding with further uranium-235 nuclei. Figure 11.14 shows the chain reaction that can occur once the fission process is started.

The process of nuclear fission—or "splitting the atom," as it is called in popularized science—can be carried out in both an uncontrolled and a controlled manner. The key to this control lies in what happens to the neutrons produced during fission. Do they react further, causing further fission, or do they escape into the surroundings? If the majority of produced neutrons react further (Figure 11.14), an uncontrolled nuclear reaction (an atomic bomb) results (see Figure 11.15). When only a few neutrons react further (on the average, one per fission), the fission reaction self-propagates in a controlled manner.

The process of *controlled* nuclear fission is the basis for the operation of nuclear power plants that are used to produce electricity (see Figure 11.16). The reaction is controlled with rods that absorb excess neutrons (so that they cannot cause unwanted fissions) and with moderating substances that decrease the speed of the neutrons. The energy produced during the fission process, which appears as heat, is used to operate steam-powered electricity-generating equipment.

Fusion Reactions

Another type of nuclear reaction, nuclear fusion, produces even more energy than nuclear fission. **Nuclear fusion** *is a nuclear reaction in which two small nuclei are collided together to produce a larger nucleus and a large amount of energy.* This process is essentially

Figure 11.14 A fission chain reaction is caused by further reaction of the neutrons produced during fission.

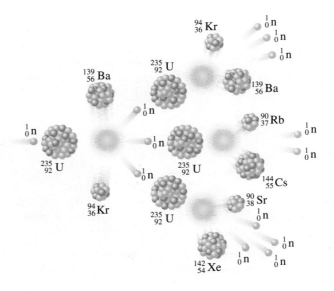

Figure 11.15 Enormous amounts of energy are released in the explosion of a nuclear fission bomb.

the opposite of nuclear fission. In order for fusion to occur, a very high temperature (several hundred million degrees) is required.

Nuclear fusion is the process by which the sun generates its energy (see Figure 11.17). Within the sun hydrogen-1 nuclei are converted to helium-4 nuclei with the release of extraordinarily large amounts of energy.

The use of nuclear fusion on Earth might seem impossible because of the high temperatures required. It has, however, been accomplished in a hydrogen bomb. In such a weapon, a *fission* bomb is used to achieve the high temperatures needed to start the following process:

$$_1^3H + _1^2H \longrightarrow _2^4He + _0^1n$$

The use of nuclear fusion as a *controlled* (peaceful) energy source is a very active area of current scientific research. "Harnessing" this type of nuclear reaction would have numerous advantages:

1. Unlike the by-products of fission reactions, the by-products of fusion reactions are stable (nonradioactive) nuclides. Thus the problem of storing radioactive wastes does not arise.
2. The major fuel under study for controlled fusion is $_1^2H$ (called deuterium), a hydrogen isotope that can be readily extracted from ocean water (0.015% of all hydrogen

At the high temperature of fusion reactions, electrons completely separate from nuclei. Neutral atoms cannot exist. This high-temperature, gaslike mixture of nuclei and electrons is called a *plasma* and is considered by some scientists to represent a fourth state of matter.

Figure 11.16 A nuclear power plant. The cooling tower at the Trojan nuclear power plant in Oregon dominates the landscape. The nuclear reactor is housed in the dome-shaped enclosure.

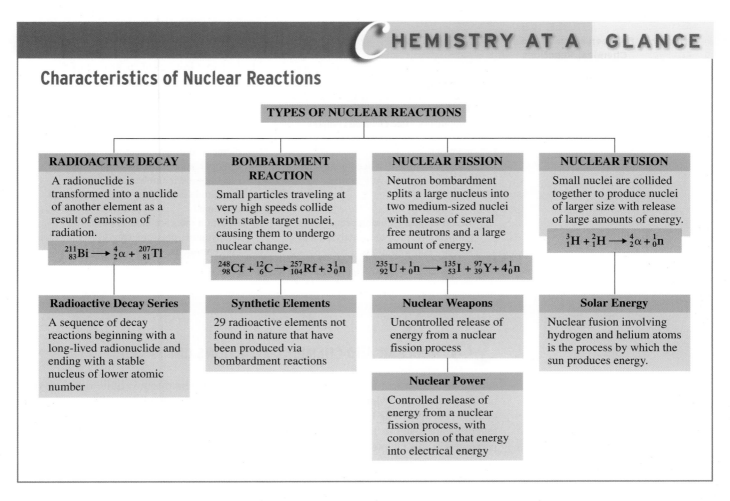

CHEMISTRY AT A GLANCE

Characteristics of Nuclear Reactions

TYPES OF NUCLEAR REACTIONS

RADIOACTIVE DECAY

A radionuclide is transformed into a nuclide of another element as a result of emission of radiation.

$$^{211}_{83}\text{Bi} \longrightarrow {}^{4}_{2}\alpha + {}^{207}_{81}\text{Tl}$$

Radioactive Decay Series

A sequence of decay reactions beginning with a long-lived radionuclide and ending with a stable nucleus of lower atomic number

BOMBARDMENT REACTION

Small particles traveling at very high speeds collide with stable target nuclei, causing them to undergo nuclear change.

$$^{248}_{98}\text{Cf} + {}^{12}_{6}\text{C} \longrightarrow {}^{257}_{104}\text{Rf} + 3{}^{1}_{0}\text{n}$$

Synthetic Elements

29 radioactive elements not found in nature that have been produced via bombardment reactions

NUCLEAR FISSION

Neutron bombardment splits a large nucleus into two medium-sized nuclei with release of several free neutrons and a large amount of energy.

$$^{235}_{92}\text{U} + {}^{1}_{0}\text{n} \longrightarrow {}^{135}_{53}\text{I} + {}^{97}_{39}\text{Y} + 4{}^{1}_{0}\text{n}$$

Nuclear Weapons

Uncontrolled release of energy from a nuclear fission process

Nuclear Power

Controlled release of energy from a nuclear fission process, with conversion of that energy into electrical energy

NUCLEAR FUSION

Small nuclei are collided together to produce nuclei of larger size with release of large amounts of energy.

$$^{3}_{1}\text{H} + {}^{2}_{1}\text{H} \longrightarrow {}^{4}_{2}\alpha + {}^{1}_{0}\text{n}$$

Solar Energy

Nuclear fusion involving hydrogen and helium atoms is the process by which the sun produces energy.

---

atoms are ${}^{2}_{1}\text{H}$). Just 0.005 km^3 of ocean water contains enough ${}^{2}_{1}\text{H}$ to supply the United States with all the energy it needs for 1 year!

However, difficult scientific and engineering problems still remain to be solved before controlled fusion is a reality.

The Chemistry at a Glance feature above summarizes important concepts about the major types of nuclear reactions that have been considered in this chapter.

Figure 11.17 A close-up view of the sun. The process of nuclear fusion maintains the interior of the sun at a temperature of approximately 15 million degrees.

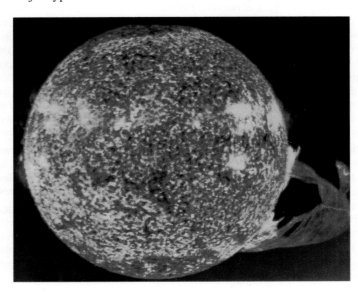

TABLE 11.6
Differences Between Nuclear and Chemical Reactions

Chemical Reaction	Nuclear Reaction
1. Different isotopes of an element have identical chemical properties.	1. Different isotopes of an element have different nuclear properties.
2. The chemical reactivity of an element depends on the element's state of combination (free element, compound, etc.).	2. The nuclear reactivity of an element is independent of the state of chemical combination.
3. Elements retain their identity in chemical reactions.	3. Elements may be changed into other elements during nuclear reactions.
4. Energy changes that accompany chemical reactions are relatively small.	4. Energy changes that accompany nuclear reactions are a number of orders of magnitude larger than those in chemical reactions.
5. Reaction rates are influenced by temperature, pressure, catalysts, and reactant concentrations.	5. Reaction rates are independent of temperature, pressure, catalysts, and reactant concentrations.

 NUCLEAR AND CHEMICAL REACTIONS COMPARED

As the discussions in this chapter have shown, nuclear chemistry is quite different from ordinary chemistry. Many of the laws of chemistry must be modified when we consider nuclear reactions. The major differences between nuclear reactions and ordinary chemical reactions are listed in Table 11.6. This table serves as a summary of many of the concepts presented in this chapter.

CONCEPTS TO REMEMBER

Radioactivity. Some atoms possess nuclei that are unstable. To achieve stability, these unstable nuclei spontaneously emit energy (radiation). Such atoms are said to be radioactive (Section 11.1).

Emissions from radioactive nuclei. The types of radiation emitted by naturally occurring radioactive nuclei are alpha, beta, and gamma. These radiations can be characterized by mass and charge values. Alpha particles carry a positive charge, beta particles carry a negative charge, and gamma radiation has no charge (Section 11.2).

Balanced nuclear equations. The procedures for balancing nuclear equations are different from those for balancing ordinary chemical equations. In nuclear equations, mass numbers and atomic numbers (rather than atoms) balance on both sides (Section 11.3).

Half-life. Every radionuclide decays at a characteristic rate given by its half-life. One half-life is the time required for half of any given quantity of a radioactive substance to undergo decay (Section 11.4).

Bombardment reactions. A bombardment reaction is a nuclear reaction in which small particles traveling at very high speeds are collided with stable nuclei; this causes these nuclei to undergo nuclear change (become unstable). Over 2000 synthetically produced radionuclides that do not occur naturally have been produced by using bombardment reactions (Section 11.5).

Radioactive decay series. The product of the radioactive decay of an unstable nuclide is a nuclide of another element, which may or may not be stable. If it is not stable, it will decay and produce still another nuclide. Further decay will continue until a stable nuclide is formed. Such a sequence of reactions is called a radioactive decay series (Section 11.6).

Chemical effects of radiation. Radiation from radioactive decay is ionizing radiation—radiation with enough energy to remove an electron from an atom or molecule. Interaction of ionizing radiation with matter produces ion pairs, with many ion pairs being produced by a single "particle" of radiation. Free-radical formation usually accompanies ion pair formation. A free radical is a chemical species that contains an unpaired electron. Free radicals, very reactive species, rapidly interact with other chemical species nearby, causing many of the undesirable effects that, in living organisms, are associated with radiation exposure (Section 11.7).

Biochemical effects of radiation. The biochemical effects of radiation depend on the energy, ionizing ability, and penetrating ability of the radiation. Alpha particles exhibit the greatest ionizing effect, and gamma rays have the greatest penetrating ability (Section 11.8).

Detection of radiation. Radiation can be detected by making use of the fact that radiation ionizes atoms and molecules. The Geiger counter operates on this principle. Radiation also affects photographic film in the same way as ordinary light; the film is exposed. Hence film badges are used to record the extent of radiation exposure (Section 11.9).

Sources of radiation exposure. Both natural and human-generated sources of low-level radiation exposure exist, with natural sources accounting for almost 80% of exposure (on the average). Current dosage levels received by the general population are very small compared with those known to cause serious radiation sickness (Section 11.10).

Nuclear medicine. Radionuclides are used in medicine for both diagnosis and therapy. The choice of radionuclide is dictated by the purpose as well as the target organ. Bombardment reactions are used to produce the nuclides used in medicine; such nuclides all have short half-lives (Section 11.11).

Nuclear fission. Nuclear fission occurs when fissionable nuclides are bombarded with neutrons. The nuclides split into two fragments of about the same size. Also, more neutrons and large amounts of energy are produced. Nuclear fission is the process by which nuclear power plants generate energy (Section 11.12).

Nuclear fusion. In nuclear fusion, small nuclei fuse to make heavier nuclei. Nuclear fusion is the process by which the sun generates its energy (Section 11.12).

KEY REACTIONS AND EQUATIONS

1. General equation for alpha decay (Section 11.3)
$$\begin{matrix} A \\ Z \end{matrix}X \longrightarrow \begin{matrix} A-4 \\ Z-2 \end{matrix}Y + {}^{4}_{2}\alpha$$

2. General equation for beta decay (Section 11.3)
$$\begin{matrix} A \\ Z \end{matrix}X \longrightarrow {}^{A}_{Z+1}Y + {}^{0}_{-1}\beta$$

3. Half-life and amount of undecayed radionuclide (Section 11.4)
$$\left(\begin{matrix}\text{Amount of radionuclide} \\ \text{undecayed after } n \text{ half-lives}\end{matrix}\right) = \left(\begin{matrix}\text{original amount} \\ \text{of radionuclide}\end{matrix}\right) \times \left(\frac{1}{2^n}\right)$$

EXERCISES *and* PROBLEMS

The members of each pair of problems in this section test the same material.

⬤ **Notation for Nuclides (Section 11.1)**

11.1   Use two different notations to denote each of the following nuclides.
   a. Contains 4 protons, 4 electrons, and 6 neutrons
   b. Contains 11 protons, 11 electrons, and 14 neutrons
   c. Contains 41 protons, 41 electrons, and 55 neutrons
   d. Contains 103 protons, 103 electrons, and 154 neutrons

11.2   Use two different notations to denote each of the following nuclides.
   a. Contains 20 protons, 20 electrons, and 18 neutrons
   b. Contains 37 protons, 37 electrons, and 43 neutrons
   c. Contains 51 protons, 51 electrons, and 74 neutrons
   d. Contains 99 protons, 99 electrons, and 157 neutrons

11.3   Use a notation different from that given to designate each of the following nuclides.
   a. Nitrogen-14   b. Gold-197   c. ${}^{121}_{50}Sn$ d. ${}^{10}_{5}B$

11.4   Use a notation different from that given to designate each of the following nuclides.
   a. Oxygen-17   b. Lead-212   c. ${}^{92}_{37}Rb$ d. ${}^{201}_{83}Bi$

11.5   What physical manifestation indicates that an atom possesses an unstable nucleus?

11.6   What is the limit for nuclear stability in terms of number of nucleons present in a nucleus?

11.7   How do the neutron-to-proton ratios compare for stable nuclei of low atomic number and stable nuclei of high atomic number?

11.8   Indicate whether or not all isotopes would be radioactive for each of the following elements.
   a. ${}_{76}Os$ b. ${}_{86}Rn$ c. ${}_{96}Cm$ d. ${}_{104}Sg$

⬤ **The Nature of Radioactive Emissions (Section 11.2)**

11.9   Supply a complete symbol, with superscript and subscript, for each of the following types of radiation.
   a. Alpha particle   b. Beta particle   c. Gamma ray

11.10 Give the charge and mass (in amu) of each of the following types of radiation.
   a. Alpha particle   b. Beta particle   c. Gamma ray

11.11 State the composition of an alpha particle in terms of protons and neutrons.

11.12 What is the relationship between a beta particle and an electron?

⬤ **Equations for Radioactive Decay (Section 11.3)**

11.13 Write balanced nuclear equations for the alpha decay of each of the following nuclides.
   a. ${}^{200}_{84}Po$ b. Curium-240   c. ${}^{244}_{96}Cm$ d. Uranium-238

11.14 Write balanced nuclear equations for the alpha decay of each of the following nuclides.
   a. ${}^{229}_{90}Th$ b. Bismuth-210   c. ${}^{152}_{64}Gd$ d. Americium-243

11.15 Write balanced nuclear equations for the beta decay of each of the following nuclides.
   a. ${}^{10}_{4}Be$ b. Carbon-14   c. ${}^{21}_{9}F$ d. Sodium-25

11.16 Write balanced nuclear equations for the beta decay of each of the following nuclides.
   a. ${}^{77}_{32}Ge$ b. Uranium-235   c. ${}^{16}_{7}N$ d. Iron-60

11.17 What is the effect on the mass number and atomic number of the parent nuclide when alpha particle decay occurs?

11.18 What is the effect on the mass number and atomic number of the parent nuclide when beta particle decay occurs?

11.19 Supply the missing symbol in each of the following nuclear equations.
   a. ${}^{34}_{14}Si \rightarrow {}^{34}_{15}P + ?$ b. $? \rightarrow {}^{28}_{13}Al + {}^{0}_{-1}\beta$
   c. ${}^{252}_{99}Es \rightarrow {}^{248}_{97}Bk + ?$ d. ${}^{204}_{82}Pb \rightarrow ? + {}^{4}_{2}\alpha$

11.20 Supply the missing symbol in each of the following nuclear equations.
   a. $? \rightarrow {}^{230}_{92}U + {}^{4}_{2}\alpha$ b. ${}^{192}_{78}Pt \rightarrow ? + {}^{4}_{2}\alpha$
   c. ${}^{84}_{35}Br \rightarrow ? + {}^{0}_{-1}\beta$ d. ${}^{10}_{4}Be \rightarrow {}^{10}_{5}B + ?$

11.21 Identify the mode of decay for each of the following parent radionuclides, given the identity of the daughter nuclide.
a. Parent = platinum-190; daughter = osmium-186
b. Parent = oxygen-19; daughter = fluorine-19

11.22 Identify the mode of decay for each of the following parent radionuclides, given the identity of the daughter nuclide.
a. Parent = uranium-238; daughter = thorium-234
b. Parent = rhodium-104; daughter = palladium-104

11.23 Which of the following diagrams is a correct representation for the process of beta particle decay?

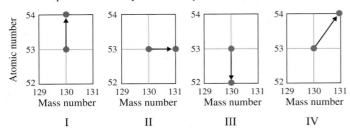

I   II   III   IV

11.24 Which of the following diagrams is a correct representation for the process of alpha particle decay?

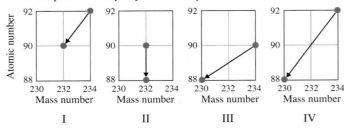

I   II   III   IV

⬤ Rate of Radioactive Decay (Section 11.4)

11.25 Technetium-99 has a half-life of 6.0 hr. What fraction of the technetium-99 atoms in a sample will remain undecayed after the following times?
a. 12 hr   b. 36 hr   c. 3 half-lives   d. 6 half-lives

11.26 Copper-66 has a half-life of 5.0 min. What fraction of the copper-66 atoms in a sample will remain undecayed after the following times?
a. 20 min   b. 30 min   c. 3 half-lives   d. 8 half-lives

11.27 Determine the half-life of a radionuclide if after 5.4 days the fraction of undecayed nuclides present is
a. 1/16   b. 1/64   c. 1/256   d. 1/1024

11.28 Determine the half-life of a radionuclide if after 3.2 days the fraction of undecayed nuclides present is
a. 1/8   b. 1/128   c. 1/32   d. 1/512

11.29 The half-life of sodium-24 is 15.0 hr. How many grams of this nuclide in a 4.00-g sample will remain after 60.0 hr?

11.30 The half-life of strontium-90 is 28 years. How many grams of this nuclide in a 4.00-g sample will remain after 112 years?

11.31 Silicon-34 decays to phosphorus-34 through beta particle emission. Based on the following diagram, in which yellow spheres represent silicon-34 nuclides and red spheres represent phosphorus-34 nuclides, how many half-lives have elapsed for the sample?

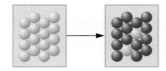

11.32 Platinum-192 decays to osmium-188 through alpha particle emission. Based on the following diagram, in which yellow spheres represent platinum-192 nuclides and red spheres represent osmium-188 nuclides, how many half-lives have elapsed for the sample?

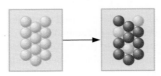

⬤ Bombardment Reactions (Section 11.5)

11.33 Approximately how many laboratory-produced radionuclides are known?

11.34 How does the number of laboratory-produced radionuclides compare with the number of naturally occurring nuclides?

11.35 What is the highest-atomic-numbered naturally occurring element?

11.36 What is the highest-atomic-numbered element for which nonradioactive isotopes exist?

11.37 Supply the missing nuclear symbol in each of the following equations for bombardment reactions.
a. $^{24}_{12}\text{Mg} + ? \rightarrow {}^{27}_{14}\text{Si} + {}^{1}_{0}\text{n}$ b. $^{27}_{13}\text{Al} + {}^{2}_{1}\text{H} \rightarrow ? + {}^{4}_{2}\alpha$
c. $^{9}_{4}\text{Be} + ? \rightarrow {}^{12}_{6}\text{C} + {}^{1}_{0}\text{n}$ d. $^{6}_{3}\text{Li} + ? \rightarrow {}^{4}_{2}\text{He} + {}^{3}_{2}\text{He}$

11.38 Supply the missing nuclear symbol in each of the following equations for bombardment reactions.
a. $? + {}^{4}_{2}\alpha \rightarrow {}^{250}_{99}\text{Es} + {}^{3}_{1}\text{H}$ b. $^{14}_{7}\text{N} + {}^{4}_{2}\alpha \rightarrow ? + {}^{1}_{1}\text{H}$
c. $^{12}_{6}\text{C} + {}^{2}_{1}\text{H} \rightarrow {}^{13}_{7}\text{N} + ?$ d. $^{27}_{13}\text{Al} + ? \rightarrow {}^{30}_{15}\text{P} + {}^{1}_{0}\text{n}$

⬤ Radioactive Decay Series (Section 11.6)

11.39 The uranium-235 decay series terminates with lead-207. Would you expect lead-207 to be a stable or an unstable nuclide? Explain your answer.

11.40 A textbook erroneously indicates that the uranium-235 decay series terminates with radon-222. Explain why such a situation cannot be.

11.41 In the thorium-232 natural decay series, the thorium-232 initially undergoes alpha decay, the resulting daughter emits a beta particle, and the succeeding daughters emit a beta and an alpha particle in that order. Write four nuclear equations, one to represent each of the first four steps in the thorium-232 decay series.

11.42 In the uranium-235 natural decay series, the uranium-235 initially undergoes alpha decay, the resulting daughter emits a beta particle, and the succeeding daughters emit an alpha and a beta particle in that order. Write four nuclear equations, one to represent each of the first four steps in the uranium-235 decay series.

⬤ Chemical Effects of Radiation (Section 11.7)

11.43 What is an ion pair?

11.44 What is a free radical?

11.45 Indicate whether each of the following species is a free radical.
a. H_2O^+ b. H_3O^+ c. OH d. OH^-

11.46 Write a chemical equation that involves water as a reactant for the formation of the
a. water free radical   b. hydroxyl free radical

11.47 What is the fate of a radiation "particle" that is involved in an ion pair formation reaction?

11.48 What are two possible fates for a free radical produced from a molecule–radiation interaction?

Biochemical Effects of Radiation (Section 11.8)

11.49 Contrast the abilities of alpha, beta, and gamma radiations to penetrate a thick sheet of paper.

11.50 Contrast the abilities of alpha, beta, and gamma radiations to penetrate human skin.

11.51 Contrast the velocities with which alpha, beta, and gamma radiations are emitted by nuclei.

11.52 Contrast the following for alpha and beta particles.
a. The distance that they can travel in air before their excess energy is dissipated
b. The number of ion pairs they can produce in traveling the distance noted in part **a**

11.53 What would be the expected effect of each of the following short-term, whole-body radiation exposures?
a. 10 rems   b. 150 rems

11.54 What would be the expected effect of each of the following short-term, whole-body radiation exposures?
a. 50 rems   b. 250 rems

Detection of Radiation (Section 11.9)

11.55 Why do technicians who work around radiation usually wear film badges?

11.56 Explain the principle of operation of a Geiger counter.

Sources of Radiation Exposure (Section 11.10)

11.57 What is background radiation?

11.58 Contrast the radiation exposure that an average American receives from natural sources with the radiation exposure an average American receives from human-activity sources.

11.59 List the four major sources of low-level exposure to background radiation for an average American.

11.60 List major sources of low-level exposure to radiation generated from human activities for an average American.

Nuclear Medicine (Section 11.11)

11.61 Why are the radionuclides used for diagnostic procedures usually gamma emitters?

11.62 Why do the radionuclides used in diagnostic procedures nearly always have short half-lives?

11.63 Explain how each of the following radionuclides is used in diagnostic medicine.
a. Gallium-67   b. Sodium-24
c. Thallium-201   d. Chromium-51

11.64 The radionuclide technetium-99m has numerous uses in diagnostic medicine.
a. What determines the target location for technetium-99m in its uses?
b. What is the significance of the *m* in the notation technetium-99*m*?

11.65 How do the radionuclides used for therapeutic purposes differ from the radionuclides used for diagnostic purposes?

11.66 Contrast the different ways in which cobalt-60 and yttrium-90 are used in radiation therapy.

Nuclear Fission and Nuclear Fusion (Section 11.12)

11.67 How many neutrons are produced in each of the following uranium-235 fission reactions?
a. $^{235}_{92}U + ^{1}_{0}n \rightarrow ^{135}_{53}I + ^{97}_{39}Y + $ neutron(s)
b. $^{235}_{92}U + ^{1}_{0}n \rightarrow ^{72}_{30}Zn + ^{160}_{62}Sm + $ neutron(s)

c. U-235 + neutron $\rightarrow$ Rb-90 + Cs-144 + neutron(s)
d. U-235 + neutron $\rightarrow$ Ba-142 + Kr-91 + neutron(s)

11.68 How many neutrons are produced in each of the following uranium-235 fission reactions?
a. $^{235}_{92}U + ^{1}_{0}n \rightarrow ^{87}_{35}Br + ^{146}_{57}La + $ neutron(s)
b. $^{235}_{92}U + ^{1}_{0}n \rightarrow ^{139}_{56}Ba + ^{94}_{36}Kr + $ neutron(s)
c. U-235 + neutron $\rightarrow$ Xe-139 + Sr-95 + neutron(s)
d. U-235 + neutron $\rightarrow$ Sn-131 + Mo-103 + neutron(s)

11.69 What nuclide undergoes nuclear fission to give barium-143, rubidium-94, and three neutrons?

11.70 What nuclide undergoes nuclear fission to give cesium-143, rubidium-89, and two neutrons?

11.71 Identify particle X in each of the following nuclear fusion reactions.
a. Two helium-3 nuclei fuse to give two protons and particle X.
b. Lithium-7 and particle X fuse to give two alpha particles and a neutron.

11.72 Identify particle X in each of the following nuclear fusion reactions.
a. Hydrogen-3 and particle X fuse to give helium-4 and a neutron.
b. Two identical X particles fuse to give helium-4 and two protons.

11.73 Identify which of the following characteristics apply to the fission process, which to the fusion process, and which to both processes.
a. An extremely high temperature is required to start the process.
b. An example of the process occurs on the sun.
c. Transmutation of elements occurs.
d. Neutrons are needed to start the process.

11.74 Identify which of the following characteristics apply to the fission process, which to the fusion process, and which to both processes.
a. Large amounts of energy are released in the process.
b. Energy released in the process is called *nuclear energy*.
c. The process is now used to generate some electrical power in the United States.
d. A fourth state of matter called *plasma* is encountered in studying this process.

11.75 Identify each of the following nuclear reactions as fission, fusion, or neither.
a. $^{3}_{2}He + ^{3}_{2}He \rightarrow ^{4}_{2}He + 2\,^{1}_{1}H$
b. $^{235}_{92}U + ^{1}_{0}n \rightarrow ^{144}_{55}Cs + ^{90}_{37}Rb + 2\,^{1}_{0}n$
c. $^{209}_{83}Bi + ^{4}_{2}\alpha \rightarrow ^{210}_{85}At + 3\,^{1}_{0}n$
d. $^{238}_{92}U \rightarrow ^{234}_{90}Th + ^{4}_{2}\alpha$

11.76 Identify each of the following nuclear reactions as fission, fusion, or neither.
a. $^{239}_{92}U \rightarrow ^{239}_{93}Np + ^{0}_{-1}\beta$
b. $^{230}_{90}Th + ^{1}_{1}p \rightarrow ^{223}_{87}Fr + 2\,^{4}_{2}\alpha$
c. $^{3}_{1}H + ^{2}_{1}H \rightarrow ^{4}_{2}He + ^{1}_{0}n$
d. $^{235}_{92}U + ^{1}_{0}n \rightarrow ^{139}_{54}Xe + ^{95}_{38}Sr + 2\,^{1}_{0}n$

Nuclear and Chemical Reactions Compared (Section 11.13)

11.77 Contrast the behavior of different isotopes of an element in nuclear and chemical reactions.

11.78 Contrast the energy changes that accompany nuclear and chemical reactions.

11.79 Contrast how factors such as temperature, pressure, and catalysts affect reaction rate for nuclear and chemical reactions.

11.80 Contrast how the state of chemical combination of an element affects reactivity in nuclear and chemical reactions.

ADDITIONAL PROBLEMS

11.81 Write nuclear equations for each of the following radioactive-decay processes.
   a. Thallium-206 is formed by beta emission.
   b. Palladium-109 undergoes beta emission.
   c. Plutonium-241 is formed by alpha emission.
   d. Fermium-249 undergoes alpha emission.

11.82 Cobalt-55 has a half-life of 18 hours. How long will it take, in hours, for the following fractions of nuclides in a cobalt-55 sample to decay?
   a. 7/8   b. 31/32
   c. 63/64   d. 127/128

11.83 Write equations for the following nuclear bombardment processes.
   a. Bombardment of a radionuclide with an alpha particle produces curium-242 and one neutron.
   b. Bombardment of curium-246 with a small particle produces nobelium-254 and four neutrons.
   c. Aluminum-27 is bombarded with an alpha particle and produces a neutron.
   d. Bombardment of sodium-23 with hydrogen-2 produces neon-21.

11.84 The second artificially produced element was promethium. Write the equation for the production of ^{143}Pm by the bombardment of ^{142}Nd with neutrons.

11.85 Using Table 11.2 as your source of information, determine for how many of the transuranium elements the most stable isotope has a half-life that is less than 1.0 day.

11.86 How many neutrons are produced as a result of the following fission reaction?

$$\text{Neutron} + \text{U-235} \longrightarrow \text{I-135} + \text{Y-97} + \text{? neutrons}$$

11.87 Consider the decay series

$$E \longrightarrow F \longrightarrow G \longrightarrow H$$

where E, F, and G are radioactive, with half-lives of 10.0 sec, 1.2 min, and 12.5 days, respectively, and H is nonradioactive. Starting with 1000 atoms of E, and none of F, G, and H, estimate the numbers of atoms of E, F, G, and H that are present after 50 days.

11.88 Fill in the blanks in the following segment of a radioactive decay series.

$$\underline{\ ?\ } \xrightarrow{\ \beta\ } \underline{\ ?\ } \xrightarrow{\ \alpha\ } {}^{224}\text{Ra} \xrightarrow{\ \beta\ } \underline{\ ?\ }$$

𝓜ultiple-Choice Practice Test

11.89 Which of the following statements concerning the nature of emissions from naturally occurring radioactive materials is *incorrect*?
   a. Alpha particles carry a positive charge.
   b. Beta particles have a mass less than that of alpha particles.
   c. Gamma particles have a mass of 1 amu.
   d. Alpha particles have a mass of 4 amu.

11.90 The loss of an alpha particle by a radiounuclide causes which of the following?
   a. Both its atomic number and its mass number increase.
   b. Both its atomic number and its mass number decrease.
   c. Its atomic number increases, and its mass number decreases.
   d. Its atomic number decreases, and its mass number increases.

11.91 Which is the daughter nuclide for the alpha decay of polonium-212?
   a. Lead-208   b. Lead-216
   c. Radon-208   d. Radon-216

11.92 The beta decay of a nuclide of $^{234}_{90}$Th produces a nuclide of which of the following?
   a. Element 88   b. Element 89
   c. Element 91   d. Element 92

11.93 After three half-lives have elapsed, the amount of a radioactive sample which has not decayed is which of the following?
   a. 1/3 the original amount   b. 1/9 the original amount
   c. 1/4 the original amount   d. 1/8 the original amount

11.94 The half-life of cobalt-60 is 5.2 years. Thus what happens to a sample of cobalt-60 after 5.2 years?
   a. It breaks in half.
   b. It turns into cobalt-30.
   c. It contains one-half as many cobalt-60 atoms as it did originally.
   d. It contains twice as many cobalt-60 atoms as it did originally.

11.95 The general production process for synthetic elements is which of the following?
   a. Both a transmutation reaction and a bombardment reaction
   b. A transmutation reaction but not a bombardment reaction
   c. A bombardment reaction but not a transmutation reaction
   d. Neither a transmutation reaction nor a bombardment reaction

11.96 Forms of ionizing radiation include all of the following except one. Which is the exception?
   a. Visible light
   b. X rays
   c. Alpha particles
   d. Ultraviolet light

11.97 Which of the following statements about natural radioactive emissions is *incorrect*?
   a. Beta particles are emitted from nuclei at speeds of up to 0.9 the speed of light.
   b. Alpha particles cannot penetrate the body's outer layers of skin.
   c. Alpha and beta particles and gamma radiation are all capable of knocking electrons off atoms with which they collide.
   d. A piece of aluminum foil will stop both alpha particles and gamma radiation.

11.98 Generation of electricity in a nuclear power plant and generation of energy within the sun involve which of the following processes, respectively?
   a. Nuclear fission and nuclear fusion
   b. Nuclear fusion and nuclear fission
   c. Nuclear fission and nuclear fission (both the same process)
   d. Nuclear fusion and nuclear fusion (both the same process)

Saturated Hydrocarbons

Chapter Outline

Crude oil (petroleum) constitutes the largest and most important natural source for saturated hydrocarbons, the simplest type of organic compound. Here is shown a pump and towers associated with obtaining crude oil from underground deposits.

This chapter is the first of six that deal with the subject of organic chemistry and organic compounds. Organic compounds are the chemical basis for life itself, as well as an important component of the basis for our current high standard of living. Proteins, carbohydrates, enzymes, and hormones are organic molecules. Organic compounds also include natural gas, petroleum, coal, gasoline, and many synthetic materials such as dyes, plastics, and clothing fibers.

12.1 ORGANIC AND INORGANIC COMPOUNDS

During the latter part of the eighteenth century and the early part of the nineteenth century, chemists began to categorize compounds into two types: organic and inorganic. Compounds obtained from living organisms were called *organic* compounds, and compounds obtained from mineral constituents of the earth were called *inorganic* compounds.

During this early period, chemists believed that a special "vital force" supplied by a living organism was necessary for the formation of an organic compound. This concept was proved incorrect in 1828 by the German chemist Friedrich Wöhler. Wöhler heated an aqueous solution of two inorganic compounds, ammonium chloride and silver cyanate, and obtained urea (a component of urine).

$$NH_4Cl + AgNCO \longrightarrow (NH_2)_2CO + AgCl$$
Urea

The historical origins of the terms *organic* and *inorganic* involve the following conceptual pairings:

*org*anic—living *org*anisms
*in*organic—*in*animate materials

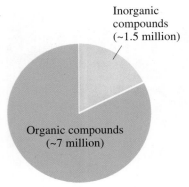

Inorganic compounds (~1.5 million)

Organic compounds (~7 million)

Figure 12.1 Sheer numbers is one reason why organic chemistry is a separate field of chemical study. Approximately 7 million organic compounds are known, compared to "just" 1.5 million inorganic compounds.

Some textbooks define organic chemistry as the study of carbon-containing compounds. Almost all carbon-containing compounds qualify as organic compounds. However, the oxides of carbon, carbonates, cyanides, and metallic carbides are classified as inorganic rather than organic compounds. Inorganic carbon compounds involve carbon atoms that are not bonded to hydrogen atoms (CO, CO_2, Na_2CO_3, and so on).

Carbon atoms in organic compounds, in accordance with the octet rule, always form four covalent bonds.

Soon other chemists had successfully synthesized organic compounds from inorganic starting materials. As a result, the vital-force theory was completely abandoned.

The terms *organic* and *inorganic* continue to be used in classifying compounds, but the definitions of these terms no longer reflect their historical origins. **Organic chemistry** *is the study of hydrocarbons (compounds of carbon and hydrogen) and their derivatives.* Nearly all compounds found in living organisms are still classified as organic compounds, as are many compounds that have been synthesized in the laboratory and have never been found in a living organism. **Inorganic chemistry** *is the study of all substances other than hydrocarbons and their derivatives.*

In essence, organic chemistry is the study of the compounds of one element (carbon), and inorganic chemistry is the study of the compounds of the other 116 elements. This unequal partitioning occurs because there are approximately 7 million organic compounds and only an estimated 1.5 million inorganic compounds (Figure 12.1). This is an approximately 5:1 ratio between organic and inorganic compounds.

12.2 BONDING CHARACTERISTICS OF THE CARBON ATOM

Why does the element carbon form five times as many compounds as all the other elements combined? The answer is that carbon atoms have the unique ability to bond to each other in a wide variety of ways that involve long chains of carbon atoms or cyclic arrangements (rings) of carbon atoms. Sometimes both chains and rings of carbon atoms are present in the same molecule.

The variety of covalent bonding "behaviors" possible for carbon atoms is related to carbon's electron configuration. Carbon is a member of Group IVA of the periodic table, so carbon atoms possess four valence electrons (Section 4.2). In compound formation, four additional valence electrons are needed to give carbon atoms an octet of valence electrons (the octet rule, Section 4.3). These additional electrons are obtained by electron sharing (covalent bond formation). The sharing of *four* valence electrons requires the formation of *four* covalent bonds.

Carbon can meet this four-bond requirement in three different ways:

1. *By bonding to four other atoms.* This situation requires the presence of four single bonds.

Four single bonds

2. *By bonding to three other atoms.* This situation requires the presence of two single bonds and one double bond.

$$-\overset{|}{C}=$$

Two single bonds and one double bond

3. *By bonding to two other atoms.* This situation requires the presence of either two double bonds or a triple bond and a single bond.

Two double bonds   One triple bond and one single bond

12.3 HYDROCARBONS AND HYDROCARBON DERIVATIVES

The field of organic chemistry encompasses the study of hydrocarbons and hydrocarbon derivatives (Section 12.1). A **hydrocarbon** *is a compound that contains only carbon atoms and hydrogen atoms.* Thousands of hydrocarbons are known. A **hydrocarbon derivative** *is a compound that contains carbon and hydrogen and one or more additional*

Figure 12.2 A summary of classification terms for organic compounds.

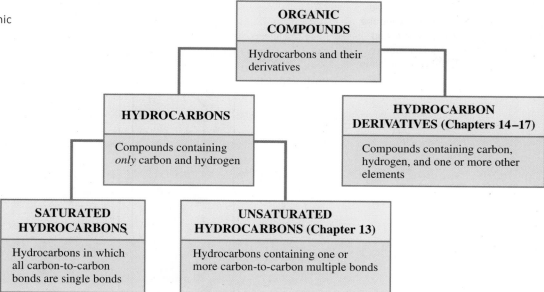

elements. Additional elements commonly found in hydrocarbon derivatives include O, N, S, P, F, Cl, and Br. Millions of hydrocarbon derivatives are known.

Hydrocarbons may be divided into two large classes: saturated and unsaturated. A **saturated hydrocarbon** *is a hydrocarbon in which all carbon–carbon bonds are single bonds.* Saturated hydrocarbons are the simplest type of organic compound. An **unsaturated hydrocarbon** *is a hydrocarbon in which one or more carbon–carbon multiple bonds (double bonds, triple bonds, or both) are present.* In general, saturated and unsaturated hydrocarbons undergo distinctly different chemical reactions.

Saturated hydrocarbons are the subject of this chapter. Unsaturated hydrocarbons are considered in the next chapter. Figure 12.2 summarizes the terminology presented in this section.

Two categories of saturated hydrocarbons exist, those with *acyclic* carbon atom arrangements and those with *cyclic* carbon atom arrangements. The term *acyclic* means "not cyclic." The following notations contrast simple acyclic and cyclic arrangements of six-carbon atoms.

We first consider saturated hydrocarbons with *acyclic* carbon atom arrangements (Sections 12.4 through 12.11) and then saturated hydrocarbons with *cyclic* carbon atom arrangements (Sections 12.12 through 12.14).

12.4 ALKANES: ACYCLIC SATURATED HYDROCARBONS

An **alkane** *is a saturated hydrocarbon in which the carbon atom arrangement is acyclic.* Thus an alkane is a hydrocarbon that contains only carbon–carbon single bonds (saturated) and has no rings of carbon atoms (acyclic).

The molecular formulas of all alkanes fit the general formula C_nH_{2n+2}, where n is the number of carbon atoms present. The number of hydrogen atoms present in an alkane is always twice the number of carbon atoms plus two more, as in C_4H_{10}, C_5H_{12}, and C_8H_{18}.

The three simplest alkanes are methane (CH_4), ethane (C_2H_6), and propane (C_3H_8). Three different methods for showing the three-dimensional structures of these simplest of all alkanes

> The term *saturated* has the general meaning that there is no more room for something. Its use with hydrocarbons comes from early studies in which chemists tried to add hydrogen atoms to various hydrocarbon molecules. Compounds to which no more hydrogen atoms could be added (because they already contained the maximum number) were called saturated, and those to which hydrogen could be added were called unsaturated.

Figure 12.3 Three different three-dimensional ways of representing the structures of methane, ethane, and propane: dash-wedge-line structure, ball-and-stick model, and space-filling model.

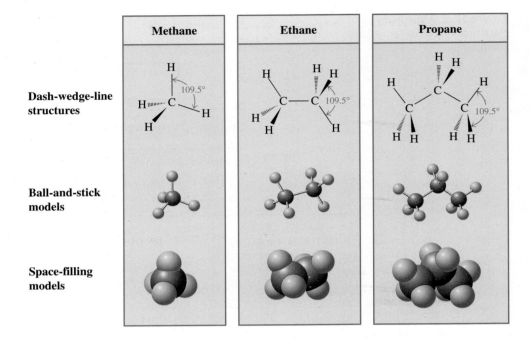

are given in Figure 12.3. They are dash-wedge-line structures, ball-and-stick models, and space-filling models. Note how each carbon atom in each of the models participates in four bonds (Section 12.2). Note also that the geometrical arrangement of atoms about each carbon atom is tetrahedral, an arrangement consistent with the principles of VSEPR theory (Section 5.8). The tetrahedral arrangement of the atoms bonded to alkane carbon atoms is fundamental to understanding the structural aspects of organic chemistry.

12.5 STRUCTURAL FORMULAS

The structures of alkanes, as well as other types of organic compounds, are generally represented in two dimensions rather than three (Figure 12.3) because of the difficulty in drawing the latter. These two-dimensional structural representations make no attempt to portray accurately the bond angles or molecular geometry of molecules. Their purpose is to convey information about which atoms in a molecule are bonded to which other atoms.

Two-dimensional structural representations for organic molecules are called structural formulas. A **structural formula** *is a two-dimensional structural representation that shows how the various atoms in a molecule are bonded to each other.* Structural formulas are of two types: expanded structural formulas and condensed structural formulas. An **expanded structural formula** *is a structural formula that shows all atoms in a molecule and all bonds connecting the atoms.* When written out, expanded structural formulas generally occupy a lot of space, and condensed structural formulas represent a shorthand method for conveying the same information. A **condensed structural formula** *is a structural formula that uses groupings of atoms, in which central atoms and the atoms connected to them are written as a group, to convey molecular structural information.* The expanded and condensed structural formulas for methane, ethane, and propane follow.

Structural formulas, whether expanded or condensed, do not show the geometry (shape) of the molecule. That information can be conveyed only by 3-D drawings or models such as those in Figure 12.3.

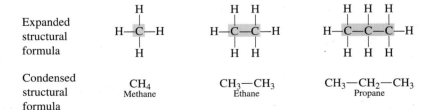

The Occurrence of Methane

Methane (CH_4), the simplest of all hydrocarbons, is a major component of the atmospheres of Jupiter, Saturn, Uranus, and Neptune but only a minor component of Earth's atmosphere (see the accompanying table). Earth's gravitational field, being weaker than that of the large outer planets, cannot retain enough hydrogen (H_2) in its atmosphere to permit the formation of large amounts of methane; H_2 molecules (the smallest and fastest-moving of all molecules) escape from it into outer space.

The small amount of methane present in Earth's atmosphere comes from terrestrial sources. The decomposition of animal and plant matter in an oxygen-deficient environment—swamps, marshes, bogs, and the sediments of lakes—produces methane. A common name for methane, marsh gas, refers to the production of methane in this manner.

Composition of Earth's Atmosphere (in parts per million by volume)

Major Components		Minor Components	
nitrogen	780,800	argon	9340
oxygen	209,500	carbon dioxide	314
		neon	18
		helium	5
		methane	2
		krypton	1

Bacteria that live in termites and in the digestive tracts of plant-eating animals have the ability to produce methane from plant materials (cellulose). The methane output of a large cow (belching and flatulence) can reach 20 liters per day.

Methane entering the atmosphere from terrestrial sources presents an environmental problem. Methane is a "greenhouse gas" that contributes to global warming. Methane is 15 to 30 times more efficient than carbon dioxide (the primary greenhouse gas) in trapping heat radiated from Earth. Fortunately, its atmospheric level of 2.0 ppm by volume is much lower than that of carbon dioxide (over 300 ppm).

Methane gas is also found associated with coal and petroleum deposits. Methane associated with coal mines is considered a hazard. If left to accumulate, it can form pockets where air is not present, and asphyxiation of miners can occur. When mixed with air in certain ratios, it can also present an explosion hazard. Methane associated with petroleum deposits is most often recovered, processed, and marketed as *natural gas*. The processed natural gas used in the heating of homes is 85% to 95% methane by volume. Because methane is odorless, an odorant (smelly compound) must be added to the processed natural gas used in home heating. Otherwise, natural gas leaks could not be detected.

Decomposition of plant and animal matter in marshes is a source of methane gas.

The condensed structural formula for propane, $CH_3—CH_2—CH_3$, is interpreted in the following manner: The first carbon atom is bonded to three hydrogen atoms, and its fourth bond is to the middle carbon atom. The middle carbon atom, besides its bond to the first carbon atom, is also bonded to two hydrogen atoms and to the last carbon atom. The last carbon atom has bonds to three hydrogen atoms in addition to its bond to the middle carbon atom. As is always the case, each carbon atom has four bonds (Section 12.2).

The condensed structural formulas of hydrocarbons in which a long chain of carbon atoms is present are often condensed even more. The formula

$$CH_3—CH_2—CH_2—CH_2—CH_2—CH_2—CH_2—CH_3$$

can be further abbreviated as

$$CH_3—(CH_2)_6—CH_3$$

where parentheses and a subscript are used to denote the number of $—CH_2—$ groups in the chain.

It is important to note that expanded structural formulas show all bonds within a molecule and that condensed structural formulas show only certain bonds—the bonds between carbon atoms. Specifically, the bond line in the condensed structural formula

$$CH_3—CH_3$$

denotes the bond between the first carbon atom and the second carbon atom; it is not a bond between hydrogen atoms and the second carbon atom.

In situations where the focus is solely on the arrangement of carbon atoms in an alkane, *skeletal structural formulas* that omit the hydrogen atoms are often used. A **skeletal structural formula** *is a structural formula that shows the arrangement and bonding of carbon atoms present in an organic molecule but does not show the hydrogen atoms attached to the carbon atoms.*

C—C—C—C—C   means the same as $CH_3—CH_2—CH_2—CH_2—CH_3$
Skeletal structural formula   Condensed structural formula

The skeletal structural formula still represents a unique alkane because we know that each carbon atom shown must have enough hydrogen atoms attached to it to give the carbon four bonds.

ALKANE ISOMERISM

The molecular formulas CH_4, C_2H_6, and C_3H_8 represent the alkanes methane, ethane, and propane, respectively. Next in the alkane molecular formula sequence (C_nH_{2n+2}) is C_4H_{10}, which would be expected to be the molecular formula of the four-carbon alkane. A new phenomenon arises, however, when an alkane has four or more carbon atoms. There is more than one structural formula that is consistent with the molecular formula. Consequently, more than one compound exists with that molecular formula. This situation brings us to the topic of *isomerism*.

Isomers *are compounds that have the same molecular formula (that is, the same numbers and kinds of atoms) but that differ in the way the atoms are arranged.* Isomers, even though they have the same molecular formula, are always different compounds with different properties.

There are two four-carbon alkane isomers, the compounds *butane* and *isobutane*. Both have the molecular formula C_4H_{10}.

$CH_3—CH_2—CH_2—CH_3$ $CH_3—CH—CH_3$
                                                                 |
                                                              CH_3

Butane   Isobutane

Butane and isobutane are different compounds with different properties. Butane has a boiling point of −1°C and a melting point of −138°C, whereas the corresponding values for isobutane are −12°C and −159°C.

Contrasting the two C_4H_{10} isomers structurally, note that butane has a chain of four carbon atoms. It is an example of a continuous-chain alkane. A **continuous-chain alkane** *is an alkane in which all carbon atoms are connected in a continuous nonbranching chain.* The other C_4H_{10} isomer, isobutane, has a chain of three carbon atoms with the fourth carbon attached as a branch on the middle carbon of the three-carbon chain. It is an example of a branched-chain alkane. A **branched-chain alkane** *is an alkane in which one or more branches (of carbon atoms) are attached to a continuous chain of carbon atoms.*

There are three isomers for alkanes with five carbon atoms (C_5H_{12}):

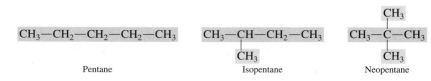

Pentane   Isopentane   Neopentane

The word *isomer* comes from the Greek *isos,* which means "the same," and *meros,* which means "parts." Isomers have the same parts put together in different ways.

The existence of isomers necessitates the use of structural formulas in organic chemistry. Isomers always have the same molecular formula and different structural formulas.

Figure 12.4 Space-filling models for the three isomeric C_5H_{12} alkanes: (a) pentane, (b) isopentane, and (c) neopentane.

(a) Pentane	(b) Isopentane	(c) Neopentane
Boiling point = 36.1°C	Boiling point = 27.8°C	Boiling point = 9.5°C
Density = 0.626 g/mL	Density = 0.620 g/mL	Density = 0.614 g/mL

Constitutional isomers are also frequently called *structural isomers*. The general characteristics of such isomers, independent of which name is used, are the same molecular formula and different structural formulas.

TABLE 12.1

Number of Isomers Possible for Alkanes of Various Carbon Chain Lengths

Molecular Formula	Possible Number of Isomers
CH_4	1
C_2H_6	1
C_3H_8	1
C_4H_{10}	2
C_5H_{12}	3
C_6H_{14}	5
C_7H_{16}	9
C_8H_{18}	18
C_9H_{20}	35
$C_{10}H_{22}$	75
$C_{15}H_{32}$	4,347
$C_{20}H_{42}$	336,319
$C_{25}H_{52}$	36,797,588
$C_{30}H_{62}$	4,111,846,763

Figure 12.4 shows space-filling models for the three isometric C_5 alkanes. Note how neopentane, the most branched isomer, has the most compact, most spherical three-dimensional shape.

The number of possible alkane isomers increases dramatically with increasing numbers of carbon atoms in the alkane, as shown in Table 12.1. Such isomerism is one of the major reasons for the existence of so many organic compounds.

Several different types of isomerism exist. The alkane isomerism examples discussed in this section are examples of *constitutional isomerism*. **Constitutional isomers** *are isomers that differ in the connectivity of atoms, that is, in the order in which atoms are attached to each other within molecules.* We will see shortly (Section 12.14) and in later chapters that other types of isomers are also possible, even among compounds whose atoms are connected in the same order. In the biochemistry portion of the text, where carbohydrates, lipids, and proteins are considered, we will find that different isomers elucidate different responses within the human body. Often, when many isomers are possible with the same molecular formula, only one isomer will be physiologically active.

12.7 CONFORMATIONS OF ALKANES

Rotation about carbon–carbon single bonds is an important property of alkane molecules. Two groups of atoms in an alkane connected by a carbon–carbon single bond can rotate with respect to one another around that bond, much as a wheel rotates around an axle.

As a result of rotation around single bonds, alkane molecules (except for methane) can exist in infinite numbers of orientations, or conformations. A **conformation** *is the specific three-dimensional arrangement of atoms in an organic molecule at a given instant that results from rotations about carbon–carbon single bonds.*

The following skeletal formulas represent four different conformations for a continuous-chain six-carbon alkane molecule.

C—C—C—C—C—C   C—C—C—C—C   C—C—C—C   C—C—C—C
                       | | | |
                       C   C   C   C

All four skeletal formulas represent the same molecule; that is, they are different conformations of the same molecule. In all four cases, a continuous chain of six carbon atoms is present. In all except the first case, the chain is "bent," but bends do not disrupt the continuity of the chain.

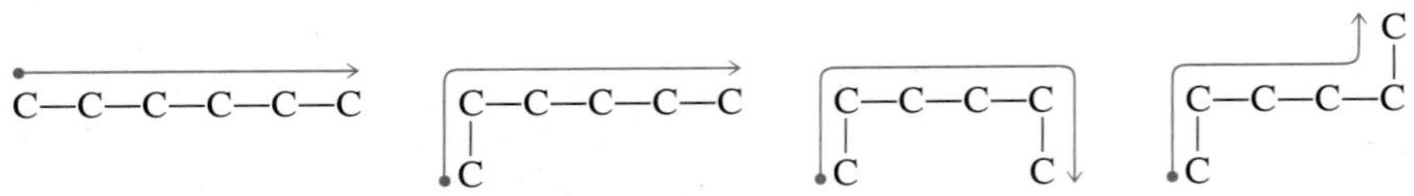

C—C—C—C—C—C

Note that the structures

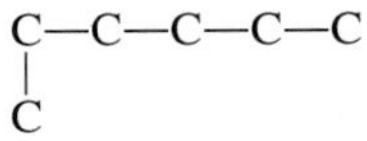

C—C—C—C—C   and   C—C—C—C—C
| |
C   C

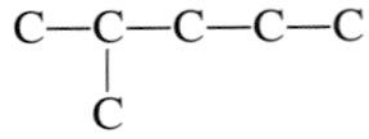

are not two conformations of the same alkane but, rather, represent two different alkanes. The first structure involves a continuous chain of six carbon atoms, and the second structure involves a continuous chain of five carbon atoms to which a branch is attached. There is no way that you can get a continuous chain of six carbon atoms out of the second structure without "back-tracking," and "back-tracking" is not allowed.

You should learn to recognize molecules drawn in several different ways (conformations). Like friends, they can be recognized whether they are sitting, reclining, or standing.

EXAMPLE 12.1

Recognizing Different Conformations of a Molecule and Constitutional Isomers

Determine whether the members of each of the following pairs of structural formulas represent (1) different conformations of the same molecule, (2) different compounds that are constitutional isomers, or (3) different compounds that are not constitutional isomers.

a. CH_3—CH_2—CH_2—CH_3 and CH_2—CH_2
                                              | |
                                           CH_3 CH_3

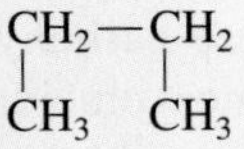

b. 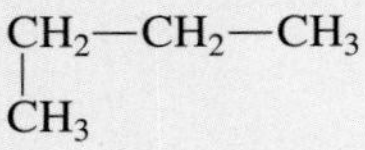 CH_2—CH_2—CH_3 and CH_2—CH_2—CH_2—CH_3
        | |
     CH_3 CH_3

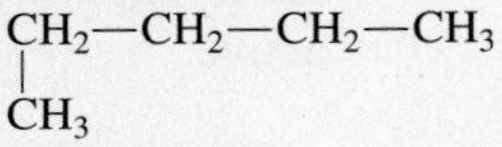

c. 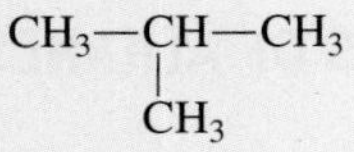 CH_3—CH—CH_3 and CH_3—CH_2—CH_2
             | |
          CH_3 CH_3

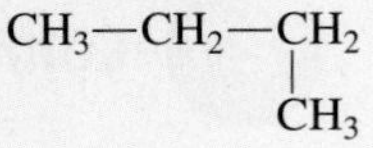

Solution
a. Both molecules have the molecular formula C_4H_{10}. The connectivity of carbon atoms is the same for both molecules: a continuous chain of four carbon atoms. For the second structural formula, we need to go around two corners to get a four-carbon-atom chain, which is fine because of the free rotation associated with single bonds in alkanes.

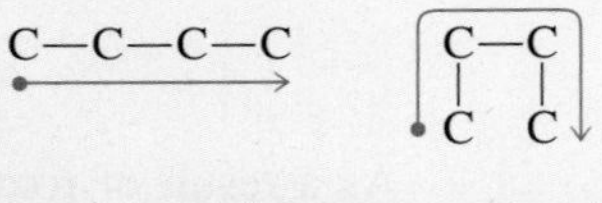

With the same molecular formula and the same connectivity of atoms, these two structural formulas are conformations of the same molecule.

b. The molecular formula of the first compound is C_4H_{10}, and that of the second compound is C_5H_{12}. Thus the two structural formulas represent different compounds that are not constitutional isomers. Constitutional isomers must have the same molecular formula.

c. Both molecules have the same molecular formula, C_4H_{10}. The connectivity of atoms is different. In the first case, we have a chain of three carbon atoms with a branch off the chain. In the second case, a continuous chain of four carbon atoms is present.

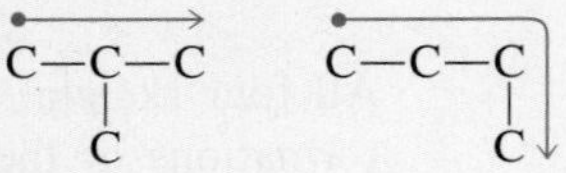

These two structural formulas are those of constitutional isomers.

Practice Exercise 12.1

Determine whether the members of each of the following pairs of structural formulas represent (1) different conformations of the same molecule, (2) different compounds that are constitutional isomers, or (3) different compounds that are not constitutional isomers.

a. CH_3—CH_2—CH_2—CH_2—CH_3 and CH_3—CH_2
CH_2—CH_2
CH_3

b. CH_3—CH—CH_2—CH_3 and CH_3—CH—CH_2
CH_3 CH_3 CH_3

c. CH_3—CH—CH_2—CH_3 and CH_2—CH_2—CH_2
CH_3 CH_3 CH_3

Answers: a. Different conformations; **b.** Different conformations; **c.** Constitutional isomers

The condensed structural formulas for branched-chain alkanes can be further condensed to give linear (straight-line) condensed structural formulas. The linear condensed structural formula for the alkane

$$CH_3—CH—CH_2—CH—CH_3$$
$$CH_3 \qquad CH_3$$

is

$$CH_3—CH—(CH_3)—CH_2—CH—(CH_3)—CH_3 \quad \text{or} \quad (CH_3)_2—CH—CH_2—CH—(CH_3)_2$$

Groups in parentheses in such formulas are understood to be attached to the carbon atom that *precedes* the group in the structural formula, unless the parenthesized group starts the formula. In that case the group is attached to the carbon atom that *follows*. Writing structural formulas in this format is done primarily to reduce the vertical space that the structural formula takes.

12.8 IUPAC NOMENCLATURE FOR ALKANES

When relatively few organic compounds were known, chemists arbitrarily named them using what today are called *common names*. These common names gave no information about the structures of the compounds they described. However, as more organic compounds became known, this nonsystematic approach to naming compounds became unwieldy.

Today, formal systematic rules exist for generating names for organic compounds. These rules, which were formulated and are updated periodically by the International Union of Pure and Applied Chemistry (IUPAC), are known as *IUPAC rules*. The advantage of the IUPAC naming system is that it assigns each compound a name that not only identifies it but also enables one to draw its structural formula.

IUPAC names for the first ten *continuous-chain* alkanes are given in Table 12.2. Note that all of these names end in -*ane,* the characteristic ending for all alkane names. Note also that beginning with the five-carbon alkane, Greek numerical prefixes are used to denote the actual number of carbon atoms in the continuous chain.

To name *branched-chain* alkanes, we must be able to name the branch or branches that are attached to the main carbon chain. These branches are formally called substituents. A **substituent** *is an atom or group of atoms attached to a chain (or ring) of*

IUPAC is pronounced "eye-you-pack."

Continuous-chain alkanes are also frequently called *straight-chain alkanes* and *normal-chain alkanes.*

You need to memorize the prefixes in column two of Table 12.2. This is the way to count from 1 to 10 in "organic chemistry language."

TABLE 12.2 IUPAC Names for the First Ten Continuous-Chain Alkanes*

Molecular Formula	IUPAC Prefix	IUPAC Name	Condensed Structural Formula
CH_4	meth-	methane	CH_4
C_2H_6	eth-	ethane	$CH_3—CH_3$
C_3H_8	prop-	propane	$CH_3—CH_2—CH_3$
C_4H_{10}	but-	butane	$CH_3—CH_2—CH_2—CH_3$
C_5H_{12}	pent-	pentane	$CH_3—CH_2—CH_2—CH_2—CH_3$
C_6H_{14}	hex-	hexane	$CH_3—CH_2—CH_2—CH_2—CH_2—CH_3$
C_7H_{16}	hept-	heptane	$CH_3—CH_2—CH_2—CH_2—CH_2—CH_2—CH_3$
C_8H_{18}	oct-	octane	$CH_3—CH_2—CH_2—CH_2—CH_2—CH_2—CH_2—CH_3$
C_9H_{20}	non-	nonane	$CH_3—CH_2—CH_2—CH_2—CH_2—CH_2—CH_2—CH_2—CH_3$
$C_{10}H_{22}$	dec-	decane	$CH_3—CH_2—CH_2—CH_2—CH_2—CH_2—CH_2—CH_2—CH_2—CH_3$

*The IUPAC naming system also includes prefixes for naming continuous-chain alkanes that have more than 10 carbon atoms, but we will not consider them in this text.

carbon atoms. Note that *substituent* is a general term that applies to carbon-chain attachments in all organic molecules, not just alkanes.

For branched-chain alkanes, the substituents are specifically called *alkyl groups*. An **alkyl group** *is the group of atoms that would be obtained by removing a hydrogen atom from an alkane.*

The two most commonly encountered alkyl groups are the two simplest: the one-carbon and two-carbon alkyl groups. Their formulas and names are

$$——CH_3 \qquad ——CH_2—CH_3$$
Methyl group $\qquad$ Ethyl group

The extra long bond in these formulas (on the left) denotes the point of attachment to the carbon chain. Note that alkyl groups do not lead a stable, independent existence; that is, they are not molecules. They are always found attached to another entity (usually a carbon chain).

Alkyl groups are named by taking the stem of the name of the alkane that contains the same number of carbon atoms and adding the ending *-yl.* Table 12.3 gives the names for small continuous-chain alkyl groups.

We are now ready for the IUPAC rules for naming branched-chain alkanes.

Rule 1: *Identify the longest continuous carbon chain (the parent chain), which may or may not be shown in a straight line, and name the chain.*

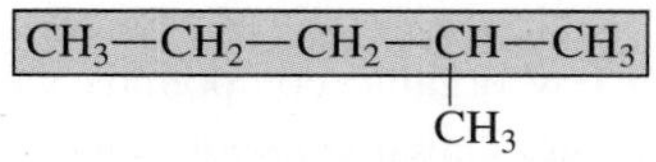

> An additional guideline for identifying the longest continuous carbon chain: If two different carbon chains in a molecule have the same largest number of carbon atoms, select as the parent chain the one with the larger number of substituents (alkyl groups) attached to the chain.

$$CH_3—CH_2—CH_2—CH—CH_3$$
$$|$$
$$CH_3$$

The parent chain name is *pentane,* because it has five carbon atoms.

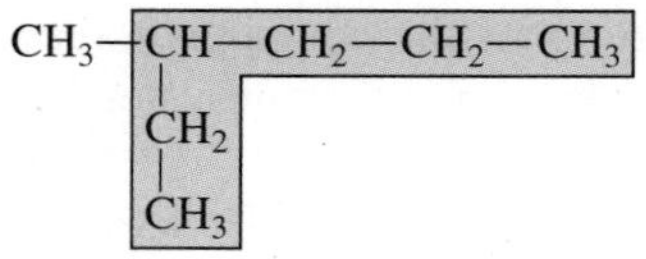

$$CH_3—CH—CH_2—CH_2—CH_3$$
$$|$$
$$CH_2$$
$$|$$
$$CH_3$$

The parent chain name is *hexane,* because it has six carbon atoms.

TABLE 12.3
Names for the First Six Continuous-Chain Alkyl Groups

The ending *-yl,* as in meth*yl,* eth*yl,* prop*yl,* and but*yl,* appears in the names of all alkyl groups.

Number of Carbons	Structural Formula	Stem of Alkane Name	Suffix	Alkyl Group Name
1	$—CH_3$	meth-	$—yl$	methyl
2	$—CH_2—CH_3$	eth-	$—yl$	ethyl
3	$—CH_2—CH_2—CH_3$	prop-	$—yl$	propyl
4	$—CH_2—CH_2—CH_2—CH_3$	but-	$—yl$	butyl
5	$—CH_2—CH_2—CH_2—CH_2—CH_3$	pent-	$—yl$	pentyl
6	$—CH_2—CH_2—CH_2—CH_2—CH_2—CH_3$	hex-	$—yl$	hexyl

Additional guidelines for numbering carbon atom chains:

1. If both ends of the chain have a substituent the same distance in, number from the end closest to the second-encountered substituent.
2. If there are substituents equidistant from each end of the chain and there is no third substituent to use as the "tie-breaker," begin numbering at the end nearest the substituent that has alphabetical priority—that is, the substituent whose name occurs first in the alphabet.

Rule 2: *Number the carbon atoms in the parent chain from the end of the chain nearest a substituent (alkyl group).*

There always are two ways to number the chain (either from left to right or from right to left). This rule gives the first-encountered alkyl group the lowest possible number.

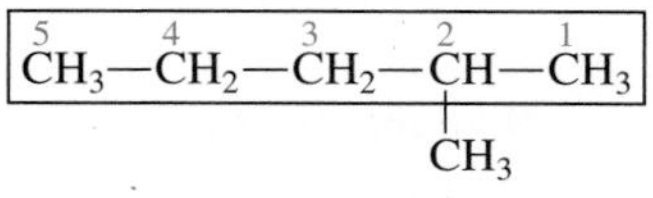

$$\overset{5}{C}H_3-\overset{4}{C}H_2-\overset{3}{C}H_2-\overset{2}{C}H-\overset{1}{C}H_3$$
$$\underset{CH_3}{|}$$

Right-to-left numbering system

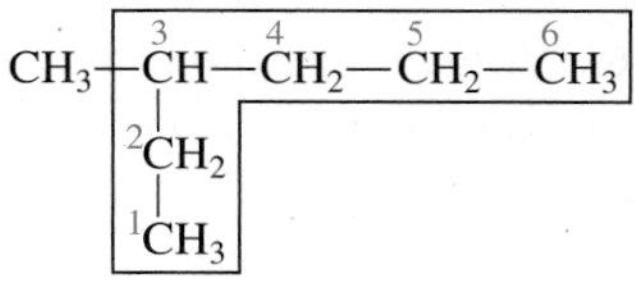

$$CH_3-\overset{3}{C}H-\overset{4}{C}H_2-\overset{5}{C}H_2-\overset{6}{C}H_3$$
$$\overset{2}{|}{C}H_2$$
$$\overset{1}{|}{C}H_3$$

Left-to-right numbering system

Rule 3: *If only one alkyl group is present, name and locate it (by number), and prefix the number and name to that of the parent carbon chain.*

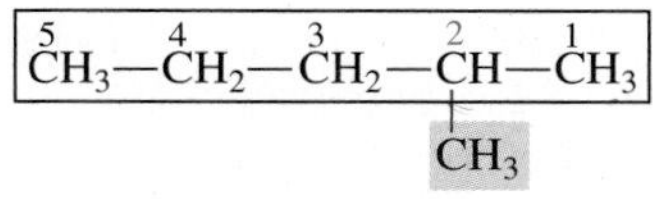

$$\overset{5}{C}H_3-\overset{4}{C}H_2-\overset{3}{C}H_2-\overset{2}{C}H-\overset{1}{C}H_3$$
$$\underset{CH_3}{|}$$

2-Methylpentane

$$CH_3-\overset{3}{C}H-\overset{4}{C}H_2-\overset{5}{C}H_2-\overset{6}{C}H_3$$
$$\overset{2}{|}{C}H_2$$
$$\overset{1}{|}{C}H_3$$

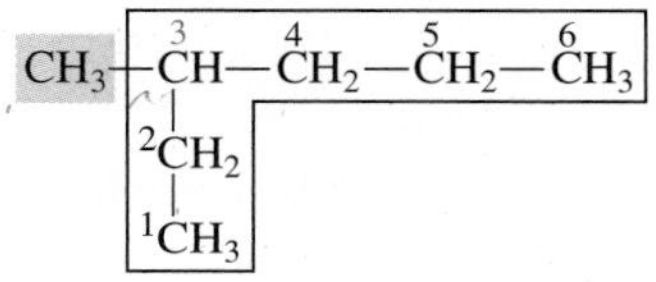

3-Methylhexane

Note that the name is written as one word, with a hyphen between the number (location) and the name of the alkyl group.

Rule 4: *If two or more of the same kind of alkyl group are present in a molecule, indicate the number with a Greek numerical prefix (di-, tri-, tetra-, penta-, and so forth). In addition, a number specifying the location of each identical group must be included. These position numbers, separated by commas, precede the numerical prefix. Numbers are separated from words by hyphens.*

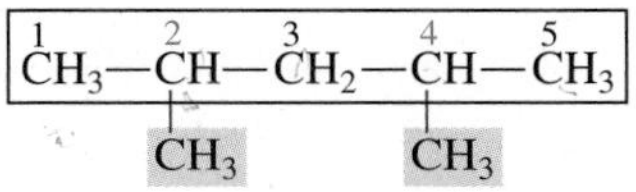

$$\overset{1}{C}H_3-\overset{2}{C}H-\overset{3}{C}H_2-\overset{4}{C}H-\overset{5}{C}H_3$$
$$\underset{CH_3}{|}\quad\quad\underset{CH_3}{|}$$

2,4-Dimethylpentane

$$\overset{1}{C}H_3-\overset{2}{C}H_2-\overset{3}{\underset{|}{C}}-\overset{4}{C}H_2-\overset{5}{C}H_3$$
$$\overset{CH_3}{|}$$
$$\underset{CH_3}{}$$

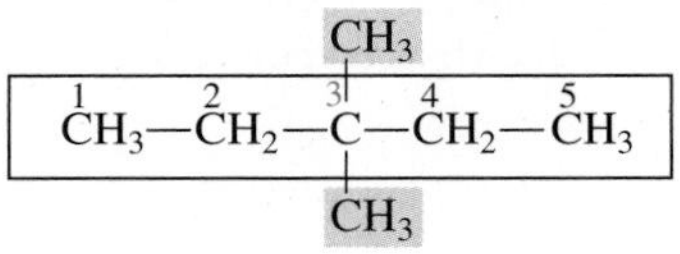

3,3-Dimethylpentane

There must be as many numbers as there are alkyl groups in the IUPAC name of a branched-chain alkane.

Note that the numerical prefix *di-* must always be accompanied by two numbers, *tri-* by three, and so on, even if the same number is written twice, as in 3,3-dimethylpentane.

Rule 5: *When two kinds of alkyl groups are present on the same carbon chain, number each group separately, and list the names of the alkyl groups in alphabetical order.*

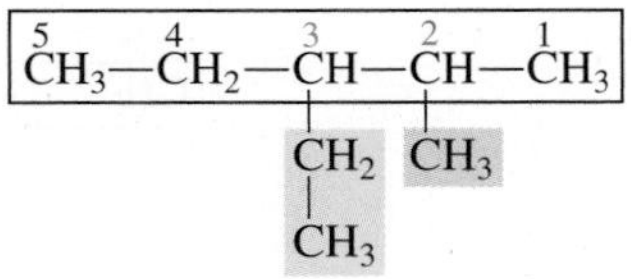

$$\overset{5}{C}H_3-\overset{4}{C}H_2-\overset{3}{C}H-\overset{2}{C}H-\overset{1}{C}H_3$$
$$\underset{CH_2}{|}\quad\underset{CH_3}{|}$$
$$\underset{CH_3}{|}$$

3-Ethyl-2-methylpentane

Note that ethyl is named first in accordance with the alphabetical rule.

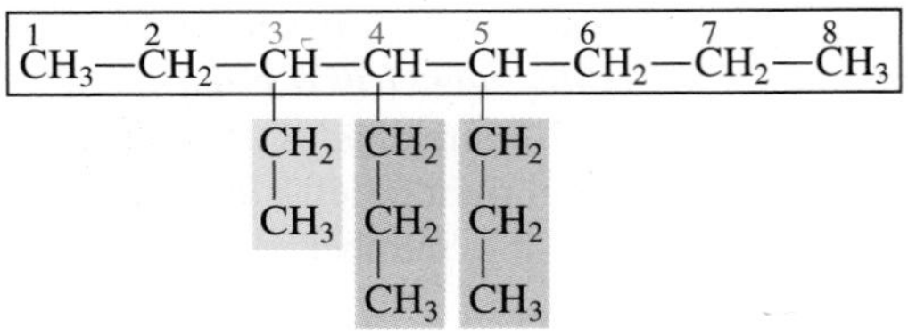

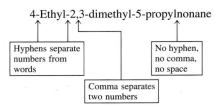

3-Ethyl-4,5-dipropyloctane

Numerical prefixes that designate numbers of alkyl groups, such as *di-*, *tri-*, and *tetra-*, are not considered when determining alphabetical priority for alkyl groups.

Note that the prefix *di-* does not affect the alphabetical order; *ethyl* precedes *propyl*.

Rule 6: *Follow IUPAC punctuation rules, which include the following: (1) Separate numbers from each other by commas. (2) Separate numbers from letters by hyphens. (3) Do not add a hyphen or a space between the last-named substituent and the name of the parent alkane that follows.*

4-Ethyl-2,3-dimethyl-5-propylnonane

Hyphens separate numbers from words

No hyphen, no comma, no space

Comma separates two numbers

EXAMPLE 12.2

Determining IUPAC Names for Branched-Chain Alkanes

Give the IUPAC name for each of the following branched-chain alkanes.

a. 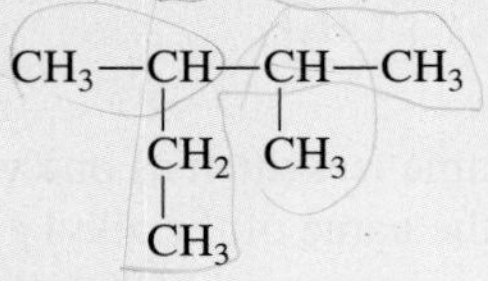 $CH_3-CH-CH-CH_3$
        $\quad\quad CH_2\ \ CH_3$
        $\quad\quad CH_3$

b. 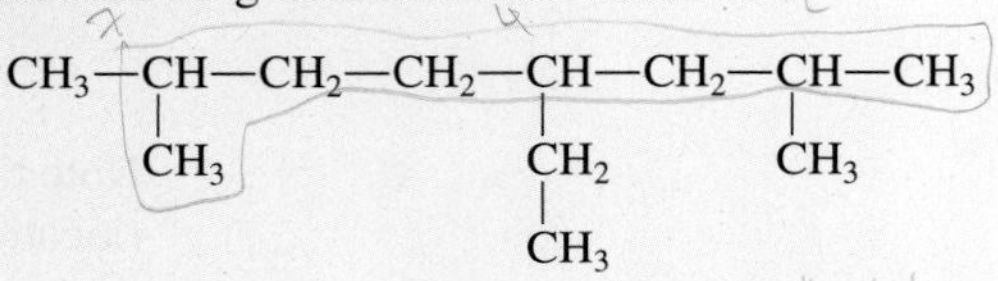 $CH_3-CH-CH_2-CH_2-CH-CH_2-CH-CH_3$
        $\quad\quad\quad CH_3 \quad\quad\quad\quad CH_2 \quad\quad CH_3$
        $\quad\quad\quad\quad\quad\quad\quad\quad\quad\quad CH_3$

v-c- 2,7-dimethyloctane

Solution

a. The longest carbon chain possesses five carbon atoms. Thus the parent-chain name is pentane.

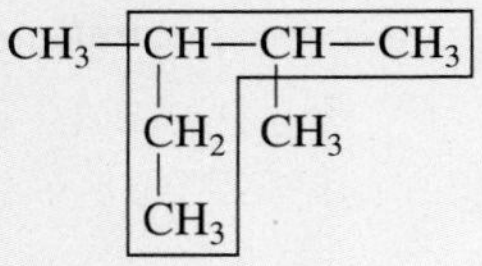

$CH_3-CH-CH-CH_3$
$\quad\quad CH_2\ \ CH_3$
$\quad\quad CH_3$

This parent chain is numbered from right to left because an alkyl substituent is closer to the right end of the chain than to the left end.

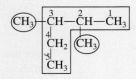

There are two methyl group substituents (circled). One methyl group is located on carbon 2 and the other on carbon 3. The IUPAC name for the compound is 2,3-dimethylpentane.

b. There are eight carbon atoms in the longest carbon chain, so the parent name is octane. There are three alkyl groups present (circled).

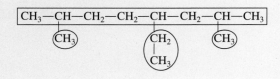

Selection of the numbering system to be used cannot be made based on the "first-encountered-alkyl-group rule" because an alkyl group is equidistant from each end of the chain. Thus the second-encountered alkyl group is used as the "tie-breaker." It is closer to the right end of the parent chain (carbon 4) than to the left end (carbon 5). Thus we use the right-to-left numbering system.

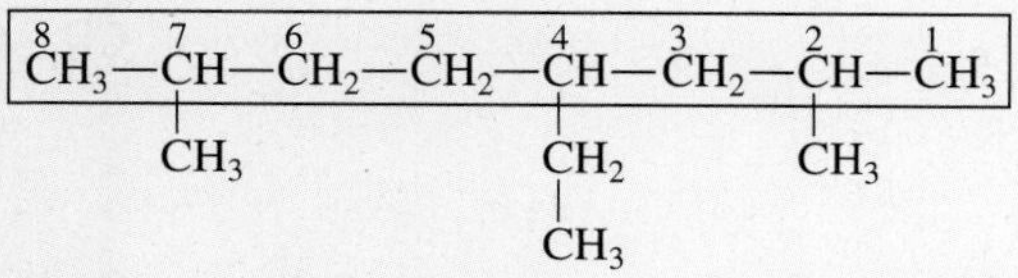

$$\underset{CH_3}{\overset{8}{C}H_3}-\overset{7}{C}H-\overset{6}{C}H_2-\overset{5}{C}H_2-\overset{4}{C}H-\overset{3}{C}H_2-\overset{2}{C}H-\overset{1}{C}H_3$$

Two different kinds of alkyl groups are present: ethyl and methyl. Ethyl has alphabetical priority over methyl and precedes methyl in the IUPAC name. The IUPAC name is 4-ethyl-2,7-dimethyloctane.

> Always compare the total number of carbon atoms in the name with the number of carbon atoms in the structure to make sure they match. The name 4-ethyl-2,7-dimethyloctane indicates the presence of $2 + 2(1) + 8 = 12$ carbon atoms. The structure does have 12 carbon atoms.

Practice Exercise 12.2

Give the IUPAC name for each of the following alkanes.

a. 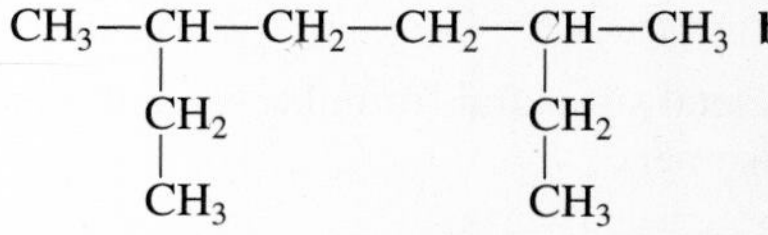 $CH_3-CH-CH_2-CH_2-CH-CH_3$ with CH_2 and CH_3 branches **b.** 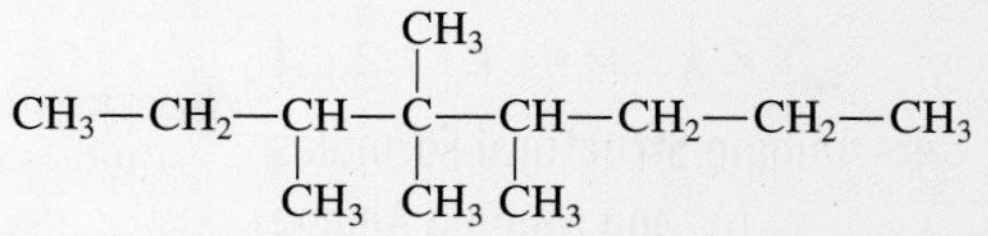 $CH_3-CH_2-CH-C-CH-CH_2-CH_2-CH_3$ with CH_3, CH_3, CH_3, CH_3 branches

Answers: a. 3,6-Dimethyloctane; **b.** 3,4,4,5-Tetramethyloctane

After you learn the rules for naming alkanes, it is relatively easy to reverse the procedure and translate the name of an alkane into a structural formula. Example 12.3 shows how this is done.

● EXAMPLE 12.3

Generating the Structural Formula of an Alkane from Its IUPAC Name

> A few smaller branched alkanes have common names—that is, non-IUPAC names—that still have widespread use. They make use of the prefixes *iso* and *neo,* as in isobutane, isopentane, and neohexane. These prefixes denote particular end-of-chain carbon atom arrangements. 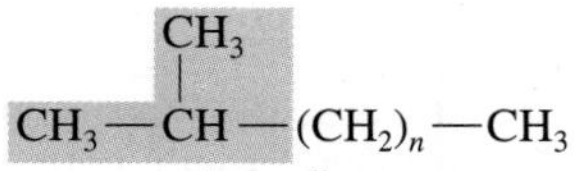
>
> $$\underset{\text{An isoalkane}}{CH_3-\underset{\overset{|}{CH_3}}{CH}-(CH_2)_n-CH_3}$$
> (e.g., $n = 1$, Isopentane) 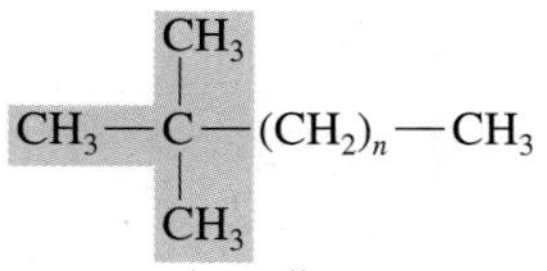
>
> $$\underset{\text{A neoalkane}}{CH_3-\underset{\overset{|}{CH_3}}{\overset{\overset{CH_3}{|}}{C}}-(CH_2)_n-CH_3}$$
> (e.g., $n = 1$, Neohexane)

▶ Draw the condensed structural formula for 3-ethyl-2,3-dimethylpentane.

Solution

Step 1: The name of this compound ends in *pentane,* so the longest continuous chain has five carbon atoms. Draw this chain of five carbon atoms and number it.

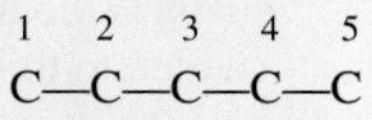

$$\overset{1}{C}-\overset{2}{C}-\overset{3}{C}-\overset{4}{C}-\overset{5}{C}$$

Step 2: Complete the carbon skeleton by attaching alkyl groups as they are specified in the name. An ethyl group goes on carbon 3, and methyl groups are attached to carbons 2 and 3.

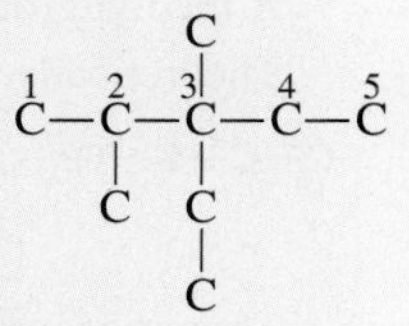

$$\overset{1}{C}-\overset{2}{C}-\overset{3}{C}-\overset{4}{C}-\overset{5}{C}$$ with C above carbon 3, C below carbon 2 and carbon 3, and C below carbon 3

Step 3: Add hydrogen atoms to the carbon skeleton so that each carbon atom has four bonds.

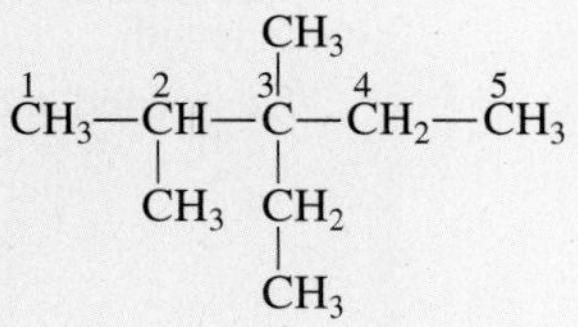

$$\overset{1}{C}H_3-\underset{\overset{|}{CH_3}}{\overset{2}{C}H}-\underset{\overset{|}{\underset{\overset{|}{CH_3}}{CH_2}}}{\overset{\overset{CH_3}{|}}{\overset{3}{C}}}-\overset{4}{C}H_2-\overset{5}{C}H_3$$

(continued)

Practice Exercise 12.3

Draw the condensed structural formula for 4,5-diethyl-3,4,5-trimethyloctane.

Answer:

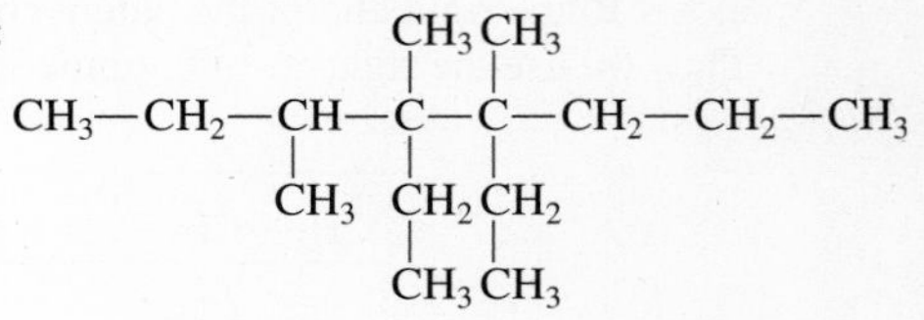

$$CH_3-CH_2-CH-\overset{\overset{\displaystyle CH_3}{|}}{C}-\overset{\overset{\displaystyle CH_3}{|}}{C}-CH_2-CH_2-CH_3$$

with CH_3 and CH_2CH_2 substituents below, ending in $CH_3 CH_3$.

The following example, which involves determining the structural formulas for and naming of alkane constitutional isomers, serves as a good review of the structural and naming concepts for alkanes considered so far in this chapter.

● **EXAMPLE 12.4**

Determining Structural Formulas for and Naming Alkane Constitutional Isomers

▶ Draw skeletal structural formulas for, and assign IUPAC names to, all C_6H_{14} alkane constitutional isomers.

Solution

Table 12.1 indicates that there are five constitutional isomers with the chemical formula C_6H_{14}. Part of the purpose of this example is to consider the "thinking pattern" needed to identify these five isomers. There are two concepts embedded in the thinking pattern.

1. Carbon chains of varying length are examined for isomerism possibilities, starting with the chain of maximum length and then examining increasingly shorter chain lengths.
2. Substituents are added to the various carbon chains, with the number of added carbons determined by the chain length. Various location possibilities for the substituents are examined.

Step 1: A C_6 carbon chain is the longest chain possible; it contains all available carbon atoms.

$$C-C-C-C-C-C$$

This is the molecule hexane, the first of the five constitutional isomers. No substituents are added to this chain, as that would increase the carbon count beyond six.

Step 2: Decreasing the carbon-chain length by one gives a C_5 chain.

$$C-C-C-C-C$$

A methyl group must be added to the chain to bring the carbon count back up to six. Theoretically, there are five possible positions for the methyl group:

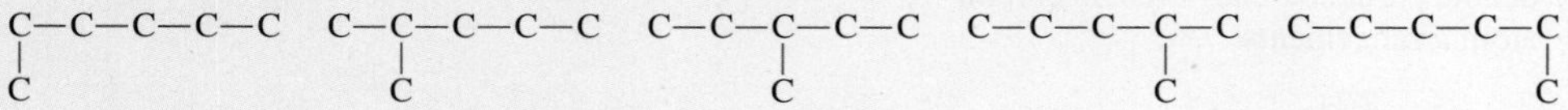

C—C—C—C—C C—C—C—C—C C—C—C—C—C C—C—C—C—C C—C—C—C—C
| | | | |
C   C   C   C   C

These five structures do not represent five new isomers. The first and last structures represent two alternate ways of drawing the molecule hexane, the first isomer. A methyl group (or any alkyl group) added to the end carbons of a carbon chain will always increase the chain length.

The second and third structures do represent new isomers:

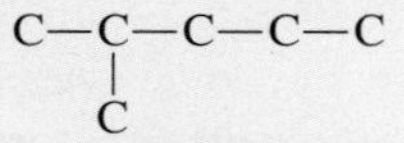

C—C—C—C—C   C—C—C—C—C
    | |
    C   C
2-methylpentane 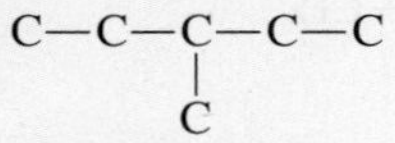   3-methylpentane

The fourth of the five structures is not a new isomer. Numbering its carbon chain from the right end shows that it is 2-methylpentane rather than 4-methylpentane. Thus the second and fourth structures are two representations of the same molecule.

Step 3: Decreasing the chain length to four carbon atoms is the next consideration. Two carbon atoms must now be added as attachments. This can be done in two ways—dimethyl and ethyl.

Examining dimethyl possibilities first, eliminating structures that have methyl groups on terminal carbon atoms gives the following possibilities.

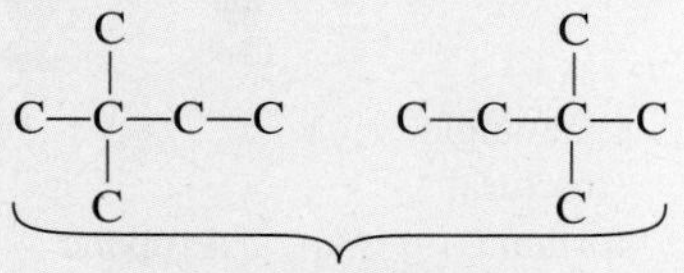

2,2-Dimethylbutane 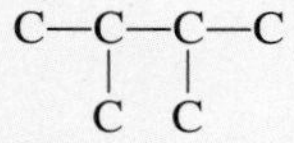   2,3-Dimethylbutane

The first and second structures are the same; both represent the molecule 2,2-dimethylbutane, a fourth isomer.

The third structure, 2,3-dimethylbutane, is different from the other two. It is the fifth isomer.

What about ethyl butanes?

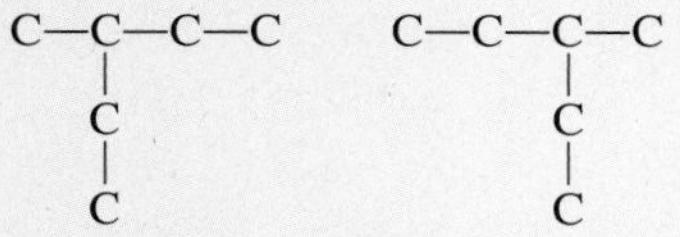

Neither of these structures is a new isomer because both have a five-carbon chain. Both structures are actually depictions of 3-methylpentane, one of the isomers previously identified.

Step 4: A chain length of three does not generate any new isomers. A trimethyl structure is impossible, as the middle carbon atom, the only carbon to which substituents can be attached, would have five bonds. An ethyl methyl structure extends the carbon chain length, as does a single three-carbon attachment.

Thus, there are five constitutional isomers: *hexane, 2-methylpentane, 3-methylpentane, 2,2-dimethylbutane, and 2,3-dimethylbutane.*

Practice Exercise 12.4

Draw skeletal structural formulas for, and assign IUPAC names to, all C_5H_{12} alkane constitutional isomers.

Answer:

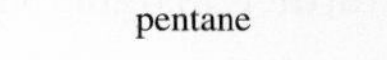

pentane 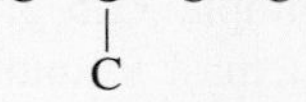   2-methylbutane 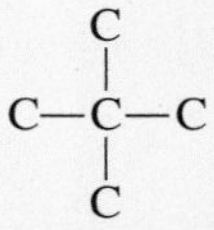   2, 2-dimethylpropane

12.9 LINE-ANGLE STRUCTURAL FORMULAS FOR ALKANES

Three two-dimensional methods for denoting alkane structures have been used in previous sections of this chapter. They are expanded structural formulas, condensed structural formulas, and skeletal structural formulas. An even more concise method for denoting molecular structure of alkanes (and other hydrocarbons and their derivatives) exists. This method, *line-angle structural formulas,* is particularly useful for molecules in which several carbon atoms are present.

A **line-angle structural formula** *is a structural representation in which a line represents a carbon–carbon bond and a carbon atom is understood to be present at every point where two lines meet and at the ends of lines.* Ball-and-stick-models and line-angle structural formulas for the alkanes propane, butane, and pentane are as follows:

Ball-and-stick
model

Line-angle
structural
formula

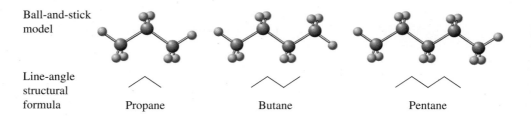

Propane   Butane   Pentane

Note that the zigzag (sawtooth) pattern used in line-angle structural formulas has a relationship to the three-dimensional shape of the molecules that are represented.

The line-angle structural formula for an unbranched chain of eight carbon atoms would be

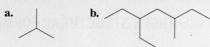

Octane

The structures of branched-chain alkanes can also be designated using line-angle structural formulas. The five constitutional alkane isomers in which six carbon atoms are present (C_6H_{14}) have the following line-angle formulas:

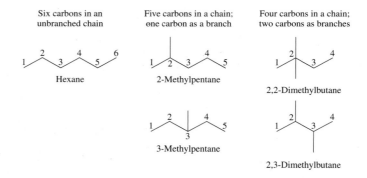

Example 12.5 gives further insights concerning the use and interpretation of line-angle structural formulas.

EXAMPLE 12.5

Generating Condensed Structural Formulas from Line-Angle Structural Formulas for Alkanes

For each of the following alkanes, determine the number of hydrogen atoms present on each carbon atom and then write the condensed structural formula for the alkane.

a.   b.

Solution

a. Each carbon atom in an alkane must be bonded to four atoms. Thus, carbon atoms bonded to only one carbon atom have three hydrogen atoms attached; those bonded to two other carbon atoms have two hydrogen atoms attached; those bonded to three other carbon atoms have only one atom attached; and those bonded to four other carbon atoms bear no hydrogen atoms. For this alkane, each carbon atom's hydrogen content is indicated by circled numbers as follows.

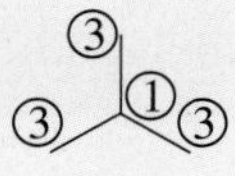

With this information on hydrogen content, the condensed structural formula is written as

$$CH_3-\overset{\overset{\textstyle CH_3}{|}}{CH}-CH_3$$

b. Using the methods of part **a,** the hydrogen content of this alkane is

and the condensed structural formula becomes

$$CH_3-CH_2-\overset{\overset{\textstyle CH_2}{|}}{\underset{\underset{\textstyle CH_3}{|}}{CH}}-CH_2-\overset{\overset{\textstyle CH_3}{|}}{CH}-CH_2-CH_3$$

Practice Exercise 12.5

For each of the following alkanes, determine the number of hydrogen atoms present on each carbon atom and then write the condensed structural formula for the alkane.

a. **b.**

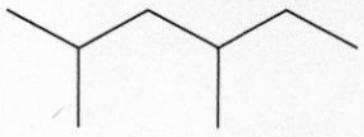

Answers: a.

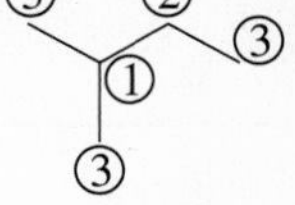

$$CH_3-\overset{\overset{\textstyle CH_3}{|}}{CH}-CH_2-CH_3$$

b.

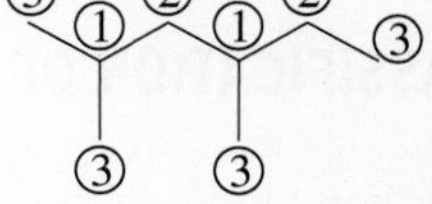

$$CH_3-\overset{\overset{\textstyle CH_3}{|}}{CH}-CH_2-\overset{\overset{\textstyle CH_3}{|}}{CH}-CH_2-CH_3$$

The Chemistry at a Glance feature on page 338 contrasts the line-angle structural formula notation for alkanes with all other structural formula notations for alkanes encountered so far in this chapter.

Structural Representations for Alkane Molecules

THREE-DIMENSIONAL STRUCTURAL REPRESENTATIONS

DASH-WEDGE-LINE STRUCTURE

Dashes represent bonds receding behind the page, wedges bonds coming out of the page, and solid lines bonds in the plane of the page.

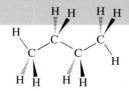

BALL-AND-STICK MODEL

This type of model emphasizes the connections (bonds) among the atoms and shows the tetrahedral arrangement of bonds about carbon atoms.

SPACE-FILLING MODEL

This type of model emphasizes the overall shape of the molecule and shows the tetrahedral arrangement of bonds about carbon atoms.

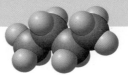

TWO-DIMENSIONAL STRUCTURAL REPRESENTATIONS

EXPANDED STRUCTURAL FORMULA

A structural formula that shows all atoms in a molecule and all bonds connecting the atoms.

$$H-\overset{\overset{\textstyle H}{|}}{\underset{\underset{\textstyle H}{|}}{C}}-\overset{\overset{\textstyle H}{|}}{\underset{\underset{\textstyle H}{|}}{C}}-\overset{\overset{\textstyle H}{|}}{\underset{\underset{\textstyle H}{|}}{C}}-\overset{\overset{\textstyle H}{|}}{\underset{\underset{\textstyle H}{|}}{C}}-H$$

CONDENSED STRUCTURAL FORMULA

A structural formula that uses grouping of atoms, in which central atoms and the atoms connected to them are written as a group.

$$CH_3-CH_2-CH_2-CH_3$$

SKELETAL STRUCTURAL FORMULA

A structural formula that shows the arrangement and bonding of carbon atoms present but does not show the hydrogen atoms attached to the carbon atoms.

$$C-C-C-C$$

LINE-ANGLE STRUCTURAL FORMULA

A structural formula in which a line represents a carbon–carbon bond and a carbon atom is understood to be present at every point where lines meet and at the ends of lines.

12.10 CLASSIFICATION OF CARBON ATOMS

Each of the carbon atoms within a hydrocarbon structure can be classified as a primary, secondary, tertiary, or quaternary carbon atom. A **primary carbon atom** *is a carbon atom in an organic molecule that is bonded to only one other carbon atom.* Each of the "end" carbon atoms in the three-carbon propane structure is a primary carbon atom, whereas the middle carbon atom of propane is a secondary carbon atom. A **secondary carbon atom** *is a carbon atom in an organic molecule that is bonded to two other carbon atoms.*

$$CH_3-CH_2-CH_3$$

Primary carbon atom   Secondary carbon atom   Primary carbon atom

The notations 1°, 2°, 3°, and 4° are often used as designations for the terms *primary, secondary, tertiary,* and *quaternary.* Thus we can write

1° carbon atom
2° carbon atom
3° carbon atom
4° carbon atom

A **tertiary carbon atom** *is a carbon atom in an organic molecule that is bonded to three other carbon atoms.* The molecule 2-methylpropane contains a tertiary carbon atom.

$$CH_3-\overset{\overset{\displaystyle CH_3}{|}}{CH}-CH_3$$

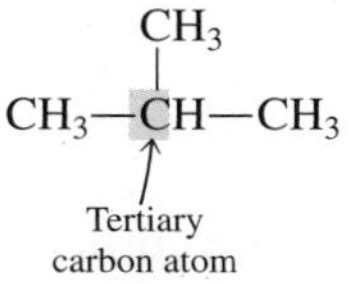

Tertiary
carbon atom

A **quaternary carbon atom** *is a carbon atom in an organic molecule that is bonded to four other carbon atoms.* The molecule 2,2-dimethylpropane contains a quaternary carbon atom.

$$CH_3-\overset{\overset{\displaystyle CH_3}{|}}{\underset{\underset{\displaystyle CH_3}{|}}{C}}-CH_3$$

Quaternary
carbon atom

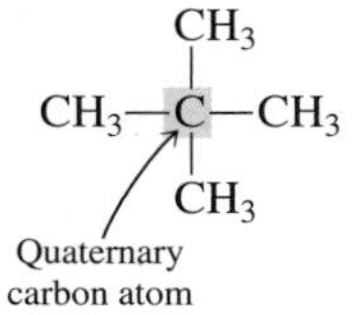

12.11 BRANCHED-CHAIN ALKYL GROUPS

To this point in the chapter, all alkyl groups encountered in structures have been continuous-chain alkyl groups (Table 12.3), the simplest type of alkyl group. Just as there are continuous-chain and branched-chain alkanes, there are continuous-chain and branched-chain alkyl groups. Four branched-chain alkyl groups, shown in Figure 12.5, are so common that you should know their names and structures.

For the two groups whose names contain the prefix *iso-* the common structural feature is an end-of-chain arrangement that contains two methyl groups.

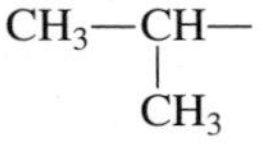

$$CH_3-\overset{\overset{\displaystyle }{|}}{\underset{\underset{\displaystyle CH_3}{|}}{CH}}-$$

You need to be able to recognize various conformations of branched-chain alkyl groups. For example, these structures all represent an isopropyl group:

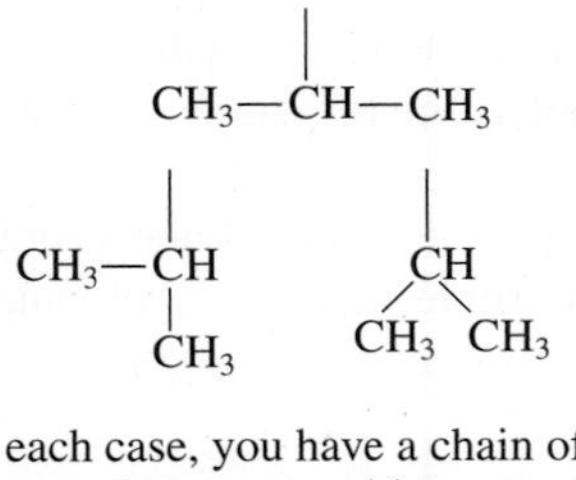

$$CH_3-\overset{\overset{\displaystyle |}{|}}{CH}-CH_3$$

$$CH_3-\overset{\overset{\displaystyle }{}}{\underset{\underset{\displaystyle CH_3}{|}}{CH}} \qquad \overset{\overset{\displaystyle |}{}}{\underset{\underset{\displaystyle CH_3\ \ CH_3}{}}{CH}}$$

In each case, you have a chain of three carbon atoms with an attachment point (the long bond) involving the middle carbon atom of the chain.

For the *sec*-butyl group, the point of attachment of the group to the main carbon chain involves a *secondary* carbon atom. For the *tert*-butyl group, the point of attachment of the group to the main carbon chain involves a *tertiary* carbon atom.

Two examples of alkanes containing branched-chain alkyl groups follow.

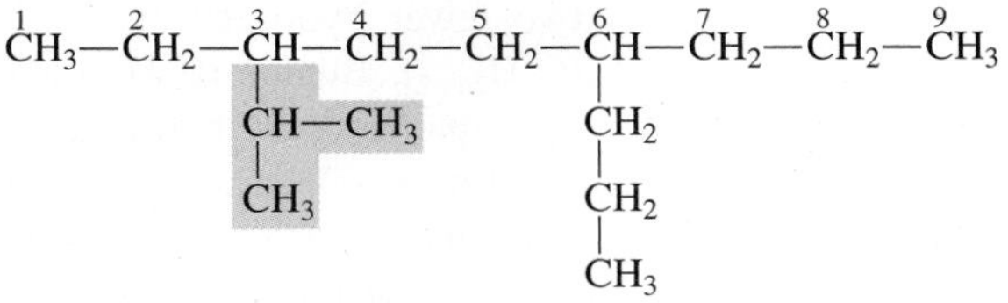

$$\overset{1}{C}H_3-\overset{2}{C}H_2-\overset{3}{C}H-\overset{4}{C}H_2-\overset{5}{C}H_2-\overset{6}{C}H-\overset{7}{C}H_2-\overset{8}{C}H_2-\overset{9}{C}H_3$$

3-Isopropyl-6-propylnonane

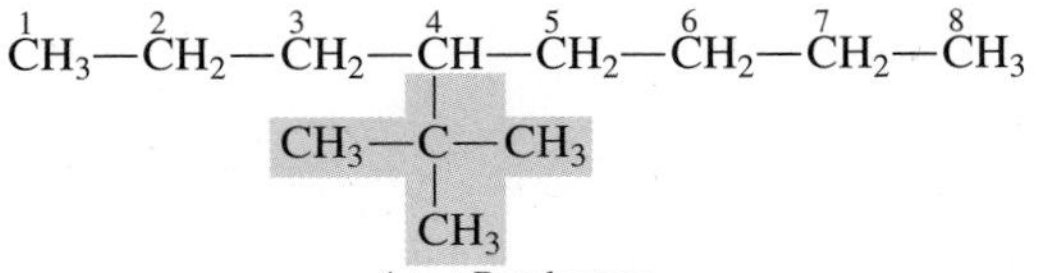

$$\overset{1}{C}H_3-\overset{2}{C}H_2-\overset{3}{C}H_2-\overset{4}{C}H-\overset{5}{C}H_2-\overset{6}{C}H_2-\overset{7}{C}H_2-\overset{8}{C}H_3$$

4-*tert*-Butyloctane

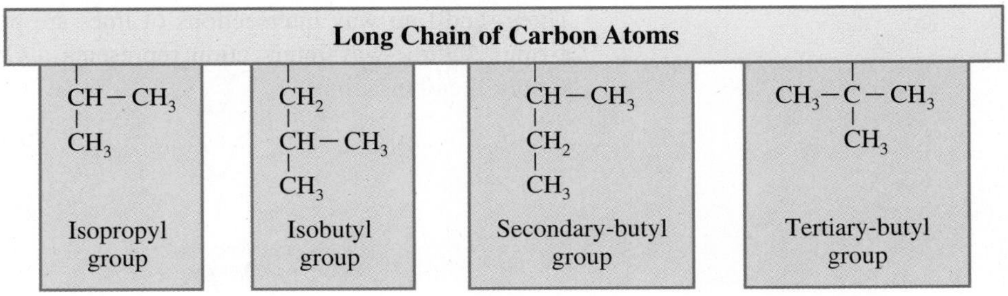

Figure 12.5 The four most common branched-chain alkyl groups and their IUPAC names.

Long Chain of Carbon Atoms			
CH—CH₃ CH₃ Isopropyl group	CH₂ CH—CH₃ CH₃ Isobutyl group	CH—CH₃ CH₂ CH₃ Secondary-butyl group	CH₃—C—CH₃ CH₃ Tertiary-butyl group

In IUPAC naming, hyphenated prefixes, such as *sec*- and *tert*-, are not considered when alphabetizing. The prefixes *iso* and *neo* are not hyphenated prefixes and are included when alphabetizing. The following IUPAC name is thus correct:

5-*sec*-Butyl-4-isopropyl-3-methyldecane

Complex Branched Alkyl Groups

Complex branched alkyl groups, for which no "simple" name is available (Figure 12.5), are occasionally encountered. The IUPAC system provision for such groups involves naming them as though they were themselves compounds. Select the *longest alkyl chain* in the complex substituent as the base alkyl group. The base alkyl group is then numbered beginning with the carbon atom attached to the main carbon chain. The substituents on the base alkyl group are listed with appropriate numbers, and parentheses are used to set off the name of the complex alkyl group. Two examples of such nomenclature follow.

(1,1-Dimethylpropyl) group   (1,1,3-Trimethylbutyl) group

12.12 CYCLOALKANES

It takes a minimum of three carbon atoms to form a cyclic arrangement of carbon atoms.

A **cycloalkane** *is a saturated hydrocarbon in which carbon atoms connected to one another in a cyclic (ring) arrangement are present.* The simplest cycloalkane is cyclopropane, which contains a cyclic arrangement of three carbon atoms. Figure 12.6 shows a three-dimensional model of cyclopropane's structure and those of the four-, five-, and six-carbon cycloalkanes.

Cyclopropane's three carbon atoms lie in a flat ring. In all other cycloalkane molecules, some puckering of the ring occurs; that is, the ring systems are nonplanar, as shown in Figure 12.6.

The general formula for cycloalkanes is C_nH_{2n}. Thus a given cycloalkane contains two fewer hydrogen atoms than an alkane with the same number of hydrogen atoms (C_nH_{2n+2}). Butane (C_4H_{10}) and cyclobutane (C_4H_8) are not isomers; isomers must have the same molecular formula (Section 12.6).

Line-angle structural formulas are generally used to represent cycloalkane structures. The line-angle structural formula for cyclopropane is a triangle, that for cyclobutane a square, that for cyclopentane a pentagon, and that for cyclohexane a hexagon.

Cyclopropane   Cyclobutane   Cyclopentane   Cyclohexane

Note that, in such structures, the intersection of two lines represents a CH_2 group. Three- and four-way intersections of lines are possible when substituents are present on a ring. A three-way intersection represents a CH group and a four-way intersection is simply a carbon atom.

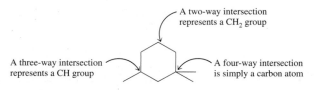

A two-way intersection represents a CH_2 group

A three-way intersection represents a CH group   A four-way intersection is simply a carbon atom

(a) Cyclopropane, C_3H_6   **(b) Cyclobutane, C_4H_8**   **(c) Cyclopentane, C_5H_{10}**   **(d) Cyclohexane, C_6H_{12}**

Figure 12.6 Three-dimensional representations of the structures of simple cycloalkanes.

● **E XAMPLE 12.6**

Generating Condensed Structural Formulas from Line-Angle Structural Formulas for Cycloalkanes

▶ Generate the condensed structural formula for each of the following cycloalkanes.

a.   **b.**

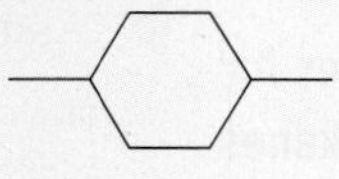

Solution

a. First replace each angle and line terminus with a carbon atom, and then add hydrogens as necessary to give each carbon four bonds. The molecular formula of this compound is C_8H_{16}.

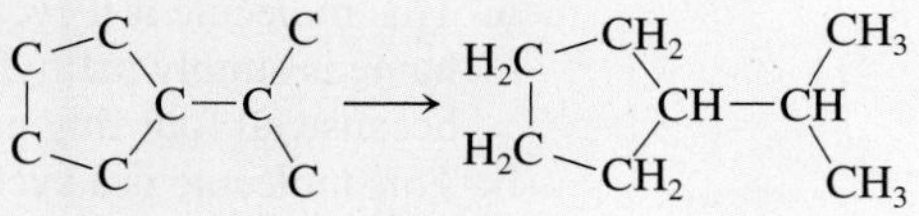

$$\underset{\substack{C-C}}{\overset{C-C}{\underset{|}{C}}}\!\!\!\!\!C-C\!\!\underset{C}{\overset{C}{<}} \longrightarrow \underset{\substack{H_2C-CH_2}}{\overset{H_2C-CH_2}{\underset{|}{CH_2}}}\!\!\!CH-CH\!\!\underset{CH_3}{\overset{CH_3}{<}}$$

b. Similarly, we have

$$\underset{\substack{C-C}}{\overset{C-C}{C-C}}\!\!\!C-C \longrightarrow CH_3-CH\underset{H_2C-CH_2}{\overset{H_2C-CH_2}{<}}CH-CH_3$$

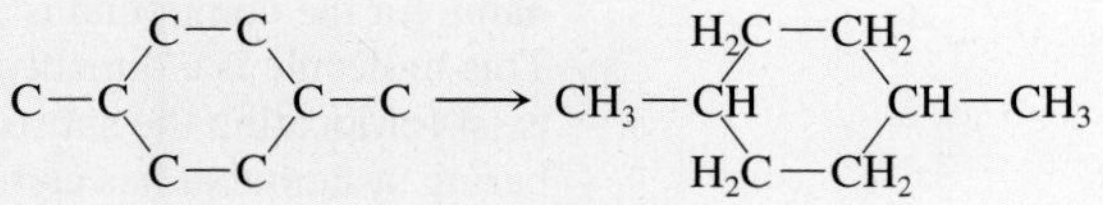

Practice Exercise 12.6

Generate the condensed structural formula for each of the following cycloalkanes.

a. 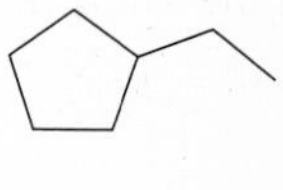   **b.**

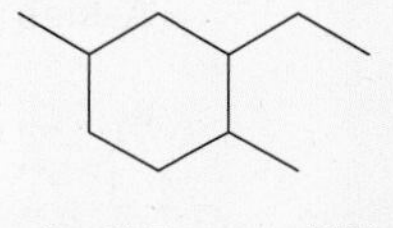

Answers: **a.** 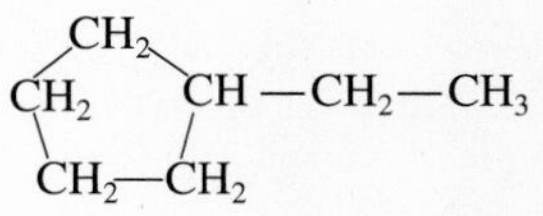 $\underset{\substack{CH_2-CH_2}}{\overset{CH_2}{\underset{CH_2}{<}}}CH-CH_2-CH_3$

b. 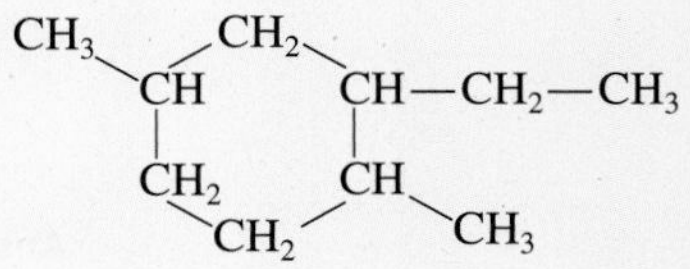 $\underset{\substack{CH_2\\CH_2}}{\overset{CH_3}{\underset{CH}{<}}}CH\underset{CH_3}{\overset{CH_2}{<}}CH-CH_2-CH_3$

The observed C—C—C bond angles in cyclopropane are 60°, and those in cyclobutane are 90°, values that are considerably smaller than the 109° angle associated with a tetrahedral arrangement of bonds about a carbon atom (Section 5.8). Consequently, cyclopropane and cyclobutane are relatively unstable compounds. Five- and six-membered cycloalkane structures are much more stable, and these structural entities are encountered in many organic molecules.

12.13 IUPAC NOMENCLATURE FOR CYCLOALKANES

IUPAC naming procedures for cycloalkanes are similar to those for alkanes. The ring portion of a cycloalkane molecule serves as the name base, and the prefix *cyclo-* is used to indicate the presence of the ring. Alkyl substituents are named in the same manner as in alkanes. Numbering conventions used in locating substituents on the ring include the following:

1. If there is just one ring substituent, it is not necessary to locate it by number.
2. When two ring substituents are present, the carbon atoms in the ring are numbered beginning with the substituent of higher alphabetical priority and proceeding in the direction (clockwise or counterclockwise) that gives the other substituent the lower number.
3. When three or more ring substituents are present, ring numbering begins at the substituent that leads to the lowest set of location numbers. When two or more equivalent numbering sets exist, alphabetical priority among substituents determines the set used.

Example 12.7 illustrates the use of the ring-numbering guidelines.

Cycloalkanes of ring sizes ranging from 3 to over 30 are found in nature, and in principle, there is no limit to ring size. Five-membered rings (cyclopentanes) and six-membered rings (cyclohexanes) are especially abundant in nature.

● EXAMPLE 12.7

Determining IUPAC Names for Cycloalkanes

► Assign IUPAC names to each of the following cycloalkanes.

a.   b. 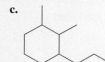   c.

Solution

a. This molecule is a cyclobutane (four-carbon ring) with a methyl substituent. The IUPAC name is simply methylcyclobutane. No number is needed to locate the methyl group, because all four ring positions are equivalent.
b. This molecule is a cyclopentane with ethyl and methyl substituents. The numbers for the carbon atoms that bear the substituents are 1 and 2. On the basis of alphabetical priority, the number 1 is assigned to the carbon atom that bears the ethyl group. The IUPAC name for the compound is 1-ethyl-2-methylcyclopentane.
c. This molecule is a dimethylpropylcyclohexane. Two different 1,2,3 numbering systems exist for locating the substituents. On the basis of alphabetical priority, we use the numbering system that has carbon 1 bearing a methyl group; methyl has alphabetical priority over propyl. Thus the compound name is 1,2-dimethyl-3-propylcyclohexane.

Practice Exercise 12.7

Assign IUPAC names to each of the following cycloalkanes.

a.   b.   c.

Answers: **a.** Methylcyclopropane; **b.** 1-Ethyl-4-methylcyclohexane; **c.** 4-Ethyl-1,2-dimethylcyclopentane

12.14 ISOMERISM IN CYCLOALKANES

Constitutional isomers are possible for cycloalkanes that contain four or more carbon atoms. For example, there are five cycloalkane constitutional isomers that have the formula C_5H_{10}: one based on a five-membered ring, one based on a four-membered ring, and three based on a three-membered ring. These isomers are

Cyclopentane   Methylcyclobutane   1,2-Dimethyl-cyclopropane   1,1-Dimethyl-cyclopropane   Ethylcyclopropane

A second type of isomerism, called *stereoisomerism,* is possible for some *substituted* cycloalkanes. Whereas constitutional isomerism results from differences in *connectivity,* stereoisomerism results from differences in *configuration.* **Stereoisomers** *are isomers that have the same molecular and structural formulas but different orientations of atoms in space.* Several forms of stereoisomerism exist. The form associated with cycloalkanes is called *cis–trans isomerism.* **Cis-trans isomers** *are isomers that have the same molecular and structural formulas but different orientations of atoms in space because of restricted rotation about bonds.*

In alkanes, there is free rotation about all carbon–carbon bonds (Section 12.7). In cycloalkanes, the ring structure restricts rotation for the carbon atoms in the ring. The consequence of this lack of rotation in a cycloalkane is the creation of "top" and "bottom" positions for the two attachments on each of the ring carbon atoms. This "top–bottom" situation leads to *cis–trans* isomerism in cycloalkanes in which each of two ring carbon atoms bears two different attachments.

Consider the following two structures for the molecule 1,2-dimethylcyclopentane.

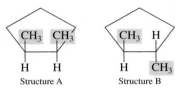

In structure A, both methyl groups are above the plane of the ring (the "top" side). In structure B, one methyl group is above the plane of the ring (the "top" side) and the other below it (the "bottom" side). Structure A cannot be converted into structure B without breaking bonds. Hence structures A and B are isomers; there are two 1,2-dimethylcyclopentanes. The first isomer is called *cis*-1,2-dimethylcyclopentane and the second *trans*-1,2-dimethylcyclopentane.

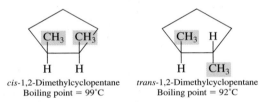

cis-1,2-Dimethylcyclopentane
Boiling point = 99°C

trans-1,2-Dimethylcyclopentane
Boiling point = 92°C

Cis- *is a prefix that means "on the same side."* In *cis*-1,2-dimethylcyclopentane, the two methyl groups are on the same side of the ring. **Trans-** *is a prefix that means "across from."* In *trans*-1,2-dimethylcyclopentane, the two methyl groups are on opposite sides of the ring.

Cis–trans isomerism can occur in rings of all sizes. The presence of a substituent on each of two carbon atoms in the ring is the requirement for its occurrence. In biochemistry, we will find that the human body often selectively distinguishes between the *cis* and *trans* isomers of a compound. One isomer will be active in the body and the other inactive.

Cis–trans isomers have the same molecular formula and the same structural formula. The only difference between them is the orientation of atoms in space. *Constitutional isomers* have the same molecular formula but different structural formulas.

The Latin *cis* means "on the same side," and the Latin *trans* means "across from." Consider the use of the prefix *trans-* in the phrase "transatlantic voyage."

Cis–trans isomerism will also be encountered in the next chapter (Section 13.6), where the required restricted rotation barrier will be a carbon–carbon double bond rather than a ring of carbon atoms. Another type of stereoisomerism called enantiomerism (left- and right-handed forms of a molecule) will be considered in the discussion of carbohydrates in Chapter 18.

● **EXAMPLE 12.8**

Identifying and Naming Cycloalkane *Cis-Trans* Isomers

Determine whether *cis–trans* isomerism is possible for each of the following cycloalkanes. If so, then draw structural formulas for the *cis* and *trans* isomers.

a. Methylcyclohexane
b. 1,1-Dimethylcyclohexane
c. 1,3-Dimethylcyclobutane
d. 1-Ethyl-2-methylcyclobutane

Solution

a. *Cis–trans* isomerism is not possible because we do not have two substituents on the ring.
b. *Cis–trans* isomerism is not possible. We have two substituents on the ring, but they are on the same carbon atom. Each of two different carbons must bear substituents.

(continued)

In cycloalkanes, *cis–trans* isomerism can also be denoted by using wedges and dotted lines. A heavy wedge-shaped bond to a ring structure indicates a bond *above* the plane of the ring, and a broken dotted line indicates a bond *below* the plane of the ring.

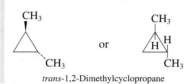

cis-1,2-Dimethylcyclopropane

trans-1,2-Dimethylcyclopropane

c. *Cis–trans* isomerism does exist.

cis-1,3-Dimethylcyclobutane   *trans*-1,3-Dimethylcyclobutane

d. *Cis–trans* isomerism does exist.

cis-1-Ethyl-2-methylcyclobutane   *trans*-1-Ethyl-2-methylcyclobutane

Practice Exercise 12.8

Determine whether *cis–trans* isomerism is possible for each of the following cycloalkanes. If so, then draw structural formulas for the *cis* and *trans* isomers.

a. 1-Ethyl-1-methylcyclopentane   **b.** Ethylcyclohexane
c. 1,3-Dimethylcyclopentane   **d.** 1,1-Dimethylcyclooctane

Answers: a. Not possible; **b.** Not possible; **c.**
d. Not possible

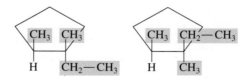

cis isomer   *trans* isomer

Use of the terms *cis-* and *trans-* in designating stereoisomers in cycloalkanes is limited to substituted cycloalkanes in which the two substituted carbon atoms each have one hydrogen atom and one substituent other than hydrogen. The designations *cis-* and *trans-* become ambiguous in situations where either or both of the substituted carbons have two different substituents but no hydrogen atoms. Following is an example of such a situation in substituted cycloalkanes.

The first structure is the *cis-* isomer if the focus is on H and the ethyl group; the second structure is the *cis-* isomer if the focus is on H and the methyl group. A different nomenclature system, called the *E,Z* nomenclature system (which is not covered in this textbook), must be used to distinguish such isomerism.

12.15 SOURCES OF ALKANES AND CYCLOALKANES

The word *petroleum* comes from the Latin *petra,* which means "rock," and *oleum,* which means "oil."

Alkanes and cycloalkanes are not "laboratory curiosities" but rather two families of extremely important naturally occurring compounds. Natural gas and petroleum (crude oil) constitute their largest and most important natural source. Deposits of these resources are usually associated with underground dome-shaped rock formations (Figure 12.7). When a hole is drilled into such a rock formation, it is possible to recover some of the trapped

Figure 12.7 A rock formation such as this is necessary for the accumulation of petroleum and natural gas.

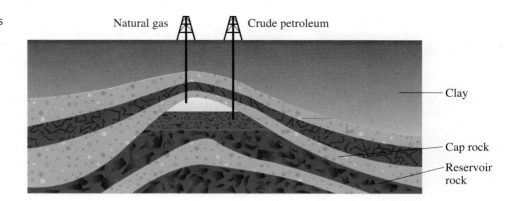

Natural gas   Crude petroleum

Clay

Cap rock

Reservoir rock

Figure 12.8 An oil rig pumping oil from an underground rock formation.

hydrocarbons—that is, the natural gas and/or petroleum (Figure 12.8). Note that petroleum and natural gas do not occur in the earth in the form of "liquid pools" but rather are dispersed throughout a porous rock formation.

Unprocessed natural gas contains 50%–90% methane, 1%–10% ethane, and up to 8% higher-molecular-mass alkanes (predominantly propane and butanes). The higher alkanes found in crude natural gas are removed prior to release of the gas into the pipeline distribution systems. Because the removed alkanes can be liquefied by the use of moderate pressure, they are stored as liquids under pressure in steel cylinders and are marketed as bottled gas.

Crude petroleum is a complex mixture of hydrocarbons (both cyclic and acyclic) that can be separated into useful fractions through refining. During refining, the physical separation of the crude into component fractions is accomplished by fractional distillation, a process that takes advantage of boiling-point differences between the components of the crude petroleum. Each fraction contains hydrocarbons within a specific boiling-point range. The gasoline fraction consists primarily of alkanes and cycloalkanes with 5 to 12 carbon atoms present. The fractions obtained from a typical fractionation process are shown in Figure 12.9.

Figure 12.9 The complex hydrocarbon mixture present in petroleum is separated into simpler mixtures by means of a fractionating column.

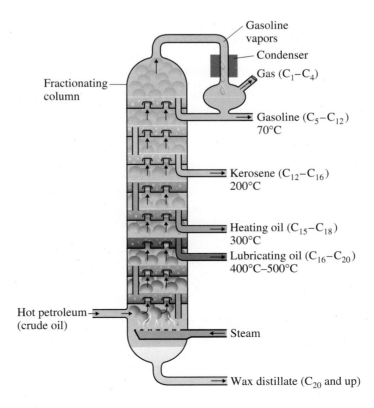

Gasoline vapors

Condenser

Gas (C_1–C_4)

Fractionating column

Gasoline (C_5–C_{12})
70°C

Kerosene (C_{12}–C_{16})
200°C

Heating oil (C_{15}–C_{18})
300°C

Lubricating oil (C_{16}–C_{20})
400°C–500°C

Hot petroleum (crude oil)

Steam

Wax distillate (C_{20} and up)

Figure 12.10 The insolubility of alkanes in water is used to advantage by many plants, which produce unbranched long-chain alkanes that serve as protective coatings on leaves and fruits. Such protective coatings minimize water loss for plants. Apples can be "polished" because of the long-chain alkane coating on their skin, which involves the unbranched alkanes $C_{27}H_{56}$ and $C_{29}H_{60}$. The leaf wax of cabbage and broccoli is mainly unbranched $C_{29}H_{60}$.

12.16 PHYSICAL PROPERTIES OF ALKANES AND CYCLOALKANES

In this section, we consider a number of generalizations about the physical properties of alkanes and cycloalkanes.

1. *Alkanes and cycloalkanes are insoluble in water.* Water molecules are polar, and alkane and cycloalkane molecules are nonpolar. Molecules of unlike polarity have limited solubility in one another (Section 8.4). The water insolubility of alkanes makes them good preservatives for metals. They prevent water from reaching the metal surface and causing corrosion. They also have biological functions as protective coatings (see Figure 12.10).

2. *Alkanes and cycloalkanes have densities lower than that of water.* Alkane and cycloalkane densities fall in the range 0.6 g/mL to 0.8 g/mL, compared with water's density of 1.0 g/mL. When alkanes and cycloalkanes are mixed with water, two layers form (because of insolubility), with the hydrocarbon layer on top (because of its lower density). This density difference between alkanes/cycloalkanes and water explains why oil spills in aqueous environments spread so quickly. The *floating* oil follows the movement of the water.

3. *The boiling points of continuous-chain alkanes and cycloalkanes increase with an increase in carbon chain length or ring size.* For continuous-chain alkanes, the boiling point increases roughly 30°C for every carbon atom added to the chain. This trend, shown in Figure 12.11, is the result of increasing London force strength (Section 7.13). London forces become stronger as molecular surface area increases. Short, continuous-chain alkanes (1 to 4 carbon atoms) are gases at room temperature. Continuous-chain alkanes containing 5 to 17 carbon atoms are liquids, and alkanes that have carbon chains longer than this are solids at room temperature.

    Branching on a carbon chain lowers the boiling point of an alkane. A comparison of the boiling points of unbranched alkanes and their 2-methyl-branched isomers is given in Figure 12.11. Branched alkanes are more compact, with smaller surface areas than their straight-chain isomers.

    Cycloalkanes have higher boiling points than their noncyclic counterparts with the same number of carbon atoms (Figure 12.11). These differences are due in large part to cyclic systems having more rigid and more symmetrical structures.

    Cyclopropane and cyclobutane are gases at room temperature, and cyclopentane through cyclooctane are liquids at room temperature. Figure 12.12 is a physical-state summary for unbranched alkanes or unsubstituted cycloalkanes with 8 or fewer carbon atoms.

The alkanes and cycloalkanes whose boiling points are compared in Figure 12.11 constitute *homologous series* of organic compounds. In a homologous series, the members

Figure 12.11 Trends in normal boiling points for continuous-chain alkanes, 2-methyl branched alkanes, and unsubstituted cycloalkanes as a function of the number of carbon atoms present. For a series of alkanes or cycloalkanes, melting point increases as carbon chain length increases.

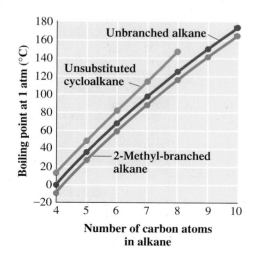

Unbranched Alkanes			
C_1	C_3	C_5	C_7
C_2	C_4	C_6	C_8

Unsubstituted Cycloalkanes			
✕	C_3	C_5	C_7
✕	C_4	C_6	C_8

☐ Gas   ☐ Liquid

Figure 12.12 A physical-state summary for unbranched alkanes and unsubstituted cycloalkanes at room temperature and pressure.

The term *paraffins* is an older name for the alkane family of compounds. This name comes from the Latin *parum affinis*, which means "little activity." That is a good summary of the general chemical properties of alkanes.

Figure 12.13 Propane fuel tank on a home barbecue unit.

differ structurally only in the number of —CH_2— groups present. Members exhibit gradually changing physical properties and usually have very similar chemical properties.

The existence of homologous series of organic compounds gives organization to organic chemistry in the same way that the periodic table gives organization to the chemistry of the elements. Knowing something about a few members of a homologous series usually enables us to deduce the properties of other members in the series.

12.17 CHEMICAL PROPERTIES OF ALKANES AND CYCLOALKANES

Alkanes are the least reactive type of organic compound. They can be heated for long periods of time in strong acids and bases with no appreciable reaction. Strong oxidizing agents and reducing agents have little effect on alkanes.

Alkanes are not absolutely unreactive. Two important reactions that they undergo are combustion, which is reaction with oxygen, and halogenation, which is reaction with halogens.

Combustion

A **combustion reaction** *is a chemical reaction between a substance and oxygen (usually from air) that proceeds with the evolution of heat and light (usually as a flame).* Alkanes readily undergo combustion when ignited. When sufficient oxygen is present to support total combustion, carbon dioxide and water are the products.

$$CH_4 + 2O_2 \longrightarrow CO_2 + 2H_2O + \text{heat energy}$$
$$2C_6H_{14} + 19O_2 \longrightarrow 12CO_2 + 14H_2O + \text{heat energy}$$

The exothermic nature (Section 9.5) of alkane combustion reactions explains the extensive use of alkanes as fuels. Natural gas, used in home heating, is predominantly methane. Propane is used in home heating in rural areas and in gas barbecue units (see Figure 12.13). Butane fuels portable camping stoves. Gasoline is a complex mixture of many alkanes and other types of hydrocarbons.

Incomplete combustion can occur if insufficient oxygen is present during the combustion process. When this is the case, some carbon monoxide (CO) and/or elemental carbon are reaction products along with carbon dioxide (CO_2). In a chemical laboratory setting, incomplete combustion is often observed. The appearance of deposits of carbon black (soot) on the bottom of glassware is physical evidence that incomplete combustion is occurring. The problem is that the air-to-fuel ratio for the Bunsen burner is not correct. It is too rich; it contains too much fuel and not enough oxygen (air).

Halogenation

The halogens are the elements in Group VIIA of the periodic table: fluorine (F_2), chlorine (Cl_2), bromine (Br_2), and iodine (I_2) (Section 3.4). A **halogenation reaction** *is a chemical reaction between a substance and a halogen in which one or more halogen atoms are incorporated into molecules of the substance.*

Halogenation of an alkane produces a hydrocarbon derivative in which one or more halogen atoms have been substituted for hydrogen atoms. An example of an alkane halogenation reaction is

$$
\underset{\substack{| \quad |\\ H \quad H}}{\overset{\substack{H \quad H\\ | \quad |}}{H-C-C-H}} + Br_2 \xrightarrow[\text{light}]{\text{Heat or}} \underset{\substack{| \quad |\\ H \quad H}}{\overset{\substack{H \quad H\\ | \quad |}}{H-C-C-Br}} + HBr
$$

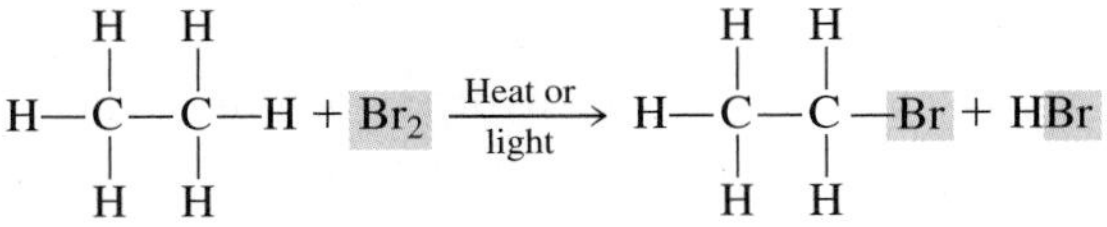

Alkane halogenation is an example of a substitution reaction, a type of reaction that occurs often in organic chemistry. A **substitution reaction** *is a chemical reaction in which part of a small reacting molecule replaces an atom or a group of atoms on a*

CHEMICAL Connections

The Physiological Effects of Alkanes

The simplest alkanes (methane, ethane, propane, and butane) are gases at room temperature and pressure. Methane and ethane are difficult to liquefy, so they are usually handled as compressed gases. Propane and butane are easily liquefied at room temperature under a moderate pressure. They are stored in low-pressure cylinders in a liquefied form. These four gases are colorless, odorless, and nontoxic, and they have limited physiological effects. The danger in inhaling them lies in potential suffocation due to lack of oxygen. The major immediate danger associated with a natural gas leak is the potential formation of an explosive air–alkane mixture rather than the formation of a toxic air–alkane mixture.

The C_5 to C_8 alkanes, of which there are many isomeric forms, are free-flowing, nonpolar, volatile liquids. They are the primary constituents of gasoline. These compounds are not particularly toxic, but gasoline should not be swallowed because (1) some of the additives present are harmful and (2) liquid alkanes can damage lung tissue because of physical rather than chemical effects. Physical effects include the dissolving of lipid molecules of cell membranes (see Chapter 19), causing pneumonia-like symptoms. Liquid alkanes can also affect the skin for related reasons. These alkanes dissolve natural body oils, causing the skin to dry out. (This "drying out" effect is easily noticed when paint thinner, a mixture of hydrocarbons, is used to remove paint from the hands.)

In direct contrast to liquid alkanes, solid alkanes are used to protect the skin. Pharmaceutical-grade *petrolatum* and *mineral oil* (also called liquid petrolatum), obtained as products from petroleum distillation, have such a function. Petrolatum is a mixture of C_{25} to C_{30} alkanes, and mineral oil involves alkanes in the C_{18} to C_{24} range.

Petrolatum (Vaseline is a well-known brand name) is a semi-solid hydrocarbon mixture that is useful both as a skin softener and as a skin protector. Many moisturizing hand lotions and some medicated salves contain petrolatum. Neither water nor

A semi-solid alkane mixture, such as Vaseline, is useful as a skin protector because neither water nor water solutions will penetrate a coating of it. Here, Vaseline is applied to a baby's bottom as a protection against diaper rash.

water solutions (for example, urine) will penetrate protective petrolatum coatings. This explains why petrolatum products protect a baby's bottom from diaper rash.

Mineral oil is often used to replace natural skin oils washed away by frequent bathing and swimming. Too much mineral oil, however, can be detrimental; it will dissolve nonpolar skin materials. Mineral oil has some use as a laxative; it effectively softens and lubricates hard stools. When taken by mouth, it passes through the gastrointestinal tract unchanged and is excreted chemically intact. Loss of fat-soluble vitamins (A, D, E, and K) can occur if mineral oil is consumed while these vitamins are in the digestive tract. Using a mineral oil enema instead avoids this drawback.

---

hydrocarbon or hydrocarbon derivative. A diagrammatic representation of a substitution reaction is shown in Figure 12.14.

A *general* equation for the substitution of a single halogen atom for one of the hydrogen atoms of an alkane is

$$\underset{\text{Alkane}}{R-H} + \underset{\text{Halogen}}{X_2} \xrightarrow{\text{Heat or light}} \underset{\text{Halogenated alkane}}{R-X} + \underset{\text{Hydrogen halide}}{H-X}$$

Occasionally, it is useful to represent alkyl groups in a nonspecific way. The symbol R is used for this purpose. Just as *city* is a generic term for Chicago, New York, or San Francisco, the symbol R is a generic designation for any alkyl group. The symbol R comes from the German word *radikal,* which means, in a chemical context, "molecular fragment."

Note the following features of this general equation:

1. The notation R—H is a general formula for an alkane. R— in this case represents an alkyl group. Addition of a hydrogen atom to an alkyl group produces the parent hydrocarbon of the alkyl group.
2. The notation R—X on the product side is the general formula for a halogenated alkane. X is the general symbol for a halogen atom.
3. Reaction conditions are noted by placing these conditions on the equation arrow that separates reactants from products. Halogenation of an alkane requires the presence of heat or light.

Figure 12.14 In an alkane substitution reaction, an incoming atom or group of atoms (represented by the orange sphere) replaces a hydrogen atom in the alkane molecule.

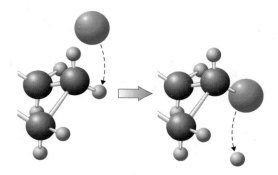

(The symbol R is used frequently in organic chemistry and will be encountered in numerous generalized formulas in subsequent chapters; it always represents a generalized organic group in a structural formula. An R group can be an alkyl group—methyl, ethyl, propyl, etc.—or any number of other organic groups. Consider the symbol R to represent the Rest of an organic molecule, which is not specifically specified because it is not the focal point of the discussion occurring at that time.)

In halogenation of an alkane, the alkane is said to undergo *fluorination, chlorination, bromination,* or *iodination,* depending on the identity of the halogen reactant. Chlorination and bromination are the two widely used alkane halogenation reactions. Fluorination reactions generally proceed too quickly to be useful, and iodination reactions go too slowly.

Halogenation usually results in the formation of a mixture of products rather than a single product. More than one product results because more than one hydrogen atom on an alkane can be replaced with halogen atoms. To illustrate this concept, let us consider the chlorination of methane, the simplest alkane.

Methane and chlorine, when heated to a high temperature or in the presence of light, react as follows:

$$CH_4 + Cl_2 \xrightarrow[\text{light}]{\text{Heat or}} CH_3Cl + HCl$$

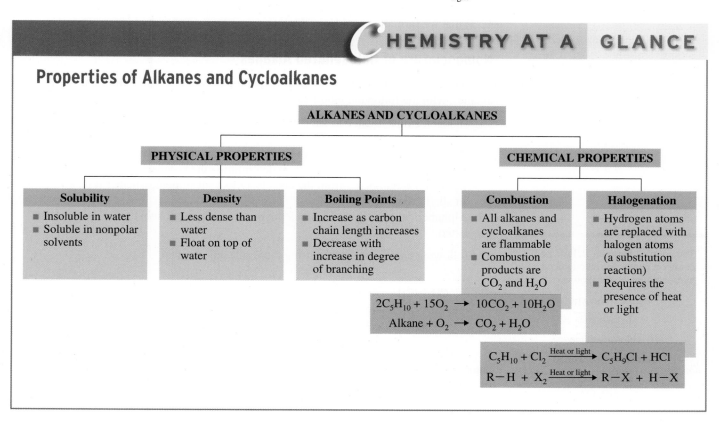

CHEMISTRY AT A GLANCE

Properties of Alkanes and Cycloalkanes

ALKANES AND CYCLOALKANES

PHYSICAL PROPERTIES

CHEMICAL PROPERTIES

Solubility
- Insoluble in water
- Soluble in nonpolar solvents

Density
- Less dense than water
- Float on top of water

Boiling Points
- Increase as carbon chain length increases
- Decrease with increase in degree of branching

Combustion
- All alkanes and cycloalkanes are flammable
- Combustion products are CO_2 and H_2O

Halogenation
- Hydrogen atoms are replaced with halogen atoms (a substitution reaction)
- Requires the presence of heat or light

$$2C_5H_{10} + 15O_2 \rightarrow 10CO_2 + 10H_2O$$
$$\text{Alkane} + O_2 \rightarrow CO_2 + H_2O$$

$$C_5H_{10} + Cl_2 \xrightarrow{\text{Heat or light}} C_5H_9Cl + HCl$$
$$R-H + X_2 \xrightarrow{\text{Heat or light}} R-X + H-X$$

The reaction does not stop at this stage, however, because the chlorinated methane product can react with additional chlorine to produce polychlorinated products.

$$CH_3Cl + Cl_2 \xrightarrow[\text{light}]{\text{Heat or}} CH_2Cl_2 + HCl$$

$$CH_2Cl_2 + Cl_2 \xrightarrow[\text{light}]{\text{Heat or}} CHCl_3 + HCl$$

$$CHCl_3 + Cl_2 \xrightarrow[\text{light}]{\text{Heat or}} CCl_4 + HCl$$

By controlling the reaction conditions and the ratio of chlorine to methane, it is possible to *favor* formation of one or another of the possible chlorinated methane products.

The chemical properties of cycloalkanes are similar to those of alkanes. Cycloalkanes readily undergo combustion as well as chlorination and bromination. With unsubstituted cycloalkanes, monohalogenation produces a single product because all hydrogen atoms present in the cycloalkane are equivalent to one another.

The Chemistry at a Glance feature on page 349 summarizes the physical properties and chemical reactions of alkanes and cycloalkanes.

12.18 NOMENCLATURE AND PROPERTIES OF HALOGENATED ALKANES

A **halogenated alkane** *is an alkane derivative in which one or more halogen atoms are present.* Similarly, a **halogenated cycloalkane** *is a cycloalkane derivative in which one or more halogen atoms are present.* Produced by halogenation reactions (Section 12.17), these two types of compounds represent the first class of hydrocarbon derivatives (Section 12.3) that we formally consider in this text.

Nomenclature of Halogenated Alkanes

The IUPAC rules for naming halogenated alkanes are similar to those for naming branched alkanes, with the following modifications:

1. Halogen atoms, treated as substituents on a carbon chain, are called *fluoro-, chloro-, bromo-,* and *iodo-.*
2. When a carbon chain bears both a halogen and an alkyl substituent, the two substituents are considered of equal rank in determining the numbering system for the chain. The chain is numbered from the end closer to a substituent, whether it be a *halo-* or an alkyl group.
3. Alphabetical priority determines the order in which all substituents present are listed.

The following names are derived using these rule adjustments.

CH₃—CH—CH—CH₃
     | |
    Cl CH₃
2-Chloro-3-methylbutane

CH₃—CH—CH₂—CH₂
     | |
    Br   Cl
3-Bromo-1-chlorobutane

1-Ethyl-2-fluorocyclohexane

> The contrast between IUPAC and common names for halogenated hydrocarbons is as follows:
>
> IUPAC (one word)
> ┌──────────┐
> │ haloalkane │
> └──────────┘
> chloromethane
>
> Common (two words)
> ┌────────────┐
> │ alkyl halide │
> └────────────┘
> methyl chloride

Simple halogenated alkanes can also be named as *alkyl halides*. These non-IUPAC names have two parts. The first part is the name of the hydrocarbon portion of the molecule (the alkyl group). The second part (as a separate word) identifies the halogen portion, which is named as if it were an ion (chloride, bromide, and so on), even though no ions are present

CHEMICAL Connections

Chlorofluorocarbons and the Ozone Layer

Chlorofluorocarbons (CFCs) are compounds composed of the elements chlorine, fluorine, and carbon. CFCs are synthetic compounds that have been developed primarily for use as refrigerants. The two most widely used of the CFCs are trichlorofluoromethane and dichlorodifluoromethane. Both of these compounds are marketed under the trade name Freon.

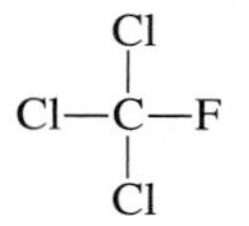

Trichlorofluoromethane
(Freon-11)

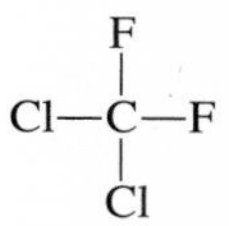

Dichlorodifluoromethane
(Freon-12)

Freon-11 and Freon-12 possess ideal properties for use as a refrigerant gas. Both are inert, nontoxic, and easily compressible. Prior to their development, ammonia was used in refrigeration. Ammonia is toxic, and leaking ammonia-based refrigeration units have been fatal.

We now know that CFCs contribute to a serious environmental problem: destruction of the stratospheric (high-altitude) ozone that we commonly call the ozone layer. Once released into the atmosphere, CFCs persist for long periods without reaction. Consequently, they slowly drift upward in the atmosphere, finally reaching the stratosphere.

It is in the stratosphere, the location of the "ozone layer," that environmental problems occur. At these high altitudes, the CFCs are exposed to ultraviolet light (from the sun), which activates them. The ultraviolet light breaks carbon–chlorine bonds within the CFCs, releasing chlorine atoms.

$$CCl_2F_2 + \text{ultraviolet light} \longrightarrow CClF_2 + Cl$$

The Cl atoms so produced (called atomic chlorine) are extremely reactive species. One of the molecules with which they react is ozone (O_3).

$$Cl + O_3 \longrightarrow ClO + O_2$$

A reaction such as this upsets the O_3–O_2 equilibrium in the stratosphere (Section 9.8).

The Montreal Protocol of 1987 (an international agreement on substances that deplete the ozone layer), and later amendments to this agreement, limit—and in some cases ban—future production and use of CFCs. The following graph shows the effects of the implementation of this agreement.

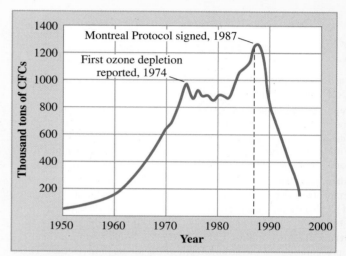

Worldwide Production of CFCs (1950–1996)

Replacements for the phased-out CFCs are HFCs (hydrogen-fluorocarbons) such as

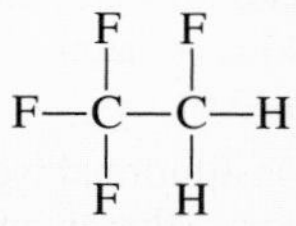

1,1,1,2-Tetrafluoroethane

Haloalkanes with some carbon–hydrogen bonds are more reactive than CFCs and are generally destroyed at lower altitudes before they reach the stratosphere. Unfortunately, however, their refrigeration properties are not as good as those of the CFCs.

(all bonds are covalent bonds). The following examples contrast the IUPAC names and the common names (in parentheses) of selected halogenated alkanes.

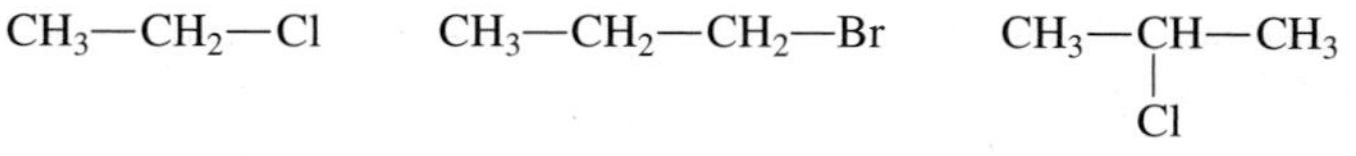

$$CH_3—CH_2—Cl \qquad CH_3—CH_2—CH_2—Br \qquad CH_3—CH—CH_3$$
$$\qquad\qquad\qquad\qquad\qquad\qquad\qquad\qquad\qquad\qquad | $$
$$\qquad\qquad\qquad\qquad\qquad\qquad\qquad\qquad\qquad\qquad Cl$$

Chloroethane
(ethyl chloride)

1-Bromopropane
(propyl bromide)

2-Chloropropane
(isopropyl chloride)

An alternative designation for a halogenated alkane is *alkyl halide*.

Several polyhalogenated methanes have acquired common names that are not clearly related to their structures. Five important examples of this additional nomenclature are CH_2Cl_2 (methylene chloride), $CHCl_3$ (chloroform), CCl_4 (carbon tetrachloride), CCl_3F (Freon-11), and CCl_2F_2 (Freon-12). The compounds Freon-11 and Freon-12 are examples of chlorofluorocarbons (CFCs). CFCs are synthetic compounds that have been heavily used as refrigerants and as air conditioning chemicals. We now know that CFCs are factors in the destruction of stratospheric (high-altitude) ozone, as discussed in the Chemical Connections feature "Chlorofluorocarbons and the Ozone Layer" on page 351.

Physical Properties of Halogenated Alkanes

Halogenated alkane boiling points are generally higher than those of the corresponding alkane. An important factor contributing to this effect is the polarity of carbon–halogen bonds, which results in increased dipole–dipole interactions.

Some halogenated alkanes have densities that are greater than that of water, a situation not common for organic compounds. Chloroalkanes containing two or more chlorine atoms, bromoalkanes, and iodoalkanes are all more dense than water.

CONCEPTS TO REMEMBER

Carbon atom bonding characteristics. Carbon atoms in organic compounds must have four bonds (Section 12.2).

Types of hydrocarbons. Hydrocarbons, binary compounds of carbon and hydrogen, are of two types: saturated and unsaturated. In saturated hydrocarbons, all carbon–carbon bonds are single bonds. Unsaturated hydrocarbons have one or more carbon–carbon multiple bonds—double bonds, triple bonds, or both (Section 12.3).

Alkanes. Alkanes are saturated hydrocarbons in which the carbon atom arrangement is that of an unbranched or branched chain. The formulas of all alkanes can be represented by the general formula C_nH_{2n+2}, where n is the number of carbon atoms present (Section 12.4).

Structural formulas. Structural formulas are two-dimensional representations of the arrangement of the atoms in molecules. These formulas give complete information about the arrangement of the atoms in a molecule but not the spatial orientation of the atoms. Two types of structural formulas are commonly encountered: expanded and condensed (Section 12.5).

Isomers. Isomers are compounds that have the same molecular formula, (that is, the same numbers and kinds of atoms) but that differ in the way the atoms are arranged (Section 12.6).

Constitutional isomers. Constitutional isomers are isomers that differ in the connectivity of atoms, that is, in the order in which atoms are attached to each other within molecules (Section 12.6).

Conformations. Conformations are differing orientations of the same molecule made possible by free rotation about single bonds in the molecule (Section 12.7).

Alkane nomenclature. The IUPAC name for an alkane is based on the longest continuous chain of carbon atoms in the molecule. A group of carbon atoms attached to the chain is an alkyl group. Both the position and the identity of the alkyl group are prefixed to the name of the longest carbon chain (Section 12.8).

Line-angle structural formulas. A line-angle structural formula is a structural representation in which a line represents a carbon–carbon bond and a carbon atom is understood to be present at every point where two lines meet and at the ends of the line. Line-angle structural formulas are the most concise method for representing the structure of a hydrocarbon or hydrocarbon derivative (Section 12.9).

Cycloalkanes. Cycloalkanes are saturated hydrocarbons in which at least one cyclic arrangement of carbon atoms is present. The formulas of all cycloalkanes can be represented by the general formula C_nH_{2n}, where n is the number of carbon atoms present (Section 12.12).

Cycloalkane nomenclature. The IUPAC name for a cycloalkane is obtained by placing the prefix *cyclo-* before the alkane name that corresponds to the number of carbon atoms in the ring. Alkyl groups attached to the ring are located by using a ring-numbering system (Section 12.13).

Cis-trans isomerism. For certain disubstituted cycloalkanes, *cis–trans* isomers exist. *Cis–trans* isomers are compounds that have the same molecular and structural formulas but different arrangements of atoms in space because of restricted rotation about bonds (Section 12.14).

Natural sources of saturated hydrocarbons. Natural gas and petroleum are the largest and most important natural sources of both alkanes and cycloalkanes (Section 12.15).

Physical properties of saturated hydrocarbons. Saturated hydrocarbons are not soluble in water and have lower densities than water. Melting and boiling points increase with increasing carbon chain length or ring size (Section 12.16).

Chemical properties of saturated hydrocarbons. Two important reactions that saturated hydrocarbons undergo are combustion and halogenation. In combustion, saturated hydrocarbons burn in air to produce CO_2 and H_2O. Halogenation is a substitution reaction in which one or more hydrogen atoms of the hydrocarbon are replaced by halogen atoms (Section 12.17).

Halogenated alkanes. Halogenated alkanes are hydrocarbon derivatives in which one or more halogen atoms have replaced hydrogen atoms of the alkane (Section 12.18).

Halogenated alkane nomenclature. Halogenated alkanes are named by using the rules that apply to branched-chain alkanes, with halogen substituents being treated the same as alkyl groups (Section 12.18).

KEY REACTIONS AND EQUATIONS

1. Combustion (rapid reaction with O_2) of alkanes (Section 12.17)

$$\text{Alkane} + O_2 \rightarrow CO_2 + H_2O$$

2. Halogenation of alkanes (Section 12.17)

$$R\text{—}H + X_2 \xrightarrow[\text{light}]{\text{Heat or}} R\text{—}X + H\text{—}X$$

EXERCISES *and* PROBLEMS

The members of each pair of problems in this section test similar material.

Organic and Inorganic Compounds (Section 12.1)

12.1 Indicate whether each of the following statements is true or false.
   a. The number of organic compounds exceeds the number of inorganic compounds by a factor of about 2.
   b. Chemists now believe that a special "vital force" is needed to form an organic compound.
   c. Historically, the *org-* of the term *organic* was conceptually paired with the *org-* in the term *living organism.*
   d. Most but not all compounds found in living organisms are organic compounds.

12.2 Indicate whether each of the following statements is true or false.
   a. Over 7 million organic compounds have been characterized.
   b. The number of known organic compounds and the number of known inorganic compounds are approximately the same.
   c. In essence, organic chemistry is the study of the compounds of one element.
   d. Numerous organic compounds are known that do not occur in living organisms.

Bonding Characteristics of the Carbon Atom (Section 12.2)

12.3 Indicate whether each of the following situations meet or do not meet the "bonding requirement" for carbon atoms.
   a. Two single bonds and a double bond
   b. A single bond and two double bonds
   c. Three single bonds and a triple bond
   d. A double bond and a triple bond

12.4 Indicate whether each of the following situations meet or do not meet the "bonding requirement" for carbon atoms.
   a. Four single bonds
   b. Three single bonds and a double bond
   c. Two double bonds and two single bonds
   d. Two double bonds

Hydrocarbons and Hydrocarbon Derivatives (Section 12.3)

12.5 What is the difference between a *hydrocarbon* and a *hydrocarbon derivative*?

12.6 Contrast hydrocarbons and hydrocarbon derivatives in terms of number of compounds that are known.

12.7 What is the difference between a *saturated hydrocarbon* and an *unsaturated hydrocarbon*?

12.8 What structural feature is present in an unsaturated hydrocarbon that is not present in a saturated hydrocarbon?

12.9 Classify each of the following hydrocarbons as saturated or unsaturated.

12.10 Classify each of the following hydrocarbons as saturated or unsaturated.

General Formulas for Alkanes (Section 12.4)

12.11 Using the general formula for an alkane, derive the following for specific alkanes.
   a. Number of hydrogen atoms present when 8 carbon atoms are present
   b. Number of carbon atoms present when 10 hydrogen atoms are present
   c. Number of carbon atoms present when 41 total atoms are present
   d. Total number of covalent bonds present in the molecule when 7 carbon atoms are present

12.12 Using the general formula for an alkane, derive the following for specific alkanes.
   a. Number of carbon atoms present when 14 hydrogen atoms are present
   b. Number of hydrogen atoms present when 6 carbon atoms are present
   c. Number of hydrogen atoms present when 32 total atoms are present
   d. Total number of covalent bonds present in the molecule when 16 hydrogen atoms are present

Structural Formulas (Section 12.5)

12.13 Convert each of the following expanded structural formulas into a condensed structural formula.

d.

12.14 Convert each of the following expanded structural formulas into a condensed structural formula.

a.

b.

c.

d.

12.15 The following skeletal structural formulas for alkanes are incomplete in that the hydrogen atoms attached to each carbon are not shown. Complete each of these formulas by writing in the correct number of hydrogen atoms attached to each carbon atom. That is, rewrite each of these formulas as a condensed structural formula such as CH_3—CH_2—CH_3.

a. C—C—C—C
      |
      C

b. C—C—C—C—C—C
      | | |
      C   C   C

c. C—C—C—C—C—C

d.
      C
      |
   C—C—C—C
      |
      C

12.16 The following skeletal structural formulas for alkanes are incomplete in that the hydrogen atoms attached to each carbon are not shown. Complete each of these formulas by writing in the correct number of hydrogen atoms attached to each carbon atom. That is, rewrite each of these formulas as a condensed structural formula such as CH_3—CH_2—CH_3.

a. C—C—C—C—C
      | |
      C   C

b.
      C
      |
   C—C—C
      |
      C

c. C—C—C—C—C

d. C—C—C—C—C
      | |
      C   C
      |
      C

12.17 Draw the indicated type of formula for the following alkanes.
   a. The expanded structural formula for a continuous-chain alkane with the formula C_5H_{12}
   b. The expanded structural formula for CH_3—$(CH_2)_6$—CH_3
   c. The condensed structural formula, using parentheses for the —CH_2— groups, for the continuous-chain alkane $C_{10}H_{22}$
   d. The molecular formula for the alkane CH_3—$(CH_2)_4$—CH_3

12.18 Draw the indicated type of formula for the following alkanes.
   a. The expanded structural formula for a continuous-chain alkane with the molecular formula C_6H_{14}
   b. The condensed structural formula, using parentheses for the —CH_2— groups, for the straight-chain alkane $C_{12}H_{26}$
   c. The molecular formula for the alkane CH_3—$(CH_2)_6$—CH_3
   d. The expanded structural formula for CH_3—$(CH_2)_3$—CH_3

● **Alkane Isomerism (Section 12.6)**

12.19 What general requirement must be met before two compounds can be isomers?

12.20 Explain why two alkanes with the molecular formulas C_5H_{12} and C_6H_{14} could not be constitutional isomers.

12.21 Indicate whether each of the following would be expected to be the same or different for two alkane constitutional isomers.
   a. Number of hydrogen atoms present in a molecule
   b. Condensed structural formula
   c. Boiling point
   d. Melting point

12.22 Indicate whether each of the following would be expected to be the same or different for two alkane constitutional isomers.
   a. Number of carbon atoms present in a molecule
   b. Shape of molecule
   c. Density
   d. Molecular formula

12.23 What is the difference between a continuous-chain alkane and a branched-chain alkane?

12.24 The general formula for a continuous-chain alkane is C_nH_{2n+2}. What is the general formula for a branched-chain alkane?

12.25 With the help of Table 12.1, indicate how many constitutional isomers exist for each of the following.
   a. Four-carbon alkanes   b. Six-carbon alkanes
   c. Eight-carbon alkanes   d. Ten-carbon alkanes

12.26 With the help of Table 12.1, indicate how many constitutional isomers exist for each of the following.
   a. Three-carbon alkanes   b. Five-carbon alkanes
   c. Seven-carbon alkanes   d. Nine-carbon alkanes

12.27 How many of the numerous eight-carbon alkane constitutional isomers are continuous-chain alkanes?

12.28 How many of the numerous seven-carbon alkane constitutional isomers are continuous-chain alkanes?

● **Conformations of Alkanes (Section 12.7)**

12.29 For each of the following pairs of structures, determine whether they are
   1. Different conformations of the same molecule
   2. Different compounds that are constitutional isomers
   3. Different compounds that are not constitutional isomers

a. $CH_3-CH_2-CH_2-CH-CH_3$
    $\quad\quad\quad\quad\quad\quad\quad\quad\quad |$
    $\quad\quad\quad\quad\quad\quad\quad\quad CH_3$

 and $CH_3-CH-CH_2-CH_3$
    $\quad\quad\quad\quad\quad |$
    $\quad\quad\quad\quad CH_3$

b. $CH_3-CH_2-CH_2-CH_2-CH_3$

 and $CH_3-CH-CH_3$
    $\quad\quad\quad\quad\quad |$
    $\quad\quad\quad\quad CH_2$
    $\quad\quad\quad\quad\quad |$
    $\quad\quad\quad\quad CH_3$

c. $CH_3-CH_2-CH_2$ and CH_3-CH_2
    $\quad\quad\quad\quad\quad\quad\quad |$ $\quad\quad\quad\quad\quad\quad\quad\quad |$
    $\quad\quad\quad\quad\quad\quad CH_3$ $\quad\quad\quad\quad\quad\quad CH_2-CH_3$

d. $\quad\quad\quad CH_3$
    $\quad\quad\quad\quad |$
   CH_3-C-CH_3 and $CH_3-CH-CH_2-CH_3$
    $\quad\quad\quad\quad |$ $\quad\quad\quad\quad\quad\quad\quad\quad |$
    $\quad\quad\quad CH_3$ $\quad\quad\quad\quad\quad\quad\quad CH_3$

12.30 For each of the following pairs of structures, determine whether they are
1. Different conformations of the same molecule
2. Different compounds that are constitutional isomers
3. Different compounds that are not constitutional isomers

a. $CH_3-CH-CH_3$
    $\quad\quad\quad\quad |$
    $\quad\quad\quad CH_2-CH_3$

 and $CH_3-CH-CH_2-CH_3$
    $\quad\quad\quad\quad\quad |$
    $\quad\quad\quad\quad CH_3$

b. $CH_3-CH-CH_2-CH_3$
    $\quad\quad\quad\quad |$
    $\quad\quad\quad CH_3$

 and $CH_3-CH_2-CH-CH_3$
    $\quad\quad\quad\quad\quad\quad\quad |$
    $\quad\quad\quad\quad\quad\quad CH_3$

c. $CH_3-CH-CH_2-CH_3$
    $\quad\quad\quad\quad |$
    $\quad\quad\quad CH_2$
    $\quad\quad\quad\quad |$
    $\quad\quad\quad CH_3$

 and $CH_3-CH-CH-CH_3$
    $\quad\quad\quad\quad\quad |\quad\quad |$
    $\quad\quad\quad\quad CH_3\ CH_3$

d. $CH_3-CH_2-CH_2-CH_2-CH_2-CH_3$
    $\quad\quad\quad\quad\quad\quad\quad\quad CH_3$
    $\quad\quad\quad\quad\quad\quad\quad\quad\quad |$
 and CH_3-C-CH_3
    $\quad\quad\quad\quad\quad\quad\quad |$
    $\quad\quad\quad\quad\quad CH_3$

12.31 Convert each of the following linear condensed structural formulas into "regular" condensed structural formulas.
a. $CH_3-CH_2-CH-(CH_3)-CH_2-CH_3$
b. $(CH_3)_2-CH-CH_2-CH-(CH_3)_2$
c. $CH_3-CH-(CH_3)-CH_3$
d. $CH_3-CH_2-CH-(CH_2-CH_3)-CH_2-CH_3$

12.32 Convert each of the following linear condensed structural formulas into "regular" condensed structural formulas.
a. $CH_3-CH-(CH_3)-CH_2-CH_3$
b. $CH_3-C-(CH_3)_2-CH_3$
c. $(CH_3)_2-CH-CH_3$
c. $CH_2-CH_2-CH-(CH_3)-CH-(CH_3)-CH_3$

IUPAC Nomenclature for Alkanes (Section 12.8)

12.33 The first step in naming an alkane is to identify the longest continuous chain of carbon atoms. For each of the following skeletal structural formulas, how many carbon atoms are present in the longest continuous chain?

a. $\quad\quad\quad\quad\quad\quad\quad C$
    $\quad\quad\quad\quad\quad\quad\quad |$
   $C-C-C-C-C-C-C$
    $\quad\quad\quad\quad |$
    $\quad\quad\quad\quad C$
    $\quad\quad\quad\quad |$
    $\quad\quad\quad\quad C$

b. $C-C-C-C-C-C$
    $\quad\quad |\quad\quad\quad\quad\quad |$
    $\quad\quad C\quad\quad\quad\quad\quad C$
    $\quad\quad |$
    $\quad\quad C$

c. $\quad\quad\quad\quad\quad\quad C-C-C$
    $\quad\quad\quad\quad\quad\quad |$
   $C-C-C-C$
    $\quad\quad\quad\quad\quad |$
    $\quad\quad\quad\quad C-C-C-C$

d. $C-C-C$
    $\quad\quad\quad |$
   $C-C-C-C$
    $\quad\quad |$
    $\quad\quad C$
    $\quad\quad |$
    $\quad\quad C-C$

12.34 The first step in naming an alkane is to identify the longest continuous chain of carbon atoms. For each of the following skeletal structural formulas, how many carbon atoms are present in the longest continuous chain?

a. $\quad\quad\quad\quad\quad\quad\quad\quad C$
    $\quad\quad\quad\quad\quad\quad\quad\quad |$
   $C-C-C-C-C-C$
    $\quad\quad\quad\quad |\quad\quad\quad |$
    $\quad\quad\quad\quad C\quad\quad\quad C$
    $\quad\quad\quad\quad |$
    $\quad\quad\quad\quad C$

b. $C-C-C-C-C$
    $\quad |\quad\quad\quad |\quad\quad\quad |$
    $\quad C\quad\quad\quad C\quad\quad\quad C$
    $\quad\quad\quad\quad\quad |$
    $\quad\quad\quad\quad\quad C$

c. $\quad\quad\quad\quad\quad\quad C-C-C-C$
    $\quad\quad\quad\quad\quad\quad |$
   $C-C-C-C$
    $\quad\quad\quad\quad |$
   $C-C-C-C$

d. $C-C-C-C-C-C-C$
    $\quad\quad\quad |\quad\quad\quad\quad |$
    $\quad\quad\quad C\quad\quad\quad\quad C$
    $\quad\quad\quad |\quad\quad\quad\quad |$
    $\quad\quad\quad C\quad\quad\quad\quad C$
    $\quad\quad\quad |\quad\quad\quad\quad |$
    $\quad\quad\quad C\quad\quad\quad\quad C$

12.35 Give the IUPAC name for each of the following alkanes.
a. $CH_3-CH_2-CH-CH_2-CH_3$
    $\quad\quad\quad\quad\quad\quad |$
    $\quad\quad\quad\quad\quad CH_3$

b. $CH_3-CH_2-CH_2-CH_2-CH-CH_3$
    $\quad\quad\quad\quad\quad\quad\quad\quad\quad\quad\quad |$
    $\quad\quad\quad\quad\quad\quad\quad\quad\quad CH_3$

c. $CH_2-CH_2-CH_2-CH-CH_3$
    $\quad |\quad\quad\quad\quad\quad\quad\quad\quad |$
    $CH_3\quad\quad\quad\quad\quad\quad CH_3$

d. 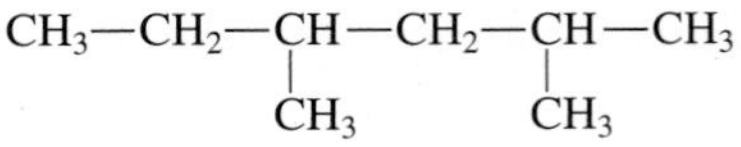 $CH_3-CH_2-CH-CH_2-CH-CH_3$
              | |
             CH_3 CH_3

12.36 Give the IUPAC name for each of the following alkanes.

a. 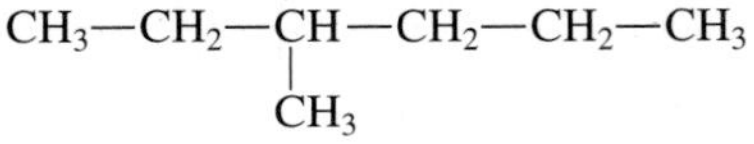 $CH_3-CH_2-CH-CH_2-CH_2-CH_3$
              |
             CH_3

b. 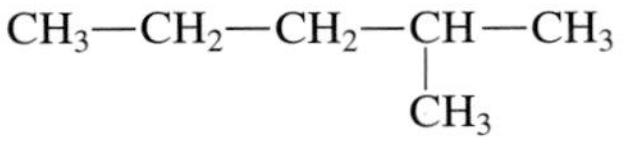 $CH_3-CH_2-CH_2-CH-CH_3$
                    |
                   CH_3

c. 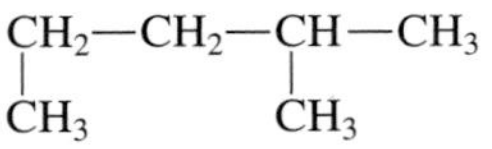 $CH_2-CH_2-CH-CH_3$
   | |
  CH_3 CH_3

d. 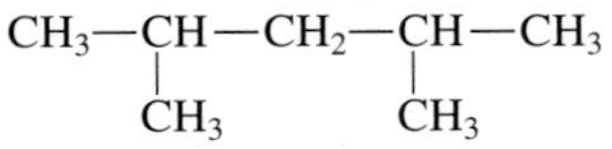 $CH_3-CH-CH_2-CH-CH_3$
        | |
     CH_3 CH_3

12.37 Give the IUPAC name for each of the following alkanes.

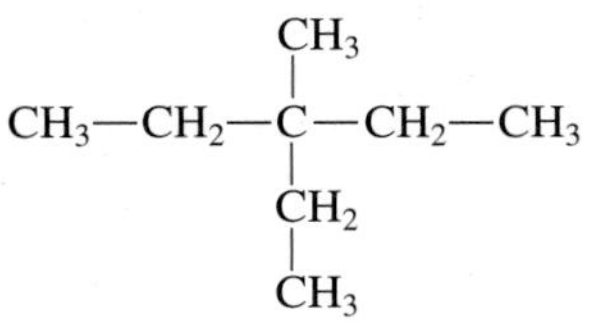

a. 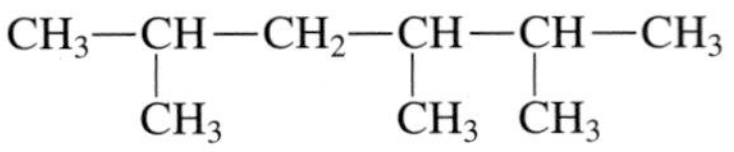 $CH_3-CH-CH_2-CH-CH-CH_3$
       | | |
    CH_3 CH_3 CH_3

(handwritten: 235 trimethyl hexane)

          CH_3
          |
b. 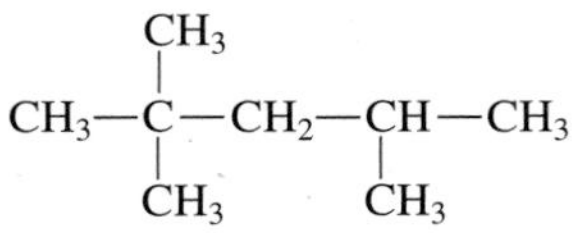 $CH_3-C-CH_2-CH-CH_3$
        | |
     CH_3 CH_3

          CH_3
          |
c. $CH_3-CH_2-C-CH_2-CH_3$
          |
          CH_2
          |
          CH_3

          CH_3
          |
d. 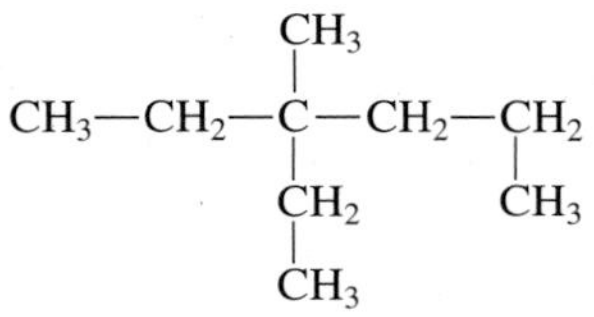 $CH_3-CH_2-C-CH_2-CH_2$
          | |
          CH_2 CH_3
          |
          CH_3

12.38 Give the IUPAC name for each of the following alkanes.

a. 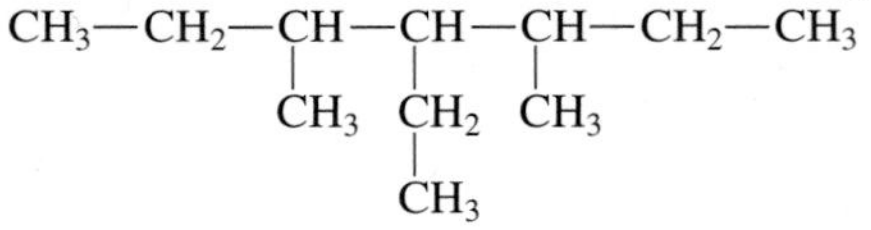 $CH_3-CH_2-CH-CH-CH-CH_2-CH_3$
               | | |
              CH_3 CH_2 CH_3
                      |
                      CH_3

                        CH_3
                        |
b. 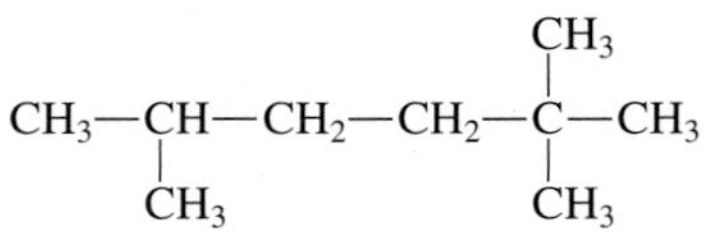 $CH_3-CH-CH_2-CH_2-C-CH_3$
        | |
     CH_3 CH_3

c. $CH_2-CH-CH_2-CH_2-CH_3$
   | |
  CH_3 CH_3

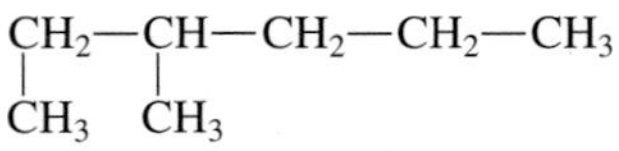

d. 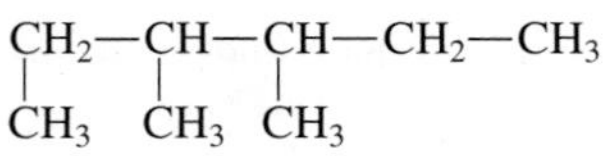 $CH_2-CH-CH-CH_2-CH_3$
   | | |
  CH_3 CH_3 CH_3

12.39 Two different carbon chains of eight atoms can be located in the following alkane.

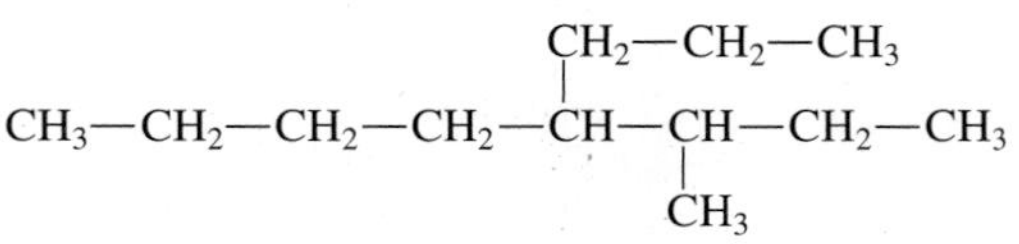

                     $CH_2-CH_2-CH_3$
                     |
$CH_3-CH_2-CH_2-CH_2-CH-CH-CH_2-CH_3$
                            |
                          CH_3

Which of these chains should be used in determining the IUPAC name for the alkane? Explain your answer.

12.40 Two different carbon chains of seven atoms can be located in the following alkane.

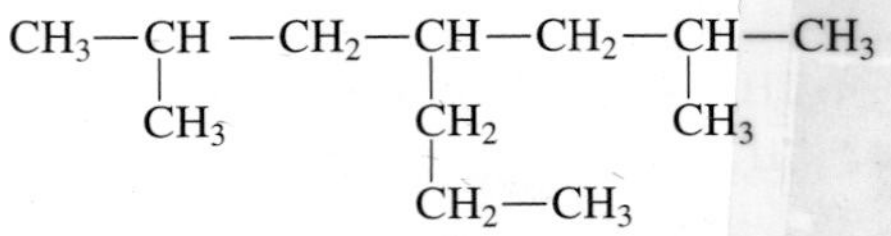

$CH_3-CH-CH_2-CH-CH_2-CH-CH_3$
        | | |
     CH_3 CH_2 CH_3
                   |
                  CH_2-CH_3

Which of these chains should be used in determining the IUPAC name for the alkane? Explain your answer.

12.41 Draw a condensed structural formula for each of the following alkanes.

a. 3,4-Dimethylhexane   b. 3-Ethyl-3-methylpentane
c. 3,5-Diethyloctane   d. 4-Propylnonane

12.42 Draw a condensed structural formula for each of the following alkanes.

a. 2,4-Dimethylhexane   b. 5-Propyldecane
c. 2,3,4-Trimethyloctane   d. 3-Ethyl-3-methylheptane

12.43 For each of the alkanes in Problem 12.41 determine (a) the number of alkyl groups present and (b) the number of substituents present.

12.44 For each of the alkanes in Problem 12.42 determine (a) the number of alkyl groups present and (b) the number of substituents present.

12.45 Explain why the name given for each of the following alkanes is not the correct IUPAC name. Then give the correct IUPAC name for the compound.

a. 2-Ethyl-2-methylpropane   b. 2,3,3-Trimethylbutane
c. 3-Methyl-4-ethylhexane   d. 2-Methyl-4-methylhexane

12.46 Explain why the name given for each of the following alkanes is not the correct IUPAC name. Then give the correct IUPAC name for the compound.

a. 2-Ethylpentane   b. 3,3,4-Trimethylpentane
c. 4-Ethyl-3-methylhexane   d. 3-Ethyl-4-ethylhexane

Line-Angle Structural Formulas for Alkanes (Section 12.9)

12.47 Convert each of the following line-angle structural formulas to a skeletal structural formula.

a. 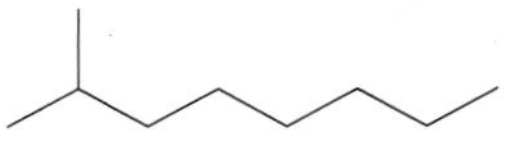   b.

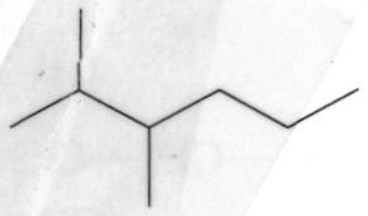

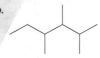

c.    d.

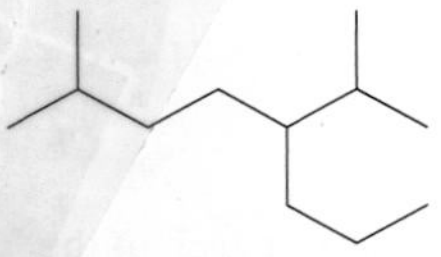

12.48 Convert each of the following line-angle structural formulas to a skeletal structural formula.

a.    b.

c.   d.

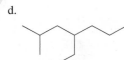

12.49 Convert each of the following line-angle structural formulas to a condensed structural formula.

a.   b.

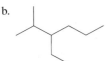

c.   d.

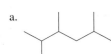

12.50 Convert each of the following line-angle structural formulas to a condensed structural formula.

a.   b.

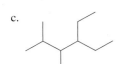

c.   d.

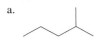

12.51 Do the line-angle structural formulas in each of the following sets represent (1) the same compound, (2) constitutional isomers, or (3) different compounds that are not constitutional isomers?

a.

and

b.

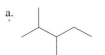

and

12.52 Do the line-angle structural formulas in each of the following sets represent (1) the same compound, (2) constitutional isomers, or (3) different compounds that are not constitutional isomers?

a.

and

b.

and

12.53 Convert each of the condensed structural formulas in Problem 12.37 to a line-angle structural formula.

12.54 Convert each of the condensed structural formulas in Problem 12.38 to a line-angle structural formula.

12.55 Assign an IUPAC name to each of the compounds in Problem 12.47.

12.56 Assign an IUPAC name to each of the compounds in Problem 12.48.

12.57 Determine the molecular formula for each of the compounds in Problem 12.49.

12.58 Determine the molecular formula for each of the compounds in Problem 12.50.

● **Classification of Carbon Atoms (Section 12.10)**

12.59 For each of the alkane structures in Problem 12.37, give the number of (a) primary, (b) secondary, (c) tertiary, and (d) quaternary carbon atoms present.

12.60 For each of the alkane structures in Problem 12.38, give the number of (a) primary, (b) secondary, (c) tertiary, and (d) quaternary carbon atoms present.

● **Branched-Chain Alkyl Groups (Section 12.11)**

12.61 Give the name of the branched alkyl group attached to each of the following carbon chains, where the carbon chain is denoted by a horizontal line.

a. ———————   b. ———————

        $CH—CH_3$ CH_2

         CH_3 $CH_3—CH—CH_3$

c. ———————   d. ———————

    $CH_3—CH—CH_3$ $CH—CH_3$

                                CH_2

                                CH_3

12.62 Give the name of the branched alkyl group attached to each of the following carbon chains, where the carbon chain is denoted by a horizontal line.

a. ———————   b. ———————

      CH_2 $CH_3—C—CH_3$

      $CH—CH_3$ CH_3

      CH_3

c. ———————   d. ———————

  $CH_3—CH_2—CH—CH_3$ CH

                           CH_3 CH_3

12.63 Draw condensed structural formulas for the following branched alkanes.
   a. 5-(sec-Butyl)decane
   b. 4,4-Diisopropyloctane
   c. 5-Isobutyl-2,3-dimethylnonane
   d. 4-(1,1-Dimethylethyl)octane

12.64 Draw condensed structural formulas for the following branched alkanes.
   a. 5-Isobutylnonane
   b. 4,4-Di(sec-butyl)decane
   c. 4-(tert-Butyl)-3,3-diethylheptane
   d. 5-(2-Methylpropyl)nonane

12.65 To which carbon atoms in a hexane molecule can each of the following alkyl groups be attached without extending the longest carbon chain beyond six carbons?
   a. Ethyl   b. Isopropyl   c. Isobutyl   d. tert-Butyl

12.66 To which carbon atoms in a heptane molecule can each of the following alkyl groups be attached without extending the longest carbon chain beyond seven carbons?
   a. Ethyl   b. Isopropyl   c. sec-Butyl   d. tert-Butyl

12.67 Using IUPAC rules, name the following "complex" five-carbon branched alkyl groups.

a.
$$\text{———CH}_2\text{—CH—CH}_2\text{—CH}_3$$
with CH₃ above the CH

$$\text{a. ———CH}_2\text{—}\overset{\overset{\displaystyle CH_3}{|}}{CH}\text{—CH}_2\text{—CH}_3$$

$$\text{b. ———}\overset{\overset{\displaystyle CH_3}{|}}{\underset{\underset{\displaystyle CH_3}{|}}{C}}\text{—CH}_2\text{—CH}_3$$

12.68 Using IUPAC rules, name the following "complex" five-carbon branched alkyl groups.

$$\text{a. ———CH}_2\text{—}\overset{\overset{\displaystyle CH_3}{|}}{\underset{\underset{\displaystyle CH_3}{|}}{CH}}\text{—CH}_3 \qquad \text{b. ———}\overset{\overset{\displaystyle CH_3}{|}}{CH}\text{—}\overset{\overset{\displaystyle CH_3}{|}}{CH}\text{—CH}_3$$

● **Cycloalkanes (Section 12.12)**

12.69 Using the general formula for a cycloalkane, derive the following for specific cycloalkanes.
a. Number of hydrogen atoms present when 8 carbon atoms are present
b. Number of carbon atoms present when 12 hydrogen atoms are present
c. Number of carbon atoms present when a total of 15 atoms are present
d. Number of covalent bonds present when 5 carbon atoms are present

12.70 Using the general formula for a cycloalkane, derive the following for specific cycloalkanes.
a. Number of hydrogen atoms present when 4 carbon atoms are present
b. Number of carbon atoms present when 6 hydrogen atoms are present
c. Number of hydrogen atoms present when a total of 18 atoms are present
d. Number of covalent bonds present when 8 hydrogen atoms are present

12.71 What is the molecular formula for each of the following cycloalkane molecules?

12.72 What is the molecular formula for each of the following cycloalkane molecules?

● **IUPAC Nomenclature for Cycloalkanes (Section 12.13)**

12.73 Assign an IUPAC name to each of the cycloalkanes in Problem 12.71.

12.74 Assign an IUPAC name to each of the cycloalkanes in Problem 12.72.

12.75 What is wrong with each of the following attempts to name a cycloalkane using IUPAC rules?
a. Dimethylcyclohexane   b. 3,4-Dimethylcyclohexane
c. 1-Ethylcyclobutane   d. 2-Ethyl-1-methylcyclopentane

12.76 What is wrong with each of the following attempts to name a cycloalkane using IUPAC rules?
a. Dimethylcyclopropane
b. 1-Methylcyclohexane
c. 2,5-Dimethylcyclobutane
d. 1-Propyl-2-ethylcyclohexane

12.77 Draw line-angle structural formulas for the following cycloalkanes.
a. Propylcyclobutane
b. Isopropylcyclobutane
c. *cis*-1,2-Diethylcyclohexane
d. *trans*-1-Ethyl-3-propylcyclopentane

12.78 Draw line-angle structural formulas for the following cycloalkanes.
a. Butylcyclopentane
b. Isobutylcyclopentane
c. *cis*-1,3-Diethylcyclopentane
d. *trans*-1-Ethyl-4-methylcyclohexane

● **Isomerism in Cycloalkanes (Section 12.14)**

12.79 Determine the number of constitutional isomers that are possible for each of the following situations.
a. Four-carbon cycloalkanes
b. Five-carbon cycloalkanes where the ring has three carbon atoms
c. Six-carbon cycloalkanes where the ring has five carbon atoms
d. Seven-carbon cycloalkanes where the ring has five carbon atoms

12.80 Determine the number of constitutional isomers that are possible for each of the following situations.
a. Five-carbon cycloalkanes
b. Six-carbon cycloalkanes where the ring has four carbon atoms
c. Six-carbon cycloalkanes where the ring has three carbon atoms
d. Seven-carbon cycloalkanes where the ring has four carbon atoms

12.81 Determine whether *cis–trans* isomerism is possible for each of the following cycloalkanes. If it is, then draw structural formulas for the *cis* and *trans* isomers.
a. Isopropylcyclobutane
b. 1,2-Diethylcyclopropane
c. 1-Ethyl-1-propylcyclopentane
d. 1,3-Dimethylcyclohexane

12.82 Determine whether *cis–trans* isomerism is possible for each of the following cycloalkanes. If it is, then draw structural formulas for the *cis* and *trans* isomers.
a. *sec*-Butylcyclohexane
b. 1-Ethyl-3-methylcyclobutane
c. 1,1-Dimethylcyclohexane
d. 1,3-Dipropylcyclopentane

● **Sources of Alkanes and Cycloalkanes (Section 12.15)**

12.83 In terms of the types of hydrocarbons present, what is the composition of unprocessed natural gas?

12.84 In terms of the types of hydrocarbons present, what is the composition of the gasoline fraction obtained by processing crude petroleum?

12.85 What physical property of hydrocarbons is the basis for the fractional distillation process for separating hydrocarbons?

12.86 Describe the process by which crude petroleum is separated into simpler mixtures (fractions).

Physical Properties of Alkanes and Cycloalkanes (Section 12.16)

12.87 Which member in each of the following pairs of compounds has the higher boiling point?
a. Hexane and octane
b. Cyclobutane and cyclopentane
c. Pentane and 1-methylbutane
d. Pentane and cyclopentane

12.88 Which member in each of the following pairs of compounds has the higher boiling point?
a. Methane and ethane
b. Cyclohexane and hexane
c. Butane and methylpropane
d. Pentane and 2,2-dimethylpropane

12.89 With the help of Figures 12.11 and 12.12, determine in which of the following pairs of compounds both members of the pair have the same physical state (solid, liquid, or gas) at room temperature and pressure.
a. Ethane and hexane
b. Cyclopropane and butane
c. Octane and 3-methyloctane
d. Pentane and decane

12.90 With the help of Figures 12.11 and 12.12, determine in which of the following pairs of compounds both members of the pair have the same physical state (solid, liquid, or gas) at room temperature and pressure.
a. Methane and butane
b. Cyclobutane and cyclopentane
c. Hexane and 2,3-dimethylbutane
d. Pentane and octane

Chemical Properties of Alkanes and Cycloalkanes (Section 12.17)

12.91 Write the formulas of the products from the complete combustion of each of the following alkanes or cycloalkanes.
a. C_3H_8
b. Butane
c. Cyclobutane
d. $CH_3-(CH_2)_{15}-CH_3$

12.92 Write the formulas of the products from the complete combustion of each of the following alkanes or cycloalkanes.
a. C_4H_{10}
b. 2-Methylpentane
c. Cyclopentane
d. $CH_3-(CH_2)_7-CH_3$

12.93 Write molecular formulas for all the possible halogenated hydrocarbon products from the bromination of methane.

12.94 Write molecular formulas for all the possible halogenated hydrocarbon products from the fluorination of methane.

12.95 Write structural formulas for all the possible halogenated hydrocarbon products from the monochlorination of the following alkanes or cycloalkanes.
a. Ethane
b. Butane
c. 2-Methylpropane
d. Cyclopentane

12.96 Write structural formulas for all the possible halogenated hydrocarbon products from the monobromination of the following alkanes or cycloalkanes.
a. Propane
b. Pentane
c. 2-Methylbutane
d. Cyclohexane

Nomenclature of Halogenated Alkanes (Section 12.18)

12.97 Give both IUPAC and common names to each of the following halogenated hydrocarbons.
a. CH_3-I
b. $CH_3-CH_2-CH_2-Cl$
c. $CH_3-CH-CH_2-CH_3$ with F below the CH
d. cyclobutane with Cl

12.98 Give both IUPAC and common names to each of the following halogenated hydrocarbons.
a. $CH_3-CH_2-CH_2-CH_2-Br$
b. $CH_3-CH-Cl$ with CH_3 below
c. $CH_3-C(CH_3)(Cl)-CH_3$
d. cyclohexane with Br

12.99 Draw structural formulas for the following halogenated hydrocarbons.
a. Trichloromethane
b. 1,2-Dichloro-1,1,2,2-tetrafluoroethane
c. Isopropyl bromide
d. trans-1-Bromo-3-chlorocyclopentane

12.100 Draw structural formulas for the following halogenated hydrocarbons.
a. Trifluorochloromethane
b. Pentafluoroethane
c. Isobutyl chloride
d. cis-1,2-Dichlorocyclohexane

ADDITIONAL PROBLEMS

12.101 Answer the following questions about the unbranched alkane with six carbon atoms.
a. How many hydrogen atoms are present?
b. How many carbon–carbon bonds are present?
c. How many carbon atoms have two hydrogen atoms bonded to them?
d. How many total covalent bonds are present?

12.102 Answer the following questions about the unbranched alkane with seven carbon atoms.
a. Is the alkane a solid, a liquid, or a gas at room temperature?
b. Is the alkane less dense or more dense than water?
c. Is the alkane soluble or insoluble in water?
d. Is the alkane flammable or nonflammable in air?

12.103 Indicate whether the members of each of the following pairs of compounds are constitutional isomers.
a. Hexane and 2-methylhexane
b. Hexane and 2,2-dimethylbutane
c. Hexane and methylcyclopentane
d. Hexane and cyclohexane

12.104 Draw structural formulas for the following compounds.
a. trans-1,4-Difluorocyclohexane
b. cis-1-Chloro-2-methylcyclobutane
c. tert-Butyl bromide
d. Isobutyl iodide

12.105 Give the molecular formula for each of the following compounds.
a. An 18-carbon alkane
b. A 7-carbon cycloalkane

c. A 7-carbon difluorinated alkane

d. A 6-carbon dibrominated cycloalkane

12.106 Classify each of the following molecular formulas as representing an alkane, a cycloalkane, a halogenated alkane, or a halogenated cycloalkane.

   a. C_6H_{14} b. $C_6H_{10}Cl_2$ c. $C_4H_8Cl_2$ d. C_4H_8

12.107 There are eighteen eight-carbon noncyclic alkane constitutional isomers. How many of these isomers have names that end in the following?

   a. Octane

   b. Heptane

   c. Dimethylhexane

   d. Ethylhexane

12.108 Write skeletal structural formulas and assign names to all saturated hydrocarbon constitutional isomers (ignore *cis–trans* isomers) with the following molecular formulas.

   a. C_7H_{16} (9 isomers)   b. C_6H_{12} (12 isomers)

   c. $C_5H_{11}Cl$ (8 isomers)   d. $C_3H_6Br_2$ (4 isomers)

12.109 Assign an IUPAC name to each of the following hydrocarbons, whose line-angle structural formulas are

a.   b.

c. 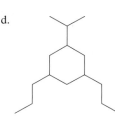   d.

12.110 How many different alkyl groups exist that contain the following?

   a. Two carbon atoms   b. Three carbon atoms

   c. Four carbon atoms   d. Five carbon atoms

$m$ultiple-Choice Practice Test

12.111 Which of the following statements concerning saturated hydrocarbons is *incorrect*?

   a. Every carbon atom present must have four bonds.

   b. All bonds present must be single bonds.

   c. Every carbon atom present must be bonded to at least one hydrogen atom.

   d. This classification includes both alkanes and cycloalkanes.

12.112 Which of the following gives the generalized molecular formulas for alkanes and cycloalkanes, respectively?

   a. C_nH_{2n+2} and C_nH_{2n} b. C_nH_{2n+2} and C_nH_{2n+4}

   c. C_nH_{2n} and C_nH_{2n-2} d. C_nH_{2n} and C_nH_{2n+2}

12.113 The formula CH_3—CH_2—CH_2—CH_2—CH_3 is an example of which of the following?

   a. An expanded structural formula

   b. A condensed structural formula

   c. A skeletal structural formula

   d. A line-angle structural formula

12.114 Which of the following compounds is a constitutional isomer of CH_3—CH_2—CH_2—CH_2—CH_3?

   a. 2-Methylpentane   b. 2-Methylbutane

   c. 2,2-Dimethylpentane   d. 2,2-Dimethylbutane

12.115 One of the three five-carbon alkane constitutional isomers has the molecular formula C_5H_{12}. Which of the following gives the molecular formulas for the other two isomers respectively?

   a. C_5H_{11} and C_5H_{10} b. C_5H_{13} and C_5H_{14}

   c. C_4H_{12} and C_6H_{12} d. C_5H_{12} and C_5H_{12}

12.116 Which of the following statements concerning alkanes and alkyl groups is *incorrect*?

   a. Isobutane and 2-methylpropane are two names for the same compound.

   b. 2-Methylpentane and 2-methylbutane contain the same number of alkyl groups.

   c. Butane and cyclobutane contain the same number of hydrogen atoms.

   d. Secondary-butyl group and (1,1-dimethylethyl) group are two names for the same alkyl group.

12.117 In which of the following alkanes are both secondary and tertiary carbon atoms present?

   a. CH_3—CH_2—CH_2—CH_3

   b. CH_3—CH—CH_3
$$\qquad\qquad |$$
$$\qquad\quad\; CH_3$$

   c. CH_3—CH_2—CH—CH_3
$$\qquad\qquad\qquad\; |$$
$$\qquad\qquad\qquad CH_3$$

   d. CH_3—CH—CH—CH_3
$$\qquad\qquad |\quad\; |$$
$$\qquad\quad\; CH_3\;\; CH_3$$

12.118 For which of the following halogenated cycloalkanes is *cis–trans* isomerism possible?

   a. 1,1-Dibromocyclobutane

   b. 1-Bromo-1-chlorocyclobutane

   c. 1-Bromo-2-chlorocyclobutane

   d. 1,1-Dichlorocyclobutane

12.119 Which of the following statements concerning the boiling points of specific alkanes is *correct*?

   a. Hexane has a higher boiling point than heptane.

   b. Pentane has a higher boiling point than 2-methylbutane.

   c. Butane has a higher boiling point than cyclobutane.

   d. Butane and pentane have approximately the same boiling point.

12.120 Which of the following statements concerning the chemical properties of alkanes and cycloalkanes is *correct*?

   a. Alkanes undergo combustion reactions but cycloalkanes do not.

   b. Neither alkanes nor cycloalkanes undergo combustion reactions.

   c. Both alkanes and cycloalkanes undergo combustion reactions.

   d. Alkanes undergo combustion reactions but do not undergo halogenation reactions.

Unsaturated Hydrocarbons

The unsaturated hydrocarbon ethene is used to stimulate the ripening process in fruit that has been picked while still green, such as bananas.

Two general types of hydrocarbons exist, *saturated* and *unsaturated*. Saturated hydrocarbons, discussed in the previous chapter, include the alkanes and cycloalkanes. All bonds in saturated hydrocarbons are single bonds. Unsaturated hydrocarbons, the topic of this chapter, contain one or more carbon–carbon multiple bonds. There are three classes of unsaturated hydrocarbons: the *alkenes*, the *alkynes*, and the *aromatic hydrocarbons*, all of which we will consider.

13.1 UNSATURATED HYDROCARBONS

An **unsaturated hydrocarbon** *is a hydrocarbon in which one or more carbon–carbon multiple bonds (double bonds, triple bonds, or both) are present.* Unsaturated hydrocarbons have *physical* properties similar to those of saturated hydrocarbons. However, their *chemical* properties are much different. Unsaturated hydrocarbons are chemically more reactive than their saturated counterparts. The increased reactivity of unsaturated hydrocarbons is related to the presence of the carbon–carbon multiple bond(s) in such compounds. These multiple bonds serve as locations where chemical reactions can occur.

Whenever a specific portion of a molecule governs its chemical properties, that portion of the molecule is called a functional group. A **functional group** *is the part of an organic molecule where most of its chemical reactions occur.* Carbon–carbon multiple bonds are the functional group for an unsaturated hydrocarbon.

The study of various functional groups and their respective reactions provides the organizational structure for organic chemistry. Each of the organic chemistry chapters

Alkanes and cycloalkanes (Chapter 12) lack functional groups; as a result, they are relatively unreactive.

The general molecular formula for an alkene with one double bond, C_nH_{2n}, is the same as that for a cycloalkane (Section 12.12). Thus such alkenes and cycloalkanes with the same number of carbon atoms are isomeric with one another.

An older but still widely used name for alkenes is *olefins*, pronounced "oh-la-fins." The term *olefin* means "oil-forming." Many alkenes react with Cl_2 to form "oily" compounds.

that follow introduces new functional groups that characterize families of hydrocarbon derivatives.

Unsaturated hydrocarbons are subdivided into three groups on the basis of the type of multiple bond(s) present: (1) *alkenes,* which contain one or more carbon–carbon double bonds, (2) *alkynes,* which contain one or more carbon–carbon triple bonds, and (3) *aromatic hydrocarbons,* which exhibit a special type of "delocalized" bonding that involves a six-membered carbon ring (to be discussed in Section 13.12).

We begin our consideration of unsaturated hydrocarbons with a discussion of alkenes. Information about alkynes and aromatic hydrocarbons then follows.

13.2 CHARACTERISTICS OF ALKENES AND CYCLOALKENES

An **alkene** *is an acyclic unsaturated hydrocarbon that contains one or more carbon–carbon double bonds.* The alkene functional group is, thus, a $C{=}C$ group. Note the close similarity between the family names *alkene* and *alkane* (Section 12.4); they differ only in their endings: *-ene* versus *-ane.* The *-ene* ending means a double bond is present.

The simplest type of alkene contains only one carbon–carbon double bond. Such compounds have the general molecular formula C_nH_{2n}. Thus alkenes with one double bond have two fewer hydrogen atoms than do alkanes (C_nH_{2n+2}).

The two simplest alkenes are ethene (C_2H_4) and propene (C_3H_6).

$$CH_2{=}CH_2 \qquad CH_2{=}CH{-}CH_3$$
Ethene   Propene

Comparing the geometrical shape of ethene with that of methane (the simplest alkane) reveals a major difference. The arrangement of bonds about the carbon atom in methane is tetrahedral (Section 12.4), whereas the carbon atoms in ethene have a trigonal planar arrangement of bonds; that is, they form a flat, triangle-shaped arrangement (see Figure 13.1). The two carbon atoms participating in a double bond and the four other atoms attached to these two carbon atoms always lie in a plane with a trigonal planar arrangement of atoms about each carbon atom of the double bond. Such an arrangement of atoms is consistent with the principles of VSEPR theory (Section 5.8).

A **cycloalkene** *is a cyclic unsaturated hydrocarbon that contains one or more carbon–carbon double bonds within the ring system.* Cycloalkenes in which there is only one double bond have the general molecular formula C_nH_{2n-2}. This general formula reflects the loss of four hydrogen atoms from that of an alkane (C_nH_{2n+2}). Note that two hydrogen atoms are lost because of the double bond and two because of the ring structure.

The simplest cycloalkene is the compound cyclopropene (C_3H_4), a three-membered carbon ring system containing one double bond.

Figure 13.1 Three-dimensional representations of the structures of ethene and methane. In ethene, the atoms are in a flat (planar) rather than a tetrahedral arrangement. Bond angles are 120°.

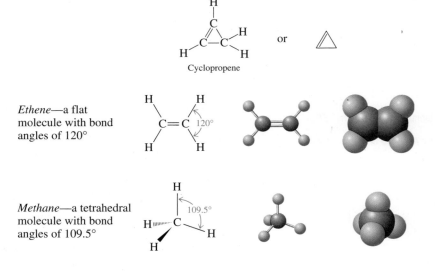

Alkenes with more than one carbon–carbon double bond are relatively common. When two double bonds are present, the compounds are often called *dienes;* for three double bonds the designation *trienes* is used. Cycloalkenes that contain more than one double bond are possible but are not common.

13.3 IUPAC NOMENCLATURE FOR ALKENES AND CYCLOALKENES

The IUPAC rules previously presented for naming alkanes and cycloalkanes (Sections 12.8 and 12.13) can be used, with some modification, to name alkenes and cycloalkenes.

Rule 1. *Replace the alkane suffix -ane with the suffix -ene, which is used to indicate the presence of a carbon–carbon double bond.*

Rule 2. *Select as the parent carbon chain the longest continuous chain of carbon atoms that contains both carbon atoms of the double bond.* For example, select

$$CH_2=C-CH_2-CH_2-CH_3 \qquad \text{not} \qquad CH_2=C-CH_2-CH_2-CH_3$$
$$\quad\;\; | \qquad\qquad\qquad\qquad\qquad\qquad\qquad\qquad | $$
$$\quad\;\; CH_2 \qquad\qquad\qquad\qquad\qquad\qquad\qquad\; CH_2$$
$$\quad\;\; | \qquad\qquad\qquad\qquad\qquad\qquad\qquad\qquad | $$
$$\quad\;\; CH_3 \qquad\qquad\qquad\qquad\qquad\qquad\qquad\; CH_3$$

Longest carbon chain containing both carbon atoms of the double bond

Carbon chain that does not contain both carbon atoms of the double bond

Rule 3. *Number the parent carbon chain beginning at the end nearest the double bond.*

$$\overset{1}{C}H_3-\overset{2}{C}H=\overset{3}{C}H-\overset{4}{C}H_2-\overset{5}{C}H_3 \qquad \text{not} \qquad \overset{5}{C}H_3-\overset{4}{C}H=\overset{3}{C}H-\overset{2}{C}H_2-\overset{1}{C}H_3$$

If the double bond is equidistant from both ends of the parent chain, begin numbering from the end closer to a substituent.

$$\overset{4}{C}H_3-\overset{3}{C}H=\overset{2}{C}H-\overset{1}{C}H_2 \qquad \text{not} \qquad \overset{1}{C}H_3-\overset{2}{C}H=\overset{3}{C}H-\overset{4}{C}H_2$$
$$\qquad\qquad\qquad\quad | \qquad\qquad\qquad\qquad\qquad\qquad\qquad\quad |$$
$$\qquad\qquad\qquad\quad Cl \qquad\qquad\qquad\qquad\qquad\qquad\qquad\quad Cl$$

Rule 4. *Give the position of the double bond in the chain as a single number, which is the lower-numbered carbon atom participating in the double bond.* This number is placed immediately before the name of the parent carbon chain.

$$\overset{1}{C}H_3-\overset{2}{C}H=\overset{3}{C}H-\overset{4}{C}H_3 \qquad\qquad \overset{1}{C}H_2=\overset{2}{C}H-\overset{3}{C}H-\overset{4}{C}H_3$$
$$\qquad\qquad\qquad\qquad\qquad\qquad\qquad\qquad\qquad\qquad\qquad | $$
$$\qquad\qquad\qquad\qquad\qquad\qquad\qquad\qquad\qquad\qquad\; CH_3$$

2-Butene   3-Methyl-1-butene

Rule 5. *Use the suffixes -diene, -triene, -tetrene, and so on when more than one double bond is present in the molecule.* A separate number must be used to locate each double bond.

$$\overset{1}{C}H_2=\overset{2}{C}H-\overset{3}{C}H=\overset{4}{C}H_2 \qquad\qquad \overset{1}{C}H_2=\overset{2}{C}H-\overset{3}{C}H-\overset{4}{C}H=\overset{5}{C}H_2$$
$$\qquad\qquad\qquad\qquad\qquad\qquad\qquad\qquad\qquad\qquad\qquad | $$
$$\qquad\qquad\qquad\qquad\qquad\qquad\qquad\qquad\qquad\qquad\; CH_3$$

1,3-Butadiene   3-Methyl-1,4-pentadiene

Rule 6. *Do not use a number to locate the double bond in unsubstituted cycloalkenes with only one double bond because that bond is assumed to be between carbons 1 and 2.*

Rule 7. *In substituted cycloalkenes with only one double bond, the double-bonded carbon atoms are numbered 1 and 2 in the direction (clockwise or counterclockwise) that gives the first-encountered substituent the lower number.* Again, no number is used in the name to locate the double bond.

Cyclohexene   4-Methylcyclohexene

Carbon–carbon double bonds take precedence over alkyl groups and halogen atoms in determining the direction in which the parent carbon chain is numbered.

A number is not needed to specify double bond position in ethene and propene because there is only one way of positioning the double bond in these molecules.

In 1993, IUPAC rules were revised relative to the positioning of numbers used in names. Instead of placing the number immediately in front of the base name (2-pentene) IUPAC recommended that the number be placed immediately before the part of the name that the number serves as a locator for (pent-2-ene). Transition to this new number positioning has been slow. In this textbook, the newer system will be used whenever it helps to clarify a name; otherwise the older system will be used. Two additional examples comparing the two systems are:

2-methyl-1-butene (old)
2-methylbut-1-ene (new)

1,3-butadiene (old)
buta-1,3-diene (new)

Rule 8. *In cycloalkenes with more than one double bond within the ring, assign one double bond the numbers 1 and 2 and the other double bonds the lowest numbers possible.*

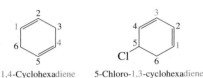

1,4-Cyclohexadiene   5-Chloro-1,3-cyclohexadiene

● **EXAMPLE 13.1**

Assigning IUPAC Names to Alkenes and Cycloalkenes

Assign IUPAC names to the following alkenes and cycloalkenes.

a. $CH_3-CH=CH-CH_2-CH_2-CH_3$

b. $CH_3-CH_2-C=CH_3$
               |
              CH_2
               |
              CH_3

c.

d.

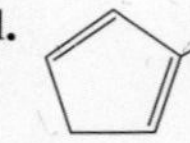

Solution

a. The carbon chain in this hexene is numbered from the end closest to the double bond.

$$\overset{1}{C}H_3-\overset{2}{C}H=\overset{3}{C}H-\overset{4}{C}H_2-\overset{5}{C}H_2-\overset{6}{C}H_3$$

The complete IUPAC name is 2-hexene.

b. The longest carbon chain containing *both* carbons of the double bond has four carbon atoms. Thus we have a butene.

$$\boxed{CH_3-CH_2-C=CH_2}$$
               |
              CH_2
               |
              CH_3

The chain is numbered from the end closest to the double bond. The complete IUPAC name is 2-ethyl-1-butene.

c. This compound is a methylcyclobutene. The numbers 1 and 2 are assigned to the carbon atoms of the double bond, and the ring is numbered clockwise, which results in a carbon 3 location for the methyl group. (Counterclockwise numbering would have placed the methyl group on carbon 4.) The complete IUPAC name of the cycloalkene is 3-methylcyclobutene. The double bond is understood to involve carbons 1 and 2.

d. A ring system containing five carbon atoms, two double bonds, and a methyl substituent on the ring is called a methylcyclopentadiene. Two different numbering systems produce the same locations (carbons 1 and 3) for the double bonds.

The counterclockwise numbering system assigns the lower number to the methyl group. The complete IUPAC name of the compound is 2-methyl-1,3-cyclopentadiene.

Practice Exercise 13.1

Assign IUPAC names to the following alkenes and cycloalkenes.

a. $CH_3—CH{=}CH—CH_2—\underset{\underset{\displaystyle CH_3}{|}}{CH}—CH_3$

b.

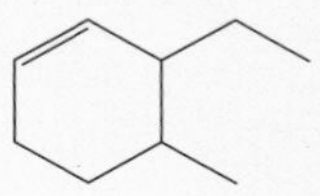

c. $CH_2{=}CH—CH{=}CH_2$

d.

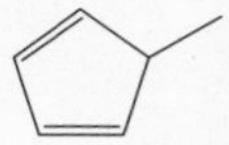

Answers: a. 5-Methyl-2-hexene; **b.** 3-Ethyl-4-methylcyclohexene; **c.** 1,3-Butadiene; **d.** 5-Methyl-1,3-pentadiene

Common Names (Non-IUPAC Names)

The simpler members of most families of organic compounds, including alkenes, have common names in addition to IUPAC names. In many cases these common (non-IUPAC) names are used almost exclusively for the compounds. It would be nice if such common names did not exist, but they do. We have no choice but to memorize these names; fortunately, there are not many of them.

The two simplest alkenes, ethene and propene, have common names you should be familiar with. They are ethylene and propylene, respectively.

$$CH_2{=}CH_2 \qquad CH_2{=}CH—CH_3$$
Ethylene $\qquad$ Propylene

Alkenes as Substituents

Just as there are *alkanes* and *alkyl groups* (Section 12.8), there are *alkenes* and *alkenyl groups*. An **alkenyl group** *is a noncyclic hydrocarbon substituent in which a carbon–carbon double bond is present.* The three most frequently encountered alkenyl groups are the one-, two-, and three-carbon entities, which may be named using IUPAC nomenclature (methylidene, ethenyl, and 2-propenyl) or with common names (methylene, vinyl, and allyl).

$$CH_2{=} \qquad\qquad CH_2{=}CH— \qquad\qquad CH_2{=}CH—CH_2—$$
Methylene group $\qquad$ Vinyl group $\qquad$ Allyl group
(IUPAC name: methylidene group) $\quad$ (IUPAC name: ethenyl group) $\quad$ (IUPAC name: 2-propenyl group)

The use of these alkenyl group names in actual compound nomenclature is illustrated in the following examples.

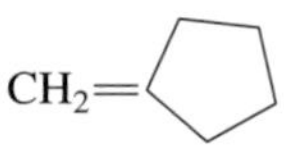

$$CH_2{=}\text{⬠} \qquad\qquad CH_2{=}CH—Cl \qquad\qquad CH_2{=}CH—CH_2—Br$$
Methylene cyclopentane $\qquad$ Vinyl chloride $\qquad$ Allyl bromide
(IUPAC name: methylidenecyclopentane) $\quad$ (IUPAC name: chloroethene) $\quad$ (IUPAC name: 3-bromopropene)

13.4 LINE-ANGLE STRUCTURAL FORMULAS FOR ALKENES

Line-angle formulas for the three- to six-carbon acyclic 1-alkenes are as follows.

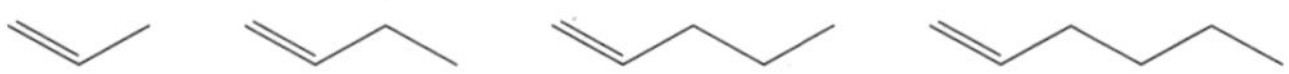

Propene $\qquad$ 1-Butene $\qquad$ 1-Pentene $\qquad$ 1-Hexene

---

Despite the universal acceptance and precision of the IUPAC nomenclature system, some alkenes (those of low molecular mass) are known almost exclusively by common names.

Representative line-angle structural formulas for substituent-bearing alkenes include

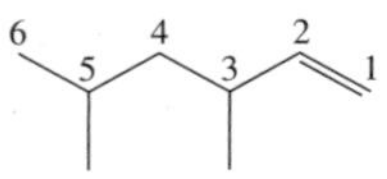

3,5-Dimethyl-1-hexene 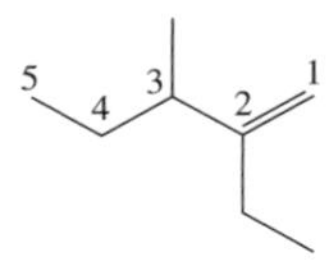   2-Ethyl-3-methyl-1-pentene

Diene representations in terms of line-angle structural formulas include

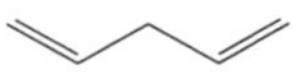

1,4-Pentadiene 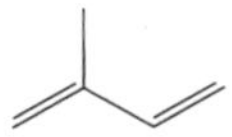   2-Methyl-1,3-butadiene

13.5 CONSTITUTIONAL ISOMERISM IN ALKENES

Constitutional isomerism is possible for alkenes, just as it was for alkanes (Section 12.6). In general, there are more alkene isomers for a given number of carbon atoms than there are alkane isomers. This is because there is more than one location where a double bond can be placed in systems containing four or more carbon atoms. Figure 13.2 compares constitutional isomer possibilities for C_4 and C_5 alkanes and their counterpart alkenes with one double bond.

Two different subtypes of constitutional isomerism are represented among the alkene isomers shown in Figure 13.2: *positional* isomers and *skeletal* isomers. **Positional isomers** *are constitutional isomers with the same carbon-chain arrangement but different hydrogen atom arrangements as the result of differing location of the functional group present.* Positional isomer sets found in Figure 13.2 are:

1-butene and 2-butene

1-pentene and 2-pentene

2-methyl-1-butene, 3-methyl-1-butene, and 2-methyl-2-butene

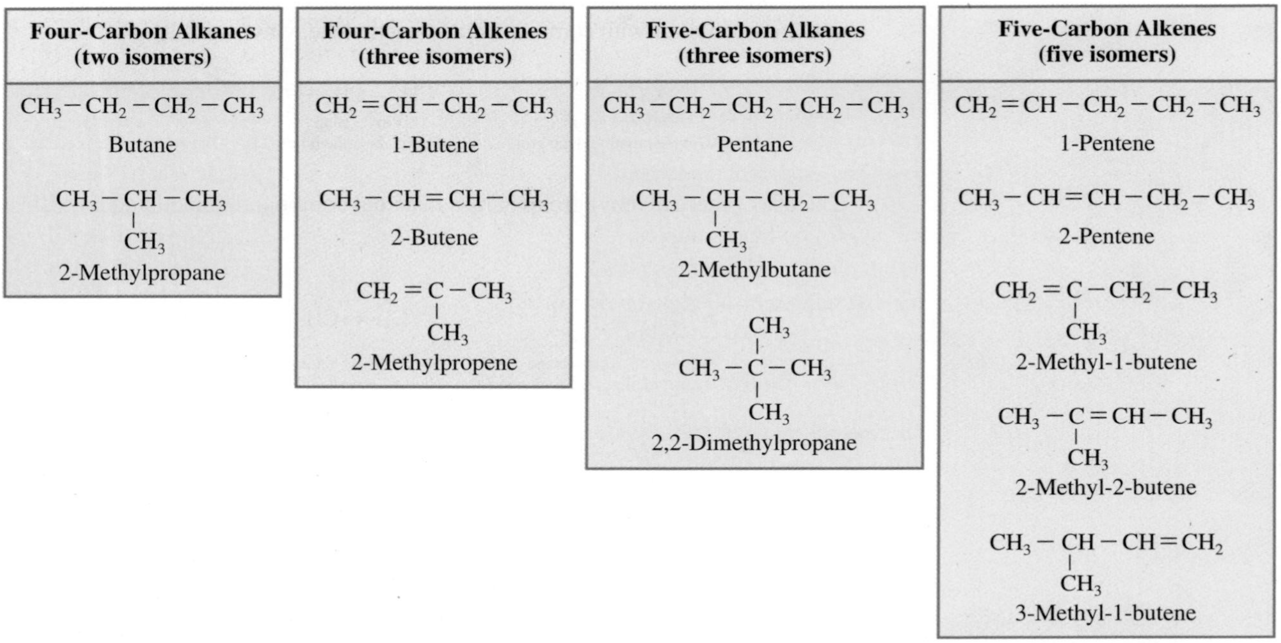

Four-Carbon Alkanes (two isomers)	Four-Carbon Alkenes (three isomers)	Five-Carbon Alkanes (three isomers)	Five-Carbon Alkenes (five isomers)
$CH_3-CH_2-CH_2-CH_3$ Butane	$CH_2=CH-CH_2-CH_3$ 1-Butene	$CH_3-CH_2-CH_2-CH_2-CH_3$ Pentane	$CH_2=CH-CH_2-CH_2-CH_3$ 1-Pentene

$CH_3-CH-CH_3$
        $\vert$
      CH_3
 2-Methylpropane
	$CH_3-CH=CH-CH_3$
	 2-Butene
		$CH_3-CH-CH_2-CH_3$
		       $\vert$
		     CH_3
		 2-Methylbutane
			$CH_3-CH=CH-CH_2-CH_3$
			 2-Pentene

	$CH_2=C-CH_3$
	       $\vert$
	     CH_3
	 2-Methylpropene
		CH_3
		   $\vert$
		 CH_3-C-CH_3
		   $\vert$
		 CH_3
		 2,2-Dimethylpropane
			$CH_2=C-CH_2-CH_3$
			       $\vert$
			     CH_3
			 2-Methyl-1-butene

			$CH_3-C=CH-CH_3$
			      $\vert$
			    CH_3
			 2-Methyl-2-butene

			$CH_3-CH-CH=CH_2$
			      $\vert$
			    CH_3
			 3-Methyl-1-butene

Figure 13.2 A comparison of structural isomerism possibilities for four- and five-carbon alkane and alkene systems

CHEMICAL Connections

Ethene: A Plant Hormone and High-Volume Industrial Chemical

Ethene (ethylene), the simplest unsaturated hydrocarbon (C_2H_4), is a colorless, flammable gas with a slightly sweet odor. It occurs naturally in *small* amounts in plants, where it functions as a plant hormone. A few parts per million ethene (less than 10 parts per million) stimulates the fruit-ripening process.

The commercial fruit industry uses ethene's ripening property to advantage. Bananas, tomatoes, and some citrus fruits are picked green to prevent spoiling and bruising during transportation to markets. At their destinations, the fruits are exposed to

Ethene is the hormone that causes tomatoes to ripen.

small amounts of ethene gas, which stimulates the ripening process.

Despite having no large natural source, ethene is an exceedingly important industrial chemical. Indeed, industrial production of ethene exceeds that of every other organic compound. Petrochemicals (substances found in natural gas and petroleum) are the starting materials for ethene production.

In one process, ethane (from natural gas) is *dehydrogenated* at a high temperature to produce ethene.

$$CH_3-CH_3 \xrightarrow{750\ C} CH_2=CH_2 + H_2$$

Ethane   Ethene

In another process, called *thermal cracking,* hydrocarbons from petroleum are heated to a high temperature in the absence of air (to prevent combustion), which causes the cleavage of carbon–carbon bonds. Ethene is one of the smaller molecules produced by this process.

Industrially produced ethene serves as a starting material for the production of many plastics and fibers. Almost one-half of ethene production is used in the production of the well-known plastic polyethylene (Section 13.10). Polyvinyl chloride (PVC) and polystyrene (Styrofoam) are two other important ethene-based materials. About one-sixth of ethene production is converted to ethylene glycol, the principal component of most brands of antifreeze for automobile radiators (Section 14.5).

Skeletal isomers *are constitutional isomers that have different carbon-chain arrangements as well as different hydrogen atom arrangements.* The C_4 alkenes 1-butene and 2-methylpropene are skeletal isomers. All alkane isomers discussed in the previous chapter were skeletal isomers; positional isomerism is not possible for alkanes because they lack a functional group.

● **EXAMPLE 13.2**

Determining Structural Formulas for Alkene Constitutional Isomers

▶ Draw condensed structural formulas for all alkene constitutional isomers that have the molecular formula C_5H_{10}.

Solution

The answers for this problem have already been considered. The structures of the five C_5H_{10} alkene constitutional isomers are given in Figure 13.2. The purpose of this example is to consider the "thinking pattern" used to obtain the given answers.

There are two concepts in the thinking pattern.

1. The different carbon skeletons (both unbranched and branched) that are possible using five carbon atoms are determined.
2. For each of the carbon skeletons determined, different positions for placement of the double bond are then considered.

(continued)

Step 1: There are three possible arrangements for five carbon atoms:

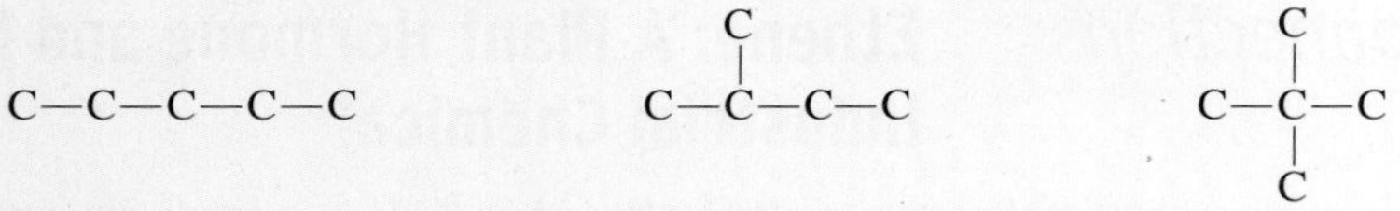

$$C-C-C-C-C \qquad C-\overset{\overset{\displaystyle C}{|}}{C}-C-C \qquad C-\overset{\overset{\displaystyle C}{|}}{\underset{\underset{\displaystyle C}{|}}{C}}-C$$

(These arrangements are the constitutional isomers for a 5-carbon alk*ane,* the situation considered in Section 12.6 of the previous chapter.)

Step 2: For the first carbon skeleton (the unbranched chain), there are two possible locations for the double bond; that is, there are two positional isomers:

$$CH_2\!\!=\!\!CH-CH_2-CH_2-CH_3 \qquad CH_3-CH\!\!=\!\!CH-CH_2-CH_3$$
1-Pentene   2-Pentene

Moving the double bond farther to the right than in the second structure does not produce new isomers but rather duplicates of the two given structures. A double bond between carbons 3 and 4 (numbering from the left side) is the same as having the double bond between carbons 2 and 3 (numbering from the right side).

For the second carbon skeleton there are three positional isomers, that is, three different positions for the double bond:

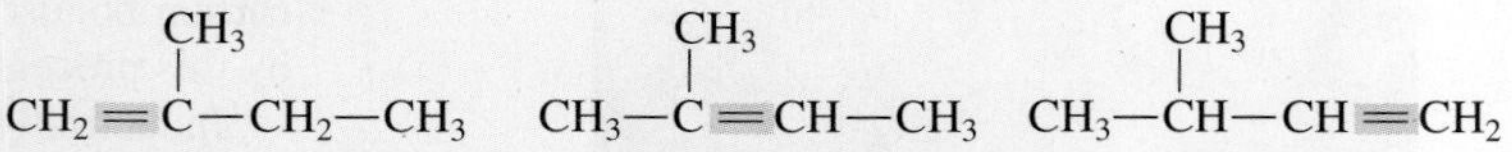

$$CH_2\!\!=\!\!\overset{\overset{\displaystyle CH_3}{|}}{C}-CH_2-CH_3 \qquad CH_3-\overset{\overset{\displaystyle CH_3}{|}}{C}\!\!=\!\!CH-CH_3 \qquad CH_3-\overset{\overset{\displaystyle CH_3}{|}}{CH}-CH\!\!=\!\!CH_2$$
2-Methyl-1-butene   2-Methyl-2-butene   3-Methyl-1-butene

For the third carbon skeleton there are no isomers. Placing a double bond at any location within the structure creates a situation where the central carbon atom has five bonds. Thus, there are five alkene constitutional isomers with the molecular formula C_5H_{10}.

Practice Exercise 13.2

Draw condensed structural formulas for all alkene constitutional isomers that have the molecular formula C_4H_8.

Answers: $CH_2\!\!=\!\!CH-CH_2-CH_3;$ $CH_3-CH\!\!=\!\!CH-CH_3;$ $CH_2\!\!=\!\!\overset{\overset{\displaystyle }{}}{C}-CH_3$
1-Butene   2-Butene   CH_3
2-Methylpropene

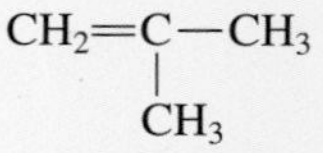

 ## 13.6 *CIS–TRANS* ISOMERISM IN ALKENES

Cis–trans isomerism (Section 12.14) is possible for some alkenes. Such isomerism results from the structural rigidity associated with carbon–carbon double bonds: Unlike the situation in alkanes, where free rotation about carbon–carbon single bonds is possible (Section 12.7), no rotation about carbon–carbon double bonds (or carbon–carbon triple bonds) can occur.

To determine whether an alkene has *cis* and *trans* isomers, draw the alkene structure in a manner that emphasizes the four attachments to the double-bonded carbon atoms.

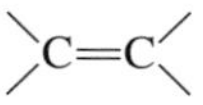

$$\overset{\diagdown}{\underset{\diagup}{}}C\!\!=\!\!C\overset{\diagup}{\underset{\diagdown}{}}$$

> The double bond of alkenes, like the ring of cycloalkanes, imposes rotational restrictions.

If *each* of the two carbons of the double bond has two *different* groups attached to it, *cis* and *trans* isomers exist.

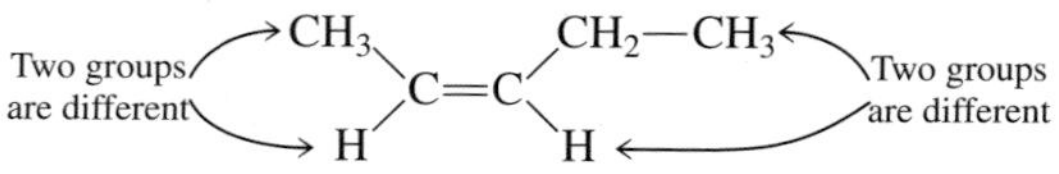

Two groups   CH_3   CH_2—CH_3   Two groups
are different   C=C   are different
H   H

Figure 13.3 *Cis-trans* isomers: Different representations of the *cis* and *trans* isomers of 2-butene.

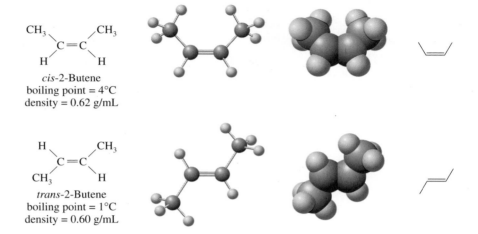

cis-2-Butene
boiling point = 4°C
density = 0.62 g/mL

trans-2-Butene
boiling point = 1°C
density = 0.60 g/mL

The simplest alkene for which *cis* and *trans* isomers exist is 2-butene.

Structure A
(*cis*-2-butene)

Structure B
(*trans*-2-butene)

Recall from Section 12.14 that *cis* means "on the same side" and *trans* means "across from." Structure A is the *cis* isomer; both methyl groups are on the same side of the double bond. Structure B is the *trans* isomer; the methyl groups are on opposite sides of the double bond. The only way to convert structure A to structure B is to break the double bond. At room temperature, such bond breaking does not occur. Hence these two structures represent two different compounds (*cis–trans* isomers) that differ in boiling point, density, and so on. Figure 13.3 shows three-dimensional representations of the *cis* and *trans* isomers of 2-butene.

Cis–trans isomerism is not possible when one of the double-bonded carbons bears two identical groups. Thus neither 1-butene nor 2-methylpropene is capable of existing in *cis* and *trans* forms.

Two identical groups

1-Butene

Two identical groups

Two identical groups

2-Methylpropene

When alkenes contain more than one double bond, *cis–trans* considerations are more complicated. Orientation about each double bond must be considered independently of that at other sites. For example, for the molecule 2,4-heptadiene (two double bonds) there are four different *cis–trans* isomers (*trans–trans, trans–cis, cis–trans,* and *cis–cis*). The structures of two of these isomers are

trans-trans-2,4-Heptadiene

trans-cis-2,4-Heptadiene

EXAMPLE 13.3

Determining Whether *Cis–Trans* Isomerism Is Possible in Substituted Alkenes

Determine whether each of the following substituted alkenes can exist in *cis–trans* isomeric forms.

a. 1-Bromo-1-chloroethene　　　**b.** 2-Chloro-2-butene

Solution

a. The condensed structural formula for this compound is

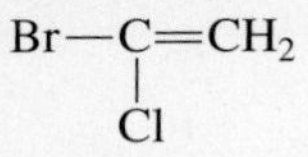

$$Br—C=CH_2$$
$$|$$
$$Cl$$

Redrawing this formula to emphasize the four attachments to the double-bonded carbon atoms gives

Br　　　H
　C=C
Cl　　　H

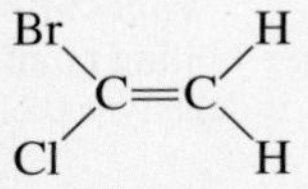

The carbon atom on the right has two identical attachments. Hence *cis–trans* isomerism is not possible.

b. The condensed structural formula for this compound is

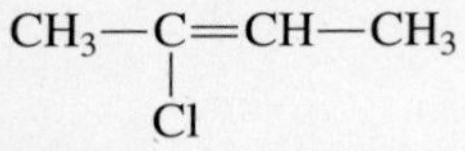

$$CH_3—C=CH—CH_3$$
$$|$$
$$Cl$$

Redrawing this formula to emphasize the four attachments to the double-bonded carbon atoms gives

CH₃　　　CH₃
　　C=C
Cl　　　H

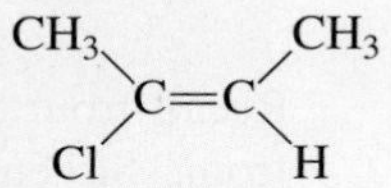

Because both carbon atoms of the double bond bear two different attachments, *cis–trans* isomers are possible.

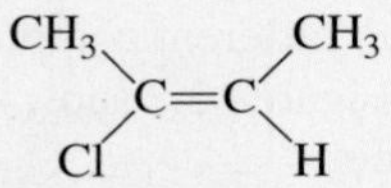

CH₃　　CH₃　　　　Cl　　　CH₃
　C=C　　　　　　　C=C
Cl　　H　　　　　CH₃　　H

cis-2-Chloro-2-butene 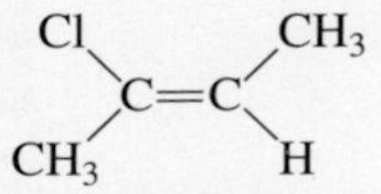　　　*trans*-2-Chloro-2-butene

Practice Exercise 13.3

Determine whether each of the following substituted alkenes can exist in *cis–trans* isomeric forms.

a. 1-Chloropropene　　　**b.** 2-Chloropropene

Answers: a. Yes; **b.** No

 13.7 NATURALLY OCCURRING ALKENES

Alkenes are abundant in nature. Many important biological molecules are characterized by the presence of carbon–carbon double bonds within their structure. Two important types of naturally occurring substances to which alkenes contribute are pheromones and terpenes.

Pheromones

A **pheromone** *is a compound used by insects (and some animals) to transmit a message to other members of the same species.* Pheromones are often alkenes or alkene derivatives. The biological activity of alkene-type pheromones is usually highly dependent on whether the double bonds present are in a *cis* or a *trans* arrangement (Section 13.6).

Chemical Connections — *Cis-Trans* Isomerism and Vision

Cis–trans isomerism plays an important role in many biochemical processes, including the reception of light by the retina of the eye. Within the retina, microscopic structures called rods and cones contain a compound called *retinal,* which absorbs light. Retinal contains a carbon chain with five carbon–carbon double bonds, four in a *trans* configuration and one in a *cis* configuration. This arrangement of double bonds gives retinal a shape that fits the protein *opsin,* to which it is attached, as shown in the accompanying diagram.

When light strikes retinal, the *cis* double bond is converted to a *trans* double bond. The resulting *trans*-retinal no longer fits the protein opsin and is subsequently released. Accompanying this release is an electrical impulse, which is sent to the brain. Receipt of such impulses by the brain is what enables us to see.

In order to trigger nerve impulses again, *trans*-retinal must be converted back to *cis*-retinal. This occurs in the membranes of the rods and cones, where enzymes change *trans*-retinal back into *cis*-retinal.

cis double bond

This double bond is now *trans.*

Opsin (protein)

Opsin (protein)

The sex attractant of the female silkworm is a 16-carbon alkene derivative containing an —OH group. Two double bonds are present, *trans* at carbon 10 and *cis* at carbon 12.

$$CH_3—(CH_2)_2 \quad \overset{H}{\underset{\;}{\underset{13}{C}}} \quad \overset{(11)\;(10)}{\underset{12}{C=C}} \quad \overset{(CH_2)_8—CH_2—OH}{\underset{H}{H}}$$

This compound is 10 billion times more effective in eliciting a response from the male silkworm than the 10-*cis*–12-*trans* isomer and 10 trillion times more effective than the isomer wherein both bonds are in a *trans* configuration.

Insect sex pheromones are useful in insect control. A small amount of synthetically produced sex pheromone is used to lure male insects of a single species into a trap (see Figure 13.4). The trapped males are either killed or sterilized. Releasing sterilized males has proved effective in some situations. A sterile male can mate many times, preventing fertilization in many females, who usually mate only once. Sex attractant pheromones are now used to control the gypsy moth and the Mediterranean fruit fly.

Terpenes

A **terpene** *is an organic compound whose carbon skeleton is composed of two or more 5-carbon isoprene structural units.* Isoprene (2-methyl-1,3-butadiene) is a five-carbon diene.

Figure 13.4 The application of sex pheromones in insect control involves using a small amount of synthetically produced pheromone to lure a particular insect into a trap. This is accomplished without harming other "beneficial" insects.

$$\overset{1}{CH_2}=\overset{2}{\underset{\underset{CH_3}{|}}{C}}-\overset{3}{CH}=\overset{4}{CH_2}$$

2-Methyl-1,3-butadiene
(isoprene)

Figure 13.5 Selected terpenes containing two, three, and eight isoprene units. Dashed lines in the structures separate the individual isoprene units.

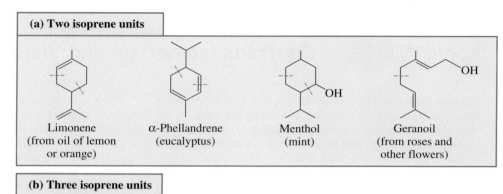

(a) Two isoprene units

Limonene
(from oil of lemon
or orange)

α-Phellandrene
(eucalyptus)

Menthol
(mint)

Geranoil
(from roses and
other flowers)

(b) Three isoprene units

Zingiberene
(from oil of ginger)

α-Farnesene
(from natural coating of apples)

(c) Eight isoprene units

β-Carotene
(present in carrots and other vegetables)

In later chapters, we will encounter additional isoprene-based molecules important in the functioning of the human body. They include vitamin K (Section 21.13), coenzyme Q (Section 23.7), and cholesterol (Section 19.9).

Terpenes are formed by joining the tail of one isoprene structural unit to the head of another unit.

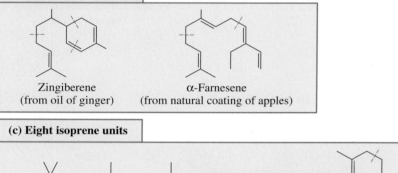

The isoprene structural unit maintains its isopentyl structure (Section 12.8) in a terpene, usually with modification of the isoprene double bonds.

Terpenes are among the most widely distributed compounds in the biological world, with over 22,000 structures known. Such compounds are responsible for the odors of many trees and for many characteristic plant fragrances.

The number of carbon atoms present in a terpene is always a multiple of the number 5 (10, 15, and so on). Parts (a) and (b) of Figure 13.5 give the structures of selected 10- and 15-carbon terpenes found in plants. Beta-carotene is a terpene whose structure has 40 carbon atoms present in 8 isoprene units (Figure 13.5c).

In the human body, dietary beta-carotene (obtained by eating yellow-colored vegetables) serves as a precursor for vitamin A (see Figure 13.6); splitting of a beta-carotene molecule produces two vitamin A molecules (Section 21.13). An additional role of beta-carotene in the body, independent of its vitamin A function, is that of antioxidant. An antioxidant is a substance that helps protect cells from damage from reactive oxygen-derived species called free radicals (Section 23.11).

Figure 13.6 The molecule β-carotene is responsible for the yellow-orange color of carrots, apricots, and yams.

Unbranched 1-Alkenes		
C_3	C_5	C_7
C_2	C_4	C_6	C_8

Unsubstituted Cycloalkenes		
C_3*	C_5	C_7
C_4*	C_6	C_8

☐ Gas   ☐ Liquid

*Cyclopropene and cyclobutene are relatively unstable compounds, readily converting to other hydrocarbons because of the severe bond angle strain associated with a small ring containing a double bond.

Figure 13.7 A physical-state summary for unbranched 1-alkenes and unsubstituted cycloalkenes with one double bond at room temperature and pressure.

13.8 PHYSICAL PROPERTIES OF ALKENES AND CYCLOALKENES

The general physical properties of alkenes and cycloalkenes include insolubility in water, solubility in nonpolar solvents, and densities lower than that of water. Thus they have physical properties similar to those of alkanes (Section 12.16). The melting point of an alkene is usually lower than that of the alkane with the same number of carbon atoms.

Alkenes with 2 to 4 carbon atoms are gases at room temperature. Unsubstituted alkenes with 5 to 17 carbon atoms and one double bond are liquids, and those with still more carbon atoms are solids. Figure 13.7 is a physical-state summary for unbranched 1-alkenes and unsubstituted cycloalkenes with 8 or fewer carbon atoms.

13.9 CHEMICAL REACTIONS OF ALKENES

Alkenes, like alkanes, are very flammable. The combustion products, as with any hydrocarbon, are carbon dioxide and water.

$$C_2H_4 + 3O_2 \longrightarrow 2CO_2 + 2H_2O$$
Ethene

Pure alkenes are, however, too expensive to be used as fuel.

Aside from combustion, nearly all other reactions of alkenes take place at the carbon–carbon double bond(s). These reactions are called *addition reactions* because a substance is *added* to the double bond. This behavior contrasts with that of alkanes, where the most common reaction type, aside from combustion, is *substitution* (Section 12.17).

An **addition reaction** *is a reaction in which atoms or groups of atoms are added to each carbon atom of a carbon–carbon multiple bond in a hydrocarbon or hydrocarbon derivative.* A general equation for an alkene addition reaction is

$$\diagdown C=C \diagup \; + \; A\!-\!B \longrightarrow -\overset{|}{\underset{A}{C}}-\overset{|}{\underset{B}{C}}-$$

In this reaction, the A part of the reactant A—B becomes attached to one carbon atom of the double bond, and the B part to the other carbon atom (see Figure 13.8). As this occurs, the carbon–carbon double bond simultaneously becomes a carbon–carbon single bond.

Addition reactions can be classified as symmetrical or unsymmetrical. A **symmetrical addition reaction** *is an addition reaction in which identical atoms (or groups of atoms) are added to each carbon of a carbon–carbon multiple bond.* An **unsymmetrical addition reaction** *is an addition reaction in which different atoms (or groups of atoms) are added to the carbon atoms of a carbon–carbon multiple bond.*

Figure 13.8 In an alkene addition reaction, the atoms provided by an incoming molecule are attached to the carbon atoms originally joined by a double bond. In the process, the double bond becomes a single bond.

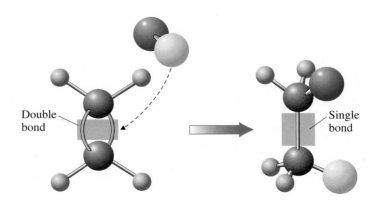

Double bond

Single bond

CHEMICAL Connections

Carotenoids: A Source of Color

Carotenoids are the most widely distributed of the substances that give color to our world; they occur in flowers, fruits, plants, insects, and animals. These compounds are terpenes (Section 13.7) in which eight isoprene units are present. Structural formulas for two members of the carotenoid family, β-carotene and lycopene, follow.

β-Carotene

Lycopene

Present in both of these carotenoid structures is a *conjugated* system of 11 double bonds. (Conjugated double bonds are double bonds separated from each other by one single bond.) Color is frequently caused by the presence of compounds that contain extended conjugated-double-bond systems. When *visible* light strikes these compounds, certain wavelengths of the visible light are absorbed by the electrons in the conjugated-bond system. The unabsorbed wavelengths of visible light are reflected and are perceived as color.

The molecule β-carotene is responsible for the yellow-orange color in carrots, apricots, and yams. The yellow-orange color of autumn leaves comes from β-carotene. Leaves contain chlorophyll (green pigment) and β-carotene (yellow-pigment) in a ratio of approximately 3 to 1. The yellow-orange β-carotene color is masked by the chlorophyll until autumn, when the chlorophyll molecules decompose as a result of lower temperatures and less sunlight and are not replaced.

The molecule lycopene is the red pigment in tomatoes, paprika, and watermelon. Lycopene's structure differs from that of β-carotene in that the two rings in β-carotene have been opened. The ripening of a green tomato involves the gradual decomposition of chlorophyll with an associated unmasking of the red color of the lycopene present. A green pepper becomes red after ripening for the same reason.

Research studies indicate that lycopene has anticancer properties. One study comparing a group of men on a lycopene-rich diet with another group on a low-lycopene diet showed the incidence of prostate cancer was one-third lower in the lycopene-rich diet group.

Heat-processed tomatoes are a good source of dietary lycopene, with concentrated juice containing the highest levels of this substance. The lycopene in cooked tomatoes is absorbed more readily during digestion than is the lycopene in raw tomatoes. Recent studies indicate that red seedless watermelon contains as much lycopene as cooked tomatoes.

The anticancer properties of lycopene relate to its ability to react with highly reactive oxygen-containing molecules in the body, thereby preventing these molecules from oxidizing cellular components and creating new substances that might negatively affect cellular activity. Thus, lycopene has *antioxidant* properties. (See Section 14.14 for further information about antioxidants.)

Carotenoids such as β-carotene and lycopene are synthesized only by plants. They can, however, reach animal tissues via feed and can be modified and deposited therein. The yellowish tint of animal fat comes from β-carotene present in animal diets. The chicken egg yolk is another example of color imparted by dietary carotenoids.

Carotenoids, molecules that contain eight isoprene units, are responsible for the yellow-orange color of autumn leaves.

Symmetrical Addition Reactions

The two most common examples of symmetrical addition reactions are hydrogenation and halogenation.

A **hydrogenation reaction** *is an addition reaction in which H₂ is incorporated into molecules of an organic compound.* In alkene hydrogenation a hydrogen atom is added to each carbon atom of a double bond. It is accomplished by heating the alkene and H_2 in the presence of a catalyst (usually Ni or Pt).

The following word associations are important to remember:

   alkane—substitution reaction
   alkene—addition reaction

An analogy can be drawn to a basketball team. When a *substitution* is made, one player leaves the game as another enters. The number of players on the court remains at five per team. If *addition* were allowed during a basketball game, two players could enter the game and no one would leave; there would be seven players per team on the court rather than five.

Hydrogenation of an alkene requires a catalyst. No reaction occurs if the catalyst is not present.

The Chemical Connections feature "*Trans* Fatty Acids and Blood Cholesterol Levels" in Chapter 19 addresses health issues relative to consumption of partially hydrogenated products.

$$CH_2{=}CH{-}CH_3 + H_2 \xrightarrow[\substack{150°C \\ 12{-}15\ atm \\ pressure}]{Ni\ or\ Pt} \overset{\overset{\displaystyle H\ \ \ H}{|\ \ \ |}}{CH_2{-}CH{-}CH_3}$$
Propene   Propane

The identity of the catalyst used in hydrogenation is specified by writing it above the arrow in the chemical equation for the hydrogenation. In general terms, hydrogenation of an alkene can be written as

$$\overset{\diagdown}{\underset{\diagup}{}}C{=}C\overset{\diagup}{\underset{\diagdown}{}} + H_2 \xrightarrow[\substack{Heat, \\ pressure}]{Ni\ or\ Pt} \overset{\overset{\displaystyle H\ \ \ H}{|\ \ \ |}}{{-}C{-}C{-}}$$
Alkene   Alkane

The hydrogenation of vegetable oils is a very important commercial process today. Vegetable oils from sources such as soybeans and cottonseeds are composed of long-chain organic molecules that contain several double bonds. When these oils are hydrogenated, they are converted to low-melting solids that are used in margarines and shortenings.

A **halogenation reaction** *is an addition reaction in which a halogen is incorporated into molecules of an organic compound.* In alkene halogenation a halogen atom is added to each carbon atom of a double bond. Chlorination (Cl_2) and bromination (Br_2) are the two halogenation processes most commonly encountered. No catalyst is needed.

$$CH_3{-}CH{=}CH{-}CH_3 + Cl_2 \longrightarrow CH_3{-}\overset{\overset{\displaystyle Cl}{|}}{CH}{-}\overset{\overset{\displaystyle Cl}{|}}{CH}{-}CH_3$$
2-Butene   2,3-Dichlorobutane

In general terms, halogenation of an alkene can be written as

$$\overset{\diagdown}{\underset{\diagup}{}}C{=}C\overset{\diagup}{\underset{\diagdown}{}} + X_2 \longrightarrow \overset{\overset{\displaystyle X\ \ \ X}{|\ \ \ |}}{{-}C{-}C{-}} \qquad (X = Cl, Br)$$
Alkene   Halogen   Dihalogenated alkane

Bromination is often used to test for the presence of carbon–carbon double bonds in organic substances. Bromine in water or carbon tetrachloride is reddish brown. The dibromo compound(s) formed from the symmetrical addition of bromine to an organic compound is(are) colorless. Thus the decolorization of a Br_2 solution indicates the presence of carbon–carbon double bonds (see Figure 13.9).

Unsymmetrical Addition Reactions

Two important types of unsymmetrical addition reactions are hydrohalogenation and hydration.

A **hydrohalogenation reaction** *is an addition reaction in which a hydrogen halide (HCl, HBr, or HI) is incorporated into molecules of an organic compound.* In alkene hydrohalogenation one carbon atom of a double bond receives a halogen atom

Figure 13.9 A bromine in water solution is reddish brown (left). When a small amount of such a solution is added to an unsaturated hydrocarbon, the added solution is decolorized as the bromine adds to the hydrocarbon to form colorless dibromo compounds (right).

and the other carbon atom receives a hydrogen atom. Hydrohalogenation reactions require no catalyst. For *symmetrical* alkenes, such as ethene, only one product results from hydrohalogenation.

$$CH_2\text{=}CH_2 + H\text{—}Cl \longrightarrow \overset{\displaystyle H \quad\; Cl}{\underset{\text{Chloroethane}}{CH_2\text{—}CH_2}}$$
Ethene

A **hydration reaction** *is an addition reaction in which H_2O is incorporated into molecules of an organic compound.* In alkene hydration one carbon atom of a double bond receives a hydrogen atom and the other carbon atom receives an —OH group. Alkene hydration requires a small amount of H_2SO_4 (sulfuric acid) as a catalyst. For *symmetrical* alkenes, only one product results from hydration.

$$CH_2\text{=}CH_2 + H\text{—}OH \xrightarrow{\;H_2SO_4\;} \overset{\displaystyle H \quad\; OH}{\underset{\text{An alcohol}}{CH_2\text{—}CH_2}}$$
Ethene

In this equation, the water (H_2O) is written as H—OH to emphasize how this molecule adds to the double bond. Note also that the product of this hydration reaction contains an —OH group. Hydrocarbon derivatives of this type are called *alcohols*. Such compounds are the subject of Chapter 14.

When the alkene involved in a hydrohalogenation or hydration reaction is itself *unsymmetrical,* more than one product is possible. (An unsymmetrical alkene is one in which the two carbon atoms of the double bond are not equivalently substituted.) For example, the addition of HCl to propene (an unsymmetrical alkene) could produce either 1-chloropropane or 2-chloropropane, depending on whether the H from the HCl attaches itself to carbon 2 or carbon 1.

$$CH_2\text{=}CH\text{—}CH_3 + HCl \longrightarrow \overset{\displaystyle Cl \quad\;\; H}{\underset{\text{1-Chloropropane}}{CH_2\text{—}CH\text{—}CH_3}}$$
Propene

or

$$CH_2\text{=}CH\text{—}CH_3 + HCl \longrightarrow \overset{\displaystyle H \quad\;\; Cl}{\underset{\text{2-Chloropropane}}{CH_2\text{—}CH\text{—}CH_3}}$$
Propene

When two isomeric products are possible, one product usually predominates. The dominant product can be predicted by using Markovnikov's rule, named after the Russian chemist Vladimir V. Markovnikov (see Figure 13.10). **Markovnikov's rule** states that *when an unsymmetrical molecule of the form HQ adds to an unsymmetrical alkene, the hydrogen atom from the HQ becomes attached to the unsaturated carbon atom that already has the most hydrogen atoms.* Thus the major product in our example involving propene is 2-chloropropane.

The addition of water to carbon–carbon double bonds occurs in many biochemical reactions that take place in the human body—for example, in the citric acid cycle (Section 23.6) and in the oxidation of fatty acids (Section 25.4).

Figure 13.10 Vladimir Vasilevich Markovnikov (1837-1904). A professor of chemistry at several Russian universities, Markovnikov (pronounced Mar-cove-na-coff) synthesized rings containing four carbon atoms and seven carbon atoms, thereby disproving the notion of the day that carbon could form only five- and six-membered rings.

Two catchy summaries of Markovnikov's rule are "hydrogen goes where hydrogen is" and "the rich get richer" (in terms of hydrogen).

● **EXAMPLE 13.4**

Predicting Products in Alkene Addition Reactions Using Markovnikov's Rule

▶ Using Markovnikov's rule, predict the predominant product in each of the following addition reactions.

a. $CH_3\text{—}CH_2\text{—}CH_2\text{—}CH\text{=}CH_2 + HBr \longrightarrow$

b. ⬠ $+ HCl \longrightarrow$

c. $CH_3\text{—}CH\text{=}CH\text{—}CH_2\text{—}CH_3 + HBr \longrightarrow$

Solution

a. The hydrogen atom will add to carbon 1, because carbon 1 already contains more hydrogen atoms than carbon 2. The predominant product of the addition will be 2-bromopentane.

$$CH_3-CH_2-CH_2-\overset{②}{C}H=\overset{①}{C}H_2 + HBr \longrightarrow CH_3-CH_2-CH_2-\overset{\overset{Br}{|}}{C}H-\overset{\overset{H}{|}}{C}H_2$$

b. Carbon 1 of the double bond does not have any H atoms directly attached to it. Carbon 2 of the double bond has one H atom (H atoms are not shown in the structure but are implied) attached to it. The H atom from the HCl will add to carbon 2, giving 1-chloro-1-methylcyclopentane as the product.

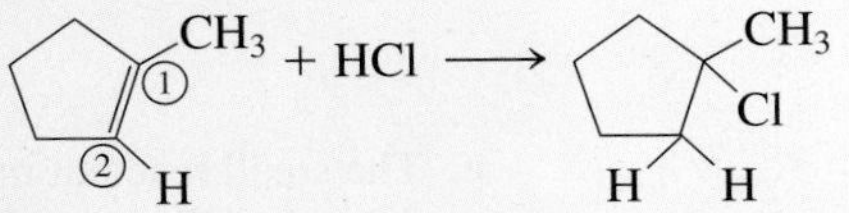

c. Each carbon atom of the double bond in this molecule has one hydrogen atom. Thus Markovnikov's rule does not favor either carbon atom. The result is two isomeric products that are formed in almost equal quantities.

$$CH_3-\overset{\overset{}{|}}{C}H-CH_2-CH_2-CH_3 \quad \text{and} \quad CH_3-CH_2-\overset{\overset{}{|}}{C}H-CH_2-CH_3$$
$$Br Br$$

2-Bromopentane   3-Bromopentane

Practice Exercise 13.4

Using Markovnikov's rule, predict the predominant product in each of the following addition reactions.

a. $CH_2=CH-CH_2-CH_3 + HCl \rightarrow$

b. 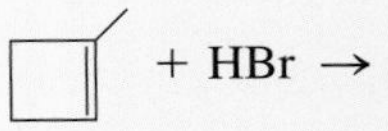 + HBr $\rightarrow$

Answers: a. $CH_3-\overset{\overset{}{|}}{C}H-CH_2-CH_3$
Cl

b.

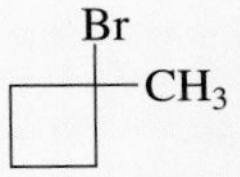

In compounds that contain more than one carbon–carbon double bond, such as dienes and trienes, addition can occur at each of the double bonds. In the complete hydrogenation of a diene and in that of a triene, the amounts of hydrogen needed are twice as much and three times as much, respectively, as that needed for the hydrogenation of an alkene with one double bond.

$$CH_2=CH-CH_2-CH_2-CH_2-CH_3 + H_2 \xrightarrow{Ni} CH_3-(CH_2)_4-CH_3$$

$$CH_2=CH-CH=CH-CH_2-CH_3 + 2H_2 \xrightarrow{Ni} CH_3-(CH_2)_4-CH_3$$

$$CH_2=CH-CH=CH-CH=CH_2 + 3H_2 \xrightarrow{Ni} CH_3-(CH_2)_4-CH_3$$

● **E X A M P L E 13.5**

Predicting Reactants and Products in Alkene Addition Reactions

Supply the structural formula of the missing substance in each of the following addition reactions.

a. $CH_3-CH_2-CH=CH_2 + H_2O \xrightarrow{H_2SO_4}$?

b.

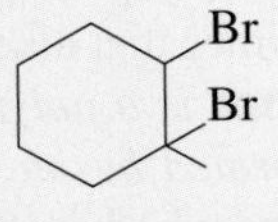

c.

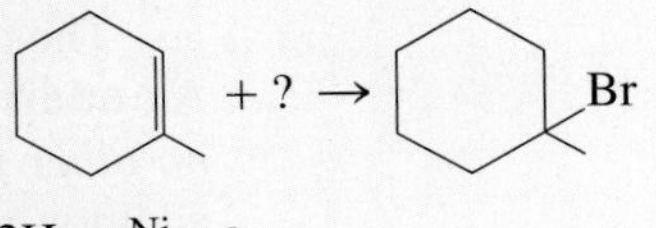

d. $CH_3-CH=CH-CH=CH_2 + 2H_2 \xrightarrow{Ni}$?

(continued)

Solution

a. This is a hydration reaction. Using Markovnikov's rule, we determine that the H will become attached to carbon 1, which has more hydrogen atoms than carbon 2, and that the —OH group will be attached to carbon 2.

$$CH_3—CH_2—\overset{\displaystyle OH}{\underset{\displaystyle |}{CH}}—CH_3$$

b. The reactant alkene will have to have a double bond between the two carbon atoms that bromine atoms are attached to in the product.

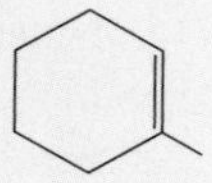

c. The small reactant molecule that adds to the double bond is HBr. The added Br atom from the HBr is explicity shown in the product's structural formula, but the added H atom is not shown.

d. Hydrogen will add at each of the double bonds. The product hydrocarbon is pentane.

$$CH_3—CH_2—CH_2—CH_2—CH_3$$

Practice Exercise 13.5

Supply the structural formula of the missing substance in each of the following addition reactions.

a. $CH_3—CH_2—CH{=}CH_2 + HBr \rightarrow$?

b. $? + H_2O \xrightarrow{H_2SO_4}$

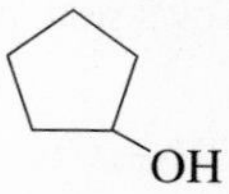

c.

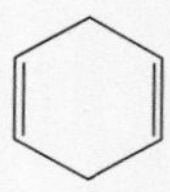

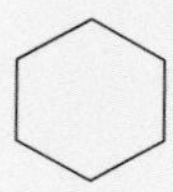

d. $CH_3—\overset{\displaystyle |}{\underset{\displaystyle |}{\underset{\displaystyle CH_3}{C}}}{=}CH—CH_3 + Cl_2 \longrightarrow$?

Answers: a. $CH_3—CH_2—\overset{\displaystyle |}{\underset{\displaystyle |}{\underset{\displaystyle Br}{CH}}}—CH_3$; **b.**  **c.** H_2 ; **d.** $CH_3—\overset{\displaystyle Cl}{\underset{\displaystyle |}{\underset{\displaystyle CH_3}{\overset{\displaystyle |}{C}}}}—\overset{\displaystyle Cl}{\underset{\displaystyle |}{CH}}—CH_3$

13.10 POLYMERIZATION OF ALKENES: ADDITION POLYMERS

The word *polymer* comes from the Greek *poly*, which means "many," and *meros*, which means "parts."

A **polymer** *is a large molecule formed by the repetitive bonding together of many smaller molecules.* The smaller repeating units of a polymer are called *monomers.* A **monomer** *is the small molecule that is the structural repeating unit in a polymer.* The process by which a polymer is made is called *polymerization.* A **polymerization reaction** *is a chemical reaction in which the repetitious combining of many small molecules (monomers) produces a very large molecule (the polymer).* With appropriate catalysts, simple alkenes and simple substituted alkenes readily undergo polymerization.

The type of polymer that alkenes and substituted alkenes form is an *addition polymer.* An **addition polymer** *is a polymer in which the monomers simply "add together" with no other products formed besides the polymer.* Addition polymerization is similar to the addition reactions described in Section 13.9 except that there is no reactant other than the alkene or substituted alkene.

We will consider polymer types other than addition polymers in Sections 15.12 and 16.18.

The simplest alkene addition polymer has ethylene (ethene) as the monomer. With appropriate catalysts, ethylene readily adds to itself to produce polyethylene.

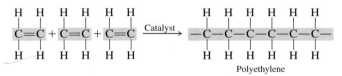

Polyethylene

An *exact* formula for a polymer such as polyethylene cannot be written because the length of the carbon chain varies from polymer molecule to polymer molecule. In recognition of this "inexactness" of formula, the notation used for denoting polymer formulas is independent of carbon chain length. We write the formula of the simplest repeating unit (the monomer with the double bond changed to a single bond) in parentheses and then add the subscript n after the parentheses, with n being understood to represent a very large number. Using this notation, we have, for the formula of polyethylene,

$$\left(\begin{array}{cc} H & H \\ | & | \\ C - C \\ | & | \\ H & H \end{array}\right)_n$$

This notation clearly identifies the basic repeating unit found in the polymer.

Substituted-Ethene Addition Polymers

Many substituted alkenes undergo polymerization similar to that of ethene when they are treated with the proper catalyst. For a monosubstituted-ethene monomer, the general polymerization equation is

$$H_2C=\overset{\overset{\displaystyle Z}{|}}{CH} \xrightarrow{\text{Polymerization}} \left(CH_2-\overset{\overset{\displaystyle Z}{|}}{CH}\right)_n$$

Variation in the substituent group Z can change polymer properties dramatically, as is shown by the entries in Table 13.1, a listing of monomers for the five ethene-based polymers *polyethylene, polypropylene, poly(vinyl chloride) (PVC), Teflon,* and *polystyrene,* along with several uses of each. Figure 13.11 depicts the preparation of polystyrene. Figure 13.12 contrasts the structures of polyethylene, polypropylene, and poly (vinyl chloride) as depicted in space-filling models.

The properties of an ethene-based polymer depend not only on monomer identity but also on the average size (length) of polymer molecules and on the extent of polymer branching. For example, there are three major types of polyethylene: high-density polyethylene (HDPE), low-density polyethylene (LDPE), and linear low-density polyethylene (LLDPE). The major difference among these three materials is the degree of branching

Figure 13.11 Preparation of polystyrene. When styrene, $C_6H_5 — CH = CH_2$, is heated with a catalyst (benzoyl peroxide), it yields a viscous liquid. After some time, this liquid sets to a hard plastic (sample shown at left).

Figure 13.12 Line-angle structural formulas and space-filling models of segments of the ethene-based polymers (a) polyethylene, (b) polypropylene, and (c) poly(vinyl chloride).

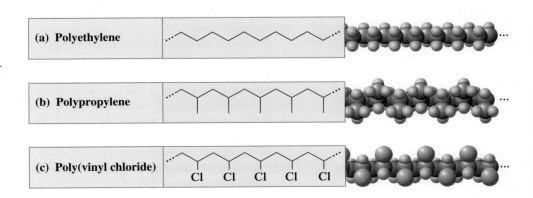

TABLE 13.1
Some Common Polymers Obtained from Ethene-Based Monomers

Polymer Formula and Name	Monomer Formula and Name	Uses of Polymer
polyethylene	ethylene	bottles, plastic bags, toys, electrical insulation
polypropylene	propylene	indoor–outdoor carpeting, bottles, molded parts (including heart valves)
poly(vinyl chloride) (PVC)	vinyl chloride	plastic wrap, bags for intravenous drugs, garden hose, plastic pipe, simulated leather (Naugahyde)
Teflon	tetrafluoroethylene	cooking utensil coverings, electrical insulation, component of artificial joints in body parts replacement
polystyrene	styrene	toys, Styrofoam packaging, cups, simulated wood furniture

of the polymer chain. HDPE and LLDPE are composed of linear, unbranched carbon chains, while LDPE chains are branched. The strong and thick plastic bags from a shopping mall are LLDPE, the thin and flimsy grocery store plastic bags are HDPE, and the very wispy garment bags dry cleaners use are LDPE.

In general, HDPE materials are rigid or semi-rigid with uses such as threaded bottle caps, toys, bottles, and milk jugs whereas LDPE materials are more flexible with uses such as plastic film and squeeze bottles (see Figure 13.13). Objects made of HDPE hold their shape in boiling water, whereas objects made of LDPE become severely deformed at this temperature.

With their alkane-like structures, ethene-based addition polymers are very unreactive, as are alkanes. This unreactivity means that these polymers do not readily decompose when they are deposited into landfill sites.

Decreasing availability of space at landfills has led to an increasingly important plastics recycling industry. Because of the wide diversity of addition polymers in consumer products, recycling requires sorting the polymers into various subtypes, which are then recycled separately.

To standardize the recycling process manufacturers now place recycling symbols on their products, identifying the type of addition polymer present. Table 13.2 lists the recycling symbols/codes now in use. Figure 13.14 shows a portion of a recycling collection point where collection organization is based on these recycling codes.

Figure 13.13 Examples of objects made of polyethylene. Polyethylene objects that are strong and rigid (bottles, toys, covering for wire) contain HDPE (high-density polyethylene). Polyethylene objects that are very flexible (plastic bags and packaging materials) contain LDPE (low-density polyethylene).

Code	Type	Name	Examples
PET	1	Polyethylene terephthalate	soft drink bottles, peanut butter jars, vegetable oil bottles
HDPE	2	High-density polyethylene	milk, water, and juice containers, squeezable ketchup and syrup bottles
PVC	3	Polyvinyl chloride	shampoo bottles, plastic pipe, shower curtains
LDPE	4	Low-density polyethylene	six-pack rings, shrink-wrap, sandwich bags, grocery bags
PP	5	Polypropylene	margarine tubs, straws, diaper linings, toys
PS	6	Polystyrene	egg cartons, disposable utensils, packing peanuts, foam cups
Other	7	Multilayer plastics	various flexible items

Butadiene-Based Addition Polymers

When dienes such as 1,3-butadiene are used as the monomers in addition polymerization reactions, the resulting polymers contain double bonds and are thus still unsaturated.

$$CH_2{=}CH{-}CH{=}CH_2 \xrightarrow{\text{Polymerization}} \left(CH_2{-}CH{=}CH{-}CH_2\right)_n$$

1,3-Butadiene   Polybutadiene

In general, unsaturated polymers are much more flexible than the ethene-based saturated polymers listed in Table 13.1. Natural rubber is a flexible addition polymer whose repeating unit is isoprene (Section 13.7)—that is, 2-methyl-1,3-butadiene (see Figure 13.15).

$$CH_2{=}\underset{\underset{CH_3}{|}}{C}{-}CH{=}CH_2 \xrightarrow{\text{Polymerization}} \left(CH_2{-}\underset{\underset{CH_3}{|}}{C}{=}CH{-}CH_2\right)_n$$

Isoprene
(2-methyl-1,3-butadiene)

Polyisoprene
(natural rubber)

Addition Copolymers

Saran Wrap is a polymer in which two different monomers are present: chloroethene (vinyl chloride) and 1,1-dichloroethene.

Figure 13.14 Sorting of ethene-based polymers into various subtypes often occurs at recycling collection points.

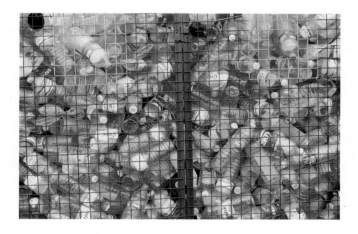

Figure 13.15 Natural rubber being harvested in Malaysia.

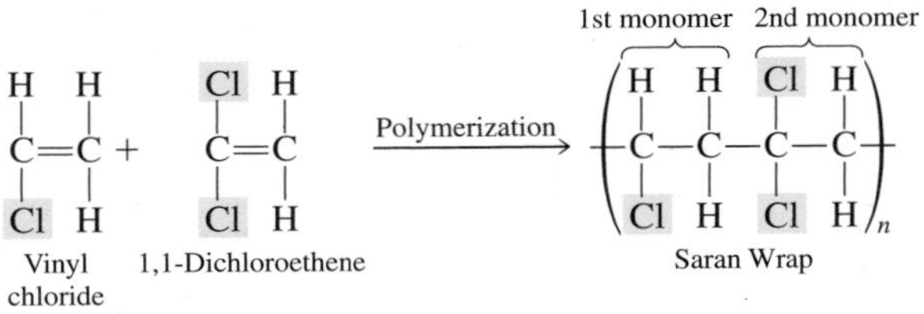

Very few biological molecules are known that contain a carbon–carbon triple bond.

$$CH_2{=}CHCl + CCl_2{=}CH_2 \xrightarrow{\text{Polymerization}} \text{Saran Wrap}$$

Vinyl chloride   1,1-Dichloroethene   1st monomer   2nd monomer   Saran Wrap

Such a polymer is an example of a *copolymer*. A **copolymer** *is a polymer in which two different monomers are present.* Another important copolymer is styrene–butadiene rubber, the leading synthetic rubber in use today. It contains the monomers 1,3-butadiene and styrene in a 3:1 ratio. It is a major ingredient in automobile tires

The Chemistry at a Glance feature on page 383 summarizes the reaction chemistry of alkenes presented in this and the previous section.

13.11 ALKYNES

Alkynes represent the second class of unsaturated hydrocarbons (Section 13.1). An **alkyne** *is an acyclic unsaturated hydrocarbon that contains one or more carbon–carbon triple bonds.* The alkyne functional group is, thus, a $C{\equiv}C$ group. As the family name *alkyne* indicates, the characteristic "ending" associated with a triple bond is *-yne*.

The general formula for an alkyne with one triple bond is C_nH_{2n-2}. Thus the simplest member of this type of alkyne has the formula C_2H_2, and the next member, with $n = 3$, has the formula C_3H_4.

$$CH{\equiv}CH \qquad CH{\equiv}C{-}CH_3$$
Ethyne   Propyne

The presence of a carbon–carbon triple bond in a molecule always results in a linear arrangement for the two atoms attached to the carbons of the triple bond. Thus ethyne is a linear molecule (see Figure 13.16).

The simplest alkyne, ethyne (C_2H_2), is the most important alkyne from an industrial standpoint. A colorless gas, it goes by the common name *acetylene* and is used in oxyacetylene torches, high-temperature torches used for cutting and welding materials.

IUPAC Nomenclature for Alkynes

Cycloalkynes, molecules that contain a triple bond as part of a ring structure, are known, but they are not common. Because of the 180° angle associated with a triple bond, a ring system containing a triple bond has to be quite large. The smallest cycloalkyne that has been isolated is cyclooctyne.

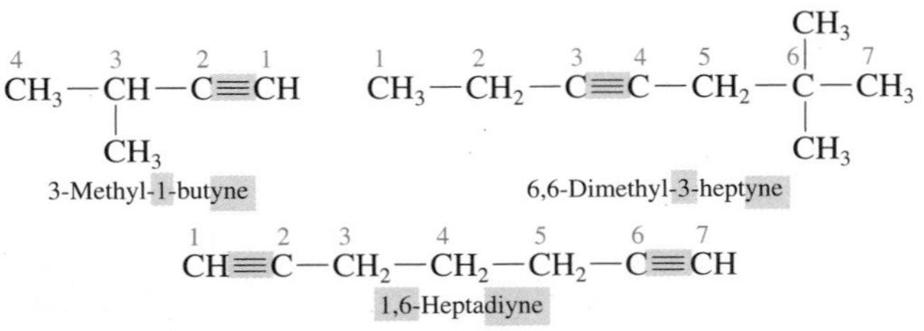

The rules for naming alkynes are identical to those used to name alkenes (Section 13.3), except the ending *-yne* is used instead of *-ene*. Consider the following structures and their IUPAC names.

$$\overset{4}{C}H_3{-}\overset{3}{C}H{-}\overset{2}{C}{\equiv}\overset{1}{C}H$$
$$\underset{CH_3}{|}$$
3-Methyl-1-butyne

$$\overset{1}{C}H_3{-}\overset{2}{C}H_2{-}\overset{3}{C}{\equiv}\overset{4}{C}{-}\overset{5}{C}H_2{-}\overset{6}{C}{-}\overset{7}{C}H_3$$
6,6-Dimethyl-3-heptyne

$$\overset{1}{C}H{\equiv}\overset{2}{C}{-}\overset{3}{C}H_2{-}\overset{4}{C}H_2{-}\overset{5}{C}H_2{-}\overset{6}{C}{\equiv}\overset{7}{C}H$$
1,6-Heptadiyne

Common names for simple alkynes are based on the name *acetylene,* as shown in the following examples.

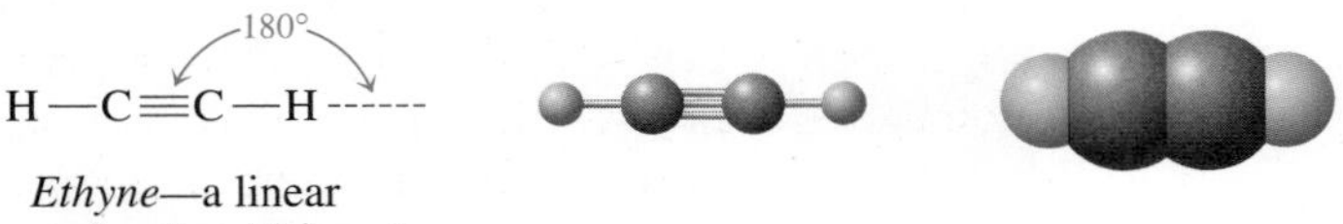

Figure 13.16 Structural representations of ethyne (acetylene), the simplest alkyne. The molecule is linear—that is, the bond angles are 180°.

$$H{-}C{\equiv}C{-}H \xrightarrow{180°}$$

Ethyne—a linear molecule with bond angles of 180°

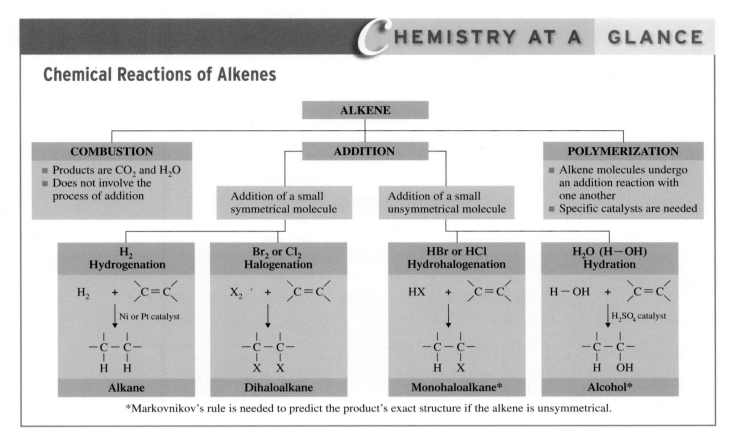

CHEMISTRY AT A GLANCE

Chemical Reactions of Alkenes

ALKENE

COMBUSTION
- Products are CO_2 and H_2O
- Does not involve the process of addition

ADDITION

Addition of a small symmetrical molecule

Addition of a small unsymmetrical molecule

POLYMERIZATION
- Alkene molecules undergo an addition reaction with one another
- Specific catalysts are needed

H_2 Hydrogenation

$$H_2 \ + \ \text{C=C}$$
↓ Ni or Pt catalyst
$$-\text{C}-\text{C}- \quad \text{H H}$$

Alkane

Br_2 or Cl_2 Halogenation

$$X_2 \ + \ \text{C=C}$$
↓
$$-\text{C}-\text{C}- \quad \text{X X}$$

Dihaloalkane

HBr or HCl Hydrohalogenation

$$HX \ + \ \text{C=C}$$
↓
$$-\text{C}-\text{C}- \quad \text{H X}$$

Monohaloalkane*

H_2O (H—OH) Hydration

$$H-OH \ + \ \text{C=C}$$
↓ H_2SO_4 catalyst
$$-\text{C}-\text{C}- \quad \text{H OH}$$

Alcohol*

*Markovnikov's rule is needed to predict the product's exact structure if the alkene is unsymmetrical.

Early cars had carbide headlights that produced acetylene by the action of slowly dripping water on calcium carbide. This same type of lamp, which was also used by miners, is still often used by spelunkers (cave explorers).

$$CH \equiv CH \qquad CH_3 - C \equiv CH \qquad CH_3 - C \equiv C - CH_3$$
Acetylene   Methylacetylene   Dimethylacetylene

Isomerism and Alkynes

Because of the linearity (180° angles) about an alkyne's triple bond, *cis–trans* isomerism, such as that found in alkenes, is not possible for alkynes because there are no "up" and "down" positions. However, constitutional isomers are possible—both relative to the carbon chain (skeletal isomers) and to the position of the triple bond (positional isomers).

Skeletal isomers:

1-Pentyne   and   3-Methyl-1-butyne

Positional isomers:

1-Hexyne   and   3-Hexyne

Physical and Chemical Properties of Alkynes

Unbranched 1-Alkynes			
✕	C_3	C_5	C_7
C_2	C_4	C_6	C_8

☐ Gas ☐ Liquid

Figure 13.17 A physical-state summary for unbranched 1-alkynes at room temperature and pressure.

The physical properties of alkynes are similar to those of alkenes and alkanes. In general, alkynes are insoluble in water but soluble in organic solvents, have densities less than that of water, and have boiling points that increase with molecular mass. Low-molecular-mass alkynes are gases at room temperature. Figure 13.17 is a physical-state summary for unbranched 1-alkynes with eight or fewer carbon atoms.

The triple-bond functional group of alkynes behaves chemically quite similarly to the double-bond functional group of alkenes. Thus there are many parallels between alkene chemistry and alkyne chemistry. The same substances that add to double bonds

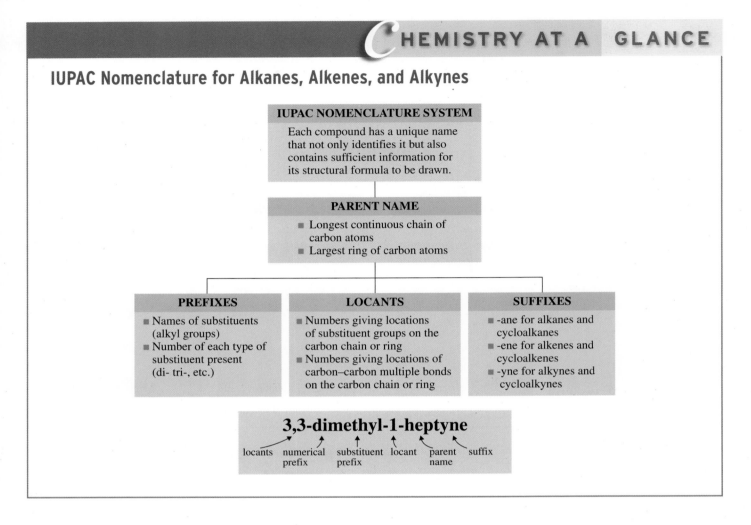

CHEMISTRY AT A GLANCE

IUPAC Nomenclature for Alkanes, Alkenes, and Alkynes

IUPAC NOMENCLATURE SYSTEM

Each compound has a unique name that not only identifies it but also contains sufficient information for its structural formula to be drawn.

PARENT NAME

■ Longest continuous chain of carbon atoms
■ Largest ring of carbon atoms

PREFIXES

■ Names of substituents (alkyl groups)
■ Number of each type of substituent present (di- tri-, etc.)

LOCANTS

■ Numbers giving locations of substituent groups on the carbon chain or ring
■ Numbers giving locations of carbon–carbon multiple bonds on the carbon chain or ring

SUFFIXES

■ -ane for alkanes and cycloalkanes
■ -ene for alkenes and cycloalkenes
■ -yne for alkynes and cycloalkynes

3,3-dimethyl-1-heptyne

locants   numerical   substituent   locant   parent   suffix
          prefix   prefix   name

(H$_2$, HCl, Cl$_2$, and so on) also add to triple bonds. However, two molecules of a specific reactant can add to a triple bond, as contrasted to the addition of one molecule of reactant to a double bond. In triple-bond addition, the first molecule converts the triple bond into a double bond, and the second molecule then converts the double bond into a single bond. For example, propyne reacts with H$_2$ to form propene first and then to form propane.

$$CH \equiv C - CH_3 \xrightarrow[Ni]{H_2} CH_2 = CH - CH_3 \xrightarrow[Ni]{H_2} CH_3 = CH_2 - CH_3$$

An alkyne (propyne)   An alkene (propene)   An alkane (propane)

Alkynes, like alkenes and alkanes, are flammable, that is, they readily undergo combustion reactions.

13.12 AROMATIC HYDROCARBONS

Aromatic hydrocarbons are the third class of unsaturated hydrocarbons; the alkenes and alkynes (previously considered) are the other two classes. An **aromatic hydrocarbon** *is an unsaturated cyclic hydrocarbon that does not readily undergo addition reactions.* This reaction behavior, which is very different from that of alkenes and alkynes, explains the separate classification for aromatic hydrocarbons.

The fact that, even though they are unsaturated compounds, aromatic hydrocarbons do not readily undergo addition reactions suggests that the bonding present in this type of compound must differ significantly from that in alkenes and alkynes. Such is indeed the case.

Students often ask whether it is possible to have hydrocarbons in which both double and triple bonds are present. The answer is yes. Immediately, another question is asked. How do we name such compounds? Such compounds are called *alkenynes.* An example is

1-Buten-3-yne

A double bond has priority over a triple bond in numbering the chain when numbering systems are equivalent. Otherwise, the chain is numbered from the end closest to a multiple bond.

Figure 13.18 Space-filling and ball-and-stick models for the structure of benzene.

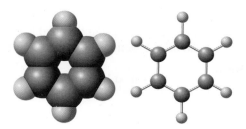

Let's look at the bonding present in *benzene,* the simplest aromatic hydrocarbon, to explore this new type of bonding situation and to also characterize the aromatic hydrocarbon functional group. Benzene, a flat, symmetrical molecule with a molecular formula of C_6H_6 (see Figure 13.18), has a structural formula that is often formalized as that of a cyclohexatriene—in other words, as a structural formula that involves a six-membered carbon ring in which three double bonds are present.

This structure is one of two equivalent structures that can be drawn for benzene that differ only in the locations of the double bonds (1,3,5 positions versus 2,4,6 positions):

Neither of these conventional structures, however, is totally correct. Experimental evidence indicates that all of the carbon–carbon bonds in benzene are equivalent (identical), and these structures imply three bonds of one type (double bonds) and three bonds of a different type (single bonds).

The equivalent nature of the carbon–carbon bonds in benzene is addressed by considering the correct bonding structure for benzene to be an *average* of the two "triene" structures. Related to this "average"-structure situation is the concept that electrons associated with the ring double bonds are not held between specific carbon atoms; instead, they are free to move "around" the carbon ring. Thus the true structure for benzene, an intermediate between that represented by the two "triene" structures, is a situation in which all carbon–carbon bonds are equivalent; they are neither single nor double bonds but something in between. Placing a double-headed arrow between the conventional structures that are averaged to obtain the true structure is one way to denote the average structure.

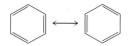

An alternative notation for denoting the bonding in benzene—a notation that involves a single structure—is

In this "circle-in-the-ring" structure for benzene, the circle denotes the electrons associated with the double bonds that move "around" the ring. Each carbon atom in the ring can be considered to participate in three conventional (localized) bonds (two C—C bonds and one C—H bond) and in one *delocalized bond* (the circle) that involves all six carbon atoms. A **delocalized bond** *is a covalent bond in which electrons are shared among more than two atoms.* This delocalized bond is what causes benzene and its derivatives to be resistant to addition reactions, a property normally associated with unsaturation in a molecule.

The structure represented by the notation

is called an *aromatic ring system,* and it is the functional group present in aromatic compounds. An **aromatic ring system** *is a highly unsaturated carbon ring system in which both localized and delocalized bonds are present.*

13.13 NAMES FOR AROMATIC HYDROCARBONS

Replacement of one or more of the hydrogen atoms on benzene with other groups produces benzene derivatives. Compounds with alkyl groups or halogen atoms attached to the benzene ring are commonly encountered. We consider first the naming of benzene derivatives with one substituent, then the naming of those with two substituents, and finally the naming of those with three or more substituents.

Benzene Derivatives with One Substituent

The IUPAC system of naming monosubstituted benzene derivatives uses the name of the substituent as a prefix to the name benzene. Examples of this type of nomenclature include

A few monosubstituted benzenes have names wherein the substituent and the benzene ring taken together constitute a new parent name. Two important examples of such nomenclature with hydrocarbon substituents are

Both of these compounds are industrially important chemicals.

Monosubstituted benzene structures are often drawn with the substituent at the "12 o'clock" position, as in the previous structures. However, because all the hydrogen atoms in benzene are equivalent, it does not matter at which carbon of the ring the substituted group is located. Each of the following formulas represents chlorobenzene.

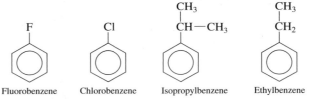

For monosubstituted benzene rings that have a group attached that is not easily named as a substituent, the benzene ring is often treated as a group attached to this substituent. In this reversed approach, the benzene ring attachment is called a *phenyl* group, and the compound is named according to the rules for naming alkanes, alkenes, and alkynes.

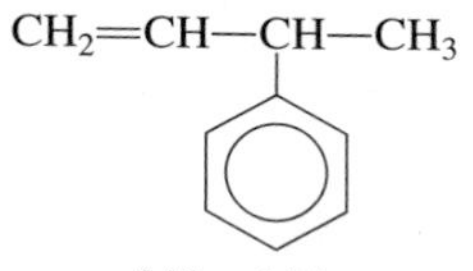

$$CH_2=CH-CH-CH_3$$

3-Phenyl-1-butene

Benzene Derivatives with Two Substituents

When two substituents, either the same or different, are attached to a benzene ring, three isomeric structures are possible.

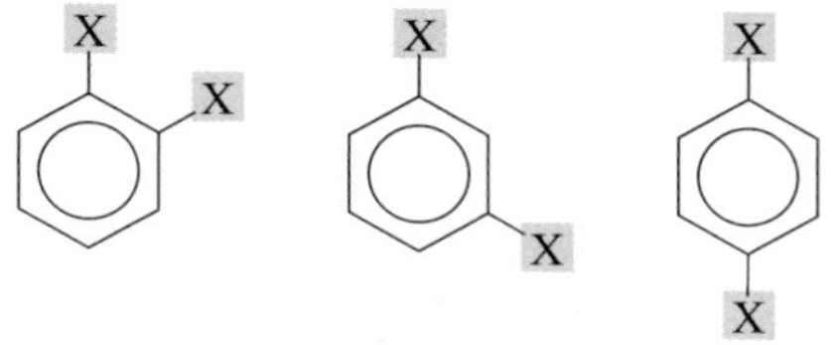

To distinguish among these three isomers, we must specify the positions of the substituents relative to one another. This can be done in either of two ways: by using numbers or by using nonnumerical prefixes.

When numbers are used, the three isomeric dichlorobenzenes have the first-listed set of names:

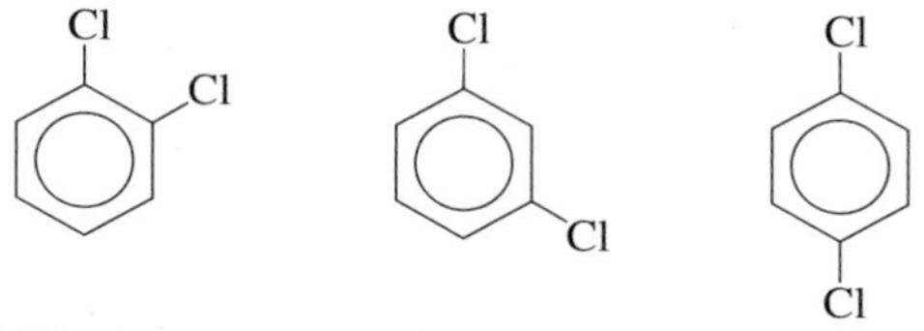

1,2-Dichlorobenzene   1,3-Dichlorobenzene   1,4-Dichlorobenzene
(*ortho*-dichlorobenzene)   (*meta*-dichlorobenzene)   (*para*-dichlorobenzene)

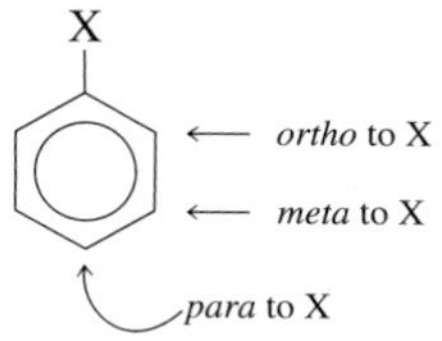

The prefix system uses the prefixes *ortho-*, *meta-*, and *para-* (abbreviated *o-*, *m-*, and *p-*).

   Ortho- means 1,2 disubstitution; the substituents are on adjacent carbon atoms.
   Meta- means 1,3 disubstitution; the substituents are one carbon removed from each other.
   Para- means 1,4 disubstitution; the substituents are two carbons removed from each other (on opposite sides of the ring).

When prefixes are used, the three isomeric dichlorobenzenes have the second-listed set of names above.

When one of the two substituents in a disubstituted benzene imparts a special name to the compound (as, for example, toluene), the compound is named as a derivative of that parent molecule. The special substituent is assumed to be at ring position 1.

4-Bromotoluene   2-Ethyltoluene
(not 1-bromo-4-methylbenzene)   (not 1-ethyl-2-methylbenzene)

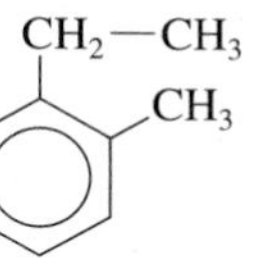

The word *phenyl* comes from "phene," a European term used during the 1800s for benzene. The word is pronounced *fen*-nil.

Cis–trans isomerism is not possible for disubstituted benzenes. All 12 atoms of benzene are in the same plane—that is, benzene is a flat molecule. When a substituent group replaces an H atom, the atom that bonds the group to the ring is also in the plane of the ring.

Learn the meaning of the prefixes *ortho-*, *meta-*, and *para-*. These prefixes are extensively used in naming disubstituted benzenes.

   ⟵ *ortho* to X
   ⟵ *meta* to X
   para to X

The use of *ortho-*, *meta-*, and *para-* in place of position numbers is reserved for disubstituted benzenes. The system is never used with cyclohexanes or other ring systems.

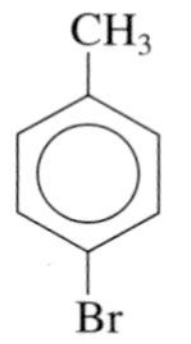

When neither substituent group imparts a special name, the substituents are cited in alphabetical order before the ending -*benzene*. The carbon of the benzene ring bearing the substituent with alphabetical priority becomes carbon 1.

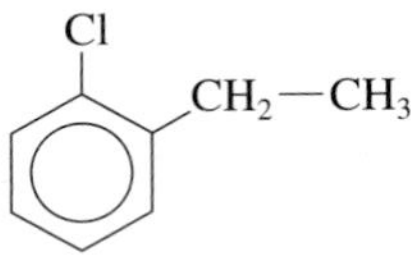

1-Chloro-2-ethylbenzene
(not 2-chloro-1-ethylbenzene)

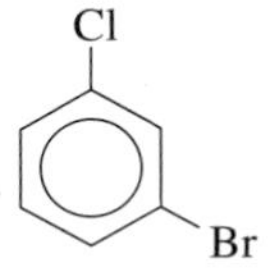

1-Bromo-3-chlorobenzene
(not 3-bromo-1-chlorobenzene)

A benzene ring bearing two methyl groups is a situation that generates a new special base name. Such compounds (there are three isomers) are not named as dimethylbenzenes or as methyl toluenes. They are called xylenes.

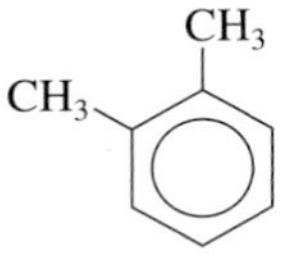

o-Xylene 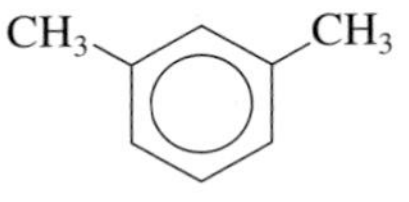   *m*-Xylene 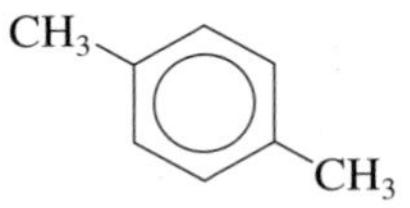   *p*-Xylene

The xylenes are good solvents for grease and oil and are used for cleaning microscope slides and optical lenses and for removing wax from skis.

Benzene Derivatives with Three or More Substituents

When more than two groups are present on the benzene ring, their positions are indicated with *numbers*. The ring is numbered in such a way as to obtain the lowest possible numbers for the carbon atoms that have substituents. If there is a choice of numbering systems (two systems give the same lowest set), then the group that comes first alphabetically is given the lower number.

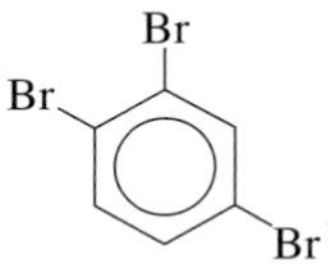

1,2,4-Tribromobenzene 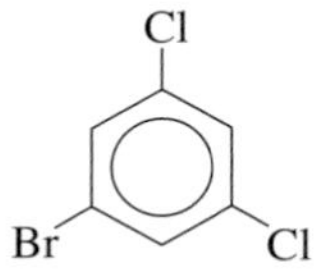   1-Bromo-3,5-dichlorobenzene

> When parent names such as *toluene* and *xylene* are used, additional substituents present cannot be the same as those included in the parent name. If such is the case, name the compound as a substituted benzene. The compound
>
>
> is named as a trimethylbenzene and not as a methylxylene or a dimethyltoluene.

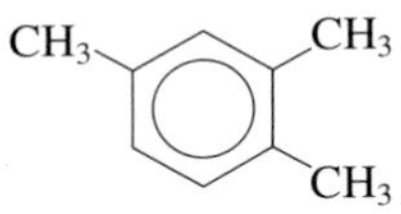

EXAMPLE 13.6

Assigning IUPAC Names to Benzene Derivatives

Assign IUPAC names to the following benzene derivatives.

a.

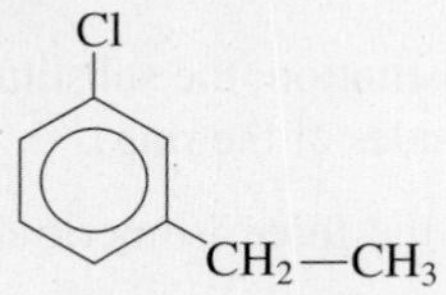

b.

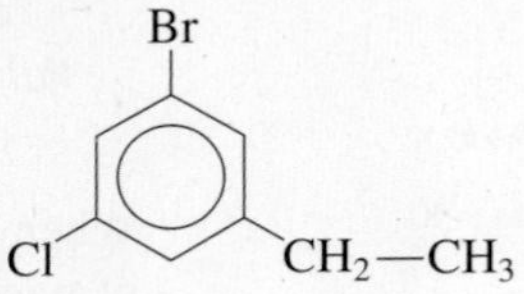

c.

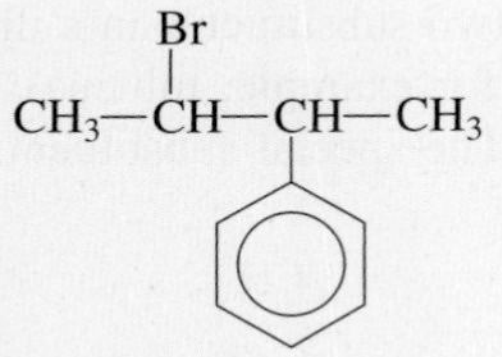

d.

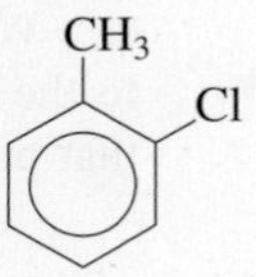

Solution

a. No substituents that will change the parent name from benzene are present on the ring. Alphabetical priority dictates that the chloro group is on carbon 1 and the ethyl group on carbon 3. The compound is named 1-chloro-3-ethylbenzene (or *m*-chloroethylbenzene).

Figure 13.19 Space-filling model for the compound 2-chlorotoluene.

b. Again, no substituents that will change the parent name from benzene are present on the ring. Alphabetical priority among substituents dictates that the bromo group is on carbon 1, the chloro group on carbon 3, and the ethyl group on carbon 5. The compound is named 1-bromo-3-chloro-5-ethylbenzene.

c. This compound is named with the benzene ring treated as a substituent—that is, as a phenyl group. The compound is named 2-bromo-3-phenylbutane.

d. The methyl group present on the benzene ring changes the parent name from benzene to toluene. Carbon 1 bears the methyl group. Numbering clockwise, we obtain the name 2-chlorotoluene. (See Figure 13.19).

Practice Exercise 13.6

Assign IUPAC names to the following benzene derivatives.

a. Br—⬡—CH$_2$—CH$_3$

b. CH$_2$—CH$_2$—CH$_3$ (on ring with Cl para)

c. CH$_3$—CH$_2$—CH$_2$—CH—CH$_2$—CH$_3$ (with phenyl group)

d. ⬡—Br with Cl, Cl

Answers: a. 1-Bromo-3-ethylbenzene or *m*-bromoethylbenzene; **b.** 1-Chloro-4-propylbenzene or *o*-chloropropylbenzene; **c.** 3-Phenylhexane; **d.** 4-Bromo-1,2-dichlorobenzene

13.14 AROMATIC HYDROCARBONS: PHYSICAL PROPERTIES AND SOURCES

In general, aromatic hydrocarbons resemble other hydrocarbons in physical properties. They are insoluble in water, are good solvents for other nonpolar materials, and are less dense than water.

Benzene, monosubstituted benzenes, and many disubstituted benzenes are liquids at room temperature. Benzene itself is a colorless, flammable liquid that burns with a sooty flame because of incomplete combustion.

At one time, coal tar was the main source of aromatic hydrocarbons. Petroleum is now the primary source of such compounds. At high temperatures, with special catalysts, saturated hydrocarbons obtained from petroleum can be converted to aromatic hydrocarbons. The production of toluene from heptane is representative of such a conversion.

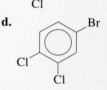

$$CH_3-CH_2-CH_2-CH_2-CH_2-CH_2-CH_3 \xrightarrow[\text{High temperature}]{\text{Catalyst}} \text{[toluene]} + 4H_2$$

Benzene was once widely used as an organic solvent. Such use has been discontinued because benzene's short- and long-term toxic effects are now recognized. Benzene inhalation can cause nausea and respiratory problems.

Two common situations in which a person can be exposed to low-level benzene vapors are

1. Inhaling gasoline vapors while refueling an automobile. Gasoline contains about 2% (v/v) benzene.
2. Being around a cigarette smoker. Benzene is a combustion product present in cigarette smoke. For smokers themselves, inhaled cigarette smoke is a serious benzene-exposure source.

13.15 CHEMICAL REACTIONS OF AROMATIC HYDROCARBONS

We have noted that aromatic hydrocarbons do not readily undergo the addition reactions characteristic of other unsaturated hydrocarbons. An addition reaction would require breaking up the delocalized bonding (Section 13.12) present in the ring system.

If benzene is so unresponsive to addition reactions, what reactions does it undergo? Benzene undergoes *substitution* reactions. As you recall from Section 12.17, substitution reactions are characterized by different atoms or groups of atoms replacing hydrogen atoms in a hydrocarbon molecule. Two important types of substitution reactions for benzene and other aromatic hydrocarbons are alkylation and halogenation.

1. *Alkylation:* An alkyl group (R—) from an alkyl chloride (R—Cl) substitutes for a hydrogen atom on the benzene ring. A catalyst, $AlCl_3$, is needed for alkylation.

Benzene Chloroethane Ethylbenzene

In general terms, the alkylation of benzene can be written as

Alkylation is the most important industrial reaction of benzene.

2. *Halogenation* (bromination or chlorination): A hydrogen atom on a benzene ring can be replaced by bromine or chlorine if benzene is treated with Br_2 or Cl_2 in the presence of a catalyst. The catalyst is usually $FeBr_3$ for bromination and $FeCl_3$ for chlorination.

Aromatic halogenation differs from alkane halogenation (Section 12.17) in that light is not required to initiate aromatic halogenation.

> Alkylation, the reaction that attaches an alkyl group to an aromatic ring, is also known as a *Friedel–Crafts reaction,* named after Charles Friedel and James Mason Crafts, the French and American chemists responsible for its discovery in 1877.

13.16 FUSED-RING AROMATIC HYDROCARBONS

Benzene and its substituted derivatives are not the only type of aromatic hydrocarbon that exists. Another large class of aromatic hydrocarbons is the fused-ring aromatic hydrocarbons. A **fused-ring aromatic hydrocarbon** *is an aromatic hydrocarbon whose structure contains two or more rings fused together.* Two carbon rings that share a pair of carbon atoms are said to be *fused.*

The three simplest fused-ring aromatic compounds are naphthalene, anthracene, and phenanthrene. All three are solids at room temperature.

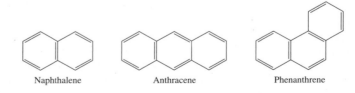

Naphthalene   Anthracene   Phenanthrene

CHEMICAL Connections   Fused-Ring Aromatic Hydrocarbons and Cancer

A number of fused-ring aromatic hydrocarbons are known to be carcinogens—that is, to cause cancer. Three of the most potent carcinogens are

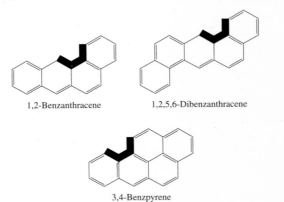

1,2-Benzanthracene

1,2,5,6-Dibenzanthracene

3,4-Benzpyrene

Very small amounts of these substances, when applied to the skin of mice, cause cancer.

Carcinogenic fused-ring aromatic hydrocarbons share some structural features. They all contain four or more fused rings, and they all have the same "angle" in the series of rings (the dark area in the structures shown).

Fused-ring aromatic hydrocarbons are often formed when hydrocarbon materials are heated to high temperatures. These resultant compounds are present in low concentrations in tobacco smoke, in automobile exhaust, and sometimes in burned (charred) food. The charred portions of a well-done steak cooked over charcoal are a likely source.

Angular, fused-ring hydrocarbon systems are believed to be partially responsible for the high incidence of lung and lip cancer among cigarette smokers because tobacco smoke contains 3,4-benzpyrene. The more a person smokes, the greater his or her risk of developing cancer.

We now know that the high incidence of lung cancer in British chimney sweeps (documented over 200 years ago) was caused by fused-ring hydrocarbon compounds present in the chimney soot that the sweeps inhaled regularly.

CONCEPTS TO REMEMBER

Unsaturated hydrocarbons. An unsaturated hydrocarbon is a hydrocarbon that contains one or more carbon–carbon multiple bonds. Three main classes of unsaturated hydrocarbons exist: alkenes, alkynes, and aromatic hydrocarbons (Section 13.1).

Alkenes and cycloalkenes. An alkene is an acyclic unsaturated hydrocarbon in which one or more carbon–carbon double bonds are present. A cycloalkene is a cyclic unsaturated hydrocarbon that contains one or more carbon–carbon double bonds within the ring system (Section 13.2).

Alkene nomenclature. Alkenes and cycloalkenes are given IUPAC names using rules similar to those for alkanes and cycloalkanes, except that the ending -ene is used. Also, the double bond takes precedence both in selecting and in numbering the main chain or ring (Section 13.3).

Isomerism in alkenes. Two subtypes of constitutional isomers are possible for alkenes: skeletal isomers and positional isomers. Positional isomers differ in the location of the functional group (double bond) present (Section 13.5).

Cis-trans isomerism in alkenes. Cis–trans isomerism is possible for some alkenes because there is restricted rotation about a carbon–carbon double bond (Section 13.6).

Physical properties of alkenes. Alkenes and alkanes have similar physical properties. They are nonpolar, insoluble in water, less dense than water, and soluble in nonpolar solvents (Section 13.8).

Addition reactions of alkanes. Numerous substances, including H_2, Cl_2, Br_2, HCl, HBr, and H_2O, add to an alkene carbon–carbon double bond. When both the alkene and the substance to be added are unsym-

metrical, the addition proceeds according to Markovnikov's rule: The carbon atom of the double bond that already has the greater number of H atoms gets one more (Section 13.9).

Addition polymers. Addition polymers are formed from alkene monomers that undergo repeated addition reactions with each other. Many familiar and widely used materials, such as fibers and plastics, are addition polymers (Section 13.10).

Alkynes and cycloalkynes. Alkynes and cycloalkynes are unsaturated hydrocarbons that contain one or more carbon–carbon triple bonds. They are named in the same way as alkenes and cycloalkenes, except that their parent names end in -yne. Like alkenes, alkynes undergo addition reactions. These occur in two steps, an alkene forming first and then an alkane (Section 13.11).

Aromatic hydrocarbons. Benzene, the simplest aromatic hydrocarbon, and other members of this family of compounds contain a six-membered ring with a cyclic, delocalized bond. This aromatic ring is often drawn as a hexagon containing a circle, which represents six electrons that move freely around the ring (Section 13.12).

Nomenclature of aromatic hydrocarbons. Monosubstituted benzene compounds are named by adding the substituent name to the word benzene. Positions of substituents in disubstituted benzenes are indicated by using a numbering system or the ortho- (1,2), meta- (1,3), and para- (1,4) prefix system (Section 13.13).

Chemical reactions of aromatic hydrocarbons. Aromatic hydrocarbons undergo substitution reactions rather than addition reactions. Important substitution reactions are alkylation and halogenation (Section 13.15).

KEY REACTIONS AND EQUATIONS

1. Hydrogenation of an alkene (Section 13.9)

$$\text{C}=\text{C} + \text{H--H} \xrightarrow{\text{Ni or Pt}} -\overset{|}{\underset{H}{C}}-\overset{|}{\underset{H}{C}}-$$

2. Halogenation of an alkene (Section 13.9)

$$\text{C}=\text{C} + \text{Br--Br} \longrightarrow -\overset{|}{\underset{Br}{C}}-\overset{|}{\underset{Br}{C}}-$$

3. Hydrohalogenation of an alkene (Section 13.9)

$$\text{C}=\text{C} + \text{H--Cl} \longrightarrow -\overset{|}{\underset{H}{C}}-\overset{|}{\underset{Cl}{C}}-$$

4. Hydration of an alkene (Section 13.9)

$$\text{C}=\text{C} + \text{H--OH} \xrightarrow{\text{H}_2\text{SO}_4} -\overset{|}{\underset{H}{C}}-\overset{|}{\underset{OH}{C}}-$$

5. Hydrogenation of an alkyne (Section 13.11)

$$-\text{C}\equiv\text{C}- + \text{H}_2 \xrightarrow{\text{Ni}} -\overset{H}{\underset{}{C}}=\overset{H}{\underset{}{C}}- \xrightarrow[\text{Ni}]{\text{H}_2} -\overset{H}{\underset{H}{C}}-\overset{H}{\underset{H}{C}}-$$

6. Halogenation of an alkyne (Section 13.11)

$$-\text{C}\equiv\text{C}- + \text{Br}_2 \longrightarrow -\overset{Br}{\underset{}{C}}=\overset{Br}{\underset{}{C}}- \xrightarrow{\text{Br}_2} -\overset{Br}{\underset{Br}{C}}-\overset{Br}{\underset{Br}{C}}-$$

7. Hydrohalogenation of an alkyne (Section 13.11)

$$-\text{C}\equiv\text{C}- + \text{HBr} \longrightarrow -\overset{H}{\underset{}{C}}=\overset{Br}{\underset{}{C}}- \xrightarrow{\text{HBr}} -\overset{H}{\underset{H}{C}}-\overset{Br}{\underset{Br}{C}}-$$

8. Alkylation of benzene (Section 13.15)

$$\bigcirc + \text{R--Cl} \xrightarrow{\text{AlCl}_3} \bigcirc^{\text{R}} + \text{HCl}$$

9. Halogenation of benzene (Section 13.15)

$$\bigcirc + \text{Br}_2 \xrightarrow{\text{FeBr}_3} \bigcirc^{\text{Br}} + \text{HBr}$$

EXERCISES and PROBLEMS

The members of each pair of problems in this section test similar material.

● Unsaturated Hydrocarbons (Section 13.1)

13.1 How does an unsaturated hydrocarbon differ structurally from a saturated hydrocarbon?

13.2 What type of functional group is present in an unsaturated hydrocarbon?

13.3 What is the functional group present in an alkene?

13.4 What is the functional group present in an alkyne?

13.5 In general terms, compare the physical properties of unsaturated and saturated hydrocarbons.

13.6 In general terms, compare the chemical properties of unsaturated and saturated hydrocarbons.

13.7 Classify each of the following hydrocarbons as saturated or unsaturated.

a. $CH_3-CH_2-CH=CH-CH_3$

b.
$$CH_2=C-\overset{\overset{\displaystyle CH_3}{|}}{C}-CH_3$$
$$\underset{\overset{|}{CH_3}\overset{|}{CH_3}}{}$$

c. $CH_2=CH-CH_2-\overset{\overset{\displaystyle }{}}{C}-CH_3$
$$\underset{\overset{||}{CH_2}}{}$$

d. $CH_2=CH-CH=CH-CH=CH_2$

13.8 Classify each of the following hydrocarbons as saturated or unsaturated.

a. $CH_3-CH=CH-CH=CH_2$

b. $CH_3-CH=CH-CH_3$

c. $CH_2=C-CH_2-CH_3$
$$\underset{\overset{|}{CH_3}}{}$$

d. $CH_2=C-CH=CH-CH=CH_2$
$$\underset{\overset{|}{CH_3}}{}$$

● Characteristics of Alkenes and Cycloalkenes (Section 13.2)

13.9 Write the *molecular formula* for hydrocarbons with each of the following structural features.
a. Acyclic, four carbon atoms, no multiple bonds
b. Acyclic, five carbon atoms, one double bond
c. Cyclic, five carbon atoms, one double bond
d. Cyclic, seven carbon atoms, two double bonds

13.10 Write the *molecular formula* for hydrocarbons with each of the following structural features.
a. Acyclic, six carbon atoms, two double bonds
b. Acyclic, six carbon atoms, three double bonds
c. Cyclic, five carbon atoms, no multiple bonds
d. Cyclic, eight carbon atoms, four double bonds

13.11 Write the *general* molecular formula (C_nH_{2n} and so on) for each of the following families of compounds.
a. Cycloalkene with one double bond
b. Alkadiene
c. Diene
d. Cycloalkatriene

13.12 Write the *general* molecular formula (C_nH_{2n} and so on) for each of the following families of compounds.
a. Cycloalkadiene
b. Alkene with one double bond
c. Triene
d. Alkatriene

13.13 Classify each of the compounds in Problem 13.7 as an alkene with one double bond, as a diene, or as a triene.

13.14 Classify each of the compounds in Problem 13.8 as an alkene with one double bond, as a diene, or as a triene.

IUPAC Nomenclature for Alkenes and Cycloalkenes (Section 13.3)

13.15 Assign an IUPAC name to each of the following unsaturated hydrocarbons.
a. $CH_3{-}CH{=}CH{-}CH_3$
b. $CH_3{-}\underset{\underset{CH_3}{|}}{C}{=}CH{-}\underset{\underset{CH_3}{|}}{CH}{-}CH_3$

c.

d.

13.16 Assign an IUPAC name to each of the following unsaturated hydrocarbons.
a. $CH_3{-}CH_2{-}CH{=}CH{-}CH_3$
b. $CH_3{-}CH_2{-}\underset{\underset{CH_3}{|}}{C}{=}CH{-}CH_3$

c.

d.

13.17 Assign an IUPAC name to each of the hydrocarbons in Problem 13.7.

13.18 Assign an IUPAC name to each of the hydrocarbons in Problem 13.8.

13.19 Draw a condensed structural formula for each of the following unsaturated hydrocarbons.
a. 3-Methyl-1-pentene
b. 3-Methylcyclopentene
c. 1,3-Butadiene
d. 3-Ethyl-1,4-pentadiene

13.20 Draw a condensed structural formula for each of the following unsaturated hydrocarbons.
a. 4-Methyl-1-hexene
b. 4-Methylcyclohexene
c. 1,3-Pentadiene .
d. 2-Ethyl-1,4-pentadiene

13.21 The following names are *incorrect* by IUPAC rules. Determine the correct IUPAC name for each compound.
a. 2-Ethyl-2-pentene
b. 4,5-Dimethyl-4-hexene
c. 3,5-Cyclopentadiene
d. 1,2-Dimethyl-4-cyclohexene

13.22 The following names are *incorrect* by IUPAC rules. Determine the correct IUPAC name for each compound.
a. 2-Methyl-4-pentene
b. 3-Methyl-2,4-pentadiene
c. 3-Methyl-3-cyclopentene
d. 1,2-Dimethyl-3-cyclohexene

13.23 Draw a condensed structural formula for each of the following unsaturated hydrocarbons.
a. Ethylene
b. Methylenecyclobutane
c. Vinyl bromide
d. Allyl iodide

13.24 Draw a condensed structural formula for each of the following unsaturated hydrocarbons.
a. Propylene
b. Methylenecyclopentane
c. Vinyl iodide
d. Allyl chloride

Line-Angle Structural Formulas for Alkenes (Section 13.4)

13.25 Draw a line-angle structural formula for each of the following unsaturated hydrocarbons.
a. 1-Butene
b. 2-Butene
c. 1,3-Butadiene
d. 3-Methyl-1-butene

13.26 Draw a line-angle structural formula for each of the following unsaturated hydrocarbons.
a. 1-Pentene
b. 2-Pentene
c. 1,4-Pentadiene
d. 4-Methyl-2-pentene

13.27 Name, using IUPAC rules, each of the following unsaturated hydrocarbons.
a.
b.
c.
d.

13.28 Name, using IUPAC rules, each of the following unsaturated hydrocarbons.
a.
b.
c.
d.

Constitutional Isomerism in Alkenes (Section 13.5)

13.29 For each of the following pairs of alkenes, indicate whether the members of each pair are *positional* constitutional isomers or *skeletal* constitutional isomers.
a. 1-Hexene and 2-hexene
b. 2-Methyl-1-pentene and 3-methyl-1-pentene
c. 3-Hexene and 2-ethyl-1-butene
d. 3-Methyl-2-pentene and 3-methyl-1-pentene

13.30 For each of the following pairs of alkenes, indicate whether the members of each pair are *positional* constitutional isomers or *skeletal* constitutional isomers.
a. 2-Hexene and 3-hexene
b. 4-Methyl-1-pentene and 4-methyl-2-pentene
c. 2,3-Dimethyl-2-butene and 3,3-dimethyl-1-butene
d. 2-Methyl-2-pentene and 2-methyl-1-pentene

13.31 How many constitutional isomers exist that fit each of the following specifications?
   a. Unbranched chain of five carbon atoms; one carbon–carbon double bond
   b. Unbranched chain of five carbon atoms; two carbon–carbon double bonds
   c. Five carbon atoms; one methyl group; one carbon–carbon double bond
   d. Five carbon atoms; two methyl groups; one carbon–carbon double bond

13.32 How many constitutional isomers exist that fit each of the following specifications?
   a. Unbranched chain of six carbon atoms; one carbon–carbon double bond
   b. Unbranched chain of six carbon atoms; two carbon–carbon double bonds
   c. Six carbon atoms; one methyl group; one carbon–carbon double bond
   d. Six carbon atoms; two methyl groups; one carbon–carbon double bond

13.33 Draw skeletal structural formulas and give the IUPAC names for the 13 possible alkene constitutional isomers with the formula C_6H_{12}. (Three of the constitutional isomers are hexenes, six are methyl-pentenes, three are dimethylbutenes, and one is an ethylbutene.)

13.34 Draw skeletal structural formulas and give the IUPAC names for the 16 possible alkadiene constitutional isomers with the formula C_6H_{10}. (Six of the constitutional isomers are hexa-dienes, eight are methylpentadienes, one is a dimethylbutadi-ene, and one is an ethylbutadiene.)

Cis-Trans Isomerism in Alkenes (Section 13.6)

13.35 For each molecule, tell whether *cis–trans* isomers exist. If they do, draw the two isomers and label them as *cis* and *trans*.
   a. $CH_2{=}CH{-}CH_3$ b. $CH_3{-}\underset{\underset{CH_3}{|}}{C}{=}CH{-}CH_3$
   c. 3-Hexene   d. 4-Methyl-2-pentene

13.36 For each molecule, tell whether *cis–trans* isomers exist. If they do, draw the two isomers and label them as *cis* and *trans*.
   a. $CH_3{-}CH_2{-}CH{=}CH_2$ b. $CH_3{-}CH_2{-}CH{=}\underset{\underset{Cl}{|}}{CH}$
   c. 2-Pentene   d. 1,2-Dichloroethene

13.37 Assign an IUPAC name to each of the following molecules. Include the prefix *cis-* or *trans-* when appropriate.

   a. b.
   c.   d.

13.38 Assign an IUPAC name to each of the following molecules. Include the prefix *cis-* or *trans-* when appropriate.
   a. H, Cl / H, H   b. Br, Br / H, H
   c. H, CH₃ / CH₃, H   d. H, Br / H, Br

13.39 Draw a structural formula for each of the following compounds.
   a. *trans*-3-Methyl-3-hexene   b. *cis*-2-Pentene
   c. *trans*-5-Methyl-2-heptene   d. *trans*-1,3-Pentadiene

13.40 Draw a structural formula for each of the following compounds.
   a. *trans*-2-Hexene   b. *cis*-4-Methyl-2-pentene
   c. *cis*-1-Chloro-1-pentene   d. *cis*-1,3-Pentadiene

Naturally Occurring Alkenes (Section 13.7)

13.41 What is a pheromone?

13.42 What is a terpene?

13.43 Why is the number of carbon atoms in a terpene always a multiple of the number 5?

13.44 What is the structural relationship between β-carotene and vitamin A?

Physical Properties of Alkenes and Cycloalkenes (Section 13.8)

13.45 With the help of Figure 13.7, indicate whether each of the following alkenes would be expected to be a solid, a liquid, or a gas at room temperature and pressure.
   a. Propene   b. 1-Pentene
   c. 1-Octene   d. Cyclopentene

13.46 With the help of Figure 13.7, indicate whether each of the following statements is true or false.
   a. 1-Butene has a density greater than that of water.
   b. 1-Butene has a higher boiling point than 1-hexene.
   c. 1-Butene is flammable but 1-hexene is not.
   d. Both 1-pentene and cyclopentene are gases at room tempera-ture and pressure.

Chemical Reactions of Alkenes (Section 13.9)

13.47 Which of the following chemical reactions are addition reactions?
   a. $C_4H_8 + Cl_2 \rightarrow C_4H_8Cl_2$
   b. $C_6H_6 + Cl_2 \rightarrow C_6H_5Cl + HCl$
   c. $C_3H_6 + HCl \rightarrow C_3H_7Cl$
   d. $C_7H_{16} \rightarrow C_7H_8 + 4H_2$

13.48 Which of the following chemical reactions are addition reactions?
   a. $C_3H_6 + Cl_2 \rightarrow C_3H_6Cl_2$
   b. $C_8H_{10} \rightarrow C_8H_8 + H_2$
   c. $C_6H_6 + C_2H_5Cl \rightarrow C_8H_{10} + HCl$
   d. $C_4H_8 + HCl \rightarrow C_4H_9Cl$

13.49 Write a chemical equation showing reactants, products, and catalysts needed (if any) for the reaction of ethene with each of the following substances.
   a. Cl_2 b. HCl   c. H_2 d. HBr

13.50 Write a chemical equation showing reactants, products, and catalysts needed (if any) for the reaction of ethene with each of the following substances.
   a. H_2O b. Br_2 c. HI   d. I_2

13.51 Write a chemical equation showing reactants, products, and catalysts needed (if any) for the reaction of propene with each of the reactants in Problem 13.49. Use Markovnikov's rule as needed.

13.52 Write a chemical equation showing reactants, products, and catalysts needed (if any) for the reaction of propene with each of the reactants in Problem 13.50. Use Markovnikov's rule as needed.

13.53 Supply the structural formula of the product in each of the following alkene addition reactions.

a. $CH_3-CH=CH-CH_3 + Cl_2 \rightarrow$?

b. $CH_3-CH_2-CH=CH_2 + HCl \rightarrow$?

c. $+ H_2 \xrightarrow[\text{catalyst}]{\text{Ni}}$?

d. □ $+ H_2O \xrightarrow{H_2SO_4}$?

13.54 Supply the structural formula of the product in each of the following alkene addition reactions.

a. $CH_3-CH_2-CH=CH_2 + Cl_2 \rightarrow$?

b. $CH_3-\underset{\underset{CH_3}{|}}{CH}-CH=CH_2 + HBr \rightarrow$?

c. $+ H_2 \xrightarrow[\text{catalyst}]{\text{Ni}}$?

d. $CH_3-CH=CH_2 + H_2O \xrightarrow{H_2SO_4}$?

13.55 What reactant would you use to prepare each of the following compounds from cyclohexene?

a. [structure: cyclohexane with Br and Br]

b. [structure: cyclohexane]

c. [structure: cyclohexane with Cl]

d. [structure: cyclohexane with OH]

13.56 What reactant would you use to prepare each of the following compounds from cyclopentene?

a. [structure: cyclopentane]

b. [structure: cyclopentane with OH]

c. [structure: cyclopentane with Cl and Cl]

d. [structure: cyclopentane with Br]

13.57 How many molecules of H_2 gas will react with 1 molecule of each of the following unsaturated hydrocarbons?

a. $CH_3-CH=CH-CH=CH-CH_3$

b. [structure: bicyclic with double bond]

c. 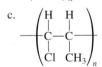 [structure with CH=CH₂]

d. $CH_3-CH=C=\underset{\underset{CH_3}{|}}{C}-CH=CH_2$

13.58 How many molecules of H_2 gas will react with 1 molecule of each of the following unsaturated hydrocarbons?

a. $CH_3-CH=CH-CH_3$

b. [structure: cyclobutane with substituent]

c. 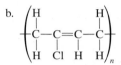 [structure with CH=CH₂]

d. $CH_2=CH-\underset{\underset{CH_2}{||}}{C}-CH=CH_2$

Polymerization of Alkenes (Section 13.10)

13.59 What is meant by the term *polymer*?

13.60 What is meant by the term *monomer*?

13.61 What is meant by the term *addition polymer*?

13.62 What is meant by the term *copolymer*?

13.63 Draw the structural formula of the monomer(s) from which each of the following polymers was made.

a. $\left(\begin{array}{c}\underset{|}{\overset{|}{F}}\ \underset{|}{\overset{|}{F}}\\ C-C\\ \underset{}{\overset{|}{F}}\ \underset{}{\overset{|}{F}}\end{array}\right)_n$

b. $\left(\begin{array}{c}H\quad\quad H\\ C-C=C-C\\ H\ Cl\ H\ H\end{array}\right)_n$

c. $\left(\begin{array}{c}H\ H\\ C-C\\ H\ Cl\end{array}\right)_n$

d. $\left(\begin{array}{c}H\ H\\ C-C\\ H\ \phi\end{array}\right)_n$

13.64 Draw the structural formula of the monomer(s) from which each of the following polymers was made.

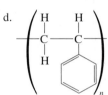

a. $\left(\begin{array}{c}H\ F\\ C-C\\ H\ F\end{array}\right)_n$

b. $\left(\begin{array}{c}H\quad\quad H\\ C-C=C-C\\ H\ Cl\ Cl\ H\end{array}\right)_n$

c. $\left(\begin{array}{c}H\ H\\ C-C\\ Cl\ CH_3\end{array}\right)_n$

d. $\left(\begin{array}{c}H\ H\\ C-C\\ Cl\ \phi\end{array}\right)_n$

13.65 Draw the "start" (the first three repeating units) of the structural formula of the addition polymers made from the following monomers.

a. Ethylene   b. Vinyl chloride

c. 1,2-Dichloroethene   d. 1-Chloroethene

13.66 Draw the "start" (the first three repeating units) of the structural formula of the addition polymers made from the following monomers.

a. Propylene   b. 1,1,2,2-Tetrafluoroethene

c. 2-Methyl-1-propene   d. 1,2-Dichloroethylene

Alkynes (Section 13.11)

13.67 What is the general molecular formula for an alkyne in which two carbon–carbon triple bonds are present?

13.68 What is the general molecular formula for a cycloalkyne in which one carbon–carbon triple bond is present?

13.69 Assign an IUPAC name to each of the following unsaturated hydrocarbons.

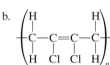

a. $CH_3-CH_2-CH_2-CH_2-C\equiv CH$

b. $CH_3-C\equiv C-\underset{\underset{CH_3}{|}}{CH}-CH_3$

c.

$CH_3-\underset{\underset{CH_3}{|}}{\overset{\overset{CH_3}{|}}{C}}-C\equiv C-CH_2-CH_2-CH_3$

d.

13.70 Assign an IUPAC name to each of the following unsaturated hydrocarbons.

a. $CH_3-\underset{\underset{CH_3}{|}}{CH}-C\equiv CH$

b.

c. $CH_3-\underset{\underset{CH_3}{|}}{CH}-C\equiv C-\underset{\underset{CH_3}{|}}{CH}-CH_3$

d. $CH_3-\underset{\underset{CH_2}{|}}{\underset{\underset{CH_3}{|}}{CH}}-CH_2-\underset{\overset{||}{CH}}{C}$

13.71 Draw skeletal structural formulas and give the IUPAC names for the three possible alkyne isomers with the molecular formula C_5H_8.

13.72 Draw skeletal structural formulas and give the IUPAC names for the seven possible alkyne isomers with the molecular formula C_6H_{10}. (Three of the constitutional isomers are hexynes, three are pentynes, and one is a butyne.)

13.73 Why is *cis–trans* isomerism not possible for an alkyne?

13.74 What are the bond angles about the triple bond in an alkyne?

13.75 Contrast alkynes and alkenes in terms of general physical properties.

13.76 Contrast alkynes and alkenes in terms of general chemical properties.

13.77 Supply the condensed structural formula of the product in each of the following alkyne addition reactions.

a. $CH\equiv CH + 2H_2 \xrightarrow{Ni} ?$

b. $CH_3-C\equiv CH + 2Br_2 \longrightarrow ?$

c. $CH_3-C\equiv CH + 2HBr \longrightarrow ?$

d. $CH\equiv CH + 1HCl \longrightarrow ?$

13.78 Supply the condensed structural formula of the product in each of the following alkyne addition reactions.

a. $CH_3-C\equiv C-CH_3 + 2Br_2 \longrightarrow ?$

b. $CH_3-C\equiv C-CH_3 + 2HBr \longrightarrow ?$

c. $CH\equiv C-CH_2-CH_3 + 1H_2 \xrightarrow{Ni} ?$

d. $CH\equiv C-CH_3 + 1HCl \longrightarrow ?$

Aromatic Hydrocarbons (Section 13.12)

13.79 Draw a structural representation for the functional group present in an aromatic hydrocarbon.

13.80 A circle (ring) within a hexagon is often used to represent an aromatic carbon ring. What does the circle represent?

13.81 What is the shortcoming of representing the bonding in a benzene ring using alternating single and double carbon–carbon bonds?

13.82 What is a *delocalized* bond?

Nomenclature for Aromatic Hydrocarbons (Section 13.13)

13.83 Assign an IUPAC name to each of the following disubstituted benzenes. Use numbers rather than prefixes to locate the substituents on the benzene ring.

13.84 Assign an IUPAC name to each of the following disubstituted benzenes. Use numbers rather than prefixes to locate the substituents on the benzene ring.

13.85 Assign each of the compounds in Problem 13.83 an IUPAC name in which the substituents on the benzene ring are located using the *ortho-*, *meta-*, *para-* prefix system.

13.86 Assign each of the compounds in Problem 13.84 an IUPAC name in which the substituents on the benzene ring are located using the *ortho-*, *meta-*, *para-* prefix system.

13.87 Assign an IUPAC name to each of the following substituted benzenes.

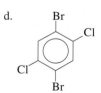

13.88 Assign an IUPAC name to each of the following substituted benzenes.

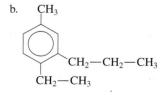

c. 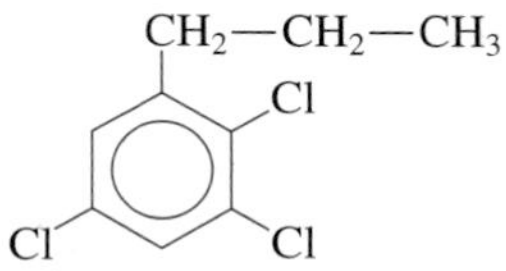 $CH_2-CH_2-CH_3$ d.

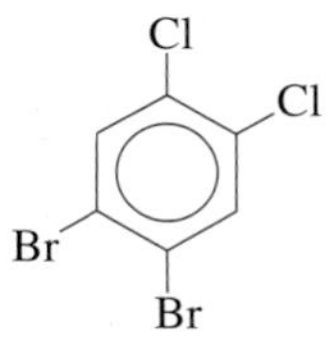

13.89 Assign an IUPAC name to each of the following compounds in which the benzene ring is treated as a substituent.

a. 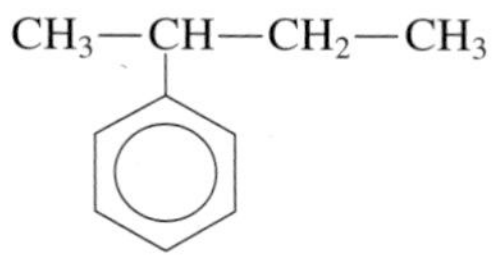 $CH_3-CH-CH_2-CH_3$

b. 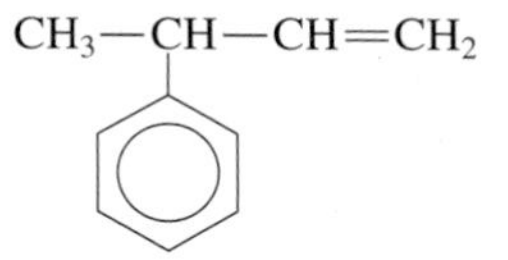 $CH_3-CH-CH=CH_2$

c. 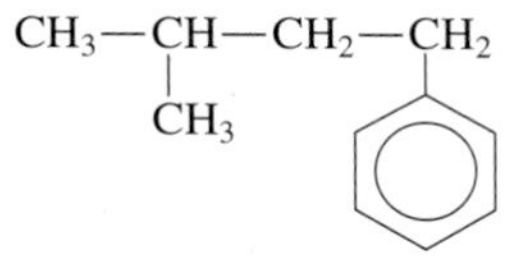 $CH_3-CH-CH_2-CH_2$
  $\quad\;\; CH_3$

d. 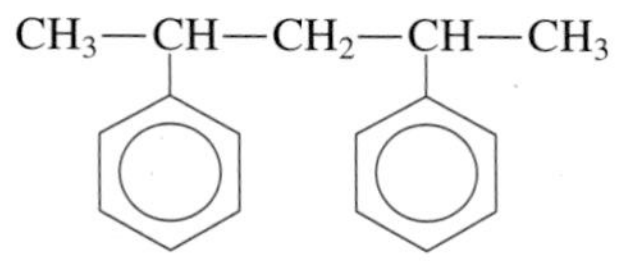 $CH_3-CH-CH_2-CH-CH_3$

13.90 Assign an IUPAC name to each of the following compounds in which the benzene ring is treated as a substituent.

a. 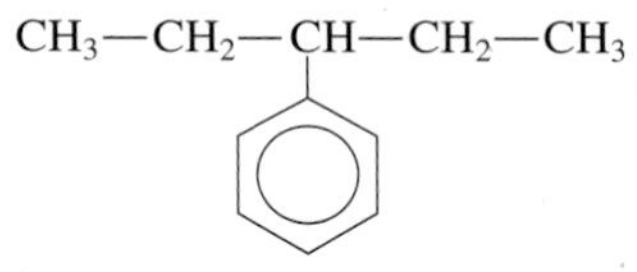 $CH_3-CH_2-CH-CH_2-CH_3$

b. 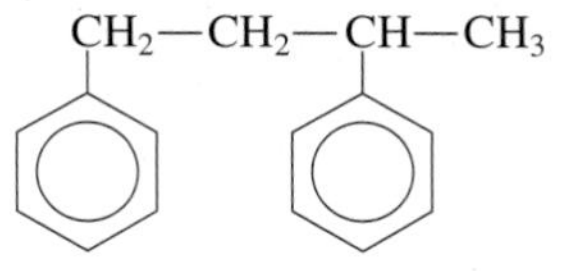 $CH_2-CH_2-CH-CH_3$

c. 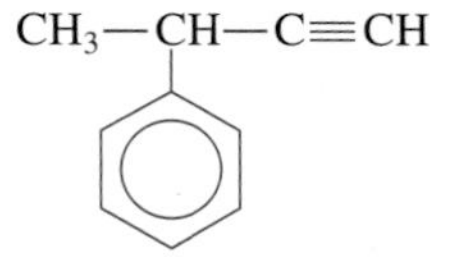 $CH_3-CH-C\equiv CH$

d. 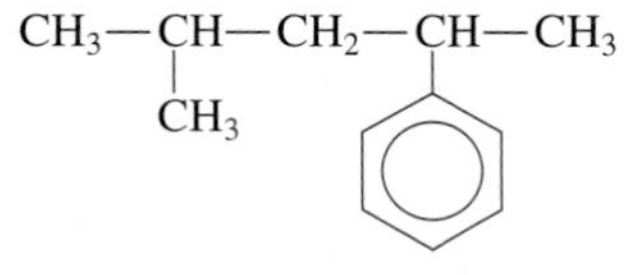 $CH_3-CH-CH_2-CH-CH_3$
  $\quad\;\; CH_3$

13.91 Write a structural formula for each of the following compounds.
  a. 1,3-Diethylbenzene   b. *o*-Xylene
  c. *p*-Ethyltoluene   d. Phenylbenzene

13.92 Write a structural formula for each of the following compounds.
  a. *o*-Ethylpropylbenzene   b. *m*-Xylene
  c. 2-Bromotoluene   d. 2-Phenylpropane

● **Aromatic Hydrocarbons: Physical Properties and Sources (Section 13.14)**

13.93 What is the physical state at room temperature for benzene, monosubstituted benzenes, and many disubstituted benzenes?

13.94 Indicate whether or not each of the following is a general physical property of aromatic hydrocarbons.
  a. Soluble in water
  b. Less dense than water
  c. Good solvent for nonpolar substances
  d. All solids at room temperatures

13.95 What is currently the primary source for aromatic hydrocarbons?

13.96 What used to be the primary source for aromatic hydrocarbons?

● **Chemical Reactions of Aromatic Hydrocarbons (Section 13.15)**

13.97 For each of the following classes of compounds, indicate whether addition or substitution is the most characteristic reaction for the class.
  a. Alkanes   b. Dienes
  c. Alkylbenzenes   d. Cycloalkenes

13.98 For each of the following classes of compounds, indicate whether addition or substitution is the most characteristic reaction for the class.
  a. Alkynes
  b. Cycloalkanes
  c. Aromatic hydrocarbons
  d. Saturated hydrocarbons

13.99 Complete the following reaction equations by supplying the formula of the missing reactant or product.
  a.

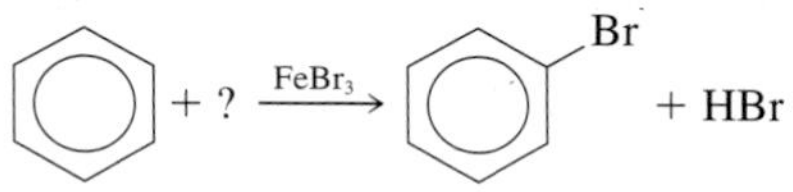

  b.

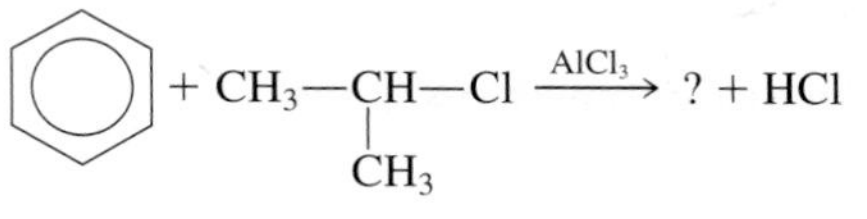

  c.

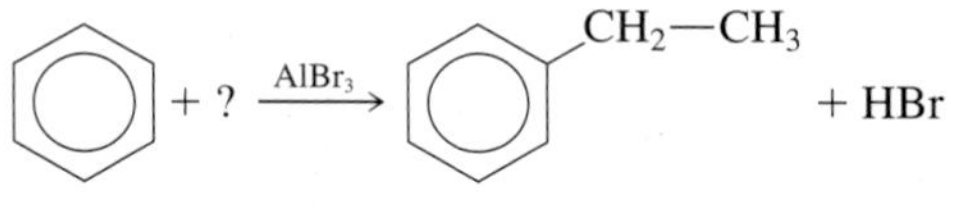

13.100 Complete the following reaction equations by supplying the formula of the missing reactant, product, or catalyst.
  a.

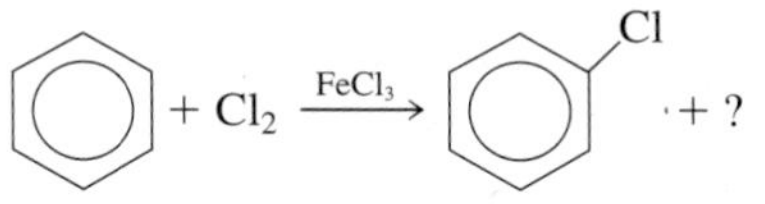

  b.

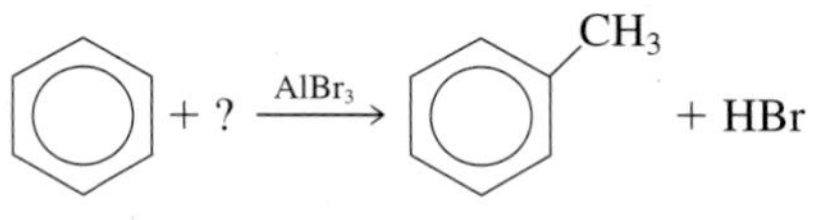

  c.

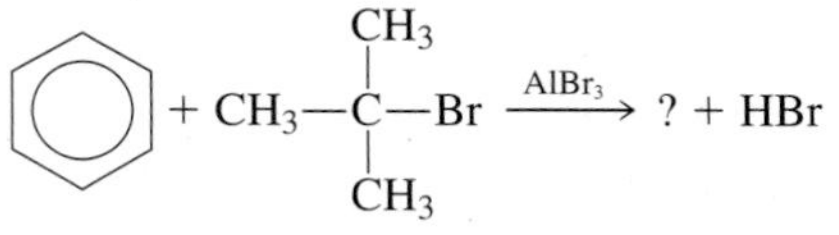

● **Fused-Ring Aromatic Hydrocarbons (Section 13.16)**

13.101 What is the general characteristic associated with two carbon rings that are *fused* together?

13.102 What is the structural formula for naphthalene, the simplest fused-ring aromatic hydrocarbon?

ADDITIONAL PROBLEMS

13.103 What is the molecular formula for the simplest compound of each of the following types?
a. Alkene with one multiple bond
b. Cycloalkene with one multiple bond
c. Alkyne with one multiple bond
d. Alkane

13.104 Indicate whether the hydrocarbon listed first in each of the following pairs of hydrocarbons contains (1) more hydrogen atoms, (2) the same number of hydrogen atoms, or (3) fewer hydrogen atoms than the hydrocarbon listed second.
a. Propane and propene
b. Propene and propyne
c. Propene and cyclopropene
d. Propyne and cyclopropene

13.105 Indicate whether each of the following pairs of hydrocarbons are *constitutional* isomers.
a. Propene and cyclopropene
b. 1-Pentene and 2-pentene
c. *cis*-2-Butene and *trans*-2-butene
d. Cyclobutene and 2-butyne

13.106 Contrast the compounds cyclohexane, cyclohexene, and benzene in terms of each of the following:
a. Number of carbon atoms present
b. Number of hydrogen atoms present
c. Whether they undergo substitution or addition reactions
d. Whether they are a solid, a liquid, or a gas at room temperature and pressure

13.107 Draw a condensed structural formula for each of the following unsaturated hydrocarbons or hydrocarbon derivatives.
a. 5-Methyl-2-hexyne
b. 1-Chloro-2-butene
c. 1,6-Heptadiene
d. 3-Methyl-1,4-pentadiene

13.108 How many molecules of H_2 will react with 1 molecule of each of the compounds in Problem 13.107 when the appropriate catalyst is present?

13.109 Draw a condensed structural formula for each of the following compounds.
a. Vinylbenzene
b. Allyl chloride
c. Propylacetylene
d. Dipropylacetylene

13.110 The compound 2-methyl-1-propene is a well-known substance. The compound 2,2-dimethyl-1-propene does not exist. Explain why this is so.

13.111 The compound 1,2-dichlorocyclohexane exists in *cis–trans* forms. However, *cis–trans* isomerism is not possible for the compound 1,2-dichlorobenzene. Explain why this is so.

13.112 Hydrocarbons with the formula C_5H_{10} can be either alkenes or cycloalkanes. Draw the ten constitutional isomers that fit this formula; five are alkenes and five are cycloalkanes. Then indicate which of these ten isomers can exist in *cis–trans* forms.

13.113 There are eight isomeric substituted benzenes that have the formula C_9H_{12}. What are the IUPAC names for these eight constitutional isomers?

13.114 How many different compounds are there that fit each of the following descriptions?
a. Bromochlorobenzenes
b. Trichlorobenzenes
c. Dibromodichlorobenzenes
d. Monobromoanthracenes

*M*ultiple-Choice Practice Test

13.115 All of the following compounds are unsaturated hydrocarbons except which one?
a. 2-Butene
b. 3-Heptyne
c. Cyclopropane
d. 1,3-Dimethylbenzene

13.116 What is the correct IUPAC name for the compound
$$CH_3{-}CH{-}CH{=}CH_2$$
$$\underset{CH_3}{\big|}$$
a. 2-Methylbutene
b. 2-Methyl-3,4-butene
c. 2-Methyl-3-butene
d. 3-Methyl-1-butene

13.117 What is the number of carbon atoms present in a vinyl group?
a. One   b. Two   c. Three   d. Four

13.118 Which of the following types of unsaturated hydrocarbons does *not* have the general formula C_nH_{2n-2}?
a. Alkenes with one double bond
b. Cycloalkenes with one double bond
c. Alkenes with two double bonds
d. Alkenes with one triple bond

13.119 For which of the following halogenated hydrocarbons is *cis–trans* isomerism possible?
a. 1,1-Dichloro-1-propene
b. 1,3-Dichloro-1-propene
c. 2,3-Dichloro-1-propene
d. 3,3-Dichloro-1-propene

13.120 Which of the following reactions can be used to convert an alkene to an alkane?
a. Hydrogenation
b. Halogenation
c. Hydrohalogenation
d. Hydration

13.121 In which of the following addition polymers are methyl groups present as attachments to the carbon chain?
a. Polyethylene
b. Polypropylene
c. Teflon
d. PVC

13.122 Which of the following statements concerning alkynes is *incorrect*?
a. Alkynes are generally insoluble in water.
b. Alkynes generally have densities less than that of water.
c. Alkynes do not undergo halogenation reactions.
d. Alkynes undergo combustion reactions.

13.123 In naming aromatic hydrocarbons using IUPAC rules, which of the following is a correct pairing of prefix and numbers for substituent locations?
a. *Para*- and 1,2-
b. *Ortho*- and 1,4-
c. *Meta*- and 1,3-
d. *Iso*- and 2,3-

13.124 Which of the following aromatic compounds contains 7 carbon atoms?
a. Toluene
b. 1,2-Dichlorobenzene
c. *o*-Xylene
d. 3-Phenylbutane

Alcohols, Phenols, and Ethers

14

Chapter Outline

Two alcohols, 1-octanol and 3-octanol, contribute to the distinctive flavor of mushrooms.

This chapter is the first of three that consider hydrocarbon derivatives with *oxygen-containing functional groups*. Many biochemically important molecules contain carbon atoms bonded to oxygen atoms.

In this chapter we consider hydrocarbon derivatives whose functional groups contain one oxygen atom participating in two single bonds (alcohols, phenols, and ethers). Chapter 15 focuses on derivatives whose functional groups have one oxygen atom participating in a double bond (aldehydes and ketones), and in Chapter 16 we examine functional groups that contain two oxygen atoms, one participating in single bonds and the other in a double bond (carboxylic acids, esters, and other acid derivatives).

14.1 BONDING CHARACTERISTICS OF OXYGEN ATOMS IN ORGANIC COMPOUNDS

An understanding of the bonding characteristics of the oxygen atom is a prerequisite to our study of compounds with oxygen-containing functional groups. Normal bonding behavior for oxygen atoms in such functional groups is the formation of two covalent bonds. Oxygen is a member of Group VIA of the periodic table and thus possesses six valence electrons. To complete its octet by electron sharing, an oxygen atom can form either two single bonds or a double bond.

Two single bonds One double bond

Figure 14.1 Space-filling models for the three simplest unbranched-chain alcohols: methyl alcohol, ethyl alcohol, and propyl alcohol.

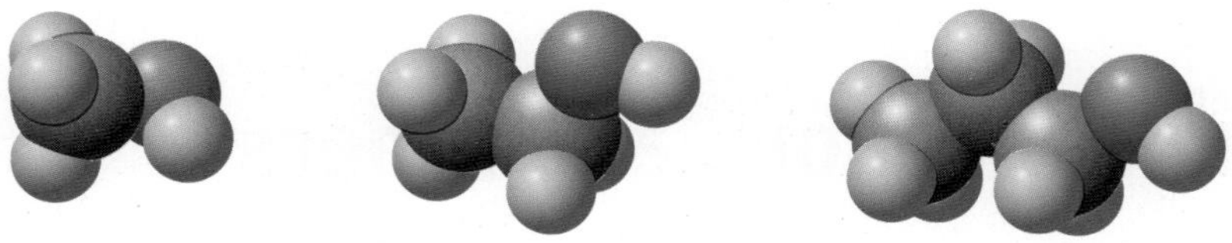

CH_3— OH CH_3— CH_2— OH CH_3— CH_2— CH_2— OH
One-carbon alcohol   **Two-carbon alcohol**   **Three-carbon alcohol**

Thus, in organic chemistry, carbon forms four bonds, hydrogen forms one bond, and oxygen forms two bonds.

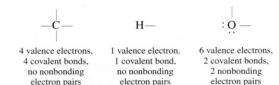

| 4 valence electrons, 4 covalent bonds, no nonbonding electron pairs | 1 valence electron, 1 covalent bond, no nonbonding electron pairs | 6 valence electrons, 2 covalent bonds, 2 nonbonding electron pairs |

14.2 STRUCTURAL CHARACTERISTICS OF ALCOHOLS

We begin our discussion of hydrocarbon derivatives containing a single oxygen atom by considering *alcohols,* substances with the generalized formula

R—OH

An **alcohol** *is an organic compound in which an —OH group is bonded to a saturated carbon atom.* A *saturated* carbon atom is a carbon atom that is bonded to four other atoms.

The —OH group, the functional group that is characteristic of an alcohol, is called a *hydroxyl group.* A **hydroxyl group** *is the —OH functional group.*

Examples of structural formulas for alcohols include

CH_3—OH CH_3—CH_2—OH CH_3—CH_2—CH_2—OH

Space-filling models for these three alcohols, the simplest alcohols possible that have unbranched carbon chains, are given in Figure 14.1.

Alcohols may be viewed structurally as being alkyl derivatives of water in which a hydrogen atom has been replaced by an alkyl group.

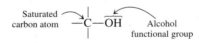

Figure 14.2 shows the similarity in oxygen bond angles for water and CH_3—OH, the simplest alcohol.

Alcohols may also be viewed structurally as hydroxyl derivatives of alkanes in which a hydrogen atom has been replaced by a hydroxyl group.

R—H   R—OH
An alkane   An alcohol

The hydroxyl group (—OH) should not be confused with the *hydroxide* ion (OH⁻) that we have encountered previously. Alcohols are not *hydroxides.* Hydroxides are ionic compounds that contain the OH⁻ polyatomic ion (Section 4.10). Alcohols are not ionic compounds. In an alcohol, the —OH group, which is not an ion, is *covalently* bonded to a saturated carbon atom.

Water (HOH)

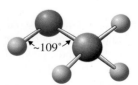

Methyl alcohol (CH₃OH)

Figure 14.2 The similar shapes of water and methanol. Methyl alcohol may be viewed structurally as an alkyl derivative of water.

Line-angle structural formulas for selected simple alcohols:

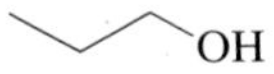

Propyl alcohol
(1-propanol)

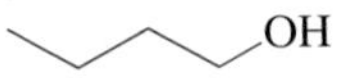

Butyl alcohol
(1-butanol)

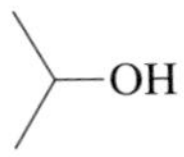

Isopropyl alcohol
(2-propanol)

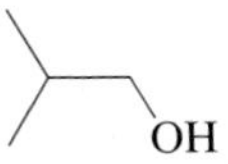

Isobutyl alcohol
(2-methyl-1-propanol)

14.3 NOMENCLATURE FOR ALCOHOLS

Common names exist for alcohols with simple (generally C_1 through C_4) alkyl groups. To assign a common name:

Rule 1: Name all of the carbon atoms of the molecule as a single *alkyl group.*

Rule 2: Add the word *alcohol,* separating the words with a space.

CH_3—OH
Methyl alcohol

CH_3—CH_2—OH
Ethyl alcohol

CH_3—CH_2—CH_2—OH
Propyl alcohol

CH_3—CH—OH
|
CH_3
Isopropyl alcohol

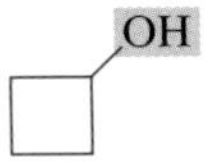

OH
Cyclobutyl alcohol

IUPAC rules for naming alcohols that contain a single hydroxyl group follow.

Rule 1: *Name the longest carbon chain to which the hydroxyl group is attached.* The chain name is obtained by dropping the final -*e* from the alkane name and adding the suffix -*ol.*

Rule 2: *Number the chain starting at the end nearest the hydroxyl group, and use the appropriate number to indicate the position of the —OH group.* (In numbering of the longest carbon chain, the hydroxyl group has priority over double and triple bonds, as well as over alkyl, cycloalkyl, and halogen substituents.)

Rule 3: *Name and locate any other substituents present.*

Rule 4: *In alcohols where the —OH group is attached to a carbon atom in a ring, the hydroxyl group is assumed to be on carbon 1.*

Table 14.1 gives both IUPAC and common names for monohydroxy alcohols that contain four or fewer carbon atoms.

TABLE 14.1
IUPAC and Common Names of Monohydroxy Alcohols That Contain Up to Four Carbon Atoms

Formula	IUPAC Name	Common Name
One carbon atom (CH_3OH)		
CH_3—OH	methanol	methyl alcohol
Two carbon atoms (C_2H_5OH)		
CH_3—CH_2—OH	ethanol	ethyl alcohol
Three carbon atoms (C_3H_7OH); two constitutional isomers exist		
CH_3—CH_2—CH_2—OH	1-propanol	propyl alcohol
CH_3—CH—CH_3 | OH	2-propanol	isopropyl alcohol
Four carbon atoms (C_4H_9OH); four constitutional isomers exist		
CH_3—CH_2—CH_2—CH_2—OH	1-butanol	butyl alcohol
CH_3—CH—CH_2—OH | CH_3	2-methyl-1-propanol	isobutyl alcohol
CH_3—CH_2—CH—OH | CH_3	2-butanol	*sec*-butyl alcohol
CH_3 | CH_3—C—OH | CH_3	2-methyl-2-propanol	*tert*-butyl alcohol

EXAMPLE 14.1

Determining IUPAC Names
for Alcohols

Name the following alcohols, utilizing IUPAC nomenclature rules.

3-m-3-

a.

$$CH_3-CH_2-\underset{\underset{OH}{|}}{\overset{\overset{CH_3}{|}}{C}}-CH_2-CH_2-CH_3$$

b. $CH_3-CH_2-\underset{\underset{CH_2-OH}{|}}{CH}-CH_2-CH_3$

c.

$$CH_3-\underset{\overset{CH_3}{|}}{\bigcirc}-OH$$

d.

(structure) OH

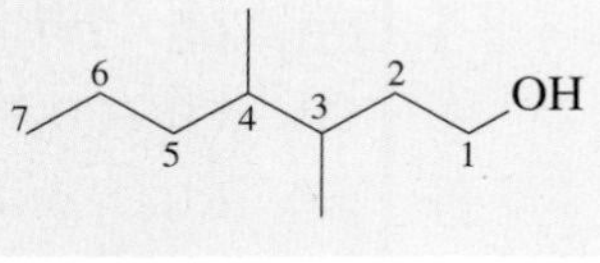

In the naming of alcohols with *un-saturated* carbon chains, two endings are needed: one for the double or triple bond and one for the hydroxyl group. The *-ol* suffix always comes last in the name; that is, unsaturated alcohols are named as *alkenols* or *alkynols*.

$$\overset{3}{C}H_2=\overset{2}{C}H-\overset{1}{C}H_2-OH$$

2-Propen-1-ol
(common name: allyl alcohol)

The contrast between IUPAC and common names for alcohols is as follows:

IUPAC (one word)

[alkanol]

ethanol

Common (two words)

[alkyl alcohol]

ethyl alcohol

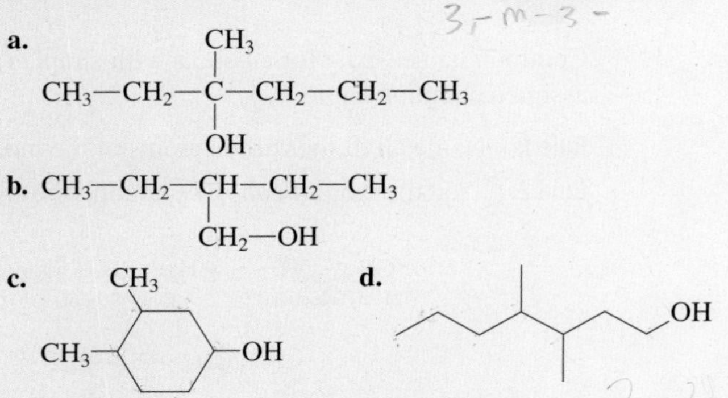

Solution

a. The longest carbon chain that contains the alcohol functional group has six carbons. When we change the *-e* to *-ol*, hexane becomes *hexanol*. Numbering the chain from the end nearest the —OH group identifies carbon number 3 as the location of both the —OH group and a methyl group. The complete name is 3-methyl-3-hexanol.

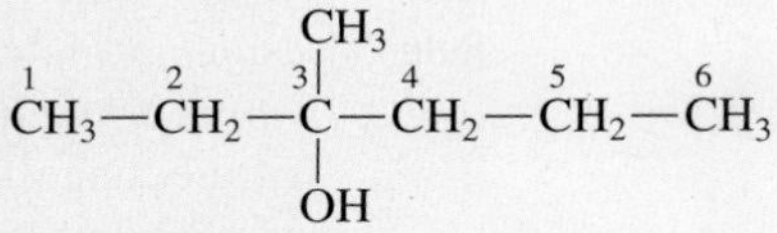

$$\overset{1}{C}H_3-\overset{2}{C}H_2-\overset{3}{\underset{\underset{OH}{|}}{\overset{\overset{CH_3}{|}}{C}}}-\overset{4}{C}H_2-\overset{5}{C}H_2-\overset{6}{C}H_3$$

b. The longest carbon chain containing the —OH group has four carbon atoms. It is numbered from the end closest to the —OH group as follows:

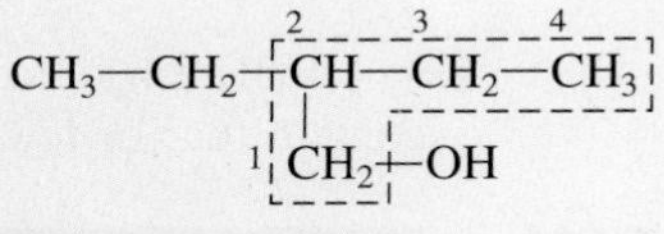

$$CH_3-CH_2\overset{2}{\underset{\underset{1}{|}}{\overset{3}{C}H}}-\overset{4}{C}H_2-CH_3$$
$$\overset{1}{C}H_2-OH$$

The base name is 1-butanol. The complete name is 2-ethyl-1-butanol.

c. This alcohol is a cyclohexanol. The carbon to which the —OH group is attached is assigned the number 1. The complete name for this alcohol is 3,4-dimethylcyclohexanol. Note that the number 1 is not part of the name.

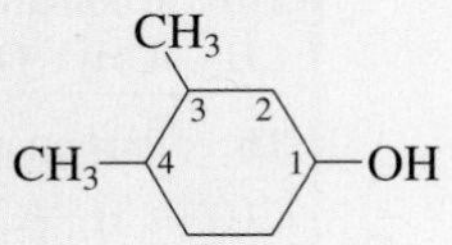

$$CH_3-\overset{\overset{CH_3}{|}}{\underset{4}{\bigcirc}}\overset{3\ \ 2}{\underset{1}{}}-OH$$

d. This alcohol is a dimethylheptanol. Numbering from right to left, the location of the hydroxyl group is 1, and locants for the methyl groups are 3 and 4. The complete IUPAC name is 3,4-dimethyl-1-heptanol.

(structure with carbons numbered 7,6,5,4,3,2,1 and OH)

Practice Exercise 14.1

Name the following alcohols utilizing IUPAC nomenclature rules.

a. $CH_3-\underset{\underset{CH_3}{|}}{CH}-\underset{\underset{OH}{|}}{CH}-CH_2-\underset{\underset{CH_3}{|}}{CH}-CH_3$

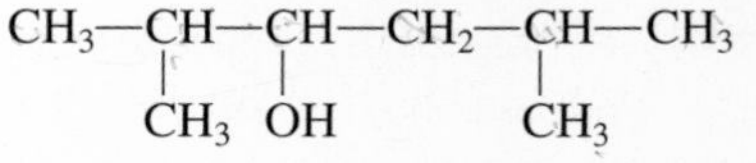

b. $CH_3-CH_2-\underset{\underset{CH_2-CH_2-OH}{|}}{CH}-CH_3$

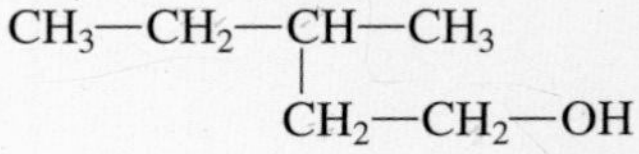

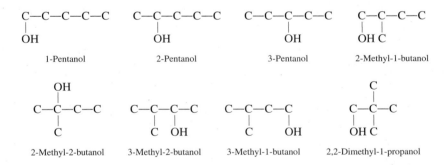

c.

d.

Answers: a. 2,5-Dimethyl-3-hexanol; **b.** 3-Methyl-1-pentanol; **c.** 1,2-Dimethylcyclopentanol; **d.** 5-Methyl-4-octanol

Alcohols with More Than One Hydroxyl Group

A hydroxyl group as a substituent in a molecule is called a hydroxy group; an *-oxy* rather than an *-oxyl* ending is used.

Polyhydroxy alcohols—alcohols that possess more than one hydroxyl group—can be named with only a slight modification of the preceding IUPAC rules. An alcohol in which two hydroxyl groups are present is named as a *diol,* one containing three hydroxyl groups is named as a *triol,* and so on. In these names for diols, triols, and so forth, the final *-e* of the parent alkane name is retained for pronunciation reasons.

$$CH_2-CH_2 \qquad CH_3-CH-CH_2 \qquad CH_2-CH-CH_2$$
$$|\quad\quad| \qquad\qquad\quad |\quad\quad | \qquad\qquad | \quad\quad |\quad\quad |$$
$$OH \quad OH \qquad\qquad OH \quad OH \qquad\quad OH \quad OH \quad OH$$

1,2-Ethanediol   1,2-Propanediol   1,2,3-Propanetriol

The first two of the preceding compounds have the common names *ethylene glycol* and *propylene glycol.* These two alcohols are synthesized, respectively, from the alkenes ethylene and propylene (Section 13.3); hence the common names.

14.4 ISOMERISM FOR ALCOHOLS

Addition of a functional group greatly increases constitutional isomer possibilities. There are 75 alkane isomers with the formula $C_{10}H_{22}$ and 507 alcohol isomers with the formula $C_{10}H_{21}OH$.

Constitutional isomerism is possible for alcohols containing three or more carbon atoms. As with alkenes (Section 13.5), both *skeletal* isomers and *positional* isomers are possible. For monohydroxy saturated alcohols, there are two C_3 isomers, four C_4 isomers, and eight C_5 isomers. Structures for the C_3 and C_4 isomers are found in Table 14.1. The C_5 isomers are

C—C—C—C—C   C—C—C—C—C   C—C—C—C—C   C—C—C—C
| | | | |
OH   OH   OH   OH C

1-Pentanol   2-Pentanol   3-Pentanol   2-Methyl-1-butanol

OH   C
| |
C—C—C—C   C—C—C—C   C—C—C—C   C—C—C
| | | | |
C   C OH   C OH   OH C

2-Methyl-2-butanol   3-Methyl-2-butanol   3-Methyl-1-butanol   2,2-Dimethyl-1-propanol

The three pentanols are positional isomers as are the four methylbutanols.

14.5 IMPORTANT COMMONLY ENCOUNTERED ALCOHOLS

In this section we consider the properties and uses of six commonly encountered alcohols: methyl, ethyl, and isopropyl alcohols (all monohydroxy alcohols), ethylene glycol and propylene glycol (both diols), and glycerol (a triol).

Figure 14.3 Racing cars at the Indianapolis Speedway are fueled with methyl alcohol.

Methyl alcohol poisoning is treated with ethyl alcohol, which ties up the enzyme that oxidizes methyl alcohol to its toxic metabolites. Ethyl alcohol has 10 times the affinity for the alcohol dehydrogenase enzyme that methyl alcohol has. This situation is considered further in Section 21.7.

Many people imagine ethanol to be relatively nontoxic and methanol to be extremely toxic. Actually, their toxicities differ by a factor of only 2. Typical fatal doses for adults are about 100 mL for methanol and about 200 mL for ethanol, although smaller doses of methanol may damage the optic nerve.

The alcohol content of strong alcoholic beverages is often stated in terms of proof. *Proof* is twice the percentage of alcohol. This system dates back to the seventeenth century and is based on the fact that a 50% (v/v) alcohol–water mixture will burn. Its flammability was *proof* that a liquor had not been watered down.

Methyl Alcohol (Methanol)

Methyl alcohol, with one carbon atom and one —OH group, is the simplest alcohol. This colorless liquid is a good fuel for internal combustion engines. Since 1965 all racing cars at the Indianapolis Speedway have been fueled with methyl alcohol (Figure 14.3). (Methyl alcohol fires are easier to put out than gasoline fires because water mixes with and dilutes methyl alcohol.) Methyl alcohol also has excellent solvent properties, and it is the solvent of choice for paints, shellacs, and varnishes.

Methyl alcohol is sometimes called *wood alcohol,* terminology that draws attention to an early method for its preparation—the heating of wood to a high temperature in the absence of air. Today, nearly all methyl alcohol is produced via the reaction between H_2 and CO.

$$CO + 2H_2 \xrightarrow[300°C - 400°C, \, 200 \, atm]{ZnO - Cr_2O_3} CH_3—OH$$

Drinking methyl alcohol is very dangerous. Within the human body, methyl alcohol is oxidized by the liver enzyme *alcohol dehydrogenase* to the toxic metabolites formaldehyde and formic acid.

$$CH_3—OH \xrightarrow[\text{dehydrogenase}]{\text{Alcohol}} \underset{\text{Formaldehyde}}{H—\overset{\overset{\textstyle O}{\|}}{C}—H} \xrightarrow[\text{oxidation}]{\text{Further}} \underset{\text{Formic acid}}{H—\overset{\overset{\textstyle O}{\|}}{C}—OH}$$

Formaldehyde can cause blindness (temporary or permanent). Formic acid causes acidosis (see the Chemical Connections feature on page 279). Ingesting as little as 1 oz (30 mL) of methyl alcohol can cause optic nerve damage.

Ethyl Alcohol (Ethanol)

Ethyl alcohol, the two-carbon monohydroxy alcohol, is the alcohol present in alcoholic beverages and is commonly referred to simply as alcohol or *drinking alcohol.* Like methyl alcohol, ethyl alcohol is oxidized in the human body by the liver enzyme *alcohol dehydrogenase.*

$$CH_3—CH_2—OH \xrightarrow[\text{dehydrogenase}]{\text{Alcohol}} \underset{\text{Acetaldehyde}}{CH_3—\overset{\overset{\textstyle O}{\|}}{C}—H} \xrightarrow[\text{oxidation}]{\text{Further}} \underset{\text{Acetic acid}}{CH_3—\overset{\overset{\textstyle O}{\|}}{C}—OH}$$

Acetaldehyde, the first oxidation product, is largely responsible for the symptoms of hangover. The odors of both acetaldehyde and acetic acid are detected on the breath of someone who has consumed a large amount of alcohol. Ethyl alcohol oxidation products are less toxic than those of methyl alcohol.

Long-term excessive use of ethyl alcohol may cause undesirable effects such as cirrhosis of the liver, loss of memory, and strong physiological addiction. Links have also been established between certain birth defects and the ingestion of ethyl alcohol by women during pregnancy (fetal alcohol syndrome).

Ethyl alcohol can be produced by yeast fermentation of sugars found in plant extracts (see Figure 14.4). The synthesis of ethyl alcohol in this manner, from grains such as corn, rice, and barley, is the reason why ethyl alcohol is often called *grain alcohol.*

$$\underset{\underset{\text{(glucose)}}{\text{Sugar}}}{C_6H_{12}O_6} \xrightarrow[\text{Fermentation}]{\text{Yeast}} \underset{\text{Ethyl alcohol}}{2CH_3—CH_2—OH} + 2CO_2$$

Fermentation is the process by which ethyl alcohol for alcoholic beverages is produced. The maximum concentration of ethyl alcohol obtainable by fermentation is about 18% (v/v), because yeast enzymes cannot function in stronger alcohol solutions. Alcoholic beverages with a higher concentration of alcohol than this are prepared by either distillation

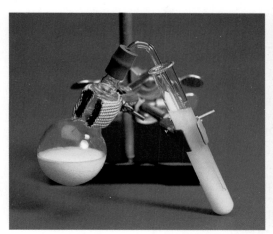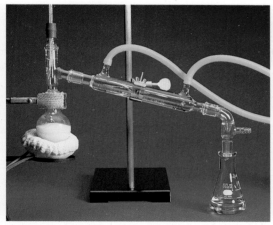

Figure 14.4 An experimental setup for preparing ethyl alcohol by fermentation. (a) A small amount of yeast has been added to the aqueous sugar solution in the flask. Yeast enzymes catalyze the decomposition of sugar to ethanol and carbon dioxide, CO_2. The CO_2 is bubbling through lime water, $Ca(OH)_2$, producing calcium carbonate, $CaCO_3$. (b) More concentrated ethanol is produced from the solution in the flask by collecting the fraction that boils at about 78°C. (c) Concentrated ethanol (50% v/v) burns when it is ignited.

or fortification with alcohol obtained by the distillation of another fermentation product. Table 14.2 lists the alcohol content of common alcoholic beverages and of selected common household products and over-the-counter drug products.

Denatured alcohol is ethyl alcohol that has been rendered unfit to drink by the addition of small amounts of toxic substances (denaturing agents). Almost all of the ethyl alcohol used for industrial purposes is denatured alcohol.

Most ethyl alcohol used in industry is prepared from ethene via a hydration reaction (Section 13.9).

$$CH_2{=}CH_2 + H_2O \xrightarrow{\text{Catalyst}} CH_3{-}CH_2{-}OH$$

The reaction produces a product that is 95% alcohol and 5% water. In applications where water does interfere with its use, the mixture is treated with a dehydrating agent to produce 100% ethyl alcohol. Such alcohol, with all traces of water removed, is called *absolute alcohol*.

Isopropyl Alcohol (2-Propanol)

The "medicinal" odor associated with doctors' offices is usually that of isopropyl alcohol.

Isopropyl alcohol is one of two three-carbon monohydroxy alcohols; the other is propyl alcohol. A 70% isopropyl alcohol–30% water solution is marketed as *rubbing alcohol*. Isopropyl alcohol's rapid evaporation rate creates a dramatic cooling effect when it is applied to the skin, hence its use for alcohol rubs to combat high body temperature. It also finds use in cosmetics formulations such as after-shave lotion and hand lotions.

Isopropyl alcohol has a bitter taste. Its toxicity is twice that of ethyl alcohol, but it causes few fatalities because it often induces vomiting and thus doesn't stay down long enough to be fatal. In the body it is oxidized to acetone.

$$\underset{\text{Isopropyl alcohol}}{CH_3{-}\underset{\underset{OH}{|}}{CH}{-}CH_3} \xrightarrow[\text{dehydrogenase}]{\text{Alcohol}} \underset{\text{Acetone}}{CH_3{-}\overset{\overset{O}{\|}}{C}{-}CH_3}$$

Large amounts (about 150 mL) of ingested isopropyl alcohol can be fatal; death occurs from paralysis of the central nervous system.

TABLE 14.2
Ethyl Alcohol Content (volume percent) of Common Alcoholic Beverages, Household Products, and Over-the-Counter Drugs

Product Type	Product	Volume Percent Ethyl Alcohol
Alcoholic Beverages	beer	3.2–9
	wine (unfortified)	12
	brandy	40–45
	whiskey	45–55
	rum	45
Flavorings	vanilla extract	35
	almond extract	50
Cough and Cold Remedies	Pertussin Plus	25
	Nyquil	25
	Dristan	12
	Vicks 44	10
	Robitussin, DM	1.4
Mouthwashes	Listerine	25
	Scope	18
	Colgate 100	17
	Cepacol	14
	Lavoris	5

Ethylene Glycol (1,2-Ethanediol) and Propylene Glycol (1,2-Propanediol)

Ethylene glycol and propylene glycol are the two simplest alcohols possessing two —OH groups. Besides being diols, they are also classified as glycols. A **glycol** *is a diol in which the two —OH groups are on adjacent carbon atoms.*

$$CH_2-CH_2 \qquad CH_3-CH-CH_2$$
$$\;\;|\;\;\;\;\;\;|\qquad\qquad\qquad\;\;\;|\;\;\;\;\;\;|$$
$$\;\;OH\;\;\;OH\qquad\qquad\quad\;OH\;\;\;OH$$

Ethylene glycol   Propylene glycol

> The ethylene glycol and propylene glycol used in antifreeze formulations are colorless and odorless; the color and odor of antifreezes come from additives for rust protection and the like.

Both of these glycols are colorless, odorless, high-boiling liquids that are completely miscible with water. Their major uses are as the main ingredient in automobile "year-round" antifreeze and airplane "de-icers" (Figure 14.5) and as a starting material for the manufacture of polyester fibers (Section 16.18).

Ethylene glycol is extremely toxic when ingested. In the body, liver enzymes oxidize it to oxalic acid.

> Ethylene glycol and propylene glycol are synthesized from ethylene and propylene, respectively, hence their common names.

Figure 14.5 Ethylene glycol is the major ingredient in airplane "de-icers."

$$HO-CH_2-CH_2-OH \xrightarrow[\text{enzymes}]{\text{Liver}} HO-\underset{\displaystyle O}{\overset{\displaystyle O}{C}}-\underset{\displaystyle O}{\overset{\displaystyle O}{C}}-OH$$

Ethylene glycol   Oxalic acid

Oxalic acid, as a calcium salt, crystallizes in the kidneys, which leads to renal problems.

Propylene glycol, on the other hand, is essentially nontoxic and has been used as a solvent for drugs. Like ethylene glycol, it is oxidized by liver enzymes; however, pyruvic acid, its oxidation product, is a compound normally found in the human body, being an intermediate in carbohydrate metabolism (Chapter 24).

$$CH_3-CH-CH_2 \xrightarrow[\text{enzymes}]{\text{Liver}} CH_3-\overset{\displaystyle O}{C}-\overset{\displaystyle O}{C}-OH$$
$$\;\;\;\;\;\;|\;\;\;\;\;\;|$$
$$\;\;\;\;\;OH\;\;\;OH$$

Propylene glycol   Pyruvic acid

Figure 14.6 Glycerol is often called biological antifreeze. For survival in Arctic and northern winters, many fish and insects, including the common housefly, produce large amounts of glycerol that dissolve in their blood, thereby lowering the freezing point of the blood.

Glycerol (1,2,3-Propanetriol)

Glycerol, which is often also called glycerin, is a clear, thick liquid that has the consistency of honey. Its molecular structure involves three —OH groups on three different carbon atoms.

Glycerol is normally present in the human body because it is a product of fat metabolism. It is present, in combined form, in all animal fats and vegetable oils (Section 19.4). In some Arctic species, glycerol functions as a "biological antifreeze" (see Figure 14.6).

Because glycerol has a great affinity for water vapor (moisture), it is often added to pharmaceutical preparations such as skin lotions and soap. Florists sometimes use glycerol on cut flowers to help retain water and maintain freshness. Its lubricative properties also make it useful in shaving creams and in applications such as glycerol suppositories for rectal administration of medicines. It is used in candies and icings as a retardant for preventing sugar crystallization.

14.6 PHYSICAL PROPERTIES OF ALCOHOLS

Alcohol molecules have both polar and nonpolar character. The hydroxyl groups present are polar, and the alkyl (R) group present is nonpolar.

Nonpolar portion↘   ↙Polar portion
$$\overline{CH_3—CH_2—CH_2}—\overline{OH}$$

The physical properties of an alcohol depend on whether the polar or the nonpolar portion of its structure "dominates." Factors that determine this include the *length* of the nonpolar carbon chain present and the *number* of polar hydroxyl groups present (see Figure 14.7).

Boiling Points and Water Solubilities

Figure 14.8a shows that the boiling point for 1-alcohols, unbranched-chain alcohols with an —OH group on an end carbon, increases as the length of the carbon chain increases. This trend results from increasing London forces (Sections 7.13 and 12.16) with increasing carbon chain length. Alcohols with more than one hydroxyl group present have significantly higher boiling points (bp) than their monohydroxy counterparts.

$$CH_3—CH_2—CH_2 \atop OH \qquad CH_3—CH—CH_2 \atop OH \quad OH \qquad CH_2—CH—CH_2 \atop OH \quad OH \quad OH$$

bp = 97°C   bp = 188°C   bp = 290°C

Figure 14.7 Space-filling molecular models showing the nonpolar (green) and polar (pink) parts of methanol and 1-octanol. (a) The polar hydroxyl functional group dominates the physical properties of methanol. The molecule is completely soluble in water (polar) but only partially so in hexane (nonpolar). (b) Conversely, the nonpolar portion of 1-octanol dominates its physical properties; it is infinitely soluble in hexane and has limited solubility in water.

CH₃ — OH
Nonpolar   Polar
(a) Methanol

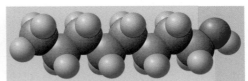

CH₃CH₂CH₂CH₂CH₂CH₂CH₂CH₂ — OH
Nonpolar   Polar
(b) 1- Octanol

CHEMICAL Connections

Menthol: A Useful Naturally Occurring Terpene Alcohol

Menthol is a naturally occurring terpene (Section 13.7) alcohol with a pleasant, minty odor. Its IUPAC name is 2-isopropyl-5-methylcyclohexanol.

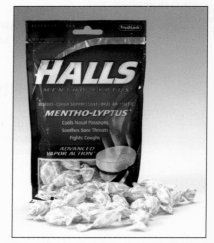

In the pure state, menthol is a white crystalline solid with a melting point of 41°C to 43°C. It can be obtained from peppermint oil and can also be prepared synthetically.

Topical application of menthol to the skin causes a refreshing, cooling sensation followed by a slight burning-and-prickling sensation. Its mode of action is that of a *differential* anesthetic. It stimulates the receptor cells in the skin that normally respond to cold to give a sensation of coolness that is unrelated to body temperature. (This cooling sensation is particularly noticeable in the respiratory tract when low concentrations of menthol are inhaled.) At the same time as cooling is perceived, menthol can depress the nerves for pain reception.

Numerous products contain menthol.

■ Throat sprays and lozenges containing menthol temporarily soothe inflamed mucous surfaces of the nose and throat. Lozenges contain 2–20 milligrams of menthol per wafer.

■ Cough drops and cigarettes of the "mentholated" type use menthol for its counterirritant effect.

■ Pre-electric shave preparations and aftershave lotions often contain menthol. A concentration of only 0.1% (m/v) gives ample cooling to allay the irritation of a "close" shave.

■ Many dermatologic preparations contain menthol as an anti-pruritic (anti-itching agent).

■ Chest-rub preparations containing menthol include Ben Gay [7% (m/v)] and Mentholatum [6% (m/v)].

■ Artificial mint flavors have menthol as an ingredient. Several toothpastes and mouthwashes use menthol as a flavoring agent.

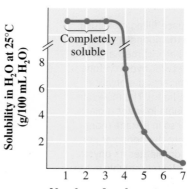

Many kinds of cough drops contain menthol as a counterirritant.

This boiling-point trend is related to increased hydrogen bonding between alcohol molecules (to be discussed shortly). Figure 14.9 is a physical-state summary for unbranched 1-alcohols and unsubstituted cycloalcohols with eight or fewer carbon atoms.

Small monohydroxy alcohols are soluble in water in all proportions. As carbon chain length increases beyond three carbons, solubility in water rapidly decreases (Figure 14.8b) because of the increasingly nonpolar character of the alcohol. Alcohols with two —OH groups present are more soluble in water than their counterparts with only one —OH

Figure 14.8 (a) Boiling points and (b) solubilities in water of selected 1-alcohols.

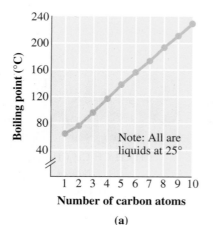

Note: All are liquids at 25°

(a)

Completely soluble

(b)

TABLE 14.3
A Comparison of Selected Physical Properties of Alcohols with Alkane Counterparts of Similar Molecular Mass

Type of Compound	Compound	Structure	Molecular Mass (amu)	Boiling Point (°C)	Solubility in Water
alkane	ethane	$CH_3—CH_3$	30	−89	slight solubility
alcohol	methanol	$CH_3—OH$	32	65	unlimited solubility
alkane	propane	$CH_3—CH_2—CH_3$	44	−42	slight solubility
alcohol	ethanol	$CH_3—CH_2—OH$	46	78	unlimited solubility
alkane	butane	$CH_3—CH_2—CH_2—CH_3$	58	−1	slight solubility
alcohol	1-propanol	$CH_3—CH_2—CH_2—OH$	60	97	unlimited solubility
alcohol	2-propanol	$CH_3—CH—OH$ $\quad\quad\;\;	$ $\quad\quad CH_3$	60	83	unlimited solubility

group. Increased hydrogen bonding is responsible for this. Diols containing as many as seven carbon atoms show appreciable solubility in water.

Alcohols and Hydrogen Bonding

A comparison of the properties of alcohols with their alkane counterparts (Table 14.3) shows that

1. Alcohols have *higher* boiling points than alkanes of similar molecular mass.
2. Alcohols have much *higher* solubility in water than alkanes of similar molecular mass.

The differences in physical properties between alcohols and alkanes are related to hydrogen bonding. Because of their hydroxyl group(s), alcohols can participate in hydrogen bonding, whereas alkanes cannot. Hydrogen bonding between alcohol molecules (see Figure 14.10) is similar to that which occurs between water molecules (Section 7.13).

Extra energy is needed to overcome alcohol–alcohol hydrogen bonds before alcohol molecules can enter the vapor phase. Hence alcohol boiling points are higher than those for the corresponding alkanes (where no hydrogen bonds are present).

Alcohol molecules can also hydrogen-bond to water molecules (see Figure 14.11). The formation of such hydrogen bonds explains the solubility of small alcohol molecules

Unbranched 1-Alcohols			
C_1	C_3	C_5	C_7
C_2	C_4	C_6	C_8

Unsubstituted Cycloalcohols			
⨉	C_3	C_5	C_7
⨉	C_4	C_6	C_8

☐ Liquid

Figure 14.9 A physical-state summary for unbranched 1-alcohols and unsubstituted cycloalcohols at room temperature and pressure.

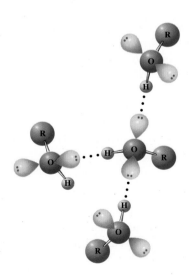

Figure 14.10 Alcohol boiling points are higher than those of the corresponding alkanes because of alcohol-alcohol hydrogen bonding.

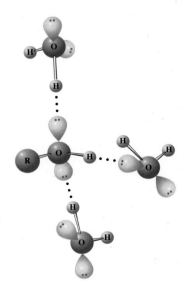

Figure 14.11 Because of hydrogen bonding between alcohol molecules and water molecules, alcohols of small molecular mass have unlimited solubility in water.

in water. As the alcohol chain length increases, alcohols become more alkane-like (nonpolar), and solubility decreases.

14.7 PREPARATION OF ALCOHOLS

Alcohols are intermediate products in the metabolism of both carbohydrates (Chapter 24) and fats (Chapter 25). In these metabolic processes, both addition of water to a carbon–carbon double bond and addition of hydrogen to a carbon–oxygen double bond lead to the introduction of the alcohol functional group into a biomolecule.

A general method for preparing alcohols—the hydration of alkenes—was discussed in the previous chapter (Section 13.9). Alkenes react with water (an unsymmetrical addition agent) in the presence of sulfuric acid (the catalyst) to form an alcohol. Markovnikov's rule is used to determine the predominant alcohol product.

Another method of synthesizing alcohols involves the addition of H_2 to a carbon–oxygen double bond (a carbonyl group, $\diagdown C{=}O$). (The carbonyl group is a functional group that will be discussed in detail in Chapter 15.) A carbonyl group behaves very much like a carbon–carbon double bond when it reacts with H_2 under the proper conditions. As a result of H_2 addition, the oxygen of the carbonyl group is converted to an —OH group.

Aldehyde (Section 15.3)   Alcohol

Ketone (Section 15.3)   Alcohol

14.8 CLASSIFICATION OF ALCOHOLS

Prior to considering chemical reactions of alcohols (Section 14.9), we consider a classification system for alcohols that is often needed when predicting the products in a chemical reaction that involves an alcohol.

Pronounce 1° as "primary," 2° as "secondary," and 3° as "tertiary."

Methyl alcohol, CH_3—OH, an alcohol in which the hydroxyl-bearing carbon atom is attached to three hydrogen atoms, does not fit any of the alcohol classification definitions. It is usually grouped with the primary alcohols because its reactions are similar to theirs.

Alcohols are classified as primary (1°), secondary (2°), or tertiary (3°) depending on the number of carbon atoms bonded to the carbon atom that bears the hydroxyl group. A **primary alcohol** *is an alcohol in which the hydroxyl-bearing carbon atom is bonded to only one other carbon atom.* A **secondary alcohol** *is an alcohol in which the hydroxyl-bearing carbon atom is bonded to two other carbon atoms.* A **tertiary alcohol** *is an alcohol in which the hydroxyl-bearing carbon atom is bonded to three other carbon atoms.* Chemical reactions of alcohols often depend on alcohol class (1°, 2°, or 3°).

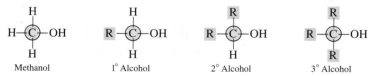

Methanol   1° Alcohol   2° Alcohol   3° Alcohol

Although all alcohols are able to participate in hydrogen bonding (Section 14.6), increasing the number of R groups around the carbon atom bearing the OH group decreases the extent of hydrogen bonding. This effect, called stearic hindrance, becomes particularly important when the R groups are large. Thus, 1° alcohols are best able to hydrogen-bond and 3° alcohols are least able to hydrogen-bond.

● EXAMPLE 14.2

Classifying Alcohols as Primary, Secondary, or Tertiary Alcohols

▶ Classify each of the following alcohols as a primary, secondary, or tertiary alcohol.

$$\text{a. } CH_3-CH_2-CH_2-OH \qquad \text{b. } CH_3-CH_2-\underset{\underset{CH_3}{|}}{\overset{\overset{CH_3}{|}}{C}}-OH$$

$$\text{c. } CH_3-\overset{\overset{CH_3}{|}}{CH}-\underset{\underset{OH}{|}}{CH}-\overset{\overset{CH_3}{|}}{CH}-CH_3 \qquad \text{d. } \underset{}{\overset{}{\bigcirc}}-OH$$

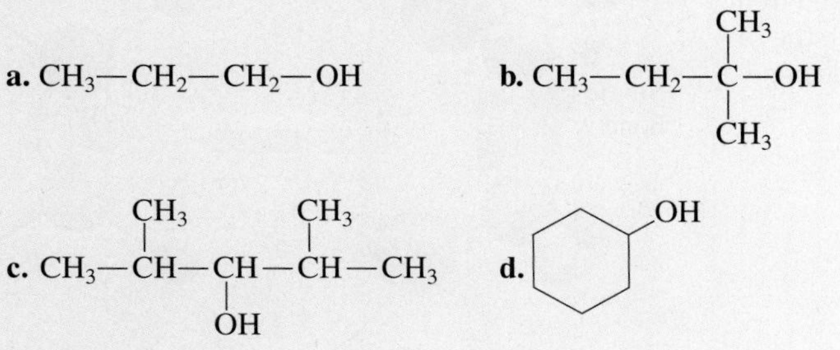

Solution

a. This is a primary alcohol. The carbon atom to which the —OH group is attached is bonded to only one other carbon atom.
b. This is a tertiary alcohol. The carbon atom bearing the —OH group is bonded to three other carbon atoms.
c. This is a secondary alcohol. The hydroxyl-bearing carbon atom is bonded to two other carbon atoms.
d. This is a secondary alcohol. The ring carbon atom to which the —OH group is attached is bonded to two other ring carbon atoms.

Practice Exercise 14.2

Classify each of the following alcohols as a primary, secondary, or tertiary alcohol.

$$\text{a. } CH_3-\underset{\underset{OH}{|}}{CH}-CH_3 \qquad \text{b. } CH_3-\underset{\underset{CH_3}{|}}{\overset{\overset{CH_3}{|}}{C}}-CH_2-OH$$

$$\text{c. } CH_3-\underset{\underset{CH_3}{|}}{CH}-\underset{\underset{CH_3}{|}}{CH}-OH \qquad \text{d. } \underset{}{\overset{OH}{\bigcirc}}-CH_3$$

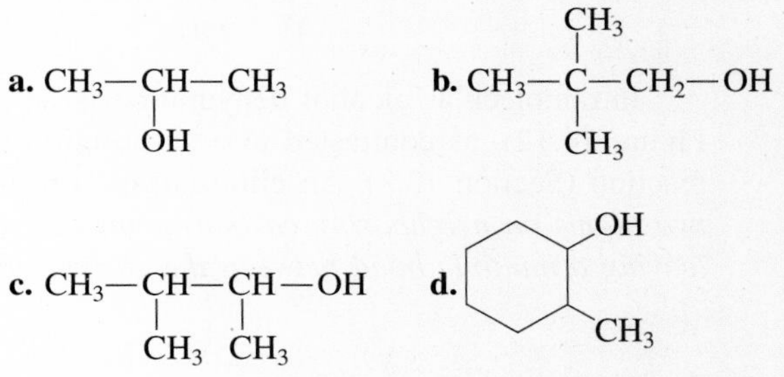

Answers: a. Secondary; **b.** Primary; **c.** Secondary; **d.** Secondary

14.9 CHEMICAL REACTIONS OF ALCOHOLS

Of the many chemical reactions that alcohols undergo, we consider four in this section: (1) combustion, (2) dehydration, (3) oxidation, and (4) halogenation.

Combustion

As we have seen in the previous two chapters, hydrocarbons of all types undergo combustion in air to produce carbon dioxide and water. Alcohols are also flammable; as with hydrocarbons, the combustion products are carbon dioxide and water. Methyl alcohol is the fuel of choice for racing cars (Section 14.5). Oxygenated gasoline, which is used in winter in many areas of the United States because it burns "cleaner," contains ethyl alcohol as one of the "oxygenates."

Intramolecular Alcohol Dehydration

A **dehydration reaction** is *a chemical reaction in which the components of water (H and OH) are removed from a single reactant or from two reactants (H from one and OH from*

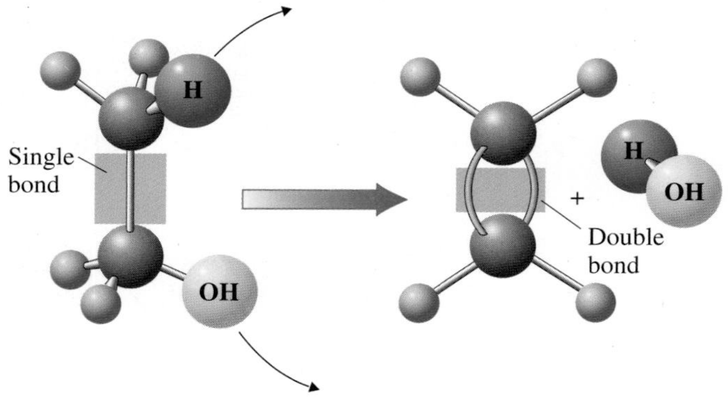

Figure 14.12 In an intramolecular alcohol dehydration, the components of water (H and OH) are removed from neighboring carbon atoms with the resultant introduction of a double bond into the molecule.

the other). In *intramolecular* dehydration, both water components are removed from the same molecule.

Reaction conditions for the intramolecular dehydration of an alcohol are a temperature of 180°C and the presence of sulfuric acid (H_2SO_4) as a catalyst. The dehydration product is an alkene.

Ease of alcohol dehydration depends on alcohol classification. Primary alcohols are the most difficult to dehydrate, requiring temperatures of around 180°C. Secondary alcohol dehydration occurs at lower temperatures and tertiary alcohols dehydrate at temperatures slightly above room temperature.

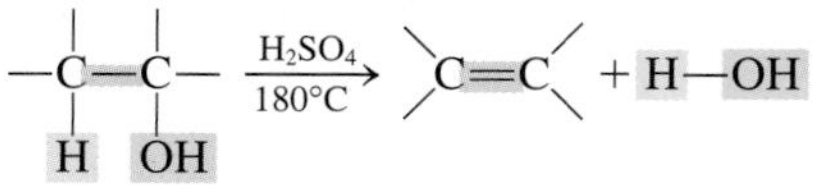

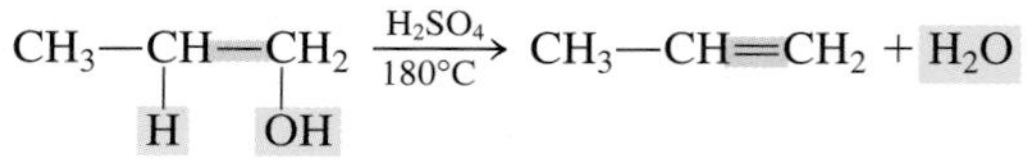

$$-\overset{|}{\underset{H}{C}}-\overset{|}{\underset{OH}{C}}- \xrightarrow[180°C]{H_2SO_4} \hspace{0.5em} \diagup\!\!\!\!C\!\!=\!\!C\diagdown\!\!\!\! + H-OH$$

$$CH_3-\overset{|}{\underset{H}{CH}}-\overset{|}{\underset{OH}{CH_2}} \xrightarrow[180°C]{H_2SO_4} CH_3-CH\!\!=\!\!CH_2 + H_2O$$

Intramolecular alcohol dehydration is an example of an *elimination reaction* (see Figure 14.12), as contrasted to a substitution reaction (Section 12.17) and an addition reaction (Section 13.9). An **elimination reaction** *is a reaction in which two groups or two atoms on neighboring carbon atoms are removed, or eliminated, from a molecule, leaving a multiple bond between the carbon atoms.*

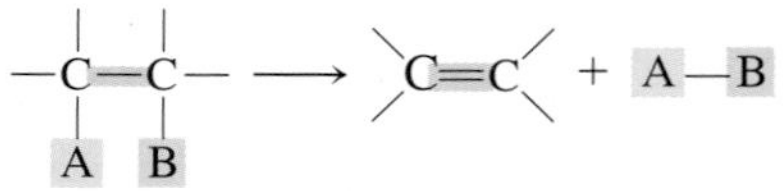

$$-\overset{|}{\underset{A}{C}}-\overset{|}{\underset{B}{C}}- \longrightarrow \hspace{0.5em} \diagup\!\!\!\!C\!\!=\!\!C\diagdown\!\!\!\! + A-B$$

What occurs in an elimination reaction is the reverse of what occurs in an addition reaction.

Dehydration of an alcohol can result in the production of more than one alkene product. This happens when there is more than one neighboring carbon atom from which hydrogen loss can occur. Dehydration of 2-butanol produces two alkenes.

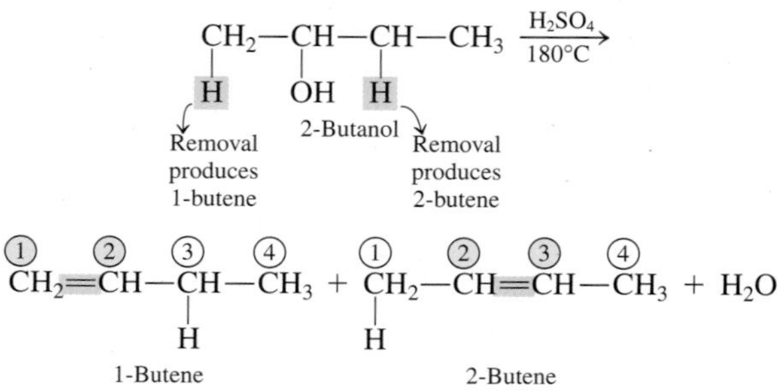

Dehydration of alcohols to form carbon–carbon double bonds occurs in several metabolic pathways in living systems, such as the citric acid cycle (Section 23.6) and the β oxidation pathway (Section 25.4). In these biochemical dehydrations, enzymes serve as catalysts instead of acids, and the reaction temperature is 37°C instead of the elevated temperatures required in the laboratory.

$$CH_2-\overset{|}{\underset{H}{CH}}-\overset{|}{\underset{OH}{CH}}-CH_3 \xrightarrow[180°C]{H_2SO_4}$$

Removal produces 1-butene   2-Butanol   Removal produces 2-butene

$$\overset{①}{CH_2}\!\!=\!\!\overset{②}{CH}-\overset{③}{\underset{H}{CH}}-\overset{④}{CH_3} + \overset{①}{CH_2}-\overset{②}{\underset{H}{CH}}\!\!=\!\!\overset{③}{CH}-\overset{④}{CH_3} + H_2O$$

1-Butene   2-Butene

Alexander Zaitsev (1841–1910), a nineteenth-century Russian chemist, studied at the University of Paris and then returned to his native Russia to become a professor of chemistry at the University of Kazan. His surname is pronounced "zait-zeff."

The dominant product can be predicted using Zaitsev's rule, named after the Russian chemist Alexander Zaitsev. **Zaitsev's rule** states that *the major product in an intramolecular alcohol dehydration reaction is the alkene that has the greatest number of alkyl groups attached to the carbon atoms of the double bond.* In the preceding reaction, 2-butene (with two alkyl groups) is favored over 1-butene (with one alkyl group).

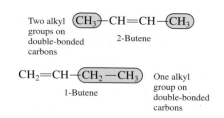

Two alkyl groups on double-bonded carbons

CH_3—CH=CH—CH_3 2-Butene

CH_2=CH—CH_2—CH_3 One alkyl group on double-bonded carbons

1-Butene

Alkene formation via intramolecular alcohol *dehydration* is the "reverse reaction" to the reaction for preparing an alcohol through *hydration* of an alkene (Section 13.9). This relationship can be diagrammed as follows:

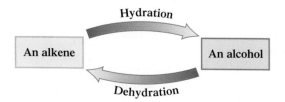

Hydration

An alkene   **An alcohol**

Dehydration

This "reverse reaction" situation illustrates the fact that many organic reactions can go both forward or backward, depending on reaction conditions. Noting relationships such as this helps in keeping track of the numerous reactions that hydrocarbon derivatives undergo. These two "reverse reactions" actually involve an equilibrium situation.

$$C=C + H_2O \underset{\text{Dehydration}}{\overset{\text{Hydration}}{\rightleftharpoons}} -\underset{H}{\overset{|}{C}}-\underset{OH}{\overset{|}{C}}-$$

An alkene   An alcohol

Whether the forward reaction (alcohol formation) or the reverse reaction (alkene formation) is favored depends on experimental conditions. The favored direction for the reaction can be predicted using Le Châtelier's principle (Section 9.9).

1. The addition of water favors alcohol formation.
2. The removal of water favors alkene formation.

Experimental conditions for alcohol formation involve the use of a *dilute* sulfuric acid solution as a catalyst. *Concentrated* sulfuric acid (a dehydrating agent) as well as higher temperatures are used for alkene formation. Dilute acid solutions are mainly water; concentrated acid solutions have less water and heat also removes water.

Intermolecular Alcohol Dehydration

At a lower temperature (140°C) than that required for alkene formation (180°C), an *inter*molecular rather than an *intra*molecular alcohol dehydration process can occur to produce an ether—a compound with the general structure R—O—R. In such ether formation, two alcohol molecules interact, an H atom being lost from one and an —OH group from the other. The resulting "leftover" portions of the two alcohol molecules join to form the ether. This reaction, which gives useful yields only for primary alcohol reactants (2° and 3° alcohols yield predominantly alkenes), can be written as

$$-\overset{|}{\underset{|}{C}}-O-H + H-O-\overset{|}{\underset{|}{C}}- \xrightarrow[140°C]{H_2SO_4} -\overset{|}{\underset{|}{C}}-O-\overset{|}{\underset{|}{C}}- + H-O-H$$

$$CH_3-CH_2-O-H + H-O-CH_2-CH_3 \xrightarrow[140°C]{H_2SO_4}$$

Ethanol   Ethanol

$$CH_3-CH_2-O-CH_2-CH_3 + H_2O$$

An alternative way of expressing Zaitsev's rule is "Hydrogen atom loss, during intramolecular alcohol dehydration to form an alkene, will occur preferentially from the carbon atom (adjacent to the hydroxyl-bearing carbon) that already has the fewest hydrogen atoms."

Ethers, like alcohols, constitute a many-membered important class of oxygen-containing hydrocarbon derivatives. An extended discussion of ethers, whose general properties are much different from those of alcohols, is found in Sections 14.15 through 14.19 of this chapter.

The preceding reaction is an example of *condensation*. A **condensation reaction** *is a chemical reaction in which two molecules combine to form a larger one while liberating a small molecule, usually water.* In this case, two alcohol molecules combine to give an ether and water.

EXAMPLE 14.3

Predicting the Reactant in an Alcohol Dehydration Reaction When Given the Product

Identify the alcohol reactant needed to produce each of the following compounds as the *major product* of an alcohol dehydration reaction.

a. Alcohol $\xrightarrow[180°C]{H_2SO_4}$ $CH_3-CH{=}CH-CH_3$

b. Alcohol $\xrightarrow[180°C]{H_2SO_4}$ $CH_2{=}CH-CH-CH_3$
$\qquad\qquad\qquad\qquad\qquad\quad |$
$\qquad\qquad\qquad\qquad\qquad\ CH_3$

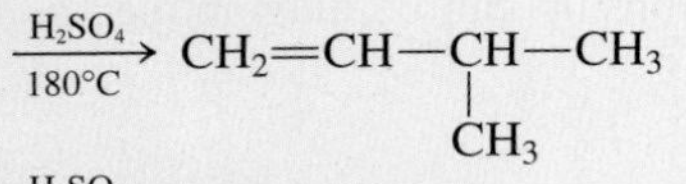

c. Alcohol $\xrightarrow[140°C]{H_2SO_4}$ $CH_3-CH-CH_2-O-CH_2-CH-CH_3$
$\qquad\qquad\qquad\qquad\qquad\quad |\qquad\qquad\qquad\qquad\ |$
$\qquad\qquad\qquad\qquad\qquad\ CH_3\qquad\qquad\qquad\quad CH_3$

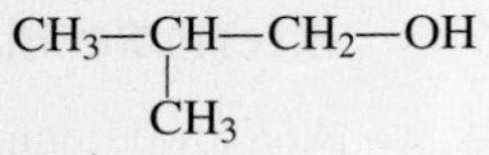

The following is a summary of products obtained from alcohol dehydration reactions using H_2SO_4 as a catalyst.

Primary alcohol $\xrightarrow[140°C]{180°C}$ alkene / ether

Secondary alcohol $\xrightarrow[140°C]{180°C}$ alkene / alkene

Tertiary alcohol $\xrightarrow[140°C]{180°C}$ alkene / alkene

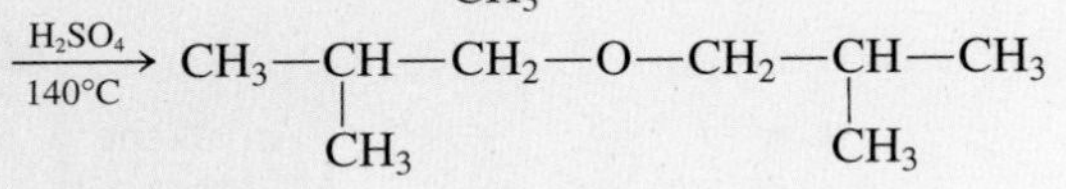

Solution

a. Both carbon atoms of the double bond are equivalent to each other. Add an H atom to one carbon atom of the double bond and an OH group to the other carbon atom of the double bond. It does not matter which goes where; you get the same molecule either way.

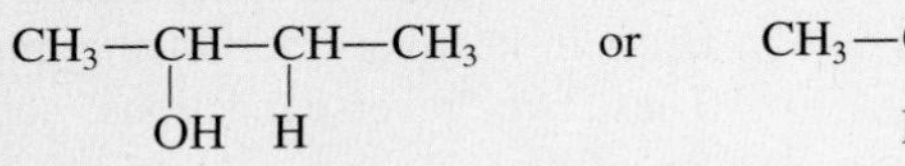

$CH_3-CH-CH-CH_3$ or $CH_3-CH-CH-CH_3$
$\qquad\ \ |\quad\ |$ $\qquad\qquad\qquad\qquad\ |\quad\ \ |$
$\qquad\ OH\ H$ $\qquad\qquad\qquad\qquad\ H\quad OH$

b. There are two possible parent alcohols: one with an —OH group on carbon 1 and the other with an —OH group on carbon 2.

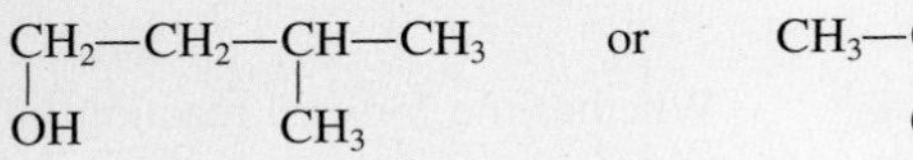

$CH_2-CH_2-CH-CH_3$ or $CH_3-CH-CH-CH_3$
$\ |\qquad\qquad\quad |$ $\qquad\qquad\quad\ \ |\quad\ \ |$
$OH\qquad\qquad\ CH_3$ $\qquad\qquad\ OH\ \ CH_3$

Using the reverse of Zaitsev's rule, we find that the hydrogen atom will go back on the double-bonded carbon that bears the most alkyl groups.

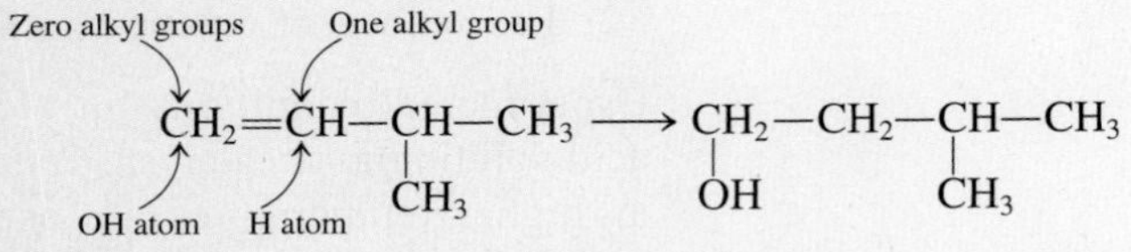

Zero alkyl groups   One alkyl group

$CH_2{=}CH-CH-CH_3 \longrightarrow CH_2-CH_2-CH-CH_3$
$\qquad\qquad\quad\ |$ $\qquad\qquad\quad |\qquad\qquad\ |$
$\qquad\qquad\ CH_3$ $\qquad\qquad\ OH\qquad\ CH_3$
OH atom   H atom

c. This is an ether. The primary alcohol from which the ether was formed will have the same alkyl group present as is in the ether. Thus the alcohol is

$CH_3-CH-CH_2-OH$
$\qquad\quad\ |$
$\qquad\ CH_3$

Practice Exercise 14.3

Identify the starting alcohol from which each of the following products was obtained by an alcohol dehydration reaction.

a. Alcohol $\xrightarrow[180°C]{H_2SO_4}$ $CH_2{=}CH-CH_2-CH_3$

b. Alcohol $\xrightarrow[180°C]{H_2SO_4}$ $CH_3-C{=}C-CH_3$
$\qquad\qquad\qquad\qquad\qquad\qquad |\quad\ |$
$\qquad\qquad\qquad\qquad\qquad\ CH_3\ CH_3$

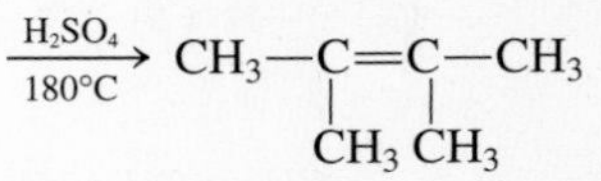

c. Alcohol $\xrightarrow[140°C]{H_2SO_4}$ $CH_3-CH_2-CH_2-O-CH_2-CH_2-CH_3$

Answers: **a.** 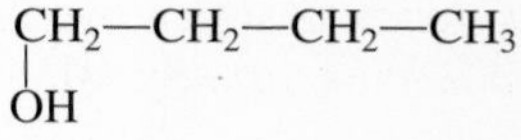 $CH_2-CH_2-CH_2-CH_3$ **b.** 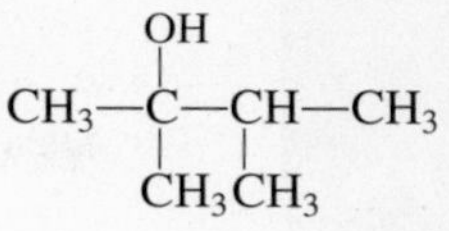 $CH_3-\overset{\overset{\displaystyle OH}{|}}{\underset{\underset{\displaystyle CH_3}{|}}{C}}-\overset{\overset{}{}}{\underset{\underset{\displaystyle CH_3}{|}}{CH}}-CH_3$ **c.** $CH_3-CH_2-CH_2-OH$

Oxidation

Before discussing alcohol oxidation reactions, we consider a new method for recognizing when oxidation and reduction have occurred in a chemical reaction.

The processes of oxidation and reduction were considered in Section 9.2 in the context of inorganic, rather than organic, reactions. Oxidation numbers were used to characterize oxidation–reduction processes. This same technique could be used in characterizing oxidation–reduction processes involving organic compounds, but it is not. Formal use of the oxidation number rules with organic compounds is usually cumbersome because of the many carbon and hydrogen atoms present; often, fractional oxidation numbers for carbon result.

A better approach for organic redox reactions is to use the following set of operational rules instead of oxidation numbers.

1. An *organic oxidation* is an oxidation that increases the number of C—O bonds and/or decreases the number of C—H bonds.
2. An *organic reduction* is a reduction that decreases the number of C—O bonds and/or increases the number of C—H bonds.

Note that these operational definitions for oxidation and reduction are "opposites." This is just as it should be; oxidation and reduction are "opposite" processes.

Some alcohols readily undergo oxidation with mild oxidizing agents; others are resistant to oxidation with these same oxidizing agents. Primary and secondary alcohols, but not tertiary alcohols, readily undergo oxidation in the presence of mild oxidizing agents to produce compounds that contain a carbon–oxygen double bond (aldehydes, ketones, and carboxylic acids). A number of different oxidizing agents can be used for the oxidation, including potassium permanganate ($KMnO_4$), potassium dichromate ($K_2Cr_2O_7$), and chromic acid (H_2CrO_4).

The net effect of the action of a mild oxidizing agent on a primary or secondary alcohol is the removal of two hydrogen atoms from the alcohol. One hydrogen comes from the —OH group, the other from the carbon atom to which the —OH group is attached. This H removal generates a carbon–oxygen double bond.

> This chemical reaction is consistent with the operational definition given earlier in this section for an organic oxidation. A new C—O bond is formed and a C—H bond is broken.

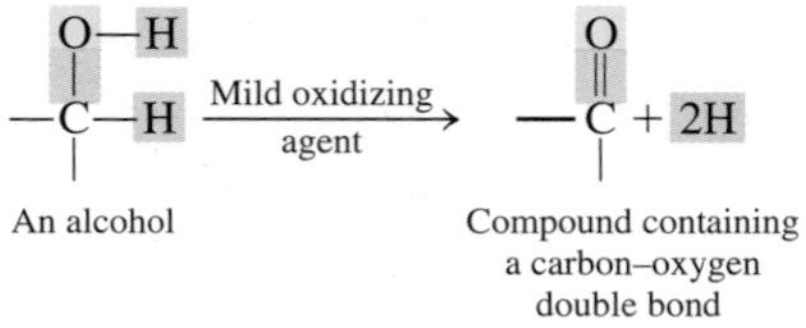

An alcohol $\xrightarrow[\text{agent}]{\text{Mild oxidizing}}$ Compound containing a carbon–oxygen double bond $+ 2H$

The two "removed" hydrogen atoms combine with oxygen supplied by the oxidizing agent to give H_2O.

Primary and secondary alcohols, the two types of oxidizable alcohols, yield different products upon oxidation. A 1° alcohol produces an *aldehyde* that is often then further oxidized to a *carboxylic acid,* and a 2° alcohol produces a *ketone.*

$$\text{Primary alcohol} \xrightarrow[\text{ox. agent}]{\text{Mild}} \text{aldehyde} \xrightarrow[\text{ox. agent}]{\text{Mild}} \text{carboxylic acid}$$

$$\text{Secondary alcohol} \xrightarrow[\text{ox. agent}]{\text{Mild}} \text{ketone}$$

$$\text{Tertiary alcohol} \xrightarrow[\text{ox. agent}]{\text{Mild}} \text{no reaction}$$

Figure 14.13 The oxidation of ethanol is the basis for the "breathalyzer test" that law enforcement officers use to determine whether an individual suspected of driving under the influence (DUI) has a blood alcohol level exceeding legal limits.

The DUI suspect is required to breathe into an apparatus containing a solution of potassium dichromate ($K_2Cr_2O_7$). The unmetabolized alcohol in the person's breath is oxidized by the dichromate ion ($Cr_2O_7^{2-}$), and the extent of the reaction gives a measure of the amount of alcohol present.

The dichromate ion is a yellow-orange color in solution. As oxidation of the alcohol proceeds, the dichromate ions are converted to Cr^{3+} ions, which have a green color in solution. The intensity of the green color that develops is measured and is proportional to the amount of ethanol in the suspect's breath, which in turn has been shown to be proportional to the person's blood alcohol level.

The general reaction for the oxidation of a primary alcohol is

$$R-\overset{\overset{\text{O}-\text{H}}{|}}{\underset{\text{H}}{C}}-H \xrightarrow{[O]} R-\overset{\overset{\text{O}}{||}}{C}-H \xrightarrow{[O]} R-\overset{\overset{\text{O}}{||}}{C}-OH$$

1° Alcohol   Aldehyde   Carboxylic acid

In this equation, the symbol [O] represents the mild oxidizing agent. The immediate product of the oxidation of a primary alcohol is an aldehyde. Because aldehydes themselves are readily oxidized by the same oxidizing agents that oxidize alcohols, aldehydes are further converted to carboxylic acids. A specific example of a primary alcohol oxidation reaction is

$$CH_3-CH_2-OH \xrightarrow{[O]} CH_3-\overset{\overset{\text{O}}{||}}{C}-H \xrightarrow{[O]} CH_3-\overset{\overset{\text{O}}{||}}{C}-OH$$

Ethanol

This specific oxidation reaction—that of ethanol—is the basis for the "breathalyzer test" used by law enforcement officers to determine whether an automobile driver is "drunk" (see Figure 14.13).

The general reaction for the oxidation of a secondary alcohol is

$$R-\overset{\overset{\text{O}-\text{H}}{|}}{\underset{\text{H}}{C}}-R \xrightarrow{[O]} R-\overset{\overset{\text{O}}{||}}{C}-R$$

2° Alcohol   Ketone

As with primary alcohols, oxidation involves the removal of two hydrogen atoms. Unlike aldehydes, ketones are resistant to further oxidation. A specific example of the oxidation of a secondary alcohol is

$$CH_3-\overset{\overset{\text{OH}}{|}}{CH}-CH_3 \xrightarrow{[O]} CH_3-\overset{\overset{\text{O}}{||}}{C}-CH_3$$

Tertiary alcohols do not undergo oxidation with mild oxidizing agents. This is because they do not have hydrogen on the —OH-bearing carbon atom.

$$R-\overset{\overset{\text{OH}}{|}}{\underset{\text{R}}{C}}-R \xrightarrow{[O]} \text{no reaction}$$

3° Alcohol

● EXAMPLE 14.4

Predicting Products in Alcohol Oxidation Reactions

▶ Draw the structural formula(s) for the product(s) formed by oxidation of the following alcohols with a mild oxidizing agent. If no reaction occurs, write "no reaction."

a. $CH_3-CH_2-CH_2-\overset{\overset{}{\underset{\text{OH}}{|}}}{CH}-CH_3$ **b.** $CH_3-\overset{\overset{}{\underset{\text{CH}_3}{|}}}{CH}-CH_2-OH$

c. $CH_3-CH_2-\overset{\overset{}{\underset{\text{CH}_3}{|}}}{CH}-OH$ **d.** (cyclohexane ring with OH and CH₃)

Solution

a. The oxidation product will be a ketone, as this is a 2° alcohol.

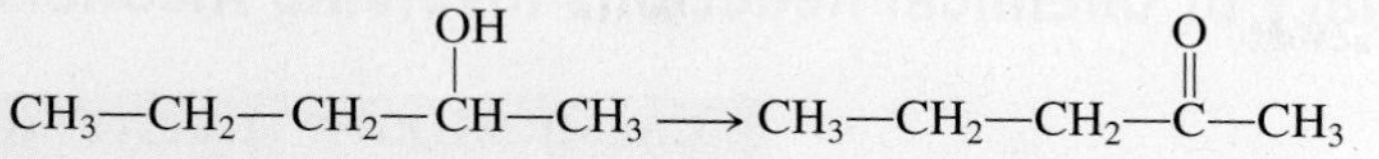

$$\text{CH}_3\text{—CH}_2\text{—CH}_2\overset{\overset{\displaystyle \text{OH}}{|}}{\text{—CH}}\text{—CH}_3 \longrightarrow \text{CH}_3\text{—CH}_2\text{—CH}_2\overset{\overset{\displaystyle \text{O}}{\|}}{\text{—C}}\text{—CH}_3$$

b. A 1° alcohol undergoes oxidation first to an aldehyde and then to a carboxylic acid.

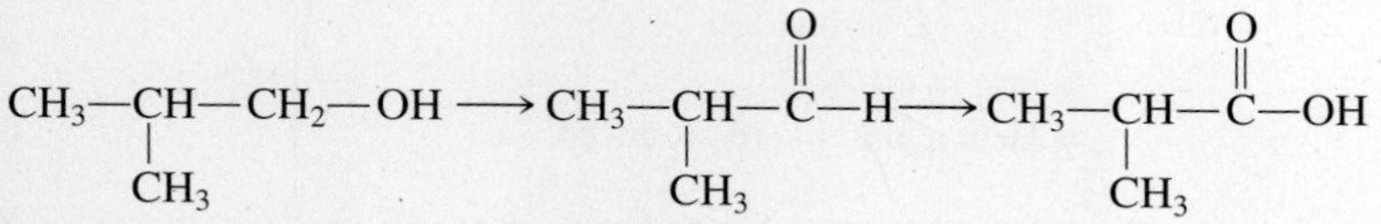

$$\text{CH}_3\overset{\overset{\displaystyle }{|}}{\underset{\underset{\displaystyle \text{CH}_3}{|}}{\text{—CH}}}\text{—CH}_2\text{—OH} \longrightarrow \text{CH}_3\underset{\underset{\displaystyle \text{CH}_3}{|}}{\text{—CH}}\overset{\overset{\displaystyle \text{O}}{\|}}{\text{—C}}\text{—H} \longrightarrow \text{CH}_3\underset{\underset{\displaystyle \text{CH}_3}{|}}{\text{—CH}}\overset{\overset{\displaystyle \text{O}}{\|}}{\text{—C}}\text{—OH}$$

c. A ketone is the product from the oxidation of a 2° alcohol.

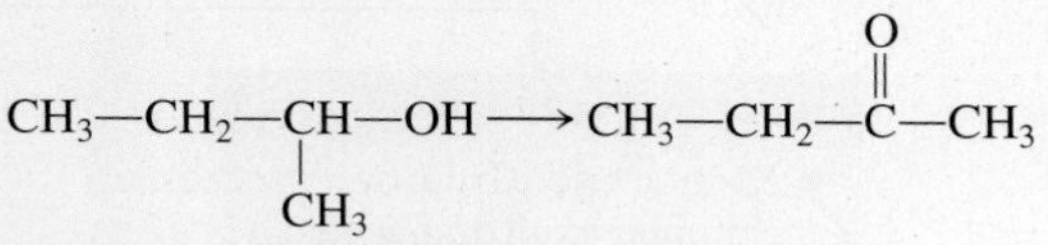

$$\text{CH}_3\text{—CH}_2\underset{\underset{\displaystyle \text{CH}_3}{|}}{\text{—CH}}\text{—OH} \longrightarrow \text{CH}_3\text{—CH}_2\overset{\overset{\displaystyle \text{O}}{\|}}{\text{—C}}\text{—CH}_3$$

d. This cyclic alcohol is a tertiary alcohol. The hydroxyl-bearing carbon atom is attached to two ring carbon atoms and a methyl group. Tertiary alcohols do not undergo oxidation with mild oxidizing agents. Therefore, "no reaction."

Practice Exercise 14.4

Draw the structural formula(s) for the product(s) formed by oxidation of the following alcohols with a mild oxidizing agent. If no reaction occurs, write "no reaction."

a. $\text{CH}_3\text{—CH}_2\text{—CH}_2\text{—OH}$ **b.** 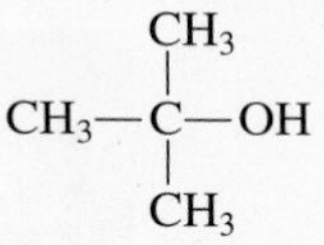 $\text{CH}_3\overset{\overset{\displaystyle \text{CH}_3}{|}}{\underset{\underset{\displaystyle \text{CH}_3}{|}}{\text{—C}}}\text{—OH}$

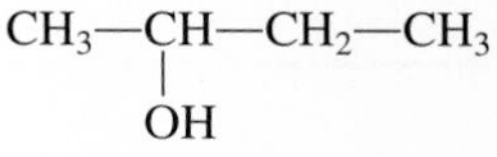

c. $\text{CH}_3\underset{\underset{\displaystyle \text{OH}}{|}}{\text{—CH}}\text{—CH}_2\text{—CH}_3$ **d.**

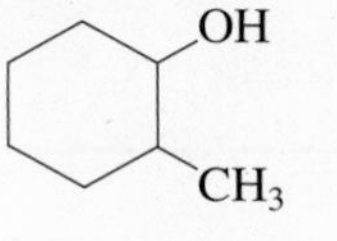

Answers:

a. 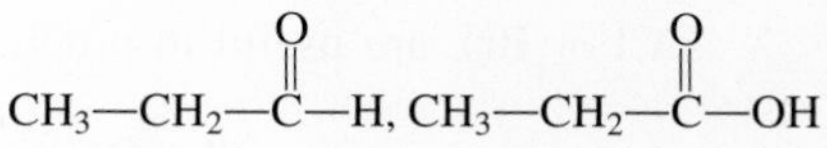 $\text{CH}_3\text{—CH}_2\overset{\overset{\displaystyle \text{O}}{\|}}{\text{—C}}\text{—H},\ \text{CH}_3\text{—CH}_2\overset{\overset{\displaystyle \text{O}}{\|}}{\text{—C}}\text{—OH}$ **b.** No reaction

c. 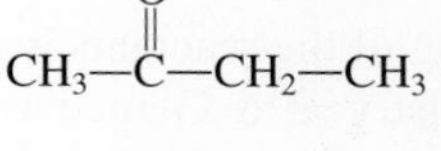 $\text{CH}_3\overset{\overset{\displaystyle \text{O}}{\|}}{\text{—C}}\text{—CH}_2\text{—CH}_3$ **d.**

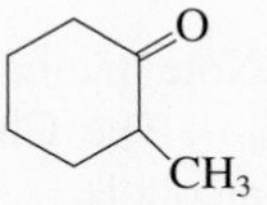

Halogenation

Alcohols undergo halogenation reactions in which a halogen atom is substituted for the hydroxyl group, producing an alkyl halide. Alkyl halide production in this manner is superior to alkyl halide production through halogenation of an alkane (Section 12.18) because mixtures of products are *not* obtained. A single product is produced in which the halogen atom is found only where the —OH group was originally located.

CHEMISTRY AT A GLANCE

Summary of Chemical Reactions Involving Alcohols

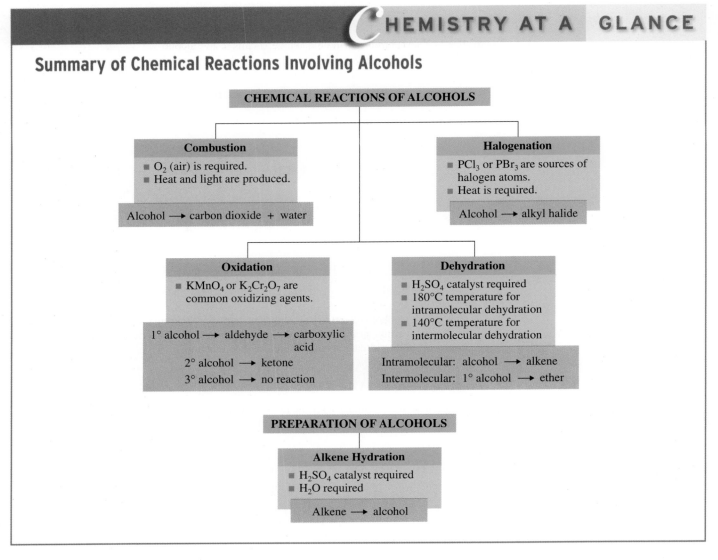

CHEMICAL REACTIONS OF ALCOHOLS

Combustion
- O_2 (air) is required.
- Heat and light are produced.

Alcohol ⟶ carbon dioxide + water

Halogenation
- PCl_3 or PBr_3 are sources of halogen atoms.
- Heat is required.

Alcohol ⟶ alkyl halide

Oxidation
- $KMnO_4$ or $K_2Cr_2O_7$ are common oxidizing agents.

1° alcohol ⟶ aldehyde ⟶ carboxylic acid
2° alcohol ⟶ ketone
3° alcohol ⟶ no reaction

Dehydration
- H_2SO_4 catalyst required
- 180°C temperature for intramolecular dehydration
- 140°C temperature for intermolecular dehydration

Intramolecular: alcohol ⟶ alkene
Intermolecular: 1° alcohol ⟶ ether

PREPARATION OF ALCOHOLS

Alkene Hydration
- H_2SO_4 catalyst required
- H_2O required

Alkene ⟶ alcohol

Several different halogen-containing reactants, including phosphorus trihalides (PX_3; X is Cl or Br), are useful in producing alkyl halides from alcohols.

$$3R{-}OH + PX_3 \xrightarrow{\text{heat}} 3R{-}X + H_3PO_3$$

Note that heating of the reactants is required.

The Chemistry at a Glance feature above summarizes the reaction chemistry of alcohols.

14.10 POLYMERIC ALCOHOLS

It is possible to synthesize polymeric alcohols with structures similar to those of substituted polyethylenes (Section 13.9). One of the simplest such compounds is poly(vinyl alcohol) (PVA).

PVA

Poly(vinyl alcohol) is a tough, whitish polymer that can be formed into strong films, tubes, and fibers that are highly resistant to hydrocarbon solvents. Unlike most organic polymers, PVA is water-soluble. Water-soluble films and sheetings are important PVA products. PVA has oxygen-barrier properties under dry conditions that are superior to those of any other polymer. PVA can be rendered insoluble in water, if needed, by use of chemical agents that cross-link individual polymer strands.

14.11 STRUCTURAL CHARACTERISTICS OF PHENOLS

A **phenol** *is an organic compound in which an —OH group is attached to a carbon atom that is part of an aromatic carbon ring system.*

The general formula for phenols is Ar–OH, where Ar represents an *aryl group*. An **aryl group** *is an aromatic carbon ring system from which one hydrogen atom has been removed.*

A hydroxyl group is thus the functional group for both phenols and alcohols. The reaction chemistry for phenols is sufficiently different from that for nonaromatic alcohols (Section 14.9) to justify discussing these compounds separately. Remember that phenols contain a "benzene ring" and that the chemistry of benzene is much different from that of other unsaturated hydrocarbons (Section 13.14).

The following are examples of compounds classified as phenols.

> The generic term *aryl group* (Ar) is the aromatic counterpart of the nonaromatic generic term *alkyl group* (R).

14.12 NOMENCLATURE FOR PHENOLS

Besides being the name for a family of compounds, *phenol* is also the IUPAC-approved name for the simplest member of the phenol family of compounds.

Phenol

A space-filling model for the compound *phenol* is shown in Figure 14.14. The name *phenol* is derived from a combination of the terms *phen*yl and alcoh*ol*.

The IUPAC rules for naming phenols are simply extensions of the rules used to name benzene derivatives with hydrocarbon or halogen substituents (Section 13.12). The parent name is phenol. Ring numbering always begins with the hydroxyl group and proceeds in the direction that gives the lower number to the next carbon atom bearing a substituent. The numerical position of the hydroxyl group is not specified in the name because it is 1 by definition.

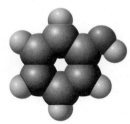

Figure 14.14 A space-filling model for *phenol*, a compound that has an —OH group bonded directly to a benzene (aromatic) ring.

3-Chlorophenol
(or *meta*-Chlorophenol)

4-Ethyl-2-methylphenol

2,5-Dibromophenol

Methyl and hydroxy derivatives of phenol have IUPAC-accepted common names. Methylphenols are called cresols. The name *cresol* applies to all three isomeric methylphenols.

ortho-Cresol   *meta*-Cresol   *para*-Cresol

For hydroxyphenols, each of the three isomers has a different common name.

Catechol   Resorcinol   Hydroquinone

Several neurotransmitters in the human body (Section 17.10), including norepinephrine, epinephrine (adrenaline), and dopamine, are catechol derivatives.

14.13 PHYSICAL AND CHEMICAL PROPERTIES OF PHENOLS

Phenols are generally low-melting solids or oily liquids at room temperature. Most of them are only slightly soluble in water. Many phenols have antiseptic and disinfectant properties. The simplest phenol, phenol itself, is a colorless solid with a medicinal odor. Its melting point is 41°C, and it is more soluble in water than are most other phenols.

We have previously noted that the chemical properties of phenols are significantly different from those of alcohols (Section 14.11). The similarities and differences between these two reaction chemistries are as follows:

1. Both alcohols and phenols are flammable.
2. Dehydration is a reaction of alcohols but not of phenols; phenols cannot be dehydrated.
3. Both 1° and 2° alcohols are oxidized by mild oxidizing agents. Tertiary (3°) alcohols and phenols do not react with the oxidizing agents that cause 1° and 2° alcohol oxidation. Phenols can be oxidized by stronger oxidizing agents.
4. Both alcohols and phenols undergo halogenation in which the hydroxyl group is replaced by a halogen atom in a substitution reaction.

Acidity of Phenols

One of the most important properties of phenols is their acidity. Unlike alcohols, phenols are weak acids in solution. As acids, phenols have K_a values (Section 10.5) of about 10^{-10}. Such K_a values are lower than those of most weak inorganic acids (10^{-5} to 10^{-10}). The acid ionization reaction for phenol itself is

Phenol   Phenoxide ion

$$+ H_2O \rightleftharpoons + H_3O^+$$

Note that the negative ion produced from the ionization is called the phenoxide ion. When phenol itself is reacted with sodium hydroxide (a base), the salt sodium phenoxide is produced.

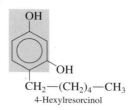

14.14 OCCURRENCE OF AND USES FOR PHENOLS

Dilute (2%) solutions of phenol have long been used as antiseptics. Concentrated phenol solutions, however, can cause severe skin burns. Today, phenol has been largely replaced by more effective phenol derivatives such as 4-hexylresorcinol. The compound 4-hexylresorcinol is an ingredient in many mouthwashes and throat lozenges.

An *antiseptic* is a substance that kills microorganisms on living tissue. A *disinfectant* is a substance that kills microorganisms on inanimate objects.

The "parent" name for a benzene ring bearing two hydroxyl groups "meta" to each other is resorcinol (Section 14.12).

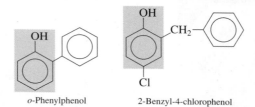

4-Hexylresorcinol

The phenol derivatives *o*-phenylphenol and 2-benzyl-4-chlorophenol are the active ingredients in Lysol, a disinfectant for walls, floors, and furniture in homes and hospitals.

Figure 14.15 Many commercially baked goods contain the antioxidants BHA and BHT to help prevent spoilage.

o-Phenylphenol   2-Benzyl-4-chlorophenol

A number of phenols possess antioxidant activity. An **antioxidant** *is a substance that protects other substances from being oxidized by being oxidized itself in preference to the other substances.* An antioxidant has a greater affinity for a particular oxidizing agent than do the substances the antioxidant is "protecting"; the antioxidant, therefore, reacts with the oxidizing agent first. Many foods sensitive to air are protected from oxidation through the use of phenolic antioxidants. Two commercial phenolic antioxidant food additives are BHA (butylated hydroxy anisole) and BHT (butylated hydroxy toluene) (see Figure 14.15).

Within the human body, natural dietary antioxidants also offer protection against undesirable oxidizing agents. They include vitamin C (Section 21.12), beta-carotene (Section 21.13), vitamin E (Section 21.14), and flavonoids (Section 23.11).

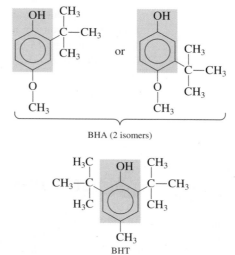

BHA (2 isomers)

BHT

A naturally occurring phenolic antioxidant that is important in the functioning of the human body is vitamin E (Section 21.14).

Vitamin E

A number of phenols found in plants are used as flavoring agents and/or antibacterials. Included among these phenols are

Thymol

Eugenol

Isoeugenol

Vanillin

Thymol, obtained from the herb thyme, possesses both flavorant and antibacterial properties. It is used as an ingredient in several mouthwash formulations.

Eugenol is responsible for the flavor of cloves. Dentists traditionally used clove oil as an antiseptic because of eugenol's presence; they use it to a limited extent even today.

Isoeugenol, which differs in structure from eugenol only in the location of the double bond in the hydrocarbon side chain, is responsible for the odor associated with nutmeg (see Figure 14.16).

Vanillin, which gives vanilla its flavor, is extracted from the dried seed pods of the vanilla orchid. Natural supplies of vanillin are inadequate to meet demand for this flavoring agent. Synthetic vanillin is produced by oxidation of eugenol. Vanillin is an unusual substance in that even though its odor can be perceived at extremely low concentrations, the strength of its odor does not increase greatly as its concentration is increased.

Certain phenols exert profound physiological effects. For example, the irritating constituents of poison ivy and poison oak are derivatives of catechol (Section 14.12). These skin irritants have 15-carbon alkyl side chains with varying degrees of unsaturation (zero to three double bonds).

Catechol

Poison ivy irritants

Figure 14.16 Nutmeg tree fruit. A phenolic compound, isoeugenol, is responsible for the odor associated with nutmeg.

A diisopropyl phenol, with the medical name *Propofol*, is a short-acting intravenous agent used for the induction of general anesthesia and for sedation in several current medical contexts.

Propofol
(2,6-diisopropylphenol)

It is extensively used in medical contexts such as intensive care unit (ICU) sedation for intubated, mechanically ventilated adults and in procedures such as a colonoscopy. It provides no analgesia (pain relief).

14.15 STRUCTURAL CHARACTERISTICS OF ETHERS

An **ether** *is an organic compound in which an oxygen atom is bonded to two carbon atoms by single bonds.* In an ether, the carbon atoms that are attached to the oxygen atom can be part of alkyl, cycloalkyl, or aryl groups. Examples of ethers include

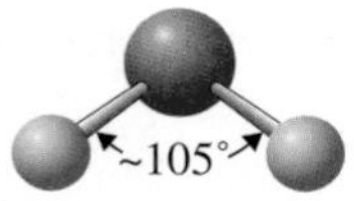

Water (HOH)

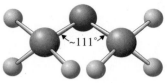

Dimethyl ether (CH₃OCH₃)

Figure 14.17 The similar shapes of water and dimethyl ether molecules. Dimethyl ether may be viewed structurally as a dialkyl derivative of water.

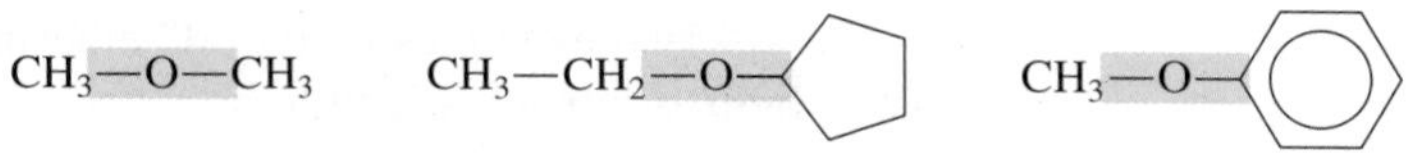

$$CH_3\text{—}O\text{—}CH_3 \qquad CH_3\text{—}CH_2\text{—}O\text{—}\bigpentagon \qquad CH_3\text{—}O\text{—}\bigcirc$$

The two groups attached to the oxygen atom of an ether can be the same (first structure), but they need not be so (second and third structures).

All ethers contain a C—O—C unit, which is the ether functional group.

Ether functional group

$$\text{---}\overset{\frown}{C\text{—}O\text{—}C}\text{---}$$

Generalized formulas for ethers, which depend on the types of groups attached to the oxygen atom (alkyl or aryl), include R—O—R, R—O—R′ (where R′ is an alkyl group different from R), R—O—Ar, and Ar—O—Ar.

Structurally, an ether can be visualized as a derivative of water in which both hydrogen atoms have been replaced by hydrocarbon groups (see Figure 14.17). Note that unlike alcohols and phenols, ethers do not possess a hydroxyl (—OH) group.

$$H\text{—}\overset{..}{\underset{..}{O}}\text{—}H \qquad\qquad R\text{—}\overset{..}{\underset{..}{O}}\text{—}R$$

Water   An ether

14.16 NOMENCLATURE FOR ETHERS

Common names are almost always used for ethers whose alkyl groups contain four or fewer carbon atoms. There are two rules, one for unsymmetrical ethers (two different alkyl/aryl groups) and one for symmetrical ethers (both alkyl/aryl groups the same).

Line-angle structural formulas for selected simple ethers:

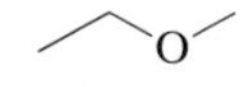

Ethyl methyl ether
(methoxyethane)

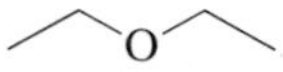

Diethyl ether
(ethoxyethane)

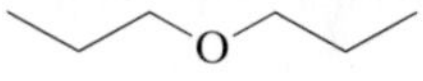

Dipropyl ether
(1-propoxypropane)

Rule 1: For unsymmetrical ethers, name both hydrocarbon groups bonded to the oxygen atom in alphabetical order and add the word *ether,* separating the words with a space. Such ether names have three separate words within them.

$$CH_3\text{—}O\text{—}CH_2\text{—}CH_3 \qquad\qquad CH_3\text{—}CH_2\text{—}O\text{—}\bigcirc$$

Ethyl methyl ether   Ethyl phenyl ether

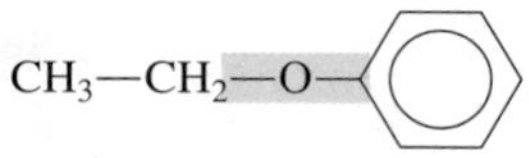

Rule 2: For symmetrical ethers, name the alkyl group, add the prefix *di-*, and then add the word *ether,* separating the words with a space. Such ether names have two separate words within them.

$$CH_3\text{—}O\text{—}CH_3 \qquad\qquad CH_3\text{—}CH_2\text{—}O\text{—}CH_2\text{—}CH_3$$

Dimethyl ether   Diethyl ether

Ethers with more complex alkyl/aryl groups are named using the IUPAC system. In this system, ethers are named as substituted hydrocarbons. The smaller hydrocarbon attachment and the oxygen atom are called an *alkoxy group,* and this group is considered a substituent on the larger hydrocarbon group. An **alkoxy group** *is an —OR group, an alkyl (or aryl) group attached to an oxygen atom.* Simple alkoxy groups include the following:

It is possible to have compounds that contain both ether and alcohol functional groups such as

$$CH_3\text{—}\underset{\underset{OH}{|}}{CH}\text{—}CH_2\text{—}CH_2\text{—}O\text{—}CH_3$$

4-Methoxy-2-butanol

(The alcohol functional group has higher priority in IUPAC nomenclature, so the compound is named as an alcohol rather than as an ether.)

$$CH_3\text{—}O\text{—} \qquad CH_3\text{—}CH_2\text{—}O\text{—} \qquad CH_3\text{—}CH_2\text{—}CH_2\text{—}O\text{—}$$

Methoxy group   Ethoxy group   Propoxy group

The general symbol for an alkoxy group is —O—R (or —OR).

The rules for naming an ether using the IUPAC system are

Rule 1: Select the longest carbon chain and use its name as the base name.

Rule 2: Change the -yl ending of the other hydrocarbon group to -oxy to obtain the alkoxy group name; *methyl* becomes *methoxy, ethyl* becomes *ethoxy,* etc.

Rule 3: Place the alkoxy name, with a locator number, in front of the base chain name.

Two examples of IUPAC ether nomenclature, with the alkoxy groups present highlighted in each structure are:

$$\boxed{CH_3-O}-CH_2-CH_2-CH_2-CH_3$$

1-Methoxybutane

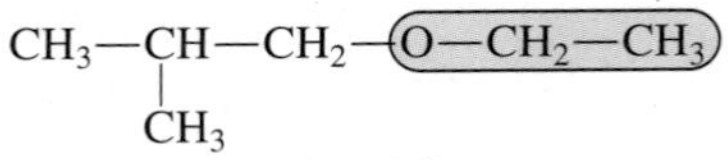

$$CH_3-CH-CH_2-\boxed{O-CH_2-CH_3}$$
$$\quad\quad |$$
$$\quad\quad CH_3$$

1-Ethoxy-2-methylpropane

The simplest aromatic ether involves a methoxy group attached to a benzene ring. This ether goes by the common name *anisole*.

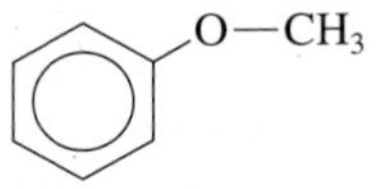

Anisole

Derivatives of anisole are named as substituted anisoles, in a manner similar to that for substituted phenols (Section 14.12). Anisole derivatives were encountered in Section 14.14 when considering antioxidant food additives: BHAs are both a phenol and an anisole.

The compound responsible for the characteristic odor of anise and fennel is *anethole*, an allyl derivative of anisole.

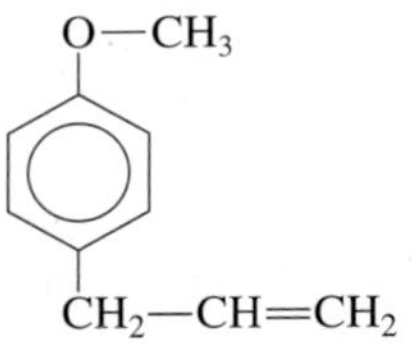

O—CH₃

CH₂—CH=CH₂

● **E X A M P L E 1 4 . 5**

Determining IUPAC Names for Ethers

▶ Name the following ethers utilizing IUPAC nomenclature rules.

a. $CH_3-CH_2-O-CH_2-CH_2-CH_3$ **b.** 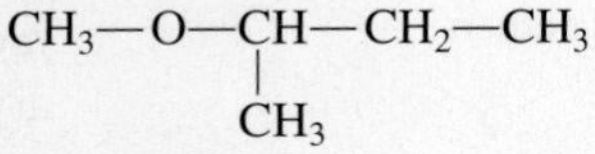 $CH_3-O-CH-CH_2-CH_3$
$$\quad\quad\quad\quad\quad\quad\quad\quad\quad\quad\quad\quad\quad\quad |$$
$$\quad\quad\quad\quad\quad\quad\quad\quad\quad\quad\quad\quad\quad\quad CH_3$$

c. 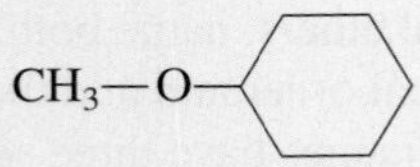 CH_3-O-⬡   **d.** Ethyl methyl ether

Solution

a. The base name is propane. An ethoxy group is attached to carbon-1 of the propane chain.

$$CH_3-CH_2-O-\overset{①}{CH_2}-\overset{②}{CH_2}-\overset{③}{CH_3}$$

The IUPAC name is 1-ethoxypropane.

b. The base name is butane, as the longest carbon chain contains four carbon atoms.

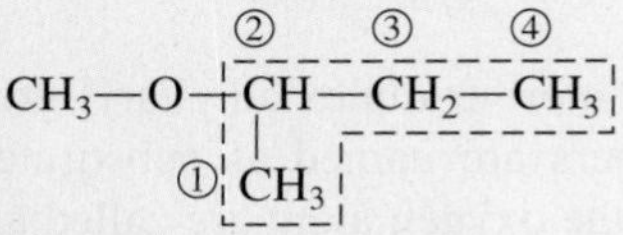

$$CH_3-O-\overset{②}{CH}-\overset{③}{CH_2}-\overset{④}{CH_3}$$
$$\quad\quad\quad\overset{①}{|}$$
$$\quad\quad\quad CH_3$$

The IUPAC name is 2-methoxybutane.

c. The base name is cyclohexane. The complete IUPAC name is methoxycyclohexane. No number is needed to locate the methoxy group since all ring carbon atoms are equivalent to each other.

d. The ether structure is $CH_3-CH_2-O-CH_3$, and the IUPAC name is methoxyethane.

The contrast between IUPAC and common names for ethers is as follows:

IUPAC (one word)

| alkoxyalkane |

2-methoxybutane

Common (three or two words)

| alkyl alkyl ether |

ethyl methyl ether

or

| dialkyl ether |

dipropyl ether

Practice Exercise 14.5

Name the following ethers utilizing IUPAC nomenclature rules.

a. $CH_3-CH_2-CH_2-O-CH_2-CH_2-CH_3$ **b.** 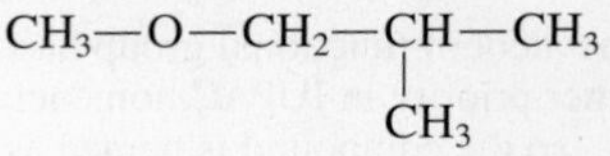 $CH_3-O-CH_2-CH-CH_3$
$$\quad\quad\quad\quad\quad\quad\quad\quad\quad\quad\quad\quad\quad\quad\quad\quad\quad |$$
$$\quad\quad\quad\quad\quad\quad\quad\quad\quad\quad\quad\quad\quad\quad\quad\quad\quad CH_3$$

c.

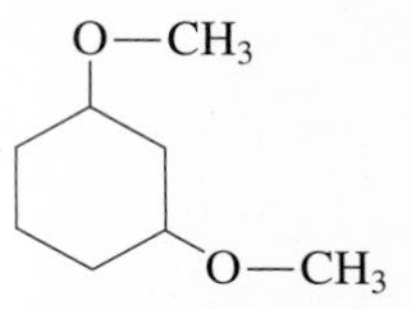

O—CH₃

O—CH₃

d. Dimethyl ether

Answers: a. 1-Propoxypropane; **b.** 1-Methoxy-2-methylpropane; **c.** 1, 3-Dimethoxycyclohexane; **d.** Methoxymethane

---

CHEMICAL Connections

Ethers as General Anesthetics

For many people, the word *ether* evokes thoughts of hospital operating rooms and anesthesia. This response derives from the *former* large-scale use of diethyl ether as a general anesthetic. In 1846, the Boston dentist William Morton was the first to demonstrate publicly the use of diethyl ether as a surgical anesthetic.

In many ways, diethyl ether is an ideal general anesthetic. It is relatively easy to administer, it is readily made in pure form, and it causes excellent muscle relaxation. There is less danger of an overdose with diethyl ether than with almost any other anesthetic because there is a large gap between the effective level for anesthesia and the lethal dose.

Despite these ideal properties, diethyl ether is rarely used today because of two drawbacks: (1) It causes nausea and irritation of the respiratory passage, and (2) it is a highly flammable substance, forming explosive mixtures with air, which can be set off by a spark.

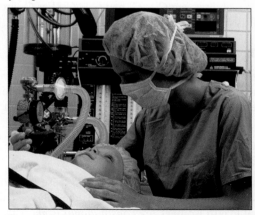

Preparing to administer an anesthetic to a child.

By the 1930s, nonether anesthetics had been developed that solved the problems of nausea and irritation. They also, however, were extremely flammable compounds. The simple hydrocarbon cyclopropane was the most widely used of these newer compounds.

It was not until the late 1950s and early 1960s that nonflammable general anesthetics became available. Anesthetic nonflammability was achieved by incorporating halogen atoms into anesthetic molecules. Three of the most used of these "halogenated" anesthetics are enflurane, isoflurane, and halothane.

Enflurane and isoflurane, which are constitutional isomers, are hexahalogenated ethers.

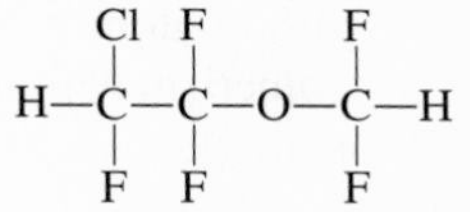

Enflurane

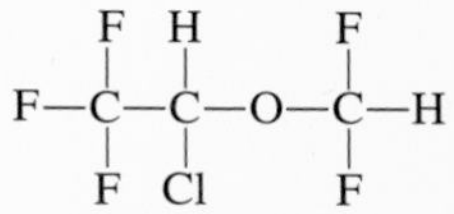

Isoflurane

With these compounds, induction of anesthesia can be achieved in less than 10 minutes with an inhaled concentration of 3% in oxygen.

Halothane, which is potent at relatively low doses, and whose effects wear off quickly, is a pentahalogenated alkane derivative rather than an ether.

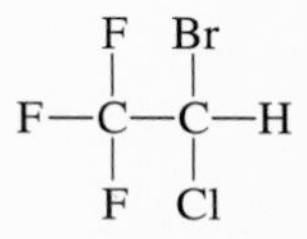

Halothane

It is the only inhalation anesthetic that contains a bromine atom.

Phase-out of the use of the preceding three compounds began in the late 1980s and early 1990s. They have largely been replaced by a "second generation" of similar halogenated ethers with even better anesthetic properties. In general, these new compounds have more fluorine atoms and fewer chlorine atoms. Two of the most prominent of these second-generation anesthetic agents now in use are Sevoflurane and Desflurane.

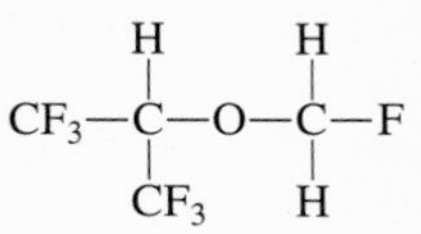

Sevoflurane

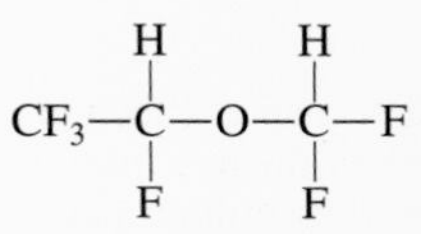

Desflurane

The ether MTBE (methyl *tert*-butyl ether) has been a widely used gasoline additive since the early 1980s.

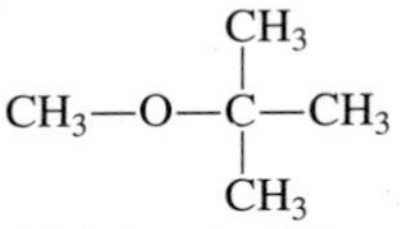

$$CH_3-O-\underset{\underset{CH_3}{|}}{\overset{\overset{CH_3}{|}}{C}}-CH_3$$

Methyl *tert*-butyl ether
(MTBE)

As an additive, MTBE not only raises octane levels but also functions as a clean-burning "oxygenate" in EPA-mandated reformulated gasolines used to improve air quality in polluted areas. The amount of MTBE used in gasoline is now decreasing in response to a growing problem: contamination of water supplies by small amounts of MTBE from leaking gasoline tanks and from spills. MTBE in the water supplies is not a health-and-safety issue at this time, but its presence does affect taste and odor in contaminated supplies.

Compounds with ether functional groups occur in a variety of plants. The phenolic flavoring agents eugenol, isoeugenol, and vanillin (Section 14.14) are also ethers; each has a methoxy substituent on the ring.

Technically, the name methyl *tert*-butyl ether (MTBE) is incorrect because the convention for naming ethers dictates an alphabetical ordering of alkyl groups (*tert*-butyl methyl ether). However, the compound is called MTBE rather than TBME by those in the petroleum industry and by environmental scientists.

14.17 ISOMERISM FOR ETHERS

Ethers contain two carbon chains (two alkyl groups), unlike the one carbon chain found in alcohols. Constitutional isomerism possibilities in ethers depend on (1) the partitioning of carbon atoms between the two alkyl groups and (2) isomerism possibilities for the individual alkyl groups present. Isomerism is not possible for a C_2 ether (two methyl groups) or a C_3 ether (a methyl and an ethyl group). For C_4 ethers, isomerism arises not only from carbon atom partitioning between the alkyl groups (C_1-C_3 and C_2-C_2) but also from isomerism within a C_3 group (propyl and isopropyl). There are three C_4 ether constitutional isomers.

$$CH_3-CH_2-O-CH_2-CH_3 \qquad CH_3-O-CH_2-CH_2-CH_3 \qquad CH_3-O-\underset{\underset{CH_3}{|}}{CH}-CH_3$$

Diethyl ether   Methyl propyl ether   Isopropyl methyl ether

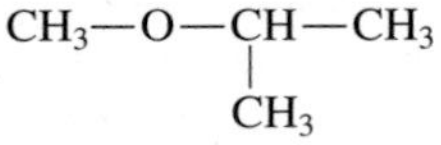

For C_5 ethers, carbon partitioning possibilities are C_2-C_3 and C_1-C_4. For C_4 groups there are four isomeric variations: butyl, isobutyl, *sec*-butyl, and *tert*-butyl (Section 12.11).

Functional Group Isomerism

Ethers and alcohols with the same number of carbon atoms and the same degree of saturation have the same molecular formula. The simplest manifestation of this phenomenon involves dimethyl ether, the C_2 ether, and ethyl alcohol, the C_2 alcohol. Both have the molecular formula C_2H_6O.

$$CH_3-O-CH_3 \qquad CH_3-CH_2-OH$$

Dimethyl ether   Ethyl alcohol

With the same molecular formula and different structural formulas, these two compounds are constitutional isomers. This type of constitutional isomerism is the subtype called *functional group isomerism*. **Functional group isomers** *are constitutional isomers that contain different functional groups.* When three carbon atoms are present the ether–alcohol functional group isomerism possibilities are

Later in this chapter (Section 14.21) and in each of the next two chapters we will encounter other pairs of functional groups for which functional group isomerism is possible.

$$CH_3-CH_2-O-CH_3 \qquad CH_3-CH_2-CH_2-OH \qquad CH_3-\underset{\underset{CH_3}{|}}{CH}-OH$$

Ethyl methyl ether   Propyl alcohol   Isopropyl alcohol

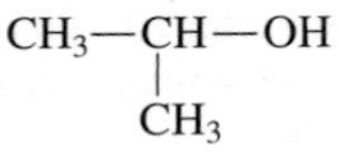

Figure 14.18 Alcohols and ethers with the same number of carbon atoms and the same degree of saturation are functional group isomers, as is illustrated here for propyl alcohol and ethyl methyl ether.

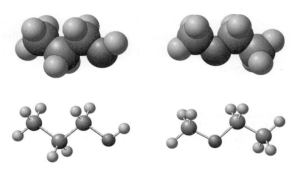

Propyl alcohol (C_3H_8O)   **Ethyl methyl ether (C_3H_8O)**

All three compounds have the molecular formula C_3H_8O. Figure 14.18 shows molecular models for the isomeric propyl alcohol and ethyl methyl ether molecules.

Unbranched Alkyl Alkyl Ethers			
C_1–C_1			
C_1–C_2	C_2–C_2		
C_1–C_3	C_2–C_3	C_3–C_3	
C_1–C_4	C_2–C_4	C_3–C_4	C_4–C_4

☐ Gas   ☐ Liquid

Figure 14.19 A physical-state summary for unbranched alkyl alkyl ethers at room temperature and pressure.

The term *ether* comes from the Latin *aether*, which means "to ignite." This name is given to these compounds because of their high vapor pressure at room temperature, which makes them very flammable.

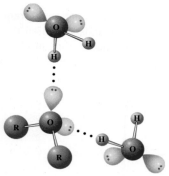

Figure 14.20 Although ether molecules cannot hydrogen-bond to one another, they can hydrogen-bond to water molecules. Such hydrogen bonding causes ethers to be more soluble in water than alkanes of similar molecular mass.

14.18 PHYSICAL AND CHEMICAL PROPERTIES OF ETHERS

The boiling points of ethers are similar to those of alkanes of comparable molecular mass and are much lower than those of alcohols of comparable molecular mass.

Alkane	$CH_3-CH_2-CH_2-CH_2-CH_3$	Mol. mass = 72 amu bp = 36°C
Ether	$CH_3-CH_2-O-CH_2-CH_3$	Mol. mass = 74 amu bp = 35°C
Alcohol	$CH_3-CH_2-CH_2-CH_2-OH$	Mol. mass = 74 amu bp = 117°C

The much higher boiling point of the alcohol results from hydrogen bonding between alcohol molecules. Ether molecules, like alkanes, cannot hydrogen-bond to one another. Ether oxygen atoms have no hydrogen atom attached directly to them. Figure 14.19 is a physical-state summary for unbranched alkyl alkyl ethers where the alkyl groups range in size from C_1 to C_4.

Ethers, in general, are more soluble in water than are alkanes of similar molecular mass because ether molecules are able to form hydrogen bonds with water (Figure 14.20).

Ethers have water solubilities similar to those of alcohols of the same molecular mass. For example, diethyl ether and butyl alcohol have the same solubility in water. Because ethers can also hydrogen-bond to alcohols, alcohols and ethers tend to be mutually soluble. Nonpolar substances tend to be more soluble in ethers than in alcohols because ethers have no hydrogen-bonding network that has to be broken up for solubility to occur.

Two chemical properties of ethers are especially important.

1. *Ethers are flammable.* Special care must be exercised in laboratories where ethers are used. Diethyl ether, whose boiling point of 35°C is only a few degrees above room temperature, is a particular flash-fire hazard.

2. *Ethers react slowly with oxygen from the air to form unstable hydroperoxides and peroxides.*

R—O—O—H	R—O—O—R
Hydroperoxide	Peroxide

Such compounds, when concentrated, represent an explosion hazard and must be removed before *stored* ethers are used.

Like alkanes, ethers are unreactive toward acids, bases, and oxidizing agents. Like alkanes, they do undergo combustion and halogenation reactions.

The general chemical unreactivity of ethers, coupled with the fact that most organic compounds are ether-soluble, makes ethers excellent solvents in which to carry out organic reactions. Their relatively low boiling points simplify their separation from the reaction products.

A chemical reaction for the preparation of ethers has been previously considered. In Section 14.9 we noted that the intermolecular dehydration of a primary alcohol will produce an ether.

$$R_1\text{—OH} + R_2\text{—OH} \xrightarrow[\text{Heat}]{H^+} R_1\text{—O—}R_2 + H_2O$$

Alcohol   Alcohol   Ether   Water

Although additional methods exist for ether preparation, we will not consider them in this text.

14.19 CYCLIC ETHERS

Cyclic ethers contain ether functional groups as part of a ring system. Some examples of such cyclic ethers, along with their common names, follow.

Ethylene   Tetrahydrofuran   Furan   Pyran
oxide   (THF)

Ethylene oxide has few direct uses. Its importance is as a starting material for the production of ethylene glycol (Section 14.5), a major component of automobile antifreeze. THF is a particularly useful solvent in that it dissolves many organic compounds and yet is miscible with water. In carbohydrate chemistry (Chapter 18), we will encounter many cyclic structures that are polyhydroxy derivatives of the five-membered (furan) and six-membered (pyran) cyclic ether systems. These carbohydrate derivatives are called *furanoses* and *pyranoses*, respectively (Section 18.10).

Vitamin E (Section 14.14) and THC, the active ingredient in marijuana (see page 429), have structures in which a cyclic ether component is present.

Cyclic ethers are our first encounter with heterocyclic organic compounds. A **heterocyclic organic compound** *is a cyclic organic compound in which one or more of the carbon atoms in the ring have been replaced with atoms of other elements.* The hetero atom is usually oxygen or nitrogen.

We have just seen that *cyclic ethers*—compounds in which the ether functional group is part of a ring system—exist. In contrast, cyclic alcohols—compounds in which the alcohol functional group is part of a ring system—do not exist. To incorporate an alcohol functional group into a ring system would require an oxygen atom with three bonds, and oxygen atoms form only two bonds.

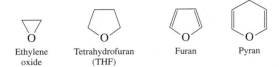

Compounds such as

OH   and   OH

which do exist, are not cyclic alcohols in the sense in which we are using the term because the alcohol functional group is attached to a ring system rather than part of it.

CHEMICAL Connections   Marijuana: The Most Commonly Used Illicit Drug

Prepared from the leaves, flowers, seeds, and small stems of a hemp plant called *Cannabis sativa,* marijuana, which is also called pot or grass, is the most commonly used illicit drug in the United States. The most active ingredient of the many in marijuana is the molecule *tetrahydrocannabinol,* called THC for short. Three different functional groups are present in a THC molecule; it is a phenol, a cyclic ether, and a cycloalkene. The THC content of marijuana varies considerably. Most marijuana sold in the North American illegal drug market has a THC content of 1% to 2%.

Marijuana has a pharmacology unlike that of any other drug. A marijuana "high" is a combination of sedation, tran-

quilization, and mild hallucination. THC readily penetrates the brain. The portions of the brain that involve memory and motor control contain the receptor sites where THC molecules interact. Even moderate doses of marijuana cause short-term memory loss. Marijuana unquestionably impairs driving ability, even after ordinary social use. THC readily crosses the placental barrier and reaches the fetus. Heavy marijuana users experience inflammation of the bronchi, sore throat, and inflamed sinuses. Increased heart rate, to as high as a dangerous 160 beats per minute, can occur with marijuana use.

The onset of action of THC is usually within minutes after smoking begins, and peak concentration in plasma occurs in 10 to 30 minutes. Unless more is smoked, the effects seldom last longer than 2 to 3 hours. Because THC is only slightly soluble in water, it tends to be deposited in fatty tissues. Unlike alcohol, THC persists in the bloodstream for several days, and the products of its breakdown remain in the blood for as long as 8 days.

New research indicates that physical dependence on THC can develop. Drug withdrawal symptoms are seen in some individuals who have been exposed repeatedly to high doses.

Tetrahydrocannabinol

14.20 SULFUR ANALOGS OF ALCOHOLS

Many organic compounds containing oxygen have sulfur analogs, in which a sulfur atom has replaced an oxygen atom. Sulfur is in the same group of the periodic table as oxygen, so the two elements have similar electron configurations (Section 3.8).

Thiols, the sulfur analogs of alcohols, contain —SH functional groups instead of —OH functional groups. The thiol functional group is called a *sulfhydryl group.* A **sulfhydryl group** *is the —SH functional group.* A **thiol** *is an organic compound in which a sulfhydryl group is bonded to a saturated carbon atom.* An older term used for thiols is *mercaptans.* Contrasting the general structures for alcohols and thiols, we have

$$\text{R—OH} \qquad \text{and} \qquad \text{R—SH}$$

An alcohol   A thiol

Hydroxyl group   Sulfhydryl group

Nomenclature for Thiols

The root *thio-* indicates that a sulfur atom has replaced an oxygen atom in a compound. It originates from the Greek *theion,* meaning "brimstone," which is an older name for the element sulfur.

Thiols are named in the same way as alcohols in the IUPAC system, except that the *-ol* becomes *-thiol.* The prefix *thio-* indicates the substitution of a sulfur atom for an oxygen atom in a compound.

$$\text{CH}_3\text{—CH—CH}_2\text{—CH}_3 \qquad \text{CH}_3\text{—CH—CH}_2\text{—CH}_3$$
$$\qquad\quad |\qquad\qquad\qquad\qquad\qquad |$$
$$\qquad\quad\text{OH}\qquad\qquad\qquad\qquad\quad\text{SH}$$

2-Butanol   2-Butanethiol

As in the case of diols and triols, the *-e* at the end of the alkane name is also retained for thiols.

Even though thiols have a higher molecular mass than alcohols with the same number of carbon atoms, they have much lower boiling points because they do not exhibit hydrogen bonding as alcohols do.

Common names for thiols are based on use of the term *mercaptan*, the older name for thiols. The name of the alkyl group present (as a separate word) precedes the word *mercaptan*.

$$CH_3—CH_2—SH \qquad CH_3—\overset{\displaystyle |}{\underset{\displaystyle CH_3}{CH}}—SH$$

Ethyl mercaptan   Isopropyl mercaptan

● **EXAMPLE 14.6**

Determining IUPAC and Common Names for Thiols

Convert each of the following common names for thiols to IUPAC names or vice versa.

a. Propyl mercaptan   **b.** Isobutyl mercaptan
c. 1-Butanethiol   **d.** 2-Propanethiol

Solution

a. The structural formula for propyl mercaptan is $CH_3—CH_2—CH_2—SH$. In the IUPAC system the name base is propane; the complete name is 1-propanethiol.
b. The structural formula for isobutyl mercaptan is

$$CH_3—\overset{\displaystyle |}{\underset{\displaystyle CH_3}{CH}}—CH_2—SH$$

The longest carbon chain has three carbon atoms (propane), and both a methyl group and a sulfhydryl group are attached to the chain. The IUPAC name is 2-methyl-1-propanethiol.
c. The structure of this thiol is $CH_3—CH_2—CH_2—CH_2—SH$. The alkyl group is a butyl group, giving a common name of butyl mercaptan for this thiol.
d. The thiol structural formula is

$$CH_3—\overset{\displaystyle |}{\underset{\displaystyle SH}{CH}}—CH_3$$

The sulfhydryl group is attached to an isopropyl group; the common name is isopropyl mercaptan.

Practice Exercise 14.6

Convert each of the following common names for thiols to IUPAC names or vice versa.

a. Methyl mercaptan   **b.** *sec*-Butyl mercaptan
c. 2-Methyl-2-propanethiol   **d.** 1-Pentanethiol

Answers: a. Methanethiol; **b.** 2-Butanethiol; **c.** *tert*-Butyl mercaptan; **d.** Pentyl mercaptan

Figure 14.21 Thiols are responsible for the strong odor of "essence of skunk." Their odor is an effective defense mechanism.

Properties of Thiols

Two important properties of thiols are lower boiling points than alcohols of similar size (because of lack of hydrogen bonding) and a strong, disagreeable odor. The familiar odor of natural gas results from the addition of a low concentration of methanethiol ($CH_3—SH$) to the gas. The exceptionally low threshold of detection for this thiol enables consumers to smell a gas leak long before the gas, which is itself odorless, reaches dangerous levels. The scent of skunks (Figure 14.21) is due primarily to two thiols.

$$\underset{\displaystyle SH}{\overset{\displaystyle |}{CH_2}}—CH_2—\underset{\displaystyle CH_3}{\overset{\displaystyle |}{CH}}—CH_3 \qquad \underset{\displaystyle SH—CH_2}{}\overset{\displaystyle H}{\diagup}C=C\overset{\displaystyle CH_3}{\diagdown}\underset{\displaystyle H}{}$$

3-Methyl-1-butanethiol   *trans*-2-Butene-1-thiol

A major contributor to the typical smell of the human armpit is a compound that contains both an alcohol and a thiol functional group.

3-Methyl-3-sulfanyl-1-hexanol

Although hundreds of substances contribute to the aroma of freshly brewed coffee, the one most responsible for this characteristic odor is 2-(sulfhydrylmethyl)furan. Structurally this compound is both a thiol and a cyclic ether.

2-(Sulfhydrylmethyl)furan

Thiols are easily oxidized but yield different products than their alcohol analogs. Thiols form *disulfides*. Each of two thiol groups loses a hydrogen atom, thus linking the two sulfur atoms together via a disulfide group, —S—S—.

$$R—SH + HS—R \xrightarrow{\text{Oxidation}} R—S—S—R + 2H$$

A disulfide

Reversal of this reaction, a reduction process, is also readily accomplished. Breaking of the disulfide bond regenerates two thiol molecules.

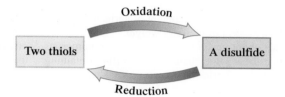

These two "opposite reactions" are of biological importance in the area of protein chemistry. Disulfide bonds formed from the interaction of two —SH groups contribute in a major way to protein structure (Chapter 20).

14.21 SULFUR ANALOGS OF ETHERS

Sulfur analogs of ethers are known as thioethers (or sulfides). A **thioether** *is an organic compound in which a sulfur atom is bonded to two carbon atoms by single bonds*. The generalized formula for a thioether is R—S—R. Like thiols, thioethers (or sulfides) have strong characteristic odors.

Thioethers are named in the same way as ethers, with *sulfide* used in place of *ether* in common names and *alkylthio* used in place of *alkoxy* in IUPAC names.

$CH_3—S—CH_3$
Dimethyl sulfide
(methylthiomethane)

Methyl phenyl sulfide
(methylthiobenzene)

4-(ethylthio)-2-Methyl-2-pentene

Bacteria in the mouth interact with saliva and leftover food to produce such compounds as hydrogen sulfide, methanethiol (a thioalcohol), and dimethyl sulfide (a thioether). These compounds, which have odors detectable in air at concentrations of parts per billion, are responsible for "morning breath."

In general, thiols are more reactive than their alcohol counterparts, and thioethers are more reactive than their corresponding ethers. The larger size of a sulfur atom compared to an oxygen atom (see Figure 14.22) results in a carbon–sulfur covalent bond that is weaker than a carbon–oxygen covalent bond. An added factor is that sulfur's electronegativity (2.5) is significantly lower than that of oxygen (3.5). Dimethyl sulfide is a gas at room temperature and ethyl methyl sulfide is a liquid.

Thiols and thioethers are functional group isomers in the same manner that alcohols and ethers are functional group isomers (Section 14.17). For example, the

Figure 14.22 A comparison involving space-filling models for dimethyl ether and dimethyl sulfide (dimethyl thioether). A sulfur atom is much larger than an oxygen atom. This results in a carbon-sulfur bond being weaker than a carbon-oxygen bond.

Dimethyl ether   **Dimethyl sulfide**

Chemical Connections

Garlic and Onions: Odiferous Medicinal Plants

Garlic and onions, which botanically belong to the same plant genus, are vegetables known for the bad breath—and perspiration odors—associated with their consumption. These effects are caused by organic sulfur-containing compounds, produced when garlic and onions are cut, that reach the lungs and sweat glands via the bloodstream. The total sulfur content of garlic and onions amounts to about one percent of their dry weight.

Less well known about garlic and onions are the numerous studies showing that these same "bad breath" sulfur-containing compounds are health-promoting substances that have the capacity to prevent or at least ameliorate a host of ailments in humans and animals. The list of beneficial effects associated with garlic use is longer than that for any other medicinal plant. Only onions come close to having the same kind of efficacy. Garlic has been shown to function as an antibacterial, antiviral, antifungal, antiprotozal, and antiparasitic agent. In the area of heart and circulatory problems, garlic contains vasodilative compounds that improve blood fluidity and reduce platelet aggregation. The health-promoting role of onions has not been explored as thoroughly as that of garlic, but the studies undertaken so far seem to confirm that onions are second only to garlic in their "healing powers."

Whole garlic bulbs and whole onions that remain undisturbed and intact do not contain any strongly odiferous compounds and display virtually no physiological activity. The act of cutting or crushing these vegetables causes a cascade of reactions to occur in damaged plant cells. Exposure to oxygen in the air is an important facet of these reactions. Over one hundred sulfur-containing organic compounds are formed in garlic and probably a similar number are produced in the less-studied onion. Many of the compounds so produced are common to both garlic and onions. The compounds associated with garlic ingestion that contribute to bad breath include allyl methyl sulfide, allyl methyl disulfide, diallyl sulfide, and diallyl disulfide. Their structures are given in the accompanying table.

Not all of the strongly odiferous compounds associated with garlic and onions elicit negative responses from the human olfactory system. For example, the smell of fried onions is considered a pleasant odor by most people. Compounds contributing to the "fried onion smell" include methyl propyl disulfide, methyl propyl trisulfide, allyl propyl disulfide, and

dipropyl trisulfide. Structures for these compounds are also given in the accompanying table.

In addition to physiologically active sulfur compounds, garlic and onions also contain a variety of other healthful ingredients. Among these are the B vitamins thiamine and riboflavin and vitamin C. Almost all of the trace elements are also present, including manganese, iron, phosphorus, selenium, and chromium. The actual amount of a given trace element depends on the soil in which the garlic or onion was grown.

Garlic Breath

$CH_2\!=\!CH\!-\!CH_2\!-\!S\!-\!CH_3$
Allyl methyl sulfide

$CH_2\!=\!CH\!-\!CH_2\!-\!S\!-\!S\!-\!CH_3$
Allyl methyl disulfide

$CH_2\!=\!CH\!-\!CH_2\!-\!S\!-\!CH_2\!-\!CH\!=\!CH_2$
Diallyl sulfide

$CH_2\!=\!CH\!-\!CH_2\!-\!S\!-\!S\!-\!CH_2\!-\!CH\!=\!CH_2$
Diallyl disulfide

Fried Onions

$CH_3\!-\!S\!-\!S\!-\!CH_2\!-\!CH_2\!-\!CH_3$
Methyl propyl disulfide

$CH_3\!-\!S\!-\!S\!-\!S\!-\!CH_2\!-\!CH_2\!-\!CH_3$
Methyl propyl trisulfide

$CH_2\!=\!CH\!-\!CH_2\!-\!S\!-\!S\!-\!CH_2\!-\!CH_2\!-\!CH_3$
Allyl propyl disulfide

$CH_3\!-\!CH_2\!-\!CH_2\!-\!S\!-\!S\!-\!S\!-\!CH_2\!-\!CH_2\!-\!CH_3$
Dipropyl trisulfide

thiol 1-propanethiol and the thioether methylthioethane both have the molecular formula C_3H_8S.

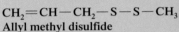

$CH_3\!-\!CH_2\!-\!CH_2\!-\!SH$ $CH_3\!-\!S\!-\!CH_2\!-\!CH_3$

1-Propanethiol   Methylthioethane

The Chemistry at a Glance feature on page 433 contrasts the four major types of compounds considered in this chapter—alcohols, thiols, ethers, and thioethers—in terms of structure, hydrogen bonding characteristics, and nomenclature.

CHEMISTRY AT A GLANCE

Alcohols, Thiols, Ethers, and Thioethers

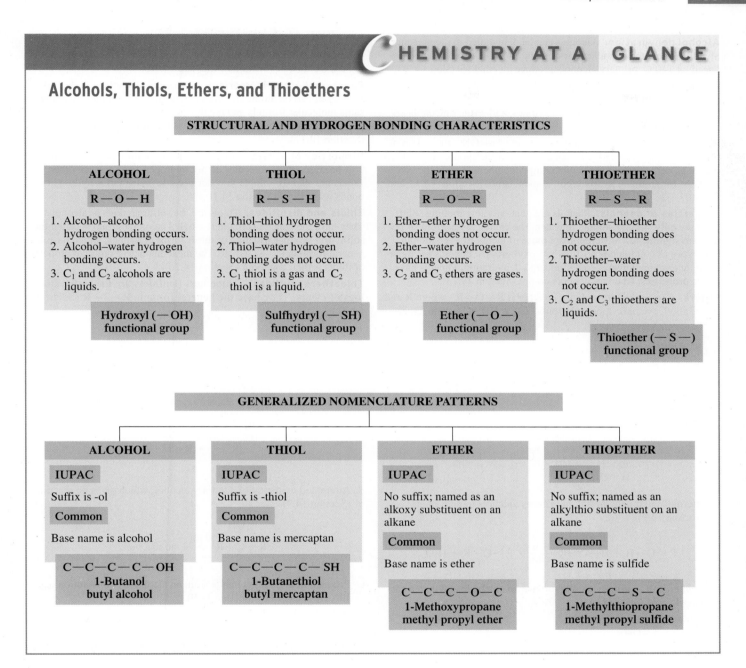

STRUCTURAL AND HYDROGEN BONDING CHARACTERISTICS

ALCOHOL

R—O—H

1. Alcohol–alcohol hydrogen bonding occurs.
2. Alcohol–water hydrogen bonding occurs.
3. C_1 and C_2 alcohols are liquids.

Hydroxyl (—OH) functional group

THIOL

R—S—H

1. Thiol–thiol hydrogen bonding does not occur.
2. Thiol–water hydrogen bonding does not occur.
3. C_1 thiol is a gas and C_2 thiol is a liquid.

Sulfhydryl (—SH) functional group

ETHER

R—O—R

1. Ether–ether hydrogen bonding does not occur.
2. Ether–water hydrogen bonding occurs.
3. C_2 and C_3 ethers are gases.

Ether (—O—) functional group

THIOETHER

R—S—R

1. Thioether–thioether hydrogen bonding does not occur.
2. Thioether–water hydrogen bonding does not occur.
3. C_2 and C_3 thioethers are liquids.

Thioether (—S—) functional group

GENERALIZED NOMENCLATURE PATTERNS

ALCOHOL

IUPAC

Suffix is -ol

Common

Base name is alcohol

C—C—C—C—OH
1-Butanol
butyl alcohol

THIOL

IUPAC

Suffix is -thiol

Common

Base name is mercaptan

C—C—C—C—SH
1-Butanethiol
butyl mercaptan

ETHER

IUPAC

No suffix; named as an alkoxy substituent on an alkane

Common

Base name is ether

C—C—C—O—C
1-Methoxypropane
methyl propyl ether

THIOETHER

IUPAC

No suffix; named as an alkylthio substituent on an alkane

Common

Base name is sulfide

C—C—C—S—C
1-Methylthiopropane
methyl propyl sulfide

CONCEPTS TO REMEMBER

Alcohols. Alcohols are organic compounds that contain an —OH group attached to a saturated carbon atom. The general formula for an alcohol is R—OH, where R is an alkyl group (Section 14.2).

Nomenclature of alcohols. The IUPAC names of simple alcohols end in *-ol*, and their carbon chains are numbered to give precedence to the location of the —OH group. Alcohol common names contain the word *alcohol* preceded by the name of the alkyl group (Section 14.3).

Isomerism for alcohols. Constitutional isomerism is possible for alcohols containing three or more carbon atoms. Both skeletal and positional isomers are possible (Section 14.4).

Physical properties of alcohols. Alcohol molecules hydrogen-bond to each other and to water molecules. They thus have higher-than-normal boiling points, and the low-molecular-mass alcohols are soluble in water (Section 14.6).

Classification of alcohols. Alcohols are classified on the basis of the number of carbon atoms bonded to the carbon attached to the —OH group. In primary alcohols, the —OH group is bonded to a carbon atom bonded to only one other C atom. In secondary alcohols, the —OH-containing C atom is attached to two other C atoms. In tertiary alcohols, it is attached to three other C atoms (Section 14.8).

Alcohol dehydration. Alcohols can be dehydrated in the presence of sulfuric acid to form alkenes or ethers. At 180°C, an alkene is produced; at 140°C, primary alcohols produce an ether (Section 14.9).

Alcohol oxidation. Oxidation of primary alcohols first produces an aldehyde, which is then further oxidized to a carboxylic acid. Secondary alcohols are oxidized to ketones, and tertiary alcohols are resistant to oxidation (Section 14.9).

Phenols. Phenols have the general formula Ar—OH, where Ar represents an aryl group derived from an aromatic compound. Phenols are named as derivatives of the parent compound phenol, using the conventions for aromatic hydrocarbon nomenclature (Sections 14.11 and 14.12).

Properties of phenols. Phenols are generally low-melting solids; most are only slightly soluble in water. The chemical reactions of phenols are significantly different from those of alcohols, even though both types of compounds possess hydroxyl groups. Phenols are more resistant to oxidation and do not undergo dehydration. Phenols have acidic properties, whereas alcohols do not (Section 14.13).

Ethers. The general formula for an ether is R—O—R′, where R and R′ are alkyl, cycloalkyl, or aryl groups. In the IUPAC system, ethers are named as alkoxy derivatives of alkanes. Common names are obtained by giving the R group names in alphabetical order and adding the word *ether* (Sections 14.15 and 14.16).

Functional group isomerism. Ethers and alcohols with the same number of carbon atoms and the same degree of saturation have the same molecular formula and are thus isomers of each other. This type of constitutional isomerism is known as functional group isomerism (Section 14.17).

Properties of ethers. Ethers have lower boiling points than alcohols because ether molecules do not hydrogen-bond to each other. Ethers are slightly soluble in water because water forms hydrogen bonds with ethers (Section 14.18).

Thiols and disulfides. Thiols are the sulfur analogs of alcohols. They have the general formula R—SH. The —SH group is called the sulfhydryl group. Oxidation of thiols forms disulfides, which have the general formula R—S—S—R. The most distinctive physical property of thiols is their foul odor (Section 14.20).

Thioethers (sulfides). Thioethers (sulfides) are the sulfur analogs of ethers. They have the general formula R—S—R (Section 14.21).

KEY REACTIONS AND EQUATIONS

1. Intramolecular dehydration of alcohols to give alkenes (Section 14.9)

$$-\overset{|}{\underset{H}{C}}-\overset{|}{\underset{OH}{C}}- \xrightarrow[180°C]{H_2SO_4} \ \ \diagup C = C \diagdown \ + H_2O$$

2. Intermolecular dehydration of primary alcohols to give ethers (Section 14.9)

$$R\!-\!O\!-\!H + H\!-\!O\!-\!R \xrightarrow[140°C]{H_2SO_4} R\!-\!O\!-\!R + H_2O$$

3. Oxidation of a primary alcohol to give an aldehyde and then a carboxylic acid (Section 14.9)

$$R-\overset{OH}{\underset{H}{\overset{|}{C}}}-H \xrightarrow{[O]} R-\overset{O}{\overset{\|}{C}}-H + H_2O$$

Aldehyde

$$\xrightarrow[\text{oxidation}]{[O] \ \text{Further}} R-\overset{O}{\overset{\|}{C}}-OH$$

Carboxylic acid

4. Oxidation of a secondary alcohol to give a ketone (Section 14.9)

$$R-\overset{OH}{\underset{H}{\overset{|}{C}}}-R' \xrightarrow{[O]} R-\overset{O}{\overset{\|}{C}}-R' + 2H$$

5. Attempted oxidation of a tertiary alcohol, which gives no reaction (Section 14.9)

$$R-\overset{OH}{\underset{R''}{\overset{|}{C}}}-R' \xrightarrow{[O]} \text{no reaction}$$

6. Production of an alkyl halide from an alcohol by substitution using PX₃ (X is Cl or Br) (Section 14.9)

$$R-OH \xrightarrow[\text{heat}]{PX_3} R-X$$

7. Oxidation of a thiol to give a disulfide (Section 14.20)

$$R-SH + HS-R \xrightarrow{[O]} R-S-S-R + 2H$$

8. Reduction of a disulfide to give a thiol (Section 14.20)

$$R-S-S-R + 2H \longrightarrow R-SH + HS-R$$

EXERCISES *and* PROBLEMS

The members of each pair of problems in this section test similar material.

⬤ Bonding Characteristics of Oxygen (Section 14.1)

14.1 In organic compounds, how many covalent bonds does each of the following types of atoms form?
   a. Oxygen   b. Hydrogen   c. Carbon   d. A halogen

14.2 Indicate whether or not each of the following covalent bonding behaviors are possible for an oxygen atom in an organic compound.
   a. One single bond   b. Two single bonds
   c. One double bond   d. Two double bonds

Structural Characteristics of Alcohols (Section 14.2)

14.3 What is the generalized formula for an alcohol?

14.4 What is the name of the functional group that characterizes an alcohol?

14.5 Contrast, in general terms, the structures of an alcohol and water.

14.6 Contrast, in general terms, the structures of an alcohol and an alkane.

Nomenclature for Alcohols (Section 14.3)

14.7 Assign an IUPAC name to each of the following alcohols.

a.
$$CH_3—CH_2—CH_2—\overset{\overset{\displaystyle OH}{|}}{CH}—CH_3$$

b.
$$CH_3—\overset{\overset{\displaystyle CH_3}{|}}{CH}—\overset{\overset{\displaystyle OH}{|}}{CH}—CH_3$$

c.
$$CH_3—CH_2—CH_2—\overset{|}{CH}—CH_2—CH_3$$
$$\overset{|}{CH_2}$$
$$\overset{|}{OH}$$

d.
$$CH_3—CH_2—\overset{\overset{\displaystyle |}{CH}}{}—OH$$
$$\overset{|}{CH_3}$$

14.8 Assign an IUPAC name to each of the following alcohols.

a.
$$CH_3—CH_2—\overset{\overset{\displaystyle |}{CH}}{}—CH_2—CH_3$$
$$\overset{|}{OH}$$

b.
$$CH_3—CH_2—\overset{|}{CH}—\overset{|}{CH}—CH_3$$
$$\overset{|}{OH}\quad\overset{|}{CH_3}$$

c.
$$CH_3—CH_2—\overset{|}{CH}—CH_2—CH_2—CH_3$$
$$\overset{|}{CH_2}—CH_2—CH_2—OH$$

d.
$$CH_3—\overset{\overset{\displaystyle OH}{|}}{\underset{\underset{\displaystyle CH_3}{|}}{C}}—CH_2—CH_3$$

14.9 Assign an IUPAC name to each of the following alcohols.

a. b. c. d.

14.10 Assign an IUPAC name to each of the following alcohols.

a. b. c. d.

14.11 Write a condensed structural formula for each of the following alcohols.

a. 2-Methyl-1-propanol b. 4-Methyl-2-pentanol
c. 2-Phenyl-2-propanol d. 2-Methylcyclobutanol

14.12 Write a condensed structural formula for each of the following alcohols.

a. 2-Methyl-2-heptanol b. 3-Ethyl-2-pentanol
c. 3-Phenyl-1-butanol d. 3,5-Dimethylcyclohexanol

14.13 Write a condensed structural formula for, and assign an IUPAC name to, each of the following alcohols.

a. Pentyl alcohol b. Propyl alcohol
c. Isobutyl alcohol d. *sec*-Butyl alcohol

14.14 Write a condensed structural formula for, and assign an IUPAC name to, each of the following alcohols.

a. Butyl alcohol b. Hexyl alcohol
c. Isopropyl alcohol d. *tert*-Butyl alcohol

14.15 Assign an IUPAC name to each of the following polyhydroxy alcohols.

a.
$$CH_2—CH—CH_3$$
$$\overset{|}{OH}\quad\overset{|}{OH}$$

b.
$$CH_2—CH_2—CH_2—CH—CH_3$$
$$\overset{|}{OH}\qquad\qquad\overset{|}{OH}$$

c.
$$CH_3—CH_2—CH—CH_2—CH_2$$
$$\overset{|}{OH}\qquad\overset{|}{OH}$$

d.
$$CH_2—CH—CH—CH_2$$
$$\overset{|}{OH}\quad\overset{|}{OH}\quad\overset{|}{CH_3}\ \overset{|}{OH}$$

14.16 Assign an IUPAC name to each of the following polyhydroxy alcohols.

a.
$$CH_2—CH—CH_2$$
$$\overset{|}{OH}\quad\overset{|}{CH_3}\ \overset{|}{OH}$$

b.
$$CH_2—CH—CH_2$$
$$\overset{|}{OH}\quad\overset{|}{OH}\quad\overset{|}{CH_3}$$

c.
$$CH_3—CH_2—CH—CH—CH_3$$
$$\overset{|}{OH}\quad\overset{|}{OH}$$

d.
$$CH_2—CH—CH—CH_3$$
$$\overset{|}{OH}\quad\overset{|}{OH}\quad\overset{|}{OH}$$

14.17 Utilizing IUPAC rules, name each of the following compounds. Don't forget to use *cis*- and *trans*- prefixes (Section 12.14) where needed.

a. b. c. d.

14.18 Utilizing IUPAC rules, name each of the following compounds. Don't forget to use *cis*- and *trans*- prefixes (Section 12.14) where needed.

a. b. c. d.

14.19 Write a condensed structural formula for each of the following *unsaturated* alcohols.
 a. 4-Penten-2-ol
 b. 1-Pentyn-3-ol
 c. 3-Methyl-3-buten-2-ol
 d. *cis*-2-Buten-1-ol

14.20 Write a condensed structural formula for each of the following *unsaturated* alcohols.
 a. 1-Penten-3-ol
 b. 3-Butyn-1-ol
 c. 2-Methyl-3-buten-1-ol
 d. *trans*-3-Penten-1-ol

14.21 Each of the following alcohols is named incorrectly. However, the names give correct structural formulas. Draw structural formulas for the compounds, and then write the correct IUPAC name for each alcohol.
 a. 2-Ethyl-1-propanol
 b. 2,4-Butanediol
 c. 2-Methyl-3-butanol
 d. 1,4-Cyclopentanediol

14.22 Each of the following alcohols is named incorrectly. However, the names give correct structural formulas. Draw structural formulas for the compounds, and then write the correct IUPAC name for each alcohol.
 a. 3-Ethyl-2-butanol
 b. 3,4-Pentanediol
 c. 3-Methyl-3-butanol
 d. 1,1-Dimethyl-1-butanol

Isomerism for Alcohols (Section 14.4)

14.23 Indicate whether each of the following compounds is or is not a constitutional isomer of 1-hexanol.

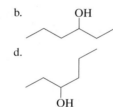
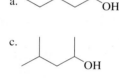

a.
b.
c.
d.

14.24 Indicate whether each of the following compounds is or is not a constitutional isomer of 2-pentanol.

a.
b.
c.
d.

14.25 Give IUPAC names for all isomeric C_7 monohydroxy alcohols in which the carbon chain is unbranched.

14.26 Give IUPAC names for all isomeric C_8 monohydroxy alcohols in which the carbon chain is unbranched.

14.27 For which values of x is the alcohol name 2-methyl-x-pentanol a correct IUPAC name?

14.28 For which values of x is the alcohol name 3-methyl-x-pentanol a correct IUPAC name?

Important Common Alcohols (Section 14.5)

14.29 What does each of the following terms mean?
 a. Absolute alcohol
 b. Grain alcohol
 c. Rubbing alcohol
 d. Drinking alcohol

14.30 What does each of the following terms mean?
 a. Wood alcohol
 b. Denaturated alcohol
 c. 70-Proof alcohol
 d. "Alcohol"

14.31 Give the IUPAC name of the alcohol that fits each of the following descriptions.
 a. Thick liquid that has the consistency of honey
 b. Often produced via a fermentation process
 c. Used as a race car fuel
 d. Industrially produced from CO and H_2

14.32 Give the IUPAC name of the alcohol that fits each of the following descriptions.
 a. Sometimes used as a skin coolant for the human body
 b. Antifreeze ingredient
 c. Active ingredient in alcoholic beverages
 d. Moistening agent in many cosmetics

Physical Properties of Alcohols (Section 14.6)

14.33 Explain why the boiling points of alcohols are much higher than those of alkanes with similar molecular masses.

14.34 Explain why the water solubilities of alcohols are much higher than those of alkanes with similar molecular masses.

14.35 Which member of each of the following pairs of compounds would you expect to have the higher boiling point?
 a. 1-Butanol and 1-heptanol
 b. Butane and 1-propanol
 c. Ethanol and 1,2-ethanediol

14.36 Which member of each of the following pairs of compounds would you expect to have the higher boiling point?
 a. 1-Octanol and 1-pentanol
 b. Pentane and 1-butanol
 c. 1,3-Propanediol and 1-propanol

14.37 Which member of each of the following pairs of compounds would you expect to be more soluble in water?
 a. Butane and 1-butanol
 b. 1-Octanol and 1-pentanol
 c. 1,2-Butanediol and 1-butanol

14.38 Which member of each of the following pairs of compounds would you expect to be more soluble in water?
 a. 1-Pentanol and 1-butanol
 b. 1-Propanol and 1-hexanol
 c. 1,2,3-Propanetriol and 1-hexanol

14.39 Determine the maximum number of hydrogen bonds that can form between an ethanol molecule and
 a. other ethanol molecules
 b. water molecules
 c. methanol molecules
 d. 1-propanol molecules

14.40 Determine the maximum number of hydrogen bonds that can form between a methanol molecule and
 a. other methanol molecules
 b. water molecules
 c. 1-propanol molecules
 d. 2-propanol molecules

Preparation of Alcohols (Section 14.7)

14.41 Write the structure of the expected predominant organic product formed in each of the following reactions.
 a. $CH_2{=}CH_2 + H_2O \xrightarrow{H_2SO_4}$
 b.
$$CH_3{-}CH_2{-}\overset{\displaystyle O}{\overset{\|}{C}}{-}H + H_2 \xrightarrow{\text{Catalyst}}$$

c. $CH_3-CH_2-C=CH_2 + H_2O \xrightarrow{H_2SO_4}$
                 |
                CH_3

d.
              O
              ||
$CH_3-CH_2-C-CH_2-CH_3 + H_2 \xrightarrow{Catalyst}$

14.42 Write the structure of the expected predominant organic product formed in each of the following reactions.

a. $CH_3-CH=CH-CH_3 + H_2O \xrightarrow{H_2SO_4}$

b.
              O
              ||
$CH_3-CH_2-C-CH_3 + H_2 \xrightarrow{Catalyst}$

c.
    O
    ||
$CH_3-C-H + H_2 \xrightarrow{Catalyst}$

d. $CH_3-CH-CH=CH-CH_3 + H_2O \xrightarrow{H_2SO_4}$
        |
       CH_3

Classification of Alcohols (Section 14.8)

14.43 Classify each of the alcohols in Problem 14.7 as a primary, secondary, or tertiary alcohol.

14.44 Classify each of the alcohols in Problem 14.8 as a primary, secondary, or tertiary alcohol.

Chemical Reactions of Alcohols (Section 14.9)

14.45 Draw the structure of the organic product expected to be predominant when each of the following alcohols is dehydrated using sulfuric acid at the temperature indicated.

a. $CH_3-CH-CH_3 \xrightarrow[180°C]{H_2SO_4}$ $CH_2=CH-CH_3$
        |
       OH

b. $CH_3-CH_2-CH-CH_2-OH \xrightarrow[180°C]{H_2SO_4}$
              |
            CH_3

c. $CH_3-CH-OH \xrightarrow[140°C]{H_2SO_4}$ $CH_3-CH=CH_2$
        |
       CH_3

d. $CH_3-CH_2-CH_2-OH \xrightarrow[140°C]{H_2SO_4}$

14.46 Draw the structure of the organic product expected to be predominant when each of the following alcohols is dehydrated using sulfuric acid at the temperature indicated.

a. $CH_3-CH-CH_2-OH \xrightarrow[140°C]{H_2SO_4}$
       |
      CH_3

b. $CH_3-CH-CH_2-OH \xrightarrow[180°C]{H_2SO_4}$
       |
      CH_3

c. $CH_3-CH-CH_2-CH_3 \xrightarrow[180°C]{H_2SO_4}$
       |
      OH

d. $CH_3-CH-CH_2-CH_3 \xrightarrow[140°C]{H_2SO_4}$
       |
      OH

14.47 Identify the alcohol reactant from which each of the following products was obtained by an alcohol dehydration reaction.

a. Alcohol $\xrightarrow[180°C]{H_2SO_4}$ $CH_3-CH=C-CH_3$
                             |
                           CH_3

b. Alcohol $\xrightarrow[180°C]{H_2SO_4}$ $CH_3-CH=CH_2$

c. Alcohol $\xrightarrow[140°C]{H_2SO_4}$ $CH_3-CH_2-O-CH_2-CH_3$

d. Alcohol $\xrightarrow[140°C]{H_2SO_4}$ $CH_3-CH-CH_2-O-CH_2-CH-CH_3$
                         | |
                        CH_3 CH_3

14.48 Identify the alcohol reactant from which each of the following products was obtained by an alcohol dehydration reaction.

a. Alcohol $\xrightarrow[180°C]{H_2SO_4}$ $CH_2=C-CH_2-CH_3$
                               |
                            CH_3

b. Alcohol $\xrightarrow[180°C]{H_2SO_4}$ $CH_3-CH_2-CH=CH_2$

c. Alcohol $\xrightarrow[140°C]{H_2SO_4}$ CH_3-O-CH_3

d. Alcohol $\xrightarrow[140°C]{H_2SO_4}$ $CH_3-CH_2-O-CH_2-CH_3$

14.49 Draw the structure of the alcohol that could be used to prepare each of the following compounds in an oxidation reaction.

a.
        OH
        ||
$CH_3-CH_2-CH-CH_3$

b.
        O
        ||
$CH_3-CH_2-CH-OH$

c.
        O
        ||
CH_3-CH_2-C-H

d.
      O   OH
      || |
       C-H
(cyclopentane ring)

14.50 Draw the structure of the alcohol that could be used to prepare each of the following compounds in an oxidation reaction.

a.
          O
          ||
$CH_3-CH-C-H$
     |
   CH_3

b.
          O
          ||
$CH_3-CH-C-CH_3$
     |
   CH_3

c.
               O
               ||
$CH_3-CH-CH_2-C-OH$
     |
   CH_3

d.
        O
        ||
        C-H
(cyclohexane ring)

14.51 Draw the structure of the expected predominant organic product formed in each of the following reactions.

a. $CH_3-CH_2-CH_2-OH \xrightarrow[heat]{PCl_3}$

b.
     CH_3
(cyclopentane ring with CH₃ and OH) $\xrightarrow[180°C]{H_2SO_4}$
     OH

c. $CH_3-CH-CH_2-CH_3 \xrightarrow{K_2Cr_2O_7}$
       |
      OH

d. $CH_3-CH_2-OH \xrightarrow[140°C]{H_2SO_4}$

14.52 Draw the structure of the expected predominant organic product formed in each of the following reactions.

a.

![structure: cyclopentane with CH3 and OH, PCl3, heat]

b. $CH_2{-}CH_2{-}CH_2{-}CH_3$ with OH on first carbon, $\xrightarrow[140°C]{H_2SO_4}$

c. $CH_3{-}CH_2{-}CH{-}CH_2{-}CH_3$ with OH below, $\xrightarrow{K_2Cr_2O_7}$

d. $CH_3{-}CH{-}OH$ with CH_3 below, $\xrightarrow[180°C]{H_2SO_4}$

Polymeric Alcohols (Section 14.10)

14.53 Describe the physical and chemical characteristics of the polymeric alcohol PVA [poly(vinyl alcohol)].

14.54 Draw a structural representation for the polymeric alcohol PVA [poly(vinyl alcohol)].

Structural Characteristics of Phenols (Section 14.11)

14.55 Explain why the first of the following two compounds is a phenol, and the second is not.

$CH_3{-}$[benzene ring]${-}OH$ $CH_3{-}$[benzene ring]${-}CH_2{-}OH$

14.56 Explain why the first of the following two compounds is a phenol, and the second is not.

$CH_3{-}$[benzene ring]${-}OH$ $CH_3{-}$[cyclohexane ring]${-}OH$

Nomenclature for Phenols (Section 14.12)

14.57 Name the following phenols.

a. [benzene ring with OH and CH2—CH3]

b. [benzene ring with Cl and OH]

c. [benzene ring with OH and CH3]

d. [benzene ring with OH top and OH bottom]

14.58 Name the following phenols.

a. [benzene ring with OH and CH2—CH2—CH3]

b. [benzene ring with OH and OH]

c. [benzene ring with OH and CH3]

d. [benzene ring with OH, Cl, and CH—CH3 with CH3]

14.59 Draw a structural formula for each of the following phenols.
a. 2-Ethylphenol
b. 2,4-Dibromophenol
c. *m*-Cresol
d. Resorcinol

14.60 Draw a structural formula for each of the following phenols.
a. 3-Bromophenol
b. *o*-Cresol
c. Catechol
d. 2,6-Dichlorophenol

Physical and Chemical Properties of Phenols (Section 14.13)

14.61 Characterize phenols in terms of their physical state at room temperature.

14.62 Characterize phenols in terms of their solubility in water.

14.63 What is the difference, if any, between phenols and alcohols in terms of the following?
a. Flammability
b. Ability to undergo halogenation reactions

14.64 What is the difference, if any, between phenols and alcohols in terms of the following?
a. Reaction with weak oxidizing agents
b. Ability to undergo dehydration reactions

14.65 Phenols are weak acids. Write an equation for the acid ionization of the compound phenol.

14.66 How does the acidity of phenols compare with that of inorganic weak acids?

Occurrence and Uses for Phenols (Section 14.14)

14.67 Phenolic compounds are frequently used as antiseptics and disinfectants. What is the difference between an antiseptic and a disinfectant?

14.68 Phenolic compounds are frequently used as antioxidants. What is an antioxidant?

14.69 Describe the structures of the phenolic antioxidants BHA and BHT in terms of the group attached to the phenolic ring system.

14.70 Identity the flavor/odor and uses associated with each of the following phenolic flavoring agents.
a. Thymol   b. Eugenol
c. Isoeugenol   d. Vanillin

Structural Characteristics of Ethers (Section 14.15)

14.71 Indicate whether each of the following structural notations denotes an ether.
a. R—O—R   b. R—O—H
c. Ar—O—R   d. Ar—O—Ar

14.72 What is the difference in meaning associated with each of the following pairs of notations?
a. R—O—R and R—O—H
b. Ar—O—R and Ar—O—Ar
c. Ar—O—H and Ar—O—R
d. R—O—R and Ar—O—Ar

Nomenclature for Ethers (Section 14.16)

14.73 Assign an IUPAC name to each of the following ethers.
a. $CH_3{-}CH_2{-}CH_2{-}O{-}CH_2{-}CH_3$
b. $CH_3{-}CH{-}CH_3$ with $O{-}CH_3$ below

c.

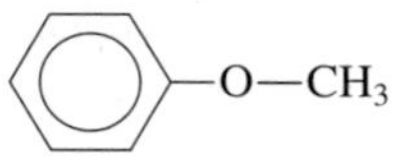

—O—CH₃

d.

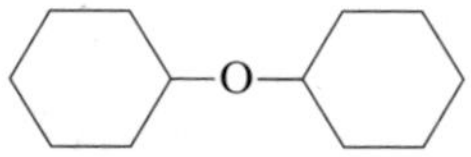

14.74 Assign an IUPAC name to each of the following ethers.

a. CH₃—CH₂—O—CH₂—CH₃

b. CH₃—CH—O—CH₃
      |
      CH₃

c. CH₃—CH₂—CH—CH₃
           |
           O—CH₂—CH₃

d.

O—CH₂—CH₃

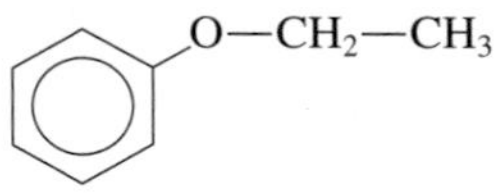

14.75 Assign a common name to each of the ethers in Problem 14.73.

14.76 Assign a common name to each of the ethers in Problem 14.74.

14.77 Assign an IUPAC name to each of the following ethers.

a.    b.

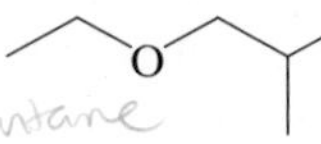

c. 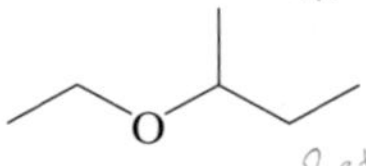   d.

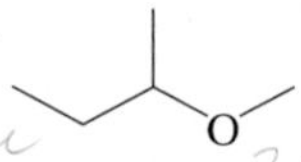

14.78 Assign an IUPAC name to each of the following ethers.

a. 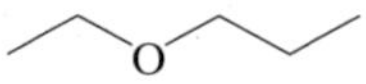   b.

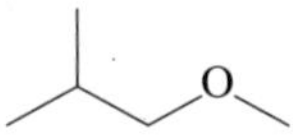

c. 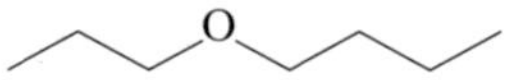   d.

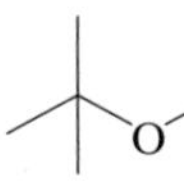

14.79 Draw the structure of each of the following ethers.

a. Isopropyl propyl ether   b. Ethyl phenyl ether
c. 3-Methylanisole   d. Ethoxycyclobutane

14.80 Draw the structure of each of the following ethers.

a. Butyl methyl ether   b. Anisole
c. Phenyl propyl ether   d. 1,3-Dimethoxybenzene

Isomerism for Ethers (Section 14.17)

14.81 Indicate whether each of the following ethers is or is not a constitutional isomer of ethyl propyl ether.

a.    b.

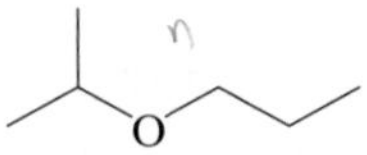

c. 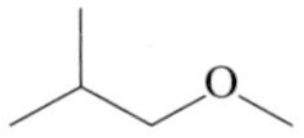   d.

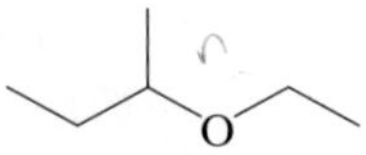

14.82 Indicate whether each of the following ethers is or is not a constitutional isomer of dipropyl ether.

a. 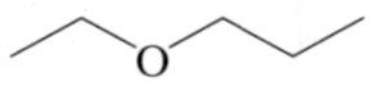   b.

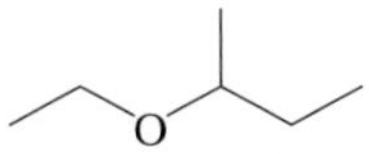

c. 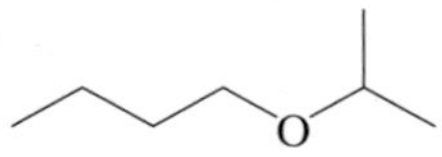   d.

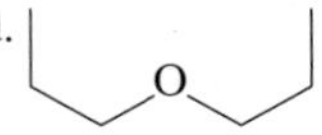

14.83 Give common names for all ethers that are constitutional isomers of ethyl propyl ether.

14.84 Give common names for all ethers that are constitutional isomers of butyl methyl ether.

14.85 Draw condensed structural formulas for the following.
a. All ethers that are functional group isomers of 1-butanol
b. All alcohols that are functional group isomers of 2-methoxypropane

14.86 Draw condensed structural formulas for the following.
a. All ethers that are functional group isomers of 2-methyl-1-propanol
b. All alcohols that are functional group isomers of 1-ethoxyethane

14.87 For which values of x is the ether name x-methoxy-3-methylpentane a correct IUPAC name?

14.88 For which values of x is the ether name x-methoxy-2-methylpentane a correct IUPAC name?

Physical and Chemical Properties of Ethers (Section 14.18)

14.89 Dimethyl ether and ethanol have the same molecular mass. Dimethyl ether is a gas at room temperature, and ethanol is a liquid at room temperature. Explain these observations.

14.90 Compare the solubility in water of ethers and alcohols that have similar molecular masses.

14.91 What are the two chemical hazards associated with ether use?

14.92 How do the chemical reactivities of ethers compare with those of
a. alkanes   b. alcohols

14.93 Explain why ether molecules cannot hydrogen-bond to each other.

14.94 How many hydrogen bonds can form between a single ether molecule and water molecules?

Cyclic Ethers (Section 14.19)

14.95 Classify each of the following molecular structures as that of a cyclic ether, a noncyclic ether, or a nonether.

a. 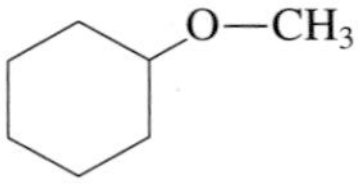
O—CH₃   b.

c. 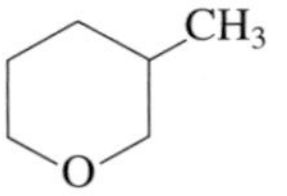
CH₃   d. 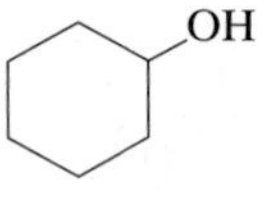
OH

14.96 Classify each of the following molecular structures as that of a cyclic ether, a noncyclic ether, or a nonether.

a. 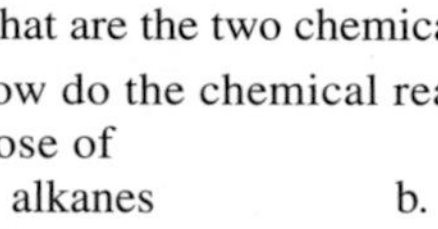   b.

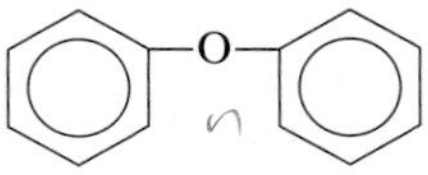

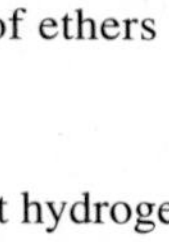

c. 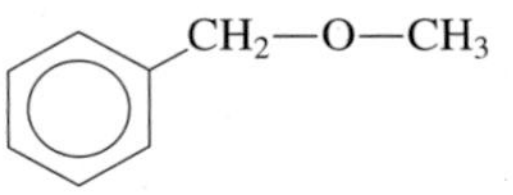
CH₂—O—CH₃   d. 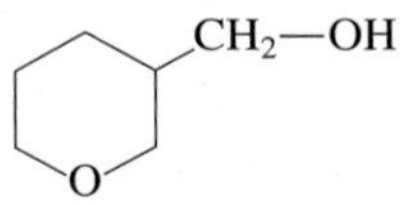
CH₂—OH

● **Sulfur Analogs of Alcohols (Section 14.20)**

14.97 Contrast the general structural formulas for a thioalcohol and an alcohol.

14.98 What is the generalized structure for and name of the functional group present in a thiol?

14.99 Draw a condensed structural formula for each of the following thiols.
   a. 1-Butanethiol
   b. 3-Methyl-1-pentanethiol
   c. Cyclopentanethiol
   d. 1,2-Ethanedithiol

14.100 Draw a condensed structural formula for each of the following thiols.
   a. 1-Propanethiol   b. 1,3-Pentanedithiol
   c. 3-Methyl-3-pentanethiol   d. 2-Methylcyclopentanethiol

14.101 Assign a common name to each of the following thiols.
   a. CH_3—SH
   b. CH_3—CH_2—CH_2—SH
   c. CH_3—CH_2—CH—SH
            |
            CH_3
   d. CH_3—CH—CH_2—SH
            |
            CH_3

14.102 Assign a common name to each of the following thiols.
   a. CH_3—CH_2—SH
   b. CH_3—CH—SH
            |
            CH_3
   c. CH_3—CH_2—CH_2—CH_2—SH
            CH_3
            |
   d. CH_3—C—SH
            |
            CH_3

14.103 Contrast the products that result from the oxidation of an alcohol and the oxidation of a thiol.

14.104 Write the formulas for the sulfur-containing organic products of the following reactions.
   a. $2CH_3$—CH_2—SH $\xrightarrow{\text{Oxidizing agent}}$
   b. CH_3—CH_2—S—S—CH_2—CH_3 $\xrightarrow{\text{Reducing agent}}$

● **Sulfur Analogs of Ethers (Section 14.21)**

14.105 Contrast the general structural formulas for a thioether and an ether.

14.106 What is the generalized structure for the functional group present in a thioether?

14.107 Assign both an IUPAC name and a common name to each of the following thioethers.
   a. CH_3—CH_2—S—CH_3
   b. CH_3—CH—S—CH_3
            |
            CH_3
   c. ⬡—S—CH_3
   d. CH_2=CH—CH_2—S—CH_3

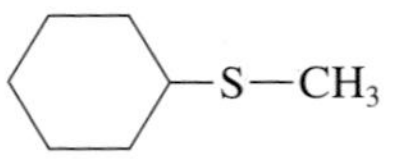

14.108 Assign both an IUPAC name and a common name to each of the following thioethers.
   a. CH_3—CH_2—S—CH_2—CH_3
   b. CH_3—CH—S—CH_2—CH_3
            |
            CH_3
   c. ⬠—S—CH_3
   d. CH_2=CH—CH_2—S—CH_2—CH_3

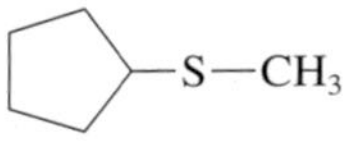

ADDITIONAL PROBLEMS

14.109 Assign an IUPAC name to each of the following compounds.
   a.
   b. (structure)
   c. (structure)
   d. (structure)

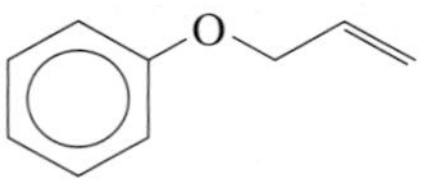

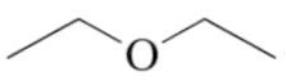

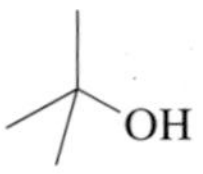

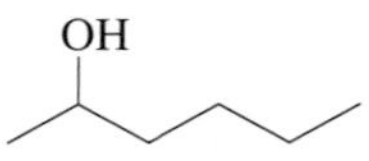

14.110 Draw structural formulas for the eight isomeric alcohols and six isomeric ethers that have the molecular formula $C_5H_{10}O$.

14.111 Three isomeric pentanols with unbranched carbon chains exist. Which of these, upon dehydration at 180°C, yields only 1-pentene as a product?

14.112 A mixture of methanol, 1-propanol, and H_2SO_4 (catalyst) is heated to 140°C. After reaction, the solution contains three different ethers. Draw a structural formula for each of the ethers.

14.113 Which of the terms *ether, alcohol, diol, thiol, thioether, thioalcohol, disulfide, sulfide,* and *peroxide* characterize(s) each of the following compounds? Note that more than one term may apply to a given compound.
   a. CH_3—S—S—CH_3
   b. CH_3—CH_2—O—O—CH_3
   c. HO—CH_2—CH_2—SH
   d. CH_3—O—CH_2—S—CH_3

14.114 Assign IUPAC names to the following compounds.
   a. CH_3—O—CH_2—CH_2—CH_2—OH
   b. CH_3—CH_2—O—CH_2—CH_2—O—CH_2—CH_3
   c. CH_3—S—CH_2—CH_3
   d. CH_3—O—CH_2—CH_2—S—CH_2—CH_3

*M*ultiple-Choice Practice Test

14.115 What is the correct IUPAC name for the alcohol whose structural formula is

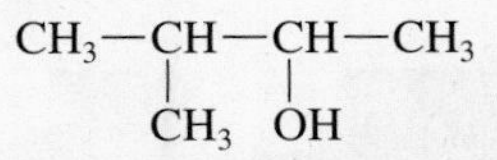

$$CH_3-CH-CH-CH_3$$
$$\quad\quad | \quad\quad |$$
$$\quad CH_3 \quad OH$$

   a. 2-Methylbutanol   b. 3-Methylbutanol
   c. 2-Methyl-3-butanol   d. 3-Methyl-2-butanol

14.116 Which of the following statements concerning common alcohols is *incorrect*?
   a. Wood alcohol and methyl alcohol are two names for the same compound.
   b. Denatured alcohol is drinking alcohol rendered unfit to drink.
   c. Rubbing alcohol is a 70% solution of ethyl alcohol.
   d. Glycerin and ethylene glycol are both polyhydroxy alcohols.

14.117 What is the organic product formed by the oxidation of a secondary alcohol?
   a. Aldehyde   b. Ketone
   c. Carboxylic acid   d. Alkene

14.118 How many constitutional isomeric alcohols are there that have the molecular formula $C_4H_{10}O$?
   a. Two   b. Three
   c. Four   d. Five

14.119 Which of the following statements concerning the physical properties of alcohols is *incorrect*?
   a. Alcohol solubility in water decreases as the carbon chain length increases.
   b. Alcohol solubility in water decreases as the number of —OH groups present increases.

   c. Alcohol boiling points increase as carbon chain length increases.
   d. C_1 to C_4 straight-chain alcohols are liquids at room temperature.

14.120 What is the molecular formula for the compound called phenol?
   a. C_6H_6O b. $C_6H_{12}O$
   c. $C_6H_6O_2$ d. $C_6H_{12}O_2$

14.121 Simple ethers may be viewed as derivatives of water in which both hydrogen atoms have been replaced with which of the following?
   a. Alkyl groups   b. Alkoxy groups
   c. Hydroxyl groups   d. Sulfhydryl groups

14.122 What is the common name for the compound 2-ethoxypropane?
   a. Diethyl ether
   b. Diisopropyl ether
   c. Ethyl propyl ether
   d. Ethyl isopropyl ether

14.123 Which of the following is a characteristic property of thiols?
   a. Extremely strong odors
   b. Abnormally high boiling points
   c. Extensive intermolecular hydrogen bonding
   d. Strong resistance to oxidation

14.124 In which of the following pairs of compounds are the two members of the pair constitutional isomers?
   a. Methoxymethane and methoxyethane
   b. 2-Propanol and isopropyl alcohol
   c. Ethanol and ethanediol
   d. Propyl alcohol and ethyl methyl ether

Aldehydes and Ketones

Chapter Outline

Benzaldehyde is the main flavor component in almonds. Aldehydes and ketones are responsible for the odor and taste of numerous nuts and spices.

The word *carbonyl* is pronounced "carbon-EEL."

The difference in electronegativity between oxygen and carbon causes a carbon–oxygen double bond to be polar.

In this chapter, we continue our discussion of hydrocarbon derivatives that contain the element oxygen. The functional groups we considered in the previous chapter (alcohols, phenols, and ethers) have the common feature of carbon–oxygen *single* bonds. Carbon–oxygen *double* bonds are also possible in hydrocarbon derivatives. We will now consider the simplest types of compounds that contain this structural feature: aldehydes and ketones.

15.1 THE CARBONYL GROUP

Both aldehydes and ketones contain a carbonyl functional group. A **carbonyl group** *is a carbon atom double-bonded to an oxygen atom.* The structural representation for a carbonyl group is

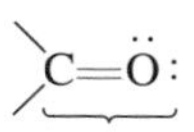

$$\text{\textbackslash C=\textddot{O}:}$$

Carbonyl group

Carbon–oxygen and carbon–carbon double bonds differ in a major way. A carbon–oxygen double bond is *polar,* and a carbon–carbon double bond is *nonpolar.* The electronegativity (Section 5.9) of oxygen (3.5) is much greater than that of carbon (2.5). Hence the carbon–oxygen double bond is polarized, the oxygen atom acquiring a fractional negative charge (δ^-) and the carbon atom acquiring a fractional positive charge (δ^+).

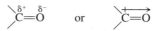

Polar nature of carbon–oxygen double bond

All carbonyl groups have a trigonal planar structure. The bond angles between the three atoms attached to the carbonyl carbon atom are 120°, as would be predicted using VSEPR theory (Section 5.8).

 A trigonal planar structure

15.2 COMPOUNDS CONTAINING A CARBONYL GROUP

The carbon atom of a carbonyl group must form two other bonds in addition to the carbon–oxygen double bond in order to have four bonds. The nature of these two additional bonds determines the type of carbonyl-containing compound it is. There are five major classes of carbonyl-containing hydrocarbon derivatives that we consider in this text.

1. **Aldehydes.** In an aldehyde, one of the two additional bonds that the carbonyl carbon atom forms must be to hydrogen atom. The other may be to a hydrogen atom, an alkyl or cycloalkyl group, or an aromatic ring system.

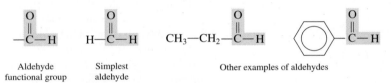

    Aldehyde functional group   Simplest aldehyde   Other examples of aldehydes

2. **Ketones.** In a ketone, both of the additional bonds of the carbonyl carbon atom must be to another carbon atom that is part of an alkyl, cycloalkyl, or aromatic group.

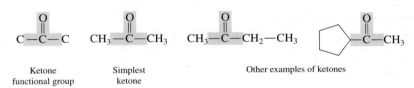

    Ketone functional group   Simplest ketone   Other examples of ketones

3. **Carboxylic acids.** In a carboxylic acid, one of the two additional bonds of the carbonyl carbon atom must be to a hydroxyl group, and the other may be to a hydrogen atom, an alkyl or cycloalkyl group, or an aromatic ring system. The structural parameters for a carboxylic acid are the same as those for an aldehyde except that the mandatory hydroxyl group replaces the mandatory hydrogen atom of an aldehyde.

    Carboxylic acid functional group   Simplest carboxylic acid   Other examples of carboxylic acids

4. **Esters.** In an ester, one of the two additional bonds of the carbonyl carbon atom must be to an oxygen atom, which in turn is bonded to an alkyl, cycloalkyl, or aromatic group. The other bond may be to a hydrogen atom, alkyl or cycloalkyl group, or an aromatic ring system. The structural parameters for an ester differ from those for a carboxylic acid only in that an —OH group has become an —O—R or —O—Ar group.

$$\text{Ester functional group}\qquad\text{Simplest ester}$$

—C—O—C—
‖
O
Ester
functional group

H—C—O—CH₃
‖
O
Simplest
ester

CH₃—C—O—CH₂—CH₃
‖
O

CH₃—C—O—⬠
‖
O

Other examples of esters

5. **Amides.** The previous four types of carbonyl compounds contain the elements carbon, hydrogen, and oxygen. Amides are different from these compounds in that the element nitrogen, in addition to carbon, hydrogen, and oxygen, is present. In an amide, an amino group (—NH₂) or substituted amino group replaces the —OH group of a carboxylic acid.

—C—NH₂
‖
O
Amide
functional group

H—C—NH₂
‖
O
Simplest
amide

CH₃—CH₂—C—NH₂
‖
O

⬡—C—NH₂
‖
O

Other examples of amides

In the remainder of this chapter we consider aldehydes and ketones, the first two of the five major classes of carbonyl compounds. They share the common feature of having only one oxygen atom present, the oxygen atom of the carbonyl group.

In Chapter 16, the third and fourth of the carbonyl classes, carboxylic acids and esters, are considered. They share the common feature of having an additional oxygen atom present besides the one in the carbonyl group.

In Chapter 17, the fifth of the carbonyl classes is considered. Here the element nitrogen is present in addition to oxygen, carbon, and hydrogen.

15.3 THE ALDEHYDE AND KETONE FUNCTIONAL GROUPS

The word *aldehyde* is pronounced "AL-da-hide."

In interpreting general *condensed* functional group structures such as RCHO, remember that carbon always has four bonds and hydrogen always has only one. In RCHO, you know one of carbon's bonds goes to the R group and one to H; therefore, two bonds must go to O.

The word *ketone* is pronounced "KEY-tone."

In an aldehyde, the carbonyl group is always located at the end of a hydrocarbon chain.

CH₃—CH₂—CH₂—CH₂—C—H
‖
O

In a ketone, the carbonyl group is always at a nonterminal (interior) position on the hydrocarbon chain.

CH₃—CH₂—CH₂—C—CH₂—CH₃
‖
O

An **aldehyde** *is a carbonyl-containing organic compound in which the carbonyl carbon atom has at least one hydrogen atom directly attached to it.* The remaining group attached to the carbonyl carbon atom can be hydrogen, an alkyl group (R), a cycloalkyl group, or an aryl group (Ar). The aldehyde functional group is

Linear notations for an aldehyde functional group and for an aldehyde itself are —CHO and RCHO, respectively. Note that the ordering of the symbols H and O in these notations is HO, not OH (which denotes a hydroxyl group).

A **ketone** *is a carbonyl-containing organic compound in which the carbonyl carbon atom has two other carbon atoms directly attached to it.* The groups containing these bonded carbon atoms may be alkyl, cycloalkyl, or aryl.

The ketone functional group is

The general condensed formula for a ketone is RCOR, in which the oxygen atom is understood to be double-bonded to the carbonyl carbon at the left of it in the formula.

An aldehyde functional group can be bonded to only one carbon atom because three of the four bonds from an aldehyde carbonyl carbon must go to oxygen and hydrogen. Thus, an aldehyde functional group is always found at the end of a carbon chain. A ketone functional group, by contrast, is always found within a carbon chain, as it must be bonded to two other carbon atoms.

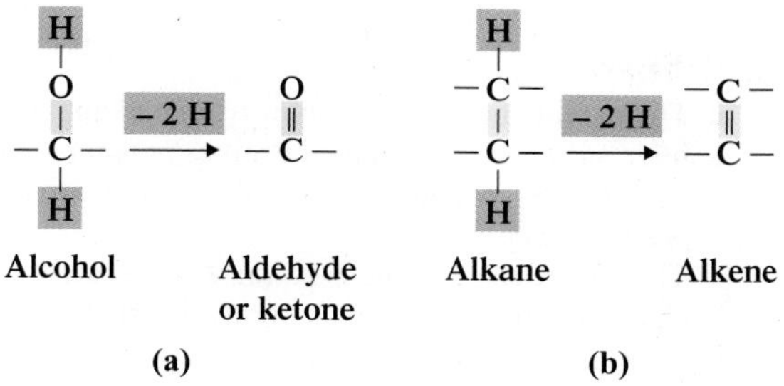

Figure 15.1 Aldehydes and ketones are related to alcohols in the same manner that alkenes are related to alkanes; removal of two hydrogen atoms produces a double bond.

Cyclic aldehydes are not possible. For an aldehyde carbonyl carbon atom to be part of a ring system it would have to form two bonds to ring atoms, which would give it five bonds. Unlike aldehydes, ketones can form cyclic structures, such as

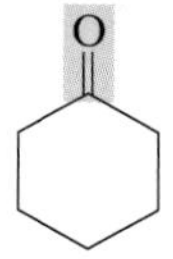

Six-membered ring, one ketone group

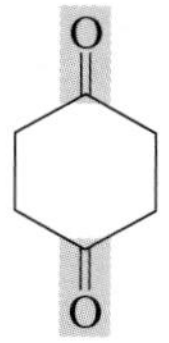

Six-membered ring, two ketone groups

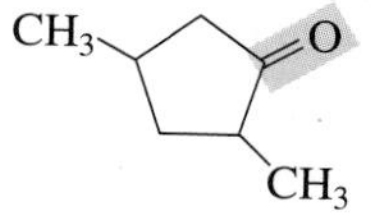

Five-membered ring, one ketone group, two alkyl groups

Cyclic ketones are *not* heterocyclic ring systems as were cyclic ethers (Section 14.19).

Aldehydes and ketones are related to alcohols in the same manner that alkenes are related to alkanes. Removal of hydrogen atoms from each of two adjacent carbon atoms in an alkane produces an alkene. In a like manner, removal of a hydrogen atom from the —OH group of an alcohol and from the carbon atom to which the hydroxyl group is attached produces a carbonyl group (see Figure 15.1).

15.4 NOMENCLATURE FOR ALDEHYDES

The IUPAC rules for naming aldehydes are as follows:

Rule 1: Select as the parent carbon chain the longest chain that *includes* the carbon atom of the carbonyl group.

Rule 2: Name the parent chain by changing the -*e* ending of the corresponding alkane name to -*al*.

Rule 3: Number the parent chain by assigning the number 1 to the carbonyl carbon atom of the aldehyde group.

Rule 4: Determine the identity and location of any substituents, and append this information to the front of the parent chain name.

The carbonyl carbon atom in an aldehyde cannot have any number but 1, so we do not have to include this number in the aldehyde's IUPAC name.

● **EXAMPLE 15.1**
Determining IUPAC Names for Aldehydes

Assign IUPAC names to the following aldehydes.

a.

$$CH_3-CH_2-CH_2-CH_2-\overset{\displaystyle O}{\overset{\displaystyle \|}{C}}-H$$

b.

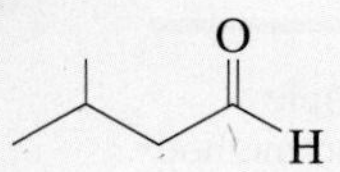

c.

$$CH_3-CH_2-CH_2-\underset{\underset{\displaystyle CH_3}{\overset{\displaystyle |}{CH_2}}}{\overset{\displaystyle |}{CH}}-\overset{\displaystyle O}{\overset{\displaystyle \|}{C}}-H$$

d.

$$CH_3-CH_2-\underset{\underset{\displaystyle OH}{\overset{\displaystyle |}{CH}}}{\overset{}{}}-CH_2-\overset{\displaystyle O}{\overset{\displaystyle \|}{C}}-H$$

(continued)

Line-angle structural formulas for the simpler unbranched-chain aldehydes:

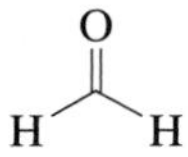

Methanal 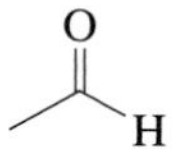   Ethanal

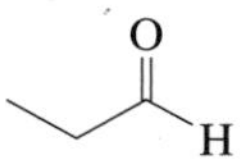

Propanal 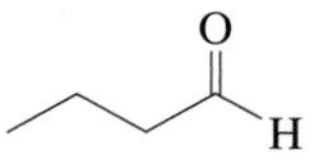   Butanal

Be careful about the endings -al and -ol. They are easily confused. The suffix -al (pronounced like the man's name Al) denotes an aldehyde; the suffix -ol (pronounced like the ol in old) denotes an alcohol.

When a compound contains more than one type of functional group, the suffix for only one of them can be used as the ending of the name. The IUPAC rules establish priorities that specify which suffix is used. For the functional groups we have discussed up to this point in the text, the IUPAC priority system is

Increasing priority ↑

aldehyde
ketone
alcohol
alkene
alkyne
alkoxy ⎤ Equal-priority
alkyl ⎬ substituents (listed in alphabetical
halogen ⎦ order)

aldehyde–ether

$$CH_3-O-CH_2-\overset{\overset{O}{\|}}{C}-H$$

2-methoxyethanal

aldehyde–alkene

$$CH_2{=}CH-\overset{\overset{O}{\|}}{C}-H$$

2-propenal

aldehyde–alcohol

$$HO-CH_2-CH_2-\overset{\overset{O}{\|}}{C}-H$$

3-hydroxypropanal

The common names for simple aldehydes illustrate a second method for counting from one to four: *form-*, *acet-*, *propion-*, and *butyr-*. We will use this method again in the next chapter when we consider the common names for carboxylic acids and esters. (The first method for counting from one to four, with which you are now thoroughly familiar, is *meth-*, *eth-*, *prop-*, and *but-*, as in methane, ethane, propane, and butane.)

Solution

a. The parent chain name comes from pentane. Remove the *-e* ending and add the aldehyde suffix *-al*. The name becomes *pentanal*. The location of the carbonyl carbon atom need not be specified because this carbon atom is always number 1. The complete name is simply *pentanal*.

b. The parent chain name is *butanal*. To locate the methyl group, we number the carbon chain beginning with the carbonyl carbon atom. The complete name of the aldehyde is *3-methylbutanal*.

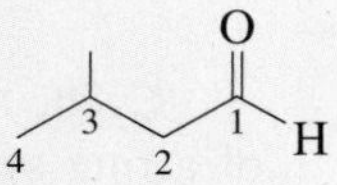

c. The longest chain containing the carbonyl atom is five carbons long, giving a parent chain name of *pentanal*. An ethyl group is present on carbon 2. Thus the complete name is *2-ethylpentanal*.

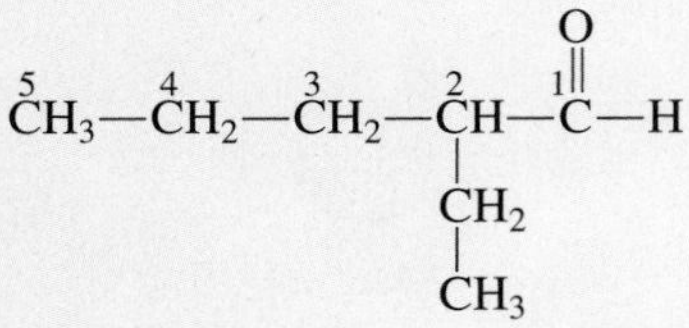

$$\overset{5}{C}H_3-\overset{4}{C}H_2-\overset{3}{C}H_2-\overset{2}{C}H-\overset{1}{\overset{\overset{O}{\|}}{C}}-H$$
$$\underset{\underset{CH_3}{|}}{CH_2}$$

d. This is a hydroxyaldehyde, with the hydroxyl group located on carbon 3.

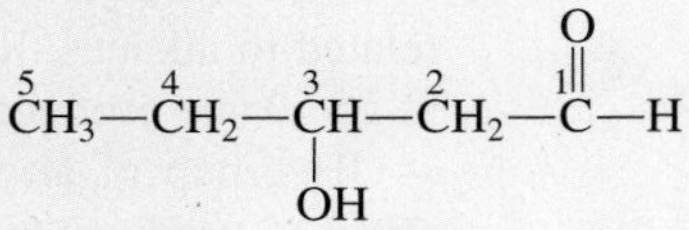

$$\overset{5}{C}H_3-\overset{4}{C}H_2-\overset{3}{C}H-\overset{2}{C}H_2-\overset{1}{\overset{\overset{O}{\|}}{C}}-H$$
$$\underset{OH}{|}$$

The complete name of the compound is *3-hydroxypentanal*. An aldehyde functional group has priority over an alcohol functional group in IUPAC nomenclature. An alcohol group named as a substituent is a *hydroxy* group.

Practice Exercise 15.1

Assign IUPAC names to the following aldehydes.

a.
$$CH_3-\underset{\underset{CH_3}{|}}{CH}-\overset{\overset{O}{\|}}{C}-H$$

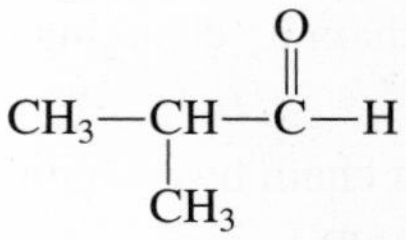

b.
$$CH_3-CH_2-\underset{\underset{CH_3-CH_2-CH_2}{|}}{CH}-\overset{\overset{O}{\|}}{C}-H$$

c.
$$CH_3-\underset{\underset{Cl}{|}}{CH}-\underset{\underset{Cl}{|}}{CH}-\overset{\overset{O}{\|}}{C}-H$$

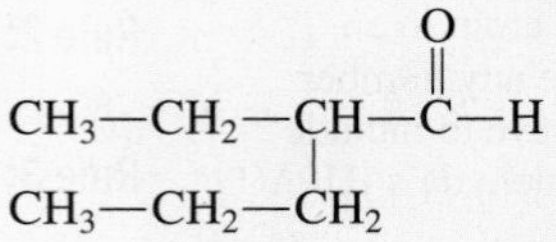

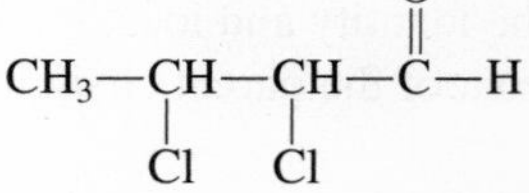

d.

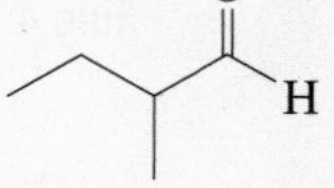

Answers: a. 2-Methylpropanal; **b.** 2-Ethylpentanal; **c.** 2,3-Dichlorobutanal; **d.** 2-Methylbutanal

Unbranched aldehydes with a small number of carbon atoms have common names:

$$H-\overset{\overset{O}{\|}}{C}-H$$
Formaldehyde

$$CH_3-\overset{\overset{O}{\|}}{C}-H$$
Acetaldehyde

$$CH_3-CH_2-\overset{\overset{O}{\|}}{C}-H$$
Propionaldehyde

$$CH_3-CH_2-CH_2-\overset{\overset{O}{\|}}{C}-H$$
Butyraldehyde

The contrast between IUPAC names and common names for aldehydes is as follows:

IUPAC (one word)

> alkanal

pentanal
Common (one word)

> (prefix) aldehyde*

butyraldehyde.

*The common-name prefixes are related to natural sources for carboxylic acids with the same number of carbon atoms (see Section 16.3).

Aromatic aldehydes are not cyclic aldehydes (which do not exist). The carbonyl carbon atom in an aromatic aldehyde is not part of the ring system.

Figure 15.2 Space-filling model for benzaldehyde, the simplest aromatic aldehyde.

Unlike the common names for alcohols and ethers, the common names for aldehydes are *one* word rather than two or three.

In the IUPAC system, aromatic aldehydes—compounds in which an aldehyde group is attached to a benzene ring—are named as derivatives of benzaldehyde, the parent compound (see Figure 15.2).

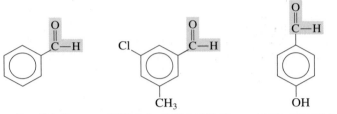

Benzaldehyde   3-Chloro-5-methylbenzaldehyde   4-Hydroxybenzaldehyde

The last of these compounds is named as a benzaldehyde rather than as a phenol because the aldehyde group has priority over the hydroxyl group in the IUPAC naming system.

15.5 NOMENCLATURE FOR KETONES

Assigning IUPAC names to ketones is similar to naming aldehydes except that the ending -*one* is used instead of -*al*. The rules for IUPAC ketone nomenclature follow.

Rule 1: Select as the parent carbon chain the longest carbon chain that *includes* the carbon atom of the carbonyl group.

Rule 2: Name the parent chain by changing the -*e* ending of the corresponding alkane name to -*one*. This ending, -one, is pronounced "own."

Rule 3: Number the carbon chain such that the carbonyl carbon atom receives the lowest possible number. The position of the carbonyl carbon atom is noted by placing a number immediately before the name of the parent chain.

Rule 4: Determine the identity and location of any substituents, and append this information to the front of the parent chain name.

Rule 5: Cyclic ketones are named by assigning the number 1 to the carbon atom of the carbonyl group. The ring is then numbered to give the lowest number(s) to the atom(s) bearing substituents.

● **EXAMPLE 15.2**

Determining IUPAC Names for Ketones

Assign IUPAC names to the following ketones.

a.

$$CH_3-C-CH_2-CH_2-CH_3$$

b.

c.

d.

Solution

a. The parent chain name is *pentanone*. We number the chain from the end closest to the carbonyl carbon atom. Locating the carbonyl carbon at carbon 2 completes the name, *2-pentanone*.

b. The longest carbon chain of which the carbonyl carbon is a member is four carbons long. The parent chain name is *butanone*. There is one methyl group attached, and the numbering system is from right to left.

(continued)

Propanone is the simplest possible ketone. One- and two-carbon ketones cannot exist. A minimum of three carbon atoms is required for a ketone: one C atom for the carbonyl group and one C atom for each of the groups attached to the carbonyl carbon atom. No locator number is needed in the name propanone, because there is only one possible location for the double bond.

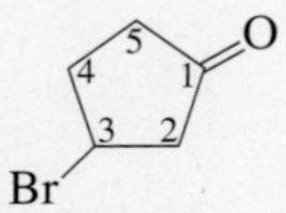

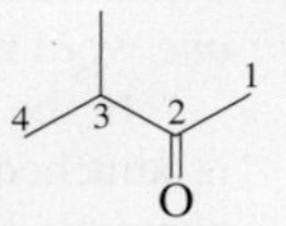

In IUPAC nomenclature, the ketone functional group has precedence over all groups we have discussed so far except the aldehyde group. When both aldehyde and ketone groups are present in the same molecule, the ketone group is named as a substituent (the *oxo-* group).

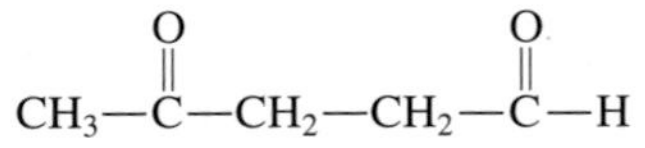

$$CH_3-\overset{\overset{O}{\|}}{C}-CH_2-CH_2-\overset{\overset{O}{\|}}{C}-H$$

4-Oxopentanal

Line-angle structural formulas for the simpler unbranched-chain 2-ketones:

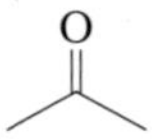

2-Propanone

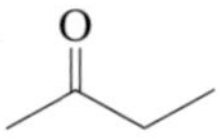

2-Butanone

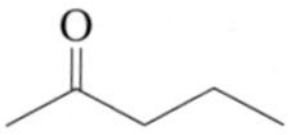

2-Pentanone

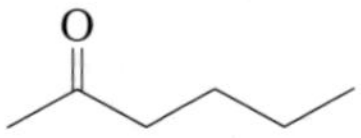

2-Hexanone

The complete name for the compound is *3-methyl-2-butanone.*

c. The base name is *cyclohexanone*. The methyl group is bonded to carbon 2 because we begin numbering at the carbonyl carbon. The name is *2-methylcyclohexanone.*

d. This ketone has a base name of *cyclopentanone*. Numbering clockwise from the carbonyl carbon atom locates the bromo group on carbon 3. The complete name is *3-bromocyclopentanone.*

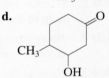

Practice Exercise 15.2

Assign IUPAC names to the following ketones.

a.

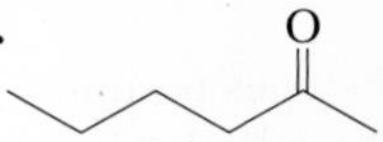

b.

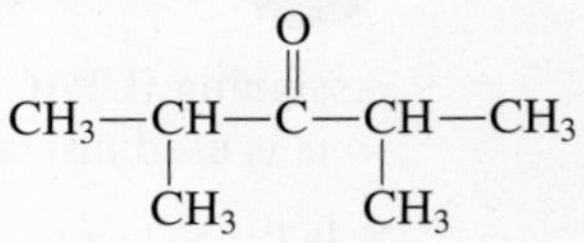

$$CH_3-\underset{\underset{CH_3}{|}}{CH}-\overset{\overset{O}{\|}}{C}-\underset{\underset{CH_3}{|}}{CH}-CH_3$$

c.

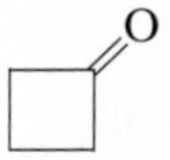

d.

Answers: a. 2-Hexanone; **b.** 2,4-Dimethyl-3-pentanone; **c.** Cyclobutanone;
d. 3-Hydroxy-4-methylcyclohexanone

The procedure for coining common names for ketones is the same as that used for ether common names (Section 14.16). They are constructed by giving, in alphabetical order, the names of the alkyl or aryl groups attached to the carbonyl functional group and then adding the word *ketone*. Unlike aldehyde common names, which are one word, those for ketones are two or three words.

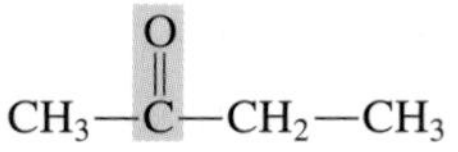

The contrast between IUPAC names and common names for ketones is as follows:

IUPAC (one word)

alkanone

2-butanone

Common (three or two words)

alkyl alkyl ketone

ethyl methyl ketone

or

dialkyl ketone

dipropyl ketone

$$CH_3-\overset{\overset{O}{\|}}{C}-CH_2-CH_3$$

Ethyl methyl ketone

$$CH_3-CH_2-\overset{\overset{O}{\|}}{C}-\langle\rangle$$

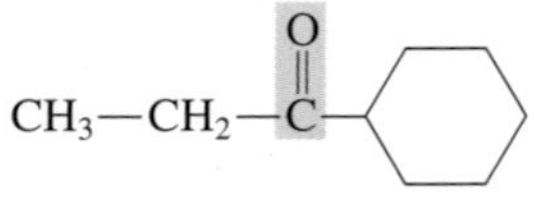

Cyclohexyl ethyl ketone

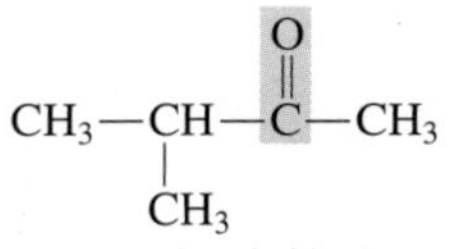

$$CH_3-\underset{\underset{CH_3}{|}}{CH}-\overset{\overset{O}{\|}}{C}-CH_3$$

Isopropyl methyl ketone

Three ketones have additional common names besides those obtained with the preceding procedures. These three ketones are

Acetone
(dimethyl ketone)

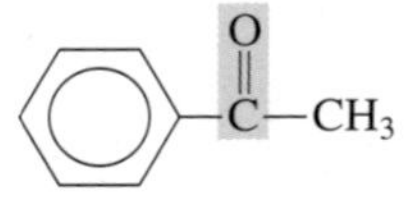

Acetophenone
(methyl phenyl ketone)

Benzophenone
(diphenyl ketone)

Acetophenone is the simplest aromatic ketone.

15.6 ISOMERISM FOR ALDEHYDES AND KETONES

Like the classes of organic compounds previously discussed (alkanes, alkenes, alkynes, alcohols, ethers, etc.), constitutional isomers exist for aldehydes and for ketones, and *between* aldehydes and ketones (functional group isomerism). The compounds butanal and 2-methylpropanal are examples of skeletal aldehyde isomers; the compounds 2-pentanone and 3-pentanone are examples of positional ketone isomers.

CHEMICAL Connections

Lachrymatory Aldehydes and Ketones

A lachrymator, pronounced "lack-ra-mater," is a compound that causes the production of tears. A number of aldehydes and ketones have lachrymatic properties.

Two lachrymal ketones are 2-chloroacetophenone and bromo-acetone.

2-Chloroacetophenone
(2-chloro-1-phenylethanone)

$$CH_3-\overset{\overset{\displaystyle O}{\|}}{C}-CH_2-Br$$

Bromoacetone
(1-bromopropanone)

2-Chloroacetophenone is a component of the tear gas used by police and the military. It is also the active ingredient in MACE canisters now marketed for use by individuals to protect themselves from attackers. The compound bromoacetone has been used as a chemical war gas.

Smoke contains compounds that cause the eyes to tear. A predominant lachrymator in wood smoke is formaldehyde, the one-carbon aldehyde. The smoke associated with an outdoor barbecue contains the unsaturated aldehyde *acrolein*.

$$CH_2=CH-\overset{\overset{\displaystyle O}{\|}}{C}-H$$

Acrolein
(propenal)

Acrolein is formed as fats that are present in meat break down when heated. (Besides being a lachrymator, acrolein is responsible for the "pleasant" odor associated with the process of barbecuing meat.)

The lachrymatory compound associated with onions is a derivative of thiopropionaldehyde.

$$CH_3-CH_2-\overset{\overset{\displaystyle S}{\|}}{C}-H$$

Thiopropionaldehyde
(propanethial)

$$CH_3-CH_2-\overset{\overset{\displaystyle S}{\|}}{C}-H \quad \overset{\displaystyle O}{}$$

Thiopropionaldehyde-S-oxide
(propanethial-S-oxide)
[lachrymator in chopped onions]

Onions do not cause tear production until they are chopped or sliced. The onion cells damaged by these actions release the

The smoke generated from outdoor barbecuing contains the lachrymatory aldehyde *acrolein.*

enzyme *allinase,* which converts an odorless compound naturally present in onions to the lachrymatic compound.

Scientists are not sure why thiopropionaldehyde-S-oxide causes tear production, but it is known that it is an unstable molecule that is readily broken up by water into propanal, hydrogen sulfide, and sulfuric acid.

$$CH_3-CH_2-\overset{\overset{\displaystyle S}{\|}}{C}-H \overset{\displaystyle O}{} \xrightarrow{H_2O}$$

$$CH_3-CH_2-\overset{\overset{\displaystyle O}{\|}}{C}-H + H_2S + H_2SO_4$$

Sulfuric acid may be responsible for the eye irritation.

Many people can peel onions under water, or peel cold ones from the refrigerator, without crying. Water washes away the soluble lachrymator and also breaks it down chemically. If the onion is cold, the enzymatic reaction making the lachrymator is slower, so less is formed. Also the vapor pressure of the lachrymator is greatly reduced at the lower temperature, so its concentration in air is reduced.

Figure 15.3 Aldehydes and ketones with the same number of carbon atoms and the same degree of saturation are functional group isomers, as is illustrated here for the three-carbon aldehyde (propanal) and the three-carbon ketone (propanone). Both have the molecular formula C_3H_6O.

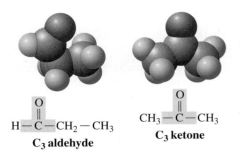

$$\underset{\textbf{C}_3\ \textbf{aldehyde}}{H-\overset{\overset{\displaystyle O}{\|}}{C}-CH_2-CH_3} \qquad \underset{\textbf{C}_3\ \textbf{ketone}}{CH_3-\overset{\overset{\displaystyle O}{\|}}{C}-CH_3}$$

Aldehyde skeletal isomers: $CH_3-CH_2-CH_2-\overset{\overset{\displaystyle O}{\|}}{C}-H$ and $CH_3-\overset{\overset{\displaystyle CH_3}{|}}{CH}-\overset{\overset{\displaystyle O}{\|}}{C}-H$

Ketone positional isomers: $CH_3-\overset{\overset{\displaystyle O}{\|}}{C}-CH_2-CH_2-CH_3$ and $CH_3-CH_2-\overset{\overset{\displaystyle O}{\|}}{C}-CH_2-CH_3$

This is the third time we have encountered functional group isomerism. Alcohols and ethers (Section 14.17) and thiols and thioethers (Section 14.21) also exhibit this type of isomerism.

Aldehydes and ketones with the same number of carbon atoms and the same degree of saturation are functional group isomers. Molecular models for the isomeric C_3 compounds propanal and propanone, which both have the molecular formula C_3H_6O, are shown in Figure 15.3.

15.7 SELECTED COMMON ALDEHYDES AND KETONES

Formaldehyde, the simplest aldehyde, with only one carbon atom, is manufactured on a large scale by the oxidation of methanol.

$$CH_3-OH \xrightarrow[600°C-700°C]{Ag\ catalyst} H-\overset{\overset{\displaystyle O}{\|}}{C}-H + H_2$$

Its major use is in the manufacture of polymers (Section 15.12). At room temperature and pressure, formaldehyde is an irritating gas. Bubbling this gas through water produces *formalin*, an aqueous solution containing 37% formaldehyde by mass or 40% by volume. (This represents the solubility limit of formaldehyde gas in water.) Very little free formaldehyde gas is actually present in formalin; most of it reacts with water, producing methylene glycol.

$$H-\overset{\overset{\displaystyle O}{\|}}{C}-H + H-O-H \longrightarrow HO-CH_2-OH$$

Figure 15.4 Formalin is used to preserve biological specimens. This coelacanth, a "prehistoric" fish, is preserved in formaldehyde.

Formalin is used for preserving biological specimens (see Figure 15.4); anyone who has experience in a biology laboratory is familiar with the pungent odor of formalin. Formalin is also the most widely used preservative chemical in embalming fluids used by morticians. Its mode of action involves reaction with protein molecules in a manner that links the protein molecules together; the result is a "hardening" of the protein.

Acetone, a colorless, volatile liquid with a pleasant, mildly "sweet" odor, is the simplest ketone and is also the ketone used in largest volume in industry. Acetone is an excellent solvent because it is miscible with both water and nonpolar solvents. Acetone is the main ingredient in gasoline treatments that are designed to solubilize water in the gas tank and allow it to pass through the engine in miscible form. Acetone can also be used to remove water from glassware in the laboratory. And it is a major component of some nail polish removers.

The oxidation of methanol to formaldehyde has been previously mentioned (Section 14.5). Ingested methanol is oxidized in the human body to formaldehyde, and it is the formaldehyde that causes blindness.

Small amounts of acetone are produced in the human body in reactions related to obtaining energy from fats. Normally, such acetone is degraded to CO_2 and H_2O. Diabetic people produce larger amounts of acetone, not all of which can be degraded. The presence of acetone in urine is a sign of diabetes. In severe diabetes, the odor of acetone can be detected on the person's breath.

TABLE 15.1
Selected Aldehydes and Ketones Whose Uses Are Based on Their Odor or Flavor

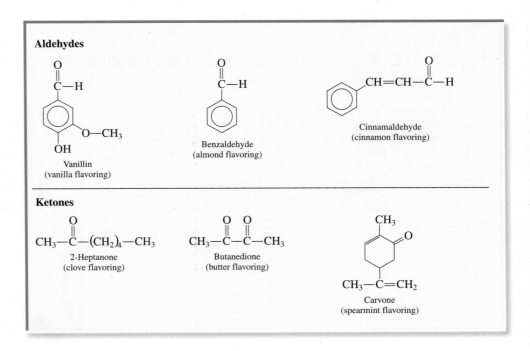

Aldehydes

Vanillin
(vanilla flavoring)

Benzaldehyde
(almond flavoring)

Cinnamaldehyde
(cinnamon flavoring)

Ketones

2-Heptanone
(clove flavoring)

Butanedione
(butter flavoring)

Carvone
(spearmint flavoring)

Naturally Occurring Aldehydes and Ketones

Aldehydes and ketones occur widely in nature. Naturally occurring compounds of these types, with higher molecular masses, usually have pleasant odors and flavors and are often used for these properties in consumer products (perfumes, air fresheners, and the like). Table 15.1 gives the structures and uses for selected naturally occurring aldehydes and ketones. The unmistakable odor of melted butter is largely due to the four-carbon diketone butanedione (see Figure 15.5).

Many important steroid hormones (Section 19.12) are ketones, including testosterone, the hormone that controls the development of male sex characteristics; progesterone, the hormone secreted at the time of ovulation in females; and cortisone, a hormone from the adrenal glands that is used medicinally to relieve inflammation.

Figure 15.5 The delightful odor of melted butter is largely due to butanedione, whose structure is given in Table 15.1.

Testosterone

Progesterone

Cortisone

15.8 PHYSICAL PROPERTIES OF ALDEHYDES AND KETONES

The C_1 and C_2 aldehydes are gases at room temperature (Figure 15.6). The C_3 through C_{11} straight-chain saturated aldehydes are liquids, and the higher aldehydes are solids. The presence of alkyl groups tends to lower both boiling points and melting points, as does the presence of unsaturation in the carbon chain. Lower-molecular-mass ketones are colorless liquids at room temperature (Figure 15.6).

The boiling points of aldehydes and ketones are intermediate between those of alcohols and alkanes of similar molecular mass. Aldehydes and ketones have higher boiling points than alkanes because of dipole–dipole attractions between molecules. Carbonyl group polarity (Section 15.1) makes these dipole–dipole interactions possible.

Unbranched Aldehydes			
C_1	C_3	C_5	C_7
C_2	C_4	C_6	C_8

Unbranched 2-Ketones			
	C_3	C_5	C_7
	C_4	C_6	C_8

☐ Gas ☐ Liquid

Figure 15.6 A physical-state summary for unbranched aldehydes and unbranched 2-ketones at room temperature and pressure.

CHEMICAL Connections

Melanin: A Hair and Skin Pigment

Human hair as well as human skin is colored naturally by the pigment melanin—a polymeric substance involving many interconnected *cyclic ketone* units. The more melanin a person produces, which is genetically controlled, the darker his or her hair and skin. The number of melanin-producing cells is essentially the same in dark-skinned and light-skinned people, but they are more active in dark-skinned people. The following is a representation of a portion of the structure of polymeric melanin pigment.

The interaction of melanin pigments with ultraviolet light is what produces a "suntan."

Hair pigmentation (hair color) results from biosynthesis of melanin within hair follicles. The melanin molecules so produced are incorporated into the growing hair shaft and distributed throughout the hair cortex. The melanin tends to accumulate within hair protein as granules. Hair, once it exits the scalp, is no longer alive and any damage to the melanin or the hair itself cannot be repaired by the body.

For most people, starting sometime in their thirties, the production of melanin in hair follicles begins to gradually decrease. Once a hair follicle stops producing melanin, the hair will be colorless but will appear white due to light scattering. A proportion of white hair in colored hair will make a head of hair appear gray.

Suntan and sunburn both involve melanin (as a skin pigment) and ultraviolet (UV) radiation. Sudden high-level exposure to UV radiation causes the skin to burn, and steady low-level exposure to UV radiation can cause tanning. Melanin molecules in the skin constitute a built-in defense system to protect the skin

against UV radiation. The melanin molecules act as a protective barrier by absorbing and scattering the UV radiation. A dark-skinned person has more melanin molecules in the upper layers of the skin (and more protection against sunburn) than a light-skinned person.

When melanin-producing cells deep in the skin are exposed to UV radiation, melanin production increases. The presence of this extra melanin in the skin gives the skin an appearance that we call a "tan." The larger the melanin molecules so produced, the deeper the tan. People who tan readily have skin that can produce a large amount of melanin.

When a person experiences a sunburn, the skin peels. When peeling occurs, any tan that has been built up (excess melanin) sloughs off with the dead skin. Thus the tanning process must begin anew.

TABLE 15.2
Boiling Points of Some Alkanes, Aldehydes, and Alcohols of Similar Molecular Mass

Type of Compound	Compound	Structure	Molecular Mass	Boiling Point (8°C)
alkane	ethane	CH_3—CH_3	30	−89
aldehyde	methanal	H—CHO	30	−21
alcohol	methanol	CH_3—OH	32	65
alkane	propane	CH_3—CH_2—CH_3	44	−42
aldehyde	ethanal	CH_3—CHO	44	20
alcohol	ethanol	CH_3—CH_2—OH	46	78
alkane	butane	CH_3—CH_2—CH_2—CH_3	58	−1
aldehyde	propanal	CH_3—CH_2—CHO	58	49
alcohol	1-propanol	CH_3—CH_3—CH_2—OH	60	97

Figure 15.7 Low-molecular-mass aldehydes and ketones are soluble in water because of hydrogen bonding.

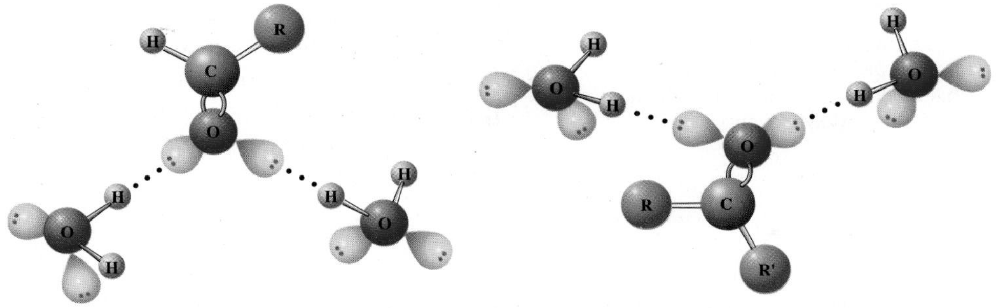

(a) Aldehyde–Water Hydrogen Bonding

(b) Ketone–Water Hydrogen Bonding

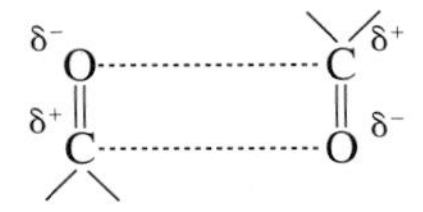

Dipole–dipole attraction

Aldehydes and ketones have lower boiling points than the corresponding alcohols because no hydrogen bonding occurs as it does with alcohols. Dipole–dipole attractions are weaker forces than hydrogen bonds (Section 7.13). Table 15.2 provides boiling-point information for selected aldehydes, alcohols, and alkanes.

Water molecules can hydrogen-bond with aldehyde and ketone molecules (Figure 15.7). This hydrogen bonding causes low-molecular-mass aldehydes and ketones to be water soluble. As the hydrocarbon portions get larger, the water solubility of aldehydes and ketones decreases. Table 15.3 gives data on solubility in water for selected aldehydes and ketones.

Although low-molecular-mass aldehydes have pungent, penetrating, unpleasant odors, higher-molecular-mass aldehydes (above C_8) are more fragrant, especially benzaldehyde derivatives. Ketones generally have pleasant odors, and several are used in perfumes and air fresheners.

The ordering of boiling points for carbonyl compounds (aldehydes and ketones), alcohols, and alkanes of similar molecular mass is

$$\text{Alcohols} > \frac{\text{carbonyl}}{\text{compounds}} > \text{alkanes}$$

15.9 PREPARATION OF ALDEHYDES AND KETONES

Aldehydes and ketones can be produced by the oxidation of primary and secondary alcohols, respectively, using mild oxidizing agents such as $KMnO_4$ or $K_2Cr_2O_7$ (Section 14.9).

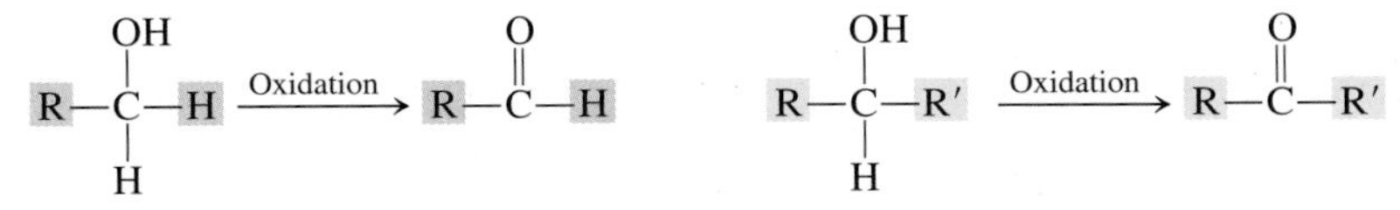

Primary alcohol   Aldehyde   Secondary alcohol   Ketone

TABLE 15.3
Water Solubility (g/100 g H_2O) for Various Aldehydes and Ketones

Number of Carbon Atoms	Aldehyde	Water Solubility of Aldehyde	Ketone	Water Solubility of Ketone
1	methanal	very soluble		
2	ethanal	infinite		
3	propanal	16	propanone	infinite
4	butanal	7	2-butanone	26
5	pentanal	4	2-pentanone	5
6	hexanal	1	2-hexanone	1.6
7	heptanal	0.1	2-heptanone	0.4
8	octanal	insoluble	2-octanone	insoluble

The term *aldehyde* stems from *al*cohol *dehyd*rogenation, indicating that aldehydes are related to alcohols by the loss of hydrogen.

When this type of reaction is used for aldehyde preparation, reaction conditions must be sufficiently mild to avoid further oxidation of the aldehyde to a carboxylic acid (Section 14.9). Ketones do not undergo the further oxidation that aldehydes do.

In the oxidation of an alcohol to an aldehyde or a ketone, the alcohol molecule loses H atoms. Recall that a *decrease in the number of C—H bonds* in an organic molecule is one of the operational definitions for the process of *oxidation* (Section 14.9).

● **EXAMPLE 15.3**

Predicting Products in Alcohol Oxidation Reactions

▶ Draw the structure of the aldehyde or ketone formed from the oxidation of each of the following alcohols. Assume that reaction conditions are sufficiently mild that any aldehydes produced are not oxidized further.

a. $CH_3—CH_2—CH_2—OH$

b. 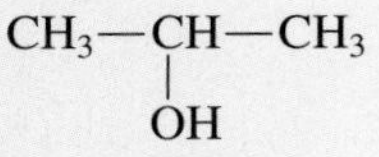 $CH_3—CH—CH_3$ with OH below the CH

c.

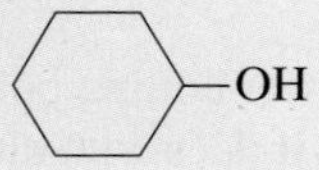

—OH

d.

CH_3
|
$CH_3—C—OH$
|
CH_3

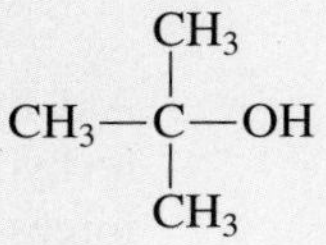

Solution

a. This is a primary alcohol that will give the aldehyde *propanal* as the oxidation product.

$$CH_3—CH_2—\overset{\overset{\displaystyle O}{\|}}{C}—H$$

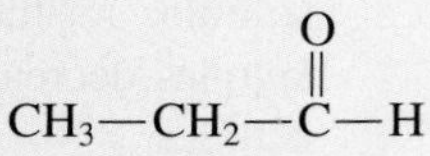

Propanal

b. This is a secondary alcohol. Upon oxidation, secondary alcohols are converted to ketones.

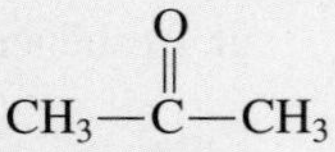

$$CH_3—\overset{\overset{\displaystyle O}{\|}}{C}—CH_3$$

Propanone

c. This cyclic alcohol is a secondary alcohol; hence a ketone is the oxidation product.

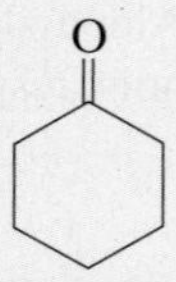

Cyclohexanone

d. This is a tertiary alcohol. Tertiary alcohols do not undergo oxidation (Section 14.9).

Practice Exercise 15.3

Draw the structure of the aldehyde or ketone formed from the oxidation of each of the following alcohols. Assume that reaction conditions are sufficiently mild that any aldehydes produced are not oxidized further to carboxylic acids.

a. 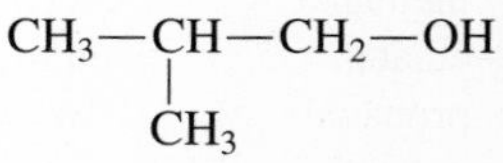 $CH_3—CH—CH_2—OH$ with CH_3 below the CH

b. 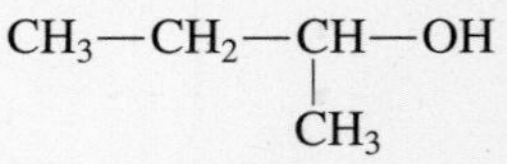 $CH_3—CH_2—CH—OH$ with CH_3 below the CH

c.

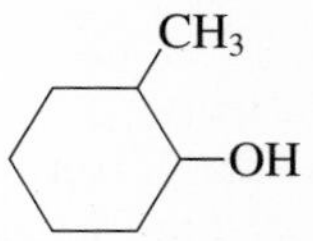

CH_3
—OH

d.

CH_3
|
$CH_3—C—CH_2—OH$
|
CH_3

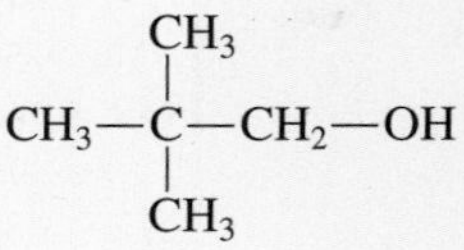

Answers: a.

$$CH_3-CH-\overset{\overset{\displaystyle O}{\|}}{C}-H$$
$$\underset{\displaystyle CH_3}{|}$$

b.

$$CH_3-CH_2-\overset{\overset{\displaystyle O}{\|}}{C}-CH_3$$

c.

(cyclohexanone ring with CH_3 substituent and =O)

d.

$$CH_3-\overset{\overset{\displaystyle CH_3}{|}}{\underset{\displaystyle CH_3}{C}}-\overset{\overset{\displaystyle O}{\|}}{C}-H$$

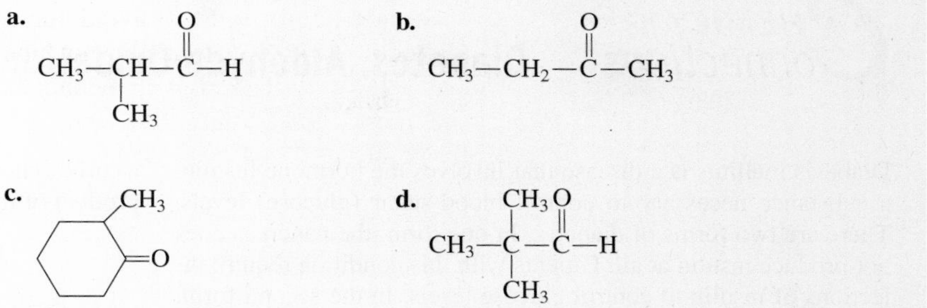

15.10 OXIDATION AND REDUCTION OF ALDEHYDES AND KETONES

Oxidation of Aldehydes and Ketones

Aldehydes readily undergo oxidation to carboxylic acids (Section 15.9), and ketones are resistant to oxidation.

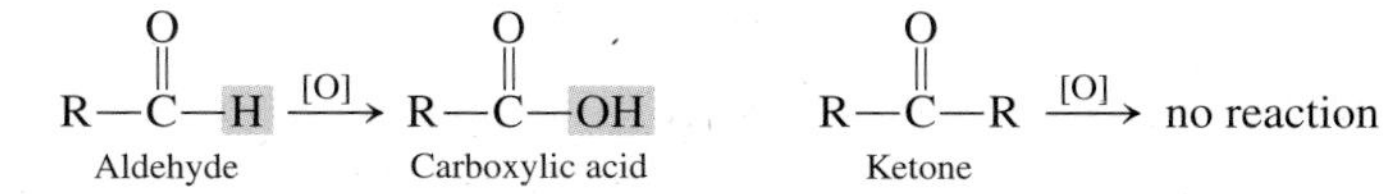

$$\underset{\text{Aldehyde}}{R-\overset{\overset{\displaystyle O}{\|}}{C}-H} \xrightarrow{[O]} \underset{\text{Carboxylic acid}}{R-\overset{\overset{\displaystyle O}{\|}}{C}-OH} \qquad \underset{\text{Ketone}}{R-\overset{\overset{\displaystyle O}{\|}}{C}-R} \xrightarrow{[O]} \text{no reaction}$$

In aldehyde oxidation, the aldehyde gains an oxygen atom (supplied by the oxidizing agent). An *increase in the number of C—O bonds* is one of the operational definitions for the process of *oxidation* (Section 14.9).

Among the mild oxidizing agents that convert aldehydes into carboxylic acids is oxygen in air. Thus aldehydes must be protected from air. When an aldehyde is prepared from oxidation of a primary alcohol (Section 15.9), it is usually removed from the reaction mixture immediately to prevent it from being further oxidized to a carboxylic acid.

Because both aldehydes and ketones contain carbonyl groups, we might expect similar oxidation reactions for the two types of compounds. Oxidation of an aldehyde involves breaking a carbon–hydrogen bond, and oxidation of a ketone involves breaking a carbon–carbon bond. The former is much easier to accomplish than the latter. For ketones to be oxidized, strenuous reaction conditions must be employed.

Several tests, based on the ease with which aldehydes are oxidized, have been developed for distinguishing between aldehydes and ketones, for detecting the presence of aldehyde groups in sugars (carbohydrates), and for measuring the amounts of sugars present in a solution. The most widely used of these tests are the Tollens test and Benedict's test.

The Tollens test, also called the silver mirror test, involves a solution that contains silver nitrate ($AgNO_3$) and ammonia (NH_3) in water. When Tollens solution is added to an aldehyde, Ag^+ ion (the oxidizing agent) is reduced to silver metal, which deposits on the inside of the test tube, forming a silver mirror. The appearance of this silver mirror (see Figure 15.8) is a positive test for the presence of the aldehyde group.

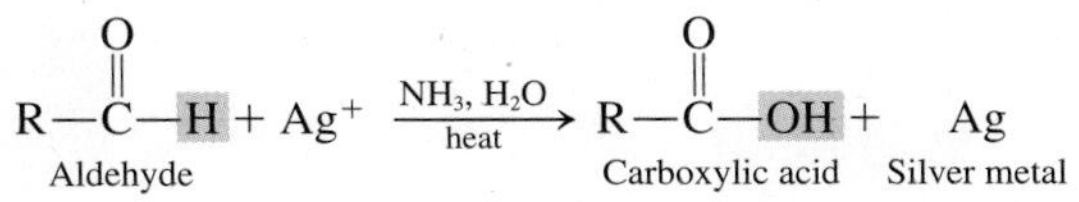

$$\underset{\text{Aldehyde}}{R-\overset{\overset{\displaystyle O}{\|}}{C}-H} + Ag^+ \xrightarrow[\text{heat}]{NH_3, H_2O} \underset{\text{Carboxylic acid}}{R-\overset{\overset{\displaystyle O}{\|}}{C}-OH} + \underset{\text{Silver metal}}{Ag}$$

The Ag^+ ion will not oxidize ketones.

Benedict's test is similar to the Tollens test in that a metal ion is the oxidizing agent. With this test, Cu^{2+} ion is reduced to Cu^+ ion, which precipitates from solution as Cu_2O (a brick-red solid; Figure 15.9).

CHEMICAL Connections — Diabetes, Aldehyde Oxidation, and Glucose Testing

Diabetes mellitus is a disease that involves the hormone insulin, a substance necessary to control blood-sugar (glucose) levels. There are two forms of diabetes. In one form, the pancreas does not produce insulin at all. Patients with this condition require injections of insulin to control glucose levels. In the second form, the body cannot make proper use of insulin. Patients with this form of diabetes can often control glucose levels through their diet but may require medication. If the blood-sugar level of a diabetic becomes too high, serious kidney damage can result.

Normal urine does not contain glucose. When the kidneys become overloaded with glucose (the blood-glucose level is too high), glucose is excreted in the urine. Benedict's test (Section 15.10) can be used to detect glucose in urine, because glucose has an aldehyde group present in its structure.

$$CH_2-CH-CH-CH-CH-\overset{\overset{O}{\|}}{C}-H$$
$$\ \ \ |\ \ \ \ |\ \ \ \ |\ \ \ \ |\ \ \ \ |$$
$$OH\ \ OH\ \ OH\ \ OH\ \ OH$$
Glucose

For many years diabetes monitoring involved testing urine for the presence of glucose with a Benedict's solution. The test was carried out using either plastic test strips coated with Benedict's solution or Clinitest tablets (a convenient solid form of Benedict's reagent). A few drops of urine were added to the plastic strip or tablet, and the degree of coloration was used to estimate the blood-glucose level. The solution turned greenish at a low glucose level, then yellow-orange, and finally a dark orange-red.

Today, use of Benedict's solution for urine glucose testing has been replaced largely by other chemical tests that provide better results. Blood glucose testing has been found to detect rising glucose levels earlier than urine glucose testing. Blood glucose tests are now the preferred method for monitoring glucose levels.

Blood glucose testing involves placing a drop of blood (from a finger prick) on a plastic strip containing a dye and an enzyme that will oxidize glucose's aldehyde group. A two-step reaction sequence occurs. First, the enzyme causes glucose oxidation to

a carboxylic acid, with hydrogen peroxide (H_2O_2) being another product of the reaction.

$$R-\overset{\overset{O}{\|}}{C}-H + O_2 \xrightarrow{\text{Enzyme}} R-\overset{\overset{O}{\|}}{C}-OH + H_2O_2$$
Aldehyde (glucose) → Carboxylic acid + Hydrogen peroxide

Then the H_2O_2 reacts with the dye to produce a colored product. The amount of color produced, measured by comparison with a color chart or by an electronic monitor, is proportional to the blood-glucose concentration.

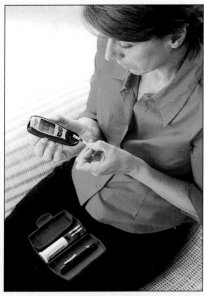

In a blood glucose test, a small drop of blood obtained from a finger prick is absorbed on a test strip present within the blood glucose monitor. The results of the blood–strip interaction, which involves aldehyde oxidation, are processed electronically and displayed on an LCD readout.

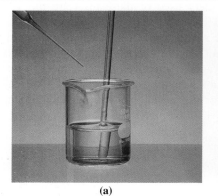  
(a) (b) (c)

Figure 15.8 A positive Tollens test for aldehydes involves the formation of a silver mirror. (a) An aqueous solution of ethanal is added to a solution of silver nitrate in aqueous ammonia and stirred. (b) The solution darkens as ethanal is oxidized to ethanoic acid, and Ag+ ion is reduced to silver. (c) The inside of the beaker becomes coated with metallic silver.

$$\underset{\text{Aldehyde}}{R-\overset{\overset{\displaystyle O}{\|}}{C}-H} + Cu^{2+} \longrightarrow \underset{\text{Carboxylic acid}}{R-\overset{\overset{\displaystyle O}{\|}}{C}-OH} + \underset{\text{Brick-red solid}}{Cu_2O}$$

Benedict's solution is made by dissolving copper sulfate, sodium citrate, and sodium carbonate in water.

Reduction of Aldehydes and Ketones

Aldehydes and ketones are easily reduced by hydrogen gas (H_2), in the presence of a catalyst (Ni, Pt, or Cu), to form alcohols. The reduction of aldehydes produces primary alcohols, and the reduction of ketones yields secondary alcohols.

$$\text{Aldehyde} \xrightarrow{\text{Reduction}} \text{primary alcohol}$$

$$\text{Ketone} \xrightarrow{\text{Reduction}} \text{secondary alcohol}$$

Specific examples of such reactions follow.

Aldehyde reduction:
$$\underset{\text{Ethanal}}{CH_3-\overset{\overset{\displaystyle O}{\|}}{C}-H} + H_2 \xrightarrow{\text{Ni}} \underset{\text{Ethanol}}{CH_3-\overset{\overset{\displaystyle OH}{|}}{\underset{\underset{\displaystyle H}{|}}{C}}-H}$$

Ketone reduction:
$$\underset{\text{Propanone}}{CH_3-\overset{\overset{\displaystyle O}{\|}}{C}-CH_3} + H_2 \xrightarrow{\text{Ni}} \underset{\text{2-Propanol}}{CH_3-\overset{\overset{\displaystyle OH}{|}}{\underset{\underset{\displaystyle H}{|}}{C}}-CH_3}$$

It is the addition of hydrogen atoms to the carbon–oxygen double bond that produces the alcohol in each of these reactions.

$$\underset{}{\diagdown}C{=}O + H_2 \xrightarrow{\text{Catalyst}} \underset{\underset{\displaystyle H\ \ H}{|\ \ |}}{\diagdown}C{-}O$$

This hydrogen addition process is very similar to the addition of hydrogen to the carbon–carbon double bond of an alkene to produce an alkane, which we encountered in Section 13.9.

$$\underset{}{\diagdown}C{=}C\diagup + H_2 \xrightarrow{\text{Catalyst}} \underset{\underset{\displaystyle H\ \ H}{|\ \ |}}{\diagdown}C{-}C\diagup$$

Aldehyde reduction and ketone reduction to produce alcohols are the "opposite" of the oxidation of alcohols to produce aldehydes and ketones (Section 15.9). These "opposite" relationships can be diagrammed as follows:

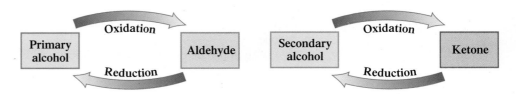

As we noted in Section 14.9, "keeping track" of such relationships is an aid in remembering organic chemistry reaction schemes.

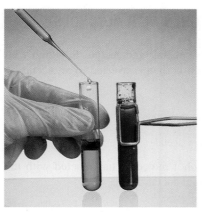

Figure 15.9 Benedict's solution, which is blue in color, turns brick red when an aldehyde reacts with it.

15.11 REACTION OF ALDEHYDES AND KETONES WITH ALCOHOLS

Aldehydes and ketones react with alcohols to form hemiacetals and acetals. Reaction with *one* molecule of alcohol produces a hemiacetal, which is then converted to an acetal by reaction with a second alcohol molecule.

$$\text{Aldehyde or ketone} + \text{alcohol} \xrightarrow[\text{catalyst}]{\text{Acid}} \text{hemiacetal}$$

$$\text{Hemiacetal} + \text{alcohol} \xrightarrow[\text{catalyst}]{\text{Acid}} \text{acetal}$$

The Greek prefix *hemi-* means "half." When one alcohol molecule has reacted with the aldehyde or ketone, the compound is halfway to the final acetal.

Further information about these two reactions follows.

> Hemiacetal and acetal formation are very important biochemical reactions; they are crucial to understanding the chemistry of carbohydrates, which is considered in Chapter 18.

Hemiacetal Formation

Hemiacetal formation is an addition reaction in which a molecule of alcohol adds to the carbonyl group of an aldehyde or ketone. The H portion of the alcohol adds to the carbonyl oxygen atom, and the R—O portion of the alcohol adds to the carbonyl carbon atom.

$$\underset{\text{An aldehyde}}{\overset{\displaystyle O}{\underset{R_1 \quad H}{\|}{C}}} + \underset{\text{An alcohol}}{\boxed{H}\,\boxed{O{-}R_2}} \rightleftharpoons \underset{\text{A hemiacetal}}{R_1{-}\overset{\displaystyle O{-}H}{\underset{H}{C}}{-}O{-}R_2}$$

$$\underset{\text{A ketone}}{\overset{\displaystyle O}{\underset{R_1 \quad R_2}{\|}{C}}} + \underset{\text{An alcohol}}{\boxed{H}\,\boxed{O{-}R_3}} \rightleftharpoons \underset{\text{A hemiacetal}}{R_1{-}\overset{\displaystyle O{-}H}{\underset{R_2}{C}}{-}O{-}R_3}$$

> Hemiacetals contain an alcohol group (hydroxyl group) and an ether group (alkoxy group) on the same carbon atom.

Formally defined, a **hemiacetal** *is an organic compound in which a carbon atom is bonded to both a hydroxyl group (—OH) and an alkoxy group (—OR).* The functional group for a hemiacetal is thus

The carbon atom of the hemiacetal functional group is often referred to as the *hemiacetal carbon atom;* it was the carbonyl carbon atom of the aldehyde or ketone that reacted.

A reaction mixture containing a hemiacetal is always in equilibrium with the alcohol and carbonyl compound from which it was made, and the equilibrium lies to the carbonyl compound side of the reaction (Section 9.9).

$$\text{Alcohol} + \text{aldehyde} \rightleftharpoons \text{hemiacetal}$$

$$\text{Alcohol} + \text{ketone} \rightleftharpoons \text{hemiacetal}$$

This situation makes isolation of the hemiacetal difficult; in practice, it usually cannot be done.

An important exception to this difficulty with isolation is the case where the —OH and $\overset{\diagdown}{\underset{\diagup}{C}}{=}O$ functional groups that react to form the hemiacetal come from the *same* molecule. This produces a *cyclic* hemiacetal rather than a noncyclic one, and cyclic acetals are more stable than the noncyclic ones and can be isolated.

Illustrative of intramolecular hemiacetal formation is the reaction

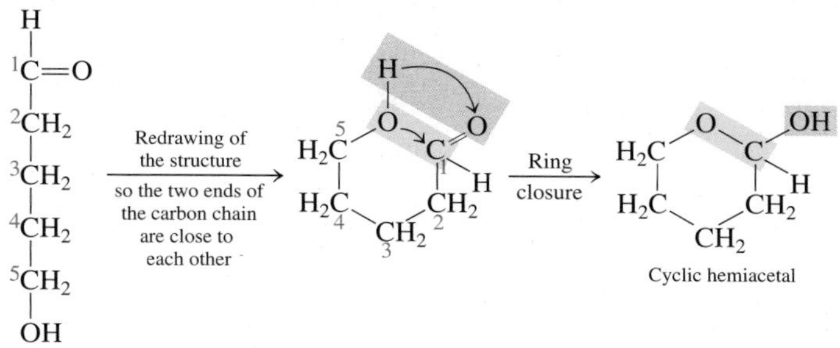

Cyclic hemiacetals are very important compounds in carbohydrate chemistry, the topic of Chapter 18.

EXAMPLE 15.4

Recognizing Hemiacetal Structures

Indicate whether each of the following compounds is a hemiacetal.

a. 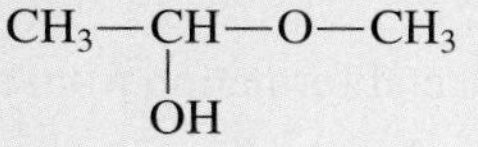 $CH_3-CH-O-CH_3$
                 $|$
                OH

b. 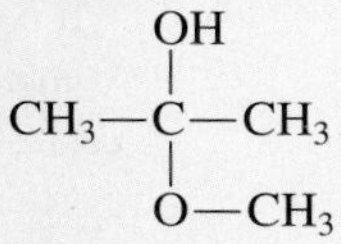
$CH_3-\overset{\displaystyle OH}{\underset{\displaystyle O-CH_3}{C}}-CH_3$

c.
$CH_3-CH-\overset{\displaystyle CH_3}{\underset{\displaystyle OH}{CH}}-O-CH_3$

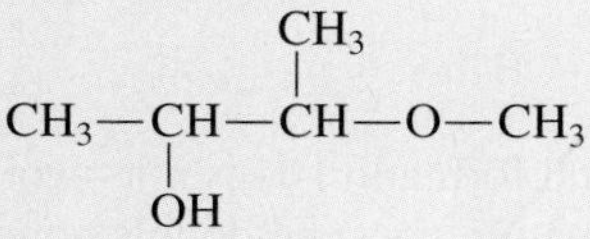

d.

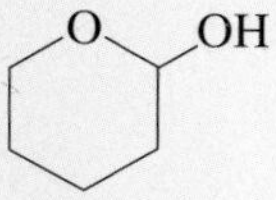

Solution

In each part, we will be looking for the following structural feature: the presence of an —OH group and an —OR group attached to the same carbon atom.

a. We have an —OH group and an —OR group attached to the same carbon atom. The compound is a *hemiacetal*.
b. We have an —OH group and an —OR group attached to the same carbon atom. The compound is a *hemiacetal*.
c. The —OH and —OR groups present in this molecule are attached to *different* carbon atoms. Therefore, the molecule is *not a hemiacetal*.
d. We have a ring carbon atom bonded to two oxygen atoms: one oxygen atom in an —OH substituent and the other oxygen atom bonded to the rest of the ring (the same as an R group). This is a *hemiacetal*.

Practice Exercise 15.4

Indicate whether each of the following compounds is a hemiacetal.

a. 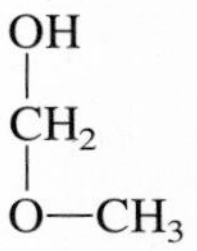
$\overset{\displaystyle OH}{\underset{\displaystyle \underset{\displaystyle O-CH_3}{CH_2}}{|}}$

b. 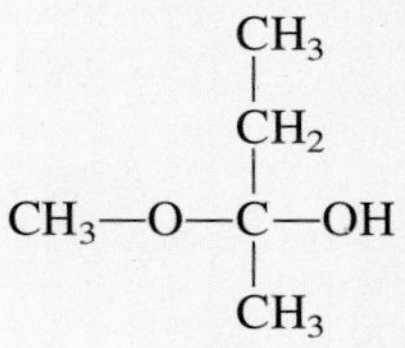
$CH_3-O-\overset{\displaystyle CH_2}{\underset{\displaystyle CH_3}{C}}-OH$
with CH_3 on top

c.
$CH_3-O-\overset{\displaystyle CH_3}{\underset{\displaystyle HO-CH}{CH}}$
$\underset{\displaystyle CH_3}{|}$

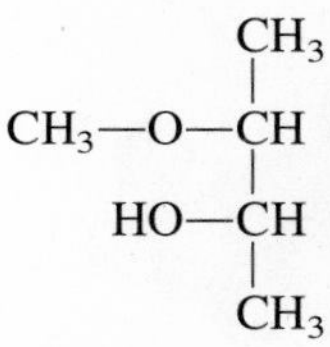

d.

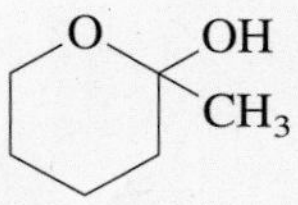

Answers: a. Yes; **b.** Yes; **c.** No; **d.** Yes

This is our second encounter with condensation reactions. The first encounter involved intermolecular alcohol dehydration (Section 14.9).

Acetals have two alkoxy groups (—OR) attached to the same carbon atom.

Acetal Formation

If a small amount of acid catalyst is added to a hemiacetal reaction mixture, the hemiacetal reacts with a second alcohol molecule, in a condensation reaction, to form an acetal.

$$
\underset{\text{A hemiacetal}}{R_1\!-\!\overset{\displaystyle OH}{\underset{\displaystyle H}{\overset{|}{\underset{|}{C}}}}\!-\!OR_2} + \underset{\text{An alcohol}}{R_3\!-\!OH} \overset{H^+}{\rightleftharpoons} \underset{\text{An acetal}}{R_1\!-\!\overset{\displaystyle OR_3}{\underset{\displaystyle H}{\overset{|}{\underset{|}{C}}}}\!-\!OR_2} + H\!-\!OH
$$

An **acetal** *is an organic compound in which a carbon atom is bonded to two alkoxy groups (—OR).* The functional group for an acetal is thus

$$
-\overset{\displaystyle OR}{\underset{\displaystyle |}{\overset{|}{C}}}\!-\!OR
$$

A specific example of acetal formation from a hemiacetal is

$$
\underset{}{CH_3\!-\!\overset{\displaystyle OH}{\underset{\displaystyle O-CH_3}{\overset{|}{\underset{|}{CH}}}}} + CH_3\!-\!CH_2\!-\!OH \overset{H^+}{\rightleftharpoons} CH_3\!-\!\overset{\displaystyle O-CH_2-CH_3}{\underset{\displaystyle O-CH_3}{\overset{|}{\underset{|}{CH}}}} + H\!-\!OH
$$

Note that acetal formation does not involve addition to a carbon–oxygen double bond as hemiacetal formation does; no double bond is present in either of the reactants involved in acetal formation. Acetal formation involves a substitution reaction; the —OR group of the alcohol replaces the —OH group on the hemiacetal.

Figure 15.10 shows molecular models for acetaldehyde (the two-carbon aldehyde) and the hemiacetal and acetal formed when this aldehyde reacts with ethyl alcohol.

Acetal Hydrolysis

In Section 24.1, we will find that the enzyme-catalyzed hydrolysis of acetals is an important step in the digestion of carbohydrates.

Acetals, unlike hemiacetals, are easily isolated from reaction mixtures. They are stable in basic solution but undergo *hydrolysis* in acidic solution. A **hydrolysis reaction** *is the reaction of a compound with H_2O, in which the compound splits into two or more fragments as the elements of water (H— and —OH) are added to the compound.* The products of acetal hydrolysis are the aldehyde or ketone and alcohols that originally reacted to form the acetal.

$$
\underset{\text{Acetal}}{-\overset{\displaystyle O-R_1}{\underset{\displaystyle |}{\overset{|}{C}}}\!-\!O\!-\!R_2} + H\!-\!OH \overset{\text{Acid catalyst}}{\rightleftharpoons} \underset{\substack{\text{Aldehyde or}\\\text{ketone}}}{\overset{\displaystyle O}{\overset{\|}{C}}} + R_1\!-\!OH + R_2\!-\!O\!-\!H
$$

Figure 15.10 Molecular models for acetaldehyde and its hemiacetal and acetal formed by reaction with ethyl alcohol.

Acetaldehyde **Acetaldehyde hemiacetal with ethyl alcohol** **Acetaldehyde acetal with ethyl alcohol**

For example,

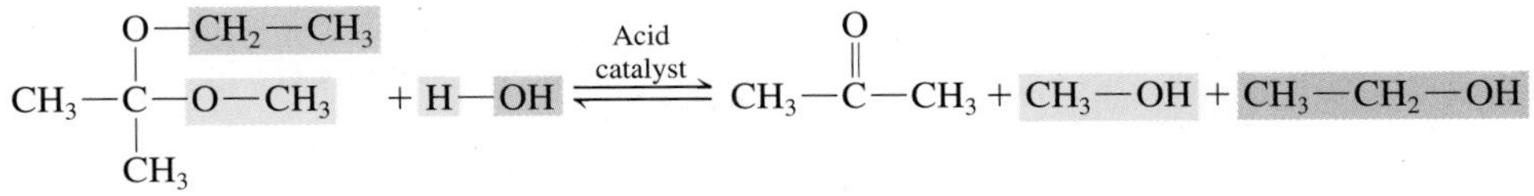

$$CH_3-\underset{\underset{CH_3}{|}}{\overset{\overset{O-CH_2-CH_3}{|}}{C}}-O-CH_3 \;+\; H-OH \underset{\text{catalyst}}{\overset{\text{Acid}}{\rightleftharpoons}} CH_3-\overset{\overset{O}{\|}}{C}-CH_3 + CH_3-OH + CH_3-CH_2-OH$$

The carbonyl hydrolysis product is an aldehyde if the acetal carbon atom has a hydrogen atom attached directly to it, and it is a ketone if no hydrogen attachment is present. In the preceding example, the carbonyl product is a ketone because the two additional acetal carbon atom attachments are methyl groups.

EXAMPLE 15.5

Predicting Products in Acetal Hydrolysis Reactions

Draw the structures of the aldehyde (or ketone) and the two alcohols produced when the following acetals undergo hydrolysis in acidic solution.

a.

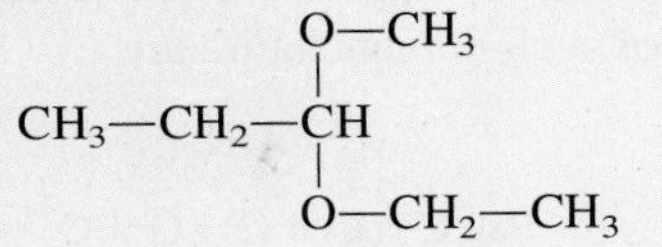

$$CH_3-CH_2-\underset{\underset{O-CH_2-CH_3}{|}}{\overset{\overset{O-CH_3}{|}}{CH}}$$

b.

$$CH_3-\underset{\underset{\underset{\underset{CH_3}{|}}{C}}{|}}{\overset{\overset{CH_3}{|}}{C}}-O-\underset{\underset{CH_3}{|}}{CH}-CH_3$$

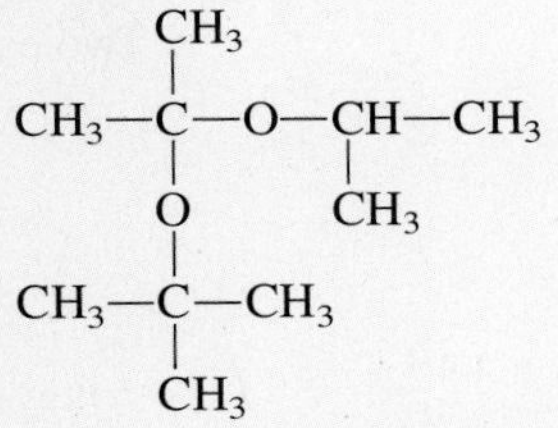

Solution

a. Each of the alkoxy (—OR) groups present will be converted into an alcohol during the hydrolysis. Because the acetal carbon atom has a H attachment, the remainder of the molecule becomes an aldehyde, with the carbon atom to which the alkoxy groups were attached becoming the carbonyl carbon atom.

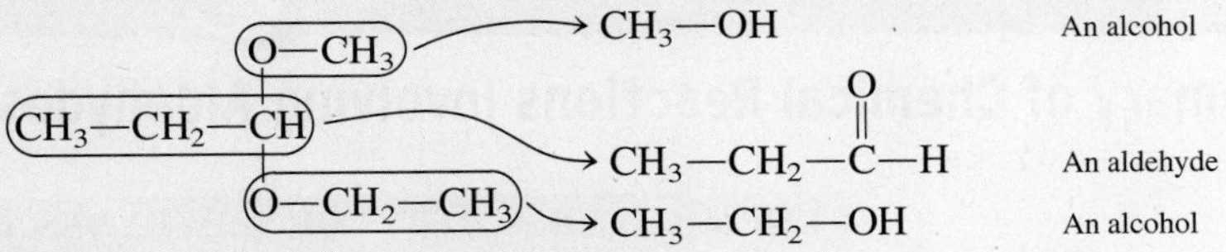

$$
\begin{array}{ll}
\boxed{O-CH_3} & \longrightarrow CH_3-OH & \text{An alcohol} \\
\boxed{CH_3-CH_2-CH} & \longrightarrow CH_3-CH_2-\overset{\overset{O}{\|}}{C}-H & \text{An aldehyde} \\
\boxed{O-CH_2-CH_3} & \longrightarrow CH_3-CH_2-OH & \text{An alcohol}
\end{array}
$$

b. Again, each of the alkoxy groups present will be converted into an alcohol during the hydrolysis. Because the acetal carbon atoms lacks a H attachment, the remainder of the molecule becomes a ketone.

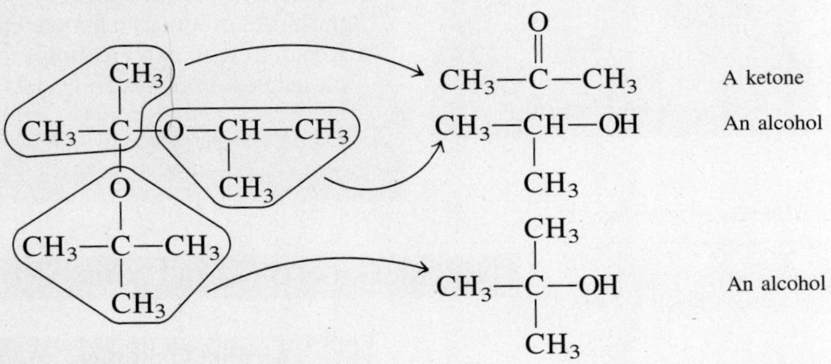

$$
\begin{array}{ll}
\boxed{CH_3-\underset{\underset{O}{|}}{C}-O-\underset{\underset{CH_3}{|}}{CH}-CH_3} & \longrightarrow CH_3-\overset{\overset{O}{\|}}{C}-CH_3 & \text{A ketone} \\
& \longrightarrow CH_3-\underset{\underset{CH_3}{|}}{CH}-OH & \text{An alcohol} \\
\boxed{CH_3-\underset{\underset{CH_3}{|}}{C}-CH_3} & \longrightarrow CH_3-\underset{\underset{CH_3}{|}}{\overset{\overset{CH_3}{|}}{C}}-OH & \text{An alcohol}
\end{array}
$$

Practice Exercise 15.5

Draw the structures of the aldehyde (or ketone) and the two alcohols produced when the following acetal undergoes hydrolysis in acidic solution.

(continued)

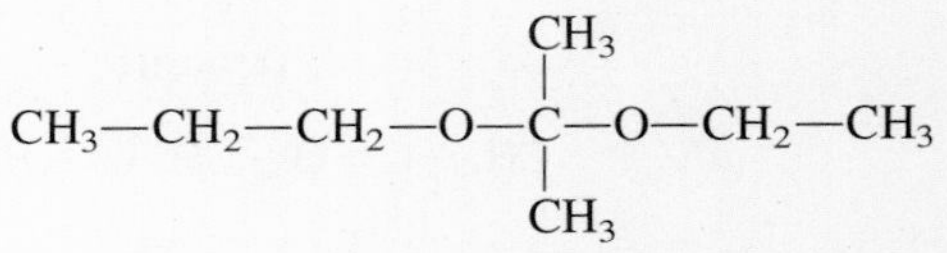

$$CH_3-CH_2-CH_2-O-\overset{\displaystyle CH_3}{\underset{\displaystyle CH_3}{C}}-O-CH_2-CH_3$$

Answers: $CH_3-CH_2-CH_2-OH$ (alcohol), CH_3-CH_2-OH (alcohol),

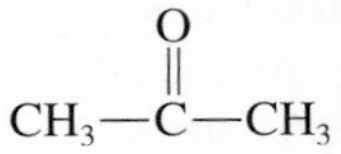

$$CH_3-\overset{\displaystyle O}{\overset{\|}{C}}-CH_3 \text{ (ketone),}$$

Nomenclature for Hemiacetals and Acetals

A "descriptive" type of common nomenclature that includes the terms *hemiacetal* and *acetal* as well as the name of the carbonyl compound (aldehyde or ketone) produced in the hydrolysis of the hemiacetal or acetal is commonly used in describing such compounds. Two examples of such nomenclature are

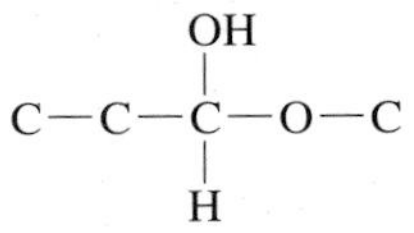

$$C-C-\overset{\displaystyle OH}{\underset{\displaystyle H}{C}}-O-C \qquad C-\overset{\displaystyle O-C-C}{\underset{\displaystyle C}{C}}-O-C-C$$

Methyl hemiacetal of propanal 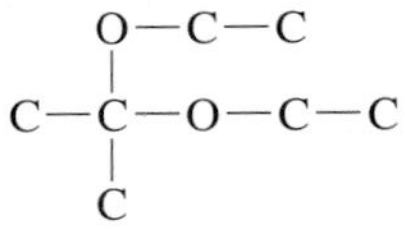   Diethyl acetal of propanone

The Chemistry at a Glance feature below summarizes reactions that involve aldehydes and ketones.

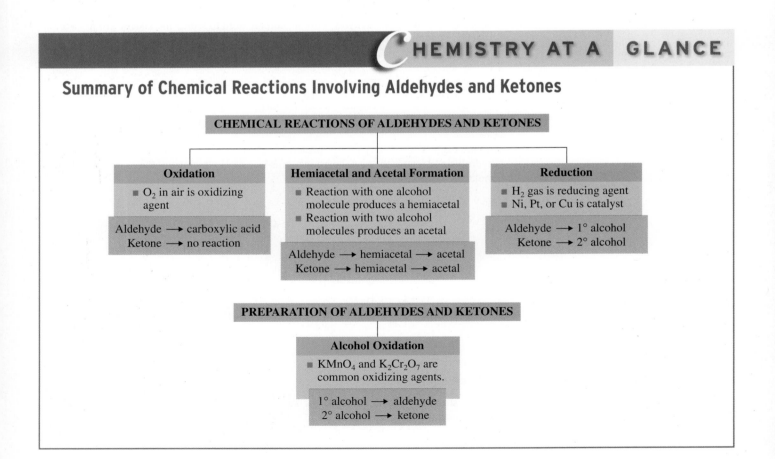

CHEMISTRY AT A GLANCE

Summary of Chemical Reactions Involving Aldehydes and Ketones

CHEMICAL REACTIONS OF ALDEHYDES AND KETONES

Oxidation
- O_2 in air is oxidizing agent

Aldehyde ⟶ carboxylic acid
Ketone ⟶ no reaction

Hemiacetal and Acetal Formation
- Reaction with one alcohol molecule produces a hemiacetal
- Reaction with two alcohol molecules produces an acetal

Aldehyde ⟶ hemiacetal ⟶ acetal
Ketone ⟶ hemiacetal ⟶ acetal

Reduction
- H_2 gas is reducing agent
- Ni, Pt, or Cu is catalyst

Aldehyde ⟶ 1° alcohol
Ketone ⟶ 2° alcohol

PREPARATION OF ALDEHYDES AND KETONES

Alcohol Oxidation
- $KMnO_4$ and $K_2Cr_2O_7$ are common oxidizing agents.

1° alcohol ⟶ aldehyde
2° alcohol ⟶ ketone

Figure 15.11 When a mixture of phenol and formaldehyde dissolved in acetic acid is treated with concentrated hydrochloric acid, a cross-linked phenol–formaldehyde network polymer is formed.

15.12 FORMALDEHYDE-BASED POLYMERS

Many types of organic compounds can serve as reactants (monomers) for polymerization reactions, including ethylenes (Section 13.10), alcohols (Section 14.10), and carbonyl compounds.

Formaldehyde, the simplest aldehyde, is a prolific "polymer former." As representative of its polymer reactions, let us consider the reaction between formaldehyde and phenol, under acidic conditions, to form a phenol–formaldehyde network polymer (see Figure 15.11). A **network polymer** *is a polymer in which monomers are connected in a three-dimensional cross-linked network.*

When excess formaldehyde is present, the polymerization proceeds via mono-, di-, and trisubstituted phenols that are formed as intermediates in the reaction between phenol and formaldehyde.

The substituted phenols then interact with each other by splitting out water molecules. The final product is a complex, large, three-dimensional network polymer in which monomer units are linked via methylene (—CH_2—) bridges.

Figure 15.12 Bakelite jewelry in use during the 1930–1950 time period.

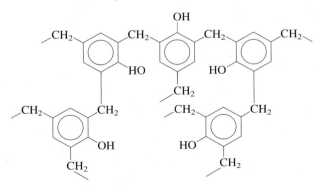

The first synthetic plastic, Bakelite, produced in 1907, was a phenol–formaldehyde polymer. Early uses of Bakelite were in the manufacture of billiard balls and "plastic" jewelry (Figure 15.12). Modern phenol–formaldehyde polymers, called phenolics, are adhesives used in the production of plywood and particle board.

15.13 SULFUR-CONTAINING CARBONYL GROUPS

The introduction of sulfur into a carbonyl group produces two different classes of compounds depending on whether the sulfur atom replaces the carbonyl oxygen atom or the carbonyl carbon atom.

Replacement of the carbonyl *oxygen* atom with sulfur produces *thiocarbonyl compounds*—thioaldehydes (thials) and thioketones (thiones)—the simplest of which are

$$
\underset{\substack{\text{Thioformaldehyde}\\\text{(Methanethial)}}}{H-\overset{\overset{\textstyle S}{\|}}{C}-H}
\qquad
\underset{\substack{\text{Thioacetone}\\\text{(Propanethione)}}}{CH_3-\overset{\overset{\textstyle S}{\|}}{C}-CH_3}
$$

Thiocarbonyl compounds such as these are unstable and readily decompose.

Replacement of the carbonyl *carbon* atom with sulfur produces *sulfoxides,* compounds that are much more stable than thiocarbonyl compounds. The oxidation of a thioether (sulfide) [Section 14.21] constitutes the most common route to a sulfoxide.

$$
\underset{\text{Thioether}}{R-S-R} \xrightarrow{\text{[O]}} \underset{\text{Sulfoxide}}{R-\overset{\overset{\textstyle O}{\|}}{S}-R}
$$

A highly interesting sulfoxide is DMSO (dimethyl sulfoxide), a sulfur analog of acetone, the simplest ketone.

$$
\underset{\text{DMSO}}{CH_3-\overset{\overset{\textstyle O}{\|}}{S}-CH_3}
\qquad
\underset{\text{Acetone}}{CH_3-\overset{\overset{\textstyle O}{\|}}{C}-CH_3}
$$

DMSO is an odorless liquid with unusual properties. Because of the presence of the polar sulfur–oxide bond, DMSO is miscible with water and also quite soluble in less polar organic solvents. When rubbed on the skin, DMSO has remarkable penetrating power and is quickly absorbed into the body, where it relieves pain and inflammation. For many years it has been heralded as a "miracle drug" for arthritis, sprains, burns, herpes, infections, and high blood pressure. However, the FDA has steadfastly refused to approve it for general medical use. For example, the FDA says that DMSO's powerful penetrating action could cause an insecticide on a gardener's skin to be carried accidentally into his or her bloodstream. Another complication is that DMSO is reduced in the body to dimethyl sulfide, a compound with a strong garlic-like odor that soon appears on the breath.

$$
\underset{}{CH_3-\overset{\overset{\textstyle O}{\|}}{S}-CH_3} \xrightarrow{\text{Reduction}} CH_3-S-CH_3
$$

The FDA has approved DMSO for use in certain bladder conditions and as a veterinary drug for topical use in nonbreeding dogs and horses. For example, DMSO is used as an anti-inflammatory rub for race horses.

*C*ONCEPTS TO REMEMBER

The carbonyl group. A carbonyl group consists of a carbon atom bonded to an oxygen atom through a double bond. Aldehydes and ketones are compounds that contain a carbonyl functional group. The carbonyl carbon in an aldehyde has at least one hydrogen attached to it, and the carbonyl carbon in a ketone has no hydrogens attached to it (Sections 15.1 through 15.3).

Nomenclature of aldehydes and ketones. The IUPAC names of aldehydes and ketones are based on the longest carbon chain that contains the carbonyl group. The chain numbering is done from the end that results in the lowest number for the carbonyl group. The names of aldehydes end in *-al,* those of ketones in *-one* (Sections 15.4 and 15.5).

Isomerism for aldehydes and ketones. Constitutional isomerism is possible for aldehydes and for ketones when four or more carbon atoms are present. Aldehydes and ketones with the same number of carbon atoms and the same degree of saturation have the same molecular formula and thus are functional group isomers of each other (Section 15.6).

Physical properties of aldehydes and ketones. The boiling points of aldehydes and ketones are intermediate between those of alcohols and alkanes. The polarity of the carbonyl groups enables aldehyde and ketone molecules to interact with each other through dipole–dipole interactions. They cannot, however, hydrogen-bond to each other. Lower-molecular-mass aldehydes and ketones are soluble in water (Section 15.8).

Preparation of aldehydes and ketones. Oxidation of primary and secondary alcohols, using mild oxidizing agents, produces aldehydes and ketones, respectively (Section 15.9).

Oxidation and reduction of aldehydes and ketones. Aldehydes are easily oxidized to carboxylic acids; ketones do not readily undergo oxidation. Reduction of aldehydes and ketones produces primary and secondary alcohols, respectively (Section 15.10).

Hemiacetals and acetals. A characteristic reaction of aldehydes and ketones is the addition of an alcohol across the carbonyl double bond to produce hemiacetals. The reaction of a second alcohol molecule with a hemiacetal produces an acetal (Section 15.11).

KEY REACTIONS AND EQUATIONS

1. Oxidation of an aldehyde to give a carboxylic acid (Section 15.10)

$$R-\overset{\overset{\displaystyle O}{\|}}{C}-H \xrightarrow{[O]} R-\overset{\overset{\displaystyle O}{\|}}{C}-OH$$

2. Attempted oxidation of a ketone (Section 15.10)

$$R-\overset{\overset{\displaystyle O}{\|}}{C}-R' \xrightarrow{[O]} \text{no reaction}$$

3. Reduction of an aldehyde to give a primary alcohol (Section 15.10)

$$R-\overset{\overset{\displaystyle O}{\|}}{C}-H + H_2 \xrightarrow{\text{Catalyst}} R-\overset{\overset{\displaystyle OH}{|}}{\underset{\underset{\displaystyle H}{|}}{C}}-H$$

4. Reduction of a ketone to give a secondary alcohol (Section 15.10)

$$R-\overset{\overset{\displaystyle O}{\|}}{C}-R' + H_2 \xrightarrow{\text{Catalyst}} R-\overset{\overset{\displaystyle OH}{|}}{\underset{\underset{\displaystyle H}{|}}{C}}-R'$$

5. Addition of an alcohol to an aldehyde to form a hemiacetal and then an acetal (Section 15.11)

$$R_1-\overset{\overset{\displaystyle O}{\|}}{C}-H + R_2-O-H \underset{}{\overset{H^+}{\rightleftharpoons}} R_1-\overset{\overset{\displaystyle OH}{|}}{\underset{\underset{\displaystyle H}{|}}{C}}-OR_2$$

Aldehyde   Hemiacetal

$$R_1-\overset{\overset{\displaystyle OH}{|}}{\underset{\underset{\displaystyle H}{|}}{C}}-OR_2 + R_3-OH \underset{}{\overset{H^+}{\rightleftharpoons}} R_1-\overset{\overset{\displaystyle OR_3}{|}}{\underset{\underset{\displaystyle H}{|}}{C}}-OR_2 + H_2O$$

Hemiacetal   Acetal

6. Hydrolysis of an acetal to yield an aldehyde and two alcohols (Section 15.11)

$$R_1-\overset{\overset{\displaystyle OR_3}{|}}{\underset{\underset{\displaystyle H}{|}}{C}}-OR_2 + H_2O \underset{}{\overset{H^+}{\rightleftharpoons}} R_1-\overset{\overset{\displaystyle O}{\|}}{C}-H + R_2-OH + R_3-OH$$

EXERCISES *and* PROBLEMS

The members of each pair of problems in this section test similar material.

● The Carbonyl Group (Section 15.1)

15.1 Indicate which of the following compounds contain a carbonyl group.

a.

$$CH_3-CH_2-CH_2-\overset{\overset{\displaystyle O}{\|}}{C}-H$$

b. CH_3-O-CH_3

c.

$$CH_3-\overset{\overset{\displaystyle O}{\|}}{C}-CH_2-CH_3$$

d.

$$\overset{\overset{\displaystyle CH_3}{|}}{O=C-H}$$

15.2 Indicate which of the following compounds contain a carbonyl group.

a.

$$CH_3-CH_2-\overset{\overset{\displaystyle O}{\|}}{C}-CH_2-CH_3$$

b. $CH_3-CH_2-O-CH_2-CH_3$

c.

$$CH_3-CH_2-\overset{\overset{\displaystyle O}{\|}}{C}-H$$

d.

$$CH_3-CH_2-\overset{\overset{\displaystyle CH_3}{|}}{C}=O$$

15.3 What are the similarities and differences between the bonding in a carbon–oxygen double bond and that in a carbon–carbon double bond?

15.4 Use δ^+ and δ^- notation to show the polarity in a carbon–oxygen double bond.

15.5 What are the approximate bond angles between the atoms attached to the carbonyl carbon atom of a carbonyl group?

15.6 What is the geometrical arrangement for the atoms directly attached to the carbonyl carbon atom in a carbonyl compound?

● **Compounds Containing a Carbonyl Group (Section 15.2)**

15.7 Indicate whether each of the following types of compounds contain a carbonyl group.
a. Aldehyde   b. Ester
c. Alcohol   d. Carboxylic acid

15.8 What elements are present in each of the following types of hydrocarbon derivatives?
a. Carboxylic acid   b. Amide
c. Aldehyde   d. Ketone

15.9 Identify the type of hydrocarbon derivative associated with each of the following functional group designations.

a. $R-\overset{\overset{\displaystyle O}{\|}}{C}-H$ b. $R-\overset{\overset{\displaystyle O}{\|}}{C}-R$

c. $R-\overset{\overset{\displaystyle O}{\|}}{C}-NH_2$ d. $R-\overset{\overset{\displaystyle O}{\|}}{C}-OH$

15.10 Write the structural formula for the simplest member of each of the following types of hydrocarbon derivatives.
a. Aldehyde   b. Ester
c. Ketone   d. Carboxylic acid

● **The Aldehyde and Ketone Functional Groups (Section 15.3)**

15.11 Classify each of the following structures as an aldehyde, a ketone, or neither.

a. $CH_3-CH_2-CH_2-\overset{\overset{\displaystyle O}{\|}}{C}-OH$

b. $CH_3-\overset{\overset{\displaystyle O}{\|}}{C}-CH_3$

c. $CH_3-O-CH_2-CH_3$

d. CH_3-CHO

15.12 Classify each of the following structures as an aldehyde, a ketone, or neither.

a. $CH_3-\overset{\overset{\displaystyle O}{\|}}{C}-CH_2-CH_3$

b. $CH_3-CH_2-CH_2-\overset{\overset{\displaystyle O}{\|}}{C}-O-CH_3$

c. $CH_3-CH_2-\underset{\underset{\displaystyle CH_3}{|}}{CH}-\overset{\overset{\displaystyle O}{\|}}{C}-H$

d. $CH_3-\underset{\underset{\displaystyle CH_3}{|}}{\overset{\overset{\displaystyle CH_3}{|}}{C}}-CH_2-CHO$

15.13 Draw the structures of the two simplest aldehydes and the two simplest ketones.

15.14 One- and two-carbon ketones do not exist. Explain why.

15.15 Classify each of the following structures as an aldehyde, a ketone, or neither.

a. 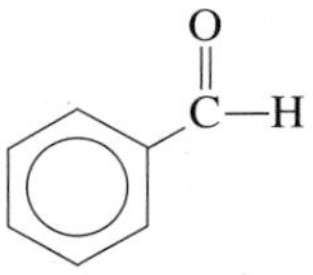   b.

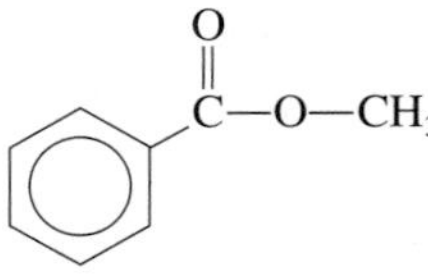

c. 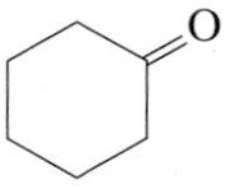   d.

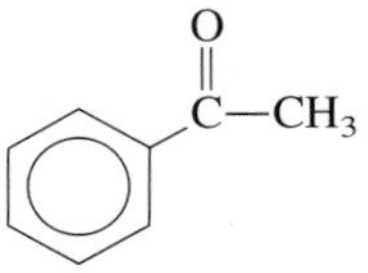

15.16 Classify each of the following structures as an aldehyde, a ketone, or neither.

a. 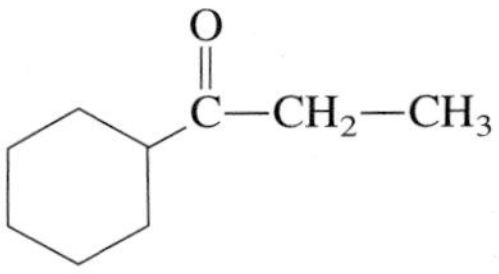   b.

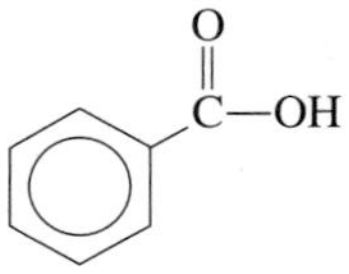

c. 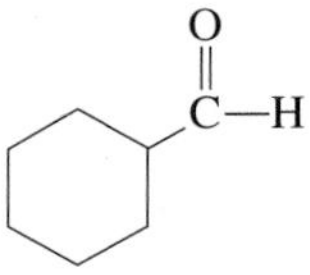   d.

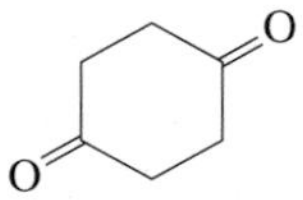

● **Nomenclature for Aldehydes (Section 15.4)**

15.17 Assign an IUPAC name to each of the following aldehydes.

a. $CH_3-CH_2-\underset{\underset{\displaystyle CH_3}{|}}{CH}-\overset{\overset{\displaystyle O}{\|}}{C}-H$

b. $CH_3-\underset{\underset{\displaystyle CH_2-CH_2-CH_3}{|}}{CH}-CH_2-CH_2-\overset{\overset{\displaystyle O}{\|}}{C}-H$

c.

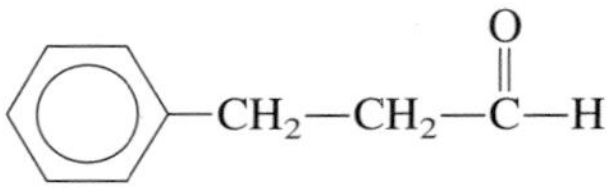

d. CH_3-CH_2-CHO

15.18 Assign an IUPAC name to each of the following aldehydes.

a. $CH_3-\underset{\underset{\displaystyle CH_3}{|}}{CH}-CH_2-CH_2-\overset{\overset{\displaystyle O}{\|}}{C}-H$

b. $CH_3-CH_2-\underset{\underset{\underset{\displaystyle CH_3}{|}}{\underset{\displaystyle CH_2}{|}}}{CH}-\overset{\overset{\displaystyle O}{\|}}{C}-H$

c.

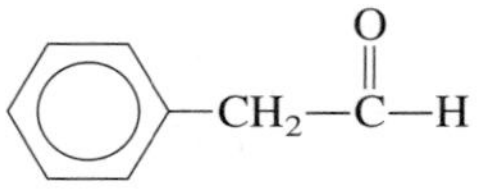

d. $CH_3-CH_2-CH_2-CHO$

15.19 Assign an IUPAC name to each of the following aldehydes.

a. 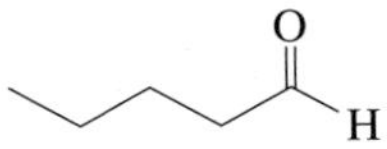   b.

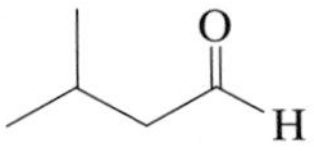

c. 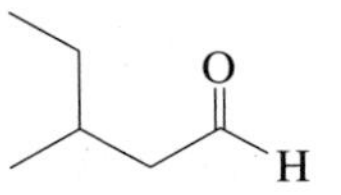   d.

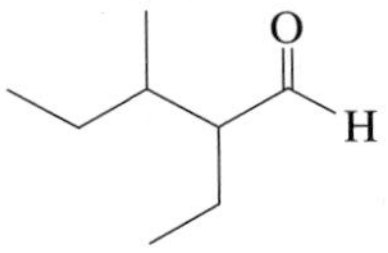

15.20 Assign an IUPAC name to each of the following aldehydes.

a. 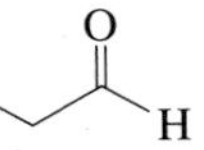   b.

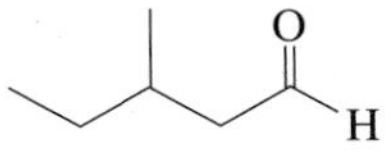

c. 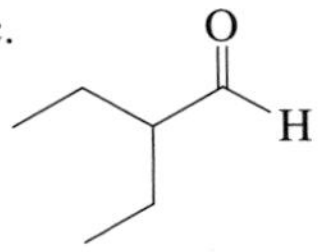   d.

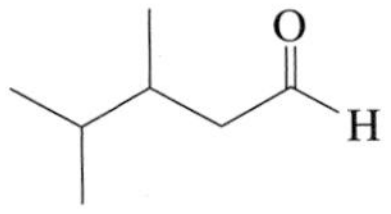

15.21 Draw a structural formula for each of the following aldehydes.
a. 3-Methylpentanal   b. 2-Ethylhexanal
c. 2,2-Dichloropropanal d. 4-Hydroxy-2-methyloctanal

15.22 Draw a structural formula for each of the following aldehydes.
a. 2-Methylpentanal   b. 4-Ethylhexanal
c. 3,3-Dimethylhexanal   d. 2,3-Dibromopropanal

15.23 Draw a structural formula for each of the following aldehydes.
a. Formaldehyde   b. Propionaldehyde
c. 2-Chlorobenzaldehyde d. 2,4-Dimethylbenzaldehyde

15.24 Draw a structural formula for each of the following aldehydes.
a. Acetaldehyde   b. Butyraldehyde
c. Dichloroacetaldehyde d. 2-Methylbenzaldehyde

15.25 Assign a common name to each of the following aldehydes.

a. 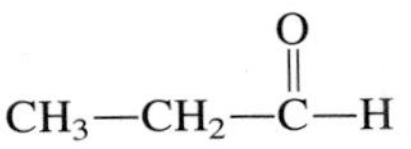 $CH_3-CH_2-\overset{\displaystyle O}{\overset{\|}{C}}-H$ b. CH_3-CH_2-CHO

c. 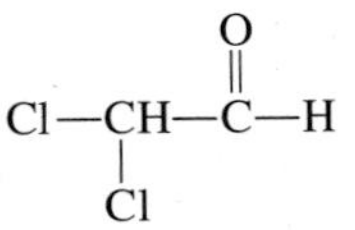 $Cl-\underset{\underset{\displaystyle Cl}{|}}{CH}-\overset{\displaystyle O}{\overset{\|}{C}}-H$ d.

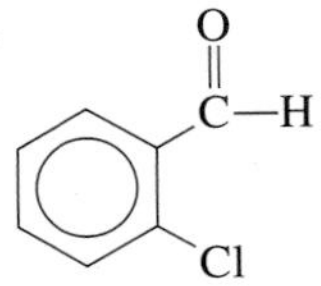

15.26 Assign a common name to each of the following aldehydes.

a. 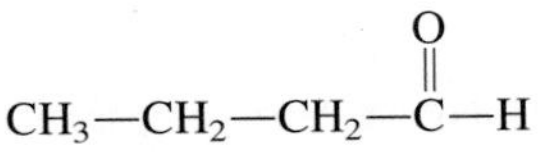 $CH_3-CH_2-CH_2-\overset{\displaystyle O}{\overset{\|}{C}}-H$

b. $CH_3-CH_2-CH_2-CHO$

c. 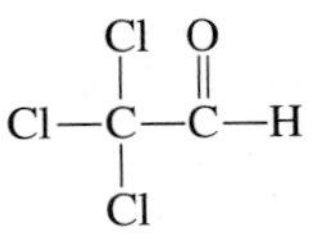 $Cl-\underset{\underset{\displaystyle Cl}{|}}{\overset{\overset{\displaystyle Cl}{|}}{C}}-\overset{\displaystyle O}{\overset{\|}{C}}-H$

d.

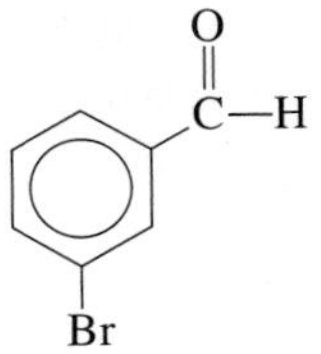

● **Nomenclature for Ketones (Section 15.5)**

15.27 Using IUPAC nomenclature, name each of the following ketones.

a. 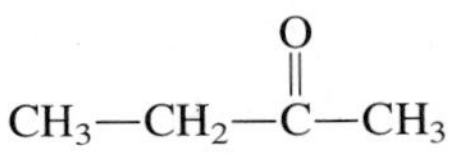 $CH_3-CH_2-\overset{\displaystyle O}{\overset{\|}{C}}-CH_3$

b. $CH_3-\underset{\underset{\displaystyle CH_3}{|}}{CH}-\underset{\underset{\displaystyle CH_3}{|}}{CH}-\overset{\displaystyle O}{\overset{\|}{C}}-\underset{\underset{\displaystyle CH_3}{|}}{CH}-CH_3$

c. $CH_3-\underset{\underset{\displaystyle CH_3}{|}}{CH}-CH_2-CH_2-\overset{\displaystyle O}{\overset{\|}{C}}-CH_2-CH_3$

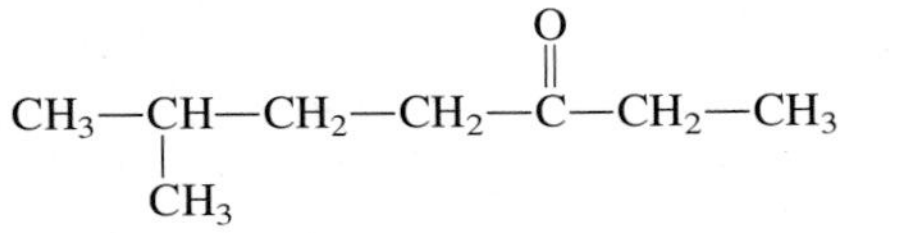

d. 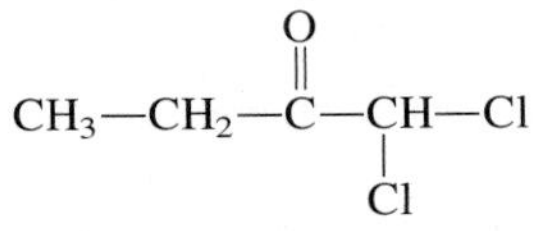 $CH_3-CH_2-\overset{\displaystyle O}{\overset{\|}{C}}-\underset{\underset{\displaystyle Cl}{|}}{CH}-Cl$

15.28 Using IUPAC nomenclature, name each of the following ketones.

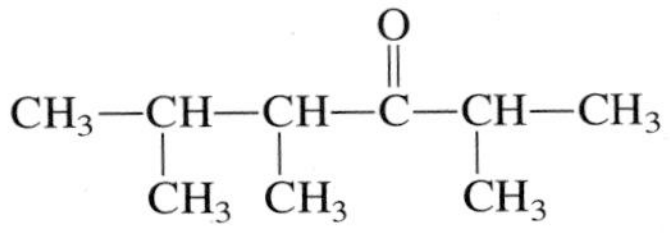

a. 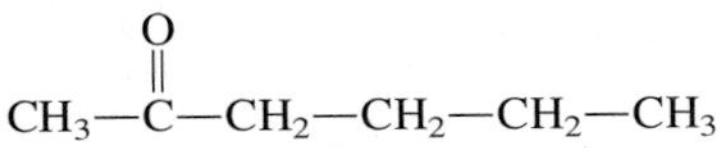 $CH_3-\overset{\displaystyle O}{\overset{\|}{C}}-CH_2-CH_2-CH_2-CH_3$

b. $CH_3-CH_2-\overset{\displaystyle O}{\overset{\|}{C}}-CH_2-\underset{\underset{\underset{\displaystyle CH_3}{|}}{\overset{\displaystyle CH_2}{|}}}{CH}-CH_3$

c. $CH_3-\underset{\underset{\displaystyle Cl}{|}}{CH}-\overset{\displaystyle O}{\overset{\|}{C}}-\underset{\underset{\displaystyle Br}{|}}{CH}-CH_3$

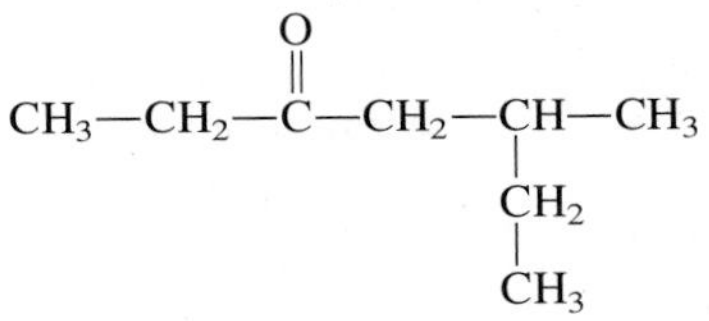

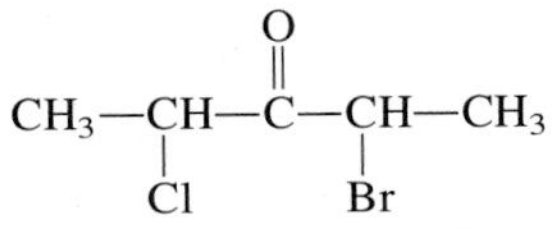

d. 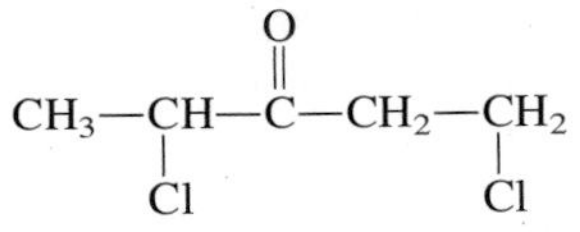 $CH_3-\underset{\underset{\displaystyle Cl}{|}}{CH}-\overset{\displaystyle O}{\overset{\|}{C}}-CH_2-\underset{\underset{\displaystyle Cl}{|}}{CH_2}$

15.29 Assign an IUPAC name to each of the following ketones.

a. 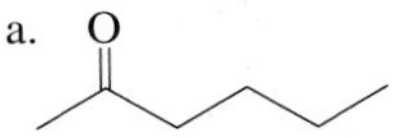   b.

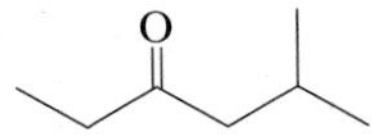

c. 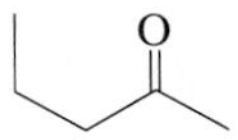   d.

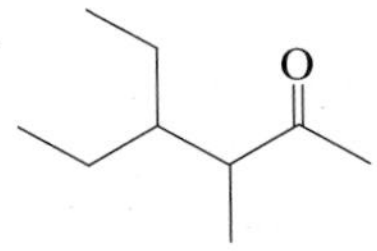

15.30 Assign an IUPAC name to each of the following ketones.

a. 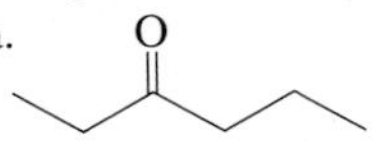   b.

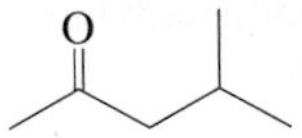

c. 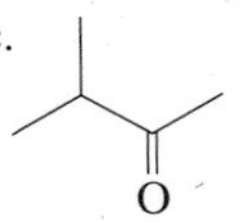   d.

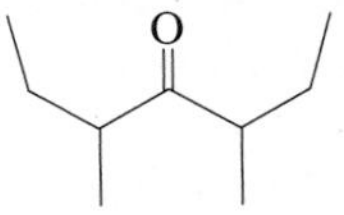

15.31 Using IUPAC nomenclature, name each of the following ketones.

a. 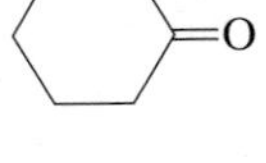   b.

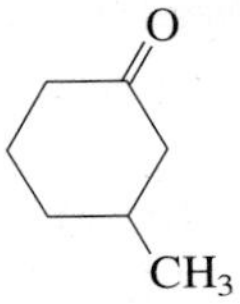

c. 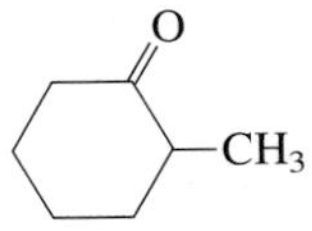   d.

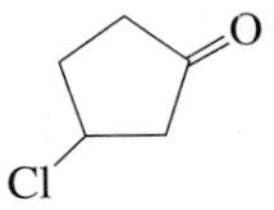

15.32 Using IUPAC nomenclature, name each of the following ketones.

a.

b.

CH₃

c.

CH₃

d.

Br

15.33 Draw a structural formula for each of the following ketones.
- a. 3-Methyl-2-pentanone
- b. 3-Hexanone
- c. Cyclobutanone
- d. Chloropropanone

15.34 Draw a structural formula for each of the following ketones.
- a. 2-Methyl-3-pentanone
- b. 2-Pentanone
- c. Bromopropanone
- d. Cyclopentanone

15.35 Draw a structural formula for each of the following ketones.
- a. Isopropyl propyl ketone
- b. Chloromethyl methyl ketone
- c. Acetophenone
- d. Methyl phenyl ketone

15.36 Draw a structural formula for each of the following ketones.
- a. Methyl *tert*-butyl ketone
- b. Dichloromethyl ethyl ketone
- c. Benzophenone
- d. Diphenyl ketone

● **Isomerism for Aldehydes and Ketones (Section 15.6)**

15.37 Give IUPAC names for all saturated unbranched-chain compounds that are named as the following.
- a. Heptanals
- b. Heptanones

15.38 Give IUPAC names for all saturated unbranched-chain compounds that are named as the following.
- a. Hexanals
- b. Hexanones

15.39 How many aldehydes and how many ketones exist with each of the following molecular formulas?
- a. CH_2O
- b. C_3H_6O

15.40 How many aldehydes and how many ketones exist with each of the following molecular formulas?
- a. C_2H_4O
- b. C_4H_8O

15.41 For which values of x is the ketone name x-methyl-3-hexanone a correct IUPAC name?

15.42 For which values of x is the ketone name x-methyl-3-pentanone a correct IUPAC name?

15.43 Draw skeletal structural formulas for the four aldehydes and three ketones that have the molecular formula $C_5H_{10}O$.

15.44 Draw skeletal structural formulas for the eight aldehydes and six ketones that have the molecular formula $C_6H_{12}O$.

● **Selected Common Aldehydes and Ketones (Section 15.7)**

15.45 What is the difference between the substances formaldehyde and formalin?

15.46 What is the odor associated with and what are the uses for the substance formalin?

15.47 What are the general properties of and uses for the substance acetone?

15.48 What is the significance of the odor of acetone on the breath of a person?

● **Physical Properties of Aldehydes and Ketones (Section 15.8)**

15.49 Aldehydes and ketones have higher boiling points than alkanes of similar molecular mass. Explain why.

15.50 Aldehydes and ketones have lower boiling points than alcohols of similar molecular mass. Explain why.

15.51 How many hydrogen bonds can form between an acetone molecule and water molecules?

15.52 How many hydrogen bonds can form between an acetaldehyde molecule and water molecules?

15.53 Would you expect ethanal or octanal to be more soluble in water? Explain your answer.

15.54 Would you expect ethanal or octanal to have the more fragrant odor? Explain your answer.

● **Preparation of Aldehydes and Ketones (Section 15.9)**

15.55 Draw the structure of the aldehyde or ketone formed from oxidation of each of the following alcohols. Assume that reaction conditions are sufficiently mild that any aldehydes produced are not oxidized further to carboxylic acids.
- a. $CH_3-CH_2-CH_2-CH_2-CH_2-OH$
- b. $CH_3-CH_2-CH-OH$
  $\ \ \ \ \ \ \ \ \ \ \ \ \ \ \ CH_3$
- c. $\ \ \ \ \ \ \ \ \ CH_3$
  $\ \ \ \ \ \ \ \ \ \ |$
  $CH_3-C-CH_2-CH_2-OH$
  $\ \ \ \ \ \ \ \ \ \ |$
  $\ \ \ \ \ \ \ \ \ CH_3$
- d. $CH_3-\!\!\bigcirc\!\!-OH$

15.56 Draw the structure of the aldehyde or ketone formed from oxidation of each of the following alcohols. Assume that reaction conditions are sufficiently mild that any aldehydes formed are not oxidized further to carboxylic acids.
- a. $CH_3-CH_2-CH-CH_2-OH$
  $\ \ \ \ \ \ \ \ \ \ \ \ \ \ CH_3$
- b. $CH_3-CH-CH-OH$
  $\ \ \ \ \ \ \ \ CH_3\ \ CH_3$
- c. $\ \ \ \ \ \ \ \ \ CH_3$
  $\ \ \ \ \ \ \ \ \ \ |$
  CH_3-C-OH
  $\ \ \ \ \ \ \ \ \ \ |$
  $\ \ \ \ \ \ \ \ \ CH_3$
- d. $\ \ \ \ \ \ \ \ \ CH_2-CH_3$
  $\bigcirc\!\!-OH$

15.57 Draw the structure of the alcohol needed to prepare each of the following aldehydes or ketones by alcohol oxidation.
- a. Diethyl ketone
- b. Phenylpropanone
- c. Acetaldehyde
- d. 2-Ethylhexanal

15.58 Draw the structure of the alcohol needed to prepare each of the following aldehydes or ketones by alcohol oxidation.
- a. Propanal
- b. Dipropyl ketone
- c. 3-Phenyl-2-butanone
- d. Cyclohexanone

● **Oxidation and Reduction of Aldehydes and Ketones (Section 15.10)**

15.59 Draw the structural formula of the organic product when each of the following aldehydes is oxidized to a carboxylic acid.
- a. Ethanal
- b. Pentanal
- c. Formaldehyde
- d. 3,4-Dichlorohexanal

15.60 Draw the structural formula of the organic product when each of the following aldehydes is oxidized to a carboxylic acid.
  a. Butanal
  b. 2-Methylpentanal
  c. Acetaldehyde
  d. Benzaldehyde

15.61 What are the characteristics of a positive Tollens test for aldehydes?

15.62 What are the characteristics of a positive Benedict's test for aldehydes?

15.63 What is the oxidizing agent in Benedict's solution?

15.64 What is the oxidizing agent in Tollens solution?

15.65 Which of the following compounds would react with Tollens solution?

a.
$$CH_3-CH_2-CH_2-\overset{\overset{\displaystyle O}{\|}}{C}-CH_3$$

b.
$$CH_3-CH_2-CH_2-\overset{\overset{\displaystyle O}{\|}}{C}-H$$

c.
$$CH_3-\underset{\underset{\displaystyle OH}{|}}{CH}-CH_2-\overset{\overset{\displaystyle O}{\|}}{C}-H$$

d.
$$CH_3-\underset{\underset{\displaystyle OH}{|}}{CH}-\overset{\overset{\displaystyle O}{\|}}{C}-CH_3$$

15.66 Which of the following compounds would react with Benedict's solution?

a.
$$CH_3-CH_2-\overset{\overset{\displaystyle O}{\|}}{C}-H$$

b.
$$CH_3-\overset{\overset{\displaystyle O}{\|}}{C}-CH_3$$

c.
$$CH_3-CH_2-\underset{\underset{\displaystyle OH}{|}}{CH}-\overset{\overset{\displaystyle O}{\|}}{C}-CH_2-CH_3$$

d.
$$CH_3-\underset{\underset{\displaystyle CH_3}{|}}{CH}-\underset{\underset{\displaystyle CH_3}{|}}{CH}-CH_2-\overset{\overset{\displaystyle O}{\|}}{C}-H$$

15.67 Draw the structure of the major organic compound produced when each of the following compounds is reduced using molecular H_2 and a Ni catalyst.

a.
$$CH_3-CH_2-CH_2-\overset{\overset{\displaystyle O}{\|}}{C}-H$$

b.
$$CH_3-CH_2-\overset{\overset{\displaystyle O}{\|}}{C}-CH_2-CH_3$$

c.
$$CH_3-\underset{\underset{\displaystyle CH_3}{|}}{CH}-CH_2-\overset{\overset{\displaystyle O}{\|}}{C}-H$$

d.
$$CH_3-\underset{\underset{\displaystyle CH_3}{|}}{CH}-\overset{\overset{\displaystyle O}{\|}}{C}-CH_2-CH_2-CH_3$$

15.68 Draw the structure of the major organic compound produced when each of the following compounds is reduced using molecular H_2 and a Ni catalyst.

a.
$$CH_3-CH_2-CH_2-\overset{\overset{\displaystyle O}{\|}}{C}-CH_3$$

b.
$$CH_3-CH_2-CH_2-CH_2-\overset{\overset{\displaystyle O}{\|}}{C}-H$$

c.
$$CH_3-\underset{\underset{\displaystyle CH_3}{|}}{\overset{\overset{\displaystyle CH_3}{|}}{C}}-CH_2-CH_2-\overset{\overset{\displaystyle O}{\|}}{C}-H$$

d.
$$CH_3-CH_2-\overset{\overset{\displaystyle O}{\|}}{C}-CH_3$$

Hemiacetal Formation (Section 15.11)

15.69 When an alcohol molecule (R—O—H) adds across a carbon–oxygen double bond, into what "fragments" is the alcohol split?

15.70 When an alcohol molecule (R—O—H) adds across a carbon–oxygen double bond, which part of the alcohol molecule adds to the carbonyl oxygen atom?

15.71 Indicate whether each of the following compounds is a hemiacetal.
  a. $CH_3-CH_2-O-CH_3$

b.
$$CH_3-\underset{\underset{\displaystyle O-CH_3}{|}}{\overset{\overset{\displaystyle OH}{|}}{C}}-CH_3$$

c.   d.

15.72 Indicate whether each of the following compounds is a hemiacetal.

a.
$$CH_3-CH_2-\underset{\underset{\displaystyle OH}{|}}{\overset{\overset{\displaystyle OH}{|}}{C}}-CH_3$$

b.
$$CH_3-CH_2-\underset{\underset{\displaystyle O-CH_3}{|}}{\overset{\overset{\displaystyle OH}{|}}{C}}-CH_3$$

c.   d.

15.73 Draw the structural formula of the hemiacetal formed from each of the following pairs of reactants.
  a. Acetaldehyde and ethyl alcohol
  b. 2-Pentanone and methanol
  c. Butanal and ethanol
  d. Acetone and isopropyl alcohol

15.74 Draw the structural formula of the hemiacetal formed from each of the following pairs of reactants.
  a. Acetaldehyde and methanol
  b. 2-Pentanone and ethyl alcohol
  c. Butanal and isopropyl alcohol
  d. Acetone and ethanol

15.75 Draw the structural formula of the missing compound in each of the following reactions.

a.
$$CH_3-(CH_2)_2-\overset{\overset{\displaystyle O}{\|}}{C}-H + CH_3-CH_2-OH \overset{H^+}{\rightleftharpoons} ?$$

b.
$$? + CH_3-OH \overset{H^+}{\rightleftharpoons} CH_3-CH_2-\underset{\underset{\displaystyle O-CH_3}{|}}{\overset{\overset{\displaystyle OH}{|}}{CH}}$$

c.
$$CH_3-CH_2-\overset{\overset{\displaystyle O}{\|}}{C}-CH_3 + CH_3-OH \overset{H^+}{\rightleftharpoons} ?$$

d.

15.76 Draw the structural formula of the missing compound in each of the following reactions.

a.

$$CH_3-CH_2-\overset{\overset{\displaystyle O}{\|}}{C}-H + CH_3-OH \underset{\longleftarrow}{\overset{H^+}{\longrightarrow}} ?$$

b.

$$? + CH_3-CH_2-OH \underset{\longleftarrow}{\overset{H^+}{\longrightarrow}} CH_3-CH_2-\underset{\underset{\displaystyle O-CH_2-CH_3}{|}}{\overset{\overset{\displaystyle OH}{|}}{CH}}$$

c.

$$CH_3-CH_2-CH_2-\overset{\overset{\displaystyle O}{\|}}{C}-CH_3 + CH_3-CH_2-OH \underset{\longleftarrow}{\overset{H^+}{\longrightarrow}} ?$$

d.

Acetal Formation (Section 15.11)

15.77 Indicate whether each of the following compounds is an acetal.

a.
$$CH_3-CH_2-CH_2-\underset{\underset{\displaystyle O-CH_3}{|}}{\overset{\overset{\displaystyle O-CH_3}{|}}{CH}}$$

b.
$$CH_3-CH_2-CH_2-O-\underset{\underset{\displaystyle CH_3}{|}}{\overset{\overset{\displaystyle O-CH_3}{|}}{C}}-CH_3$$

c.
$$CH_3-CH_2-CH_2-O-\underset{\underset{\displaystyle CH_3}{|}}{\overset{\overset{\displaystyle CH_3}{|}}{C}}-OH$$

d.
$$CH_3-\underset{\underset{\displaystyle O-CH_2-CH_2-CH_3}{|}}{\overset{\overset{\displaystyle O-CH_2-CH_2-CH_3}{|}}{C}}-CH_2-CH_2-CH_3$$

15.78 Indicate whether each of the following compounds is an acetal.

a.
$$CH_3-CH_2-\underset{\underset{\displaystyle O-CH_3}{|}}{\overset{\overset{\displaystyle O-CH_2-CH_3}{|}}{CH}}$$

b.
$$CH_3-CH_2-O-\underset{\underset{\displaystyle CH_3}{|}}{\overset{\overset{\displaystyle O-CH_3}{|}}{C}}-CH_3$$

c.
$$CH_3-CH_2-CH_2-\underset{\underset{\displaystyle OH}{|}}{\overset{\overset{\displaystyle H}{|}}{C}}-O-CH_3$$

d.
$$CH_3-\underset{\underset{\displaystyle O-CH_3}{|}}{\overset{\overset{\displaystyle CH_3}{|}}{C}}-O-CH_3$$

15.79 Draw the structural formula of the missing compound(s) in each of the following reactions.

a.
$$CH_3-\underset{\underset{\displaystyle CH_3}{|}}{\overset{\overset{\displaystyle O-CH_3}{|}}{C}}-OH + ? \overset{H^+}{\longrightarrow} CH_3-\underset{\underset{\displaystyle CH_3}{|}}{\overset{\overset{\displaystyle O-CH_3}{|}}{C}}-O-CH_3 + H_2O$$

b.
$$? + CH_3-CH_2-OH \overset{H^+}{\longrightarrow} CH_3-\underset{\underset{\displaystyle O-CH_2-CH_3}{|}}{\overset{\overset{\displaystyle H}{|}}{C}}-O-CH_3 + H_2O$$

c.
$$CH_3-CH_2-\underset{\underset{\displaystyle H}{|}}{\overset{\overset{\displaystyle OH}{|}}{C}}-O-CH_3 + CH_3-\underset{\underset{\displaystyle CH_3}{|}}{CH}-OH \overset{H^+}{\longrightarrow}$$
$$? + H_2O$$

d.
$$\underset{\text{Hemiacetal}}{?} + \underset{\text{Alcohol}}{?} \overset{H^+}{\longrightarrow} CH_3-\underset{\underset{\displaystyle O-CH_3}{|}}{CH}-O-CH_3 + H_2O$$

15.80 Draw the structural formula of the missing compound(s) in each of the following reactions.

a.
$$CH_3-CH_2-\underset{\underset{\displaystyle CH_3}{|}}{\overset{\overset{\displaystyle O-CH_3}{|}}{C}}-OH + ? \overset{H^+}{\longrightarrow}$$
$$CH_3-CH_2-\underset{\underset{\displaystyle CH_3}{|}}{\overset{\overset{\displaystyle O-CH_3}{|}}{C}}-O-CH_3 + H_2O$$

b. $? + CH_3-CH_2-OH \overset{H^+}{\longrightarrow}$
$$CH_3-CH_2-\underset{\underset{\displaystyle O-CH_2-CH_3}{|}}{\overset{\overset{\displaystyle H}{|}}{C}}-O-CH_3 + H_2O$$

c.
$$CH_3-CH_2-\underset{\underset{\displaystyle H}{|}}{\overset{\overset{\displaystyle OH}{|}}{C}}-O-CH_3 + CH_3-\underset{\underset{\displaystyle CH_3}{|}}{CH}-OH \overset{H^+}{\longrightarrow}$$
$$? + H_2O$$

d.
$$\underset{\text{Hemiacetal}}{?} + \underset{\text{Alcohol}}{?} \overset{H^+}{\longrightarrow}$$
$$CH_3-CH_2-\underset{\underset{\displaystyle O-CH_3}{|}}{CH}-O-CH_3 + H_2O$$

15.81 Draw the structural formulas of the aldehyde (or ketone) and the two alcohols produced when the following acetals undergo hydrolysis in acidic solution.

a.
$$CH_3-\underset{\underset{\displaystyle O-CH_3}{|}}{\overset{\overset{\displaystyle O-CH_3}{|}}{CH}}$$

b.
$$CH_3-\underset{\underset{\displaystyle O-CH_3}{|}}{\overset{\overset{\displaystyle O-CH_3}{|}}{C}}-CH_3$$

c.
$$CH_3-O-\underset{\underset{\displaystyle CH_2-CH_3}{|}}{\overset{\overset{\displaystyle CH_2-CH_3}{|}}{C}}-O-CH_2-CH_3$$

d.
$$CH_3-CH_2-CH_2-CH_2-\underset{\underset{\displaystyle H}{|}}{\overset{\overset{\displaystyle O-CH_3}{|}}{C}}-O-CH_3$$

15.82 Draw the structural formulas of the aldehyde (or ketone) and the two alcohols produced when the following acetals undergo hydrolysis in acidic solution.

a.
$$CH_3-CH_2-\overset{\overset{\displaystyle O-CH_3}{|}}{\underset{\underset{\displaystyle O-CH_3}{|}}{CH}}$$

b.
$$CH_3-CH_2-\overset{\overset{\displaystyle O-CH_3}{|}}{\underset{\underset{\displaystyle O-CH_3}{|}}{C}}-CH_3$$

c.
$$CH_3-CH_2-O-\overset{\overset{\displaystyle H}{|}}{\underset{\underset{\displaystyle CH_3}{|}}{C}}-O-CH_2-CH_3$$

d.
$$CH_3-CH_2-\overset{\overset{\displaystyle O-CH_2-CH_3}{|}}{\underset{\underset{\displaystyle O-CH_2-CH_3}{|}}{C}}-CH_2-CH_2-CH_3$$

15.83 Name each of the compounds in Problem 15.81 in the manner described in Section 15.11.

15.84 Name each of the compounds in Problem 15.82 in the manner described in Section 15.11.

ADDITIONAL PROBLEMS

15.93 Explain each of the following.
  a. The IUPAC name for the three-carbon aldehyde is propanal rather than 1-propanal.
  b. The IUPAC name for the three-carbon ketone is propanone rather than 2-propanone.

15.94 Each of the following compound names represents an impossible structure. In each case, explain why.
  a. Methanone   b. 1-Chlorobutanal
  c. 3-Methyl-3-pentanone   d. Cyclohexanal

15.95 What is the characteristic structural feature of each of the following?
  a. Hemiacetal   b. Acetal

15.96 Draw the structural formula of the hemiacetal formed and then the acetal formed when each of the following compounds reacts with an excess of the reactant alcohol.
  a. Propanal and ethanol   b. Cyclohexanone and methanol

15.97 The compound 4-hydroxybutanal can form an intramolecular cyclic hemiacetal. Draw the structural formula of this cyclic hemiacetal.

15.98 Name the functional groups present in each of the following polyfunctional compounds.

● **Formaldehyde-Based Polymers (Section 15.12)**

15.85 What are the structural characteristics associated with a network polymer?

15.86 What is a major current use for phenol–formaldehyde network polymers?

15.87 In phenol–formaldehyde polymer formation what are the intermediate compounds that are formed?

15.88 In a phenol–formaldehyde network polymer what type of "bridges" cross-link the various substituted phenols?

● **Sulfur-Containing Carbonyl Groups (Section 15.13)**

15.89 What type of compound is formed by replacement of the carbonyl oxygen atom with a sulfur atom?

15.90 What type of compound is formed by replacement of the carbonyl carbon atom with a sulfur atom?

15.91 Draw structural formulas for the following compounds.
  a. Thioformaldehyde   b. Methanethial
  c. Thioacetone   d. Propanethione

15.92 Dimethyl sulfoxide (DMSO) is a sulfur analog of acetone in which the sulfur has substituted for the carbonyl carbon atom.
  a. Draw the structural formula for DMSO.
  b. Describe the "unusual" solubility properties of DMSO.

  a. 4-Octen-2-one
  b. 2-Methoxy-4-hydroxypentanal
  c. 3-Hexyn-2-one
  d. 4-Oxohexanal

15.99 Indicate whether each of the following compounds would be named as an alcohol, an aldehyde, or a ketone.

a.

b.

c.

d.

𝑀ultiple-Choice Practice Test

15.100 Which of the following statements concerning aldehydes and ketones is correct?
  a. Aldehydes contain a carbonyl group but ketones do not.
  b. Ketones contain a carbonyl group but aldehydes do not.
  c. Both aldehydes and ketones contain a carbonyl group.
  d. Neither aldehydes nor ketones contain a carbonyl group.

15.101 Which is the IUPAC name for the ketone *ethyl propyl ketone*?
  a. 3-Pentanone
  b. 4-Pentanone
  c. 3-Hexanone
  d. 4-Hexanone

15.102 Which of the following compounds is a constitutional isomer of acetone?
a. Formaldehyde   b. Acetaldehyde
c. Propionaldehyde   d. Butyraldehyde

15.103 The physical state, at room temperature and pressure, for the simplest aldehyde and the simplest ketone is, respectively, which of the following?
a. Gas and gas   b. Gas and liquid
c. Liquid and gas   d. Liquid and liquid

15.104 For which of the following molecular combinations is hydrogen bonding possible?
a. Aldehyde–aldehyde   b. Ketone–ketone
c. Aldehyde–ketone   d. Water–ketone

15.105 A general method for the preparation of ketones is oxidation of which of the following?
a. 1° alcohols   b. 2° alcohols
c. 3° alcohols   d. Aldehydes

15.106 Which of the following reactions is classified as a reduction reaction?
a. Alcohol to ketone   b. Alcohol to aldehyde
c. Aldehyde to alcohol   d. Aldehyde to carboxylic acid

15.107 In a hemiacetal, the hemiacetal carbon atom is bonded to which of the following?
a. Two hydroxyl groups
b. Two alkoxy groups
c. One hydroxyl group and one alkoxy group
d. Two hydroxyl groups and one alkoxy group

15.108 To produce an acetal from a ketone, the ketone must react with which of the following?
a. One alcohol molecule
b. Two identical alcohol molecules
c. Two different alcohol molecules
d. Two alcohol molecules, which may or may not be identical

15.109 What is the number of organic product molecules produced from the complete hydrolysis of an acetal molecule?
a. Two
b. Three
c. Four
d. Five

Carboxylic Acids, Esters, and Other Acid Derivatives

Chapter Outline

Esters, a type of carboxylic acid derivative, are largely responsible for the flavors and fragrances of ripe fruits such as red raspberries.

I n Chapter 15, we discussed the carbonyl group and two families of compounds—aldehydes and ketones—that contain this group. In this chapter, we discuss four more families of compounds in which the carbonyl group is present: carboxylic acids, esters, acid chlorides, and acid anhydrides.

16.1 STRUCTURE OF CARBOXYLIC ACIDS AND THEIR DERIVATIVES

A **carboxylic acid** *is an organic compound whose functional group is the carboxyl group.* What is a carboxyl group? A **carboxyl group** *is a carbonyl group (C=O) with a hydroxyl group (—OH) bonded to the carbonyl carbon atom.* A general structural representation for a carboxyl group is

Abbreviated linear designations for the carboxyl group are

$$-\text{COOH} \quad \text{and} \quad -\text{CO}_2\text{H}$$

Although we see within a carboxyl group both a carbonyl group (C=O) and a hydroxyl group (—OH), the carboxyl group does not show characteristic behavior of

The term *carboxyl* is a contraction of the words *carb*onyl and hydr*oxyl*.

either an alcohol or a carbonyl compound (aldehyde or ketone). Rather, it is a unique functional group with a set of characteristics different from those of its component parts.

The simplest carboxylic acid has a hydrogen atom attached to the carboxyl group carbon atom.

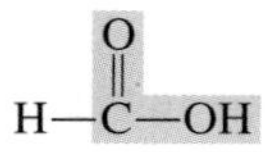

$$H-\overset{\overset{\displaystyle O}{\|}}{C}-OH$$

Structures for the next two simplest carboxylic acids, those with methyl and ethyl alkyl groups, are

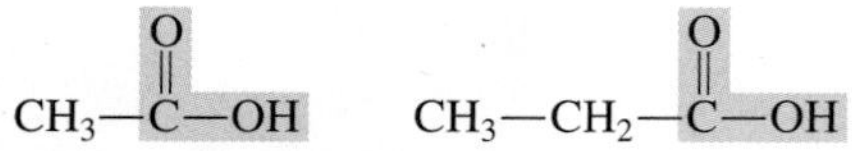

$$CH_3-\overset{\overset{\displaystyle O}{\|}}{C}-OH \qquad CH_3-CH_2-\overset{\overset{\displaystyle O}{\|}}{C}-OH$$

The structure of the simplest aromatic carboxylic acid involves a benzene ring to which a carboxyl group is attached.

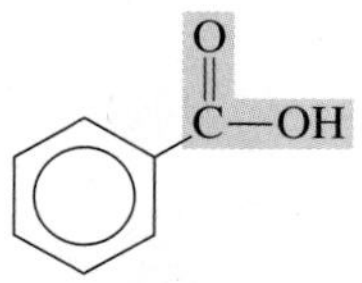

General formulas for carboxylic acids containing alkyl and aryl groups, respectively, are

R—COOH   and   Ar—COOH

Cyclic carboxylic acids do not exist; having the carboxyl carbon atom as part of a ring system creates a situation where the carboxyl carbon atom would have five bonds. The nonexistence of cyclic carboxylic acids parallels the nonexistence of cyclic aldehydes (Section 15.3).

A **carboxylic acid derivative** *is an organic compound that can be synthesized from or converted into a carboxylic acid.* Four important families of carboxylic acid derivatives are esters, acid chlorides, acid anhydrides, and amides. The group attached to the carbonyl carbon atom distinguishes these derivative types from each other and also from carboxylic acids.

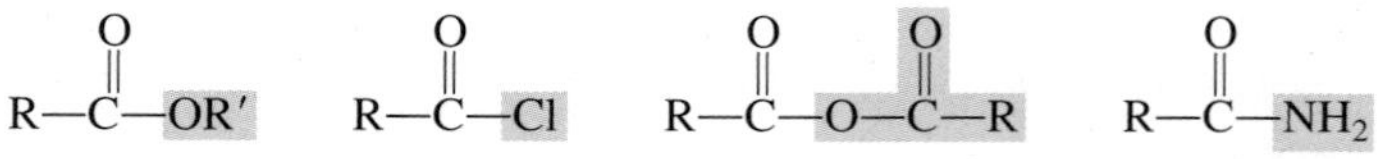

$$\underset{\text{Ester}}{R-\overset{\overset{\displaystyle O}{\|}}{C}-OR'} \qquad \underset{\text{Acid chloride}}{R-\overset{\overset{\displaystyle O}{\|}}{C}-Cl} \qquad \underset{\text{Acid anhydride}}{R-\overset{\overset{\displaystyle O}{\|}}{C}-O-\overset{\overset{\displaystyle O}{\|}}{C}-R} \qquad \underset{\text{Amide}}{R-\overset{\overset{\displaystyle O}{\|}}{C}-NH_2}$$

Further information about the first three of these four families of carboxylic acid derivatives is found in later sections of this chapter. Consideration of amides, which are nitrogen-containing compounds, will be found in Chapter 17.

16.2 IUPAC NOMENCLATURE FOR CARBOXYLIC ACIDS

IUPAC rules for naming carboxylic acids resemble those for naming aldehydes (Section 15.4).

Monocarboxylic Acids

A **monocarboxylic acid** *is a carboxylic acid in which one carboxyl group is present.* IUPAC rules for naming such compounds are:

Rule 1: Select as the parent carbon chain the longest carbon chain that *includes* the carbon atom of the carboxyl group.

Rule 2: Name the parent chain by changing the *-e* ending of the corresponding alkane to *-oic acid.*

Rule 3: Number the parent chain by assigning the number 1 to the carboxyl carbon atom.

Rule 4: Determine the identity and location of any substituents in the usual manner, and append this information to the front of the parent chain name.

A carboxyl group must occupy a terminal (end) position in a carbon chain because there can be only one other bond to it.

Space-filling models for the three simplest carboxylic acids—methanoic acid, ethanoic acid, and propanoic acid—are shown in Figure 16.1.

Figure 16.1 Space-filling models for the three simplest carboxylic acids: methanoic acid, ethanoic acid, and propanoic acid.

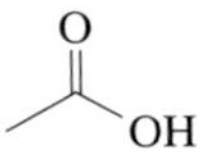

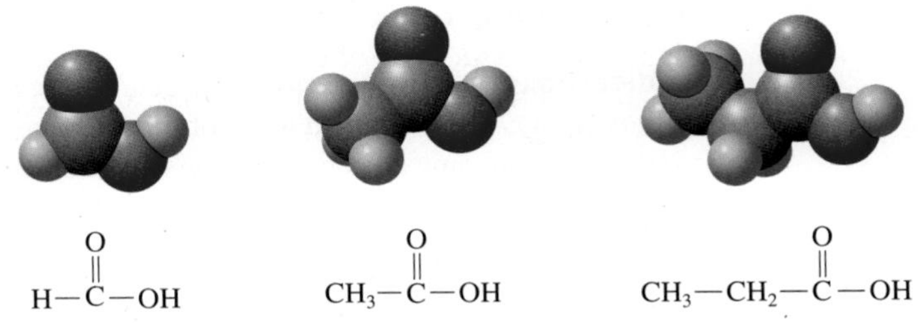

$$H-\overset{\overset{\displaystyle O}{\|}}{C}-OH \qquad CH_3-\overset{\overset{\displaystyle O}{\|}}{C}-OH \qquad CH_3-CH_2-\overset{\overset{\displaystyle O}{\|}}{C}-OH$$

Methanoic acid   **Ethanoic acid**   **Propanoic acid**

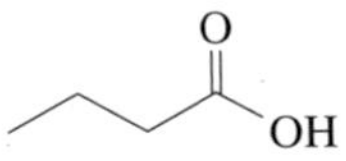

⬤ **EXAMPLE 16.1**

Determining IUPAC Names for Carboxylic Acids

Line-angle structural formulas for the simpler unbranched-chain carboxylic acids:

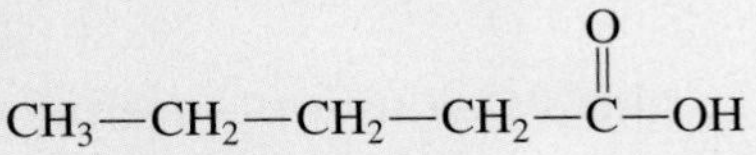

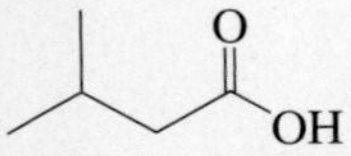

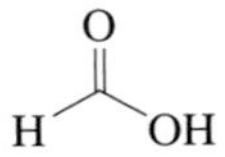

H   OH
Methanoic acid

O
OH
Ethanoic acid

O
OH
Propanoic acid

O
OH
Butanoic acid

Assign IUPAC names to the following carboxylic acids.

a. $CH_3-CH_2-CH_2-CH_2-\overset{\overset{\displaystyle O}{\|}}{C}-OH$

b. (line-angle structure with O and OH)

c. $CH_3-\underset{\underset{\underset{CH_3}{|}}{\overset{|}{CH_2}}}{\overset{|}{CH}}-\overset{|}{CH}-\overset{\overset{\displaystyle O}{\|}}{C}-OH$, with Br on the first CH

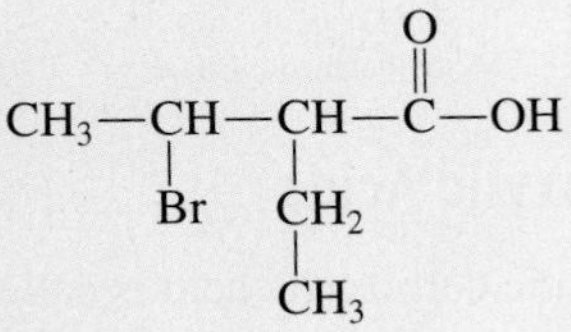

d. CH_3-CH_2-COOH

Solution

a. The parent chain name is based on pentane. Removing the -*e* ending from pentane and replacing it with the ending -*oic acid* gives *pentanoic acid*. The location of the carboxyl group need not be specified, because by definition the carboxyl carbon atom is always carbon 1.

b. The parent chain name is *butanoic acid*. To locate the methyl group substituent, we number the carbon chain beginning with the carboxyl carbon atom. The complete name of the acid is *3-methylbutanoic acid*.

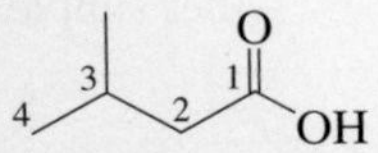

(line-angle structure numbered 4, 3, 2, 1 with O and OH)

c. The longest carboxyl-carbon-containing chain has four carbon atoms. The parent chain name is thus *butanoic acid*. There are two substituents present, an ethyl group on carbon 2 and a bromo group on carbon 3. The complete name is *3-bromo-2-ethylbutanoic acid*.

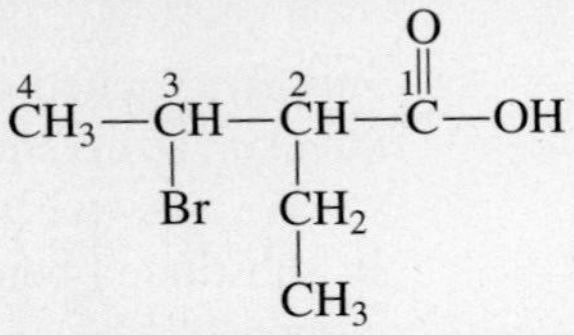

$$\overset{4}{C}H_3-\overset{3}{C}H-\overset{2}{C}H-\overset{\overset{\displaystyle O}{\|}}{C}-OH$$
with Br and CH₂—CH₃ substituents

d. This is the three-carbon, unsubstituted carboxylic acid with the carboxyl group designated using "linear" notation. The acid's name is simply *propanoic acid*.

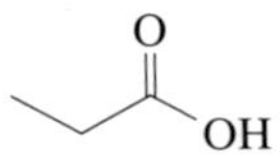

Practice Exercise 16.1

Assign IUPAC names to the following carboxylic acids.

a. (line-angle structure with O and OH)

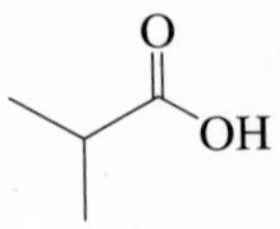

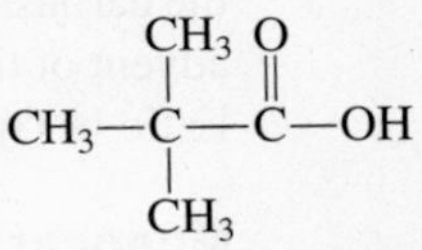

b. $CH_3-\underset{\underset{CH_3}{|}}{\overset{\overset{CH_3}{|}}{C}}-\overset{\overset{\displaystyle O}{\|}}{C}-OH$

c. $CH_3-CH_2-\underset{\underset{CH_3-CH_2-CH_2}{|}}{CH}-\overset{\overset{\displaystyle O}{\|}}{C}-OH$

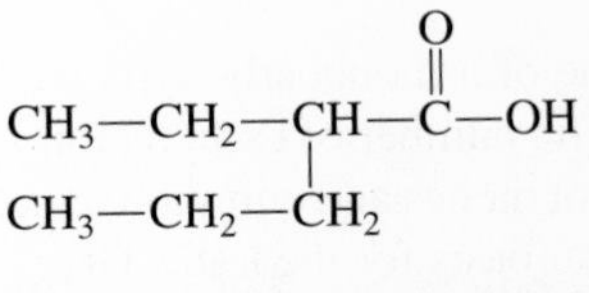

d. $CH_3-CH_2-CH_2-COOH$

Answers: a. 2-Methylpropanoic acid; **b.** 2,2-Dimethylpropanoic acid; **c.** 2-Ethylpentanoic acid; **d.** Butanoic acid

The carboxyl functional group has the highest priority in the IUPAC naming system of all functional groups considered so far. When both a carboxyl group and a carbonyl group (aldehyde, ketone) are present in the same molecule, the prefix *oxo-* is used to denote the carbonyl group.

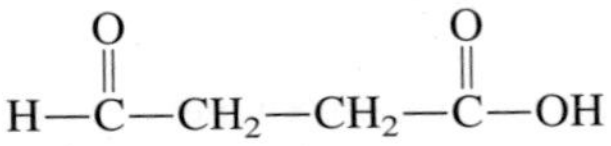

$$H-\overset{\overset{\displaystyle O}{\|}}{C}-CH_2-CH_2-\overset{\overset{\displaystyle O}{\|}}{C}-OH$$

4-Oxobutanoic acid

Dicarboxylic Acids

A **dicarboxylic acid** *is a carboxylic acid that contains two carboxyl groups, one at each end of a carbon chain.* Saturated acids of this type are named by appending the suffix *-dioic acid* to the corresponding alkane name (the *-e* is retained to facilitate pronunciation). Both carboxyl carbon atoms must be part of the parent carbon chain, and the carboxyl locations need not be specified with numbers because they will always be at the two ends of the chain.

Pentanedioic acid

2-Methylbutanedioic acid

Aromatic Carboxylic Acids

The simplest aromatic carboxylic acid is called benzoic acid (Figure 16.2).

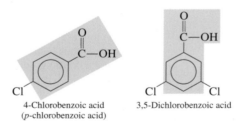

Benzoic acid

Other simple aromatic acids are named as derivatives of benzoic acid.

4-Chlorobenzoic acid
(*p*-chlorobenzoic acid)

3,5-Dichlorobenzoic acid

In substituted benzoic acids, the ring carbon atom bearing the carboxyl group is always carbon 1.

Figure 16.2 Space-filling model for benzoic acid, the simplest aromatic carboxylic acid.

Methyl benzoic acids go by the name *toluic acid*. (This situation parallels methyl benzene being called toluene.)

o-Toluic acid

The common names of monocarboxylic acids are the basis for aldehyde common names (Section 15.4).

C_1:   formic acid and formaldehyde
C_2:   acetic acid and acetaldehyde
C_3:   propionic acid and propionaldehyde
C_4:   butyric acid and butyraldehyde

16.3 COMMON NAMES FOR CARBOXYLIC ACIDS

The use of common names is more prevalent for carboxylic acids than for any other family of organic compounds. Because of their abundance in nature, carboxylic acids were among the earliest classes of organic compounds to be studied, and they acquired names before the advent of the IUPAC naming system. These common names are usually derived from some Latin or Greek word that is related to a source for the acid.

Monocarboxylic Acids

The common name of a monocarboxylic acid is formed by taking the Latin or Greek root name for the specific number of carbon atoms and appending the suffix *–ic acid*. Table 16.1 gives the parent root names and common names for the first six unbranched monocarboxylic acids. The historical basis for the Latin-Greek root name system is as follows.

The stinging sensation associated with red ant bites is due in part to formic acid (Latin, *formica,* "ant"). Acetic acid gives vinegar its tartness (sour taste); vinegar contains

TABLE 16.1
Common Names for the First Six Unbranched Monocarboxylic Acids

Length of Carbon Chain	Structural Formula	Latin or Greek Root	Common Name*
C_1 monoacid	H—COOH	form-	formic acid
C_2 monoacid	CH_3—COOH	acet-	acetic acid
C_3 monoacid	CH_3—CH_2—COOH	propion-	propionic acid
C_4 monoacid	CH_3—$(CH_2)_2$—COOH	butyr-	butyric acid
C_5 monoacid	CH_3—$(CH_2)_3$—COOH	valer-	valeric acid
C_6 monoacid	CH_3—$(CH_2)_4$—COOH	capro-	caproic acid

*The mnemonic "*F*rogs *a*re *p*olite, *b*eing *v*ery *c*ourteous" is helpful in remembering, in order, the first letters of the common names of these six simple saturated monocarboxylic acids.

There is a connection between acetic acid and sourdough bread. The yeast used in leavening the dough for this bread is a type that cannot metabolize the sugar maltose as most yeasts do. Consequently, bacteria that thrive on maltose become abundant in the dough. These bacteria produce acetic acid and lactic acid from the maltose, and the dough becomes *sour* (acidic); hence the name *sourdough* bread.

Note that the common and IUPAC naming systems for carboxylic acids differ in three ways:

1. The base name for the carbon chain differs.
2. The suffix that ends the name differs, being -*ic acid* in the common system and -*oic acid* in the IUPAC system.
3. The carbon numbering system differs, involving Greek letters in the common system and Arabic numbers in the IUPAC system.

Figure 16.3 "Drug-sniffing" dogs used by narcotics agents can find hidden heroin by detecting the odor of acetic acid (vinegar odor). Acetic acid is a by-product of the final step in illicit heroin production, and trace amounts remain in the heroin.

Ingesting a fatal dose of oxalic acid from eating spinach would require that you eat nine pounds of spinach at one sitting, a slightly above-average serving of spinach!

small amounts of acetic acid (Latin, *acetum,* "sour"). Propionic acid is the smallest acid that can be obtained from fats (Greek, *protos,* "first," and *pion,* "fat"). Rancid butter contains butyric acid (Latin, *butyrum,* "butter"). Valeric acid, found in valerian root (an herb), has a strong odor (Latin, *valere,* "to be strong"). The skin secretions of goats contain caproic acid, which contributes to the odor associated with these animals (Latin, *caper,* "goats").

Acetic acid is the most widely used of all carboxylic acids. Its primary use is as an *acidulant*—a substance that gives the proper acidic conditions for a chemical reaction. In the pure state, acetic acid is a colorless liquid with a sharp odor (see Figure 16.3). Vinegar is a 4%–8% (v/v) acetic acid solution; its characteristic odor comes from the acetic acid present. Pure acetic acid is often called *glacial* acetic acid because it freezes on a moderately cold day (f.p. = 17°C), producing icy-looking crystals.

When using common names for carboxylic acids, the positions (locations) of substituents are denoted by using letters of the Greek alphabet rather than numbers. The first four letters of the Greek alphabet are alpha (α), beta (β), gamma (γ), and delta (δ). The alpha-carbon atom is carbon 2, the beta-carbon atom is carbon 3, and so on.

$$
\cdots \cdots \overset{}{C}-C-C-C-\overset{\displaystyle O}{\overset{\|}{C}}-OH
$$

IUPAC: 5   4   3   2   1

Greek letter: δ γ β α

With the Greek-letter system, the compound

$$
CH_3-CH_2-\underset{\underset{\displaystyle CH_3}{|}}{CH}-CH_2-\overset{\displaystyle O}{\overset{\|}{C}}-OH
$$

β carbon α carbon

would be called β-*methylvaleric acid.*

Figure 16.4 contrasts the different carbon-atom numbering systems in IUPAC and common-name nomenclature for carboxylic acids.

Dicarboxylic Acids

Common names for the first six dicarboxylic acids are given in Table 16.2. Oxalic acid, the simplest dicarboxylic acid, is found in plants of the genus *Oxalis,* which includes rhubarb and spinach, and in cabbage (see Figure 16.5). This acid and its salts are poisonous in *high* concentrations. The amount of oxalic acid present in spinach, cabbage, and rhubarb is not harmful. Oxalic acid is used to remove rust, bleach straw and leather, and remove ink stains. Succinic and glutaric acid and their derivatives play important roles in biochemical reactions that occur in the human body (Section 23.6).

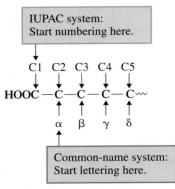

IUPAC system:
Start numbering here.

C1 C2 C3 C4 C5

HOOC—C—C—C—C~~~

α   β   γ   δ

Common-name system:
Start lettering here.

Figure 16.4 A Greek-letter numbering system is used in common-name nomenclature for carboxylic acids.

TABLE 16.2
Common Names for the First Six Unbranched Dicarboxylic Acids

Length of Carbon Chain	Structural Formula	Latin or Greek Root	Common Name*
C_2 diacid	HOOC—COOH	oxal-	oxalic acid
C_3 diacid	HOOC—CH_2—COOH	malon-	malonic acid
C_4 diacid	HOOC—$(CH_2)_2$—COOH	succin-	succinic acid
C_5 diacid	HOOC—$(CH_2)_3$—COOH	glutar-	glutaric acid
C_6 diacid	HOOC—$(CH_2)_4$—COOH	adip-	adipic acid
C_7 diacid	HOOC—$(CH_2)_5$—COOH	pimel-	pimelic acid

*The mnemonic "*Oh my, such good apple pie*" is helpful in remembering, in order, the first letters of the common names of these six simple dicarboxylic acids.

● **EXAMPLE 16.2**

Generating the Structural Formulas of Carboxylic Acids from Their Common Names

The contrast between IUPAC names and common names for mono- and dicarboxylic acids is as follows:

Monocarboxylic Acids
   IUPAC (two words)
      alkanoic acid
   Common (two words)
      (prefix)ic acid*

Dicarboxylic Acids
   IUPAC (two words)
      alkanedioic acid
   Common (two words)
      (prefix)ic acid*

*The common-name prefixes are related to natural sources for the acids.

Draw a structural formula for each of the following carboxylic acids.

a. Caproic acid   **b.** Glutaric acid
c. α-Phenylsuccinic acid   **d.** β-Chlorobutyric acid

Solution

a. Caproic acid is the six-carbon unsubstituted monocarboxylic acid. Its structural formula is

$$CH_3—CH_2—CH_2—CH_2—CH_2—\overset{\overset{\displaystyle O}{\|}}{C}—OH$$

b. Glutaric acid is the five-carbon unsubstituted dicarboxylic acid, with a carboxyl group at each end of the carbon chain.

$$HO—\overset{\overset{\displaystyle O}{\|}}{C}—CH_2—CH_2—CH_2—\overset{\overset{\displaystyle O}{\|}}{C}—OH$$

c. Succinic acid is the four-carbon unsubstituted dicarboxylic acid. A phenyl group (Section 13.12) is present on the alpha-carbon atom.

$$HO—\overset{\overset{\displaystyle O}{\|}}{C}—\overset{\alpha}{CH}—\overset{\beta}{CH_2}—\overset{\overset{\displaystyle O}{\|}}{C}—OH$$

d. Butyric acid is the four-carbon unsubstituted monocarboxylic acid. A chloro group is attached to the beta-carbon atom (carbon 3).

$$\overset{\gamma}{CH_3}—\overset{\beta}{\underset{\underset{\displaystyle Cl}{|}}{CH}}—\overset{\alpha}{CH_2}—\overset{\overset{\displaystyle O}{\|}}{C}—OH$$

Practice Exercise 16.2

Draw a structural formula for each of the following carboxylic acids.

a. Adipic acid   **b.** β-Chlorovaleric acid   **c.** Malonic acid   **d.** Phenylacetic acid

Answers:

a. $HO—\overset{\overset{\displaystyle O}{\|}}{C}—CH_2—CH_2—CH_2—CH_2—\overset{\overset{\displaystyle O}{\|}}{C}—OH$

b. $CH_3-CH_2-\overset{}{\underset{\underset{Cl}{|}}{CH}}-CH_2-\overset{\overset{O}{\|}}{C}-OH$

c. $HO-\overset{\overset{O}{\|}}{C}-CH_2-\overset{\overset{O}{\|}}{C}-OH$

d. $CH_2-\overset{\overset{O}{\|}}{C}-OH$ (attached to benzene ring)

$HO-\overset{\overset{O}{\|}}{C}-\overset{\overset{O}{\|}}{C}-OH$

Figure 16.5 The C_2 dicarboxylic acid, oxalic acid, contributes to the tart taste of rhubarb stalks.

An unsaturated monocarboxylic acid with the structure

$CH_3-CH_2-CH_2-\overset{}{\underset{\underset{CH_3}{|}}{C}}=CH-COOH$

3-Methyl-2-hexenoic acid

has been found to be largely responsible for "body odor." It is produced by skin bacteria, particularly those found in armpits.

16.4 POLYFUNCTIONAL CARBOXYLIC ACIDS

A **polyfunctional carboxylic acid** *is a carboxylic acid that contains one or more additional functional groups besides one or more carboxyl groups.* Such acids occur naturally in many fruits, are important in the normal functioning of the human body (metabolism), and find use in over-the-counter skin-care products and in prescription drugs. Three commonly encountered types of polyfunctional carboxylic acids are *unsaturated* acids, *hydroxy* acids, and *keto* acids.

$C-C=C-COOH$
An unsaturated acid

$C-C-\overset{\overset{OH}{|}}{C}-COOH$
A hydroxy acid

$C-\overset{\overset{O}{\|}}{C}-C-COOH$
A keto acid

More information about these types of polyfunctional acids follows.

Unsaturated Acids

The simplest *unsaturated mono*carboxylic acid is propenoic acid (acrylic acid), a substance used in the manufacture of several polymeric materials. Two isomers exist for the simplest *unsaturated di*carboxylic acid, butenedioic acid. The two isomers have separate common names, fumaric acid (*trans*) and maleic acid (*cis*), a naming procedure seldom encountered.

$CH_2=CH-COOH$
Acrylic acid

$\underset{H}{\overset{HOOC}{}}C=C\underset{H}{\overset{COOH}{}}$
Maleic acid
(*cis* isomer)

$\underset{HOOC}{\overset{H}{}}C=C\underset{H}{\overset{COOH}{}}$
Fumaric acid
(*trans* isomer)

Some antihistamines (Section 17.10) are salts of maleic acid. The addition of small amounts of maleic acid to fats and oils prevents them from becoming rancid. Fumaric acid is a *metabolic acid*. Metabolic acids are intermediate compounds in the metabolic reactions (Section 23.1) that occur in the human body. More information about metabolic acids is presented in the next section.

Hydroxy Acids

Four of the simpler *hydroxy* acids are

$\underset{OH}{\overset{CH_2}{|}}-COOH$
Glycolic acid

$CH_3-\underset{OH}{\overset{CH}{|}}-COOH$
Lactic acid

$HOOC-\underset{OH}{\overset{CH}{|}}-CH_2-COOH$
Malic acid

$HOOC-\underset{OH}{\overset{CH}{|}}-\underset{OH}{\overset{CH}{|}}-COOH$
Tartaric acid

Malic and tartaric acids are derivatives of succinic acid, the four-carbon unsubstituted diacid (Section 16.3).

Hydroxy acids occur naturally in many foods. Glycolic acid is present in the juice from sugar cane and sugar beets. Lactic acid is present in sour milk, sauerkraut, and dill

CHEMICAL Connections

Nonprescription Pain Relievers Derived from Propanoic Acid

Consumers are faced with a shelf-full of choices when looking for an over-the-counter medicine to treat aches, pains, and fever. The vast majority of brands available, however, represent only four chemical formulations. Besides the long-available aspirin and acetaminophen, consumers can now purchase products that contain ibuprofen and naproxen.

These two newer entrants into the over-the-counter pain-reliever market are derivatives of propanoic acid, the three-carbon monocarboxylic acid.

Ibuprofen, marketed under the brand names Advil, Motrin-IB, and Nuprin, was cleared by the FDA in 1984 for nonprescription sales. Numerous studies have shown that nonprescription-strength ibuprofen relieves minor pain and fever as well as aspirin or acetaminophen. Like aspirin, ibuprofen reduces inflammation. (Prescription-strength ibuprofen has extensive use as an anti-inflammatory agent for the treatment of rheumatoid arthritis.) There is evidence that ibuprofen is more effective than either aspirin or acetaminophen in reducing dental pain and menstrual pain. Both aspirin and ibuprofen can cause stomach bleeding in some people, although ibuprofen seems to cause fewer problems. Ibuprofen is more expensive than either aspirin or acetaminophen.

Naproxen, marketed under the brand names Aleve and Anaprox, was cleared by the FDA in 1994 for nonprescription use. The effects of naproxen last longer in the body (8–12 hr per

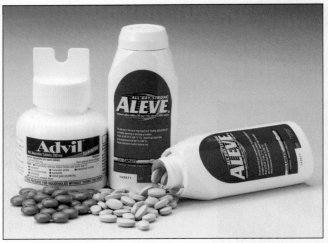

The active ingredients in Aleve and Advil are derivatives of propanoic acid.

dose) than the effects of ibuprofen (4–6 hr per dose) and of aspirin and acetaminophen (4 hr per dose). Naproxen is more likely to cause slight intestinal bleeding and stomach upset than is ibuprofen. It is also not recommended for use by children under 12.

Propanoic acid | Ibuprofen | Naproxen

Figure 16.6 Tartaric acid, the dihydroxy derivative of succinic acid, is particularly abundant in ripe grapes.

The IUPAC name for citric acid is 2-hydroxy-1,2,3-propanetrioic acid.

pickles. Both malic acid and tartaric acid occur naturally in fruits. The sharp taste of apples (fruit of trees of the genus *Malus*) is due to malic acid. Tartaric acid is particularly abundant in grapes (Figure 16.6). It is also a component of tartar sauce and an acidic ingredient in many baking powders. Lactic and malic acids are also *metabolic acids* (Section 16.5).

Citric acid, perhaps the best known of all carboxylic acids, is a hydroxy acid with a structural feature we have not previously encountered. It is a hydroxy *tri*carboxylic acid. Besides there being acid groups at both ends of a carbon chain, a third acid group is present as a substituent on the chain. An acid group as a substituent is called a *carboxy* group. Thus citric acid is a hydroxycarboxy diacid.

$$\underset{\text{Citric acid}}{\text{HOOC}-CH_2-\overset{\overset{\displaystyle OH}{|}}{\underset{\underset{\displaystyle COOH}{|}}{C}}-CH_2-COOH}$$

Citric acid gives citrus fruits their "sharp" taste; lemon juice contains 4%–8% citric acid, and orange juice is about 1% citric acid. Citric acid is used widely in beverages and in foods. In jams, jellies, and preserves, it produces tartness and pH adjustment to optimize conditions for gelation. In fresh salads, citric acid prevents enzymatic browning reactions, and in frozen fruits it prevents deterioration of color and flavor. Addition of citric acid to seafood retards microbial growth by lowering pH. Citric acid is also a metabolic acid (Section 16.5).

Keto Acids

Keto acids, as the designation implies, contain a carbonyl group within a carbon chain. Pyruvic acid, with three carbon atoms, is the simplest keto acid that can exist.

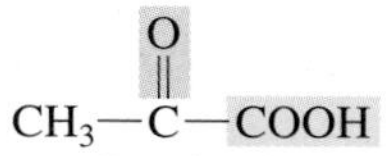

$$CH_3-\overset{\overset{\displaystyle O}{\|}}{C}-COOH$$

Pyruvic acid

In the pure state, pyruvic acid is a liquid with an odor resembling that of vinegar (acetic acid; Section 16.3). Pyruvic acid is a metabolic acid (Section 16.5).

CHEMICAL Connections

Carboxylic Acids and Skin Care

A number of carboxylic acids are used as "skin-care acids." Heavily advertised at present are cosmetic products that contain *alpha-hydroxy* acids, carboxylic acids in which a hydroxyl group is attached to the acid's alpha-carbon atom. Such cosmetic products address problems such as dryness, flaking, and itchiness of the skin and are highly promoted for removing wrinkles.

Alpha-hydroxy acids "work" by loosening the cells of the outer layer of skin (the epidermis) and by accelerating the flaking off of dead skin. The result is healthier-looking skin.

The alpha-hydroxy acids most commonly found in cosmetic products are glycolic acid and lactic acid, the two simplest alpha-hydroxy acids.

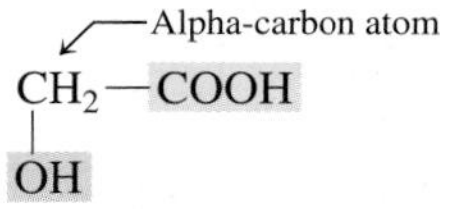

Alpha-carbon atom
CH_2-COOH
$|$
OH
Glycolic acid

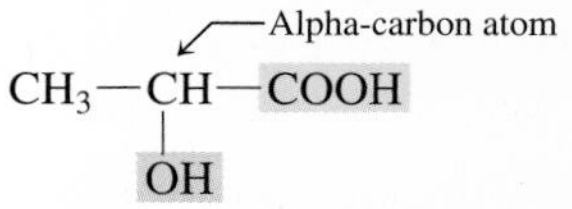

Alpha-carbon atom
$CH_3-CH-COOH$
$|$
OH
Lactic acid

Both acids are naturally occurring substances. Glycolic acid occurs in sugar cane and sugar beets, and lactic acid occurs in sour milk.

The use of alpha-hydroxy acids in cosmetics is considered safe at acid concentrations of less than 10%; higher concentrations can cause skin irritation, burning, and stinging. (Lactic acid becomes a prescription drug at concentrations of 12% or more.) One drawback of the cosmetic use of alpha-hydroxy products is that such use can increase the skin's sensitivity to the ultraviolet light component of sunlight; it is this component that causes sunburn. Individuals who are using "alpha-hydroxys" should apply a sunscreen whenever they go outside for an extended period of time.

Glycolic acid, at higher concentrations than that found in cosmetics, is used by dermatologists for the "spot" removal of *keratoses* (precancerous lesions and/or patches of darker, thickened skin).

Two skin-care products containing alpha-hydroxy acids.

Polyunsaturated carboxylic acids are used extensively in the treatment of severe acne. The prescription drugs Tretinoin and Accutane are such compounds.

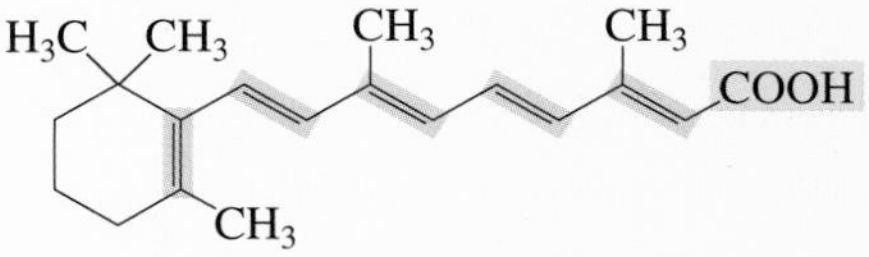

Tretinoin (5 *trans*-double bonds)

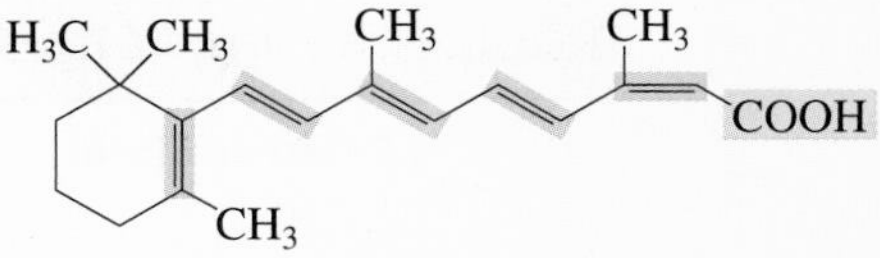

Accutane (4 *trans*- and 1 *cis*-double bonds)

16.5 METABOLIC CARBOXYLIC ACIDS

Numerous polyfunctional acids, including some mentioned in the previous section, are intermediates in the metabolic reactions that occur in the human body as food is processed. There are eight such acids that will appear repeatedly in the biochemistry chapters of this text.

Interestingly, these eight key metabolic intermediates are derived from only three of the simple carboxylic acids. These three simple acids and the metabolic acids related to them are

1. Propionic acid (3-carbon monoacid): lactic, glyceric, and pyruvic acids
2. Succinic acid (4-carbon diacid): fumaric, oxaloacetic, and malic acids
3. Glutaric acid (5-carbon diacid): α-ketoglutaric and citric acids

Metabolic acids derived from the diacids succinic and glutaric are encountered in the citric acid cycle (Section 23.6), a series of reactions in which C_2 units obtained from all types of foods are further processed for the purpose of obtaining energy. Glyceric and pyruvic acid (propionic acid derivatives) are encountered in glycolysis (Section 24.2), a series of reactions in which glucose is processed. Lactic acid (a propionic derivative) is a by-product of strenuous exercise (Section 24.3). Figure 16.7 gives further details about the eight metabolic acids.

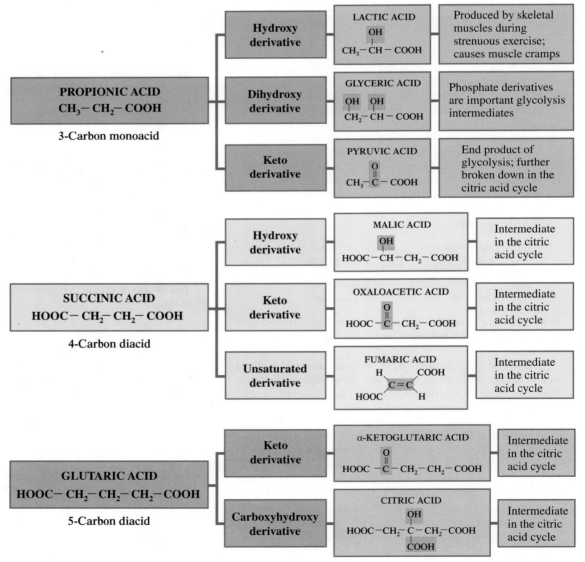

Figure 16.7 Structural characteristics and functions of several polyfunctional carboxylic acids that are important in metabolic reactions in the human body.

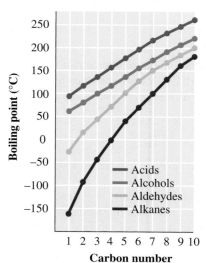

Figure 16.8 The boiling points of monocarboxylic acids compared to those of other types of compounds. All compounds in the comparison have unbranched carbon chains.

Unbranched Monocarboxylic Acids			
C_1	C_3	C_5	C_7
C_2	C_4	C_6	C_8

Unbranched Dicarboxylic Acids			
✕	C_3	C_5	C_7
C_2	C_4	C_6	C_8

☐ Liquid ☐ Solid

Figure 16.9 A physical-state summary for unbranched mono- and dicarboxylic acids at room temperature and pressure.

 PHYSICAL PROPERTIES OF CARBOXYLIC ACIDS

Carboxylic acids are the most *polar* organic compounds we have discussed so far. Both the carbonyl part ($\diagdown$C$=$O) and the hydroxyl part (—OH) of the carboxyl functional group are polar. The result is very high melting and boiling points for carboxylic acids, the highest of any type of organic compound yet considered (Figure 16.8).

Unsubstituted saturated monocarboxylic acids containing up to nine carbon atoms are liquids that have strong, sharp odors (Figure 16.9). Acids with 10 or more carbon atoms in an unbranched chain are waxy solids that are odorless (because of low volatility). Aromatic carboxylic acids, as well as dicarboxylic acids, are also odorless solids.

The high boiling points of carboxylic acids indicate the presence of strong intermolecular attractive forces. A unique hydrogen-bonding arrangement, shown in Figure 16.10, contributes to these attractive forces. A given carboxylic acid molecule forms two hydrogen bonds to another carboxylic acid molecule, producing a "complex" known as a *dimer*. Because dimers have twice the mass of a single molecule, a higher temperature is needed to boil a carboxylic acid than would be needed for similarly sized aldehyde and alcohol molecules where dimerization does not occur.

Carboxylic acids readily hydrogen-bond to water molecules. Such hydrogen bonding contributes to water solubility for short-chain carboxylic acids. The unsubstituted C_1 to C_4 monocarboxylic acids are completely miscible with water. Solubility then rapidly decreases with carbon number, as shown in Figure 16.11. Short-chain dicarboxylic acids are also water-soluble. In general, aromatic acids are not water-soluble.

 PREPARATION OF CARBOXYLIC ACIDS

Oxidation of primary alcohols or aldehydes, using an oxidizing agent such as CrO_3 or $K_2Cr_2O_7$, produces carboxylic acids, a process that we examined in Sections 14.9 and 15.10.

$$\text{Primary alcohol} \xrightarrow{[O]} \text{aldehyde} \xrightarrow{[O]} \text{carboxylic acid}$$

Aromatic acids can be prepared by oxidizing a carbon side chain (alkyl group) on a benzene derivative. In this process, all the carbon atoms of the alkyl group except the one attached to the ring are lost. The remaining carbon becomes part of a carboxyl group.

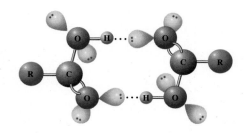

Figure 16.10 A given carboxylic acid molecule can form two hydrogen bonds to another carboxylic acid molecule, producing a "dimer," a complex with a mass twice that of a single molecule.

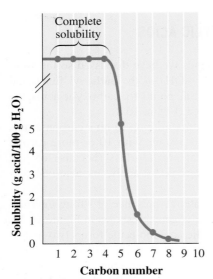

Figure 16.11 The solubility in water of saturated unbranched-chain carboxylic acids.

16.8 ACIDITY OF CARBOXYLIC ACIDS

Carboxylic acids, as the name implies, are *acidic*. When a carboxylic acid is placed in water, hydrogen ion transfer (proton transfer; Section 10.2) occurs to produce hydronium ion (the acidic species in water; Section 10.2) and carboxylate ion.

$$R-COOH + H_2O \longrightarrow H_3O^+ + R-COO^-$$
Hydronium ion   Carboxylate ion

A **carboxylate ion** *is the negative ion produced when a carboxylic acid loses one or more acidic hydrogen atoms.*

Carboxylate ions formed from monocarboxylic acids always carry a -1 charge; only one acidic hydrogen atom is present in such molecules. Dicarboxylic acids, which possess two acidic hydrogen atoms (one in each carboxyl group), can produce carboxylate ions bearing a -2 charge.

Carboxylate ions are named by dropping the *-ic acid* ending from the name of the parent acid and replacing it with *-ate*.

$$CH_3-\overset{\overset{\displaystyle O}{\|}}{C}-OH + H_2O \longrightarrow H_3O^+ + CH_3-\overset{\overset{\displaystyle O}{\|}}{C}-O^-$$
Acetic acid (ethanoic acid)   Acetate ion (ethanoate ion)

$$HO-\overset{\overset{\displaystyle O}{\|}}{C}-\overset{\overset{\displaystyle O}{\|}}{C}-OH + 2H_2O \longrightarrow 2H_3O^+ + {}^-O-\overset{\overset{\displaystyle O}{\|}}{C}-\overset{\overset{\displaystyle O}{\|}}{C}-O^-$$
Oxalic acid (ethanedioic acid)   Oxalate ion (ethanedioate ion)

Carboxylic acids are weak acids (Section 10.4). The extent of proton transfer is usually less than 5%; that is, an equilibrium situation exists in which the equilibrium lies far to the left.

$$R-COOH + H_2O \rightleftharpoons H_3O^+ + R-COO^-$$
More than 95% of molecules in this form   Less than 5% of molecules in this form

Table 16.3 gives K_a values and percent ionization values in 0.100 M solution (topics previously discussed in Section 10.5) and pK_a values (Section 10.10) for selected monocarboxylic acids.

16.9 CARBOXYLIC ACID SALTS

In a manner similar to that of inorganic acids (Section 10.6), carboxylic acids react completely with strong bases to produce water and a carboxylic acid salt.

$$CH_3-\overset{\overset{\displaystyle O}{\|}}{C}-OH + NaOH \longrightarrow CH_3-\overset{\overset{\displaystyle O}{\|}}{C}-O^- Na^+ + H_2O$$
Carboxylic acid   Strong base   Carboxylic acid salt   Water

At normal human body pH values (pH = 7.35 to 7.45), most carboxylic acids exist as carboxylate ions. Acetic acid is in the form of acetate ion, pyruvic acid is in the form of pyruvate ion, lactic acid is in the form of lactate ion, and so on.

TABLE 16.3
Acid Strength for Selected Monocarboxylic Acids

Acid	K_a	Percent Ionization (0.100 M Solution)	pK_a
formic	1.8×10^{-4}	4.2%	3.75
acetic	1.8×10^{-5}	1.3%	4.75
propionic	1.3×10^{-5}	1.2%	4.89
butyric	1.5×10^{-5}	1.2%	4.82
valeric	1.5×10^{-5}	1.2%	4.82
caproic	1.4×10^{-5}	1.2%	4.85

Carboxylic acid salt formation involves an acid–base neutralization reaction (Section 10.7).

A **carboxylic acid salt** *is an ionic compound in which the negative ion is a carboxylate ion.*
Carboxylic acid salts are named similarly to other ionic compounds (Section 4.9): *The positive ion is named first, followed by a separate word giving the name of the negative ion.* The salt formed in the preceding reaction contains sodium ions and acetate ions (from acetic acid); hence the salt's name is sodium acetate.

EXAMPLE 16.3

Writing Equations for the Formation of Carboxylic Acid Salts

Using an acid–base neutralization reaction, write a chemical equation for the formation of each of the following carboxylic acid salts.

a. Sodium propionate   **b.** Potassium oxalate

Solution

a. This salt contains sodium ion (Na^+) and propionate ion, the three-carbon monocarboxylate ion.

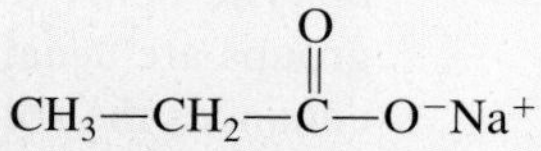

$$CH_3-CH_2-\overset{\displaystyle O}{\overset{\|}{C}}-O^-Na^+$$

From a neutralization standpoint, the sodium ion's source is the base sodium hydroxide, NaOH, and the negative ion's source is the acid propanoic acid. The acid–base neutralization equation is

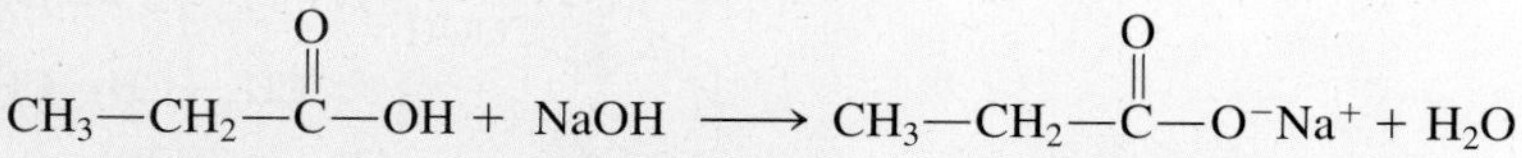

$$CH_3-CH_2-\overset{\displaystyle O}{\overset{\|}{C}}-OH + NaOH \longrightarrow CH_3-CH_2-\overset{\displaystyle O}{\overset{\|}{C}}-O^-Na^+ + H_2O$$

Propionic acid   Sodium hydroxide   Sodium propionate   Water

b. This salt contains potassium ions (K^+) whose source would be the base potassium hydroxide, KOH. The salt also contains oxalate ions, whose source would be the acid oxalic acid.

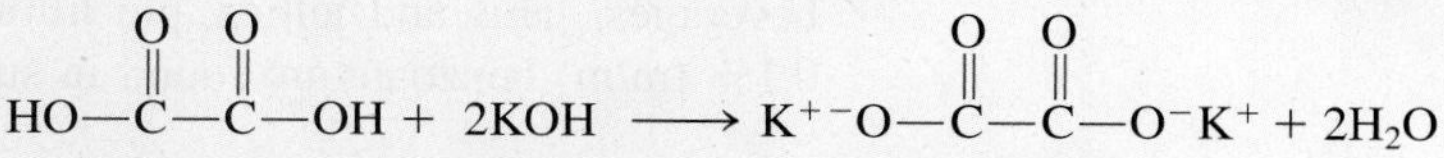

$$HO-\overset{\displaystyle O}{\overset{\|}{C}}-\overset{\displaystyle O}{\overset{\|}{C}}-OH + 2KOH \longrightarrow K^{+-}O-\overset{\displaystyle O}{\overset{\|}{C}}-\overset{\displaystyle O}{\overset{\|}{C}}-O^-K^+ + 2H_2O$$

Oxalic acid   Potassium hydroxide   Potassium oxalate   Water

Note that two molecules of base are needed to react completely with one molecule of acid because the acid is a dicarboxylic acid.

Practice Exercise 16.3

Using an acid–base neutralization reaction, write a chemical equation for the formation of each of the following carboxylic acid salts.

a. Sodium formate   **b.** Potassium malonate

Answers: **a.** $H-\overset{\displaystyle O}{\overset{\|}{C}}-OH + NaOH \longrightarrow H-\overset{\displaystyle O}{\overset{\|}{C}}-O^-Na^+ + H_2O$;

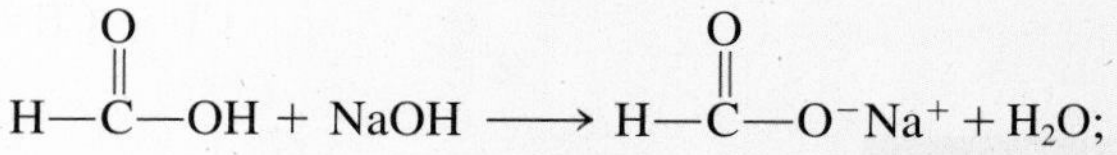

b. $HO-\overset{\displaystyle O}{\overset{\|}{C}}-CH_2-\overset{\displaystyle O}{\overset{\|}{C}}-OH + 2KOH \longrightarrow$

$$K^{+-}O-\overset{\displaystyle O}{\overset{\|}{C}}-CH_2-\overset{\displaystyle O}{\overset{\|}{C}}-O^-K^+ + 2H_2O$$

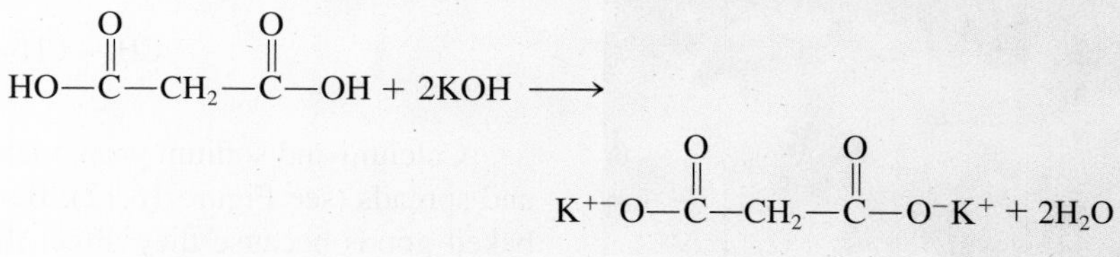

Converting a carboxylic acid salt back to a carboxylic acid is very simple. React the salt with a solution of a strong acid such as hydrochloric acid (HCl) or sulfuric acid (H_2SO_4).

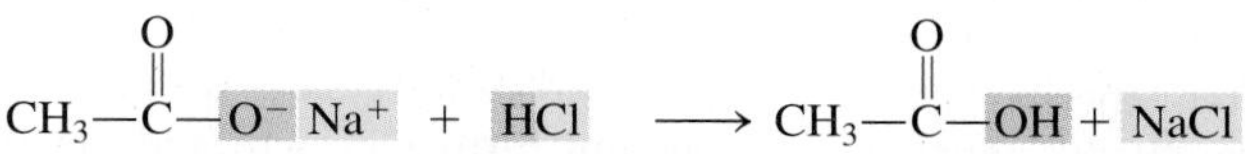

$$CH_3-\overset{\displaystyle O}{\overset{\|}{C}}-O^-Na^+ + HCl \longrightarrow CH_3-\overset{\displaystyle O}{\overset{\|}{C}}-OH + NaCl$$

Sodium acetate   Hydrochloric acid   Acetic acid   Sodium chloride

The interconversion reactions between carboxylic acid salts and their "parent" carboxylic acids are so easy to carry out that organic chemists consider these two types of compounds interchangeable.

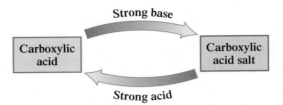

Uses for Carboxylic Acid Salts

The solubility of carboxylic acid salts in water is much greater than that of the carboxylic acids from which they are derived. Drugs and medicines that contain acid groups are usually marketed as the sodium or potassium salt of the acid. This greatly enhances the solubility of the medication, increasing the ease of its absorption by the body.

Many *antimicrobials,* compounds used as food preservatives, are carboxylic acid salts. Particularly important are the salts of benzoic, sorbic, and propionic acids.

$$CH_3-CH=CH-CH=CH-COOH \qquad CH_3-CH_2-COOH$$

Benzoic acid

Sorbic acid
(2,4-hexadienoic acid)

Propionic acid

The benzoate salts of sodium and potassium are effective against yeast and mold in beverages, jams and jellies, pie fillings, ketchup, and syrups. Concentrations of up to 0.1% (m/m) benzoate are found in such products.

Sodium benzoate   Potassium benzoate

The solubility of benzoic acid in water at 25°C is 3.4 g/L. The solubility of sodium benzoate, the sodium salt of benzoic acid in water at 25°C is 550 g/L.

Sodium and potassium sorbates inhibit mold and yeast growth in dairy products, dried fruits, sauerkraut, and some meat and fish products. Sorbate preservative concentrations range from 0.02% to 0.2% (m/m).

$$CH_3-CH=CH-CH=CH-\overset{\overset{\displaystyle O}{\|}}{C}-O^-\,Na^+$$
Sodium sorbate

$$CH_3-CH=CH-CH=CH-\overset{\overset{\displaystyle O}{\|}}{C}-O^-\,K^+$$
Potassium sorbate

Calcium and sodium propionates are used in baked products and also in cheese foods and spreads (see Figure 16.12). Benzoates and sorbates cannot be used in yeast-leavened baked goods because they affect the activity of the yeast.

$$\left(CH_3-CH_2-\overset{\overset{\displaystyle O}{\|}}{C}-O^-\right)_2 Ca^{2+} \qquad CH_3-CH_2-\overset{\overset{\displaystyle O}{\|}}{C}-O^-\,Na^+$$
Calcium propionate   Sodium propionate

Figure 16.12 Propionates, salts of propionic acid, extend the shelf life of bread by preventing the formation of mold.

Carboxylate salts do not directly kill microorganisms present in food. Rather, they prevent further growth and proliferation of these organisms by increasing the pH of the foods in which they are used.

16.10 STRUCTURE OF ESTERS

An **ester** *is a carboxylic acid derivative in which the —OH portion of the carboxyl group has been replaced with an —OR group.*

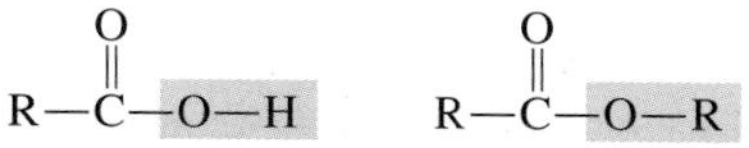

$$R-\overset{\overset{\displaystyle O}{\|}}{C}-O-H \qquad R-\overset{\overset{\displaystyle O}{\|}}{C}-O-R$$

Carboxylic acid   Ester

The ester functional group is thus

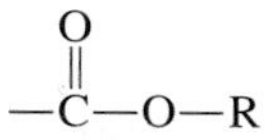

$$-\overset{\overset{\displaystyle O}{\|}}{C}-O-R$$

In linear form, the ester functional group can be represented as —COOR or —CO$_2$R.

The simplest ester, which has two carbon atoms, has a hydrogen atom attached to the ester functional group.

$$H-\overset{\overset{\displaystyle O}{\|}}{C}-O-CH_3$$

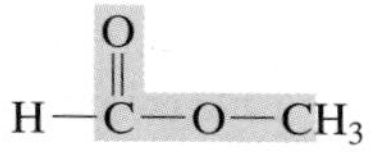

Note that the two carbon atoms present are not bonded to each other.

There are two three-carbon esters.

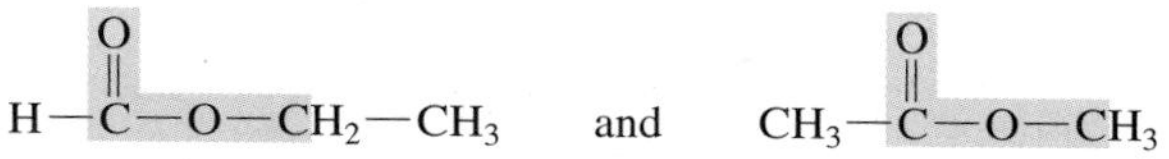

$$H-\overset{\overset{\displaystyle O}{\|}}{C}-O-CH_2-CH_3 \quad \text{and} \quad CH_3-\overset{\overset{\displaystyle O}{\|}}{C}-O-CH_3$$

The structure of the simplest aromatic ester is derived from the structure of benzoic acid, the simplest aromatic carboxylic acid.

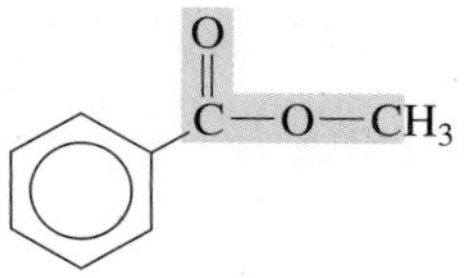

$$\overset{\overset{\displaystyle O}{\|}}{C}-O-CH_3$$

Note that the difference between a carboxylic acid and an ester is a "H versus R" relationship.

$$R-\overset{\overset{\displaystyle O}{\|}}{C}-O-H \quad \text{and} \quad R-\overset{\overset{\displaystyle O}{\|}}{C}-O-R$$

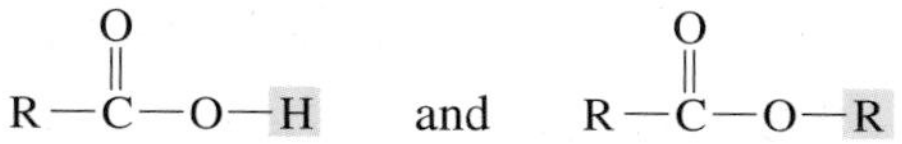

Acid   Ester

We have encountered this "H versus R" relationship several times before in our study of hydrocarbon derivatives. The Chemistry at a Glance feature on page 488 summarizes the "H versus R" relationships we have encountered so far.

16.11 PREPARATION OF ESTERS

Esters are produced through *esterification.* An **esterification reaction** *is the reaction of a carboxylic acid with an alcohol (or phenol) to produce an ester.* A strong acid catalyst (generally H$_2$SO$_4$) is needed for esterification.

$$R-\overset{\overset{\displaystyle O}{\|}}{C}-O-H + H-O-R' \underset{}{\overset{H^+}{\rightleftharpoons}} R-\overset{\overset{\displaystyle O}{\|}}{C}-O-R' + H_2O$$

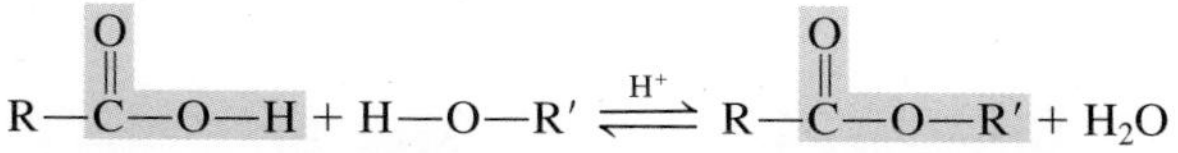

Carboxylic acid   Alcohol   Ester   Water

In the esterification process, a —OH group is lost from the carboxylic acid, a —H atom is lost from the alcohol, and water is formed as a by-product. The net effect of

Esterification is a *condensation reaction.* This is the third time we have encountered this type of reaction. The first encounter involved intermolecular alcohol dehydration (Section 14.9) and the second encounter involved the preparation of acetals (Section 15.11).

this reaction is substitution of the —OR group of the alcohol for the —OH group of the acid.

$$R-\overset{\overset{\displaystyle O}{\|}}{C}-O-H + H-O-R' \overset{H^+}{\rightleftharpoons} R-\overset{\overset{\displaystyle O}{\|}}{C}-O-R' + H_2O$$

A specific example of esterification is the reaction of acetic acid with methyl alcohol.

$$CH_3-\overset{\overset{\displaystyle O}{\|}}{C}-O-H + H-O-CH_3 \overset{H^+}{\rightleftharpoons} CH_3-\overset{\overset{\displaystyle O}{\|}}{C}-O-CH_3 + H_2O$$

> Studies show that in ester formation, the hydroxyl group of the acid (not of the alcohol) becomes part of the water molecule.

Esterification reactions are equilibrium processes, with the position of equilibrium (Section 9.8) usually favoring products only slightly. That is, at equilibrium, substantial amounts of both reactants and products are present. The amount of ester formed can be increased by using an excess of alcohol or by constantly removing one of the products. According to Le Châtelier's principle (Section 9.9), either of these techniques will shift the position of equilibrium to the right (the product side of the equation). This equilibrium problem explains the use of the "double-arrow notation" in all the esterification equations in this section.

It is often useful to think of the structure of an ester in terms of its "parent" alcohol and acid molecules; the ester has an acid part and an alcohol part.

Acid part   Alcohol part

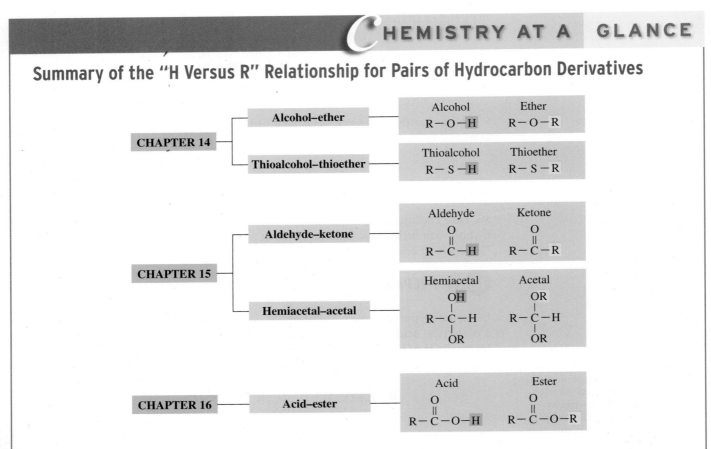

CHEMISTRY AT A GLANCE

Summary of the "H Versus R" Relationship for Pairs of Hydrocarbon Derivatives

CHAPTER 14

Alcohol–ether

Alcohol	Ether
R—O—H	R—O—R

Thioalcohol–thioether

Thioalcohol	Thioether
R—S—H	R—S—R

CHAPTER 15

Aldehyde–ketone

Aldehyde	Ketone
$R-\overset{\overset{\displaystyle O}{\|}}{C}-H$	$R-\overset{\overset{\displaystyle O}{\|}}{C}-R$

Hemiacetal–acetal

Hemiacetal	Acetal
$R-\overset{\overset{\displaystyle OH}{\|}}{\underset{\underset{\displaystyle OR}{\|}}{C}}-H$	$R-\overset{\overset{\displaystyle OR}{\|}}{\underset{\underset{\displaystyle OR}{\|}}{C}}-H$

CHAPTER 16

Acid–ester

Acid	Ester
$R-\overset{\overset{\displaystyle O}{\|}}{C}-O-H$	$R-\overset{\overset{\displaystyle O}{\|}}{C}-O-R$

In this context, it is easy to identify the acid and alcohol from which a given ester can be produced; just add a —OH group to the acid part of the ester and a —H atom to the alcohol part to generate the parent molecules.

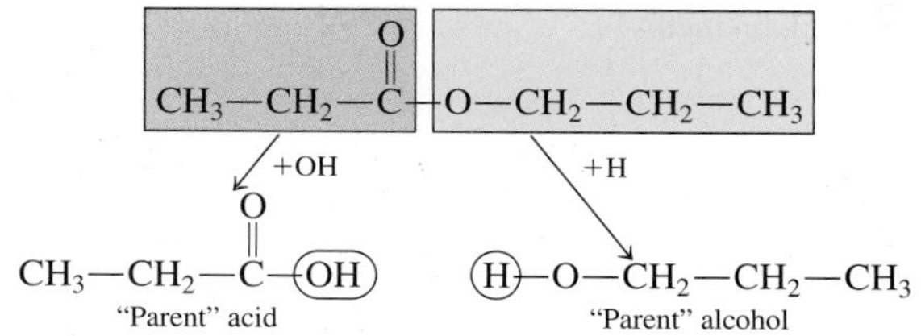

$$CH_3—CH_2—\overset{\overset{\displaystyle O}{\|}}{C}—O—CH_2—CH_2—CH_3$$

$+OH$

$+H$

$$CH_3—CH_2—\overset{\overset{\displaystyle O}{\|}}{C}—(OH)$$

"Parent" acid

$$(H)—O—CH_2—CH_2—CH_3$$

"Parent" alcohol

Cyclic esters formed from hydroxy-acids are called *lactones*.

Cyclic Esters (Lactones)

Hydroxy acids—compounds which contain both a hydroxyl and a carboxyl group (Section 16.4)—have the capacity to undergo intermolecular esterification to form cyclic esters. Such internal esterification easily takes place in situations where a five- or six-membered ring can be formed.

$$\overset{4}{C}H_2—\overset{3}{C}H_2—\overset{2}{C}H_2—\overset{\overset{\displaystyle O}{\|}}{\overset{1}{C}}—OH \longrightarrow$$

$$\underset{OH}{}$$

$$\longrightarrow \quad + H_2O$$

Cyclic ester

Salts and esters of carboxylic acids are named in the same way. The name of the positive ion (in the case of a salt) or the name of the organic group attached to the single-bonded oxygen of the carbonyl group (in the case of an ester) precedes the name of the acid. The *-ic acid* part of the name of the acid is converted to *-ate*.

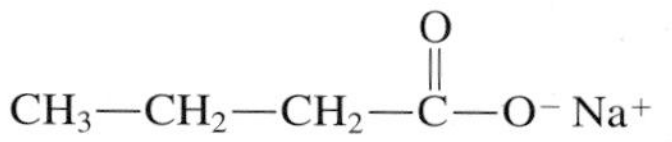

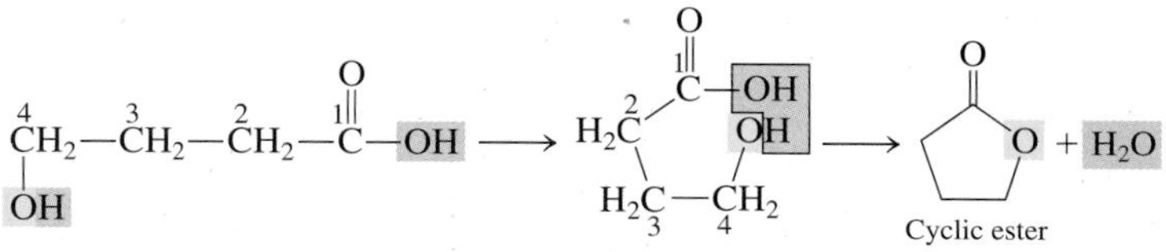

Cyclic esters are formally called *lactones*. A **lactone** *is a cyclic ester*. The ring size in a lactone is indicated using a Greek letter. A lactone with a five-membered ring is a γ-lactone and one with a six-membered ring is a δ-lactone.

$$CH_3—CH_2—CH_2—\overset{\overset{\displaystyle O}{\|}}{C}—O^- \, Na^+$$

IUPAC: Sodium butanoate
Common: Sodium butyrate

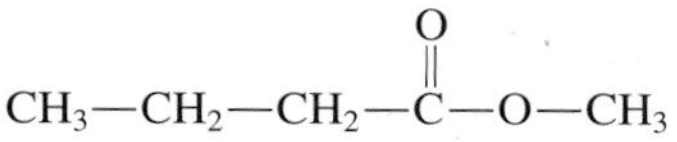

$$CH_3—CH_2—CH_2—\overset{\overset{\displaystyle O}{\|}}{C}—O—CH_3$$

IUPAC: Methyl butanoate
Common: Methyl butyrate

γ-lactone

δ-lactone

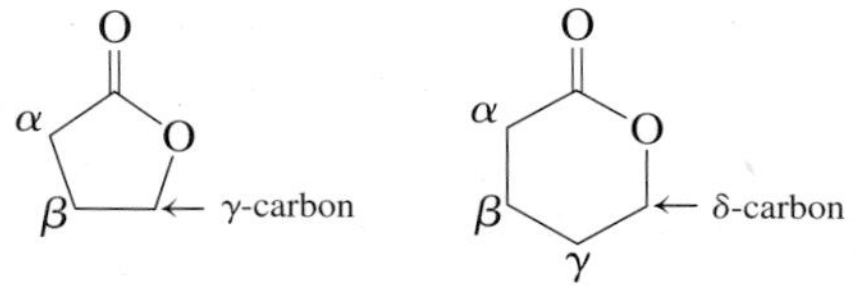

Chemical reactions that are expected to produce a hydroxy carboxylic acid often yield a lactone instead if a five- or six-membered ring can be formed.

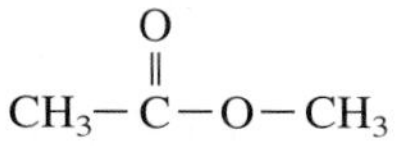

$$CH_3—\overset{\overset{\displaystyle O}{\|}}{C}—O—CH_3$$

Methyl acetate

16.12 NOMENCLATURE FOR ESTERS

Visualizing esters as having an "alcohol part" and an "acid part" (Section 16.11) is the key to naming them in both the common and the IUPAC systems of nomenclature. The rules are as follows:

Rule 1: The name for the alcohol part of the ester appears first and is followed by a *separate word* giving the name for the acid part of the ester.

Rule 2: The name for the alcohol part of the ester is simply the name of the R group (alkyl, cycloalkyl, or aryl) present in the —OR portion of the ester.

Rule 3: The name for the acid part of the ester is obtained by dropping the *-ic acid* ending for the acid's name and adding the suffix *-ate*.

Consider the ester derived from ethanoic acid (acetic acid) and methanol (methyl alcohol). Its name will be *methyl ethanoate* (IUPAC) or *methyl acetate* (common); see Figure 16.13.

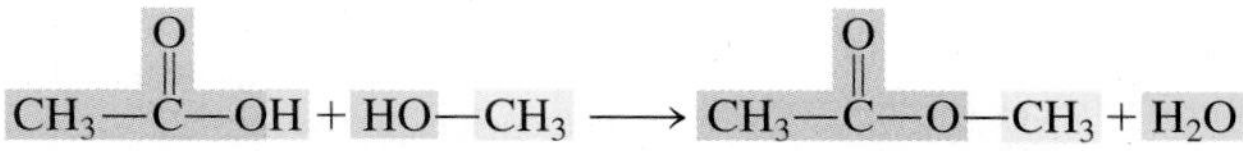

$$CH_3—\overset{\overset{\displaystyle O}{\|}}{C}—O—CH_2—CH_3$$

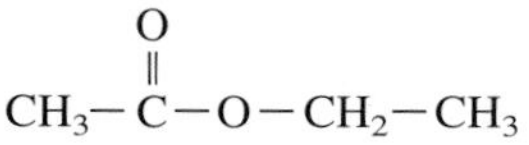

Ethyl acetate

Figure 16.13 Space-filling models for the methyl and ethyl esters of acetic acid.

$$CH_3—\overset{\overset{\displaystyle O}{\|}}{C}—OH + HO—CH_3 \longrightarrow CH_3—\overset{\overset{\displaystyle O}{\|}}{C}—O—CH_3 + H_2O$$

IUPAC: Ethanoic acid   Methanol   Methyl ethanoate
Common: Acetic acid   Methyl alcohol   Methyl acetate

Dicarboxylic acids can form diesters, with each of the carboxyl groups undergoing esterification. An example of such a molecule and how it is named is

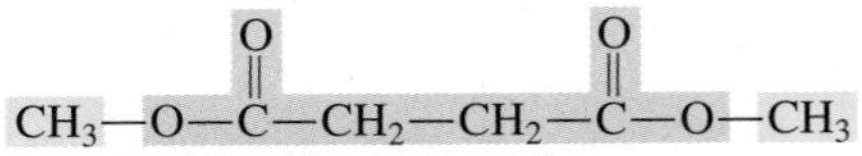

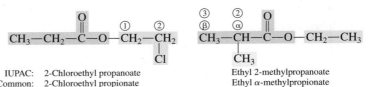

IUPAC: Dimethyl butanedioate
Common: Dimethyl succinate

Further examples of ester nomenclature, for compounds in which substituents are present, are

$$CH_3-CH_2-\overset{\overset{\displaystyle O}{\|}}{C}-O-\underset{1}{CH_2}-\underset{2}{CH_2}$$
$$|$$
$$Cl$$

IUPAC: 2-Chloroethyl propanoate
Common: 2-Chloroethyl propionate

$$\underset{\beta}{\overset{3}{C}}H_3-\underset{\alpha}{\overset{2}{C}}H-\overset{\overset{\displaystyle O}{\|}}{C}-O-CH_2-CH_3$$
$$|$$
$$CH_3$$

Ethyl 2-methylpropanoate
Ethyl α-methylpropionate

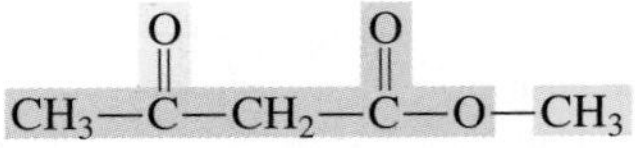

$$CH_3-\overset{\overset{\displaystyle O}{\|}}{C}-CH_2-\overset{\overset{\displaystyle O}{\|}}{C}-O-CH_3$$

IUPAC: Methyl 3-oxobutanoate
Common: Methyl β-oxobutyrate

Line-angle structural formulas for the simpler unbranched-chain methyl esters:

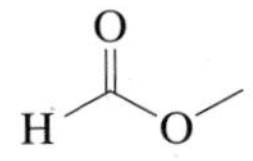

Methyl methanoate

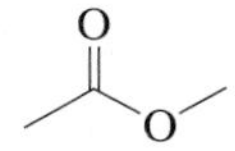

Methyl ethanoate

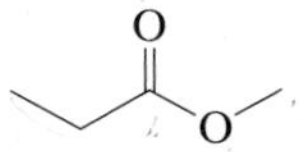

Methyl propanoate

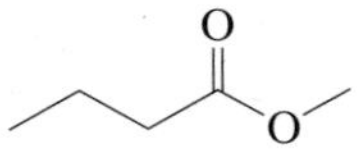

Methyl butanoate

● EXAMPLE 16.4

Determining IUPAC and Common Names for Esters

▶ Assign both IUPAC and common names to the following esters.

a.
$$CH_3-CH_2-\overset{\overset{\displaystyle O}{\|}}{C}-O-CH_2-CH_3$$

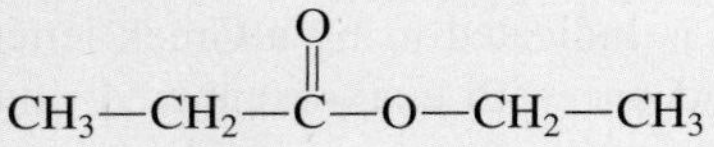

b.

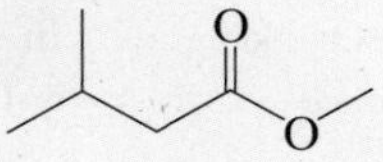

c.
$$\overset{\overset{\displaystyle O}{\|}}{C}-O-CH_2-CH_2-CH_3$$

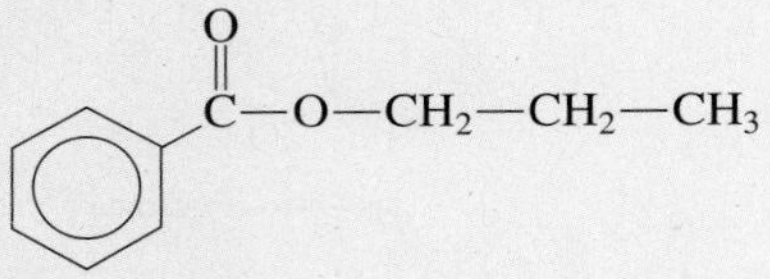

Solution

a. The name *ethyl* characterizes the alcohol part of the molecule. The name of the acid is propanoic acid (IUPAC) or propionic acid (common). Deleting the *-ic acid* ending and adding *-ate* gives the name *ethyl propanoate* (IUPAC) or *ethyl propionate* (common).
b. The name of the alcohol part of the molecule is methyl (from methanol or methyl alcohol). The name of the five-carbon acid is 3-methylbutanoic acid or β-methylbutyric acid. Hence the ester name is *methyl 3-methylbutanoate* (IUPAC) or *methyl β-methylbutyrate* (common).
c. The name *propyl* characterizes the alcohol part of the molecule. The acid part of the molecule is derived from benzoic acid (both IUPAC and common name). Hence the ester name in both systems is *propyl benzoate.*

The contrast between IUPAC names and common names for unbranched esters of carboxylic acids is as follows:

IUPAC (two words)

alkyl alkanoate

methyl propanoate

Common (two words)

alkyl (prefix)ate*

methyl acetate

*The common-name prefixes are related to natural sources for the "parent" carboxylic acids.

Practice Exercise 16.4

Assign both IUPAC and common names to the following esters.

a.
$$CH_3-\overset{\overset{\displaystyle O}{\|}}{C}-O-CH_2-CH_3$$

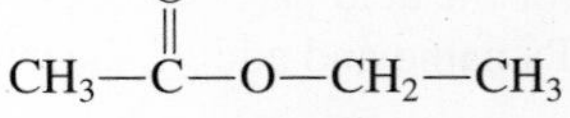

b.

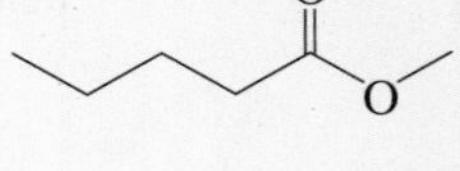

c.
$$H-\overset{\overset{\displaystyle O}{\|}}{C}-O-CH_2-CH_2-CH_3$$

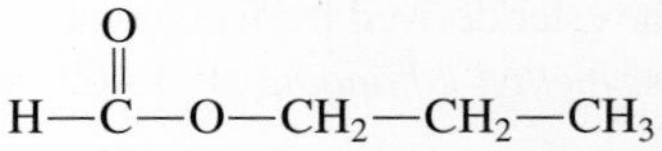

Answers: a. Ethyl ethanoate, ethyl acetate; **b.** Methyl pentanoate, methyl valerate; **c.** Propyl methanoate, propyl formate

IUPAC names for lactones are generated by replacing the *-oic* ending of the parent hydroxycarboxylic acid name with *-olide* and identifying the hydroxyl-bearing carbon by number.

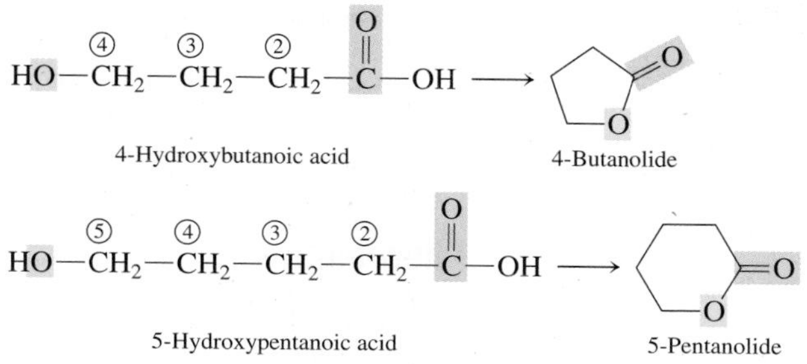

4-Hydroxybutanoic acid   4-Butanolide

5-Hydroxypentanoic acid   5-Pentanolide

16.13 SELECTED COMMON ESTERS

In this section we consider selected esters that function as flavoring agents, pheromones, and medications.

Flavor/Fragrance Agents

Esters are largely responsible for the flavor and fragrance of fruits and flowers. Generally, a natural flavor or odor is caused by a mixture of esters, with one particular compound being dominant. The synthetic production of these "dominant" compounds is the basis for the flavoring agents used in ice cream, gelatins, soft drinks, and so on. Table 16.4 gives the structures of selected esters used as flavoring agents. What is surprising about the structures in Table 16.4 is how closely some of them resemble each other. For example, the apple and pineapple flavoring agents differ by one carbon atom (methyl versus ethyl); a five-carbon chain versus an eight-carbon chain makes the difference between banana and orange flavor.

Fats and oils, substances that are part of our dietary intake, are triesters—molecules containing three ester functional groups. Such compounds are considered in Chapter 19.

TABLE 16.4
Selected Esters That Are Used as Flavoring Agents

IUPAC Name	Structural Formula	Characteristic Flavor and Odor
isobutyl methanoate	$H-\overset{O}{\overset{\|}{C}}-O-CH_2-\overset{CH_3}{\overset{\|}{CH}}-CH_3$	raspberry
propyl ethanoate	$CH_3-\overset{O}{\overset{\|}{C}}-O-(CH_2)_2-CH_3$	pear
pentyl ethanoate	$CH_3-\overset{O}{\overset{\|}{C}}-O-(CH_2)_4-CH_3$	banana
octyl ethanoate	$CH_3-\overset{O}{\overset{\|}{C}}-O-(CH_2)_7-CH_3$	orange
pentyl propanoate	$CH_3-CH_2-\overset{O}{\overset{\|}{C}}-O-(CH_2)_4-CH_3$	apricot
methyl butanoate	$CH_3-(CH_2)_2-\overset{O}{\overset{\|}{C}}-O-CH_3$	apple
ethyl butanoate	$CH_3-(CH_2)_2-\overset{O}{\overset{\|}{C}}-O-CH_2-CH_3$	pineapple

Numerous lactones are common in plants. Two examples are 4-decanolide, a compound partially responsible for the taste and odor of ripe peaches, and coumarin (common name), the compound responsible for the pleasant odor of newly mown hay.

$$CH_3—CH_2—CH_2—CH_2—CH_2—CH_2$$

4-Decanolide
(peach odor)

Coumarin
(newly mown hay odor)

Pheromones

A number of pheromones (Section 13.7) contain ester functional groups. The compound isoamyl acetate,

$$CH_3—\overset{\overset{\displaystyle O}{\|}}{C}—O—CH_2—CH_2—\overset{\overset{\displaystyle CH_3}{|}}{CH}—CH_3$$

is an alarm pheromone for the honey bee. The compound methyl *p*-hydroxybenzoate,

$$HO—\langle\ \rangle—\overset{\overset{\displaystyle O}{\|}}{C}—O—CH_3$$

is a sexual attractant for canine species. It is secreted by female dogs in heat and evokes attraction and sexual arousal in male dogs.

The compound nepetalactone, a lactone present in the catnip plant, is an attractant for cats of all types. It is not considered a pheromone, however, because different species are involved (see Figure 16.14).

Medications

Numerous esters have medicinal value, including benzocaine (a local anesthetic), aspirin, and oil of wintergreen (a counterirritant). The structure of benzocaine is

$$H_2N—\langle\ \rangle—\overset{\overset{\displaystyle O}{\|}}{C}—O—CH_2—CH_3$$

Both aspirin and oil of wintergreen are esters of salicylic acid, an aromatic hydroxyacid.

$$\overset{\overset{\displaystyle O}{\|}}{C}—OH$$

OH

Salicylic acid

Figure 16.14 Cats of all types (from lions to house cats) are strongly attracted to the catnip plant. The attractant in the catnip plant is nepetalactone, a cyclic ester.

Nepetalactone

Because this acid has both an acid group and a hydroxyl group, it can form two different types of esters: one by reaction of its acid group with an alcohol, the other by reaction of its alcohol group with a carboxylic acid.

Reaction of acetic acid with the alcohol group of salicylic acid produces aspirin.

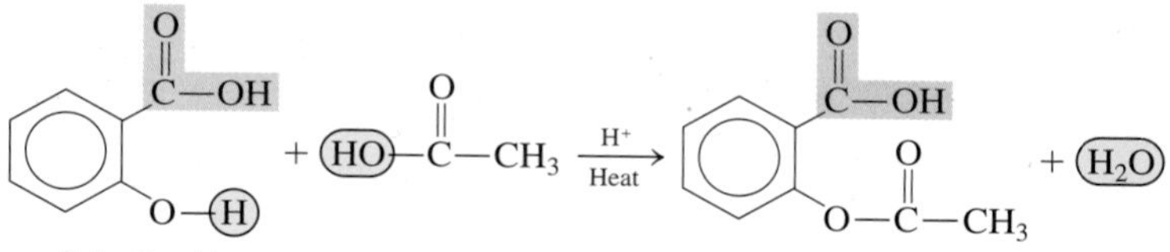

| Salicylic acid | Acetic acid | Aspirin |

Aspirin's mode of action in the human body is considered in the Chemical Connections feature on page 494.

Reaction of methanol with the acid group of salicylic acid produces oil of wintergreen.

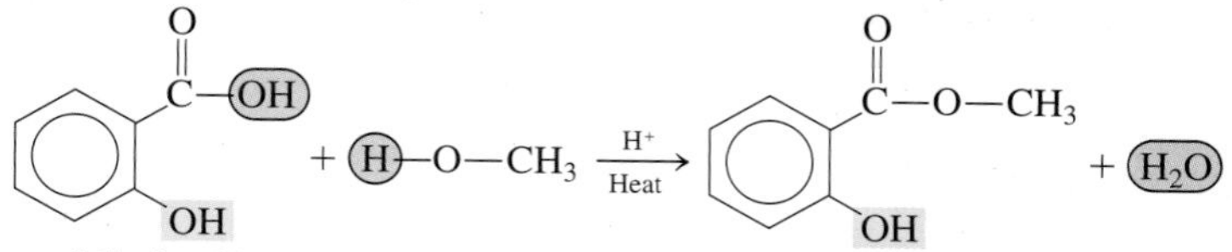

| Salicylic acid | Methanol | Oil of wintergreen |

Oil of wintergreen, also called methyl salicylate, is used in skin rubs and liniments to help decrease the pain of sore muscles. It is absorbed through the skin, where it is hydrolyzed to produce salicylic acid. Salicylic acid, as with aspirin, is the actual pain reliever.

The *macrolide antibiotics* are a family of large-ring lactones. Erythromycin, the best known member of this antibiotic family, has an antimicrobial spectrum similar to that of penicillin (Section 21.10) and is often used for people who have an allergy to penicillins. Structurally, this antibiotic contains a 14-membered lactone ring.

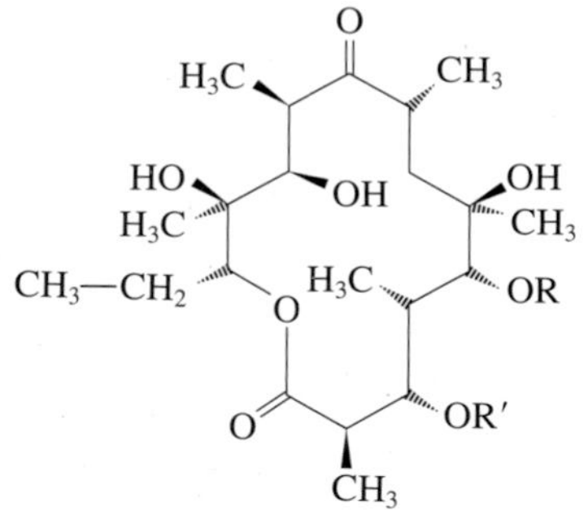

Erythromycin (R and R′ are
carbohydrate units)

Erythromycin is a naturally occurring substance first isolated from a red-pigmented soil bacterium. The laboratory synthesis of this compound has now been achieved. Its chemical formula is $C_{37}H_{67}NO_{13}$.

16.14 ISOMERISM FOR CARBOXYLIC ACIDS AND ESTERS

As with the other families of organic compounds previously discussed, constitutional isomers based on different carbon skeletons and on different positions for the functional group are possible for carboxylic acids and esters as well as other types of carboxylic acid derivatives. The following two examples illustrate carboxylic acid skeletal isomerism and ester positional isomerism.

Carboxylic acid
skeletal isomers:

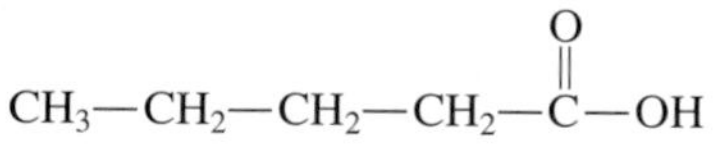

Pentanoic acid   and   2-Methylbutanoic acid

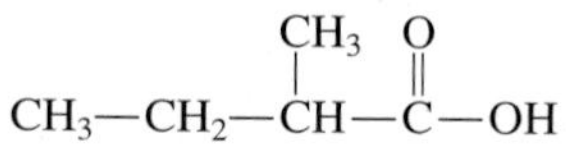

Ester positional
isomers:

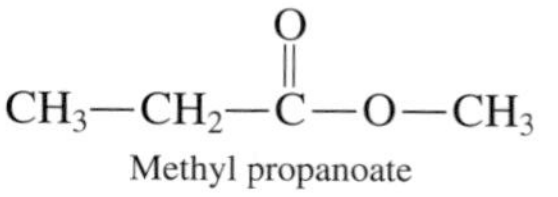

Methyl propanoate   and   Ethyl ethanoate

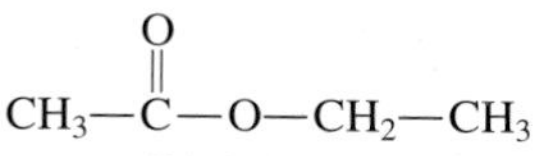

CHEMICAL Connections   Aspirin

Aspirin, an ester of salicylic acid (Section 16.13), is a drug that has the ability to decrease pain (analgesic properties), to lower body temperature (antipyretic properties), and to reduce inflammation (anti-inflammatory properties). It is most frequently taken in tablet form, and the tablet usually contains 325 mg of aspirin held together with an inert starch binder.

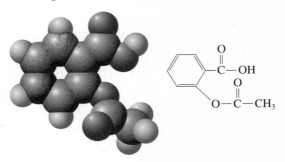

After ingestion, aspirin undergoes hydrolysis to produce salicylic acid and acetic acid. Salicylic acid is the active ingredient of aspirin—the substance that has analgesic, antipyretic, and anti-inflammatory effects.

Salicylic acid is capable of irritating the lining of the stomach, inducing a small amount of bleeding. Breaking (or chewing) an aspirin tablet, rather than taking it whole, reduces the chance of bleeding by eliminating drug concentration on one part of the stomach lining. Buffered aspirin products contain alkaline chemicals (such as aluminum glycinate or aluminum hydroxide) to neutralize the acidity of the aspirin when it contacts the stomach lining.

Aspirin—that is, salicylic acid—inhibits the synthesis of a class of hormones called prostaglandins (Section 19.13), molecules that cause pain, fever, and inflammation when present in the bloodstream in higher-than-normal levels. Salicylic acid's mode of action is irreversible inhibition (Section 21.7) of

cyclooxygenase, an enzyme necessary for the production of prostaglandins.

Recent studies show that aspirin also increases the time it takes blood to coagulate (clot). For blood to coagulate, platelets must first be able to aggregate, and prostaglandins (which aspirin inhibits) appear to be necessary for platelet aggregation to occur. One study suggests that healthy men can cut their risk of heart attacks nearly in half by taking one baby aspirin per day (81 mg compared to the 325 mg in a regular tablet). Aspirin acts by making the blood less likely to clot. Heart attacks usually occur when clots form in the coronary arteries, cutting off blood supply to the heart.

Aspirin manufacturers indicate that "low dose" (81 mg) aspirin tablet use is rapidly increasing among adults in the United States. In 2007 more than 50 million people (36% of the adult population) took aspirin regularly for cardiovascular health reasons.

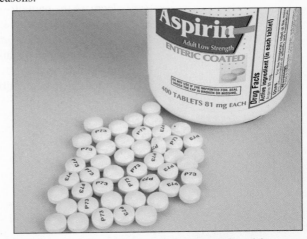

Low-dose aspirin tablets contain 81 mg of aspirin.

Esters and carboxylic acid functional group isomerism represents the fourth time we have encountered this type of isomerism. Previous examples are alcohol–ether, thiol–thioether, and aldehyde–ketone isomers.

Carboxylic acids and esters with the same number of carbon atoms and the same degree of saturation are functional group isomers. The ester ethyl propanoate and the carboxylic acid pentanoic acid both have the molecular formula, $C_5H_{10}O_2$, and are thus functional group isomers.

Carboxylic acid–ester functional group isomers:

$$CH_3{-}CH_2{-}\overset{\displaystyle O}{\overset{\|}{C}}{-}O{-}CH_2{-}CH_3 \quad \text{and} \quad CH_3{-}CH_2{-}CH_2{-}CH_2{-}\overset{\displaystyle O}{\overset{\|}{C}}{-}OH$$

Ethyl propanoate   Pentanoic acid

Name	Functional-Group Class	Molecular Mass (amu)	Boiling Point (°C)
diethyl ether	ether	74	34
ethyl formate	ester	74	54
methyl acetate	ester	74	57
butanal	aldehyde	72	76
1-butanol	alcohol	74	118
propionic acid	acid	74	141

16.15 PHYSICAL PROPERTIES OF ESTERS

Ester molecules cannot form hydrogen bonds to each other because they do not have a hydrogen atom bonded to an oxygen atom. Consequently, the boiling points of esters are much lower than those of alcohols and carboxylic acids of comparable molecular mass. Esters are more like ethers in their physical properties. Table 16.5 gives boiling-point data for compounds of similar molecular mass that contain different functional groups.

Water molecules can hydrogen-bond to esters through the oxygen atoms present in the ester functional group (Figure 16.15). Because of such hydrogen bonding, low-molecular-mass esters are soluble in water. Solubility rapidly decreases with increasing carbon chain length; borderline solubility situations are reached when three to five carbon atoms are in a chain.

Low- and intermediate-molecular-mass esters are usually colorless liquids at room temperature (see Figure 16.16). Most have pleasant odors (Section 16.13).

16.16 CHEMICAL REACTIONS OF ESTERS

The most important reaction of esters involves breaking the carbon–oxygen single bond that holds the "alcohol part" and the "acid part" of the ester together. This reaction process is called either ester hydrolysis or ester saponification, depending on reaction conditions.

Ester Hydrolysis

This is our second encounter with hydrolysis reactions. The first encounter involved the hydrolysis of acetals (Section 15.11).

The breaking of a bond within a molecule and the attachment of the components of water to the fragments are characteristics of all hydrolysis reactions.

In ester hydrolysis, an ester reacts with water, producing the carboxylic acid and alcohol from which the ester was formed.

$$R-\overset{\overset{\displaystyle O}{\|}}{C}-O-R' + H-OH \xrightarrow{H^+} R-\overset{\overset{\displaystyle O}{\|}}{C}-OH + R'-O-H$$

$$CH_3-\overset{\overset{\displaystyle O}{\|}}{C}-O-CH_3 + H-OH \xrightarrow{H^+} CH_3-\overset{\overset{\displaystyle O}{\|}}{C}-OH + CH_3-O-H$$

Methyl acetate   Water   Acetic acid   Methyl alcohol

Figure 16.15 Low-molecular-mass esters are soluble in water because of ester-water hydrogen bonding.

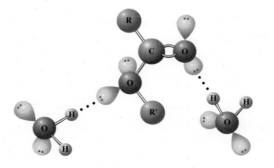

Methyl Esters		
	C_3	C_5	C_7
C_2	C_4	C_6	C_8

Ethyl Esters		
	C_3	C_5	C_7
	C_4	C_6	C_8

☐ Liquid

Figure 16.16 A physical-state summary for methyl and ethyl esters of unbranched-chain carboxylic acids at room temperature and pressure.

Ester hydrolysis requires the presence of a strong-acid catalyst or enzymes. Ester hydrolysis is the reverse of esterification (Section 16.11), the formation of an ester from a carboxylic acid and an alcohol.

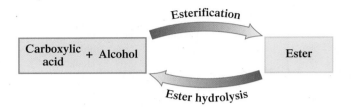

Ester Saponification

A **saponification reaction** is *the hydrolysis of an organic compound, under basic conditions, in which a carboxylic acid salt is one of the products.* Esters, amides (Section 17.17), and fats and oils (Section 19.6) all undergo saponification reactions.

In ester saponification either NaOH or KOH is used as the base and the saponification products are an alcohol and a carboxylic acid salt. (Any carboxylic acid product formed is converted to its salt because of the basic reaction conditions.)

$$\underset{\text{An ester}}{R-\overset{\overset{\displaystyle O}{\|}}{C}-O-R'} + \underset{\text{A strong base}}{NaOH} \xrightarrow{H_2O} \underset{\text{A carboxylate salt}}{R-\overset{\overset{\displaystyle O}{\|}}{C}-O^-\ Na^+} + \underset{\text{An alcohol}}{R'-OH}$$

A specific example of ester saponification is

$$\underset{\text{Methyl benzoate}}{C_6H_5-\overset{\overset{\displaystyle O}{\|}}{C}-O-CH_3} + \underset{\text{Sodium hydroxide}}{NaOH} \xrightarrow{H_2O} \underset{\text{Sodium benzoate}}{C_6H_5-\overset{\overset{\displaystyle O}{\|}}{C}-O^-\ Na^+} + \underset{\text{Methyl alcohol}}{CH_3-OH}$$

In both ester hydrolysis and ester saponification, an alcohol is produced. Under acidic conditions (ester hydrolysis), the other product is a carboxylic acid. Under basic conditions (ester saponification), the other product is a carboxylic acid salt.

EXAMPLE 16.5

Structural Equations for Reactions That Involve Esters

Write structural equations for each of the following reactions.

a. Hydrolysis, with an acidic catalyst, of ethyl acetate
b. Saponification, with NaOH, of methyl formate
c. Esterification of propionic acid using isopropyl alcohol

Solution

a. Hydrolysis, under acidic conditions, cleaves an ester to produce its "parent" carboxylic acid and alcohol.

$$\underset{\text{Ethyl acetate}}{CH_3-\overset{\overset{\displaystyle O}{\|}}{C}-O-CH_2-CH_3} + H_2O \xrightarrow{H^+} \underset{\text{Acetic acid}}{CH_3-\overset{\overset{\displaystyle O}{\|}}{C}-OH} + \underset{\text{Ethyl alcohol}}{CH_3-CH_2-OH}$$

b. Saponification cleaves an ester to produce its "parent" alcohol and the *salt* of its "parent" carboxylic acid.

$$\underset{\text{Methyl formate}}{H-\overset{\overset{\displaystyle O}{\|}}{C}-O-CH_3} + \underset{\text{Sodium hydroxide}}{NaOH} \xrightarrow{H_2O} \underset{\text{Sodium formate}}{H-\overset{\overset{\displaystyle O}{\|}}{C}-O^-\ Na^+} + \underset{\text{Methyl alcohol}}{CH_3-OH}$$

c. Esterification is the reaction in which a carboxylic acid and an alcohol react to produce an ester.

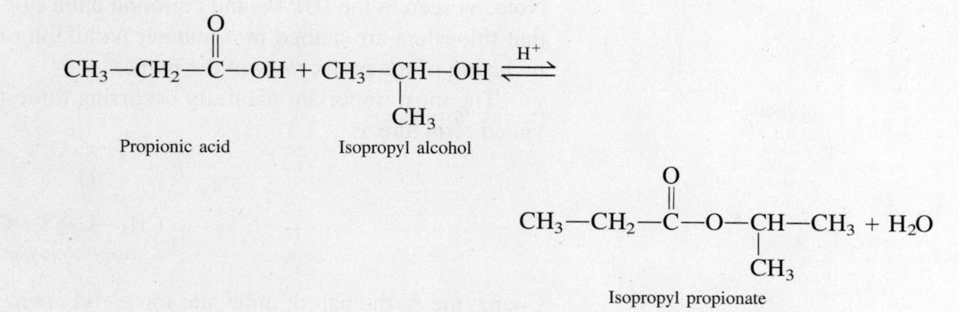

$$CH_3-CH_2-\overset{\overset{\displaystyle O}{\|}}{C}-OH + CH_3-\underset{\underset{\displaystyle CH_3}{|}}{CH}-OH \underset{\longleftarrow}{\overset{H^+}{\longrightarrow}}$$

Propionic acid   Isopropyl alcohol

$$CH_3-CH_2-\overset{\overset{\displaystyle O}{\|}}{C}-O-\underset{\underset{\displaystyle CH_3}{|}}{CH}-CH_3 + H_2O$$

Isopropyl propionate

Practice Exercise 16.5

Write structural equations for each of the following reactions.
a. Hydrolysis, with an acidic catalyst, of propyl propanoate
b. Saponification, with KOH, of ethyl propanoate
c. Esterification of acetic acid with propyl alcohol

Answers:

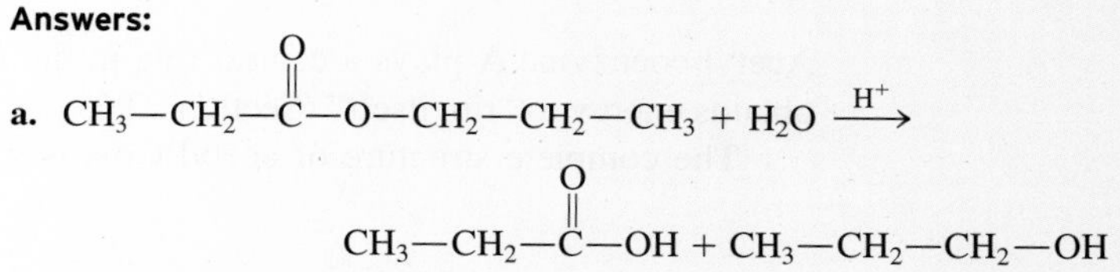

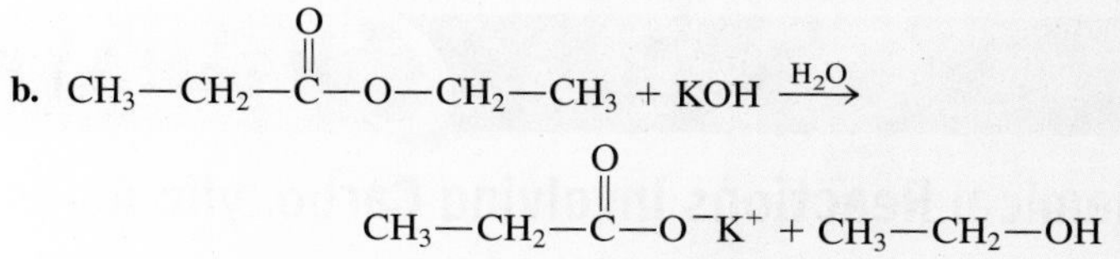

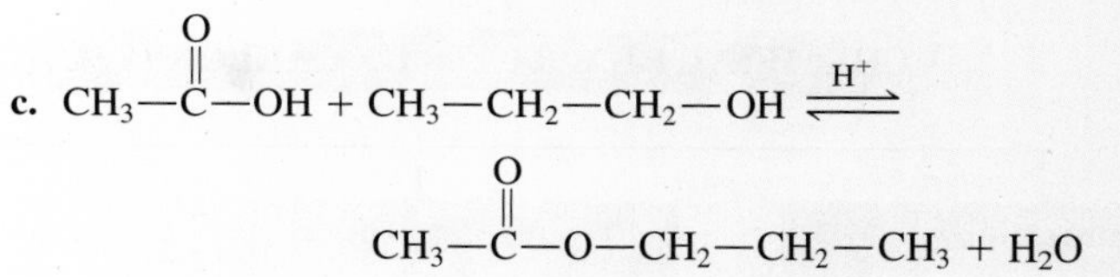

a. $CH_3-CH_2-\overset{\overset{\displaystyle O}{\|}}{C}-O-CH_2-CH_2-CH_3 + H_2O \overset{H^+}{\longrightarrow}$

$CH_3-CH_2-\overset{\overset{\displaystyle O}{\|}}{C}-OH + CH_3-CH_2-CH_2-OH$

b. $CH_3-CH_2-\overset{\overset{\displaystyle O}{\|}}{C}-O-CH_2-CH_3 + KOH \overset{H_2O}{\longrightarrow}$

$CH_3-CH_2-\overset{\overset{\displaystyle O}{\|}}{C}-O^-K^+ + CH_3-CH_2-OH$

c. $CH_3-\overset{\overset{\displaystyle O}{\|}}{C}-OH + CH_3-CH_2-CH_2-OH \rightleftharpoons$

$CH_3-\overset{\overset{\displaystyle O}{\|}}{C}-O-CH_2-CH_2-CH_3 + H_2O$

The Chemistry at a Glance feature on page 498 summarizes reactions that involve carboxylic acids and esters.

16.17 SULFUR ANALOGS OF ESTERS

Just as alcohols react with carboxylic acids to produce esters, thiols (Section 14.20) react with carboxylic acids to produce thioesters. A **thioester** *is a sulfur-containing analog of an ester in which an —SR group has replaced the —OR group.*

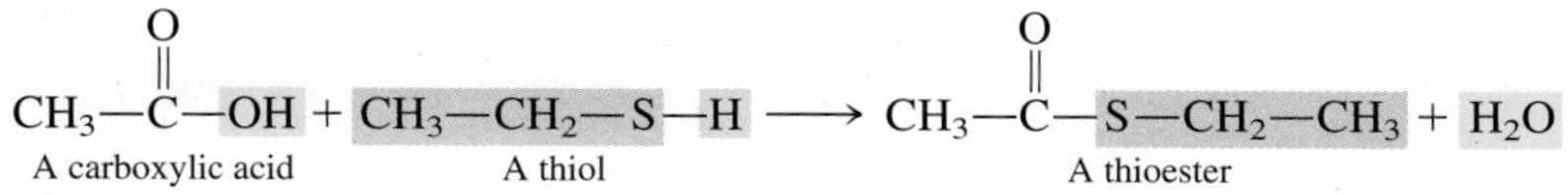

$$CH_3-\overset{\overset{\displaystyle O}{\|}}{C}-OH + CH_3-CH_2-S-H \longrightarrow CH_3-\overset{\overset{\displaystyle O}{\|}}{C}-S-CH_2-CH_3 + H_2O$$

A carboxylic acid   A thiol   A thioester

The thioester methyl thiobutanoate is used as an artificial flavoring agent. It generates the taste we call strawberry.

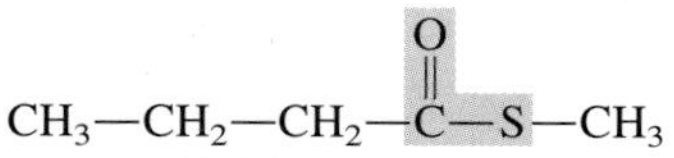

$$CH_3-CH_2-CH_2-\overset{\overset{\displaystyle O}{\|}}{C}-S-CH_3$$

Methyl thiobutanoate
(methyl thiobutyrate)

The ester functional group is

$$-\overset{\overset{\displaystyle O}{\|}}{C}-O-R$$

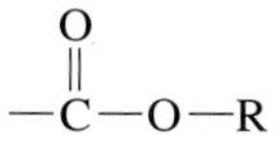

The thioester functional group is

$$-\overset{\overset{\displaystyle O}{\|}}{C}-S-R$$

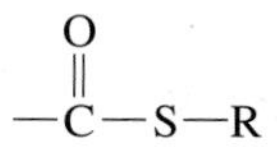

Note, as seen in the IUPAC and common names of the preceding "strawberry" compound, that thioesters are named in a manner paralleling that for esters (Section 16.12) with the inclusion of the prefix *thio-* in the name.

The most important naturally occurring thioester is acetyl coenzyme A, whose abbreviated structure is

$$CH_3-\overset{\displaystyle O}{\overset{\displaystyle \|}{C}}-S-CoA$$
Acetyl coenzyme A

Coenzyme A, the parent molecule for acetyl coenzyme A, is a large, complex *thiol* whose structure, for simplicity, is usually abbreviated as CoA—S—H. The formation of acetyl coenzyme A (acetyl CoA) from coenzyme A can be envisioned as a thioesterification reaction between acetic acid and coenzyme A.

$$CH_3-\overset{\displaystyle O}{\overset{\displaystyle \|}{C}}-OH + CoA-S-H \longrightarrow CH_3-\overset{\displaystyle O}{\overset{\displaystyle \|}{C}}-S-CoA + H_2O$$

Acetic acid   Coenzyme A   Acetyl coenzyme A
                   (a thiol)   (acetyl CoA)

Acetyl coenzyme A plays a central role in the metabolic cycles through which the body obtains energy to "run itself" (Section 23.6).

The complete structure of acetyl CoA is given in Section 23.3.

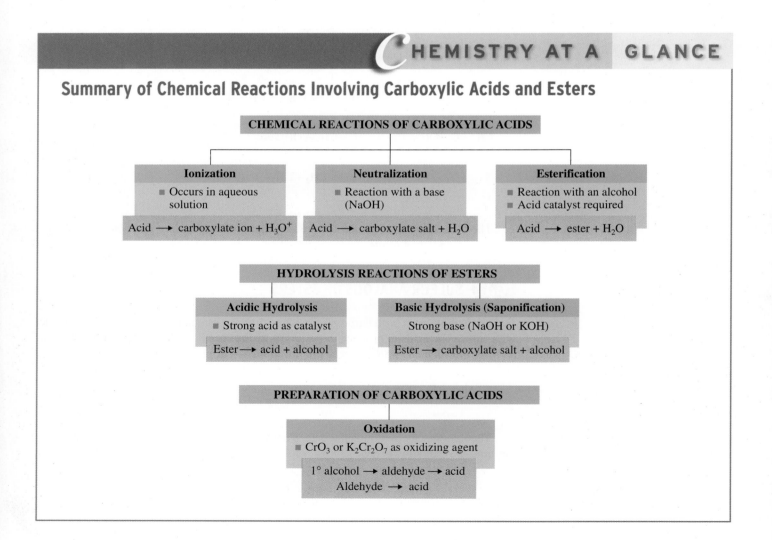

CHEMISTRY AT A GLANCE

Summary of Chemical Reactions Involving Carboxylic Acids and Esters

CHEMICAL REACTIONS OF CARBOXYLIC ACIDS

Ionization
- Occurs in aqueous solution

Acid ⟶ carboxylate ion + H_3O^+

Neutralization
- Reaction with a base (NaOH)

Acid ⟶ carboxylate salt + H_2O

Esterification
- Reaction with an alcohol
- Acid catalyst required

Acid ⟶ ester + H_2O

HYDROLYSIS REACTIONS OF ESTERS

Acidic Hydrolysis
- Strong acid as catalyst

Ester ⟶ acid + alcohol

Basic Hydrolysis (Saponification)
Strong base (NaOH or KOH)

Ester ⟶ carboxylate salt + alcohol

PREPARATION OF CARBOXYLIC ACIDS

Oxidation
- CrO_3 or $K_2Cr_2O_7$ as oxidizing agent

1° alcohol ⟶ aldehyde ⟶ acid
Aldehyde ⟶ acid

Figure 16.17 Space-filling model of a segment of the polyester condensation polymer known as poly(ethylene terephthalate), or PET.

16.18 POLYESTERS

A **condensation polymer** *is a polymer formed by reacting difunctional monomers to give a polymer and some small molecule (such as water) as a by-product of the process. Polyesters* are an important type of condensation polymer. A **polyester** *is a condensation polymer in which the monomers are joined through ester linkages.* Dicarboxylic acids and dialcohols are the monomers generally used in forming polyesters.

The best known of the many polyesters now marketed is *poly(ethylene terephthalate)*, which is also known by the acronym *PET.* The monomers used to produce PET are terephthalic acid (a diacid) and ethylene glycol (a dialcohol).

Condensation polymerization reactions produce two products: the polymer and a small molecule. This contrasts with addition polymerization reactions (Section 13.9) where the polymer is the only product.

Terephthalic acid   Ethylene glycol

The reaction of one acid group of the diacid with one alcohol group of the dialcohol initially produces an ester molecule, with an acid group left over on one end and an alcohol group left over on the other end.

Leftover acid group that can react further

Ester linkage

Leftover alcohol group that can react further

This species can react further. The remaining acid group can react with an alcohol group from another monomer, and the alcohol group can react with an acid group from another monomer. This process continues until an extremely long polymer molecule called a *polyester* is produced (see Figure 16.17).

Ester linkage   Ester linkage

Poly(ethylene terephthalate), a polyester

About 50% of PET production goes into textile products, including clothing fibers, curtain and upholstery materials, and tire cord. The trade name for PET as a clothing fiber is *Dacron.* The other 50% of PET production goes into plastics applications. As a film-like material, it is called *Mylar.* Mylar products include the plastic backing for audio and video tapes and computer diskettes. Its chemical name PET is applied when this polyester is used in clear, flexible soft-drink bottles and as the wrapping material for frozen foods and boil-in-bag foods.

PET is also used in medicine. Because it is physiologically inert, PET is used in the form of a mesh to replace diseased sections of arteries. It has also been used in synthetic heart valves (see Figure 16.18).

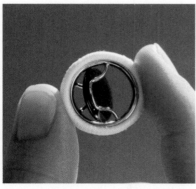

Figure 16.18 The polyester PET, as a fabric, is used in the construction of artificial heart valves.

Plastic bottles made of PET cannot be reused because they cannot withstand the high temperatures needed to sterilize them for reuse. Also such bottles cannot be used for any food items that must be packaged at high temperatures, such as jams and jellies. For these uses, the polyester PEN (polyethylene naphthalate), a polymer that can withstand higher temperatures, is available. The monomers for this polymer are ethylene glycol and one or more naphthalene dicarboxylic acids.

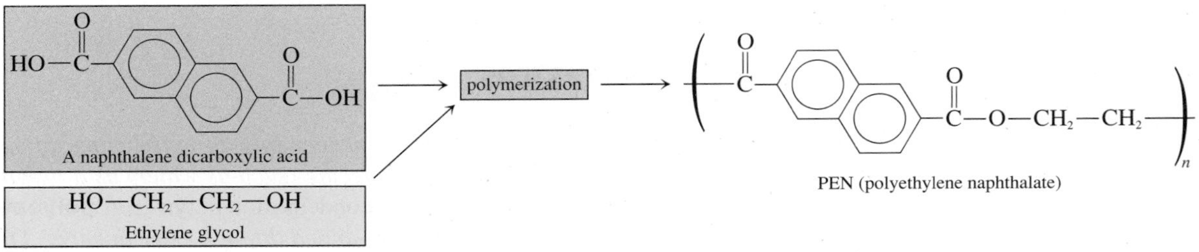

A naphthalene dicarboxylic acid

Ethylene glycol

polymerization

PEN (polyethylene naphthalate)

A variation of the diacid–dialcohol monomer formulation for polyesters involves using hydroxyacids as monomers. In this situation, both of the functional groups required are present in the same molecule.

A polymerization reaction in which lactic acid and glycolic acid (both hydroxyacids, Section 16.4) are monomers produces a biodegradable material (trade name *Lactomer*) that is used as surgical staples in several types of surgery. Traditional suture materials must be removed later on, after they have served their purpose. Lactomer staples start to dissolve (hydrolyze) after a period of several weeks. The hydrolysis products are the starting monomers, lactic acid and glycolic acid, both of which are normally present in the human body. By the time the tissue has fully healed, the staples have fully degraded.

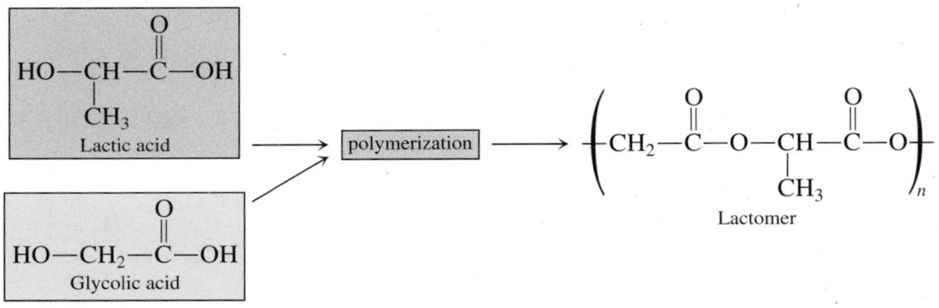

Lactic acid

Glycolic acid

polymerization

Lactomer

Another commercially available biodegradable polyester is PHBV, a substance that finds use in specialty packaging, orthopedic devices, and controlled drug-release formulations. The monomers for this polymer are 3-hydroxybutanoic acid (β-hydroxybutyrate) and 3-hydroxypentanoic acid (β-hydroxyvalerate). The properties of PHBV vary according to the reacting ratio for the two monomers. With more of the butanoic acid present a stiffer polymer is produced while more of the pentanoic acid imparts flexibility to the plastic. A nonconventional method for producing the polymer is used; the two monomers are produced by bacterial fermentation of mixtures containing acetic and propionic acids. The generalized formula for the polymer is written as

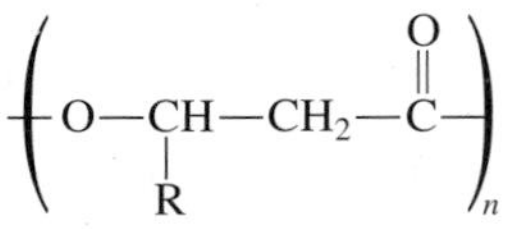

$(R = -CH_3$ and $-CH_2-CH_3)$

PHBV [poly(β-hydroxybutyrate-co-β-hydroxyvalerate)]

ACID CHLORIDES AND ACID ANHYDRIDES

Sections 16.10 through 16.18 have focused on the carboxylic acid derivatives called esters. We now consider two more of the carboxylic acid derivative types listed in Section 16.1, namely carboxylic acid chlorides and carboxylic acid anhydrides.

Acid Chlorides

An **acid chloride** *is a carboxylic acid derivative in which the —OH portion of the carboxyl group has been replaced with a —Cl atom.* Thus, acid chlorides have the general formula

$$\text{R}-\overset{\displaystyle\overset{O}{\|}}{\text{C}}-\boxed{\text{Cl}}$$

Acid chlorides are named in either of two ways:

Rule 1: Replace the *-ic acid* ending of the common name of the parent carboxylic acid with *-yl chloride.*

$$\text{CH}_3-\text{CH}_2-\text{CH}_2-\text{CH}_2-\overset{\displaystyle\overset{O}{\|}}{\text{C}}-\text{Cl}$$

Buty*ric acid* becomes buty*ryl chloride.*

Rule 2: Replace the *-oic acid* ending of the IUPAC name of the parent carboxylic acid with *-oyl chloride.*

$$\text{CH}_3-\text{CH}_2-\overset{\displaystyle\overset{\text{CH}_3}{|}}{\text{CH}}-\text{CH}_2-\overset{\displaystyle\overset{O}{\|}}{\text{C}}-\text{Cl}$$

3-Methylpentan*oic acid* becomes 3-methylpentan*oyl chloride.*

Preparation of an acid chloride from its parent carboxylic acid involves reacting the acid with one of several inorganic chlorides (PCl_3, PCl_5, or $SOCl_2$). The general reaction is

$$\text{R}-\overset{\displaystyle\overset{O}{\|}}{\text{C}}-\text{OH} \xrightarrow[\text{chloride}]{\text{Inorganic}} \text{R}-\overset{\displaystyle\overset{O}{\|}}{\text{C}}-\text{Cl} + \text{Inorganic products}$$

Acid chlorides react rapidly with water, in a hydrolysis reaction, to regenerate the parent carboxylic acid.

$$\text{R}-\overset{\displaystyle\overset{O}{\|}}{\text{C}}-\boxed{\text{Cl}} + \text{H}_2\text{O} \longrightarrow \text{R}-\overset{\displaystyle\overset{O}{\|}}{\text{C}}-\boxed{\text{OH}} + \text{HCl}$$

This reactivity with water means that acid chlorides cannot exist in biological systems.

Acid chlorides are useful starting materials for the synthesis of other carboxylic acid derivatives, particularly esters and amides. Synthesis of esters and amides using acid chlorides is a more efficient process than ester and amide synthesis using a carboxylic acid.

Acid Anhydrides

An **acid anhydride** *is a carboxylic acid derivative in which the —OH portion of the carboxyl group has been replaced with a* $-\text{O}-\overset{\displaystyle\overset{O}{\|}}{\text{C}}-\text{R}$ *group.* Thus, acid anhydrides have the general formula

$$\text{R}-\overset{\displaystyle\overset{O}{\|}}{\text{C}}-\text{O}-\overset{\displaystyle\overset{O}{\|}}{\text{C}}-\text{R}'$$

The word *anhydride* means "without water." Structurally, acid anhydrides can be visualized as two carboxylic acid molecules bonded together after removal of a water molecule from the acid molecules.

$$\text{R}-\overset{\displaystyle\overset{O}{\|}}{\text{C}}-\boxed{\text{O}-\text{H}} + \boxed{\text{H}-\text{O}}-\overset{\displaystyle\overset{O}{\|}}{\text{C}}-\text{R}' \longrightarrow \text{R}-\overset{\displaystyle\overset{O}{\|}}{\text{C}}-\text{O}-\overset{\displaystyle\overset{O}{\|}}{\text{C}}-\text{R}' + \boxed{\text{H}_2\text{O}}$$

Symmetrical acid anhydrides (both R groups are the same) are named by replacing the *acid* ending of the parent carboxylic acid name with the word *anhydride*.

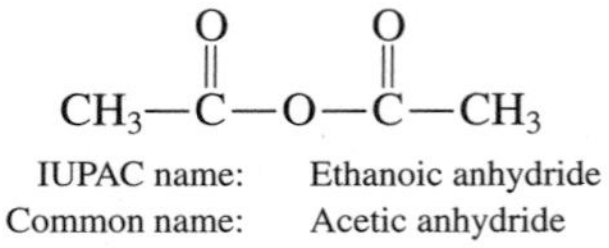

$$CH_3-\overset{\overset{\displaystyle O}{\|}}{C}-O-\overset{\overset{\displaystyle O}{\|}}{C}-CH_3$$

IUPAC name: Ethanoic anhydride
Common name: Acetic anhydride

Mixed acid anhydrides (different R groups present) are named by using the names of the individual parent carboxylic acids (in alphabetic order) followed by the word *anhydride*.

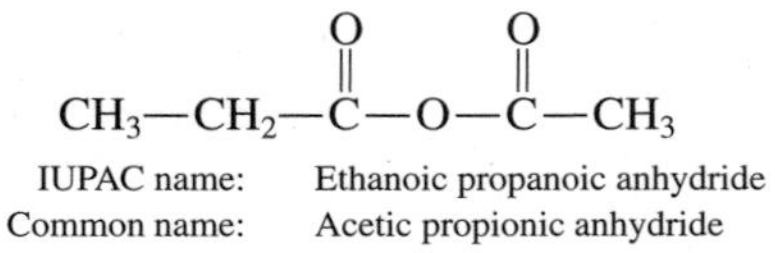

$$CH_3-CH_2-\overset{\overset{\displaystyle O}{\|}}{C}-O-\overset{\overset{\displaystyle O}{\|}}{C}-CH_3$$

IUPAC name: Ethanoic propanoic anhydride
Common name: Acetic propionic anhydride

In general, acid anhydrides cannot be formed by directly reacting the parent carboxylic acids together. Instead, an acid chloride is reacted with a carboxylate ion to produce the acid anhydride.

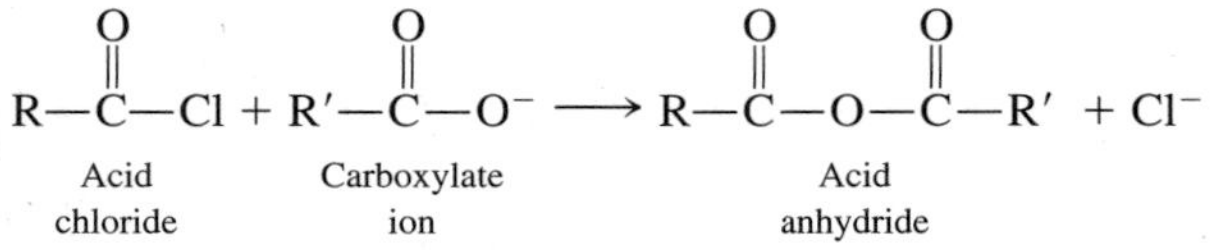

$$R-\overset{\overset{\displaystyle O}{\|}}{C}-Cl + R'-\overset{\overset{\displaystyle O}{\|}}{C}-O^- \longrightarrow R-\overset{\overset{\displaystyle O}{\|}}{C}-O-\overset{\overset{\displaystyle O}{\|}}{C}-R' + Cl^-$$

  Acid   Carboxylate   Acid
chloride   ion   anhydride

Acid anhydrides are very reactive compounds, although generally not as reactive as the acid chlorides. Like acid chlorides, they cannot exist in biological systems, as they undergo hydrolysis to regenerate the parent carboxylic acids.

$$R-\overset{\overset{\displaystyle O}{\|}}{C}-O-\overset{\overset{\displaystyle O}{\|}}{C}-R' + H_2O \xrightarrow{\text{Heat}} R-\overset{\overset{\displaystyle O}{\|}}{C}-OH + R'-\overset{\overset{\displaystyle O}{\|}}{C}-OH$$

     Acid   Acid   Acid
anhydride

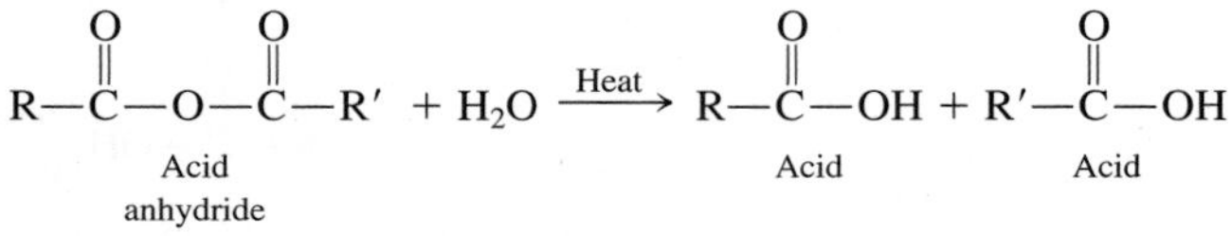

Reaction of an alcohol with an acid anhydride is a useful method for synthesizing esters.

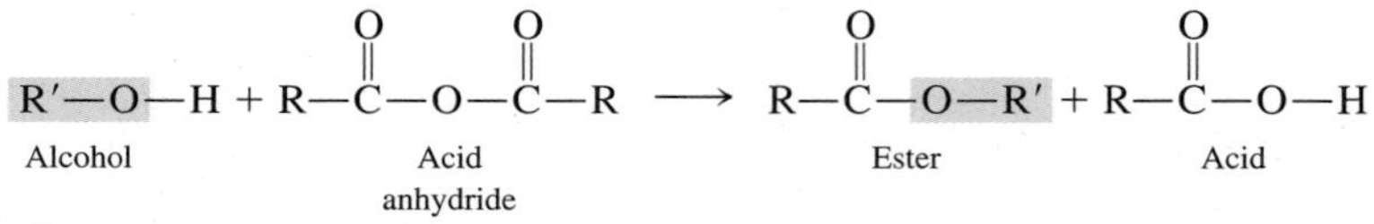

$$R'-O-H + R-\overset{\overset{\displaystyle O}{\|}}{C}-O-\overset{\overset{\displaystyle O}{\|}}{C}-R \longrightarrow R-\overset{\overset{\displaystyle O}{\|}}{C}-O-R' + R-\overset{\overset{\displaystyle O}{\|}}{C}-O-H$$

Alcohol   Acid   Ester   Acid
anhydride

Acyl Transfer Reactions

Carboxylic acids contain acyl groups. An **acyl group** *is that portion of a carboxylic acid that remains after the —OH group is removed from the carboxyl carbon atom.* Acid chlorides and acid anhydrides also contain acyl groups.

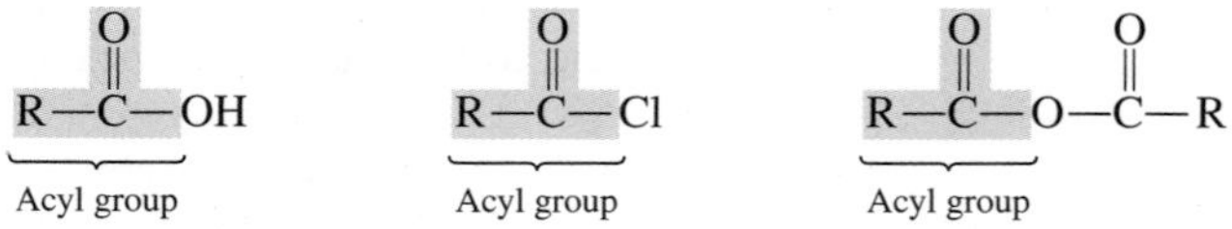

$$R-\overset{\overset{\displaystyle O}{\|}}{C}-OH \qquad R-\overset{\overset{\displaystyle O}{\|}}{C}-Cl \qquad R-\overset{\overset{\displaystyle O}{\|}}{C}-O-\overset{\overset{\displaystyle O}{\|}}{C}-R$$

Acyl group   Acyl group   Acyl group

Any compound with the generalized formula

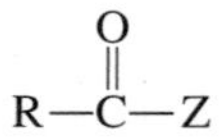

$$R-\overset{\overset{\displaystyle O}{\|}}{C}-Z$$

contains an acyl group, which is the portion of the molecule that remains after the Z entity is removed.

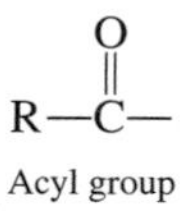

$$R-\overset{\overset{\displaystyle O}{\|}}{C}-$$

Acyl group

Compounds that contain acyl groups, when they react with an alcohol or phenol, transfer the acyl group to the oxygen atom of the alcohol or phenol.

$$
\underset{\text{Carboxylic acid}}{R-\overset{\overset{\displaystyle O}{\|}}{C}-OH} + R'-O-H \longrightarrow R-\overset{\overset{\displaystyle O}{\|}}{C}-O-R' + H_2O
$$

$$
\underset{\text{Acid chloride}}{R-\overset{\overset{\displaystyle O}{\|}}{C}-Cl} + R'-O-H \longrightarrow R-\overset{\overset{\displaystyle O}{\|}}{C}-O-R' + HCl
$$

$$
\underset{\text{Acid anhydride}}{R-\overset{\overset{\displaystyle O}{\|}}{C}-O-\overset{\overset{\displaystyle O}{\|}}{C}-R} + R'-O-H \longrightarrow R-\overset{\overset{\displaystyle O}{\|}}{C}-O-R' + R-\overset{\overset{\displaystyle O}{\|}}{C}-OH
$$

Chemical reactions such as these are called *acyl transfer reactions*. An **acyl transfer reaction** *is a chemical reaction in which an acyl group is transferred from one molecule to another.* Acyl transfer reactions occur frequently in biochemical systems. The process of protein synthesis (Section 22.11) is dependent upon acyl transfer reactions, as are many metabolic reactions. Often in metabolic reactions the thioester acetyl coenzyme A serves as an acyl transfer agent (Section 23.6).

The acyl group present in carboxylic acids and carboxylic acid derivatives is named by replacing the *-ic acid* ending of the acid name with the suffix *-yl*.

Common names: *-ic acid* become *-yl*
IUPAC names: *-oic acid* becomes *-oyl*

Thus, the two- and three-carbonyl acyl groups are named as follows:

$$
\overset{\overset{\displaystyle O}{\|}}{CH_3-C-} \qquad\qquad CH_3-CH_2-\overset{\overset{\displaystyle O}{\|}}{C-}
$$

IUPAC name:	Ethanoyl group (from ethanoic acid)	Propanoyl group (from propanoic acid)
Common name:	Acetyl group (from acetic acid)	Propionyl group (from propionic acid)

16.20 ESTERS AND ANHYDRIDES OF INORGANIC ACIDS

Inorganic acids such as sulfuric, phosphoric, and nitric acids react with alcohols to form esters in a manner similar to that for carboxylic acids.

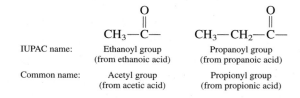

$$
\underset{\substack{\text{Sulfuric acid}\\(H_2SO_4)}}{HO-\overset{\overset{\displaystyle O}{\|}}{\underset{\underset{\displaystyle O}{\|}}{S}}-OH} + CH_3-OH \longrightarrow \underset{\substack{\text{Methyl ester of}\\\text{sulfuric acid}}}{HO-\overset{\overset{\displaystyle O}{\|}}{\underset{\underset{\displaystyle O}{\|}}{S}}-O-CH_3} + H_2O
$$

$$
\underset{\substack{\text{Phosphoric acid}\\(H_3PO_4)}}{HO-\overset{\overset{\displaystyle O}{\|}}{\underset{\underset{\displaystyle OH}{|}}{P}}-OH} + CH_3-OH \longrightarrow \underset{\substack{\text{Methyl ester of}\\\text{phosphoric acid}}}{HO-\overset{\overset{\displaystyle O}{\|}}{\underset{\underset{\displaystyle OH}{|}}{P}}-O-CH_3} + H_2O
$$

$$
\underset{\substack{\text{Nitric acid}\\(HNO_3)}}{\overset{\overset{\displaystyle O}{\|}}{\underset{\underset{\displaystyle O}{|}}{N}}-OH} + CH_3-OH \longrightarrow \underset{\substack{\text{Methyl ester of}\\\text{nitric acid}}}{\overset{\overset{\displaystyle O}{\|}}{\underset{\underset{\displaystyle O}{|}}{N}}-O-CH_3} + H_2O
$$

Esters of inorganic acids undergo hydrolysis reactions in a manner similar to that for esters of carboxylic acids (Section 16.16).

The most important inorganic esters, from a biochemical standpoint, are those of phosphoric acid—that is, phosphate esters. A **phosphate ester** *is an organic compound formed by reaction of an alcohol with phosphoric acid.* Because phosphoric acid has three hydroxyl groups, it can form mono-, di-, and triesters by reaction with one, two, and three molecules of alcohol, respectively.

Phosphoric Acid Anhydrides

Three biologically important phosphoric acids exist: phosphoric acid, diphosphoric acid, and triphosphoric acid. Phosphoric acid, the simplest of the three acids, undergoes intermolecular dehydration to produce diphosphoric acid.

Another intermolecular dehydration, involving diphosphoric acid and phosphoric acid, produces triphosphoric acid.

In the same manner that carboxylic acids are acidic (Section 16.8), phosphoric acid, diphosphoric acid, and triphosphoric acid are also acidic. The phosphoric acids are, however, polyprotic rather than monoprotic acids. The hydrogen atom in each of the —OH groups possesses acidic properties. All three phosphoric acids undergo esterification reactions with alcohols, producing species such as

We will encounter esters of these types in Chapter 23 when we consider the biochemical production of energy in the human body. Adenosine diphosphate (ADP) and adenosine triphosphate (ATP) are important examples of such compounds.

Diphosphoric acid and triphosphoric acid are phosphoric acid anhydrides as well as acids. Note the structural similarities between a carboxylic acid anhydride and diphosphoric acid.

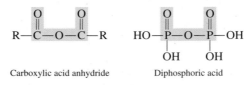

CHEMICAL Connections

Nitroglycerin: An Inorganic Triester

The reaction of one molecule of glycerol (a trihydroxy-alcohol) with three molecules of nitric acid produces the trinitrate ester called nitroglycerin.

$$
\begin{array}{l}
CH_2-OH \\
| \\
CH-OH \quad + \; 3HO-NO_2 \longrightarrow \\
| \\
CH_2-OH
\end{array}
\begin{array}{l}
CH_2-O-NO_2 \\
| \\
CH-O-NO_2 \quad + \; 3H_2O \\
| \\
CH_2-O-NO_2
\end{array}
$$

Besides being a component of dynamite explosives, nitroglycerin has medicinal value. It is used in treating patients with angina pectoris—sharp chest pains caused by an insufficient supply of oxygen reaching heart muscle. Its effect on the human body is that of a vasodilator, a substance that increases blood flow by relaxing constricted muscles around blood vessels.

Nitroglycerin medication is available in several forms: (1) as a liquid diluted with alcohol to render it nonexplosive, (2) as a liquid adsorbed to a tablet for convenience of sublingual (under the tongue) administration, (3) in ointments for topical use, and (4) as "skin patches" that release the drug continuously through the skin over a 24-hr period. Nitroglycerin is rapidly absorbed through the skin, enters the bloodstream, and finds its way to heart muscle within seconds.

In the pure state, nitroglycerin is a shock-sensitive liquid that can decompose to produce large volumes of gases (N_2, CO_2, H_2O, and O_2). When used in dynamite, it is adsorbed on clay-like materials, giving products that will not explode without a formal ignition system.

Another compound used for the same medicinal purposes as nitroglycerin is isopentyl nitrite. It is a monoester involving nitrous acid (HNO_2) and isopentyl alcohol (3-methyl-1-butanol).

$$
\begin{array}{l}
CH_3-CH-CH_2-CH_2-OH \; + \; HO-NO \longrightarrow \\
\qquad\quad | \\
\qquad\quad CH_3
\end{array}
$$

$$
\begin{array}{l}
CH_3-CH-CH_2-CH_2-O-NO \; + \; H_2O \\
\qquad\quad | \\
\qquad\quad CH_3
\end{array}
$$

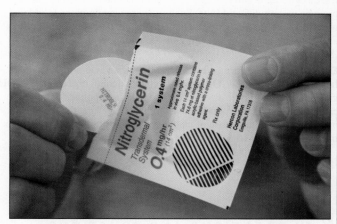

Nitroglycerin is used in treating chest pains related to angina pectoris.

Phosphoric acid anhydride systems play important roles in cellular processes through which biochemical energy is produced. The presence of phosphoric anhydride systems in biological settings contrasts markedly with carboxylic acid anhydride systems (Section 16.19), which are not found in biological settings because of their reactivity with water.

CONCEPTS TO REMEMBER

The carboxyl group. The functional group present in carboxylic acids is the carboxyl group. A carboxyl group is composed of a hydroxyl group bonded to a carbonyl carbon atom. It thus contains two oxygen atoms directly bonded to the same carbon atom (Section 16.1).

Carboxylic acid derivatives. Four important families of carboxylic acid derivatives are esters, acid chlorides, acid anhydrides, and amides. The group attached to the carbonyl carbon atom distinguishes these derivatives from each other and also from carboxylic acids (Section 16.1).

Nomenclature of carboxylic acids. The IUPAC name for a monocarboxylic acid is formed by replacing the final -e of the

hydrocarbon parent name with -oic acid. As with previous IUPAC nomenclature, the longest carbon chain containing the functional group is identified, and it is numbered starting with the carboxyl carbon atom. Common-name usage is more prevalent for carboxylic acids than for any other type of organic compound (Sections 16.2 and 16.3).

Types of carboxylic acids. Carboxylic acids are classified by the number of carboxyl groups present (monocarboxylic, dicarboxylic, etc.), by the degree of saturation (saturated, unsaturated, aromatic), and by additional functional groups present (hydroxy, keto, etc.) (Sections 16.4 and 16.5).

Physical properties of carboxylic acids. Low-molecular-mass carboxylic acids are liquids at room temperature and have sharp or unpleasant odors. Long-chain acids are waxlike solids. The carboxyl group is polar and forms hydrogen bonds to other carboxyl groups or other molecules. Thus carboxylic acids have relatively high boiling points, and those with lower molecular masses are soluble in water (Section 16.6).

Preparation of carboxylic acids. Carboxylic acids are synthesized through oxidation of primary alcohols or aldehydes using strong oxidizing agents. Aromatic carboxylic acids can be prepared by oxidizing a carbon side chain on a benzene derivative using a strong oxidizing agent (Section 16.7).

Acidity of carboxylic acids. Soluble carboxylic acids behave as weak acids, donating protons to water molecules. The portion of the acid molecule left after proton loss is called a carboxylate ion (Section 16.8).

Carboxylic acid salts. Carboxylic acids are neutralized by bases to produce carboxylic acid salts. Such salts are usually more soluble in water than are the acids from which they were derived. Carboxylic acid salts are named by changing the *-ic* ending of the acid to *-ate* (Section 16.9).

Esters. Esters are formed by the reaction of an acid with an alcohol. In such reactions, the —OR group from the alcohol replaces the —OH group in the carboxylic acid. Esters are polar compounds, but they cannot form hydrogen bonds to each other. Therefore, their boiling points are lower than those of alcohols and acids of similar molecular mass (Sections 16.10 and 16.11).

Nomenclature of esters. An ester is named as an alkyl (from the name of the alcohol reactant) carboxylate (from the name of the acid reactant) (Section 16.12).

Chemical reactions of esters. Esters can be converted back to carboxylic acids and alcohols under either acidic or basic conditions. Under acidic conditions, the process is called hydrolysis, and the products are the acid and alcohol. Under basic conditions, the process is called saponification, and the products are the acid salt and alcohol (Section 16.16).

Thioesters. Thioesters are sulfur-containing analogs of esters in which a —SR group has replaced the —OR group (Section 16.17).

Polyesters. Polyesters are polymers in which the monomers (diacids and dialcohols) are joined through ester linkages (Section 16.18).

Acid chlorides and acid anhydrides. An acid chloride is a carboxylic acid derivative in which the —OH portion of the carboxyl group has been replaced with a —Cl atom. An acid anhydride involves two carboxylic acid molecules bonded together after intermolecular dehydration has occurred. Both acid chlorides and acid anhydrides are very reactive molecules (Section 16.19).

Esters and anhydrides of inorganic acids. Alcohols can react with inorganic acids, such as nitric, sulfuric, and phosphoric acids, to form esters. Phosphate esters are an important class of biochemical compounds. Anhydrides of phosphoric acid (diphosphoric acid and triphosphoric acid) and their esters are also important types of biochemical molecules (Section 16.20).

KEY REACTIONS AND EQUATIONS

1. Oxidation of a primary alcohol to a carboxylic acid (Section 16.7)

$$R-CH_2-OH \xrightarrow{[O]} R-\overset{\overset{\displaystyle O}{\|}}{C}-H \xrightarrow{[O]} R-\overset{\overset{\displaystyle O}{\|}}{C}-OH$$

2. Oxidation of an alkylbenzene to a carboxylic acid (Section 16.7)

3. Ionization of a carboxylic acid to give a carboxylate ion and a hydronium ion (Section 16.8)

$$R-\overset{\overset{\displaystyle O}{\|}}{C}-OH + H_2O \rightleftharpoons R-\overset{\overset{\displaystyle O}{\|}}{C}-O^- + H_3O^+$$

4. Reaction of a carboxylic acid with a base to produce a carboxylic acid salt plus water (Section 16.9)

$$R-\overset{\overset{\displaystyle O}{\|}}{C}-OH + NaOH \longrightarrow R-\overset{\overset{\displaystyle O}{\|}}{C}-O^-Na^+ + H_2O$$

5. Preparation of an ester from a carboxylic acid and an alcohol (Section 16.11)

$$R-\overset{\overset{\displaystyle O}{\|}}{C}-OH + R'-OH \underset{}{\overset{H^+}{\rightleftharpoons}} R-\overset{\overset{\displaystyle O}{\|}}{C}-O-R' + H_2O$$

6. Ester hydrolysis to produce a carboxylic acid and an alcohol (Section 16.16)

$$R-\overset{\overset{\displaystyle O}{\|}}{C}-O-R' + H-OH \underset{}{\overset{H^+}{\rightleftharpoons}} R-\overset{\overset{\displaystyle O}{\|}}{C}-OH + R'-OH$$

7. Ester saponification to give a carboxylic acid salt and alcohol (Section 16.16)

$$R-\overset{\overset{\displaystyle O}{\|}}{C}-O-R' + NaOH \xrightarrow{H_2O} R-\overset{\overset{\displaystyle O}{\|}}{C}-O^-Na^+ + R'-OH$$

8. Preparation of a thioester (Section 16.17)

$$R-\overset{\overset{\displaystyle O}{\|}}{C}-OH + R'-SH \xrightarrow{H^+} R-\overset{\overset{\displaystyle O}{\|}}{C}-S-R' + H_2O$$

9. Phosphate ester formation (Section 16.20)

$$R-OH + HO-\overset{\overset{\displaystyle O}{\|}}{\underset{\underset{\displaystyle OH}{|}}{P}}-OH \longrightarrow R-O-\overset{\overset{\displaystyle O}{\|}}{\underset{\underset{\displaystyle OH}{|}}{P}}-OH + H_2O$$

EXERCISES *and* PROBLEMS

The members of each pair of problems in this section test similar material.

● **Structure of Carboxylic Acids and Their Derivatives (Section 16.1)**

16.1 In which of the following compounds is a carboxyl group present?

a. $CH_3-CH_2-\overset{\overset{\displaystyle O}{\|}}{C}-OH$

b. $CH_3-CH_2-CH_2-\overset{\overset{\displaystyle O}{\|}}{C}-O-CH_3$

c. $CH_3-\overset{\overset{\displaystyle OH}{|}}{CH}-\overset{\overset{\displaystyle O}{\|}}{C}-CH_3$

d. CH_3-CH_2-COOH

16.2 In which of the following compounds is a carboxyl group present?

a. $CH_3-CH_2-\overset{\overset{\displaystyle O}{\|}}{C}-O-CH_3$

b. $CH_3-CH_2-\overset{\overset{\displaystyle O}{\|}}{C}-OH$

c.   (benzaldehyde) $\overset{\overset{\displaystyle O}{\|}}{C}-H$ attached to benzene ring

d. $CH_2-CH-CH_2-CH_2-CO_2H$ with CH_3 on the CH

16.3 Classify each of the structures in Problem 16.1 as a carboxylic acid, a carboxylic acid derivative, or neither.

16.4 Classify each of the structures in Problem 16.2 as a carboxylic acid, a carboxylic acid derivative, or neither.

● **IUPAC Nomenclature for Carboxylic Acids (Section 16.2)**

16.5 Give the IUPAC name for each of the following carboxylic acids.

a. $CH_3-CH_2-CH_2-\overset{\overset{\displaystyle O}{\|}}{C}-OH$

b. $CH_3-\overset{\overset{\displaystyle Br}{|}}{CH}-CH_2-CH_2-\overset{\overset{\displaystyle O}{\|}}{C}-OH$

c. $CH_3-\underset{\underset{\displaystyle CH_2-CH_3}{|}}{CH}-CH_2-COOH$

d. $Cl-CH_2-\overset{\overset{\displaystyle O}{\|}}{C}-OH$

16.6 Give the IUPAC name for each of the following carboxylic acids.

a. $CH_3-CH_2-CH_2-CH_2-CH_2-\overset{\overset{\displaystyle O}{\|}}{C}-OH$

b. $\underset{\underset{\displaystyle CH_3}{|}}{CH_2}-\underset{\underset{\displaystyle CH_3}{|}}{CH}-\overset{\overset{\displaystyle O}{\|}}{C}-OH$

c. $CH_3-\overset{\overset{\displaystyle CH_3}{|}}{\underset{\underset{\displaystyle Cl}{|}}{C}}-\overset{\overset{\displaystyle O}{\|}}{C}-OH$

d. $HOOC-CH_3$

16.7 Assign an IUPAC name to each of the following carboxylic acids.

a. [skeletal structure, carboxylic acid]   b. [skeletal structure, carboxylic acid]

c. [skeletal structure, carboxylic acid]   d. [skeletal structure, carboxylic acid]

16.8 Assign an IUPAC name to each of the following carboxylic acids.

a. [skeletal structure]   b. [skeletal structure]

c. [skeletal structure]   d. [skeletal structure]

16.9 Draw a condensed structural formula that corresponds to each of the following carboxylic acids.
a. 2-Ethylbutanoic acid   b. 2,5-Dimethylhexanoic acid
c. Methylpropanoic acid   d. Dichloroethanoic acid

16.10 Draw a condensed structural formula that corresponds to each of the following carboxylic acids.
a. 3,3-Dimethylheptanoic acid   b. 4-Methylpentanoic acid
c. 3-Chloropropanoic acid   d. Trichloroethanoic acid

16.11 Give the IUPAC name for each of the following carboxylic acids.

a. $HO-\overset{\overset{\displaystyle O}{\|}}{C}-CH_2-\overset{\overset{\displaystyle O}{\|}}{C}-OH$

b. $HO-\overset{\overset{\displaystyle O}{\|}}{C}-CH_2-\overset{\overset{\displaystyle CH_3}{|}}{CH}-CH_2-\overset{\overset{\displaystyle O}{\|}}{C}-OH$

c. [benzene ring with COOH and Cl]   d. [benzene ring with COOH, Cl, Br]

16.12 Give the IUPAC name for each of the following carboxylic acids.

a. $HO-\underset{\underset{\displaystyle O}{\|}}{\overset{\overset{\displaystyle O}{\|}}{C}}-OH$

b. $HO-\overset{\overset{\displaystyle O}{\|}}{C}-\overset{\overset{\displaystyle Cl}{|}}{CH}-CH_2-CH_2-\overset{\overset{\displaystyle O}{\|}}{C}-OH$

c. [benzene ring with COOH and CH_2-CH_3]   d. [benzene ring with COOH and CH_3]

16.13 Draw a condensed structural formula that corresponds to each of the following carboxylic acids.
a. 2,2-Dimethylbutanoic acid
b. 2,2-Dimethylbutanedioic acid
c. 2,2-Dimethylpentanedioic acid
d. 2,4-Dichlorobenzoic acid

16.14 Draw a condensed structural formula that corresponds to each of the following carboxylic acids.
a. 2,3-Dichlorohexanoic acid
b. 2,3-Dichlorohexanedioic acid
c. 2,3-Dichloroheptanedioic acid
d. 3,5-Dichlorobenzoic acid

● **Common Names for Carboxylic Acids (Section 16.3)**

16.15 Draw a condensed structural formula that corresponds to each of the following carboxylic acids.
a. Valeric acid   b. Acetic acid
c. α-Chlorobutyric acid   d. β-Bromocaproic acid

16.16 Draw a condensed structural formula that corresponds to each of the following carboxylic acids.
a. Butyric acid   b. Caproic acid
c. α-Methylpropionic acid   d. β-Chloro-β-iodocaproic acid

16.17 Draw a condensed structural formula that corresponds to each of the following carboxylic acids.
a. Malonic acid   b. Succinic acid
c. γ-Bromopimelic acid   d. α-Methylglutaric acid

16.18 Draw a condensed structural formula that corresponds to each of the following carboxylic acids.
a. Glutaric acid   b. Pimelic acid
c. α-Methyladipic acid   d. α,β-Dichlorosuccinic acid

16.19 Classify the two carboxylic acids in each of the following pairs as (1) both dicarboxylic acids, (2) both monocarboxylic acids, or (3) one dicarboxylic and one monocarboxylic acid.
a. Glutaric acid and valeric acid
b. Adipic acid and oxalic acid
c. Caproic acid and formic acid
d. Succinic acid and malonic acid

16.20 Classify the two carboxylic acids in each of the following pairs as (1) both dicarboxylic acids, (2) both monocarboxylic acids, or (3) one dicarboxylic and one monocarboxylic acid.
a. Formic acid and acetic acid
b. Butyric acid and succinic acid
c. Pimelic acid and caproic acid
d. Malonic acid and adipic acid

● **Polyfunctional Carboxylic Acids (Section 16.4)**

16.21 Each of the following acids contains an additional type of functional group besides the carboxyl group. For each acid, specify the noncarboxyl functional group present.
a. Acrylic acid   b. Lactic acid
c. Maleic acid   d. Glycolic acid

16.22 Each of the following acids contains an additional type of functional group besides the carboxyl group. For each acid, specify the noncarboxyl functional group present.
a. Fumaric acid   b. Pyruvic acid
c. Malic acid   d. Tartaric acid

16.23 Give the IUPAC name for each of the acids in Problem 16.21.

16.24 Give the IUPAC name for each of the acids in Problem 16.22.

16.25 Draw a structural formula for each of the following acids.
a. 3-Oxopentanoic acid   b. 2-Hydroxybutanoic acid
c. *trans*-4-Hexenoic acid   d. α,β-Dihydroxyglutaric acid

16.26 Draw a structural formula for each of the following acids.
a. 3-Hydroxypentanoic acid   b. α,γ-Dihydroxyvaleric acid
c. 2-Oxobutanoic acid   d. *cis*-3-Heptenoic acid

● **Metabolic Carboxylic Acids (Section 16.5)**

16.27 Classify each of the following polyfunctional acids as a derivative of (1) propionic acid, (2) succinic acid, or (3) glutaric acid.
a. Lactic acid   b. Glyceric acid
c. Oxaloacetic acid   d. Citric acid

16.28 Classify each of the following polyfunctional acids as a derivative of (1) propionic acid, (2) succinic acid, or (3) glutaric acid.
a. Pyruvic acid   b. Malic acid
c. Fumaric acid   d. α-Ketoglutaric acid

16.29 For each of the acids in Problem 16.27, list the functional groups that are present.

16.30 For each of the acids in Problem 16.28, list the functional groups that are present.

● **Physical Properties of Carboxylic Acids (Section 16.6)**

16.31 Determine the maximum number of hydrogen bonds that can form between an acetic acid molecule and
a. another acetic acid molecule
b. water molecules

16.32 Determine the maximum number of hydrogen bonds that can form between a butanoic acid molecule and
a. another butanoic acid molecule
b. water molecules

16.33 What is the physical state (solid, liquid, or gas) of each of the following carboxylic acids at room temperature?
a. Oxalic acid   b. Decanoic acid
c. Hexanoic acid   d. Benzoic acid

16.34 What is the physical state (solid, liquid, or gas) of each of the following carboxylic acids at room temperature?
a. Succinic acid   b. Octanoic acid
c. Pentanoic acid   d. *p*-Chlorobenzoic acid

● **Preparation of Carboxylic Acids (Section 16.7)**

16.35 Draw a structural formula for the carboxylic acid expected to be formed when each of the following substances is oxidized using a strong oxidizing agent.
a. CH₃—CH₂—OH
b.
$$CH_3-\overset{\overset{\displaystyle O}{\|}}{C}-H$$
c.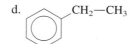
d.

16.36 Draw a structural formula for the carboxylic acid expected to be formed when each of the following substances is oxidized using a strong oxidizing agent.
a. 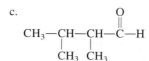   b. CH₃—CH₂—CH₂—OH
c. 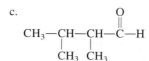   d.

Acidity of Carboxylic Acids (Section 16.8)

16.37 How many *acidic* hydrogen atoms are present in each of the following carboxylic acids?
   a. Pentanoic acid   b. Citric acid
   c. Succinic acid   d. Oxalic acid

16.38 How many *acidic* hydrogen atoms are present in each of the following carboxylic acids?
   a. Acetic acid   b. Benzoic acid
   c. Propanoic acid   d. Glutaric acid

16.39 What is the charge on the carboxylate ion formed when each of the acids in Problem 16.37 ionizes in water?

16.40 What is the charge on the carboxylate ion formed when each of the acids in Problem 16.38 ionizes in water?

16.41 What is the name of the carboxylate ion that forms when each of the acids in Problem 16.37 ionizes in water? (Use an IUPAC carboxylate name if the acid name is IUPAC; use a common name if the acid name is common.)

16.42 What is the name of the carboxylate ion that forms when each of the acids in Problem 16.38 ionizes in water? (Use an IUPAC carboxylate name if the acid name is IUPAC; use a common name if the acid name is common.)

16.43 Write a chemical equation for the formation of each of the following carboxylate ions, in aqueous solution, from its parent acid.
   a. Acetate   b. Citrate
   c. Ethanoate   d. 2-Methylbutanoate

16.44 Write a chemical equation for the formation of each of the following carboxylate ions, in aqueous solution, from its parent acid.
   a. Butanoate   b. Succinate
   c. Benzoate   d. α-Methylbutyrate

Carboxylic Acid Salts (Section 16.9)

16.45 Give the IUPAC name for each of the following carboxylic acid salts.

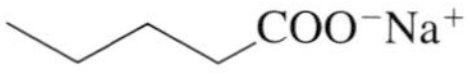

16.46 Give the IUPAC name for each of the following carboxylic acid salts.

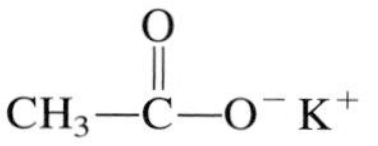

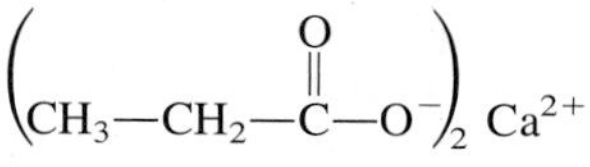

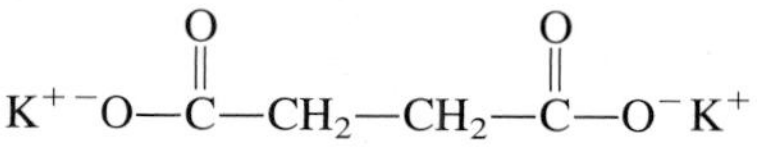

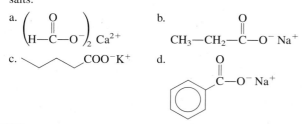

16.47 Write a chemical equation for the preparation of each of the salts in Problem 16.45 using an acid–base neutralization reaction.

16.48 Write a chemical equation for the preparation of each of the salts in Problem 16.46 using an acid–base neutralization reaction.

16.49 Write a chemical equation for the conversion of each of the following carboxylic acid salts to its parent carboxylic acid. Let hydrochloric acid (HCl) be the source of the needed hydronium ions.
   a. Sodium butanoate   b. Potassium oxalate
   c. Calcium malonate   d. Sodium benzoate

16.50 Write a chemical equation for the conversion of each of the following carboxylic acid salts to its parent carboxylic acid. Let hydrochloric acid (HCl) be the source of the needed hydronium ions.
   a. Calcium propanoate   b. Sodium lactate
   c. Magnesium succinate   d. Potassium benzoate

16.51 What is an antimicrobial?

16.52 Which three carboxylic acids have salts that are used extensively as food preservatives?

16.53 Which carboxylic acid has salts that are used to inhibit yeast and mold growth in the following?
   a. Jams and jellies   b. Dairy products
   c. Dried fruit   d. Cheese spreads

16.54 Which carboxylic acid has salts that are used to inhibit yeast and mold growth in the following?
   a. Ketchup and syrup   b. Sauerkraut
   c. Pie filling   d. Baked products

Structure of Esters (Section 16.10)

16.55 Characterize an ester functional group in terms of the following.
   a. The number of oxygen atoms that are present
   b. The minimum number of carbon atoms that must be present

16.56 Give two linear notations for designating the ester functional group.

16.57 Which of the following structures represent esters?

a. 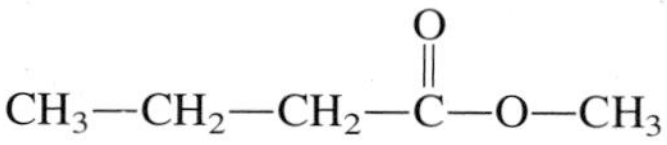
$$CH_3-CH_2-CH_2-\overset{O}{\overset{\|}{C}}-O-CH_3$$

b.
$$CH_3-O-\overset{O}{\overset{\|}{C}}-CH_3$$

c.
$$CH_3-O-CH_2-\overset{O}{\overset{\|}{C}}-CH_3$$

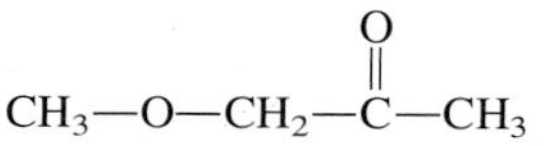

d.

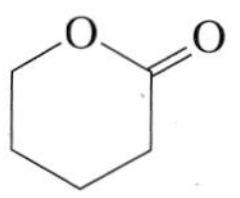

16.58 Which of the following structures represent esters?

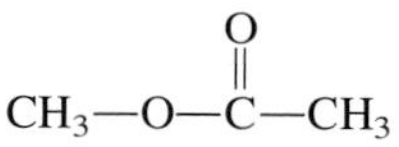

a. 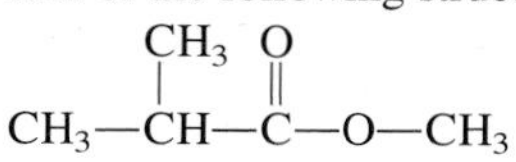
$$CH_3-\overset{CH_3}{\overset{|}{CH}}-\overset{O}{\overset{\|}{C}}-O-CH_3$$

b. 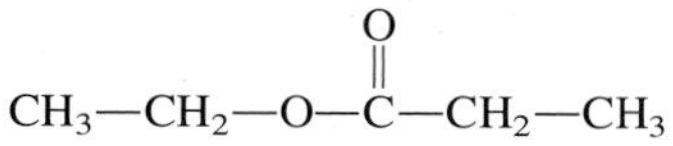
$$CH_3-CH_2-O-\overset{O}{\overset{\|}{C}}-CH_2-CH_3$$

c.

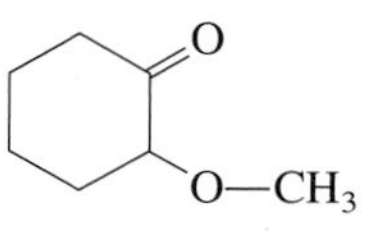

d.

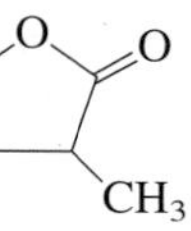

Preparation of Esters (Section 16.11)

16.59 Draw the structure of the ester produced when each of the following pairs of carboxylic acid and alcohol react.
   a. Propanoic acid and methanol
   b. Acetic acid and 1-propanol
   c. 2-Methylbutanoic acid and 2-propanol
   d. Valeric acid and *sec*-butyl alcohol

16.60 Draw the structure of the ester produced when each of the following pairs of carboxylic acid and alcohol react.
a. Methanoic acid and 1-propanol
b. Propanoic acid and ethanol
c. 2-Methylpropanoic acid and 2-butanol
d. Valeric acid and isobutyl alcohol

16.61 For each of the following esters, draw the structural formula of the "parent" acid and the "parent" alcohol.

a.

$$CH_3-CH_2-\overset{\overset{\displaystyle O}{\|}}{C}-O-CH_2-CH_3$$

b.

$$CH_3-O-\overset{\overset{\displaystyle O}{\|}}{C}-CH_2-CH_2-CH_3$$

c.

$$CH_3-\overset{\overset{\displaystyle O}{\|}}{C}-O-\bigcirc$$

d.

$$\bigcirc-\overset{\overset{\displaystyle O}{\|}}{C}-O-CH_3$$

16.62 For each of the following esters, draw the structural formula of the "parent" acid and the "parent" alcohol.

a.

$$CH_3-\overset{\overset{\displaystyle O}{\|}}{C}-O-CH_2-CH_3$$

b.

$$CH_3-O-\overset{\overset{\displaystyle O}{\|}}{C}-CH_2-CH_3$$

c.

$$\bigcirc-\overset{\overset{\displaystyle O}{\|}}{C}-O-CH_2-CH_3$$

d.

$$CH_3-\overset{\overset{\displaystyle CH_3}{|}}{CH}-CH_2-\overset{\overset{\displaystyle O}{\|}}{C}-O-\bigcirc$$

16.63 What are lactones and by what chemical reaction are they produced?

16.64 What is the difference between a γ-lactone and a δ-lactone?

● **Nomenclature for Esters (Section 16.12)**

16.65 Assign an IUPAC name to each of the following esters.

a.

$$CH_3-CH_2-\overset{\overset{\displaystyle O}{\|}}{C}-O-CH_3$$

b.

$$H-\overset{\overset{\displaystyle O}{\|}}{C}-O-CH_3$$

c.

$$CH_3-CH_2-CH_2-O-\overset{\overset{\displaystyle O}{\|}}{C}-CH_3$$

d.

$$CH_3-CH_2-\overset{\overset{\displaystyle O}{\|}}{C}-O-\overset{\overset{\displaystyle CH_3}{|}}{CH}-CH_3$$

16.66 Assign an IUPAC name to each of the following esters.

a.

$$CH_3-CH_2-CH_2-\overset{\overset{\displaystyle O}{\|}}{C}-O-CH_3$$

b.

$$CH_3-CH_2-\overset{\overset{\displaystyle O}{\|}}{C}-O-CH_2-CH_2-CH_3$$

c.

$$CH_3-CH_2-O-\overset{\overset{\displaystyle O}{\|}}{C}-H$$

d.

$$CH_3-\overset{\overset{\displaystyle CH_3}{|}}{CH}-\overset{\overset{\displaystyle CH_3}{|}}{CH}-\overset{\overset{\displaystyle O}{\|}}{C}-O-CH_2-CH_3$$

16.67 Assign a common name to each of the esters in Problem 16.65.

16.68 Assign a common name to each of the esters in Problem 16.66.

16.69 Assign an IUPAC name to each of the following esters.

a.

b.

c.

d.

16.70 Assign an IUPAC name to each of the following esters.

a.

b.

c.

d.

16.71 Draw a structural formula for each of the following esters.
a. Methyl formate   b. Ethyl phenylacetate
c. Isopropyl acetate   d. 2-Bromopropyl ethanoate

16.72 Draw a structural formula for each of the following esters.
a. Ethyl butyrate   b. Butyl ethanoate
c. 2-Methylpropyl formate   d. Ethyl α-methylpropionate

16.73 Assign IUPAC names to the esters that are produced from the reaction of the following carboxylic acids and alcohols.
a. Acetic acid and ethanol
b. Ethanoic acid and methanol
c. Butyric acid and ethyl alcohol
d. 2-Butanol and caproic acid

16.74 Assign IUPAC names to the esters that are produced from the reaction of the following carboxylic acids and alcohols.
a. Ethanoic acid and propyl alcohol
b. Acetic acid and 1-pentanol
c. Acetic acid and 2-pentanol
d. Ethanol and benzoic acid

● **Selected Common Esters (Section 16.13)**

16.75 Structurally contrast the ester flavoring agents that generate apple and pineapple flavors in terms of their "acid part" and "alcohol part."

16.76 Structurally contrast the ester flavoring agents that generate pear and banana flavors in terms of their "acid part" and "alcohol part."

16.77 How does the structure of aspirin differ from that of salicylic acid?

16.78 How does the structure of oil of wintergreen differ from that of salicylic acid?

● **Isomerism for Carboxylic Acids and Esters (Section 16.14)**

16.79 Give IUPAC names for the four isomeric C_5 monocarboxylic acids with saturated carbon chains.

16.80 Give IUPAC names for the eight isomeric C_6 monocarboxylic acids with saturated carbon chains.

16.81 Give IUPAC names for the four isomeric methyl esters that contain six carbon atoms and saturated carbon chains.

16.82 Give IUPAC names for the two isomeric ethyl esters that contain six carbon atoms and saturated carbon chains.

16.83 How many esters exist that are isomeric with 2-methylbutanoic acid?

16.84 How many esters exist that are isomeric with 2-methylpropanoic acid?

16.85 Draw condensed structural formulas for all carboxylic acids and all esters that have the molecular formula $C_3H_6O_2$.

16.86 Draw condensed structural formulas for all carboxylic acids and all esters that have the molecular formula $C_4H_8O_2$.

● **Physical Properties of Esters (Section 16.15)**

16.87 Explain why ester molecules cannot form hydrogen bonds to each other.

16.88 How many hydrogen bonds can form between a methyl acetate molecule and two water molecules?

16.89 Explain why esters have lower boiling points than carboxylic acids of comparable molecular mass.

16.90 Explain why esters are less soluble in water than carboxylic acids of comparable molecular mass.

● **Chemical Reactions of Esters (Section 16.16)**

16.91 Write the structural formulas of the reaction products when each of the following esters is hydrolyzed under acidic conditions.

a.
$$CH_3—CH_2—\overset{\overset{\displaystyle O}{\|}}{C}—O—CH_2—CH_3$$

b.
$$CH_3—\overset{\overset{\displaystyle CH_3}{|}}{CH}—\overset{\overset{\displaystyle O}{\|}}{C}—O—\bigcirc$$

c. Methyl butanoate   d. Isopropyl benzoate

16.92 Write the structural formulas of the reaction products when each of the following esters is hydrolyzed under acidic conditions.

a.
$$CH_3—\overset{\overset{\displaystyle CH_3}{|}}{CH}—\overset{\overset{\displaystyle O}{\|}}{C}—O—\overset{\overset{\displaystyle CH_3}{|}}{CH}—CH_3$$

b.
$$\bigcirc—\overset{\overset{\displaystyle O}{\|}}{C}—O—\bigcirc$$

c. Ethyl valerate   d. Pentyl benzoate

16.93 Write the structural formulas of the reaction products when each of the esters in Problem 16.91 is saponified using sodium hydroxide.

16.94 Write the structural formulas of the reaction products when each of the esters in Problem 16.92 is saponified using sodium hydroxide.

16.95 Draw structures of the reaction products in the following chemical reactions.

a.
$$CH_3—\overset{\overset{\displaystyle CH_3}{|}}{CH}—\overset{\overset{\displaystyle O}{\|}}{C}—O—CH_2—CH_3 + H_2O \xrightarrow{H^+}$$

b.
$$CH_3—\overset{\overset{\displaystyle CH_3}{|}}{CH}—\overset{\overset{\displaystyle O}{\|}}{C}—O—CH_2—CH_3 + NaOH \xrightarrow{H_2O}$$

c.
$$H—\overset{\overset{\displaystyle O}{\|}}{C}—O—CH_2—CH_2—CH_2—CH_3 + H_2O \xrightarrow{H^+}$$

d.
$$CH_3—\overset{\overset{\displaystyle O}{\|}}{C}—O—CH_2—\overset{\overset{\displaystyle CH_3}{|}}{CH}—\overset{\overset{\displaystyle CH_3}{|}}{CH}—CH_3 + NaOH \xrightarrow{H_2O}$$

16.96 Draw structures of the reaction products in the following chemical reactions.

a.
$$CH_3—\overset{\overset{\displaystyle CH_3}{|}}{CH}—CH_2—\overset{\overset{\displaystyle O}{\|}}{C}—O—CH_3 + H_2O \xrightarrow{H^+}$$

b.
$$CH_3—\overset{\overset{\displaystyle CH_3}{|}}{CH}—CH_2—\overset{\overset{\displaystyle O}{\|}}{C}—O—CH_3 + NaOH \xrightarrow{H_2O}$$

c.
$$CH_3—CH_2—\overset{\overset{\displaystyle O}{\|}}{C}—O—(CH_2)_5—CH_3 + H_2O \xrightarrow{H^+}$$

d.
$$CH_3—(CH_2)_5—\overset{\overset{\displaystyle O}{\|}}{C}—O—CH_2—CH_3 + NaOH \xrightarrow{H_2O}$$

● **Sulfur Analogs of Esters (Section 16.17)**

16.97 Draw the structures of the thioesters formed as a result of each of the following reactions between carboxylic acids and thiols.

a.
$$CH_3—\overset{\overset{\displaystyle O}{\|}}{C}—OH + CH_3—CH_2—SH \rightarrow$$

b.
$$CH_3—(CH_2)_8—\overset{\overset{\displaystyle O}{\|}}{C}—OH + CH_3—SH \rightarrow$$

c.
$$\bigcirc—COOH + CH_3—\overset{\overset{\displaystyle CH_3}{|}}{CH}—SH \rightarrow$$

d.
$$H—\overset{\overset{\displaystyle O}{\|}}{C}—OH + CH_3—CH_2—CH_2—SH \rightarrow$$

16.98 Draw the structures of the thioesters formed as a result of each of the following reactions between carboxylic acids and thiols.

a.
$$CH_3—CH_2—\overset{\overset{\displaystyle O}{\|}}{C}—OH + CH_3—CH_2—SH \rightarrow$$

b. $CH_3—CH_2—CH_2—COOH + CH_3—SH \rightarrow$

c.
$$CH_3—\overset{\overset{\displaystyle O}{\|}}{C}—OH + CH_3—CH_2—\overset{\overset{\displaystyle }{|}}{CH}—SH \rightarrow$$
$$\underset{\displaystyle CH_3}{}$$

d.
$$\bigcirc—COOH + \bigcirc—SH \rightarrow$$

16.99 Draw the structural formula for each of the following thioesters.
  a. Methyl thioethanoate   b. Methyl thioacetate
  c. Ethyl thioformate   d. Ethyl thiomethanoate

16.100 Draw the structural formula for each of the following thioesters.
  a. Methyl thiopropanoate
  b. Methyl thiopropionate
  c. Ethyl thioacetate
  d. Ethyl thioethanoate

16.101 Formation of the thioester acetyl coenzyme A can be envisioned as a thioesterification reaction involving which two substances?

16.102 What is the generalized abbreviated structural formula for the thioester acetyl CoA?

● **Polyesters (Section 16.18)**

16.103 Write the structure (two repeating units) of the polyester polymer formed from oxalic acid and 1,3-propanediol.

16.104 Write the structure (two repeating units) of the polyester polymer formed from malonic acid and ethylene glycol.

16.105 Draw the structural formulas of the monomers needed to form the following polyester.

$$\left(O-(CH_2)_3-O-\overset{\overset{\displaystyle O}{\|}}{C}-(CH_2)_2-\overset{\overset{\displaystyle O}{\|}}{C}-O \right)_n$$

16.106 Draw the structural formulas of the monomers needed to form the following polyester.

$$\left(O-\overset{\overset{\displaystyle O}{\|}}{C}-(CH_2)_3-\overset{\overset{\displaystyle O}{\|}}{C}-O-(CH_2)_2-O \right)_n$$

16.107 What are the names of the monomers used to produce the polyester whose acronym is PET?

16.108 What are the names of the monomers used to produce the polyester whose acronym is PEN?

16.109 What are some uses for the biodegradable polyester PHBV?

16.110 What is the major use for the biodegradable polyester known as *Lactomer*?

● **Acid Chlorides and Acid Anhydrides (Section 16.19)**

16.111 Draw the condensed structural formula for each of the following compounds.
  a. Propionyl chloride
  b. 3-Methylbutanoyl chloride
  c. Butyric anhydride
  d. Butanoic ethanoic anhydride

16.112 Draw the condensed structural formula for each of the following compounds.
  a. Acetyl chloride
  b. 2-Methylbutanoyl chloride
  c. Propionic anhydride
  d. Ethanoic methanoic anhydride

16.113 Assign an IUPAC name to each of the following compounds.
  a.

$$CH_3-\overset{\overset{\displaystyle O}{\|}}{C}-O-\overset{\overset{\displaystyle O}{\|}}{C}-CH_2-CH_3$$

  b.

$$CH_3-CH_2-CH_2-CH_2-\overset{\overset{\displaystyle O}{\|}}{C}-Cl$$

  c.

$$CH_3-\underset{\underset{\displaystyle CH_3}{|}}{CH}-\underset{\underset{\displaystyle CH_3}{|}}{CH}-\overset{\overset{\displaystyle O}{\|}}{C}-Cl$$

  d.

$$CH_3-CH_2-\overset{\overset{\displaystyle O}{\|}}{C}-O-\overset{\overset{\displaystyle O}{\|}}{C}-H$$

16.114 Assign an IUPAC name to each of the following compounds.
  a.

$$CH_3-CH_2-\overset{\overset{\displaystyle O}{\|}}{C}-O-\overset{\overset{\displaystyle O}{\|}}{C}-CH_2-CH_3$$

  b.

$$CH_3-CH_2-\overset{\overset{\displaystyle O}{\|}}{C}-Cl$$

  c.

$$CH_3-\underset{\underset{\displaystyle CH_3}{|}}{\overset{\overset{\displaystyle CH_3}{|}}{C}}-CH_2-\overset{\overset{\displaystyle O}{\|}}{C}-Cl$$

  d.

$$CH_3-\overset{\overset{\displaystyle O}{\|}}{C}-O-\overset{\overset{\displaystyle O}{\|}}{C}-H$$

16.115 Draw a condensed structural formula for the organic product of the reaction of each of the following compounds with water.
  a. Pentanoyl chloride   b. Pentanoic anhydride

16.116 Draw a condensed structural formula for the organic product of the reaction of each of the following compounds with water.
  a. Butanoyl chloride
  b. Butanoic anhydride

16.117 Draw the condensed structural formulas for the ester formed and the carboxylic acid formed when acetic anhydride reacts with the following alcohols.
  a. Ethyl alcohol
  b. 1-Butanol

16.118 Draw the condensed structural formulas for the ester formed and the carboxylic acid formed when ethanoic anhydride reacts with the following alcohols.
  a. Propyl alcohol
  b. 2-Butanol

16.119 What is the IUPAC name for the acyl group present in each of the following hydrocarbon derivatives?
  a. Propanoic acid
  b. Butanoic anhydride
  c. Butanoyl chloride
  d. Acetic acid

16.120 What is the IUPAC name for the acyl group present in each of the following hydrocarbon derivatives?
  a. Butanoic acid   b. Propanoic anhydride
  c. Propanoyl chloride   d. Formic acid

16.121 In general terms, what are the products obtained in an acyl transfer reaction involving an acid chloride and an alcohol?

16.122 In general terms, what are the products obtained in an acyl transfer reaction involving an acid anhydride and an alcohol?

● **Esters and Anhydrides of Inorganic Acids (Section 16.20)**

16.123 Draw the structures of the esters formed by reacting the following substances.
  a. 1 molecule methanol and 1 molecule phosphoric acid
  b. 2 molecules methanol and 1 molecule phosphoric acid
  c. 1 molecule methanol and 1 molecule nitric acid
  d. 1 molecule ethylene glycol and 2 molecules nitric acid

16.124 Draw the structures of the esters formed by reacting the following substances.
  a. 1 molecule ethanol and 1 molecule phosphoric acid
  b. 2 molecules methanol and 1 molecule sulfuric acid
  c. 1 molecule ethylene glycol and 1 molecule nitric acid
  d. 1 molecule glycerol and 3 molecules nitric acid

16.125 Phosphoric acid can form triesters but sulfuric acid cannot. Explain why.

16.126 Sulfuric acid can form diesters but nitric acid cannot. Explain why.

ADDITIONAL PROBLEMS

16.127 With the help of Figure 16.7 and IUPAC naming rules, specify the number of carbon atoms present and the number of carboxyl groups present in each of the following carboxylic acids.
  a. Oxalic acid   b. Heptanoic acid
  c. *cis*-3-Heptenoic acid   d. Citric acid

16.128 Malonic, maleic, and malic acids are dicarboxylic acids with similar-sounding names. How do the structures of these acids differ from each other?

16.129 The general molecular formula for an alkane is C_nH_{2n+2}. What is the general molecular formula for an unsaturated unsubstituted monocarboxylic acid containing one carbon–carbon double bond?

16.130 Assign IUPAC names to the following compounds.

a.   b.

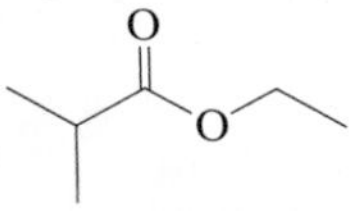

c. 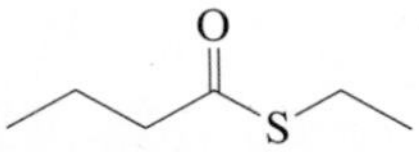   d.

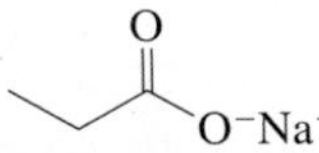

16.131 A sample of ethyl alcohol is divided into two portions. Portion A is added to an aqueous solution of a strong oxidizing agent and allowed to react. The organic product of this reaction is mixed with portion B of the ethyl alcohol. A trace of acid is added and the solution is heated. What is the structure of the final product of this reaction scheme?

16.132 For each of the following reactions, draw the structure(s) of the organic product(s).

a.
$$CH_3-CH_2-\overset{\overset{\displaystyle O}{\|}}{C}-O-CH_3 + NaOH \xrightarrow{H_2O}$$

b.
$$CH_3-CH_2-\overset{\overset{\displaystyle O}{\|}}{C}-OH + CH_3-SH \longrightarrow$$

c.
$$CH_3-\overset{\overset{\displaystyle O}{\|}}{C}-OH + NaOH \longrightarrow$$

d.

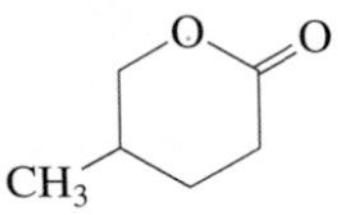

+ H_2O \xrightarrow{H^+}

Multiple-Choice Practice Test

16.133 Which of the following statements concerning the carboxylic acid functional group is *correct*?
  a. It is called a carboxylate group.
  b. It can be denoted using the notation —COOH.
  c. An oxygen–oxygen single bond is present.
  d. A carbon–hydrogen single bond is present.

16.134 What are the common names for the C_1 and C_2 monocarboxylic acids, respectively?
  a. Formic acid and acetic acid
  b. Acetic acid and formic acid
  c. Oxalic acid and acetic acid
  d. Acetic acid and oxalic acid

16.135 In which of the following pairs of carboxylic acids does the first member of the pair have more carbon atoms than the second member of the pair?
  a. Malonic acid and succinic acid
  b. Glutaric acid and succinic acid
  c. Oxalic acid and malonic acid
  d. Oxalic acid and glutaric acid

16.136 Which statement is true for the carboxyl carbon atom in the IUPAC nomenclature system for monocarboxylic acids?
  a. It is always assigned the number one.
  b. It is always assigned the highest number possible.
  c. It is always known as the alpha carbon atom.
  d. It is always known as the beta carbon atom.

16.137 Which of the following is a C_3 monohydroxy carboxylic acid?
  a. Tartaric acid   b. Lactic acid
  c. Citric acid   d. Pyruvic acid

16.138 An ester is a carboxylic acid derivative in which the —OH portion of the carboxyl group has been replaced with which of the following?
  a. —OR group   b. —OCl group
  c. —Cl atom   d. —O⁻Na⁺ group

16.139 Which of the following esters, upon hydrolysis, produces two-carbon alcohol as one of the products?
  a. Methyl methanoate   b. Propyl ethanoate
  c. Methyl propanoate   d. Ethyl methanoate

16.140 Which of the following is neither a reactant nor a produc an ester saponification reaction?
  a. A strong base   b. An alcohol
  c. A carboxylic acid   d. A carboxylic acid salt

16.141 A polyester is a condensation polymer in which the ing monomers are a dicarboxylic acid and which of following?
  a. Carboxylic acid anhydride   b. Carboxylic ac
  c. Monoalcohol   d. Dialcohol

16.142 What is the number of oxygen atoms present in a t phosphoric acid?
  a. One   b. Two   c. Three   d. Four

Amines and Amides

Chapter Outline

Parachutist with a parachute made of the polyamide nylon.

The four most abundant elements in living organisms are carbon, hydrogen, oxygen, and nitrogen. In previous chapters, we have discussed compounds containing the first three of these elements. Alkanes, alkenes, alkynes, and aromatic hydrocarbons are all carbon–hydrogen compounds. The carbon–hydrogen–oxygen compounds we have discussed include alcohols, phenols, ethers, aldehydes, ketones, carboxylic acids, and esters. We now extend our discussion to organic compounds that contain the element nitrogen.

Two types of organic nitrogen-containing compounds are the focus of this chapter: amines and amides. Amines are carbon–hydrogen–nitrogen compounds, and amides contain oxygen in addition to these elements. Amines and amides occur widely in living organisms. Many of these naturally occurring compounds are very active physiologically. In addition, numerous drugs used for the treatment of mental illness, hay fever, heart problems, and other physical disorders are amines or amides.

17.1 BONDING CHARACTERISTICS OF NITROGEN ATOMS IN ORGANIC COMPOUNDS

An understanding of the bonding characteristics of the nitrogen atom is a prerequisite to our study of amines and amides. Nitrogen is a member of Group VA of the periodic table; it has five valence electrons (Section 4.2) and will form three covalent bonds to complete its octet of electrons (Section 4.3). Thus, in organic chemistry, carbon forms four bonds (Section 12.2), nitrogen forms three bonds, and oxygen forms two bonds (Section 14.1).

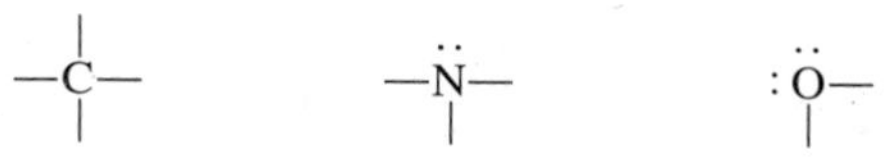

$$-\overset{|}{\underset{|}{C}}-$$

4 valence electrons
4 covalent bonds
no nonbonding
electron pairs

$$-\overset{..}{\underset{|}{N}}-$$

5 valence electrons
3 covalent bonds
1 nonbonding
electron pair

$$:\overset{..}{O}-$$

6 valence electrons
2 covalent bonds
2 nonbonding
electron pairs

17.2 STRUCTURE AND CLASSIFICATION OF AMINES

Amines bear the same relationship to ammonia that alcohols and ethers bear to water (Sections 14.2 and 14.15).

An **amine** *is an organic derivative of ammonia* (NH_3) *in which one or more alkyl, cycloalkyl, or aryl groups are attached to the nitrogen atom.* Amines are classified as primary (1°), secondary (2°), or tertiary (3°) on the basis of how many hydrocarbon groups are bonded to the ammonia nitrogen atom (see Figure 17.1). A **primary amine** *is an amine in which the nitrogen atom is bonded to one hydrocarbon group and two hydrogen atoms.* The generalized formula for a primary amine is RNH_2. A **secondary amine** *is an amine in which the nitrogen atom is bonded to two hydrocarbon groups and one hydrogen atom.* The generalized formula for a secondary amine is R_2NH. A **tertiary amine** *is an amine in which the nitrogen atom is bonded to three hydrocarbon groups and no hydrogen atoms.* The generalized formula for a tertiary amine is R_3N.

The basis for the amine primary-secondary-tertiary classification system differs from that for alcohols (Section 14.8).

1. For alcohols we look at how many R groups are on a *carbon* atom, the hydroxyl-bearing carbon atom.
2. For amines we look at how many R groups are on the *nitrogen* atom.

Tert-butyl alcohol is a *tertiary* alcohol, whereas *tert*-butylamine is a *primary* amine.

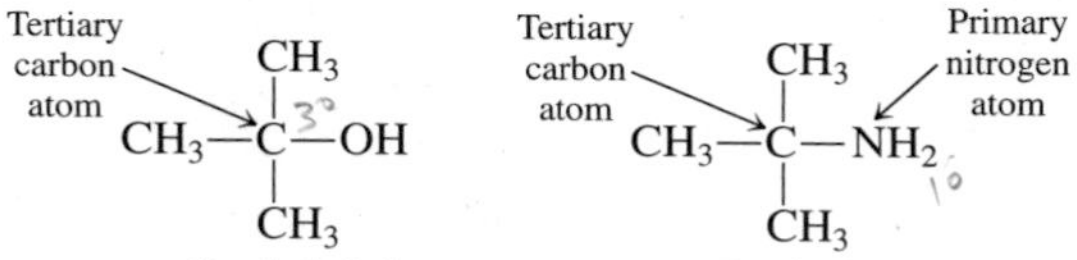

Tertiary carbon atom →

$$CH_3-\overset{\overset{\displaystyle CH_3}{|}}{\underset{\underset{\displaystyle CH_3}{|}}{C}}-OH$$

tert-Butyl alcohol
(a tertiary alcohol)

Tertiary carbon atom →
Primary nitrogen atom

$$CH_3-\overset{\overset{\displaystyle CH_3}{|}}{\underset{\underset{\displaystyle CH_3}{|}}{C}}-NH_2$$

tert-Butylamine
(a primary amine)

The functional group present in a primary amine, the $-NH_2$ group, is called an *amino* group. An **amino group** *is the* $-NH_2$ *functional group.* Secondary and tertiary amines possess substituted amino groups.

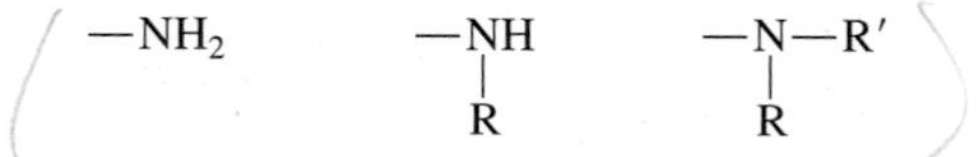

$$-NH_2$$

Amino group

$$-\overset{|}{\underset{\underset{\displaystyle R}{|}}{N}}H$$

Monosubstituted
amino group

$$-\overset{|}{\underset{\underset{\displaystyle R}{|}}{N}}-R'$$

Disubstituted
amino group

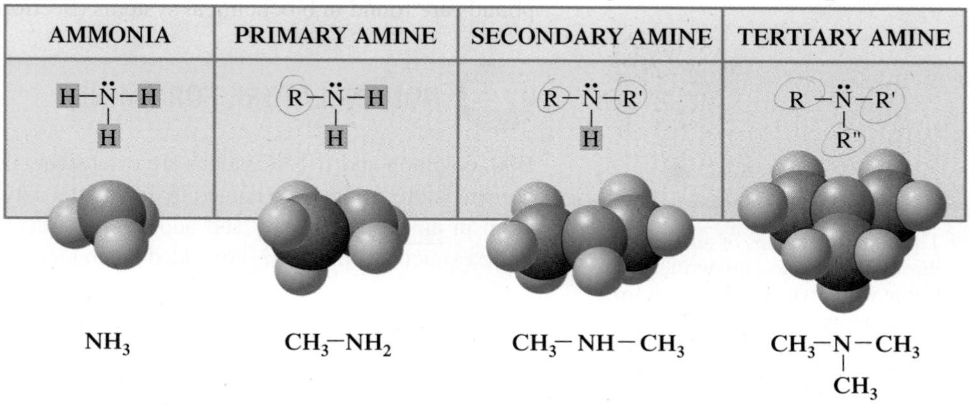

Figure 17.1 Classification of amines is related to the number of R groups attached to the nitrogen atom.

AMMONIA	PRIMARY AMINE	SECONDARY AMINE	TERTIARY AMINE
$H-\overset{..}{\underset{\underset{\displaystyle H}{\mid}}{N}}-H$	$R-\overset{..}{\underset{\underset{\displaystyle H}{\mid}}{N}}-H$	$R-\overset{..}{\underset{\underset{\displaystyle H}{\mid}}{N}}-R'$	$R-\overset{..}{\underset{\underset{\displaystyle R''}{\mid}}{N}}-R'$
NH_3	CH_3-NH_2	$CH_3-NH-CH_3$	$CH_3-\overset{}{\underset{\underset{\displaystyle CH_3}{\mid}}{N}}-CH_3$

EXAMPLE 17.1

Classifying Amines as Primary, Secondary, or Tertiary

Classify each of the following amines as a primary, secondary, or tertiary amine.

a. 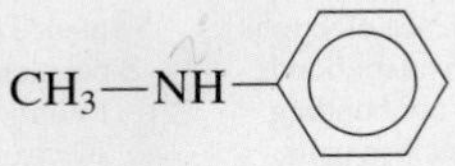 CH₃—NH—⬡

b. 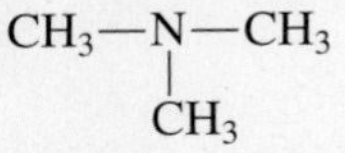 CH₃—N—CH₃ with CH₃ below N

c. 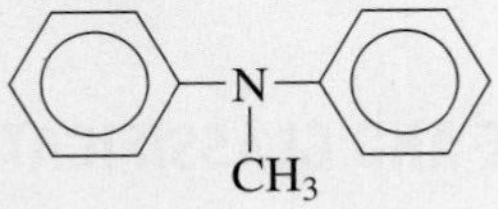 ⬡—N(CH₃)—⬡ (two phenyl groups bonded to N, CH₃ below)

d. 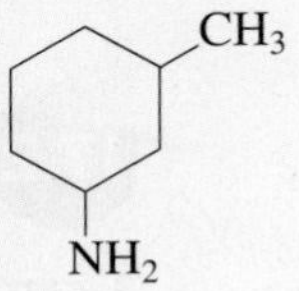 cyclohexane ring with CH₃ and NH₂

Solution

The number of carbon atoms directly bonded to the nitrogen atom determines the amine classification.

a. This is a secondary amine because the nitrogen is bonded to both a methyl group and a phenyl group.
b. Here we have a tertiary amine because the nitrogen atom is bonded to three methyl groups.
c. This is also a tertiary amine; the nitrogen atom is bonded to two phenyl groups and a methyl group.
d. This is a primary amine. The nitrogen atom is bonded to only one carbon atom.

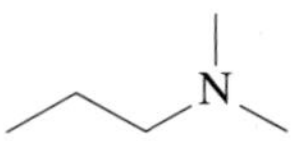

Practice Exercise 17.1

Classify each of the following amines as a primary, secondary, or tertiary amine.

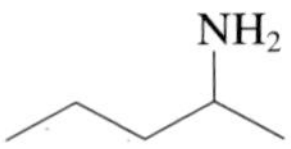

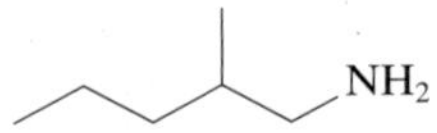

a. CH₃—CH₂—CH₂—NH₂

b. CH₃—NH—CH₂—CH₃

c. 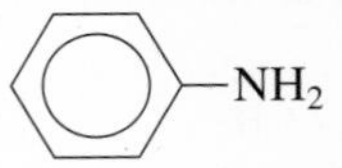 ⬡—NH₂

d. 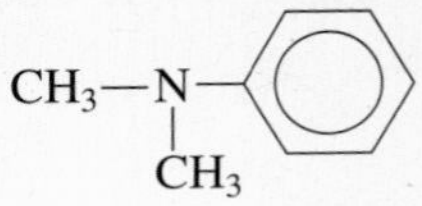 CH₃—N(CH₃)—⬡

Answers: **a.** Primary; **b.** Secondary; **c.** Primary; **d.** Tertiary

Line-angle structural formulas for selected primary, secondary, and tertiary amines.

1° ‾‾‾‾‾‾NH₂

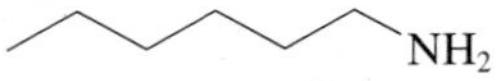

1° ‾‾‾‾ with NH₂

1° ‾‾‾‾ with NH₂

2° ‾NH‾

3° ‾N‾

Cyclic amines exist. Such compounds are always either secondary or tertiary amines.

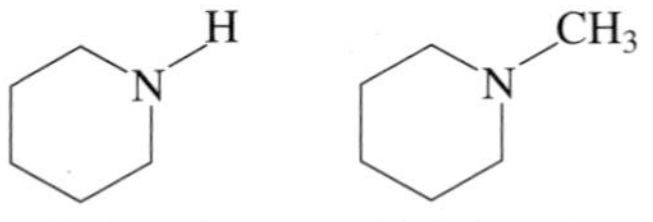

2° Cyclic amine (piperidine with N—H)   3° Cyclic amine (piperidine with N—CH₃)

Cyclic amines are heterocyclic compounds (Section 14.19). Numerous cyclic amine compounds are found in biochemical systems (Section 17.9).

17.3 NOMENCLATURE FOR AMINES

Both common and IUPAC names are extensively used for amines. In the common system of nomenclature, amines are named by listing the alkyl group or groups attached to the nitrogen atom in alphabetical order and adding the suffix -*amine;* all of this appears as one word. Prefixes such as *di-* and *tri-* are added when identical groups are bonded to the nitrogen atom.

The common names of amines, like those of aldehydes, are written as a single word, which is different from the common names of alcohols (two words), ethers (two or three words), ketones (two or three words), acids (two words), and esters (two words).

CH₃—CH₂—NH₂   CH₃—NH—CH₃   ⬡—N(CH₃)—CH₂—CH₃

Ethylamine   Dimethylamine   Ethylmethylphenylamine

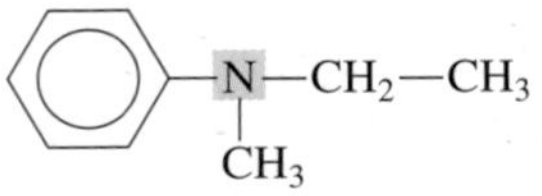

The IUPAC rules for naming amines are similar to those for alcohols (Section 14.3). Alcohols are named as *alkanols* and amines are named as *alkanamines*. IUPAC rules for naming *primary* amines are as follows:

Rule 1: Select as the parent carbon chain the longest chain to which the nitrogen atom is attached.

Rule 2: Name the parent chain by changing the *-e* ending of the corresponding alkane name to *-amine*.

Rule 3: Number the parent chain from the end nearest the nitrogen atom.

Rule 4: The position of attachment of the nitrogen atom is indicated by a number in front of the parent chain name.

Rule 5: The identity and location of any substituents are appended to the front of the parent chain name.

> IUPAC nomenclature for primary amines is similar to that for alcohols, except that the suffix is *-amine* rather than *-ol*. An —NH_2 group, like an —OH group, has priority in numbering the parent carbon chain.

$$CH_3-CH-CH_2-CH_3$$
$$\vert$$
$$NH_2$$
2-Butanamine

$$CH_3-CH-CH_2-CH_2-NH_2$$
$$\vert$$
$$CH_3$$
3-Methyl-1-butanamine

In diamines, the final *-e* of the carbon chain name is retained for ease of pronunciation. Thus the base name for a four-carbon chain bearing two amino groups is butan*e*diamine.

$$H_2N-CH_2-CH_2-CH_2-CH_2-NH_2$$
1,4-Butanediamine

Secondary and tertiary amines are named as *N*-substituted primary amines. The largest carbon group bonded to the nitrogen is used as the parent amine name. The names of the other groups attached to the nitrogen are appended to the front of the base name, and *N*- or *N,N*- prefixes are used to indicate that these groups are attached to the nitrogen atom rather than to the base carbon chain.

$$NH-CH_3$$
$$④ \quad ③ \quad ② \vert \quad ①$$
$$CH_3-CH_2-CH-CH_3$$
N-Methyl-2-butanamine

$$CH_3$$
$$\vert \quad ① \quad ② \quad ③$$
$$CH_3-N-CH_2-CH_2-CH_3$$
N,N-Dimethyl-1-propanamine

$$CH_3$$
$$\vert ① \quad ② \quad ③$$
$$CH_3-CH_2-N-CH_2-CH_2-CH_3$$
N-Ethyl-*N*-methyl-1-propanamine

$$CH_3 \quad NH-CH_3$$
$$① \quad ②\vert \quad ③\vert \quad ④ \quad ⑤$$
$$CH_3-CH-CH-CH_2-CH_3$$
2,*N*-Dimethyl-3-pentanamine

> In IUPAC nomenclature, the amino group has a priority just below that of an alcohol. The priority list for functional groups is
>
> carboxylic acid ↑
> aldehyde
> ketone   Increasing
> alcohol   priority
> amine

In amines where additional functional groups are present, the amine group is treated as a substituent. As a substituent, an —NH_2 group is called an *amino* group.

$$NH_2 \qquad O$$
$$\vert \qquad \Vert$$
$$CH_3-CH_2-CH-CH_2-C-OH$$
3-Aminopentanoic acid

$$NH_2 \qquad O$$
$$\vert \qquad \Vert$$
$$CH_3-CH-CH_2-C-CH_3$$
4-Amino-2-pentanone

$$③ \qquad ② \qquad ①$$
$$CH_3-NH-CH_2-CH_2-CH_2-OH$$
3-(*N*-Methylamino)-1-propanol

The simplest aromatic amine, a benzene ring bearing an amino group, is called *aniline* (Figure 17.2). Other simple aromatic amines are named as derivatives of aniline.

Aniline   *m*-Chloroaniline   2,3-Dichloroaniline

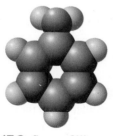

Figure 17.2 Space-filling model of aniline, the simplest aromatic amine. Aromatic amines, including aniline, are generally toxic; they are readily absorbed through the skin.

In secondary and tertiary aromatic amines, the additional group or groups attached to the nitrogen atom are located using a capital *N*-.

N-Ethylaniline   *N*,*N*-Dimethylaniline   3,*N*-Dimethylaniline

E X A M P L E 17.2

Determining IUPAC Names for Amines

Assign IUPAC names to each of the following amines.

a. $CH_3—CH_2—NH—(CH_2)_4—CH_3$

b. $Br—\langle\bigcirc\rangle—NH_2$

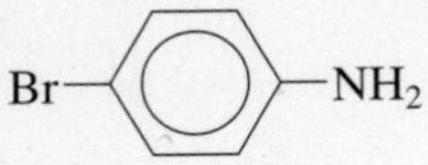

c. $H_2N—CH_2—CH_2—NH_2$

d. (structure of triethyl/ethyl-dimethyl amine)

A benzene ring with both an amino group and a methyl group as substituents is called *toluidine*. This name is a combination of the names *toluene* and *aniline*.

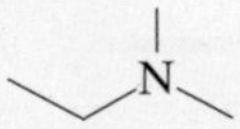

Solution

a. The longest carbon chain has five carbons. The name of the compound is *N-ethyl-1-pentanamine*.
b. This compound is named as a derivative of aniline: *4-bromoaniline* (or *p*-bromoaniline). The carbon in the ring to which the —NH_2 is attached is carbon 1.
c. Two —NH_2 groups are present in this molecule. The name is *1,2-ethanediamine*.
d. This is a tertiary amine in which the longest carbon chain has two carbons (ethane). The base name is thus *ethanamine*. We also have two methyl groups attached to the nitrogen atom. The name of the compound is *N,N-dimethylethanamine*.

The contrast between IUPAC names and common names for primary, secondary, and tertiary amines is as follows:

Primary Amines

IUPAC (one word)

    alkanamine

Common (one word)

    alkylamine

Secondary Amines

IUPAC (one word)

    N-alkylalkanamine

Common (one word)

    alkylalkylamine

Tertiary Amines

IUPAC (one word)

    N-alkyl-N-alkylalkanamine

Common (one word)

    alkylalkylalkylamine

Practice Exercise 17.2

Assign IUPAC names to each of the following amines.

a. (structure with NH_2)

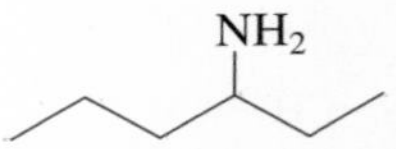

b. $CH_3—CH_2—CH_2—NH—CH_2—CH_2—CH_3$

c. $CH_3—N—CH_3$ with CH_3 below

d. $\langle\bigcirc\rangle—NH—CH_3$

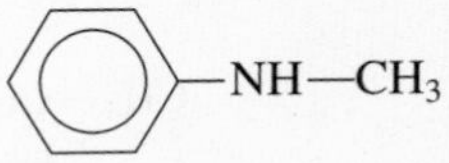

Answers: a. 3-hexanamine; **b.** *N*-propyl-1-propanamine; **c.** *N,N*-dimethylmethanamine; **d.** *N*-methylaniline

17.4 ISOMERISM FOR AMINES

Constitutional isomerism in amines can arise from several causes. Different carbon atom arrangements produce isomers, as in

$CH_3—CH_2—CH_2—CH_2—CH_2—NH_2$ and $CH_3—CH_2—CH—CH_2—NH_2$ with CH_3
        1-Pentanamine   2-Methyl-1-butanamine

Different positioning of the nitrogen atom on a carbon chain is another cause for isomerism, illustrated in the following compounds.

$CH_2—CH_2—CH_2—CH_3$ with NH_2 and $CH_3—CH—CH_2—CH_3$ with NH_2
        1-Butanamine   2-Butanamine

For secondary and tertiary amines, different partitioning of carbon atoms among the carbon chains present produces constitutional isomers. There are three C₄ secondary amines;

Unbranched Primary Amines			
C_1	C_3	C_5	C_7
C_2	C_4	C_6	C_8

☐ Gas   ☐ Liquid

Figure 17.3 A physical-state summary for unbranched primary amines at room temperature and room pressure.

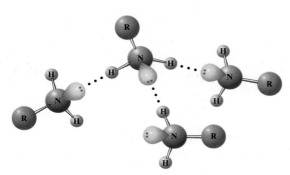

Figure 17.4 Hydrogen bonding interactions among amine molecules involves the hydrogen atoms and nitrogen atoms of amino groups.

carbon atom partitioning can be two ethyl groups, a propyl group and a methyl group, or an isopropyl group and a methyl group.

$$CH_3—CH_2—NH—CH_2—CH_3 \quad \text{and} \quad CH_3—CH_2—CH_2—NH—CH_3 \quad \text{and} \quad CH_3—\overset{\underset{|}{CH_3}}{CH}—NH—CH_3$$

N-Ethylethanamine   *N*-Methyl-1-propanamine   *N*-Methyl-2-propanamine

17.5 PHYSICAL PROPERTIES OF AMINES

The methylamines (mono-, di-, and tri-) and ethylamine are gases at room temperature and have ammonia-like odors. Most other amines are liquids (see Figure 17.3), and many have odors resembling that of raw fish. A few amines, particularly diamines, have strong, disagreeable odors. The foul odor arising from dead fish and decaying flesh is due to amines released by the bacterial decomposition of protein. Two of these "odoriferous" compounds are the diamines putrescine and cadaverine.

$$H_2N—(CH_2)_4—NH_2 \qquad H_2N—(CH_2)_5—NH_2$$
Putrescine   Cadaverine
(1,4-butanediamine)   (1,5-pentanediamine)

The simpler amines are irritating to the skin, eyes, and mucous membranes and are toxic by ingestion. Aromatic amines are generally toxic. Many are readily absorbed through the skin and affect both the blood and the nervous system.

The boiling points of amines are intermediate between those of alkanes and alcohols of similar molecular mass. They are higher than alkane boiling points, because hydrogen bonding is possible between amine molecules but not between alkane molecules. Intermolecular hydrogen bonding of amines involves the hydrogen atoms and nitrogen atoms of the amino groups (Figure 17.4).

The boiling points of amines are lower than those of corresponding alcohols (Figure 17.5), because N···H hydrogen bonds are weaker than O···H hydrogen bonds. [The difference in hydrogen-bond strength results from electronegativity differences; nitrogen is less electronegative than oxygen (Section 5.9).]

Amines with fewer than six carbon atoms are infinitely soluble in water. This solubility results from hydrogen bonding between the amines and water. Even tertiary amines are water-soluble, because the amine nitrogen atom has a nonbonding electron pair that can form a hydrogen bond with a hydrogen atom of water (Figure 17.6).

17.6 BASICITY OF AMINES

Amines, like ammonia, are weak bases. In Section 10.4 we learned that ammonia's weak-base behavior results from its accepting a proton (H^+) from water to produce ammonium ion (NH_4^+) and hydroxide ion (OH^-).

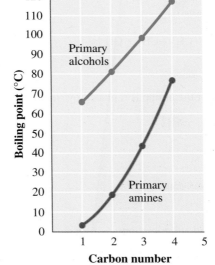

Figure 17.5 A comparison of boiling points of unbranched primary amines and unbranched primary alcohols.

Primary alcohols

Primary amines

Boiling point (°C)

Carbon number

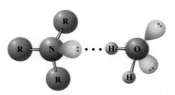

Figure 17.6 Low-molecular-mass amines are soluble in water because of amine–water hydrogen bonding interactions.

Tertiary amines have <u>lower</u> boiling points than primary and secondary amines because intermolecular hydrogen bonding is not possible in tertiary amines. Such amines have no hydrogen atoms directly bonded to the nitrogen atom.

$$\overset{..}{N}H_3 \;+\; HOH \;\rightleftharpoons\; NH_4{}^+ \;+\; OH^-$$
Ammonia Ammonium ion Hydroxide ion

Amines behave in a similar manner.

$$CH_3-\overset{..}{N}H_2 \;+\; HOH \;\rightleftharpoons\; CH_3-\overset{+}{N}H_3 \;+\; OH^-$$
Methylamine Methylammonium ion Hydroxide ion

Amines, like ammonia, have a pair of unshared electrons on the nitrogen atom present. These unshared electrons can accept a hydrogen ion from water. Thus both amines and ammonia produce basic aqueous solutions.

The result of the interaction of an amine with water is a basic solution containing substituted ammonium ions and hydroxide ions. A **substituted ammonium ion** *is an ammonium ion in which one or more alkyl, cycloalkyl, or aryl groups have been substituted for hydrogen atoms.*

Two important generalizations apply to substituted ammonium ions.

1. *Substituted ammonium ions are charged species rather than neutral molecules.*
2. *The nitrogen atom in an ammonium ion or a substituted ammonium ion participates in four bonds.* In a neutral compound, nitrogen atoms form only three bonds. Four bonds about a nitrogen atom are possible, however, when the species is a positive ion because the fourth bond is a coordinate covalent bond (Section 5.5).

Naming the positive ion that results from the interaction of an amine with water is based on the following two rules:

Rule 1: For alkylamines, the ending of the name of the amine is changed from *amine* to *ammonium ion.*

$$CH_3-CH_2-NH_2 \xrightarrow{H_2O} CH_3-CH_2-\overset{+}{N}H_3 + OH^-$$
Ethylamine Ethylammonium ion

$$CH_3-\underset{\underset{CH_3}{|}}{N}-CH_2-CH_3 \xrightarrow{H_2O} CH_3-\underset{\underset{CH_3}{|}}{\overset{+}{N}H}-CH_2-CH_3 + OH^-$$
Ethyldimethylamine Ethyldimethylammonium ion

Substituted ammonium ions always contain one more hydrogen atom than their "parent" amine. They also always carry a $+1$ charge, whereas the "parent" amine is a <u>neutral</u> molecule.

Rule 2: For aromatic amines, the final *-e* of the name of the amine is replaced by *-ium ion.*

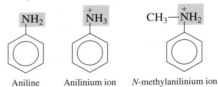
Aniline Anilinium ion *N*-methylanilinium ion

● **EXAMPLE 17.3**
Determining Names for Substituted Ammonium and Substituted Anilinium Ions

Name the following substituted ammonium or substituted anilinium ions.

a. $CH_3-CH_2-\overset{+}{N}H_2-CH_2-CH_3$

b. $CH_3-\underset{\underset{CH_3}{|}}{CH}-CH_2-\overset{+}{N}H_3$

c. $CH_3-\underset{\underset{CH_3}{|}}{\overset{+}{N}H}-CH_3$

d. $CH_3-\overset{+}{N}H-CH_3$

Solution

a. The parent amine is diethylamine. Replacing the word *amine* in the parent name with *ammonium ion* generates the name of the ion, *diethylammonium ion.*
b. The parent amine is isobutylamine. The name of the ion is *isobutylammonium ion.*
c. The parent amine is trimethylamine. The name of the ion is *trimethylammonium ion.*
d. The parent name is *N,N*-dimethylaniline. Replacing the word *aniline* in the parent name with *anilinium ion* generates the name of the ion, *N,N*-dimethylanilinium ion.

Practice Exercise 17.3

Name the following substituted ammonium or substituted anilinium ions.

a. $CH_3-CH_2-\overset{+}{N}H_2-CH_3$ **b.** $CH_3-CH-\overset{+}{N}H_3$
                                                      CH_3

c. $CH_3-CH_2-\overset{+}{N}H-CH_2-CH_3$ **d.** 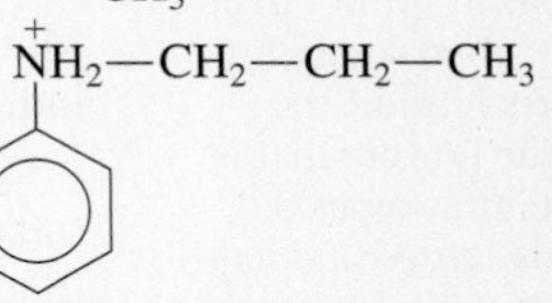 $\overset{+}{N}H_2-CH_2-CH_2-CH_3$
                      CH_3

Answers: a. Ethylmethylammonium ion; **b.** Isopropylammonium ion; **c.** Diethylmethylammonium ion; **d.** *N*-propylanilinium ion

Amines are <u>stronger proton acceptors</u> than oxygen-containing organic compounds such as alcohols and ethers; that is, they are <u>stronger bases</u> than these compounds. A 0.1 M aqueous solution of methylamine has a pH of 11.8 and a 0.1 M aqueous solution of aniline has a pH of 8.6. These solutions are sufficiently basic to turn red litmus paper blue (Section 10.1). (Carboxylic acid salts (Section 16.9) are the only other type of organic compound sufficiently basic to turn red litmus paper blue.)

17.7   AMINE SALTS

The reaction of an acid with a base (<u>neutralization</u>) produces <u>a salt</u> (Section 10.7). Because amines are <u>bases</u> (Section 17.6), their reaction with <u>an acid produces a salt</u>, an <u>amine salt</u>.

$CH_3-\overset{..}{N}H_2 + \overset{\frown}{\textcircled{H}}-Cl \longrightarrow CH_3-\overset{+}{N}H_3\ Cl^-$

Amine   Acid   Amine salt

Aromatic amines react with acids in a similar manner.

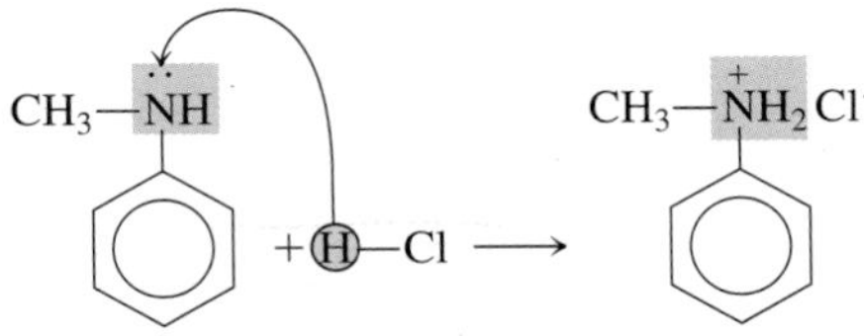

An **amine salt** *is an ionic compound in which the (positive ion is a mono-, di-, or trisubstituted ammonium ion)* $(RNH_3^+, R_2NH_2^+, or\ R_3NH^+)$ *and the <u>negative ion comes from an acid</u>.* Amine salts can be obtained in crystalline form (odorless, white crystals) by evaporating the water from the acidic solutions in which amine salts are prepared.

Amine salts are named using standard nomenclature procedures for ionic compounds (Section 4.9). The name of the <u>positive ion</u>, the substituted ammonium or anilinium ion, is given first and is followed by a <u>separate word</u> for the name of the <u>negative ion</u>.

$CH_3-CH_2-\overset{+}{N}H_3\ Cl^-$ $CH_3-\overset{+}{N}H_2-CH_3\ Br^-$

Ethylammonium chloride   Dimethylammonium bromide

An older naming system for amine salts, still used in the pharmaceutical industry, treats amine salts as amine–acid complexes rather than as ionic compounds. In this system,

The cocaine molecule is both an amine and an ester; one amine and two ester functional groups are present. As an illegal street drug, co-caine is consumed as a water-soluble amine salt and in a water-insoluble, free-base form (nonsalt form—freed of the base required to make the salt). Cocaine hydrochloride, the amine salt, is a white powder that is snorted or injected intravenously. Free-base cocaine is heated and its vapors are inhaled. Cocaine users and dealers call the water-insoluble form of the drug "crack." "Snow" and "coke" are street names for the water-soluble form of the drug.

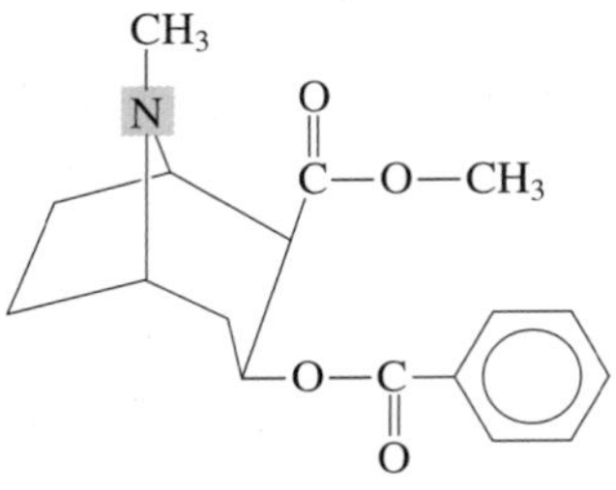

Cocaine
(an amine)

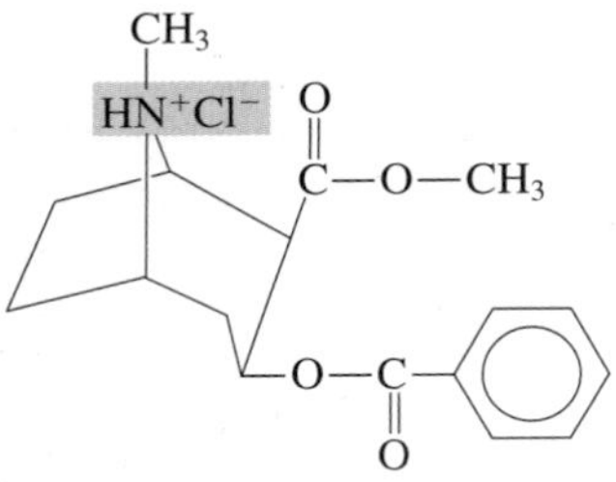

Cocaine hydrochloride
(an amine salt)

the amine salt made from dimethylamine and hydrochloric acid is named and repre-sented as

$$CH_3-NH \cdot HCl$$
$$\underset{|}{}$$
$$CH_3$$
Dimethylamine hydrochloride

rather than as

$$CH_3-\overset{+}{N}H_2 \; Cl^-$$
$$\underset{|}{}$$
$$CH_3$$
Dimethylammonium chloride

Many medication labels refer to hydrochlorides or hydrogen sulfates (from sulfuric acid), indicating that the medications are in a water-soluble ionic (salt) form.

Many higher-molecular-mass amines are water-insoluble; however, virtually all amine salts are water-soluble. Thus amine salt formation, like carboxylic acid salt formation (Section 16.9), provides a means for converting water-insoluble compounds into water-soluble compounds. Many drugs that contain amine functional groups are administered to patients in the form of amine salts because of their increased solubility in water in this form.

Many people unknowingly use acids to form amine salts when they put vinegar or lemon juice on fish. Such action converts amines in fish (often smelly compounds) to salts, which are odorless.

The process of forming amine salts with acids is an easily reversed process. Treating an amine salt with a strong base such as NaOH regenerates the "parent" amine.

$$\underset{\text{Amine salt}}{CH_3-\overset{+}{N}H_3 \; Cl^-} + \underset{\text{Base}}{NaOH} \longrightarrow \underset{\text{Amine}}{CH_3-NH_2} + NaCl + H_2O$$

The "opposite nature" of the processes of amine salt formation from an amine and the regeneration of the amine from its amine salt can be diagrammed as follows:

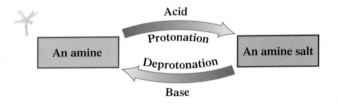

An amine gains a hydrogen ion to produce an amine salt when treated with an acid (a protonation reaction), and an amine salt loses a hydrogen ion to produce an amine when treated with a base (a deprotonation reaction).

● **E X A M P L E 17.4**

Writing Chemical Equations for Reactions That Involve Amine Salts

▶ Write the structures of the products that form when each of the following reactions involving amines or amine salts takes place.

a. $CH_3-NH-CH_3 + HCl \longrightarrow$

b.

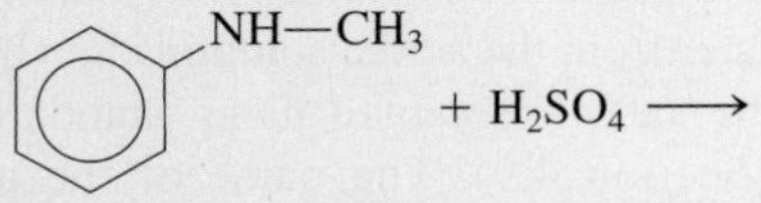

NH—CH$_3$

$+ H_2SO_4 \longrightarrow$

c. $CH_3-\overset{+}{N}H_2-CH_3 \; Cl^- + NaOH \longrightarrow$

Solution

a. The reactants are an amine and a strong acid. Their interaction produces an amine salt.

$$CH_3-NH-CH_3 + \textcircled{H}Cl \longrightarrow CH_3-\overset{+}{N}H_2-CH_3 \; Cl^-$$

b. Again, we have the reaction of an amine with a strong acid. A hydrogen ion is transferred from the acid to the amine.

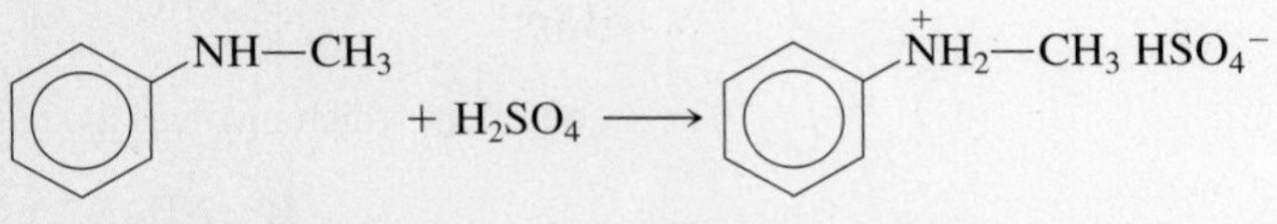

c. The reactants are an amine salt and a strong base. Their interaction regenerates the "parent" amine.

$$CH_3 \overset{+}{-NH_2} -CH_3\ Cl^- + NaOH \longrightarrow CH_3 -NH -CH_3 + NaCl + H_2O$$

Practice Exercise 17.4

Write the structures of the products formed in the following reactions.

a. $CH_3 -CH_2 -\underset{\underset{CH_3}{|}}{N} -CH_3 + HCl \longrightarrow$

b. $CH_3 -CH_2 -NH_2 + H_2SO_4 \longrightarrow$

c. $CH_3 -CH_2 -\underset{\underset{CH_3}{|}}{\overset{+}{NH}} -CH_3\ Br^- + NaOH \longrightarrow$

Answers:

a. $CH_3 -CH_2 -\underset{\underset{CH_3}{|}}{\overset{+}{NH}} -CH_3\ Cl^-$ **b.** $CH_3 -CH_2 -\overset{+}{NH_3}\ HSO_4^-$

c. $CH_3 -CH_2 -\underset{\underset{CH_3}{|}}{N} -CH_3 + NaBr + H_2O$

Because amines and their salts are so easily interconverted, the amine itself is often designated as the *free amine* or *free base* or as the *deprotonated form* of the amine, to distinguish it from the *protonated form* of the amine, which is present in the amine salt.

$$CH_3 -NH_2 \qquad\qquad CH_3 -\overset{+}{NH_3}\ Cl^-$$

Deprotonated form (free amine or free base) Protonated form = salt

Whether an amine in solution exists in its deprotonated form or its protonated form is dependent on the pH of the solution. An aqueous solution of a simple amine has a pH of about 12 and at this pH the amine is almost entirely (99%) present in its deprotonated form. In biochemical solutions such as body fluids, which are buffered solutions with a pH near 7, amines exist predominately in protonated form.

17.8 PREPARATION OF AMINES AND QUATERNARY AMMONIUM SALTS

Several methods exist for preparing amines. We consider only one: alkylation in the presence of base. Generalized equations for the alkylation process are

$$\text{Ammonia + alkyl halide} \xrightarrow{\text{Base}} 1° \text{ amine}$$

$$1° \text{ Amine + alkyl halide} \xrightarrow{\text{Base}} 2° \text{ amine}$$

$$2° \text{ Amine + alkyl halide} \xrightarrow{\text{Base}} 3° \text{ amine}$$

$$3° \text{ Amine + alkyl halide} \xrightarrow{\text{Base}} \text{quaternary ammonium salt}$$

Alkylation under basic conditions is actually a two-step process. In the first step, using a primary amine preparation as an example, an amine salt is produced.

$$NH_3 + R\!-\!X \longrightarrow R\!-\!\overset{+}{N}H_3\ X^-$$

The second step, which involves the base present (NaOH), converts the amine salt to free amine.

$$R\!-\!\overset{+}{N}H_3\ X^- + NaOH \longrightarrow RNH_2 + NaX + H_2O$$

A specific example of the production of a primary amine from ammonia is the reaction of ethyl bromide with ammonia to produce ethylamine. The chemical equation (with both steps combined) is

$$NH_3 + CH_3\!-\!CH_2\!-\!Br + NaOH \longrightarrow CH_3\!-\!CH_2\!-\!NH_2 + NaBr + H_2O$$

If the newly formed primary amine produced in an ammonia alkylation reaction is not quickly removed from the reaction mixture, (the nitrogen atom of the amine may react with further alkyl halide molecules, giving, in succession, secondary and tertiary amines.)

$$NH_3 \xrightarrow[\text{OH}^-]{\text{RX}} RNH_2 \xrightarrow[\text{OH}^-]{\text{RX}} R_2NH \xrightarrow[\text{OH}^-]{\text{RX}} R_3N$$

Primary   Secondary   Tertiary
amine   amine   amine

Examples of the production of a 2° amine and a 3° amine via alkylation are

$$CH_3\!-\!CH_2\!-\!NH_2 + CH_3\!-\!Br + NaOH \longrightarrow CH_3\!-\!CH_2\!-\!NH\!-\!CH_3 + NaBr + H_2O$$

Primary amine   Alkyl halide   Base   Secondary amine

$$CH_3\!-\!CH_2\!-\!NH\!-\!CH_3 + CH_3\!-\!Br + NaOH \longrightarrow CH_3\!-\!CH_2\!-\!\underset{\underset{\displaystyle CH_3}{|}}{N}\!-\!CH_3 + NaBr + H_2O$$

Secondary amine   Alkyl halide   Base   Tertiary amine

Tertiary amines react with alkyl halides in the presence of a strong base to produce a quaternary ammonium salt. A **quaternary ammonium salt** *is an ammonium salt in which all four groups attached to the nitrogen atom of the ammonium ion are hydrocarbon groups.*

$$R\!-\!\underset{\underset{\displaystyle R}{|}}{N}\!-\!R + R\!-\!X \xrightarrow{\text{OH}^-} R\!-\!\overset{\overset{\displaystyle R}{|}}{\underset{\underset{\displaystyle R}{|}}{N}}{}^{\!+}\!-\!R\ X^-$$

(Quaternary ammonium salts differ from amine salts in that the addition of a strong base does not convert quaternary ammonium salts back to their "parent" amines; there is no hydrogen atom on the nitrogen with which the OH⁻ can react. Quaternary ammonium salts are colorless, odorless, crystalline solids that have high melting points and are usually water-soluble.

Quaternary ammonium salts are named in the same way as amine salts (Section 17.7), taking into account that four organic groups are attached to the nitrogen atom rather than a lesser number of groups. Figure 17.7 contrasts the structures of an ammonium ion and a tetramethyl ammonium ion.

Compounds that contain quaternary ammonium ions are important in biochemical systems. Choline and acetylcholine are two important quaternary ammonium *ions* present in the human body. Choline has important roles in both fat transport and growth regulation. Acetylcholine is involved in the transmission of nerve impulses.)

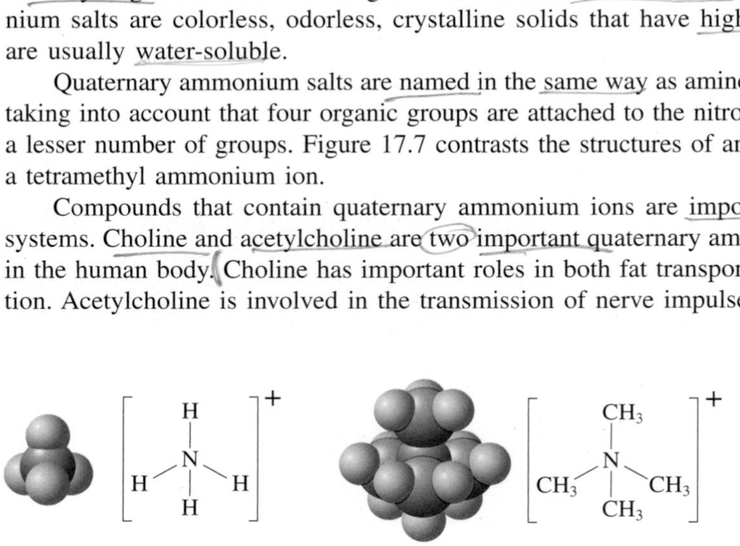

Figure 17.7 Space-filling models showing that the ammonium ion (NH_4^+) has a tetrahedral structure, as does the quaternary ammonium ion, in which four methyl groups are present [$(CH_3)_4N^+$].

$$CH_3-\overset{\overset{\displaystyle CH_3}{|}}{\underset{\underset{\displaystyle CH_3}{|}}{N^+}}-CH_2-CH_2-OH$$

Choline

$$CH_3-\overset{\overset{\displaystyle CH_3}{|}}{\underset{\underset{\displaystyle CH_3}{|}}{N^+}}-CH_2-CH_2-O-\overset{\overset{\displaystyle O}{\|}}{C}-CH_3$$

Acetylcholine

17.9 HETEROCYCLIC AMINES

A **heterocyclic amine** *is an organic compound in which nitrogen atoms of amine groups are part of either an aromatic or a nonaromatic ring system.* Heterocyclic amines are the most common type of heterocyclic organic compound (Section 14.19). Figure 17.8 gives structures for a number of "key" unsubstituted heterocyclic amines. These compounds are the "parent" compounds for numerous derivatives that are important in medicinal, agricultural, food, and industrial chemistry, as well as in the functioning of the human body.

Study of the heterocyclic amine structures in Figure 17.8 shows that (1) ring systems may be saturated, unsaturated, or aromatic, (2) more than one nitrogen atom may be present in a given ring, and (3) fused ring systems often occur.

The two most widely used central nervous system stimulants in our society, caffeine and nicotine, are heterocyclic amine derivatives. Caffeine's structure is based on a purine ring system. Nicotine's structure contains one pyridine ring and one pyrrolidine ring.

Caffeine

Nicotine

A large cyclic structure built on four pyrrole rings (Figure 17.8), called a porphyrin, is important in the chemistry of living organisms. Porphyrins form metal ion complexes in which the metal ion is located in the middle of the large ring structure. *Heme*, an iron–porphyrin complex present in the red blood pigment hemoglobin, is responsible for oxygen transport in the human body.

Porphyrin ring

Heme, a component
of hemoglobin

Heterocyclic amines are the first heterocyclic compounds we have encountered that have nitrogen heteroatoms. In previous chapters, we have discussed heterocyclic compounds with oxygen as the heteroatom: cyclic ethers (Section 14.19); cyclic ketones (Section 15.3); the cyclic forms of hemiacetals and acetals (Section 15.11); and cyclic esters (Section 16.10).

Heterocyclic amines often have strong odors, some agreeable and others disagreeable. The "pleasant" aroma of many heat-treated foods is caused by heterocyclic amines formed during the heat treatment. The compounds responsible for the pervasive odors of popped popcorn and hot roasted peanuts are heterocyclic amines.

Methyl-2-pyridyl ketone
(odor of popcorn)

2-Methoxy-5-methylpyrazine
(odor of peanuts)

Figure 17.8 Structural formulas for selected heterocyclic amines that serve as "parent" molecules for more complex amine derivatives.

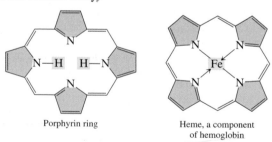

PYRROLIDINE	PYRROLE	IMIDAZOLE	INDOLE
PYRIDINE	**PYRIMIDINE**	**QUINOLINE**	**PURINE**

Caffeine: The Most Widely Used Central Nervous System Stimulant

Caffeine, which is naturally present in coffee beans and tea leaves and is added to many soft drinks, is both the most widely used and the most frequently used central nervous system (CNS) stimulant in our society. Studies indicate that 82% of the adult U.S. population drink coffee and 52% drink tea. Over one-half of the coffee drinkers consume three or more cups daily. In the soft drink market, the use of caffeine has now spread beyond cola drinks to orange drinks.

Caffeine belongs to a family of compounds called *xanthines*. Its formal chemical name is 1,3,7-trimethylxanthine and its structure is

Caffeine
(1,3,7-trimethylxanthine)

Xanthine

Caffeine is naturally present in coffee beans. Over 60 plants and trees cultivated by humans contain caffeine.

Besides its CNS system effects, caffeine also increases basal metabolic rate, increases heart rate by stimulating heart muscles, promotes secretion of stomach acid, functions as a bronchial tube dilator and increases urine production because of its diuretic properties. The overall effect that most individuals experience from caffeine consumption is interpreted as a "lift."

Used in small quantities, caffeine's effects are temporary; hence it must be consumed on a regular basis throughout the day. Its half-life in the body is 3.0–7.5 hours. Caffeine tolerance develops in regular users of the substance. Over time, larger amounts of caffeine are needed in order for an individual to achieve his or her "lift."

Caffeine is mildly addicting. People who ordinarily consume substantial amounts of caffeine-containing beverages or drugs experience withdrawal symptoms if caffeine is eliminated. Such symptoms include headache and depression for a period of several days. As a result of caffeine dependence, many people need a cup of coffee before they feel good each morning.

In large quantities, caffeine has been shown to cause significant undesirable effects including anxiety, sleeplessness, headaches, and dehydration. The latter occurs because of caffeine's diuretic effects.

Several studies indicate that caffeine has an effect on the child-bearing abilities of women. One study indicates a decreased rate of conception for women who consume more than three cups of coffee daily. Another study indicates that the consumption of this amount of caffeine increases the risk of miscarriage by 30% compared to women who do not drink coffee. Lactating mothers probably should limit their caffeine intake because it appears in mother's milk, thus affecting newborn babies.

Current scientific thought holds that caffeine's mode of action in the body is exerted through a chemical substance called cyclic adenosine monophosphate (cyclic AMP). Caffeine inhibits an enzyme that ordinarily breaks down cyclic AMP to its inactive end product. The resulting increase in cyclic AMP leads to increased glucose production within cells and thus makes available more energy to allow higher rates of cellular activity.

Coffee is the major source of caffeine for most Americans. However, substantial amounts of caffeine may be consumed in soft drinks, tea, and numerous nonprescription medications including combination pain relievers (Anacin, Midol, Empirin), cold remedies (Dristan, Triaminicin), and antisleep agents (No Doz, Vivarin).

17.10 SELECTED BIOCHEMICALLY IMPORTANT AMINES

Neurotransmitters

A **neurotransmitter** *is a chemical substance that is released at the end of a nerve, travels across the synaptic gap between the nerve and another nerve, and then bonds to a receptor site on the other nerve, triggering a nerve impulse.* Figure 17.9 shows schematically how neurotransmitters function.

The most important neurotransmitters in the human body are acetylcholine (Section 17.8) and the amines norepinephrine, dopamine, and serotonin.

The names of the amine neurotransmitters are pronounced nor-ep-in-NEFF-rin, DOPE-a-mean, and SER-oh-tone-in.

CHEMICAL Connections

Nicotine Addiction: A Widespread Example of Drug Dependence

Next to caffeine, the heterocyclic amine nicotine is the most widely used central nervous system stimulant in our society. It is obtained primarily from the tobacco plant, where it constitutes 0.3 to 0.5% by dry mass of the plant. Biosynthesis of nicotine occurs in the roots of the plant and it then is translocated to the leaves, where it accumulates. In the pure state, in its free base form, nicotine is an oily liquid that is miscible with water. It readily reacts with water to form salts that are usually solids and water soluble, behavior associated with the amine nature of this substance (Section 17.7).

Structurally, nicotine contains both a pyrrolidine and a pyridine ring system (see Figure 17.8); these rings are connected through a carbon–carbon bond rather than fused. Two amine functional groups are present in a nicotine molecule.

Nicotine

Cigarettes contain 8 to 20 milligrams of nicotine (depending on the brand) but only approximately 1 mg is actually absorbed by the body when a cigarette is smoked. This absorbed nicotine is responsible for the addictive process that develops with continued cigarette smoking. In its free-base form, nicotine will burn and its vapors will combust at 95°C in air. Thus, most of the nicotine is burned when a cigarette is smoked; however, enough is inhaled to provide the desired stimulatory effects.

Two biochemical effects of nicotine are

1. Nicotine acts on acetylcholine receptors, increasing their activity. This leads to an increased flow of adrenaline (Section 17.10), which in turn causes an increase in heart rate, blood pressure, and respiration rate, as well as higher blood glucose levels.
2. Nicotine increases the release of the brain neurotransmitter dopamine (Section 17.10), which makes a person feel "good." Getting the dopamine boost is part of the addiction process.

Inhaled nicotine is rapidly distributed throughout the body. On average, it takes 7 seconds for nicotine to reach the brain once it is inhaled. The overall stimulatory effect of nicotine is, however, rather transient. After an initial response, depression follows. The half-life of nicotine in the body is between 1 and 2 hours.

Nicotine induces both physiological and psychological dependence. Withdrawal from nicotine is accompanied by headache, stomach pain, irritability, and insomnia. Full withdrawal can take six months or longer.

In large doses, nicotine is a potent poison. Nicotine poisoning causes vomiting, diarrhea, nausea, and abdominal pain. Death occurs from respiratory paralysis. The lethal dose of the drug is considered to be approximately 60 mg, a level that is never reached with cigarette smoking.

Nicotine can influence the action of prescription drugs that the smoker is taking concurrently. For example, Propoxyphene (Darvon) can lose its painkilling effect in a smoker, and tranquilizers such as Diazepam (Valium) exert a diminished effect on the central nervous system.

The carcinogenic effects associated with cigarette smoking are caused not by nicotine but by other substances present in tobacco smoke, including fused-ring aromatic hydrocarbons (see the Chemical Connections feature on page 391) and radon-222 decay products (see the Chemical Connections feature on page 309).

There are actually three forms for a nicotine molecule, with the form adopted determined by the pH of the molecule's environment. At a low pH (less than 3.0) nicotine exists in a diprotonated form; at pH values between 3 and 8 it is monoprotonated; and at higher pH values it is in a nonprotonated (free-base) form (see Section 17.7).

Diprotonated form   Monoprotonated form   Nonprotonated form (free-base form)

Note that the protonation sites in the molecules are the nitrogen atoms of the amine functional groups (Section 17.7). The nonprotonated (free-base) form of nicotine is very volatile and evaporates (as well as combusts) at the temperatures associated with a burning cigarette.

Cigarette manufacturers purposely add ammonia to tobacco leaves, to raise the pH and make the free-base form more available to the smoker. Such pH adjustment also occurs for smokeless tobacco products. Products of lower pH are marketed for new users and higher pH products, which supply more nicotine to the user, are available for "experienced" users.

Nicotine is formed in the roots of tobacco plants and then translocates to the leaves, where it accumulates.

Figure 17.9 Neurotransmitters are chemical messengers between nerve cells. Neurotransmitters released from one nerve cell stimulate (activate) an adjacent nerve cell. (a) Before the conduction of a nerve impulse. (b) An incoming nerve impulse triggers the release of neurotransmitter molecules. (c) Neurotransmitters bind to receptor sites, activating the receptor nerve cell.

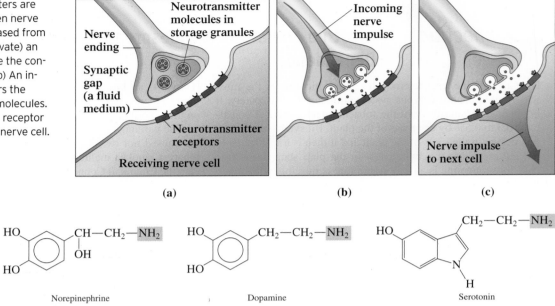

Norepinephrine, a compound secreted by the adrenal glands into the blood, helps maintain muscle tone in the blood vessels.

Dopamine is found in the brain. A deficiency of this neurotransmitter results in Parkinson's disease, a degenerative neurological disease. Administration of dopamine to a patient does not relieve the symptoms of this disease because dopamine in the blood cannot cross the blood–brain barrier. The drug L-dopa, which can pass through the blood–brain barrier, does give relief from Parkinson's symptoms. Inside brain cells, enzymes catalyze the conversion of L-dopa to dopamine.

Serotonin, also a brain chemical, is involved in sleep, sensory perception, and the regulation of body temperature. Serotonin deficiency has been implicated in mental illness. Treatment of mental depression can involve the use of drugs that help maintain serotonin at normal levels by preventing its breakdown within the brain.

Recent research shows that serotonin also regulates lactation in women. Such serotonin is produced in the mammary glands rather than in the brain. At this location it controls milk production and secretion. The concentration of serotonin builds up in mammary glands as they fill with milk. The increase in serotonin inhibits further milk synthesis and secretion by suppressing the expression of milk protein genes. After nursing, the cycle of milk and serotonin production begins again.

Epinephrine

Epinephrine, also known as adrenaline, has some neurotransmitter functions but is more important as a central nervous system stimulant. Produced by the adrenal glands, epinephrine differs in structure from norepinephrine in that a methyl group substituent is present on the amine nitrogen atom.

Prozac, the most widely prescribed drug for mental depression, inhibits the reuptake of serotonin, thus maintaining serotonin levels. Chemically, Prozac is a derivative of methyl propyl amine with three fluorine atoms present in the structure. The element fluorine is rarely encountered in biochemical molecules.

Ephedrin, pronounced "eh-FEH-drin," is a substance extracted from the Asian plant ephedra that was used in numerous dietary supplements sold as weight-loss aids and energy boosters at nutrition stores, in supermarkets, and on the Internet until it was banned in the United States by the Food and Drug Administration (FDA) in 2004. Its chemical structure resembles that of epinephrine (adrenaline).

Ephedrin

Within the body, it behaves as a stimulant to the heart and the central nervous system. Concerns about the safety of its use, particularly with people who have hypertension and other cardiovascular problems, led to its ban.

Pain, excitement, and fear trigger the release of large amounts of epinephrine into the bloodstream. The effect is increased blood glucose levels, which in turn increase blood pressure, rate and force of heart contraction, and muscular strength. These changes cause the body to function at a "higher" level. Epinephrine is often called the "fight or flight" hormone.

Histamine

The heterocyclic amine *histamine* is responsible for the unpleasant effects felt by individuals susceptible to hay fever and various pollen allergies.

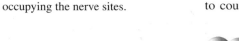

Histamine

Histamine is naturally present in the human body in a "stored" form; it is part of more complex molecules. A number of situations can trigger the release of the "stored" histamine. Activators include (1) contact with pollen, dust, and other allergens, (2) substances released from damaged cells, and (3) contact with chemicals to which an individual has become sensitized.

The presence of "free" histamine causes the symptoms associated with hay fever, such as watery eyes and stuffy nose, and many of the symptoms associated with the common cold. A group of substances called *antihistamines* can be taken as medication to counteract the effects of the histamine.

Antihistamines are drugs that counteract, to some extent, the effects of histamine release in the body. Antihistamines share a common structural feature with histamine—an ethanamine chain.

$$-CH_2-CH_2-N\big\langle$$

This structure allows antihistamines to occupy receptor sites in nerves normally occupied by histamine, thus blocking histamine from occupying the nerve sites.

17.11 ALKALOIDS

People in various parts of the world have known for centuries that physiological effects can be obtained by eating or chewing the leaves, roots, or bark of certain plants. Over 5000 different compounds that are physiologically active have been isolated from such plants. Nearly all of these compounds, which are collectively called alkaloids, contain amine functional groups. An **alkaloid** *is a nitrogen-containing organic compound extracted from plant material.*

Three well-known compounds that we have considered previously are alkaloids. They are nicotine (tobacco plant), caffeine (coffee beans and tea leaves), and cocaine (coca plant).

A number of alkaloids are currently used in medicine. Quinine, which occurs in cinchona bark, is used to treat malaria. Atropine, which is isolated from the belladonna plant, is used to dilate the pupil of the eye in patients undergoing eye examinations (Figure 17.10). Atropine is also used as a preoperative drug to relax muscles and reduce the secretion of saliva in surgical patients.

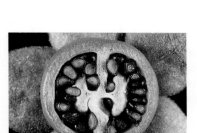

Figure 17.10 Fruit of the belladonna plant; the alkaloid atropine is obtained from this plant.

The name *alkaloid,* which means "like a base," reflects the fact that alkaloids react with acids. Such behavior is expected for substances with amine functional groups because amines are weak bases.

Quinine   Atropine

An extremely important family of alkaloids is the narcotic painkillers, a class of drugs derived from the resin (opium) of the oriental poppy plant (Figure 17.11). The most important drugs obtained from opium are morphine and codeine. Synthetic modification of morphine produces the illegal drug heroin.

These three compounds have similar chemical structures.

Morphine is one of the most effective painkillers known; its painkilling properties are about a hundred times greater than those of aspirin. Morphine acts by blocking the process in the brain that interprets pain signals coming from the peripheral nervous system. The major drawback to the use of morphine is that it is addictive.

Codeine is a methylmorphine. Almost all codeine used in modern medicine is produced by methylating the more abundant morphine. Codeine is less potent than morphine, having a painkilling effect about one-sixth that of morphine.

Heroin is a semi-synthetic compound, the diacetyl ester of morphine; it is produced from morphine. This chemical modification increases painkilling potency; heroin has more than three times the painkilling effect of morphine. However, heroin is so addictive that it has no accepted medical use in the United States.

The most widely prescribed painkillers in the United States, at present, for the relief of moderate to heavy pain contain the codeine derivatives hydrocodone or oxycodone as an active ingredient. Usually a second painkilling agent is present in addition to the codeine derivative.

Lortab and Norco are examples of hydrocodone-containing formulations in which acetaminophen is the second painkilling agent present. The painkilling effect of this two-drug combination is actually greater than that of the sum of the individual effects. The presence of the acetaminophen also limits the potential for the unsafe addictive side effects associated with higher than prescribed doses of hydrocodone because of the liver-toxicity problem associated with higher doses of the accompanying acetaminophen.

Oxycodone-containing formulations include Percodan (an aspirin-oxycodone combination) and Percocet (an acetaminophen-oxycodone combination). Oxycontin is a continuous-release form of oxycodone as a single agent. The use of two-drug combinations for oxycodone is based on the same reasons that apply to hydrocodone two-drug combinations.

There are two driving forces for the preferred use of hydrocodone formulations over oxycodone formulations. The first is cost. Hydrocodone production requires a

CHEMICAL Connections   Alkaloids Present in Chocolate

Chocolate is a food preparation made from the beans (seeds) of the tropical cacao tree. Growth conditions for cacao trees require a warm, moist climate like that found near the equator. The majority of the world's supply of cacao beans comes from the west coast of Africa—Ivory Coast, Ghana, and Nigeria. (Because of a mistake in spelling, probably made by early English importers, cacao beans are known as cocoa beans in English-speaking countries.)

All chocolate products are manufactured from ground cocoa beans. The heat from the grinding process causes the cocoa bean mixture to melt, forming a free-flowing mixture called *chocolate liquor*. Unsweetened baker's chocolate is simply cooled, hardened chocolate liquor. Semisweet chocolate has added granulated sugar. Milk chocolate has added sugar, milk solids, and vanilla flavoring.

Because of their plant origins, chocolate products contain alkaloids. The dominant alkaloid present is theobromine, with caffeine being present in a smaller amount. The name theobromine comes from the Greek term *theobroma* meaning "food of the Gods." The concentrations of these two alkaloids in cocoa beans varies depending on the origin of the beans. The following table gives theobromine and caffeine content of several finished chocolate products.

The caffeine content of a typical chocolate bar is 30 mg and that of a slice of chocolate cake 20–30 mg. By contrast, a cup of coffee contains 100–150 mg of caffeine and a twelve-ounce cola drink contains 33–52 mg.

Structurally, theobromine and caffeine differ only by a methyl group.

Theobromine
(3,7-dimethyxanthine)

Caffeine
(1,3,7-trimethylxanthine)

This close structural similarity does not, however, translate into close pharmacological properties. Theobromine's stimulant effects on the central nervous system are minimal compared to those of caffeine. A mild diuretic effect and relaxation of the smooth muscles of the bronchi in the lungs are two other effects of theobromine; caffeine has similar effects in these areas.

Theobromine has been used as a pharmaceutical drug for its diuretic effect. Because of its ability to dilate blood vessels, theobromine also has been used to treat high blood pressure. Research shows that pets, especially dogs, are sensitive to theobromine because the animals metabolize theobromine more slowly than humans. A chocolate bar, inadvertently ingested, is poisonous to dogs and can even be lethal. The same holds true for cats.

Occasionally, chocolate is touted as a "health" food because cocoa beans have relatively high levels of several kinds of antioxidant flavonoids (see the Chemical Connections feature on page 805). Studies show that people with high blood levels of flavonoids are at lower risk of developing heart disease, asthma, and type 2 diabetes. Dark chocolate contains the most cocoa and thus the most flavonoids. As a "health" food, however, chocolate should be consumed only occasionally because the downside of consumption is the high number of calories associated with chocolate.

Theobromine and Caffeine Content of Finished Chocolate Products

Product	Theobromine, %	Caffeine, %	Theobromine/ Caffeine Ratio
baking chocolate	1.386	0.164	8.45 to 1
dark sweet chocolate	0.474	0.076	6.3 to 1
milk chocolate	0.197	0.022	9.0 to 1

less-costly sequence of chemical reactions. The second is less potential for abuse of the drug. While the use of either codeine derivative can be habit-forming, and can lead to physical and psychological addiction, oxycodone effects are generally greater than those for hydrocodone at equivalent dosages. A driving force for the use of oxycodone formulations over hydrocodone formulations is that the onset of pain relief occurs faster for oxycodone.

17.12 STRUCTURE AND CLASSIFICATION OF AMIDES

An **amide** *is a carboxylic acid derivative in which the carboxyl —OH group has been replaced with an amino or a substituted amino group.* The amide functional group is thus

$$\underset{\text{}}{\overset{\displaystyle O}{\underset{\|}{-C-NH_2}}} \quad \text{or} \quad \overset{\displaystyle O}{\underset{\|}{-C-NH-R}} \quad \text{or} \quad \overset{\displaystyle O}{\underset{\underset{R}{\|}}{-C-N-R}}$$

depending on the degree of substitution.

Amides, like amines, can be classified as primary (1°), secondary (2°), or tertiary (3°), depending on how many hydrogen atoms are attached to the nitrogen atom.

$$\underset{\text{Primary amide}}{R-\overset{\displaystyle O}{\underset{\|}{C}}-NH_2} \qquad \underset{\text{Secondary amide}}{R-\overset{\displaystyle O}{\underset{\|}{C}}-NH-R'} \qquad \underset{\text{Tertiary amide}}{R-\overset{\displaystyle O}{\underset{\underset{R''}{\|}}{C}}-N-R'}$$

A **primary amide** *is an amide in which two hydrogen atoms are bonded to the amide nitrogen atom.* Such amides are also called *unsubstituted* amides. A **secondary amide** *is an amide in which an alkyl (or aryl) group and a hydrogen atom are bonded to the amide nitrogen atom.* *Monosubstituted* amide is another name for this type of amide. A **tertiary amide** *is an amide in which two alkyl (or aryl) groups and no hydrogen atoms are bonded to the amide nitrogen atom.* Such amides are *disubstituted* amides.

Note that the difference between a 1° amide and a 2° amide is "H versus R" and that the difference between a 2° amide and a 3° amide is again "H versus R." These "H versus R" relationships are the same relationships that exist between 1° and 2° amines and 2° and 3° amines (Section 17.2), as is summarized in Figure 17.12.

> Primary, secondary, and tertiary amides are also called unsubstituted, monosubstituted, and disubstituted amides, respectively.

Figure 17.12 Primary, secondary, and tertiary amines and amides and the "H versus R" relationship.

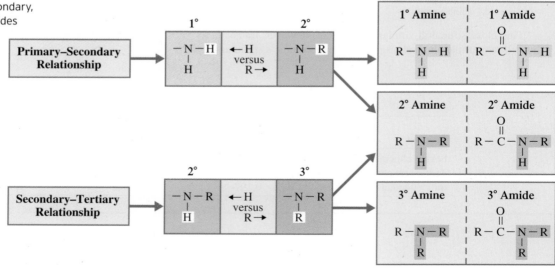

The simplest amide has a hydrogen atom attached to an unsubstituted amide functional group.

$$H-\underset{\underset{\|}{O}}{C}-NH_2$$

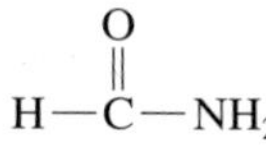

Next in complexity are amides in which a methyl group is present. There are two of them, one with the methyl group attached to the carbon atom and the other with the methyl group attached to the nitrogen atom.

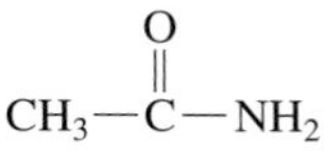

$$CH_3-\underset{\underset{\|}{O}}{C}-NH_2 \quad \text{and} \quad H-\underset{\underset{\|}{O}}{C}-NH-CH_3$$

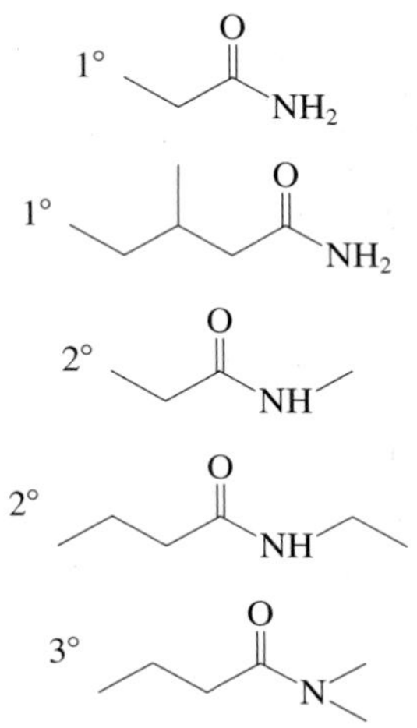

The first of these structures is a 1° amide, and the second structure is a 2° amide. The structure of the simplest aromatic amide involves a benzene ring to which an unsubstituted amide functional group is attached.

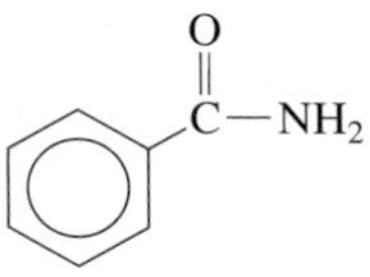

Cyclic amide structures are possible. Examples of such structures include

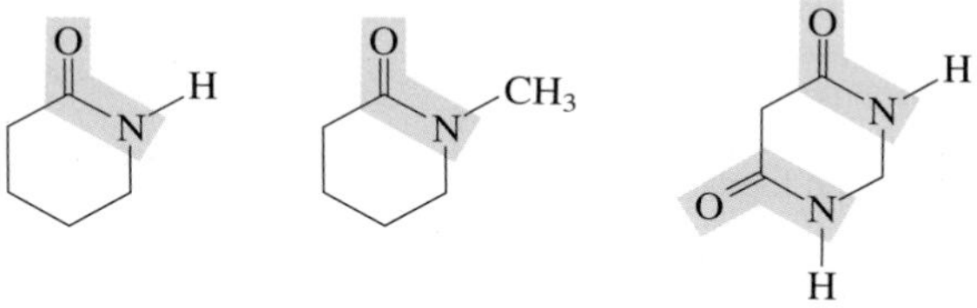

Cyclic amides are called *lactams,* a term that parallels the use of the term *lactones* for cyclic esters (Section 16.11).

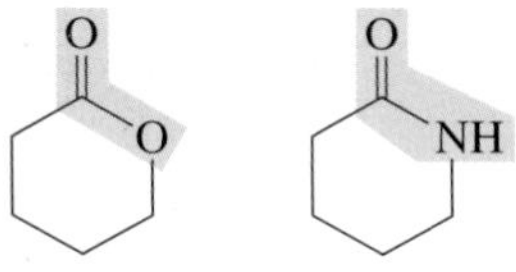

A lactone
(a cyclic ester)

A lactam
(a cyclic amide)

A **lactam** *is a cyclic amide.* The ring size in a lactam is indicated using a Greek letter. A lactam with a four-membered ring is a β-lactam because the β carbon from the carbonyl group is bonded to the heteroatom. A lactam with a five-membered ring is a γ-lactam.

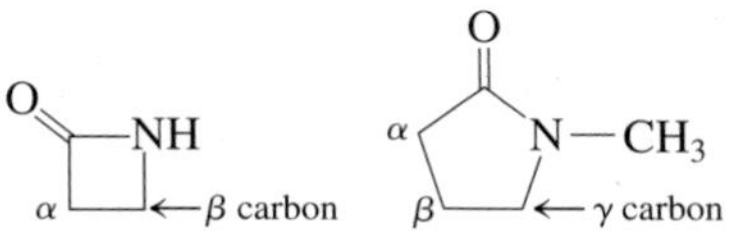

β-Lactam

γ-Lactam

The members of the penicillin family of antibiotics (Section 21.9) have structures that contain a β-lactam ring.

17.13 NOMENCLATURE FOR AMIDES

For nomenclature purposes (both IUPAC and common), amides are considered to be derivatives of carboxylic acids. Hence their names are based on the name of the parent carboxylic acid. (A similar procedure was used for naming esters; Section 16.12). The rules are as follows:

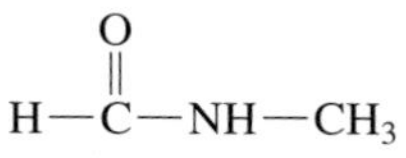

Line-angle structural formulas for selected primary, secondary, and tertiary amides:

1°

1°

2°

2°

3°

Rule 1: The ending of the name of the carboxylic acid is changed from *-ic acid* (common) or *-oic acid* (IUPAC) to *-amide*. For example, *benzoic acid* becomes *benzamide*.

Rule 2: The names of groups attached to the nitrogen (2° and 3° amides) are appended to the front of the base name, using an *N-* prefix as a locator.

Selected primary amide IUPAC names (with the common name in parentheses) are

$$H-\overset{\overset{\displaystyle O}{\|}}{C}-NH_2 \qquad CH_3-\overset{\overset{\displaystyle O}{\|}}{C}-NH_2 \qquad CH_3-CH_2-\overset{\overset{\displaystyle O}{\|}}{C}-NH_2$$

Methanamide (formamide) Ethanamide (acetamide) Propanamide (propionamide)

$$CH_3-\underset{\underset{\displaystyle CH_3}{|}}{CH}-CH_2-\overset{\overset{\displaystyle O}{\|}}{C}-NH_2 \qquad CH_3-\underset{\underset{\displaystyle Cl}{|}}{\overset{\overset{\displaystyle CH_3}{|}}{C}}-\overset{\overset{\displaystyle O}{\|}}{C}-NH_2$$

3-Methylbutanamide (*β*-methylbutyramide) 2-Chloro-2-methylpropanamide (*α*-chloro-*α*-methylpropionamide)

Nomenclature for secondary and tertiary amides, amides with substituted amino groups, involves use of the prefix *N-*, a practice we previously encountered with amine nomenclature (Section 17.3).

$$CH_3-CH_2-\overset{\overset{\displaystyle O}{\|}}{C}-NH-CH_3 \qquad CH_3-\overset{\overset{\displaystyle O}{\|}}{C}-\underset{\underset{\displaystyle CH_3}{|}}{\overset{\overset{\displaystyle CH_3}{|}}{N}}-CH_3$$

N-Methylpropanamide (*N*-methylpropionamide) *N,N*-Dimethylethanamide (*N,N*-dimethylacetamide)

Molecular models for methanamide and its *N*-methyl and *N,N*-dimethyl derivatives (the simplest 1°, 2°, and 3° amides, respectively) are given in Figure 17.13.

The simplest aromatic amide, a benzene ring bearing an unsubstituted amide group, is called *benzamide*. Other aromatic amides are named as benzamide derivatives.

Benzamide 2-Methylbenzamide *N*-Methylbenzamide

Acrylamide (2-propenamide), the simplest unsaturated amide, has the structure

$$CH_2=CH-\overset{\overset{\displaystyle O}{\|}}{C}-NH_2$$

It is a neurotoxic agent and a possible human carcinogen.

Surprisingly, low concentrations of acrylamide have been found in potato chips, french fries, and other starchy foods prepared at high temperatures (greater than 120°C). Its possible source is the reaction between the amino acid asparagine (present in food proteins; Section 20.2) and carbohydrate sugars (Section 18.8) present in food.

Human risk studies are underway concerning acrylamide presence in fried and some baked foods. No traces of acrylamide have been found in uncooked or boiled foods.

● **E X A M P L E 17.5**

Determining IUPAC and Common Names for Amides

The contrast between IUPAC names and common names for unbranched unsubstituted amides is as follows:

IUPAC (one word)

| alkanamide |

ethanamide

Common (one word)

| (prefix)amide* |

acetamide

*The common-name prefixes are re-lated to natural sources for the acids.

Assign both common and IUPAC names to each of the following amides.

a.
$$CH_3-CH_2-CH_2-\overset{\overset{\displaystyle O}{\|}}{C}-NH_2$$

b.
$$CH_3-\underset{\underset{\displaystyle Br}{|}}{CH}-\overset{\overset{\displaystyle O}{\|}}{C}-NH-CH_3$$

c.

d.

Solution

a. The parent acid for this amide is butyric acid (common) or butanoic acid (IUPAC). The common name for this amide is *butyramide,* and the IUPAC name is *butanamide*.

b. The common and IUPAC names of the acid are very similar; they are propionic acid and propanoic acid, respectively. The common name is *α-bromo-N-methylpropionamide,*

$$O \\ \parallel \\ H-C-NH_2$$

Methanamide
(a primary amide)

$$O \\ \parallel \\ H-C-NH-CH_3$$

N-Methyl methanamide
(a secondary amide)

$$O \\ \parallel \\ H-C-N-CH_3 \\ \quad\quad | \\ \quad\quad CH_3$$

N,N-Dimethyl methanamide
(a tertiary amide)

Figure 17.13 Space-filling models for the simplest primary, secondary, and tertiary amides.

and the IUPAC name is *2-bromo-N-methylpropanamide*. The prefix *N-* must be used with the methyl group to indicate that it is attached to the nitrogen atom.

c. In both the common and IUPAC systems of nomenclature, the name of the parent acid is the same: benzoic acid. The name of the amide is *N,N-diphenylbenzamide*.

d. The amide is derived from valeric acid (common name) or pentanoic acid (IUPAC name). The complete name must take into account the presence of the methyl group on the carbon chain. The amide's common name is *β-methylvaleramide* and its IUPAC name is *3-methylpentanamide*.

Practice Exercise 17.5

Assign both common and IUPAC names to each of the following amides.

a. 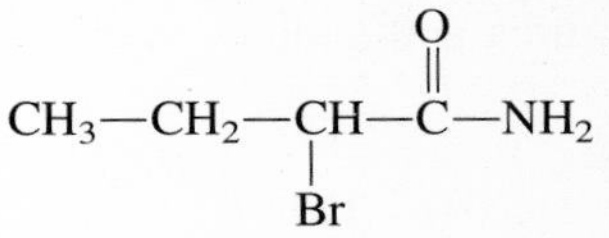
$$CH_3-CH_2-CH-C-NH_2 \\ \quad\quad\quad | \\ \quad\quad\quad Br$$
with C=O above.

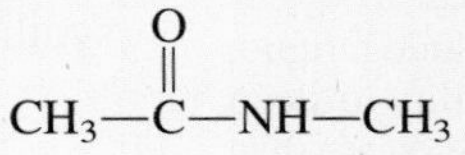

b.
$$CH_3-C-NH-CH_3$$ with C=O.

c.
benzene ring—C(=O)—N—CH₃ with CH₃ below.

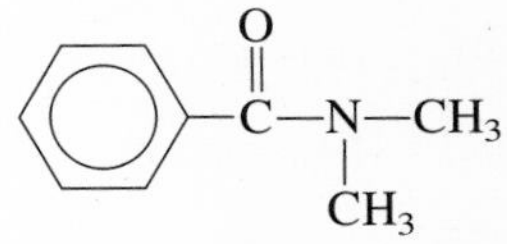

d. 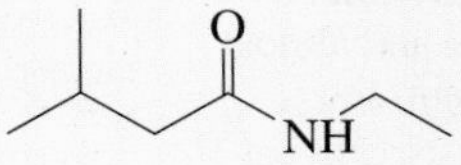
(CH₃)₂CH-CH₂-C(=O)-NH-CH₂CH₃

Answers: a. *α*-bromobutyramide, 2-bromobutanamide; **b.** *N*-methylacetamide, *N*-methylethanamide; **c.** *N,N*-dimethylbenzamide (both common and IUPAC name); **d.** *N*-ethyl-*β*-methylbutyramide, *N*-ethyl-3-methylbutanamide

17.14 SELECTED AMIDES AND THEIR USES

The simplest naturally occurring amide is urea, a water-soluble white solid produced in the human body from carbon dioxide and ammonia through a complex series of metabolic reactions (Section 26.4).

$$CO_2 + 2NH_3 \longrightarrow (H_2N)_2CO + H_2O$$

Urea is a one-carbon diamide. Its molecular structure is

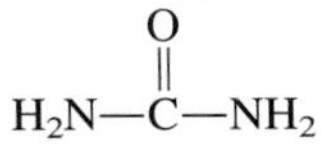

$$H_2N-C-NH_2$$ with C=O.

Urea formation is the human body's primary method for eliminating "waste" nitrogen. The kidneys remove urea from the blood and provide for its excretion in urine. With malfunctioning kidneys, urea concentrations in the body can build to toxic levels—a condition called *uremia*.

The complex amide melatonin is a hormone that is synthesized by the pineal gland and that regulates the sleep–wake cycle in humans. Melatonin levels within the body increase in evening hours and then decrease as morning approaches. High melatonin levels are associated with longer and more sound sleeping. The concentration of this hormone in the blood decreases with age; a six-year-old has a blood melatonin concentration over five times that of an 80-year-old. This is one reason why young children have less trouble sleeping than senior citizens. As a prescription drug, melatonin is used to treat insomnia and jet lag.

Structurally, melatonin is a polyfunctional amide; amine and ether groups are also present.

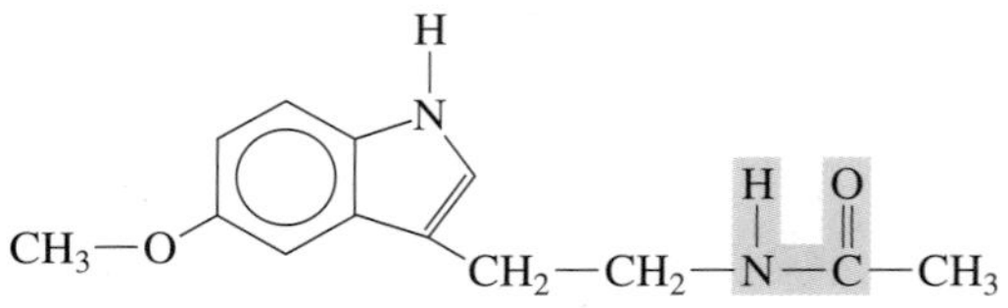

Synthetically produced melatonin is under investigation as a drug for treating jet lag. Jet lag is a condition caused by desynchronization of the biological clock. It is usually caused by drastically changing the sleep–wake cycle, as when crossing several time zones during an airline flight or when performing shift work. Symptoms of jet lag include fatigue, early awakening or inability to sleep, and headaches. Studies indicate that melatonin taken in the evening in a new time zone will usually reset a person's biological clock and almost totally alleviate (or prevent) the symptoms of jet lag.

A number of synthetic amides exhibit physiological activity and are used as drugs in the human body. Foremost among them, in terms of use, is acetaminophen, which in 1992 replaced aspirin as the top-selling over-the-counter pain reliever. Acetaminophen is a derivative of acetamide (see the Chemical Connections feature on page 538).

Barbiturates, which are cyclic amide compounds, are a heavily used group of prescription drugs that cause relaxation (tranquilizers), sleep (sedatives), and death (overdoses). All barbiturates are derivatives of barbituric acid, a cyclic amide that was first synthesized from urea and malonic acid.

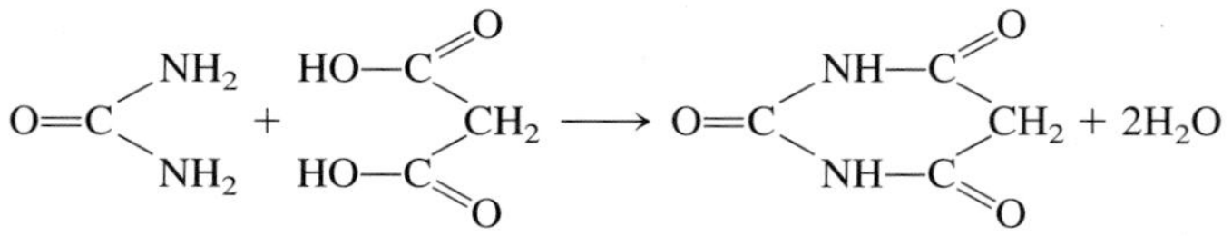

Urea Malonic acid Barbituric acid

(The researcher who first synthesized this compound named it after his girlfriend Barbara.)

Unbranched Primary Amides			
C_1	C_3	C_5	C_7
C_2	C_4	C_6	C_8

☐ Liquid ☐ Solid

Figure 17.14 A physical-state summary for unbranched primary amides at room temperature and pressure.

17.15 PHYSICAL PROPERTIES OF AMIDES

Amides do not exhibit basic properties in solution as amines do (Section 17.6). Although the nitrogen atom present in amides has a nonbonding pair of electrons, as in amines, these electrons are not available for bonding to a H^+ ion. The reason for this is related to the polarity of the carbonyl portion (—C=O) of the amide functional group.

Methanamide and its *N*-methyl and *N,N*-dimethyl derivatives (the simplest 1°, 2°, and 3° amides, respectively), are all liquids at room temperature. All unbranched primary amides, except methanamide, are solids at room temperature (Figure 17.14), as are most other amides. In many cases, the amide melting point is even higher than that of the corresponding carboxylic acid. The high melting points result from the numerous intermolecular hydrogen-bonding possibilities that exist between amide H atoms and carbonyl O atoms. Figure 17.15 shows selected hydrogen-bonding interactions that are possible among several primary amide molecules.

Fewer hydrogen-bonding possibilities exist for 2° amides because the nitrogen atom now has only one hydrogen atom; hence lower melting points are the rule for such amides. Still lower melting points are observed for 3° amides because no hydrogen bonding is possible. The disubstituted *N,N*-dimethylacetamide has a melting point of −20°C, which is about 100°C lower than that of the unsubstituted acetamide.

Amides of low molecular mass, up to five or six carbon atoms, are soluble in water. Again, numerous hydrogen-bonding possibilities exist between water and the amide. Even disubstituted amides can participate in such hydrogen bonding.

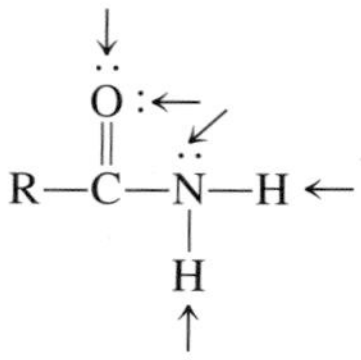

Arrows denote sites where hydrogen bonding to water can occur.

Lidocaine (xylocaine), a substance commonly administered by injection as a dental anesthetic, is a synthetic molecule that contains both amide and amine functional groups.

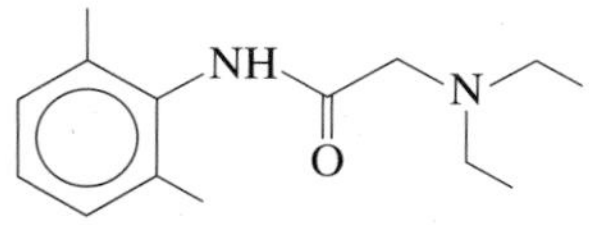

Another well known local anesthetic is procaine (novocaine). Its structure contains two amine groups and an ester group but no amide group.

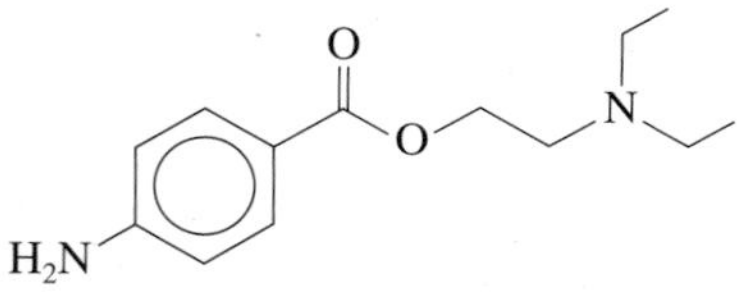

Both lidocaine and procaine share a common structural feature—the presence of a diethyl amino group (on the right side of each structure).

Figure 17.15 The high boiling points of amides are related to the numerous amide-amide hydrogen-bonding possibilities that exist.

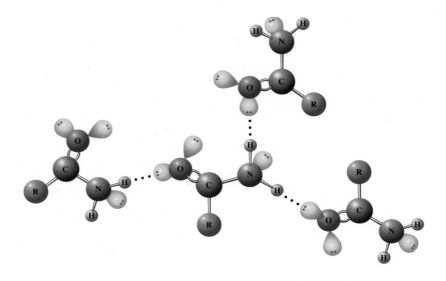

17.16 PREPARATION OF AMIDES

Amides are the least reactive of the common carboxylic acid derivatives and they can be synthesized from an acid chloride, an acid anhydride, an ester, or the carboxylic acid itself.

The reaction of a carboxylic acid with ammonia or a 1° or 2° amine produces an amide, provided that the reaction is carried out at an elevated temperature (greater than 100°C) and a dehydrating agent is present.

$$\text{Ammonia + carboxylic acid} \xrightarrow[\text{Catalyst}]{100°C} 1° \text{ amide}$$

$$1° \text{ Amine + carboxylic acid} \xrightarrow[\text{Catalyst}]{100°C} 2° \text{ amide}$$

$$2° \text{ Amine + carboxylic acid} \xrightarrow[\text{Catalyst}]{100°C} 3° \text{ amide}$$

If the preceding reactions are run at room temperature (25°C), no amide formation occurs; instead an acid–base reaction occurs in which a carboxylic acid salt is produced. This acid–base reaction when a 1° amine is the reactant is

$$\underset{\text{Acid}}{\text{R}-\overset{\text{O}}{\overset{\|}{\text{C}}}-\text{OH}} + \underset{\substack{\text{Primary} \\ \text{amine}}}{\text{H}-\overset{\text{H}}{\underset{}{\text{N}}}-\text{R}} \xrightarrow{25° \text{ C}} \underset{\text{Carboxylate salt}}{\text{H}-\overset{\text{H}}{\underset{\text{H}}{\overset{+}{\text{N}}}}-\text{R} \quad \text{R}-\overset{\text{O}}{\overset{\|}{\text{C}}}-\text{O}^-}$$

General structural equations for 1°, 2°, and 3° amide production from carboxylic acids are

> The reaction of a carboxylic acid with ammonia or an amine to produce an amide is similar to the reaction of a carboxylic acid with an alcohol. In both cases, water is formed as a by-product as the —OH part of the carboxylic acid is replaced.

$$\underset{\substack{\text{Carboxylic} \\ \text{acid}}}{\text{R}-\overset{\text{O}}{\overset{\|}{\text{C}}}-\text{OH}} + \underset{\text{Ammonia}}{\text{H}-\overset{\text{H}}{\underset{}{\text{N}}}-\text{H}} \xrightarrow[\text{catalyst}]{100°C} \underset{\text{Primary amide}}{\text{R}-\overset{\text{O}}{\overset{\|}{\text{C}}}-\overset{\text{H}}{\underset{}{\text{N}}}-\text{H}} + \text{H}_2\text{O}$$

$$\underset{\substack{\text{Carboxylic} \\ \text{acid}}}{\text{R}-\overset{\text{O}}{\overset{\|}{\text{C}}}-\text{OH}} + \underset{\substack{\text{Primary} \\ \text{amine}}}{\text{H}-\overset{\text{H}}{\underset{}{\text{N}}}-\text{R}} \xrightarrow[\text{catalyst}]{100°C} \underset{\text{Secondary amide}}{\text{R}-\overset{\text{O}}{\overset{\|}{\text{C}}}-\overset{\text{H}}{\underset{}{\text{N}}}-\text{R}} + \text{H}_2\text{O}$$

$$\underset{\substack{\text{Carboxylic} \\ \text{acid}}}{\text{R}-\overset{\text{O}}{\overset{\|}{\text{C}}}-\text{OH}} + \underset{\substack{\text{Secondary} \\ \text{amine}}}{\text{H}-\overset{\text{R}}{\underset{}{\text{N}}}-\text{R}} \xrightarrow[\text{catalyst}]{100°C} \underset{\text{Tertiary amide}}{\text{R}-\overset{\text{O}}{\overset{\|}{\text{C}}}-\overset{\text{R}}{\underset{}{\text{N}}}-\text{R}} + \text{H}_2\text{O}$$

CHEMICAL Connections

Acetaminophen: A Substituted Amide

Often called the aspirin substitute, acetaminophen is the most widely used of all nonprescription pain relievers, accounting for over half of that market. Acetaminophen is a derivative of acetamide in which a hydroxyphenyl group has replaced one of the amide hydrogens.

$$CH_3-\overset{\overset{\displaystyle O}{\|}}{C}-NH_2 \qquad\qquad CH_3-\overset{\overset{\displaystyle O}{\|}}{C}-NH-\underset{}{\bigcirc}-OH$$

Acetamide   Acetaminophen

The pharmaceutical designation APAP for this compound comes from its IUPAC name, which is *N-acetyl-p-amino*phenol.

Acetaminophen is the active ingredient in Tylenol, Datril, Tempra, Equate, and Anacin-3. Excedrin, which contains both acetaminophen and aspirin, is a combination pain reliever.

Acetaminophen is often used as an aspirin substitute because it has no irritating effect on the intestinal tract and yet has comparable analgesic and antipyretic effects. Unlike aspirin, however, it is not effective against inflammation and is of limited use for the aches and pains of arthritis. Also, acetaminophen does not inhibit platelet aggregation and therefore is not useful for preventing blood clotting.

Acetaminophen is available in a liquid form that is used extensively for small children and other patients who have difficulty taking solid tablets. The wide use of acetaminophen for children has a drawback; it is the drug most often involved in childhood poisonings.

In *large* doses, acetaminophen can cause liver and kidney damage. Such effects are not found when acetaminophen is taken as directed. For this reason, the maximum adult dosage of 4 g should not be exceeded (eight 500 mg tablets) and extra-strength formulations should be used with great caution. Analgesic abuse is a real potential with the heavily advertised extra-strength formulations.

Acetaminophen's mode of action in the body is similar to that of aspirin—inhibition of prostaglandin synthesis.

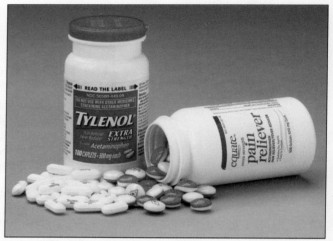

Over-the-counter pain relievers such as Tylenol contain acetaminophen as the active ingredient.

This is the fourth time we have encountered *condensation* reactions. Esterification (Section 16.11), acetal formation (Section 15.11), and intermolecular alcohol dehydration (Section 14.9) were the other three condensation situations.

These reactions are called *amidification* reactions. An **amidification reaction** *is the reaction of a carboxylic acid with an amine (or ammonia) to produce an amide.* In amidification, an —OH group is lost from the carboxylic acid, a —H atom is lost from the ammonia or amine, and water is formed as a by-product. Amidification reactions are thus *condensation* reactions.

Two specific amidification reactions, in which a 2° amide and a 3° amide are produced, respectively, are

$$CH_3-CH_2-CH_2-\overset{\overset{\displaystyle O}{\|}}{C}-OH + H-\overset{\overset{\displaystyle H}{|}}{N}-CH_3 \xrightarrow[\text{catalyst}]{100°C} CH_3-CH_2-CH_2-\overset{\overset{\displaystyle O}{\|}}{C}-\overset{\overset{\displaystyle H}{|}}{N}-CH_3 + H_2O$$

Butanoic acid   Methylamine (1° amine)   *N*-Methylbutanamide (a 2° amide)

$$\bigcirc-\overset{\overset{\displaystyle O}{\|}}{C}-OH + H-\overset{\overset{\displaystyle |}{N}}{\underset{\overset{\displaystyle |}{CH_2-CH_3}}{}}-CH_2-CH_3 \xrightarrow[\text{catalyst}]{100°C} \bigcirc-\overset{\overset{\displaystyle O}{\|}}{C}-\underset{\overset{\displaystyle |}{CH_2-CH_3}}{N}-CH_2-CH_3 + H_2O$$

Benzoic acid   Diethylamine (2° amine)   *N,N*-Diethylbenzamide (a 3° amide)

Just as it is useful to think of the structure of an ester (Section 16.10) in terms of an "acid part" and an "alcohol part," it is useful to think of an amide in terms of an "acid part" and an "amine (or ammonia) part."

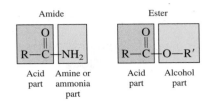

In this context, it is easy to identify the parent acid and amine from which a given amide can be produced; to generate the parent molecules, just add an —OH group to the acid part of the amide and an H atom to the amine part.

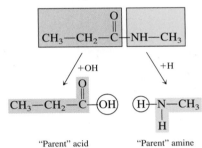

EXAMPLE 17.6

Predicting Reactants Needed to Prepare Specific Amides

What carboxylic acid and amine (or ammonia) are needed to prepare each of the following amides?

a.

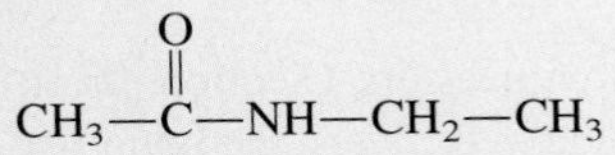

$$CH_3-\overset{\displaystyle O}{\overset{\displaystyle \|}{C}}-NH-CH_2-CH_3$$

b.

$$CH_3-CH_2-\overset{\displaystyle O}{\overset{\displaystyle \|}{C}}-NH_2$$

c.

$$CH_3-CH_2-\overset{\displaystyle O}{\overset{\displaystyle \|}{C}}-\underset{\displaystyle \underset{\displaystyle CH_3}{|}}{N}-CH_3$$

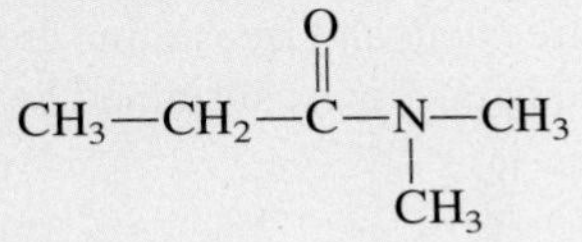

Solution

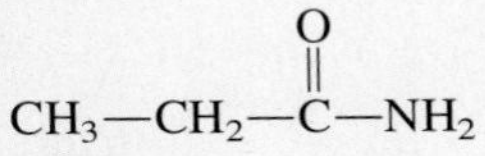

a. Viewing the molecule as having an acid part and an amine part, we obtain

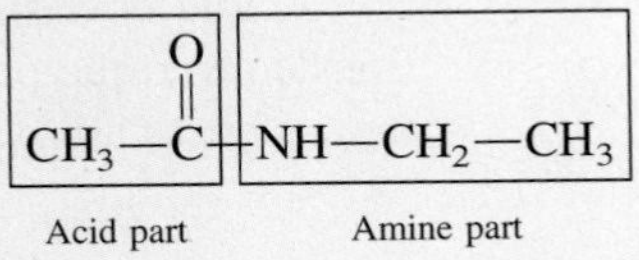

$$\boxed{CH_3-\overset{\displaystyle O}{\overset{\displaystyle \|}{C}}}\,\boxed{NH-CH_2-CH_3}$$

Acid part   Amine part

Adding an —OH group to the acid part and a H atom to the amine part, we obtain the "parent" molecules, which are

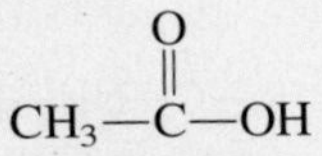

$$CH_3-\overset{\displaystyle O}{\overset{\displaystyle \|}{C}}-OH \quad \text{and} \quad CH_3-CH_2-NH_2$$

b. Proceeding as in part **a,** we find that the "parent" acid and amine molecules are, respectively,

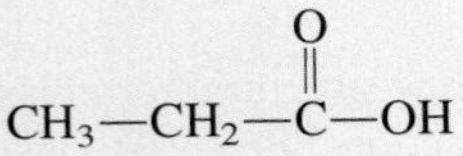

$$CH_3-CH_2-\overset{\displaystyle O}{\overset{\displaystyle \|}{C}}-OH \quad \text{and} \quad NH_3$$

c. Proceeding again as in part **a,** we find that the "parent" acid and amine molecules are, respectively,

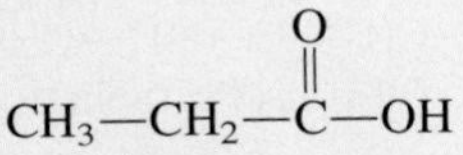

$$CH_3-CH_2-\overset{\displaystyle O}{\overset{\displaystyle \|}{C}}-OH \quad \text{and} \quad CH_3-NH-CH_3$$

(continued)

Practice Exercise 17.6

What carboxylic acid and amine (or ammonia) are needed to prepare each of the following amides?

a.

$$CH_3-\overset{\displaystyle O}{\overset{\|}{C}}-\underset{\underset{\displaystyle CH_3}{|}}{N}-CH_2-CH_3$$

b.

$$CH_3-CH_2-\overset{\displaystyle O}{\overset{\|}{C}}-NH-CH_3$$

c.

Benzene ring with $-\overset{\displaystyle O}{\overset{\|}{C}}-NH_2$

Answers:

a.

$$CH_3-\overset{\displaystyle O}{\overset{\|}{C}}-OH,\ CH_3-CH_2-NH-CH_3$$

b.

$$CH_3-CH_2-\overset{\displaystyle O}{\overset{\|}{C}}-OH,\ CH_3-NH_2$$

c.

Benzene ring with $-\overset{\displaystyle O}{\overset{\|}{C}}-OH,\ NH_3$

17.17 HYDROLYSIS OF AMIDES

As was the case with esters (Section 16.16), the most important reaction of amides is hydrolysis. In amide hydrolysis, the bond between the carbonyl carbon atom and the nitrogen is broken, and free acid and free amine are produced. Amide hydrolysis is catalyzed by acids, bases, or certain enzymes; sustained heating is also often required.

$$R-\overset{\displaystyle O}{\overset{\|}{C}}\overset{\frown}{-}NH-R' + H_2O \xrightarrow{\text{Heat}} R-\overset{\displaystyle O}{\overset{\|}{C}}-OH + R'-NH_2$$

Amide   Carboxylic   Amine
acid

Amide hydrolysis under basic conditions is also called amide saponification, just as ester hydrolysis under basic conditions is called ester saponification (Section 16.16).

Acidic or basic hydrolysis conditions have an effect on the products. *Acidic* conditions convert the product amine to an amine salt (Section 17.7). *Basic* conditions convert the product carboxylic acid to a carboxylic acid salt (Section 16.9).

$$R-\overset{\displaystyle O}{\overset{\|}{C}}-NH-R' + H_2O + \boxed{HCl} \xrightarrow{\text{Heat}} R-\overset{\displaystyle O}{\overset{\|}{C}}-OH + R'-\overset{+}{N}H_3\ Cl^-$$

Acidic hydrolysis of an amide   Carboxylic acid   Amine salt

$$R-\overset{\displaystyle O}{\overset{\|}{C}}-NH-R' + \boxed{NaOH} \xrightarrow{\text{Heat}} R-\overset{\displaystyle O}{\overset{\|}{C}}-O^-\ Na^+ + R'-NH_2$$

Basic hydrolysis of an amide   Carboxylic acid salt   Amine

● EXAMPLE 17.7

Predicting the Products of Amide Hydrolysis Reactions

Draw structural formulas for the organic products of each of the following amide hydrolysis reactions. Be sure to take into account whether the hydrolysis occurs under neutral, acidic, or basic conditions.

a.

$$CH_3-CH_2-\overset{\displaystyle O}{\overset{\|}{C}}-NH-CH_3 + H_2O \xrightarrow{\text{Heat}}$$

b.

$$CH_3-CH_2-\overset{\displaystyle O}{\overset{\|}{C}}-NH-CH_2-CH_3 + H_2O \xrightarrow[\text{HCl}]{\text{Heat}}$$

c.

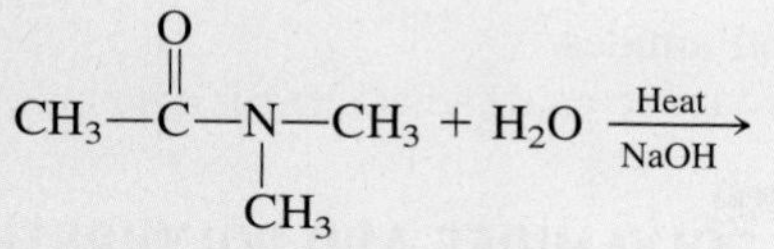

$$CH_3-\overset{\overset{\displaystyle O}{\|}}{C}-\underset{\underset{\displaystyle CH_3}{|}}{N}-CH_3 + H_2O \xrightarrow[\text{NaOH}]{\text{Heat}}$$

d.

$$CH_3-\underset{\underset{\displaystyle CH_3}{|}}{CH}-\overset{\overset{\displaystyle O}{\|}}{C}-NH_2 + H_2O \xrightarrow{\text{Heat}}$$

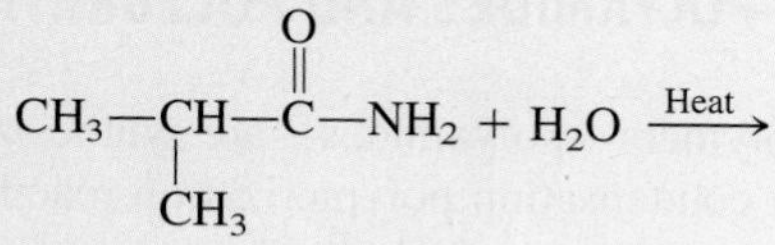

Solution

a. This reaction is hydrolysis under neutral conditions. The products will be the "parent" acid and amine for the amide. These "parents" are

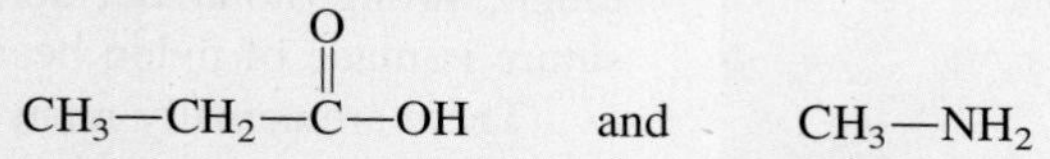

$$CH_3-CH_2-\overset{\overset{\displaystyle O}{\|}}{C}-OH \quad \text{and} \quad CH_3-NH_2$$

b. This reaction is hydrolysis under acidic conditions. The acid is hydrochloric acid (HCl). The products will be the "parent" carboxylic acid and the chloride salt of the amine. The HCl converts the amine to its chloride salt.

$$CH_3-CH_2-\overset{\overset{\displaystyle O}{\|}}{C}-OH \quad \text{and} \quad CH_3-CH_2-\overset{+}{N}H_3\ Cl^-$$

c. This reaction is hydrolysis under basic conditions. The base is sodium hydroxide (NaOH). The products will be the "parent" amine and the salt of the carboxylic acid. The NaOH converts the carboxylic acid to its sodium salt.

$$CH_3-\overset{\overset{\displaystyle O}{\|}}{C}-O^-\ Na^+ \quad \text{and} \quad CH_3-NH-CH_3$$

d. This reaction is hydrolysis under neutral conditions. The products will be the "parent" acid and amine of the amide. Because the amide is unsubstituted, the parent amine is actually ammonia.

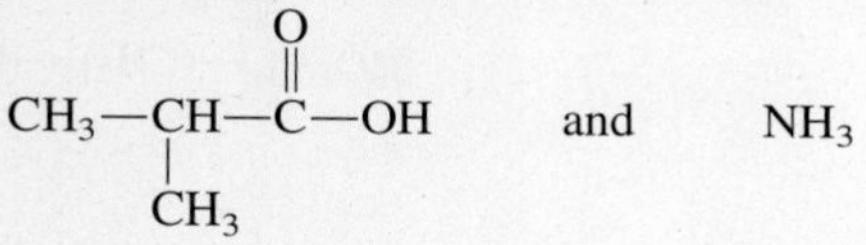

$$CH_3-\underset{\underset{\displaystyle CH_3}{|}}{CH}-\overset{\overset{\displaystyle O}{\|}}{C}-OH \quad \text{and} \quad NH_3$$

Practice Exercise 17.7

Draw structural formulas for the organic products of each of the following amide hydrolysis reactions. Be sure to take into account whether the hydrolysis occurs under neutral, acidic, or basic conditions.

a.
$$CH_3-\overset{\overset{\displaystyle O}{\|}}{C}-NH-CH_3 + H_2O \xrightarrow[\text{NaOH}]{\text{Heat}}$$

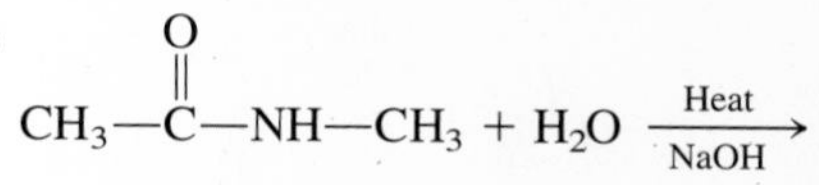

b.
$$CH_3-\overset{\overset{\displaystyle O}{\|}}{C}-NH-CH_3 + H_2O \xrightarrow[\text{HCl}]{\text{Heat}}$$

c.
$$CH_3-\overset{\overset{\displaystyle O}{\|}}{C}-NH-CH_3 + H_2O \xrightarrow{\text{Heat}}$$

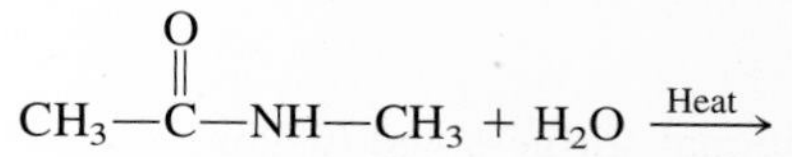

d.
$$\bigcirc\!\!\!\!\!\!-\overset{\overset{\displaystyle O}{\|}}{C}-NH_2 + H_2O \xrightarrow{\text{Heat}}$$

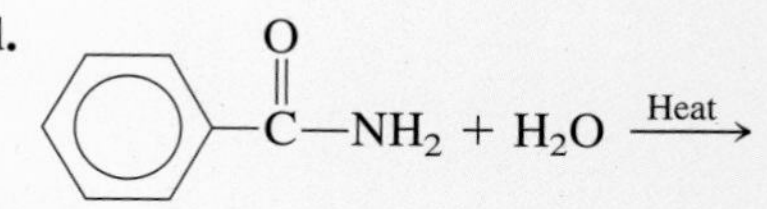

Answers:

a.
$$CH_3-\overset{\overset{\displaystyle O}{\|}}{C}-O^-\ Na^+,\ CH_3-NH_2$$

b.
$$CH_3-\overset{\overset{\displaystyle O}{\|}}{C}-OH,\ CH_3-\overset{+}{N}H_3\ Cl^-$$

c.
$$CH_3-\overset{\overset{\displaystyle O}{\|}}{C}-OH,\ CH_3-NH_2$$

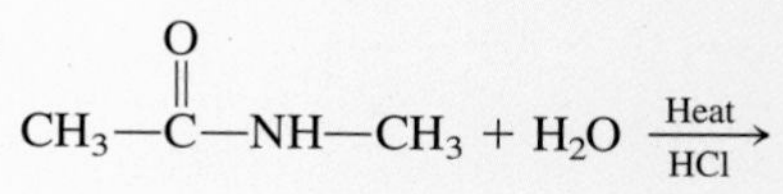

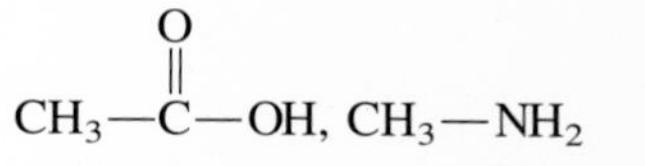

d.
$$\bigcirc\!\!\!\!\!\!-\overset{\overset{\displaystyle O}{\|}}{C}-OH,\ NH_3$$

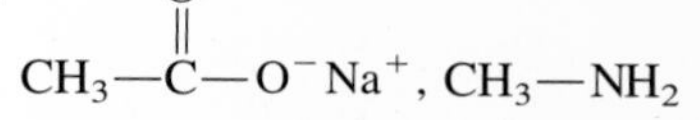

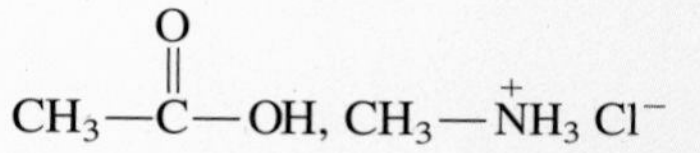

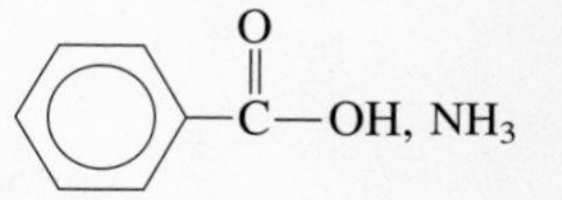

The Chemistry at a Glance feature on page 543 summarizes the reactions that involve amines and amides.

POLYAMIDES AND POLYURETHANES

Amide polymers—polyamides—are synthesized by combining diamines and dicarboxylic acids in a condensation polymerization reaction (Section 16.18). A **polyamide** *is a condensation polymer in which the monomers are joined through amide linkages.*

The most important synthetic polyamide is *nylon*. Nylon is used in clothing and hosiery, as well as in carpets, tire cord, rope, and parachutes. It also has nonfiber uses; for example, it is used in paint brushes, electrical parts, valves, and fasteners. It is a tough, strong, nontoxic, nonflammable material that is resistant to chemicals. Surgical suture is made of nylon because it is such a strong fiber.

There are actually many different types of nylon, all of which are based on diamine and diacid monomers. The most important nylon is nylon 66, which is made by using 1,6-hexanediamine and hexanedioic acid as monomers (Figure 17.16).

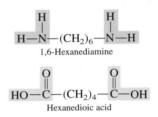

The reaction of one acid group of the diacid with one amine group of the diamine initially produces an amide molecule; an acid group is left over on one end, and an amine group is left over on the other end.

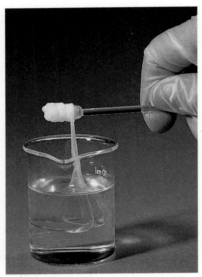

Figure 17.16 A white strand of a nylon polymer forms between the two layers of a solution containing a diacid (bottom layer) and a diamine (top layer).

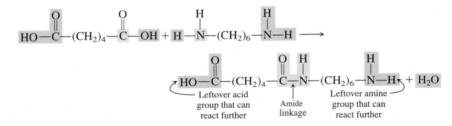

The name *nylon 66* comes from the fact that each of the monomers has six carbon atoms.

This species then reacts further, and the process continues until a long polymeric molecule, nylon, has been produced.

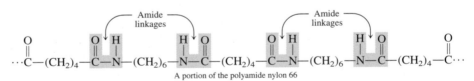

Additional stiffness and toughness are imparted to polyamides if aromatic rings are present in the polymer "backbone." The polyamide Kevlar is now used in place of steel in bullet-resistant vests. The polymeric repeating unit in Kevlar is

Kevlar

A uniform system of hydrogen bonds that holds polymer chains together accounts for the "amazing" strength of Kevlar (see Figure 17.17).

CHEMISTRY AT A GLANCE

Summary of Chemical Reactions Involving Amines and Amides

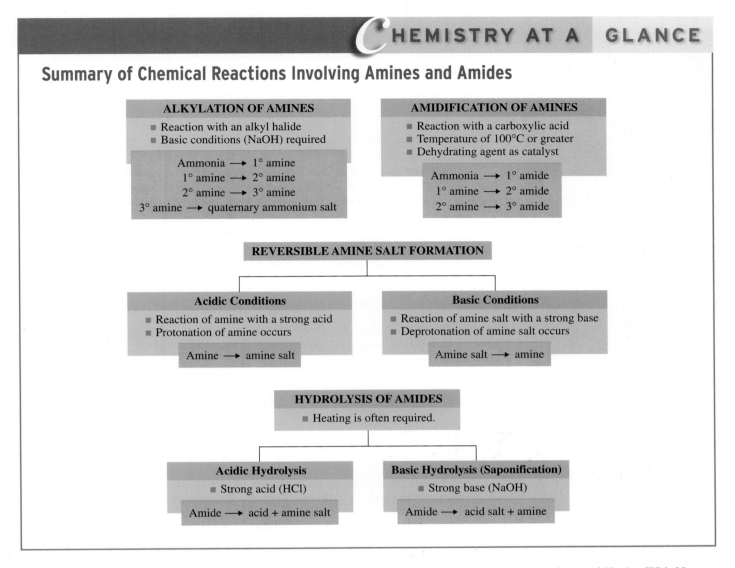

ALKYLATION OF AMINES
- Reaction with an alkyl halide
- Basic conditions (NaOH) required

Ammonia → 1° amine
1° amine → 2° amine
2° amine → 3° amine
3° amine → quaternary ammonium salt

AMIDIFICATION OF AMINES
- Reaction with a carboxylic acid
- Temperature of 100°C or greater
- Dehydrating agent as catalyst

Ammonia → 1° amide
1° amine → 2° amide
2° amine → 3° amide

REVERSIBLE AMINE SALT FORMATION

Acidic Conditions
- Reaction of amine with a strong acid
- Protonation of amine occurs

Amine → amine salt

Basic Conditions
- Reaction of amine salt with a strong base
- Deprotonation of amine salt occurs

Amine salt → amine

HYDROLYSIS OF AMIDES
- Heating is often required.

Acidic Hydrolysis
- Strong acid (HCl)

Amide → acid + amine salt

Basic Hydrolysis (Saponification)
- Strong base (NaOH)

Amide → acid salt + amine

Nomex is a polyamide whose structure is a variation of that of Kevlar. With Nomex, the monomers are *meta* isomers rather than *para* isomers. Nomex is used in flame-resistant clothing for fire fighters and race car drivers (Figure 17.18).

Silk and wool are examples of *naturally occurring* polyamide polymers. Silk and wool are proteins, and proteins are polyamide polymers. Because much of the human

Figure 17.17 A regular hydrogen-bonding pattern among Kevlar polymer strands contributes to the great strength of this polymer.

Figure 17.18 Firefighters with flame-resistant clothing containing Nomex.

body is protein material, much of the human body is polyamide polymer. The monomers for proteins are amino acids, difunctional molecules containing both amino and carboxyl groups. Here are some representative structures for amino acids, of which there are many (Section 20.2).

$$H_2N—CH_2—COOH \qquad H_2N—CH—COOH \qquad H_2N—CH—COOH$$
$$\qquad\qquad\qquad\qquad\qquad\quad |\qquad\qquad\qquad\qquad\quad |$$
$$\qquad\qquad\qquad\qquad\qquad CH_3\qquad\qquad\qquad\qquad CH—CH_3$$
$$\qquad\qquad\qquad\qquad\qquad\qquad\qquad\qquad\qquad\qquad\qquad |$$
$$\qquad\qquad\qquad\qquad\qquad\qquad\qquad\qquad\qquad\qquad\qquad CH_3$$

A **urethane** *is a hydrocarbon derivative that contains a carbonyl group bonded to both an —OR group and a —NHR (or –NR₂) group.* Such compounds are prepared by reaction of an alcohol with an isocyanate ($RN{=}C{=}O$).

$$R—N{=}C{=}O + R'OH \longrightarrow$$

Isocyanate   Urethane

A **polyurethane** *is a polymer formed from the reaction of dialcohol and diisocyanate monomers.* With the monomers benzene diisocyanate and ethylene glycol, the polymerization reaction is

Benzene 2,6-diisocyanate   Ethylene glycol   A polyurethane

Structurally, polyurethanes have aspects of the structures of both polyesters and polyamides as shown in the following segment of a polyurethane structure.

Amide | Ester

Foam rubber in furniture upholstery, packaging materials, life preservers, elastic fibers, and many other products contain polyurethane polymers (see Figure 17.19).

One of the best known polyurethanes is Spandex, a strong yet flexible polymer used in both men's and women's athletic wear. It has also been used in support hosiery. On the molecular level, it has rigid regions (for strength) that are joined together by flexible segments.

Flexible portion

Rigid segment

Spandex
Trade name: Lycra

Figure 17.19 Polyurethanes have medical applications. For example, polyurethane membranes are used as skin substitutes for severe burn victims. Because they pass only oxygen and water, these membranes help patients recover more rapidly.

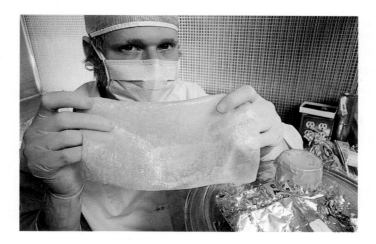

CONCEPTS TO REMEMBER

Structural characteristics of amines. Amines are derivatives of ammonia (NH_3) in which one or more hydrogen atoms have been replaced by an alkyl, a cycloalkyl, or an aryl group (Section 17.2).

Classification of amines. Amines are classified as primary, secondary or tertiary, depending on the number of hydrocarbon groups (one, two, or three) directly attached to the nitrogen atom. The functional group present in a primary amine, the —NH_2 group, is called an *amino* group. Secondary and tertiary amines contain substituted amino groups (Section 17.2).

Nomenclature for amines. Common names for amines are formed by listing the hydrocarbon groups attached to the nitrogen atom in alphabetical order, followed by the suffix *-amine*. In the IUPAC system, the *-e* ending of the name of the longest carbon chain present is changed to *-amine*, and a number is used to locate the position of the amino group. Carbon-chain substituents are given numbers to designate their locations (Section 17.3).

Properties of amines. The methylamines and ethylamine are gases at room temperature; amines of higher molecular mass are usually liquids and smell like raw fish. Primary and secondary, but not tertiary, amines can participate in hydrogen bonding to other amine molecules (Section 17.5).

Basicity of amines. Amines are weak bases because of the ability of the unshared electron pair on the amine nitrogen atom to accept a proton in acidic solution (Section 17.6).

Amine salts. The reaction of a strong acid with an amine produces an amine salt. Such salts are more soluble in water than are the parent amines (Section 17.7).

Alkylation of ammonia and amines. Alkylation of ammonia, primary amines, secondary amines, and tertiary amines produces primary amines, secondary amines, tertiary amines, and quaternary ammonium salts, respectively (Section 17.8).

Heterocyclic amines. In a heterocyclic amine, the nitrogen atoms of amino groups present are part of either an aromatic or a nonaromatic ring system. Numerous heterocyclic amines are important biochemical compounds (Section 17.9).

Structural characteristics of amides. An amide is derived from a carboxylic acid by replacing the hydroxyl group with an amino or a substituted amino group (Section 17.12).

Classification of amides. Amides, like amines, can be classified as primary, secondary, or tertiary, depending on how many nonhydrogen atoms are attached to the nitrogen atom (Section 17.12).

Nomenclature for amides. The nomenclature for amides is derived from that for carboxylic acids by changing the *-oic acid* ending to *-amide*. Groups attached to the nitrogen atom of the amide are located using the prefix *N-* (Section 17.13).

Properties of amides. Amides do not exhibit basic properties in solution. Most unbranched amides are solids at room temperature and have correspondingly high boiling points because of strong hydrogen bonds between molecules (Section 17.15).

Preparation of amides. Reaction, at elevated temperature, of carboxylic acids with ammonia, primary amines, and secondary amines produces primary, secondary, and tertiary amides, respectively (Section 17.16).

Hydrolysis of amides. In amide hydrolysis, the bond between the carbonyl carbon atom and the nitrogen is broken, and free acid and free amine are produced. Acidic hydrolysis conditions convert the product amine to an amine salt. Basic hydrolysis conditions convert the product acid to an acid salt (Section 17.17).

Polyamides. Polyamides are condensation polymers with monomers joined together by amide linkages. The monomers for polyamides are diacids and diamines (Section 17.18).

KEY REACTIONS AND EQUATIONS

1. Reaction of amines with water to give a basic solution (Section 17.6)

$$R—NH_2 + H_2O \rightleftharpoons R—\overset{+}{N}H_3 + OH^-$$

2. Reaction of amines with acids to produce amine salts (Section 17.7)

$$R—NH_2 + HCl \longrightarrow R—\overset{+}{N}H_3\ Cl^-$$

3. Conversion of an amine salt to an amine (Section 17.7)

$$R—\overset{+}{N}H_3\ Cl^- + NaOH \longrightarrow R—NH_2 + NaCl + H_2O$$

4. Alkylation of ammonia to produce a primary amine (Section 17.8)

$$NH_3 + R—X + NaOH \longrightarrow R—NH_2 + NaX + H_2O$$

5. Alkylation of primary and secondary amines to produce, respectively, secondary and tertiary amines (Section 17.8)

$$RNH_2 + R{-}X + NaOH \longrightarrow R_2NH + NaX + H_2O$$

$$R_2NH + R{-}X + NaOH \longrightarrow R_3N + NaX + H_2O$$

6. Alkylation of a tertiary amine to produce a quaternary ammonium salt (Section 17.8)

$$R_3N + R{-}X \longrightarrow R_4\overset{+}{N}\ X^-$$

7. Reaction of amines with carboxylic acids to form amides (Section 17.16)

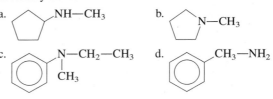

8. Acidic hydrolysis of amides to produce a carboxylic acid and an amine salt (Section 17.17)

9. Basic hydrolysis of amides to produce a carboxylic acid salt and an amine (Section 17.17)

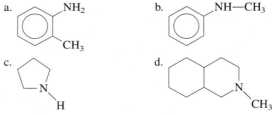

EXERCISES *and* PROBLEMS

The members of each pair of problems in this section test similar material.

● Bonding Characteristics of Nitrogen Atoms (Section 17.1)

17.1 Contrast N, O, and C atoms in terms of the number of covalent bonds they usually form in organic compounds.

17.2 Contrast N, O, and C atoms in terms of the number of nonbonding electron pairs they normally possess when they are present in organic compounds.

● Structure and Classification of Amines (Section 17.2)

17.3 What is the generalized molecular formula for the following?
a. Primary amine   b. Secondary amine   c. Tertiary amine

17.4 Draw a generalized structural formula for the functional group present in the following.
a. Primary amine   b. Secondary amine   c. Tertiary amine

17.5 In which of the following compounds is an amine functional group present?

a. $CH_3{-}CH{-}CH_3$
      $\underset{NH_2}{|}$

b. $CH_3{-}NH{-}CH_3$

c. $CH_3{-}CH_2{-}\overset{\overset{O}{\|}}{C}{-}NH_2$

d. $CH_3{-}CH_2{-}\underset{\underset{CH_3}{|}}{N}{-}CH_2{-}CH_3$

17.6 In which of the following compounds is an amine functional group present?

a. $CH_3{-}CH_2{-}CH_2{-}NH_2$

b. $CH_3{-}CH_2{-}\underset{\underset{CH_3}{|}}{N}{-}CH_3$

c. $CH_3{-}NH{-}\bigcirc$

d. $CH_3{-}CH_2{-}CH_2{-}\overset{\overset{O}{\|}}{C}{-}NH_2$

17.7 Classify each of the following amines as a primary, secondary, or tertiary amine.

a. $CH_3{-}NH_2$

b. $CH_3{-}CH{-}CH_3$
      $\underset{NH_2}{|}$

c. $CH_3{-}NH{-}CH_3$

d. $CH_3{-}CH_2{-}\underset{\underset{CH_3}{|}}{N}{-}CH_2{-}\underset{\underset{CH_3}{|}}{CH}{-}CH_3$

17.8 Classify each of the following amines as a primary, secondary, or tertiary amine.

a. $CH_3{-}CH_2{-}CH{-}CH_2{-}CH_3$
                  $\underset{NH_2}{|}$

b. $CH_3{-}\underset{\underset{NH_2}{|}}{\overset{\overset{CH_3}{|}}{C}}{-}CH_3$

c. $CH_3{-}\underset{\underset{CH_3}{|}}{N}{-}CH_3$

d. $CH_3{-}\underset{\underset{CH_3}{|}}{CH}{-}NH{-}\underset{\underset{CH_3}{|}}{CH}{-}CH_2$

17.9 Classify each of the following amines as a primary, secondary, or tertiary amine.

a. ⬠—NH—CH₃   b. ⬠ N—CH₃

c. ⬡—N—CH₂—CH₃   d. ⬡—CH₃—NH₂
        |
        CH₃

17.10 Classify each of the following amines as a primary, secondary, or tertiary amine.

a. ⬡—NH₂ with CH₃   b. ⬡—NH—CH₃

c. ⬠ N—H   d. (bicyclic) N—CH₃

● Nomenclature for Amines (Section 17.3)

17.11 Assign a common name to each of the following amines.

a. $CH_3{-}NH{-}CH_2{-}CH_3$

b. $CH_3{-}CH_2{-}CH_2{-}NH_2$

c. $CH_3{-}CH_2{-}\underset{\underset{CH_3}{|}}{N}{-}CH_2{-}CH_3$

d. $CH_3{-}\underset{\underset{CH_3}{|}}{CH}{-}NH{-}CH_3$

17.12 Assign a common name to each of the following amines.

a. $CH_3—CH—CH_3$
  $\quad\quad\;\;|$
  $\quad\quad\;NH_2$

b. $H_2N—CH_2—CH_2—CH_2—CH_3$

c. $CH_3—CH_2—N—CH_2—CH_3$
  $\quad\quad\quad\quad|$
  $\quad\quad\quad CH_2—CH_3$

d. $CH_3—CH_2—CH_2—NH—CH—CH_3$
  $\quad\quad\quad\quad\quad\quad\quad\quad|$
  $\quad\quad\quad\quad\quad\quad\quad CH_3$

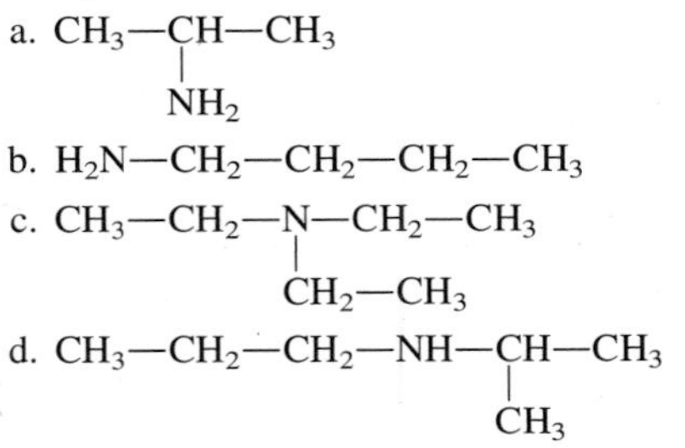

17.13 Assign an IUPAC name to each of the following amines.

a. $CH_3—CH_2—CH—CH_2—CH_3$
  $\quad\quad\quad\quad|$
  $\quad\quad\quad NH_2$

b. $CH_3—CH—CH—CH_2—CH_3$
  $\quad\quad\quad|\quad\;\;|$
  $\quad\quad CH_3\;NH_2$

c. $CH_3—CH_2—CH—CH_2—CH_3$
  $\quad\quad\quad\quad|$
  $\quad\quad\quad NH—CH_3$

d. $CH_3—CH—CH—CH_3$
  $\quad\quad\;\;|\quad\;|$
  $\quad\;\;NH_2\;NH_2$

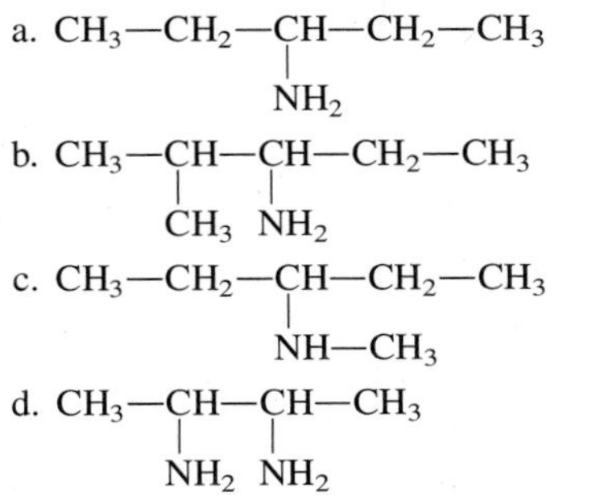

17.14 Assign an IUPAC name to each of the following amines.

a. $CH_3—CH_2—CH_2—NH_2$

b. $CH_3—CH—NH_2$
  $\quad\quad\;|$
  $\quad\;\;CH_3$

c. $CH_3—CH—CH—CH—CH_3$
  $\quad\quad|\quad\;|\quad\;|$
  $\quad NH_2\;CH_3\;NH_2$

d. $CH_3—CH_2—CH_3—NH—CH_3$

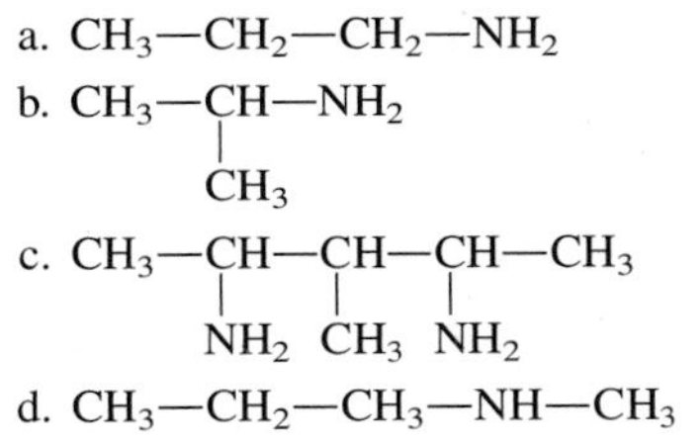

17.15 Assign an IUPAC name to each of the following amines.

a. ⌁NH₂   b. ⌁N⌁

c. ⌁NH⌁   d. ⌁NH⌁

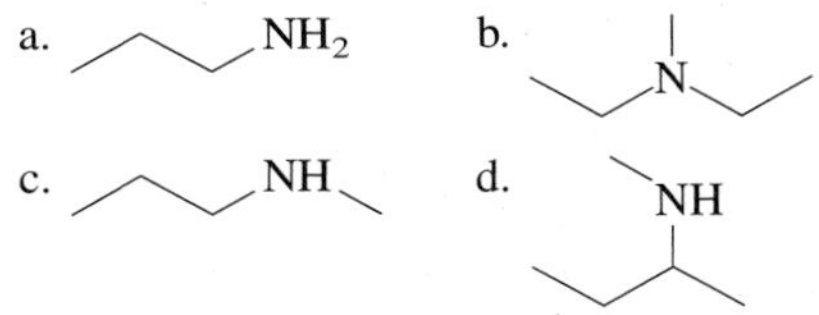

17.16 Assign an IUPAC name to each of the following amines.

a. ⌁NH₂   b. ⌁NH₂

c. ⌁NH⌁   d. ⌁N⌁

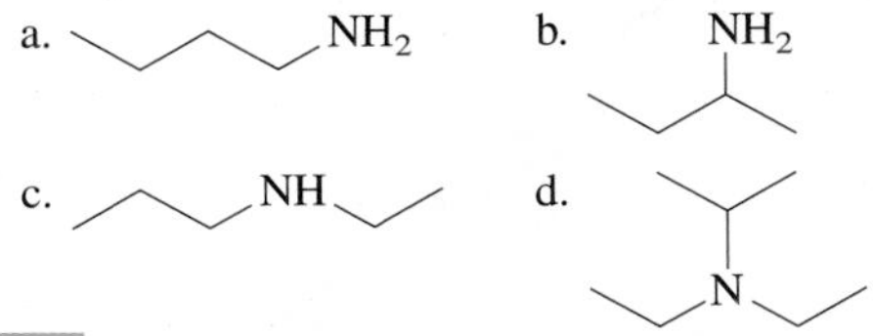

17.17 Name each of the following aromatic amines as a derivative of aniline.

a. ⬡—NH₂ (Br)

b. ⬡—NH—CH—CH₃ (CH₃)

c. ⬡—N—CH₂—CH₃ (CH₃)

d. ⬡—N—⬡ (CH₃)

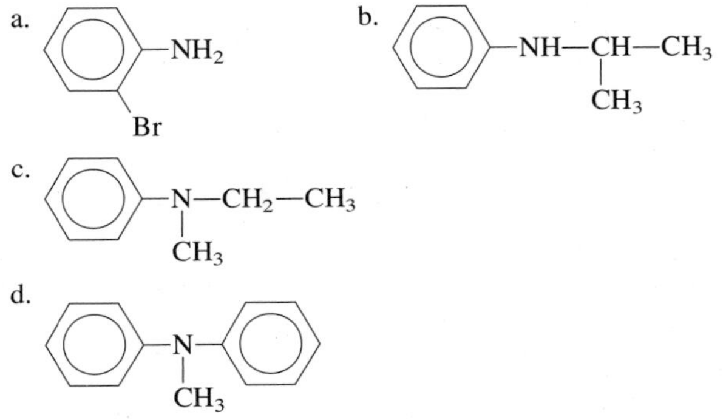

17.18 Name each of the following aromatic amines as a derivative of aniline.

a. Cl—⬡—NH₂

b. ⬡—NH—CH₂—CH₃

c. CH₃—⬡—NH—CH₃

d. ⬡ with NH₂, Br, Cl

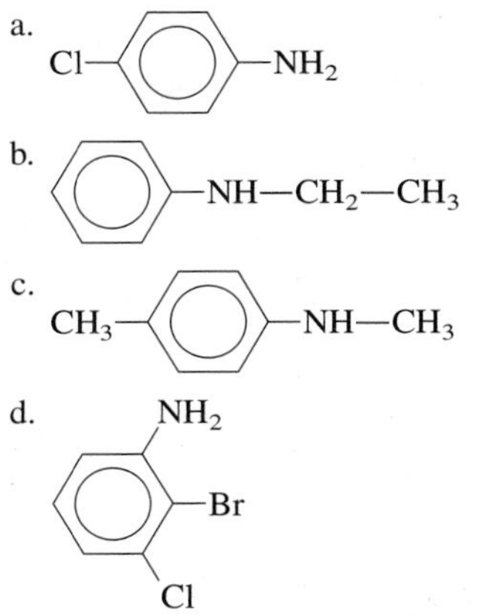

17.19 Draw a structural formula for each of the following amines.
a. 2-Methyl-2-butanamine   b. 1,6-Hexanediamine
c. 2-Amino-3-pentanone   d. 2-Aminopropanoic acid

17.20 Draw a structural formula for each of the following amines.
a. 2-Methyl-3-ethyl-1-hexanamine
b. 1,3-Pentanediamine
c. 3-Amino-2-pentanol
d. *N,N*-dimethyl-1-butanamine

⬤ **Isomerism for Amines (Section 17.4)**

17.21 Draw condensed structural formulas for the eight isomeric primary amines that have the molecular formula $C_5H_{13}N$.

17.22 Draw condensed structural formulas for the six isomeric secondary amines that have the molecular formula $C_5H_{13}N$.

17.23 Give common names for the three isomeric tertiary amines that have the molecular formula $C_5H_{13}N$.

17.24 Give common names for the seven isomeric tertiary amines that have the molecular formula $C_6H_{15}N$.

17.25 Assign an IUPAC name to each of the four isomeric amines that have the molecular formula C_3H_9N.

17.26 Assign an IUPAC name to each of the eight isomeric amines that have the molecular formula $C_4H_{11}N$.

⬤ **Physical Properties of Amines (Section 17.5)**

17.27 Indicate whether each of the following amines is a liquid or a gas at room temperature.
a. Butylamine   b. Dimethylamine
c. Ethylamine   d. Dibutylamine

17.28 Indicate whether each of the following amines is a liquid or a gas at room temperature.
a. Methylamine   b. Propylamine
c. Trimethylamine   d. Pentylamine

17.29 Determine the maximum number of hydrogen bonds that can form between a methylamine molecule and
a. other methylamine molecules
b. water molecules

17.30 Determine the maximum number of hydrogen bonds that can form between a dimethylamine molecule and
a. other dimethylamine molecules
b. water molecules

17.31 Although they have similar molecular masses (73 and 72 amu, respectively), the boiling point of butylamine is much higher (78°C) than that of pentane (36°C). Explain why.

17.32 Although they have similar molecular masses (73 and 74 amu, respectively), the boiling point of 1-butanamine is much lower (78°C) than that of 1-butanol (118°C). Explain why.

17.33 Which compound in each of the following pairs of amines would you expect to be more soluble in water? Justify each answer.
   a. CH_3—CH_2—NH_2 and
      CH_3—CH_2—CH_2—CH_2—CH_2—NH_2
   b. CH_3—CH_2—CH_2—NH_2 and
      H_2N—CH_2—CH_2—CH_2—NH_2

17.34 Which compound in each of the following pairs of amines would you expect to be more soluble in water? Justify each answer.
   a. CH_3—CH_2—CH_2—NH_2 and
      CH_3—CH_2—CH_2—CH_2—NH_2
   b. CH_3—CH_2—NH—CH_3 and CH_3—$\overset{\displaystyle |}{\underset{\displaystyle CH_3}{N}}$—$CH_3$

● **Basicity of Amines (Section 17.6)**

17.35 Show the structures of the missing substance(s) in each of the following acid–base equilibria.
   a. CH_3—CH_2—NH_2 + H_2O ⇌ ? + OH^-
   b. 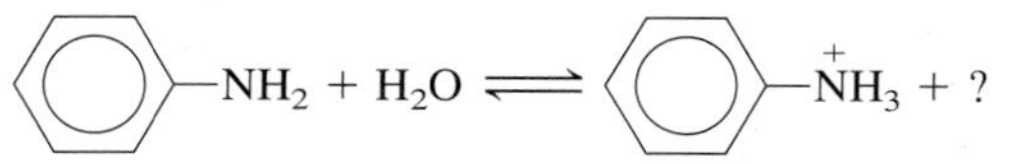
   ⬡—NH_2 + H_2O ⇌ ⬡—$\overset{+}{N}H_3$ + ?
   c. ? + H_2O ⇌ CH_3—$\overset{\displaystyle |}{\underset{\displaystyle CH_3}{CH}}$—$\overset{+}{N}H_2$—$CH_3$ + OH^-
   d. Diethylamine + H_2O ⇌ ? + ?

17.36 Show the structures of the missing substance(s) in each of the following acid–base equilibria.
   a. CH_3—CH_2—CH_2—NH_2 + H_2O ⇌
      CH_3—CH_2—CH_2—$\overset{+}{N}H_3$ + ?
   b. 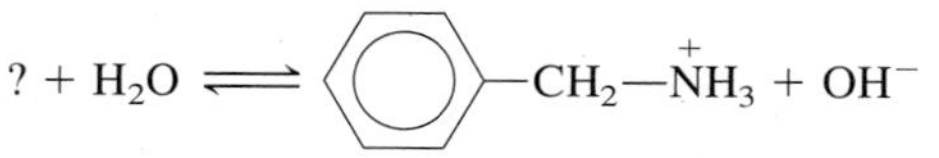
   ? + H_2O ⇌ ⬡—CH_2—$\overset{+}{N}H_3$ + OH^-
   c. CH_3—$\overset{\displaystyle |}{\underset{\displaystyle CH_3}{CH}}$—$CH_2$—$NH$—$CH_3$ + H_2O ⇌ ? + OH^-
   d. Trimethylamine + H_2O ⇌ ? + ?

17.37 Name each of the following substituted ammonium and substituted anilinium ions.
   a. CH_3—$\overset{+}{N}H_2$—CH_3
   b. CH_3—CH_2—$\overset{+}{\underset{\displaystyle CH_2-CH_3}{N}H}$—$CH_2$—$CH_3$
   c. CH_3—CH_2—$\overset{+}{N}H$—CH_2—CH_3
      ⬡
   d. $\overset{+}{N}H_2$—$\overset{\displaystyle CH_3}{\overset{\displaystyle |}{CH}}$—$CH_3$
      ⬡

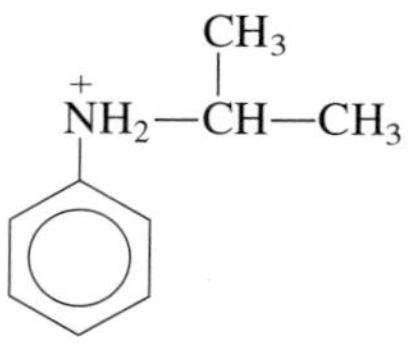

17.38 Name each of the following substituted ammonium and substituted anilinium ions.
   a. CH_3—$\overset{+}{N}H_3$
   b. CH_3—CH_2—CH_2—$\overset{+}{N}H_2$—CH_3
   c. $\overset{+}{N}H_2$—CH_2—CH_3

   d. CH_3—CH_2—$\overset{+}{N}H$—CH_2—CH_2—CH_3
      ⬡

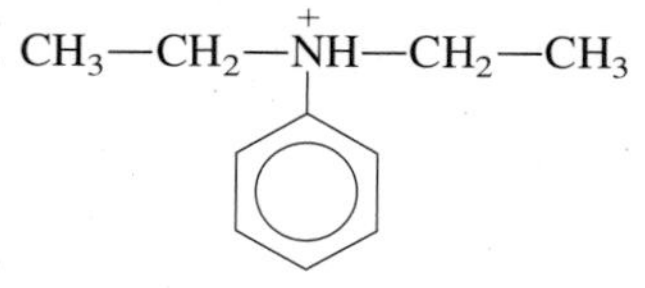

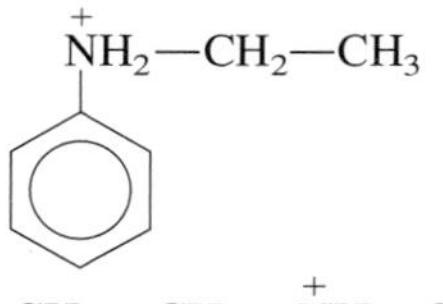

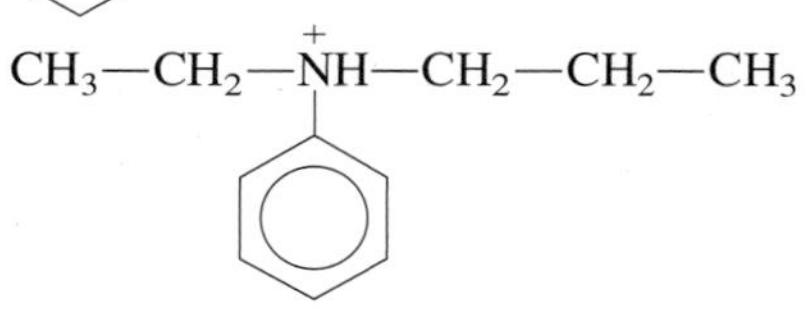

17.39 Draw a structural formula for the "parent" amine of each of the substituted ammonium and substituted anilinium ions in Problem 17.37.

17.40 Draw a structural formula for the "parent" amine of each of the substituted ammonium and substituted anilinium ions in Problem 17.38.

● **Amine Salts (Section 17.7)**

17.41 Draw the structure of the missing substance in each of the following reactions involving amine salts.
   a. CH_3—CH_2—NH_2 + HCl ⟶ ?
   b. 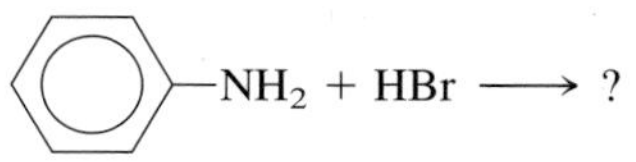
   ⬡—NH_2 + HBr ⟶ ?
   c.
   ? + HBr ⟶ CH_3—$\overset{\displaystyle CH_3}{\overset{\displaystyle |}{\underset{\displaystyle CH_3}{C}}}$—$\overset{+}{N}H_3$ Br^-
   d. CH_3—CH_2—NH—CH_3 + ? ⟶
      CH_3—CH_2—$\overset{+}{N}H_2$—CH_3 Cl^-

17.42 Draw the structure of the missing substance in each of the following reactions involving amine salts.
   a. CH_3—CH_2—NH—CH_2—CH_3 + HBr ⟶ ?
   b. CH_3—NH_2 + ? ⟶ CH_3—$\overset{+}{N}H_3$ Cl^-
   c. ? + HBr ⟶ CH_3—$\overset{\displaystyle |}{\underset{\displaystyle CH_3}{CH}}$—$\overset{+}{\underset{\displaystyle CH_3}{N}H}$—$CH_3$ Br^-
   d.
   ⬡—NH—CH_3 + HCl ⟶ ?

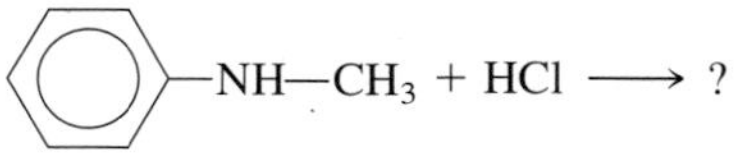

17.43 Draw the structures of the missing substance(s) in each of the following reactions involving amine salts.
   a. CH_3—$\overset{\displaystyle |}{\underset{\displaystyle CH_3}{CH}}$—$\overset{+}{N}H_3$ Cl^- + $NaOH$ ⟶ ? + $NaCl$ + H_2O
   b. ? + $NaOH$ ⟶ CH_3—NH—CH_3 + $NaCl$ + H_2O
   c. 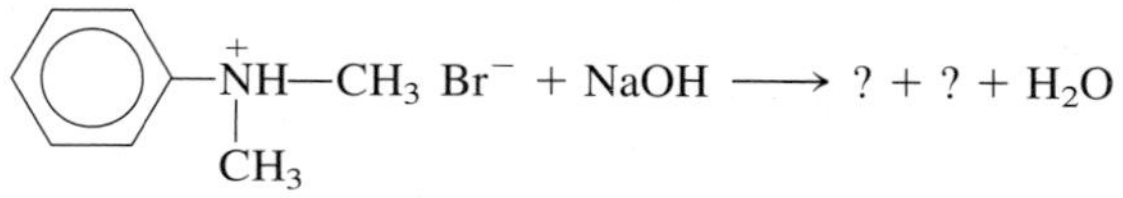
   ⬡—$\overset{+}{\underset{\displaystyle CH_3}{N}H}$—$CH_3$ Br^- + $NaOH$ ⟶ ? + ? + H_2O
   d. CH_3—$\overset{+}{N}H_2$—CH_3 Cl^- + $NaOH$ ⟶ ? + $NaCl$ + H_2O

17.44 Draw the structures of the missing substance(s) in each of the following reactions involving amine salts.

a. $CH_3-CH_2-CH_2-\overset{+}{N}H_3\ Br^- + NaOH \longrightarrow$
$? + NaBr + H_2O$

b. $? + NaOH \longrightarrow CH_3-\underset{\underset{CH_3}{|}}{N}-CH_3 + NaBr + H_2O$

c. $CH_3-\underset{\underset{CH_3}{|}}{CH}-\overset{+}{N}H_2-CH_3\ Cl^- + NaOH \longrightarrow$
$? + NaCl + H_2O$

d.
⬡$-\overset{+}{N}H_3\ Cl^- + NaOH \longrightarrow ? + NaCl + ?$

17.45 Name each of the following amine salts.

a. $CH_3-CH_2-CH_2-\overset{+}{N}H_3\ Cl^-$

b. $CH_3-CH_2-CH_2-\underset{\underset{CH_3}{|}}{\overset{+}{N}H_2}\ Cl^-$

c. $CH_3-CH_2-\underset{\underset{CH_3}{|}}{\overset{+}{N}H}-CH_3\ Br^-$

d.
⬡$-\underset{\underset{CH_3}{|}}{\overset{+}{N}H}-CH_3\ Br^-$

17.46 Name each of the following amine salts.

a. $CH_3-CH_2-\underset{\underset{CH_3}{|}}{\overset{+}{N}H_2}\ Cl^-$

b. $CH_3-CH_2-CH_2-CH_2-\overset{+}{N}H_3\ Cl^-$

c. $CH_3-\underset{\underset{CH_3}{|}}{CH}-\underset{\underset{CH_3}{|}}{\overset{+}{N}H}-CH_3\ Br^-$

d.
⬡$-\underset{\underset{CH_3}{|}}{\overset{+}{N}H_2}\ Cl^-$

17.47 Which of the four terms *free amine, free base, deprotonated base,* and *protonated base* apply to each of the following amine species? More than one term may apply in a given situation.

a. $CH_3-CH_2-NH_2$

b. $CH_3-CH_2-CH_2-NH-CH_3$

c. $CH_3-\overset{+}{N}H_3$

d. $CH_3-CH_2-\overset{+}{N}H_2-CH_3$

17.48 Which of the four terms *free amine, free base, deprotonated base,* and *protonated base* apply to each of the following amine species? More than one term may apply in a given situation.

a. $CH_3-CH_2-CH_2-NH_2$

b. $CH_3-CH_2-CH_2-\overset{+}{N}H_2-CH_3$

c. $CH_3-NH-CH_2-CH_3$

d. $CH_3-CH_2-\overset{+}{N}H_3$

17.49 Why are drugs that contain the amine functional group most often administered to patients in the form of amine chloride or hydrogen sulfate salts?

17.50 Both heptylamine and heptyl alcohol are insoluble in water. If you were given a mixture of these two liquids, how could you separate them without heating (distilling) them?

17.51 How would the structure and name of the amine salt ethyl-methylammonium chloride probably be written by someone in the pharmaceutical industry?

17.52 A student looking in an old chemistry book found the following name and structure for a compound.

$$CH_3-CH_2-NH_2\cdot HBr$$
Ethylamine hydrobromide

What are the modern name and structural representation for this compound?

⬤ **Alkylation of Ammonia and Amines (Section 17.8)**

17.53 Identify the three products in each of the following reactions.

a. $NH_3 + CH_3-CH_2-CH_2-Cl + NaOH \longrightarrow$

b. $CH_3-Br + CH_3-\underset{\underset{CH_3}{|}}{CH}-NH-CH_3 + NaOH \longrightarrow$

c. $CH_3-CH_2-NH_2 + CH_3-CH_2-Cl + NaOH \longrightarrow$

d. $CH_3-\underset{\underset{CH_3}{|}}{\overset{\overset{CH_3}{|}}{C}}-Br + NH_3 + NaOH \longrightarrow$

17.54 Identify the three products in each of the following reactions.

a. $CH_3-\underset{\underset{CH_3}{|}}{CH}-Cl + NH_3 + NaOH \longrightarrow$

b. $CH_3-NH-CH_3 + CH_3-Br + NaOH \longrightarrow$

c. $CH_3-CH_2-CH_2-NH_2 + CH_3-CH_2-Br + NaOH \longrightarrow$

d. $CH_3-CH_2-\underset{\underset{CH_3}{|}}{CH}-Cl +$
$CH_3-CH_2-NH-\underset{\underset{CH_3}{|}}{CH}-CH_3 + NaOH \longrightarrow$

17.55 List three different sets of alkyl chloride–secondary amine reactants that could be used to prepare the tertiary amine ethylmethylpropylamine.

17.56 List three different sets of alkyl chloride–secondary amine reactants that could be used to prepare the tertiary amine butylethylpropylamine.

17.57 Draw the structure of the amine or quaternary ammonium salt produced when each of the following pairs of compounds reacts in the presence of a strong base.
a. Trimethylamine and ethyl bromide
b. Diisopropylamine and methyl bromide
c. Ethylmethylpropylamine and methyl chloride
d. Ethylamine and ethyl chloride

17.58 Draw the structure of the amine or quaternary ammonium salt produced when each of the following pairs of compounds reacts in the presence of a strong base.
a. Dimethylamine and propyl bromide
b. Diethylmethylamine and isopropyl chloride
c. Methylpropylamine and ethyl chloride
d. Tripropylamine and propyl chloride

17.59 Classify each of the following salts as an amine salt or a quaternary ammonium salt.

a. $CH_3-\overset{+}{\underset{\underset{CH_3}{|}}{N}H}-CH_3 \ Br^-$

b. $CH_3-\overset{\overset{CH_3}{|}}{\underset{\underset{CH_3}{|}}{N}{}^+}-CH_3 \ Cl^-$

c. $CH_3-CH_2-\overset{+}{N}H_2-CH_3 \ Br^-$

d. $CH_3-CH_2-\overset{\overset{CH_3}{|}}{\underset{\underset{CH_3}{|}}{N}{}^+}-CH_2-CH_3 \ Cl^-$

17.60 Classify each of the following salts as an amine salt or a quaternary ammonium salt.

a. $CH_3-\overset{\overset{CH_3}{|}}{\underset{\underset{CH_3}{|}}{N}{}^+}-CH_2-CH_3 \ Cl^-$

b. $CH_3-\overset{\overset{H}{|}}{\underset{\underset{CH_3}{|}}{N}{}^+}-CH_2-CH_3 \ Cl^-$

c. $CH_3-\overset{\overset{H}{|}}{\underset{\underset{H}{|}}{N}{}^+}-CH_3 \ Br^-$

d. $CH_3-CH_2-CH_2-\overset{\overset{CH_3}{|}}{\underset{\underset{CH_3}{|}}{N}{}^+}-CH_3 \ Br^-$

17.61 Name each of the salts in Problem 17.59.

17.62 Name each of the salts in Problem 17.60.

● **Heterocyclic Amines (Section 17.9)**

17.63 Indicate whether or not each of the following compounds is a heterocyclic amine.

a. b.

c. d.

17.64 Indicate whether or not each of the following compounds is a heterocyclic amine.

a. b.

c. d.

17.65 With the help of Figure 17.8, identify the heterocyclic amine ring system or systems present in each of the following substances.

a. Caffeine b. Heme
c. Histamine d. Serotonin

17.66 With the help of Figure 17.8, identify the heterocyclic amine ring system or systems present in each of the following.

a. Nicotine b. Quinine
c. "Odor of popcorn" d. Porphyrin ring

● **Biochemically Important Amines (Section 17.10)**

17.67 Indicate whether each of the following statements about biochemically important amines is true or false.

a. Serotonin deficiency is associated with Parkinson's disease.
b. Structurally, epinephrine and norepinephrine differ by a methyl group.
c. An antihistamine is taken to counteract the effects caused by histamine.
d. L-Dopa and dopamine are two names for the same compound.

17.68 Indicate whether each of the following statements about biochemically important amines is true or false.

a. Both serotonin and dopamine have neurotransmitter functions in the human body.
b. Structurally, ephedrin and epinephrine differ by a methyl group.
c. Adrenaline is a central nervous system stimulant produced by the adrenal glands.
d. The presence of "free" histamine causes the symptoms associated with hay fever.

17.69 How many amine functional groups are present in molecules of each of the following substances?

a. Epinephrine b. Adrenaline
c. Histamine d. Serotonin

17.70 How many amine functional groups are present in molecules of each of the following substances?

a. Dopamine b. Norepinephrine
c. Ephedrin d. Prozac

● **Alkaloids (Section 17.11)**

17.71 Indicate whether or not each of the following substances is an alkaloid.

a. Nicotine b. Quinine
c. Morphine d. Cocaine

17.72 Indicate whether or not each of the following substances is an alkaloid.

a. Caffeine b. Theobromine
c. Atropine d. Codeine

17.73 Indicate whether each of the following statements about alkaloids is true or false.

a. A medicinal use for atropine is dilation of the pupil of the eye.
b. Quinine, used to treat malaria, can be obtained from the bark of a specific tree.
c. Structurally, morphine and codeine differ by a methyl group.
d. Heroin is a naturally occurring substance obtained from the poppy plant.

17.74 Indicate the number of amine functional groups present in each of the following heterocyclic amines.

a. Morphine b. Codeine
c. Nicotine d. Quinine

Structure of and Classification of Amides (Section 17.12)

17.75 What is the generalized molecular formula for each of the following?
  a. Primary amide   b. Secondary amide
  c. Tertiary amide

17.76 Draw a generalized structural formula for the functional group present in each of the following.
  a. Primary amide   b. Secondary amide
  c. Tertiary amide

17.77 Indicate whether or not each of the following compounds contain an amide functional group.

a. 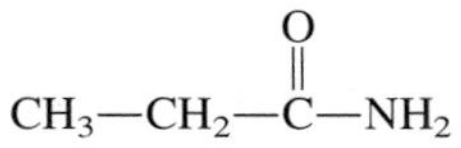
$$CH_3-CH_2-\overset{\overset{\displaystyle O}{\|}}{C}-NH_2$$

b.

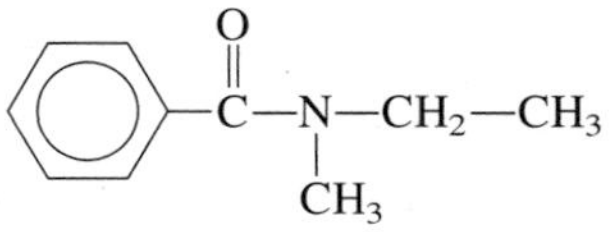

c. 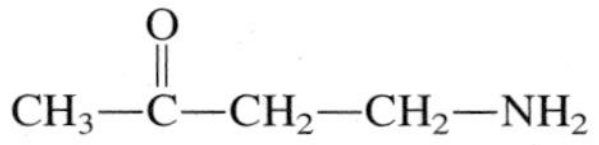
$$CH_3-\overset{\overset{\displaystyle O}{\|}}{C}-CH_2-CH_2-NH_2$$

d.

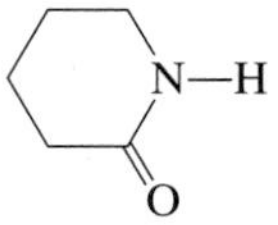

17.78 Indicate whether or not each of the following compounds contain an amide functional group.

a. 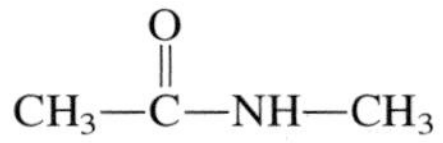
$$CH_3-\overset{\overset{\displaystyle O}{\|}}{C}-NH-CH_3$$

b.
$$CH_3-\overset{\overset{\displaystyle O}{\|}}{C}-\overset{\overset{\displaystyle CH_2-CH_3}{|}}{N}-CH_2-CH_2-CH_3$$

c.
$$H_2N-\overset{\overset{\displaystyle O}{\|}}{C}-CH_2-CH_3$$

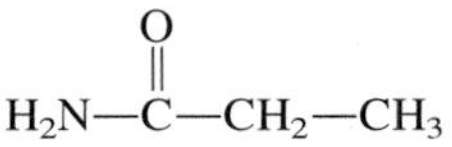

d.

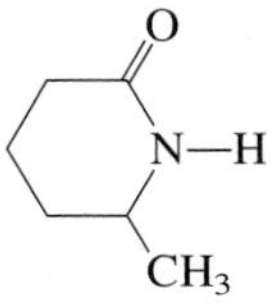

17.79 Classify each of the following amides as unsubstituted, monosubstituted, or disubstituted.

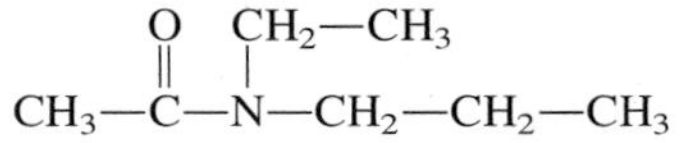

a. 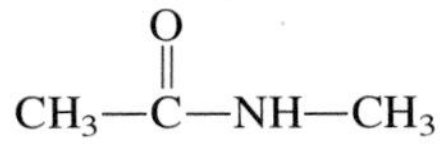
$$CH_3-\overset{\overset{\displaystyle O}{\|}}{C}-NH-CH_3$$

b.
$$CH_3-\overset{\overset{\displaystyle O}{\|}}{C}-\overset{\overset{\displaystyle CH_3}{|}}{N}-CH_2-CH_3$$

c.
$$CH_3-\overset{\overset{\displaystyle O}{\|}}{C}-NH_2$$

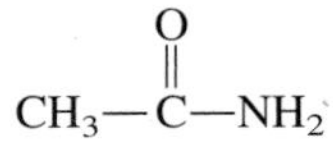

d.

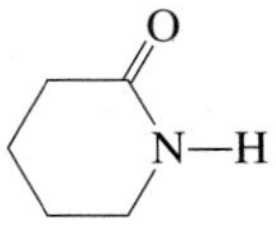

17.80 Classify each of the following amides as unsubstituted, mono-substituted, or disubstituted.

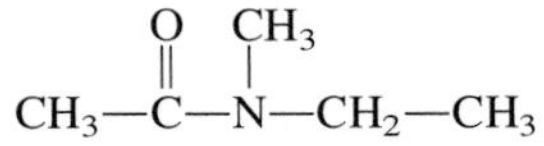

a. 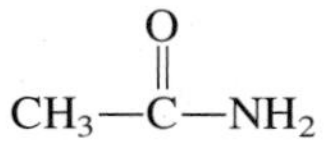
$$CH_3-\overset{\overset{\displaystyle O}{\|}}{C}-NH_2$$

b.
$$CH_3-CH_2-\overset{\overset{\displaystyle CH_3}{|}}{CH}-\overset{\overset{\displaystyle O}{\|}}{C}-NH-CH_3$$

c.
$$CH_3-\overset{\overset{\displaystyle O}{\|}}{C}-\overset{\overset{\displaystyle }{|}}{N}-CH_2-CH_2-CH_3$$
$$\overset{\displaystyle |}{CH_3-CH_2-CH_3}$$

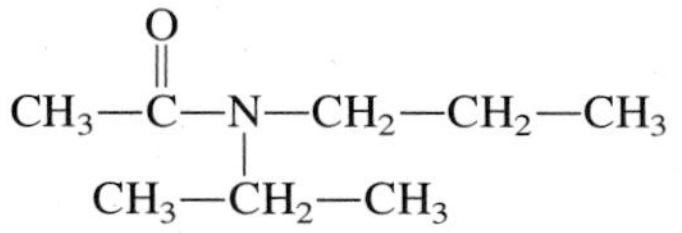

d.

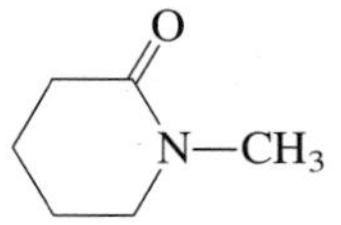

17.81 Classify each of the amides in Problem 17.79 as a primary, secondary, or tertiary amide.

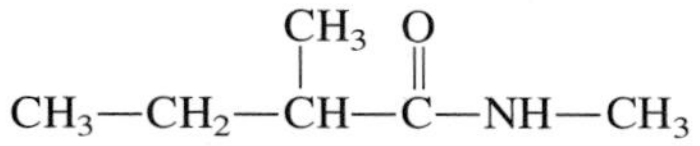

17.82 Classify each of the amides in Problem 17.80 as a primary, secondary, or tertiary amide.

Nomenclature for Amides (Section 17.13)

17.83 Assign an IUPAC name to each of the following amides.

a. 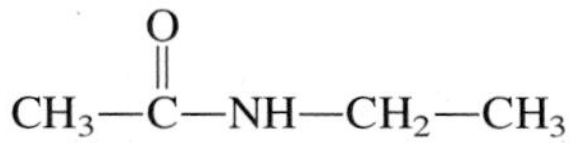
$$CH_3-\overset{\overset{\displaystyle O}{\|}}{C}-NH-CH_2-CH_3$$

b.
$$CH_3-CH_2-\overset{\overset{\displaystyle O}{\|}}{C}-\overset{\overset{\displaystyle CH_3}{|}}{N}-CH_3$$

c.
$$H_2N-\overset{\overset{\displaystyle O}{\|}}{C}-CH_2-CH_2-CH_3$$

d. 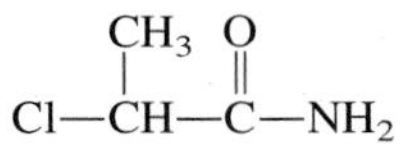
$$Cl-\overset{\overset{\displaystyle CH_3}{|}}{CH}-\overset{\overset{\displaystyle O}{\|}}{C}-NH_2$$

17.84 Assign an IUPAC name to each of the following amides.

a. 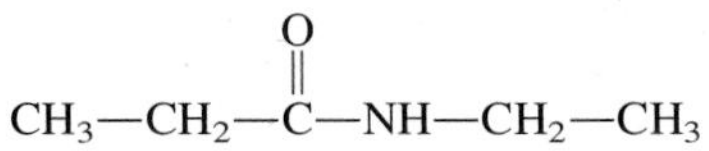
$$CH_3-CH_2-\overset{\overset{\displaystyle O}{\|}}{C}-NH-CH_2-CH_3$$

b.
$$CH_3-CH_2-CH_2-CH_2-\overset{\overset{\displaystyle O}{\|}}{C}-NH_2$$

c.
$$CH_3-CH_2-CH_2-\overset{\overset{\displaystyle O}{\|}}{C}-\overset{\overset{\displaystyle CH_3}{|}}{N}-CH_3$$

d. 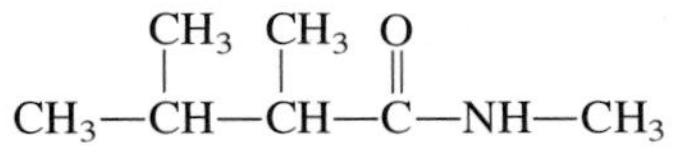
$$CH_3-\overset{\overset{\displaystyle CH_3}{|}}{CH}-\overset{\overset{\displaystyle CH_3}{|}}{CH}-\overset{\overset{\displaystyle O}{\|}}{C}-NH-CH_3$$

17.85 Assign a common name to each of the amides in Problem 17.83.

17.86 Assign a common name to each of the amides in Problem 17.84.

17.87 Assign an IUPAC name to each of the following amides.

a.

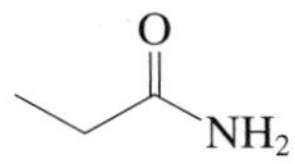

b.

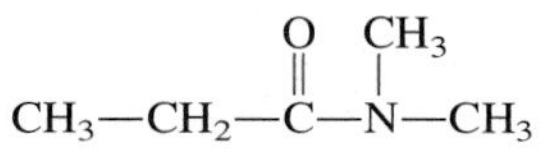

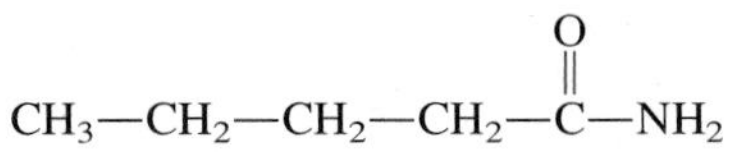

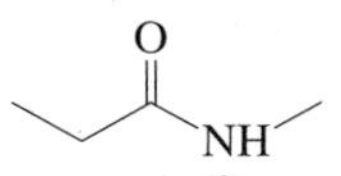

c.

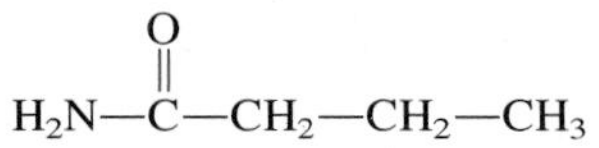

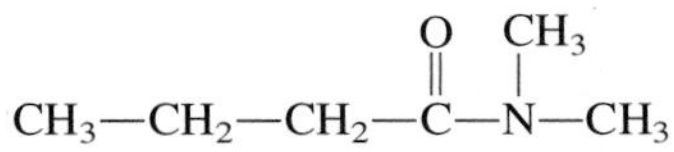

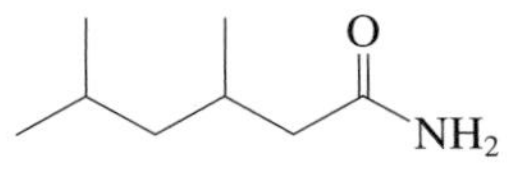

d.

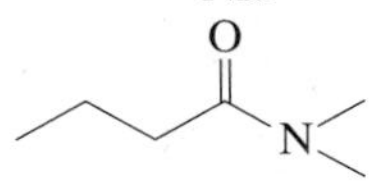

17.88 Assign an IUPAC name to each of the following amides.

a.

b.

c.

d.

17.89 Write a structural formula for each of the following amides.
a. *N,N*-dimethylacetamide   b. α-methylbutyramide
c. 3,*N*-dimethylbutanamide   d. Formamide

17.90 Write a structural formula for each of the following amides.
a. *N,N*-diethylpropanamide
b. β-methylbutyramide
c. *N*-methylbenzamide
d. β,β,*N*-trimethylbutyramide

● **Selected Amides and Their Uses (Section 17.14)**

17.91 What is the structural formula for the compound urea?

17.92 Describe the compound urea in terms of the following.
a. Physical properties
b. Method for its biochemical preparation

17.93 Contrast the structures of acetamide and acetaminophen.

17.94 What is a function for the hormone melatonin in the human body?

● **Physical Properties of Amides (Section 17.15)**

17.95 Although amides contain a nitrogen atom, they are not bases as amines are. Explain why.

17.96 Would you expect *N*-ethylacetamide or *N,N*-diethylacetamide to have the higher boiling point? Explain.

17.97 Determine the maximum number of hydrogen bonds that can form between an acetamide molecule and
a. other acetamide molecules
b. water molecules

17.98 Determine the maximum number of hydrogen bonds that can form between a propanamide molecule and
a. other propanamide molecules
b. water molecules

● **Preparation of Amides (Section 17.16)**

17.99 Draw the structures of the missing substances in each of the following reactions involving amides.

a.

$$CH_3-CH_2-\overset{\overset{\displaystyle O}{\|}}{C}-OH + ? \xrightarrow{100°C}$$

$$CH_3-CH_2-\overset{\overset{\displaystyle O}{\|}}{C}-NH-CH_3 + H_2O$$

b.

$$CH_3-\overset{\overset{\displaystyle CH_3}{|}}{\underset{\underset{\displaystyle CH_3}{|}}{C}}-\overset{\overset{\displaystyle O}{\|}}{C}-OH + CH_3-NH-CH_3 \xrightarrow{100°C}$$

$$? + H_2O$$

c.

$$CH_3-\overset{\overset{\displaystyle O}{\|}}{C}-OH + ? \xrightarrow{100°C} CH_3-\overset{\overset{\displaystyle O}{\|}}{C}-NH_2 + H_2O$$

d.

17.100 Draw the structures of the missing substances in each of the following reactions involving amides.

a.

$$CH_3-CH_2-CH_2-\overset{\overset{\displaystyle O}{\|}}{C}-OH +$$
$$CH_3-CH_2-NH_2 \xrightarrow{100°C} ? + H_2O$$

b.

$$? + NH_3 \xrightarrow{100°C} H-\overset{\overset{\displaystyle O}{\|}}{C}-NH_2 + H_2O$$

c.

$$CH_3-\overset{\overset{\displaystyle CH_3}{|}}{CH}-\overset{\overset{\displaystyle O}{\|}}{C}-OH + ? \xrightarrow{100°C}$$

$$CH_3-\overset{\overset{\displaystyle CH_3}{|}}{CH}-\overset{\overset{\displaystyle O}{\|}}{C}-\overset{\overset{\displaystyle CH_3}{|}}{N}-CH_3 + H_2O$$

d.

$$? + CH_3-NH_2 \xrightarrow{100°C}$$

17.101 Draw the structures of the carboxylic acid and the amine from which each of the following amides could be formed.

a.

$$CH_3-\overset{\overset{\displaystyle O}{\|}}{C}-\overset{\overset{\displaystyle CH_3}{|}}{N}-\overset{\overset{\displaystyle CH_3}{|}}{CH}-CH_3$$

b. *N*-methylpentanamide

c.

$$CH_3-\overset{\overset{\displaystyle CH_3}{|}}{CH}-\overset{\overset{\displaystyle O}{\|}}{C}-NH-CH_3$$

d. 2,3,*N*-trimethylbutanamide

17.102 Draw the structures of the carboxylic acid and the amine from which each of the following amides could be formed.

a.

$$CH_3-CH_2-\overset{\overset{\displaystyle O}{\|}}{C}-\overset{\overset{\displaystyle }{N}}{\underset{\underset{\displaystyle CH_3}{|}}{}}-CH_3$$

b. 2-Methylpentanamide

c.

$$CH_3-\overset{\overset{\displaystyle CH_3}{|}}{\underset{\underset{\displaystyle CH_3}{|}}{C}}-\overset{\overset{\displaystyle O}{\|}}{C}-NH-CH_2-CH_3$$

d. *N,N*-diethylacetamide

● **Hydrolysis of Amides (Section 17.17)**

17.103 Draw the structures of the organic products in each of the following hydrolysis reactions.

a.

$$CH_3-CH_2-CH_2-\overset{\overset{\displaystyle O}{\|}}{C}-NH-CH_3 + H_2O \xrightarrow{Heat}$$

b.

$$CH_3-CH_2-CH_2-\overset{\overset{\displaystyle O}{\|}}{C}-NH-CH_3 + H_2O \xrightarrow[HCl]{Heat}$$

c.

$$CH_3-CH_2-CH_2-\overset{\displaystyle O}{\overset{\|}{C}}-NH-CH_3 + H_2O \xrightarrow[\text{NaOH}]{\text{Heat}}$$

d.

$+ H_2O \xrightarrow{\text{Heat}}$

17.104 Draw the structures of the organic products in each of the following hydrolysis reactions.

a.

$$CH_3-CH_2-\overset{\displaystyle O}{\overset{\|}{C}}-NH-CH_2-CH_3 + H_2O \xrightarrow{\text{Heat}}$$

b.

$$CH_3-CH_2-\overset{\displaystyle O}{\overset{\|}{C}}-NH-CH_2-CH_3 + H_2O \xrightarrow[\text{HCl}]{\text{Heat}}$$

c.

$$CH_3-CH_2-\overset{\displaystyle O}{\overset{\|}{C}}-NH-CH_2-CH_3 + H_2O \xrightarrow[\text{NaOH}]{\text{Heat}}$$

d.

$$CH_3-\overset{\displaystyle}{\underset{\displaystyle}{CH}}-\overset{\displaystyle O}{\overset{\|}{C}}-NH_2 + H_2O \xrightarrow{\text{Heat}}$$

● **Polyamides and Polyurethanes (Section 17.18)**

17.105 List the general characteristics of the monomers needed to produce a polyamide.

17.106 Contrast the monomers needed to produce a polyamide with those needed to produce a polyester.

17.107 Draw a structural representation for the polyamide formed from the reaction of succinic acid and 1,4-butanediamine.

17.108 Draw a structural representation for the polyamide formed from the reaction of adipic acid and 1,2-ethanediamine.

17.109 Draw the generalized structural formula for a urethane.

17.110 What are the two types of monomers used to form a polyurethane polymer?

ADDITIONAL PROBLEMS

17.111 Draw structural formulas for the following compounds.
   a. 2-Methylpentanamide
   b. *N*-Isopropylethanamide
   c. Diethylammonium chloride
   d. Trimethylanilinium chloride

17.112 What is the structure of the organic product (or products) in each of the following reactions?

   a. $CH_3-CH_2-\overset{+}{N}H_2-CH_3\ Br^- + NaOH \longrightarrow$

   b.
$$CH_3-CH_2-\overset{\displaystyle CH_3}{\overset{|}{CH}}-\overset{\displaystyle O}{\overset{\|}{C}}-OH + CH_3-NH-CH_3 \xrightarrow{\text{heat}}$$

   c.
$$CH_3-\overset{\displaystyle CH_3}{\overset{|}{N}}-CH_3 + CH_3-Cl \xrightarrow{\text{NaOH}}$$

   d.
$$CH_3-CH_2-\overset{\displaystyle CH_3}{\overset{|}{CH}}-\overset{\displaystyle O}{\overset{\|}{C}}-NH_2 + H_2O \xrightarrow[\text{heat}]{\text{NaOH}}$$

17.113 What is the structure of the organic product (or products) in each of the following reactions?

   a. $CH_3-CH_2-CH_2-NH_2 + CH_3-CH_2-COOH \xrightarrow[\text{catalyst}]{100°C}$

   b. $CH_3-Br + CH_3-NH_2 \xrightarrow{\text{NaOH}}$

   c. $CH_3-CH_2-\overset{\displaystyle CH_3}{\overset{|}{CH}}-\overset{\displaystyle O}{\overset{\|}{C}}-NH-CH_3 + H_2O \xrightarrow[\text{heat}]{\text{HCl}}$

   d. $CH_3-CH_2-NH_2 + H_2O \longrightarrow$

17.114 Draw structural formulas and assign IUPAC names to the four amide constitutional isomers with the formula C_3H_7ON.

17.115 Draw the structural formula of the quaternary ammonium salt with the formula $C_5H_{14}NCl$.

17.116 Classify each of the following amines or amides as unsubstituted, monosubstituted, or disubstituted.
   a. *o*-Methylbenzamide   b. *N*-Methylbenzamide
   c. Cyclopentylmethylamine   d. Isopropylamine

17.117 Indicate whether each of the following compounds is an amine, an amide, both, or neither.

   a. $CH_3-CH_2-\overset{\displaystyle}{\underset{\displaystyle NH_2}{CH}}-\overset{\displaystyle}{\underset{\displaystyle NH_2}{CH}}-CH_3$

   b.

   c.
$$NH_2-CH_2-\overset{\displaystyle O}{\overset{\|}{C}}-OH$$

   d.

17.118 Assign IUPAC names to each of the following compounds.

   a.

   b.

   c.

   d.

Multiple-Choice Practice Test

17.119 Which of the following elements is not present in an unsubstituted amine?
    a. Carbon
    b. Hydrogen
    c. Oxygen
    d. Nitrogen

17.120 Which of the following amines is classified as a *secondary* amine?
    a. 1-Butanamine
    b. 2-Butanamine
    c. *N*-methyl-2-butanamine
    d. 3-Methyl-2-butanamine

17.121 Why are the boiling points of amines lower than those of alcohols of similar molecular mass?
    a. Amines do not contain an oxygen atom as do alcohols.
    b. Amine–amine hydrogen bonding is not possible.
    c. N···H hydrogen bonds are weaker than O···H hydrogen bonds.
    d. Amines are insoluble in water.

17.122 What is the molecular formula for the compound aniline?
    a. C_6H_6N
    b. C_6H_7N
    c. C_6H_8N
    d. $C_6H_8N_2$

17.123 Which of the following sets of reactants, under appropriate conditions, produces a secondary amine?
    a. Ammonia + alkyl halide
    b. Ammonia + carboxylic acid
    c. Primary amine + alkyl halide
    d. Primary amine + carboxylic acid

17.124 Which of the following statements concerning amines and amides is correct?
    a. Both amines and amides exhibit basic properties in aqueous solution.
    b. Amines but not amides exhibit basic properties in aqueous solution.
    c. Amides but not amines exhibit basic properties in aqueous solution.
    d. Neither amines nor amides exhibit basic properties in aqueous solution.

17.125 Which of the following statements concerning amines and amides is correct?
    a. Both amines and amides undergo hydrolysis reactions.
    b. Amines but not amides undergo hydrolysis reactions.
    c. Amides but not amines undergo hydrolysis reactions.
    d. Neither amines nor amides undergo hydrolysis reactions.

17.126 What is the name of the amide produced by the reaction of butanoic acid and methyl amine?
    a. *N*-methylbutanamide
    b. 2-Methylbutanamide
    c. Butyl amide
    d. Methyl butyl amide

17.127 What are the organic products when an amide undergoes hydrolysis under basic conditions?
    a. Carboxylic acid and amine salt
    b. Carboxylic acid salt and amine
    c. Carboxylic acid salt and amine salt
    d. Carboxylic acid and amine

17.128 Which of the following sets of monomers would produce a polyamide?
    a. Dicarboxylic acid and dialcohol
    b. Dicarboxylic acid and diamine
    c. Diamide and dialcohol
    d. Diamide and diamine

Carbohydrates

Chapter Outline

The naturally present sugars fructose, glucose, and sucrose all contribute to the sweetness of ripe peaches.

Beginning with this chapter on carbohydrates, we will focus almost exclusively on biochemistry, the chemistry of living systems. Like organic chemistry, biochemistry is a vast subject, and we can discuss only a few of its facets. Our approach to biochemistry will be similar to our approach to organic chemistry. We will devote individual chapters to each of the major classes of biochemical compounds, which are carbohydrates, lipids, proteins, and nucleic acids. Then we will examine the major types of chemical reactions in living organisms. In this first "biochapter," carbohydrates are considered.

The same functional groups found in organic compounds are also present in biochemical compounds. Usually, however, there is greater structural complexity associated with biochemical compounds as a result of polyfunctionality; several different functional groups are present. Often biochemical compounds interact with each other, within cells, to form larger structures. But the same chemical principles and chemical reactions associated with the various organic functional groups that we have studied apply to these larger biochemical structures as well.

 18.1 BIOCHEMISTRY–AN OVERVIEW

Biochemistry *is the study of the chemical substances found in living organisms and the chemical interactions of these substances with each other.* Biochemistry is a field in which new discoveries are made almost daily about how cells manufacture the molecules needed for life and how the chemical reactions by which life is maintained occur. The knowledge explosion that has occurred in the field of biochemistry during the last decades of the twentieth century and the beginning of the twenty-first is truly phenomenal.

Figure 18.1 Mass composition data for the human body in terms of major types of biochemical substances.

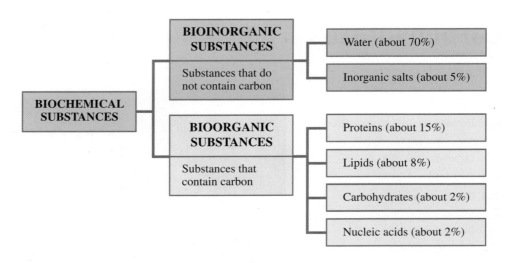

As isolated compounds, bioinorganic and bioorganic substances have no life in and of themselves. Yet when these substances are gathered together in a cell, their chemical interactions are able to sustain life.

It is estimated that more than half of all organic carbon atoms are found in the carbohydrate materials of plants.

Human uses for carbohydrates of the plant kingdom extend beyond food. Carbohydrates in the form of cotton and linen are used as clothing. Carbohydrates in the form of wood are used for shelter and heating and in making paper.

A **biochemical substance** *is a chemical substance found within a living organism.* Biochemical substances are divided into two groups: bioinorganic substances and bioorganic substances. *Bioinorganic substances* include water and inorganic salts. *Bioorganic substances* include carbohydrates, lipids, proteins, and nucleic acids. Figure 18.1 gives an approximate mass composition for the human body in terms of types of biochemical substances present.

Although we tend to think of the human body as made up of organic substances, bioorganic molecules make up only about one-fourth of body mass. The bioinorganic substance water constitutes over two-thirds of the mass of the human body, and another 4%–5% of body mass comes from inorganic salts (Section 10.6).

 18.2   OCCURRENCE AND FUNCTIONS OF CARBOHYDRATES

Carbohydrates are the most abundant class of bioorganic molecules on planet Earth. Although their abundance in the human body is relatively low (Section 18.1), carbohydrates constitute about 75% by mass of dry plant materials (see Figure 18.2).

Green (chlorophyll-containing) plants produce carbohydrates via *photosynthesis.* In this process, carbon dioxide from the air and water from the soil are the reactants, and sunlight absorbed by chlorophyll is the energy source.

$$CO_2 + H_2O + \text{solar energy} \xrightarrow[\text{Plant enzymes}]{\text{Chlorophyll}} \text{carbohydrates} + O_2$$

Plants have two main uses for the carbohydrates they produce. In the form of *cellulose,* carbohydrates serve as structural elements, and in the form of *starch,* they provide energy reserves for the plants.

Dietary intake of plant materials is the major carbohydrate source for humans and animals. The average human diet should ideally be about two-thirds carbohydrate by mass. Carbohydrates have the following functions in humans:

1. Carbohydrate oxidation provides energy.
2. Carbohydrate storage, in the form of glycogen, provides a short-term energy reserve.
3. Carbohydrates supply carbon atoms for the synthesis of other biochemical substances (proteins, lipids, and nucleic acids).
4. Carbohydrates form part of the structural framework of DNA and RNA molecules.
5. Carbohydrates linked to lipids (Chapter 19) are structural components of cell membranes.
6. Carbohydrates linked to proteins (Chapter 20) function in a variety of cell–cell and cell–molecule recognition processes.

Figure 18.2 Most of the matter in plants, except water, is carbohydrate material. Photosynthesis, the process by which carbohydrates are made, requires sunlight.

18.3 CLASSIFICATION OF CARBOHYDRATES

Most simple carbohydrates have empirical formulas that fit the general formula $C_nH_{2n}O_n$. An early observation by scientists that this general formula can also be written as $C_n(H_2O)_n$ is the basis for the term *carbohydrate*—that is, "hydrate of carbon." It is now known that this hydrate viewpoint is not correct, but the term *carbohydrate* still persists. Today the term is used to refer to an entire family of compounds, only some of which have the formula $C_nH_{2n}O_n$.

A **carbohydrate** *is a polyhydroxy aldehyde, a polyhydroxy ketone, or a compound that yields polyhydroxy aldehydes or polyhydroxy ketones upon hydrolysis.* The carbohydrate glucose is a polyhydroxy aldehyde, and the carbohydrate fructose is a polyhydroxy ketone.

Aldehyde group → CHO

Glucose (a polyhydroxy aldehyde)

Fructose (a polyhydroxy ketone)

Ketone group ← C=O

A striking structural feature of carbohydrates is the large number of functional groups present. In glucose and fructose there is a functional group attached to each carbon atom.

Carbohydrates are classified on the basis of molecular size as monosaccharides, oligosaccharides, and polysaccharides.

A **monosaccharide** *is a carbohydrate that contains a single polyhydroxy aldehyde or polyhydroxy ketone unit.* Monosaccharides cannot be broken down into simpler units by hydrolysis reactions. Both glucose and fructose are monosaccharides. Naturally occurring monosaccharides have from three to seven carbon atoms; five- and six-carbon species are especially common. Pure monosaccharides are water-soluble, white, crystalline solids.

An **oligosaccharide** *is a carbohydrate that contains two to ten monosaccharide units covalently bonded to each other.* Disaccharides are the most common type of oligosaccharide. A **disaccharide** *is a carbohydrate that contains two monosaccharide units covalently bonded to each other.* Like monosaccharides, disaccharides are crystalline, water-soluble substances. Sucrose (table sugar) and lactose (milk sugar) are disaccharides.

Within the human body, oligosaccharides are often found associated with proteins and lipids in complexes that have both structural and regulatory functions. Free oligosaccharides, other than disaccharides, are seldom encountered in biochemical systems.

Complete hydrolysis of an oligosaccharide produces monosaccharides. Upon hydrolysis, a disaccharide produces two monosaccharides, a trisaccharide three monosaccharides, a hexasaccharide six monosaccharides, and so on.

A **polysaccharide** *is a polymeric carbohydrate that contains many monosaccharide units covalently bonded to each other.* Polysaccharides often contain several thousand monosaccharide units. Both cellulose and starch are polysaccharides. We encounter these two substances everywhere. The paper on which this book is printed is mainly cellulose, as are the cotton in our clothes and the wood in our houses. Starch is a component of many types of foods, including bread, pasta, potatoes, rice, corn, beans, and peas.

18.4 CHIRALITY: HANDEDNESS IN MOLECULES

Monosaccharides are the simplest type of carbohydrate. Before considering specific structures for and specific reactions of monosaccharides, we will consider an important general structural property called *handedness,* which most monosaccharides exhibit.

The term *monosaccharide* is pronounced "mon-oh-SACK-uh-ride."

The *oligo* in the term *oligosaccharides* comes from the Greek *oligos,* which means "small" or "few." The term *oligosaccharide* is pronounced "OL-ee-go-SACK-uh-ride."

Types of carbohydrates are related to each other through hydrolysis.

Polysaccharides
↓ Hydrolysis
Oligosaccharides
↓ Hydrolysis
Monosaccharides

Figure 18.3 The mirror image of the right hand is the left hand. Conversely, the mirror image of the left hand is the right hand.

Left   **Right**

Mirror image of left hand
is in the back of the mirror

The property of handedness is not restricted to carbohydrates. It is a general phenomenon found in all classes of organic compounds.

Every object has a mirror image. The question is, "Is the mirror image the same (superimposable) or different (nonsuperimposable)?"

The term *chiral* (rhymes with *spiral*) comes from the Greek word *cheir,* which means "hand." Chiral objects are said to possess "handedness."

Most monosaccharides exist in two forms: a "left-handed" form and a "right-handed" form. These two forms are related to each other in the same way your left and right hands are related to each other. That relationship is that of *mirror images.* Figure 18.3 shows this mirror-image relationship for human hands.

Mirror Images

The concept of *mirror images* is the key to understanding molecular handedness. All objects, including all molecules, have mirror images. A **mirror image** *is the reflection of an object in a mirror.* Objects can be divided into two classes on the basis of their mirror images: objects with *superimposable* mirror images and objects with *nonsuperimposable* mirror images. **Superimposable mirror images** *are images that coincide at all points when the images are laid upon each other.* A dinner plate with no design features has superimposable mirror images. **Nonsuperimposable mirror images** *are images where not all points coincide when the images are laid upon each other.* Human hands are nonsuperimposable mirror images, as Figure 18.4 shows; note in this figure that the two thumbs point in opposite directions and that the fingers do not align correctly. Like human hands, all objects with nonsuperimposable mirror images exist in "left-handed" and "right-handed" forms.

Chirality

Of particular concern to us is the "handedness concept" as it applies to molecules. Not all molecules possess handedness. What, then, is the molecular structural feature that generates "handedness"? Any organic molecule that contains a carbon atom with four *different* groups attached to it in a tetrahedral orientation possesses handedness. Such a carbon atom is called a *chiral center.* A **chiral center** *is an atom in a molecule that has four different groups tetrahedrally bonded to it.*

A molecule that contains a chiral center is said to be *chiral.* A **chiral molecule** *is a molecule whose mirror images are not superimposable.* Chiral molecules have handedness. An **achiral molecule** *is a molecule whose mirror images are superimposable.* Achiral molecules do not possess handedness.

A trisubstituted methane molecule, such as bromochloroiodomethane, is the simplest example of a chiral organic molecule.

$$\begin{array}{c} H \\ | \\ Br-C-Cl \\ | \\ I \end{array}$$

Bromochloroiodomethane

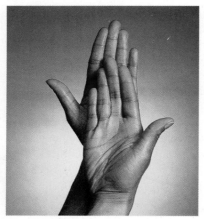

Figure 18.4   A person's left and right hands are not superimposable upon each other.

Note the four different groups attached to the carbon atom present: —H, —Br, —Cl, and —I. Figure 18.5a shows the nonsuperimposability of the two mirror-image forms of this molecule.

Figure 18.5 Examples of simple molecules that are chiral. (a) The mirror image forms of the molecule bromochloroiodomethane are nonsuperimposable. (b) The mirror-image forms of the molecule glyceraldehyde are nonsuperimposable.

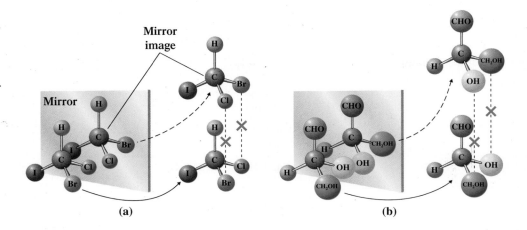

The simplest example of a chiral monosaccharide molecule is the three-carbon monosaccharide called glyceraldehyde.

$$
\begin{array}{c}
CHO \\
| \\
H-C-OH \\
| \\
CH_2OH
\end{array}
$$

Glyceraldehyde

There are a few chiral molecules known that do not have a chiral center. Such exceptions are not important for the applications of the chirality concept that we will make in this text.

The four different groups attached to the carbon atom at the chiral center in this molecule are —H, —OH, —CHO, and —CH₂OH. The nonsuperimposability of the two mirror-image forms of glyceraldehyde is shown in Figure 18.5b.

Chiral centers within molecules are often denoted by a small asterisk. Note the chiral centers in the following molecules.

$$
\begin{array}{ccc}
\underset{OH}{CH_3-CH_2-\overset{H}{\underset{|}{*C}}-CH_3,} & \underset{I}{H-\overset{Cl}{\underset{|}{*C}}-CH_3} & CH_3-CH_2-CH_2-\overset{CH_3}{\underset{\underset{H}{|}}{*C}}-CH_2-CH_3
\end{array}
$$

2-Butanol   1-Chloro-1-iodoethane   3-Methylhexane

EXAMPLE 18.1

Identifying Chiral Centers in Molecules

Indicate whether the circled carbon atom in each of the following molecules is a chiral center.

a. CH₃—ⒸH—CH₂—CH₃
　　　　　|
　　　　　Cl

b. CH₃—CH₂—Ⓒ—CH₃
　　　　　　　　‖
　　　　　　　　O

c. CH₃—CH₂—ⒸH—OH
　　　　　　　　　|
　　　　　　　　CH₂
　　　　　　　　　|
　　　　　　　　CH₃

d.
　　　　　　Br
　　　　　　|
　　　　　ⒸH
　　　H₂C　　CH₂
　　　|　　　　|
　　　HC　　　CH₂
　　　|　　　　|
　　　Br　　CH₂

Solution

a. This is a chiral center. The four different groups attached to the carbon atom are —CH₃, —Cl, —CH₂—CH₃, and —H.

b. No chiral center is present. The carbon atom is attached to only *three* groups.

c. No chiral center is present. Two of the groups attached to the carbon atom are identical.

(continued)

d. The chirality rules for ring carbon atoms are the same as those for acyclic carbon atoms. A chiral center is present. Two of the groups are —H and —Br. The third group, obtained by proceeding clockwise around the ring, is —CH_2—CH_2—CH_2. The fourth group, obtained by proceeding counterclockwise around the ring, is —CH_2—CHBr—CH_2.

Practice Exercise 18.1

Indicate whether the circled carbon atom in each of the following molecules is a chiral center.

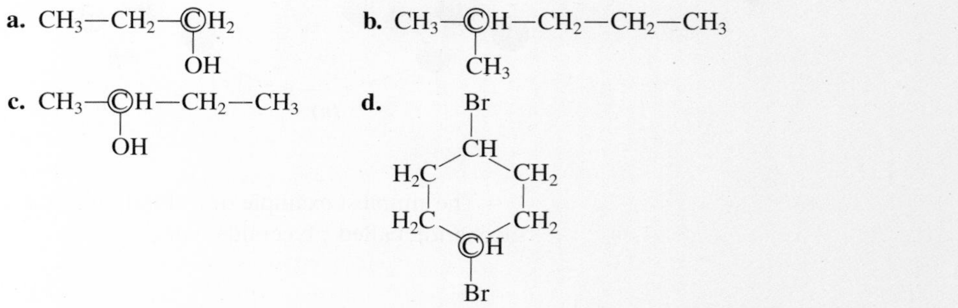

a. CH_3—CH_2—ⒸH_2
　　　　　　　　OH

b. CH_3—ⒸH—CH_2—CH_2—CH_3
　　　　　　CH$_3$

c. CH_3—ⒸH—CH_2—CH_3
　　　　OH

d.

Answers:　a. Not a chiral center;　**b.** Not a chiral center;　**c.** Chiral center;　**d.** Not a chiral center

---

> Remember the meaning of the structural notations —CHO and —CH_2OH.
>
> —CHO　means
> $$-\overset{\displaystyle O}{\underset{}{\overset{\|}{C}}}-H$$
>
> —CH_2OH　means
> $$-\overset{\displaystyle H}{\underset{\displaystyle H}{\overset{|}{\underset{|}{C}}}}-OH$$

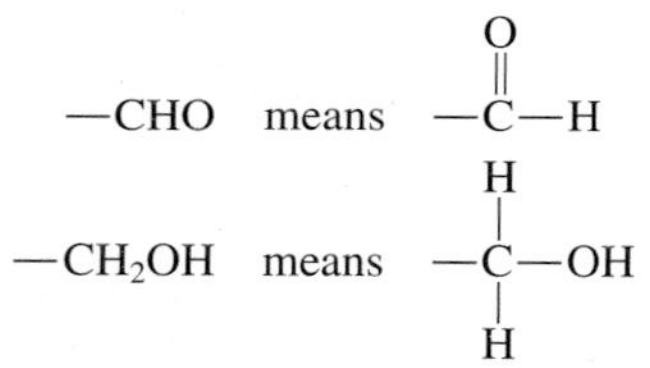

Organic molecules, especially monosaccharides, may contain more than one chiral center. For example, the following monosaccharide has two chiral centers.

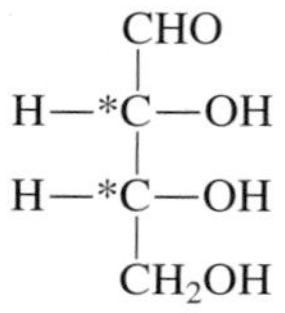

$$
\begin{array}{c}
CHO \\
| \\
H-{}^*C-OH \\
| \\
H-{}^*C-OH \\
| \\
CH_2OH
\end{array}
$$

What is the importance of the handedness that we have been discussing? In human body chemistry, right-handed and left-handed forms of a molecule often elicit different responses within the body. Sometimes both forms are biologically active, each form giving a different response; sometimes both elicit the same response, but one form's response is many times greater than that of the other; and sometimes only one of the two forms is biochemically active. For example, studies show that the body's response to the right-handed form of the hormone epinephrine (Section 17.10) is 20 times greater than its response to the left-handed form.

Naturally occurring monosaccharides are almost always "right-handed." Plants, our dietary source for carbohydrates, produce only right-handed monosaccharides. Interestingly, when we consider protein chemistry (Chapter 20), we will find that amino acids, the building blocks for proteins, are always left-handed molecules.

18.5 STEREOISOMERISM: ENANTIOMERS AND DIASTEREOMERS

The left- and right-handed forms of a chiral molecule are isomers. They are not *constitutional isomers,* the type of isomerism that we encountered repeatedly in the organic chemistry chapters of the text, but rather are *stereoisomers.* **Stereoisomers** *are isomers that have the same molecular and structural formulas but differ in the orientation of atoms in space.* By contrast, atoms are connected to each other in different ways in constitutional isomers (Section 12.6).

There are two major structural features that generate *stereoisomerism:* (1) the presence of a chiral center in a molecule and (2) the presence of "structural rigidity" in a

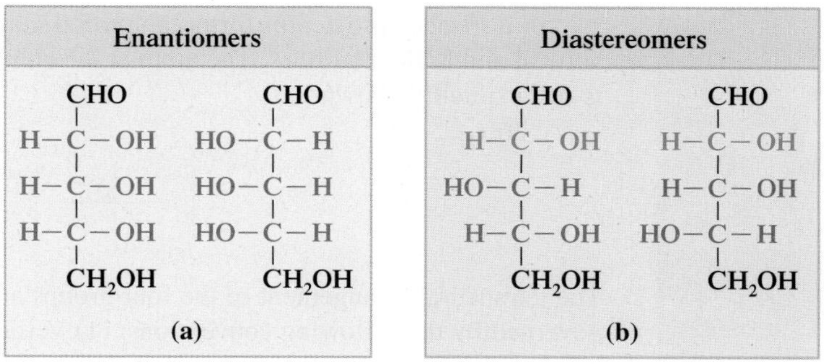

Figure 18.6 (a) Enantiomers are stereoisomers whose molecules are nonsuperimposable mirror images of each other, as in left-handed and right-handed forms of a molecule. (b) Diastereomers are stereoisomers whose molecules are not mirror images of each other.

Enantiomers	Diastereomers
CHO CHO	CHO CHO
H—C—OH HO—C—H	H—C—OH H—C—OH
H—C—OH HO—C—H	HO—C—H H—C—OH
H—C—OH HO—C—H	H—C—OH HO—C—H
CH₂OH CH₂OH	CH₂OH CH₂OH
(a)	**(b)**

molecule. Structural rigidity is caused by restricted rotation about chemical bonds. It is the basis for *cis–trans* isomerism, a phenomenon found in some substituted cycloalkanes (Section 12.14) and some alkenes (Section 13.5). Thus handedness is our second encounter with stereoisomerism. (When we discussed *cis–trans* isomerism, we did not mention that it is a form of stereoisomerism.)

Stereoisomers can be subdivided into two types: *enantiomers* and *diastereomers* (Figure 18.6). **Enantiomers** *are stereoisomers whose molecules are nonsuperimposable mirror images of each other.* Left- and right-handed forms of a molecule with a single chiral center are enantiomers.

Diastereomers *are stereoisomers whose molecules are not mirror images of each other.* Cis–trans isomers (of both the alkene and the cycloalkane types) are diastereomers. Molecules that contain more than one chiral center can also exist in diastereomeric as well as enantiomeric forms, as is shown in Figure 18.6.

Figure 18.7 shows the "thinking pattern" involved in using the terms *stereoisomers, enantiomers,* and *diastereomers.*

The term *enantiomer* comes from the Greek *enantios,* which means "opposite." It is pronounced "en-AN-tee-o-mer."

Some textbooks use the term *diastereoisomers* instead of *diastereomers.* The pronunciation for *diastereomer* is "dye-a-STEER-ee-o-mer."

18.6 DESIGNATING HANDEDNESS USING FISCHER PROJECTION FORMULAS

Drawing *three*-dimensional representations of chiral molecules can be both time-consuming and awkward. Fischer projection formulas represent a method for giving molecular chirality specifications in *two* dimensions. A **Fischer projection formula** *is a two-dimensional structural notation for showing the spatial arrangement of groups about chiral centers in molecules.*

Fischer projection formulas carry the name of their originator, the German chemist Hermann Emil Fischer (see Figure 18.8).

Figure 18.7 A summary of the "thought process" used in classifying molecules as enantiomers or diastereomers.

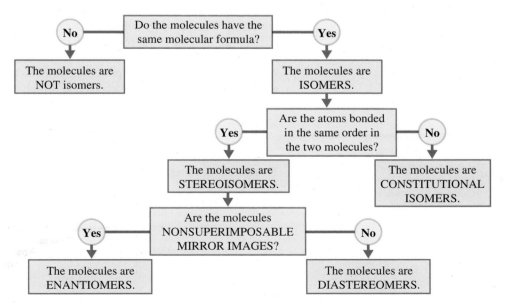

In a Fischer projection formula a chiral center is represented as the intersection of vertical and horizontal lines. The atom at the chiral center, which is almost always carbon, is not explicitly shown.

The tetrahedral arrangement of the four groups attached to the atom at the chiral center is governed by the following conventions: (1) Vertical lines from the chiral center represent bonds to groups directed into the printed page. (2) Horizontal lines from the chiral center represent bonds to groups directed out of the printed page.

In Fischer projection formulas for monosaccharides (the simplest type of carbohydrate; Section 18.2), the monosaccharide *carbon chain* is positioned vertically with the carbonyl group (aldehyde or ketone) at or near the top. The smallest monosaccharide that has a chiral center is the compound glyceraldehyde (2,3-dihydroxypropanal; Section 18.4).

$$CH_2-{}^*CH-\overset{\overset{\displaystyle O}{\|}}{C}-H$$
$$\ \ |\qquad\ |$$
$$OH\quad OH$$

The Fischer projection formulas for the two enantiomers of glyceraldehyde are

The handedness (right and left) of these two enantiomers is specified by using the designations D and L. The enantiomer with the chiral center —OH group on the right in the Fischer projection formula is by definition the right-handed isomer (D-glyceraldehyde), and the enantiomer with the chiral center —OH group on the left in the Fischer projection formula is by definition the left-handed isomer (L-glyceraldehyde).

We now consider Fischer projection formulas for the compound 2,3,4-trihydroxybutanal, a monosaccharide with four carbons and *two* chiral centers.

$$CH_2-{}^*CH-{}^*CH-\overset{\overset{\displaystyle O}{\|}}{C}-H$$
$$\ \ |\qquad\ |\qquad\ |$$
$$OH\quad OH\quad OH$$

There are four stereoisomers for this compound—two pairs of enantiomers.

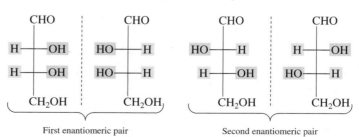

First enantiomeric pair   Second enantiomeric pair

The D and L designations for the handedness of the two members of an enantiomeric pair come from the Latin words *dextro,* which means "right," and *levo,* which means "left."

To draw the mirror image of a Fischer projection structure, keep up-and-down and front-and-back aspects of the structure the same and reverse the left-and-right aspects.

> Any given molecular structure can have only one mirror image. Hence enantiomers always come in pairs; there can never be more than two.

In the first enantiomeric pair, both chiral-center —OH groups are on the same side of the Fischer projection formula, and in the second enantiomeric pair, the chiral-center —OH groups are on opposite sides of the Fischer projection formula. These are the only —OH group arrangements possible.

The D,L system used to designate the handedness of glyceraldehyde enantiomers is extended to monosaccharides with more than one chiral center in the following manner. The carbon chain is numbered starting at the carbonyl group end of the molecule, and the highest-numbered chiral center is used to determine D or L configuration.

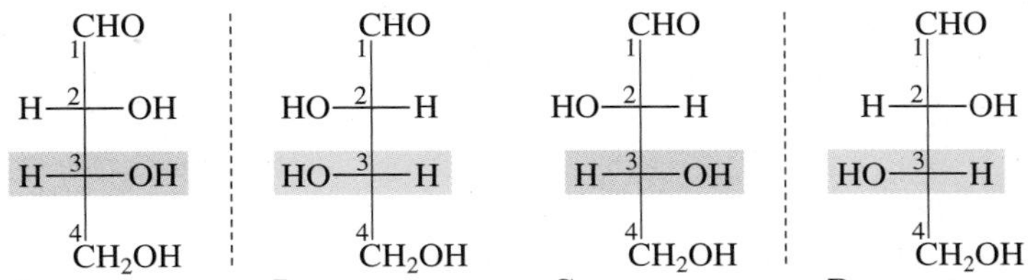

A D isomer | B L isomer | C D isomer | D L isomer

> Diastereomers that have two chiral centers must have the same handedness (both left or both right) at one chiral center and opposite handedness (one left and one right) at the other chiral center.

The D,L nomenclature gives the configuration (handedness) only at the highest-numbered chiral center. The configuration at other chiral centers in a molecule is accounted for by assigning a different common name to each pair of D,L enantiomers. In our present example, compounds A and B (the first enantiomeric pair) are D-erythrose and L-erythrose; compounds C and D (the second enantiomeric pair) are D-threose and L-threose.

What is the relationship between compounds A and C in our present example? They are diastereomers (Section 18.5), stereoisomers that are not mirror images of each other. Other diastereomeric pairs in our example are A and D, B and C, and B and D. The members of each of these four pairs are epimers. **Epimers** *are diastereomers whose molecules differ only in the configuration at one chiral center.*

EXAMPLE 18.2 Drawing Fischer Projection Formulas for Monosaccharides

Draw a Fischer projection formula for the enantiomer of each of the following monosaccharides.

a.

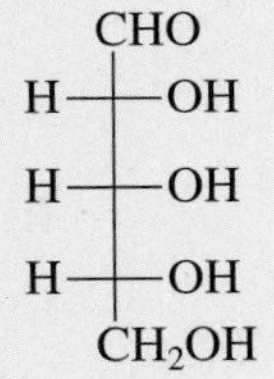

b.

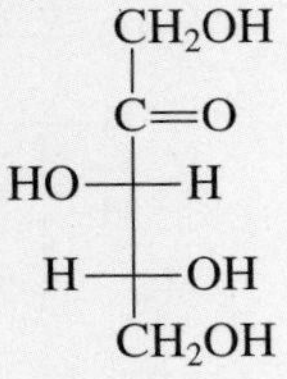

Solution

Given the Fischer projection formula of one member of an enantiomeric pair, we draw the other enantiomer's Fischer projection formula by reversing the substituents that are in horizontal positions *at each chiral center.*

a. Three chiral centers are present in this polyhydroxy aldehyde. Reversing the positions of the —H and —OH groups at each chiral center produces the Fischer projection formula of the other enantiomer.

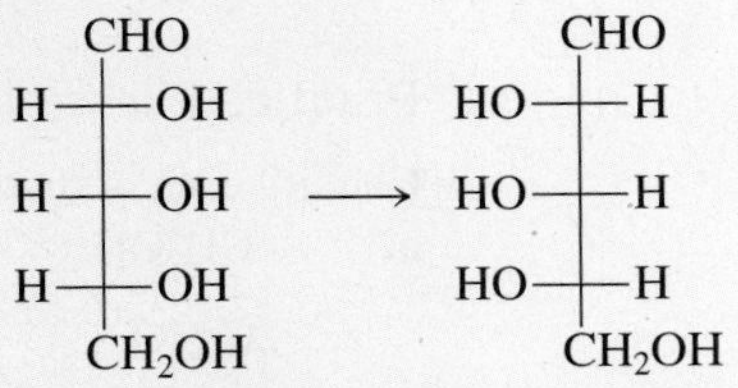

The given enantiomer — The other enantiomer

b. This monosaccharide is a polyhydroxy ketone with two chiral centers. Reversing the positions of the —H and —OH groups at both chiral centers generates the Fischer projection formula of the other enantiomer.

(continued)

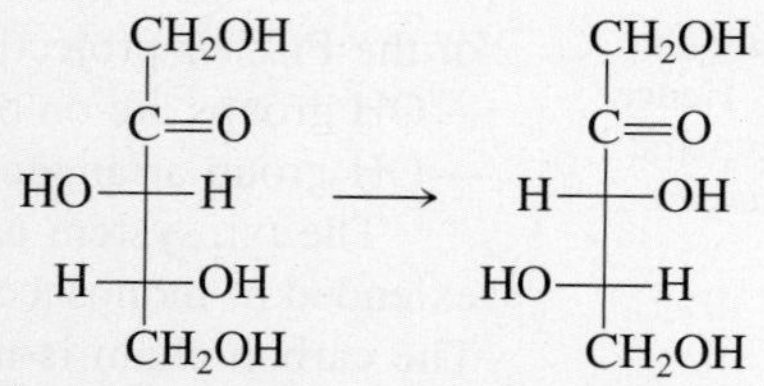

$$\begin{array}{ccc}
\text{CH}_2\text{OH} & & \text{CH}_2\text{OH} \\
\text{C}{=}\text{O} & & \text{C}{=}\text{O} \\
\text{HO}{-\!-}\text{H} & \longrightarrow & \text{H}{-\!-}\text{OH} \\
\text{H}{-\!-}\text{OH} & & \text{HO}{-\!-}\text{H} \\
\text{CH}_2\text{OH} & & \text{CH}_2\text{OH} \\
\text{The given enantiomer} & & \text{The other enantiomer}
\end{array}$$

Practice Exercise 18.2

Draw a Fischer projection formula for the enantiomer of each of the following monosaccharides.

a.
$$\begin{array}{c}
\text{CHO} \\
\text{H}{-\!-}\text{OH} \\
\text{HO}{-\!-}\text{H} \\
\text{HO}{-\!-}\text{H} \\
\text{CH}_2\text{OH}
\end{array}$$

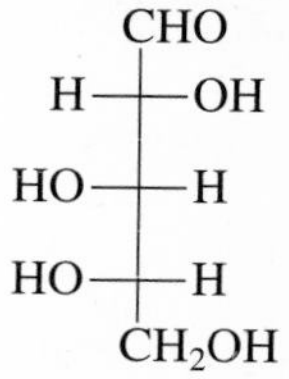

b.
$$\begin{array}{c}
\text{CH}_2\text{OH} \\
\text{C}{=}\text{O} \\
\text{H}{-\!-}\text{OH} \\
\text{H}{-\!-}\text{OH} \\
\text{CH}_2\text{OH}
\end{array}$$

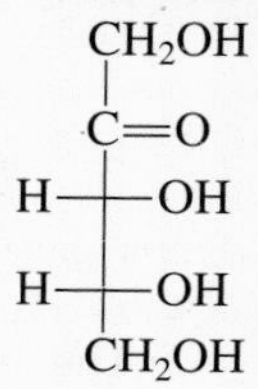

Answers:

a.
$$\begin{array}{c}
\text{CHO} \\
\text{HO}{-\!-}\text{H} \\
\text{H}{-\!-}\text{OH} \\
\text{H}{-\!-}\text{OH} \\
\text{CH}_2\text{OH}
\end{array}$$

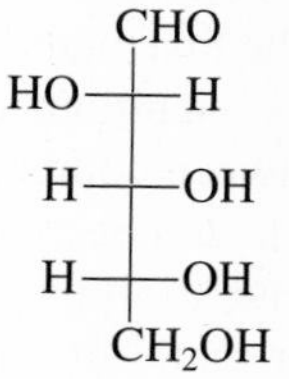

b.
$$\begin{array}{c}
\text{CH}_2\text{OH} \\
\text{C}{=}\text{O} \\
\text{HO}{-\!-}\text{H} \\
\text{HO}{-\!-}\text{H} \\
\text{CH}_2\text{OH}
\end{array}$$

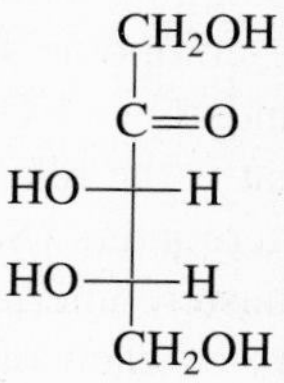

EXAMPLE 18.3

Classifying Monosaccharides as D or L Enantiomers

Classify each of the following monosaccharides as a D enantiomer or an L enantiomer.

a.
$$\begin{array}{c}
{}^1\text{CHO} \\
\text{HO}{-}{}^2{-}\text{H} \\
\text{H}{-}{}^3{-}\text{OH} \\
\text{H}{-}{}^4{-}\text{OH} \\
{}^5\text{CH}_2\text{OH}
\end{array}$$

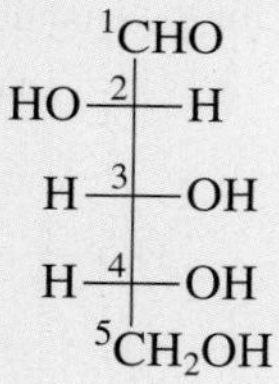

b.
$$\begin{array}{c}
{}^1\text{CH}_2\text{OH} \\
{}^2\text{C}{=}\text{O} \\
\text{H}{-}{}^3{-}\text{OH} \\
\text{HO}{-}{}^4{-}\text{H} \\
\text{HO}{-}{}^5{-}\text{H} \\
{}^6\text{CH}_2\text{OH}
\end{array}$$

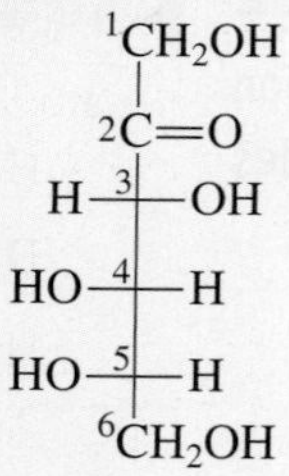

Solution

D or L configuration for a monosaccharide is determined by the highest-numbered chiral center, the one farthest from the carbonyl carbon atom.

a. The highest-numbered chiral center, which involves carbon 4, has the —OH group on the right. Thus this monosaccharide is a D enantiomer.

b. The highest-numbered chiral center, which involves carbon 5, has the —OH group on the left. Thus this monosaccharide is an L enantiomer.

Practice Exercise 18.3

Classify each of the following monosaccharides as a D enantiomer or an L enantiomer.

a.
$$\begin{array}{c}
\text{CH}_2\text{OH} \\
\text{C}{=}\text{O} \\
\text{HO}{-\!-}\text{H} \\
\text{H}{-\!-}\text{OH} \\
\text{H}{-\!-}\text{OH} \\
\text{CH}_2\text{OH}
\end{array}$$

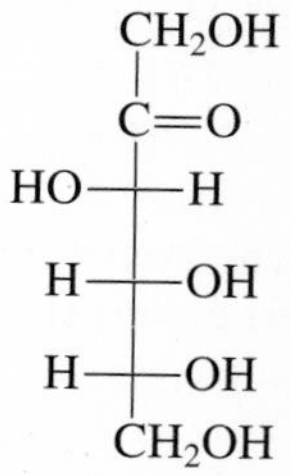

b.
$$\begin{array}{c}
\text{CHO} \\
\text{H}{-\!-}\text{OH} \\
\text{HO}{-\!-}\text{H} \\
\text{H}{-\!-}\text{OH} \\
\text{HO}{-\!-}\text{H} \\
\text{CH}_2\text{OH}
\end{array}$$

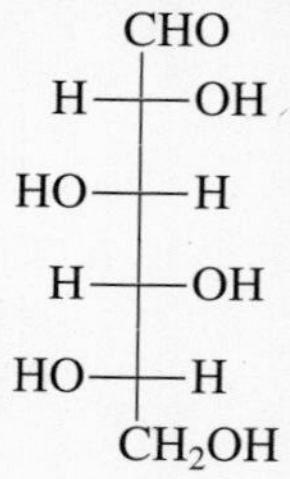

Answers: a. D enantiomer; **b.** L enantiomer

● **E X A M P L E 18.4**

Recognizing Enantiomers and Diastereomers

Characterize each of the following pairs of structures as enantiomers, diastereomers, or neither enantiomers nor diastereomers.

a.

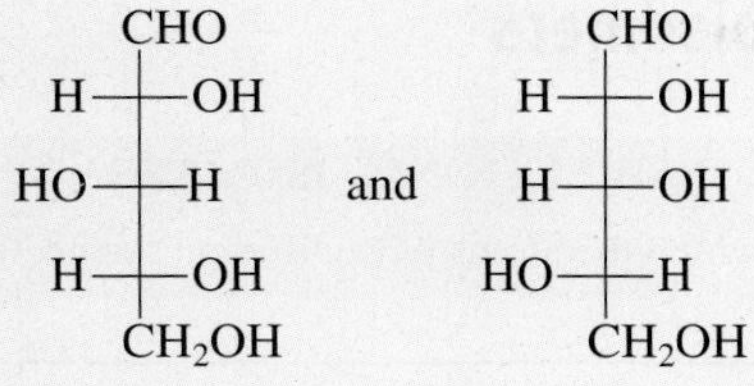

```
    CHO                CHO
H——OH            H——OH
HO——H     and    H——OH
H——OH            HO——H
   CH2OH             CH2OH
```

b.

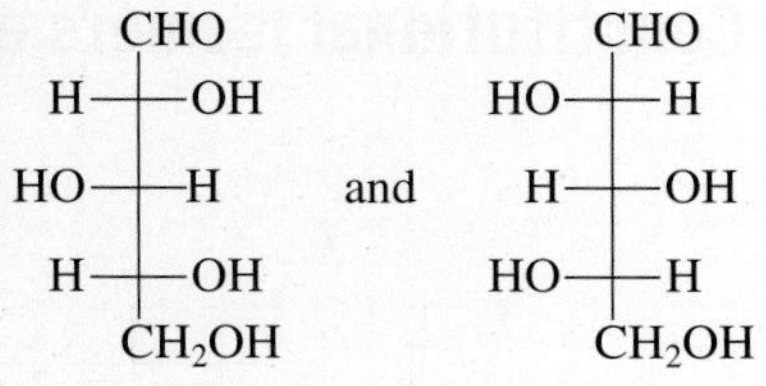

```
    CHO                CHO
H——OH            HO——H
HO——H     and    H——OH
H——OH            HO——H
   CH2OH             CH2OH
```

c.

```
    CHO                CHO
H——OH           H——C——H
HO——H     and    H——OH
H——OH            HO——H
   CH2OH             CH2OH
```

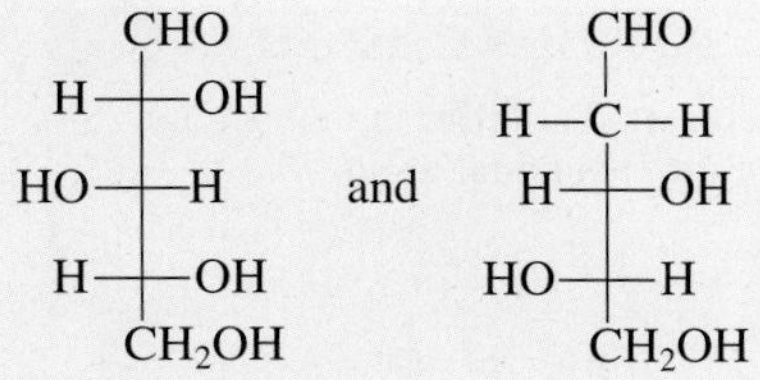

Solution

a. These two structures represent *diastereomers*—the arrangement of —H and —OH substituents is identical for at least one chiral center, whereas the arrangement of —H and —OH substituents at remaining chiral centers is that of mirror images. The —H and —OH substituent arrangement is the same at the first chiral center and is that of mirror images at the second and third chiral centers.

b. These two structures represent *enantiomers*—a mirror-image substituent relationship exists between the two isomers at *every* chiral center.

c. These two structures are *neither enantiomers nor diastereomers*. The connectivity of atoms differs in the two structures at carbon 2. Stereoisomers (enantiomers and diastereomers) must have the same connectivity throughout both structures. (The two structures are not even constitutional isomers because the first structure contains one more oxygen atom than the second.)

Practice Exercise 18.4

Characterize the following pairs of structures as enantiomers, diastereomers, or neither enantiomers nor diastereomers.

a.

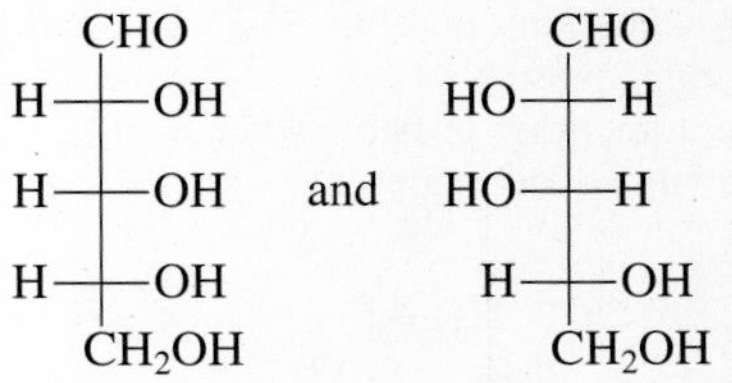

```
    CHO                CHO
H——OH            HO——H
H——OH     and    HO——H
H——OH            H——OH
   CH2OH             CH2OH
```

b.

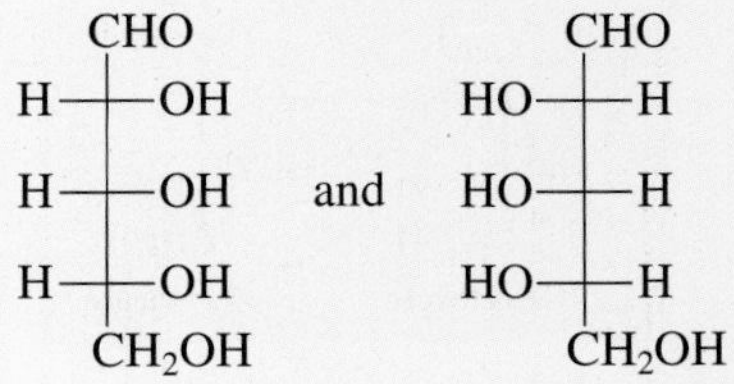

```
    CHO                CHO
H——OH            HO——H
H——OH     and    HO——H
H——OH            HO——H
   CH2OH             CH2OH
```

c.

```
    CHO                CHO
H——OH            HO——H
HO——H     and    H——OH
HO——H            HO——H
   CH2OH             CH2OH
```

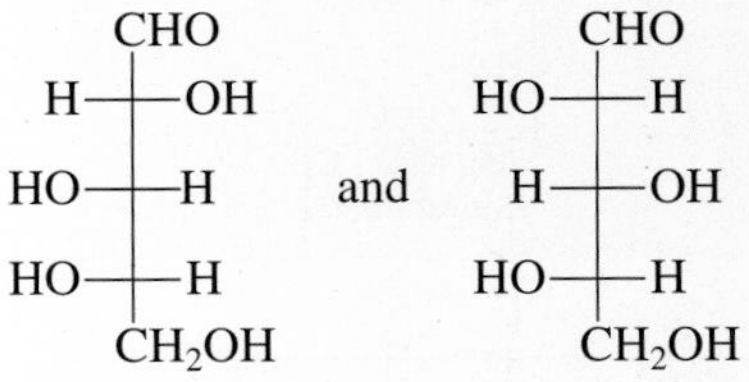

Answers: a. Diastereomers; **b.** Enantiomers; **c.** Diastereomers

We calculate 2^n to predict the maximum possible number of stereoisomers for a molecule with n chiral atoms. In a few cases, the actual number of stereoisomers is less than the maximum because of symmetry considerations that make some mirror images superimposable.

In general, a compound that has n chiral centers may exist in a *maximum* of 2^n stereoisomeric forms. For example, when three chiral centers are present, at most eight stereoisomers ($2^3 = 8$) are possible (four pairs of enantiomers).

The Chemistry at a Glance feature on page 566 summarizes information about the various types of isomers we have encountered so far in the text—the various subtypes of constitutional isomers and the various subtypes of stereoisomers.

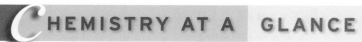

Constitutional Isomers and Stereoisomers

CONSTITUTIONAL ISOMERS

Isomers in which the atoms have different connectivity

SKELETAL ISOMERS

Isomers with different carbon atom arrangements and different hydrogen atom arrangements

$$CH_3-CH_2-CH_2-CH_3$$
Butane (C_4H_{10})

$$CH_3-CH-CH_3$$
$$|$$
$$CH_3$$
2-Methylpropane (C_4H_{10})

POSITIONAL ISOMERS

Isomers that differ in the location of the functional group

$$CH_2=CH-CH_2-CH_3$$
1-Butene (C_4H_8)

$$CH_3-CH=CH-CH_3$$
2-Butene (C_4H_8)

FUNCTIONAL GROUP ISOMERS

Isomers that contain different functional groups

$$CH_3-CH_2-CH_2-\overset{\overset{\displaystyle O}{\|}}{C}-H$$
Butanal (aldehyde, C_4H_8O)

$$CH_3-CH_2-\overset{\overset{\displaystyle O}{\|}}{C}-CH_3$$
2-Butanone (ketone, C_4H_8O)

STEREOISOMERS

Isomers with atoms of the same connectivity that differ only in the orientation of the atoms in space

---

ENANTIOMERS

- Stereoisomers that are nonsuperimposable mirror images of each other
- Handedness (D and L forms) is determined by the configuration at the high-numbered chiral center

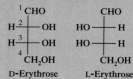

D and L Enantiomers

D-Erythrose   L-Erythrose

DIASTEREOMERS

Stereoisomers that are not mirror images of each other

CIS–TRANS ISOMERS

Stereoisomerism that results from restricted rotation about chemical bonds
- Is sometimes possible when a ring is present
- Is sometimes possible when a double bond is present

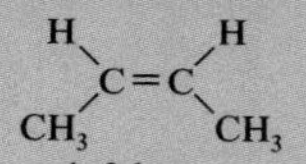

cis-2-butene

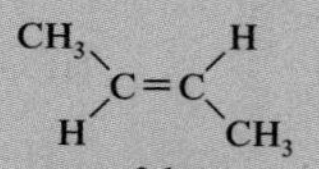

trans-2-butene

MOST OTHER DIASTEREOMERS
(two or more chiral centers)

Stereoisomerism that results from
- A mirror image relationship at one (or more) chiral centers, and
- The same configuration at one (or more) chiral centers

Three chiral centers

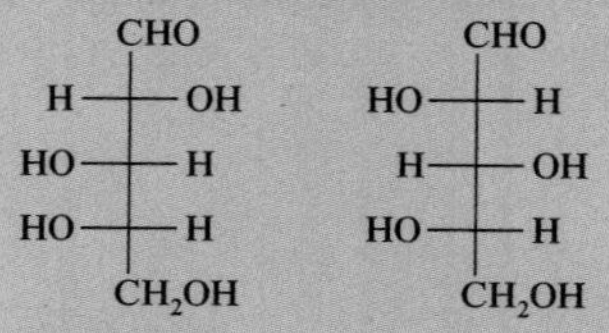

L-Arabinose   L-Xylose

  18.7   PROPERTIES OF ENANTIOMERS

Constitutional isomers differ in most chemical and physical properties. For example, constitutional isomers have different boiling points and melting points. Diastereomers also differ in most chemical and physical properties. They also have different boiling points and melting points. In contrast, nearly all the properties of a pair of enantiomers are the same;

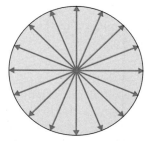

(a) Ordinary (unpolarized) light

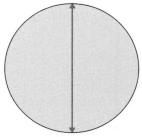

(b) Plane-polarized light

Figure 18.9 Vibrational characteristics of ordinary (unpolarized) light (a), and polarized light (b). The direction of travel of the light is toward the reader.

Achiral molecules are optically *inactive*. Chiral molecules are optically *active*.

Because of their ability to rotate the plane of polarized light, enantiomers are sometimes referred to as *optical isomers*.

In any pair of enantiomers, one, the (+)-enantiomer, always rotates the plane of polarized light to the right, and the other, the (−)-enantiomer, to the left.

for example, they have identical boiling points and melting points. Enantiomers exhibit different properties in only two areas: (1) their interaction with plane-polarized light and (2) their interaction with other chiral substances.

Interaction of Enantiomers with Plane-Polarized Light

All light moves through space with a wave motion. Ordinary light waves—that is, unpolarized light waves—vibrate in *all* planes at right angles to their direction of travel. Plane-polarized light waves, by contrast, vibrate in *only one* plane at right angles to their direction of travel. Figure 18.9 contrasts the vibrational behavior of ordinary light with that of plane-polarized light.

Ordinary light can be converted to plane-polarized light by passing it through a *polarizer,* an instrument with lenses or filters containing special types of crystals. When plane-polarized light is passed through a solution containing a *single* enantiomer, the plane of the polarized light is rotated counterclockwise (to the left) or clockwise (to the right), depending on the enantiomer. The extent of rotation depends on the concentration of the enantiomer as well as on its identity. Furthermore, the two enantiomers of a pair rotate the plane-polarized light the same number of degrees, but in opposite directions. If a 0.50 M solution of one enantiomer rotates the light 30° to the right, then a 0.50 M solution of the other enantiomer rotates the light 30° to the left.

Instruments used to measure the degree of rotation of plane-polarized light by enantiomeric compounds are called *polarimeters.* The schematic diagram in Figure 18.10 shows the basis for these instruments.

Dextrorotatory and Levorotatory Compounds

Enantiomers are said to be optically active because of the way they interact with plane-polarized light. An **optically active compound** *is a compound that rotates the plane of polarized light.*

An enantiomer that rotates plane-polarized light in a clockwise direction (to the *right*) is said to be dextrorotatory (the Latin *dextro* means "right"). A **dextrorotatory compound** *is a chiral compound that rotates the plane of polarized light in a clockwise direction.* An enantiomer that rotates plane-polarized light in a counterclockwise direction (to the *left*) is said to be levorotatory (the Latin *levo* means "left"). A **levorotatory compound** *is a chiral compound that rotates the plane of polarized light in a counterclockwise direction.* If one member of an enantiomeric pair is dextrorotatory, then the other member must be levorotatory.

A plus or minus sign inside parentheses is used to denote the direction of rotation of plane-polarized light by a chiral compound. The notation (+) means rotation to the right (clockwise), and (−) means rotation to the left (counterclockwise). Thus the dextrorotatory enantiomer of glucose is (+)-glucose.

The handedness of enantiomers (D or L, Section 18.6) and the direction of rotation of plane-polarized light by enantiomers [(+) or (−)] are not connected entities. There is no way of knowing which way an enantiomer will rotate light until it is examined with

Figure 18.10 Schematic depiction of how a polarimeter works.

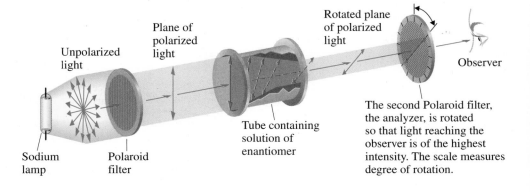

Unpolarized light · Plane of polarized light · Rotated plane of polarized light · Observer · Sodium lamp · Polaroid filter · Tube containing solution of enantiomer · The second Polaroid filter, the analyzer, is rotated so that light reaching the observer is of the highest intensity. The scale measures degree of rotation.

Both handedness and direction of rotation of plane-polarized light can be incorporated into the name of an enantiomer. For example, the notation D-(+)-mannose specifies that the right-handed isomer of the monosaccharide mannose rotates plane-polarized light in a clockwise direction (to the right).

a polarimeter. Not all D enantiomers rotate plane-polarized light in the same direction, nor do all L enantiomers rotate plane-polarized light in the same direction. Some D enantiomers are dextrorotatory; others are levorotatory.

Interactions Between Chiral Compounds

A left-handed baseball player (chiral) and a right-handed baseball player (chiral) can use the same baseball bat (achiral) or wear the same baseball hat (achiral). However, left- and right-handed baseball players (chiral) cannot use the same baseball glove (chiral). This nonchemical example illustrates that the chirality of an object becomes important when the object interacts with another chiral object.

Applying this generalization to molecules, we find that the two members of an enantiomeric pair have the same interaction with achiral molecules and different interactions with chiral molecules. We find that

1. Enantiomers have identical boiling points, melting points, and densities because such properties depend on the strength of intermolecular forces (Section 7.13), and intermolecular force strength does not depend on chirality. Intermolecular force strength is the same for both forms of a chiral molecule because both forms have identical sets of functional groups.
2. A pair of enantiomers have the same solubility in an achiral solvent, such as ethanol, but differing solubilities in a chiral solvent, such as D-2-butanol.
3. The rate and extent of reaction of enantiomers with another reactant are the same if the reactant is achiral but differ if the reactant is chiral.
4. Receptor sites for molecules within the body have chirality associated with them. Thus enantiomers always generate different responses within the human body as they interact at such sites. Sometimes the responses are only slightly different, and at other times they are very different.

Let us consider two specific examples of differing chiral–chiral interactions involving enantiomers that occur within the human body. The first example involves taste perceptions. The distinctly different natural flavors "spearmint" and "caraway" are generated by molecules that are enantiomers interacting with chiral "taste receptors" (see Figure 18.11).

The second example involves the body's response to the enantiomeric forms of the hormone epinephrine (adrenaline). The response of the body to the D isomer of the hormone is 20 times greater than its response to the L isomer of the hormone. Epinephrine binds to its cellular receptor site by means of a three-point contact, as is shown in Figure 18.12. D-Epinephrine makes a perfect three-point contact with the receptor surface, but the biochemically weaker L-epinephrine can make only a two-point contact. Because of the poorer fit, the binding of the L isomer is weaker, and less physiological response is observed.

Figure 18.11 The distinctly different natural flavors of spearmint and caraway are caused by enantiomeric molecules. Spearmint leaves contain L-carvone, and caraway seeds contain D-carvone.

Figure 18.12 D-Epinephrine binds to the receptor at three points, whereas the biochemically weaker L-epinephrine binds at only two sites.

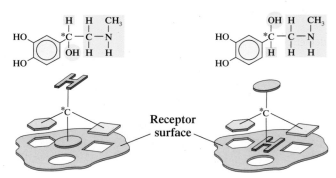

18.8 CLASSIFICATION OF MONOSACCHARIDES

Now that we have considered molecular chirality and its consequences (Sections 18.4 through 18.7), we return to the subject of carbohydrates by considering further details about monosaccharides, the simplest carbohydrates (Section 18.3).

Although there is no limit to the number of carbon atoms that can be present in a monosaccharide, only monosaccharides with three to seven carbon atoms are commonly found in nature. A three-carbon monosaccharide is called a *triose,* and those that contain four, five, and six carbon atoms are called *tetroses, pentoses,* and *hexoses,* respectively.

Monosaccharides are classified as *aldoses* or *ketoses* on the basis of type of carbonyl group (Section 15.1) present. An **aldose** *is a monosaccharide that contains an aldehyde functional group.* Aldoses are polyhydroxy aldehydes. A **ketose** *is a monosaccharide that contains a ketone functional group.* Ketoses are polyhydroxy ketones.

Monosaccharides are often classified by both their number of carbon atoms and their functional group. A six-carbon monosaccharide with an aldehyde functional group is an *aldohexose;* a five-carbon monosaccharide with a ketone functional group is a *ketopentose.*

Monosaccharides are also often called sugars. Hexoses are six-carbon sugars, pentoses five-carbon sugars, and so on. The word *sugar* is associated with "sweetness," and most (but not all) monosaccharides have a sweet taste. The designation *sugar* is also applied to disaccharides, many of which also have a sweet taste. Thus **sugar** *is a general designation for either a monosaccharide or a disaccharide.*

The term *saccharide* comes from the Latin word for "sugar," which is *saccharum.*

● **EXAMPLE 18.5**

Classifying Monosaccharides on the Basis of Structural Characteristics

Classify each of the following monosaccharides according to both the number of carbon atoms and the type of carbonyl group present.

a.

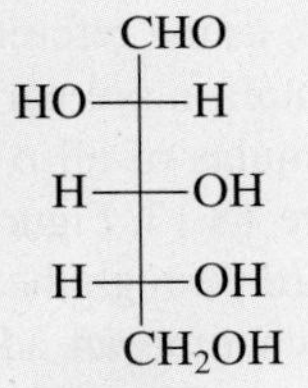

```
        CHO
   HO ——— H
    H ——— OH
    H ——— OH
       CH₂OH
```

b.

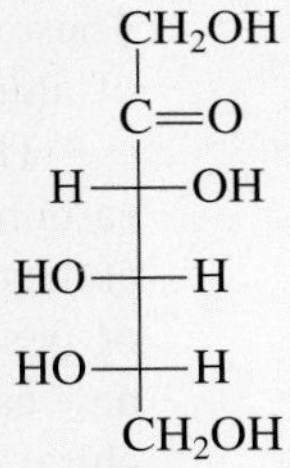

```
       CH₂OH
        |
        C == O
    H ——— OH
   HO ——— H
   HO ——— H
       CH₂OH
```

c.

```
        CHO
   HO ——— H
    H ——— OH
   HO ——— H
   HO ——— H
       CH₂OH
```

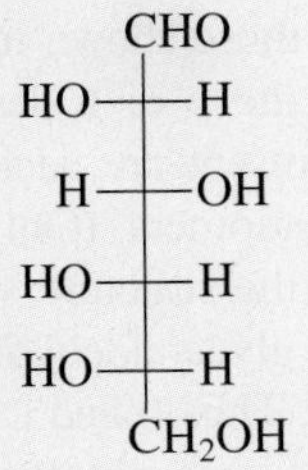

d.

```
       CH₂OH
        |
        C == O
    H ——— OH
    H ——— OH
       CH₂OH
```

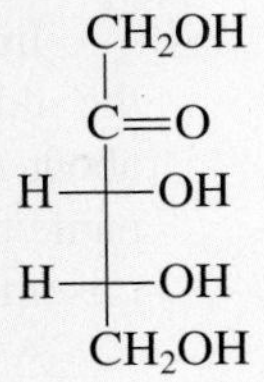

Solution

a. An aldehyde functional group is present as well as five carbon atoms. This monosaccharide is thus an *aldopentose.*

b. This monosaccharide contains a ketone group and six carbon atoms, so it is a *ketohexose.*

c. Six carbon atoms and an aldehyde group in a monosaccharide are characteristic of an *aldohexose.*

d. This monosaccharide is a *ketopentose.*

(continued)

Practice Exercise 18.5

Classify each of the following monosaccharides according to both the number of carbon atoms and the type of carbonyl group present.

a. 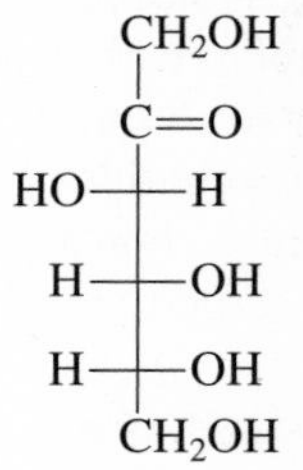
$$CH_2OH$$
$$C=O$$
$$HO-H$$
$$H-OH$$
$$H-OH$$
$$CH_2OH$$

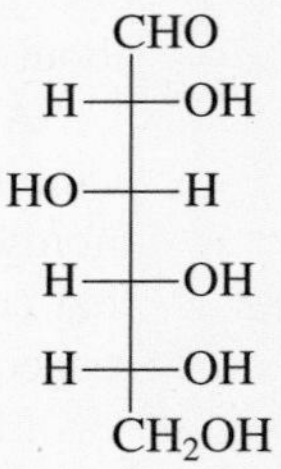

b.
$$CHO$$
$$H-OH$$
$$HO-H$$
$$H-OH$$
$$H-OH$$
$$CH_2OH$$

c.
$$CHO$$
$$H-OH$$
$$H-OH$$
$$CH_2OH$$

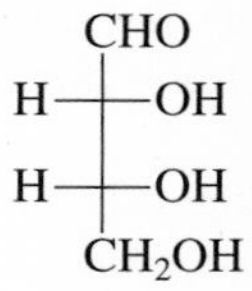

d. 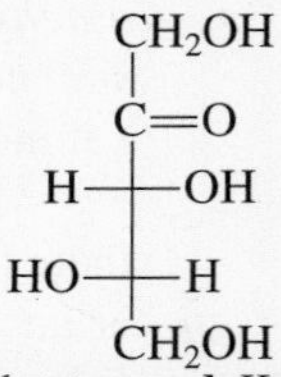
$$CH_2OH$$
$$C=O$$
$$H-OH$$
$$HO-H$$
$$CH_2OH$$

Answers: a. Ketohexose; **b.** Aldohexose; **c.** Aldotetrose; **d.** Ketopentose

---

In terms of carbon atoms, trioses are the smallest monosaccharides that can exist. There are two such compounds, one an aldose (glyceraldehyde) and the other a ketose (dihydroxyacetone).

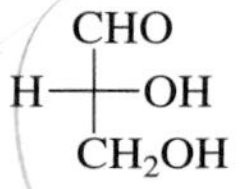

$$CHO$$
$$H-OH$$
$$CH_2OH$$
D-Glyceraldehyde

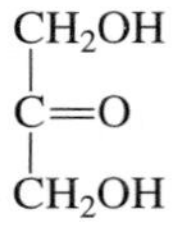

$$CH_2OH$$
$$C=O$$
$$CH_2OH$$
Dihydroxyacetone

These two triose structures serve as the reference points for consideration of the structures of aldoses and ketoses that contain more carbon atoms.

The Fischer projection formulas of all D aldoses containing three, four, five, and six carbon atoms are given in Figure 18.13. Figure 18.13 starts with the triose glyceraldehyde at the top and proceeds downward through the tetroses, pentoses, and hexoses. The number of possible aldoses doubles each time an additional carbon atom is added because the new carbon atom is a chiral center. Glyceraldehyde has one chiral center, the tetroses two chiral centers, the pentoses three chiral centers, and the hexoses four chiral centers.

In aldose structures such as those shown in Figure 18.13, the chiral center farthest from the aldehyde group determines the D or L designation for the aldose. The configurations about the other chiral centers present are accounted for by assigning a different common name to each set of D and L enantiomers. (Only the D isomer is shown in Figure 18.13; the L isomer is the mirror image of the structure shown.)

A major difference between glyceraldehyde and dihydroxyacetone is that the latter does not possess a chiral carbon atom. Thus D and L forms are not possible for dihydroxyacetone. This reduces by half (compared with aldoses) the number of stereoisomers possible for ketotetroses, ketopentoses, and ketohexoses. An aldohexose has four chiral carbon atoms, but a ketohexose has only three. Figure 18.14 gives the Fischer projection formulas and common names for the D forms of ketoses containing three, four, five, and six carbon atoms.

18.9 BIOCHEMICALLY IMPORTANT MONOSACCHARIDES

Of the many monosaccharides, six that are particularly important in the functioning of the human body are the trioses D-glyceraldehyde and dihydroxyacetone and the D forms of glucose, galactose, fructose, and ribose. Glucose and galactose are aldohexoses, fructose is

Nearly all naturally occurring monosaccharides are D isomers. These D monosaccharides are important energy sources for the human body. L monosaccharides, which can be produced in the laboratory, cannot be used by the body as energy sources. Body enzymes are specific for D isomers.

All monosaccharides have names that end in *-ose* except the trioses glyceraldehyde and dihydroxyacetone.

You should memorize the structures of the six monosaccharides considered in this section.

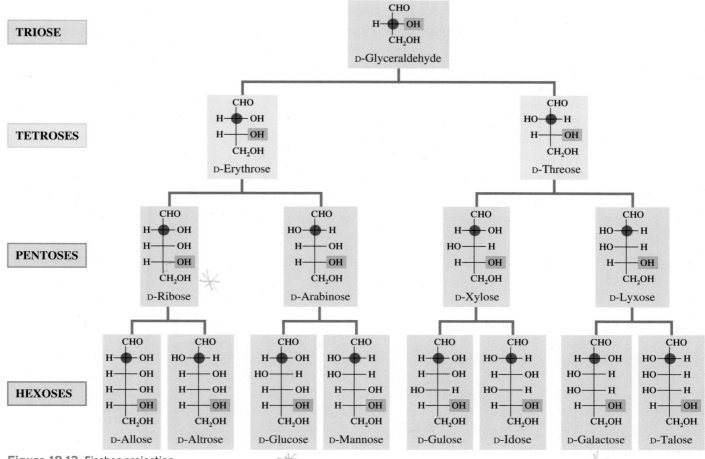

Figure 18.13 Fischer projection formulas and common names for D aldoses containing three, four, five, and six carbon atoms. The new chiral-center carbon atom added in going from triose to tetrose to pentose to hexose is marked in color. This new chiral center can have the hydroxyl group at the right or left in the Fischer projection formula, which doubles the number of stereoisomers. The hydroxyl group that specifies the D configuration is highlighted in orange.

a ketohexose, and ribose is an aldopentose. All six of these monosaccharides are water-soluble, white, crystalline solids.

D-Glyceraldehyde and Dihydroxyacetone

The simplest of the monosaccharides, these two trioses are important intermediates in the process of glycolysis (Section 24.2), a series of reactions whereby glucose is converted into two molecules of pyruvate. D-Glyceraldehyde is a chiral molecule but dihydroxyacetone is not.

D-Glucose

D-Glucose tastes sweet, is nutritious, and is an important component of the human diet. L-Glucose, on the other hand, is tasteless, and the body cannot use it.

Of all monosaccharides, D-glucose is the most abundant in nature and the most important from a human nutritional standpoint. Its Fischer projection formula is

Figure 18.14 Fischer projection formulas and common names for D ketoses containing three, four, five, and six carbon atoms. The new chiral-center carbon atom added in going from triose to tetrose to pentose to hexose is marked in color. This new chiral center can have the hydroxyl group at the right or left in the Fischer projection formula, which doubles the number of stereoisomers. The hydroxyl group that specifies the D configuration is highlighted in orange.

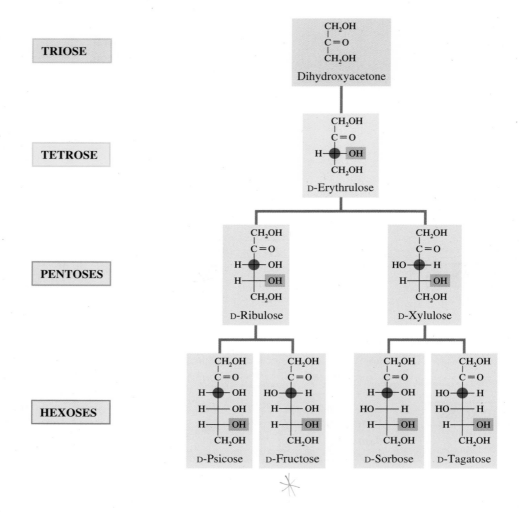

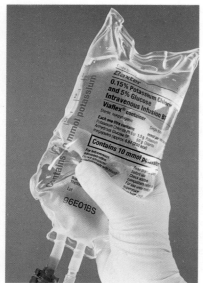

Figure 18.15 A 5% (m/v) glucose solution is often used in hospitals as an intravenous source of nourishment for patients who cannot take food by mouth. The body can use it as an energy source without digesting it.

Ripe fruits, particularly ripe grapes (20%–30% glucose by mass), are a good source of glucose, which is often referred to as *grape sugar.* Two other names for D-glucose are *dextrose* and *blood sugar.* The name *dextrose* draws attention to the fact that the optically active D-glucose, in aqueous solution, rotates plane-polarized light to the right.

The term *blood sugar* draws attention to the fact that blood contains dissolved glucose. The normal concentration of glucose in human blood is in the range of 70–100 mg/dL (1 dL = 100 mL). The actual glucose concentration in blood is dependent on the time that has elapsed since the last meal was eaten. A concentration of about 130 mg/dL occurs in the first hour after eating, and then the concentration decreases over the next 2–3 hours back to the normal range. Cells use glucose as a primary source of energy (see Figure 18.15).

Two hormones, insulin and glucagon (Section 24.9), have important roles in keeping glucose blood concentrations within the normal range, which is required for normal body function. Abnormal functioning of the hormonal control process for blood glucose levels leads to the condition known as diabetes (see Section 24.9).

D-Galactose

A comparison of the Fischer projection formulas for D-galactose and D-glucose shows that these two compounds differ only in the configuration of the —OH group and —H group on carbon 4.

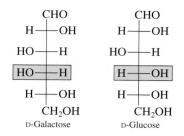

D-Galactose and D-glucose are epimers (diastereomers that differ only in the configuration at one chiral center; Section 18.6).

D-Galactose is seldom encountered as a free monosaccharide. It is, however, a component of numerous important biochemical substances. In the human body, galactose is synthesized from glucose in the mammary glands for use in lactose (milk sugar), a disaccharide consisting of a glucose unit and a galactose unit (Section 18.13). D-Galactose is sometimes called *brain sugar* because it is a component of glycoproteins (protein–carbohydrate compounds; Section 18.18) found in brain and nerve tissue. D-Galactose is also present in the chemical markers that distinguish various types of blood—A, B, AB, and O (see the Chemical Connections feature on page 580).

D-Fructose

D-Fructose is biochemically the most important ketohexose. It is also known as *levulose* and *fruit sugar*. Aqueous solutions of naturally occurring D-fructose rotate plane-polarized light to the left; hence the name *levulose*. The sweetest-tasting of all sugars, D-fructose is found in many fruits and is present in honey in equal amounts with glucose. It is sometimes used as a dietary sugar, not because it has fewer calories per gram than other sugars but because less is needed for the same amount of sweetness. The Chemical Connections feature on page 588 provides additional information about fructose use as a sweetener in the form of high fructose corn syrup (HFCS).

From the third to the sixth carbon, the structure of D-fructose is identical to that of D-glucose. Differences at carbons 1 and 2 are related to the presence of a ketone group in fructose and of an aldehyde group in glucose.

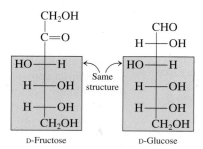

D-Ribose

The last three monosaccharides discussed in this section have all been hexoses. D-Ribose is a pentose. If carbon 3 and its accompanying —H and —OH groups were eliminated from the structure of D-glucose, the remaining structure would be that of D-ribose.

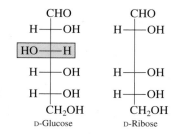

D-Ribose is a component of a variety of complex molecules, including ribonucleic acids (RNAs) and energy-rich compounds such as adenosine triphosphate (ATP). The compound 2-deoxy-D-ribose is also important in nucleic acid chemistry. This monosaccharide is a component of DNA molecules. The prefix *deoxy-* means "minus an oxygen"; the structures of ribose and 2-deoxyribose differ in that the latter compound lacks an oxygen atom at carbon 2.

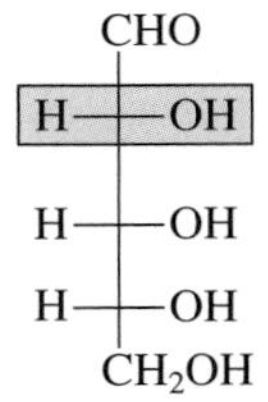

$$\begin{array}{cc}
\text{CHO} & \text{CHO} \\
\boxed{\text{H——OH}} & \boxed{\text{H——C——H}} \\
\text{H——OH} & \text{H——OH} \\
\text{H——OH} & \text{H——OH} \\
\text{CH}_2\text{OH} & \text{CH}_2\text{OH} \\
\text{D-Ribose} & \text{2-Deoxy-D-ribose}
\end{array}$$

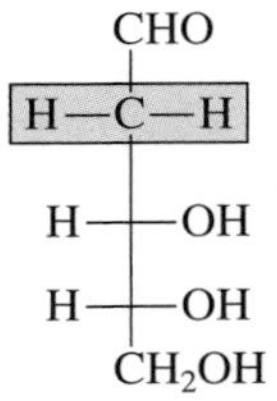

18.10 CYCLIC FORMS OF MONOSACCHARIDES

Recall from Section 15.11 that hemiacetals have both an —OH group and an —OR group attached to the same carbon atom. In the cyclic hemiacetals that monosaccharides form, it is the carbonyl carbon atom that bears the —OH and —OR groups.

So far in this chapter, the structures of monosaccharides have been depicted as open-chain polyhydroxy aldehydes or ketones. However, experimental evidence indicates that for monosaccharides containing five or more carbon atoms, such open-chain structures are actually in equilibrium with two cyclic structures, and the cyclic structures are the dominant forms at equilibrium.

The cyclic forms of monosaccharides result from the ability of their carbonyl group to react intramolecularly with a hydroxyl group. The result is a cyclic hemiacetal (Section 15.10). Such an intramolecular cyclization reaction for D-glucose is shown in Figure 18.16.

In Figure 18.16, structure 2 is a rearrangement of the projection formula for D-glucose in which the carbon atoms have locations similar to those found for carbon atoms in a six-membered ring. All hydroxyl groups drawn to the right in the original Fischer projection formula appear below the ring. Those to the left in the Fischer projection formula appear above the ring.

Structure 3 in Figure 18.16 is obtained by rotating the groups attached to carbon 5 in a counterclockwise direction so that they are in the positions where it is easiest to visualize intramolecular hemiacetal formation. The intramolecular reaction occurs between the hydroxyl group on carbon 5 and the carbonyl group (carbon 1). The —OH group adds across the carbon–oxygen double bond, producing a heterocyclic ring that contains five carbon atoms and one oxygen atom.

Cyclization of glucose (hemiacetal formation) creates a new chiral center at carbon 1, and the presence of this new chiral center produces two stereoisomers, called α and β isomers.

Addition across the carbon–oxygen double bond with its accompanying ring formation produces a chiral center at carbon 1, so two stereoisomers are possible (see Figure 18.16, structures 4–6). These two forms differ in the orientation of the —OH group on the hemiacetal carbon atom (carbon 1). In α-D-glucose, the —OH group is on the opposite side of the ring from the CH$_2$OH group attached to carbon 5. In β-D-glucose, the CH$_2$OH group on carbon 5 and the —OH group on carbon 1 are on the same side of the ring.

In an aqueous solution of D-glucose, a dynamic equilibrium exists among the α, β, and open-chain forms, and there is continual interconversion among them. For example, a freshly mixed solution of pure α-D-glucose slowly converts to a mixture of both α- and β-D-glucose by an opening and a closing of the cyclic structure. When equilibrium is established, 63% of the molecules are β-D-glucose, 37% are α-D-glucose, and less than 0.01% are in the open-chain form.

$$\alpha\text{-D-Glucose} \rightleftharpoons \text{Open-chain D-Glucose} \rightleftharpoons \beta\text{-D-Glucose}$$
$$(37\%) \qquad\qquad (\text{less than } 0.01\%) \qquad\qquad (63\%)$$

Intramolecular cyclic hemiacetal formation and the equilibrium between forms associated with it are not restricted to glucose. All aldoses with five or more carbon atoms establish similar equilibria, but with different percentages of the alpha, beta, and

Figure 18.16 The cyclic hemiacetal forms of D-glucose result from the intramolecular reaction between the carbonyl group and the hydroxyl group on carbon 5.

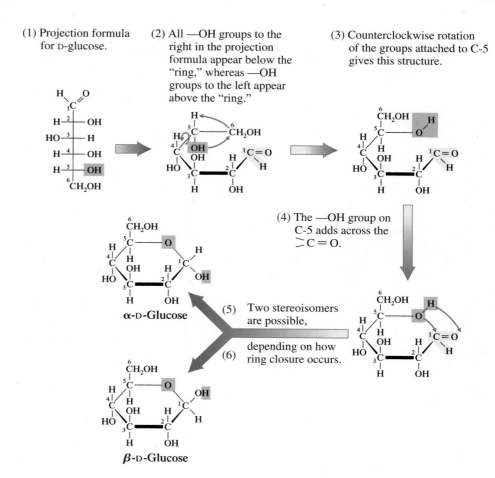

open-chain forms. Fructose and other ketoses with a sufficient number of carbon atoms also cyclize.

Galactose, like glucose, forms a six-membered ring, but both D-fructose and D-ribose form a five-membered ring.

D-Fructose cyclization involves carbon 2 (the keto group) and carbon 5, which results in two CH₂OH groups being outside the ring (carbons 1 and 6). D-Ribose cyclization involves carbon 1 (the aldehyde group) and carbon 4.

A cyclic monosaccharide containing a six-atom ring is called a *pyranose,* and one containing a five-atom ring is called *furanose* because their ring structures resemble the ring structures in the cyclic ethers *pyran* and *furan* (Section 14.19), respectively.

Such nomenclature leads to more specific names for the cyclic forms of monosaccharides—names that specify ring size. The more specific name for α-D-glucose is α-D-glucopyranose,

Figure 18.17 Walter Norman Haworth (1883–1950), the developer of Haworth projection formulas, was a British carbohydrate chemist. He helped determine the structures of the cyclic forms of glucose, was the first to synthesize vitamin C, and was a corecipient of the 1937 Nobel Prize in chemistry.

and the more specific name for α-D-fructose is α-D-fructofuranose. The last part of each of these names specifies ring size.

18.11 HAWORTH PROJECTION FORMULAS

The structural representations of the cyclic forms of monosaccharides found in the previous section are examples of Haworth projection formulas. A **Haworth projection formula** *is a two-dimensional structural notation that specifies the three-dimensional structure of a cyclic form of a monosaccharide.* Such projections carry the name of their originator, the British chemist Walter Norman Haworth (see Figure 18.17).

In a Haworth projection, the hemiacetal ring system is viewed "edge on" with the oxygen ring atom at the upper right (six-membered ring) or at the top (five-membered ring).

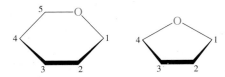

The D or L form of a monosaccharide is determined by the position of the terminal CH_2OH group on the highest-numbered ring carbon atom. In the D form, this group is positioned above the ring. In the L form, which is not usually encountered in biochemical systems, the terminal CH_2OH group is positioned below the ring.

Up = D → CH_2OH —O   Down = L ↘ —O CH_2OH

α or β configuration is determined by the position of the —OH group on carbon 1 relative to the CH_2OH group that determines D or L series. In a β configuration, both of these groups point in the same direction; in an α configuration, the two groups point in opposite directions.

CH_2OH ... OH CH_2OH ... OH CH_2OH ... OH

β-D-Monosaccharide α-D-Monosaccharide β-L-Monosaccharide

In situations where α or β configuration does not matter, the —OH group on carbon 1 is placed in a horizontal position, and a wavy line is used as the bond that connects it to the ring.

CH_2OH —O   ⁓OH

The specific identity of a monosaccharide is determined by the positioning of the other —OH groups in the Haworth projection formula. Any —OH group at a chiral center that is to the right in a Fischer projection formula points down in the Haworth projection formula. Any group to the left in a Fischer projection formula points up in the Haworth projection formula. The following is a matchup between the Haworth projection formula and a Fischer projection formula.

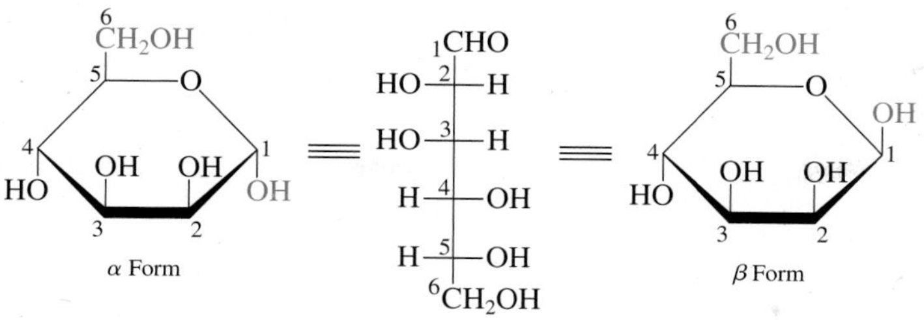

Comparison of this Fischer projection formula with those given in Figure 18.13 reveals that the monosaccharide is D-mannose.

EXAMPLE 18.6

Distinguishing Common Monosaccharides from Each Other on the Basis of Structural Characteristics

Which of the monosaccharides *glucose, fructose, galactose,* and *ribose* has each of the following structural characteristics? There may be more than one correct answer for a given characteristic.

a. It is a hexose.
b. It is an aldose.
c. Its cyclic form has a 5-membered ring.
d. Its cyclic form exists in alpha and beta forms.

Solution

a. *Glucose, fructose,* and *galactose.* The only monosaccharide of the four listed that is not a hexose is ribose, which is a pentose.
b. *Glucose, galactose,* and *ribose.* The only monosaccharide of the four listed that is not an aldose is fructose, which is a ketose.
c. *Fructose* and *ribose.* Fructose is a ketohexose and when ketoses cyclize, the number of atoms in the ring is always one less than the number of carbon atoms in the open chain form; the ring contains 4 carbon atoms and 1 oxygen atom and there are 2 carbon atoms outside the ring. Ribose is an aldopentose and when aldoses cyclize, the number of atoms in the ring is the same as the number of carbon atoms in the open chain form; the ring contains 4 carbon atoms and 1 oxygen atom and there is 1 carbon atom outside the ring.
d. *Glucose, fructose, galactose,* and *ribose.* All monosaccharides have cyclic structures that have alpha and beta forms. As ring closure occurs, the random "twisting" of the two ends of the chain can produce two orientations for the —OH group on the hemiacetal carbon atom, giving rise to alpha and beta forms.

Practice Exercise 18.6

Which of the monosaccharides *glucose, fructose, galactose,* and *ribose* has each of the following structural characteristics? There may be more than one correct answer for a given characteristic.

a. It is a pentose.
b. It is a ketose.
c. Its cyclic form has a 6-membered ring.
d. Its cyclic form has two carbon atoms outside the ring.

Answers: a. Ribose; **b.** Fructose; **c.** Glucose, galactose; **d.** Fructose

18.12 REACTIONS OF MONOSACCHARIDES

Five important reactions of monosaccharides are oxidation to acidic sugars, reduction to sugar alcohols, glycoside formation, phosphate ester formation, and amino sugar formation. In considering these reactions, we will use glucose as the monosaccharide reactant. Remember, however, that other aldoses, as well as ketoses, undergo similar reactions.

Oxidation to Produce Acidic Sugars

The redox chemistry of monosaccharides is closely linked to that of the alcohol and aldehyde functional groups. This latter redox chemistry, which we considered in Chapters 14 and 15, is summarized in the following diagram.

Monosaccharide oxidation can yield three different types of *acidic sugars*. The oxidizing agent used determines the product.

Weak oxidizing agents, such as Tollens and Benedict's solutions (Section 15.10), oxidize the aldehyde end of an aldose to give an *aldonic acid.* Oxidation of the aldehyde end of glucose produces gluconic acid, and oxidation of the aldehyde end of galactose produces galactonic acid. The structures involved in the glucose reaction are

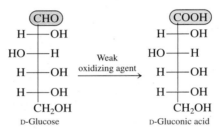

Because aldoses act as reducing agents in such reactions, they are called *reducing sugars.* With Tollens solution, glucose reduces Ag^+ ion to Ag, and with Benedict's solution, glucose reduces Cu^{2+} ion to Cu^+ ion (see Section 15.10). A **reducing sugar** *is a carbohydrate that gives a positive test with Tollens and Benedict's solutions.*

Under the basic conditions associated with Tollens and Benedict's solutions, ketoses are also reducing sugars. In this situation the ketose undergoes a structural rearrangement that produces an aldose, and the aldose then reacts. Thus all monosaccharides, both aldoses and ketoses, are reducing sugars.

Tollens and Benedict's solutions can be used to test for glucose in urine, a symptom of diabetes. For example, using Benedict's solution, we observe that if no glucose is present in the urine (a normal condition), the Benedict's solution remains blue. The presence of glucose is indicated by the formation of a red precipitate. Testing for the presence of glucose in urine is such a common laboratory procedure that much effort has been put into the development of easy-to-use test methods (Figure 18.18).

Strong oxidizing agents can oxidize both ends of a monosaccharide at the same time (the carbonyl group and the terminal primary alcohol group) to produce a dicarboxylic acid. Such polyhydroxy dicarboxylic acids are known as *aldaric acids.* For glucose, this oxidation produces glucaric acid.

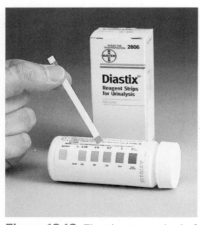

Figure 18.18 The glucose content of urine can be determined by dipping a plastic strip treated with oxidizing agents into the urine sample and comparing the color change of the strip to a color chart that indicates glucose concentration.

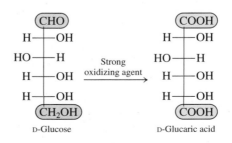

Although it is difficult to do in the laboratory, in biochemical systems *enzymes* can oxidize the primary alcohol end of an aldose such as glucose, without oxidation of the aldehyde group, to produce an *alduronic acid*. For glucose, such an oxidation produces D-glucuronic acid.

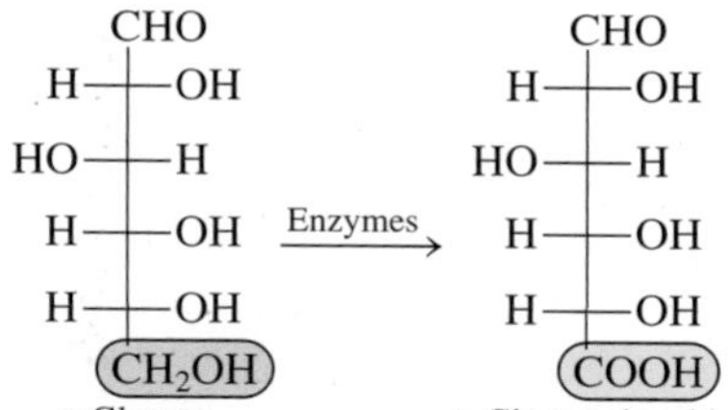

D-Glucose   D-Glucuronic acid

Reduction to Produce Sugar Alcohols

The carbonyl group present in a monosaccharide (either an aldose or a ketose) can be reduced to a hydroxyl group, using hydrogen as the reducing agent. For aldoses and ketoses, the product of the reduction is the corresponding polyhydroxy alcohol, which is sometimes called a *sugar alcohol*. For example, the reduction of D-glucose gives D-glucitol.

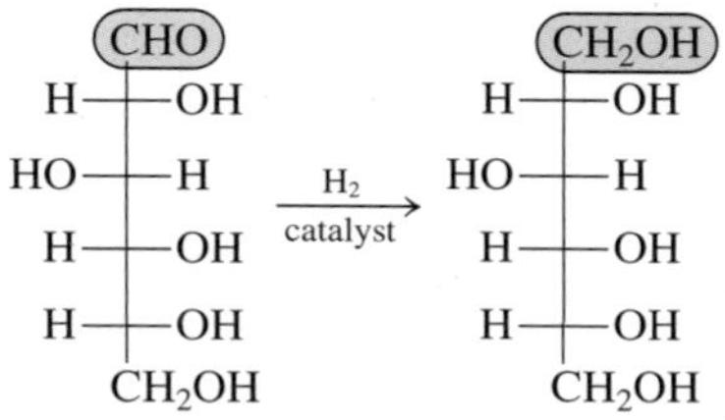

D-Glucose   D-Glucitol

d-Glucitol is also known by the common name D-sorbitol. Hexahydroxy alcohols such as D-sorbitol have properties similar to those of the trihydroxy alcohol *glycerol* (Section 14.5). These alcohols are used as moisturizing agents in foods and cosmetics because of their affinity for water. D-Sorbitol is also used as a sweetening agent in chewing gum; bacteria that cause tooth decay cannot use polyalcohols as food sources, as they can glucose and many other monosaccharides.

D-Sorbitol accumulation in the eye is a major factor in the formation of cataracts due to diabetes.

Glycoside Formation

Remember, from Section 15.11, that acetals have two —OR groups attached to the same carbon atom.

In Section 15.11 we learned that hemiacetals can react with alcohols in acid solution to produce acetals. Because the cyclic forms of monosaccharides are hemiacetals, they react with alcohols to form acetals, as is illustrated here for the reaction of β-D-glucose with methyl alcohol.

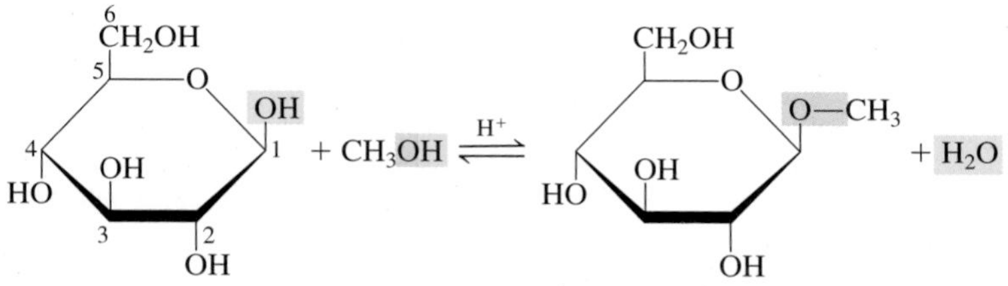

β-D-Glucose   Methyl-β-D-glucoside

The general name for monosaccharide acetals is *glycoside*. A **glycoside** *is an acetal formed from a cyclic monosaccharide by replacement of the hemiacetal carbon —OH group with an —OR group.* More specifically, a glycoside produced from glucose is called a glucoside, that from galactose is called a galactoside, and so on. Glycosides, like the hemiacetals from which they are formed, can exist in both α and β forms. Glycosides are named by listing the alkyl or aryl group attached to the

CHEMICAL Connections

Blood Types and Monosaccharides

Human blood is classified into four types: A, B, AB, and O. If a blood transfusion is necessary and the patient's own blood is not available, the donor's blood must be matched to that of the patient. Blood of one type cannot be given to a recipient with blood of another type unless the two types are compatible. A transfusion of the wrong blood type can cause the blood cells to form clumps, a potentially fatal reaction. The following table shows compatibility relationships. People with type O blood are universal donors, and those with type AB blood are universal recipients.

Human Blood Group Compatibilities

Donor blood type	Recipient Blood Type			
	A	**B**	**AB**	**O**
A	+	−	+	−
B	−	+	+	−
AB	−	−	+	−
O	+	+	+	+

+ = compatible; − = incompatible

In the United States, sampling studies show that type O is the most common type of blood, with type A the second most common. There is a definite correlation between ethnicity and blood type, as shown in the following table.

Blood Types in the United States

	Caucasian	African	Hispanic	Asian
O	45%	51%	57%	40%
A	40%	26%	33%	27%
B	11%	19%	11%	26%
AB	4%	4%	2%	7%

Percentages may not add to 100 because of rounding.

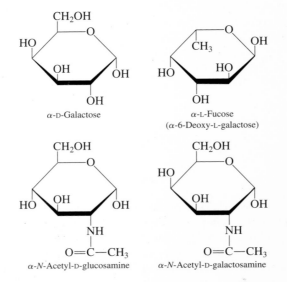

α-D-Galactose

α-L-Fucose
(α-6-Deoxy-L-galactose)

α-N-Acetyl-D-glucosamine

α-N-Acetyl-D-galactosamine

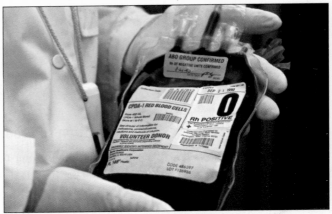

A unit of blood obtained from a blood bank.

The biochemical basis for the various blood types involves monosaccharides. The plasma membranes of red blood cells carry biochemical markers made up of monosaccharides. Four monosaccharides are involved in the "marking system." One is the simple monosaccharide D-galactose and the other three are monosaccharide derivatives. Two of these are *N*-acetyl amino derivatives (Section 18.12), those of D-glucose and D-galactose. The third is L-fucose (6-deoxy-L-galactose), an L-galactose derivative in which the oxygen atom at carbon 6 has been removed (converting the —CH$_2$OH group to a —CH$_3$ group). The L configuration of this derivative is unusual in that L-monosaccharides are seldom found in the human body. The arrangement of these monosaccharides in the biochemical marker determines blood type.

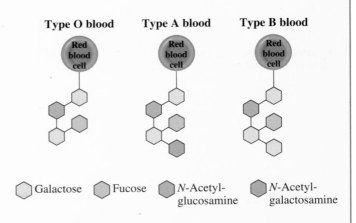

Type O blood   **Type A blood**   **Type B blood**

Galactose   Fucose   *N*-Acetyl-glucosamine   *N*-Acetyl-galactosamine

Note that all of the biochemical markers have a common structural portion that involves four monosaccharide units. Type A markers differ from type O markers in that an *N*-acetyl galactosamine unit is also present. In type B markers, a second galactose unit is present. Type AB blood contains both type A and type B markers.

oxygen, followed by the name of the monosaccharide involved, with the suffix *-ide* appended to it.

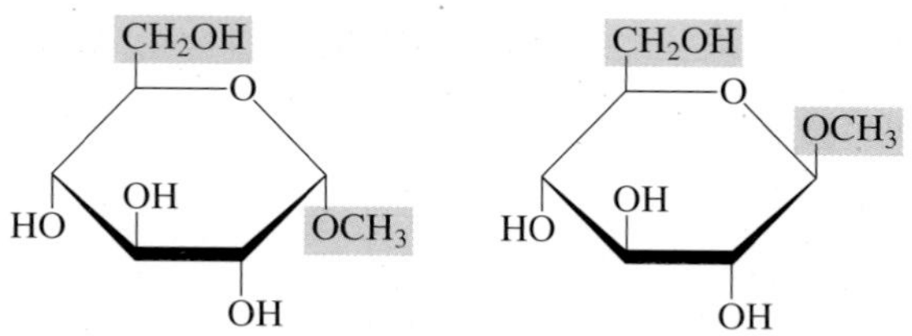

Methyl-α-D-glucoside Methyl-β-D-glucoside

Phosphate Ester Formation

The hydroxyl groups of a monosaccharide can react with inorganic oxyacids to form inorganic esters (Section 16.20). Phosphate esters, formed from phosphoric acid and various monosaccharides, are commonly encountered in biochemical systems. For example, specific enzymes in the human body catalyze the esterification of the hemiacetal group (carbon 1) and the primary alcohol group (carbon 6) in glucose to produce the compounds glucose 1-phosphate and glucose 6-phosphate, respectively.

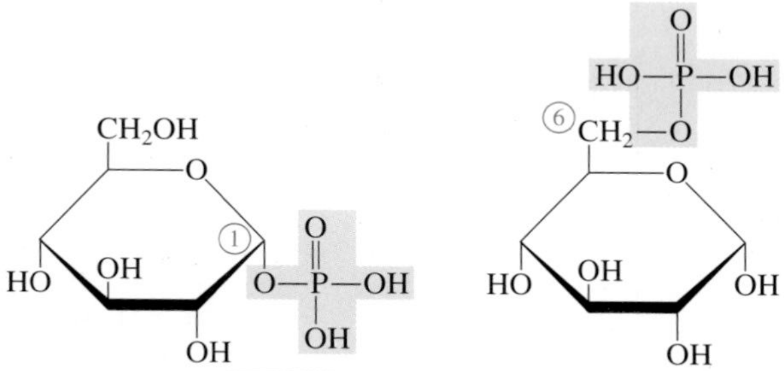

α-D-Glucose 1-phosphate α-D-Glucose 6-phosphate

These phosphate esters of glucose are stable in aqueous solution and play important roles in the metabolism of carbohydrates.

Amino Sugar Formation

If one of the hydroxyl groups of a monosaccharide is replaced with an amino group, an amino sugar is produced. In naturally occurring amino sugars, of which there are three common ones, the amino group replaces the carbon 2 hydroxyl group. The three common natural amino sugars are

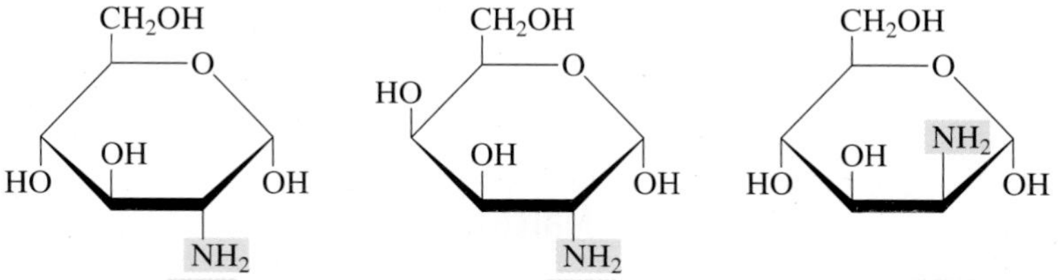

D-Glucosamine D-Galactosamine D-Mannosamine

Amino sugars and their *N*-acetyl derivatives are important building blocks of polysaccharides found in chitin (Section 18.16) and hyaluronic acid (Section 18.17). The *N*-acetyl derivatives of D-glucosamine and D-galactosamine are present in the biochemical markers on red blood cells, which distinguish the various blood types. (See the Chemical Connections feature on page 580.)

An *acetyl group* has the structure

$$CH_3-\overset{\overset{\displaystyle O}{\|}}{C}-$$

It can be considered to be derived from acetic acid by removal of the —OH portion of that structure.

$$CH_3-\overset{\overset{\displaystyle O}{\|}}{C}-\overbrace{OH}\longrightarrow$$
Acetic acid

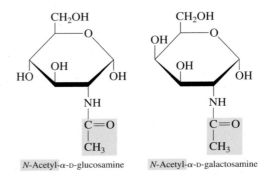

N-Acetyl-α-D-glucosamine   N-Acetyl-α-D-galactosamine

The Chemistry at a Glance feature on page 583 summarizes the "sugar terminology" associated with the common types of monosaccharides and monosaccharide derivatives that we have considered so far in this chapter.

18.13 DISACCHARIDES

A monosaccharide that has cyclic forms (hemiacetal forms) can react with an alcohol to form a glycoside (acetal), as we noted in Section 18.12. This same type of reaction can be used to produce a *disaccharide*, a carbohydrate in which two monosaccharides are bonded together (Section 18.3). In disaccharide formation, one of the monosaccharide reactants functions as a hemiacetal, and the other functions as an alcohol.

Monosaccharide + monosaccharide ⟶ disaccharide + H₂O
(Functioning as a hemiacetal)   (Functioning as an alcohol)   (Glycoside)

The bond that links the two monosaccharides of a disaccharide (glycoside) together is called a glycosidic linkage. A **glycosidic linkage** *is the bond in a disaccharide resulting from the reaction between the hemiacetal carbon atom —OH group of one monosaccharide and an —OH group on the other monosaccharide.* It is always a carbon–oxygen–carbon bond in a disaccharide.

We now examine the structures and properties of four important disaccharides: maltose, cellobiose, lactose, and sucrose. As we consider details of the structures of these compounds, we will find that the configuration (α or β) at carbon 1 of the reacting monosaccharides that functions as a hemiacetal is of prime importance.

Maltose

Maltose, often called *malt sugar,* is produced whenever the polysaccharide starch (Section 18.15) breaks down, as happens in plants when seeds germinate and in human beings during starch digestion. It is a common ingredient in baby foods and is found in malted milk. Malt (germinated barley that has been baked and ground) contains maltose; hence the name *malt sugar.*

Structurally, maltose is made up of two D-glucose units, one of which must be α-D-glucose. The formation of maltose from two glucose molecules is as follows:

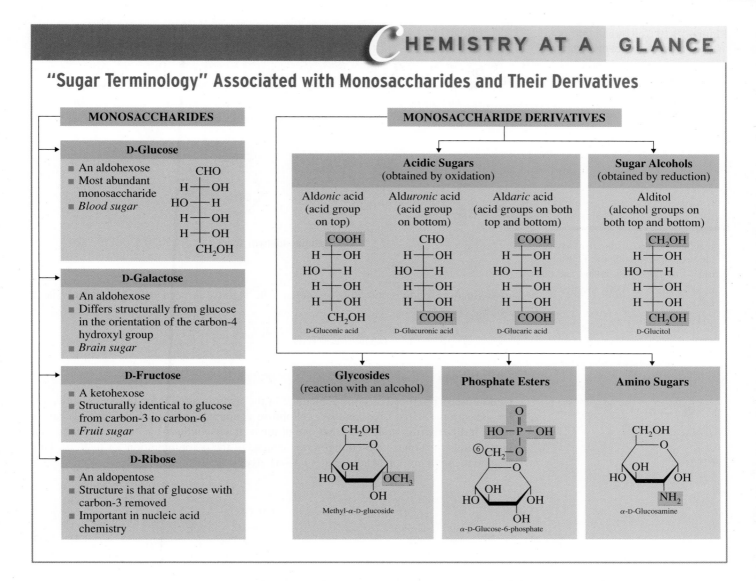

CHEMISTRY AT A GLANCE

"Sugar Terminology" Associated with Monosaccharides and Their Derivatives

The glycosidic linkage between the two glucose units is called an $\alpha(1 \rightarrow 4)$ linkage. The two —OH groups that form the linkage are attached, respectively, to carbon 1 of the first glucose unit (in an α configuration) and to carbon 4 of the second.

Maltose is a reducing sugar (Section 18.12) because the glucose unit on the right has a hemiacetal carbon atom (C-1). Thus this glucose unit can open and close; it is in equilibrium with its open-chain aldehyde form (Section 18.10). This means there are actually three forms of the maltose molecule: α-maltose, β-maltose, and the open-chain form. Structures for these three maltose forms are shown in Figure 18.19. In the solid state, the β form is dominant.

The most important chemical reaction of maltose is that of hydrolysis. Hydrolysis of D-maltose, whether in a laboratory flask or in a living organism, produces two molecules

Figure 18.19 The three forms of maltose present in aqueous solution.

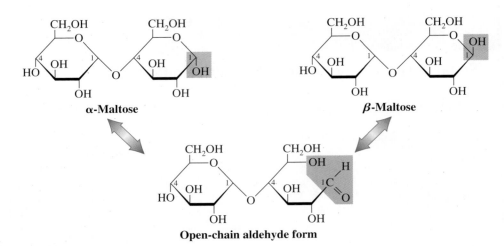

α-Maltose

β-Maltose

Open-chain aldehyde form

It is important to distinguish between the structural notation used for an α(1 → 4) glycosidic linkage and that used for a β(1 → 4) glycosidic linkage.

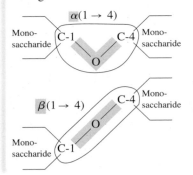

of D-glucose. Acidic conditions or the enzyme *maltase* is needed for the hydrolysis to occur.

$$\text{D-Maltose} + \text{H}_2\text{O} \xrightarrow{\text{H}^+ \text{ or maltase}} 2 \text{ D-glucose}$$

Cellobiose

Cellobiose is produced as an intermediate in the hydrolysis of the polysaccharide cellulose (Section 18.16). Like maltose, cellobiose contains two D-glucose monosaccharide units. It differs from maltose in that one of the D-glucose units—the one functioning as a hemiacetal—must have a β configuration instead of the α configuration for maltose. This change in configuration results in a β(1 → 4) glycosidic linkage.

β-D-Glucose

D-Glucose

β(1→ 4) Linkage

Like maltose, cellobiose is a reducing sugar, has three isomeric forms in aqueous solution, and upon hydrolysis produces two D-glucose molecules.

$$\text{D-Cellobiose} + \text{H}_2\text{O} \xrightarrow{\text{H}^+ \text{ or cellobiase}} 2 \text{ D-glucose}$$

Despite these similarities, maltose and cellobiose have different biochemical behaviors. These differences are related to the stereochemistry of their glycosidic linkages. Maltase, the enzyme that breaks the glucose–glucose α(1 → 4) linkage present in maltose, is found both in the human body and in yeast. Consequently, maltose is digested easily by humans and is readily fermented by yeast. Both the human body and yeast lack the enzyme cellobiase needed to break the glucose–glucose β(1 → 4) linkage of cellobiose. Thus cellobiose cannot be digested by humans or fermented by yeast.

Lactose

In both maltose and cellobiose, the monosaccharide units present are identical—two glucose units in each case. However, the two monosaccharide units in a disaccharide need not

CHEMICAL Connections

Lactose Intolerance and Galactosemia

Lactose is the principal carbohydrate in milk. Human mother's milk obtained by nursing infants contains 7%–8% lactose, almost double the 4%–5% lactose found in cow's milk.

For many people, the digestion and absorption of lactose are a problem. This problem, called *lactose intolerance,* is a condition in which people lack the enzyme *lactase,* which is needed to hydrolyze lactose to galactose and glucose.

$$\text{Lactose} + H_2O \xrightarrow{\text{Lactase}} \text{glucose} + \text{galactose}$$

Human milk contains more lactose than does cow's milk.

Deficiency of lactase can be caused by a genetic defect, by physiological decline with age, or by injuries to the mucosa lining the intestines. When lactose molecules remain in the intestine undigested, they attract water to themselves, causing fullness, discomfort, cramping, nausea, and diarrhea. Bacterial fermentation of the lactose further along the intestinal tract produces acid (lactic acid) and gas, adding to the discomfort.

The level of the enzyme lactase in humans varies with age. Most children have sufficient lactase during the early years of

their life when milk is a much-needed source of calcium in their diet. In adulthood, the enzyme level decreases, and lactose intolerance develops. This explains the change in milk-drinking habits of many adults. Some researchers estimate that as many as one of three adult Americans exhibits a degree of lactose intolerance.

The level of the enzyme lactase in humans varies widely among ethnic groups, indicating that the trait is genetically determined (inherited). The occurrence of lactose intolerance is lowest among Scandinavians and other northern Europeans and highest among native North Americans, Southeast Asians, Africans, and Greeks. The estimated prevalence of lactose intolerance is as follows:

80% Asian Americans	60% Inuits
80% Native Americans	50% Hispanics
75% African Americans	20% Caucasians
70% Mediterranean peoples	10% Northern Europeans

After lactose has been degraded into glucose and galactose, the galactose has to be converted into glucose before it can be used by cells. In humans, the genetic condition called *galactosemia* is caused by the absence of one or more of the enzymes needed for this conversion. In people with this condition, galactose and its toxic metabolic derivative galactitol (dulcitol) accumulate in the blood.

CHO		CH₂OH	
H	OH	H	OH
HO	H	HO	H
HO	H	HO	H
H	OH	H	OH
CH₂OH		CH₂OH	
D-Galactose		D-Galactitol (D-Dulcitol)	

If not treated, galactosemia can cause mental retardation in infants and even death. Treatment involves exclusion of milk and milk products from the diet.

---

be identical. *Lactose* is made up of a β-D-galactose unit and a D-glucose unit joined by a β(1 → 4) glycosidic linkage.

The α form of lactose is sweeter to the taste and more soluble in water than the β form. The β form can be found in ice cream that has been stored for a long time; it crystallizes and gives the ice cream a gritty texture.

β(1 → 4) Glycosidic linkage

β-D-Galactose · D-Glucose

Lactose

The glucose hemiacetal center is unaffected when galactose bonds to glucose in the formation of lactose, so lactose is a reducing sugar (the glucose ring can open to give an aldehyde).

Lactose is the major sugar found in milk. This accounts for its common name, *milk sugar.* Enzymes in mammalian mammary glands take glucose from the bloodstream and synthesize lactose in a four-step process. Epimerization (Section 18.6) of glucose yields galactose, and then the $\beta(1 \rightarrow 4)$ linkage forms between a galactose and a glucose unit. Lactose is an important ingredient in commercially produced infant formulas that are designed to simulate mother's milk. Souring of milk is caused by the conversion of lactose to lactic acid by bacteria in the milk. Pasteurization of milk is a quick-heating process that kills most of the bacteria and retards the souring process.

Lactose can be hydrolyzed by acid or by the enzyme *lactase,* forming an equimolar mixture of galactose and glucose.

$$\text{D-Lactose} + H_2O \xrightarrow{H^+ \text{ or lactase}} \text{D-galactose} + \text{D-glucose}$$

In the human body, the galactose so produced is then converted to glucose by other enzymes. The genetic condition *lactose intolerance,* an inability of the human digestive system to hydrolyze lactose, is considered in the Chemical Connections feature on page 585.

Sucrose

Sucrose, common *table sugar,* is the most abundant of all disaccharides and occurs throughout the plant kingdom. It is produced commercially from the juice of sugar cane and sugar beets. Sugar cane contains up to 20% by mass sucrose, and sugar beets contain up to 17% by mass sucrose. Figure 18.20 shows a molecular model for sucrose.

The two monosaccharide units present in a D-sucrose molecule are α-D-glucose and β-D-fructose. The glycosidic linkage is not a $(1 \rightarrow 4)$ linkage, as was the case for maltose, cellobiose, and lactose. It is instead an $\alpha, \beta(1 \rightarrow 2)$ glycosidic linkage. The —OH group on carbon 2 of D-fructose (the hemiacetal carbon) reacts with the —OH group on carbon 1 of D-glucose (the hemiacetal carbon).

Sucrose, unlike maltose, cellobiose, and lactose, is a *nonreducing sugar.* No hemiacetal is present in the molecule, because the glycosidic linkage involves the reducing ends of both monosaccharides. Sucrose, in the solid state and in solution, exists in only one form—there are no α and β isomers, and an open-chain form is not possible.

Sucrase, the enzyme needed to break the $\alpha, \beta(1 \rightarrow 2)$ linkage in sucrose, is present in the human body. Hence sucrose is an easily digested substance. Sucrose hydrolysis (digestion) produces an equimolar mixture of glucose and fructose called *invert sugar* (see Figure 18.21).

$$\text{D-Sucrose} + H_2O \xrightarrow{H^+ \text{ or sucrase}} \text{D-glucose} + \text{D-fructose}$$

The enzyme needed to break the $\beta(1 \rightarrow 4)$ linkage in lactose is different from the one needed to break the $\beta(1 \rightarrow 4)$ linkage in cellobiose. Because the two disaccharides have slightly different structures, different enzymes are required—*lactase* for lactose and *cellobiase* for cellobiose.

The glycosidic linkage in sucrose is very different from that in maltose, cellobiose, and lactose. The linkages in the latter three compounds can be characterized as "head-to-tail" linkages—that is, the front end (carbon 1) of one monosaccharide is linked to the back end (carbon 4) of the other monosaccharide. Sucrose has a "head-to-head" glycosidic linkage; the front ends of the two monosaccharides (carbon 1 for glucose and carbon 2 for fructose) are linked.

The term *invert sugar* comes from the observation that the direction of rotation of plane-polarized light (Section 18.7) changes from positive (clockwise) to negative (counterclockwise) when sucrose is hydrolyzed to invert sugar. The rotation is +66° for sucrose. The *net* rotation for the invert sugar mixture of fructose (−92°) and glucose (+52°) is −40°.

Figure 18.20 Space-filling model of the disaccharide sucrose.

When sucrose is cooked with acid-containing foods such as fruits or berries, partial hydrolysis takes place, forming some invert sugar. Jams and jellies prepared in this manner are actually sweeter than the pure sucrose added to the original mixture because one-to-one mixtures of glucose and fructose taste sweeter than sucrose. The Chemical Connections feature on page 588 discusses the increasing trend of using high-fructose corn syrup (HFCS) instead of sucrose as a sweetener in beverages and processed foods. The topic of artificial sweeteners is considered in the Chemical Connections feature on page 590.

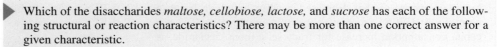

● EXAMPLE 18.7

Distinguishing Common Disaccharides from Each Other on the Basis of Structural and Reaction Characteristics

▶ Which of the disaccharides *maltose, cellobiose, lactose,* and *sucrose* has each of the following structural or reaction characteristics? There may be more than one correct answer for a given characteristic.

a. Both monosaccharide units present are glucose.
b. It is a reducing sugar.
c. It can exist in alpha and beta forms.
d. Its glycoside linkage is an $\alpha(1 \rightarrow 4)$ linkage.

Solution

a. *Maltose* and *cellobiose*. Both of these disaccharides contain two glucose units. They differ from each other in the type of glycosidic linkage present; maltose contains an $\alpha(1 \rightarrow 4)$ linkage and cellobiose contains a $\beta(1 \rightarrow 4)$ glycosidic linkage.
b. *Maltose, cellobiose,* and *lactose.* A reducing sugar has one "free" hemiacetal carbon atom. Such is the case any time two monosaccharides are bonded through a $(1 \rightarrow 4)$ glycosidic linkage. Sucrose is the only one of the four disaccharides that does not have a $(1 \rightarrow 4)$ glycosidic linkage; its glycosidic linkage is of the $(1 \rightarrow 2)$ variety.
c. *Maltose, cellobiose,* and *lactose.* A "free" hemiacetal carbon atom is a prerequisite for alpha and beta cyclic forms. Sucrose is the only one of the four disaccharides for which this is not the case. Reducing sugars (part **b**) and alpha-beta forms are based on the same structural feature.
d. *Maltose.* Maltose, cellobiose, and lactose all have $(1 \rightarrow 4)$ glycosidic linkages. However, only maltose has an $\alpha(1 \rightarrow 4)$ linkage; the other two monosaccharides have $\beta(1 \rightarrow 4)$ linkages.

Practice Exercise 18.7

Which of the disaccharides *maltose, cellobiose, lactose,* and *sucrose* has each of the following structural or reaction characteristics? There may be more than one correct answer for a given characteristic.

a. Two different monosaccharide units are present.
b. Hydrolysis produces only monosaccharides.
c. Its glycosidic linkage is a "head-to-head" linkage.
d. It is not a reducing sugar.

Answers: **a.** Lactose, sucrose; **b.** Maltose, cellobiose, lactose, sucrose; **c.** Sucrose; **d.** Sucrose

Figure 18.21 Honeybees and many other insects possess an enzyme called *invertase* that hydrolyzes sucrose to invert sugar. Thus honey is predominantly a mixture of D-glucose and D-fructose with some unhydrolyzed sucrose. Honey also contains flavoring agents obtained from the particular flowers whose nectars are collected. Whether a person eats monosaccharides individually, as in honey, or linked together, as in sucrose, they end up the same way in the human body: as glucose and fructose.

18.14 GENERAL CHARACTERISTICS OF POLYSACCHARIDES

A **polysaccharide** *is a polymer that contains many monosaccharide units bonded to each other by glycosidic linkages.* Polysaccharides are often also called *glycans.* **Glycan** *is an alternate name for a polysaccharide.*

Important parameters that distinguish various polysaccharides (or glycans) from each other are:

1. *The identity of the monosaccharide repeating unit(s) in the polymer chain.* The more abundant polysaccharides in nature contain only one type of monosaccharide repeating unit. Such polysaccharides, including starch, glycogen, cellulose, and chitin, are examples of *homopolysaccharides.* A **homopolysaccharide** *is a polysaccharide in*

Changing Sugar Patterns: Decreased Sucrose, Increased Fructose

For many decades, until the mid-1970s, the disaccharide sucrose (table sugar) was the main sweetener added to foods consumed by humans, particularly sweetened beverages. In the last 30 years, however, sucrose use has decreased steadily, with high-fructose corn syrup (HFCS) progressively displacing sucrose. The per capita consumption increase for fructose, a monosaccharide, is matched with an almost equal decline in sucrose use, as shown in the accompanying graph.

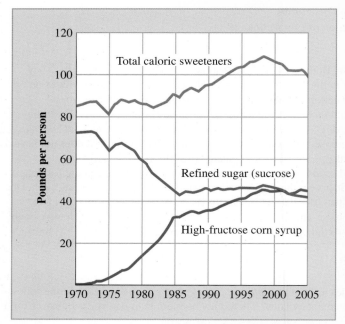

Per capita consumption in the United States, in pounds, of selected sweeteners (1970–2005).

Three major reasons for the switch from sucrose to high-fructose corn syrup are

1. HFCS is cheaper to produce because of the relative abundance of corn, farm subsidies, and sugar import tariffs in the United States.
2. HFCS is easier to blend and transport because it is a liquid.
3. HFCS usage leads to products with a much longer shelf life.

The HFCS production process, first introduced in 1957, involves milling corn to produce corn starch and then hydrolyzing the starch (a glucose polymer; Section 18.15) to glucose using the enzymes α-amylase and glucoamylase (Section 21.3). The glucose so produced is then treated with the enzyme glucose isomerase to produce a mixture of glucose and fructose. The isomerase enzymatic process produces HFCS whose composition is 42% fructose, 50% glucose, and 8% other sugars (HFCS-42). Concentration procedures are then used to produce a syrup that is 90% fructose (HFCS-90). HFCS-42 and HFCS-90 are then blended to produce HFCS-55, which is 55% fructose, 41% glucose, and 4% other sugars. HFCS-55 is the preferred sweetener used by the soft drink industry, although HFCS-42 is commonly used as a sweetener in many processed food products. HFCS-55 has properties, including sweetness, that are very similar to sucrose, which is 50% glucose and 50% fructose; remember that sucrose is a disaccharide in which the monosaccharides are glucose and fructose.

Concentrated fruit juices, particularly apple juice, are also used to sweeten beverages. The first listed ingredient in most fruit juice blends is apple juice, which actually provides more fructose relative to glucose than either sucrose or HFCS-55. A one-cup serving of apple juice contains 4.22 g of sucrose, 6.20 g of glucose, and 13.89 g of fructose. Pear juice and grape juice also have very high fructose levels relative to glucose.

Fruits are much better natural sources of fructose than are vegetables; only a few vegetables have higher fructose levels than glucose levels. The following two tables, one for fruits and one for vegetables, shows sucrose, glucose, and fructose amounts present per serving.

Sucrose, Glucose, and Fructose Amounts, per Serving, in Selected Fruits

Food Item	Serving Size	Sucrose (g)	Glucose (g)	Fructose (g)
apple	1 medium	2.86	3.35	8.14
avocado	1 fruit	0.10	0.14	0.14
banana	1 medium	2.82	5.88	5.72
blueberries	1 cup	0.16	7.08	7.21
cantaloupe	1/8 melon	3.00	1.06	1.29
cherries	1 cup	0.18	7.71	6.28
grapes	1 cup	0.24	11.52	13.01
orange	1 medium	5.99	2.76	3.15
peach	1 medium	4.66	1.91	1.50
pear	1 medium	1.29	4.58	10.34
pineapple	1 cup diced	8.48	2.70	3.18
plum	1 medium	1.04	3.35	2.03
raspberries	1 cup	0.25	2.29	2.89
strawberries	1 cup	0.20	3.39	4.15

Food Item	Serving Size	Sucrose (g)	Glucose (g)	Fructose (g)
watermelon	1/16 melon	3.46	4.52	9.61
broccoli	1 cup	0.09	0.43	0.60
carrots, baby	10 small	2.70	1.00	1.00
corn, sweet	1 ear	1.85	0.45	0.43
cucumber	1 cup	0.00	0.75	0.89
onions	1 slice	0.44	0.74	0.44
peas, green	1 cup	7.24	0.17	0.57
potatoes	1 medium	0.36	0.70	0.58
spinach	1 cup	0.02	0.03	0.04
sweet potato	1 medium	3.28	1.25	0.91
tomatoes	1 medium	0.00	1.54	1.69

which only one type of monosaccharide monomer is present. Polysaccharides whose structures contain two or more types of monosaccharide monomers, including hyaluronic acid and heparin, are called *heteropolysaccharides*. A **heteropolysaccharide** *is a polysaccharide in which more than one (usually two) type of monosaccharide monomer is present.*

2. *The length of the polymer chain.* Polysaccharide chain length can vary from less than a hundred monomer units to up to a million monomer units.

3. *The type of glycosidic linkage between monomer units.* As with disaccharides (Section 18.13), several different types of glycosidic linkages are encountered in polysaccharide structures.

4. *The degree of branching of the polymer chain.* The ability to form *branched-chain* structures distinguishes polysaccharides from the other two major types of biochemical polymers: proteins (Chapter 20) and nucleic acids (Chapter 22), which occur only as linear (unbranched) polymers.

Figure 18.22 illustrates important general structural considerations relative to polysaccharides.

In nutrition discussions, monosaccharides and disaccharides are called *simple carbohydrates,* and polysaccharides are called *complex carbohydrates.*

Figure 18.22 The polymer chain of a polysaccharide may be unbranched or branched. The monosaccharide monomers in the polymer chain may all be identical or two or more kinds of monomers may be present.

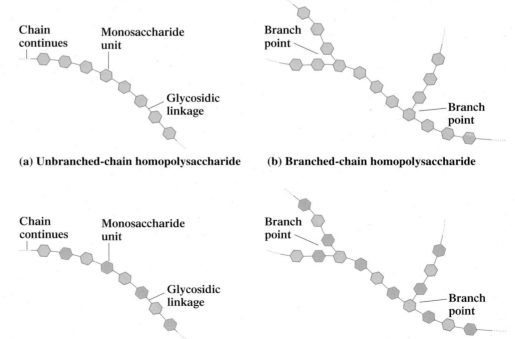

(a) Unbranched-chain homopolysaccharide

(b) Branched-chain homopolysaccharide

(c) Unbranched-chain heteropolysaccharide

(d) Branched-chain heteropolysaccharide

CHEMICAL Connections   Artificial Sweeteners

Because of the high caloric value of sucrose, it is often difficult to satisfy a demanding "sweet tooth" with sucrose without adding pounds to the body frame or inches to the waistline. Artificial sweeteners, which provide virtually no calories, are now used extensively as a solution to the "sucrose problem."

Three artificial sweeteners that have been widely used are saccharin, sodium cyclamate, and aspartame.

Saccharin

Sodium cyclamate

Aspartame

All three of these artificial sweeteners have received much publicity because of concern about their safety.

Saccharin is the oldest of the artificial sweeteners, having been in use for over 100 years. Questions about its safety arose in 1977 from a study that suggested that large doses of saccharin caused bladder tumors in rats. As a result, the FDA proposed banning saccharin, but public support for its use caused Congress to impose a moratorium on the ban. In 1991, on the basis of many further studies, the FDA withdrew its proposal to ban saccharin.

Sodium cyclamate, approved by the FDA in 1949, dominated the artificial sweetener market for 20 years. In 1969, principally on the basis of one study suggesting that it caused cancer in laboratory animals, the FDA banned its use. Further studies have shown that neither sodium cyclamate nor its metabolites cause cancer in animals. Reapproval of sodium cyclamate has been suggested, but there has been little action by the FDA. Interestingly, Canada has approved sodium cyclamate use but banned the use of saccharin.

Aspartame (Nutra-Sweet), approved by the FDA in 1981, is used in both the United States and Canada and accounts for

three-fourths of current artificial sweetener use. It tastes like sucrose but is 150 times sweeter. It provides 4 kcal/g, as does sucrose, but because so little is used, its calorie contribution is negligible. Aspartame has quickly found its way into almost every diet food on the market today.

The safety of aspartame lies with its hydrolysis products: the amino acids aspartic acid and phenylalanine. These amino acids are identical to those obtained from digestion of proteins. The only danger aspartame poses is that it contains phenylalanine, an amino acid that can lead to mental retardation among young children suffering from PKU (phenylketonuria). Labels on all products containing aspartame warn phenylketonurics of this potential danger.

Sucralose, a derivative of sucrose, is a new low-calorie entry into the artificial sweetener market. It is synthesized from sucrose by substitution of three chlorine atoms for hydroxyl groups.

An advantage of sucralose over aspartame is that it is heat-stable and can therefore be used in cooked food. Aspartame loses its sweetness when heated. Sucralose is 600 times sweeter than sucrose and has a similar taste. It is calorie-free because it cannot be hydrolyzed as it passes through the digestive tract.

Name	Type	Sweetness*
lactose	disaccharide	16
glucose	monosaccharide	74
sucrose	disaccharide	100
fructose	monosaccharide	173
sodium cyclamate	noncarbohydrate	3,000
aspartame	noncarbohydrate	15,000
saccharin	noncarbohydrate	35,000
sucralose	disaccharide derivative	60,000

*Sweetness is compared to table sugar (sucrose), which is 100 on the scale.

Unlike monosaccharides and most disaccharides, polysaccharides are not sweet and do not test positive in Tollens and Benedict's solutions. They have limited water solubility because of their size. However, the —OH groups present can individually become hydrated by water molecules. The result is usually a thick colloidal suspension of the polysaccharide in water. Polysaccharides, such as flour and cornstarch, are often used as thickening agents in sauces, desserts, and gravy.

Although there are many naturally occurring polysaccharides of biochemical importance, we will focus on only six of them: starch, glycogen, cellulose, chitin, hyaluronic acid, and heparin. Starch and glycogen are examples of *storage polysaccharides,* cellulose and chitin are *structural polysaccharides,* and hyaluronic acid and heparin are *acidic polysaccharides.*

18.15 STORAGE POLYSACCHARIDES

A **storage polysaccharide** *is a polysaccharide that is a storage form for monosaccharides and is used as an energy source in cells.* In cells, monosaccharides are stored in the form of polysaccharides rather than as individual monosaccharides in order to lower the osmotic pressure within cells. Osmotic pressure depends on the number of *individual* molecules present (Section 8.9). Incorporating many monosaccharide molecules into a single polysaccharide molecule results in a dramatic reduction in molecular numbers. The most important storage polysaccharides are starch (in plant cells) and glycogen (in animal and human cells).

Starch

Starch is a homopolysaccharide containing only glucose monosaccharide units. It is the energy-storage polysaccharide in plants. If excess glucose enters a plant cell, it is converted to starch and stored for later use. When the cell cannot get enough glucose from outside the cell, it hydrolyzes starch to release glucose.

Two different polyglucose polysaccharides can be isolated from most starches: amylose and amylopectin. *Amylose,* a straight-chain glucose polymer, usually accounts for 15%–20% of the starch; *amylopectin,* a branched glucose polymer, accounts for the remaining 80%–85% of the starch.

In amylose's non-branched structure, the glucose units are connected by $\alpha(1 \rightarrow 4)$ glycosidic linkages.

> Amylose and cellulose are both linear chains of D-glucose molecules. They are stereoisomers that differ in the configuration at carbon 1 of each D-glucose unit. In amylose, α-D-glucose is present; in cellulose, β-D-glucose.

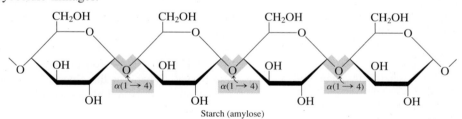

Starch (amylose)

The number of glucose units present in an amylose chain depends on the source of the starch; 300–500 monomer units are usually present.

Amylopectin, the other polysaccharide in starch, has a high degree of branching in its polyglucose structure. A branch occurs about once every 25–30 glucose units. The branch points involve $\alpha(1 \rightarrow 6)$ linkages (Figure 18.23). Because of the branching, amylopectin has a larger average molecular mass than the linear amylose. Up to 100,000 glucose units may be present in an amylopectin polymer chain.

> The glucose polymers amylose, amylopectin, and glycogen compare as follows in molecular size and degree of branching.
>
> Amylose: Up to 1000 glucose units; no branching
>
> Amylopectin: Up to 100,000 glucose units; branch points every 25–30 glucose units
>
> Glycogen: Up to 1,000,000 glucose units; branch points every 8–12 glucose units

All of the glycosidic linkages in starch (both amylose and amylopectin) are of the α type. In amylose, they are all $(1 \rightarrow 4)$; in amylopectin, both $(1 \rightarrow 4)$ and $(1 \rightarrow 6)$ linkages are present. Because both types of α linkages can be broken through hydrolysis within the human digestive tract (with the help of enzymes), starch has nutritional value for humans. The starches present in potatoes and cereal grains (wheat, rice, corn, etc.) account for approximately two-thirds of the world's food consumption.

Iodine is often used to test for the presence of starch in solution. Starch-containing solutions turn a dark blue-black when iodine is added (see Figure 18.24). As starch is broken down through acid or enzymatic hydrolysis to glucose monomers, the blue-black color disappears.

(a)

Figure 18.23 Two perspectives on the structure of the polysaccharide amylopectin. (a) Molecular structure of amylopectin. (b) An overview of the branching that occurs in the amylopectin structure. Each dot is a glucose unit.

An $\alpha(1 \rightarrow 6)$ linkage is present in the amylopectin structure at each branch point.

(b)

Glycogen

Glycogen, like starch, is a polysaccharide containing only glucose units. It is the glucose storage polysaccharide in humans and animals. Its function is thus similar to that of starch in plants, and it is sometimes referred to as *animal starch*. Liver cells and muscle cells are the storage sites for glycogen in humans.

Glycogen has a structure similar to that of amylopectin; all glycosidic linkages are of the α type, and both $(1 \rightarrow 4)$ and $(1 \rightarrow 6)$ linkages are present. Glycogen and amylopectin differ in the number of glucose units between branches and in the total number of glucose units present in a molecule. Glycogen is about three times more highly branched than amylopectin, and it is much larger, with up to 1,000,000 glucose units present.

When excess glucose is present in the blood (normally from eating too much starch), the liver and muscle tissue convert the excess glucose to glycogen, which is then stored in these tissues. Whenever the glucose blood level drops (from exercise, fasting, or normal activities), some stored glycogen is hydrolyzed back to glucose. These two opposing processes are called *glycogenesis* and *glycogenolysis,* the formation and decomposition of glycogen, respectively.

$$\text{Glucose} \underset{\text{Glycogenolysis}}{\overset{\text{Glycogenesis}}{\rightleftarrows}} \text{glycogen}$$

Glycogen is an ideal storage form for glucose. The large size of these macromolecules prevents them from diffusing out of cells. Also, conversion of glucose to glycogen reduces

The amount of stored glycogen in the human body is relatively small. Muscle tissue is approximately 1% glycogen, liver tissue 2%–3%. However, this amount is sufficient to take care of normal-activity glucose demands for about 15 hours. During strenuous exercise, glycogen supplies can be exhausted rapidly. At this point, the body begins to oxidize fat as a source of energy.

Many marathon runners eat large quantities of starch foods the day before a race. This practice, called *carbohydrate loading,* maximizes body glycogen reserves.

Figure 18.24 Use of iodine to test for starch. Starch-containing solutions and foods turn dark blue-black when iodine is added.

Figure 18.25 The small, dense particles within this electron micrograph of a liver cell are glycogen granules.

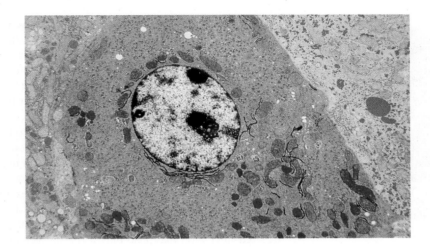

osmotic pressure (Section 8.9). Cells would burst because of increased osmotic pressure if all of the glucose in glycogen were present in cells in free form. High concentrations of glycogen in a cell sometimes precipitate or crystallize into *glycogen granules.* These granules are discernible in photographs of cells under electron microscope magnification (Figure 18.25).

18.16 STRUCTURAL POLYSACCHARIDES

A **structural polysaccharide** *is a polysaccharide that serves as a structural element in plant cell walls and animal exoskeletons.* Two of the most important structural polysaccharides are cellulose and chitin. Both are homopolysaccharides.

Cellulose

Cellulose, the structural component of plant cell walls, is the most abundant naturally occurring polysaccharide. The "woody" portions of plants—stems, stalks, and trunks—have particularly high concentrations of this fibrous, water-insoluble substance.

Like amylose, cellulose is an unbranched glucose polymer. The structural difference between cellulose and amylose, which gives them completely different properties, is that the glucose residues present in cellulose have a beta-configuration whereas the glucose residues in amylose have an alpha-configuration. The glycosidic linkages in cellulose are therefore $\beta(1 \rightarrow 4)$ linkages rather than $\alpha(1 \rightarrow 4)$ linkages.

This difference in glycosidic linkage type causes cellulose and amylose to have different molecular shapes. Amylose molecules tend to have spiral-like structures whereas cellulose molecules tend to have linear structures. The linear (straight-chain) cellulose molecules, when aligned side by side, become water-insoluble fibers because of interchain hydrogen bonding involving the numerous hydroxyl groups present.

Typically, cellulose chains contain about 5000 glucose units, which gives macromolecules with molecular masses of about 900,000 amu. Cotton is almost pure cellulose (95%), and wood is about 50% cellulose.

Even though it is a glucose polymer, cellulose is not a source of nutrition for human beings. Humans lack the enzymes capable of catalyzing the hydrolysis of $\beta(1 \rightarrow 4)$ linkages in cellulose. Even grazing animals lack the enzymes necessary for cellulose digestion. However, the intestinal tracts of animals such as horses, cows, and sheep contain bacteria that produce *cellulase,* an enzyme that can hydrolyze cellulose $\beta(1 \rightarrow 4)$ linkages and produce free glucose from cellulose. Thus grasses and other plant materials are a source of nutrition for grazing animals. The intestinal tracts of termites contain the same microorganisms, which enable termites to use wood as their source of food. Microorganisms in the soil can also metabolize cellulose, which makes possible the biodegradation of dead plants.

Despite its nondigestibility, cellulose is still an important component of a balanced diet. It serves as dietary fiber. Dietary fiber provides the digestive tract with "bulk" that helps move food through the intestinal tract and facilitates the excretion of solid wastes. Cellulose readily absorbs water, leading to softer stools and frequent bowel action. Links have been found between the length of time stools spend in the colon and possible colon problems.

High-fiber food may also play a role in weight control. Obesity is not seen in parts of the world where people eat large amounts of fiber-rich foods (see Figure 18.26). Many of the weight-loss products on the market are composed of bulk-inducing fibers such as methylcellulose.

Some dietary fibers bind lipids such as cholesterol (Section 19.9) and carry them out of the body with the feces. This lowers blood lipid concentrations and, possibly, the risk of heart and artery disease.

About 25–35 grams of dietary fiber daily is a desirable intake. This is two to three times higher than the average intake of Americans.

Chitin

Chitin is a polysaccharide that is similar to cellulose in both function and structure. Its function is to give rigidity to the exoskeletons of crabs, lobsters, shrimp, insects, and other arthropods (see Figure 18.27). It also occurs in the cell walls of fungi.

Structurally, chitin is a linear polymer (no branching) with all $\beta(1 \rightarrow 4)$ glycosidic linkages, as is cellulose. Chitin differs from cellulose in that the monosaccharide present is an *N*-acetyl amino derivative of D-glucose (Section 18.12). Figure 18.28 contrasts the structures of chitin and cellulose.

Figure 18.26 A sandwich such as this is high in dietary fiber; that is, it is a cellulose-rich "meal."

The word *chitin* is pronounced "kye-ten"; it rhymes with *Titan.*

Figure 18.27 Chitin, a linear $\beta(1 \rightarrow 4)$ polysaccharide, produces the rigidity in the exoskeletons of crabs and other arthropods.

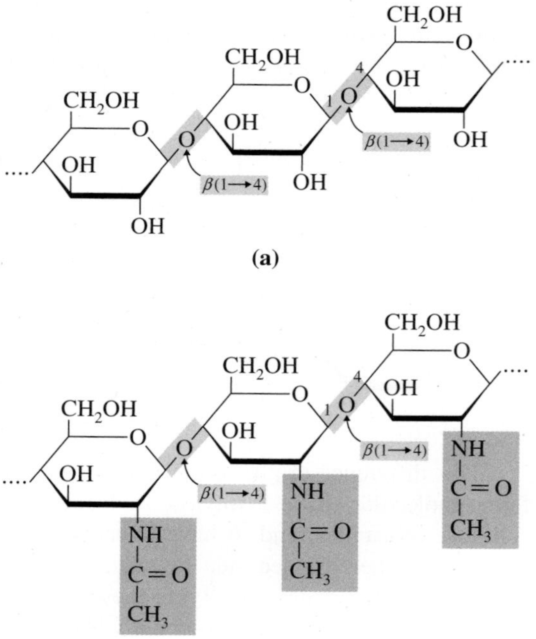

Figure 18.28 The structures of cellulose (a) and chitin (b). In both substances, all glycosidic linkages are of the $\beta(1 \rightarrow 4)$ type.

(a)

(b)

CHEMISTRY AT A GLANCE

Types of Glycosidic Linkages for Common Glucose-Containing Di- and Polysaccharides

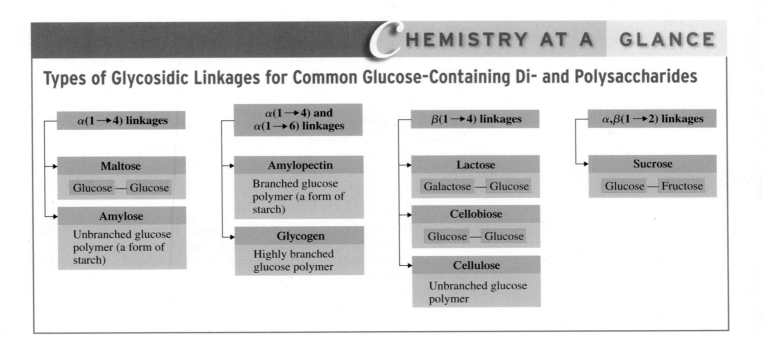

$\alpha(1 \rightarrow 4)$ linkages

Maltose
Glucose — Glucose

Amylose
Unbranched glucose polymer (a form of starch)

$\alpha(1 \rightarrow 4)$ and $\alpha(1 \rightarrow 6)$ linkages

Amylopectin
Branched glucose polymer (a form of starch)

Glycogen
Highly branched glucose polymer

$\beta(1 \rightarrow 4)$ linkages

Lactose
Galactose — Glucose

Cellobiose
Glucose — Glucose

Cellulose
Unbranched glucose polymer

$\alpha,\beta(1 \rightarrow 2)$ linkages

Sucrose
Glucose — Fructose

The Chemistry at a Glance feature above summarizes the types of glycosidic linkages present in commonly encountered glucose-containing di- and polysaccharides.

18.17 ACIDIC POLYSACCHARIDES

An **acidic polysaccharide** *is a polysaccharide with a disaccharide repeating unit in which one of the disaccharide components is an amino sugar and one or both disaccharide components has a negative charge due to a sulfate group or a carboxyl group.* Unlike the polysaccharides discussed in the previous two sections, acidic polysaccharides are *heteropolysaccharides;* two different monosaccharides are present in an alternating pattern. Acidic polysaccharides are involved in a variety of cellular functions and tissues (Figure 18.29). Two of the most well-known acidic polysaccharides are hyaluronic acid and heparin, both of which have unbranched-chain structures.

Hyaluronic Acid

The structure of hyaluronic acid contains alternating residues of *N*-acetyl-β-D-glucosamine and D-glucuronic acid.

Figure 18.29 Acidic polysaccharides associated with the connective tissue of joints give hurdlers such as these the flexibility needed to accomplish their task.

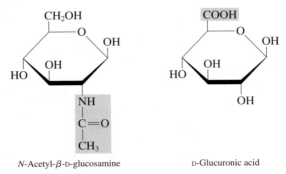

N-Acetyl-β-D-glucosamine   D-Glucuronic acid

Both of these monosaccharide derivatives have been encountered previously. *N*-acetyl-β-D-glucosamine is the repeating unit in chitin (see Section 18.16). Glucuronic acid is derived from glucose by oxidation of the —OH group at carbon 6 to an acid group (see Section 18.12). A section of the polymeric structure for hyaluronic acid is

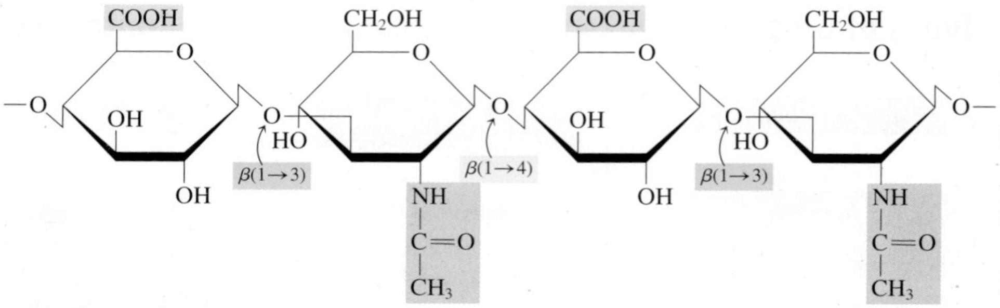

In this structure, note the alternating pattern of glycosidic bond types, $\beta(1 \rightarrow 3)$ and $\beta(1 \rightarrow 4)$. There are approximately 50,000 disaccharide units per chain.

Highly viscous hyaluronic acid solutions serve as lubricants in the fluid of joints and they are also associated with the jelly-like consistency of the vitreous humor of the eye. (The Greek word *hyalos* means "glass"; hyaluronic acid solutions have a glass-like appearance.)

Heparin

The best known of heparin's biochemical functions is that of an anticoagulant; it helps prevent blood clots. It binds strongly to a protein involved in terminating the process of blood clotting, thus inhibiting blood clotting.

The monosaccharides present in the disaccharide repeating unit for heparin are D-glucuronate-2-sulfate and *N*-sulfo-D-glucosamine-6-sulfate, both of which contain two negatively charged acidic groups.

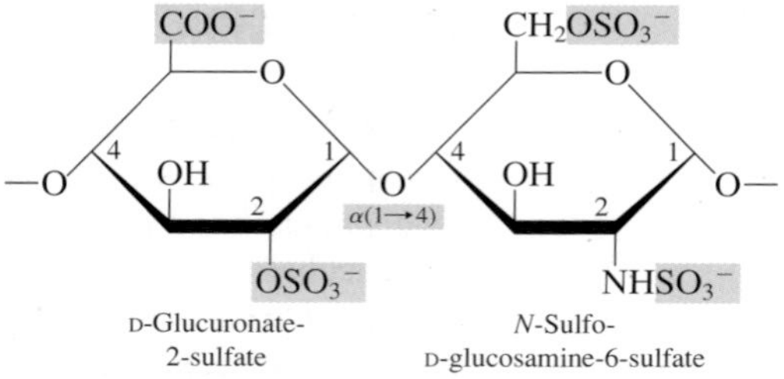

Heparin is a small polysaccharide with only 15–90 disaccharide residues per chain.

18.18 GLYCOLIPIDS AND GLYCOPROTEINS: CELL RECOGNITION

Prior to 1960, the biochemistry of carbohydrates was thought to be rather simple. These compounds served (1) as energy sources for plants, humans, and animals and (2) as structural materials for plants and arthropods.

Research since then has shown that oligosaccharides (Section 18.2) attached through glycosidic linkages to lipid molecules (Chapter 19) or to protein molecules (Chapter 20) have a wide variety of cellular functions including the process of cell recognition. Such molecules, called *glycolipids* and *glycoproteins,* respectively, often govern how individual cells of differing function within a biochemical system recognize each other and how cells interact with invading bacteria and viruses. A **glycolipid** *is a lipid molecule that has one or more carbohydrate (or carbohydrate derivative) units covalently bonded to it.* Similarly, a **glycoprotein** *is a protein molecule that has one or more carbohydrate (or carbohydrate derivative) units covalently bonded to it.*

The lipid or protein part of the glycolipid or glycoprotein is incorporated into the cell membrane structure and the carbohydrate (oligosaccharide) part functions as a marker on the outer cell membrane surface. Cell recognition generally involves the interaction

The prefix *glyco-*, used in the terms *glycolipid* and *glycoprotein,* is derived from the Greek word *glykys,* which means "sweet." Most monosaccharides and disaccharides have a sweet taste.

"Good and Bad Carbs": The Glycemic Index

The glycemic index (GI) is a dietary carbohydrate rating system that indicates how fast a particular carbohydrate is broken down into glucose (through hydrolysis) and the level of blood glucose that results. Its focal point is thus blood glucose levels.

Slow generation of glucose, a modest rise in blood glucose, and a smooth return to normal blood glucose levels are desirable. A rapid increase (surge) in blood glucose levels, with a resulting overcorrection (from excess insulin production) that drops glucose levels below normal, is undesirable. Low-GI foods promote the first of these two effects, and high-GI foods

promote the latter. Selected examples of GI ratings for foods are given in the accompanying tables.

Considerations relative to use of GI values such as these include the following:

1. At least one low-GI food should be part of each meal.
2. Fruits, vegetables, and legumes tend to have low-GI ratings.
3. Whole-grain foods, substances high in fiber, tend to have slower digestion rates.
4. High-GI rated foods should still be consumed, but as part of meals that also contain low-GI foods.

Low-Glycemic Foods (less than 55)		Intermediate-Glycemic Foods (55 to 70)		High-Glycemic Foods (over 70)	
low-fat yogurt, artificially sweetened	14	brown rice	55	corn chips	72
grapefruit	25	popcorn, sweet corn	55	watermelon†	72
kidney beans	27	long-grain white rice	56	Cheerios†	74
low-fat yogurt, sugar sweetened	33	mini shredded wheats	58	french fries	76
apple, pear	38	cheese pizza	60	rice cakes	82
spaghetti	41	Coca-Cola*	63	pretzels	83
orange	44	raisins	64	cornflakes	84
low-fat ice cream*	50	table sugar*	65	baked potato†	85
potato chips*	54	white bread	70	dried dates†	103

*High in a lot of "empty calories," so eat sparingly. Otherwise, they'll crowd out more nutritious foods.
†Don't avoid these healthful foods. Instead, combine them with low-glycemic choices.

between the carbohydrate marker of one cell and a protein imbedded into the cell membrane of another cell.

In the human reproductive process, fertilization involves a binding interaction between oligosaccharide markers on the outer membrane surface of an ovulated egg and protein receptor sites on a sperm cell membrane. This binding process is followed by release of enzymes by the sperm cell which dissolve the egg cell membrane allowing for entry of the sperm and the ensuing egg–sperm fertilization process.

The oligosaccharide markers on cell surfaces that are the basis for blood types were considered in the Chemical Connections feature on page 580. Additional aspects of glycoprotein chemistry are considered in Section 20.17.

DIETARY CONSIDERATIONS AND CARBOHYDRATES

Foods high in carbohydrate content constitute over 50% of the diet of most people of the world—rice in Asia, corn in South America, cassava (a starchy root vegetable) in parts of Africa, the potato and wheat in North America, and so on. Current nutritional recommendations support such a situation; a balanced diet should ideally be about 60% carbohydrate.

Nutritionists usually subdivide dietary carbohydrates into the categories *simple* and *complex*. A **simple carbohydrate** *is a dietary monosaccharide or disaccharide*. Simple carbohydrates are usually sweet to the taste and are commonly referred to as sugars (Section 18.8). A **complex carbohydrate** *is a dietary polysaccharide*. The main complex carbohydrates are starch and cellulose, substances not generally sweet to the taste.

Simple carbohydrates provide 20% of the energy in the U.S. diet. Half of this energy content comes from *natural sugars* and the other half from *refined sugars* added to foods. A **natural sugar** *is a sugar naturally present in whole foods.* Milk and fresh fruit are two important sources of natural sugars. A **refined sugar** *is a sugar that has been separated from its plant source.* Sugar beets and sugar cane are major sources for refined sugars. Despite claims to the contrary, refined sugars are chemically and structurally no different from the sugars naturally present in foods. The only difference is that the refined sugar is in a pure form, whereas natural sugars are part of mixtures of substances obtained from a plant source.

Refined sugars are often said to provide *empty calories* because they provide energy but few other nutrients. Natural sugars, on the other hand, are accompanied by nutrients. A tablespoon of sucrose (table sugar) provides 50 calories of energy just as a small orange does. The small orange, however, also supplies vitamin C, potassium, calcium, and fiber; table sugar provides no other nutrients.

The major dietary source for complex carbohydrates in the U.S. diet is grains, a source of both starch and fiber as well as of protein, vitamins, and minerals. The pulp of a potato provides starch, and the skin provides fiber. Vegetables such as broccoli and green beans are low in starch but high in fiber.

A developing concern about dietary intake of carbohydrates involves *how fast* a given dietary carbohydrate is broken down to generate glucose within the human body. The term *glycemic effect* refers to how quickly carbohydrates are digested (broken down into glucose), how high blood glucose levels rise, and how quickly blood glucose levels return to normal. A measurement system called the *glycemic index (GI)* has been developed for rating foods in terms of their glycemic effect. The Chemical Connections feature on page 597 discusses this topic further.

CONCEPTS TO REMEMBER

Biochemistry. Biochemistry is the study of the chemical substances found in living systems and the chemical interactions of these substances with each other (Section 18.1).

Carbohydrates. Carbohydrates are polyhydroxy aldehydes, polyhydroxy ketones, or compounds that yield such substances upon hydrolysis. Plants contain large quantities of carbohydrates produced via photosynthesis (Section 18.2).

Carbohydrate classification. Carbohydrates are classified into three groups: monosaccharides, oligosaccharides, and polysaccharides (Section 18.3).

Chirality and achirality. A chiral object is not identical to its mirror image. An achiral object is identical to its mirror image (Section 18.4).

Chiral center. A chiral center is an atom in a molecule that has four different groups tetrahedrally bonded to it. Molecules that contain a single chiral center exist in a left-handed and a right-handed form (Section 18.4).

Stereoisomerism. The atoms of stereoisomers are connected in the same way but are arranged differently in space. The major causes of stereoisomerism in molecules are structural rigidity and the presence of a chiral center (Section 18.5).

Enantiomers and diastereomers. Two types of stereoisomers exist: enantiomers and diastereomers. Enantiomers have structures that are nonsuperimposable mirror images of each other. Enantiomers have identical achiral properties but different chiral properties. Diastereomers have structures that are not mirror images of each other (Section 18.5).

Fischer projection formulas. Fischer projection formulas are two-dimensional structural formulas used to depict the three-dimensional shapes of molecules with chiral centers (Section 18.6).

Chirality of monosaccharides. Monosaccharides are classified as D or L stereoisomers on the basis of the configuration of the chiral center farthest from the carbonyl group (Section 18.6).

Optical activity. Chiral compounds are optically active—that is, they rotate the plane of polarized light. Enantiomers rotate the plane of polarized light in opposite directions. The prefix $(+)$ indicates that the compound rotates the plane of polarized light in a clockwise direction, whereas compounds that rotate the plane of polarized light in a counterclockwise direction have the prefix $(-)$ (Section 18.7).

Classification of monosaccharides. Monosaccharides are classified as aldoses or ketoses on the basis of the type of carbonyl group present. They are further classified as trioses, tetroses, pentoses, etc. on the basis of the number of carbon atoms present (Section 18.8).

Important monosaccharides. Important monosaccharides include glucose, galactose, fructose, and ribose. Glucose and galactose are aldohexoses, fructose is a ketohexose, and ribose is an aldopentose (Section 18.9).

Cyclic monosaccharides. Cyclic monosaccharides form through an intramolecular reaction between the carbonyl group and an alcohol group of an open-chain monosaccharide. These cyclic forms predominate in solution (Section 18.10).

Reactions of monosaccharides. Five important reactions of monosaccharides are (1) oxidation to an acidic sugar, (2) reduction to a sugar alcohol, (3) glycoside formation, (4) phosphate ester formation, and (5) amino sugar formation (Section 18.12).

Disaccharides. Disaccharides are glycosides formed from the linkage of two monosaccharides. The most important disaccharides are maltose, cellobiose, lactose, and sucrose. Each of these has at least one glucose unit in its structure (Section 18.13).

Polysaccharides. Polysaccharides are polymers in which monosaccharides are the monomers. In homopolysaccharides only one type of monomer is present. Two or more monosaccharide monomers are present in heteropolysaccharides. Storage polysaccharides (starch, glycogen) are storage molecules for monosaccharides. Structural polysaccharides (cellulose, chitin) serve as structural elements in plant cell walls and animal exoskeletons (Sections 18.14 to 18.17).

Glycolipids and glycoproteins. Glycolipids and glycoproteins are molecules in which oligosaccharides are attached through glycosidic linkages to lipids and proteins, respectively. Such molecules often govern how cells of differing function interact with each other (Section 18.18).

KEY REACTIONS AND EQUATIONS

1. Monosaccharide oxidation (Section 18.12)

   Aldose or ketose + weak oxidizing agent → acidic sugar

2. Monosaccharide reduction (Section 18.12)

   $$\text{Aldose or ketose} + H_2 \xrightarrow{\text{Catalyst}} \text{sugar alcohol}$$

3. Glycoside (acetal) formation (Section 18.12)

   Cyclic monosaccharide + alcohol → glycoside (acetal) + H_2O

4. Monosaccharide ester formation (Section 18.12)

   Monosaccharide + oxyacid → ester + H_2O

5. Hydrolysis of disaccharide (Section 18.13)

   $$\text{Disaccharide} + H_2O \xrightarrow{\text{Catalyst}} \text{two monosaccharides}$$

6. Hydrolysis of maltose (Section 18.13)

   $$\text{D-Maltose} + H_2O \xrightarrow{H^+ \text{ or maltase}} 2 \text{ D-glucose}$$

7. Hydrolysis of cellobiose (Section 18.13)

   $$\text{D-Cellobiose} + H_2O \xrightarrow{H^+ \text{ or cellobiase}} 2 \text{ D-glucose}$$

8. Hydrolysis of lactose (Section 18.13)

   $$\text{D-Lactose} + H_2O \xrightarrow{H^+ \text{ or lactase}} \text{D-galactose} + \text{D-glucose}$$

9. Hydrolysis of sucrose (Section 18.13)

   $$\text{D-Sucrose} + H_2O \xrightarrow{H^+ \text{ or sucrase}} \text{D-fructose} + \text{D-glucose}$$

10. Complete hydrolysis of starch (Section 18.15)

    $$\text{Starch} + H_2O \xrightarrow{H^+ \text{ or enzymes}} \text{many D-glucose}$$

11. Complete hydrolysis of glycogen (Section 18.15)

    $$\text{Glycogen} + H_2O \xrightarrow{H^+ \text{ or enzymes}} \text{many D-glucose}$$

EXERCISES and PROBLEMS

The members of each pair of problems in this section test similar material.

Biochemical Substances (Section 18.1)

18.1 Define each of the following terms.
   a. Biochemistry   b. Biochemical substance

18.2 Contrast the relative amounts, by mass, of bioorganic and bioinorganic substances present in the human body.

18.3 What are the four major types of bioorganic substances?

18.4 For each of the following pairs of bioorganic substances, indicate which member of the pair is more abundant in the human body.
   a. Proteins and nucleic acids   b. Proteins and carbohydrates
   c. Lipids and carbohydrates   d. Lipids and nucleic acids

Occurrence of Carbohydrates (Section 18.2)

18.5 Write a general chemical equation for photosynthesis.

18.6 What role does chlorophyll play in photosynthesis?

18.7 What are the two major functions of carbohydrates in the plant kingdom?

18.8 What are the six major functions of carbohydrates in the human body?

Structural Characteristics of Carbohydrates (Section 18.3)

18.9 Define the term *carbohydrate*.

18.10 What functional group is present in all carbohydrates?

18.11 Indicate how many monosaccharide units are present in each of the following.
   a. Disaccharide   b. Tetrasaccharide
   c. Oligosaccharide   d. Polysaccharide

18.12 Identify, in general terms, the product produced from the complete hydrolysis of each of the following types of carbohydrates.
   a. Disaccharide   b. Tetrasaccharide
   c. Oligosaccharide   d. Polysaccharide

Chirality (Section 18.4)

18.13 Explain what the term *superimposable* means.

18.14 Explain what the term *nonsuperimposable* means.

18.15 In each of the following lists of objects, identify those objects that are chiral.
   a. Nail, hammer, screwdriver, drill bit
   b. Your hand, your foot, your ear, your nose
   c. The words TOT, TOOT, POP, PEEP

18.16 In each of the following lists of objects, identify those objects that are chiral.
   a. Baseball cap, glove, shoe, scarf
   b. Pliers, scissors, spoon, fork
   c. The words MOM, DAD, AHA, WAX

18.17 Indicate whether the circled carbon atom in each of the following molecules is a chiral center.

a. CH₃—ⒸH₂—OH

b. CH₃—ⒸH—OH
       |
       CH₃

c. CH₃—ⒸH—OH
       |
       Cl

d. CH₃—CH₂—ⒸH—OH
              |
              CH₃

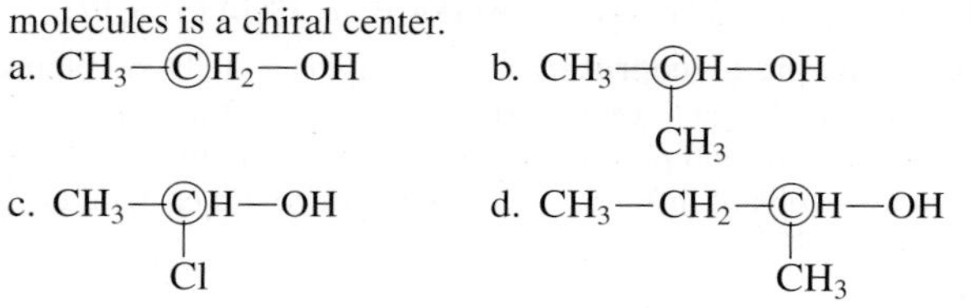

18.18 Indicate whether the circled carbon atom in each of the following molecules is a chiral center.

a. CH₃—ⒸH₂—NH₂

b. CH₃—ⒸH—CH₃
       |
       NH₂

c. CH₃—ⒸH—NH₂
       |
       CH₃

d. CH₃—ⒸH—NH₂
       |
       Cl

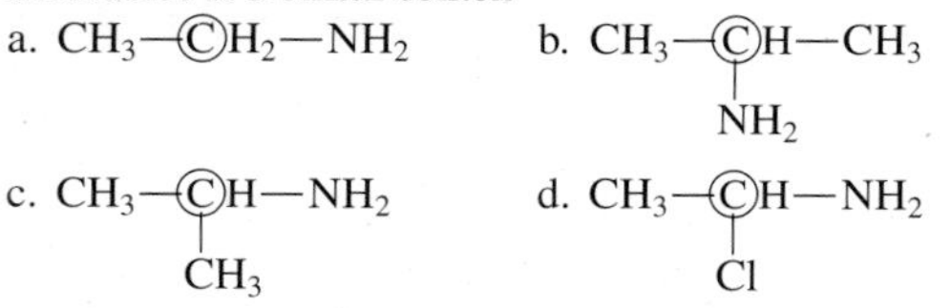

18.19 Use asterisks to show the chiral center(s), if any, in the following structures.

a.
     H Cl
     | |
Cl—C—C—Br
     | |
     H Cl

b.
     H Cl H
     | | |
Br—C—C—C—Cl
     | | |
     H Br Br

c.
                        O
                        ||
CH₂—CH—CH—CH—C—H
 | | | |
 OH OH OH OH

d. CH₂—CH—CH—CH—CH—CH₂
    | | | | | |
    OH OH OH OH OH OH

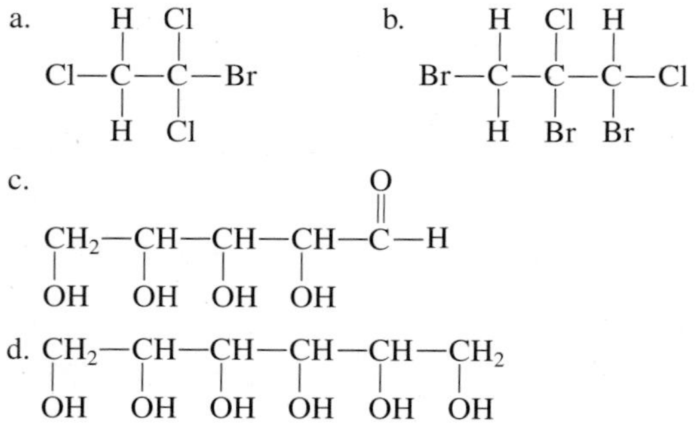

18.20 Use asterisks to show the chiral center(s), if any, in the following structures.

a.
     Br Cl
     | |
Cl—C—C—Cl
     | |
     Br Br

b.
     H H H
     | | |
Cl—C—C—C—Cl
     | | |
     Br OH Br

c.
                   O
                   ||
CH₃—CH—CH—CH—C—H
     | | |
     OH OH OH

d. CH₂—CH—CH—CH—CH₂
    | | | | |
    OH OH OH OH OH

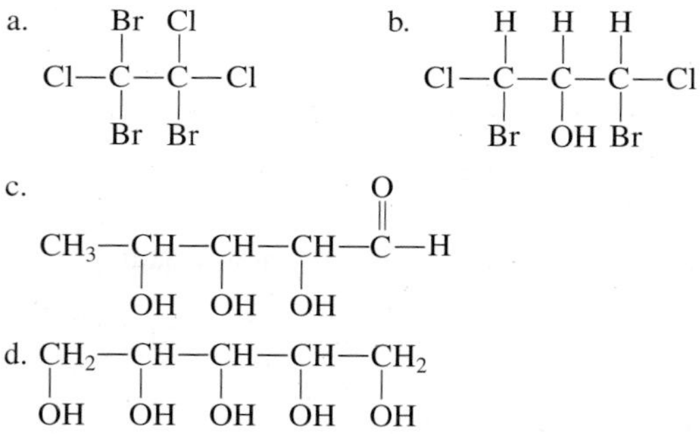

18.21 How many chiral centers are present in each of the following molecular structures?

a. <cyclohexane with Cl and Cl>

b. <cyclohexane with Cl and Cl>

c. <cyclohexane with OH>

d. <benzene ring with OH>

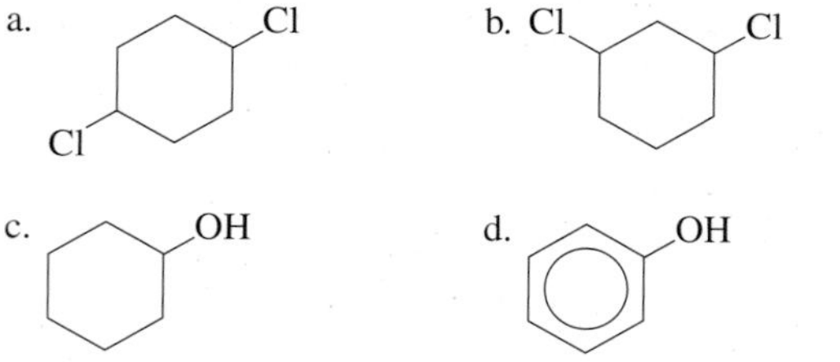

18.22 How many chiral centers are present in each of the following molecular structures?

a. <cyclohexane with Cl and Br>

b. <cyclohexane with OH and CH₃>

18.23 Classify each of the molecules in Problem 18.19 as chiral or achiral.

18.24 Classify each of the molecules in Problem 18.20 as chiral or achiral.

Stereoisomerism: Enantiomers and Diastereomers (Section 18.5)

18.25 What is the difference between constitutional isomers and stereoisomers?

18.26 Both enantiomers and diastereomers are stereoisomers. How do they differ?

18.27 What are two major structural features that can generate stereoisomerism?

18.28 Explain why *cis–trans* isomers are diastereomers rather than enantiomers.

Fischer Projection Formulas (Section 18.6)

18.29 Draw the Fischer projection formula for each of the following molecules.

a.
        H
        |
   Br—C—Cl
        |
       CH₃

b.
        CH₃
        |
   Br—C—Cl
        |
        H

c.
        CH₃
        |
   Br—C—H
        |
        Cl

d.
        CH₃
        |
    H—C—Br
        |
        Cl

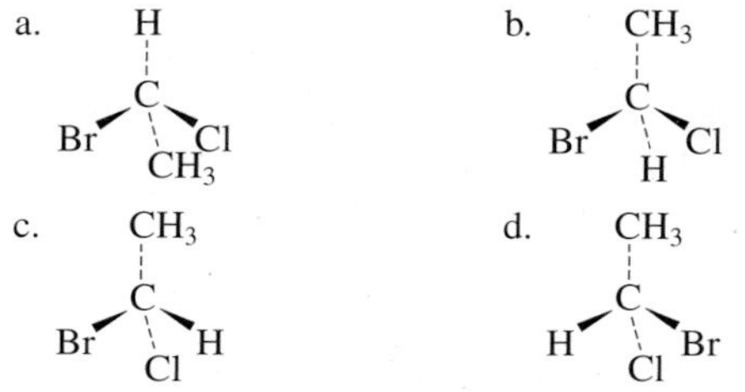

18.30 Draw the Fischer projection formula for each of the following molecules.

a.
        Cl
        |
   OH—C—CH₃
        |
        H

b.
        OH
        |
   Cl—C—CH₃
        |
        H

c.
        CH₃
        |
   HO—C—Cl
        |
        H

d.
        CH₃
        |
    H—C—OH
        |
        Cl

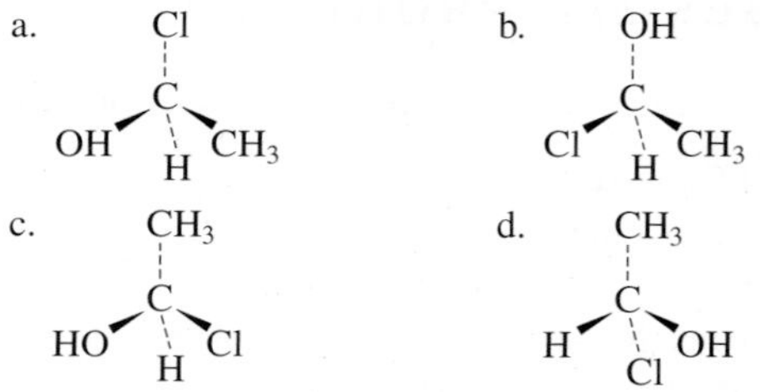

18.31 Draw a Fischer projection formula for the enantiomer of each of the following monosaccharides.

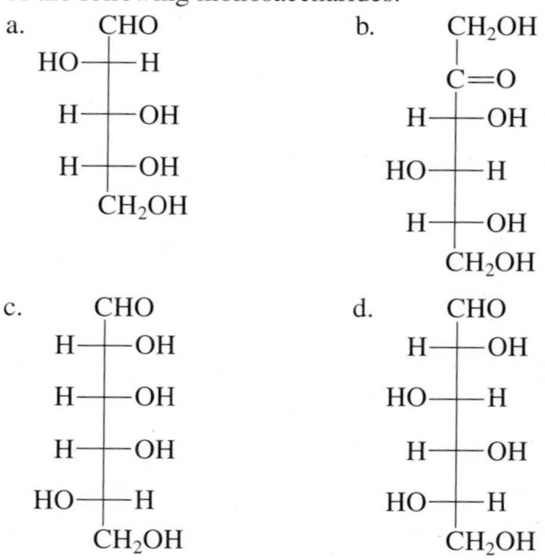

a.
      CHO
   HO—+—H
    H—+—OH
    H—+—OH
     CH₂OH

b.
     CH₂OH
      |
      C=O
    H—+—OH
   HO—+—H
    H—+—OH
     CH₂OH

c.
      CHO
    H—+—OH
    H—+—OH
    H—+—OH
   HO—+—H
     CH₂OH

d.
      CHO
    H—+—OH
   HO—+—H
    H—+—OH
   HO—+—H
     CH₂OH

18.32 Draw a Fischer projection formula for the enantiomer of each of the following monosaccharides.

a.
```
      CHO
 HO ──┼── H
 HO ──┼── H
 HO ──┼── H
     CH₂OH
```

b.
```
     CH₂OH
      C=O
 HO ──┼── H
 HO ──┼── H
     CH₂OH
```

c.
```
      CHO
 HO ──┼── H
 HO ──┼── H
  H ──┼── OH
  H ──┼── OH
     CH₂OH
```

d.
```
     CH₂OH
      C=O
  H ──┼── OH
  H ──┼── OH
 HO ──┼── H
     CH₂OH
```

18.33 Classify each of the molecules in Problem 18.31 as a D enantiomer or an L enantiomer.

18.34 Classify each of the molecules in Problem 18.32 as a D enantiomer or an L enantiomer.

18.35 Characterize the members of each of the following pairs of structures as (1) enantiomers, (2) diastereomers, or (3) neither enantiomers nor disastereomers.

a.
```
      CHO                    CHO
  H ──┼── OH             H ──┼── OH
 HO ──┼── H    and       H ──┼── OH
  H ──┼── OH             H ──┼── OH
     CH₂OH                  CH₂OH
```

b.
```
      CHO                    CHO
  H ──┼── OH             H ──┼── H
 HO ──┼── H    and      HO ──┼── H
     CH₂OH                  CH₂OH
```

c.
```
      CHO                    CHO
  H ──┼── OH            HO ──┼── H
 HO ──┼── H    and       H ──┼── OH
  H ──┼── OH            HO ──┼── H
 HO ──┼── H             H ──┼── OH
     CH₂OH                  CH₂OH
```

d.
```
     CH₂OH                  CH₂OH
      C=O                    C=O
 HO ──┼── H    and      HO ──┼── H
 HO ──┼── H             H ──┼── OH
     CH₂OH                  CH₂OH
```

18.36 Characterize the members of each of the following pairs of structures as (1) enantiomers, (2) diastereomers, or (3) neither enantiomers nor disastereomers.

a.
```
      CHO                    CHO
  H ──┼── OH            HO ──┼── H
  H ──┼── OH    and     HO ──┼── H
 HO ──┼── H             H ──┼── OH
     CH₂OH                  CH₂OH
```

b.
```
      CHO                    CHO
  H ──┼── OH             H ──┼── OH
  H ──┼── OH    and     HO ──┼── H
     CH₂OH                  CH₂OH
```

c.
```
      CHO                    CHO
  H ──┼── OH            HO ──┼── H
 HO ──┼── H    and      H ──┼── OH
  H ──┼── OH            H ──┼── OH
  H ──┼── OH            HO ──┼── H
     CH₂OH                  CH₂OH
```

d.
```
     CH₂OH                  CH₂OH
      C=O                    C=O
  H ──┼── OH    and      H ──┼── H
  H ──┼── OH             H ──┼── OH
     CH₂OH                  CH₂OH
```

Properties of Enantiomers (Section 18.7)

18.37 D-glucose and L-glucose would be expected to show differences in which of the following properties?
a. Solubility in an achiral solvent
b. Density
c. Melting point
d. Effect on plane-polarized light

18.38 D-glucose and L-glucose would be expected to show differences in which of the following properties?
a. Solubility in a chiral solvent
b. Freezing point
c. Reaction with ethanol
d. Reaction with (+)-lactic acid

18.39 Compare (+)-lactic acid and (−)-lactic acid with respect to each of the following properties.
a. Boiling point   b. Optical activity
c. Solubility in water   d. Reaction with (+)-2,3-butanediol

18.40 Compare (+)-glyceraldehyde and (−)-glyceraldehyde with respect to each of the following properties.
a. Freezing point   b. Rotation of plane-polarized light
c. Reaction with ethanol   d. Reaction with (−)-2,3-butanediol

Classification of Monosaccharides (Section 18.8)

18.41 Classify each of the following monosaccharides as an aldose or a ketose.

a.
```
      CHO
  H ─ C ─ OH
 HO ─ C ─ H
 HO ─ C ─ H
  H ─ C ─ OH
     CH₂OH
```

b.
```
     CH₂OH
      C=O
  H ─ C ─ OH
  H ─ C ─ OH
  H ─ C ─ OH
     CH₂OH
```

c.
```
 CH₂OH
  C=O
 CH₂OH
```

d.
```
     CH₂OH
      C=O
 HO ─ C ─ H
     CH₂OH
```

18.42 Classify each of the following monosaccharides as an aldose or a ketose.

a.
$$
\begin{array}{c}
\text{CHO} \\
| \\
\text{HO}-\text{C}-\text{H} \\
| \\
\text{HO}-\text{C}-\text{H} \\
| \\
\text{HO}-\text{C}-\text{H} \\
| \\
\text{H}-\text{C}-\text{OH} \\
| \\
\text{CH}_2\text{OH}
\end{array}
$$

b.
$$
\begin{array}{c}
\text{CH}_2\text{OH} \\
| \\
\text{C}=\text{O} \\
| \\
\text{HO}-\text{C}-\text{H} \\
| \\
\text{HO}-\text{C}-\text{H} \\
| \\
\text{CH}_2\text{OH}
\end{array}
$$

c.
$$
\begin{array}{c}
\text{CHO} \\
| \\
\text{HO}-\text{C}-\text{H} \\
| \\
\text{H}-\text{C}-\text{OH} \\
| \\
\text{CH}_2\text{OH}
\end{array}
$$

d.
$$
\begin{array}{c}
\text{CH}_2\text{OH} \\
| \\
\text{C}=\text{O} \\
| \\
\text{HO}-\text{C}-\text{H} \\
| \\
\text{HO}-\text{C}-\text{H} \\
| \\
\text{H}-\text{C}-\text{OH} \\
| \\
\text{CH}_2\text{OH}
\end{array}
$$

18.43 Classify each monosaccharide in Problem 18.41 by its number of carbon atoms and its type of carbonyl group.

18.44 Classify each monosaccharide in Problem 18.42 by its number of carbon atoms and its type of carbonyl group.

18.45 Using the information in Figures 18.13 and 18.14, assign a name to each of the monosaccharides in Problem 18.41.

18.46 Using the information in Figures 18.13 and 18.14, assign a name to each of the monosaccharides in Problem 18.42.

● **Biochemically Important Monosaccharides (Section 18.9)**

18.47 Indicate at what carbon atom(s) the structures of each of the following pairs of monosaccharides differ.
a. D-Glucose and D-galactose
b. D-Glucose and D-fructose
c. D-Glyceraldehyde and dihydroxyacetone
d. D-Ribose and 2-deoxy-D-ribose

18.48 Indicate whether the members of each of the following pairs of monosaccharides have the same molecular formula.
a. D-Glucose and D-galactose
b. D-Glucose and D-fructose
c. D-Glyceraldehyde and dihydroxyacetone
d. D-Ribose and 2-deoxy-D-ribose

18.49 Indicate which of the terms *aldoses, ketoses, hexoses,* and *aldohexoses* apply to both members of each of the following pairs of monosaccharides. More than one term may apply in a given situation.
a. D-Glucose and D-galactose
b. D-Glucose and D-fructose
c. D-Galactose and D-fructose
d. D-Glyceraldehyde and D-ribose

18.50 Indicate which of the terms *aldoses, ketoses, trioses,* and *aldohexoses* apply to both members of each of the following pairs of monosaccharides. More than one term may apply in a given situation.
a. D-Glucose and D-ribose
b. D-Fructose and dihydroxyacetone
c. D-Glyceraldehyde and dihydroxyacetone
d. D-Galactose and D-ribose

18.51 Draw the Fischer projection formula for each of the following monosaccharides.
a. D-Glucose   b. D-Glyceraldehyde
c. D-Fructose   d. L-Galactose

18.52 Draw the Fischer projection formula for each of the following monosaccharides.
a. D-Galactose
b. D-Ribose
c. Dihydroxyacetone
d. L-Glucose

18.53 To which of the common monosaccharides does each of the following terms apply?
a. Levulose   b. Grape sugar   c. Brain sugar

18.54 To which of the common monosaccharides does each of the following terms apply?
a. Dextrose   b. Fruit sugar   c. Blood sugar

● **Cyclic Forms of Monosaccharides (Section 18.10)**

18.55 The intermolecular reaction that produces the cyclic forms of monosaccharides involves functional groups on which two carbon atoms in the case of each of the following?
a. D-Glucose
b. D-Galactose
c. D-Fructose
d. D-Ribose

18.56 How many carbon atoms and how many oxygen atoms are present in the ring portion of the cyclic forms of each of the following monosaccharides?
a. D-Glucose
b. D-Galactose
c. D-Fructose
d. D-Ribose

18.57 What is the structural difference between the alpha and beta forms of D-glucose?

18.58 What is the structural difference between the alpha forms of D-glucose and D-galactose?

18.59 Fructose contains six carbon atoms, and ribose has only five carbon atoms. Why do both of these monosaccharides have cyclic forms that involve a five-membered ring?

18.60 Fructose and glucose both contain six carbon atoms. Why do the cyclic forms of fructose have a five-membered ring instead of the six-membered ring found in the cyclic forms of glucose?

18.61 The structure of glucose is sometimes written in an open-chain form and sometimes as a cyclic hemiacetal structure. Explain why either form is acceptable.

18.62 When pure α-D-glucose is dissolved in water, β-D-glucose and α-D-glucose are both soon present. Explain how this is possible.

● **Haworth Projection Formulas (Section 18.11)**

18.63 Identify each of the following structures as an α-D-monosaccharide or a β-D-monosaccharide.

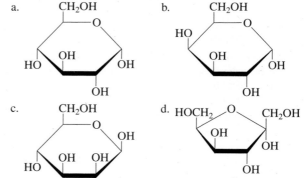

18.64 Identify each of the following structures as an α-D-monosaccharide or a β-D-monosaccharide.

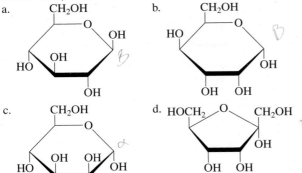

a.

b.

c.

d.

18.65 Identify whether each of the structures in Problem 18.63 is that of a hemiacetal.

18.66 Identify whether each of the structures in Problem 18.64 is that of a hemiacetal.

18.67 Draw the open-chain form for each of the monosaccharides in Problem 18.63.

18.68 Draw the open-chain form for each of the monosaccharides in Problem 18.64.

18.69 Using the information in Figures 18.13 and 18.14, assign a name to each of the monosaccharides in Problem 18.63.

18.70 Using the information in Figures 18.13 and 18.14, assign a name to each of the monosaccharides in Problem 18.64.

18.71 Draw the Haworth projection formula for each of the following monosaccharides.
   a. α-D-Galactose
   b. β-D-Galactose
   c. α-L-Galactose
   d. β-L-Galactose

18.72 Draw the Haworth projection formula for each of the following monosaccharides.
   a. α-D-Mannose
   b. β-D-Mannose
   c. α-L-Mannose
   d. β-L-Mannose

● **Reactions of Monosaccharides (Section 18.12)**

18.73 Which of the following monosaccharides is a *reducing sugar*?
   a. D-Glucose   b. D-Galactose
   c. D-Fructose   d. D-Ribose

18.74 Which of the following monosaccharides will give a positive test with Benedict's solution?
   a. D-Glucose   b. D-Galactose
   c. D-Fructose   d. D-Ribose

18.75 In terms of oxidation and reduction, explain what occurs to both D-glucose and Tollens solution when they react with each other.

18.76 Describe the chemical reaction used to detect glucose in urine that involves Benedict's solution.

18.77 Draw structures for the following compounds.
   a. Galactonic acid   b. Galactaric acid
   c. Galacturonic acid   d. Galactitol

18.78 Draw structures for the following compounds.
   a. Mannonic acid   b. Mannaric acid
   c. Mannuronic acid   d. Mannitol

18.79 Indicate whether each of the following structures is that of a glycoside.

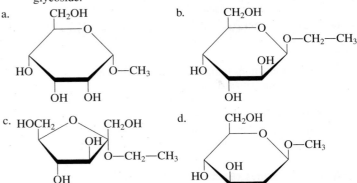

a.

b.

c.

d.

18.80 Indicate whether each of the following structures is that of a glycoside.

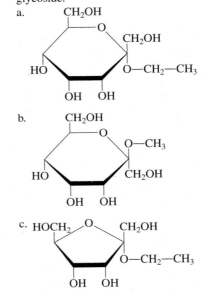

a.

b.

c.

d.

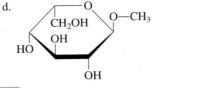

18.81 For each structure in Problem 18.79, identify the configuration at the acetal carbon atom as α or β.

18.82 For each structure in Problem 18.80, identify the configuration at the acetal carbon atom as α or β.

18.83 Identify the alcohol needed to produce each of the compounds in Problem 18.79 by reaction of the alcohol with the appropriate monosaccharide.

18.84 Identify the alcohol needed to produce each of the compounds in Problem 18.80 by reaction of the alcohol with the appropriate monosaccharide.

18.85 What is the difference in meaning between the terms *glycoside* and *glucoside*?

18.86 What is the difference in meaning between the terms *glycoside* and *galactoside*?

18.87 Draw structures for the following compounds.
a. Ethyl-β-D-glucoside
b. Methyl-α-D-galactoside

18.88 Draw structures for the following compounds.
a. Ethyl-α-D-galactoside
b. Methyl-β-D-glucoside

18.89 Draw structures for the following compounds.
a. α-D-Galactose 6-phosphate
b. N-Acetyl-α-D-galactosamine

18.90 Draw structures for the following compounds.
a. α-D-Mannose 6-phosphate
b. N-Acetyl-α-D-mannosamine

● Disaccharides (Section 18.13)

18.91 What monosaccharides are produced from the hydrolysis of the following disaccharides?
a. Sucrose
b. Maltose
c. Lactose
d. Cellobiose

18.92 What type of glycosidic linkage [α(1 → 4), etc.] is present in each of the following disaccharides?
a. Sucrose
b. Maltose
c. Lactose
d. Cellobiose

18.93 Explain why lactose is a reducing sugar.

18.94 Explain why sucrose is not a reducing sugar.

18.95 Indicate whether each of the following disaccharides gives a positive or a negative Benedict's test.
a. Sucrose   b. Maltose
c. Lactose   d. Cellobiose

18.96 Indicate whether each of the following disaccharides gives a positive or a negative Tollens test.
a. Maltose   b. Lactose
c. Cellobiose   d. Sucrose

18.97 What type of glycosidic linkage [α(1 → 4), etc.] is present in each of the following disaccharides?
a.

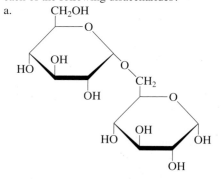

b.

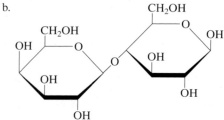

c.

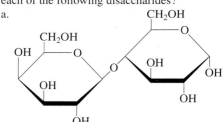

d.

18.98 What type of glycosidic linkage [α(1 → 4), etc.] is present in each of the following disaccharides?
a.

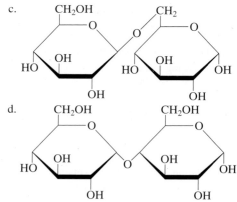

b.

c.

d.

18.99 For each of the structures in Problem 18.97, specify whether the disaccharide is in an α configuration or a β configuration, or neither.

18.100 For each of the structures in Problem 18.98, specify whether the disaccharide is in an α configuration or a β configuration, or neither.

18.101 Identify each of the structures in Problem 18.97 as a reducing sugar or a nonreducing sugar.

18.102 Identify each of the structures in Problem 18.98 as a reducing sugar or a nonreducing sugar.

18.103 Using the information in Figures 18.13 and 18.14, assign a name to each monosaccharide present in each of the structures in Problem 18.97.

18.104 Using the information in Figures 18.13 and 18.14, assign a name to each monosaccharide present in each of the structures in Problem 18.98.

General Characteristics of Polysaccharides (Section 18.14)

18.105 What is the difference, if any, between a polysaccharide and a glycan?

18.106 What is the difference, if any, between a homopolysaccharide and a heteropolysaccharide?

18.107 What is the range for the polymer chain length in a polysaccharide?

18.108 Contrast polysaccharides with mono- and disaccharides in terms of general property differences.

Storage Polysaccharides (Section 18.15)

18.109 Indicate whether or not each of the following is a correct characterization for the amylose form of starch.
a. It is a homopolysaccharide.
b. It contains two different types of monosaccharide molecules.
c. It is a branched-chain glucose polymer.
d. All glycosidic linkages present are $\alpha(1 \rightarrow 4)$.

18.110 Indicate whether or not each of the following is a correct characterization for glycogen.
a. It is a homopolysaccharide.
b. It contains two different types of monosaccharide molecules.
c. It is a branched-chain glucose polymer.
d. All glycosidic linkages present are $\alpha(1 \rightarrow 4)$.

18.111 What is the difference, if any, between the amylose and amylopectin forms of starch in terms of the following?
a. Relative abundance
b. Length of polymer chain
c. Type of glycosidic linkages present
d. Type of monosaccharide monomers present

18.112 Which of the characterizations *homopolysaccharide, heteropolysaccharide, straight-chain polysaccharide,* and *storage polysaccharide* applies to both members of each of the following pairs of substances? More than one characterization may apply in a given situation.
a. Glycogen and starch
b. Amylose and amylopectin
c. Glycogen and amylose
d. Starch and amylopectin

Structural Polysaccharides (Section 18.16)

18.113 Indicate whether or not each of the following is a correct characterization for cellulose.
a. It is an unbranched glucose polymer.
b. Its glycosidic linkages are of the same type as those in starch.
c. It is a source of nutrition for humans.
d. One of its biochemical functions is that of dietary fiber.

18.114 Indicate whether or not each of the following is a correct characterization for chitin.
a. It is an unbranched polymer.
b. Two different types of monomers are present.
c. Glycosidic linkages present are the same as those in cellulose.
d. The monomers present are glucose derivatives rather than glucose itself.

18.115 Indicate whether or not each of the following characterizations applies to (1) both cellulose and chitin, (2) to cellulose only, (3) to chitin only, or (4) to neither cellulose nor chitin.
a. Storage polysaccharide
b. Monomers are glucose units
c. Glycosidic linkages are all $\alpha(1 \rightarrow 4)$
d. An unbranched polymer

18.116 Indicate whether or not each of the following characterizations applies to (1) both cellulose and chitin, (2) to cellulose only, (3) to chitin only, or (4) to neither cellulose nor chitin.
a. Structural polysaccharide
b. Monomers are glucose derivatives
c. Glycosidic linkages are all $(1 \rightarrow 4)$
d. Homopolysaccharide

Acidic Polysaccharides (Section 18.17)

18.117 Indicate whether or not each of the following is a correct characterization for hyaluronic acid.
a. Heteropolysaccharide
b. Monomer repeating units are a disaccharide
c. Two types of glycosidic linkages are present
d. A very small polysaccharide

18.118 Indicate whether or not each of the following is a correct characterization for heparin.
a. Heteropolysaccharide
b. Monomer repeating units are a disaccharide
c. Two types of glycosidic linkages are present
d. A very small polysaccharide

18.119 What is the biological function for heparin?

18.120 What is the biological function for hyaluronic acid?

Glycolipids and Glycoproteins (Section 18.18)

18.121 In terms of structure, what is a glycolipid?

18.122 In terms of structure, what is a glycoprotein?

18.123 Describe the general features of the cell recognition process in which glycoproteins participate.

18.124 Describe the general features of the cell recognition process in which glycolipids participate.

Dietary Considerations and Carbohydrates (Section 18.19)

18.125 In a dietary context, what is the difference between a *simple* carbohydrate and a *complex* carbohydrate?

18.126 In a dietary context, what is the difference between a *natural* sugar and a *refined* sugar?

18.127 In a dietary context, what are *empty* calories?

18.128 In a dietary context, what is the *glycemic effect*?

ADDITIONAL PROBLEMS

18.129 Indicate whether each of the following compounds is chiral or achiral.

a. 1-Chloro-2-methylpentane   b. 2-Chloro-2-methylpentane
c. 2-Chloro-3-methylpentane   d. 3-Chloro-2-methylpentane

18.130 Indicate whether each of the following compounds is optically active or optically inactive.

a.
$$Cl-\overset{\overset{\displaystyle H}{|}}{\underset{\underset{\displaystyle Cl}{|}}{C}}-Cl$$

b.
$$Cl-\overset{\overset{\displaystyle H}{|}}{\underset{\underset{\displaystyle H}{|}}{C}}-Cl$$

c.
$$H-\overset{\overset{\displaystyle COOH}{|}}{\underset{\underset{\displaystyle COOH}{|}}{C}}-OH$$

d.
$$F-\overset{\overset{\displaystyle CH_3}{|}}{\underset{\underset{\displaystyle H}{|}}{C}}-Cl$$

18.131 In which of the following pairs of monosaccharides do both members of the pair contain the same number of carbon atoms?

a. Glyceraldehyde and glucose
b. Dihydroxyacetone and ribose
c. Ribose and deoxyribose
d. Glyceraldehyde and dihydroxyacetone

18.132 Draw Fischer projection formulas for the four stereoisomers of the molecule

$$CH_2-\underset{\underset{\displaystyle OH}{|}}{CH}-\underset{\underset{\displaystyle OH}{|}}{CH}-\overset{\overset{\displaystyle O}{\|}}{C}-CH_2$$
$$\underset{\displaystyle OH}{|} \qquad\qquad\qquad \underset{\displaystyle OH}{|}$$

18.133 What is the alkane of lowest molecular mass that is a chiral compound?

18.134 What monosaccharide(s) is (are) obtained from the hydrolysis of each of the following?

a. Sucrose   b. Glycogen   c. Starch   d. Amylose

18.135 Classify each of the following carbohydrates as a glucose polymer or a glucose-derivative polymer.

a. Chitin   b. Amylopectin
c. Hyaluronic acid   d. Glycogen

18.136 List the reactant(s) necessary to effect the following chemical changes.

a.
$$H-\overset{\overset{\displaystyle CHO}{|}}{C}-OH \atop H-\underset{\underset{\displaystyle CH_2OH}{|}}{C}-OH \quad\overset{?}{\longrightarrow}\quad H-\overset{\overset{\displaystyle COOH}{|}}{C}-OH \atop H-\underset{\underset{\displaystyle COOH}{|}}{C}-OH$$

18.137 Which of the characterizations *homopolysaccharides, heteropolysaccharides, branched polysaccharides,* and *unbranched polysaccharides* applies to both members of each of the following pairs of substances? More than one characterization may apply in a given case.

a. Glycogen and cellulose
b. Glycogen and amylopectin
c. Amylose and chitin
d. Heparin and hyaluronic acid

18.138 Match each of the following structural characteristics to the polysaccharides *amylopectin, amylose, glycogen, cellulose,* and *chitin.* A specific characteristic may apply to more than one of the polysaccharides.

a. Contains both $\alpha(1\rightarrow4)$ and $\alpha(1\rightarrow6)$ glycosidic linkages
b. Polymer chain is unbranched
c. Glucose derivatives are present in the polymer chain
d. Contains only $\beta(1\rightarrow4)$ glycosidic linkages

b.

c.

d.
$$H-\overset{\overset{\displaystyle CHO}{|}}{C}-OH \atop H-\underset{\underset{\displaystyle CH_2OH}{|}}{C}-OH \quad\overset{?}{\longrightarrow}\quad H-\overset{\overset{\displaystyle CHO}{|}}{C}-OH \atop H-\underset{\underset{\displaystyle COOH}{|}}{C}-OH$$

Multiple-Choice Practice Test

18.139 Which of the following statements relating to chirality is *incorrect*?

a. A chiral center is an atom in a molecule that has four different groups tetrahedrally bonded to it.
b. A chiral molecule is a molecule whose mirror images are superimposable.
c. Naturally occurring monosaccharides are almost always "right-handed."
d. The simplest example of a chiral monosaccharide is glyceraldehyde.

18.140 Which of the following statements concerning the D and L forms of a monosaccharide is *incorrect*?

a. Structurally they are nonsuperimposable mirror images of each other.
b. They must contain the same number of chiral centers.
c. They are enantiomers.
d. They are diastereomers.

18.141 Which of the following is a correct characterization for the monosaccharide glucose?
a. Aldopentose
b. Aldohexose
c. Ketopentose
d. Ketohexose

18.142 The structures of D-glucose and D-fructose differ at which carbon atom(s)?
a. Carbon 1 only
b. Carbon 2 only
c. Carbon 1 and carbon 2
d. Carbon 1 and carbon 6

18.143 How many different forms of a D-monosaccharide are present, at equilibrium, in an aqueous solution of the monosaccharide?
a. One
b. Two
c. Three
d. Four

18.144 Which of the following disaccharides produces both D-glucose and D-fructose upon hydrolysis?
a. Sucrose
b. Lactose
c. Maltose
d. Cellobiose

18.145 In which of the following pairs of disaccharides do both members of the pair have the same type of glycosidic linkage?
a. Sucrose and lactose
b. Cellobiose and maltose
c. Lactose and cellobiose
d. Sucrose and maltose

18.146 In which of the following pairs of carbohydrates are both members of the pair heteropolysaccharides?
a. Cellulose and amylose
b. Starch and chitin
c. Hyaluronic acid and heparin
d. Glycogen and amylopectin

18.147 In which of the following pairs of polysaccharides are both members of the pair structural polysaccharides?
a. Glycogen and cellulose
b. Starch and chitin
c. Glycogen and starch
d. Cellulose and chitin

18.148 The carbohydrate portion of glycolipids and glycoproteins that are involved in cell recognition processes is which of the following?
a. Monosaccharide
b. Glucose molecule
c. Oligosaccharide
d. Polysaccharide

Lipids

Chapter Outline

Fats and oils are the most widely occurring types of lipids. Thick layers of fat help insulate polar bears against the effects of low temperatures.

There are four major classes of bioorganic substances: carbohydrates, lipids, proteins, and nucleic acids (Section 18.1). In the previous chapter we considered the first of these classes, carbohydrates. We now turn our attention to the second of the bioorganic classes, the compounds we call lipids.

Lipids known as fats provide a major way of storing chemical energy and carbon atoms in the body. Fats also surround and insulate vital body organs, providing protection from mechanical shock and preventing excessive loss of heat energy. Phospholipids, glycolipids, and cholesterol (a lipid) are the basic components of cell membranes. Several cholesterol derivatives function as chemical messengers (hormones) within the body.

STRUCTURE AND CLASSIFICATION OF LIPIDS

Unlike carbohydrates and most other classes of compounds, lipids do not have a common structural feature that serves as the basis for defining such compounds. Instead, their characterization is based on solubility characteristics. A **lipid** *is an organic compound found in living organisms that is insoluble (or only sparingly soluble) in water but soluble in nonpolar organic solvents.* When a biochemical material (human, animal, or plant tissue) is homogenized in a blender and mixed with a nonpolar organic solvent, the substances that dissolve in the solvent are the lipids.

Figure 19.1 shows the structural diversity that is associated with lipid molecules. Some are esters, some are amides, and some are alcohols; some are acyclic, some are cyclic, and some are polycyclic. The common thread that ties all of the compounds of Figure 19.1 together is solubility rather than structure. All are insoluble in water.

Figure 19.1 The structural formulas of these types of lipids illustrate the great structural diversity among lipids. The defining parameter for lipids is solubility rather than structure.

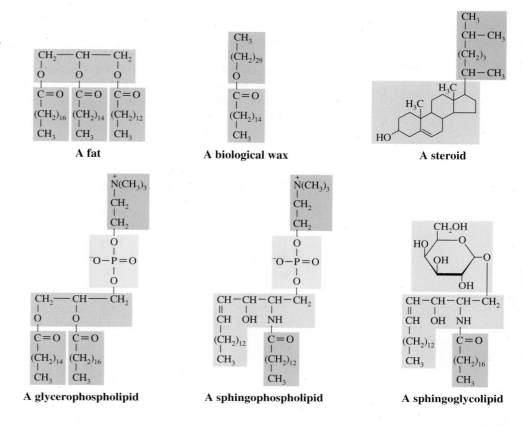

For purposes of study, we will divide lipids into five categories on the basis of lipid function:

1. **Energy-storage lipids** (triacylglycerols)
2. **Membrane lipids** (phospholipids, sphingoglycolipids, and cholesterol)
3. **Emulsification lipids** (bile acids)
4. **Messenger lipids** (steroid hormones and eicosanoids)
5. **Protective-coating lipids** (biological waxes)

Our entry point into a discussion of these five general types of lipids is a consideration of molecules called *fatty acids.* Fatty acids are structural components of all the lipids that we consider in this chapter except cholesterol, bile acids, and steroid hormones. Familiarity with the structural characteristics and physical properties of fatty acids makes it easier to understand the behavior of the many fatty-acid-containing lipids found in the human body.

19.2 TYPES OF FATTY ACIDS

A **fatty acid** *is a naturally occurring monocarboxylic acid.* Because of the pathway by which they are biosynthesized (Section 25.7), fatty acids nearly always contain an even number of carbon atoms and have a carbon chain that is unbranched. In terms of carbon chain length, fatty acids are characterized as *long-chain fatty acids* (C_{12} to C_{26}), *medium-chain fatty acids* (C_8 and C_{10}), or *short-chain fatty acids* (C_4 and C_6). Fatty acids are rarely found free in nature but rather occur as part of the structure of more complex lipid molecules.

Fatty acids are acids that were first isolated from naturally occurring fats; hence the designation *fatty acids.*

Saturated and Unsaturated Fatty Acids

The carbon chain of a fatty acid may or may not contain carbon–carbon double bonds. On the basis of this consideration, fatty acids are classified as saturated fatty acids (SFAs), monounsaturated fatty acids (MUFAs), or polyunsaturated fatty acids (PUFAs).

A **saturated fatty acid** *is a fatty acid with a carbon chain in which all carbon–carbon bonds are single bonds.* The structural formula for the 16-carbon SFA is

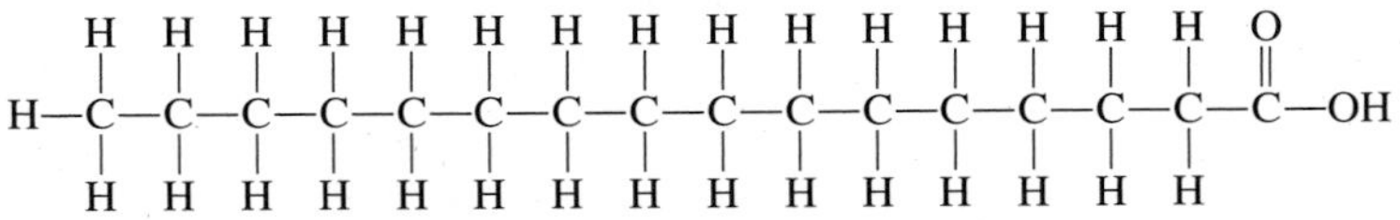

IUPAC name: hexadecanoic acid
Common name: palmitic acid

The structural formula for a fatty acid is usually written in a more condensed form than the preceding structural formula. Two alternative structural notations for palmitic acid are

$$CH_3-(CH_2)_{14}-\overset{\overset{\displaystyle O}{\|}}{C}-OH$$

and

(We first encountered line-angle structural formulas in Section 12.9.)

A **monounsaturated fatty acid** *is a fatty acid with a carbon chain in which one carbon–carbon double bond is present.* In biochemically important MUFAs, the configuration about the double bond is nearly always *cis* (Section 13.5). Different ways of depicting the structure of a MUFA follow.

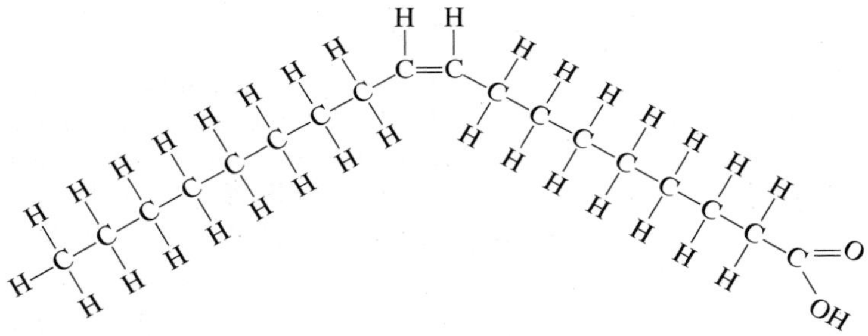

$$CH_3-(CH_2)_7-CH=CH-(CH_2)_7-\overset{\overset{\displaystyle O}{\|}}{C}-OH$$

IUPAC name: *cis*-9-octadecenoic acid
Common name: oleic acid

The first of these structures correctly emphasizes that the presence of a *cis* double bond in the carbon chain puts a rigid 30° bend in the chain. Such a bend affects the physical properties of a fatty acid, as we will see in Section 19.3.

A **polyunsaturated fatty acid** *is a fatty acid with a carbon chain in which two or more carbon–carbon double bonds are present.* Up to six double bonds are found in biochemically important PUFAs.

Fatty acids are nearly always referred to using their common names. IUPAC names for fatty acids, although easily constructed, are usually quite long. These two types of

More than 500 different fatty acids have been isolated from the lipids of microorganisms, plants, animals, and humans. These fatty acids differ from one another in the length of their carbon chains, their degree of unsaturation (number of double bonds), and the positions of the double bonds in the chains.

The fatty acids present in naturally occurring lipids almost always have the following three characteristics:

1. An unbranched carbon chain
2. An even number of carbon atoms in the carbon chain
3. Double bonds, when present in the carbon chain, in a *cis* configuration

names for an 18-carbon PUFA containing *cis* double bonds in the 9 and 12 positions are as follows:

IUPAC name: *cis,cis*-9,12-octadecadienoic acid
Common name: linoleic acid

Unsaturated Fatty Acids and Double-Bond Position

A numerically based shorthand system exists for specifying key structural parameters for fatty acids. In this system, two numbers separated by a colon are used to specify the number of carbon atoms and the number of carbon–carbon double bonds present. The notation 18:0 denotes a C_{18} fatty acid with no double bonds, whereas the notation 18:2 signifies a C_{18} fatty acid in which two double bonds are present.

To specify double-bond positioning within the carbon chain of an unsaturated fatty acid, the preceding notation is expanded by adding the Greek capital letter delta (Δ) followed by one or more superscript numbers. The notation $18:3(\Delta^{9,12,15})$ denotes a C_{18} PUFA with three double bonds at locations between carbons 9 and 10, 12 and 13, and 15 and 16.

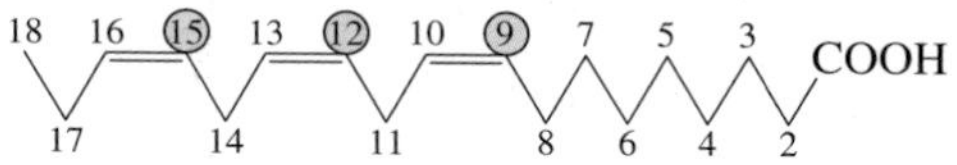

MUFAs are usually Δ^9 acids, and the first two additional double bonds in PUFAs are generally at the Δ^{12} and Δ^{15} locations. [A notable exception to this generalization is the biochemically important arachidonic acid, a PUFA with the structural parameters $20:4(\Delta^{5,8,11,14})$]. Denoting double-bond locations using this "delta notation" always assumes a numbering system in which the carboxyl carbon atom is C-1.

Several different "families" of unsaturated fatty acids exist. These family relationships become apparent when double-bond position is specified relative to the methyl (noncarboxyl) end of the fatty acid carbon chain. Double-bond positioning determined in this manner is denoted by using the Greek lower-case letter omega (ω). An **omega-3 fatty acid** *is an unsaturated fatty acid with its endmost double bond three carbon atoms away from its methyl end.* An example of an omega-3 fatty acid is

An **omega-6 fatty acid** *is an unsaturated fatty acid with its endmost double bond six carbon atoms away from its methyl end.*

The following three acids all belong to the omega-6 fatty acid family.

The structural feature common to these omega-6 fatty acids is highlighted with color in the preceding structural formulas. All the members of an omega family of fatty acids have structures in which the same "methyl end" is present.

Table 19.1 gives the names and structures of the fatty acids most commonly encountered as building blocks in biochemically important lipid structures, as well as the "delta" and "omega" notations for the acids.

TABLE 19.1
Selected Fatty Acids of Biological Importance

Structural Notation			Common Name	Structure
Saturated Fatty Acids				
12:0			lauric acid	
14:0			myristic acid	
16:0			palmitic acid	
18:0			stearic acid	
20:0			arachidic acid	
Monounsaturated Fatty Acids				
16:1	Δ^9	ω-7	palmitoleic acid	
18:1	Δ^9	ω-9	oleic acid	
Polyunsaturated Fatty Acids				
18:2	$\Delta^{9,12}$	ω-6	linoleic acid	
18:3	$\Delta^{9,12,15}$	ω-3	linolenic acid	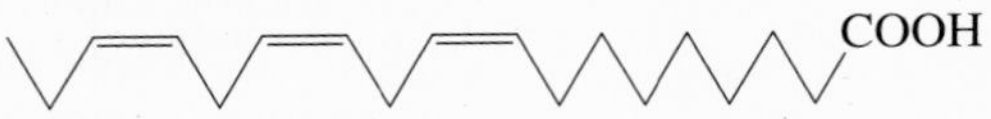
20:4	$\Delta^{5,8,11,14}$	ω-6	arachidonic acid	
20:5	$\Delta^{5,8,11,14,17}$	ω-3	EPA (eicosapentaenoic acid)	
22:6	$\Delta^{4,7,10,13,16,19}$	ω-3	DHA (docosahexaenoic acid)	

⬤ **E X A M P L E   19.1**

Classifying Fatty Acids on the Basis of Structural Characteristics

▶ Classify the fatty acid with the following structural formula in the ways indicated.

a. What is the type designation (SFA, MUFA, or PUFA) for this fatty acid?
b. On the basis of carbon chain length and degree of unsaturation, what is the numerical shorthand designation for this fatty acid?
c. To which "omega" family of fatty acids does this fatty acid belong?
d. What is the "delta" designation for the carbon chain double-bond locations for this fatty acid?

Solution

a. Two carbon–carbon double bonds are present in this molecule, which makes it a *polyunsaturated fatty acid (PUFA)*.
b. Eighteen carbon atoms and two carbon–carbon double bonds are present. The shorthand numerical designation for this fatty acid is thus *18:2*.
c. Counting from the methyl end of the carbon chain, the first double bond encountered involves carbons 6 and 7. This fatty acid belongs to the *omega-6* family of fatty acids.
d. Counting from the carboxyl end of the carbon chain, with C-1 being the carboxyl group, the double-bond locations are 9 and 12. This is a $\Delta^{9,12}$ *fatty acid.*

Practice Exercise 19.1

Classify the fatty acid with the following structural formula in the ways indicated.

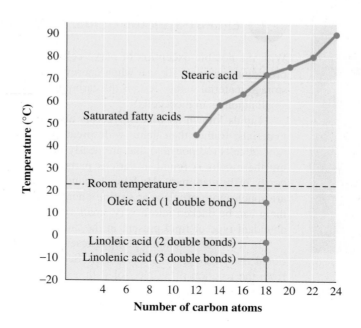

a. What is the type designation (SFA, MUFA, or PUFA) for this fatty acid?
b. On the basis of carbon chain length and degree of unsaturation, what is the numerical shorthand designation for this fatty acid?
c. To which "omega" family of fatty acids does this fatty acid belong?
d. What is the "delta" designation for the carbon chain double-bond location for this fatty acid?

Answers: a. MUFA (monounsaturated fatty acid); **b.** 12:1 fatty acid; **c.** omega-3 fatty acid (ω-3); **d.** delta-9 fatty acid (Δ^9)

19.3 PHYSICAL PROPERTIES OF FATTY ACIDS

The physical properties of fatty acids, and of lipids that contain them, are largely determined by the length and degree of unsaturation of the fatty acid carbon chain.

Water solubility for fatty acids is a direct function of carbon chain length; solubility decreases as carbon chain length increases. Short-chain fatty acids have a slight solubility in water. Long-chain fatty acids are essentially insoluble in water. The slight solubility of short-chain fatty acids is related to the polarity of the carboxyl group present. In longer-chain fatty acids, the nonpolar nature of the hydrocarbon chain completely dominates solubility considerations.

Melting points for fatty acids are strongly influenced by both carbon chain length and degree of unsaturation (number of double bonds present). Figure 19.2 shows melting-point variation as a function of both of these variables. As carbon chain length increases, melting point increases. This trend is related to the greater surface area associated with a longer carbon chain and to the increased opportunities that this greater surface area affords for intermolecular attractions between fatty acid molecules.

A trend of particular significance is that saturated fatty acids have higher melting points than unsaturated fatty acids with the same number of carbon atoms. The greater

Fatty acids have low water solubilities, which decrease with increasing carbon chain length; at 30°C, lauric acid (12:0) has a water solubility of 0.063 g/L and stearic acid (18:0) a solubility of 0.0034 g/L. Contrast this with glucose's solubility in water at the same temperature, 1100 g/L.

Figure 19.2 The melting point of a fatty acid depends on the length of the carbon chain and on the number of double bonds present in the carbon chain.

Figure 19.3 Space-filling models of four 18-carbon fatty acids, which differ in the number of double bonds present. Note how the presence of double bonds changes the shape of the molecule.

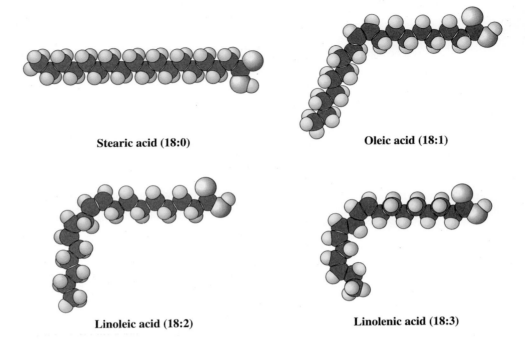

Stearic acid (18:0)

Oleic acid (18:1)

Linoleic acid (18:2)

Linolenic acid (18:3)

the degree of unsaturation, the greater the reduction in melting points. Figure 19.2 shows this effect for the 18-carbon acids with zero, one, two, and three double bonds. Long-chain saturated fatty acids tend to be solids at room temperature, whereas long-chain unsaturated fatty acids tend to be liquids at room temperature.

The decreasing melting point associated with increasing degree of unsaturation in fatty acids is explained by decreased molecular attractions between carbon chains. The double bonds in unsaturated fatty acids, which generally have the *cis* configuration, produce "bends" in the carbon chains of these molecules (see Figure 19.3). These "bends" prevent unsaturated fatty acids from packing together as tightly as saturated fatty acids. The greater the number of double bonds, the less efficient the packing. As a result, unsaturated fatty acids always have fewer intermolecular attractions, and therefore lower melting points, than their saturated counterparts.

 19.4 ENERGY-STORAGE LIPIDS: TRIACYLGLYCEROLS

With the notable exception of nerve cells, human cells store small amounts of energy-providing materials for use when energy demand is high. The most widespread energy-storage material within cells is the carbohydrate glycogen (Section 18.15); it is present in small amounts in most cells.

Lipids known as triacylglycerols also function within the body as energy-storage materials. Rather than being widespread, triacylglycerols are concentrated primarily in special cells (adipocytes) that are nearly filled with the material. Adipose tissue containing these cells is found in various parts of the body: under the skin, in the abdominal cavity, in the mammary glands, and around various organs (see Figure 19.4). Triacylglycerols are much more efficient at storing energy than is glycogen because large quantities of them can be packed into a very small volume. These energy-storage lipids are the most abundant type of lipid present in the human body.

In terms of functional groups present, triacylglycerols are triesters; three ester functional groups are present. Recall from Section 16.11 that an ester is a compound produced from the reaction of an alcohol with a carboxylic acid. The alcohol involved in triacyl-glycerol formation is always glycerol, a three-carbon alcohol with three hydroxyl groups.

Figure 19.4 An electron micrograph of adipocytes, the body's triacylglycerol-storing cells. Note the bulging spherical shape.

$$CH_2-OH$$
$$CH-OH$$
$$CH_2-OH$$
Glycerol

Fatty acids are the carboxylic acids involved in triacylglycerol formation. In the esterification reaction producing a triacylglycerol, a single molecule of glycerol reacts with three fatty acid molecules; each of the three hydroxyl groups present is esterified. Figure 19.5 shows the triple esterification reaction that occurs between glycerol and three molecules of stearic acid (18:0); note the production of three molecules of water as a by-product of the reaction.

Two general ways to represent the structure of a triacylglycerol are

Triacylglycerols do not actually contain glycerol and three fatty acids, as the block diagram for a triacylglycerol implies. They actually contain a glycerol *residue* and three fatty acid *residues*. In the formation of the triacylglycerol, three molecules of water have been removed from the structural components of the tri-acylglycerol, leaving residues of the reacting molecules.

The first representation, a block diagram, shows the four subunits present in the structure: glycerol and three fatty acids. The second representation, a general structural formula, shows the three ester linkages present in a triacylglycerol. Each of the fatty acids is attached to glycerol through an ester linkage.

Formally defined, a **triacylglycerol** *is a lipid formed by esterification of three fatty acids to a glycerol molecule.* Within the name *triacylglycerol* is the term *acyl.* An *acyl group,* previously defined and considered in Section 16.19, is the portion of a carboxylic acid that remains after the —OH group is removed from the carboxyl carbon atom. The structural representation for an acyl group is

$$O$$
$$R-C-$$
An acyl group

Thus, as the name implies, triacylglycerol molecules contain three fatty acid residues (three acyl groups) attached to a glycerol residue. An older name that is still frequently used for a triacylglycerol is *triglyceride.*

The triacylglycerol produced from glycerol and three molecules of stearic acid (as in Figure 19.5) is an example of a simple triacylglycerol. A **simple triacylglycerol**

Figure 19.5 Structure of the simple triacylglycerol produced from the triple esterification reaction between glycerol and three molecules of stearic acid (18:0 acid). Three molecules of water are a by-product of this reaction.

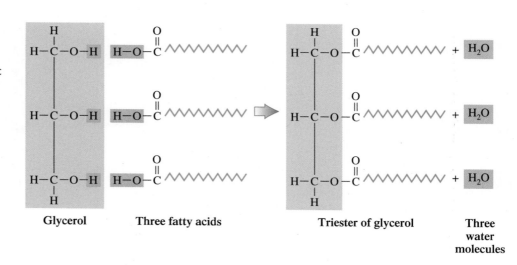

Glycerol   Three fatty acids   Triester of glycerol   Three water molecules

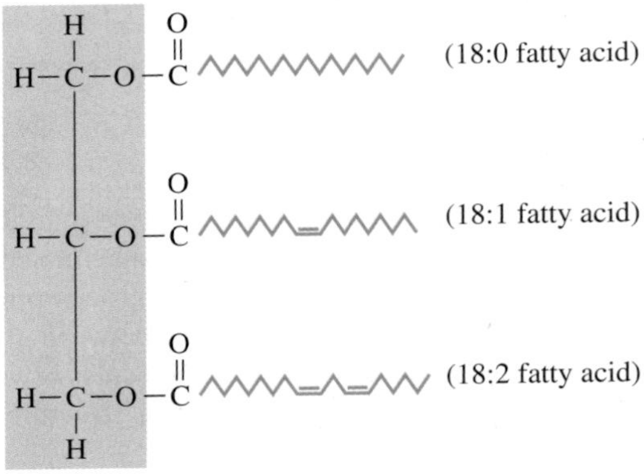

Figure 19.6 Structure of a mixed triacylglycerol in which three different fatty acid residues are present.

(18:0 fatty acid)

(18:1 fatty acid)

(18:2 fatty acid)

is a triester formed from the esterification of glycerol with three identical fatty acid molecules. If the reacting fatty acid molecules are not all identical, then the result is a mixed triacylglycerol. A **mixed triacylglycerol** *is a triester formed from the esterification of glycerol with more than one kind of fatty acid molecule.* Figure 19.6 shows the structure of a mixed triacylglycerol in which one fatty acid is saturated, another monounsaturated, and the third polyunsaturated. Naturally occurring *simple* triacylglycerols are rare. Most biochemically important triacylglycerols are *mixed* triacylglycerols.

● **EXAMPLE 19.2**

Drawing the Structural Formula of a Triacylglycerol

▶ Draw the structural formula of the triacylglycerol produced from the reaction between glycerol and three molecules of myristic acid.

Solution

From Table 19.1 we note that myristic acid is the 14:0 fatty acid. Draw the structure of glycerol and then place three molecules of myristic acid alongside the glycerol. The fatty acid placements should be such that their carboxylic groups are lined up alongside the hydroxyl groups of glycerol. Form an ester linkage between each carboxylic group and a glycerol hydroxyl group with the accompanying production of a water molecule.

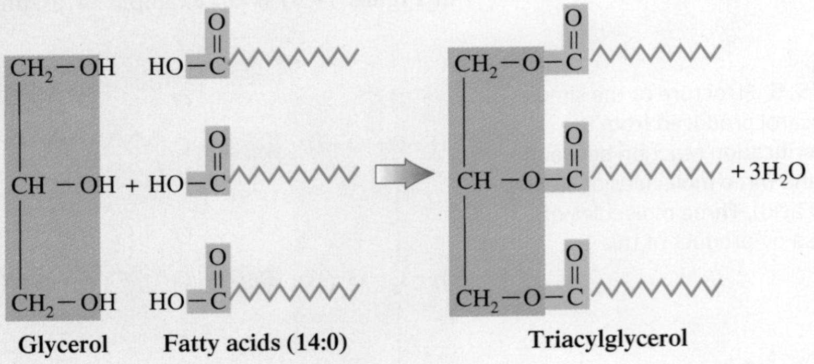

Glycerol   Fatty acids (14:0)   Triacylglycerol

Practice Exercise 19.2

Draw the structural formula of the triacylglycerol produced from the reaction between glycerol and three molecules of lauric acid.

Answer:

$$CH_2-OH \quad HO-C\text{(12:0)} + \quad \rightarrow \quad Triacylglycerol + 3H_2O$$

Glycerol　　Fatty acids (12:0)　　　　　　　Triacylglycerol

Fats and Oils

Fats are naturally occurring complex mixtures of triacylglycerol molecules in which many different kinds of triacylglycerol molecules are present. *Oils* are also naturally occurring complex mixtures of triacylglycerol molecules in which there are many different kinds of triacylglycerol molecules present. Given that both are triacylglycerol mixtures, what distinguishes a fat from an oil? The answer is physical state at room temperature. A **fat** *is a triacylglycerol mixture that is a solid or a semi-solid at room temperature (25°C)*. Generally, fats are obtained from animal sources. An **oil** *is a triacylglycerol mixture that is a liquid at room temperature (25°C)*. Generally, oils are obtained from plant sources. Because they are mixtures, no fat or oil can be represented by a single specific chemical formula. Many different fatty acids are represented in the triacylglycerol molecules present in the mixture. The actual composition of a fat or oil varies even for the species from which it is obtained. Composition depends on both dietary and climatic factors. For example, fat obtained from corn-fed hogs has a different overall composition than fat obtained from peanut-fed hogs. Flax seed grown in warm climates gives oil with a different composition from that obtained from flax seed grown in colder climates.

Additional generalizations and comparisons between fats and oils follow.

1. Fats are composed largely of triacylglycerols in which saturated fatty acids predominate, although some unsaturated fatty acids are present. Such triacylglycerols can pack closely together because of the "linearity" of their fatty acid chains (Figure 19.7a), thus causing the higher melting points associated with fats. Oils contain triacylglycerols with larger amounts of mono- and polyunsaturated fatty acids than those in fats. Such triacylglycerols cannot pack as tightly together because of "bends" in their fatty acid chains (Figure 19.7b). The result is lower melting points.
2. Fats are generally obtained from animals; hence the term *animal fat*. Although fats are solids at room temperature, the warmer body temperature of the living animal

Petroleum oils (Section 12.15) are structurally different from *lipid oils*. The former are mixtures of alkanes and cycloalkanes. The latter are mixtures of triesters of glycerol.

Figure 19.7 Representative triacylglycerols from (a) a fat and (b) an oil.

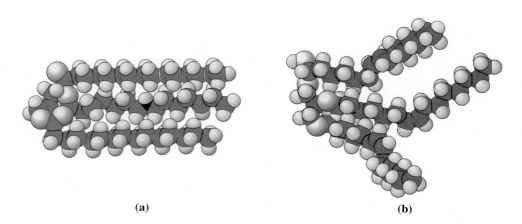

(a)　　　　　　　　　　(b)

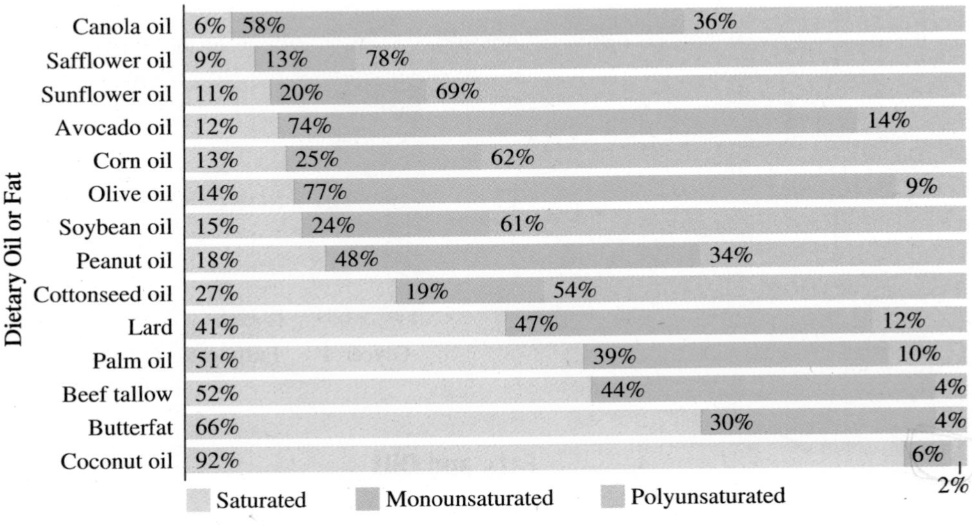

Figure 19.8 Percentages of saturated, monounsaturated, and polyunsaturated fatty acids in the triacylglycerols of various dietary fats and oils.

Dietary Oil or Fat	Saturated	Monounsaturated	Polyunsaturated
Canola oil	6%	58%	36%
Safflower oil	9%	13%	78%
Sunflower oil	11%	20%	69%
Avocado oil	12%	74%	14%
Corn oil	13%	25%	62%
Olive oil	14%	77%	9%
Soybean oil	15%	24%	61%
Peanut oil	18%	48%	34%
Cottonseed oil	27%	19%	54%
Lard	41%	47%	12%
Palm oil	51%	39%	10%
Beef tallow	52%	44%	4%
Butterfat	66%	30%	4%
Coconut oil	92%	6%	2%

■ Saturated   ■ Monounsaturated   ■ Polyunsaturated

All oils, even polyunsaturated oils, contain some *saturated* fatty acids. All fats, even highly saturated fats, contain some *unsaturated* fatty acids.

keeps the fat somewhat liquid (semi-solid) and thus allows for movement. Oils typically come from plants, although there are also fish oils. A fish would have some serious problems if its triacylglycerols "solidified" when it encountered cold water.

3. Pure fats and pure oils are colorless, odorless, and tasteless. The tastes, odors, and colors associated with dietary plant oils are caused by small amounts of other naturally occurring substances present in the plant that have been carried along during processing. The presence of these "other" compounds is usually considered desirable.

Figure 19.8 gives the percentages of saturated, monounsaturated, and polyunsaturated fatty acids found in common dietary oils and fats. In general, a higher degree of fatty acid unsaturation is associated with oils than with fats. A notable exception to this generalization is coconut oil, which is highly saturated. This oil is a liquid not because it contains many double bonds within the fatty acids but because it is rich in *shorter-chain* fatty acids, particularly lauric acid (12:0).

 DIETARY CONSIDERATIONS AND TRIACYLGLYCEROLS

In the past two decades, considerable research has been carried out concerning the role of dietary factors as a cause of disease (obesity, diabetes, cancer, hypertension, and atherosclerosis). Numerous studies have shown that, *in general*, nations whose citizens have high dietary intakes of triacylglycerols (fats and oils) tend to have higher incidences of heart disease and certain types of cancers. This is the reason for concern that the typical American diet contains too much fat and the call for Americans to reduce their total dietary fat intake.

Contrary to the general trend, however, there are several areas of the world where high dietary fat intake does not translate into high risks for cardiovascular disease, obesity, and certain types of cancers. These exceptions, which include some Mediterranean countries and the Inuit people of Greenland, suggest that relationships between dietary triacylglycerol intake and risk factors for disease involve more than simply the *total amount* of triacylglycerols consumed.

A grain- and vegetable-rich diet that contains small amounts of extra-virgin olive oil (three to four teaspoons daily) has been found to help people with high blood pressure reduce the amount of blood pressure medication they require, on average, by 48%. Substitution of sunflower oil for the olive oil resulted in only a 4% reduction in medication dosage.

The blood-pressure-reduction benefits of olive oil do not relate to the triacylglycerols present but rather come from "other" compounds naturally present, namely from antioxidant polyphenols olive oil contains. These antioxidants help promote the relaxation of blood vessels.

"Good Fats" Versus "Bad Fats"

In dietary discussions, the term *fat* is used as a substitute for the term *triacylglycerol*. Thus a dietary fat can be either a "fat" or an "oil." Ongoing studies indicate that both the *type of dietary fat* consumed and the *amount of dietary fat* consumed are important factors in determining human body responses to dietary fat. Current dietary fat recommendations are that people limit their total fat intake to 30% of total calories—with up to 15% coming from monounsaturated fat, up to 10% from polyunsaturated fat, and less than 10% from saturated fats.

Freshly pressed extra-virgin olive oil contains a compound that has the same pharmacological activity as the over-the-counter pain reliever ibuprofen (Section 16.4). This finding suggests a possible explanation for some of the various health benefits attributed to a Mediterranean diet that typically is rich in olive oil. It is estimated that this olive oil compound, called oleocanthal, is present in a typical Mediterranean diet in an amount equivalent to about 10% of the ibuprofen dose recommended for headache relief.

Oleocanthal

These recommendations imply correctly that different types of dietary fat have different effects. In simplified terms, research studies indicate that saturated fats are "bad fat," monounsaturated fats are "good fat," and polyunsaturated fats can be both "good fat" and "bad fat." In the latter case, fatty acid omega type (Section 19.2) becomes important, a situation addressed later in this section. (Studies indicate that saturated fat can increase heart disease risk, that monounsaturated fat can decrease both heart disease and breast cancer risk, and that polyunsaturated fat can reduce heart disease risk but promote the risk of certain types of cancers.)

Referring to Figure 19.8, note the wide variance in the three general types of fatty acids (SFAs, MUFAs, and PUFAs) present in various kinds of dietary fats. Dietary fats high in "good" monounsaturated fatty acids include olive, avocado, and canola oils. Monounsaturated fatty acids help reduce the stickiness of blood platelets. This helps prevent the formation of blood clots and may also dissolve clots once they form.

Many people do not realize that most tree nuts and peanuts are good sources of MUFAs. The Chemical Connections feature on page 620 looks at recent research on the fat content of nuts.

Omega-3 and Omega-6 Fatty Acids

In the 1980s, researchers found that the Inuit people of Greenland exhibit a low incidence of heart disease despite having a diet very high in fat. This contrasts markedly with studies on the U.S. population, which show a correlation between a high-fat diet and a high incidence of heart disease. What accounts for the difference between the two peoples? The Inuit diet is high in omega-3 fatty acids (from fish), and the U.S. diet is high in omega-6 fatty acids (from plant oils). An American consumes about double the amount of omega-6 fatty acids and half the amount of omega-3 fatty acids that an Inuit consumes.

Several large studies now confirm that benefits can be derived from eating several servings of fish each week. The choice of fish is important, however. Not all fish are equal in omega-3 fatty acid content. Cold-water fish, also called fatty fish because of the extra amounts of fat they have for insulation against the cold, contain more omega-3 acids than leaner, warm-water fish. Fatty fish include albacore tuna, salmon, and mackerel (see Figure 19.9). Leaner, warm-water fish, which include cod, catfish, halibut, sole, and snapper, do not appear to offer as great a positive effect on heart health as do their "fatter" counterparts. (Note that most of the fish used in fish and chips (e.g., cod, halibut) is on the low end of the omega-3 scale.) Table 19.2 gives the actual omega-3 fatty acid concentrations associated with various kinds of cold-water fish.

Figure 19.9 Fish that live in deep, cold water—mackerel, herring, tuna, and salmon—are better sources of omega-3 fatty acids than other fish.

TABLE 19.2
Omega-3 Fatty Acid Amounts Associated with Various Kinds of Cold-Water Fish

Per 3.5-oz. Serving (raw)	Omega-3s (grams)*
mackerel	2.3
albacore tuna	2.1
herring, Atlantic	1.6
anchovy	1.5
salmon, wild king (Chinook)	1.4
salmon, wild sockeye (red)	1.2
tuna, bluefin	1.2
salmon, wild pink	1.0
salmon, wild Coho (silver)	0.8
oysters, Pacific	0.7
salmon, farm-raised Atlantic	0.6
swordfish	0.6
trout, rainbow	0.6

*Omega-3 content of fish can vary depending on harvest location and time of year.

CHEMICAL Connections

The Fat Content of Tree Nuts and Peanuts

People who bypass the nut tray at holiday parties usually believe in a myth—that nuts are *unhealthful* high-fat foods. Indeed, nuts are high-fat food. However, the fat is "good fat" rather than "bad fat" (Section 19.5); that is, the fatty acids present are MUFAs and PUFAs rather than SFAs. In most cases, a handful of nuts is better for you than a cookie or bagel.

Numerous studies now indicate that eating nuts can have a strong protective effect against coronary heart disease. The most improvement comes from adding small amounts of nuts—an ounce (3–4 teaspoons)—to the diet five or more times a week. Raw, dry-roasted, or lightly salted varieties are best.

The recommendation of only one ounce of nuts per day relates to the high calorie content of nuts, which is 160 to 200 calories per ounce. The number of nuts and number of calories per ounce for common types of nuts is as follows:

Nuts	Calories
18 cashews	160
20 peanuts	160
47 pistachios	160
24 almonds	166
14 walnut halves	180
8 Brazil nuts	186
12 hazelnuts	188
15 pecan halves	190
12 macadamias	200

The amount of fat present in nuts ranges from 74% in the macadamia nut, 68% in pecans, and 63% in hazelnuts to around 50% in nuts such as the almond, cashew, peanut, and pistachio, as is shown in the table below.

The different fatty acid fractions (SFAs, MUFAs, and PUFAs) present in nuts also vary, but with definite trends. Unsaturated fatty acids always significantly dominate saturated fatty acids. The unsaturation/saturation ratio is highest for hazelnuts (11.9), pecans (10.9), walnuts (9.0), and almonds (9.0) and is lowest for cashews (3.9).

Their low amounts of saturated fatty acids are not the only reason why nuts help reduce the risk of coronary heart disease. Nuts also offer valuable antioxidant vitamins, minerals, and plant fiber protein. The protein content is highest (18%–26%) in the cashew, pistachio, almond, and peanut; here the amount of protein is about the same as in meat, fish, and cheese. The carbohydrate content of nuts is relatively low, less than 10% in most cases.

An unexpected discovery involving the anticancer drug Taxol and hazelnuts was made in the year 2000. The active chemical component in this drug, paclitaxel, was found in hazelnuts. It was the first report of this potent chemical being found in a plant other than in the bark of the Pacific yew tree, a slow-growing plant found in limited quantities in the Pacific Northwest. Although the amount of the chemical found in a hazelnut tree is about one-tenth that found in yew bark, the effort required to extract paclitaxel from these sources is comparable. Because hazelnut trees are more common, this finding could reduce the cost of the commercial drug and make it more readily available.

Fat and Fatty Acid Composition of Selected Nuts

	Total Fat (percentage of weight)	SFA	MUFA	PUFA	UFA/SFA Ratio
		(percentage of total fat)			
almonds	52	10	68	22	9.0
cashews	46	20	62	18	3.9
hazelnuts	63	8	82	10	11.9
macadamias	74	16	82	2	5.4
peanuts	49	15	51	34	5.7
pecans	68	8	66	26	10.9
pistachios	48	13	72	15	6.6
walnuts	62	10	24	66	9.0

Essential Fatty Acids

An **essential fatty acid** *is a fatty acid needed in the human body that must be obtained from dietary sources because it cannot be synthesized within the body, in adequate amounts, from other substances.* There are two essential fatty acids: *linoleic acid* and *linolenic acid.* Linoleic acid (18:2) is the primary member of the omega-6 acid family, and linolenic acid (18:3) is the primary member of the omega-3 acid family. Their structures were given in Table 19.1.

These two acids (1) are needed for proper membrane structure and (2) serve as starting materials for the production of several nutritionally important longer-chain omega-6

TABLE 19.3
Biochemically Important Omega-3 and Omega-6 Fatty Acids

Omega-3 Acids	Omega-6-Acids
linolenic acid (18:3) (lin-oh-LEN-ic)	linoleic acid (18:2) (lin-oh-LAY-ic)
eicosapentaenoic acid (20:5) (EYE-cossa-PENTA-ee-NO-ic)	arachidonic acid (20:4) (a-RACK-ih-DON-ic)
docosahexaenoic acid (20:6) (DOE-cossa-HEXA-ee-NO-ic)	

In 2001 the FDA gave approval for manufacturers of baby formula to add the fatty acids DHA (docosahexaenoic acid) and AA (arachidonic acid) to infant formulas. Human breast milk naturally contains these acids, which are important in brain and vision development. Because not all mothers can breast-feed, health officials regulate the ingredients in infant formula so that formula-fed babies get the next best thing to mother's milk.

and omega-3 acids. When these two acids are missing from the diet, the skin reddens and becomes irritated, infections and dehydration are likely to occur, and the liver may develop abnormalities. If the fatty acids are restored, then the conditions reverse themselves. Infants are especially in need of these acids for their growth. Human breast milk has a much higher percentage of the essential fatty acids than cow's milk.

Linoleic acid is the starting material for the biosynthesis of arachidonic acid.

Linoleic acid (18:2) ⟶ arachidonic acid (20:4)
Omega-6 fatty acids

Arachidonic acid is the major starting material for eicosanoids (Section 19.13), substances that help regulate blood pressure, clotting, and several other important body functions.

Linolenic acid is the starting material for the biosynthesis of two additional omega-3 fatty acids.

Linolenic acid (18:3) ⟶ EPA (20:5) ⟶ DHA (22:6)
Omega-3 fatty acids

EPA (eicosapentaenoic acid) and DHA (docosahexaenoic acid) are important constituents of the communication membranes of the brain and are necessary for normal brain development. EPA and DHA are also active in the retina of the eye.

Table 19.3 gives pronunciation guidelines for the names of the two essential fatty acids and of the other acids mentioned that are biosynthesized from them.

Fat Substitutes (Artificial Fats)

In response to consumer demand for low-fat, low-calorie foods, food scientists have developed several types of "artificial fats." Such substances replicate the taste, texture, and cooking properties of fats but are themselves not lipids. See the Chemical Connections feature on page 624 for further discussion of this topic.

 19.6 CHEMICAL REACTIONS OF TRIACYLGLYCEROLS

The chemical properties of triacylglycerols (fats and oils) are typical of esters and alkenes because these are the two functional groups present in triacylglycerols. Four important triacylglycerol reactions are hydrolysis, saponification, hydrogenation, and oxidation.

Hydrolysis

Naturally occurring mono- and diacylglycerols are seldom encountered. *Synthetic* mono- and diacylglycerols are used as emulsifiers in many food products. Emulsifiers prevent suspended particles in colloidal solutions (Section 8.7) from coalescing and settling. Emulsifiers are usually present in so-called fat-free cakes and other fat-free products.

Hydrolysis of a triacylglycerol is the reverse of the esterification reaction by which it was formed (see Figure 19.5). Triacylglyercol hydrolysis, when carried out in a laboratory setting, requires the presence of an acid or a base. Under acidic conditions, the hydrolysis products are glycerol and fatty acids. Under basic conditions, the hydrolysis products are glycerols and fatty acid salts.

Within the human body, triacylglycerol hydrolysis occurs during the process of digestion. Such hydrolysis requires the help of enzymes (protein catalysts; Section 21.1) produced by the pancreas. These enzymes cause the triacylglycerol to be hydrolyzed in a

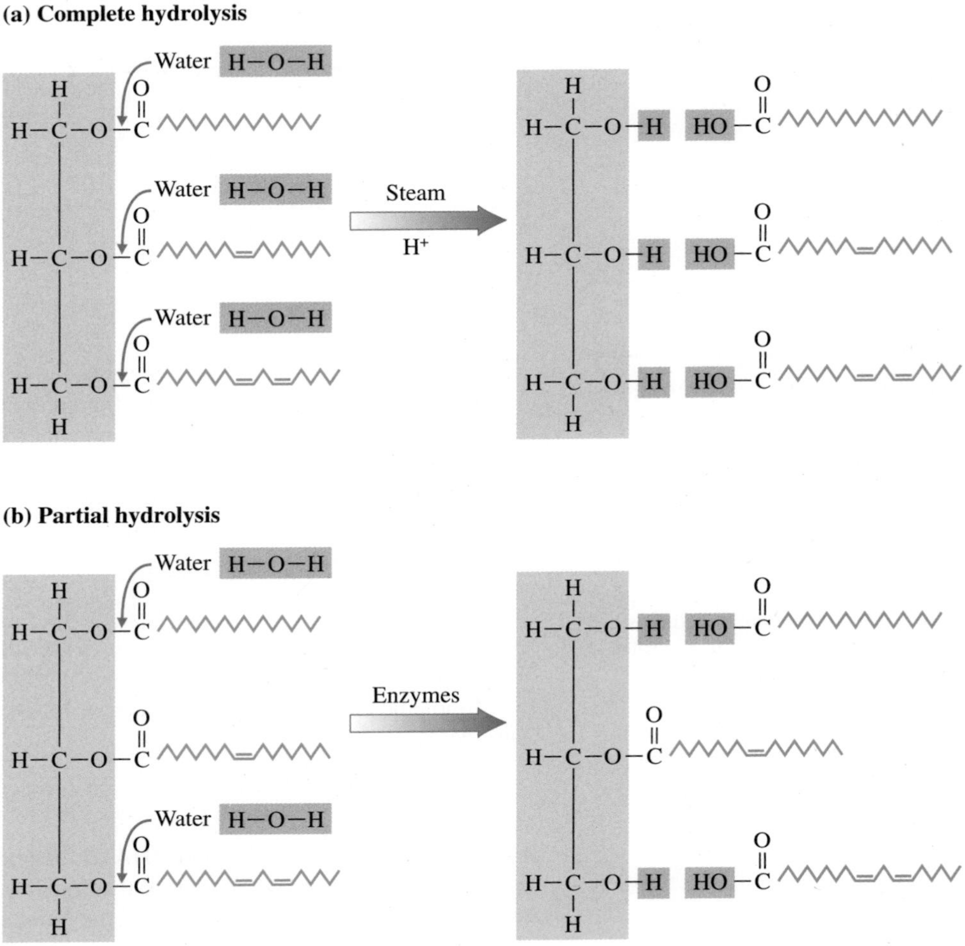

Figure 19.10 Complete and partial hydrolysis of a triacylglycerol. (a) Complete hydrolysis of a triacylglycerol produces glycerol and three fatty acid molecules. (b) Partial hydrolysis (during digestion) of a triacylglycerol produces a monoacylglycerol and two fatty acid molecules.

(a) Complete hydrolysis

(b) Partial hydrolysis

stepwise fashion. First, one of the outer fatty acids is removed, then the other outer one, leaving a monoacylglycerol. In most cases this is the end product of the initial digestion (hydrolysis) of the triacylglycerol. Sometimes, enzymes remove all three fatty acids, leaving a free molecule of glycerol.

In situations where all three fatty acids are removed the hydrolysis process is referred to as *complete hydrolysis*, which is depicted in Figure 19.10a. If one or more of the fatty acid residues remains attached to the glycerol, the hydrolysis process is called *partial hydrolysis* (see Figure 19.10b).

● **EXAMPLE 19.3**

Writing a Structural Equation for the Hydrolysis of a Triacylglycerol

Write an equation for the acid-catalyzed hydrolysis of the following triacylglycerol.

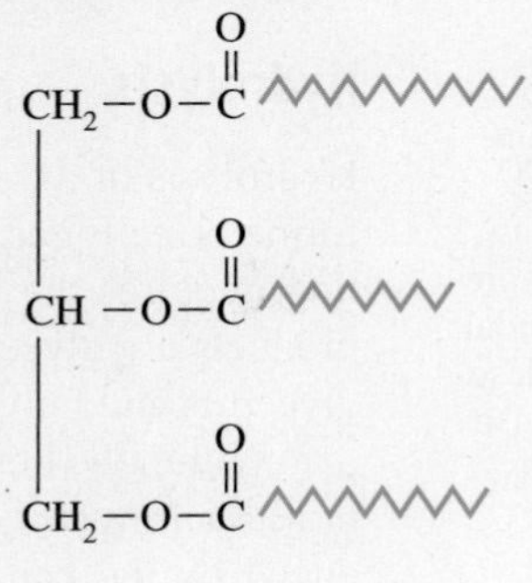

Solution

Three water molecules are required for the hydrolysis, one to interact with each of the ester linkages present in the triacylglycerol. Breaking of the three ester linkages produces four product molecules: glycerol and three fatty acids.

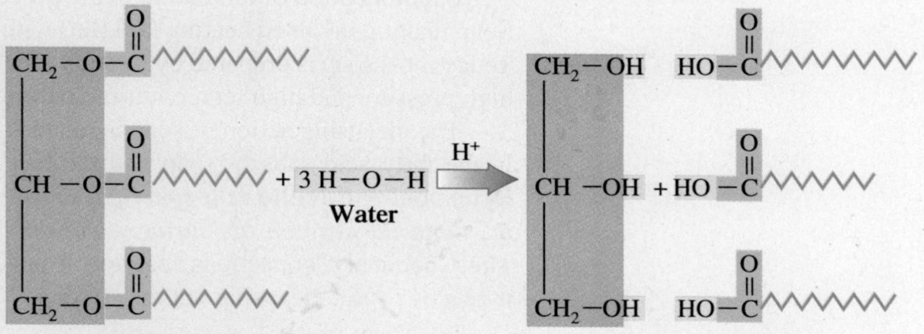

Triacylglycerol　　　　　　　　**Glycerol**　　**Fatty acids**

Practice Exercise 19.3

Write a structural equation for the acid-catalyzed hydrolysis of the following triacylglycerol.

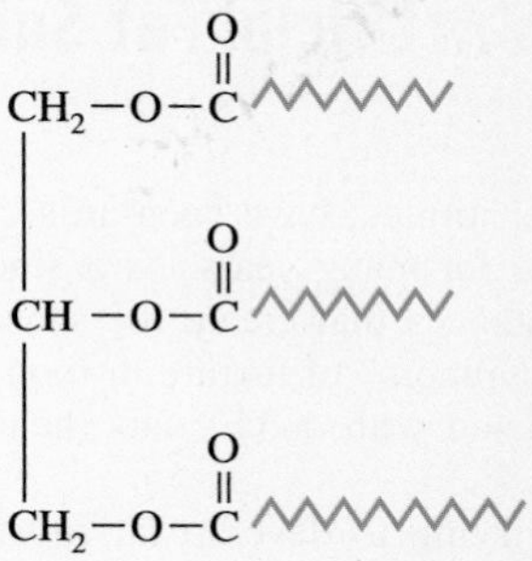

Answer:

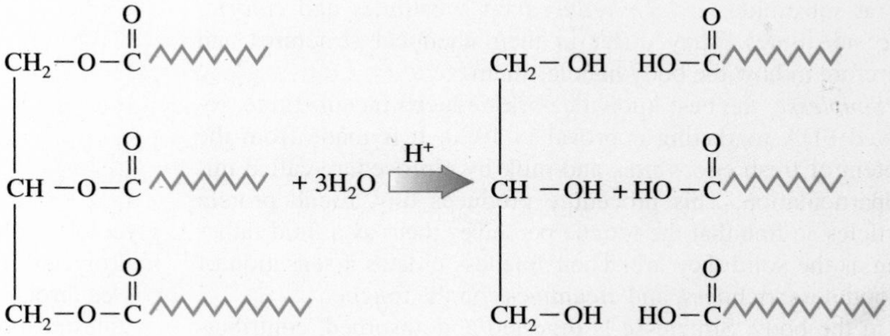

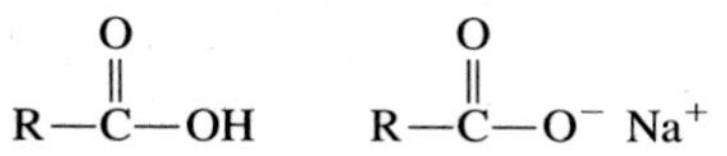

Saponification

Recall from Section 16.9 the structural difference between a carboxylic acid and a carboxylic acid salt.

Carboxylic　　　　Carboxylic
acid　　　　　　　acid salt

Saponification (Section 16.16) is a reaction carried out in an alkaline (basic) solution. For fats and oils, the products of saponification are glycerol and fatty acid *salts*.

The overall reaction of triacylglycerol saponification can be thought of as occurring in two steps. The first step is the hydrolysis of the ester linkages to produce glycerol and three fatty acid molecules:

$$\text{Fat or oil} + 3H_2O \longrightarrow 3 \text{ fatty acids} + \text{glycerol}$$

The second step involves a reaction between the fatty acid molecules and the base (usually NaOH) in the alkaline solution. (This is an acid–base reaction that produces water plus salts:)

$$3 \text{ Fatty acids} + 3\text{NaOH} \longrightarrow 3 \text{ fatty acid salts} + 3\text{H}_2\text{O}$$

Saponification of animal fat is the process by which soap was made in pioneer times. Soap making involved heating lard (fat) with lye (ashes of wood, an impure form of KOH). Today most soap is prepared by hydrolyzing fats and oils (animal fat and coconut oil) under high pressure and high temperature. Sodium carbonate is used as the base.

The cleansing action of soap is related to the structure of the carboxylate ions present in the fatty acid salts of soap and the fact that these ions readily participate in micelle formation. A **micelle** is a spherical cluster of molecules in which the polar portions of the molecules are on the surface, and the nonpolar portions are located in the interior. The Chemical Connections feature on page 625 discusses micelle formation further as it relates to the cleansing action of soap.

Artificial Fat Substitutes

Artificial sweeteners (sugar substitutes) have been an accepted part of the diet of most people for many years. New since the 1990s are artificial fats—substances that create the sensations of "richness" of taste and "creaminess" of texture in food without the negative effects associated with dietary fats (heart disease and obesity).

Food scientists have been trying to develop fat substitutes since the 1960s. Now available for consumer use are two types of fat substitutes: *calorie-reduced* fat substitutes and *calorie-free* substitutes. They differ in their chemical structures and therefore in how the body handles them.

Simplesse, the best-known calorie-reduced fat substitute, received FDA marketing approval in 1990. It is made from the protein of fresh egg whites and milk by a procedure called microparticulation. This procedure produces tiny, round protein particles so fine that the tongue perceives them as a fluid rather than as the solid they are. Their fineness creates a sensation of smoothness, richness, and creaminess on the tongue.

In the body, Simplesse is digested and absorbed, contributing to energy intake. But 1 g of Simplesse provides 1.3 cal, compared with the 9 cal provided by 1 g of fat. Simplesse is used only to replace fats in *formulated* foods such as salad dressings, cheeses, sour creams, and other dairy products. Simplesse is unsuitable for frying or baking because it turns rubbery or rigid (gels) when heated. Consequently, it is not available for home use.

Olestra, the best-known calorie-free fat substitute, received FDA marketing approval in 1996. It is produced by heating cottonseed and/or soybean oil with sucrose in the presence of methyl alcohol. Chemically, olestra has a structure somewhat similar to that of a triacylglycerol; sucrose takes the place of the glycerol molecule, and six to eight fatty acids are attached by ester linkages to it rather than the three fatty acids in a triacyl-

Olestra

glycerol. Unlike triacylglycerols, however, olestra cannot be hydrolyzed by the body's digestive enzymes and therefore passes through the digestive tract undigested.

Olestra looks, feels, and tastes like dietary fat and can substitute for fats and oils in foods such as shortenings, oils, margarines, snacks, ice creams, and other desserts. It has the same cooking properties as fats and oils.

In the digestive tract, Olestra interferes with the absorption of both dietary and body-produced cholesterol; thus it may lower total cholesterol levels. A problem with its use is that it also reduces the absorption of the fat-soluble vitamins A, D, E, and K. To avoid such depletion, Olestra is fortified with these vitamins. Another problem with Olestra use is that in some individuals it can cause gastrointestinal irritation and/or diarrhea. All products containing Olestra must carry the following label: "Olestra may cause abdominal cramping and loose stools. Olestra inhibits the absorption of some vitamins and other nutrients. Vitamins A, D, E, and K have been added."

Hydrogenation

Hydrogenation is a chemical reaction we first encountered in Section 13.9. It involves hydrogen addition across carbon–carbon multiple bonds, which increases the degree of saturation as some double bonds are converted to single bonds. With this change, there is a corresponding increase in the melting point of the substance.

Hydrogenation involving just one carbon–carbon bond within a fatty acid residue of a triacylglycerol can be diagrammed as follows:

$$---CH_2-CH_2-CH{=}CH-CH_2-CH_2--- \; + \; H_2 \longrightarrow \; ---CH_2-CH_2-CH_2-CH_2-CH_2-CH_2---$$

Portion of an unsaturated fatty acid residue in a triacylglycerol containing one double bond	The double bond has been converted to a single bond; the degree of saturation has increased

The structural equation for the complete hydrogenation of a triacylglycerol in which all three fatty acid residues are oleic acid (18:1) is shown in Figure 19.11.

Many food products are produced via partial hydrogenation. In partial hydrogenation some, but not all, of the double bonds present are converted into single bonds. In this manner, liquids (usually plant oils) are converted into semi-solid materials.

Peanut butter is produced from peanut oil through partial hydrogenation. Solid cooking shortenings and stick margarine are produced from liquid plant oils through partial hydrogenation. Soft-spread margarines are also partial-hydrogenation products. Here, the extent of hydrogenation is carefully controlled to make the margarine soft at refrigerator temperatures (4°C). Concern has arisen about food products obtained from hydrogenation processes because the hydrogenation process itself converts some *cis* double bonds within fatty acid residues into *trans* double bonds producing *trans* unsaturated fatty acids. The Chemical Connections feature on page 626 explores this issue further.

CHEMICAL Connections

The Cleansing Action of Soap

The cleansing action of soap is directly related to the structure of the carboxylate ions present in soap within fatty acid salts. Their structure is such that they exhibit a "dual polarity." The hydrocarbon portion of the carboxylate ion is nonpolar, and the carboxyl portion is polar. This dual polarity for the fatty acid salt *sodium stearate,* which is representative of all fatty acid salts present in soap, is as follows:

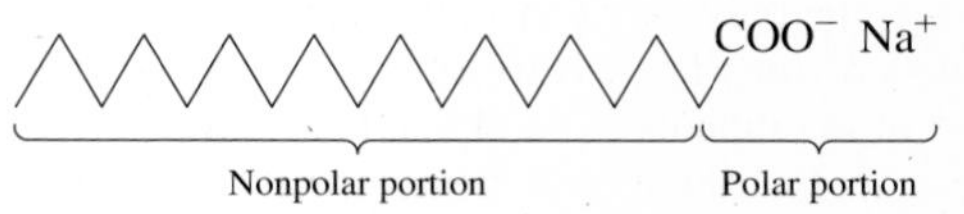

$$\bigwedge\!\!\bigwedge\!\!\bigwedge\!\!\bigwedge\!\!\bigwedge\; COO^- \, Na^+$$

Nonpolar portion   Polar portion

Soap solubilizes oily and greasy materials in the following manner: The nonpolar portion of the carboxylate ion dissolves in the nonpolar oil or grease, and the polar carboxyl portion maintains its solubility in the polar water.

The penetration of the oil or grease by the nonpolar end of the carboxylate ion is followed by the formation of micelles (see the accompanying diagram). The carboxyl groups (the micelle exterior) and water molecules are attracted to each other, causing the solubilizing of the micelle.

The micelles do not combine into larger drops because their surfaces are all negatively charged, and like charges repel each other. The water-soluble micelles are subsequently rinsed away, leaving a material devoid of oil and grease.

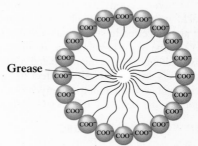

Fatty acid micelle

For most cleansing purposes, synthetic detergents have largely replaced soaps. The basis of the cleansing action of synthetic detergents is very similar to that of soaps because their structures are very similar. The structure of the sodium salt of a benzene sulfonic acid is typical of the types of molecules used in detergents.

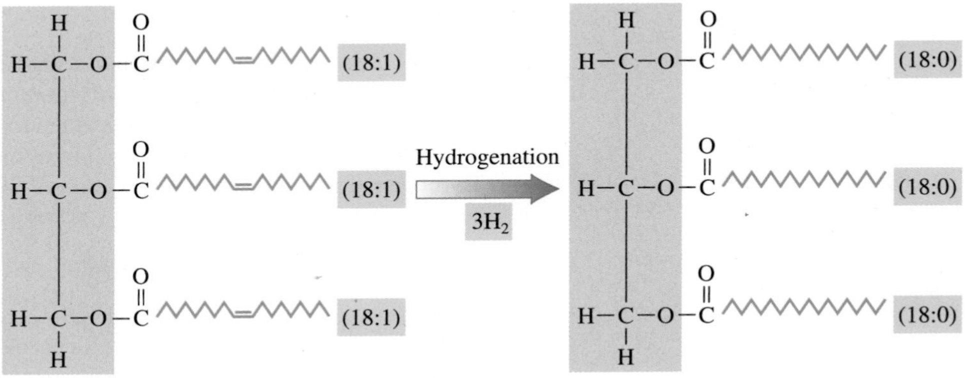

Figure 19.11 Structural equation for the complete hydrogenation of a triacylglycerol with oleic acid (18:1) fatty acid residues.

Chemical Connections

Trans Fatty Acids and Blood Cholesterol Levels

All current dietary recommendations stress reducing saturated fat intake. In accordance with such recommendations, many people have switched from butter to margarine and now use partially hydrogenated vegetable oils rather than animal fat for cooking. However, recent studies suggest that partially hydrogenated products also play a role in raising blood cholesterol levels. Why would this be so?

It is now known that when triacylglycerols are subjected to partial hydrogenation (Section 19.6) two types of changes occur in the fatty acid residues present: (1) some of the *cis* double bonds present are converted to single bonds (the objective of the process), and (2) some of the remaining *cis* double bonds are converted to *trans* double bonds (an unanticipated result of the process). These latter *cis–trans* conversions affect the general shape of the fatty acid residues present in triacylglycerols, which in turn affects the biochemical behavior of the triacylglycerols. In the

preceding diagram, note how conversion of a *cis,cis*-18:2 fatty acid to a *trans,trans*-18:2 fatty acid affects molecular shape. The *trans,trans*-18:2 fatty acid has a shape very much like that of an 18:0 saturated fatty acid (the structure on the right).

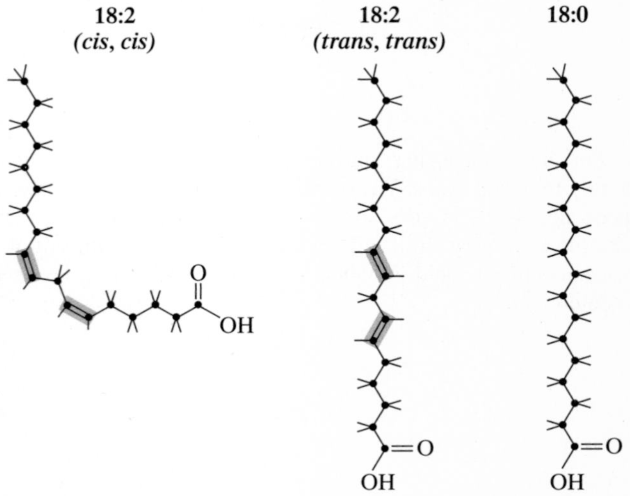

Studies show that fatty acids with *trans* double bonds affect blood cholesterol levels in a manner similar to saturated fatty acids.

Trans fatty acids (*trans* fat) make up approximately 5% of the fat intake in the typical diet in the United States, and the amount of *trans* fat a person consumes depends on the amount of fat eaten and on the types of foods selected. The best example of a *trans* fat food may be stick margarine, but it is also found in crackers, cookies, pastries, and deep-fried fast foods. Spreadable margarine in tubs, though, contains little if any *trans* fat.

Beginning in 2006, the U.S. Food and Drug Administration (FDA) requires that the *trans* fat content of a food be included in the nutrition facts panel found on all food products. Prior to this rule change, the only way consumers could determine whether a food included *trans* fat was to look for the word *hydrogenated* on the list of ingredients. A food that lists partially hydrogenated oils among its first three ingredients usually contains substantial amounts of *trans* fatty acids as well as some saturated fat.

A label of "zero grams trans fat per serving" on a product does not mean the product is absolutely *trans*-fat free. FDA regulations allow *trans* fat levels of less than 0.5 gram per serving to be labeled as 0 grams per serving.

The health implications of *trans* fatty acids is an area of active research; many answers are yet to be found. Preliminary studies indicate that *trans* fat raises bad (LDL) cholesterol, but it does not raise good (HDL) cholesterol. Saturated fat, on the other hand, raises both bad and good cholesterol. Thus, just as too much saturated fat isn't healthy, too much *trans* fat is also not healthy. Recommendations are that total fat intake be limited to 30% of daily calories and that combined saturated fat and *trans* fat intake should be limited to 10% or less of daily calories.

Oxidation

The carbon–carbon double bonds present in the fatty acid residues of a triacylglycerol are subject to oxidation with molecular oxygen (from air) as the oxidizing agent. Such oxidation breaks the carbon–carbon bonds, producing both aldehyde and carboxylic acid products.

$$-CH=CH- \xrightarrow{\text{Oxidation}} \underset{\substack{\text{Short-chain}\\\text{aldehydes}}}{-\overset{\overset{\displaystyle O}{\|}}{C}-H + H-\overset{\overset{\displaystyle O}{\|}}{C}-} \xrightarrow{\text{Oxidation}} \underset{\substack{\text{Short-chain}\\\text{carboxylic acids}}}{-\overset{\overset{\displaystyle O}{\|}}{C}-OH + HO-\overset{\overset{\displaystyle O}{\|}}{C}-}$$

Unsaturated fatty acids

Antioxidants are compounds that are easily oxidized. When added to foods, they are more easily oxidized than the food. Thus they prevent the food from being oxidized (see Section 14.14).

The short-chain aldehydes and carboxylic acids so produced often have objectionable odors, and fats and oils containing them are said to have become *rancid.* To avoid this unwanted oxidation process, commercially prepared foods containing fats and oils nearly always contain *antioxidants*—substances that are more easily oxidized than the food. Two naturally occurring antioxidants are vitamin C (Section 21.12) and vitamin E (Section 21.13). Two synthetic oxidation inhibitors are BHA and BHT (Section 14.14). In the presence of air, antioxidants, rather than food, are oxidized.

● EXAMPLE 19.4

Determining the Products for Reactions That Triacylglycerols Undergo

▶ Using words rather than structural formulas, characterize the products formed when the following triacylglycerol undergoes the reactions listed.

$$
\begin{array}{l}
\overset{\displaystyle H}{\underset{\displaystyle |}{}} \quad \overset{\displaystyle O}{\|} \\
H-\overset{|}{\underset{|}{C}}-O-C\!\!\!\diagup\!\!\!\diagdown\!\!\!\diagup\!\!\!\diagdown\!\!\!\diagup\!\!\!\diagdown \quad \text{(18:0 fatty acid residue)}
\end{array}
$$

a. Complete hydrolysis   b. Complete saponification using NaOH
c. Complete hydrogenation

Solution

a. When a triacylglycerol undergoes complete hydrolysis, there are four organic products: glycerol and three fatty acids. For the given triacylglycerol the products are *glycerol, an 18:0 fatty acid, an 18:1 fatty acid,* and *an 18:2 fatty acid.* Three molecules of water are also consumed.

b. When a triacylglycerol undergoes complete saponification, there are four organic products: glycerol and three fatty acid salts. For the given triacylglycerol, with NaOH as the base involved in the saponification, the products are *glycerol, the sodium salt of the 18:0 fatty acid, the sodium salt of the 18:1 fatty acid,* and *the sodium salt of the 18:2 fatty acid.*

c. Complete hydrogenation will change the given triacylglycerol into a *triacylglycerol in which all three fatty acid residues are 18:0 fatty acid residues.* That is, all of the fatty acid residues are completely saturated (there are no carbon–carbon double bonds).

Practice Exercise 19.4

Using words rather than structural formulas, characterize the organic products formed when the following triacylglycerol undergoes the reactions listed.

(continued)

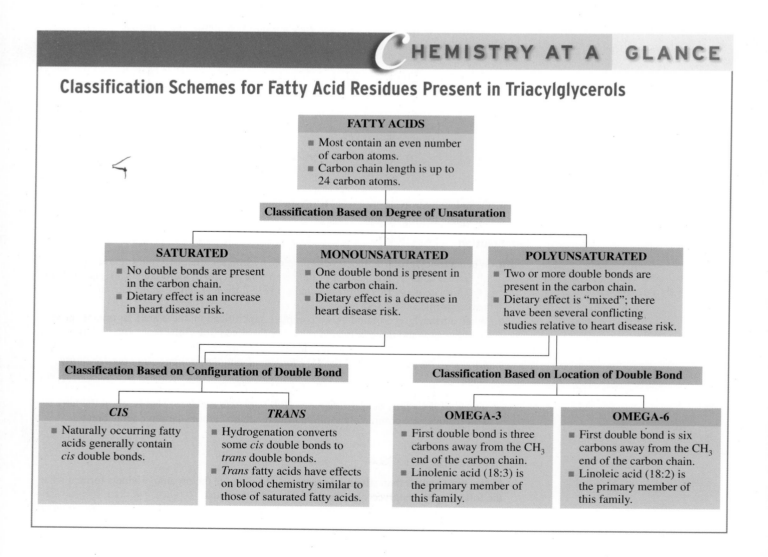

a. Complete hydrolysis
b. Complete saponification using NaOH
c. Complete hydrogenation

Answers: a. Four products: glycerol and three fatty acids; **b.** Four products: glycerol and three fatty acid salts; **c.** One product: a triacylglycerol in which all fatty acid residues are saturated (18:0) residues

CHEMISTRY AT A GLANCE

Classification Schemes for Fatty Acid Residues Present in Triacylglycerols

FATTY ACIDS
- Most contain an even number of carbon atoms.
- Carbon chain length is up to 24 carbon atoms.

Classification Based on Degree of Unsaturation

SATURATED
- No double bonds are present in the carbon chain.
- Dietary effect is an increase in heart disease risk.

MONOUNSATURATED
- One double bond is present in the carbon chain.
- Dietary effect is a decrease in heart disease risk.

POLYUNSATURATED
- Two or more double bonds are present in the carbon chain.
- Dietary effect is "mixed"; there have been several conflicting studies relative to heart disease risk.

Classification Based on Configuration of Double Bond

CIS
- Naturally occurring fatty acids generally contain *cis* double bonds.

TRANS
- Hydrogenation converts some *cis* double bonds to *trans* double bonds.
- *Trans* fatty acids have effects on blood chemistry similar to those of saturated fatty acids.

Classification Based on Location of Double Bond

OMEGA-3
- First double bond is three carbons away from the CH_3 end of the carbon chain.
- Linolenic acid (18:3) is the primary member of this family.

OMEGA-6
- First double bond is six carbons away from the CH_3 end of the carbon chain.
- Linoleic acid (18:2) is the primary member of this family.

Figure 19.12 The oils (triacylglycerols) present in skin perspiration rapidly undergo oxidation. The oxidation products, short-chain aldehydes and short-chain carboxylic acids, often have strong odors.

An aminodialcohol contains two hydroxyl groups, —OH, and an amino group, —NH$_2$.

Perspiration generated by strenuous exercise or by "hot and muggy" climatic conditions contains numerous triacylglycerols (oils). Rapid oxidation of these oils, promoted by microorganisms on the skin, generates the body odor that accompanies most "sweaty" people (see Figure 19.12).

The Chemistry at a Glance on page 628 contains a summary of the terminology used in characterizing the properties of the fatty acid residues that are part of the structure of triacylglycerols (fats and oils).

19.7 MEMBRANE LIPIDS: PHOSPHOLIPIDS

All cells are surrounded by a membrane that confines their contents. Up to 80% of the mass of a cell membrane can be lipid materials; the rest is primarily protein. It is membranes that give cells their individuality by separating them from their environment.

There are three common types of membrane lipids: phospholipids, sphingoglycolipids, and cholesterol. We consider phospholipids in this section and the other two types of membrane lipids in the next two sections.

Phospholipids are the most abundant type of membrane lipid. A **phospholipid** *is a lipid that contains one or more fatty acids, a phosphate group, a platform molecule to which the fatty acid(s) and the phosphate group are attached, and an alcohol that is attached to the phosphate group.* The platform molecule on which a phospholipid is built may be the 3-carbon alcohol *glycerol* or a more complex C$_{18}$ aminodialcohol called *sphingosine*. Glycerol-based phospholipids are called *glycerophospholipids,* and those based on sphingosine are called *sphingophospholipids.* The general block diagrams for a glycerophospholipid and a sphingophospholipid are as follows:

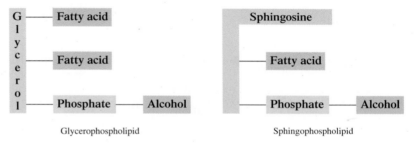

Glycerophospholipid   Sphingophospholipid

Glycerophospholipids

A **glycerophospholipid** *is a lipid that contains two fatty acids and a phosphate group esterified to a glycerol molecule and an alcohol esterified to the phosphate group.* All attachments (bonds) between groups in a glycerophospholipid are ester linkages, a situation similar to that in triacylglycerols (Section 19.4). However, glycerophospholipids have four ester linkages as contrasted to three ester linkages in triacylglycerols.

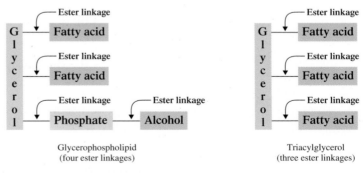

Glycerophospholipid   Triacylglycerol
(four ester linkages)   (three ester linkages)

Because of the ester linkages present, glycerophospholipids undergo hydrolysis and saponification reactions in a manner similar to that for triacylglycerols (Section 19.6). There will be five reaction products, however, instead of the four for triacylglycerols.

Phosphoric acid is the parent source for the minus one charged phosphate group used in the formation of glycerophospholipids. The structures of these two entities are

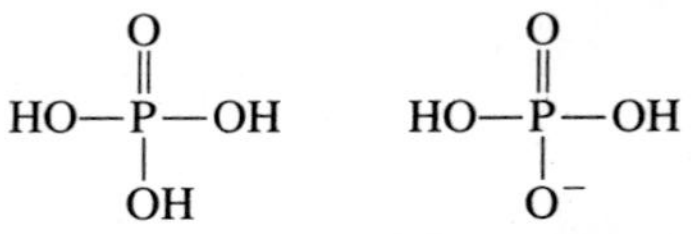

$$\underset{\substack{\text{Phosphoric acid}}}{\overset{\displaystyle O \atop \displaystyle \|}{HO\!-\!\overset{\displaystyle |}{\underset{\displaystyle OH}{P}}\!-\!OH}} \qquad \underset{\substack{\text{Minus one charged} \\ \text{phosphate group}}}{\overset{\displaystyle O \atop \displaystyle \|}{HO\!-\!\overset{\displaystyle |}{\underset{\displaystyle O^-}{P}}\!-\!OH}}$$

Phosphoric acid structures were previously considered in Section 16.20 when esters of inorganic acids were considered.

The alcohol attached to the phosphate group in a glycophospholipid is usually one of three amino alcohols: choline, ethanolamine, or serine. The structures of these three amino alcohols, given in terms of the charged forms (Sections 17.8 and 20.4) that they adopt in neutral solution, are

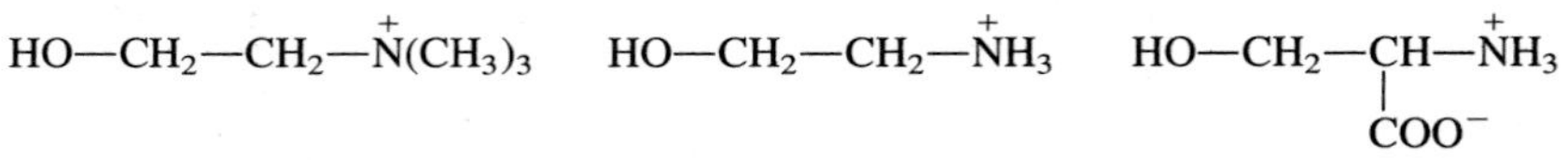

$HO{-}CH_2{-}CH_2{-}\overset{+}{N}(CH_3)_3$ $HO{-}CH_2{-}CH_2{-}\overset{+}{N}H_3$ $HO{-}CH_2{-}\underset{\underset{\displaystyle COO^-}{|}}{CH}{-}\overset{+}{N}H_3$

Choline   Ethanolamine   Serine
(a quaternary ammonium ion)   (positive-ion form)   (two ionic groups present)

Glycerophospholipids containing these three amino alcohols are respectively known as phosphatidylcholines, phosphatidylethanolamines, and phosphatidylserines. The fatty acid, glycerol, and phosphate portions of a glycerophospholipid structure constitute a *phosphatidyl group*.

EXAMPLE 19.5

Drawing the Structural Formula for a Glycerophospholipid

Draw the structural formula for the glycerophospholipid that produces, upon hydrolysis, equimolar amounts of glycerol, phosphoric acid, and ethanolamine, and twice that molar amount of stearic acid, the 18:0 fatty acid.

Solution

The fact that the molar amount of stearic acid produced is twice that of the other products indicates that both fatty acid residues present in the glycerophospholipid are stearic acid residues.

To draw the structural formula for this lipid, arrange the component parts in the following manner. Draw on the left the structure of glycerol, the "backbone" of the structure. Then place alongside it the two stearic acid and one phosphoric acid molecules. The acid placements should be such that their acid groups are lined up alongside the hydroxyl groups of glycerol. Then place the ion form of ethanolamine alongside the phosphoric acid molecule.

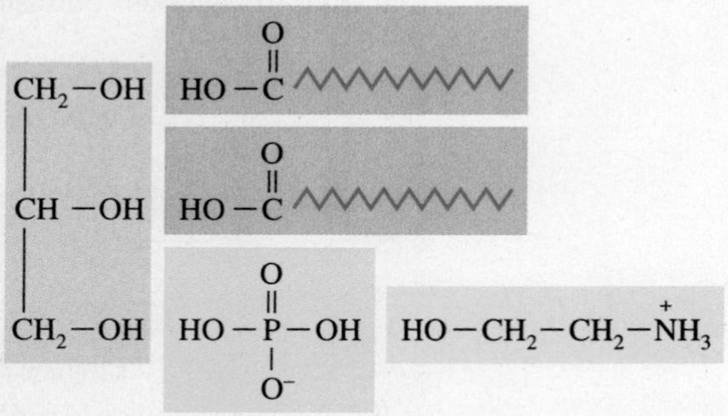

The five components of the overall structure are then bonded together via ester linkages. The product of the formation of each ester linkage is a molecule of water. The atoms involved in this water formation are highlighted in color in the structures at the left and the ester linkages formed are highlighted in color in the structure at the right.

$$CH_2-OH \quad HO-\overset{\overset{\displaystyle O}{\|}}{C}\wwww \qquad CH_2-O-\overset{\overset{\displaystyle O}{\|}}{C}\wwww$$

$$CH-OH \quad HO-\overset{\overset{\displaystyle O}{\|}}{C}\wwww \quad \Longrightarrow \quad 4H_2O \;+\; CH-O-\overset{\overset{\displaystyle O}{\|}}{C}\wwww$$

$$CH_2-OH \quad HO-\underset{\underset{\displaystyle O^-}{|}}{\overset{\overset{\displaystyle O}{\|}}{P}}-OH \quad HO-CH_2-CH_2-\overset{+}{N}H_3 \qquad CH_2-O-\underset{\underset{\displaystyle O^-}{|}}{\overset{\overset{\displaystyle O}{\|}}{P}}-O-CH_2-CH_2-\overset{+}{N}H_3$$

Practice Exercise 19.5

Draw the structure formula for the glycerophospholipid that produces, upon hydrolysis, equimolar amounts of glycerol, phosphoric acid, and choline, and twice that molar amount of lauric acid, the 12:0 fatty acid.

Answer:

$$CH_2-O-\overset{\overset{\displaystyle O}{\|}}{C}\wwww$$

$$CH-O-\overset{\overset{\displaystyle O}{\|}}{C}\wwww$$

$$CH_2-O-\underset{\underset{\displaystyle O^-}{|}}{\overset{\overset{\displaystyle O}{\|}}{P}}-O-CH_2-CH_2-\overset{+}{N}(CH_3)_3$$

Although the general structural features of glycerophospholipids are similar in many respects to those of triacylglycerols, these two types of lipids have quite different biochemical functions. Triacylglycerols serve as storage molecules for metabolic fuel. Glycerophospholipids function almost exclusively as components of cell membranes (Section 19.10) and are not stored. A major structural difference between the two types of lipids, that of polarity, is related to their differing biochemical functions. Triacylglycerols are a nonpolar class of lipids, whereas glycerophospholipids are polar. In general, membrane lipids have polarity associated with their structures.

Further consideration of general glycerophospholipid structure reveals an additional structural characteristic of most membrane lipids. Let us consider a phosphatidylcholine containing stearic and oleic acids to illustrate this additional feature. The chemical structure of this molecule is shown in Figure 19.13a.

A molecular model for this compound, which gives the orientation of groups in space, is illustrated in Figure 19.13b. There are two important things to notice about this model: (1) There is a "head" part, the choline and phosphate, and (2) there are two "tails," the two fatty acid carbon chains. The head part is polar. The two tails, the carbon chains, are nonpolar.

All glycerophospholipids have structures similar to that shown in Figure 19.13. All have a "head" and two "tails." A simplified representation for this structure uses a circle to represent the polar head and two wavy lines to represent the nonpolar tails.

Glycerophospholipids have a *hydrophobic* ("water-hating") portion, the nonpolar fatty acid groups, and a *hydrophilic* ("water-loving") portion, the polar head group.

Polar head group   Nonpolar tails

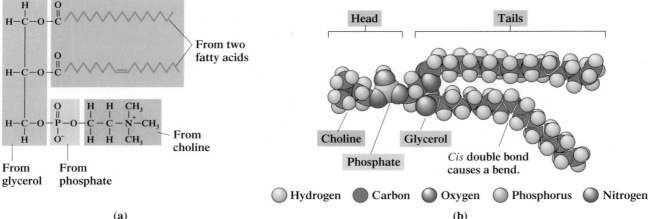

Figure 19.13 (a) Structural formula and (b) molecular model showing the "head and two tails" structure of a phosphatidylcholine molecule containing stearic acid (18:0) and oleic acid (18:1).

The amino alcohol in phosphatidyl-cholines (pronounced fahs-fuh-TIDE-ul-KOH-leens) is choline.

The polar head group of a glycerophospholipid is soluble in water. The nonpolar tail chains are insoluble in water but soluble in nonpolar substances. This dual polarity, which we previously encountered when we discussed soaps (see the Chemical Connections feature on page 625), is a structural characteristic of most membrane lipids.

Phosphatidylcholines are also known as _lecithins_. There are a number of different phosphatidylcholines because different fatty acids may be bonded to the glycerol portion of the phosphatidylcholine structure. In general, phosphatidylcholines are waxy solids that form colloidal suspensions in water. Egg yolks and soybeans are good dietary sources of these lipids. Within the body, phosphatidylcholines are prevalent in cell membranes.

Periodically, claims arise that phosphatidylcholine should be taken as a nutritive supplement; some even maintain it will improve memory. There is no evidence that these supplements are useful. The enzyme _lecithinase_ in the intestine hydrolyzes most of the phosphatidylcholine taken orally before it passes into body fluids, so it does not reach body tissues. The phosphatidylcholine present in cell membranes is made by the liver; thus phosphatidylcholines are not essential nutrients.

The food industry uses phosphatidylcholines as emulsifiers to promote the mixing of otherwise immiscible materials. Mayonnaise, ice cream, and custards are some of the products they are found in. It is the polar–nonpolar (head–tail) structure of phosphatidylcholines that enables them to function as emulsifiers.

Phosphatidylethanolamines and phosphatidylserines are also known as _cephalins._ These compounds are found in heart and liver tissue and in high concentrations in the brain. They are important in blood clotting. Much is yet to be learned about how these compounds function within the human body.

Sphingophospholipids

Sphingophospholipids have structures based on the 18-carbon monounsaturated aminodi-alcohol _sphingosine_. A **sphingophospholipid** _is a lipid that contains one fatty acid and one phosphate group attached to a sphingosine molecule and an alcohol attached to the phosphate group._

The structure of sphingosine, the platform molecule for a sphingophospholipid, is

$$CH_3{-}(CH_2)_{12}{-}CH{=}CH{-}CH{-}CH{-}CH_2$$

OH   NH$_2$ OH

Sphingosine

When sphingolipids were discovered over a century ago by the physician–chemist Johann Thudichum (1829–1901), their biochemical role seemed as enigmatic as the Sphinx, for which he named them.

All phospholipids derived from sphingosine have (1) the fatty acid attached to the sphingosine —NH$_2$ group via an _amide linkage,_ (2) the phosphate group attached to the sphingosine

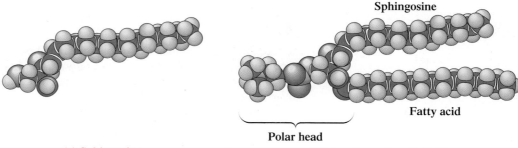

(a) Sphingosine   **(b) A sphingophospholipid**

Figure 19.14 Molecular models for (a) sphingosine and (b) a sphingophospholipid. The particular sphingophospholipid shown has choline as the alcohol esterified to the phosphate group. Note the "head and two tails" structure for the sphingophospholipid.

terminal —OH group via an *ester linkage,* and (3) an additional alcohol esterified to the phosphate group. The general block diagram for a sphingophospholipid is

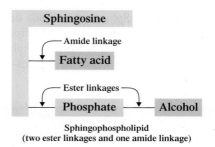

Sphingophospholipid
(two ester linkages and one amide linkage)

> In sphingophospholipids, the first three carbon atoms at the polar end of sphingosine are analogous to the three carbon atoms of glycerol in glycerophospholipids.

Molecular models showing orientation of atoms in space for sphingosine itself and for a sphingophospholipid are given in Figure 19.14. Note that, as in glycerophospholipids, the "head and two tails" structure is present in sphingophospholipids. For sphingophospholipids, the fatty acid is one of the tails, and the long carbon chain of sphingosine itself is the other tail. The polar head is the phosphate group with its esterified alcohol.

Like glycerophospholipids, sphingophospholipids participate in saponification reactions. Amide linkages behave much as ester linkages do in this type of reaction.

Sphingophospholipids in which the alcohol esterified to the phosphate group is *choline* are called *sphingomyelins*. Sphingomyelins are found in all cell membranes and are important structural components of the myelin sheath, the protective and insulating coating that surrounds nerves. The molecule depicted in Figure 19.14b is a sphingomyelin. The structural formula for a sphingomyelin in which stearic acid (18:0) is the fatty acid is

Sphingosine

$$HO-CH-CH=CH-(CH_2)_{12}-CH_3$$

$$CH-NH-\overset{\overset{\displaystyle O}{\|}}{C}-(CH_2)_{16}-CH_3$$

Stearic acid (18:0)

$$CH_2-O-\overset{\overset{\displaystyle O}{\|}}{\underset{\underset{\displaystyle O^-}{|}}{P}}-O-CH_2-CH_2-\overset{\overset{\displaystyle CH_3}{|+}}{\underset{\underset{\displaystyle CH_3}{|}}{N}}-CH_3$$

Phosphate   Choline

19.8 MEMBRANE LIPIDS: SPHINGOGLYCOLIPIDS

The second of the three major types of membrane lipids is *sphingoglycolipids*. A **sphingoglycolipid** *is a lipid that contains both a fatty acid and a carbohydrate component attached to a sphingosine molecule.* A fatty acid is attached to the sphingosine through an amide linkage, and a monosaccharide or oligosaccharide (Section 18.3) is attached to the sphingosine at the terminal —OH carbon atom through a glycosidic linkage (Section 18.13). The generalized block diagram for a sphingoglycolipid is

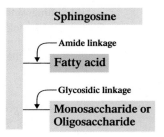

Sphingoglycolipids undergo saponification reactions; both the amide and the glycosidic linkages can be hydrolyzed.

The simplest sphingoglycolipids, which are called *cerebrosides,* contain a single monosaccharide unit—either glucose or galactose. As the name suggests, cerebrosides

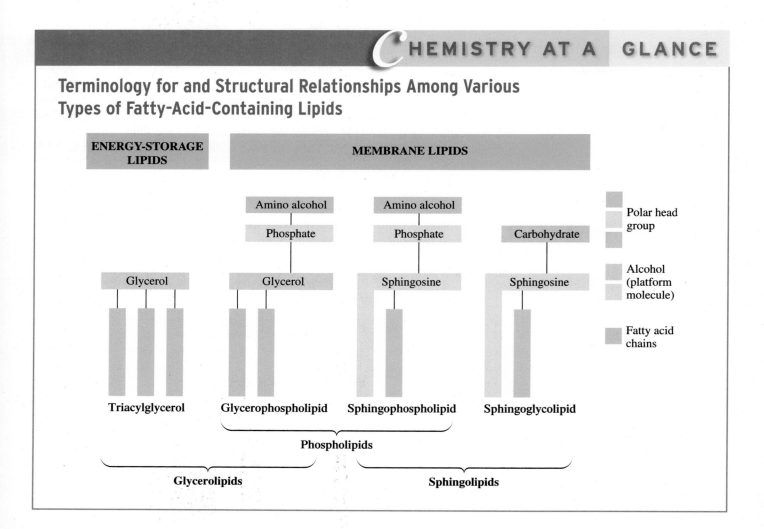

CHEMISTRY AT A GLANCE

Terminology for and Structural Relationships Among Various Types of Fatty-Acid-Containing Lipids

occur primarily in the brain (7% of dry mass). They are also present in the myelin sheath of nerves. The specific structure for a cerebroside in which stearic acid (18:0) is the fatty acid and galactose is the monosaccharide is

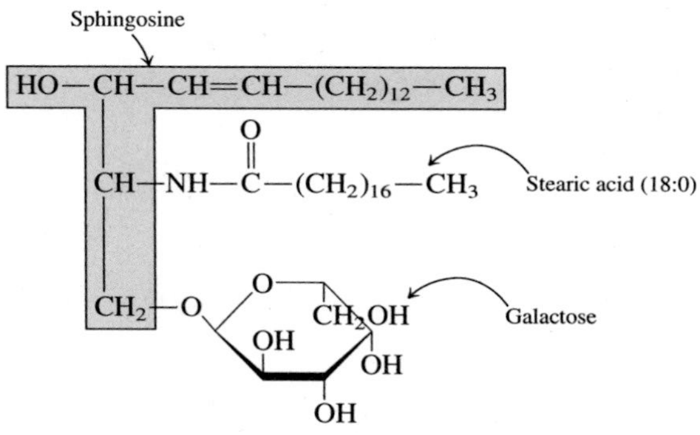

More complex sphingoglycolipids, called *gangliosides,* contain a branched chain of up to seven monosaccharide residues. These substances occur in the gray matter of the brain as well as in the myelin sheath.

The Chemistry at a Glance feature on page 634 summarizes terminology and structural relationships among the types of lipids that we have considered up to this point. The common thread among all of the structures is the presence of at least one fatty acid residue.

19.9 MEMBRANE LIPIDS: CHOLESTEROL

Cholesterol, the third of the three major types of membrane lipids, is a specific compound rather than a family of compounds like the phospholipids (Section 19.7) and sphingoglycolipids (Section 19.8). Cholesterol's structure differs markedly from that of other membrane lipids in that (1) there are no fatty acid residues present and (2) neither glycerol nor sphingosine is present as the platform molecule.

Cholesterol is a *steroid.* A **steroid** *is a lipid whose structure is based on a fused-ring system that involves three 6-membered rings and one 5-membered ring.* This steroid fused-ring system, which is called the *steroid nucleus,* has the following structure:

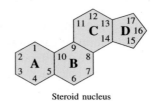

Steroid nucleus

Note that each of the rings of the steroid nucleus carries a letter designation and that a "consecutive" numbering system is used to denote individual carbon atoms.

Numerous steroids have been isolated from plants, animals, and human beings. Location of double bonds within the fused-ring system and the nature and location of substituents distinguish one steroid from another. Most steroids have an oxygen functional group (=O or —OH) at carbon 3 and some kind of side chain at carbon 17. Many also have a double bond from carbon 5 to either carbon 4 or carbon 6.

Cholesterol *is a C_{27} steroid molecule that is a component of cell membranes and a precursor for other steroid-based lipids.* It is the most abundant steroid in the human body. The *-ol* ending in the name cholester*ol* conveys the information that an alcohol functional group is present in this molecule; it is located on carbon 3 of the steroid nucleus. In addition, cholesterol has methyl group attachments at carbons 10 and 13, a carbon–carbon double bond between carbons 5 and 6, and an eight-carbon branched side chain at carbon 17. Figure 19.15 gives both the structural formula and a molecular model

Besides being an important molecule in and of itself, cholesterol serves as a precursor for several other important steroid molecules including bile acids (Section 19.11), steroid hormones (Section 19.12), and vitamin D (Section 21.13).

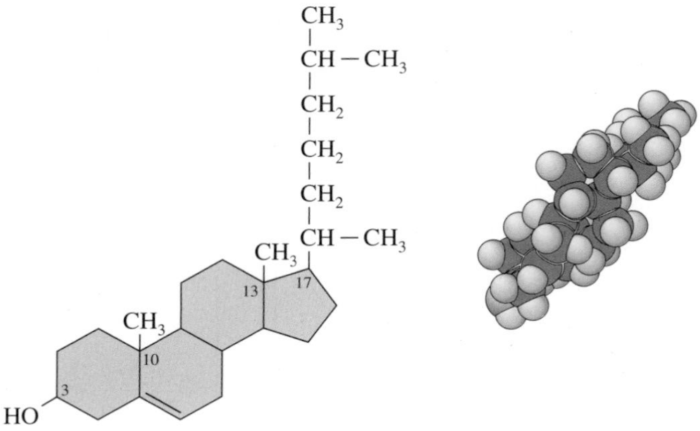

Figure 19.15 Structural formula and molecular model for the cholesterol molecule.

for cholesterol. The molecular model shows the rather compact nature of the cholesterol molecule. The "head and two tails" arrangement found in other membrane lipids is not present. The lack of a *large* polar head group causes cholesterol to have limited water solubility. The —OH group on carbon 3 is considered the head of the molecule.

Within the human body, cholesterol is found in cell membranes (up to 25% by mass), in nerve tissue, in brain tissue (about 10% by dry mass), and in virtually all fluids. Every 100 mL of human blood plasma contains about 50 mg of free cholesterol and about 170 mg of cholesterol esterified with various fatty acids.

Although a portion of the body's cholesterol is obtained from dietary intake, most of it is biosynthesized by the liver and (to a lesser extent) the intestine. Typically, 800–1000 mg are biosynthesized each day. Ingested cholesterol decreases biosynthetic cholesterol production. However, the reduction is less than the amount ingested. Therefore, total body cholesterol levels increase with increased dietary intake of cholesterol.

Biosynthetic cholesterol is distributed to cells throughout the body for various uses via the bloodstream. Because cholesterol is only sparingly soluble in water (blood), a protein carrier system is used for its distribution. These cholesterol–protein combinations are called *lipoproteins.*

The lipoproteins that carry cholesterol *from the liver* to various tissues are called LDLs (low-density lipoproteins), and those that carry excess cholesterol from tissues *back to the liver* are called HDLs (high-density lipoproteins). If too much cholesterol is being transported by LDLs or too little by HDLs, the imbalance results in an increase in blood cholesterol levels. High blood cholesterol levels contribute to atherosclerosis, a form of cardiovascular disease characterized by the buildup of plaque along the inner walls of arteries. Plaque is a mound of lipid material mixed with smooth muscle cells and calcium. Much of the lipid material in plaque is cholesterol. Plaque deposits in the arteries that serve the heart reduce blood flow to the heart muscle and can lead to a heart attack. Figure 19.16 shows the occlusion that can occur in an artery as a result of plaque buildup.

The cholesterol associated with LDLs is often called "bad cholesterol" because it contributes to increased blood cholesterol levels, and the cholesterol associated with HDLs is often called "good cholesterol" because it contributes to reduced blood cholesterol levels. The Chemical Connections feature titled "Lipoproteins and Heart Disease Risk" on page 690 in the next chapter considers this topic in further detail.

Much still needs to be learned concerning the actual role played by serum cholesterol in plaque buildup within arteries. Current knowledge suggests that it makes good sense to reduce the amount of cholesterol (as well as saturated fats) taken into the body through dietary intake. People who want to reduce dietary cholesterol intake should reduce the amount of animal products they eat (meat, dairy products, etc.) and eat more fruit and vegetables. Plant foods contain negligible amounts of cholesterol; cholesterol is found primarily in foods of animal origin. Table 19.4 gives cholesterol amounts associated with selected foods.

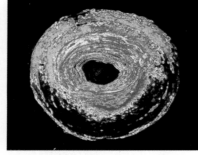

Figure 19.16 A severely occluded artery—the result of the buildup of cholesterol-containing plaque deposits.

TABLE 19.4
The Amount of Cholesterol Found in Various Foods

Food	Cholesterol (mg)
liver (3 oz)	410
egg (1 large)	213
shrimp (3 oz)	166
pork chop (3 oz)	83
chicken (3 oz)	75
beef steak (3 oz)	70
fish fillet (3 oz)	54
whole milk (1 cup)	33
cheddar cheese (1 oz)	30
Swiss cheese (1 oz)	26
low-fat milk (1 cup)	22

19.10 CELL MEMBRANES

Cell membranes are also commonly called *plasma membranes* because they separate the cytoplasm (aqueous contents) of a cell from its surroundings.

The percentage of lipid and protein components in a cell membrane is related to the function of the cell. The lipid/protein ratio ranges from about 80% lipid/20% protein by mass in the myelin sheath of nerve cells to the unique 20% lipid/80% protein ratio for the inner mitochondrial membrane (Section 23.2). Red blood cell membranes contain approximately equal amounts of lipid and protein. A typical membrane also has a carbohydrate content that varies between 2% and 10% by mass.

Prior to discussing additional types of lipid molecules—emulsification lipids (Section 19.11), messenger lipids (Sections 19.12 and 19.13), and protective-coating lipids (Section 19.14)—we will extend our discussion of membrane lipids to include how these types of lipids interact with each other to form cell membranes.

Living cells contain an estimated 10,000 different kinds of molecules in an aqueous environment confined by a *cell membrane*. A **cell membrane** *is a lipid-based structure that separates a cell's aqueous-based interior from the aqueous environment surrounding the cell.* Besides its "separation" function, a cell membrane also controls the movement of substances into and out of the cell. Up to 80% of the mass of a cell membrane is lipid material consisting primarily of the three types of membrane lipids we have just discussed: phospholipids, glycolipids, and cholesterol.

The keys to understanding the structural basis for a cell membrane are (1) the virtually insoluble nature of membrane lipids in water and (2) the "head and two tails" structure (Section 19.7) of phospholipids and sphingoglycolipids. When these lipids are placed in water, the polar heads of phospholipids and sphingoglycolipids favor contact with water, whereas their nonpolar tails interact with one another rather than with water. The result is a remarkable bit of molecular architecture called a *lipid bilayer*. A **lipid bilayer** *is a two-layer-thick structure of phospholipids and glycolipids in which the nonpolar tails of the lipids are in the middle of the structure and the polar heads are on the outside surfaces of the structure.* Such a bilayer is six-billionths to nine-billionths of a meter thick—that is, 6 to 9 nanometers thick. There are three distinct parts to the bilayer: the exterior polar "heads," the interior polar "heads," and the central nonpolar "tails," as shown in Figure 19.17.

Figure 19.17 Cross section of a lipid bilayer. The circles represent the polar heads of the lipid components, and the wavy lines represent the nonpolar tails of the lipid components. The heads occupy "surface" positions, and the tails occupy "internal" positions.

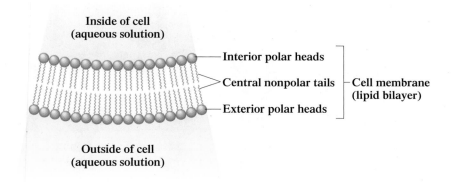

Figure 19.18 Space-filling model of a section of a lipid bilayer. The key to the structure is the "head and two tails" structure of the membrane lipids that constitute the bilayer.

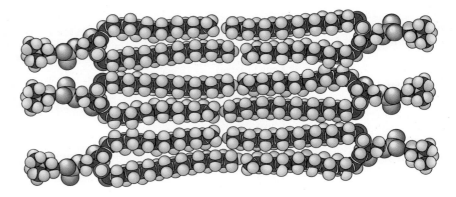

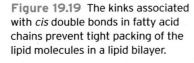

The glycerophospholipids and sphingoglycolipids found in a lipid bilayer are chiral molecules with the chiral center (Section 18.4) at carbon 2 of the glycerol or sphingosine components of the molecules. The stereoconfiguration of these chiral molecules is always "left-handed," that is, L isomers. The fact that they all have the same configuration enhances their ability to aggregate together in the lipid bilayer.

Figure 19.18, which is based on space-filling models for phospholipids, gives a "close-up" view of the arrangement of lipid molecules in a section of a lipid bilayer. Note the "exterior" nature of the polar heads of these membrane lipids.

A lipid bilayer is held together by intermolecular interactions, not by covalent bonds. This means each phospholipid or sphingolipid is free to diffuse laterally within the lipid bilayer. Most lipid molecules in the bilayer contain at least one unsaturated fatty acid. The presence of such acids, with the kinks in their carbon chains (Section 19.3), prevents tight packing of fatty acid chains (Figure 19.19). The open packing imparts a liquidlike character to the membrane—a necessity because numerous types of biochemicals must pass into and out of a cell.

Cholesterol molecules are also components of cell membranes. They regulate membrane fluidity. Because of their compact shape (Section 19.9; Figure 19.15), cholesterol molecules fit between the fatty acid chains of the lipid bilayer (Figure 19.20), restricting movement of the fatty acid chains. Within the membrane, the cholesterol molecule orientation is "head" to the outside (the hydroxyl group) and "tail" to the inside (the steroid ring structure with its attached alkyl groups).

Proteins are also components of lipid bilayers. The proteins are responsible for moving substances such as nutrients and electrolytes across the membrane, and they also act as receptors that bind hormones and neurotransmitters.

There are two general types of membrane proteins: *integral* and *peripheral*. An **integral membrane protein** *is a membrane protein that penetrates the cell membrane.* Some membrane proteins penetrate only partially through the lipid bilayer while others go completely from one side to the other side of the lipid bilayer. A **peripheral membrane protein** *is a nonpenetrating membrane protein located on the surface of the cell membrane.*

Figure 19.19 The kinks associated with *cis* double bonds in fatty acid chains prevent tight packing of the lipid molecules in a lipid bilayer.

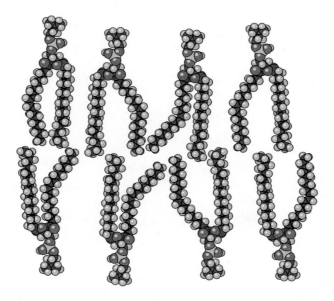

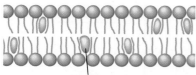

Cholesterol

Figure 19.20 Cholesterol molecules fit between fatty acid chains in a lipid bilayer.

Intermolecular forces rather than chemical bonds govern the interactions between membrane proteins and the lipid bilayer. Figure 19.21 shows diagrammatically the relationship between membrane proteins and the overall structure of a cell membrane.

Small carbohydrate molecules are also components of cell membranes. They are found on the *outer* membrane surface covalently bonded to protein molecules (a glycoprotein) or lipid molecules (a glycolipid). The carbohydrate portions of glycoproteins and glycolipids function as *markers,* substances that play key roles in the process by which different cells recognize each other.

Transport Across Cell Membranes

In order for cellular processes to be maintained, molecules of various types must be able to cross cell membranes. Three common transport mechanisms exist by which molecules can enter and leave cells. They are *passive* transport, *facilitated* transport, and *active* transport.

Passive transport *is the transport process in which a substance moves across a cell membrane by diffusion from a region of higher concentration to a region of lower concentration without the expenditure of any cellular energy.* Only a few types of molecules, including O_2, N_2, H_2O, urea, and ethanol, can cross membranes in this manner. Passive transport is closely related to the process of osmosis (Section 8.9).

Facilitated transport *is the transport process in which a substance moves across a cell membrane, with the aid of membrane proteins, from a region of higher concentration to a region of lower concentration without the expenditure of cellular energy.* The specific protein molecules involved in the process are called *carriers* or *transporters.* A carrier protein forms a complex with a specific molecule at one surface of the membrane. Formation of the complex induces a conformational change in the protein that allows the molecule to move through a "gate" to the other side of the membrane. Once the molecule is released, the protein returns to its original conformation. Glucose, chloride ion, and bicarbonate ion cross membranes in this manner.

Active transport *is the transport process in which a substance moves across a cell membrane, with the aid of membrane proteins, against a concentration gradient with the expenditure of cellular energy.* Proteins involved in active transport are called "pumps," because they require energy much as a water pump requires energy in order to function. The needed energy is supplied by molecules such as ATP (Section 23.3). The need for energy expenditure is related to the molecules moving against a concentration gradient— from lower to higher concentration. It is essential to life processes to have some solutes "permanently" at different concentrations on the two sides of a membrane, a situation contrary to the natural tendency (osmosis) to establish equal concentrations on both sides

Figure 19.21 Proteins are important structural components of cell membranes.

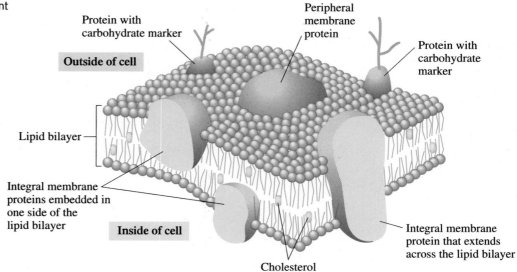

Protein with carbohydrate marker

Peripheral membrane protein

Protein with carbohydrate marker

Outside of cell

Lipid bilayer

Integral membrane proteins embedded in one side of the lipid bilayer

Inside of cell

Cholesterol

Integral membrane protein that extends across the lipid bilayer

Figure 19.22 Three processes by which substances can cross plasma membranes: (a) passive transport, (b) facilitated transport, and (c) active transport.

Gate	Pump

Concentration gradient: movement with the gradient; from high to low concentration

Cellular energy expenditure: none required

Protein help: none required

(a) Passive Transport

Concentration gradient: movement with the gradient; from high to low concentration

Cellular energy expenditure: none required

Protein help: proteins serve as "gates"

(b) Facilitated Transport

Concentration gradient: movement against the gradient; from low to high concentration

Cellular energy expenditure: energy input required

Protein help: proteins serve as "pumps"

(c) Active Transport

of a membrane. Hence the need for active transport. Sodium, potassium, and hydronium ions cross membranes through active transport.

Figure 19.22 contrasts the processes of passive transport, facilitated transport, and active transport.

19.11 EMULSIFICATION LIPIDS: BILE ACIDS

An **emulsifier** *is a substance that can disperse and stabilize water-insoluble substances as colloidal particles in an aqueous solution.* Cholesterol derivatives called *bile acids* function as emulsifying agents that facilitate the absorption of dietary lipids in the intestine. Their mode of action is much like that of soap during washing (see the Chemical Connections feature on page 625).

A **bile acid** *is a cholesterol derivative that functions as a lipid-emulsifying agent in the aqueous environment of the digestive tract.* Approximately one-third of the daily production of cholesterol by the liver is converted to bile acids. Obtained by oxidation of cholesterol, bile acids differ structurally from cholesterol in three respects:

1. They are tri- or dihydroxy cholesterol derivatives.
2. The carbon 17 side chain of cholesterol has been oxidized to a carboxylic acid.
3. The oxidized acid side chain is bonded to an amino acid (either glycine or taurine) through an amide linkage.

Figure 19.23 gives structural formulas for the three major types of bile acids produced from cholesterol by biochemical oxidation: cholic acid, 7-deoxycholic acid, and 12-deoxycholic

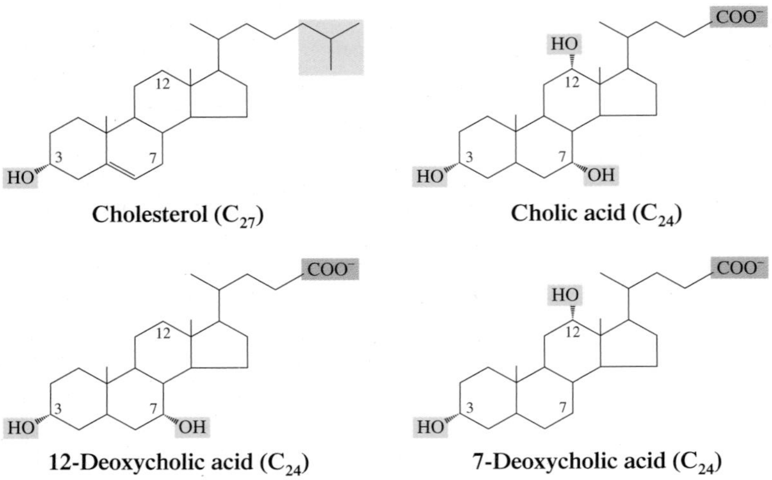

Figure 19.23 Structural formulas for cholesterol, cholic acid, and two deoxycholic acids.

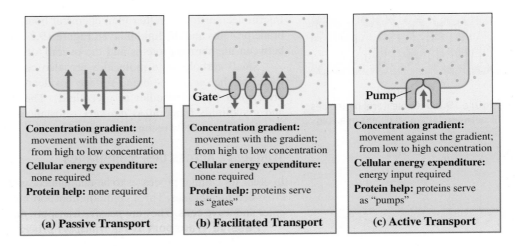

Cholesterol (C_{27})

Cholic acid (C_{24})

12-Deoxycholic acid (C_{24})

7-Deoxycholic acid (C_{24})

Figure 19.24 The structures of glycocholic acid and taurocholic acid.

Cholic acid

Taurine

Glycine

Taurocholic acid

Glycocholic acid

Figure 19.25 A large percentage of gallstones, the causative agent for many "gallbladder attacks," are almost pure crystallized cholesterol that has precipitated from bile solution.

The average bile acid composition in normal human adult bile is 38% cholic acid derivatives, 34% 7-deoxycholic acid derivatives, and 28% 12-deoxycholic acid derivatives. Glycine-containing derivatives predominate over taurine-containing derivatives by a 3:1 to 4:1 ratio. Uncomplexed (free) bile acids are not present in bile.

acid. The structural formulas are those for these bile acids prior to the attachment of the amino acid to the carbon 17 side chain.

Bile acids always carry an amino acid (either glycine or taurine) attached to the side-chain carboxyl group via an amide linkage. The presence of this amino acid attachment increases both the polarity of the bile acid and its water solubility. Figure 19.24 shows the structures of glycocholic acid (glycine is the amino acid) and taurocholic acid (taurine is the amino acid).

The medium through which bile acids are supplied to the small intestine is *bile*. **Bile** *is a fluid containing emulsifying agents that is secreted by the liver, stored in the gallbladder, and released into the small intestine during digestion.* Besides bile acids, bile also contains bile pigments (breakdown products of hemoglobin; Section 26.7), cholesterol itself, and electrolytes such as bicarbonate ion. The bile acids that are present increase the solubility of the cholesterol in the bile fluid.

A number of factors, including increased secretion of cholesterol and a decrease in the size of the bile pool, can upset the balance between the cholesterol present in bile and the bile acid derivatives needed to maintain cholesterol's solubility in the bile. The result is the precipitation of crystallized cholesterol from the bile and the resulting formation of gallstones in the gallbladder. In Western countries, approximately 80% of gallstones are almost pure cholesterol (see Figure 19.25).

19.12 MESSENGER LIPIDS: STEROID HORMONES

We have previously considered lipids that function as energy-storage molecules (triacylglycerols; Section 19.4), as components of cell membranes (phospholipids, sphingoglycolipids, and cholesterol; Sections 19.7 through 19.9), and as emulsifying agents (bile acids; Section 19.11). An additional role played by lipids is that of "chemical messenger." *Steroid hormones* and *eicosanoids* are two large families of lipids that have messenger functions. In this section we consider steroid hormones, which are cholesterol derivatives. In Section 19.13 we consider eicosanoids, which are fatty acid derivatives.

A **hormone** *is a biochemical substance, produced by a ductless gland, that has a messenger function.* Hormones serve as a means of communication *between* various tissues. Some hormones, though not all, are lipids.

A **steroid hormone** *is a hormone that is a cholesterol derivative.* There are two major classes of steroid hormones: (1) sex hormones, which control reproduction and secondary sex characteristics, and (2) adrenocorticoid hormones, which regulate numerous biochemical processes in the body.

Sex Hormones

The sex hormones can be classified into three major groups:

1. Estrogens—the female sex hormones
2. Androgens—the male sex hormones
3. Progestins—the pregnancy hormones

Estrogens are a class of molecules rather than a single molecule. Statements like "the estrogen level is high" should be rephrased as "there is a high level of estrogens."

Estrogens are synthesized in the ovaries and adrenal cortex and are responsible for the development of female secondary sex characteristics at the onset of puberty and for regulation of the menstrual cycle. They also stimulate the development of the mammary glands during pregnancy and induce estrus (heat) in animals.

Androgens are synthesized in the testes and adrenal cortex and promote the development of secondary male characteristics. They also promote muscle growth.

Progestins are synthesized in the ovaries and the placenta and prepare the lining of the uterus for implantation of the fertilized ovum. They also suppress ovulation.

Figure 19.26a gives the structure of the primary hormone in each of the three subclasses of sex hormones. Other members of these hormone families are metabolized forms of the primary hormone.

Note, in Figure 19.26a how similar the structures are for these principal hormones, and yet how different their functions. The fact that seemingly minor changes in structure effect great changes in biofunction points out, again, the extreme specificity (Section 21.5) of the enzymes that control biochemical reactions.

Increased knowledge of the structures and functions of sex hormones has led to the development of a number of *synthetic* steroids whose actions often mimic those of the natural steroid hormones. The best known types of synthetic steroids are oral contraceptives and anabolic agents.

Figure 19.26 Structures of selected sex hormones and synthetic compounds that have similar actions.

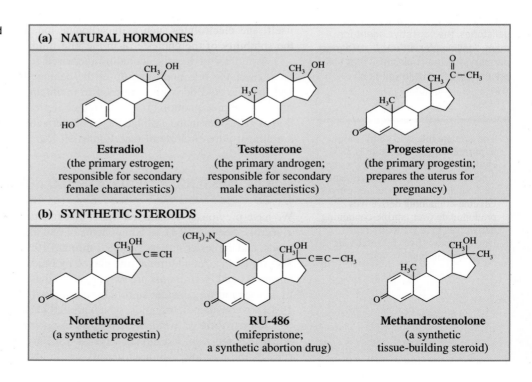

(a) NATURAL HORMONES

Estradiol
(the primary estrogen; responsible for secondary female characteristics)

Testosterone
(the primary androgen; responsible for secondary male characteristics)

Progesterone
(the primary progestin; prepares the uterus for pregnancy)

(b) SYNTHETIC STEROIDS

Norethynodrel
(a synthetic progestin)

RU-486
(mifepristone; a synthetic abortion drug)

Methandrostenolone
(a synthetic tissue-building steroid)

Oral contraceptives are used to suppress ovulation as a method of birth control. Generally, a mixture of a synthetic estrogen and a synthetic progestin is used. The synthetic estrogen regulates the menstrual cycle, and the synthetic progestin prevents ovulation, thus creating a false state of pregnancy. The structure of norethynodrel (Enovid), a synthetic progestin, is given in Figure 19.26b. Compare its structure to that of progesterone (the real hormone); the structures are very similar.

Interestingly, the controversial "morning after" pill developed in France and known as RU-486, is also similar in structure to progesterone. RU-486 interferes with gestation of a fertilized egg and terminates a pregnancy within the first 9 weeks of gestation more effectively and safely than surgical methods. The structure of RU-486 appears next to that of norethynodrel in Figure 19.26b.

Anabolic agents include the illegal steroid drugs used by some athletes to build up muscle strength and enhance endurance. Anabolic agents are now known to have serious side effects on the user. The Chemical Connections feature on page 644 focuses on the use of anabolic steroids. The structure of one of the more commonly used anabolic agents, methandrostenolone, is given in Figure 19.26b. Note the similarities between its structure and that of the naturally occurring testosterone.

Adrenocorticoid Hormones

The second major group of steroid hormones consists of the adrenocorticoid hormones. Produced by the adrenal glands, small organs located on top of each kidney, at least 28 different hormones have been isolated from the adrenal cortex (the outer part of the glands).

There are two types of adrenocorticoid hormones.

1. *Mineralocorticoids* control the balance of Na^+ and K^+ ions in cells and body fluids.
2. *Glucocorticoids* control glucose metabolism and counteract inflammation.

The major mineralocorticoid is aldosterone, and the major glucocorticoid is cortisol (hydrocortisone). Cortisol is the hormone synthesized in the largest amount by the adrenal glands. Cortisol and its synthetic ketone derivative cortisone exert powerful anti-inflammatory effects in the body. Both cortisone and prednisolone, a similar synthetic derivative, are used as prescription drugs to control inflammatory diseases such as rheumatoid arthritis. Figure 19.27 gives the structures of these adrenocorticoid hormones.

The C≡C functional group, which occurs in both norethynodrel (Enovid) and RU-486, is rarely found in biomolecules.

Figure 19.27 Structures of selected adrenocorticoid hormones and related synthetic compounds.

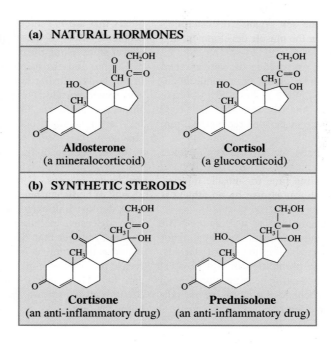

(a) NATURAL HORMONES

Aldosterone
(a mineralocorticoid)

Cortisol
(a glucocorticoid)

(b) SYNTHETIC STEROIDS

Cortisone
(an anti-inflammatory drug)

Prednisolone
(an anti-inflammatory drug)

19.13 MESSENGER LIPIDS: EICOSANOIDS

An **eicosanoid** *is an oxygenated C$_{20}$ fatty acid derivative that functions as a messenger lipid.* The term *eicosanoid* is derived from the Greek word *eikos,* which means "twenty." The metabolic precursor for most eicosanoids is arachidonic acid, the 20:4 fatty acid.

Almost all cells, except red blood cells, produce eicosanoids. These substances, like hormones, have profound physiological effects at extremely low concentrations. Eicosanoids are hormonelike molecules rather than true hormones because they are not transported in the bloodstream to their site of action as true hormones are. Instead, they exert their effects in the tissues where they are synthesized. Eicosanoids usually have a very short "life," being broken down, often within seconds of their synthesis, to inactive residues (which are eliminated in urine). For this reason, they are difficult to study and monitor within cells.

Eicosanoids exert their effects at very low concentrations, sometimes less than one part in a billion (10^9).

The physiological effects of eicosanoids include mediation of

1. The inflammatory response, a normal response to tissue damage
2. The production of pain and fever
3. The regulation of blood pressure
4. The induction of blood clotting
5. The control of reproductive functions, such as induction of labor
6. The regulation of the sleep/wake cycle

There are three principal types of eicosanoids: prostaglandins, thromboxanes, and leukotrienes.

CHEMICAL Connections — Steroid Drugs in Sports

The steroid hormone testosterone is the principal male sex hormone. It has masculinizing (androgenic) effects and muscle-building (anabolic) effects. Masculinizing effects of testosterone include the growth of facial and body hair, deepening of the voice, and maturation of the male sex organs. Testosterone's anabolic effects are responsible for the muscle development that boys experience at puberty.

Some of the many synthetic testosterone derivatives exert primarily androgenic effects, whereas others exert primarily anabolic effects. Androgenic compounds can be used to correct hormonal imbalances in the body. Anabolic steroids can be used to prevent the withering of muscle in persons recovering from major surgery or serious injuries.

Anabolic steroids have also been "discovered" by athletes, who have found that these compounds can be used to help build muscle mass and reduce the healing time for muscle injuries. Often the net result of anabolic hormone use by an athlete is enhanced athletic performance.

The International Olympic Committee, as well as the NBA and NFL, prohibits steroid use and tests for it. Recently, major league baseball adopted a policy regarding the use of steroids. There are two reasons for steroid prohibition in athletics: (1) Their use is considered a form of cheating because it confers an unfair advantage, and (2) their "beneficial effects" are far outweighed by serious negative side effects.

Current medical evidence indicates that using anabolic steroids is dangerous. Steroid abuse is associated with a wide range of adverse side effects ranging from some that are physically unattractive, such as acne and breast development in men, to others that are life-threatening, such as heart attacks and liver problems. Most of these alarming effects are reversible if the abuser stops taking the drugs, but some are permanent.

Steroid abuse disrupts the normal production of hormones in the body, causing both reversible and irreversible change. The male reproductive system is altered, causing testicular shrinkage and decreased sperm production. Both of these effects are reversible, but breast development is an irreversible change.

In the female body, steroid abuse causes masculinization. Breast size and body fat decrease, the skin becomes coarse, and the voice deepens. Some women may experience excessive growth of body hair but lose hair from the scalp.

A definite link exists between steroid abuse and cardiovascular diseases, including heart attacks and strokes, even in persons younger than 30. Steroids, particularly the oral types, increase the level of low-density lipoprotein (LDL) and decrease the level of high-density lipoprotein (HDL). High LDL and low HDL levels increase the risk of atherosclerosis. Steroids also increase the risk that blood clots will form in blood vessels.

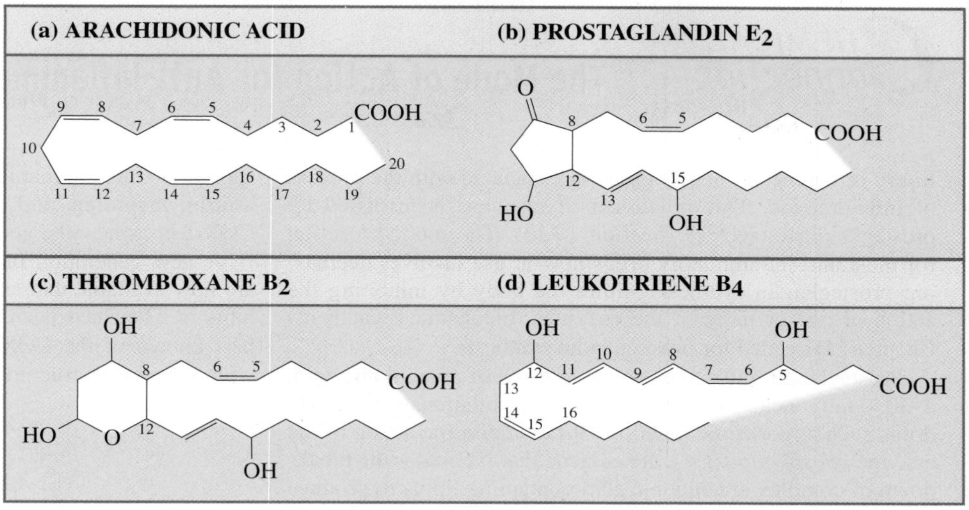

Figure 19.28 Relationship of the structures of various eicosanoids to their precursor, arachidonic acid.

(a) ARACHIDONIC ACID

(b) PROSTAGLANDIN E$_2$

(c) THROMBOXANE B$_2$

(d) LEUKOTRIENE B$_4$

The capital letter–numerical subscript designations for individual eicosanoids is based on selected structural characteristics of the molecules. The numerical subscript indicates the number of carbon–carbon double bonds present. The letters denote subgroups of molecules. The prostaglandin E group, for example, has a carbonyl group on carbon 9.

Naturally occurring fatty acids are normally found in the form of fatty acid residues in such molecules as triacylglycerols, phospholipids, and sphingolipids, substances previously considered in this chapter. In eicosanoids the fatty acids are not esterified and carry out their biochemical roles in the form of the free acids, that is, free fatty acid derivatives.

Prostaglandins

A **prostaglandin** *is a messenger lipid that is a C$_{20}$-fatty-acid derivative that contains a cyclopentane ring and oxygen-containing functional groups.* Twenty-carbon fatty acids are converted into a prostaglandin structure when the eighth and twelfth carbon atoms of the fatty acid become connected to form a five-membered ring (Figure 19.28b).

Prostaglandins are named after the prostate gland, which was first thought to be their only source. Today, more than 20 prostaglandins have been discovered in a variety of tissues in both males and females.

Within the human body, prostaglandins are involved in many regulatory functions, including raising body temperature, inhibiting the secretion of gastric juices, increasing the secretion of a protective mucus layer into the stomach, relaxing and contracting smooth muscle, directing water and electrolyte balance, intensifying pain, and enhancing inflammation responses. Aspirin reduces inflammation and fever because it inactivates the enzyme needed for prostaglandin synthesis.

Thromboxanes

A **thromboxane** *is a messenger lipid that is a C$_{20}$-fatty-acid derivative that contains a cyclic ether ring and oxygen-containing functional groups.* As with prostaglandins, the cyclic structure involves a bond between carbons 8 and 12 (Figure 19.28c). An important function of thromboxanes is to promote the formation of blood clots. Thromboxanes are produced by blood platelets and promote platelet aggregation.

Leukotrienes

A **leukotriene** *is a messenger lipid that is a C$_{20}$-fatty-acid derivative that contains three conjugated double bonds and hydroxy groups.* Fatty acids and their derivatives do not normally contain *conjugated* double bonds (see the Chemical Connections feature "Carotenoids: A Source of Color" on page 374 in Chapter 13), as is the case in leukotrienes (Figure 19.28d). Leukotrienes are found in leukocytes (white blood cells). Their source and the presence of the three conjugated double bonds account for their name. Various inflammatory and hypersensitivity (allergy) responses are associated with elevated levels of leukotrienes. The development of drugs that inhibit leukotriene synthesis has been an active area of research.

CHEMICAL Connections   The Mode of Action for Anti-Inflammatory Drugs

Injury or damage to bodily tissue is associated with the process of inflammation. This inflammation response is mediated by prostaglandin molecules (Section 19.13). The mode of action for most anti-inflammatory drugs now in use involves decreasing prostaglandin synthesis within the body by inhibiting the action of one or more of the enzymes (biochemical catalysts; Chapter 21) needed for prostaglandin synthesis.

Prostaglandin molecules are derivatives of arachidonic acid, a 20:4 fatty acid (Section 19.13). Anti-inflammatory steroid drugs such as cortisone (Section 19.12) inhibit the action of the enzyme *phospholipase A₂*, the enzyme that facilitates the breakdown of complex arachidonic acid-containing lipids to produce free arachidonic acid. Inhibiting arachidonic acid release stops the prostaglandin synthesis process, which in turn prevents (or diminishes) inflammation.

Besides anti-inflammatory steroid drugs, many nonsteroidal anti-inflammatory drugs (NSAIDs) are also available for inflammation control. The most frequently used NSAIDs are the over-the-counter pain relievers aspirin, ibuprofen (Advil), and naproxen (Aleve). These substances, which have anti-pain, anti-fever, and anti-inflammatory properties, prevent prostaglandin synthesis by inhibiting the enzyme needed for the ring closure reaction at carbons 8 and 12 in arachidonic acid, a necessary step in prostaglandin synthesis (Section 19.13). The enzyme that NSAIDs inhibit is called *cyclooxygenase*, an enzyme known by the acronym *COX*.

There are actually two forms of the COX enzyme: COX-1 and COX-2. The COX-1 enzyme is involved in the normal physiological production of prostaglandin molecules (a desirable situation) and the COX-2 enzyme is responsible for the prostaglandin production associated with the inflammation re-

sponse (a situation that is desirable to control). NSAIDs such as aspirin, ibuprofen, and naproxen inhibit both the COX-1 and COX-2 enzymes (the good and the bad).

A new generation of prescription anti-inflammatory agents are now available that are COX-2 inhibitors but not COX-1 inhibitors. They have been touted by some as "super aspirins." The best known of the COX-2 inhibitors are Vioxx and Celebrex, whose chemical structures are

Vioxx   **Celebrex**

Like almost all anti-inflammatory drugs, these drugs have ulcer-causing side effects.

In 2004, Vioxx was withdrawn from the market because of concerns relative to heart attacks and strokes. A study indicated that patients taking Vioxx were twice as likely to suffer a heart attack or stroke as a control group involved in the study who were taking a placebo. The actual risk was 3.5% in the Vioxx group, compared with 1.9% in the control group, according to the FDA. The difference became apparent after 18 months of Vioxx use. The use of COX-2 inhibitors is now a topic under intense scrutiny.

19.14 PROTECTIVE-COATING LIPIDS: BIOLOGICAL WAXES

A **biological wax** *is a lipid that is a monoester of a long-chain fatty acid and a long-chain alcohol.* Biological waxes are *monoesters,* unlike fats and oils (Section 19.4), which are *triesters.* The fatty acids found in biological waxes generally are saturated and contain from 14 to 36 carbon atoms. The alcohols found in biological waxes may be saturated or unsaturated and may contain from 16 to 30 carbon atoms.

The block diagram for a biological wax is

| Long-chain fatty acid | Long-chain alcohol |

with the fatty acid and alcohol linked through an ester linkage. An actual structural formula for a biological wax that bees secrete and use as a structural material is

The term *wax* derives from the old English word *weax,* which means "the material of the honeycomb."

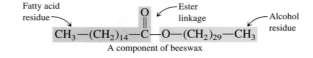

A component of beeswax

Figure 19.29 A biological wax has a structure with a small, weakly polar "head" and two long, nonpolar "tails." The polarity of the small "head" is not sufficient to impart any degree of water solubility to the molecule.

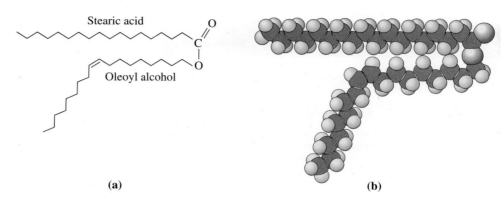

Stearic acid

Oleoyl alcohol

(a)   **(b)**

Figure 19.30 Plant leaves often have a biological wax coating to prevent excessive loss of water.

Note that the general structural formula for a biological wax is the same as that for a simple ester (Section 16.10).

$$R-\overset{\overset{\displaystyle O}{\|}}{C}-O-R'$$

However, for waxes both R and R′ must be long carbon chains (usually 20–30 carbon atoms).

The water-insoluble, water-repellent properties of biological waxes result from the complete dominance of the nonpolar nature of the long hydrocarbon chains present (from the alcohol and the fatty acid) over the weakly polar nature of the ester functional group that links the two carbon chains together (see Figure 19.29).

In living organisms biological waxes have numerous functions, all of which are related directly or indirectly to their water-repellent properties. Both humans and animals possess skin glands that secrete biological waxes to protect hair and skin and to keep it pliable and lubricated. With animal fur, waxes impart water repellency to the fur. Birds, particularly aquatic birds, rely on waxes secreted from preen glands to keep their feathers water repellent. Such wax coatings also help minimize loss of body heat when the bird is in cold water. Many plants, particularly those that grow in arid regions, have leaves that are coated with a thin layer of biological waxes, which serve to prevent excessive evaporation of water and to protect against parasite attack (see Figure 19.30). Similarly, insects with a high surface-area-to-volume ratio are often coated with a protective biological wax.

Biological waxes find use in the pharmaceutical, cosmetics, and "polishing" industries. Carnauba wax (obtained from a species of Brazilian palm tree) is a particularly hard wax whose uses involve high-gloss finishes: automobile wax, boat wax, floor wax, and shoe wax.

$$CH_3-(CH_2)_{28}-\overset{\overset{\displaystyle O}{\|}}{C}-O-(CH_2)_{31}-CH_3$$
A component of carnauba wax

When aquatic birds are caught in an oil spill, the oil dissolves the wax coating on their feathers. This causes the birds to lose their buoyancy (they cannot swim properly) and compromises their protection against the effects of cold water.

Naturally occuring waxes, such as beeswax, are usually mixtures of several monoesters rather than being a single monoester. This parallels the situation for fats and oils, which are mixtures of numerous triesters (triacylglycerols).

Lanolin, a mixture of waxes obtained from sheep wool, is used as a base for skin creams and ointments intended to enhance retention of water (which softens the skin).

Many synthetic materials are now available with properties that closely match—and even improve on—the properties of biological waxes. Such synthetic materials, which are generally polymers, have now replaced biological waxes in many cosmetics, ointments, and the like. The synthetic *carbowax,* for example, is a polyether.

Throughout this discussion, we have used the term *biological wax* rather than just *wax*. This is because the everyday meaning of the term *wax* is broader in scope than the chemical definition of the term *biological wax*. In general discussions, a **wax** *is a pliable, water-repelling substance used particularly in protecting surfaces and producing polished surfaces.* This broadened definition for waxes includes not only *biological waxes*

CHEMISTRY AT A GLANCE

Types of Lipids in Terms of How They Function

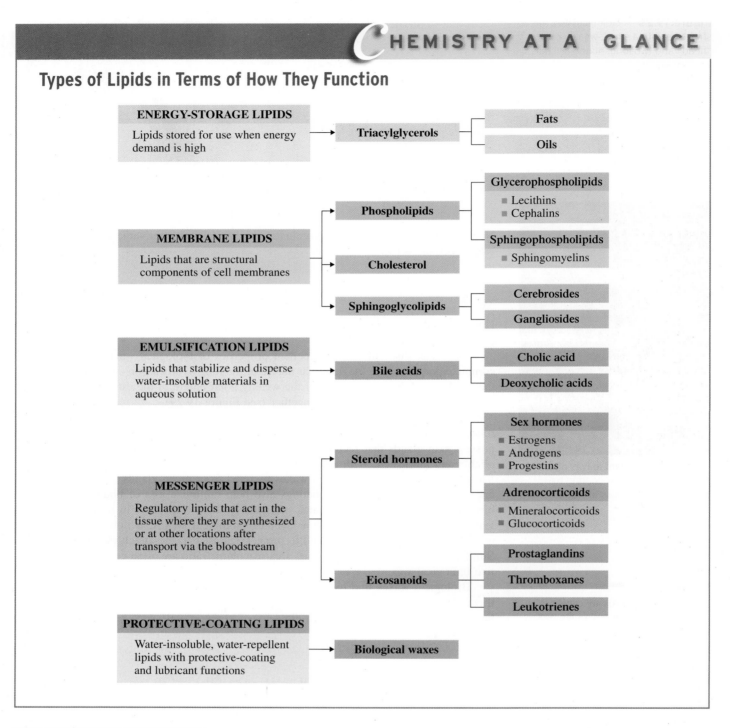

ENERGY-STORAGE LIPIDS

Lipids stored for use when energy demand is high

→ Triacylglycerols

- Fats
- Oils

MEMBRANE LIPIDS

Lipids that are structural components of cell membranes

→ Phospholipids

- Glycerophospholipids
  - Lecithins
  - Cephalins
- Sphingophospholipids
  - Sphingomyelins

→ Cholesterol

→ Sphingoglycolipids

- Cerebrosides
- Gangliosides

EMULSIFICATION LIPIDS

Lipids that stabilize and disperse water-insoluble materials in aqueous solution

→ Bile acids

- Cholic acid
- Deoxycholic acids

MESSENGER LIPIDS

Regulatory lipids that act in the tissue where they are synthesized or at other locations after transport via the bloodstream

→ Steroid hormones

- Sex hormones
  - Estrogens
  - Androgens
  - Progestins
- Adrenocorticoids
  - Mineralocorticoids
  - Glucocorticoids

→ Eicosanoids

- Prostaglandins
- Thromboxanes
- Leukotrienes

PROTECTIVE-COATING LIPIDS

Water-insoluble, water-repellent lipids with protective-coating and lubricant functions

→ Biological waxes

Human ear wax, which acts as a protective barrier against infection by capturing airborne particles, is not a true biological wax—that is, it is not a mixture of simple esters. Human ear wax is a yellow waxy secretion that is a mixture of triacylglycerols, phospholipids, and esters of cholesterol. Its medical name is *cerumen.*

but also *mineral waxes.* A **mineral wax** *is a mixture of long-chain alkanes obtained from the processing of petroleum.* (How mineral waxes are obtained from petroleum was considered in Section 12.15.) Mineral waxes, which are also called *paraffin waxes,* resist moisture and chemicals and have no odor or taste. They serve as a waterproof coating for such paper products as milk cartons and waxed paper. Most candles are made from mineral waxes. Some "wax products" are a blend of biological and mineral waxes. For example, beeswax is sometimes a component of candle wax.

The Chemistry at a Glance feature above summarizes the function-based lipid classifications we have considered in this chapter. Subclassifications within each function classification are also given in this summary.

CONCEPTS TO REMEMBER

Lipids. Lipids are a structurally heterogeneous group of compounds of biochemical origin that are soluble in nonpolar organic solvents and insoluble in water. Lipids are divided into five major types on the basis of biochemical function: energy-storage lipids, membrane lipids, emulsification lipids, messenger lipids, and protective-coating lipids (Section 19.1).

Types of fatty acids. Fatty acids are monocarboxylic acids that contain long, unbranched carbon chains. The carbon chain may be saturated, monounsaturated, or polyunsaturated. Length of carbon chain, degree of unsaturation, and location of the unsaturation influence the properties of fatty acids. Omega-3 and omega-6 fatty acids are unsaturated fatty acids with the endmost double bond three and six carbons, respectively, away from the methyl end of the carbon chain (Section 19.2).

Triacylglycerols. Triacylglycerols are energy-storage lipids formed by esterification of three fatty acids to a glycerol molecule. Fats are triacylglycerol mixtures that are solids or semi-solids at room temperature; they contain a relatively high percentage of saturated fatty acid residues. Oils are triacylglycerol mixtures that are liquids at room temperature; they contain a relatively high percentage of unsaturated fatty acid residues (Section 19.4).

Phospholipids. Phospholipids are membrane lipids that contain one or more fatty acids, a phosphate group, a platform molecule to which the fatty acid(s) and phosphate group are attached, and an alcohol attached to the phosphate group. The platform molecule is either glycerol (glycerophospholipids) or sphingosine (sphingophospholipids). Phospholipids have a "head and two tails" structure. Lecithins, cephalins, and sphingomyelins are types of phospholipids (Section 19.7).

Sphingoglycolipids. Sphingoglycolipids are membrane lipids in which a fatty acid and a mono- or oligosaccharide are attached to the platform molecule sphingosine. Cerebrosides and gangliosides are types of sphingoglycolipids (Section 19.8).

Cholesterol. Cholesterol is a membrane lipid whose structure contains a steroid nucleus. It is the most abundant type of steroid. Besides its membrane functions, it also serves as a precursor for several other types of lipids (Section 19.9).

Lipid bilayer. A lipid bilayer is the fundamental structure associated with a cell membrane. It is a two-layer structure of lipid molecules (mostly phospholipids and glycolipids) in which the nonpolar tails of the lipids are in the interior and the polar heads are on the outside surfaces (Section 19.10).

Membrane transport mechanisms. The transport mechanisms by which molecules enter and leave cells include *passive* transport, *facilitated* transport, and *active* transport. Passive and facilitated transport follow a concentration gradient and do not involve cellular energy expenditure. Active transport involves movement against a concentration gradient and requires the expenditure of cellular energy (Section 19.10).

Bile acids. Bile acids are cholesterol derivatives that function as emulsification lipids. They cause dietary lipids to be soluble in the aqueous environment of the digestive tract. Cholic acid and deoxycholic acids are the major types of bile acids (Section 19.11).

Steroid hormones. Steroid hormones are cholesterol derivatives that function as messenger lipids. The two major types of steroid hormones are sex hormones and adrenocorticoid hormones (Section 19.12).

Eicosanoids. Eicosanoids are fatty acid derivatives that function as messenger lipids. The major classes of eicosanoids are prostaglandins, thromboxanes, and leukotrienes (Section 19.13).

Biological waxes. Biological waxes are protective-coating lipids formed through the esterification of a long-chain fatty acid to a long-chain alcohol (Section 19.14).

KEY REACTIONS AND EQUATIONS

1. Formation of a triacylglycerol (Section 19.6)

Glycerol + 3 fatty acids $\xrightarrow{\text{Enzymes}}$ Glycerol[— Fatty acid, — Fatty acid, — Fatty acid] + $3H_2O$

2. Hydrolysis of a triacylglycerol to produce glycerol and fatty acids (Section 19.6)

Glycerol[— Fatty acid, — Fatty acid, — Fatty acid] + $3H_2O \xrightarrow{\text{H}^+ \text{ or enzymes}}$ Glycerol + 3 fatty acids

3. Saponification of a triacylglycerol to produce glycerol and fatty acid salts (Section 19.6)

Glycerol[— Fatty acid, — Fatty acid, — Fatty acid] + $3OH^- \longrightarrow$ Glycerol + 3 fatty acid salts

4. Hydrogenation of a triacylglycerol to reduce the unsaturation of its fatty acid components (Section 19.6)

Glycerol[— Monounsaturated fatty acid, — Saturated fatty acid, — Monounsaturated fatty acid] + $2H_2 \longrightarrow$ Glycerol[— Saturated fatty acid, — Saturated fatty acid, — Saturated fatty acid]

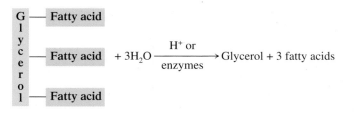

EXERCISES *and* PROBLEMS

The members of each pair of problems in this section test similar material.

● Structure and Classification of Lipids (Section 19.1)

19.1 What characteristic do all lipids have in common?

19.2 What structural feature, if any, do all lipid molecules have in common? Explain your answer.

19.3 Would you expect lipids to be soluble or insoluble in each of the following solvents?
   a. H_2O (polar)
   b. $CH_3-CH_2-O-CH_2-CH_3$ (nonpolar)
   c. CH_3-OH (polar)
   d. $CH_3-CH_2-CH_2-CH_2-CH_3$ (nonpolar)

19.4 Would you expect lipids to be soluble or insoluble in each of the following solvents?
   a. $CH_3-(CH_2)_7-CH_3$ (nonpolar)
   b. CH_3-Cl (polar)
   c. CCl_4 (nonpolar)
   d. CH_3-CH_2-OH (polar)

19.5 In terms of biochemical function, what are the five major categories of lipids?

19.6 What is the biochemical function of each of the following types of lipids?
   a. Triacylglycerols   b. Bile acids
   c. Sphingoglycolipids   d. Eicosanoids

● Types of Fatty Acids (Section 19.2)

19.7 Classify each of the following fatty acids as long-chain, medium-chain, or short-chain.
   a. Myristic (14:0)   b. Caproic (6:0)
   c. Arachidic (20:0)   d. Capric (10:0)

19.8 Classify each of the following fatty acids as long-chain, medium-chain, or short-chain.
   a. Lauric (12:0)   b. Oleic (18:1)
   c. Butyric (4:0)   d. Stearic (18:0)

19.9 Classify each of the following fatty acids as saturated, mono-unsaturated, or polyunsaturated.
   a. Stearic (18:0)   b. Linolenic (18:3)
   c. Docosahexaenoic (22:6) d. Oleic (18:1)

19.10 Classify each of the following fatty acids as saturated, mono-unsaturated, or polyunsaturated.
   a. Palmitic (16:0)   b. Linoleic (18:2)
   c. Arachidonic (20:4)   d. Palmitoleic (16:1)

19.11 Structurally, what is the difference between a SFA and a MUFA?

19.12 Structurally, what is the difference between a MUFA and a PUFA?

19.13 With the help of Table 19.1, classify each of the acids in Problem 19.9 as an omega-3 acid, an omega-6 acid, or neither an omega-3 nor an omega-6 acid.

19.14 With the help of Table 19.1, classify each of the acids in Problem 19.10 as an omega-3 acid, an omega-6 acid, or neither an omega-3 nor an omega-6 acid.

19.15 Draw the condensed structural formula for the fatty acid whose numerical shorthand designation is 18:2 ($\Delta^{9,12}$).

19.16 Draw the condensed structural formula for the fatty acid whose numerical shorthand designation is 20:4 ($\Delta^{5,8,11,14}$).

19.17 Using the structural information given in Table 19.1, assign an IUPAC name to each of the following fatty acids.
   a. Myristic acid   b. Palmitoleic acid

19.18 Using the structural information given in Table 19.1, assign an IUPAC name to each of the following fatty acids.
   a. Stearic acid   b. Linolenic acid

● Physical Properties of Fatty Acids (Section 19.3)

19.19 What is the relationship between carbon chain length and melting point for fatty acids?

19.20 What is the relationship between degree of unsaturation and melting point for fatty acids?

19.21 Why does the introduction of a *cis* double bond into a fatty acid lower its melting point?

19.22 Why does increasing carbon chain length decrease water solubility for fatty acids?

19.23 In each of the following pairs of fatty acids, select the fatty acid that has the lower melting point.
   a. 18:0 acid and 18:1 acid   b. 18:2 acid and 18:3 acid
   c. 14:0 acid and 16:0 acid   d. 18:1 acid and 20:0 acid

19.24 In each of the following pairs of fatty acids, select the fatty acid that has the higher melting point.
   a. 14:0 acid and 18:0 acid   b. 20:4 acid and 20:5 acid
   c. 18:3 acid and 20:3 acid   d. 16:0 acid and 16:1 acid

● Triacylglycerols (Section 19.4)

19.25 What are the four structural subunits that contribute to the structure of a triacylglycerol?

19.26 Draw the general block diagram for a triacylglycerol.

19.27 How many different kinds of functional groups are present in a triacylglycerol in which all three fatty acid residues come from saturated fatty acids?

19.28 How many different kinds of functional groups are present in a triacylglycerol in which all three fatty acid residues come from unsaturated fatty acids?

19.29 Draw the condensed structural formula of a triacylglycerol formed from glycerol and three molecules of palmitic acid.

19.30 Draw the condensed structural formula of a triacylglycerol formed from glycerol and three molecules of stearic acid.

19.31 Draw block diagram structures for the four different triacylglycerols that can be produced from glycerol, stearic acid, and linolenic acid.

19.32 Draw block diagram structures for the three different triacylglycerols that can be produced from glycerol, palmitic acid, stearic acid, and linolenic acid.

19.33 Identify the fatty acids present in each of the following triacylglycerols.

a.
$$CH_2-O-\overset{\overset{\textstyle O}{\|}}{C}-(CH_2)_{14}-CH_3$$
$$CH-O-\overset{\overset{\textstyle O}{\|}}{C}-(CH_2)_{12}-CH_3$$
$$CH_2-O-\overset{\overset{\textstyle O}{\|}}{C}-(CH_2)_7-CH=CH-(CH_2)_7-CH_3$$

b.

$CH_2-O-\overset{O}{\overset{\|}{C}}$ /\/\/\/\/\=\/\/\

$CH-O-\overset{O}{\overset{\|}{C}}$ /\/\/\/\/\/\/\

$CH_2-O-\overset{O}{\overset{\|}{C}}$ /\/\/\=\/\/\

19.34 Identify the fatty acids present in each of the following triacylglycerols.

a.

$CH_2-O-\overset{O}{\overset{\|}{C}}-(CH_2)_{16}-CH_3$

$CH-O-\overset{O}{\overset{\|}{C}}-(CH_2)_7-CH=CH-(CH_2)_7-CH_3$

$CH_2-O-\overset{O}{\overset{\|}{C}}-(CH_2)_{12}-CH_3$

b.

$CH_2-O-\overset{O}{\overset{\|}{C}}$ /\/\/\/\=\/\/\

$CH-O-\overset{O}{\overset{\|}{C}}$ /\/\/\/\=\/\/\

$CH_2-O-\overset{O}{\overset{\|}{C}}$ /\/\/\/\/\/\/\

19.35 For each of the acyl groups present in the triacylglycerol of Problem 19.33a, indicate how many carbon atoms are present and how many oxygen atoms are present.

19.36 For each of the acyl groups present in the triacylglycerol of Problem 19.34a, indicate how many carbon atoms are present and how many oxygen atoms are present.

19.37 What is the difference in meaning, if any, between the members of each of the following pairs of terms?
a. Triacylglycerol and triglyceride
b. Triacylglycerol and fat
c. Triacylglycerol and mixed triacylglycerol
d. Fat and oil

19.38 What is the difference in meaning, if any, between the members of each of the following pairs of terms?
a. Triacylglycerol and oil
b. Triacylglycerol and simple triacylglycerol
c. Simple triacylglycerol and mixed triacylglycerol
d. Triglyceride and fat

● **Dietary Considerations and Triacylglycerols (Section 19.5)**

19.39 In a dietary context, indicate whether each of the following pairings of concepts is correct.
a. "Saturated fat" and "good fat"
b. "Polyunsaturated fat" and "bad fat"

19.40 In a dietary context, indicate whether each of the following pairings of concepts is correct.
a. "Monounsaturated fat" and "good fat"
b. "Saturated fat" and "good and bad fat"

19.41 In a dietary context, which of the following pairings of concepts is correct?
a. "Cold-water fish" and "high in omega-3 fatty acids"
b. "Fatty fish" and "low in omega-3 fatty acids"

19.42 In a dietary context, which of the following pairings of concepts is correct?
a. "Warm-water fish" and "low in omega-3 fatty acids"
b. "Fish and chips" and "high in omega-3 fatty acids"

19.43 In a dietary context, classify each of the following fatty acids as an essential fatty acid or as a nonessential fatty acid.
a. Lauric acid (12:0)
b. Linoleic acid (18:2)
c. Myristic acid (14:0)
d. Palmitoleic (16:1)

19.44 In a dietary context, classify each of the following fatty acids as an essential fatty acid or as a nonessential fatty acid.
a. Stearic acid (18:0)
b. Linolenic acid (18:3)
c. Oleic acid (18:1)
d. Arachidic acid (20:0)

● **Chemical Reactions of Triacylglycerols (Section 19.6)**

19.45 Name, in general terms, the organic products of the complete
a. hydrolysis of a fat
b. saponification of an oil

19.46 Name, in general terms, the organic products of the complete
a. saponification of a fat
b. hydrolysis of an oil

19.47 Draw condensed structural formulas for all products you would obtain from the complete hydrolysis of the following triacylglycerol.

$CH_2-O-\overset{O}{\overset{\|}{C}}-(CH_2)_{14}-CH_3$

$CH-O-\overset{O}{\overset{\|}{C}}-(CH_2)_{12}-CH_3$

$CH_2-O-\overset{O}{\overset{\|}{C}}-(CH_2)_7-CH=CH-(CH_2)_7-CH_3$

19.48 Draw condensed structural formulas for all products you would obtain from the complete hydrolysis of the following triacylglycerol.

$CH_2-O-\overset{O}{\overset{\|}{C}}-(CH_2)_{16}-CH_3$

$CH-O-\overset{O}{\overset{\|}{C}}-(CH_2)_{12}-CH_3$

$CH_2-O-\overset{O}{\overset{\|}{C}}-(CH_2)_6-(CH_2-CH=CH)_3-CH_2-CH_3$

19.49 With the help of Table 19.1, determine the names of each of the organic products obtained in Problem 19.47.

19.50 With the help of Table 19.1, determine the names of each of the organic products obtained in Problem 19.48.

19.51 Draw condensed structural formulas for all products you would obtain from the saponification with NaOH of the triacylglycerol in Problem 19.47.

19.52 Draw condensed structural formulas for all products you would obtain from the saponification with KOH of the triacylglycerol in Problem 19.48.

19.53 With the help of Table 19.1, name each of the products obtained in Problem 19.51.

19.54 With the help of Table 19.1, name each of the products obtained in Problem 19.52.

19.55 Why can only unsaturated triacylglycerols undergo hydrogenation?

19.56 A food package label lists an oil as "partially hydrogenated." What does this mean?

19.57 How many molecules of H_2 will react with one molecule of the following triacylglycerol?

$$CH_2-O-\overset{\overset{O}{\|}}{C}-(CH_2)_6-(CH_2-CH=CH)_2-(CH_2)_4-CH_3$$
$$CH-O-\overset{\overset{O}{\|}}{C}-(CH_2)_7-CH=CH-(CH_2)_7-CH_3$$
$$CH_2-O-\overset{\overset{O}{\|}}{C}-(CH_2)_6-(CH_2-CH=CH)_3-CH_2-CH_3$$

19.58 How many molecules of H_2 will react with one molecule of the following triacylglycerol?

$$CH_2-O-\overset{\overset{O}{\|}}{C}-(CH_2)_7-CH=CH-(CH_2)_7-CH_3$$
$$CH-O-\overset{\overset{O}{\|}}{C}-(CH_2)_{16}-CH_3$$
$$CH_2-O-\overset{\overset{O}{\|}}{C}-(CH_2)_6-(CH_2-CH=CH)_3-CH_2-CH_3$$

19.59 Draw block diagram structures for all possible products of the partial hydrogenation, with two molecules of H_2, of the following molecules.

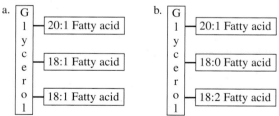

a.
Glycerol	18:1 Fatty acid
	18:1 Fatty acid
	16:1 Fatty acid

b.
Glycerol	18:0 Fatty acid
	18:2 Fatty acid
	16:1 Fatty acid

19.60 Draw block diagram structures for all possible products of the partial hydrogenation, with two molecules of H_2, of the following molecules.

a.
Glycerol	20:1 Fatty acid
	18:1 Fatty acid
	18:1 Fatty acid

b.
Glycerol	20:1 Fatty acid
	18:0 Fatty acid
	18:2 Fatty acid

19.61 Why do animal fats and vegetable oils become rancid when exposed to moist, warm air?

19.62 Why are the compounds BHA and BHT often added to foods that contain fats and oils?

Phospholipids (Section 19.7)

19.63 What are the two common types of platform molecules for a phospholipid?

19.64 How many fatty acid residues are present in a phospholipid?

19.65 Draw the general block diagram for a glycerophospholipid.

19.66 Draw the general block diagram for a sphingophospholipid.

19.67 Draw the structures of the three amino alcohols commonly esterified to the phosphate group in a glycerophospholipid.

19.68 What structural subunits are present in a phosphatidyl group?

19.69 Sphingophospholipids have a "head and two tails" structure. Give the chemical identity of the head and of each of the two tails.

19.70 Glycerophospholipids have a "head and two tails" structure. Give the chemical identity of the head and of each of the two tails.

19.71 Which portion of the structure of a phospholipid has hydrophobic characteristics?

19.72 Which portion of the structure of a phospholipid has hydrophilic characteristics?

19.73 Indicate how many ester linkages are present in the structure of a
a. glycerophospholipid   b. sphingophospholipid

19.74 Indicate how many amide linkages are present in the structure of a
a. glycerophospholipid   b. sphingophospholipid

19.75 Structurally, what is the difference between a lecithin and a phosphatidylserine?

19.76 Structurally, what is the difference between a lecithin and a sphingomyelin?

Sphingoglycolipids (Section 19.8)

19.77 Draw the general block diagram for a sphingoglycolipid.

19.78 How many of each of the following types of linkages are present in a sphingoglycolipid?
a. Ester linkages   b. Amide linkages
c. Glycosidic linkages

19.79 How does the general structure of a sphingoglycolipid differ from that of a sphingophospholipid?

19.80 Structurally, what is the difference between a cerebroside and a ganglioside?

Cholesterol (Section 19.9)

19.81 Draw and number the fused hydrocarbon ring system characteristic of all steroids.

19.82 What positions in the steroid nucleus are particularly likely to bear substituents?

19.83 Describe the structure of cholesterol in terms of substituents attached to the steroid nucleus.

19.84 Structurally, what is considered the "head" of a cholesterol molecule?

19.85 In a dietary context, what is the difference between "good cholesterol" and "bad cholesterol"?

19.86 In a dietary context, how do HDL and LDL differ in function?

Cell Membranes (Section 19.10)

19.87 What are the three major types of lipids present in cell membranes?

19.88 What is the structural characteristic common to all the nonsteroid lipids present in cell membranes?

19.89 What is a lipid bilayer?

19.90 What is the basic structure of a cell membrane?

19.91 What is the function of unsaturation in the hydrocarbon tails of membrane lipids?

19.92 What function does cholesterol serve when it is present in cell membranes?

19.93 What is an integral membrane protein?

19.94 What is a peripheral membrane protein?

19.95 What is the difference between *passive transport* and *facilitated transport*?

19.96 What is the difference between *facilitated transport* and *active transport*?

19.97 Match each of the following statements related to membrane transport processes to the appropriate term: *passive transport, facilitated transport, active transport*. More than one term may apply in a given situation.
a. Movement across the membrane is against the concentration gradient.
b. Proteins serve as "gates."
c. Expenditure of cellular energy is required.
d. Movement across the membrane is from a high to a low concentration.

19.98 Match each of the following statements related to membrane transport processes to the appropriate term: *passive transport, facilitated transport, active transport*. More than one term may apply in a given situation.
a. Movement across the membrane is with the concentration gradient.
b. Proteins serve as "pumps."
c. Expenditure of cellular energy is not required.
d. Movement across the membrane is from a low to a high concentration.

Bile Acids (Section 19.11)

19.99 What is an emulsifier?

19.100 How do bile acids aid in the digestion of lipids?

19.101 Describe the structural differences between a bile acid and cholesterol.

19.102 Describe the structural differences between cholic acid and a deoxycholic acid.

19.103 Describe the structural differences between glycocholic acid and taurocholic acid.

19.104 Describe the structural differences between glycocholic acid and glyco-7-deoxycholic acid.

19.105 What is the medium through which bile acids are supplied to the small intestine?

19.106 What is the chemical composition of bile?

19.107 At what location in the body are bile acids stored until needed?

19.108 What is the chemical composition of the majority of gallstones?

Steroid Hormones (Section 19.12)

19.109 What are the two major classes of steroid hormones?

19.110 Describe the general function of each of the following types of steroid hormones.
a. Estrogens
b. Androgens
c. Progestins
d. Mineralocorticoids

19.111 How do the sex hormones estradiol and testosterone differ in structure?

19.112 What functional groups are present in each of the following steroid hormones or synthetic steroids?
a. Estradiol
b. Testosterone
c. Progesterone
d. Cortisone

19.113 Indicate whether or not each of the following is an adrenal corticoid hormone or a sex hormone.
a. Aldosterone
b. Testosterone
c. Cortisol
d. Estradiol

19.114 Indicate whether each of the following is a natural steroid hormone or a synthetic steroid.
a. Cortisone
b. Progesterone
c. Prednisolone
d. Norethynodrel

19.115 What is the biochemical function for each of the steroids listed in Problem 19.113?

19.116 What is the biochemical function for each of the steroids listed in Problem 19.114?

Eicosanoids (Section 19.13)

19.117 What is the major structural difference between a prostaglandin and its parent fatty acid?

19.118 What is the major structural difference between a leukotriene and its parent fatty acid?

19.119 What structural feature distinguishes a prostaglandin from a leukotriene?

19.120 What structural feature distinguishes a thromboxane from a leukotriene?

19.121 List six physiological processes that are regulated by eicosanoids.

19.122 What is the biochemical basis for the effectiveness of aspirin in decreasing inflammation?

Biological Waxes (Section 19.14)

19.123 Draw the general block diagram for a biological wax.

19.124 Draw the condensed structural formula of a wax formed from palmitic acid (Table 19.1) and cetyl alcohol, $CH_3-(CH_2)_{14}-CH_2-OH$.

19.125 What is the difference between a biological wax and a mineral wax?

19.126 Biological waxes have a "head and two tails" structure. Give the chemical identity of the head and of the two tails.

ADDITIONAL PROBLEMS

19.127 Classify each of the following types of lipids as (1) glycerol-based, (2) sphingosine-based, or (3) neither glycerol-based nor sphingosine-based.
a. Bile acids
b. Fats
c. Waxes
d. Leukotrienes

19.128 Indicate whether each of the lipid types in Problem 19.127 has a "head and two tails" structure.

19.129 Identify the type of lipid that fits each of the following "structural component" characterizations.
a. Sphingosine + fatty acid + phosphoric acid + choline
b. Fused-ring system with three 6-membered rings and one 5-membered ring
c. 20-carbon fatty acid + cyclopentane ring
d. Sphingosine + fatty acid + monosaccharide

19.130 Classify each of the following types of lipids as (1) an energy-storage lipid, (2) a membrane lipid, (3) an emulsification lipid, (4) a messenger lipid, or (5) a protective-coating lipid.
a. Fats
b. Estrogens
c. Sphingomyelins
d. Prostaglandins

19.131 Which of the terms *glycerolipid, sphingolipid,* and *phospholipid* apply to each of the following lipids? More than one term may apply in a given situation.
a. Triacylglycerol
b. Sphingoglycolipid
c. Glycerophospholipid
d. Sphingophospholipid

19.132 Specify the numbers of ester linkages, amide linkages, and glycosidic linkages present in each of the following types of lipids.
a. Oils
b. Sphinogomyelins
c. Cerebrosides
d. Phosphatidylcholines

19.133 Indicate whether each of the following types of lipids contain a "steroid nucleus" as part of its structure.
a. Prostaglandins
b. Cortisone
c. Bile acids
d. Leukotrienes

*M*ultiple-Choice Practice Test

19.134 Which of the following statements concerning fatty acids is *correct*?
a. They are naturally occurring dicarboxylic acids.
b. They are rarely found in the free state in nature.
c. Their carbon chains always contain at least two double bonds.
d. They almost always contain an odd number of carbon atoms.

19.135 Which of the following is a distinguishing characteristic between fats and oils?
a. Physical state at room temperature
b. Identity of the alcohol component present
c. Number of structural subunits present
d. Number of fatty acid residues present

19.136 *Partial* hydrogenation of a fat or an oil does which of the following?
a. Produces fatty acid salts
b. Decreases the degree of fatty acid unsaturation
c. Decreases the melting point
d. Increases the number of fatty acid residues present

19.137 In the oxidation of fats and oils, which part of the molecule is attacked by the oxidizing agent?
a. Carbon–carbon double bonds
b. Ester linkages
c. Hydroxyl groups
d. Carboxyl groups

19.138 In which of the following pairs of lipids are both members of the pair *membrane* lipids?
a. Triacylglycerols and cholesterol
b. Triacylglycerols and sphingophospholipids
c. Sphingophospholipids and sphingoglycolipids
d. Eicosanoids and bile salts

19.139 Which of the following types of lipids does *not* have a "head and two tails" structure?
a. Glycerophospholipids
b. Sphingophospholipids
c. Sphingoglycolipids
d. Triacylglycerols

19.140 The "steroid nucleus" of steroid lipids involves a fused-ring system that has how many rings?
a. Two
b. Three
c. Four
d. Five

19.141 Which of the following polarity-based descriptions is correct for a lipid bilayer?
a. Both the outer and inner surfaces contain polar "heads."
b. Both the outer and inner surfaces contain nonpolar "heads."
c. Both the outer and the inner surfaces contain polar "tails."
d. Both the outer and the inner surfaces contain nonpolar "tails."

19.142 Based on function, eicosanoids are classified as which of the following?
a. Membrane lipids
b. Emulsification lipids
c. Messenger lipids
d. Protective-coating lipids

19.143 How many structural subunits are present in the "block diagram" for a biological wax?
a. Two
b. Three
c. Four
d. Five

Proteins

The fibrous protein alpha-keratin is the major structural element present in sheep's wool.

In this chapter we consider the third of the bioorganic classes of molecules (Section 18.1), the compounds called proteins. An extraordinary number of different proteins, each with a different function, exist in the human body. A typical human cell contains about 9000 different kinds of proteins, and the human body contains about 100,000 different proteins. Proteins are needed for the synthesis of enzymes, certain hormones, and some blood components; for the maintenance and repair of existing tissues; for the synthesis of new tissue; and sometimes for energy.

20.1 CHARACTERISTICS OF PROTEINS

Next to water, proteins are the most abundant substances in nearly all cells—they account for about 15% of a cell's overall mass (Section 18.1) and for almost half of a cell's dry mass. All proteins contain the elements carbon, hydrogen, oxygen, and nitrogen; most also contain sulfur. The presence of nitrogen in proteins sets them apart from carbohydrates and lipids, which most often do not contain nitrogen. The average nitrogen content of proteins is 15.4% by mass. Other elements, such as phosphorus and iron, are essential constituents of certain specialized proteins. Casein, the main protein of milk, contains phosphorus, an element very important in the diet of infants and children. Hemoglobin, the oxygen-transporting protein of blood, contains iron.

A **protein** *is a naturally occurring, unbranched polymer in which the monomer units are amino acids.* Thus the starting point for a discussion of proteins is an understanding of the structures and chemical properties of amino acids.

The word protein comes from the Greek *proteios*, which means "of first importance." This reflects the key role that proteins play in life processes.

In an α-amino acid, the carboxyl group and the amino group are attached to the same carbon atom.

The nature of the side chain (R group) distinguishes α-amino acids from each other, both physically and chemically.

The nonpolar amino acid *proline* has a structural feature not found in any other standard amino acid. Its side chain, a propyl group, is bonded to both the α-carbon atom and the amino nitrogen atom, giving a cyclic side chain.

Proline

A variety of functional groups are present in the side chains of the 20 standard amino acids: six have alkyl groups (Section 12.8), three have aromatic groups (Section 13.11), two have sulfur-containing groups (Section 14.20), two have hydroxyl (alcohol) groups (Section 14.2), three have amino groups (Section 17.2), two have carboxyl groups (Section 16.1), and two have amide groups (Section 17.12).

20.2 AMINO ACIDS: THE BUILDING BLOCKS FOR PROTEINS

An **amino acid** *is an organic compound that contains both an amino* (—NH₂) *group and a carboxyl* (—COOH) *group.* The amino acids found in proteins are always α-amino acids. An **α-amino acid** *is an amino acid in which the amino group and the carboxyl group are attached to the α-carbon atom.* The general structural formula for an α-amino acid is

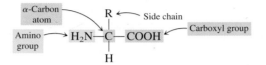

The R group present in an α-amino acid is called the amino acid *side chain*. The nature of this side chain distinguishes α-amino acids from each other. Side chains vary in size, shape, charge, acidity, functional groups present, hydrogen-bonding ability, and chemical reactivity.

Over 700 different naturally occurring amino acids are known, but only 20 of them, called standard amino acids, are normally present in proteins. A **standard amino acid** *is one of the 20 α-amino acids normally found in proteins.* The structures of the 20 standard amino acids are given in Table 20.1. Within Table 20.1, amino acids are grouped according to side-chain polarity. In this system there are four categories: (1) nonpolar amino acids, (2) polar neutral amino acids, (3) polar acidic amino acids, and (4) polar basic amino acids. This classification system gives insights into how various types of amino acid side chains help determine the properties of proteins (Section 20.11).

A **nonpolar amino acid** *is an amino acid that contains one amino group, one carboxyl group, and a nonpolar side chain.* When incorporated into a protein, such amino acids are *hydrophobic* ("water-fearing"); that is, they are not attracted to water molecules. They are generally found in the interior of proteins, where there is limited contact with water. There are nine nonpolar amino acids. Tryptophan is a borderline member of this group because water can weakly interact through hydrogen bonding with the NH ring location on tryptophan's side-chain ring structure. Thus, some textbooks list tryptophan as a polar neutral amino acid.

The three types of polar amino acids have varying degrees of affinity for water. Within a protein, such amino acids are said to be *hydrophilic* ("water-loving"). Hydrophilic amino acids are often found on the surfaces of proteins.

A **polar neutral amino acid** *is an amino acid that contains one amino group, one carboxyl group, and a side chain that is polar but neutral.* In solution at physiological pH, the side chain of a polar neutral amino acid is neither acidic nor basic. There are six polar neutral amino acids. These amino acids are more soluble in water than the nonpolar amino acids as, in each case, the R group present can hydrogen bond to water.

A **polar acidic amino acid** *is an amino acid that contains one amino group and two carboxyl groups, the second carboxyl group being part of the side chain.* In solution at physiological pH, the side chain of a polar acidic amino acid bears a negative charge; the side-chain carboxyl group has lost its acidic hydrogen atom. There are two polar acidic amino acids: aspartic acid and glutamic acid.

A **polar basic amino acid** *is an amino acid that contains two amino groups and one carboxyl group, the second amino group being part of the side chain.* In solution at physiological pH, the side chain of a polar basic amino acid bears a positive charge; the nitrogen atom of the amino group has accepted a proton (basic behavior; Section 17.6). There are three polar basic amino acids: lysine, arginine, and histidine.

The names of the standard amino acids are often abbreviated using three-letter codes. Except in four cases, these abbreviations are the first three letters of the amino acid's name. In addition, a new one-letter code for amino acid names is currently gaining popularity (particularly in computer applications). Both sets of abbreviations are used extensively in specifying the amino acid make-up of protein. Both sets of abbreviations are given in Table 20.1.

TABLE 20.1
The 20 Standard Amino Acids, Grouped According to Side-Chain Polarity

Below each amino acid's structure are its name (with pronunciation), its three-letter abbreviation, and its one-letter abbreviation.

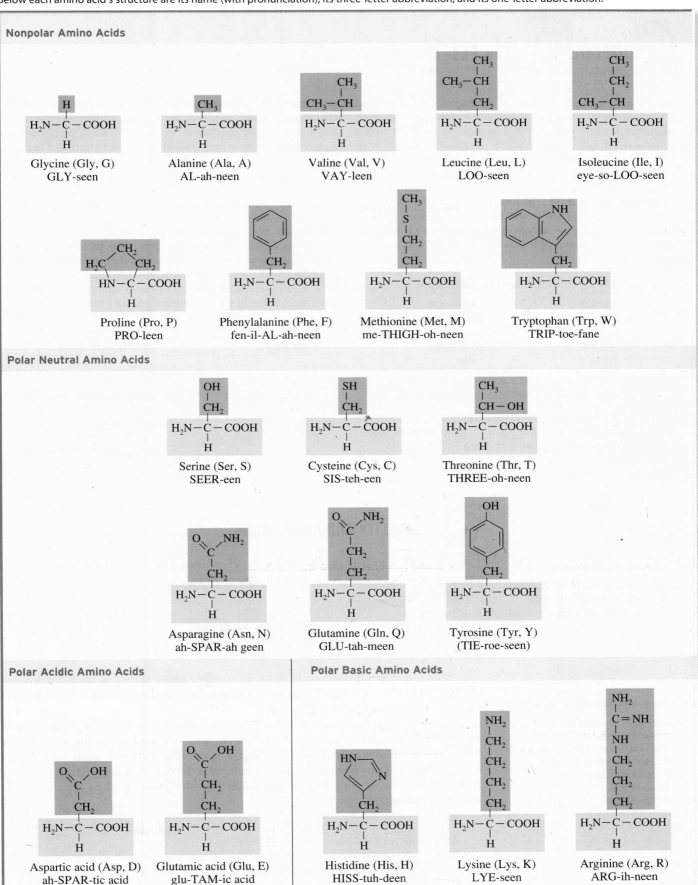

Nonpolar Amino Acids

Glycine (Gly, G)
GLY-seen

Alanine (Ala, A)
AL-ah-neen

Valine (Val, V)
VAY-leen

Leucine (Leu, L)
LOO-seen

Isoleucine (Ile, I)
eye-so-LOO-seen

Proline (Pro, P)
PRO-leen

Phenylalanine (Phe, F)
fen-il-AL-ah-neen

Methionine (Met, M)
me-THIGH-oh-neen

Tryptophan (Trp, W)
TRIP-toe-fane

Polar Neutral Amino Acids

Serine (Ser, S)
SEER-een

Cysteine (Cys, C)
SIS-teh-een

Threonine (Thr, T)
THREE-oh-neen

Asparagine (Asn, N)
ah-SPAR-ah geen

Glutamine (Gln, Q)
GLU-tah-meen

Tyrosine (Tyr, Y)
(TIE-roe-seen)

Polar Acidic Amino Acids

Aspartic acid (Asp, D)
ah-SPAR-tic acid

Glutamic acid (Glu, E)
glu-TAM-ic acid

Polar Basic Amino Acids

Histidine (His, H)
HISS-tuh-deen

Lysine (Lys, K)
LYE-seen

Arginine (Arg, R)
ARG-ih-neen

CHEMICAL Connections

The Essential Amino Acids

All of the amino acids in Table 20.1 are necessary constituents of human protein. Adequate amounts of 11 of the 20 amino acids can be synthesized from carbohydrates and lipids in the body if a source of nitrogen is also available. Because the human body is incapable of producing 9 of these 20 acids fast enough or in sufficient quantities to sustain normal growth, these 9 amino acids, called essential amino acids, must be obtained from food. An **essential amino acid** *is an amino acid needed in the human body that must be obtained from dietary sources because it cannot be synthesized within the body from other substances in adequate amounts.* The following table lists the *essential* amino acids for humans.

The Essential Amino Acids for Humans

arginine*	methionine
histidine	phenylalanine
isoleucine	threonine
leucine	tryptophan
lysine	valine

*Arginine is required for growth in children but is not required by adults.

The human body can synthesize small amounts of some of the essential amino acids, but not enough to meet its needs, especially in the case of growing children.

A **complete dietary protein** *is a protein that contains all the essential amino acids in approximately the same relative amounts in which the human body needs them.* A complete dietary protein may or may not contain all the nonessential amino acids. Most animal proteins, including casein from milk and proteins found in meat, fish, and eggs, are complete proteins, although gelatin is an exception (it lacks tryptophan). Proteins from plants (vegetables, grains, and legumes) have quite diverse amino acid patterns and some tend to be limited in one or more essential amino acids. Some plant proteins (for example, corn protein) are far from complete. Others (for example, soy protein) are complete. Thus vegetarians must eat a variety of plant foods to obtain all of the essential amino acids in appropriate quantities.

The following table lists the essential amino acid deficiencies associated with selected vegetables and grains.

Amino Acids Missing in Selected Vegetables and Grains

Food Source	Amino Acid Deficiency
soy	none
wheat, rice, oats	lysine
corn	lysine, tryptophan
beans	methionine, tryptophan
peas	methionine
almonds, walnuts	lysine, tryptophan

20.3 CHIRALITY AND AMINO ACIDS

Glycine, the simplest of the standard amino acids, is achiral. All of the other standard amino acids are chiral.

Four different groups are attached to the α-carbon atom in all of the standard amino acids except glycine, where the R group is a hydrogen atom.

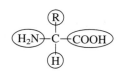

This means that the structures of 19 of the 20 standard amino acids possess a chiral center (Section 18.4) at this location, so enantiomeric forms (left- and right-handed forms; Section 18.5) exist for each of these amino acids.

Because only L amino acids are constituents of proteins, the enantiomer designation of L or D will be omitted in subsequent amino acid and protein discussions. It is understood that it is the L isomer that is always present.

With few exceptions (in some bacteria), the amino acids found in nature and in proteins are L isomers. Thus, as is the case with monosaccharides (Section 18.8), nature favors one mirror-image form over the other. Interestingly, for amino acids the L isomer is the preferred form, whereas for monosaccharides the D isomer is preferred.

The rules for drawing Fischer projection formulas (Section 18.6) for amino acid structures follow.

1. The —COOH group is put at the top of the projection, the R group at the bottom. This positions the carbon chain vertically.
2. The —NH₂ group is in a horizontal position. Positioning it on the left denotes the L isomer, and positioning it on the right denotes the D isomer.

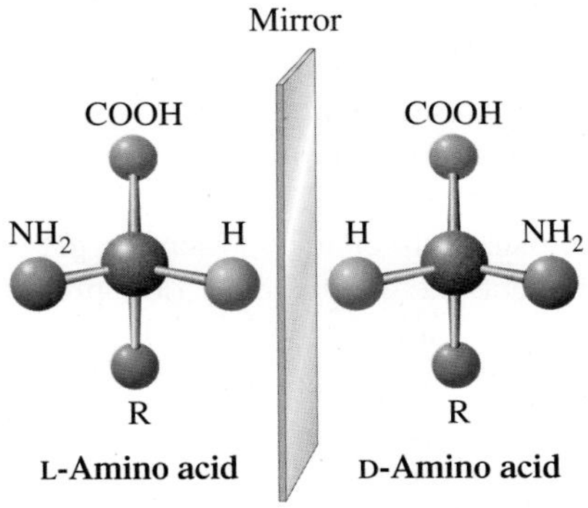

Figure 20.1 Designation of handedness in standard amino acid structures involves aligning the carbon chain vertically and looking at the position of the horizontally aligned —NH₂ group. The L form has the —NH₂ group on the left, and the D form has the —NH₂ group on the right.

Mirror

L-Amino acid   D-Amino acid

Figure 20.1 shows molecular models that illustrate the use of these rules. Fischer projection formulas for both enantiomers of the amino acids alanine and serine follow.

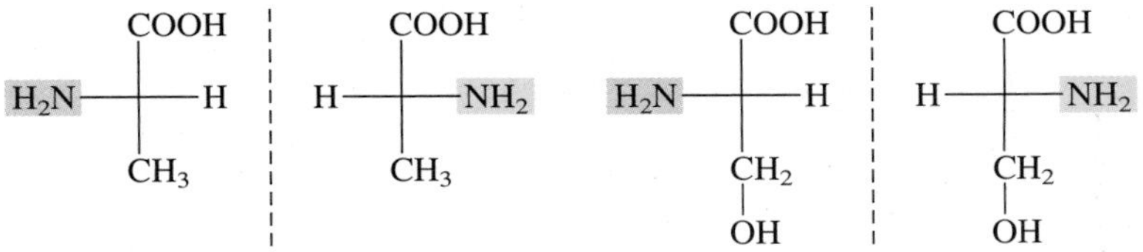

$$
\begin{array}{cccc}
\text{COOH} & \text{COOH} & \text{COOH} & \text{COOH} \\
H_2N \!-\!\!-\! H & H \!-\!\!-\! NH_2 & H_2N \!-\!\!-\! H & H \!-\!\!-\! NH_2 \\
\text{CH}_3 & \text{CH}_3 & \text{CH}_2 & \text{CH}_2 \\
 & & \text{OH} & \text{OH} \\
\text{L-Alanine} & \text{D-Alanine} & \text{L-Serine} & \text{D-Serine}
\end{array}
$$

Two of the 19 chiral standard amino acids, isoleucine and threonine, possess two chiral centers (see Table 20.1). With two chiral centers present, four stereoisomers are possible for these amino acids. However, only one of the L isomers is found in proteins.

20.4 ACID-BASE PROPERTIES OF AMINO ACIDS

In pure form, amino acids are white crystalline solids with relatively high decomposition points. (Most amino acids decompose before they melt.) Also, most amino acids are *not* very soluble in water because of strong intermolecular forces within their crystal structures. Such properties are those often exhibited by compounds in which charged species are present. Studies of amino acids confirm that they are charged species both in the solid state and in solution. Why is this so?

Both an acidic group (—COOH) and a basic group (—NH₂) are present on the same carbon in an α-amino acid.

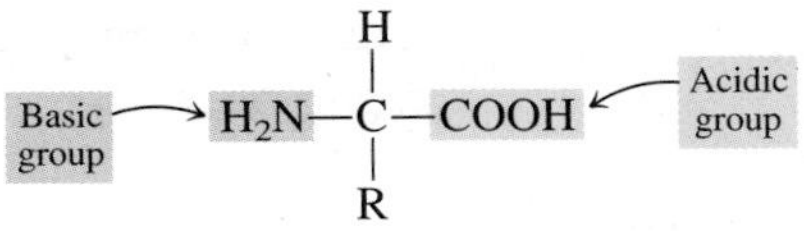

In Section 16.8, we learned that in neutral solution, carboxyl groups have a tendency to lose protons (H⁺), producing a negatively charged species:

$$-\text{COOH} \longrightarrow -\text{COO}^- + \text{H}^+$$

In drawing amino acid structures, where handedness designation is not required, the placement of the four groups about the α-carbon atom is arbitrary. From this point on in the text, we will draw amino acid structures such that the —COOH group is on the right, the —NH₂ group is on the left, the R group points down, and the H atom points up. Drawing amino acids in this "arrangement" makes it easier to draw structures where amino acids are linked together to form longer amino acid chains.

In Section 17.6, we learned that in neutral solution, amino groups have a tendency to accept protons (H⁺), producing a positively charged species:

$$-\text{NH}_2 + \text{H}^+ \longrightarrow -\overset{+}{\text{N}}\text{H}_3$$

Consistent with the behavior of these groups, in neutral solution, the —COOH group of an amino acid donates a proton to the —NH₂ of the same amino acid. We can characterize this behavior as an *internal* acid–base reaction. The net result is that in neutral solution, amino acid molecules have the structure

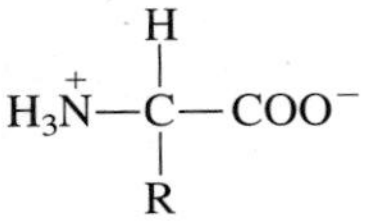

$$
\begin{array}{c}
\text{H} \\
| \\
H_3\overset{+}{\text{N}}\!-\!\text{C}\!-\!\text{COO}^- \\
| \\
\text{R}
\end{array}
$$

Strong intermolecular forces between the positive and negative centers of zwitterions are the cause of the high melting points of amino acids.

From this point on in the text, the structures of amino acids will be drawn in their zwitterion form unless information given about the pH of the solution indicates otherwise.

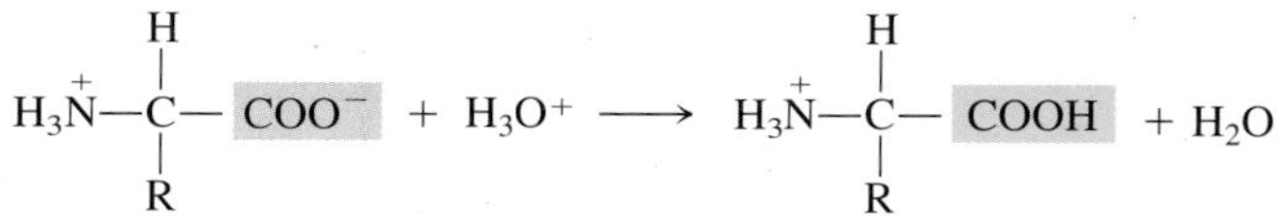

The ability of amino acids to react with both H_3O^+ and OH^- ions means that amino acid solutions can function as buffers (Section 10.12). The same is true for proteins, which are amino acid polymers (Section 20.1). The buffering action of proteins present in blood is a major function of such proteins.

Such a molecule is known as a zwitterion, from the German term meaning "double ion." A **zwitterion** *is a molecule that has a positive charge on one atom and a negative charge on another atom, but which has no net charge.* Note that the net charge on a zwitterion is zero even though parts of the molecule carry charges. In solution and also in the solid state, α-amino acids exist as zwitterions.

Zwitterion structure changes when the pH of a solution containing an amino acid is changed from neutral either to acidic (low pH) by adding an acid such as HCl or to basic (high pH) by adding a base such as NaOH. In an acidic solution, the zwitterion accepts a proton (H^+) to form a positively charged ion.

$$H_3\overset{+}{N}-\underset{R}{\overset{H}{C}}-COO^- + H_3O^+ \longrightarrow H_3\overset{+}{N}-\underset{R}{\overset{H}{C}}-COOH + H_2O$$

Zwitterion (no net charge) Positively charged ion

In basic solution, the $-\overset{+}{N}H_3$ of the zwitterion loses a proton, and a negatively charged species is formed.

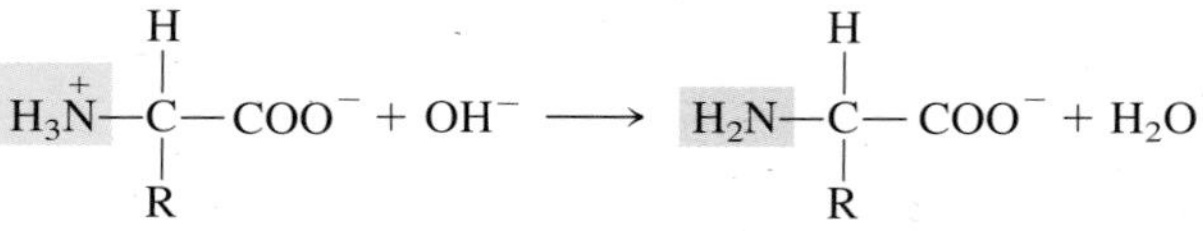

$$H_3\overset{+}{N}-\underset{R}{\overset{H}{C}}-COO^- + OH^- \longrightarrow H_2N-\underset{R}{\overset{H}{C}}-COO^- + H_2O$$

Zwitterion (no net charge) Negatively charged ion

Thus, in solution, three different amino acid forms can exist (zwitterion, negative ion, and positive ion). The three species are actually in equilibrium with each other, and the equilibrium shifts with pH change. The overall equilibrium process can be represented as follows:

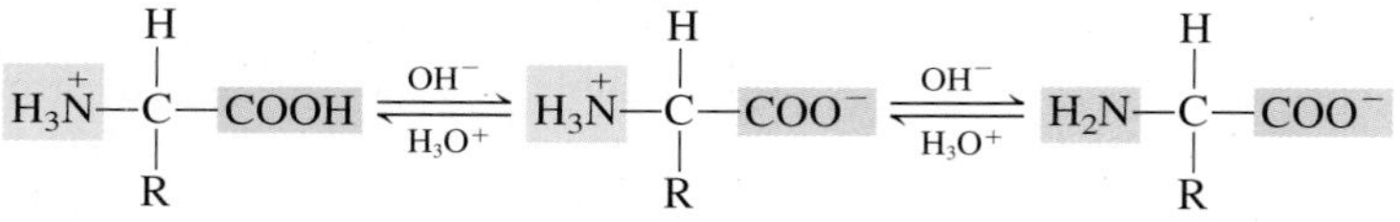

$$H_3\overset{+}{N}-\underset{R}{\overset{H}{C}}-COOH \underset{H_3O^+}{\overset{OH^-}{\rightleftharpoons}} H_3\overset{+}{N}-\underset{R}{\overset{H}{C}}-COO^- \underset{H_3O^+}{\overset{OH^-}{\rightleftharpoons}} H_2N-\underset{R}{\overset{H}{C}}-COO^-$$

Acidic solution (low pH) Neutral solution (pH = 7.0) Basic solution (high pH)

In acidic solution, the positively charged species on the left predominates; nearly neutral solutions have the middle species (the zwitterion) as the dominant species; in basic solution, the negatively charged species on the right predominates.

EXAMPLE 20.1

Determining Amino Acid Form in Solutions of Various pH

Draw the structural form of the amino acid alanine that predominates in solution at each of the following pH values.

a. pH = 1.0
b. pH = 7.0
c. pH = 11.0

Solution

At low pH, both amino and carboxyl groups are protonated. At high pH, both groups have lost their protons. At neutral pH, the zwitterion is present.

a. $H_3\overset{+}{N}-\underset{CH_3}{\overset{H}{C}}-COOH$
pH = 1.0 (net charge of +1)

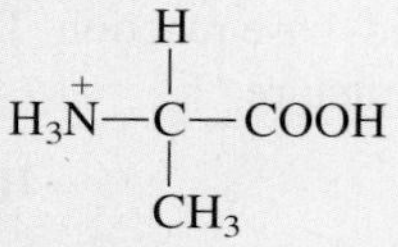

b. $H_3\overset{+}{N}-\underset{CH_3}{\overset{H}{C}}-COO^-$
pH = 7.0 (no net charge)

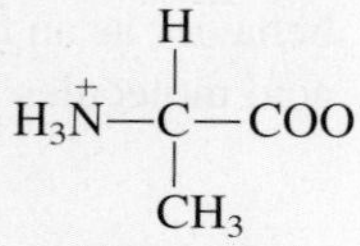

c. $H_2N-\underset{CH_3}{\overset{H}{C}}-COO^-$
pH = 11.0 (net charge of -1)

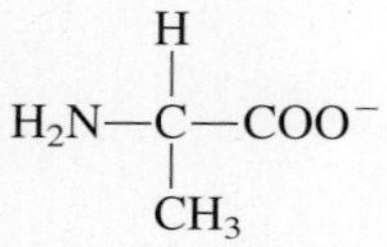

Guidelines for amino acid form as a function of solution pH follow.

Low pH: All acid groups are protonated (—COOH). All amino groups are protonated (—$\overset{+}{N}H_3$).

High pH: All acid groups are deprotonated (—COO⁻). All amino groups are deprotonated (—NH₂).

Neutral pH: All acid groups are deprotonated (—COO⁻). All amino groups are protonated (—$\overset{+}{N}H_3$).

The term *protonated* denotes gain of a H⁺ ion, and the term *deprotonated* denotes loss of a H⁺ ion.

Side-chain carboxyl groups are weaker acids than α-carbon carboxyl groups.

Practice Exercise 20.1

Draw the structural form of the amino acid valine that predominates in solution at each of the following pH values.

a. pH = 7.0   **b.** pH = 12.0   **c.** pH = 2.0

Answers: a.

$$H_3\overset{+}{N}-\underset{\underset{\displaystyle CH_3}{|}}{\overset{\overset{\displaystyle H}{|}}{\underset{\displaystyle CH-CH_3}{C}}}-COO^-$$

b.

$$H_2N-\underset{\underset{\displaystyle CH_3}{|}}{\overset{\overset{\displaystyle H}{|}}{\underset{\displaystyle CH-CH_3}{C}}}-COO^-$$

c.

$$H_3\overset{+}{N}-\underset{\underset{\displaystyle CH_3}{|}}{\overset{\overset{\displaystyle H}{|}}{\underset{\displaystyle CH-CH_3}{C}}}-COOH$$

The previous discussion assumed that the side chain (R group) of an amino acid remains unchanged in solution as the pH is varied. This is the case for neutral amino acids but not for acidic or basic ones. For these latter compounds, the side chain can also acquire a charge because it contains an amino or a carboxyl group that can, respectively, gain or lose a proton.

Because of the extra site that can be protonated or deprotonated, acidic and basic amino acids have four charged forms in solution. These four forms for aspartic acid, one of the acidic amino acids, are

$$H_3\overset{+}{N}-\underset{\underset{\displaystyle COOH}{|}}{\overset{\overset{\displaystyle H}{|}}{\underset{\displaystyle CH_2}{C}}}-COOH \underset{H_3O^+}{\overset{OH^-}{\rightleftharpoons}} H_3\overset{+}{N}-\underset{\underset{\displaystyle COOH}{|}}{\overset{\overset{\displaystyle H}{|}}{\underset{\displaystyle CH_2}{C}}}-COO^- \underset{H_3O^+}{\overset{OH^-}{\rightleftharpoons}} H_3\overset{+}{N}-\underset{\underset{\displaystyle COO^-}{|}}{\overset{\overset{\displaystyle H}{|}}{\underset{\displaystyle CH_2}{C}}}-COO^- \underset{H_3O^+}{\overset{OH^-}{\rightleftharpoons}} H_2N-\underset{\underset{\displaystyle COO^-}{|}}{\overset{\overset{\displaystyle H}{|}}{\underset{\displaystyle CH_2}{C}}}-COO^-$$

Low-pH form (+1 charge) | Moderately-low-pH form (no net charge) (zwitterion) | Intermediate-pH form (−1 net charge) | High-pH form (−2 net charge)

The existence of two low-pH forms for aspartic acid results from the two carboxyl groups being deprotonated at different pH values. For basic amino acids, two high-pH forms exist because deprotonation of the amino groups does not occur simultaneously. The side-chain amino group deprotonates before the α-amino group for histidine, but the opposite is true for lysine and arginine.

Isoelectric Points and Electrophoresis

The amounts of the various forms of an amino acid—zwitterion, negative ion(s), and positive ion(s)—that are present in an aqueous solution of the amino acid vary with solution pH. There is no pH at which ionic amino acid forms are absent, but there is a pH at which there is an equal number of positive and negative charges present, which produces a "no net charge" situation. The "no net charge" pH value for an amino acid solution is called its *isoelectric point*. An **isoelectric point** *is the pH at which an amino acid solution has no net charge because an equal number of positive and negative charges are present.* At the isoelectric point, almost all amino acid molecules in a solution (more than 99%) are present in their zwitterion form.

Every amino acid has a different isoelectric point. Fifteen of the 20 amino acids, those with nonpolar or polar neutral side chains (Table 20.1), have isoelectric points in the range of 4.8–6.3. The three basic amino acids have higher isoelectric points and the two acidic amino acids have lower ones. Table 20.2 lists the isoelectric points for the 20 standard amino acids.

The isoelectric point of an amino acid is measured by observing its behavior in an electric field. In an electric field, a charged molecule is attracted to (migrates toward) the electrode of opposite charge. At a high pH, an amino acid has a net negative charge and migrates toward the positive electrode. At a low pH, the opposite is true; with a net positive charge, the amino acid migrates toward the negative electrode. At the isoelectric point, migration does not occur because the zwitterions present have no net charge.

TABLE 20.2

Isoelectric Points for the 20 Amino Acids Commonly Found in Proteins

Name	Isoelectric Point
alanine	6.01
arginine	10.76
asparagine	5.41
aspartic acid	2.77
cysteine	5.07
glutamic acid	3.22
glutamine	5.65
glycine	5.97
histidine	7.59
isoleucine	6.02
leucine	5.98
lysine	9.74
methionine	5.74
phenylalanine	5.48
proline	6.48
serine	5.68
threonine	5.87
tryptophan	5.88
tyrosine	5.66
valine	5.97

Figure 20.2 Separation, at a pH of 5.5, of the three amino acids Lys, Phe, and Glu using electrophoresis.

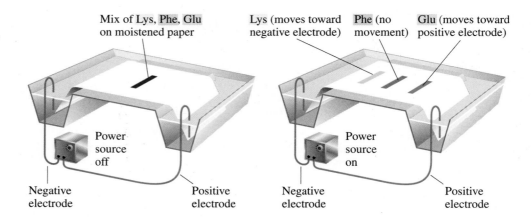

Mixtures of amino acids in solution can be separated by using their different migration patterns at various pH values. This type of analytical separation is called electrophoresis. **Electrophoresis** *is the process of separating charged molecules on the basis of their migration toward charged electrodes associated with an electric field.*

Figure 20.2 schematically shows the separation of the amino acids Lys, Phe, and Glu by electrophoresis. At a pH of 5.5, these amino acids exist in the following forms:

Lys
(+1 charge)

Phe
(no net charge)

Glu
(−1 charge)

When a current is applied, Phe does not move (it has no net charge); Lys, because of its positive charge, migrates toward the negative electrode; and Glu, with a negative charge, moves toward the positive electrode.

Proteins, which are amino acid polymers (Section 20.8), also have isoelectric points and also can be separated via electrophoresis techniques.

EXAMPLE 20.2

Migration Patterns of Amino Acids at Various pH Values in an Electric Field

Predict the direction of migration (if any) toward the positively or negatively charged electrode for the following amino acids in solutions of the specified pH. Write "isoelectric" if no migration occurs.

a. Lysine at pH = 7.0
b. Glutamic acid at pH = 7.0
c. Serine at pH = 1.0

Solution

a. Lysine at pH = 7.0

$$H_3\overset{+}{N}-\overset{\overset{\displaystyle H}{|}}{\underset{\underset{\underset{\underset{\displaystyle ^+NH_3}{|}}{\underset{\displaystyle CH_2}{|}}}{\underset{\displaystyle CH_2}{|}}}{\underset{\displaystyle CH_2}{\underset{|}{C}}}-COO^-$$

Net positive charge (2 "+" and 1 "−")

Migrates toward negatively charged electrode

b. Glutamic acid at pH = 7.0

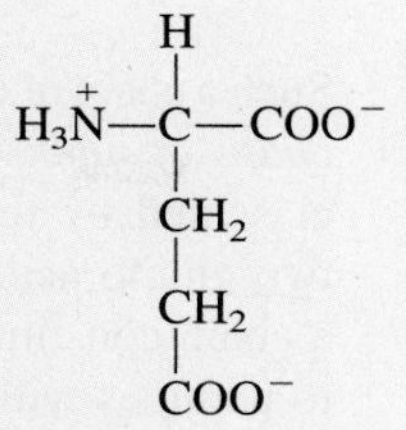

$$H_3\overset{+}{N}-\overset{\displaystyle H}{\underset{\displaystyle CH_2}{\overset{|}{C}}}-COO^-$$

CH_2

CH_2

COO^-

Net negative charge
(1 "+" and 2 "−")

Migrates toward
positively charged
electrode

c. Serine at pH = 1.0

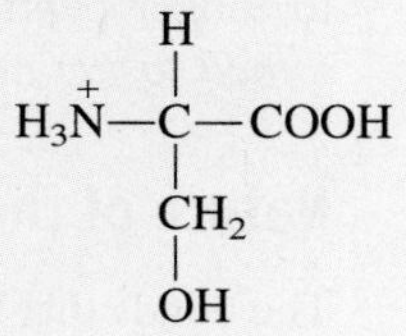

$$H_3\overset{+}{N}-\overset{\displaystyle H}{\underset{\displaystyle CH_2}{\overset{|}{C}}}-COOH$$

CH_2

OH

One positive charge

Migrates toward nega-
tively charged electrode

Practice Exercise 20.2

Predict the direction of migration (if any) toward the positively or negatively charged electrode for the following amino acids in solutions of the specified pH. Write "isoelectric" if no
migration occurs.

a. Lysine at pH = 12.0   **b.** Glutamic acid at pH = 2.0   **c.** Serine at pH = 5.68

Answers: a. Toward positively charged electrode; **b.** Toward negatively charged electrode; **c.** Isoelectric

20.5 CYSTEINE: A CHEMICALLY UNIQUE AMINO ACID

Cysteine is the only standard amino acid (Table 20.1) that has a side chain that contains a
sulfhydryl group (—SH group; Section 14.20). The presence of this sulfhydryl group imparts
to cysteine a chemical property that is unique among the standard amino acids. Cysteine, in
the presence of mild oxidizing agents, readily *dimerizes,* that is, reacts with another cysteine
molecule to form a cystine molecule. (A *dimer* is a molecule that is made up of two like
subunits.) In cystine, the two cysteine residues are linked via a covalent disulfide bond.

Cystine contains two *cysteine*
residues linked by a disulfide bond.

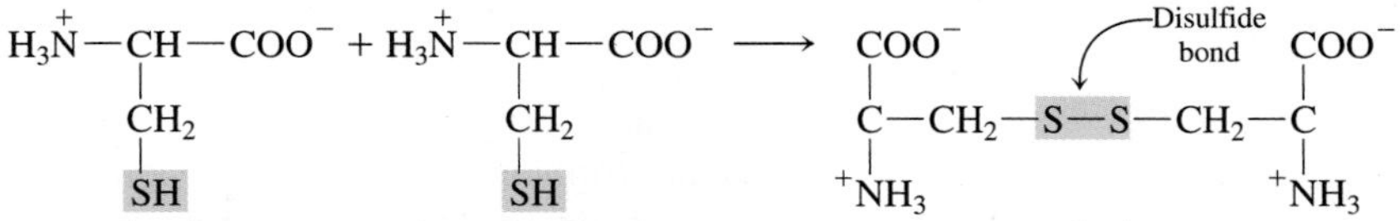

$$H_3\overset{+}{N}-\underset{\underset{SH}{\overset{|}{\underset{|}{CH_2}}}}{\overset{|}{CH}}-COO^- + H_3\overset{+}{N}-\underset{\underset{SH}{\overset{|}{\underset{|}{CH_2}}}}{\overset{|}{CH}}-COO^- \longrightarrow$$

Cysteine   Cysteine

$$\underset{^+NH_3}{\overset{COO^-}{\overset{|}{\underset{|}{C}}}}-CH_2-S-S-CH_2-\underset{^+NH_3}{\overset{COO^-}{\overset{|}{\underset{|}{C}}}}$$

Disulfide
bond

Cystine

The covalent disulfide bond of cystine is readily broken, using reducing agents, to
regenerate two cysteine molecules. This oxidation–reduction behavior involving sulfhydryl
groups and disulfide bonds was previously encountered in Section 14.20 when the reactions of thioalcohols were considered.

$$-SH + HS-\ \underset{\text{Reduction}}{\overset{\text{Oxidation}}{\rightleftharpoons}}\ -S-S-\ +\ 2H$$

As we shall see in Section 20.10, the formation of disulfide bonds between cysteine
residues present in protein molecules has important consequences relative to protein
structure and protein shape.

 PEPTIDES

Under proper conditions, amino acids can bond together to produce an unbranched chain
of amino acids. The length of the amino acid chain can vary from a few amino acids to
many amino acids. Representative of such chains is the following five-amino-acid chain.

Amino acid — Amino acid — Amino acid — Amino acid — Amino acid

Such a chain of covalently linked amino acids is called a *peptide*. A **peptide** *is an unbranched chain of amino acids, each joined to the next by a peptide bond.* Peptides are further classified by the number of amino acids present in the chain. A compound containing two amino acids is specifically called a *dipeptide;* three amino acids joined together in a chain constitute a *tripeptide;* and so on. The name *oligopeptide* is loosely used to refer to peptides with 10 to 20 amino acid residues, and the name *polypeptide* is used to refer to longer peptides. A **polypeptide** *is a long unbranched chain of amino acids, each joined to the next by a peptide bond.*

Nature of the Peptide Bond

The bonds that link amino acids together in a peptide chain are called peptide bonds. There are four peptide bonds present in a pentapeptide.

Amino acid — Amino acid — Amino acid — Amino acid — Amino acid

| Peptide bond | Peptide bond | Peptide bond | Peptide bond |

The nature of the peptide bond becomes apparent by reconsidering a chemical reaction previously encountered. In Section 17.16, we learned that a carboxylic acid and an amine can react to produce an amide. The general equation for this reaction is

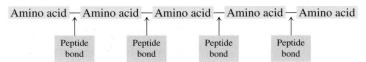

$$R-\overset{\overset{\text{O}}{\|}}{C}-OH + H-\overset{\overset{\text{H}}{|}}{N}-R \longrightarrow R-\overset{\overset{\text{O}}{\|}}{C}-\overset{\overset{\text{H}}{|}}{N}-R + H_2O$$

Acid   Amine   Amide

Two amino acids can combine in a similar way—the carboxyl group of one amino acid interacts with the amino group of the other amino acid. The products are a molecule of water and a molecule containing the two amino acids linked by an amide bond.

> Amide bond formation is an example of a condensation reaction.

$$H_3\overset{+}{N}-\overset{\overset{\text{H}}{|}}{\underset{\text{R}_1}{C}}-COO^- + H_3\overset{+}{N}-\overset{\overset{\text{H}}{|}}{\underset{\text{R}_2}{C}}-COO^- \longrightarrow H_3\overset{+}{N}-\overset{\overset{\text{H}}{|}}{\underset{\text{R}_1}{C}}-\overset{\overset{\text{O}}{\|}}{C}-\overset{\overset{\text{H}}{|}}{N}-\overset{\overset{\text{H}}{|}}{\underset{\text{R}_2}{C}}-COO^- + H_2O$$

Amide bond

Removal of the elements of water from the reacting carboxyl and amino groups and the ensuing formation of the amide bond are better visualized when expanded structural formulas for the reacting groups are used.

$$-\overset{\overset{\text{O}}{\|}}{C}-O + H-\overset{\overset{\text{H}}{|}}{\underset{\text{H}}{N}}- \longrightarrow -\overset{\overset{\text{O}}{\|}}{C}-\overset{\overset{\text{H}}{|}}{N}- + H_2O$$

Carboxyl group ($-COO^-$)   Amino group ($H_3\overset{+}{N}-$)   (Amide bond ≈ peptide bond)

> A peptide chain has *directionality* because its two ends are different. There is an N-terminal end and a C-terminal end. By convention, the direction of the peptide chain is always
>
> N-terminal end → C-terminal end
>
> The N-terminal end is always on the left, and the C-terminal end is always on the right.

In amino acid chemistry, amide bonds that link amino acids together are given the specific name of peptide bond. A **peptide bond** *is a covalent bond between the carboxyl group of one amino acid and the amino group of another amino acid.*

In all peptides, long or short, the amino acid at one end of the amino acid sequence has a free $H_3\overset{+}{N}$ group, and the amino acid at the other end of the sequence has a free COO^- group. The end with the free $H_3\overset{+}{N}$ group is called the *N-terminal end,* and the end with the free COO^- group is called the *C-terminal end.* By convention, the sequence of amino acids in a peptide is written with the N-terminal end amino acid on the left. The individual amino acids within a peptide chain are called *amino acid residues.* An

amino acid residue *is the portion of an amino acid structure that remains, after the release of H_2O, when an amino acid participates in peptide bond formation as it becomes part of a peptide chain.*

The structural formula for a peptide may be written out in full, or the sequence of amino acids present may be indicated by using the standard three-letter amino acid abbreviations. The abbreviated formula for the tripeptide

Glycine   Alanine   Serine

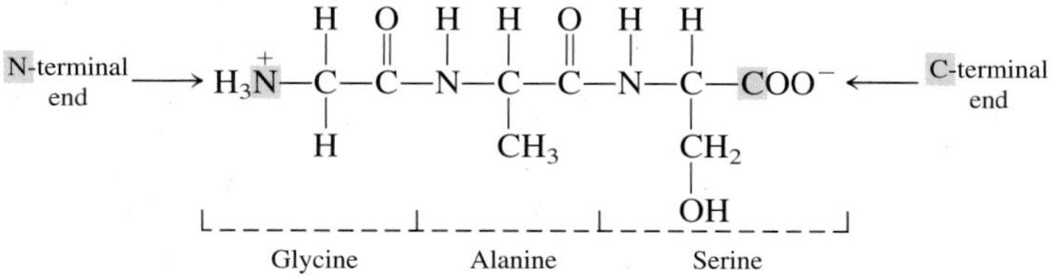

which contains the amino acids glycine, alanine, and serine, is Gly–Ala–Ser. When we use this abbreviated notation, by convention, the amino acid at the N-terminal end of the peptide is always written on the left.

The repeating sequence of peptide bonds and α-carbon (—CH groups in a peptide is referred to as the *backbone* of the peptide.)

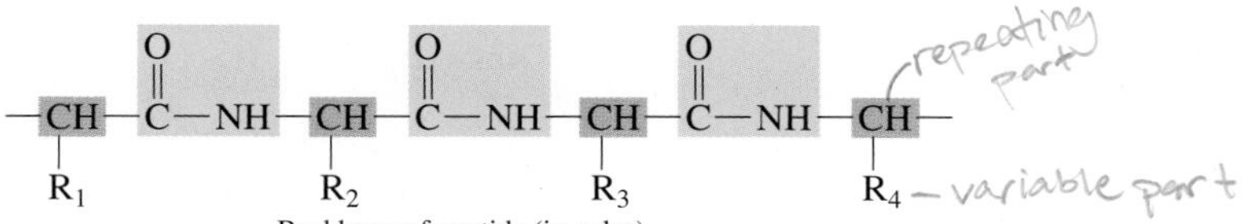

Backbone of peptide (in color)

The R group side chains are considered substituents on the backbone rather than part of the backbone.

Thus, structurally, a peptide has a regularly repeating part (the backbone) and a variable part (the sequence of R groups). (It is the variable R group sequence that distinguishes one peptide from another.)

EXAMPLE 20.3

Converting an Abbreviated Peptide Formula to a Structural Peptide Formula

Draw the structural formula for the tripeptide Ala–Gly–Val.

Solution

Step 1: The N-terminal end of the peptide involves alanine. Its structure is written first.

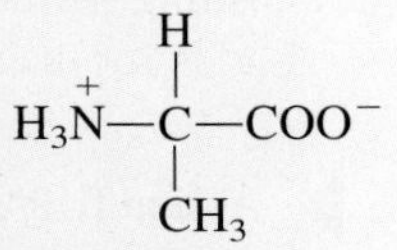

Step 2: The structure of glycine is written to the right of the alanine structure, and a peptide bond is formed between the two amino acids by removing the elements of H_2O and bonding the N of glycine to the carboxyl C of alanine.

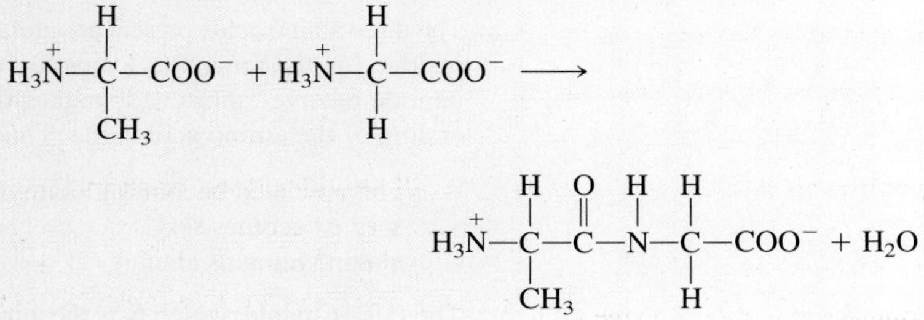

(continued)

Step 3:   To the right of the just-formed dipeptide, draw the structure of valine. Then repeat Step 2 to form the desired tripeptide.

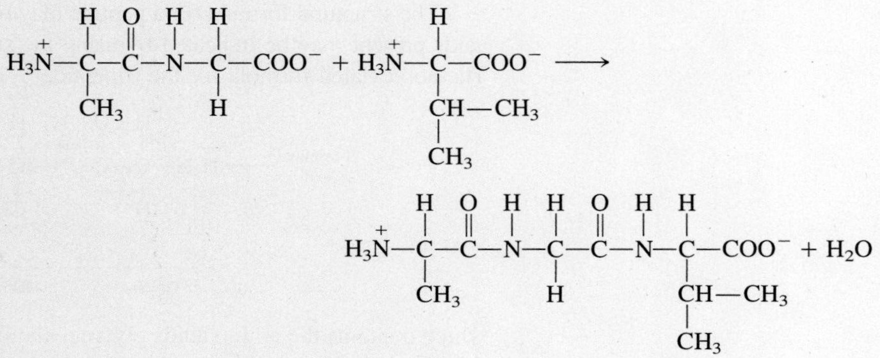

$$
\overset{+}{H_3N}-\underset{\underset{CH_3}{|}}{\overset{\overset{H}{|}}{C}}-\overset{\overset{O}{\|}}{C}-\underset{\overset{|}{H}}{\overset{\overset{H}{|}}{N}}-\overset{\overset{H}{|}}{C}-COO^- \; + \; \overset{+}{H_3N}-\underset{\underset{CH_3}{\underset{|}{CH}}}{\overset{\overset{H}{|}}{C}}-COO^- \longrightarrow
$$

$$
\overset{+}{H_3N}-\underset{\underset{CH_3}{|}}{\overset{\overset{H}{|}}{C}}-\overset{\overset{O}{\|}}{C}-\underset{\overset{|}{H}}{N}-\underset{\overset{|}{H}}{\overset{\overset{H}{|}}{C}}-\overset{\overset{O}{\|}}{C}-\overset{\overset{H}{|}}{N}-\underset{\underset{CH_3}{\underset{|}{CH}}}{\overset{\overset{H}{|}}{C}}-COO^- \; + \; H_2O
$$

Practice Exercise 20.3

Draw the structural formula for the tripeptide Cys–Ala–Gly.

Answer:

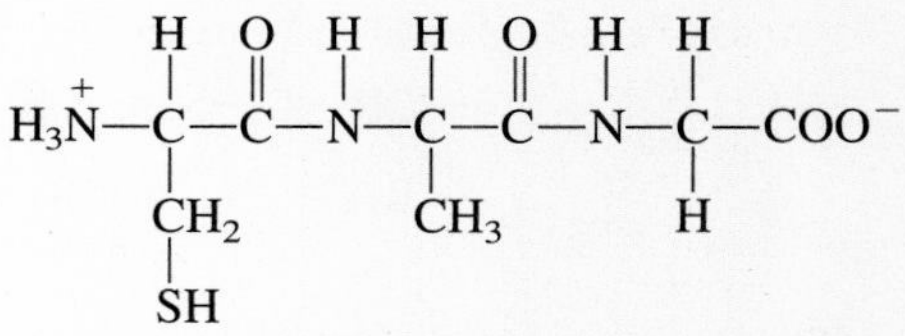

$$
\overset{+}{H_3N}-\underset{\underset{\underset{SH}{|}}{\underset{CH_2}{|}}}{\overset{\overset{H}{|}}{C}}-\overset{\overset{O}{\|}}{C}-\underset{\overset{|}{H}}{N}-\underset{\underset{CH_3}{|}}{\overset{\overset{H}{|}}{C}}-\overset{\overset{O}{\|}}{C}-\underset{\overset{|}{H}}{N}-\underset{\overset{|}{H}}{\overset{\overset{H}{|}}{C}}-COO^-
$$

Peptide Nomenclature

IUPAC rules for naming small peptides are as follows:

Rule 1:   The C-terminal amino acid residue (located at the far right of the structure) keeps its full amino acid name.

Rule 2:   All of the other amino acid residues have names that end in *-yl.* The *-yl* suffix replaces the *-ine* or *-ic acid* ending of the amino acid name, except for tryptophan (tryptophyl), cysteine (cysteinyl), glutamine (glutaminyl), and asparagine (asparaginyl).

Rule 3:   The amino acid naming sequence begins at the N-terminal amino acid residue.

⬤ **E X A M P L E   2 0 . 4**

Determining IUPAC Names for Small Peptides

Assign IUPAC names to each of the following small peptides.

a. Glu–Ser–Ala
b. Gly–Tyr–Leu–Val

Solution

a. The three amino acids present are glutamic acid, serine, and alanine. Alanine, the C-terminal residue (on the far right), keeps its full name. The other amino acid residues in the peptide receive "shortened" names that end in *-yl*. The *-yl* replaces the *-ine* or *-ic acid* ending of the amino acid name. Thus

glutamic acid becomes glutamyl
serine becomes seryl
alanine remains alanine

The IUPAC name, which lists the amino acids in the sequence from N-terminal residue to C-terminal residue, becomes *glutamylserylalanine.*

b. The four amino acids present are glycine, tyrosine, leucine, and valine. Proceeding as in part **a,** we note that

> glycine becomes glycyl
> tyrosine becomes tyrosyl
> leucine becomes leucyl
> valine remains valine

Combining these individual names gives the IUPAC name *glycyltyrosylleucylvaline.*

The complete name for a small peptide should actually include a handedness designation (L-designation; Section 20.3) before the name of each residue. For example the dipeptide serylglycine (Ser–Gly) should actually be L-seryl–L-glycine (L-Ser–L-Gly). The L-handedness designation is, however, usually not included because it is understood that all amino acids present in a peptide, unless otherwise noted, are L enantiomers.

Practice Exercise 20.4

Assign IUPAC names to each of the following small peptides.

a. Gly–Ala–Leu   **b.** Gly–Tyr–Ser–Ser

Answers: **a.** Glycylalanylleucine; **b.** Glycyltyrosylserylserine

Isomeric Peptides

Peptides that contain the same amino acids but in different order are different molecules (constitutional isomers) with different properties. For example, two different dipeptides can be formed from one molecule of alanine and one molecule of glycine.

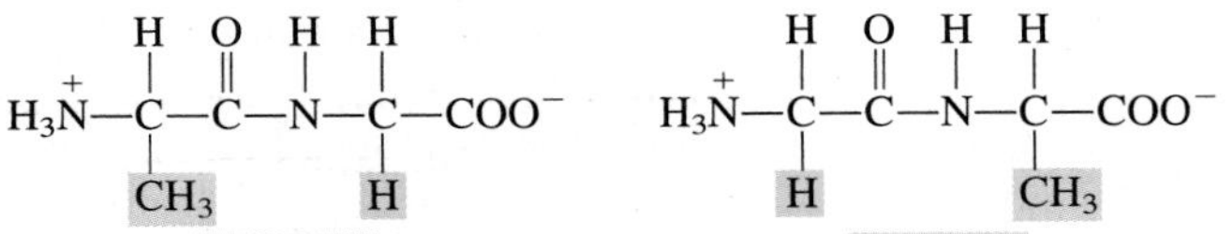

Ala–Gly   Gly–Ala

In the first dipeptide, the alanine is the N-terminal residue, and in the second molecule, it is the C-terminal residue. These two compounds are isomers with different chemical and physical properties.

The number of isomeric peptides possible increases rapidly as the length of the peptide chain increases. Let us consider the tripeptide Ala–Ser–Cys as another example. In addition to this sequence, five other arrangements of these three components are possible, each representing another isomeric tripeptide: Ala–Cys–Ser, Ser–Ala–Cys, Ser–Cys–Ala, Cys–Ala–Ser, and Cys–Ser–Ala. For a pentapeptide containing 5 different amino acids, 120 isomers are possible.

> Amino acid sequence in a peptide has biochemical importance. Isomeric peptides give different biochemical responses; that is, they have different biochemical specificities.

> For a peptide containing one each of *n* different kinds of amino acids, the number of constitutional isomers is given by *n!* (*n* factorial).
>
> $5! = 5 \times 4 \times 3 \times 2 \times 1 = 120$

20.7 BIOCHEMICALLY IMPORTANT SMALL PEPTIDES

Many relatively small peptides have been shown to be biochemically active. Functions for them include hormonal action, neurotransmission, and antioxidant activity.

Small Peptide Hormones

The two best-known peptide hormones, both produced by the pituitary gland, are *oxytocin* and *vasopressin.* Each hormone is a nonapeptide (nine amino acid residues) with six of the residues held in the form of a loop by a disulfide bond formed from the interaction of two

Oxytocin plays a role in stimulating the flow of milk in a nursing mother. The baby's suckling action sends nerve signals to the mother's brain, triggering the release of oxytocin, via the blood, to the mammary glands. The oxytocin causes muscle contraction in the mammary gland, forcing out milk. As suckling continues, more oxytocin is released and more milk is available for the baby.

cysteine residues (Section 20.5). Structurally, these nonapeptides differ in the amino acid present in positions 3 and 8 of the peptide chain. In both structures an amide group replaces the C terminal single-bonded oxygen atom.

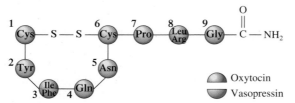

(Oxytocin regulates uterine contractions and lactation. Vasopressin regulates the excretion of water by the kidneys; it also affects blood pressure.)

Small Peptide Neurotransmitters

Enkephalins are pentapeptide neurotransmitters produced by the brain itself that bind at receptor sites in the brain to reduce pain. The two best-known enkephalins are Met-enkephalin and Leu-enkephalin, whose structures are

Met-enkephalin: Tyr–Gly–Gly–Phe–Met

Leu-enkephalin: Tyr–Gly–Gly–Phe–Leu

The two enkephalins differ structurally only in the amino acid at the end of the chain.

The pain-reducing effects of enkephalin action play a role in the "high" reported by long-distance runners, in the competitive athlete's managing to finish the game despite being injured, and in the pain-relieving effects of acupuncture.

The action of the prescription painkillers morphine and codeine is based on their binding at the same receptor sites in the brain as the naturally occurring enkephalins.

Small Peptide Antioxidants

Other antioxidants previously considered are BHA and BHT (Section 14.13) and β-carotene (Section 13.7).

The tripeptide *glutathione* (Glu–Cys–Gly) is present in significant concentrations in most cells and is of considerable physiological importance as a regulator of oxidation–reduction reactions. Specifically, glutathione functions as an antioxidant (Section 14.13), protecting cellular contents from oxidizing agents such as peroxides and superoxides (highly reactive forms of oxygen often generated within the cell in response to bacterial invasion) (Section 23.11).

The tripeptide structure of glutathione has an unusual feature. The amino acid Glu, an acidic amino acid, is bonded to Cys through the side-chain carboxyl group rather than through its α-carbon carboxyl group.

20.8 GENERAL STRUCTURAL CHARACTERISTICS OF PROTEINS

Proteins are the second type of biochemical polymer we have encountered; the other was polysaccharides (Section 18.14). Protein monomers are amino acids, whereas polysaccharide monomers are monosaccharides.

In Section 20.1, we defined a protein simply as a naturally occurring, unbranched polymer in which the monomer units are amino acids. A more specific protein definition is now in order. A **protein** *is a peptide in which at least 40 amino acid residues are present.* The defining line governing the use of the term *protein*—40 amino acid residues—is an arbitrary line. (The terms *polypeptide* and *protein* are often used interchangeably; a protein is a relatively long polypeptide. The key point is that the term *protein* is reserved for peptides with a large number of amino acids; it is not correct to call a tripeptide a protein. Over 10,000 amino acid residues are present in several proteins; 400–500 amino acid residues are common in proteins; small proteins contain 40–100 amino acid residues.

More than one peptide chain may be present in a protein. On this basis, proteins are classified as *monomeric* or *multimeric*. A **monomeric protein** *is a protein in which only one peptide chain is present*. Large proteins, those with many amino acid residues, usually are multimeric. A **multimeric protein** *is a protein in which more than one peptide chain is present*. The peptide chains present in multimeric proteins are called *protein subunits*. The protein subunits within a multimeric protein may all be identical to each other or different kinds of subunits may be present. Proteins with up to 12 subunits are known. The small protein insulin, which functions as a hormone in the human body, is a multimeric protein with two protein subunits; one subunit contains 21 amino acid residues and the other 30 amino acid residues. The structure of insulin is considered in more detail in Section 20.11.

Proteins, on the basis of chemical composition, are classified as *simple* or *complex*. A **simple protein** *is a protein in which only amino acid residues are present*. More than one protein subunit may be present in a simple protein, but all subunits contain only amino acids. A **conjugated protein** *is a protein that has one or more non-amino acid entities present in its structure in addition to one or more peptide chains*. These non-amino acid components, which may be organic or inorganic, are called *prosthetic groups*. A **prosthetic group** *is a non-amino acid group present in a conjugated protein*.

Conjugated proteins may be further classified according to the nature of the prosthetic group(s) present. *Lipoproteins* contain lipid prosthetic groups, *glycoproteins* contain carbohydrate groups, *metalloproteins* contain a specific metal, and so on (see Table 20.3). Some proteins contain more than one type of prosthetic group. In general, prosthetic groups have important roles in the biochemical functions for conjugated proteins. Several examples of glycoproteins and lipoproteins are discussed in Sections 20.17 and 20.18, respectively.

In general, the three-dimensional structures of proteins, even those with just a single peptide chain, are more complex than those of carbohydrates and lipids—the biomolecules discussed in the two previous chapters. Our approach to describing and understanding this complexity in protein structure involves considering this structure at four levels. These four protein structural levels, listed in order of increasing complexity, are *primary* structure, *secondary* structure, *tertiary* structure, and *quaternary* structure. They are the subject matter for the next four sections of this chapter.

TABLE 20.3
Types of Conjugated Proteins

Class	Prosthetic Group	Specific Example	Function of Example
hemoproteins	heme unit	hemoglobin	carrier of O_2 in blood
		myoglobin	oxygen binder in muscles
lipoproteins	lipid	low-density lipoprotein (LDL)	lipid carrier
		high-density lipoprotein (HDL)	lipid carrier
glycoproteins	carbohydrate	gamma globulin	antibody
		mucin	lubricant in mucous secretions
		interferon	antiviral protection
phosphoproteins	phosphate group	glycogen phosphorylase	enzyme in glycogen phosphorylation
nucleoproteins	nucleic acid	ribosomes	site for protein synthesis in cells
		viruses	self-replicating, infectious complex
metalloproteins	metal ion	iron–ferritin	storage complex for iron
		zinc–alcohol dehydrogenase	enzyme in alcohol oxidation

 PRIMARY STRUCTURE OF PROTEINS

Primary protein structure *is the order in which amino acids are linked together in a protein.* Every protein has its own unique amino acid sequence. Primary protein structure always involves more than just the numbers and kinds of amino acids present; it also involves the *order of attachment* of the amino acids to each other through peptide bonds.

Insulin, the hormone that regulates blood-glucose levels, was the first protein for which primary structure was determined; the "sequencing" of its 51 amino acids was completed in 1953, after 8 years of work by the British biochemist Frederick Sanger (see Figure 20.3). Today, primary structures are known for many thousands of proteins, and the sequencing procedures involve automated methods that require relatively short periods of time (days). Figure 20.4 shows the primary structure of myoglobin, a protein involved in oxygen transport in muscles; it contains 153 amino acids assembled in the particular, definite order shown in this diagram.

The primary structure of a specific protein is always the same regardless of where the protein is found within an organism. The structures of certain proteins are even similar

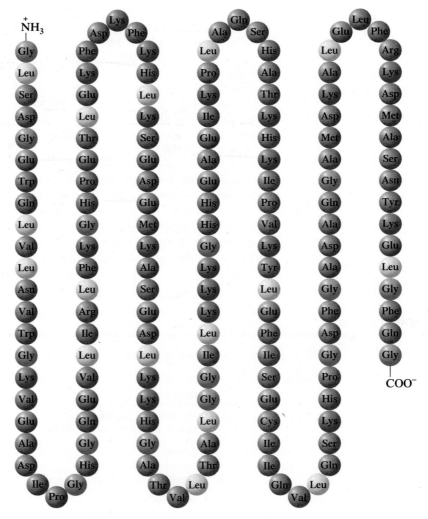

Figure 20.4 The primary structure of human myoglobin. This diagram gives only the sequence of the amino acids present and conveys no information about the actual three-dimensional shape of the protein. The "wavy" pattern for the 153 amino acid sequence was chosen to minimize the space used to present the needed information. The actual shape of the protein is determined by secondary and tertiary levels of protein structure, levels yet to be discussed.

CHEMICAL Connections

Substitutes for Human Insulin

In humans, an insufficient production of insulin results in the disease *diabetes mellitus*. Treatment of this disease often involves giving the patient extra insulin via subcutaneous injection. For many years, because of the limited availability of human insulin, most insulin used by diabetics was obtained from the pancreases of slaughter-house animals. Such animal insulin, primarily from cows and pigs, was used by most diabetics without serious side effects because it is structurally very similar to human insulin. Immunological reactions gradually do increase over time, however, because the animal insulin is foreign to the human body.

A comparison of the primary structure of human insulin with pig and cow insulins shows differences at only 4 of the 51 amino acid positions: positions 8, 9, and 10 on chain A and position 30 on chain B (see Figure 20.11 and the following table).

Species	Chain A			Chain B
	#8	**#9**	**#10**	**#30**
human	Thr	Ser	Ile	Thr
pig (porcine)	Thr	Ser	Ile	Ala
cow (bovine)	Ala	Ser	Val	Ala

The dependence of diabetics on animal insulin has declined because of the availability of human insulin produced by genetically engineered bacteria (Section 22.14). These bacteria carry a gene that directs the synthesis of human insulin. Such bacteria-produced insulin is fully functional. All diabetics now have the choice of using human insulin or using animal insulin. Many still continue to use the animal insulin because it is cheaper.

The *primary structure* of a protein is the *sequence* of amino acids in a protein chain—that is, the order in which the amino acids are connected to each other.

among different species of animals. For example, the primary structures of insulin in cows, pigs, sheep, and horses are very similar both to each other and to human insulin. Until recently, this similarity was particularly important for diabetics who required supplemental injections of insulin. (See the Chemical Connections feature above.)

An analogy is often drawn between the primary structure of proteins and words. Words, which convey information, are formed when the 26 letters of the English alphabet are properly sequenced. Proteins are formed from proper sequences of the 20 standard amino acids. Just as the proper sequence of letters in a word is necessary for it to make sense, the proper sequence of amino acids is necessary to make biochemically active protein. Furthermore, the letters that form a word are written from left to right, as are amino acids in protein formulas. As any dictionary of the English language will document, a tremendous variety of words can be formed by different letter sequences. Imagine the number of amino acid sequences possible for a large protein. There are 1.55×10^{66} sequences possible for the 51 amino acids found in insulin! From these possibilities, the body reliably produces only *one,* illustrating the remarkable precision of life processes. From the simplest bacterium to the human brain cell, only those amino acid sequences needed by the cell are produced. The fascinating process of protein biosynthesis and the way in which genes in DNA direct this process will be discussed in Chapter 22.

SECONDARY STRUCTURE OF PROTEINS

Secondary protein structure *is the arrangement in space adopted by the backbone portion of a protein.* The two most common types of secondary structure are the *alpha helix (α helix)* and the *beta pleated sheet (β pleated sheet).* The type of interaction responsible for both of these types of secondary structure is hydrogen bonding (Section 7.13) between a carbonyl oxygen atom of a peptide linkage and the hydrogen atom of an amino group of another peptide linkage farther along the backbone. Information about the geometry associated with these peptide linkages is helpful in understanding how hydrogen bonding interactions occur between peptide linkages of a protein backbone. Important geometrical considerations are:

1. The peptide linkages are essentially planar. This means that for two amino acids linked through a peptide linkage, six atoms lie in the same plane: the α-carbon atom and the C=O group from the first amino acid and the N–H group and the α-carbon atom from the second amino acid.

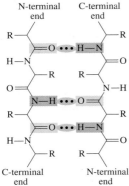

Figure 20.5 The hydrogen bonding between the carbonyl oxygen atom of one peptide linkage and the amide hydrogen atom of another peptide linkage.

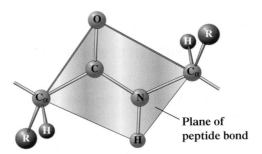

Plane of peptide bond

2. The planar peptide linkage structure has considerable rigidity, which means that rotation of groups about the C—N bond is hindered, and *cis–trans* isomerism is possible about this bond. The *trans* isomer orientation is the preferred orientation, as shown in the preceding diagram. The O atom of the C=O group and the H atom of the N—H group are positioned *trans* to each other.

Figure 20.5 shows the hydrogen bonding possibilities that exist between carbonyl oxygen atoms and amide hydrogen atoms associated with different peptide linkages in a protein backbone. The protein backbone segments shown can be two segments of the same backbone or two segments from different backbones. We consider both of these situations in further detail in this section.

The Alpha Helix

An **alpha helix structure** *is a protein secondary structure in which a single protein chain adopts a shape that resembles a coiled spring (helix), with the coil configuration maintained by hydrogen bonds.* The hydrogen bonds are between $\rangle$N—H and $\rangle$C=O groups of every fourth amino acid, as is shown diagrammatically in Figure 20.6.

Proteins have varying amounts of α-helical secondary structure, ranging from a few percent to nearly 100%. In an α helix, all of the amino acid side chains (R groups) lie outside the helix; there is not enough room for them in the interior. Figure 20.6d illustrates this situation.

The Beta Pleated Sheet

A **beta pleated sheet structure** *is a protein secondary structure in which two fully extended protein chain segments in the same or different molecules are held together by*

The hydrogen bonding present in an α helix is *intra*molecular. In a β pleated sheet, the hydrogen bonding can be *inter*molecular (between two different chains) or *intra*molecular (a single chain folding back on itself).

Figure 20.6 Four representations of the α helix protein secondary structure. (a) Arrangement of protein backbone with no detail shown. (b) Backbone arrangement with hydrogen-bonding interactions shown. (c) Backbone atomic detail shown, as well as hydrogen-bonding interactions. (d) Top view of an α helix showing that amino acid side chains (R groups) point away from the long axis of the helix.

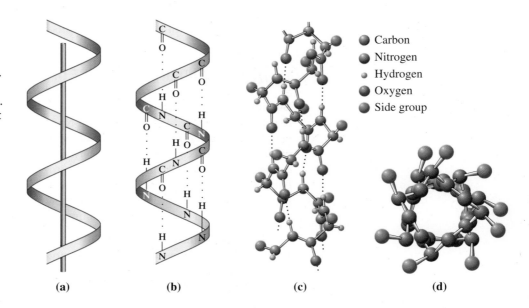

- Carbon
- Nitrogen
- Hydrogen
- Oxygen
- Side group

(a) (b) (c) (d)

hydrogen bonds. Hydrogen bonds form between oxygen and hydrogen peptide linkage atoms that are either in different parts of a single chain that folds back on itself (intrachain bonds) or between atoms in different peptide chains in those proteins that contain more than one chain (interchain bonds). In molecules where the β pleated sheet involves a single molecule, several U-turns in the protein chain arrangement are needed in order to form the structure.

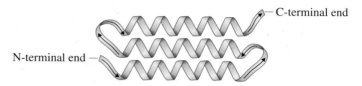

This "U-turn structure" is the most frequently encountered type of β pleated sheet structure.

Figure 20.7a shows a representation of the β pleated sheet structure that occurs when portions of two different peptide chains are aligned parallel to each other (interchain bonds). The term *pleated sheet* arises from the repeated zigzag pattern in the structure (Figure 20.7b). Note how in a pleated sheet structure the amino acid side chains are positioned above and below the plane of the sheet.

Very few proteins have entirely α helix or β pleated sheet structures. Instead, most proteins have only certain portions of their molecules in these conformations. The rest of the molecule consists of "unstructured segments." It is possible to have both α helix and β pleated sheet structures within the same protein. Figure 20.8 is a diagram of a protein chain where both helical and pleated sheet segments, as well as unstructured segments, are present within a single peptide chain. The β pleated sheet segment involves a single peptide chain folding back on itself (intrachain bonds). Helical structure and pleated sheet structure are found only in portions of a protein where the amino acid R groups present are relatively small; large R groups tend to disrupt both of these types of secondary structure. The term *unstructured* used in describing portions of a protein structure is somewhat of a misnomer, because all molecules of a given protein exhibit *identical* unstructured segments.

An active area of protein research at present involves learning more about the biochemical functions for the unstructured portions of proteins. A growing number of researchers now believe that some "unstructure" is essential to the functioning of many proteins. It confers flexibility to proteins, thereby allowing them to interact with several different substances, an important mechanism for rapid response to changing cellular

The β pleated sheet is found extensively in the protein of silk. Because such proteins are already fully extended, silk fibers cannot be stretched. When wool, which has an α helix structure, becomes wet, it stretches as hydrogen bonds of the helix are broken. The wool returns to its original shape as it dries. Wet stretched wool, dried under tension, maintains its stretched length because it has assumed a β pleated sheet configuration.

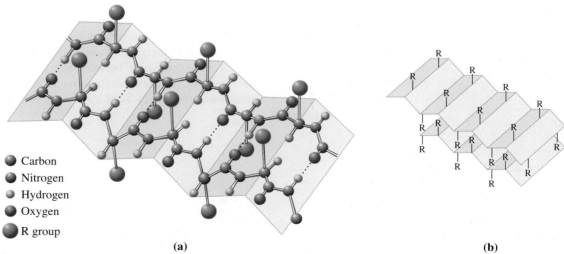

Carbon
Nitrogen
Hydrogen
Oxygen
R group

(a) **(b)**

Figure 20.7 Two representations of the β pleated sheet protein structure. (a) A representation emphasizing the hydrogen bonds between protein chains. (b) A representation emphasizing the pleats and the location of the R groups.

Figure 20.8 The secondary structure of a single protein often shows areas of α helix and β pleated sheet configurations, as well as areas of "unstructure."

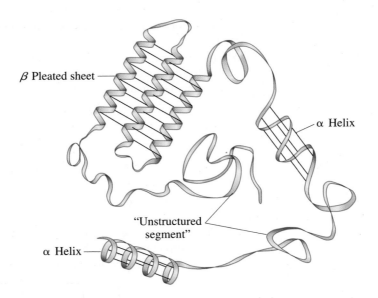

β Pleated sheet

α Helix

"Unstructured segment"

α Helix

conditions. Often unstructured regions of a protein have the flexibility to bind to several different protein partners, allowing the same amino acid sequence to multitask. The act of binding to another protein does bring added structure to the unstructured portion of the protein, but the added structure is lost when the binding interaction ceases.

TERTIARY STRUCTURE OF PROTEINS

Tertiary protein structure *is the overall three-dimensional shape of a protein that results from the interactions between amino acid side chains (R groups) that are widely separated from each other within a peptide chain.*

A good analogy for the relationships among the primary, secondary, and tertiary structures of a protein is that of a telephone cord (Figure 20.9). The primary structure is the long, straight cord. The coiling of the cord into a helical arrangement gives the secondary structure. The supercoiling arrangement the cord adopts after you hang up the receiver is the tertiary structure.

Interactions Responsible for Tertiary Structure

Four types of attractive interactions contribute to the tertiary structure of a protein: (1) covalent disulfide bonds, (2) electrostatic attractions (salt bridges), (3) hydrogen bonds, and (4) hydrophobic attractions. All four of these interactions are interactions between amino acid R groups. This is a major distinction between tertiary-structure interactions and secondary-structure interactions. Tertiary-structure interactions involve the R groups of amino acids; secondary-structure interactions involve the peptide linkages between amino acid residues.

Figure 20.9 A telephone cord has three levels of structure. These structural levels are a good analogy for the first three levels of protein structure.

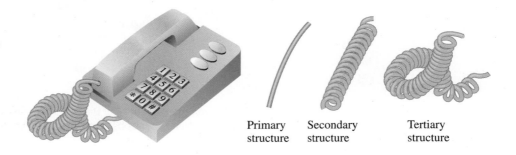

Primary structure

Secondary structure

Tertiary structure

Figure 20.10 Disulfide bonds involving cysteine residues can form in two different ways: (a) between two —SH groups on the same chain or (b) between two —SH groups on different chains.

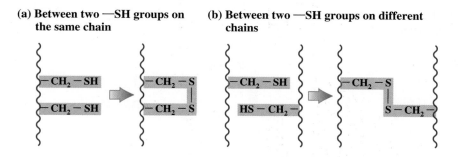

(a) Between two —SH groups on the same chain

(b) Between two —SH groups on different chains

Cysteine is the only α-amino acid that contains a sulfhydryl group (—SH).

Disulfide bonds, the strongest of the tertiary-structure interactions, result from the —SH groups of two cysteine residues reacting with each other to form a *covalent* disulfide bond (Section 20.5). This type of interaction is the only one of the four tertiary-structure interactions that involves a covalent bond. Disulfide bond formation may involve two cysteine units in the same peptide chain (an intramolecular disulfide bond; see Figure 20.10a) or two cysteine units in different chains (an intermolecular disulfide bond; see Figure 20.10b). Figure 20.11 gives the structure of the protein hormone insulin, a protein that has two peptide chains and a total of 51 amino acid residues; both inter- and intramolecular disulfide bonds are present in its structure.

Electrostatic interactions, also called *salt bridges,* always involve the interaction between an acidic side chain (R group) and a basic side chain (R group). The side chains of acidic and basic amino acids, at the appropriate pH, carry charges, with the acidic side chain being negatively charged and the basic side chain being positively charged. Such side chain charges occur when a —COOH group becomes a —COO$^-$ group and when a —NH$_2$ group becomes a — $\overset{+}{\text{N}}$H$_3$ group. The interaction that occurs between the two types of side chains is a positive–negative ion–ion attraction. Figure 20.12b shows an electrostatic interaction.

Hydrogen bonds can occur between amino acids with polar R groups. A variety of polar side chains can be involved, especially those that possess the following functional groups:

$$-\text{OH} \qquad -\text{NH}_2 \qquad \overset{\displaystyle O}{\overset{\|}{-\text{C}}}-\text{OH} \qquad \overset{\displaystyle O}{\overset{\|}{-\text{C}}}-\text{NH}_2$$

Hydrogen bonds are relatively weak and are easily disrupted by changes in pH and temperature. Figure 20.12c shows the hydrogen-bonding interactions between the R groups of glutamine and serine.

Figure 20.11 Human insulin, a small two-chain protein, has both intrachain and interchain disulfide linkages as part of its tertiary structure.

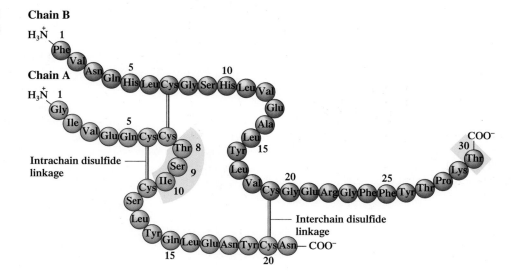

Chain B

H$_3$$\overset{+}{\text{N}}$ 1

Chain A

H$_3$$\overset{+}{\text{N}}$ 1

Intrachain disulfide linkage

Interchain disulfide linkage

Figure 20.12 Four types of interactions between amino acid R groups produce the tertiary structure of a protein. (a) Disulfide bonds. (b) Electrostatic interactions (salt bridges). (c) Hydrogen bonds. (d) Hydrophobic interactions.

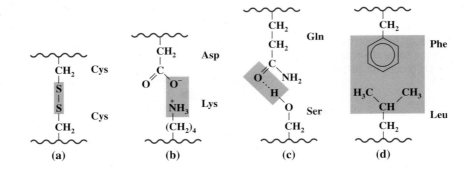

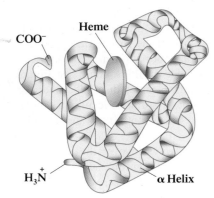

Figure 20.13 A schematic diagram showing the tertiary structure of the single-chain protein myoglobin.

Hydrophobic interactions result when two nonpolar side chains are close to each other. In aqueous solution, many proteins have their polar R groups outward, toward the aqueous solvent (which is also polar), and their nonpolar R groups inward (away from the polar water molecules). The nonpolar R groups then interact with each other. The attractive forces are London forces (Section 7.13) resulting from the momentary uneven distribution of electrons within the side chains. Hydrophobic interactions are common between phenyl rings and alkyl side chains. Although hydrophobic interactions are weaker than hydrogen bonds or electrostatic interactions, they are a significant force in some proteins because there are so many of them; their cumulative effect can be greater in magnitude than the effects of hydrogen bonding. Figure 20.12d shows the hydrophobic interactions between the R groups of phenylalanine and leucine.

In 1959, a protein tertiary structure was determined for the first time. The determination involved myoglobin, a conjugated protein (Section 20.8) whose function is oxygen storage in muscle tissue. Figure 20.13 shows myoglobin's tertiary structure. It involves

● E X A M P L E 2 0 . 5

Drawing Structural Representations for Amino Acid Side Chain Interactions Associated with Protein Tertiary Structure

▶ Identify the type of noncovalent interaction that occurs between the side chains of the following amino acids and show the interaction using structural representations for the side chains.

a. Serine and asparagine
b. Glutamic acid and lysine

Solution

a. Both serine and asparagine are polar neutral amino acids (see Table 20.1). The side chains of such amino acids interact through *hydrogen bonding*. A structural representation for this hydrogen bonding interaction is:

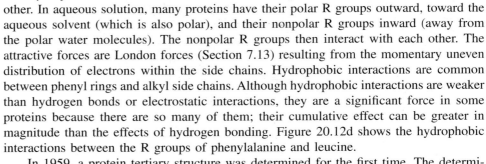

b. Glutamic acid and lysine have, respectively, acidic and basic side chains (see Table 20.1). Such side chains carry a charge, in solution, as the result of proton transfer. The interaction between the negatively charged acidic side chain and the positively charged basic side chain is an *electrostatic interaction*. A structural representation for this electrostatic interaction is:

Practice Exercise 20.5

Identify the type of noncovalent interaction that occurs between the side chains of the following amino acids and show the interaction using structural representations for the side chains.

a. Threonine and asparagine
b. Aspartic acid and arginine

Answers:

a. Hydrogen bonding;

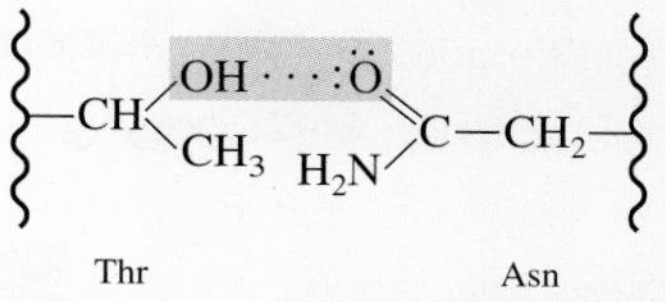

Thr   Asn

b. Electrostatic interaction;

Asp   Arg

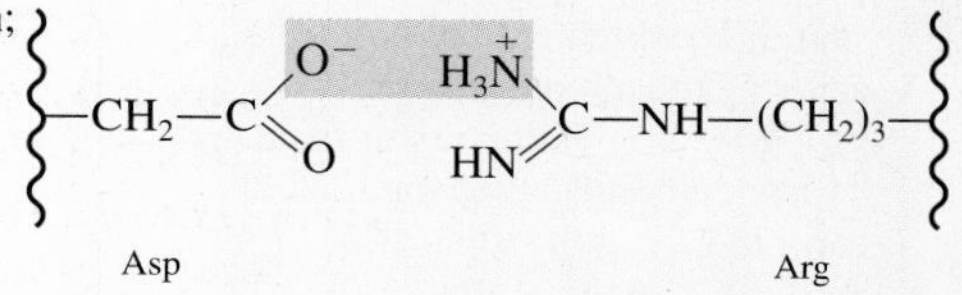

a single peptide chain of 153 amino acids with numerous α helix segments within the chain. The structure also contains a prosthetic heme group, an iron-containing group with the ability to bind molecular oxygen.

A comparison of Figure 20.13 (myoglobin's tertiary structure) with Figure 20.4 (myoglobin's primary structure) shows how different the perspectives of primary and tertiary structure are for a protein.

20.12 QUATERNARY STRUCTURE OF PROTEINS

Quaternary structure is the highest level of protein organization. It is found only in multimeric proteins (Section 20.8). Such proteins have structures involving two or more peptide chains that are independent of each other—that is, are not covalently bonded to each other. **Quaternary protein structure** *is the organization among the various peptide chains in a multimeric protein.*

Most multimeric proteins contain an even number of subunits (two subunits = a dimer, four subunits = a tetramer, and so on). The subunits are held together mainly by hydrophobic interactions between amino acid R groups.

The noncovalent interactions that contribute to tertiary structure (electrostatic interactions, hydrogen bonds, and hydrophobic interactions) are also responsible for the maintenance of quaternary structure. The noncovalent interactions that contribute to quaternary structure are, however, more easily disrupted. For example, only small changes in cellular conditions can cause a tetrameric protein to fall apart, dissociating into dimers or perhaps four separate subunits, with a resulting temporary loss of protein activity. As original cellular conditions are restored, the tertiary structure automatically re-forms, and normal protein function is restored.

An example of a protein with quaternary structure is hemoglobin, the oxygen-carrying protein in blood (Figure 20.14). It is a tetramer in which there are two identical α chains and two identical β chains. Each chain enfolds a heme group, the site where oxygen binds to the protein.

The Chemistry at a Glance feature on page 678 reviews what we have said about protein structural levels.

CHEMISTRY AT A GLANCE

Protein Structure

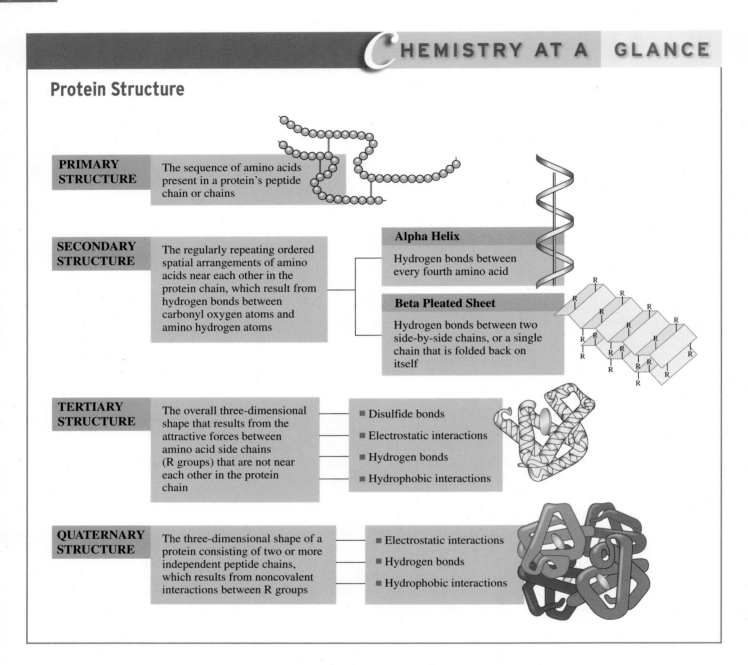

PRIMARY STRUCTURE
The sequence of amino acids present in a protein's peptide chain or chains

SECONDARY STRUCTURE
The regularly repeating ordered spatial arrangements of amino acids near each other in the protein chain, which result from hydrogen bonds between carbonyl oxygen atoms and amino hydrogen atoms

Alpha Helix
Hydrogen bonds between every fourth amino acid

Beta Pleated Sheet
Hydrogen bonds between two side-by-side chains, or a single chain that is folded back on itself

TERTIARY STRUCTURE
The overall three-dimensional shape that results from the attractive forces between amino acid side chains (R groups) that are not near each other in the protein chain

- Disulfide bonds
- Electrostatic interactions
- Hydrogen bonds
- Hydrophobic interactions

QUATERNARY STRUCTURE
The three-dimensional shape of a protein consisting of two or more independent peptide chains, which results from noncovalent interactions between R groups

- Electrostatic interactions
- Hydrogen bonds
- Hydrophobic interactions

Figure 20.14 A schematic diagram showing the tertiary and quaternary structure of the oxygen-carrying protein hemoglobin.

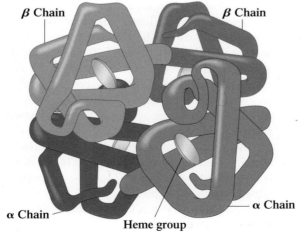

β Chain   β Chain

α Chain   α Chain

Heme group

PROTEIN CLASSIFICATION BASED ON SHAPE

Based on molecular shape, which is determined primarily by tertiary and quaternary structural features, there are three main types of proteins: *fibrous, globular,* and *membrane.* A **fibrous protein** *is a protein whose molecules have an elongated shape with one dimension much longer than the others.* Fibrous proteins tend to have simple, regular, linear structures. There is a tendency for such proteins to aggregate together to form macromolecular structures. A **globular protein** *is a protein whose molecules have peptide chains that are folded into spherical or globular shapes.* The folding in such proteins is such that most of the amino acids with hydrophobic side chains (nonpolar R groups) are in the interior of the molecule and most of the hydrophilic side chains (polar R groups) are on the outside of the molecule. Generally, globular proteins are water-soluble substances. A **membrane protein** *is a protein that is found associated with a membrane system of a cell.* Membrane protein structure is somewhat opposite that of globular proteins, with most of the hydrophobic amino acid side chains oriented outward. Thus, such proteins tend to be water-insoluble and they usually have fewer hydrophobic amino acids than globular proteins.

Further discussion of proteins in this section is limited to fibrous and globular proteins, as membrane proteins were considered in the previous chapter in relation to cell membrane structure (Section 19.10).

Table 20.4 gives examples of selected fibrous and globular proteins. Comparison of fibrous and globular proteins in terms of general properties shows the following differences.

1. Fibrous proteins are generally water-insoluble, whereas globular proteins dissolve in water. This enables globular proteins to travel through the blood and other body fluids to sites where their activity is needed.
2. Fibrous proteins usually have a single type of secondary structure, whereas globular proteins often contain several types of secondary structure.
3. Fibrous proteins generally have structural functions that provide support and external protection, whereas globular proteins are involved in metabolic chemistry, performing functions such as catalysis, transport, and regulation.
4. The number of different kinds of globular protein far exceeds the number of different kinds of fibrous protein. However, because the most abundant proteins in the human body are fibrous proteins rather than globular proteins, the total mass of fibrous proteins present exceeds the total mass of globular proteins present.

We will now more closely examine the characteristics of two fibrous proteins (α keratin and collagen) and two globular proteins (hemoglobin and myoglobin) as representatives of their types.

TABLE 20.4
Some Common Fibrous and Globular Proteins

Name	Occurrence and Function
Fibrous proteins (insoluble)	
keratins	found in wool, feathers, hooves, silk, and fingernails
collagens	found in tendons, bone, and other connective tissue
elastins	found in blood vessels and ligaments
myosins	found in muscle tissue
fibrin	found in blood clots
Globular proteins (soluble)	
insulin	regulatory hormone for controlling glucose metabolism
myoglobin	involved in oxygen storage in muscles
hemoglobin	involved in oxygen transport in blood
transferrin	involved in iron transport in blood
immunoglobulins	involved in immune system responses

Figure 20.15 The tail feathers of a peacock contain the fibrous protein α keratin.

Natural silk (silkworm silk) and spider silk (spider webs) are made of *fibroin,* a fibrous protein that exists mainly in a beta pleated sheet form. The great strength and toughness of silk fibers, which exceed those of many synthetic fibers, is related to the *close* stacking of the beta sheets. A high percentage of the amino acid residues (primary structure) in silk are either glycine (R = H) or alanine (R = CH_3). It is the smallness of these two R groups that makes the close stacking possible.

Collagen, pronounced "KAHL-uh-jen," is the most abundant protein in the human body.

α Keratin

The fibrous protein α *keratin* is particularly abundant in nature, where it is found in protective coatings for organisms. It is the major protein constituent of hair, feathers (Figure 20.15), wool, fingernails and toenails, claws, scales, horns, turtle shells, quills, and hooves.

The structure of a typical α keratin, that of hair, is depicted in Figure 20.16. The individual molecules are almost wholly α helical (Figure 20.16a). Pairs of these helices twine about one another to produce a coiled coil (Figure 20.16b). In hair, two of the coiled coils then further twist together to form a protofilament (Figure 20.16c). Protofilaments then coil together in groups of four to form microfilaments (Figure 20.16d), which become the "core" unit in the structure of the α keratin of hair. These microfilaments in turn coil at even higher levels. This coiling at higher and higher levels is what produces the strength associated with α-keratin-containing proteins. All levels of coiling organization are stabilized by attractive forces of the types previously considered in the discussion of generalized secondary and tertiary protein structure (Sections 20.10 and 20.11). Particularly important are *inter*coil disulfide bridges that form between cysteine residues.

Introduction of disulfide bridges within the several levels of coiling structure determines the "hardness" of an α keratin. "Hard" keratins, such as those found in horns and nails, have considerably more disulfide bridges than their softer counterparts found in hair, wool, and feathers.

Collagen

Collagen, the most abundant of all proteins in humans (30% of total body protein), is a major structural material in tendons, ligaments, blood vessels, and skin; it is also the organic component of bones and teeth. Table 20.5 gives the collagen content of selected body tissues. The predominant structural feature within collagen molecules is a *triple helix* formed when three chains of amino acids wrap around each other to give a ropelike arrangement of polypeptide chains (see Figure 20.17).

The rich content of the amino acid proline (up to 20%) in collagen is one reason why it has a triple-helix conformation rather than the simpler α helix structure (Section 20.10). Proline amino acid residues do not fit into regular α helices because of the cyclic nature of the side chain present and its accompanying different "geometry."

Portion of a collagen chain

Figure 20.16 The coiled-coil structure of the fibrous protein α keratin.

(a) α Helix

(b) Coiled coil of two α helices

(c) Protofilament (pair of coiled coils)

(d) Microfilament (four coiled protofilaments)

Figure 20.17 A schematic diagram emphasizing how three helical polypeptide chains intertwine to form a triple helix. The chains are partially unwound and cut away to show their structure.

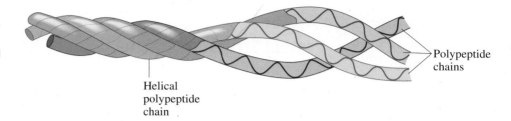

Helical polypeptide chain

Polypeptide chains

TABLE 20.5
The Collagen Content of Selected Body Tissues

Tissue	Collagen (% dry mass)
Achilles tendon	86
aorta	12–24
bone (mineral-free)	88
cartilage	46–63
cornea	68
ligament	17
skin	72

The hemoglobin of a fetus is slightly different in structure from adult hemoglobin. Called *fetal hemoglobin,* this hemoglobin has a greater affinity for oxygen than the mother's hemoglobin. This ensures a steady flow of oxygen to the fetus. Shortly after birth, a baby's body ceases to produce fetal hemoglobin, and its production of "adult" hemoglobin begins.

Figure 20.18 Electron micrograph of collagen fibers.

The function of hemoglobin is oxygen *transfer,* and the function of myoglobin is oxygen *storage.*

Collagen molecules (triple helices) are very long, thin, and rigid. Many such molecules, lined up alongside each other, combine to make collagen fibrils. Cross-linking between helices gives the fibrils extra strength. The greater the number of cross links, the more rigid the fibril is. The stiffening of skin and other tissues associated with aging is thought to result, at least in part, from an increasing amount of cross-linking between collagen molecules. The process of tanning, which converts animal hides to leather, involves increasing the degree of cross-linking.

Figure 20.18 shows an electron micrograph of collagen fibers.

Hemoglobin

The globular protein *hemoglobin* transports oxygen from the lungs to tissue. Its tertiary structure was shown in Figure 20.14. It is a tetramer (four peptide chains) with each subunit also containing a *heme group,* the entity that binds oxygen. With four heme groups present, a hemoglobin molecule can transport four oxygen molecules at the same time.

The structure of a heme group is

$$\text{Heme}$$

It is the iron atom at the center of the heme molecule that actually interacts with the O_2.

Myoglobin

The globular protein *myoglobin* functions as an oxygen storage molecule in muscles. Its tertiary structure was shown in Figure 20.13. Myoglobin is a monomer, whereas hemoglobin is a tetramer. That is, myoglobin consists of a single peptide chain and a heme unit, and hemoglobin has four peptide chains and four heme units. Thus only one O_2 molecule can be carried by a myoglobin molecule. The tertiary structure of the single peptide chain of myoglobin is almost identical to the tertiary structure of each of the subunits of hemoglobin.

Myoglobin has a higher affinity for oxygen than does hemoglobin. Thus the transfer of oxygen from hemoglobin to myoglobin occurs readily. Oxygen stored in myoglobin molecules serves as a reserve oxygen source for working muscles when their demand for oxygen exceeds that which can be supplied by hemoglobin.

The Chemical Connections feature on page 682 considers how the amount of myoglobin present in muscle tissue is related to the color of the meats that humans eat.

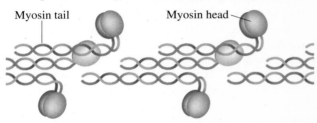

CHEMICAL Connections

Protein Structure and the Color of Meat

The meat that humans eat is composed primarily of muscle tissue. The major proteins present in such muscle tissue are *myosin* and *actin*, which lie in alternating layers and which slide past each other during muscle contraction. Contraction is temporarily maintained through interactions between these two types of proteins.

Structurally, myosin consists of a rodlike coil of two alpha helices (fibrous protein) with two globular protein heads. It is the "head portions" of myosin that interact with the actin.

Myosin tail   Myosin head

Structurally, actin has the appearance of two filaments spiraling about one another (see diagram below). Each circle in this structural diagram represents a monomeric unit of actin (called globular actin). The monomeric actin units associate to form a long polymer (called fibrous actin). Each identical monomeric actin unit is a globular protein containing many amino acid residues.

The chemical process associated with muscle contraction (interaction between myosin and actin) requires molecular oxygen. The oxygen storage protein *myoglobin* (Section 20.13) is the oxygen source. The amount of myoglobin present in a muscle is determined by how the muscle is used. Heavily used muscles require larger amounts of myoglobin than infrequently used muscles require.

The amount of myoglobin present in muscle tissue is a major determiner of the color of the muscle tissue. Myoglobin molecules have a red color when oxygenated and a purple color when deoxygenated. Thus, heavily worked muscles have a darker color than infrequently used muscles.

The different colors of meat reflect the concentration of myoglobin in the muscle tissue. In turkeys and chickens, which walk around a lot but rarely fly, the leg meat is dark, the breast meat is white. On the other hand, game birds that do fly a lot have dark breast meat. In general, game animals (which use all of their muscles regularly) tend to have darker meat than domesticated animals.

All land animals and birds need to support their own weight. Fish, on the other hand, are supported by water as they swim, which reduces the need for myoglobin oxygen support. Hence fish tend to have lighter flesh. Fish that spend most of their time lying at the bottom of a body of water have the lightest (whitest) flesh of all. Salmon flesh contains additional pigments that give it its characteristic "orange-pink" color.

Meat, when cooked, turns brown as the result of changes in myoglobin structure caused by the heat; the iron atom in the heme unit of myoglobin (Section 20.13) becomes oxidized. When meat is heavily salted with preservatives (NaCl, NaNO$_2$, or the like), as in the preparation of ham, the myoglobin picks up nitrite ions, and its color changes to pink.

20.14 PROTEIN CLASSIFICATION BASED ON FUNCTION

Proteins play crucial roles in almost all biochemical processes. The diversity of functions exhibited by proteins far exceeds that of other major types of biochemical molecules. The functional versatility of proteins stems from (1) their ability to bind small molecules specifically and strongly to themselves, (2) their ability to bind other proteins, often other like proteins, to form fiber-like structures, and (3) their ability to bind to, and often become integrated into, cell membranes.

The following list gives a number of the major categories of proteins based on function. The order is not to be taken as an indication of relative functional importance. The list is also to be considered a selective list with a number of functions not included because of space considerations.

1. **Catalytic proteins.** Proteins are probably best known for their role as catalysts. Proteins with the role of biochemical catalyst are called *enzymes*. An entire chapter in this text, Chapter 21, is devoted to this most important role for proteins. Enzymes participate in almost all of the metabolic reactions that occur in cells. The chemistry of human genetics, to be discussed in detail in Chapter 22, is very dependent on the presence of enzymes.

2. **Defense proteins.** These proteins, also called *immunoglobulins* or *antibodies* (Section 20.17), are central to the functioning of the body's immune system. They bind to foreign substances, such as bacteria and viruses, to help combat invasion of the body by foreign particles.

3. **Transport proteins.** These proteins bind to particular small biomolecules and transport them to other locations in the body and then release the small molecules as needed at the destination location. The most well-known example of a transport protein is *hemoglobin* (Section 20.12), which carries oxygen from the lungs to other organs and tissues. Another transport protein is *transferrin,* which carries iron from the liver to the bone marrow. *High-* and *low-density lipoproteins* (Section 20.18) are carriers of cholesterol in the bloodstream.

4. **Messenger proteins.** These proteins transmit signals to coordinate biochemical processes between different cells, tissues, and organs. A number of hormones (Section 19.12) that regulate body processes are messenger proteins, including *insulin* and *glucagon* (Section 24.9). *Human growth hormone* is another example of a messenger protein.

5. **Contractile proteins.** These proteins are necessary for all forms of movement. Muscles are composed of filament-like contractile proteins that, in response to nerve stimuli, undergo conformation changes that involve contraction and extension. *Actin* and *myosin* (Section 20.13) are examples of such proteins. Human reproduction depends on the movement of sperm. Sperm can "swim" because of long flagella made up of contractile proteins.

6. **Structural proteins.** These proteins confer stiffness and rigidity to otherwise fluid-like biochemical systems. *Collagen* (Section 20.13) is a component of cartilage and *α keratin* (Section 20.13) gives mechanical strength as well as protective covering to hair, fingernails, feathers, hooves, and some animal shells.

7. **Transmembrane proteins.** These proteins, which span a cell membrane (Section 19.10) help control the movement of small molecules and ions through the cell membrane. Many such proteins have channels through which molecules can enter and exit a cell. Such protein channels are very selective, often allowing passage to just one type of molecule or ion.

8. **Storage proteins.** These proteins bind (and store) small molecules for future use. During degradation of hemoglobin (Section 26.7) the iron atoms present are released and become part of *ferritin,* an iron-storage protein, which saves the iron for use in the biosynthesis of new hemoglobin molecules. *Myoglobin* (Section 20.13) is an oxygen-storage protein present in muscle; the oxygen so stored is a reserve oxygen source for working muscle.

9. **Regulatory proteins.** These proteins are often found "embedded" in the exterior surface of cell membranes. They act as sites at which messenger molecules, including messenger proteins such as insulin, can bind and thereby initiate the effect that the messenger "carries." Regulatory proteins are often the molecules that bind to enzymes (catalytic proteins), thereby turning them "on" and "off," and thus controlling enzymatic action (Section 21.8).

10. **Nutrient proteins.** These proteins are particularly important in the early stages of life, from embryo to infant. *Casein,* found in milk, and *ovalbumin,* found in egg white, are two examples of such proteins. The role of milk in nature is to nourish and provide immunological protection for mammalian young. Three-fourths of the protein in milk is casein. Over 50% of the protein in egg white is *ovalbumin.*

20.15 PROTEIN HYDROLYSIS

When a protein or smaller peptide in a solution of strong acid or strong base is heated, the peptide bonds of the amino acid chain are hydrolyzed and free amino acids are produced. The hydrolysis reaction is the reverse of the formation reaction for a peptide bond. Amine and carboxylic acid functional groups are regenerated.

Let us consider the hydrolysis of the tripeptide Ala–Gly–Cys under acidic conditions. Complete hydrolysis produces one unit each of the amino acids alanine, glycine, and cysteine. The equation for the hydrolysis is

Ala–Gly–Cys $\xrightarrow[\text{heat}]{\text{H}_2\text{O, H}^+}$ Ala   Gly   Cys

Note that the product amino acids in this reaction are written in positive-ion form because of the acidic reaction conditions.

Protein digestion (Section 26.1) is simply enzyme-catalyzed hydrolysis of ingested protein. The free amino acids produced from this process are absorbed through the intestinal wall into the bloodstream and transported to the liver. Here they become the raw materials for the synthesis of new protein. Also, the hydrolysis of cellular proteins to amino acids is an ongoing process, as the body resynthesizes needed molecules and tissue.

> Protein hydrolysis produces free amino acids. This process is the reverse of protein synthesis, where free amino acids are combined.

20.16 PROTEIN DENATURATION

> A consequence of protein denaturation, the partial or complete loss of a protein's three-dimensional structure, is loss of biochemical activity for the protein.

Protein denaturation *is the partial or complete disorganization of a protein's characteristic three-dimensional shape as a result of disruption of its secondary, tertiary, and quaternary structural interactions.* Because the biochemical function of a protein depends on its three-dimensional shape, the result of denaturation is loss of biochemical activity. Protein denaturation does not affect the primary structure of a protein.

Although some proteins lose all of their three-dimensional structural characteristics upon denaturation (Figure 20.19), most proteins maintain some three-dimensional structure. Often, for limited denaturation changes, it is possible to find conditions under which the effects of denaturation can be reversed; this restoration process, in which the protein is "refolded," is called *renaturation*. However, for extensive denaturation changes, the process is usually irreversible.

Loss of water solubility is a frequent physical consequence of protein denaturation. The precipitation out of biochemical solution of denatured protein is called *coagulation.*

A most dramatic example of protein denaturation occurs when egg white (a concentrated solution of the protein albumin) is poured onto a hot surface. The clear albumin solution immediately changes into a white solid with a jelly-like consistency (see Figure 20.20). A similar process occurs when hamburger juices encounter a hot surface. A brown jelly-like solid forms.

Figure 20.19 Protein denaturation involves loss of the protein's three-dimensional structure. Complete loss of such structure produces an "unstructured" protein strand.

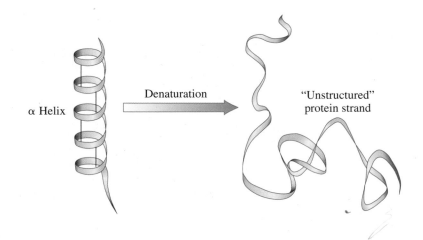

α Helix   Denaturation   "Unstructured" protein strand

Figure 20.20 Heat denatures the protein in egg white, producing a white jellylike solid. The primary structure of the protein remains intact, but all higher levels of protein structure are disrupted.

When protein-containing foods are cooked, protein denaturation occurs. Such "cooked" protein is more easily digested because it is easier for digestive enzymes to "work on" denatured (unraveled) protein. Cooking foods also kills microorganisms through protein denaturation. For example, ham and bacon can harbor parasites that cause trichinosis. Cooking the ham or bacon denatures parasite protein.

In surgery, heat is often used to seal small blood vessels. This process is called *cauterization.* Small wounds can also be sealed by cauterization. Heat-induced denaturation is used in sterilizing surgical instruments and in canning foods; bacteria are destroyed when the heat denatures their protein.

The body temperature of a patient with fever may rise to 102°F, 103°F, or even 104° without serious consequences. A temperature above 106°F (41°C) is extremely dangerous, for at this level the enzymes of the body begin to be inactivated. Enzymes, which function as catalysts for almost all body reactions, are protein. Inactivation of enzymes, through denaturation, can have lethal effects on body chemistry.

The effect of ultraviolet radiation from the sun, an ionizing radiation (Section 11.7), is similar to that of heat. Denatured skin proteins cause most of the problems associated with sunburn.

A curdy precipitate of casein, the principal protein in milk, is formed in the stomach when the hydrochloric acid of gastric juice denatures the casein. The curdling of milk that takes place when milk sours or cheese is made (see Figure 20.21) results from the presence of lactic acid, a by-product of bacterial growth. Yogurt is prepared by growing lactic-acid-producing bacteria in skim milk. The coagulated denatured protein gives yogurt its semi-solid consistency.

Serious eye damage can result from eye tissue contact with acids or bases, when irreversibly denatured and coagulated protein causes a clouded cornea. This reaction is part of the basis for the rule that students wear protective eyewear in the chemistry laboratory.

Globular proteins denature more readily than fibrous proteins because of weaker secondary and tertiary attractive forces.

Alcohols are an important type of denaturing agent. Denaturation of bacterial protein takes place when isopropyl or ethyl alcohol is used as a disinfectant—hence the common practice of swabbing the skin with alcohol before giving an injection. Interestingly, pure isopropyl or ethyl alcohol is less effective than the commonly used 70% alcohol solution. Pure alcohol quickly denatures and coagulates the bacterial surface, thereby forming an effective barrier to further penetration by the alcohol. The 70% solution denatures more slowly and allows complete penetration to be achieved before coagulation of the surface proteins takes place.

The process of giving a person a "hair permanent" involves protein denaturation through the use of reducing agents and oxidizing agents (see the Chemical Connections feature on page 686).

Table 20.6 is a listing of selected physical and chemical agents that cause protein denaturation. The effectiveness of a given denaturing agent depends on the type of protein upon which it is acting.

Figure 20.21 Storage room for cheese; during storage cheese "matures" as bacteria and enzymes ferment the cheese, giving it a stronger flavor.

GLYCOPROTEINS

A glycoprotein is a protein that contains carbohydrates or carbohydrate derivatives in addition to amino acids (Section 18.18). The carbohydrate content of glycoproteins is variable (from a few percent up to 85%), but it is fixed for any specific glycoprotein.

Glycoproteins include a number of very important substances; two of these, collagen and immunoglobulins, are considered in this section. Many of the proteins in cell membranes (lipid bilayers; see Section 19.10) are actually glycoproteins. The blood group markers of the ABO system (see the Chemical Connections feature on page 580 in Chapter 18) are also glycoproteins in which the carbohydrate content can reach 85%.

Chemical Connections

Denaturation and Human Hair

The process used in waving hair—that is, in a hair permanent—involves reversible denaturation. Hair is protein in which many disulfide (—S—S—) linkages occur as part of its tertiary structure; 16%–18% of hair is the amino acid cysteine. It is these disulfide linkages that give hair protein its overall shape. When a permanent is administered, hair is first treated with a reducing agent (ammonium thioglycolate) that breaks the disulfide linkages in the hair, producing two sulfhydryl (—SH) groups:

$$\text{Disulfide bridges} \xrightarrow[\text{agents}]{\text{Reducing}} \text{sulfhydryl groups}$$

The "reduced" hair, whose tertiary structure has been disrupted, is then wound on curlers to give it a new configuration.

Finally, the reduced and rearranged hair is treated with an oxidizing agent (potassium bromate) to form disulfide linkages at new locations within the hair:

$$\text{Sulfhydryl groups} \xrightarrow[\text{agents}]{\text{Oxidizing}} \text{re-formed disulfide bridges}$$

The new shape and curl of the hair are maintained by the newly formed disulfide bonds and the resulting new tertiary structure accompanying their formation. Of course, as new hair grows in, the "permanent" process has to be repeated.

Disulfide bridges, which involve covalent bonds, impart considerable resistance to denaturation because they are much stronger than the noncovalent interactions otherwise present.

TABLE 20.6
Selected Physical and Chemical Denaturing Agents

Denaturing Agent	Mode of Action
heat	disrupts hydrogen bonds by making molecules vibrate too violently; produces coagulation, as in the frying of an egg
microwave radiation	causes violent vibrations of molecules that disrupt hydrogen bonds
ultraviolet radiation	operates very similarly to the action of heat (e.g., sunburning)
violent whipping or shaking	causes molecules in globular shapes to extend to longer lengths, which then entangle (e.g., beating egg white into meringue)
detergent	affects R-group interactions
organic solvents (e.g., ethanol, 2-propanol, acetone)	interfere with R-group interactions because these solvents also can form hydrogen bonds; quickly denature proteins in bacteria, killing them (e.g., the disinfectant action of 70% ethanol)
strong acids and bases	disrupt hydrogen bonds and salt bridges; prolonged action leads to actual hydrolysis of peptide bonds
salts of heavy metals (e.g., salts of Hg^{2+}, Ag^+, Pb^{2+})	metal ions combine with —SH groups and form poisonous salts
reducing agents	reduce disulfide linkages to produce —SH groups

Collagen

The fibrous protein collagen, whose structure was first considered in Section 20.13, qualifies as a *glycoprotein* because carbohydrate units are present in its structure. This structural feature of collagen, not considered previously, involves the presence of the *nonstandard* amino acids 4-hydroxyproline (5%) and 5-hydroxylysine (1%)—derivatives of the standard amino acids proline and lysine (Table 20.1).

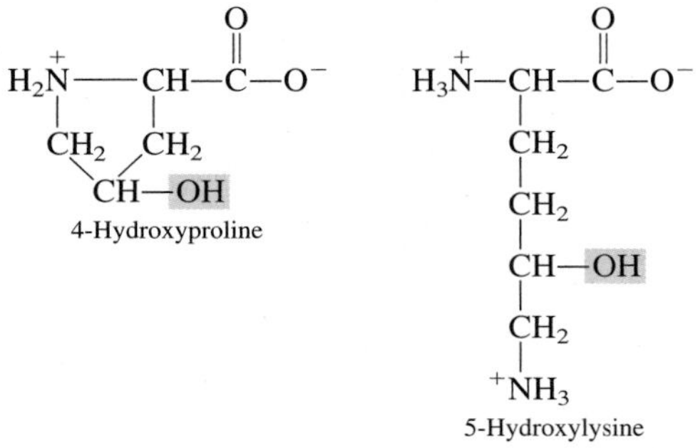

Nonstandard amino acids consist of amino acid residues that have been chemically modified after their incorporation into a protein (as is the case with 4-hydroxyproline and 5-hydroxylysine) and amino acids that occur in living organisms but are not found in proteins.

4-Hydroxyproline   5-Hydroxylysine

The presence of carbohydrate units (mostly glucose, galactose, and their disaccharides) attached by glycosidic linkages (Section 18.13) to collagen at its 5-hydroxylysine residues causes collagen to be classified as a *glycoprotein*. The function of the carbohydrate groups in collagen is related to cross-linking; they direct the assembly of collagen triple helices into more complex aggregations called *collagen fibrils*.

A primary biochemical function of vitamin C involves the hydroxylation of proline and lysine during collagen formation. These hydroxylation processes require the enzymes *proline hydroxylase* and *lysine hydroxylase*. These enzymes can function only in the presence of vitamin C.

When collagen is boiled in water, under basic conditions, it is converted to the water-soluble protein gelatin. This process involves both denaturation (Section 20.15) and hydrolysis (Section 20.14). Heat acts as a denaturant, causing rupture of the hydrogen bonds supporting collagen's triple-helix structure. Regions in the amino acid chains where proline and hydroxyproline concentrations are high are particularly susceptible to hydrolysis, which breaks up the polypeptide chains. Meats become more tender when cooked because of the conversion of some collagen to gelatin. Tougher cuts of meat (more cross-linking), such as stew meat, need longer cooking times.

Immunoglobulins

Immunoglobulins are among the most important and interesting of the soluble proteins in the human body. An **immunoglobulin** *is a glycoprotein produced by an organism as a protective response to the invasion of microorganisms or foreign molecules.* Different classes of immunoglobulins, identified by differing carbohydrate content and molecular mass, exist.

Immunoglobulins serve as *antibodies* to combat invasion of the body by *antigens*. An **antigen** *is a foreign substance, such as a bacterium or virus, that invades the human body.* An **antibody** *is a biochemical molecule that counteracts a specific antigen.* The immune system of the human body has the capability to produce immunoglobulins that respond to several million different antigens.

All types of immunoglobulin molecules have much the same basic structure, which includes the following features:

1. Four polypeptide chains are present: two identical heavy (H) chains and two identical light (L) chains.
2. The H chains, which usually contain 400–500 amino acid residues, are approximately twice as long as the L chains.
3. Both the H and L chains have constant and variable regions. The constant regions have the same amino acid sequence from immunoglobulin to immunoglobulin, and the variable regions have a different amino acid sequence in each immunoglobulin.
4. The carbohydrate content of various immunoglobulins varies from 1% to 12% by mass.

CHEMICAL Connections

Cyclosporine: An Antirejection Drug

The survival rate for patients undergoing human organ transplant operations such as heart, liver, or kidney replacement has risen dramatically since the late 1980s. This increased success coincides with the introduction of new drugs for controlling transplant rejection by a patient's own immune system. One of the best known of these immunosuppressive agents (antirejection drugs) is cyclosporine, a substance obtained from a particular type of soil fungus.

The primary structure of cyclosporine is that of a cyclic peptide containing 11 amino acid units. Ten of these are amino acids, which have simple side chains (four or fewer carbon atoms). The eleventh amino acid, which is the key to cyclosporine's pharmacological activity, is very unusual, having not been previously encountered. It has a 7-carbon branched, unsaturated, hydroxylated side chain with the structure

$$-CH-CH-CH_2-CH=CH-CH_3$$
$$\quad\ |\quad\ \ |$$
$$\quad OH\ \ CH_3$$

The following diagram shows the amino acid sequence within the cyclosporine ring. Seven of the amino acid units,

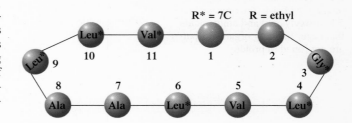

denoted by asterisks, have their nitrogen atom methylated; that is, a methyl group has replaced the hydrogen atom. This unique structural feature makes cyclosporine water-insoluble but fat-soluble.

The fat solubility of cyclosporine allows it to cross cell membranes readily and to be widely distributed in the body. It is administered either intravenously or orally. Because of its low water solubility, the drug is supplied in olive oil for oral administration. Cyclosporine has a narrow therapeutic index. When the blood concentration is too low, inadequate immunosuppression occurs. On the other hand, a high cyclosporine concentration can lead to kidney problems.

5. The secondary and tertiary structures are similar for all immunoglobulins. They involve a Y-shaped conformation (Figure 20.22) with disulfide linkages between H and L chains stabilizing the structure.

The interaction of an immunoglobulin molecule with an antigen occurs at the "tips" (uppermost part) of the Y structure. These tips are the variable-composition region of the immunoglobulin structure. It is here that the antigen binds specifically, and it is here that the amino acid sequence differs from one immunoglobulin to another.

Each immunoglobulin has two identical active sites and can thus bind to two molecules of the antigen it is "designed for." The action of many such immunoglobulins of a given type in concert with each other creates an *antigen–antibody complex*

Figure 20.22 This schematic diagram shows the structure of an immunoglobulin. Two heavy (H) polypeptide chains and two light (L) polypeptide chains are cross-linked by disulfide bridges. The purple areas are the constant amino acid regions, and the areas shown in red are the variable amino acid regions of each chain. Carbohydrate molecules attached to the heavy chains aid in determining the destinations of immunoglobulins in the tissues.

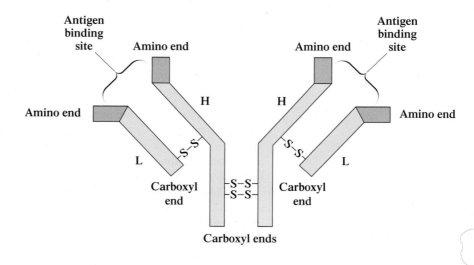

Figure 20.23 In this immunoglobulin-antigen complex, note that more than one immunoglobulin molecule can attach itself to a given antigen. Also, any given immunoglobulin has only two sites where antigen can bind.

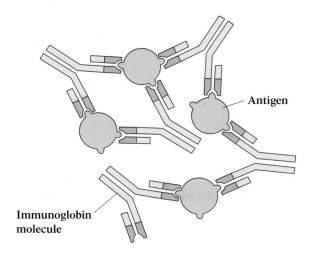

Antigen

Immunoglobin molecule

that precipitates from solution (Figure 20.23). Eventually, an invading antigen can be eliminated from the body through such precipitation. The bonding of an antigen to the variable region of an immunoglobulin occurs through hydrophobic interactions, dipole–dipole interactions, and hydrogen bonds rather than covalent bonds.

The importance of immunoglobulins is amply and tragically demonstrated by the effects of AIDS (acquired immunodeficiency syndrome). The AIDS virus upsets the body's normal production of immunoglobulins and leaves the body susceptible to what would otherwise not be debilitating and deadly infections.

Individuals who receive organ transplants must be given drugs to suppress the production of immunoglobulins against foreign proteins in the new organ, thus preventing rejection of the organ. The major reason for the increasing importance of organ transplants is the successful development of drugs that can properly manipulate the body's immune system (see the Chemical Connections feature on cyclosporine on page 688).

Many reasons exist for a mother to breast-feed a newborn infant. One of the most important is immunoglobulins. During the first two or three days of lactation, the breasts produce *colostrum,* a premilk substance containing immunoglobulins from the mother's blood. Colostrum helps protect the newborn infant from those infections to which the mother has developed immunity. These diseases are the ones in her environment— precisely those the infant needs protection from. Breast milk, once it is produced, is a source of immunoglobulins for the infant for a short time. (After the first week of nursing, immunoglobulin concentrations in the milk decrease rapidly.) Infant formula used as a substitute for breast milk is almost always nutritionally equivalent, but it does not contain immunoglobulins.

20.18 LIPOPROTEINS

A **lipoprotein** *is a conjugated protein that contains lipids in addition to amino acids.* The major function of such proteins is to help suspend lipids and transport them through the bloodstream. Lipids, in general, are insoluble in blood (an aqueous medium) because of their nonpolar nature (Section 19.1).

A **plasma lipoprotein** *is a lipoprotein that is involved in the transport system for lipids in the bloodstream.* These proteins have a spherical structure that involves a central core of lipid material (triacylglycerols and cholesterol esters) surrounded by a shell (membrane structure) of phospholipids, cholesterol, and proteins. In the blood, cholesterol exists primarily in the form of cholesterol esters formed from the esterification of cholesterol's hydroxyl group with a fatty acid.

Fatty acid   Ester   Cholesterol
linkage

Figure 20.24 is a depiction of such a spherical lipoprotein structure.

There are four major classes of plasma lipoproteins:

1. **Chylomicrons.** Their function is to transport dietary triacylglycerols from the intestine to the liver and to adipose tissue.
2. **Very-low-density lipoproteins (VLDL).** Their function is to transport triacylglycerols synthesized in the liver to adipose tissue.
3. **Low-density lipoproteins (LDL).** Their function is to transport cholesterol synthesized in the liver to cells throughout the body.
4. **High-density lipoproteins (HDL).** Their function is to collect excess cholesterol from body tissues and transport it back to the liver for degradation to bile acids.

CHEMICAL Connections

Lipoproteins and Heart Disease Risk

In the United States, heart disease is the number-one health problem for both men and women. There are many risk factors associated with heart disease, some of which are controllable and some of which are not. You can't change your age, race, or family history (genetics). But you can control (manage) factors such as your weight, whether or not you smoke, and your cholesterol level.

Relative to high cholesterol as a risk factor for heart disease, studies now indicate that just focusing on the numerical value of cholesterol present in the blood is an oversimplification of the problem. Consideration of the "form" in which the cholesterol is present is as important as the total amount. Much of the research concerning the "form" of cholesterol present relates to the lipoproteins LDL and HDL present in the blood.

Both LDLs and HDLs are involved in cholesterol transport (Section 20.18). LDLs carry approximately 80% of the cholesterol and HDLs the remainder. Of significance, LDLs and HDLs carry cholesterol for different purposes. LDLs carry cholesterol to cells for their use, whereas HDLs carry excess cholesterol away from cells to the liver for processing and excretion from the body.

Studies show that LDL levels correlate *directly* with heart disease, whereas HDL levels correlate *inversely* with heart disease risk. Thus HDL is sometimes referred to as "good" cholesterol (*H*DL = *H*ealthy) and LDL as "bad" cholesterol (*L*DL = *L*ess healthy).

The goal of dietary measures to slow the advance of atherosclerosis is to reduce LDL cholesterol levels. Reduction in the dietary intake of saturated fat appears to be a key action (Section 19.5).

High HDL levels are desirable because they give the body an efficient means of removing excess cholesterol. Low HDL levels can result in excess cholesterol depositing within the circulatory system.

In general, women have higher HDL levels than men—an average of 55 mg per 100 mL of blood serum versus 45 mg per 100 mL. This may explain in part why proportionately fewer women have heart attacks than men. Nonsmokers have uniformly higher HDL levels than smokers. Exercise on a regular basis tends to increase HDL levels. This discovery has increased the popularity of walking and running exercise. Genetics also plays a role in establishing HDL as well as other lipoprotein concentrations in the blood.

A person's *total* blood cholesterol level does not necessarily correlate with that individual's real risk for heart and blood vessel disease. A better measure is the *cholesterol ratio,* which is defined as

$$\text{Cholesterol ratio} = \frac{\text{total cholesterol}}{\text{HDL cholesterol}}$$

For example, if a person's total cholesterol is 200 and his or her HDL is 45, then the cholesterol ratio would be 4.4. According to the accompanying guidelines for interpreting cholesterol ratio values, this indicates an average risk for heart disease.

What Your Cholesterol Ratio Means

Ratio	Heart Disease Risk
6.0	high
5.0	above average
4.5	average
4.0	below average
3.0	low

Figure 20.24 A model for the structure of a plasma lipoprotein.

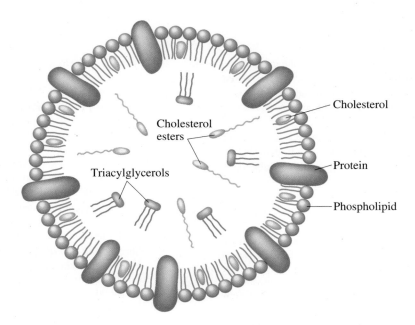

The density of a lipoprotein is related to the fractions of protein and lipid material present. The greater the amount of protein in the lipoprotein, the higher the density. Figure 20.25 characterizes the major plasma lipoprotein types in terms of density as well as lipid–protein composition.

The presence or absence of various types of lipoproteins in the blood appears to have implications for the health of the heart and blood vessels. Lipoprotein levels in the blood are now used as an indicator of heart disease risk (see the Chemical Connections feature on page 689).

Figure 20.25 Densities and relative amounts of proteins, phospholipids, cholesterol, and triacylglycerols present in the four major classes of plasma lipoproteins.

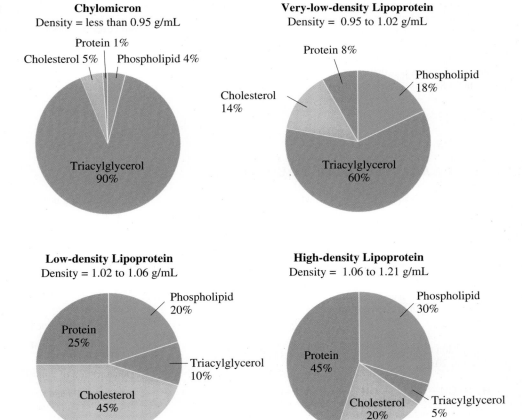

CONCEPTS TO REMEMBER

Protein. A protein is a polymer in which the monomer units are amino acids (Section 20.1).

α-Amino acid. An α-amino acid is an amino acid in which the amino group and the carboxyl group are both attached to the α-carbon atom (Section 20.2).

Standard amino acid. A standard amino acid is one of the 20 α-amino acids that are normally present in protein (Section 20.2).

Amino acid classifications. Amino acids are classified as nonpolar, polar neutral, polar basic, or polar acidic depending on the nature of the side chain (R group) present (Section 20.2).

Chirality of amino acids. Amino acids found in proteins are always left-handed (L isomer) (Section 20.3).

Zwitterion. A zwitterion is a molecule that has a positive charge on one atom and a negative charge on another atom. In neutral solution and in the solid state, amino acids exist as zwitterions. For amino acids in solution, the isoelectric point is the pH at which the solution has no net charge because an equal number of positive and negative charges are present (Section 20.4).

Disulfide bond formation. The amino acid cysteine readily dimerizes; the —SH groups of two cysteine molecules interact to form a covalent disulfide bond (Section 20.5).

Peptide bond. A peptide bond is an amide bond involving the carboxyl group of one amino acid and the amino group of another amino acid. In a protein, the amino acids are linked to each other through peptide bonds (Section 20.6).

Biochemically important peptides. Numerous small peptides are biochemically active. Their functions include hormonal action, neurotransmission functions, and antioxidant activity (Section 20.7).

General characteristics of proteins. Proteins are peptides with at least 40 amino acid residues. A single peptide chain is present in a monomeric protein and two or more peptide chains are present in a multimeric protein. A simple protein contains only one or more peptide chains. A conjugated protein contains one or more additional chemical components, called prosthetic groups, in addition to peptide chains (Section 20.8).

Primary protein structure. The primary structure of a protein is the sequence of amino acids present in the peptide chain or chains of the protein (Section 20.9).

Secondary protein structure. The secondary structure of a protein is the arrangement in space of the backbone portion of the protein. The two major types of protein secondary structure are the α helix and the β pleated sheet (Section 20.10).

Tertiary protein structure. The tertiary structure of a protein is the overall three-dimensional shape that results from the attractive forces among amino acid side chains (R groups) (Section 20.11).

Quaternary protein structure. The quaternary structure of a protein involves the associations among the peptide chains present in a multimeric protein (Section 20.12).

Fibrous and globular proteins. Fibrous proteins are generally insoluble in water and have a long, thin, fibrous shape. α Keratin and collagen are important fibrous proteins. Globular proteins are generally soluble in water and have a roughly spherical or globular overall shape. Hemoglobin and myoglobin are important globular proteins (Section 20.13).

Protein hydrolysis. Protein hydrolysis is a chemical reaction in which peptide bonds within a protein are broken through reaction with water. Complete hydrolysis produces free amino acids (Section 20.15).

Protein denaturation. Protein denaturation is the partial or complete disorganization of a protein's characteristic three-dimensional shape as a result of disruption of its secondary, tertiary, and quaternary structural interactions (Section 20.16).

Glycoproteins. Glycoproteins are conjugated proteins that contain carbohydrates or carbohydrate derivatives in addition to amino acids. Collagen and immunoglobulins are important glycoproteins (Section 20.17).

Lipoproteins. Lipoproteins are conjugated proteins that are composed of both lipids and amino acids. Lipoproteins are classified on the basis of their density (Section 20.18).

KEY REACTIONS AND EQUATIONS

1. Formation of a zwitterion at pH 7 (Section 20.4)

$$\underset{R}{H_2N-CH-COOH} \longrightarrow \underset{R}{\overset{+}{H_3}N-CH-COO^-}$$

2. Conversion of a zwitterion to a positive ion in acidic solution (Section 20.4)

$$\underset{R}{\overset{+}{H_3}N-CH-COO^-} + H_3O^+ \longrightarrow \underset{R}{\overset{+}{H_3}N-CH-COOH} + H_2O$$

3. Conversion of a zwitterion to a negative ion in basic solution (Section 20.4)

$$\underset{R}{\overset{+}{H_3}N-CH-COO^-} + OH^- \longrightarrow \underset{R}{H_2N-CH-COO^-} + H_2O$$

4. Formation of a peptide bond (Section 20.6)

$$\underset{R}{\overset{+}{H_3}N-CH-COO^-} + \underset{R'}{\overset{+}{H_3}N-CH-COO^-} \longrightarrow$$

$$\underset{R}{\overset{+}{H_3}N-CH}-\overset{O}{\overset{\|}{C}}-\overset{H}{\underset{}{N}}-\underset{R'}{CH-COO^-} + H_2O$$

5. Hydrolysis of a protein in acidic solution (Section 20.15)

$$\text{Protein} + H_2O \xrightarrow{H^+} \text{smaller peptides} \xrightarrow{H^+} \text{amino acids}$$

6. Denaturation of a protein (Section 20.16)

Protein with 1°, 2°, and 3° structure $\xrightarrow[\text{agent}]{\text{Denaturing}}$ Protein with 1° structure only

EXERCISES *and* PROBLEMS

The members of each pair of problems in this section test similar material.

🔵 **Characteristics of Proteins (Section 20.1)**

20.1 What is the general name for the building blocks (monomers) from which a protein (polymer) is made?

20.2 What element is always present in proteins that is seldom present in carbohydrates and lipids?

20.3 What percent of a cell's overall mass is accounted for by proteins?

20.4 Approximately how many different proteins are present in a typical human cell?

🔵 **Amino Acid Structural Characteristics (Section 20.2)**

20.5 Which of the following structures represent α-amino acids?

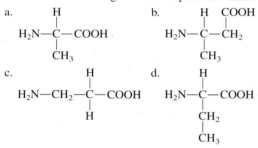

20.6 What is the significance of the prefix α in the designation α-amino acid?

20.7 What is the major structural difference among the various standard amino acids?

20.8 On the basis of polarity, what are the four types of side chains found in the standard amino acids?

20.9 With the help of Table 20.1, determine which of the standard amino acids have a side chain with the following characteristics.
  a. Contains an aromatic group
  b. Contains the element sulfur
  c. Contains a carboxyl group
  d. Contains a hydroxyl group

20.10 With the help of Table 20.1, determine which of the standard amino acids have a side chain with the following characteristics.
  a. Contains only carbon and hydrogen
  b. Contains an amino group
  c. Contains an amide group
  d. Contains more than four carbon atoms

20.11 What is the distinguishing characteristic of a polar basic amino acid?

20.12 What is the distinguishing characteristic of a polar acidic amino acid?

20.13 In what way is the structure of the amino acid proline different from that of the other 19 standard amino acids?

20.14 Which two of the standard amino acids are constitutional isomers?

20.15 What amino acids do these abbreviations stand for?
  a. Ala   b. Leu   c. Met   d. Trp

20.16 What amino acids do these abbreviations stand for?
  a. Asp   b. Cys   c. Phe   d. Val

20.17 Which four standard amino acids have three-letter abbreviations that are not the first three letters of their common names?

20.18 What are the three-letter abbreviations for the three polar basic amino acids?

20.19 Classify each of the following amino acids as nonpolar, polar neutral, polar acidic, or polar basic.
  a. Asn   b. Glu   c. Pro   d. Ser

20.20 Classify each of the following amino acids as nonpolar, polar neutral, polar acidic, or polar basic.
  a. Gly   b. Thr   c. Tyr   d. His

🔵 **Chirality and Amino Acids (Section 20.3)**

20.21 To which family of mirror-image isomers do nearly all naturally occurring amino acids belong?

20.22 In what way is the structure of glycine different from that of the other 19 common amino acids?

20.23 Draw Fischer projection formulas for the following amino acids.
  a. L-Serine   b. D-Serine
  c. D-Alanine   d. L-Leucine

20.24 Draw Fischer projection formulas for the following amino acids.
  a. L-Cysteine   b. D-Cysteine
  c. D-Alanine   d. L-Valine

🔵 **Acid-Base Properties of Amino Acids (Section 20.4)**

20.25 At room temperature, amino acids are solids with relatively high decomposition points. Explain why.

20.26 Amino acids exist as zwitterions in the solid state. Explain why.

20.27 Draw the zwitterion structure for each of the following amino acids.
  a. Leucine   b. Isoleucine
  c. Cysteine   d. Glycine

20.28 Draw the zwitterion structure for each of the following amino acids.
  a. Serine   b. Methionine
  c. Threonine   d. Phenylalanine

20.29 Draw the structure of serine at each of the following pH values.
  a. 5.68   b. 1.0   c. 12.0   d. 3.0

20.30 Draw the structure of glycine at each of the following pH values.
  a. 5.97   b. 13.0   c. 2.0   d. 11.0

20.31 Explain what is meant by the term *isoelectric point*.

20.32 Most amino acids have isoelectric points between 5.0 and 6.0, but the isoelectric point of lysine is 9.7. Explain why lysine has such a high value for its isoelectric point.

20.33 Glutamic acid exists in two low-pH forms instead of the usual one. Explain why.

20.34 Arginine exists in two high-pH forms instead of the usual one. Explain why.

20.35 Predict the direction of movement of each of the following amino acids in a solution at the pH value specified under the influence of an electric field. Indicate the direction as toward the positive electrode or toward the negative electrode. Write "isoelectric" if no net movement occurs.
  a. Alanine at pH = 12.0   b. Valine at pH = 5.97
  c. Aspartic acid at pH = 1.0   d. Arginine at pH = 13.0

20.36 Predict the direction of movement of each of the following amino acids in a solution at the pH value specified under the influence of an electric field. Indicate the direction as toward the positive electrode or toward the negative electrode. Write "isoelectric" if no net movement occurs.
a. Alanine at pH = 2.0   b. Valine at pH = 12.0
c. Aspartic acid at pH = 13.0   d. Arginine at pH = 1.0

20.37 A direct current was passed through a solution containing valine, histidine, and aspartic acid at a pH of 6.0. One amino acid migrated to the positive electrode, one migrated to the negative electrode, and one did not migrate to either electrode. Which amino acids went where?

20.38 A direct current was passed through a solution containing alanine, arginine, and glutamic acid at a pH of 6.0. One amino acid migrated to the positive electrode, one migrated to the negative electrode, and one did not migrate to either electrode. Which amino acids went where?

● **Cysteine and Disulfide Bonds (Section 20.5)**

20.39 When two cysteine molecules dimerize, what happens to the R groups present?

20.40 What chemical reaction involving the cysteine molecule produces a disulfide bond?

● **Peptides (Section 20.6)**

20.41 What two functional groups are involved in the formation of a peptide bond?

20.42 What is meant by the N-terminal end and the C-terminal end of a peptide?

20.43 Write out the full structure of the tripeptide Val–Phe–Cys.

20.44 Write out the full structure of the tripeptide Glu–Ala–Leu.

20.45 Explain why the notations Ser–Cys and Cys–Ser represent two different molecules rather than the same molecule.

20.46 Explain why the notations Ala–Gly–Val–Ala and Ala–Val–Gly–Ala represent two different molecules rather than the same molecule.

20.47 There are a total of six different amino acid sequences for a tripeptide containing one molecule each of serine, valine, and glycine. Using three-letter abbreviations for the amino acids, draw the six possible sequences of amino acids.

20.48 There are a total of six different amino acid sequences for a tetrapeptide containing two molecules each of serine and valine. Using three-letter abbreviations for the amino acids, draw the six possible sequences of amino acids.

20.49 Identify the amino acids contained in each of the following tripeptides.

a.

$$H_3\overset{+}{N}-CH-\overset{O}{\overset{\|}{C}}-\overset{H}{\overset{|}{N}}-CH-\overset{O}{\overset{\|}{C}}-\overset{H}{\overset{|}{N}}-CH-COO^-$$
with side chains CH_2-OH, CH_3, CH_2-SH

b.

$$H_3\overset{+}{N}-CH-\overset{O}{\overset{\|}{C}}-\overset{H}{\overset{|}{N}}-CH-\overset{O}{\overset{\|}{C}}-\overset{H}{\overset{|}{N}}-CH-COO^-$$
with side chains CH_2-COO^-, $CH-OH-CH_3$, CH_2-C-NH_2 (with $\|O$)

20.50 Identify the amino acids contained in each of the following tripeptides.

a.

$$H_3\overset{+}{N}-CH_2-\overset{O}{\overset{\|}{C}}-\overset{H}{\overset{|}{N}}-CH-\overset{O}{\overset{\|}{C}}-\overset{H}{\overset{|}{N}}-CH_2-COO^-$$
with side chain $CH-CH_3$ / CH_3

b.

$$H_3\overset{+}{N}-CH-\overset{O}{\overset{\|}{C}}-\overset{H}{\overset{|}{N}}-CH-\overset{O}{\overset{\|}{C}}-\overset{H}{\overset{|}{N}}-CH-COO^-$$
with side chains CH_2-OH, $CH_2-CH-CH_3-CH_3$, $CH_2-CH_2-COO^-$

20.51 How many peptide bonds are present in each of the molecules in Problem 20.49?

20.52 How many peptide bonds are present in each of the molecules in Problem 20.50?

20.53 With the help of Table 20.1, assign an IUPAC name to each of the following small peptides.
a. Ser–Cys   b. Gly–Ala–Val
c. Tyr–Asp–Gln   d. Leu–Lys–Trp–Met

20.54 With the help of Table 20.1, assign an IUPAC name to each of the following small peptides.
a. Cys–Ser   b. Val–Ala–Gly
c. Tyr–Gln–Asp   d. Phe–Met–Tyr–Asn

20.55 What are the two repeating units present in the "backbone" of a peptide?

20.56 For a peptide, describe
a. the regularly repeating part of its structure.
b. the variable part of its structure.

● **Biochemically Important Small Peptides (Section 20.7)**

20.57 Contrast the structures of the protein hormones oxytocin and vasopressin in terms of
a. what they have in common.
b. how they differ.

20.58 Contrast the protein hormones oxytocin and vasopressin in terms of their biochemical functions.

20.59 Contrast the binding-site locations in the brain for enkephalins and the prescription painkillers morphine and codeine.

20.60 Contrast the structures of the peptide neurotransmitters Met-enkephalin and Leu-enkephalin in terms of
a. what they have in common.
b. how they differ.

20.61 What is the unusual structural feature present in the molecule glutathione?

20.62 What is the major biochemical function of glutathione?

● **General Structural Characteristics of Proteins (Section 20.8)**

20.63 What is the major difference between a monomeric protein and a multimeric protein?

20.64 What is the major difference between a simple protein and a conjugated protein?

20.65 Indicate whether each of the following statements about proteins is true or false.

a. Two or more peptide chains are always present in a multimeric protein.

b. A simple protein contains only one type of amino acid.

c. A conjugated protein can also be a monomeric protein.

d. The prosthetic group(s) present in a glycoprotein are carbohydrate groups.

20.66 Indicate whether each of the following statements about proteins is true or false.

a. Conjugated proteins always have only one peptide chain.

b. All peptide chains in a multimeric protein must be identical to each other.

c. A simple protein can also be a multimeric protein.

d. Both monomeric proteins and multimeric proteins can contain prosthetic groups.

● Primary Protein Structure (Section 20.9)

20.67 What is meant by the *primary structure* of a protein?

20.68 Two proteins with the same amino acid composition do not have to have the same primary structure. Explain why.

20.69 What type of bond is responsible for the primary structure of a protein?

20.70 A segment of a protein contains two alanine and two glycine amino acid residues. How many different primary structures are possible for this segment of the protein?

● Secondary Protein Structure (Section 20.10)

20.71 What are the two common types of secondary protein structure?

20.72 Hydrogen bonding between which functional groups stabilizes protein secondary structure arrangements?

20.73 The β pleated sheet secondary structure can be formed through either intramolecular hydrogen bonding or intermolecular hydrogen bonding. Explain why.

20.74 The α helix secondary structure always involves intramolecular hydrogen bonding and never involves intermolecular hydrogen bonding. Explain why.

20.75 Can more than one type of secondary structure be present in the same protein molecule? Explain your answer.

20.76 What is meant by the statement that the secondary structure of a segment of a protein is "unstructured"?

20.77 What is the function of the "unstructured" secondary structure portions of a protein?

20.78 Why is the term "unstructured segment" of a protein somewhat of a misnomer?

● Tertiary Protein Structure (Section 20.11)

20.79 State the four types of attractive forces that give rise to tertiary protein structure.

20.80 What is the difference between the types of hydrogen bonding that occur in secondary and tertiary protein structure.

20.81 Give the type of amino acid R group that is involved in each of the following interactions that contribute to tertiary protein structure.

a. Hydrophobic interaction   b. Hydrogen bond

c. Disulfide bond   d. Electrostatic interaction

20.82 Classify each of the interactions listed in Problem 20.81 as a covalent bond or as a noncovalent interaction.

20.83 With the help of Table 20.1, specify the nature of each of the following tertiary-structure interactions, using the choices

hydrophobic, electrostatic, hydrogen bonding, and disulfide bond.

a. Phenylalanine and leucine   b. Arginine and glutamic acid

c. Two cysteines   d. Serine and tyrosine

20.84 With the help of Table 20.1, specify the nature of each of the following tertiary-structure interactions using the choices hydrophobic, electrostatic, hydrogen bonding, and disulfide bond.

a. Lysine and aspartic acid   b. Threonine and tyrosine

c. Alanine and valine   d. Leucine and isoleucine

● Quaternary Protein Structure (Section 20.12)

20.85 What is meant by the *quaternary structure* of a protein?

20.86 Not all proteins have quaternary structure. Explain why this is so.

20.87 Compare the types of noncovalent interactions that contribute to tertiary protein structure with those that contribute to quaternary protein structure.

20.88 Quaternary protein structure is more easily disrupted than tertiary protein structure. Explain why this is so.

● Protein Classification Based on Shape (Section 20.13)

20.89 Contrast fibrous and globular proteins in terms of

a. solubility characteristics in water.

b. general biochemical function.

20.90 Contrast fibrous and globular proteins in terms of

a. general secondary structure.

b. relative abundance within the human body.

20.91 Classify each of the following proteins as a globular protein or a fibrous protein.

a. α-Keratin   b. Collagen

c. Hemoglobin   d. Myoglobin

20.92 What is the major biochemical function for each of the following proteins?

a. α-Keratin   b. Collagen

c. Hemoglobin   d. Myoglobin

20.93 Contrast the structures of the proteins α keratin and collagen.

20.94 Contrast the structures of the proteins myoglobin and hemoglobin.

● Protein Classification Based on Function (Section 20.14)

20.95 Using the list in Section 20.14, characterize each of the following proteins in terms of its function classification.

a. Actin   b. Myoglobin

c. Transferrin   d. Insulin

20.96 Using the list in Section 20.14, characterize each of the following proteins in terms of its function classification.

a. Collagen   b. Hemoglobin

c. Ferritin   d. Casein

● Protein Hydrolysis (Section 20.15)

20.97 Will hydrolysis of the dipeptides Ala–Val and Val–Ala yield the same products? Explain your answer.

20.98 A shampoo bottle lists "partially hydrolyzed protein" as one of its ingredients. What is the difference between partially hydrolyzed protein and completely hydrolyzed protein?

20.99 Drugs that are proteins, such as insulin, must always be injected rather than taken by mouth. Explain why.

20.100 Which structural levels of a protein are affected by hydrolysis?

20.101 Identify the primary structure of a hexapeptide containing six different amino acids if the following smaller peptides are among the partial-hydrolysis products: Ala–Gly, His–Val–Arg, Ala–Gly–Met, and Gly–Met–His.

20.102 Identify the primary structure of a hexapeptide containing five different amino acids if the following smaller peptides are among the partial-hydrolysis products: Gly–Cys, Ala–Ser, Ala–Gly, and Cys–Val–Ala.

20.103 How many different di- and tripeptides could be present in a solution of partially hydrolyzed Ala–Gly–Ser–Tyr?

20.104 How many different di- and tripeptides could be present in a solution of partially hydrolyzed Ala–Gly–Ala–Gly?

Protein Denaturation (Section 20.16)

20.105 Which structural levels of a protein are affected by denaturation?

20.106 Suppose a sample of protein is completely hydrolyzed and another sample of the same protein is denatured. Compare the final products of these processes.

20.107 In what way is the protein in a cooked egg the same as that in a raw egg?

20.108 Why is 70% ethanol rather than pure ethanol preferred for use as an antiseptic agent?

Glycoproteins (Section 20.17)

20.109 What two nonstandard amino acids are present in collagen?

20.110 Where are the carbohydrate units located in collagen?

20.111 What is the function of the carbohydrate groups present in collagen?

20.112 What is the role of vitamin C in the biosynthesis of collagen?

20.113 What is the difference between an antigen and an antibody?

20.114 What is an immunoglobulin?

20.115 Describe the structural features of a typical immunoglobulin molecule.

20.116 Describe the process by which blood immunoglobulins help protect the body from invading bacteria and viruses.

Lipoproteins (Section 20.18)

20.117 Describe the general overall structure of a plasma lipoprotein.

20.118 In what chemical form does cholesterol usually exist in the bloodstream?

20.119 What are the four major classes of plasma lipoproteins?

20.120 What do the designations VLDL, LDL, and HDL stand for?

20.121 What factor determines the density of a plasma lipoprotein?

20.122 As the lipid content of a plasma lipoprotein increases does its density increase or decrease?

20.123 What is the biochemical function for the following?
a. Chylomicrons
b. Low-density lipoproteins

20.124 What is the biochemical function for the following?
a. Very-low-density lipoproteins
b. High-density lipoproteins

ADDITIONAL PROBLEMS

20.125 State whether each of the following statements applies to primary, secondary, tertiary, or quaternary protein structure.
a. A disulfide bond forms between two cysteine residues in the same protein chain.
b. A salt bridge forms between amino acids with acidic and basic side chains.
c. Hydrogen bonding between carbonyl oxygen atoms and nitrogen atoms of amino groups causes a peptide to coil into a helix.
d. Peptide linkages hold amino acids together in a polypeptide chain.

20.126 What is the common name for each of the following IUPAC-named standard amino acids?
a. 2-Aminopropanoic acid
b. 2-Amino-4-methylpentanoic acid
c. 2-Amino-3-hydroxybutanoic acid
d. 2-Aminobutanedioic acid

20.127 What is the net charge at a pH of 1.0 for each of the following peptides?
a. Val–Ala–Leu
b. Tyr–Trp–Thr
c. Asp–Asp–Glu–Gly
d. His–Arg–Ser–Ser

20.128 What is the net charge at a pH of 13.0 for each of the peptides in Problem 20.127?

20.129 The amino acid isoleucine possesses two chiral centers. Draw Fischer projection formulas for the four stereoisomers that are possible for this amino acid.

20.130 Indicate how many structurally isomeric tetrapeptides are possible for a tetrapeptide in which
a. four different amino acids are present.
b. three different amino acids are present.
c. two different amino acids are present.

20.131 Draw the structures of the three hydrolysis products obtained when the tripeptide in part a of Problem 20.49 undergoes hydrolysis under
a. low-pH (acidic) conditions.
b. high-pH (basic) conditions.

20.132 Classify each of the following proteins as a simple protein, a conjugated protein, a glycoprotein, a lipoprotein, a fibrous protein, or a globular protein. More than one classification may apply to a given protein.
a. α Keratin
b. Hemoglobin
c. Myoglobin
d. Collagen

*M*ultiple-Choice Practice Test

20.133 Which of the following sets of four elements are found in all amino acids?
  a. C, H, O, S   b. C, H, S, N
  c. C, H, O, N   d. C, O, S, N

20.134 Which of the following statements concerning the structure of α-amino acids is *correct*?
  a. The amino group is attached to the carbon atom of the carboxyl group.
  b. The amino group and the carboxyl group are directly bonded to the same carbon atom.
  c. The amino acid contains only two carbon atoms.
  d. The amino acid contains only one carbon atom.

20.135 Which of the following is an *incorrect* statement about glycine, the amino acid with the simplest structure?
  a. It does not contain a chiral center.
  b. It has a side chain that does not contain the element carbon.
  c. It is one of the 20 standard amino acids.
  d. Its amino group and carboxyl group are directly bonded to each other.

20.136 In a solution of high pH, all of the acidic and basic sites in an amino acid are which of the following?
  a. Protonated   b. Deprotonated
  c. Positively charged   d. Negatively charged

20.137 Which of the standard amino acids exist as zwitterions in the solid state?
  a. All of them
  b. Only those that have nonpolar side chains

  c. Only those that are polar neutral
  d. Only those that are acidic or basic

20.138 Which of the following statements concerning the tripeptide Val–Ala–Gly is *correct*?
  a. The C-terminal amino acid residue is Val.
  b. The N-terminal amino acid residue is Gly.
  c. Its structure can also be written as Gly–Ala–Val.
  d. It is constitutionally isomeric with five other tripeptides.

20.139 Which of the following types of bonding is responsible for protein secondary structure?
  a. Peptide linkages   b. Amide linkages
  c. Hydrogen bonds   d. Bonds involving R groups

20.140 R-group interaction between which of the following pairs of amino acids produces a covalent bond?
  a. Cysteine–cysteine   b. Proline–proline
  c. Alanine–glycine   d. Valine–lysine

20.141 Which of the following levels of protein structure is not disrupted when protein denaturation occurs?
  a. Primary structure   b. Secondary structure
  c. Tertiary structure   d. Quaternary structure

20.142 In which of the following pairs of proteins are both members of the pair *fibrous* proteins?
  a. α-Keratin and collagen
  b. Collagen and hemoglobin
  c. Hemoglobin and myoglobin
  d. α-Keratin and hemoglobin

Enzymes and Vitamins

The microbial life present in hot springs and thermal vents possesses enzymes that are specially adapted to higher temperature conditions.

In this chapter we consider two topics: enzymes and vitamins. Enzymes govern all chemical reactions in living organisms. They are specialized proteins that, with fascinating precision and selectivity, catalyze biochemical reactions that store and release energy, make pigments in our hair and eyes, digest the food we eat, synthesize cellular building materials, and protect us by repairing cellular damage and clotting our blood. Enzymes are sensitive to their environment, responding quickly to changes in the cell. The deficiency or excess of particular enzymes can cause certain diseases or signal problems such as heart attacks and other organ damage. Our knowledge of protein structure (Chapter 20) can help us appreciate and better understand how enzymes function in living cells.

Vitamins, which are necessary components of a healthful diet, play important roles in cellular metabolism. In most cases, they function as enzyme cofactors or carriers of functional groups during biosynthesis.

 21.1 GENERAL CHARACTERISTICS OF ENZYMES

An **enzyme** *is an organic compound that acts as a catalyst for a biochemical reaction.* Each cell in the human body contains thousands of different enzymes because almost every reaction in a cell requires its own specific enzyme. Enzymes cause cellular reactions to occur millions of times faster than corresponding uncatalyzed reactions. As catalysts (Section 9.6), enzymes are not consumed during the reaction but merely help the reaction occur more rapidly.

The word *enzyme* comes from the Greek words *en,* which means "in," and *zyme,* which means "yeast." Long before their chemical nature was understood, yeast enzymes were used in the production of bread and alcoholic beverages. The action of yeast on sugars

Figure 21.1 Bread dough rises as the result of carbon dioxide production resulting from the action of yeast enzymes on sugars present in the dough.

Enzymes, the most efficient catalysts known, increase the rates of biochemical reactions by factors of up to 10^{20} over uncatalyzed reactions. Nonenzymatic catalysts, on the other hand, typically enhance the rate of a reaction by factors of 10^2 to 10^4.

produces the carbon dioxide gas that causes bread to rise (see Figure 21.1). Fermentation of sugars in fruit juices with the same yeast enzymes produces alcoholic beverages.

Most enzymes are globular proteins (Section 20.13). Some are simple proteins, consisting entirely of amino acid chains. Others are conjugated proteins, containing additional chemical components (Section 21.2). Until the 1980s, it was thought that *all* enzymes were proteins. A few enzymes are now known that are made of ribonucleic acids (RNA; Section 22.7) and that catalyze cellular reactions involving nucleic acids. In this chapter, we will consider only enzymes that are proteins.

Enzymes undergo all the reactions of proteins, including *denaturation* (Section 20.16). Slight alterations in pH, temperature, or other protein denaturants affect enzyme activity dramatically. Good cooks realize that overheating yeast kills the action of the yeast. A person suffering from a high fever (greater than 106°F) runs the risk of denaturing certain enzymes. The biochemist must exercise extreme caution in handling enzymes to avoid the loss of their activity. Even vigorous shaking of an enzyme solution can destroy enzyme activity.

Enzymes differ from nonbiochemical (laboratory) catalysts not only in size, being much larger, but also in that their activity is usually regulated by other substances present in the cell in which they are found. Most laboratory catalysts need to be removed from a reaction mixture to stop their catalytic action; this is not so with enzymes. In some cases, if a certain chemical is needed in the cell, the enzyme responsible for its production is activated by other cellular components. When a sufficient quantity has been produced, the enzyme is then deactivated. In other situations, the cell may produce more or less enzyme as required. Because different enzymes are required for nearly all cellular reactions, certain necessary reactions can be accelerated or decelerated without affecting the rest of the cellular chemistry.

ENZYME STRUCTURE

Enzymes can be divided into two general structural classes: simple enzymes and conjugated enzymes. A **simple enzyme** *is an enzyme composed only of protein (amino acid chains).* A **conjugated enzyme** *is an enzyme that has a nonprotein part in addition to a protein part.* By itself, neither the protein part nor the nonprotein portion of a conjugated enzyme has catalytic properties. An **apoenzyme** *is the protein part of a conjugated enzyme.* A **cofactor** *is the nonprotein part of a conjugated enzyme.* It is the combination of apoenzyme with cofactor that produces a biochemically active enzyme. A **holoenzyme** *is the biochemically active conjugated enzyme produced from an apoenzyme and a cofactor.*

$$\text{Apoenzyme} + \text{cofactor} = \text{holoenzyme}$$

Why do apoenzymes need cofactors? Cofactors provide additional chemically reactive functional groups besides those present in the amino acid side chains of apoenzymes.

A cofactor is generally either a small organic molecule or an inorganic ion (usually a metal ion). A **coenzyme** *is a small organic molecule that serves as a cofactor in a conjugated enzyme.* Many vitamins (Section 21.11) have coenzyme functions in the human body.

Typical inorganic ion cofactors include Zn^{2+}, Mg^{2+}, Mn^{2+}, and Fe^{2+}. The nonmetallic Cl^- ion occasionally acts as a cofactor. Dietary minerals are an important source of inorganic ion cofactors.

NOMENCLATURE AND CLASSIFICATION OF ENZYMES

Enzymes are most commonly named by using a system that attempts to provide information about the *function* (rather than the structure) of the enzyme. Type of reaction catalyzed and *substrate* identity are focal points for the nomenclature. A **substrate** *is the reactant in an enzyme-catalyzed reaction.* The substrate is the substance upon which the enzyme "acts."

Three important aspects of the enzyme-naming process are the following:

1. The suffix -*ase* identifies a substance as an enzyme. Thus ure*ase*, sucr*ase*, and lip*ase* are all enzyme designations. The suffix -*in* is still found in the names of some of the first enzymes studied, many of which are digestive enzymes. Such names include *trypsin, chymotrypsin,* and *pepsin.*
2. The type of reaction catalyzed by an enzyme is often noted with a prefix. An *oxidase* enzyme catalyzes an oxidation reaction, and a *hydrolase* enzyme catalyzes a hydrolysis reaction.
3. The identity of the substrate is often noted in addition to the type of reaction. Enzyme names of this type include *glucose oxidase, pyruvate carboxylase,* and *succinate dehydrogenase.* Infrequently, the substrate but not the reaction type is given, as in the names *urease* and *lactase.* In such names, the reaction involved is hydrolysis; *urease* catalyzes the hydrolysis of urea, *lactase* the hydrolysis of lactose.

● **EXAMPLE 21.1**

Predicting Enzyme Function from an Enzyme's Name

▶ Predict the function of the following enzymes.

a. Cellulase   **b.** Sucrase
c. L-Amino acid oxidase   **d.** Aspartate aminotransferase

Solution

a. Cellulase catalyzes the hydrolysis of cellulose.
b. Sucrase catalyzes the hydrolysis of the disaccharide sucrose.
c. L-Amino acid oxidase catalyzes the oxidation of L-amino acids.
d. Aspartate aminotransferase catalyzes the transfer of an amino group from aspartate to a different molecule.

Practice Exercise 21.1

Predict the function of the following enzymes.

a. Maltase
b. Lactate dehydrogenase
c. Fructose oxidase
d. Maleate isomerase

Answers: a. Hydrolysis of maltose; **b.** Removal of hydrogen from lactate ion; **c.** Oxidation of fructose; **d.** Rearrangement (isomerization) of maleate ion

Ⓑ
Ⓕ
Recall from Section 14.9 that for organic redox reactions, the following two operational rules are used instead of oxidation numbers (Section 9.2) to characterize oxidation and reduction processes:

1. An *organic oxidation reaction* is an oxidation that *increases* the number of C—O bonds and/or *decreases* the number of C—H bonds.
2. An *organic reduction reaction* is a reduction that *decreases* the number of C—O bonds and/or *increases* the number of C—H bonds.

Enzymes are grouped into six major classes on the basis of the types of reactions they catalyze.

1. An **oxidoreductase** is *an enzyme that catalyzes an oxidation–reduction reaction.* Because oxidation and reduction are not independent processes but linked processes that must occur together (Section 9.3), an oxidoreductase requires a coenzyme that is oxidized or reduced as the substrate is reduced or oxidized. *Lactate dehydrogenase* is an oxidoreductase that removes hydrogen atoms from a molecule.

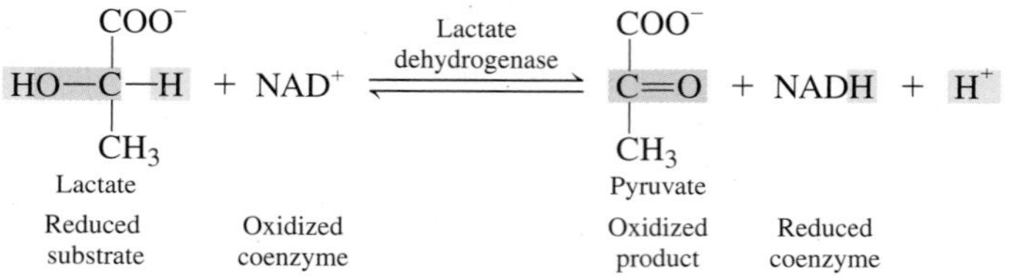

$$\underset{\substack{\text{Lactate}\\\text{Reduced}\\\text{substrate}}}{\text{HO}-\overset{\displaystyle\text{COO}^-}{\underset{\displaystyle\text{CH}_3}{\overset{|}{\underset{|}{\text{C}}}}-\text{H}} + \underset{\substack{\text{Oxidized}\\\text{coenzyme}}}{\text{NAD}^+} \underset{\text{dehydrogenase}}{\overset{\text{Lactate}}{\rightleftharpoons}} \underset{\substack{\text{Pyruvate}\\\text{Oxidized}\\\text{product}}}{\overset{\displaystyle\text{COO}^-}{\underset{\displaystyle\text{CH}_3}{\overset{|}{\underset{|}{\text{C}}=\text{O}}}} + \underset{\substack{\text{Reduced}\\\text{coenzyme}}}{\text{NADH}} + \text{H}^+$$

2. A **transferase** *is an enzyme that catalyzes the transfer of a functional group from one molecule to another.* Two major subtypes of transferases are *transaminases* and *kinases.* (A transaminase catalyzes the transfer of an amino group from one molecule to another. Kinases, which play a major role in metabolic energy-production reactions (see Section 24.2), catalyze the transfer of a phosphate group from adenosine triphosphate (ATP) to give adenosine diphosphate (ADP) and a phosphorylated product (a product containing an additional phosphate group).

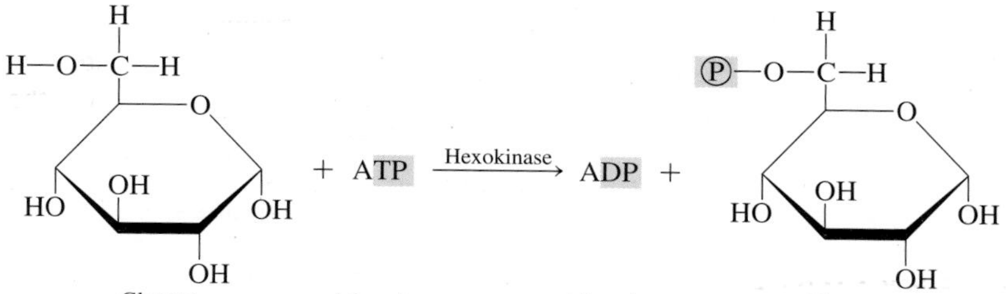

Glucose + ATP —(Hexokinase)→ ADP + Glucose 6-phosphate

Glucose

Adenosine triphosphate
(3 phosphate groups present)

Adenosine diphosphate
(2 phosphate groups present)

Glucose 6-phosphate

The symbol ℗ is a shorthand notation for a PO_3^{2-} unit.

3. A **hydrolase** *is an enzyme that catalyzes a hydrolysis reaction in which the addition of a water molecule to a bond causes the bond to break.* Hydrolysis reactions are central to the process of digestion. *Carbohydrases* effect the breaking of glycosidic bonds in oligo- and polysaccharides, *proteases* effect the breaking of peptide linkages in proteins, and *lipases* effect the breaking of ester linkages in triacylglycerols.

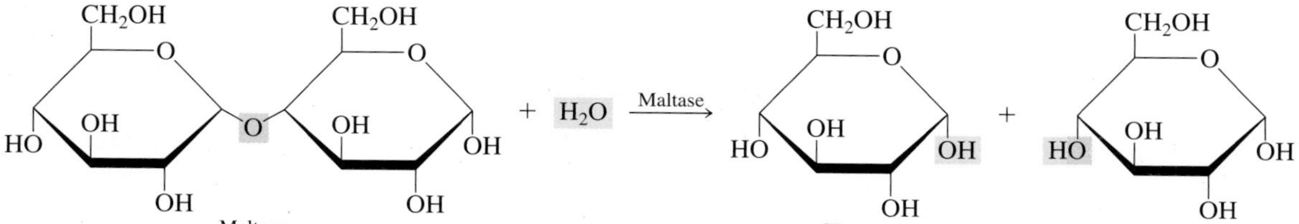

Maltose + H_2O —(Maltase)→ Glucose + Glucose

Maltose

Glucose

Glucose

4. A **lyase** *is an enzyme that catalyzes the addition of a group to a double bond or the removal of a group to form a double bond in a manner that does not involve hydrolysis or oxidation.* A *dehydratase* effects the removal of the components of water from a double bond and a *hydratase* effects the addition of the components of water to a double bond.

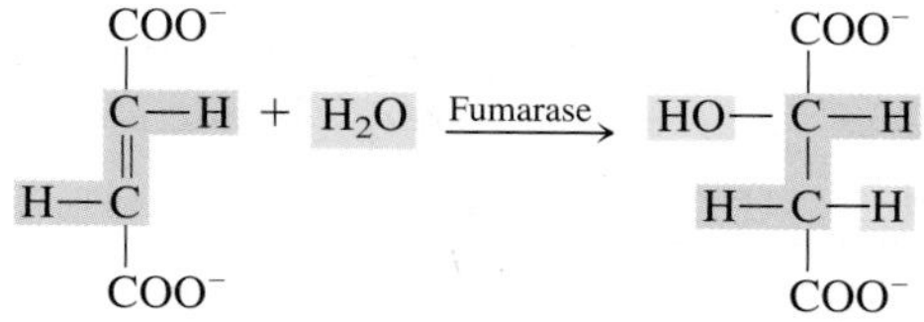

Fumarate + H_2O —(Fumarase)→ L-Malate

Fumarate

L-Malate

5. An **isomerase** *is an enzyme that catalyzes the isomerization (rearrangement of atoms) of a substrate in a reaction, converting it into a molecule isomeric with itself.* There is only one reactant and one product in reactions where isomerases are operative.

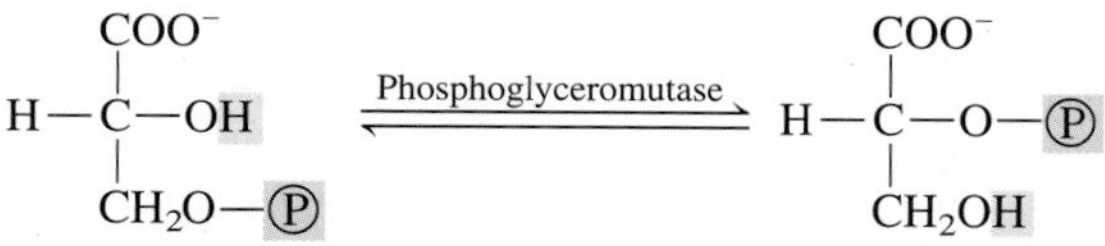

3-Phosphoglycerate ⇌(Phosphoglyceromutase) 2-Phosphoglycerate

3-Phosphoglycerate

2-Phosphoglycerate

TABLE 21.1
**Main Classes and Subclasses
of Enzymes**

Main Classes	Selected Subclasses	Type of Reaction Catalyzed
oxidoreductases	oxidases	oxidation of a substrate
	reductases	reduction of a substrate
	dehydrogenases	introduction of double bond (oxidation) by formal removal of two H atoms from substrate, the H being accepted by a coenzyme
transferases	transaminases	transfer of an amino group between substrates
	kinases	transfer of a phosphate group between substrates
hydrolases	lipases	hydrolysis of ester linkages in lipids
	proteases	hydrolysis of amide linkages in proteins
	nucleases	hydrolysis of sugar–phosphate ester bonds in nucleic acids
	carbohydrases	hydrolysis of glycosidic bonds in carbohydrates
	phosphatases	hydrolysis of phosphate–ester bonds
lyases	dehydratases	removal of H_2O from substrate
	decarboxylases	removal of CO_2 from substrate
	deaminases	removal of NH_3 from substrate
	hydratases	addition of H_2O to a substrate
isomerases	racemases	conversion of D to L isomer, or vice versa
	mutases	transfer of a functional group from one position to another in the same molecule
ligases	synthetases	formation of new bond between two substrates, with participation of ATP
	carboxylases	formation of new bond between a substrate and CO_2, with participation of ATP

6. A **ligase** *is an enzyme that catalyzes the bonding together of two molecules into one with the participation of ATP.* ATP involvement is required because such reactions are generally energetically unfavorable and they require the simultaneous input of energy obtained by a hydrolysis reaction in which ATP is converted to ADP (such energy release is considered in Section 23.3).

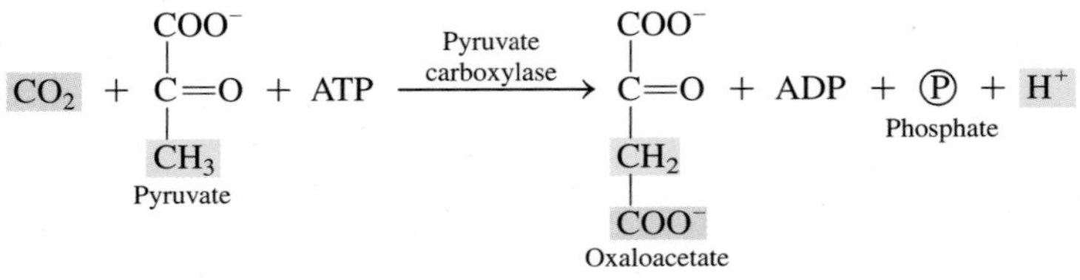

Within each of these six main classes of enzymes there are enzyme subclasses. Table 21.1 gives further information about enzyme subclass terminology, some of which was used in the preceding discussion of the main enzyme classes. For example, dehydratases and hydratases are subclass designations; both of these enzyme subtypes are lyases (see Table 21.1).

EXAMPLE 21.2

Classifying Enzymes by the Type of Chemical Reaction They Catalyze

To what main enzyme class do the enzymes that catalyze the following chemical reactions belong?

a.

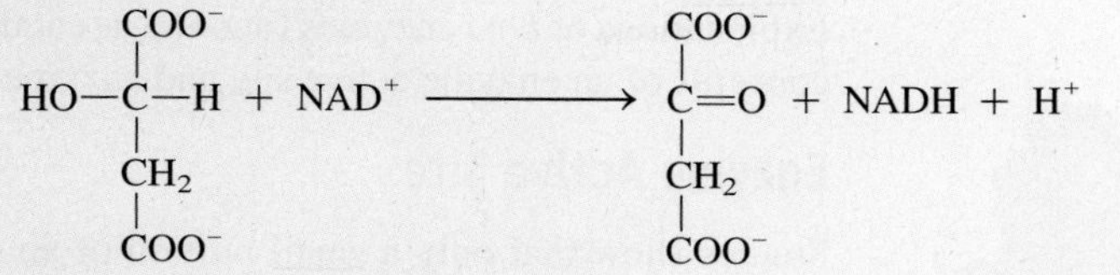

b.

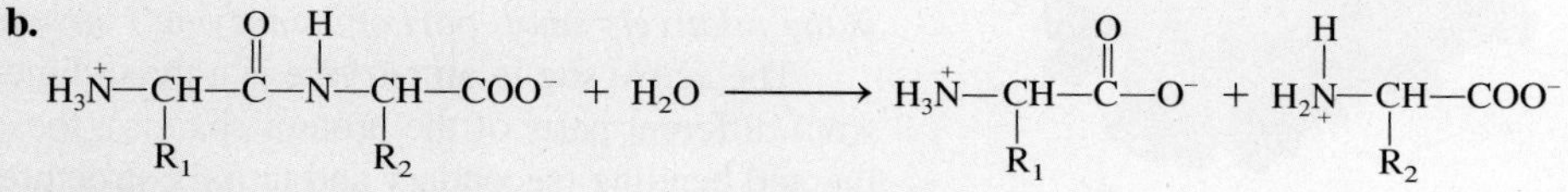

Solution

a. In this reaction two H atoms are removed from the substrate and a carbon–oxygen double bond is formed. Loss of two hydrogen atoms by a molecule is an indication of oxidation. This is an oxidation–reduction reaction and the enzyme needed to effect the change will be an *oxidoreductase*.

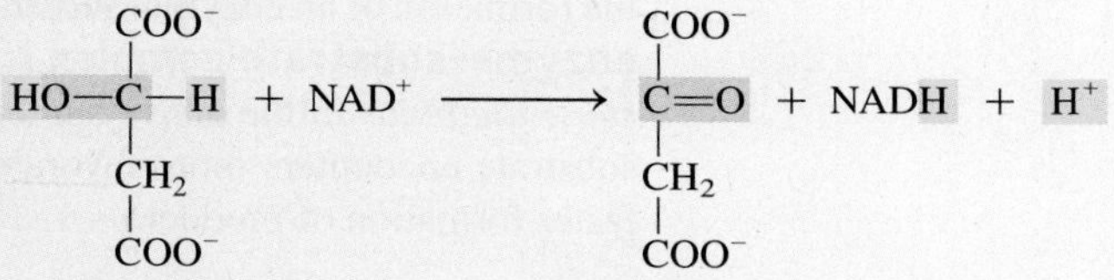

Malate   Oxaloacetate

b. In this reaction the components of a molecule of H_2O are added to the substrate (a dipeptide) with the resulting breaking of the peptide bond to produce two amino acids. This is an example of a hydrolysis reaction. An enzyme that effects such a change is called a *hydrolase*.

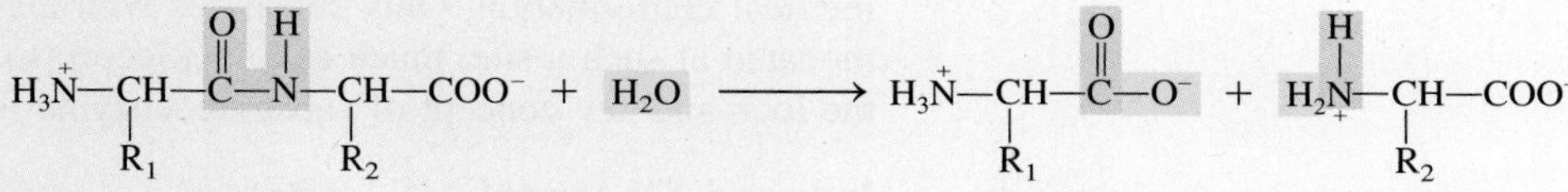

Dipeptide   Amino acid   Amino acid

Practice Exercise 21.2

To what main enzyme class do the enzymes that catalyze the following chemical reactions belong?

a.

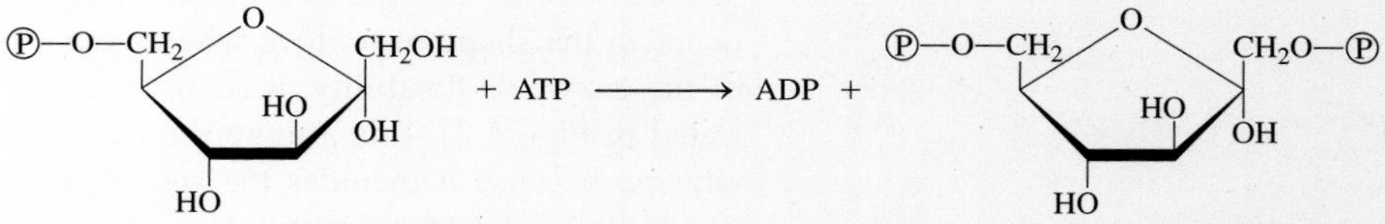

Fructose 6-phosphate   Fructose 1,6-bisphosphate

b.

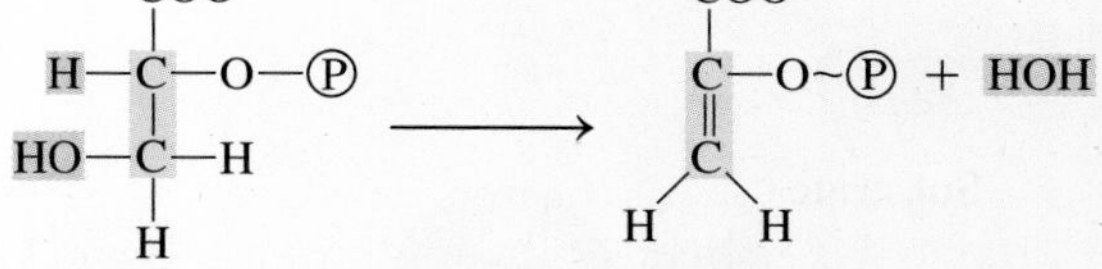

2-Phosphoglycerate   Phosphoenolpyruvate

Answers: a. Transferase; **b.** Lyase

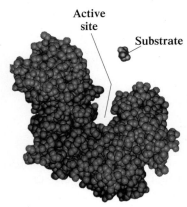

Figure 21.2 The active site of an enzyme is usually a crevicelike region formed as a result of the protein's secondary and tertiary structural characteristics.

The lock-and-key model is more than just a "shape fit." In addition, there are weak binding forces (R group interactions) between parts.

 21.4 **MODELS OF ENZYME ACTION**

Explanations of *how* enzymes function as catalysts in biochemical systems are based on the concepts of an enzyme active site and enzyme–substrate complex formation.

Enzyme Active Site

Studies show that only a small portion of an enzyme molecule called the active site participates in the interaction with a substrate or substrates during a reaction. The **active site** *is the relatively small part of an enzyme's structure that is actually involved in catalysis.*

The active site in an enzyme is a three-dimensional entity formed by groups that come from different parts of the protein chain(s); these groups are brought together by the folding and bending (secondary and tertiary structure; Sections 20.10 and 20.11) of the protein. The active site is usually a "crevicelike" location in the enzyme (see Figure 21.2).

Enzyme–Substrate Complex

Catalysts offer an alternative pathway with lower activation energy through which a reaction can occur (Section 9.6). In enzyme-controlled reactions, this alternative pathway involves the formation of an enzyme–substrate complex as an intermediate species in the reaction. An **enzyme-substrate complex** *is the intermediate reaction species that is formed when a substrate binds to the active site of an enzyme.* Within the enzyme–substrate complex, the substrate encounters more favorable reaction conditions than if it were free. The result is faster formation of product.

Lock-and-Key Model

To account for the highly specific way an enzyme recognizes a substrate and binds it to the active site, researchers have proposed several models. The simplest of these models is the lock-and-key model.

In the lock-and-key model, the active site in the enzyme has a fixed, rigid geometrical conformation. Only substrates with a complementary geometry can be accommodated at such a site, much as a lock accepts only certain keys. Figure 21.3 illustrates the lock-and-key concept of substrate–enzyme interaction.

Induced-Fit Model

The lock-and-key model explains the action of numerous enzymes. It is, however, too restrictive for the action of many other enzymes. Experimental evidence indicates that many enzymes have flexibility in their shapes. They are not rigid and static; there is constant change in their shape. The induced-fit model is used for this type of situation.

The induced-fit model allows for small changes in the shape or geometry of the active site of an enzyme to accommodate a substrate. A good analogy is the changes that occur in the shape of a glove when a hand is inserted into it. The induced fit is a result of the enzyme's flexibility; it adapts to accept the incoming substrate. This model, illustrated in Figure 21.4, is a more thorough explanation for the active-site properties of an enzyme because it includes the specificity of the lock-and-key model coupled with the flexibility of the enzyme protein.

Figure 21.3 The lock-and-key model for enzyme activity. Only a substrate whose shape and chemical nature are complementary to those of the active site can interact with the enzyme.

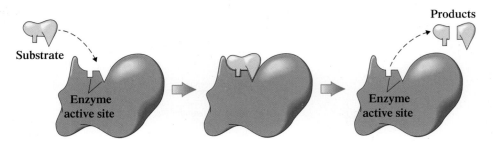

Figure 21.4 The induced-fit model for enzyme activity. The enzyme active site, although not exactly complementary in shape to that of the substrate, is flexible enough that it can adapt to the shape of the substrate.

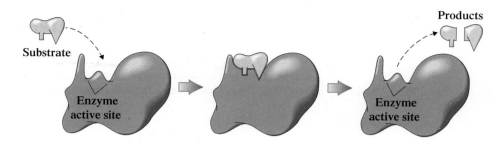

The forces that draw the substrate into the active site are many of the same forces that maintain tertiary structure in the folding of peptide chains. Electrostatic interactions, hydrogen bonds, and hydrophobic interactions all help attract and bind substrate molecules. For example, a protonated (positively charged) amino group in a substrate could be attracted and held at the active site by a negatively charged aspartate or glutamate residue. Alternatively, cofactors such as positively charged metal ions often help bind substrate molecules. Figure 21.5 is a schematic representation of the amino acid R group interactions that bind a substrate to an enzyme active site.

21.5 ENZYME SPECIFICITY

Enzymes exhibit different levels of selectivity, or specificity, for substrates. The degree of enzyme specificity is determined by the active site. Some active sites accommodate only one particular compound, whereas others can accommodate a "family" of closely related compounds. Types of enzyme specificity include

1. *Absolute Specificity.* Such specificity means an enzyme will catalyze a particular reaction for *only one* substrate. This most restrictive of all specificities is not common. Urease is an enzyme with absolute specificity.
2. *Stereochemical Specificity.* Such specificity means an enzyme can distinguish between stereoisomers. Chirality is inherent in an active site, because amino acids are chiral compounds. L-Amino-acid oxidase will catalyze reactions of L-amino acids but not of D-amino acids.
3. *Group Specificity.* Such specificity involves structurally similar compounds that have the same functional groups. Carboxypeptidase is group-specific; it cleaves amino acids, one at a time, from the carboxyl end of the peptide chain.
4. *Linkage Specificity.* Such specificity involves a particular type of bond, irrespective of the structural features in the vicinity of the bond. Phosphatases hydrolyze phosphate–ester bonds in all types of phosphate esters. Linkage specificity is the most general of the specificities considered.

Figure 21.5 A schematic diagram representing amino acid R group interactions that bind a substrate to an enzyme active site. The R group interactions that maintain the three-dimensional structure of the enzyme (secondary and tertiary structure) are also shown.

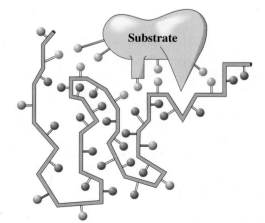

- R group interactions that bind the substrate to the enzyme active site
- R group interactions that maintain the three-dimensional structure of the enzyme
- Noninteracting R groups that help determine the solubility of the enzyme

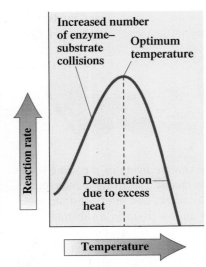

Figure 21.6 A graph showing the effect of temperature on the rate of an enzymatic reaction.

The upper temperature limit for life now stands at 121°C as the result of the discovery, in 2004, of a new "heat-loving" microbe. The microbe was found in a water sample from a hydrothermal vent deep in the Northeast Pacific Ocean. Its method of respiration involves reduction of Fe(III) to Fe(II) to produce energy.

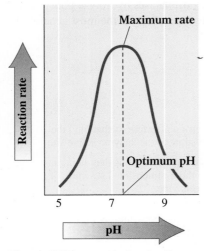

Figure 21.7 A graph showing the effect of pH on the rate of an enzymatic reaction.

21.6 FACTORS THAT AFFECT ENZYME ACTIVITY

Enzyme activity *is a measure of the rate at which an enzyme converts substrate to products in a biochemical reaction.* Four factors affect enzyme activity: temperature, pH, substrate concentration, and enzyme concentration.

Temperature

Temperature is a measure of the kinetic energy (energy of motion) of molecules. Higher temperatures mean molecules are moving faster and colliding more frequently. This concept applies to collisions between substrate molecules and enzymes. As the temperature of an enzymatically catalyzed reaction increases, so does the rate (velocity) of the reaction.

However, when the temperature increases beyond a certain point, the increased energy begins to cause disruptions in the tertiary structure of the enzyme; denaturation is occurring. Change in tertiary structure at the active site impedes catalytic action, and the enzyme activity quickly decreases as the temperature climbs past this point (Figure 21.6). The temperature that produces maximum activity for an enzyme is known as the optimum temperature for that enzyme. **Optimum temperature** *is the temperature at which an enzyme exhibits maximum activity.*

For human enzymes, the optimum temperature is around 37°C, normal body temperature. A person who has a fever where body core temperature exceeds 40°C can be in a life-threatening situation because such a temperature is sufficient to initiate enzyme denaturation. The loss of function of critical enzymes, particularly those of the central nervous system, can result in dysfunction sufficient to cause death.

The "destroying" effect of temperature on bacterial enzymes is used in a hospital setting to sterilize medical instruments and laundry. In high-temperature high-pressure vessels called *autoclaves* super-heated steam is used to produce a temperature sufficient to denature bacterial enzymes.

Not all enzymes have optimal temperatures around the physiological temperature of 37°C. This is particularly true for enzymes found in microbes associated with hydrothermal areas such as those in Yellowstone National Park and hydrothermal vents on the ocean floor, where temperature and pressure can be extremely high. The ability of microbial enzymes to survive under such harsh conditions is related to the amino acid sequences in their protein structures, sequences that are stable under such extraordinary conditions. Microbial enzymes that survive in such extreme environments are collectively called *extremozymes.*

The study of extremozymes is an area of special interest for industrial chemists. Enzymes can function as catalysts for industrial processes, just as they do for biochemical reactions, provided they can survive the conditions associated with the process. Because industrial processes usually require higher temperature and pressure than physiological processes, extremozymes can be useful. The enzymes present in some detergent formulations, which must function in hot water, are the result of research associated with high-temperature microbial enzymes.

pH

The pH of an enzyme's environment can affect its activity. This is not surprising because the *charge* on acidic and basic amino acids (Section 20.2) located at the active site depends on pH. Small changes in pH (less than one unit) can result in enzyme denaturation (Section 20.16) and subsequent loss of catalytic activity.

Most enzymes exhibit maximum activity over a very narrow pH range. Only within this narrow pH range do the enzyme's amino acids exist in properly charged forms (Section 20.4). **Optimum pH** *is the pH at which an enzyme exhibits maximum activity.* Figure 21.7 shows the effect of pH on an enzyme's activity. Biochemical buffers help maintain the optimum pH for an enzyme.

Each enzyme has a characteristic optimum pH, which usually falls within the physiological pH range of 7.0–7.5. Notable exceptions to this generalization are the digestive

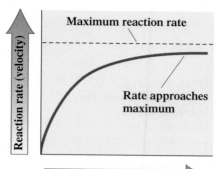

Figure 21.8 A graph showing the change in enzyme activity with a change in substrate concentration at constant temperature, pH, and enzyme concentration. Enzyme activity remains constant after a certain substrate concentration is reached.

enzymes pepsin and trypsin. Pepsin, which is active in the stomach, functions best at a pH of 2.0. On the other hand, trypsin, which operates in the small intestine, functions best at a pH of 8.0. The amino acid sequences present in pepsin and trypsin are those needed such that the R groups present can maintain protein tertiary structure (Section 20.11) at low (2.0) and high (8.0) pH values, respectively.

A variation from normal pH can also affect substrates, causing either protonation or deprotonation of groups on the substrate. The interaction between the altered substrate and the enzyme active site may be less efficient than normal—or even impossible.

Substrate Concentration

When the concentration of an enzyme is kept constant and the concentration of substrate is increased, the enzyme activity pattern shown in Figure 21.8 is obtained. This activity pattern is called a *saturation curve*. Enzyme activity increases up to a certain substrate concentration and thereafter remains constant.

What limits enzymatic activity to a certain maximum value? As substrate concentration increases, the point is eventually reached where enzyme capabilities are used to their maximum extent. The rate remains constant from this point on (Figure 21.8). Each substrate must occupy an enzyme active site for a finite amount of time, and the products must leave the site before the cycle can be repeated. When each enzyme molecule is

---

CHEMICAL Connections

H. pylori and Stomach Ulcers

Helicobacter pylori, commonly called *H. pylori*, is a bacterium that can function in the highly acidic environment of the stomach. The discovery in 1982 of the existence of this bacterium in the stomach was startling to the medical profession because conventional thought at the time was that bacteria could not survive at the stomach's pH of about 1.4.

It is now known that *H. pylori* causes more than 90% of duodenal ulcers and up to 80% of gastric ulcers. Before this discovery, it was thought that most ulcers were caused by excess stomach acid eating the stomach lining. Contributory causes were thought to be spicy food and stress. Conventional treatment involved acid-suppression or acid-neutralization medications. Now, treatment regimens involve antibiotics. The medical profession was slow to accept the concept of a bacterial cause for most ulcers, and it was not until the mid-1990s that antibiotic treatment became common.

How the enzymes present in the *H. pylori* bacterium can function in the acidic environment of the stomach (where they should be denatured) is now known. Present on the surface of the bacterium is the enzyme *urease*, an enzyme that converts urea to the basic substance ammonia. The ammonia then neutralizes acid present in its immediate vicinity; a protective barrier is thus created. The *urease* itself is protected from denaturation by its complex quaternary structure.

H. pylori causes ulcers by weakening the protective mucous coating of the stomach and duodenum, which allows acid to get through to the sensitive lining beneath. Both the acid and the bacteria irritate the lining and cause a sore—the ulcer. Ultimately the *H. pylori* themselves burrow into the lining to an acid-safe area within the lining.

Approximately two-thirds of the world's population is infected with *H. pylori*. In the United States 30% of the adult population is infected, with the infection most prevalent among older adults. About 20% of people under the age of 40 and half of those over 60 have it. Only one out of every six people infected with *H. pylori* ever suffer symptoms related to ulcers. Why *H. pylori* does not cause ulcers in every infected person is not known.

H. pylori bacteria are most likely spread from person to person through fecal–oral or oral–oral routes. Possible environmental sources include contaminated water sources. The infection is more common in crowded living conditions with poor sanitation. In countries with poor sanitation, 90% of the adult population can be infected.

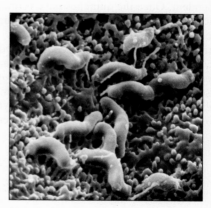

H. pylori bacteria.

TABLE 21.2
Turnover Numbers for Selected Enzymes

Enzyme	Turnover Number (per minute)	Reaction Catalyzed
carbonic anhydrase	36,000,000	$CO_2 + H_2O \rightleftharpoons H_2CO_3$
catalase	5,600,000	$2H_2O_2 \rightleftharpoons 2H_2O + O_2$
cholinesterase	1,500,000	hydrolysis of acetylcholine
penicillinase	120,000	hydrolysis of penicillin
lactate dehydrogenase	60,000	conversion of pyruvate to lactate
DNA polymerase I	900	addition of nucleotides to DNA chains

working at full capacity, the incoming substrate molecules must "wait their turn" for an empty active site. At this point, the enzyme is said to be under saturation conditions.

The rate at which an enzyme accepts substrate molecules and releases product molecules at substrate saturation is given by its turnover number. An enzyme's **turnover number** *is the number of substrate molecules transformed per minute by one molecule of enzyme under optimum conditions of temperature, pH, and saturation.* Table 21.2 gives turnover numbers for selected enzymes. Some enzymes have a much faster mode of operation than others.

CHEMICAL Connections

Enzymatic Browning: Discoloration of Fruits and Vegetables

Everyone is familiar with the way fruits such as apples, pears, peaches, apricots, and bananas, and vegetables such as potatoes quickly turn brown when their tissue is exposed to oxygen. Such oxygen exposure occurs when the food is sliced or bitten into or when it has sustained bruises, cuts, or other injury to the peel. This "browning reaction" is related to the work of an enzyme called phenolase (or polyphenoloxidase), a conjugated enzyme in which copper is present.

Phenolase is classified as an oxidoreductase. The substrates for phenolase are phenolic compounds present in the tissues of the fruits and vegetables. Phenolase hydroxylates monophenols to *o*-diphenols and oxidizes *o*-diphenols to *o*-quinones (see chemical equations below). The *o*-quinones then enter into a number of other reactions, which produce the "undesirable" brown discolorations. Quinone formation is enzyme- and oxygen-dependent. Once the quinones have formed, the subsequent reactions occur spontaneously and no longer depend on the presence of phenolase or oxygen.

Enzymatic browning can be prevented or slowed in several ways. Immersing the "injured" food (for example, apple slices) in cold water slows the browning process. The lower temperature decreases enzyme activity, and the water limits the enzyme's access to oxygen. Refrigeration slows enzyme activity even more, and boiling temperatures destroy (denature) the

At left, a freshly cut apple. Brownish oxidation products form in a few minutes (at right).

enzyme. A long-used method for preventing browning involves lemon juice. Phenolase works very slowly in the acidic environment created by the lemon juice's presence. In addition, the vitamin C (ascorbic acid) present in lemon juice functions as an antioxidant. It is more easily oxidized than the phenolic-derived compounds, and its oxidation products are colorless.

Monophenol derivatives   *o*-Diphenol derivatives   *o*-Quinone derivatives

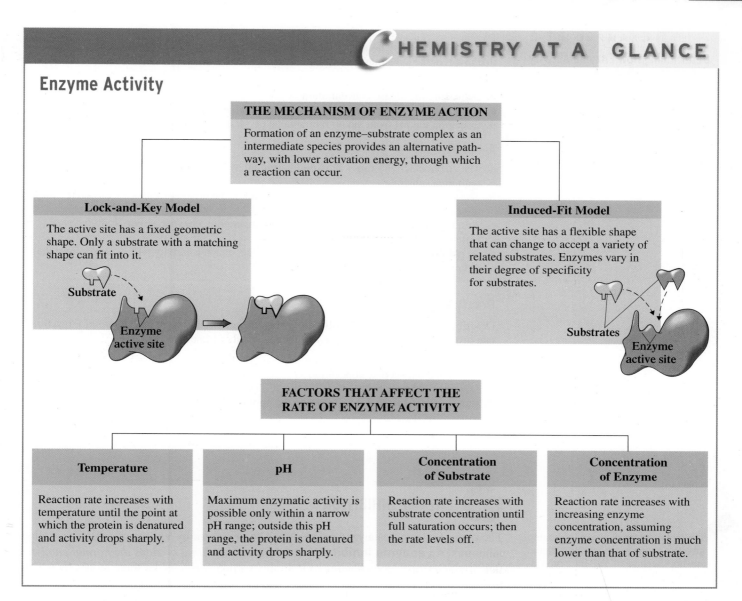

CHEMISTRY AT A GLANCE

Enzyme Activity

THE MECHANISM OF ENZYME ACTION

Formation of an enzyme–substrate complex as an intermediate species provides an alternative pathway, with lower activation energy, through which a reaction can occur.

Lock-and-Key Model

The active site has a fixed geometric shape. Only a substrate with a matching shape can fit into it.

Substrate

Enzyme active site

Induced-Fit Model

The active site has a flexible shape that can change to accept a variety of related substrates. Enzymes vary in their degree of specificity for substrates.

Substrates

Enzyme active site

FACTORS THAT AFFECT THE RATE OF ENZYME ACTIVITY

Temperature	pH	Concentration of Substrate	Concentration of Enzyme
Reaction rate increases with temperature until the point at which the protein is denatured and activity drops sharply.	Maximum enzymatic activity is possible only within a narrow pH range; outside this pH range, the protein is denatured and activity drops sharply.	Reaction rate increases with substrate concentration until full saturation occurs; then the rate levels off.	Reaction rate increases with increasing enzyme concentration, assuming enzyme concentration is much lower than that of substrate.

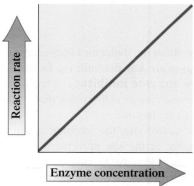

Figure 21.9 A graph showing the change in reaction rate with a change in enzyme concentration for an enzymatic reaction. Temperature, pH, and substrate concentration are constant. The substrate concentration is high relative to enzyme concentration.

Enzyme Concentration

Because enzymes are not consumed in the reactions they catalyze, the cell usually keeps the number of enzymes low compared with the number of substrate molecules. This is efficient; the cell avoids paying the energy costs of synthesizing and maintaining a large work force of enzyme molecules. Thus, in general, the concentration of substrate in a reaction is much higher than that of the enzyme.

If the amount of substrate present is kept constant and the enzyme concentration is increased, the reaction rate increases because more substrate molecules can be accommodated in a given amount of time. A plot of enzyme activity versus enzyme concentration, at a constant substrate concentration that is high relative to enzyme concentration, is shown in Figure 21.9. The greater the enzyme concentration, the greater the reaction rate.

The Chemistry at a Glance feature above reviews what we have said about enzyme activity.

 **E X A M P L E 21.3**

Determining How Enzyme Activity Is Affected by Various Changes

▶ Describe the effect that each of the following changes would have on the rate of a reaction that involves the substrate urea and the liver enzyme urease.

a. Increasing the urea concentration
b. Increasing the urease concentration
c. Increasing the temperature from its optimum value to a value 10° higher than this value
d. Lowering the pH from the optimum value of 5.0 to a value of 3.0

Solution

a. The enzyme activity rate will *increase* until all of the enzyme molecules are engaged with urea substrate.
b. The enzyme activity rate will *increase* until all of the urea molecules are engaged with urease enzymes.
c. At temperatures higher than the optimum temperature, enzyme activity will *decrease* from that at the optimum temperature.
d. At pH values lower than the optimum pH value, enzyme activity will *decrease* from that at the optimum pH.

Practice Exercise 21.3

Describe the effect that each of the following changes would have on the rate of a reaction that involves the substrate sucrose and the intestinal enzyme sucrase.

a. Decreasing the sucrase concentration
b. Increasing the sucrose concentration
c. Lowering the temperature to 10° C
d. Raising the pH from 6.0 to 8.0 when the optimum pH is 6.2

Answers: a. Decrease rate; **b.** Increase rate; **c.** Decrease rate; **d.** Decrease rate

 ENZYME INHIBITION

The rates of enzyme-catalyzed reactions can be *decreased* by a group of substances called inhibitors. An **enzyme inhibitor** *is a substance that slows or stops the normal catalytic function of an enzyme by binding to it.* In this section, we consider three modes by which inhibition takes place: reversible competitive inhibition, reversible noncompetitive inhibition, and irreversible inhibition.

Reversible Competitive Inhibition

In Section 21.5 we noted that enzymes are quite specific about the molecules they accept at their active sites. Molecular shape and charge distribution are key determining factors in whether an enzyme accepts a molecule. A **competitive enzyme inhibitor** *is a molecule that sufficiently resembles an enzyme substrate in shape and charge distribution that it can compete with the substrate for occupancy of the enzyme's active site.*

When a competitive inhibitor binds to an enzyme active site, the inhibitor remains unchanged (no reaction occurs), but its physical presence at the site prevents a normal substrate molecule from occupying the site. The result is a decrease in enzyme activity.

The formation of an enzyme–competitive inhibitor complex is a reversible process because it is maintained by weak interactions (hydrogen bonds, etc.). With time (fractions of a second), the complex breaks up. The empty active site is then available for a new occupant. Substrate and inhibitor again compete for the empty active site. Thus the active site of an enzyme binds either inhibitor or normal substrate on a random basis. If inhibitor concentration is greater than substrate concentration, the inhibitor dominates the occupancy process. The reverse is also true. Competitive inhibition can be reduced by simply increasing the concentration of the substrate.

 The treatment for methanol poisoning involves giving a patient intravenous ethanol (Section 14.5). This action is based on the principle of competitive enzyme inhibition. The same enzyme, *alcohol dehydrogenase,* detoxifies both methanol and ethanol. Ethanol has 10 times the affinity for the enzyme that methanol has, keeping the enzyme busy with ethanol as the substrate gives the body time to excrete the methanol before it is oxidized to the potentially deadly formaldehyde (Section 14.5).

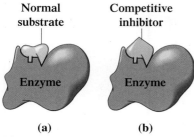

Figure 21.10 A comparison of an enzyme with a substrate at its active site (a) and an enzyme with a competitive inhibitor at its active site (b).

Figure 21.10 compares the binding of a normal substrate and that of a competitive inhibitor at an enzyme's active site. Note that the portions of these two molecules that bind to the active site have the same shape but that the two molecules differ in *overall* shape. It is because of this overall difference in shape that the substrate reacts at the active site but the inhibitor does not.

Numerous drugs act by means of competitive inhibition. For example, antihistamines are competitive inhibitors of histidine decarboxylation, the enzymatic reaction that converts histidine to histamine. Histamine causes the usual allergy and cold symptoms: watery eyes and runny nose.

Reversible Noncompetitive Inhibition

A **noncompetitive enzyme inhibitor** *is a molecule that decreases enzyme activity by binding to a site on an enzyme other than the active site.* The substrate can still occupy the active site, but the presence of the inhibitor causes a change in the structure of the enzyme sufficient to prevent the catalytic groups at the active site from properly effecting their catalyzing action. Figure 21.11 contrasts the processes of reversible competitive inhibition and reversible noncompetitive inhibition.

Unlike the situation in competitive inhibition, increasing the concentration of substrate does not completely overcome the inhibitory effect in this case. However, lowering the concentration of a noncompetitive inhibitor sufficiently does free up many enzymes, which then return to normal activity.

Examples of noncompetitive inhibitors include the heavy metal ions Pb^{2+}, Ag^+, and Hg^{2+}. The binding sites for these ions are sulfhydryl (—SH) groups located away from the active site. Metal sulfide linkages are formed, an effect that disrupts secondary and tertiary structure.

Irreversible Inhibition

An **irreversible enzyme inhibitor** *is a molecule that inactivates enzymes by forming a strong covalent bond to an amino acid side-chain group at the enzyme's active site.* In general, such inhibitors do *not* have structures similar to that of the enzyme's normal substrate. The inhibitor–active site bond is sufficiently strong that addition of excess substrate does not reverse the inhibition process. Thus the enzyme is permanently deactivated. The actions of chemical warfare agents (nerve gases) and organophosphate insecticides are based on irreversible inhibition.

The Chemistry at a Glance feature on page 712 summarizes what we have considered concerning enzyme inhibition.

Figure 21.11 The difference between a reversible competitive inhibitor and a reversible noncompetitive inhibitor.

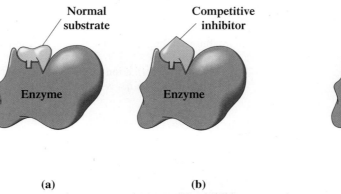

(a)
An enzyme–substrate complex in absence of an inhibitor.

(b)
A competitive inhibitor binds to the active site, which prevents the normal substrate from binding to the site.

(c)
A noncompetitive inhibitor binds to a site other than the active site; the normal substrate still binds to the active site but the enzyme cannot catalyze the reaction due to the presence of the inhibitor.

CHEMISTRY AT A GLANCE

Enzyme Inhibition

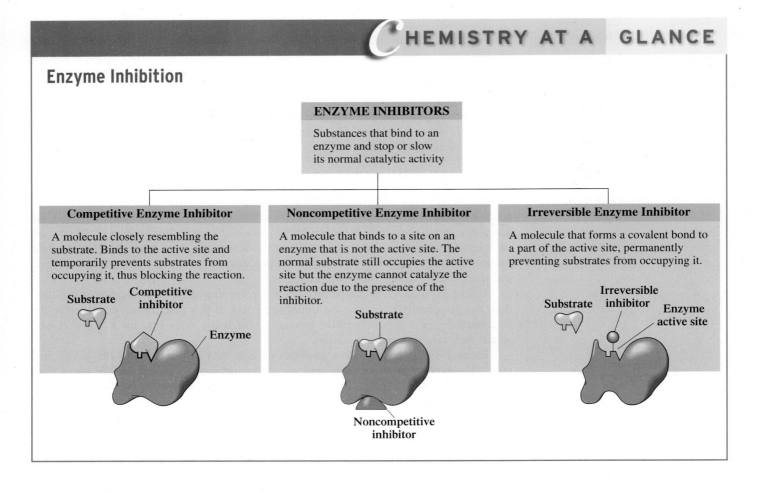

ENZYME INHIBITORS

Substances that bind to an enzyme and stop or slow its normal catalytic activity

Competitive Enzyme Inhibitor

A molecule closely resembling the substrate. Binds to the active site and temporarily prevents substrates from occupying it, thus blocking the reaction.

Substrate   Competitive inhibitor

Enzyme

Noncompetitive Enzyme Inhibitor

A molecule that binds to a site on an enzyme that is not the active site. The normal substrate still occupies the active site but the enzyme cannot catalyze the reaction due to the presence of the inhibitor.

Substrate

Noncompetitive inhibitor

Irreversible Enzyme Inhibitor

A molecule that forms a covalent bond to a part of the active site, permanently preventing substrates from occupying it.

Substrate   Irreversible inhibitor   Enzyme active site

EXERCISE 21.4

Identifying the Type of Enzyme Inhibition from Inhibitor Characteristics

Identify the type of enzyme inhibition each of the following inhibitor characteristics is associated with.

a. An inhibitor that decreases enzyme activity by binding to a site on the enzyme other than the active site

b. An inhibitor that inactivates enzymes by forming a strong covalent bond at the enzyme active site

Solution

a. Inhibitor binding at a nonactive site location is a characteristic of a *reversible noncompetitive inhibitor.*

b. Covalent bond formation at the active site, with amino acid residues located there, is a characteristic of an *irreversible inhibitor.*

Practice Exercise 21.4

Identify the type of enzyme inhibition each of the following inhibitor characteristics is associated with.

a. An inhibitor that has a shape and charge distribution similar to that of the enzyme's normal substrate

b. An inhibitor whose effect can be reduced by simply increasing the concentrate of normal substrate present

Answers: a. Reversible competitive inhibitor; **b.** Reversible competitive inhibitor

21.8 REGULATION OF ENZYME ACTIVITY

Regulation of enzyme activity within a cell is a necessity for many reasons. Illustrative of this need are the following two situations, both of which involve the concept of energy conservation.

1. A cell that continually produces large amounts of an enzyme for which substrate concentration is always very low is wasting energy. The production of the enzyme needs to be "turned off."
2. A product of an enzyme-catalyzed reaction that is present in plentiful (more than needed) amounts in a cell, is a waste of energy if the enzyme continues to catalyze the reaction that produces the product. The enzyme needs to be "turned off."

Many mechanisms exist by which enzymes within a cell can be "turned on" and "turned off." In this section we consider, in general terms, three such mechanisms: (1) feedback control associated with allosteric enzymes, (2) proteolytic enzymes and zymogens, and (3) covalent modification.

Allosteric Enzymes

Many, but not all, of the enzymes responsible for regulating cellular processes are *allosteric enzymes.* Characteristics of allosteric enzymes are as follows:

1. All allosteric enzymes have quaternary structure; that is, they are composed of two or more protein chains.
2. All allosteric enzymes have two kinds of binding sites: those for substrate and those for regulators.
3. Active and regulatory binding sites are distinct from each other in both location and shape. Often the regulatory site is on one protein chain and the active site is on another.
4. Binding of a molecule at the regulatory site causes changes in the overall three-dimensional structure of the enzyme, including structural changes at the active site.

The term *allosteric* comes from the Greek *allo,* which means "other," and *stereos,* which means "site or space."

Thus an **allosteric enzyme** *is an enzyme with two or more protein chains (quaternary structure) and two kinds of binding sites (substrate and regulator).*

Substances that bind at regulatory sites of allosteric enzymes are called *regulators.* The binding of a *positive regulator* increases enzyme activity; the shape of the active site is changed such that it can more readily accept substrate. The binding of a *negative regulator* (a noncompetitive inhibitor) decreases enzyme activity; changes to the active site are such that substrate is less readily accepted. Figure 21.12 contrasts the different effects on an allosteric enzyme that positive and negative regulators produce.

Some regulators of allosteric enzyme function are inhibitors (negative regulators), and some increase enzyme activity (positive regulators).

Feedback Control

One of the mechanisms by which allosteric enzyme activity is regulated is feedback control. **Feedback control** *is a process in which activation or inhibition of the first reaction in a reaction sequence is controlled by a product of the reaction sequence.*

To illustrate the feedback control mechanism, let us consider a biochemical process within a cell that occurs in several steps, each step catalyzed by a different enzyme.

Most biochemical processes within cells take place in several steps rather than in a single step. A different enzyme is required for each step of the process.

$$A \xrightarrow{\text{Enzyme 1}} B \xrightarrow{\text{Enzyme 2}} C \xrightarrow{\text{Enzyme 3}} D$$

The product of each step is the substrate for the next enzyme.

What will happen in this reaction series if the final product (D) is a negative regulator of the first enzyme (enzyme 1)? At low concentrations of D, the reaction sequence proceeds rapidly. At higher concentrations of D, the activity of enzyme 1 becomes inhibited (by feedback), and eventually the activity stops. At the stopping point, there is sufficient D present in the cell to meet its needs. Later, when the concentration of D decreases through use in other cell reactions, the activity of enzyme 1 increases and more D is produced.

Feedback control is operative in devices that maintain a constant temperature, such as a furnace or an oven. If you set the control thermostat at 68°F, the furnace or oven produces heat until that temperature is reached, and then, through electronic feedback, the furnace or oven is shut off.

Figure 21.12 The differing effects of positive and negative regulators on an allosteric enzyme.

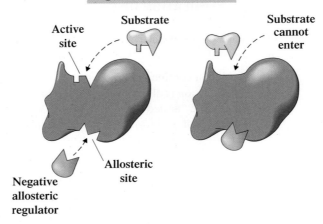

Negative Allosteric Control

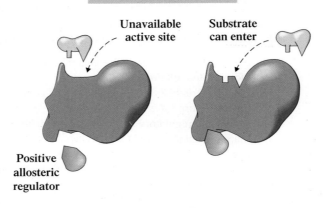

Positive Allosteric Control

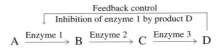

The general term *allosteric control* is often used to describe a process in which a regulatory molecule that binds at one site in an enzyme influences substrate binding at the active site in the enzyme.

Feedback control is not the only mechanism by which an allosteric enzyme can be regulated; it is just one of the more common ways. Regulators of a particular allosteric enzyme may be products of entirely different pathways of reaction within the cell, or they may even be compounds produced outside the cell (hormones).

Proteolytic Enzymes and Zymogens

A second mechanism for regulating cellular enzyme activity is based on the production of enzymes in an inactive form. These inactive enzyme precursors are then "turned on" at the appropriate time. Such a mechanism for control is often encountered in the production of *proteolytic enzymes*. A **proteolytic enzyme** *is an enzyme that catalyzes the breaking of peptide bonds that maintain the primary structure of a protein.* Because they would otherwise destroy the tissues that produce them, proteolytic enzymes are generated in an inactive form and then later, when they are needed, are converted to their active form. Most digestive and blood-clotting enzymes are proteolytic enzymes. The inactive forms of proteolytic enzymes are called *zymogens*. A **zymogen** *is the inactive precursor of a proteolytic enzyme.* (An alternative, but less often used, name for a zymogen is *proenzyme.*)

The names of zymogens can be recognized by the suffix *-ogen* or the prefix *pre-* or *pro-*.

Activation of a zymogen requires an enzyme-controlled reaction that removes some part of the zymogen structure. Such modification changes the three-dimensional structure (secondary and tertiary structure) of the zymogen, which affects active site conformation. For example, the zymogen pepsinogen is converted to the active enzyme pepsin in the stomach, where it then functions as a digestive enzyme. Pepsin would digest

Figure 21.13 Conversion of a zymogen (the inactive form of a proteolytic enzyme) to a proteolytic enzyme (the active form of the enzyme) often involves removal of a peptide chain segment from the zymogen structure.

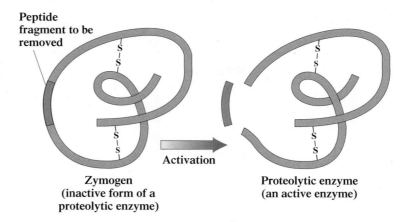

Peptide fragment to be removed

Activation

Zymogen (inactive form of a proteolytic enzyme)

Proteolytic enzyme (an active enzyme)

the tissues of the stomach wall if it were prematurely generated in active form. Pepsinogen activation involves removal of a peptide fragment from its structure (Figure 21.13).

Covalent Modification of Enzymes

A third mechanism for regulation of enzyme activity within a cell, called *covalent modification,* involves adding or removing a group from an enzyme through the forming or breaking of a covalent bond. **Covalent modification** *is a process in which enzyme activity is altered by covalently modifying the structure of the enzyme through attachment of a chemical group to or removal of a chemical group from a particular amino acid within the enzyme's structure.*

The most commonly encountered type of covalent modification involves the processes by which a phosphate group is added to or removed from an enzyme. The source of the added phosphate group is often an ATP molecule. The process of addition of the phosphate group to the enzyme is called *phosphorylation* and the removal of the phosphate group from the enzyme is called *dephosphorylation.* This phosphorylation/dephosphorylation process is the off/on or on/off switch for the enzyme. For some enzymes the active ("turned-on" form) is the phosphorylated version of the enzyme; however, for other enzymes it is the dephosphorylated version that is active.

The preceding covalent modification processes are governed by other enzymes. *Protein kinases* effect the addition of phosphate groups and *phosphatases* catalyze removal of the phosphate groups. Usually, the phosphate group is added to (or removed from) the R group of a serine, tyrosine, or threonine amino acid residue present in the protein (enzyme). The R groups of these three amino acids have a common structural feature, the presence of a free —OH group (see Table 21.1). The hydroxyl group is the site where phosphorylation or dephosphorylation occurs.

Glycogen phosphorylase, an enzyme involved in the breakdown of glycogen to glucose (Section 24.5) is activated by the addition of a phosphate group. *Glycogen synthase,* an enzyme involved in the synthesis of glycogen (Section 24.5) is deactivated by phosphorylation.

ANTIBIOTICS THAT INHIBIT ENZYME ACTIVITY

An **antibiotic** *is a substance that kills bacteria or inhibits their growth.* Antibiotics exert their action selectively on bacteria and do not affect the normal metabolism of the host organism. Antibiotics usually inhibit specific enzymes essential to the life processes of bacteria. In this section we consider the actions of two families of antibiotics, sulfa drugs and penicillins, as well as the specific antibiotic *Cipro.*

Sulfa Drugs

The 1932 discovery of the antibacterial activity of the compound sulfanilamide by the German bacteriologist Gerhard Domagk (1895–1964) led to the characterization of a whole family of sulfanilamide derivatives collectively called sulfa drugs—the first "antibiotics" in the medical field.

Figure 21.14 Structures of selected sulfa drugs in use today as antibiotics.

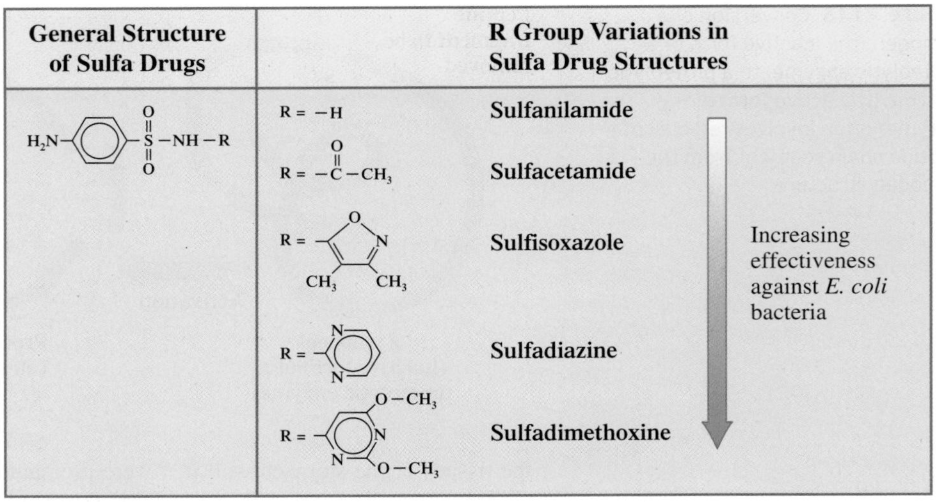

General Structure of Sulfa Drugs	R Group Variations in Sulfa Drug Structures	
H_2N—⟨benzene ring⟩—$\overset{O}{\underset{O}{\overset{\|}{\underset{\|}{S}}}}$—$NH$—$R$	$R = -H$	Sulfanilamide
	$R = -\overset{O}{\overset{\|}{C}}-CH_3$	Sulfacetamide
	$R = $ ⟨isoxazole, CH_3 CH_3⟩	Sulfisoxazole
	$R = $ ⟨pyrimidine⟩	Sulfadiazine
	$R = $ ⟨pyrimidine, $O-CH_3$ $O-CH_3$⟩	Sulfadimethoxine

Increasing effectiveness against *E. coli* bacteria

Sulfanilamide inhibits bacterial growth because it is structurally similar to PABA (*p*-aminobenzoic acid).

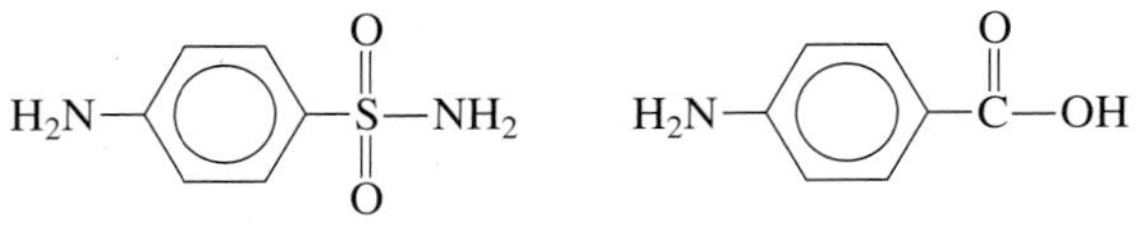

H_2N—⟨benzene ring⟩—$\overset{O}{\underset{O}{\overset{\|}{\underset{\|}{S}}}}$—$NH_2$ H_2N—⟨benzene ring⟩—$\overset{O}{\overset{\|}{C}}$—$OH$

Sulfanilamide   *p*-Aminobenzoic acid
                       (PABA)

Many bacteria need PABA in order to produce an important coenzyme, folic acid. Sulfanilamide acts as a competitive inhibitor to enzymes in the biosynthetic pathway for converting PABA into folic acid in these bacteria. Folic acid deficiency retards growth of the bacteria and can eventually kill them.

Sulfa drugs selectively inhibit only bacteria metabolism and growth because humans absorb folic acid from their diet and thus do not use PABA for its synthesis. A few of the most common sulfa drugs and their structures are shown in Figure 21.14.

Penicillins

Penicillin, one of the most widely used antibiotics, was accidentally discovered by Alexander Fleming in 1928 while he was working with cultures of an infectious staphylococcus bacterium. A decade later, the scientists Howard Flory and Ernst Chain isolated penicillin in pure form and proved its effectiveness as an antibiotic.

Several naturally occurring penicillins have now been isolated, and numerous derivatives of these substances have been synthetically produced. All have structures containing a four-membered β-lactam ring (Section 17.12) fused with a five-membered thiazolidine ring (Figure 21.15). As with sulfa drugs, derivatives of the basic structure differ from each other in the identity of a particular R group.

Penicillins inhibit *transpeptidase,* an enzyme that catalyzes the formation of peptide cross links between polysaccharide strands in bacterial cell walls. These cross links strengthen cell walls. A strong cell wall is necessary to protect the bacterium from lysis (breaking open). By inhibiting transpeptidase, penicillin prevents the formation of a strong cell wall. Any osmotic or mechanical shock then causes lysis, killing the bacterium.

Penicillin's unique action depends on two aspects of enzyme deactivation that we have discussed before: structural similarity to the enzyme's natural substrate and irreversible inhibition. Penicillin is *highly specific* in binding to the active site of transpeptidase. In this sense, it acts as a very selective competitive inhibitor. However, unlike a normal competitive inhibitor, once bound to the active site, the β-lactam ring opens as the highly reactive amide bond forms a covalent bond to a critical serine residue required for normal catalytic action. The result is an irreversibly inhibited transpeptidase enzyme (Figure 21.16).

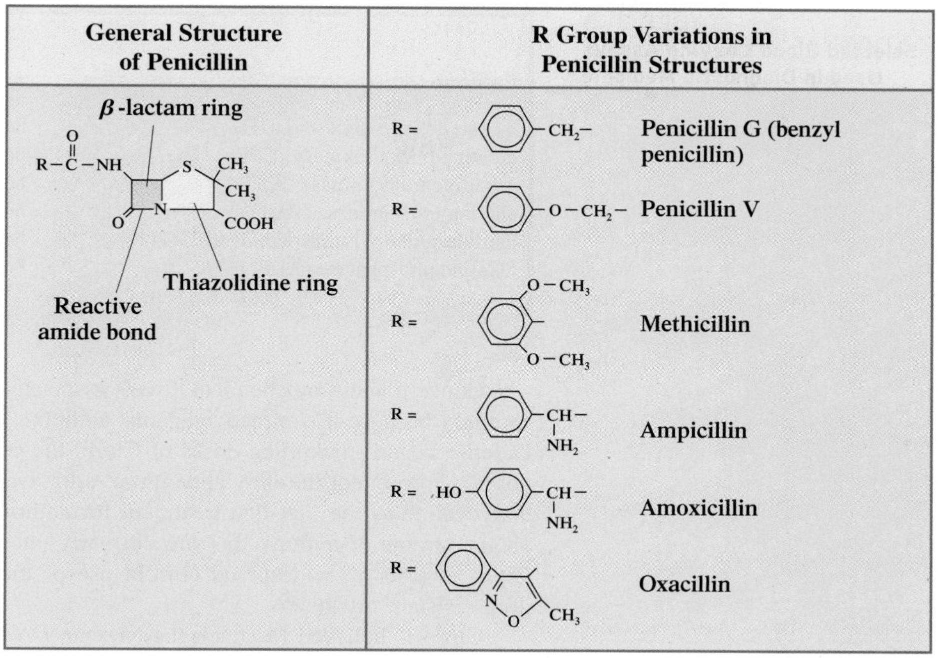

Figure 21.15 Structures of selected penicillins in use today as antibiotics.

General Structure of Penicillin	R Group Variations in Penicillin Structures

Some bacteria produce the enzyme *penicillinase,* which protects them from penicillin. Penicillinase selectively binds penicillin and catalyzes the opening of the β-lactam ring before penicillin can form a covalent bond to the enzyme. Once the ring is opened, the penicillin is no longer capable of inactivating transpeptidase.

Certain semi-synthetic penicillins such as methicillin and amoxicillin have been produced that are resistant to penicillinase activity and are thus clinically important.

Penicillin does not usually interfere with normal metabolism in humans because of its highly selective binding to bacterial transpeptidase. This selectivity makes penicillin an extremely useful antibiotic.

Cipro

The antibiotic ciprofloxacin hydrochloride (marketed as "Cipro") is an effective agent against bacterial infections in many different parts of the body. It is effective against skin and bone infections as well as against infections involving the urinary, gastrointestinal, and respiratory systems. It is the drug of choice for treatment of traveler's diarrhea. It is considered one of the best broad-spectrum antibiotics available. Bacteria are slow to acquire resistance to Cipro.

Structurally, Cipro contains several common functional groups (carboxylic acid, ketone, and amine), as well as two seldom-encountered groups (fluoro and cyclopropyl).

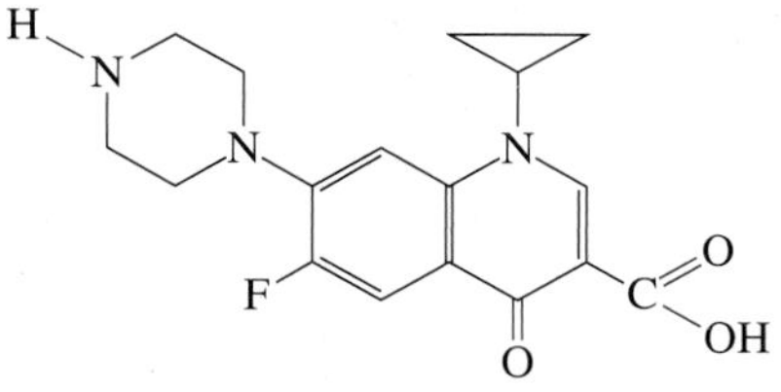

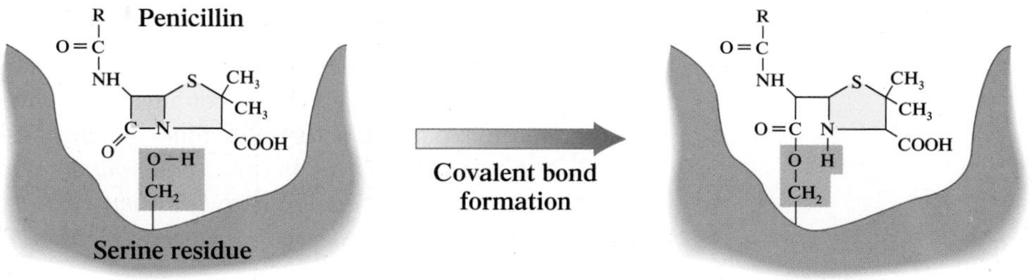

Figure 21.16 The selective binding of penicillin to the active site of transpeptidase. Subsequent irreversible inhibition through formation of a covalent bond to a serine residue permanently blocks the active site.

TABLE 21.3
Selected Blood Enzyme Assays
Used in Diagnostic Medicine

Enzyme	Condition Indicated by Abnormal Level
lactate dehydrogenase (LDH)	heart disease, liver disease
creatine phosphokinase (CPK)	heart disease
aspartate transaminase (AST)	heart disease, liver disease, muscle damage
alanine transaminase (ALT)	heart disease, liver disease, muscle damage
gamma-glutamyl transpeptidase (GGTP)	heart disease, liver disease
alkaline phosphatase (ALP)	bone disease, liver disease

Concern about biochemical threats associated with terrorism has thrust Cipro into the spotlight because it is effective against anthrax. As early as 1990, the U.S. Department of Defense began stockpiling doses of Cipro; the government's 2001 order was 100 million doses. Cipro is not the only antianthrax drug available. The FDA recommends the use of doxycycline as the first-line treatment for anthrax because this compound can deal with all the strains of anthrax that are currently encountered. Authorities would rather keep Cipro in reserve; widespread current use of the drug could speed up the evolution of drug-resistant organisms.

Cipro is believed to attack the enzyme *DNA gyrase,* which controls how DNA in a bacterial chromosome coils into its tertiary structure. When tertiary structure is disrupted, replication and transcription of the DNA cannot occur.

21.10 MEDICAL USES OF ENZYMES

Enzymes can be used to diagnose certain diseases. Although blood serum contains many enzymes, some enzymes are not normally found in the blood but are produced only inside cells of certain organs and tissues. The appearance of these enzymes in the blood often indicates that there is tissue damage in an organ and that cellular contents are spilling out (leaking) into the bloodstream (see Figure 21.17). Assays of abnormal enzyme activity in blood serum can be used to diagnose many disease states, some of which are listed in Table 21.3. The Chemical Connections feature on page 719 examines the use of enzymes to diagnose heart attacks (myocardial infarctions).

Enzymes can also be used in the treatment of diseases. A recent advance in treating heart attacks is the use of tissue plasminogen activator (TPA), which activates the enzyme plasminogen. When so activated, this enzyme dissolves blood clots in the heart and often provides immediate relief.

Another medical use for enzymes is in clinical laboratory chemical analysis. For example, no simple direct test for the measurement of urea in the blood is available. However, if the urea in the blood is converted to ammonia via the enzyme *urease,* the ammonia produced, which is easily measured, becomes an indicator of urea. This *blood urea nitrogen (BUN) test* is a common clinical laboratory procedure. High urea levels in the blood indicate kidney malfunction.

21.11 GENERAL CHARACTERISTICS OF VITAMINS

This section and the two that follow deal with vitamins. Vitamins are considered in conjunction with enzymes because many enzymes contain vitamins as part of their structure. Recall from Section 21.2 that conjugated enzymes have a protein part (apoenzyme) and a nonprotein part (cofactor). Vitamins, in many cases, are cofactors in conjugated enzymes.

A **vitamin** *is an organic compound, essential in small amounts for the proper functioning of the human body, that must be obtained from dietary sources because the body cannot synthesize it.*

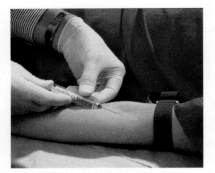

Figure 21.17 Drawing of a blood sample. Determination of enzyme concentrations in blood provides important information about the "state" of various organs within the human body.

CHEMICAL Connections

Heart Attacks and Enzyme Analysis

The symptoms of a heart attack—that is, a *myocardial infarction (MI)*—include irregular breathing and pain in the left chest that may radiate to the neck, left shoulder, and arm. An initial diagnosis of an MI is based on these and other physical symptoms, and treatment is initiated on this basis.

Physicians then use enzyme analysis to confirm the diagnosis and to monitor the course of treatment. The blood levels of three enzymes are commonly assayed in MI situations: creatine phosphokinase (CPK), aspartate transaminase (AST), and lactate dehydrogenase (LDH). The CPK level rises and falls relatively rapidly after a heart attack, reaching a maximum after about 30 hours at a level approximately six times normal. The AST level triples after about 40 hours. LDH, whose concentrations rise slowly, is used to monitor the later stages of the MI and to assess the extent of heart damage. The accompanying graph shows blood levels of these three enzymes as a function of time in an MI situation.

Further information about the seriousness of an MI is obtained by studying isoenzymes. **Isoenzymes** *are forms of the same enzyme with slightly different amino acid sequences.* The use of the prefix "iso" in the term *isoenzymes* implies that such enzymes catalyze the "same" reaction and not that they have isomeric structures. Lactate dehydrogenase is a mixture of five isoenzymes denoted LDH_1 to LDH_5. Creatine phosphokinase is a mixture of three isoenzymes denoted CK-MM, CK-MB, and CK-BB.

Consideration of further details about the LDH isoenzymes shows how isoenzymes give information about whether a heart attack has occurred or not. Note from the following table that heart tissue is particularly high in LDH_1 and that liver and skeletal muscle are particularly high in LDH_5.

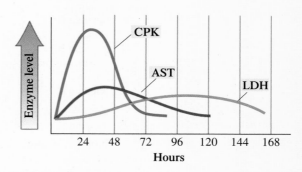

When a heart attack occurs, some heart muscle cells are damaged (destroyed), and their enzymes "leak" into the bloodstream. This changes the ratios among the various LDHs present in blood, because heart muscle is particularly high in LDH_1. The accompanying graph of an LDH isoenzyme assay for a heart attack victim shows that the LDH_1/LDH_2 ratio, which is normally less than one in blood serum, is now greater than one.

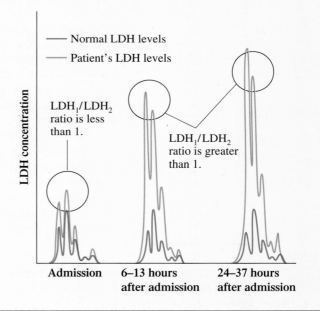

Tissue	LDH_1	LDH_2	LDH_3	LDH_4	LDH_5
brain	23	34	30	10	3
heart	50	36	9	3	2
kidney	28	34	21	11	6
liver	4	6	17	16	57
lung	10	20	30	25	15
serum	28	41	19	7	5
skeletal muscle	5	5	10	22	58

The spelling of the term *vitamin* was originally *vitamine,* a word derived from the Latin *vita,* meaning "life," and from the fact that these substances were all thought to contain the *amine* functional group. When this supposition was found to be false, the final *e* was dropped from *vitamine,* and the term *vitamin* came into use. Some vitamins contain amine functional groups, but others do not.

Vitamins differ from the major classes of foods (carbohydrates, lipids, and proteins) in the amount required; for vitamins it is *micro*gram or *milli*gram quantities per day compared with 50–200 grams per day for the major food categories. To illustrate the small amount of vitamins needed by the human body, consider the recommended daily allowance (RDA) of vitamin B_{12}, which is 2.0 micrograms per day for an adult. Just 1.0 gram of this vitamin could theoretically supply the daily needs of 500,000 people.

A well-balanced diet usually meets all the body's vitamin requirements. However, supplemental vitamins are often required for women during pregnancy and for people recovering from certain illnesses.

One of the most common myths associated with the nutritional aspects of vitamins is that vitamins from natural sources are superior to synthetic vitamins. In truth, synthetic

TABLE 21.4
The Water-Soluble and
Fat-Soluble Vitamins

Water-Soluble Vitamins	Fat-Soluble Vitamins
Thiamin	Vitamin A
Riboflavin	Vitamin D
Niacin	Vitamin E
Vitamin B_6	Vitamin K
Folate	
Vitamin B_{12}	
Pantothenic acid	
Biotin	
Vitamin C	

vitamins, manufactured in the laboratory, are identical to the vitamins found in foods. The body cannot tell the difference and gets the same benefits from either source.

There are 13 known vitamins, and scientists believe that the discovery of additional vitamins is unlikely. Despite searches for new vitamins, it has been over 50 years since the last of the known vitamins (B_{12}) was discovered. Strong evidence that the vitamin family is complete comes from the fact that many people have lived for years being fed, intravenously, solutions containing the known vitamins and nutrients, and they have not developed any known vitamin deficiency disease.

Solubility characteristics divide the vitamins into two major classes: the water-soluble vitamins and the fat (lipid)-soluble vitamins. There are nine water-soluble vitamins and four fat-soluble vitamins. Table 21.4 gives the identity of the vitamins that fall in each of these solubility categories. Water-soluble vitamins must be constantly replenished in the body because they are rapidly eliminated from the body in the urine. They are carried in the bloodstream, are needed in frequent, small doses, and are unlikely to be toxic except when taken in unusually large doses. The fat-soluble vitamins are found dissolved in lipid materials. They are, in general, carried in the blood by protein carriers, are stored in fat tissues, are needed in periodic doses, and are more likely to be toxic when consumed in excess of need.

An important difference exists, in terms of function, between water-soluble and fat-soluble vitamins. Water-soluble vitamins function as coenzymes for a number of important biochemical reactions in humans, animals, and microorganisms. Fat-soluble vitamins generally do not function as coenzymes in humans and animals and are rarely utilized in any manner by microorganisms. Other differences between the two categories of vitamins are summarized in Table 21.5. A few exceptions occur, but the differences shown in this table are generally valid.

	Water-Soluble Vitamins (B vitamins and vitamin C)	Fat-Soluble Vitamins (vitamins A, D, E, and K)
absorption	directly into the blood	first enter into the lymph system
transport	travel without carriers	many require protein carriers
storage	circulate in the water-filled parts of the body	found in the cells associated with fat
excretion	kidneys remove excess in urine	tend to remain in fat-storage sites
toxicity	not likely to reach toxic levels when consumed from supplements	likely to reach toxic levels when consumed from supplements
requirements	needed in frequent doses	needed in periodic doses
relationship to coenzymes	function as coenzymes	do not function as coenzymes

21.12 WATER-SOLUBLE VITAMINS

The nine water-soluble vitamins, vitamin C and eight B vitamins, got their names from the labels B and C on the test tubes in which they were first collected. Later, test tube B was found to contain more than one vitamin.

Vitamin C

Vitamin C, which has the simplest structure of the 13 vitamins, exists in two active forms in the human body: an oxidized form and a reduced form.

Ascorbic acid (reduced) ⇌ (Oxidation / Reduction) Dehydroascorbic acid (oxidized) + 2H

Humans, monkeys, apes, and guinea pigs are among the relatively few species that require dietary sources of vitamin C. Other species synthesize vitamin C from carbohydrates. Vitamin C's biosynthesis involves L-gulonic acid, an acid derivative of the monosaccharide L-gulose (see Figure 18.13). L-Gulonic acid is changed by the enzyme *lactonase* into a cyclic ester (lactone, Section 16.11); ring closure involves carbons 1 and 4. An *oxidase* then introduces a double bond into the ring, producing L-ascorbic acid.

L-Gulonic acid →(Lactonase)→ γ-L-Gulonolactone →(Oxidase)→ L-Ascorbic acid

The four —OH groups present in vitamin C's reduced form are suggestive of its biosynthetic monosaccharide (*polyhydroxy* aldehyde) origins. Its chemical name, L-ascorbic acid, correctly indicates that vitamin C is a weak acid. Although no carboxyl group is present, the carbon 3 hydroxyl group hydrogen atom exhibits acidic behavior as a result of its attachment to an unsaturated carbon atom.

The most completely characterized role of vitamin C is its function as a cosubstrate in the formation of the structural protein collagen (Section 20.17), which makes up much of the skin, ligaments, and tendons and also serves as the matrix on which bone and teeth are formed. Specifically, biosynthesis of the amino acids hydroxyproline and hydroxylysine (important in binding collagen fibers together) from proline and lysine requires the presence of both vitamin C and iron. Iron serves as a cofactor in the reaction, and vitamin C maintains iron in the oxidation state that allows it to function. In this role, vitamin C is functioning as a *specific* antioxidant.

Vitamin C also functions as a *general* antioxidant (Section 14.14) for water-soluble substances in the blood and other body fluids. Its antioxidant properties are also beneficial for several other vitamins. The active form of vitamin E is regenerated by vitamin C, and it also helps keep the active form of folate (a B vitamin) in its reduced state. Because of its antioxidant properties, vitamin C is often added to foods as a preservative.

Vitamin C is also involved in the metabolism of several amino acids that end up being converted to the hormones norepinephrine and thyroxine. The adrenal glands contain a higher concentration of vitamin C than any other organ in the body.

Vitamin C, the best known of all vitamins, was the first vitamin to be discovered (1928), the first to be structurally characterized (1933), and the first to be synthesized in the laboratory (1933). Laboratory production of vitamin C, which exceeds 80 million pounds per year, is greater than the combined production of all the other vitamins. In addition to its use as a vitamin supplement, synthetic vitamin C is used as a food additive (preservative), a flour additive, and an animal feed additive.

Why is vitamin C called ascorbic *acid* when there is no carboxyl group (acid group) present in its structure? Vitamin C is a cyclic ester in which a carbon 1 carboxyl group has reacted with a carbon 4 hydroxyl group, forming the ring structure.

Other naturally occurring dietary antioxidants include glutathione (Section 20.7), vitamin E (Section 21.13), beta-carotene (Section 21.13), and flavonoids (Section 23.11).

An intake of 100 mg/day of vitamin C saturates all body tissues with the compound. After the tissues are saturated, all additional vitamin C is excreted. The RDA for vitamin C varies from country to country. It is 30 mg/day in Great Britain, 60 mg/day in the United States and Canada, and 75 mg/day in Germany. A variety of fruits and vegetables have a relatively high vitamin C content (see Figure 21.18).

Vitamin B

There are eight B vitamins. Our discussion of them involves four topics: nomenclature, function, structural characteristics, and dietary sources.

Much confusion exists about the B vitamins' names. Many have "number" names as well as "word" names (often several). The *preferred* names for the B vitamins (alternative names in parentheses) are

1. **Thiamin** (vitamin B_1)
2. **Riboflavin** (vitamin B_2)
3. **Niacin** (nicotinic acid, nicotinamide, vitamin B_3)
4. **Vitamin B_6** (pyridoxine, pyridoxal, pyridoxamine)
5. **Folate** (folic acid)
6. **Vitamin B_{12}** (cobalamin)
7. **Pantothenic acid** (vitamin B_5)
8. **Biotin**

B vitamin structure is very diverse. The only common thread among structures is that all structures, except that of pantothenic acid, involve heterocyclic nitrogen ring systems. The element sulfur is present in two structures (thiamin and biotin), and vitamin B_{12} contains a metal atom (cobalt). (Biotin does not contain a tin atom, as the name might imply.) Table 21.6 gives structural forms for the eight B vitamins. Note that for two B vitamins (niacin and vitamin B_6), more than one form of the vitamin exists.

The major function of B vitamins within the human body is as components of coenzymes (Section 21.2). Unlike vitamin C, all of the B vitamins must be chemically modified before they become functional within the coenzymes. For example, thiamine is converted to thiamine pyrophosphate (TPP), which then serves as the coenzyme in several reactions involving carbohydrate metabolism.

Figure 21.18 Rows of cabbage plants. Although many people think citrus fruits (50 mg per 100 g) are the best source of vitamin C, peppers (128 mg per 100 g), cauliflower (70 mg per 100 g), strawberries (60 mg per 100 g), and spinach or cabbage (60 mg per 100 g) are all richer in vitamin C.

Pronunciation guidelines for the standard names of the B vitamins:

THIGH-ah-min
RYE-boh-flay-vin
NIGH-a-sin
FOLL-ate
PAN-toe-THEN-ick acid
BY-oh-tin

Thiamine (vitamin B_1)

Thiamine pyrophosphate (TPP)

Another example of chemical modification for a B vitamin is the conversion of folate to tetrahydrofolate (THF).

Folate

Tetrahydrofolate (THF)

Table 21.7 lists selected important coenzymes that involve B vitamins and indicates how these coenzymes function. In general, coenzymes serve as temporary carriers of atoms or functional groups in redox and group transfer reactions.

TABLE 21.6
Structures of the Eight B Vitamins

Thiamine

Riboflavin

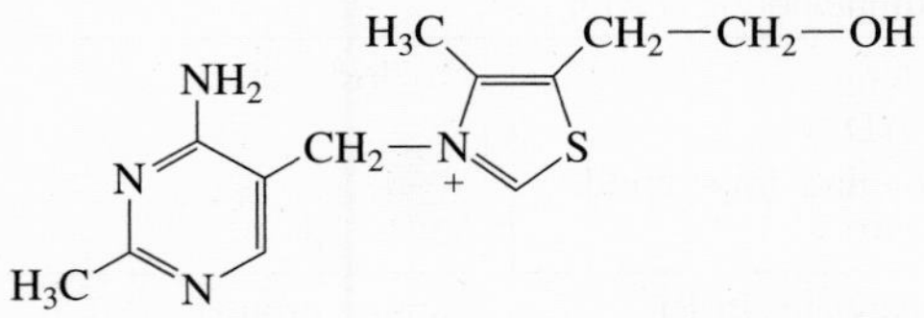

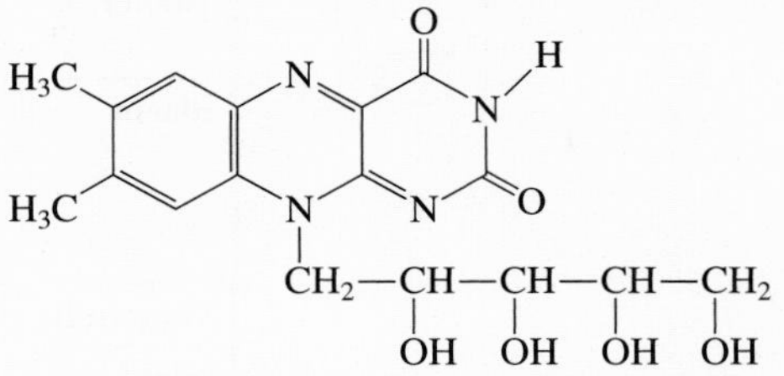

Niacin (two forms)

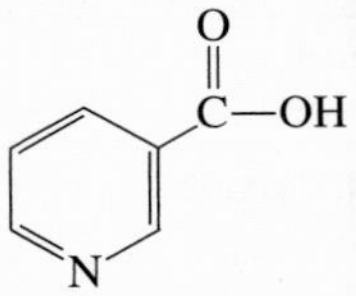

Nicotinic acid   Nicotinamide

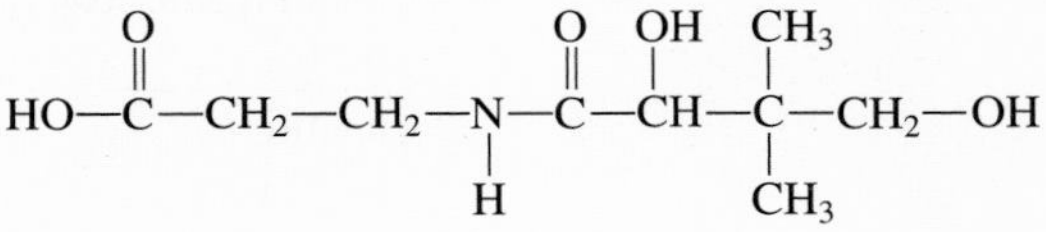

Pantothenic Acid

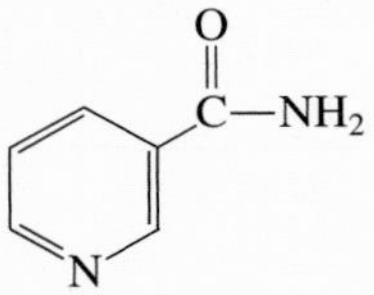

Vitamin B₆ (three forms)

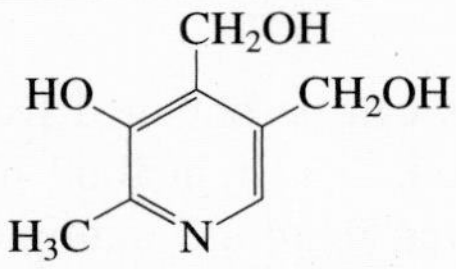

Pyridoxine 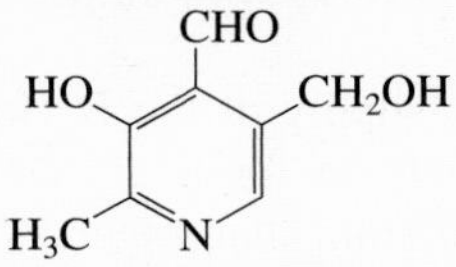   Pyridoxal 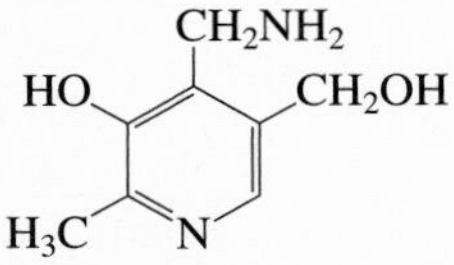   Pyridoxamine

Folate

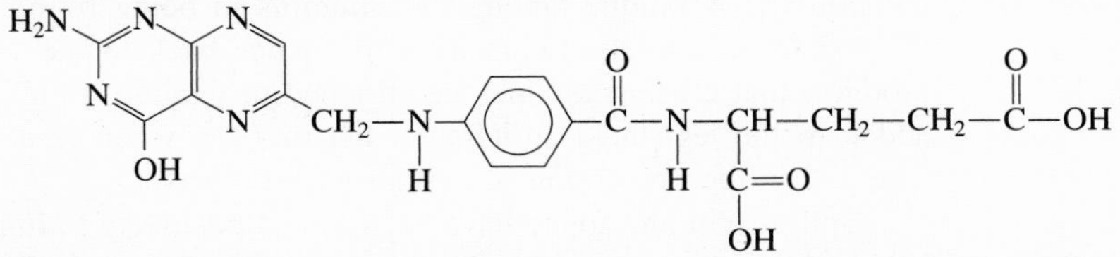

Vitamin B₁₂

Biotin

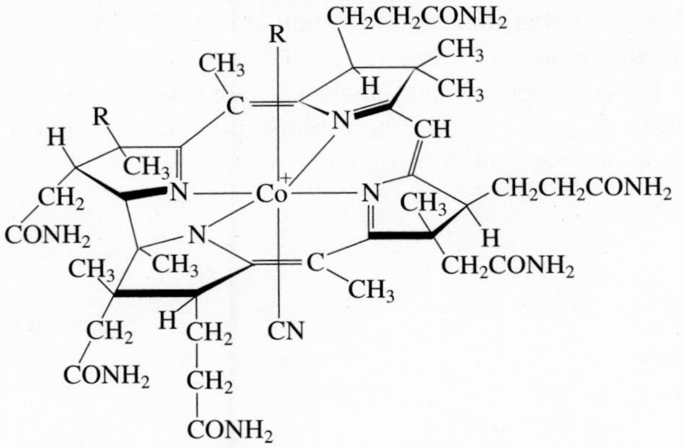

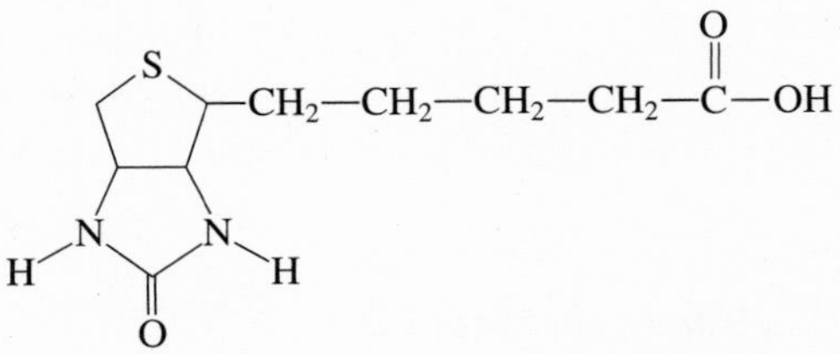

In their function as coenzymes, some of the B vitamins do not remain permanently bonded to the apoenzyme (Section 21.2) that they are associated with. This means that they can be repeatedly used by various enzymes. This reuse (recycling) diminishes the need for large amounts of the B vitamins in biochemical systems. Figure 21.19 shows diagrammatically how a vitamin is used and then released in an enzyme-catalyzed reaction.

TABLE 21.7
Selected Important Coenzymes in Which B Vitamins Are Present

B Vitamin	Coenzymes	Groups Transferred
thiamine	thiamine pyrophosphate (TPP)	aldehydes
riboflavin	flavin mononucleotide (FMN) flavin adenine dinucleotide (FAD)	hydrogen atoms
niacin	nicotinamide adenine dinucleotide (NAD^+) nicotinamide adenine dinucleotide phosphate ($NADP^+$)	hydride ion (H^-)
vitamin B_6	pyridoxal-5-phosphate (PLP)	amino groups
folate	tetrahydrofolate (THF)	one-carbon groups other than CO_2
vitamin B_{12}	5′-deoxyadenosylcobalamin	alkyl groups, hydrogen atoms
pantothenic acid	coenzyme A (CoA) acyl carrier protein (ACP)	acyl groups
biotin	biocytin	carbon dioxide

An ample supply of the B vitamins can be obtained from normal dietary intake as long as a variety of foods are consumed. A certain food may be a better source of a particular B vitamin than others; however, there are multiple sources for each of the B vitamins as Table 21.8 shows. Note from Table 21.8 that fruits, in general, are very poor sources of B vitamins and that only certain vegetables are good B vitamin sources. Vitamin B_{12} is unique among the vitamins in being found almost exclusively in food derived from animals. Legislation that dates back to the 1940s requires that all grain products that cross state lines be enriched in thiamin, riboflavin, and niacin. Folate was added to the legislated enrichment list in 1996 when research showed that folate was essential in the prevention of certain birth defects.

Both niacin and folate have been linked positively to improvement in cardiovascular health. Adding prescription-strength, extended-release niacin to cholesterol-lowering statin medications slows the progression of atherosclerosis among people with coronary heart disease and low HDL levels better than statin therapy alone. Additionally, prescription niacin is the most effective treatment currently available to increase low levels of HDL. Another study shows that younger women (26–46 years old) who consume 800 μg of folate per day reduce the risk of developing high blood pressure by almost a third compared to those who consume less than 200 μg/day.

Figure 21.19 Many enzymes require a vitamin-based coenzyme in order to be active. After the catalytic action, the vitamin is released and can be reused.

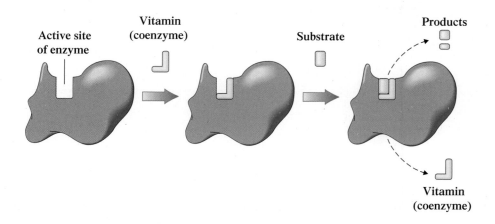

TABLE 21.8
A Summary of Dietary Sources of B Vitamins

Vitamin	Vegetable Group	Fruit Group	Bread, Cereal, Rice, and Pasta Group	Milk, Yogurt, and Cheese Group	Meat, Poultry, Fish, Dry Beans, Eggs, and Nuts Group			
					Meats	Dry Beans	Eggs	Nuts and Seeds
thiamine		watermelon	whole and enriched grains		pork, organ meats	legumes		sunflower seeds
riboflavin	mushrooms, asparagus, broccoli, leafy greens		whole and enriched grains	milk, cheeses	liver, red meat, poultry, fish	legumes	eggs	
niacin	mushrooms, asparagus, potato		whole and enriched grains, wheat bran		tuna, chicken, beef, turkey	legumes, peanuts		sunflower seeds
vitamin B$_6$	broccoli, spinach, potato, squash	bananas, watermelon	whole wheat, brown rice		chicken, fish, pork, organ meats	soybeans		sunflower seeds
folate	mushrooms, leafy greens, broccoli, asparagus, corn	oranges	fortified grains		organ meats (muscle meats are poor sources)	legumes		sunflower seeds, nuts
vitamin B$_{12}$				milk products	beef, poultry, fish, shellfish		egg yolk	
pantothenic acid	mushrooms, broccoli, avocados		whole grains		meat	legumes	egg yolk	
biotin			fortified cereals	yogurt	liver (muscle meats are poor sources)	soybeans	egg yolk	nuts

---

Lipids are substances that are not soluble in water but are soluble in nonpolar substances such as fat (Section 19.1). Thus, the four fat-soluble vitamins are a type of lipid. They could have been, but were not, discussed in Chapter 19.

Beta-carotene is a deep yellow (almost orange) compound. If a plant food is white or colorless, it possesses little or no vitamin A activity. Potatoes, pasta, and rice are foods in this category.

21.13 FAT-SOLUBLE VITAMINS

The four fat-soluble vitamins are designated using the letters A, D, E, and K. Many of the functions of the fat-soluble vitamins involve processes that occur in cell membranes. The structures of the fat-soluble vitamins are more hydrocarbon-like, with fewer functional groups than the water-soluble vitamins. Their structures as a whole are nonpolar, which enhances their solubility in cell membranes.

Vitamin A

Normal dietary intake provides a person with both *preformed* and *precursor forms* (provitamin forms) of vitamin A. Preformed vitamin A forms are called *retinoids*. The retinoids include retin*al*, retin*ol*, and retin*oic acid*.

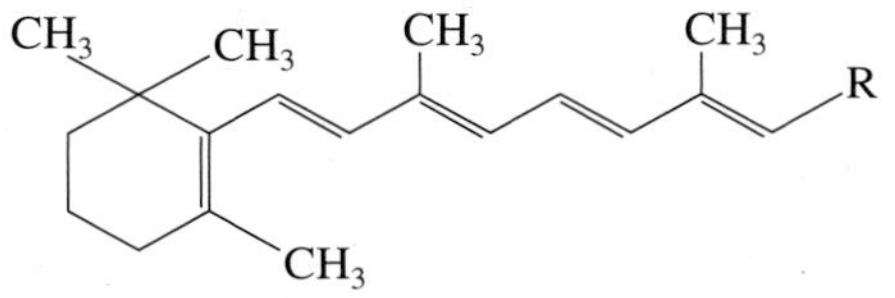

R = CH$_2$OH (Retinol)
R = CHO (Retinal)
R = COOH (Retinoic acid)

Retinoids

Foods derived from animals, including egg yolks and dairy products, provide compounds (retinyl esters) that are easily hydrolyzed to retinoids in the intestine.

The retinoids are terpenes (Section 13.7) in which four isoprene units are present. Beta-carotene has an eight-unit terpene structure.

Beta-carotene cleavage does not always occur in the "middle" of the molecule, so only one molecule of vitamin A is produced. Furthermore, not all beta-carotene is converted to vitamin A, and its absorption is not as efficient as that of vitamin A itself. It is estimated that 6 mg of beta-carotene is needed to produce 1 mg of retinol. Unconverted beta-carotene serves as an antioxidant (see Section 13.7), a role independent of its conversion to vitamin A.

Foods derived from plants provide carotenoids (see the Chemical Connections feature "Carotenoids: A Source of Color" on page 374 in Chapter 13), which serve as precursor forms of vitamin A. The major carotenoid with vitamin A activity is beta-carotene (β-carotene), which can be cleaved to yield two molecules of vitamin A.

Cleavage at this point can yield two molecules of vitamin A

Beta-carotene, a precursor for vitamin A

Beta-carotene is a yellow to red-orange pigment plentiful in carrots, squash, cantaloupe, apricots, and other yellow vegetables and fruits, as well as in leafy green vegetables (where the yellow pigment is masked by green chlorophyll).

Vitamin A has four major functions in the body.

1. *Vision.* In the eye, vitamin A (as retinal) combines with the protein opsin to form the visual pigment rhodopsin (see the Chemical Connections feature "Cis–Trans Isomerism and Vision" on page 371 in Chapter 13). Rhodopsin participates in the conversion of light energy into nerve impulses that are sent to the brain. Although vitamin A's involvement in the process of vision is its best-known function ("Eat your carrots and you'll see better"), only 0.1% of the body's vitamin A is found in the eyes.

2. *Regulating Cell Differentiation.* Cell differentiation is the process whereby immature cells change in structure and function to become specialized cells. For example, some immature bone marrow cells differentiate into white blood cells and others into red blood cells. In the cellular differentiation process, vitamin A (as retinoic acid) binds to protein receptors; these vitamin A–protein receptor complexes then bind to regulatory regions of DNA molecules.

3. *Maintenance of the Health of Epithelial Tissues.* Epithelial tissue covers outer body surfaces as well as lining internal cavities and tubes. It includes skin and the linings of the mouth, stomach, lungs, vagina, and bladder. Lack of vitamin A (as retinoic acid) causes such surfaces to become drier and harder than normal. Vitamin A's role here is related to cellular differentiation involving mucus-secreting cells.

4. *Reproduction and Growth.* In men, vitamin A participates in sperm development. In women, normal fetal development during pregnancy requires vitamin A. In both cases, it is the retinoic acid form of vitamin A that is needed. Again vitamin A's role is related to cellular differentiation processes.

Vitamin D

The two most important members of the vitamin D family of molecules are vitamin D_3 (cholecalciferol) and vitamin D_2 (ergocalciferol). Vitamin D_3 is produced in the skin of humans and animals by the action of sunlight (ultraviolet light) on its precursor molecule, the cholesterol derivative 7-dehydrocholesterol (a normal metabolite of cholesterol found in the skin). Absorption of light energy induces breakage of the 9, 10 carbon bond; a spontaneous isomerization (shifting of double bonds) then occurs (see Figure 21.20).

Figure 21.20 The quantity of vitamin D synthesized by exposure of the skin to sunlight (ultraviolet radiation) varies with latitude, the length of exposure time, and skin pigmentation. (Darker-skinned people synthesize less vitamin D because the pigmentation filters out ultraviolet light.)

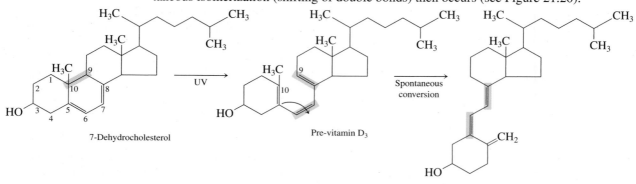

Vitamin D_3 (cholecalciferol) is sometimes called the "sunshine vitamin" because of its synthesis in the skin by sunlight irradiation.

Vitamin D_2 (ergocalciferol) differs from vitamin D_3 only in the side-chain structure. It is produced from the plant sterol ergosterol through the action of light.

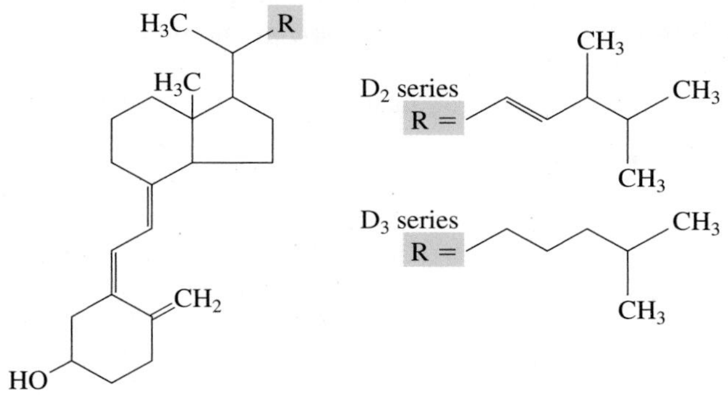

Vitamin D

Both the cholecalciferol and the ergocalciferol forms of vitamin D must undergo two further hydroxylation steps before the vitamin D becomes fully functional. The first step, which occurs in the liver, adds a —OH group to carbon 25. The second step, which occurs in the kidneys, adds a —OH group to carbon 1.

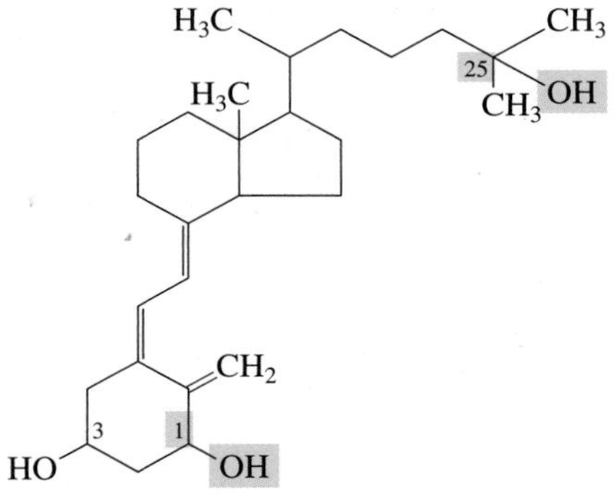

1,25-Dihydroxyvitamin D_3

Milk is enriched in vitamin D by exposure to ultraviolet light. Cholesterol in milk is converted to cholecalciferol (vitamin D) by ultraviolet light.

When it comes to strong bones, calcium won't do you a lot of good unless you are also getting enough vitamin D. In one study, women consuming 500 IU of vitamin D a day had a 37% lower risk of hip fracture than women consuming only 140 IU daily of vitamin D. Some researchers now recommend a standard of 800–1000 IU per day instead of the long-established standard recommendation for vitamin D of 400 IU per day.

Only a few foods, including liver, fatty fish (such as salmon), and egg yolks, are good natural sources of vitamin D. Such vitamin D is vitamin D_3. Foods fortified with vitamin D include milk and margarine. The rest of the body's vitamin D supplies are made within the body (skin) with the help of sunlight.

The principal function of vitamin D is to maintain normal blood levels of calcium ion and phosphate ion so that bones can absorb these ions. Vitamin D stimulates absorption of these ions from the gastrointestinal tract and aids in their retention by the kidneys. Vitamin D triggers the deposition of calcium salts into the organic matrix of bones by activating the biosynthesis of calcium-binding proteins.

Vitamin E

There are four forms of vitamin E: alpha-, beta-, delta-, and gamma-tocopherol. These forms differ from each other structurally in what substituents (—CH_3 or —H) are present at two positions on an aromatic ring.

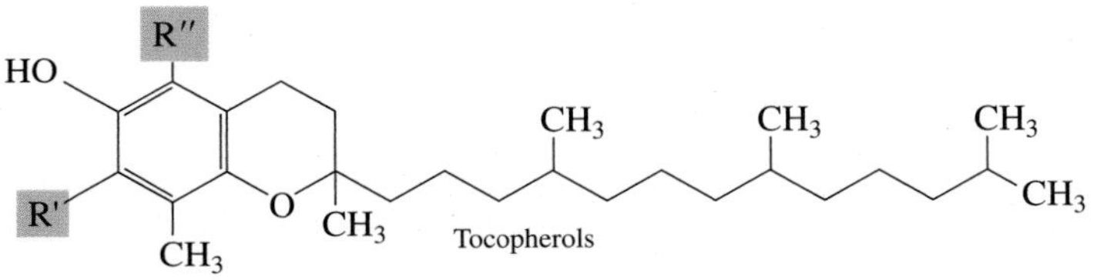

Tocopherols

	R′	R″
α	CH_3	CH_3
β	CH_3	H
γ	H	CH_3
δ	H	H

The word *tocopherol* is pronounced "tuh-KOFF-er-ol."

The tocopherol form with the greatest biochemical activity is alpha-tocopherol, the vitamin E form in which methyl groups are present at both the R and R′ positions on the aromatic ring. Gamma-tocopherol is the main form of vitamin E in vitamin-E rich foods.

Plant oils (margarine, salad dressings, and shortenings), green and leafy vegetables, and whole-grain products are sources of vitamin E.

The primary function of vitamin E in the body is as an antioxidant—a compound that protects other compounds from oxidation by being oxidized itself. Vitamin E is particularly important in preventing the oxidation of polyunsaturated fatty acids (Section 19.2) in membrane lipids. It also protects vitamin A from oxidation. Vitamin E's antioxidant action involves it giving up the hydrogen present on its —OH group to oxygen-containing free radicals. After vitamin E is "spent" as an antioxidant, reaction with vitamin C restores the hydrogen atom previously lost by the vitamin E.

A most important location in the human body where vitamin E exerts its antioxidant effect is the lungs, where exposure of cells to oxygen (and air pollutants) is greatest. Both red and white blood cells that pass through the lungs, as well as the cells of the lung tissue itself, benefit from vitamin E's protective effect.

Infants, particularly premature infants, do not have a lot of vitamin E, which is passed from the mother to the infant only in the last weeks of pregnancy. Often, premature infants require oxygen supplementation for the purpose of controlling respiratory distress. In such situations, vitamin E is administered to the infant along with oxygen to give antioxidant protection.

Vitamin E has also been found to be involved in the conversion of arachidonic acid (20:4) to prostaglandins (Section 19.13).

> Vitamin E is unique among the vitamins in that antioxidant activity is its *principal* biochemical role.

Vitamin K

Like the other fat-soluble vitamins, vitamin K has more than one form. Structurally, all forms have a methylated napthoquinone structure to which a long side chain of carbon atoms is attached. The various forms differ structurally in the length and degree of unsaturation of the side chain.

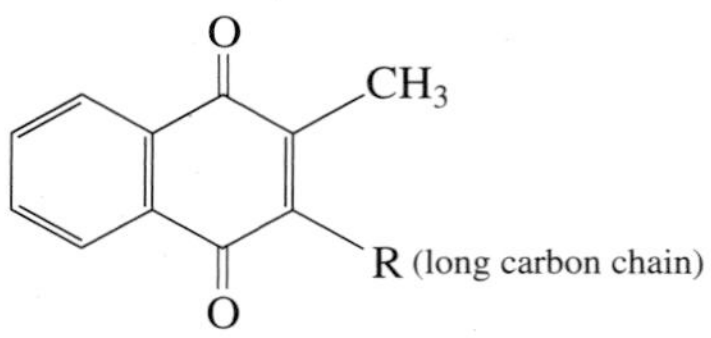

Vitamin K

Vitamin K_1, also called phylloquinone, has a side chain that is predominantly saturated; only one carbon–carbon double bond is present. It is a substance found in plants. Vitamin K_2 has several forms, called menaquinones, with the various forms differing in the length of the side chain. Menaquinone side chains have several carbon–carbon double bonds, in contrast to the one carbon–carbon double bond present in phylloquinone. Vitamin K_2 is found in animals and humans and can be synthesized by bacteria, including those found in the human intestinal tract.

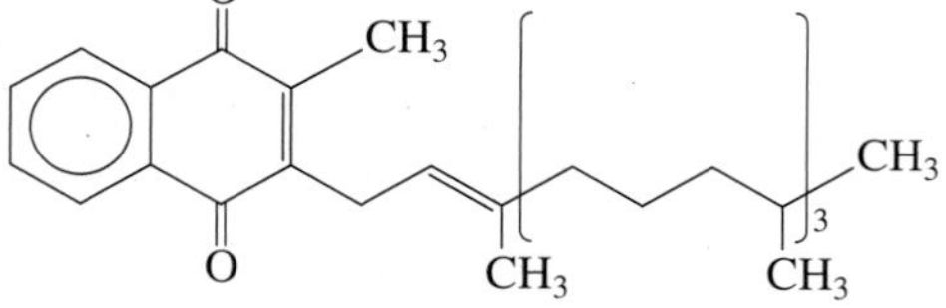

Vitamin K_1 (phylloquinone)

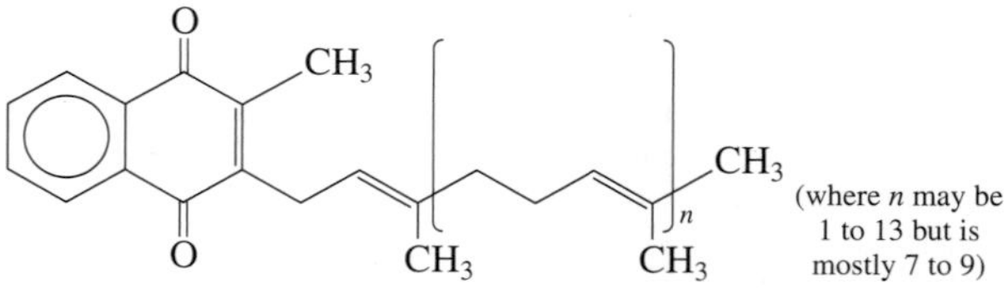

Vitamin K_2 (menaquinone)

(where n may be 1 to 13 but is mostly 7 to 9)

Typically, about half of the human body's vitamin K is synthesized by intestinal bacteria and half comes from the diet. Menaquinones are the form of vitamin K found in vitamin K supplements. Only leafy green vegetables such as spinach and cabbage are particularly rich in vitamin K. Other vegetables such as peas and tomatoes and animal tissues, including liver, contain lesser amounts.

All of the fat-soluble vitamins share a common structural feature; they all have terpenelike structures. That is, they are all made up of five-carbon isoprene units (Section 13.5). No common structural pattern exists for the water-soluble vitamins. On the other hand, the water-soluble vitamins have *functional* uniformity, whereas the fat-soluble vitamins have diverse functions.

Vitamin K is essential to the blood-clotting process. Over a dozen different proteins and the mineral calcium are involved in the formation of a blood clot. Vitamin K is essential for the formation of prothrombin and at least five other proteins involved in the regulation of blood clotting. Vitamin K is sometimes given to presurgical patients to ensure adequate prothrombin levels and prevent hemorrhaging.

Vitamin K is also required for the biosynthesis of several other proteins found in the plasma, bone, and kidney.

CONCEPTS TO REMEMBER

Enzymes. Enzymes are highly specialized protein molecules that act as biochemical catalysts. Enzymes have common names that provide information about their function rather than their structure. The suffix *-ase* is characteristic of most enzyme names (Section 21.1).

Enzyme structure. Simple enzymes are composed only of protein (amino acids). Conjugated enzymes have a nonprotein portion (cofactor) in addition to a protein portion (apoenzyme). Cofactors may be small organic molecules (coenzymes) or inorganic ions (Section 21.2).

Enzyme classification. There are six classes of enzymes based on function: oxidoreductases, transferases, hydrolases, lyases, isomerases, and ligases (Section 21.3).

Enzyme active site. An enzyme active site is the relatively small part of the enzyme that is actually involved in catalysis. It is where substrate binds to the enzyme (Section 21.4).

Lock-and-key model of enzyme activity. The active site in an enzyme has a fixed, rigid geometrical conformation. Only substrates with a complementary geometry can be accommodated at the active site (Section 21.4).

Induced-fit model of enzyme activity. The active site in an enzyme can undergo small changes in geometry in order to accommodate a series of related substrates (Section 21.4).

Enzyme activity. Enzyme activity is a measure of the rate at which an enzyme converts substrate to products. Four factors that affect enzyme activity are temperature, pH, substrate concentration, and enzyme concentration (Section 21.6).

Enzyme inhibition. An enzyme inhibitor slows or stops the normal catalytic function of an enzyme by binding to it. Three modes of inhibition are reversible competitive inhibition, reversible noncompetitive inhibition, and irreversible inhibition (Section 21.7).

Allosteric enzyme. An allosteric enzyme is an enzyme with two or more protein chains and two kinds of binding sites (for substrate and regulator) (Section 21.8).

Zymogen. A zymogen is an inactive precursor of a proteolytic enzyme; the zymogen is activated by a chemical reaction that removes to part of its structure (Section 21.8).

Covalent modification. Covalent modification is a cellular process for regulation of enzyme activity in which the structure of an enzyme is modified through formation of, or breaking of, a covalent bond. The most commonly encountered type of covalent modification involves a phosphate group being added to, or removed from, an enzyme (Section 21.8).

Vitamins. A vitamin is an organic compound necessary in small amounts for the normal growth of humans and some animals. Vitamins must be obtained from dietary sources because they cannot be synthesized in the body (Section 21.11).

Water-soluble vitamins. Vitamin C and the eight B vitamins are the water-soluble vitamins. Vitamin C is essential for the proper formation of bones and teeth and is also an important antioxidant. All eight B vitamins function as coenzymes (Section 21.12).

Fat-soluble vitamins. The four fat-soluble vitamins are vitamins A, D, E, and K. The best-known function of vitamin A is its role in vision. Vitamin D is essential for the proper use of calcium and phosphorus to form bones and teeth. The primary function of vitamin E is as an antioxidant. Vitamin K is essential in the regulation of blood clotting (Section 21.13).

KEY REACTIONS AND EQUATIONS

1. Conversion of an apoenzyme to an active enzyme (Section 21.2)

   Apoenzyme + cofactor $\longrightarrow$ holoenzyme (active enzyme)

2. Mechanism of enzyme action (Section 21.4)

   Enzyme + substrate $\longrightarrow$ enzyme–substrate complex

   Enzyme–substrate complex $\longrightarrow$ enzyme + product

3. Enzyme inhibition (Section 21.7)

   Enzyme + inhibitor $\longrightarrow$ inactive enzyme

EXERCISES *and* PROBLEMS

The members of each pair of problems in this section test similar material.

● **Importance of Enzymes (Section 21.1)**

21.1 What is the general role of enzymes in the human body?

21.2 Why does the body need so many different enzymes?

21.3 List two ways in which enzymes differ from inorganic laboratory catalysts.

21.4 Occasionally we refer to the "delicate" nature of enzymes. Explain why this adjective is appropriate.

● **Enzyme Structure (Section 21.2)**

21.5 Indicate whether each of the following phrases describes a simple or a conjugated enzyme.
   a. An enzyme that has both a protein and a nonprotein portion
   b. An enzyme that requires Mg^{2+} ion for activity
   c. An enzyme in which only amino acids are present
   d. An enzyme in which a cofactor is present

21.6 Indicate whether each of the following phrases describes a simple or a conjugated enzyme.
   a. An enzyme that contains a carbohydrate portion
   b. An enzyme that contains only protein
   c. A holoenzyme
   d. An enzyme that has a vitamin as part of its structure

21.7 What is the difference between a cofactor and a coenzyme?

21.8 All coenzymes are cofactors, but not all cofactors are coenzymes. Explain this statement.

21.9 Why are cofactors present in most enzymes?

21.10 What is the difference between an apoenzyme and a holoenzyme?

● **Enzyme Nomenclature (Section 21.3)**

21.11 Which of the following substances are enzymes?
   a. Sucrase   b. Galactose
   c. Trypsin   d. Xylulose reductase

21.12 Which of the following substances are enzymes?
   a. Sucrose   b. Pepsin
   c. Glutamine synthetase   d. Cellulase

21.13 Predict the function of each of the following enzymes.
   a. Pyruvate carboxylase   b. Alcohol dehydrogenase
   c. L-Amino acid reductase   d. Maltase

21.14 Predict the function of each of the following enzymes.
   a. Cytochrome oxidase   b. *Cis–trans* isomerase
   c. Succinate dehydrogenase   d. Lactase

21.15 Suggest a name for an enzyme that catalyzes each of the following reactions.
   a. Hydrolysis of sucrose
   b. Decarboxylation of pyruvate
   c. Isomerization of glucose
   d. Removal of hydrogen from lactate

21.16 Suggest a name for an enzyme that catalyzes each of the following reactions.
   a. Hydrolysis of lactose   b. Oxidation of nitrite
   c. Decarboxylation of citrate   d. Reduction of oxalate

21.17 Give the name of the substrate on which each of the following enzymes acts.
   a. Pyruvate carboxylase   b. Galactase
   c. Alcohol dehydrogenase   d. L-Amino acid reductase

21.18 Give the name of the substrate on which each of the following enzymes acts.
   a. Cytochrome oxidase   b. Lactase
   c. Succinate dehydrogenase   d. Tyrosine kinase

21.19 To which of the six major classes of enzymes does each of the following belong?
   a. Mutase   b. Dehydratase
   c. Carboxylase   d. Kinase

21.20 To which of the six major classes of enzymes does each of the following belong?
   a. Protease   b. Racemase
   c. Dehydrogenase   d. Synthetase

21.21 To which of the six major classes of enzymes does the enzyme that catalyzes each of the following reactions belong?
   a. A *cis* double bond is converted to a *trans* double bond.
   b. An alcohol is dehydrated to form a compound with a double bond.
   c. An amino group is transferred from one substrate to another.
   d. An ester linkage is hydrolyzed.

21.22 To which of the six major classes of enzymes does the enzyme that catalyzes each of the following reactions belong?
   a. An L isomer is converted to a D isomer.
   b. A phosphate group is transferred from one substrate to another.
   c. An amide linkage is hydrolyzed.
   d. Hydrolysis of a carbohydrate to monosaccharides occurs.

21.23 Identify the enzyme needed in each of the following reactions as an isomerase, a decarboxylase, a dehydrogenase, a lipase, or a phosphatase.

a.
$$CH_3-\overset{O}{\overset{\|}{C}}-COOH \rightarrow CH_3-\overset{O}{\overset{\|}{C}}-H + CO_2$$

b.
$$\begin{array}{l} CH_2-O-\overset{O}{\overset{\|}{C}}-R \\ | \\ CH-O-\overset{O}{\overset{\|}{C}}-R + 3H_2O \\ | \\ CH_2-O-\overset{O}{\overset{\|}{C}}-R \end{array} \rightarrow \begin{array}{l} CH_2-OH \\ | \\ CH-OH \\ | \\ CH_2-OH \end{array} + 3R-COOH$$

c.
$$\overset{+}{H_3}N-CH-COO^- + H_2O \rightarrow$$
$$| \quad CH_2$$
$$| \quad OPO_3^{2-}$$

$$\overset{+}{H_3}N-CH-COO^- + HPO_4^{2-}$$
$$| \quad CH_2$$
$$| \quad OH$$

d.
$$CH_3-\overset{OH}{\overset{|}{C}}H-COOH + NAD^+ \rightarrow$$
$$CH_3-\overset{O}{\overset{\|}{C}}-COOH + NADH + H^+$$

21.24 Identify the enzyme needed in each of the following reactions as an isomerase, a decarboxylase, a dehydrogenase, a protease, or a phosphatase.

a.
$$CH_3-\overset{\overset{\displaystyle O}{\|}}{C}-\overset{\overset{\displaystyle O}{\|}}{C}-OH \rightarrow CH_3-\overset{\overset{\displaystyle O}{\|}}{C}-H + CO_2$$

b.
$$\begin{array}{c} CHO \\ | \\ HC-OH \\ | \\ HO-CH \\ | \\ HC-OH \\ | \\ HO-CH \\ | \\ CH_2OPO_3{}^{2-} \end{array} \rightarrow \begin{array}{c} CH_2OH \\ | \\ C=O \\ | \\ HO-CH \\ | \\ HC-OH \\ | \\ HO-CH \\ | \\ CH_2OPO_3{}^{2-} \end{array}$$

c.
$$HO-\overset{\overset{\displaystyle O}{\|}}{C}-CH_2-CH_2-\overset{\overset{\displaystyle O}{\|}}{C}-OH \rightarrow$$
$$HO-\overset{\overset{\displaystyle O}{\|}}{C}-CH=CH-\overset{\overset{\displaystyle O}{\|}}{C}-OH + 2H$$

d.
$$\overset{+}{H_3}N-CH-\overset{\overset{\displaystyle O}{\|}}{C}-NH-CH-COO^- + H_2O \rightarrow$$
$$\begin{array}{cc} | & | \\ CH_3 & CH_3 \end{array}$$
$$2\ \overset{+}{H_3}N-CH-COO^-$$
$$| \atop CH_3$$

Models of Enzyme Action (Section 21.4)

21.25 What is an enzyme active site?

21.26 What is an enzyme–substrate complex?

21.27 How does the lock-and-key model of enzyme action explain the highly specific way some enzymes select a substrate?

21.28 How does the induced-fit model of enzyme action explain the broad specificities of some enzymes?

21.29 What types of forces hold a substrate at an enzyme active site?

21.30 The forces that hold a substrate at an enzyme active site are not covalent bonds. Explain why not.

Enzyme Specificity (Section 21.5)

21.31 Define the following terms dealing with enzyme specificity.
a. Absolute specificity   b. Linkage specificity

21.32 Define the following terms dealing with enzyme specificity.
a. Group specificity   b. Stereochemical specificity

21.33 Which type(s) of enzyme specificity are best accounted for by the lock-and-key model of enzyme action?

21.34 Which type(s) of enzyme specificity are best accounted for by the induced-fit model of enzyme action?

21.35 Which enzyme in each of the following pairs would be more limited in its catalytic scope?
a. An enzyme that exhibits absolute specificity or an enzyme that exhibits group specificity
b. An enzyme that exhibits stereochemical specificity or an enzyme that exhibits linkage specificity

21.36 Which enzyme in each of the following pairs would be more limited in its catalytic scope?
a. An enzyme that exhibits linkage specificity or an enzyme that exhibits absolute specificity
b. An enzyme that exhibits group specificity or an enzyme that exhibits stereochemical specificity

Factors That Affect Enzyme Activity (Section 21.6)

21.37 Temperature affects enzymatic reaction rates in two ways. An increase in temperature can accelerate the rate of a reaction or it can stop the reaction. Explain each of these effects.

21.38 Define the optimum temperature for an enzyme.

21.39 Explain why all enzymes do not possess the same optimum pH.

21.40 Why does an enzyme lose activity when the pH is drastically changed from the optimum pH?

21.41 Draw a graph that shows the effect of increasing substrate concentration on the rate of an enzyme-catalyzed reaction (at constant temperature, pH, and enzyme concentration).

21.42 Draw a graph that shows the effect of increasing enzyme concentration on the rate of an enzyme-catalyzed reaction (at constant temperature, pH, and substrate concentration).

21.43 In an enzyme-catalyzed reaction, all of the enzyme active sites are saturated by substrate molecules at a certain substrate concentration. What happens to the rate of the reaction when the substrate concentration is doubled?

21.44 What is an enzyme turnover number?

21.45 What is an *extremozyme*?

21.46 Why are *extremozymes* of commercial importance?

Enzyme Inhibition (Section 21.7)

21.47 In competitive inhibition, can both the inhibitor and the substrate bind to an enzyme at the same time? Explain your answer.

21.48 Compare the sites where competitive and noncompetitive inhibitors bind to enzymes.

21.49 Indicate whether each of the following statements describes a reversible competitive inhibitor, a reversible noncompetitive inhibitor, or an irreversible inhibitor. More than one answer may apply.
a. Both inhibitor and substrate bind at the active site on a random basis.
b. The inhibitor effect cannot be reversed by the addition of more substrate.
c. Inhibitor structure does not have to resemble substrate structure.
d. The inhibitor can bind to the enzyme at the same time as substrate.

21.50 Indicate whether each of the following statements describes a reversible competitive inhibitor, a reversible noncompetitive inhibitor, or an irreversible inhibitor. More than one answer may apply.
a. It bonds covalently to the enzyme active site.
b. The inhibitor effect can be reversed by the addition of more substrate.
c. Inhibitor structure must be somewhat similar to that of substrate.
d. The inhibitor cannot bind to the enzyme at the same time as substrate.

Regulation of Enzyme Activity (Section 21.8)

21.51 What is an allosteric enzyme?

21.52 What is a regulator molecule?

21.53 What is feedback control?

21.54 What is the difference between positive and negative feedback to an allosteric enzyme?

21.55 What is the general relationship between zymogens and proteolytic enzymes?

21.56 What, if any, is the difference in meaning between the terms *zymogen* and *proenzyme*?

21.57 Why are proteolytic enzymes always produced in an inactive form?

21.58 What is the mechanism by which most zymogens are activated?

21.59 What is covalent modification?

21.60 What are the two most commonly encountered covalent modification processes?

21.61 What is the most common source for the phosphate group involved in phosphorylation?

21.62 Is the phosphorylated version of an enzyme the "turned-on" form or the "turned-off" form of the enzyme?

21.63 What is the general name for enzymes that effect the phosphorylation of another enzyme?

21.64 What is the general name for enzymes that effect the dephosphorylation of another enzyme?

Antibiotics That Inhibit Enzyme Activity (Section 21.9)

21.65 By what mechanism do sulfa drugs kill bacteria?

21.66 By what mechanism do penicillins kill bacteria?

21.67 Why is penicillin toxic to bacteria but not to higher organisms?

21.68 What amino acid in transpeptidase forms a covalent bond to penicillin?

21.69 What situation has made Cipro a prominent antibiotic?

21.70 Describe the structure of Cipro in terms of common and "unusual" functional groups present.

Medical Uses of Enzymes (Section 21.10)

21.71 What does the acronym TPA stand for and how is TPA used in diagnostic medicine?

21.72 What does the acronym BUN test stand for and how is this test used in clinical laboratory analysis?

21.73 What are *isoenzymes*?

21.74 How are *isoenzymes* used in the diagnosis of a heart attack?

General Characteristics of Vitamins (Section 21.11)

21.75 What is a vitamin?

21.76 List a way in which vitamins differ from carbohydrates, fats, and proteins (the major classes of food).

21.77 Indicate whether each of the following is a fat-soluble or a water-soluble vitamin.
a. Vitamin K
b. Vitamin B_{12}
c. Vitamin C
d. Thiamin

21.78 Indicate whether each of the following is a fat-soluble or a water-soluble vitamin.
a. Vitamin A
b. Vitamin B_6
c. Vitamin E
d. Riboflavin

21.79 Indicate whether each of the vitamins in Problem 21.77 would be likely or unlikely to be toxic when consumed in excess.

21.80 Indicate whether each of the vitamins in Problem 21.78 would be likely or unlikely to be toxic when consumed in excess.

Water-Soluble Vitamins (Section 21.12)

21.81 Describe the two most completely characterized roles of vitamin C in the body.

21.82 Structurally, how do the oxidized and reduced forms of vitamin C differ?

21.83 What is the dominant function within the human body of the B vitamins as a group?

21.84 With the help of Table 21.7, identify the B vitamin or vitamins to which each of the following characterizations applies.
a. Is part of the coenzymes NAD and NADP
b. Is part of coenzyme A
c. Is part of THF
d. Is part of TPP

21.85 With the help of Table 21.6, indicate whether each of the following B vitamins exists in more than one structural form.
a. Folate
b. Niacin
c. Vitamin B_6
d. Biotin

21.86 With the help of Table 21.6, identify the B vitamin or vitamins to which each of the following characterizations applies.
a. Contains a heterocyclic nitrogen ring system
b. Has the most complex structure
c. Contains a metal atom as part of its structure
d. Contains sulfur as part of its structure

Fat-Soluble Vitamins (Section 21.13)

21.87 Describe the structural differences among the three retinoid forms of vitamin A.

21.88 What is the relationship between the plant pigment beta-carotene and vitamin A?

21.89 What is *cell differentiation* and how does vitamin A participate in this process?

21.90 List four major functions for vitamin A in the human body.

21.91 How do vitamin D_2 and vitamin D_3 differ in structure?

21.92 In terms of source, how do vitamin D_2 and vitamin D_3 differ?

21.93 What is the principal function of vitamin D in the human body?

21.94 Why is vitamin D often called the sunshine vitamin?

21.95 Which form of tocopherol (vitamin E) exhibits the greatest biochemical activity?

21.96 How do the various forms of tocopherol differ in structure?

21.97 What is the principal function of vitamin E in the human body?

21.98 Why is vitamin E often given to premature infants that are on oxygen therapy?

21.99 How do vitamin K_1 and vitamin K_2 differ in structure?

21.100 In terms of source, how do vitamin K_1 and vitamin K_2 differ?

21.101 How are *menaquinones, phylloquinones,* and vitamin K related?

21.102 What is the principal function of vitamin K in the human body?

ADDITIONAL PROBLEMS

21.103 Explain the difference, if any, between the following types of enzymes.
  a. Apoenzyme and proenzyme
  b. Simple enzyme and allosteric enzyme
  c. Coenzyme and isoenzyme
  d. Conjugated enzyme and holoenzyme

21.104 Identify the functional groups present in a molecule of each of the following vitamins.
  a. Vitamin C
  b. Vitamin A (retinol)
  c. Vitamin D
  d. Vitamin K

21.105 Indicate whether each of the following vitamins functions as a coenzyme.
  a. Vitamin C　　b. Vitamin A
  c. Riboflavin　　d. Biotin

21.106 Which vitamin has each of the following functions?
  a. Water-soluble antioxidant
  b. Fat-soluble antioxidant
  c. Involved in the process of calcium deposition in bone
  d. Involved in the blood-clotting process

21.107 Which vitamin has each of the following functions?
  a. Involved in cell differentiation
  b. Involved in vision
  c. Involved in collagen formation
  d. Involved in prostaglandin formation

21.108 What general kinds of reactions do the following types of enzymes catalyze?
  a. Lyases　　　　b. Ligases
  c. Hydrolases　　d. Transferases

21.109 Alcohol dehydrogenase catalyzes the conversion of ethanol to acetaldehyde. This enzyme, in its active state, consists of a protein molecule and a zinc ion. On the basis of this information, identify the following for this chemical system.
  a. Substrate　　　b. Cofactor
  c. Apoenzyme　　d. Holoenzyme

21.110 Each of the following is an abbreviation for an enzyme used in the diagnosis and/or treatment of heart attacks. What does each abbreviation stand for?
  a. TPA　　　　　b. LDH
  c. CPK　　　　　d. AST

*M*ultiple-Choice Practice Test

21.111 Which are the two most common endings for the name of an enzyme?
  a. -ase and -ose　　b. -ase and -in
  c. -in and -ogen　　d. -in and -ine

21.112 Which of the following pairings of enzyme type and enzyme function is *incorrect*?
  a. Kinase and transfer of a phosphate group between substrates
  b. Mutase and introduction of a double bond within a molecule
  c. Protease and hydrolysis of amide linkages in proteins
  d. Decarboxylase and removal of CO_2 from a substrate

21.113 Which of the following is true for a conjugated enzyme?
  a. It contains only protein.
  b. It does not contain protein.
  c. It has a nonprotein part.
  d. It always contains a metal ion.

21.114 Which of the following statements concerning *cofactors* is *incorrect*?
  a. All conjugated enzymes contain cofactors.
  b. Some cofactors are metal ions.
  c. A cofactor is the nonprotein portion of an enzyme.
  d. Vitamins cannot be cofactors.

21.115 What happens to substrate molecules at an enzyme active site?
  a. They always react with O_2.
  b. They become covalently bonded to the enzyme.
  c. They become catalysts.
  d. They undergo change to a desired product.

21.116 Which of the following statements about a reversible non-competitive inhibitor is *correct*?
  a. It prevents substrate from occupying the enzyme active site.
  b. It must resemble the substrate in general shape.
  c. It and the substrate can simultaneously occupy the active site.
  d. It binds to the enzyme at a location other than the active site.

21.117 What is the shape of a plot of enzyme activity (*y*-axis) versus temperature (*x*-axis) with other variables constant?
  a. Straight line with an upward slope
  b. Line with an upward slope and a long flat top
  c. Line with an upward slope followed by a downward slope
  d. Straight horizontal line

21.118 Which of the following statements concerning the B vitamins is *correct*?
  a. Structurally, they are all very similar.
  b. All except two of them are water soluble.
  c. In chemically modified form they often serve as cofactors in enzymes.
  d. Fruits, in general, are very good sources of these vitamins.

21.119 Cholesterol is a precursor for which of the following vitamins?
  a. Vitamin A　　b. Vitamin C
  c. Vitamin D　　d. Vitamin E

21.120 Beta-carotene is a precursor for which of the following vitamins?
  a. Vitamin A　　b. Vitamin D
  c. Vitamin E　　d. Vitamin K

Nucleic Acids

Chapter Outline

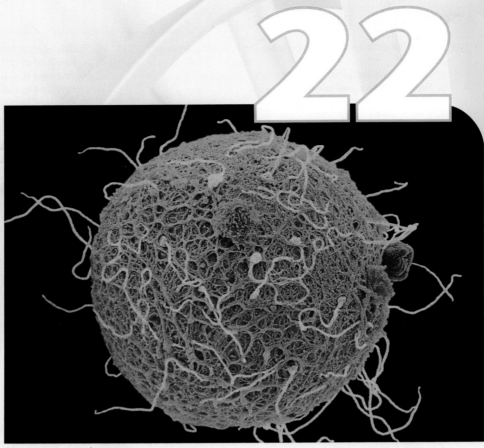

Human egg and sperm.

A most remarkable property of living cells is their ability to produce exact replicas of themselves. Furthermore, cells contain all the instructions needed for making the complete organism of which they are a part. The molecules within a cell that are responsible for these amazing capabilities are nucleic acids.

The Swiss physiologist Friedrich Miescher (1844–1895) discovered nucleic acids in 1869 while studying the nuclei of white blood cells. The fact that they were initially found in cell nuclei and are acidic accounts for the name *nucleic acid.* Although we now know that nucleic acids are found throughout a cell, not just in the nucleus, the name is still used for such materials.

TYPES OF NUCLEIC ACIDS

Two types of nucleic acids are found within cells of higher organisms: *deoxyribonucleic acid* (DNA) and *ribonucleic acid* (RNA). Nearly all the DNA is found within the cell nucleus. Its primary function is the storage and transfer of genetic information. This information is used (indirectly) to control many functions of a living cell. In addition, DNA is passed from existing cells to new cells during cell division. RNA occurs in all parts of a cell. It functions primarily in synthesis of proteins, the molecules that carry out essential cellular functions. The structural distinctions between DNA and RNA molecules are considered in Section 22.3.

All nucleic acid molecules are polymers. A **nucleic acid** *is a polymer in which the monomer units are nucleotides.* Thus the starting point for a discussion of nucleic acids is an understanding of the structures and chemical properties of nucleotides.

It was not until 1944, 75 years after the discovery of nucleic acids, that scientists obtained the first evidence that these molecules are responsible for the storage and transfer of genetic information.

22.2 NUCLEOTIDES: BUILDING BLOCKS OF NUCLEIC ACIDS

A **nucleotide** *is a three-subunit molecule in which a pentose sugar is bonded to both a phosphate group and a nitrogen-containing heterocyclic base.* With a three-subunit structure, nucleotides are more complex monomers than the monosaccharides of polysaccharides (Section 18.8) and the amino acids of proteins (Section 20.2). A block structural diagram for a nucleotide is

Pentose Sugars

The sugar unit of a nucleotide is either the pentose *ribose* or the pentose *2′-deoxyribose.*

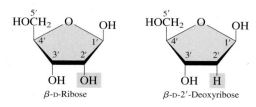

β-D-Ribose β-D-2′-Deoxyribose

Structurally, the only difference between these two sugars occurs at carbon 2′. The —OH group present on this carbon in ribose becomes a —H atom in 2′-deoxyribose. (The prefix *deoxy-* means "without oxygen.")

RNA and DNA differ in the identity of the sugar unit in their nucleotides. In RNA the sugar unit is *ribose*—hence the *R* in RNA. In DNA the sugar unit is *2′-deoxyribose*—hence the *D* in DNA.

Nitrogen-Containing Heterocyclic Bases

Five nitrogen-containing heterocyclic bases are nucleotide components. Three of them are derivatives of pyrimidine (Section 17.9), a monocyclic base with a six-membered ring, and two are derivatives of purine (Section 17.9), a bicyclic base with fused five- and six-membered rings.

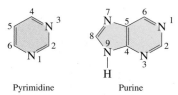

Pyrimidine   Purine

Both of these heterocyclic compounds are bases because they contain amine functional groups (secondary or tertiary), and amine functional groups exhibit basic behavior (proton acceptors; Section 17.6).

The three pyrimidine derivatives found in nucleotides are thymine (T), cytosine (C), and uracil (U).

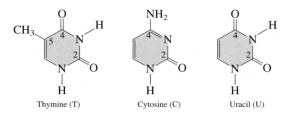

Thymine (T)   Cytosine (C)   Uracil (U)

Thymine is the 5-methyl-2,4-dioxo derivative, cytosine the 4-amino-2-oxo derivative, and uracil the 2,4-dioxo derivative of pyrimidine.

Proteins are polypeptides, many carbohydrates are polysaccharides, and nucleic acids are polynucleotides.

The systems for numbering the atoms in the pentose and nitrogen-containing base subunits of a nucleotide are important and will be used extensively in later sections of this chapter. The convention is that

1. Pentose ring atoms are designated with *primed* numbers.
2. Nitrogen-containing base ring atoms are designated with *unprimed* numbers.

A pyrimidine derivative that we have encountered previously is the B vitamin thiamine (see Section 21.12).

Caffeine, the most widely used nonprescription central nervous system stimulant, is the 1,3,7-trimethyl-2,6-dioxo derivative of purine (Section 17.9).

Figure 22.1 Space-filling model of the molecule adenine, a nitrogen-containing heterocyclic base present in both DNA and RNA.

The two purine derivatives found in nucleotides are adenine (A) and guanine (G).

Adenine (A)   Guanine (G)

Adenine is the 6-amino derivative of purine, and guanine is the 2-amino-6-oxo purine derivative. A space-filling model for adenine is shown in Figure 22.1.

Adenine, guanine, and cytosine are found in both DNA and RNA. Uracil is found only in RNA, and thymine usually occurs only in DNA. Figure 22.2 summarizes the occurrences of nitrogen-containing heterocyclic bases in nucleic acids.

Phosphate

Phosphate, the third component of a nucleotide, is derived from phosphoric acid (H_3PO_4). Under cellular pH conditions, the phosphoric acid loses two of its hydrogen atoms to give a hydrogen phosphate ion (HPO_4^{2-}).

Phosphoric acid   Hydrogen phosphate ion

Nucleotide Formation

The formation of a nucleotide from sugar, base, and phosphate can be visualized as occurring in the following manner:

Base

Phosphate   Sugar   Nucleotide

Important characteristics of this combining of three molecules into one molecule (the nucleotide) are that

1. Condensation, with formation of a water molecule, occurs at two locations: between sugar and base and between sugar and phosphate.
2. The base is always attached at the C-1′ position of the sugar. For purine bases, attachment is through N-9; for pyrimidine bases, N-1 is involved. The C-1′ carbon atom of the ribose unit is always in a β configuration (Section 18.10), and the bond connecting the sugar and base is a β-N-glycosidic linkage (Section 18.13).
3. The phosphate group is attached to the sugar at the C-5′ position through a phosphate–ester linkage.

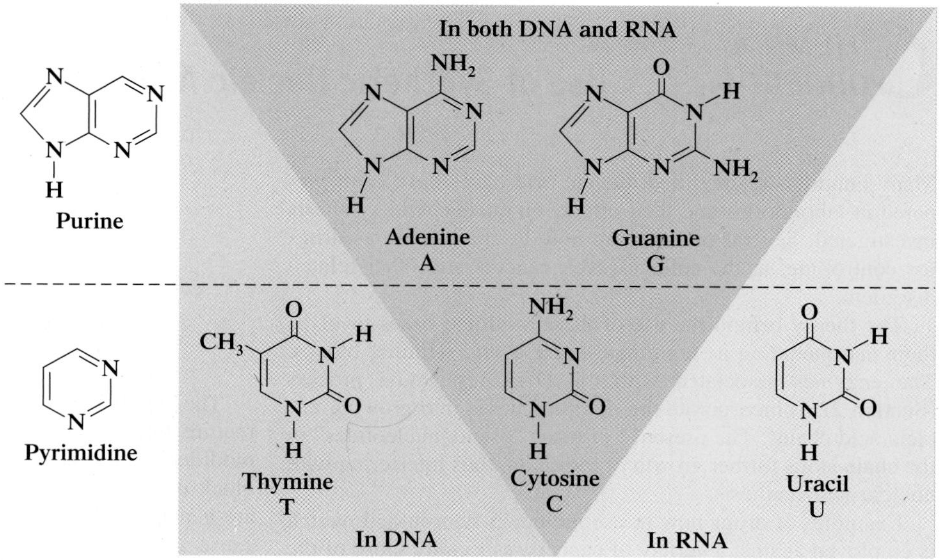

Figure 22.2 Two purine bases and three pyrimidine bases are found in the nucleotides present in nucleic acids.

Figure 22.2 Two purine bases and three pyrimidine bases are found in the nucleotides present in nucleic acids.

To remember which two of the five nucleotide bases are the purine derivatives (fused rings), use the phrase "pure silver" and the chemical symbol for silver, which is Ag.

pure Ag
purine A and G

There are four possible RNA nucleotides, differing in the base present (A, C, G, or U), and four possible DNA nucleotides, differing in the base present (A, C, G, or T).

Nucleotide Nomenclature

The common names and abbreviations for the eight nucleotides of DNA and RNA molecules are given in Table 22.1. It is important to be familiar with them because they are frequently encountered in biochemistry.

We can make several generalizations about the nomenclature given in Table 22.1.

1. All of the names end in 5'-monophosphate, which signifies the presence of a phosphate group attached to the 5' carbon atom of ribose or deoxyribose. (In Chapter 23 we will encounter nucleotides that contain two or three phosphate groups—diphosphates and triphosphates.)
2. Preceding the monophosphate ending is the name of the base present in a modified form. The suffix -*osine* is used with purine bases, the suffix -*idine* with pyrimidine bases.
3. The prefix *deoxy-* at the start of the name signifies that the sugar present is deoxyribose. When no prefix is present, the sugar is ribose.

TABLE 22.1
The Names of the Eight Nucleotides Found in DNA and RNA

Base	Sugar	Nucleotide Name	Nucleotide Abbreviation
DNA Nucleotides			
adenine	deoxyribose	deoxyadenosine 5'-monophosphate	dAMP
guanine	deoxyribose	deoxyguanosine 5'-monophosphate	dGMP
cytosine	deoxyribose	deoxycytidine 5'-monophosphate	dCMP
thymine	deoxyribose	deoxythymidine 5'-monophosphate	dTMP
RNA Nucleotides			
adenine	ribose	adenosine 5'-monophosphate	AMP
guanine	ribose	guanosine 5'-monophosphate	GMP
cytosine	ribose	cytidine 5'-monophosphate	CMP
uracil	ribose	uridine 5'-monophosphate	UMP

CHEMICAL Connections

Use of Synthetic Nucleic Acid Bases in Medicine

Many hundreds of modified nucleic acid bases have been prepared in laboratories and their effects on nucleic acid synthesis investigated. Several of them are now in clinical use as drugs for controlling, at the cellular level, cancers and other related disorders.

The theory behind the use of these modified bases involves their masquerading as legitimate nucleic acid building blocks. The enzymes associated with the DNA replication process (Section 22.5) incorporate the modified bases into growing nucleic acid chains. The presence of these "pseudonucleotides" in the chain stops further growth of the chain, thus interfering with nucleic acid synthesis.

Examples of drugs now in use include 5-fluorouracil, which is employed against a variety of cancers, especially those of the breast and digestive tract, and 6-mercaptopurine, which is used in the treatment of leukemia.

6-Mercaptopurine
(a modified adenine)

Adenine

The rapidly dividing cells that are characteristic of cancer require large quantities of DNA. Anticancer drugs based on modified nucleic acid bases block DNA synthesis and therefore block the increase in the number of cancer cells. Cancer cells are generally affected to a greater extent than normal cells because of this rapid growth. Eventually, the normal cells are affected to such a degree that use of the drugs must be discontinued. 5-Fluorouracil inhibits the formation of thymine-containing nucleotides required for DNA synthesis. 6-Mercaptopurine, which substitutes for adenine, inhibits the synthesis of nucleotides that incorporate adenine and guanine.

5-Fluorouracil
(a modified thymine)

Thymine

4. The abbreviations in Table 22.1 for the nucleotides come from the one-letter symbols for the bases (A, C, G, T, and U), the use of MP for monophosphate, and a lower-case *d* at the start of the abbreviation whenever deoxyribose is the sugar.

The use of synthetic nucleic acid bases in medicine is considered in the Chemical Connections feature on this page.

22.3 PRIMARY NUCLEIC ACID STRUCTURE

Nucleic acids are polymers in which the repeating units, the monomers, are nucleotides (Section 22.2). The nucleotide units within a nucleic acid molecule are linked to each other through sugar–phosphate bonds. The resulting molecular structure (Figure 22.3) involves a chain of alternating sugar and phosphate groups with a base group protruding from the chain at regular intervals.

We can now define, in terms of structure, the two major types of nucleic acids: ribonucleic acids and deoxyribonucleic acids (Section 22.1). A **ribonucleic acid (RNA)** *is a nucleotide polymer in which each of the monomers contains ribose, a phosphate group, and one of the heterocyclic bases adenine, cytosine, guanine, or uracil.* Two changes to this definition generate the deoxyribonucleic acid definition; deoxyribose replaces ribose and thymine replaces uracil. A **deoxyribonucleic acid (DNA)** *is a*

Nucleotides are related to nucleic acids in the same way that amino acids are related to proteins.

Figure 22.3 The general structure of a nucleic acid in terms of nucleotide subunits.

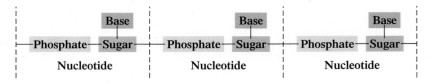

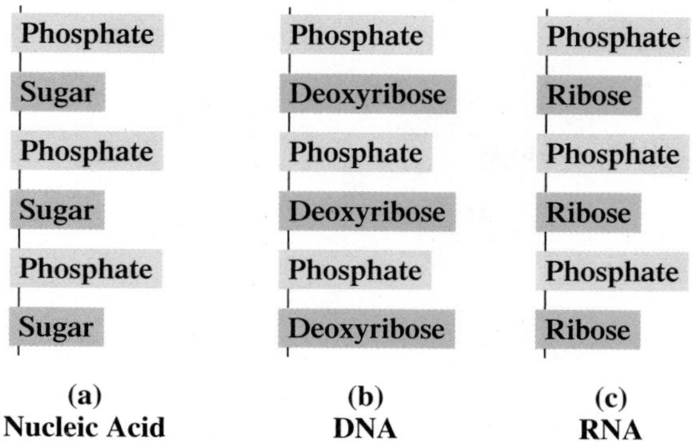

Figure 22.4 (a) The generalized backbone structure of a nucleic acid. (b) The specific backbone structure for a deoxyribonucleic acid (DNA). (c) The specific backbone structure for a ribonucleic acid (RNA).

Phosphate	Phosphate	Phosphate
Sugar	Deoxyribose	Ribose
Phosphate	Phosphate	Phosphate
Sugar	Deoxyribose	Ribose
Phosphate	Phosphate	Phosphate
Sugar	Deoxyribose	Ribose
(a)	**(b)**	**(c)**
Nucleic Acid	**DNA**	**RNA**

The backbone of a nucleic acid structure is always an alternating sequence of phosphate and sugar groups. The sugar is ribose in RNA and deoxyribose in DNA.

nucleotide polymer in which each of the monomers contains deoxyribose, a phosphate group, and one of the heterocyclic bases adenine, cytosine, guanine, or thymine.

The alternating sugar–phosphate chain in a nucleic acid structure is often called the *nucleic acid backbone*. This backbone is constant throughout the entire nucleic acid structure. For DNA molecules, the backbone consists of alternating phosphate and *deoxyribose* sugar units; for RNA molecules, the backbone consists of alternating phosphate and *ribose* sugar units. Figure 22.4 contrasts the generalized backbone structure for a nucleic acid with the specific backbone structures of DNAs and RNAs.

The variable portion of nucleic acid structure is the sequence of bases attached to the sugar units of the backbone. The sequence of these base side chains distinguishes various DNAs from each other and various RNAs from each other. Only four types of bases are found in any given nucleic acid structure. This situation is much simpler than that for proteins, where 20 side-chain entities (amino acids) are available (Section 20.2). In both RNA and DNA, adenine, guanine, and cytosine are encountered as side-chain components; thymine is found mainly in DNA, and uracil is found only in RNA (Figure 22.2).

Just as the order of amino acid side chains determines the primary structure of a protein (Section 20.9), the order of nucleotide bases determines the primary structure of a nucleic acid.

Primary nucleic acid structure *is the sequence in which nucleotides are linked together in a nucleic acid.* Because the sugar–phosphate backbone of a given nucleic acid does not vary, the primary structure of the nucleic acid depends only on the sequence of bases present. Further information about nucleic acid structure can be obtained by considering the detailed four-nucleotide segment of a DNA molecule shown in Figure 22.5.

The following list describes some important points about nucleic acid structure that are illustrated in Figure 22.5.

1. Each nonterminal phosphate group of the sugar–phosphate backbone is bonded to two sugar molecules through a *3′,5′-phosphodiester linkage.* There is a phosphoester bond to the 5′ carbon of one sugar unit and a phosphoester bond to the 3′ carbon of the other sugar.
2. A nucleotide chain has *directionality.* One end of the nucleotide chain, the *5′ end,* normally carries a free phosphate group attached to the 5′ carbon atom. The other end of the nucleotide chain, the *3′ end,* normally has a free hydroxyl group attached to the 3′ carbon atom. By convention, the sequence of bases of a nucleic acid strand is read from the 5′ end to the 3′ end.
3. Each nonterminal phosphate group in the backbone of a nucleic acid carries a -1 charge. The parent phosphoric acid molecule from which the phosphate was derived originally had three —OH groups (Section 22.2). Two of these become involved in the 3′,5′-phosphodiester linkage. The remaining —OH group is free to exhibit acidic behavior—that is, to produce a H^+ ion.

$$\cdots O-\underset{\underset{\displaystyle OH}{|}}{\overset{\overset{\displaystyle O}{\parallel}}{P}}-O\cdots \;\rightleftharpoons\; \cdots O-\underset{\underset{\displaystyle O^-}{|}}{\overset{\overset{\displaystyle O}{\parallel}}{P}}-O\cdots \;+\; H^+$$

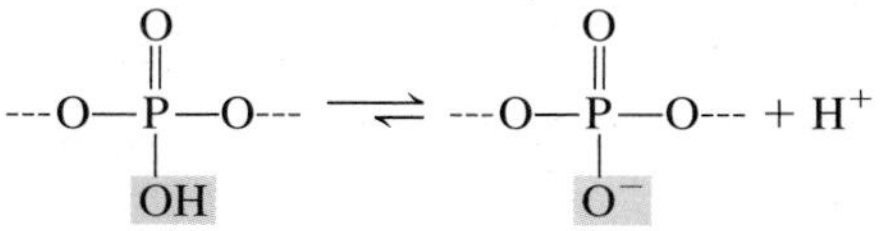

This behavior by the many phosphate groups in a nucleic acid backbone gives nucleic acids their acidic properties.

For both nucleic acids and proteins, a distinction is made between the two ends of the polymer chain. For nucleic acids there is a 5′ end and a 3′ end; for proteins there is an N-terminal end and a C-terminal end (Section 20.6).

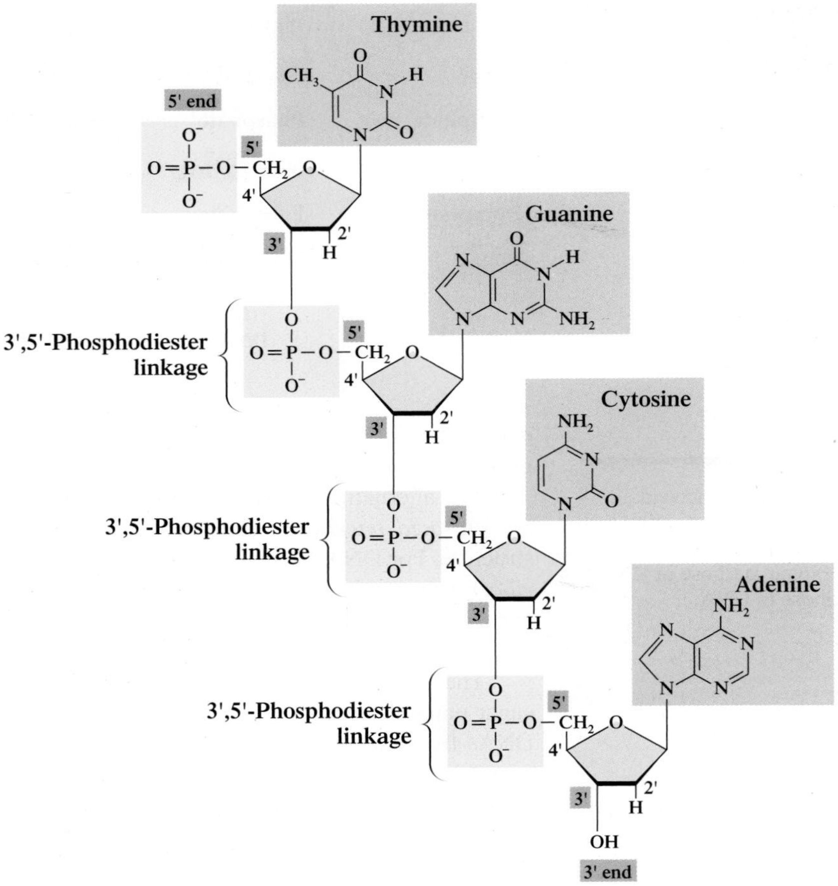

Figure 22.5 A four-nucleotide-long segment of DNA. (The choice of bases was arbitrary.)

Three parallels between primary nucleic acid structure and primary protein structure (Section 20.9) are worth noting.

1. DNAs, RNAs, and proteins all have backbones that do not vary in structure (see Figure 22.6).
2. The sequence of attachments to the backbones (nitrogen bases in nucleic acids and amino acid R groups in proteins) distinguishes one DNA from another, one RNA from another, and one protein from another (see Figure 22.6).
3. Both nucleic acid polymer chains and protein polymer chains have directionality; for nucleic acids there is a 5′ end and a 3′ end, and for proteins there is an N-terminal end and a C-terminal end.

The Chemistry at a Glance feature on page 741 summarizes important concepts relative to the makeup of the nucleotide building blocks (monomers) present in polymeric DNA and RNA molecules.

Figure 22.6 A comparison of the general primary structures of nucleic acids and proteins.

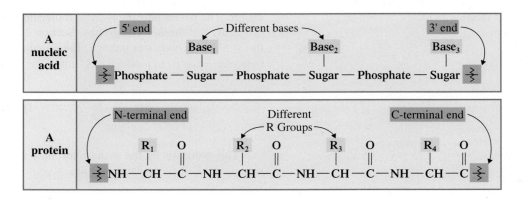

CHEMISTRY AT A GLANCE

Nucleic Acid Structure

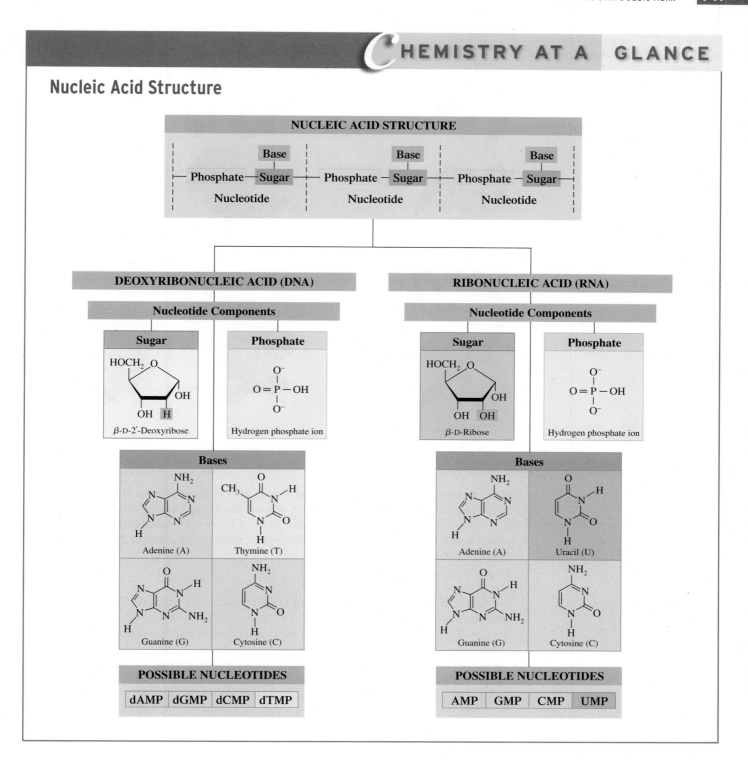

THE DNA DOUBLE HELIX

Like proteins, nucleic acids have secondary, or three-dimensional, structure as well as primary structure. The secondary structures of DNAs and RNAs differ, and we will discuss them separately.

The amounts of the bases A, T, G, and C present in DNA molecules were the key to determination of the general three-dimensional structure of DNA molecules. Base composition data for DNA molecules from many different organisms revealed a definite pattern of

Figure 22.7 Three views of the DNA double helix. (a) A schematic drawing that emphasizes the hydrogen bonding between bases on the two chains. (b) A space-filling model in which one DNA strand is blue and the other strand is orange. The bases are shown in lighter shades of blue and orange. (c) A top view of the double helix.

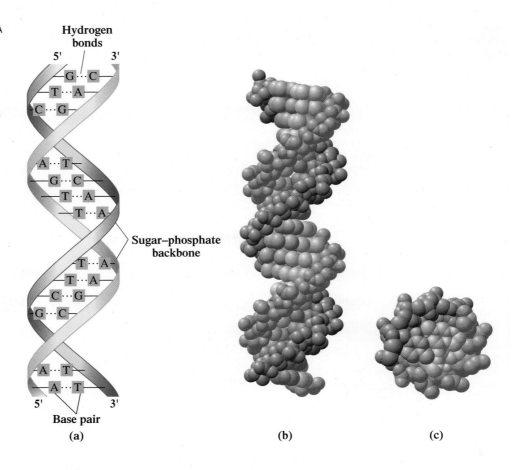

(a)   (b)   (c)

The α-helix secondary structure of proteins (Section 20.10) involves *one* polypeptide chain; the double-helix secondary structure of DNA involves *two* polynucleotide chains. In the α-helix of proteins, the R groups are on the *outside* of the helix; in the double helix of DNA, the bases are on the *inside* of the double helix.

base occurrence. The amounts of A and T were always equal, and the amounts of C and G were always equal, as were the amounts of total purines and total pyrimidines.

The relative amounts of these base pairs in DNA vary depending on the life form from which the DNA is obtained. (Each animal or plant has a unique base composition.) However, the relationships

$$\%A = \%T \qquad \text{and} \qquad \%C = \%G$$

always hold true. For example, human DNA contains 30% adenine, 30% thymine, 20% guanine, and 20% cytosine.

In 1953, an explanation for the base composition patterns associated with DNA molecules was proposed by the American microbiologist James Watson and the English biophysicist Francis Crick. Their model, which has now been validated in numerous ways, involves a double-helix structure that accounts for the equality of bases present, as well as for other known DNA structural data.

The DNA double helix involves two polynucleotide strands coiled around each other in a manner somewhat like a spiral staircase. The sugar–phosphate backbones of the two polynucleotide strands can be thought of as being the outside banisters of the spiral staircase (see Figure 22.7). The bases (side chains) of each backbone extend inward toward the bases of the other strand. The two strands are connected by *hydrogen bonds* (Section 7.13) between their bases. Additionally, the two strands of the double helix are *antiparallel*—that is, they run in opposite directions. One strand runs in the 5′-to-3′ direction, and the other is oriented in the 3′-to-5′ direction.

Base Pairing

A physical restriction, the size of the interior of the DNA double helix, limits the base pairs that can hydrogen-bond to one another. Only pairs involving one small base (a

Figure 22.8 Hydrogen-bonding possibilities are more favorable when A-T and G-C base pairing occurs than when A-C and G-T base pairing occurs. (a) Two and three hydrogen bonds can form, respectively, between A-T and G-C base pairs. These combinations are present in DNA molecules. (b) Only one hydrogen bond can form between G-T and A-C base pairs. These combinations are not present in DNA molecules.

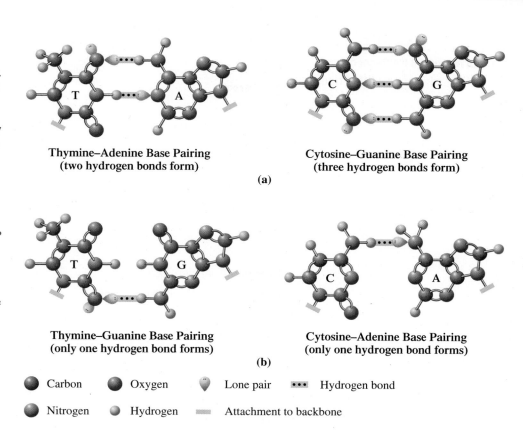

Thymine–Adenine Base Pairing
(two hydrogen bonds form)

Cytosine–Guanine Base Pairing
(three hydrogen bonds form)

(a)

Thymine–Guanine Base Pairing
(only one hydrogen bond forms)

Cytosine–Adenine Base Pairing
(only one hydrogen bond forms)

(b)

● Carbon ● Oxygen ◉ Lone pair ••• Hydrogen bond

● Nitrogen ● Hydrogen ══ Attachment to backbone

A mnemonic device for recalling base-pairing combinations in DNA involves listing the base abbreviations in alphabetical order. Then the first and last bases pair, and so do the middle two bases.

DNA: A C G T

Another way to remember these base-pairing combinations is to note that AT spells a word and that C and G look very much alike.

important DNA Replication

pyrimidine) and one large base (a purine) correctly "fit" within the helix interior. There is not enough room for two large purine bases to fit opposite each other (they overlap), and two small pyrimidine bases are too far apart to hydrogen-bond to one another effectively. Of the four possible purine–pyrimidine combinations (A–T, A–C, G–T, and G–C), hydrogen-bonding possibilities are *most favorable* for the A–T and G–C pairings, and these two combinations are the *only two* that normally occur in DNA. Figure 22.8 shows the specific hydrogen-bonding interactions for the four possible purine–pyrimidine base-pairing combinations.

The pairing of A with T and that of G with C are said to be *complementary*. A and T are complementary bases, as are G and C. **Complementary bases** *are pairs of bases in a nucleic acid structure that can hydrogen-bond to each other.* The fact that complementary base pairing occurs in DNA molecules explains, very simply, why the amounts of the bases A and T present are always equal, as are the amounts of G and C.

The two strands of DNA in a double helix are *not identical*—they are complementary. **Complementary DNA strands** *are strands of DNA in a double helix with base pairing such that each base is located opposite its complementary base.* Wherever G occurs in one strand, there is a C in the other strand; wherever T occurs in one strand, there is an A in the other strand. An important ramification of this complementary relationship is that knowing the base sequence of one strand of DNA enables us to predict the base sequence of the complementary strand.

In specifying the base sequence of a segment of a strand of DNA (or RNA), we list the bases in sequential order (using their one-letter abbreviations) in the direction from the 5′ end to the 3′ end of the segment.

The *antiparallel* nature of the two polynucleotide chains in the DNA double helix means that there is a 5′ end and a 3′ end at both ends of the double helix.

The two strands of DNA in a double helix are complementary. This means that if you know the order of bases in one strand, you can predict the order of bases in the other strand.

5′ A–A–G–C–T–A–G–C–T–T–A–C–T 3′

EXAMPLE 22.1

Predicting Base Sequence in a Complementary DNA Strand

Predict the sequence of bases in the DNA strand that is complementary to the single DNA strand shown.

$$5'\ C–G–A–A–T–C–C–T–A\ 3'$$

Solution

Because only A forms a complementary base pair with T, and only G with C, the complementary strand is as follows:

Given: $5'\ C–G–A–A–T–C–C–T–A\ 3'$

Complementary strand: $3'\ G–C–T–T–A–G–G–A–T\ 5'$

Note the reversal of the numbering of the ends of the complementary strand compared to the given strand. This is due to the antiparallel nature of the two strands in a DNA double helix.

Practice Exercise 22.1

Predict the sequence of bases in the DNA strand complementary to the single DNA strand shown.

$$5'\ A–A–T–G–C–A–G–C–T\ 3'$$

Answer: $3'\ T–T–A–C–G–T–C–G–A\ 5'$

 Hydrogen bonding is responsible for the secondary structure (double helix) of DNA. Hydrogen bonding is also responsible for secondary structure in proteins (Section 20.10).

Hydrogen bonding between base pairs is an important factor in stabilizing the DNA double helix structure. Although hydrogen bonds are relatively weak forces, each DNA molecule has so many base pairs that collectively these hydrogen bonds are a force of significant strength. In addition to hydrogen bonding, base-stacking interactions also contribute to DNA double-helix stabilization.

Base-Stacking Interactions

The bases in a DNA double helix are positioned with the planes of their rings parallel (like a stack of coins). Stacking interactions involving a given base and the parallel bases directly above it and below it also contribute to the stabilization of the DNA double helix. These stacking interactions are as important in their stabilization effects as is the hydrogen bonding associated with base pairing—perhaps even more important. Purine and pyrimidine bases are hydrophobic in nature, so their stacking interactions are those associated with hydrophobic molecules—mainly London forces (Section 7.13). The concept of hydrophobic interactions has been encountered twice previously. Hydrophobic interactions involving the nonpolar tails of membrane lipids contribute to the structural stability of cell membranes (Section 19.10), and hydrophobic interactions involving nonpolar R groups of amino acids contribute to protein tertiary structure stability (Section 20.11).

Use of the Term "DNA Molecule"

The term *DNA molecule* is actually a misnomer, even though general usage of the term is common in news reports, in textbooks, and even in the vocabulary of scientists. It is technically a misnomer for two reasons.

1. Cellular solutions have pH values such that the phosphate groups present in the DNA backbone structure are negatively charged. This means DNA is actually a multicharged ionic species rather than a neutral molecule.
2. The two strands of DNA in a double-helix structure are not held together by covalent bonds but rather by hydrogen bonds, which are noncovalent interactions. Thus, double-helix DNA is an entity that involves two intertwined ionic species rather than a single molecule.

Despite these considerations, usage of the term *DNA molecule* is accepted by most scientists and it is used through the remainder of this textbook.

22.5 REPLICATION OF DNA MOLECULES

DNA molecules are the carriers of genetic information within a cell; that is, they are the molecules of heredity. Each time a cell divides, an exact copy of the DNA of the parent cell is needed for the new daughter cell. The process by which new DNA molecules are generated is DNA replication. **DNA replication** *is the biochemical process by which DNA molecules produce exact duplicates of themselves.* The key concept in understanding DNA replication is the base pairing associated with the DNA double helix.

DNA Replication Overview

To understand DNA replication, we must regard the two strands of the DNA double helix as a pair of *templates,* or patterns. During replication, the strands separate. Each can then act as a template for the synthesis of a new, complementary strand. The result is two daughter DNA molecules with base sequences identical to those of the parent double helix. Let us consider details of this replication.

Under the influence of the enzyme *DNA helicase,* the DNA double helix unwinds, and the hydrogen bonds between complementary bases are broken. This unwinding process, as shown in Figure 22.9, is somewhat like opening a zipper.

The bases of the separated strands are no longer connected by hydrogen bonds. They can pair with *free* individual nucleotides present in the cell's nucleus. As shown in Figure 22.9, the base pairing always involves C pairing with G and A pairing with T. The pairing process occurs one nucleotide at a time. After a free nucleotide has formed hydrogen bonds with a base of the old strand (the template), the enzyme *DNA polymerase* verifies that the base pairing is correct and then catalyzes the formation of a new phosphodiester linkage between the nucleotide and the growing strand (represented by the darker blue ribbons in Figure 22.9). The *DNA polymerase* then slides down the strand to the next unpaired base of the template, and the same process is repeated.

Each of the two daughter molecules of double-stranded DNA formed in the DNA replication process contains one strand from the original parent molecule and one newly formed strand.

The Replication Process in Finer Detail

Though simple in principle, the DNA replication process has many intricacies.

1. The enzyme *DNA polymerase* can operate on a forming DNA daughter strand only in the 5′-to-3′ direction. Because the two strands of parent DNA run in opposite directions

Figure 22.9 In DNA replication, the two strands of the DNA double helix unwind, the separated strands serving as templates for the formation of new DNA strands. Free nucleotides pair with the complementary bases on the separated strands of DNA. This process ultimately results in the complete replication of the DNA molecule.

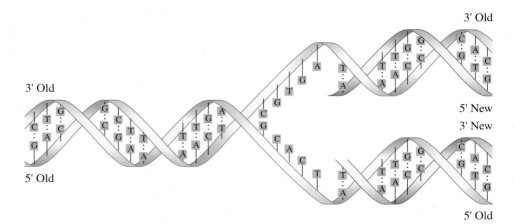

Figure 22.10 Because the enzyme *DNA polymerase* can act only in the 5′-to-3′ direction, one strand (top) grows continuously in the direction of the unwinding, and the other strand grows in segments in the opposite direction. The segments in this latter chain are then connected by a different enzyme, *DNA ligase.*

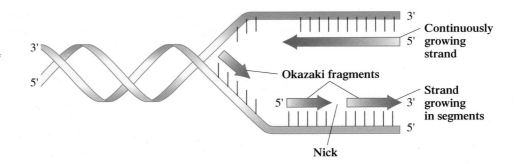

(one is 5′ to 3′ and the other 3′ to 5′; Section 22.4), only one strand can grow continuously in the 5′-to-3′ direction. The other strand must be formed in short segments, called *Okazaki fragments* (after their discoverer, Reiji Okazaki), as the DNA unwinds (see Figure 22.10). The breaks or gaps in this daughter strand are called *nicks.* To complete the formation of this strand, the Okazaki fragments are connected by action of the enzyme *DNA ligase.*

2. The process of DNA unwinding does not have to begin at an end of the DNA molecule. It may occur at any location within the molecule. Indeed, studies show that unwinding usually occurs at several interior locations simultaneously and that DNA replication is bidirectional for these locations; that is, it proceeds in both directions from the unwinding sites. As shown in Figure 22.11, the result of this multiple-site replication process is formation of "bubbles" of newly synthesized DNA. The bubbles grow larger and eventually coalesce, giving rise to two complete daughter DNAs. Multiple-site replication enables large DNA molecules to be replicated rapidly.

Chromosomes

Once the DNA within a cell has been replicated, it interacts with specific proteins in the cell called *histones* to form structural units that provide the most stable arrangement for the long DNA molecules. These histone–DNA complexes are called *chromosomes.* A **chromosome** *is an individual DNA molecule bound to a group of proteins.* Typically, a chromosome is about 15% by mass DNA and 85% by mass protein.

Cells from different kinds of organisms have different numbers of chromosomes. A normal human has 46 chromosomes per cell, a mosquito 6, a frog 26, a dog 78, and a turkey 82.

Chromosomes occur in matched (*homologous*) pairs. The 46 chromosomes of a human cell constitute 23 homologous pairs. One member of each homologous pair is

Figure 22.11 DNA replication usually occurs at multiple sites within a molecule, and the replication is bidirectional from these sites.

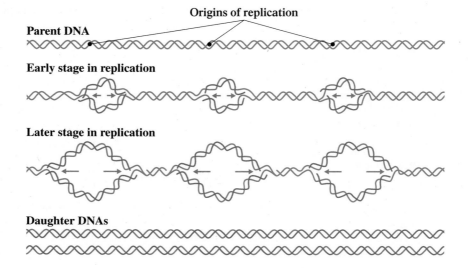

Figure 22.12 Identical twins share identical physical characteristics because they received identical DNA from their parents.

Chromosomes are *nucleoproteins.* They are a combination of nucleic acid (DNA) and various proteins.

The bases thymine (T) and uracil (U) have similar structures. Thymine is a methyluracil (Section 22.2). The hydrogen-bonding patterns (Figure 22.7) for the A–U base pair (RNA) and the A–T base pair (DNA) are identical.

Heterogeneous nuclear RNA (hnRNA) also goes by the name *primary transcript RNA* (ptRNA).

derived from a chromosome inherited from the father, and the other is a copy of one of the chromosomes inherited from the mother. Homologous chromosomes have similar, but not identical, DNA base sequences; both code for the same traits but for different forms of the trait (for example, blue eyes versus brown eyes). Offspring are like their parents, but they are different as well; part of their DNA came from one parent and part from the other parent. Occasionally, identical twins are born (see Figure 22.12). Such twins have received identical DNA from their parents.

The Chemistry at a Glance feature on page 748 summarizes the steps in DNA replication.

22.6 OVERVIEW OF PROTEIN SYNTHESIS

We saw in the previous section how the replication of DNA makes it possible for a new cell to contain the same genetic information as its parent cell. We will now consider how the genetic information contained in a cell is expressed in cell operation. This brings us to the topic of protein synthesis. The synthesis of proteins (skin, hair, enzymes, hormones, and so on) is under the direction of DNA molecules. It is this role of DNA that establishes the similarities between parent and offspring that we regard as hereditary characteristics.

We can divide the overall process of protein synthesis into two phases. The first phase is called *transcription* and the second *translation.* The following diagram summarizes the relationship between transcription and translation.

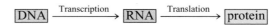

$$\boxed{\text{DNA}} \xrightarrow{\text{Transcription}} \boxed{\text{RNA}} \xrightarrow{\text{Translation}} \boxed{\text{protein}}$$

Before discussing the details of transcription and translation, we need to learn more about RNA molecules. They are involved in both transcription, as the end products, and translation, as the starting materials. We will be particularly concerned with differences between RNA and DNA and among various types of RNA molecules.

22.7 RIBONUCLEIC ACIDS

Four major differences exist between RNA molecules and DNA molecules.

1. The sugar unit in the backbone of RNA is ribose; it is deoxyribose in DNA.
2. The base thymine found in DNA is replaced by uracil in RNA (Figure 22.2). In RNA, uracil, instead of thymine, pairs with (forms hydrogen bonds with) adenine.
3. RNA is a single-stranded molecule; DNA is double-stranded (double helix). Thus RNA, unlike DNA, does not contain equal amounts of specific bases.
4. RNA molecules are much smaller than DNA molecules, ranging from 75 nucleotides to a few thousand nucleotides.

We should note that the single-stranded nature of RNA does not prevent *portions* of an RNA molecule from folding back upon itself and forming double-helical regions. If the base sequences along two portions of an RNA strand are complementary, a structure with a hairpin loop results, as shown in Figure 22.13. The amount of double-helical structure present in an RNA varies with RNA type, but a value of 50% is not atypical.

Types of RNA Molecules

RNA molecules found in human cells are categorized into five major types, distinguished by their function. These five RNA types are heterogeneous nuclear RNA (hnRNA), messenger RNA (mRNA), small nuclear RNA (snRNA), ribosomal RNA (rRNA), and transfer RNA (tRNA).

CHEMISTRY AT A GLANCE

DNA Replication

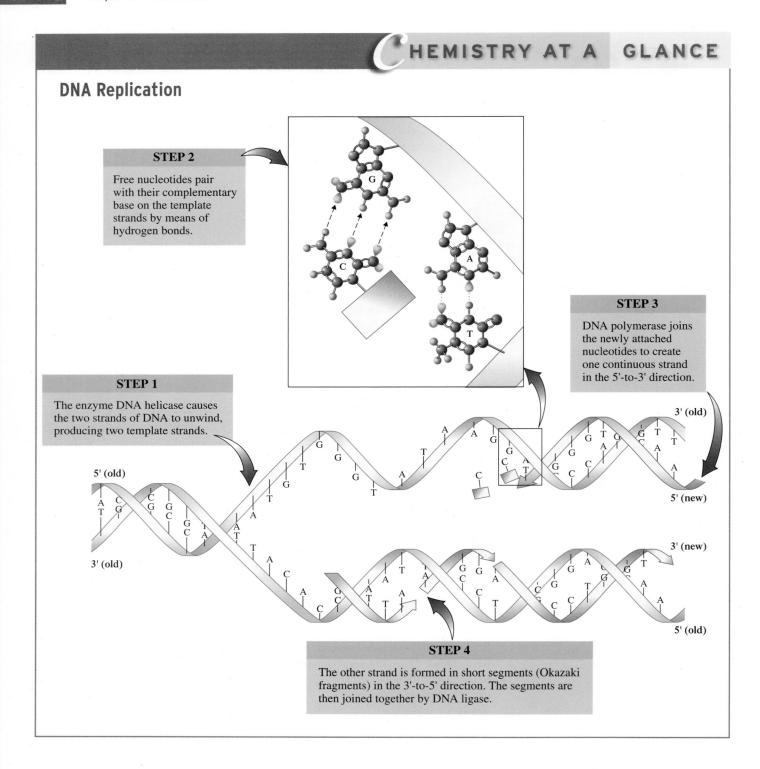

STEP 2

Free nucleotides pair with their complementary base on the template strands by means of hydrogen bonds.

STEP 1

The enzyme DNA helicase causes the two strands of DNA to unwind, producing two template strands.

STEP 3

DNA polymerase joins the newly attached nucleotides to create one continuous strand in the 5'-to-3' direction.

STEP 4

The other strand is formed in short segments (Okazaki fragments) in the 3'-to-5' direction. The segments are then joined together by DNA ligase.

The most abundant type of RNA in a cell is ribosomal RNA (75% to 80% by mass). Transfer RNA constitutes 10%–15% of cellular RNA; messenger RNA and its precursor, heterogeneous nuclear RNA, make up 5%–10% of RNA material in the cell.

Heterogeneous nuclear RNA (hnRNA) *is RNA formed directly by DNA transcription.* Post-transcription processing converts the heterogeneous nuclear RNA to messenger RNA.

Messenger RNA (mRNA) *is RNA that carries instructions for protein synthesis (genetic information) to the sites for protein synthesis.* The molecular mass of messenger RNA varies with the length of the protein whose synthesis it will direct.

Small nuclear RNA (snRNA) *is RNA that facilitates the conversion of heterogeneous nuclear RNA to messenger RNA.* It contains from 100 to 200 nucleotides.

Figure 22.13 A hairpin loop is produced when single-stranded RNA doubles back on itself and complementary base pairing occurs.

Hairpin loop

Hydrogen bonds

Ribosomal RNA (rRNA) *is RNA that combines with specific proteins to form ribosomes, the physical sites for protein synthesis.* Ribosomes have molecular masses on the order of 3 million amu. The rRNA present in ribosomes has no informational function.

Transfer RNA (tRNA) *is RNA that delivers amino acids to the sites for protein synthesis.* Transfer RNAs are the smallest of the RNAs, possessing only 75–90 nucleotide units.

At a nondetail level, a cell consists of a nucleus and an extranuclear region called the cytoplasm. The process of DNA transcription occurs in the nucleus, as does the processing of hnRNA to mRNA. [DNA replication (Section 22.5) also occurs in the nucleus.] The mRNA formed in the nucleus travels to the cytoplasm where translation (protein synthesis) occurs. Figure 22.14 summarizes the transcription and translation processes in terms of the types of RNA involved and the cellular locations where the processes occur.

An additional role for RNA, besides its major involvement in protein synthesis, has been recently discovered by scientists. It also plays a part in the process of blood coagulation near a wound. RNA that is released from damaged cells associated with the wound helps to activate two enzymes needed for the blood coagulation process.

A detailed look at cellular structure is found in Section 23.2.

22.8 TRANSCRIPTION: RNA SYNTHESIS

Transcription *is the process by which DNA directs the synthesis of hnRNA/mRNA molecules that carry the coded information needed for protein synthesis.* Messenger RNA production via transcription is actually a "two-step" process in which an hnRNA molecule is initially produced and then is "edited" to yield the desired mRNA molecule. The mRNA molecule so produced then functions as the carrier of the information needed to direct protein synthesis.

Within a strand of a DNA molecule are instructions for the synthesis of numerous hnRNA/mRNA molecules. During transcription, a DNA molecule unwinds, under enzyme influence, at the particular location where the appropriate base sequence is found for the hnRNA/mRNA of concern, and the "exposed" base sequence is transcribed. A short segment of a DNA strand so transcribed, which contains instructions for the formation of a particular hnRNA/mRNA, is called a *gene*. A **gene** *is a segment of a DNA strand that contains the base sequence for the production of a specific hnRNA/mRNA molecule.*

In humans, most genes are composed of 1000–3500 nucleotide units. Hundreds of genes can exist along a DNA strand. Obtaining information concerning the total number of genes and the total number of nucleotide base pairs present in human DNA has been an area of intense research activity for the last two decades. The central activity in this research has been the *Human Genome Project,* a decade-long internationally based research project to determine the location and base sequence of each of the genes in the human *genome.* A **genome** *is all of the genetic material (the total DNA) contained in the chromosomes of an organism.*

Figure 22.14 An overview of types of RNA in terms of cellular locations where they are encountered and processes in which they are involved.

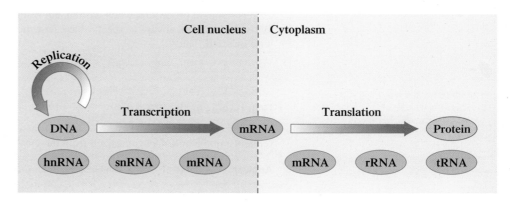

Before the Human Genome Project began, current biochemical thought predicted the presence of about 100,000 genes in the human genome. Initial results of the human genome project, announced in 2001, pared this number down to 30,000–40,000 genes and also indicated that the base pairs present in these genes constitute only a very small percentage (2%) of the 2.9 billion base pairs present in the chromosomes of the human genome. In 2004, based on reanalysis of human genome project information, the human gene count was pared down further to 20,000–25,000 genes. (Later in this section, the significance and ramifications of this dramatic decrease in estimates of the human gene count are considered.)

Steps in the Transcription Process

The mechanics of transcription are in many ways similar to those of DNA replication. Four steps are involved.

1. A *portion* of the DNA double helix unwinds, exposing some bases (a gene). The unwinding process is governed by the enzyme *RNA polymerase* rather than by *DNA helicase* (replication enzyme).
2. Free *ribo*nucleotides, one nucleotide at a time, align along *one* of the exposed strands of DNA bases, the *template* strand, forming new base pairs. In this process, U rather than T aligns with A in the base-pairing process. Because ribonucleotides rather than deoxyribonucleotides are involved in the base pairing, ribose, rather than deoxyribose, becomes incorporated into the new nucleic acid backbone.
3. *RNA polymerase* is involved in the linkage of ribonucleotides, one by one, to the growing hnRNA molecule.
4. Transcription ends when the *RNA polymerase* enzyme encounters a sequence of bases that is "read" as a stop signal. The newly formed hnRNA molecule and the *RNA polymerase* enzyme are released, and the DNA then rewinds to re-form the original double helix.

Figure 22.15 shows the overall process of transcription of DNA to form hnRNA.

In DNA–RNA base pairing, the complementary base pairs are

DNA RNA

A — U

G — C

C — G

T — A

RNA molecules contain the base U instead of the base T.

● EXAMPLE 22.2

Base Pairing Associated with the Transcription Process

▶ From the base sequence 5′ A–T–G–C–C–A 3′ in a DNA template strand, determine the base sequence in the hnRNA synthesized from the DNA template strand.

Solution

An RNA molecule cannot contain the base T. The base U is present instead. Therefore, U–A base pairing will occur instead of T–A base pairing. The other base-pairing combination, G–C, remains the same. The hnRNA product of the transcription process will therefore be

DNA template: 5′ A–T–G–C–C–A 3′

⋮ ⋮ ⋮ ⋮ ⋮ ⋮

hnRNA molecule: 3′ U–A–C–G–G–U 5′

Note that the direction of the hnRNA strand is antiparallel to that of the DNA template. This will always be the case during transcription.

It is standard procedure, when writing and reading base sequences for nucleic acids (both DNAs and RNAs), always to specify base sequence in the 5′ ⟶ 3′ direction unless otherwise directed. Thus

3′ U–A–C–G–G–U 5′   becomes   5′ U–G–G–C–A–U 3′

Practice Exercise 22.2

From the base sequence 5′ T–A–A–C–C–T 3′ in a DNA template strand, determine the base sequence in the hnRNA synthesized from the DNA template strand.

Answer: 3′ A–U–U–G–G–A 5′ which becomes 5′ A–G–G–U–U–A 3′

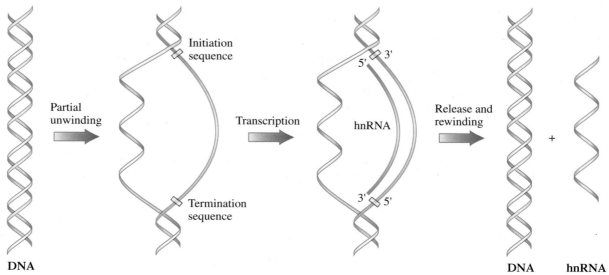

Figure 22.15 The transcription of DNA to form hnRNA involves an unwinding of a portion of the DNA double helix. Only one strand of the DNA is copied during transcription.

Post-Transcription Processing: Formation of mRNA

The RNA produced from a gene through transcription is hnRNA, the precursor for mRNA. The conversion of hnRNA to mRNA involves *post-transcription processing* of the hnRNA. In this processing, certain portions of the hnRNA are deleted and the retained parts are then spliced together. This process leads us to the concepts of *exons* and *introns*.

It is now known that not all bases in a gene convey genetic information. Instead, a gene is *segmented;* it has portions called *exons* that contain genetic information and portions called *introns* that do not convey genetic information.

An **exon** *is a gene segment that conveys (codes for) genetic information. Ex*ons are DNA segments that help *ex*press a genetic message. An **intron** *is a gene segment that does not convey (code for) genetic information. In*trons are DNA segments that *in*terrupt a genetic message. A gene consists of alternating exon and intron segments (Figure 22.16).

Both the exons and the introns of a gene are transcribed during production of hnRNA. The hnRNA is then "edited," under enzyme direction, to remove the introns, and the remaining exons are joined together to form a shortened RNA strand that carries the genetic information of the transcribed gene. The removal of the introns and joining together of the exons takes place simultaneously in a single process. The "edited" RNA so produced is the messenger RNA (mRNA) that serves as a blueprint for protein assembly. Much is yet to be learned about introns and why they are present in genes; investigating their function is an active area of biochemical research.

Splicing *is the process of removing introns from an hnRNA molecule and joining the remaining exons together to form an mRNA molecule.* The splicing process involves snRNA

Figure 22.16 Heterogeneous nuclear RNA contains both exons and introns. Messenger RNA is heterogeneous nuclear RNA from which the introns have been excised.

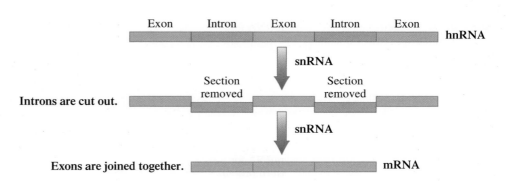

molecules, the most recent of the RNA types to be discovered. This type of RNA is never found "free" in a cell. An snRNA molecule is always found complexed with proteins in particles called *small nuclear ribonucleoprotein particles,* which are usually called *snRNPs* (pronounced "snurps"). A **small nuclear ribonucleoprotein particle** *is a complex formed from an snRNA molecule and several proteins.* "Snurps" always further collect together into larger complexes called *spliceosomes.* A **spliceosome** *is a large assembly of snRNA molecules and proteins involved in the conversion of hnRNA molecules to mRNA molecules.*

Alternative Splicing

Prior to the announcement of the Human Genome Project's results, biochemistry had largely embraced the "one-protein-one-gene" concept. It was generally assumed that each type of protein had "its own" gene that carried the instructions for its synthesis. This is no longer plausible because the estimated number of different proteins present in the human body now significantly exceeds the estimated number of genes.

The concept of *alternative splicing* bridges the gap between the larger estimated number of proteins and the now-lower estimated number of genes. **Alternative splicing** *is a process by which several different proteins that are variations of a basic structural motif can be produced from a single gene.* In alternative splicing, an hnRNA molecule with multiple exons present is spliced in several different ways. Figure 22.17 shows the four alternative splicing patterns that can occur when an hnRNA contains four exons, two of which are *alternative exons.*

The Human Transcriptome

As biochemists were mapping the human genome, they anticipated that they were close to unlocking many secrets of the human body and that the results of the project would provide a list of human genes for which the function of each could be quickly investigated. The results of the project, however, only complicated the situation.

Results indicate that the total number of genes present in a genome is not as important in understanding human cell behavior as was previously thought, while mRNA transcripts obtained from the genes are *more* important in understanding human cell behavior than was previously thought. Because of alternative splicing, different cell types interpret the information encoded on a DNA molecule differently and so produce a different number of mRNA molecules and ultimately a different number of proteins. For each cell type, the number of mRNA transcripts generated varies in response to complex signals within a cell and between cells.

Research now shows that the information-bearing sections of DNA within a gene can be spliced together an average of eight different ways. There could turn out to be around 200,000 relevant mRNA molecules as compared to 20,000–25,000 genes within the human genome. Collectively, the total number of mRNA molecules for an organism is known as its *transcriptome.* A **transcriptome** *is all of the mRNA molecules that can be generated from the genetic material in a genome.* A transcriptome differs from a genome in that it acknowledges the biochemical complexity created by splice variants obtained from hnRNA. Transcriptome research is now a developing biochemical frontier.

Completion of the Human Genome Project is not the end to high-profile cooperative international study of the human genome. A new multiyear, perhaps multidecade, project called ENCODE (encyclopedia of DNA elements) is in its initial stages. The focus of this new project is identification of functional elements within the genome. Functional elements not only include base sequences that code for proteins, but also regulatory sequences that control DNA transcription, and sequences that control the packaging of the genome.

Figure 22.17 An hnRNA molecule containing four exons, two of which (B and C) are alternative exons, can be spliced in four different ways, producing four different proteins. Proteins can be produced with neither, either, or both of the alternative exons present.

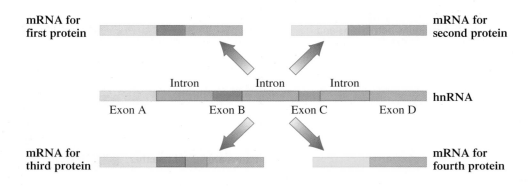

mRNA for first protein — mRNA for second protein — Intron — Intron — Intron — hnRNA — Exon A — Exon B — Exon C — Exon D — mRNA for third protein — mRNA for fourth protein

There is a rough correlation between the number of codons for a particular amino acid and that amino acid's frequency of occurrence in proteins. For example, the two amino acids that have a single codon, Met and Trp, are two of the least common amino acids in proteins.

22.9 THE GENETIC CODE

The nucleotide (base) sequence of an mRNA molecule is the informational part of such a molecule. This base sequence in a given mRNA determines the amino acid sequence for the protein synthesized under that mRNA's direction.

How can the base sequence of an mRNA molecule (which involves only *4* different bases—A, C, G, and U) encode enough information to direct proper sequencing of *20* amino acids in proteins? If each base encoded for a particular standard amino acid, then only 4 amino acids would be specified out of the 20 needed for protein synthesis, a clearly inadequate number. If two-base sequences were used to code amino acids, then there would be $4^2 = 16$ possible combinations, so 16 amino acids could be represented uniquely. This is still an inadequate number. If three-base sequences were used to code for amino acids, there would be $4^3 = 64$ possible combinations, which is more than enough combinations for uniquely specifying each of the 20 standard amino acids found in proteins.

Research has verified that sequences of three nucleotides in mRNA molecules specify the amino acids that go into synthesis of a protein. Such three-nucleotide sequences are called codons. A **codon** *is a three-nucleotide sequence in an mRNA molecule that codes for a specific amino acid.*

Which amino acid is specified by which codon? (We have 64 codons to choose from.) Researchers deciphered codon–amino acid relationships by adding different *synthetic* mRNA molecules (whose base sequences were known) to cell extracts and then determining the structure of any newly formed protein. After many such experiments, researchers finally matched all 64 possible codons with their functions in protein synthesis. It was found that 61 of the 64 codons formed by various combinations of the bases A, C, G, and U were related to specific amino acids; the other 3 combinations were termination codons ("stop" signals) for protein synthesis. Collectively, these relationships between three-nucleotide sequences in mRNA and amino acid identities are known as the genetic code. The **genetic code** *is the assignment of the 64 mRNA codons to specific amino acids (or stop signals).* The determination of this code during the early 1960s is one of the most remarkable of twentieth-century scientific achievements. The 1968 Nobel Prize in chemistry was awarded to Marshall Nirenberg and Har Gobind Khorana for their work in illuminating how mRNA encodes for proteins.

The complete genetic code is given in Table 22.2. Examination of this table indicates that the genetic code has several remarkable features.

1. *The genetic code is highly degenerate; that is, many amino acids are designated by more than one codon.* Three amino acids (Arg, Leu, and Ser) are represented by six codons. Two or more codons exist for all other amino acids except Met and Trp, which have only a single codon. Codons that specify the same amino acid are called *synonyms.*

2. *There is a pattern to the arrangement of synonyms in the genetic code table.* All synonyms for an amino acid fall within a single box in Table 22.2, unless there are more than four synonyms, where two boxes are needed. The significance of the "single box" pattern is that with synonyms, the first two bases of the codon are the same—they differ only in the third base. For example, the four synonyms for the amino acid proline (Pro) are CCU, CCC, CCA, and CCG.

3. *The genetic code is almost universal.* Although Table 22.2 does not show this feature, studies of many organisms indicate that with minor exceptions, the code is the same in all of them. The same codon specifies the same amino acid whether the cell is a bacterial cell, a corn plant cell, or a human cell.

4. *An initiation codon exists.* The existence of "stop" codons (UAG, UAA, and UGA) suggests the existence of "start" codons. There is one initiation codon. Besides coding for the amino acid methionine, the codon AUG functions as an initiator of protein synthesis when it occurs as the first codon in an amino acid sequence.

TABLE 22.2

The Universal Genetic Code

The code is composed of 64 three-nucleotide sequences (codons), which can be read from the table. The left-hand column indicates the nucleotide base found in the first (5′) position of the codon. The nucleotide in the second (middle) position of the codon is given by the base listing at the top of the table. The right-hand column indicates the nucleotide found in the third (3′) position. Thus the codon ACG encodes for the amino acid Thr, and the codon GGG encodes for the amino acid Gly.

Second letter

		U		C		A		G	
U	UUU UUC	Phenylalanine Phe	UCU UCC	Serine	UAU UAC	Tyrosine Tyr	UGU UGC	Cysteine Cys	U C
	UUA UUG	Leucine Leu	UCA UCG	Ser	UAA UAG	Stop codon Stop codon	UGA UGG	Stop codon Tryptophan Trp	A G
C	CUU CUC	Leucine	CCU CCC	Proline	CAU CAC	Histidine His	CGU CGC	Arginine	U C
	CUA CUG	Leu	CCA CCG	Pro	CAA CAG	Glutamine Gln	CGA CGG	Arg	A G
A	AUU AUC	Isoleucine Ile	ACU ACC	Threonine	AAU AAC	Asparagine Asn	AGU AGC	Serine Ser	U C
	AUA AUG	Methionine Met Initiation codon	ACA ACG	Thr	AAA AAG	Lysine Lys	AGA AGG	Arginine Arg	A G
G	GUU GUC	Valine	GCU GCC	Alanine	GAU GAC	Aspartic acid Asp	GGU GGC	Glycine	U C
	GUA GUG	Val	GCA GCG	Ala	GAA GAG	Glutamic acid Glu	GGA GGG	Gly	A G

First letter (5′ end)

Third letter (3′ end)

EXAMPLE 22.3

Using the Genetic Code and mRNA Codons to Predict Amino Acid Sequences

Using the genetic code in Table 22.2, determine the sequence of amino acids encoded by the mRNA codon sequence

5′ GCC–AUG–GUA–AAA–UGC–GAC–CCA 3′

Solution

Matching the codons with the amino acids, using Table 22.2, yields

mRNA: 5′ GCC–AUG–GUA–AAA–UGC–GAC–CCA 3′

Peptide: Ala   Met   Val   Lys   Cys   Asp   Pro

Practice Exercise 22.3

Using the genetic code in Table 22.2, determine the sequence of amino acids encoded by the mRNA codon sequence

5′ CAU–CCU–CAC–ACU–GUU–UGU–UGG 3′

Answer: His–Pro–His–Thr–Val–Cys–Trp

● EXAMPLE 22.4

Relating Exons and Introns to hnRNA and mRNA Structures

Introns and exons are actually never as short as those given in this simplified example.

Sections A, C, and E of the following base sequence section of a DNA template strand are exons, and sections B and D are introns.

DNA   5′ ATT – CGT – TGT – TTT – CCC – AGT – GCC 3′
         A   B   C   D   E

a. What is the structure of the hnRNA transcribed from this template?
b. What is the structure of the mRNA obtained by splicing the hnRNA?

Solution

a. The base sequence in the hnRNA will be complementary to that of the template DNA, except that U is used in the RNA instead of T. The hnRNA will have a directionality antiparallel to that of the DNA sequence.

hnRNA   3′ UAA – GCA – ACA – AAA – GGG – UCA – CGG 5′

b. In the splicing process, introns are removed and the exons combined to give the mRNA.

mRNA   3′ UAA – ACA – AAA – CGG 5′

Practice Exercise 22.4

Sections A, C, and E of the following base sequence section of a DNA template strand are exons, and sections B and D are introns.

DNA   5′ CGC – CGT – AGT – TGG – CCC – GGA – GGA 3′
         A   B   C   D   E

a. What is the structure of the hnRNA transcribed from this template?
b. What is the structure of the mRNA obtained by splicing the hnRNA?

Answers: a. 3′ GCG–GCA–UCA–ACC–GGG–CCU–CCU 5′;   **b.** 3′ GCG–ACC–CCU–CCU 5′

● EXAMPLE 22.5

Relating Protein Amino Acid Sequence to the Directionality of an mRNA Segment

The structure of an mRNA segment obtained from a DNA template strand is

mRNA   3′ AUU – CCG – UAC – GAC 5′

What polypeptide amino acid sequence will be synthesized using this mRNA?

Solution

The directionality of an mRNA segment obtained from template DNA is 3′-to-5′ because the two segments must be antiparallel to each other (Section 22.5). The codons in an mRNA must be read in the 5′-to-3′ direction to correctly use genetic code relationships to determine the sequence of amino acids in the peptide. Rewriting the given mRNA with reversed directionality (5′-to-3′ direction) gives

mRNA   5′ CAG – CAU – GCC – UUA 3′

Note that in reversing the directionality from 3′-to-5′ to 5′-to-3′ the sequence of bases in a codon is also reversed; for example, GAC becomes CAG.

Using the genetic code relationships between codon and amino acid (Table 22.2) shows that this mRNA codon sequence codes for the amino acid sequence

Gln–His–Ala–Leu

(continued)

Practice Exercise 22.5

The structure of an mRNA segment obtained from a DNA template strand is

mRNA   3′ ACG – AGC – CCU – CUU 5′

What polypeptide amino acid sequence will be synthesized using this mRNA?

Answer: Phe–Ser–Arg–Ala

22.10 ANTICODONS AND tRNA MOLECULES

The amino acids used in protein synthesis do not directly interact with the codons of an mRNA molecule. Instead, tRNA molecules function as intermediaries that deliver amino acids to the mRNA. At least one type of tRNA molecule exists for each of the 20 amino acids found in proteins.

All tRNA molecules have the same general shape, and this shape is crucial to how they function. Figure 22.18a shows the general *two-dimensional* "cloverleaf" shape of a tRNA molecule, a shape produced by the molecule's folding and twisting into regions of parallel strands and regions of hairpin loops. (The actual three-dimensional shape of a tRNA molecule involves considerable additional twisting of the "cloverleaf" shape— Figure 22.18b.)

Two features of the tRNA structure are of particular importance.

1. The 3′ end of the open part of the cloverleaf structure is where an amino acid becomes *covalently* bonded to the tRNA molecule through an ester bond. Each of the different tRNA molecules is specifically recognized by an *aminoacyl tRNA synthetase* enzyme. These enzymes also recognize the one kind of amino acid that "belongs" with the particular tRNA and facilitates its bonding to the tRNA (see Figure 22.19).
2. The loop *opposite* the open end of the cloverleaf is the site for a sequence of three bases called an anticodon. An **anticodon** *is a three-nucleotide sequence on a tRNA molecule that is complementary to a codon on an mRNA molecule.*

Figure 22.18 A tRNA molecule. The amino acid attachment site is at the open end of the cloverleaf (the 3′ end), and the anticodon is located in the hairpin loop opposite the open end.

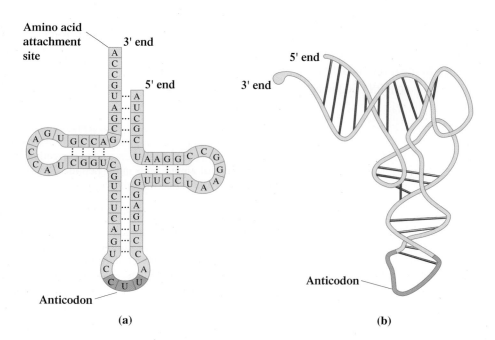

(a)   (b)

Figure 22.19 An aminoacyl-tRNA synthetase has an active site for tRNA and a binding site for the particular amino acid that is to be attached to that tRNA.

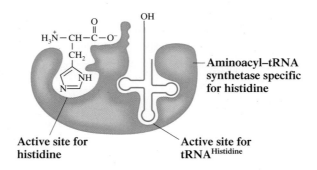

Aminoacyl–tRNA synthetase specific for histidine

Active site for histidine

Active site for tRNA^{Histidine}

The interaction between the anticodon of the tRNA and the codon of the mRNA leads to the proper placement of an amino acid into a growing peptide chain during protein synthesis. This interaction, which involves complementary base pairing, is shown in Figure 22.20.

● EXAMPLE 22.6

Determining Codon–Anticodon Relationships

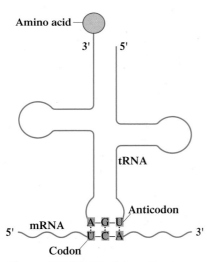

Figure 22.20 The interaction between anticodon (tRNA) and codon (mRNA), which involves complementary base pairing, governs the proper placement of amino acids in a protein.

A tRNA molecule has the anticodon 5′ AAG 3′. With which mRNA codon will this anticodon interact?

Solution

The interaction between a codon and an anticodon has directionality (antiparallel) considerations, that is,

	3′	Anticodon	5′
		⋮ ⋮ ⋮	
	5′	Codon	3′

The given anticodon, with reversed directionality, is

3′ GAA 5′

and its codon interaction, which involves complementary base pairing, is

Anticodon	3′	GAA	5′
		⋮ ⋮ ⋮	
Codon	5′	CUU	3′

The codon is, thus, 5′ CUU 3′.

Practice Exercise 22.6

A tRNA molecule has the anticodon 5′ ACG 3′. With which mRNA codon will this anticodon interact?

Answer: 5′ CGU 3′

● EXAMPLE 22.7

Determining Anticodon–tRNA Amino Acid Relationships

A tRNA molecule possesses the anticodon 5′ CGU 3′. Which amino acid will this tRNA molecule carry?

Solution

Codons, rather than anticodons, are involved in the genetic code relationships. The codon with which this anticodon 5′ CGU 3′ base pairs is 5′ ACG 3′. Remember, as shown in

(continued)

Example 22.6, that the codon–anticodon pairing involves the anticodon 3′ UGC 5′ (the anticodon written in the 3′-to-5′ direction).

Anticodon	3′	UGC	5′
		⋮ ⋮ ⋮	
Codon	5′	ACG	3′

The amino acid that the codon ACG codes for (see Table 22.2) is Thr (Threonine).

Practice Exercise 22.7

A tRNA molecule possesses the anticodon 5′ UGA 3′. Which amino acid will this tRNA molecule carry?

Answer: Ser (Serine)

22.11 TRANSLATION: PROTEIN SYNTHESIS

Translation *is the process by which mRNA codons are deciphered and a particular protein molecule is synthesized.* The substances needed for the translation phase of protein synthesis are mRNA molecules, tRNA molecules, amino acids, ribosomes, and a number of different enzymes. A **ribosome** *is an rRNA–protein complex that serves as the site for the translation phase of protein synthesis.*

The number of ribosomes present in a cell for higher organisms varies from hundreds of thousands to even a few million. Recent research concerning ribosome structure suggests the following for such structures:

1. They contain four rRNA molecules and about 80 proteins that are packed into two rRNA-protein subunits, one small subunit and one large subunit (see Figure 22.21).
2. Each subunit contains approximately 65% rRNA and 35% protein by mass.
3. A ribosome's active site, the location where proteins are synthesized by one-at-a-time addition of amino acids to a growing peptide chain, is located in the large ribosomal subunit.
4. The active site is mostly rRNA, with only one of the ribosome's many protein components being present.
5. Because rRNA is so predominant at the active site, the ribosome is thought to be a RNA enzyme (Section 21.1), that is, a *ribozyme.*
6. The mRNA involved in the translation phase of protein synthesis binds to the small subunit of the ribosome.

There are five general steps to the translation process: (1) activation of tRNA, (2) initiation, (3) elongation, (4) termination, and (5) post-translational processing.

Figure 22.21 Ribosomes, which contain both rRNA and protein, have structures that contain two subunits. One subunit is much larger than the other.

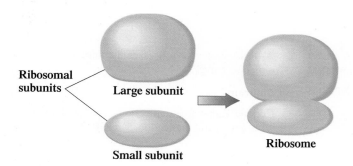

Ribosomal subunits

Large subunit

Small subunit

Ribosome

Activation of tRNA

There are two steps involved in tRNA activation. First, an amino acid interacts with an activator molecule (ATP; Section 23.3) to form a highly energetic complex. This complex then reacts with the appropriate tRNA molecule to produce an *activated tRNA molecule,* a tRNA molecule that has an amino acid covalently bonded to it at its 3′ end through an ester linkage.

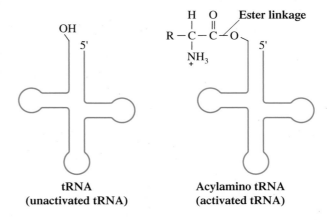

tRNA
(unactivated tRNA)

Acylamino tRNA
(activated tRNA)

Initiation

The initiation of protein synthesis in human cells begins when mRNA attaches itself to the surface of a small ribosomal subunit such that its first codon, which is always the initiating codon AUG, occupies a site called the P site (peptidyl site). (See Figure 22.22a.) An activated tRNA molecule with anticodon complementary to the codon AUG attaches itself, through complementary base pairing, to the AUG codon (Figure 22.22b). The resulting complex then interacts with a large ribosomal subunit to complete the formation of an initiation complex (Figure 22.22c). (Since the initiating codon AUG codes for the amino acid methionine, the first amino acid in a developing human protein chain will always be methionine.)

Elongation

Next to the P site in an mRNA–ribosome complex is a second binding site called the A site (aminoacyl site). (See Figure 22.23a.) At this second site the next mRNA codon is exposed, and a tRNA with the appropriate anticodon binds to it (Figure 22.23b). With amino acids in place at both the P and the A sites, the enzyme *peptidyl transferase* effects the linking of

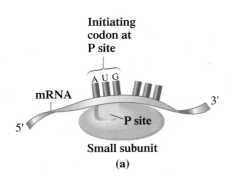

Figure 22.22 Initiation of protein synthesis begins with the formation of an initiation complex.

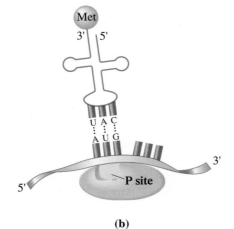

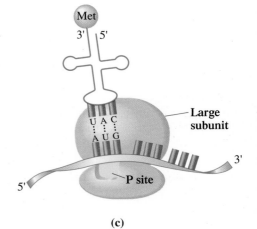

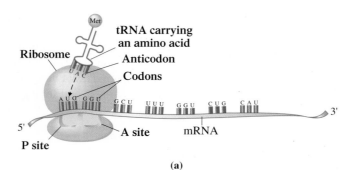

(a)

The initiation tRNA carrying the amino acid Met binds at the P site.

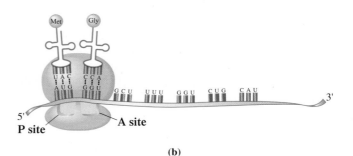

(b)

A tRNA with amino acid 2 binds at the A site.

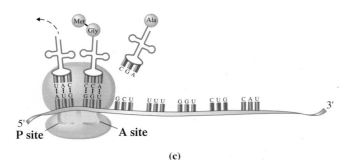

(c)

A peptide bond forms between amino acid 1 and amino acid 2 as amino acid 1 moves from the P site to the A site.

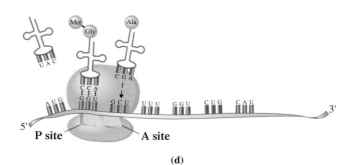

(d)

The first tRNA is released, the ribosome moves one codon to the right, translocating the dipeptide to the P site, and the tRNA with amino acid 3 occupies the A site.

Figure 22.23 The process of translation that occurs during protein synthesis. The anticodons of tRNA molecules are paired with the codons of an mRNA molecule to bring the appropriate amino acids into sequence for protein formation.

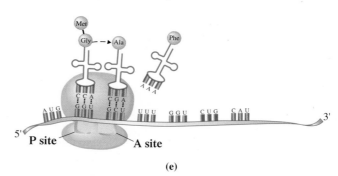

(e)

Elongation continues as the dipeptide at the P site is bonded to the amino acid at the A site to form a tripeptide.

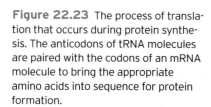

In elongation, the polypeptide chain grows one amino acid at a time.

The process of *translocation* occurs at approximately 50-millisecond intervals, that is, 20 times per second.

the P site amino acid to the A site amino acid to form a dipeptide. Such peptide bond formation leaves the tRNA at the P site empty and the tRNA at the A site bearing the dipeptide (Figure 22.23c).

The empty tRNA at the P site now leaves that site and is free to pick up another molecule of its specific amino acid. Simultaneously with the release of tRNA from the P site, the ribosome shifts along the mRNA. This shift puts the newly formed dipeptide at the P site, and the third codon of mRNA is now available, at site A, to accept a tRNA molecule whose anticodon complements this codon (see Figure 22.23d). The movement of a ribosome along an mRNA molecule is called *translocation*. **Translocation** *is the part of translation in which a ribosome moves down an mRNA molecule three base positions (one codon) so that a new codon can occupy the ribosomal A site.*

CHEMICAL Connections

Antibiotics That Inhibit Bacterial Protein Synthesis

Some antibiotics work because they inhibit protein synthesis in bacteria but not in humans. They inhibit one specific enzyme or another in the bacterial ribosomes. These antibiotics are useful in treating disease and in studying protein synthesis mecha- nisms in bacteria. The accompanying table lists a few of the most commonly encountered antibiotics and their modes of action relative to protein synthesis.

Antibiotic	Biological Action	Antibiotic	Biological Action
chloramphenicol	inhibits an important enzyme (peptidyl transferase) in the large ribosomal subunit	streptomycin	inhibits initiation of protein synthesis and also causes the mRNA codons to be read incorrectly
erythromycin	binds to the large subunit and stops the ribosome from moving along the mRNA from one codon to the next	tetracycline	binds to the small ribosomal subunit and inhibits the binding of incoming tRNA molecules
puromycin	induces premature polypeptide chain termination		

Now a repetitious process begins. The third codon, now at the A site, accepts an incoming tRNA with its accompanying amino acid; and then the entire dipeptide at the P site is transferred and bonded to the A site amino acid to give a tripeptide (see Figure 22.23e). The empty tRNA at the P site is released, the ribosome shifts along the mRNA, and the process continues.

The transfer of the growing peptide chain from the P site to the A site is an exam- ple of an *acyl transfer reaction,* a reaction type first introduced in Section 16.19. Figure 22.24 shows the structural detail for such transfer when Met is the amino acid at the P site and Gly is the amino acid at the A site.

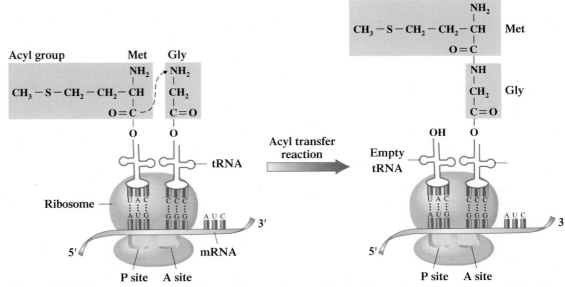

Figure 22.24 The transfer of an amino acid (or growing peptide chain) from the ribosomal P site to the ribo- somal A site during translation is an example of an acyl transfer reaction.

Termination

The polypeptide continues to grow by way of translocation until all necessary amino acids are in place and bonded to each other. Appearance in the mRNA codon sequence of one of the three stop codons (UAA, UAG, or UGA) terminates the process. No tRNA has an anti-codon that can base-pair with these stop codons. The polypeptide is then cleaved from the tRNA through hydrolysis.

Post-Translation Processing

Some modification of proteins usually occurs after translation. This *post-translation processing* gives the protein the final form it needs to be fully functional. Some of the aspects of post-translation processing are the following.

1. In most proteins, the methionine (Met) residue that initiated protein synthesis is removed by a specialized enzyme in a hydrolysis reaction. A second hydrolysis reaction releases the polypeptide chain from its tRNA carrier.
2. Some covalent modification of a protein can occur, such as the formation of disulfide bridges between cysteine residues (Section 20.5).
3. Completion of the folding of polypeptides into their active conformations occurs. Protein folding actually begins as the polypeptide chain is elongated on the ribosome. For protein with quaternary structure (Section 20.12), the various components are assembled together.

Recent research indicates that there may be a connection between synonymous codons within the genetic code (Section 22.9) and protein folding. It now appears that synonymous codons, even though they translate into the same amino acids during protein synthesis, have an effect on the way emerging proteins fold into their three-dimensional shapes (tertiary structure; Section 20.11) as they elongate and then leave a ribosome. This means that two stretches of mRNA that differ only in synonymous codons can produce proteins with identical amino acid sequences but different folding patterns. Two differently folded proteins would be expected to produce different biochemical responses within a cell when interacting with other substances; there is some evidence, now, that this is the case.

Efficiency of mRNA Utilization

Many ribosomes can move simultaneously along a single mRNA molecule (Figure 22.25). In this highly efficient arrangement, many identical protein chains can be synthesized almost at the same time from a single strand of mRNA. This multiple use of mRNA molecules reduces the amount of resources and energy that the cell expends to synthesize needed protein. Such complexes of several ribosomes and mRNA are called polyribosomes or polysomes. A **polyribosome** *is a complex of mRNA and several ribosomes.*

The Chemistry at a Glance feature on page 764 summarizes the steps in protein synthesis.

Figure 22.25 Several ribosomes can simultaneously proceed along a single strand of mRNA one after another. Such a complex of mRNA and ribosomes is called a polyribosome or polysome.

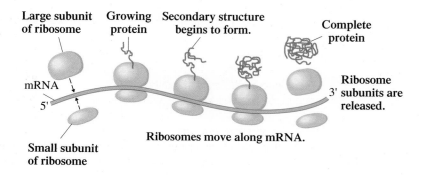

22.12 MUTATIONS

A **mutation** *is an error in base sequence in a gene that is reproduced during DNA replication.* Such errors alter the genetic information that is passed on during transcription. The altered information can cause changes in amino acid sequence during protein synthesis. Sometimes, such changes have a profound effect on an organism.

A **mutagen** *is a substance or agent that causes a change in the structure of a gene.* Radiation and chemical agents are two important types of mutagens. Radiation, in the form of ultraviolet light, X rays, radioactivity (Chapter 11), and cosmic rays, has the potential to be mutagenic. Ultraviolet light from the sun is the radiation that causes sunburn and can induce changes in the DNA of the skin cells. Sustained exposure to ultraviolet light can lead to serious problems such as skin cancer.

Chemical agents can also have mutagenic effects. Nitrous acid (HNO_2) is a mutagen that causes deamination of heterocyclic nitrogen bases. For example, HNO_2 can convert cytosine to uracil.

$$\text{Cytosine} \xrightarrow{HNO_2} \text{Uracil}$$

Deamination of a cytosine that was part of an mRNA codon would change the codon; for example, CGG would become UGG.

A variety of chemicals—including nitrites, nitrates, and nitrosamines—can form nitrous acid in the body. The use of nitrates and nitrites as preservatives in foods such as bologna and hot dogs is a cause of concern because of their conversion to nitrous acid in the body and possible damage to DNA.

Fortunately, the body has *repair enzymes* that recognize and replace altered bases. Normally, the vast majority of altered DNA bases are repaired, and mutations are avoided. Occasionally, however, the damage is not repaired, and the mutation persists.

22.13 NUCLEIC ACIDS AND VIRUSES

Viruses are very small disease-causing agents that are considered the lowest order of life. Indeed, their structure is so simple that some scientists do not consider them truly alive because they are unable to reproduce in the absence of other organisms. Figure 22.26 shows an electron microscope image of an influenza virus.

A **virus** *is a small particle that contains DNA or RNA (but not both) surrounded by a coat of protein and that cannot reproduce without the aid of a host cell.* Viruses do not possess the nucleotides, enzymes, amino acids, and other molecules necessary to replicate their nucleic acid or to synthesize proteins. To reproduce, viruses must invade the cells of another organism and cause these host cells to carry out the reproduction of the virus. Such an invasion disrupts the normal operation of cells, causing diseases within the host organism. The only function of a virus is reproduction; viruses do not generate energy.

There is no known form of life that is not subject to attack by viruses. Viruses attack bacteria, plants, animals, and humans. Many human diseases are of viral origin. Among them are the common cold, mumps, measles, smallpox, rabies, influenza, infectious mononucleosis, hepatitis, and AIDS.

Viruses most often attach themselves to the outside of specific cells in a host organism. An enzyme within the protein overcoat of the virus catalyzes the breakdown of the cell membrane, opening a hole in the membrane. The virus then injects its DNA or RNA into the cell. Once inside, this nucleic acid material is mistaken by the host cell for its own, whereupon that cell begins to translate and/or transcribe the viral nucleic acid. When all

Figure 22.26 An electron microscope image of an influenza virus.

CHEMISTRY AT A GLANCE

Protein Synthesis: Transcription and Translation

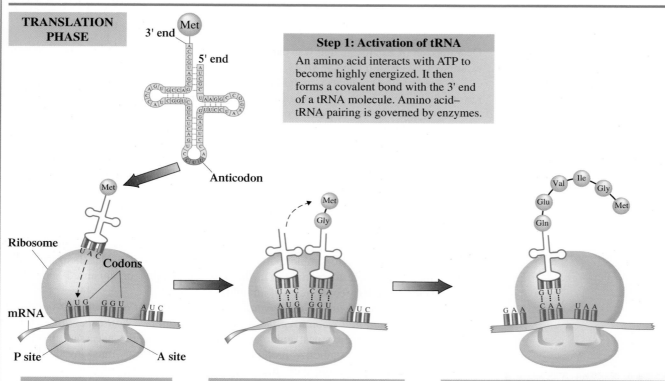

TRANSCRIPTION PHASE

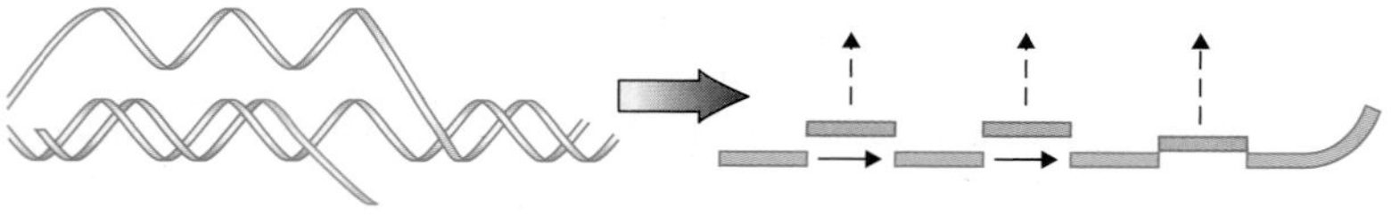

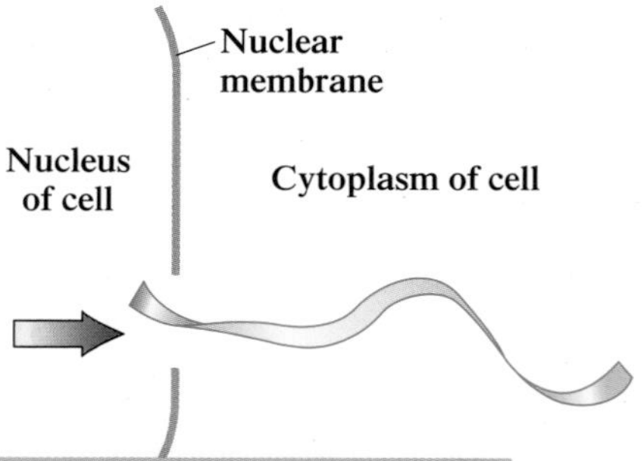

Nuclear membrane

Nucleus of cell

Cytoplasm of cell

Step 1: Formation of hnRNA

DNA in the nucleus partially unwinds to allow a strand of hnRNA to be made.

Step 2: Formation of mRNA

Introns are removed from the hnRNA strand.

Step 3: mRNA Enters the Cytoplasm

The mRNA leaves the nucleus and enters the cytoplasm.

TRANSLATION PHASE

Met

3' end

5' end

Anticodon

Step 1: Activation of tRNA

An amino acid interacts with ATP to become highly energized. It then forms a covalent bond with the 3' end of a tRNA molecule. Amino acid–tRNA pairing is governed by enzymes.

Ribosome

Codons

mRNA

P site

A site

Met

Gly

Val Ile Gly

Glu   Met

Gln

Step 2: Initiation

The mRNA attaches to a ribosome so that the first codon (AUG) is at the P site. A tRNA carrying methionine attaches to the first codon.

Step 3: Elongation

Another tRNA with the second amino acid binds at the A site. The methionine transfers from the P site to the A site. The ribosome shifts to the next codon, making its A site available for the tRNA carrying the third amino acid.

Steps 4 and 5: Termination and Post-Translation Processing

The polypeptide chain continues to lengthen until a stop codon appears on the mRNA. The new protein is cleaved from the last tRNA.

During post-translation processing, cleavage of Met (the initiation codon) usually occurs. S—S bonds between Cys units also can form.

the virus components have been synthesized by the host cell, they assemble automatically to form many new virus particles. Within 20 to 30 minutes after a single molecule of viral nucleic acid enters the host cell, hundreds of new virus particles have formed. So many are formed that they eventually burst the host cell and are free to infect other cells.

If a virus contains DNA, the host cell replicates the viral DNA in a manner similar to the way it replicates its own DNA. The newly produced viral DNA then proceeds to make the proteins needed for the production of protein coats for additional viruses.

An RNA-containing virus is called a *retrovirus*. Once inside a host, such viruses first make viral DNA. This *reverse* synthesis is governed by the enzyme *reverse transcriptase*. The template is the viral RNA rather than DNA. The viral DNA so produced then produces additional viral DNA and the proteins necessary for the protein coats.

The AIDS (acquired immunodeficiency syndrome) virus is an example of a retrovirus. This virus has an affinity for a specific type of white blood cell called a *helper T cell,* which is an important part of the body's immune system. When helper T cells are unable to perform their normal functions as a result of such viral infection, the body becomes more susceptible to infection and disease.

A **vaccine** is *a preparation containing an inactive or weakened form of a virus or bacterium.* The antibodies produced by the body against these specially modified viruses or bacteria effectively act against the naturally occurring active forms as well. Thanks to vaccination programs, many diseases, such as polio and mumps (caused by RNA-containing viruses) and smallpox and yellow fever (caused by DNA-containing viruses), are now seldom encountered.

> Viral infections are more difficult to treat than bacterial infections because viruses, unlike bacteria, replicate inside cells. It is difficult to design drugs that prevent the replication of the virus that do not also affect the normal activities of the host cells.

RECOMBINANT DNA AND GENETIC ENGINEERING

Increased knowledge about how DNA molecules function under various chemical conditions has opened the door to the field of technology called *genetic engineering* or *biotechnology.* Techniques now exist whereby a "foreign" gene can be added to an organism, and the organism will produce the protein associated with the added gene.

As an example of benefits that can come from genetic engineering, consider the case of human insulin. For many years, because of the very limited availability of human insulin, the insulin used by diabetics was obtained from the pancreases of slaughterhouse animals. Such insulin is structurally very similar to human insulin (see the Chemical Connections feature "Substitutes for Human Insulin" on page 671) and can be substituted for it. Today, diabetics can also choose to use "real" human insulin produced by genetically altered bacteria. Such "genetically engineered" bacteria are grown in large numbers, and the insulin they produce is harvested in a manner similar to the way some antibiotics are obtained from cultured microorganisms. Human growth hormone is another substance that is now produced by genetically altered bacteria. Table 22.3 gives additional

TABLE 22.3
Selected Human Proteins Produced Using Recombinant DNA Technology and Their Uses

Protein	Treatment
insulin	diabetes
erythropoietin (EPO)	anemia
human growth hormone (HGH)	stimulate growth
interleukins	stimulate immune system
interferons	leukemia and other cancers
lung surfactant protein	respiratory distress
serum albumin	plasma supplement
tumor necrosis factor (TNF)	cancers
tissue plasminogen activator (TPA)	heart attacks
epidermal growth factor	healing of wounds and burns
fibroblast growth factor	ulcers

Figure 22.27 Recombinant DNA is made by inserting a gene obtained from DNA of one organism into the DNA from another kind of organism.

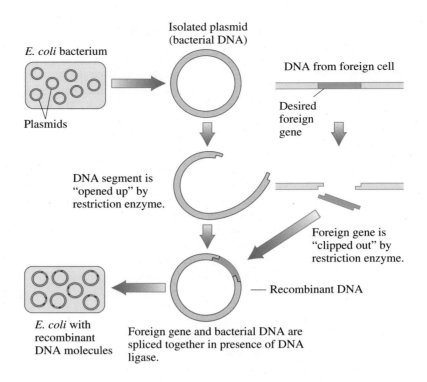

examples of human proteins that are used in therapeutic medicine that have become available through the use of recombinant DNA technology.

Genetic engineering procedures involve a type of DNA called *recombinant DNA*. **Recombinant DNA (rDNA)** *is DNA that contains genetic material from two different organisms.* Let us examine the theory and procedures used in obtaining recombinant DNA through genetic engineering.

The notation rDNA is often used to designate recombinant DNA.

The bacterium *E. coli,* which is found in the intestinal tract of humans and animals, is the organism most often used in recombinant DNA experiments. Yeast cells are also used, with increasing frequency, in this research.

In addition to their chromosomal DNA, *E. coli* (and other bacteria) contain DNA in the form of small, circular, double-stranded molecules called *plasmids*. These plasmids, which carry only a few genes, replicate independently of the chromosome. Also, they are transferred relatively easily from one cell to another. Plasmids from *E. coli* are used in recombinant DNA work.

The procedure used to obtain *E. coli* cells that contain recombinant DNA involves the following steps (see Figure 22.27).

Step 1: **Cell membrane dissolution.** *E. coli* cells of a specific strain are placed in a solution that dissolves cell membranes, thus releasing the contents of the cells.

Step 2: **Isolation of plasmid fraction.** The released cell components are separated into fractions, one fraction being the plasmids. The isolated plasmid fraction is the material used in further steps.

Step 3: **Cleavage of plasmid DNA.** A special enzyme, called a *restriction enzyme,* is used to cleave the double-stranded DNA of a circular plasmid. The result is a linear (noncircular) DNA molecule.

Step 4: **Gene removal from another organism.** The same restriction enzyme is then used to remove a desired gene from a chromosome of another organism.

Step 5: **Gene–plasmid splicing.** The gene (from Step 4) and the opened plasmid (from Step 3) are mixed in the presence of the enzyme *DNA ligase,* which splices the two together. This splicing, which attaches one end of the gene to one end of the

Figure 22.28 Cleavage pattern resulting from the use of a restriction enzyme that cleaves DNA between G and A bases in the 5′-to-3′ direction in the sequence G–A–A–T–T–C. The double-helix structure is not cut straight across.

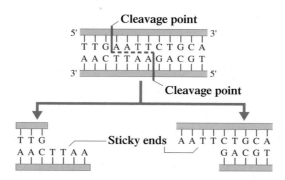

opened plasmid and attaches the other end of the gene to the other end of the plasmid, results in an altered circular plasmid (the recombinant DNA).

Step 6: **Uptake of recombinant DNA.** The altered plasmids (recombinant DNA) are placed in a live *E. coli* culture, where they are taken up by the *E. coli* bacteria. The *E. coli* culture into which the plasmids are placed need not be identical to that from which the plasmids were originally obtained.

We noted in Step 3 that the conversion of a circular plasmid into a linear DNA molecule requires a restriction enzyme. A **restriction enzyme** *is an enzyme that recognizes specific base sequences in DNA and cleaves the DNA in a predictable manner at these sequences.* The discovery of restriction enzymes made genetic engineering possible.

Restriction enzymes occur naturally in numerous types of bacterial cells. Their function is to protect the bacteria from invasion by foreign DNA by catalyzing the cleavage of the invading DNA. The term *restriction* relates to such enzymes placing a "restriction" on the type of DNA allowed into the bacterial cells.

To understand how a restriction enzyme works, let us consider one that cleaves DNA between G and A bases in the 5′-to-3′ direction in the sequence G–A–A–T–T–C. This enzyme will cleave the double-helix structure of a DNA molecule in the manner shown in Figure 22.28.

Note that the double helix is not cut straight across; the individual strands are cut at different points, giving a staircase cut. (Both cuts must be between G and A in the 5′-to-3′ direction.) This staircase cut leaves unpaired bases on each cut strand. These ends with unpaired bases are called "sticky ends" because they are ready to "stick to" (pair up with) a complementary section of DNA if they can find one.

If the same restriction enzyme used to cut a plasmid is also used to cut a gene from another DNA molecule, the sticky ends of the gene will be complementary to those of the plasmid. This enables the plasmid and gene to combine readily, forming a new, modified plasmid molecule. This modified plasmid molecule is called recombinant DNA. In addition to the newly spliced gene, the recombinant DNA plasmid contains all of the genes and characteristics of the original plasmid. Figure 22.29 shows diagrammatically the match between sticky ends that occurs when plasmid and gene combine.

Step 6 involves inserting the recombinant DNA (modified plasmids) back into *E. coli* cells. The process is called transformation. **Transformation** *is the process of incorporating recombinant DNA into a host cell.*

The transformed cells then reproduce, resulting in large numbers of identical cells called clones. **Clones** *are cells with identical DNA that have descended from a single cell.* Within a few hours, a single genetically altered bacterial cell can give rise to thousands of clones. Each clone has the capacity to synthesize the protein directed by the foreign gene it carries.

Researchers are not limited to selection of naturally occurring genes for transforming bacteria. Chemists have developed nonenzymatic methods of linking nucleotides

Figure 22.29 The "sticky ends" of the cut plasmid and the cut gene are complementary and combine to form recombinant DNA.

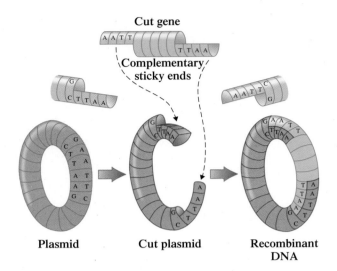

together such that they can construct artificial genes of any sequence they desire. In fact, benchtop instruments are now available that can be programmed by a microprocessor to synthesize any DNA base sequence *automatically*. The operator merely enters a sequence of desired bases, starts the instrument, and returns later to obtain the product. Such flexibility in manufacturing DNA has opened many doors, accelerated the pace of recombinant DNA research, and redefined the term *designer genes*!

22.15 THE POLYMERASE CHAIN REACTION

The **polymerase chain reaction (PCR)** *is a method for rapidly producing multiple copies of a DNA nucleotide sequence.* Billions of copies of a specific DNA sequence (gene) can be produced in a few hours via this reaction. The PCR is easy to carry out, requiring only a few chemicals, a container, and a source of heat. (In actuality, the PCR process is now completely automated.)

By means of the PCR process, DNA that is available only in very small quantities can be amplified to quantities large enough to analyze. The PCR process, devised in 1983, has become a valuable tool for diagnosing diseases and detecting pathogens in the body. It is now used in the prenatal diagnosis of a number of genetic disorders, including muscular dystrophy and cystic fibrosis, and in the identification of bacterial pathogens. It is also the definitive way to detect the AIDS virus.

The PCR process has also proved useful in certain types of forensic investigations. A DNA sample may be obtained from a single drop of blood or semen or a single strand of hair at a crime scene and amplified by the PCR process. A forensic chemist can then compare the amplified samples with DNA samples taken from suspects. Work with DNA in the forensic area is often referred to as *DNA fingerprinting.*

DNA polymerase, an enzyme present in all living organisms, is a key substance in the PCR process. It can attach additional nucleotides to a short starter nucleotide chain, called a *primer,* when the primer is bound to a complementary strand of DNA that functions as a template. The original DNA is heated to separate its strands, and then primers, DNA polymerase, and deoxyribonucleotides are added so that the *DNA polymerase* can replicate the original strand. The process is repeated until, in a short time, millions of copies of the original DNA have been made.

Figure 22.30 shows diagrammatically, in very simplified terms, the basic steps in the PCR process.

PCR temperature conditions are higher than those in the human body. This is possible because the DNA polymerase used was isolated from an organism that lives in the "hot pots" of Yellowstone National Park at temperatures of 70°C–75°C.

After n cycles of the PCR process, the amount of DNA will have increased 2^n times.

2^{10} is approximately 1000.
2^{20} is approximately 1,000,000.

Twenty-five cycles of the PCR can be carried out in an hour in a process that is fully automated.

Figure 22.30 The basic steps, in simplified terms, of the polymerase chain reaction process. Each cycle of the polymerase chain reaction doubles the number of copies of the target DNA sequence.

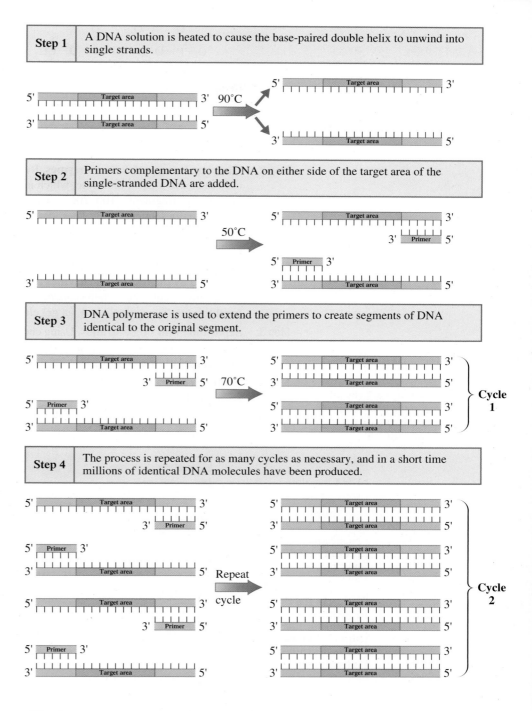

| Step 1 | A DNA solution is heated to cause the base-paired double helix to unwind into single strands. |

| Step 2 | Primers complementary to the DNA on either side of the target area of the single-stranded DNA are added. |

| Step 3 | DNA polymerase is used to extend the primers to create segments of DNA identical to the original segment. |

| Step 4 | The process is repeated for as many cycles as necessary, and in a short time millions of identical DNA molecules have been produced. |

22.16 DNA SEQUENCING

DNA sequencing *is a method by which the base sequence in a DNA molecule (or a portion of it) is determined.* Discovered in 1977, this is the process that made the Human Genome Project (Section 22.8) possible. Today, thanks to computer technology, sequencing a nucleic acid is a fairly routine, fully automated process.

The key concept in DNA sequencing is the *selective interruption* of polynucleotide synthesis. This interruption of synthesis, which is caused to occur at every possible nucleotide site, depends on the presence of 2′,3′-*dideoxy*ribonucleotide triphosphates (ddNTPs) in the synthesis mixture. Such compounds are synthetic analogs of the standard deoxyribonucleotide triphosphates in which both the 2′ and the 3′ hydroxy groups of deoxyribose have been replaced by hydrogen substituents.

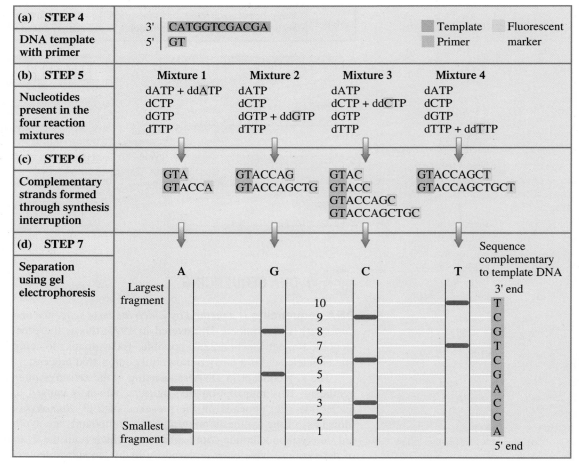

The basic steps involved in the DNA sequencing process are as follows:

Step 1: **Cleavage using restriction enzymes.** Restriction enzymes (Section 22.14) are used to cleave a DNA molecule, which is too large to be sequenced as a whole, into smaller fragments (100–200 base pairs). These smaller fragments are the DNA actually sequenced. By later identifying the points of overlap among the fragments sequenced, it is possible to determine the base sequence of the entire original DNA molecule.

Step 2: **Separation into individual components.** The mixture of small DNA fragments generated by the restriction enzymes is separated into individual components. Each component type is then sequenced independently. Separation of the fragment mixture is accomplished via gel electrophoresis techniques (Section 20.4).

Step 3: **Separation into single strands.** Using chemical methods, a given DNA fragment is separated into its two strands, and one strand is then used as a template to create complementary strands of varying lengths via the "interruption of synthesis" process.

Step 4: **Addition of primer to the single strands.** The single-stranded DNA to be sequenced (the template) is mixed with short polynucleotides that serve as a primer for the complementary strands (see Figure 22.31a).

Figure 22.31 A schematic diagram of selected steps in the DNA sequencing procedure for the 10-base DNA segment 5′ AGCAGCTGGT 3′.

Step 5: **Separation of the reaction mixture into four parts.** The mixture of template DNA and primer is divided into four portions, and four parallel synthesis reactions are carried out. Each reaction mixture contains all four deoxyribonucleotide triphosphates: dATP, dCTP, dGTP, and dTTP (Section 22.2). Each test tube also contains a unique ingredient—one of the ddNTPs that has been labeled with a fluorescent material that can be detected using instrumentation (see Figure 22.31b).

Step 6: **Polynucleotide synthesis with interruption.** As DNA complementary strand synthesis proceeds, nucleotides from the solution are added to the growing polynucleotide chain. Elongation of the growing chain takes place without complication until a ddNTP is incorporated into the chain. Synthesis stops at this point because a ddNTP lacks a hydroxyl group at carbon 3 and hence cannot participate in a 3′-to-5′ phosphodiester linkage, a necessary requirement for chain elongation (Section 22.3). Thus the portion of the reaction mixture that contains ddATP will be a mixture of all possible lengths of DNA complementary strands that terminate in ddA. Similarly, all of the complementary strands in the portion that contains ddGTP will terminate in ddG, and so on (see Figure 22.31c).

Step 7: **Identification of the reaction mixture components.** The newly synthesized complementary DNA strands of the four portions of the reaction mixture are then subjected to gel electrophoresis. Smaller DNA fragments move more rapidly through the gel than do larger ones, which is the basis for the separation. Fluorescence from the four differently marked ddNTPs present in the complementary strands is the basis for identification; the labeling pattern observed indicates the sequence of bases. Figure 22.31d shows the gel separation of the complementary nucleotide strands for the 10-base DNA segment shown in Figure 22.31a.

CONCEPTS TO REMEMBER

Nucleic acids. Nucleic acids are polymeric molecules in which the repeating units are nucleotides. Cells contain two kinds of nucleic acids—deoxyribonucleic acids (DNA) and ribonucleic acids (RNA). The major biochemical functions of DNA and RNA are, respectively, transfer of genetic information and synthesis of proteins (Section 22.1).

Nucleotides. Nucleotides, the monomers of nucleic acid polymers, are molecules composed of a pentose sugar bonded to both a phosphate group and a nitrogen-containing heterocyclic base. The pentose sugar must be either ribose or deoxyribose. Five nitrogen-containing bases are found in nucleotides: adenine (A), guanine (G), cytosine (C), thymine (T), and uracil (U) (Section 22.2).

Primary nucleic acid structure. The "backbone" of a nucleic acid molecule is a constant alternating sequence of sugar and phosphate groups. Each sugar unit has a nitrogen-containing base attached to it (Section 22.3).

Complementary bases. Complementary bases are specific pairs of bases in nucleic acid structures that hydrogen-bond to each other (Section 22.4).

Secondary DNA structure. A DNA molecule exists as two polynucleotide chains coiled around each other in a double-helix arrangement. The double helix is held together by hydrogen bonding between complementary pairs of bases. Only two base-pairing combinations occur: A with T, and C with G (Section 22.4).

DNA replication. DNA replication occurs when the two strands of a parent DNA double helix separate and act as templates for the synthesis of new chains using the principle of complementary base pairing (Section 22.5).

Chromosome. A chromosome is a cell structure that consists of an individual DNA molecule bound to a group of proteins (Section 22.5).

RNA molecules. Five important types of RNA molecules, distinguished by their function, are ribosomal RNA (rRNA), messenger RNA (mRNA), heterogeneous nuclear RNA (hnRNA), transfer RNA (tRNA), and small nuclear RNA (snRNA) (Section 22.7).

Transcription. Transcription is the process in which the genetic information encoded in the base sequence of DNA is copied into hnRNA/mRNA molecules (Section 22.8).

Gene. A gene is a portion of a DNA molecule that contains the base sequences needed for the production of a specific hnRNA/mRNA molecule. Genes are segmented, with portions called exons that contain genetic information and portions called introns that do not convey genetic information (Section 22.8).

Codon. A codon is a three-nucleotide sequence in mRNA that codes for a specific amino acid needed during the process of protein synthesis (Section 22.9).

Genetic code. The genetic code consists of all the mRNA codons that specify either a particular amino acid or the termination of protein synthesis (Section 22.9).

Anticodon. An anticodon is a three-nucleotide sequence in tRNA that binds to a complementary sequence (a codon) in mRNA (Section 22.10).

Translation. Translation is the stage of protein synthesis in which the codons in mRNA are translated into amino acid sequences of new proteins. Translation involves interactions between the codons of mRNA and the anticodons of tRNA (Section 22.11).

Mutations. Mutations are changes in the base sequence in DNA molecules (Section 22.12).

Recombinant DNA. Recombinant DNA molecules are synthesized by splicing a segment of DNA, usually a gene, from one organism into the DNA of another organism (Section 22.14).

Polymerase chain reaction. The polymerase chain reaction is a method for rapidly producing many copies of a DNA sequence (Section 22.15).

DNA sequencing. DNA sequencing is a multistep process for determining the sequence of bases in a DNA segment (Section 22.16).

KEY REACTIONS AND EQUATIONS

1. Formation of a nucleotide (Section 22.2)

   Pentose sugar (ribose or deoxyribose) + phosphate ion
       + nitrogen-containing heterocyclic base ⟶

   Phosphate — Sugar
              |
             Base   + 2H₂O

2. Formation of a nucleic acid (Section 22.3)

   Many deoxyribose-containing nucleotides ⟶ DNA

   Many ribose-containing nucleotides ⟶ RNA

3. Protein synthesis (Section 22.6)

   DNA $\xrightarrow{\text{Transcription}}$ RNA $\xrightarrow{\text{Translation}}$ protein

EXERCISES and PROBLEMS

The members of each pair of problems in this section test similar material.

● Types of Nucleic Acids (Section 22.1)

22.1 What does the designation DNA stand for?

22.2 What does the designation RNA stand for?

22.3 What is the primary function within cells for RNA?

22.4 What is the primary function within cells for DNA?

22.5 Within human cells, where is the DNA located?

22.6 Within human cells, where is the RNA located?

22.7 What is the general name for the building blocks (monomers) from which an RNA molecule is made?

22.8 What is the general name for the building blocks (monomers) from which a DNA molecule is made?

● Nucleotides (Section 22.2)

22.9 How many subunits are present within a nucleotide?

22.10 What are the name of the subunits present within a nucleotide?

22.11 What is the structural difference between the pentose sugars ribose and 2-deoxyribose?

22.12 What are the names of the pentose sugars present, respectively, in DNA and RNA molecules?

22.13 Characterize each of the following nitrogen-containing bases as a purine derivative or a pyrimidine derivative.
   a. Thymine   b. Cytosine
   c. Adenine   d. Guanine

22.14 Characterize each of the following nitrogen-containing bases as a component of (1) both DNA and RNA, (2) DNA but not RNA, or (3) RNA but not DNA.
   a. Adenine   b. Thymine
   c. Uracil   d. Cytosine

22.15 How many different choices are there for each of the following subunits in the specified type of nucleotide?
   a. Pentose sugar subunit in DNA nucleotides
   b. Nitrogen-containing base subunit in RNA nucleotides
   c. Phosphate subunit in DNA nucleotides

22.16 How many different choices are there for each of the following subunits in the specified type of nucleotide?
   a. Pentose sugar subunit in RNA nucleotides
   b. Nitrogen-containing base subunit in DNA nucleotides
   c. Phosphate subunit in RNA nucleotides

22.17 Which nitrogen-containing base is present in each of the following nucleotides?
   a. AMP   b. dGMP   c. dTMP   d. UMP

22.18 Which nitrogen-containing base is present in each of the following nucleotides?
   a. GMP   b. dAMP   c. CMP   d. dCMP

22.19 Which pentose sugar is present in each of the nucleotides in Problem 22.17?

22.20 Which pentose sugar is present in each of the nucleotides in Problem 22.18?

22.21 Characterize as true or false each of the following statements about the given nucleotide.

   a. The nitrogen-containing base is a purine derivative.
   b. The phosphate group is attached to the sugar unit at carbon 3′.
   c. The sugar unit is ribose.
   d. The nucleotide could be a component of both DNA and RNA.

22.22 Characterize as true or false each of the following statements about the given nucleotide.

   a. The sugar unit is 2-deoxyribose.
   b. The sugar unit is attached to the nitrogen-containing base at nitrogen 3.
   c. The nitrogen-containing base is a pyrimidine derivative.
   d. The nucleotide could be a component of both DNA and RNA.

22.23 Draw the structures of the three products produced when the nucleotide in Problem 22.21 undergoes hydrolysis.

22.24 Draw the structures of the three products produced when the nucleotide in Problem 22.22 undergoes hydrolysis.

Primary Nucleic Acid Structure (Section 22.3)

22.25 What are the two repeating subunits present in the *backbone* portion of a nucleic acid?

22.26 To which type of subunit in a nucleic acid *backbone* are the nitrogen-containing bases attached?

22.27 What distinguishes various DNA molecules from each other?

22.28 What distinguishes various RNA molecules from each other?

22.29 What is the difference between a nucleic acid's 3′ end and its 5′ end?

22.30 In the lengthening of a polynucleotide chain, which type of nucleotide subunit would bond to the 3′ end of the polynucleotide chain?

22.31 What are the nucleotide subunits that participate in a nucleic acid 3′,5′-phosphodiester linkage?

22.32 How many 3′,5′-phosphodiester linkages are present in a tetranucleotide segment of a nucleic acid?

22.33 Draw the structure of the dinucleotide product obtained by combining the nucleotides of Problems 22.21 and 22.22 such that the Problem 22.21 nucleotide is the 5′ end of the dinucleotide.

22.34 Draw the structure of the dinucleotide product obtained by combining the nucleotides of Problems 22.21 and 22.22 such that the Problem 22.21 nucleotide is the 3′ end of the dinucleotide.

The DNA Double Helix (Section 22.4)

22.35 Describe the DNA double helix in terms of
a. general shape.
b. what is on the outside of the helix and what is within the interior of the helix.

22.36 Describe the DNA double helix in terms of
a. the directionality of the two polynucleotide chains present.
b. a comparison of the total number of nitrogen-containing bases present in each of the two polynucleotide chains.

22.37 The base content of a particular DNA molecule is 36% thymine. What is the percentage of each of the following bases in the molecule?
a. Adenine   b. Guanine   c. Cytosine

22.38 The base content of a particular DNA molecule is 24% guanine. What is the percentage of each of the following bases in the molecule?
a. Adenine   b. Cytosine   c. Thymine

22.39 In terms of hydrogen bonding, a G–C base pair is more stable than an A–T base pair. Explain why this is so.

22.40 What structural consideration prevents the following bases from forming complementary base pairs?
a. A and G   b. T and C

22.41 What is the relationship between the total number of purine bases (A and G) and the total number of pyrimidine bases (C and T) present in a DNA double helix?

22.42 The base composition for one of the strands of a DNA double helix is 19% A, 34% C, 28% G, and 19% T. What is the percent base composition for the other strand of the DNA double helix?

22.43 Identify the 3′ and 5′ ends of the DNA base sequence TAGCC.

22.44 The two-base DNA sequences TA and AT represent different dinucleotides. Explain why this is so.

22.45 Using the concept of complementary base pairing, write the complementary DNA strands, with their 5′ and 3′ ends labeled, for each of the following DNA base sequences.
a. 5′ ACGTAT 3′   b. 5′ TTACCG 3′
c. 3′ GCATAA 5′   d. AACTGG

22.46 Using the concept of complementary base pairing, write the complementary DNA strands, with their 5′ and 3′ ends labeled, for each of the following DNA base sequences.
a. 5′ CCGGTA 3′   b. 5′ CACAGA 3′
c. 3′ TTTAGA 5′   d. CATTAC

22.47 How many total hydrogen bonds would exist between the DNA strand 5′ AGTCCTCA 3′ and its complementary strand?

22.48 How many total hydrogen bonds would exist between the DNA strand 5′ CCTAGGAT 3′ and its complementary strand?

Replication of DNA Molecules (Section 22.5)

22.49 What is the function of the enzyme *DNA helicase* in the DNA replication process?

22.50 What are two functions of the enzyme *DNA polymerase* in the DNA replication process?

22.51 In the replication of a DNA molecule, two daughter molecules, Q and R, are formed. The following base sequence is part of the newly formed strand in daughter molecule Q.

5′ ACTTAG 3′

Indicate the corresponding base sequence in
a. the newly formed strand in daughter molecule R.
b. the "parent" strand in daughter molecule Q.
c. the "parent" strand in daughter molecule R.

22.52 In the replication of a DNA molecule, two daughter molecules, S and T, are formed. The following base sequence is part of the "parent" strand in daughter molecule S.

5′ TTCAGAG 3′

Indicate the corresponding base sequence in
a. the newly formed strand in daughter molecule T.
b. the newly formed strand in daughter molecule S.
c. the "parent" strand in daughter molecule T.

22.53 During DNA replication, one of the newly formed strands grows continuously, whereas the other grows in segments that are later connected together. Explain why this is so.

22.54 DNA replication is most often a bidirectional process. Explain why this is so.

22.55 What is a chromosome?

22.56 Chromosomes are nucleoproteins. Explain.

Overview of Protein Synthesis (Section 22.6)

22.57 What type of molecule is the starting material for the transcription phase of protein synthesis?

22.58 What type of molecule is the end product for the transcription phase of protein synthesis?

22.59 What type of molecule is the starting material for the translation phase of protein synthesis?

22.60 What type of molecule is the end product for the translation phase of protein synthesis?

RNA Molecules (Section 22.7)

22.61 What are the four major differences between RNA molecules and DNA molecules?

22.62 What are the names and abbreviations for the five major types of RNA molecules?

22.63 State whether each of the following phrases applies to hnRNA, mRNA, tRNA, rRNA, or snRNA.
a. Material from which messenger RNA is made
b. Delivers amino acids to protein synthesis sites
c. Smallest of the RNAs in terms of nucleotide units present
d. Also goes by the designation ptRNA

22.64 State whether each of the following phrases applies to hnRNA, mRNA, tRNA, rRNA, or snRNA.
a. Associated with a series of proteins in a complex structure
b. Contains genetic information needed for protein synthesis
c. Most abundant type of RNA in a cell
d. Involved in the editing of hnRNA molecules

22.65 For each of the following types of RNA, indicate whether the predominant cellular location for the RNA is the nuclear region, the extranuclear region, or both the nuclear and the extranuclear regions.
a. hnRNA   b. tRNA   c. rRNA   d. mRNA

22.66 Indicate whether each of the following processes occurs in the nuclear or the extranuclear region of a cell.
a. DNA transcription
b. Processing of hnRNA to mRNA
c. mRNA translation (protein synthesis)
d. DNA replication

● **Transcription: RNA Synthesis (Section 22.8)**

22.67 What serves as a template in the process of *transcription*?

22.68 What is the initial product of the *transcription* process?

22.69 What are two functions of the enzyme *RNA polymerase* in the transcription process?

22.70 What is a *gene*?

22.71 What are the complementary base pairs in DNA–RNA interactions?

22.72 In DNA–DNA interactions there are two complementary base pairs, and in DNA–RNA interactions there are three complementary base pairs. Explain.

22.73 Write the base sequence of the hnRNA formed by transcription of the following DNA base sequence.

5′ ATGCTTA 3′

22.74 Write the base sequence of the hnRNA formed by transcription of the following DNA base sequence.

5′ TAGTGAT 3′

22.75 From what DNA base sequence was the following hnRNA sequence transcribed?

5′ UUCGCAG 3′

22.76 From what DNA base sequence was the following hnRNA sequence transcribed?

5′ GCUUAUC 3′

22.77 What is the relationship between an exon and a gene?

22.78 What is the relationship between an intron and a gene?

22.79 What mRNA base sequence would be obtained from the following portion of a gene?

| exon | intron | exon |
5′ TCAG–TAGC–TTCA 3′

22.80 What mRNA base sequence would be obtained from the following portion of a gene?

| intron | exon | intron |
5′ TTAC–AACG–GCAT 3′

22.81 In the process of splicing, which type of RNA
a. undergoes the splicing?
b. is present in the spliceosomes?

22.82 What is the difference between snRNA and snRNPs?

22.83 What is *alternative splicing*?

22.84 How many different mRNAs can be produced from an hnRNA that contains three exons, one of which is an "alternative" exon?

● **The Genetic Code (Section 22.9)**

22.85 What is a codon?

22.86 On what type of RNA molecule are codons found?

22.87 Using the information in Table 22.2, determine what amino acid is coded for by each of the following codons.
a. CUU   b. AAU   c. AGU   d. GGG

22.88 Using the information in Table 22.2, determine what amino acid is coded for by each of the following codons.
a. GUA   b. CCC   c. CAC   d. CCA

22.89 Using the information in Table 22.2, determine the synonyms, if any, of each of the codons in Problem 22.87.

22.90 Using the information in Table 22.2, determine the synonyms, if any, of each of the codons in Problem 22.88.

22.91 Explain why the base sequence ATC could not be a codon.

22.92 Explain why the base sequence AGAC could not be a codon.

22.93 Predict the sequence of amino acids coded by the mRNA sequence

5′ AUG–AAA–GAA–GAC–CUA 3′

22.94 Predict the sequence of amino acids coded by the mRNA sequence

5′ GGA–GGC–ACA–UGG–GAA 3′

● **Anticodons and tRNA Molecules (Section 22.10)**

22.95 Describe the general structure of a tRNA molecule.

22.96 Where is the anticodon site on a tRNA molecule?

22.97 By what type of bond is an amino acid attached to a tRNA molecule?

22.98 What principle governs the codon–anticodon interaction that leads to proper placement of amino acids in proteins?

22.99 What is the anticodon that would interact with each of the following codons?
a. AGA   b. CGU   c. UUU   d. CAA

22.100 What is the anticodon that would interact with each of the following codons?
a. CCU   b. GUA   c. AUC   d. GCA

22.101 Which amino acid will a tRNA molecule be carrying if its anticodon is the following?
a. CCC   b. CAG   c. UGC   d. GAG

22.102 Which amino acid will a tRNA molecule be carrying if its anticodon is the following?
a. GGG   b. GAC   c. AUA   d. CGA

22.103 What are the possible codons that a tRNA molecule could react with if it is carrying each of the following amino acids? (Note that a given tRNA will be specific for one or more of the possible codons.)
a. Serine   b. Leucine   c. Isoleucine   d. Glycine

22.104 What are the possible codons that a tRNA molecule could interact with if it is carrying each of the following amino acids? (Note that a given tRNA will be specific for one or more of the possible codons.)
a. Valine
b. Cysteine
c. Alanine
d. Tyrosine

● Translation: Protein Synthesis (Section 22.11)

22.105 What is a ribosome?

22.106 Approximately how many ribosomes are present in cells of higher organisms?

22.107 What is the chemical composition of a ribosome's active site?

22.108 To which subunit of a ribosome does mRNA bind?

22.109 What are the two steps involved in the activation of a tRNA molecule?

22.110 Why is the first amino acid in a developing human protein always the amino acid Met?

22.111 In the elongation phase of translation, at which site in the ribosome does new peptide bond formation actually take place?

22.112 What two changes occur at a ribosome during protein synthesis immediately after peptide bond formation?

22.113 What types of events occur during post-translation processing of a protein?

22.114 At what stage of protein formation does folding of the protein into its active conformation occur?

22.115 Write a possible mRNA base sequence that would lead to the production of this pentapeptide. (There is more than one correct answer.)

Gly–Ala–Cys–Val–Tyr

22.116 Write a possible mRNA base sequence that would lead to the production of this pentapeptide. (There is more than one correct answer.)

Lys–Met–Thr–His–Phe

● Mutations (Section 22.12)

22.117 For the codon sequence

5′ GGC–UAU–AGU–AGC–CCC 3′

write the amino acid sequence produced in each of the following ways.
a. Translation proceeds in a normal manner.
b. A mutation changes CCC to CCU.
c. A mutation changes CCC to ACC.

22.118 For the codon sequence

5′ GGA–AUA–UGG–UUC–CUA 3′

write the amino acid sequence produced in each of the following ways.
a. Translation proceeds in a normal manner.
b. A mutation changes GGA to GGG.
c. A mutation changes GGA to CGA.

● Viruses and Vaccines (Section 22.13)

22.119 Describe the general structure of a virus.

22.120 What is the only function of a virus?

22.121 What is the most common method by which viruses invade cells?

22.122 Why must a virus infect another organism in order to reproduce?

● Recombinant DNA and Genetic Engineering (Section 22.14)

22.123 How does recombinant DNA differ from normal DNA?

22.124 Give two reasons why bacterial cells are used for recombinant DNA procedures.

22.125 What role do plasmids play in recombinant DNA procedures?

22.126 Describe what occurs when a particular restriction enzyme operates on a segment of double-stranded DNA.

22.127 Describe what happens during transformation.

22.128 How are plasmids obtained from *E. coli* bacteria?

22.129 A particular restriction enzyme will cleave DNA between A and A in the sequence AAGCTT in the 5′-to-3′ direction. Draw a diagram showing the structural details of the "sticky ends" that result from cleavage of the following DNA segment.

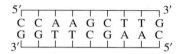

22.130 A particular restriction enzyme will cleave DNA between A and A in the sequence AAGCTT in the 5′-to-3′ direction. Draw a diagram showing the structural details of the "sticky ends" that result from cleavage of the following DNA segment.

● Polymerase Chain Reaction (Section 22.15)

22.131 What is the function of the polymerase chain reaction?

22.132 What is the function of the enzyme *DNA polymerase* in the PCR process?

22.133 What is a *primer* and what is its function in the PCR process?

22.134 What are the four types of substances needed to carry out the PCR process?

● DNA Sequencing (Section 22.16)

22.135 How do the notations dATP and ddATP differ in meaning?

22.136 What role do dideoxynucleotides play in the DNA sequencing process?

22.137 Assume that the red lines in the first and second columns of Figure 22.31d are interchanged, while the other labels stay the same.
a. Given this change, what would be the sequence of bases in the DNA fragment under study?
b. Given this change, what would be the sequence of bases in the original DNA fragment that was to be sequenced?

22.138 Assume that the "red lines" in the second and fourth columns of Figure 22.31d are interchanged, while the other labels stay the same.
a. Given this change, what would be the sequence of bases in the DNA fragment under study?
b. Given this change, what would be the sequence of bases in the original DNA fragment that was to be sequenced?

ADDITIONAL PROBLEMS

22.139 With the help of the structures given in Section 22.2, describe the structural differences between the following pairs of nucleotide bases.
a. Thymine and uracil
b. Adenine and guanine

22.140 The following is a sequence of bases for an exon portion of a strand of a gene.

5′ CATACAGCCTGGAAGCTC 3′

a. What is the sequence of bases on the strand of DNA complementary to this segment?
b. What is the sequence of bases on the mRNA molecule synthesized from this strand?
c. What codons are present on the mRNA molecule from part b?
d. What anticodons will be found on the tRNA molecules that interact with the codons from part c?
e. What is the sequence of amino acids in the peptide formed using these protein synthesis instructions?

22.141 Which of these RNA types, (1) mRNA, (2) hnRNA, (3) rRNA, or (4) tRNA, is most closely associated with each of the following terms?
a. Codon
b. Anticodon
c. Intron
d. Amino acid carrier

22.142 Which of these processes, (1) translation phase of protein synthesis, (2) transcription phase of protein synthesis, (3) replication of DNA, or (4) formation of recombinant DNA, is associated with each of the following events?
a. Complete unwinding of a DNA molecule occurs.
b. Partial unwinding of a DNA molecule occurs.
c. An mRNA–ribosome complex is formed.
d. Okazaki fragments are formed.

22.143 Which of these base-pairing situations, (1) between two DNA segments, (2) between two RNA segments, (3) between a DNA segment and an RNA segment, or (4) between a codon and an anticodon, fits each of the following base-pairing sequences? More than one response may apply to a given base-pairing situation.
a. A G T
⋮ ⋮ ⋮
U C A
b. A C T
⋮ ⋮ ⋮
T G A
c. A G U
⋮ ⋮ ⋮
U C A
d. C C G
⋮ ⋮ ⋮
G G C

22.144 Which of these characterizations, (1) found in DNA but not RNA, (2) found in RNA but not DNA, (3) found in both DNA and RNA, or (4) not found in DNA or RNA, fits each of the following mono-, di-, or trinucleotides?
a. 5′ dAMP–dAMP 3′
b. 5′ AMP–AMP–CMP 3′
c. 5′ dAMP–CMP 3′
d. 5′ GGA 3′

22.145 Suppose that 28% of the nucleotides of a DNA molecule are deoxythymidine 5′-monophosphate, and during replication the relative amounts of available nucleotide bases are 22% A, 28% T, 22% C, and 28% G. What base would be depleted first in the replication process?

22.146 On the basis of the most recent results of the Human Genome Project concerning the DNA present in a human cell
a. How many base-pairs are present in the DNA?
b. How many genes are present in the DNA?
c. What percentage of the base-pairs are accounted for by the genes?

$\mathcal{M}$ultiple-Choice Practice Test

22.147 Which of the following is not a structural subunit of a nucleotide?
a. A nitrogen-containing heterocyclic base
b. A pentose sugar
c. An amino acid
d. A phosphate

22.148 The number of kinds of RNA nucleotides is which of the following?
a. The same as the number of kinds of DNA nucleotides
b. Double the number of kinds of DNA nucleotides
c. Less than the number of kinds of DNA nucleotides
d. Greater than the number of kinds of DNA nucleotides

22.149 In which of the following pairs of nucleotide bases are both members of the pair "single-ring" bases?
a. A and C
b. G and T
c. T and U
d. A and G

22.150 Which of the following elements is not present in the "backbone" of a nucleic acid molecule?
a. Nitrogen
b. Carbon
c. Oxygen
d. Hydrogen

22.151 In a DNA double helix, the base pairs are which of the following?
a. Part of the backbone structure
b. Located inside the helix
c. Located outside the helix
d. Covalently bonded to each other

22.152 Which of the following types of RNA has a "cloverleaf shape" with three hairpin loops?
a. mRNA
b. rRNA
c. hnRNA
d. tRNA

22.153 Which of the following types of RNA contains introns and exons?
a. hnRNA
b. mRNA
c. tRNA
d. rRNA

22.154 Which of the following events occurs during the *translation* phase of protein synthesis?
a. mRNA is converted to hnRNA.
b. tRNAs carry amino acids to the site for protein synthesis.
c. A partial unwinding of a DNA double helix occurs.
d. tRNAs interact with ribosomes.

22.155 The genetic code is a listing that gives the relationships between which of the following?
a. Codons and anticodons
b. Codons and amino acids
c. Anticodons and amino acids
d. Codons and genes

22.156 What is the complementary hnRNA sequence to the DNA sequence CTA–TAC?
a. TCG–CGT
b. GAU–AUG
c. UGC–GCU
d. TCG–CGT

Biochemical Energy Production

23

Chapter Outline

The energy consumed by these scarlet ibises in flight is generated by numerous sequences of biochemical reactions that occur within their bodies.

This chapter is the first of four dealing with the chemical reactions that occur in a living organism. In this first chapter, we consider those molecules that are repeatedly encountered in biological reactions, as well as those reactions that are common to the processing of carbohydrates, lipids, and proteins. The three following chapters consider the reactions associated uniquely with carbohydrate, lipid, and protein processing, respectively.

 23.1 METABOLISM

Metabolism *is the sum total of all the biochemical reactions that take place in a living organism.* Human metabolism is quite remarkable. An average human adult whose weight remains the same for 40 years processes about 6 *tons* of solid food and 10,000 gallons of water, during which time the composition of the body is essentially constant. Just as we must put gasoline in a car to make it go or plug in a kitchen appliance to make it run, we also need a source of energy to think, breathe, exercise, or work. As we have seen in previous chapters, even the simplest living cell is continually carrying on energy-demanding processes such as protein synthesis, DNA replication, RNA transcription, and membrane transport.

Metabolic reactions fall into one of two subtypes: catabolism and anabolism. **Catabolism** *is all metabolic reactions in which large biochemical molecules are broken down to smaller ones.* Catabolic reactions usually release energy. The reactions involved in the oxidation of glucose are catabolic. **Anabolism** *is all metabolic reactions in which small biochemical molecules are joined together to form larger ones.* Anabolic reactions usually require energy in order to proceed. The synthesis of proteins

Catabolism is pronounced ca-TAB-o-lism, and *anabolism* is pronounced an-ABB-o-lism. *Catabolic* is pronounced CAT-a-bol-ic, and *anabolic* is pronounced AN-a-bol-ic.

Figure 23.1 The processes of catabolism and anabolism are opposite in nature. The first usually produces energy, and the second usually consumes energy.

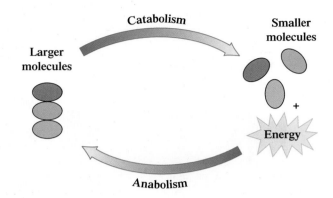

from amino acids is an anabolic process. Figure 23.1 contrasts catabolic and anabolic processes.

The metabolic reactions that occur in a cell are usually organized into sequences called *metabolic pathways.* A **metabolic pathway** *is a series of consecutive biochemical reactions used to convert a starting material into an end product.* Such pathways may be *linear,* in which a series of reactions generates a final product, or *cyclic,* in which a series of reactions regenerates the first reactant.

Linear metabolic pathway: $A \xrightarrow{\text{Enzyme 1}} B \xrightarrow{\text{Enzyme 2}} C \xrightarrow{\text{Enzyme 3}} D$

Cyclic metabolic pathway:

$$A \longrightarrow B$$
$$D \longleftarrow C$$

The major metabolic pathways for all life forms are similar. This enables scientists to study metabolic reactions in simpler life forms and use the results to help understand the corresponding metabolic reactions in more complex organisms, including humans.

● **E X A M P L E   2 3 . 1**

Distinguishing Between Anabolic and Catabolic Processes

▶ Classify each of the following chemical processes as anabolic or catabolic.

   a. Synthesis of a polysaccharide from monosaccharides
   b. Hydrolysis of a pentasaccharide to monosaccharides
   c. Formation of a nucleotide from phosphate, nitrogenous base, and pentose sugar
   d. Hydrolysis of a triacylglycerol to glycerol and fatty acids

Solution

   a. *Anabolic.* A large molecule is formed from many small molecules.
   b. *Catabolic.* Five monosaccharides are produced from a larger molecule.
   c. *Anabolic.* Three subunits are combined to give a larger unit.
   d. *Catabolic.* A four-subunit molecule is broken down into four smaller molecules.

Practice Exercise 23.1

Classify each of the following chemical processes as anabolic or catabolic.

   a. Synthesis of a protein from amino acids
   b. Formation of a triacylglycerol from glycerol and fatty acids
   c. Hydrolysis of a polysaccharide to monosaccharides
   d. Formation of a nucleic acid from nucleotides

Answers: a. Anabolic; **b.** Anabolic; **c.** Catabolic; **d.** Anabolic

23.2 METABOLISM AND CELL STRUCTURE

Knowledge of the major structural features of a cell is a prerequisite to understanding *where* metabolic reactions take place.

Cells are of two types: prokaryotic and eukaryotic. *Prokaryotic cells* have no nucleus and are found only in bacteria. The DNA that governs the reproduction of prokaryotic cells is usually a single circular molecule found near the center of the cell in a region called the *nucleoid*. A **eukaryotic cell** *is a cell in which the DNA is found in a membrane-enclosed nucleus.* Cells of this type, which are found in all higher organisms, are about 1000 times larger than bacterial cells. Our focus in the remainder of this section will be on eukaryotic cells, the type present in humans. Figure 23.2 shows the general internal structure of a eukaryotic cell. Note the key components shown: the outer membrane, nucleus, cytosol, ribosomes, lysosomes, and mitochondria.

The **cytoplasm** *is the water-based material of a eukaryotic cell that lies between the nucleus and the outer membrane of the cell.* Within the cytoplasm are several kinds of small structures called *organelles*. An **organelle** *is a minute structure within the cytoplasm of a cell that carries out a specific cellular function.* The organelles are surrounded by the *cytosol*. The **cytosol** *is the water-based fluid part of the cytoplasm of a cell.*

Three important types of organelles are ribosomes, lysosomes, and mitochondria. We have considered ribosomes before; they are the sites where protein synthesis occurs (Section 22.11). A **lysosome** *is an organelle that contains hydrolytic enzymes needed for cellular rebuilding, repair, and degradation.* Some lysosome enzymes hydrolyze proteins to amino acids; others hydrolyze polysaccharides to monosaccharides. Bacteria and viruses "trapped" by the body's immune system (Section 20.16) are degraded and destroyed by enzymes from lysosomes.

A **mitochondrion** *is an organelle that is responsible for the generation of most of the energy for a cell.* Much of the discussion of this chapter deals with the energy-producing chemical reactions that occur within mitochondria. Further details of mitochondrion structure will help us understand more about how these reactions occur.

Mitochondria are sausage-shaped organelles containing both an *outer membrane* and a *multifolded inner membrane* (see Figure 23.3). The outer membrane, which is about 50% lipid and 50% protein, is freely permeable to small molecules. The inner membrane, which is about 20% lipid and 80% protein, is highly impermeable to most substances. The nonpermeable nature of the inner membrane divides a mitochondrion into two separate compartments—an interior region called the *matrix* and the region between the

---

The term *eukaryotic*, pronounced you-KAHR-ee-ah-tic, is from the Greek *eu*, meaning "true," and *karyon*, meaning "nucleus." The term *prokaryotic*, which contains the Greek *pro*, meaning "before," literally means "before the nucleus."

Eukaryotic and prokaryotic cells differ in that the former contain a well-defined nucleus, set off from the rest of the cell by a membrane.

A protective mechanism exists to prevent lysosome enzymes from destroying the cell in which they are found if they should be accidently released (via membrane rupture or leakage). The optimum pH (Section 21.6) for lysosome enzyme activity is 4.8. The cytoplasmic pH of 7.0–7.3 renders them inactive.

Mitochondria, pronounced my-toe-KON-dree-ah, is plural. The singular form of the term is *mitochondrion*. The threadlike shape of the inner membrane of the mitochondria is responsible for this organelle's name; *mitos* is Greek for "thread," and *chondrion* is Greek for "granule."

Figure 23.2 A schematic representation of a eukaryotic cell with selected internal components identified.

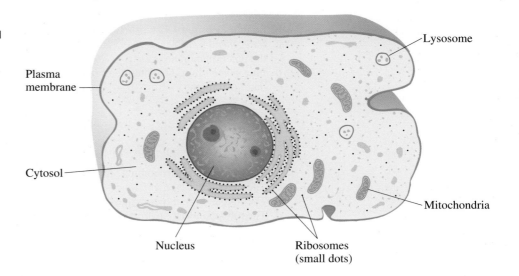

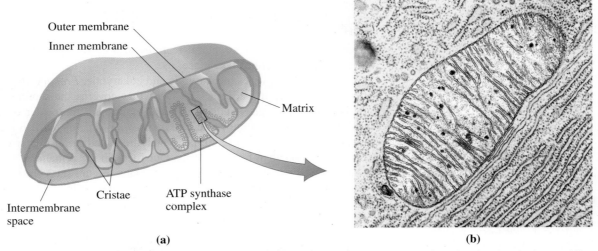

Outer membrane

Inner membrane

Matrix

Cristae

ATP synthase complex

Intermembrane space

(a)

(b)

Figure 23.3 (a) A schematic representation of a mitochondrion, showing key features of its internal structure. (b) An electron micrograph showing the ATP synthase knobs extending into the matrix.

inner and outer membranes, called the *intermembrane space*. The folds of the inner membrane that protrude into the matrix are called *cristae*.

The invention of high-resolution electron microscopes allowed researchers to see the interior structure of the mitochondrion more clearly and led to the discovery, in 1962, of small spherical knobs attached to the cristae called *ATP synthase complexes*. As their name implies, these relatively small knobs, which are located on the matrix side of the inner membrane, are responsible for ATP synthesis, and their association with the inner membrane is critically important for this task. More will be said about ATP in the next section.

 **EXAMPLE 23.2**

Recognizing Structural Characteristics of a Mitochondrion

Identify each of the following structural features of a mitochondrion.

a. The more permeable of the two mitochondrial membranes
b. The mitochondrial membrane that has cristae
c. The mitochondrial membrane that determines the size of the matrix
d. The mitochondrial membrane that is interior to the intermembrane space

Solution

a. *Outer membrane.* The outer membrane is more permeable.
b. *Inner membrane.* Cristae are "folds" present on the inner membrane.
c. *Inner membrane.* The matrix is the center part of a mitochondrion and is surrounded by the inner membrane.
d. *Inner membrane.* The intermembrane space separates the outer membrane from the inner membrane.

Practice Exercise 23.2

Identify each of the following structural features of a mitochondrion.

a. The mitochondrial membrane that is highly folded.
b. The mitochondrial membrane with the higher protein content.
c. The mitochondrial membrane that is exterior to the intermembrane space.
d. The mitochondrial membrane with which ATP synthase complexes are associated.

Answers: a. Inner membrane; **b.** Inner membrane; **c.** Outer membrane; **d.** Inner membrane

23.3 IMPORTANT INTERMEDIATE COMPOUNDS IN METABOLIC PATHWAYS

Nucleotides, besides being the monomer units from which nucleic acids are made, are also present in several *nonpolymeric* molecules that are important in energy production in living things.

As a prelude to an overview presentation (Section 23.5) of the metabolic processes by which our food is converted to energy, we now consider several compounds that repeatedly function as key intermediates in these metabolic pathways. Knowing about these compounds will make it easier to understand the details of metabolic pathways. The compounds to be discussed all have *nucleotides* (Section 22.2) as part of their structures.

Adenosine Phosphates (ATP, ADP, and AMP)

Several adenosine phosphates exist. Of importance in metabolism are adenosine *mono*phosphate (AMP), adenosine *di*phosphate (ADP), and adenosine *tri*phosphate (ATP). AMP is not a new molecule to us; it is one of the nucleotides present in RNA molecules (Section 22.2). ADP and ATP differ structurally from AMP only in the number of phosphate groups present. Block structural diagrams for these three adenosine phosphates follow.

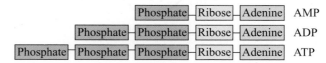

Figure 23.4 shows actual structural formulas for these three adenosine phospates. In ATP, a phosphoester bond joins the first phosphoryl group to the pentose sugar ribose. The other two phosporyl groups are joined to one another by phosphoanhydride bonds, as shown in the figure. A **phosphoryl group** *is the functional group derived from a phosphate ion that is part of another molecule.* ATP contains three phosphoryl groups, ADP two phosphoryl groups, and AMP one phosphoryl group. When two phosphate groups react with one another, a water molecule is produced (Section 16.20); hence the use of the word *anhydride* in describing the chemical bonds present between phosphoryl groups in ATP and ADP. A **phosphoanhydride bond** *is the chemical bond formed when two phosphate groups react with each other and a water molecule is produced.*

ATP and ADP molecules readily undergo hydrolysis reactions in which phosphate groups (P_i, inorganic phosphate) are released

$$ATP + H_2O \longrightarrow ADP + P_i + energy$$
$$ADP + H_2O \longrightarrow AMP + P_i + energy$$
$$ATP + 2H_2O \longrightarrow AMP + 2P_i + energy$$

Figure 23.4 Structures of the various phosphate forms of adenosine.

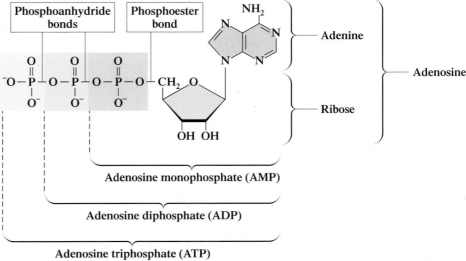

In metabolic pathways in which they are involved, the adenosine phosphates continually change back and forth among the various forms:

$$ATP \rightleftharpoons ADP \rightleftharpoons AMP$$

These hydrolyses are energy-producing reactions that are used to drive cellular processes that require energy input. The phosphoanhydride bonds in ATP and ADP are *very reactive* bonds that require less energy than normal to break. The presence of such reactive bonds, which are often called *strained bonds* (see Section 23.4), is the basis for the net energy production that accompanies hydrolysis. Greater-than-normal electron–electron repulsive forces at specific locations within a molecule are the cause for bond strain; in ATP and ADP, it is the highly electronegative oxygen atoms in the additional phosphate groups that cause the increased repulsive strain.

A typical cellular reaction in which ATP functions as both a source of a phosphate group and a source of energy is the conversion of glucose to glucose-6-phosphate, a reaction that is the first step in the process of glycolysis (Section 24.2).

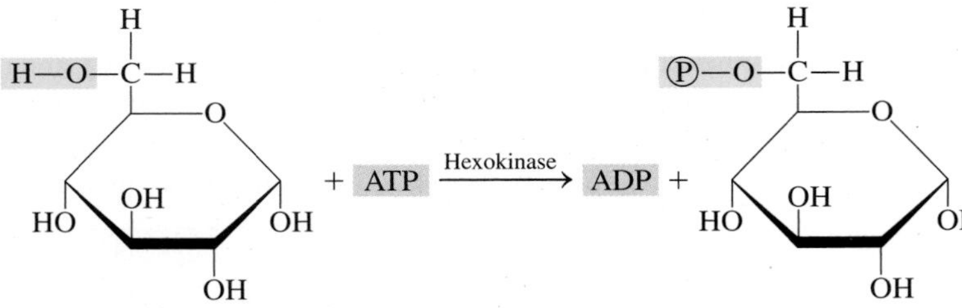

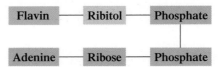

The symbol ⒫ is a shorthand notation for a PO_3^{2-} unit.

Glucose   Glucose 6-phosphate

ATP is not the only nucleotide triphosphate present in cells, although it is the most prevalent. The other nitrogen-containing bases associated with nucleotides (Section 22.2) are also present in triphosphate form. Uridine triphosphate (UTP) is involved in carbohydrate metabolism, guanosine triphosphate (GTP) participates in protein and carbohydrate metabolism, and cytidine triphosphate (CTP) is involved in lipid metabolism.

Flavin Adenine Dinucleotide (FAD, FADH$_2$)

Flavin adenine dinucleotide (FAD) is a coenzyme (Section 21.2) required in numerous metabolic redox reactions. Structurally, FAD can be visualized as containing either three subunits or six subunits. A block diagram of FAD from the three-subunit viewpoint is

Flavin	Ribitol	ADP

Flavin and ribitol, the two components attached to the ADP unit, together constitute the B vitamin riboflavin (Section 21.12). The block diagram for FAD from the six-subunit viewpoint is

Flavin	Ribitol	Phosphate
Adenine	Ribose	Phosphate

This block diagram shows the basis for the name *f*lavin *a*denine *d*inucleotide. Ribitol is a reduced form of ribose; a —CH$_2$OH group is present in place of the —CHO group (Section 18.12).

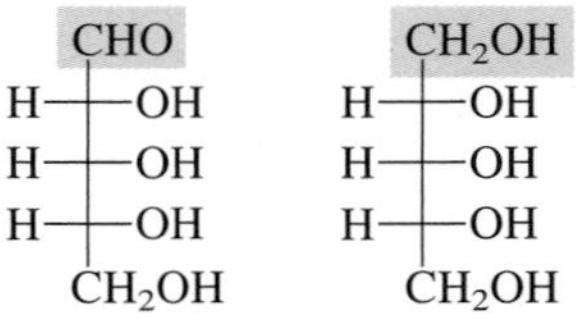

$$
\begin{array}{cc}
\text{CHO} & \text{CH}_2\text{OH} \\
\text{H}-\!\!\!-\text{OH} & \text{H}-\!\!\!-\text{OH} \\
\text{H}-\!\!\!-\text{OH} & \text{H}-\!\!\!-\text{OH} \\
\text{H}-\!\!\!-\text{OH} & \text{H}-\!\!\!-\text{OH} \\
\text{CH}_2\text{OH} & \text{CH}_2\text{OH} \\
\text{D-Ribose} & \text{D-Ribitol}
\end{array}
$$

The complete structural formula of FAD is given in Figure 23.5a.

The active portion of FAD in metabolic redox reactions is the flavin subunit of the molecule. The flavin is reduced, converting the FAD to FADH$_2$, a molecule with two additional hydrogen atoms. Thus FAD is the *oxidized* form of the molecule, and FADH$_2$ is the *reduced* form.

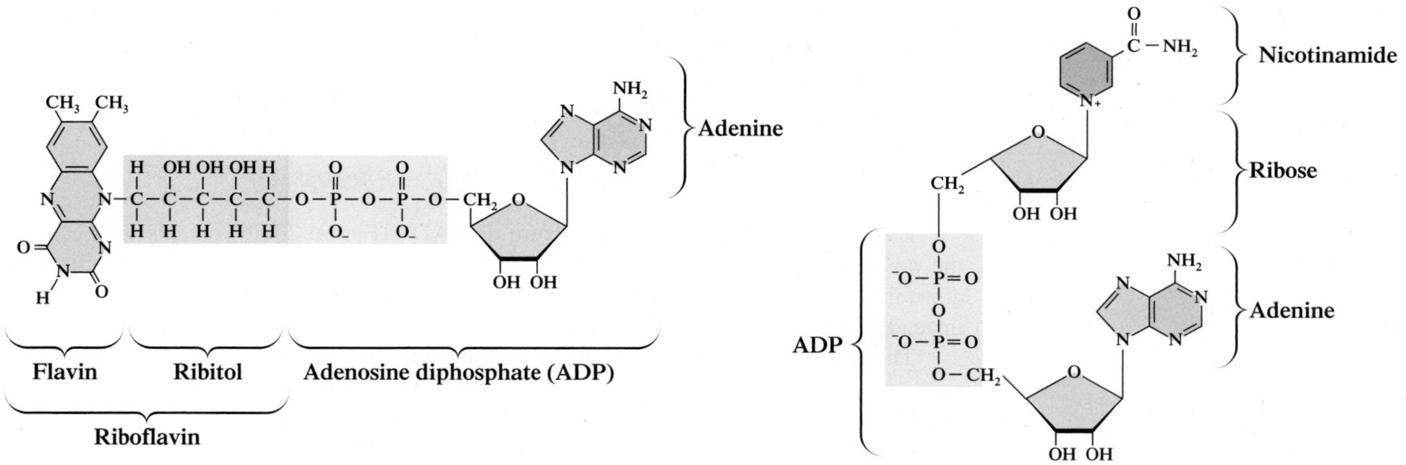

(a) Flavin adenine dinucleotide (FAD)   **(b) Nicotinamide adenine dinucleotide (NAD⁺)**

Figure 23.5 Structural formulas of the molecules flavin adenine dinucleotide, FAD (a) and nicotinamide adenine dinucleotide, NAD^+ (b).

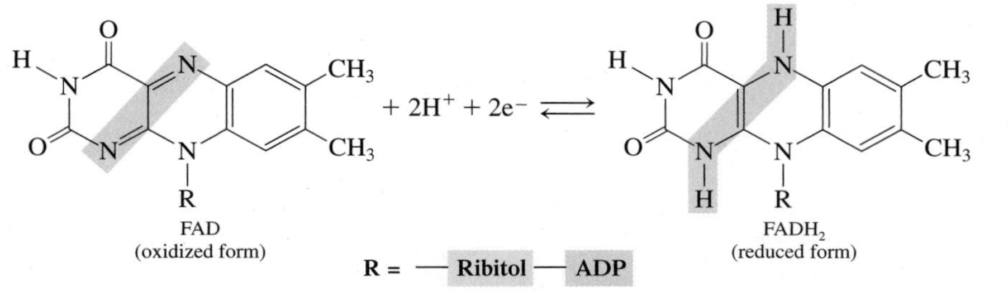

A typical cellular reaction in which FAD serves as the oxidizing agent involves a $-CH_2-CH_2-$ portion of a substrate being oxidized to produce a carbon–carbon double bond.

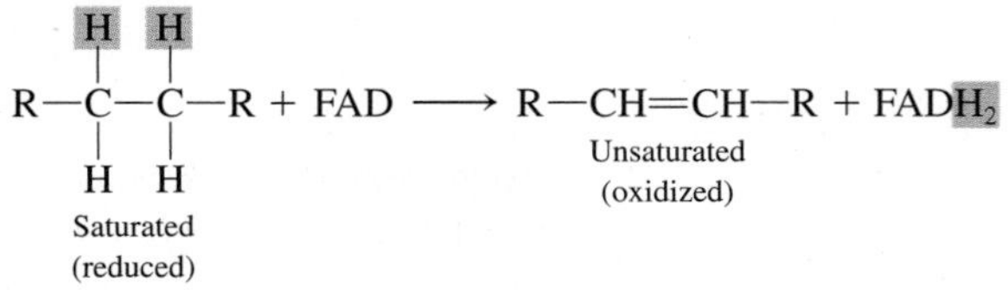

In metabolic pathways in which it is involved, flavin adenine dinucleotide continually changes back and forth between its oxidized form and its reduced form.

$$2H^+ + 2e^- + FAD \rightleftharpoons FADH_2$$

For an enzyme-catalyzed redox reaction involving removal of two hydrogen atoms, such as this, each removed hydrogen atom is equivalent to a hydrogen *ion,* H^+, plus an electron, e^-.

$$2 \text{ H atoms (removed)} \longrightarrow 2H^+ + 2e^-$$

On the basis of this equivalency, the summary equation relating the oxidized and reduced forms of flavin adenine dinucleotide is usually written as

$$\underbrace{2H^+ + 2e^-}_{\text{2 H atoms}} + FAD \rightleftharpoons FADH_2$$

Nicotinamide Adenine Dinucleotide (NAD⁺, NADH)

Several parallels exist between the characteristics of nicotinamide adenine dinucleotide (NAD^+) and those of FAD. Both have coenzyme functions in metabolic redox pathways, both have a B vitamin as a structural component, and both can be represented structurally by using a three-subunit or a six-subunit formulation. In the case of NAD^+, the B vitamin present is nicotinamide (Section 21.12).

The three-subunit block diagram for the structure of NAD^+ is

| Nicotinamide | Ribose | ADP |

The six-subunit block diagram, which emphasizes the dinucleotide nature of the coenzyme, as well as the origin of its name, is

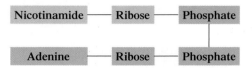

Examination of the detailed structure of NAD^+ (Figure 23.5b) reveals the basis for the positive electrical charge. The $+$ sign refers to the positive charge on the nitrogen atom in the nicotinamide component of the structure; this nitrogen atom has four bonds instead of the usual three (Section 17.6).

The active portion of NAD^+ in metabolic redox reactions is the nicotinamide subunit of the molecule. The nicotinamide is reduced, converting the NAD^+ to NADH, a molecule with one additional hydrogen atom and two additional electrons. Thus NAD^+ is the *oxidized* form of the molecule, and NADH is the *reduced* form.

$$\text{NAD}^+ \text{ (oxidized form)} \quad + \text{ H}^+ + 2e^- \rightleftharpoons \quad \text{NADH (reduced form)}$$

$$R = \quad \boxed{\text{Ribose}} - \boxed{\text{ADP}}$$

A typical cellular reaction in which NAD^+ serves as the oxidizing agent is the oxidation of a secondary alcohol to give a ketone.

> In metabolic pathways, nicotinamide adenine dinucleotide continually changes back and forth between its oxidized form and its reduced form.
>
> $$2H^+ + 2e^- + NAD^+ \rightleftharpoons NADH + H^+$$

$$\underset{\text{2° alcohol}}{R-\overset{OH}{\underset{H}{C}}-R} + NAD^+ \longrightarrow \underset{\text{Ketone}}{R-\overset{O}{C}-R} + NADH + H^+$$

In this reaction, one hydrogen atom of the alcohol substrate is directly transferred to NAD^+, whereas the other appears in solution as H^+ ion. Both electrons lost by the alcohol go to the nicotinamide ring in NADH. (Two electrons are required, rather than one, because of the original positive charge on NAD^+.) Thus the summary equation relating the oxidized and reduced forms of nicotinamide adenine dinucleotide is written as

$$\underbrace{2H^+ + 2e^-}_{\text{2 H atoms}} + NAD^+ \rightleftharpoons NADH + H^+$$

Coenzyme A (CoA–SH)

Another important coenzyme in metabolic pathways is coenzyme A, a derivative of the B vitamin pantothenic acid (Section 21.12). The three-subunit and six-subunit block diagrams for coenzyme A are

and

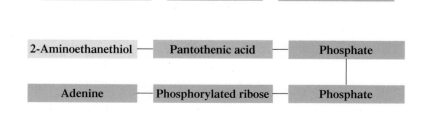

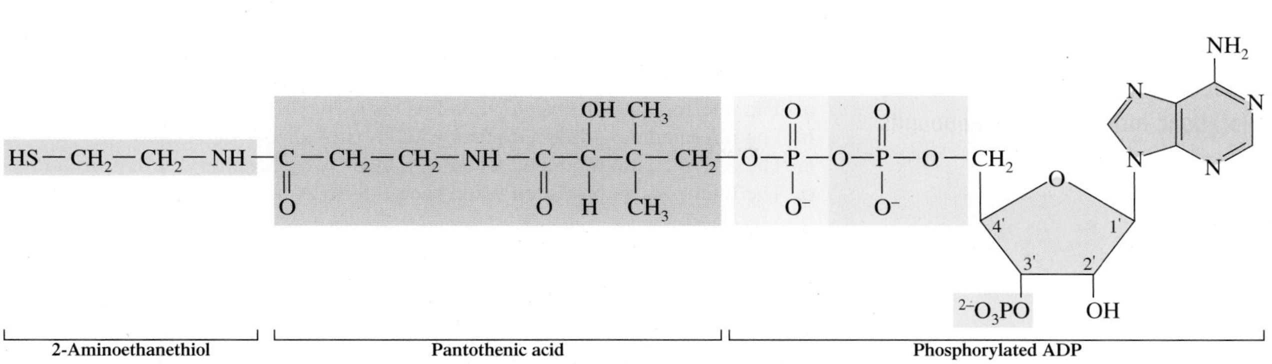

Figure 23.6 Structural formula for coenzyme A (CoA-SH).

2-Aminoethanethiol | Pantothenic acid | Phosphorylated ADP

In metabolic pathways, coenzyme A is continually changing back and forth between its CoA form and its acetyl CoA form.

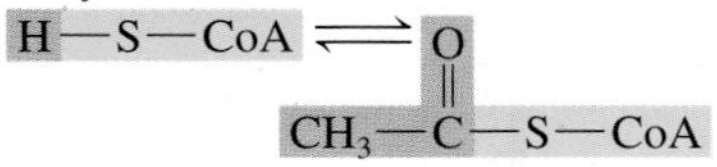

Note, in the three-subunit block diagram, that the ADP subunit present is phosphorylated. As shown in Figure 23.6, the complete structural formula for coenzyme A, the phosphorylated version of ADP carries an extra phosphate group attached to carbon 3′ of its ribose.

The active portion of coenzyme A is the sulfhydryl group (—SH group; Section 14.20) in the ethanethiol subunit of the coenzyme. For this reason, the abbreviation CoA–SH is used for coenzyme A.

Think of the letter A in the name *coenzyme A* as reflecting a general metabolic function of this substance; it is the transfer of *acetyl* groups in metabolic pathways. An **acetyl group** *is the portion of an acetic acid molecule* (CH₃–COOH) *that remains after the* —OH *group is removed from the carboxyl carbon atom.* An acetyl group bonds to CoA–SH through a thioester bond (Section 16.17) to give acetyl CoA.

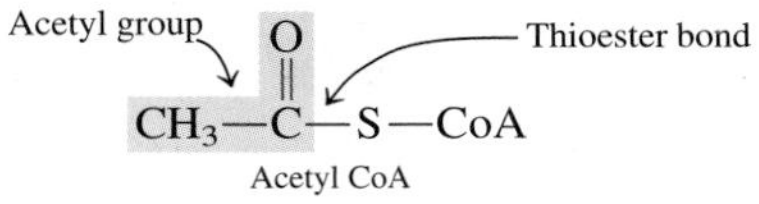

An acetyl group, which can be considered to be derived from acetic acid, has the structure

Classification of Metabolic Intermediate Compounds

The metabolic intermediate compounds considered in this section can be classified into three groups based on function. The classifications are:

1. Intermediates for the storage of energy and transfer of phosphate groups
2. Intermediates for the transfer of electrons in metabolic redox reactions
3. Intermediates for the transfer of acetyl groups

Figure 23.7 shows the category assignment for the intermediates previously considered.

Figure 23.7 Classification of metabolic intermediate compounds in terms of function.

Intermediates for the storage of energy and transfer of phosphate groups	ATP ⇌ ADP ⇌ AMP
Intermediates for the transfer of electrons in metabolic redox reactions	FAD ⇌ FADH₂ NAD⁺ ⇌ NADH
Intermediates for the transfer of acetyl groups	H–S–CoA ⇌ acetyl–S–CoA

EXAMPLE 23.3

Recognizing Relationships Among Metabolic Intermediate Compounds

▶ Give the abbreviated formula for the following metabolic intermediate compounds.

a. The intermediate produced when FAD is reduced
b. The intermediate produced when FADH$_2$ is oxidized
c. The intermediate produced when ATP loses two phosphoryl groups as a PP$_i$
d. The intermediate produced when acetyl–S–CoA transfers an acetyl group

Solution

a. In FAD reduction two hydrogen ions and two electrons are acquired by the FAD to produce *FADH$_2$*.
b. In FADH$_2$ oxidation two hydrogen ions and two electrons are released by the FADH$_2$ to produce *FAD*. FADH$_2$ oxidation and FAD reduction (part a) are reverse processes.
c. Loss of one phosphoryl group by ATP as P$_i$ produces ADP. Loss of two phosphoryl groups by ATP as PP$_i$ produces *AMP*.
d. Release of the acetyl group from acetyl CoA (acetyl–S–CoA) produces *coenzyme A (H–S–CoA)* itself.

Practice Exercise 23.3

Give the abbreviated formula for the following metabolic intermediate compounds.

a. The intermediate produced when NADH is oxidized
b. The intermediate produced when NAD$^+$ is reduced
c. The intermediate produced when a phosphate group is added to AMP
d. The intermediate produced when CoA–S–H bonds to an acetyl group

Answers: a. NAD$^+$; **b.** NADH; **c.** ADP; **d.** Acetyl–S–CoA

23.4 HIGH-ENERGY PHOSPHATE COMPOUNDS

In the previous section, we noted that knowing about several key intermediate compounds in metabolic reactions makes it easier to understand the yet-to-come details of metabolic processes. In like manner, knowing about a particular type of bond present in certain phosphate-containing metabolic intermediates makes the details of metabolic processes easier to understand.

Several phosphate-containing compounds found in metabolic pathways are known as high-energy compounds. A **high-energy compound** *is a compound that has a greater free energy of hydrolysis than that of a typical compound*. High-energy compounds differ from other compounds in that they contain one or more *very reactive* bonds, often called *strained bonds*. The energy required to break these strained bonds during hydrolysis is less than that generally required to break a chemical bond. Consequently, the balance between the energy needed to break bonds in the reactants and that released by bond formation in the products is such that more than the typical amount of free energy is released during the hydrolysis reaction.

Greater-than-normal electron–electron repulsive forces at specific locations within a molecule are the cause of bond strain. Highly electronegative atoms and/or highly charged atoms occurring together in a molecule cause increased repulsive forces and thus increase bond strain.

Let us specifically consider bond strain as it is related to phosphate-containing organic molecules involved in metabolic pathways. The parent molecule for phosphate groups is phosphoric acid, H_3PO_4, a weak triprotic inorganic acid (Section 10.4). This acid exists in aqueous solution in several forms, the dominant form at cellular pH being HPO_4^{2-} ion.

$$\begin{array}{c} O \\ \parallel \\ HO-P-O^- \\ | \\ O^- \end{array}$$

Diphosphate and triphosphate ions can also exist in cellular fluids.

In the definition for a high-energy compound, the term *free energy* rather than simply *energy* was used. Free energy is the amount of energy released by a chemical reaction that is actually available for further use at a given temperature and pressure. In reality, the energy released in a chemical reaction is divisible into two parts. One part, lost as heat, is not available for further use. The other part, the free energy, is available for further use; in cells, it can be used to "drive" reactions that require energy.

In a chemical reaction, the energy balance between bond breaking among reactants (energy input) and new bond formation among products (energy release) determines whether there is a net loss or a net gain of energy (Section 9.5).

The designation *high-energy compound* does not mean that a compound is different from other compounds in terms of bonding. High-energy compounds obey the normal rules for chemical bonding. The only difference between such compounds and other compounds is the presence of one or more *strained bonds*. The breaking of such bonds requires lower-than-normal amounts of energy.

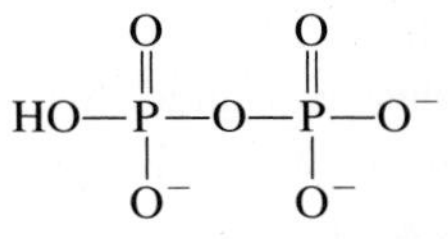

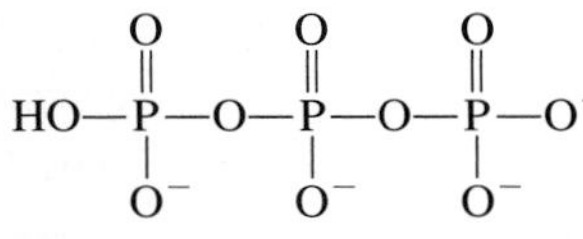

Note the presence in these three phosphate structures of highly electronegative oxygen atoms, many of which bear negative charges. The factors that can produce bond strain are present when phosphates (mono-, di-, and tri-) are bonded to certain organic molecules.

Table 23.1 gives the structures of commonly encountered phosphate-containing compounds, as well as a numerical parameter—the free energy of hydrolysis—that can be considered a measure of the extent of bond strain in the molecules. The more negative the free energy of

TABLE 23.1
Free Energies of Hydrolysis of Common Phosphate-Containing Metabolic Compounds

The —PO_3^{2-} group as part of a larger organic phosphate molecule is referred to as a *phosphoryl group*.

Type	Example	Free Energy of Hydrolysis (kcal/mole)
enol phosphates 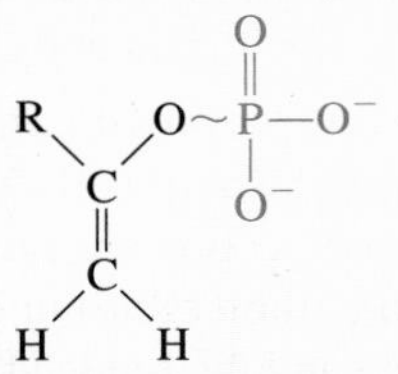	phosphoenolpyruvate	−14.8
acyl phosphates	1,3-bisphosphoglycerate	−11.8
	acetyl phosphate	−11.3
guanidine phosphates	creatine phosphate	−10.3
	arginine phosphate	−9.1
triphosphates	ATP ⟶ AMP + PP$_i$*	−7.7
	ATP ⟶ ADP + P$_i$*	−7.5
diphosphates	PP$_i$ ⟶ 2P$_i$	−7.8
	ADP ⟶ AMP + P$_i$	−7.5
sugar phosphates	glucose 1-phosphate	−5.0
	fructose 6-phosphate	−3.8
	AMP ⟶ adenosine + P$_i$	−3.4
	glucose 6-phosphate	−3.3
	glycerol 3-phosphate	−2.2

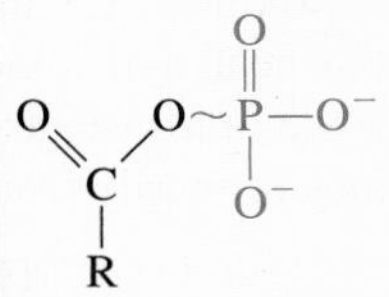

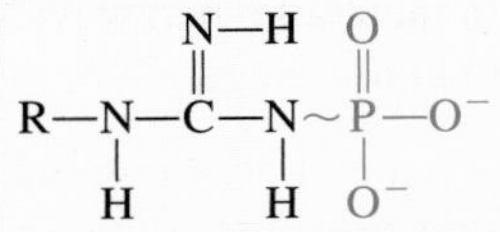

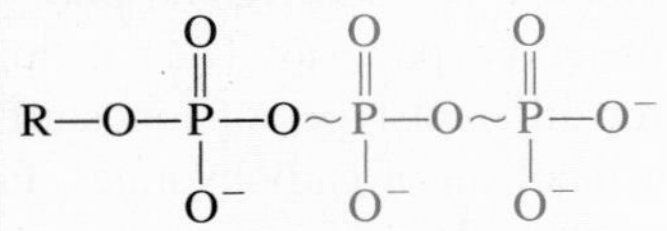

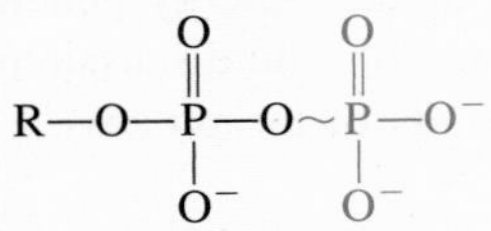

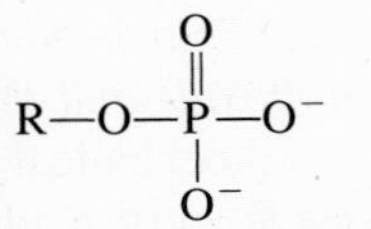

*The notation P$_i$ is used as a general designation for any free monophosphate species present in cellular fluid. Free diphosphate ions are designated as PP$_i$ ("i" stands for *inorganic*).

hydrolysis, the greater the bond strain. A free-energy release greater than 6.0 kcal/mole is generally considered indicative of bond strain. In Table 23.1, strained bonds within the molecules are noted with a squiggle (~), a notation often employed to denote strained bonds.

AN OVERVIEW OF BIOCHEMICAL ENERGY PRODUCTION

The energy needed to run the human body is obtained from ingested food through a multistep process that involves several different catabolic pathways. There are four general stages in the biochemical energy production process, and numerous reactions are associated with each stage.

> The first stage of biochemical energy production, digestion, is not considered part of metabolism because it is extracellular. Metabolic processes are intracellular.

Stage 1: The first stage, *digestion,* begins in the mouth (saliva contains starch-digesting enzymes), continues in the stomach (gastric juices), and is completed in the small intestine (the majority of digestive enzymes and bile salts). The end products of digestion—glucose and other monosaccharides from carbohydrates, amino acids from proteins, and fatty acids and glycerol from fats and oils—are small enough to pass across intestinal membranes and into the blood, where they are transported to the body's cells.

Stage 2: The second stage, *acetyl group formation,* involves numerous reactions, some of which occur in the cytosol of cells and some in cellular mitochondria. The small molecules from digestion are further oxidized during this stage. Primary products include two-carbon acetyl units (which become attached to coenzyme A to give acetyl CoA) and the reduced coenzyme NADH.

Stage 3: The third stage, the *citric acid cycle,* occurs inside mitochondria. Here acetyl groups are oxidized to produce CO_2 and energy. Some of the energy released by these reactions is lost as heat, and some is carried by the reduced coenzymes NADH and $FADH_2$ to the fourth stage. The CO_2 that we exhale as part of the breathing process comes primarily from this stage.

Stage 4: The fourth stage, the *electron transport chain and oxidative phosphorylation,* also occurs inside mitochondria. NADH and $FADH_2$ supply the "fuel" (hydrogen ions and electrons) needed for the production of ATP molecules, the primary energy carriers in metabolic pathways. Molecular O_2, inhaled via breathing, is converted to H_2O in this stage.

The reactions in stages 3 and 4 are the same for all types of foods (carbohydrates, fats, proteins). These reactions constitute the common metabolic pathway. The **common metabolic pathway** *is the sum total of the biochemical reactions of the citric acid cycle, the electron transport chain, and oxidative phosphorylation.* The remainder of this chapter deals with the common metabolic pathway. The reactions of stages 1 and 2 of biochemical energy production differ for different types of foodstuffs. They are discussed in Chapters 24–26, which cover the metabolism of carbohydrates, fats (lipids), and proteins, respectively.

The Chemistry at a Glance feature on page 790 summarizes the four general stages in the process of production of biochemical energy from ingested food. This diagram is a *very simplified* version of the "energy generation" process that occurs in the human body, as will become clear from the discussions presented in later sections of this chapter, which give further details of the process.

THE CITRIC ACID CYCLE

The **citric acid cycle** *is the series of biochemical reactions in which the acetyl portion of acetyl CoA is oxidized to carbon dioxide and the reduced coenzymes $FADH_2$ and NADH are produced.* This cycle, stage 3 of biochemical energy production, gets its name from the first intermediate product in the cycle, citric acid. It is also known as the *Krebs cycle,* after its discoverer Hans Adolf Krebs (see Figure 23.8), and as the *tricarboxylic acid cycle,* in reference to the three carboxylate groups present in citric acid. Figure 23.9 lists the compounds produced in all eight steps of the citric acid cycle.

Figure 23.8 Hans Adolf Krebs (1900-1981), a German-born British biochemist, received the 1953 Nobel Prize in medicine for establishing the relationships among the different compounds in the cycle that carries his name, the Krebs cycle.

Figure 23.9 The citric acid cycle. Details of the numbered steps are given in the text.

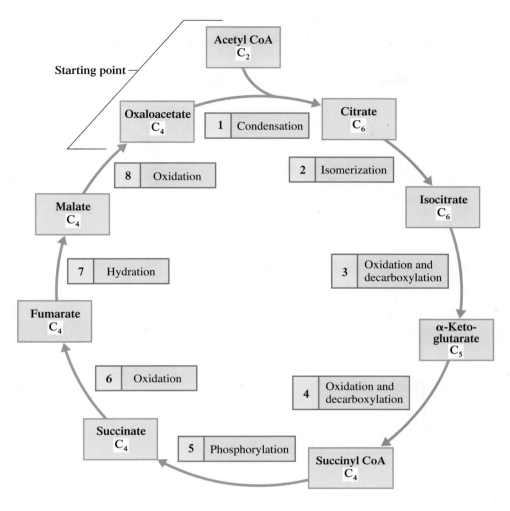

The formation of citryl CoA involves addition of acetyl CoA to the carbon–oxygen double bond. A hydrogen atom of the acetyl —CH$_3$ group adds to the oxygen atom of the double bond, and the remainder of the acetyl CoA adds to the carbon atom of the double bond (Section 15.10).

The chemical reactions of the citric acid cycle take place in the mitochondrial matrix where the needed enzymes are found, except the succinate dehydrogenase reaction that involves FAD. The enzyme that catalyzes this reaction is an integral part of the inner mitochondrial membrane.

We shall now consider the individual steps of the cycle in detail. As we go through these steps, we will observe two important types of reactions: (1) oxidation, which produces NADH and FADH$_2$, and (2) decarboxylation, wherein a carbon chain is shortened by the removal of a carbon atom as CO$_2$.

Reactions of the Citric Acid Cycle

At cellular pH, citric acid is actually present as citrate ion. Despite this, the name of the cycle is the citric acid cycle, which references the molecular, rather than ionic, form of the substance.

Step 1: *Formation of Citrate.* Acetyl CoA, which carries the two-carbon degradation product of carbohydrates, fats, and proteins (Section 23.5), enters the cycle by combining with the four-carbon keto dicarboxylate species oxaloacetate. This results in the transfer of the acetyl group from coenzyme A to oxaloacetate, producing the C$_6$ citrate species and free coenzyme A.

CHEMISTRY AT A GLANCE

Simplified Summary of the Four Stages of Biochemical Energy Production

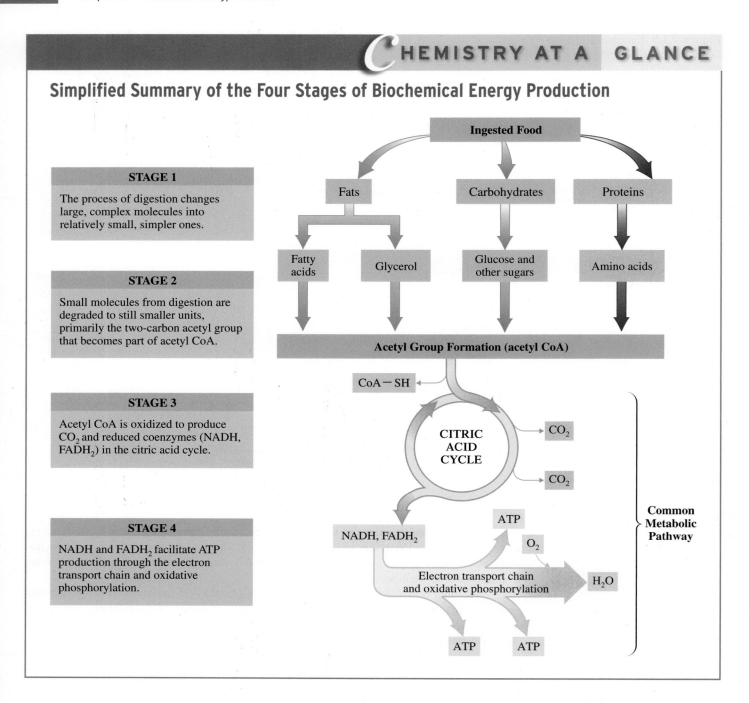

STAGE 1

The process of digestion changes large, complex molecules into relatively small, simpler ones.

STAGE 2

Small molecules from digestion are degraded to still smaller units, primarily the two-carbon acetyl group that becomes part of acetyl CoA.

STAGE 3

Acetyl CoA is oxidized to produce CO_2 and reduced coenzymes (NADH, $FADH_2$) in the citric acid cycle.

STAGE 4

NADH and $FADH_2$ facilitate ATP production through the electron transport chain and oxidative phosphorylation.

A *synthase* is an enzyme that makes a new covalent bond during a reaction without the direct involvement of an ATP molecule.

There are two parts to the reaction: (1) the condensation of acetyl CoA and oxaloacetate to form citryl CoA, a process catalyzed by the enzyme *citrate synthase,* and (2) hydrolysis of the thioester bond in citryl CoA to produce CoA—SH and citrate, also catalyzed by the enzyme *citrate synthase.*

Step 2: *Formation of Isocitrate.* Citrate is converted to its less symmetrical isomer isocitrate in an isomerization process that involves a dehydration followed by a hydration, both catalyzed by the enzyme *aconitase.* The net result of these reactions is that the —OH group from citrate is moved to a different carbon atom.

All acids found in the citric acid cycle exist as carboxylate ions (Section 16.8) at cellular pH.

The key happenings in Step 3 are the following:

1. The hydroxyl group of isocitrate is oxidized to a ketone group.
2. NAD$^+$ is converted to its reduced form, NADH.
3. A carboxyl group from the original oxaloacetate is removed as CO_2.

The CO_2 molecules produced in Steps 3 and 4 of the citric acid cycle are the CO_2 molecules we exhale in the process of respiration.

The key happenings in Step 4 are the following:

1. A second NAD$^+$ is converted to its reduced form, NADH.
2. A second carboxyl group is removed as CO_2.
3. Coenzyme A reacts with the decarboxylation product succinate to produce succinyl CoA, a compound with a high-energy thioester bond. This is the second involvement of a coenzyme A molecule in the cycle, the other instance occurring in Step 1.

The thioester bond in succinyl CoA is a strained bond. Its hydrolysis releases energy, which is trapped by GTP formation. The function of the GTP produced is similar to that of ATP: to store energy in the form of a high-energy phosphate bond (Section 23.4).

Citrate is a tertiary alcohol and isocitrate a secondary alcohol. Tertiary alcohols are not readily oxidized; secondary alcohols are easier to oxidize (Section 14.9). The next step in the cycle involves oxidation.

Step 3: *Oxidation of Isocitrate and Formation of CO_2.* This step involves oxidation–reduction (the first of four redox reactions in the citric acid cycle) and decarboxylation. The reactants are a NAD$^+$ molecule and isocitrate. The reaction, catalyzed by *isocitrate dehydrogenase,* is complex: (1) Isocitrate is oxidized to a ketone (oxalosuccinate) by NAD$^+$, releasing two hydrogens. (2) One hydrogen and two electrons are transferred to NAD$^+$ to form NADH; the remaining hydrogen ion (H$^+$) is released. (3) The oxalosuccinate remains bound to the enzyme and undergoes decarboxylation (loses CO_2), which produces the C_5 α-ketoglutarate (a keto dicarboxylate species).

This step yields the first molecules of CO_2 and NADH in the cycle.

Step 4: *Oxidation of α-Ketoglutarate and Formation of CO_2.* This second redox reaction of the cycle involves one molecule each of NAD$^+$, CoA—SH, and α-ketoglutarate. The catalyst is an aggregate of three enzymes called the *α-ketoglutarate dehydrogenase* complex. As in Step 3, both oxidation and decarboxylation occur. There are three products: CO_2, NADH, and the C_4 species succinyl CoA.

Step 5: *Thioester bond cleavage in Succinyl CoA and Phosphorylation of GDP.* Two molecules react with succinyl CoA—a molecule of GDP (similar to ADP; Section 23.3) and a free phosphate group (P$_i$). The enzyme *succinyl CoA synthetase* removes coenzyme A by thioester bond cleavage. The energy released is used to combine GDP and P$_i$ to form GTP. Succinyl CoA has been converted to succinate.

Steps 6 through 8 of the citric acid cycle involve a sequence of functional group changes that we have encountered several times in the organic sections of the text. The reaction sequence is

$$\text{Alkane} \xrightarrow[\text{(dehydrogenation)}]{\overset{①}{\text{Oxidation}}} \text{alkene} \xrightarrow{\overset{②}{\text{Hydration}}} \begin{array}{c}\text{secondary}\\\text{alcohol}\end{array} \xrightarrow[\text{(dehydrogenation)}]{\overset{③}{\text{Oxidation}}} \text{ketone}$$

Step 6: *Oxidation of Succinate.* This is the third redox reaction of the cycle. The enzyme involved is *succinate dehydrogenase,* and the oxidizing agent is FAD rather than NAD^+. Two hydrogen atoms are removed from the succinate to produce fumarate, a C_4 species with a *trans* double bond. FAD is reduced to $FADH_2$ in the process.

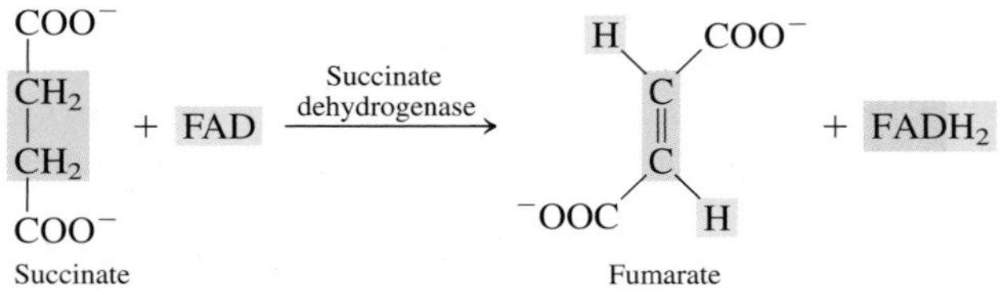

Fumarate, with its *trans* double bond, is an essential metabolic intermediate in both plants and animals. Its isomer, with a *cis* double bond, is called maleate, and it is toxic and irritating to tissues. *Succinate dehydrogenase* produces only the *trans* isomer of this unsaturated diacid.

Succinate + FAD →(Succinate dehydrogenase) Fumarate + FADH₂

Step 7: *Hydration of Fumarate.* The enzyme *fumarase* catalyzes the addition of water to the double bond of fumarate. The enzyme is stereospecific, so only the L isomer of the product malate is produced.

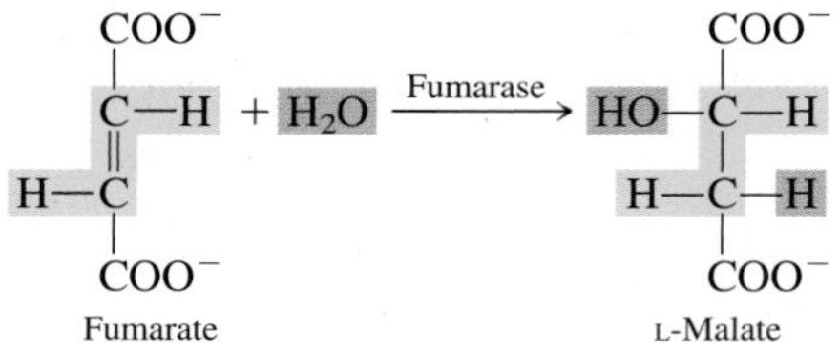

Fumarate + H₂O →(Fumarase) L-Malate

Step 8: *Oxidation of L-Malate to Regenerate Oxaloacetate.* In the fourth oxidation–reduction reaction of the cycle, a molecule of NAD^+ reacts with malate, picking up two hydrogen atoms with their associated energy to form $NADH + H^+$. The product of this reaction is oxaloacetate, so we are back where we started. The oxaloacetate formed in this step can combine with another molecule of acetyl CoA (Step 1), and the cycle can begin again.

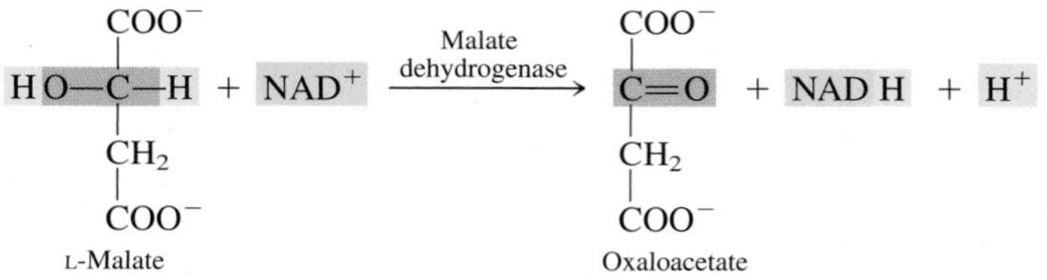

L-Malate + NAD⁺ →(Malate dehydrogenase) Oxaloacetate + NAD H + H⁺

Summary of the Citric Acid Cycle

An overall summary equation for the citric acid cycle is obtained by adding together the individual reactions of the cycle:

$$\text{Acetyl CoA} + 3NAD^+ + FAD + GDP + P_i + 2H_2O \longrightarrow$$
$$2CO_2 + \text{CoA—SH} + 3NADH + 2H^+ + FADH_2 + GTP$$

Important features of the cycle include the following:

1. The "fuel" for the cycle is acetyl CoA, obtained from the breakdown of carbohydrates, fats, and proteins.
2. Four of the cycle reactions involve oxidation and reduction. The oxidizing agent is either NAD^+ (three times) or FAD (once). The operation of the cycle depends on the availability of these oxidizing agents.

The eight B vitamins and their structures were discussed in Section 21.12.

3. In redox reactions, NAD^+ is the oxidizing agent when a carbon–oxygen double bond is formed; FAD is the oxidizing agent when a carbon–carbon double bond is formed.

4. The three NADH and one $FADH_2$ that are formed during the cycle carry electrons and H^+ to the electron transport chain (Section 23.7) through which ATP is synthesized.

5. Two carbon atoms enter the cycle as the acetyl unit of acetyl CoA, and two carbon atoms leave the cycle as two molecules of CO_2. The carbon atoms that enter and leave are not the same ones. The carbon atoms that leave during one turn of the cycle are carbon atoms that entered during the previous turn of the cycle.

6. Four B vitamins are necessary for the proper functioning of the cycle: riboflavin (in both FAD and the α-ketoglutarate dehydrogenase complex), nicotinamide (in NAD^+), pantothenic acid (in CoA—SH), and thiamine (in the α-ketoglutarate dehydrogenase complex).

7. One high-energy GTP molecule is produced by phosphorylation.

The Chemistry at a Glance feature on page 794 gives a detailed diagrammatic summary of the reactions that occur in the citric acid cycle.

EXAMPLE 23.4

Recognizing the Reactants and Products of Various Steps in the Citric Acid Cycle

When one acetyl CoA is processed through the citric acid cycle, how many times does each of the following events occur?

a. A secondary alcohol group is oxidized to a ketone group.
b. A NADH molecule is produced.
c. A decarboxylation reaction occurs.
d. A C_6 molecule is produced.

Solution

a. *Two.* Both isocitrate (Step 3) and malate (Step 8) are secondary alcohols that are oxidized by NAD^+ to ketones.
b. *Three.* The product of the use of NAD^+ as an oxidizing agent is NADH. Oxidation using NAD^+ occurs in Steps 3, 4, and 8.
c. *Two.* In Step 3 isocitrate is decarboxylated and in Step 4 α-ketoglutarate is decarboxylated. In each case the product is a CO_2 molecule.
d. *Two.* Citrate and isocitrate are C_6 molecules. Citrate is produced in Step 1 and isocitrate is produced in Step 2.

Practice Exercise 23.4

When one acetyl CoA is processed through the citric acid cycle, how many times does each of the following events occur?

a. A FAD molecule is a reactant.
b. A CoA–S–H molecule is produced.
c. A dehydrogenase enzyme is needed for the reaction to occur.
d. A C_5 molecule is produced.

Answers: a. One (Step 6); **b.** Two (Steps 1 and 5); **c.** Four (Steps 3, 4, 6, and 8); **d.** One (Step 3)

Regulation of the Citric Acid Cycle

The rate at which the citric acid cycle operates is controlled by the body's need for energy (ATP). When the body's ATP supply is high, the ATP present inhibits the activity of citrate synthase, the enzyme in Step 1 of the cycle. When energy is being used at a high rate, a state of low ATP and high ADP concentrations, the ADP activates citrate synthase and the cycle speeds up. A similar control mechanism exists at Step 3, which involves isocitrate dehydrogenase; here NADH acts as an inhibitor and ADP as an activator.

$\mathcal{C}$HEMISTRY AT A GLANCE

Summary of the Reactions of the Citric Acid Cycle

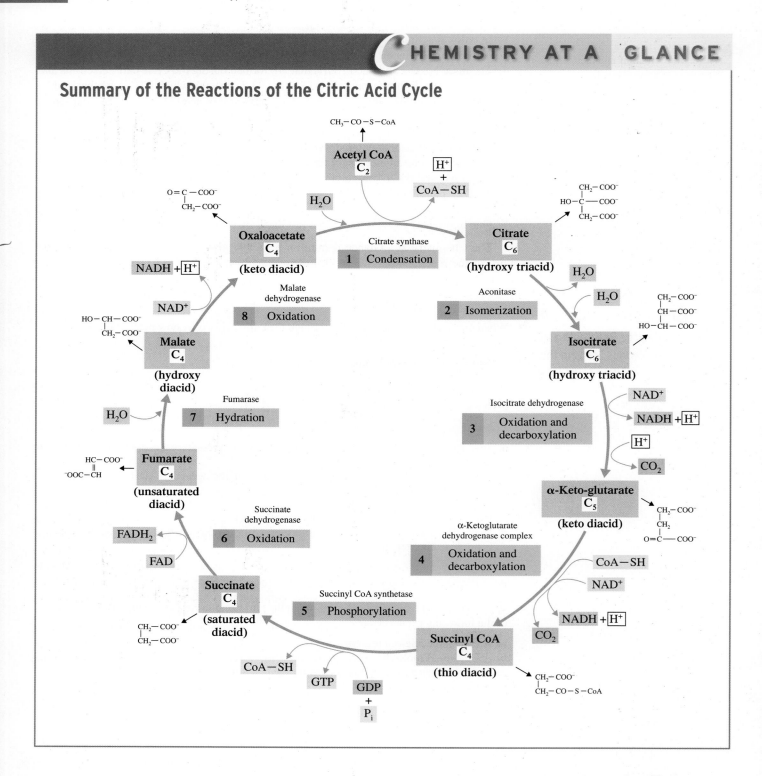

23.7 THE ELECTRON TRANSPORT CHAIN

The *electron transport chain* is also frequently called the *respiratory chain*.

The NADH and FADH$_2$ produced in the citric acid cycle pass to the electron transport chain. The **electron transport chain** *is a series of biochemical reactions in which electrons and hydrogen ions from NADH and FADH$_2$ are passed to intermediate carriers and* *then ultimately react with molecular oxygen to produce water.* NADH and $FADH_2$ are oxidized in this process.

$$NADH + H^+ \longrightarrow NAD^+ + 2H^+ + 2e^-$$
$$FADH_2 \longrightarrow FAD + 2H^+ + 2e^-$$

Water is formed when the electrons and hydrogen ions that originate from these reactions react with molecular oxygen.

$$O_2 + 4e^- + 4H^+ \longrightarrow 2H_2O$$

The electrons that pass through the various steps of the electron transport chain (ETC) lose some energy with each transfer along the chain. Some of this "lost" energy is used to make ATP from ADP (oxidative phosphorylation), as we will see in Section 23.8.

The enzymes and electron carriers needed for the ETC are located along the inner mitochondrial membrane. Within this membrane are four distinct protein complexes, each containing some of the molecules needed for the ETC process to occur. These four protein complexes, which are tightly bound to the membrane, are

Complex I: NADH–coenzyme Q reductase
Complex II: Succinate–coenzyme Q reductase
Complex III: Coenzyme Q–cytochrome c reductase
Complex IV: Cytochrome c oxidase

Two electron carriers, coenzyme Q and cytochrome c, which are not tightly associated with any of the four complexes, serve as mobile electron carriers that shuttle electrons between the various complexes.

Our discussion of the individual reactions that occur in the ETC is divided into four parts, each part dealing with the reactions associated with one of the four protein complexes.

Complex I: NADH–Coenzyme Q Reductase

NADH, from the citric acid cycle, is the source for the electrons that are processed through complex I, the largest of the four protein complexes. Complex I contains over 40 subunits, including flavin mononucleotide (FMN) and several iron–sulfur proteins (FeSP). The net result of electron movement through complex I is the transfer of electrons from NADH to coenzyme Q (CoQ), a result implied by the name of complex I: *NADH–coenzyme Q reductase*. The actual electron transfer process is not, however, a single-step direct transfer of electrons from NADH to CoQ; several intermediate carriers are involved.

The first electron transfer step that occurs in complex I involves the interaction of NADH with flavin mononucleotide (FMN). The NADH is oxidized to NAD^+ (which can again participate in the citric acid cycle) as it passes two hydrogen ions and two electrons to FMN, which is reduced to $FMNH_2$.

$$NADH + H^+ \xrightarrow{\text{Oxidation}} NAD^+ + 2H^+ + 2e^-$$

$$2H^+ + 2e^- + FMN \xrightarrow{\text{Reduction}} FMNH_2$$

NADH supplies both electrons and one of the H^+ ions that are transferred; the other H^+ ion comes from the matrix solution. The actual changes that occur within the structure of FMN as it accepts the two electrons and two H^+ ions are shown in Figure 23.10a.

The next steps involve transfer of electrons from the reduced $FMNH_2$ through a series of iron/sulfur proteins (FeSPs). The iron present in these FeSPs is Fe^{3+}, which is reduced to Fe^{2+}. The two H atoms of $FMNH_2$ are released to solution as two H^+ ions. Two FeSP molecules are needed to accommodate the two electrons released by $FMNH_2$ because an Fe^{3+}/Fe^{2+} reduction involves only one electron.

The oxygen involved in the water formation associated with the electron transport chain is the oxygen we breathe.

The $FMN/FMNH_2$ pair is the third biochemical situation we have encountered in which a flavin molecule is present. The other two are the $FAD/FADH_2$ pair and the B vitamin riboflavin. FMN differs from FAD in not having an adenine nucleotide. Both FMN and FAD are synthesized within the body from riboflavin.

Figure 23.10 Structural characteristics of the electron carriers flavin mononucleotide and coenzyme Q. (a) The oxidized form (FMN) and reduced form (FMNH$_2$) of the electron carrier flavin mononucleotide. (b) The oxidized form (CoQ) and reduced form (CoQH$_2$) of the electron carrier coenzyme Q.

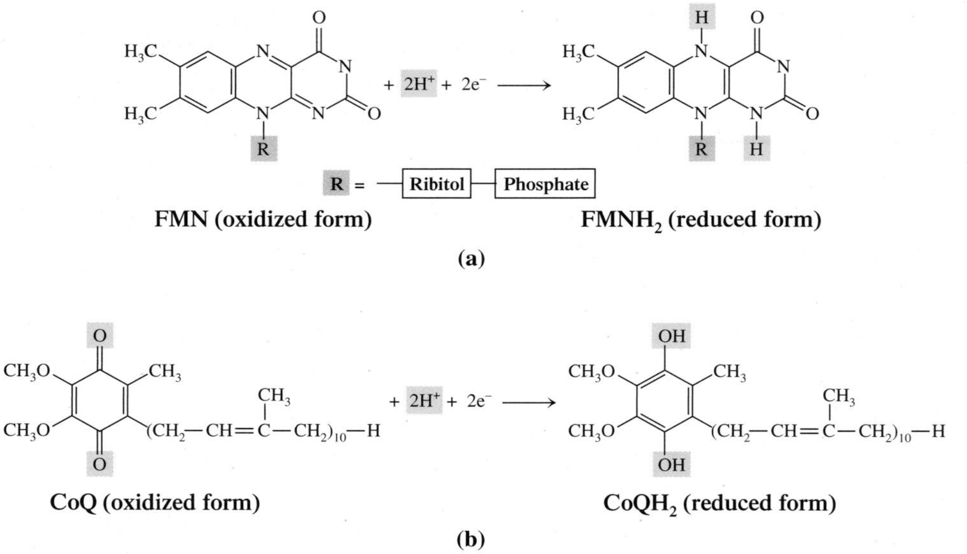

FMN (oxidized form) $\xrightarrow{+\ 2H^+ +\ 2e^-}$ **FMNH$_2$ (reduced form)**

R = — Ribitol — Phosphate

(a)

CoQ (oxidized form) $\xrightarrow{+\ 2H^+ +\ 2e^-}$ **CoQH$_2$ (reduced form)**

(b)

$$FMNH_2 \xrightarrow{\text{Oxidation}} FMN + 2H^+ + 2e^-$$

$$2e^- + 2Fe(III)SP \xrightarrow{\text{Reduction}} 2Fe(II)SP$$

In the final complex I reaction, Fe(II)SP is reconverted into Fe(III)SP as each of two Fe(II)SP units passes an electron to CoQ, changing it from its oxidized form (CoQ) to its reduced form (CoQH$_2$).

$$2Fe(II)SP \xrightarrow{\text{Oxidation}} 2Fe(III)SP + 2e^-$$

$$2e^- + 2H^+ + CoQ \xrightarrow{\text{Reduction}} CoQH_2$$

Coenzyme Q, in both its oxidized and reduced forms, is lipid soluble and can move laterally within the mitochondrial membrane. Its function is to shuttle its newly acquired electrons to complex III, where it becomes the initial substrate for reactions at this complex.

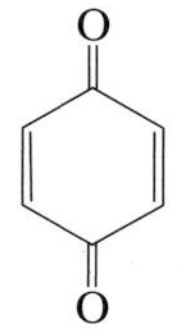

The Q in the designation coenzyme Q comes from the molecule quinone. Structurally, coenzyme Q is a quinone derivative. In its most common form, coenzyme Q has a long carbon chain containing 10 isoprene units (Section 13.6) attached to its quinone unit. The actual changes that occur within the structure of CoQ as it accepts the two electrons and the two H$^+$ ions involve the quinone part of its structure, as is shown in Figure 23.10b. The two H$^+$ ions that CoQ picks up in forming CoQH$_2$ come from solution.

Complex II: Succinate–Coenzyme Q Reductase

Complex II, which is much smaller than complex I, contains only 4 subunits, including two FeSPs. This complex is used to process the FADH$_2$ that is generated in the citric acid cycle when succinate is converted to fumarate. (Thus the use of the term *succinate* in the name of complex II.)

CoQ is associated with the operations in complex II in a manner similar to its actions in complex I. It is the final recipient of the electrons from FADH$_2$, with iron–sulfur proteins serving as intermediaries.

Thus complexes I and II produce a common product, the reduced form of coenzyme Q (CoQH$_2$). As was the case with complex I, the reduced CoQH$_2$ shuttles electrons to complex III.

The molecule quinone, a cyclic ketone (Section 15.3), has the structure.

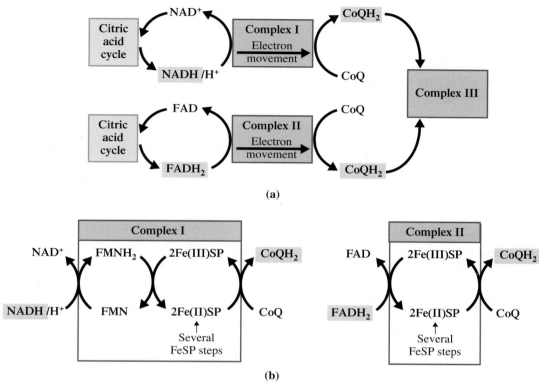

Figure 23.11 An overview of electron movement through complexes I and II of the electron transport chain. (a) $CoQH_2$ carries electrons from both complexes I and II to complex III. (b) NADH is the substrate for complex I and $FADH_2$ is the substrate for complex II. $CoQH_2$ is the common product from both electron transfer processes.

$$FADH_2 \xrightarrow{\text{Oxidation}} FAD + 2H^+ + 2e^-$$

$$2e^- + 2Fe(III)SP \xrightarrow{\text{Reduction}} 2Fe(II)SP$$

$$2Fe(II)SP \xrightarrow{\text{Oxidation}} 2Fe(III)SP + 2e^-$$

$$2e^- + 2H^+ + CoQ \xrightarrow{\text{Reduction}} CoQH_2$$

Figure 23.11 summarizes the electron transport chain reactions associated with complexes I and II. In Figure 23.11a the net process is shown with only starting and end products shown. In Figure 23.11b individual reaction detail is shown. Note the general pattern that is developing for the electron carriers. They are reduced in one step (accept electrons) and then regenerated (oxidized; lose electrons) in the next step so that they can again participate in electron transport chain reactions.

Complex III: Coenzyme Q–Cytochrome c Reductase

Complex III contains 11 different subunits. Electron carriers present include several iron–sulfur proteins as well as several cytochromes. A **cytochrome** *is a heme-containing protein in which reversible oxidation and reduction of an iron atom occur.* Heme, a compound also present in hemoglobin and myoglobin (Section 20.13), has the structure

All H^+ ions required for the reactions of NADH, CoQ, and O_2 in the ETC come from the matrix side of the inner mitochondrial membrane.

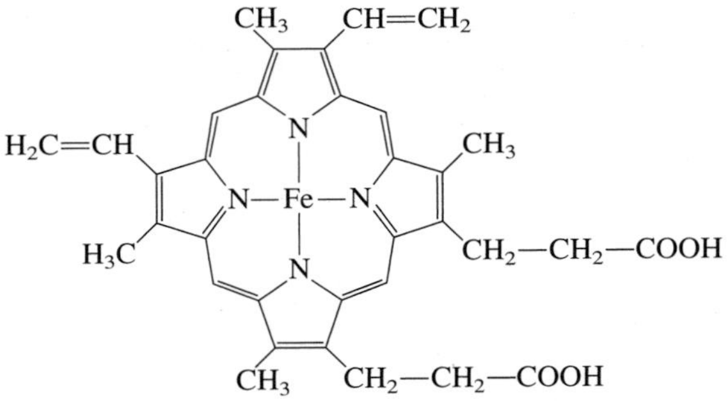

> In cytochromes the iron of the heme is involved in redox reactions in which the iron changes back and forth between the +2 and +3 oxidation states.

> In cytochromes the heme present is bound to protein in such a way as to prevent the heme from combining with oxygen as it does when it is present in hemoglobin.

> Iron/sulfur protein (FeSP) is a *non-heme iron protein*. Most proteins of this type contain sulfur, as is the case with FeSP. Often the iron is bound to the sulfur atom in the amino acid cysteine.

> A feature that all steps in the ETC share is that as each electron carrier passes electrons along the chain, it becomes reoxidized and thus able to accept more electrons.

Heme-containing proteins function similarly to FeSP; iron changes back and forth between the +3 and +2 oxidation states.

Various cytochromes, abbreviated cyt a, cyt b, cyt c, and so on, differ from each other in (1) their protein constituents, (2) the manner in which the heme is bound to the protein, and (3) attachments to the heme ring. Again, because the Fe^{3+}/Fe^{2+} system involves only a one-electron change, two cytochrome molecules are needed to move two electrons along the chain.

The initial substrate for complex III is $CoQH_2$ molecules carrying the electrons that have been processed through complex I (from NADH) and also those processed through complex II (from $FADH_2$). The electron transfer process proceeds from $CoQH_2$ to an FeSP, then to cyt b, then to another FeSP, then to cyt c_1, and finally to cyt c. Cyt c can move laterally in the intermembrane space; it delivers its electrons to complex IV. Cyt c is the only one of the cytochromes that is water soluble.

The initial oxidation–reduction reaction at complex III is between $CoQH_2$ and an iron–sulfur protein (FeSP).

$$CoQH_2 \xrightarrow{\text{Oxidation}} CoQ + 2e^- + 2H^+$$

$$2e^- + 2Fe(III)SP \xrightarrow{\text{Reduction}} 2Fe(II)SP$$

The H^+ ions produced from the oxidation of $CoQH_2$ go into cellular solution. All further redox reactions at complex III involve only electrons, which are conveyed further down the enzyme complex chain. Figure 23.12 shows diagrammatically the electron transfer steps associated with complex III.

Complex IV: Cytochrome c Oxidase

Complex IV contains 13 subunits, including two cytochromes. The electron movement flows from cyt c (carrying electrons from complex III) to cyt a to cyt a_3. In the final step of electron transfer, the electrons from cyt a_3 and hydrogen ions from cellular solution combine with oxygen (O_2) to form water.

$$O_2 + 4H^+ + 4e^- \longrightarrow 2H_2O$$

It is estimated that 95% of the oxygen used by cells serves as the final electron acceptor for the ETC.

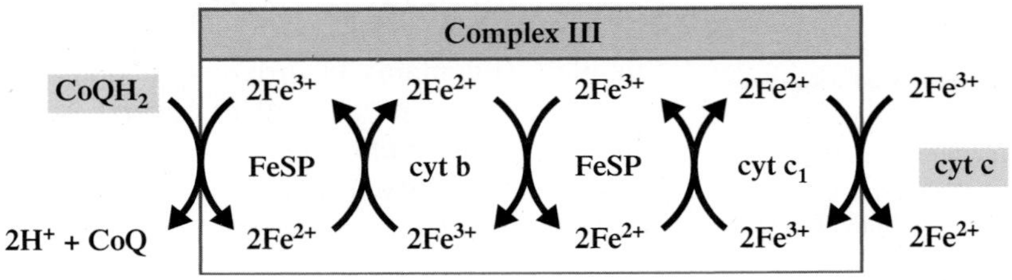

Figure 23.12 Electron movement through complex III is initiated by the electron carrier $CoQH_2$. In several steps the electrons are passed to cyt c.

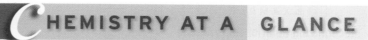

CHEMISTRY AT A GLANCE

Summary of the Flow of Electrons Through the Four Complexes of the Electron Transport Chain

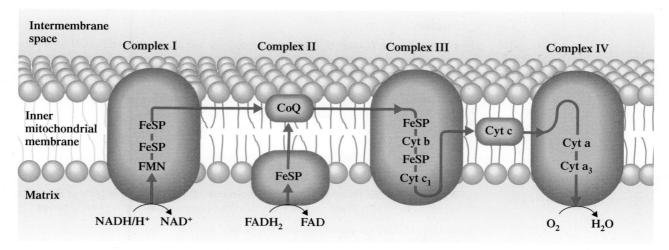

Flow of electrons ⟶

Both hydrogen and electrons from NADH and FADH$_2$ participate in the reactions involving enzyme complexes I and II. Following the formation of CoQH$_2$, hydrogen ions no longer directly participate in enzyme complex reactions, that is, they are not passed further down the enzyme complex chain. Instead they become part of the cellular solution, where they participate in several reactions including the process by which ATP is synthesized (Section 23.8).

The two cytochromes present in cytochrome c oxidase (a and a$_3$) differ from previously encountered cytochromes in that each has a copper atom associated with it in addition to its iron center. The copper atom sites participate in the electron transfer process as do the iron atom sites, with the copper atoms going back and forth between the reduced Cu$^+$ state and the oxidized Cu^{2+} state. Figure 23.13 shows the electron transfer sequence through these copper and iron sites.

The Chemistry at a Glance feature above is a schematic diagram summarizing the flow of electrons through the four complexes of the electron transport chain.

Figure 23.13 The electron transfer pathway through complex IV (cytochrome c oxidase). Electrons pass through both copper and iron centers and in the last step interact with molecular O$_2$. Reduction of one O$_2$ molecule requires the passage of four electrons through complex IV, one at a time.

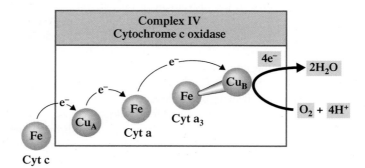

E XAMPLE 23.5

Recognizing Relationships Among
Electron Carriers and Enzyme
Complexes in the Electron
Transport Chain

With which of the four complexes in the electron transport chain is each of the following events associated? (There may be more than one correct answer in a given situation.)

a. Iron–sulfur proteins (FeSPs) are needed as reactants.
b. The mobile electron carrier CoQ serves as a "shuttle molecule."
c. Molecular O_2 is needed as a reactant.
d. FAD is a product.

Solution

a. *Complexes I, II, and III.* Iron sulfur proteins accept electrons from NADH (complex I), $FADH_2$ (complex II), and $CoQH_2$ (complex III).
b. *Complexes I, II, and III.* CoQ functions as a shuttle molecule between complex I and complex III and also between complex II and complex III.
c. *Complex IV.* The final electron acceptor in the ETC is molecular O_2. It combines with electrons and H^+ ions to produce H_2O.
d. *Complex II.* $FADH_2$ is converted to FAD at complex II.

Practice Exercise 23.5

With which of the four complexes in the electron transport chains is each of the following events associated? (There may be more than one correct answer in a given situation.)

a. The metal iron is present in the form of Fe^{2+} and Fe^{3+} ions.
b. $FADH_2$ is needed as a reactant.
c. The metal copper is present in the form of Cu^+ and Cu^{2+} ions.
d. Cytochromes are needed as reactants.

Answers: a. Complexes I, II, III, and IV; **b.** Complex II; **c.** Complex IV; **d.** Complexes III and IV

23.8 OXIDATIVE PHOSPHORYLATION

Oxidative phosphorylation *is the biochemical process by which ATP is synthesized from ADP as a result of the transfer of electrons and hydrogen ions from NADH or $FADH_2$ to O_2 through the electron carriers involved in the electron transport chain.* Oxidative phosphorylation is conceptually simple but mechanistically complex. Learning the "details" of oxidative phosphorylation has been—and still is—one of the most challenging research areas in biochemistry.

One concept central to the oxidative phosphorylation process is that of *coupled* reactions. **Coupled reactions** *are pairs of biochemical reactions that occur concurrently in which energy released by one reaction is used in the other reaction.* Oxidative phosphorylation and the oxidation reactions of the electron transport chain are coupled systems.

The interdependence (coupling) of ATP synthesis with the reactions of the ETC is related to the movement of protons (H^+ ions) across the inner mitochondrial membrane. Three of the four protein complexes involved in the ETC chain (I, III, and IV) have a second function besides electron transfer down the chain. They also serve as "proton pumps," transferring protons from the matrix side of the inner mitochondrial membrane to the intermembrane space (Figure 23.14).

Some of the H^+ ions crossing the inner mitochondrial membrane come from the reduced electron carriers, and some come from the matrix; the details of how the H^+ ions cross the inner mitochondrial membrane are not fully understood.

For every two electrons passed through the ETC, four protons cross the inner mitochondrial membrane through complex I, four through complex III, and two more through complex IV. This proton flow causes a buildup of H^+ ions (protons) in the

Oxidative phosphorylation is not the only process by which ATP is produced in cells. A second process, substrate phosphorylation (Section 24.2), can also be an ATP source. However, the amount of ATP produced by this second process is much less than that produced by oxidative phosphorylation.

Figure 23.14 A second function for protein complexes I, III, and IV involved in the electron transport chain is that of proton pump. For every two electrons passed through the ETC, 10 H⁺ ions are transferred from the mitochondrial matrix to the intermembrane space through these complexes.

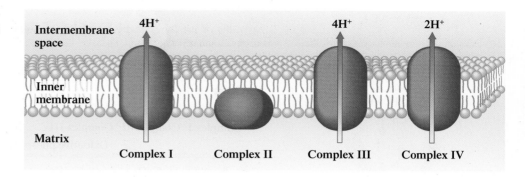

The difference in H⁺ ion concentration between the two sides of the inner mitochondrial membrane causes a pH difference of about 1.4 units. A pH difference of 1.4 units means that the intermembrane space, the more acidic region, has 25 times more protons than the matrix.

intermembrane space; this high concentration of protons becomes the basis for ATP synthesis (Figure 23.15).

The "proton flow" explanation for ATP–ETC coupling is formally called chemiosmotic coupling. **Chemiosmotic coupling** *is an explanation for the coupling of ATP synthesis with electron transport chain reactions that requires a proton gradient across the inner mitochondrial membrane.* The main concepts in this explanation for coupling follow.

1. The result of the pumping of protons from the mitochondrial matrix across the inner mitochondrial membrane is a higher concentration of protons in the intermembrane space than in the matrix. This concentration difference constitutes an *electrochemical (proton) gradient.* A chemical gradient exists whenever a substance has a higher concentration in one region than in another. Because the proton has an electrical

CHEMICAL Connections

Cyanide Poisoning

Inhalation of hydrogen cyanide gas (HCN) or ingestion of solid potassium cyanide (KCN) rapidly inhibits the electron transport chain in all tissues, making cyanide one of the most potent and rapidly acting poisons known. The attack point for the cyanide ion (CN^-) is cytochrome c oxidase, the last of the four protein complexes in the electron transport chain. Cyanide inactivates this complex by bonding itself to the Fe^{3+} in the complex's heme portions. As a result, Fe^{3+} is unable to transfer electrons to oxygen, blocking the cell's use of oxygen. Death results from

tissue asphyxiation, particularly of the central nervous system. Cyanide also binds to the heme group in hemoglobin, blocking oxygen transport in the bloodstream.

One treatment for cyanide poisoning is to administer various nitrites, NO_2^-, which oxidize the iron atoms of hemoglobin to Fe^{3+}. This form of hemoglobin helps draw CN^- back into the bloodstream, where it can be converted to thiocyanate (SCN^-) by thiosulfate ($S_2O_3^{2-}$), which is administered along with the nitrite (see the accompanying figure).

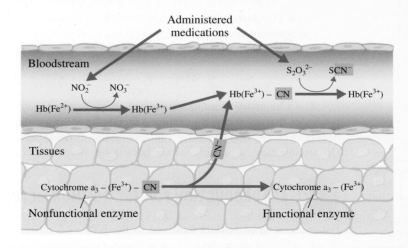

CHEMISTRY AT A GLANCE

Summary of the Common Metabolic Pathway

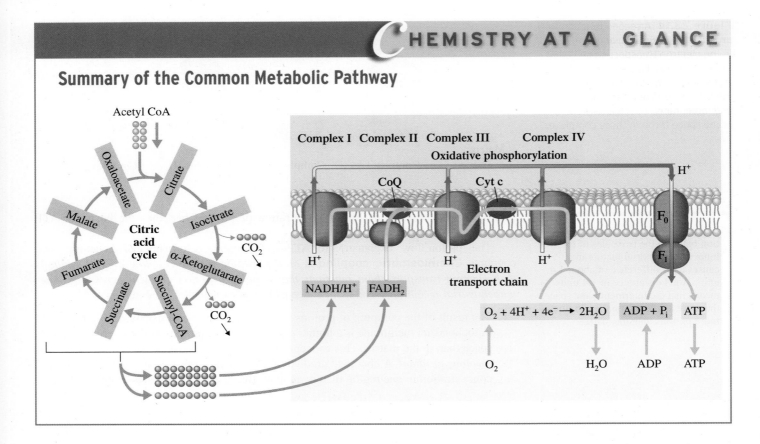

Some of the energy released at each of the protein complexes I, III, and IV is consumed in the movement of H^+ ions across the inner membrane from the matrix into the intermembrane space. Movement of ions from a region of lower concentration (the matrix) to one of higher concentration (the intermembrane space) requires the expenditure of energy because it opposes the natural tendency, as exhibited in the process of osmosis (Section 8.9), to equalize concentrations.

charge (H^+ ion), an electrical gradient also exists. Potential energy (Section 7.2) is always associated with an electrochemical gradient.

2. A spontaneous flow of protons from the region of high concentration to the region of low concentration occurs because of the electrochemical gradient. This proton flow is not through the membrane itself (it is not permeable to H^+ ions) but rather through enzyme complexes called *ATP synthases* located on the inner mitochondrial membrane (Section 23.2). This proton flow through the ATP synthases "powers" the synthesis of ATP. ATP synthases are thus the *coupling factors* that link the processes of oxidative phosphorylation and the electron transport chain.

3. ATP synthase has two subunits, the F_0 and F_1 subunits (Figure 23.15). The F_0 part of the synthase is the channel for proton flow, whereas the formation of ATP takes place in the F_1 subunit. As protons return to the mitochondrial matrix through the F_0 subunit, the potential energy associated with the electrochemical gradient is released and used in the F_1 subunit for the synthesis of ATP.

$$ADP + P_i \xrightarrow{\text{ATP synthase}} ATP + H_2O$$

The Chemistry at a Glance feature above brings together into one diagram the three processes that constitute the common metabolic pathway: the citric acid cycle, the electron transport chain, and oxidative phosphorylation. These three processes operate together. Discussing them separately, as we have done, is a matter of convenience only.

Without oxygen, the biochemical systems of the human body quickly shut down and death occurs. Why? Without oxygen as the final electron acceptor in the ETC, the ETC chain shuts down and ATP production stops. Without ATP to power life's processes (Chapters 24–26), these processes stop.

23.9 ATP PRODUCTION FOR THE COMMON METABOLIC PATHWAY

For each mole of NADH oxidized in the ETC, 2.5 moles of ATP are formed. $FADH_2$, which does not enter the ETC at its start, produces only 1.5 moles of ATP per mole of $FADH_2$ oxidized. $FADH_2$'s entrance point into the chain, complex II, is beyond the first

Figure 23.15 Formation of ATP accompanies the flow of protons from the intermembrane space back into the mitochondrial matrix. The proton flow results from an electrochemical gradient across the inner mitochondrial membrane.

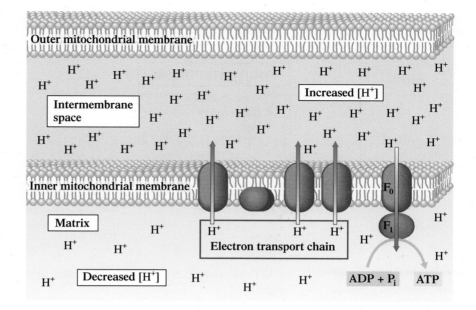

"proton-pumping" site, complex I. Hence fewer ATP molecules are produced from $FADH_2$ than from NADH.

The energy yield, in terms of ATP production, can now be totaled for the common metabolic pathway (Section 23.5). Every acetyl CoA entering the citric acid cycle (CAC) produces three NADH, one $FADH_2$, and one GTP (which is equivalent in energy

CHEMICAL Connections   Brown Fat, Newborn Babies, and Hibernating Animals

Ordinarily, metabolic processes generate enough heat to maintain normal body temperature. In certain cases, however, including newborn infants and hibernating animals, normal metabolism is not sufficient to meet the body's heat requirements. In these cases, a supplemental method of heat generation, which involves *brown fat tissue*, occurs.

Brown fat tissue, as the name implies, is darker in color than ordinary fat tissue, which is white. Brown fat is specialized for heat production. It contains many more blood vessels and mitochondria than white fat. (The increased number of mitochondria gives brown fat its color.)

Another difference between the two types of fat is that the mitochondria in brown fat cells contain a protein called *thermogenin*, which functions as an *uncoupling agent*. This protein "uncouples" the ATP production associated with the electron transport chain. The ETC reactions still take place, but the energy that would ordinarily be used for ATP synthesis is simply released as heat.

Brown fat tissue is of major importance for newborn infants. Newborns are immediately faced with a temperature regulation problem. They leave an environment of constant 37°C temperature and enter a much colder environment (25°C). A supply of *active* brown fat, present at birth, helps the baby adapt to the cooler environment.

Very limited amounts of brown fat are present in most adults. However, stores of brown fat increase in adults who are regularly exposed to cold environs. Thus the production of

Hibernating bears rely on brown fat tissue to help meet their bodies' heat requirements.

brown fat is one of the body's mechanisms for adaptation to cold.

Thermogenin, the uncoupling agent in brown fat, is a protein bound to the inner mitochondrial membrane. When activated, it functions as a proton channel through the inner membrane. The proton gradient produced by the electron transport chain is dissipated through this "new" proton channel, and less ATP synthesis occurs because the normal proton channel, ATP synthase, has been bypassed. The energy of the proton gradient, no longer useful for ATP synthesis, is released as heat.

Figure 23.16 The interconversion of ATP and ADP is the principal medium for energy exchange in biochemical processes.

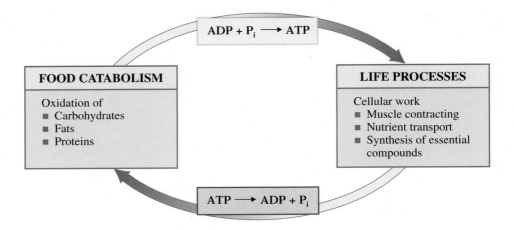

to ATP; Section 23.6). Thus 10 molecules of ATP are produced for each acetyl CoA catabolized.

$$3 \text{ NADH} \longrightarrow 7.5 \text{ ATP}$$
$$1 \text{ FADH}_2 \longrightarrow 1.5 \text{ ATP}$$
$$\underline{1 \text{ GTP} \longrightarrow 1 \quad \text{ATP}}$$
$$10 \quad \text{ATP}$$

Biochemistry textbooks published before the mid-1990s make the following statements:

1 NADH produces 3 ATP in the ETC.

1 FADH$_2$ produces 2 ATP in the ETC.

As more has been learned about the electron transport chain and oxidative phosphorylation, these numbers have had to be reduced. The overall conversion process is more complex than was originally thought, and not as much ATP is produced.

 ### THE IMPORTANCE OF ATP

The cycling of ATP and ADP in metabolic processes is the principal medium for energy exchange in biochemical processes. The conversion

$$\text{ATP} \longrightarrow \text{ADP} + \text{P}_i$$

powers life processes (the biosynthesis of essential compounds, muscle contraction, nutrient transport, and so on). The conversion

$$\text{P}_i + \text{ADP} \longrightarrow \text{ATP}$$

which occurs in food catabolism cycles, regenerates the ATP expended in cell operation. Figure 23.16 summarizes the ATP–ADP cycling process.

ATP is a high-energy phosphate compound (Section 23.4). Its hydrolysis to ADP produces an *intermediate* amount of free energy (-7.5 kcal/mole; Table 23.1) compared with hydrolysis energies for other organophosphate compounds (Table 23.1). Of major importance, the energy derived from ATP hydrolysis is a *biochemically useful* amount of energy. It is larger than the amount of energy needed by compounds to which ATP donates energy, and yet it is smaller than that available in compounds used to form ATP. If the ATP hydrolysis energy were *unusually high,* the body would not be able to convert ADP back to ATP because ATP synthesis requires an energy input equal to or greater than the hydrolysis energy, and such an unusually high amount of energy would not be available.

ATP molecules in cells have a high turnover rate. Normally, a given ATP molecule in a cell does not last more than a minute before it is converted to ADP. The concentration of ATP in a cell varies from 0.5 to 2.5 milligram per milliliter of cellular fluid.

 ### NON-ETC OXYGEN-CONSUMING REACTIONS

The electron transport chain/oxidative phosphorylation phase of metabolism consumes more than 90% of the oxygen taken into the human body via respiration. What happens to the remainder of the inspired O$_2$?

As a normal part of metabolic chemistry, significant amounts of this remaining O_2 are converted into several highly reactive oxygen species (ROS). Among these ROSs are hydrogen peroxide (H_2O_2), superoxide ion (O_2^-), and hydroxyl radical (OH). The latter two of these substances are free radicals, substances that contain an unpaired electron (Section 11.7). Reactive oxygen species have beneficial functions within the body, but they can also cause problems if they are not eliminated when they are no longer needed.

White blood cells have a significant concentration of superoxide free radicals. Here, these free radicals aid in the destruction of invading bacteria and viruses. Their formation reaction is

$$2O_2 + NADPH \longrightarrow 2O_2^- + NADP^+ + H^+$$

(NADP is a phosphorylated version of the coenzyme NADH; see Section 24.8.)

CHEMICAL Connections

Flavonoids: An Important Class of Dietary Antioxidants

Numerous studies indicate that diets high in fruits and vegetables are associated with a "healthy lifestyle." One reason for this is that fruits and vegetables contain compounds called *phytochemicals.* Phytochemicals are compounds found in plants that have biochemical activity in the human body even though they have no nutritional value. The functions that phytochemicals perform in the human body include antioxidant activity, cancer inhibition, cholesterol regulation, and anti-inflammatory activity.

Each fruit and vegetable is a unique package of phytochemicals, so consuming a wide variety of fruits and vegetables provides the body with the broadest spectrum of benefits. In such a situation, many phytochemicals are consumed in *small* amounts. This approach is much safer than taking supplemental doses of particular phytochemicals; in larger doses, some phytochemicals are toxic.

A major group of phytochemicals are the *flavonoids,* of which over 4000 individual compounds are known. All flavonoids are antioxidants (Section 14.14), but some are stronger antioxidants than others, depending on their molecular structure. About 50 flavonoids are present in foods and in beverages derived from plants (tea leaves, grapes, oranges, and so on).

The core *flavonoid* structure is

Both aromatic and cyclic ether ring systems are present. Of particular importance as antioxidants in foods are flavonoids known as *flavones* and *flavonols,* flavonoids whose core structures are enhanced by the presence of ketone and/or hydroxyl groups and a double bond in the oxygen-containing ring system.

Flavones   Flavonols

The formation of flavone and flavonol compounds depends normally on the action of light, so in general the highest concentration of these compounds occurs in leaves or in the skins of fruits, whereas only traces are found in parts of plants that grow below the ground. The common onion is, however, a well-known exception to this generalization.

The most widespread flavonoid in food is the flavonol *quercetin.*

It is predominant in fruits, vegetables, and the leaves of various vegetables. In fruits, apples contain the highest amounts of quercetin, the majority of it being found in the outer tissues (skin, peel). A small peeled apple contains about 5.7 mg of the antioxidant vitamin C. But the same amount of apple *with the skin* contains flavonoids and other phytochemicals that have the effect of 1500 mg of vitamin C. Onions are also major dietary sources of quercetin.

In addition to their antioxidant benefits, flavonoids may also help fight bacterial infections. Recent studies indicate that flavonoids can stop the growth of some strains of drug-resistant bacteria.

Superoxide ion that is not needed is eliminated from cells in a two-step process governed by the enzymes *superoxide dismutase* and *catalase,* two of the most rapidly working enzymes known (see Table 21.2). In the first step, superoxide ion is converted to hydrogen peroxide, which is then, in the second step, converted to H_2O.

$$2O_2^- + 2H^+ \xrightarrow{\text{superoxide dismutase}} H_2O_2 + O_2$$

$$2H_2O_2 \xrightarrow{\text{catalase}} 2H_2O + O_2$$

Immediate destruction of the hydrogen peroxide produced in the first of these two steps is critical, because if it persisted, then unwanted production of hydroxyl radical would occur via hydrogen peroxide's reaction with superoxide ion.

$$H_2O_2 + O_2^- + H^+ \longrightarrow H_2O + O_2 + OH$$

Hydroxyl radicals quickly react with other substances by taking an electron from them. Such action usually causes bond breaking. Lipids in cell membranes are particularly vulnerable to such attack by hydroxyl radicals.

It is estimated that 5% of the ROSs escape destruction through normal channels (superoxide dimutase and catalase). Operating within a cell is a backup system—a network of antioxidants—to deal with this problem. Participating in this antioxidant network are glutathione (Section 20.7), vitamin C (Section 21.12), and beta-carotene and vitamin E (Section 21.13), as well as other compounds obtained from plants through dietary intake. Particularly important in this latter category are compounds called *flavonoids* (see the Chemical Connections feature on page 805). The vitamin antioxidants as well as the other antioxidants present prevent oxidative damage by reacting with the harmful ROS oxidizing agents before they can react with other biologically important substances.

Reactive oxygen species can also be formed in the body as the result of external influences such as polluted air, cigarette smoke, and radiation exposure (including solar radiation). Vitamin C is particularly active against such free-radical damage.

> Antioxidant molecules provide electrons to convert free radicals and other ROSs into less-reactive substances.

CONCEPTS TO REMEMBER

Metabolism. Metabolism is the sum total of all the biochemical reactions that take place in a living organism. Metabolism consists of catabolism and anabolism. Catabolic biochemical reactions involve the breakdown of large molecules into smaller fragments. Anabolic biochemical reactions synthesize large molecules from smaller ones (Section 23.1).

Mitochondria. Mitochondria are membrane-enclosed subcellular structures that are the site of energy production in the form of ATP molecules. Enzymes for both the citric acid cycle and the electron transport chain are housed in the mitochondria (Section 23.2).

Important coenzymes. Three very important coenzymes involved in catabolism are NAD^+, FAD, and CoA. NAD^+ and FAD are oxidizing agents that participate in the oxidation reactions of the citric acid cycle. They transport hydrogen atoms and electrons from the citric acid cycle to the electron transport chain. CoA interacts with acetyl groups produced from food degradation to form acetyl CoA. Acetyl CoA is the "fuel" for the citric acid cycle (Section 23.3).

High-energy compounds. A high-energy compound liberates a larger-than-normal amount of free energy upon hydrolysis because structural features in the molecule contribute to repulsive strain in one

or more bonds. Most high-energy biochemical molecules contain phosphate groups (Section 23.4).

Common metabolic pathway. The common metabolic pathway includes the reactions of the citric acid cycle and those of the electron transport chain and oxidative phosphorylation. The degradation products from all types of foods (carbohydrates, fats, and proteins) participate in the reactions of the common metabolic pathway (Section 23.5).

Citric acid cycle. The citric acid cycle is a cyclic series of eight reactions that oxidize the acetyl portion of acetyl CoA, resulting in the production of two molecules of CO_2. The complete oxidation of one acetyl group produces three molecules of NADH, one of $FADH_2$, and one of GTP besides the CO_2 (Section 23.6).

Electron transport chain. The electron transport chain is a series of reactions that passes electrons from NADH and $FADH_2$ to molecular oxygen. Each electron carrier that participates in the chain has an increasing affinity for electrons. Upon accepting the electrons and hydrogen ions, the O_2 is reduced to H_2O (Section 23.7).

Oxidative phosphorylation. Oxidative phosphorylation is the biochemical process by which ATP is synthesized from ADP as the result of a proton gradient across the inner mitochondrial membrane.

Oxidative phosphorylation is coupled to the reactions of the electron transport chain (Section 23.8).

Chemiosmotic coupling. Chemiosmotic coupling explains how the energy needed for ATP synthesis is obtained. Synthesis takes place because of a flow of protons across the inner mitochondrial membrane (Section 23.8).

Importance of ATP. ATP is the link between energy production and energy use in cells. The conversion of ATP to ADP powers life processes, and the conversion of ADP back to ATP regenerates the energy expended in cell operation (Section 23.10).

KEY REACTIONS AND EQUATIONS

1. Oxidation by FAD (Section 23.3)

$$FAD + 2H^+ + 2e^- \longrightarrow FADH_2$$

2. Oxidation by NAD^+ (Section 23.3)

$$NAD^+ + 2H^+ + 2e^- \longrightarrow NADH + H^+$$

3. The citric acid cycle (Section 23.6)

$$Acetyl\ CoA + 3NAD^+ + FAD + GDP + P_i + 2H_2O \longrightarrow$$
$$2CO_2 + CoA-SH + 3NADH + 2H^+ + FADH_2 + GTP$$

4. The electron transport chain (Section 23.7)

$$NADH + H^+ \longrightarrow NAD^+ + 2H^+ + 2e^-$$
$$FADH_2 \longrightarrow FAD + 2H^+ + 2e^-$$
$$O_2 + 4H^+ + 4e^- \longrightarrow 2H_2O$$

5. Oxidative phosphorylation (Section 23.8)

$$ADP + P_i \xrightarrow[\text{from ETC}]{\text{Energy}} ATP + H_2O$$

EXERCISES and PROBLEMS

The members of each pair of problems in this section test similar material.

Metabolism (Section 23.1)

23.1 Classify anabolism and catabolism as synthetic or degradative processes.

23.2 Classify anabolism and catabolism as energy-producing or energy-consuming processes.

23.3 What is a metabolic pathway?

23.4 What is the difference between a linear and a cyclic metabolic pathway?

23.5 What general characteristics are associated with a catabolic pathway?

23.6 What general characteristics are associated with an anabolic pathway?

Cell Structure (Section 23.2)

23.7 List several differences between prokaryotic cells and eukaryotic cells.

23.8 What kinds of organisms have prokaryotic cells and what kinds have eukaryotic cells?

23.9 What is an organelle?

23.10 What is the general function of each of the following types of organelles?
a. Ribosome   b. Lysosome   c. Mitochondrion

23.11 In a mitochondrion, what separates the matrix from the intermembrane space?

23.12 In what major way do the inner and outer mitochondrial membranes differ?

23.13 What is the intermembrane space of a mitochondrion?

23.14 Where are ATP synthase complexes located in a mitochondrion?

Intermediate Compounds in Metabolic Pathways (Section 23.3)

23.15 What does each letter in ATP stand for?

23.16 What does each letter in ADP stand for?

23.17 Draw a block diagram structure for ATP.

23.18 Draw a block diagram structure for ADP.

23.19 What is the structural difference between ATP and AMP?

23.20 What is the structural difference between ADP and AMP?

23.21 What is the structural difference between ATP and GTP?

23.22 What is the structural difference between ATP and CTP?

23.23 In terms of hydrolysis, what is the relationship between ATP and ADP?

23.24 In terms of hydrolysis, what is the relationship between ADP and AMP?

23.25 What does each letter in FAD stand for?

23.26 What does each letter in NAD^+ stand for?

23.27 Draw a block diagram structure for FAD based on the presence of an ADP core (three-block diagram).

23.28 Draw a block diagram structure for FAD based on the presence of two nucleotides (six-block diagram).

23.29 Draw a block diagram structure for NAD^+ based on the presence of two nucleotides (six-block diagram).

23.30 Draw a block diagram structure for NAD^+ based on the presence of an ADP core (three-block diagram).

23.31 Which part of an NAD^+ molecule is the active participant in redox reactions?

23.32 Which part of an FAD molecule is the active participant in redox reactions?

23.33 Give the letter designation for
a. the reduced form of FAD.
b. the oxidized form of NADH.

23.34 Give the letter designation for
a. the oxidized form of $FADH_2$.
b. the reduced form of NAD^+.

23.35 Name the vitamin B molecule that is part of the structure of
a. NAD^+.   b. FAD.

23.36 Indicate whether or not the vitamin B portion of the following molecules is the "active" portion of the molecule in redox processes.
a. NAD^+.   b. FAD.

23.37 Draw the three-block diagram structure for coenzyme A.

23.38 Which part of a coenzyme A molecule is the active participant in an acetyl transfer reaction?

● High-Energy Phosphate Compounds (Section 23.4)

23.39 What is a high-energy compound?

23.40 What factors contribute to a strained bond in high-energy phosphate compounds?

23.41 What does the designation P_i denote?

23.42 What does the designation PP_i denote?

23.43 With the help of Table 23.1, determine which compound in each of the following pairs of phosphate-containing compounds releases more free energy upon hydrolysis.
a. ATP and phosphoenolpyruvate
b. Creatine phosphate and ADP
c. Glucose 1-phosphate and 1,3-bisphosphoglycerate
d. AMP and glycerol 3-phosphate

23.44 With the help of Table 23.1, determine which compound in each of the following pairs of phosphate-containing compounds releases more free energy upon hydrolysis.
a. ATP and creatine phosphate
b. Glucose 1-phosphate and glucose 6-phosphate
c. ADP and AMP
d. Phosphoenolpyruvate and PP_i

● Biochemical Energy Production (Section 23.5)

23.45 Describe the four general stages of the process by which biochemical energy is obtained from food.

23.46 Of the four general stages of biochemical energy production from food, which are part of the common metabolic pathway?

● The Citric Acid Cycle (Section 23.6)

23.47 What are two other names for the citric acid cycle?

23.48 What is the basis for the name *citric acid cycle*?

23.49 What is the "fuel" for the citric acid cycle?

23.50 What are the products of the citric acid cycle?

23.51 Consider the reactions that occur during *one turn* of the citric acid cycle in answering each of the following questions.
a. How many CO_2 molecules are formed?
b. How many molecules of $FADH_2$ are formed?
c. How many times is a secondary alcohol oxidized?
d. How many times does water add to a carbon–carbon double bond?

23.52 Consider the reactions that occur during *one turn* of the citric acid cycle in answering each of the following questions.
a. How many molecules of NADH are formed?
b. How many GTP molecules are formed?

c. How many decarboxylation reactions occur?
d. How many oxidation–reduction reactions occur?

23.53 There are eight steps in the citric acid cycle. List those steps that involve
a. oxidation.
b. isomerization.
c. hydration.

23.54 There are eight steps in the citric acid cycle. List those steps that involve
a. oxidation and decarboxylation.
b. phosphorylation.
c. condensation.

23.55 There are four C_4 dicarboxylic acid species in the citric acid cycle. What are their names and structures?

23.56 There are two keto carboxylic acid species in the citric acid cycle. What are their names and structures?

23.57 What type of reaction occurs in the citric acid cycle whereby a C_6 compound is converted to a C_5 compound?

23.58 What type of reaction occurs in the citric acid cycle whereby a C_5 compound is converted to a C_4 compound?

23.59 Identify the oxidized coenzyme (NAD^+ or FAD) that participates in each of the following citric acid cycle reactions.
a. Isocitrate $\longrightarrow$ α-ketoglutarate
b. Succinate $\longrightarrow$ fumarate

23.60 Identify the oxidized coenzyme (NAD^+ or FAD) that participates in each of the following citric acid cycle reactions.
a. Malate $\longrightarrow$ oxaloacetate
b. α-Ketoglutarate $\longrightarrow$ succinyl CoA

23.61 List the two citric acid cycle intermediates involved in the reaction governed by each of the following enzymes. List the reactant first.
a. Isocitrate dehydrogenase
b. Fumarase
c. Malate dehydrogenase
d. Aconitase

23.62 List the two citric acid cycle intermediates involved in the reaction governed by each of the following enzymes. List the reactant first.
a. α-Ketoglutarate dehydrogenase
b. Succinate dehydrogenase
c. Citrate synthase
d. Succinyl CoA synthetase

● The Electron Transport Chain (Section 23.7)

23.63 By what other name is the electron transport chain known?

23.64 Give a one-sentence summary of what occurs during the reactions known as the electron transport chain.

23.65 What is the final electron acceptor of the electron transport chain?

23.66 Which substances generated in the citric acid cycle participate in the electron transport chain?

23.67 Give the abbreviation for each of the following electron carriers.
a. The oxidized form of flavin mononucleotide
b. The reduced form of coenzyme Q

23.68 Give the abbreviation for each of the following electron carriers.
a. The reduced form of flavin mononucleotide
b. The oxidized form of coenzyme Q

23.69 Indicate whether each of the following electron carriers is in its oxidized form or its reduced form.
   a. Fe(III)SP   b. Cyt b (Fe^{3+})
   c. NADH   d. FAD

23.70 Indicate whether each of the following electron carriers is in its oxidized form or its reduced form.
   a. $FMNH_2$ b. Fe(II)SP
   c. Cyt c_1 (Fe^{2+})   d. NAD^+

23.71 Indicate whether each of the following changes represents oxidation or reduction.
   a. $CoQH_2 \longrightarrow CoQ$
   b. $NAD^+ \longrightarrow NADH$
   c. Cyt c (Fe^{2+}) $\longrightarrow$ cyt c (Fe^{3+})
   d. Cyt b (Fe^{3+}) $\longrightarrow$ cyt b (Fe^{2+})

23.72 Indicate whether each of the following changes represents oxidation or reduction.
   a. $FADH_2 \longrightarrow FAD$
   b. $FMN \longrightarrow FMNH_2$
   c. Fe(III)SP $\longrightarrow$ Fe(II)SP
   d. Cyt c_1 (Fe^{3+}) $\longrightarrow$ cyt c_1 (Fe^{2+})

23.73 With which of the protein complexes (I, II, III, and IV) of the ETC is each of the following electron carriers associated? More than one answer may apply in a given situation.
   a. NADH   b. CoQ   c. Cyt b   d. Cyt a

23.74 With which of the protein complexes (I, II, III, and IV) of the ETC is each of the following electron carriers associated? More than one answer may apply in a given situation.
   a. $FADH_2$ b. FeSP   c. Cyt c   d. Cyt c_1

23.75 Which electron carrier shuttles electrons between protein complexes I and III?

23.76 Which electron carrier shuttles electrons between protein complexes II and III?

23.77 How many electrons does the electron carrier between complexes II and III carry per "trip"?

23.78 How many electrons does the electron carrier between complexes III and IV carry per "trip"?

23.79 Fill in the missing substances in the following electron transport chain reaction sequences.

   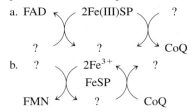

23.80 Fill in the missing substances in the following electron transport chain reaction sequences.

   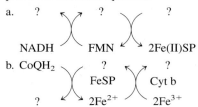

Oxidative Phosphorylation (Section 23.8)

23.81 What is oxidative phosphorylation?

23.82 What are coupled reactions?

23.83 The coupling of ATP synthesis with the reactions of the ETC is related to the movement of what chemical species across the inner mitochrondial membrane?

23.84 At what protein complex location(s) in the electron transport chain does proton pumping occur?

23.85 At what mitochondrial location does H^+ ion buildup occur as the result of proton pumping?

23.86 How many protons cross the inner mitochondrial membrane for every two electrons that are passed through the electron transport chain?

23.87 What is the name of the enzyme that catalyzes ATP production during oxidative phosphorylation?

23.88 What is the location of the enzyme that uses stored energy in a proton gradient to drive the reaction that produces ATP?

23.89 How is the proton gradient associated with chemiosmotic coupling dissipated during ATP synthesis?

23.90 What are the "starting materials" from which ATP is synthesized as the proton gradient associated with chemiosmotic coupling is dissipated?

ATP Production (Section 23.9)

23.91 How many ATP molecules are formed for each NADH molecule that enters the electron transport chain?

23.92 How many ATP molecules are formed for each $FADH_2$ molecule that enters the electron transport chain?

23.93 NADH and $FADH_2$ molecules do not yield the same number of ATP molecules. Explain why.

23.94 What is the energy yield, in terms of ATP molecules, from one turn of the citric acid cycle, assuming that the products of the cycle enter the electron transport chain?

The Importance of ATP (Section 23.10)

23.95 How does the free-energy release associated with the hydrolysis of ATP to ADP compare with the hydrolysis energies of other high-energy phosphate compounds?

23.96 What would the biochemical consequences be if the free-energy release associated with the hydrolysis of ATP to ADP had an unusually high value?

23.97 Indicate whether each of the following processes would be expected to involve the conversion of ATP to ADP or the conversion of ADP to ATP.
   a. Heart muscle contraction
   b. Transport of nutrients to various locations in the body

23.98 Indicate whether each of the following processes would be expected to involve the conversion of ATP to ADP or the conversion of ADP to ATP.
   a. Degradation of dietary carbohydrates
   b. Synthesis of protein from amino acids

Non-ETC Oxygen-Consuming Reactions (Section 23.11)

23.99 What does the designation ROS stand for?

23.100 Give the chemical formula for each of the following.
   a. Superoxide ion   b. Hydroxyl radical

23.101 Give the chemical equation for the reaction by which
   a. superoxide ion is generated within cells.
   b. superoxide ion is converted to hydrogen peroxide within cells.

23.102 Give the chemical equation for the reaction by which
   a. hydrogen peroxide is converted to desirable products within cells.
   b. hydrogen peroxide is converted to an undesirable product within cells.

ADDITIONAL PROBLEMS

23.103 Classify each of the following substances as (1) a reactant in the citric acid cycle, (2) a reactant in the electron transport chain, or (3) a reactant in both the CAC and the ETC.
a. NADH
b. O_2
c. Fumarate
d. Cytochrome a

23.104 Classify each of the following substances as (1) a product in the citric acid cycle, (2) a product in the electron transport chain, or (3) a product in both the CAC and the ETC.
a. FAD
b. CO_2
c. Malate
d. Flavin mononucleotide

23.105 Which of these substances, (1) ATP, (2) CoA, (3) FAD, and (4) NAD^+, contain the following subunits of structure? More than one choice may apply in a given situation.
a. Contains two ribose subunits
b. Contains two phosphate subunits
c. Contains one adenine subunit
d. Contains one ribitol subunit

23.106 Characterize, in terms of number of carbon atoms present, each of the following citric acid cycle changes as (1) a C_6 to C_6 change, (2) a C_6 to C_5 change, (3) a C_5 to C_4 change, or (4) a C_4 to C_4 change.
a. Citrate to isocitrate
b. Succinate to fumarate
c. Malate to oxaloacetate
d. Isocitrate to α-ketoglutarate

23.107 In what way are the processes of the citric acid cycle and the electron transport chain interrelated?

23.108 Where within a cell does each of the following take place?
a. Citric acid cycle
b. Electron transport chain and oxidative phosphorylation

23.109 One of the oxidation steps that occurs when lipids are metabolized is

$$R-CH_2-CH_2-\overset{\overset{\displaystyle O}{\|}}{C}-S-CoA \longrightarrow R-CH{=}CH-\overset{\overset{\displaystyle O}{\|}}{C}-S-CoA$$

Would you expect this reaction to require FAD or NAD^+ as the oxidizing agent?

23.110 In oxidative phosphorylation, what is oxidized and what is phosphorylated?

*M*ultiple-Choice Practice Test

23.111 Which of the following is true for a mitochondrion?
a. The inner membrane separates the matrix from the intermembrane space.
b. The inner membrane is more permeable than the outer membrane.
c. The outer membrane has ATP-synthase complexes attached to it.
d. The outer membrane has a highly folded structure.

23.112 Which of the following molecules has two unsubstituted ribose molecules as structural subunits?
a. FAD
b. NAD^+
c. CoA
d. ATP

23.113 Which of the following are products of the citric acid cycle?
a. Acetyl CoA and NADH
b. Acetyl CoA and CO_2
c. CO_2 and H_2O
d. CO_2 and $FADH_2$

23.114 Which of the following citric acid cycle intermediates is not a C_4 species?
a. Fumarate
b. Citrate
c. Malate
d. Oxaloacetate

23.115 Which are the first two intermediates, respectively, in the citric acid cycle?
a. Isocitrate and α-ketoglutarate
b. Citrate and α-ketoglutarate
c. Citrate and isocitrate
d. Isocitrate and succinate

23.116 Which of the following is an electron carrier that shuttles electrons between various protein complexes in the electron transport chain?
a. FMN
b. NADH
c. Cyt c
d. Cyt a_3

23.117 What is the substrate that interacts with protein complex III in the electron transport chain?
a. $CoQH_2$
b. Cyt c_1
c. FMN
d. FeSP

23.118 Which of the following is both a reactant and a product in the operation of the electron transport chain?
a. O_2
b. H_2O
c. $FADH_2$
d. Cyt b

23.119 At how many protein complex sites in the electron transport chain does proton pumping occur?
a. One
b. Two
c. Three
d. Four

23.120 How many moles of ATP result from the entry of one mole of NADH into the electron transport chain?
a. 1 mole ATP
b. 1.5 moles ATP
c. 2 moles ATP
d. 2.5 moles ATP

Carbohydrate Metabolism

24

Chapter Outline

Carbohydrates are the major energy source for human beings.

I n this chapter we explore the relationship between carbohydrate metabolism and energy production in cells. The molecule glucose is the focal point of carbohydrate metabolism. Commonly called blood sugar, glucose is supplied to the body via the circulatory system and, after being absorbed by a cell, can be either oxidized to yield energy or stored as glycogen for future use. When sufficient oxygen is present, glucose is totally oxidized to CO_2 and H_2O. However, in the absence of oxygen, glucose is only partially oxidized to lactic acid. Besides supplying energy needs, glucose and other six-carbon sugars can be converted into a variety of different sugars (C_3, C_4, C_5, and C_7) needed for biosynthesis. Some of the oxidative steps in carbohydrate metabolism also produce NADH and NADPH, sources of reductive power in cells.

24.1 DIGESTION AND ABSORPTION OF CARBOHYDRATES

Digestion *is the biochemical process by which food molecules, through hydrolysis, are broken down into simpler chemical units that can be used by cells for their metabolic needs.* Digestion is the first stage in the processing of food products.

The digestion of carbohydrates begins in the mouth, where the enzyme *salivary α-amylase* catalyzes the hydrolysis of α-glycosidic linkages (Section 18.13) in starch from plants and glycogen from meats to produce smaller polysaccharides and the disaccharide maltose.

Only a small amount of carbohydrate digestion occurs in the mouth because food is swallowed so quickly. Although the food mass remains longer in the stomach, very little further carbohydrate digestion occurs there either, because *salivary α-amylase* is

Salivary α-amylase is a constituent of saliva, the fluid secreted by the salivary glands. Saliva is 99% water plus small amounts of several inorganic ions and organic molecules. Saliva secretion can be triggered by the taste, smell, sight, and even thought of food. Average saliva output is about 1.5 L per day.

Figure 24.1 A section of the small intestine, showing its folds and the villi that cover the inner surface of the folds. Villi greatly increase the inner intestinal surface area.

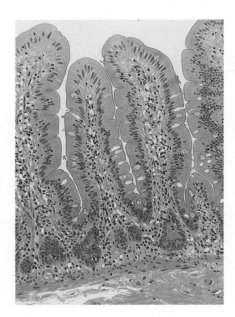

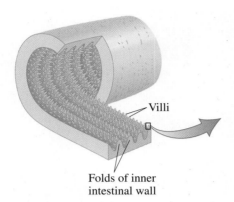

Villi

Folds of inner
intestinal wall

inactivated by the acidic environment of the stomach, and the stomach's own secretions do not contain any carbohydrate-digesting enzymes.

The primary site for carbohydrate digestion is within the small intestine, where α-amylase, this time secreted by the pancreas, again begins to function. The *pancreatic α-amylase* breaks down polysaccharide chains into shorter and shorter segments until the disaccharide maltose (two glucose units; Section 18.13) and glucose itself are the dominant species.

The final step in carbohydrate digestion occurs on the outer membranes of intestinal mucosal cells, where the enzymes that convert disaccharides to monosaccharides are located. The important disaccharidase enzymes are *maltase, sucrase,* and *lactase.* These enzymes convert, respectively, maltose to two glucose units, sucrose to one glucose and one fructose unit, and lactose to one glucose and one galactose unit (Section 18.13). (The disaccharides sucrose and lactose present in food are not digested until they reach this point.)

The three major breakdown products from carbohydrate digestion are thus glucose, galactose, and fructose. These monosaccharides are absorbed into the bloodstream through the intestinal wall. The folds of the intestinal wall are lined with fingerlike projections called *villi,* which are rich in blood capillaries (Figure 24.1). Absorption is by *active transport* (Section 19.10), which, unlike passive transport, is an energy-requiring process. In this case, ATP is needed. Protein carriers mediate the passage of the monosaccharides through cell membranes. Figure 24.2 summarizes the different phases in the digestive process for carbohydrates.

After their absorption into the bloodstream, monosaccharides are transported to the liver, where fructose and galactose are rapidly converted into compounds that are metabolized by the same pathway as glucose. Thus the central focus of carbohydrate metabolism is the pathway by which glucose is further processed, a pathway called *glycolysis* (Section 24.2)—a series of ten reactions, each of which involves a different enzyme.

 **EXAMPLE 24.1**

Determining the Sites Where Various Aspects of Carbohydrate Digestion Occur

Based on the information in Figure 24.2, determine the location within the human body where each of the following aspects of carbohydrate digestion occurs.

a. The enzyme sucrase is active.
b. Hydrolysis reactions converting polysaccharides to disaccharides occur.
c. First site where breaking of glycosidic linkages occurs.
d. The monosaccharides glucose, fructose, and galactose are produced.

(continued)

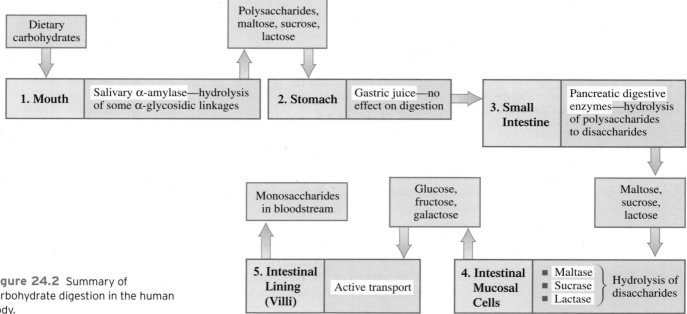

Figure 24.2 Summary of carbohydrate digestion in the human body.

The term *glycolysis,* pronounced "gligh-KOLL-ih-sis," comes from the Greek *glyco,* meaning "sweet," and *lysis,* meaning "breakdown."

Pyruvate, pronounced "PIE-roo-vate," is the carboxylate ion (Section 16.8) produced when pyruvic acid (a three-carbon keto acid) loses its acidic hydrogen atom.

Pyruvic acid   Pyruvate ion

24.2 GLYCOLYSIS

Glycolysis *is the metabolic pathway by which glucose (a C_6 molecule) is converted into two molecules of pyruvate (a C_3 molecule), chemical energy in the form of ATP is produced, and NADH-reduced coenzymes are produced.* This metabolic pathway functions in almost all cells.

The conversion of glucose to pyruvate is an oxidation process in which no molecular oxygen is utilized. The oxidizing agent is the coenzyme NAD^+. Metabolic pathways in which molecular oxygen is not a participant are called *anaerobic* pathways. Pathways that require molecular oxygen are called *aerobic* pathways. Glycolysis is an anaerobic pathway.

Glycolysis is a ten-step process (compared to the eight steps of the citric acid cycle; Section 23.6) in which every step is enzyme-catalyzed. Figure 24.3 gives an overview of glycolysis. There are two stages in the overall process, a *six-carbon stage* (Steps 1–3) and a *three-carbon stage* (Steps 4–10). All of the enzymes needed for glycolysis are present in the cell cytosol (Section 23.2), which is where glycolysis takes place. Details of the individual steps within the glycolysis pathway are now considered.

Six-Carbon Stage of Glycolysis (Steps 1–3)

The six-carbon stage of glycolysis is an *energy-consuming stage.* The energy release associated with the conversion of two ATP molecules to two ADP molecules is used to transform monosaccharides into monosaccharide phosphates. The intermediates of the six-carbon stage of glycolysis are all either *glucose* or *fructose* derivatives in which phosphate groups are present.

Step 1: **Phosphorylation:** *Formation of Glucose 6-Phosphate.* Glycolysis begins with the phosphorylation of glucose to yield glucose 6-phosphate, a glucose molecule with a phosphate group attached to the hydroxyl oxygen on carbon 6 (the carbon atom outside the ring). The phosphate group is from an ATP molecule. *Hexokinase,* an enzyme that requires Mg^{2+} ion for its activity, catalyzes the reaction.

The symbol Ⓟ is a shorthand notation for a PO_3^{2-} unit.

Glucose → Hexokinase (ATP → ADP) → Glucose 6-phosphate

This reaction requires energy, which is provided by the breakdown of an ATP molecule. This energy expenditure will be recouped later in the cycle. Phosphorylation of glucose provides a way of "trapping" glucose within a cell. Glucose can cross cell membranes, but glucose 6-phosphate cannot.

Step 2: **Isomerization:** *Formation of Fructose 6-Phosphate.* Glucose 6-phosphate is isomerized to fructose 6-phosphate by *phosphoglucoisomerase.*

Glucose 6-phosphate → Phosphoglucoisomerase → Fructose 6-phosphate

The net result of this change is that carbon 1 of glucose is no longer part of the ring structure. [Glucose, an aldose, forms a six-membered ring, and fructose, a ketose, forms a five-membered ring (Section 18.10); both sugars, however, contain six carbon atoms.]

Step 3: **Phosphorylation:** *Formation of Fructose 1,6-Bisphosphate.* This step, like Step 1, is a phosphorylation reaction and therefore requires the expenditure of energy. ATP is the source of the phosphate and the energy. The enzyme involved, *phosphofructokinase,* is another enzyme that requires Mg^{2+} ion for its activity. The fructose molecule now contains two phosphate groups.

Anaerobic is pronounced "AN-air-ROE-bic." *Aerobic* is pronounced "air-ROE-bic."

Glycolysis is also called the *Embden–Meyerhof pathway* after the German chemists Gustav Embden (1874–1933) and Otto Meyerhof (1884–1951), who discovered many of the details of the pathway in the early 1930s.

A *kinase* is an enzyme that catalyzes the transfer of a phosphoryl group (PO_3^{2-}) from ATP (or some other high-energy phosphate compound) to a substrate (Section 21.3).

Step 1 of glycolysis is the first of two steps in which an ATP molecule is converted to an ADP with the energy released used to effect a phosphorylation. The other ATP-phosphorylation reaction occurs in Step 3.

Step 3 of glycolysis commits the original glucose molecule to the glycolysis pathway. Glucose 6-phosphate (Step 1) and fructose 6-phosphate (Step 2) can enter other metabolic pathways, but fructose 1,6-bisphosphate can enter only glycolysis.

Figure 24.3 An overview of glycolysis.

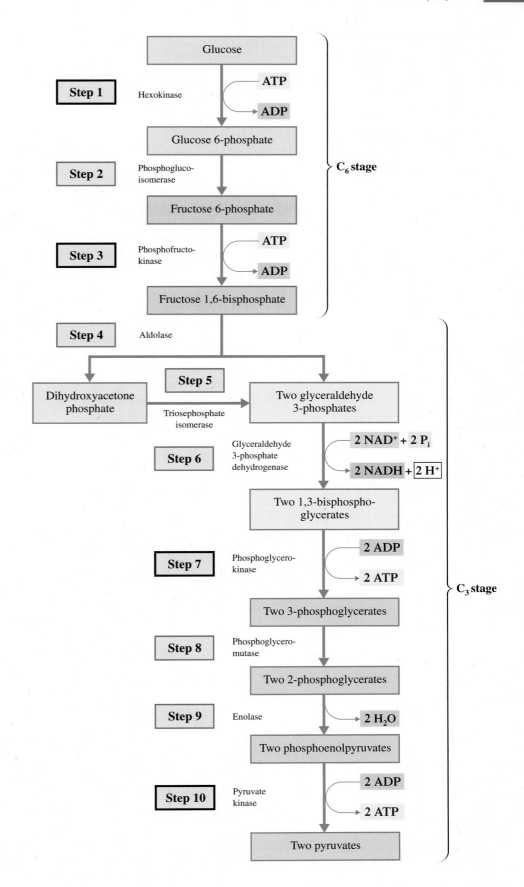

Step 3 of glycolysis, like Step 1, is a step in which phosphorylation is effected as an ATP molecule is converted to an ADP molecule.

The term *bisphosphate* is used instead of *diphosphate* to indicate the two phosphates are on different carbon atoms in fructose and not connected to each other.

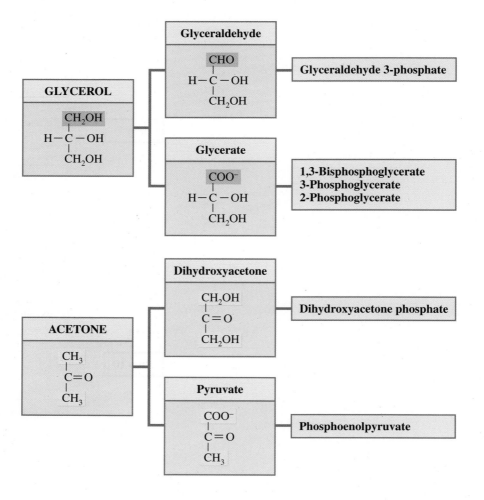

Fructose 6-phosphate   Fructose 1,6-bisphosphate

Three-Carbon Stage of Glycolysis (Steps 4–10)

The three-carbon stage of glycolysis is an *energy-generating stage* rather than an energy-consuming stage. All of the intermediates in this stage are C_3-phosphates, two of which are high-energy phosphate species (Section 23.4). Loss of a phosphate from these high-energy species effects the conversion of ADP molecules to ATP molecules.

The C_3 intermediates in this stage of glycolysis are all phosphorylated derivatives of *dihydroxyacetone, glyceraldehyde, glycerate,* or *pyruvate,* which in turn are derivatives of either glycerol or acetone. Figure 24.4 shows the structural relationships among these molecules.

Step 4: **Cleavage:** *Formation of Two Triose Phosphates.* In this step, the reacting C_6 species is split into two C_3 (triose) species. Because fructose 1,6-bisphosphate, the molecule being split, is unsymmetrical, the two trioses produced are not identical. One product is dihydroxyacetone phosphate, and the other is glyceraldehyde 3-phosphate. *Aldolase* is the enzyme that catalyzes this reaction. A better understanding of the structural relationships between reactant and products is obtained if the fructose 1,6-bisphosphate is written in its open-chain form (Section 18.10) rather than in its cyclic form.

Figure 24.4 Structural relationships among glycerol and acetone and the C_3 intermediates in the process of glycolysis.

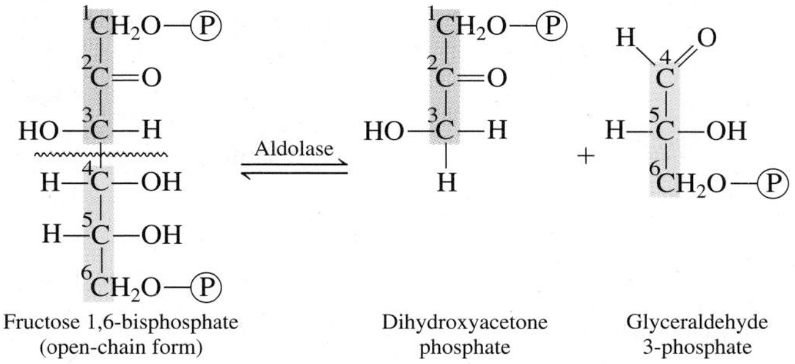

Fructose 1,6-bisphosphate (open-chain form) — Aldolase — Dihydroxyacetone phosphate + Glyceraldehyde 3-phosphate

Step 5: **Isomerization:** *Formation of Glyceraldehyde 3-Phosphate.* Only one of the two trioses produced in Step 4, glyceraldehyde 3-phosphate, is a glycolysis intermediate. Dihydroxyacetone phosphate, the other triose, can, however, be readily converted into glyceraldehyde 3-phosphate. Dihydroxyacetone phosphate (a ketose) and glyceraldehyde 3-phosphate (an aldose) are isomers, and the isomerization process from ketose to aldose is catalyzed by the enzyme *triosephosphate isomerase.*

Dihydroxyacetone phosphate — Triosephosphate isomerase — Glyceraldehyde 3-phosphate

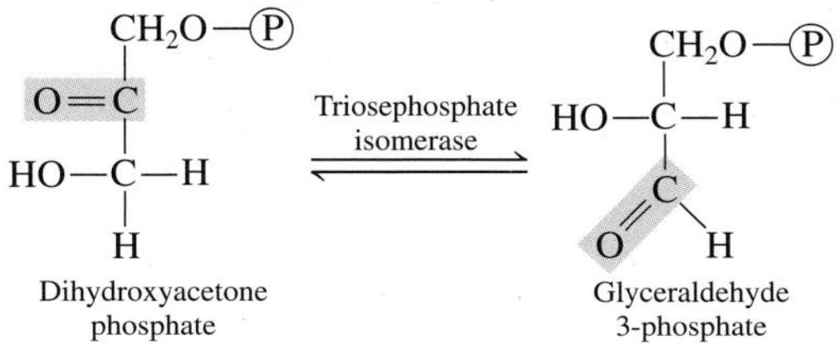

Step 6: **Oxidation and Phosphorylation:** *Formation of 1,3-Bisphosphoglycerate.* In a reaction catalyzed by *glyceraldehyde 3-phosphate dehydrogenase,* a phosphate group is added to glyceraldehyde 3-phosphate to produce 1,3-bisphosphoglycerate. The hydrogen of the aldehyde group becomes part of NADH.

Step 6 is the first of two glycolysis steps in which a high-energy phosphate compound, that is an "energy-rich" compound, is formed. The other high-energy phosphate compound is formed in Step 9. In the step that follows each of these two situations, energy from the high-energy phosphate compound is used to convert ADP to ATP.

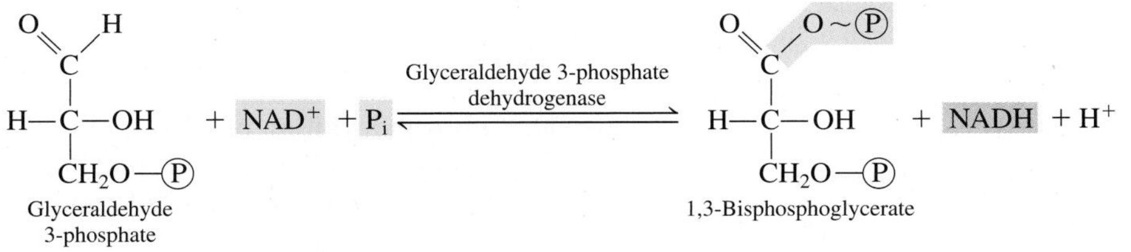

Glyceraldehyde 3-phosphate + NAD^+ + P_i — Glyceraldehyde 3-phosphate dehydrogenase — 1,3-Bisphosphoglycerate + NADH + H^+

The newly added phosphate group in 1,3-bisphosphoglycerate is a high-energy phosphate group (Section 23.4). A high-energy phosphate group is produced when a phosphate group is attached to a carbon atom that is also participating in a carbon–carbon or carbon–oxygen double bond.

Note that a molecule of the reduced coenzyme NADH is a product of this reaction and also that the source of the added phosphate is inorganic phosphate (P_i).

Keep in mind that from Step 6 onward, two molecules of each of the C_3 compounds take part in every reaction for each original C_6 glucose molecule.

Step 7: **Phosphorylation of ADP:** *Formation of 3-Phosphoglycerate.* In this step, the diphosphate species just formed is converted back to a monophosphate species. This is an ATP-producing step in which the C-1 phosphate group of 1,3-bisphosphoglycerate (the high-energy phosphate) is transferred to an ADP molecule to form the ATP. The enzyme involved is *phosphoglycerokinase.*

Step 7 is the first of two steps in which ATP is formed from ADP. This process, called *substrate-level phosphorylation,* always involves a high-energy phosphate compound. This same process also occurs in Step 10.

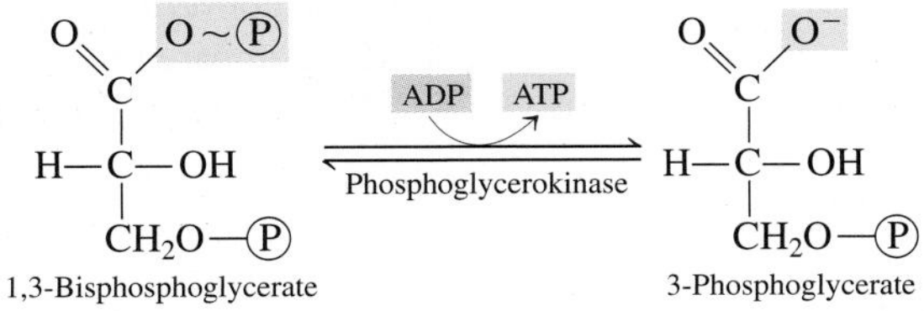

1,3-Bisphosphoglycerate — ADP ATP — Phosphoglycerokinase — 3-Phosphoglycerate

B-F

A *mutase* is an enzyme that effects the shift of a functional group from one position to another within a molecule (Section 21.2).

Remember that two ATP molecules are produced for each original glucose molecule because both C_3 molecules produced from the glucose react.

ATP production in this step involves substrate-level phosphorylation. **Substrate-level phosphorylation** *is the biochemical process by which a high-energy phosphate group from an intermediate compound (substrate) is directly transferred to ADP to produce ATP.* Substrate-level phosphorylation differs from oxidative phosphorylation (Section 23.8) in that the latter process involves the transfer of free phosphate ions in solution (P_i) to ADP molecules to form ATP.

Step 8: **Isomerization:** *Formation of 2-Phosphoglycerate.* In this isomerization step, the phosphate group of 3-phosphoglycerate is moved from carbon 3 to carbon 2. The enzyme *phosphoglyceromutase* catalyzes the exchange of the phosphate group between the two carbons.

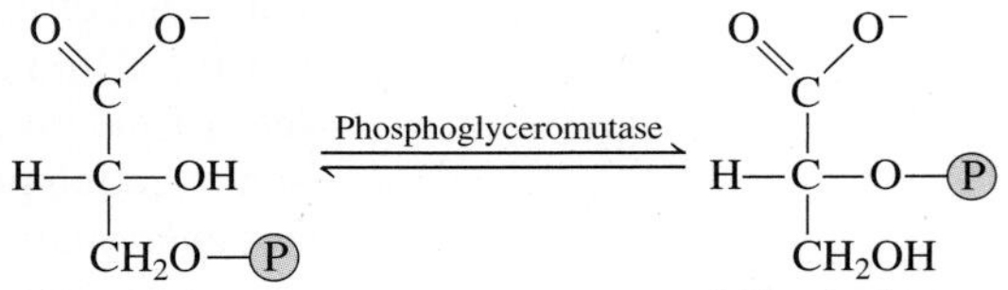

3-Phosphoglycerate → (Phosphoglyceromutase) → 2-Phosphoglycerate

An *enol* (from *ene* + *ol*), as in phospho*enol*pyruvate, is a compound in which an —OH group is attached to a carbon atom involved in a carbon–carbon double bond. Note that in phosphoenolpyruvate, the —OH group has been phosphorylated.

Step 9: **Dehydration:** *Formation of Phosphoenolpyruvate.* This is an alcohol dehydration reaction that proceeds with the enzyme *enolase,* another Mg^{2+}-requiring enzyme. The result is another compound containing a high-energy phosphate group; the phosphate group is attached to a carbon atom that is involved in a carbon–carbon double bond.

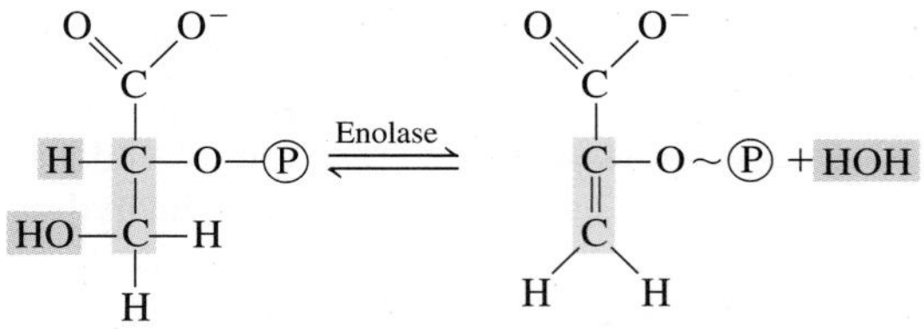

2-Phosphoglycerate → (Enolase) → Phosphoenolpyruvate + HOH

Step 9 is the second of two glycolytic steps in which a high-energy phosphate compound, that is, "energy-rich" compound, is formed; the other step was Step 6. In the next step, energy from the high-energy phosphate compound will be used to convert ADP and ATP.

Step 10 is the second of two steps in which ATP is formed from ADP. This same process also occurred in Step 7.

Step 10: **Phosphorylation of ADP:** *Formation of Pyruvate.* In this step, substrate-level phosphorylation again occurs. Phosphoenolpyruvate transfers its high-energy phosphate group to an ADP molecule to produce ATP and pyruvate.

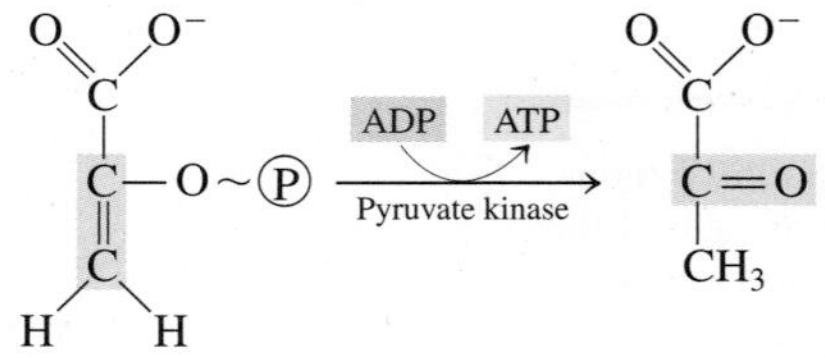

Phosphoenolpyruvate → (ADP, ATP, Pyruvate kinase) → Pyruvate

The enzyme involved, *pyruvate kinase,* requires both Mg^{2+} and K^+ ions for its activity. Again, because two C_3 molecules are reacting, two ATP molecules are produced.

ATP molecules are involved in Steps 1, 3, 7, and 10 of glycolysis. Considering these steps collectively shows that there is a net gain of two ATP molecules for every glucose molecule converted into two pyruvates (Table 24.1). Though useful, this is a small amount of ATP compared to that generated in oxidative phosphorylation (Section 23.8).

The net overall equation for the process of glycolysis is

$$\text{Glucose} + 2NAD^+ + 2ADP + 2P_i \longrightarrow$$

$$2 \text{ pyruvate} + 2NADH + 2ATP + 2H^+ + 2H_2O$$

TABLE 24.1
**ATP Production and Consumption
During Glycolysis**

Step	Reaction	ATP Change per Glucose
1	Glucose → glucose 6-phosphate	−1
3	Fructose 6-phosphate → fructose 1,6-bisphosphate	−1
7	2(1,3-Bisphosphoglycerate → 3-phosphoglycerate)	+2
10	2(Phosphoenolpyruvate → pyruvate)	+2
		Net +2

 **EXAMPLE 24.2**

Recognizing Structural Characteristics of Glycolysis Intermediates

▶ Specify the number of carbon atoms present and the number of phosphate groups present in each of the following glycolysis intermediates.

a. Glucose 6-phosphate
b. Fructose 1,6-bisphosphate
c. Glyceraldehyde 3-phosphate
d. 3-Phosphoglycerate

Solution

a. Glucose is a hexose. It is a C_6 *molecule* that carries *one phosphate group*, attached to carbon 6.
b. Fructose is also a hexose. It is a C_6 *molecule* that carries *two phosphate groups*, one attached to carbon 1 and the other attached to carbon 6.
c. Glyceraldehyde is a C_3 *molecule* with *one phosphate group*, attached to carbon 3.
d. Glycerate, like glyceraldehyde, is a C_3 *molecule* with *one phosphate group*, attached to carbon 3.

Practice Exercise 24.2

Specify the number of carbon atoms present and the number of phosphate groups present in each of the following glycolysis intermediates.

a. Fructose 6-phosphate
b. 1,3-Bisphosphoglycerate
c. 2-Phosphoglycerate
d. Phosphoenolpyruvate

Answers: a. C_6 with one phosphate; **b.** C_3 with two phosphates; **c.** C_3 with one phosphate; **d.** C_3 with one phosphate

EXAMPLE 24.3

Recognizing Reaction Types and Events That Occur in the Various Steps of Glycolysis

▶ Indicate at what step in the glycolysis pathway each of the following events occurs.

a. First phosphorylation of ADP occurs
b. First "energy-rich" compound is produced
c. Second "energy-rich" compound undergoes reaction
d. First isomerization reaction occurs

Solution

a. Phosphorylation of ADP produces ATP with the added phosphate coming from a glycolysis intermediate. This process occurs for the first time in *Step 7*, where 1,3-bisphosphoglycerate is converted to 3-phosphoglycerate. It occurs a second time in Step 10.
b. In carbohydrate metabolism, an "energy-rich" compound is a high-energy phosphate. Two such compounds are produced during glycolysis, the first in *Step 6* and the second in Step 9. The Step 6 compound is 1,3-bisphosphoglycerate.
c. The second "energy-rich" compound is phosphoenolpyruvate, which is produced in Step 9 and undergoes reaction in *Step 10*.

(continued)

d. There are three isomerization reactions in glycolysis. They occur in Steps 2, 5, and 8. In the first occurrence, *Step 2*, glucose 6-phosphate is converted to fructose 6-phosphate. Glucose and fructose are isomeric hexose (Section 18.8) molecules.

Practice Exercise 24.3

Indicate at what step in the glycolysis pathway each of the following events occurs.

a. Second formation of ATP occurs.
b. Second "energy-rich" compound is produced.
c. Second time ATP is converted to ADP.
d. A dehydration reaction occurs.

Answers: a. Step 10; **b.** Step 9; **c.** Step 3; **d.** Step 9

EXAMPLE 24.4

Relating Enzyme Names and Functions to Steps in the Process of Glycolysis

Relate the names and functions of the following glycolytic enzymes to steps in the process of glycolysis.

a. Phosphofructokinase
b. Phosphoglyceromutase
c. Triosephosphate isomerase
d. Enolase

Solution

a. A kinase is an enzyme that is involved with phosphate group transfer. The phospho-fructo portion of the enzyme name refers to a fructose phosphate. Transfer of a phosphate group to a fructose phosphate occurs in *Step 3*. Fructose 6-phosphate is converted to fructose 1,6-bisphosphate.
b. A mutase is an enzyme that shifts the position of a functional group within a molecule. Such a functional group shift occurs in *Step 8*, where 3-phosphoglycerate is isomerized to 2-phosphoglycerate. The phosphoglycero portion of the enzyme name indicates that a glycerate phosphate is the substrate for the enzyme.
c. A triosephosphate isomerase effects the isomerization of a triose. Such a process occurs in *Step 5*, where dihydroxyacetone phosphate (a ketone) is converted to glyceraldehyde 3-phosphate (an aldehyde). Ketones and aldehydes with the same number of carbon atoms are often isomers.
d. The only enol species in glycolysis is the compound phosphoenolpyruvate produced in *Step 9* through a dehydration reaction that introduces a carbon–carbon double bond in the molecule. An enolase effects such a change.

Practice Exercise 24.4

Relate the names and functions of the following glycolytic enzymes to steps in the process of glycolysis.

a. Hexokinase
b. Pyruvate kinase
c. Glyceraldehyde 3-phosphate dehydrogenase
d. Phosphoglycerokinase

Answers: a. Step 1; **b.** Step 10; **c.** Step 6; **d.** Step 7

Entry of Galactose and Fructose into Glycolysis

The breakdown products from carbohydrate digestion are glucose, fructose, and galactose (Section 24.1). Both fructose and galactose are converted, in the liver, to intermediates that enter into the glycolysis pathway.

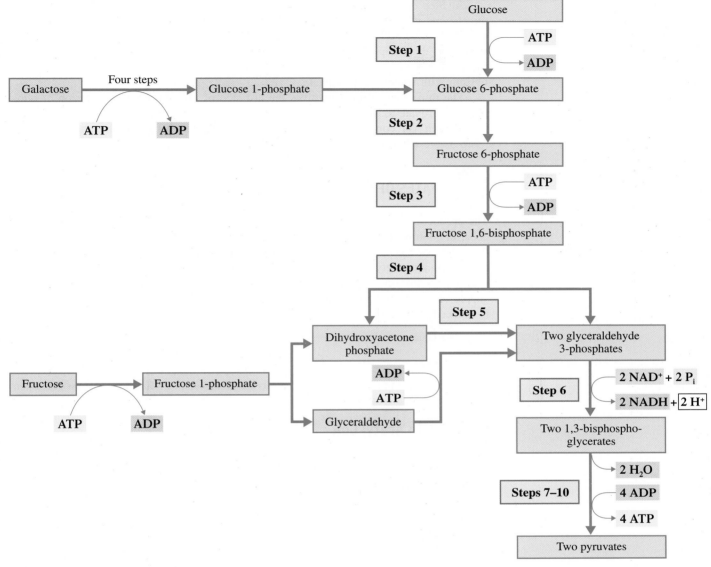

Figure 24.5 Entry points for fructose and galactose into the glycolysis pathway.

The entry of fructose into the glycolytic pathway involves phosphorylation by ATP to produce fructose 1-phosphate, which is then split into two trioses—glyceraldehyde and dihydroxyacetone phosphate. Dihydroxyacetone phosphate enters glycolysis directly; glyceraldehyde must be phosphorylated by ATP to glyceraldehyde 3-phosphate before it enters the pathway (see Figure 24.5).

The entry of galactose into the glycolytic pathway begins with its conversion to glucose 1-phosphate (a four-step sequence), which is then converted to glucose 6-phosphate, a glycolysis intermediate (see Figure 24.5).

Regulation of Glycolysis

Glycolysis, like all metabolic pathways, must have control mechanisms associated with it. In glycolysis, the control points are Steps 1, 3, and 10 (see Figure 24.3).

Step 1, the conversion of glucose to glucose 6-phosphate, involves the enzyme *hexokinase*. This particular enzyme is inhibited by glucose 6-phosphate, the substance produced by its action (feedback inhibition; Section 21.8).

At Step 3, where fructose 6-phosphate is converted to fructose 1,6-bisphosphate by the enzyme *phosphofructokinase,* high concentrations of ATP and citrate inhibit enzyme activity. A high ATP concentration, which is characteristic of a state of low energy consumption, thus stops glycolysis at the fructose 6-phosphate stage. This stoppage also causes increases in glucose 6-phosphate stores because glucose 6-phosphate is in equilibrium with fructose 6-phosphate.

The third control point involves the last step of glycolysis, the conversion of phosphoenolpyruvate to pyruvate. *Pyruvate kinase,* the enzyme needed at this point, is inhibited by high ATP concentrations. Both pyruvate kinase (Step 10) and phosphofructokinase (Step 3) are allosteric enzymes (Section 21.8).

24.3 FATES OF PYRUVATE

The production of pyruvate from glucose (glycolysis) occurs in a similar manner in most cells. In contrast, the fate of the pyruvate so produced varies with cellular conditions and the nature of the organism. Three common fates for pyruvate are of prime importance: conversion into acetyl CoA, into lactate, and into ethanol (see Figure 24.6).

A key concept in considering these fates of pyruvate is the need for a continuous supply of NAD^+ for glycolysis. As glucose is oxidized to pyruvate in glycolysis, NAD^+ is reduced to NADH.

$$\text{Glucose} + 2NAD^+ \longrightarrow 2 \text{ pyruvate} + 2NADH + 2H^+$$
$$2ADP + 2P_i \qquad 2ATP$$

It is significant that each pathway of pyruvate metabolism includes provisions for regeneration of NAD^+ from NADH so that glycolysis can continue.

Oxidation to Acetyl CoA

Under *aerobic* (oxygen-rich) conditions, pyruvate is oxidized to acetyl CoA. Pyruvate formed in the cytosol through glycolysis crosses the two mitochondrial membranes and enters the mitochondrial matrix, where the oxidation takes place. The overall reaction, in simplified terms, is

$$\underset{\text{Pyruvate}}{CH_3-\overset{\overset{O}{\|}}{C}-COO^-} + CoA-SH + NAD^+ \xrightarrow{\underset{\text{complex}}{\overset{\text{Pyruvate}}{\text{dehydrogenase}}}} \underset{\text{Acetyl CoA}}{CH_3-\overset{\overset{O}{\|}}{C}-S-CoA} + NADH + CO_2$$

Figure 24.6 The three common fates of pyruvate generated by glycolysis.

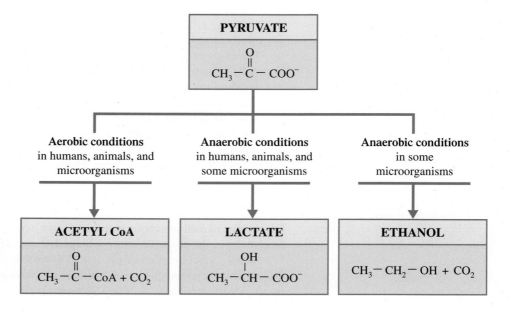

An additional fate for pyruvate is conversion to oxaloacetate. This fate for pyruvate, which occurs during the process called *gluconeogenesis,* is discussed in Section 24.6.

This reaction, which involves both oxidation and decarboxylation (CO_2 is produced), is far more complex than the simple stoichiometry of the equation suggests. The enzyme complex involved contains three different enzymes, each with numerous subunits. The overall reaction process involves four separate steps and requires NAD^+, CoA—SH, FAD, and two other coenzymes (lipoic acid and thiamine pyrophosphate, the latter derived from the B vitamin thiamine).

Most acetyl CoA molecules produced from pyruvate enter the citric acid cycle. Citric acid cycle operations change more NAD^+ to its reduced form, NADH. The NADH from glycolysis, from the conversion of pyruvate to acetyl CoA, and from the citric acid cycle enters the electron transport chain directly (Section 23.7) or indirectly (Section 24.4). In the ETC, electrons from NADH are transferred to O_2, and the NADH is changed back to NAD^+. The NAD^+ needed for glycolysis, pyruvate–acetyl CoA conversion, and the citric acid cycle is regenerated.

The net overall reaction for processing one glucose molecule to two molecules of acetyl CoA is

$$\text{Glucose} + 2ADP + 2P_i + 4NAD^+ + 2CoA-SH \longrightarrow$$
$$2 \text{ acetyl CoA} + 2CO_2 + 2ATP + 4NADH + 4H^+ + 2H_2O$$

Fermentation Processes

When the body becomes oxygen deficient (anaerobic conditions), such as during strenuous exercise, the electron transport chain process slows down because its last step is dependent on oxygen. The result of this "slowing down" is a buildup in NADH concentration (it is not being consumed so fast) and a decreased amount of available NAD^+ (it is not being produced so fast). Decreased NAD^+ concentration then negatively affects the rate of glycolysis. An alternative method for conversion of NADH to NAD^+—a method that does not require oxygen—is needed if glycolysis is to continue, it being the only available source of *new* ATP under these conditions.

Fermentation processes solve this problem. **Fermentation** *is a biochemical process by which NADH is oxidized to NAD^+ without the need for oxygen.* We consider here two fermentation processes: lactate fermentation and ethanol fermentation.

Lactate Fermentation

Lactate fermentation *is the enzymatic anaerobic reduction of pyruvate to lactate.* The sole purpose of this process is the conversion of NADH to NAD^+. The lactate so formed is converted back to pyruvate when aerobic conditions are again established in a cell (Section 24.6).

Working muscles often produce lactate. If strenuous exercise is continued too long, the buildup of lactate in the muscles reaches a point beyond which fermentation cannot continue. This slows glycolysis and new ATP production, and the muscle action can no longer continue (fatigue and exhaustion; see the accompanying Chemical Connections feature on page 825) until oxygen supplies are re-established.

The equation for lactate formation from pyruvate is

$$\underset{\text{Pyruvate}}{CH_3-\overset{\overset{\displaystyle O}{\|}}{C}-COO^-} + NAD\,H + H^+ \xrightarrow[\text{dehydrogenase}]{\text{Lactate}} \underset{\text{Lactate}}{CH_3-\overset{\overset{\displaystyle O\;H}{|}}{C}H-COO^-} + NAD^+$$

When the reaction for conversion of pyruvate to lactate is added to the net glycolysis reaction (Section 24.2), an overall reaction for the conversion of glucose to lactate is obtained.

$$\text{Glucose} + 2ADP + 2P_i \longrightarrow 2 \text{ lactate} + 2ATP + 2H_2O$$

Note that NADH and NAD^+ do not appear in this equation, even though the process cannot proceed without them. The NADH generated during glycolysis (Step 6) is consumed

---

B

Not all acetyl CoA produced from pyruvate enters the citric acid cycle. Particularly when high levels of acetyl CoA are produced (from excess ingestion of dietary carbohydrates), some acetyl CoA is used as the starting material for the production of the fatty acids needed for fat (triacylglycerol) formation (Section 25.7).

Red blood cells have no mitochondria and therefore always form lactate as the end product of glycolysis.

in the conversion of pyruvate to lactate. Thus there is no net oxidation–reduction in the conversion of glucose to lactate.

Ethanol Fermentation

With bread and other related products obtained using yeast, the ethanol produced by fermentation evaporates during baking.

Under anaerobic conditions, several simple organisms, including yeast, possess the ability to regenerate NAD^+ through ethanol, rather than lactate, production. Such a process is called ethanol fermentation. **Ethanol fermentation** *is the enzymatic anaerobic conversion of pyruvate to ethanol and carbon dioxide.* Ethanol fermentation involving yeast causes bread and related products to rise as a result of CO_2 bubbles being released during baking. Beer, wine, and other alcoholic drinks are produced by ethanol fermentation of the sugars in grain and fruit products.

The first step in conversion of pyruvate to ethanol is a decarboxylation reaction to produce acetaldehyde.

$$CH_3-\overset{\overset{\displaystyle O}{\|}}{C}-COO^- + H^+ \xrightarrow[\text{decarboxylase}]{\text{Pyruvate}} CH_3-\overset{\overset{\displaystyle O}{\|}}{C}-H + CO_2$$

Pyruvate   Acetaldehyde

The second step involves acetaldehyde reduction to produce ethanol.

$$CH_3-\overset{\overset{\displaystyle O}{\|}}{C}-H + NAD\,H + H^+ \xrightarrow[\text{dehydrogenase}]{\text{Alcohol}} CH_3-\overset{\overset{\displaystyle OH}{|}}{\underset{\underset{\displaystyle H}{|}}{C}}-H + NAD^+$$

Acetaldehyde   Ethanol

The overall equation for the conversion of pyruvate to ethanol (the sum of the two steps) is

$$CH_3-\overset{\overset{\displaystyle O}{\|}}{C}-COO^- + 2H^+ + NADH \xrightarrow{\text{Two steps}} CH_3-CH_2-OH + NAD^+ + CO_2$$

Pyruvate   Ethanol

An overall reaction for the production of ethanol from glucose is obtained by combining the reaction for the conversion of pyruvate with the net reaction for glycolysis (Section 24.2).

$$\text{Glucose} + 2ADP + 2P_i \longrightarrow 2 \text{ ethanol} + 2CO_2 + 2ATP + 2H_2O$$

Again note that NADH and NAD^+ do not appear in the final equation; they are both generated and consumed.

Figure 24.7 summarizes the relationship between the fates of pyruvate and the regeneration of NAD^+ from NADH.

Figure 24.7 All three of the common fates of pyruvate from glycolysis provide for the regeneration of NAD^+ from NADH.

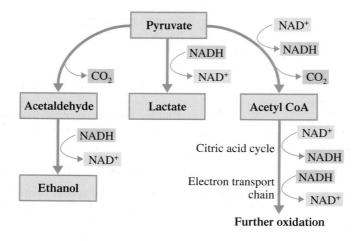

CHEMICAL Connections

Lactate Accumulation

During strenuous exercise, conditions in muscle cells can change from aerobic to anaerobic as the oxygen supply becomes inadequate to meet demand. Such conditions cause pyruvate to be converted to lactate rather than acetyl CoA. (Lactate production can also be high at the start of strenuous exercise before the delivery of oxygen is stepped up via an increased respiration rate.)

The resulting lactate begins to accumulate in the cytosol of cells where it is produced. Some lactate diffuses out of the cells into the blood, where it contributes to a slight decrease in blood pH. This lower pH triggers fast breathing, which helps supply more oxygen to the cells.

Lactate accumulation is the cause of muscle pain and cramping during prolonged, strenuous exercise. As a result of such cramping, muscles may be stiff and sore the next day. Regular, hard exercise increases the efficiency with which oxygen is delivered to the body. Thus athletes can function longer than non-athletes under aerobic conditions without lactate production.

Lactate accumulation can also occur in heart muscle if it experiences decreased oxygen supply (from artery blockage). The heart muscle experiences cramps and stops beating (cardiac arrest). Massage of heart muscle often reduces such cramps, just as it does for skeletal muscle, and it is sometimes possible to start the heart beating again by using such a technique.

Premature infants born with underdeveloped lungs are often given increased amounts of oxygen to minimize lactate accumulation. They are also given bicarbonate (HCO_3^-) solution to counteract the acidity change in blood that accompanies lactate buildup.

Rapid breathing, which is also called hyperventilation, raises slightly the pH of blood. The CO_2 loss associated with the rapid breathing causes carbonic acid (H_2CO_3) present in the blood to dissociate in CO_2 and H_2O to replace the lost CO_2.

$$H_2CO_3 \rightleftharpoons CO_2 + H_2O$$

A decreased amount of carbonic acid causes blood pH to rise, which makes the blood slightly more basic. Athletes who compete in short-distance races, such as 50 meters and 100 meters, have learned how to use this phenomenon of hyperventilation to their advantage. A few seconds before the start of the race, through hyperventilation, they decrease the amount of CO_2 in their lungs, making their blood a little bit more basic. This slight increase in basicity means the runner can absorb slightly more lactic acid before the blood pH drops to the point where cramping becomes a problem. Having such an advantage for only a few seconds in a short race can be helpful.

Strenuous muscular activity can result in lactate accumulation.

 EXAMPLE 24.5

Identifying Characteristics of the Three Common Pathways by Which Pyruvate Is Converted to Other Products

▶ Which of the three common metabolic pathways for pyruvate is compatible with each of the following characterizations concerning the reactions that pyruvate undergoes?

a. CO_2 is not a product for this pathway.
b. A C_3 molecule is a product for this pathway.
c. NAD^+ is needed as an oxidizing agent for this pathway.
d. A C_2 molecule is a product under anaerobic reaction conditions for this pathway.

Solution

a. CO_2 is a product for both acetyl CoA formation and ethanol fermentation. In both of these processes the C_3 pyruvate is converted to a C_2 species and CO_2. No CO_2 production occurs during *lactate fermentation*, as C_3 pyruvate is converted to C_3 lactate.

b. The only pathway that produces a C_3 molecule is *lactate fermentation*. In the other two pathways a C_2 molecule and CO_2 are produced.

c. NAD^+, as an oxidizing agent, is needed in pathways that function under aerobic conditions. NADH, as a reducing agent, is needed in pathways that function under anaerobic conditions. The only pathway that involves aerobic conditions is *acetyl CoA formation*.

d. *Alcohol fermentation*, an anaerobic process, produces ethanol, a C_2 molecule. The other anaerobic process, lactate fermentation, produces the C_3 molecule lactate.

(continued)

Practice Exercise 24.5

Which of the three common metabolic pathways for pyruvate is compatible with each of the following characterizations concerning the reactions that pyruvate undergoes?

a. Acetaldehyde is an intermediate in this pathway.
b. An anaerobic pathway that does not function in humans.
c. An anaerobic pathway that does function in humans.
d. A C_2 molecule is a product under aerobic reaction conditions for this pathway.

Answers: a. Ethanol fermentation; **b.** Ethanol fermentation; **c.** Lactate fermentation; **d.** Acetyl CoA formation

ATP PRODUCTION FOR THE COMPLETE OXIDATION OF GLUCOSE

We now assemble energy production figures for glycolysis, oxidation of pyruvate to acetyl CoA, the citric acid cycle, and the electron transport chain. The result, with one added piece of information, gives the ATP yield for the *complete* oxidation of one molecule of glucose.

The new piece of information involves the NADH produced during Step 6 of glycolysis. This NADH, produced in the cytosol, cannot *directly* participate in the electron transport chain because mitochondria are impermeable to NADH (and NAD^+). A transport system shuttles the electrons from NADH, but not NADH itself, across the membrane. This shuttle involves dihydroxyacetone phosphate (a glycolysis intermediate) and glycerol 3-phosphate.

The first step in the shuttle is the cytosolic reduction of dihydroxyacetone phosphate by NADH to produce glycerol 3-phosphate and NAD^+ (see Figure 24.8). Glycerol 3-phosphate then crosses the outer mitochondrial membrane, where it is reoxidized to dihydroxyacetone phosphate. The oxidizing agent is FAD rather than NAD^+. The regenerated dihydroxyacetone phosphate diffuses out of the mitochondrion and returns to the cytosol for participation in another "turn" of the shuttle. The $FADH_2$ coproduced in the mitochondrial reaction can participate in the electron transport chain reactions. The net reaction of this shuttle process is

Figure 24.8 The dihydroxyacetone phosphate–glycerol 3-phosphate shuttle.

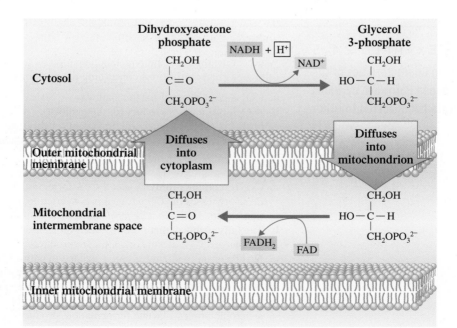

TABLE 24.2
Production of ATP from the Complete Oxidation of One Glucose Molecule in a Skeletal Muscle Cell

Reaction	Comments	Yield of ATP
Glycolysis		
glucose → glucose 6-phosphate	consumes 1 ATP	−1
glucose 6-phosphate → fructose 1,6-bisphosphate	consumes 1 ATP	−1
2(glyceraldehyde 3-phosphate → 1,3-bisphosphoglycerate)	each produces 1 cytosolic NADH	—
2(1,3-bisphosphoglycerate → 3-phosphoglycerate)	each produces 1 ATP	+2
2(phosphoenolpyruvate → pyruvate)	each produces 1 ATP	+2
Oxidation of Pyruvate		
2(pyruvate → acetyl CoA + CO_2)	each produces 1 NADH	—
Citric Acid Cycle		
2(isocitrate → α-ketoglutarate + CO_2)	each produces 1 NADH	—
2(α-ketoglutarate → succinyl CoA + CO_2)	each produces 1 NADH	—
2(succinyl CoA → succinate)	each produces 1 GTP	+2
2(succinate → fumarate)	each produces 1 $FADH_2$	—
2(malate → oxaloacetate)	each produces 1 NADH	—
Electron Transport Chain and Oxidative Phosphorylation		
2 cytosolic NADH formed in glycolysis	each produces 1.5 ATP	+3
2 NADH formed in the oxidation of pyruvate	each produces 2.5 ATP	+5
2 $FADH_2$ formed in the citric acid cycle	each produces 1.5 ATP	+3
6 NADH formed in the citric acid cycle	each produces 2.5 ATP	+15
Net production of ATP		**+30**

$$\underset{\text{(cytosolic)}}{NADH} + H^+ + \underset{\text{(mitochondrial)}}{FAD} \longrightarrow \underset{\text{(cytosolic)}}{NAD^+} + \underset{\text{(mitochondrial)}}{FADH_2}$$

The consequence of this reaction is that only 1.5 rather than 2.5 molecules of ATP are formed for each cytosolic NADH, because $FADH_2$ yields one less ATP than does NADH in the electron transport chain.

Table 24.2 shows ATP production for the complete oxidation of a molecule of glucose. The final number is 30 ATP, 26 of which come from the oxidative phosphorylation associated with the electron transport chain. This total of 30 ATP for complete oxidation contrasts markedly with a total of 2 ATP for oxidation of glucose to lactate and 2 ATP for oxidation of glucose to ethanol. Neither of these latter processes involves the citric acid cycle or the electron transport chain. Thus the aerobic oxidation of glucose is 15 times more efficient in the production of ATP than the anaerobic lactate and ethanol processes.

The production of 30 ATP molecules per glucose (Table 24.2) is for those cells where the dihydroxyacetone phosphate–glycerol 3-phosphate shuttle operates (skeletal muscle and nerve cells). In certain other cells, particularly heart and liver cells, a more complex shuttle system called the malate–aspartate shuttle functions. In this shuttle, 2.5 ATP molecules result from 1 cytosolic NADH, which changes the total ATP production to 32 molecules per glucose.

The net overall reaction for the *complete* metabolism (oxidation) of a glucose molecule is the simple equation

$$\text{Glucose} + 6O_2 + 30ADP + 30P_i \longrightarrow 6CO_2 + 6H_2O + 30 \text{ ATP}$$

Note that substances such as NADH, NAD^+, and $FADH_2$ are not part of this equation. Why? They cancel out—that is, they are consumed in one step (reactant) and regenerated in another step (product). Note also what the net equation does not acknowledge: the many dozens of reactions that are needed to generate the 30 molecules of ATP.

GLYCOGEN SYNTHESIS AND DEGRADATION

Glycogen, a branched polymeric form of glucose (Section 18.15), is the storage form of carbohydrates in humans and animals. It is found primarily in muscle and liver tissue. In muscles it is the source of glucose needed for glycolysis. In the liver, it is the source of glucose needed to maintain normal glucose levels in the blood.

Glycogenesis

Glycogenesis *is the metabolic pathway by which glycogen is synthesized from glucose 6-phosphate.* Glycogenesis involves three reactions (steps).

Step 1:   *Formation of Glucose 1-phosphate.* The starting material for this step is not glucose itself but rather glucose 6-phosphate (available from the first step of glycolysis). The enzyme *phosphoglucomutase* effects the change from a 6-phosphate to a 1-phosphate.

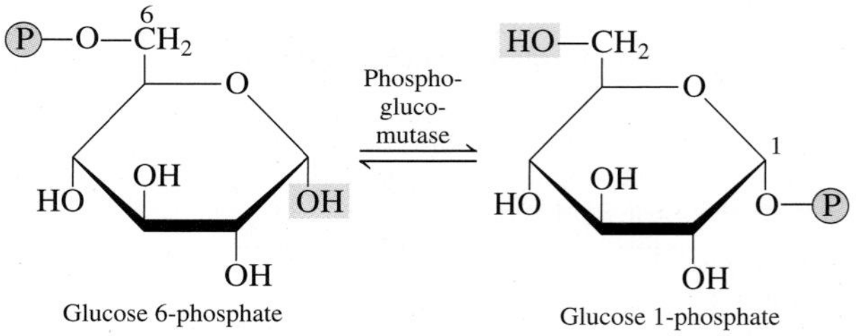

Glucose 6-phosphate   Glucose 1-phosphate

Step 2:   *Formation of UDP-glucose.* Glucose 1-phosphate from Step 1 must be activated before it can be added to a growing glycogen chain. The activator is the high-energy compound UTP (uridine triphosphate). A UMP is transferred to glucose 1-phosphate and the resulting PP_i is hydrolyzed to $2P_i$.

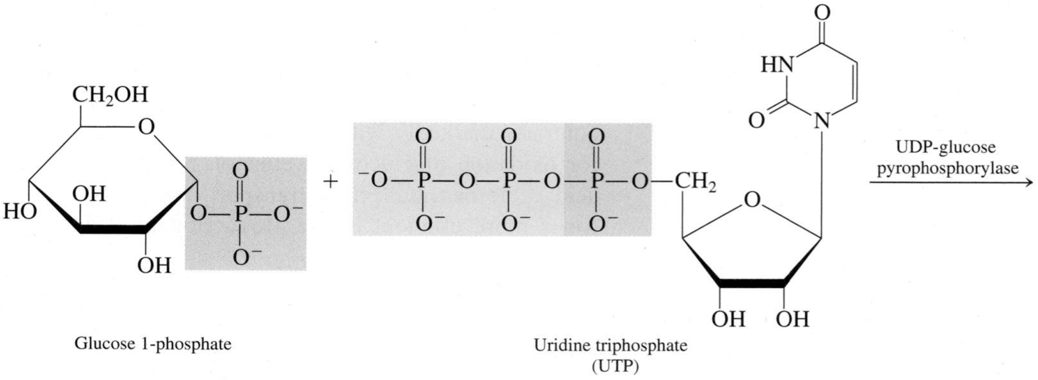

Glucose 1-phosphate

Uridine triphosphate
(UTP)

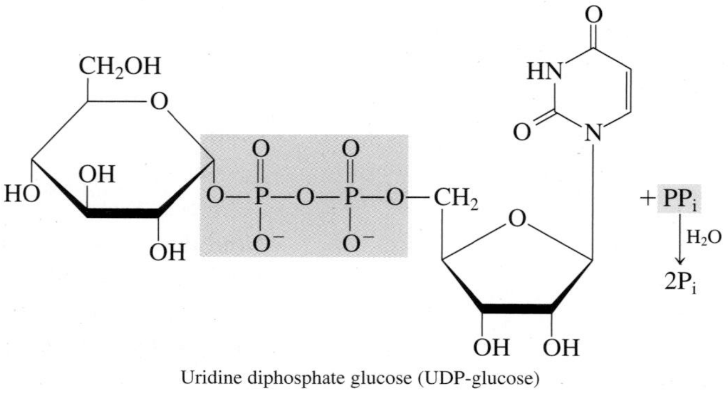

UDP-glucose
pyrophosphorylase

$+ PP_i$

H_2O

$2P_i$

Uridine diphosphate glucose (UDP-glucose)

Step 3: *Glucose Transfer to a Glycogen Chain.* The glucose unit of UDP-glucose is then attached to the end of a glycogen chain.

$$\text{UDP-glucose} + (\text{glucose})_n \xrightarrow[\substack{\text{Glycogen} \\ \text{chain}}]{\substack{\text{Glycogen} \\ \text{synthase}}} (\text{glucose})_{n+1} \underset{\substack{\text{Glycogen with an} \\ \text{additional glucose unit}}}{+ \text{UDP}}$$

In a subsequent reaction, the UDP produced in Step 3 is converted back to UTP, which can then react with another glucose 1-phosphate (Step 2). The conversion reaction requires ATP.

$$\text{UDP} + \text{ATP} \longrightarrow \text{UTP} + \text{ADP}$$

Adding a single glucose unit to a growing glycogen chain requires the investment of two ATP molecules: one in the formation of glucose 6-phosphate and one in the regeneration of UTP.

Glycogenolysis

Glycogenolysis *is the metabolic pathway by which glucose 6-phosphate is produced from glycogen.* This process is not simply the reverse of glycogen synthesis (glycogenesis), because it does not require UTP or UDP molecules. Glycogenolysis is a two-step process rather than a three-step process.

Step 1: *Phosphorylation of a Glucose Residue.* The enzyme *glycogen phosphorylase* effects the removal of an end glucose unit from a glycogen molecule as glucose 1-phosphate.

> A *phosphorylase* is an enzyme that catalyzes the cleavage of a bond by P_i (in contrast to hydrolysis, which refers to bond cleavage by water), such as removal of a glucose unit from glycogen to give glucose 1-phosphate.

$$(\text{Glucose})_n + P_i \xrightarrow[\substack{\text{Glycogen}}]{\substack{\text{Glycogen} \\ \text{Phosphorylase}}} (\text{glucose})_{n-1} + \text{glucose 1-phosphate}$$
$$\underset{\substack{\text{Glycogen with one} \\ \text{fewer glucose unit}}}{}$$

Step 2: *Glucose 1-phosphate Isomerization.* The enzyme *phosphoglucomutase* catalyzes the isomerization process whereby the phosphate group of glucose 1-phosphate is moved to the carbon 6 position.

$$\text{Glucose 1-phosphate} \underset{\substack{\text{Phosphogluco-} \\ \text{mutase}}}{\rightleftharpoons} \text{glucose 6-phosphate}$$

This process is the reverse of the first step of glycogenesis.

> A *phosphatase* is an enzyme that effects the removal of a phosphate group (P_i) from a molecule, such as converting glucose 6-phosphate to glucose, with H_2O as the attacking species.

In muscle and brain cells an immediate need for energy is the stimulus that initiates glycogenolysis. The glucose 6-phosphate that is produced directly enters the glycolysis pathway at Step 1 (Figure 24.3) and its multistep conversion to pyruvate begins. A low level of glucose is the stimulus that initiates glycogenolysis in liver cells. Here, the glucose 6-phosphate produced must be converted to free glucose before it can enter the bloodstream, as glucose 6-phosphate cannot cross cell membranes (Section 24.2). This change is effected by the enzyme *glucose 6-phosphatose,* an enzyme found in liver cells but not in muscle cells or brain cells.

> The fact that glycogen synthesis (glycogenesis) and glycogen degradation (glycogenolysis) are not totally reverse processes has significance. In fact, it is almost always the case in biochemistry that "opposite" biosynthetic and degradative pathways differ in some steps. This allows for separate control of the pathways.

$$\text{Glucose 6-phosphate} + H_2O \xrightarrow[\text{6-phosphatase}]{\text{Glucose}} \text{glucose} + P_i$$

Because muscle and brain cells lack glucose 6-phosphatose, they cannot form free glucose from glucose 6-phosphate. Thus, muscle and brain cells can use glucose 6-phosphate from glycogen for energy production only. The liver, however, with this enzyme present, has the capacity to use glucose 6-phosphate obtained from glycogen to supply additional glucose to the blood.

Figure 24.9 contrasts the "opposite" processes of glycogenesis and glycogenolysis. Both processes involve the glycolysis intermediate glucose 6-phosphate. UDP–glucose is unique to glycogenesis.

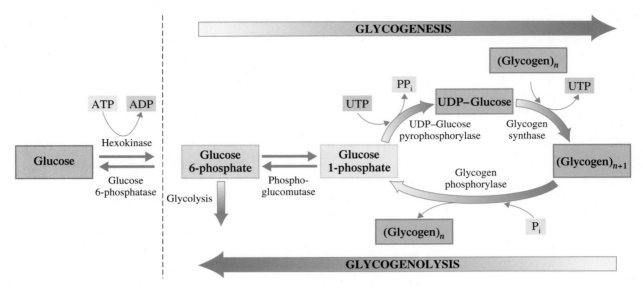

GLYCOGENESIS

GLYCOGENOLYSIS

Figure 24.9 The processes of glycogenesis and glycogenolysis contrasted. The intermediate UDP–glucose is part of glycogenesis but not of glycogenolysis.

24.6 GLUCONEOGENESIS

Gluconeogenesis *is the metabolic pathway by which glucose is synthesized from non-carbohydrate materials.* Glycogen stores in muscle and liver tissue are depleted within 12–18 hours from fasting or in even less time from heavy work or strenuous exercise. Without gluconeogenesis, the brain, which is dependent on glucose as a fuel, would have problems functioning if food intake were restricted for even one day.

The noncarbohydrate starting materials for gluconeogenesis are lactate (from hard-working muscles and from red blood cells), glycerol (from triacylglycerol hydrolysis), and certain amino acids (from dietary protein hydrolysis or from muscle protein during starvation). About 90% of gluconeogenesis takes place in the liver. Hence gluconeogenesis helps to maintain normal blood-glucose levels in times of inadequate dietary carbohydrate intake (such as between meals).

The processes of gluconeogenesis (pyruvate to glucose) and glycolysis (glucose to pyruvate) are not exact opposites. The most obvious difference between these two processes is that 12 compounds are involved in gluconeogenesis and only 11 in glycolysis. Why the difference? The last step of glycolysis is the conversion of the high-energy compound phosphoenolpyruvate to pyruvate. The reverse of this process, which is the beginning of gluconeogenesis, cannot be accomplished in a single step because of the large energy difference between the two compounds and the slow rate of the reaction. Instead, a two-step process by way of oxaloacetate is required to effect the change, and this adds an extra compound to the gluconeogenesis pathway (see Figure 24.10). Both an ATP molecule and a GTP molecule are needed to drive this two-step process.

Figure 24.10 The "opposite" processes of gluconeogenesis (pyruvate to glucose) and glycolysis (glucose to pyruvate) are not exact opposites. The reversal of the last step of glycolysis requires two steps in gluconeogenesis. Therefore, gluconeogenesis has 11 steps, whereas glycolysis has only 10 steps.

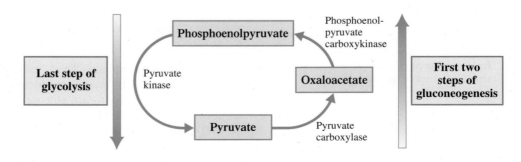

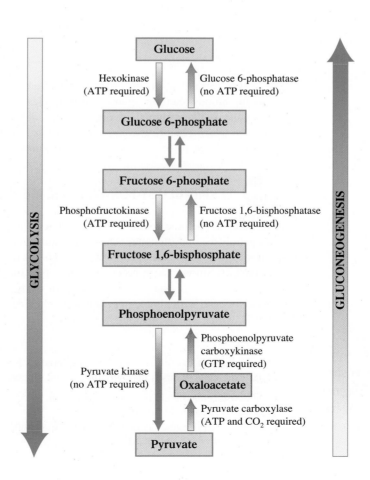

The oxaloacetate intermediate in this two-step process provides a connection to the citric acid cycle. In the first step of this cycle, oxaloacetate combines with acetyl CoA. If energy rather than glucose is needed, then oxaloacetate can go directly into the citric acid cycle.

As is shown in Figure 24.11, there are two other locations where gluconeogenesis and glycolysis differ. In Steps 9 and 11 of gluconeogenesis (Steps 1 and 3 of glycolysis), the reactant–product combinations match between pathways. However, different enzymes are required for the forward and reverse processes. The new enzymes for gluconeogenesis are *fructose 1,6-bisphosphatase* and *glucose 6-phosphatase*.

The overall net reaction for gluconeogenesis is

$$2 \text{ Pyruvate} + 4ATP + 2GTP + 2NADH + 2H_2O \longrightarrow$$
$$\text{glucose} + 4ADP + 2GDP + 6P_i + 2NAD^+$$

Glycolysis has a net production of 2 ATP (Section 24.2). Gluconeogenesis has a net expenditure of 4 ATP and 2 GTP, which is equivalent to the expenditure of 6 ATP.

Figure 24.11 The pathway for gluconeogenesis is similar, but not identical, to the pathway for glycolysis.

Figure 24.12 The Cori cycle. Lactate, formed from glucose under anaerobic conditions in muscle cells, is transferred to the liver, where it is reconverted to glucose, which is then transferred back to the muscle cells.

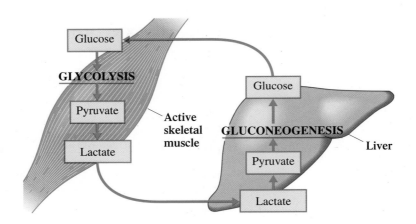

Thus to reconvert pyruvate to glucose requires the expenditure of 4 ATP and 2 GTP. Whenever gluconeogenesis occurs, it is at the expense of other ATP-producing metabolic processes.

The Cori Cycle

Gluconeogenesis using lactate as a source of pyruvate is particularly important because of lactate formation during strenuous exercise. The lactate so produced (Section 24.3) diffuses from muscle cells into the blood, where it is transported to the liver. Here the enzyme lactate dehydrogenase (the same enzyme that catalyzes lactate formation in muscle) converts lactate back to pyruvate.

$$
\underset{\text{Lactate}}{
\begin{array}{c}
\text{COO}^- \\
| \\
\text{H---C---OH} \\
| \\
\text{CH}_3
\end{array}}
+ \text{NAD}^+ \;\xrightarrow[\text{dehydrogenase}]{\text{Lactate}}\;
\underset{\text{Pyruvate}}{
\begin{array}{c}
\text{COO}^- \\
| \\
\text{C}=\text{O} \\
| \\
\text{CH}_3
\end{array}}
+ \text{NADH} + \text{H}^+
$$

The newly formed pyruvate is then converted via gluconeogenesis to glucose, which enters the bloodstream and goes to the muscles. This cyclic process, which is called the Cori cycle, is diagrammed in Figure 24.12. The **Cori cycle** *is a cyclic biochemical process in which glucose is converted to lactate in muscle tissue, the lactate is reconverted to glucose in the liver, and the glucose is returned to the muscle tissue.*

> The Cori cycle is named in honor of Gerty Radnitz Cori (1896–1957) and Carl Cori (1896–1984), the husband-and-wife team who discovered it. They were awarded a Nobel Prize in 1947, the third husband-and-wife team to be so recognized. Marie and Pierre Curie were the first, Irene and Frederic Joliot-Curie the second.

24.7 TERMINOLOGY FOR GLUCOSE METABOLIC PATHWAYS

In the preceding three sections we considered the processes of glycolysis, glycogenesis, glycogenolysis, and gluconeogenesis. Because of their like-sounding names, keeping the terminology for these four processes "straight" is often a problem. Figure 24.13 shows the relationships among these processes. Note that the glycogen degradation pathways (left side of Figure 24.13) have names ending in -*lysis,* which means "breakdown." The pathways associated with glycogen synthesis (right side of Figure 24.13) have names ending in -*genesis,* which means "making."

 EXAMPLE 24.6

Recognizing Characteristics of Glycolysis, Glycogenesis, Glycogenolysis, and Gluconeogenesis

▶ Identify each of the following as a characteristic of one or more of the following processes: glycolysis, glycogenesis, glycogenolysis, and gluconeogenesis.

a. Glucose 6-phosphate is the initial reactant.
b. Glucose is the final product.
c. Glucose 6-phosphate is produced in the first step.
d. UTP is involved in this process.

Figure 24.13 The relationships among four common metabolic pathways that involve glucose.

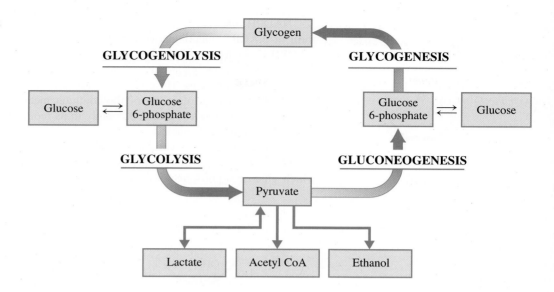

Solution

a. *Glycogenesis.* In this 3-step process glucose 6-phosphate units are added to a growing glycogen molecule.
b. *Gluconeogenesis.* This 11-step process converts pyruvate into glucose.
c. *Glycolysis.* The first step of glycolysis is the production of glucose 6-phosphate from glucose. The glucose 6-phosphate produced can then be further processed in the glycolysis pathway or processed to glycogen using the glycogenesis pathway.
d. *Glycogenesis.* The second step of glycogenesis is the conversion of glucose 1-phosphate to UDP-glucose.

Practice Exercise 24.6

Identify each of the following as a characteristic of one or more of the following processes: glycolysis, glycogenesis, glycogenolysis, and gluconeogenesis.

a. Glycogen is the final product.
b. Glucose is the initial reactant.
c. Glucose 1-phosphate is produced in the first step.
d. ADP is converted to ATP in this process.

Answers: **a.** Glycogenesis; **b.** Glycolysis; **c.** Glycogenesis; **d.** Glycolysis

 24.8 THE PENTOSE PHOSPHATE PATHWAY

Glycolysis is not the only pathway by which glucose may be degraded. Depending on the type of cell, various amounts of glucose are degraded by the pentose phosphate pathway, a pathway whose main focus is *not* subsequent ATP production as is the case for glycolysis. Major functions of this alternative pathway are (1) synthesis of the coenzyme NADPH needed in lipid biosynthesis (Section 25.7), and (2) production of ribose 5-phosphate, a pentose derivative needed for the synthesis of nucleic acids and many coenzymes. The **pentose phosphate pathway** *is the metabolic pathway by which glucose is used to produce NADPH, ribose 5-phosphate (a pentose phosphate), and numerous other sugar phosphates.* The operation of the pentose phosphate pathway is significant in cells that produce lipids: fatty tissue, the liver, mammary glands, and the adrenal cortex (an active producer of steroid lipids).

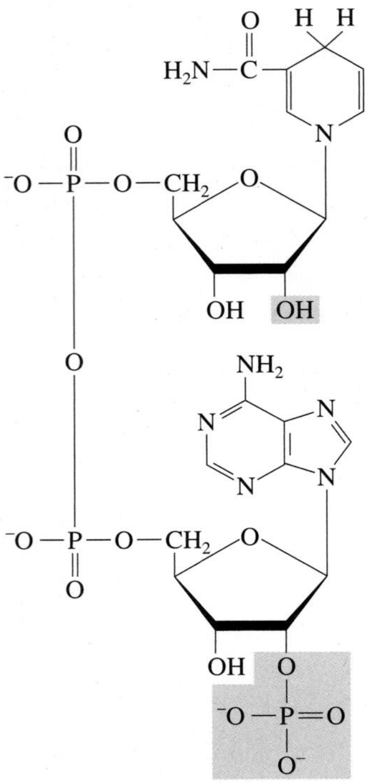

Figure 24.14 The Structure of NADPH. The phosphate group shown in color is the structural feature that distinguishes NADPH from NADH.

NADPH, the coenzyme produced in the pentose phosphate pathway, is the reduced form of $NADP^+$ (nicotinamide adenine dinucleotide phosphate). Structurally, $NADP^+$/NADPH is a phosphorylated version of NAD^+/NADH (see Figure 24.14).

The nonphosphorylated and phosphorylated versions of this coenzyme have significantly different functions. The nonphosphorylated version is involved, mainly in its oxidized form (NAD^+), in the reactions of the common metabolic pathway (Section 23.5). The phosphorylated version is involved, mainly in its reduced form (NADPH), in biosynthetic reactions of lipids and nucleic acids.

There are two stages within the pentose phosphate pathway—an oxidative stage and a nonoxidative stage. The oxidative stage, which occurs first, involves three steps through which glucose 6-phosphate is converted to ribulose 5-phosphate and CO_2.

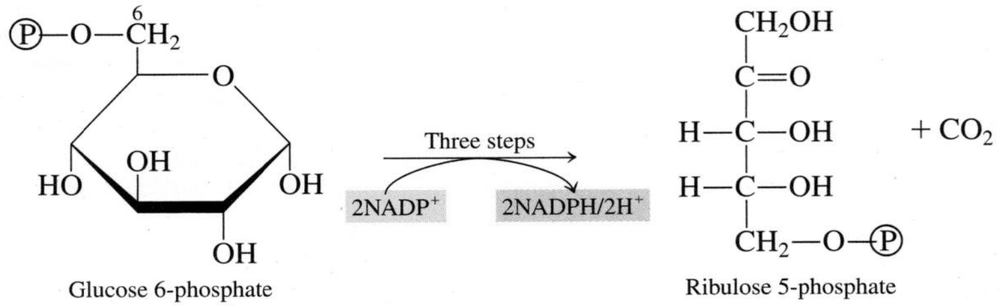

The net equation for the oxidative stage of the pentose phosphate pathway is

Glucose 6-phosphate + $2NADP^+$ + H_2O $\longrightarrow$
$$\text{ribulose 5-phosphate} + CO_2 + 2NADPH + 2H^+$$

Note the production of two NADPH molecules per glucose 6-phosphate processed during this stage.

In the first step of the nonoxidative stage of the pentose phosphate pathway, ribulose 5-phosphate (a ketose) is isomerized to ribose 5-phosphate (an aldose).

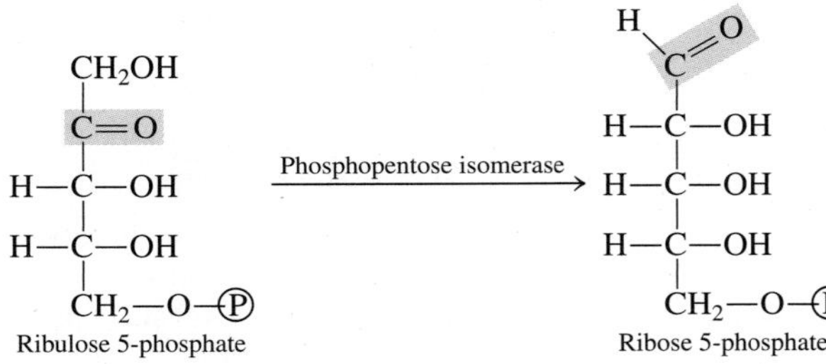

The pentose ribose is a component of ATP, GTP, UTP, CoA, NAD^+/NADH, FAD/$FADH_2$, and RNA. Further steps in the nonoxidative stage contain provision for the conversion of ribose 5-phosphate to numerous other sugar phosphates. Ultimately, glyceraldehyde 3-phosphate and fructose 6-phosphate (both glycolysis intermediates) are formed. The overall net reaction for the pentose phosphate pathway is

3 Glucose 6-phosphate + $6NADP^+$ + $3H_2O$ $\longrightarrow$
$$2 \text{ fructose 6-phosphate} + 3CO_2 + \text{glyceraldehyde 3-phosphate} + 6NADPH + 6H^+$$

The pentose phosphate pathway, with its many intermediates, helps meet cellular needs in numerous ways.

1. When ATP demand is high, the pathway continues to its end products, which enter glycolysis.
2. When NADPH demand is high, intermediates are recycled to glucose 6-phosphate (the start of the pathway), and further NADPH is produced.
3. When ribose 5-phosphate demand is high, for nucleic acid and coenzyme production, most of the nonoxidative stage is nonfunctional, leaving ribose 5-phosphate as a major product.

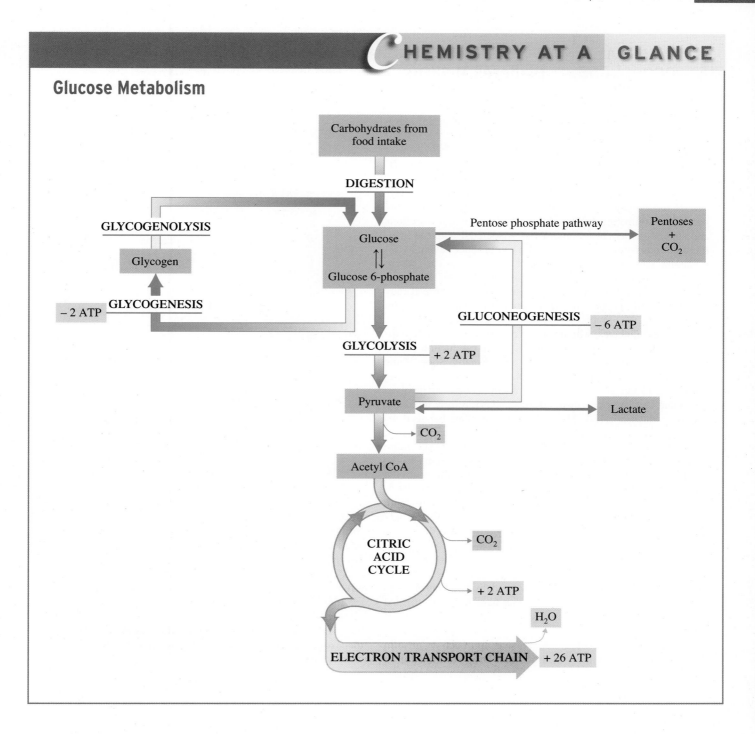

$\mathcal{C}$HEMISTRY AT A GLANCE

Glucose Metabolism

The Chemistry at a Glance feature above shows how the pentose phosphate pathway is related to the other major pathways of glucose metabolism that we have considered.

 24.9 HORMONAL CONTROL OF CARBOHYDRATE METABOLISM

A second major method for regulating carbohydrate metabolism, besides enzyme inhibition by metabolites (Section 24.2), is hormonal control. Among others, three hormones—insulin, glucagon, and epinephrine—affect carbohydrate metabolism.

CHEMICAL Connections   Diabetes Mellitus

Diabetes mellitus is the best-known and most prevalent metabolic disease in humans, affecting approximately 4% of the population. There are two major forms of this disease: insulin-dependent (type I) and non–insulin-dependent (type II) diabetes.

Type I diabetes, which often appears in children, is the result of inadequate insulin production by the beta cells of the pancreas. Control of this condition involves insulin injections and special dietary programs. A risk associated with the insulin injections is that too much insulin can produce severe hypoglycemia (insulin shock); blackout or a coma can result. Treatment involves a quick infusion of glucose. Diabetics often carry candy bars (quick glucose sources) for use if they feel any of the symptoms that signal the onset of insulin shock.

In Type II diabetes, which usually occurs in overweight individuals more than 40 years old, body insulin production can be normal, but the cells do not respond to it normally. Some of the insulin receptors on the cell membranes are not functioning properly and fail to recognize the insulin. Treatment involves drugs that increase body insulin levels and a carefully regulated diet (to reduce obesity). More efficient use of undamaged receptors occurs at increased insulin levels.

About 10% of all cases of diabetes are type I. The more common non–insulin-dependent type II diabetes occurs in the other 90% of cases. The effects of both types of diabetes are the same—inadequate glucose uptake by cells. The result is blood-glucose levels much higher than normal (hyperglycemia). With an inadequate glucose intake, cells must resort to other procedures for energy production, procedures that involve the breakdown of fats and protein.

Insulin was discovered in 1921, the first human body hormone to be discovered. Since that time, researchers have been searching for noninvasive (needle-free) ways to deliver the hormone to people who have diabetes. Avenues that have been investigated include oral delivery, nasal delivery, and dermal delivery using a patch.

The first approval of a noninvasive approach to insulin therapy was given by the U.S. Food and Drug Administration

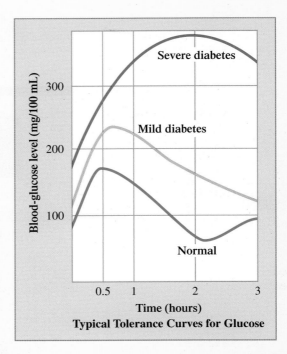

Typical Tolerance Curves for Glucose

in 2006. It involves inhalable insulin, delivered with an inhaler, much the same way medicine to treat asthma is delivered. The insulin can be packaged as a powder or liquid. Tests show that inhalable insulin is as effective as injections in controlling blood sugar. The inhalable insulin reaches the deep lung alveoli structures, which offer the largest surface area for absorption.

A concern about inhalable insulin is its cost, which is three to four times more than injectable insulin. It is still an important development for people who sometimes refuse insulin treatment because of a fear of needles. Another concern is convenience. The inhaler used to dispense the insulin is about 6 inches long (the size of an eyeglass case) when stored and about a foot long when it is in use.

Insulin

Insulin, a 51-amino-acid protein whose structure we considered in Section 20.11, is a hormone produced by the beta cells of the pancreas. Insulin promotes the uptake and utilization of glucose by cells. Thus its function is to lower blood glucose levels. It is also involved in lipid metabolism.

The release of insulin is triggered by *high* blood-glucose levels. The mechanism for insulin action involves insulin binding to protein receptors on the outer surfaces of cells, which facilitates entry of glucose into the cells. Insulin also produces an increase in the rates of glycogen synthesis, glycolysis, and fatty acid synthesis.

Figure 24.15 The series of events by which the hormone epinephrine stimulates glucose production.

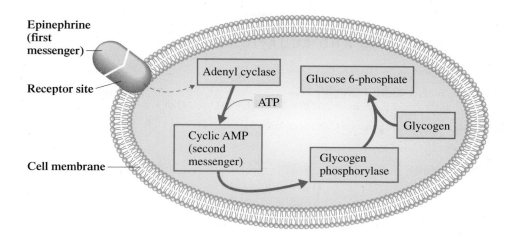

Glucagon

Glucagon is a polypeptide hormone (29 amino acids) produced in the pancreas by alpha cells. It is released when blood-glucose levels are *low*. Its principal function is to increase blood-glucose concentrations by speeding up the conversion of glycogen to glucose (glycogenolysis) and gluconeogenesis in the liver. Thus glucagon's effects are opposite to those of insulin.

Epinephrine

Epinephrine (Section 17.10), also called adrenaline, is released by the adrenal glands in response to anger, fear, or excitement. Its function is similar to that of glucagon—stimulation of glycogenolysis, the release of glucose from glycogen. Its primary target is muscle cells, where energy is needed for quick action. It also functions in lipid metabolism.

Epinephrine acts by binding to a receptor site on the outside of the cell membrane, stimulating the enzyme *adenyl cyclase* to begin production of a *second messenger,* cyclic AMP (cAMP) from ATP. The cAMP is released in the cell interior, where, in a series of reactions, it activates *glycogen phosphorylase,* the enzyme that initiates glycogenolysis. The glucose 6-phosphate that is produced from the glycogen breakdown provides a source of quick energy. Figure 24.15 shows the series of events initiated by the release of the hormone epinephrine. Cyclic AMP also inhibits glycogenesis, thus preventing glycogen production at the same time.

CONCEPTS TO REMEMBER

Glycolysis. Glycolysis, a series of ten reactions that occur in the cytosol, is a process in which one glucose molecule is converted into two molecules of pyruvate. A net gain of two molecules of ATP and two molecules of NADH results from the metabolizing of glucose to pyruvate (Section 24.2).

Fates of pyruvate. With respect to energy-yielding metabolism, the pyruvate produced by glycolysis can be converted to acetyl CoA under aerobic conditions or to lactate under anaerobic conditions. Some microorganisms convert pyruvate to ethanol, an anaerobic process (Section 24.3).

Glycogenesis. Glycogenesis is the process whereby excess glucose 6-phosphate is converted into glycogen. The glycogen is stored in the liver and in muscle tissue (Section 24.5).

Glycogenolysis. Glycogenolysis is the breakdown of glycogen into glucose 6-phosphate. This process occurs when muscles need energy and when the liver is restoring a low blood-sugar level to normal (Section 24.5).

Gluconeogenesis. Gluconeogenesis is the formation of glucose from pyruvate, lactate, and certain other substances. This process takes place in the liver when glycogen supplies are being depleted and when carbohydrate intake is low (Section 24.6).

Cori cycle. The Cori cycle is the cyclic process involving the transport of lactate from muscle tissue to the liver, the resynthesis of glucose by gluconeogenesis, and the return of glucose to muscle tissue (Section 24.6).

Pentose phosphate pathway. The pentose phosphate pathway metabolizes glucose to produce ribose (a pentose), NADPH, and other sugars needed for biosynthesis (Section 24.8).

Carbohydrate metabolism and hormones. Insulin decreases blood-glucose levels by promoting the uptake of glucose by cells. Glucagon increases blood-glucose levels by promoting the conversion of glycogen to glucose. Epinephrine stimulates the release of glucose from glycogen in muscle cells (Section 24.9).

*K*EY REACTIONS AND EQUATIONS

1. Glycolysis (Section 24.2)

$$\text{Glucose} + 2P_i + 2ADP + 2NAD^+ \longrightarrow$$
$$2 \text{ pyruvate} + 2ATP + 2NADH + 2H^+ + 2H_2O$$

2. Oxidation of pyruvate to acetyl CoA (Section 24.3)

$$CH_3-\overset{\overset{\displaystyle O}{\|}}{C}-COO^- + \text{CoA—SH} + NAD^+ \xrightarrow{\text{Four steps}}$$

$$CH_3-\overset{\overset{\displaystyle O}{\|}}{C}-S-CoA + NADH + CO_2$$

3. Reduction of pyruvate to lactate (Section 24.3)

$$CH_3-\overset{\overset{\displaystyle O}{\|}}{C}-COO^- + NADH + H^+ \longrightarrow$$

$$CH_3-\overset{\overset{\displaystyle OH}{|}}{CH}-COO^- + NAD^+$$

4. Reduction of pyruvate to ethanol (Section 24.3)

$$CH_3-\overset{\overset{\displaystyle O}{\|}}{C}-COO^- + 2H^+ + NADH \xrightarrow{\text{Two steps}}$$

$$CH_3-CH_2-OH + NAD^+ + CO_2$$

5. Complete oxidation of glucose (Section 24.4)

$$\text{Glucose} + 6O_2 + 30ADP + 30P_i \longrightarrow$$
$$6CO_2 + 6H_2O + 30ATP$$

6. Glycogenesis (Section 24.5)

$$\text{Glucose 6-phosphate} \xrightarrow[\text{steps}]{\text{Three}} \text{glycogen}$$

7. Glycogenolysis (Section 24.5)

$$\text{Glycogen} \xrightarrow[\text{steps}]{\text{Two}} \text{glucose 6-phosphate}$$

8. Gluconeogenesis (Section 24.6)

$$\left.\begin{array}{c}\text{Lactate, certain}\\\text{amino acids,}\\\text{citric acid cycle}\\\text{intermediates}\end{array}\right\} \longrightarrow \text{pyruvate} \xrightarrow[\text{steps}]{\text{Eleven}} \text{glucose}$$

EXERCISES *and* PROBLEMS

The members of each pair of problems in this section test similar material.

● Carbohydrate Digestion (Section 24.1)

24.1 Where does carbohydrate digestion begin in the body, and what is the name of the enzyme involved in this initial digestive process?

24.2 Very little digestion of carbohydrates occurs in the stomach. Why?

24.3 What is the primary site for carbohydrate digestion, and what organ produces the enzymes that are active at this location?

24.4 Where does the final step in carbohydrate digestion take place, and in what form are carbohydrates as they enter this final step?

24.5 Where does the digestion of sucrose begin, and what is the reaction that occurs?

24.6 Where does the digestion of lactose begin, and what is the reaction that occurs?

24.7 Identify the three major monosaccharides produced by digestion of carbohydrates.

24.8 The various stages of carbohydrate digestion all involve the same general type of reaction. What is this reaction type?

● Glycolysis (Section 24.2)

24.9 What is the starting material for glycolysis?

24.10 What is the end product from glycolysis?

24.11 What coenzyme functions as the oxidizing agent in glycolysis?

24.12 What is meant by the statement that glycolysis is an anaerobic pathway?

24.13 What is the first step of glycolysis, and why is it important in retaining glucose inside the cell?

24.14 Step 3 of glycolysis is the commitment step. Explain.

24.15 What two C_3 fragments are formed by the splitting of a fructose 1,6-bisphosphate molecule?

24.16 In one step of the glycolysis pathway, a C_6 chain is broken into two C_3 fragments, only one of which can be further degraded. What happens to the other C_3 fragment?

24.17 How many pyruvate molecules are produced per glucose molecule during glycolysis?

24.18 How many molecules of ATP and NADH are produced per glucose molecule during glycolysis?

24.19 How many steps in the glycolysis pathway produce ATP?

24.20 How many steps in the glycolysis pathway consume ATP?

24.21 Of the 10 steps of glycolysis, which ones involve phosphorylation?

24.22 Of the 10 steps of glycolysis, which ones involve oxidation?

24.23 Where in a cell does glycolysis occur?

24.24 Do the reactions of glycolysis and the citric acid cycle occur at the same location in a cell? Explain.

24.25 Replace the question mark in each of the following word equations with the name of a substance.

a. Glucose + ATP $\xrightarrow{\text{Hexokinase}}$? + ADP

b. ? $\xrightarrow{\text{Enolase}}$ phosphoenolpyruvate + water

c. 3-Phosphoglycerate $\xrightarrow{?}$ 2-phosphoglycerate

d. 1,3-Bisphosphoglycerate + ? $\xrightarrow[\text{kinase}]{\text{Phosphoglycero-}}$ 3-phosphoglycerate + ATP

24.26 Replace the question mark in each of the following word equations with the name of a substance.

a. Glucose 6-phosphate $\xrightarrow[\text{isomerase}]{\text{Phosphogluco-}}$?

b. ? $\xrightarrow{\text{Aldolase}}$ dihydroxyacetone phosphate + glyceraldehyde 3-phosphate

c. Phosphoenolpyruvate + ? $\xrightarrow[\text{kinase}]{\text{Pyruvate}}$ pyruvate + ATP

d. Dihydroxyacetone phosphate $\xrightarrow{?}$ glyceraldehyde 3-phosphate

24.27 In which step of glycolysis does each of the following occur?
a. Second substrate-level phosphorylation reaction
b. First ATP-consuming reaction
c. Third isomerization reaction
d. Use of NAD^+ as an oxidizing agent

24.28 In which step of glycolysis does each of the following occur?
a. First energy-producing reaction
b. First ATP-producing reaction
c. A dehydration reaction
d. First isomerization reaction

24.29 What is the net ATP production when each of the following molecules is processed through the glycolysis pathway?
a. One glucose molecule
b. One sucrose molecule

24.30 What is the net ATP production when each of the following molecules is processed through the glycolysis pathway?
a. One lactose molecule
b. One maltose molecule

24.31 Draw structural formulas for each of the following pairs of molecules.
a. Pyruvic acid and pyruvate
b. Dihydroxyacetone and dihydroxyacetone phosphate
c. Fructose 6-phosphate and fructose 1,6-bisphosphate
d. Glyceric acid and glyceraldehyde

24.32 Draw structural formulas for each of the following pairs of molecules.
a. Glyceric acid and glycerate
b. Glycerate and pyruvate
c. Glucose 6-phosphate and fructose 6-phosphate
d. Dihydroxyacetone and glyceric acid

24.33 Number the carbon atoms of fructose 1,6-bisphosphate 1 through 6, and show the location of each carbon in the two trioses produced during Step 4 of glycolysis.

24.34 Number the carbon atoms of glucose 1 through 6, and show the location of each carbon in the two molecules of pyruvate produced by glycolysis.

Fates of Pyruvate (Section 24.3)

24.35 What are the three common possible fates for pyruvate produced from glycolysis?

24.36 Compare the fates of pyruvate in the body under aerobic and anaerobic conditions.

24.37 What is the overall reaction equation for the conversion of pyruvate to acetyl CoA?

24.38 What is the overall reaction equation for the conversion of pyruvate to lactate?

24.39 Explain how lactate fermentation allows glycolysis to continue under anaerobic conditions.

24.40 How is the ethanol fermentation in yeast similar to lactate fermentation in skeletal muscle?

24.41 In ethanol fermentation, a C_3 pyruvate molecule is changed to a C_2 ethanol molecule. What is the fate of the third pyruvate carbon?

24.42 What are the structural differences between pyruvate and lactate ions?

24.43 What is the net reaction for the conversion of one glucose molecule to two lactate molecules?

24.44 What is the net reaction for the conversion of one glucose molecule to two ethanol molecules?

Complete Oxidation of Glucose (Section 24.4)

24.45 How does the fact that cytosolic $NADH/H^+$ cannot cross the mitochondrial membranes affect ATP production from cytosolic $NADH/H^+$?

24.46 What is the net reaction for the shuttle mechanism involving glycerol 3-phosphate by which NADH electrons are shuttled across the mitochondrial membrane?

24.47 Contrast, in terms of ATP production, the oxidation of glucose to CO_2 and H_2O with the oxidation of glucose to pyruvate.

24.48 Contrast, in terms of ATP production, the oxidation of glucose to CO_2 and H_2O with the oxidation of glucose to ethanol.

24.49 How many of the 30 ATP molecules produced from the complete oxidation of 1 glucose molecule are produced during glycolysis?

24.50 How many of the 30 ATP molecules produced from the complete oxidation of 1 glucose molecule are produced during the oxidation of pyruvate to acetyl CoA?

Glycogen Metabolism (Section 24.5)

24.51 Compare the meanings of the terms *glycogenesis* and *glycogenolysis*.

24.52 Where is most of the body's glycogen stored?

24.53 Glucose 1-phosphate is the product of the first step of glycogenesis. What is the reactant?

24.54 Glucose 1-phosphate is the product of the first step of glycogenolysis. What are the reactants?

24.55 What is the source of the PP_i produced during the second step of glycogenesis?

24.56 What is the function of the PP_i produced during the second step of glycogenesis?

24.57 How is ATP involved in glycogenesis?

24.58 How many ATP molecules are needed to attach a single glucose molecule to a growing glycogen chain?

24.59 Which step of glycogenolysis is the reverse of Step 1 of glycogenesis?

24.60 What reaction determines whether glucose 6-phosphate formed by glycogenolysis can leave a cell?

24.61 What is the difference between the processing of glucose 6-phosphate in liver cells and that in muscle cells?

24.62 The liver, but not the brain or muscle cells, has the capacity to supply free glucose to the blood. Explain.

24.63 In what form does glycogen enter the glycolysis pathway?

24.64 Explain why one more ATP is produced when glucose is obtained from glycogen than when it is obtained directly from the blood.

Gluconeogenesis (Section 24.6)

24.65 What organ is primarily responsible for gluconeogenesis?

24.66 What is the physiological function of gluconeogenesis?

24.67 How does gluconeogenesis get around the three irreversible steps of glycolysis?

24.68 Although gluconeogenesis and glycolysis are "reverse" processes, there are 11 steps in gluconeogenesis and only 10 steps in glycolysis. Explain.

24.69 What intermediate in gluconeogenesis is also an intermediate in the citric acid cycle?

24.70 What are the sources of high-energy bonds in gluconeogenesis?

24.71 What is the fate of lactate formed by muscular activity?

24.72 What is the physiological function of the Cori cycle?

Terminology for Glucose Metabolic Pathways (Section 24.7)

24.73 Indicate in which of the four processes *glycolysis, glycogenesis, glycogenolysis,* and *gluconeogenesis* each of the following compounds is encountered. There may be more than one correct answer for a given compound.
a. Glucose 6-phosphate
b. Dihydroxyacetone phosphate
c. Oxaloacetate
d. UDP-glucose

24.74 Indicate in which of the four processes *glycolysis, glycogenesis, glycogenolysis,* and *gluconeogenesis* each of the following compounds is encountered. There may be more than one correct answer for a given compound.
a. Glucose 1-phosphate
b. Glycogen
c. Pyruvate
d. Fructose 6-phosphate

24.75 Indicate in which of the four processes *glycolysis, glycogenesis, glycogenolysis,* and *gluconeogenesis* each of the following situations is encountered. There may be more than one correct answer for each.
a. NAD^+ is consumed.
b. ATP is produced.
c. UDP is involved.
d. ADP is consumed.

24.76 Indicate in which of the four processes *glycolysis, glycogenesis, glycogenolysis,* and *gluconeogenesis* each of the following

situations is encountered. There may be more than one correct answer for each.
a. NADH is consumed.
b. ATP is consumed.
c. CO_2 is involved.
d. H_2O is a product.

24.77 Indicate in which of the four processes *glycolysis, glycogenesis, glycogenolysis,* and *gluconeogenesis* each of the following enzymes is needed. There may be more than one correct answer for each.
a. Fructose 1,6-bisphosphatase.
b. Pyruvate kinase.
c. Glycogen synthase.
d. Phosphoglucomutase.

24.78 Indicate in which of the four processes *glycolysis, glycogenesis, glycogenolysis,* and *gluconeogenesis* each of the following enzymes is needed. There may be more than one correct answer for each.
a. Pyruvate carboxylase.
b. Glycogen phosphorylase.
c. Hexokinase.
d. Glucose 6-phosphatase.

The Pentose Phosphate Pathway (Section 24.8)

24.79 What is the starting material for the pentose phosphate pathway?

24.80 What are two major functions of the pentose phosphate pathway?

24.81 How do the biochemical functions of NADH and NADPH differ?

24.82 How do the structures of NADH and NADPH differ?

24.83 Write a general equation for the oxidative stage of the pentose phosphate pathway.

24.84 Write a general equation for the entire pentose phosphate pathway.

24.85 What compound contains the carbon atom lost from glucose (a hexose) in its conversion to ribose (a pentose)?

24.86 How many molecules of NADPH are produced per glucose 6-phosphate in the pentose phosphate pathway?

Control of Carbohydrate Metabolism (Section 24.9)

24.87 What effect does insulin have on glycogen metabolism?

24.88 What effect does insulin have on blood-glucose levels?

24.89 What effect does glucagon have on blood-glucose levels?

24.90 What effect does glucagon have on glycogen metabolism?

24.91 What organ is the source of insulin?

24.92 What organ is the source of glucagon?

24.93 The hormone epinephrine generates a "second messenger." Explain.

24.94 What is the relationship between cAMP and the hormone epinephrine?

24.95 Compare the target tissues for glucagon and epinephrine.

24.96 Compare the biological functions of glucagon and epinephrine.

ADDITIONAL PROBLEMS

24.97 What is the ATP yield *per glucose molecule* in each of the following processes?
a. Glycolysis
b. Glycolysis, acetyl CoA formation, and the common metabolic pathway
c. Glycolysis plus oxidation of pyruvate to acetyl CoA
d. Glycolysis plus reduction of pyruvate to lactate

24.98 Which one of these characterizations, (1) Cori cycle, (2) an anaerobic process, (3) oxidative stage of pentose phosphate pathway, or (4) nonoxidative stage of pentose phosphate pathway, applies to each of the following chemical changes?
a. Pyruvate to lactate
b. Pyruvate to ethanol
c. Glucose 6-phosphate to ribulose 5-phosphate
d. Ribulose 5-phosphate to ribose 5-phosphate

24.99 What condition or conditions determine that pyruvate is involved in each of the following?
a. Gluconeogenesis
b. Converted to lactate

c. Citric acid cycle
d. Converted to ethanol

24.100 In the complete metabolism of 1 mole of sucrose, how many moles of each of the following are produced?
a. CO_2
b. Pyruvate
c. Acetyl CoA
d. ATP

24.101 Under what conditions does glucose 6-phosphate enter each of the following processes.
a. Glycogenesis
b. Glycolysis
c. Pentose phosphate pathway
d. Hydrolysis to free glucose

*M*ultiple-Choice Practice Test

24.102 Which of the following statements concerning glycolysis is *correct*?
a. It is an oxidation process in which molecular oxygen is used.
b. All reactions take place in the cytosol of a cell.
c. There are two stages, each of which involves a series of five reactions.
d. The overall process converts a C_6 molecule into three C_2 molecules.

24.103 What are the two steps in glycolysis in which ATP is converted to ADP?
a. 1 and 2
b. 1 and 3
c. 2 and 3
d. 7 and 10

24.104 Intermediates in the glycolysis pathway include two derivatives of which of the following?
a. Glucose
b. Fructose
c. Pyruvate
d. Glyceraldehyde

24.105 What are the total number of steps in the C_6 stage and C_3 stage of glycolysis, respectively?
a. 10 and 10
b. 5 and 5
c. 4 and 6
d. 3 and 7

24.106 During the overall process of glycolysis, which of the following occurs for each glucose molecule processed?
a. Net loss of two ATP molecules
b. Net loss of four ATP molecules

c. Net gain of two ATP molecules
d. Net gain of four ATP molecules

24.107 Lactate fermentation can occur in which of the following?
a. Humans, animals, and microorganisms
b. Humans and animals but not in microorganisms
c. Microorganisms but not in humans and animals
d. Microorganisms and animals but not in humans

24.108 What is the name of the process in which glycogen is converted to glucose 6-phosphate?
a. Glycolysis
b. Glycogenolysis
c. Glycogenesis
d. Glyconeogenesis

24.109 The compound oxaloacetate is an intermediate in which of the following conversions?
a. Glycogen to glucose
b. Glucose to glycogen
c. Pyruvate to glucose
d. Pyruvate to acetyl CoA

24.110 As part of the Cori cycle, which of the following occurs in liver cells?
a. Glucose is converted to pyruvate.
b. Glucose is converted to lactate.
c. Pyruvate is converted to lactate.
d. Lactate is converted to pyruvate.

24.111 Which of the following are products of the first stage of the pentose phosphate pathway?
a. Ribose 5-phosphate and ribulose 5-phosphate
b. Ribose 5-phosphate and carbon dioxide
c. Ribulose 5-phosphate and carbon dioxide
d. Ribose and carbon dioxide

Lipid Metabolism

Chapter Outline

The fat stored in a camel's hump serves not only as a source of energy but also as a source of water. Water is one of the products of fat (triacylglycerol) metabolism.

Certain classes of lipids play an extremely important role in cellular metabolism, because they represent an energy-rich "fuel" that can be stored in large amounts in adipose (fat) tissue. Between one-third and one-half of the calories present in the diet of the average U.S. resident are supplied by lipids. Furthermore, excess energy derived from carbohydrates and proteins beyond normal daily needs is stored in lipid molecules (in adipose tissue), later to be mobilized and used when needed.

 25.1 DIGESTION AND ABSORPTION OF LIPIDS

Because 98% of total *dietary* lipids are triacylglycerols (fats and oils; Section 19.4), this chapter focuses on triacylglycerol metabolism. Like all lipids, triacylglycerols (TAGs) are insoluble in water. Hence, water-based salivary enzymes in the mouth have little effect on them. The *major* change that TAGs undergo in the stomach is physical rather than chemical. The churning action of the stomach breaks up triacylglycerol materials into small globules, or droplets, which float as a layer above the other components of swallowed food. The resulting material is called *chyme.*

High-fat foods remain in the stomach longer than low-fat foods. The conversion of high-fat materials into chyme takes longer than the breakup of low-fat materials. This is why a high-fat meal causes a person to feel "full" for a longer period of time.

Lipid digestion also begins in the stomach. Under the action of *gastric lipase* enzymes, hydrolysis of TAGs occurs. Normally, about 10% of TAGs undergo hydrolysis in the

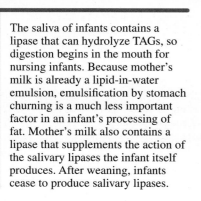

The saliva of infants contains a lipase that can hydrolyze TAGs, so digestion begins in the mouth for nursing infants. Because mother's milk is already a lipid-in-water emulsion, emulsification by stomach churning is a much less important factor in an infant's processing of fat. Mother's milk also contains a lipase that supplements the action of the salivary lipases the infant itself produces. After weaning, infants cease to produce salivary lipases.

stomach, but regular consumption of a high-fat diet can induce the production of higher levels of gastric lipases.

The arrival of chyme from the stomach triggers in the small intestine, through the action of the hormone *cholecystokinin,* the release of bile stored in the gallbladder. The bile (Section 19.11), which contains no enzymes, acts as an emulsifier (Section 19.11). Colloid particle formation (Section 8.7) through bile emulsification "solubilizes" the triacylglycerol globules, and digestion of the TAGs resumes. The major enzymes involved at this point are the *pancreatic lipases,* which hydrolyze ester linkages between the glycerol and fatty acid units of the TAGs. *Complete* hydrolysis does not usually occur; only two of the three fatty acid units are liberated, producing a monoacylglycerol and two free fatty acids. Occasionally, enzymes remove all three fatty acid units, leaving a free glycerol molecule.

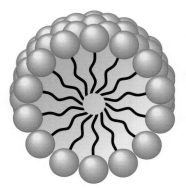

Figure 25.1 In a fatty acid micelle, the hydrophobic chains of the fatty acids and monoacylglycerols are in the interior of the micelle.

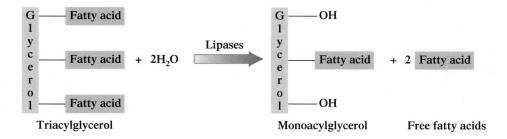

Triacylglycerol $\qquad$ Monoacylglycerol $\qquad$ Free fatty acids

When freed of the triacylglycerol molecules they "transport" during digestion, bile acids are mostly recycled. Small amounts are excreted.

Chylomicron is pronounced "kye-lo-MY-cron."

With the help of bile, the free fatty acids and monoacylglycerols produced from hydrolysis are combined into tiny spherical droplets called micelles (Section 19.6). A **fatty acid micelle** *is a micelle in which fatty acids and/or monoacylglycerols and some bile are present.* Fatty acid micelles are very small compared to the original triacylglycerol globules, which contain thousands of triacylglycerol molecules. Figure 25.1 shows a cross section of the three-dimensional structure of a fatty acid micelle.

Micelles, containing free fatty acid and monoacylglycerol components, are small enough to be readily absorbed through the membranes of intestinal cells. Within the intestinal cells, a "repackaging" occurs in which the free fatty acids and monoacylglycerols are reassembled into triacylglycerols. The newly formed triacylglycerols are then combined with membrane lipids (phospholipids and cholesterol) and water-soluble proteins to produce a type of lipoprotein (Section 20.18) called a *chylomicron* (see Figure 25.2). A **chylomicron** *is a lipoprotein that transports triacylglycerols from intestinal cells, via the lymphatic system, to the bloodstream.* Triacylglycerols constitute 95% of the core lipids present in a chylomicron.

Chylomicrons are too large to pass through capillary walls directly into the bloodstream. Consequently, delivery of the chylomicrons to the bloodstream is accomplished through the body's lymphatic system. Chylomicrons enter the lymphatic system through tiny lymphatic vessels in the intestinal lining. They enter the bloodstream through the thoracic duct (a large lymphatic vessel just below the collarbone), where the fluid of the lymphatic system flows into a vein, joining the bloodstream.

Once the chylomicrons reach the bloodstream, the TAGs they carry are again hydrolyzed to produce glycerol and free fatty acids. TAG release from chylomicrons and their ensuing hydrolysis is mediated by *lipoprotein lipases.* These enzymes are located on the lining of blood vessels in muscle and other tissues that use fatty acids for fuel and in fat synthesis. The fatty acid and glycerol hydrolysis products from TAG hydrolysis are absorbed by the cells of the body and are either broken down to acetyl CoA for energy or stored as lipids (they are again repackaged as TAGs). Figure 25.3 summarizes the events that must occur before triacylglycerols can reach the bloodstream through the digestive process.

Soon after a meal heavily laden with TAGs is ingested, the chylomicron content of both blood and lymph increases dramatically. Chylomicron concentrations usually begin to rise within 2 hours after a meal, reach a peak in 4–6 hours, and then drop

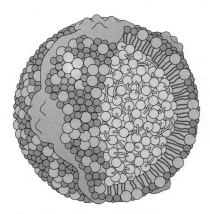

 Triacylglycerols (TAGs)

■ **Protein**

Membrane lipids

Figure 25.2 A three-dimensional model of a chylomicron, a type of lipoprotein. Chylomicrons are the form in which TAGs are delivered to the bloodstream via the lymphatic system.

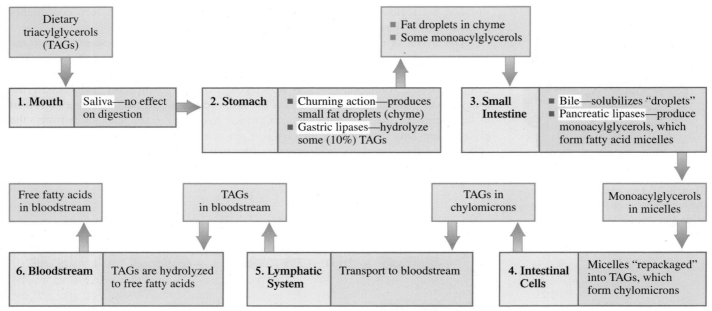

Figure 25.3 A summary of the events that must occur before triacylglycerols (TAGs) can reach the bloodstream through the digestive process.

rather rapidly to a normal level as they move into adipose cells (Section 25.2) or into the liver.

EXAMPLE 25.1

Determining the Sites Where Various Aspects of Lipid Digestion Occur

Based on the information in Figure 25.3, determine the location within the human body where each of the following aspects of lipid digestion occurs.

a. Interaction with bile occurs.
b. Monoacylglycerols are produced.
c. Chyme is produced.
d. Gastric lipases are active.

Solution

a. The *small intestine* is the location where bile released from the gallbladder interacts with the fat droplets present in chyme. The bile functions as an emulsifying agent.
b. Pancreatic lipases present in the *small intestine* convert triacylglycerols to monoacylglycerols.
c. The churning action of the *stomach* breaks triacylglycerol materials into small droplets that float on top of other ingested food materials; the resulting mixture is called chyme.
d. Gastric lipases present in the *stomach* begin the triacylglycerol hydrolysis process; about 10% of triacylglycerols undergo hydrolysis in the stomach.

Practice Exercise 25.1

Based on the information in Figure 25.3, determine the location within the human body where each of the following aspects of lipid digestion occurs.

a. Pancreatic lipases are active.
b. Fatty acid micelles are produced.
c. Chylomicrons are produced.
d. Monoacylglycerols are converted back to triacylglycerols.

Answers: a. Small intestine; b. Small intestine; c. Intestinal cells; d. Intestinal cells

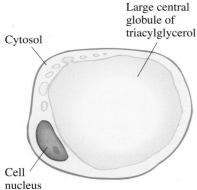

Large central globule of triacylglycerol

Cytosol

Cell nucleus

Figure 25.4 Structural characteristics of an adipose cell.

25.2 TRIACYLGLYCEROL STORAGE AND MOBILIZATION

Most cells in the body have limited capability for storage of TAGs. However, this activity is the major function of specialized cells called adipocytes, found in adipose tissue. *An* **adipocyte** *is a triacylglycerol-storing cell.* **Adipose tissue** *is tissue that contains large numbers of adipocyte cells.*

Adipose tissue is located primarily directly beneath the skin (subcutaneous), particularly in the abdominal region, and in areas around vital organs. Besides its function as a storage location for the chemical energy inherent in TAGs, subcutaneous adipose tissue also serves as an insulator against excessive heat loss to the environment and provides organs with protection against physical shock.

Adipose cells are among the largest cells in the body. They differ from other cells in that most of the cytoplasm has been replaced with a large triacylglycerol droplet (Figure 25.4). This droplet accounts for nearly the entire volume of the cell. As newly formed TAGs are imported into an adipose cell, they form small droplets at the periphery of the cell that later merge with the large central droplet.

Use of the TAGs stored in adipose tissue for energy production is triggered by several hormones, including epinephrine and glucagon. Hormonal interaction with adipose cell membrane receptors stimulates production of cAMP from ATP inside the adipose cell. In a series of enzymatic reactions, the cAMP activates *hormone-sensitive lipase (HSL)* through phosphorylation. HSL is the lipase needed for triacylglycerol hydrolysis, a prerequisite for fatty acids to enter the bloodstream from an adipose cell. This cAMP activation process is illustrated in Figure 25.5.

The overall process of tapping the body's triacylglycerol energy reserves (adipose tissue) for energy is called triacylglycerol mobilization. **Triacylglycerol mobilization** *is the hydrolysis of triacylglycerols stored in adipose tissue, followed by release into the bloodstream of the fatty acids and glycerol so produced.* Triacylglycerol mobilization is an ongoing process. On the average, about 10% of the TAGs in adipose tissue are replaced daily by new triacylglycerol molecules.

Triacylglycerol energy reserves (fat reserves) are the human body's major source of stored energy. Energy reserves associated with protein, glycogen, and glucose are small to very small when compared to fat reserves. Table 25.1 shows relative amounts of stored energy associated with the various types of energy reserves present in the human body.

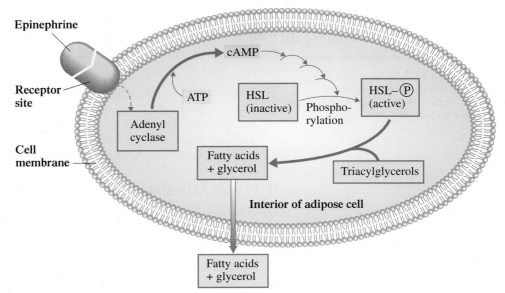

Figure 25.5 Hydrolysis of stored triacylglycerols in adipose tissue is triggered by hormones that stimulate cAMP production within adipose cells.

TABLE 25.1
Stored Energy Reserves of Various Types for a 150-lb (70-kg) Person

Type of Energy Reserve	Amount of Stored Energy	Percent of Total Stored Energy
triacylglycerol	135,000 kcal	84.3%
protein	24,000 kcal	15%
glycogen	720 kcal	0.45%
blood glucose	80 kcal	0.05%

Triacylglycerol reserves would enable the average person to survive starvation for about 30 days, given sufficient water. Glycogen reserves (stored glucose) would be depleted within 1 day.

GLYCEROL METABOLISM

During triacylglycerol mobilization, one molecule of glycerol is produced for each triacylglycerol completely hydrolyzed. Glycerol metabolism primarily involves processes considered in the previous chapter. After entering the bloodstream, glycerol travels to the liver or kidneys, where it is converted, in a two-step process, to dihydroxyacetone phosphate.

$$
\begin{array}{ccc}
\text{H}_2\text{C}\!-\!\text{OH} & & \text{H}_2\text{C}\!-\!\text{OH} \\
| & \xrightarrow[\text{kinase}]{\text{Glycerol}} & | \\
\text{HC}\!-\!\text{OH} & & \text{HC}\!-\!\text{OH} \\
| & & | \\
\text{H}_2\text{C}\!-\!\text{OH} \quad \text{ATP} \quad \text{ADP} & & \text{H}_2\text{C}\!-\!\text{O}\!-\!\textcircled{P} \quad \text{NAD}^+ \quad \text{NADH/H}^+
\end{array}
$$

Glycerol   Glycerol 3-phosphate   Dihydroxyacetone phosphate

Glycerol 3-phosphate dehydrogenase

$$
\begin{array}{c}
\text{H}_2\text{C}\!-\!\text{OH} \\
| \\
\text{C}\!=\!\text{O} \\
| \\
\text{H}_2\text{C}\!-\!\text{O}\!-\!\textcircled{P}
\end{array}
$$

The first step involves phosphorylation of a primary hydroxyl group of the glycerol. In the second step, glycerol's secondary alcohol group (C-2) is oxidized to a ketone.

The following equation represents the overall reaction for the metabolism of glycerol.

$$\text{Glycerol} + \text{ATP} + \text{NAD}^+ \longrightarrow \text{Dihydroxyacetone phosphate} + \text{ADP} + \text{NADH} + \text{H}^+$$

Dihydroxyacetone phosphate is an intermediate in both glycolysis (Section 24.2) and gluconeogenesis (Section 24.6). It can be converted to pyruvate, then acetyl CoA, and finally carbon dioxide, or it can be used to form glucose. Dihydroxyacetone phosphate formation from glycerol represents the first of several situations we will consider wherein carbohydrate and lipid metabolism are connected.

OXIDATION OF FATTY ACIDS

There are three parts to the process by which fatty acids are broken down to obtain energy.

1. The fatty acid must be *activated* by bonding to coenzyme A.
2. The fatty acid must be *transported* into the mitochondrial matrix by a shuttle mechanism.
3. The fatty acid must be repeatedly *oxidized,* cycling through a series of four reactions, to produce acetyl CoA, FADH$_2$, and NADH.

Fatty Acid Activation

The outer mitochondrial membrane is the site of fatty acid *activation,* the first stage of fatty acid oxidation. Here the fatty acid is converted to a high-energy derivative of coenzyme A. Reactants are the fatty acid, coenzyme A, and a molecule of ATP.

$$
\underset{\text{Free fatty acid}}{\text{R}\!-\!\overset{\overset{\textstyle O}{\|}}{\text{C}}\!-\!\text{O}^-} + \text{HS}\!-\!\text{CoA} \xrightarrow[\text{synthetase}]{\text{Acyl CoA}} \underset{\text{Acyl CoA}}{\text{R}\!-\!\overset{\overset{\textstyle O}{\|}}{\text{C}}\!-\!\text{S}\!-\!\text{CoA}}
$$

ATP   AMP + 2P$_i$

The stored TAGs in adipose tissue supply approximately 60% of the body's energy needs when the body is in a resting state.

Recall, from Section 16.19, that *acyl* is a generic term for

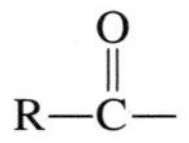

$$R-\overset{\overset{\displaystyle O}{\parallel}}{C}-$$

which is the species formed when the carboxyl —OH is removed from a carboxylic acid. The R group can involve a carbon chain of any length.

This reaction requires the expenditure of two high-energy phosphate bonds from a single ATP molecule; the ATP is converted to AMP rather than ADP, and the resulting pyrophosphate (PP_i) is hydrolyzed to $2P_i$.

The activated fatty acid–CoA molecule is called *acyl* CoA. The difference between the designations *acyl* CoA and *acetyl* CoA is that *acyl* refers to a random-length fatty acid carbon chain that is covalently bonded to coenzyme A, whereas *acetyl* refers to a two-carbon chain covalently bonded to coenzyme A.

$$R-\overset{\overset{\displaystyle O}{\parallel}}{C}-S-CoA \qquad CH_3-\overset{\overset{\displaystyle O}{\parallel}}{C}-S-CoA$$

Acyl CoA
R = carbon chain of any length

Acetyl CoA
R = CH$_3$ group

Fatty Acid Transport

Acyl CoA is too large to pass through the inner mitochondrial membrane to the mitochondrial matrix, where the enzymes needed for fatty acid oxidation are located. A shuttle mechanism involving the molecule carnitine effects the entry of acyl CoA into the matrix (see Figure 25.6). The acyl group is transferred to a carnitine molecule, which carries it through the membrane. The acyl group is then transferred from the carnitine back to a CoA molecule.

Reactions of the β-Oxidation Pathway

In the mitochondrial matrix, a sequence of four reactions *repeatedly* cleaves two-carbon units from the carboxyl end of the acyl CoA molecule. This repetitive four-reaction sequence is called the *β-oxidation pathway* because the second carbon from the carboxyl end of the chain, the beta carbon, is the carbon atom that is oxidized. The **β-oxidation pathway** *is a repetitive series of four biochemical reactions that degrades acyl CoA to acetyl CoA by removing two carbon atoms at a time, with FADH₂ and NADH also being produced.* Each repetition of the four-reaction sequence generates an acetyl CoA molecule and an acyl CoA molecule that has two fewer carbon atoms.

For a *saturated* fatty acid, the β-oxidation pathway involves the following functional group changes at the β carbon and the following reaction types.

We have encountered an identical set of functional group changes before, in the back side of the citric acid cycle (Section 23.6), Steps 6–8 of this cycle.

$$\textbf{Alkane} \xrightarrow[\text{(dehydrogenation)}]{\textcircled{1}\ \text{Oxidation}} \textbf{alkene} \xrightarrow{\textcircled{2}\ \text{Hydration}} \textbf{secondary alcohol} \xrightarrow[\text{(dehydrogenation)}]{\textcircled{3}\ \text{Oxidation}} \textbf{ketone} \xrightarrow{\textcircled{4}\ \text{Chain cleavage}}$$

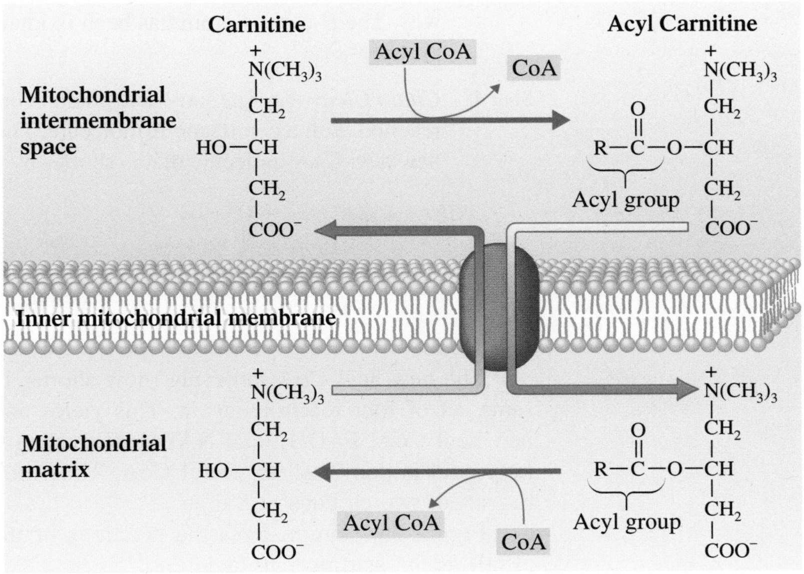

Figure 25.6 Fatty acids are transported across the inner mitochondrial membrane in the form of acyl carnitine.

Details about Steps 1–4 of the β-oxidation pathway follow.

Step 1: *Oxidation (dehydrogenation).* Hydrogen atoms are removed from the α and β carbons, creating a double bond between these two carbon atoms. FAD is the oxidizing agent, and a $FADH_2$ molecule is a product.

The enzyme involved is stereospecific in that only *trans* double bonds are produced.

Step 2: *Hydration.* A molecule of water is added across the *trans* double bond, producing a secondary alcohol at the β-carbon position. Again, the enzyme involved is stereospecific in that only the L-hydroxy isomer is produced from the *trans* double bond.

The enzyme involved in this hydration will also hydrate a *cis* double bond, but the product then is the D isomer. We shall return to this point later in considering how unsaturated fatty acids are oxidized.

The reaction sequence dehydrogenation–hydration–dehydrogenation in the fatty acid spiral has a parallel in Steps 6–8 of the citric acid cycle (Section 23.6), where succinate is dehydrogenated to fumarate, which is hydrated to malate, which is dehydrogenated to oxaloacetate.

Step 3: *Oxidation (dehydrogenation).* The β-hydroxy group is oxidized to a ketone functional group with NAD^+ serving as the oxidizing agent. The required enzyme exhibits absolute stereospecificity for the L isomer.

It is now apparent why the name for this series of reactions is β-oxidation pathway. The β-carbon atom has been oxidized from a —CH_2— group to a ketone group.

Step 4: *Chain Cleavage.* The fatty acid chain is broken between the α and β carbons by reaction with a coenzyme A molecule. The result is an acetyl CoA molecule and a new acyl CoA molecule that is shorter by two carbon atoms than its predecessor.

The new acyl CoA molecule (now shorter by two carbons) is *recycled* through the same set of four reactions again. This yields another acetyl CoA, a two-carbon-shorter new acyl CoA, $FADH_2$, and NADH. Recycling occurs again and again, until the entire fatty acid is converted to acetyl CoA. Thus the fatty acid carbon chain is sequentially degraded, two carbons at a time.

Figure 25.7 summarizes the reactions of the β-oxidation pathway for stearic acid (18:0) as the starting fatty acid.

Figure 25.7 Reactions of the β-oxidation pathway for an 18:0 fatty acid (stearic acid).

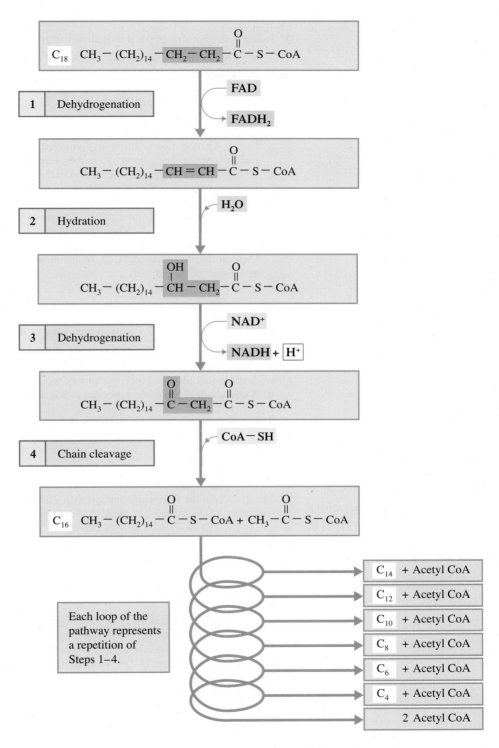

The fatty acids normally found in dietary triacylglycerols contain an *even* number of carbon atoms. Thus the number of acetyl CoA molecules produced in the β-oxidation pathway is equal to half the number of carbon atoms in the fatty acid. The number of *repetitions* of the β-oxidation pathway that are needed to produce the acetyl CoA is always one less than the number of acetyl CoA molecules produced because the last repetition produces two acetyl CoA molecules as a C_4 unit splits into two C_2 units.

C_{18} fatty acid $\longrightarrow$ 9 acetyl CoA (8 repetitive sequences)

C_{14} fatty acid $\longrightarrow$ 7 acetyl CoA (6 repetitive sequences)

This sequence of reactions is called the β-oxidation *pathway* rather than the β-oxidation *cycle* because a different product results from each repetition.

EXAMPLE 25.2

Recognizing Reaction Types and Events That Occur in the β-Oxidation Pathway

Indicate at what step in the β-oxidation pathway each of the following events occurs.

a. A carbon–carbon single bond is converted to a carbon–carbon double bond.
b. NAD⁺ is reduced to NADH.
c. A hydration reaction occurs.
d. An acetyl CoA molecule is produced.

Solution

a. *Step 1.* An oxidation reaction involving removal of two hydrogen atoms changes the carbon–carbon single bond to a carbon–carbon double bond.
b. *Step 3.* This step is the second of two oxidation reactions that occur. The oxidizing agent for this oxidation is NAD⁺, which is converted to NADH. Note that the oxidizing agent for the first oxidation reaction (step 1) is not NAD⁺ but rather FAD.
c. *Step 2.* A molecule of water is added to the carbon–carbon double bond producing a secondary alcohol at the β-carbon position.
d. *Step 4.* Breakage of the bond between the α- and β-carbon atoms produces an acetyl CoA molecule and an acyl CoA molecule whose carbon chain is two atoms shorter than at the start of the pathway.

Practice Exercise 25.2

Indicate at what step in the β-oxidation pathway each of the following events occurs.

a. A carbon–carbon double bond is converted to a carbon–carbon single bond.
b. FAD is reduced to FADH₂.
c. A secondary alcohol group is oxidized to a ketone group.
d. A coenzyme A molecule is needed as a reactant.

Answers: a. Step 2; **b.** Step 1; **c.** Step 3; **d.** Step 4

EXAMPLE 25.3

Relating Characteristics of the β-Oxidation Pathway to Types of Reactions That Occur

Match each of the following characteristics of the β-oxidation pathway to the terms (1) first oxidation, (2) hydration, (3) second oxidation, and (4) chain cleavage.

a. The enzyme needed is thiolase.
b. The enzyme needed is acyl CoA dehydrogenase.
c. The substance *trans*-enoyl CoA is a product.
d. The substance β-ketoacyl CoA is a reactant.

Solution

a. *Chain cleavage* (Step 4). The prefix *thio* in the enzyme name *thiolase* indicates a reaction that involves an S bond. When coenzyme A reacts with the acyl group produced by chain cleavage to form acyl CoA a new C—S bond is formed.
b. *First oxidation* (Step 1). This oxidation step, as well as the second one, involves the removal of two hydrogen atoms with the resulting creation of a double bond. A dehydrogenase enzyme effects such a change.
c. *First oxidation* (Step 1). *Trans*-enoyl CoA is produced in Step 1 and becomes the reactant for Step 2.
d. *Chain cleavage* (Step 4). β-ketoacyl CoA is produced in Step 3 and becomes the reactant for Step 4.

Practice Exercise 25.3

Match each of the following characteristics of the β-oxidation pathway to the terms (1) first oxidation, (2) hydration, (3) second oxidation, and (4) chain cleavage.

a. The enzyme enoyl CoA hydratase is needed.

b. A stereospecific enzyme that produces *trans*-carbon–carbon double bonds is needed.
c. The substance acyl CoA is a reactant.
d. The substance acyl CoA is a product.

Answers: a. Hydration; **b.** First oxidation; **c.** First oxidation; **d.** Chain cleavage

Unsaturated Fatty Acids

Unsaturated fatty acids are common components of dietary triacylglycerols. Their oxidation through the β-oxidation pathway requires two additional enzymes besides those needed for oxidation of saturated fatty acids. These two—an epimerase that can change a D configuration to an L configuration and a *cis–trans* isomerase—are needed for two reasons. First, the double bonds in naturally occurring unsaturated fatty acids are nearly always *cis* double bonds, which yield on hydration a D-hydroxy product rather than the L-hydroxy product needed for Step 3 of the pathway. The epimerase enzyme effects a configuration change from the D form to the L form.

$$
\underset{\text{D-}\beta\text{-Hydroxyacyl CoA}}{R-\overset{\overset{\displaystyle H}{|}}{\underset{\underset{\displaystyle OH}{|}}{C}}-\overset{\overset{\displaystyle H}{|}}{\underset{\underset{\displaystyle H}{|}}{C}}-\overset{\overset{\displaystyle O}{\|}}{C}-S-CoA} \xrightarrow{\text{Epimerase}} \underset{\text{L-}\beta\text{-Hydroxyacyl CoA}}{R-\overset{\overset{\displaystyle OH}{|}}{\underset{\underset{\displaystyle H}{|}}{C}}-\overset{\overset{\displaystyle H}{|}}{\underset{\underset{\displaystyle H}{|}}{C}}-\overset{\overset{\displaystyle O}{\|}}{C}-S-CoA}
$$

Second, the double bonds in naturally occurring unsaturated fatty acids often occupy odd-numbered positions (Section 19.2). The hydratase in Step 2 of the pathway can effect hydration of only an even-numbered double bond. The *cis–trans* isomerase produces a *trans*-(2,3) double bond from a *cis*-(3,4) double bond.

cis-(3,4)　　　　　　　　　　　*trans*-(2,3)

The Step 2 hydratase can then work on the *trans*-(2,3) double bond in the normal fashion.

25.5 ATP PRODUCTION FROM FATTY ACID OXIDATION

How does the total energy output from fatty acid oxidation compare to that of glucose oxidation? Let us calculate ATP production for the oxidation of a specific fatty acid molecule, stearic acid (18:0), and compare it with that from glucose.

Figure 25.7 shows that for each four-reaction sequence except the last one, one $FADH_2$ molecule, one NADH molecule, and one acetyl CoA molecule are produced. In the final four-reaction sequence, two acetyl CoA molecules are produced in addition to the $FADH_2$ and NADH molecules.

Eight repetitions of the β-oxidation pathway are required for the oxidation of stearic acid, an 18-carbon acid. These eight repetitions of the pathway produce 9 acetyl CoA molecules, 8 $FADH_2$ molecules, and 8 NADH molecules. Further processing of these products through the common metabolic pathway (citric acid cycle, electron transport chain, and oxidative phosphorylation) leads to ATP production as follows:

$$9 \text{ acetyl CoA} \times \frac{10 \text{ ATP}}{1 \text{ acetyl CoA}} = \quad 90 \text{ ATP}$$

$$8 \text{ FADH}_2 \times \frac{1.5 \text{ ATP}}{1 \text{ FADH}_2} = \quad 12 \text{ ATP}$$

$$8 \text{ NADH} \times \frac{2.5 \text{ ATP}}{1 \text{ NADH}} = \quad \underline{20 \text{ ATP}}$$

$$122 \text{ ATP}$$

The conversion factors used in this calculation were first presented in Section 23.9.

This *gross* production of 122 ATP must be decreased by the ATP needed to activate the fatty acid before it enters the β-oxidation pathway. The activation consumes two high-energy phosphate bonds of an ATP molecule. For accounting purposes, this is equivalent to hydrolyzing 2 ATP molecules to ADP. Thus the *net* ATP production from oxidation of stearic acid is 120 ATP (122 minus 2).

 **E X A M P L E   25.4**

Calculating Net ATP Production from the Complete Oxidation of a Fatty Acid

▶ What is the net ATP production for the complete oxidation of lauric acid, the C_{12} saturated fatty acid, to CO_2 and H_2O?

Solution

The total ATP production for oxidation of a fatty acid is the sum of the ATP produced from the metabolic products $FADH_2$, NADH, and acetyl CoA. In obtaining the *net* ATP production, it must be remembered that the activation of a fatty acid molecule prior to its oxidation requires ATP consumption equivalent to two ATP molecules (Section 25.4).

Oxidation of a C_{12} fatty acid requires five passages through the four-reaction sequence of the β-oxidation pathway. These five repetitions produce a total of 6 acetyl CoA, 5 $FADH_2$, and 5 NADH. The number of acetyl CoA produced is always equal to the number of carbon atoms present in the fatty acid divided by 2, which is $12/2 = 6$ in this case. The number of $FADH_2$ and NADH produced will always be one less than the number of acetyl CoA produced.

Further processing of these metabolic products through the citric acid cycle, electron transport chain, and oxidative phosphorylation leads to ATP production as follows:

Fatty Acid Activation: -2 ATP

Acetyl CoA Production:

$$6 \text{ acetyl CoA} \times \frac{10 \text{ ATP}}{1 \text{ acetyl CoA}} \qquad +60 \text{ ATP}$$

FADH₂ Production:

$$5 \text{ FADH}_2 \times \frac{1.5 \text{ ATP}}{1 \text{ FADH}_2} \qquad +7.5 \text{ ATP}$$

NADH Production:

$$5 \text{ NADH} \times \frac{2.5 \text{ ATP}}{1 \text{ NADH}} \qquad +12.5 \text{ ATP}$$

Net ATP Production: $+78$ ATP

The ATP equivalents for acetyl CoA, $FADH_2$, and NADH used in this calculation were first presented in Section 23.9.

Practice Exercise 25.4

What is the net ATP production for the complete oxidation of palmitic acid, the C_{16} saturated fatty acid, to CO_2 and H_2O?

Answer: 106 ATP

The comparison between complete fatty acid oxidation and complete glucose oxidation (Section 24.4) shows that a stearic acid molecule produces four times as much ATP as a glucose molecule.

$$1 \text{ glucose} \longrightarrow \boxed{30 \text{ ATP}}$$
$$1 \text{ stearic acid} \longrightarrow \boxed{120 \text{ ATP}}$$

Taking into account the fact that glucose has only 6 carbon atoms and stearic acid has 18 carbon atoms still shows more ATP production from the fatty acid.

$$3 \text{ glucose (18 C)} \longrightarrow \boxed{90 \text{ ATP}}$$
$$1 \text{ stearic acid (18 C)} \longrightarrow \boxed{120 \text{ ATP}}$$

Thus, on the basis of equal numbers of carbon atoms, lipids are 33% more efficient than carbohydrates as energy-storage systems.

High-Intensity Versus Low-Intensity Workouts

In a resting state, the human body burns more fat than carbohydrate. The fuel consumed is about one-third carbohydrate and two-thirds fat.

Information about fuel consumption ratios is obtainable from respiratory gas measurements, specifically from the respiratory exchange ratio (RER). The RER is the ratio of carbon dioxide to oxygen inhaled divided by the ratio of carbon dioxide to oxygen exhaled. For 100% fat burning, the RER would be 0.7; for 100% carbohydrate burning, the RER would be 1.0.

When a person at rest begins exercising, his or her body suddenly needs energy at a greater rate—more fuel and more oxygen are needed. It takes 0.7 L of oxygen to burn 1 gram of carbohydrate and 1.0 L of oxygen to burn 1 gram of fat. At the onset of exercise, the body is immediately short of oxygen. Also, there is a time delay in triacylglycerol mobilization. Triacylglycerols have to be broken down to fatty acids, which have to be

attached to protein carriers before they can be carried in the bloodstream to working muscles. At their destination, they must be released from the carriers and then undergo energy-producing reactions. By contrast, glycogen is already present in muscle cells, and it can release glucose 6-phosphate as an instant fuel.

Consequently, the initial stages of exercise are fueled primarily by glucose—it requires less oxygen and can even be burned anaerobically (to lactate). During the first few minutes of exercise, up to 80% of the fuel used comes from glycogen.

With time, increased breathing rates increase oxygen supplies to muscles, and triacylglycerol use increases. Continued activity for three-quarters of an hour achieves a 50–50 balance of triacylglycerol and glucose use. Beyond an hour, triacylglycerol use may be as high as 80%.

Suppose a person is exercising at a moderate rate and decides to speed up. Immediately, body fuel and oxygen needs are increased. The response is increased use of glycogen supplies.

The accompanying table compares exercise on a stationary cycle at 45% and 70% of maximum oxygen uptake sufficient to burn 300 calories.

The initial stages of exercise are fueled primarily by glucose; in later stages, triacylglycerols become the primary fuel.

	Low-Intensity Exercise	High-Intensity Exercise
percent of maximum oxygen uptake	45%	70%
time required to burn 300 calories	48 min	30 min
calories obtained from fat	133 cal	65 cal
percent of calories from fat	44%	22%
rate of fat burning per minute	2.8 cal/min	2.1 cal/min

On an equal-mass basis, fatty acids produce 2.5 times as much energy per gram as carbohydrates (glucose); this is shown by the following calculation involving 1.00 gram of stearic acid and 1.00 gram of glucose.

$$1.00 \text{ g stearic acid} \times \left(\frac{1 \text{ mole stearic acid}}{284 \text{ g stearic acid}} \right) \times \left(\frac{120 \text{ moles ATP}}{1 \text{ mole stearic acid}} \right) = 0.423 \text{ mole ATP}$$

$$1.00 \text{ g glucose} \times \left(\frac{1 \text{ mole glucose}}{180 \text{ g glucose}} \right) \times \left(\frac{30 \text{ moles ATP}}{1 \text{ mole glucose}} \right) = 0.167 \text{ mole ATP}$$

The fact that fatty acids (stearic acid) yield 2.5 times as much energy per gram as carbohydrates (glucose) means that, in terms of calories consumed, the former "do 2.5 times as much damage" to a person on a diet.

In dietary considerations, nutritionists say that 1 gram of carbohydrate equals 4 kcal and that 1 gram of fat equals 9 kcal. We now know the basis for these numbers. The value of 9 kcal for fat takes into account the fact that not all fatty acids present in fat contain 18 carbon atoms (the basis for our preceding calculations) and also the fact that fats contain glycerol, which produces ATP when degraded.

Is the preferred fuel for "running" the human body fatty acids, which yield 2.5 times as much energy per gram as glucose, or is it glucose? In a normally functioning human body, certain organs use both fuels, others prefer glucose, and still others prefer fatty acids. Here are some generalizations about "fuel" use:

1. Skeletal muscle uses glucose (from glycogen) when in an active state. In a resting state, it uses fatty acids.
2. Cardiac muscle depends first on fatty acids and secondarily on ketone bodies (Section 25.6), glucose, and lactate.
3. The liver uses fatty acids as the preferred fuel.
4. Brain function is maintained by glucose and ketone bodies (Section 25.6). Fatty acids cannot cross the blood–brain barrier and thus are unavailable.

25.6   KETONE BODIES

Ordinarily, when there is adequate balance between lipid and carbohydrate metabolism, most of the acetyl CoA produced from the β-oxidation pathway is further processed through the citric acid cycle. The first step of the citric acid cycle (Section 23.6) involves the reaction between oxaloacetate and acetyl CoA. Sufficient oxaloacetate must be present for the acetyl CoA to react with. Oxaloacetate concentration depends on pyruvate produced from glycolysis (Section 24.2); pyruvate can be converted to oxaloacetate by *pyruvate carboxylase* (Section 24.6).

Certain body conditions upset the lipid–carbohydrate balance required for acetyl CoA generated by fatty acids to be processed by the citric acid cycle. These conditions include (1) dietary intake high in fat and low in carbohydrates, (2) diabetic conditions where the body cannot adequately process glucose even though it is present, and (3) *prolonged* fasting conditions, including starvation, where glycogen supplies are exhausted. Under these conditions, the problem of inadequate oxaloacetate supplies arises, which is compounded by the body's using oxaloacetate that is present to produce glucose through gluconeogenesis (Section 24.6).

What happens when oxaloacetate supplies are too low for all acetyl CoA present to be processed through the citric acid cycle? The excess acetyl CoA is diverted to the formation of ketone bodies. A **ketone body** *is one of three substances (acetoacetate, β-hydroxybutyrate, and acetone) produced from acetyl CoA when an excess of acetyl CoA from fatty acid degradation accumulates because of triacylglycerol–carbohydrate metabolic imbalances.* The structural formulas for the three ketone bodies, two of which are C_4 molecules and the other a C_3 molecule, are

Ketone bodies are produced when the amount of acetyl CoA is excessive compared with the amount of oxaloacetate available to react with it (Step 1 of the citric acid cycle).

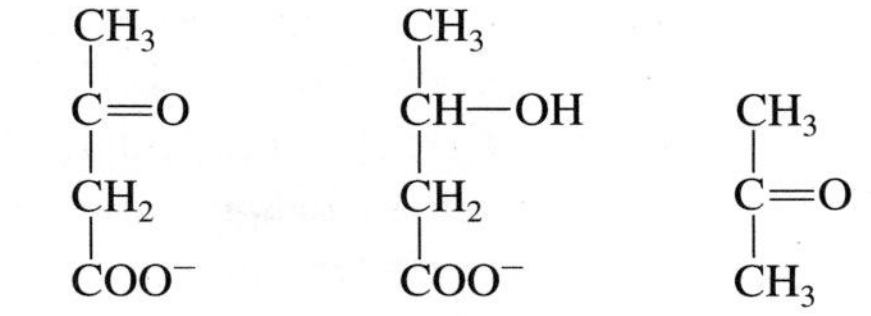

$$
\begin{array}{ccc}
\text{CH}_3 & \text{CH}_3 & \\
| & | & \text{CH}_3 \\
\text{C}=\text{O} & \text{CH}-\text{OH} & | \\
| & | & \text{C}=\text{O} \\
\text{CH}_2 & \text{CH}_2 & | \\
| & | & \text{CH}_3 \\
\text{COO}^- & \text{COO}^- & \\
\text{Acetoacetate} & \beta\text{-Hydroxybutyrate} & \text{Acetone} \\
\boxed{\text{C}_4 \text{ ketoacid}} & \boxed{\text{C}_4 \text{ hydroxyacid}} & \boxed{\text{C}_3 \text{ ketone}}
\end{array}
$$

Chemically, these three structures are closely related. The relationships are most easily seen if the focus starts with the molecule acetoacetate.

1. Reduction of the ketone group present in acetoacetate to a secondary alcohol produces β-hydroxybutyrate. Such a reduction process was initially considered in Section 15.10.

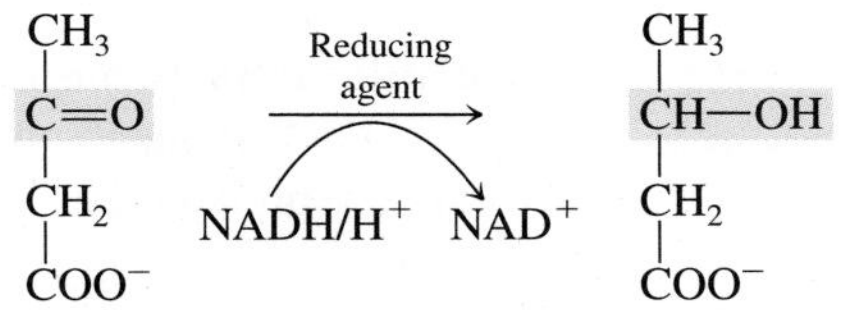

$$
\begin{array}{ccc}
\text{CH}_3 & \xrightarrow[\text{NADH/H}^+ \quad \text{NAD}^+]{\text{Reducing agent}} & \text{CH}_3 \\
| & & | \\
\boxed{\text{C}=\text{O}} & & \text{CH}-\text{OH} \\
| & & | \\
\text{CH}_2 & & \text{CH}_2 \\
| & & | \\
\text{COO}^- & & \text{COO}^- \\
\text{Acetoacetate} & & \beta\text{-Hydroxybutyrate}
\end{array}
$$

Structurally, there is no ketone functional group present in β-hydroxybutyrate. Despite it not being a ketone, it is still called a ketone body because of its structural relationship to acetoacetate.

2. Decarboxylation of acetoacetate produces acetone.

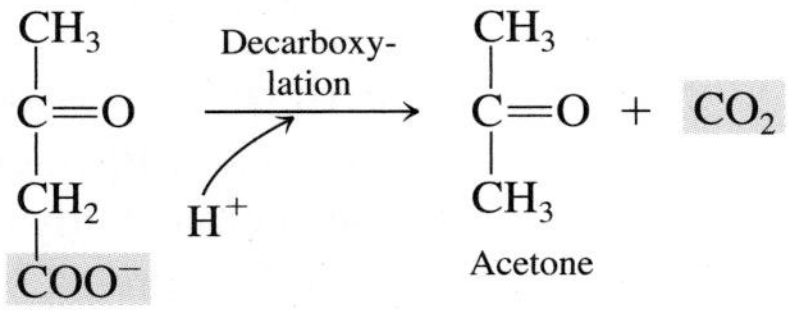

$$
\begin{array}{ccc}
\text{CH}_3 & \xrightarrow[\text{H}^+]{\text{Decarboxy-lation}} & \text{CH}_3 \\
| & & | \\
\text{C}=\text{O} & & \text{C}=\text{O} \quad + \quad \boxed{\text{CO}_2} \\
| & & | \\
\text{CH}_2 & & \text{CH}_3 \\
| & & \text{Acetone} \\
\boxed{\text{COO}^-} & & \\
\text{Acetoacetate} & &
\end{array}
$$

For a number of years, ketone bodies were thought of as degradation products that had little physiological significance. It is now known that ketone bodies can serve as sources of energy for various tissues and are very important energy sources in heart muscle and the renal cortex. Even the brain, which requires glucose, can adapt to obtain a portion of its energy from ketone bodies in dieting situations that involve a properly constructed low-carbohydrate diet.

● **EXAMPLE 25.5**

Recognizing Structural Characteristics of Ketone Bodies

▶ For each of the following structural characterizations for ketone bodies identify the ketone body to which it applies. There may be more than one correct answer for a given characterization.

a. It is a C_4 molecule.
b. It is a ketoacid.
c. It can be produced by reduction of acetoacetate.
d. Its structure contains a ketone functional group.

Solution

a. *Acetoacetate* and *β-hydroxybutyrate.* Acetoacetate is a C_4 ketoacid and β-hydroxy-butyrate is a C_4 hydroxyacid.
b. *Acetoacetate.* Both acetoacetate and β-hydroxybutyrate are acids; the first is a ketoacid and the second is a hydroxyacid.
c. *β-Hydroxybutyrate.* Reduction of the ketone functional group in acetoacetate to a secondary alcohol group produces the molecule β-hydroxybutyrate.
d. *Acetoacetate* and *acetone.* Acetoacetate is a ketoacid and acetone is a simple ketone.

(continued)

Practice Exercise 25.5

For each of the following structural characterizations for ketone bodies identify the ketone body to which it applies. There may be more than one correct answer for a given characterization.

a. It is a C_3 molecule.
b. It can be produced by decarboxylation of acetoacetate.
c. Its structure contains a carboxyl functional group.
d. It is classified as a ketone body but it is not a ketone.

Answers: **a.** Acetone; **b.** Acetone; **c.** Acetoacetate and β-hydroxybutyrate; **d.** β-Hydroxybutyrate

Ketogenesis

Even when ketogenic conditions are not present in the human body, the liver produces a *small amount* of ketone bodies.

Ketogenesis *is the metabolic pathway by which ketone bodies are synthesized from acetyl CoA.* Items to consider about this process prior to looking at the actual steps in this four-step process are:

1. The primary site for the process is liver mitochondria.
2. The first ketone body to be produced is acetoacetate. This production occurs in Step 3 of ketogenesis.
3. Some of the acetoacetate produced in Step 3 is converted to the second ketone body, β-hydroxybutyrate, in Step 4 of ketogenesis.
4. The acetoacetate and β-hydroxybutyrate synthesized by ketogenesis in the liver are released to the bloodstream where acetone, the third ketone body, is produced.
5. Acetoacetate is somewhat unstable and can spontaneously or enzymatically lose its carboxyl group to form acetone. Thus, the ketone body acetone is not actually a product of the metabolic pathway ketogenesis.
6. The ketone body acetone present in the bloodstream is a volatile substance that is mainly excreted by exhalation. Its sweet odor is detectible in the breath of a diabetic.
7. The amount of acetone present is usually small compared to the concentrations of the other two ketone bodies.

The actual reaction steps in the process of ketogenesis are:

Step 1: *First condensation.* Ketogenesis begins as two acetyl CoA molecules combine to produce acetoacetyl CoA, a reversal of the last step of the β-oxidation pathway (Section 25.4).

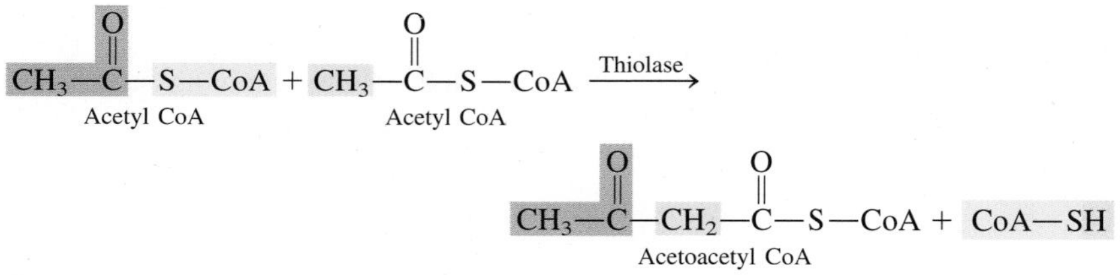

Step 2: *Second condensation.* Acetoacetyl CoA then reacts with a third acetyl CoA and water to produce 3-hydroxy-3-methylglutaryl CoA (HMG-CoA) and CoA—SH.

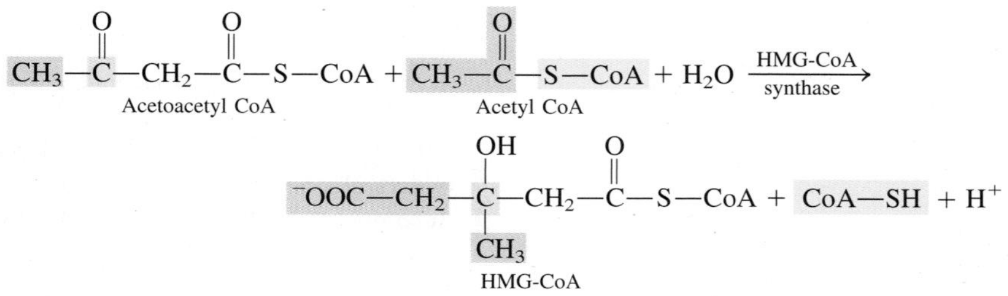

Step 3: *Chain cleavage.* HMG-CoA is then cleaved to acetyl CoA and acetoacetate.

$$^-OOC-CH_2-\underset{\underset{CH_3}{|}}{\overset{\overset{OH}{|}}{C}}-CH_2-\overset{\overset{O}{\|}}{C}-S-CoA \xrightarrow[\text{lyase}]{\text{HMG-CoA}}$$

HMG-CoA

$$^-OOC-CH_2-\overset{\overset{O}{\|}}{C}-CH_3 + CH_3-\overset{\overset{O}{\|}}{C}-S-CoA$$

Acetoacetate   Acetyl CoA

Step 4: *Reduction.* Acetoacetate is reduced to β-hydroxybutyrate. The reducing agent is NADH.

$$^-OOC-CH_2-\overset{\overset{O}{\|}}{C}-CH_3 \xrightarrow[\text{NADH/H}^+ \quad \text{NAD}^+]{\substack{\beta\text{-Hydroxybutyrate} \\ \text{dehydrogenase}}} {}^-OOC-CH_2-\overset{\overset{OH}{|}}{C}H-CH_3$$

Acetoacetate β-Hydroxybutyrate

The four chemical reactions associated with ketogenesis are summarized diagrammatically in Figure 25.8.

Energy Production from Acetoacetate

The ketone bodies β-hydroxybutyrate and acetoacetate are connected by a reversible reaction. Using a different enzyme than used in ketogenesis, β-hydroxybutyrate can be converted back to acetoacetate when cellular energy needs require it.

For acetoacetate to be used as a fuel—in heart muscle, for example—it must first be activated. Acetoacetate is activated by transfer of a CoA group from succinyl CoA (a citric acid cycle intermediate). The resulting acetoacetyl CoA is then cleaved to give two acetyl CoA molecules that can enter the citric acid cycle (see Figure 25.9). In effect, acetoacetate is a water-soluble, transportable form of acetyl units.

Ketosis

Under normal metabolic conditions (an appropriate glucose–fatty acid balance), the concentration of ketone bodies in the blood is very low—about 1 mg/100 mL. Abnormal

> Heart muscle and the renal cortex use acetoacetate in preference to glucose. The brain adapts to the utilization of acetoacetate with starvation or diabetes. 75% of the fuel needs of the brain are obtained from acetoacetate during prolonged starvation.

Figure 25.8 Ketogenesis involves the production of ketone bodies from acetyl CoA.

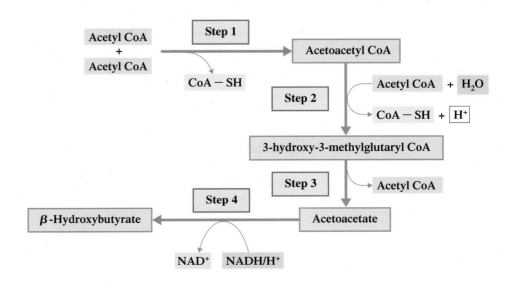

Figure 25.9 The pathway for utilization of acetoacetate as a fuel. The required succinyl CoA comes from the citric acid cycle.

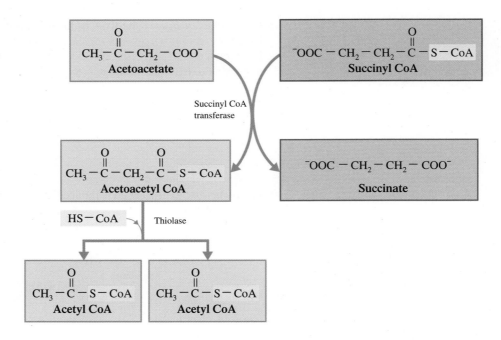

metabolic conditions, such as those mentioned at the start of this section, produce elevated blood ketone levels, levels 50–100 times greater than normal. Excess accumulation of ketone bodies in blood (20 mg/100 mL) is called *ketonemia.* At a level of 70 mg/100 mL, the renal threshold is exceeded, and ketone bodies are excreted in the urine, a condition called *ketonuria.* The overall accumulation of ketone bodies in the blood and urine is called *ketosis.* Ketosis is often detectable by the smell of acetone on a person's breath; acetone is very volatile and is excreted through the lungs.

For the vast majority of persons following a low-carbohydrate diet, the effects of ketosis appear to be harmless or nearly so. The symptoms of the *mild* ketosis that occurs as the result of such dieting include headache, dry mouth, and sometimes foul-smelling breath.

Two of the three ketone bodies—acetoacetate and β-hydroxybutyrate—are acids. Their presence in blood causes a slight but significant decrease in blood pH. This can result in acidosis (Section 10.12) in *severe* ketosis situations. Symptoms include heavy breathing (because acidic blood can carry less oxygen) and increased urine output that can lead to dehydration. Ultimately, the condition can cause coma and death.

Acidosis from elevated ketone body levels is often called *keto* acidosis or *metabolic* acidosis to distinguish it from *respiratory* acidosis (Section 10.12), which is not linked to ketone bodies.

25.7 BIOSYNTHESIS OF FATTY ACIDS: LIPOGENESIS

Lipogenesis *is the metabolic pathway by which fatty acids are synthesized from acetyl CoA.* As was the case for the opposing processes of glycolysis and gluconeogenesis, lipogenesis is not simply a reversal of the steps for degradation of fatty acids (the β-oxidation pathway). Before we look at the details of fatty acid synthesis, we will consider some differences between the synthesis and degradation of fatty acids.

1. Lipogenesis occurs in the cell cytosol, whereas degradation of fatty acids occurs in the mitochondrial matrix. Because they have different reaction sites, these two opposing processes can occur at the same time when necessary.

2. Different enzymes are involved in the two processes. Lipogenesis enzymes are collected into a multienzyme complex called *fatty acid synthase*. This enzyme complex ties the reaction steps of lipogenesis closely together. The enzymes involved in fatty acid degradation are not physically associated, so the reaction steps are independent.
3. Intermediates of the two processes are covalently bonded to different carriers. The carrier for fatty acid degradation intermediates is CoA. Lipogenesis intermediates are bonded to ACP (acyl carrier protein).
4. Fatty acid synthesis is dependent on the reducing agent NADPH. Fatty acid degradation is dependent on the oxidizing agents FAD and NAD^+.
5. Fatty acids are built up two carbons at a time during synthesis and are broken down two carbons at a time during degradation. The source of the two carbon units differs between the two processes. In lipogenesis, acetyl CoA is used to form malonyl ACP, which becomes the carrier of the two carbon units. CoA derivatives are involved in all steps of fatty acid degradation.

In general, fatty acid biosynthesis (lipogenesis) occurs any time dietary intake provides more nutrients than are needed for energy requirements. The primary lipogenesis sites are the liver, adipose tissue, and mammary glands. The mammary glands show increased synthetic activity during periods of lactation.

The Citrate–Malate Shuttle System

Acetyl CoA is the starting material for lipogenesis. Because acetyl CoA is generated in mitochondria and lipogenesis occurs in the cytosol, the acetyl CoA must first be transported to the cytosol. It exits the mitochondria through a transport system that involves citrate ion.

The outer mitochondrial matrix is freely permeable to acetyl CoA, as well as many other substances such as citrate, malate, and pyruvate. The inner mitochondrial membrane, however, is not permeable to acetyl CoA. An indirect shuttle system involving citrate solves this problem.

This shuttle system, which is diagrammed in Figure 25.10, functions as follows. Mitochondrial acetyl CoA reacts with oxaloacetate (the first step of the citric acid cycle; Section 23.6) to produce citrate, which is then transported through the inner mitochondrial membrane by a citrate transporter (a membrane protein structure). Once in the cytosol, the citrate undergoes the reverse reaction to its formation to regenerate the acetyl CoA and oxaloacetate, with ATP involved in the process. The acetyl CoA so generated becomes the "fuel" for lipogenesis; the oxaloacetate so generated reacts further to produce malate, in an NADH dependent change. The malate reenters the mitochondrial matrix through a malate transporter, and is then converted to oxaloacetate, which can then react with another acetyl CoA molecule to form citrate and the shuttle process repeats itself. Additional particulars about the shuttle system are found in Figure 25.10.

Compared to the carnitine shuttle system for long-chain fatty acid groups (acyl groups; Figure 25.6), the citrate–malate shuttle system is more complex. However, the intermediates in the shuttle have been encountered before, in glycolysis and the citric acid cycle.

ACP Complex Formation

Studies show that all intermediates in fatty acid biosynthesis (lipogenesis) are bound to acyl carrier proteins (ACP—SH) rather than coenzyme A (CoA—SH). This applies even to the C_2 acetyl group. An acyl carrier protein can be regarded as a "giant coenzyme A molecule." Involved in the ACP structure are the 2-ethanethiol and pantothenic acid components present in CoA—SH (Section 23.3), which are attached to a polypeptide chain containing 77 amino acid residues.

Figure 25.10 The citrate-malate shuttle system for transferring acetyl CoA from a mitochondrion to the cytosol.

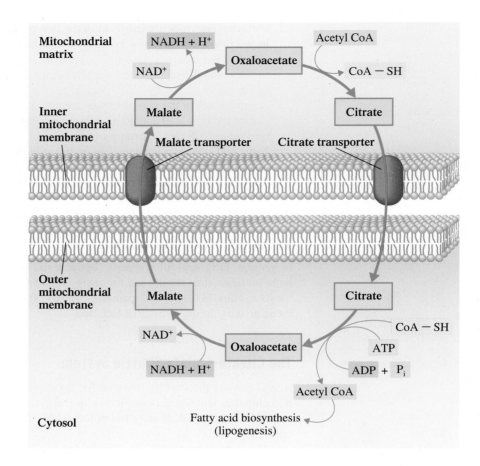

Two simple ACP complexes are needed to start the lipogenesis process. They are acetyl ACP, a C_2—ACP, and malonyl ACP, a C_3—ACP. Additional malonyl ACP molecules are needed as the lipogenesis process proceeds.

Cytosolic acetyl CoA is the starting material for the production of both of these simple ACP complexes. Acetyl ACP is produced by the direct reaction of acetyl CoA with an ACP molecule.

$$
\underset{\text{Acetyl CoA}}{CH_3-\overset{\overset{\textstyle O}{\|}}{C}-S-CoA} + \underset{\text{ACP}}{ACP-S-H} \xrightarrow{\underset{\text{transferase}}{\text{Acetyl}}} \underset{\text{Acetyl ACP}}{CH_3-\overset{\overset{\textstyle O}{\|}}{C}-S-ACP} + \underset{\text{CoA}}{CoA-S-H}
$$

The reaction to produce malonyl ACP requires two steps. The first step is a carboxylation reaction with ATP involvement.

$$
\underset{\text{Acetyl CoA}}{CH_3-\overset{\overset{\textstyle O}{\|}}{C}-S-CoA} + CO_2 \xrightarrow[\underset{ATP \quad ADP + P_i}{}]{\underset{\text{carboxylase}}{\text{Acetyl CoA}}} \underset{\text{Malonyl CoA}}{{}^-O-\overset{\overset{\textstyle O}{\|}}{C}-CH_2-\overset{\overset{\textstyle O}{\|}}{C}-S-CoA}
$$

This reaction occurs only when cellular ATP levels are high. It is catalyzed by *acetyl CoA carboxylase complex,* which requires both Mn^{2+} ion and the B vitamin biotin for its activity. The malonyl CoA so produced then reacts with ACP to produce malonyl ACP.

$$
\underset{\text{Malonyl CoA}}{{}^-O-\overset{\overset{\textstyle O}{\|}}{C}-CH_2-\overset{\overset{\textstyle O}{\|}}{C}-S-CoA} + \underset{\text{ACP}}{ACP-S-H} \xrightarrow{\underset{\text{transferase}}{\text{Malonyl}}} \underset{\text{Malonyl ACP}}{{}^-O-\overset{\overset{\textstyle O}{\|}}{C}-CH_2-\overset{\overset{\textstyle O}{\|}}{C}-S-ACP} + \underset{\text{CoA}}{CoA-S-H}
$$

The parent compound for the malonyl group is malonic acid, the C_3 dicarboxylic acid.

$$
\underset{}{HO-\overset{\overset{\textstyle O}{\|}}{C}-CH_2-\overset{\overset{\textstyle O}{\|}}{C}-OH}
$$

Figure 25.11 In the first "turn" of the fatty acid biosynthetic pathway, acetyl ACP is converted to butyryl ACP. In the next cycle (not shown), the butyryl ACP reacts with another malonyl ACP to produce a 6-carbon acid. Continued cycles produce acids with 8, 10, 12, 14, and 16 carbon atoms.

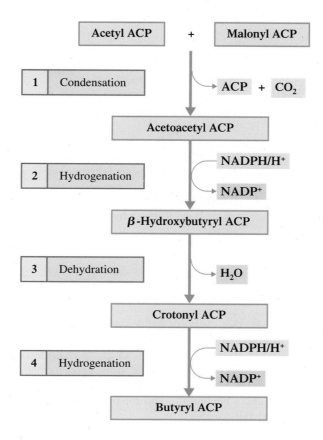

Chain Elongation

Four reactions that occur in a cyclic pattern within the multienzyme *fatty acid synthase complex* constitute the chain elongation process used for fatty acid synthesis. The reactions of the *first* turn of the cycle, in general terms, are shown in Figure 25.11. Specific details about this series of reactions follow.

Step 1:   *Condensation.* Acetyl ACP and malonyl ACP condense together to form acetoacetyl ACP.

$$CH_3-\overset{O}{\overset{\|}{C}}-S-ACP + {}^-O-\overset{O}{\overset{\|}{C}}-CH_2-\overset{O}{\overset{\|}{C}}-S-ACP \longrightarrow CH_3-\overset{O}{\overset{\|}{C}}-CH_2-\overset{O}{\overset{\|}{C}}-S-ACP + CO_2 + ACP-SH$$

Acetyl ACP   Malonyl ACP   Acetoacetyl ACP

Note that a C_2 species (acetyl) and a C_3 species (malonyl) react to produce a C_4 species (acetoacetyl) rather than a C_5 species. One carbon atom leaves the reaction in the form of a CO_2 molecule.

Steps 2 through 4 involve a sequence of functional group changes that we have encountered twice before—in fatty acid degradation (Section 25.4) and in the citric acid cycle (Section 23.6). This time, however, the changes occur in the reverse sequence to that previously encountered. The functional group changes are

$$\textbf{Ketone} \xrightarrow[\text{(hydrogenation)}]{\overset{②}{\text{Reduction}}} \textbf{secondary alcohol} \xrightarrow{\overset{③}{\text{Dehydration}}} \textbf{alkene} \xrightarrow[\text{(hydrogenation)}]{\overset{④}{\text{Reduction}}} \textbf{alkane}$$

Step 2:   *First hydrogenation.* The keto group of the acetoacetyl complex, which involves the β-carbon atom, is reduced to the corresponding alcohol by NADPH.

$$CH_3-\underset{\substack{\| \\ O}}{C}-CH_2-\underset{\substack{\| \\ O}}{C}-S-ACP \longrightarrow CH_3-\underset{\substack{| \\ OH}}{CH}-CH_2-\underset{\substack{\| \\ O}}{C}-S-ACP$$

Acetoacetyl ACP   NADPH/H$^+$ NADP$^+$ β-Hydroxybutyryl ACP

Step 3: *Dehydration.* The alcohol produced in Step 2 is dehydrated to introduce a double bond into the molecule (between the α and β carbons).

$$CH_3-\underset{\substack{| \\ OH}}{CH}-CH_2-\underset{\substack{\| \\ O}}{C}-S-ACP \longrightarrow CH_3-\overset{trans}{CH}=CH-\underset{\substack{\| \\ O}}{C}-S-ACP$$

β-Hydroxybutyryl ACP   H$_2$O   Crotonyl ACP

Step 4: *Second hydrogenation.* The double bond introduced in Step 3 is converted to a single bond through hydrogenation. As in Step 2, NADPH is the reducing agent.

$$CH_3-\overset{trans}{CH}=CH-\underset{\substack{\| \\ O}}{C}-S-ACP \longrightarrow CH_3-CH_2-CH_2-\underset{\substack{\| \\ O}}{C}-S-ACP$$

Crotonyl ACP   NADPH/H$^+$ NADP$^+$ Butyryl ACP

Further cycles of the preceding four-step process convert the four-carbon acyl group to a six-carbon acyl group, then to an eight-carbon acyl group, and so on (see Figure 25.12). Elongation of the acyl group chain through this procedure, which is tied to the fatty acid synthase complex, stops upon formation of the C_{16} acyl group (palmitic acid). Different enzyme systems and different cellular locations are required for elongation of the chain beyond C_{16} and for introduction of double bonds into the acyl group (unsaturated fatty acids).

A relatively large input of energy is needed to biosynthesize a fatty acid molecule, as can be seen from the data in Table 25.2, which gives a net summary of the reactants and products involved in the synthesis of one molecule of palmitic acid, the 16:0 fatty acid.

Figure 25.12 The sequence of cycles needed to produce a C_{16} fatty acid from acetyl ACP. Each loop represents one cycle.

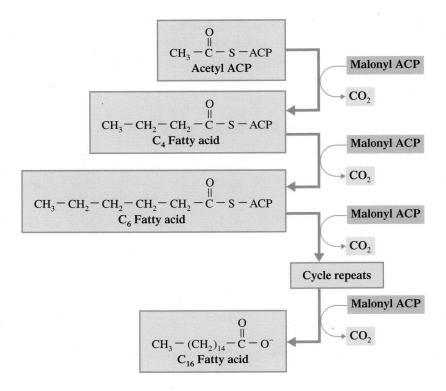

● **EXAMPLE 25.6**

Characterizing Intermediates as to When They Are Produced in the Lipogenesis Process

▶ In which step (of Steps 1 through 4) and in which cycle (first or second turn) of fatty acid biosynthesis (lipogenesis) is each of the following compounds encountered as a product?

a.

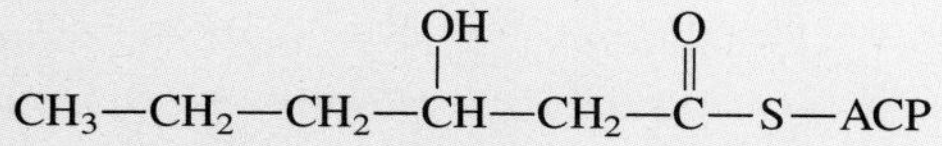

$$CH_3-CH_2-CH_2-\overset{\overset{\textstyle OH}{|}}{CH}-CH_2-\overset{\overset{\textstyle O}{\|}}{C}-S-ACP$$

b.

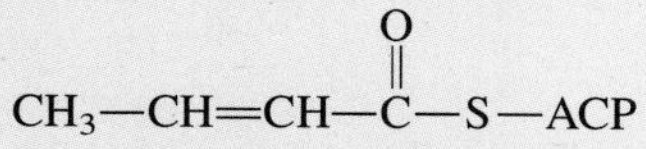

$$CH_3-CH=CH-\overset{\overset{\textstyle O}{\|}}{C}-S-ACP$$

c.

$$CH_3-\overset{\overset{\textstyle O}{\|}}{C}-CH_2-\overset{\overset{\textstyle O}{\|}}{C}-S-ACP$$

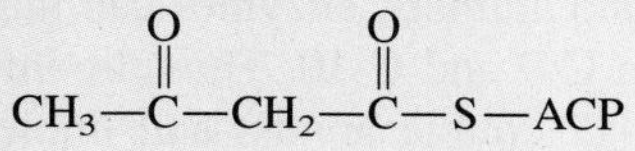

d.

$$CH_3-CH_2-CH_2-CH_2-CH_2-\overset{\overset{\textstyle O}{\|}}{C}-S-ACP$$

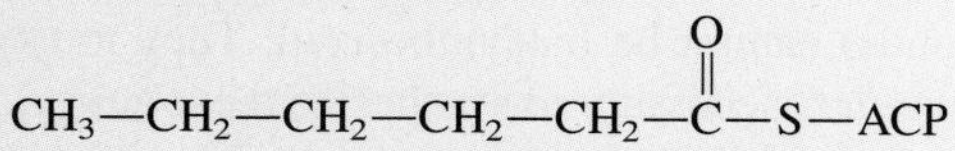

Solution

All intermediates produced during the first turn of the cycle will have a C_4 carbon chain, those produced during the second turn a C_6 carbon chain, those during the third turn a C_8 carbon chain, and so on.

Determination of which step in a turn is involved is based on the functional group present in the carbon chain. Step 1 produces a keto acid chain, Step 2 a hydroxy acid chain, Step 3 an unsaturated acid chain, and Step 4 a saturated acid chain.

a. The characterization for this molecule's carbon chain is C_6 hydroxyacid. The hydroxy-acid designation indicates *Step 2* and the C_6 chain length indicates *second turn*.
b. The carbon chain is a C_4 unsaturated acid; this indicates *Step 3* and *first turn*.
c. The carbon chain is a C_4 keto acid; this indicates *Step 1* and *first turn*.
d. The carbon chain is a C_6 saturated acid; this indicates *Step 4* and *second turn*.

Practice Exercise 25.6

In which step (of Steps 1 through 4) and in which cycle (first or second turn) of fatty acid biosynthesis (lipogenesis) is each of the following compounds encountered as a product?

a.

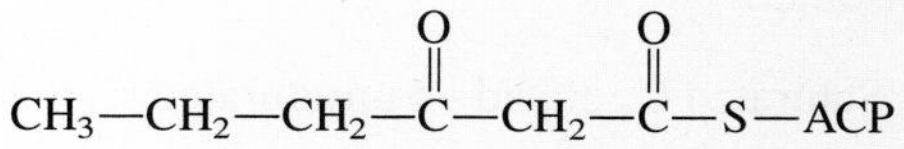

$$CH_3-CH_2-CH_2-\overset{\overset{\textstyle O}{\|}}{C}-CH_2-\overset{\overset{\textstyle O}{\|}}{C}-S-ACP$$

b.

$$CH_3-CH_2-CH_2-\overset{\overset{\textstyle O}{\|}}{C}-S-ACP$$

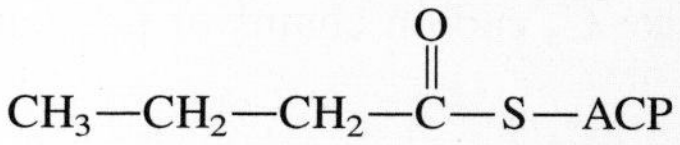

c.

$$CH_3-CH_2-CH_2-CH=CH-\overset{\overset{\textstyle O}{\|}}{C}-S-ACP$$

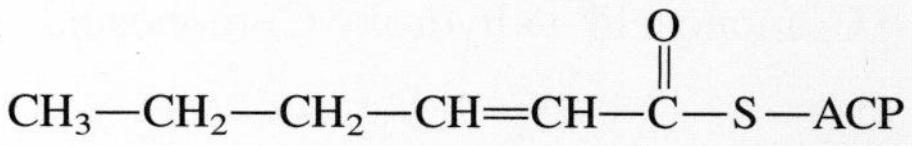

d.

$$CH_3-\overset{\overset{\textstyle OH}{|}}{CH}-CH_2-\overset{\overset{\textstyle O}{\|}}{C}-S-ACP$$

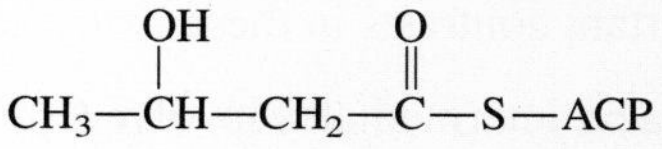

Answers: a. Step 1, second turn; **b.** Step 4, first turn; **c.** Step 3, second turn; **d.** Step 2, first turn

TABLE 25.2
Reactants and Products in the Biosynthesis of One Molecule of Palmitic Acid, the 16:0 Fatty Acid

Reactants	Products
8 acetyl CoA	1 palmitate
7 ATP	8 CoA
14 NADPH	7 ADP
6 H$^+$	7 P$_i$
	14 NADP$^+$
	6 H$_2$O

Unsaturated Fatty Acid Biosynthesis

Production of unsaturated fatty acids (insertion of double bonds) requires molecular oxygen (O_2). In an oxidation step, hydrogen is removed and combined with the O_2 to form water.

$$R-\overset{\overset{\displaystyle H}{|}}{\underset{\underset{\displaystyle H}{|}}{C}}-\overset{\overset{\displaystyle H}{|}}{\underset{\underset{\displaystyle H}{|}}{C}}-(CH_2)_n-\overset{\overset{\displaystyle O}{\|}}{C}-O^- + O_2 \xrightarrow{\quad NADPH/H^+ \quad NADP^+ \quad}$$

$$R-\overset{\overset{\displaystyle }{|}}{C}=\overset{\overset{\displaystyle }{|}}{\underset{\underset{\displaystyle H}{|}}{C}}-(CH_2)_n-\overset{\overset{\displaystyle O}{\|}}{C}-O^- + 2H_2O$$

In humans and animals, enzymes can introduce double bonds only between C-4 and C-5 and between C-9 and C-10. Thus the important unsaturated fatty acids linoleic (C_{18} with C-9 and C-12 double bonds) and linolenic (C_{18} with C-9, C-12, and C-15 double bonds) cannot be biosynthesized. They must be obtained from the diet. (Plants have the enzymes necessary to synthesize these acids.) Acids such as linoleic and linolenic (Section 19.2), which cannot be synthesized by the body but are necessary for its proper functioning, are called *essential fatty acids*.

Lipogenesis can be used to convert glucose to fatty acids via acetyl CoA. The reverse process, conversion of fatty acids to glucose, is not possible within the human body. Fatty acids can be broken down to acetyl CoA, but there is no enzyme present for the conversion of acetyl CoA to pyruvate or oxaloacetate, starting materials for gluconeogenesis (Section 24.6). Plants and some bacteria do possess the needed enzymes and thus can convert fatty acids to carbohydrates.

25.8 RELATIONSHIPS BETWEEN LIPOGENESIS AND CITRIC ACID CYCLE INTERMEDIATES

The intermediates in the last four steps of the citric acid cycle are all C_4 molecules (Section 23.6). In the first cycle of the four repetitive reactions in lipogenesis all of the carbon chains attached to ACP are C_4 chains. Several relationships exist between these two sets of C_4 entities.

The last four intermediates of the citric acid cycle bear the following relationship to each other.

Saturated C_4 diacid → unsaturated C_4 diacid → hydroxy C_4 diacid → keto C_4 diacid

The intermediate C_4 carbon chains of lipogenesis bear the following relationship to each other.

Keto C_4 monoacid → hydroxy C_4 monoacid → unsaturated C_4 monoacid

→ saturated C_4 monoacid

Note two important contrasts in these compound sequences:

1. The citric acid intermediates involve C_4 *di*acids and the lipogenesis intermediates involve C_4 *mono*acids.
2. The order in which the various acid derivative types are encountered in lipogenesis is the reverse of the order in which they are encountered in the citric acid cycle.

Figure 25.13 contrasts the structures of the various C_4 diacid intermediates from the citric acid cycle with the various C_4 monoacid intermediates encountered in lipogenesis.

25.9 BIOSYNTHESIS OF CHOLESTEROL

So far in this chapter, our discussion of lipid metabolism has focused on fats and oils (triacylglycerols) and their hydrolysis products, fatty acids and glycerol. We now consider another very important lipid—cholesterol.

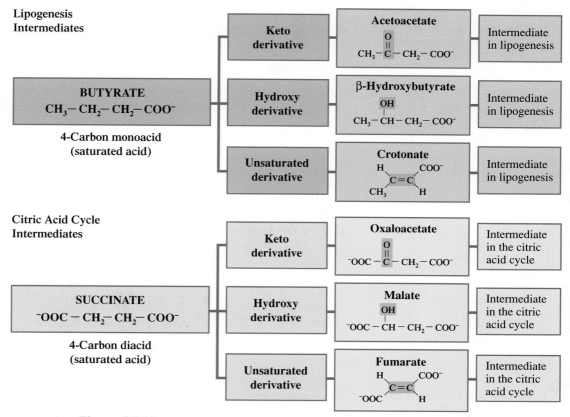

Figure 25.13 Structural relationships between C_4 citric acid cycle intermediates and C_4 lipogenesis intermediates. C_4 citric acid cycle intermediates are diacid derivatives and C_4 lipogenesis intermediates are monoacid derivatives. The same derivative types are present in both series of intermediates.

Every membrane of every cell in the body has cholesterol as a necessary component. This substance is also the precursor for bile salts, sex hormones, and adrenal hormones (Sections 19.11 and 19.12).

In today's health-conscious world, dietary intake of cholesterol is of great interest because of correlations between high serum cholesterol levels and coronary heart disease. Average daily dietary intake of cholesterol is approximately 0.3 gram. This amount, though important, is small compared to the 1.5–2.0 grams of cholesterol that the body synthesizes every day from acetyl CoA units.

The biosynthesis of cholesterol, a C_{27} molecule, occurs primarily in the liver. Its production consumes 18 molecules of acetyl CoA and involves at least 27 separate enzymatic steps. An overview of cholesterol synthesis is given in Figure 25.14.

Figure 25.14 An overview of the biosynthetic pathway for cholesterol synthesis.

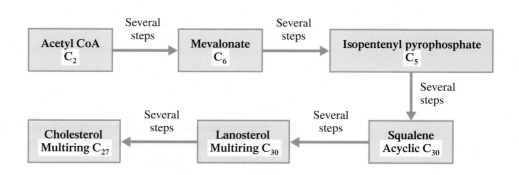

The "parent" compound for mevalonate ion is mevalonic acid (3,5-dihydroxy-3-methylpentanoic acid).

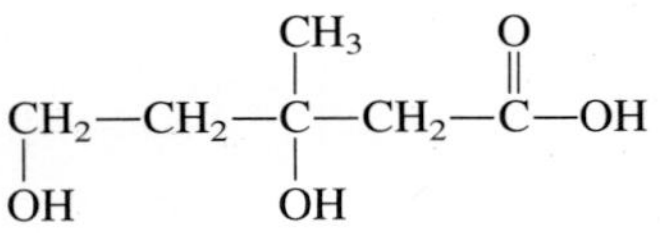

$$\underset{\text{OH}}{\overset{}{\text{CH}_2}}-\text{CH}_2-\underset{\text{OH}}{\overset{\text{CH}_3}{\text{C}}}-\text{CH}_2-\overset{\text{O}}{\overset{\|}{\text{C}}}-\text{OH}$$

In the first phase of cholesterol synthesis, three molecules of acetyl CoA are condensed into a C_6 mevalonate ion.

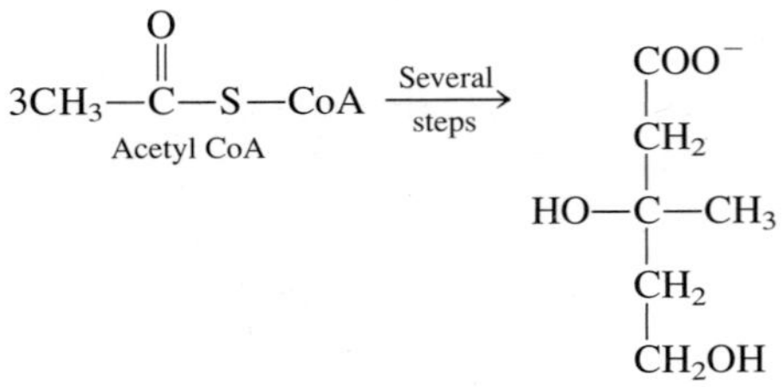

$$3\text{CH}_3-\overset{\text{O}}{\overset{\|}{\text{C}}}-\text{S}-\text{CoA} \xrightarrow{\text{Several steps}} \begin{array}{c}\text{COO}^-\\|\\\text{CH}_2\\|\\\text{HO}-\text{C}-\text{CH}_3\\|\\\text{CH}_2\\|\\\text{CH}_2\text{OH}\end{array}$$

Acetyl CoA   Mevalonate

The C_6 mevalonate undergoes a decarboxylation to yield a C_5 isoprene derivative called isopentenyl pyrophosphate and CO_2. Three ATP molecules are needed in accomplishing this process.

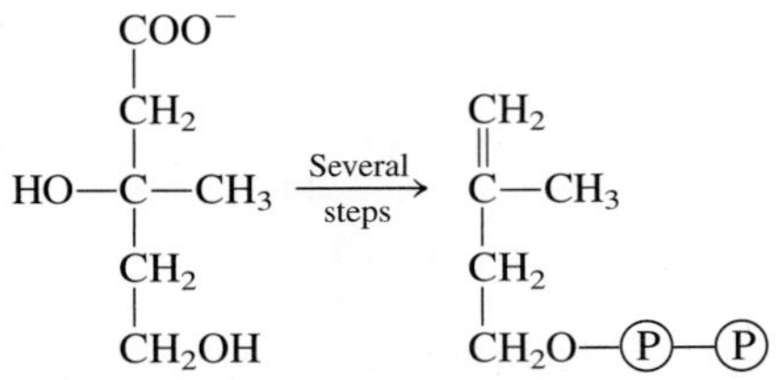

$$\begin{array}{c}\text{COO}^-\\|\\\text{CH}_2\\|\\\text{HO}-\text{C}-\text{CH}_3\\|\\\text{CH}_2\\|\\\text{CH}_2\text{OH}\end{array} \xrightarrow{\text{Several steps}} \begin{array}{c}\text{CH}_2\\\|\\\text{C}-\text{CH}_3\\|\\\text{CH}_2\\|\\\text{CH}_2\text{O}-\text{(P)}-\text{(P)}\end{array}$$

Mevalonate   Isopentenyl pyrophosphate

Compounds whose structures are based on the five-carbon isoprene unit are called *terpenes* (Section 13.7).

The isoprene structural unit (Section 13.7), present in isoprene derivatives in a modified form, is a commonly used five-carbon building block in biosynthetic processes.

The next stage of cholesterol biosynthesis involves the condensation of six isoprene units to give the C_{30} squalene molecule.

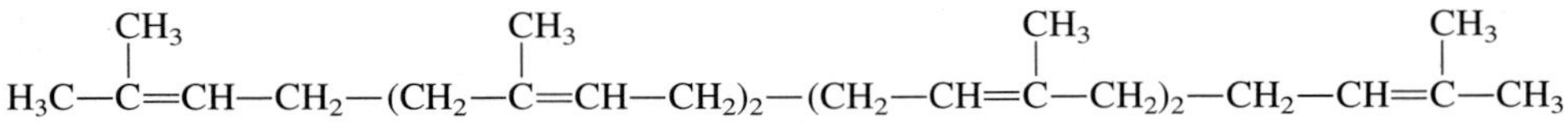

$$\text{H}_3\text{C}-\overset{\text{CH}_3}{\overset{|}{\text{C}}}=\text{CH}-\text{CH}_2-(\text{CH}_2-\overset{\text{CH}_3}{\overset{|}{\text{C}}}=\text{CH}-\text{CH}_2)_2-(\text{CH}_2-\text{CH}=\overset{\text{CH}_3}{\overset{|}{\text{C}}}-\text{CH}_2)_2-\text{CH}_2-\text{CH}=\overset{\text{CH}_3}{\overset{|}{\text{C}}}-\text{CH}_3$$

Squalene

A "redrawing" of the squalene structure, with numerous twists and bends in it, is helpful in visualizing the next stage of cholesterol biosynthesis, the formation of the four-ring steroid nucleus (Section 19.12) associated with lanosterol (and cholesterol).

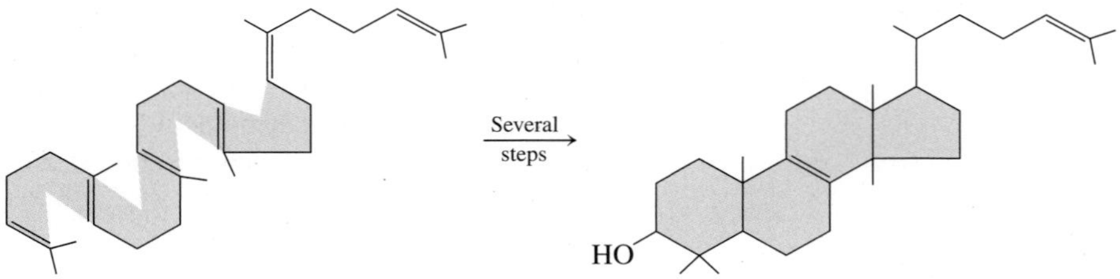

Squalene   Lanosterol

The multistep squalene-to-lanosterol transition involves the formation of four ring systems, a decrease in double bonds from six to two, the migration of two methyl groups to new locations, and the addition of an —OH group to the C_{30} system. Addition of the —OH group requires the use of molecular oxygen; the O of the —OH group comes from the molecular O_2.

The transition from lanosterol to cholesterol involves removal of three methyl groups (C_{30} to C_{27}), reduction of the double bond in the side chain, and migration of the other double bond to a new location.

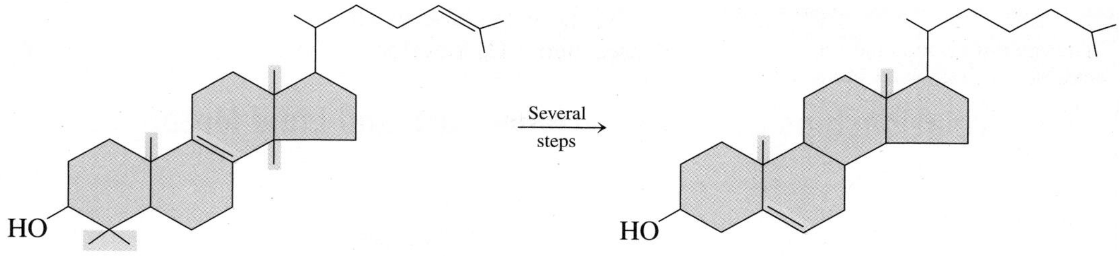

Lanosterol (C_{30})   Cholesterol (C_{27})

CHEMICAL Connections

Statins: Drugs That Lower Plasma Levels of Cholesterol

Over half of all deaths in the United States are directly or indirectly related to heart disease, in particular to atherosclerosis. Atherosclerosis results from the buildup of plaque (fatty deposits) on the inner walls of arteries. Cholesterol, obtained from low-density-lipoproteins (LDL) that circulate in blood plasma, is also a major component of plaque.

Because most of the cholesterol in the human body is synthesized in the liver, from acetyl CoA, much research has focused on finding ways to inhibit its biosynthesis. The rate-determining step in cholesterol biosynthesis involves the conversion of 3-hydroxy-3-methylglutaryl CoA (HMG-CoA) to mevalonate, a process catalyzed by the enzyme HMG-CoA reductase.

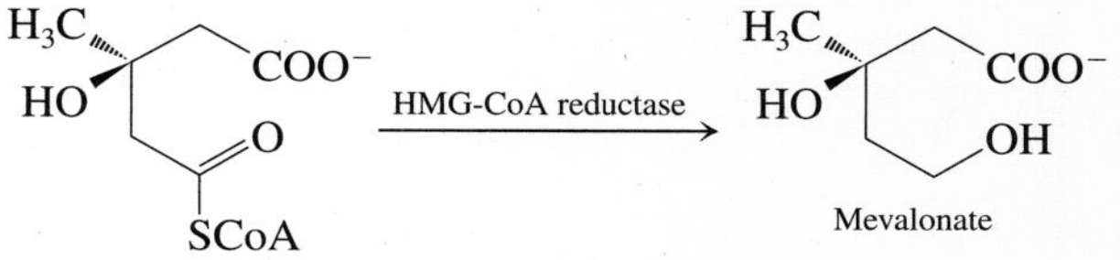

3-Hydroxy-3-methylglutaryl-CoA
(HMG-CoA)   Mevalonate

In 1976, as the result of screening more than 8000 strains of microorganisms, a compound now called *mevastatin*—a potent inhibitor of HMG-CoA reductase—was isolated from culture broths of a fungus. Soon thereafter, a second, more active compound called *lovastatin* was isolated.

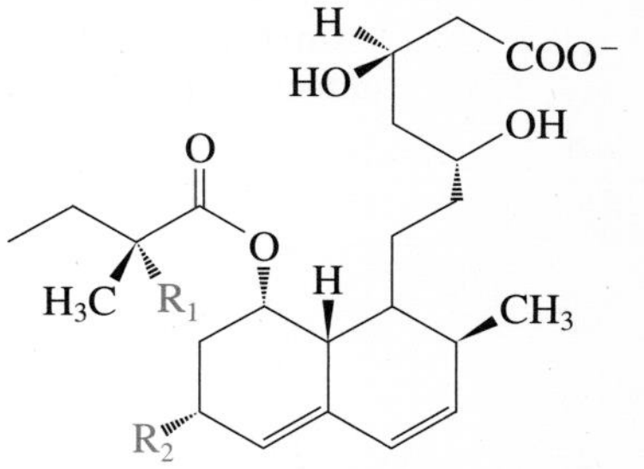

$R_1 = R_2 = H$, mevastatin
$R_1 = H$, $R_2 = CH_3$, lovastatin (Mevacor)
$R_1 = R_2 = CH_3$, simvastatin (Zocor)

These "statins" are very effective in lowering plasma concentrations of LDL by functioning as competitive inhibitors of HMG-CoA reductase.

After years of testing, the statins are now available as prescription drugs for lowering blood cholesterol levels. Clinical studies indicate that use of these drugs lowers the incidence of heart disease in individuals with mildly elevated blood cholesterol levels. A later-generation statin with a ring structure distinctly different from that of earlier statins—atorvastatin (Lipitor)—became the most prescribed medication in the United States in the year 2000. Note the structural resemblance between part of the structure of Lipitor and that of mevalonate.

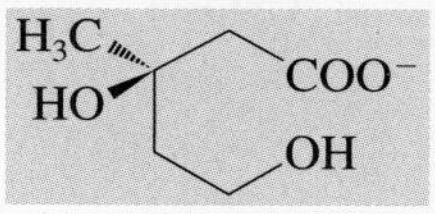

Mevalonate

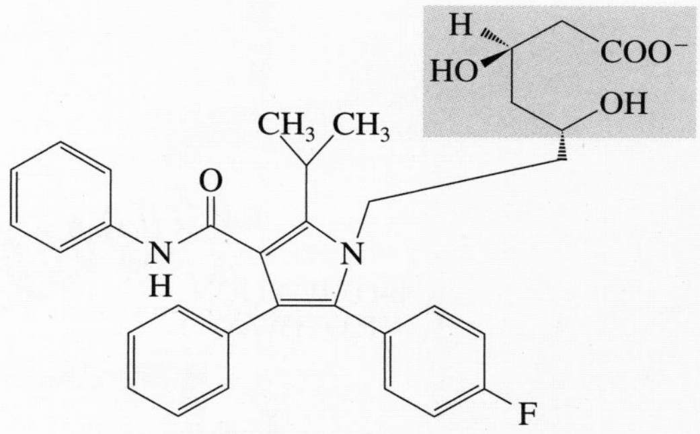

Atorvastatin (Lipitor)

Recent research studies have unexpectedly shown that the cholesterol-lowering statins have two added benefits.

Laboratory studies with animals indicate that statins prompt growth of cells to build new bone, replacing bone that has been leached away by osteoporosis ("brittle-bone disease"). A retrospective study of osteoporosis patients who also took statins shows evidence that their bones became more dense than did bones of osteoporosis patients who did not take the drugs.

Statins have also been shown to function as antiinflammatory agents that counteract the effects of a common virus, cytomegalovirus, which is now believed to contribute to the development of coronary heart disease. Researchers believe that by age 65, more than 70% of all people have been exposed to this virus. The virus, along with other infecting agents in blood, may actually trigger the inflammation mechanism for heart disease.

Interrelationships Between Carbohydrate and Lipid Metabolism

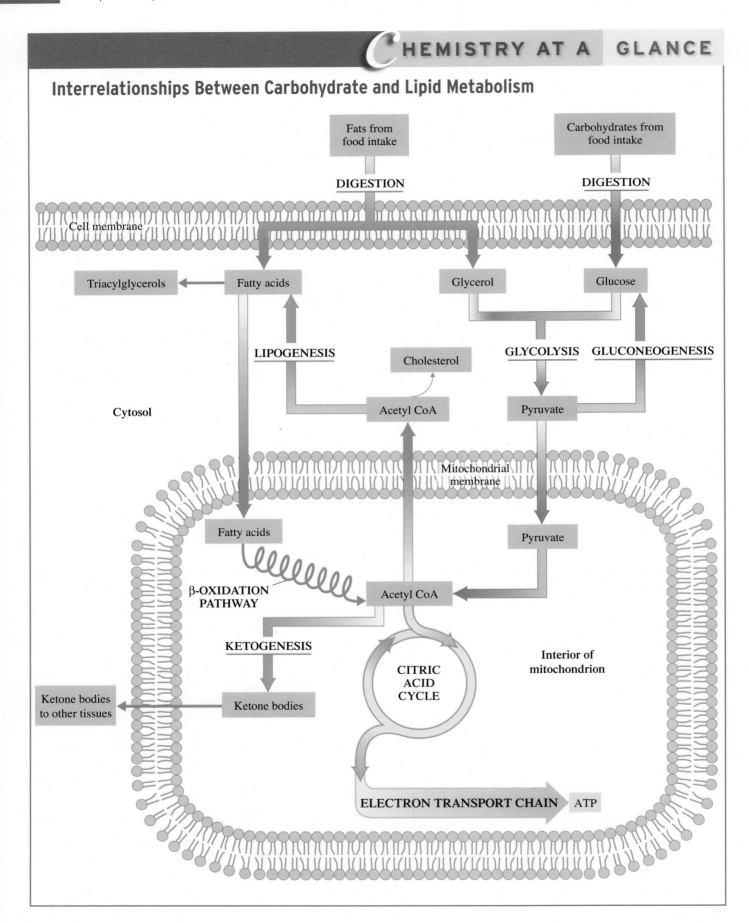

Figure 25.15 Biosynthetic relationships among steroid hormones.

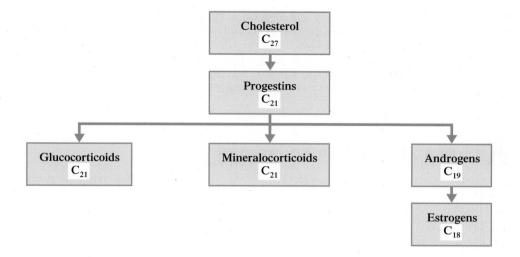

Once cholesterol has been formed, biosynthetic pathways are available to convert it to each of the five major classes of steroid hormones: progestins, androgens, estrogens, glucocorticoids, and mineralocorticoids (see Figure 25.15), as well as to bile acids and vitamin D (Section 21.13).

25.10 RELATIONSHIPS BETWEEN LIPID AND CARBOHYDRATE METABOLISM

Acetyl CoA is the primary link between lipid and carbohydrate metabolism. As shown in the Chemistry at a Glance feature on page 868, acetyl CoA is the degradation product for glucose, glycerol, and fatty acids, and it is also the starting material for the biosynthesis of fatty acids, cholesterol, and ketone bodies.

Note the four possible fates of acetyl CoA produced from fatty acid, glycerol, and glucose degradation processes.

1. *Oxidation in the citric acid cycle.* Both lipids (fatty acids and glycerol) and carbohydrates (glucose) supply acetyl CoA for the operation of this cycle.
2. *Ketone body formation.* This process is of major importance when there is imbalance between lipid and carbohydrate metabolic processes. The imbalance is caused by inadequate glucose metabolism during times of adequate lipid metabolism.
3. *Fatty acid biosynthesis.* The buildup of excess acetyl CoA when dietary intake exceeds energy needs leads to accelerated fatty acid biosynthesis.
4. *Cholesterol biosynthesis.* As with fatty acid biosynthesis, cholesterol biosynthesis occurs primarily when the body is in an acetyl CoA–rich state.

CONCEPTS TO REMEMBER

Triacylglycerol digestion and absorption. Triacylglycerols are digested (hydrolyzed) in the intestine and then reassembled after passage into the intestinal wall. Chylomicrons transport the reassembled triacylglycerols from intestinal cells to the bloodstream (Section 25.1).

Triacylglycerol storage and mobilization. Triacylglycerols are stored as fat droplets in adipose tissue. When they are needed for energy, enzyme-controlled hydrolysis reactions liberate the fatty acids, which then enter the bloodstream and travel to tissues where they are utilized (Section 25.2).

Glycerol metabolism. Glycerol is first phosphorylated and then oxidized to dihydroxyacetone phosphate, a glycolysis pathway intermediate.

Through glycolysis and the common metabolic pathway, the glycerol can be converted to CO_2 and H_2O (Section 25.3).

Fatty acid degradation. Fatty acid degradation is accomplished through the β-oxidation pathway. The degradation process involves removal of carbon atoms, two at a time, from the carboxyl end of the fatty acid. There are four repeating reactions that accompany the removal of each two-carbon unit. A turn of the cycle also produces one molecule each of acetyl CoA, NADH, and $FADH_2$ (Section 25.4).

Ketone bodies. Acetoacetate, β-hydroxybutyrate, and acetone are known as ketone bodies. Synthesis occurs mainly in the liver from acetyl CoA as a result of excessive fatty acid degradation. During starvation and in unchecked diabetes, the level of ketone bodies in the blood becomes very high (Section 25.6).

Fatty acid biosynthesis. Fatty acid biosynthesis, lipogenesis, occurs through the addition of two-carbon units to a growing acyl chain. The added two-carbon units come from malonyl CoA. A multienzyme complex, an acyl carrier protein (ACP), and NADPH are important parts of the biosynthetic process (Section 25.7).

Biosynthesis of cholesterol. Cholesterol is biosynthesized from acetyl CoA in a complex series of reactions in which isoprene units are key intermediates. Cholesterol is the precursor for the various classes of steroid hormones (Section 25.9).

KEY REACTIONS AND EQUATIONS

1. Digestion of triacylglycerols (Section 25.1)

$$\text{Triacylglycerol} + H_2O \xrightarrow{\text{Lipase}}$$
$$\text{fatty acids} + \text{glycerol} + \text{monoacylglycerols}$$

2. Mobilization of triacylglycerols (Section 25.2)

$$\text{Triacylglycerol} + 3H_2O \xrightarrow{\text{Lipase}} 3 \text{ fatty acids} + \text{glycerol}$$

3. Glycerol metabolism (Section 25.3)

$$\text{Glycerol} + \boxed{\text{ATP}} + \boxed{\text{NAD}^+} \xrightarrow{\text{Two steps}}$$
$$\text{dihydroxyacetone} + \text{ADP} + \text{NADH} + H^+$$
$$\text{phosphate}$$

4. One cycle of the β-oxidation pathway (Section 25.4)

$$R-CH_2-CH_2-\overset{\overset{\displaystyle O}{\|}}{C}-S-CoA + \boxed{\text{NAD}^+} + \boxed{\text{FAD}} + CoA-SH \longrightarrow$$
$$R-\overset{\overset{\displaystyle O}{\|}}{C}-S-CoA + CH_3-\overset{\overset{\displaystyle O}{\|}}{C}-S-CoA + \boxed{\text{NADH}} + \boxed{\text{FADH}_2}$$

5. First turn of lipogenesis (Section 25.7)

$$\text{Acetyl ACP} + \text{malonyl ACP} \xrightarrow{\text{2NADPH/H}^+ \quad \text{2NADP}^+}$$
$$\text{butyryl ACP} + \text{ACP} + CO_2 + H_2O$$

EXERCISES and PROBLEMS

The members of each pair of problems in this section test similar material.

● Digestion and Absorption of Lipids (Section 25.1)

25.1 What percent of dietary lipids are triacylglycerols?

25.2 What are the solubility characteristics of triacylglycerols?

25.3 What effect do salivary enzymes have on triacylglycerols?

25.4 What effect do stomach fluids have on triacylglycerols?

25.5 The process of lipid digestion occurs primarily at two sites within the human body.
   a. What are the identities of these two sites?
   b. What is the relative amount of TAG digestion that occurs at each site?
   c. What type of digestive enzyme functions at each site?

25.6 Why does ingestion of lipids make one feel "full" for a long time?

25.7 What function does bile serve in lipid digestion?

25.8 What are the major products of triacylglycerol digestion?

25.9 *Complete* hydrolysis of triacylglycerols during digestion is unusual. Explain.

25.10 What is a fatty acid micelle?

25.11 What happens to the products of triacylglycerol digestion after they pass through the intestinal wall?

25.12 What is a chylomicron, and what is its function?

● Triacylglycerol Storage and Mobilization (Section 25.2)

25.13 What is the distinctive structural feature of *adipocytes*?

25.14 What is the major metabolic function of adipose tissue?

25.15 What is triacylglycerol mobilization?

25.16 What situation signals the need for mobilization of triacyl-glycerols from adipose tissue?

25.17 What role does cAMP play in triacylglycerol mobilization?

25.18 Triacylglycerols in adipose tissue do not enter the bloodstream as triacylglycerols. Explain.

● Glycerol Metabolism (Section 25.3)

25.19 In which step of glycerol metabolism does each of the following events occur?
   a. ATP is consumed.
   b. A phosphorylation reaction occurs.
   c. Dihydroxyacetone phosphate is produced.
   d. Glycerol kinase is active.

25.20 In which step of glycerol metabolism does each of the following events occur?
   a. The oxidizing agent NAD^+ is needed.
   b. ADP is produced.
   c. Glycerol 3-phosphate is a reactant.
   d. A secondary alcohol group is oxidized.

25.21 How many ATP molecules are expended in the conversion of glycerol to a glycolysis intermediate?

25.22 What are the two fates of glycerol after it has been converted to a glycolysis intermediate?

● Oxidation of Fatty Acids (Section 25.4)

25.23 Indicate whether each of the following statements concerning the fatty acid activation process that occurs prior to β-oxidation is true or false.
   a. CoA—SH is a reactant.
   b. A molecule of ADP is produced.
   c. The site for the process is the outer mitochondrial membrane.
   d. Two ATP molecules are consumed.

25.24 Indicate whether each of the following statements concerning the fatty acid activation process that occurs prior to β-oxidation is true or false.
a. Acetyl CoA is a product.
b. A phosphorylation enzyme is needed.
c. A molecule of AMP is produced.
d. The activation product is acyl CoA.

25.25 Indicate whether each of the following aspects of the carnitine shuttle system associated with the process of β-oxidation occurs in the mitochondrial matrix or in the mitochondrial intermembrane space.
a. Acyl CoA is a reactant.
b. Carnitine enters the inner mitochondrial membrane.
c. Carnitine is converted to acyl carnitine.
d. Free CoA is a reactant.

25.26 Indicate whether each of the following aspects of the carnitine shuttle system associated with the process of β-oxidation occurs in the mitochondrial matrix or in the mitochondrial intermembrane space.
a. Acyl CoA is a product.
b. Acyl carnitine enters the inner mitochondrial membrane.
c. Acyl carnitine is converted to carnitine.
d. Coenzyme A is a product.

25.27 Only one molecule of ATP is needed to activate fatty acids before oxidation occurs, yet in "bookkeeping" on this energy expenditure it is counted as the loss of two ATP molecules. Explain why.

25.28 What is the difference between an acetyl CoA molecule and an acyl CoA molecule?

25.29 Explain what functional group change occurs, during one turn of the β-oxidation pathway in
a. Step 1.
b. Step 2.
c. Step 3.

25.30 For one turn of the β-oxidation pathway, arrange the following β-carbon functional groups in the order in which they are encountered: secondary alcohol, ketone, alkane, and alkene.

25.31 What is the configuration of the unsaturated enoyl CoA formed by dehydrogenation during a turn of the β-oxidation pathway?

25.32 What is the configuration of the β-hydroxyacyl CoA formed by hydration during a turn of the β-oxidation pathway?

25.33 In which step (of Steps 1 through 4) and in which turn (second or third) of the β-oxidation pathway is each of the following compounds encountered as a reactant if the fatty acid to be degraded is decanoic acid?

a.
$$CH_3—(CH_2)_4—\overset{OH}{\underset{|}{CH}}—CH_2—\overset{O}{\overset{||}{C}}—S—CoA$$

b.
$$CH_3—(CH_2)_2—CH=CH—\overset{O}{\overset{||}{C}}—S—CoA$$

c.
$$CH_3—(CH_2)_2—\overset{O}{\overset{||}{C}}—CH_2—\overset{O}{\overset{||}{C}}—S—CoA$$

d.
$$CH_3—(CH_2)_6—\overset{O}{\overset{||}{C}}—S—CoA$$

25.34 In which step (of Steps 1 through 4) and in which turn (second or third) of the β-oxidation pathway is each of the following compounds encountered as a reactant if the fatty acid to be degraded is decanoic acid?

a.
$$CH_3—(CH_2)_4—\overset{O}{\overset{||}{C}}—CH_2—\overset{O}{\overset{||}{C}}—S—CoA$$

b.
$$CH_3—(CH_2)_4—\overset{O}{\overset{||}{C}}—S—CoA$$

c.
$$CH_3—(CH_2)_4—CH=CH—\overset{O}{\overset{||}{C}}—S—CoA$$

d.
$$CH_3—(CH_2)_2—\overset{OH}{\underset{|}{CH}}—CH_2—\overset{O}{\overset{||}{C}}—S—CoA$$

25.35 Which compound(s) in Problem 25.33 undergo(es) a dehydrogenation reaction during a turn of the β-oxidation pathway?

25.36 Which compound(s) in Problem 25.34 undergo(es) a chain-cleavage reaction during a turn of the β-oxidation pathway?

25.37 How many turns of the β-oxidation pathway would be needed to degrade each of the following fatty acids to acetyl CoA?
a. 16:0 fatty acid
b. 12:0 fatty acid

25.38 How many turns of the β-oxidation pathway would be needed to degrade each of the following fatty acids to acetyl CoA?
a. 20:0 fatty acid
b. 10:0 fatty acid

25.39 The degradation of *cis*-3-hexenoic acid, a 6:1 acid, requires one more step than the degradation of hexanoic acid, a 6:0 acid. Describe the nature of the extra step.

25.40 The degradation of *cis*-4-hexenoic acid, a 6:1 acid, requires one more step than the degradation of hexanoic acid, a 6:0 acid. Describe the nature of this extra step.

⬤ ATP Production from Fatty Acid Oxidation (Section 25.5)

25.41 Identify the major fuel for skeletal muscle in
a. an active state.
b. a resting state.

25.42 Explain why fatty acids cannot serve as fuel for the brain.

25.43 Consider the conversion of a C_{10} saturated acid entirely to acetyl CoA.
a. How many turns of the β-oxidation pathway are required?
b. What is the yield of acetyl CoA?
c. What is the yield of NADH?
d. What is the yield of FADH$_2$?
e. How many high-energy ATP bonds are consumed?

25.44 Consider the conversion of a C_{14} saturated acid entirely to acetyl CoA.
a. How many turns of the β-oxidation pathway are required?
b. What is the yield of acetyl CoA?
c. What is the yield of NADH?
d. What is the yield of FADH$_2$?
e. How many high-energy ATP bonds are consumed?

25.45 What is the net ATP production for the complete oxidation to CO_2 and H_2O of the fatty acid in Problem 25.43?

25.46 What is the net ATP production for the complete oxidation to CO_2 and H_2O of the fatty acid in Problem 25.44?

25.47 Which yield more FADH$_2$, saturated or unsaturated fatty acids? Explain.

25.48 Which yield more NADH, saturated or unsaturated fatty acids? Explain.

25.49 Compare the energy released when 1 g of carbohydrate and 1 g of lipid are completely degraded in the body.

25.50 Compare the net ATP produced from 1 molecule of glucose and 1 molecule of hexanoic acid when they are completely degraded in the body.

● **Ketone Bodies (Section 25.6)**

25.51 What three body conditions are conducive to ketone body formation?

25.52 Why does a deficiency of carbohydrates in the diet lead to ketone body formation?

25.53 What is the relationship between oxaloacetate concentration and ketone body formation?

25.54 What is the relationship between pyruvate concentration and ketone body formation?

25.55 Draw the structures of the three compounds classified as ketone bodies.

25.56 Two of the three ketone bodies can be formed from the third one. Write equations for the formation of these two compounds.

25.57 Identify the ketone body (or bodies) to which each of the following statements applies.
   a. Its structure contains a ketone functional group.
   b. Its structure contains a hydroxyl functional group.
   c. It is produced in Step 3 of ketogenesis.
   d. It is produced from the decomposition of acetoacetate.

25.58 Identify the ketone body (or bodies) to which each of the following statements applies.
   a. Its structure contains two methyl groups.
   b. Its structure contains a carboxyl functional group.
   c. It is produced in step 4 of ketogenesis.
   d. It is produced from the reduction of acetoacetate.

25.59 Identify the step in ketogenesis to which each of the following statements applies.
   a. The first condensation reaction occurs.
   b. A reduction reaction occurs.
   c. A C_4—CoA molecule is produced.
   d. A C_2—CoA molecule is produced.

25.60 Identify the step in ketogenesis to which each of the following statements applies.
   a. The second condensation reaction occurs.
   b. A chain-cleavage reaction occurs.
   c. A C_6—CoA molecule is produced.
   d. A C_4 molecule is converted into another C_4 molecule.

25.61 In which step or steps of ketogenesis is/are the following molecules encountered?
   a. Acetyl CoA as a reactant
   b. Acetoacetyl CoA as a reactant
   c. HMG-CoA as a product
   d. Acetoacetate as a product

25.62 In which step or steps of ketogenesis is/are the following molecules encountered?
   a. Acetyl CoA as a product
   b. Acetoacetyl CoA as a product
   c. HMG-CoA as a reactant
   d. Acetoacetate as a reactant

25.63 Before acetoacetate can be used as a "fuel" it must first be activated. What are the names of the reactants and products in the activation reaction?

25.64 What are the names of the reactants and products in the reaction by which "usable fuel" is produced from the activated form of acetoacetate?

25.65 What is ketosis?

25.66 Severe ketosis situations produce acidosis. Explain.

● **Biosynthesis of Fatty Acids (Section 25.7)**

25.67 Compare the intracellular locations where fatty acid biosynthesis and fatty acid degradation take place.

25.68 How does the structure of fatty acid synthase differ from that of the enzymes that degrade fatty acids?

25.69 Coenzyme A plays an important role in fatty acid degradation. What is its counterpart in fatty acid biosynthesis, and how does its structure differ from that of coenzyme A?

25.70 What does the designation ACP stand for?

25.71 Indicate whether each of the following aspects of the citrate–malate shuttle system associated with the process of lipogenesis occurs in the mitochondrial matrix or in the cytosol.
   a. Citrate is produced from oxaloacetate and acetyl CoA.
   b. ATP is converted to ADP.
   c. Acetyl CoA and oxaloacetate are produced from citrate.
   d. NADH is used as a reducing agent.

25.72 Indicate whether each of the following aspects of the citrate–malate shuttle system associated with the process of lipogenesis occurs in the mitochondrial matrix or in the cytosol.
   a. Citrate enters the inner mitochondrial membrane.
   b. Malate is produced from oxaloacetate.
   c. NAD^+ is used to convert malate to oxaloacetate.
   d. CoA—SH is generated.

25.73 Indicate whether each of the following lipogenesis events associated with ACP complex formation applies to (1) acetyl CoA, (2) acetyl ACP, (3) malonyl CoA, or (4) malonyl ACP.
   a. A carboxylation occurs in its production.
   b. The enzyme acetyl transferase is needed in its production.
   c. The B vitamin biotin is involved in its production.
   d. Acetyl CoA and ACP are the reactants in its production.

25.74 Indicate whether each of the following lipogenesis events associated with ACP complex formation applies to (1) acetyl CoA, (2) acetyl ACP, (3) malonyl CoA, or (4) malonyl ACP.
   a. Acetyl CoA and CO_2 are the reactants in its production.
   b. The enzyme malonyl transferase is needed in its production.
   c. ATP is consumed in its production.
   d. Malonyl CoA and ACP are the reactants in its production.

25.75 For the first cycle of the lipogenesis pathway, identify the step to which each of the following statements apply.
   a. First hydrogenation reaction occurs.
   b. A dehydration reaction occurs.
   c. Acetoacetyl ACP is a product.
   d. Crotonyl ACP is a reactant.

25.76 For the first cycle of the lipogenesis pathway, identify the step to which each of the following statements apply.
   a. Second hydrogenation reaction occurs.
   b. A condensation reaction occurs.
   c. Acetoacetyl ACP is a reactant.
   d. Crotonyl ACP is a product.

25.77 What is the name of the intermediate compound generated in the first cycle of the lipogenesis pathway that has each of the following characteristis?
   a. Has a carbon chain that is derived from a C_4 keto monoacid
   b. Has a carbon chain that is derived from a C_4 unsaturated monoacid
   c. Produced by a dehydration reaction
   d. Produced in a reaction that also produces CO_2

25.78 What is the name of the intermediate compound generated in the first cycle of the lipogenesis pathway that has each of the following characteristics?
   a. Has a carbon chain that is derived from a C_4 hydroxy monoacid
   b. Has a carbon chain that is derived from a C_4 saturated monoacid
   c. Undergoes a dehydration reaction
   d. Produced by a condensation reaction

25.79 Indicate whether each of the following intermediate compounds generated in the first or second cycle of the lipogenesis pathway is produced by (1) a dehydration reaction, (2) a hydrogenation reaction, or (3) a condensation reaction.
   a.
$$CH_3-\overset{\overset{\displaystyle O}{\|}}{C}-CH_2-\overset{\overset{\displaystyle O}{\|}}{C}-S-ACP$$
   b.
$$CH_3-CH_2-CH_2-\overset{\overset{\displaystyle OH}{|}}{CH}-CH_2-\overset{\overset{\displaystyle O}{\|}}{C}-S-ACP$$
   c.
$$CH_3-CH_2-CH_2-CH=CH-\overset{\overset{\displaystyle O}{\|}}{C}-S-ACP$$
   d.
$$CH_3-CH_2-CH_2-\overset{\overset{\displaystyle O}{\|}}{C}-S-ACP$$

25.80 Indicate whether each of the following intermediate compounds generated in the first or second cycle of the lipogenesis pathway is produced by (1) a dehydration reaction, (2) a hydrogenation reaction, or (3) a condensation reaction.
   a.
$$CH_3-CH_2-CH_2-CH_2-CH_2-\overset{\overset{\displaystyle O}{\|}}{C}-S-ACP$$
   b.
$$CH_3-CH=CH-\overset{\overset{\displaystyle O}{\|}}{C}-S-ACP$$
   c.
$$CH_3-\overset{\overset{\displaystyle OH}{|}}{CH}-CH_2-\overset{\overset{\displaystyle O}{\|}}{C}-S-ACP$$
   d.
$$CH_3-CH_2-CH_2-\overset{\overset{\displaystyle O}{\|}}{C}-CH_2-\overset{\overset{\displaystyle O}{\|}}{C}-S-ACP$$

25.81 What role does molecular oxygen, O_2, play in fatty acid biosynthesis?

25.82 What is the characteristic structural feature of an essential fatty acid?

25.83 Consider the biosynthesis of a C_{14} saturated fatty acid from acetyl CoA molecules.
   a. How many turns of the fatty acid biosynthetic pathway are needed?
   b. How many molecules of malonyl ACP must be formed?
   c. How many high-energy ATP bonds are consumed?
   d. How many NADPH molecules are needed?

25.84 Consider the biosynthesis of a C_{16} saturated fatty acid from acetyl CoA molecules.
   a. How many turns of the fatty acid biosynthetic pathway are needed?
   b. How many molecules of malonyl ACP must be formed?
   c. How many high-energy ATP bonds are consumed?
   d. How many NADPH molecules are needed?

● **Relationships Between Lipogenesis and Citric Acid Cycle Intermediates (Section 25.8)**

25.85 What is the citric acid cycle diacid intermediate counterpart for each of the following lipogenesis C_4—ACP monoacid intermediates?
   a. Butyrate   b. Acetoacetate
   c. β-Hydroxybutyrate   d. Crotonate

25.86 What is the lipogenesis C_4—ACP monoacid intermediate counterpart for each of the following citric acid cycle C_4 diacid intermediates?
   a. Succinate   b. Malate
   c. Oxaloacetate   d. Fumarate

25.87 Identify each of the following C_4 species as a (1) hydroxy acid, (2) keto acid, (3) saturated acid, or (4) unsaturated acid.
   a. Crotonate   b. Oxaloacetate
   c. Acetoacetate   d. Malate

25.88 Identify each of the following C_4 species as a monocarboxylic acid or dicarboxylic acid.
   a. Succinate   b. Butyrate
   c. β-Hydroxybutyrate   d. Fumarate

● **Biosynthesis of Cholesterol (Section 25.9)**

25.89 Approximately what percent of the total amount of cholesterol in your body is derived from the following?
   a. Your diet
   b. Biosynthesis

25.90 What is the starting material for the biosynthesis of cholesterol?

25.91 In each of the following pairs of intermediates in the biosynthetic pathway for cholesterol, specify which one is encountered first in the pathway.
   a. Mevalonate and squalene
   b. Isopentenyl pyrophosphate and lanosterol
   c. Lanosterol and squalene

25.92 In each of the following pairs of intermediates in the biosynthetic pathway for cholesterol, specify which one is encountered first in the pathway.
   a. Mevalonate and lanosterol
   b. Isopentenyl pyrophosphate and squalene
   c. Mevalonate and isopentenyl pyrophosphate

25.93 For each pair of compounds in Problem 25.91, tell whether the number of carbon atoms in the first compound is less than, the same as, or greater than the number of carbon atoms in the second compound.

25.94 For each pair of compounds in Problem 25.92, tell whether the number of carbon atoms in the first compound is less than, the same as, or greater than the number of carbon atoms in the second compound.

● **Relationships Between Lipid and Carbohydrate Metabolism (Section 25.10)**

25.95 Name three types of compounds for which acetyl CoA, the primary link between lipid and carbohydrate metabolism, is the following.
   a. Biosynthetic starting material
   b. Degradation product

25.96 Name four possible fates for the acetyl CoA produced through degradation of carbohydrate and lipid dietary sources.

ADDITIONAL PROBLEMS

25.97 With which of these processes, (1) glycerol catabolism, (2) β-oxidation pathway, (3) lipogenesis, or (4) ketogenesis, is each of the following molecules associated?
a. Enoyl CoA
b. Malonyl ACP
c. Dihydroxyacetone phosphate
d. β-Hydroxybutyrate

25.98 With which of these processes, (1) fatty acid catabolism, (2) lipogenesis, (3) ketogenesis, or (4) consumption of molecular O_2, is each of the following situations associated?
a. Carnitine shuttle system
b. Citrate shuttle system
c. Conversion of squalene to cholesterol
d. Conversion of a saturated fatty acid to an unsaturated fatty acid

25.99 Identify the step (among Steps 1 through 4) of the fatty acid chain elongation process in lipogenesis to which each of the following characterizations applies.
a. Malonyl ACP is a reactant.
b. CO_2 is a product.
c. A dehydration reaction occurs.
d. A carbon–carbon double bond is converted to a carbon–carbon single bond.

25.100 Indicate in what order the following events occur in the digestion of triacylglycerols (TAGs). (1) Bile emulsifies TAG "droplets." (2) TAGs incorporated into chylomicrons enter the lymph system. (3) TAGs are hydrolyzed to monoacylglycerols. (4) Free fatty acids are "repackaged" into TAGs.

25.101 Indicate whether each of the following statements is true or false.
a. Chylomicrons are lipoproteins.
b. Acetoacetate is an intermediate in the conversion of glycerol to dihydroxyacetone phosphate.
c. The molecule carnitine is involved in fatty acid activation.
d. One turn of the β-oxidation pathway produces two molecules of ATP.

25.102 Indicate whether each of the following pairings of terms is correct or incorrect as a reactant/product pair for reactions in the β-oxidation pathway.
a. Alkene functional group; dehydrogenation
b. Ketone functional group; chain cleavage
c. Alkane functional group; hydration
d. Secondary alcohol functional group; oxidation

25.103 Indicate whether each of the following pairings of terms is correct or incorrect as a reactant/product pair for reactions in the chain elongation phase of lipogenesis.
a. Alkene functional group; hydrogenation
b. Secondary alcohol group; dehydration
c. Ketone group; reduction
d. Ketone group; hydrogenation

25.104 Arrange the four molecules (1) glucose, (2) sucrose, (3) C_8 unsaturated fatty acid, and (4) C_{14} unsaturated fatty acid in order of increasing biochemical energy content (ATP production) per mole.

𝓜ultiple-Choice Practice Test

25.105 Which of the following statements concerning digestion of dietary triacylglycerols in adults is *correct*?
a. It begins in the mouth.
b. It occurs to a small extent (10%) in the stomach.
c. It occurs to a large extent (90%) in the stomach.
d. It occurs only in the small intestine.

25.106 Monoacylglycerols are the predominant constituent in which of the following?
a. Fatty acid micelles   b. Chylomicrons
c. Adipocytes   d. Bile

25.107 The first stage of glycerol metabolism is the two-step conversion of glycerol to dihydroxyacetone phosphate. What is the intermediate in this process?
a. Dihydroxyacetone
b. Monohydroxyacetone phosphate
c. Glycerol 3-phosphate
d. 3-phosphoglycerate

25.108 In the oxidation of fatty acids, what is the molecule that shuttles the activated fatty acid across the inner mitochondrial membrane?
a. CoA   b. Acetyl CoA
c. Carnitine   d. Citrate

25.109 What is the first functional group change that occurs in the β-oxidation pathway?
a. Alkane to alkene   b. Alkene to 2° alcohol
c. Alkane to 2° alcohol   d. 2° alcohol to ketone

25.110 Which of the following pairings of terms is *correct* as a reactant/product pair for reactions in the β-oxidation pathway?
a. Alkene functional group; dehydrogenation
b. Ketone functional group; chain cleavage
c. Alkane functional group; hydration
d. 2° alcohol functional group; hydrogenation

25.111 How many turns of the β-oxidation pathway are needed to "process" a C_{16} fatty acid molecule?
a. Seven   b. Eight
c. Fourteen   d. Sixteen

25.112 Which of the following compounds is a *ketone body*?
a. Carnitine   b. Oxaloacetate
c. Acetoacetate   d. Acetyl CoA

25.113 What are the starting materials for the processes of ketogenesis and lipogenesis, respectively?
a. Acetyl CoA and a fatty acid
b. A fatty acid and acetyl CoA
c. Acetyl CoA and acetyl CoA
d. A fatty acid and a fatty acid

25.114 Which of the following is an intermediate in the process of lipogenesis?
a. Isopentyl pyrophosphate
b. Malonyl ACP
c. Oxaloacetate
d. Acetoacetate

Protein Metabolism

Chapter Outline

Fish, such as the Atlantic salmon, and other aquatic species process (eliminate) the nitrogen from protein in a manner different from that which occurs in human beings.

From an energy production standpoint, proteins supply only a small portion of the body's needs. With a normal diet, carbohydrates and fats supply 90% of the body's energy, and only 10% comes from proteins. However, despite its minor role in energy production, protein metabolism plays an important role in maintaining good health. The amino acids obtained from proteins are needed for both protein synthesis and synthesis of other nitrogen-containing compounds in the cell. In this chapter, we examine protein digestion, the oxidative degradation of amino acids, and amino acid biosynthesis.

26.1 PROTEIN DIGESTION AND ABSORPTION

Protein digestion begins in the stomach rather than in the mouth because saliva contains no enzymes that affect proteins. Both protein denaturation (Section 20.16) and protein hydrolysis (Section 20.15) occur in the stomach. The partially digested protein (large polypeptides) passes from the stomach into the small intestine, where digestion is completed (Figure 26.1).

Proteins are denatured in the stomach by the hydrochloric acid present in gastric juice. The acid gives gastric juice a pH of between 1.5 and 2.0. The enzyme *pepsin* effects the hydrolysis of about 10% of peptide bonds in proteins, producing a variety of polypeptides. In the small intestine, *trypsin, chymotrypsin,* and *carboxypeptidase* in pancreatic juice attack peptide bonds. The pH of pancreatic juice is between 7.0 and 8.0, and it neutralizes the acidity of the material from the stomach. *Aminopeptidase,* secreted by intestinal mucosal cells, also attacks peptide bonds. Pepsin, trypsin, chymotrypsin,

A very small number of people are unable to synthesize enough stomach acid, and these individuals must ingest capsules of dilute hydrochloric acid with every meal.

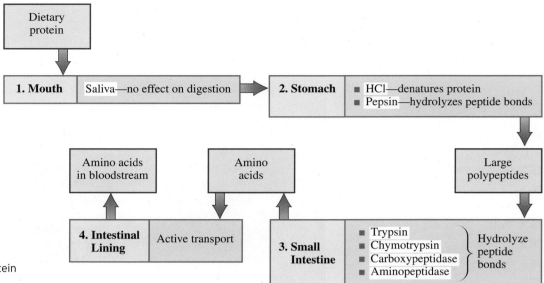

Figure 26.1 Summary of protein digestion in the human body.

The passage of polypeptide chains and small proteins across the intestinal wall is uncommon in adults. In infants, however, such transport allows the passage of antibodies (proteins) in colostral milk from a mother to a nursing infant to build up immunologic protection in the infant.

carboxypeptidase, and aminopeptidase are all examples of *proteolytic* enzymes (Section 21.8). Enzymes of this type are produced in inactive forms called *zymogens* that are activated at their site of action (Section 21.9).

The net result of protein digestion is the release of the protein's constituent amino acids. Absorption of these "free" amino acids through the intestinal wall requires active transport with the expenditure of energy (Section 19.10). Different transport systems exist for the various kinds of amino acids. After passing through the intestinal wall, the free amino acids enter the bloodstream, which distributes them throughout the body.

● **EXAMPLE 26.1**

Determining the Sites Where Various Aspects of Protein Digestion Occur

▶ Based on the information in Figure 26.1, determine the location within the human body where each of the following aspects of protein digestion occurs.

a. The enzyme trypsin is active.
b. Large polypeptides are produced.
c. Protein is denatured by HCl.
d. Active transport moves amino acids into the bloodstream.

Solution

a. Trypsin is one of several proteolytic enzymes, found in the *small intestine,* that hydrolyze peptide bonds.
b. Large polypeptides are the product of enzymatic action that occurs in the *stomach.*
c. Protein denaturation, effected by HCl (stomach acid), occurs in the *stomach.*
d. Active transport processes effect the passage of amino acids through the *intestinal lining* into the bloodstream.

Practice Exercise 26.1

Based on the information in Figure 26.1, determine the location within the human body where each of the following aspects of protein digestion occurs.

a. The enzyme aminopeptidase is active.
b. Peptide bonds are hydrolyzed under the action of pepsin.
c. Individual amino acids are produced.
d. The enzyme carboxypeptidase is active.

Answers: a. Small intestine; **b.** Stomach; **c.** Small intestine; **d.** Small intestine

26.2 AMINO ACID UTILIZATION

Amino acids produced from the digestion of proteins enter the amino acid pool of the body. The **amino acid pool** *is the total supply of free amino acids available for use in the human body.* Dietary protein is one of three sources that contributes amino acids to the amino acid pool. The other two sources are *protein turnover* and *biosynthesis* of amino acids in the liver.

Within the human body, proteins are continually being degraded (hydrolyzed) to amino acids and resynthesized. Disease, injury, and "wear and tear" are all causes of degradation. The degradation–resynthesis process is called protein turnover. **Protein turnover** *is the repetitive process in which proteins are degraded and resynthesized within the human body.*

Biosynthesis of amino acids by the liver also supplies the amino acid pool with amino acids. However, only the *nonessential* amino acids (Sections 20.2 and 26.6) can be produced in this manner.

In a healthy adult, the amount of nitrogen taken into the body each day (dietary proteins) equals the amount of nitrogen excreted from the body. Such a person is said to be in a state of nitrogen balance. **Nitrogen balance** *is the state that results when the amount of nitrogen taken into the human body as protein equals the amount of nitrogen excreted from the body in waste materials.*

Two types of nitrogen imbalance can occur. When protein degradation exceeds protein synthesis, the amount of nitrogen in the urine exceeds the amount of nitrogen ingested (dietary protein). This condition of *negative nitrogen balance* accompanies a state of "tissue wasting," because more tissue proteins are being catabolized than are being replaced by protein synthesis. Protein-poor diets, starvation, and wasting illnesses produce a negative nitrogen balance.

A *positive nitrogen balance* (nitrogen intake exceeds nitrogen output) indicates that the rate of protein anabolism (synthesis) exceeds that of protein catabolism. This state indicates that large amounts of tissue are being synthesized, such as during growth, pregnancy, and convalescence from an emaciating illness.

Although the overall nitrogen balance in the body often varies, the relative concentrations of amino acids within the amino acid pool remain essentially constant. No specialized storage forms for amino acids exist in the body, as is the case for glucose (glycogen) and fatty acids (triacylglycerols). Therefore, the body needs a relatively constant source of amino acids to maintain normal metabolism. During negative nitrogen balance, the body must resort to degradation of proteins that were synthesized for other functions.

The amino acids from the body's amino acid pool are used in four different ways.

1. *Protein synthesis.* It is estimated that about 75% of the free amino acids in a healthy, well-nourished adult go into protein synthesis. Proteins are continually needed to replace old tissue (protein turnover) and also to build new tissue (growth). The subject of protein synthesis was considered in Section 22.11.

2. *Synthesis of nonprotein nitrogen-containing compounds.* Amino acids are regularly withdrawn from the amino acid pool for the synthesis of nonprotein nitrogen-containing compounds. Such molecules include the purines and pyrimidines of nucleic acids, the heme of hemoglobin, neurotransmitters such as acetylcholine and serotonin, the choline and ethanolamine of phosphoglycerides, and hormones such as epinephrine.

3. *Synthesis of nonessential amino acids.* When required, the body draws on the amino acid pool for raw materials for the production of *nonessential* amino acids that are in short supply. The "roadblock" preventing the synthesis of the *essential* amino acids is not lack of nitrogen but lack of a correct carbon skeleton upon which enzymes can work. In general, the essential amino acids contain carbon chains or aromatic rings not present in other amino acids or the intermediates of carbohydrate or lipid metabolism. Table 26.1 lists the essential amino acids and the nonessential amino acids with the precursors needed to form the latter.

4. *Production of energy.* Because excess amino acids cannot be stored for later use, the body's response is to degrade them. The degradation process is complex because each of the 20 standard amino acids has a different degradation pathway.

The rate of protein turnover varies from a few minutes to several hours. Proteins with short turnover rates include many enzymes and regulatory hormones. In a healthy adult, about 2% of the body's protein is broken down and resynthesized every day.

Higher plants and certain microorganisms are capable of synthesizing all the protein amino acids from carbon dioxide, water, and inorganic salts.

There are approximately 100 grams of free amino acids present in the amino acid pool. Glutamine is the most abundant, followed closely by alamine, glycine, and valine. The essential amino acids constitute approximately 10 grams of the pool.

TABLE 26.1
Essential and Nonessential Amino Acids

Nutritionally Essential Amino Acids	Nutritionally Nonessential Amino Acids	
	Amino Acid	Precursor
histidine	alanine	pyruvate
isoleucine	arginine	glutamate
leucine	asparagine	aspartate
lysine	aspartic acid	oxaloacetate
methionine	cysteine	serine
phenylalanine	glutamic acid	α-ketoglutarate
threonine	glutamine	glutamate
tryptophan	glycine	serine
valine	proline	glutamate
	serine	3-phosphoglycerate
	tyrosine	phenylalanine

In all the degradation pathways, the amino nitrogen atom is removed and converted to ammonium ion, which ultimately is excreted from the body as urea. The remaining carbon skeleton is then converted to pyruvate, acetyl CoA, or a citric acid cycle intermediate, depending on its makeup, with the resulting energy production or energy storage. Figure 26.2 shows the various pathways available for the further use of amino acid catabolism products. Subsequent sections of this chapter give further details about these processes.

 TRANSAMINATION AND OXIDATIVE DEAMINATION

Degradation of an amino acid has two stages: (1) the removal of the α-amino group and (2) the degradation of the remaining carbon skeleton. In this section and the next, we consider what happens to the amino group; in Section 26.5, the fate of the carbon skeleton is considered.

The release of an amino group from most amino acids requires a two-step process involving *transamination* followed by *oxidative deamination*. The following two procedures will make these processes easier to visualize.

Figure 26.2 Possible fates for amino acid degradation products.

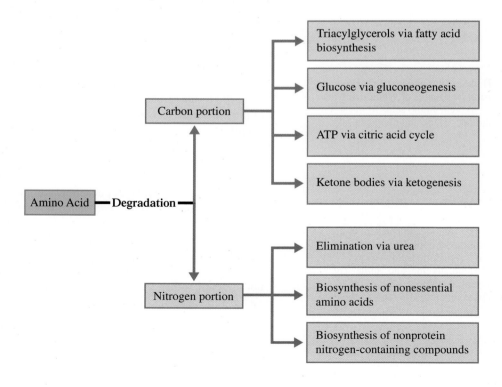

1. Draw amino acid structures in the general format

$$\overset{\overset{+}{N}H_3}{\underset{|}{R-CH-COO^-}}$$

Remember that the ordering of the four groups attached to the carbon in an amino acid is not critical except in stereochemical considerations (Fischer projection formulas; Section 20.3).

2. Review the structural relationships among six molecules—three pairs of keto/amino acids. In Section 16.5, we noted that the derivatives of three carboxylic acids—propionic (a three-carbon monoacid), succinic (a four-carbon diacid) and glutaric (a five-carbon diacid)—are particularly important in metabolic reactions. It is the α-keto and α-amino derivatives of these three acids that are the "key players" in the transamination/oxidative deamination process. Figure 26.3 gives the structural relationships among these compounds.

Transamination

A **transamination reaction** *is a biochemical reaction that involves the interchange of the amino group of an α-amino acid with the keto group of an α-keto acid.* The general equation for a transamination reaction is

$$\underset{\text{α-Amino acid}}{\overset{\overset{+}{N}H_3}{\underset{|}{R-CH-COO^-}}} + \underset{\text{α-Keto acid}}{\overset{O}{\underset{\|}{R'-C-COO^-}}} \longrightarrow \underset{\text{New α-keto acid}}{\overset{O}{\underset{\|}{R-C-COO^-}}} + \underset{\text{New α-amino acid}}{\overset{\overset{+}{N}H_3}{\underset{|}{R'-CH-COO^-}}}$$

Figure 26.3 Key compounds in the transamination/oxidative deamination process include three keto acid/amino acid pairs.

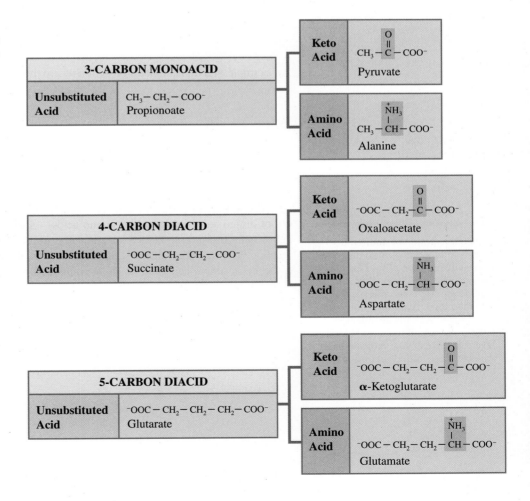

The purpose of transamination is to remove amino groups from the various α-amino acids and collect them in a single amino acid, glutamate. Glutamate then acts as the source of amino groups for continued nitrogen metabolism (excretion or biosynthesis).

There are at least 50 aminotransferase enzymes associated with transamination reactions. Most are specific for α-ketoglutarate as the amino group acceptor but are less specific for the amino acid. Glutamate is the amino acid produced from the action of α-ketoglutarate specific aminotransferases.

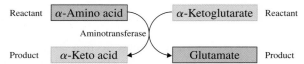

An important exception to the α-ketoglutarate specificity of aminotransferases is found in muscle cells. Here, many aminotransferases have specificity for the keto acid pyruvate. Alanine is the amino acid produced from the action of such pyruvate specific enzymes.

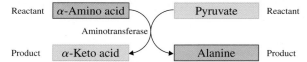

An example of a transamination reaction where the aminotransferase is α-ketoglutarate specific is

$$
\underset{\text{Glycine}}{\overset{\text{COO}^-}{\underset{\text{H}}{\overset{|}{\underset{|}{\text{H}_3\overset{+}{\text{N}}-\text{C}-\text{H}}}}}}
\;+\;
\underset{\substack{\text{α-Ketoglutarate}}}{\overset{\text{COO}^-}{\underset{\text{COO}^-}{\overset{|}{\underset{|}{\overset{\text{C}=\text{O}}{\underset{\text{CH}_2}{\text{CH}_2}}}}}}}
\;\longrightarrow\;
\underset{\text{Ketoacetate}}{\overset{\text{COO}^-}{\underset{\text{H}}{\overset{|}{\underset{|}{\text{C}=\text{O}}}}}}
\;+\;
\underset{\text{Glutamate}}{\overset{\text{COO}^-}{\underset{\text{COO}^-}{\overset{|}{\underset{|}{\overset{\text{H}_3\overset{+}{\text{N}}-\text{C}-\text{H}}{\underset{\text{CH}_2}{\text{CH}_2}}}}}}}
$$

Glycine — Old amino group carrier

Glutamate — New amino group carrier

An example of a transamination reaction where the aminotransferase is pyruvate specific is

$$
\underset{\text{Glycine}}{\overset{\text{COO}^-}{\underset{\text{H}}{\text{H}_3\overset{+}{\text{N}}-\text{C}-\text{H}}}}
\;+\;
\underset{\text{Pyruvate}}{\overset{\text{COO}^-}{\underset{\text{CH}_3}{\text{C}=\text{O}}}}
\;\longrightarrow\;
\underset{\text{Ketoacetate}}{\overset{\text{COO}^-}{\underset{\text{H}}{\text{C}=\text{O}}}}
\;+\;
\underset{\text{Analine}}{\overset{\text{COO}^-}{\underset{\text{CH}_3}{\text{H}_3\overset{+}{\text{N}}-\text{C}-\text{H}}}}
$$

Glycine — Old amino group carrier

Analine — New amino group carrier

The concentration of aminotransferases in blood is used to diagnose liver and heart disorders. Liver damage releases the enzyme alanine aminotransferase (ALT) into the blood. Aspartate aminotransferase (AST) is abundant in heart muscle, and increased blood levels of this enzyme indicate heart damage (myocardial infarction).

The initial effect of transamination reactions is to collect the amino acids from a variety of amino acids into just two amino acids—glutamate (most cells) and alanine (muscle cells). After transport to the liver, alanine is converted to glutamate through a transamination reaction. The net effect of transamination is, thus, the collection of the amino groups from a variety of amino acids into a single compound—the amino acid glutamate.

Although the transamination reaction appears to involve the simple transfer of an $-\overset{+}{\text{N}}\text{H}_3$ group between two molecules, the reaction involves several steps and requires the presence of pyridoxal phosphate, a coenzyme produced from pyridoxine (vitamin B_6).

Pyridoxine (vitamin B_6)

Pyridoxal phosphate (coenzyme)

Figure 26.4 The role of pyridoxal phosphate in the process of transamination.

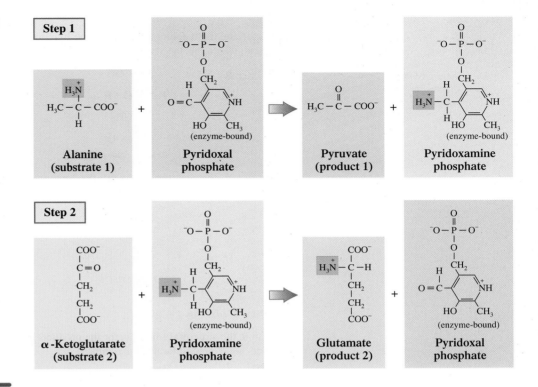

Transamination reactions are reversible and can go easily in either direction, depending on the reactant concentrations. This reversibility is the basis for regulation of amino acid concentrations in the body.

This coenzyme is an integral part of the transamination process. The amino group of the amino acid is transferred first to the pyridoxal phosphate and then from the pyridoxal phosphate to the α-keto acid. Figure 26.4 shows the role of this coenzyme in the transamination process, where alanine is the amino acid and α-ketoglutarate is the α-keto acid.

● E X A M P L E 2 6 . 2

Determining the Products in a Transamination Reaction When Given the Reactants

▶ Determine the structural formulas for the products in a transamination reaction in which the reactants are aspartate and pyruvate.

Solution

Use Figure 26.3 to obtain the structures for the named reactants. Aspartate is an amino acid and pyruvate is a keto acid.

Aspartate (amino acid)   Pyruvate (keto acid)

The interchange of functional groups associated with transamination is such that the new keto acid produced will have the same carbon skeleton as the reacting amino acid and the new amino acid produced will have the same carbon skeleton as the reacting keto acid.

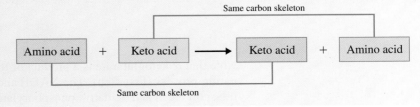

For aspartate, the reacting amino acid, the functional group change produces the following keto acid, which has the same carbon skeleton as the aspartate.

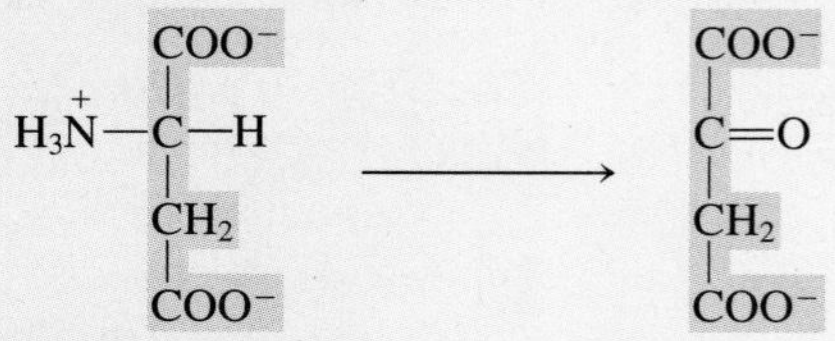

Reacting amino acid   Product keto acid

For pyruvate, the reacting keto acid, the functional group change produces the following amino acid, which has the same carbon skeleton as pyruvate.

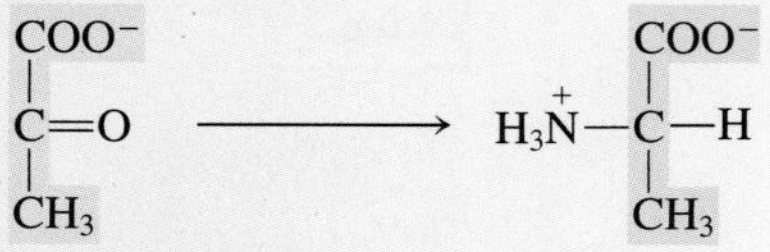

Reacting keto acid   Product amino acid

Practice Exercise 26.2

Determine the structural formulas for the products in a transamination reaction in which the reactants are α-ketoglutarate and alanine.

Answer:

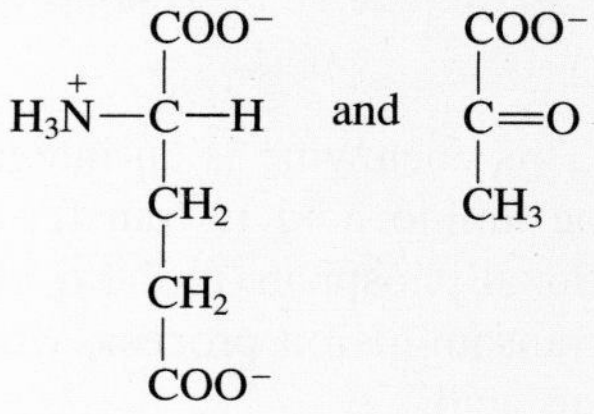

Oxidative Deamination

In the second step of amino acid degradation, ammonium ion (NH_4^+) is liberated from the glutamate formed by transamination. This step involves oxidative deamination. An **oxidative deamination reaction** *is a biochemical reaction in which an α-amino acid is converted into an α-keto acid with release of an ammonium ion.* Oxidative deamination occurs primarily in liver and kidney mitochondria.

Oxidative deamination of glutamate requires the enzyme *glutamate dehydrogenase.* This enzyme is unusual in that it can function with either $NADP^+$ or NAD^+ as a coenzyme. With NAD^+ as the coenzyme, the reaction is

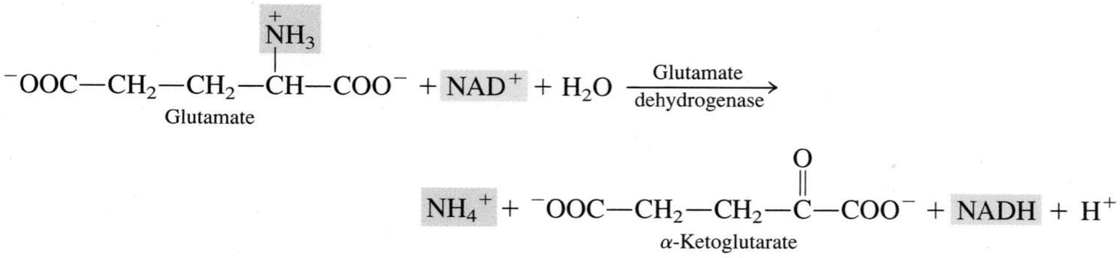

Note that α-ketoglutarate is a product of this process. It can be reused in the transamination process (first step). The NADH and H^+ formed can participate in the electron transport chain and oxidative phosphorylation to produce ATP molecules (Sections 23.7 and 23.8).

The sum of the transamination and deamination steps of the degradation of amino acids is

$$\alpha\text{-Amino acid} + NAD^+ + H_2O \longrightarrow \alpha\text{-keto acid} + NH_4^+ + NADH + H^+$$

The NH_4^+ so produced, a toxic substance if left to accumulate in the body, is then converted to urea in the urea cycle (Section 26.4).

Two amino acids, serine and threonine, exhibit different behavior from the other amino acids. They undergo *direct deamination* by a dehydration–hydration process rather than *oxidative deamination*. This different behavior results from the presence of a side-chain β-hydroxyl group, a feature unique to these two acids. The direct deamination reaction for serine is

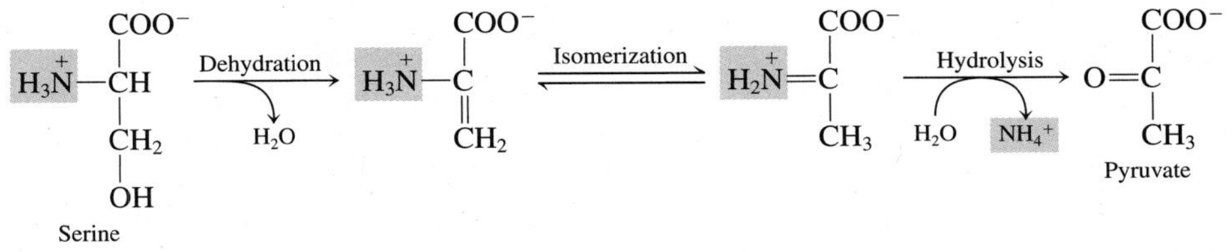

Threonine goes through a similar series of steps.

EXAMPLE 26.3

Distinguishing Between Characteristics of Oxidative Deamination and Transamination

Indicate whether each of the following reaction characteristics is associated with the process of *transamination* or with the process of *oxidative deamination*.

a. The enzyme glutamate dehydrogenase is active.
b. The coenzyme pyridoxal phosphate is needed.
c. An amino acid is converted into a keto acid with release of ammonium ion.
d. One of the reactants is an amino acid and one of the products is an amino acid.

Solution

a. *Oxidative deamination.* The substrate for oxidative deamination is glutamate and the process is an oxidation reaction (removal of H atoms) with NAD^+ serving as the oxidizing agent.
b. *Transamination.* The coenzyme pyridoxal phosphate, a derivative of vitamin B_6, is an intermediate in the amino group transfer process.
c. *Oxidative deamination.* Ammonium ion production is a characteristic of oxidative deamination; no ammonium ion is produced during transamination.
d. *Transamination.* In transamination there are two reactants (an amino acid and a keto acid) and two products (a new amino acid and a new keto acid).

Practice Exercise 26.3

Indicate whether each of the following reaction characteristics is associated with the process of *transamination* or with the process of *oxidative deamination*.

a. One of the reactants is a keto acid and one of the products is a keto acid.
b. Enzymes with a specificity toward α-ketoglutarate are often active.
c. NAD^+ is used as an oxidizing agent.
d. An aminotransferase enzyme is active.

Answers: a. Transamination; **b.** Transamination; **c.** Oxidative deamination; **d.** Transamination

26.4 THE UREA CYCLE

From a nitrogen standpoint, the net effect of amino acid degradation is the production of ammonium ion. The accumulation of this ion in the body has potential toxic effects. Consequently, the ammonium ions are converted to urea, a less toxic nitrogen-containing

---

The *transamination* process previously discussed is not a net *deamination* (loss of an amino group) process because the reacting α-keto acid becomes *aminated* as the reacting α-amino acid becomes *deaminated*.

The toxicity of ammonium ion is related to the oxidative deamination reaction by which it is formed, the conversion of glutamate to α-ketoglutarate. This reaction, which is an equilibrium situation, is shifted to the glutamate side by increased ammonium ion levels. This shift decreases α-ketoglutarate levels significantly, which affects the citric acid cycle of which α-ketoglutarate is an intermediate. Cellular ATP production drops, and the lack of ATP causes central nervous system problems.

compound, in the liver by a series of metabolic reactions called the urea cycle. The **urea cycle** *is the series of biochemical reactions in which urea is produced from ammonium ions and carbon dioxide.* The urea so produced is then transported in the blood from the liver to the kidneys and eliminated from the body in urine.

In the pure state, urea is a white solid with a melting point of 133°C. Its structure is

$$H_2N-\overset{\overset{\displaystyle O}{\|}}{C}-NH_2$$

Urea is very soluble in water (1 g per 1 mL), is odorless and colorless, and has a salty taste. (Urea does not contribute to the odor or color of urine.) With normal metabolism, an adult excretes about 30 g of urea daily in urine, although the exact amount varies with the protein content of the diet.

Three amino acids are involved as intermediates in the conversion of ammonium ions to urea through the urea cycle. These acids are arginine, ornithine, and citrulline, the latter two of which are nonstandard amino acids—that is, amino acids not found in protein. Structurally, all three of these amino acids have the same carbon chain.

Arginine is the most nitrogen-rich of the standard amino acids. It contains four nitrogen atoms.

$$H_2N-\overset{\overset{\displaystyle \overset{+}{N}H_2}{\|}}{C}-NH-CH_2-CH_2-CH_2-\overset{\overset{\displaystyle \overset{+}{N}H_3}{|}}{CH}-COO^-$$

Arginine
(standard amino acid)

$$H_3\overset{+}{N}-CH_2-CH_2-CH_2-\overset{\overset{\displaystyle \overset{+}{N}H_3}{|}}{CH}-COO^-$$

Ornithine
(nonstandard amino acid)

$$H_2N-\overset{\overset{\displaystyle O}{\|}}{C}-NH-CH_2-CH_2-CH_2-\overset{\overset{\displaystyle \overset{+}{N}H_3}{|}}{CH}-COO^-$$

Citrulline
(nonstandard amino acid)

Carbamoyl Phosphate

The "fuel" for the urea cycle is the compound *carbamoyl phosphate.* This fuel is formed from ammonium ion (from oxidative deamination; Section 26.3), carbon dioxide (from the citric acid cycle), water, and two ATP molecules. The formation equation for carbamoyl phosphate is

The functional group attached to the phosphate in carbamoyl phosphate is the simple amide functional group

$$-\overset{\overset{\displaystyle O}{\|}}{C}-NH_2$$

The term *carbamoyl* is the *prefix* that denotes an amide group. Most often, amide groups are named by using the *suffix* system, in which case the suffix is *amide.*

$$NH_4^+ + CO_2 + H_2O + \boxed{2ATP} \longrightarrow H_2N-\overset{\overset{\displaystyle O}{\|}}{C}-O\sim\overset{\overset{\displaystyle O}{\|}}{\underset{\underset{\displaystyle O^-}{|}}{P}}-O^- + \boxed{2ADP} + P_i + 3H^+$$

Carbamoyl phosphate

Note that two ATP molecules are expended in the formation of one carbamoyl phosphate molecule and that carbamoyl phosphate contains a high-energy phosphate bond. The carbamoyl phosphate formation reaction, like the reactions of the citric acid cycle, takes place in the mitochondrial matrix.

Steps of the Urea Cycle

Figure 26.5 shows the four-step urea cycle in outline form. Note that the urea cycle occurs partially in the mitochondria and partially in the cytosol and that ornithine and citrulline must be transported across the inner mitochondrial membrane. We will now consider in detail the individual steps of the urea cycle.

Figure 26.5 The four-step urea cycle in which carbamoyl phosphate is converted to urea.

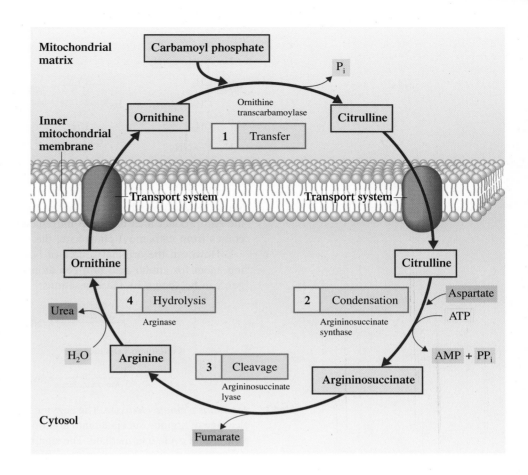

Step 1: *Carbamoyl group transfer.* The carbamoyl group of carbamoyl phosphate is transferred to ornithine to form citrulline in a reaction catalyzed by *ornithine transcarbamoylase.*

The breaking of the high-energy phosphate bond in carbamoyl phosphate drives the transfer process. With the carbamoyl transfer, the first of the two nitrogen atoms and the carbon atom needed for the formation of urea have been introduced into the cycle.

Step 2: *Citrulline–aspartate condensation.* Citrulline is transported into the cytosol where a condensation reaction between citrulline and aspartate (a standard amino acid) produces argininosuccinate. This condensation, catalyzed by *argininosuccinate synthase,* is driven by the expenditure of ATP.

The standard amino acid lysine and the nonstandard amino acid ornithine, both basic amino acids, have closely related structures.

$$\overset{\overset{\displaystyle +}{NH_3}}{\underset{}{H_3\overset{+}{N}-(CH_2)_4-CH-COO^-}}$$

Lysine

$$\overset{\overset{\displaystyle +}{NH_3}}{\underset{}{H_3\overset{+}{N}-(CH_2)_3-CH-COO^-}}$$

Ornithine

Lysine has one more CH_2 group than does ornithine.

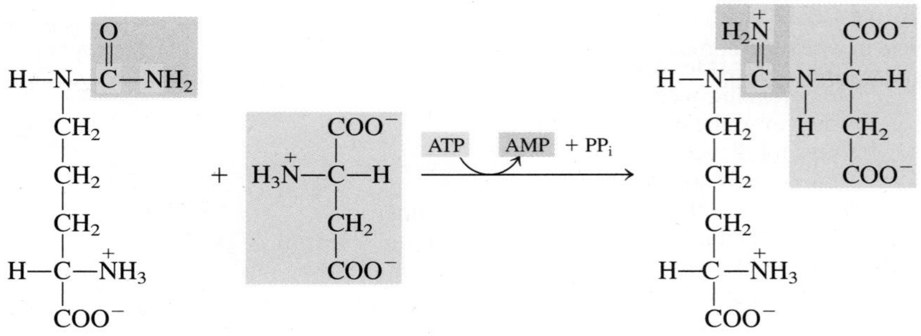

Citrulline   Aspartate   Argininosuccinate

With this reaction, the second of the two nitrogen atoms that will be part of the end-product urea has been introduced into the cycle. One nitrogen atom comes from carbamoyl phosphate, the other from aspartate.

However, the original source of both nitrogens is glutamate, the collecting agent for amino acid nitrogen atoms (Section 26.3). The flow of the nitrogen can be shown by these reactions:

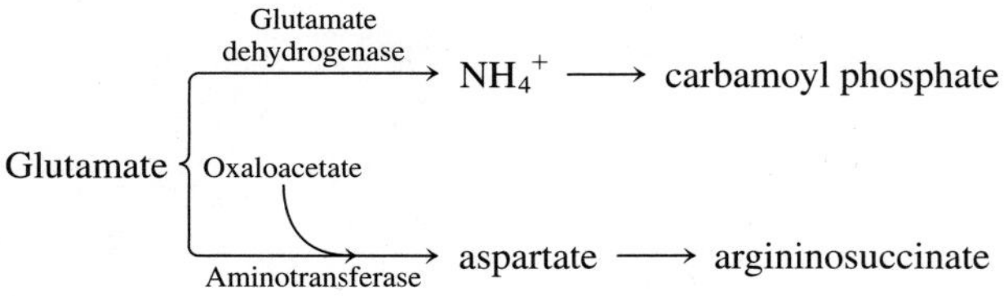

Step 3: *Argininosuccinate cleavage.* The enzyme *argininosuccinate lyase* catalyzes the cleavage of argininosuccinate into arginine, a standard amino acid, and fumarate, a citric acid cycle intermediate. The significance of this will be considered shortly.

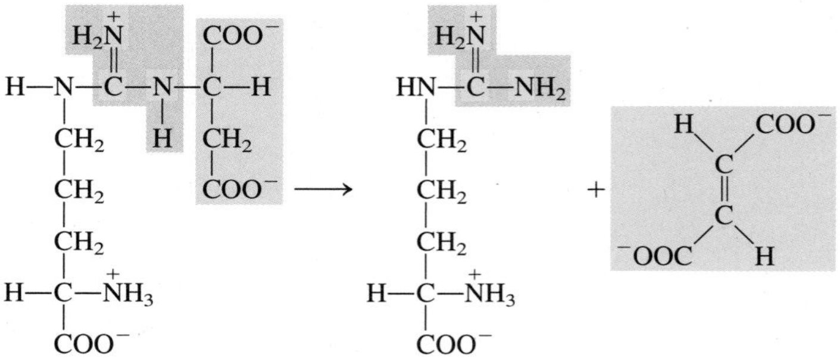

Argininosuccinate   Arginine   Fumarate

Step 4: *Urea from arginine hydrolysis.* Hydrolysis of arginine produces urea and regenerates ornithine, one of the cycle's starting materials. The enzyme involved is *arginase.*

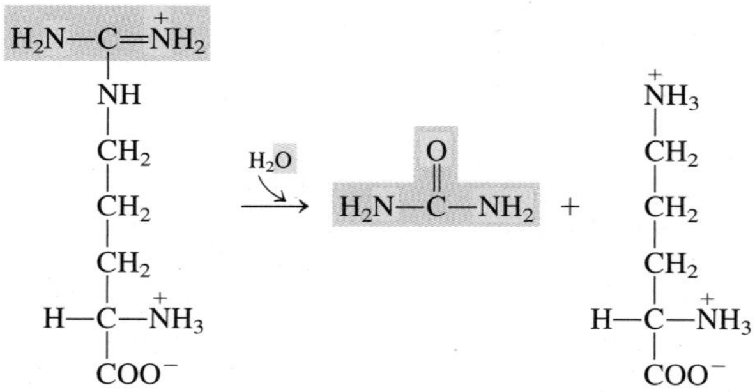

Arginine   Urea   Ornithine

The oxygen atom present in the urea comes from the water involved in the hydrolysis. The ornithine is transported back into the mitochondria, where it becomes available to participate in the urea cycle again.

Humans and most terrestrial animals excrete excess nitrogen as urea. Urea is not, however, the only biochemical means for disposing of excess nitrogen. Aquatic species (bacteria and fish) release ammonia directly into the surrounding water. Birds, terrestrial reptiles, and many insects secrete nitrogen as uric acid; it is the familar white solid in bird droppings. The structure of uric acid, a compound with a purine ring system (Section 17.9), is

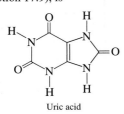

Uric acid

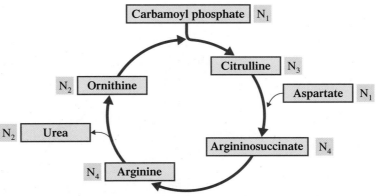

Figure 26.6 The nitrogen content of the various compounds that participate in the urea cycle.

Figure 26.6 analyzes the urea cycle in terms of the nitrogen content of the various compounds that participate in it. The fuel, the N_1 carbamoyl phosphate, condenses with the N_2 ornithine to produce the N_3 citrulline. Next come two N_4 compounds, argininosuccinate and arginine. The N_4 arginine undergoes hydrolysis to produce the N_2 urea and regenerate the N_2 ornithine.

 EXAMPLE 26.4

Relating Urea Cycle Intermediates to Events That Occur in the Urea Cycle

For each of the following urea cycle events identify the urea cycle intermediate that is involved. The urea cycle intermediates are ornithine, citrulline, argininosuccinate, and arginine.

a. Intermediate participates in a condensation reaction.
b. Intermediate is produced at the same time that urea is formed.
c. Intermediate must be transported from the mitochondrial matrix to the cytosol.
d. Intermediate is the product in Step 3 of the urea cycle.

Solution

a. *Citrulline.* The reaction between citrulline and aspartate in Step 2 of the urea cycle is a condensation reaction.
b. *Ornithine.* In Step 4 of the urea cycle arginine undergoes hydrolysis to produce ornithine and urea.
c. *Citrulline.* Citrulline is formed in the mitochondrial matrix, transported across the inner mitochondrial membrane, and reacts with aspartate in the cytosol.
d. *Arginine.* In Step 3 of the urea cycle, in a cleavage reaction, the large molecule argininosuccinate is split into two pieces—the smaller arginine and fumarate molecules.

Practice Exercise 26.4

For each of the following urea cycle events identify the urea cycle intermediate that is involved. The urea cycle intermediates are ornithine, citrulline, argininosuccinate, and arginine.

a. Intermediate reacts with carbamoyl phosphate.
b. Intermediate must be transported from the cytosol to the mitochondrial matrix.
c. Intermediate is a reactant in a hydrolysis reaction.
d. Intermediate is the reactant in Step 2 of the urea cycle.

Answers: a. Ornithine; **b.** Ornithine; **c.** Arginine; **d.** Citrulline

Figure 26.7 Fumarate from the urea cycle enters the citric acid cycle, and aspartate produced from oxalo-acetate of the citric acid cycle enters the urea cycle.

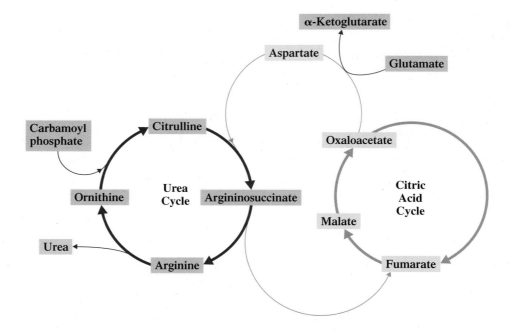

Sulfur-containing amino acids (cysteine and methionine) contain both sulfur and nitrogen. The nitrogen-containing group is lost through transamination and oxidative deamination and processed to urea. The sulfur-containing group is pro-cessed to sulfur dioxide (SO_2), which is then oxidized to sulfate (SO_4^{2-}). The SO_4^{2-} ion, the negative ion from sulfuric acid, is eliminated in urine.

Urea Cycle Net Reaction

The net reaction for urea formation, in which all of the urea cycle intermediates cancel out of the equation, is

$$NH_4^+ + CO_2 + 3ATP + 2H_2O + aspartate \longrightarrow$$
$$urea + 2ADP + AMP + PP_i + 2P_i + fumarate$$

The equivalent of a total of four ATP molecules is expended in the production of one urea molecule. Two ATP molecules are consumed in the production of carbamoyl phosphate, and the equivalent of two ATP molecules is consumed in Step 2 of the urea cycle, where an ATP is hydrolyzed to AMP and PP_i and the PP_i is then further hydrolyzed to two P_i.

Linkage Between the Urea and Citric Acid Cycles

The net equation for urea formation shows fumarate, a citric acid cycle intermediate, as a product. This fumarate enters the citric acid cycle, where it is converted to malate and then to oxaloacetate, which can then be converted to aspartate through transamination. The aspartate then re-enters the urea cycle at Step 2 (see Figure 26.7).

Besides undergoing transamination, the oxaloacetate produced from fumarate of the urea cycle can be (1) converted to glucose via gluconeogenesis, (2) condensed with acetyl CoA to form citrate, or (3) converted to pyruvate.

 26.5 AMINO ACID CARBON SKELETONS

The removal of the amino group of an amino acid by transamination/oxidative deamina-tion (Section 26.3) produces an α-keto acid that contains the carbon skeleton from the amino acid. Each of the 20 amino acid carbon skeletons undergoes a different degradation process. For alanine and serine, the degradation requires a single step. For most carbon arrangements, however, multistep reaction sequences are required. We will not consider the details of these various degradation procedures in this text. It is important, however, to consider the products of these 20 degradation sequences. There are only seven, and each is a compound that we have previously encountered in our discussions of metabolism. The seven degradation products are pyruvate, acetyl CoA, acetoacetyl CoA, α-ketoglutarate, succinyl CoA, fumarate, and oxaloacetate. The last four products are intermediates in the

The 20 standard amino acids are de-graded by 20 different pathways that converge to produce just 7 products (metabolic products).

The Chemical Composition of Urine

Urine is a dilute aqueous solution containing many solutes whose concentrations are dependent on the diet and state of health of the individual. On average, about 4 g of solutes are present in a 100-g urine sample; thus urine is an approximately 4%-by-mass aqueous solution of materials eliminated from the body.

The solutes present in urine are of two general types: organic compounds and inorganic ions. Generally, the organic compounds are more abundant because of the dominance of urea, as shown in the following composition data.

Major Constituents of Urine (for a 1400-mL specimen obtained over a 24-hour period)

Organic Constituents		Inorganic Constituents	
urea	25.0 g	chloride (Cl^-)	6.3 g
creatinine	1.5 g	sodium (Na^+)	3.0 g
amino acids	0.8 g	potassium (K^+)	1.7 g
uric acid	0.7 g	sulfate (SO_4^{2-})	1.4 g
		dihydrogen phosphate ($H_2PO_4^-$)	1.2 g
		ammonium (NH_4^+)	0.8 g
		calcium (Ca^{2+})	0.2 g
		magnesium (Mg^{2+})	0.2 g

Urea, the solute present in the greatest quantity in urine, is odorless and colorless in solution (Section 26.4). (The pale yellow color of urine is due to small amounts of urobilin and related compounds, as discussed in Section 26.7.) Urea is the principal nitrogen-containing end product of protein metabolism.

Creatinine, the second most abundant organic product in urine, is produced from the amino acids arginine, methionine, and glycine. Uric acid is a product of the metabolism of purines from nucleic acids.

The most abundant inorganic constituent of urine is chloride ion. Its primary source is dietary table salt (NaCl). Correspondingly, the second most abundant ion present is sodium ion, the positive ion in table salt. The sulfate ion present in urine comes primarily from the metabolism of sulfur-containing amino acids. Ammonium ions come primarily from the hydrolysis of urea.

Urine is normally slightly acidic, having an average pH value of 6.6. However, the pH range is wide—from 4.5 to 8.0. Fruits and vegetables in the diet tend to raise urine pH, and high-protein foods tend to lower urine pH.

A normal adult excretes 1000–1500 mL of urine daily. Actual urine volume depends on liquid intake and weather. During hot weather, urine volume decreases as a result of increased water loss through perspiration.

Arginine, Citrulline, and the Chemical Messenger Nitric Oxide

A somewhat startling biochemical discovery made during the early 1990s was the existence within the human body of a *gaseous* chemical messenger, the simple diatomic molecule nitric oxide (NO). Its production involves two of the amino acid intermediates of the urea cycle—arginine and citrulline. Arginine reacts with oxygen to produce citrulline and NO. The reaction requires NADPH and the enzyme *nitric oxide synthase* (NOS).

$$O_2 + \underset{\text{Arginine}}{\begin{array}{c} NH_2 \\ | \\ C=\overset{+}{N}H_2 \\ | \\ NH \\ | \\ (CH_2)_3 \\ | \\ HC-\overset{+}{N}H_3 \\ | \\ COO^- \end{array}} \xrightarrow[\substack{\boxed{NADPH} \quad \boxed{NADP^+}}]{\substack{\text{Nitric oxide} \\ \text{synthase}}} \underset{\text{Citrulline}}{\begin{array}{c} NH_2 \\ | \\ C=O \\ | \\ NH \\ | \\ (CH_2)_3 \\ | \\ HC-\overset{+}{N}H_3 \\ | \\ COO^- \end{array}} + \boxed{NO}$$

Even though this reaction involves urea cycle intermediates, it is completely independent of the urea cycle.

Nitric oxide affects many kinds of cells and has particularly striking effects in the following areas:

1. NO helps maintain blood pressure by dilating blood vessels.
2. NO is a chemical messenger in the central nervous system.
3. NO is involved in the immune system's response to invasion by foreign organisms or materials.
4. NO is found in the brain and may be a major biochemical component of long-term memory.

In humans, nitric oxide is the first known biochemical messenger compound that is a gas. It can easily pass through cell membranes by diffusion. No specific receptor or transport system is needed. Because of its extreme reactivity, NO exists for less than 10 seconds before undergoing reaction. This high reactivity prevents it from getting more than 1 millimeter from its site of synthesis.

The action of nitroglycerin, when it is used as a heart medication (for angina pectoris), is now known to be related to NO. Nitric oxide is the active metabolite from nitroglycerin.

Before the discovery of nitric oxide's role as a biochemical messenger, this gas was thought of mainly as a noxious atmospheric gas found in cigarette smoke and smog, as a destroyer of ozone, and as a precursor of acid rain. The contrast between nitric oxide's role in environmental pollution and its function in the human body as a chemical messenger is indeed startling.

citric acid cycle. Figure 26.8 relates these seven degradation products to the amino acids from which they are obtained. Some amino acids appear in more than one box in Figure 26.8. This means either that there is more than one pathway for degradation or that some of the carbon atoms of the skeleton emerge as one product and others as another product.

Amino acids that are degraded to citric acid cycle intermediates can serve as glucose precursors and are called glucogenic. A **glucogenic amino acid** *is an amino acid that has a carbon-containing degradation product that can be used to produce glucose via gluconeogenesis.*

Amino acids that are degraded to acetyl CoA or acetoacetyl CoA can contribute to the formation of fatty acids or ketone bodies and are called ketogenic. A **ketogenic amino acid** *is an amino acid that has a carbon-containing degradation product that can be used to produce ketone bodies.* Even though acetyl CoA can enter the citric acid cycle, there can be no *net* production of glucose from it. Acetyl groups are C_2 species, and such species only maintain the carbon count in the cycle because two CO_2 molecules exit the cycle (Section 23.6). Thus amino acids that are degraded to acetyl CoA (or acetoacetyl CoA) are not glucogenic.

Amino acids that are degraded to pyruvate can be either glucogenic or ketogenic. Pyruvate can be metabolized to either oxaloacetate (glucogenic) or acetyl CoA (ketogenic).

Only two amino acids are purely ketogenic: leucine and lysine. Nine amino acids are both glucogenic and ketogenic: those degraded to pyruvate (see Figure 26.8), as well as tyrosine, phenylalanine, and isoleucine (which have two degradation products). The remaining nine amino acids are purely glucogenic.

Our discussion of glucogenicity and ketogenicity for amino acids points out that ATP production (common metabolic pathway) is not the only fate for amino acid degradation products. They can also be converted to glucose, ketone bodies, or fatty acids (via acetyl CoA).

> Only amino acids that can replenish, either directly or indirectly, oxaloacetate supplies, are glucogenic, that is, can be used to produce glucose through gluconeogenesis.

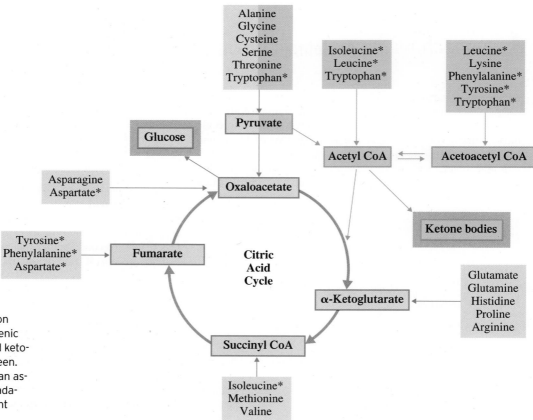

Figure 26.8 Fates of the carbon skeletons of amino acids. Glucogenic amino acids are shaded blue, and ketogenic amino acids are shaded green. Some amino acids (marked with an asterisk) have more than one degradation pathway, and thus are present more than once in the diagram.

26.6 AMINO ACID BIOSYNTHESIS

The classification of amino acids as essential or nonessential for humans (Section 20.2) roughly parallels the number of steps in their biosynthetic pathways and the energy required for their synthesis. The nonessential amino acids can be made in 1–3 steps. The essential ones have biosynthetic pathways that require 7–10 steps, judging on the basis of observations of their synthesis in microorganisms. Most bacteria and plants can synthesize all the amino acids by pathways not present in humans. Plants, consumed as food, are the major source of the essential amino acids in humans and animals.

There is considerable variation in biosynthetic pathways for amino acids among different species. By contrast, the basic pathways of carbohydrate and lipid metabolism are almost universal.

The starting materials for the biosynthesis of the 11 nonessential amino acids are the glycolysis intermediates 3-phosphoglycerate and pyruvate and the citric acid cycle intermediates oxaloacetate and α-ketoglutarate (see Figure 26.9).

Three of the nonessential amino acids—alanine, aspartate, and glutamate—are biosynthesized by transamination (Section 26.3) of the appropriate α-keto acid starting material.

$$CH_3-\overset{\overset{O}{\|}}{C}-COO^- \xrightarrow{\text{Transamination}} CH_3-\overset{\overset{+}{\underset{|}{NH_3}}}{CH}-COO^-$$

Pyruvate → Alanine

$$^-OOC-CH_2-\overset{\overset{O}{\|}}{C}-COO^- \xrightarrow{\text{Transamination}} {}^-OOC-CH_2-\overset{\overset{+}{\underset{|}{NH_3}}}{CH}-COO^-$$

Oxaloacetate → Aspartate

$$^-OOC-CH_2-CH_2-\overset{\overset{O}{\|}}{C}-COO^- \xrightarrow{\text{Transamination}} {}^-OOC-CH_2-CH_2-\overset{\overset{+}{\underset{|}{NH_3}}}{CH}-COO^-$$

α-Ketoglutarate → Glutamate

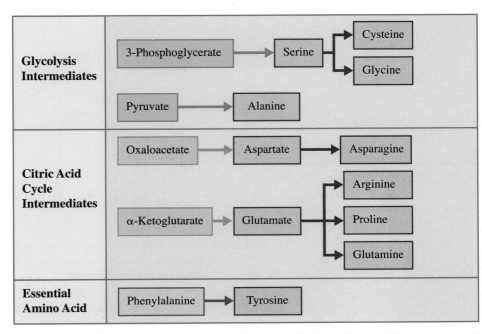

Glycolysis Intermediates	3-Phosphoglycerate → Serine → Cysteine / Glycine
	Pyruvate → Alanine
Citric Acid Cycle Intermediates	Oxaloacetate → Aspartate → Asparagine
	α-Ketoglutarate → Glutamate → Arginine / Proline / Glutamine
Essential Amino Acid	Phenylalanine → Tyrosine

Figure 26.9 A summary of the starting materials for the biosynthesis of the 11 nonessential amino acids.

PKU is characterized by elevated blood levels of phenylalanine and phenylpyruvate. The physical consequence of PKU is damage to *developing* brain cells. In children up to six years old, PKU leads to retarded mental development. The major defense against PKU is mandatory screening of newborns to identify the one in every 20,000 who is afflicted and then restricting those children's dietary phenylalanine intake to that needed for protein synthesis until they are six years old. After that age, brain cells are not so susceptible to the toxic effect of phenylpyruvate.

The nonessential amino acid tyrosine is obtained from the essential amino acid phenylalanine in a one-step oxidation that involves molecular O_2, NADPH, and the enzyme *phenylalanine hydroxylase.* Lack of this enzyme causes the metabolic disease phenylketonuria (PKU).

 26.7 HEMOGLOBIN CATABOLISM

Red blood cells are highly specialized cells whose primary function is to deliver oxygen to, and remove carbon dioxide from, body tissues. Mature red blood cells have no nucleus or DNA. Instead, they are filled with the red pigment hemoglobin. Red blood cell formation occurs in the bone marrow, and approximately 200 billion new red blood cells are formed daily. The life span of a red blood cell is about 4 months.

The oxygen-carrying ability of red blood cells is due to the protein hemoglobin present in such cells (see Figure 26.10). Hemoglobin is a conjugated protein (Section 20.8); the protein portion is called *globin,* and the prosthetic group (nonprotein portion) is *heme.* Heme contains four pyrrole groups (Section 17.9) joined together with an iron atom in the center.

Figure 26.10 A molecular model of the protein hemoglobin.

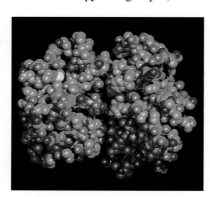

Heme

The tetrapyrrole heme ring is the only component of hemoglobin that is not reused by the body.

It is the iron atom in heme that interacts with O_2, forming a reversible complex with it. This complexation increases the amount of O_2 that the blood can carrry by a factor of 80 over that which simply "dissolves" in the blood.

Old red blood cells are broken down in the spleen (primary site) and liver (secondary site). Part of this process is degradation of hemoglobin. The globin protein is hydrolyzed to amino acids, which become part of the amino acid pool (Section 26.2). The iron atom of heme becomes part of *ferritin,* an iron-storage protein, which saves the iron for use in the biosynthesis of new hemoglobin molecules. The tetrapyrrole carbon arrangement of

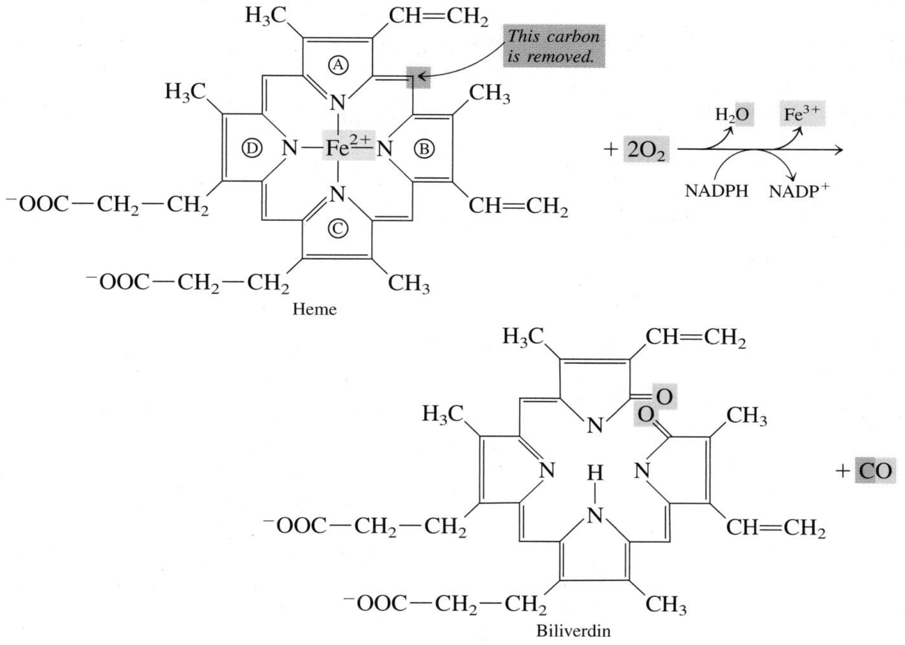

Heme

Biliverdin

heme is degraded to *bile pigments* that are eliminated in feces and to a lesser extent in urine.

Degradation of heme begins with a ring-opening reaction in which a single carbon atom is removed. The product is called *biliverdin*.

This reaction has several important characteristics. (1) Molecular oxygen, O_2, is required as a reactant. (2) Ring opening releases the iron atom to be incorporated into ferritin. (3) The product containing the excised carbon atom is *carbon monoxide* (a substance toxic to the human body). The carbon monoxide so produced reacts with functioning hemoglobin, forming a CO–hemoglobin complex; this decreases the oxygen-carrying ability of the blood. CO–hemoglobin complexes are very stable; CO release to the lungs is a slow process. An alternative rendering of the structure of biliverdin is

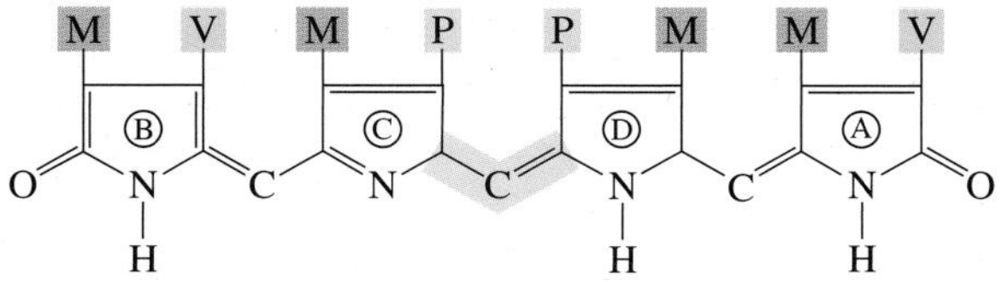

> The level of carbon monoxide produced in the first step of hemoglobin degradation is sufficient to complex 1% of the oxygen-binding sites of the blood's hemoglobin.

Biliverdin

M = —CH₃ (methyl)
V = —CH=CH₂ (vinyl)
P = —CH₂—CH₂—COO⁻ (propionate)

This structure employs a notation, common in heme chemistry, in which letters are used to denote attachments to the pyrrole rings; such notation easily distinguishes the attachments. The structure's linear arrangement of pyrrole rings also saves space compared to the heme-like representation of the rings. However, the linear structure incorrectly implies that the arrangement of the pyrrole rings that results from the ring opening is linear (straight-line); rather, the pyrrole rings actually have a hemi-like arrangement.

In the second step of heme degradation, biliverdin is converted to bilirubin. This change involves reduction of the central methylene bridge of biliverdin.

In 2002 it was discovered that bilirubin has antioxidant properties. It protects against peroxyl radicals (Section 23.11) by being oxidized back to biliverdin. Its antioxidant properties are significantly better than those of glutathione (Section 20.7), the molecule believed for 80 years to be the most important cellular antioxidant.

Bilirubin is found only in low concentrations in cells but in higher concentrations in blood. This new research suggests that bilirubin is probably the major antioxidant protector for cell membranes, while glutathione protects components inside cells.

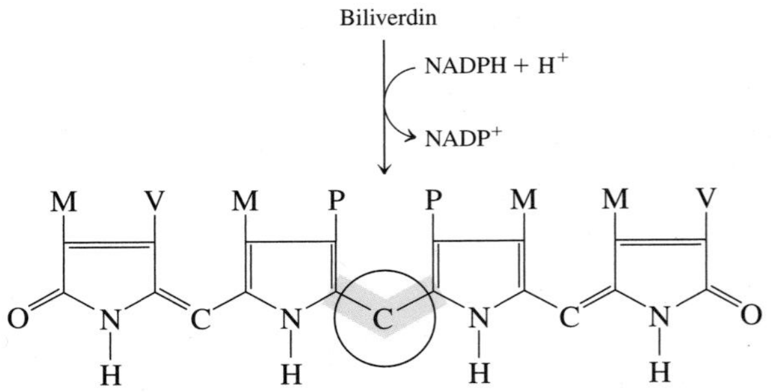

The change from heme to biliverdin to bilirubin usually occurs in the spleen. The bilirubin is then transported by serum albumin to the liver, where it is rendered more water-soluble by the attachment of sugar residues to its propionate side chains (P side chains). The solubilizing sugar is *glucuronate* (glucose with a —COO⁻ group on C-6 instead of a —CH₂OH group).

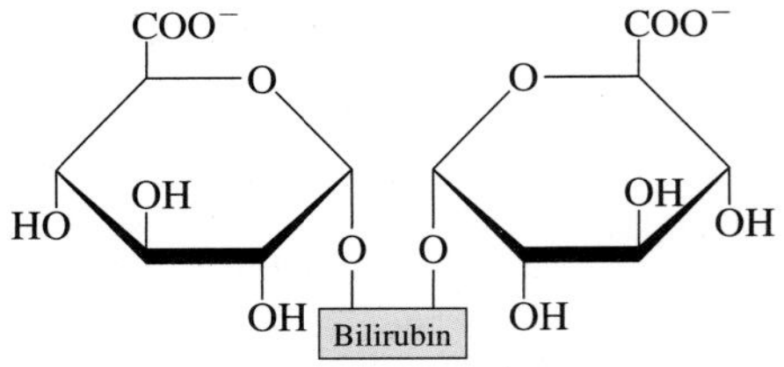

Bilirubin diglucuronide

The solubilized bilirubin is excreted from the liver in bile, which flows into the small intestine. Here the bilirubin diglucuronide is changed, in a multistep process, to either stercobilin for excretion in feces or urobilin for excretion in urine. Both stercobilin and urobilin still have tetrapyrrole structures (Figure 26.11). Intestinal bacteria are primarily responsible for the changes that produce stercobilin and urobilin.

The first part of the names *biliverdin* and *bilirubin* and the last part of the names *stercobilin* and *urobilin* all come from the Latin *bilis,* which means "bile." As for the other parts of the names:

1. Latin *virdis* means "green"; biliverdin = "green bile."
2. Latin *rubin* means "red"; bilirubin = "red bile."
3. Latin *urina* means "urine"; urobilin = "urine bile."
4. Latin *sterco* means "dung"; stercobilin = "dung bile."

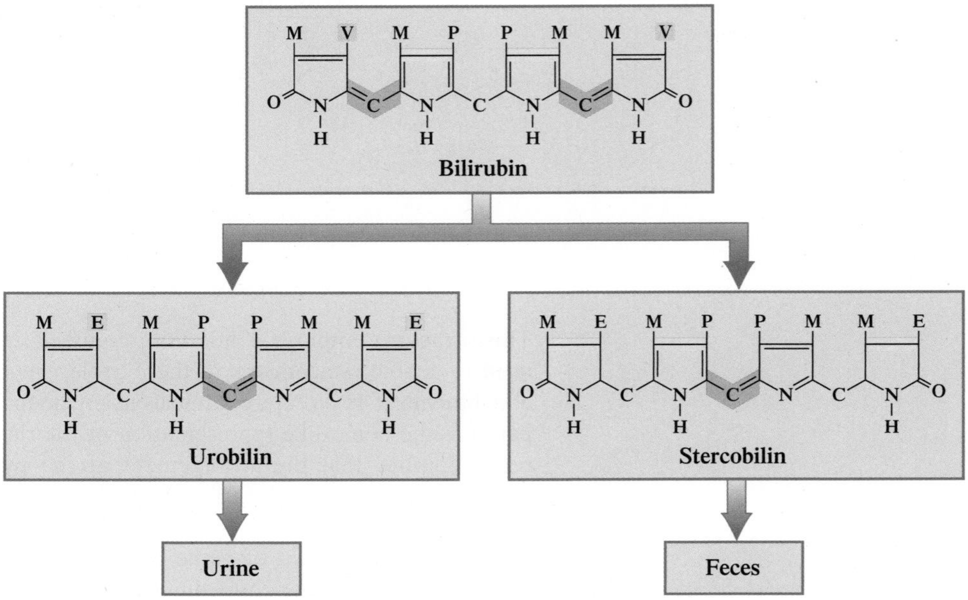

Figure 26.11 Stercobilin and urobilin have structures closely resembling that of bilirubin. Changes include reduction of vinyl (V) groups to ethyl (E) groups and reduction of —CH₂— bridges.

The word *jaundice* comes from the French *jaune,* which means "yellow."

A mild form of jaundice is common among premature infants because of underdeveloped liver function. Treatment involves the use of white or ultraviolet light, which breaks the bilirubin down to simpler compounds that are more easily excreted.

Bile Pigments

The tetrapyrrole degradation products obtained from heme are known as bile *pigments* because they are secreted with the bile (Section 25.1), and most of them are highly colored. A **bile pigment** *is a colored tetrapyrrole degradation product present in bile.* Biliverdin and bilirubin are, respectively, green and reddish orange in color. Stercobilin has a brownish hue and is the compound that gives feces their characteristic color. Urobilin is the pigment that gives urine its characteristic yellow color. Normally, the body excretes 1–2 mg of bile pigments in urine daily and 250–350 mg of bile pigments in feces daily.

When the body is functioning properly, the degradation of heme in the spleen to bilirubin and the removal of bilirubin from the blood by the liver balance each other.

Jaundice is the condition that occurs when this balance is upset such that bilirubin concentrations in the blood become higher than normal. The skin and the white of the

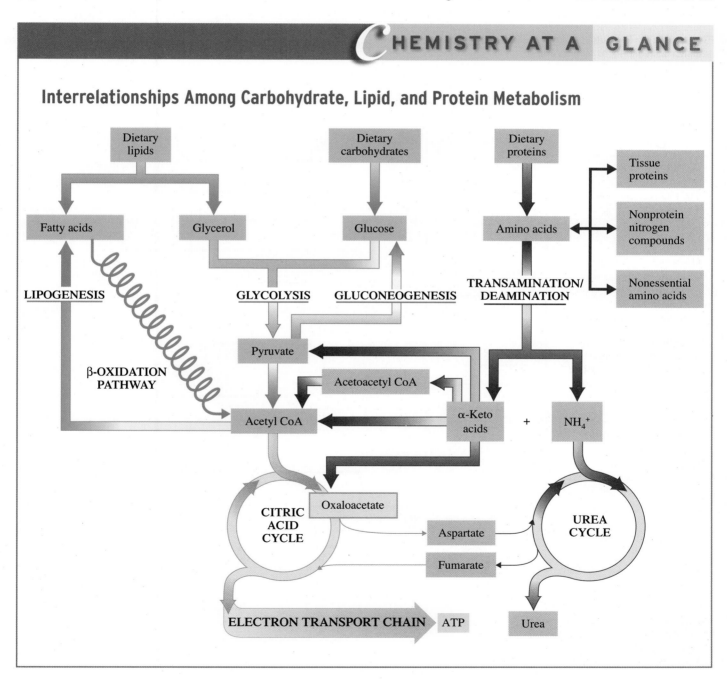

eyes acquire a yellowish tint because of the excess bilirubin in the blood. Jaundice can occur as a result of liver diseases, such as infectious hepatitis and cirrhosis, that decrease the liver's ability to process bilirubin; from spleen malfunction, in which heme is degraded more rapidly than it can be absorbed by the liver; and from gallbladder malfunction, usually from an obstruction of the bile duct.

The local coloration associated with a deep bruise is also related to the pigmentation associated with heme, biliverdin, and bilirubin. The changing color of the bruise as it heals reflects the dominant degradation product present at the time as the tissue repairs itself.

 INTERRELATIONSHIPS AMONG METABOLIC PATHWAYS

In this chapter and the previous two chapters, we have considered metabolic pathways of carbohydrates, lipids, and proteins. These pathways are not independent of each other but rather are integrally linked, as shown in the Chemistry at a Glance feature on page 895.

Figure 26.12 The human body's response to feasting, to fasting, and to starvation.

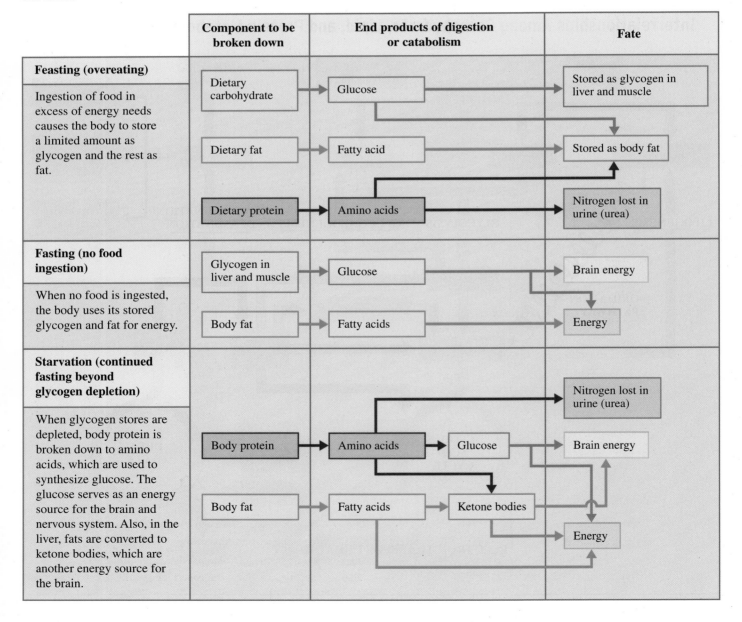

The numerous connections among pathways mean that a change in one pathway can affect many other pathways.

A good illustration of the interrelationships among pathways emerges from comparing the processes of eating (feasting), not eating for a short period (fasting), and not eating for a prolonged period (starvation). Figure 26.12 shows how the body responds to each of these situations.

CONCEPTS TO REMEMBER

Protein digestion and absorption. Digestion of proteins involves the hydrolysis of the peptide bonds that link amino acids to each other. This process begins in the stomach and is completed in the small intestine. The amino acids released by digestion are absorbed through the intestinal wall into the bloodstream (Section 26.1).

Amino acid pool. The amino acid pool within cells consists of varying amounts of each of the 20 standard amino acids found in proteins (Section 26.2).

Amino acid utilization. Amino acids from the amino acid pool are used for protein synthesis, synthesis of nonprotein nitrogen compounds, synthesis of nonessential amino acids, and energy production (Section 26.2).

Transamination. A transamination reaction is an enzyme-catalyzed transfer of an amino group from an α-amino acid to an α-keto acid. Transamination is a step in obtaining energy from amino acids (Section 26.3).

Oxidative deamination. An oxidative deamination reaction is a reaction in which an α-amino acid is converted into an α-keto acid, accompanied by the release of a free ammonium ion. Oxidative deamination is a step in obtaining energy from amino acids (Section 26.3).

Urea cycle. The urea cycle is the metabolic pathway that converts ammonium ions into urea. This cycle processes the ammonium ions in the form of carbamoyl phosphate, a compound formed from CO_2, $NH_4{}^+$, ATP, and H_2O (Section 26.4).

Amino acid carbon skeletons. Amino acid carbon skeletons (keto acids) are classified as glucogenic or ketogenic on the basis of their catabolic pathways. Glucogenic amino acids are degraded to intermediates of the citric acid cycle and can be used for glucose synthesis. Ketogenic amino acids are degraded into acetoacetyl CoA or acetyl CoA and can be used to make ketone bodies (Section 26.5).

Amino acid biosynthesis. Amino acid biosynthesis is the process in which the body synthesizes amino acids from intermediates of the glycolysis pathway and the citric acid cycle. Eleven amino acids can be synthesized by the body. The other nine amino acids, called essential amino acids, must be obtained from the diet (Section 26.6).

Hemoglobin catabolism. Hemoglobin from red blood cells undergoes a stepwise degradation to biliverdin, to bilirubin, and then to bile pigments that are excreted from the body (Section 26.7).

KEY REACTIONS AND EQUATIONS

1. Digestion of protein (Section 26.1)

$$\text{Protein} \xrightarrow[\text{and HCl}]{\text{Pancreatic enzymes}} \text{amino acids}$$

2. Transamination (Section 26.3)

$$R_1\!-\!\overset{\overset{+}{N}H_3}{\underset{}{C}H}\!-\!COO^- + R_2\!-\!\overset{O}{\overset{\|}{C}}\!-\!COO^- \longrightarrow$$
$$R_1\!-\!\overset{O}{\overset{\|}{C}}\!-\!COO^- + R_2\!-\!\overset{\overset{+}{N}H_3}{\underset{}{C}H}\!-\!COO^-$$

3. Oxidative deamination (Section 26.3)

$$R\!-\!\overset{\overset{+}{N}H_3}{\underset{}{C}H}\!-\!COO^- + NAD^+ + H_2O \longrightarrow$$
$$R\!-\!\overset{O}{\overset{\|}{C}}\!-\!COO^- + NH_4{}^+ + NADH + H^+$$

4. Formation of urea (Section 26.4)

$$NH_4{}^+ + CO_2 + 3ATP + 2H_2O + \text{aspartate} \longrightarrow$$
$$\text{urea} + \text{fumarate} + 2ADP + AMP + PP_i + 2P_i$$

EXERCISES and PROBLEMS

The members of each pair of problems in this section test similar material.

● **Protein Digestion and Absorption (Section 26.1)**

26.1 The first step in protein digestion is denaturation. Where does denaturation occur in the body, and what is the denaturant?

26.2 What is the first digestive enzyme that protein encounters, and where does this encounter take place?

26.3 What is the relationship between pepsinogen and pepsin?

26.4 What is the relationship between trypsinogen and trypsin?

26.5 Contrast gastric juice and pancreatic juice in terms of pH.

26.6 Contrast gastric juice and pancreatic juice in terms of enzymes present.

26.7 Absorption of amino acids through the intestinal wall requires a transport system. Explain.

26.8 The passage of small polypeptides through the intestinal wall is particularly important in infants. Explain.

● Amino Acid Utilization (Section 26.2)

26.9 What is the amino acid pool?

26.10 What are the three major sources of amino acids for the amino acid pool?

26.11 What is protein turnover?

26.12 The protein turnover rate is not the same for all proteins. Explain.

26.13 What is the difference between a positive nitrogen balance and a negative nitrogen balance?

26.14 What happens to the nitrogen balance during a period of fasting?

26.15 What happens to the nitrogen balance when the diet is lacking in one of the essential amino acids?

26.16 What happens to the nitrogen balance of a pregnant woman?

26.17 What four types of processes draw amino acids out of the amino acid pool?

26.18 What percent of amino acid utilization from the amino acid pool is for protein synthesis?

26.19 Classify each of the following amino acids as essential or nonessential.
  a. Lysine   b. Arginine
  c. Serine   d. Tryptophan

26.20 Classify each of the following amino acids as essential or nonessential.
  a. Proline   b. Asparagine
  c. Glutamic acid d. Tyrosine

26.21 Which of the amino acids in Problem 26.19 can be biosynthesized in the human body?

26.22 Which of the amino acids in Problem 26.20 can be biosynthesized in the human body?

● Transamination and Oxidative Deamination (Section 26.3)

26.23 In general terms, what are the two reactants in a transamination reaction?

26.24 In general terms, what are the two products in a transamination reaction?

26.25 Write structural equations for the transamination reactions that involve the following pairs of reactants.
  a. Threonine and pyruvate
  b. Alanine and oxaloacetate
  c. Glycine and α-ketoglutarate
  d. Threonine and α-ketoglutarate

26.26 Write structural equations for the transamination reactions that involve the following pairs of reactants.
  a. Threonine and oxaloacetate
  b. Glycine and pyruvate
  c. Alanine and oxaloacetate
  d. Isoleucine and α-ketoglutarate

26.27 What are the two α-keto acids that are most often reactants in transamination reactions?

26.28 The net effect of transamination is to collect the amino groups from a variety of α-amino acids into the compound glutamate. Explain.

26.29 What is the function of pyridoxal phosphate in transamination processes?

26.30 Which one of the B vitamins is important in the process of transamination?

26.31 Describe the process of oxidative deamination.

26.32 What coenzyme is required for an oxidative deamination reaction?

26.33 How does oxidative deamination differ from transamination?

26.34 What do the processes of oxidative deamination and transamination have in common?

26.35 Draw the structure of the α-keto acid produced from the oxidative deamination of each of the following amino acids.
  a. Glutamate   b. Cysteine
  c. Alanine   d. Phenylalanine

26.36 Draw the structure of the α-keto acid produced from the oxidative deamination of each of the following amino acids.
  a. Glycine   b. Leucine
  c. Aspartate   d. Tyrosine

26.37 The following α-keto acid can be used as a substitute for a particular *essential* amino acid in the diet. Explain how this is possible, and draw the structure of the essential amino acid.

$$CH_3-CH-CH_2-\overset{\overset{\displaystyle O}{\|}}{C}-COO^-$$
$$|$$
$$CH_3$$

26.38 The following α-keto acid can be used as a substitute for a particular *essential* amino acid in the diet. Explain how this is possible, and draw the structure of the essential amino acid.

$$CH_3-CH_2-CH-\overset{\overset{\displaystyle O}{\|}}{C}-COO^-$$
$$|$$
$$CH_3$$

26.39 Give the *name* of the compound produced from each reactant or the reactant needed to produce each product using transamination.
  a. Oxaloacetate $\longrightarrow$?
  b. ? $\longrightarrow$ α-ketoglutarate
  c. Alanine $\longrightarrow$?
  d. ? $\longrightarrow$ glutamate

26.40 Give the *name* of the compound produced from each reactant or the reactant needed to produce each product using transamination.
  a. Pyruvate $\longrightarrow$?   b. ? $\longrightarrow$ oxaloacetate
  c. Aspartate $\longrightarrow$?   d. ? $\longrightarrow$ alanine

● The Urea Cycle (Section 26.4)

26.41 Draw the chemical structure of urea.

26.42 What are some of the physical characteristics of urea?

26.43 In what chemical form do ammonium ions enter the urea cycle?

26.44 What are the chemical reactants for the formation of carbamoyl phosphate?

26.45 What is a carbamoyl group?

26.46 Draw the structure of the molecule carbamoyl phosphate.

26.47 How do the structures of the three amino acids involved as intermediates in the urea cycle differ from each other?

26.48 Three amino acids are involved as intermediates in the urea cycle. Name them and classify them as standard or nonstandard amino acids.

26.49 Name the compound that enters the urea cycle by combining with
a. ornithine.   b. citrulline.

26.50 Identify the first reaction of the urea cycle that occurs in the
a. mitochondrial matrix.
b. cytosol.

26.51 What substance is the "fuel" for the urea cycle?

26.52 If the urea cycle were named in the same way as the citric acid cycle, what would the cycle's name be?

26.53 Characterize each of the following "urea cycle compounds" in terms of its nitrogen content (N_1, N_2, N_3, or N_4).
a. Ornithine   b. Citrulline
c. Aspartate   d. Argininosuccinate

26.54 Characterize each of the following "urea cycle compounds" in terms of its nitrogen content (N_1, N_2, N_3, or N_4).
a. Carbamoyl phosphate
b. Ammonium ion
c. Aspartate
d. Urea

26.55 In each of the following pairs of compounds associated with the urea cycle, specify which one is encountered first in the cycle.
a. Citrulline and arginine
b. Ornithine and aspartate
c. Argininosuccinate and fumarate
d. Carbamoyl phosphate and citrulline

26.56 In each of the following pairs of compounds associated with the urea cycle, specify which one is encountered first in the cycle.
a. Carbamoyl phosphate and fumarate
b. Argininosuccinate and arginine
c. Ornithine and aspartate
d. Citrulline and ATP

26.57 How much energy, in terms of ATP, is expended in the synthesis of a molecule of urea?

26.58 What are the sources of the carbon atom and the two nitrogen atoms in urea?

26.59 What is the fate of the fumarate formed in the urea cycle?

26.60 Explain how the urea cycle is linked to the citric acid cycle.

● **Amino Acid Carbon Skeletons (Section 26.5)**

26.61 What are the four possible degradation products of the carbon skeletons of amino acids that are citric acid cycle intermediates?

26.62 What are the three possible degradation products of the carbon skeletons of amino acids that are not citric acid cycle intermediates?

26.63 With the help of Figure 26.8, write the name of the compound (or compounds) to which each of the following amino acid carbon skeletons is metabolized.
a. Leucine   b. Isoleucine
c. Aspartate   d. Arginine

26.64 With the help of Figure 26.8, write the name of the compound (or compounds) to which each of the following amino acid carbon skeletons is metabolized.
a. Serine   b. Tyrosine
c. Tryptophan   d. Histidine

26.65 What degradation characteristics do all purely glucogenic amino acids share?

26.66 What degradation characteristics do all purely ketogenic amino acids share?

● **Amino Acid Biosynthesis (Section 26.6)**

26.67 What compound is a major source of amino groups in amino acid biosynthesis?

26.68 How does transamination play a role in both catabolism and anabolism of amino acids?

26.69 What are the five starting materials for the biosynthesis of the 11 nonessential amino acids?

26.70 What is a major difference between the biosynthetic pathways for the essential and the nonessential amino acids?

● **Hemoglobin Catabolism (Section 26.7)**

26.71 What happens to the globin produced from the breakdown of hemoglobin?

26.72 What happens to the iron (Fe^{2+}) produced from the breakdown of hemoglobin?

26.73 What are the structural differences between heme and biliverdin?

26.74 What are the structural differences between biliverdin and bilirubin?

26.75 Arrange the following substances in the order in which they are encountered during the catabolism of heme: bilirubin, urobilin, biliverdin, and bilirubin diglucuronide.

26.76 Carbon monoxide is a by-product of the degradation of heme. At what point in the degradation process is it formed, and what happens to it once it is formed?

26.77 Which bile pigment is responsible for the yellow color of urine?

26.78 Which bile pigment is responsible for the brownish-red color of feces?

26.79 What chemical condition is responsible for jaundice?

26.80 What physical conditions cause jaundice?

● **Interrelationships of Metabolic Pathways (Section 26.8)**

26.81 Briefly explain how the carbon atoms from amino acids can end up in ketone bodies.

26.82 Briefly explain how the carbon atoms from amino acids can end up in glucose.

26.83 How are the amino acids from protein "processed" when they are present in amounts that exceed the body's needs?

26.84 How are the amino acids from protein "processed" when an individual is in a state of starvation?

ADDITIONAL PROBLEMS

26.85 In which of the processes (1) urea cycle, (2) hemoglobin catabolism, and (3) transamination reactions would each of the following molecules be encountered?
a. Citrulline
b. Carbon monoxide
c. Pyruvate
d. Urobilin

26.86 Characterize each of the following molecules as a possible reactant, product, or enzyme of (1) transamination, (2) oxidative deamination, or (3) both transamination and oxidative deamination.
a. Arginine
b. Glutamate
c. α-Ketoglutarate
d. Ammonium ion

26.87 Arrange the following events in the order in which they occur in the digestive process for proteins.
(1) Peptide bonds are hydrolyzed with the help of pepsin.
(2) Peptide bonds are hydrolyzed with the help of trypsin.
(3) Large polypeptides pass from the stomach into the small intestine.
(4) Amino acids pass through the intestinal wall into the bloodstream.

26.88 With the help of Figure 26.8 and the given conversion information, classify each of the following amino acids as (1) ketogenic but not glucogenic, (2) glucogenic but not ketogenic, (3) both ketogenic and glucogenic, or (4) neither ketogenic nor glucogenic.
a. Alanine is converted to pyruvate.
b. Aspartate is converted to either fumarate or oxaloacetate.
c. Lysine is converted to acetoacetyl CoA.
d. Isoleucine is converted to either succinyl CoA or acetyl CoA.

26.89 Indicate whether each of the following statements refers to *transamination* or *deamination*.
a. Both an amino acid and a keto acid are reactants.
b. An amino acid and water are reactants.
c. The ammonium ion is a product.
d. An amino acid is produced from a keto acid.

26.90 Which of the compounds (1) ornithine, (2) citrulline, (3) argininosuccinate, and (4) arginine is associated with each of the following urea cycle "occurrences"?
a. Reacts with carbamoyl phosphate
b. Reacts with water to produce urea
c. Reacts with aspartate
d. Fumarate is a product of its "breakup."

26.91 Indicate whether each of the following statements is true or false.
a. Glutamate is a very abundant amino acid in the amino acid pool.
b. Pyruvate is a compound that participates in both the urea cycle and the citric acid cycle.
c. Citrulline, a participant in the urea cycle, is a nonstandard amino acid.
d. Glutamate is a reactant in oxidative deamination.

26.92 Which of the heme degradation products (1) bilirubin, (2) biliverdin, (3) stercobilin, and (4) urobilin is associated with each of the following heme degradation characterizations?
a. CO is produced at the same time as this substance.
b. The buildup of this substance in the blood produces jaundice.
c. Molecular O_2 is involved in the reaction that produces this substance.
d. This degradation product gives feces its characteristic color.

*M*ultiple-Choice Practice Test

26.93 Amino acid metabolism differs from that of carbohydrates and triacylglycerols in what way?
a. There is no storage form for amino acids in the body.
b. Amino acids cannot be used for energy production.
c. Amino acids cannot be converted to acetyl CoA.
d. All metabolic intermediates contain the element nitrogen.

26.94 Which of the following is not a use for the amino acids present in the body's amino acid pool?
a. Synthesis of proteins
b. Synthesis of nonprotein nitrogen-containing substances
c. Synthesis of nonessential amino acids
d. Synthesis of essential amino acids

26.95 Which of the following is always one of the reactants in a transamination reaction?
a. A keto acid
b. Glycerol
c. Ammonia
d. Ammonium ion

26.96 Which of the following is always a reactant in an oxidative deamination reaction?
a. Ammonium ion
b. Water
c. NADH
d. A keto acid

26.97 Which of the following statements concerning the compound urea is *incorrect*?
a. It is a white solid in the pure state.
b. It is very soluble in water.
c. It gives urine its odor and color.
d. Two —NH_2 groups are present in its structure.

26.98 Which of the following compounds is not a reactant in the formation of carbamoyl phosphate?
a. Carbon dioxide
b. Urea
c. Water
d. Ammonium ion

26.99 In which of the following pairs of amino acids are both members of the pair nonstandard amino acids?
a. Arginine and citrulline
b. Arginine and ornithine
c. Citrulline and ornithine
d. Aspartate and glutamate

26.100 Which of the following statements concerning amino acid "carbon skeleton" degradation is correct?
a. Each amino acid is degraded to a different product.
b. All amino acids are degraded to the same product.

c. Each amino acid is degraded by a different metabolic pathway.

d. All glucogenic amino acids are degraded to the same product.

26.101 Which of the following is produced in the first step of the degradation of the heme portion of hemoglobin?

a. Molecular O_2

b. Carbon monoxide

c. Individual pyrrole groups

d. Bilirubin

26.102 In which of the following is the compound pyridoxal phosphate encountered?

a. The urea cycle

b. Hemoglobin catabolism

c. Transamination reactions

d. Oxidative deamination reactions

Answers to Selected Exercises

Chapter 1 **1.1** has mass, occupies space **1.3** (a) matter (b) matter (c) energy (d) energy (e) matter (f) matter **1.5** (a) shape (b) volume **1.7** (a) does not, yes (b) does, no (c) does, yes (d) does, yes **1.9** (a) physical (b) chemical (c) chemical (d) physical **1.11** (a) chemical (b) physical (c) chemical (d) physical **1.13** (a) physical (b) physical (c) chemical (d) physical **1.15** (a) chemical (b) physical (c) chemical (d) physical **1.17** (a) physical (b) physical (c) chemical (d) physical **1.19** (a) false (b) true (c) false (d) true **1.21** (a) heterogeneous mixture (b) homogeneous mixture (c) pure substance (d) heterogeneous mixture **1.23** (a) homogeneous mixture, one phase (b) heterogeneous mixture, two phases (c) heterogeneous mixture, three phases (d) heterogeneous mixture, three phases **1.25** (a) compound (b) compound (c) classification not possible (d) classification not possible **1.27** (a) A, classification not possible; B, classification not possible; C, compound (b) D, compound; E, classification not possible; F, classification not possible; G, classification not possible **1.29** (a) true (b) false (c) false (d) false **1.31** (a) true (b) false (c) false (d) false **1.33** (a) true (b) true (c) false (d) true **1.35** (a) more abundant (b) less abundant (c) less abundant (d) more abundant **1.37** (a) nitrogen (b) nickel (c) lead (d) tin **1.39** (a) Al (b) Ne (c) H (d) U **1.41** (a) Na, S (b) Mg, Mn (c) Ca, Cd (d) As, Ar **1.43** (a) no (b) yes (c) yes (d) no **1.45** (a) heteroatomic molecules (b) heteroatomic molecules (c) homoatomic molecules (d) heteroatomic molecules **1.47** (a) triatomic molecules (b) triatomic molecules (c) diatomic molecules (d) diatomic molecules **1.49** (a) compound (b) compound (c) element (d) compound **1.51** (a) true (b) false; Triatomic molecules must contain at least one kind of atom. (c) true (d) false; Both homoatomic and heteroatomic molecules may contain three or more atoms. **1.53** (a) ⬤⬤ (b) ⬤⬤⬤ (c) ⬤⬤
(d) ⬤⬤ **1.55** (a) H_2O (b) CO_2 (c) O_2 (d) CO **1.57** (a) compound (b) compound (c) element (d) element **1.59** (a) lithium, chlorine, oxygen (b) carbon, oxygen (c) cobalt (d) sulfur **1.61** (a) $C_{12}H_{22}O_{11}$ (b) $C_8H_{10}N_4O_2$ **1.63** (a) HCN (b) H_2SO_4 **1.65** (a) $BaCl_2$ (b) HNO_3 (c) Na_3PO_4 (d) $Mg(OH)_2$ **1.67** (a) diagram I (b) diagram III (c) diagram II (d) diagrams I and IV **1.68** (a) diagram III (b) diagrams I and IV (c) diagram I (d) diagram III **1.69** (a) element (b) compound (c) mixture (d) compound **1.70** (a) homogeneous mixture (b) heterogeneous mixture (c) compound (d) compound **1.71** (a) element (b) mixture (c) mixture (d) compound **1.72** (a) changes 3 and 4 (b) changes 1 and 2 (c) change 2 (d) change 1 **1.73** (a) B-Ar-Ba-Ra (b) Eu-Ge-Ne (c) He-At-H-Er (d) Al-La-N **1.74** (a) same, both 4 (b) more, 6 and 5 (c) same, both 5 (d) fewer, 13 and 15 **1.75** (a) $1 + 2 + x = 6; x = 3$ (b) $2 + 3 + 3x = 17; x = 4$ (c) $1 + x + x = 5; x = 2$ (d) $x + 2x + x = 8; x = 2$ **1.76** (a) 2 (N_2, NH_3) (b) 4 (N, H, C, Cl) (c) 110; 5(2 + 6 + 4 + 5 + 5) (d) 56; 4(4 + 3 + 4 + 3) **1.77** b **1.78** a **1.79** c **1.80** c **1.81** d **1.82** d **1.83** c **1.84** b **1.85** d **1.86** c

Chapter 2 **2.1** It is easier to use because it is decimal based. **2.3** (a) kilo (b) milli (c) micro (d) deci **2.5** (a) centimeter (b) kiloliter (c) microliter (d) nanogram **2.7** (a) nanogram, milligram, centigram (b) kilometer, megameter, gigameter (c) picoliter, microliter, deciliter (d) microgram, milligram, kilogram **2.9** 60 minutes is a defined (exact) number and 60 feet is a measured (inexact) number. **2.11** (a) exact (b) exact (c) inexact (d) exact **2.13** (a) inexact (b) exact (c) exact (d) inexact **2.15** (a) 0.001 (b) 1 (c) 0.0001 (d) 10 **2.17** (a) 0.1°C (b) 0.01 mL (c) 1 mL (d) 0.1 mm **2.19** (a) between 40,000 and 60,000

(b) between 49,000 and 51,000 (c) between 49,900 and 50,100 (d) between 49,990 and 50,010 **2.21** (a) 0.1 (b) 0.1 **2.23** (a) 2.70 cm (b) 27 cm **2.25** (a) 4 (b) 1 or 4 (c) 2 (d) 3 **2.27** (a) 4 (b) 2 (c) 4 (d) 3 (e) 5 (f) 4 **2.29** (a) same (b) different (c) same (d) same **2.31** (a) the last zero (b) the 2 (c) the last 1 (d) the 4 (e) the last zero (f) the last zero **2.33** (a) ± 0.001 (b) ± 0.0001 (c) ± 0.00001 (d) ± 100 (e) ± 0.001 (f) ± 0.00001 **2.35** (a) 0.351 (b) 653,900 (c) 22.556 (d) 0.2777 **2.37** (a) 2 (b) 2 (c) 2 (d) 2 **2.39** (a) 0.0080 (b) 0.0143 (c) 14 (d) 0.182 (e) 1.1 (f) 5720 **2.41** (a) 162 (b) 9.3 (c) 1261 (d) 20.0 **2.43** (a) 1.207×10^2 (b) 3.4×10^{-3} (c) 2.3100×10^2 (d) 2.3×10^4 (e) 2.00×10^{-1} (f) 1.011×10^{-1} **2.45** (a) 1.0×10^{-3} (b) 1.0×10^3 (c) 6.3×10^4 (d) 6.3×10^{-4} **2.47** (a) two (b) three (c) three (d) four **2.49** (a) 5.50×10^{12} (b) 4.14×10^{-2} (c) 1.5×10^4 (d) 2.0×10^{-7} (e) 1.5×10^{11} (f) 1.2×10^6 **2.51** (a) 10^2 (b) 10^4 (c) 10^4 (d) 10^{-4} **2.53** (a) 1 day/24 hours, 24 hours/1 day (b) 10 decades/1 century, 1 century/10 decades (c) 3 feet/1 yard, 1 yard/3 feet (d) 4 quarts/1 gallon, 1 gallon/4 quarts **2.55** (a) 1 kL/10^3 L, 10^3 L/1 kL (b) 1 mg/10^{-3} g, 10^{-3} g/1 mg (c) 10^{-2} m/1 cm, 1 cm/10^{-2} m (d) 1 μsec/10^{-6} sec, 10^{-6} sec/1 μsec **2.57** (a) exact (b) inexact (c) exact (d) exact **2.59** (a) 1.6×10^2 m (b) 2.4×10^{-8} m (c) 3 m (d) 3.0×10^5 m **2.61** 2.5 L **2.63** 3.41 lb **2.65** 0.0066 gal **2.67** 183 lb, 6.30 ft **2.69** 13.55 g/cm³ **2.71** 25.3 mL **2.73** 243 g **2.75** (a) float (b) sink **2.77** 274°C **2.79** −38.0°F **2.81** −10°C **2.83** 2.6 J/g·°C **2.85** 0.155 cal/g·°C **2.87** (a) 48 cal (b) 8.40×10^2 cal (c) 180 cal **2.89** The first 12 is an exact number (no uncertainty), and the second 12 is an inexact number (contains uncertainty). **2.90** (a) 4.720506 (b) 4.7205 (c) 4.721 (d) 4.7 **2.91** (a) 3.00×10^{-3} (b) 9.4×10^5 (c) 2.35×10^1 (d) 4.50000×10^8 **2.92** (a) smaller, 10^3 (b) larger, 10^9 (c) smaller, 10^8 (d) smaller, 10^8 **2.93** (a) four significant figures (b) four significant figures (c) three significant figures (d) exact **2.94** (a) 5.0×10^{-1} g/cm³ (b) 5.00×10^{-1} g/cm³ (c) 5.0000×10^{-1} g/cm³ (d) 5.000×10^{-1} g/cm³ **2.95** (a) 1.3×10^2 mL (b) 81 mL (c) 9.88×10^4 mL (d) 5.51 mL **2.96** (a) 2.0 calories (b) 1.0 kilocalorie (c) 100 Calories (d) 1000 kilocalories **2.97** (a) 4.5×10^3 mg/L (b) 4.5×10^9 pg/mL (c) 4.5 g/L (d) 4.5 kg/m³ **2.98** a **2.99** c **2.100** b **2.101** d **2.102** d **2.103** d **2.104** c **2.105** d **2.106** b **2.107** c

Chapter 3 **3.1** (a) electron (b) neutron (c) proton (d) proton **3.3** (a) false (b) false (c) false (d) true **3.5** (a) 2 and 4 (b) 4 and 9 (c) 5 and 9 (d) 28 and 58 **3.7** (a) 8, 8, and 8 (b) 8, 10, and 8 (c) 20, 24, and 20 (d) 100, 157, and 100 **3.9** (a) atomic number (b) both atomic number and mass number (c) mass number (d) both atomic number and mass number **3.11** (a) 19 (b) 39 (c) 19 (d) 19 **3.13** (a) nitrogen (b) aluminum (c) barium (d) gold **3.15** (a) $^{15}_{7}$N (b) $^{28}_{14}$Si (c) $^{40}_{18}$Ar (d) $^{48}_{22}$Ti **3.17** (a) S, Cl, Ar, and K (b) Ar, K, Cl, and S (c) S, Cl, Ar, and K (d) S, (Cl, K), Ar **3.19** (a) 34 (b) 23 (c) 23 (d) +11 **3.21** $^{12}_{6}$C, $^{13}_{6}$C, $^{14}_{6}$C **3.23** (a) not isotopes (b) not isotopes (c) isotopes **3.25** (a) false (b) false (c) true (d) true **3.27** (a) not the same (b) same (c) not the same (d) same **3.29** (a) 6.95 amu (b) 24.31 amu **3.31** (a) 55.85 (b) 14.01 (c) calcium (d) iodine **3.33** (a) Ca (b) Mo (c) Li (d) Sn **3.35** (a) 6 (b) 28.09 amu (c) 39 (d) 9.01 amu **3.37** (a) K and Rb (b) P and As (c) F and I (d) Na and Cs **3.39** (a) group (b) periodic law (c) periodic law (d) group **3.41** (a) fluorine (b) sodium (c) krypton (d) strontium **3.43** (a) 3 (b) 4 (c) 4 (d) 4 **3.45** (a) blue element (b) yellow element (c) yellow element (d) green element **3.47** (a) no (b) no (c) yes (d) yes **3.49** (a) S (b) P (c) I (d) Cl **3.51** (a) metal (b) nonmetal (c) metal (d) nonmetal

3.53 (a) metal (b) nonmetal (c) poor conductor of electricity (d) good conductor of heat   **3.55** (a) orbital (b) orbital (c) shell (d) shell   **3.57** (a) true (b) true (c) false (d) true   **3.59** (a) 2 (b) 2 (c) 6 (d) 18   **3.61** (a) $1s^2 2s^2 2p^2$ (b) $1s^2 2s^2 2p^6 3s^1$ (c) $1s^2 2s^2 2p^6 3s^2 3p^4$ (d) $1s^2 2s^2 2p^6 3s^2 3p^6$ **3.63** (a) oxygen (b) neon (c) aluminum (d) calcium   **3.65** (a) $1s^2 2s^2 2p^6 3s^2 3p^5$ (b) $1s^2 2s^2 2p^6 3s^2 3p^6 4s^2 3d^{10} 4p^6 5s^2 4d^7$ (c) $1s^2 2s^2 2p^6 3s^2 3p^6 4s^2$ (d) $1s^2 2s^2 2p^6 3s^2 3p^6 4s^2 3d^1$

3.67

(a) ⊥ ⊥ ⊥ ⊥ ⊥

(b) ⊥ ⊥ ⊥ ⊥ ⊥ ⊥

(c) ⊥ ⊥ ⊥ ⊥ ⊥ ⊥ ⊥ ⊥ ⊥

(d) ⊥ ⊥ ⊥ ⊥ ⊥ ⊥ ⊥ ⊥ ⊥ ⊥ ⊥ ⊥ ⊥ ⊥

3.69 (a) 3 (b) 0 (c) 1 (d) 5   **3.71** (a) no (b) yes (c) no (d) yes   **3.73** (a) s area (b) d area (c) p area (d) d area   **3.75** (a) p^1 (b) d^3 (c) s^2 (d) p^6 **3.77** (a) s area (b) d area (c) p^4 element (d) s^2 element   **3.79** (a) representative element (b) noble gas (c) transition element (d) inner transition element   **3.81** (a) noble gas (b) representative element (c) transition element (d) representative element   **3.83** (a) 4 (b) 1 (c) 2 (d) 6   **3.85** (a) helium, 2, 3, 2, 1 (b) $^{60}_{28}$Ni, 28, 28, 32 (c) argon, $^{37}_{18}$Ar, 18, 19 (d) strontium, $^{90}_{38}$Sr, 38, 38 (e) uranium, 92, 235, 92, 143 (f) chlorine, $^{37}_{17}$Cl, 17, 37 (g) plutonium, $^{232}_{94}$Pu, 94, 138 (h) sulfur, 16, 32, 16, 16 (i) $^{56}_{26}$Fe, 26, 26, 30 (j) $^{44}_{20}$Ca, 20, 40, 20   **3.86** (a) $^{40}_{20}$Ca (b) $^{211}_{86}$Rn (c) $^{110}_{47}$Ag (d) $^{9}_{4}$Be   **3.87** (a) same number of neutrons, 7 (b) same number of neutrons, 10 (c) same total number of subatomic particles, 54 (d) same number of electrons, 17   **3.88** (a) same (b) different (c) different (d) same   **3.89** (a) $^{57}_{24}$Cr (b) $^{50}_{24}$Cr (c) $^{55}_{24}$Cr (d) $^{65}_{24}$Cr   **3.90** 1638 electrons   **3.91** (a) Be, Al (b) Be, Al, Ag, Au (the metals) (c) N, Be, Ar, Al, Ag, Au (d) Ag, Au   **3.92** the same   **3.93** (a) $1s^2 2s^2 2p^1$ (B) (b) $1s^2 2s^2 2p^6 3s^2 3p^1$ (Al) (c) $1s^2 2s^1$ (Li) (d) $1s^2 2s^2 2p^6 3s^2 3p^3$ (P)   **3.94** (a) $_8$O (b) $_{10}$Ne (c) $_{30}$Zn (d) $_{12}$Mg   **3.95** a   **3.96** c   **3.97** b   **3.98** d   **3.99** a   **3.100** c   **3.101** b   **3.102** b   **3.103** d   **3.104** d

Chapter 4   **4.1** The mechanism for ionic bond formation is electron transfer and that for covalent bond formation is electron sharing.   **4.3** (a) 2 (b) 2 (c) 3 (d) 4   **4.5** (a) Group IA, 1 valence electron (b) Group VIIIA, 8 valence electrons (c) Group IIA, 2 valence electrons (d) Group VIIA, 7 valence electrons   **4.7** (a) $1s^2 2s^2 2p^2$ (b) $1s^2 2s^2 2p^5$ (c) $1s^2 2s^2 2p^6 3s^2$ (d) $1s^2 2s^2 2p^6 3s^2 3p^3$

4.9 (a) Mg· (b) K· (c) ·P· (d) :Kr: **4.11** (a) Li (b) F (c) Be (d) N   **4.13** They are the most unreactive of all elements.   **4.15** They lose, gain, or share electrons in such a way that they achieve a noble-gas electron configuration.   **4.17** (a) O^{2-} (b) Mg^{2+} (c) F^- (d) Al^{3+} **4.19** (a) Ca^{2+} (b) O^{2-} (c) Na^+ (d) Al^{3+} **4.21** (a) 15p, 18e (b) 7p, 10e (c) 12p, 10e (d) 3p, 2e   **4.23** line 1: 18 electrons, 20 protons; line 2: Be, 4 protons; line 3: I, I^-; line 4: Al^{3+}, 10 electrons   **4.25** (a) $^{14}_6$C (b) $^{27}_{13}$Al^{3+} (c) $^{35}_{17}$Cl$^-$ (d) $^{52}_{24}$Cr   **4.27** (a) 2+ (b) 3− (c) 1+ (d) 1−   **4.29** (a) 2 lost (b) 1 gained (c) 2 lost (d) 2 gained   **4.31** (a) Ne (b) Ar (c) Ar (d) Ar   **4.33** (a) Ne (b) Ar (c) Ar (d) Ar   **4.35** (a) Group IIA (b) Group VIA (c) Group VA (d) Group IA   **4.37** (a) $1s^2 2s^2 2p^6 3s^2 3p^1$ (b) $1s^2 2s^2 2p^6$

4.39

(a) Be → :O: (b) Mg → :S:

(c) K → :N: (d) Ca → :F: (×2)

4.41 (a) 2 extra electrons, 2− charge (b) 1 extra electron, 1− charge (c) 3 extra electrons, 3− charge (d) 2 extra electrons, 2− charge   **4.43** (a) $BaCl_2$ (b) $BaBr_2$ (c) Ba_3N_2 (d) BaO   **4.45** (a) MgF_2 (b) BeF_2 (c) LiF (d) AlF_3 **4.47** (a) Na_2S (b) CaI_2 (c) Li_3N (d) $AlBr_3$ **4.49** (a) $BeCl_2$ (b) BaI_2 (c) Na_2O (d) AlN   **4.51** an extended array of alternating positive and negative ions   **4.53** smallest whole-number repeating ratio of ions present   **4.55** (a) diagram III (2:3 ratio) (b) diagram IV (1:2 ratio) (c) diagram I (3:1 ratio) (d) diagram II (1:1 ratio)

4.57 all pairs except nitrogen and chlorine   **4.59** Al_2O_3 and K_2S **4.61** (a) potassium iodide (b) beryllium oxide (c) aluminum fluoride (d) sodium phosphide   **4.63** (a) +1 (b) +2 (c) +4 (d) +2   **4.65** (a) iron(II) oxide (b) gold(III) oxide (c) copper(II) sulfide (d) cobalt(II) bromide   **4.67** (a) gold(I) chloride (b) potassium chloride (c) silver chloride (d) copper(II) chloride   **4.69** (a) KBr (b) Ag_2O (c) BeF_2 (d) Ba_3P_2 **4.71** (a) CoS (b) Co_2S_3 (c) SnI_4 (d) Pb_3N_2 **4.73** (a) SO_4^{2-} (b) ClO_3^- (c) OH^- (d) CN^- **4.75** (a) PO_4^{3-} and HPO_4^{2-} (b) NO_3^- and NO_2^- (c) H_3O^+ and OH^- (d) CrO_4^{2-} and $Cr_2O_7^{2-}$ **4.77** (a) $NaClO_4$ (b) $Fe(OH)_3$ (c) $Ba(NO_3)_2$ (d) $Al_2(CO_3)_3$ **4.79** line 1: NH_4CN, NH_4NO_3, NH_4HCO_3, $(NH_4)_2SO_4$; line 2: $Al(CN)_3$, $Al(NO_3)_3$, $Al(HCO_3)_3$, $Al_2(SO_4)_3$; line 3: AgCN, $AgHCO_3$, Ag_2SO_4; line 4: $Ca(CN)_2$, $Ca(NO_3)_2$, $Ca(HCO_3)_2$, $CaSO_4$ **4.81** (a) magnesium carbonate (b) zinc sulfate (c) beryllium nitrate (d) silver phosphate   **4.83** (a) iron(II) hydroxide (b) copper(II) carbonate (c) gold(I) cyanide (d) manganese(II) phosphate   **4.85** (a) $KHCO_3$ (b) $Au_2(SO_4)_3$ (c) $AgNO_3$ (d) $Cu_3(PO_4)_2$ **4.87** line 1: Mg^{2+}, OH^-, $Mg(OH)_2$; line 2: Ba^{2+}, Br^-, barium bromide; line 3: $Zn(NO_3)_2$, zinc nitrate; line 4: Fe^{3+}, ClO_3^-, $Fe(ClO_3)_3$; line 5: Pb^{4+}, O^{2-}, lead(IV) oxide; line 6: $Co_3(PO_4)_2$, cobalt(II) phosphate; line 7: KI, potassium iodide; line 8: Cu^+, SO_4^{2-}, copper(I) sulfate; line 9: Li^+, N^{3-}, Li_3N; line 10: Al_2S_3, aluminum sulfide   **4.88** line 1: 26, 33, 23, 3+; line 2: $^{59}_{28}$Ni^{3+}, 3+; line 3: $^{197}_{77}$Ir^{2+}, 75; line 4: $^{18}_9$F$^-$, 9   **4.89** (a) Na^+ (b) F^- (c) S^{2-} (d) Ca^{2+} **4.90** (a) XZ_2 (b) X_2Z (c) XZ (d) XZ   **4.91** (a) S (b) Mg (c) P (d) Al   **4.92** (a) only monatomic ions (b) both monatomic and polyatomic ions (c) only polyatomic ions (d) only monatomic ions   **4.93** (a) K^+, Cl^- (b) Ca^{2+}, S^{2-} (c) Be^{2+}, two F^- (d) two Al^{3+}, three S^{2-} **4.94** (a) Na_3N, sodium nitride (b) KNO_3, potassium nitrate (c) MgO, magnesium oxide (d) $(NH_4)_3PO_4$, ammonium phosphate   **4.95** (a) tin(IV) chloride, tin(II) chloride (b) iron(II) sulfide, iron(III) sulfide (c) copper(I) nitride, copper(II) nitride (d) nickel(II) iodide, nickel(III) iodide   **4.96** (a) same (3+) (b) different (1+ and 2+) (c) different (1+ and 3+) (d) different (2+ and 1+)   **4.97** (a) copper(I) nitrate, copper(II) nitrate (b) lead(II) phosphate, lead(IV) phosphate (c) manganese(III) cyanide, manganese(II) cyanide (d) cobalt(II) chlorate, cobalt(III) chlorate   **4.98** (a) Na_2S (b) Na_2SO_4 (c) Na_2SO_3 (d) $Na_2S_2O_3$ **4.99** a   **4.100** d   **4.101** d   **4.102** c   **4.103** a   **4.104** b   **4.105** a   **4.106** d   **4.107** c   **4.108** a

Chapter 5   **5.1** ionic: metal and nonmetal; covalent: two or more nonmetals   **5.3** ionic: extended array of alternating positive and negative ions; covalent: molecules   **5.5** (a) :Br:Br: (b) H:I: (c) :I:Br: (d) :Br:F: **5.7** (a) 8 (b) 4 (c) 0 (d) 6   **5.9** (a) NF_3 (b) Cl_2O (c) H_2S (d) CH_4 **5.11** (a) one triple bond (b) three single bonds (c) one double and two single bonds (d) one double and four single bonds

5.13 (a) :N≡N: (b) H—C=C—H (c) H—C—H with H H below and :O: above

(d) H—O—O—H   **5.15** (a) normal (b) normal (c) not normal (d) normal   **5.17** (a) N (b) C (c) N (d) C   **5.19** Oxygen forms three bonds instead of the normal two.   **5.21** (a) nitrogen–oxygen (b) none present (c) oxygen–chlorine (d) two (oxygen–bromine)   **5.23** (a) 20 (b) 8 (c) 8 (d) 24

5.25 (a) H:P:H with H below (b) :Cl:P:Cl: with :Cl: below (c) :Br:Si:Br: with :Br: above and :Br: below (d) :F:O:F: with :Br: above

5.27 (a) :F:S:F: (b) :I:C:I: with :I: above and below (c) :Br:N:Br: with :Br: below (d) H:Se:H

5.29 (a) H:C::C:H (b) :F:N::N:F: (c) H:C:C:::N: with H H below and H below

(d) $H : \overset{..}{C} : C ::: C : H$ **5.31** (a) $\left[: \overset{..}{O} : H\right]^{-}$ (b) $\left[H : \overset{..}{Be} : H\right]^{2-}$

(c) $\left[: \overset{..}{\underset{..}{Cl}} : \overset{..}{\underset{..}{Al}} : \overset{..}{\underset{..}{Cl}} :\right]^{-}$ (d) $\left[: \overset{..}{O} : N : \overset{..}{O} : \right]^{-}$ **5.33** (a) $Na^{+} \left[: C ::: N :\right]^{-}$

(b) $3\left[K\right]^{+} \left[: \overset{..}{O} : \overset{..}{\underset{..}{P}} : \overset{..}{O} : \right]^{3-}$ **5.35** (a) linear (b) angular

(c) tetrahedral (d) linear   **5.37** (a) angular (b) angular (c) angular (d) linear   **5.39** (a) trigonal pyramidal (b) trigonal planar (c) tetrahedral (d) tetrahedral   **5.41** (a) trigonal pyramidal (b) tetrahedral (c) angular (d) angular   **5.43** (a) trigonal planar about each carbon atom (b) tetrahedral about carbon atom and angular about oxygen atom   **5.45** (a) Na, Mg, Al, P (b) I, Br, Cl, F (c) Al, P, S, O (d) Ca, Mg, C, O   **5.47** (a) N, O, Cl, F, Br (b) Na, K, Rb, (c) F, O, Cl, N (d) 0.5 units

5.49 (a) $\overset{\delta^{+}}{B} — \overset{\delta^{-}}{N}$ (b) $\overset{\delta^{+}}{Cl} — \overset{\delta^{-}}{F}$ (c) $\overset{\delta^{-}}{N} — \overset{\delta^{+}}{C}$ (d) $\overset{\delta^{+}}{F} — \overset{\delta^{-}}{O}$ **5.51** (a) H—Br, H—Cl, H—O (b) O—F, P—O, Al—O (c) Br—Br, H—Cl, B—N (d) P—N, S—O, Br—F   **5.53** (a) polar covalent (b) ionic (c) nonpolar covalent (d) ionic   **5.55** (a) nonpolar (b) polar (c) polar (d) polar   **5.57** (a) nonpolar (b) polar (c) polar (d) polar   **5.59** (a) polar (b) polar (c) nonpolar (d) polar   **5.61** (a) sulfur tetrafluoride (b) tetraphosphorus hexoxide (c) chlorine dioxide (d) hydrogen sulfide   **5.63** (a) diagram I (b) diagram II (c) diagram IV (d) diagram III   **5.65** (a) ICl (b) N_2O (c) NCl_3 (d) HBr   **5.67** (a) H_2O_2 (b) CH_4 (c) NH_3 (d) PH_3 **5.69** (a) 26 (b) 24 (c) 14 (d) 8   **5.70** (a) same, both single (b) different, double and single (c) same, both triple (d) same, both single   **5.71** (a) not enough electron dots (b) not enough electron dots (c) improper placement of a correct number of electron dots (d) too many electron dots   **5.72** (a) tetrahedral; tetrahedral (b) tetrahedral; tetrahedral (c) trigonal planar; angular (d) trigonal planar; trigonal planar   **5.73** (a) can't classify (b) nonpolar (c) can't classify (d) polar   **5.74** (a) BrI (b) SO_2 (c) NF_3 (d) H_3CF **5.75** BA, CA, DB, DA

5.76 (a)

$$H-\underset{\underset{H}{|}}{\overset{\overset{H}{|}}{C}}-H, \quad H-\underset{\underset{H}{|}}{\overset{\overset{H}{|}}{C}}-\overset{..}{\underset{..}{F}}:, \quad H-\underset{\underset{:\overset{..}{F}:}{|}}{\overset{\overset{H}{|}}{C}}-\overset{..}{\underset{..}{F}}:, \quad H-\underset{\underset{:\overset{..}{F}:}{|}}{\overset{\overset{:\overset{..}{F}:}{|}}{C}}-\overset{..}{\underset{..}{F}}:, \quad :\overset{..}{\underset{..}{F}}-\underset{\underset{:\overset{..}{F}:}{|}}{\overset{\overset{:\overset{..}{F}:}{|}}{C}}-\overset{..}{\underset{..}{F}}:$$

(b) all are tetrahedral (c) nonpolar, polar, polar, polar, nonpolar   **5.77** Chemical formulas for molecular compounds are written with the least electronegative atom first; N is the least electronegative atom   **5.78** $NaNO_3$ is an ionic compound; the ions present are sodium ion and nitrate ion.   **5.79** (a) sodium chloride (b) bromine monochloride (c) potassium sulfide (d) dichlorine monoxide   **5.80** d   **5.81** c   **5.82** b   **5.83** b   **5.84** d   **5.85** b   **5.86** c   **5.87** c   **5.88** c   **5.89** d

Chapter 6   **6.1** (a) 342.34 amu (b) 100.23 amu (c) 183.20 amu (d) 132.17 amu   **6.3** (a) 6.02×10^{23} apples (b) 6.02×10^{23} elephants (c) 6.02×10^{23} Zn atoms (d) 6.02×10^{23} CO_2 molecules   **6.5** (a) 9.03×10^{23} atoms Fe (b) 9.03×10^{23} atoms Ni (c) 9.03×10^{23} atoms C (d) 9.03×10^{23} atoms Ne   **6.7** (a) 0.200 mole (b) Avogadro's number (c) 1.50 moles (d) 6.50×10^{23} atoms   **6.9** (a) 28.01 g (b) 44.01 g (c) 58.44 g (d) 342.34 g   **6.11** (a) 6.7 g (b) 3.7 g (c) 48.0 g (d) 96.0 g   **6.13** (a) 0.179 mole (b) 0.114 mole (c) 0.0937 mole (d) 0.0210 mole

6.15 (a) $\dfrac{2 \text{ moles H}}{1 \text{ mole H}_2SO_4}$ $\dfrac{1 \text{ mole H}_2SO_4}{2 \text{ moles H}}$ $\dfrac{1 \text{ mole S}}{1 \text{ mole H}_2SO_4}$

$\dfrac{1 \text{ mole H}_2SO_4}{1 \text{ mole S}}$ $\dfrac{4 \text{ moles O}}{1 \text{ mole H}_2SO_4}$ $\dfrac{1 \text{ mole H}_2SO_4}{4 \text{ moles O}}$

(b) $\dfrac{1 \text{ mole P}}{1 \text{ mole POCl}_3}$ $\dfrac{1 \text{ mole POCl}_3}{1 \text{ mole P}}$ $\dfrac{1 \text{ mole O}}{1 \text{ mole POCl}_3}$

$\dfrac{1 \text{ mole POCl}_3}{1 \text{ mole O}}$ $\dfrac{3 \text{ moles Cl}}{1 \text{ mole POCl}_3}$ $\dfrac{1 \text{ mole POCl}_3}{3 \text{ moles Cl}}$

6.17 (a) 2.00 moles S, 4.00 moles O (b) 2.00 moles S, 6.00 moles O (c) 3.00 moles N, 9.00 moles H (d) 6.00 moles N, 12.0 moles H   **6.19** (a) 16.0 moles of atoms (b) 14.0 moles of atoms (c) 45.0 moles of atoms (d) 15.0 moles of atoms

6.21 (a) $\dfrac{3 \text{ moles H}}{1 \text{ mole H}_3PO_4}$ (b) $\dfrac{4 \text{ moles O}}{1 \text{ mole H}_3PO_4}$

(c) $\dfrac{8 \text{ moles atoms}}{1 \text{ mole H}_3PO_4}$ (d) $\dfrac{4 \text{ moles O}}{1 \text{ mole P}}$

6.23 (a) 5.57×10^{23} atoms (b) 4.81×10^{23} atoms (c) 6.0×10^{22} atoms (d) 3.0×10^{23} atoms   **6.25** (a) 63.6 g (b) 31.8 g (c) 5.88×10^{-20} g (d) 1.06×10^{-22} g   **6.27** (a) 2.50 moles (b) 0.227 mole (c) 6.6×10^{-14} mole (d) 6.6×10^{-14} mole   **6.29** (a) 6.14×10^{22} atoms S (b) 1.50×10^{23} atoms S (c) 3.61×10^{23} atoms S (d) 2.41×10^{24} atoms S   **6.31** (a) 32.1 g S (b) 6.39×10^{-22} g S (c) 64.1 g S (d) 1150 g S   **6.33** (a) balanced (b) balanced (c) not balanced (d) balanced   **6.35** (a) 4 N, 6 O (b) 10 N, 12 H, 6 O (c) 1 P, 3 Cl, 6 H (d) 2 Al, 3 O, 6 H, 6 Cl   **6.37** (a) $2Na + 2H_2O \rightarrow 2NaOH + H_2$ (b) $2Na + ZnSO_4 \rightarrow Na_2SO_4 + Zn$ (c) $2NaBr + Cl_2 \rightarrow 2NaCl + Br_2$ (d) $2ZnS + 3O_2 \rightarrow 2ZnO + 2SO_2$ **6.39** (a) $CH_4 + 2O_2 \rightarrow CO_2 + 2H_2O$ (b) $2C_6H_6 + 15O_2 \rightarrow 12CO_2 + 6H_2O$ (c) $C_4H_8O_2 + 5O_2 \rightarrow 4CO_2 + 4H_2O$ (d) $C_5H_{10}O + 7O_2 \rightarrow 5CO_2 + 5H_2O$ **6.41** (a) $3PbO + 2NH_3 \rightarrow 3Pb + N_2 + 3H_2O$ (b) $2Fe(OH)_3 + 3H_2SO_4 \rightarrow Fe_2(SO_4)_3 + 6H_2O$ **6.43** (a) $A_2 + 3B_2 \rightarrow 2AB_3$ (b) $A_2 + B_2 \rightarrow 2AB$ **6.45** diagram III

6.47 $\dfrac{2 \text{ moles Ag}_2CO_3}{4 \text{ moles Ag}}$ $\dfrac{2 \text{ moles Ag}_2CO_3}{2 \text{ moles CO}_2}$ $\dfrac{2 \text{ moles Ag}_2CO_3}{1 \text{ mole O}_2}$

$\dfrac{4 \text{ moles Ag}}{2 \text{ moles CO}_2}$ $\dfrac{4 \text{ moles Ag}}{1 \text{ mole O}_2}$ $\dfrac{2 \text{ moles CO}_2}{1 \text{ mole O}_2}$

The other six are the reciprocals of these six factors.   **6.49** (a) 14.0 moles CO_2 (b) 1.00 mole CO_2 (c) 4.00 moles CO_2 (d) 2.00 moles CO_2

6.51 (a) $\dfrac{3 \text{ moles H}_2S}{2 \text{ moles SbCl}_3}$ (b) $\dfrac{6 \text{ moles HCl}}{1 \text{ mole Sb}_2S_3}$

(c) $\dfrac{6 \text{ moles HCl}}{3 \text{ moles H}_2S}$ (d) $\dfrac{2 \text{ moles SbCl}_3}{1 \text{ mole Sb}_2S_3}$

6.53 8 water molecules   **6.55** (a) CO_2 and H_2O (b) 6.0 moles CO_2 and 12 moles H_2O **6.57** (a) 24.3 g NH_3 (b) 1.80×10^2 g $(NH_4)_2Cr_2O_7$ (c) 22.9 g N_2H_4 (d) 24.3 g NH_3 **6.59** 5.09 g O_2 **6.61** 14.3 g O_2 **6.63** 5.63 g H_2O **6.65** $y = 8$ **6.66** (a) 1.00 mole S_8 (b) 28.0 g Al (c) 30.0 g Mg (d) 6.02×10^{23} atoms He   **6.67** (a) 0.03560 mole SiH_4 (b) 2.139 g SiO_2 (c) 2.144×10^{22} molecules $(CH_3)_3SiCl$ (d) 2.144×10^{22} atoms Si   **6.68** 59.0 g Si   **6.69** C_4H_6 **6.70** 8.33 g N_2, 21.4 g H_2O, and 45.2 g Cr_2O_3 **6.71** 109 g Ag and 16.2 g S   **6.72** 43.4 g Be   **6.73** d   **6.74** b   **6.75** b   **6.76** d   **6.77** a   **6.78** b   **6.79** b   **6.80** b   **6.81** b   **6.82** c

Chapter 7   **7.1** (a) potential energy (b) magnitude increases as temperature increases (c) cause order within the system (d) electrostatic attractions   **7.3** (a) liquid state (b) solid state (c) gaseous state (d) solid state   **7.5** (a) gaseous state (b) solid state (c) liquid state (d) solid and liquid states   **7.7** (a) amount (b) volume (c) pressure (d) temperature   **7.9** (a) 0.967 atm (b) 403 mm Hg (c) 403 torr (d) 0.816 atm   **7.11** 7.2 atm   **7.13** 2.71 L   **7.15** diagram II   **7.17** 3.64 L   **7.19** 144°C   **7.21** diagram II

7.23 (a) $T_1 = \dfrac{P_1 V_1 T_2}{P_2 V_2}$ (b) $P_2 = \dfrac{P_1 V_1 T_2}{V_2 T_1}$ (c) $V_1 = \dfrac{P_2 V_2 T_1}{P_1 T_2}$

7.25 (a) 5.90 L (b) 2.11 atm (c) −171°C (d) 3.70×10^3 mL   **7.27** diagram II   **7.29** −209°C   **7.31** 1.12 L   **7.33** (a) 4.11 L (b) 3.16 atm (c) −98°C (d) 16,300 mL   **7.35** 0.42 atm   **7.37** 98 mm Hg   **7.39** (a) 2.4 atm (b) 2.4 atm (c) 1.2 atm   **7.41** (a) endothermic (b) endothermic (c) exothermic   **7.43** (a) no (b) yes (c) yes   **7.45** amount of liquid decreases; temperature of liquid decreases   **7.47** rate increases   **7.49** (a) true (b) true   **7.51** the higher the temperature, the higher the vapor pressure   **7.53** volatile   **7.55** (a) true (b) true

(c) false (d) false   **7.57** the higher the external pressure, the higher the boiling point   **7.59** Molecules must be polar.   **7.61** Boiling point increases as intermolecular force strength increases.   **7.63** (a) London (b) hydrogen bonding (c) dipole–dipole (d) London   **7.65** (a) no (b) yes (c) yes (d) no   **7.67** four (see Figure 7.21)   **7.69** (a) 0.871 atm (b) 298°C (c) 869°C   **7.70** (a) 915°C (b) −199°C (c) 24°C (d) 172°C   **7.71** (a) Boyle's law, Charles's law and combined gas law (b) Charles's law (c) Boyle's law (d) Boyle's law   **7.72** 8.08×10^6 L He   **7.73** (a) 4.6 atm (b) 29 atm (c) 0.14 atm (d) 0.90 atm   **7.74** 5.37×10^{22} molecules H_2S **7.75** 24.7 L   **7.76** 0.22 g N_2 **7.77** (a) 1.00 atm (b) 1.00 atm (c) 1.50 atm (He), 1.00 atm (Ne), 0.50 atm (Ar) (d) 0.50 atm (He), 1.00 atm (Ne), 1.50 atm (Ar)   **7.78** (a) boils (b) does not boil (c) does not boil (d) does not boil   **7.79** (a) PBr_3 (b) PI_3 (c) PI_3 **7.80** (a) Br_2, larger mass (b) H_2O, hydrogen bonding (c) CO, dipole–dipole (d) C_3H_8, larger size   **7.81** a   **7.82** d   **7.83** d   **7.84** a   **7.85** b   **7.86** a   **7.87** d   **7.88** d   **7.89** d   **7.90** a

Chapter 8   **8.1** (a) true (b) true (c) true (d) false   **8.3** (a) solute: sodium chloride; solvent: water (b) solute: sucrose; solvent: water (c) solute: water; solvent: ethyl alcohol (d) solute: ethyl alcohol; solvent: methyl alcohol   **8.5** (a) first solution (b) first solution (c) first solution (d) second solution   **8.7** (a) supersaturated (b) saturated (c) saturated (d) unsaturated   **8.9** (a) saturated (b) unsaturated (c) unsaturated (d) saturated   **8.11** (a) dilute (b) concentrated (c) dilute (d) concentrated   **8.13** (a) hydrated ion (b) hydrated ion (c) oxygen atom (d) hydrogen atom   **8.15** (a) decrease (b) increase (c) increase (d) increase   **8.17** (a) slightly soluble (b) very soluble (c) slightly soluble (d) slightly soluble   **8.19** (a) ethanol (b) carbon tetrachloride (c) ethanol (d) ethanol   **8.21** (a) soluble with exceptions (b) soluble (c) insoluble with exceptions (d) soluble   **8.23** (a) all are soluble (b) all are soluble (c) $CaBr_2$, $Ca(OH)_2$, $CaCl_2$ (d) $NiSO_4$ **8.25** (a) diagram IV (b) diagrams I and III   **8.27** (a) 7.10%(m/m) (b) 6.19%(m/m) (c) 9.06%(m/m) (d) 0.27%(m/m)   **8.29** (a) 3.62 g (b) 14.5 g (c) 68.8 g (d) 124 g   **8.31** 0.6400 g   **8.33** 276 g   **8.35** (a) 4.21%(v/v) (b) 4.60%(v/v)   **8.37** 18%(v/v)   **8.39** (a) 2.0%(m/v) (b) 15%(m/v)   **8.41** 0.500 g   **8.43** 3.75 g   **8.45** (a) 6.0 M (b) 0.456 M (c) 0.342 M (d) 0.500 M   **8.47** (a) 273 g (b) 0.373 g (c) 136 g (d) 88 g   **8.49** (a) 85.6 mL (b) 2.64 mL (c) 9180 mL (d) 0.24 mL   **8.51** (a) 0.183 M (b) 0.0733 M (c) 0.0120 M (d) 0.00275 M   **8.53** (a) 1450 mL (b) 18.0 mL (c) 85,600 mL (d) 7.5 mL   **8.55** (a) 3.0 M (b) 3.0 M (c) 4.5 M (d) 1.5 M   **8.57** diagram II   **8.59** (a) suspension (b) suspension (c) true solution and colloidal dispersion (d) true solution   **8.61** (a) false (b) false (c) false (d) true   **8.63** The presence of solute molecules decreases the ability of solvent molecules to escape.   **8.65** It is a more concentrated solution and thus has a lower vapor pressure.   **8.67** diagram III   **8.69** (a) same as (b) greater than (c) less than (d) greater than   **8.71** 2 to 1   **8.73** (a) swell (b) remain the same (c) swell (d) shrink   **8.75** (a) hemolyze (b) remain unaffected (c) hemolyze (d) crenate   **8.77** (a) hypotonic (b) isotonic (c) hypotonic (d) hypertonic   **8.79** (a) decrease (b) increase (c) not change (d) decrease   **8.81** (a) K^+ and Cl^- leave the bag. (b) K^+, Cl^-, and glucose leave the bag.   **8.83** (a) like (both soluble) (b) unlike (c) unlike (d) like (both insoluble)   **8.84** (a) 4.02 g (b) 7.303 g (c) 12.6 g (d) 0.148 g   **8.85** 0.0700 qt   **8.86** (a) 7.1 L (b) 9.5 L (c) 11 L (d) 14 L   **8.87** (a) 0.472 M (b) 0.708 M (c) 1.04 M (d) 1.60 M   **8.88** (a) 37.5%(m/v) (b) 2.23 M   **8.89** (a) 4.00 M (b) 3.22 M   **8.90** (a) NaCl (b) $MgCl_2$ **8.91** c   **8.92** b   **8.93** d   **8.94** b   **8.95** b   **8.96** d   **8.97** b   **8.98** d   **8.99** c   **8.100** b

Chapter 9   **9.1** (a) X + YZ ⟶ Y + XZ (b) X + Y ⟶ XY   **9.3** (a) single-replacement (b) decomposition (c) double-replacement (d) combination   **9.5** (a) combination, single-replacement, combustion (b) decomposition, single-replacement (c) combination, decomposition, single-replacement, double-replacement, combustion (d) combination, decomposition, single-replacement, double-replacement, combustion   **9.7** (a) +2 (b) +6 (c) 0 (d) +5   **9.9** (a) +3 (b) +4 (c) +6 (d) +6 (e) +6 (f) +6 (g) +6 (h) +5   **9.11** (a) +3P, −1F (b) +1Na, −2O, +1H (c) +1Na, +6S, −2O (d) +4C, −2O   **9.13** (a) redox (b) nonredox (c) redox (d) redox   **9.15** (a) H_2 oxidized, N_2 reduced (b) KI oxidized, Cl_2 reduced (c) Fe oxidized, Sb_2O_3 reduced (d) H_2SO_3 oxidized, HNO_3 reduced   **9.17** (a) N_2 oxidizing agent, H_2 reducing agent (b) Cl_2 oxidizing agent, KI reducing agent (c) Sb_2O_3 oxidizing agent, Fe reducing agent (d) HNO_3 oxidizing agent, H_2SO_3 reducing agent   **9.19** molecular collisions, activation energy, and collision orientation   **9.21** total kinetic energy of colliding reactants; collision orientation   **9.23** (a) exothermic (b) endothermic (c) endothermic (d) exothermic   **9.25** (a) exothermic (b) released

9.27

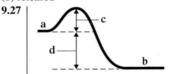

9.29 (a) As temperature increases, so does the number of collisions per second. (b) A catalyst lowers the activation energy.   **9.31** The concentration of O_2 has increased from 21% to 100%.   **9.33** (a) increase (b) decrease (c) increase (d) decrease

9.35

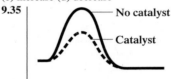

9.37 (a) 1 (b) 3 (c) 4 (d) 3   **9.39** rate of forward reaction = rate of reverse reaction   **9.41** (a) $N_2(g) + O_2(g) \longrightarrow 2NO(g)$ (b) $2NO(g) \longrightarrow N_2(g) + O_2(g)$

9.43

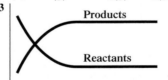

9.45 Yes, the concentrations of molecules in diagrams III and IV are the same.   **9.47** Diagrams II and IV; diagram III cannot be produced from the original mixture.

9.49 (a) $K_{eq} = \dfrac{[NO_2]^2}{[N_2O_4]}$ (b) $K_{eq} = \dfrac{[Cl_2][CO]}{[COCl_2]}$

(c) $K_{eq} = \dfrac{[CH_4][H_2S]^2}{[CS_2][H_2]^4}$ (d) $K_{eq} = \dfrac{[SO_3]^2}{[O_2][SO_2]^2}$

9.51 (a) $K_{eq} = [SO_3]$ (b) $K_{eq} = \dfrac{1}{[Cl_2]}$

(c) $K_{eq} = \dfrac{[NaCl]^2}{[Na_2SO_4][BaCl_2]}$ (d) $K_{eq} = [O_2]$

9.53 4.8×10^{-5} **9.55** (a) more products than reactants (b) essentially all reactants (c) significant amounts of both reactants and products (d) significant amounts of both reactants and products **9.57** diagram IV **9.59** diagram IV **9.61** (a) right (b) left (c) left (d) right **9.63** (a) left (b) left (c) left (d) left **9.65** (a) left (b) no effect (c) right (d) no effect **9.67** endothermic reaction **9.69** (a) redox, single replacement (b) redox, combustion (c) redox, decomposition (d) nonredox, double replacement **9.70** (a) redox (b) redox (c) redox (d) can't classify **9.71** (a) gain (b) reduction (c) decrease (d) increase **9.72** (a) gains (b) loses (c) loses (d) gains **9.73** (a) decrease (b) increase (c) increase (d) decrease **9.74** (a) no (b) no (c) yes (d) no **9.75** $CS_2(g) + 4H_2(g) \rightleftharpoons CH_4(g) + 2H_2S(g)$ **9.76** (a) yes (b) yes (c) no (d) yes **9.77** (a) no effect (b) right (c) right (d) right **9.78** c **9.79** c **9.80** a **9.81** b **9.82** c **9.83** b **9.84** c **9.85** d **9.86** b **9.87** d

Chapter 10 **10.1** (a) H^+ (b) OH^- **10.3** (a) Arrhenius acid (b) Arrhenius base **10.5** (a) $HI \xrightarrow{H_2O} H^+ + I^-$ (b) $HClO \xrightarrow{H_2O} H^+ + ClO^-$ (c) $LiOH^- \xrightarrow{H_2O} Li^+ + OH^-$ (d) $CsOH \xrightarrow{H_2O} Cs^+ + OH^-$ **10.7** (a) acid (b) base (c) acid (d) acid **10.9** (a) $HClO + H_2O \rightarrow H_3O^+ + ClO^-$ (b) $HClO_4 + NH_3 \rightarrow NH_4^+ + ClO_4^-$ (c) $H_3O^+ + OH^- \rightarrow H_2O + H_2O$ (d) $H_3O^+ + NH_2^- \rightarrow H_2O + NH_3$ **10.11** (a) HSO_3^- (b) HCN (c) $C_2O_4^{2-}$ (d) $H_2PO_4^-$ **10.13** (a) $HS^- + H_2O \rightarrow H_3O^+ + S^{2-}$, $HS^- + H_2O \rightarrow H_2S + OH^-$ (b) $HPO_4^{2-} + H_2O \rightarrow H_3O^+ + PO_4^{3-}$, $HPO_4^{2-} + H_2O \rightarrow H_2PO_4^- + OH^-$ (c) $NH_3 + H_2O \rightarrow H_3O^+ + NH_2^-$, $NH_3 + H_2O \rightarrow NH_4^+ + OH^-$ (d) $OH^- + H_2O \rightarrow H_3O^+ + O^{2-}$, $OH^- + H_2O \rightarrow H_2O + OH^-$ **10.15** (a) monoprotic (b) diprotic (c) monoprotic (d) diprotic **10.17** $H_3C_6H_5O_7 + H_2O \rightarrow H_3O^+ + H_2C_6H_5O_7^-$, $H_2C_6H_5O_7^- + H_2O \rightarrow H_3O^+ + HC_6H_5O_7^{2-}$, $HC_6H_5O_7^{2-} + H_2O \rightarrow H_3O^+ + C_6H_5O_7^{3-}$ **10.19** (a) 1, 0 (b) 2, 4 (c) 1, 7 (d) 0, 4 **10.21** To show that it is a monoprotic acid **10.23** Monoprotic; only one H atom is involved in a polar bond. **10.25** (a) strong (b) weak (c) weak (d) strong **10.27** The equilibrium position is far to the right for strong acids and far to the left for weak acids. **10.29** 0.10 M in both H_3O^+ and Cl^- ions and zero in HCl **10.31** acid in diagram IV

10.33 (a) $K_a = \dfrac{[H^+][F^-]}{[HF]}$ (b) $K_a = \dfrac{[H^+][C_2H_3O_2^-]}{[HC_2H_3O_2]}$

10.35 (a) $K_b = \dfrac{[NH_4^+][OH^-]}{[NH_3]}$ (b) $K_b = \dfrac{[C_6H_5NH_3^+][OH^-]}{[C_6H_5NH_2]}$

10.37 (a) H_3PO_4 (b) HF (c) H_2CO_3 (d) HNO_2 **10.39** 4.9×10^{-5} **10.41** (a) acid (b) salt (c) salt (d) base **10.43** (a) base (b) salt (c) acid (d) salt **10.45** (a) $Ba(NO_3)_2 \xrightarrow{H_2O} Ba^{2+} + 2NO_3^-$ (b) $Na_2SO_4 \xrightarrow{H_2O} 2Na^+ + SO_4^{2-}$ (c) $CaBr_2 \xrightarrow{H_2O} Ca^{2+} + 2Br^-$ (d) $K_2CO_3 \xrightarrow{H_2O} 2K^+ + CO_3^{2-}$ **10.47** (a) no (b) yes (c) yes (d) no **10.49** (a) 1 to 1 (b) 1 to 2 (c) 1 to 1 (d) 2 to 1 **10.51** (a) $HCl + NaOH \rightarrow NaCl + H_2O$ (b) $HNO_3 + KOH \rightarrow KNO_3 + H_2O$ (c) $H_2SO_4 + 2LiOH \rightarrow Li_2SO_4 + 2H_2O$ (d) $2H_3PO_4 + 3Ba(OH)_2 \rightarrow Ba_3(PO_4)_2 + 6H_2O$ **10.53** (a) $H_2SO_4 + 2LiOH \rightarrow Li_2SO_4 + 2H_2O$ (b) $HCl + NaOH \rightarrow NaCl + H_2O$ (c) $HNO_3 + KOH \rightarrow KNO_3 + H_2O$ (d) $2H_3PO_4 + 3Ba(OH)_2 \rightarrow Ba_3(PO_4)_2 + 6H_2O$ **10.55** (a) 3.3×10^{-12} M (b) 1.5×10^{-9} M (c) 1.1×10^{-7} M (d) 8.3×10^{-4} M **10.57** (a) acidic (b) basic (c) basic (d) acidic **10.59** line 1: 4.5×10^{-13}, acidic; line 2: 3.0×10^{-12}, basic; line 3: 1.5×10^{-7}, basic; line 4: 1.4×10^{-7}, acidic **10.61** (a) 4.00 (b) 11.00 (c) 11.00 (d) 7.00 **10.63** (a) 7.68 (b) 7.40 (c) 3.85 (d) 11.85 **10.65** (a) 1×10^{-2} (b) 1×10^{-6} (c) 1×10^{-8} (d) 1×10^{-10} **10.67** (a) 2.1×10^{-4} M (b) 8.1×10^{-6} M (c) 4.5×10^{-8} M (d) 3.6×10^{-13} M **10.69** line 1: 6.2×10^{-8}, 1.6×10^{-7}, basic; line 2: 1.4×10^{-5}, 9.14, basic; line 3: 5.0×10^{-6}, 2.0×10^{-9}, acidic; line 4: 1.4×10^{-5}, 4.85, acidic **10.71** (a) 3.35 (b) 6.37 (c) 7.21 (d) 1.82 **10.73** acid B **10.75** (a) strong acid–strong base salt (b) weak acid–strong base salt (c) strong acid–weak base salt (d) strong acid–strong base salt **10.77** (a) none (b) $C_2H_3O_2^-$ (c) NH_4^+ (d) none **10.79** (a) neutral (b) basic (c) acidic (d) neutral **10.81** (a) no (b) yes (c) no (d) yes **10.83** (a) HCN and CN^- (b) H_3PO_4 and $H_2PO_4^-$ (c) H_2CO_3 and HCO_3^- (d) HCO_3^- and CO_3^{2-} **10.85** (a) $F^- + H_3O^+ \rightarrow HF + H_2O$ (b) $H_2CO_3 + OH^- \rightarrow HCO_3^- + H_2O$ (c) $CO_3^{2-} + H_3O^+ \rightarrow HCO_3^- + H_2O$ (d) $H_3PO_4 + OH^- \rightarrow H_2PO_4^- + H_2O$ **10.87** all four diagrams **10.89** 7.06 **10.91** 5.17 **10.93** (a) weak (b) strong (c) strong (d) strong **10.95** (a) both (b) molecules (c) ions (d) both **10.97** (a) 2 (b) 3 (c) 3 (d) 2 **10.99** (a) $NaCl \rightarrow Na^+ + Cl^-$ (b) $Mg(NO_3)_2 \rightarrow Mg^{2+} + 2NO_3^-$ (c) $K_2S \rightarrow 2K^+ + S^{2-}$ (d) $NH_4CN \rightarrow NH_4^+ + CN^-$ **10.101** diagram 3 **10.103** (a) 1 Eq (b) 1 Eq (c) 2 Eq (d) 1 Eq **10.105** (a) 2 Eq (b) 3 Eq (c) 4 Eq (d) 14 Eq **10.107** 0.094 mole Cl^- ion **10.109** 21 mg Ca^{2+} ion **10.111** (a) 0.0500 M (b) 0.800 M (c) 0.950 M (d) 0.120 M **10.113** (a) yes (b) no (c) no (d) yes **10.114** (a) no (b) yes, both strong (c) no (d) yes, both weak **10.115** (a) solution A, solution D, solution C, solution B (b) solution B, solution C, solution D, solution A (c) solution B, solution C, solution D, solution A (d) solution A, solution D, solution C, solution B **10.116** 7.0 **10.117** HCl, HCN, KCl, NaOH **10.118** (a) HCN and KCN (b) HF and NaF **10.119** $H_3PO_4/H_2PO_4^-$ and $H_2PO_4^-/HPO_4^{2-}$ **10.120** CN^- ion undergoes hydrolysis to a greater extent than NH_4^+ ion, resulting in a basic solution; NH_4^+ ion and $C_2H_3O_2^-$ ion hydrolyze to an equal extent, resulting in a neutral solution. **10.121** 0.0035 g NaOH **10.122** a **10.123** d **10.124** b **10.125** c **10.126** b **10.127** a **10.128** d **10.129** b **10.130** a **10.131** d

Chapter 11 **11.1** (a) $^{10}_4Be$, Be-10 (b) $^{25}_{11}Na$, Na-25 (c) $^{96}_{41}Nb$, Nb-96 (d) $^{257}_{103}Lr$, Lr-257 **11.3** (a) ^{14}N (b) $^{197}_{79}Au$ (c) Sn-121 (d) B-10 **11.5** the spontaneous emission of radiation from the nucleus of the atom **11.7** approximately 1-to-1 ratio for low-atomic-numbered stable nuclei and 3-to-2 ratio for higher-atomic-numbered stable nuclei **11.9** (a) $^4_2\alpha$ (b) $^0_{-1}\beta$ (c) $^0_0\gamma$ **11.11** 2 protons and 2 neutrons **11.13** (a) $^{200}_{84}Po \rightarrow ^4_2\alpha + ^{196}_{82}Pb$ (b) $^{240}_{96}Cm \rightarrow ^4_2\alpha + ^{236}_{94}Pu$ (c) $^{244}_{96}Cm \rightarrow ^4_2\alpha + ^{240}_{94}Pu$ (d) $^{238}_{92}U \rightarrow ^4_2\alpha + ^{234}_{90}Th$ **11.15** (a) $^{10}_4Be \rightarrow ^0_{-1}\beta + ^{10}_5B$ (b) $^{14}_6C \rightarrow ^0_{-1}\beta + ^{14}_7N$ (c) $^{21}_9F \rightarrow ^0_{-1}\beta + ^{21}_{10}Ne$ (d) $^{25}_{11}Na \rightarrow ^0_{-1}\beta + ^{25}_{12}Mg$ **11.17** $A \rightarrow A - 4$, $Z \rightarrow Z - 2$ **11.19** (a) $^0_{-1}\beta$ (b) $^{25}_{12}Mg$ (c) $^4_2\alpha$ (d) $^{200}_{80}Hg$ **11.21** (a) $^4_2\alpha$ (b) $^0_{-1}\beta$ **11.23** diagram I **11.25** (a) $\frac{1}{4}$ (b) $\frac{1}{64}$ (c) $\frac{1}{8}$ (d) $\frac{1}{64}$ **11.27** (a) 1.4 days (b) 0.90 day (c) 0.68 day (d) 0.54 day **11.29** 0.250 g **11.31** three half-lives **11.33** 2000 **11.35** 92 **11.37** (a) $^4_2\alpha$ (b) $^{25}_{12}Mg$ (c) $^4_2\alpha$ (d) ^1_1p **11.39** Termination of a decay series requires a stable nuclide. **11.41** $^{232}_{90}Th \rightarrow ^4_2\alpha + ^{228}_{88}Ra$, $^{228}_{88}Ra \rightarrow ^0_{-1}\beta + ^{228}_{89}Ac$, $^{228}_{89}Ac \rightarrow ^0_{-1}\beta + ^{228}_{90}Th$, $^{228}_{90}Th \rightarrow ^4_2\alpha + ^{224}_{88}Ra$ **11.43** the electron and positive ion that are produced during an ionizing interaction between a molecule (or atom) and radiation **11.45** (a) yes (b) no (c) yes (d) no **11.47** It continues on, interacting with other atoms and forming more ion pairs. **11.49** Alpha is stopped; beta and gamma go through. **11.51** alpha, 0.1 the speed of light; beta, up to 0.9 the speed of light; gamma, the speed of light **11.53** (a) no detectable effects (b) nausea, fatigue, lowered blood cell count **11.55** to record the extent of their exposure to radiation **11.57** naturally occurring ionizing radiation **11.59** radon seepage, cosmic radiation, rocks and minerals, food and drink **11.61** so radiation can be detected externally **11.63** (a) tagged to white blood cells, which migrate to sites of infection thus locating the infection site (b) injected into the blood, used to locate blood flow blockages (c) injected into the blood, has affinity for healthy heart muscle tissue (d) tagged to red blood cells, used to determine blood volume based on concentration differences **11.65** They are usually α or β emitters. **11.67** (a) 4 (b) 4 (c) 2 (d) 3 **11.69** neptunium-239 **11.71** (a) $^4_2\alpha$ (b) ^2_1H **11.73** (a) fusion (b) fusion (c) both (d) fission **11.75** (a) fusion (b) fission (c) neither (d) neither **11.77** Different isotopes of an element have the same chemical properties but different nuclear properties. **11.79** Temperature, pressure, and catalysts affect chemical reaction rates but do not affect nuclear reaction rates. **11.81** (a) $^{206}_{80}Hg \rightarrow ^0_{-1}\beta + ^{206}_{81}Tl$ (b) $^{109}_{46}Pd \rightarrow ^0_{-1}\beta + ^{109}_{47}Ag$ (c) $^{245}_{96}\rightarrow ^4_2\alpha + ^{241}_{94}Pu$ (d) $^{249}_{100}Fm \rightarrow ^4_2\alpha + ^{245}_{98}Cf$ **11.82** (a) 54 hr (b) 90 hr (c) 108 hr (d) 126 hr **11.83** (a) $^{239}_{94}Pu + ^4_2\alpha \rightarrow ^{242}_{96}Cm + ^1_0n$ (b) $^{246}_{96} + ^{12}_6C \rightarrow ^{254}_{102}No + 4^1_0n$ (c) $^{27}_{13}Al + ^4_2\alpha \rightarrow ^1_0n + ^{30}_{15}P$ (d) $^{23}_{11}Na + ^2_1H \rightarrow ^{21}_{10}Ne + ^4_2\alpha$

11.84 $^{142}_{60}\text{Nd} + ^{1}_{0}\text{n} \rightarrow ^{143}_{61}\text{Pm} + ^{0}_{-1}\beta$ **11.85** 12 elements
11.86 4 neutrons   **11.87** E = 0 (negligible amount), F = 0 (negligible amount), G = 63 atoms, H = 937 atoms   **11.88** ^{228}Ac, ^{228}Th, ^{224}Ac
11.89 c   **11.90** b   **11.91** a   **11.92** c   **11.93** d   **11.94** c   **11.95** a
11.96 a   **11.97** d   **11.98** a

Chapter 12   12.1 (a) false (b) false (c) true (d) true   **12.3** (a) meets
(b) does not meet (c) does not meet (d) does not meet   **12.5** Hydrocarbons
contain C and H, and hydrocarbon derivatives contain at least one additional
element besides C and H.   **12.7** All bonds are single bonds in a saturated
hydrocarbon, and at least one carbon–carbon multiple bond is present
in an unsaturated hydrocarbon.   **12.9** (a) saturated (b) unsaturated
(c) unsaturated (d) unsaturated   **12.11** (a) 18 (b) 4 (c) 13 (d) 22
12.13 (a) $\text{CH}_3\text{—CH}_2\text{—CH}_2\text{—CH}_3$
(b) $\text{CH}_3\text{—CH}_2\text{—CH—CH}_2\text{—CH}_3$
        |
        CH_3
(c) $\text{CH}_3\text{—CH}_2\text{—CH—CH}_2\text{—CH—CH}_3$
        | |
        CH_3 CH_3
(d) $\text{CH}_3\text{—CH}_2\text{—CH—CH}_2\text{—CH}_3$
        |
        CH_2
        |
        CH_3
12.15 (a) $\text{CH}_3\text{—CH—CH}_2\text{—CH}_3$
        |
        CH_3
(b) $\text{CH}_3\text{—CH—CH—CH—CH}_2\text{—CH}_3$
        | | |
        CH_3 CH_3 CH_3
(c) $\text{CH}_3\text{—CH}_2\text{—CH}_2\text{—CH}_2\text{—CH}_2\text{—CH}_3$
(d) CH_3
            |
    $\text{CH}_3\text{—C—CH}_2\text{—CH}_3$
            |
            CH_3
12.17 (a)   H   H   H   H   H
                 | | | | |
            H—C—C—C—C—C—H
                 | | | | |
                 H   H   H   H   H
(b)   H   H   H   H   H   H   H   H
       | | | | | | | |
   H—C—C—C—C—C—C—C—C—H
       | | | | | | | |
       H   H   H   H   H   H   H   H
(c) $\text{CH}_3\text{—(CH}_2)_8\text{—CH}_3$ (d) C_6H_{14}
12.19 the same molecular formula   **12.21** (a) the same (b) different
(c) different (d) different   **12.23** a continuous chain of carbon atoms versus a continuous chain of carbon atoms to which one or more branches of
carbon atoms are attached   **12.25** (a) 2 (b) 5 (c) 18 (d) 75   **12.27** one
12.29 (a) different compounds that are not constitutional isomers
(b) different compounds that are constitutional isomers (c) different
conformations of the same molecule (d) different compounds that are
constitutional isomers
12.31
(a) $\text{CH}_3\text{—CH}_2\text{—CH—CH}_2\text{—CH}_3$ (b) $\text{CH}_3\text{—CH—CH}_2\text{—CH—CH}_3$
            | | |
            CH_3 CH_3 CH_3
(c) $\text{CH}_3\text{—CH—CH}_3$ (d) $\text{CH}_3\text{—CH}_2\text{—CH—CH}_2\text{—CH}_3$
        | |
        CH_3 CH_2
                                                         |
                                                         CH_3

12.33 (a) seven-carbon chain (b) eight-carbon chain (c) eight-
carbon chain (d) seven-carbon chain   **12.35** (a) 3-methylpentane
(b) 2-methylhexane (c) 2-methylhexane (d) 2,4-dimethylhexane

12.37 (a) 2,3,5-trimethylhexane (b) 2,2,4-trimethylpentane
(c) 3-ethyl-3-methylpentane (d) 3-ethyl-3-methylhexane   **12.39** horizontal
chain, because it has more substituents (two)
12.41 (a) $\text{CH}_3\text{—CH}_2\text{—CH—CH—CH}_2\text{—CH}_3$
                         | |
                         CH_3 CH_3
            CH_3
            |
(b) $\text{CH}_3\text{—CH}_2\text{—C—CH}_2\text{—CH}_3$
            |
            CH_2
            |
            CH_3
(c) $\text{CH}_3\text{—CH}_2\text{—CH—CH}_2\text{—CH—CH}_2\text{—CH}_2\text{—CH}_3$
                     | |
                     CH_2 CH_2
                     | |
                     CH_3 CH_3
(d) $\text{CH}_3\text{—CH}_2\text{—CH}_2\text{—CH—CH}_2\text{—CH}_2\text{—CH}_2\text{—CH}_2\text{—CH}_3$
                             |
                             CH_2
                             |
                             CH_2
                             |
                             CH_3
12.43 (a) 2, 2 (b) 2, 2 (c) 2, 2 (d) 1, 1   **12.45** (a) not based on longest
carbon chain; 2,2-dimethylbutane (b) carbon chain numbered from wrong
end; 2,2,3-trimethylbutane (c) carbon chain numbered from wrong end and
alkyl groups not listed alphabetically; 3-ethyl-4-methylhexane (d) like alkyl
groups listed separately; 2,4-dimethylhexane
12.47 (a) C—C—C—C—C—C—C—C
                          |
                          C
(b) C—C—C—C—C—C
            | |
            C   C
(c) C—C—C—C—C
            |
            C
(d) C—C—C—C—C—C—C—C
        | |
        C   C—C
                     |
                     C
12.49 (a) $\text{CH}_3\text{—CH—CH—CH—CH}_3$
                     | | |
                     CH_3 CH_3 CH_3
(b) $\text{CH}_3\text{—CH—CH—CH}_2\text{—CH}_3$
            | |
            CH_3 CH_2
                   |
                   CH_3
(c) $\text{CH}_3\text{—CH}_2\text{—CH—CH—CH}_2\text{—CH}_2\text{—CH}_3$
                     | |
                     CH_3 CH_2
                            |
                            CH_3
(d) $\text{CH}_3\text{—CH—CH}_2\text{—CH}_2\text{—CH—CH}_2\text{—CH}_2\text{—CH}_3$
            | |
            CH_3 CH_2
                                     |
                                     CH_3
12.51 (a) constitutional isomers (b) same compound
12.53 (a)   (b)

(c)   (d)

12.55 (a) 2-methyloctane (b) 2,3-dimethylhexane (c) 3-methylpentane (d) 5-isopropyl-2-methyloctane   **12.57** (a) C_8H_{18} (b) C_9H_{20} (c) $C_{10}H_{22}$ (d) $C_{11}H_{24}$ **12. 59** (a) 5, 1, 3, 0 (b) 5, 1, 1, 1 (c) 4, 3, 0, 1 (d) 4, 4, 0, 1
12.61 (a) isopropyl (b) isobutyl (c) isopropyl (d) *sec*-butyl
12.63

(a) $CH_3-CH_2-CH_2-CH_2-CH-CH_2-CH_2-CH_2-CH_2-CH_3$
$\qquad\qquad\qquad\qquad\quad CH-CH_3$
$\qquad\qquad\qquad\qquad\quad CH_2$
$\qquad\qquad\qquad\qquad\quad CH_3$

(b) $CH_3-CH_2-CH_2-\overset{\displaystyle CH_3}{\underset{\displaystyle CH-CH_3}{\overset{\displaystyle CH-CH_3}{C}}}-CH_2-CH_2-CH_2-CH_3$
$\qquad\qquad\qquad\qquad\qquad CH_3$

(c) $CH_3-CH-CH-CH_2-CH-CH_2-CH_2-CH_2-CH_3$
$\qquad\quad CH_3 \ \ CH_3 \qquad CH_2$
$\qquad\qquad\qquad\qquad\qquad CH-CH_3$
$\qquad\qquad\qquad\qquad\qquad CH_3$

(d) $CH_3-CH_2-CH_2-CH-CH_2-CH_2-CH_2-CH_3$
$\qquad\qquad\qquad\quad CH_3-C-CH_3$
$\qquad\qquad\qquad\qquad\quad CH_3$

12.65 (a) 3 or 4 (b) 3 or 4 (c) none of them (d) 3 or 4
12.67 (a) (2-methylbutyl) group (b) (1,1-dimethylpropyl) group
12.69 (a) 16 (b) 6 (c) 5 (d) 15   **12.71** (a) C_6H_{12} (b) C_6H_{12} (c) C_4H_8
(d) C_7H_{14} **12.73** (a) cyclohexane (b) 1,2-dimethylcyclobutane
(c) methylcyclopropane (d) 1,2-dimethylcyclopentane   **12.75** (a) must locate methyl groups with numbers (b) wrong numbering system for ring (c) no number needed (d) wrong numbering system for ring
12.77

(a)   (b)   (c)   (d)

12.79 (a) two (cyclobutane, methylcyclopropane) (b) three: dimethyl (1,1; 1,2); ethyl (c) one (methyl) (d) four: dimethyl (1,1; 1,2; 1,3); ethyl
12.81 (a) not possible

(b) $CH_3-CH_2 \triangle CH_2-CH_3$ $CH_3-CH_2 \triangle H$
$\qquad\qquad H \ \ H$ $\qquad\quad H \ \ CH_2-CH_3$
$\qquad\qquad\quad cis$ $\qquad\qquad\quad trans$

(c) not possible   (d)
$\qquad\qquad\qquad\qquad cis \qquad\qquad trans$

12.83 50–90% methane, 1–10% ethane, up to 8% propane and butanes
12.85 boiling point   **12.87** (a) octane (b) cyclopentane (c) pentane (d) cyclopentane   **12.89** (a) different states (b) same state (c) same state (d) same state   **12.91** (a) CO_2 and H_2O (b) CO_2 and H_2O (c) CO_2 and H_2O (d) CO_2 and H_2O **12.93** CH_3Br, CH_2Br_2, $CHBr_3$, CBr_4
12.95 (a) CH_3-CH_2
$\qquad\qquad\quad Cl$

(b) $CH_2-CH_2-CH_2-CH_3$ $CH_3-CH-CH_2-CH_3$
$\quad Cl$ Cl
(c) Cl (d)
$CH_2-CH-CH_3$ CH_3-C-CH_3
$\quad Cl \quad CH_3$ CH_3

12.97 (a) iodomethane, methyl iodide (b) 1-chloropropane, propyl chloride (c) 2-fluorobutane, *sec*-butyl fluoride (d) chlorocyclobutane, cyclobutyl chloride
12.99 (a) Cl (b) F F
$\qquad H-C-Cl$ $F-C-C-F$
$\qquad\qquad Cl$ $Cl \ \ Cl$
(c) $CH_3-CH-Br$ (d) Br
$\qquad\qquad CH_3$ H
$\qquad\qquad\qquad\qquad\qquad\qquad H$
$\qquad\qquad\qquad\qquad\qquad Cl$

12.101 (a) 14 (b) 5 (c) 4 (d) 19   **12.102** (a) liquid (b) less dense (c) insoluble (d) flammable   **12.103** (a) no (b) yes (c) no (d) no
12.104 (a) F H (b) Cl CH_3
$\qquad\qquad H \qquad F \qquad\qquad\quad H \quad H$
(c) CH_3 (d) $CH_3-CH-CH_2-I$
$\qquad CH_3-C-Br \qquad\qquad\quad CH_3$
$\qquad\qquad CH_3$

12.105 (a) $C_{18}H_{38}$ (b) C_7H_{14} (c) $C_7H_{14}F_2$ (d) $C_6H_{10}Br_2$ **12.106** (a) alkane (b) halogenated cycloalkane (c) halogenated alkane (d) cycloalkane
12.107 (a) 1 (b) (2-methyl, 3-methyl, 4-methyl) (c) 6 (2,2; 2,3; 2,4; 2,5; 3,3; 3,4) (d) 1 (3-ethyl)
12.108
(a) $C-C-C-C-C-C-C$ $C-C-C-C-C$
$\qquad\qquad heptane$ C
$\qquad\qquad\qquad\qquad\qquad\qquad 2\text{-methylhexane}$

$\qquad\qquad\qquad\qquad\qquad\qquad C$
$C-C-C-C-C-C$ $C-C-C-C-C$
$\qquad C$ C
$\quad 3\text{-methylhexane}$ $2,2\text{-dimethylpentane}$

$C-C-C-C-C$ $C-C-C-C-C$
$\quad\ C \ \ C$ $C \qquad C$
$2,3\text{-dimethylpentane}$ $2,4\text{-dimethylpentane}$

$\qquad\quad C \qquad\qquad\qquad\qquad\qquad\qquad\qquad C$
$C-C-C-C-C$ $C-C-C-C-C$ $C-C-C-C$
$\qquad\quad C \qquad\qquad\qquad\qquad C \qquad\qquad\qquad C \ \ C$
$3,3\text{-dimethylpentane}$ C $2,2,3\text{-trimethylbutane}$
$\qquad\qquad\qquad\qquad\qquad 3\text{-ethylpentane}$

(b)

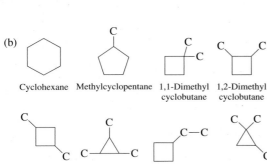

Cyclohexane Methylcyclopentane 1,1-Dimethyl cyclobutane 1,2-Dimethyl cyclobutane

1,3-Dimethyl cyclobutane 1,2,3-Trimethyl cyclopropane Ethylcyclobutane 1,1,2-Trimethyl cyclopropane

1-Ethyl-2-methyl cyclopropane 1-Ethyl-1-methyl cyclopropane Propylcyclopropane

Isopropylcyclopropane

(c)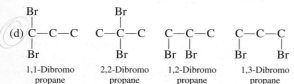

1-chloropentane   2-chloropentane   3-chloropentane

1-chloro-2-methylbutane   2-chloro-2-methylbutane   2-chloro-3-methylbutane

1-chloro-3-methylbutane   1-chloro-2,2-dimethylpropane

(d)

1,1-Dibromo propane 2,2-Dibromo propane 1,2-Dibromo propane 1,3-Dibromo propane

12.109 (a) 1,2-diethylcyclohexane (b) 3-methylhexane (c) 2,3-dimethyl-4-propylnonane (d) 1-isopropyl-3,5-dipropylcyclohexane   **12.110** (a) one (b) two (c) four (d) eight   **12.111** c   **12.112** a   **12.113** b   **12.114** b   **12.115** d   **12.116** c   **12.117** c   **12.118** c   **12.119** b   **12.120** c

Chapter 13 **13.1** one or more carbon–carbon multiple bonds are present
13.3 carbon–carbon double bond   **13.5** they are very similar
13.7 (a) unsaturated, alkene with one double bond (b) unsaturated, alkene with one double bond (c) unsaturated, diene (d) unsaturated, triene
13.9 (a) C_4H_{10} (b) C_5H_{10} (c) C_5H_8 (d) C_7H_{10} **13.11** (a) C_nH_{2n-2} (b) C_nH_{2n-2} (c) C_nH_{2n-2} (d) C_nH_{2n-6} **13.13** (a) alkene with one double bond (b) alkene with one double bond (c) diene (d) triene
13.15 (a) 2-butene (b) 2,4-dimethyl-2-pentene (c) 3-methylcyclohexane (d) 1,3-cyclopentadiene   **13.17** (a) 2-pentene (b) 2,3,3-trimethyl-1-butene (c) 2-methyl-1,4-pentadiene (d) 1,3,5-hexatriene

13.19 (a) $CH_2{=}CH{-}CH{-}CH_2{-}CH_3$
                                          $|$
                                        CH_3

(b)

(c) $CH_2{=}CH{-}CH{=}CH_2$

(d) $CH_2{=}CH{-}CH{-}CH{=}CH_2$
                          $|$
                        CH_2
                          $|$
                        CH_3

13.21 (a) 3-methyl-3-hexene (b) 2,3-dimethyl-2-hexene (c) 1,3-cyclopentadiene (d) 4,5-dimethylcyclohexene
13.23 (a) $CH_2{=}CH_2$ (b)   (c) $CH_2{=}CH{-}Br$

(d) $CH_2{=}CH{-}CH_2{-}I$

13.25 (a)   (b)

(c)   (d)

13.27 (a) 3-octene (b) 3-octene (c) 1,3-octadiene (d) 1,5-octadiene
13.29 (a) positional (b) skeletal (c) skeletal (d) positional
13.31 (a) 2 (b) 4 (c) 3 (d) zero
13.33 $C{=}C{-}C{-}C{-}C{-}C$ $C{-}C{=}C{-}C{-}C{-}C$
              1-Hexene   2-Hexene

$C{-}C{=}C{-}C{-}C$ $C{=}C{-}C{-}C{-}C$ $C{=}C{-}C{-}C{-}C$
      3-Hexene $|$ $|$
                                        C C
                          2-Methyl-1-pentene   3-Methyl-1-pentene

$C{=}C{-}C{-}C{-}C$ $C{-}C{=}C{-}C{-}C$ $C{-}C{=}C{-}C{-}C$
            $|$ $|$ $|$
          C C C
4-Methyl-1-pentene   2-Methyl-2-pentene   3-Methyl-2-pentene

$C{-}C{=}C{-}C{-}C$ $C{=}C{-}C{-}C$ C
            $|$ $|$ $|$ $|$
          C C C $C{=}C{-}C{-}C$
4-Methyl-2-pentene   2,3-Dimethyl-1-butene $|$
                                                                          C
                                                          3,3-Dimethyl-1-butene

$C{-}C{=}C{-}C$ $C{=}C{-}C{-}C$
        $|$ $|$ $|$
      C C C
2,3-Dimethyl-2-butene $|$
                                            C
                              2-Ethyl-1-butene

13.35 (a) no (b) no

(c)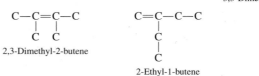

(d)

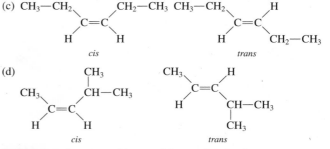

13.37 (a) *cis*-2-pentene (b) *trans*-1-bromo-2-iodoethene
(c) tetrafluoroethene (d) 2-methyl-2-butene

13.39 (a)

CH_3—CH_2, H / C=C / CH_3 CH_2—CH_3

(b) CH_3, CH_2—CH_3 / C=C / H H

(c) CH_3, H / C=C / H CH_2—CH—CH_2—CH_3 with CH_3

(d) CH_2=CH, H / C=C / H CH_3

13.41 a compound used by insects (and some animals) to transmit messages to other members of the same species   **13.43** Isoprene, the building block for terpenes, contains 5 carbon atoms.   **13.45** (a) gas (b) liquid (c) liquid (d) liquid   **13.47** (a) yes (b) no (c) yes (d) no

13.49 (a) CH_2=CH_2 + Cl_2 ⟶ CH_2—CH_2 with Cl Cl

(b) CH_2=CH_2 + HCl ⟶ CH_3—CH_2 with Cl

(c) CH_2=CH_2 + H_2 $\xrightarrow{Ni}$ CH_3—CH_3

(d) CH_2=CH_2 + HBr ⟶ CH_3—CH_2 with Br

13.51 (a) CH_2=CH—CH_3 + Cl_2 ⟶ CH_2—CH—CH_3 with Cl Cl

(b) CH_2=CH—CH_3 + HCl ⟶ CH_3—CH—CH_3 with Cl

(c) CH_2=CH—CH_3 + H_2 $\xrightarrow{Ni}$ CH_3—CH_2—CH_3

(d) CH_2=CH—CH_3 + HBr ⟶ CH_3—CH—CH_3 with Br

13.53 (a) CH_3—CH—CH—CH_3 with Cl Cl

(b) CH_3—CH_2—CH—CH_3 with Cl (c) (pentagon ring)   (d) HO—(square ring)

13.55 (a) Br_2 (b) H_2 + Ni catalyst (c) HCl (d) H_2O + H_2SO_4 catalyst   **13.57** (a) 2 (b) 2 (c) 2 (d) 3   **13.59** a large molecule formed by the repetitive bonding together of many smaller molecules   **13.61** a polymer in which the monomers add together to give the polymer as the only product

13.63 (a) CF_2=CF_2

(b) CH_2=C—CH=CH_2 with Cl (c) CH_2=CH with Cl (d) CH_2=CH with (benzene ring)

13.65 (a) —CH_2—CH_2—CH_2—CH_2—CH_2—CH_2—

(b) —CH_2—CH—CH_2—CH—CH_2—CH— with Cl Cl Cl

(c) —CH—CH—CH—CH—CH—CH— with Cl Cl Cl Cl Cl Cl

(d) —CH_2—CH—CH_2—CH—CH_2—CH— with Cl Cl Cl

13.67 C_nH_{2n-6} **13.69** (a) 1-hexyne (b) 4-methyl-2-pentyne (c) 2,2-dimethyl-3-heptyne (d) 1-butyne

13.71 C≡C—C—C—C (1-pentyne)

C—C≡C—C—C (2-pentyne)

C≡C—C—C (3-methyl-1-butyne) with C

13.73 because of the linearity (180° angles) about an alkyne's carbon–carbon triple bond   **13.75** Their physical properties are very similar.   **13.77** (a) CH_3—CH_3 (b) Br Br / CH_3—C—CH / Br Br   (c) Br / CH_3—C—CH_3 / Br

(d) CH_2=CH with Cl

13.79 (benzene ring)

13.81 implies that there are two types of carbon–carbon bonds present which is not the case   **13.83** (a) 1,3-dibromobenzene (b) 1-chloro-2-fluorobenzene (c) 1-chloro-4-fluorobenzene (d) 3-chlorotoluene

13.85 (a) *m*-dibromobenzene (b) *o*-chlorofluorobenzene (c) *p*-chlorofluorobenzene (d) *m*-chlorotoluene   **13.87** (a) 2,4-dibromo-1-chlorotoluene (b) 3-bromo-5-chlorotoluene (c) 1-bromo-3-chloro-2-fluorobenzene (d) 1,4-dibromo-2,5-dichlorobenzene

13.89 (a) 2-phenylbutane (b) 3-phenyl-1-butene (c) 3-methyl-1-phenylbutane (d) 2,4-diphenylpentane

13.91 (a) (benzene ring with CH_2—CH_3 and CH_2—CH_3) (b) (benzene ring with CH_3 and CH_3) (c) (benzene ring with CH_3 and CH_2—CH_3)

(d) (biphenyl rings)

13.93 liquid state   **13.95** petroleum   **13.97** (a) substitution (b) addition (c) substitution (d) addition

13.99 (a) Br_2 (b) (benzene ring with CH_3 / CH—CH_3)   (c) CH_3—CH_2—Br

13.101 carbon atoms are shared between rings   **13.103** (a) C_2H_4 (b) C_3H_4 (c) C_2H_2 (d) CH_4 **13.104** (a) more (b) more (c) more (d) the same number   **13.105** (a) no (b) yes (c) no (d) yes   **13.106** (a) All have six carbon atoms. (b) Cyclohexane has 12 H atoms, cyclohexene 10, and benzene 6. (c) Cyclohexane and benzene undergo substitution; cyclohexene undergoes addition. (d) All are liquids.

13.107 (a) CH_3—C≡C—CH_2—CH—CH_3 with CH_3

(b) CH_2—CH=CH—CH_3 with Cl

(c) CH_2=CH—CH_2—CH_2—CH_2—CH=CH_2

(d) CH_2=CH—CH—CH=CH_2 with CH_3

13.108 (a) two (b) one (c) two (d) two

13.109 (a) CH_2=CH—(benzene ring)

(b) $CH_2=CH-CH_2-Cl$ (c) $CH_3-CH_2-CH_2-C\equiv CH$
(d) $CH_3-CH_2-CH_2-C\equiv C-CH_2-CH_2-CH_3$
13.110 It would require a carbon atom that formed five bonds.
13.111 The substituted carbon atoms in 1,2-dichlorobenzene have only one substituent.
13.112 $CH_2=CH-CH_2-CH_2-CH_3$ $CH_3-CH=CH-CH_2-CH_3$
(*cis–trans* forms)

$CH_2=\underset{\underset{CH_3}{|}}{C}-CH_2-CH_3$ $CH_2=CH-\underset{\underset{CH_3}{|}}{CH}-CH_3$

$CH_3-\underset{\underset{CH_3}{|}}{C}=CH-CH_3$

(*cis–trans* forms)

13.113 1,2,3-trimethylbenzene; 1,2,4-trimethylbenzene; 1,3,5-trimethylbenzene; 2-ethyltoluene; 3-ethyltoluene; 4-ethyltoluene; propylbenzene; isopropylbenzene   **13.114** (a) 3 (b) 3 (c) 11
(d) 3   **13.115** c   **13.116** d   **13.117** b   **13.118** a   **13.119** b
13.120 a   **13.121** b   **13.122** c   **13.123** c   **13.124** a

Chapter 14   **14.1** (a) 2 (b) 1 (c) 4 (d) 1   **14.3** R—OH
14.5 R—O—H versus H—O—H   **14.7** (a) 2-pentanol
(b) 3-methyl-2-butanol (c) 2-ethyl-1-pentanol (d) 2-butanol
14.9 (a) 1-hexanol (b) 3-hexanol (c) 5,6-dimethyl-2-heptanol
(d) 2-methyl-3-pentanol
14.11 (a) $CH_2-\underset{\underset{CH_3}{|}}{CH}-CH_3$ (b) $CH_3-\underset{\underset{OH}{|}}{CH}-CH_2-\underset{\underset{CH_3}{|}}{CH}-CH_3$
$\overset{|}{OH}$

(c) $CH_3-\underset{\underset{C_6H_5}{|}}{\overset{\overset{OH}{|}}{C}}-CH_3$ (d) cyclobutane with OH and CH₃ substituents

14.13 (a) $CH_2-CH_2-CH_2-CH_2-CH_3$
$\overset{|}{OH}$
1-Pentanol
(b) $CH_2-CH_2-CH_3$ (c) $CH_3-\underset{\underset{CH_3}{|}}{CH}-CH_2-OH$
$\overset{|}{OH}$
1-Propanol   2-Methyl-1-Propanol
(d) $CH_3-CH_2-\underset{\underset{CH_3}{|}}{CH}-OH$
2-Butanol
14.15 (a) 1,2-propanediol (b) 1,4-pentanediol (c) 1,3-pentanediol
(d) 3-methyl-1,2,4-butanetriol   **14.17** (a) cyclohexanol
(b) *trans*-3-chlorocyclohexanol (c) *cis*-2-methylcyclohexanol
(d) 1-methylcyclobutanol
14.19 (a) $CH_3-\underset{\underset{OH}{|}}{CH}-CH_2-CH=CH_2$
(b) $CH\equiv C-\underset{\underset{OH}{|}}{CH}-CH_2-CH_3$ (c) $CH_3-\underset{\underset{OH}{|}}{CH}-\underset{\underset{CH_3}{|}}{C}=CH_2$
(d) $HO-CH_2$ CH_3
$\underset{\underset{H}{|}}{C}=\underset{\underset{H}{|}}{C}$

14.21 (a) $CH_2-\underset{\underset{CH_2}{|}}{CH}-CH_3$ (b) $CH_3-\underset{\underset{OH}{|}}{CH}-CH_2-\underset{\underset{OH}{|}}{CH_2}$
$\overset{|}{OH}$ $\underset{\underset{CH_3}{|}}{}$
2-Methyl-1-butanol   1,3-Butanediol
(c) $CH_3-\underset{\underset{CH_3}{|}}{CH}-\underset{\underset{OH}{|}}{CH}-CH_3$ (d) cyclopentane with OH and HO (1,3-Cyclopentanediol)
3-Methyl-2-butanol

14.23 (a) no (b) yes (c) yes (d) yes   **14.25** 1-heptanol, 2-heptanol, 3-heptanol, 4-heptanol   **14.27** $x = 1, 2, 3$ **14.29** (a) ethanol with all traces of H_2O removed (b) ethanol (c) 70% solution of isopropyl alcohol (d) ethanol   **14.31** (a) 1,2,3-propanetriol (b) ethanol (c) methanol (d) methanol   **14.33** Alcohol molecules can hydrogen-bond to each other; alkane molecules cannot.   **14.35** (a) 1-heptanol (b) 1-propanol (c) 1,2-ethanediol   **14.37** (a) 1-butanol (b) 1-pentanol (c) 1,2-butanediol   **14.39** (a) 3 (b) 3 (c) 3 (d) 3
14.41 (a) CH_2-CH_3 (b) $CH_3-CH_2-CH_2$
$\overset{|}{OH}$ $\overset{|}{OH}$
(c) $CH_3-CH_2-\underset{\underset{CH_3}{|}}{\overset{\overset{OH}{|}}{C}}-CH_3$ (d) $CH_3-CH_2-\underset{\underset{OH}{|}}{CH}-CH_2-CH_3$

14.43 (a) 2° (b) 2° (c) 1° (d) 2°
14.45 (a) $CH_2=CH-CH_3$ (b) $CH_3-CH_2-\underset{\underset{CH_3}{|}}{C}=CH_2$
(c) $CH_3-CH=CH_2$
(d) $CH_3-CH_2-CH_2-O-CH_2-CH_2-CH_3$
14.47
(a) $CH_3-\underset{\underset{OH}{|}}{CH}-\underset{\underset{CH_3}{|}}{CH}-CH_3$ (b) $CH_3-CH_2-CH_2$ or $CH_3-\underset{\underset{OH}{|}}{CH}-CH_3$
$\overset{|}{OH}$
(c) CH_3-CH_2-OH (d) $CH_3-\underset{\underset{CH_3}{|}}{CH}-CH_2-OH$
14.49 (a) $CH_3-CH_2-\underset{\underset{OH}{|}}{CH}-CH_3$ (b) $CH_3-CH_2-\underset{\underset{OH}{|}}{CH_2}$
(c) $CH_3-CH_2-\underset{\underset{OH}{|}}{CH_2}$ (d) cyclopentane with CH_2-OH
14.51 (a) $CH_3-CH_2-CH_2-Cl$ (b) cyclopentene with CH_3
(c) $CH_3-\overset{\overset{O}{||}}{C}-CH_2-CH_3$ (d) $CH_3-CH_2-O-CH_2-CH_3$

14.53 white solid, water soluble, hydrocarbon insoluble, does not absorb oxygen   **14.55** Phenols require the —OH groups to be attached directly to the benzene ring.   **14.57** (a) 3-ethylphenol (b) 2-chlorophenol (c) *o*-cresol (d) hydroquinone
14.59 (a) phenol ring with CH_2-CH_3 (b) phenol ring with Br and Br

(c)

(d)

14.61 low-melting solids or oily liquids
14.63 (a) both are flammable (b) both undergo halogenation
14.65

14.67 An antiseptic kills microorganisms on living tissue; a disinfectant kills microorganisms on inanimate objects.
14.69 BHA has methoxy and *tert*-butyl groups; BHT has a methyl group and two *tert*-butyl groups.   **14.71** (a) yes (b) no (c) yes (d) yes
14.73 (a) 1-ethoxypropane (b) 2-methoxypropane (c) methoxybenzene (d) cyclohexoxycyclohexane   **14.75** (a) ethyl propyl ether (b) isopropyl methyl ether (c) methyl phenyl ether (d) dicyclohexyl ether
14.77 (a) 1-methoxypentane (b) 1-ethoxy-2-methylpropane (c) 2-ethoxybutane (d) 2-methoxybutane
14.79

(a) $CH_3-CH-O-CH_2-CH_2-CH_3$ with CH_3 branch   (b) CH_3-CH_2-O- phenyl

(c) phenyl with $O-CH_3$ and CH_3

(d) cyclobutyl $O-CH_2-CH_3$

14.81 (a) no (b) no (c) yes (d) no   **14.83** butyl methyl ether, *sec*-butyl methyl ether, isobutyl methyl ether, *tert*-butyl methyl ether, ethyl isopropyl ether
14.85
(a) $CH_3-O-CH_2-CH_2-CH_3$, $CH_3-O-CH-CH_3$ (with CH_3 branch),

$CH_3-CH_2-O-CH_2-CH_3$

(b) $CH_3-CH_2-CH_2-CH_2-OH$, $CH_3-CH_2-CH-OH$ (with CH_3 branch),

$CH_3-CH-CH_2-OH$ (with CH_3 branch), CH_3-C-OH (with two CH_3 branches)

14.87 $x = 1, 2,$ and 3   **14.89** Dimethyl ether molecules cannot hydrogen-bond to each other; ethanol molecules can.   **14.91** flammability and peroxide formation   **14.93** No oxygen–hydrogen bonds are present.
14.95 (a) noncyclic ether (b) noncyclic ether (c) cyclic ether (d) nonether
14.97 R—S—H versus R—O—H
14.99 (a) $CH_3-CH_2-CH_2-CH_2$ (with SH branch)   (b) $CH_3-CH_2-CH-CH_2-CH_2$ (with CH_3 and SH branches)

(c) cyclopentyl—SH   (d) CH_2-CH_2 (with SH and SH)

14.101 (a) methyl mercaptan (b) propyl mercaptan (c) *sec*-butyl mercaptan (d) isobutyl mercaptan   **14.103** Alcohol oxidation produces aldehydes and ketones; thioalcohol oxidation produces disulfides.
14.105 R—S—R versus R—O—R   **14.107** (a) methylthioethane; ethyl methyl sulfide (b) 2-methylthiopropane; isopropyl methyl sulfide (c) methylthiocyclohexane; cyclohexyl methyl sulfide (d) 3-(methylthio)-

1-propene; allyl methyl sulfide   **14.109** (a) 2-hexanol (b) 3-phenyoxy-1-propene (c) 2-methyl-2-propanol (d) ethoxyethane
14.110

$CH_2-CH_2-CH_2-CH_2-CH_3$ (with OH)

$CH_3-C-CH_2-CH_3$ (with CH_3 and OH)

$CH_2-CH-CH_2-CH_3$ (with OH and CH_3)

$CH_3-CH-CH_2-CH_2-CH_3$ (with OH)

$CH_2-CH_2-CH-CH_3$ (with OH and CH_3)

CH_3-C-CH_2 (with CH_3 and OH)

$CH_3-CH_2-CH-CH_2-CH_3$ (with OH)

$CH_3-CH-CH-CH_3$ (with OH and CH_3)

$CH_3-CH_2-CH_2-CH_2-O-CH_3$

$CH_3-CH-CH_2-O-CH_3$ (with CH_3)

$CH_3-CH_2-CH-O-CH_3$ (with CH_3)

$CH_3-C-O-CH_3$ (with two CH_3)

$CH_3-CH_2-CH_2-O-CH_2-CH_3$

$CH_3-CH-O-CH_2-CH_3$ (with CH_3)

14.111 1-pentanol   **14.112** CH_3-O-CH_3, $CH_3-CH_2-CH_2-O-CH_2-CH_2-CH_3$, and $CH_3-O-CH_2-CH_2-CH_3$ **14.113** (a) disulfide (b) peroxide (c) alcohol, thiol, thioalcohol (d) ether, sulfide, thioether
14.114 (a) 3-methoxy-1-propanol (b) 1,2-dimethoxyethane (c) methylthioethane (d) 1-ethylthio-2-methoxyethane   **14.115** d
14.116 c   **14.117** b   **14.118** c   **14.119** b   **14.120** a   **14.121** a
14.122 d   **14.123** a   **14.124** d

Chapter 15   **15.1** (a) yes (b) no (c) yes (d) yes   **15.3** similarity: both have bonds involving four shared electrons; difference: C=O is polar, C=C is not polar   **15.5** 120°   **15.7** (a) yes (b) yes (c) no (d) yes
15.9 (a) aldehyde (b) ketone (c) amide (d) carboxylic acid
15.11 (a) neither (b) ketone (c) neither (d) aldehyde
15.13

$H-\overset{O}{\overset{\|}{C}}-H$, $CH_3-\overset{O}{\overset{\|}{C}}-H$, $CH_3-\overset{O}{\overset{\|}{C}}-CH_3$, $CH_3-CH_2-\overset{O}{\overset{\|}{C}}-CH_3$

15.15 (a) aldehyde (b) neither (c) ketone (d) ketone
15.17 (a) 2-methylbutanal (b) 4-methylheptanal (c) 3-phenylpropanal (d) propanal   **15.19** (a) pentanal (b) 3-methylbutanal (c) 3-methylpentanal (d) 2-ethyl-3-methylpentanal
15.21

(a)

$CH_3-CH_2-\overset{CH_3}{\underset{}{CH}}-CH_2-\overset{O}{\overset{\|}{C}}-H$

(b)

$CH_3-CH_2-CH_2-CH_2-\underset{\underset{CH_3}{\overset{|}{CH_2}}}{CH}-\overset{O}{\overset{\|}{C}}-H$

(c)

$$CH_3-\underset{\underset{Cl}{|}}{\overset{\overset{Cl}{|}}{C}}-\overset{\overset{O}{||}}{C}-H$$

(d)

$$CH_3-CH_2-CH_2-CH_2-\underset{\underset{OH}{|}}{CH}-CH_2-\underset{\underset{CH_3}{|}}{CH}-\overset{\overset{O}{||}}{C}-H$$

15.23 (a)

$$H-\overset{\overset{O}{||}}{C}-H$$

(b)

$$CH_3-CH_2-\overset{\overset{O}{||}}{C}-H$$

(c)

benzaldehyde with Cl (ortho-chlorobenzaldehyde structure)

(d)

benzaldehyde with two CH$_3$ groups

15.25 (a) propionaldehyde (b) propionaldehyde (c) dichloroacetaldehyde
(d) *o*-chlorobenzaldehyde   **15.27** (a) 2-butanone (b) 2,4,5-trimethyl-3-
hexanone (c) 6-methyl-3-heptanone (d) 1,1-dichloro-2-butanone
15.29 (a) 2-hexanone (b) 5-methyl-3-hexanone (c) 2-pentanone (d) 4-ethyl-
3-methyl-2-hexanone   **15.31** (a) cyclohexanone (b) 3-methylcyclohexanone
(c) 2-methylcyclohexanone (d) 3-chlorocyclopentanone
15.33
(a)

$$CH_3-\overset{\overset{O}{||}}{C}-\underset{\underset{CH_3}{|}}{CH}-CH_2-CH_3$$

(b)

$$CH_3-CH_2-\overset{\overset{O}{||}}{C}-CH_2-CH_2-CH_3$$

(c)

cyclobutanone

(d)

$$Cl-CH_2-\overset{\overset{O}{||}}{C}-CH_3$$

15.35 (a)

$$CH_3-\underset{\underset{CH_3}{|}}{CH}-\overset{\overset{O}{||}}{C}-CH_2-CH_2-CH_3$$

(b)

$$Cl-CH_2-\overset{\overset{O}{||}}{C}-CH_3$$

(c)

$$CH_3-\overset{\overset{O}{||}}{C}-\text{(phenyl)}$$

(d)

$$CH_3-\overset{\overset{O}{||}}{C}-\text{(phenyl)}$$

15.37 (a) heptanal (b) 2-heptanone, 3-heptanone, 4-heptanone
15.39 (a) 1 aldehyde, no ketones (b) 1 aldehyde, 1 ketone   **15.41** $x = 2, 4, 5$

15.43

$$C=C-C-\overset{\overset{O}{||}}{C}-H, \quad C-C-\overset{\overset{O}{||}}{\underset{\underset{C}{|}}{C}}-H, \quad C-C-\overset{\overset{O}{||}}{\underset{\underset{C}{|}}{C}}-H,$$

$$\underset{\underset{C}{|}}{\overset{\overset{C}{|}}{C}}-\overset{\overset{O}{||}}{C}-H, \quad C-\overset{\overset{O}{||}}{C}-C-C-C, \quad C-\overset{\overset{O}{||}}{C}-\underset{\underset{C}{|}}{C}-C, \quad C-C-\overset{\overset{O}{||}}{C}-C-C$$

15.45 formaldehyde is an irritating gas at room temperature; formalin is
an aqueous solution containing 37% formaldehyde by mass
15.47 a colorless, volatile liquid that is miscible with both water and
nonpolar solvents; its main use is as a solvent   **15.49** Dipole–dipole
attractions between molecules raise the boiling point.   **15.51** 2
15.53 ethanal, because it has a shorter carbon chain
15.55 (a)

$$CH_3-CH_2-CH_2-CH_2-\overset{\overset{O}{||}}{C}-H$$

(b)

$$CH_3-CH_2-\overset{\overset{O}{||}}{C}-CH_3$$

(c)

$$CH_3-\underset{\underset{CH_3}{|}}{\overset{\overset{CH_3}{|}}{C}}-CH_2-\overset{\overset{O}{||}}{C}-H$$

(d)

methylcyclohexanone (CH$_3$ substituted cyclohexanone with =O)

15.57 (a)

$$CH_3-CH_2-\underset{\underset{OH}{|}}{CH}-CH_2-CH_3$$

(b) CH_3-CH_2-OH

(c)

$$CH_3-\underset{\underset{OH}{|}}{CH}-CH_3$$

(d)

$$CH_3-CH_2-CH_2-CH_2-\underset{\underset{CH_3-CH_2}{|}}{CH}-\underset{\underset{OH}{|}}{CH_2}$$

15.59 (a)

$$CH_3-\overset{\overset{O}{||}}{C}-OH$$

(b)

$$CH_3-CH_2-CH_2-CH_2-\overset{\overset{O}{||}}{C}-OH$$

(c)

$$H-\overset{\overset{O}{||}}{C}-OH$$

(d)

$$CH_3-CH_2-\underset{\underset{Cl}{|}}{CH}-\underset{\underset{Cl}{|}}{CH}-CH_2-\overset{\overset{O}{||}}{C}-OH$$

15.61 appearance of a silver mirror   **15.63** Cu^{2+} ion   **15.65** (a) no (b)
yes (c) yes (d) no
15.67
(a) $CH_3-CH_2-CH_2-\underset{\underset{OH}{|}}{CH_2}$ (b) $CH_3-CH_2-\underset{\underset{OH}{|}}{CH}-CH_2-CH_3$

(c) $CH_3-\underset{\underset{CH_3}{|}}{CH}-CH_2-\underset{\underset{OH}{|}}{CH_2}$ (d) $CH_3-\underset{\underset{CH_3}{|}}{CH}-\underset{\underset{OH}{|}}{CH}-CH_2-CH_2-CH_3$

15.69 R—O— and H—   **15.71** (a) no (b) yes (c) yes (d) no
15.73
(a) $CH_3-\underset{\underset{OH}{|}}{CH}-O-CH_2-CH_3$ (b)

$$CH_3-\underset{\underset{O-CH_3}{|}}{\overset{\overset{OH}{|}}{C}}-CH_2-CH_2-CH_3$$

(c)

$$CH_3-CH_2-CH_2-\underset{\underset{O-CH_2-CH_3}{|}}{\overset{\overset{OH}{|}}{CH}}$$

(d)

$$CH_3-\underset{\underset{O-CH-CH_3}{|}}{\overset{\overset{OH}{|}}{C}}-CH_3$$
$$\underset{\underset{CH_3}{}}{}$$

15.75
(a)

$$CH_3-CH_2-CH_2-\underset{\underset{O-CH_2-CH_3}{|}}{\overset{\overset{OH}{|}}{CH}}$$

(b)

$$CH_3-CH_2-\overset{\overset{O}{||}}{C}-H$$

(c)

$$CH_3-CH_2-\underset{\underset{O-CH_3}{|}}{\overset{\overset{OH}{|}}{C}}-CH_3$$

(d)

cyclic structure with CH$_2$OH, O, OH, OH

15.77 (a) yes (b) yes (c) no (d) yes
15.79
(a) CH_3-OH (b) $CH_3-\underset{\underset{OH}{|}}{CH}-O-CH_3$

(c) $CH_3-CH_2-\underset{\underset{O-CH-CH_3}{|}}{\overset{}{CH}}-O-CH_3$ (d) $CH_3-\underset{\underset{OH}{|}}{CH}-O-CH_3, \; CH_3-OH$
$$\underset{\underset{CH_3}{}}{}$$

15.81
(a)

$$CH_3-\overset{\overset{O}{||}}{C}-H, \quad 2\,CH_3-OH$$

(b)

$$CH_3-\overset{\overset{O}{||}}{C}-CH_3, \quad 2\,CH_3-OH$$

(c)

$$CH_3-CH_2-\overset{\overset{O}{||}}{C}-CH_2-CH_3, \; CH_3-OH, \; CH_3-CH_2-OH$$

(d)

$$CH_3-CH_2-CH_2-CH_2-\overset{\overset{O}{||}}{C}-H, \; 2\,CH_3-OH$$

15.83 (a) dimethyl acetal of ethanal (b) dimethyl acetal of propanone
(c) ethyl methyl acetal of 3-pentanone (d) dimethyl acetal of pentanal
15.85 monomers are connected in a three-dimensional cross-linked
network   **15.87** mono-, di-, and tri-substituted phenols
15.89 thiocarbonyl compound

15.91

(a) H—C(=S)—H (b) H—C(=S)—H

(c) CH₃—C(=S)—CH₃ (d) CH₃—C(=S)—CH₃

15.93 (a) By definition, the carbonyl carbon atom is number 1 in an aldehyde; therefore, the number does not have to be specified in the name. (b) There is only one possible location for the carbonyl group in propanone; therefore, its location does not have to be specified.
15.94 (a) A ketone carbonyl group cannot be on a terminal carbon atom. (b) It requires a carbon atom with five bonds. (c) It requires a carbon atom with five bonds. (d) It requires a carbon atom with five bonds.
15.95 (a) a carbon atom bonded to both a hydroxyl group and an alkoxy group (b) a carbon atom bonded to two alkoxy groups
15.96
(a) CH₃—CH₂—CH(OH)—O—CH₂—CH₃   CH₃—CH₂—CH(O—CH₂—CH₃)—O—CH₂—CH₃
(b) cyclohexane with OH and O—CH₃   cyclohexane with O—CH₃ and O—CH₃
15.97 ring: CH₂, CH, O; CH with OH; CH₂—CH₂

15.98 (a) ketone, alkene (b) aldehyde, alcohol, ether (c) ketone, alkyne
(d) aldehyde, ketone **15.99** (a) ketone (b) aldehyde (c) aldehyde
(d) aldehyde **15.100** c **15.101** c **15.102** c **15.103** b **15.104** d
15.105 b **15.106** c **15.107** c **15.108** d **15.109** b

Chapter 16 **16.1** (a) yes (b) no (c) no (d) yes
16.3 (a) carboxylic acid (b) carboxylic acid derivative (c) neither
(d) carboxylic acid **16.5** (a) butanoic acid (b) 4-bromopentanoic acid
(c) 3-methylpentanoic acid (d) chloroethanoic acid **16.7** (a) hexanoic
acid (b) 3-methylpentanoic acid (c) 2,3-dimethylbutanoic acid
(d) 4,5-dimethylhexanoic acid
16.9
(a) CH₃—CH₂—CH(CH₃—CH₂)—C(=O)—OH
(b) CH₃—CH(CH₃)—CH₂—CH₂—CH(CH₃)—C(=O)—OH
(c) CH₃—CH(CH₃)—C(=O)—OH
(d) Cl—CH(Cl)—C(=O)—OH

16.11 (a) propanedioic acid (b) 3-methylpentanedioic acid
(c) 2-chlorobenzoic acid (d) 2-bromo-4-chlorobenzoic acid
16.13
(a) CH₃—CH₂—C(CH₃)(CH₃)—C(=O)—OH
(b) HO—C(=O)—C(CH₃)(CH₃)—CH₂—C(=O)—OH
(c) HO—C(=O)—C(CH₃)(CH₃)—CH₂—CH₂—C(=O)—OH
(d) benzene ring with COOH, Cl (ortho), Cl (para)

16.15
(a) CH₃—CH₂—CH₂—CH₂—C(=O)—OH
(b) CH₃—C(=O)—OH
(c) CH₃—CH₂—CH(Cl)—C(=O)—OH
(d) CH₃—CH₂—CH₂—CH(Br)—CH₂—C(=O)—OH
16.17
(a) HO—C(=O)—CH₂—C(=O)—OH
(b) HO—C(=O)—CH₂—CH₂—C(=O)—OH
(c) HO—C(=O)—CH₂—CH₂—CH(Br)—CH₂—CH₂—C(=O)—OH
(d) HO—C(=O)—CH(CH₃)—CH₂—CH₂—C(=O)—OH

16.19 (a) 3 (b) 1 (c) 2 (d) 1 **16.21** (a) carbon–carbon double bond
(b) hydroxyl group (c) carbon–carbon double bond (d) hydroxyl group
16.23 (a) propenoic acid (b) 2-hydroxypropanoic acid (c) *cis*-
butenedioic acid (d) 2-hydroxyethanoic acid
16.25
(a) CH₃—CH₂—C(=O)—CH₂—C(=O)—OH
(b) CH₃—CH₂—CH(OH)—C(=O)—OH
(c) CH₃—CH=CH—CH₂—CH₂—C(=O)—OH
(d) HO—C(=O)—CH(OH)—CH(OH)—CH₂—C(=O)—OH

16.27 (a) propionic acid (b) propionic acid (c) succinic acid (d) glutaric
acid **16.29** (a) hydroxy, carboxy (b) hydroxy, carboxy (c) keto, carboxy
(d) hydroxy, carboxy **16.31** (a) 2 (b) 5 **16.33** (a) solid (b) solid
(c) liquid (d) solid
16.35
(a) CH₃—C(=O)—OH (b) CH₃—C(=O)—OH
(c) CH₃—CH₂—CH(CH₃)—CH₂—C(=O)—OH (d) benzene ring with C(=O)—OH

16.37 (a) 1 (b) 3 (c) 2 (d) 2 **16.39** (a) −1 (b) −3 (c) −2 (d) −2
16.41 (a) pentanoate ion (b) citrate ion (c) succinate ion (d) oxalate ion

16.43

(a)

$$CH_3-\overset{O}{\overset{\|}{C}}-OH + H_2O \rightarrow H_3O^+ + CH_3-\overset{O}{\overset{\|}{C}}-O^-$$

(b)

$$HO-\overset{O}{\overset{\|}{C}}-CH_2-\overset{\overset{OH}{|}}{\underset{\overset{|}{\underset{O}{C}}-OH}{C}}-CH_2-\overset{O}{\overset{\|}{C}}-OH + 3H_2O \rightarrow$$

$$3H_3O^+ + {}^-O-\overset{O}{\overset{\|}{C}}-CH_2-\overset{\overset{OH}{|}}{\underset{\overset{|}{\underset{O}{C}}-O^-}{C}}-CH_2-\overset{O}{\overset{\|}{C}}-O^-$$

(c)

$$CH_3-\overset{O}{\overset{\|}{C}}-OH + H_2O \rightarrow H_3O^+ + CH_3-\overset{O}{\overset{\|}{C}}-O^-$$

(d)

$$CH_3-CH_2-\underset{\overset{|}{CH_3}}{CH}-\overset{O}{\overset{\|}{C}}-OH + H_2O \rightarrow$$

$$H_3O^+ + CH_3-CH_2-\underset{\overset{|}{CH_3}}{CH}-\overset{O}{\overset{\|}{C}}-O^-$$

16.45 (a) potassium ethanoate (b) calcium propanoate (c) potassium butanedioate (d) sodium pentanoate

16.47

(a)

$$CH_3-\overset{O}{\overset{\|}{C}}-OH + KOH \rightarrow CH_3-\overset{O}{\overset{\|}{C}}-O^- K^+ + H_2O$$

(b)

$$2\, CH_3-CH_2-\overset{O}{\overset{\|}{C}}-OH + Ca(OH)_2 \rightarrow$$

$$\left(CH_3-CH_2-\overset{O}{\overset{\|}{C}}-O^-\right)_2 Ca^{2+} + 2H_2O$$

(c)

$$HO-\overset{O}{\overset{\|}{C}}-CH_2-CH_2-\overset{O}{\overset{\|}{C}}-OH + 2KOH \rightarrow$$

$$K^+\ {}^-O-\overset{O}{\overset{\|}{C}}-CH_2-CH_2-\overset{O}{\overset{\|}{C}}-O^- K^+ + 2H_2O$$

(d)

$$CH_3-CH_2-CH_2-CH_2-\overset{O}{\overset{\|}{C}}-OH + NaOH \rightarrow$$

$$CH_3-CH_2-CH_2-CH_2-\overset{O}{\overset{\|}{C}}-O^- Na^+ + H_2O$$

16.49

(a)

$$CH_3-CH_2-CH_2-\overset{O}{\overset{\|}{C}}-O^- Na^+ + HCl \rightarrow$$

$$CH_3-CH_2-CH_2-\overset{O}{\overset{\|}{C}}-OH + NaCl$$

(b)

$$K^+\ {}^-O-\overset{O}{\overset{\|}{C}}-\overset{O}{\overset{\|}{C}}-O^- K^+ + 2HCl \rightarrow$$

$$HO-\overset{O}{\overset{\|}{C}}-\overset{O}{\overset{\|}{C}}-OH + 2KCl$$

(c)

$$\left({}^-O-\overset{O}{\overset{\|}{C}}-CH_2-\overset{O}{\overset{\|}{C}}-O^-\right)_2 Ca^{2+} + 2HCl \rightarrow$$

$$HO-\overset{O}{\overset{\|}{C}}-CH_2-\overset{O}{\overset{\|}{C}}-OH + CaCl_2$$

(d)

$$\text{(benzene ring)}-\overset{O}{\overset{\|}{C}}-O^-\ Na^+ + HCl \rightarrow \text{(benzene ring)}-\overset{O}{\overset{\|}{C}}-OH + NaCl$$

16.51 a compound used as a food preservative **16.53** (a) benzoic acid (b) sorbic acid (c) sorbic acid (d) propionic acid **16.55** (a) two oxygen atoms (b) two carbon atoms **16.57** (a) yes (b) yes (c) no (d) yes

16.59

(a)

$$CH_3-CH_2-\overset{O}{\overset{\|}{C}}-O-CH_3$$

(b)

$$CH_3-\overset{O}{\overset{\|}{C}}-O-CH_2-CH_2-CH_3$$

(c)

$$CH_3-CH_2-\underset{\overset{|}{CH_3}}{CH}-\overset{O}{\overset{\|}{C}}-O-\underset{\overset{|}{CH_3}}{CH}-CH_3$$

(d)

$$CH_3-CH_2-CH_2-CH_2-\overset{O}{\overset{\|}{C}}-O-\underset{\overset{|}{CH_3}}{CH}-CH_2-CH_3$$

16.61

(a)

$$CH_3-CH_2-\overset{O}{\overset{\|}{C}}-OH;\ CH_3-CH_2-OH$$

(b)

$$CH_3-CH_2-CH_2-\overset{O}{\overset{\|}{C}}-OH;\ CH_3-OH$$

(c)

$$CH_3-\overset{O}{\overset{\|}{C}}-OH;\ \text{(benzene ring)}-OH$$

(d)

$$\text{(benzene ring)}-\overset{O}{\overset{\|}{C}}-OH;\ CH_3-OH$$

16.63 a cyclic ester; intermolecular esterification of a hydroxyacid **16.65** (a) methyl propanoate (b) methyl methanoate (c) propyl ethanoate (d) isopropyl propanoate **16.67** (a) methyl propionate (b) methyl formate (c) propyl acetate (d) isopropyl propionate **16.69** (a) ethyl butanoate (b) propyl pentanoate (c) methyl 3-methylpropanoate (d) ethyl propanoate

16.71

(a)

$$H-\overset{O}{\overset{\|}{C}}-O-CH_3$$

(b)

$$\text{(benzene ring)}-CH_2-\overset{O}{\overset{\|}{C}}-O-CH_2-CH_3$$

(c)

$$CH_3-\overset{O}{\overset{\|}{C}}-O-\underset{\overset{|}{CH_3}}{CH}-CH_3$$

(d)

$$CH_3-\overset{O}{\overset{\|}{C}}-O-CH_2-\underset{\overset{|}{Br}}{CH}-CH_3$$

16.73 (a) ethyl ethanoate (b) methyl ethanoate (c) ethyl butanoate (d) 1-methylpropyl hexanoate **16.75** acid part is the same; alcohol part is methyl (apple) versus ethyl (pineapple) **16.77** the —OH group of salicylic acid has been esterified (methyl ester) in aspirin **16.79** pentanoic acid, 2-methylbutanoic acid, 3-methylbutanoic acid, 2,2-dimethylpropanoic acid **16.81** methyl pentanoate, methyl 2-methylbutanoate, methyl 3-methylbutanoate methyl 2,2-dimethylpropanoate **16.83** nine (methyl butanoate, methyl 2-methylpropanoate, ethyl propanoate, propyl ethanoate, isopropyl ethanoate, butyl methanoate, *sec*-butyl methanoate, isobutyl methanoate, *tert*-butyl methanoate)

16.85

$$CH_3-CH_2-\overset{O}{\overset{\|}{C}}-OH,\ CH_3-\overset{O}{\overset{\|}{C}}-O-CH_3,\ H-\overset{O}{\overset{\|}{C}}-O-CH_2-CH_3$$

16.87 No oxygen–hydrogen bonds are present. **16.89** There is no hydrogen bonding between ester molecules.

16.91

(a)

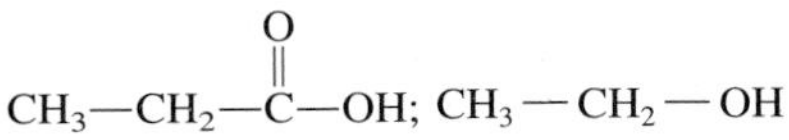

CH₃—CH₂—C(=O)—OH; CH₃—CH₂—OH

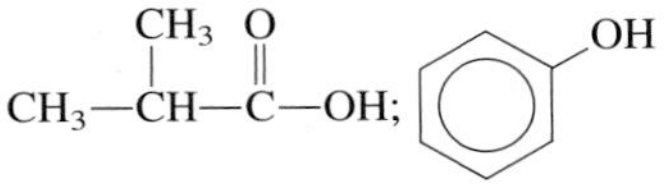

(b)

CH₃—CH(CH₃)—C(=O)—OH; (phenol with OH)

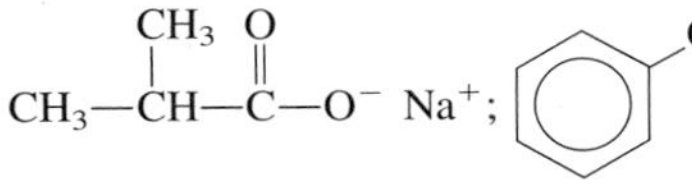

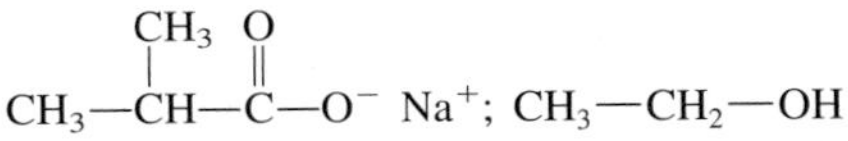

(c)

CH₃—CH₂—CH₂—C(=O)—OH; CH₃—OH

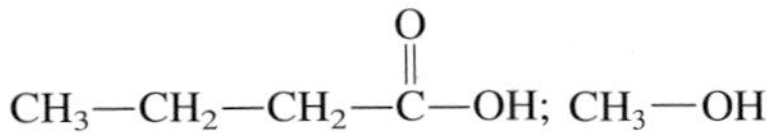

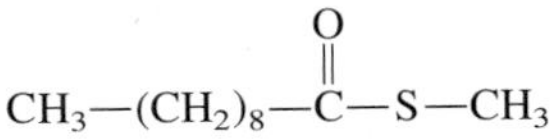

(d)

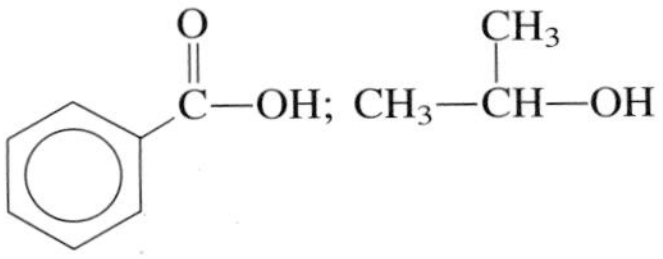

(benzene)—C(=O)—OH; CH₃—CH(OH)—H

16.93

(a)

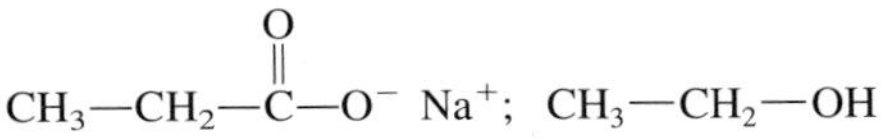

CH₃—CH₂—C(=O)—O⁻ Na⁺; CH₃—CH₂—OH

(b)

CH₃—CH(CH₃)—C(=O)—O⁻ Na⁺; (phenol with OH)

(c)

CH₃—CH₂—CH₂—C(=O)—O⁻ Na⁺; CH₃—OH

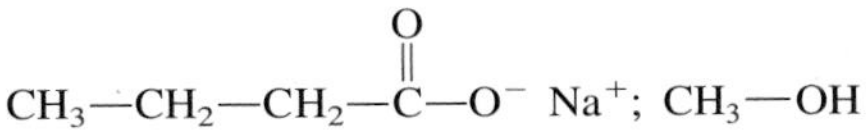

(d)

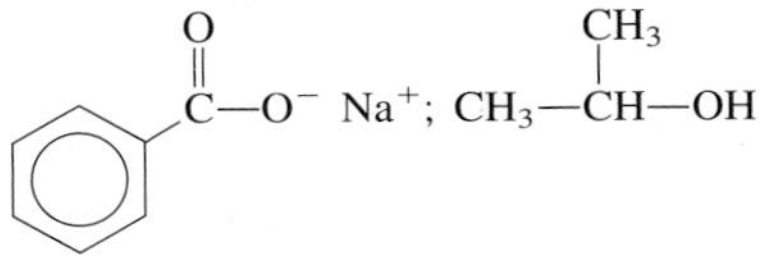

(benzene)—C(=O)—O⁻ Na⁺; CH₃—CH(OH)—H

16.95

(a)

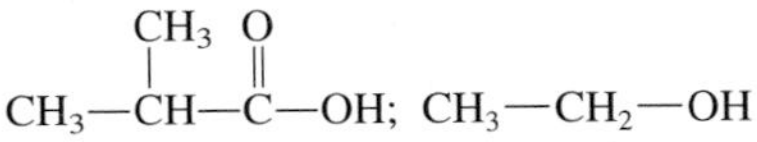

CH₃—CH(CH₃)—C(=O)—OH; CH₃—CH₂—OH

(b)

CH₃—CH(CH₃)—C(=O)—O⁻ Na⁺; CH₃—CH₂—OH

(c)

H—C(=O)—OH; CH₃—CH₂—CH₂—CH₂—OH

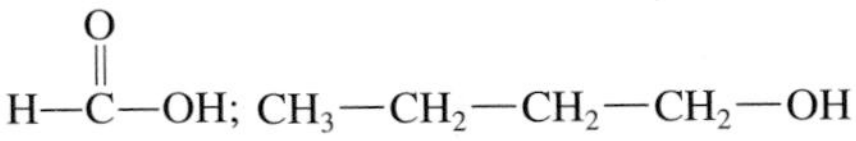

(d)

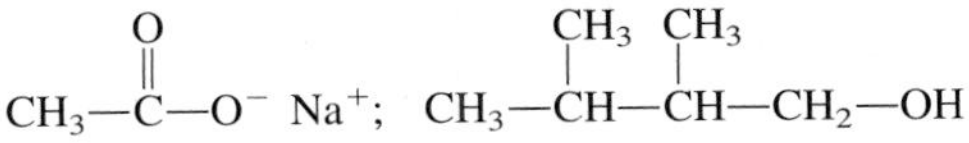

CH₃—C(=O)—O⁻ Na⁺; CH₃—CH(CH₃)—CH(CH₃)—CH₂—OH

16.97

(a)

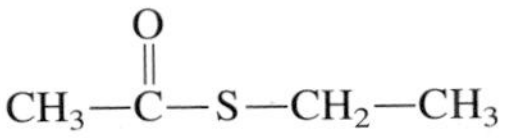

CH₃—C(=O)—S—CH₂—CH₃

(b)

CH₃—(CH₂)₈—C(=O)—S—CH₃

(c)

(benzene)—C(=O)—S—CH(CH₃)—CH₃

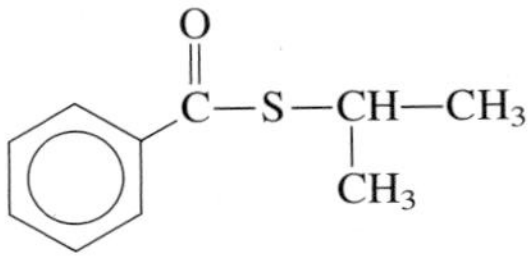

(d)

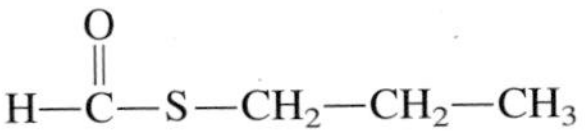

H—C(=O)—S—CH₂—CH₂—CH₃

16.99

(a)

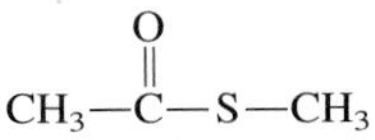

CH₃—C(=O)—S—CH₃

(b) same as part a.

(c)

H—C(=O)—S—CH₂—CH₃

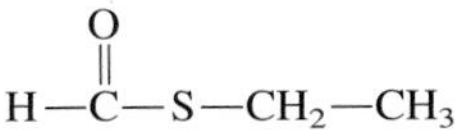

(d) same as part c.

16.101 acetic acid and coenzyme A

16.103

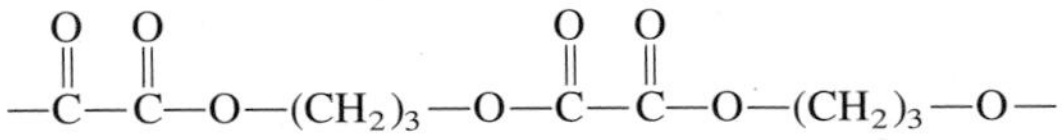

—C(=O)—C(=O)—O—(CH₂)₃—O—C(=O)—C(=O)—O—(CH₂)₃—O—

16.105

HO—C(=O)—CH₂—CH₂—C(=O)—OH;   HO—CH₂—CH₂—CH₂—OH

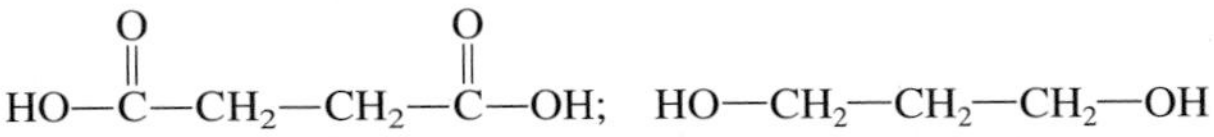

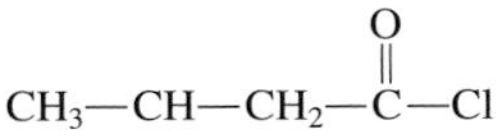

16.107 ethylene glycol and terephthalic acid   **16.109** specialty packaging, orthopedic devices, and controlled drug-release formulations

16.111

(a)

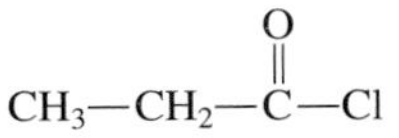

CH₃—CH₂—C(=O)—Cl

(b) CH₃—CH—CH₂—C(=O)—Cl

(c)

CH₃—CH₂—CH₂—C(=O)—O—C(=O)—CH₂—CH₂—CH₃

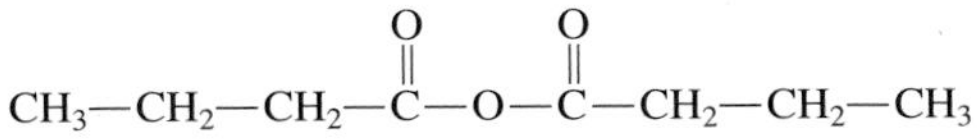

(d) 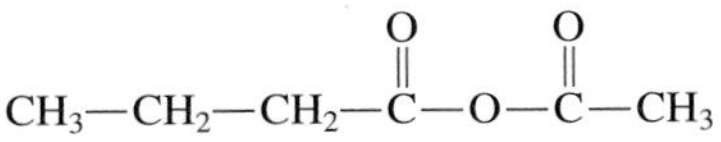 CH₃—CH₂—CH₂—C(=O)—O—C(=O)—CH₃

16.113 (a) ethanoic propanoic anhydride (b) pentanoyl chloride (c) 2,3-dimethylbutanoyl chloride (d) methanoic propanoic anhydride

16.115

(a)

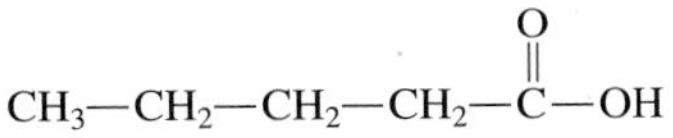

CH₃—CH₂—CH₂—CH₂—C(=O)—OH

(b) 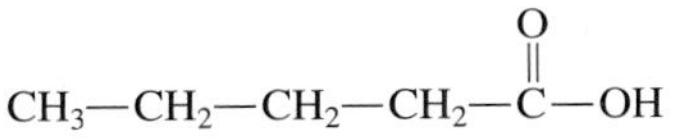 CH₃—CH₂—CH₂—CH₂—C(=O)—OH

16.117

(a)

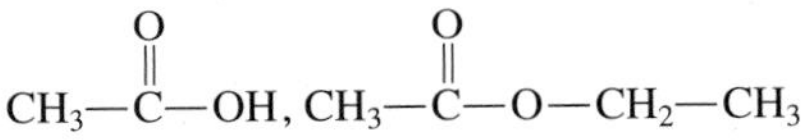

CH₃—C(=O)—OH, CH₃—C(=O)—O—CH₂—CH₃

(b)

CH₃—C(=O)—OH, CH₃—C(=O)—O—CH₂—CH₂—CH₂—CH₃

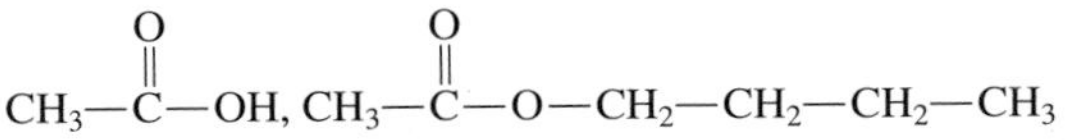

16.119 (a) propanoyl group (b) butanoyl group (c) butanoyl group (d) ethanoyl group   **16.121** an ester and HCl

16.123

(a)

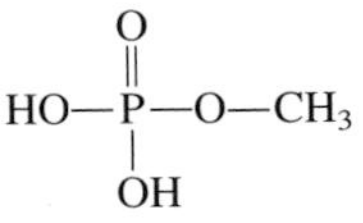

HO—P(=O)(OH)—O—CH₃

(b)

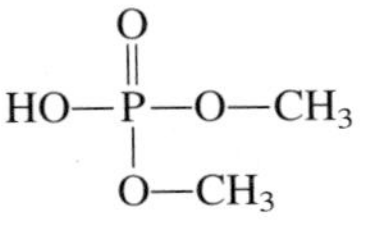

HO—P(=O)(O—CH₃)—O—CH₃

(c)

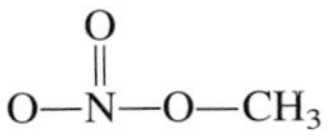

O—N(=O)—O—CH₃

(d)

O—N(=O)—O—CH₂—CH₂—O—N(=O)—O

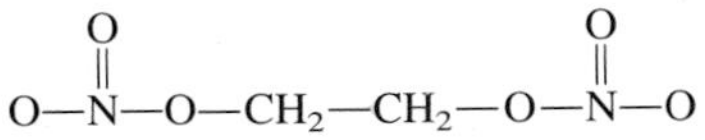

16.125 H₃PO₄ is a triprotic acid, and H₂SO₄ is a diprotic acid.

16.127 (a) 2,2 (b) 7,1 (c) 7,1 (d) 6,3

16.128

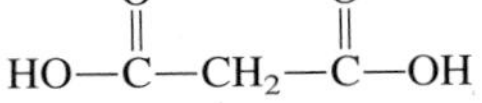

HO—C(=O)—CH₂—C(=O)—OH
Malonic acid

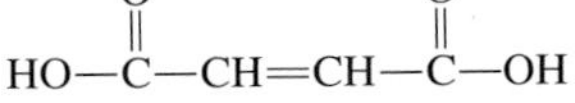

HO—C(=O)—CH=CH—C(=O)—OH
Maleic acid (*cis* isomer)

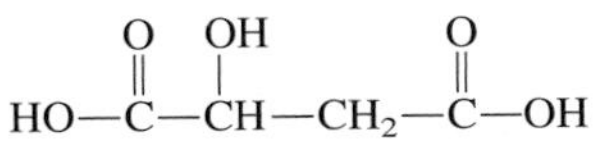

HO—C(=O)—CH(OH)—CH₂—C(=O)—OH
Malic acid

16.129 $C_nH_{2n-2}O_2$ **16.130** (a) ethyl 2-methylpropanoate (b) 2-methylbutanoic acid (c) ethyl thiobutanoate (d) sodium propanoate

16.131

$$CH_3-\overset{\overset{\displaystyle O}{\|}}{C}-O-CH_2-CH_3$$

16.132

(a)

$$CH_3-CH_2-\overset{\overset{\displaystyle O}{\|}}{C}-O^-\ Na^+;\ CH_3-OH$$

(b)

$$CH_3-CH_2-\overset{\overset{\displaystyle O}{\|}}{C}-S-CH_3$$

(c)

$$CH_3-\overset{\overset{\displaystyle O}{\|}}{C}-O^-\ Na^+$$

(d)

$$HO-\overset{\overset{\displaystyle O}{\|}}{C}-CH_2-CH_2-\underset{\underset{\displaystyle CH_3}{|}}{CH}-CH_2-CH_2\overset{\displaystyle OH}{}$$

16.133 b **16.134** a **16.135** b **16.136** a **16.137** b **16.138** a
16.139 d **16.140** c **16.141** d **16.142** d

Chapter 17   **17.1** 3 (N), 2 (O), and 4 (C)   **17.3** (a) R—NH$_2$
(b) R—NH—R (c) R—N—R
    |
    R

17.5 (a) yes (b) yes (c) no (d) yes   **17.7** (a) 1° (b) 1° (c) 2° (d) 3°
17.9 (a) 2° (b) 3° (c) 3° (d) 1°   **17.11** (a) ethylmethylamine
(b) propylamine (c) diethylmethylamine (d) isopropylmethylamine
17.13 (a) 3-pentanamine (b) 2-methyl-3-pentanamine (c) N-methyl-3-pentanamine (d) 2,3-butanediamine   **17.15** (a) 1-propanamine
(b) N-ethyl-N-methylethanamine (c) N-methyl-1-propanamine (d) N-methyl-2-butanamine   **17.17** (a) 2-bromoaniline (b) N-isopropylaniline
(c) N-ethyl-N-methylaniline (d) N-methyl-N-phenylaniline
17.19
(a)

$$CH_3-\overset{\overset{\displaystyle NH_2}{|}}{\underset{\underset{\displaystyle CH_3}{|}}{C}}-CH_2-CH_3$$

(b) H$_2$N—CH$_2$—CH$_2$—CH$_2$—CH$_2$—CH$_2$—CH$_2$—NH$_2$

(c)

$$CH_3-\overset{\overset{\displaystyle NH_2}{|}}{CH}-\overset{\overset{\displaystyle O}{\|}}{C}-CH_2-CH_3$$

(d)

$$CH_3-\underset{\underset{\displaystyle NH_2}{|}}{CH}-\overset{\overset{\displaystyle O}{\|}}{C}-OH$$

17.21
CH$_2$—CH$_2$—CH$_2$—CH$_2$—CH$_3$, CH$_3$—CH—CH$_2$—CH$_2$—CH$_3$,
| |
NH$_2$ NH$_2$

CH$_3$—CH$_2$—CH—CH$_2$—CH$_3$, CH$_2$—CH—CH$_2$—CH$_3$,
            | | |
            NH$_2$ NH$_2$ CH$_3$

$$CH_3-\overset{\overset{\displaystyle CH_3}{|}}{\underset{\underset{\displaystyle NH_2}{|}}{C}}-CH_2-CH_3,\ CH_3-\underset{\underset{\displaystyle CH_3}{|}}{CH}-\underset{\underset{\displaystyle NH_2}{|}}{CH}-CH_3,$$

$$CH_3-\underset{\underset{\displaystyle CH_3}{|}}{CH}-CH_2-CH_2,\ CH_3-\overset{\overset{\displaystyle CH_3}{|}}{\underset{\underset{\displaystyle CH_3}{|}}{C}}-CH_2-NH_2$$
                              NH$_2$

17.23 dimethylpropylamine, isopropyldimethylamine, diethylmethylamine
17.25 1-propanamine, 2-propanamine, N-methylethanamine,
N,N-dimethylmethanamine   **17.27** (a) liquid (b) gas (c) gas (d) liquid
17.29 (a) 3 (b) 3   **17.31** Hydrogen bonding is possible for the amine.
17.33 (a) CH$_3$—CH$_2$—NH$_2$; it has fewer carbon atoms. (b) H$_2$N—CH$_2$—CH$_2$—CH$_2$—NH$_2$; it has two amino groups rather than one.

17.35
(a) CH$_3$—CH$_2$—NH$_3$ (b) OH$^-$
(c) CH$_3$—CH—NH—CH$_3$
            |
            CH$_3$

(d) CH$_3$—CH$_2$—$\overset{+}{N}$H$_2$—CH$_2$—CH$_3$; OH$^-$

17.37 (a) dimethylammonium ion (b) triethylammonium ion
(c) N,N-diethylanilinium ion (d) N-isopropylanilinium ion
17.39
(a) CH$_3$—NH—CH$_3$

(b) CH$_3$—CH$_2$—N—CH$_2$—CH$_3$
                |
                CH$_2$—CH$_3$

(c) CH$_3$—CH$_2$—N—CH$_2$—CH$_3$
                |
            (benzene ring)

(d)
NH—CH—CH$_3$
    |
    CH$_3$
(benzene ring)

17.41
(a) CH$_3$—CH$_2$—$\overset{+}{N}$H$_3$ Cl$^-$

(b) (benzene ring)—$\overset{+}{N}$H$_3$ Br$^-$

(c)
$$CH_3-\overset{\overset{\displaystyle CH_3}{|}}{\underset{\underset{\displaystyle CH_3}{|}}{C}}-NH_2$$

(d) HCl

17.43
(a) CH$_3$—CH—NH$_2$ (b) CH$_3$—$\overset{+}{N}$H$_2$ Cl$^-$
        | |
        CH$_3$ CH$_3$

(c)
(benzene ring)—N—CH$_3$
                |
                CH$_3$
                         (d) CH$_3$—NH—CH$_3$

17.45 (a) propylammonium chloride (b) methylpropylammonium
chloride (c) ethyldimethylammonium bromide (d) N,N-dimethylanilinium
bromide
17.47 (a) free amine, free base, deprotonated base (b) free amine, free
base, deprotonated base (c) protonated base (d) protonated base
17.49 to increase water solubility   **17.51** ethylmethylamine
hydrochloride
17.53
(a) CH$_3$—CH$_2$—CH$_2$—NH$_2$, NaCl, H$_2$O

(b) CH$_3$—CH—N—CH$_3$, NaBr, H$_2$O
            | |
            CH$_3$ CH$_3$

(c) CH$_3$—CH$_2$—NH—CH$_2$—CH$_3$, NaCl, H$_2$O

(d)
$$CH_3-\overset{\overset{\displaystyle CH_3}{|}}{\underset{\underset{\displaystyle CH_3}{|}}{C}}-NH_2,$$
                    NaBr, H$_2$O

17.55 ethylmethylamine and propyl chloride, ethylpropylamine and
methyl chloride, methylpropylamine and ethyl chloride

17.57

(a)

$$CH_3-\overset{\overset{\displaystyle CH_3}{|}}{\underset{\underset{\displaystyle CH_3}{|}}{N^+}}-CH_2-CH_3 \ \ Br^-$$

(b)

$$CH_3-\overset{\overset{\displaystyle CH_3}{|}}{\underset{\underset{\displaystyle CH_3}{|}}{CH}}-N-\overset{\overset{\displaystyle }{}}{\underset{\underset{\displaystyle CH_3}{|}}{CH}}-CH_3$$

(c)

$$CH_3-CH_2-\overset{\overset{\displaystyle CH_3}{|}}{\underset{\underset{\displaystyle CH_3}{|}}{N^+}}-CH_2-CH_2-CH_3 \ \ Cl^-$$

(d) $CH_3-CH_2-NH-CH_2-CH_3$

17.59 (a) amine salt (b) quaternary ammonium salt (c) amine salt (d) quaternary ammonium salt **17.61** (a) trimethylammonium bromide (b) tetramethylammonium chloride (c) ethylmethylammonium bromide (d) diethyldimethylammonium chloride **17.63** (a) yes (b) yes (c) no (d) yes **17.65** (a) purine (b) pyrrole (c) imidazole (d) indole **17.67** (a) false (b) true (c) true (d) false **17.69** (a) one (b) one (c) three (d) two **17.71** (a) yes (b) yes (c) yes (d) yes **17.73** (a) true (b) true (c) true (d) false

17.75

(a) $R-\overset{\overset{\displaystyle O}{||}}{C}-NH_2$ (b) $R-\overset{\overset{\displaystyle O}{||}}{C}-NH-R$ (c) $R-\overset{\overset{\displaystyle O}{||}}{C}-\underset{\underset{\displaystyle R}{|}}{N}-R$

17.77 (a) yes (b) yes (c) no (d) yes **17.79** (a) monosubstituted (b) disubstituted (c) unsubstituted (d) monosubstituted **17.81** (a) secondary amide (b) tertiary amide (c) primary amide (d) secondary amide **17.83** (a) N-ethylethanamide (b) N,N-dimethylpropanamide (c) butanamide (d) 2-chloropropanamide **17.85** (a) N-ethylacetamide (b) N,N-dimethylpropionamide (c) butyramide (d) 2-chloropropionamide **17.87** (a) propanamide (b) N-methylpropanamide (c) 3,5-dimethylhexanamide (d) N,N-dimethylbutanamide

17.89

(a) $CH_3-\overset{\overset{\displaystyle O}{||}}{C}-\underset{\underset{\displaystyle CH_3}{|}}{N}-CH_3$

(b) $CH_3-CH_2-\overset{\overset{\displaystyle CH_3}{|}}{CH}-\overset{\overset{\displaystyle O}{||}}{C}-NH_2$

(c) $CH_3-\overset{\overset{\displaystyle CH_3}{|}}{CH}-CH_2-\overset{\overset{\displaystyle O}{||}}{C}-NH-CH_3$

(d) $H-\overset{\overset{\displaystyle O}{||}}{C}-NH_2$

17.91 $H_2N-\overset{\overset{\displaystyle }{}}{C}-NH_2$ **17.93** one of acetamide's amide hydrogen atoms has been replaced with a hydroxyphenyl group in acetaminophen **17.95** An electronegativity effect induced by the carbonyl oxygen atom makes the lone pair of electrons on the nitrogen atom unavailable. **17.97** (a) 5 (b) 5

17.99

(a) CH_3-NH_2 (b)

$$CH_3-\overset{\overset{\displaystyle CH_3}{|}}{\underset{\underset{\displaystyle CH_3}{|}}{C}}-\overset{\overset{\displaystyle O}{||}}{C}-\underset{\underset{\displaystyle CH_3}{|}}{N}-CH_3$$

(c) NH_3

(d)

$$\text{(benzene ring)}-\overset{\overset{\displaystyle O}{||}}{C}-OH$$

17.101

(a) $CH_3-\overset{\overset{\displaystyle O}{||}}{C}-OH, \ CH_3-NH-\overset{\overset{\displaystyle CH_3}{|}}{CH}-CH_3$

(b) $CH_3-CH_2-CH_2-CH_2-\overset{\overset{\displaystyle O}{||}}{C}-OH, \ CH_3-NH_2$

(c) $CH_3-\overset{\overset{\displaystyle CH_3}{|}}{CH}-\overset{\overset{\displaystyle O}{||}}{C}-OH, \ CH_3-NH_2$

(d) $CH_3-\overset{\overset{\displaystyle CH_3}{|}}{CH}-\overset{\overset{\displaystyle CH_3}{|}}{CH}-\overset{\overset{\displaystyle O}{||}}{C}-OH, \ CH_3-NH_2$

17.103

(a) $CH_3-CH_2-CH_2-\overset{\overset{\displaystyle O}{||}}{C}-OH, \ CH_3-NH_2$

(b) $CH_3-CH_2-CH_2-\overset{\overset{\displaystyle O}{||}}{C}-OH, \ CH_3-\overset{+}{N}H_3 \ Cl^-$

(c) $CH_3-CH_2-CH_2-\overset{\overset{\displaystyle O}{||}}{C}-O^- \ Na^+, \ CH_3-NH_2$

(d) $\text{(benzene ring)}-\overset{\overset{\displaystyle O}{||}}{C}-OH, \ \text{(benzene ring)}-NH-CH_3$

17.105 diacid and diamine

17.107

$$\left(-\overset{\overset{\displaystyle O}{||}}{C}-CH_2-CH_2-\overset{\overset{\displaystyle O}{||}}{C}-\overset{\overset{\displaystyle H}{|}}{N}-CH_2-CH_2-CH_2-CH_2-\overset{\overset{\displaystyle H}{|}}{N}-\right)_n$$

17.109

$$R-\overset{\overset{\displaystyle H}{|}}{N}-\overset{\overset{\displaystyle O}{||}}{C}-O-R$$

17.111

(a) $CH_3-CH_2-CH_2-\overset{\overset{\displaystyle CH_3}{|}}{CH}-\overset{\overset{\displaystyle O}{||}}{C}-NH_2$

(b) $CH_3-\overset{\overset{\displaystyle O}{||}}{C}-NH-\overset{\overset{\displaystyle CH_3}{|}}{CH}-CH_3$

(c) $CH_3-CH_2-\overset{+}{N}H_2-CH_2-CH_3 \ Cl^-$

(d) $CH_3-\overset{\overset{\displaystyle }{}}{\underset{\underset{\displaystyle \text{(benzene ring)}}{}}{\overset{+}{N}H}}\overset{\overset{\displaystyle CH_3}{\diagup}}{\underset{\underset{\displaystyle CH_3}{\diagdown}}{}} \ Cl^-$

17.112

(a) $CH_3-CH_2-NH-CH_3$

(b) $CH_3-CH_2-\overset{\overset{\displaystyle CH_3}{|}}{CH}-\overset{\overset{\displaystyle O}{||}}{C}-\underset{\underset{\displaystyle CH_3}{|}}{N}-CH_3$

(c) $CH_3-\overset{\overset{\displaystyle CH_3}{|}}{\underset{\underset{\displaystyle CH_3}{|}}{N^+}}-CH_3 \ Cl^-$

(d) $CH_3-CH_2-\overset{\overset{\displaystyle CH_3}{|}}{CH}-\overset{\overset{\displaystyle O}{||}}{C}-O^- \ Na^+$

17.113

(a)

$$CH_3-CH_2-\overset{\overset{\displaystyle O}{\|}}{C}-NH-CH_2-CH_2-CH_3$$

(b) $CH_3-NH-CH_3$

(c)

$$CH_3-CH_2-\overset{\overset{\displaystyle CH_3}{|}}{CH}-\overset{\overset{\displaystyle O}{\|}}{C}-OH, \quad CH_3-\overset{+}{N}H_3\ Cl^-$$

(d) $CH_3-CH_2-\overset{+}{N}H_3,\ OH^-$

17.114

$$CH_3-CH_2-\overset{\overset{\displaystyle O}{\|}}{C}-NH_2 \qquad CH_3-\overset{\overset{\displaystyle O}{\|}}{C}-NH-CH_3$$

Propanamide   *N*-methylethanamide

$$H-\overset{\overset{\displaystyle O}{\|}}{C}-NH-CH_2-CH_3 \qquad H-\overset{\overset{\displaystyle O}{\|}}{C}-\overset{\overset{\displaystyle CH_3}{|}}{N}-CH_3$$

N-ethylmethanamide   *N,N*-dimethylmethanamide

17.115

$$CH_3-\overset{\overset{\displaystyle CH_3}{|}}{\underset{\underset{\displaystyle CH_3}{|}}{\overset{+}{N}}}-CH_2-CH_3\ Cl^-$$

17.116 (a) unsubstituted (b) monosubstituted (c) monosubstituted (d) unsubstituted   **17.117** (a) amine (b) amide (c) amine (d) amide **17.118** (a) 2,*N*-dimethylpentanamine (b) 3-methylpentanamide (c) 1,4-pentanediamine (d) 4-bromo-*N*-ethyl-*N*-methylpentanamide **17.119** c   **17.120** c   **17.121** c   **17.122** b   **17.123** c   **17.124** b **17.125** c   **17.126** a   **17.127** b   **17.128** b

Chapter 18   18.1 (a) study of the chemical substances in living organisms and the interactions of these substances with each other (b) chemical substance found within a living organism   **18.3** proteins, lipids, carbohydrates, and nucleic acids

18.5 $CO_2 + H_2O + \text{solar energy} \xrightarrow[\text{Plant enzymes}]{\text{Chlorophyll}} \text{carbohydrates} + O_2$

18.7 serve as structural elements, provide energy reserves **18.9** Carbohydrates are polyhydroxy aldehydes, polyhydroxy ketones, or compounds that yield such substances upon hydrolysis.   **18.11** (a) 2 (b) 4 (c) 2 to 10 (d) many (several thousand usually)   **18.13** Superimposable objects have parts that coincide exactly at all points when the objects are laid upon each other.   **18.15** (a) drill bit (b) hand, foot, ear (c) PEEP, POP **18.17** (a) no (b) no (c) yes (d) yes

18.19

(a) no chiral center   (b)

$$\underset{\underset{\displaystyle Br}{|}}{CH_2}-\overset{*}{\underset{\underset{\displaystyle Br}{|}}{C}}-\overset{*}{\underset{\underset{\displaystyle Br}{|}}{CH}}$$

(c)

$$\underset{\underset{\displaystyle OH}{|}}{CH_2}-\overset{*}{\underset{\underset{\displaystyle OH}{|}}{CH}}-\overset{*}{\underset{\underset{\displaystyle OH}{|}}{CH}}-\overset{*}{\underset{\underset{\displaystyle OH}{|}}{CH}}-\overset{\overset{\displaystyle O}{\|}}{C}-H$$

(d) $\underset{\underset{\displaystyle OH}{|}}{CH_2}-\overset{*}{\underset{\underset{\displaystyle OH}{|}}{CH}}-\overset{*}{\underset{\underset{\displaystyle OH}{|}}{CH}}-\overset{*}{\underset{\underset{\displaystyle OH}{|}}{CH}}-\overset{*}{\underset{\underset{\displaystyle OH}{|}}{CH}}-\underset{\underset{\displaystyle OH}{|}}{CH_2}$

18.21 (a) zero (b) two (c) zero (d) zero   **18.23** (a) achiral (b) chiral (c) chiral (d) chiral   **18.25** Constitutional isomers have a different connectivity of atoms. Stereoisomers have the same connectivity of atoms with different arrangements of the atoms in space.   **18.27** the presence of a chiral center and the presence of "structural rigidity"

18.29

(a)

$$Br-\overset{\overset{\displaystyle H}{|}}{\underset{\underset{\displaystyle CH_3}{|}}{C}}-Cl$$

(b)

$$Br-\overset{\overset{\displaystyle CH_3}{|}}{\underset{\underset{\displaystyle H}{|}}{C}}-Cl$$

(c)

$$Br-\overset{\overset{\displaystyle CH_3}{|}}{\underset{\underset{\displaystyle Cl}{|}}{C}}-H$$

(d)

$$H-\overset{\overset{\displaystyle CH_3}{|}}{\underset{\underset{\displaystyle Cl}{|}}{C}}-Br$$

18.31

(a)

$$\begin{array}{c} CHO \\ H-\!\!-OH \\ HO-\!\!-H \\ HO-\!\!-H \\ CH_2OH \end{array}$$

(b)

$$\begin{array}{c} CH_2OH \\ C\!=\!O \\ HO-\!\!-H \\ H-\!\!-OH \\ HO-\!\!-H \\ CH_2OH \end{array}$$

(c)

$$\begin{array}{c} CHO \\ HO-\!\!-H \\ HO-\!\!-H \\ HO-\!\!-H \\ H-\!\!-OH \\ CH_2OH \end{array}$$

(d)

$$\begin{array}{c} CHO \\ HO-\!\!-H \\ H-\!\!-OH \\ HO-\!\!-H \\ H-\!\!-OH \\ CH_2OH \end{array}$$

18.33 (a) D enantiomer (b) D enantiomer (c) L enantiomer (d) L enantiomer   **18.35** (a) diastereomers (b) neither enantiomers nor diastereomers (c) enantiomers (d) diastereomers   **18.37** (d) effect on plane-polarized light   **18.39** (a) same (b) different (c) same (d) different   **18.41** (a) aldose (b) ketose (c) ketose (d) ketose **18.43** (a) aldohexose (b) ketohexose (c) ketotriose (d) ketotetrose **18.45** (a) D-galactose (b) D-psicose (c) dihydroxyacetone (d) L-erythrulose   **18.47** (a) carbon 4 (b) carbons 1 and 2 (c) carbons 1 and 2 (d) carbon 2   **18.49** (a) aldoses, hexoses, aldohexoses (b) hexoses (c) hexoses (d) aldoses **18.51**

(a)

$$\begin{array}{c} CHO \\ H-\!\!-OH \\ HO-\!\!-H \\ H-\!\!-OH \\ H-\!\!-OH \\ CH_2OH \end{array}$$

(b)

$$\begin{array}{c} CHO \\ H-\!\!-OH \\ CH_2OH \end{array}$$

(c)

$$\begin{array}{c} CH_2OH \\ C\!=\!O \\ HO-\!\!-H \\ H-\!\!-OH \\ H-\!\!-OH \\ CH_2OH \end{array}$$

(d)

$$\begin{array}{c} CHO \\ HO-\!\!-H \\ H-\!\!-OH \\ H-\!\!-OH \\ HO-\!\!-H \\ CH_2OH \end{array}$$

18.53 (a) D-fructose (b) D-glucose (c) D-galactose   **18.55** (a) carbons 1 and 5 (b) carbons 1 and 5 (c) carbons 2 and 5 (d) carbons 1 and 4 **18.57** the hydroxyl group orientation on carbon 1   **18.59** In fructose the cyclization involves carbons 2 and 5, and in ribose the cyclization involves carbons 1 and 4; both processes give five-membered rings. **18.61** The cyclic and noncyclic forms interconvert; an equilibrium exists between the forms.   **18.63** (a) α-D-monosaccharide (b) α-D-monosaccharide (c) β-D-monosaccharide (d) α-D-monosaccharide   **18.65** All four structures are hemiacetals.

18.67

(a)
```
    CHO
H —— OH
HO —— H
H —— OH
H —— OH
   CH₂OH
```
(b)
```
    CHO
H —— OH
HO —— H
HO —— H
H —— OH
   CH₂OH
```
(c)
```
    CHO
HO —— H
HO —— H
H —— OH
H —— OH
   CH₂OH
```
(d)
```
   CH₂OH
   C=O
H —— OH
HO —— H
H —— OH
   CH₂OH
```

18.69 (a) α-D-glucose (b) α-D-galactose (c) β-D-mannose (d) α-D-sorbose

18.71

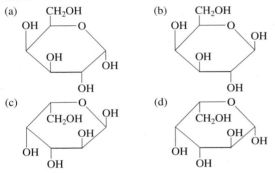

18.73 (a) reducing sugar (b) reducing sugar (c) reducing sugar (d) reducing sugar   **18.75** The aldehyde group in glucose is oxidized to an acid group. The Ag⁺ in Tollens solution is reduced to Ag.

18.77

(a)
```
   COOH
H —— OH
HO —— H
HO —— H
H —— OH
  CH₂OH
```
(b)
```
   COOH
H —— OH
HO —— H
HO —— H
H —— OH
   COOH
```
(c)
```
    CHO
H —— OH
HO —— H
HO —— H
H —— OH
   COOH
```
(d)
```
   CH₂OH
H —— OH
HO —— H
HO —— H
H —— OH
   CH₂OH
```

18.79 (a) yes (b) yes (c) yes (d) yes   **18.81** (a) alpha (b) beta (c) alpha (d) beta   **18.83** (a) methyl alcohol (b) ethyl alcohol (c) ethyl alcohol (d) methyl alcohol   **18.85** A glycoside is an acetal formed from a cyclic monosaccharide. A glucoside is a glycoside in which the monosaccharide is glucose.

18.87

a.

b.

18.89

a.

b.

18.91 (a) glucose and fructose (b) glucose (c) glucose and galactose (d) glucose   **18.93** The glucose part of the lactose structure has a hemiacetal carbon atom.   **18.95** (a) negative (b) positive (c) positive (d) positive   **18.97** (a) α(1 → 6) (b) β(1 → 4) (c) α(1 → 4) (d) α(1 → 4)   **18.99** (a) alpha (b) beta (c) alpha (d) beta **18.101** (a) reducing sugar (b) reducing sugar (c) reducing sugar (d) reducing sugar   **18.103** (a) glucose (b) galactose and glucose (c) glucose and altrose (d) glucose   **18.105** they are two names for the same thing   **18.107** less than 100 monomer units up to a million monomer units   **18.109** (a) correct (b) incorrect (c) incorrect (d) correct **18.111** (a) amylopectin is more abundant (b) amylopectin has the longer polymer chain (c) amylopectin has two types of glycosidic linkages and amylose one type of glycosidic linkage (d) same for both **18.113** (a) correct (b) incorrect (c) incorrect (d) correct   **18.115** (a) to neither (b) to cellulose only (c) to neither (d) to both   **18.117** (a) correct (b) correct (c) correct (d) incorrect   **18.119** anticoagulant for blood **18.121** a lipid molecule that has a carbohydrate unit covalently bonded to it **18.123** interaction between the carbohydrate unit of one cell and a protein imbedded into the cell membrane of another cell.   **18.125** Simple carbohydrates are the mono- and disaccharides, and complex carbohydrates are the polysaccharides.   **18.127** carbohydrates that provide energy but few other nutrients   **18.129** (a) chiral (b) achiral (c) chiral (d) chiral   **18.130** (a) no (b) no (c) no (d) yes   **18.131** (a) no (b) no (c) yes (d) yes

18.132
```
   CH₂OH        CH₂OH        CH₂OH        CH₂OH
   C=O          C=O          C=O          C=O
H —— OH      HO —— H      H —— OH      HO —— H
H —— OH      HO —— H      HO —— H      H —— OH
   CH₂OH        CH₂OH        CH₂OH        CH₂OH
```

18.133 3-methylhexane (hydrogen, methyl, ethyl, and propyl groups attached to a carbon atom)   **18.134** (a) glucose, fructose (b) glucose (c) glucose (d) glucose   **18.135** (a) glucose-derivative (b) glucose (c) glucose-derivative (d) glucose   **18.136** (a) strong oxidizing agent (b) ethyl alcohol, H⁺ ion (c) water (H⁺ ion or enzymes) (d) enzymes **18.137** (a) homopolysaccharides (b) homopolysaccharides, branched polysaccharides (c) homopolysaccharides, unbranched polysaccharides (d) heteropolysaccharides, unbranched polysaccharides **18.138** (a) amylopectin, glycogen (b) amylose, cellulose, chitin (c) chitin (d) cellulose, chitin   **18.139** b   **18.140** d   **18.141** b **18.142** c   **18.143** c   **18.144** a   **18.145** c   **18.146** c   **18.147** d **18.148** c

Chapter 19   **19.1** All lipids are insoluble or only sparingly soluble in water.   **19.3** (a) insoluble (b) soluble (c) insoluble (d) soluble **19.5** energy-storage lipids, membrane lipids, emulsification lipids, messenger lipids, and protective-coating lipids   **19.7** (a) long-chain (b) short-chain (c) long-chain (d) medium-chain   **19.9** (a) saturated (b) polyunsaturated (c) polyunsaturated (d) monounsaturated **19.11** In a SFA there are no double bonds in the carbon chain; in a MUFA there is one carbon–carbon double bond in the carbon chain.

19.13 (a) neither (b) omega-3 (c) omega-3 (d) neither

19.15 $CH_3—(CH_2)_4—CH=CH—CH_2—CH=CH—(CH_2)_7—COOH$

19.17 (a) tetradecanoic acid (b) *cis*-9-hexadecenoic acid

19.19 as carbon chain length increases melting point increases

19.21 *cis* double bonds in unsaturated fatty acids bend the carbon chain, which decreases the strength of molecular attractions **19.23** (a) 18:1 acid (b) 18:3 acid (c) 14:0 acid (d) 18:1 acid **19.25** a glycerol molecule and three fatty acid molecules **19.27** one, ester

19.29

$$H_2C—O—\overset{\overset{O}{\|}}{C}—(CH_2)_{14}—CH_3$$
$$HC—O—\overset{\overset{O}{\|}}{C}—(CH_2)_{14}—CH_3$$
$$H_2C—O—\overset{\overset{O}{\|}}{C}—(CH_2)_{14}—CH_3$$

19.31

S	L
L	S
L	L

S	S
S	L
L	S

19.33 (a) palmitic, myristic, oleic (b) oleic, palmitic, palmitoleic

19.35 (top) 16 carbon atoms and 1 oxygen atom; (middle) 14 carbon atoms and 1 oxygen atom; (bottom) 18 carbon atoms and 1 oxygen atom

19.37 (a) no difference (b) A triacylglycerol may be a solid or a liquid; a fat is a triacylglycerol that is a solid. (c) A triacylglycerol can have fatty acid residues that are all the same, or two or more different kinds may be present. In a mixed triacylglycerol, two or more different fatty acid residues must be present. (d) A fat is a triacylglycerol that is a solid; an oil is a triacylglycerol that is a liquid. **19.39** (a) not correct (b) not correct

19.41 (a) correct (b) not correct **19.43** (a) nonessential fatty acid (b) essential fatty acid (c) nonessential fatty acid (d) nonessential fatty acid

19.45 (a) glycerol and three fatty acids (b) glycerol and three fatty acid salts

19.47

$$\underset{OH}{CH_2}—\underset{OH}{CH}—\underset{OH}{CH_2}$$
$$CH_3—(CH_2)_{12}—COOH$$
$$CH_3—(CH_2)_{14}—COOH$$
$$CH_3—(CH_2)_7—CH=CH—(CH_2)_7—COOH$$

19.49 glycerol, palmitic acid, myristic acid, oleic acid

19.51

$$\underset{OH}{CH_2}—\underset{OH}{CH}—\underset{OH}{CH_2}$$
$$CH_3—(CH_2)_{12}—COO^-\,Na^+$$
$$CH_3—(CH_2)_{14}—COO^-\,Na^+$$
$$CH_3—(CH_2)_7—CH=CH—(CH_2)_7—COO^-\,Na^+$$

19.53 glycerol, sodium palmitate, sodium myristate, sodium oleate

19.55 Carbon–carbon double bond(s) must be present. **19.57** six

19.59

(a)

18:0		18:0		18:1
18:0		18:1		18:0
16:1		16:0		16:0

(b)

18:0		18:0
18:1A		18:1B
16:0		16:0

18:0
18:0
16:1

There are two possibilities for converting the 18:2 acid to 18:1 acid depending on which double bond is hydrogenated (denoted as 18:1A and 18:1B).

19.61 Rancidity results from hydrolysis of ester linkages and oxidation of carbon–carbon double bonds. **19.63** glycerol and sphingosine

19.65

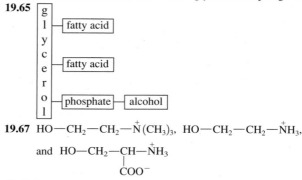

19.67 $HO—CH_2—CH_2—\overset{+}{N}(CH_3)_3,$ $HO—CH_2—CH_2—\overset{+}{N}H_3,$

and $HO—CH_2—\underset{COO^-}{CH}—\overset{+}{N}H_3$

19.69 The two tails are the carbon chain of sphingosine and the fatty acid carbon chain; the head is the phosphate–alcohol portion of the molecule. **19.71** the two tails **19.73** (a) four (b) two **19.75** They differ in the identity of the amino alcohol group; it is choline in a lecithin and serine in a phosphatidylserine.

19.77

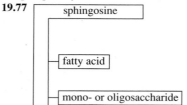

19.79 carbohydrate group versus a phosphate–alcohol group

19.81

(structure with numbered carbons: 1, 2, 3, 4, 5, 6, 7, 8, 9, 10, 11, 12, 13, 14, 15, 16, 17)

19.83 —OH on carbon 3, —CH_3 on carbons 10 and 13, and a hydrocarbon chain on carbon 17 **19.85** "Good cholesterol" is that present in HDLs, and "bad cholesterol" is that present in LDLs. **19.87** phospholipids, sphingoglycolipids, and cholesterol **19.89** a two-layer structure of lipid molecules with nonpolar "tails" in the interior and polar "heads" on the exterior **19.91** creates "open" areas in the bilayer

19.93 a membrane protein that penetrates the interior of the lipid bilayer (cell membrane) **19.95** Protein help is required in facilitated transport but not in passive transport. **19.97** (a) active transport (b) facilitated transport (c) active transport (d) passive transport and facilitated transport

19.99 a substance that can disperse and stabilize water-insoluble substances as colloidal particles in an aqueous solution **19.101** tri- or dihydroxy versus monohydroxy; oxidized side chain amidified to an amino acid versus nonoxidized side chain **19.103** amino acid glycine versus amino acid taurine **19.105** bile fluid **19.107** gall bladder **19.109** sex hormones, adrenocortical hormones **19.111** estradiol has an —OH on carbon 3, while testosterone has a ketone group at this location; testosterone has an extra —CH_3 group at carbon 10 **19.113** (a) adrenocorticoid hormone (b) sex hormone (c) adrenocorticoid hormone (d) sex hormone **19.115** (a) control Na^+/K^+ ion balance in cells and body fluids (b) responsible for secondary male characteristics (c) controls glucose metabolism and is an anti-inflammatory agent (d) responsible for secondary female characteristics **19.117** Prostaglandins have a bond between carbons 8 and 12 that creates a cyclopentane ring structural feature. **19.119** a prostaglandin is similar to a leukotriene but has a cyclopentane ring formed by a bond between C8 and C12 **19.121** inflammatory response, production of pain and fever, blood pressure regulation, induction of blood clotting, control of some reproductive functions, regulation of sleep/wake cycle

19.123 long-chain alchohol — long-chain fatty acid

19.125 mixture of esters involving a long-chain fatty acid and a long-chain alcohol versus a long-chain alkane mixture   **19.127** (a) neither (b) glycerol-based (c) neither (d) neither   **19.128** (a) no (b) no (c) yes (d) no   **19.129** (a) sphingomyelins (b) steroids (c) prostaglandins (d) cerebrosides   **19.130** (a) energy-storage lipids (b) messenger lipids (c) membrane lipids (d) messenger lipids   **19.131** (a) glycerolipid (b) sphingolipid (c) glycerolipid and phospholipid (d) sphingolipid and phospholipid   **19.132** (a) 3, 0, 0 (b) 2, 1, 0 (c) 0, 1, 1 (d) 4, 0, 0   **19.133** (a) no (b) yes (c) yes (d) no   **19.134** b   **19.135** a   **19.136** b   **19.137** a   **19.138** c   **19.139** d   **19.140** c   **19.141** a   **19.142** c   **19.143** a

Chapter 20   **20.1** amino acids   **20.3** 15% by mass   **20.5** (a) yes (b) no (c) no (d) yes   **20.7** the identity of the R group (side chain)   **20.9** (a) phenylalanine, tyrosine, tryptophan (b) methionine, cysteine (c) aspartic acid, glutamic acid (d) serine, threonine, tyrosine   **20.11** An amino group is part of the side chain.   **20.13** The side chain covalently bonds to the amino acid's amino group, producing a cyclic structure.   **20.15** (a) alanine (b) leucine (c) methionine (d) tryptophan   **20.17** asparagine, glutamine, isoleucine, tryptophan   **20.19** (a) polar neutral (b) polar acidic (c) nonpolar (d) polar neutral   **20.21** L family

20.23

(a) (b)

(c) (d)

20.25 They exist as zwitterions.

20.27

(a) (b)

(c) (d)

20.29

(a) (b)

(c) (d)

20.31 the pH at which zwitterion concentration in a solution is maximized   **20.33** Two —COOH groups are present, which deprotonate at different pH values.   **20.35** (a) toward positive electrode (b) isoelectric (c) toward negative electrode (d) toward positive electrode   **20.37** Aspartic acid migrates toward the positive electrode, histidine migrates toward the negative electrode, and valine does not migrate.   **20.39** They react with each other to produce a covalent disulfide bond.   **20.41** —COOH and —NH$_2$

20.43

20.45 Ser is the N-terminal end of Ser–Cys and Cys is the N-terminal end of Cys–Ser.   **20.47** Ser–Val–Gly, Val–Ser–Gly, Gly–Ser–Val, Ser–Gly–Val, Val–Gly–Ser, Gly–Val–Ser   **20.49** (a) Ser–Ala–Cys (b) Asp–Thr–Asn   **20.51** two in each   **20.53** (a) serylcysteine (b) glycylalanylvaline (c) tyrosylaspartylglutamine (d) leucyllysyltryptophylmethionine   **20.55** peptide bonds and α-carbon —CH groups   **20.57** (a) Both are nonapeptides with six of the residues held in the form of a loop by a disulfide bond. (b) They differ in the identity of the amino acid present at two positions in the nonapeptide.   **20.59** They bind at the same sites.   **20.61** Glu is bonded to Cys through the side-chain carboxyl group rather than through the α-carbon carboxyl group.   **20.63** Monomeric proteins contain a single peptide chain and multimeric proteins have two or more peptide chains.   **20.65** (a) true (b) false (c) true (d) true   **20.67** the order in which the amino acids are bonded to each other   **20.69** peptide bond   **20.71** α helix, β pleated sheet   **20.73** Intermolecular involves two separate chains and intramolecular involves a single chain bending back on itself.   **20.75** Yes, both α helix and β pleated sheet can occur at different regions in the same chain.   **20.77** It confers flexibility to the protein, allowing it to interact with several different substances.   **20.79** disulfide bonds, electrostatic interactions, hydrogen bonds, and hydrophobic interactions   **20.81** (a) nonpolar (b) polar neutral R groups (c) —SH groups (d) acidic and basic R groups   **20.83** (a) hydrophobic (b) electrostatic (c) disulfide bond (d) hydrogen bonding   **20.85** the organization among the various peptide chains in a multimeric protein   **20.87** They are the same.   **20.89** (a) fibrous: generally water-insoluble; globular: generally water-soluble (b) fibrous: support and external protection; globular: involvement in metabolic reactions   **20.91** (a) fibrous (b) fibrous (c) globular (d) globular   **20.93** α-Keratin has a double-helix structure and collagen a triple-helix structure.   **20.95** (a) contractile protein (b) storage protein (c) transport protein (d) messenger protein   **20.97** Yes, both Ala and Val are products in each case.   **20.99** Drug hydrolysis would occur in the stomach.   **20.101** Ala–Gly–Met–His–Val–Arg   **20.103** five: Ala–Gly–Ser, Gly–Ser–Tyr, Ala–Gly, Gly–Ser, Ser–Tyr   **20.105** secondary, tertiary, and quaternary   **20.107** same primary structure   **20.109** 4-hydroxyproline and 5-hydroxylysine   **20.111** They are involved with cross-linking.   **20.113** An antigen is a substance foreign to the human body, and an antibody is a substance that defends against an invading antigen.   **20.115** four polypeptide chains that have constant and variable amino acid regions; two chains are longer than the other two; 1%–12% carbohydrates present; long and short chains are connected through disulfide linkages   **20.117** a spherical structure with an inner core of lipid material

surrounded by a shell of phospholipids, cholesterol, and proteins **20.119** chylomicrons, very-low-density lipoproteins, low-density lipoproteins, and high-density lipoproteins **20.121** the lipid/protein mass ratio **20.123** (a) transport dietary triacylglycerols from the intestine to various locations (b) transport cholesterol from the liver to cells throughout the body **20.125** (a) tertiary (b) tertiary (c) secondary (d) primary **20.126** (a) alanine (b) leucine (c) threonine (d) aspartic acid **20.127** (a) +1 (b) +1 (c) +1 (d) +3 **20.128** (a) −1 (b) −1 (c) −4 (d) −1
20.129

COOH	COOH	COOH	COOH
H₂N—H	H—NH₂	H₂N—H	H—NH₂
H₃C—H	H—CH₃	H—CH₃	H₃C—H
CH₂	CH₂	CH₂	CH₂
CH₃	CH₃	CH₃	CH₃

20.130 (a) 24 (b) 12 (c) 6
20.131
(a) $H_3\overset{+}{N}$—CH—COOH; $H_3\overset{+}{N}$—CH—COOH; $H_3\overset{+}{N}$—CH—COOH
with side chains CH₂—OH; CH₃; CH₂—SH

(b) H_2N—CH—COO⁻; H_2N—CH—COO⁻; H_2N—CH—COO⁻
with side chains CH₂—OH; CH₃; CH₂—SH

20.132 (a) simple protein, fibrous protein (b) conjugated protein, globular protein (c) conjugated protein, globular protein (d) conjugated protein, fibrous protein, glycoprotein **20.133** c **20.134** b **20.135** d **20.136** b **20.137** a **20.138** d **20.139** c **20.140** a **20.141** a **20.142** a

Chapter 21 **21.1** catalyst **21.3** larger molecular size, activity regulated by other substances **21.5** (a) conjugated (b) conjugated (c) simple (d) conjugated **21.7** A coenzyme is a cofactor that is an organic substance. A cofactor can be an inorganic or an organic substance. **21.9** to provide additional functional groups **21.11** (a) yes (b) no (c) yes (d) yes **21.13** (a) add a carboxylate group to pyruvate (b) remove H₂ from an alcohol (c) reduce an L-amino acid (d) hydrolyze maltose **21.15** (a) sucrase (or sucrose hydrolase) (b) pyruvate decarboxylase (c) glucose isomerase (d) lactate dehydrogenase **21.17** (a) pyruvate (b) galactose (c) an alcohol (d) an L-amino acid **21.19** (a) isomerase (b) lyase (c) ligase (d) transferase **21.21** (a) isomerase (b) lyase (c) transferase (d) hydrolase **21.23** (a) decarboxylase (b) lipase (c) phosphatase (d) dehydrogenase **21.25** the portion of an enzyme actually involved in the catalysis process **21.27** The substrate must have the same shape as the active site. **21.29** interactions with amino acid R groups **21.31** (a) accepts only one substrate (b) accepts substrate with a particular type of bond **21.33** absolute specificity and stereochemical specificity **21.35** (a) absolute (b) stereochemical **21.37** Rate increases until enzyme denaturation occurs. **21.39** Enzymes vary in the number of acidic and basic amino acids present.
21.41

21.43 nothing; the rate remains constant **21.45** microbial enzymes that survive under extreme temperature and pressure conditions **21.47** no; only one molecule may occupy the active site at a given time **21.49** (a) reversible competitive (b) reversible noncompetitive,

irreversible (c) reversible noncompetitive, irreversible (d) reversible noncompetitive **21.51** enzyme that has quaternary structure and more than one binding site **21.53** The product of a subsequent reaction in a series of reactions inhibits a prior reaction. **21.55** A zymogen is an inactive precursor for a proteolytic enzyme. **21.57** so that they will not destroy the tissues that produce them **21.59** process in which enzyme activity is altered by covalently modifying the structure of the enzyme **21.61** ATP molecules **21.63** protein kinases **21.65** competitive inhibition of the conversion of PABA to folic acid **21.67** has absolute specificity for bacterial transpeptidase **21.69** the threat of biological weapon use by terrorists **21.71** tissue plasminogen activator, which activates an enzyme that dissolves blood clots **21.73** forms of the same enzyme with slightly different amino acid sequences **21.75** dietary organic compound needed by the body in trace amounts **21.77** (a) fat-soluble (b) water-soluble (c) water-soluble (d) water-soluble **21.79** (a) likely (b) unlikely (c) unlikely (d) unlikely **21.81** serves as a cosubstrate in the formation of collagen; general antioxidant for water-soluble substances in body fluids **21.83** coenzymes **21.85** (a) no (b) yes (c) yes (d) no **21.87** alcohol, aldehyde, acid **21.89** Cell differentiation is the process whereby immature cells change in structure and function to become specialized cells. Vitamin A binds to protein receptors in the process. **21.91** They differ only in the identity of the side chain present. **21.93** to maintain normal blood levels of calcium and phosphate ion so that bones can absorb these minerals **21.95** α-tocopherol **21.97** antioxidant effect **21.99** in the length and degree of unsaturation of the side chain present **21.101** phylloquinone is an alternate name for vitamin K₁ and menaquinones is an alternate name for the various forms of vitamin K₂ **21.103** (a) An apoenzyme is the protein portion of a conjugated enzyme; a proenzyme is an inactive precursor of an enzyme. (b) A simple enzyme is pure protein; an allosteric enzyme has two or more protein chains and two binding sites. (c) A coenzyme is an organic cofactor, and an isoenzyme is one of several similar forms of an enzyme. (d) A conjugated enzyme has both a protein and a nonprotein portion; holoenzyme is just another name for a conjugated enzyme. **21.104** (a) alcohol, C=C ester (b) double bond, alcohol (c) double bond, alcohol (d) double bond, ketone **21.105** (a) no (b) no (c) yes (d) yes **21.106** (a) vitamin C (b) vitamin E (c) vitamin D (d) vitamin K **21.107** (a) vitamin A (b) vitamin A (c) vitamin C (d) vitamin E **21.108** (a) addition of a group to, or removal of a group from, a double bond in a manner that does not involve hydrolysis or oxidation–reduction (b) bonding together of two molecules with the involvement of ATP (c) hydrolysis reactions (d) transfer of functional groups between two molecules **21.109** (a) ethanol (b) zinc ion (c) protein molecule (d) alcohol dehydrogenase **21.110** (a) tissue plasminogen activator (b) lactate dehydrogenase (c) creatine phosphokinase (d) aspartate transaminase **21.111** b **21.112** b **21.113** c **21.114** d **21.115** d **21.116** d **21.117** c **21.118** c **21.119** c **21.120** a

Chapter 22 **22.1** deoxyribonucleic acid **22.3** need for the synthesis of proteins **22.5** within the cell nucleus **22.7** nucleotides **22.9** three **22.11** Ribose has both an —H group and an —OH group on carbon 2; deoxyribose has 2 —H atoms on carbon 2. **22.13** (a) pyrimidine (b) pyrimidine (c) purine (d) purine **22.15** (a) one (b) four (c) one **22.17** (a) adenine (b) guanine (c) thymine (d) uracil **22.19** (a) ribose (b) deoxyribose (c) deoxyribose (d) ribose **22.21** (a) false (b) false (c) false (d) false
22.23

22.25 a pentose sugar and a phosphate **22.27** base sequence **22.29** 5′ end has a phosphate group attached to the 5′ carbon; 3′ end has

a hydroxyl group attached to the 3′ carbon   **22.31** a phosphate group and two pentose sugars
22.33

22.35 (a) two polynucleotide chains coiled around each other in a helical fashion (b) The nucleic acid backbones are the outside, and the nitrogen-containing bases are on the inside.   **22.37** (a) 36% (b) 14% (c) 14%
22.39 A G–C pairing involves 3 hydrogen bonds, and an A–T pairing involves 2 hydrogen bonds.   **22.41** They are the same.
22.43 5′ TAGCC 3′   **22.45** (a) 3′ TGCATA 5′ (b) 3′ AATGGC 5′ (c) 5′ CGTATT 3′ (d) 3′ TTGACC 5′   **22.47** 20 hydrogen bonds
22.49 catalyzes the unwinding of the double helix structure
22.51 (a) 3′ TGAATC 5′ (b) 3′ TGAATC 5′ (c) 5′ ACTTAG 3′
22.53 The unwound strands are antiparallel (5′ → 3′ and 3′ → 5′). Only the 5′ → 3′ strand can grow continuously.   **22.55** a DNA molecule bound to a group of small proteins   **22.57** DNA   **22.59** RNA
22.61 (1) RNA contains ribose instead of deoxyribose, (2) RNA contains the base U instead of T, (3) RNA is single-stranded rather than double-stranded, and (4) RNA has a lower molecular mass.
22.63 (a) hnRNA (b) tRNA (c) tRNA (d) hnRNA   **22.65** (a) nuclear region (b) extranuclear region (c) extranuclear region (d) both nuclear and extranuclear regions   **22.67** a strand of DNA   **22.69** causes a DNA helix to unwind; links aligned ribonucleotides together   **22.71** T–A, A–U, G–C, C–G   **22.73** 3′ UACGAAU 5′   **22.75** 3′ AAGCGTC 5′
22.77 Exons convey genetic information whereas introns do not.
22.79 3′ AGUCAAGU 5′   **22.81** (a) hnRNA (b) snRNA
22.83 a mechanism by which a number of proteins that are variations of a basic structural motif can be produced from a single gene
22.85 A three-nucleotide sequence in mRNA that codes for a specific amino acid   **22.87** (a) Leu (b) Asn (c) Ser (d) Gly   **22.89** (a) CUC, CUA, CUG, UUA, UUG (b) AAC (c) AGC, UCU, UCC, UCA, UCG (d) GGU, GGC, GGA   **22.91** The base T cannot be present in a codon.
22.93 Met–Lys–Glu–Asp–Leu   **22.95** A cloverleaf shape with three hairpin loops and one open side.   **22.97** covalent bond (ester)
22.99 anticodons written in the 5′-to-3′direction: (a) UCU (b) ACG (c) AAA (d) UUG   **22.101** (a) Gly (b) Leu (c) Ala (d) Leu
22.103 (a) UCU, UCC, UCA, UCG, AGU, AGC (b) UUA, UUG, CUU, CUC, CUA, CUG (c) AUU, AUC, AUA (d) GGU, GGC, GGA, GGG
22.105 an rRNA-protein complex that serves as the site for the translation phase of protein synthesis   **22.107** mostly rRNA with some protein
22.109 an amino acid reacts with ATP, the resulting complex reacts with tRNA   **22.111** A site   **22.113** initial Met residue is removed; covalent modification occurs if needed; completion of folding of protein occurs
22.115 Gly: GGU, GGC, GGA or GGG; Ala: GCU, GCC, GCA or GCG; Cys: UGU or UGC; Val: GUU, GUC, GUA or GUG; Tyr: UAU or UAC
22.117 (a) Gly–Tyr–Ser–Ser–Pro (b) Gly–Tyr–Ser–Ser–Pro (c) Gly–Tyr–Ser–Ser–Thr   **22.119** a DNA or an RNA molecule with a protein coating   **22.121** (1) attaches itself to cell membrane, (2) opens a hole in the membrane, and (3) injects itself into the cell   **22.123** contains a "foreign" gene   **22.125** host for a "foreign" gene

22.127 Recombinant DNA is incorporated into a host cell.
22.129

22.131 to produce many copies of a specific DNA sequence in a relatively short time   **22.133** a short nucleotide chain bound to the template DNA strand to which new nucleotides can be attached   **22.135** dATP stands for an ATP in which deoxyribose is present; ddATP stands for an ATP in which dideoxyribose is present.   **22.137** (a) 5′ GCCGACTACT 3′ (b) 5′ AGTAGTCGGC 3′   **22.139** (a) Thymine has a methyl group on carbon-5 that uracil lacks. (b) Adenine is 6-aminopurine, and guanine is 2-amino-6-oxopurine.   **22.140** (a) 3′ GTATGTCGGACCTTCGAG 5′ (b) 3′ GUAUGUCGGACCUUCGAG 5′ (c) 3′ GUA–UGU–CGG–ACC–UUC–GAG 5′ (d) 5′ CAU–ACA–GCC–UGG–AAG–CUC 3′ (e) Glu–Leu–Pro–Gly–Lys–Met   **22.141** (a) 1 (b) 4 (c) 2 (d) 4
22.142 (a) 3 (b) 2 (c) 1 (d) 3   **22.143** (a) 3 (b) 1 (c) 2 and 4 (d) 1, 2, 3, and 4   **22.144** (a) 1 (b) 2 (c) 4 (d) 3   **22.145** A   **22.146** (a) 2.9 billion base pairs (b) 20,000–25,000 genes (c) 2% of base pairs   **22.147** c
22.148 a   **22.149** c   **22.150** a   **22.151** b   **22.152** d   **22.153** a
22.154 b   **22.155** b   **22.156** b

Chapter 23   23.1 anabolism—synthetic; catabolism—degradative
23.3 a series of consecutive biochemical reactions   **23.5** Large molecules are broken down to smaller ones; energy is released.
23.7 Prokaryotic cells have no nucleus, and the DNA is usually a single circular molecule. Eukaryotic cells have their DNA in a membrane-enclosed nucleus.   **23.9** An organelle is a small structure within the cell cytosol that carries out a specific cellular function.   **23.11** inner membrane
23.13 region between inner and outer membranes   **23.15** adenosine triphosphate

23.17 phosphate—phosphate—phosphate—ribose—adenine

23.19 three phosphates versus one phosphate   **23.21** adenine versus guanine   **23.23** ADP is produced from the hydrolysis of ATP.
23.25 flavin adenine dinucleotide   **23.27** flavin—ribitol—ADP
23.29 nicotinamide—ribose—phosphate
adenine—ribose—phosphate

23.31 nicotinamide subunit   **23.33** (a) FADH$_2$ (b) NAD$^+$
23.35 (a) nicotinamide (b) riboflavin

23.37 2-aminoethanethiol—pantothenic acid—phosphorylated ADP

23.39 a compound with a greater free energy of hydrolysis than is typical for a compound   **23.41** free monophosphate species
23.43 (a) phosphoenolpyruvate (b) creatine phosphate (c) 1,3-bisphosphoglycerate (d) AMP   **23.45** (1) digestion, (2) acetyl group formation, (3) citric acid cycle, (4) electron transport chain and oxidative phosphorylation   **23.47** tricarboxylic acid cycle, Krebs cycle   **23.49** acetyl CoA   **23.51** (a) 2 (b) 1 (c) 2 (Steps 3, 8) (d) 2 (Steps 2, 7)
23.53 (a) Steps 3, 4, 6, 8 (b) Step 2 (c) Step 7
23.55 succinate ($^-$OOC—CH$_2$—CH$_2$—COO$^-$);
fumarate ($^-$OOC—CH=CH—COO$^-$);
malate $\left(\begin{array}{c}^-\text{OOC—CH}_2\text{—CH—COO}^- \\ | \\ \text{OH}\end{array}\right)$;
oxaloacetate $\left(^-\text{OOC—CH}_2\text{—}\overset{\overset{\text{O}}{\|}}{\text{C}}\text{—COO}^-\right)$
23.57 oxidation and decarboxylation   **23.59** (a) NAD$^+$ (b) FAD
23.61 (a) isocitrate, α-ketoglutarate (b) fumarate, malate (c) malate, oxaloacetate (d) citrate, isocitrate   **23.63** respiratory chain   **23.65** O$_2$
23.67 (a) FMN (b) CoQH$_2$ **23.69** (a) oxidized (b) oxidized (c) reduced (d) oxidized   **23.71** (a) oxidation (b) reduction (c) oxidation

(d) reduction **23.73** (a) I (b) I, II, III (c) III (d) IV **23.75** $CoQH_2$ **23.77** two **23.79** (a) $FADH_2$, 2 Fe(II)SP, $CoQH_2$ (b) $FMNH_2$, 2 Fe^{2+}, $CoQH_2$ **23.81** ATP synthesis from ADP Pi using energy from the electron transport chain **23.83** protons (H^+ ions) **23.85** intermembrane space **23.87** ATP synthase **23.89** Protons flow through ATP synthase complex. **23.91** 2.5 ATP molecules **23.93** They enter the ETC at different stages. **23.95** It is an intermediate amount of free energy, higher than some reactions and lower than others. **23.97** (a) ATP to ADP (b) ATP to ADP **23.99** reactive oxygen species

23.101 (a) $2O_2$ + NADPH $\longrightarrow$ $2O_2^-$ + $NADP^+$ + H^+ (b) $2O_2^-$ + $2H^+$ $\longrightarrow$ H_2O_2 + O_2 **23.103** (a) 2 (b) 2 (c) 1 (d) 2 **23.104** (a) 2 (b) 1 (c) 1 (d) 2 **23.105** (a) 4 (b) 3 and 4 (c) 1, 2, 3, and 4 (d) 3 **23.106** (a) 1 (b) 4 (c) 4 (d) 2 **23.107** The products from the CAC, which are $FADH_2$ and NADH, are the starting reactants for the ETC. **23.108** (a) inside mitochondria matrix (b) inside mitochondria (inner membrane) **23.109** FAD is the oxidizing agent for carbon–carbon double bond formation. **23.110** Reduced coenzymes are oxidized, and ADP is phosphorylated. **23.111** a **23.112** b **23.113** d **23.114** b **23.115** c **23.116** c **23.117** a **23.118** d **23.119** c **23.120** d

Chapter 24 **24.1** mouth, salivary α-amylase **24.3** small intestine, pancreas **24.5** outer membranes of intestinal mucosal cells, sucrose hydrolysis **24.7** glucose, galactose, fructose **24.9** glucose **24.11** NAD^+ **24.13** formation of glucose 6-phosphate, a species that cannot cross cell membranes **24.15** dihydroxyacetone phosphate, glyceraldehyde 3-phosphate **24.17** two **24.19** two **24.21** Steps 1, 3, and 6 **24.23** cytosol **24.25** (a) glucose 6-phosphate (b) 2-phosphoglycerate (c) phosphoglyceromutase (d) ADP **24.27** (a) Step 10 (b) Step 1 (c) Step 8 (d) Step 6 **24.29** (a) +2 (b) +4
24.31

and

24.33

24.35 acetyl CoA, lactate, ethanol **24.37** pyruvate + CoA + NAD^+ $\rightarrow$ acetyl CoA + NADH + CO_2 **24.39** NADH is oxidized to NAD^+, a substance needed for glycolysis. **24.41** CO_2 **24.43** glucose + 2ADP + $2P_i$ $\rightarrow$ 2lactate + 2ATP + $2H_2O$ **24.45** decreases ATP production by 2 **24.47** 30 ATP versus 2 ATP **24.49** two **24.51** Glycogenesis converts glucose 6-phosphate to glycogen and glycogenolysis is the reverse process. **24.53** glucose 6-phosphate **24.55** UTP **24.57** UDP + ATP $\rightarrow$ UTP + ADP **24.59** Step 2 **24.61** Glucose 6-phosphate is converted to glucose in liver cells and used as is in muscle cells. **24.63** as glucose 6-phosphate **24.65** the liver **24.67** two-step pathway for Step 10; different enzymes for Steps 1 and 3 **24.69** oxaloacetate **24.71** goes to the liver, where it is converted to glucose **24.73** (a) all four processes (b) glycolysis and gluconeogenesis (c) gluconeogenesis

(d) glycogenesis **24.75** (a) glycolysis (b) glycolysis (c) glycogenesis (d) glycolysis **24.77** (a) gluconeogenesis (b) glycolysis (c) glycogenesis (d) glycogenesis and glycogenolysis **24.79** glucose 6-phosphate **24.81** NADPH is consumed in its reduced form; NADH is consumed in its oxidized form (NAD^+). **24.83** Glucose 6-phosphate + $2NADP^+$ + H_2O $\rightarrow$ ribulose 5-phosphate + CO_2 + 2NADPH + $2H^+$ **24.85** CO_2 **24.87** increases rate of glyco-gen synthesis **24.89** increases blood glucose levels **24.91** pancreas **24.93** Epinephrine attaches to cell membrane and stimulates the production of cAMP, which activates glycogen phosphorylase. **24.95** glucagon (liver cells) and epinephrine (muscle cells) **24.97** (a) 2 ATP (b) 30 ATP (c) 2 ATP (d) 2 ATP **24.98** (a) 1,2 (b) 2 (c) 3 (d) 4 **24.99** (a) when the body requires free glucose (b) anaerobic conditions in muscle; red blood cells (c) when the body requires energy (d) anaerobic conditions in yeast **24.100** (a) 12 moles (b) 4 moles (c) 4 moles (d) 60 moles **24.101** (a) The glucose supply is adequate, and the body does not need energy. (b) The glucose supply is adequate, and the body needs energy. (c) Ribose 5-phosphate or NADPH is needed. (d) The free glucose supply is not adequate. **24.102** b **24.103** b **24.104** b **24.105** d **24.106** c **24.107** a **24.108** b **24.109** c **24.110** d **24.111** c

Chapter 25 **25.1** 98% **25.3** no effect **25.5** (a) stomach and small intestine (b) stomach (10%) and small intestine (90%) (c) stomach (gastric lipases), small intestine (pancreatic lipases) **25.7** acts as an emulsifier **25.9** monoacylglycerols are the major product **25.11** reassembled into triacylglycerols; converted to chylomicrons **25.13** They have a large storage capacity for triacylglycerols. **25.15** hydrolysis of triacylglycerols in adipose tissue; entry of hydrolysis products into bloodstream **25.17** activates hormone-sensitive lipase **25.19** (a) Step 1 (b) Step 1 (c) Step 2 (d) Step 1 **25.21** one **25.23** (a) true (b) false (c) true 2 (d) false **25.25** (a) intermembrane space (b) matrix (c) intermembrane space (d) matrix **25.27** the ATP is converted to AMP rather than ADP **25.29** (a) alkane to alkene (b) alkene to 2° alcohol (c) 2° alcohol to ketone **25.31** *trans* isomer **25.33** (a) Step 3, turn 2 (b) Step 2, turn 3 (c) Step 4, turn 3 (d) Step 1, turn 2 **25.35** compounds a and d **25.37** (a) 7 turns (b) 5 turns **25.39** A *cis–trans* isomerase converts a *cis* bond to a *trans* bond **25.41** (a) glucose (b) fatty acids **25.43** (a) 4 turns (b) 5 acetyl CoA (c) 4 NADH (d) 4 $FADH_2$ (e) 2 high-energy bonds **25.45** 64 ATP **25.47** unsaturated fatty acid produces less $FADH_2$ **25.49** 4 kcal versus 9 kcal **25.51** (1) dietary intakes high in fat and low in carbohydrates, (2) inadequate processing of glucose present, and (3) prolonged fasting **25.53** Ketone body formation occurs when oxaloacetate concentrations are low.
25.55

25.57 (a) acetoacetate, acetone (b) β-hydroxybutyrate (c) acetoacetate (d) acetone **25.59** (a) Step 1 (b) Step 4 (c) Step 1 (d) Step 3 **25.61** (a) Steps 1 and 2 (b) Step 2 (c) Step 2 (d) Step 3 **25.63** acetoacetate and succinyl CoA are reactants; acetoacetyl CoA and succinate are products **25.65** accumulation of ketone bodies in blood and urine **25.67** cytosol versus mitochondrial matrix **25.69** acyl carrier protein; polypeptide chain replaces phosphorylated AMP **25.71** (a) matrix (b) cytosol (c) cytosol (d) cytosol **25.73** (a) malonyl CoA (b) acetyl ACP (c) malonyl CoA (d) acetyl ACP **25.75** (a) Step 2 (b) Step 3 (c) Step 1 (d) Step 4 **25.77** (a) acetoacetyl ACP (b) crotonyl ACP (c) crotonyl ACP (d) acetoacetyl ACP **25.79** (a) condensation (b) hydrogenation (c) dehydration (d) hydrogenation **25.81** needed to convert saturated

fatty acids to unsaturated fatty acids **25.83** (a) 6 turns
(b) 6 malonyl ACP (c) 6 ATP bonds (d) 12 NADPH **25.85** (a) succinate
(b) oxaloacetate (c) malate (d) fumarate **25.87** (a) unsaturated acid
(b) ketoacid (c) ketoacid (d) hydroxyacid **25.89** (a) 13–17%
(b) 83–87% **25.91** (a) mevalonate (b) isopentenyl pyrophosphate
(c) squalene **25.93** (a) fewer than (b) fewer than (c) same as
25.95 (a) fatty acids, cholesterol, and ketone bodies (b) glucose, glycerol,
and fatty acids **25.97** (a) β-oxidation pathway (b) lipogenesis
(c) glycerol catabolism (d) ketogenesis **25.98** (a) fatty acid catabolism
(b) lipogenesis (c) consumption of molecular O_2 (d) consumption of
molecular O_2 **25.99** (a) Step 1 (b) Step 1 (c) Step 3 (d) Step
4 **25.100** (1), (3), (4), and (2) **25.101** (a) true (b) false (c) false
(d) false **25.102** (a) incorrect (b) correct (c) incorrect (d) correct
25.103 (a) correct (b) correct (c) correct (d) correct **25.104** glucose,
C_8 fatty acid, sucrose, C_{14} fatty acid **25.105** b **25.106** a **25.107** c
25.108 c **25.109** a **25.110** b **25.111** a **25.112** c **25.113** c
25.114 b

Chapter 26 **26.1** Denaturation occurs in the stomach with gastric
juice as the denaturant. **26.3** Pepsinogen is the inactive precursor of
pepsin. **26.5** Gastric juice is acidic (1.5–2.0 pH) and pancreatic juice is
basic (7.0–8.0 pH). **26.7** Membrane protein molecules facilitate the
passage of amino acids through the intestinal wall. **26.9** total supply of
free amino acids available for use **26.11** cyclic process of protein degradation and resynthesis **26.13** A positive nitrogen balance has nitrogen
intake exceeding nitrogen output; a negative nitrogen balance has nitrogen output exceeding nitrogen intake. **26.15** negative balance; proteins
are degraded to get the needed amino acid **26.17** protein synthesis;
synthesis of nonprotein nitrogen-containing compounds; nonessential
amino acid synthesis; energy production **26.19** (a) essential
(b) nonessential (c) nonessential (d) essential **26.21** b and c
26.23 an amino acid and an α-keto acid
26.25

(a)
$$\overset{\overset{+}{N}H_3}{\underset{CH_3}{HO-CH-\underset{|}{CH}-COO^-}} + CH_3-\overset{O}{\overset{\|}{C}}-COO^- \longrightarrow$$

$$\overset{O}{\underset{CH_3}{HO-CH-\underset{|}{C}-COO^-}} + CH_3-\overset{\overset{+}{N}H_3}{\underset{}{CH}-COO^-}$$

(b)
$$\overset{\overset{+}{N}H_3}{CH_3-\underset{|}{CH}-COO^-} + {}^-OOC-CH_2-\overset{O}{\overset{\|}{C}}-COO^- \longrightarrow$$

$$CH_3-\overset{O}{\overset{\|}{C}}-COO^- + {}^-OOC-CH_2-\overset{\overset{+}{N}H_3}{\underset{}{CH}-COO^-}$$

(c)
$$\overset{\overset{+}{N}H_3}{H-\underset{|}{CH}-COO^-} + {}^-OOC-CH_2-CH_2-\overset{O}{\overset{\|}{C}}-COO^- \longrightarrow$$

$$H-\overset{O}{\overset{\|}{C}}-COO^- + {}^-OOC-CH_2-CH_2-\overset{\overset{+}{N}H_3}{\underset{}{CH}-COO^-}$$

(d)
$$\overset{\overset{+}{N}H_3}{\underset{CH_3}{HO-CH-\underset{|}{CH}-COO^-}} + {}^-OOC-CH_2-CH_2-\overset{O}{\overset{\|}{C}}-COO^- \longrightarrow$$

$$\overset{O}{\underset{CH_3}{HO-CH-\underset{|}{C}-COO^-}} + {}^-OOC-CH_2-CH_2-\overset{\overset{+}{N}H_3}{\underset{}{CH}-COO^-}$$

26.27 pyruvate, α-ketoglutarate **26.29** coenzyme that participates in
the amino group transfer **26.31** conversion of an amino acid into a keto
acid with the release of ammonium ion **26.33** Oxidative deamination
produces ammonium ion, and transamination produces an amino acid.
26.35

(a)
$${}^-OOC-CH_2-CH_2-\overset{O}{\overset{\|}{C}}-COO^-$$

(b) $$HS-CH_2-\overset{O}{\overset{\|}{C}}-COO^-$$ (c) $$CH_3-\overset{O}{\overset{\|}{C}}-COO^-$$

(d)
$$\langle\!\!\!\!\bigcirc\!\!\!\!\rangle-CH_2-\overset{O}{\overset{\|}{C}}-COO^-$$

26.37 Transamination of the α-keto acid produces the amino acid.

$$\overset{CH_3}{\underset{}{CH_3-\underset{|}{CH}-CH_2-\overset{\overset{+}{N}H_3}{\underset{}{CH}-COO^-}}}$$

26.39 (a) aspartate (b) glutamate (c) pyruvate (d) α-ketoglutarate
26.41
$$H_2N-\overset{O}{\overset{\|}{C}}-NH_2$$

26.43 carbamoyl phosphate **26.45** an amide group,
$$-\overset{O}{\overset{\|}{C}}-NH_2$$

26.47
$$H_3\overset{+}{N}-, \quad H_2N-\overset{O}{\overset{\|}{C}}-NH-, \quad H_2N-\overset{\overset{+}{N}H_2}{\overset{\|}{C}}-NH-$$

26.49 (a) carbamoyl phosphate (b) aspartate **26.51** carbamoyl phosphate **26.53** (a) N_2 (b) N_3 (c) N_1 (d) N_4 **26.55** (a) citrulline
(b) ornithine (c) argininosuccinate (d) carbamoyl phosphate
26.57 equivalent of four ATP molecules **26.59** goes to the citric acid
cycle where it is converted to oxaloacetate, which is then converted to
aspartate **26.61** α-ketoglutarate, succinyl CoA, fumarate, oxaloacetate
26.63 (a) acetoacetyl CoA and acetyl CoA (b) succinyl CoA and acetyl
CoA (c) fumarate and oxaloacetate (d) α-ketoglutarate
26.65 Degradation products can be used to make glucose.
26.67 glutamate **26.69** pyruvate, α-ketoglutarate, 3-phosphoglycerate,
oxaloacetate, and phenylalanine **26.71** hydrolyzed to amino acids
26.73 In biliverdin the heme ring has been opened and one carbon
atom has been lost (as CO). **26.75** biliverdin, bilirubin, bilirubin
diglucuronide, urobilin **26.77** urobilin **26.79** excess bilirubin
26.81 Amino acid carbon skeletons are degraded to acetyl CoA or
acetoacetyl CoA; ketogenesis converts these degradation products to
ketone bodies. **26.83** converted to body fat stores **26.85** (a) 1 (b) 2
(c) 3 (d) 2 **26.86** (a) 1 (b) 3 (c) 3 (d) 2 **26.87** (1), (3), (2), and (4)
26.88 (a) 3 (b) 2 (c) 1 (d) 3 **26.89** (a) transamination (b) deamination
(c) deamination (d) transamination **26.90** (a) 1 (b) 4 (c) 2 (d) 3
26.91 (a) false (b) false (c) true (d) true **26.92** (a) 2 (b) 1 (c) 2 (d) 3
26.93 a **26.94** d **26.95** a **26.96** b **26.97** c **26.98** b **26.99** c
26.100 c **26.101** b **26.102** c

Photo Credits

Index/Glossary

Glycosidic linkage *The bond in a disaccha-ride resulting from the reaction between the hemiacetal carbon atom —OH group of one monosaccharide and a —OH group on the other monosaccharide,* 582
disaccharides and, 582–587
summary diagram for, 595
Gold, lack of corrosion of, 232
Grain alcohol, 404
Gram *The base unit of mass in the metric system,* 24
compared to English system units, 24
Greenhouse gases, global warming and, 227
Group *A vertical column of elements in the periodic table,* 60
in periodic table, notation for, 60
Guanine
nucleotide subunit, 735–737
structure of, 735–737

Hair
denaturation of, 686
permanent for, 686
pigmentation of, melanin and, 452
Half life ($t_{1/2}$) *The time required for one-half of a given quantity of a radioactive substance to undergo decay,* 297
magnitude, interpretation of, 297–300
selected values for, table of, 298
use of, in calculations, 298–300
Halogen *A general name for any element in Group VIIA of the periodic table,* 60
London forces and, 186
periodic table location of, 60, 75
Halogenated alkane *An alkane derivative in which one or more halogen atoms are present,* 350
IUPAC-common name contrast for, 350
nomenclature for, 350–351
physical properties of, 352
Halogenated cycloalkane *A cycloalkane derivative in which one or more halogen atoms are present,* 350
preparation of, from alcohols, 417–418
Halogenation reaction *A chemical reaction between a substance and a halogen in which one or more halogen atoms are incorporated into molecules of the sub-stance; an addition reaction in which a halogen is incorporated into molecules of an organic compound,* 347, 375
alcohol, 417–418
alkane, 347–350
alkene, 375
aromatic hydrocarbon, 390
cycloalkane, 350
Halothane, 425
Handedness
molecular, notation for, 561–565
molecular, recognition of, 558–560
Haworth, Walter Norman, 576
Haworth projection formula *A two-dimensional structural notation that specifies the three-dimensional structure of a cyclic form of a monosaccharide,* 576
conventions for drawing, 576
monosaccharides and, 576–577

HDL
biochemical functions of, 690–691
structural characteristics of, 690–691
Heart attack, enzyme analysis and, 719
Heat energy, units for, 43
Helicobacter pylori
conditions for existence of, 707
enzymes present in, 707
stomach ulcers and, 707
Heme, structure of, 892
Hemiacetal *An organic compound in which a carbon atom is bonded to both a hydroxyl group (—OH) and an alkoxy group (—OR),* 458
cyclic, formation of from monosaccharides, 574–577
cyclic, structure of, 458–459
nomenclature for, 462
preparation of, from aldehydes, 458–459
preparation of, from ketones, 458–459
reaction of, with alcohols, 460
Hemoglobin
biochemical function of, 681
catabolism of, 892–895
tertiary structure of, 678
Hemoglobin catabolism
bile pigments and, 893–896
carbon monoxide production and, 893
heme degradation products and, 892–896
oxygen consumption and, 893
Hemolysis, hypertonic solutions and, 213–214
Henderson–Hasselbalch equation, buffer sys-tems and, 277–278
Henry, William, 194
Henry's law *The amount of gas that will dis-solve in a liquid at a given temperature is directly proportional to the partial pres-sure of the gas above the liquid,* 194
Heparin
properties of, 596
structure of, 596
Heroin, 530
Heteroatomic molecule *A molecule in which two or more kinds of atoms are present,* 13
examples of, 13–14
Heterocyclic amine *An organic compound in which nitrogen atoms of amino groups are part of either an aromatic or a nonaro-matic ring system,* 525
ring systems in, 525
Heterocyclic organic compound *A cyclic organic compound in which one or more of the carbon atoms in the ring have been replaced with atoms of other elements,* 428
amides as, 533
amines as, 516, 525
esters as, 489
ethers as, 428
lactams as, 533
lactones as, 489
Heterogeneous mixture *A mixture that con-tains visibly different phases (parts), each of which has different properties,* 6
characteristics of, 6–9
Heterogeneous nuclear RNA (hnRNA) *RNA formed directly by DNA transcription,* 748

exons and introns and, 751–752
post-transcription processing of, 751–752
Heteropolysaccharide *A polysaccharide in which more than one type of monosaccha-ride monomer (usually two) are present,* 588
examples of, 589, 595–596
HFCS. *See* High fructose corn syrup
High fructose corn syrup
production of, 588
types of, 588
uses for, 588
High-energy compound *A compound that has a greater free energy of hydrolysis than that of a typical compound,* 786
phosphate containing, 786–788
strained bonds and, 786–788
Histamine, allergy response from, 529
Histone, 746
HMG-CoA. *See* 3-Hydroxy-3-methylglutaryl CoA
hnRNA. *See* Heterogeneous nuclear RNA
Holoenzyme *The biochemically active conju-gated enzyme produced from an apoen-zyme and a cofactor,* 699
Homoatomic molecule *A molecule in which all atoms present are of the same kind,* 13
examples of, 13–14
Homogeneous mixture *A mixture that contains only one visibly distinct phase (part), which has uniform properties throughout,* 6
characteristics of, 6–9
Homopolysaccharide *A polysaccharide in which only one type of monosaccharide monomer is present,* 587–588
examples of, 588, 591–594
Hormone *A biochemical substance, produced by a ductless gland, that has a messenger function,* 641
adrenocortical, 643
carbohydrate metabolism control and, 835–837
peptide, 667–668
sex, 642–643
Human body
density measurement for, 40
elemental composition of, 11
normal temperature for, 44
percent body fat, determination of, 40
Human genome, compared to human transcriptome, 752
Human genome project, results from, 749–750
Human transcriptome, compared to human genome, 752
Hyaluronic acid
properties of, 596
structure of, 595–596
Hydration reaction *An addition reaction in which H_2O is incorporated into molecules of an organic compound,* 376
alkenes, 376
β oxidation pathway and, 848
Hydrocarbon *A compound that contains only carbon atoms and hydrogen atoms,* 322
alkane, 323–340
alkene, 361–382
alkyne, 382–384

Common Functional Groups

Name of Class	Structural Feature

Alkane

$$-\overset{\displaystyle |}{\underset{\displaystyle |}{C}}-$$

Alkene

$$\overset{\diagdown}{\diagup}C{=}C\overset{\diagup}{\diagdown}$$

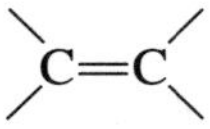

Alkyne

$$-C{\equiv}C-$$

Aromatic hydrocarbon   or

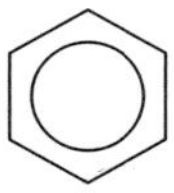

Alcohol

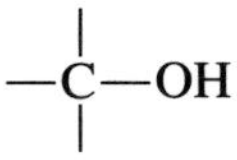

$$-\overset{\displaystyle |}{\underset{\displaystyle |}{C}}-OH$$

Phenol

OH

Ether

$$-\overset{\displaystyle |}{\underset{\displaystyle |}{C}}-O-\overset{\displaystyle |}{\underset{\displaystyle |}{C}}-$$

Thiol

$$-\overset{\displaystyle |}{\underset{\displaystyle |}{C}}-SH$$

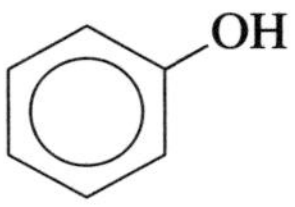

Aldehyde

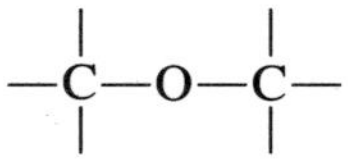

$$\overset{\displaystyle O}{\overset{\displaystyle \|}{-C}}-H\;(-CHO)$$

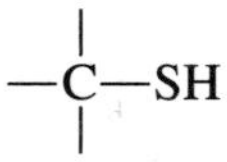

Ketone

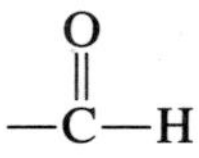

$$-\overset{\displaystyle |}{\underset{\displaystyle |}{C}}-\overset{\displaystyle O}{\overset{\displaystyle \|}{C}}-\overset{\displaystyle |}{\underset{\displaystyle |}{C}}-$$

Carboxylic acid

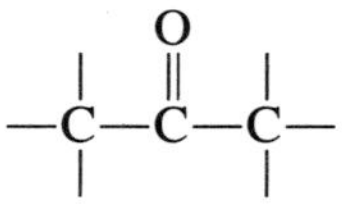

$$\overset{\displaystyle O}{\overset{\displaystyle \|}{-C}}-OH\;(-COOH\text{ or }-CO_2H)$$

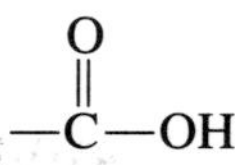

Ester

$$\overset{\displaystyle O}{\overset{\displaystyle \|}{-C}}-O-\overset{\displaystyle |}{\underset{\displaystyle |}{C}}-\;(-COOR\text{ or }-CO_2R)$$

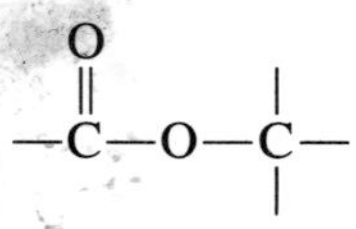

Amine

$$-\overset{\displaystyle |}{\underset{\displaystyle |}{C}}-NH_2$$

Amide

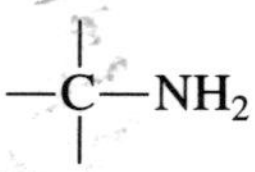

$$\overset{\displaystyle O}{\overset{\displaystyle \|}{-C}}-NH_2$$

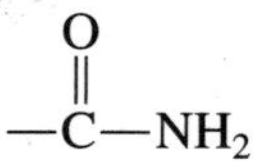